Harper's Illustrated Biochemistry

THIRTY-SECOND EDITION

Peter J. Kennelly, PhD

Professor
Department of Biochemistry
Virginia Tech
Blacksburg, Virginia

Kathleen M. Botham, PhD, DSc

Emeritus Professor of Biochemistry
Department of Comparative Biomedical Sciences
Royal Veterinary College
University of London
London, United Kingdom

Owen P. McGuinness, PhD

Professor
Department of Molecular Physiology & Biophysics
Vanderbilt University
School of Medicine
Nashville, Tennessee

Victor W. Rodwell, PhD

Professor (Emeritus) of Biochemistry
Purdue University
West Lafayette, Indiana

P. Anthony Weil, PhD

Professor Emeritus of Molecular Physiology & Biophysics
Vanderbilt University
Nashville, Tennessee

New York   Chicago   San Francisco   Athens   London   Madrid   Mexico City
Milan   New Delhi   Singapore   Sydney   Toronto

Harper's Illustrated Biochemistry, Thirty-second Edition

1 2 3 4 5 6 7 8 9   LWI   27 26 25 24 23 22

ISBN 978-1-264-79567-3
MHID 1-264-79567-X
ISSN 1043-9811

Notice

Medicine is an ever-changing science. As new research and clinical experience broaden our knowledge, changes in treatment and drug therapy are required. The authors and the publisher of this work have checked with sources believed to be reliable in their efforts to provide information that is complete and generally in accord with the standards accepted at the time of publication. However, in view of the possibility of human error or changes in medical sciences, neither the authors nor the publisher nor any other party who has been involved in the preparation or publication of this work warrants that the information contained herein is in every respect accurate or complete, and they disclaim all responsibility for any errors or omissions or for the results obtained from use of the information contained in this work. Readers are encouraged to confirm the information contained herein with other sources. For example and in particular, readers are advised to check the product information sheet included in the package of each drug they plan to administer to be certain that the information contained in this work is accurate and that changes have not been made in the recommended dose or in the contraindications for administration. This recommendation is of particular importance in connection with new or infrequently used drugs.

This book was set in Minion Pro by KnowledgeWorks Global Ltd.
The editors were Michael Weitz and Peter J. Boyle.
The production supervisor was Catherine H. Saggese.
Project management was provided by Tasneem Kauser, KnowledgeWorks Global Ltd.

Co-Authors

David A. Bender, PhD
Professor (Emeritus) of Nutritional Biochemistry
University College London
London, United Kingdom

Peter L. Gross, MD, MSc, FRCP(C)
Associate Professor
Department of Medicine
McMaster University
Hamilton, Ontario, Canada

January D. Haile, PhD
Associate Professor of Biochemistry and Molecular Biology
Centre College
Danville, Kentucky

Molly Jacob, MD, PhD, MNASc, FRCPath
Professor of Biochemistry
Christian Medical College
Vellore, Tamil Nadu, India

Peter A. Mayes, PhD, DSc
Emeritus Professor of Veterinary Biochemistry
Royal Veterinary College
University of London
London, United Kingdom

Robert K. Murray, MD, PhD
Emeritus Professor of Biochemistry
University of Toronto
Toronto, Ontario, Canada

Margaret L. Rand, PhD
Senior Associate Scientist
Division of Hematology/Oncology
Hospital for Sick Children, Toronto
Professor, Department of Biochemistry
University of Toronto, Toronto, Canada

Joe Varghese, MD, PhD
Professor and Head of Biochemistry
Christian Medical College
Vellore, Tamil Nadu, India

Contents

Preface ix

SECTION I
Structures & Functions of Proteins & Enzymes 1

1 Biochemistry & Medicine 1
Victor W. Rodwell, PhD

2 Water & pH 6
Peter J. Kennelly, PhD, & Victor W. Rodwell, PhD

3 Amino Acids & Peptides 15
Peter J. Kennelly, PhD, & Victor W. Rodwell, PhD

4 Proteins: Determination of Primary Structure 24
Peter J. Kennelly, PhD, & Victor W. Rodwell, PhD

5 Proteins: Higher Orders of Structure 34
Peter J. Kennelly, PhD, & Victor W. Rodwell, PhD

SECTION II
Enzymes: Kinetics, Mechanism, Regulation, & Role of Transition Metals 49

6 Proteins: Myoglobin & Hemoglobin 49
Peter J. Kennelly, PhD, & Victor W. Rodwell, PhD

7 Enzymes: Mechanism of Action 59
Peter J. Kennelly, PhD, & Victor W. Rodwell, PhD

8 Enzymes: Kinetics 71
Victor W. Rodwell, PhD

9 Enzymes: Regulation of Activities 85
Peter J. Kennelly, PhD, & Victor W. Rodwell, PhD

10 The Biochemical Roles of Transition Metals 96
Peter J. Kennelly, PhD

SECTION III
Bioenergetics 109

11 Bioenergetics: The Role of ATP 109
Kathleen M. Botham, PhD, DSc, & Peter A. Mayes, PhD, DSc

12 Biologic Oxidation 115
Kathleen M. Botham, PhD, DSc, & Peter A. Mayes, PhD, DSc

13 The Respiratory Chain & Oxidative Phosphorylation 121
Kathleen M. Botham, PhD, DSc, & Peter A. Mayes, PhD, DSc

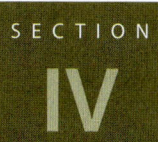

SECTION IV
Metabolism of Carbohydrates 133

14 Overview of Metabolism & the Provision of Metabolic Fuels 133
Owen P. McGuinness, PhD

15 Saccharides (ie, Carbohydrates) of Physiological Significance 147
Owen P. McGuinness, PhD

16 The Citric Acid Cycle: A Pathway Central to Carbohydrate, Lipid, & Amino Acid Metabolism 156
Owen P. McGuinness, PhD

17 Glycolysis & the Oxidation of Pyruvate 163
Owen P. McGuinness, PhD

18 Metabolism of Glycogen 171
Owen P. McGuinness, PhD

19 Gluconeogenesis & the Control of
Blood Glucose 180
Owen P. McGuinness, PhD

20 The Pentose Phosphate Pathway & Other
Pathways of Hexose Metabolism 191
Owen P. McGuinness, PhD

SECTION V Metabolism of Lipids 205

21 Lipids of Physiologic Significance 205
Kathleen M. Botham, PhD, DSc, & Peter A. Mayes, PhD, DSc

22 Oxidation of Fatty Acids: Ketogenesis 217
Kathleen M. Botham, PhD, DSc, & Peter A. Mayes, PhD, DSc

23 Biosynthesis of Fatty Acids &
Eicosanoids 226
Kathleen M. Botham, PhD, DSc, & Peter A. Mayes, PhD, DSc

24 Metabolism of Acylglycerols &
Sphingolipids 239
Kathleen M. Botham, PhD, DSc, & Peter A. Mayes, PhD, DSc

25 Lipid Transport & Storage 247
Kathleen M. Botham, PhD, DSc, & Peter A. Mayes, PhD, DSc

26 Cholesterol Synthesis, Transport, &
Excretion 259
Kathleen M. Botham, PhD, DSc, & Peter A. Mayes, PhD, DSc

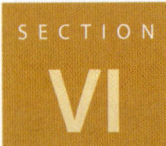

SECTION VI Metabolism of Proteins & Amino Acids 273

27 Biosynthesis of the Nutritionally
Nonessential Amino Acids 273
Victor W. Rodwell, PhD

28 Catabolism of Proteins & of Amino Acid
Nitrogen 279
Victor W. Rodwell, PhD

29 Catabolism of the Carbon Skeletons of
Amino Acids 290
Victor W. Rodwell, PhD

30 Conversion of Amino Acids to Specialized
Products 306
Victor W. Rodwell, PhD

31 Porphyrins & Bile Pigments 315
Victor W. Rodwell, PhD

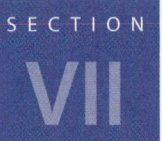

SECTION VII Structure, Function, & Replication of Informational Macromolecules 329

32 Nucleotides 329
Victor W. Rodwell, PhD

33 Metabolism of Purine & Pyrimidine
Nucleotides 337
Victor W. Rodwell, PhD

34 Nucleic Acid Structure & Function 348
P. Anthony Weil, PhD

35 DNA Organization, Replication, &
Repair 360
P. Anthony Weil, PhD

36 RNA Synthesis, Processing, &
Modification 384
P. Anthony Weil, PhD

37 Protein Synthesis & the Genetic Code 404
P. Anthony Weil, PhD

38 Regulation of Gene Expression 420
P. Anthony Weil, PhD

39 Molecular Genetics, Recombinant DNA, &
Genomic Technology 444
P. Anthony Weil, PhD

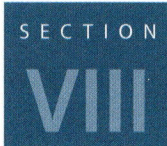

SECTION VIII

Biochemistry of Extracellular & Intracellular Communication 467

40 Membranes: Structure & Function 467
P. Anthony Weil, PhD

41 The Diversity of the Endocrine System 488
P. Anthony Weil, PhD

42 Hormone Action & Signal Transduction 508
P. Anthony Weil, PhD

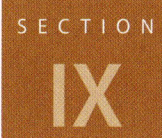

SECTION IX

Special Topics (A) 527

43 Nutrition, Digestion, & Absorption 527
David A. Bender, PhD

44 Micronutrients: Vitamins & Minerals 535
David A. Bender, PhD

45 Free Radicals & Antioxidant Nutrients 549
David A. Bender, PhD

46 Glycoproteins 554
David A. Bender, PhD

47 Metabolism of Xenobiotics 564
David A. Bender, PhD & Robert K. Murray, MD, PhD

48 Clinical Biochemistry 568
David A. Bender, PhD

SECTION X

Special Topics (B) 581

49 Intracellular Traffic & Sorting of Proteins 581
Kathleen M. Botham, PhD, DSc, & Robert K. Murray, MD, PhD

50 The Extracellular Matrix 599
Kathleen M. Botham, PhD, DSc, & Robert K. Murray, MD, PhD

51 Muscle & the Cytoskeleton 618
January D. Haile, PhD, & Peter J. Kennelly, PhD

52 Plasma Proteins & Immunoglobulins 634
Peter J. Kennelly, PhD

53 Red Blood Cells 653
Peter J. Kennelly, PhD

54 White Blood Cells 664
Peter J. Kennelly, PhD

SECTION XI

Special Topics (C) 677

55 Hemostasis & Thrombosis 677
*Peter L. Gross, MD, MSc, FRCP(C),
P. Anthony Weil, PhD, & Margaret L. Rand, PhD*

56 Cancer: An Overview 689
*Molly Jacob, MD, PhD, MNASc, FRCPath,
Joe Varghese, MD, PhD, & P. Anthony Weil, PhD*

57 The Biochemistry of Aging 717
Peter J. Kennelly, PhD

58 Biochemical Case Histories 729
David A. Bender, PhD

The Answer Bank 741
Index 745

Preface

The authors and publishers are pleased to present the thirty-second edition of *Harper's Illustrated Biochemistry*. The first edition, entitled *Harper's Biochemistry*, was published in 1939 under the sole authorship of Dr Harold Harper at the University of California School of Medicine, San Francisco, California. Presently entitled *Harper's Illustrated Biochemistry*, the book continues, as originally intended, to provide a concise survey of aspects of biochemistry most relevant to the study of medicine. Various authors have contributed to subsequent editions of this medically oriented biochemistry text, which is now observing its 83rd year.

Cover Illustration for the Thirty-Second Edition

The global COVID-19 pandemic has provided a dramatic, face-to-face demonstration of both the power and limitations of molecular medicine and epidemiology. The rapid development of highly effective vaccines was made possible by the adaptation of novel RNA-based approaches in which the patient's immune response is activated via the endogenous expression of genetically-encoded antigens, rather than the physical injection of a non-infectious antigen. Utilizing the patient's own cells as the bioreactor for generating antigens, rather than some animal or culture, enabled scientists to use the self-amplifying capacity of polynucleotides to accelerate both the speed of vaccine development and subsequent large-scale manufacture. The illustration on the cover of the thirty-second edition depicts a neutralizing antibody, in blue, bound to the spike protein on the surface of the SARS-CoV-2 coronavirus, better known as COVID-19, which is shown in red. The epitope to which the antibody binds overlaps that at which the virus binds to the ACE-2 receptor, the membrane protein by which the pathogen recognizes, binds to, and subsequently invades human cells. Therapeutic antibodies thus protect by physically blocking association of the Spike protein with the ACE-2 receptor.

Changes in the Thirty-Second Edition

As always, *Harper's Illustrated Biochemistry* continues to emphasize the close relationship of biochemistry to the understanding of diseases, their pathology, and the practice of medicine. With the retirement of long-time contributor David A. Bender, Prof. Owen P. McGuinness of Vanderbilt University has joined as a new coauthor. In addition to the fresh perspectives and novel insights provided by Prof. McGuinness, the contents of most chapters have been updated and provide the reader with the most current and pertinent information.

For example, in Chapter 6 the description of the Bohr effect's contributions to CO_2 transport and release from the lungs has been reorganized and expanded, while Chapter 9 has been updated and reorganized to include expanded coverage of zymogen activation in enzyme regulation.

Organization of the Book

All 58 chapters of the thirty-second edition place major emphasis on the medical relevance of biochemistry. Topics are organized under 11 major headings. In order to assist study and to facilitate retention of the contained information, Questions follow each Section. An Answer Bank follows Chapter 58.

Section I includes a brief history of biochemistry and emphasizes the interrelationships between biochemistry and medicine. Water and the importance of homeostasis of intracellular pH are reviewed, and the various orders of proteins structure are addressed.

Section II begins with a chapter on hemoglobin. The next four chapters address the mechanism of action, kinetics, metabolic regulation of enzymes, and the role of metal ions in multiple aspects of intermediary metabolism.

Section III addresses bioenergetics and the role of high-energy phosphates in energy capture and transfer, the oxidation–reduction reactions involved in biologic oxidation, and metabolic details of energy capture via the respiratory chain and oxidative phosphorylation.

Section IV considers the metabolism of carbohydrates via glycolysis, the citric acid cycle, the pentose phosphate pathway, glycogen metabolism, gluconeogenesis, and the control of blood glucose.

Section V outlines the nature of simple and complex lipids, lipid transport and storage, the biosynthesis and degradation of fatty acids and more complex lipids, and the reactions and metabolic regulation of cholesterol biosynthesis and transport in human subjects.

Section VI discusses protein catabolism, urea biosynthesis, and the catabolism of amino acids, and stresses the medically significant metabolic disorders associated with their incomplete catabolism. The final chapter in this section considers the biochemistry of the porphyrins and bile pigments.

Section VII first outlines the structure and function of nucleotides and nucleic acids, and then details DNA replication and repair, RNA synthesis and modification, protein synthesis, the principles of recombinant DNA technology, and the regulation of gene expression.

Section VIII considers aspects of extracellular and intracellular communication. Specific topics include membrane structure and function, the molecular bases of the actions of hormones, and signal transduction.

Sections IX, X, and XI address many topics of significant medical importance.

Section IX discusses nutrition, digestion, and absorption, micronutrients including, vitamins, free radicals and antioxidants, glycoproteins, the metabolism of xenobiotics, and clinical biochemistry.

Section X addresses intracellular traffic and the sorting of proteins, the extracellular matrix, muscle and the cytoskeleton, plasma proteins and immunoglobulins, and the biochemistry of red cells and of white cells.

Section XI includes hemostasis and thrombosis, an overview of cancer, the biochemistry of aging, and a selection of case histories.

Acknowledgments

The authors thank Michael Weitz for his role in the planning of this edition and Peter Boyle for overseeing its preparation for publication. We also thank Tasneem Kauser and her colleagues at KnowledgeWorks Global Ltd. for their efforts in managing editing, typesetting, and artwork. We gratefully acknowledge numerous suggestions and corrections received from students and colleagues from around the world.

Peter J. Kennelly
Kathleen M. Botham
Owen P. McGuinness
Victor W. Rodwell
P. Anthony Weil

C H A P T E R

1

Biochemistry & Medicine

Victor W. Rodwell, PhD

OBJECTIVES

After studying this chapter, you should be able to:

- Understand the importance of the ability of cell-free extracts of yeast to ferment sugars, an observation that enabled discovery of the intermediates of fermentation, glycolysis, and other metabolic pathways.
- Appreciate the scope of biochemistry and its central role in the life sciences, and that biochemistry and medicine are intimately related disciplines.
- Appreciate that biochemistry integrates knowledge of the chemical processes in living cells with strategies to maintain health, understand disease, identify potential therapies, and enhance our understanding of the origins of life on earth.
- Describe how genetic approaches have been critical for elucidating many areas of biochemistry, and how the Human Genome Project has furthered advances in numerous aspects of biology and medicine.

BIOMEDICAL IMPORTANCE

Biochemistry and medicine enjoy a mutually cooperative relationship. Biochemical studies have illuminated many aspects of health and disease, and the study of various aspects of health and disease has opened up new areas of biochemistry. The medical relevance of biochemistry both in normal and abnormal situations is emphasized throughout this book. Biochemistry makes significant contributions to the fields of cell biology, physiology, immunology, microbiology, pharmacology, toxicology, and epidemiology, as well as the fields of inflammation, cell injury, and cancer. These close relationships emphasize that life, as we know it, depends on biochemical reactions and processes.

DISCOVERY THAT A CELL-FREE EXTRACT OF YEAST CAN FERMENT SUGAR

Although the ability of yeast to "ferment" various sugars to ethyl alcohol has been known for millennia, only comparatively recently did this process initiate the science of biochemistry. The great French microbiologist Louis Pasteur maintained that fermentation could only occur in intact cells. However, in 1899, the brothers Büchner discovered that fermentation could occur in the *absence* of intact cells when they stored a yeast extract in a crock of concentrated sugar solution, added as a preservative. Overnight, the contents of the crock fermented, spilled over the laboratory bench and floor, and dramatically demonstrated that fermentation can proceed in the absence of an intact cell. This discovery unleashed an avalanche of research that initiated the science of biochemistry. Investigations revealed the vital roles of inorganic phosphate, ADP, ATP, and NAD(H), and ultimately identified the phosphorylated sugars and the chemical reactions and enzymes that convert glucose to pyruvate (glycolysis) or to ethanol and CO_2 (fermentation). Research beginning in the 1930s identified the intermediates of the citric acid cycle and of urea biosynthesis, and revealed the essential roles of certain vitamin-derived cofactors or "coenzymes" such as thiamin pyrophosphate, riboflavin, and ultimately coenzyme A, coenzyme Q, and cobamide coenzyme. The 1950s revealed how complex carbohydrates are synthesized from, and broken down into simple sugars, and the pathways for biosynthesis of pentoses, and the catabolism of amino acids and fatty acids.

Investigators employed animal models, perfused intact organs, tissue slices, cell homogenates and their subfractions, and subsequently purified enzymes. Advances were enhanced by the development of analytical ultracentrifugation, paper and other forms of chromatography, and the post-World War II availability of radioisotopes, principally ^{14}C, ^{3}H, and ^{32}P, as "tracers" to identify the intermediates in complex pathways such as that of cholesterol biosynthesis. X-ray crystallography was then used to solve the three-dimensional structures of numerous proteins, polynucleotides, enzymes, and viruses. Genetic advances that followed the realization that DNA was a double helix include the polymerase chain reaction, and transgenic animals or those with gene knockouts. The methods used to prepare, analyze, purify, and identify metabolites and the activities of natural and recombinant enzymes and their three-dimensional structures are discussed in the following chapters.

BIOCHEMISTRY & MEDICINE HAVE PROVIDED MUTUAL ADVANCES

The two major concerns for workers in the health sciences—and particularly physicians—are the understanding and maintenance of health and effective treatment of disease. Biochemistry impacts both of these fundamental concerns, and

the interrelationship of biochemistry and medicine is a wide, two-way street. Biochemical studies have illuminated many aspects of health and disease, and conversely, the study of various aspects of health and disease has opened up new areas of biochemistry (**Figure 1–1**). An early example of how investigation of protein structure and function revealed the single difference in amino acid sequence between normal hemoglobin and sickle cell hemoglobin. Subsequent analysis of numerous variant sickle cell and other hemoglobins has contributed significantly to our understanding of the structure and function both of hemoglobin and of other proteins. During the early 1900s, the English physician Archibald Garrod studied patients with the relatively rare disorders of alkaptonuria, albinism, cystinuria, and pentosuria, and established that these conditions were genetically determined. Garrod designated these conditions as **inborn errors of metabolism**. His insights provided a foundation for the development of the field of human biochemical genetics. A more recent example was investigation of the genetic and molecular basis of familial hypercholesterolemia, a disease that results in early-onset atherosclerosis. In addition to clarifying different genetic mutations responsible for this disease, this provided a deeper understanding of cell receptors and mechanisms of uptake, not only of cholesterol but also of how other molecules cross cell membranes. Studies of **oncogenes** and **tumor suppressor genes** in cancer cells have directed attention to the molecular mechanisms involved in the control of normal cell growth. These examples illustrate how the study of disease can open up areas of basic biochemical research. Science provides physicians and other workers in health care and biology with a foundation that impacts practice, stimulates curiosity, and promotes the adoption of scientific approaches for continued learning.

BIOCHEMICAL PROCESSES UNDERLIE HUMAN HEALTH

Biochemical Research Impacts Nutrition & Preventive Medicine

The World Health Organization (WHO) defines health as a state of "complete physical, mental, and social well-being and not merely the absence of disease and infirmity." From a biochemical

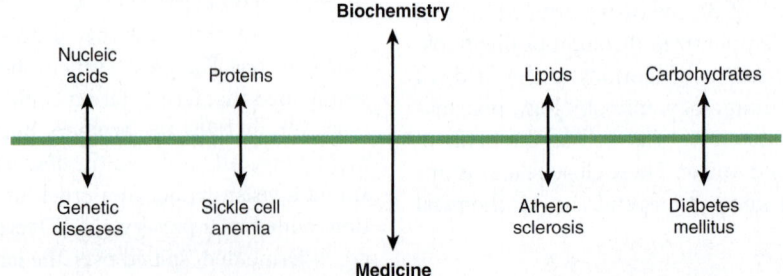

FIGURE 1–1 **A two-way street connects biochemistry and medicine.** Knowledge of the biochemical topics listed above the green line of the diagram has clarified our understanding of the diseases shown below the green line. Conversely, analyses of the diseases have cast light on many areas of biochemistry. Note that sickle cell anemia is a genetic disease, and that both atherosclerosis and diabetes mellitus have genetic components.

viewpoint, health may be considered that situation in which all of the many thousands of intra- and extracellular reactions that occur in the body are proceeding at rates commensurate with the organism's survival under pressure from both internal and external challenges. The maintenance of health requires optimal dietary intake of **vitamins**, certain **amino acids** and **fatty acids**, various **minerals**, and **water**. Understanding nutrition depends to a great extent on knowledge of biochemistry, and the sciences of biochemistry and nutrition share a focus on these chemicals. Recent increasing emphasis on systematic attempts to maintain health and forestall disease, or **preventive medicine**, includes nutritional approaches to the prevention of diseases such as atherosclerosis and cancer.

Most Diseases Have a Biochemical Basis

Apart from infectious organisms and environmental pollutants, many diseases are manifestations of abnormalities in genes, proteins, chemical reactions, or biochemical processes, each of which can adversely affect one or more critical biochemical functions. Examples of disturbances in human biochemistry responsible for diseases or other debilitating conditions include electrolyte imbalance, defective nutrient ingestion or absorption, hormonal imbalances, toxic chemicals or biologic agents, and DNA-based genetic disorders. To address these challenges, biochemical research continues to be interwoven with studies in disciplines such as genetics, cell biology, immunology, nutrition, pathology, and pharmacology. In addition, many biochemists are vitally interested in contributing to solutions to key issues such as the ultimate survival of mankind, and educating the public to support use of the scientific method in solving environmental and other major problems that confront our civilization.

Impact of the Human Genome Project on Biochemistry, Biology, & Medicine

Initially unanticipated rapid progress in the late 1990s in sequencing the human genome led in the mid-2000s to the announcement that over 90% of the genome had been sequenced. This effort was headed by the International Human Genome Sequencing Consortium and by Celera Genomics. Except for a few gaps, the sequence of the entire human genome was completed in 2003, just 50 years after the description of the double-helical nature of DNA by Watson and Crick. The implications for biochemistry, medicine, and indeed for all of biology, are virtually unlimited. For example, the ability to isolate and sequence a gene and to investigate its structure and function by sequencing and "gene knockout" experiments have revealed previously unknown genes and their products, and new insights have been gained concerning human evolution and procedures for identifying disease-related genes.

Major advances in biochemistry and understanding human health and disease continue to be made by mutation of the genomes of model organisms such as yeast, the fruit fly *Drosophila melanogaster*, the roundworm *Caenorhabditis elegans*, and the zebra fish; all organisms that can be genetically manipulated to provide insight into the functions of individual genes. These advances can potentially provide clues to curing human diseases such as cancer and Alzheimer disease. **Figure 1–2** highlights areas that have developed or accelerated as a direct result of progress made in the Human Genome Project (HGP). New "**-omics**" fields focus on comprehensive study of the structures and functions of the molecules with which each is concerned. The products of genes (RNA molecules and proteins) are being studied using the techniques of **transcriptomics** and **proteomics**. A spectacular example of the speed of progress in transcriptomics is the explosion of knowledge about small RNA molecules as regulators of gene activity. Other -omics fields include **glycomics**, **lipidomics**, **metabolomics**, **nutrigenomics**, and **pharmacogenomics**. To keep pace with the information generated, **bioinformatics** has received much attention. Other related fields to which the impetus from the HGP has carried over are **biotechnology**, **bioengineering**, **biophysics**, and **bioethics**.

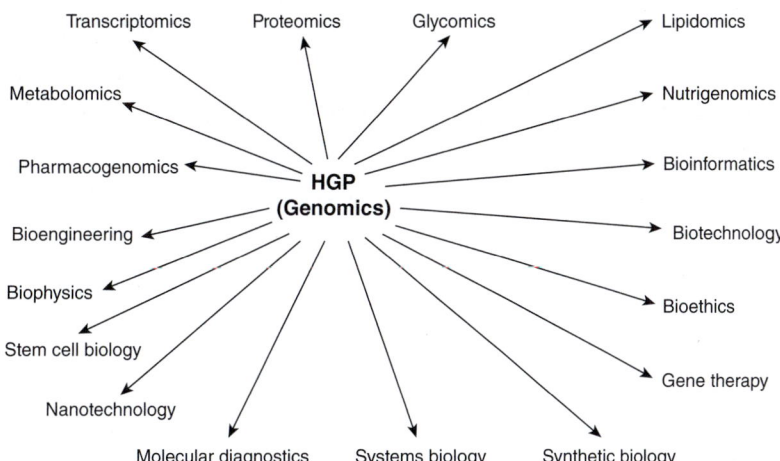

FIGURE 1–2 **The Human Genome Project (HGP) has influenced many disciplines and areas of research.** Biochemistry is not listed since it predates commencement of the HGP, but disciplines such as bioinformatics, genomics, glycomics, lipidomics, metabolomics, molecular diagnostics, proteomics, and transcriptomics are nevertheless active areas of biochemical research.

Definitions of these -omics fields and other terms appear in the Glossary of this chapter. **Nanotechnology** is an active area, which, for example, may provide novel methods of diagnosis and treatment for cancer and other disorders. **Stem cell biology** is at the center of much current research. **Gene therapy** has yet to deliver the promise that it appears to offer, but it seems probable that ultimately will occur. Many new **molecular diagnostic tests** have developed in areas such as genetic, microbiologic, and immunologic testing and diagnosis. **Systems biology** is also burgeoning. The outcomes of research in the various areas mentioned above will impact tremendously the future of biology, medicine, and the health sciences. **Synthetic biology** offers the potential for creating living organisms, initially small bacteria, from genetic material in vitro that might carry out specific tasks such as cleansing petroleum spills. All of the above make the 21st century an exhilarating time to be directly involved in biology and medicine.

SUMMARY

- Biochemistry is the science concerned with the molecules present in living organisms, individual chemical reactions and their enzyme catalysts, and the expression and regulation of each metabolic process. Biochemistry has become the basic language of all biologic sciences.

- Despite the focus on human biochemistry in this text, biochemistry concerns the entire spectrum of life forms, from viruses, bacteria, and plants to complex eukaryotes such as human beings.

- Biochemistry, medicine, and other health care disciplines are intimately related. Health in all species depends on a harmonious balance of the biochemical reactions occurring in the body, while disease reflects abnormalities in biomolecules, biochemical reactions, or biochemical processes.

- Advances in biochemical knowledge have illuminated many areas of medicine, and the study of diseases has often revealed previously unsuspected aspects of biochemistry.

- Biochemical approaches are often fundamental in illuminating the causes of diseases and in designing appropriate therapy. Biochemical laboratory tests also represent an integral component of diagnosis and monitoring of treatment.

- A sound knowledge of biochemistry and of other related basic disciplines is essential for the rational practice of medicine and related health sciences.

- Results of the HGP and of research in related areas will have a profound influence on the future of biology, medicine, and other health sciences.

- Genomic research on model organisms such as yeast, the fruit fly *D. melanogaster*, the roundworm *C. elegans*, and the zebra fish provides insight into understanding human diseases.

GLOSSARY

Bioengineering: The application of engineering to biology and medicine.

Bioethics: The area of ethics that is concerned with the application of moral and ethical principles to biology and medicine.

Bioinformatics: The discipline concerned with the collection, storage, and analysis of biologic data, for example, DNA, RNA, and protein sequences.

Biophysics: The application of physics and its techniques to biology and medicine.

Biotechnology: The field in which biochemical, engineering, and other approaches are combined to develop biologic products of use in medicine and industry.

Gene Therapy: Applies to the use of genetically engineered genes to treat various diseases.

Genomics: The genome is the complete set of genes of an organism, and genomics is the in-depth study of the structures and functions of genomes.

Glycomics: The glycome is the total complement of simple and complex carbohydrates in an organism. Glycomics is the systematic study of the structures and functions of glycomes such as the human glycome.

Lipidomics: The lipidome is the complete complement of lipids found in an organism. Lipidomics is the in-depth study of the structures and functions of all members of the lipidome and their interactions, in both health and disease.

Metabolomics: The metabolome is the complete complement of metabolites (small molecules involved in metabolism) present in an organism. Metabolomics is the in-depth study of their structures, functions, and changes in various metabolic states.

Molecular Diagnostics: Refers to the use of molecular approaches such as DNA probes to assist in the diagnosis of various biochemical, genetic, immunologic, microbiologic, and other medical conditions.

Nanotechnology: The development and application to medicine and to other areas of devices such as nanoshells, which are only a few nanometers in size (10^{-9} m = 1 nm).

Nutrigenomics: The systematic study of the effects of nutrients on genetic expression and of the effects of genetic variations on the metabolism of nutrients.

Pharmacogenomics: The use of genomic information and technologies to optimize the discovery and development of new drugs and drug targets.

Proteomics: The proteome is the complete complement of proteins of an organism. Proteomics is the systematic study of the structures and functions of proteomes and their variations in health and disease.

Stem Cell Biology: Stem cells are undifferentiated cells that have the potential to self-renew and to differentiate into any of the adult cells of an organism. Stem cell biology concerns the biology of stem cells and their potential for treating various diseases.

Synthetic Biology: The field that combines biomolecular techniques with engineering approaches to build new biologic functions and systems.

Systems Biology: The field concerns complex biologic systems studied as integrated entities.

Transcriptomics: The comprehensive study of the transcriptome, the complete set of RNA transcripts produced by the genome during a fixed period of time.

APPENDIX

Shown are selected examples of databases that assemble, annotate, and analyze data of biomedical importance.

ENCODE: ENCyclopedia Of DNA Elements. A collaborative effort that combines laboratory and computational approaches to identify every functional element in the human genome.

GenBank: Protein sequence database of the National Institutes of Health (NIH) stores all known biologic nucleotide sequences and their translations in a searchable form.

HapMap: Haplotype **Map**, an international effort to identify single nucleotide polymorphisms (SNPs) associated with common human diseases and differential responses to pharmaceuticals.

ISDB: International Sequence DataBase that incorporates DNA databases of Japan and of the European Molecular Biology Laboratory (EMBL).

PDB: Protein **D**ata**B**ase. Three-dimensional structures of proteins, polynucleotides, and other macromolecules, including proteins bound to substrates, inhibitors, or other proteins.

Water & pH

Peter J. Kennelly, PhD, & Victor W. Rodwell, PhD

O B J E C T I V E S

After studying this chapter, you should be able to:

- Describe the properties of water that account for its surface tension, viscosity, liquid state at ambient temperature, and solvent power.
- Represent the structures of organic compounds that can serve as hydrogen bond donors or acceptors.
- Explain the role played by entropy in the association and orientation, in an aqueous environment, of hydrophobic and amphipathic molecules.
- Indicate the quantitative contributions of salt bridges, hydrophobic interactions, and van der Waals forces to stabilizing the 3-D conformation of macromolecules.
- Explain the relationship of pH to acidity, alkalinity, and the quantitative determinants that characterize weak and strong acids.
- Calculate the shift in pH that accompanies the addition of a given quantity of acid or base to a buffered solution.
- Describe what buffers do, how they do it, and the conditions under which a buffer is most effective under physiologic or other conditions.
- Use the Henderson-Hasselbalch equation to calculate the net charge on a polyelectrolyte at a given pH.

BIOMEDICAL IMPORTANCE

Water is the predominant chemical component of living organisms. Its unique physical properties, which include the ability to solvate a wide range of organic and inorganic molecules, derive from water's dipolar structure and exceptional capacity for forming hydrogen bonds. The manner in which water interacts with a solvated biomolecule influences the structure both of the biomolecule and of water itself. An excellent nucleophile, water is a reactant or product in many metabolic reactions. Regulation of water balance depends on hypothalamic mechanisms that control thirst, on antidiuretic hormone (ADH), on retention or excretion of water by the kidneys, and on evaporative loss. Nephrogenic diabetes insipidus, which involves the inability to concentrate urine or adjust to subtle changes in extracellular fluid osmolarity, results from the unresponsiveness of renal tubular osmoreceptors to ADH.

Water has a slight propensity to dissociate into hydroxide ions and protons. The concentration of protons, or **acidity**, of aqueous solutions is generally reported using the logarithmic pH scale. Bicarbonate and other buffers normally maintain the pH of extracellular fluid between 7.35 and 7.45. Suspected disturbances of acid-base balance are verified by measuring the pH of arterial blood and the CO_2 content of venous blood. Causes of acidosis (blood pH <7.35) include diabetic ketosis and lactic acidosis. Alkalosis (pH >7.45) may follow vomiting of acidic gastric contents.

WATER IS AN IDEAL BIOLOGIC SOLVENT

Water Molecules Form Dipoles

A water molecule is an irregular, slightly skewed tetrahedron with oxygen at its center (**Figure 2–1**). The corners are occupied by the two hydrogens and the unshared electrons of the remaining two sp^3-hybridized orbitals of oxygen. The 105° angle between the two hydrogen atoms differs slightly from the ideal tetrahedral angle, 109.5°. The strongly electronegative oxygen atom in a water molecule attracts electrons away from the hydrogen nuclei, leaving them with a partial

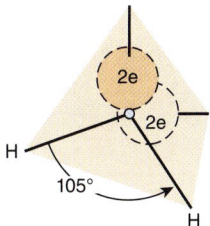

FIGURE 2–1 The water molecule has tetrahedral geometry.

positive charge, while its two unshared electron pairs constitute a region of local negative charge. This asymmetric charge distribution is referred to as a **dipole**.

Water's strong dipole is responsible for its high **dielectric constant**. As described quantitatively by Coulomb's law, the strength of interaction F between oppositely charged particles is inversely proportionate to the dielectric constant ε of the surrounding medium. The dielectric constant for a vacuum is essentially unity; for hexane it is 1.9; for ethanol, 24.3; and for water at 25°C, 78.5. When dissolved in water, the force of attraction between charged and polar species is greatly decreased relative to solvents with lower dielectric constants. Its strong dipole and high dielectric constant enable water to dissolve large quantities of charged compounds such as salts.

Water Molecules Form Hydrogen Bonds

A partially unshielded hydrogen nucleus covalently bound to an electron-withdrawing oxygen or nitrogen atom can interact with an unshared electron pair on another oxygen or nitrogen atom to form a **hydrogen bond**. Since water molecules contain both of these features, hydrogen bonding favors the self-association of water molecules into ordered arrays (**Figure 2–2**). On average, each molecule in liquid water associates through hydrogen bonds with 3.5 others. These bonds are both relatively weak and transient, with a half-life of a few picoseconds. Rupture of a hydrogen bond in liquid water requires only about 4.5 kcal/mol, less than 5% of the energy required to rupture a covalent O—H bond. The exceptional capacity of this relatively small, 18 g/mol, molecule to form hydrogen bonds profoundly influences the physical properties of water and accounts for its high viscosity, surface tension, and boiling point.

Hydrogen bonding enables water to dissolve many organic biomolecules that contain functional groups which

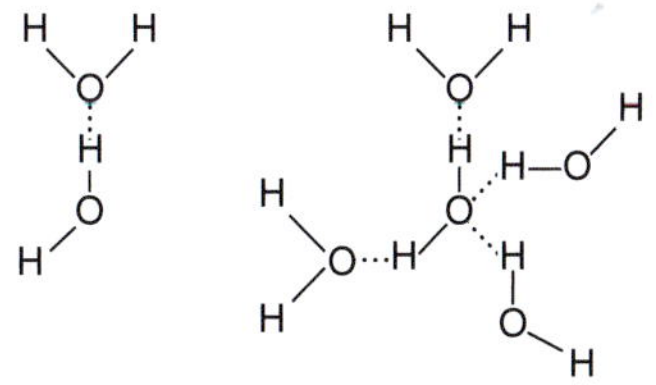

FIGURE 2–2 Water molecules self-associate via hydrogen bonds. Shown are the association of two water molecules (**left**) and a hydrogen-bonded cluster of four water molecules (**right**). Notice that water can serve simultaneously both as a hydrogen donor and as a hydrogen acceptor.

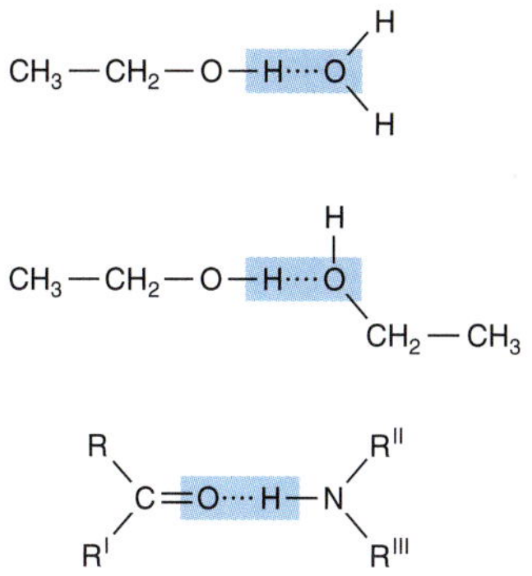

FIGURE 2–3 Additional polar groups participate in hydrogen bonding. Shown are hydrogen bonds formed between alcohol and water, between two molecules of ethanol, and between the peptide carbonyl oxygen and the peptide nitrogen hydrogen of an adjacent amino acid.

can participate in hydrogen bonding. The oxygen atoms of aldehydes, ketones, and amides, for example, provide lone pairs of electrons that can serve as hydrogen acceptors. Alcohols, carboxylic acids, and amines can serve both as hydrogen acceptors and as donors of unshielded hydrogen atoms for formation of hydrogen bonds (**Figure 2–3**).

INTERACTION WITH WATER INFLUENCES THE STRUCTURE OF BIOMOLECULES

Covalent & Noncovalent Bonds Stabilize Biologic Molecules

The covalent bond is the strongest force that holds molecules together (**Table 2–1**). Noncovalent forces, while of lesser magnitude, predominate in stabilizing the folding of the polypeptides and other macromolecules into the complex three-dimensional conformations essential to their functional competence (see Chapter 5) as well as the association of biomolecules into multicomponent complexes. Examples of the latter include the coalescence of the polypeptide subunits that form the hemoglobin tetramer (see Chapter 6); the association

TABLE 2–1 Bond Energies for Atoms of Biologic Significance

Bond Type	Energy (kcal/mol)	Bond Type	Energy (kcal/mol)
O—O	34	O=O	96
S—S	51	C—H	99
C—N	70	C=S	108
S—H	81	O—H	110
C—C	82	C=C	147
C—O	84	C=N	147
N—H	94	C=O	164

of the two polynucleotide strands that comprise a DNA double helix (see Chapter 34); and the coalescence of billions of phospholipid, glycosphingolipid, cholesterol, and other molecules into the bilayer that constitutes the foundation of the plasma membrane of an animal cell (see Chapter 40). These forces, which can be either attractive or repulsive, involve interactions both within the biomolecule and, most importantly, between it and the water that forms the principal component of the surrounding environment.

In Water, Biomolecules Fold to Position Hydrophobic Groups Within Their Interior

Most biomolecules are **amphipathic**; that is, they possess regions rich in charged or polar functional groups as well as regions with hydrophobic character. Proteins tend to fold with the R-groups of amino acids with hydrophobic side chains in the interior. Amino acids with charged or polar amino acid side chains (eg, arginine, glutamate, serine; see Table 3–1) generally are present on the surface in contact with water. A similar pattern prevails in a phospholipid bilayer where the charged "head groups" of phosphatidylserine or phosphatidylethanolamine contact water while their hydrophobic fatty acyl side chains cluster together, excluding water (see Figure 40–5). This pattern minimizes energetically unfavorable contacts between water and hydrophobic groups. It also maximizes the opportunities for the formation of energetically favorable charge-dipole, dipole-dipole, and hydrogen bonding interactions between polar groups on the biomolecule and water.

Hydrophobic Interactions

Hydrophobic interaction refers to the tendency of nonpolar compounds to self-associate in an aqueous environment. This self-association is driven neither by mutual attraction nor by what are sometimes incorrectly referred to as "hydrophobic bonds." Self-association minimizes the disruption of energetically favorable interactions between and is therefore driven by the surrounding water molecules.

While the hydrogen atoms of nonpolar groups such as the methylene groups of hydrocarbons do not form hydrogen bonds, they do affect the structure of the water with which they are in contact. Water molecules adjacent to a hydrophobic group are restricted in the number of orientations (degrees of freedom) that permit them to participate in the maximum number of energetically favorable hydrogen bonds. Maximal formation of multiple hydrogen bonds, which maximizes enthalpy, can be maintained only by increasing the order of the adjacent water molecules, with an accompanying decrease in entropy.

It follows from the second law of thermodynamics that the optimal free energy of a hydrocarbon-water mixture is a function of both maximal enthalpy (from hydrogen bonding) and highest entropy (maximum degrees of freedom). Thus, nonpolar molecules tend to form droplets that minimize exposed

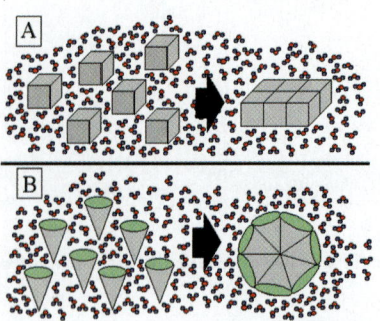

FIGURE 2–4 **Hydrophobic interactions are driven by the surrounding water molecules.** Water molecules are represented by one red (oxygen) and two blue (hydrogen) circles. The hydrophobic surfaces of solute molecules are colored gray and, where present, hydrophilic ones are colored green. **A.** When the six hydrophobic cubes shown are dispersed in water (**left**), the surrounding water molecules (red oxygens and blue hydrogens) are forced to engage in entropically unfavorable interactions with all 36 faces of the cubes. However, when the six hydrophobic cubes aggregate together (**right**), the number of exposed faces is reduced to 22. The aggregate forms and its stability is maintained, not by some attractive force, but because aggregation reduces the number of water molecules that are unfavorably affected by nearly 40%. **B.** Amphipathic molecules associate together for the same reason. However, the structure of the resulting complex (eg, micelle or bilayer) is determined by the geometries of the hydrophobic (gray) and hydrophilic (green) regions.

surface area and reduce the number of water molecules whose motional freedom becomes restricted (**Figure 2–4**). Similarly, in the aqueous environment of the living cell the hydrophobic portions of amphipathic biopolymers tend to be buried inside the structure of the molecule, or within a lipid bilayer, minimizing contact with water.

Electrostatic Interactions

Electrostatic interactions between oppositely charged groups within or between biomolecules are termed **salt bridges**. Salt bridges are comparable in strength to hydrogen bonds but act over larger distances. They therefore often facilitate the binding of charged molecules and ions to proteins and nucleic acids.

van der Waals Forces

van der Waals forces arise from attractions between transient dipoles generated by the rapid movement of electrons in all neutral atoms. Significantly weaker than hydrogen bonds but potentially extremely numerous, van der Waals forces decrease as the sixth power of the distance separating atoms (**Figure 2–5**). Thus, they act over very short distances, typically 2 to 4 Å.

Multiple Forces Stabilize Biomolecules

The DNA double helix illustrates the contribution of multiple forces to the structure of biomolecules. While each individual DNA strand is held together by covalent bonds, the two strands of the helix are held together exclusively by noncovalent

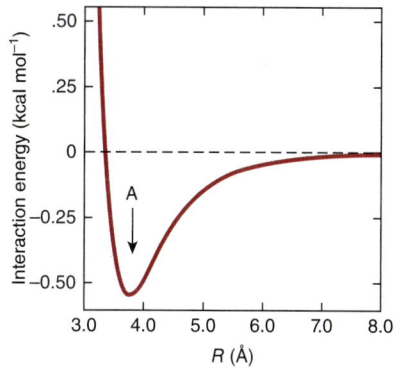

FIGURE 2–5 **The strength of van der Waals interactions varies with the distance, *R*, between interacting species.** The force of interaction between interacting species increases with decreasing distance between them until they are separated by the van der Waals contact distance (see arrow marked A). Repulsion due to interaction between the electron clouds of each atom or molecule then supervenes. While individual van der Waals interactions are extremely weak, their cumulative effect is nevertheless substantial for macromolecules such as DNA and proteins which have many atoms in close contact.

interactions such as hydrogen bonds between nucleotide bases (Watson-Crick base pairing) and van der Waals interactions between the stacked purine and pyrimidine bases. The double helix presents the charged phosphate groups and polar hydroxyl groups from the ribose sugars of the DNA backbone to water while burying the relatively hydrophobic nucleotide bases inside. The extended backbone maximizes the distance between negatively charged phosphates, minimizing unfavorable electrostatic interactions (see Figure 34–2).

WATER IS AN EXCELLENT NUCLEOPHILE

Metabolic reactions often involve the attack by lone pairs of electrons residing on electron-rich molecules termed **nucleophiles** upon electron-poor atoms called **electrophiles**. Nucleophiles and electrophiles do not necessarily possess a formal negative or positive charge. Water, whose two lone pairs of sp^3 electrons bear a partial negative charge (see Figure 2–1), is an excellent nucleophile. Other nucleophiles of biologic importance include the oxygen atoms of phosphates, alcohols, and carboxylic acids; the sulfur of thiols; and the nitrogen atoms of amines and of the imidazole ring of histidine. Common electrophiles include the carbonyl carbons in amides, esters, aldehydes, and ketones and the phosphorus atoms of phosphoesters.

Nucleophilic attack by water typically results in the cleavage of the amide, glycoside, or ester bonds that hold biopolymers together. This process is termed **hydrolysis**. Conversely, when monomer units such as amino acids or monosaccharides are joined or **condensed** together to form biopolymers, such as proteins or starch, water is a product.

Hydrolysis typically is a thermodynamically favored reaction. Yet, the amide and phosphoester bonds of polypeptides

and oligonucleotides are stable in the aqueous environment of the cell. This seemingly paradoxical behavior reflects the fact that the thermodynamics that govern the equilibrium point of a reaction do not determine the *rate* at which it will proceed toward its equilibrium point. In the cell, macromolecular catalysts called **enzymes** accelerate the rate of hydrolytic and other chemical reactions when needed. **Proteases** catalyze the hydrolysis of proteins into their component amino acids, while **nucleases** catalyze the hydrolysis of the phosphoester bonds in DNA and RNA. Precise and differential control of enzyme activity, including the sequestration of enzymes in specific organelles, enables cells to determine the physiologic circumstances under which a given biopolymer will be synthesized or degraded.

Many Metabolic Reactions Involve Group Transfer

Many of the enzymic reactions responsible for synthesis and breakdown of biomolecules involve the transfer of a chemical group G from a donor D to an acceptor A to form an acceptor group complex, A—G:

$$D-G+A \rightleftarrows A-G+D$$

The hydrolysis and phosphorolysis of glycogen, for example, involve the transfer of glucosyl groups to water or to orthophosphate. Since the equilibrium constants for these hydrolysis reactions strongly favor the formation of split products, it follows that many of the group transfer reactions responsible for the biosynthesis of macromolecules are, in and of themselves, thermodynamically unfavored. Enzyme catalysts play a critical role in surmounting these barriers by virtue of their capacity to directly link two normally separate reactions together. For example, by linking an energetically unfavorable group transfer reaction to a thermodynamically favorable one such as the hydrolysis of ATP, a new enzyme-catalyzed reaction can be generated. The free energy change of this coupled reaction will be the sum of the individual values for the two that were linked, one whose net *overall* change in free energy favors the formation of the covalent bonds required for biopolymer synthesis.

Water Molecules Exhibit a Slight but Important Tendency to Dissociate

The ability of water to ionize, while slight, is of central importance for life. Since water can act both as an acid and as a base, its ionization may be represented as an intermolecular proton transfer that forms a hydronium ion (H_3O^+) and a hydroxide ion (OH^-):

$$H_2O + H_2O \rightleftarrows H_3O + OH^-$$

The transferred proton is actually associated with a cluster of water molecules. Protons exist in solution not only as H_3O^+ but also as multimers such as $H_5O_2^+$ and $H_7O_3^+$. The proton is nevertheless routinely represented as H^+, even though it is in fact highly hydrated.

Since hydronium and hydroxide ions continuously recombine to form water molecules, an *individual* hydrogen or oxygen cannot be stated to be present as an ion or as part of a water molecule. At one instant it is an ion; an instant later it is part of a water molecule. Individual ions or molecules are therefore not considered. We refer instead to the *probability* that at any instant in time, a given hydrogen will be present as an ion or as part of a water molecule. Since 1 g of water contains 3.35×10^{22} molecules, the ionization of water can be described statistically. To state that the probability that a hydrogen exists as an ion is 0.01 means that at any given moment in time, a hydrogen atom has 1 chance in 100 of being an ion and 99 chances out of 100 of being part of a water molecule. The actual probability of a hydrogen atom in pure water existing as a hydrogen ion is approximately 1.8×10^{-9}. The probability of its being part of a water molecule thus is almost unity. Stated another way, for every hydrogen ion or hydroxide ion in pure water, there are 0.56 billion or 0.56×10^9 water molecules. Hydrogen ions and hydroxide ions nevertheless contribute significantly to the properties of water.

For dissociation of water,

$$K = \frac{[H^+][OH^-]}{[H_2O]}$$

where the brackets represent molar concentrations (strictly speaking, molar activities) and K is the **dissociation constant**. Since 1 mole (mol) of water weighs 18 g, 1 liter (L) (1000 g) of water contains $1000 \div 18 = 55.56$ mol. Pure water thus is 55.56 molar. Since the probability that a hydrogen in pure water will exist as a hydrogen ion is 1.8×10^{-9}, the molar concentration of H^+ ions (or of OH^- ions) in pure water is the product of the probability, 1.8×10^{-9}, times the molar concentration of water, 55.56 mol/L. The result is 1.0×10^{-7} mol/L.

We can now calculate the dissociation constant K for pure water:

$$K = \frac{[H^+][OH^-]}{[H_2O]} = \frac{[10^{-7}][10^{-7}]}{[55.56]}$$

$$= 0.018 \times 10^{-14} = 1.8 \times 10^{-16} \, \text{mol/L}$$

The molar concentration of water, 55.56 mol/L, is too great to be significantly affected by dissociation. It is therefore considered to be essentially constant. The concentration of pure water may therefore be incorporated into the dissociation constant K to provide a useful new constant K_w termed the **ion product** for water:

$$K = \frac{[H^+][OH^-]}{[H_2O]} = 1.8 \times 10^{-16} \, \text{mol/L}$$

$$K_w = (K)[H_2O] = [H^+][OH^-]$$

$$= (1.8 \times 10^{-16} \, \text{mol/L})(55.56 \, \text{mol/L})$$

$$= 1.00 \times 10^{-14} \, (\text{mol/L})^2$$

Note that the dimensions of K are moles per liter and those of K_w are moles^2 per liter^2. As its name suggests, the ion product

K_w is numerically equal to the product of the molar concentrations of H^+ and OH^-:

$$K_w = [H^+][OH^-]$$

At 25°C, $K_w = (10^{-7})^2$, or $10^{-14} \, (\text{mol/L})^2$. At temperatures below 25°C, K_w is somewhat less than 10^{-14}, and at temperatures above 25°C it is somewhat greater than 10^{-14}. Within the stated limitations of temperature, K_w equals $10^{-14} \, (\text{mol/L})^2$ for all aqueous solutions, even solutions containing acids or bases. We can therefore use K_w to calculate the pH of any aqueous solution.

pH IS THE NEGATIVE LOG OF THE HYDROGEN ION CONCENTRATION

The term **pH** was introduced in 1909 by Sörensen, who defined it as the negative log of the hydrogen ion concentration:

$$pH = -\log[H^+]$$

This definition, while not rigorous, suffices for most biochemical purposes. To calculate the pH of a solution:

1. Calculate the hydrogen ion concentration $[H^+]$.

2. Calculate the base 10 logarithm of $[H^+]$.

3. pH is the negative of the value found in step 2.

For example, for pure water at 25°C,

$$pH = -\log[H^+] = -\log 10^{-7} = -(-7) = 7.0$$

This value is also known as the *power* (English), *puissant* (French), or *potennz* (German) of the exponent, hence the use of the term "p."

Low pH values correspond to high concentrations of H^+ and high pH values correspond to low concentrations of H^+.

Acids are **proton donors** and bases are **proton acceptors**. **Strong acids** (eg, HCl, H_2SO_4) completely dissociate into anions and protons even in strongly acidic solutions (low pH). **Weak acids** dissociate only partially in acidic solutions. Similarly, **strong bases** (eg, KOH, NaOH), but not **weak bases** like $Ca(OH)_2$, are completely dissociated even at high pH. Many biochemicals are weak acids. Exceptions include phosphorylated intermediates, whose phosphoryl group contains two dissociable protons, the first of which is strongly acidic.

The following examples illustrate how to calculate the pH of acidic and basic solutions.

Example 1: What is the pH of a solution whose hydrogen ion concentration is 3.2×10^{-4} mol/L?

$$pH = -\log[H^+]$$

$$= -\log(3.2 \times 10^{-4})$$

$$= -\log(3.2) - \log(10^{-4})$$

$$= -0.5 + 4.0$$

$$= 3.5$$

Example 2: What is the pH of a solution whose hydroxide ion concentration is 4.0×10^{-4} mol/L? We first define a quantity **pOH** that is equal to $-\log[OH^-]$ and that may be derived from the definition of K_w:

$$K_w = [H^+][OH^-] = 10^{-14}$$

Therefore,

$$\log[H^+] + \log[OH^-] = \log 10^{-14}$$

or

$$pH + pOH = 14$$

To solve the problem by this approach:

$$[OH^-] = 4.0 \times 10^{-4}$$

$$pOH = -\log[OH^-]$$
$$= -\log(4.0 \times 10^{-4})$$
$$= -\log(4.0) - \log(10^{-4})$$
$$= -0.60 + 4.0$$
$$= 3.4$$

Now

$$pH = 14 - pOH = 14 - 3.4$$
$$= 10.6$$

Examples 1 and 2 illustrate how the logarithmic pH scale facilitates recording and comparing hydrogen ion concentrations that differ by orders of magnitude from one another, 0.00032 M (pH 3.5) and 0.000000000025 M (pH 10.6).

Example 3: What are the pH values of (a) 2.0×10^{-2} mol/L KOH and of (b) 2.0×10^{-6} mol/L KOH? The OH^- arises from two sources, KOH and water. Since pH is determined by the total $[H^+]$ (and pOH by the total $[OH^-]$), both sources must be considered. In the first case (a), the contribution of water to the total $[OH^-]$ is negligible. The same cannot be said for the second case (b):

	Concentration (mol/L)	
	(a)	**(b)**
Molarity of KOH	2.0×10^{-2}	2.0×10^{-6}
$[OH^-]$ from KOH	2.0×10^{-2}	2.0×10^{-6}
$[OH^-]$ from water	1.0×10^{-7}	1.0×10^{-7}
Total $[OH^-]$	2.00001×10^{-2}	2.1×10^{-6}

Once a decision has been reached about the significance of the contribution of water, pH may be calculated as shown in Example 3.

The above examples assume that the strong base KOH is completely dissociated in solution and that the concentration of OH^- ions was thus equal to that due to the KOH plus that present initially in the water. This assumption is valid for dilute solutions of strong bases or acids, but not for weak bases or acids. Since weak electrolytes dissociate only slightly in solution, we must use the **dissociation constant** to calculate the concentration of $[H^+]$ (or $[OH^-]$) produced by a given molarity of a weak acid (or base) before calculating total $[H^+]$ (or total $[OH^-]$) and subsequently pH.

Functional Groups That Are Weak Acids Have Great Physiologic Significance

Many biomolecules contain functional groups that are weak acids or bases. Carboxyl groups, amino groups, and phosphate esters, whose second dissociation falls within the physiologic range, are present in proteins and nucleic acids, most coenzymes, and most intermediary metabolites. Knowledge of the dissociation of weak acids and bases thus is basic to understanding the influence of intracellular pH on structure and biologic activity. Charge-based separations such as electrophoresis and ion exchange chromatography are also best understood in terms of the dissociation behavior of functional groups.

When discussing weak acids, we often refer to the protonated species (HA or R—SH) as the **acid** and the unprotonated species (A^- or R—S$^-$) as its **conjugate base**. Similarly, we may refer to the deprotonated form as the **base** (A^- or R—COO$^-$) and the protonated form as its **conjugate acid** (HA or R—COOH).

We express the relative strengths of weak acids in terms of the dissociation constants of the protonated form. Following are the expressions for the dissociation constant (K_a) for a representative weak acid, R—COOH, as well as the conjugate acid, R—NH$_3^+$, of the weak base R—NH$_2$.

$$R—COOH \rightleftarrows R—COO^- + H^+$$

$$K_a = \frac{[R—COO^-][H^+]}{[R—COOH]}$$

$$R—NH_3^+ \rightleftarrows R—NH_2 + H^+$$

$$K_a = \frac{[R—NH_2][H^+]}{[R—NH_3^+]}$$

Since the numeric values of K_a for weak acids are negative exponential numbers, we express K_a as pK_a, where

$$pK_a = -\log K_a$$

Note that pK_a is related to K_a as pH is to $[H^+]$. The stronger the acid, the lower is its pK_a value.

Representative weak acids (left), their conjugate bases (center), and pK_a values (right) include the following:

R—CH$_2$—COOH	R—CH$_2$COO$^-$	$pK_a = 4-5$
R—CH$_2$—NH$_3^+$	R—CH$_2$—NH$_2$	$pK_a = 9-10$
H$_2$CO$_3$	HCO$_3^-$	$pK_a = 6.4$
H$_2$PO$_4^-$	HPO$_4^{-2}$	$pK_a = 7.2$

pK_a is used to express the relative strengths of both weak acids and weak bases using a single, unified scale. Under this convention, **the relative strengths of bases are expressed in terms of the pK_a of their conjugate acids**. For polyprotic compounds containing more than one dissociable proton, a numerical subscript is assigned to each dissociation, numbered starting from unity in decreasing order of relative acidity. For a dissociation of the type

$$R—NH_3^+ \rightarrow R—NH_2 + H^+$$

the pK_a is the pH at which the concentration of the acid R—NH$_3^+$ equals that of the base R—NH$_2$.

From the above equations that relate K_a to [H$^+$] and to the concentrations of undissociated acid and its conjugate base, when

$$[R—COO^-] = [R—COOH]$$

or when

$$[R—NH_2] = [R—NH_3^+]$$

then

$$K_a = [H^+]$$

Thus, when the associated (protonated) and dissociated (conjugate base) species are present at equal concentrations, the prevailing hydrogen ion concentration [H$^+$] is numerically equal to the dissociation constant, K_a. If the logarithms of both sides of the above equation are taken and both sides are multiplied by −1, the expressions would be as follows:

$$K_a = [H^+]$$

$$-\log K_a = -\log[H^+]$$

Since $-\log K_a$ is defined as pK_a, and $-\log$ [H$^+$] defines pH, the equation may be rewritten as

$$pK_a = pH$$

that is, **the pK_a of an acid group is the pH at which the protonated and unprotonated species are present at equal concentrations**. The pK_a for an acid may be determined by adding 0.5 equivalent of alkali per equivalent of acid. The resulting pH will equal the pK_a of the acid.

The Henderson-Hasselbalch Equation Describes the Behavior of Weak Acids & Buffers

The Henderson-Hasselbalch equation is derived below.
A weak acid, HA, ionizes as follows:

$$HA \rightleftharpoons H^+ + A^-$$

The equilibrium constant for this dissociation is

$$K_a = \frac{[H^+][A^-]}{[HA]}$$

Cross-multiplication gives

$$[H^+][A^-] = K_a[HA]$$

Divide both sides by [A$^-$]:

$$[H^+] = K_a \frac{[HA]}{[A^-]}$$

Take the log of both sides:

$$\log[H^+] = \log\left(K_a \frac{[HA]}{[A^-]}\right)$$

$$= \log K_a + \log\frac{[HA]}{[A^-]}$$

Multiply through by −1:

$$-\log[H^+] = -\log K_a - \log\frac{[HA]}{[A^-]}$$

Substitute pH and pK_a for $-\log$ [H$^+$] and $-\log K_a$, respectively; then

$$pH = pK_a - \log\frac{[HA]}{[A^-]}$$

Inversion of the last term removes the minus sign and gives the **Henderson-Hasselbalch equation**

$$\mathbf{pH = pK_a + \log\frac{[A^-]}{[HA]}}$$

The Henderson-Hasselbalch equation has great predictive value in protonic equilibria. For example,

1. When an acid is exactly half-neutralized, [A$^-$] = [HA]. Under these conditions,

$$pH = pK_a + \log\frac{[A^-]}{[HA]} = pK_a + \log\left(\frac{1}{1}\right) = pK_a + 0$$

Therefore, at half-neutralization, pH = pK_a.

2. When the ratio [A$^-$]/[HA] = 100:1,

$$pH = pK_a + \log\frac{[A^-]}{[HA]}$$

$$pH = pK_a + \log(100/1) = pK_a + 2$$

3. When the ratio [A$^-$]/[HA] = 1:10,

$$pH = pK_a + \log(1/10) = pK_a + (-1)$$

If the equation is evaluated at ratios of [A$^-$]/[HA] ranging from 10^3 to 10^{-3} and the calculated pH values are plotted, the resulting graph describes the titration curve for a weak acid (**Figure 2–6**).

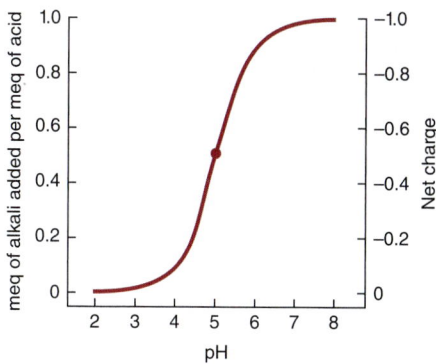

FIGURE 2–6 **Titration curve for an acid of the type HA.** The heavy dot in the center of the curve indicates the pK_a, 5.0.

Weak Acids Can Be Used to Establish & Maintain the pH of an Aqueous Solution

Solutions of weak acids or bases and their conjugates exhibit **buffering**, the ability to resist a change in pH following addition of strong acid or base. Many metabolic reactions are accompanied by the release or uptake of protons. Oxidative metabolism produces CO_2, the anhydride of carbonic acid, which if not buffered would produce severe acidosis. Biologic maintenance of a constant pH involves buffering by phosphate, bicarbonate, and proteins, which accept or release protons to resist a change in pH. For laboratory experiments using tissue extracts or enzymes, constant pH is maintained by the addition of buffers such as MES ([2-N-morpholino]-ethanesulfonic acid, pK_a 6.1), inorganic orthophosphate (pK_{a2} 7.2), HEPES (N-hydroxyethylpiperazine-N′-2-ethanesulfonic acid, pK_a 6.8), or Tris (tris[hydroxymethyl]aminomethane, pK_a 8.3). The value of pK_a relative to the desired pH is the major determinant of which buffer is selected.

Buffering can be observed by using a pH meter while titrating a weak acid or base (see Figure 2–6). We can also calculate the pH shift that accompanies addition of acid or base to a buffered solution. In the following example, the buffered solution (a weak acid, pK_a = 5.0, and its conjugate base) is initially at one of four pH values. We will calculate the pH shift that results when 0.1 meq of KOH is added to 1 meq of each solution:

Initial pH	5.00	5.37	5.60	5.86
$[A^-]_{initial}$	0.50	0.70	0.80	0.88
$[HA]_{initial}$	0.50	0.30	0.20	0.12
$([A^-]/[HA])_{initial}$	1.00	2.33	4.00	7.33
Addition of 0.1 meq of KOH Produces				
$[A^-]_{final}$	0.60	0.80	0.90	0.98
$[HA]_{final}$	0.40	0.20	0.10	0.02
$([A^-]/[HA])_{final}$	1.50	4.00	9.00	49.0
$\log ([A^-]/[HA])_{final}$	0.18	0.60	0.95	1.69
Final pH	5.18	5.60	5.95	6.69
ΔpH	**0.18**	**0.60**	**0.95**	**1.69**

Notice that ΔpH, the change in pH per milliequivalent of OH^- added, depends on the initial pH, with highest resistance to change at pH values close to the weak acid's pK_a. **Indeed, such weak acid-conjugate base combinations, called buffers, resist change most effectively when the desired pH falls within, ± 1.0 unit or less of their pK_a.**

Figure 2–6 also illustrates how the net charge on one molecule of a weak acid varies with pH. A fractional charge of −0.5 does not mean that an individual molecule bears a fractional charge but that the *probability* is 0.5 that a given molecule has a unit negative charge at any given moment in time. Consideration of the net charge on macromolecules as a function of pH provides the basis for separatory techniques such as ion exchange chromatography and electrophoresis (see Chapter 4).

The Propensity of a Proton to Dissociate Depends on Molecular Structure

Many acids of biologic interest possess more than one dissociating group. The presence of local negative charge hinders proton release from nearby acidic groups, raising their pK_a. This is illustrated by the pK_a values of the three dissociating groups of phosphoric acid and citric acid (**Table 2–2**). The effect of adjacent charge decreases with distance. The second pK_a for succinic acid, which has two methylene groups between its carboxyl groups, is 5.6, whereas the second pK_a for glutaric acid, which has one additional methylene group, is 5.4.

pK_a Values Depend on the Properties of the Medium

The pK_a of a functional group is also profoundly influenced by the surrounding medium. The medium may either raise or lower the pK_a relative to its value in water, depending on whether the undissociated acid or its conjugate base is the charged species. The effect of dielectric constant on pK_a may be observed by adding ethanol to water. The pK_a of a carboxylic acid *increases,* whereas that of an amine *decreases* on addition of ethanol because ethanol decreases the ability of water to solvate a charged species. The pK_a values of dissociating

TABLE 2–2 **Relative Strengths of Monoprotic, Diprotic, and Triprotic Acids**

Lactic acid	pK = 3.86		
Acetic acid	pK = 4.76		
Ammonium ion	pK = 9.25		
Carbonic acid	pK_1 = 6.37;	pK_2 = 10.25	
Succinic acid	pK_1 = 4.21;	pK_2 = 5.64	
Glutaric acid	pK_1 = 4.34;	pK_2 = 5.41	
Phosphoric acid	pK_1 = 2.15;	pK_2 = 6.82;	pK_3 = 12.38
Citric acid	pK_1 = 3.08;	pK_2 = 4.74;	pK_3 = 5.40

Note: Tabulated values are the pK_a values (−log of the dissociation constant).

groups in the interiors of proteins thus are profoundly affected by their local environment, including the presence or absence of water.

SUMMARY

- Water forms hydrogen-bonded clusters with itself and with other proton donors or acceptors. These extensive networks of hydrogen bonds account for the surface tension, viscosity, liquid state at room temperature, and solvent power of water.

- Compounds that contain O or N can serve as hydrogen bond donors and/or acceptors.

- Entropic forces dictate that amphipathic macromolecules bury nonpolar regions away from water.

- Salt bridges, hydrophobic interactions, and van der Waals forces participate in the formation of biomolecular complexes and maintenance of molecular conformation.

- pH is the negative log of $[H^+]$. A low pH characterizes an acidic solution, and a high pH denotes a basic solution.

- The strength of weak acids is expressed by pK_a, the negative log of the acid dissociation constant. Strong acids have low pK_a values and weak acids have high pK_a values.

- Buffers resist a change in pH when protons are produced or consumed. Maximum buffering capacity occurs within 1 pH unit on either side of pK_a. Physiologic buffers include bicarbonate, orthophosphate, and proteins.

REFERENCES

Reese KM: Whence came the symbol pH. Chem & Eng News 2004;82:64.

Segel IM: *Biochemical Calculations*. Wiley, 1968.

Skinner JL: Following the motions of water molecules in aqueous solutions. Science 2010;328:985.

Stillinger FH: Water revisited. Science 1980;209:451.

Suresh SJ, Naik VM: Hydrogen bond thermodynamic properties of water from dielectric constant data. J Chem Phys 2000;113:9727.

Wiggins PM: Role of water in some biological processes. Microbiol Rev 1990;54:432.

Amino Acids & Peptides

Peter J. Kennelly, PhD, & Victor W. Rodwell, PhD

O B J E C T I V E S

After studying this chapter, you should be able to:

- Diagram the structures and write the three- and one-letter designations for each of the amino acids present in proteins.
- Provide examples of how each type of R group of the protein amino acids contributes to their chemical properties.
- List additional important functions of amino acids and explain how certain amino acids in plant seeds can severely impact human health.
- Name the ionizable groups of the protein amino acids and list their approximate pK_a values as free amino acids in aqueous solution.
- Calculate the pH of an unbuffered aqueous solution of a polyfunctional amino acid and the change in pH that occurs following the addition of a given quantity of strong acid or base.
- Define pI and explain its relationship to the net charge on a polyfunctional electrolyte.
- Explain how pK_a and pI can be used to predict the mobility at a given pH of a polyelectrolyte, such as an amino acid, in a direct-current electrical field.
- Describe the directionality, nomenclature, and primary structure of peptides.
- Describe the conformational consequences of the partial double-bond character of the peptide bond.

BIOMEDICAL IMPORTANCE

L-α-Amino acids provide the monomer units of the long polypeptide chains of proteins. In addition, these amino acids and their derivatives participate in such diverse cellular functions as nerve transmission and the biosynthesis of porphyrins, purines, pyrimidines, and urea. The neuroendocrine system employs short polymers of amino acids called *peptides* as hormones, hormone-releasing factors, neuromodulators, and neurotransmitters. Humans and other higher animals cannot synthesize 10 of the L-α-amino acids present in proteins in amounts adequate to support infant growth or to maintain adult health. Consequently, the human diet must contain adequate quantities of these *nutritionally essential* amino acids. Each day the kidneys filter over 50 g of free amino acids from the arterial renal blood. However, only traces of free amino

acids normally appear in the urine because amino acids are almost totally reabsorbed in the proximal tubule, conserving them for protein synthesis and other vital functions.

Certain microorganisms secrete free D-amino acids, or peptides that may contain both D- and L-α-amino acids. Several of these bacterial peptides are of therapeutic value, including the antibiotics bacitracin and gramicidin A, and the antitumor agent bleomycin. Certain other microbial peptides are, however, toxic. The cyanobacterial peptides microcystin and nodularin are lethal in large doses, while small quantities promote the formation of hepatic tumors. The ingestion of certain amino acids present in the seeds of legumes of the genus *Lathyrus* can result in lathyrism, a tragic irreversible disease in which individuals lose control of their limbs. Certain other plant seed amino acids have also been implicated in a neurodegenerative disease afflicting natives of Guam.

PROPERTIES OF AMINO ACIDS

The Genetic Code Specifies 20 ʟ-α-Amino Acids

Although more than 300 amino acids occur in nature, proteins are synthesized almost exclusively from the set of 20 ʟ-α-amino acids encoded by nucleotide triplets called **codons** (see Table 37–1). While the three-letter genetic code could potentially accommodate more than 20 amino acids, the genetic code is *redundant* since several amino acids are specified by multiple codons. Scientists frequently represent the sequences of peptides and proteins using one- and three-letter abbreviations for each amino acid (**Table 3–1**). The R groups of amino acids may be either hydrophilic or hydrophobic (**Table 3–2**); properties that affect their location in a protein's mature folded conformation (see Chapter 5). Some proteins contain additional amino acids that arise by the **posttranslational** modification of an amino acid already present in a peptide. Examples include the conversion of peptidyl proline and peptidyl lysine to 4-hydroxyproline and 5-hydroxylysine; the conversion of peptidyl glutamate to γ-carboxyglutamate; and the methylation, formylation, acetylation, prenylation, and phosphorylation of certain aminoacyl residues. These modifications significantly extend the functional diversity of proteins by altering their solubility, stability, catalytic activity, and interaction with other proteins.

TABLE 3–1 ʟ-α-Amino Acids Present in Proteins

Name	Symbol	Structural Formula	pK_1 α-COOH	pK_2 α-NH$_2^+$	pK_3 R Group
With Aliphatic Side Chains					
Glycine	Gly[G]	H—CH—COO⁻, NH$_3^+$	2.4	9.8	
Alanine	Ala[A]	CH$_3$—CH—COO⁻, NH$_3^+$	2.4	9.9	
Valine	Val[V]	H$_3$C, CH—CH—COO⁻, H$_3$C NH$_3^+$	2.2	9.7	
Leucine	Leu[L]	H$_3$C, CH—CH$_2$—CH—COO⁻, H$_3$C NH$_3^+$	2.3	9.7	
Isoleucine	Ile[T]	CH$_3$, CH$_2$, CH—CH—COO⁻, CH$_3$ NH$_3^+$	2.3	9.8	
With Side Chains Containing Hydroxylic(OH) Groups					
Serine	Ser[S]	CH$_2$—CH—COO⁻, OH NH$_3^+$	2.2	9.2	about 13
Threonine	Thr[T]	CH$_3$—CH—CH—COO⁻, OH NH$_3^+$ See below.	2.1	9.1	about 13
Tyrosine	Tyr[Y]	See below.			
With Side Chains Containing Sulfur Atoms					
Cysteine	Cys[C]	CH$_2$—CH—COO⁻, SH NH$_3^+$	1.9	10.8	8.3

(Continued)

TABLE 3–1 L-α-Amino Acids Present in Proteins (Continued)

Name	Symbol	Structural Formula	pK₁	pK₂	pK₃
With Side Chains Containing Sulfur Atoms					
Methionine	Met[M]	$CH_2-CH_2-CH-COO^-$ $S-CH_3 \quad NH_3^+$	2.1	9.3	
With Side Chains Containing Acidic Groups or Their Amides					
Aspartic Acid	Asp[D]	$^-OOC-CH_2-CH-COO^-$ NH_3^+	2.1	9.9	3.9
Asparagine	Asn[N]	$H_2N-C-CH_2-CH-COO^-$ $O \qquad NH_3^+$	2.1	8.8	
Glutamic Acide	Glu[E]	$^-OOC-CH_2-CH_2-CH-COO^-$ NH_3^+	2.1	9.5	4.1
Glutamine	Gin[Q]	$H_2N-C-CH_2-CH_2-CH-COO^-$ $O \qquad\qquad NH_3^+$	2.2	9.1	
With Side Chains Containing Basic Groups					
Argine	Arg[R]	$H-N-CH_2-CH_2-CH_2-CH-COO^-$ $C=NH_2^+ \qquad\qquad NH_3^+$ NH_2	1.8	9.0	12.5
Lysine	Lys[K]	$CH_2-CH_2-CH_2-CH_2-CH-COO^-$ $NH_3^+ \qquad\qquad NH_3^+$	2.2	9.2	10.8
Histidine	His[H]	$CH_2-CH-COO^-$ $HN\diagdown N \qquad NH_3^+$	1.8	9.3	6.0
Containing Aromatic Rings					
Histidine	His[H]	See above			
Phenylalanine	Phe[F]	See above. $\bigcirc-CH_2-CH-COO^-$ NH_3^+	2.2	9.2	
Tyrosine	Try[Y]	$HO-\bigcirc-CH_2-CH-COO^-$ NH_3^+	2.2	9.1	10.1
Tryptophan	Trp[W]	$CH_2-CH-COO^-$ NH_3^+ N H	2.4	9.4	
Imino Acid					
Proline	Pro[P]	$\overset{+}{N}H_2\diagup COO^-$	2.0	10.6	

TABLE 3–2 Hydrophilic & Hydrophobic Amino Acids

Hydrophilic	Hydrophobic
Arginine	Alanine
Asparagine	Isoleucine
Aspartic acid	Leucine
Cysteine	Methionine
Glutamic acid	Phenylalanine
Glutamine	Proline
Glycine	Tryptophan
Histidine	Tyrosine
Lysine	Valine
Serine	
Threonine	

The distinction is based on the tendency of their R groups to associate with, or to minimize contact with, an aqueous environment.

Selenocysteine, the 21st Protein L-α-Amino Acid

Selenocysteine (**Figure 3–1**) is an L-α-amino acid present in proteins from every domain of life. Humans contain approximately two dozen selenoproteins that include certain peroxidases and reductases, selenoprotein P, which circulates in the plasma, and the iodothyronine deiodinases responsible for converting the prohormone thyroxine (T_4) to the thyroid hormone 3,3′, 5-triiodothyronine (T_3) (see Chapter 41). Since peptidyl selenocysteine is inserted directly into a growing polypeptide during *translation,* it is commonly termed the "21st amino acid." However, unlike the other 20 protein amino acids, incorporation of selenocysteine is specified by a large and complex genetic element for the unusual tRNA called tRNA^Sec which utilizes the UGA anticodon that normally signals STOP rather than a simple triplet codon. Rather, the protein synthetic apparatus recognizes a UGA codon as coding for selenocysteine when it is accompanied by a stem-loop structure, the selenocysteine insertion element, in the untranslated region of the mRNA (see Chapter 27).

Stereochemistry of the Protein Amino Acids

With the sole exception of glycine, the α-carbon of every amino acid is chiral. Although some protein amino acids are dextrorotatory and some levorotatory, all share the absolute

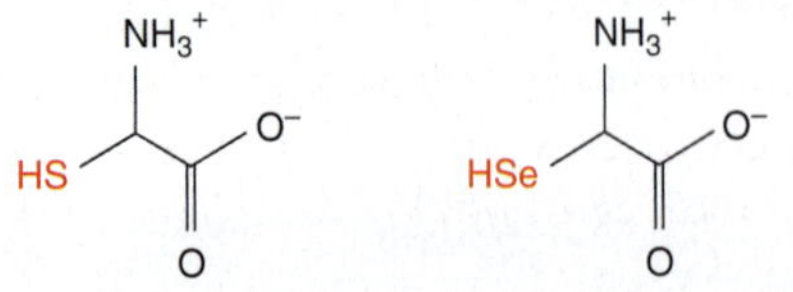

FIGURE 3–1 Cysteine (left) and selenocysteine (right). pK_3, for the selenyl proton of selenocysteine is 5.2. Since this is 3 pH units lower than that of cysteine, selenocysteine represents a better nucleophile at or below pH 7.4.

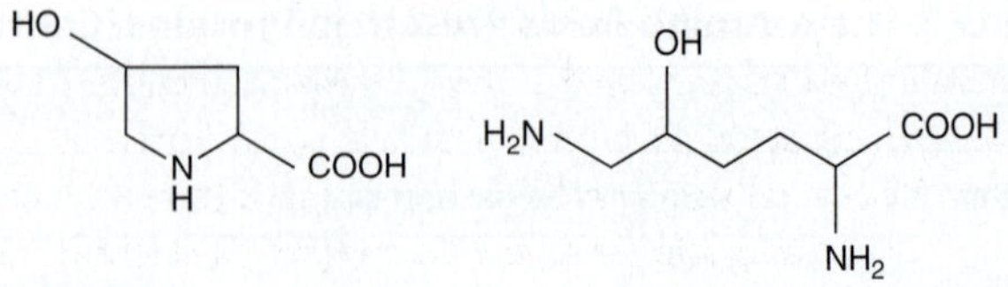

FIGURE 3–2 4-Hydroxyproline and 5-hydroxylysine.

configuration of L-glyceraldehyde and thus are defined as L-α-amino acids. Even though almost all protein amino acids are (S), the failure to use (R) or (S) to express *absolute* stereochemistry is no mere historical aberration. L-Cysteine is (R) since the atomic mass of the sulfur atom on C3 exceeds that of the amino group on C2. More significantly, in mammals the biochemical reactions of L-α-amino acids, their precursors, and their catabolites are catalyzed by enzymes that act exclusively on L-isomers, irrespective of their absolute configuration.

Posttranslational Modifications Confer Additional Properties

While some prokaryotes incorporate pyrrolysine into proteins, and plants can incorporate azetidine-2-carboxylic acid, an analog of proline, a set of just 21 L-α-amino acids clearly suffices for the formation of most proteins. Posttranslational modifications can, however, generate novel R groups that impart further properties. In collagen, protein-bound proline and lysine residues are converted to 4-hydroxyproline and 5-hydroxylysine (**Figure 3–2**). The carboxylation of glutamyl residues of proteins of the blood coagulation cascade to γ-carboxyglutamyl residues (**Figure 3–3**) completes a site for chelating the calcium ion essential for blood coagulation. The amino acid side chains of histones are subject to numerous modifications, including acetylation and methylation of lysine and methylation and deimination of arginine (see Chapters 35 and 38). It is also now possible in the laboratory to genetically introduce many different unnatural amino acids into proteins, generating proteins via recombinant gene expression with new or enhanced properties and providing a new way to explore protein structure–function relationships.

Extraterrestrial Amino Acids Have Been Detected in Meteorites

The existence of extraterrestrial amino acids was first reported in 1969 following analysis of the famous Murchison meteorite from southeastern Australia. The presence of amino acids in other meteorites, including some pristine examples from Antarctica, has now been amply confirmed. Unlike the amino acids synthesized by terrestrial organisms, these meteorites

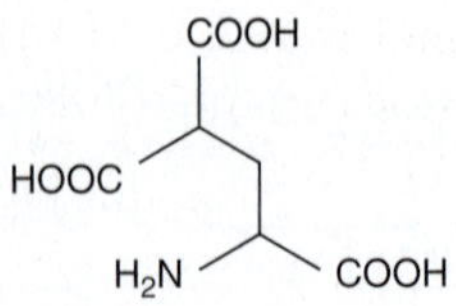

FIGURE 3–3 γ-Carboxyglutamic acid.

contain racemic mixtures of D- and L-isomers of multiple protein amino acids as well as biologically important nonprotein α-amino acids such as N-methylglycine (sarcosine) and β-alanine. Several novel amino acids that lack terrestrial counterparts of biotic origin were also discovered. Nucleobases, activated phosphates, and molecules related to sugars have also been detected in meteorites. These findings offer potential insights into the prebiotic chemistry of earth, and impact the search for extraterrestrial life. Some speculate that meteorites may have contributed to the origin of life on our planet, a conjecture entitled Panspermia, by delivering extraterrestrially generated organic molecules or even intact microorganisms to our earth.

L-α-Amino Acids Serve Additional Metabolic Roles

L-α-Amino acids fulfill vital metabolic roles in addition to serving as the "building blocks" of proteins. For example, ornithine and citrulline are key intermediates in the urea cycle (see Figure 28–16), while S-adenosyl-methionine serves as a methyl-group donor for many enzyme-catalyzed reactions. Tyrosine is a precursor of thyroid hormone, while both tyrosine and phenylalanine are metabolized to produce epinephrine, norepinephrine, and dihydroxyphenylalanine (DOPA). Glutamate is both a neurotransmitter as well as a precursor of a second neurotransmitter, γ-aminobutyric acid (GABA).

Certain Plant L-α-Amino Acids Can Adversely Impact Human Health

The consumption of plants that contain certain nonprotein amino acids can adversely impact human health. The seeds and seed products of three species of the legume *Lathyrus* have been implicated in the genesis of **neurolathyrism**, a profound neurologic disorder characterized by progressive and irreversible spastic paralysis of the legs. Lathyrism occurs widely during famines, when *Lathyrus* seeds may become a major part of the diet. L-α-Amino acids that have been implicated in human neurologic disorders, notably neurolathyrisms, include L-homoarginine and β-N-oxalyl-L-α,β-diaminopropionic acid (β-ODAP **Table 3–3**). The seeds of another *Lathyrus* legume, the "sweet pea," contain the osteolathyrogen γ-glutamyl-β-aminopropionitrile (BAPN), a glutamine derivative of β-aminopropionitrile (structure not shown). The seeds of certain *Lathyrus* species also contain α,γ-diaminobutyric acid, an analog of ornithine that inhibits the hepatic urea cycle enzyme ornithine transcarbamylase, leading to ammonia toxicity. Finally, L-β-methylaminoalanine, a neurotoxic amino acid that is present in *Cycad* seeds, has been implicated as a risk factor for neurodegenerative diseases including amyotrophic lateral sclerosis–Parkinson dementia complex in natives of Guam who consume either fruit bats that feed on cycad fruit, or flour made from cycad seeds.

D-Amino Acids

D-Amino acids occur naturally throughout the biosphere, including free D-serine and D-aspartate in human brain tissue,

TABLE 3–3 Potentially Toxic L-α-Amino Acids

Nonprotein L-α-Amino Acid	Medical Relevance
Homoarginine	Cleaved by arginase to L-lysine and urea. Implicated in human neurolathyrism.
β-N-Oxalyl diaminopropionic acid (β-ODAP)	A neurotoxin. Implicated in human neurolathyrism.
β-N-Glutamylamino-propiononitrile (BAPN)	An osteolathyrogen.
2,4-Diaminobutyric acid	Inhibits ornithine transcarbamylase, resulting in ammonia toxicity.
β-Methylaminoalanine	Possible risk factor for neurodegenerative diseases.

D-alanine and D-glutamate in the cell walls of gram-positive bacteria, and D-amino acids in certain peptides and antibiotics produced by bacteria, fungi, reptiles, and amphibians. *Bacillus subtilis* excretes D-methionine, D-tyrosine, D-leucine, and D-tryptophan to trigger biofilm disassembly, and *Vibrio cholerae* incorporates D-leucine and D-methionine into the peptide component of its peptidoglycan layer.

PROPERTIES OF THE FUNCTIONAL GROUPS OF AMINO ACIDS

Amino Acids May Have Positive, Negative, or Zero Net Charge

In aqueous solution, the charged and uncharged forms of the ionizable weak acid groups —COOH and —NH_3^+ exist in dynamic protonic equilibrium:

$$R—COOH \rightleftarrows R—COO^- + H^+$$
$$R—NH_3^+ \rightleftarrows R—NH_2 + H^+$$

While both R—COOH and R—NH_3^+ are weak acids, R—COOH is a far stronger acid than R—NH_3^+. Thus, at physiologic pH

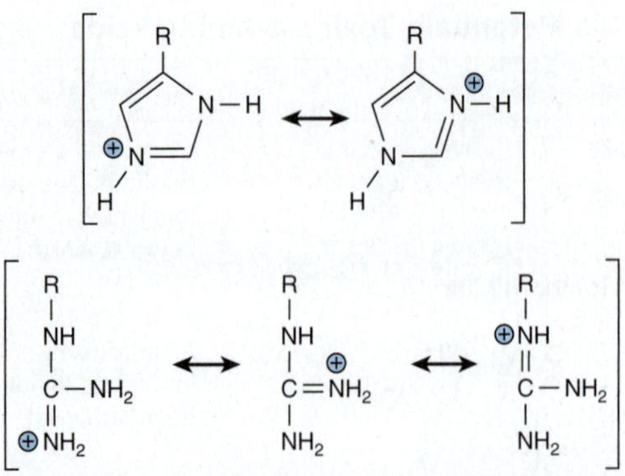

FIGURE 3-4 The protonated R groups of histidine (top) and arginine (bottom) are stabilized by electronic resonance.

(pH 7.4), the α-carboxyl groups of free amino acids exist almost entirely as R—COO⁻ and α-amino groups predominantly as R—NH_3^+. The same is true for the side chain carboxyl groups of aspartic acid and glutamic acid as well as the ε-amino group of lysine. Other ionizable groups include the sulfhydryl group of cysteine, the imidazole ring of histidine and the guanidino group of arginine (**Figure 3-4**). **Figure 3-5** illustrates the effect that the pH of the aqueous environment has on the charged state of aspartic acid.

Amino acids in blood and most tissues thus should be represented as in **A**, in the following figure. Note that these ionized yet *net* neutral molecules, a consequence of their equal numbers of positively and negatively charged groups, are termed **zwitterions**.

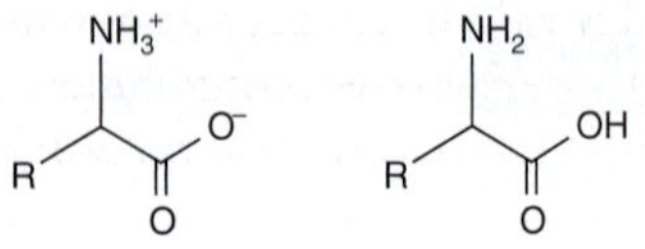

Structure **B** cannot exist in aqueous solution because at any pH low enough to protonate the carboxyl group, the amino group would also be protonated. Similarly, at any pH sufficiently high for an uncharged amino group to predominate, a carboxyl group will be present as R—COO⁻. The uncharged representation **B** is, however, often used when diagramming reactions that do not involve protonic equilibria.

pK_a Values Express the Strengths of Weak Acids & Bases

The strengths of weak acids are expressed as their **pK_a**. For molecules with multiple dissociable protons, the pK_a for each acidic group is designated by replacing the subscript "a" with a number. The strength of weak bases is generally expressed as the pK_a of their protonated, or conjugate acid, form. This enables the relative strengths of all groups capable of dissociably binding proteins to be compared on a single, continuous scale. The terms "weak acid" and "weak base" thus constitute arbitrary labels since in any protonic equilibrium both acidic and basic forms must participate. The net charge on an amino acid—the algebraic sum of all the positively and negatively charged groups present—depends on the pK_a values of its functional groups and the pH of the surrounding medium. In the laboratory, altering the charge on amino acids and their derivatives by varying the pH facilitates the physical separation of amino acids, peptides, and proteins (see Chapter 4).

At Its Isoelectric pH (pI), an Amino Acid Bears No Net Charge

Zwitterions are one example of an **isoelectric** species—the form of a molecule that has an equal number of positive and negative charges and thus is electrically neutral. The isoelectric pH, also called the isoelectric point or pI, is the pH midway between pK_a values for the ionizations on either side of the isoelectric species. For an amino acid such as alanine that has only two dissociating groups, there is no ambiguity. The first pK_a (R—COOH) is 2.35 and the second pK_a (R—NH_3^+) is 9.69. The isoelectric pH (pI) of alanine thus is

$$pI = \frac{pK_1 + pK_2}{2} = \frac{2.35 + 9.69}{2} = 6.02$$

For polyprotic acids, one must first identify the isoionic species. For example, the pI for aspartic acid is

$$pI = \frac{pK_1 + pK_2}{2} = \frac{2.09 + 3.96}{2} = 3.02$$

For lysine, Pi is calculated from:

$$pI = \frac{pK_2 + pK_3}{2}$$

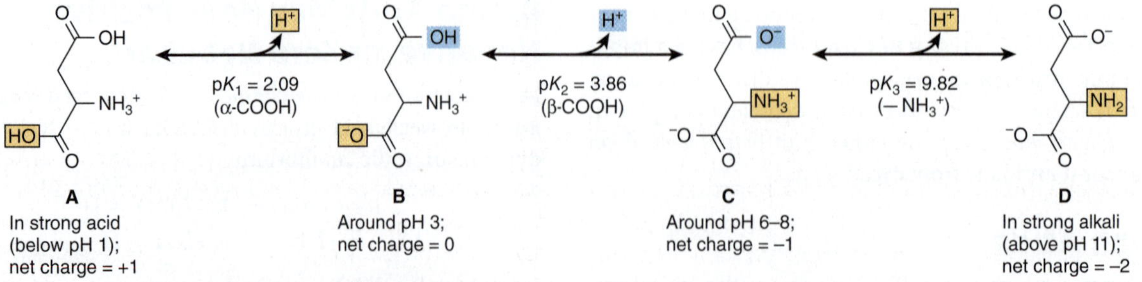

A	B	C	D
In strong acid (below pH 1); net charge = +1	Around pH 3; net charge = 0	Around pH 6–8; net charge = −1	In strong alkali (above pH 11); net charge = −2

FIGURE 3-5 Protonic equilibria of aspartic acid.

TABLE 3–4 Typical Range of pK Values for Ionizable Groups in Proteins

Dissociating Group	pK_a Range
α-Carboxyl	3.5–4.0
Non-α COOH of Asp or Glu	4.0–4.8
Imidazole of His	6.5–7.4
SH of Cys	8.5–9.0
OH of Tyr	9.5–10.5
α-Amino	8.0–9.0
ε-Amino of Lys	9.8–10.4
Guanidinium of Arg	~12.0

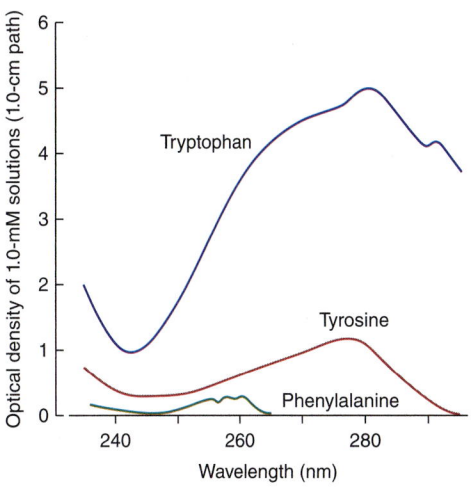

FIGURE 3–6 Ultraviolet absorption spectra of tryptophan, tyrosine, and phenylalanine.

Similar considerations apply to all polyprotic acids (eg, proteins), regardless of the number of dissociable groups present. In the clinical laboratory, knowledge of the pI guides selection of conditions for electrophoretic separations. Electrophoresis at pH 7.0 will separate two molecules with pI values of 6.0 and 8.0, because the molecule with a pI of 6.0 will have a net *negative* charge, and that with a pI of 8.0 a net *positive* charge. Similar considerations underlie chromatographic separations on ionic supports such as diethylaminoethyl (DEAE) cellulose (see Chapter 4).

pK_a Values Vary With the Environment

The environment of a dissociable group affects its pK_a (**Table 3–4**). A nonpolar environment, which possesses less capacity than water for stabilizing charged species, thus *raises* the pK_a of a carboxyl group making it a *weaker* acid, but *lowers* the pK_a of an amino group, making it a *stronger* acid. Similarly, the presence of an adjacent *oppositely* charged group can *stabilize*, or of a *similarly* charged group can *destabilize*, a developing charge. Therefore, the pK_a values of the R groups of *free* amino acids in aqueous solution (see Table 3–1) provide only an approximate guide to their pK_a values when present in proteins. The pK_a of a dissociable R group will depend on its location within a protein. For example, pK_a values that diverge from aqueous solution by as much as 3 pH units are common at the active sites of enzymes, while a buried aspartic acid of thioredoxin, has a pK_a above 9—a shift of more than 6 pH units!

The Solubility of Amino Acids Reflects Their Ionic Character

The charges conferred by the dissociable functional groups of amino acids ensure that they are readily solvated by—and thus soluble in polar solvents such as water and ethanol, but insoluble in nonpolar solvents such as benzene, hexane, or ether.

Amino acids do not absorb visible light and thus are colorless. However, tyrosine, phenylalanine, and tryptophan absorb high-wavelength (250-290 nm) ultraviolet light. Because it absorbs ultraviolet light about 10 times more efficiently than either phenylalanine or tyrosine, tryptophan makes the major contribution to the ability of most proteins to absorb light in the region of 280 nm (**Figure 3–6**).

THE α-R GROUPS DETERMINE THE PROPERTIES OF AMINO ACIDS

Each functional group of an amino acid exhibits all of its characteristic chemical reactions. For carboxylic acid groups, these reactions include the formation of esters, amides, and acid anhydrides; for amino groups, acylation, amidation, and esterification; and for —OH and —SH groups, oxidation and esterification. Since glycine, the smallest amino acid, can be accommodated in places inaccessible to other amino acids, it often occurs where peptides bend sharply. The hydrophobic R groups of alanine, valine, leucine, and isoleucine and the aromatic R groups of phenylalanine, tyrosine, and tryptophan occur primarily in the interior of cytosolic proteins. The charged R groups of basic and acidic amino acids stabilize specific protein conformations via ionic interactions, or salt bridges. These interactions also function in "charge relay" systems during enzymatic catalysis and electron transport in respiring mitochondria. Histidine plays unique roles in enzymatic catalysis. The pK_a of its imidazole proton permits histidine to function at neutral pH as either a base or an acid catalyst without the need for any environmentally induced shift. The primary thioalcohol (—SH) group of cysteine, whose pK_a is 8.3, is an excellent nucleophile and often functions as such during enzymatic catalysis. For selenocysteine, its pK_3 of 5.2 is 3 pH units lower than that of cysteine. At a distinctly acidic pH, selenocysteine thus should be the better nucleophile. The primary alcohol of serine functions as the active site nucleophile in trypsin and other serine proteases. However, the secondary alcohol group of threonine is not known to serve this role in catalysis. The —OH groups of serine, tyrosine, and threonine frequently serve as the points of covalent attachment for phosphoryl groups that regulate protein function (see Chapter 9).

Amino Acid Sequence Determines Primary Structure

Amino acids are linked together by peptide bonds.

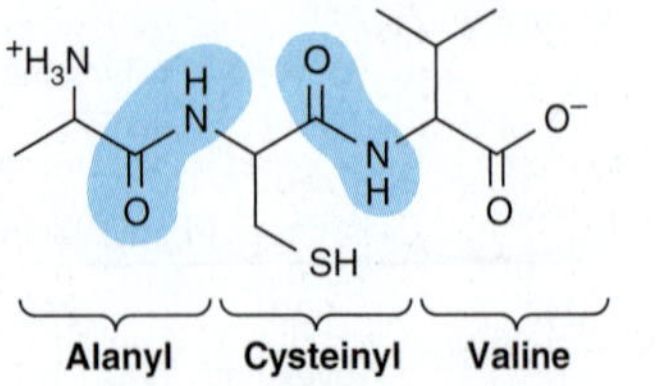

The *number* and *order* of the amino acid residues in a polypeptide constitute its *primary* **structure**. Amino acids present in peptides, called *aminoacyl residues*, are referred to by replacing the *ate* or *ine* suffixes of free amino acids with *yl* (eg, alan*yl*, aspart*yl*, tyros*yl*). Peptides are then named as derivatives of the *carboxy* terminal aminoacyl residue. For example, Lys-Leu-Tyr-Gln is called lys*yl*-leuc*yl*-tyros*yl*-gluta-m*ine*. The *ine* ending on the carboxy-terminal residue (eg, gluta*mine*) indicates that its α-carboxyl group is *not* involved in a peptide bond. Three-letter abbreviations linked by straight lines represent an unambiguous primary structure. Lines are omitted when using single-letter abbreviations.

$$\text{Glu—Ala—Lys—Gly—Tyr—Ala}$$
$$\text{E   A   K   G   Y   A}$$

Prefixes like *tri-* or *octa-* denote peptides with three or eight *residues*, respectively. By convention, peptides are written with the residue that bears the free α-amino group at the left. This convention was adopted long before it was discovered that peptides are synthesized in vivo starting from the amino-terminal residue.

Peptide Structures Are Easy to Draw

To draw a peptide, use a zigzag to represent the main chain or backbone. Add the main chain atoms, which occur in the repeating order: α-nitrogen, α-carbon, carbonyl carbon. Now add a hydrogen atom to each α-carbon and to each peptide nitrogen, then add an oxygen to the carbonyl carbon. Finally, add the appropriate R groups (shaded) to each α-carbon atom.

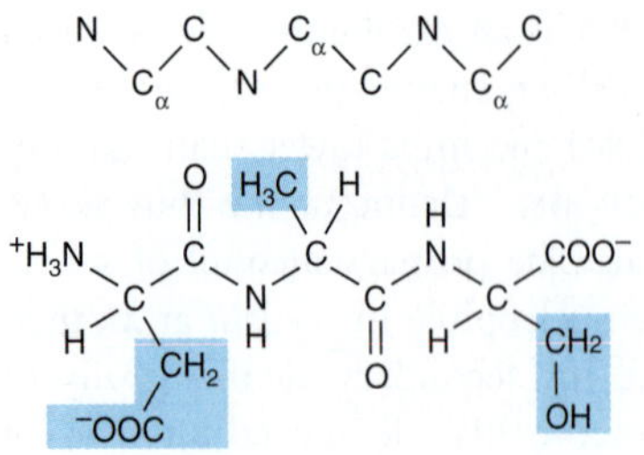

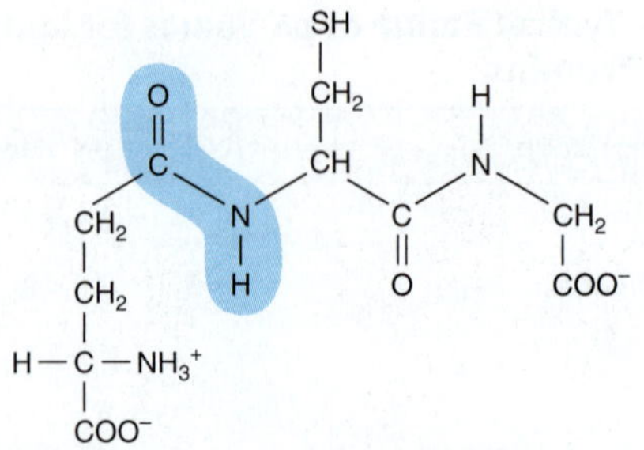

FIGURE 3–7 **Glutathione (γ-glutamyl-cysteinyl-glycine).** Note the non–α peptide bond that links Glu to Cys.

Some Peptides Contain Unusual Amino Acids

In mammals, peptide hormones typically contain only the 20 codon-specified α-amino acids linked by standard peptide bonds. Other peptides may, however, contain nonprotein amino acids, derivatives of the protein amino acids, or amino acids linked by an atypical peptide bond. For example, the amino terminal glutamate of glutathione, a tripeptide that participates in the metabolism of xenobiotics (see Chapter 47) and the reduction of disulfide bonds, is linked to cysteine by a non-α peptide bond that utilizes the side chain, rather than the α-, carboxylic acid group (**Figure 3–7**). The amino terminal glutamate of thyrotropin-releasing hormone (TRH) is cyclized to pyroglutamic acid, and the carboxyl group of the carboxyl terminal prolyl residue is amidated. The nonprotein amino acids D-phenylalanine and ornithine are present in the cyclic peptide antibiotics tyrocidine and gramicidin S, while the heptapeptide opioids dermorphin and deltophorin in the skin of South American tree frogs contain D-tyrosine and D-alanine.

The Peptide Bond Has Partial Double-Bond Character

Although peptide structures are written as if a single bond linked the α-carboxyl and α-nitrogen atoms, this bond in fact exhibits partial double-bond character:

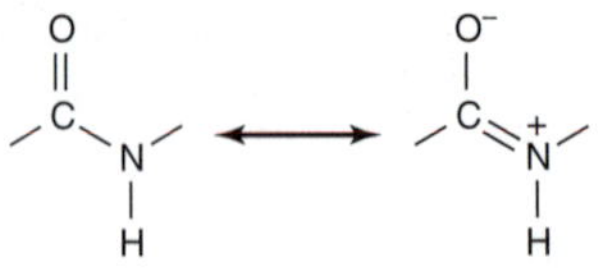

The bond that connects a carbonyl carbon to the α-nitrogen therefore cannot rotate, as this would require breaking the partial double bond. Consequently, the O, C, N, and H atoms of a peptide bond are *coplanar*. The imposed semirigidity of the peptide bond has important consequences for the manner in which peptides and proteins fold to generate higher orders of structure. In **Figure 3–8**, encircling brown arrows indicate free rotation about the remaining bonds of the polypeptide backbone.

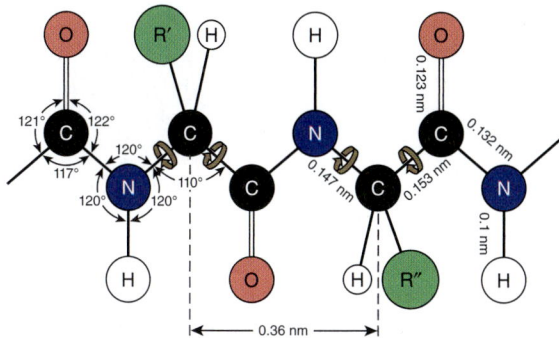

FIGURE 3–8 **Dimensions of a fully extended polypeptide chain.** The four atoms of the peptide bond are coplanar. Free rotation can occur about the bonds that connect the α-carbon with the α-nitrogen and with the α-carbonyl carbon (**brown arrows**). The extended polypeptide chain is thus a semirigid structure with two-thirds of the atoms of the backbone held in a fixed planar relationship to one another. The distance between adjacent α-carbon atoms is 0.36 nm (3.6 Å). The interatomic distances and bond angles, which are not equivalent, are also shown. (Reproduced with permission from Pauling L, Corey LP, Branson HR: The structure of proteins: two hydrogen-bonded helical configurations of the polypeptide chain. Proc Natl Acad Sci USA. 1951;37(4):205-211.)

Noncovalent Forces Constrain Peptide Conformations

Folding of a peptide probably begins coincident with its biosynthesis (see Chapter 37). The mature, physiologically active conformation reflects the collective contributions of the amino acid sequence, noncovalent interactions (eg, hydrogen bonding, hydrophobic interactions), and the minimization of steric hindrance between residues. Common repeating conformations include α-helices and β-pleated sheets (see Chapter 5).

Peptides Are Polyelectrolytes

The peptide bond is uncharged at any pH of physiologic interest. Formation of peptides from amino acids is therefore accompanied by a net loss of one positive and one negative charge per peptide bond formed at physiologic pH. Peptides nevertheless are charged owing to their terminal carboxyl and amino groups and, where present, their acidic or basic R groups. As for amino acids, the net charge on a peptide depends on the pH of its environment and on the pK_a values of its dissociating groups.

SUMMARY

- Both D-amino acids and non–α-amino acids occur in nature, but proteins are synthesized using only L-α-amino acids. D-Amino acids do, however, serve metabolic roles, not only in bacteria, but also in humans.

- L-α-Amino acids serve vital metabolic functions in addition to protein synthesis. Examples include the biosynthesis of urea, heme, nucleic acids, hormones such as epinephrine and neurotransmitters such as DOPA.

- The presence in meteorites of trace quantities of many of the protein amino acids lends credence to the hypothesis that asteroid strikes might have contributed to the development of life on earth.

- Certain of the L-α-amino acids present in plants and plant seeds can have deleterious effects on human health, for example, in lathyrism.

- The R groups of amino acids determine their particular biochemical functions. Amino acids are classified as basic, acidic, aromatic, aliphatic, or sulfur-containing based on the composition and properties of their R groups.

- The partial double-bond character of the bond that links the carbonyl carbon and the nitrogen of a peptide render the four atoms of the peptide bond *coplanar*, and hence restrict the number of possible peptide conformations.

- Peptides are often classified based on the number of amino acid residues present, and are named as derivatives of the carboxyl terminal residue. The primary structure of a peptide is its amino acid sequence, starting from the amino-terminal residue, a direction in which peptides actually are synthesized *in vivo*.

- All amino acids possess at least two weakly acidic functional groups, $R—NH_3^+$ and $R—COOH$. Many also possess additional weakly acidic functional groups such as the phenolic —OH of tyrosine or the —SH, guanidino, or imidazole moieties of cysteine, histidine, and arginine, respectively.

- The pK_a values of all functional groups of an amino acid or of a peptide dictate its net charge at a given pH. pI, the isoelectric pH, is the pH at which a molecule bears no net charge and thus does not move in a direct current electrical field.

- The pK_a values of free amino acids at best only approximate their pK_a values when present in a protein, and can differ widely due to the influence of their surroundings in a protein.

REFERENCES

Bender DA: *Amino Acid Metabolism*, 3rd ed. Wiley, 2012.

Diaz-Parga P, Goto JJ, Krishnan VV: Chemistry and chemical equilibrium dynamics of BMAA and its carbamate adducts. Neurotox Res 2018;33:76.

Koga T, Naraoka H: A new family of extraterrestrial amino acids in the Murchison meteorite. Sci Rep 2017;7:636. https://doi.org/10.1038/s41598-017-00693-9.

Osinski GR, Cockell CS, Pontefract A, et al.: The role of meteorite impacts in the origin of life. Astrobiology 2020;20:1121.

Seckler JM, Lewis SJ: Advances in D-amino acids in neurological research. Int J Mol Sci 2020;21:7325.

Wu G ed.: *Amino Acids in Nutrition and Health.* Springer, 2020.

Yoshimura T, Nishikawa T, Homma H eds.: *D-Amino Acids: Physiology, Metabolism, and Application.* Springer, 2016.

Proteins: Determination of Primary Structure

4

Peter J. Kennelly, PhD, & Victor W. Rodwell, PhD

- Cite three examples of posttranslational modifications that commonly occur during the maturation of a newly synthesized polypeptide.
- Name four chromatographic methods commonly employed for the isolation of proteins from biologic materials.
- Describe how electrophoresis in polyacrylamide gels can be used to determine the purity, subunit composition, relative mass, and isoelectric point of a protein.
- Describe the basis on which quadrupole and time-of-flight (TOF) spectrometers determine molecular mass.
- Describe how the availability of genome sequences has facilitated the determination of protein primary structure.
- Explain what is meant by "the proteome" and cite examples of its potential significance.
- Describe the advantages and limitations of gene chips as a tool for monitoring protein expression.
- Outline three strategies for resolving individual proteins and peptides from complex biologic samples to facilitate their identification by mass spectrometry (MS).
- Comment on the contributions of genomics, computer algorithms, and databases to the identification of the open reading frames (ORFs) that encode a given protein.

BIOMEDICAL IMPORTANCE

Proteins are physically and functionally complex macromolecules that perform multiple critically important roles. An internal protein network, the cytoskeleton (see Chapter 51) maintains a cell's shape and physical integrity. Actin and myosin filaments form the contractile machinery of muscle (see Chapter 51). Hemoglobin transports oxygen (see Chapter 6), while circulating antibodies defend against foreign invaders (see Chapter 52). Enzymes catalyze reactions that generate energy, synthesize and degrade biomolecules, replicate and transcribe genes, process mRNAs, etc (see Chapter 7). Receptors enable cells to sense and respond to hormones and other extracellular cues (see Chapters 41 and 42). Proteins are subject to physical and functional changes that mirror the life cycle of the organisms in which they reside. A typical protein is "born" at translation (see Chapter 37), matures through posttranslational processing events such as selective proteolysis (see Chapters 9 and 37), alternates between working and resting states through the intervention of regulatory factors (see Chapter 9), ages through oxidation, deamidation, etc (see Chapter 57), and "dies" when degraded to its component amino acids (see Chapter 28). An important goal of molecular medicine is to identify biomarkers such as proteins and/or modifications to proteins whose presence, absence, or deficiency is associated with specific physiologic states or diseases (**Figure 4–1**).

PROTEINS & PEPTIDES MUST BE PURIFIED PRIOR TO ANALYSIS

While new techniques have emerged that permit certain proteins be studied within a living cell, in general the acquisition of highly purified protein remains essential for the detailed examination of its physical and functional properties. The isolation of

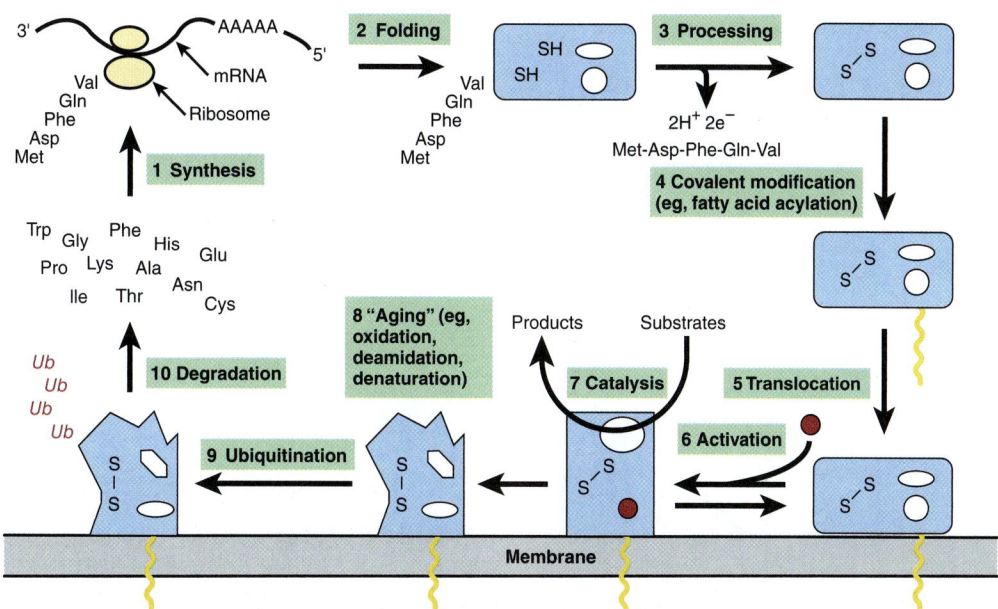

FIGURE 4–1 **Diagrammatic representation of the life cycle of a hypothetical protein.** (1) The life cycle begins with the synthesis on a ribosome of a polypeptide chain, whose primary structure is dictated by an mRNA. (2) As synthesis proceeds, the polypeptide begins to fold into its native conformation (blue). (3) Folding may be accompanied by processing events such as proteolytic cleavage of an *N*-terminal leader sequence (Met-Asp-Phe-Gln-Val) or the formation of disulfide bonds (S—S). (4) Subsequent covalent modifications may, for example, attach a fatty acid molecule (yellow) for (5) translocation of the modified protein to a membrane. (6) Binding an allosteric effector (red) may trigger the adoption of a catalytically active conformation. (7) Over time, proteins get damaged by chemical attack, deamidation, or denaturation, and (8) may be "labeled" by the covalent attachment of several ubiquitin molecules (*Ub*). (9) The ubiquitinated protein is subsequently degraded to its component amino acids, which become available for the synthesis of new proteins.

a specific protein from a natural source in quantities sufficient for analysis presents a formidable challenge as living organisms contain thousands of different proteins, each in widely varying amounts. Thus, obtaining pure protein may require successive application of multiple separation techniques. Selective precipitation exploits differences in relative solubility of individual proteins as a function of pH (isoelectric precipitation), polarity (precipitation with ethanol or acetone), or salt concentration (salting out with ammonium sulfate). Chromatographic techniques separate one protein from another based on the difference in their size and shape (size-exclusion chromatography), net charge (ion-exchange chromatography), hydrophobicity (hydrophobic interaction chromatography), or ability to bind a specific ligand (affinity chromatography).

Column Chromatography

In column chromatography, the stationary phase matrix consists of small beads loaded into a cylindrical container of glass, plastic, or steel called a column. Liquid-permeable frits confine the beads within the column while allowing the mobile-phase liquid to flow or percolate through. The stationary phase matrix can be chemically derivatized to coat each bead's surface with the acidic, basic, hydrophobic, or ligand-like groups required for ion exchange, hydrophobic interaction, or affinity chromatography. As the mobile-phase liquid emerges

from the column, it is automatically collected as a series of small portions called fractions. **Figure 4–2** depicts the basic arrangement of a simple bench-top chromatography system.

HPLC—High-Pressure Liquid Chromatography

First-generation column chromatography matrices consisted of long, intertwined oligosaccharide polymers shaped into spherical beads roughly a tenth of a millimeter in diameter. Unfortunately, their relatively large size perturbed mobile-phase flow. In theory, resolution could be increased by reducing particle size. However, the greater pressures required to overcome the increased resistance of a more tightly packed matrix crushed the soft and spongy polysaccharide or, later, polyacrylamide beads. Eventually, small spherical silicon particles became available that were physically strong and porous, with a high surface area for attaching various chemical groups. These particles were then packed into stainless steel columns capable of withstanding high pressures necessary to force liquid through the more constricted passages formed by the smaller, tightly packed beads. The high surface area and greater uniformity of these silica beads imbue high pressure liquid chromatography, or HPLC, systems with a resolving power that is orders of magnitude higher than that of the once familiar glass columns and their polysaccharide matrices.

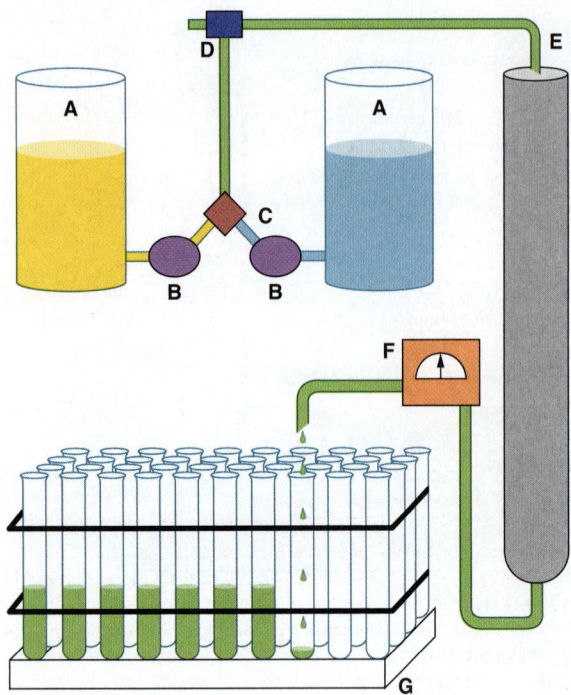

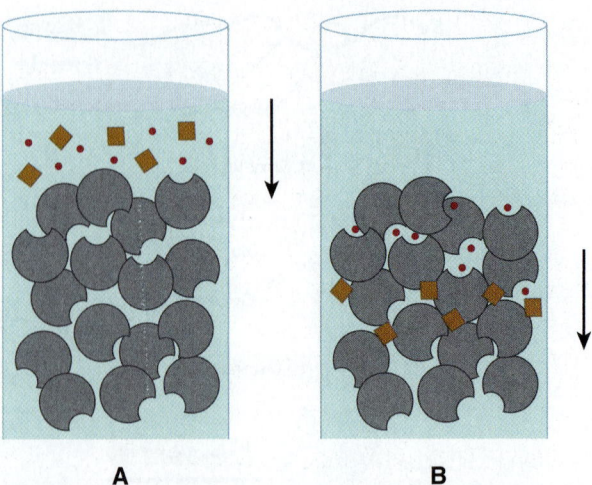

FIGURE 4–3 **Size-exclusion chromatography.** A: A mixture of large molecules (brown) and small molecules (red) is applied to the top of a gel filtration column. B: Upon entering the column, the small molecules enter pores in the stationary phase matrix (gray). As the mobile phase (blue) flows down the column, they lag behind from the large molecules, which are excluded.

FIGURE 4–2 **Components of a typical liquid chromatography apparatus.** Shown are the key components of a programmable liquid chromatography system consisting of A: reservoirs of mobile-phase liquids (yellow, light blue), B: microprocessor-controlled pumps (purple), C: mixing chamber (red), D: injection port for loading analyte (dark blue); E: glass, metal, or plastic column containing stationary phase matrix (gray), F: spectrophotometric, fluorometric, refractive index, or electrochemical detector (orange), and G: fraction collector for collecting portions, called fractions, of the eluent liquid (green) in a series of separate test tubes, vials, or wells in a microtiter plate. The microprocessor can be programmed to pump liquid from only one reservoir (isocratic elution), to switch reservoirs at some predetermined point to generate a step gradient, or to mix liquids from the two reservoirs in proportions that vary over time to generate either a multistep or a continuous gradient.

Size-Exclusion Chromatography

Size-exclusion or, as it is sometimes still referred to, gel-filtration chromatography separates proteins on the basis of their **Stokes radii**, which is a function of both molecular mass and shape. As it rapidly tumbles, an elongated protein effectively occupies, like a spinning propeller, a larger effective volume than would a globular protein of the same mass. Size-exclusion chromatography employs porous beads (**Figure 4–3**) whose openings are analogous to indentations in a river bank. If an object floating downstream drifts into an indentation in the riverbank, its downstream motion is retarded until it drifts back into the current. Similarly, proteins with Stokes radii too large to enter the pores (excluded proteins) continuously remain in the flowing mobile phase, and emerge *before* proteins able to enter some or all of the pores, where their motion is retarded. The fraction of the pores accessible to a given protein, and thus the degree to which their motion is retarded, increases with decreasing size. Proteins thus emerge from a gel filtration column in *descending* order of their Stokes radii.

Ion-Exchange Chromatography

In ion-exchange chromatography, proteins interact with the stationary phase by charge-charge interactions. Proteins with a net positive charge at a given pH will adhere to beads derivatized with negatively charged functional groups such as carboxylates or sulfates (cation exchangers). Similarly, proteins with a net negative charge will adhere to beads with positively charged functional groups, typically tertiary or quaternary amines (anion exchangers). Nonadherent proteins flow through the matrix and are washed away. Bound proteins then can be selectively displaced from the matrix by gradually raising the ionic strength of the mobile phase, thereby weakening charge-charge interactions. Purification can be enhanced by manipulating the pH of the mobile phase, which in turn will alter the magnitude and even the sign of the net charge of each component of a protein mixture.

Hydrophobic Interaction Chromatography

Hydrophobic interaction chromatography separates proteins based on their tendency to associate with a stationary phase matrix coated with hydrophobic groups (eg, phenyl Sepharose, octyl Sephadex). Proteins with exposed hydrophobic surfaces adhere to the matrix via hydrophobic interactions that are enhanced by employing a mobile phase of high ionic strength. After nonadherent proteins are washed away, the polarity of the mobile phase is decreased by gradually lowering its salt concentration. If the interaction between protein and stationary phase is particularly strong, ethanol or glycerol may be added to the mobile phase to decrease its polarity and further weaken hydrophobic interactions.

Affinity Chromatography

Affinity chromatography exploits the high selectivity displayed by enzymes and other proteins for ligands such as substrates, products, cofactors, inhibitors, or analogues thereof. In theory, when a protein mixture is applied to a matrix derivatized with a particular ligand, only those proteins that bind that ligand will adhere. Bound proteins are then eluted either by competition with free, soluble ligand or, less selectively, by disrupting protein-ligand interactions using high salt concentrations or other agents. Recombinantly expressed proteins are usually purified by fusing to their gene an additional segment of DNA that encodes a convenient ligand-binding domain (see Chapter 7).

Protein Purity Is Assessed by Polyacrylamide Gel Electrophoresis (PAGE)

The most widely used method for determining a sample's protein composition is SDS-PAGE—polyacrylamide gel electrophoresis (PAGE) in the presence of the anionic detergent sodium dodecyl sulfate (SDS). Electrophoresis separates charged biomolecules based on the rates at which they migrate through a porous matrix in an applied electrical field. For SDS-PAGE, proteins migrate as SDS-polypeptide complexes through a polyacrylamide matrix. Binding of SDS causes most polypeptides to unfold or denature. When used in conjunction with 2-mercaptoethanol or dithiothreitol to reduce and break disulfide bonds (**Figure 4–4**), SDS-PAGE separates the component polypeptides of multimeric proteins. On average, each SDS-polypeptide complex contains one molecule of SDS for every two peptide bonds. The large number of anionic SDS

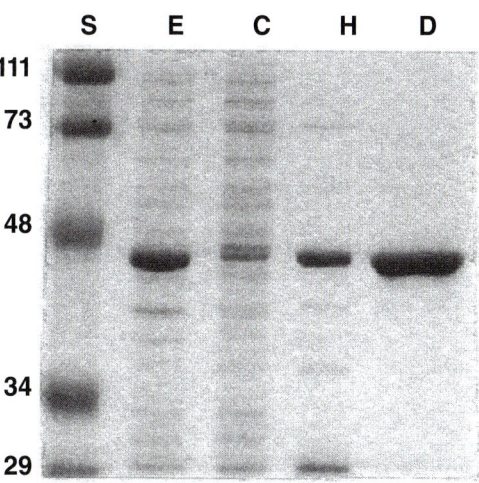

FIGURE 4–5 **Use of SDS-PAGE to observe successive purification of a recombinant protein.** The gel was stained with Coomassie Blue. Shown are protein standards (lane S) of the indicated M_r, in kDa, crude cell extract (E), cytosol (C), high-speed supernatant liquid (H), and the DEAE-Sepharose fraction (D). The recombinant protein has a mass of about 45 kDa.

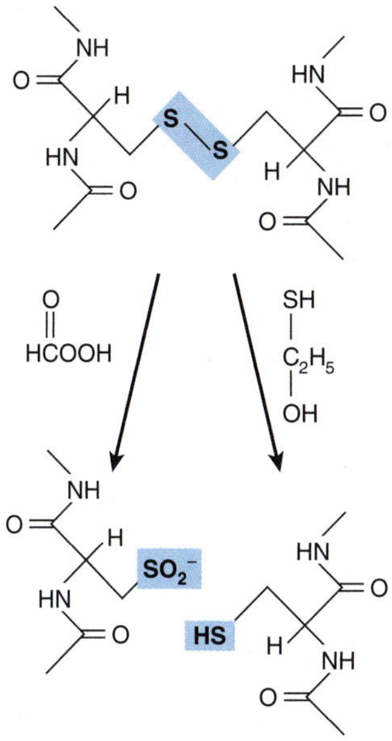

molecules, each bearing a charge of –1, overwhelms the charge contributions of the amino acid functional groups endogenous to a typical polypeptide, rendering the charge-to-mass ratio of each SDS-polypeptide complex approximately equal. Under these circumstances, the larger the polypeptide, the greater the physical resistance its SDS-polypeptide complex encounters as it moves through the acrylamide matrix. Consequently, SDS-PAGE separates most polypeptides based on their relative molecular mass (M_r). Upon completion, the individual polypeptides trapped in the polyacrylamide gel are visualized by staining with dyes such as Coomassie Blue (**Figure 4–5**).

Isoelectric Focusing (IEF)

Using polyionic buffers called ampholytes and an applied electric field, a pH gradient can be established within a polyacrylamide matrix. Applied proteins migrate until they reach the region of the matrix where the pH matches their isoelectric point (pI), the pH at which a molecule's net charge is 0. IEF frequently is used in conjunction with SDS-PAGE for two-dimensional electrophoresis, which separates polypeptides based on pI in one dimension and on M_r in the second (**Figure 4–6**). Two-dimensional electrophoresis is particularly well suited for separating the components within complex mixtures of proteins.

SANGER WAS THE FIRST TO DETERMINE THE SEQUENCE OF A POLYPEPTIDE

While scientists found it relatively simple to determine the amino acid *composition* of purified proteins, the proportion of each type of amino acid in the polypeptide, deriving the order or sequence of the amino acids proved difficult. The first

FIGURE 4–4 **Oxidative cleavage of adjacent polypeptide chains linked by disulfide bonds (highlighted in blue) by performic acid (left) or reductive cleavage by β-mercaptoethanol (right) forms two peptides that contain cysteic acid residues or cysteinyl residues, respectively.**

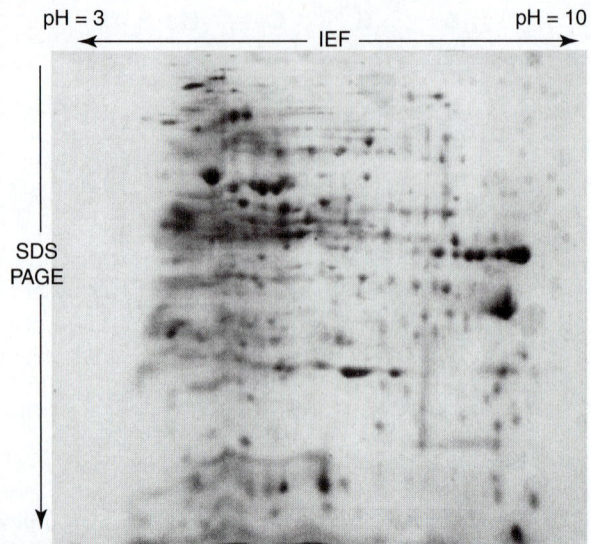

FIGURE 4–6 **Two-dimensional IEF-SDS-PAGE.** The gel was stained with Coomassie Blue. A crude bacterial extract was first subjected to isoelectric focusing (IEF) in a pH 3–10 gradient. The IEF gel was then placed horizontally on the top of an SDS-PAGE gel, and the proteins then further resolved by SDS-PAGE. Notice the greatly improved resolution of distinct polypeptides relative to ordinary SDS-PAGE gel (see Figure 4–5).

protein to have its sequence determined was the peptide hormone insulin, an achievement that earned Frederick Sanger a Nobel Prize in 1958. Mature insulin consists of the 21-residue A chain and the 30-residue B chain linked by disulfide bonds. Sanger reduced the disulfide bonds (see Figure 4–4), separated the A and B chains, and cleaved each chain into smaller and smaller peptides using the proteolytic enzymes trypsin, chymotrypsin, and pepsin followed by incubation with hydrochloric acid. Each peptide in the mixture was isolated, treated with 1-fluoro-2,4-dinitrobenzene (Sanger reagent), and their amino acid composition determined. Sanger reagent reacts with the exposed α-amino groups of the amino-terminal residues, allowing the amino-terminal amino acid of each peptide to be identified. The ε-amino group of lysine also reacts with Sanger reagent; but since an amino-terminal lysine reacts with 2 molecules of Sanger reagent, it is readily distinguished from a lysine from the interior of a peptide. Working from di- and tripeptides up through progressively larger fragments, Sanger was able to reconstruct the complete sequence of insulin. Sanger would later develop the dideoxy method for sequencing DNA, an accomplishment for which he was awarded a second Nobel Prize in 1980.

EDMAN DEVISED THE FIRST PRACTICAL METHOD FOR PEPTIDE SEQUENCING

Sanger's labor-intensive approach rendered it prohibitively difficult to apply to any but the smallest polypeptides. Pehr Edman discovered a reagent, phenyl isothiocyanate (Edman reagent),

Phenylisothiocyanate (Edman reagent) and a peptide

A phenylthiohydantoic acid

H^+, nitro-methane H_2O

A phenylthiohydantoin and a peptide shorter by one residue

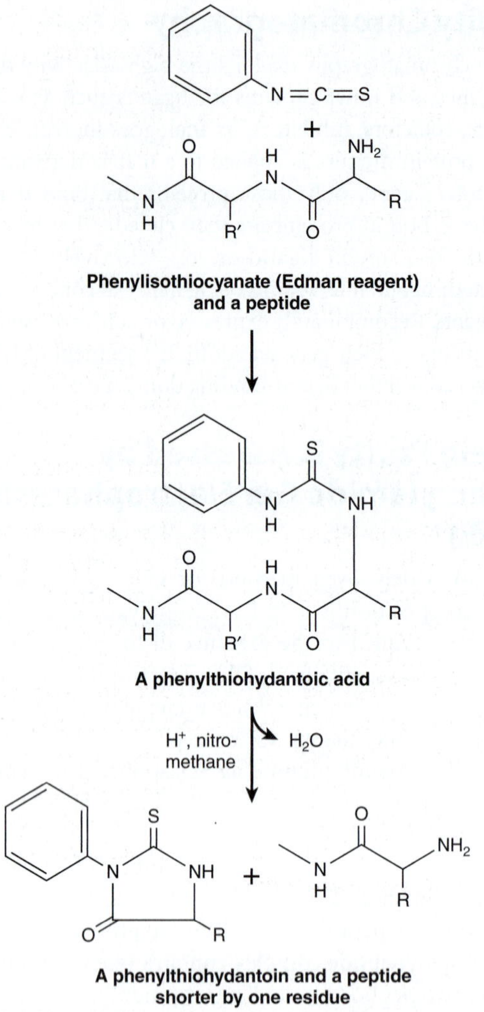

FIGURE 4–7 **The Edman reaction.** Phenyl isothiocyanate derivatizes the amino-terminal residue of a peptide as a phenylthiohydantoic acid. Treatment with acid in a nonhydroxylic solvent releases a phenylthiohydantoin, which is subsequently identified by its chromatographic mobility, and a peptide one residue shorter. The process is then repeated.

that not only could selectively label the amino-terminal residue of a peptide as its phenylthiohydantoin (PTH) derivative but, in contrast to Sanger reagent, permitted the PTH derivative to be removed under mild conditions (**Figure 4–7**). The new amino terminal residue could then be treated with Edman reagent and the process repeated to permit each successive residue in a peptide to be derivatized.

While, in theory, one could determine the entire sequence of a polypeptide using Edman reagent, the heterogeneous chemical properties of the amino acids meant that every step in the procedure represented a compromise between efficiency for any particular amino acid or set of amino acids and the flexibility needed to accommodate all 20. Consequently, each step in the process operates at less than 100% efficiency, which leads to the accumulation of polypeptide fragments with varying N-termini that eventually renders it impossible to distinguish the correct PTH amino acid for that position in the peptide from the out-of-phase contaminants. As a result, the read length for Edman sequencing varies from 5 to 30 amino acid residues depending on the quantity and purity of the peptide, hardly enough to determine the sequence of a typical protein.

To determine the complete sequence of a polypeptide several hundred residues in length, a protein must be cleaved into smaller peptides. These were then purified and analyzed by Edman sequencing. This yielded multiple segments of sequence whose location within the protein, with the exception of the amino terminus, was unknown. In order to assemble these short peptide sequences into the complete sequence of the intact polypeptide, it was necessary to generate and analyze additional peptides in search of some whose sequences overlapped with one another, thereby allowing larger segments of sequence to be pieced together. While the development of automated Edman sequencers and sophisticated HPLC systems for purifying peptides often rendered initial acquisition of a fragmentary, or partial, sequence relatively quick, finding enough "overlap" peptides to reconstruct the complete sequence of a typical protein via the Edman technique generally required large quantities of purified protein and, more importantly, several months or years of additional work. Consequently, the determination of a complete amino acid sequence via the Edman method generally was restricted to highly abundant, readily purified proteins.

MOLECULAR BIOLOGY REVOLUTIONIZED THE DETERMINATION OF PRIMARY STRUCTURE

The sequence for each and every protein in an organism is encoded in the latter's genome. By enabling scientists to decode the sequences of proteins directly from their structural genes, molecular biology revolutionized amino acid sequence analysis. The reasons for this were twofold. First, recombinant techniques permit researchers to manufacture a virtually infinite supply of DNA from even minute quantities of template present in the original sample (see Chapter 39), regardless of the encoded protein product's natural abundance or ease of isolation. Second, DNA sequencing is far more efficient than the Edman technique. Today, automated sequenators routinely "read" DNA sequences several thousand deoxyribonucleotides in length. Consequently, if one could determine just a portion of the sequence of a polypeptide of interest by Edman analysis, one could synthesize a complementary oligonucleotide probe with which to identify the DNA clone containing the gene of interest. The result was a hybrid approach in which Edman chemistry was employed to obtain a partial sequence that was subsequently exploited to determine the complete amino acid sequence by DNA cloning and sequencing.

GENOMICS ENABLES PROTEINS TO BE IDENTIFIED FROM SMALL AMOUNTS OF SEQUENCE DATA

Today the number of organisms for which the complete DNA sequence of their genomes has been determined numbers in the hundreds of thousands. Thus, for most research scientists the genetically encoded sequence of the protein(s) with which they are working has already been determined and can be accessed in a database such as GenBank. To make an unambiguous identification of the amino acid sequence for a protein, sometimes all that is required is the identities of as few as five or six consecutive residues. Today mass spectrometry (MS) has largely replaced the Edman technique as the method of choice for protein identification.

MASS SPECTROMETRY CAN DETECT COVALENT MODIFICATIONS

The development of mass spectrometric (MS) methods that offer superior sensitivity, speed, and capacity to detect posttranslational modifications (**Table 4–1**) has led to its replacement of the Edman technique as the method of choice for protein identification.

MASS SPECTROMETERS COME IN VARIOUS CONFIGURATIONS

In a simple, single quadrupole mass spectrometer, a sample is placed under vacuum and allowed to vaporize in the presence of a proton donor to impart a positive charge. An electrical field then propels the cations toward a curved flight tube where they encounter a magnetic field, which deflects them at a right angle to their original direction of flight (**Figure 4–8**). The current powering the electromagnet is gradually increased until the path of each ion is bent sufficiently to strike a detector mounted at the end of the flight tube. **For ions of identical**

TABLE 4–1 Mass Increases Resulting From Common Posttranslational Modifications

Modification	Mass Increase (Da)
Phosphorylation	80
Hydroxylation	16
Methylation	14
Acetylation	42
Myristylation	210
Palmitoylation	238
Glycosylation	162

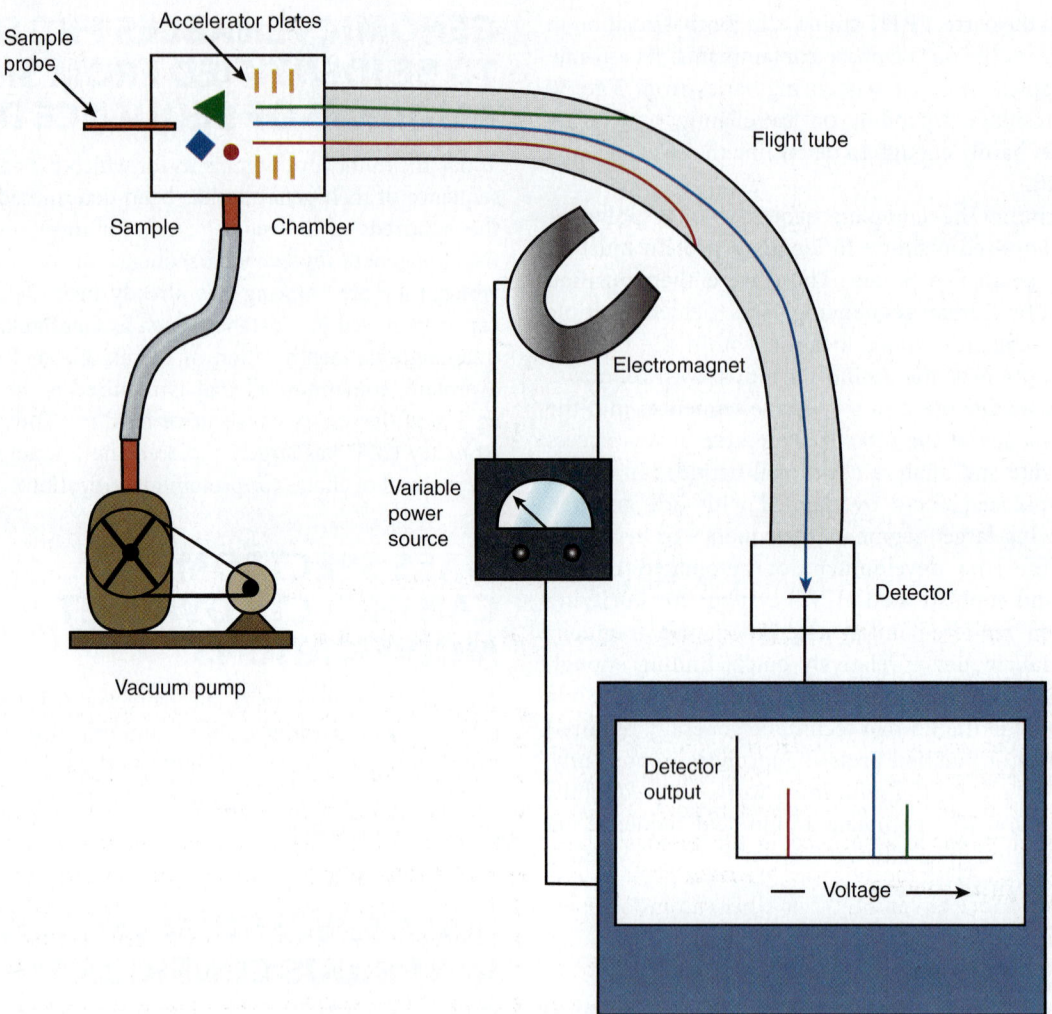

FIGURE 4–8   **Basic components of a simple mass spectrometer.** A mixture of molecules, represented by a red circle, green triangle, and blue diamond, is vaporized in an ionized state in the sample chamber. These molecules are then accelerated down the flight tube by an electrical potential applied to the accelerator grid (yellow). An adjustable field strength electromagnet applies a magnetic field that deflects the flight of the individual ions until they strike the detector. The greater the mass of the ion, the higher the magnetic field required to focus it onto the detector.

net charge, the force required to bend their path to the same extent is proportionate to their mass.

Time-of-flight (TOF) mass spectrometers employ a linear flight tube. Following vaporization of the sample in the presence of a proton donor, an electric field is briefly applied to accelerate the ions toward a detector at the end of the flight tube. **For molecules of identical charge, the velocity to which they are accelerated, and hence the time required to reach the detector, is inversely proportional to their mass.**

Quadrupole mass spectrometers are generally used to determine the masses of molecules of 4000 Da or less, whereas TOF mass spectrometers are used to determine the large masses of complete proteins. Various combinations of multiple quadrupoles, or reflection of ions back down the linear flight tube of a TOF mass spectrometer, are used to create more sophisticated instruments.

Peptides Can Be Volatilized for Analysis by Electrospray Ionization or Matrix-Assisted Laser Desorption

For many years, the analysis of peptides and proteins by MS initially was hindered by difficulties in volatilizing these large organic molecules. While small organic molecules could be readily vaporized by heating in a vacuum (**Figure 4–9**), proteins, oligonucleotides, etc. decomposed on heating. Only when reliable techniques were devised for dispersing peptides, proteins, and other large biomolecules into the vapor phase was it possible to apply MS for their structural analysis and sequence determination. Three commonly used methods for dispersion into the vapor phase are **electrospray ionization**, **matrix-assisted laser desorption and ionization (MALDI)**, and **fast atom bombardment (FAB)**. In electrospray ionization, the molecules to be analyzed are dissolved in a volatile

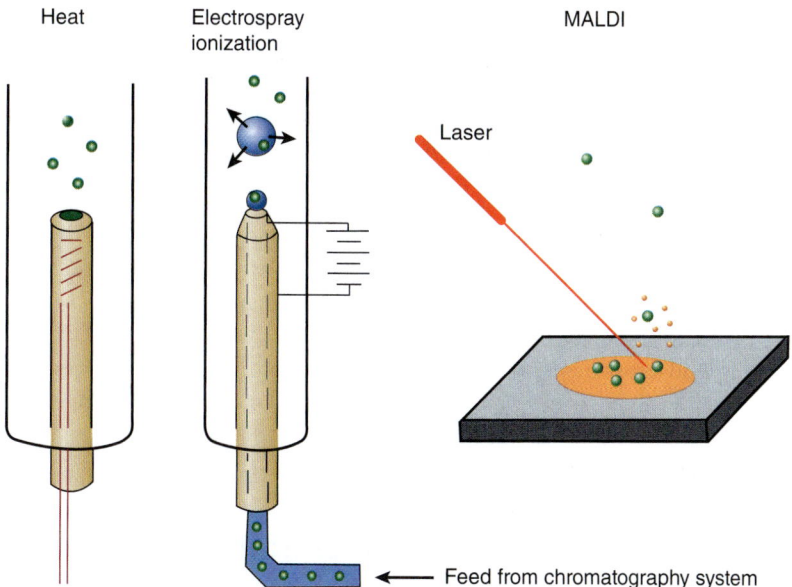

FIGURE 4–9 Three common methods for vaporizing molecules in the sample chamber of a mass spectrometer.

solvent and introduced into the sample chamber in a minute stream through a charged capillary probe (see Figure 4–9). As the droplet of liquid emerges into the sample chamber, the charged probe ionizes the sample while the solvent rapidly disperses, leaving the macromolecule suspended in the gaseous phase. Electrospray ionization is frequently used to analyze peptides and proteins as they elute from an HPLC or other chromatography column, already dissolved in a volatile solvent. In MALDI, the sample is mixed with a liquid matrix containing a light-absorbing dye and a source of protons. In the sample chamber, the mixture is excited using a laser, causing the surrounding matrix to disperse into the vapor phase so rapidly as to avoid heating embedded peptides or proteins (see Figure 4–9). In FAB, large macromolecules dispersed in glycerol or another protonic matrix are bombarded by a stream of neutral atoms, for example, xenon, that have been accelerated to a high velocity. "Soft" ionization by FAB is frequently applied to volatilize large macromolecules intact.

Peptides inside the mass spectrometer can be broken down into smaller units by collisions with neutral helium or argon atoms (collision-induced dissociation) and the masses of the individual fragments determined. Fortunately, since peptide bonds are more vulnerable to rupture than carbon-carbon bonds, the most abundant fragments will differ from one another by increments of one or two amino acids. Since—with the exceptions of (1) leucine and isoleucine and (2) glutamine and lysine—the molecular mass of each amino acid is unique, the sequence of the peptide can be reconstructed from the masses of its fragments.

Tandem Mass Spectrometry

Complex peptide mixtures can be analyzed, without prior purification, by tandem MS, which employs the equivalent of two mass spectrometers linked in series. For this reason, analysis by tandem instruments is often referred to as **MS–MS**, or MS^2. The first mass spectrometer separates individual peptides based on their differences in mass. By adjusting the field strength of the first magnet, a single peptide can be directed into the second mass spectrometer, where fragments are generated and their masses are determined. Alternatively, they can be held in an electromagnetic **ion trap** located between the two quadrupoles and selectively delivered to the second quadrupole instead of being lost when the first quadrupole is set to select ions of a different mass.

Tandem MS can be used to screen blood samples from newborns for the presence and concentrations of amino acids, fatty acids, and other metabolites. Abnormalities in metabolite levels can serve as diagnostic indicators for a variety of genetic disorders, such as phenylketonuria, ethylmalonic encephalopathy, and glutaric acidemia type 1. In recent years a new generation of tandem mass spectrometers has appeared, called Q-TOF-MS, that couple quadrupole (Q) and time-of-flight (TOF) technologies together in order to harness the best aspects of each.

PROTEOMICS & THE PROTEOME

The Goal of Proteomics Is to Identify the Entire Complement of Proteins Elaborated by a Cell Under Diverse Conditions

While the sequence of the human genome is known, the picture it provides is both static and incomplete. As genes are switched on and off and mRNA molecules customized via alternative splicing (see Chapter 36), the spectrum of proteins synthesized varies by particular cell type, stages of growth or differentiation, and in response to external stimuli. Muscle cells express proteins not expressed by neural cells, and the type of subunits present in the hemoglobin tetramer undergo change pre- and postpartum. Many proteins undergo posttranslational modifications during maturation into functionally competent forms or as a means of regulating their properties. In order to obtain a more complete and dynamic molecular description of living organisms, scientists are working to determine the **proteome**, a term that refers to the identity, abundance, and state of modification of the entire suite of proteins expressed by an individual cell at a particular time. Since the proteome for each component cell of an organism is distinct and changes with time and circumstances, the ultimate, comprehensive human proteome constitutes a target of formidable size and complexity.

Simultaneous Determination of Hundreds of Proteins Is Technically Challenging

A key goal of proteomics is the identification of proteins whose levels of expression or modification correlate with medically significant events. In addition to their potential as diagnostic indicators, these protein biomarkers may provide important clues concerning the root causes and mechanisms of a specific physiologic condition or disease. First-generation proteomics employed SDS-PAGE or two-dimensional electrophoresis to resolve the proteins in a biologic sample one from another, followed by determination of the amino acid sequence of their amino terminus by the Edman method. Identities were determined by searching available polypeptide sequences for proteins that contained a matching N-terminal sequence as well as a similar M_r and, for 2D gels, pI.

These early efforts were constrained by the limited number of polypeptide sequences available and the difficulties in isolating polypeptides from the gels in sufficient quantities for Edman analysis. Attempts to increase resolving power and sample yield by increasing the size of the gels were only marginally successful. Eventually, the development of mass spectrometric techniques provided a means for protein sequence determination whose sensitivity was compatible with electrophoretic separation approaches.

Knowledge of the genome sequence of the organism in question greatly facilitated identification by providing a comprehensive set of DNA-encoded polypeptide sequences. It also provided the nucleotide sequence data from which to construct **gene arrays, sometimes called DNA chips**, containing hundreds of distinct oligonucleotide probes. These chips could then be used to detect the presence of mRNAs containing complementary nucleotide sequences. While changes in the expression of the mRNA encoding a protein do not necessarily reflect comparable changes in the level of the corresponding protein, gene arrays were both less technically demanding and more sensitive than first-generation proteomic approaches, particularly with respect to low abundance proteins.

Second-generation proteomics coupled newly developed nanoscale chromatographic techniques with MS. The proteins in a biologic sample are first treated with a protease to hydrolyze them into smaller peptides that are then subject to reversed-phase, ion-exchange, or size-exclusion chromatography to apportion the vast number of peptides into smaller subsets more amenable to analysis. These subsets are analyzed by injecting the column eluent directly into a double quadrupole or TOF mass spectrometer. **Multidimensional protein identification technology (MudPIT)** employs successive rounds of chromatography to resolve the peptides produced from the digestion of a complex biologic sample into several simpler fractions that can be analyzed separately by MS.

Today, advances in the capability and sensitivity of tandem mass spectrometers allow them to directly analyze complex samples. The elimination of prior proteolytic or chromatographic preparation will soon render it feasible to analyze the proteome of an individual cell.

Bioinformatics Assists Identification of Protein Functions

The functions of a large proportion of the proteins encoded by the human genome are presently unknown. Efforts continue to develop protein arrays or chips for directly testing the potential functions of proteins on a mass scale. However, while some protein functions are relatively easy to assay, such as protease or esterase activity, others are much less tractable. Data mining via bioinformatics permits researchers to compare amino acid sequences of unknown proteins with those whose functions have been determined. This provides a means to uncover clues to their potential properties, physiologic roles, and mechanisms of action. Algorithms exploit the tendency of nature to employ variations of a structural theme to perform similar functions in several proteins (eg, the Rossmann nucleotide binding fold to bind NAD(P)H, nuclear targeting sequences, and EF hands to bind Ca^{2+}). These domains generally are detected in the primary structure by conservation of particular amino acids at key positions. Insights into the properties and physiologic role of a newly discovered protein thus may be inferred by comparing its primary structure with that of known proteins.

SUMMARY

- Long amino acid polymers or polypeptides constitute the basic structural unit of proteins, and the structure of a protein provides insights into how it fulfills its functions.

- Proteins undergo posttranslational alterations during their lifetime that influence their function and determine their fate.

- By generating a new amino terminus, Edman reagent permitted the determination of lengthy segments of amino acid sequence. However, physical and chemical factors generally limited the number of amino acids that could be reliably identified to 30 or less, which rendered the complete determination of a protein's sequence via the Edman technique a time- and effort- intensive process.

- Polyacrylamide gels provide a porous matrix for separating proteins on the basis of their mobility in an applied direct current electrical field.

- The nearly constant ratio at which the anionic detergent SDS binds proteins enables SDS-PAGE to separate polypeptides predominantly on the basis of relative size.

- Because mass is a universal property of all biomolecules and their derivatives, MS has emerged as a versatile technique applicable to the determination of primary structure, identification of posttranslational modifications, and the detection of metabolic abnormalities.

- DNA cloning coupled with protein chemistry provided a hybrid approach that greatly increased the speed and efficiency for determination of primary structures of proteins.

- Genomics, the determination of entire polynucleotide sequences, provides researchers with a blueprint for every genetically encoded macromolecule in an organism.

- Proteomic analysis utilizes genomic data to identify the entire complement of proteins in a biologic sample from partial amino acid sequence data obtained by coupling protein and peptide separation methods with sequencing by MS.

- A major goal of proteomics is the identification of proteins and their posttranslational modifications whose appearance or disappearance correlates with physiologic phenomena, aging, or specific diseases.

- Bioinformatics refers to the development of computer algorithms designed to infer the functional properties of macromolecules through comparison of sequences of novel proteins with others whose properties are known.

REFERENCES

Aslam B, Basit M, Nisar MA, et al: Proteomics: Technologies and their applications. J Chromatog Sci 2017;55:182.

Biemann K: Laying the groundwork for proteomics: Mass spectrometry from 1958 to 1988. J Proteomics 2014;107:62.

Bonner P: *Protein Purification,* 2nd ed., CRC Press, 2019.

Brown KA, Melby JA, Roberts DS, Ge Y: Top-down proteomics: Challenges, innovations, and applications in basic and clinical research. Exp Rev Proteomics 2020;17:719.

Burgess RR, Deutscher MP eds.: *Guide to Protein Purification.* 2nd ed., Methods Enzymol, vol. 463, Elsevier, 2009 (Entire volume).

Duarte TT, Spencer CT: Personalized proteomics: The future of precision medicine. Proteomes 2016;4:29.

Jiang Z, Zhou X, Li R, et al: Whole transcriptome analysis with sequencing: Methods, challenges and potential solutions. Cellular Molec Life Sci 2015;72:3425.

Kelly RT: Single-cell proteomics: Progress and prospects. Mol Cell Proteomics 2020;19:739.

Syu GD, Dunn S, Zhu H: Developments and applications of functional protein microarrays. Mol Cell Proteomics 2020;6:916.

Van Riper SK, de Jong EP, Carlis JV, et al: Mass spectrometry-based proteomics: Basic principles and emerging technologies and directions. Adv Exp Med Biol 2013;990:1.

Wood DW: New trends and affinity tag designs for recombinant protein purification. Curr Opin Struct Biol 2014;26:54.

Proteins: Higher Orders of Structure

5

Peter J. Kennelly, PhD, & Victor W. Rodwell, PhD

O B J E C T I V E S

After studying this chapter, you should be able to:

- List the advantages and drawbacks of several common approaches to classifying proteins.
- Explain and illustrate the primary, secondary, tertiary, and quaternary structure of proteins.
- Identify the major recognized types of protein secondary structures and explain supersecondary motifs.
- Describe the kind and relative strengths of the forces that stabilize each order of protein structure.
- Describe the information summarized by a Ramachandran plot.
- Summarize the basic operating principles underlying three key methods for determining protein structure: X-ray crystallography, nuclear magnetic resonance spectroscopy, and cryo-electron microscopy.
- Describe the stepwise process by which proteins fold to attain their native conformation.
- Identify the physiologic roles in protein maturation of chaperones, protein disulfide isomerase, and peptidylproline *cis–trans* isomerase.
- Describe the principal biophysical techniques used to study tertiary and quaternary structure of proteins.
- Explain how genetic and nutritional disorders of collagen maturation illustrate the close linkage between protein structure and function.
- Describe the basic events that underly the molecular pathology of prion diseases.

BIOMEDICAL IMPORTANCE

In nature, form follows function. In order for a newly synthesized polypeptide to mature into a biologically functional protein capable of catalyzing a metabolic reaction, powering cellular motion, or forming the macromolecular rods and cables that provide structural integrity to hair, bones, tendons, and teeth, it must fold into a specific three-dimensional arrangement, or **conformation**. In addition, during maturation, **posttranslational modifications** may add new chemical groups or remove transiently needed peptide segments. Genetic or nutritional deficiencies that impede protein maturation are deleterious to health. Examples of the former include Creutzfeldt-Jakob disease, scrapie, Alzheimer disease, and bovine spongiform encephalopathy ("mad cow disease"). Examples of the latter include scurvy (ascorbic acid) and Menkes syndrome (Cu).

Conversely, many next-generation direct-acting antiviral therapeutics for viral diseases such as hepatitis C or HIV act by inhibiting the activity of proteases, glycosidases, and peptidyl protein *cis–trans* isomerases that catalyze key steps in the maturation of essential viral proteins.

CONFORMATION VERSUS CONFIGURATION

The terms configuration and conformation are often confused. **Configuration** refers to the geometric relationship between a given set of atoms, for example, those that distinguish L- from D-amino acids. Interconversion of *configurational* alternatives requires breaking (and reforming) covalent bonds. **Conformation** refers to the spatial relationship of every atom in a molecule.

Interconversion between *conformers* occurs with retention of configuration, generally via rotation about single bonds.

PROTEINS WERE INITIALLY CLASSIFIED BY THEIR GROSS CHARACTERISTICS

Scientists initially approached the elucidation of structure–function relationships in proteins by separating them into classes based on properties such as solubility, shape, or the presence of nonprotein groups. For example, the proteins that can be extracted from cells using aqueous solutions of physiologic pH and ionic strength are classified as **soluble**. Extraction of **integral membrane proteins** requires dissolution of the membrane with detergents. **Globular proteins** are compact, roughly spherical molecules that have **axial ratios** (the ratio of their shortest to longest dimensions) of not over three. Most enzymes are globular proteins. By contrast, many structural proteins adopt highly extended conformations. These **fibrous proteins** may possess axial ratios of 10 or more.

Lipoproteins and glycoproteins contain covalently bound lipid and carbohydrate, respectively. Myoglobin, hemoglobin, cytochromes, and many other **metalloproteins** contain tightly associated metal ions. While more precise classification schemes have emerged based on similarity, or **homology**, in amino acid sequence and three-dimensional structure, many early classification terms remain in use.

PROTEINS ARE CONSTRUCTED USING MODULAR PRINCIPLES

Proteins perform complex physical and catalytic functions by positioning specific chemical groups in a precise three-dimensional arrangement. The polypeptide scaffold containing these groups must adopt a conformation that is both functionally efficient and physically strong. At first glance, the biosynthesis of polypeptides comprised of tens of thousands of individual atoms would appear to be extremely challenging. When one considers that a typical polypeptide can potentially adopt more than or equal to 10^{50} distinct conformations, folding into the conformation appropriate to their biologic function would appear to be even more difficult. As described in Chapters 3 and 4, synthesis of the polypeptide backbones of proteins employs a small set of common building blocks or modules, the amino acids, joined by a common linkage, the peptide bond. Similarly, a stepwise modular pathway simplifies the folding and processing of newly synthesized polypeptides into mature proteins.

FOUR ORDERS OF PROTEIN STRUCTURE

The modular nature of protein synthesis and folding are embodied in the concept of orders of protein structure: **primary structure**—the sequence of amino acids in a polypeptide chain; **secondary structure**—the folding of short (3-30 residue), contiguous segments of polypeptide into geometrically ordered units; **tertiary structure**—the assembly of secondary structural units into larger functional units such as the mature polypeptide and its component domains; and **quaternary structure**—the number and types of polypeptide units of oligomeric proteins and their spatial arrangement.

SECONDARY STRUCTURE

Peptide Bonds Restrict Possible Secondary Conformations

Free rotation is possible about only two of the three types of covalent bonds comprising the polypeptide backbone: the bond linking the α-carbon (Cα) to the carbonyl carbon (Co) and the bond linking Cα to nitrogen (see Figure 3–8). The partial double-bond character of the peptide bond that links Co to the α-nitrogen requires that the carbonyl carbon, carbonyl oxygen, and α-nitrogen remain coplanar, thus preventing rotation. The angle about the Cα—N bond is termed the phi (φ) angle, and that about the Co—Cα bond the psi (ψ) angle. In peptides, for amino acids other than glycine, most combinations of phi and psi angles are disallowed because of steric hindrance (**Figure 5–1**). The conformations of proline are even more restricted as its cyclic structure prevents free rotation of its N—Cα bond.

The aminoacyl residues within extended segments of polypeptide (eg, loops) adopt a variety of phi and psi angles. Regions of ordered secondary structure arise when a series

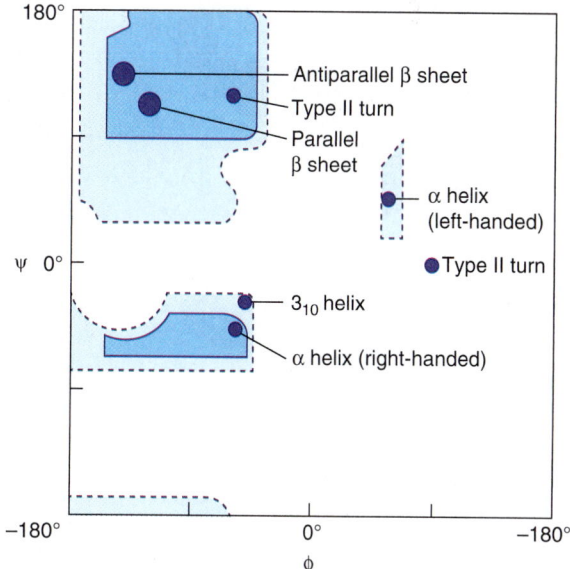

FIGURE 5–1 Ramachandran plot. The blue regions indicate sterically permissible combinations of phi–psi angles for nonglycine and nonproline amino acids in a polypeptide chain. The deeper the blue, the more thermodynamically favorable the phi–psi combination. Phi–psi angles corresponding to specific types of secondary structures are labeled.

of aminoacyl residues adopt similar phi and psi angles. The angles that define the two most common types of secondary structure, the **α helix** and the **β sheet**, fall within the lower and upper left-hand quadrants of a Ramachandran plot, respectively (see Figure 5–1).

Alpha Helix

The polypeptide backbone of an α helix is twisted by an equal amount about each α-carbon with a phi angle of approximately −57° and a psi angle of approximately −47°. A complete turn of the helix contains an average of 3.6 aminoacyl residues, and its *pitch* or rise per turn is 0.54 nm (**Figure 5–2**). The R groups of each aminoacyl residue in an α helix face outward (**Figure 5–3**). Since proteins are comprised solely of L-amino acids, the only stable α helices they can form are right-handed. Schematic diagrams of proteins often represent α helices as coils or cylinders.

The stability of an α helix arises primarily from hydrogen bonds formed between the oxygen of the peptide bond carbonyl and the hydrogen atom of the peptide bond nitrogen of the fourth residue down the polypeptide chain (**Figure 5–4**). The ability to form the maximum number of hydrogen bonds, supplemented by van der Waals interactions in the core of this tightly packed structure, provides the thermodynamic driving force for the formation of an α helix. Since the peptide bond nitrogen of proline lacks a hydrogen atom, it is incapable of

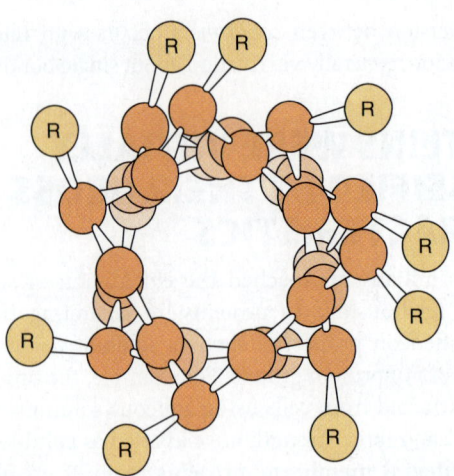

FIGURE 5–3 **View down the axis of a polypeptide α helix.** The side chains (R) are on the outside of the helix. The van der Waals radii of the atoms are larger than shown here; hence, there is almost no free space inside the helix.

forming a hydrogen bond with a carbonyl oxygen. Consequently, proline can only be stably accommodated within the first turn of an α helix. When present elsewhere, proline disrupts the conformation of the helix, producing a bend. Because it possesses

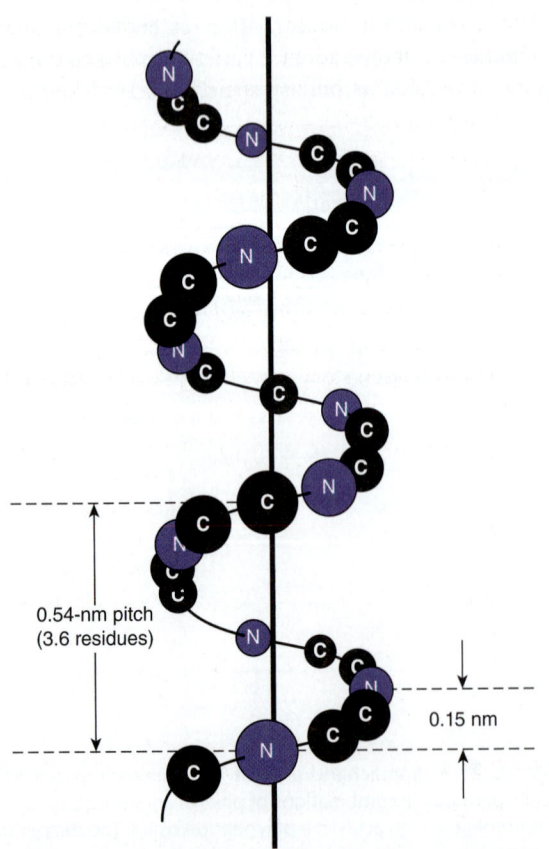

0.54-nm pitch
(3.6 residues)

0.15 nm

FIGURE 5–2 **Orientation of the main chain atoms of a peptide about the axis of an α helix.**

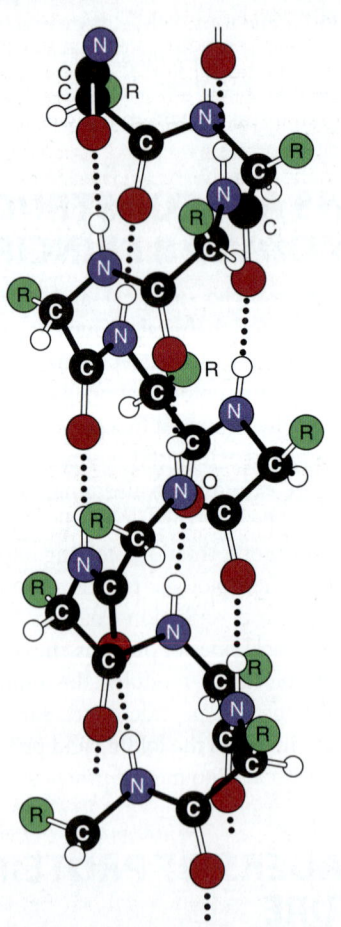

FIGURE 5–4 **Hydrogen bonds (dotted lines) formed between H and O atoms stabilize a polypeptide in an α-helical conformation.**

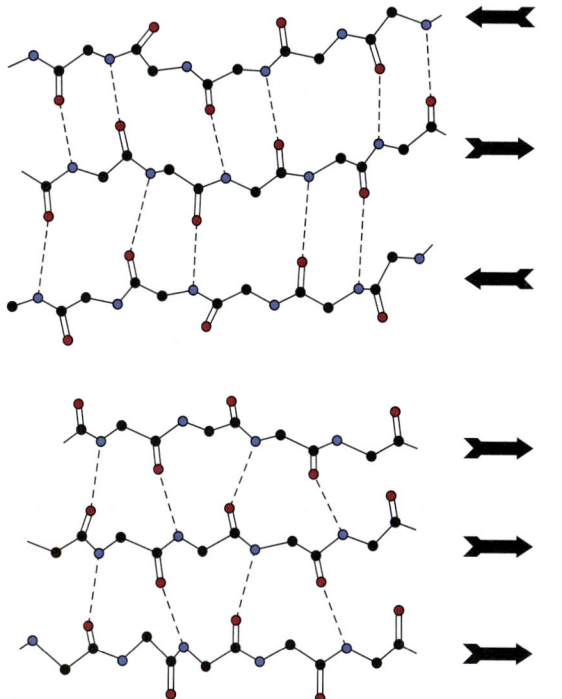

FIGURE 5–5 **Spacing and bond angles of the hydrogen bonds of antiparallel and parallel pleated β sheets.** Arrows indicate the direction of each strand. Hydrogen bonds are indicated by dotted lines with the participating α-nitrogen atoms (hydrogen donors) and oxygen atoms (hydrogen acceptors) shown in blue and red, respectively. Backbone carbon atoms are shown in black. For clarity in presentation, R groups and hydrogen atoms are omitted. **Top:** Antiparallel β sheet. Pairs of hydrogen bonds alternate between being close together and wide apart and are oriented approximately perpendicular to the polypeptide backbone. **Bottom:** Parallel β sheet. The hydrogen bonds are evenly spaced but slant in alternate directions.

such a small R group, glycine can disrupt packing, that may introduce a bend within an α helix.

Many α helices have predominantly hydrophobic R groups projecting from one side along their central axis and predominantly hydrophilic R groups projecting from the other side. These **amphipathic helices** are well adapted to the formation of interfaces between polar and nonpolar regions such as the hydrophobic interior of a protein and its aqueous environment. Clusters of amphipathic helices can create *channels*, or pores, through hydrophobic cell membranes that permit specific polar molecules to pass.

Beta Sheet

The second (hence "beta") recognizable regular secondary structure in proteins is the β sheet. Unlike the compact backbone of the α helix, the peptide backbone of the β sheet is highly extended. Viewed edge-on, the backbone follows a zigzag or pleated pattern from which the R groups of adjacent amino acid residues project in opposite directions. Like the α helix, β sheets derive much of their stability from hydrogen bonds between the carbonyl oxygens and amide hydrogens of peptide bonds. However, in contrast to the α helix, these bonds are formed between adjacent segments of the β sheet (**Figure 5–5**).

Interacting β sheets can be arranged such that the adjacent segments of the polypeptide chain proceed in the same direction amino to carboxyl to form **parallel β sheets** or in opposite directions to form **antiparallel β sheets** (see Figure 5–5). Both configurations are stabilized by hydrogen bonds between segments, or strands, of the sheet. Most β sheets are not perfectly flat but tend to have a right-handed twist. Clusters of twisted strands of β sheet, sometimes referred to as β barrels, form the core of many globular proteins (**Figure 5–6**). Schematic diagrams often represent β sheets as arrows whose points indicate the amino to the carboxyl terminal direction.

Loops & Bends

Roughly half of the residues in a "typical" globular protein reside in α helices or β sheets, and half in loops, turns, bends, and other extended conformational features. Turns and bends refer to short segments of amino acids that join two units of the secondary structure, such as two adjacent strands of an antiparallel β sheet. A β turn involves four aminoacyl residues, in which the first residue is hydrogen-bonded to the fourth, resulting in a tight 180° turn (**Figure 5–7**). Proline and glycine often are present in β turns.

Loops are regions that contain residues beyond the minimum number necessary to connect adjacent regions of secondary structure. Irregular in conformation, loops nevertheless serve key biologic roles. For many enzymes, catalytically critical aminoacyl residues are often located in the loops that bridge the domains responsible for binding substrates. **Helix-loop-helix motifs** provide the oligonucleotide-binding portion of many DNA-binding proteins such as repressors and transcription factors. Structural motifs such as the helix-loop-helix motif or the E-F hands of calmodulin (see Chapter 42) that are intermediate in scale between secondary and tertiary structures are often termed **supersecondary structures**. Since many loops and bends reside on the surface of proteins, and are thus exposed to solvent, they constitute readily accessible sites, or **epitopes**, for recognition and binding of antibodies.

While loops lack apparent structural regularity, many adopt a specific conformation stabilized through hydrogen bonding, salt bridges, and hydrophobic interactions with other portions of the protein. However, not all portions of proteins are necessarily ordered. Proteins may contain "disordered" regions, often at the extreme amino or carboxyl terminal, characterized by high conformational flexibility. In many instances, these disordered regions assume an ordered conformation upon binding of a ligand. This structural flexibility enables such regions to act as ligand-controlled switches that affect protein structure and function.

Tertiary & Quaternary Structure

The term "tertiary structure" refers to the entire three-dimensional conformation of an individual polypeptide. It indicates, in three-dimensional space, how secondary structural features—helices, sheets, bends, turns, and loops—assemble to form domains and how these domains relate spatially to

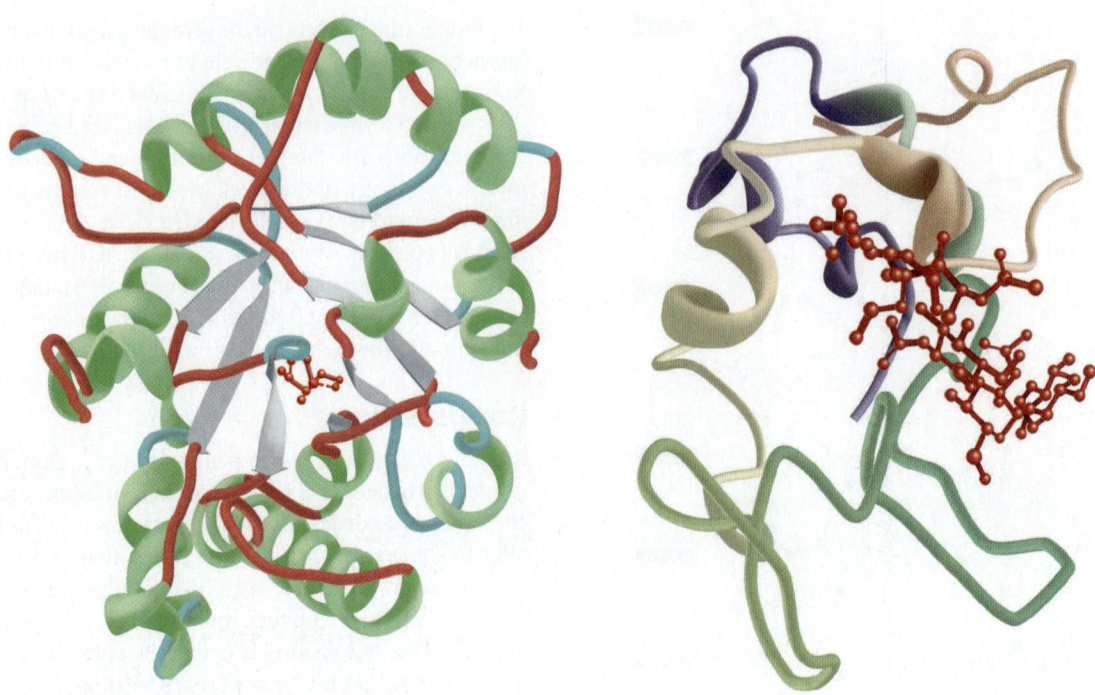

FIGURE 5–6 **Examples of the tertiary structure of proteins. Left:** The enzyme triose phosphate isomerase complexed with the substrate analog 2-phosphoglycerate (red). Note the elegant and symmetrical arrangement of alternating β sheets (gray) and α helices (green), with the β sheets forming a β-barrel core surrounded by the helices. (Adapted from Protein Data Bank ID no. 1o5x.) **Right:** Lysozyme complexed with the substrate analog penta-*N*-acetyl chitopentaose (red). The color of the polypeptide chain is graded along the visible spectrum from purple (N-terminal) to tan (C-terminal). Note, the concave shape of the domain forms a binding pocket for the pentasaccharide, the lack of β sheet, and the high proportion of loops and bends. (Adapted with permission from Protein Data Bank ID no. 1sfb.)

one another. A **domain** is a section of the protein structure sufficient to perform a particular chemical or physical task such as binding of a substrate or other ligand. Most domains are modular in nature, that is, contiguous in both primary sequence and three-dimensional space (**Figure 5–8**). Simple proteins, particularly those that interact with a single substrate or other ligand, such as lysozyme, triose phosphate isomerase (see Figure 5–6), or the oxygen storage protein myoglobin (see Chapter 6), often consist of a single domain. By contrast,

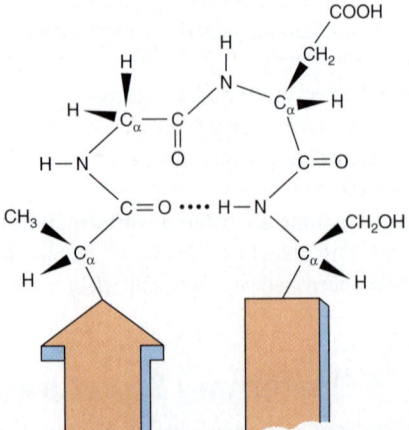

FIGURE 5–7 **A β turn that links two segments of antiparallel β sheet.** The dotted line indicates the hydrogen bond between the first and fourth amino acids of the four-residue segment Ala-Gly-Asp-Ser.

lactate dehydrogenase is comprised of two domains, an N-terminal NAD$^+$-binding domain and a C-terminal binding domain for the second substrate, pyruvate (see Figure 5–8). Lactate dehydrogenase is one of the family of oxidoreductases that share a common N-terminal NAD(P)$^+$-binding domain known as the **Rossmann fold**. By fusing a segment of DNA coding for a Rossmann fold domain to that coding for a variety of C-terminal domains, a large family of oxidoreductases have evolved that utilize NAD(P)$^+$/NAD(P)H for the oxidation and reduction of a wide range of metabolites. Examples include alcohol dehydrogenase, glyceraldehyde-3-phosphate dehydrogenase, malate dehydrogenase, quinone oxidoreductase, 6-phosphogluconate dehydrogenase, D-glycerate dehydrogenase, and formate dehydrogenase.

Not all domains bind substrates. Hydrophobic domains anchor proteins to membranes or enable them to span membranes. Localization sequences target proteins to specific subcellular or extracellular locations such as the nucleus, mitochondria, secretory vesicles, etc. Regulatory domains trigger changes in protein function in response to the binding of allosteric effectors or covalent modifications (see Chapter 9). Combining the genetic material coding for individual domain modules provides a facile route for generating proteins of great structural complexity and functional sophistication (**Figure 5–9**).

Proteins containing multiple domains can also be assembled through the association of multiple polypeptides, or protomers. Quaternary structure defines the polypeptide composition of a protein and, for an oligomeric protein, the spatial relationships

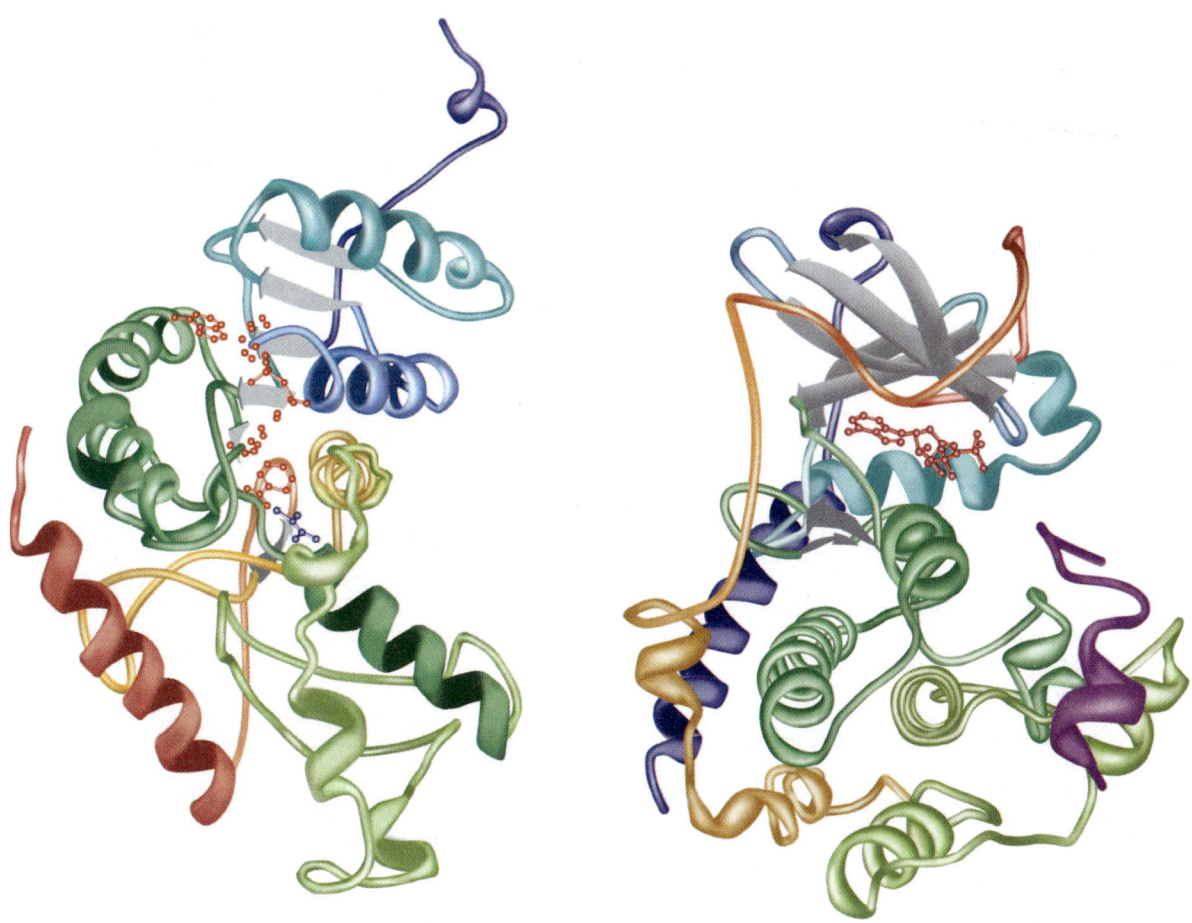

FIGURE 5–8 **Polypeptides containing two domains. Left:** Shown is the three-dimensional structure of a monomer unit of the tetrameric enzyme lactate dehydrogenase with the substrates NADH (red) and pyruvate (blue) bound. Not all bonds in NADH are shown. The color of the polypeptide chain is graded along the visible spectrum from blue (N-terminal) to orange (C-terminal). Note how the N-terminal portion of the polypeptide forms a contiguous domain, encompassing the upper portion of the enzyme, responsible for binding NADH. Similarly, the C-terminal portion forms a contiguous domain responsible for binding pyruvate. **Right:** Shown is the three-dimensional structure of the catalytic subunit of the cAMP-dependent protein kinase (see Chapter 42) with the substrate analogs ADP (red) and peptide (purple ribbon) bound. The color of the polypeptide chain is graded along the visible spectrum from blue (N-terminal) to orange (C-terminal). Protein kinases transfer the γ-phosphoryl group of ATP to protein and peptide substrates (see Chapter 9). Note how the N-terminal portion of the polypeptide forms a contiguous domain rich in β sheet that binds ADP. Similarly, the C-terminal portion forms a contiguous, α–helix–rich domain responsible for binding the peptide substrate. (Left, Adapted with permission from Protein Data Bank ID no. 3ldh; Right, Adapted with permission from Protein Data Bank ID no. 1jbp.)

between its protomers or subunits. **Monomeric** proteins consist of a single polypeptide chain. **Dimeric** proteins contain two polypeptide chains. **Homodimers** contain two copies of the same polypeptide chain, while in a **heterodimer** the polypeptides differ. Greek letters (α, β, γ, etc.) are used to distinguish different subunits of a hetero-oligomeric protein, and subscripts indicate the number of each subunit type. For example, α_4 designates a homotetrameric protein, and $\alpha_2\beta_2\gamma$, a protein with five subunits of three different types.

Schematic Diagrams Highlight Specific Structural Features

The three-dimensional structure for literally thousands of proteins can be accessed through the Protein Data Bank (http://www.rcsb.org/pdb/home/home.do) and other repositories. While one can obtain the images that indicate the position of every atom, the many thousands of atoms within even a small protein generally render such depictions too complex to be readily interpreted. Therefore, textbooks, journals, websites, etc., oftentimes utilize simplified schematic diagrams that highlight specific features of a protein's three-dimensional structure. Ribbon diagrams (see Figures 5–6 and 5–8) trace the conformation of the polypeptide backbone, with spirals (or, sometimes, cylinders) and arrows indicating regions of α helix and β sheet, respectively. In an even simpler representation, the path of the polypeptide backbone is traced by indicating the locations of the α–carbon in each amino acid residue linked order by line segments. In order to emphasize specific structure–function relationships, these schematic diagrams often depict the side chains of selected amino acids.

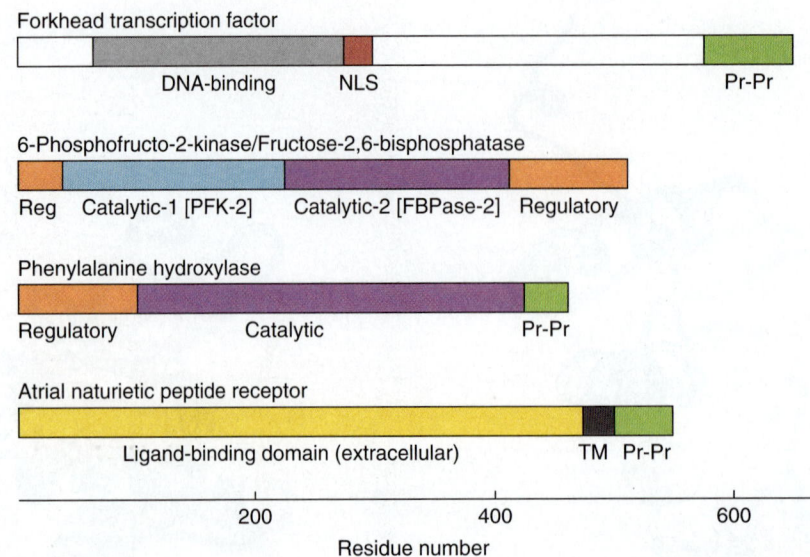

FIGURE 5–9 **Some multidomain proteins.** The rectangles represent the polypeptide sequences of a forkhead transcription factor; 6-phosphofructo-2-kinase/fructose-2,6-bisphosphatase, a bifunctional enzyme whose activities are controlled in a reciprocal fashion by allosteric effectors and covalent modification (see Chapter 19); phenylalanine hydroxylase (see Chapters 27 and 29), whose activity is stimulated by phosphorylation of its regulatory domain; and a receptor for atrial natriuretic peptide receptor, whose intracellular domain transmits signals through protein–protein interactions with heterotrimeric GTP-binding proteins (see Chapter 42). Regulatory domains are colored orange, catalytic domains in blue or purple, protein–protein interaction domains (Pr-Pr) in green, DNA-binding domains in gray, nuclear localization sequences in red, ligand-binding domains in light yellow, and transmembrane domains in black. The kinase and bisphosphatase activities of 6-phosphofructo-2-kinase/fructose-2,6-bisphosphatase are catalyzed by the N- (PFK-2) and C-terminal (FBP2-ase) proximate catalytic domains, respectively.

MULTIPLE FACTORS STABILIZE TERTIARY & QUATERNARY STRUCTURE

Higher orders of protein structure are stabilized primarily—and often exclusively—by noncovalent interactions. Principal among these are hydrophobic interactions that drive most hydrophobic amino acid side chains into the interior of the protein or at the interface where two subunits come into contact, thereby shielding them from the surrounding water molecules. Other significant contributors include hydrogen bonds and salt bridges between the carboxylates of aspartic and glutamic acid and the oppositely charged side chains of protonated lysyl, arginyl, and histidyl residues. These interactions are individually weak—1 to 5 kcal/mol relative to 80 to 120 kcal/mol for a covalent bond. However, just as a Velcro fastener harnesses the cumulative strength of a multitude of tiny plastic loops and hooks, collectively these individually weak but numerous interactions confer a high degree of stability upon the biologically functional conformation of a protein.

Some proteins contain covalent disulfide (S—S) bonds that link the sulfhydryl groups of cysteinyl residues. Formation of disulfide bonds involves oxidation of the cysteinyl sulfhydryl groups and requires oxygen. Intrapolypeptide disulfide bonds further enhance the stability of the folded conformation of a peptide, while interpolypeptide disulfide bonds stabilize the quaternary structure of certain oligomeric proteins.

BIOPHYSICAL TECHNIQUES REVEAL THREE-DIMENSIONAL STRUCTURE

X-Ray Crystallography

Following the solution of the three-dimensional structure of myoglobin by John Kendrew in 1960, x-ray crystallography has revealed the structures of thousands of biologic macromolecules ranging from proteins to oligonucleotides and viruses. For the solution of its structure by x-ray crystallography, a protein is first precipitated under conditions that form well-ordered crystals. To establish appropriate conditions, crystallization trials use a few microliters of protein solution and a matrix of variables (temperature, pH, presence of salts or organic solutes such as polyethylene glycol) to establish optimal conditions for crystal formation. Suitably sized crystals are then irradiated with a beam of x-rays and the pattern formed by those x-rays that are *diffracted* as a consequence of colliding with atomic nuclei of the protein recorded.

Early crystallographers collected the patterns formed by the diffracted x-rays as a series of spots on film. In order to calculate the position of each atom, the phases of the diffracted x-rays must first be determined. The traditional approach to solving the "phase problem" employs **isomorphous displacement**, in which an atom with a strong and distinctive x-ray "signature," such as mercury or uranium, is introduced into a crystal at known positions in the primary structure of the protein. Alternatively, recombinant expression can be used to substitute methionine with selenocysteine, whose selenium atom also possesses a distinct x-ray signature. If the protein being analyzed is homologous to one whose structure has already been solved, one can phase diffraction data without using heavy atoms by performing **molecular replacement** using the existing structure as a scaffold.

Initially, extrapolating the position of the individual atoms in the protein required manually measuring the position of each spot in the diffraction pattern followed by a series of lengthy hand calculations. Today, the patterns are recorded electronically using an area detector, then analyzed using a mathematical approach termed a *Fourier synthesis*. Do proteins inside these crystals exist in the same conformation as in free solution? The consensus is yes, a conclusion supported by the capacity of many enzymes retain the ability to catalyze chemical reactions when in the crystalline state.

Nuclear Magnetic Resonance Spectroscopy

Nuclear magnetic resonance (NMR) spectroscopy measures the absorbance of radio frequency electromagnetic energy by certain atomic nuclei. "NMR-active" isotopes of biologically relevant elements include ^{1}H, ^{13}C, ^{15}N, and ^{31}P. The frequency, or chemical shift, at which a particular nucleus absorbs energy is a function of the isotope itself as well as the presence and proximity of other NMR-active nuclei. In one-dimensional NMR a single isotope, for example ^{13}C, is monitored. The resulting chemical shift data can be used to identify the functional group in which each atom resides, for example, -CH_2- or -CHOH-, as well as the presence of any NMR-active nuclei bound to it. Two-dimensional NMR exploits the coupling of NMR-active nuclei through space, called the nuclear Overhauser effect (NOE), to determine the proximity of two different NMR-active nuclei, for example, ^{1}H and ^{13}C or ^{1}H and ^{15}N, to one another. From this information, the spatial relationships between component atoms can be used to construct a three-dimensional structure.

Since NMR spectroscopy analyzes proteins in aqueous solution, it can be used to observe changes in protein conformation that accompany ligand binding or catalysis in real time. It also provides an avenue for determining the three-dimensional structures of proteins that are difficult or impossible to crystallize. The noninvasive and nondestructive nature of NMR also offers the possibility of one day being able to observe the structure and dynamics of proteins within living cells.

Cryo–Electron Microscopy

Biochemists have long dreamed of observing proteins and other biological macromolecules in the same manner that optical microscopes enable microbiologists and cell biologists directly view living cells. However, the resolution of optical microscopes is limited by the wavelength of visible light (4 to 7×10^{-7} m) used to probe the sample. In the mid-20th century, scientists determined how to use beams of electrons as their source of electromagnetic radiation. The lower wavelength of their electron beams (1×10^{-12} m to 10×10^{-12} m) enabled **electron microscopes** (EM) to magnify objects more than a million times more than an optical microscope—sufficient to visualize large macromolecules such as ribosomes and DNA plasmids that had been coated with a coating of a dense element such as the uranium in uranium acetate. The high energy electron beam, however, tended to rapidly degrade many biomacromolecules.

In 2017, Jacques Dubochet, Joachim Frank, and Richard Henderson were awarded a Nobel Prize for the development of cryo-electron microscopy (cryo-EM). Cryo-EM employs ultracold media such as liquid nitrogen (T around −195°C), liquid ethane, or liquid helium to stabilize biomolecules in a *hydrated* state and protect them from heating when bombarded by an electron beam. This technique allows large macromolecules and macromolecular complexes to be visualized by EM. The extremely low temperatures conferred the added benefit of stabilizing macromolecules in a single *conformational state*. A survey of a set of macromolecules will often reveal the different structural conformations it can attain. Tomography takes advantage of the fact that the individual biomacromolecules and their complexes lay in multiple orientations on the grid to generate composite three-dimensional images from a set of two-dimensional images. The effect of factors that trigger changes in conformational state can thus be determined and compared. Cryo-EM is rapidly superseding x-ray crystallography as the method of choice for determining the three-dimensional structures of proteins and other biological macromolecules.

Molecular Modeling

A valuable adjunct to the empirical determination of the three-dimensional structure of proteins is the use of computer technology for molecular modeling. When the three-dimensional structure is known, **molecular dynamics** programs can be used to simulate the conformational dynamics of a protein and the manner in which factors such as temperature, pH, ionic strength, or amino acid substitutions influence these motions. **Molecular docking** programs simulate the interactions that take place when a protein encounters a substrate, inhibitor, or other ligand. Computer-aided drug design, the virtual screening for molecules likely to interact with key sites on a protein of biomedical interest, is extensively used to facilitate the discovery of potential new pharmacophores.

Molecular modeling is also employed to infer the structure of proteins for which x-ray crystallographic or NMR structures are not yet available. Secondary structure algorithms weigh the propensity of specific residues to become incorporated into α helices or β sheets in previously studied proteins to predict the secondary structure of other polypeptides. In **homology modeling**, the known three-dimensional structure of a protein is used as a template upon which to erect a model of the *probable* structure of a related protein. Thanks to recent advances in machine learning, predicting the three-dimensional conformation of a protein directly from its primary sequence may soon become routine.

PROTEIN FOLDING

Proteins are conformationally dynamic molecules that can fold into their functionally competent conformation in as little as a few milliseconds. Moreover, they often can refold if their conformation becomes disrupted, a process called renaturation. How are the remarkable speed and fidelity of protein folding attained? In nature, folding into the native state occurs too rapidly to be the product of a random, haphazard search of all possible structures. Denatured proteins are not just random coils. Native contacts are favored, and regions of the native structure persist even in the denatured state. Following are factors that facilitate and are basic mechanistic features of protein folding–refolding.

Native Conformation of a Protein Is Thermodynamically Favored

Since the biologically relevant—or native—conformation of a protein generally is the one that is most energetically favored, knowledge of the native conformation is specified in the primary sequence. However, the number of distinct combinations of phi and psi angles that can potentially be adopted by even a relatively small—100 amino acid long—polypeptide is unbelievably vast: 3^{198}. Hence, it would require billions of years for a polypeptide to arrive at its native conformation if folding occurred via the random exploration of all possible conformations. Clearly, in nature, protein folding takes place in a more orderly and guided fashion.

Folding Is Modular

Protein folding generally occurs via a stepwise process. In the first stage, as the newly synthesized polypeptide emerges from the ribosome, short segments fold into secondary structural units. Initiating folding *contranslationally*, that is, as a protein is synthesized, simplifies the folding process by allowing newly formed segments of organized structure to serve as scaffolds to direct folding of subsequent segments toward the native conformation and away from unproductive alternatives. In the second stage, the hydrophobic regions of this initial set secondary and supersecondary elements segregate together away from water to form the hydrophobic core of a "molten globule." Within the molten globule, these secondary structure modules rearrange or even morph into other types of secondary structural forms until the mature conformation of the protein is attained. This process is orderly, but not rigid. For oligomeric proteins, individual protomers tend to fold before they associate with other subunits.

Auxiliary Proteins Assist Folding

Under appropriate laboratory conditions, many proteins will spontaneously refold after being **denatured** (ie, unfolded) by mild heating or treatment with acid or base, chaotropic agents, or detergents. However, most proteins fail to spontaneously refold in vitro and, for those that do, the process is very slow—minutes to hours. In most cases, denatured proteins congeal together to form insoluble **aggregates**, disordered complexes of unfolded or partially folded polypeptides held together predominantly by hydrophobic interactions. While the native conformation may be thermodynamically favored, aggregates nonetheless generally occupy deep local energy wells that are extremely difficult to escape from. Thus, they represent unproductive and oftentimes cytotoxic dead ends in the folding process. Cells employ auxiliary proteins to speed the process of folding and to guide its trajectory away from aggregate formation and toward a productive conclusion.

Chaperones

Chaperone proteins participate in the folding of over half of all mammalian proteins. The hsp70 (70-kDa heat shock protein) family of chaperones binds short sequences of hydrophobic amino acids that emerge while a new polypeptide is being synthesized, shielding them from solvent as polypeptide synthesis continues. By preventing aggregation, chaperones provide an opportunity for appropriate secondary structural elements to form prior to their coalescence into a molten globule. The hsp60 family of chaperones, sometimes called chaperonins, differs in sequence and structure from hsp70 and its homologs. Hsp60 acts later in the folding process, often together with an hsp70 chaperone. The central cavity of the donut-shaped hsp60 oligomer provides a sheltered nonpolar environment in which a misfolded polypeptide can unfold and subsequently refold into its physiologically functional conformation.

Protein Disulfide Isomerase

Disulfide bonds between and within polypeptides stabilize tertiary and quaternary structures. The process is initiated by the enzyme protein sulfhydryl oxidase, which catalyzes the oxidation of cysteine residues to form disulfide bonds. However, disulfide bond formation is nonspecific—a given cysteine can form a disulfide bond with any accessible cysteinyl residue. By catalyzing disulfide exchange, the rupture of an S—S bond and its reformation with a different partner cysteine, protein disulfide isomerase facilitates the formation of disulfide bonds that stabilize a protein's native conformation. Since many eukaryotic sulfhydryl oxidases are flavin-dependent, dietary riboflavin deficiency often is accompanied by an increased incidence of improper folding of disulfide-containing proteins.

Proline-*cis, trans*-Isomerization

All X-Pro peptide bonds—where X represents any residue—are synthesized in the *trans* configuration. However, of the X-Pro bonds of mature proteins, approximately 6% are *cis*.

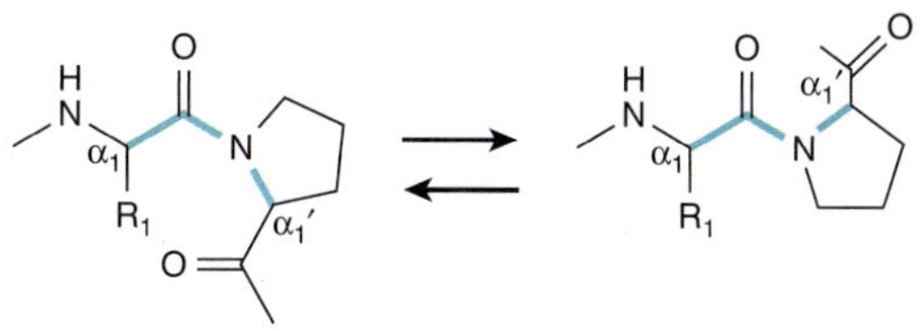

FIGURE 5–10 Isomerization of the *N*-α₁ prolyl peptide bond from a *cis* to a *trans* configuration relative to the backbone of the polypeptide.

The *cis* configuration is particularly common in β turns. Isomerization from *trans* to *cis* is catalyzed by enzymes called proline-*cis*, *trans*-isomerases, also known as cyclophilins (**Figure 5–10**). In addition to promoting the maturation of native proteins, cyclophilins also participate in the folding of proteins expressed as a consequence of viral infection. Consequently, cyclophilins are being pursued as targets for the development of drugs, such as cyclosporine and Alisporivir, for the treatment of HIV, hepatitis C, and other virally transmitted diseases.

PERTURBATION OF PROTEIN CONFORMATION MAY HAVE PATHOLOGIC CONSEQUENCES

Prions

The transmissible spongiform encephalopathies, or **prion diseases**, are fatal neurodegenerative diseases characterized by spongiform changes, astrocytic gliomas, and neuronal loss resulting from the deposition of insoluble protein aggregates in neural cells. They include Creutzfeldt-Jakob disease in humans, scrapie in sheep, and bovine spongiform encephalopathy (mad cow disease) in cattle. A variant form of Creutzfeldt-Jacob disease (vCJD) that afflicts younger patients is associated with early-onset psychiatric and behavioral disorders. Prion diseases may manifest themselves as infectious, genetic, or sporadic disorders. Because no viral or bacterial source for pathologic prion proteins could be identified, the cause and mechanism of transmission of prion diseases long remained elusive.

Today it is recognized that **prion diseases are protein conformation diseases** transmitted by altering the conformation, and hence the physical properties, of proteins endogenous to the host such as human prion-related protein (PrP). Prp, a glycoprotein encoded on the short arm of chromosome 20, normally exists as a monomer rich in α helix. Pathologic prion proteins serve as the templates for the conformational transformation of normal PrP, known as PrPc, into PrPsc. PrPsc is rich in β sheet with many hydrophobic aminoacyl side chains exposed to solvent. As each new PrPsc molecule is formed, it triggers the production of yet more pathologic variants in a conformational chain reaction. Because PrPsc molecules associate strongly with one another through their exposed hydrophobic regions, as they accumulate multiple

PrPsc molecules coalesce to form insoluble protease-resistant aggregates. Since one pathologic prion or prion-related protein can serve as template for the conformational transformation of many times its number of PrPc molecules, prions act in a completely DNA- and RNA-independent manner.

Alzheimer Disease

Senile plaques and neurofibrillary bundles comprised predominantly of aggregates of β-amyloid, are the characteristic feature of the Alzheimer disease. β-amyloid is a 4.3-kDa polypeptide produced by the proteolytic cleavage of a larger protein endogenous to brain tissue known as amyloid precursor protein. In Alzheimer disease patients, levels of β-amyloid become elevated, and this protein undergoes a conformational transformation from a soluble α helix–rich state to a state rich in β sheet and prone to self-aggregation. While the root cause of Alzheimer disease remains elusive, apolipoprotein E has been implicated as a potential mediator of this conformational transformation.

Beta-Thalassemias

Thalassemias are caused by genetic defects that impair the synthesis of one of the polypeptide subunits of hemoglobin (see Chapter 6). During the burst of hemoglobin synthesis that occurs during erythrocyte development, a specific chaperone called α-hemoglobin–stabilizing protein (AHSP) binds to free hemoglobin α-subunits as they await incorporation into the hemoglobin multimer. In the absence of this chaperone, free α-hemoglobin subunits aggregate, and the resulting precipitate has cytotoxic effects on the developing erythrocyte. Investigations using genetically modified mice suggest a role for AHSP in modulating the severity of β-thalassemia in human subjects.

COLLAGEN ILLUSTRATES THE ROLE OF POSTTRANSLATIONAL PROCESSING IN PROTEIN MATURATION

Protein Maturation Often Involves Making & Breaking of Covalent Bonds

The maturation of proteins into their final structural state often involves the cleavage or formation (or both) of covalent bonds, a process of **posttranslational modification**. Many polypeptides are initially synthesized as larger precursors called **proproteins**. The "extra" polypeptide segments in these proproteins often serve as leader sequences that target a polypeptide to a particular organelle or facilitate its passage through a membrane or to provide secondary structural scaffolds to guide folding of the polypeptide. Other segments ensure that any potentially harmful activity, such as that of the proteases trypsin and chymotrypsin, remains dormant until they reach their intended destination. However, once these transient requirements are fulfilled, the now superfluous peptide regions are removed by selective proteolysis. Other covalent modifications may add new chemical functionalities to a protein. The maturation of collagen illustrates both of these processes.

Collagen Is a Fibrous Protein

The fibrous proteins collagen, keratin, and myosin account for more than 25% of the protein mass in the human body. These fibrous proteins represent a primary source of structural strength for cells (ie, the cytoskeleton) and tissues (eg, the extracellular matrix). Skin derives its strength and flexibility from an intertwined mesh of collagen and keratin fibers, while bones and teeth are buttressed by an underlying network of collagen fibers analogous to steel strands in reinforced concrete. Collagen also is present in connective tissues such as ligaments and tendons. The high degree of tensile strength required to fulfill these structural roles derives from collagen's elongated form and high degree of crosslinking.

Collagen Forms a Unique Triple Helix

A mature collagen fiber forms an elongated rod with an axial ratio of about 200. Each fiber is comprised of three intertwined polypeptide strands, which twist to the left, wrapping around one another in a right-handed fashion to form the unique collagen triple helix (**Figure 5–11**). The opposing handedness of this superhelix and its component polypeptides makes the collagen triple helix highly resistant to unwinding—a principle also applied to the steel cables of suspension bridges. A collagen triple helix has 3.3 residues per turn and a rise per residue nearly twice that of an α helix. The R groups of each polypeptide strand of the triple helix pack so closely that, in order to fit, one of the three must be H. Thus, every third amino acid residue in collagen is a glycine residue. Staggering of the three strands provides appropriate positioning of the requisite glycines throughout the helix. Collagen is also rich in proline and hydroxyproline, yielding a repetitive Gly-X-Y pattern (see Figure 5–11) in which Y generally is proline or hydroxyproline.

Collagen triple helices are stabilized by hydrogen bonds between residues in *different* polypeptide chains, a process helped by the hydroxyl groups of hydroxyprolyl residues. Additional stability is provided by covalent cross-links formed between modified lysyl residues both within and between polypeptide chains.

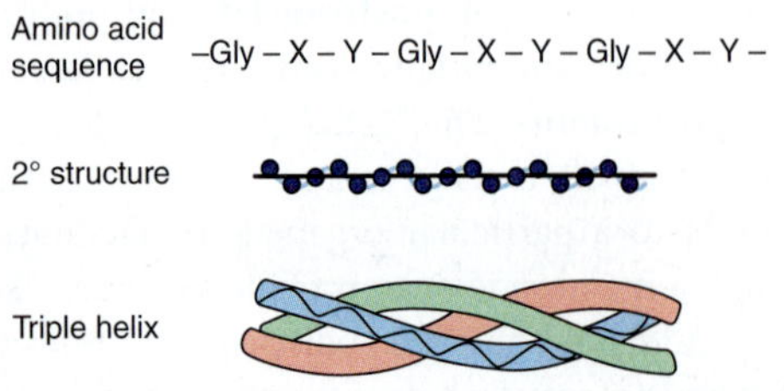

Amino acid sequence –Gly – X – Y – Gly – X – Y – Gly – X – Y –

2° structure

Triple helix

FIGURE 5–11   **Primary, secondary, and tertiary structures of collagen.**

Collagen Is Synthesized as a Larger Precursor

Collagen is initially synthesized as a larger precursor polypeptide, procollagen. Numerous prolyl and lysyl residues of procollagen are hydroxylated by prolyl hydroxylase and lysyl hydroxylase, enzymes that require ascorbic acid (vitamin C; see Chapters 27 and 44). Hydroxyprolyl and hydroxylysyl residues provide additional hydrogen-bonding capability that stabilizes the mature protein. In addition, glucosyl and galactosyl transferases attach glucosyl or galactosyl residues to the hydroxyl groups of specific hydroxylysyl residues.

The central portion of the precursor polypeptide then associates with other molecules to form the characteristic triple helix. This process is accompanied by the removal of the globular amino terminal and carboxyl terminal extensions of the precursor polypeptide by selective proteolysis. Certain lysyl residues are modified by lysyl oxidase, a copper-containing protein that converts ε-amino groups to aldehydes. The aldehydes can either undergo an aldol condensation to form a C=C double bond or form a Schiff base (eneimine) with the ε-amino group of an unmodified lysyl residue, which is subsequently reduced to form a C—N single bond. These covalent bonds cross-link the individual polypeptides and imbue the fiber with exceptional strength and rigidity.

Nutritional & Genetic Disorders Can Impair Collagen Maturation

The complex series of events in collagen maturation illustrate the consequences of incomplete polypeptide maturation. The best-known defect in collagen biosynthesis is **scurvy**, which results from a dietary deficiency of the vitamin C required by prolyl and lysyl hydroxylases. The consequent deficit in the number of hydroxyproline and hydroxylysine residues undermines the conformational stability of collagen fibers, leading to bleeding gums, swelling joints, poor wound healing, and ultimately death. **Menkes syndrome**, characterized by kinky hair and growth retardation, is brought on by a dietary deficiency of the copper required by lysyl oxidase, which catalyzes a key step in the formation of the covalent cross-links that strengthen collagen fibers.

Genetic disorders of collagen biosynthesis include several forms of osteogenesis imperfecta, characterized by fragile bones. In the Ehlers-Danlos syndrome, a group of connective tissue disorders that involve impaired integrity of supporting structures, defects in the genes that encode α collagen-1, procollagen *N*-peptidase, or lysyl hydroxylase result in mobile joints and skin abnormalities (see Chapter 50).

SUMMARY

- Proteins may be classified based on their solubility, shape, or function or on the presence of a prosthetic group, such as heme.

- The gene-encoded primary structure of a polypeptide is the sequence of its amino acids. Secondary structure results from folding of short, contiguous segments of a polypeptide into hydrogen-bonded motifs such as the α helix, the β–pleated sheet, β bends, and loops. Combinations of these motifs can form supersecondary motifs.

- Tertiary structure concerns the relationships between secondary and other small structural units to form functional protein domains and polypeptide monomers. Quaternary structure concerns the spatial relationships between various types of polypeptides in oligomeric proteins, those compromised of more than one polypeptide subunit or protomer.

- Primary structures are stabilized by covalent peptide bonds, whereas higher orders of structure are stabilized by weak forces—multiple hydrogen bonds, salt (electrostatic) bonds, and association of hydrophobic R groups.

- The phi (ϕ) angle of a polypeptide is the angle about the C_α—N bond; the psi (ψ) angle is that about the C_α—C_o bond. Most combinations of phi–psi angles are disallowed due to steric hindrance. The phi–psi angles that form the α helix and the β sheet fall within the lower and upper left-hand quadrants of a Ramachandran plot, respectively.

- Protein folding is a complex, only partially understood process. Broadly speaking, short segments of newly synthesized polypeptide fold into secondary structural units. Forces that bury hydrophobic regions from solvent then drive the partially folded polypeptide into a "molten globule" in which the modules of the secondary structure are rearranged to give the native conformation of the protein.

- Proteins that assist folding include protein disulfide isomerase, proline-*cis*, *trans*-isomerase, and the chaperones that participate in the folding of over half of mammalian proteins. Chaperones shield newly synthesized polypeptides from solvent and provide an environment for elements of secondary structure to emerge and coalesce into molten globules.

- Biomedical researchers are currently working to develop agents that interfere with the folding of viral proteins and prions as drugs for the treatment of hepatitis C and a range of neurodegenerative disorders.

- X-ray crystallography and NMR are key techniques used to study higher orders of protein structure.

- While lacking the atomic-level resolution of x-ray crystallography or NMR, cryo-EM has emerged as a powerful tool for analyzing the macromolecular dynamics of biologic macromolecules in heterogeneous samples.

- Prions proteins act autonomously, without need for DNA or RNA, to cause fatal transmissible spongiform encephalopathies such as Creutzfeldt-Jakob disease, scrapie, and bovine spongiform encephalopathy. Prion diseases involve an altered secondary–tertiary structure of a naturally occurring protein, PrPc. When PrPc interacts with its pathologic isoform PrPsc, its conformation is transformed from a predominantly α-helical structure to the β-sheet structure characteristic of PrPsc.

- Collagen illustrates the close linkage between protein structure and biologic function. Diseases of collagen maturation include Ehlers-Danlos syndrome and the vitamin C deficiency disease scurvy.

REFERENCES

Aguzzi A, de Cecco E: Shifts and drifts in prion science. Important questions remain unanswered since prions were discovered four decades ago. Science 2020;370:32.

Bryan-Marrugo OL, Ramos-Jimenez J, Barrera-Saldana H, et al: History and progress of antiviral drugs: From acyclovir to direct-acting antiviral agents (DAAs) for hepatitis C. Medicina Universitaria 2015;17:165.

Callaway E: Revolutionary cryo-EM is taking over structural biology. Nature 2020;578:201.

Gianni S, Jemth P: Protein folding: Vexing debates on a fundamental problem. Biophys Chem 2016;212:17.

Ito S, Nagata K: Quality control of procollagen in cells. Annu Rev Biochem 2021;90:631.

Jiang Y, Kalodimos CG: NMR studies of large proteins. J Mol Biol 2017;428:2667.

Luengo TM, Mayer MP, Rudiger SGD: The Hsp70-Hsp90 chaperone cascade in protein folding, Trends Cell Biol 2019;29:164.

Papageorgiou AC, Mattson S: Protein structure validation and analysis with X-ray crystallography. Meth Mol Biol 2014;1129:397.

Pastore A, Temussi PA: The Emperor's new clothes: Myths and truths of in-cell NMR. Arch Biochem Biophys 2017;628:114.

Pennisi E: Protein structure prediction now easier, faster. Science 2021;373:262.

Rein T: Peptidylprolylisomerases, protein folders, or scaffolders? The example of FKBP51 and FKBP52. Bioessays 2020;42:1900250.

Rosenzweig R, Nillegoda NB, Mayer MP, Bukau B: The Hsp70 chaperone network. Nature Rev Mol Cell Biol 2019;20:665.

Yamaguchi K, Kuwata K. Formation and properties of amyloid fibrils of prion protein. Biophys Rev 2018;10:517.

Exam Questions

Section I – Proteins: Structure & Function

1. Select the one of the following statements that is NOT CORRECT.
   A. Fermentation and glycolysis share many common biochemical features.
   B. Louis Pasteur first discovered that cell-free yeast preparations could convert sugars to ethanol and carbon dioxide.
   C. Organic orthophosphate (P_i) is essential for glycolysis.
   D. ^{14}C is an important tool for detecting metabolic intermediates.
   E. Medicine and biochemistry provide mutual insights to one another.

2. Select the one of the following statements that is NOT CORRECT.
   A. The vitamin derivative NAD is essential for conversion of glucose to pyruvate.
   B. The term "Inborn errors of metabolism" was coined by the physician Archibald Garrod.
   C. Mammalian tissue slices can incorporate inorganic ammonia into urea.
   D. Realization that DNA is a double helix permitted Watson & Crick to describe the polymerase chain reaction (PCR).
   E. Mutation of the genome of a "model organism" can provide insight into biochemical processes.

3. Explain how the Büchner's observation in the early part of the 20th century led to the discovery of the details of fermentation.

4. Name some of the earliest discoveries that followed the realization that a cell-free preparation of yeast cells could catalyze the process of fermentation.

5. Name some of the kinds of tissue preparations that early 20th century biochemists employed to study glycolysis and urea biosynthesis, and to discover the roles of vitamin derivatives.

6. Describe how the availability of radioactive isotopes facilitated the identification of metabolic intermediates.

7. Name several of the "inborn errors of metabolism" identified by the physician Archibald Garrod.

8. Cite an example in lipid metabolism for which the linking of biochemical and genetic approaches has contributed to the advance of medicine and biochemistry.

9. Name several of the intact "model organisms" whose genomes can be selectively altered to provide insight into biochemical processes.

10. Select the one of the following statements that is NOT CORRECT.

    The propensity of water molecules to form hydrogen bonds with one another is the primary factor responsible for all of the following properties of water EXCEPT:
    A. Its atypically high boiling point.
    B. Its high heat of vaporization.
    C. Its high surface tension.
    D. Its ability to dissolve hydrocarbons.
    E. Its expansion upon freezing.

11. Select the one of the following statements that is NOT CORRECT.
    A. The side-chains of the amino acids cysteine and methionine absorb light at 280 nm.
    B. Glycine is often present in regions where a polypeptide forms a sharp bend, reversing the direction of a polypeptide.
    C. Polypeptides are named as derivatives of the C-terminal aminoacyl residue.
    D. The C, N, O, and H atoms of a peptide bond are coplanar.
    E. A linear pentapeptide contains four peptide bonds.

12. Select the one of the following statements that is NOT CORRECT.
    A. Buffers of human tissue include bicarbonate, proteins, and orthophosphate.
    B. A weak acid or a weak base exhibits its greatest buffering capacity when the pH is equal to its pK_a plus or minus one pH unit.
    C. The isoelectric pH (pI) of lysine can be calculated using the formula $(pK_2 + pK_3)/2$.
    D. The mobility of a monofunctional weak acid in a direct current electrical field reaches its maximum when the pH of its surrounding environment is equal to its pK_a.
    E. For simplicity, the strengths of weak bases are generally expressed as the pK_a of their conjugate acids.

13. Select the one of the following statements that is NOT CORRECT.
    A. If the pK_a of a weak acid is 4.0, 50% of the molecules will be in the dissociated state when the pH of the surrounding environment is 4.0.
    B. A weak acid with a pK_a of 4.0 will be a more effective buffer at pH 3.8 than at pH 5.7.
    C. At a pH equal to its pI, a polypeptide carries no charged groups.
    D. Strong acids and bases are so named because they undergo complete dissociation when dissolved in water.
    E. The pK_a of an ionizable group can be influenced by the physical and chemical properties of its surrounding environment.

14. Select the one of the following statements that is NOT CORRECT.
    A. A major objective of proteomics is to identify all of the proteins present in a cell under different conditions as well as their states of modification.
    B. Mass spectrometry has largely replaced the Edman method for sequencing of peptides and proteins.
    C. Sanger reagent was an improvement on Edman's because the former generates a new amino terminus, allowing several consecutive cycles of sequencing to take place.
    D. Since mass is a universal property of all atoms and molecules, mass spectrometry is ideally suited to the detection of posttranslational modifications in proteins.
    E. Time-of-flight mass spectrometers take advantage of the relationship F = ma.

15. Why does olive oil added to water tend to form large droplets?

16. What distinguishes a strong base from a weak base?

17. Select the one of the following statements that is NOT CORRECT.
   A. Ion-exchange chromatography separates proteins based upon the sign and magnitude of their charge at a given pH.
   B. Two-dimensional gel electrophoresis separates proteins first on the basis of their pI values and second on their charge-to-mass ratio using SDS-PAGE.
   C. Affinity chromatography exploits the selectivity of protein-ligand interactions to isolate a specific protein from a complex mixture.
   D. Many recombinant proteins are expressed with an additional domain fused to their N- or C-terminus. One common component of these fusion domains is a ligand-binding site designed expressly to facilitate purification by affinity chromatography.
   E. Following purification by classical techniques, tandem mass spectrometry typically is used to analyze individual homogeneous peptides derived from a complex protein mixture.

18. Select the one of the following statements that is NOT CORRECT.
   A. Protein folding is assisted by intervention of specialized auxiliary proteins called chaperones.
   B. Protein folding tends to be modular, with areas of local secondary structure forming first, then coalescing into a molten globule.
   C. Protein folding is driven first and foremost by the thermodynamics of the water molecules surrounding the nascent polypeptide.
   D. The formation of S-S bonds in a mature protein is facilitated by the enzyme protein disulfide isomerase.
   E. Only a few unusual proteins, such as collagen, require posttranslational processing by partial proteolysis to attain their mature conformation.

19. Estimate pI for a polyelectrolyte that contains three carboxyl groups and three amino groups whose pK$_a$ values are 4.0, 4.6, 6.3, 7.7, 8.9, and 10.2.

20. State one drawback of the categorization of the protein amino acids simply as "essential" or "nonessential"?

21. Select the one of the following statements that is NOT CORRECT.
   A. Posttranslational modifications of proteins can affect both their function and their metabolic fate.
   B. The native conformational state generally is that which is thermodynamically favored.
   C. The complex three-dimensional structures of most proteins are formed and stabilized by the cumulative effects of a large number of weak interactions.
   D. Research scientists employ gene arrays for the high-throughput analysis of the presence and expression level of proteins.
   E. Examples of weak interactions that stabilize protein folding include hydrogen bonds, salt bridges, and van der Waals forces.

22. Select the one of the following statements that is NOT CORRECT.
   A. Changes in configuration involve the rupture of covalent bonds.
   B. Changes in conformation involve the rotation of one or more single bonds.
   C. The Ramachandran plot illustrates the degree to which steric hindrance limits the permissible angles of the single bonds in the backbone of a peptide or protein.
   D. Formation of an α helix is stabilized by the hydrogen bonds between each peptide bond carboxyl oxygen and the N-H group of the next peptide bond.
   E. In a β sheet the R groups of adjacent residues point in opposite directions relative to the plane of the sheet.

23. Select the one of the following statements that is NOT CORRECT.
   A. The descriptor α$_2$β$_2$γ$_3$ denotes a protein with seven subunits of three different types.
   B. Loops are extended regions that connect adjacent regions of secondary structure.
   C. More than half of the residues in a typical protein reside in either α helices or β sheets.
   D. Most β sheets have a right-handed twist.
   E. Prions are viruses that cause protein-folding diseases that attack the brain.

24. What advantage does the acidic group of phosphoric acid that is associated with pK$_2$ offer for buffering in human tissues?

25. The dissociation constants for a previously uncharacterized racemic amino acid discovered in a meteor have been determined to be pK$_1$ = 2.0, pK$_2$ = 3.5, pK$_3$ = 6.3, pK$_4$ = 8.0, pK$_5$ = 9.8, and pK$_7$ = 10.9:
   A. What carboxyl or amino functional group would you expect to be associated with each dissociation?
   B. What would be the approximate net charge on this amino acid at pH 2?
   C. What would be its approximate net charge at pH 6.3?
   D. During direct current electrophoresis at pH 8.5, toward which electrode would this amino acid be likely to move?

26. A biochemical buffer is a compound which tends to resist changes in pH even when acids or bases are added. What two properties are required of an effective physiologic buffer? In addition to phosphate, what other physiologic compounds meet these criteria?

27. Name two amino acids whose posttranslational modification confers significant new properties on a protein.

28. Explain why diets deficient in (a) copper (Cu) or (b) ascorbic acid lead to incomplete posttranslational processing of collagen.

29. Describe the role of N-terminal signal sequences in the biosynthesis of certain proteins.

SECTION II

Enzymes: Kinetics, Mechanism, Regulation, & Role of Transition Metals

CHAPTER 6

Proteins: Myoglobin & Hemoglobin

Peter J. Kennelly, PhD, & Victor W. Rodwell, PhD

OBJECTIVES

After studying this chapter, you should be able to:

- Describe the most important structural similarities and differences between myoglobin and hemoglobin.
- Sketch binding curves for the oxygenation of myoglobin and hemoglobin.
- Explain why hemoglobin's sigmoidal O_2-binding curve renders it a better vehicle for transporting oxygen than myoglobin's hyperbolic binding curve.
- Explain how the presence of the distal histidine affects the ability of hemoglobin to bind carbon monoxide (CO).
- Define P_{50} and indicate its significance in oxygen transport and delivery.
- Describe the structural and conformational changes in hemoglobin that accompany its oxygenation and subsequent deoxygenation. What role does the proximal histidine play in this transition?
- Explain the role of 2,3-bisphosphoglycerate (BPG) in oxygen binding and delivery.
- Explain how the Bohr effect a) enhances the ability of red blood cells to absorb CO_2 from peripheral tissues and b) promotes the release of transported CO_2 in the O_2 rich environment of the lungs.
- Describe the structural consequences to hemoglobin S (HbS) of lowering Po_2.
- Identify the metabolic defect that occurs as a consequence of α and β thalassemias.

BIOMEDICAL IMPORTANCE

The efficient delivery of oxygen from the lungs to the peripheral tissues and the maintenance of tissue reserves to protect against anoxic episodes are essential to health. In mammals, these functions are performed by the homologous heme proteins hemoglobin and myoglobin, respectively. Myoglobin, a monomeric protein of red muscle, binds oxygen tightly as a reserve against oxygen deprivation. The multiple subunits of hemoglobin, a tetrameric protein of erythrocytes, interact in a cooperative fashion that enables this transporter to offload a high proportion of bound O_2 in peripheral tissues while simultaneously retaining the capacity to bind it efficiently in the lungs. The offloading of oxygen is enhanced by the binding of 2,3-bisphosphoglycerate (BPG), which stabilizes the quaternary structure of deoxyhemoglobin. Cyanide and carbon monoxide (CO) kill because they disrupt the physiologic function of the heme proteins cytochrome oxidase and hemoglobin, respectively.

In addition to delivering O_2, hemoglobin plays a critical role in the transport of the waste product of respiration, CO_2, to the lungs for disposal. Proton binding to hemoglobin enhances the conversion of CO_2, a major product of respiration, into water-soluble carbonic acid and its conjugate base, bicarbonate by the enzyme carbonic anhydrase. In the lungs, O_2-induced proton release from hemoglobin reverses this process to facilitate disposal as gaseous CO_2. Additional CO_2 is carried in the form of covalently bound carbamates. Hemoglobin and myoglobin illustrate both protein structure–function relationships and the molecular basis of genetic disorders such as sickle cell disease and the thalassemias.

HEME & FERROUS IRON CONFER THE ABILITY TO STORE & TRANSPORT OXYGEN

Myoglobin and hemoglobin contain **heme**, an iron-containing cyclic tetrapyrrole consisting of four molecules of pyrrole linked by **methyne** bridges. This planar network of conjugated double bonds absorbs visible light and colors heme deep red. The substituents at the β-positions of heme are methyl (M), vinyl (V), and propionate (Pr) groups arranged in the order M, V, M, V, M, Pr, Pr, M (**Figure 6–1**). An atom of ferrous iron (Fe^{2+}) resides at the center of the planar tetrapyrrole. Oxidation of the Fe^{2+} of myoglobin or hemoglobin to Fe^{3+} destroys their biologic activity. Other proteins with metal-containing tetrapyrrole prosthetic groups include the cytochromes (Fe and Cu) and chlorophyll (Mg) (see Chapter 31).

Myoglobin Is Rich in α Helix

Oxygen stored in red muscle myoglobin is released during O_2 deprivation (eg, severe exercise) for use in muscle mitochondria for aerobic synthesis of ATP (see Chapter 13). A 153-aminoacyl residue polypeptide (MW 17,000), the compactly folded myoglobin molecule measures 4.5 × 3.5 × 2.5 nm (**Figure 6–2**). An unusually high proportion, about 75%, of the residues are present in eight right-handed 7 to 20 residue α helices. Starting at the amino terminal, these are termed helices A through H. Typical of globular proteins, the surface of myoglobin is rich in amino acids bearing polar and potentially charged side chains, while—with two exceptions—the interior contains residues that possess nonpolar R groups (eg, Leu, Val, Phe, and Met). The exceptions are the seventh and eighth residues in helices E and F, His E7 and His F8, which lie close to the heme iron, the site of O_2 binding.

Histidines F8 & E7 Perform Unique Roles in Oxygen Binding

The heme of myoglobin lies in a crevice between helices E and F oriented with its polar propionate groups facing the surface of the globin (see Figure 6–2). The remainder resides in the nonpolar interior. It is held in place mainly by hydrophobic interactions. The fifth coordination position of the iron is occupied by a nitrogen from the imidazole ring of the **proximal histidine**, His F8. The **distal histidine**, His E7, lies on the side of the heme ring opposite to His F8.

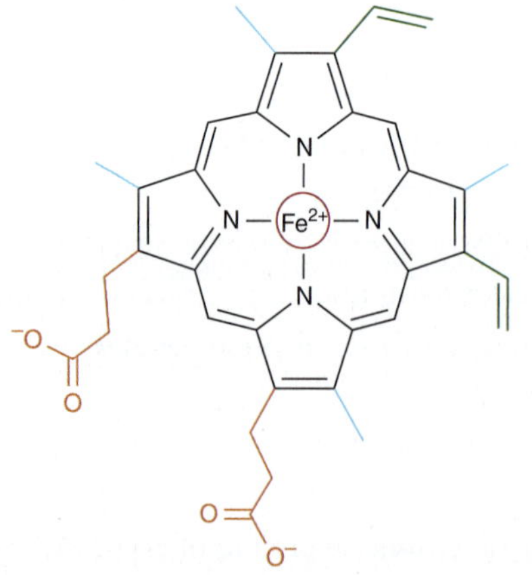

FIGURE 6–1 **Heme.** The pyrrole rings and methyne bridge carbons are coplanar, and the iron atom (Fe^{2+}) resides in almost the same plane. The fifth and sixth coordination positions of Fe^{2+} are directly perpendicular to—and directly above and below—the plane of the heme ring. Observe the nature of the methyl (blue), vinyl (green), and propionate (orange) substituent groups on the β carbons of the pyrrole rings, the central iron atom (red), and the location of the polar side of the heme ring (at about 7 o'clock) that faces the surface of the myoglobin molecule.

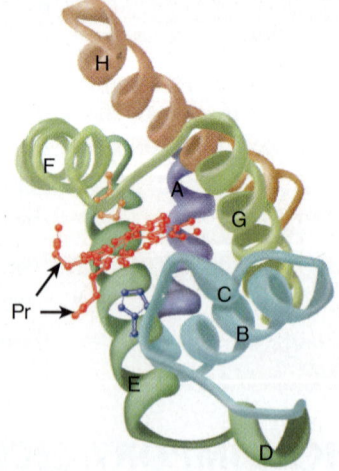

FIGURE 6–2 **Three-dimensional structure of myoglobin.** Shown is a ribbon diagram tracing the polypeptide backbone of myoglobin. The color of the polypeptide chain is graded along the visible spectrum from blue (N-terminal) to tan (C-terminal). The heme prosthetic group is red. The α-helical regions are designated A through H. The distal (E7) and proximal (F8) histidine residues are highlighted in blue and orange, respectively. Note how the polar propionate substituents (Pr) project out of the heme toward solvent. (Adapted with permission from Protein Data Bank ID no. 1a6n.)

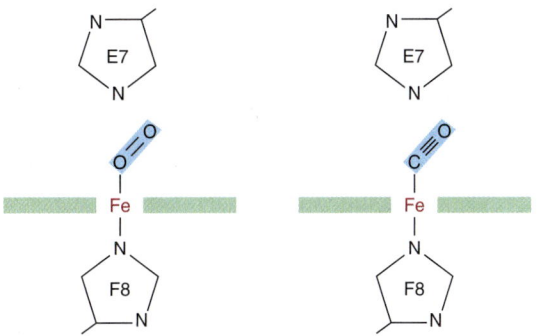

FIGURE 6–3 **Angles for bonding of oxygen and carbon monoxide (CO) to the heme iron of myoglobin.** The distal E7 histidine hinders bonding of CO at the preferred (90°) angle to the plane of the heme ring.

The Iron Moves Toward the Plane of the Heme When Oxygen Is Bound

The iron of unoxygenated myoglobin lies 0.03 nm (0.3 Å) outside the plane of the heme ring, toward the proximal histidine, His F8. Consequently, the heme "puckers" slightly. When O_2 occupies the sixth coordination position, the iron moves to within 0.01 nm (0.1 Å) of the plane of the heme ring. Oxygenation of myoglobin thus is accompanied by motion of the iron, of His F8, and of residues linked to His F8.

Apomyoglobin Provides a Hindered Environment for the Heme Iron

Minute quantities of CO arise from a variety of biologic sources, including the catabolism of red blood cells within the human body. CO is also present in the atmosphere, mainly due to the incomplete combustion of fossil fuels. CO binds to the iron in a free heme group 25,000 times more strongly than oxygen. However, the heme groups in the apoproteins of myoglobin and hemoglobin reside in a **hindered environment** for their gaseous ligands (**Figure 6–3**). When CO binds to free heme, all three atoms (Fe, C, and O) lie **perpendicular** to the plane of the heme. This geometry maximizes the overlap between the lone pair of electrons on the sp hybridized carbon of the CO molecule and the Fe^{2+} iron (**Figure 6–4**, right). By contrast, O_2 is sp^2 hybridized. Consequently, the electrons that bind to the iron lie at an angle of roughly 120° to the axis of the O=O double bond (see Figure 6–4, left). In myoglobin and hemoglobin the distal histidine sterically precludes or hinders the preferred, high-affinity orientation of CO while still permitting O_2 to attain its most favorable orientation (see Figure 6–3). Binding at a less favored angle reduces the strength of the heme-CO bond to about 200 times that of the heme-O_2 bond (see Figure 6–3, right). Therefore O_2, which typically is present in great excess over CO, normally dominates. Nevertheless, about 1% of human myoglobin typically is present combined with CO.

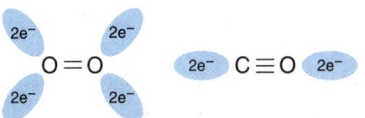

FIGURE 6–4 **Orientation of the lone pairs of electrons relative to the O=O and C≡O bonds of oxygen and carbon monoxide.** In molecular oxygen, formation of the double bond between the two oxygen atoms is facilitated by the adoption of an sp^2 hybridization state by the valence electron of each oxygen atom. As a consequence, the two atoms of the oxygen molecule and each lone pair of electrons are coplanar and separated by an angle of roughly 120° (**left**). By contrast, the two atoms of carbon monoxide are joined by a triple bond, which requires that the carbon and oxygen atoms adopt an sp hybridization state. In this state, the lone pairs of electrons and triple bonds are arranged in a linear fashion, where they are separated by an angle of 180° (**right**).

THE OXYGEN DISSOCIATION CURVES FOR MYOGLOBIN & HEMOGLOBIN SUIT THEIR PHYSIOLOGIC ROLES

The hyperbolic relationship between the concentration, or partial pressure, of O_2 (Po_2) and the quantity of O_2 bound, expressed as an O_2 saturation isotherm (**Figure 6–5**), reveals why myoglobin is more suitable for O_2 storage than for O_2 transport. Myoglobin loads O_2 readily at the Po_2 of the lung capillary bed (100 mm Hg). However, since myoglobin releases only a small fraction of its bound O_2 at the Po_2 values typically encountered in active muscle (20 mm Hg) or other tissues (40 mm Hg), it represents an ineffective vehicle for delivery of O_2. However, when strenuous exercise lowers the Po_2 of muscle tissue to about 5 mm of Hg, dissociation of O_2 from myoglobin permits mitochondrial synthesis of ATP, and hence muscular activity, to continue.

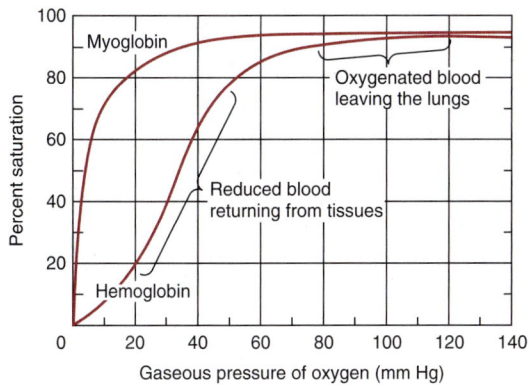

FIGURE 6–5 **Oxygen-binding curves of both hemoglobin and myoglobin.** Arterial oxygen tension is about 100 mm Hg; mixed venous oxygen tension is about 40 mm Hg; capillary (active muscle) oxygen tension is about 20 mm Hg; and the minimum oxygen tension required for cytochrome oxidase is about 5 mm Hg. Association of chains into a tetrameric structure (hemoglobin) results in much greater oxygen delivery than would be possible with single chains. (Modified with permission from Scriver CR, Beaudet AL, Sly WS: *The Molecular and Metabolic Bases of Inherited Disease*, 7th ed. New York, NY: McGraw Hill; 1995.)

By contrast, hemoglobin behaves as if it were two proteins. At the high Po_2 of the lungs, 100 mm Hg and above, it displays a high affinity for oxygen that enables it to bind oxygen to nearly every available heme iron. This form of the protein is commonly referred to as **R**, for **relaxed, state** hemoglobin. At the lower Po_2 values encountered in peripheral tissues, 40 mm Hg and below, hemoglobin exhibits a much lower apparent affinity for oxygen. Transition of hemoglobin to this low affinity, **taut** or **T state** enables it to release a large proportion of the oxygen previously picked up in the lungs. This dynamic interchange between the high and low affinity R and T states serves as the foundation for hemoglobin's sigmoidal O_2-binding curve.

THE ALLOSTERIC PROPERTIES OF HEMOGLOBINS RESULT FROM THEIR QUATERNARY STRUCTURES

The quaternary structure of hemoglobin confers striking additional properties, absent from monomeric myoglobin, which adapts this tetrameric protein to its unique biologic roles in the reciprocal transport of O_2 and CO_2 between the lungs and peripheral tissues.

Hemoglobin Is Tetrameric

Hemoglobins are tetramers composed of pairs of two similar, but distinct, polypeptide subunits (**Figure 6–6**). Greek letters are used to designate each subunit type. The subunit composition of the principal hemoglobins are $\alpha_2\beta_2$ (HbA; normal adult hemoglobin), $\alpha_2\gamma_2$ (HbF; fetal hemoglobin), $\alpha_2\beta^S_2$ (HbS; sickle cell hemoglobin), and $\alpha_2\delta_2$ (HbA$_2$; a minor adult hemoglobin). The primary structures of the β, γ, and δ chains of human hemoglobin are highly conserved.

Myoglobin & the β Subunits of Hemoglobin Share Almost Identical Secondary & Tertiary Structures

Despite differences in the kind and number of amino acids present, myoglobin and the β polypeptide of hemoglobin A share almost identical secondary and tertiary structures. Similarities include the location of the heme and the helical

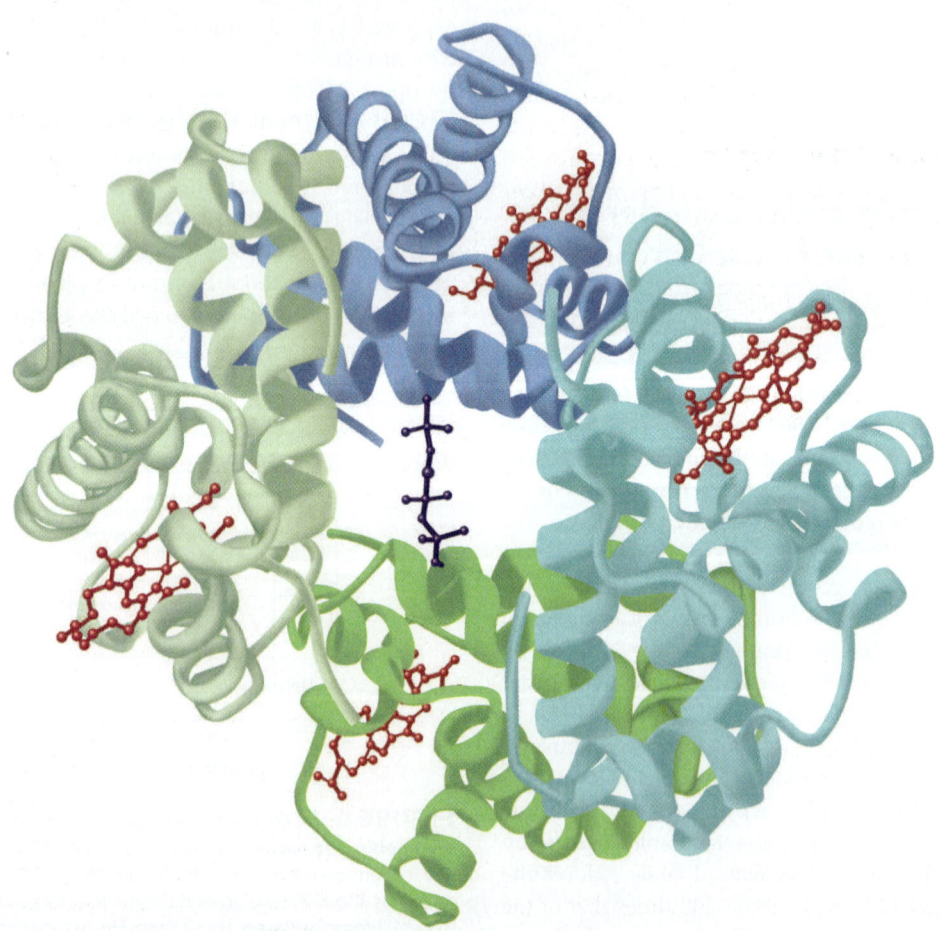

FIGURE 6–6 **Hemoglobin.** Shown is the three-dimensional structure of deoxyhemoglobin with a molecule of 2,3-bisphosphoglycerate (dark blue) bound. The two α subunits are colored in the darker shades of green and blue, the two β subunits in the lighter shades of green and blue, and the heme prosthetic groups in red. (Adapted with permission from Protein Data Bank ID no. 1b86.)

regions, and the presence of amino acids with similar properties at comparable locations. Although it possesses seven rather than eight helical regions, the α polypeptide of hemoglobin also closely resembles myoglobin.

Oxygenation of Hemoglobin Triggers Conformational Changes in the Apoprotein

Hemoglobins can bind up to four molecules of O_2 per tetramer, one per heme. However, once the first molecule of O_2 becomes bound, the affinity of the other three subunits increases (see Figure 6–5). Termed **cooperative binding**, this behavior permits hemoglobin to maximize both the quantity of O_2 loaded at the Po_2 of the lungs and the quantity of O_2 released at the Po_2 of the peripheral tissues.

P_{50} Expresses the Relative Affinities of Different Hemoglobins for Oxygen

The quantity P_{50}, a measure of O_2 concentration, is the partial pressure of O_2 at which a given hemoglobin reaches half-saturation. Depending on the organism, P_{50} can vary widely, but in all instances, it exceeds the normal Po_2 of the peripheral tissues. For example, the values of P_{50} for HbA and HbF are 26 and 20 mm Hg, respectively. In the placenta, this difference enables HbF to extract oxygen from the HbA in the mother's blood. However, HbF is suboptimal postpartum since its higher affinity for O_2 limits the quantity of O_2 delivered to the tissues.

The subunit composition of hemoglobin tetramers undergoes complex changes during development. The human fetus initially synthesizes a $\xi_2\epsilon_2$ tetramer. By the end of the first trimester, ξ and ϵ subunits have been replaced by α and γ subunits, forming HbF ($\alpha_2\gamma_2$), the hemoglobin of late fetal life. While synthesis of β subunits begins in the third trimester, the replacement of γ subunits by β subunits to yield adult HbA ($\alpha_2\beta_2$) does not reach completion until several weeks postpartum (**Figure 6–7**).

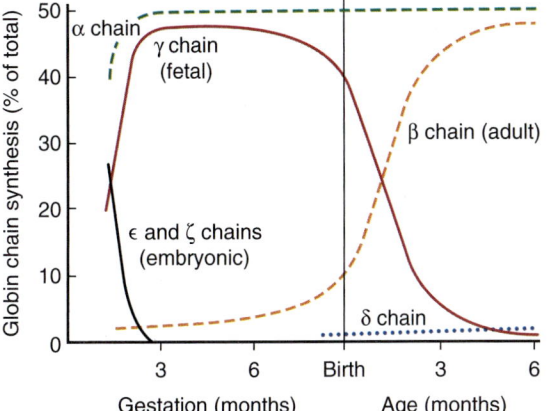

FIGURE 6–7 Developmental pattern of the quaternary structure of fetal and newborn hemoglobins. (Reproduced with permission from Ganong WF: *Review of Medical Physiology*, 20th ed. New York, NY: McGraw Hill; 2001.)

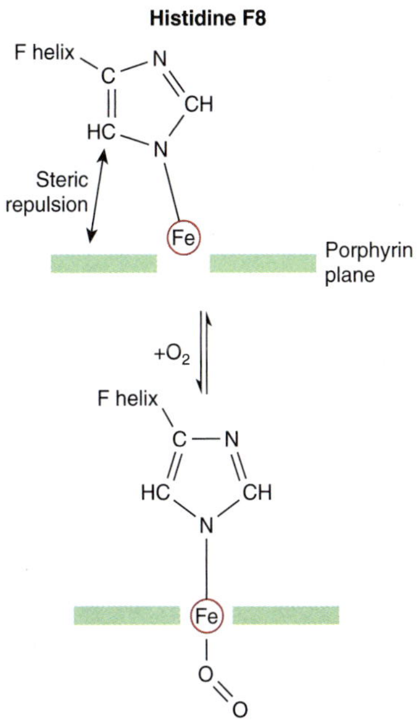

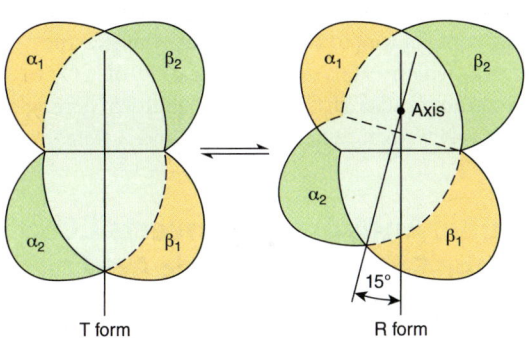

FIGURE 6–8 On oxygenation of hemoglobin the iron atom moves into the plane of the heme. Histidine F8 and its associated aminoacyl residues are pulled along with the iron atom. For a representation of this motion, see https://pdb101.rcsb.org/learn/videos/oxygen-binding-in-hemoglobin.

Oxygenation of Hemoglobin Is Accompanied by Large Conformational Changes

The binding of the first molecule of O_2 to deoxyHb shifts the heme iron toward the plane of the heme ring (**Figure 6–8**). This motion is transmitted through the proximal (F8) histidine and the residues attached thereto to the entire tetramer, triggering the rupture of salt bridges formed by the carboxyl terminal residues of all four subunits. As a result, one pair of α/β subunits rotates 15° with respect to the other, compacting the tetramer (**Figure 6–9**) and triggering other, profound

FIGURE 6–9 During transition of the T form to the R form of hemoglobin, the $\alpha_2\beta_2$ pair of subunits (green) rotates through 15° relative to the pair of $\alpha_1\beta_1$ subunits (yellow). The axis of rotation is eccentric, and the $\alpha_2\beta_2$ pair also shifts toward the axis somewhat. In the representation, the tan $\alpha_1\beta_1$ pair is shown fixed while the green $\alpha_2\beta_2$ pair of subunits both shifts and rotates.

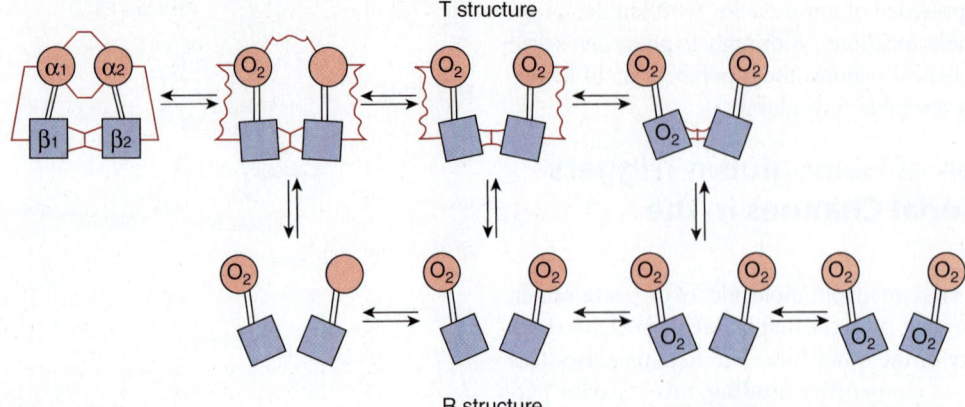

FIGURE 6–10 **Transition from the T structure to the R structure.** In this model, salt bridges (red lines) linking the subunits in the T structure break progressively as oxygen is added, and even those salt bridges that have not yet ruptured are progressively weakened (wavy red lines). The transition from T to R does not take place after a fixed number of oxygen molecules have been bound but becomes more probable as each successive oxygen binds. The transition between the two structures is influenced by protons, carbon dioxide, chloride, and 2,3-bisphosphoglycerate (BPG); the higher their concentration, the more oxygen must be bound to trigger the transition. Fully oxygenated molecules in the T structure and fully deoxygenated molecules in the R structure are not shown because they are unstable. (Modified with permission from Perutz MF. Hemoglobin structure and respiratory transport. Sci Am. 1978;239(6):92-125.)

changes in secondary, tertiary, and quaternary structures that accompany the transition of hemoglobin from the low-affinity T state to the high-affinity R state. These changes significantly increase the affinity of the remaining unoxygenated hemes for O_2, as subsequent binding events require the rupture of fewer salt bridges (**Figure 6–10**). The cooperativity of hemoglobin constitutes a form of **allosteric** (Gk *allos* "other," *steros* "space") behavior (see Chapter 19), since binding of O_2 to one heme affects the binding affinity of the others.

Hemoglobin Assists in the Transport of CO_2 to the Lungs

For every molecule of O_2 consumed during the course of respiration, one molecule of CO_2 is produced. Consequently, red blood cells must be able to efficiently absorb CO_2 from respiring tissues and subsequently release it for disposal in the lungs. Two key contributors to this process are the enzyme carbonic anhydrase, which catalyzes the hydration of CO_2 to water-soluble carbonic acid, H_2CO_3, and the ability of hemoglobin to cooperatively transition between the T and R states.

Carbonic Anhydrase Converts CO_2 to Water-Soluble Carbonic Acid & Bicarbonate

As is the case for O_2, the limited water solubility of CO_2 at neutral pH renders simply dissolving this gas in the bloodstream grossly insufficient to meet the body's transport needs. The enzyme carbonic anhydrase, present in large quantities in red blood cells, catalyzes the hydration of CO_2 molecules to form water soluble carbonic acid, H_2CO_3.

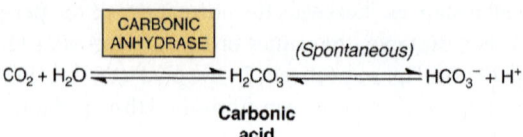

H_2CO_3 is a weak acid that, in turn, can dissociate to form bicarbonate ion, HCO_3^-, and a proton. Since the pK_{a1} of H_2CO_3, 6.35, falls below physiologic pH, in red blood cells the majority of absorbed CO_2 is carried as water-soluble bicarbonate.

Binding of Protons to T-State Hemoglobin Increases CO_2 Uptake From Respiring Tissues

When hemoglobin transitions from the R to the T state, conformational changes take place in each of the two β–chains that lead to the formation of salt bridges between the side chains of Asp 94 and His 146. Formation of each salt bridge requires T-state hemoglobin to bind a proton from the surrounding environment. As R-state hemoglobin gives up its bound O_2 to respiring tissues and subsequently transitions to the T state, the absorption of these two protons both buffers the pH of the acidifying red blood cell and shifts or pulls the equilibrium between H_2CO_3 and HCO_3^- in favor of bicarbonate, further increasing the quantity of carbon dioxide absorbed by the red blood cells from peripheral tissues.

T-state hemoglobin binds two protons per tetramer. In actively respiring tissues, the proton concentration inside the red blood cells will increase as H_2CO_3 accumulates. The greater availability of H^+, in turn, favors the formation of T-state hemoglobin, thereby enhancing the release of O_2.

Proton binding by T-state hemoglobin enables the high levels of CO_2 in actively respiring tissues to drive the release of O_2 from hemoglobin while concomitantly enhancing the quantity of CO_2 absorbed by promoting the conversion of carbonic acid to bicarbonate. As a result, transition to the T state helps buffer or moderate the CO_2-mediated decrease in the pH of the red blood cells in venous blood.

Proton Release From R-state Hemoglobin Enhances CO_2 Release in the Lungs

On reaching the lungs, the dramatic increase in the partial pressure of oxygen drives the binding of O_2 to deoxyhemoglobin. O_2 binding, in turn, triggers the transition of hemoglobin from the T to the R state. The resulting conformational change causes the rupture of salt bridges in the two β–chains and the release of the two protons formerly chelated between Asp 94 and His 146. Once freed, the protons combine with bicarbonate to increase the concentration of carbonic acid, which in turns favors the carbonic anhydrase catalyzed dehydration of H_2CO_3 to form CO_2, which can then be disposed by exhalation (**Figure 6–11**). This proton-mediated coupling of the transition of hemoglobin between T and R states with the equilibria between CO_2, H_2CO_3, and HCO_3^- is known as the **Bohr effect**.

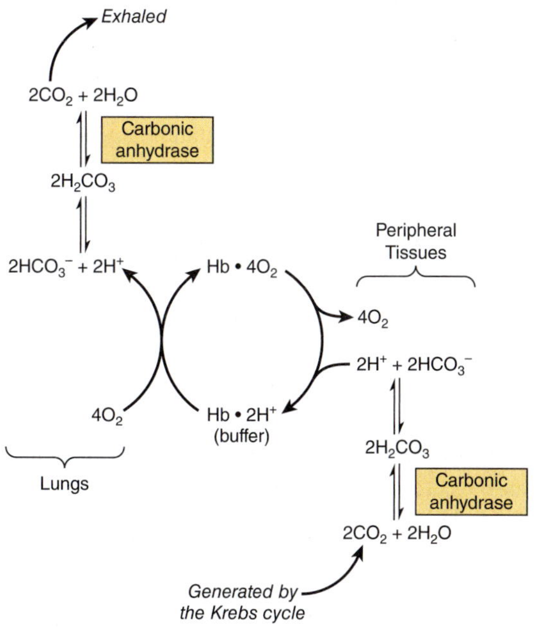

Additional CO_2 Is Transported as Carbamates of Hemoglobin

About 15% of the CO_2 in venous blood is carried by hemoglobin as carbamates formed with the amino terminal nitrogens of its polypeptide chains:

$$CO_2 + Hb{-}NH_3^+ \rightleftharpoons 2H^+ + Hb{-}\underset{H}{N}{-}\overset{\overset{\textstyle O}{\|}}{C}{-}O^-$$

Carbamate formation changes the charge on amino terminals from positive to negative, favoring salt bridge formation between α and β chains. As part of the Bohr effect, proton binding by T-state hemoglobin favors carbamate formation while proton release on transition to the R state favors carbamate breakdown and release of CO_2.

2,3-BPG Stabilizes the T Structure of Hemoglobin

The transition of hemoglobin to the T state opens a central cavity at the interface of its four subunits capable of binding one molecule of 2,3-bisphosphoglycerate, 2,3-BPG (see Figure 6–6). Binding of 2,3-BPG thus favors the shift from the R to the T state and the further release of bound O_2, as conversion back to the R state requires the rupture of the additional salt bridges formed between 2,3-BPG and Lys EF6, His H21, and the terminal amino groups of Val NA1 from both β chains (**Figure 6–12**).

2,3-BPG is synthesized from the glycolytic intermediate 1,3-BPG, a reaction catalyzed by the bifunctional enzyme **2,3-bisphosphogylcerate synthase/2-phosphatase** (BPGM). BPG is hydrolyzed to 3-phosphoglycerate by the 2-phosphatase

FIGURE 6–11 **The Bohr effect.** Carbon dioxide generated in peripheral tissues combines with water to form carbonic acid, which dissociates into protons and bicarbonate ions. Deoxyhemoglobin acts as a buffer by binding protons and delivering them to the lungs. In the lungs, the uptake of oxygen by hemoglobin releases protons that combine with bicarbonate ion, forming carbonic acid, which when dehydrated by carbonic anhydrase becomes carbon dioxide, which then is exhaled.

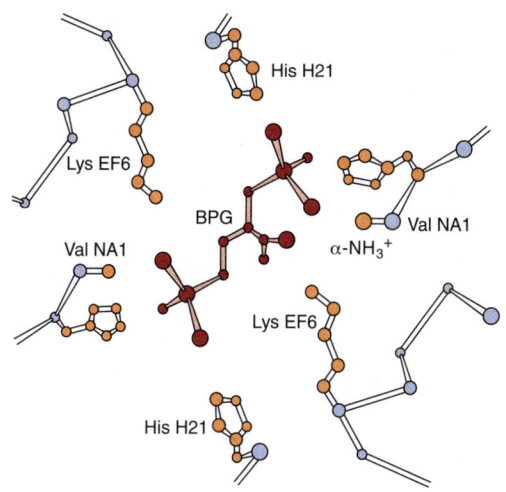

FIGURE 6–12 **Mode of binding of 2,3-bisphosphoglycerate (BPG) to human deoxyhemoglobin.** BPG interacts with three positively charged groups on each β chain. (Adapted with permission from Arnone A. X-ray diffraction study of binding of 2,3-diphosphoglycerate to human deoxyhaemoglobin. Nature. 1972;237(5351):146-149.)

activity of BPGM and to 2-phosphoglycerate by a second enzyme, multiple inositol polyphosphate phosphatase (MIPP). The activities of these enzymes, and hence the level of BPG in erythrocytes, are sensitive to pH. The CO_2-induced acidification of the red blood cells triggers production of 2,3-BPG, which in turn reinforces the impact of carbonic acid–derived protons in shifting the R-T equilibrium in favor of the T state, increasing the quantity of O_2 released in peripheral tissues.

In the fetal hemoglobin, residue H21 of the γ subunit is Ser rather than His. Since Ser cannot form a salt bridge, BPG binds more weakly to HbF than to HbA. The lower stabilization afforded to the T state by BPG helps account for HbF having a higher affinity for O_2 than HbA.

Adaptation to High Altitude

Physiologic changes that accompany prolonged exposure to high altitude include increases in the number of erythrocytes, the concentration of hemoglobin within them, and the synthesis of 2,3-BPG. Elevated 2,3-BPG lowers the affinity of HbA for O_2 (increases P_{50}), which enhances the release of O_2 at peripheral tissues.

NUMEROUS MUTATIONS AFFECTING HUMAN HEMOGLOBINS HAVE BEEN IDENTIFIED

Mutations in the genes that encode the α or β subunits of hemoglobin potentially can affect its biologic function. However, of more than 1100 known genetic mutations affecting human hemoglobins, most are both rare and benign, presenting no clinical abnormalities. When a mutation does compromise biologic function, the condition is termed a **hemoglobinopathy**. It is estimated that more than 7% of the globe's population are carriers for hemoglobin disorders. The URL http://globin.cse.psu.edu/ (Globin Gene Server) provides information about—and links for—normal and mutant hemoglobins. Selected examples are as follows.

Methemoglobin & Hemoglobin M

In methemoglobinemia, the heme iron is ferric rather than ferrous, rendering methemoglobin unable to bind or transport O_2. Normally, the Fe^{3+} of methemoglobin is returned to Fe^{2+} state through the action of the enzyme methemoglobin reductase. Methemoglobin levels can rise to pathophysiologically significant levels from a number of causes: oxidation of Fe^{2+} to Fe^{3+} as a side effect of agents such as sulfonamides, reductions in the activity of methemoglobin reductase, or inheritance of the gene for a mutationally altered form of hemoglobin called HbM.

In hemoglobin M, histidine F8 (His F8) has been replaced by tyrosine. The iron of HbM forms a tight ionic complex with the phenolate anion of tyrosine that stabilizes the Fe^{3+} form. In α-chain hemoglobin M variants, the R-T equilibrium favors the T state. Oxygen affinity is reduced, and the Bohr effect is absent. β-Chain hemoglobin M variants exhibit R-T switching, and the Bohr effect is therefore present.

Mutations that favor the R state (eg, hemoglobin Chesapeake) increase O_2 affinity. These hemoglobins therefore fail to deliver adequate O_2 to peripheral tissues. The resulting tissue hypoxia leads to **polycythemia**, an increased concentration of erythrocytes.

Hemoglobin S

In HbS, the nonpolar amino acid valine has replaced the polar surface residue Glu6 of the β subunit, generating a hydrophobic **"sticky patch"** on the surface of the β subunit of both oxyHbS and deoxyHbS. Both HbA and HbS contain a complementary sticky patch on their surfaces that is exposed only in the deoxygenated T state. Thus, at low Po_2, deoxyHbS can polymerize to form long, twisted helical fibers. Binding of deoxyHbA terminates fiber polymerization, since HbA lacks the second sticky patch necessary to bind another Hb molecule (**Figure 6–13**). These insoluble fibers distort the erythrocyte into a characteristic sickle shape, rendering it vulnerable to lysis in the interstices of the splenic sinusoids. They also cause multiple secondary clinical effects. A low Po_2, such as that at high altitudes, exacerbates the tendency to polymerize. The terms sickle cell trait and sickle cell disease refer to persons in whom the gene for either one or both β subunits are mutated, respectively. Emerging treatments for sickle cell disease include inducing HbF expression to inhibit the polymerization of HbS, stem cell transplantation, and, in the future, gene therapy.

BIOMEDICAL IMPLICATIONS

Myoglobinuria

Following massive crush injury to skeletal muscle followed by renal damage, released myoglobin may appear in the urine. Myoglobin can be detected in plasma following a myocardial infarction, but assay of serum troponin, lactate dehydrogenase isozymes, or creatine kinase (see Chapter 7) provides a more sensitive index of myocardial injury.

Anemias

Anemias, reductions in the number of red blood cells or concentration of hemoglobin in the blood, can reflect impaired synthesis of hemoglobin (eg, in iron deficiency; see Chapter 52) or impaired production of erythrocytes (eg, in folic acid or vitamin B_{12} deficiency; see Chapter 44). Diagnosis of anemias begins with spectroscopic measurement of blood hemoglobin levels.

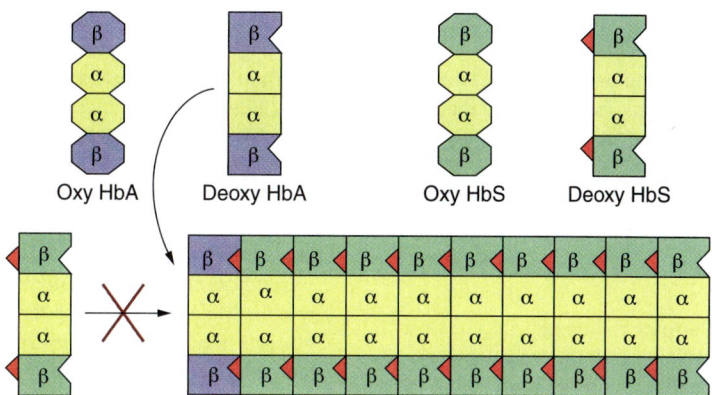

FIGURE 6–13 Polymerization of deoxyhemoglobin S. The dissociation of oxygen from hemoglobin S (HbS) unmasks a sticky patch (red triangle) on the surface of its β subunits (green) that can adhere to a complementary site on the β subunits of other molecules of deoxyHbS. Polymerization to a fibrous polymer is interrupted deoxyHbA, whose β subunits (lavender) lack the sticky patch required for binding additional HbS subunits.

Thalassemias

The genetic defects known as thalassemias result from the partial or total absence of one or more α or β chains of hemoglobin. Over 750 different mutations have been identified, but only three are common. Either the α chain (α thalassemias) or β chain (β thalassemias) can be affected. A superscript indicates whether a subunit is completely absent (α^0 or β^0) or whether its synthesis is reduced (α^- or β^-). Apart from marrow transplantation, treatment is symptomatic.

Certain mutant hemoglobins are common in many populations, and a patient may inherit more than one type. Hemoglobin disorders thus present a complex pattern of clinical phenotypes. The use of DNA probes for their diagnosis is considered in Chapter 39.

GLYCATED HEMOGLOBIN (HBA$_{1c}$)

Blood glucose that enters the erythrocytes can form a covalent adduct with the ε-amino groups of lysyl residues and the α-amino group of the N-terminal valines of hemoglobin β chains, a process referred to as **glycation**. Unlike glycosylation (see Chapter 46), glycation is not enzyme-catalyzed. The fraction of hemoglobin glycated, normally about 5%, is proportionate to blood glucose concentration. Since the life-span of an erythrocyte is typically 120 days, the level of glycated hemoglobin (HbA$_{1c}$) reflects the mean blood glucose concentration over the preceding 8 to 12 weeks. Measurement of HbA$_{1c}$ therefore provides valuable information for management of diabetes mellitus.

SUMMARY

- Myoglobin is monomeric; hemoglobin is a tetramer of two subunit types ($\alpha_2\beta_2$ in HbA). Myoglobin and the subunits of hemoglobin are homologs of one another that, despite differences in primary structure, exhibit nearly identical secondary and tertiary structures.

- Heme is an essentially planar, slightly puckered, cyclic tetrapyrrole whose four nitrogen atoms bind a central Fe^{2+} that is also linked to histidine F8, and, in oxyMb and oxyHb, to O_2.

- The O_2-binding curve for myoglobin is hyperbolic, but for hemoglobin it is sigmoidal, a consequence of cooperative interactions in the tetramer.

- Cooperativity arises from the ability of hemoglobin to exist in two different conformational states, a relaxed or R state in which all four subunits exhibit a high affinity for oxygen and a taut or T state where all four subunits display a low affinity for oxygen.

- The high levels of O_2 in the lungs drive the R-T equilibrium in favor of the R state, while acidification of the red blood cells generated from the catalytic hydration of CO_2 in the peripheral tissues favors the T state. Cooperativity thus maximizes the ability of hemoglobin both to load O_2 at the P_{O_2} of the lungs and to deliver O_2 at the P_{O_2} of the tissues.

- Relative affinities of different hemoglobins for oxygen are expressed as P_{50}, the P_{O_2} at which they become half-saturated with O_2. Hemoglobin isoforms saturate at the partial pressures of their respective respiratory organ, for example, the lung or placenta.

- On oxygenation of hemoglobin, the iron and histidine F8 move toward the heme ring. The resulting conformational changes in the hemoglobin tetramer induce the rupture of several salt bonds, the release of bound protons, and loosening of the quaternary structure that increases the affinity for the other three subunits for O_2.

- Binding of 2,3-BPG in the central cavity of T-state hemoglobin favors O_2 release in actively respiring tissues. Transition to the R state results in the closure of this cavity and extrusion of 2,3-BPG.

- Hemoglobin assists in CO_2 transport from peripheral tissues to the lungs via the formation of carbamates and the **Bohr effect**, a consequence of the binding of protons to T-state, but not R-state, hemoglobin. Proton binding enhances the conversion of CO_2 to water-soluble carbonic acid and bicarbonate. In the lungs, the release of protons from oxygenated R-state hemoglobin favors the conversion of bicarbonate and carbonic acid to CO_2, which is exhaled.

- In sickle cell hemoglobin (HbS), Val replaces Glu_6 of the β subunit of HbA, creating a "sticky patch" that has a complement on deoxyHb (but not on oxyHb). DeoxyHbS polymerizes at low O_2 concentrations, forming fibers that distort erythrocytes into sickle shapes.

- α and β Thalassemias are anemias that result from reduced production of α and β subunits of HbA, respectively.

REFERENCES

Cho J, King JS, Qian X, et al: Dephosphorylation of 2,3-bisphosphogylcerate by MIPP expands the regulatory capacity of the Rapoport-Luebering glycolytic shunt. Proc Natl Acad Sci USA 2008;105:5998.

Costa FF, Conrad N (editors): *Sickle Cell Anemia. From Basic Science to Clinical Practice.* Springer, 2016.

Gros G, Wittenberg BA, Jue T: Myoglobin's old and new clothes: from molecular structure to function in living cells. J Exp Biol 2010;213:2713.

Henry ER, Cellmer T, Dunkelberger EB, et al: Allosteric control of hemoglobin S fiber formation by oxygen and its relation to the pathophysiology of sickle cell disease. Proc Natl Acad Sci USA 2020;117:15018.

Piel FB: The present and future global burden of the inherited disorders of hemoglobin. Hematol Oncol Clin North Am 2016;30:327.

Sigler PF (editor): *The Molecular Basis of Human Hemoglobin Dysfunction.* Elsevier, 2017.

Storz JF: *Hemoglobin: Insights into Protein Structure, Function, and Evolution.* Oxford University Press, 2018 (Entire volume).

Thein SL: Molecular basis of β thalassemia and potential therapeutic targets. Blood Cells Mol Dis 2018;70:54.

Umbreit J: Methemoglobin—it's not just blue: A concise review. Am J Hematol 2007;82:134.

Yuan Y, Tam MF, Simplaceanu V, Ho C: New look at hemoglobin allostery. Chem Rev 2015;115:1702.

Enzymes: Mechanism of Action

Peter J. Kennelly, PhD, & Victor W. Rodwell, PhD

OBJECTIVES

After studying this chapter, you should be able to:

- Describe the structural relationships between specific B vitamins and certain coenzymes.
- Outline the four principal mechanistic strategies employed by enzymes to catalyze chemical reactions.
- Explain the concepts of the "lock and key" and "induced fit" models for enzyme–substrate interaction and how the latter accounts for the dynamic nature of enzyme catalysis.
- Outline the underlying principles of enzyme-linked immunoassays.
- Describe how detecting and measuring the activity of many enzymes can be facilitated by coupling with an appropriate dehydrogenase.
- Identify proteins whose plasma levels are used as biomarkers for diagnosis and prognosis.
- Describe the application of restriction endonucleases and of restriction fragment length polymorphisms in the detection of genetic diseases.
- Illustrate the utility of site-directed mutagenesis for the identification of aminoacyl residues that are involved in the recognition of substrates or allosteric effectors, or in the mechanism of catalysis.
- Describe how "affinity tags" can facilitate purification of a protein expressed from its cloned gene.
- Discuss the events that led to the discovery that RNAs can act as enzymes, and briefly describe the evolutionary concept of an "RNA world."

BIOMEDICAL IMPORTANCE

Nobel laureate (1946) James Sumner defined an enzyme as "a catalyst of high molecular weight and biological origin." Enzymes catalyze the chemical reactions that make life on the earth possible, such as the breakdown of nutrients to supply energy and biomolecular building blocks; the assembly of those building blocks into proteins, DNA, membranes, cells, and tissues; and the harnessing of energy to power cell motility, neural function, and muscle contraction. Almost all enzymes are proteins. Notable exceptions include ribosomal RNAs and a handful of RNA molecules imbued with endonuclease or nucleotide ligase activity known collectively as ribozymes. Many pathologic conditions are the direct consequence of changes in the quantity or in the catalytic activity of key enzymes that result from genetic defects, nutritional deficits, tissue damage, toxins, or infection by viral or bacterial pathogens (eg, *Vibrio cholerae*). Thus, the ability to detect and to quantify the activity of specific enzymes in blood, other tissue fluids, or cell extracts provides information that enhances the physician's ability to diagnose many diseases.

In addition to serving as the catalysts for all metabolic processes, the impressive catalytic activity and substrate specificity of enzymes enables them to fulfill unique roles in human health and well-being. The protease rennin, for example, is utilized in the production of cheeses, while lactase is employed to remove lactose from milk to benefit lactose-intolerant individuals. Proteases and amylases augment the capacity of detergents to remove dirt and stains, while other enzymes can participate in the stereospecific synthesis of complex drugs or antibiotics.

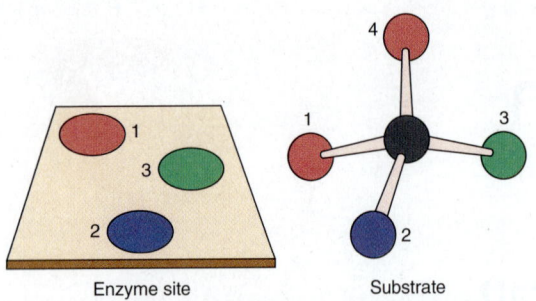

FIGURE 7–1   **Planar representation of the "three-point attachment" of a substrate to the active site of an enzyme.** Although atoms 1 and 4 are identical, once atoms 2 and 3 are bound to their complementary sites on the enzyme, only atom 1 can bind. Once bound to an enzyme, apparently identical atoms thus may be distinguishable, permitting a stereospecific chemical change.

ENZYMES ARE EFFICIENT & HIGHLY SPECIFIC CATALYSTS

The enzymes that catalyze the conversion of one or more compounds (**substrates**) into one or more different compounds (**products**) generally enhance the rates of the corresponding noncatalyzed reaction by factors of 10^6 to 10^9 or even more. Although some enzymes become modified during catalysis, these *transient* alterations are resolved in the course of the reaction being catalyzed. In addition to being highly-efficient, enzymes are also extremely *selective*. Typically, the catalysts used in synthetic chemistry are specific for the type of reaction catalyzed and will act on any compound containing the relevant functional group. Enzymes, however, are generally specific for a single substrate or even a single stereoisomer thereof—for example, D- but not L-sugars, L- but not D-amino acids—or a small set of closely related substrates. Since they bind substrates through at least "three points of attachment" (**Figure 7–1**), enzymes also can produce chiral products from nonchiral substrates. For example, reduction of the nonchiral substrate pyruvate by lactate dehydrogenase produces exclusively L-lactate, not a racemic mixture of D- and L-lactate. The exquisite specificity of enzyme catalysts imbues living cells with the ability to simultaneously conduct and independently control a broad spectrum of biochemical processes.

ENZYMES ARE CLASSIFIED BY REACTION TYPE

Some of the names for enzymes first described in the earliest days of biochemistry persist in use to this day. Examples include pepsin, trypsin, and amylase. Early biochemists generally named newly discovered enzymes by adding the suffix –*ase* to a descriptor for the type of reaction catalyzed. For example, enzymes that remove the elements of hydrogen, H_2 or H^- plus H^+, generally are referred to as dehydrogen*ases*, enzymes that hydrolyze proteins as prote*ases*, and enzymes that catalyze rearrangements in configuration as isomer*ases*. In many cases, these general descriptors were supplemented with terms

indicating the particular substrate on which the enzyme acts (*xanthine* oxidase), its source (*pancreatic* ribonuclease), its mode of regulation (*hormone-sensitive* lipase), or a characteristic feature of its mechanism of action (*cysteine* protease). Where needed, alphanumeric designators can be added to identify multiple forms, or isozymes, of an enzyme (eg, RNA polymerase *III*; protein kinase *Cβ*).

As more enzymes were discovered, these early naming conventions increasingly resulted in the inadvertent designation of some enzymes by multiple names or the assignment of duplicate names to enzymes exhibiting similar catalytic capabilities. To address these problems, the International Union of Biochemistry (IUB) developed a system of enzyme nomenclature in which each enzyme has a unique name and code number that identify the *type* of reaction catalyzed and the *substrates* involved. Enzymes are grouped into the following six classes:

1. **Oxidoreductases**—enzymes that catalyze oxidations and reductions

2. **Transferases**—enzymes that catalyze transfer of moieties such as glycosyl, methyl, or phosphoryl groups

3. **Hydrolases**—enzymes that catalyze *hydrolytic* cleavage of C—C, C—O, C—N, and other covalent bonds

4. **Lyases**—enzymes that catalyze cleavage of C—C, C—O, C—N, and other covalent bonds by *atom elimination*, generating double bonds

5. **Isomerases**—enzymes that catalyze geometric or structural changes *within* a molecule

6. **Ligases**—enzymes that catalyze the joining together (ligation) of two molecules in reactions coupled to the hydrolysis of ATP

The IUB name of hexokinase is ATP:D-hexose 6-phosphotransferase E.C. 2.7.1.1. This name identifies hexokinase as a member of class 2 (transferases), subclass 7 (transfer of a phosphoryl group), sub-subclass 1 (alcohol is the phosphoryl acceptor), and "hexose-6" indicates that the alcohol phosphorylated is on carbon six of a hexose. While EC numbers have proven particularly useful to differentiate enzymes with similar functions or similar catalytic activities, IUB names tend to be lengthy and cumbersome. Consequently, hexokinase and many other enzymes commonly are referred to using their traditional, albeit sometimes ambiguous names.

PROSTHETIC GROUPS, COFACTORS, & COENZYMES PLAY IMPORTANT ROLES IN CATALYSIS

Many enzymes contain small molecules or metal ions that participate directly in substrate binding or in catalysis. Termed **prosthetic groups**, **cofactors**, and **coenzymes**, they extend the repertoire of mechanistic capabilities beyond those afforded by the functional groups present on the aminoacyl side chains of peptides.

Prosthetic Groups

Prosthetic groups are tightly and stably incorporated into a protein's structure by covalent bonds or noncovalent forces. Examples include pyridoxal phosphate, flavin mononucleotide (FMN), flavin adenine dinucleotide (FAD), thiamin pyrophosphate, lipoic acid, biotin, and transition metals such as Fe, Co, Cu, Mg, Mn, and Zn. Metal ions that participate in redox reactions generally are bound as organometallic complexes such as the prosthetic groups heme or iron–sulfur clusters (see Chapter 10). Metals may facilitate the binding and orientation of substrates, the formation of covalent bonds with reaction intermediates (Co^{2+} in coenzyme B_{12}, see Chapter 44), or by acting as Lewis acids or bases to render substrates more **electrophilic** (electron-poor) or **nucleophilic** (electron-rich), and hence more reactive (see Chapter 10).

Cofactors Associate Reversibly With Enzymes or Substrates

Cofactors serve functions similar to those of prosthetic groups and overlap with them. The major difference between the two is operational, not chemical. Cofactors bind weakly and transiently to their cognate enzymes or substrates, forming dissociable complexes. Therefore, unlike associated prosthetic groups, cofactors must be present in the surrounding environment to promote complex formation in order for catalysis to occur. Metal ions form the most numerous class of cofactors. Enzymes that require a metal ion cofactor are termed **metal-activated enzymes** to distinguish them from the **metalloenzymes** for which bound metal ions serve as prosthetic groups. It is estimated that one-third of all enzymes fall into one of these two groups.

Many Coenzymes, Cofactors, & Prosthetic Groups Are Derivatives of B Vitamins

The water-soluble B vitamins supply important components of numerous coenzymes. **Nicotinamide** is a component of the redox coenzymes NAD and NADP (**Figure 7–2**); **riboflavin** is a component of the redox coenzymes FMN and FAD; **pantothenic acid** is a component of the acyl group carrier **coenzyme A**. As its pyrophosphate **thiamin** participates in the decarboxylation of α-keto acids while folic acid and cobamide coenzymes function in one-carbon metabolism. In addition, several coenzymes contain the adenine, ribose, and phosphoryl moieties of AMP or ADP (see Figure 7–2).

Coenzymes Serve as Substrate Shuttles

Coenzymes serve as recyclable shuttles that transport many substrates from one point within the cell to another. The function of these shuttles is twofold. First, they stabilize species such as hydrogen atoms ($FADH_2$) or hydride ions (NADH) that are too reactive to persist for any significant time in the presence of the water, oxygen, or the organic molecules that

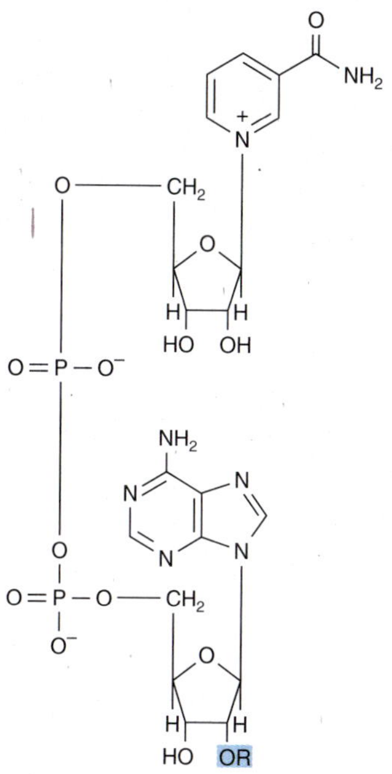

FIGURE 7–2 **Structure of NAD⁺ and NADP⁺.** For NAD⁺, OR = —OH. For NADP⁺, —OR = —OPO_3^{2-}.

permeate cells. Second, they increase the number of points of contact between substrate and enzyme, which increases the affinity and specificity with which small chemical groups such as acetate (coenzyme A), glucose (UDP), or hydride (NAD⁺) are bound by their target enzymes. Other chemical moieties transported by coenzymes include methyl groups (folates) and oligosaccharides (dolichol).

CATALYSIS OCCURS AT THE ACTIVE SITE

An important early 20th-century insight into enzymic catalysis sprang from the observation that the presence of substrates renders enzymes more resistant to the denaturing effects of elevated temperatures. This observation led Emil Fischer to propose that enzymes (E) and their substrates (S) form an enzyme–substrate (ES) complex whose thermal stability is greater than that of the enzyme itself. This insight profoundly shaped our understanding of both the chemical nature and kinetic behavior of enzymic catalysis.

Fischer reasoned that the exquisitely high specificity with which enzymes discriminate their substrates when forming an ES complex was analogous to the manner in which a mechanical lock distinguishes the proper key. The analogy to enzymes is that the "lock" is formed by a surface-accessible cleft or pocket in the enzyme called the **active site** (see Figures 5–6 and 5–8). As implied by the adjective "active," the active site is much more than simply a recognition site for binding substrates; it provides the environment wherein chemical transformation takes place. Within the active site, substrates are brought into close proximity with one another in optimal alignment with

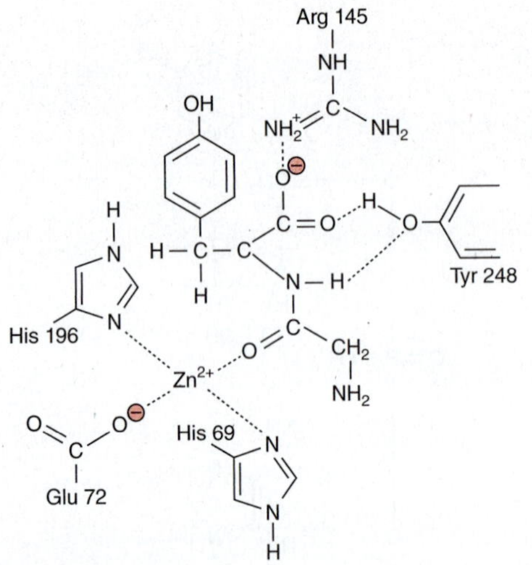

FIGURE 7–3 Two-dimensional representation of a dipeptide substrate, glycyl-tyrosine, bound within the active site of carboxypeptidase A.

the cofactors, prosthetic groups, and aminoacyl side chains that participate in catalyzing the transformation of substrates into products (**Figure 7–3**). Catalysis is further enhanced by the capacity of the active site to shield substrates from water and generate an environment whose polarity, hydrophobicity, acidity, or alkalinity can differ markedly from that of the surrounding cytoplasm.

ENZYMES EMPLOY MULTIPLE MECHANISTIC STRATEGIES TO FACILITATE CATALYSIS

Enzymes use combinations of four general mechanistic strategies to achieve dramatic enhancements of the rates of chemical reactions.

Catalysis by Proximity

In order to chemically interact, substrate molecules must come within bond-forming distance of one another. The higher their concentration, the more frequently they will encounter one another, and the greater will be the rate at which reaction products appear. When an enzyme binds substrate molecules at its active site, it creates a region of high local substrate concentration, one in which they are oriented in an ideal position to chemically interact. In and of itself, this proximity results in rate enhancements of at least a 1000-fold over that observed in the absence of an enzyme.

Acid–Base Catalysis

In addition to contributing to the ability of the active site to bind substrates, the ionizable functional groups of aminoacyl side chains and, where present, of prosthetic groups, can contribute to catalysis by acting as acids or bases. We distinguish two types of acid–base catalysis. **Specific acid or base catalysis** refers to reactions for which the only *participating* acids or bases are protons or hydroxide ions. The rate of reaction thus is sensitive to changes in the concentration of protons or hydroxide ions, but is *independent* of the concentrations of other acids (proton donors) or bases (proton acceptors) present in the solution or at the active site. Reactions whose rates are responsive to *all* the acids or bases present are said to be subject to **general acid catalysis** or **general base catalysis.**

Catalysis by Strain

For catalysis of lytic reactions, which involve breaking a covalent bond, enzymes typically bind their substrates in a conformation that weakens the bond targeted for cleavage through physical distortion and electronic polarization. This strained conformation mimics that of the **transition state intermediate**, a transient species that represents the midway point in the transformation of substrates to products. Nobel Laureate Linus Pauling was the first to suggest a role for **transition state stabilization** as a general mechanism by which enzymes accelerate the rates of chemical reactions. Knowledge of the transition state of an enzyme-catalyzed reaction is frequently exploited by chemists to design and synthesize more effective enzyme inhibitors, called **transition state analogs**, as potential pharmacophores.

Covalent Catalysis

The process of **covalent catalysis** involves the formation of a covalent bond between the enzyme and one or more substrates. **The modified enzyme thus becomes a reactant**. Covalent catalysis provides a new reaction pathway whose activation energy is lower—and rate of reaction therefore faster—than the pathways available in homogeneous solution. The chemically modified state of the enzyme is, however, transient. Completion of the reaction returns the enzyme to its original, unmodified state. Its role thus remains catalytic. Covalent catalysis is particularly common among enzymes that catalyze **group transfer reactions**. Residues on the enzyme that participate in covalent catalysis generally are cysteine or serine, and occasionally histidine. Covalent catalysis often follows a "ping-pong" mechanism—one in which the first substrate is bound and its product released prior to the binding of the second substrate (**Figure 7–4**).

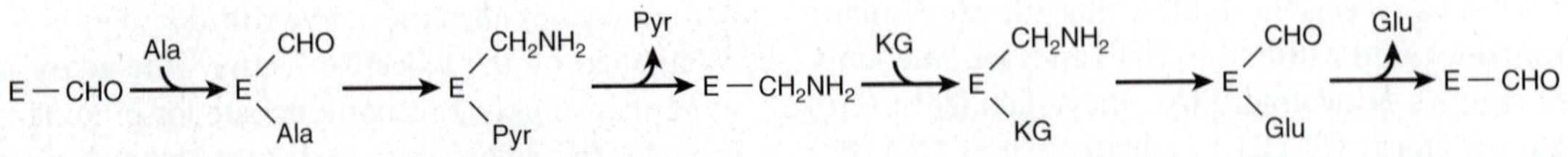

FIGURE 7–4 "Ping-pong" mechanism for transamination. E—CHO and E—CH$_2$NH$_2$ represent the enzyme-pyridoxal phosphate and enzyme-pyridoxamine complexes, respectively. (Ala, alanine; Glu, glutamate; KG, α-ketoglutarate; Pyr, pyruvate.)

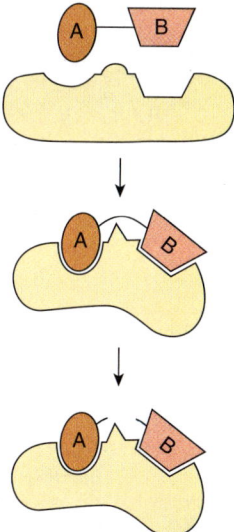

FIGURE 7–5 **Two-dimensional representation of Koshland's induced fit model of the active site of a lyase.** Binding of the substrate A—B induces conformational changes in the enzyme that align catalytic residues which participate in catalysis and strain the bond between A and B, facilitating its cleavage.

SUBSTRATES INDUCE CONFORMATIONAL CHANGES IN ENZYMES

While Fischer's "lock and key model" accounted for the exquisite specificity of enzyme–substrate interactions, the implied rigidity of the enzyme's active site failed to account for the dynamic changes that accompany catalytic transformations. This drawback was addressed by Daniel Koshland's **induced fit model**, which states that as substrates bind to an enzyme, they induce a conformational change that is analogous to placing a hand (substrate) into a glove (enzyme) (**Figure 7–5**). The enzyme in turn induces reciprocal changes in its substrates, harnessing the energy of binding to facilitate the transformation of substrates into products. The induced fit model has been amply confirmed by biophysical studies of enzyme motion during substrate binding.

HIV PROTEASE ILLUSTRATES ACID–BASE CATALYSIS

Enzymes of the **aspartic protease family**, which includes the digestive enzyme pepsin, the lysosomal cathepsins, and the protease produced by the human immunodeficiency virus (HIV), share a common mechanism that employs two conserved aspartyl residues as acid–base catalysts. In the first stage of the reaction, one aspartate functions as a general base (Asp X, **Figure 7–6**) that extracts a proton from a water molecule to make it more nucleophilic. The resulting nucleophile then attacks the electrophilic carbonyl carbon of the peptide bond targeted for hydrolysis, forming a **tetrahedral transition state intermediate**. A second aspartate (Asp Y, Figure 7–6)

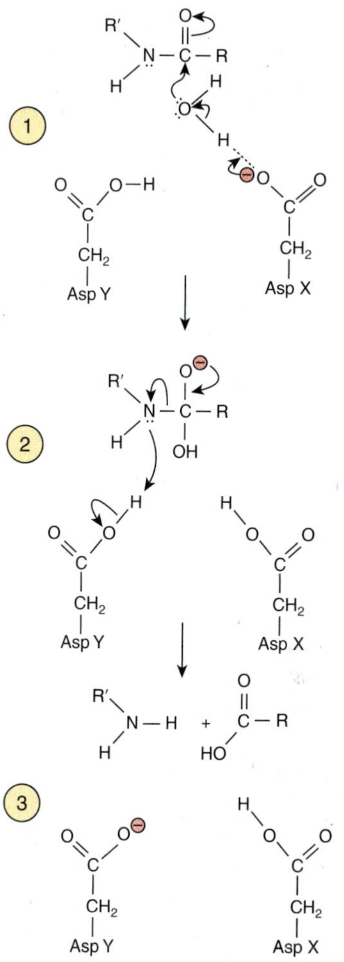

FIGURE 7–6 **Mechanism for catalysis by an aspartic protease such as HIV protease.** Curved arrows indicate directions of electron movement. ① Aspartate X acts as a base to activate a water molecule by abstracting a proton. ② The activated water molecule attacks the peptide bond, forming a transient tetrahedral intermediate. ③ Aspartate Y acts as an acid to facilitate breakdown of the tetrahedral intermediate and release of the split products by donating a proton to the newly formed amino group. Subsequent shuttling of the proton on Asp X to Asp Y restores the protease to its initial state.

then facilitates the decomposition of this tetrahedral intermediate by donating a proton to the amino group produced by rupture of the peptide bond. The two active site aspartates can act simultaneously as a general base or as a general acid because their immediate environment favors ionization of one, but not the other.

CHYMOTRYPSIN & FRUCTOSE-2, 6-BISPHOSPHATASE ILLUSTRATE COVALENT CATALYSIS

Chymotrypsin

While catalysis by aspartic proteases involves the direct hydrolytic attack of water on a peptide bond, catalysis by the **serine protease** chymotrypsin involves formation of a covalent acyl-enzyme intermediate. A conserved seryl residue, serine 195 in the bovine form, is activated via interactions with histidine 57 and aspartate 102. While these three residues are far apart in primary structure, in the active site of the mature, folded protein they reside within bond-forming distance of one another. Aligned in the order Asp 102-His 57-Ser 195, this trio forms a linked **charge-relay network** that acts as a **"proton shuttle."**

Binding of substrate initiates proton shifts that in effect transfer the hydroxyl proton of Ser 195 to Asp 102 (**Figure 7–7**). The enhanced nucleophilicity of the seryl oxygen facilitates its attack on the carbonyl carbon of the peptide bond of the substrate, forming a covalent **acyl-enzyme intermediate**. The proton on Asp 102 then shuttles via His 57 to the amino group liberated when the peptide bond is cleaved. The portion of the original peptide with a free amino group then leaves the active site and is replaced by a water molecule. The charge-relay network now activates the water molecule by withdrawing a proton through His 57 to Asp 102. The resulting hydroxide ion attacks the acyl-enzyme intermediate, and a reverse proton shuttle returns a proton to Ser 195, restoring its original state. While modified during the process of catalysis, chymotrypsin emerges unchanged on completion of the reaction. The proteases trypsin and elastase employ a similar catalytic mechanism, but the numbering of the residues in their Ser-His-Asp proton shuttles differ.

Fructose-2,6-Bisphosphatase

Fructose-2,6-bisphosphatase, a regulatory enzyme of gluconeogenesis (see Chapter 19), catalyzes the hydrolytic release of the phosphate on carbon 2 of fructose-2,6-bisphosphate. **Figure 7–8** illustrates the roles of seven active site residues. Catalysis involves a "catalytic triad" consisting of one Glu and two His residues, of which one His forms a covalent phosphohistidyl intermediate.

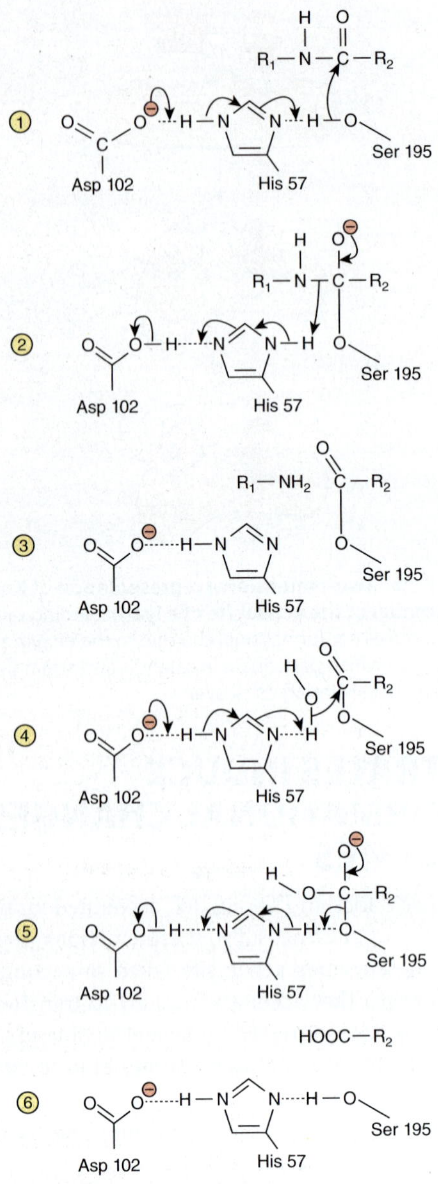

CATALYTIC RESIDUES ARE HIGHLY CONSERVED

Members of an enzyme family such as the aspartic or serine proteases employ a similar mechanism to catalyze a common reaction type, but act on different substrates. Most enzyme families appear to have arisen through gene duplication events that created a second copy of the gene that encodes a particular enzyme. The two genes, and consequently their encoded proteins, can then evolve independently, forming divergent **homologs** that recognize different substrates. The result is illustrated by chymotrypsin, which cleaves peptide bonds on the carboxyl terminal side of large hydrophobic amino acids, and trypsin, which cleaves peptide bonds on the carboxyl terminal side of basic amino acids. Proteins that diverged from a common ancestor are said to be **homologous** to one another. The common ancestry of enzymes can be inferred from the presence of specific amino acids in the same relative position in each family member. These residues are said to be **evolutionarily conserved**.

FIGURE 7–7 Catalysis by chymotrypsin. ① The charge-relay system removes a proton from Ser 195, making it a stronger nucleophile. ② Activated Ser 195 attacks the peptide bond, forming a transient tetrahedral intermediate. ③ Release of the amino terminal peptide is facilitated by donation of a proton to the newly formed amino group by His 57 of the charge-relay system, yielding an acyl-Ser 195 intermediate. ④ His 57 and Asp 102 collaborate to activate a water molecule, which attacks the acyl-Ser 195, forming a second tetrahedral intermediate. ⑤ The charge-relay system donates a proton to Ser 195, facilitating breakdown of the tetrahedral intermediate to release the carboxyl terminal peptide ⑥.

ISOZYMES ARE DISTINCT ENZYME FORMS THAT CATALYZE THE SAME REACTION

Higher organisms often elaborate several physically distinct versions of a given enzyme, each of which catalyzes the same reaction. These protein catalysts or **isozymes** can arise through gene duplication or, in higher eukaryotes, alternative mRNA

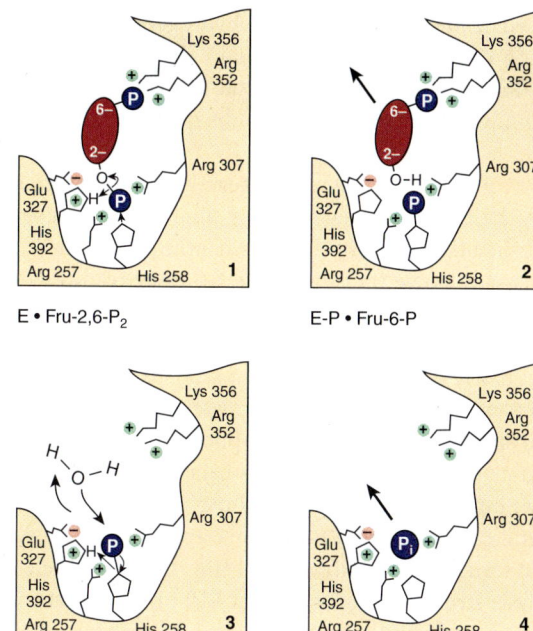

FIGURE 7–8 **Catalysis by fructose-2,6-bisphosphatase.** (1) Lys 356 and Arg 257, 307, and 352 stabilize the quadruple negative charge of the substrate by charge–charge interactions. Glu 327 stabilizes the positive charge on His 392. (2) The nucleophile His 392 attacks the C-2 phosphoryl group and transfers it to His 258, forming a phosphoryl-enzyme intermediate. Fructose-6-phosphate now leaves the enzyme. (3) Nucleophilic attack by a water molecule, possibly assisted by Glu 327 acting as a base, forms inorganic phosphate. (4) Inorganic orthophosphate is released from Arg 257 and Arg 307. (Reproduced with permission from Pilkis SJ, Claus TH, Kurland IJ, et al: 6-Phosphofructo-2-kinase/fructose-2,6-bisphosphatase: a metabolic signaling enzyme, Annu Rev Biochem. 1995;64:799-835.)

splicing (see Chapter 36). While the homologous proteases described act on different substrates, isozymes also may differ in auxiliary features such as sensitivity to particular regulatory factors (see Chapter 9) or subcellular location that adapt them to specific tissues or circumstances rather than distinct substrates. Isozymes that catalyze the identical reaction may also enhance survival by providing a "backup" copy of an essential enzyme.

THE CATALYTIC ACTIVITY OF ENZYMES FACILITATES THEIR DETECTION

The relatively small quantities of enzymes typically contained in cells hamper determination of their presence and abundance. However, the ability to rapidly transform thousands of molecules of a specific substrate into products imbues each enzyme with the ability to amplify its presence. Under appropriate conditions (see Chapter 8), the rate of the catalytic reaction being monitored is proportionate to the amount of enzyme present, which allows its concentration to be inferred.

Assays of the catalytic activity of enzymes are frequently used both in research and clinical laboratories.

Single-Molecule Enzymology

The limited sensitivity of traditional enzyme assays necessitates the use of a large group, or ensemble, of enzyme molecules in order to produce measurable quantities of product. The data obtained thus reflect the *average* activity of individual enzymes across multiple cycles of catalysis. Recent advances in **nanotechnology** and imaging have made it possible to observe catalytic events involving discrete enzyme and substrate molecules. Consequently, scientists can now measure the rate of individual catalytic events, and sometimes a specific step in catalysis, by a process called **single-molecule enzymology**, an example of which is illustrated in **Figure 7–9**.

Drug Discovery Requires Enzyme Assays Suitable for High-Throughput Screening

Enzymes are frequent targets for the development of drugs and other therapeutic agents. These generally take the form of enzyme inhibitors (see Chapter 8). The discovery of new drugs is greatly facilitated when a large number of potential pharmacophores can be simultaneously assayed in a rapid, automated fashion—a process referred to as **high-throughput screening (HTS)**. HTS employs robotics, optics, data processing, and microfluidics to simultaneously conduct and monitor thousands of parallel enzyme assays. It thus serves as a perfect complement to **combinatorial chemistry**, a method for

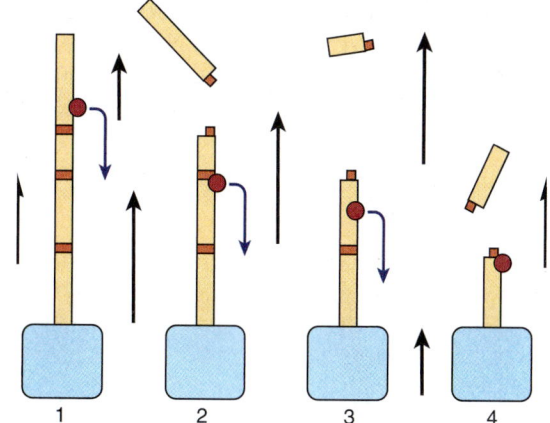

FIGURE 7–9 **Direct observation of single DNA cleavage events catalyzed by a restriction endonuclease.** DNA molecules immobilized to beads (blue) are placed in a flowing stream of buffer (black arrows), which causes them to assume an extended conformation. Cleavage at one of the restriction sites (orange) by an endonuclease leads to a shortening of the DNA molecule, which can be observed directly in a microscope since the nucleotide bases in DNA are fluorescent. Although the endonuclease (red) does not fluoresce, and hence is invisible, the progressive manner in which the DNA molecule is shortened (1→4) reveals that the endonuclease binds to the free end of the DNA molecule and moves along it from site to site.

generating large libraries of chemical compounds spanning all possible combinations of a given set of chemical precursors.

Enzyme-Linked Immunoassays

The sensitivity of enzyme assays can be exploited to detect proteins that lack catalytic activity. **Enzyme-linked immunosorbent assays** (ELISAs) use antibodies covalently linked to a "reporter enzyme", such as alkaline phosphatase or horseradish peroxidase, that can be used to produce easily-detected chromogenic or fluorescent products *in vitro*. Serum or other biologic samples to be tested are first placed in multiwell microtiter plates. Most proteins adhere to the plastic surface and thus are immobilized. Any exposed plastic that remains is subsequently "blocked" by adding a nonantigenic protein such as bovine serum albumin. A solution of antibody covalently linked to a reporter enzyme is then added and the antibodies allowed to adhere to any immobilized antigen molecules. Excess free antibody molecules are then removed by washing. The presence and quantity of bound antibody is then determined by assaying the activity of the reporter enzyme, a quantity that should be proportional to the number of antibody molecules present and, presumably, the antigen molecules to which they are bound.

NAD(P)⁺-Dependent Dehydrogenases Are Assayed Spectrophotometrically

The physicochemical properties of the reactants in an enzyme-catalyzed reaction dictate the options for the assay of enzyme activity. Spectrophotometric assays exploit the ability of a substrate or product to absorb light. The reduced coenzymes NADH and NADPH, written as NAD(P)H, absorb light at a wavelength of 340 nm, whereas their oxidized forms NAD(P)⁺ do not (**Figure 7–10**). When NAD(P)⁺ is reduced, the absorbance at 340 nm therefore increases in proportion to—and at a rate determined by—the quantity of NAD(P)H

produced. Conversely, when a dehydrogenase catalyzes the oxidation of NAD(P)H, a decrease in absorbance at 340 nm will be observed. In each case, the rate of change in absorbance at 340 nm will be proportionate to the quantity of the enzyme present.

The assay of enzymes whose reactions are not accompanied by a change in absorbance or fluorescence is generally more difficult. In some instances, either the product or remaining substrate can be transformed into a more readily detected compound, although the reaction product may have to be separated from unreacted substrate prior to measurement. An alternative strategy is to devise a synthetic substrate whose product absorbs light or fluoresces. For example, hydrolysis of the phosphoester bond in *p*-nitrophenyl phosphate (*p*NPP), an artificial substrate molecule, is catalyzed at a measurable rate by numerous phosphatases, phosphodiesterases, and serine proteases. While *p*NPP does not absorb visible light, the anionic form of the *p*-nitrophenol (pK_a 6.7) generated on its hydrolysis strongly absorbs light at 419 nm, and thus can be quantified.

Many Enzymes May Be Assayed by Coupling to a Dehydrogenase

Another quite general approach is to employ a "coupled" assay (**Figure 7–11**). Typically, a dehydrogenase whose substrate is the product of the enzyme of interest is added in catalytic excess. The rate of appearance or disappearance of NAD(P)H then depends on the rate of the enzyme reaction to which the dehydrogenase has been coupled.

THE ANALYSIS OF CERTAIN ENZYMES AIDS DIAGNOSIS

The analysis of enzymes in blood plasma plays a central role in the diagnosis of several disease processes. Many enzymes are functional constituents of blood. Examples include pseudocholinesterase, lipoprotein lipase, and components of the

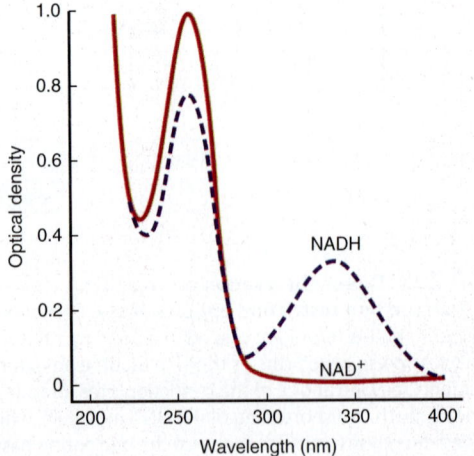

FIGURE 7–10 **Absorption spectra of NAD⁺ and NADH.** Densities are for a 44-mg/L solution in a cell with a 1-cm light path. NADP⁺ and NADPH have spectra analogous to NAD⁺ and NADH, respectively.

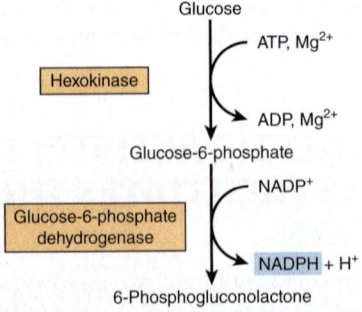

FIGURE 7–11 **Coupled enzyme assay for hexokinase activity.** The production of glucose-6-phosphate by hexokinase is coupled to the oxidation of this product by glucose-6-phosphate dehydrogenase in the presence of added enzyme and NADP⁺. When an excess of glucose-6-phosphate dehydrogenase is present, the rate of formation of NADPH, which can be measured at 340 nm, is governed by the rate of formation of glucose-6-phosphate by hexokinase.

cascades that trigger blood clotting, clot dissolution, and opsonization of invading microbes. Several enzymes are released into plasma following cell death or injury. While these latter enzymes perform no physiologic function in plasma, they can serve as **biomarkers**, molecules whose appearance or levels can assist in the diagnosis and prognosis of diseases and injuries affecting specific tissues. The plasma concentration of an enzyme or other protein released consequent to injury may rise early or late, and may decline rapidly or slowly. Cytoplasmic proteins tend to appear more rapidly than those from subcellular organelles.

Quantitative analysis of the activity of released enzymes or other proteins, typically in plasma or serum but also in urine or various cells, provides information concerning diagnosis, prognosis, and response to treatment. Assays of enzyme *activity* typically employ standard kinetic assays of initial reaction rates. **Table 7–1** lists several enzymes of value in clinical diagnosis. Note that these enzymes are not absolutely specific for the indicated disease. For example, elevated blood levels of prostatic acid phosphatase are associated typically with prostate cancer, but also may occur with certain other cancers and noncancerous conditions. Interpretation of enzyme assay data must make due allowance for the sensitivity and the diagnostic specificity of the enzyme test, together with other factors elicited through a comprehensive clinical examination that includes patient's age, sex, prior history, and possible drug use.

Analysis of Serum Enzymes Following Tissue Injury

An enzyme useful for diagnostic enzymology should be relatively specific for the tissue or organ under study, and should appear in the plasma or other fluid at a time useful for diagnosis (the "diagnostic window"). In the case of a myocardial infarction (MI), detection must be possible within a few hours

TABLE 7–1 Principal Serum Enzymes Used in Clinical Diagnosis

Serum Enzyme	Major Diagnostic Use
Alanine aminotransferase (ALT)	Viral hepatitis
Amylase	Acute pancreatitis
Ceruloplasmin	Hepatolenticular degeneration (Wilson disease)
Creatine kinase	Muscle disorders
γ-Glutamyl transferase	Various liver diseases
Lactate dehydrogenase isozyme 5	Liver diseases
Lipase	Acute pancreatitis
β-Glucocerebrosidase	Gaucher disease
Phosphatase, acid	Prostate disease, including cancer
Phosphatase, alkaline (isozymes)	Various bone disorders, obstructive liver diseases

ALT, alanine aminotransferase.
Note: Many of the above enzymes are not specific to the disease listed.

or less of a preliminary diagnosis to permit initiation of appropriate therapy. The first enzymes used to diagnose MI were aspartate aminotransferase (AST), alanine aminotransferase (ALT), and lactate dehydrogenase (LDH). Diagnosis using LDH exploits the tissue-specific variations in its quaternary structure (**Figure 7–12**). However, it is released relatively slowly following injury. Creatine kinase (CK) has three tissue-specific isozymes: CK-MM (skeletal muscle), CK-BB (brain), and CK-MB (heart and skeletal muscle), along with a more optimal diagnostic window. As with LDH, individual CK isozymes are separable by electrophoresis. Today, assays of

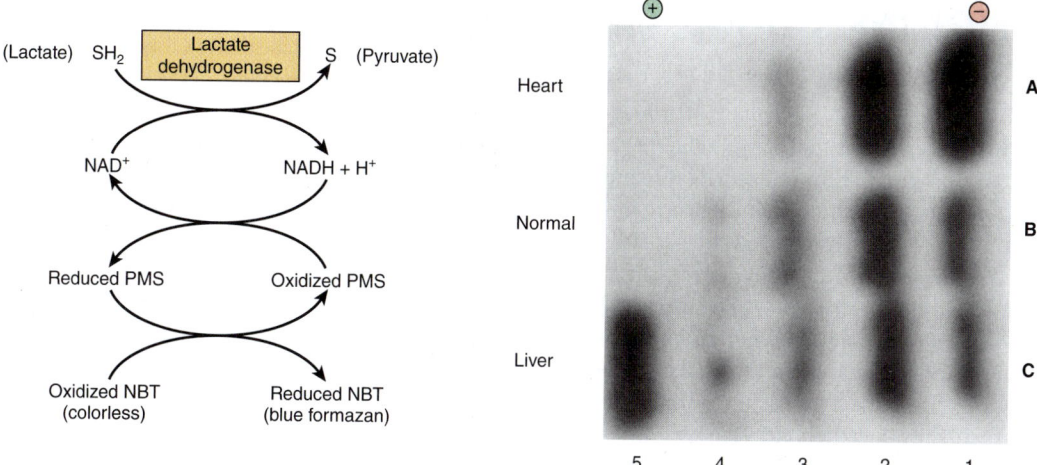

FIGURE 7–12 Normal and pathologic patterns of lactate dehydrogenase (LDH) isozymes in human serum. Samples of serum were separated by electrophoresis. LDH isozymes were then visualized using a dye-coupled reaction-specific for LDH. Pattern **A** is serum from a patient with a myocardial infarct; **B** is normal serum; and **C** is serum from a patient with liver disease. Arabic numerals identify LDH isozymes 1 through 5. Electrophoresis and a specific detection technique thus can be used to visualize isozymes of enzymes other than LDH.

plasma CK levels are primarily used to assess skeletal muscle disorders such as Duchene muscular dystrophy.

Plasma Troponin Constitutes the Currently Preferred Diagnostic Marker for an MI

Troponin is a complex of three proteins present in the contractile apparatus of *skeletal* and *cardiac muscle* but not in *smooth muscle* (see Chapter 51). Troponin levels in plasma typically rise for 2 to 6 hours after an MI, and remain elevated for 4 to 10 days. Immunological measurements of plasma levels of cardiac troponins I and T thus provide sensitive and specific indicators of damage to heart muscle. Since other sources of heart muscle damage also elevate serum troponin levels, cardiac troponins provide a general marker of cardiac injury.

Additional Clinical Uses of Enzymes

Enzymes are employed in the clinical laboratory to determine the presence and the concentration of critical metabolites. For example, glucose oxidase frequently is utilized to measure plasma glucose concentration. Enzymes also are employed with increasing frequency for the treatment of injury and disease. Examples include tissue plasminogen activator (tPA) or streptokinase for treatment of acute MI, and trypsin for treatment of cystic fibrosis. Intravenous infusion of recombinantly produced glycosylases can be used to treat lysosomal storage syndromes such as Gaucher disease (β-glucosidase), Pompe disease (α-glucosidase), Fabry disease (α-galactosidase A), Sly disease (β-glucuronidase), and mucopolysaccharidosis types I, II, and VI—also known as Hurler syndrome (α-L-iduronidase), Hunter syndrome (iduronate 2-sulfatase), and Maroteaux-Lamy syndrome (arylsulfatase B), respectively.

ENZYMES FACILITATE DIAGNOSIS OF GENETIC & INFECTIOUS DISEASES

Many diagnostic techniques take advantage of the specificity and efficiency of the enzymes that act on oligonucleotides such as DNA. Early on **restriction endonucleases** emerged as an important tool for both clinical and forensic analyses. Restriction endonucleases, often referred to as restriction enzymes, cleave double-stranded DNA at sites specified by a sequence of four, six, or more base pairs called *restriction sites* (see Chapter 39). If an individual bears a mutation or *polymorphism* that either eliminates or generates the restriction site for one of the endonucleases being used, the number and size of the DNA fragments produced on cleavage will differ from that of another individual. Such **restriction fragment length polymorphisms (RFLPs)** can be used for prenatal detection of deleterious mutations characteristic of hereditary disorders such as sickle cell trait, β-thalassemia, infant phenylketonuria, and Huntington disease. They also can be used as molecular

biological "fingerprints" to help trace the source of a sample of DNA to a specific individual. RFLP suffers from several important limitations, the most critical of which was the requirement for relatively high quantities of DNA to employ as substrates for the restriction enzymes. Today, a dramatically more sensitive set of molecular diagnostic tools based on the **polymerase chain reaction (PCR)** have largely supplanted RFLP analysis.

Medical Applications of the Polymerase Chain Reaction

As described in Chapter 39, the polymerase chain reaction (PCR) employs a thermostable DNA polymerase and appropriate oligonucleotide primers to produce thousands of copies of a defined segment of DNA from a minute quantity of starting material. PCR enables medical, biological, and forensic scientists to detect and characterize DNA present initially at levels too low for direct detection. In addition to screening for genetic mutations, PCR can be used to detect pathogens and parasites such as *Trypanosoma cruzi*, the causative agent of Chagas disease, and *Neisseria meningitides*, the causative agent of bacterial meningitis, through the selective amplification of their DNA.

RECOMBINANT DNA PROVIDES AN IMPORTANT TOOL FOR STUDYING ENZYMES

Highly purified samples of enzymes are essential for the study of their structure and function. The isolation of an individual enzyme, particularly one present in low concentration, from among the thousands of proteins present in a cell can be extremely difficult. By cloning the gene for the enzyme of interest, it generally is possible to produce large quantities of its encoded protein in *Escherichia coli* or yeast. However, not all animal proteins can be expressed in their appropriately folded, functionally competent form in microbial cells as these organisms cannot perform certain posttranslational processing tasks specific to higher organisms. In these instances, options include expression of recombinant genes in cultured animal cell systems or by employing the baculovirus expression vector of cultured insect cells. For more details concerning recombinant DNA techniques, see Chapter 39.

Recombinant Fusion Proteins Are Purified by Affinity Chromatography

Recombinant DNA technology can also be used to generate proteins specifically modified to render them readily purified by affinity chromatography. The gene of interest is linked to an additional oligonucleotide sequence that encodes a carboxyl or amino terminal extension to the protein of interest. The resulting **fusion protein** contains a new domain tailored to interact with an appropriately modified affinity support.

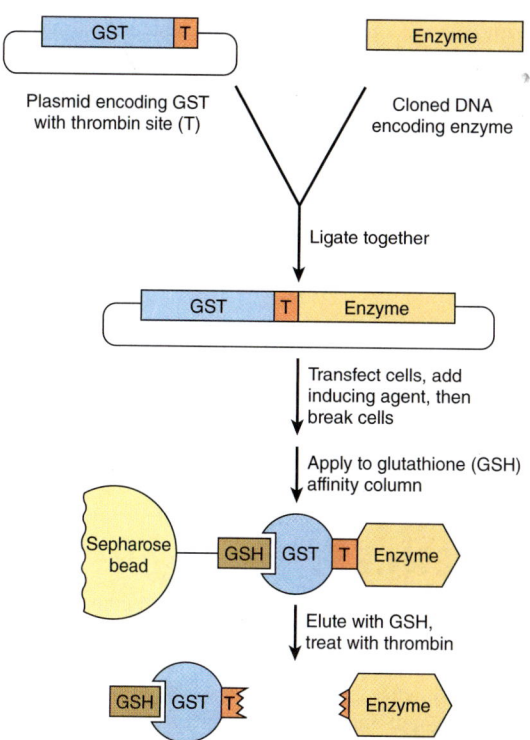

FIGURE 7–13 Use of glutathione S-transferase (GST) fusion proteins to purify recombinant proteins. (GSH, glutathione.)

One popular approach is to attach an oligonucleotide that encodes six consecutive histidine residues. The expressed "His tag" protein binds to chromatographic supports that contain an immobilized divalent metal ion such as Ni^{2+} or Cd^{2+}. This approach exploits the ability of these divalent cations to bind His residues. Once bound, contaminating proteins are washed off and the His-tagged enzyme is eluted with buffers containing high concentrations of free histidine or imidazole, which compete with the polyhistidine tails for binding to the immobilized metal ions. Alternatively, the substrate-binding domain of glutathione S-transferase (GST) can serve as a "GST tag." **Figure 7–13** illustrates the purification of a GST-fusion protein using an affinity support containing bound glutathione.

The addition of an N-terminal fusion domain may also help induce proper folding of the remainder of the recombinant polypeptide. Most fusion domains also possess a cleavage site for a highly specific protease such as thrombin in the region that links the two portions of the protein to permit its eventual removal.

Site-Directed Mutagenesis Provides Mechanistic Insights

Once the ability to express a protein from its cloned gene has been established, it is possible to employ **site-directed mutagenesis** to change specific aminoacyl residues by altering their codons. Used in combination with kinetic analyses and x-ray crystallography, this approach facilitates identification of the specific roles of given aminoacyl residues in substrate binding

and catalysis. For example, the inference that a particular aminoacyl residue functions as a general acid can be tested by replacing it with an aminoacyl residue incapable of donating a proton.

RIBOZYMES: ARTIFACTS FROM THE RNA WORLD

Cech Discovered the First Catalytic RNA Molecule

For many years after their discovery it was assumed that all enzymes were proteins. However, while examining the processing of ribosomal RNA (rRNA) molecules in the ciliated protozoan *Tetrahymena* in the early 1980s, Thomas Cech and his coworkers observed that processing of the 26S rRNA proceeded smoothly in vitro even in the total *absence* of protein. They subsequently traced the source of this splicing activity to a 413 bp catalytic segment of the RNA, which they termed a *ribozyme* (see Chapter 36)—a discovery that earned Cech a Nobel Prize.

Several other ribozymes have since been discovered. The vast majority catalyze nucleophilic displacement reactions that target the phosphodiester bonds of the RNA backbone. In small self-cleaving RNAs, such as hammerhead or hepatitis delta virus RNA, the attacking nucleophile is water and the result is hydrolysis. For the large group I intron ribozymes, the attacking nucleophile is the 3′-hydroxyl of the terminal ribose of another segment of RNA and the result is a splicing reaction.

The Ribosome—The Ultimate Ribozyme

The ribosome was the first example of a "molecular machine" to be recognized. A massive complex comprised of scores of protein subunits and several large ribosomal RNA molecules, the ribosome performs the vitally important and highly complex process of synthesizing long polypeptide chains following the instructions encoded in messenger RNA (mRNA) molecules (see Chapter 37). For many years, it was assumed that rRNAs played a passive, structural role, or perhaps assisted in the recognition of cognate mRNAs through a base pairing mechanism. It was thus somewhat surprising when it was discovered that rRNAs were both necessary and sufficient for catalyzing peptide synthesis.

The RNA World Hypothesis

The discovery of ribozymes has had a profound influence on evolutionary theory. For many years, scientists had hypothesized that the first biologic catalysts were formed when amino acids contained in the primordial soup coalesced to form simple proteins. With the realization that RNA could both carry information and catalyze chemical reactions, a new "RNA World" hypothesis emerged in which RNA constituted the first biologic macromolecule. Eventually, a more chemically stable oligonucleotide, DNA, superseded RNA for long-term information storage, while proteins, by virtue of their greater chemical functional group and conformational diversity dominated catalysis. If one assumes that some sort of RNA-protein hybrid was formed as an intermediate in the transition from ribonucleotide to polypeptide catalysts, one needs to look no further than the ribosome to find the presumed missing link.

Why did not proteins take over all catalytic functions? Presumably, in the case of the ribosome the process was both too complex and too essential to permit much opportunity for possible competitors to gain a foothold. In the case of the small self-cleaving RNAs and self-splicing introns, they may represent one of the few cases in which RNA autocatalysis is more efficient than development of a new protein catalyst.

SUMMARY

- Enzymes are efficient catalysts whose stringent specificity extends to the kind of reaction catalyzed, and typically, to the stereochemistry of their substrates.

- In many enzymes, the suite of chemical tools available to facilitate catalysis has been expanded using non-amino acid molecules. When tightly bound, these inorganic and organic molecules are referred to as prosthetic groups. If loosely bound (dissociable), they are referred to as cofactors.

- Coenzymes, many of which are derivatives of B vitamins, serve as "shuttles" for commonly used groups such as amines, electrons, and acetyl groups.

- During catalysis, enzymes redirect the conformational changes induced by substrate binding to effect complementary changes that facilitate transformation into product.

- Mechanistic strategies employed by enzymes to facilitate catalysis include the introduction of strain, approximation of reactants, acid–base catalysis, and covalent catalysis.

- Aminoacyl residues that participate in catalysis are highly conserved through the evolution of enzymes.

- The ability to change residues suspected of being important in catalysis or substrate binding by site-directed mutagenesis provides important insights into mechanisms of enzyme action.

- The catalytic activity of enzymes reveals their presence, facilitates their detection, and provides the basis for enzyme-linked immunoassays.

- Many enzymes can be assayed spectrophotometrically by coupling them to an NAD(P)H-dependent dehydrogenase.

- Assay of plasma proteins, including several enzymes, aids clinical diagnosis and prognosis.

- Restriction endonucleases facilitate diagnosis of genetic diseases by revealing restriction fragment length polymorphisms.

- The PCR amplifies DNA initially present in quantities too small for analysis.

- Attachment of a polyhistidyl, GST, or other "tag" to the N- or C-terminus of a recombinant protein facilitates its purification by affinity chromatography.

- Not all enzymes are proteins. Several ribozymes are known that can cut and resplice the phosphodiester bonds of RNA while it is the RNA components of the ribosome that are primarily responsible for catalyzing polypeptide synthesis.

REFERENCES

Apple FS, Sandoval Y, Jaffe AS, et al: Cardiac troponin assays: Guide to understanding analytical characteristics and their impact on clinical care. Clin Chem 2017;63:73.

Baumer ZT, Whitehead TA: The inner workings of an enzyme: A high-throughput mutation screen dissects the mechanistic basis of enzyme activity. Science 2021;373:391.

Bishop ML, Fody EP, Schoeff, LE: *Clinical Chemistry. Principles, Techniques, and Correlations*, 8th ed. Jones & Bartlett Learning, 2018.

Frey PA, Hegeman AD: *Enzyme Reaction Mechanisms*. Oxford University Press, 2006.

Heckman CM, Paradisi F: Looking back: A short history of the discovery of enzymes and how they became powerful chemical tool. Chem Cat Chem 2020;12:6082.

Hedstrom L: Serine protease mechanism and specificity. Chem Rev 2002;102:4501.

Knight AE: Single enzyme studies: A historical perspective. Meth Mol Biol 2011;778:1.

Rho JH, Lampe PD: High–throughput analysis of plasma hybrid markers for early detection of cancers. Proteomes 2014;2:1.

Sanabria H, Rodnin D, Hemmen K, et al: Resolving dynamics and function of transient states in single enzyme molecules. Nature Commun 2020;11:1231.

Spies M, Chemla Y (eds): *Single-Molecule Enzymology: Fluorescence-based and High-throughput Methods*. Methods Enzymol, vol. 581, Academic Press, 2016 (Entire volume).

Weinberg CA, Weinberg Z, Hammann C: Novel ribozymes: Discovery, catalytic mechanisms, and the quest to understand biological function. Nucl Acids Res 2019;47:9480.

Enzymes: Kinetics

Victor W. Rodwell, PhD

- Describe the scope and objectives of enzyme kinetic analysis.
- Indicate whether ΔG, the overall change in free energy for a reaction, is dependent on reaction mechanism.
- Indicate whether ΔG is a function of the *rates* of reactions.
- Explain the relationship between K_{eq}, concentrations of substrates and products at equilibrium, and the ratio of the rate constants k_1/k_{-1}.
- Outline how the concentration of hydrogen ions, of enzyme, and of substrate affect the rate of an enzyme-catalyzed reaction.
- Utilize collision theory to explain how temperature affects the rate of a chemical reaction.
- Define initial rate conditions and explain the advantage obtained from measuring the velocity of an enzyme-catalyzed reaction under these conditions.
- Describe the application of linear forms of the Michaelis-Menten equation to estimate K_m and V_{max}.
- Give one reason why a linear form of the Hill equation is used to evaluate how substrate-binding influences the kinetic behavior of certain multimeric enzymes.
- Contrast the effects of an increasing concentration of substrate on the kinetics of simple competitive and noncompetitive inhibition.
- Describe how substrates add to, and products depart from, an enzyme that follows a ping-pong mechanism.
- Describe how substrates add to, and products depart from, an enzyme that follows a rapid-equilibrium mechanism.
- Provide examples of the utility of enzyme kinetics in ascertaining the mode of action of drugs.

BIOMEDICAL IMPORTANCE

A complete and balanced set of enzyme activities is required for maintaining homeostasis. Enzyme kinetics, the quantitative measurement of the rates of enzyme-catalyzed reactions and the systematic study of factors that affect these rates, constitutes a central tool for the analysis, diagnosis, and treatment of the enzymic imbalances that underlie numerous human diseases. For example, kinetic analysis can reveal the number and order of the individual steps by which enzymes transform substrates into products and, in conjunction with site-directed mutagenesis, kinetic analyses can reveal details of the catalytic mechanism of a given enzyme. In the blood, the appearance or a surge in the levels of particular enzymes serves as clinical indicators for pathologies such as myocardial infarctions, prostate cancer, and damage to the liver. The involvement of enzymes in virtually all physiologic processes makes them the targets of choice for drugs that cure or ameliorate human disease.

Applied enzyme kinetics represents the principal tool by which scientists identify and characterize therapeutic agents that selectively inhibit the rates of specific enzyme-catalyzed processes. Enzyme kinetics thus plays a central and critical role in drug discovery, in comparative to pharmacodynamics, and in elucidating the mode of action of drugs.

CHEMICAL REACTIONS ARE DESCRIBED USING BALANCED EQUATIONS

A **balanced chemical equation** lists the initial chemical species (substrates) present and the new chemical species (products) formed for a particular chemical reaction, all in their respective proportions or **stoichiometry**. For example, balanced equation (1) indicates that one molecule each of substrates A and B reacts to form one molecule each of products P and Q:

$$A + B \rightleftarrows P + Q \qquad (1)$$

The double arrows indicate reversibility, an intrinsic property of all chemical reactions. Thus, for reaction (1), if A and B can form P and Q, then P and Q can also form A and B. Designation of a particular reactant as a "substrate" or "product" is therefore somewhat arbitrary since the products for a reaction written in one direction are the substrates for the reverse reaction. The term "products" is, however, often used to designate the reactants whose formation is thermodynamically favored. Reactions for which thermodynamic factors strongly favor formation of the products to which the arrow points often are represented with a single arrow as if they were "irreversible":

$$A + B \rightarrow P + Q \qquad (2)$$

Unidirectional arrows are also used to describe reactions in living cells where the products of reaction (2) are immediately consumed by a subsequent enzyme-catalyzed reaction or rapidly escape the cell, for example, CO_2. The rapid removal of product P or Q therefore effectively precludes occurrence of the reverse reaction, rendering equation (2) **functionally irreversible under physiologic conditions.**

CHANGES IN FREE ENERGY DETERMINE THE DIRECTION & EQUILIBRIUM STATE OF CHEMICAL REACTIONS

The Gibbs free-energy change ΔG (also called either free energy or Gibbs energy) describes in quantitative form both the *direction* in which a chemical reaction will tend to proceed and the concentrations of reactants and products that will be present at equilibrium. ΔG for a chemical reaction equals the sum of the free energies of formation of the reaction products ΔG_p minus the sum of the free energies of formation of the substrates ΔG_s. A similar but different quantity designated by ΔG^0 denotes the change in free energy that accompanies transition from the standard state, one-molar concentrations of substrates and products, to equilibrium. A more useful biochemical term is $\Delta G^{0\prime}$, which defines ΔG^0 at a standard state of 10^{-7} M protons, pH 7.0. If the free energy of formation of the products is *lower* than that of the substrates, the signs of ΔG^0 and $\Delta G^{0\prime}$ will be *negative*, indicating that the reaction as written is favored in the direction left to right. Such reactions are referred to as **spontaneous**. The **sign** and the **magnitude** of the free-energy change determine how far the reaction will proceed.

Equation (3) illustrates the relationship between the equilibrium constant K_{eq} and ΔG^0:

$$\Delta G^0 = -RT \ln K_{eq} \qquad (3)$$

where R is the gas constant (1.98 cal/mol°K or 8.31 J/mol°K) and T is the absolute temperature in degrees Kelvin. K_{eq} is equal to the product of the concentrations of the reaction products, each raised to the power of their stoichiometry, divided by the product of the substrates, each raised to the power of their stoichiometry:

For the reaction $A + B \rightleftarrows P + Q$

$$K_{eq} = \frac{[P][Q]}{[A][B]} \qquad (4)$$

and for reaction (5)

$$A + A \rightleftarrows P \qquad (5)$$

$$K_{eq} = \frac{[P]}{[A]^2} \qquad (6)$$

ΔG^0 may be calculated from equation (3) if the molar concentrations of substrates and products present at equilibrium are known. If ΔG^0 is a negative number, K_{eq} will be greater than unity, and the concentration of products at equilibrium will exceed that of the substrates. If ΔG^0 is positive, K_{eq} will be less than unity, and the formation of substrates will be favored.

Note that, since ΔG^0 is a function exclusively of the initial and final states of the reacting species, it can provide information only about the *direction* and *equilibrium state* of the reaction. ΔG^0 is independent of the **mechanism** of the reaction, and provides no information concerning *rates* of reactions. Consequently—and as explained below—although a reaction may have a large negative ΔG^0 or $\Delta G^{0\prime}$, it may nevertheless take place at a negligible rate.

THE RATES OF REACTIONS ARE DETERMINED BY THEIR ACTIVATION ENERGY

Reactions Proceed via Transition States

The concept of the **transition state** is fundamental to understanding the chemical and thermodynamic basis of catalysis. Equation (7) depicts a group transfer reaction in which an entering group E displaces a leaving group L, attached initially to R:

$$E + R - L \rightleftharpoons E - R + L \qquad (7)$$

The net result of this process is to transfer group R from L to E. Midway through the displacement, the bond between R and L has weakened but has not yet been completely severed, and the new bond between E and R is yet incompletely formed. This transient intermediate—in which neither free substrate nor product exists—is termed the **transition state, E⋯R⋯L.** Dotted lines represent the "partial" bonds that are undergoing formation and rupture. **Figure 8–1** provides a more detailed illustration of the transition state intermediate formed during the transfer of a phosphoryl group.

Reaction (7) can be thought of as consisting of two "partial reactions," the first corresponding to the formation (F) and the second to the subsequent decay (D) of the transition state intermediate. As for all reactions, characteristic changes in free energy, ΔG_F and ΔG_D are associated with each partial reaction:

$$E + R - L \rightleftharpoons E \cdots R \cdots L \quad \Delta G_F \qquad (8)$$

$$E \cdots R \cdots L \rightleftharpoons E - R + L \quad \Delta G_D \qquad (9)$$

$$E + R - L \rightleftharpoons E - R + L \quad \Delta G = \Delta G_F + \Delta G_D \qquad (10)$$

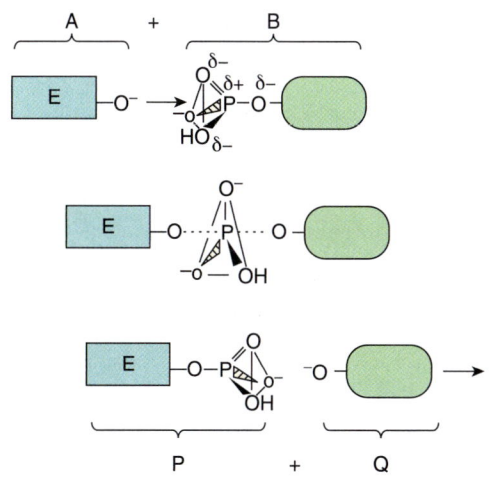

FIGURE 8–1 **Formation of a transition state intermediate during a simple chemical reaction, A + B → P + Q.** Shown are three stages of a chemical reaction in which a phosphoryl group is transferred from leaving group L (green) to entering group E (blue). **Top:** Entering group E (bracket A) approaches the other reactant, L-phosphate (bracket B). Notice how the three oxygen atoms linked by the triangular lines and the phosphorus atom of the phosphoryl group form a pyramid. **Center:** As E approaches L-phosphate, the new bond between E and the phosphoryl group begins to form (dotted line) as that linking L to the phosphoryl group weakens. These partially formed bonds are indicated by dotted lines. **Bottom:** Formation of the new product, E-phosphate (bracket P), is now complete as the leaving group L (bracket Q) exits. Notice how the geometry of the phosphoryl group differs between the transition state and the substrate or product. The phosphorus and three oxygen atoms that occupy the four corners of a pyramid in the substrate and product become coplanar, as emphasized by the triangle, in the transition state.

For the overall reaction (10), ΔG is the numeric sum of ΔG_F and ΔG_D. As for any equation of two terms, it is not possible to deduce from their resultant ΔG, either the sign or the magnitude of ΔG_F or ΔG_D.

Many reactions involve several successive transition states, each with an associated change in free energy. For these reactions, the overall ΔG represents the sum of *all* of the free-energy changes associated with the formation and decay of *all* of the transition states. **It therefore is not possible to infer from the overall ΔG the number or type of transition states through which the reaction proceeds.** Stated another way, *overall reaction thermodynamics tells us nothing about mechanism or kinetics.*

ΔG_F Defines the Activation Energy

Regardless of the sign or magnitude of ΔG, ΔG_F for the overwhelming majority of chemical reactions has a positive sign, which indicates that formation of the transition state requires surmounting one or more energy barriers. For this reason, ΔG_F for reaching a transition state is often termed the **activation energy,** E_{act}. The ease—and hence the frequency—with which this barrier is overcome is *inversely* related to E_{act}. The thermodynamic parameters that determine how *fast* a reaction proceeds thus are the ΔG_F values for formation of the transition state(s) through which the reaction proceeds. For a simple reaction, where $\propto$ means "proportionate to,"

$$\text{Rate} \propto e^{-E_{act}/RT} \qquad (11)$$

The activation energy for the reaction proceeding in the opposite direction to that drawn is equal to $-\Delta G_D$.

NUMEROUS FACTORS AFFECT REACTION RATE

The **kinetic theory**—also called the **collision theory**—of chemical kinetics states that for two molecules to react, they (1) must approach within bond-forming distance of one another, or "collide," and (2) must possess sufficient kinetic energy to overcome the energy barrier for reaching the transition state. It therefore follows that conditions that tend to increase the *frequency* or *energy* of collision between substrates will tend to increase the rate of the reaction in which they participate.

Temperature

Raising the ambient temperature increases the kinetic energy of molecules. As illustrated in **Figure 8–2**, the total number of molecules whose kinetic energy exceeds the energy barrier E_{act} (vertical bar) for formation of products increases from low (A) through intermediate (B) to high (C) temperatures. Increasing the kinetic energy of molecules also increases their rapidity of motion, and therefore the frequency with which they collide. This combination of more frequent and more highly energetic, and hence productive, collisions increases the reaction rate.

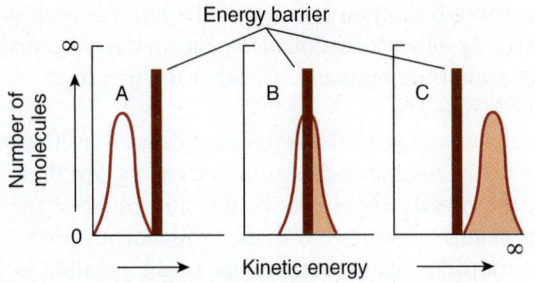

FIGURE 8–2 **Representation of the energy barrier for chemical reactions.** (See text for discussion.)

Reactant Concentration

The frequency with which molecules collide is directly proportionate to their concentrations. For two different molecules A and B, the frequency with which they collide will double if the concentration of either A or B is doubled. If the concentrations of both A and B are doubled, the probability of collision will increase fourfold.

For a chemical reaction proceeding at a constant temperature that involves one molecule each of A and B,

$$A + B \rightarrow P \tag{12}$$

the fraction of the molecules possessing a given kinetic energy will be a constant. The number of collisions between molecules whose combined kinetic energy is sufficient to produce product P therefore will be directly proportionate to the number of collisions between A and B, and thus to their molar concentrations, denoted by the square brackets:

$$\text{Rate} \propto [A]\,[B] \tag{13}$$

Similarly, for the reaction represented by

$$A + 2B \rightarrow P \tag{14}$$

which can also be written as

$$A + B + B \rightarrow P \tag{15}$$

the corresponding rate expression is

$$\text{Rate} \propto [A]\,[B]\,[B] \tag{16}$$

or

$$\text{Rate} \propto [A][B]^2 \tag{17}$$

For the general case, when n molecules of A react with m molecules of B,

$$nA + mB \rightarrow P \tag{18}$$

the rate expression is

$$\text{Rate} \propto [A]^n\,[B]^m \tag{19}$$

Replacing the proportionality sign with an equals sign by introducing a **rate constant, k,** characteristic of the reaction under study gives equations (20) and (21), in which the subscripts 1 and –1 refer to the forward and reverse reactions, respectively:

$$\text{Rate}_1 = k_1[A]^n[B]^m \tag{20}$$

$$\text{Rate}_{-1} = k_{-1}[P] \tag{21}$$

The sum of the molar ratios of the reactants defines the **kinetic order** of the reaction. Consider reaction (5). The stoichiometric coefficient for the sole reactant, A, is 2. Therefore, the rate of production of P is proportional to the square of [A] and the reaction is said to be *second order* with respect to reactant A. In this instance, the overall reaction is also *second order*. Therefore, k_1 is referred to as a *second-order rate constant*.

Reaction (12) describes a simple second-order reaction between two different reactants, A and B. The stoichiometric coefficient for each reactant is 1. Therefore, while the reaction is second order, it is said to be *first order* with respect to A and *first order* with respect to B.

In the laboratory, the kinetic order of a reaction with respect to a particular reactant, referred to as the variable reactant or substrate, can be determined by maintaining the concentration of other reactants in large excess over the variable reactant. Under these *pseudo-first-order conditions*, the concentration of the "fixed" reactant remains virtually constant. Thus, the rate of reaction will depend exclusively on the concentration of the variable reactant, sometimes also called the limiting reactant. The concepts of reaction order and pseudo-first-order conditions apply not only to simple chemical reactions but also to enzyme-catalyzed reactions.

K_{eq} Is a Ratio of Rate Constants

While all chemical reactions are to some extent reversible, at equilibrium the *overall* concentrations of reactants and products remain constant. At equilibrium, the rate of conversion of substrates to products therefore equals the rate at which products are converted to substrates:

$$\text{Rate}_1 = \text{Rate}_{-1} \tag{22}$$

Therefore,

$$k_1 = [A]^n[B]^m = k_{-1}[P] \tag{23}$$

and

$$\frac{k_1}{k_{-1}} = \frac{[P]}{[A]^n[B]^m} \tag{24}$$

The ratio of k_1 to k_{-1} is equal to the equilibrium constant, K_{eq}. The following important properties of a system at equilibrium must be kept in mind:

1. The equilibrium constant is a ratio of the reaction rate *constants* (not the reaction *rates*).

2. At equilibrium, the reaction *rates* (not the *rate constants*) of the forward and back reactions are equal.

3. The numeric value of the equilibrium constant K_{eq} can be calculated either from the concentrations of substrates and products at equilibrium or from the ratio k_1/k_{-1}.

4. Equilibrium is a *dynamic* state. Although there is no *net* change in the concentration of substrates or products, individual substrate and product molecules are continually being interconverted. Interconvertibility can be proved by adding to a system at equilibrium a trace of radioisotopic product, which can then be shown to result in the appearance of radiolabelled substrate.

THE KINETICS OF ENZYME CATALYSIS

Enzymes Lower the Activation Energy Barrier for a Reaction

All enzymes accelerate reaction rates by lowering ΔG_F for the formation of transition states. However, they may differ in the way this is achieved. While the sequence of chemical steps at the active site parallels those which occur when the substrates react in the absence of a catalyst, **the environment of the active site lowers ΔG_F** by stabilizing the transition state intermediates. To put it another way, the enzyme can be envisioned as binding to the transition state intermediate (Figure 8–1) more tightly than it does to either substrates or products. As discussed in Chapter 7, stabilization can involve (1) acid–base groups suitably positioned to transfer protons to or from the developing transition state intermediate, (2) suitably positioned charged groups or metal ions that stabilize developing charges, or (3) the imposition of steric strain on substrates so that their geometry approaches that of the transition state. HIV protease (see Figure 7–6) illustrates catalysis by an enzyme that lowers the activation barrier in part by stabilizing a transition state intermediate.

Catalysis by enzymes that proceeds via a *unique* reaction mechanism typically occurs when the transition state intermediate forms a covalent bond with the enzyme (**covalent catalysis**). The catalytic mechanism of the serine protease chymotrypsin (see Figure 7–7) illustrates how an enzyme utilizes covalent catalysis to provide a unique reaction pathway possessing a more favorable E_{act}.

ENZYMES DO NOT AFFECT K_{eq}

While enzymes undergo transient modifications during the process of catalysis, they always emerge unchanged at the completion of the reaction. **The presence of an enzyme therefore has no effect on ΔG^0 for the *overall* reaction**, which is a function solely of the **initial and final states** of the reactants. Equation (25) shows the relationship between the equilibrium constant for a reaction and the standard free-energy change for that reaction:

$$\Delta G^0 = -RT \ln K_{eq} \qquad (25)$$

This principle is perhaps most readily illustrated by including the presence of the enzyme (Enz) in the calculation of the equilibrium constant for an enzyme-catalyzed reaction:

$$A + B + Enz \rightleftarrows P + Q + Enz \qquad (26)$$

Since the enzyme on both sides of the double arrows is present in equal quantity and identical form, the expression for the equilibrium constant,

$$K_{eq} = \frac{[P][Q][Enz]}{[A][B][Enz]} \qquad (27)$$

reduces to one identical to that for the reaction in the *absence* of the enzyme:

$$K_{eq} = \frac{[P][Q]}{[A][B]} \qquad (28)$$

Enzymes therefore have no effect on K_{eq}.

MULTIPLE FACTORS AFFECT THE RATES OF ENZYME-CATALYZED REACTIONS

Temperature

Raising the temperature increases the rate of both uncatalyzed and enzyme-catalyzed reactions by increasing the kinetic energy and the collision frequency of the reacting molecules. However, heat energy can also increase the conformational flexing of the enzyme to a point that exceeds the energy barrier for disrupting the noncovalent interactions that maintain its three-dimensional structure. The polypeptide chain then begins to unfold, or **denature**, with an accompanying loss of the catalytic activity. The temperature range over which an enzyme maintains a stable, catalytically competent conformation depends on—and typically moderately exceeds—the normal temperature of the cells in which it resides. Enzymes from humans generally exhibit stability at temperatures up to 45 to 55°C. By contrast, enzymes from the thermophilic microorganisms that reside in volcanic hot springs or undersea hydrothermal vents may be stable at temperatures up to or even above 100°C.

The **temperature coefficient (Q_{10})** is the factor by which the rate of a biologic process increases for a 10°C increase in temperature. For the temperatures over which enzymes are stable, the rates of most biologic processes typically double for a 10°C rise in temperature ($Q_{10} = 2$). Changes in the rates of enzyme-catalyzed reactions that accompany a rise or fall in body temperature constitute a prominent survival feature for "cold-blooded" life forms such as lizards or fish, whose body temperatures are dictated by the external environment. However, for mammals and other homeothermic organisms, changes in enzyme reaction rates with temperature assume physiologic importance only in circumstances such as fever or hypothermia.

Hydrogen Ion Concentration

The rate of almost all enzyme-catalyzed reactions exhibits a significant dependence on hydrogen ion concentration. Most intracellular enzymes exhibit optimal activity at pH values between 5 and 9. The relationship of activity to hydrogen ion concentration (**Figure 8–3**) reflects the balance between enzyme denaturation at high or low pH and effects on the

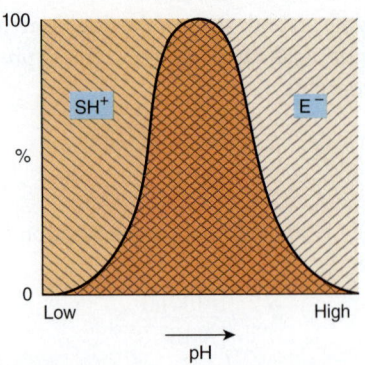

FIGURE 8–3 **Effect of pH on enzyme activity.** Consider, for example, a negatively charged enzyme (E⁻) that binds a positively charged substrate (SH⁺). Shown is the proportion (%) of SH⁺ [\\\] and of E⁻ [///] as a function of pH. Only in the cross-hatched area do both the enzyme and the substrate bear an appropriate charge.

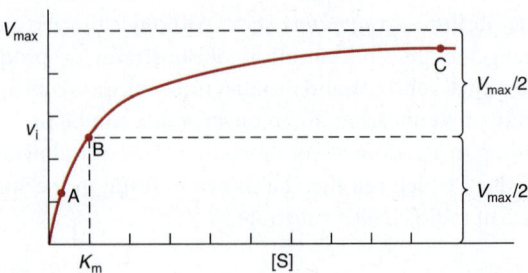

FIGURE 8–4 **Effect of substrate concentration on the initial velocity of an enzyme-catalyzed reaction.**

charged state of the enzyme, the substrates, or both. For enzymes whose mechanism involves acid–base catalysis, the residues involved must be in the appropriate state of protonation for the reaction to proceed. The binding and recognition of substrate molecules with dissociable groups also typically involves the formation of salt bridges with the enzyme. The most common charged groups are carboxylate groups (negative) and protonated amines (positive). Gain or loss of critical charged groups adversely affects substrate binding and thus will retard or abolish catalysis.

ASSAYS OF ENZYME-CATALYZED REACTIONS TYPICALLY MEASURE THE INITIAL VELOCITY

Most measurements of the rates of enzyme-catalyzed reactions employ relatively short time periods, conditions that are considered to approximate **initial rate conditions**. Under these conditions, only traces of product accumulate, rendering the rate of the reverse reaction negligible. The **initial velocity** (v_i) of the reaction thus is essentially that of the rate of the forward reaction. Assays of enzyme activity almost always employ a large (10^3-10^6) molar excess of substrate over enzyme. Under these conditions, v_i is proportionate to the concentration of enzyme, that is, it is pseudo-first-order with respect to enzyme. Measuring the initial velocity therefore permits one to estimate the quantity of enzyme present in a biologic sample.

SUBSTRATE CONCENTRATION AFFECTS THE REACTION RATE

In what follows, enzyme reactions are treated as if they had only a single substrate and a single product. For enzymes with multiple substrates, the principles discussed below apply with equal validity. Moreover, by employing pseudo-first-order conditions

(see above), scientists can study the dependence of reaction rate on an individual reactant through the appropriate choice of fixed and variable substrates. In other words, under pseudo-first-order conditions the behavior of a multisubstrate enzyme will imitate one having a single substrate. In this instance, however, the observed rate constant will be a function both of the rate constant k_i for the reaction and of the concentration of the fixed substrate.

For a typical enzyme, as substrate concentration is increased, v_i increases until it reaches a maximum value V_{max} (**Figure 8–4**). When further increases in substrate concentration fail to increase v_i, the enzyme is said to be "saturated" with the substrate. Note that the shape of the curve that relates activity to substrate concentration (Figure 8–4) is *hyperbolic*. At any given instant, only substrate molecules that are combined with the enzyme as an enzyme-substrate (ES) complex can be transformed into product. Since the equilibrium constant for the formation of the enzyme-substrate complex is not infinitely large, only a fraction of the enzyme may be present as an ES complex even when the substrate is present in considerable excess (points A and B of **Figure 8–5**). At points A or B, increasing or decreasing [S] therefore will increase or decrease the number of ES complexes with a corresponding change in v_i. At point C (Figure 8–5), however, essentially all the enzyme is present as the ES complex. Since no free enzyme remains available for forming ES, further increases in [S] cannot increase the rate of the reaction. **Under these saturating conditions, v_i depends solely on—and thus is limited by—the rapidity with which product dissociates from the enzyme so that it may combine with more substrate.**

THE MICHAELIS-MENTEN & HILL EQUATIONS MODEL THE EFFECTS OF SUBSTRATE CONCENTRATION

The Michaelis-Menten Equation

The Michaelis-Menten equation (29) illustrates in mathematical terms the relationship between initial reaction velocity v_i and substrate concentration [S], shown graphically in Figure 8–4:

$$v_i = \frac{V_{max}[S]}{K_m + [S]} \qquad (29)$$

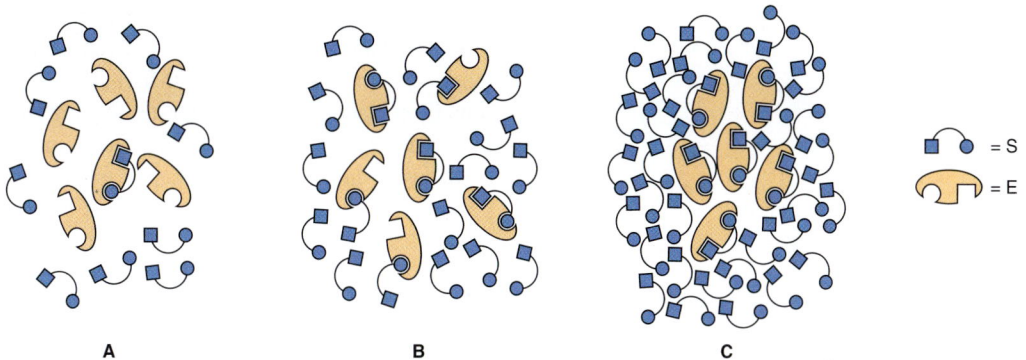

FIGURE 8–5 Representation of an enzyme in the presence of a concentration of substrate that is below K_m (A), at a concentration equal to K_m (B), and at a concentration well above K_m (C). Points A, B, and C correspond to those points in Figure 8–4.

The Michaelis constant K_m is the substrate concentration at which v_i is half the maximal velocity ($V_{max}/2$) attainable at a particular concentration of the enzyme. K_m thus has the dimensions of substrate concentration. The dependence of initial reaction velocity on [S] and K_m may be illustrated by evaluating the Michaelis-Menten equation under three conditions.

1. When [S] is much less than K_m (point A in Figures 8–4 and 8–5), the term K_m + [S] is essentially equal to K_m. Replacing K_m + [S] with K_m reduces equation (29) to

$$v_i = \frac{V_{max}[S]}{K_m+[S]} \quad v_i \approx \frac{V_{max}[S]}{K_m} \approx \left(\frac{V_{max}}{K_m}\right)[S] \quad (30)$$

where ≈ means "approximately equal to." Since V_{max} and K_m are both constants, their ratio is a constant. In other words, when [S] is considerably below K_m, v_i is proportionate to k[S]. The initial reaction velocity therefore is directly proportional to [S].

2. When [S] is much greater than K_m (point C in Figures 8–4 and 8–5), the term K_m + [S] is essentially equal to [S]. Replacing K_m + [S] with [S] reduces equation (29) to

$$v_i = \frac{V_{max}[S]}{K_m+[S]} \quad v_i \approx \frac{V_{max}[S]}{[S]} \approx V_{max} \quad (31)$$

Thus, when [S] greatly exceeds K_m, the reaction velocity is maximal (V_{max}) and unaffected by further increases in the substrate concentration.

3. When [S] = K_m (point B in Figures 8–4 and 8–5):

$$v_i = \frac{V_{max}[S]}{K_m+[S]} = \frac{V_{max}[S]}{2[S]} = \frac{V_{max}}{2} \quad (32)$$

Equation (32) states that when [S] equals K_m, the initial velocity is half-maximal. Equation (32) also reveals that K_m is—and may be determined experimentally from—the substrate concentration at which the initial velocity is half-maximal.

A Linear Form of the Michaelis-Menten Equation Is Used to Determine K_m & V_{max}

The direct measurement of the numeric value of V_{max}, and therefore the calculation of K_m, often requires impractically high concentrations of substrate to achieve saturating conditions. A linear form of the Michaelis-Menten equation circumvents this difficulty and permits V_{max} and K_m to be extrapolated from initial velocity data obtained at less than saturating concentrations of the substrate. Start with equation (29),

$$v_i = \frac{V_{max}[S]}{K_m+[S]} \quad (29)$$

invert

$$\frac{1}{v_i} = \frac{K_m+[S]}{V_{max}[S]} \quad (33)$$

factor

$$\frac{1}{v_i} = \frac{K_m}{V_{max}[S]} + \frac{[S]}{V_{max}[S]} \quad (34)$$

and simplify

$$\frac{1}{v_i} = \left(\frac{K_m}{V_{max}}\right)\frac{1}{[S]} + \frac{1}{V_{max}} \quad (35)$$

Equation (35) is the equation for a straight line, $y = ax + b$, where $y = 1/v_i$ and $x = 1/[S]$. A plot of $1/v_i$ as y as a function of $1/[S]$ as x therefore gives a straight line whose y intercept is $1/V_{max}$ and whose slope is K_m/V_{max}. Such a plot is called a **double reciprocal** or **Lineweaver-Burk plot** (**Figure 8–6**). Setting the y term of equation (36) equal to zero and solving for x reveals that the x intercept is $-1/K_m$:

$$0 = ax + b; \quad \text{therefore,} \quad x = \frac{-b}{a} = \frac{-1}{K_m} \quad (36)$$

K_m can be calculated from the slope and y intercept, but is perhaps most readily calculated from the negative x intercept.

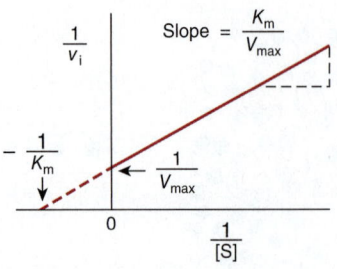

FIGURE 8–6 Double-reciprocal or Lineweaver-Burk plot of $1/v_i$ versus $1/[S]$ used to evaluate K_m and V_{max}.

The greatest virtue of the Lineweaver-Burk plot resides in the facility with which it can be used to determine the kinetic mechanism of an enzyme inhibitor (see below). However, in using a double-reciprocal plot to determine kinetic constants, it is important to avoid the introduction of bias through the clustering of data at low values of $1/[S]$. This bias can be readily avoided in the laboratory as follows. Prepare a solution of substrate whose dilution into an assay will produce the maximum desired concentration of the substrate. Now prepare dilutions of the stock solution by factors of 1:2, 1:3, 1:4, 1:5, etc. Data generated using equal volumes of these dilutions will then fall on the $1/[S]$ axis at equally spaced intervals of 1, 2, 3, 4, 5, etc. A single-reciprocal plot such as the Eadie-Hofstee (v_i vs $v_i/[S]$) or Hanes-Woolf ($[S]/v_i$ vs $[S]$) plot can also be used to minimize data clustering.

The Catalytic Constant, k_{cat}

Several parameters may be used to compare the relative activity of different enzymes or of different preparations of the same enzyme. The activity of impure enzyme preparations typically is expressed as a *specific activity* (V_{max} divided by the protein concentration). For a homogeneous enzyme, one may calculate its *turnover number* (V_{max} divided by the moles of enzyme present). However, if the number of active sites present is known, the catalytic activity of a homogeneous enzyme is best expressed as its *catalytic constant*, k_{cat} (V_{max} divided by the number of active sites, S_t):

$$k_{cat} = \frac{V_{max}}{S_t} \qquad (37)$$

Since the units of concentration cancel out, the units of k_{cat} are reciprocal time.

Catalytic Efficiency, k_{cat}/K_m

By what measure should the efficiency of different enzymes, different substrates for a given enzyme, and the efficiency with which an enzyme catalyzes a reaction in the forward and reverse directions be quantified and compared? While the maximum capacity of a given enzyme to convert substrate to product is important, the benefits of a high k_{cat} can only be realized if K_m is sufficiently low. Thus, *catalytic efficiency* of enzymes is best expressed in terms of the *ratio* of these two kinetic constants, k_{cat}/K_m.

For certain enzymes, once substrate binds to the active site, it is converted to product and released so rapidly as to render these events effectively instantaneous. For these exceptionally efficient catalysts, the rate-limiting step in catalysis is the formation of the ES complex. Such enzymes are said to be *diffusion-limited*, or catalytically perfect, since the fastest possible rate of catalysis is determined by the rate at which molecules move or diffuse through the solution. Examples of enzymes for which k_{cat}/K_m approaches the diffusion limit of 10^8-10^9 $M^{-1}s^{-1}$ include triosephosphate isomerase, carbonic anhydrase, acetylcholinesterase, and adenosine deaminase.

In living cells, the assembly of enzymes that catalyze successive reactions into multimeric complexes can circumvent the limitations imposed by diffusion. The geometric relationships of the enzymes in these complexes are such that the substrates and products do not diffuse into the bulk solution until the last step in the sequence of catalytic steps is complete. Fatty acid synthetase extends this concept one step further by covalently attaching the growing substrate fatty acid chain to a biotin tether that rotates from active site to active site within the complex until synthesis of a palmitic acid molecule is complete (see Chapter 23).

K_m May Approximate a Binding Constant

The affinity of an enzyme for its substrate is the inverse of the dissociation constant K_d for dissociation of the enzyme-substrate complex ES:

$$E + S \underset{k_{-1}}{\overset{k_1}{\rightleftharpoons}} ES \qquad (38)$$

$$K_d = \frac{k_{-1}}{k_1} \qquad (39)$$

Stated another way, the *smaller* the tendency of the enzyme and its substrate to *dissociate*, the *greater* the affinity of the enzyme for its substrate. While the Michaelis constant K_m often approximates the dissociation constant K_d, this should not be assumed, for it is by no means always the case. For a typical enzyme-catalyzed reaction:

$$E + S \underset{k_{-1}}{\overset{k_1}{\rightleftharpoons}} ES \overset{k_2}{\longrightarrow} E + P \qquad (40)$$

The value of $[S]$ that gives $v_i = V_{max}/2$ is

$$[S] = \frac{k_{-1} + k_2}{k_1} = K_m \qquad (41)$$

When $k_{-1} \gg k_2$, then

$$k_{-1} + k_2 \approx k_{-1} \qquad (42)$$

and

$$[S] \approx \frac{k_1}{k_{-1}} = K_d \qquad (43)$$

Hence, $1/K_m$ only approximates $1/K_d$ under conditions where the association and dissociation of the ES complex are rapid relative to catalysis. For the many enzyme-catalyzed reactions for which $k_{-1} + k_2$ is **not** approximately equal to k_{-1}, $1/K_m$ will underestimate $1/K_d$.

The Hill Equation Describes the Behavior of Enzymes That Exhibit Cooperative Binding of Substrate

While most enzymes display the simple **saturation kinetics** depicted in Figure 8–4 and are adequately described by the Michaelis-Menten expression, some enzymes bind their substrates in a **cooperative** fashion analogous to the binding of oxygen by hemoglobin (see Chapter 6). Cooperative behavior is an *exclusive* property of multimeric enzymes that bind substrate at multiple sites.

For enzymes that display positive cooperativity in binding the substrate, the shape of the curve that relates changes in v_i to changes in [S] is sigmoidal (**Figure 8–7**). Neither the Michaelis-Menten expression nor its derived plots can be used to evaluate cooperative kinetics. Enzymologists therefore employ a graphic representation of the **Hill equation** originally derived to describe the cooperative binding of O_2 by hemoglobin. Equation (44) represents the Hill equation arranged in a form that predicts a straight line, where k' is a complex constant:

$$\frac{\log v_i}{V_{max} - v_i} = n \log[S] - \log k' \qquad (44)$$

Equation (44) states that when [S] is low relative to k', the initial reaction velocity increases as the nth power of [S].

A graph of $\log v_i/(V_{max} - v_i)$ versus $\log[S]$ gives a straight line (**Figure 8–8**). The slope of the line, **n**, is the **Hill coefficient**, an empirical parameter whose value is a function of the number, kind, and strength of the interactions of the multiple substrate-binding sites on the enzyme. When $n = 1$, all binding sites behave independently and simple Michaelis-Menten kinetic behavior is observed. If n is greater than 1, the enzyme is said to exhibit **positive cooperativity**. Binding of substrate to one

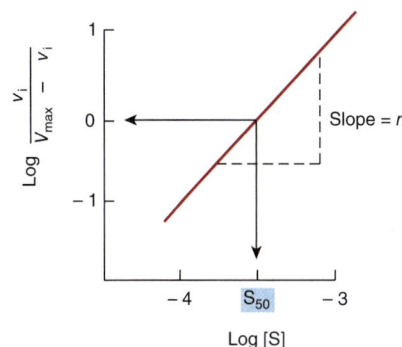

FIGURE 8–8 A graphical representation of a linear form of the Hill equation is used to evaluate S_{50}, the substrate concentration that produces half-maximal velocity, and the degree of cooperativity n.

site then enhances the affinity of the remaining sites to bind additional substrate. The greater the value for n, the higher the degree of cooperativity and the more markedly sigmoidal will be the plot of v_i versus [S]. A perpendicular dropped from the point where the y term $\log v_i/(V_{max} - v_i)$ is zero intersects the x-axis at a substrate concentration termed S_{50}, the substrate concentration that results in half-maximal velocity, S_{50}, thus is analogous to the P_{50} for oxygen binding to hemoglobin (see Chapter 6).

KINETIC ANALYSIS DISTINGUISHES COMPETITIVE FROM NONCOMPETITIVE INHIBITION

Inhibitors of the catalytic activities of enzymes provide both pharmacologic agents and research tools for the study of the mechanism of enzyme action. The strength of the interaction between an inhibitor and an enzyme depends on forces important in protein structure and ligand binding (hydrogen bonds, electrostatic interactions, hydrophobic interactions, and van der Waals forces; see Chapter 5). Inhibitors can be classified on the basis of their site of action on the enzyme, on whether they chemically modify the enzyme, or on the kinetic parameters they influence. Compounds that mimic the transition state of an enzyme-catalyzed reaction (**transition state analogs**) or that take advantage of the catalytic machinery of an enzyme (**mechanism-based inhibitors**) can be particularly potent inhibitors. Kinetically, we distinguish two classes of inhibitors based on whether raising the substrate concentration does or does not overcome the inhibition.

Competitive Inhibitors Typically Resemble Substrates

The effects of competitive inhibitors can be overcome by raising the concentration of substrate. Most frequently, in competitive inhibition, the inhibitor (**I**) binds to the substrate-binding

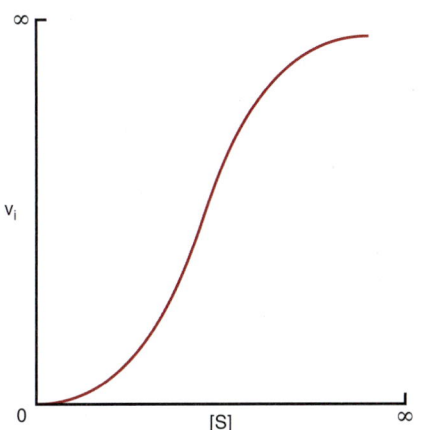

FIGURE 8–7 Representation of sigmoid substrate saturation kinetics.

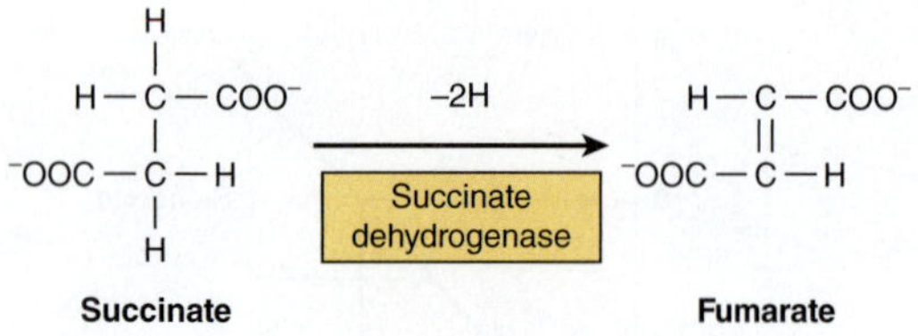

H$-$C$-$COO$^-$ $\xrightarrow{-2H}$ H$-$C$-$COO$^-$
$^-$OOC$-$C$-$H **Succinate dehydrogenase** $^-$OOC$-$C$-$H

Succinate   **Fumarate**

FIGURE 8–9   The succinate dehydrogenase reaction.

portion of the active site thereby blocking access by the substrate. The structures of most classic competitive inhibitors therefore tend to resemble the structure of a substrate, and thus are termed **substrate analogs**. Inhibition of the enzyme succinate dehydrogenase by malonate illustrates competitive inhibition by a substrate analog. Succinate dehydrogenase catalyzes the removal of one hydrogen atom from each of the two methylene carbons of succinate (**Figure 8–9**). Both succinate and its structural analog malonate ($^-$OOC—CH$_2$—COO$^-$) can bind to the active site of succinate dehydrogenase, forming an ES or an EI complex, respectively. However, since malonate contains only one methylene carbon, it cannot undergo dehydrogenation.

The formation and dissociation of the EI complex is a dynamic process described by

$$E-I \underset{k_{-1}}{\overset{k_1}{\rightleftharpoons}} E+I \tag{45}$$

for which the equilibrium constant K_i is

$$K_i = \frac{[E][I]}{[E-I]} = \frac{k_1}{k_{-1}} \tag{46}$$

In effect, **a competitive inhibitor acts by decreasing the number of free enzyme molecules available to bind substrate, that is, to form ES, and thus eventually to form product,** as described below.

A competitive inhibitor and substrate exert reciprocal effects on the concentration of the EI and ES complexes. Since the formation of ES complexes removes free enzyme available to combine with the inhibitor, increasing [S] *decreases* the concentration of the EI complex and *raises* the reaction velocity. The extent to which [S] must be increased to completely overcome the inhibition depends on the concentration of the inhibitor present, its affinity for the enzyme (K_i), and the affinity, K_m, of the enzyme for its substrate.

Double-Reciprocal Plots Facilitate the Evaluation of Inhibitors

Double-reciprocal plots typically are used both to distinguish between competitive and noncompetitive inhibitors and to simplify evaluation of inhibition constants. v_i is determined at several substrate concentrations both in the presence and in the absence of the inhibitor. For classic competitive inhibition, the lines that connect the experimental data point converge at the y-axis (**Figure 8–10**). Since the y intercept is equal to

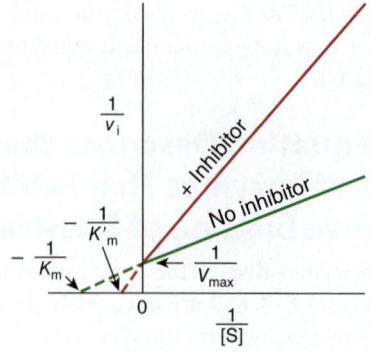

FIGURE 8–10   **Lineweaver-Burk plot of simple competitive inhibition.** Note the complete relief of inhibition at high [S] (ie, low 1/[S]).

$1/V_{max}$, this pattern indicates that **when 1/[S] approaches 0, v_i is independent of the presence of inhibitor**. Note, however, that the intercept on the x-axis *does* vary with inhibitor concentration and that, since $-1/K'_m$ is smaller than $-1/K_m$, K'_m (the "apparent K_m") becomes larger in the presence of increasing concentrations of the inhibitor. Thus, **a competitive inhibitor has no effect on V_{max} but raises K'_m, the apparent K_m for the substrate**. For a simple competitive inhibition, the intercept on the x-axis is

$$x = \frac{-1}{K_m}\left(1+\frac{[I]}{K_i}\right) \tag{47}$$

Once K_m has been determined in the absence of inhibitor, K_i can be calculated from equation (47). K_i values are used to compare different inhibitors of the same enzyme. The *lower* the value for K_i, the more effective the inhibitor. For example, the statin drugs that act as competitive inhibitors of 3-hydroxy-3-methylglutaryl coenzyme A (HMG-CoA) reductase (see Chapter 26) have K_i values several orders of magnitude lower than the K_m for the substrate, HMG-CoA.

Simple Noncompetitive Inhibitors Lower V_{max} but Do Not Affect K_m

In strict noncompetitive inhibition, binding of the inhibitor does not affect binding of the substrate. Formation of both EI and enzyme inhibitor substrate (EIS) complexes is therefore possible. However, while the enzyme-inhibitor complex can still bind the substrate, its efficiency at transforming substrate to product, reflected by V_{max}, is decreased. Noncompetitive inhibitors bind enzymes at sites distinct from the substrate-binding site and generally bear little or no structural resemblance to the substrate.

For simple noncompetitive inhibition, E and EI possess identical affinity for the substrate, and the EIS complex generates product at a negligible rate (**Figure 8–11**). More complex noncompetitive inhibition occurs when binding of the inhibitor *does* affect the apparent affinity of the enzyme for

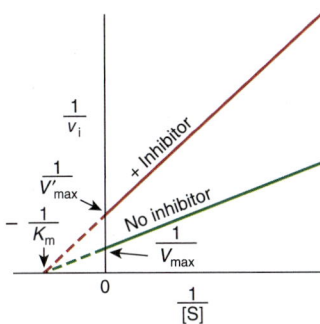

FIGURE 8–11 Lineweaver-Burk plot for simple noncompetitive inhibition.

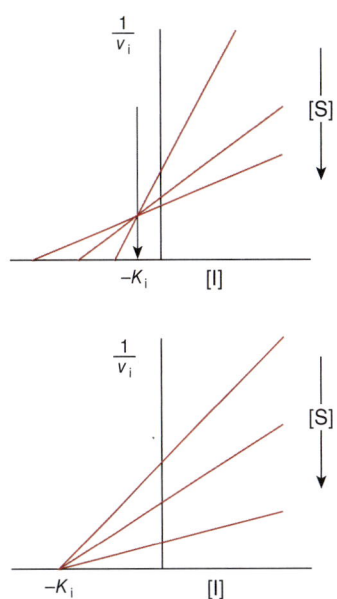

FIGURE 8–12 Applications of Dixon plots. **Top:** Competitive inhibition, estimation of K_i. **Bottom:** Noncompetitive inhibition, estimation of K_i.

the substrate, causing the lines to intercept in either the third or fourth quadrants of a double-reciprocal plot (not shown). While certain inhibitors exhibit characteristics of a mixture of competitive and noncompetitive inhibition, the evaluation of these inhibitors exceeds the scope of this chapter.

Dixon Plot

A Dixon plot is sometimes employed as an alternative to the Lineweaver-Burk plot for determining inhibition constants. The initial velocity (v_i) is measured at several concentrations of the inhibitor, but at a fixed concentration of the substrate (S). For a simple competitive or noncompetitive inhibitor, a plot of $1/v_i$ versus inhibitor concentration [I] yields a straight line. The experiment is repeated at different fixed concentrations of the substrate. The resulting set of lines intersects to the left of the y-axis. For *competitive* inhibition, a perpendicular dropped to the x-axis from the point of intersection of the lines gives $-K_i$ (**Figure 8–12**, top). For *noncompetitive* inhibition, the intercept on the x-axis is $-K_i$ (**Figure 8–12**, bottom). Pharmaceutical publications frequently employ Dixon plots to illustrate the comparative potency of competitive inhibitors.

IC$_{50}$

A less rigorous alternative to K_i as a measure of inhibitory potency is the concentration of inhibitor that produces 50% inhibition, **IC$_{50}$**. Unlike the equilibrium dissociation constant K_i, the numeric value of IC$_{50}$ varies as a function of the specific circumstances of substrate concentration, etc under which it is determined.

Tightly Bound Inhibitors

Some inhibitors bind to enzymes with such high affinity, $K_i \leq 10^{-9}$ M, that the concentration of inhibitor required to measure K_i falls below the concentration of enzyme typically present in an assay. Under these circumstances, a significant fraction of the total inhibitor may be present as an EI complex. If so, this violates the assumption, implicit in classical steady-state kinetics, that the concentration of free inhibitor is independent of the concentration of enzyme. The kinetic analysis of these tightly bound inhibitors requires specialized kinetic equations that incorporate the concentration of enzyme to estimate K_i or IC$_{50}$ and to distinguish competitive from noncompetitive tightly bound inhibitors.

Irreversible Inhibitors "Poison" Enzymes

In the above examples, the inhibitors form a dissociable, dynamic complex with the enzyme. Fully active enzyme can therefore be recovered simply by removing the inhibitor from the surrounding medium. However, a variety of other inhibitors act *irreversibly* by chemically modifying the enzyme. These modifications generally involve making or breaking covalent bonds with aminoacyl residues essential for substrate binding, catalysis, or maintenance of the enzyme's functional conformation. Since these covalent changes are relatively stable, an enzyme that has been "poisoned" by an irreversible inhibitor such as a heavy metal atom or an acylating reagent remains inhibited even after the removal of the remaining inhibitor from the surrounding medium.

Mechanism-Based Inhibition

"Mechanism-based" or "suicide" inhibitors are specialized substrate analogs that contain a chemical group that can be transformed by the catalytic machinery of the target enzyme. After binding to the active site, catalysis by the enzyme generates a highly reactive group that forms a covalent bond to and **blocks the function of a catalytically essential residue**. The specificity and persistence of suicide inhibitors, which are both enzyme-specific and unreactive outside the confines of the enzyme's active site, render them promising leads for the development of enzyme-specific drugs. The kinetic analysis of suicide inhibitors lies beyond the scope of this chapter. Neither the Lineweaver-Burk nor the Dixon approach is applicable since suicide inhibitors violate a key boundary condition common to both approaches, namely that the activity of the enzyme does not decrease during the course of the assay.

MOST ENZYME-CATALYZED REACTIONS INVOLVE TWO OR MORE SUBSTRATES

While several enzymes have a single substrate, many others have two—and sometimes more—substrates and products. The fundamental principles discussed above, while illustrated for single-substrate enzymes, apply also to multisubstrate enzymes. The mathematical expressions used to evaluate multisubstrate reactions are, however, complex. While a detailed analysis of the full range of multisubstrate reactions exceeds the scope of this chapter, some common types of kinetic behavior for two-substrate, two-product reactions (termed "Bi-Bi" reactions) are considered below.

Sequential or Single-Displacement Reactions

In **sequential reactions**, both substrates must combine with the enzyme to form a ternary complex before catalysis can proceed (**Figure 8–13**, top). Sequential reactions are sometimes referred to as single-displacement reactions because the group undergoing transfer is usually passed directly, in a single step, from one substrate to the other. Sequential Bi-Bi reactions can be further distinguished on the basis of whether the

two substrates add in a **random** or in a **compulsory** order. For random-order reactions, either substrate A or substrate B may combine first with the enzyme to form an EA or an EB complex (**Figure 8–13**, center). For compulsory-order reactions, A must first combine with E before B can combine with the EA complex. One explanation for why some enzymes follow a compulsory-order mechanism can be found in Koshland's induced fit hypothesis: the addition of A induces a conformational change in the enzyme that aligns residues that recognize and bind B.

Ping-Pong Reactions

The term "**ping-pong**" applies to mechanisms in which one or more products are released from the enzyme before all the substrates have been added. Ping-pong reactions involve covalent catalysis and a transient, modified form of the enzyme (see Figure 7–4). Ping-pong Bi-Bi reactions are often referred to as **double displacement reactions**. The group undergoing transfer is first displaced from substrate A by the enzyme to form product P and a modified form of the enzyme (F). The subsequent group transfer from F to the second substrate B, forming product Q and regenerating E, constitutes the second displacement (**Figure 8–13**, bottom).

Most Bi-Bi Reactions Conform to Michaelis-Menten Kinetics

Most Bi-Bi reactions conform to a somewhat more complex form of Michaelis-Menten kinetics in which V_{max} refers to the reaction rate attained when *both* substrates are present at saturating levels. Each substrate has its own characteristic K_m value, which corresponds to the concentration that yields half-maximal velocity when the second substrate is present at saturating levels. As for single-substrate reactions, double-reciprocal plots can be used to determine V_{max} and K_m. v_i is measured as a function of the concentration of one substrate (the variable substrate) while the concentration of the other substrate (the fixed substrate) is maintained constant. If the lines obtained for several fixed-substrate concentrations are plotted on the same graph, it is possible to distinguish a ping-pong mechanism, which yields parallel lines (**Figure 8–14**), from a sequential mechanism, which yields a pattern of intersecting lines (not shown).

    Product inhibition studies are used to complement kinetic analyses and to distinguish between ordered and random Bi-Bi reactions. For example, in a random-order Bi-Bi reaction, each product will act as a competitive inhibitor in the absence of its coproducts regardless of which substrate is designated the variable substrate. However, for a sequential mechanism (Figure 8–13, top), only product Q will give the pattern indicative of competitive inhibition when A is the variable substrate, while only product P will produce this pattern with B as the variable substrate. The other combinations of product inhibitor and variable substrate will produce forms of complex noncompetitive inhibition.

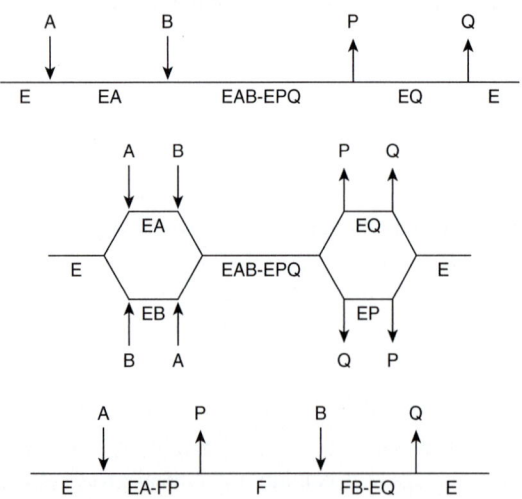

FIGURE 8–13 **Representations of three classes of Bi-Bi reaction mechanisms.** Horizontal lines represent the enzyme. Arrows indicate the addition of substrates and departure of products. **Top:** an ordered Bi-Bi reaction, characteristic of many NAD(P)H-dependent oxidoreductases. **Center:** a random Bi-Bi reaction, characteristic of many kinases and some dehydrogenases. **Bottom:** a ping-pong reaction, characteristic of aminotransferases and serine proteases.

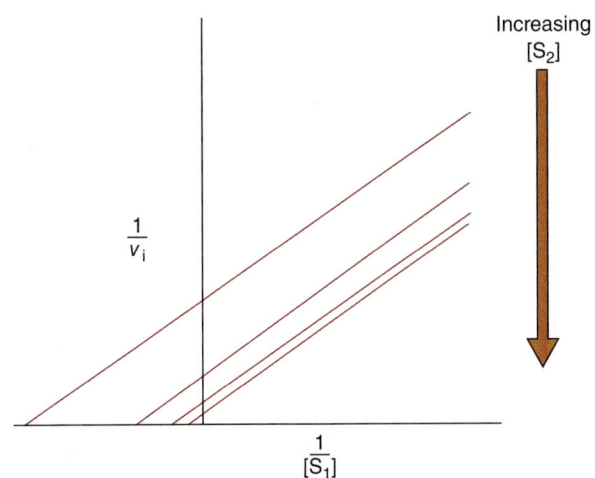

FIGURE 8–14 **Lineweaver-Burk plot for a two-substrate ping-pong reaction.** Increasing the concentration of one substrate (S_1) while maintaining that of the other substrate (S_2) constant alters both the x and y intercepts, but not the slope.

KNOWLEDGE OF ENZYME KINETICS, MECHANISM, AND INHIBITION AIDS DRUG DEVELOPMENT

Many Drugs Act as Enzyme Inhibitors

The goal of pharmacology is to identify agents that can:

1. Destroy or impair the growth, invasiveness, or development of invading pathogens.

2. Stimulate endogenous defense mechanisms.

3. Halt or impede aberrant molecular processes triggered by genetic, environmental, or biologic stimuli with minimal perturbation of the host's normal cellular functions.

By virtue of their diverse physiologic roles and high degree of substrate selectivity, enzymes constitute natural targets for the development of pharmacologic agents that are both potent and specific. Statin drugs, for example, lower cholesterol production by inhibiting the enzyme HMG-CoA reductase (see Chapter 26), while emtricitabine and tenofovir disoproxil fumarate block replication of the human immunodeficiency virus by inhibiting the viral reverse transcriptase (see Chapter 34). Pharmacologic treatment of hypertension often includes the administration of an inhibitor of angiotensin-converting enzyme, thus lowering the level of angiotensin II, a vasoconstrictor (see Chapter 41).

Enzyme Kinetics Defines Appropriate Screening Conditions

Enzyme kinetics plays a crucial role in drug discovery. Knowledge of the kinetic behavior of the enzyme of interest is necessary,

first and foremost, to select appropriate assay conditions for detecting the presence of an inhibitor. The concentration of substrate, for example, must be adjusted such that sufficient product is generated to permit facile detection of the enzyme's activity without being so high that it masks the presence of an inhibitor. Second, enzyme kinetics provides the means for quantifying and comparing the potency of different inhibitors and defining their mode of action. Noncompetitive inhibitors are particularly desirable, because—by contrast to competitive inhibitors—their effects can never be completely overcome by increases in substrate concentration.

Most Drugs Are Metabolized In Vivo

Drug development often involves more than the kinetic evaluation of the interaction of inhibitors with the target enzyme. In order to minimize its effective dosage, and hence the potential for deleterious side effects, a drug needs to be resistant to degradation by enzymes present in the patient or pathogen, a process termed **drug metabolism**. For example, penicillin and other β-lactam antibiotics block cell wall synthesis in bacteria by irreversibly inactivating the enzyme alanyl alanine carboxypeptidase-transpeptidase. Many bacteria, however, produce β-lactamases that hydrolyze the critical β-lactam function in penicillin and related drugs. One strategy for overcoming the resulting antibiotic resistance is to simultaneously administer a β-lactamase inhibitor with a β-lactam antibiotic.

Metabolic transformation is sometimes required to convert an inactive drug precursor, or **prodrug**, into its biologically active form (see Chapter 47). 2′-Deoxy-5-fluorouridylic acid, a potent inhibitor of thymidylate synthase, a common target of cancer chemotherapy, is produced from 5-fluorouracil via a series of enzymatic transformations catalyzed by a phosphoribosyl transferase and the enzymes of the deoxyribonucleoside salvage pathway (see Chapter 33). The effective design and administration of prodrugs requires knowledge of the kinetics and mechanisms of the enzymes responsible for transforming them into their biologically active forms.

SUMMARY

- The study of enzyme kinetics—the factors that affect the rates of enzyme-catalyzed reactions—reveals the individual steps by which enzymes transform substrates into products.

- ΔG, the overall change in free energy for a reaction, is independent of reaction mechanism and provides no information concerning *rates* of reactions.

- K_{eq}, a ratio of reaction *rate constants*, may be calculated from the concentrations of substrates and products at equilibrium or from the ratio k_1/k_{-1}. Enzymes do *not* affect K_{eq}.

- Reactions proceed via transition states, for whose formation the activation energy is referred to as ΔG_F. Temperature, hydrogen ion concentration, enzyme concentration, substrate concentration, and inhibitors all affect the rates of enzyme-catalyzed reactions.

- Measurement of the rate of an enzyme-catalyzed reaction generally employs initial rate conditions, for which the virtual absence of product effectively precludes the reverse reaction from taking place.

- Linear forms of the Michaelis-Menten equation simplify determination of K_m and V_{max}.

- A linear form of the Hill equation is used to evaluate the cooperative substrate-binding kinetics exhibited by some multimeric enzymes. The slope n, the Hill coefficient, reflects the number, nature, and strength of the interactions of the substrate-binding sites. A value of n greater than 1 indicates positive cooperativity.

- The effects of simple competitive inhibitors, which typically resemble substrates, are overcome by raising the concentration of the substrate. Simple noncompetitive inhibitors lower V_{max} but do not affect K_m.

- For simple competitive and noncompetitive inhibitors, the inhibitory constant K_i is equal to the dissociation constant for the relevant enzyme-inhibitor complex. A simpler and less rigorous term widely used in pharmaceutical publications for evaluating the effectiveness of an inhibitor is IC_{50}, the concentration of inhibitor that produces 50% inhibition under the particular circumstances of an experiment.

- Substrates may add in a random order (either substrate may combine first with the enzyme) or in a compulsory order (substrate A must bind before substrate B).

- In ping-pong reactions, one or more products are released from the enzyme before all the substrates have been added.

- Applied enzyme kinetics facilitate the identification, characterization, and elucidation of the mode of action of drugs that selectively inhibit specific enzymes.

- Enzyme kinetics plays a central role in the analysis and optimization of drug metabolism, a key determinant of drug efficacy.

REFERENCES

Cook PF, Cleland WW: *Enzyme Kinetics and Mechanism*. Garland Science, 2007.

Copeland RA: *Evaluation of Enzyme Inhibitors in Drug Discovery*. John Wiley & Sons, 2005.

Cornish-Bowden A: *Fundamentals of Enzyme Kinetics*. Portland Press Ltd, 2004.

Dixon M: The graphical determination of K_m and K_i. Biochem J 1972;129:197.

Fersht A: *Structure and Mechanism in Protein Science: A Guide to Enzyme Catalysis and Protein Folding*. Freeman, 1999.

Schramm, VL: Enzymatic transition-state theory and transition-state analogue design. J Biol Chem 2007;282:28297.

Segel IH: *Enzyme Kinetics*. Wiley Interscience, 1975.

Wlodawer A: Rational approach to AIDS drug design through structural biology. Annu Rev Med 2002;53:595.

Enzymes: Regulation of Activities

Peter J. Kennelly, PhD, & Victor W. Rodwell, PhD

OBJECTIVES

After studying this chapter, you should be able to:

- Explain the concept of whole-body homeostasis.
- Discuss why the cellular concentrations of substrates for most enzymes tend to be close to their K_m.
- List multiple mechanisms by which active control of metabolite flux is achieved.
- List the key characteristics that determine whether an allosteric effector is acting as a feedback regulator, indicator metabolite, or second messenger.
- State the advantages of synthesizing certain enzymes as proenzymes.
- Describe typical structural changes that accompany conversion of a proenzyme to its active form.
- Indicate two general ways in which an allosteric effector can influence catalytic activity.
- Outline the roles of protein kinases, protein phosphatases, and of regulatory and hormonal and second messengers in regulating metabolic processes.
- Explain how the substrate requirements of lysine acetyltransferases and sirtuins can trigger shifts in the degree of lysine acetylation of metabolic enzymes.
- Describe two ways by which regulatory networks can be constructed in cells.

BIOMEDICAL IMPORTANCE

The 19th-century physiologist Claude Bernard enunciated the conceptual basis for metabolic regulation. He observed that living organisms respond in ways that are both quantitatively and temporally appropriate to permit them to survive the multiple challenges posed by changes in their external and internal environments. Walter Cannon subsequently coined the term "homeostasis" to describe the ability of animals to maintain a constant internal environment despite changes in their external surroundings. At the cellular level, homeostasis is maintained by adjusting the rates of key metabolic reactions in response to internal changes by monitoring the levels of key metabolic intermediates, such as 5′-AMP and NAD⁺, or to external factors such as hormones through receptor-mediated signal transduction cascades.

Perturbations of the sensor-response machinery responsible for maintaining homeostatic balance can be deleterious to human health. Cancer, diabetes, cystic fibrosis, and Alzheimer disease, for example, are all characterized by regulatory dysfunctions triggered by the interplay between pathogenic agents, genetic mutations, nutritional inputs, and lifestyle practices. Many oncogenic viruses contribute to the initiation of cancer by elaborating protein-tyrosine kinases that modify proteins responsible for controlling patterns of gene expression. The cholera toxin produced by *Vibrio cholerae* disables sensor-response pathways in intestinal epithelial cells by catalyzing the addition of ADP-ribose to the GTP-binding proteins (G-proteins) that link cell-surface receptors to adenylyl cyclase. The ADP-ribosylation induced activation of the cyclase leads to the unrestricted flow of water into the intestines, resulting in massive diarrhea and dehydration. *Yersinia pestis*, the causative agent of plague, elaborates a protein-tyrosine phosphatase that hydrolyzes phosphoryl groups on key cytoskeletal proteins, thereby disabling the phagocytic machinery of protective macrophages. Dysfunctions in the proteolytic systems responsible for the degradation of defective or abnormal proteins are believed to play a role in neurodegenerative diseases such as Alzheimer and Parkinson.

In addition to their immediate function as regulators of enzyme activity, protein degradation, etc., covalent modifications

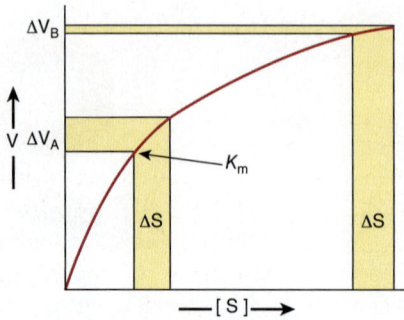

FIGURE 9–1 **Differential response of the rate of an enzyme-catalyzed reaction, ΔV, to the same incremental change in substrate concentration at a substrate concentration close to K_m (ΔV_A) or far above K_m (ΔV_B).**

such as phosphorylation, acetylation, and sumoylation provide a protein-based code for the storage and transmission of information (see Chapter 35). Such DNA-independent hereditary information is referred to as **epigenetic**. This chapter outlines the mechanisms by which metabolic processes are controlled, and provides some illustrative examples.

REGULATION OF METABOLITE FLOW CAN BE ACTIVE OR PASSIVE

While at first it might seem desirable to have metabolic enzymes normally operate at their maximal rate, this leaves little room to adjust throughput or flux to changes in substrate availability. When an enzyme is saturated by substrate, the rate of product formation cannot increase to accommodate a surge in substrate availability (see Figure 9–1). Moreover, even relatively large decreases in substrate availability will produce only disproportionately small decreases in rate. Consequently, most enzymes have evolved such that their K_m values for substrates tend to be close to the latter's average intracellular concentrations, so that changes in substrate concentration generate corresponding changes in metabolite flux (**Figure 9–1**). Responses to changes in substrate level represent an important but *passive* means for the intercellular coordination of metabolite flow. Adaptation to extracellular signals requires mechanisms for regulating enzyme efficiency in an *active* manner.

Metabolite Flow Tends to Be Unidirectional

Normally, living cells exist in a dynamic steady state in which the mean concentrations of metabolic intermediates remain relatively constant over time. This reflects the fact that, in living cells, the reaction products of one enzyme-catalyzed reaction serve as substrates for, and are rapidly removed by, other enzyme-catalyzed reactions (**Figure 9–2**). Under these circumstances flux through many nominally reversible enzyme-catalyzed reactions occurs unidirectionally. In a pathway comprised of a succession of coupled enzyme-catalyzed reactions, the continual depletion of the pathway intermediates pulls the equilibrium for even those reactions characterized by small or even slightly unfavorable changes in free energy in favor of product formation, because the *overall* change in free energy that favors unidirectional metabolite flow. This situation is somewhat analogous to the flow of water through a pipe in which one end is lower than the other. Flow of water through the pipe remains unidirectional despite the presence of bends or kinks, due to the overall change in height, which corresponds to the pathway's overall change in free energy (**Figure 9–3**).

COMPARTMENTATION ENSURES METABOLIC EFFICIENCY & SIMPLIFIES REGULATION

In eukaryotes, the anabolic and catabolic pathways that synthesize and break down common biomolecules often are physically separated from one another. Certain metabolic pathways reside only within specialized cell types or inside separate subcellular compartments. For example, fatty acid biosynthesis occurs in the cytosol, whereas fatty acid oxidation takes place within mitochondria (see Chapters 22 and 23), while many degradative enzymes are contained inside organelles called lysosomes. In addition, apparently antagonistic pathways can coexist in the absence of physical barriers provided that each proceeds with the formation of one or more *unique intermediates*. For any reaction or series of reactions, the change in free energy that takes place when metabolite

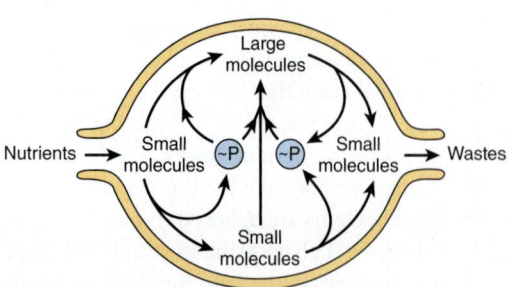

FIGURE 9–2 **An idealized cell in steady state.** Note that metabolite flow is unidirectional.

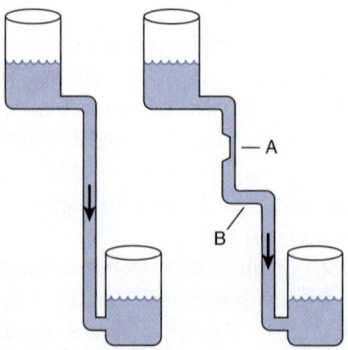

FIGURE 9–3 **Hydrostatic analogy for a pathway with a rate-limiting step (A) and a step with a ΔG value near 0 (B).**

flow proceeds in the "forward" direction is equal in magnitude *but opposite in sign* from that required to proceed in the "reverse" direction. Some enzymes within these pathways catalyze reactions, such as isomerizations, for which the difference in free energy between substrates and products is close to zero. These catalysts act bidirectionally, depending on the ratio of substrates to products. However, virtually all metabolic pathways possess one or more steps for which ΔG is significant. For example, glycolysis, the breakdown of glucose to form two molecules of pyruvate, has a favorable overall ΔG of −96 kJ/mol, a value much too large to simply operate in "reverse" in order to convert excess pyruvate to glucose. Consequently, gluconeogenesis proceeds via a pathway in which the three most energetically disfavored steps in glycolysis are circumvented using alternative, thermodynamically favorable reactions catalyzed by distinct enzymes (see Chapter 19).

The ability of enzymes to discriminate between the structurally similar coenzymes NAD^+ and $NADP^+$ also results in a form of compartmentation. The reduction potentials of both coenzymes are similar. However, most of the reactions that generate electrons destined for the electron transport chain reduce NAD^+ to NADH, while enzymes that catalyze the reductive steps in many biosynthetic pathways generally use NADPH as the electron donor.

Rate-Limiting Enzymes as Preferred Targets of Regulatory Control

While the flux of metabolites through metabolic pathways involves catalysis by numerous enzymes, homeostasis is maintained by the regulation of only a select subset of these enzymes. The ideal enzyme for regulatory intervention is one whose quantity or catalytic efficiency dictates that the reaction it catalyzes is slow relative to all others in the pathway. Decreasing the catalytic efficiency or the quantity of the catalyst participating in the "bottleneck" or **rate-limiting reaction** will immediately reduce metabolite flux through the entire pathway. Conversely, an increase in either its quantity or catalytic efficiency will elicit an increase in flux through the pathway as a whole. As natural "governors" of metabolic flux, the enzymes that catalyze rate-limiting steps also constitute promising drug targets. For example, "statin" drugs curtail synthesis of cholesterol by inhibiting HMG-CoA reductase, catalyst of the rate-limiting reaction of cholesterogenesis.

REGULATION OF ENZYME QUANTITY

The overall catalytic capacity of the rate-limiting step in a metabolic pathway is the product of the concentration of enzyme molecules and their intrinsic catalytic efficiency. It therefore follows that catalytic capacity can be controlled by changing the quantity of enzyme present, altering its intrinsic catalytic efficiency, or a combination thereof.

Proteins Are Continuously Synthesized & Degraded

By measuring the rates of incorporation and subsequent loss of ^{15}N-labeled amino acids into proteins, Schoenheimer deduced that proteins exist in a state of "dynamic equilibrium" where they are continuously synthesized and degraded—a process referred to as **protein turnover**. Even **constitutive** proteins, those whose aggregate concentrations remain essentially constant over time, are subject to continual replacement through turnover. However, the concentrations of many other enzymes are subject to dynamic shifts in response to hormonal, dietary, pathologic, and other factors that may affect the overall rate constants for their synthesis (k_s), degradation (k_{deg}), or both.

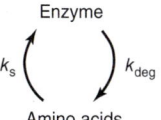

Control of Enzyme Synthesis

The synthesis of certain enzymes depends on the presence of **inducers**, typically substrates or structurally related compounds that stimulate the transcription of the gene that encodes them (see Chapter 38), or **transcription factors**. *Escherichia coli* grown on glucose will, for example, only catabolize lactose after addition of a β-galactoside, an inducer that triggers synthesis of a β-galactosidase and a galactoside permease. Inducible enzymes of humans include tryptophan pyrrolase, threonine dehydratase, tyrosine-α-ketoglutarate aminotransferase, enzymes of the urea cycle, HMG-CoA reductase, Δ-aminolevulinate synthase, and cytochrome P450. Conversely, an excess of a metabolite may curtail synthesis of its cognate enzyme via **repression**. Both induction and repression involve *cis* elements, specific DNA sequences located upstream of regulated genes, that provide binding sites for *trans*-acting regulatory proteins. The molecular mechanisms of induction and repression are discussed in Chapter 38, while detailed information on the control of protein synthesis in response to hormonal stimuli can be found in Chapter 42.

Control of Enzyme Degradation

In animals many proteins are degraded by the ubiquitin-proteasome pathway. Degradation takes place in the 26S proteasome, a large macromolecular complex made up of more than 30 polypeptide subunits arranged in the form of a hollow cylinder. The active sites of its proteolytic subunits face the interior of the cylinder, thus preventing indiscriminate degradation of cellular proteins. Proteins are targeted to the interior of the proteasome by the covalent attachment of one or more molecules of ubiquitin, a small, 76 amino acid ($\approx$ 8.5-kDa), protein that is highly conserved among eukaryotes. "Ubiquitination" is catalyzed by a large family of enzymes called E3 ligases, which attach ubiquitin to the side-chain amino group of lysyl residues on their targets.

The ubiquitin-proteasome pathway is responsible both for the regulated degradation of selected cellular proteins, for example, cyclins (see Chapter 35), and for the removal of defective or aberrant protein species. The key to the versatility and selectivity of the ubiquitin-proteasome system resides in the variety of intracellular E3 ligases and their ability to discriminate between the different physical or conformational states of target proteins. Thus, the ubiquitin-proteasome pathway can selectively degrade proteins whose physical integrity and functional competency have been compromised by the loss of or damage to a prosthetic group, oxidation of cysteine or histidine residues, partial unfolding, or deamidation of asparagine or glutamine residues (see Chapter 57). Recognition by proteolytic enzymes also can be regulated by covalent modifications such as phosphorylation; binding of substrates or allosteric effectors; or association with membranes, oligonucleotides, or other proteins. Dysfunctions of the ubiquitin-proteasome pathway sometimes contribute to the accumulation and subsequent aggregation of misfolded proteins characteristic of several neurodegenerative diseases.

MULTIPLE OPTIONS ARE AVAILABLE FOR REGULATING CATALYTIC ACTIVITY

In humans the induction of protein synthesis is a complex multistep process that typically requires hours to produce significant changes in overall enzyme level. By contrast, changes in intrinsic catalytic efficiency triggered by binding of dissociable ligands (**allosteric regulation**) or by **covalent modification** occur within a few minutes to a fraction of a second. Consequently, changes in protein level generally dominate when meeting long-term adaptive requirements, whereas changes in catalytic efficiency are favored for rapid and transient alterations in metabolite flux.

ALLOSTERIC EFFECTORS REGULATE CERTAIN ENZYMES

The activity of many enzymes is regulated by the noncovalent binding of small molecules known as **effectors**. Effector binding can modulate enzyme function by virtue of the fact that the enzyme–effector complex is a distinct molecular entity from the individual enzyme and effector molecules, and thus can manifest properties different than those of its separated components. Effector binding can increase (activate) or decrease (inhibit) the catalytic efficiency of the target enzyme, cause it to translocate to a different location within a cell, trigger its association with (or dissociation) from other proteins, or even potentiate or suppress sensitivity to the binding of a second effector or to covalent modification. Physiologic effectors include several of the end products and intermediates of key metabolic pathways, as well as specialized biomolecules whose sole function is to act as regulatory ligands.

Most effectors bear little or no resemblance to the substrates or products of the affected enzyme. For example an effector of 3-phosphoglycerate dehydrogenase, the amino acid serine, displays minimal structural overlap with either of the enzyme's substrates, 3-phosphoglycerate and NAD^+ (**Figure 9–4**). Jacques Monod reasoned that the lack of structural similarity between most physiologic effectors and the reactants targeted for catalytic transformation by their cognate enzymes indicated that these modulators bound at a site that was physically distinct and perhaps even physically distant from an enzyme's active site. Monod employed the term **allosteric**, which means to "occupy another space," to classify these modulatory sites as well as the effectors that are bound to them. The ability of allosteric ligands to influence events taking place at the active site can be attributed to their ability to induce conformational changes that affect all or large portions on an enzyme. Allosteric effectors may influence catalytic efficiency by altering the K_m for a substrate (K series allosteric enzymes), V_{max} (V-series allosteric enzymes), or both.

Allosteric Effectors Can Be Grouped Into Three Functional Classes

Allosteric effectors can be classified into three categories based on their origin and scope of action. *Feedback effectors* are pathway end products or intermediates that bind to and either activate or inhibit one or more enzymes within the pathway responsible for their synthesis. In most cases, feedback inhibitors bind to the enzyme that catalyzes the first committed step in a particular biosynthetic sequence. In the following example, the biosynthesis of D from A is catalyzed by enzymes Enz_1 through Enz_3:

$$Enz_1 \; Enz_2 \; Enz_3$$
$$A \rightarrow B \rightarrow C \rightarrow D$$

In the case of 3-phosphoglycerate dehydrogenase, serine is the end product of a pathway for which this enzyme catalyzes the first committed step (see Figure 9–4).

The regulatory function of feedback effectors is secondary in importance to their role metabolism. Their scope of action is also localized to the pathway responsible for their synthesis. In most cases, the binding of an end product or intermediate decreases the catalytic efficiency of their enzyme target, a pattern called *feedback inhibition*.

Like feedback effectors, *indicator metabolites* also are metabolic end products or intermediates with a secondary role as allosteric effectors of enzymes. They are distinguished from feedback effectors by their scope of action, which extends beyond or even lies outside of their pathway of origin. This enables indicator metabolites to coordinate flux through multiple metabolic pathways. For example, citrate—a key intermediate in the citric acid cycle (Chapter 16)—acts as an allosteric activator of the enzyme responsible for catalyzing the first committed step in fatty acid biosynthesis, acetyl-CoA carboxylase (see Figure 23–6), while alanine acts as an

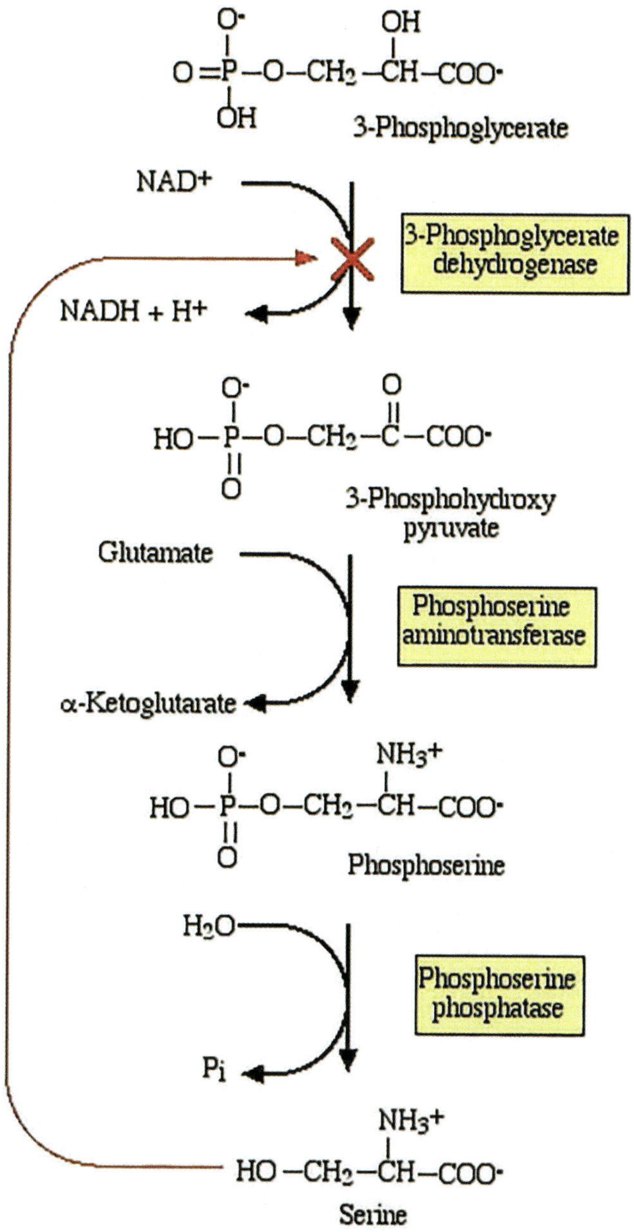

FIGURE 9–4 **Serine regulates its rate of synthesis by acting as a feedback inhibitor.** Shown are the three enzyme-catalyzed reactions by which the amino acid serine is synthesized from the glycolytic intermediate, glyceraldehyde 3-phosphate. The red arrow indicates that serine binds to 3-phosphoglycerate dehydrogenase, inhibiting the enzyme's activity (symbolized by the red "X"). Through this simple mechanism, when cellular demand for serine, the pathway end product, declines and its concentration begins to increase, serine synthesis is reduced by its action as a feedback inhibitor of the enzyme responsible for catalyzing the first committed step in the pathway.

allosteric inhibitor of the glycolytic enzyme pyruvate kinase (Chapter 19).

Unlike indicator metabolites or feedback effectors, *second messengers* are specialized allosteric ligands whose production or release is triggered in response to an external first messenger, such as a hormone or nerve impulse. Some typical second messengers include 3′, 5′-cAMP, synthesized from ATP by the enzyme adenylyl cyclase in response to the hormone epinephrine, and Ca^{2+}, which is stored inside the endoplasmic reticulum of most cells. Membrane depolarization resulting from a nerve impulse opens a membrane channel that releases calcium ions into the cytoplasm, where they bind to and activate enzymes involved in the regulation of muscle contraction and mobilize stored glucose from glycogen by binding to and activating phosphorylase kinase. Other second messengers include 3′,5′-cGMP, nitric oxide, and the polyphosphoinositols produced by the hydrolysis of inositol phospholipids by hormone-regulated phospholipases. Specific examples of the participation of second messengers in the regulation of cellular processes can be found in Chapters 18, 42, and 51.

ASPARTATE TRANSCARBAMOYLASE IS A MODEL ALLOSTERIC ENZYME

Aspartate transcarbamoylase (ATCase), the catalyst for the first reaction unique to pyrimidine biosynthesis (see Figure 33–9), is a target of allosteric regulation by two nucleotide triphosphates: cytidine triphosphate (CTP) and adenosine triphosphate. CTP, an end product of the pyrimidine biosynthetic pathway, functions as a feedback inhibitor of ATCase, whereas the purine nucleotide ATP activates it. Moreover, the binding of ATP prevents inhibition by CTP, enabling synthesis of *pyrimidine* nucleotides to proceed when *purine* nucleotide levels are elevated. The ability of the indicator metabolite ATP to prevent momentary interruptions arising from momentary, transient increases in the level of the inhibitory effector CTP ensures that pyrimidine biosynthesis keeps pace with purine biosynthesis during times of high demand, such as when DNA replication takes place.

REGULATORY COVALENT MODIFICATIONS CAN BE REVERSIBLE OR IRREVERSIBLE

Mammalian proteins are the targets of a wide range of covalent modification processes. Modifications such as prenylation, glycosylation, hydroxylation, and fatty acid acylation introduce unique structural features into newly synthesized proteins that persist for the lifetime of the protein. Some covalent modifications regulate protein function. The core principle underlying regulatory covalent modification is that, introducing a new covalent bond or cleaving an existing covalent bond alters the identity of the affected enzyme. This "new" enzyme molecule thus can possess properties distinct from those of its precursor. In mammalian cells, a wide range of regulatory covalent modifications occur. Some of these modifications, such as **partial proteolysis**, in which one or a small handful of peptide bonds within a precursor protein are hydrolyzed, are termed **irreversible** because it is impossible to reconstruct the precursor form of the protein inside the cell. Hence, once modified the activated proteins retain their new form and properties until they are damaged, disposed, or degraded.

Acetylation, ADP-ribosylation, sumoylation (the attachment of a small, ubiquitin-like modifier or SUMO protein), and phosphorylation are all examples of so-called **reversible** covalent modifications. In this context, "reversible" refers to capability to restore the modified protein to its modification-free precursor form and **not** the mechanism by which restoration takes place. Thermodynamics dictates that if the enzyme-catalyzed reaction by which the modification was introduced is thermodynamically favorable, simply reversing the process will be rendered impractical by the correspondingly unfavorable free-energy change. The phosphorylation of proteins on seryl, threonyl, or tyrosyl residues, catalyzed by protein kinases, is thermodynamically favored as a consequence of utilizing the high-energy gamma phosphoryl group of ATP. Phosphate groups are removed, not by recombining the phosphate with ADP to form ATP, but by a hydrolytic reaction catalyzed by enzymes called protein phosphatases. Similarly, acetyltransferases employ a high-energy donor substrate, NAD^+, while deacetylases catalyze a direct hydrolysis that generates free acetate.

Activation of Chymotrypsin Illustrates Control of Enzyme Activity by Selective Proteolysis

Many secreted proteins are synthesized as larger, functionally inert precursors called **proproteins**. If the latent function of the proprotein is catalytic in nature, these precursors are termed **proenzymes** or, alternatively, **zymogens**. Transformation of proproteins into their active, functionally competent forms is accomplished by selective, also known as partial, proteolysis involving one or a small handful of highly specific proteolytic clips. Activation of chymotrypsinogen, for example, involves the hydrolysis of four peptide bonds (**Figure 9–6**) that divide the original polypeptide into five smaller segments. The largest of these are designated as peptides A, B, and C. In mature, active α-chymotrypsin the A, B, and C peptides are linked together by disulfide bonds while the two small dipeptides are allowed to diffuse away. Note that while residues His 57 and Asp 102 of the catalytically essential charge relay network (see Figure 7–7) reside on the B peptide, Ser 195 resides on the C peptide, brought together in three-dimensional space by conformational changes triggered by the partial proteolysis of chymotrypsinogen.

Many Digestive Enzymes Are Stored as Functionally Dormant Proproteins

In their zymogen form, digestive proteases and lipases can be safely stored in the pancreas as they await their excretion into the stomach. On secretion into the digestive tract, these pancreatic hydrolases become activated by a series of selective proteolytic cleavages that are initiated by the conversion of trypsinogen to trypsin through the proteolytic action of **enteropeptidase**, formerly known as enterokinase, a protease found in the brush border of the duodenum. Once activated by enteropeptidase, trypsin catalyzes the subsequent conversion of numerous other pancreatic zymogens such as chymotrypsinogen, proelastase, kallikreinogen, procarboxypeptidases A and B, prophospholipase A2, and pancreatic prolipase.

Pepsinogen, which is secreted by gastric chief cells, catalyzes its own activation to pepsin. Autoproteolysis is triggered by conformational changes induced on exposure to the acidic pH of upper gastrointestinal tract. In pancreatitis, premature activation of digestive proteases and lipases leads to autodigestion of healthy tissue rather than ingested proteins.

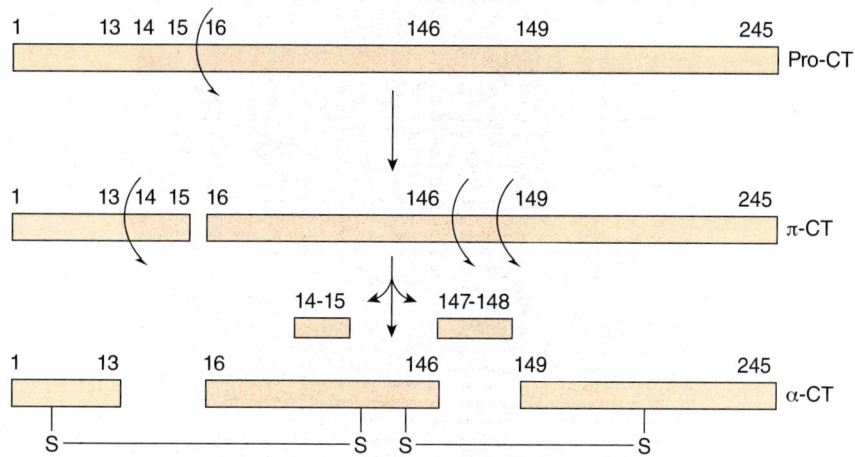

FIGURE 9–5 **Two-dimensional representation of the sequence of proteolytic events that ultimately result in formation of the catalytic site of chymotrypsin, which includes the Asp102-His57-Ser195 catalytic triad (see Figure 7–7).** Successive proteolysis forms pro-chymotrypsin (pro-CT), π-chymotrypsin (π-Ct), and ultimately α-chymotrypsin (α-CT), an active protease whose three peptides (A, B, C) remain associated by covalent interchain disulfide bonds.

Selective Proteolysis Enables Rapid Activation of the Blood Clotting & Complement Cascades

On wounding or injury, the rapidity with which blood clots are formed can be critical to survival. Synthesis and secretion of the components of the blood clotting and complement cascades as functionally dormant zymogens enable the building blocks for constructing a potentially life-saving blood clot to be prepositioned, ready to be rapidly activated via a series or *cascade* of selective proteolytic cleavage events. The components of the complement cascade that attacks invading bacteria are also prepositioned in the circulation in the form of proteolytically activated proproteins (see Chapters 52 and 55).

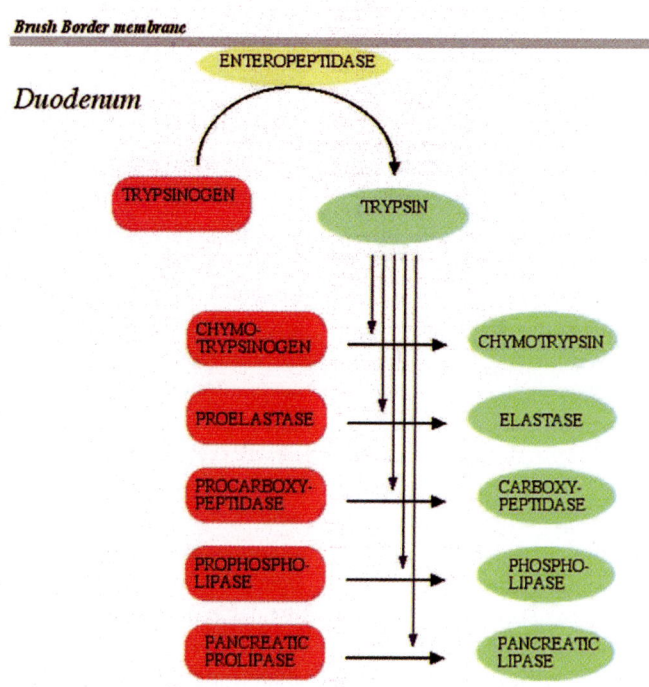

FIGURE 9–6 **Activation of pancreatic zymogens in the duodenum.**

REVERSIBLE COVALENT MODIFICATION REGULATES KEY MAMMALIAN PROTEINS

The Histone Code Is Based on Reversible Covalent Modifications

Histones and other DNA-binding proteins in chromatin are subject to extensive modification by **acetylation, methylation, ADP-ribosylation**, as well as phosphorylation and the addition of a **small ubiquitin-like modifier (SUMO)** protein, a process subbed sumoylation. These modifications modulate the manner in which the proteins within chromatin interact with each other as well as the DNA itself. The resulting changes in chromatin structure within the region affected can render genes more accessible to the proteins responsible for their transcription, thereby enhancing gene expression or, on a larger scale, facilitating replication of the entire genome (see Chapter 38). On the other hand, changes in chromatin structure that restrict the accessibility of genes to transcription factors, DNA-dependent RNA polymerases, etc., thereby inhibiting transcription, are said to **silence** gene expression.

The combination of covalent modifications that determine gene accessibility in chromatin has been termed the "histone code." This code represents a classic example of **epigenetics**, the hereditary transmission of information by the means other than the sequence of nucleotides that comprise the genome. In this instance, the pattern of gene expression within a newly formed "daughter" cell will be determined, in part, by the particular set of histone covalent modifications embodied in the chromatin proteins inherited from the "parental" cell.

Thousands of Mammalian Proteins Are Modified by Covalent Phosphorylation

Protein kinases phosphorylate proteins by catalyzing transfer of the terminal phosphoryl group of ATP to the hydroxyl groups of select seryl, threonyl, or tyrosyl residues in proteins, forming O-phosphoseryl, O-phosphothreonyl, or O-phosphotyrosyl residues, respectively (**Figure 9–7**). The unmodified form of the protein can be regenerated by hydrolytic removal of phosphoryl groups, a thermodynamically favorable reaction catalyzed by **protein phosphatases**.

A typical mammalian cell possesses thousands of phosphorylated proteins and several hundred protein kinases and protein phosphatases that catalyze their interconversion. The ease of interconversion of enzymes between their phospho and dephospho forms accounts, in part, for the frequency with which phosphorylation–dephosphorylation is utilized as a mechanism for regulatory control. Unlike structural modifications, covalent phosphorylation persists only as long as the covalently modified form of the protein serves a specific need. Once the need has passed, the enzyme can be converted back to its original form, poised to respond to the next stimulatory event. A second factor underlying the widespread use of protein phosphorylation–dephosphorylation lies in the chemical

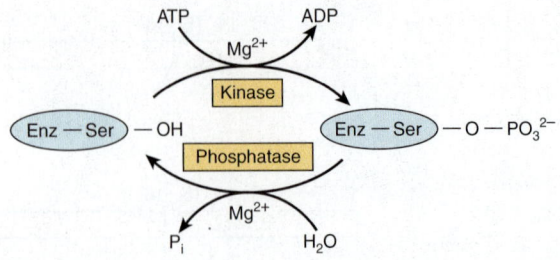

FIGURE 9–7 **Covalent modification of a regulated enzyme by phosphorylation–dephosphorylation of a seryl residue.** Shown is the protease cascade responsible for activating pancreatic zymogens (Red) by partial proteolysis into their enzymatically active (Green) forms. The cascade is triggered by the brush border enzyme enteropeptidase (Yellow), which converts trypsinogen to trypsin. Once activated, trypsin catalyzes the targeted proteolytic clips (Blue arrows) that transform chymotrypsinogen into chymotrypsin, proelastase into elastase, the procarboxypeptidases into carboxypeptidases, prophospholipase into phospholipase, and pancreatic prolipase into pancreatic lipase.

properties of the phosphoryl group itself. In order to alter an enzyme's functional properties, modifications to its chemical structure must influence the protein's three-dimensional configuration. The high-charge density of protein-bound phosphoryl groups, −1 or −2 at physiologic pH, their propensity to form strong salt bridges with arginyl and lysyl residues, and their high exceptionally high hydrogen-bonding capacity renders them potent agents for modifying protein structure and function. Phosphorylation generally influences an enzyme's intrinsic catalytic efficiency or other properties by inducing changes in its conformation. Consequently, the amino acids modified by phosphorylation can be and typically are relatively distant from the catalytic site itself.

Protein Acetylation: A Ubiquitous Modification of Metabolic Enzymes

As is the case with covalent phosphorylation, covalent acetylation possesses the twin virtues of employing the conformation altering potential of changing the charge character of the side chain which they target, in this case from the +1 of the protonated ε-amino group of lysine to neutral acetylated form, and being reversible in vivo. It is thus not surprising that the number of proteins subject to and regulated by covalent acetylation–deacetylation now numbers in the thousands. These include histones and other nuclear proteins as well as nearly every enzyme in such core metabolic pathways as glycolysis, glycogen synthesis, gluconeogenesis, the tricarboxylic acid cycle, β-oxidation of fatty acids, and the urea cycle. The potential regulatory impact of acetylation–deacetylation has been established for only a handful of these proteins. However, the latter include metabolically important enzymes such as acetyl-CoA synthetase, long-chain acyl-CoA dehydrogenase, malate dehydrogenase, isocitrate dehydrogenase, glutamate dehydrogenase, carbamoyl phosphate synthetase, phosphoenol-pyruvate carboxykinase, aconitase, and ornithine transcarbamoylase. Perturbations in the acetylation–deacetylation of lysine residues in proteins is thought to be associated with aging and neurodegeneration.

In the mitochondria, it is believed that many proteins react with acetyl-CoA directly, without the intervention of an enzyme catalyst. Their degree of acetylation thus is thought to respond to and reflect changes in the concentration of this central metabolic intermediate. The acetylation of other proteins, particularly those residing outside the mitochondria, requires the intervention of a **lysine acetyltransferase**. These enzymes catalyze the transfer of the acetyl group of acetyl-CoA to the ε-amino groups of lysyl residues, forming *N*-acetyl lysine.

All protein deacetylation is believed to be enzyme catalyzed. Two classes of protein deacetylases have been identified: **histone deacetylases** and **sirtuins**. Histone deacetylases catalyze the removal by hydrolysis of acetyl groups, regenerating the unmodified form of the protein and acetate as products. Sirtuins, on the other hand, use NAD$^+$ as substrate, which yields *O*-acetyl ADP-ribose and nicotinamide as products in addition to the unmodified protein.

Covalent Modifications Regulate Metabolite Flow

In many respects, the sites of protein phosphorylation, acetylation, and other covalent modifications can be considered another form of allosteric site. However, in this case, the "allosteric ligand" binds covalently to the protein. Phosphorylation–dephosphorylation, acetylation–deacetylation, and feedback inhibition provide short-term, readily reversible regulation of metabolite flow in response to specific physiologic signals. All three act independently of changes in gene expression. As with feedback inhibition, protein phosphorylation–dephosphorylation generally targets an early enzyme in a protracted metabolic pathway. Feedback inhibition involves a single protein that is influenced indirectly, if at all, by hormonal or neural signals. By contrast, regulation of mammalian enzymes by phosphorylation–dephosphorylation involves one or more protein kinases and protein phosphatases, and is generally under direct neural and hormonal control.

Acetylation–deacetylation, on the other hand, targets multiple proteins in a pathway. It has been hypothesized that the degree of acetylation of metabolic enzymes is modulated to a large degree by the energy status of the cell. Under this model, the high levels of acetyl-CoA (the substrate for lysine acetyltransferases and the reactant in nonenzymatic lysine acetylation) present in a well-nourished cell would promote lysine acetylation. When nutrients are lacking, acetyl-CoA levels drop and the ratio of NAD$^+$/NADH rises, favoring protein deacetylation.

PROTEIN PHOSPHORYLATION IS EXTREMELY VERSATILE

Protein phosphorylation–dephosphorylation is a highly versatile and selective process. Not all proteins are subject to phosphorylation, and of the many hydroxyl groups on a protein's

TABLE 9–1 Examples of Mammalian Enzymes Whose Catalytic Activity Is Altered by Covalent Phosphorylation–Dephosphorylation

Enzyme	Activity State	
	Low	High
Acetyl-CoA carboxylase	EP	E
Glycogen synthase	EP	E
Pyruvate dehydrogenase	EP	E
HMG-CoA reductase	EP	E
Glycogen phosphorylase	E	EP
Citrate lyase	E	EP
Phosphorylase b kinase	E	EP
HMG-CoA reductase kinase	E	EP

Abbreviations: E, dephosphoenzyme; EP, phosphoenzyme.

surface, only one or a small subset are targeted. While phosphorylation of some enzymes increases their catalytic activity, the phosphorylated form of other enzymes may be catalytically inactive (**Table 9–1**). Alternatively, phosphorylation may affect a protein's location within the cell, susceptibility to proteolytic degradation, responsiveness to regulation by allosteric ligands, or responsiveness to covalent modification on some other site.

Many proteins can be phosphorylated at multiple sites. Others are subject to regulation both by phosphorylation–dephosphorylation and by the binding of allosteric ligands, or by phosphorylation–dephosphorylation and another covalent modification. If the protein kinase that phosphorylates a particular protein responds to a signal different from that which modulates the protein phosphatase that catalyzes dephosphorylation of the resulting phosphoprotein, then even the simplest phosphoprotein becomes a *decision node*. Its functional output reflects the phosphoprotein's degree or state of phosphorylation, which reflects the relative strength of the signals that modulate the protein kinase and protein phosphatase. If the phosphorylation (or dephosphorylation) of a given site can be catalyzed by more than one protein kinase (or protein phosphatase), the phosphoprotein can then integrate and process several input signals.

The ability of many protein kinases to phosphorylate multiple proteins enables coordinate regulation of cellular activities. For example, 3-hydroxy-3-methylglutaryl-CoA reductase and acetyl-CoA carboxylase—the rate-controlling enzymes for cholesterol and fatty acid biosynthesis, respectively—both can be inactivated through phosphorylation by the AMP-activated protein kinase. When this protein kinase is activated either through phosphorylation by yet another protein kinase or in response to the binding of its allosteric activator 5′-AMP, the two major pathways responsible for the synthesis of lipids from acetyl-CoA are both inhibited.

The interplay between protein kinases and protein phosphatases, between the functional consequences of phosphorylation at different sites, between phosphorylation sites and

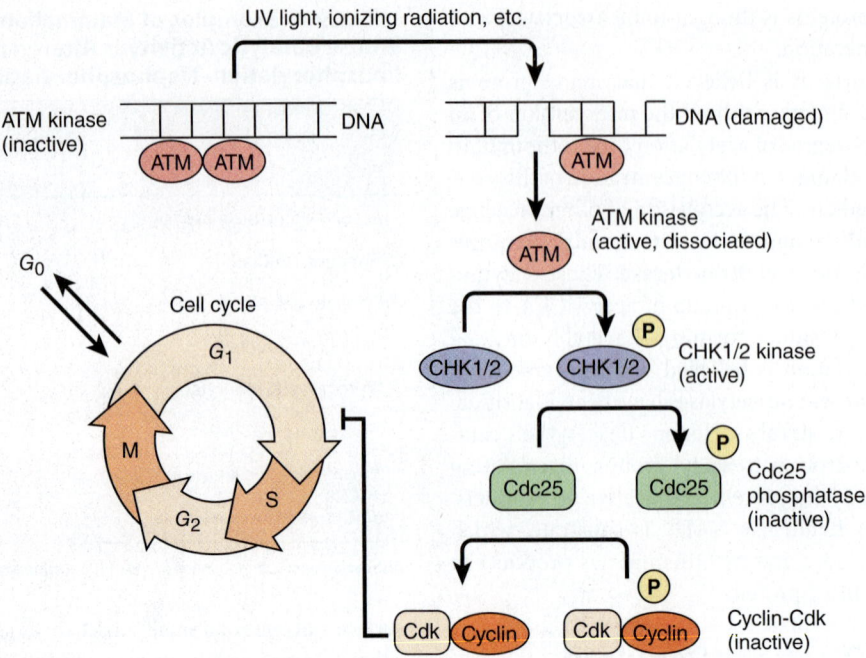

FIGURE 9–8 **A simplified representation of the G_1 to S checkpoint of the eukaryotic cell cycle.** The circle shows the various stages in the eukaryotic cell cycle. The genome is replicated during S phase, while the two copies of the genome are segregated and cell division occurs during M phase. Each of these phases is separated by a G, or growth, phase characterized by an increase in cell size and the accumulation of the precursors required for the assembly of the large macromolecular complexes formed during S and M phases.

allosteric sites, or between phosphorylation sites and other sites of covalent modification provides the basis for regulatory networks that integrate multiple inputs to evoke an appropriate coordinated cellular response. In these sophisticated regulatory networks, individual enzymes respond to different internal and environmental signals.

INDIVIDUAL REGULATORY EVENTS COMBINE TO FORM SOPHISTICATED CONTROL NETWORKS

Cells carry out a complex array of metabolic processes that must be regulated in response to a broad spectrum of internal and external factors. Hence, interconvertible enzymes and the enzymes responsible for their interconversion act, not as isolated "on" and "off" switches, but as binary elements within integrated biomolecular information processing networks.

One well-studied example of such a network is the eukaryotic cell cycle that controls cell division. Upon emergence from the G_0 or quiescent state, the extremely complex process of cell division proceeds through a series of specific phases designated G_1, S, G_2, and M (**Figure 9–8**). Elaborate monitoring systems, called **checkpoints**, assess key indicators of progress to ensure that no phase of the cycle is initiated until the prior phase is complete. Figure 9–8 outlines, in simplified form, a chromosome-associated protein kinase called ATM binding to and being activated by regions of chromatin-containing double-stranded breaks in the DNA. Upon activation, one subunit

of the activated ATM dimer dissociates and initiates a series, or cascade, of protein phosphorylation–dephosphorylation events mediated by the CHK1 and CHK2 protein kinases, the Cdc25 protein phosphatase, and finally a complex between a cyclin and a cyclin-dependent protein kinase, or Cdk. In this case, activation of the Cdk-cyclin complex blocks the G_1 to S transition, thus preventing the replication of damaged DNA. Failure at this checkpoint can lead to mutations in DNA that may lead to cancer or other diseases. Additional checkpoints and signaling cascades (not shown) interact together to control cell cycle progression in response to multiple indicators of cell status.

SUMMARY

- Homeostasis involves maintaining a relatively constant intracellular and intraorgan environment despite wide fluctuations in the external environment. This is achieved via appropriate changes in the rates of biochemical reactions in response to physiologic need.

- The substrates for most enzymes are usually present at a concentration close to their K_m. This facilitates passive adjustments to the rates of product formation in response to changes in levels of metabolic intermediates.

- Most metabolic control mechanisms target enzymes that catalyze an early, committed, and rate-limiting reaction. Control can be exerted by varying the concentration of the target protein, its functional efficiency, or some combination of the two.

- Binding of allosteric effectors to sites distinct from the catalytic sites of enzymes triggers conformational changes that may either increase or decrease catalytic efficiency.

■ Feedback regulators regulate flux through a pathway for which they are the end product or an intermediate, while indicator metabolites act beyond their pathway of origin.

■ Second messengers are allosteric effectors whose sole function is to regulate the activity of one or more proteins in response to a hormone, nerve impulse, or other extracellular input.

■ Secretion of inactive proenzymes or zymogens facilitates rapid mobilization of activity via partial proteolysis in response to injury or physiologic need while protecting the tissue of origin (eg, autodigestion by proteases).

■ Phosphorylation by protein kinases of specific seryl, threonyl, or tyrosyl residues—and subsequent dephosphorylation by protein phosphatases—regulates the activity of many human enzymes in response to hormonal and neural signals.

■ Numerous metabolic enzymes are modified by the acetylation-deacetylation of lysine residues. The degree of acetylation of these proteins is thought to be modulated by the availability of acetyl-CoA, the acetyl donor substrate for lysine acetyltransferases, and NAD^+, a substrate for the sirtuin deacetylases.

■ The capacity of protein kinases, protein phosphatases, lysine acetylases, and lysine deacetylases to target both multiple proteins and multiple sites on proteins is key to the formation of integrated regulatory networks that process complex environmental information to produce an appropriate cellular response.

REFERENCES

Bett JS: Proteostasis regulation by the ubiquitin system. Essays Biochem 2016;60:143.

Dokholyan NV: Controlling allosteric networks in proteins. Chem Rev 2016;116:6463.

Drazic A, Myklebust LM, Ree R, Arnesen T: The world of protein acetylation. Biochim Biophys Acta 2016;1864:1372.

Krauss G (ed.): *Biochemistry of Signal Transduction and Regulation,* 5th ed. Wiley VCH, Weinheim, Germany, 2014.

McGuire M (ed). *Protein Phosphorylation.* Callisto Reference, Forest Hills, NY, USA, 2015.

Narita T, Weinert BT, Choudhary C. Functions and mechanisms of non-histone protein acetylation. Nature Rev Mol Cell Biol 2019;20:156.

Traut TW. *Allosteric Regulatory Enzymes.* Springer-Verlag US, New York, New York, 2010.

Varejao N, Lascorz J, Li Y, Reverter D. Molecular mechanisms in SUMO conjugation. Biochem Soc Trans 2019;48:123.

Wolfinbarger L Jr: *Enzyme Regulation in Metabolic Pathways.* John Wiley and Sons, Hoboken, NJ, 2017.

The Biochemical Roles of Transition Metals

Peter J. Kennelly, PhD

- Explain why essential transition metals are often referred to as micronutrients.
- Explain the importance of multivalency to the ability of transition metals to participate in electron transport and oxidation–reduction reactions.
- Describe how Lewis and Bronsted-Lowry acids differ.
- Define the term complexation as it refers to metal ions.
- Provide a rationale for why zinc is a common prosthetic group in enzymes that catalyze hydrolytic reactions.
- List four benefits obtained by incorporating transition metals into organometallic complexes *in vivo*.
- Cite an example of a transition metal that functions as an electron carrier in one protein, an oxygen carrier in another, and a redox catalyst in yet another.
- Explain how the possession of multiple metal ions enables the metalloenzymes cytochrome oxidase and nitrogenase to catalyze the reduction of molecular oxygen and nitrogen, respectively.
- Describe two mechanisms by which excess levels of transition metals can be harmful to living organisms.
- Provide an operational definition of the term "heavy metal," and list three strategies for treating acute heavy metal poisoning.
- Describe the processes by which Fe, Co, Cu, and Mo are absorbed in the human gastrointestinal tract.
- Describe the metabolic role of sulfite oxidase and the pathology of sulfite oxidase deficiency.
- Describe the function of zinc finger motifs and provide an example of their role in metal ion metabolism.

BIOMEDICAL IMPORTANCE

Maintenance of human health and vitality requires the ingestion of trace levels of numerous inorganic elements, among them the transition metals iron (Fe), manganese (Mn), zinc (Zn), cobalt (Co), copper (Cu), nickel (Ni), molybdenum (Mo), vanadium (V), and chromium (Cr). In general, transition metals are sequestered within our bodies in organometallic complexes that enable their reactivity to be controlled, especially their propensity to generate harmful reactive oxygen species (ROS), and directed where needed. Transition metals are key components of numerous enzymes and electron transport proteins as well as the oxygen transport proteins hemoglobin and hemocyanin. Zinc finger motifs provide the DNA-binding domains for many transcription factors, while Fe-S clusters are found in many of the enzymes that participate in DNA replication and repair. Nutritionally or genetically induced deficiencies of these metals are associated with a variety of pathologic conditions including pernicious anemia (Fe), Menkes disease (Cu), and sulfite oxidase deficiency (Mo). When ingested in large quantities, most heavy metals, including several of the nutritionally essential transition metals, are highly toxic and nearly all are potentially carcinogenic.

TRANSITION METALS ARE ESSENTIAL FOR HEALTH

Humans Require Minute Quantities of Several Inorganic Elements

The organic elements oxygen, carbon, hydrogen, nitrogen, sulfur, and phosphorous typically account for slightly more than 97% of the mass of the human body. Calcium, the majority of which is contained in bones, teeth, and cartilage, contributes a further ≈ 2%. The remaining 0.4 to 0.5% is accounted for by numerous inorganic elements (**Table 10–1**). Many of these are essential for health, albeit in minute quantities, and thus are commonly classified as **micronutrients**. Examples of physiologically essential micronutrients include iodine, which is required for the synthesis of tri- and tetraiodothyronine (see Chapter 41); selenium, which is required for the synthesis of the amino acid selenocysteine (see Chapter 27); and vitamins (see Chapter 44). The focus of the current chapter will be on the physiologic roles of the nutritionally essential transition metals iron (Fe), manganese (Mn), zinc (Zn), cobalt (Co), copper (Cu), nickel (Ni), molybdenum (Mo), vanadium (V), and chromium (Cr).

Transition Metals Are Multivalent

One common characteristic of metals is their propensity to undergo **oxidation**, a process in which they donate one or more electrons from their outer or valence shell to an electronegative acceptor species, for example, molecular oxygen. Oxidation of an alkali or alkaline earth metal (**Figure 10–1**) results in a single ionized species, for example, Na^+, K^+, Li^+, Mg^{2+}, or Ca^{2+}. By contrast, the **oxidation of transition metals** can yield *multiple* valence states (**Table 10–2**). This capability enables transition metals to undergo dynamic transitions between valence states through the addition or donation of electrons, and therefore to function as an electron carrier during oxidation–reduction (redox) reactions. The ability of transition metals to act as *acids* further expands their biologic roles.

Transition Metal Ions Are Potent Lewis Acids

In addition to serving as electron carriers, the functional capabilities of the nutritionally essential transition metals are enhanced by their ability to act as Lewis acids. Protic (Bronsted-Lowry) acids can donate a proton (H^+) to an acceptor with a lone pair of electrons, for example, a primary amine or a molecule of water. **Lewis acids**, by contrast, are *aprotic*. Like H^+ ions, Lewis acids possess empty valence orbitals capable of noncovalently associating with or "accepting" a lone pair of electrons from a second, "donor" molecule. When the ferrous (Fe^{2+}) iron in myoglobin and hemoglobin bind oxygen or other diatomic gases such carbon monoxide, they are acting as Lewis acids (see Chapter 8). Divalent Zn^{2+} or Mn^{2+} can serve as Lewis acids during catalysis by hydrolytic enzymes, specifically by enhancing the nucleophilicity of active-site water molecules.

TOXICITY OF HEAVY METALS

Most heavy metals, a loosely defined term for metallic elements with densities greater than 5 g/cm^3 or atomic numbers greater than 20, are toxic. Some well-known examples include arsenic, antimony, lead, mercury, and cadmium, whose toxicity can arise via a number of mechanisms.

Displacement of an Essential Cation

The ability of a heavy metal to displace a functionally essential metal can readily lead to loss or impairment of enzyme function. Classic examples include the displacement of iron

TABLE 10–1 Quantities of Selected Elements in the Human Body

Element	Mass	Essential	Element	Mass	Essential	Element	Mass	Essential
Oxygen	43 kg	+	Selenium	15 µg	+	Cadmium	50 µg	−
Carbon	16 kg	+	Iron	4.2 g	+	Rubidium	680 µg	−
Hydrogen	7 kg	+	Zinc	2.3 g	+	Strontium	320 µg	−
Nitrogen	1.8 kg	+	Copper	72 µg	+	Titanium	20 µg	−
Phosphorous	780 g	+	Nickel	15 µg	+	Silver	2 µg	−
Calcium	1.0 kg	+	Chromium	14 µg	+	Niobium 1.5	µg	−
Sulfur	140 g	+	Manganese	12 µg	+	Zirconium	1 µg	−
Potassium	140 g	+	Molybdenum	5 µg	+	Tungsten	20 ng	−
Sodium	100 g	+	Cobalt	3 µg	+	Yttrium	0.6 µg	−
Chlorine	95 g	+	Vanadium	0.1 µg	+	Cerium	40 µg	−
Magnesium	19 g	+	Silicon	1.0 mg	Possibly	Bromine	260 µg	−
Iodine	20 µg	+	Fluorine*	2.6 g	−	Lead*	120 µg	−

Data for a 70-kg (150 lb) human are from Emsley J: *The Elements, 3rd ed.* New York, NY: Oxford University Press; 1998.

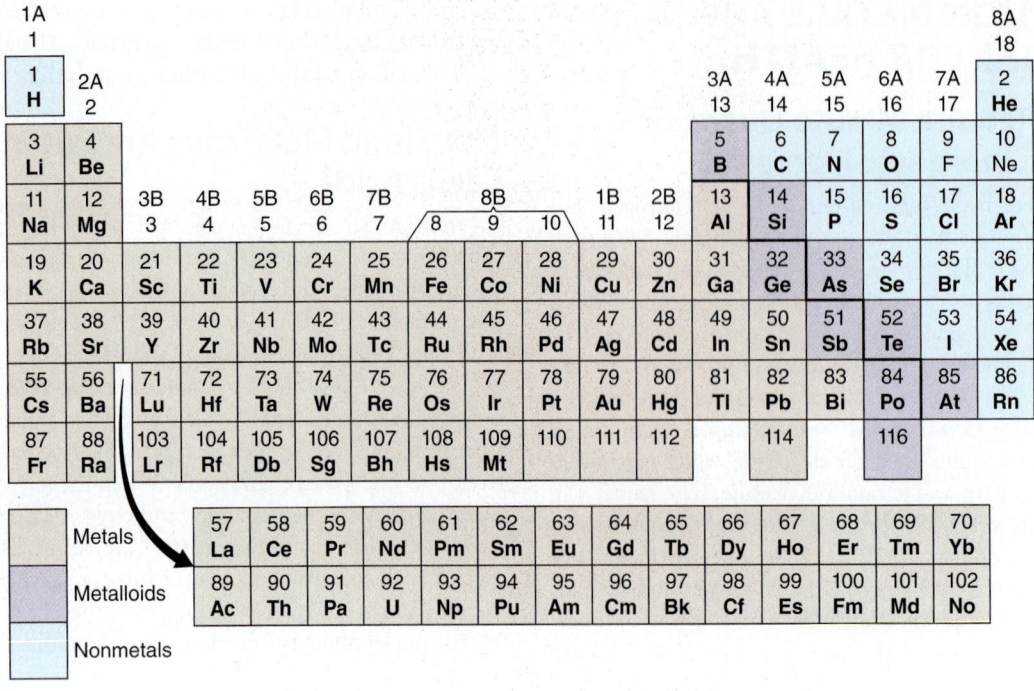

FIGURE 10–1 **Periodic table of the elements.** Transition metals occupy columns 3 to 11, also labeled 1B to 8B.

by **gallium** in the enzymes ribonucleotide reductase and Fe, Cu superoxide dismutase. Ga^{3+}, while of similar size and identical charge to Fe^{3+}, lacks the multivalence capability of iron. Replacement of Fe^{3+} by Ga^{3+} therefore renders affected enzymes catalytically inert.

Enzyme Inactivation

Heavy metals readily form adducts with free sulfhydryl groups. Enzyme inactivation can occur if the affected sulfhydryl resides

TABLE 10–2 **Valence States of Essential Transition Metals**

Transition Metal	Potential Valences
Cobalt	Co^{-1}, Co^0, **Co^+**, **Co^{2+}**, **Co^{3+}**, Co^{4+}
Chromium	Cr^{-4}, Cr^{-2}, Cr^-, Cr^0, Cr^+, Cr^{2+}, **Cr^{3+}**, Cr^{4+}, Cr^{5+}, Cr^{6+}
Copper	Cu^0, **Cu^+**, **Cu^{2+}**
Iron	Fe^0, Fe^+, **Fe^{2+}**, **Fe^{3+}**, **Fe^{4+}**, Fe^{5+}, Fe^{6+}
Manganese	Mn^{3-}, Mn^{2-}, Mn^-, Mn^0, Mn^+, **Mn^{2+}**, **Mn^{3+}**, Mn^{4+}, Mn^{5+}, Mn^{6+}, Mn^{7+}
Molybdenum	Mo^{4-}, Mo^{2-}, Mo^-, Mo^0, Mo^+, Mo^{2+}, Mo^{3+}, **Mo^{4+}**, **Mo^{5+}**, **Mo^{6+}**
Nickel	Ni^{2-}, Ni^-, Ni^0, Ni^+, **Ni^{2+}**, Ni^{4+}
Vanadium	V^-, V^0, V^+, V^{2+}, **V^{3+}**, **V^{4+}**, **V^{5+}**
Zinc	Zn^{2-}, Zn^0, Zn^+, **Zn^{2+}**

Shown are the possible valence states for each of the nutritionally essential transition states. Biochemically and physiologically relevant valence states are highlighted in red.

in the enzyme site. Alternatively, formation of heavy metal adducts with surface sulfhydryl groups can undermine a protein's structural integrity, with concomitant impairment of function. Examples include the inhibition of δ-aminolevulinate synthase by Pb (see Chapter 31) and the inactivation of the pyruvate dehydrogenase complex by arsenite or mercury. In pyruvate dehydrogenase, the heavy metals react with the sulfhydryl on the essential prosthetic group, lipoic acid (see Chapter 17), rather than with a peptidyl cysteine.

Formation of Reactive Oxygen Species

Heavy metals can induce the formation of reactive oxygen species (ROS), which can then damage DNA, membrane lipids, and other biomolecules (see Chapter 57). Oxidative damage to DNA can cause genetic mutations that may lead to cancer or other pathophysiologic conditions. ROS-mediated peroxidation of lipid molecules (see Figure 21–23) can lead to the loss of membrane integrity. The resulting dissipation of action potentials and disruption of various cross-membrane transport processes can be particularly deleterious to neurologic and neuromuscular functions. It has also been reported that rats fed with excessive levels of heavy metals are prone to develop cancerous tumors.

TOXICITY OF TRANSITION METALS

While nutritionally essential, several transition metals are nonetheless harmful if present in the body in excess (**Table 10–3**). Consequently, higher organisms exert strict control over

TABLE 10–3 Relative Toxicity of Metals

Toxicity		
Nontoxic	**Low**	**Medium to High**
Aluminum Manganese	Barium Tin	Antimony Niobium
Bismuth Molybdenum	Cerium Ytterbium	Beryllium Palladium
Calcium Potassium	Germanium Yttrium	Cadmium Platinum
Cesium Rubidium	Gold	Chromium Selenium
Iron Sodium	Rhodium	Cobalt Thorium
Lithium Strontium	Scandium	Copper Titanium
Magnesium	Terbium	Indium Tungsten
		Lead Uranium
		Mercury Vanadium
		Polonium Zirconium
		Nickel Zinc

Nutritionally essential transition metals are in red type. Data from Meade RH: Contaminants in the Mississippi River, 1987-1992. US Geological Circular 1133; 1995.

both the uptake and excretion of transition metal ions. Examples include the **hepcidin** system for regulation of iron (see Figure 52–8) to avoid its accumulation to damaging levels. These mechanisms can be circumvented to some degree when transition metals enter via inhalation or absorption through the skin or mucous membranes, and can be overwhelmed by ingestion of massive, supraphysiologic levels. Typical symptoms of acute poisoning by heavy metals or by transition metals include abdominal pain, vomiting, muscle cramps, confusion, and numbness. Treatments include administration of metal chelating agents, diuretics, or—should kidney function be compromised—hemodialysis.

LIVING ORGANISMS PACKAGE TRANSITION METALS WITHIN ORGANOMETALLIC COMPLEXES

Complexation Enhances Solubility & Controls Reactivity of Transition Metal Ions

The levels of free transition metals in the body are, under normal circumstances, extremely low. The vast majority are found either associated directly with proteins via the oxygen, nitrogen, and sulfur atoms found on the side chains of amino acids such as aspartate, glutamate, histidine, or cysteine (**Figure 10–2**); or with other organic moieties such as porphyrin (see Figure 6–1), corrin (see Figure 44–10), or pterins (**Figure 10–3**). The sequestration of transition metals into organometallic complexes confers multiple advantages that include protection against oxidation, suppression of ROS production, enhancement of solubility, control of reactivity, and assembly of multimetal units (**Figure 10–4**).

Significance of Multivalent Capability

Physiologic function of transition metal–containing cofactors and prosthetic groups is dependent on maintaining a multivalent ion's appropriate oxidation state. For example, the porphyrin ring and proximal and distal histidyl residues of the globin polypeptide chain that complex the Fe^{2+} atoms in hemoglobin protects them from oxidation to Fe^{3+}-containing methemoglobin, which is incapable of binding to and transporting oxygen (see Chapter 6). Free transition metal ions are vulnerable to oxidation by O_2 and agents inside the cell. Not only are free transition metal ions vulnerable to nonspecific oxidation, their interaction with oxidizing agents such as O_2, NO, and H_2O_2 generally results in the generation of even more potent ROS (see Figure 58–2). Incorporation into organometallic

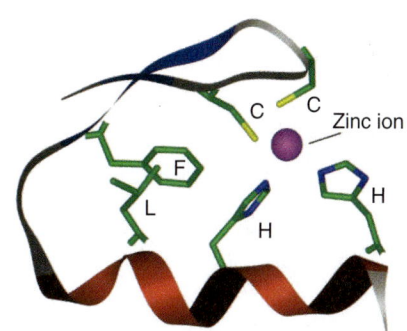

FIGURE 10–2 Ribbon diagram of a consensus C_2H_2 zinc finger domain. Shown are the bound Zn^{2+} (purple) and the R groups of the conserved phenylalanyl (F), leucyl (L) cysteinyl (C), and histidyl (H) residues with their carbon atoms in green. The polypeptide backbone is shown as a ribbon, with alpha helical portions highlighted in red. The sulfur and nitrogen atoms of the R groups of the cysteinyl and histidyl residues are shown in yellow and blue, respectively.

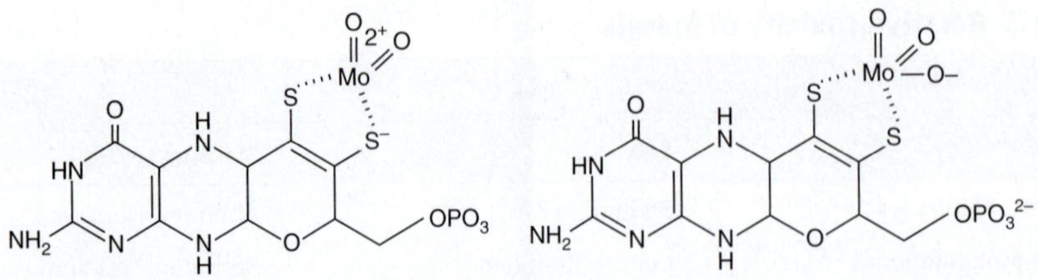

FIGURE 10-3 **Molybdopterin.** Shown are the oxidized **(left)** and reduced **(right)** forms of molybdopterin.

complexes thus protects the functionally relevant oxidation state of the transition metal and against the potential for generating harmful ROS.

Adjacent Ligands Can Modify Redox Potential

In organometallic complexes, the position and identity of the surrounding ligands can modify or tune the redox potential and Lewis acid potency of transition metal ions, thereby optimizing them for specific tasks (**Table 10-4**). For example, both cytochrome c and myoglobin are small, 12 to 17 kDa, monomeric proteins that contain a single heme iron. While the iron in myoglobin is optimized by its surroundings to maintain a constant, Fe^{2+}, valence state, for oxygen binding; the iron atom in cytochrome c is optimized to cycle between the +2 and +3 valence states so the protein can carry electrons between complexes III and IV of the electron transport chain. The superoxide dismutases (SODs) illustrate how complexation can adapt different transition metals as catalysts for a common chemical reaction, the disproportionation of H_2O_2 into H_2O and O_2. Each of the four distinct, nonhomologous SODs contain different transition metals whose atomic symbols are used to designate each family: Fe-SODs, Mn-SODs, Ni-SODs, and Cu, Zn-SODs.

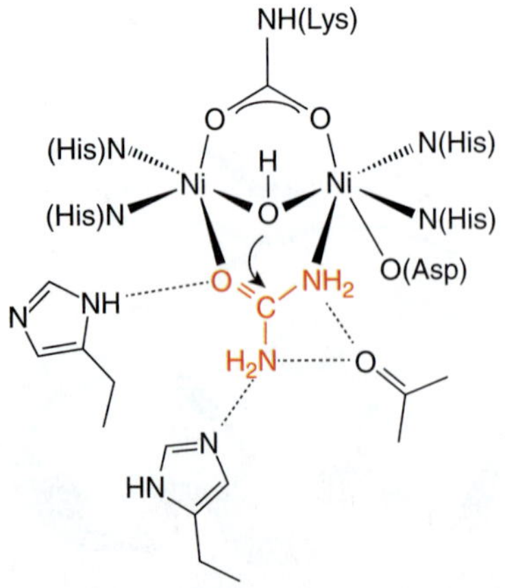

FIGURE 10-4 **Hydrolysis of urea requires the cooperative influence of two active site Ni atoms.** The figure depicts the formation of the transition state intermediate for the hydrolysis of the first C-N bond in urea (red) by the enzyme urease. Note how the Ni atoms chelate a water molecule to form a nucleophilic hydroxide and weaken the C-N bond through Lewis acid interactions with lone pairs of electrons on the O and one of the N atoms of urea.

Complexation Can Organize Multiple Metal Ions in a Single Functional Unit

The formation of organometallic complexes also allows multiple metal ions to be assembled together in a single functional unit with capabilities that lie beyond those obtainable with a single transition metal ion. In the enzyme urease, which catalyzes the hydrolysis of urea in plants, the presence of two Ni atoms within the active site enables the enzyme to simultaneously polarize electrons in the C-N bond targeted for hydrolysis and to activate the attacking water molecule (see Figure 10-4). The presence of two Fe and two Cu atoms in cytochrome oxidase enables the complex IV of the electron transport chain to accumulate the four electrons needed to carry out the reduction of oxygen to water. Similarly, the bacterial enzyme nitrogenase employs an 8Fe-7S prosthetic group called the P-cluster and a unique Fe, Mo-cofactor to carry out the eight-electron reduction of atmospheric nitrogen to ammonia.

PHYSIOLOGIC ROLES OF THE ESSENTIAL TRANSITION METALS

Iron

Iron is one of the most functionally versatile of the physiologically essential transition metals. Both in hemoglobin and myoglobin the heme-bound Fe^{2+} iron is used to bind a diatomic gas, O_2, for transport and storage, respectively (see Chapter 6). Similarly, in marine invertebrates, the iron present in the **diiron center** of hemerythrin (**Figure 10-5**) can bind and transport oxygen. By contrast, the iron atoms contained in the heme groups of the b- and c-type cytochromes and the Fe-S clusters (see Figure 13-4) and Rieske iron centers (**Figure 10-6**) of other electron transport chain components transport electrons by cycling between their ferrous (+2) and ferric state (+3).

Roles of Iron in Redox Reactions

The iron atoms of many metalloproteins facilitate the catalysis of oxidation–reduction, or redox, reactions. Stearoyl-acyl carrier protein Δ^9-desaturase and type 1 ribonucleotide reductase employ hemerythrin-like diiron centers to catalyze the reduction of carbon–carbon double bonds and an alcohol, respectively, to methylene groups. Methane monooxygenase uses a

TABLE 10–4 **Some Biologically Important Metalloproteins**

Protein	Function or Reaction Catalyzed	Metal(s)
Aconitase	Isomerization	Fe-S center
Alcohol dehydrogenase	Oxidation	Zn
Alkaline phosphatase	Hydrolysis	Zn
Arginase	Hydrolysis	Mn
Aromatase	Hydroxylation	Heme Fe
Azurin (bacteria)	e⁻ transport	Cu
Carbonic anhydrase	Hydration	Zn
Carboxypeptidase A	Hydrolysis	Zn
Cytochrome c	e⁻ transport	Heme Fe
Cytochrome oxidase	Reduction of O_2 to H_2O	Heme Fe
Cytochrome P450	Oxidation & hydroxylation	Heme Fe (2) & Cu (2)
Dopamine β-hydroxylase	Hydroxylation	Cu
Ferredoxin	e⁻ transport	Fe-S center
Galactosyl transferase	Glycoprotein synthesis	Mn
Hemoglobin	O_2 transport	Heme Fe (4)
Isocitrate dehydrogenase	Oxidation	Mn
β-Lactamase II (bacteria)	Hydrolysis	Zn
Lysyl oxidase	Oxidation	Cu
Matrix metalloprotease	Hydrolysis	Zn
Myoglobin	O_2 storage	Heme Fe
Nitric oxide synthase	Reduction	Heme Fe
Nitrogenase (bacteria)	Reduction	Fe, Mo cofactor, P-cluster (Fe), Fe-S center
Phospholipase C	Hydrolysis	Zn
Ribonucleotide reductase	Reduction	Fe (2)
Sulfite oxidase	Oxidation	Molybdopterin & Fe-S center
Superoxide dismutase (cytoplasmic)	Disproportionation	Cu, Zn
Urease (plant)	Hydrolysis	Ni
Xanthine oxidase	Oxidation	Molybdopterin & Fe-S center

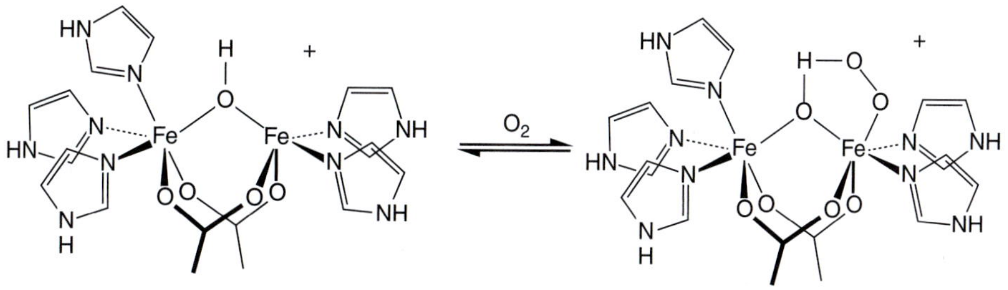

FIGURE 10–5 **Diiron center of the deoxy (left) and oxy (right) forms of hemerythrin.** Shown are the side chains of the histidine, glutamate, and aspartate residues responsible for binding the metal ions to the polypeptide.

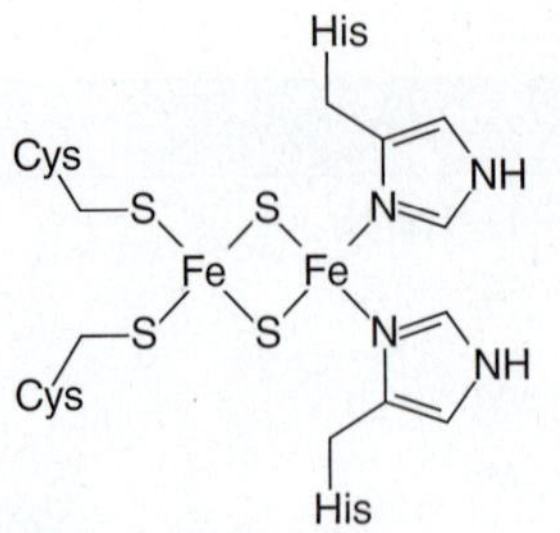

FIGURE 10–6 **Structure of a Rieske iron center.** Rieske iron centers are a type of 2Fe-2S cluster in which histidine residues replace two of the cysteine residues that normally bind the prosthetic group to the polypeptide chain.

similar diiron center for the oxidation of methane to methanol. The members of the cytochrome P450 family generate $Fe = O^{3+}$. This is a powerful oxidant that participates in the reduction and neutralization of a broad range of xenobiotics via the two-electron reduction of O_2, a complex process during which the heme iron cycles between several, that is, +2, +3, +4, and +5, oxidation states.

Participation of Iron in Non-Redox Reactions

Purple acid phosphatases, bimetallic enzymes containing one atom of iron matched with a second metal, such as Zn, Mn, Mg, or another Fe, catalyze the hydrolysis of phosphomono-esters. Myeloperoxidase employs heme iron to catalyze the condensation of H_2O_2 with Cl^- ions to generate hypochlorous acid, HOCl, a potent bacteriocide used by macrophages to kill entrapped microorganisms. It recently has been shown that many enzymes involved in DNA replication and repair, including DNA helicase, DNA primase, several DNA polymerases, some glycosylases and endonucleases, and several transcription factors contain Fe-S clusters. While their elimination generally results in a loss of protein function, the specific role(s) performed by these Fe-S centers remains cryptic. However, since most are located in the DNA binding, rather than the catalytic, domains of these proteins, it has been proposed that these Fe-S centers may function as electrochemical detectors for the identification of damaged DNA. Others speculate that these clusters serve as redox-sensitive modulators of catalytic activity or DNA binding, or simply as stabilizers of the three-dimensional structure of these proteins.

Manganese

Humans contain a handful of Mn-containing enzymes, the majority of which are located within the mitochondria. These include isocitrate dehydrogenase from the tricarboxylic acid (TCA) cycle, two key players in nitrogen metabolism: glutamate synthetase and arginase, and the gluconeogenic enzymes pyruvate carboxylase and phosphoenolpyruvate carboxykinase, isopropyl malate synthase, and the mitochondrial isozyme of superoxide dismutase. In most of these enzymes, Mn is present in the +2 oxidation state and is presumed to act as a Lewis acid. By contrast, some bacterial organisms employ Mn in several enzymes responsible for catalyzing redox reactions, where it cycles between the +2 and +3 oxidation states in, for example, Mn-superoxide dismutase (Mn-SOD), Mn-ribonucleotide reductase, and Mn-catalase.

Zinc

Unlike the divalent (+2) ions of other first-row transition metals (see Figure 10–1), the valence shell of Zn^{2+} possesses a full set of electrons. As a consequence, Zn^{2+} ions do not adopt alternate oxidation states under physiologic conditions, rendering it unsuitable to participate in electron transport processes or as a catalyst for redox reactions. On the other hand, redox inert Zn^{2+} ions also pose a minimal risk of generating harmful ROS species. Its unique status among the physiologically essential transition metals renders Zn^{2+} an ideal candidate as a ligand for stabilizing the protein conformation.

It has been estimated that the human body contains 3000 zinc-containing metalloproteins. The vast majority of these are transcription factors and other DNA- and RNA-binding proteins that contain anywhere from one to thirty copies of a Zn^{2+}-containing polynucleotide-binding domain known as the zinc finger. Zinc fingers consist of a polypeptide loop whose conformation is stabilized by the interactions between Zn^{2+} and lone pairs of electrons donated by the sulfur and nitrogen atoms contained in two conserved cysteine and two conserved histidine residues (see Figure 38–16). Zinc fingers bind polynucleotides with a high degree of site specificity that is conferred, at least in part, by variations in the sequence of amino acids that make up the remainder of the loop. Scientists are working to exploit this combination of small size and binding specificity to construct sequence-specific nucleases for use in genetic engineering and, eventually, gene therapy.

Zn^{2+} is also an essential component of several metalloenzymes, including carboxypeptidase A, carbonic anhydrase II, adenosine deaminase, alkaline phosphatase, phospholipase C, leucine aminopeptidase, the cytosolic form of superoxide dismutase, and alcohol dehydrogenase. Zn^{2+} is also a component of the type II β-lactamases used by bacteria to neutralize penicillin and other lactam antibiotics. These metalloenzymes exploit the Lewis acid properties of Zn^{2+} to stabilize the development of negatively charged intermediates, polarize the distribution of electrons in carbonyl groups, and enhance the nucleophilicity of water (**Figure 10–7**).

Cobalt

The predominant and, thus far, only known biochemical function of dietary cobalt is as the core component of 5′-deoxy-adenosylcobalamin, otherwise known as vitamin B_{12} (see Figure 44–10). The Co^{3+} in this cofactor resides at the center of a tetrapyrrole corrin ring where it acts as a Lewis base that binds to and facilitates the transfer of one-carbon methyl or methylene groups. In humans, this includes the enzyme catalyzed transfer of a $-CH_3$ group from tetrahydrofolate to homocysteine, the final step in the synthesis of the amino acid methionine (see Figure 44–13), and the rearrangement of methylmalonyl-CoA to form succinyl-CoA during the

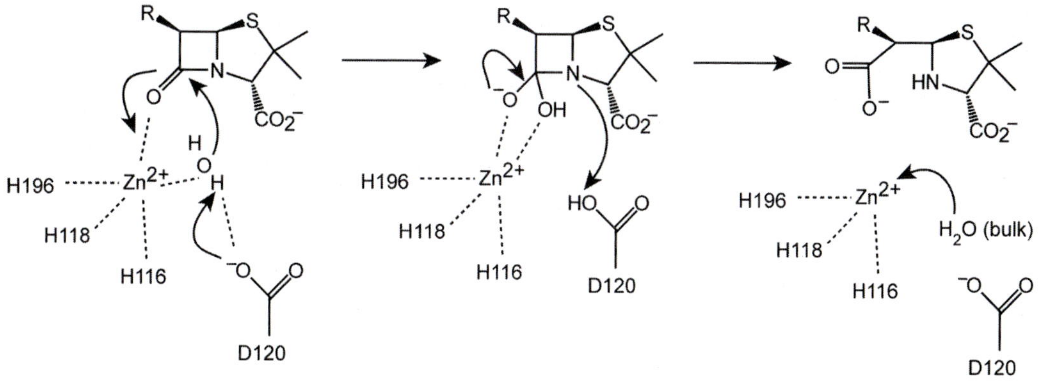

FIGURE 10–7 **Role of Zn^{2+} in catalytic mechanisms of β-lactamase II.** Zn^{2+} is bound to the enzyme via the nitrogen atoms present in the side chains of multiple histidine (H) residues. **Left:** Zn^{2+} activates a water molecule, one of whose protons is accommodated by aspartic acid (D) residue 120, which **(middle)** executes a nucleophilic attack on carbonyl C of the lactam ring of the antibiotic. **Right:** D120 then donates the bound proton to the lactam nitrogen, facilitating the cleavage of the C-N bond in the tetrahedral intermediate.

catabolism of the propionate generated from the metabolism of isoleucine and lipids containing odd numbers of amino acids (see Figure 19–2). During the latter reaction, the Co^{3+} is transiently reduced to the 2+ oxidation state by abstracting an electron to generate a reactive methylene radical, R-CH$_2^*$. More information of Co and vitamin B$_{12}$ can be found in Chapter 44.

Copper

Copper is a functionally essential component of approximately 30 different metalloenzymes in humans, including cytochrome oxidase, dopamine β-hydroxylase, tyrosinase, the cytosolic form of superoxide dismutase (Cu, Zn-SOD), and lysyl oxidase. Dopamine β-hydroxylase and tyrosinase are both catecholamine oxidases, enzymes that oxidize the ortho-position in the phenol rings of L-dopamine (see Figure 41–10) and tyrosine, respectively. The former is the final step in the pathway by which epinephrine is synthesized in the adrenal gland, while the latter is the first and rate-limiting step in the synthesis of melanin. Both dopamine β-hydroxylase and tyrosinase are members of the type-3 family of copper proteins, which share a common **dicopper** center. As shown in **Figure 10–8**, the copper atoms in catecholamine oxidases chelate a molecule of molecular oxygen, activating it for attack on a phenol ring. During this process the copper atoms cycle between the +2 and +1 oxidation states. Another type-3 copper protein is hemocyanin. Unlike the catecholamine oxidases, the dicopper center of hemocyanin serves to transport oxygen in invertebrate animals such as mollusks that lack hemoglobin.

In Cu, Zn-SOD, the Cu^{2+} atom in the bimetallic center abstracts an electron from superoxide, O$_2^-$, an extremely reactive and cytotoxic ROS, forming O$_2$ and Cu^{1+}. The Cu atom in the enzyme is then restored to its original, +2, valence state by donating an electron to a second molecule of superoxide, generating H$_2$O$_2$. While hydrogen peroxide also is a ROS, it is considerably less reactive than O$_2^-$, a radical anion. Moreover, it can be subsequently converted to water and O$_2$ through the action of a second detoxifying enzyme, catalase (see Chapter 12).

Lysyl oxidase employs a single atom of Cu^{2+} to convert the epsilon amino groups on lysine side chains in collagen or elastin to aldehydes using molecular oxygen. The aldehyde groups on the side chain of the resulting amino acid, allysine (2-amino-6-oxo-hexanoic acid), then chemically react with the side chains of other allysine or lysine residues on adjacent polypeptides to generate the chemical crosslinks essential to the exceptional tensile strength of mature collagen and elastin fibers. Another essential feature of the enzyme is the presence of a modified amino acid, 2,4,5-trihydroxyphenylalanine quinone, in the active site. This modification is generated by the autocatalytic oxidation of the side chain of a conserved tyrosine residue by lysyl oxidase itself.

Nickel

A variety of nickel-containing enzymes are present in bacterial organisms. Examples include Ni, Fe hydrogenase and methyl-coenzyme M reductase, which catalyze redox reactions; acetyl-CoA synthase, which catalyzes a transferase reaction; and superoxide dismutase, which catalyzes a disproportionation reaction. Ni is a key component of urease, an enzyme found in bacteria, fungi, and plants (see Figure 10–4). However, the molecular basis of the dietary requirement for nickel in humans and other mammals has yet to be discovered.

Molybdenum

Catalytic Roles of Molybdopterin

Molybdenum is a key component of the phylogenetically universal cofactor molybdopterin (see Figure 10–3). In animals, molybdopterin serves as a catalytically essential prosthetic group for many enzymes, including xanthine oxidase, aldehyde oxidase, and sulfite oxidase. Xanthine oxidase, which also contains flavin, catalyzes the final two oxidative steps in the pathway by which uric acid is synthesized from purine nucleotides: the oxidation of hypoxanthine to xanthine and the oxidation of xanthine to uric acid (see Chapter 33). Catalysis of this two-stage process is facilitated by the ability of the

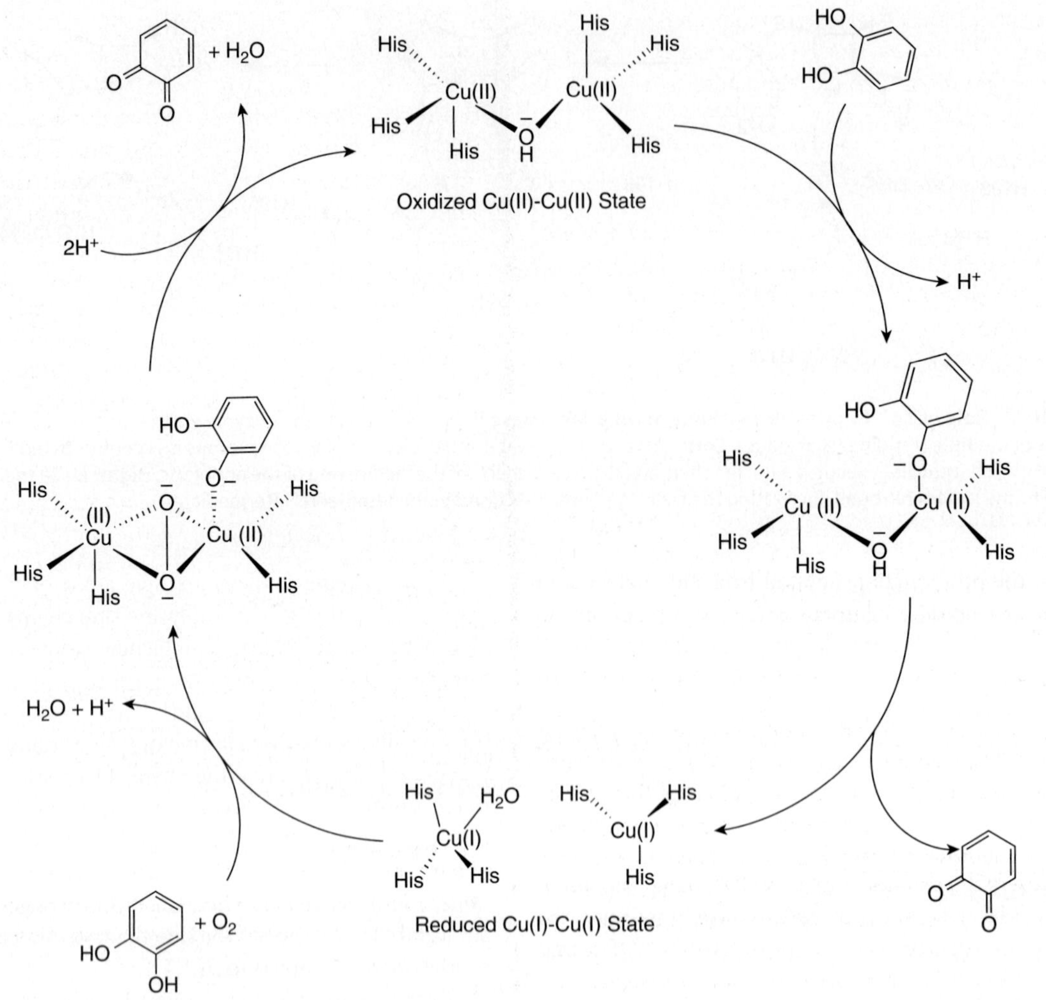

FIGURE 10-8 Reaction mechanism of the catecholamine oxidases.

bound Mo atom to cycle among the +4, +5, and +6 valence states. In addition to molybdopterin and flavin, aldehyde oxidase also contains an Fe-S cluster. Its complex suite of prosthetic groups enables the enzyme to oxidize a broad range of substrates, including many heterocyclic organic compounds. It has therefore been suggested that aldehyde oxidase participates, like the cytochrome P450 system, in the detoxification of xenobiotics (see Chapter 47).

Iron & Molybdenum Metalloenzymes

The Fe- and Mo-containing metalloenzyme sulfite oxidase is located in the mitochondria, where it catalyzes the oxidation of the sulfite (SO_3^{2-}) generated by the catabolism of sulfur-containing biomolecules to sulfate, SO_4^{2-}. As for xanthine oxidase, the ability of the molybdenum ion to transition between the +6, +5, and +4 oxidation states is critical to providing a catalytic route by which the two electrons removed from the sulfite molecule can be sequentially transferred to two molecules of cytochrome c, each of which can carry only a single electron (**Figure 10-9**). Mutations in any one of three genes—*MOCS1*, *MOCS2*, or *GPNH*—whose protein products catalyze

key steps in the synthesis of molybdopterin can lead to **sulfite oxidase deficiency**. Individuals who suffer from this autosomal inherited inborn error of metabolism are incapable of breaking down the sulfur-containing amino acids cysteine and methionine. The resulting accumulation of these amino acids

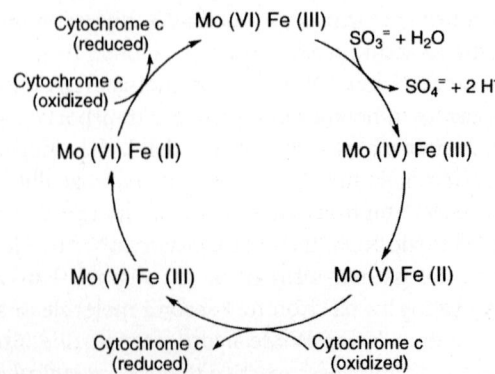

FIGURE 10-9 Reaction mechanism of sulfite oxidase showing oxidation states of enzyme-bound iron and molybdenum atoms.

and their derivatives in neonatal blood and tissues produces severe physical deformities and brain damage that leads to intractable seizures, severe mental retardation, and—in most cases—death during early childhood.

Vanadium

Although nutritionally essential, the role of vanadium in living organisms remains cryptic. No vanadium-containing cofactor has been identified to date. Vanadium is found throughout the body in both its +4, for example, HVO_4^{2-}, $H_2VO_4^-$, etc., and +5 oxidation states, for example, VO^{2+}, HVO^{3+}, etc. Various plasma proteins are known to bind oxides of vanadium, including albumin, immunoglobulin G, and transferrin. It has also recently been discovered that vanadium is a key component of the secreted "glue" with which marine mussels secure themselves to rocks, pilings, and ship's bottoms. Although vanadate, a phosphate analog, is known to inhibit protein-tyrosine phosphatases and alkaline phosphatase *in vitro*, it is unclear whether these interactions are of physiologic significance.

Chromium

The role of Cr in humans remains unknown. In the 1950s, a Cr^{3+}-containing "glucose tolerance factor" was isolated from brewer's yeast whose laboratory effects implicated this transition metal as a cofactor in the regulation of glucose metabolism. However, decades of research have failed to uncover either a Cr-containing biomolecule or a Cr-related genetic disease in animals. Nevertheless, many persons continue to ingest Cr-containing dietary supplements, such as Cr^{3+}-picolinate, for their alleged weight-loss properties.

ABSORPTION & TRANSPORT OF TRANSITION METALS

Transition Metals Are Absorbed by Diverse Mechanisms

In general, intestinal uptake of most transition metals is relatively inefficient. Only a small portion of the transition metals ingested each day is absorbed into our bodies. In addition, some transition metals, such as Ni and V, can be readily absorbed through the lungs when present in contaminated air or as a component of cigarette smoke. The perceived "inefficiency" of intestinal absorption may reflect the combination of the human body's modest requirements for these elements and the need to buffer against the accumulation of excessive amounts of these potentially toxic heavy metals. While we understand the pathways by which some transition metals, such as Fe (see Chapter 52), are taken up in great detail, in other instances, little hard evidence has been uncovered.

Fe^{2+} is absorbed directly via a transmembrane protein, the divalent metal ion transport protein (DMT-1), in the proximal duodenum. DMT-1 is also postulated to constitute the primary vehicle for the uptake of Mn^{2+}, Ni^{2+}, and—to a lesser extent—Cu^{2+}. As most of the iron in the stomach is in the ferric state, Fe^{3+}, it must be reduced to the ferrous, Fe^{2+}, state in order to be absorbed. This reaction is catalyzed by a ferric reductase that is also present on the cell surface, **duodenal cytochrome b (Dcytb)**. In addition, Dcytb is responsible for reducing Cu^{2+} to Cu^{1+} prior to transport by the high affinity Cu transport protein Ctr1. However, excess levels of Zn can inhibit Cu absorption, which can lead to a potentially lethal anemia. Molybdenum and vanadium are absorbed in the gut as the oxyanions vanadate, HVO_4^{2-}, and molybdate, $HMoO_4^{2-}$, by the same nonspecific anion transporter responsible for the absorption of their structural analogs phosphate, HPO_4^{2-}, and sulfate, SO_4^{2-}.

Cobalt is absorbed as the organometallic complex cobalamin, that is, vitamin B_{12}, via a dedicated pathway involving two secreted cobalamin-binding proteins, haptocorrin and intrinsic factor, and a cell surface receptor, cubilin. In the stomach, cobalamin released from ingested foods binds to haptocorrin, which protects the coenzyme from the extreme pH of its gastric surroundings. As the cobalamin-haptocorrin complex moves into the duodenum, the pH rises, inducing dissociation of the complex. The released cobalamin is then bound by a haptocorrin homolog known as intrinsic factor. The resulting cobalamin-intrinsic factor complex is then recognized and internalized by cubilin receptors present on the surface of intestinal epithelial cells.

SUMMARY

- Maintenance of human health and vitality requires the dietary uptake of trace quantities of several inorganic elements, including several transition metals.

- Paradoxically, many heavy metals, including excess levels of some of the nutritionally essential transition metals, are toxic and potentially carcinogenic.

- Acute heavy metal poisoning is treated by ingestion of chelating agents, the administration of diuretics along with the ingestion of water, or hemodialysis.

- Most heavy metals, including essential transition metals, can generate ROS in the presence of water and oxygen.

- The capacity of transition metals to serve as carriers for electrons and diatomic gases, as well as to facilitate catalysis of a wide range of enzymatic reactions, derives from two factors: their ability to transition between multiple valence states and their Lewis acid properties.

- In the body, transition metal ions are rarely encountered in free form. In most instances, they exist in organometallic complexes bound to proteins either directly by amino acid side chains or as part of organometallic prosthetic groups such as hemes, Fe-S clusters, or molybdopterin.

- Incorporation into organometallic complexes serves as a means of optimizing the properties of associated transition metals, prevents the collateral generation of ROS, and organizes multiple transition metals into a single functional unit.

- In the electron transport chain, many key electron transfer events rely on the ability of Fe atoms present in hemes, Fe-S clusters, and Rieske iron centers to transition between their +2 and +3 oxidation states.

- The presence of two Fe and two Cu atoms enables cytochrome oxidase to accumulate the four electrons needed to reduce molecular oxygen to water.

- Fe is commonly employed as a prosthetic group in many of the metalloenzymes that catalyze oxidation–reduction reactions.

- The majority of the nearly 3000 predicted zinc-metalloproteins encoded by the human genome contains a conserved polynucleotide-binding motif, the zinc finger.

- Fe-S clusters are present in many of the proteins involved in DNA replication and repair. It has been postulated that these prosthetic groups serve as electrochemical sensors for DNA damage.

- The Lewis acid capacity of Zn^{2+} is commonly utilized by hydrolytic enzymes to increase the nucleophilicity of water.

- The Mo atoms in xanthine oxidase and sulfite oxidase cycle between three different valence states during catalysis.

- The buildup of the S-containing amino acids methionine and cysteine in persons suffering from a deficiency in the Fe, Mo-metalloenzyme sulfite oxidase can cause serious developmental defects and death in infancy.

- In humans, the only known function of Co is as a component of 5′-deoxyadenosylcobalamin, vitamin B_{12}, a cofactor involved in the transfer of one-carbon groups. Co is absorbed into the body as its vitamin B_{12} complex.

REFERENCES

Buccella D, Lim MH, Morrow JR (eds.): Metals in biology: From metallomics to trafficking. Inorg Chem 2019;58:(Entire volume).

Cran DC, Kostenkova K: Open questions on the biological roles of first-row transition metals. Commun Chem 2020;3:104.

Fuss JO, Tsai CL, Ishida JP, Tainer JA: Emerging critical roles of Fe-S clusters in DNA replication and repair. Biochim Biophys Acta 2015;1853:1253.

Liu J, Chakraborty S, Hosseinzadeh P, et al: Metalloproteins containing cytochrome, iron-sulfur, or copper redox centers. Chem Rev 2014;114:4366.

Maret M: Zinc biochemistry: From a single zinc enzyme to a key element of life. Adv Nutr 2013;4:82.

Maret M, Wedd A (eds.): *Binding, Transport and Storage of Metal Ions in Biological Cells*. Royal Society of Chemistry, 2014:(Entire volume).

Zhang C: Essential functions of iron-requiring proteins in DNA replication, repair and cell cycle control. Protein Cell 2014;5:750.

Exam Questions

Section II – Enzymes: Kinetics, Mechanism, Regulation, & Role of Transition Metals

1. Rapid shallow breathing can lead to hyperventilation, a condition wherein carbon dioxide is exhaled from the lungs more rapidly than it is produced by the tissues. Explain how hyperventilation can lead to an increase in the pH of the blood.

2. A protein engineer desires to alter the active site of chymotrypsin so that it will cleave peptide bonds to the C-terminal side of aspartyl and glutamyl residues. The protein engineer will be most likely to succeed if he replaces the hydrophobic amino acid at the bottom of the active site pocket with:
   A. Phenylalanine
   B. Threonine
   C. Glutamine
   D. Lysine
   E. Proline

3. Select the one of the following statements that is NOT CORRECT.
   A. Many mitochondrial proteins are covalently modified by the acetylation of the epsilon amino groups of lysine residues.
   B. Protein acetylation is an example of a covalent modification that can be "reversed" under physiologic conditions.
   C. Increased levels of acetyl-CoA tend to favor protein acetylation.
   D. Acetylation increases the steric bulk of the amino acid side chains that are subject to this modification.
   E. The side chain of an acetylated lysyl residue is a stronger base than that of an unmodified lysyl residue.

4. Select the one of the following statements that is NOT CORRECT.
   A. Acid–base catalysis is a prominent feature of the catalytic mechanism of the HIV protease.
   B. Fischer's lock-and-key model explains the role of transition state stabilization in enzymic catalysis.
   C. Hydrolysis of peptide bonds by serine proteases involves the transient formation of a modified enzyme.
   D. Many enzymes employ metal ions as prosthetic groups or cofactors.
   E. In general, enzymes bind transition state analogs more tightly than substrate analogs.

5. Select the one of the following statements that is NOT CORRECT.
   A. To calculate K_{eq}, the equilibrium constant for a reaction, divide the initial rate of the forward reaction (rate 1) by the initial velocity of the reverse reaction (rate 1).
   B. The presence of an enzyme has no effect on K_{eq}.
   C. For a reaction conducted at constant temperature, the fraction of the potential reactant molecules possessing sufficient kinetic energy to exceed the activation energy of the reaction is a constant.
   D. Enzymes and other catalysts lower the activation energy of reactions.
   E. The algebraic sign of ΔG, the Gibbs free energy change for a reaction, indicates the direction in which a reaction will proceed.

6. Select the one of the following statements that is NOT CORRECT.
   A. As used in biochemistry, the standard state concentration for products and reactants other than protons is 1 molar.
   B. ΔG is a function of the logarithm of K_{eq}.
   C. As used in reaction kinetics, the term "spontaneity" refers to whether the reaction as written is favored to proceed from left to right.
   D. $\Delta G°$ denotes the change in free energy that accompanies transition from the standard state to equilibrium.
   E. On reaching equilibrium, the rates of the forward and reverse reaction both drop to zero.

7. Select the one of the following statements that is NOT CORRECT.
   A. Enzymes lower the activation energy for a reaction.
   B. Enzymes often lower the activation energy by destabilizing transition state intermediates.
   C. Active site histidyl residues frequently aid catalysis by acting as proton donors or acceptors.
   D. Covalent catalysis is employed by some enzymes to provide an alternative reaction pathway.
   E. The presence of an enzyme has no effect on $\Delta G°$.

8. Select the one of the following statements that is NOT CORRECT.
   A. For most enzymes, the initial reaction velocity, v_i, exhibits a hyperbolic dependence on [S].
   B. When [S] is much lower than K_m, the term $K_m + [S]$ in the Michaelis-Menten equation closely approaches K_m. Under these conditions, the rate of catalysis is a linear function of [S].
   C. The molar concentrations of substrates and products are equal when the rate of an enzyme-catalyzed reaction reaches half of its potential maximum value ($V_{max}/2$).
   D. An enzyme is said to have become saturated with substrate when successively raising [S] fails to produce a significant increase in v_i.
   E. When making steady-state rate measurements, the concentration of substrates should greatly exceed that of the enzyme catalyst.

9. Select the one of the following statements that is NOT CORRECT.
   A. Certain monomeric enzymes exhibit sigmoidal initial rate kinetics.
   B. The Hill equation is used to perform quantitative analysis of the cooperative behavior of enzymes or carrier proteins such as hemoglobin or calmodulin.
   C. For an enzyme that exhibits cooperative binding of substrate, a value of n (the Hill coefficient) greater than unity is said to exhibit positive cooperativity.
   D. An enzyme that catalyzes a reaction between two or more substrates is said to operate by a sequential mechanism if the substrates must bind in a fixed order.
   E. Prosthetic groups enable enzymes to add chemical groups beyond those present on amino acid side chains.

10. Select the one of the following statements that is NOT CORRECT.

    A. IC_{50} is a simple operational term for expressing the potency of an inhibitor.
    B. Lineweaver-Burk and Dixon plots employ rearranged versions of the Michaelis-Menten equation to generate linear representations of kinetic behavior and inhibition.
    C. A plot of $1/v_i$ versus $1/[S]$ can be used to evaluate the type and affinity for an inhibitor.
    D. Simple noncompetitive inhibitors lower the apparent K_m for a substrate.
    E. Noncompetitive inhibitors typically bear little or no structural resemblance to the substrate(s) of an enzyme-catalyzed reaction.

11. Select the one of the following statements that is NOT CORRECT.

    A. For a given enzyme, the intracellular concentrations of its substrates tend to be close to their K_m values.
    B. The sequestration of certain pathways within intracellular organelles facilitates the task of metabolic regulation.
    C. The earliest step in a biochemical pathway where regulatory control can be efficiently exerted is the first committed step.
    D. Feedback regulation refers to the allosteric control of an early step in a biochemical pathway by the end product(s) of that pathway.
    E. Metabolic control is most effective when one of the more rapid steps in a pathway is targeted for regulation.

12. Select the one of the following statements that is NOT CORRECT.

    A. The Bohr effect refers to the release of protons that occurs when oxygen binds to deoxyhemoglobin.
    B. Shortly after birth of a human infant, synthesis of the α-chain undergoes rapid induction until it comprises 50% of the hemoglobin tetramer.
    C. The β-chain of fetal hemoglobin is present throughout gestation.
    D. The term thalassemia refers to any genetic defect that results in partial or total absence of the α- or β-chains of hemoglobin.
    E. The taut conformation of hemoglobin is stabilized by several salt bridges that form between the subunits.

13. Select the one of the following statements that is NOT CORRECT.

    A. Steric hindrance by histidine E7 plays a critical role in weakening the affinity of hemoglobin for carbon monoxide (CO).
    B. Carbonic anhydrase plays a critical role in respiration by virtue of its capacity to break down 2,3-bisphosphoglycerate in the lungs.
    C. Hemoglobin S is distinguished by a genetic mutation that substitutes Glu6 on the β subunit with Val, creating a sticky patch on its surface.
    D. Oxidation of the heme iron from the +2 to the +3 state abolishes the ability of hemoglobin to bind oxygen.
    E. The functional differences between hemoglobin and myoglobin reflect, to a large degree, differences in their quaternary structure.

14. Select the one of the following statements that is NOT CORRECT.

    A. The charge-relay network of trypsin makes the active site serine a stronger nucleophile.
    B. The Michaelis constant is the substrate concentration at which the rate of the reaction is half-maximal.
    C. During transamination reactions, both substrates are bound to the enzyme before either product is released.
    D. Histidine residues act both as acids and as bases during catalysis by an aspartate protease.
    E. Many coenzymes and cofactors are derived from vitamins.

15. Select the one of the following statements that is NOT CORRECT.

    A. Interconvertible enzymes fulfill key roles in integrated regulatory networks.
    B. Phosphorylation of an enzyme often alters its catalytic efficiency.
    C. "Second messengers" act as intracellular extensions or surrogates for hormones and nerve impulses impinging on cell surface receptors.
    D. The ability of protein kinases to catalyze the reverse reaction that removes the phosphoryl group is key to the versatility of this molecular regulatory mechanism.
    E. Zymogen activation by partial proteolysis is irreversible under physiologic conditions.

16. Which of the following is NOT a benefit obtained by incorporating physiologically essential transition metal ions into organometallic complexes?

    A. Optimization of Lewis acid potency of the bound metal.
    B. Ability to construct complexes containing multiple transition metal ions.
    C. Attenuation of the production of reactive oxygen species.
    D. Protection against unwanted oxidation.
    E. To render the bound transition metal multivalent.

17. Which of the following is NOT a potential function of the physiologically essential transition metals?

    A. Binding diatomic gas molecules
    B. Proton carrier
    C. Stabilizing protein conformation
    D. Enhancing the nucleophilicity of water
    E. Electron carrier

18. Acute heavy metal poisoning can be treated by:

    A. Administration of diuretics
    B. Ingestion of chelating agents
    C. Hemodialysis
    D. All of the above
    E. None of the above

19. Which of the following is the name of a common organometallic DNA-binding motif?

    A. Zinc finger
    B. Molybdopterin
    C. Fe-S center
    D. All of the above
    E. None of the above

C H A P T E R

11

Bioenergetics: The Role of ATP

Kathleen M. Botham, PhD, DSc, & Peter A. Mayes, PhD, DSc

OBJECTIVES

After studying this chapter, you should be able to:

- State the first and second laws of thermodynamics and understand how they apply to biologic systems.
- Explain what is meant by the terms free energy, entropy, enthalpy, exergonic, and endergonic.
- Appreciate how reactions that are endergonic may be driven by coupling to those that are exergonic in biologic systems.
- Explain the role of group transfer potential, adenosine triphosphate (ATP), and other nucleotide triphosphates in the transfer of free energy from exergonic to endergonic processes, enabling them to act as the "energy currency" of cells.

BIOMEDICAL IMPORTANCE

Bioenergetics, or biochemical thermodynamics, is the study of the energy changes accompanying biochemical reactions. Biologic systems are essentially **isothermic** and use chemical energy to power living processes. The way in which an animal obtains suitable fuel from its food to provide this energy is basic to the understanding of normal nutrition and metabolism. Death from **starvation** occurs when available energy reserves are depleted, and certain forms of malnutrition are associated with energy imbalance (**marasmus**). Thyroid hormones control the **metabolic rate** (rate of energy release), and disease results if they malfunction. Excess storage of surplus energy causes **obesity**, an increasingly common disease of Western society which predisposes to many diseases, including cardiovascular disease and diabetes mellitus type 2, and lowers life expectancy.

FREE ENERGY IS THE USEFUL ENERGY IN A SYSTEM

Gibbs change in **free energy** (ΔG) is that portion of the total energy change in a system that is available for doing work—that is, the useful energy, also known as the chemical potential.

Biologic Systems Conform to the General Laws of Thermodynamics

The first law of thermodynamics states that **the total energy of a system, including its surroundings, remains constant**. It implies that within the total system, energy is neither lost nor gained during any change. However, energy may be transferred from one part of the system to another, or may be transformed into another form of energy. In living systems, chemical energy may be transformed into heat or into electrical, radiant, or mechanical energy.

The second law of thermodynamics states that **the total entropy of a system must increase if a process is to occur spontaneously**. **Entropy** is the extent of disorder or randomness of the system and becomes maximum as equilibrium is approached. Under conditions of constant temperature and pressure, the relationship between the free-energy change (ΔG) of a reacting system and the change in entropy (ΔS) is expressed by the following equation, which combines the two laws of thermodynamics:

$$\Delta G = \Delta H - T\Delta S$$

where ΔH is the change in **enthalpy** (heat) and T is the absolute temperature.

In biochemical reactions, since ΔH is approximately equal to the **total change in internal energy of the reaction or ΔE**, the above relationship may be expressed in the following way:

$$\Delta G = \Delta E - T\Delta S$$

If ΔG is negative, the reaction proceeds spontaneously with loss of free energy, that is, it is **exergonic**. If, in addition, ΔG is of great magnitude, the reaction goes virtually to completion and is essentially irreversible. On the other hand, if ΔG is positive, the reaction proceeds only if free energy can be gained, that is, it is **endergonic**. If, in addition, the magnitude of ΔG is great, the system is stable, with little or no tendency for a reaction to occur. If ΔG is zero, the system is at equilibrium and no net change takes place.

When the reactants are present in concentrations of 1.0 mol/L, ΔG^0 is the standard free-energy change. For biochemical reactions, a standard state is defined as having a pH of 7.0. The standard free-energy change at this standard state is denoted by $\Delta G^{0'}$.

The standard free-energy change can be calculated from the equilibrium constant K_{eq},

$$\Delta G^{0'} = -RT \ln K_{eq}$$

where R is the gas constant and T is the absolute temperature (see Chapter 8). It is important to note that the actual ΔG may be larger or smaller than $\Delta G^{0'}$ depending on the concentrations of the various reactants, including the solvent, various ions, and proteins.

In a biochemical system, an enzyme only speeds up the attainment of equilibrium; it never alters the final concentrations of the reactants at equilibrium.

ENDERGONIC PROCESSES PROCEED BY COUPLING TO EXERGONIC PROCESSES

The vital processes—for example, biosynthetic reactions, muscular contraction, nerve impulse conduction, and active transport—obtain energy by chemical linkage, or **coupling**, to oxidative reactions. In its simplest form, this type of coupling

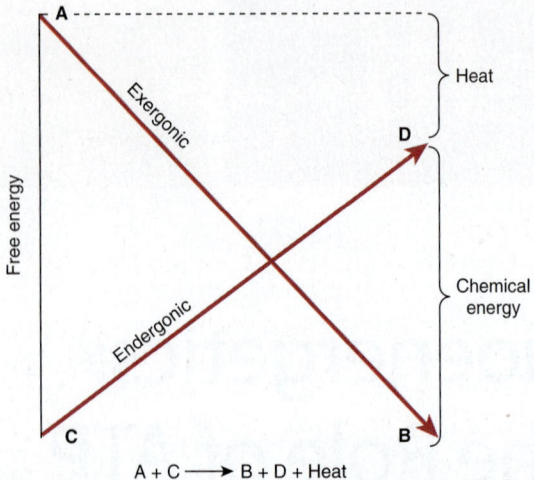

FIGURE 11–1 **Coupling of an exergonic to an endergonic reaction.**

may be represented as shown in **Figure 11–1**. The conversion of metabolite A to metabolite B occurs with release of free energy and is coupled to another reaction in which free energy is required to convert metabolite C to metabolite D. The terms **exergonic** and **endergonic**, rather than the normal chemical terms "exothermic" and "endothermic," are used to indicate that a process is accompanied by loss or gain, respectively, of free energy in any form, not necessarily as heat. In practice, an endergonic process cannot exist independently, but must be a component of a coupled exergonic–endergonic system where the overall net change is exergonic. The exergonic reactions are termed **catabolism** (generally, the breakdown or oxidation of fuel molecules), whereas the synthetic reactions that build up substances are termed **anabolism**. The combined catabolic and anabolic processes constitute **metabolism**.

If the reaction shown in Figure 11–1 is to go from left to right, then the overall process must be accompanied by loss of free energy as heat. One possible mechanism of coupling could be envisaged if a common obligatory intermediate (I) took part in both reactions, that is,

$$A + C \rightarrow I \rightarrow B + D$$

Some exergonic and endergonic reactions in biologic systems are coupled in this way. This type of system has a built-in mechanism for biologic control of the rate of oxidative processes since the common obligatory intermediate allows the rate of utilization of the product of the synthetic path (D) to determine by mass action the rate at which A is oxidized. Indeed, these relationships supply a basis for the concept of **respiratory control**, the process that prevents an organism from burning out of control. An extension of the coupling concept is provided by dehydrogenation reactions, which are coupled to hydrogenations by an intermediate carrier (**Figure 11–2**).

An alternative method of coupling an exergonic to an endergonic process is to synthesize a compound of high-energy potential in the exergonic reaction and to incorporate this

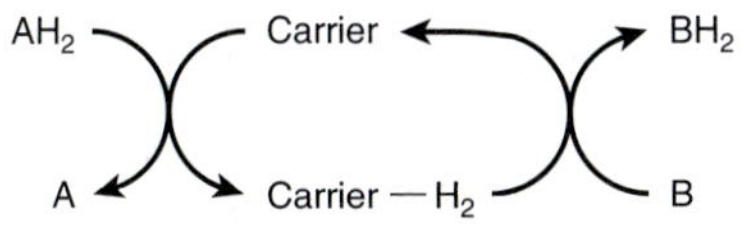

FIGURE 11–2 Coupling of dehydrogenation and hydrogenation reactions by an intermediate carrier.

new compound into the endergonic reaction, thus effecting a transference of free energy from the exergonic to the endergonic pathway. The biologic advantage of this mechanism is that the compound of high potential energy, ~Ⓔ, unlike I in the previous system, need not be structurally related to A, B, C, or D, allowing Ⓔ to serve as a transducer of energy from a wide range of exergonic reactions to an equally wide range of endergonic reactions or processes, such as biosynthesis, muscular contraction, nervous excitation, and active transport. In the living cell, the principal high-energy intermediate or carrier compound is **ATP** (**Figure 11–3**).

HIGH-ENERGY PHOSPHATES PLAY A CENTRAL ROLE IN ENERGY CAPTURE & TRANSFER

In order to maintain living processes, all organisms must obtain supplies of free energy from their environment. **Autotrophic** organisms utilize simple exergonic processes; for example, the energy of sunlight (green plants), the reaction $Fe^{2+} \rightarrow Fe^{3+}$ (some bacteria). On the other hand, **heterotrophic** organisms obtain free energy by coupling their metabolism to the breakdown of complex organic molecules in their environment. In all these organisms, ATP plays a central role in the transference of free energy from the exergonic to the endergonic processes. ATP is a nucleotide consisting of the nucleoside adenosine (adenine linked to ribose) and three phosphate groups (see Chapter 32). In its reactions in the cell, it functions as the Mg^{2+} complex (see Figure 11–3).

The importance of phosphates in intermediary metabolism became evident with the discovery of the role of ATP, adenosine diphosphate (ADP), and inorganic phosphate (P_i) in glycolysis (see Chapter 17).

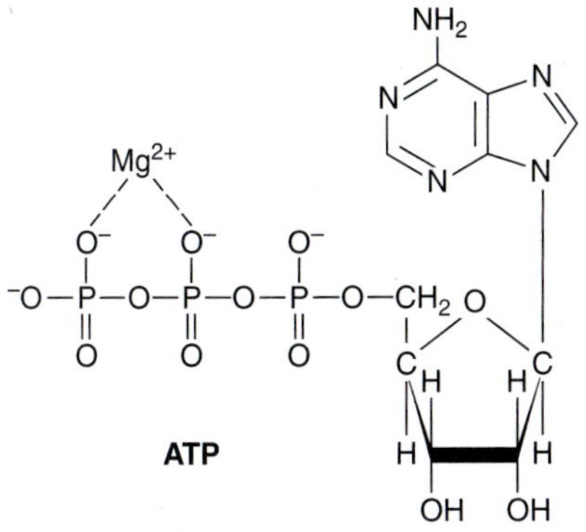

FIGURE 11–3 Adenosine triphosphate (ATP) is shown as the magnesium complex.

The Intermediate Value for the Free Energy of Hydrolysis of ATP Has Important Bioenergetic Significance

The standard free energy of hydrolysis of a number of biochemically important phosphates is shown in **Table 11–1**. An estimate of the comparative tendency of each of the phosphate groups to transfer to a suitable acceptor may be obtained from the $\Delta G^{0'}$ of hydrolysis at 37°C. This is termed the **group transfer potential**. The value for the hydrolysis of the terminal phosphate of ATP (when ATP is converted to ADP + Pi) divides the list into two groups. **Low-energy phosphates**, having a low group transfer potential, exemplified by the ester phosphates found in the intermediates of glycolysis, have $G^{0'}$ values smaller than that of ATP, while in **high-energy phosphates**, with a more negative $G^{0'}$, the value is higher than that of ATP. The components of this latter group, including ATP, are usually anhydrides (eg, the 1-phosphate of 1,3-bisphosphoglycerate), enol phosphates (eg, phosphoenolpyruvate), and phosphoguanidines (eg, creatine phosphate, arginine phosphate).

The symbol ~ Ⓟ indicates that the group attached to the bond, on transfer to an appropriate acceptor, results in transfer of the larger quantity of free energy. Thus, ATP has a high group transfer potential, whereas the phosphate in adenosine monophosphate (AMP) is of the low-energy type since it is a normal ester link (**Figure 11–4**). In energy transfer reactions, ATP may be converted to ADP and P_i or, in reactions requiring a greater energy input, to AMP + PP_i (see Table 11–1).

TABLE 11–1 Standard Free Energy of Hydrolysis of Some Organophosphates of Biochemical Importance

Compound	$\Delta G^{0'}$	
	kJ/mol	kcal/mol
Phosphoenolpyruvate	−61.9	−14.8
Carbamoyl phosphate	−51.4	−12.3
1,3-Bisphosphoglycerate (to 3-phosphoglycerate)	−49.3	−11.8
Creatine phosphate	−43.1	−10.3
ATP → AMP + PP_i	−32.2	−7.7
ATP → ADP + P_i	−30.5	−7.3
Glucose-1-phosphate	−20.9	−5.0
PP_i	−19.2	−4.6
Fructose-6-phosphate	−15.9	−3.8
Glucose-6-phosphate	−13.8	−3.3
Glycerol-3-phosphate	−9.2	−2.2

Abbreviations: PP_i, pyrophosphate; P_i, inorganic orthophosphate.
Note: All values are taken from Jencks WP: Free energies of hydrolysis and decarboxylation. *In: Handbook of Biochemistry and Molecular Biology*, vol 1. *Physical and Chemical Data.* Fasman GD (editor). CRC Press, 1976:296-304, except that for PP_i which is from Frey PA, Arabshahi A: Standard free-energy change for the hydrolysis of the alpha, beta-phosphoanhydride bridge in ATP. Biochemistry 1995;34:11307. Values differ between investigators, depending on the precise conditions under which the measurements were made.

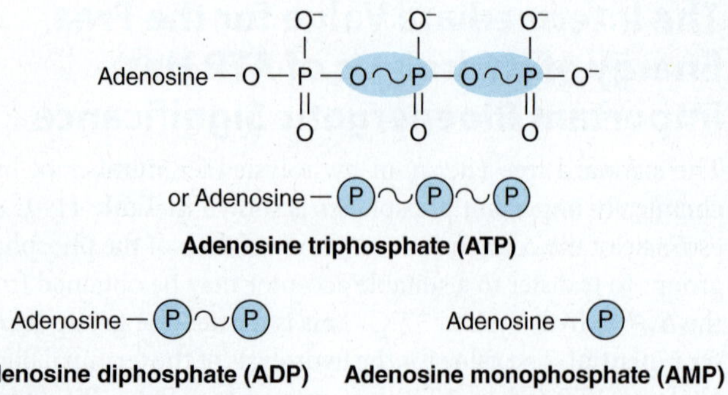

Adenosine — O — P — O∼P — O∼P — O⁻

or Adenosine — Ⓟ∼Ⓟ∼Ⓟ

Adenosine triphosphate (ATP)

Adenosine — Ⓟ∼Ⓟ   Adenosine — Ⓟ

Adenosine diphosphate (ADP)   Adenosine monophosphate (AMP)

FIGURE 11–4 Structure of ATP, ADP, and AMP showing the position and the number of high-energy phosphates (∼Ⓟ).

The intermediate position of ATP allows it to play an important role in energy transfer. The high free-energy change on hydrolysis of ATP is not in itself caused by the breaking of the P-O bond linking the terminal phosphate to the molecule (see Figure 11–4), in fact, energy is needed to bring this about. It is the consequences of this bond breakage that cause net energy to be released. Firstly, there is strong electrostatic repulsion between the negatively charged oxygen atoms in the adjacent phosphate groups of ATP (see Figure 11–4), which destabilizes the molecule and makes the removal of one phosphate group energetically favorable. Secondly, the orthophosphate produced is greatly stabilized by the formation of resonance hybrids in which the three negative charges are shared between the four oxygen atoms. Overall, therefore, the products of hydrolysis, ADP and orthophosphate, are more stable, and so lower in energy, than ATP (**Figure 11–5**). Other "high-energy compounds" are thiol esters involving coenzyme A (eg, acetyl-CoA), acyl carrier protein, amino acid esters involved in protein synthesis, S-adenosylmethionine (active methionine), uridine diphosphate glucose (UDPGlc), and 5-phosphoribosyl-1-pyrophosphate (PRPP).

ATP ACTS AS THE "ENERGY CURRENCY" OF THE CELL

The high group transfer potential of ATP enables it to act as a donor of high-energy phosphate to form those compounds below it in Table 11–1. Likewise, with the necessary enzymes, ADP can accept phosphate groups to form ATP from those compounds above ATP in the table. In effect, an **ATP/ADP cycle** connects those processes that generate ∼Ⓟ to those processes that utilize ∼Ⓟ (**Figure 11–6**), continuously consuming and regenerating ATP. This occurs at a very rapid rate since the total ATP/ADP pool is extremely small and sufficient to maintain an active tissue for only a few seconds.

There are three major sources of ∼Ⓟ taking part in **energy conservation** or **energy capture**:

1. **Oxidative phosphorylation** is the greatest quantitative source of ∼Ⓟ in aerobic organisms. ATP is generated in the mitochondrial matrix as O_2 is reduced to H_2O by electrons passing down the respiratory chain (see Chapter 13).

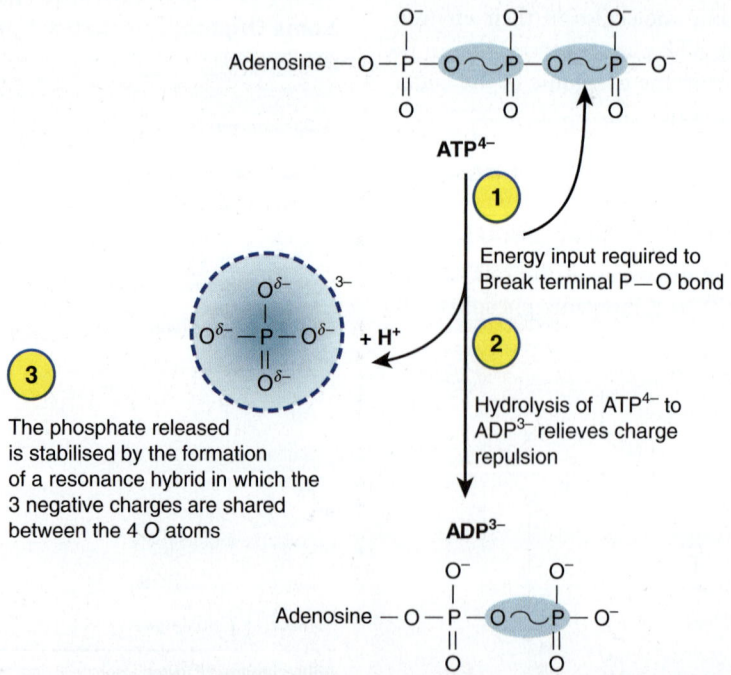

FIGURE 11–5 The free-energy change on hydrolysis of ATP to ADP. Initially, energy input is required to break the terminal P-O bond. However, the breaking of the bond relieves the strong electrostatic repulsion between the negatively charged oxygen atoms in the adjacent phosphate groups of ATP, making the removal of one phosphate group energetically favorable. In addition, the orthophosphate released is greatly stabilized by the formation of resonance hybrids in which the three negative charges are shared between the four oxygen atoms. These effects more than compensate for the initial energy input and result in the high free-energy change seen when ATP is hydrolyzed to ADP.

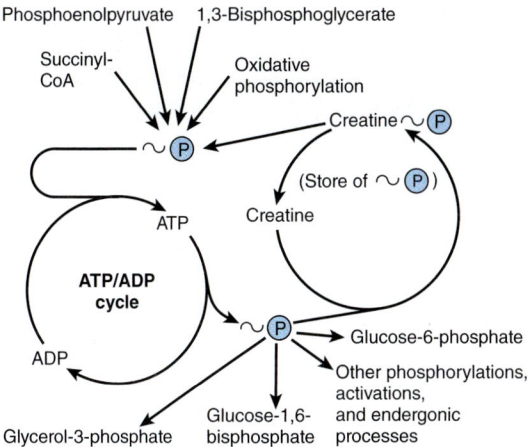

FIGURE 11–6 Role of ATP/ADP cycle in transfer of high-energy phosphate.

2. **Glycolysis**. A net formation of two ~Ⓟ results from the formation of lactate from one molecule of glucose, generated in two reactions catalyzed by phosphoglycerate kinase and pyruvate kinase, respectively (see Chapter 17).

3. **The citric acid cycle**. One ~Ⓟ is generated directly in the cycle at the succinate thiokinase step (see Figure 16–3).

   Phosphagens act as storage forms of group transfer potential and include **creatine phosphate**, which occurs in vertebrate skeletal muscle, heart, spermatozoa, and brain, and **arginine phosphate**, which occurs in invertebrate muscle. When ATP is rapidly being utilized as a source of energy for muscular contraction, phosphagens permit its concentrations to be maintained, but when the ATP/ADP ratio is high, their concentration can increase to act as an energy store (**Figure 11–7**).

   When ATP acts as a phosphate donor to form compounds of lower free energy of hydrolysis (see Table 11–1), the phosphate group is invariably converted to one of low energy. For example, the phosphorylation of glycerol to form glycerol-3-phosphate:

Glycerol + Adenosine—Ⓟ~Ⓟ~Ⓟ   **GLYCEROL KINASE** →

Glycerol—Ⓟ + Adenosine—Ⓟ~Ⓟ

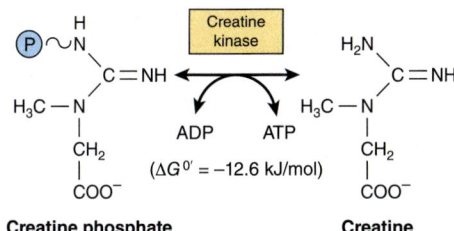

FIGURE 11–7 Transfer of high-energy phosphate between ATP and creatine.

ATP Allows the Coupling of Thermodynamically Unfavorable Reactions to Favorable Ones

Endergonic reactions cannot proceed without an input of free energy. For example, the phosphorylation of glucose to glucose-6-phosphate, the first reaction of glycolysis (see Figure 17–2):

$$Glucose + P_i \rightarrow Glucose\text{-}6\text{-}phosphate + H_2O$$

$$(\Delta G^{0\prime}) = +13.8 \text{ kJ/mol} \tag{1}$$

is highly endergonic and cannot proceed under physiologic conditions. Thus, in order to take place, the reaction must be coupled with another—more exergonic—reaction such as the hydrolysis of the terminal phosphate of ATP.

$$ATP \rightarrow ADP + P_i \ (\Delta G^{0\prime} = -30.5 \text{ kJ/mol}) \tag{2}$$

When (1) and (2) are coupled in a reaction catalyzed by hexokinase, phosphorylation of glucose readily proceeds in a highly exergonic reaction that under physiologic conditions is irreversible. Many "activation" reactions follow this pattern.

Adenylyl Kinase (Myokinase) Interconverts Adenine Nucleotides

This enzyme is present in most cells. It catalyzes the following reaction:

ADENYLYL KINASE

$$ATP + AMP \leftrightarrow 2ADP$$

Adenylyl kinase is important for the maintenance of energy homeostasis in cells because it allows:

1. The group transfer potential in ADP to be used in the synthesis of ATP.

2. The AMP formed as a consequence of activating reactions involving ATP to be rephosphorylated to ADP.

3. AMP to increase in concentration when ATP becomes depleted so that it is able to act as a metabolic (allosteric) signal to increase the rate of catabolic reactions, which in turn lead to the generation of more ATP (see Chapter 14).

When ATP Forms AMP, Inorganic Pyrophosphate (PP$_i$) Is Produced

ATP can also be hydrolyzed directly to AMP, with the release of PP$_i$ (see Table 11–1). This occurs, for example, in the activation of long-chain fatty acids (see Figure 22–3).

   This reaction is accompanied by loss of free energy as heat, which ensures that the activation reaction will go to the right, and is further aided by the hydrolytic splitting of PP$_i$, catalyzed by **inorganic pyrophosphatase**, a reaction that itself has a large $\Delta G^{0\prime}$ of -19.2 kJ/mol. Note that activations via the

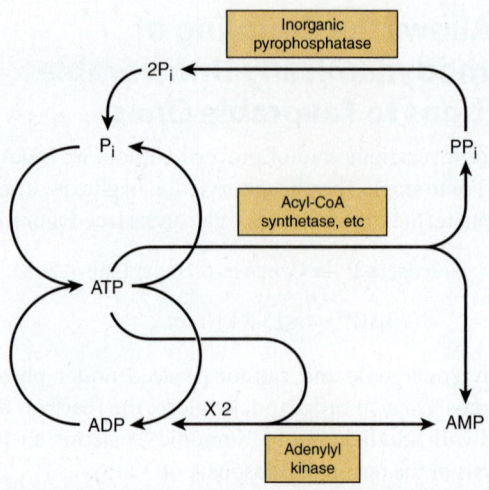

FIGURE 11–8 Phosphate cycles and interchange of adenine nucleotides.

pyrophosphate pathway result in the loss of two ~ⓟ rather than one, as occurs when ADP and P_i are formed.

$$PP_i + H_2O \xrightarrow{\text{INORGANIC PYROPHOSPHATASE}} 2P_i$$

A combination of the above reactions makes it possible for phosphate to be recycled and the adenine nucleotides to interchange (**Figure 11–8**).

Other Nucleoside Triphosphates Participate in Group Transfer Potential

By means of the **nucleoside diphosphate (NDP) kinases**, UTP, GTP, and CTP can be synthesized from their diphosphates, for example, UDP reacts with ATP to form UTP.

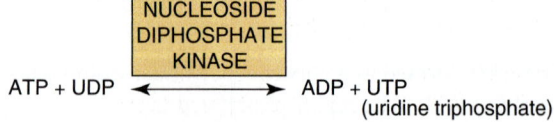
(uridine triphosphate)

All of these triphosphates take part in phosphorylations in the cell. Similarly, specific **nucleoside monophosphate (NMP) kinases** catalyze the formation of NDP from the corresponding monophosphates. Thus, adenylyl kinase is a specialized NMP kinase.

SUMMARY

- Biologic systems use chemical energy to power living processes.
- Exergonic reactions take place spontaneously with loss of free energy (ΔG is negative). Endergonic reactions require the gain of free energy (ΔG is positive) and occur only when coupled to exergonic reactions.
- ATP acts as the "energy currency" of the cell, transferring free energy derived from substances of higher energy potential to those of lower energy potential.

REFERENCES

Haynie D: *Biological Thermodynamics*. Cambridge University Press, 2008.
Nicholls DG, Ferguson S: *Bioenergetics*, 4th ed. Academic Press, 2013.

Biologic Oxidation

Kathleen M. Botham, PhD, DSc, & Peter A. Mayes, PhD, DSc

OBJECTIVES

After studying this chapter you should be able to:

- Explain the meaning of redox potential and how it can be used to predict the direction of flow of electrons in biologic systems.
- Identify the four classes of enzymes (oxidoreductases) involved in oxidation and reduction reactions.
- Describe the action of oxidases and provide examples of where they play an important role in metabolism.
- Indicate the two main functions of dehydrogenases and explain the importance of nicotinamide adenine dinucleotide (NAD)- and riboflavin-linked dehydrogenases in metabolic pathways such as glycolysis, the citric acid cycle, and the respiratory chain.
- Identify the two types of enzymes classified as hydroperoxidases; indicate the reactions they catalyze and explain why they are important.
- Give the two steps of reactions catalyzed by oxygenases and identify the two subgroups of this class of enzymes.
- Appreciate the role of cytochrome P450 in drug detoxification and steroid synthesis.
- Describe the reaction catalyzed by superoxide dismutase and explain how it protects tissues from oxygen toxicity.

BIOMEDICAL IMPORTANCE

Chemically, **oxidation** is defined as the removal of electrons and **reduction** is defined as the gain of electrons. Thus, oxidation of a molecule (the electron donor) is always accompanied by reduction of a second molecule (the electron acceptor). This principle of oxidation–reduction also applies to biochemical systems and is an important concept underlying the understanding of the nature of biologic oxidation. Note that many biologic oxidations can take place without the participation of molecular oxygen, for example, dehydrogenations. The life of higher animals is absolutely dependent on a supply of oxygen for **respiration**, the process by which cells derive energy in the form of ATP (see Chapter 11) from the controlled reaction of hydrogen with oxygen to form water. In addition, molecular oxygen is incorporated into a variety of substrates by enzymes designated as **oxygenases**; many drugs, pollutants, and chemical carcinogens (xenobiotics) are metabolized by enzymes of this

class, known as the **cytochrome P450 system**. Administration of oxygen can be lifesaving in the treatment of patients with respiratory or circulatory failure.

FREE ENERGY CHANGES CAN BE EXPRESSED IN TERMS OF REDOX POTENTIAL

In reactions involving oxidation and reduction, the free energy change is proportionate to the tendency of reactants to donate or accept electrons. Thus, in addition to expressing free energy change in terms of $\Delta G^{0'}$ (see Chapter 11), it is possible, in an analogous manner, to express it numerically as an **oxidation–reduction** or **redox potential** (E'_0). Chemically, the redox potential of a system (E_0) is usually compared with the potential of the hydrogen electrode (0.0 V at pH 0.0). However, for biologic systems, the redox potential

TABLE 12–1 Some Redox Potentials of Special Interest in Mammalian Oxidation Systems

System E'_0	Volts
H^+/H_2	−.0.42
$NAD^+/NADH$	−0.32
Lipoate; ox/red	−0.29
Acetoacetate/3-hydroxybutyrate	−0.27
Pyruvate/lactate	−0.19
Oxaloacetate/malate	−0.17
Fumarate/succinate	+0.03
Cytochrome b; Fe^{3+}/Fe^{2+}	+0.08
Ubiquinone; ox/red	+0.10
Cytochrome c_1; Fe^{3+}/Fe^{2+}	+0.22
Cytochrome a; Fe^{3+}/Fe^{2+}	+0.29
Oxygen/water	+0.82

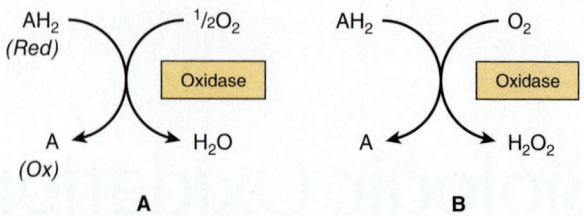

FIGURE 12–1 Oxidation of a metabolite catalyzed by an oxidase (A) forming H_2O and (B) forming H_2O_2.

(E'_0) is normally expressed at pH 7.0, at which pH of the electrode potential of the hydrogen electrode is −0.42 V. The redox potentials of some redox systems of special interest in mammalian biochemistry are shown in **Table 12–1**. The relative positions of redox systems in the table allow prediction of the direction of flow of electrons from one redox couple to another.

Enzymes involved in oxidation and reduction are called **oxidoreductases** and are classified into four groups: **oxidases**, **dehydrogenases**, **hydroperoxidases**, and **oxygenases**.

OXIDASES USE OXYGEN AS A HYDROGEN ACCEPTOR

Oxidases catalyze the removal of hydrogen from a substrate using oxygen as a hydrogen acceptor.* They form water or hydrogen peroxide as a reaction product (**Figure 12–1**).

Cytochrome Oxidase Is a Hemoprotein

Cytochrome oxidase is a hemoprotein widely distributed in many tissues, having the typical heme prosthetic group present in myoglobin, hemoglobin, and other cytochromes (see Chapter 6). It is the terminal component of the chain of respiratory carriers found in mitochondria (see Chapter 13) and it functions to transfer electrons resulting from the oxidation of substrate molecules by dehydrogenases to their final acceptor, oxygen. The action of the enzyme is blocked by **carbon monoxide**, **cyanide**, **and hydrogen sulfide**, and this causes poisoning by preventing cellular respiration. The cytochrome

---

*The term "oxidase" is sometimes used collectively to denote all enzymes that catalyze reactions involving molecular oxygen.

oxidase enzyme complex comprises heme a_3 combined with another heme, heme a, in a single protein and so is also termed **cytochrome aa_3**. It contains two molecules of heme, each having one Fe atom that oscillates between Fe^{3+} and Fe^{2+} during oxidation and reduction. Furthermore, two atoms of copper are present, one associated with each heme unit.

Other Oxidases Are Flavoproteins

Flavoprotein enzymes contain **flavin mononucleotide (FMN)** or **flavin adenine dinucleotide (FAD)** as prosthetic groups. FMN and FAD are formed in the body from the vitamin **riboflavin** (see Chapter 44). FMN and FAD are usually tightly—but not covalently—bound to their respective apoenzyme proteins. **Metalloflavoproteins** contain one or more metals as essential cofactors. Examples of flavoprotein oxidases include L-**amino acid oxidase**, an enzyme found in kidney with general specificity for the oxidative deamination of the naturally occurring L-amino acids; **xanthine oxidase**, a molybdenum-containing enzyme which plays an important role in the conversion of purine bases to uric acid (see Chapter 33), and is of particular significance in uricotelic animals (see Chapter 28); and **aldehyde dehydrogenase**, an FAD-linked enzyme present in mammalian livers, which contains molybdenum and nonheme iron and acts on aldehydes and N-heterocyclic substrates. The mechanisms of oxidation and reduction of these enzymes are complex. Evidence suggests a two-step reaction as shown in **Figure 12–2**.

DEHYDROGENASES PERFORM TWO MAIN FUNCTIONS

There are a large number of enzymes in the dehydrogenase class. Their two main functions are as follows:

1. Transfer of hydrogen from one substrate to another in a coupled oxidation–reduction reaction (**Figure 12–3**). These dehydrogenases often utilize common coenzymes or hydrogen carriers, for example, nicotinamide adenine dinucleotide (NAD^+). This type of reaction in which one substrate is oxidized/reduced at the expense of another is freely reversible, enabling reducing equivalents to be transferred within the cell and oxidative processes to occur in the absence of oxygen, such as during the anaerobic phase of glycolysis (see Figure 17–2).

2. Transfer of electrons from substrate to oxygen in the **respiratory chain** electron transport system (see Figure 13–3).

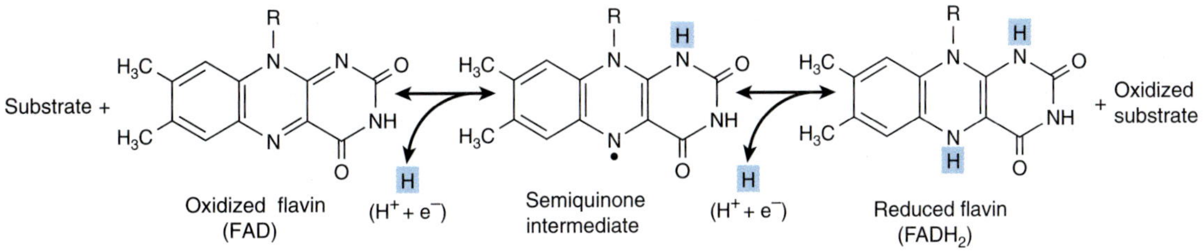

FIGURE 12–2 **Oxidoreduction of isoalloxazine ring in flavin nucleotides via a semiquinone intermediate.** In oxidation reactions, the flavin (eg, FAD) accepts two electrons and two H$^+$ in two steps, forming the semiquinone intermediate followed by the reduced flavin (eg, FADH$_2$) and the substrate is oxidized. In the reverse (reduction) reaction, the reduced flavin gives up two electrons and two H$^+$ so that it becomes oxidized (eg, to FAD) and the substrate is reduced.

Many Dehydrogenases Depend on Nicotinamide Coenzymes

These dehydrogenases use **NAD$^+$** or **nicotinamide adenine dinucleotide phosphate (NADP$^+$)**—or both—which are formed in the body from the vitamin **niacin** (see Chapter 44). The structure of NAD$^+$ is shown in **Figure 12–4**. NADP$^+$ has a phosphate group esterified to the 2′ hydroxyl of its adenosine moiety, but otherwise is identical to NAD$^+$. The oxidized forms of both nucleotides have a positive charge on the nitrogen atom of the nicotinamide moiety as indicated in Figure 12–4. The coenzymes are reduced by the specific substrate of the dehydrogenase and reoxidized by a suitable electron acceptor. They are able to freely and reversibly dissociate from their respective apoenzymes.

Generally, **NAD-linked dehydrogenases** catalyze oxidoreduction reactions of the type:

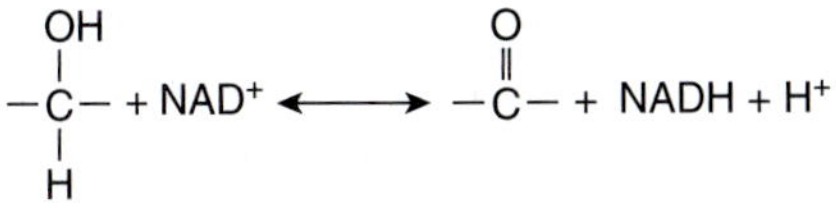

When a substrate is oxidized, it loses two hydrogen atoms and two electrons. One H$^+$ and both electrons are accepted by NAD$^+$ to form NADH and the other H$^+$ is released (see Figure 12–4). Many such reactions occur in the oxidative pathways of metabolism, particularly in glycolysis (see Chapter 17) and the citric acid cycle (see Chapter 16). NADH is generated in these pathways via the oxidation of fuel molecules, and NAD$^+$ is regenerated by the oxidation of NADH as it transfers the electrons to O$_2$ via the respiratory chain in mitochondria, a process which leads to the formation of ATP (see Chapter 13).

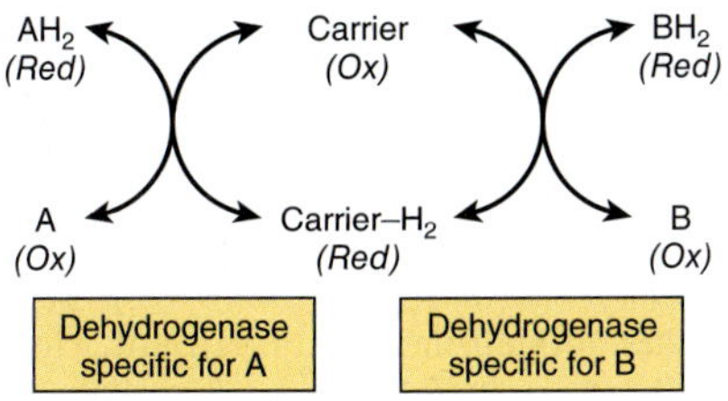

NADP-linked dehydrogenases are found characteristically in biosynthetic pathways where reductive reactions are required, as in the extramitochondrial pathway of fatty acid synthesis (see Chapter 23) and steroid synthesis (see Chapter 26)—and also in the pentose phosphate pathway (see Chapter 20).

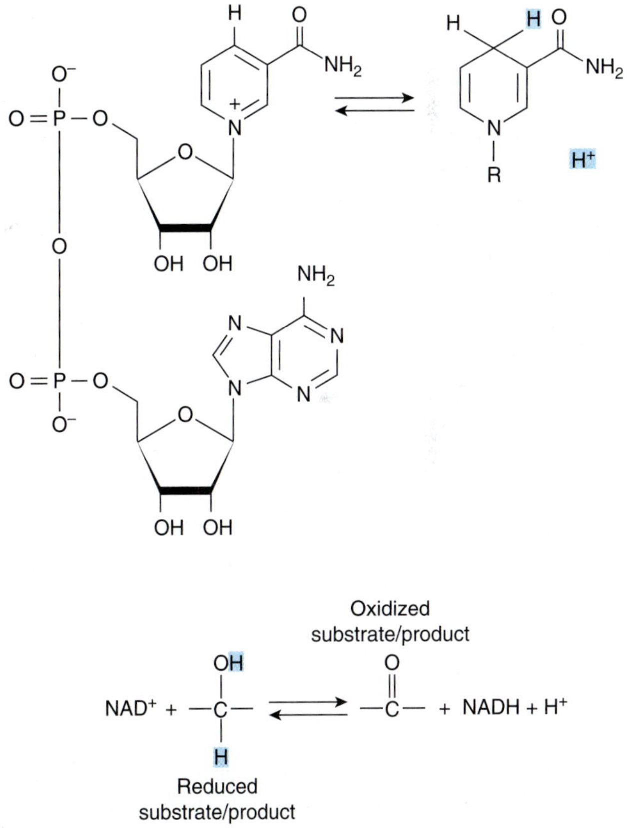

FIGURE 12–4 **Oxidation and reduction of nicotinamide coenzymes.** Nicotinamide coenzymes consist of a nicotinamide ring linked to an adenosine via a ribose and a phosphate group, forming a dinucleotide. NAD$^+$/NADH are shown, but NADP$^+$/NADPH are identical except that they have a phosphate group esterified to the 2′ OH of the adenosine. An oxidation reaction involves the transfer of two electrons and one H$^+$ from the substrate to the nicotinamide ring of NAD$^+$ forming NADH and the oxidized product. The remaining hydrogen of the hydrogen pair removed from the substrate remains free as a hydrogen ion. NADH is oxidized to NAD$^+$ by the reverse reaction. R, the part of the molecule unchanged in the oxidation/reduction reaction.

FIGURE 12–3 **Oxidation of a metabolite catalyzed by coupled dehydrogenases.**

Other Dehydrogenases Depend on Riboflavin

The **flavin groups such as FMN and FAD** are associated with dehydrogenases as well as with oxidases as described earlier. FAD is the electron acceptor in reactions of the type:

$$\underset{\underset{H}{|}}{\overset{\overset{H}{|}}{-C}}-C- + FAD \longleftrightarrow -C=C- + FADH_2$$

FAD accepts two electrons and two H^+ in the reaction (see Figure 12–2), forming **FADH$_2$**. Flavin groups are generally more tightly bound to their apoenzymes than are the nicotinamide coenzymes. Most of the **riboflavin-linked dehydrogenases** are concerned with electron transport in (or to) the respiratory chain (see Chapter 13). **NADH dehydrogenase** acts as a carrier of electrons between NADH and the components of higher redox potential (see Figure 13–3). Other dehydrogenases such as **succinate dehydrogenase, acyl-CoA dehydrogenase,** and **mitochondrial glycerol-3-phosphate dehydrogenase** transfer reducing equivalents directly from the substrate to the respiratory chain (see Figure 13–5). Another role of the flavin-dependent dehydrogenases is in the dehydrogenation (by **dihydrolipoyl dehydrogenase**) of reduced lipoate, an intermediate in the oxidative decarboxylation of pyruvate and α-ketoglutarate (see Figures 13–5 and 17–5). The **electron-transferring flavoprotein (ETF)** is an intermediary carrier of electrons between acyl-CoA dehydrogenase and the respiratory chain (see Figure 13–5).

Cytochromes May Also Be Regarded as Dehydrogenases

The **cytochromes** are iron-containing hemoproteins in which the iron atom oscillates between Fe^{3+} and Fe^{2+} during oxidation and reduction. Except for cytochrome oxidase (described earlier), they are classified as dehydrogenases. In the respiratory chain, they are involved as carriers of electrons from flavoproteins on the one hand to cytochrome oxidase on the other (see Figure 13–5). Several identifiable cytochromes occur in the respiratory chain, they are; cytochromes b, c_1, c, and cytochrome oxidase (aa_3). Cytochromes are also found in other locations, for example, the endoplasmic reticulum (cytochromes P450 and b_5), and in plant cells, bacteria, and yeasts.

HYDROPEROXIDASES USE HYDROGEN PEROXIDE OR AN ORGANIC PEROXIDE AS SUBSTRATE

Two types of enzymes found both in animals and plants fall into the **hydroperoxidase** category: **peroxidases** and **catalase**.

Hydroperoxidases play an important role in protecting the body against the harmful effects of **reactive oxygen species (ROS)**. ROS are highly reactive oxygen-containing molecules such as peroxides, which are formed during normal metabolism, but can be damaging if they accumulate. They are believed to contribute to the causation of diseases such as cancer and atherosclerosis, as well as the aging process in general (see Chapters 21, 44, 54).

Peroxidases Reduce Peroxides Using Various Electron Acceptors

Peroxidases are found in milk as well as in leukocytes, platelets, and other tissues involved in eicosanoid metabolism (see Chapter 23). Their prosthetic group is **protoheme**. In the reaction catalyzed by peroxidase, hydrogen peroxide is reduced at the expense of several substances that act as electron acceptors, such as ascorbate (vitamin C), quinones, and cytochrome c. The reaction catalyzed by peroxidase is complex, but the overall reaction is as follows:

$$H_2O_2 + AH_2 \xrightarrow{\text{PEROXIDASE}} 2H_2O + A$$

In erythrocytes and other tissues, the enzyme **glutathione peroxidase**, containing **selenium** as a prosthetic group, catalyzes the destruction of H_2O_2 and lipid hydroperoxides through the conversion of reduced glutathione to its oxidized form, protecting membrane lipids and hemoglobin against oxidation by peroxides (see Chapter 21).

Catalase Uses Hydrogen Peroxide as Electron Donor & Electron Acceptor

Catalase is a hemoprotein containing four heme groups. It can act as a peroxidase, catalyzing reactions of the type shown earlier, but it is also able to catalyze the breakdown of H_2O_2 formed by the action of oxygenases to water and oxygen:

$$2H_2O_2 \xrightarrow{\text{CATALASE}} 2H_2O + O_2$$

This reaction uses one molecule of H_2O_2 as a substrate electron donor and another molecule of H_2O_2 as an oxidant or electron acceptor. It is one of the fastest enzyme reactions known, destroying millions of potentially damaging H_2O_2 molecules per second. Under most conditions in vivo, the peroxidase activity of catalase seems to be favored. Catalase is found in blood, bone marrow, mucous membranes, kidney, and liver. **Peroxisomes** are membrane-bound organelles (see Chapter 49) found in many tissues, including liver. They are rich in oxidases and in catalase. Thus, enzymes that produce and breakdown H_2O_2 are contained within the same subcellular compartment. However, mitochondrial and microsomal electron transport systems as well as xanthine oxidase must be considered as additional sources of H_2O_2.

OXYGENASES CATALYZE THE DIRECT TRANSFER & INCORPORATION OF OXYGEN INTO A SUBSTRATE MOLECULE

Oxygenases are concerned with the synthesis or degradation of many different types of metabolites. They catalyze the incorporation of oxygen into a substrate molecule in two steps: (1) oxygen is bound to the enzyme at the active site and (2) the bound oxygen is reduced or transferred to the substrate. Oxygenases may be divided into two subgroups, dioxygenases and monooxygenases.

Dioxygenases Incorporate Both Atoms of Molecular Oxygen Into the Substrate

The basic reaction catalyzed by dioxygenases is as follows:

$$A + O_2 \rightarrow AO_2$$

Examples include the liver enzymes, **homogentisate dioxygenase** (oxidase) and **3-hydroxyanthranilate dioxygenase** (oxidase), which contain iron; and L-**tryptophan dioxygenase** (tryptophan pyrrolase) (see Chapter 29), which utilizes heme.

Monooxygenases (Mixed-Function Oxidases, Hydroxylases) Incorporate Only One Atom of Molecular Oxygen Into the Substrate

The other oxygen atom is reduced to water, an additional electron donor or cosubstrate (Z) being necessary for this purpose:

$$A - H + O_2 + ZH_2 \rightarrow A - OH + H_2O + Z$$

Cytochromes P450 Are Monooxygenases Important in Steroid Metabolism & for the Detoxification of Many Drugs

Cytochromes P450 are an important superfamily of heme-containing monooxygenases, and more than 50 such enzymes have been found in the human genome. They are located mainly in the endoplasmic reticulum in the liver and intestine, but are also found in the mitochondria in some tissues. The cytochromes participate in an electron transport chain in which both NADH and NADPH may donate reducing equivalents. Electrons are passed to cytochrome P450 in two types of reaction involving FAD or FMN. Class I systems consist of an FAD-containing reductase enzyme, an iron sulfur (Fe_2S_2) protein, and the P450 heme protein, while class II systems contain cytochrome P450 reductase, which passes electrons from $FADH_2$ to FMN (**Figure 12–5**). Class I and II systems are well characterized, but in recent years, other cytochromes P450, which do not fit into either category, have been identified. In the final step, oxygen accepts the electrons from cytochrome P450 and is reduced, with one atom being incorporated into H_2O and the other into the substrate, usually resulting in its hydroxylation. This series of enzymatic reactions, known as the **hydroxylase cycle**, is illustrated in **Figure 12–6**. In the endoplasmic reticulum of the liver, cytochromes P450 are found together with another heme-containing protein, **cytochrome b_5** (see Figure 12–5) and together they have a major role in drug metabolism and detoxification. Cytochrome b_5 also has an important role as a fatty acid desaturase. Together, cytochromes P450 and b_5 are responsible for about 75% of the modification and degradation of drugs which occurs in the body. The rate of detoxification of many medicinal drugs by

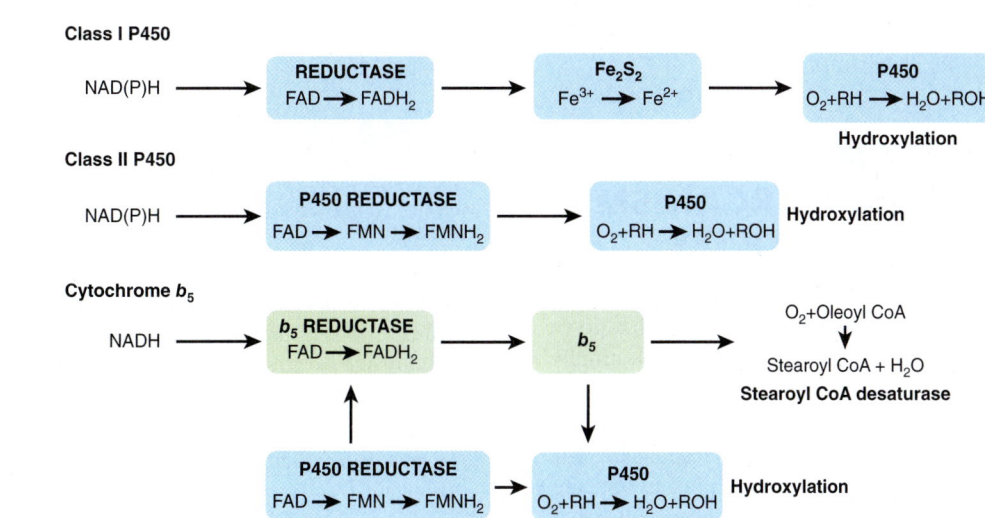

FIGURE 12–5 **Cytochromes P450 and b_5 in the endoplasmic reticulum.** Most cytochromes P450 are class I or class II. In addition to cytochrome P450, class I systems contain a small FAD-containing reductase and an iron sulfur protein, and class II contains cytochrome P450 reductase, which incorporates FAD and FMN. Cytochromes P450 catalyze many steroid hydroxylation reactions and drug detoxification steps. Cytochrome b_5 acts in conjunction with the FAD-containing cytochrome b_5 reductase in the fatty acyl-CoA desaturase (eg, stearoyl-CoA desaturase) reaction and also works together with cytochromes P450 in drug detoxification. It is able to accept electrons from cytochrome P450 reductase via cytochrome b_5 reductase and donate them to cytochrome P450.

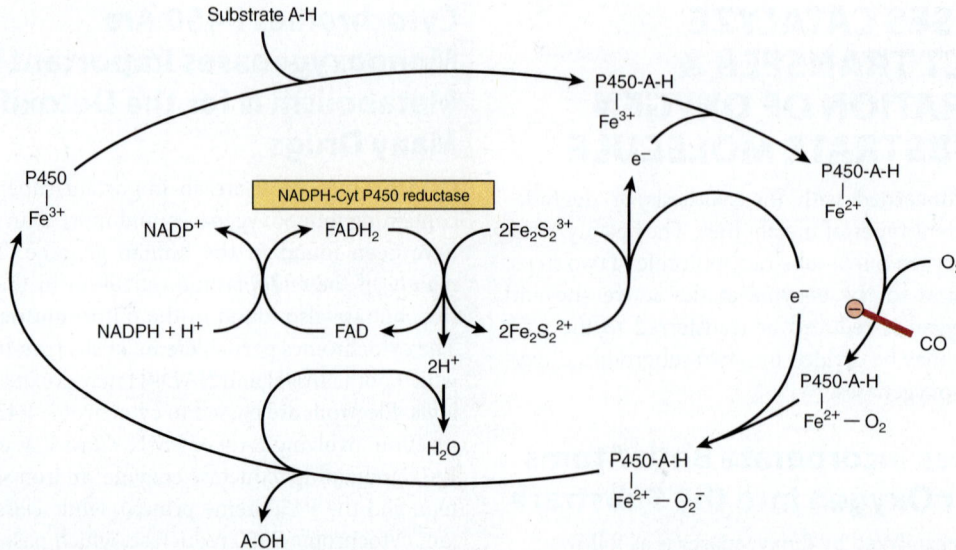

FIGURE 12–6 **Cytochrome P450 hydroxylase cycle.** The system shown is typical of steroid hydroxylases of the adrenal cortex. Liver microsomal cytochrome P450 hydroxylase does not require the iron-sulfur protein Fe_2S_2. Carbon monoxide (CO) inhibits the indicated step.

cytochromes P450 determines the duration of their action. Benzpyrene, aminopyrine, aniline, morphine, and benzphetamine are hydroxylated, increasing their solubility and aiding their excretion. Many drugs such as phenobarbital have the ability to induce the synthesis of cytochromes P450.

Mitochondrial cytochrome P450 systems are found in steroidogenic tissues such as adrenal cortex, testis, ovary, and placenta and are concerned with the biosynthesis of steroid hormones from cholesterol (hydroxylation at C_{22} and C_{20} in side chain cleavage and at the 11β and 18 positions). In addition, renal systems catalyzing 1α- and 24-hydroxylations of 25-hydroxycholecalciferol in vitamin D metabolism—and cholesterol 7α-hydroxylase and sterol 27-hydroxylase involved in bile acid biosynthesis from cholesterol in the liver (see Chapters 26, 41)—are P450 enzymes.

SUPEROXIDE DISMUTASE PROTECTS AEROBIC ORGANISMS AGAINST OXYGEN TOXICITY

Transfer of a single electron to O_2 generates the potentially damaging **superoxide anion-free radical** (O_2^-), which gives rise to free-radical chain reactions (see Chapter 21), amplifying its destructive effects. The ease with which superoxide can be formed from oxygen in tissues and the occurrence of **superoxide dismutase (SOD)**, the enzyme responsible for its removal in all aerobic organisms (although not in obligate anaerobes), indicate that the potential toxicity of oxygen is due to its conversion to superoxide.

Superoxide is formed when reduced flavins—present, for example, in xanthine oxidase—are reoxidized univalently by molecular oxygen:

$$Enz - Flavin - H_2 + O_2 \rightarrow Enz - Flavin - H + O_2^- + H^+$$

Superoxide can reduce oxidized cytochrome c

$$O_2^- + Cytc(Fe^{3+}) \rightarrow O_2 + Cytc(Fe^{2+})$$

or be removed by SOD, which catalyzes the conversion of superoxide to molecular oxygen and hydrogen peroxide.

In this reaction, superoxide acts as both oxidant and reductant. Thus, SOD protects aerobic organisms against the potential deleterious effects of superoxide. The enzyme occurs in all major aerobic tissues in the mitochondria and the cytosol. Although exposure of animals to an atmosphere of 100% oxygen causes an adaptive increase in SOD, particularly in the lungs, prolonged exposure leads to lung damage and death. Antioxidants, for example, α-tocopherol (vitamin E), act as scavengers of free radicals and reduce the toxicity of oxygen (see Chapter 44).

SUMMARY

- In biologic systems, as in chemical systems, oxidation (loss of electrons) is always accompanied by reduction of an electron acceptor.

- Oxidoreductases have a variety of functions in metabolism; oxidases and dehydrogenases play major roles in respiration; hydroperoxidases protect the body against damage by free radicals; and oxygenases mediate the hydroxylation of drugs and steroids.

- Tissues are protected from oxygen toxicity caused by the superoxide free radical by the specific enzyme superoxide dismutase.

REFERENCES

Nelson DL, Cox MM: *Lehninger Principles of Biochemistry*, 7th ed. Macmillan Education, 2017.

Nicholls DG, Ferguson SJ: *Bioenergetics*, 4th ed. Academic Press, 2013.

The Respiratory Chain & Oxidative Phosphorylation

Kathleen M. Botham, PhD, DSc, & Peter A. Mayes, PhD, DSc

O B J E C T I V E S

After studying this chapter, you should be able to:

- Describe the double membrane structure of mitochondria and indicate the location of various enzymes.

- Appreciate that energy from the oxidation of fuel substrates (fats, carbohydrates, amino acids) is almost all generated in mitochondria via a process termed electron transport in which electrons pass through a series of complexes (the respiratory chain) until they are finally reacted with oxygen to form water.

- Describe the four protein complexes involved in the transfer of electrons through the respiratory chain and explain the roles of flavoproteins, iron-sulfur proteins, and coenzyme Q.

- Explain how coenzyme Q accepts electrons from NADH via Complex I and from $FADH_2$ via Complex II.

- Indicate how electrons are passed from reduced coenzyme Q to cytochrome *c* via Complex III in the Q cycle.

- Explain the process by which reduced cytochrome *c* is oxidized and oxygen is reduced to water via Complex IV.

- Describe how electron transport generates a proton gradient across the inner mitochondrial membrane, leading to the buildup of a proton motive force that generates ATP by the process of oxidative phosphorylation.

- Describe the structure of the ATP synthase enzyme and explain how it works as a rotary motor to produce ATP from ADP and Pi.

- Explain that oxidation of reducing equivalents via the respiratory chain and oxidative phosphorylation are tightly coupled in most circumstances, so that one cannot proceed unless the other is functioning.

- Indicate examples of common poisons that block respiration or oxidative phosphorylation and identify their site of action.

- Explain, with examples, how uncouplers may act as poisons by dissociating oxidation via the respiratory chain from oxidative phosphorylation, but may also have a physiologic role in generating body heat.

- Explain the role of exchange transporters present in the inner mitochondrial membrane in allowing ions and metabolites to pass through while preserving electrochemical and osmotic equilibrium.

BIOMEDICAL IMPORTANCE

Aerobic organisms are able to capture a far greater proportion of the available free energy of respiratory substrates than anaerobic organisms. Most of this takes place inside **mitochondria**, which have been termed the "powerhouses" of the cell. Respiration is coupled to the generation of the high-energy intermediate, ATP (see Chapter 11), by **oxidative phosphorylation**. A number of drugs (eg, **amobarbital**) and poisons (eg, **cyanide, carbon monoxide**) inhibit oxidative phosphorylation, usually with fatal consequences. Several inherited defects of mitochondria involving components of the respiratory chain and oxidative phosphorylation have been reported. Patients present with **myopathy** and **encephalopathy** and often have **lactic acidosis**.

SPECIFIC ENZYMES ARE ASSOCIATED WITH COMPARTMENTS SEPARATED BY THE MITOCHONDRIAL MEMBRANES

The mitochondrial **matrix** (the internal compartment) is enclosed by a **double membrane**. The **outer membrane** is permeable to most metabolites and the **inner membrane** is selectively permeable (**Figure 13–1**). The outer membrane is characterized by the presence of various enzymes, including **acyl-CoA synthetase** (see Chapter 22) and **glycerol phosphate acyltransferase** (see Chapter 24). Other enzymes, including **adenylyl kinase** (see Chapter 11) and **creatine kinase** (see Chapter 51) are found in **the intermembrane space**. The phospholipid **cardiolipin** is concentrated in the inner membrane together with the

enzymes of **the respiratory chain**, **ATP synthase**, and various **membrane transporters**.

THE RESPIRATORY CHAIN OXIDIZES REDUCING EQUIVALENTS & ACTS AS A PROTON PUMP

Most of the energy liberated during the oxidation of carbohydrate, fatty acids, and amino acids is made available within mitochondria as reducing equivalents (—H or electrons) (**Figure 13–2**). The enzymes of the citric acid cycle and β-oxidation (see Chapters 22 and 16), **the respiratory chain complexes**, and the machinery for **oxidative phosphorylation** are all found in mitochondria. The respiratory chain collects and transports reducing equivalents, directing them to their final reaction with oxygen to form water, and oxidative phosphorylation is the process by which the liberated free energy is trapped as **high-energy phosphate (ATP)**.

Components of the Respiratory Chain Are Contained in Four Large Protein Complexes Embedded in the Inner Mitochondrial Membrane

Electrons flow through the respiratory chain through a redox span of 1.1 V from $NAD^+/NADH$ to $O_2/2H_2O$ (see Table 12–1), passing through three large protein complexes: **NADH-Q oxidoreductase (Complex I)**, where electrons are transferred from NADH to **coenzyme Q (Q)** (also called **ubiquinone**); **Q-cytochrome c oxidoreductase (Complex III)**, which passes the electrons on to **cytochrome c; and cytochrome c oxidase (Complex IV)**, which completes the chain, passing the electrons to O_2 and causing it to be reduced to H_2O (**Figure 13–3**). Some substrates with more positive redox potentials than $NAD^+/NADH$ (eg, succinate) pass electrons to Q via a fourth complex, **succinate-Q reductase (Complex II)**, rather than Complex I. The four complexes are embedded in the inner mitochondrial membrane, but Q and cytochrome c are mobile. Q diffuses rapidly within the membrane, while cytochrome c is a soluble protein.

Flavoproteins & Iron-Sulfur Proteins (Fe-S) Are Components of the Respiratory Chain Complexes

Flavoproteins (see Chapter 12) are important components of Complexes I and II. The oxidized flavin nucleotide (flavin mononucleotide [FMN] or flavin adenine dinucleotide [FAD]) can be reduced in reactions involving the transfer of two electrons (to form $FMNH_2$ or $FADH_2$), but they can also accept one electron to form the semiquinone (see Figure 12–2). **Iron-sulfur proteins (nonheme iron proteins, Fe-S)** are found in Complexes I, II, and III. These may contain one, two, or four Fe atoms linked to inorganic sulfur atoms and/or via cysteine-SH groups to the protein (**Figure 13–4**). The Fe-S take part in

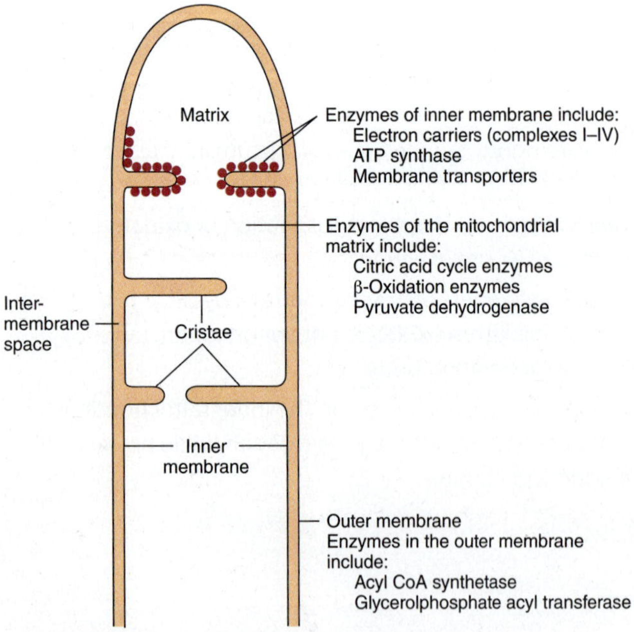

FIGURE 13–1 **Structure of the mitochondrial membranes.** Note that the inner membrane contains many folds or cristae.

Labels in figure:
- Matrix
- Enzymes of inner membrane include: Electron carriers (complexes I–IV) ATP synthase Membrane transporters
- Enzymes of the mitochondrial matrix include: Citric acid cycle enzymes β-Oxidation enzymes Pyruvate dehydrogenase
- Inter-membrane space
- Cristae
- Inner membrane
- Outer membrane Enzymes in the outer membrane include: Acyl CoA synthetase Glycerolphosphate acyl transferase

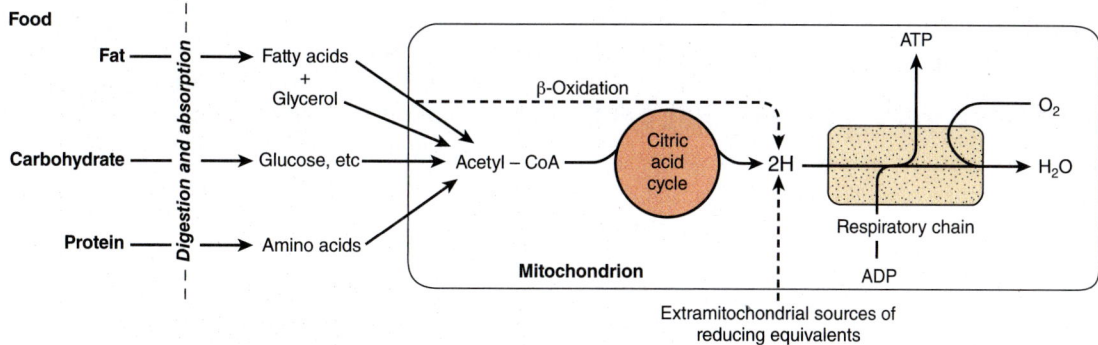

FIGURE 13–2 **Role of the respiratory chain of mitochondria in the conversion of food energy to ATP.** Oxidation of the major foodstuffs leads to the generation of reducing equivalents (2H) that are collected by the respiratory chain for oxidation and coupled generation of ATP.

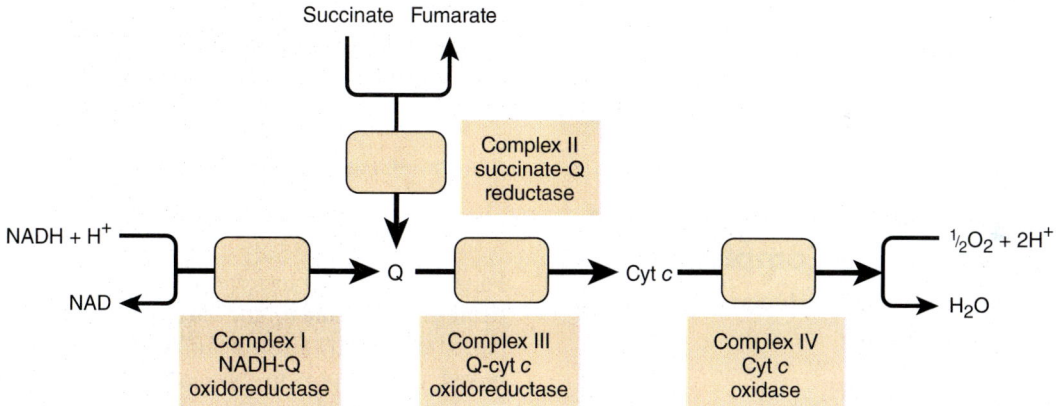

FIGURE 13–3 **Overview of electron flow through the respiratory chain.** (cyt, cytochrome; Q, coenzyme Q or ubiquinone.)

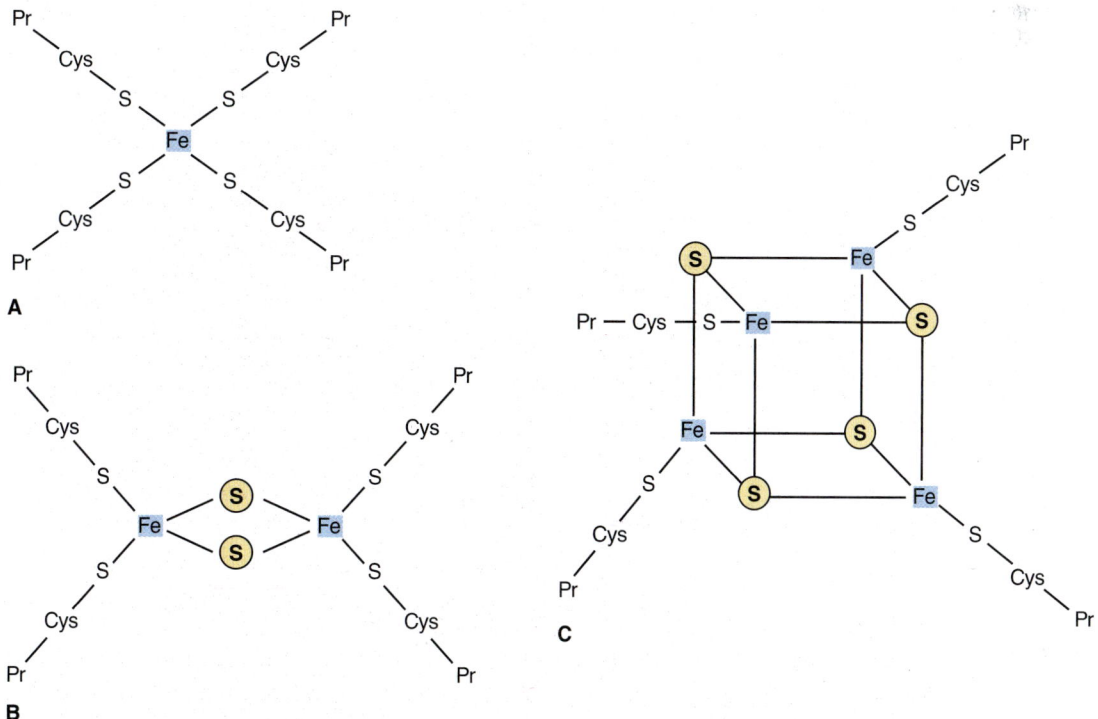

FIGURE 13–4 **Iron-sulfur proteins (Fe-S).** (**A**) The simplest Fe-S with one Fe bound by four cysteines. (**B**) 2Fe-2S center. (**C**) 4Fe-4S center. (Cys, cysteine; Pr, apoprotein; Ⓢ, inorganic sulfur.)

single electron transfer reactions in which one Fe atom undergoes oxidoreduction between Fe^{2+} and Fe^{3+}.

Q Accepts Electrons via Complexes I & II

NADH-Q oxidoreductase or Complex I is a large L-shaped multisubunit protein that catalyzes electron transfer from NADH to Q, and during the process four H^+ are transferred across the membrane into the intermembrane space:

$$NADH + Q + 5H^+_{matrix} \rightarrow NAD + QH_2 + 4H^+_{intermembrance\ space}$$

Electrons are transferred from NADH to FMN initially, then to a series of Fe-S centers, and finally to Q (**Figure 13–5**). In Complex II (succinate-Q reductase), $FADH_2$ is formed during the conversion of succinate to fumarate in the citric acid cycle (see Figure 16–3) and electrons are then passed via several Fe-S centers to Q (see Figure 13–5). Glycerol-3-phosphate (generated in the breakdown of triacylglycerols or from glycolysis; see Figure 17–2) and acyl-CoA also pass electrons to Q via different pathways involving flavoproteins (see Figure 13–5).

The Q Cycle Couples Electron Transfer to Proton Transport in Complex III

Electrons are passed from QH_2 to cytochrome c via Complex III (Q-cytochrome c oxidoreductase):

$$QH_2 + 2Cyt\,c_{oxidized} + 2H^+_{matrix} \rightarrow$$
$$Q + 2Cyt\,c_{reduced} + 4H^+_{intermembrane\ space}$$

The process is believed to involve **cytochromes c_1, b_L**, and **b_H** and **a Rieske Fe-S** (an unusual Fe-S in which one of the Fe atoms is linked to two histidine residues rather than two cysteine residues) (see Figure 13–4) and is known as the **Q cycle** (**Figure 13–6**). Q may exist in three forms: the oxidized quinone, the reduced quinol, or the semiquinone (see Figure 13–6). The semiquinone is formed transiently during the cycle, one turn of which results in the oxidation of $2QH_2$ to Q, releasing $4H^+$ into the intermembrane space, and the reduction of one Q to QH_2, causing $2H^+$ to be taken up from the matrix (see Figure 13–6). Note that while Q carries two electrons, the cytochromes carry only one, thus the oxidation of one QH_2 is coupled to the reduction of two molecules of cytochrome c via the Q cycle.

Molecular Oxygen Is Reduced to Water via Complex IV

Reduced cytochrome c is oxidized by Complex IV (cytochrome c oxidase), with the concomitant reduction of O_2 to two molecules of water:

$$4Cyt\,c_{reduced} + O_2 + 8H^+_{matrix} \rightarrow$$
$$4Cyt\,c_{oxidized} + 2H_2O + 4H^+_{intermembrane\ space}$$

Four electrons are transferred from cytochrome c to O_2 via **two heme groups, a and a_3**, and **Cu** (see Figure 13–5). Electrons are passed initially to a Cu center (Cu_A), which contains 2Cu atoms linked to two protein cysteine-SH groups (resembling an Fe-S), then in sequence to heme a, heme a_3, a second Cu center, Cu_B, which is linked to heme a_3, and finally to O_2. Eight H^+ are removed from the matrix, of which four are used to form two water molecules and four are pumped into the intermembrane space. Thus, for every pair of electrons passing down

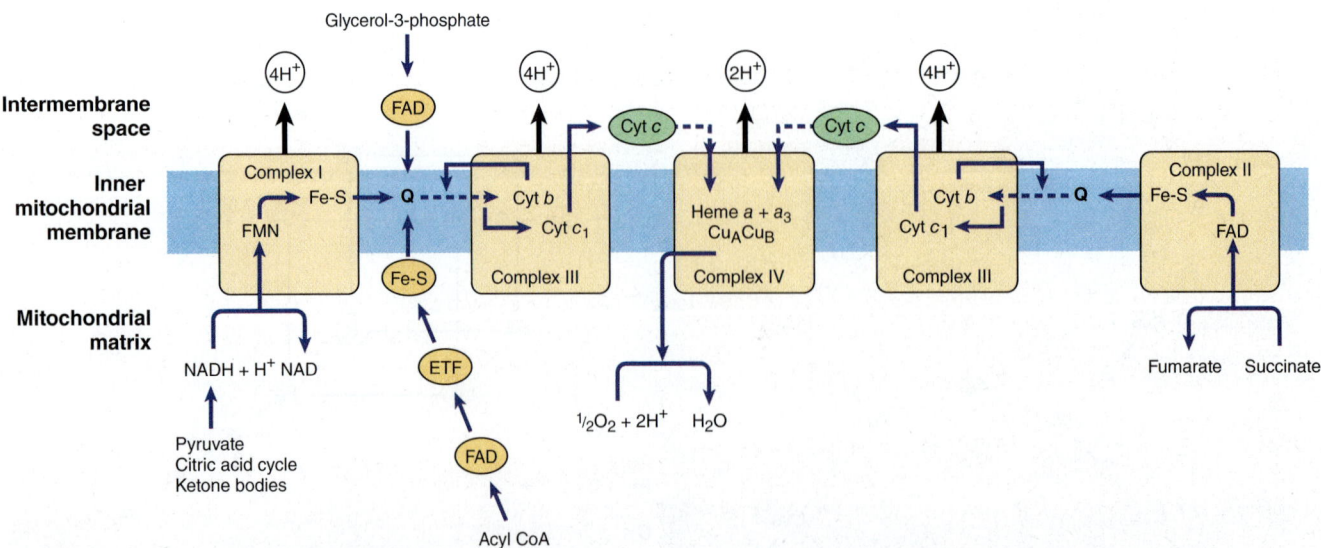

FIGURE 13–5 **Flow of electrons through the respiratory chain complexes, showing the entry points for reducing equivalents from important substrates.** Q and cyt c are mobile components of the system as indicated by the dotted arrows. The flow through Complex III (the Q cycle) is shown in more detail in Figure 13–6. (cyt, cytochrome; ETF, electron transferring flavoprotein; Fe-S, iron-sulfur protein; Q, coenzyme Q or ubiquinone.)

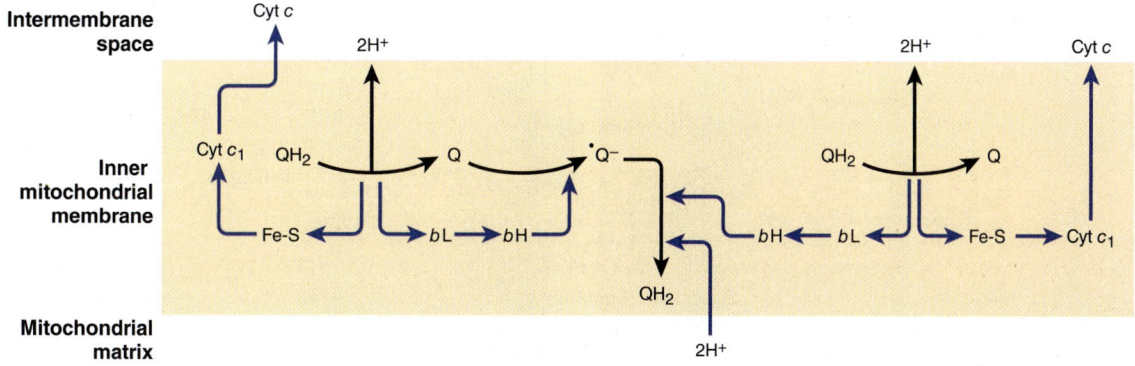

QH$_2$: Reduced (quinol) form (QH$_2$)   **Q:** Fully oxidized (quinone) form   **·Q$^-$:** Semiquinone (free radical) form

FIGURE 13–6 **The Q cycle.** During the oxidation of QH$_2$ to Q, one electron is donated to cyt c via a Rieske Fe-S and cyt c_1 and the second to a Q to form the semiquinone via cyt b_L and cyt $b_{H'}$ with 2H$^+$ being released into the intermembrane space. A similar process then occurs with a second QH$_2$, but in this case the second electron is donated to the semiquinone, reducing it to QH$_2$, and 2H$^+$ are taken up from the matrix. (cyt, cytochrome; Fe-S, iron-sulfur protein; Q, coenzyme Q or ubiquinone.)

the chain from NADH or FADH$_2$, 2H$^+$ are pumped across the membrane by Complex IV. The O$_2$ remains tightly bound to Complex IV until it is fully reduced, and this minimizes the release of potentially damaging intermediates such as superoxide anions or peroxide which are formed when O$_2$ accepts one or two electrons, respectively (see Chapter 12).

ELECTRON TRANSPORT VIA THE RESPIRATORY CHAIN CREATES A PROTON GRADIENT WHICH DRIVES THE SYNTHESIS OF ATP

The flow of electrons through the respiratory chain generates ATP by the process of **oxidative phosphorylation**. **The chemiosmotic theory**, proposed by Peter Mitchell in 1961, postulates that the two processes are coupled by a proton gradient across the inner mitochondrial membrane so that **the proton motive force** caused by the electrochemical potential difference (negative on the matrix side) drives the mechanism of ATP synthesis. As we have seen, Complexes I, III, and IV act as **proton pumps**, moving H$^+$ from the mitochondrial matrix to the intermembrane space. Since the inner mitochondrial membrane is impermeable to ions in general and particularly to protons, these accumulate in the intermembrane space, creating the proton motive force predicted by the chemiosmotic theory.

A Membrane-Located ATP Synthase Functions as a Rotary Motor to Form ATP

The proton motive force drives a membrane-located ATP synthase that forms ATP in the presence of P$_i$ + ADP. ATP synthase is embedded in the inner membrane, together with the respiratory chain complexes (**Figure 13–7**). Several subunits of the protein form a ball-like shape arranged around an axis known as F$_1$, which projects into the matrix and contains the phosphorylation mechanism (**Figure 13–8**). F$_1$ is attached to a membrane protein complex known as F$_0$, which also consists of several protein subunits. F$_0$ spans the membrane and forms a proton channel. As protons flow through F$_0$ driven by the proton gradient across the membrane it rotates, driving the production of ATP in the F$_1$ complex (see Figures 13–7 and 13–8). This is thought to occur by a **binding change mechanism** in which the conformation of the β subunits in F$_1$ is changed as the axis rotates from one that binds ATP tightly to one that releases ATP and binds ADP and P$_i$ so that the next ATP can be formed. As indicated earlier, for each NADH oxidized, Complexes I and III translocate four protons each and Complex IV translocates two.

THE RESPIRATORY CHAIN PROVIDES MOST OF THE ENERGY CAPTURED DURING CATABOLISM

ADP captures, in the form of high-energy phosphate, a significant proportion of the free energy released by catabolic processes. The resulting ATP has been called the **energy**

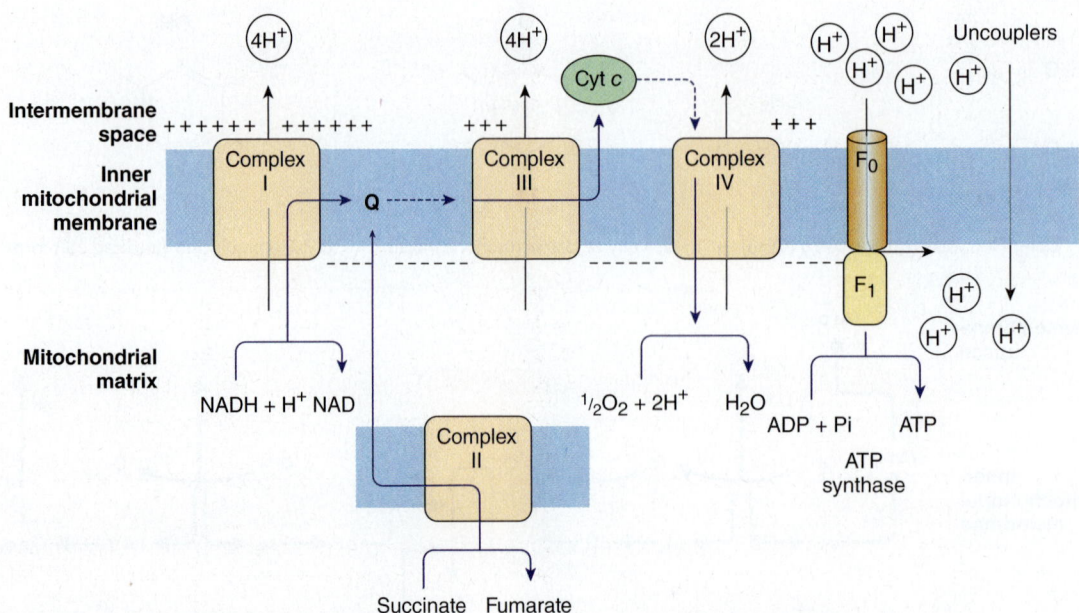

FIGURE 13–7 **The chemiosmotic theory of oxidative phosphorylation.** Complexes I, III, and IV act as proton pumps creating a proton gradient across the membrane, which is negative on the matrix side. The proton motive force generated drives the synthesis of ATP as the protons flow back into the matrix through the ATP synthase enzyme (Figure 13–8). Uncouplers increase the permeability of the membrane to ions, collapsing the proton gradient by allowing the H^+ to pass across without going through the ATP synthase, and thus uncouple electron flow through the respiratory complexes from ATP synthesis. (cyt, cytochrome; Q, coenzyme Q or ubiquinone.)

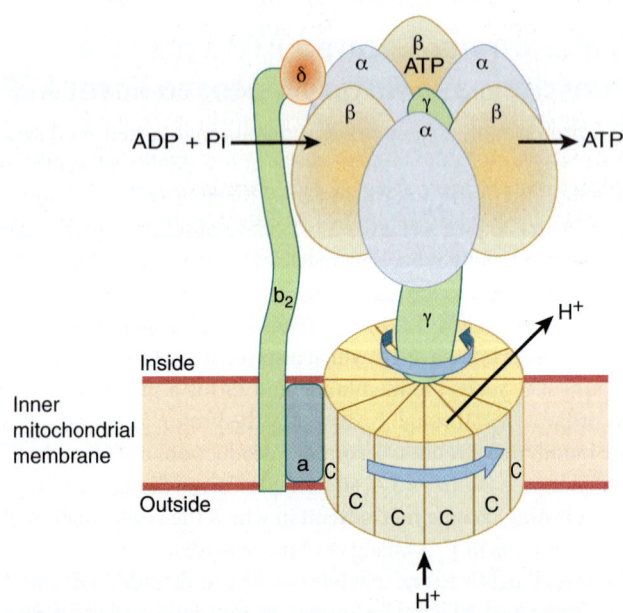

FIGURE 13–8 **Mechanism of ATP production by ATP synthase.** The enzyme complex consists of an F_0 subcomplex which is a disk of "C" protein subunits. Attached is a γ subunit in the form of a "bent axle." Protons passing through the disk of "C" units cause it and the attached γ subunit to rotate. The γ subunit fits inside the F_1 subcomplex of three α and three β subunits, which are fixed to the membrane and do not rotate. ADP and P_i are taken up sequentially by the β subunits to form ATP, which is expelled as the rotating γ subunit squeezes each β subunit in turn and changes its conformation. Thus, three ATP molecules are generated per revolution. For clarity, not all the subunits that have been identified are shown—eg, the "axle" also contains an ε subunit.

"currency" of the cell because it passes on this free energy to drive those processes requiring energy (see Figure 11–5).

There is a net direct capture of two high-energy phosphate groups in the glycolytic reactions (see Table 17–1). Two more high-energy phosphates per mole of glucose are captured in the citric acid cycle during the conversion of succinyl-CoA to succinate (see Chapter 16). All of these phosphorylations occur at the **substrate level**. For each mol of substrate oxidized via Complexes I, III, and IV in the respiratory chain (ie, via NADH), 2.5 mol of ATP are formed per 0.5 mol of O_2 consumed, that is, the P:O ratio = 2.5 (see Figure 13–7). On the other hand, when 1 mol of substrate (eg, succinate or 3-phophoglycerate) is oxidized via Complexes II, III, and IV, only 1.5 mol of ATP are formed, that is, P:O = 1.5. These reactions are known as **oxidative phosphorylation at the respiratory chain level**. Taking these values into account, it can be estimated that nearly 90% of the high-energy phosphates produced from the complete oxidation of 1 mol glucose is obtained via oxidative phosphorylation coupled to the respiratory chain (see Table 17–1).

Respiratory Control Ensures a Constant Supply of ATP

The rate of respiration of mitochondria can be controlled by the availability of ADP. This is because oxidation and phosphorylation are **tightly coupled**; that is, oxidation cannot proceed via the respiratory chain without concomitant phosphorylation of ADP. **Table 13–1** shows the five conditions controlling the rate of respiration in mitochondria. Most cells

TABLE 13–1 States of Respiratory Control

Conditions Limiting the Rate of Respiration	
State 1	Availability of ADP and substrate
State 2	Availability of substrate only
State 3	The capacity of the respiratory chain itself, when all substrates and components are present in saturating amounts
State 4	Availability of ADP only
State 5	Availability of oxygen only

in the resting state are in **state 4**, and respiration is controlled by the availability of ADP. When work is performed, ATP is converted to ADP, allowing more respiration to occur, which in turn replenishes the store of ATP. Under certain conditions, the concentration of inorganic phosphate can also affect the rate of functioning of the respiratory chain. As respiration increases (as in exercise), the cell approaches **states 3 or 5** when either the capacity of the respiratory chain becomes saturated or the PO_2 decreases below the K_m for heme a_3. There is also the possibility that the ADP/ATP transporter, which facilitates entry of cytosolic ADP into and ATP out of the mitochondrion, becomes rate limiting.

Thus, the manner in which biologic oxidative processes allow the free energy resulting from the oxidation of foodstuffs to become available and to be captured is stepwise, efficient, and controlled—rather than explosive, inefficient, and uncontrolled, as in many nonbiologic processes. The remaining free energy that is not captured as high-energy phosphate is liberated as **heat**. This need not be considered "wasted" since it ensures that the respiratory system as a whole is sufficiently exergonic to be removed from equilibrium,

allowing continuous unidirectional flow and constant provision of ATP. It also contributes to maintenance of body temperature.

MANY POISONS INHIBIT THE RESPIRATORY CHAIN

Much information about the respiratory chain has been obtained by the use of inhibitors, and, conversely, this has provided knowledge about the mechanism of action of several poisons (**Figure 13–9**). They may be classified as inhibitors of the respiratory chain, inhibitors of oxidative phosphorylation, or uncouplers of oxidative phosphorylation.

Barbiturates such as amobarbital inhibit electron transport via Complex I by blocking the transfer from Fe-S to Q. At sufficient dosage, they are fatal. **Antimycin A** and **dimercaprol** inhibit the respiratory chain at Complex III. The classic poisons **H₂S**, **carbon monoxide**, and **cyanide** inhibit Complex IV and can therefore totally arrest respiration. **Malonate** is a competitive inhibitor of Complex II.

Atractyloside inhibits oxidative phosphorylation by inhibiting the transporter of ADP into and ATP out of the mitochondrion (**Figure 13–10**). The antibiotic **oligomycin** completely blocks oxidation and phosphorylation by blocking the flow of protons through ATP synthase (see Figure 13–8).

Uncouplers dissociate oxidation in the respiratory chain from phosphorylation (see Figure 13–7). These compounds are toxic, causing respiration to become uncontrolled, since the rate is no longer limited by the concentration of ADP or P_i. The uncoupler that has been used most frequently in studies of the respiratory chain is **2,4-dinitrophenol**, but other compounds act in a similar manner. **Thermogenin (or uncoupling protein 1 [UCP1])** is a physiologic uncoupler found in brown adipose tissue that functions to generate body heat,

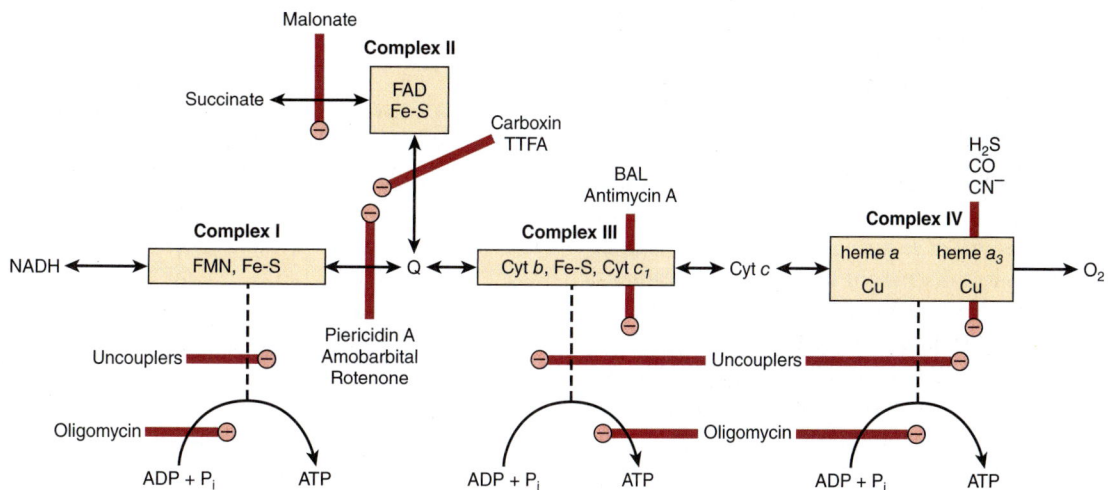

FIGURE 13–9 Sites of inhibition (⊖) of the respiratory chain by specific drugs, chemicals, and antibiotics. (BAL, dimercaprol; TTFA, an Fe-chelating agent. Other abbreviations as in Figure 13–5.)

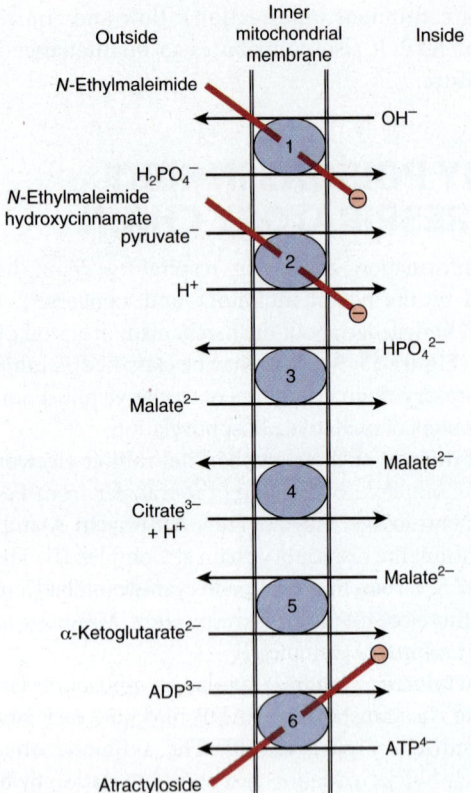

FIGURE 13–10 **Transporter systems in the inner mito-**
chondrial membrane. ① Phosphate transporter, ② pyruvate
symport, ③ dicarboxylate transporter, ④ tricarboxylate transporter,
⑤ α-ketoglutarate transporter, ⑥ adenine nucleotide transporter.
N-Ethylmaleimide, hydroxycinnamate, and atractyloside inhibit (⊖)
the indicated systems. Also present (but not shown) are transporter
systems for glutamate/aspartate (Figure 13–13), glutamine, ornithine,
neutral amino acids, and carnitine (see Figure 22–1).

particularly for the newborn and during hibernation in animals (see Chapter 25).

THE CHEMIOSMOTIC THEORY CAN ACCOUNT FOR RESPIRATORY CONTROL AND THE ACTION OF UNCOUPLERS

The electrochemical potential difference across the membrane, once established as a result of proton translocation,
inhibits further transport of reducing equivalents through the
respiratory chain unless discharged by back-translocation of
protons across the membrane through the ATP synthase. This
in turn depends on availability of ADP and P_i.

Uncouplers such as dinitrophenol are amphipathic (see
Chapter 21) and increase the permeability of the lipoid inner
mitochondrial membrane to protons, while physiologic uncouplers such as UCP1 are membrane-spanning proteins which have
a similar effect, thus the electrochemical potential is reduced and
ATP synthase is short-circuited (see Figure 13–7). In these circumstances, oxidation can proceed without phosphorylation.

THE SELECTIVE PERMEABILITY OF THE INNER MITOCHONDRIAL MEMBRANE NECESSITATES EXCHANGE TRANSPORTERS

Exchange diffusion systems involving transporter proteins that
span the membrane are present in the membrane for exchange of
anions against OH^- ions and cations against H^+ ions. Such systems are necessary for uptake and output of ionized metabolites
while preserving electrical and osmotic equilibrium. The inner
mitochondrial membrane is freely permeable to **uncharged
small molecules**, such as oxygen, water, CO_2, NH_3, and to **mono-
carboxylic acids**, such as 3-hydroxybutyric, acetoacetic, and
acetic, especially in their undissociated, more lipid soluble form.
Long-chain fatty acids are transported into mitochondria via the
carnitine system (see Figure 22–1), and there is also a special carrier for **pyruvate** involving a symport that utilizes the H^+ gradient from outside to inside the mitochondrion (see Figure 13–10).
However, **dicarboxylate and tricarboxylate anions (eg, malate,
citrate) and amino acids** require specific transporter or carrier
systems to facilitate their passage across the membrane.

The transport of di- and tricarboxylate anions is closely
linked to that of inorganic phosphate, which penetrates readily as the $H_2PO_4^-$ ion in exchange for OH^-. The net uptake
of malate by the dicarboxylate transporter requires inorganic
phosphate for exchange in the opposite direction. The net
uptake of citrate, isocitrate, or *cis*-aconitate by the tricarboxylate transporter requires malate in exchange. α-Ketoglutarate
transport also requires an exchange with malate. The adenine
nucleotide transporter allows the exchange of ATP and ADP,
but not AMP. It is vital for ATP to exit from mitochondria to the
sites of extramitochondrial utilization and for the return of ADP
for ATP production within the mitochondrion (**Figure 13–11**).
Since in this translocation four negative charges are removed
from the matrix for every three taken in, the electrochemical gradient across the membrane (the proton motive force)

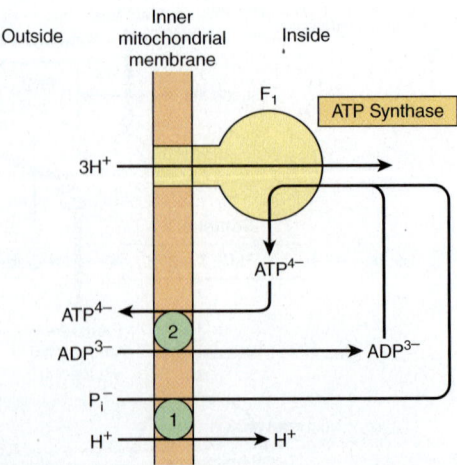

FIGURE 13–11 **Combination of phosphate transporter ① with**
the adenine nucleotide transporter ② in ATP synthesis. The H^+/P_i symport shown is equivalent to the P_i/OH^- antiport shown in Figure 13–10.

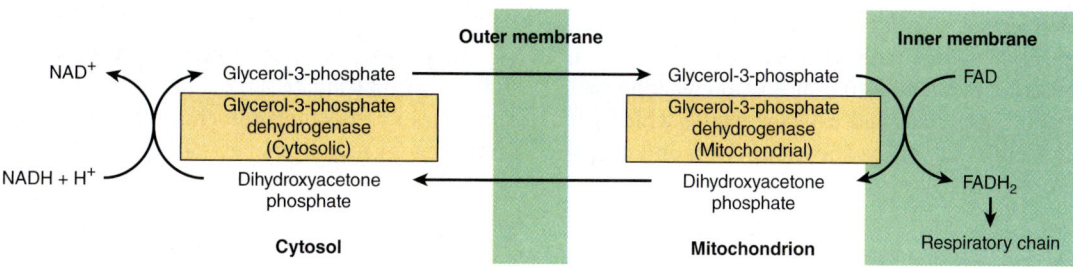

FIGURE 13–12 Glycerophosphate shuttle for transfer of reducing equivalents from the cytosol into the mitochondrion.

favors the export of ATP. Na^+ can be exchanged for H^+, driven by the proton gradient. It is believed that active uptake of Ca^{2+} by mitochondria occurs with a net charge transfer of 1 (Ca^+ uniport), possibly through a Ca^{2+}/H^+ antiport. Calcium release from mitochondria is facilitated by exchange with Na^+.

Ionophores Permit Specific Cations to Penetrate Membranes

Ionophores are lipophilic molecules that complex specific cations and facilitate their transport through biologic membranes, for example, **valinomycin** (K^+). The classic uncouplers such as dinitrophenol are, in fact, proton ionophores.

Proton-Translocating Transhydrogenase Is a Source of Intramitochondrial NADPH

Proton-translocating transhydrogenase (also called NAD(P) transhydrogenase), a protein in the inner mitochondrial membrane, couples the passage of protons down the electrochemical gradient from outside to inside the mitochondrion with the transfer of H from intramitochondrial NADH to NADP forming NADPH for intramitochondrial enzymes such as

glutamate dehydrogenase and hydroxylases involved in steroid synthesis.

Oxidation of Extramitochondrial NADH Is Mediated by Substrate Shuttles

NADH cannot penetrate the mitochondrial membrane, but it is produced continuously in the cytosol by glyceraldehyde-3-phosphate dehydrogenase, an enzyme in the glycolysis sequence (see Figure 17–2). However, under aerobic conditions, extramitochondrial NADH does not accumulate and is presumed to be oxidized by the respiratory chain in mitochondria. The transfer of reducing equivalents through the mitochondrial membrane requires **substrate pairs**, linked by suitable dehydrogenases on each side of the mitochondrial membrane. The mechanism of transfer using the **glycerophosphate shuttle** is shown in **Figure 13–12**. Since the mitochondrial enzyme is linked to the respiratory chain via a flavoprotein rather than NAD, only 1.5 mol rather than 2.5 mol of ATP are formed per atom of oxygen consumed. Although this shuttle is present in some tissues (eg, brain, white muscle), in others (eg, heart muscle) it is deficient. It is therefore believed that the **malate shuttle** system (**Figure 13–13**)

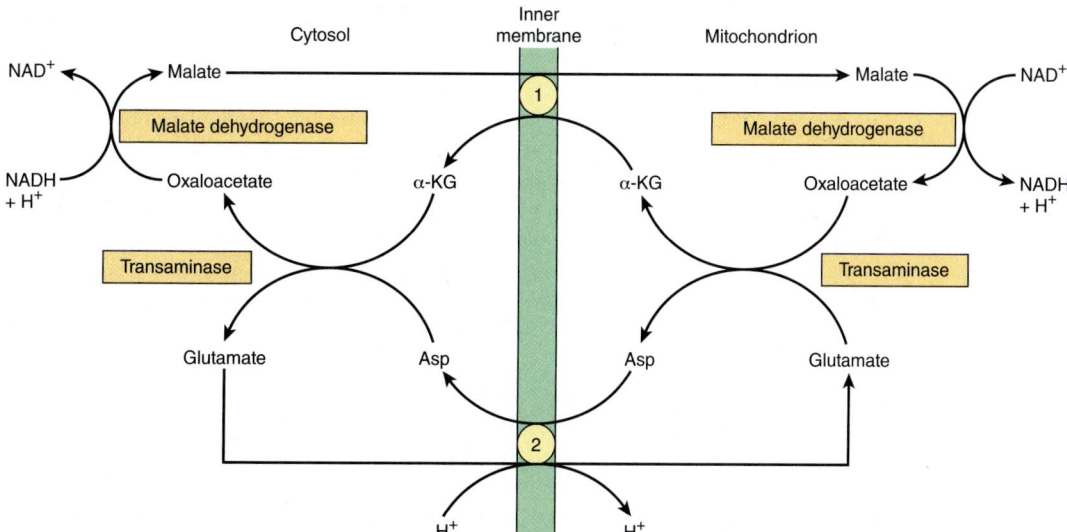

FIGURE 13–13 Malate shuttle for transfer of reducing equivalents from the cytosol into the mitochondrion. ① α-Ketoglutarate transporter and ② glutamate/aspartate transporter (note the proton symport with glutamate). α-KG, α-ketoglutarate. NAD^+ is generated from NADH outside the mitochondrion via the formation of malate from oxaloacetate. Malate crossed the inner membrane in exchange for α-KG and is converted back to oxaloacetate, releasing NADH inside the matrix. (Asp) and α-KG are then produced from oxaloacetate and glutamate by a transaminase enzyme and are transported into the cytosol, where oxaloacetate is reformed by a second transaminase and may be used to generate another NAD^+ from NADH.

is of more universal utility. The complexity of this system is due to the impermeability of the mitochondrial membrane to oxaloacetate. This is overcome by a transamination reaction with glutamate forming aspartate and α-ketoglutarate which can then cross the membrane via specific transporters and reform oxaloacetate in the cytosol.

The Creatine Phosphate Shuttle Facilitates Transport of High-Energy Phosphate From Mitochondria

The **creatine phosphate shuttle** (**Figure 13–14**) augments the functions of **creatine phosphate** as an energy buffer by acting as a dynamic system for transfer of high-energy phosphate from mitochondria in active tissues such as heart and skeletal muscle. An isoenzyme of **creatine kinase** (CK_m) is found in the mitochondrial intermembrane space, catalyzing

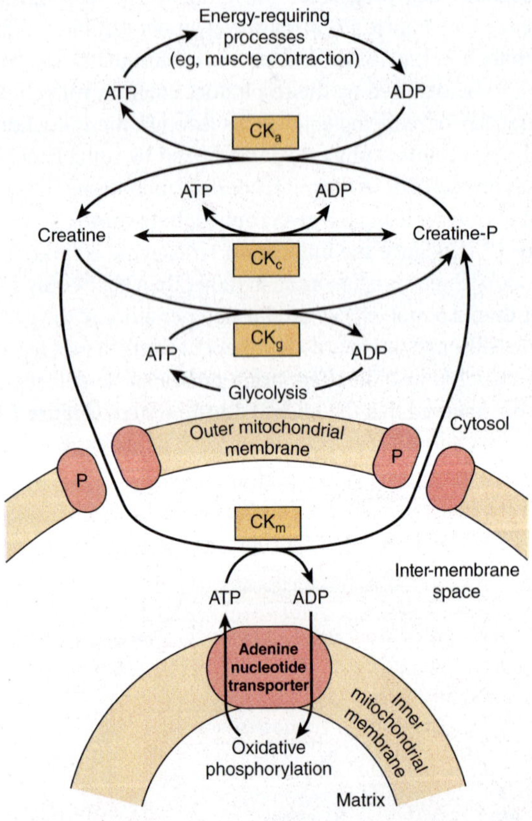

FIGURE 13–14 **The creatine phosphate shuttle of heart and skeletal muscle.** The shuttle allows rapid transport of high-energy phosphate from the mitochondrial matrix into the cytosol. (CK_a, creatine kinase concerned with large requirements for ATP, eg, muscular contraction; CK_c, creatine kinase for maintaining equilibrium between creatine and creatine phosphate and ATP/ADP; CK_g, creatine kinase coupling glycolysis to creatine phosphate synthesis; CK_m, mitochondrial creatine kinase mediating creatine phosphate production from ATP formed in oxidative phosphorylation; P, pore protein in outer mitochondrial membrane.)

the transfer of high-energy phosphate to creatine from ATP emerging from the adenine nucleotide transporter. In turn, the creatine phosphate is transported into the cytosol via protein pores in the outer mitochondrial membrane, becoming available for generation of extramitochondrial ATP.

CLINICAL ASPECTS

The condition known as **fatal infantile mitochondrial myopathy and renal dysfunction** involves severe diminution or absence of most oxidoreductases of the respiratory chain. **MELAS** (mitochondrial encephalopathy, lactic acidosis, and stroke) is an inherited condition due to NADH-Q oxidoreductase (Complex I) or cytochrome oxidase (Complex IV) deficiency. It is caused by a mutation in mitochondrial DNA and may be involved in **Alzheimer disease** and **diabetes mellitus**. A number of drugs and poisons act by inhibition of oxidative phosphorylation (see earlier).

SUMMARY

- Virtually all energy released from the oxidation of carbohydrate, fat, and protein is made available in mitochondria as reducing equivalents (—H or e⁻). These are funneled into the respiratory chain, where they are passed down a redox gradient of carriers to their final reaction with oxygen to form water.

- The redox carriers are grouped into four respiratory chain complexes in the inner mitochondrial membrane. Three of the four complexes are able to use the energy released in the redox gradient to pump protons to the outside of the membrane, creating an electrochemical potential between the matrix and the inner membrane space.

- ATP synthase spans the membrane and acts like a rotary motor using the potential energy of the proton gradient or proton motive force to synthesize ATP from ADP and P_i. In this way, oxidation is closely coupled to phosphorylation to meet the energy needs of the cell.

- Since the inner mitochondrial membrane is impermeable to protons and other ions, special exchange transporters span the membrane to allow ions such as OH⁻, ATP⁴⁻, ADP³⁻, and metabolites to pass through without discharging the electrochemical gradient across the membrane.

- Many well-known poisons such as cyanide arrest respiration by inhibition of the respiratory chain.

REFERENCES

Kocherginsky N: Acidic lipids, H(+)-ATPases, and mechanism of oxidative phosphorylation. Physico-chemical ideas 30 years after P. Mitchell's Nobel Prize award. Prog Biophys Mol Biol 2009;99:20.

Mitchell P: Keilin's respiratory chain concept and its chemiosmotic consequences. Science 1979;206:1148.

Nakamoto RK, Baylis Scanlon JA, Al-Shawi MK: The rotary mechanism of the ATP synthase. Arch Biochem Biophys 2008;476:43.

Exam Questions

Section III – Bioenergetics

1. Which one of the following statements about the free energy change (ΔG) in a biochemical reaction is CORRECT?

   A. If ΔG is negative, the reaction proceeds spontaneously with a loss of free energy.

   B. In an exergonic reaction, ΔG is positive.

   C. The standard free energy change when reactants are present in concentrations of 1.0 mol/L and the pH is 7.0 is represented as ΔG^0.

   D. In an endergonic reaction there is a loss of free energy.

   E. If a reaction is essentially irreversible, it has a high positive ΔG.

2. If the ΔG of a reaction is zero:

   A. The reaction goes virtually to completion and is essentially irreversible.

   B. The reaction is endergonic.

   C. The reaction is exergonic.

   D. The reaction proceeds only if free energy can be gained.

   E. The system is at equilibrium and no net change occurs.

3. $\Delta G^{0'}$ is defined as the standard free energy charge when:

   A. The reactants are present in concentrations of 1.0 mol/L.

   B. The reactants are present in concentrations of 1.0 mol/L at pH 7.0.

   C. The reactants are present in concentrations of 1.0 mmol/L at pH 7.0.

   D. The reactants are present in concentrations of 1.0 μmol/L.

   E. The reactants are present in concentrations of 1.0 mol/L at pH 7.4.

4. Which of the following statements about ATP is CORRECT?

   A. It contains three high-energy phosphate bonds.

   B. It is needed in the body to drive exergonic reactions.

   C. It is used as an energy store in the body.

   D. It functions in the body as a complex with Mg^{2+}.

   E. It is synthesized by ATP synthase in the presence of uncouplers such as UCP-1 (thermogenin).

5. Which one of the following enzymes uses molecular oxygen as a hydrogen acceptor?

   A. Cytochrome c oxidase

   B. Isocitrate dehydrogenase

   C. Homogentisate dioxygenase

   D. Catalase

   E. Superoxide dismutase

6. Which one of the following statement about cytochromes is INCORRECT?

   A. They are hemoproteins that take part in oxidation–reduction reactions.

   B. They contain iron which oscillates between Fe^{3+} and Fe^{2+} during the reactions they participate in.

   C. They act as electron carriers in the respiratory chain in mitochondria.

   D. They have an important role in the hydroxylation of steroids in the endoplasmic reticulum.

   E. They are all dehydrogenase enzymes.

7. Which one of the following statement about cytochromes P450 is INCORRECT?

   A. They are able to accept electrons from either NADH or NADPH.

   B. They are found only in the endoplasmic reticulum.

   C. They are monooxygenase enzymes.

   D. They play a major role in drug detoxification in the liver.

   E. In some reactions they work in conjunction with cytochrome $b5$.

8. As one molecule of NADH is oxidized via the respiratory chain:

   A. 1.5 molecules of ATP are produced in total.

   B. 1 molecule of ATP is produced as electrons pass through complex IV.

   C. 1 molecule of ATP is produced as electrons pass through complex II.

   D. 1 molecule of ATP is produced as electrons pass through complex III.

   E. 0.5 of a molecule of ATP is produced as electrons pass through complex I.

9. The number of ATP molecules produced for each molecule of $FADH_2$ oxidized via the respiratory chain is:

   A. 1

   B. 2.5

   C. 1.5

   D. 2

   E. 0.5

10. A number of compounds inhibit oxidative phosphorylation— the synthesis of ATP from ADP and inorganic phosphate linked to oxidation of substrates in mitochondria. Which of the following describes the action of oligomycin?

    A. It discharges the proton gradient across the mitochondrial inner membrane.

    B. It discharges the proton gradient across the mitochondrial outer membrane.

    C. It inhibits the electron transport chain directly by binding to one of the electron carriers in the mitochondrial inner membrane.

    D. It inhibits the transport of ADP into, and ATP out of, the mitochondrial matrix.

    E. It inhibits the transport of protons back into the mitochondrial matrix through ATP synthase.

11. A number of compounds inhibit oxidative phosphorylation— the synthesis of ATP from ADP and inorganic phosphate linked to oxidation of substrates in mitochondria. Which of the following describes the action of an uncoupler?

    A. It discharges the proton gradient across the mitochondrial inner membrane.

    B. It discharges the proton gradient across the mitochondrial outer membrane.

    C. It inhibits the electron transport chain directly by binding to one of the electron carriers in the mitochondrial inner membrane.

    D. It inhibits the transport of ADP into, and ATP out of, the mitochondrial matrix.

    E. It inhibits the transport of protons back into the mitochondrial matrix through the stalk of the primary particle.

131

12. A student takes some tablets she is offered at a disco, and without asking what they are she swallows them. A short time later she starts to hyperventilate and becomes very hot. What is the most likely action of the tablets she has taken?

    A. An inhibitor of mitochondrial ATP synthesis
    B. An inhibitor of mitochondrial electron transport
    C. An inhibitor of the transport of ADP into mitochondria to be phosphorylated
    D. An inhibitor of the transport of ATP out of mitochondria into the cytosol
    E. An uncoupler of mitochondrial electron transport and oxidative phosphorylation

13. The flow of electrons through the respiratory chain and the production of ATP are normally tightly coupled. The processes are uncoupled by which of the following?

    A. Cyanide
    B. Oligomycin
    C. Thermogenin
    D. Carbon monoxide
    E. Hydrogen sulphide

14. Which of the following statements about ATP synthase is INCORRECT?

    A. It is located in the inner mitochondrial membrane.
    B. It requires a proton motive force to form ATP in the presence of ADP and Pi.
    C. ATP is produced when part of the molecule rotates.
    D. One ATP molecule is formed for each full revolution of the molecule.
    E. The F_1 subcomplex is fixed to the membrane and does not rotate.

15. The chemiosmotic theory of Peter Mitchell proposes a mechanism for the tight coupling of electron transport via the respiratory chain to the process of oxidative phosphorylation. Which of the following options is NOT predicted by the theory?

    A. A proton gradient across the inner mitochondrial membrane generated by electron transport drives ATP synthesis.
    B. The electrochemical potential difference across the inner mitochondrial membrane caused by electron transport is positive on the matrix side.
    C. Protons are pumped across the inner mitochondrial membrane as electrons pass down the respiratory chain.
    D. An increase in the permeability of the inner mitochondrial membrane to protons uncouples the processes of electron transport and oxidative phosphorylation.
    E. ATP synthesis occurs when the electrochemical potential difference across the membrane is discharged by translocation of protons back across the inner mitochondrial membrane through an ATP synthase enzyme.

Overview of Metabolism & the Provision of Metabolic Fuels

Owen P. McGuinness, PhD

OBJECTIVES

After studying this chapter, you should be able to:

- Explain what is meant by anabolic, catabolic, and amphibolic metabolic pathways.
- Describe the movement of carbohydrates, lipids, and amino acids between organs, and metabolism within organs at the subcellular level.
- Describe the ways cells can modulate flux of metabolites through metabolic pathways.
- Describe how the movement (storage and release) of substrates is regulated in the fed and fasting states.

BIOMEDICAL IMPORTANCE

Metabolism is the term used to describe the interconversion of chemical compounds in the body. It includes the pathways taken by individual molecules in a specific cell, the interrelationships of pathways in cells and between cells in an organ as well as between organs, and the regulatory mechanisms that regulate the flow of metabolites through the pathways. Metabolic pathways fall into three categories. (1) **Anabolic pathways** are involved in the synthesis of larger and more complex compounds from smaller precursors—for example, the synthesis of protein from amino acids and the synthesis of triacylglycerol and glycogen from carbohydrates. Anabolic pathways are endothermic and thus require reducing equivalents or ATP to support the pathways. (2) **Catabolic pathways** are involved in the breakdown of larger molecules, commonly involving oxidative reactions. They are exothermic, producing reducing equivalents, and, mainly via the respiratory chain (see Chapter 13), ATP. (3) **Amphibolic pathways** occur at the

"crossroads" of metabolism. They can participate in both anabolic and catabolic functions, for example, the citric acid cycle (see Chapter 16).

Knowledge of normal metabolism is essential for an understanding of abnormalities that underlie disease. Normal metabolism requires that the metabolism in cells can not only perform their normal resting or basal metabolic functions but they can adapt to a changing environment. This includes appropriate adaptation to periods of feasting, fasting, starvation, and exercise, as well as pregnancy and lactation. Abnormal metabolism may result from nutritional deficiency, caloric excess, enzyme deficiency or inappropriate regulation, abnormal secretion of hormones, or the actions of drugs and toxins.

A 70-kg adult human being requires about 8 to 12 MJ (1920-2900 kcal) from metabolic fuels each day, depending on physical activity. Caloric requirements increase as the size of the animal increases. For any given size when animals are growing, they have a proportionally higher requirement to allow for the energy cost of growth. This energy requirement is met by our diet. For humans the caloric content of our diet is derived from carbohydrates (40-60%), lipids (mainly

This was adapted from chapter in 30th edition by David A. Bender, PhD, & Peter A. Mayes, PhD, DSc

triacylglycerol, 30-40%), and protein (10-15%), as well as alcohol. The mix of carbohydrate, lipid, and protein being oxidized varies, depending on the actual composition of the diet (ie, you burn what you eat), whether the subject is in the fed or fasting state, and on the duration and intensity of physical work.

There is a constant requirement for metabolic fuels throughout the day. The resting or basal metabolic rate accounts for ~60% of daily energy expenditure in humans. Physical activity accounts for a variable amount to total daily energy expenditure as exercise can increase the metabolic rate by 40 to 50% over the basal or resting metabolic rate. Metabolism is not constant in a given individual in a 24-hour period; we cycle between anabolism and catabolism. In a weight stable person these processes on average are equal; we are in net zero energy balance. Most people consume their daily intake of metabolic fuels in two or three meals, so there is net anabolism where we store reserves of carbohydrate (glycogen in liver and muscle), lipid (triacylglycerol in adipose tissue), and labile protein stores. During the period following a meal we mobilize and catabolize those stores when there is no intake of food.

If the intake of calories exceeds energy expenditure, the surplus is stored, either as glycogen (liver and muscle) or as triacylglycerol in adipose tissue. If this persists, it will lead to the development of **obesity** and its associated health hazards. By contrast, if the intake of metabolic fuels is consistently lower than energy expenditure (long-term energy deficit), the limited reserves of fat and carbohydrate get exhausted forcing the organism to oxidize amino acids. The amino acids arising from protein turnover that are normally reused for replacement protein synthesis are mobilized. This net protein breakdown will lead to protein wasting, **emaciation**, and, eventually, death (see Chapter 43).

OVERVIEW OF SUBSTRATE METABOLISM IN FASTING & FEASTING

After an overnight fast the liver is the main source of glucose (~9 g/hr in humans). (**Figure 14–1**). The glucose is derived from hepatic glycogen stores (via glycogenolysis) and synthesis of new glucose (via gluconeogenesis). In humans over 60% (~6 g/hr) of the glucose released by the liver is metabolized by the central nervous system (which is largely dependent on glucose) and red blood cells (which are wholly reliant on glucose). Adipose tissue releases nonesterified fatty acids by hydrolysis of stored triglycerides. These fatty acids are the primary oxidizable fuel for many tissues (heart, muscle, liver). The liver can also synthesize ketone bodies from fatty acids to export to muscle and other tissues for oxidation. As liver glycogen reserves deplete, amino acids arising from net muscle protein breakdown with prolonged fasting and lactate derived from hydrolysis of stored muscle glycogen provide carbons to support **gluconeogenesis** and thus provide glucose to the glucose-dependent tissues (see Chapter 19).

In the fed state, after a meal in which there is an ample supply of carbohydrate, the metabolic fuel for most tissues switches to glucose (see Figure 14–1). In response to a

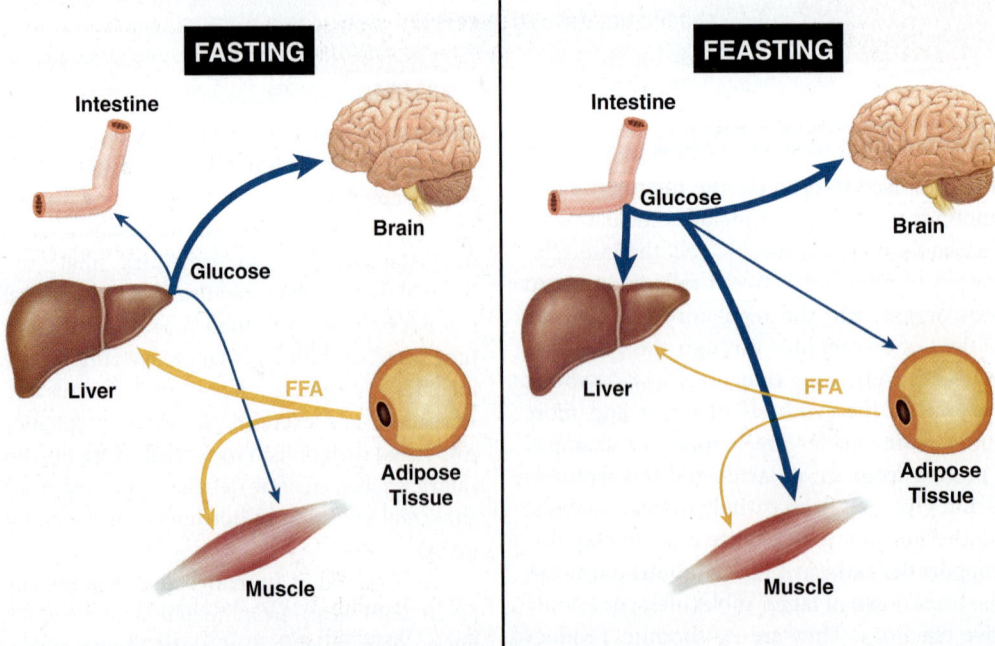

FIGURE 14–1 **Overview of Glucose and lipid flux in overnight fasted and fed humans.** In the fasted setting liver is the source of the glucose and a large fraction of the glucose production is taken up by the brain. Adipose releases nonesterified fatty acids (FFA) that are used by the liver and skeletal muscle. In the fed state, the intestine becomes the source of glucose. The liver switches to a glucose consumer, while brain glucose uptake is unaltered. Lipolysis is suppressed and muscle and adipose tissue glucose uptake is increased and muscle fatty acid uptake is decreased.

carbohydrate-rich meal the liver switches to a glucose consumer storing the majority of the glucose carbon as glycogen, with a small amount used for lipid synthesis. In contrast, glucose uptake by the brain and red blood cells is unaltered (~6 g/hr). The release of fatty acids from adipose tissue lipolysis is suppressed and tissues primarily reliant on fatty acid oxidation switch to glucose in part due to the decrease in fatty acid supply and increased glucose availability. Any dietary glucose not taken up by the liver is taken up by peripheral tissues for oxidation or storage. It is stored in muscle as glycogen or in adipose tissue as triacylglycerol.

The formation and mobilization of reserves of triacylglycerol and glycogen, and the extent to which tissues take up and oxidize glucose, are largely controlled by the hormones **insulin** and **glucagon** that are made in the endocrine pancreas. Their effects can also be modulated by other neural and/or endocrine signals (eg, sympathetic nervous system, growth hormone). Plasma glucose concentration is a tightly controlled variable. Because of the absolute dependency of the central nervous system on glucose; we have neuroendocrine systems to protect against low blood glucose (ie, hypoglycemia).

PATHWAYS TO PROCESS THE MAJOR PRODUCTS OF OUR DIET

The composition of the diet we eat dictates the general metabolism of an organism. There is a need to process the major products in the diet (carbohydrate, lipid, and protein) into their basic components. These are mainly glucose, fatty acids and glycerol, and amino acids, respectively. If the composition of the diet changes (eg, high carbohydrate vs low carbohydrate) metabolic pathways can adapt to metabolize the nutrient. In ruminants (and, to a lesser extent, other herbivores), dietary cellulose is fermented by symbiotic microorganisms to short-chain fatty acids (acetic, propionic, butyric), and metabolism in these animals is adapted to use these fatty acids as major substrates. The products of digestion when completely oxidized go to a **common product**, **acetyl-CoA**, which is then oxidized by the **citric acid cycle** (see Chapter 16) (**Figure 14–2**).

Carbohydrate Metabolism Is Centered on Oxidation & Storage of Glucose

Glucose is metabolized by all tissues and is an important energy source for many (**Figure 14–3**). Glucose is first metabolized to glucose-6-phosphate by hexokinase and from there it can go to many fates. It can be metabolized to pyruvate by the pathway of **glycolysis** (see Chapter 17). In aerobic tissues the pyruvate can be metabolized to **acetyl-CoA**, which can enter the citric acid cycle for complete oxidation to CO_2 and H_2O. The citric acid cycle is linked to the formation of ATP in the process of **oxidative phosphorylation** (see Figure 13–2). Glycolysis can also occur anaerobically (in the absence of oxygen) where pyruvate is converted to the end product lactate.

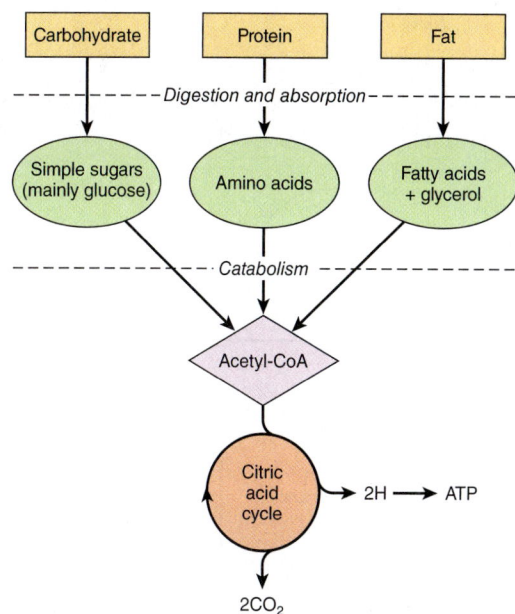

FIGURE 14–2 Outline of the pathways for the catabolism of carbohydrate, protein, and fat. All these pathways lead to the production of acetyl-CoA, which is oxidized in the citric acid cycle, ultimately yielding ATP by the process of oxidative phosphorylation.

Glucose and its metabolites also take part in other processes. Glucose can be stored as a polymer called **glycogen** in skeletal muscle and liver (see Chapter 18). It can be diverted to the **pentose phosphate pathway**, an alternative to part of the pathway of glycolysis (see Chapter 20). This pathway is a source of reducing equivalents (NADPH) for fatty acid synthesis (see Chapter 23) and the source of **ribose** for nucleotide and nucleic acid synthesis (see Chapter 33). In the glycolytic pathway triose phosphate intermediates can give rise to the **glycerol moiety** of triacylglycerols. Pyruvate can provide for the synthesis of intermediates in the citric acid cycle that can provide the carbon skeletons for the synthesis of nonessential or dispensable **amino acids** (see Chapter 27). Acetyl-CoA derived from pyruvate is the precursor for the synthesis of **fatty acids** (see Chapter 23) and **cholesterol** (see Chapter 26) and hence of all the steroid hormones synthesized in the body. Some tissues can synthesize glucose from precursors such as lactate, amino acids, and glycerol by the process of **gluconeogenesis** (see Chapter 19). This is important for suppling glucose when dietary carbohydrate is low or inadequate.

Lipid Metabolism Is Concerned Mainly With Fatty Acids & Cholesterol

The long-chain fatty acids are either derived from dietary lipid or synthesized from acetyl-CoA derived from carbohydrate or amino acids (lipogenesis). Fatty acids may be oxidized to **acetyl-CoA (β-oxidation)** or esterified with glycerol, forming **triacylglycerol** as the body's main fuel reserve. Stored triacylglycerol (adipose tissue) can be mobilized (**lipolysis**) to release nonesterified fatty acids and glycerol.

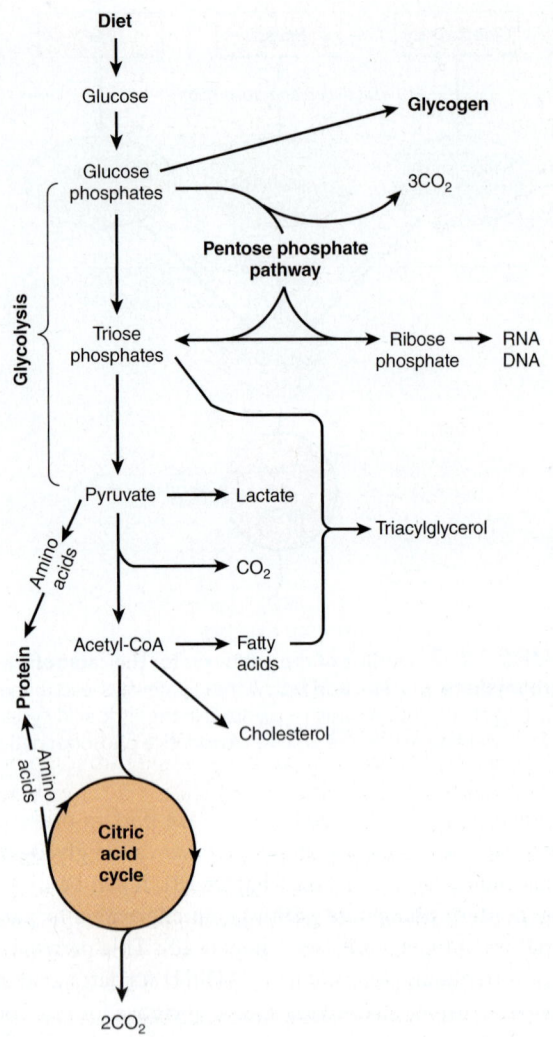

FIGURE 14–3 **Overview of carbohydrate metabolism showing the major pathways and end products.** Gluconeogenesis is not shown.

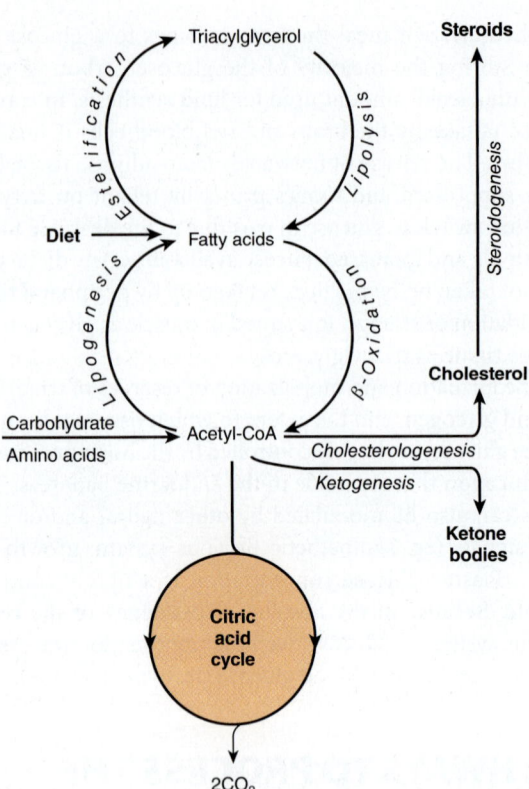

FIGURE 14–4 **Overview of fatty acid metabolism showing the major pathways and end products.** The ketone bodies are acetoacetate, 3-hydroxybutyrate, and acetone (which is formed nonenzymically by decarboxylation of acetoacetate).

Acetyl-CoA formed by β-oxidation of fatty acids may undergo three fates (**Figure 14–4**):

1. **Oxidized** to $CO_2 + H_2O$ via the citric acid cycle

2. Synthesis of **cholesterol** and other **steroids**

3. Synthesize **ketone bodies** (acetoacetate and 3-hydroxybutyrate) in the liver (see Chapter 22)

Much of Amino Acid Metabolism Involves Transamination

The amino acids are required for protein synthesis (**Figure 14–5**). The **essential or indispensable amino acids** must be supplied in the diet, since they cannot be synthesized in the body. The **nonessential or dispensable amino acids**, which are supplied in the diet, can also be formed from metabolic intermediates by **transamination** using the amino group from other amino acids (see Chapter 27). If the carbon

backbone is to be used for other processes the alpha amino nitrogen must be removed (**deamination**), metabolized in the liver to **urea**, and excreted by the kidney. The carbon skeletons that remain after transamination may (1) be oxidized to CO_2 in the citric acid cycle, (2) be used to synthesize glucose (gluconeogenesis, see Chapter 19), fatty acids (see Chapter 28), or (3) form ketone bodies.

Several amino acids are also the precursors of other compounds, for example, purines, pyrimidines, hormones such as epinephrine and thyroxine, and neurotransmitters.

METABOLIC PATHWAYS AT THE ORGAN & CELLULAR LEVEL

At the whole organism level substrates are moved between organs that can either remove or add substrates to the blood perfusing the organ. The concentrations of the substrates entering and leaving tissues and organs can be measured to help describe how substrates move between organs. Within each organ substrates can be followed as they transverse the plasma membrane and enter the metabolic pathways. Depending on the specific pathway it could all occur in the cytosol (eg, glycolysis) or be compartmentalized in subcellular organelles (eg, citric acid cycle in the mitochondrion).

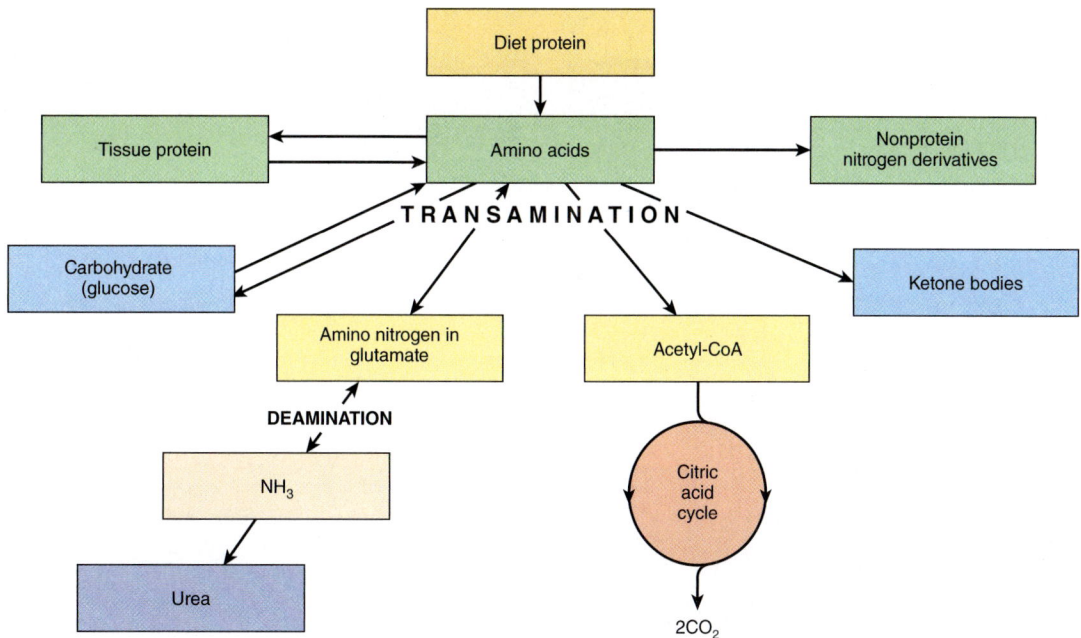

FIGURE 14–5 Overview of amino acid metabolism showing the major pathways and end products.

The Anatomical Location of an Organ & the Blood Circulation Integrates Metabolism

When food is digested in the intestine the substrates are either directly taken up and enter the portal vein or are packaged and secreted into the lymphatic system. The portal vein sends all of the absorbed substrates to the liver. Depending on the substrate, the liver can take up a small or large fraction of that which is delivered into the portal vein with the remaining allowed to pass into the systemic circulation. Substrates that enter the lymphatic circulation coalesce into a common thoracic duct which bypasses the liver and drains its contents into the systemic circulation.

Amino acids resulting from the digestion of dietary protein and **glucose** resulting from the digestion of carbohydrates are absorbed via the hepatic portal vein. The liver has the role of regulating the blood concentration of these water-soluble metabolites (**Figure 14–6**) by removing a variable portion of these substrates before they enter the systemic circulation. The uptake of glucose and amino acids is a regulated process.

In the case of glucose, in the fed state ~10 to 15% of the absorbed glucose is taken up by the liver (see Figure 14–1). The majority is used to synthesize glycogen (**glycogenesis**, see Chapter 18). A small fraction is used for fatty acid synthesis (**lipogenesis**, see Chapter 23) and the remainder is broken down by glycolysis to generate pyruvate that can be oxidized in the citric acid cycle for pyruvate oxidation. The glucose not taken up by the liver is oxidized by the brain and many other tissues including skeletal muscle. Between meals, the liver rapidly switches to a producer of glucose. It is the primary source of glucose in the fasted setting (see Figure 14–1). The glucose is derived from two sources; stored glycogen (**glycogenolysis**;

see Chapter 18) and the synthesis of glucose from metabolites such as lactate, glycerol, and amino acids (**gluconeogenesis**; see Chapter 19).

The liver is a consumer of many dietary amino acids. Like glucose only a fraction of the total absorbed amino acids are removed by the liver. The remaining are removed by peripheral tissues. In the liver they are substrates for the **synthesis of the major plasma proteins** (eg, albumin, fibrinogen). A significant fraction is **deaminated**. While the carbon backbone of amino acids can be oxidized, the nitrogen is converted to urea, transported to the kidney and excreted (see Chapter 28). The remaining amino acids are taken up by peripheral tissues primarily for protein synthesis.

The main dietary **lipids** (**Figure 14–7**) are triacylglycerols that are hydrolyzed to monoacylglycerols and fatty acids in the gut, then reesterified in the intestinal mucosa. Here they are packaged with protein (ie, apolipoproteins) and secreted into the lymphatic system and thence into the bloodstream as **chylomicrons**, the largest of the plasma **lipoproteins** (see Chapter 25). Chylomicrons also contain other lipid-soluble nutrients from the diet, including vitamins A, D, E, and K (see Chapter 44). Unlike glucose and amino acids absorbed from the small intestine, chylomicron triacylglycerol is not taken up directly by the liver. It is first metabolized by tissues that have **lipoprotein lipase**, which hydrolyzes the triacylglycerol, releasing fatty acids that are incorporated into tissue lipids or oxidized as fuel. The chylomicron remnants are cleared by the liver. The other major source of long-chain fatty acids is synthesis (**lipogenesis**) from carbohydrate, in adipose tissue and the liver (see Chapter 23).

Adipose tissue triacylglycerol is the main fuel reserve of the body. It is hydrolyzed (**lipolysis**) and glycerol and nonesterified (free) fatty acids are released into the circulation. Glycerol is used as a substrate for gluconeogenesis (see Chapter 19).

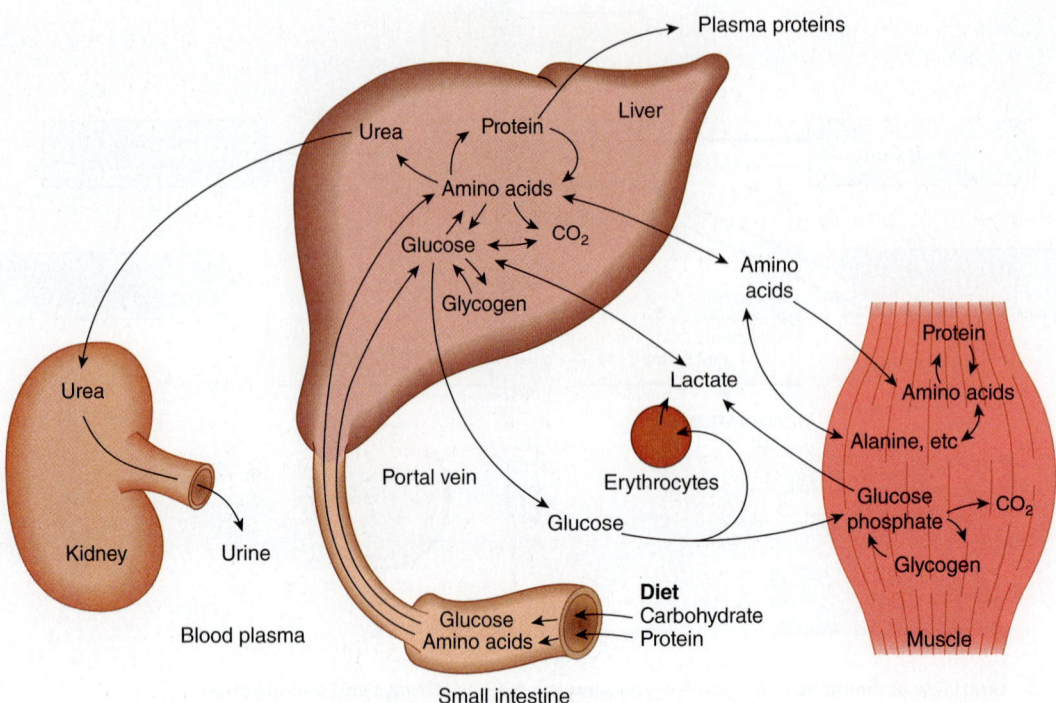

FIGURE 14–6 **Uptake and fate of major carbohydrate and amino acid substrates and metabolites.** Note: Brain and adipose tissue are not depicted.

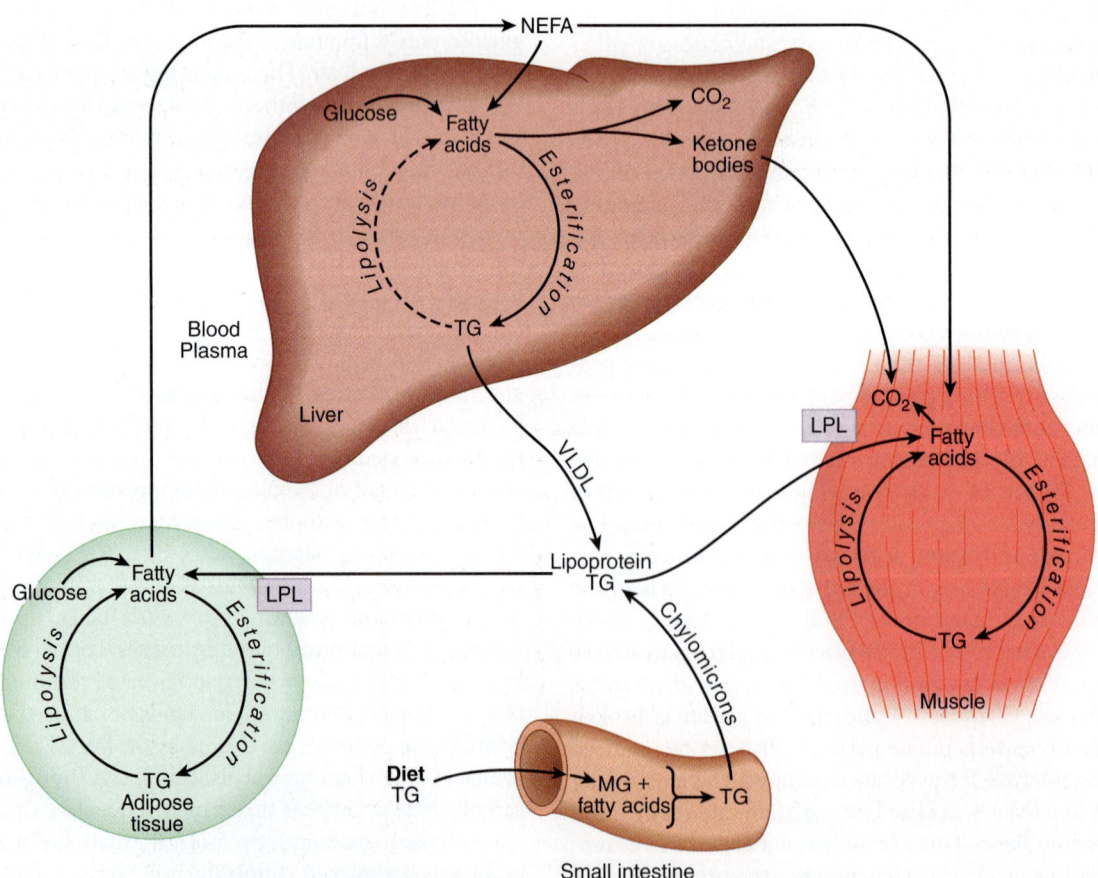

FIGURE 14–7 **Uptake and fate of major lipid substrates and metabolites.** (LPL, lipoprotein lipase; MG, monoacylglycerol; NEFA, non-esterified fatty acids; TG, triacylglycerol; VLDL, very low-density lipoprotein.)

The fatty acids are transported bound to serum albumin; they are taken up by most tissues (but not brain or erythrocytes) and either esterified to triacylglycerols for storage or oxidized as a fuel. In the liver, newly synthesized triacylglycerol, triacylglycerol from chylomicron remnants (see Figure 25–3) and nonesterified (free) fatty acids from adipose tissue are packaged in **very low-density lipoprotein (VLDL) and** secreted into the circulation. This triacylglycerol undergoes a fate similar to that of chylomicrons. Fatty acids can be partially oxidized in the liver in the fasting setting to form **ketone bodies** (**ketogenesis**, see Chapter 22). Ketone bodies are exported to extrahepatic tissues, where they provide an alternative fuel in prolonged fasting and starvation.

Skeletal muscle's primary fuel is fatty acids in the fasted setting. In the fed state muscle glucose uptake increases markedly as its preferred substrate fatty acid decreases (see Figure 14–1). The glucose is oxidized to CO_2 (aerobic) or anaerobically converted to lactate. A large fraction (>50%) is stored as glycogen in the fed state. Skeletal muscle synthesizes muscle protein from plasma amino acids. Muscle accounts for approximately 50% of body mass and consequently represents a considerable store of protein that can be drawn upon to supply amino acids for gluconeogenesis and be oxidized in skeletal muscle in starvation (see Chapter 19). In long term fasting ketones can be a significant contributor to muscle substrate oxidation.

At the Subcellular Level, Glycolysis Occurs in the Cytosol & the Citric Acid Cycle in the Mitochondria

Compartmentation of pathways in separate subcellular compartments or organelles permits integration and regulation of metabolism. Not all pathways are of equal importance in all cells. **Figure 14–8** depicts the subcellular compartmentation of metabolic pathways in a liver parenchymal cell.

The liver performs many anabolic processes simultaneously (gluconeogenesis, lipogenesis, VLDL synthesis, and protein synthesis) each of these are energy requiring (ATP, NADH, NADPH). The central role of the **mitochondrion** is immediately apparent, since it acts as the focus of carbohydrate, lipid, and amino acid metabolism as well as a site for generation of energy to support these processes. It contains the enzymes of the citric acid cycle (see Chapter 16), β-oxidation of fatty acids and ketogenesis (see Chapter 22), as well as the respiratory chain and ATP synthase (see Chapter 13).

Glycolysis (see Chapter 17), the pentose phosphate pathway (see Chapter 20), and fatty acid synthesis (see Chapter 23) all occur in the cytosol. Gluconeogenesis (see Chapter 19) requires movement of molecules between cellular compartments. Substrates such as lactate and pyruvate, which are formed in the cytosol, enter the mitochondrion to yield **oxaloacetate** that then has to be moved to the cytosol to generate phosphoenolpyruvate, which serves as a precursor for the synthesis of glucose.

The membranes of the **endoplasmic reticulum** contain the enzyme system for **triacylglycerol synthesis** (see Chapter 24), and the **ribosomes** are responsible for **protein synthesis** (see Chapter 37).

THE FLUX THROUGH METABOLIC PATHWAYS MUST BE REGULATED IN A CONCERTED MANNER

Regulation of the overall flux through a pathway is important as it allows a cell to respond to a changing environment. The regulation could be dictated by changes in overall substrates available to the cell as well as by endocrine signals that stimulate or inhibit specific metabolic pathways to subserve the needs of the organism. For example, storing glycogen in the liver in the fed state and mobilizing it in the fasted setting. This control is achieved by one or more key reactions in the pathway catalyzed by regulatory enzymes and/or transport systems that shuttle metabolites across the plasma membrane or between intracellular compartments. The physicochemical factors that control the rate of an enzyme-catalyzed reaction, such as substrate concentration, are of primary importance in the control of the overall rate of a metabolic pathway (see Chapter 9).

Nonequilibrium Reactions Are Potential Control Points

In a reaction at equilibrium, the forward and reverse reactions occur at equal rates, and there is therefore no net flux in either direction.

$$A \leftrightarrow C \leftrightarrow D$$

If this were a closed system and we added a fixed quantity of "A", the reaction would proceed to the right to make "C and D" until a new equilibrium was reached where the forward and reverse reactions are equal. The final concentration of A, C, and D would be determined by the absolute amount of A added and the properties of the enzymes. *In vivo*, there is a net flux from left to right because there is a continuous supply of substrate A and continuous removal of product D. The *in vivo* pathway is in a "steady state" if the rates of the reactions are constant and the concentration of the substrates, products, and intermediates are constant. In practice, there are normally one or more **nonequilibrium** reactions in a metabolic pathway, where the reactants are present in concentrations that are far from equilibrium. In attempting to reach equilibrium, large losses of free energy occur, making this type of reaction essentially irreversible. Such a pathway has both flow and direction. The enzymes catalyzing nonequilibrium reactions are usually present in low concentration and are subject to a variety of regulatory mechanisms. However, most reactions in metabolic pathways cannot be classified as equilibrium or nonequilibrium, but fall somewhere between the two extremes.

The Control of Flux Through Many Pathways Is Distributed

The flux-generating reaction can be identified as a nonequilibrium reaction in which the K_m of the enzyme is considerably lower than the normal concentration of substrate. The first reaction in glycolysis, catalyzed by hexokinase (see Figure 17–2), would be considered such a flux-generating step

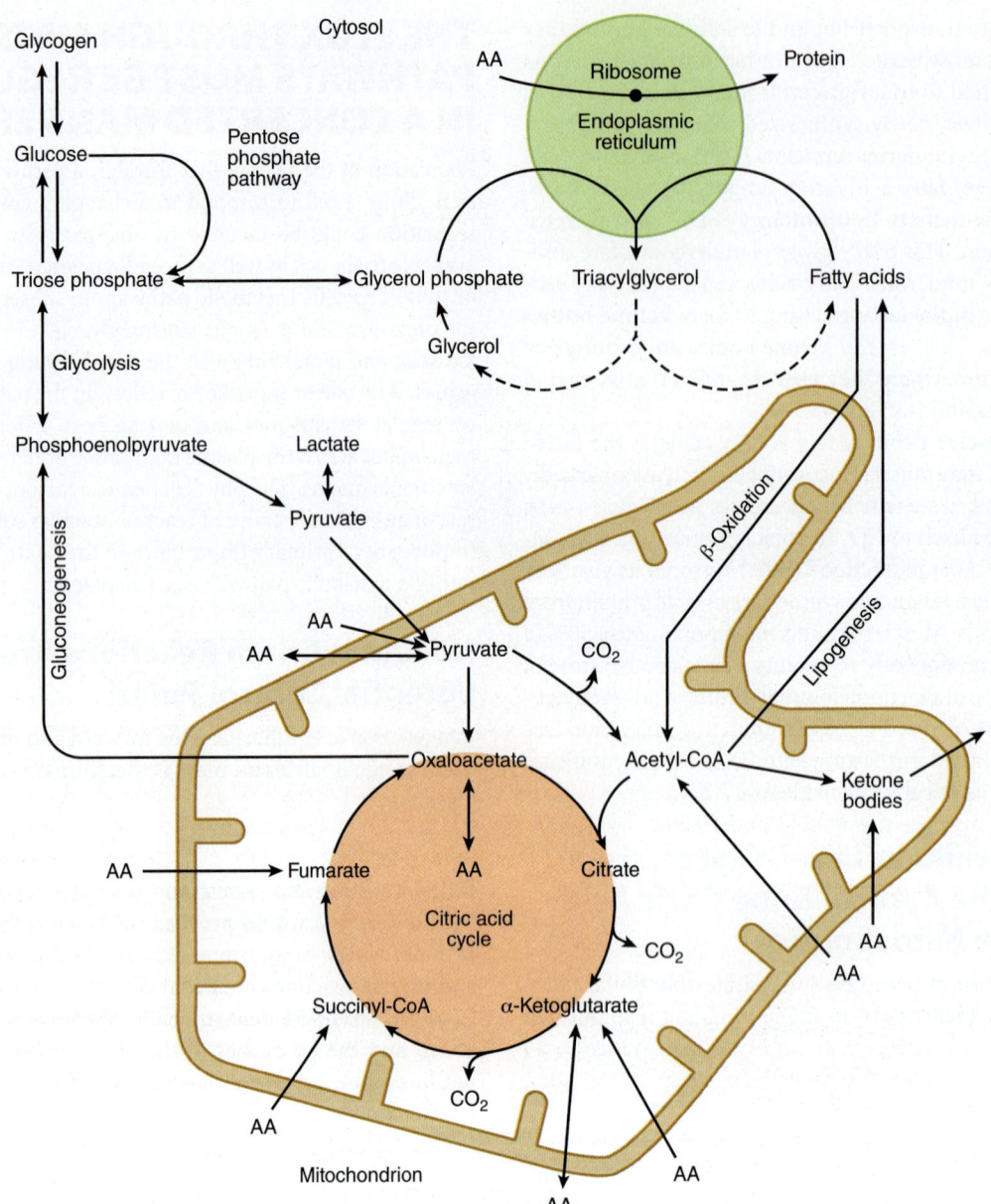

FIGURE 14–8 **Intracellular location and overview of major metabolic pathways in a liver parenchymal cell.** (AA →, metabolism of one or more essential amino acids; AA ↔, metabolism of one or more nonessential amino acids.)

because its K_m for glucose of 0.05 mmol/L is well below the normal blood glucose concentration of 3 to 5 mmol/L. However, glucose must first be transported into the cell by transporters. In some tissues, transport activity is very low in the resting state relative to the activity of hexokinase. Thus intracellular glucose concentration is kept relatively low because of the high affinity of hexokinase and the relatively low rate of glucose uptake allowed by the transport system. Thus in this setting transport activity is an important (it could be considered rate limiting) determinant of glucose uptake and subsequent metabolism. In the presence of insulin (a hormone made by the endocrine pancreas) transport activity increases so transport is no longer a significant barrier to glucose uptake. Thus,

the control of glucose uptake then shifts to hexokinase. The product of the hexokinase reaction is glucose-6-phosphate. Glucose-6-phosphate is an allosteric inhibitor of hexokinase. If the downstream pathway does not have the capacity to efficiently metabolize the additional glucose-6-phosphate when transport activity is increased, glucose-6-phosphate will increase and serve as a brake on hexokinase. Then hexokinase activity will limit how much glucose uptake is increased even though transport activity is markedly enhanced. The presence of allosteric inhibition allows downstream reactions to indirectly serve an important controlling influence on the flux through the pathway. Thus, there is rarely one enzyme controlling flux through a pathway. Rather the control is distributed;

this distribution of control can vary depending on the physiologic setting. This distribution of control allows for fine-tuning of metabolic flux under differing physiologic states.

ALLOSTERIC & HORMONAL SIGNALS CONTROL OF ENZYME-CATALYZED REACTIONS

In the metabolic pathway shown in **Figure 14–9**,

$$A \leftrightarrow B \rightarrow C \leftrightarrow D$$

reactions $A \leftrightarrow B$ and $C \leftrightarrow D$ are equilibrium reactions and $B \rightarrow C$ is a nonequilibrium reaction. The flux through this pathway can be regulated by the availability of substrate A. This depends on its supply from the blood, which in turn depends on either food intake or key reactions that release substrates from tissue reserves (glycogen and triglycerides) into the bloodstream. For example, glycogen phosphorylase in liver (see Figure 18–1) can mobilize liver glycogen and hormone-sensitive lipase in adipose tissue can mobilize adipose tissue triglycerides (see Figure 25–8). It also depends on the transport of substrate A into the cell. Muscle and adipose tissue glucose transport from the bloodstream increases in response to the hormone insulin.

Flux is also determined by removal of the end product D and the availability of cosubstrates or cofactors. Enzymes catalyzing nonequilibrium reactions are often allosteric proteins subject to the rapid actions of "feed-back" or "feed-forward" control by **allosteric modifiers**, in immediate response to the needs of the cell (see Chapter 9). Frequently, the end product of a biosynthetic pathway inhibits the enzyme catalyzing the first reaction in the pathway. Other control mechanisms depend on the action of **hormones** responding to the needs of the body as a whole; they may act rapidly by altering the activity or cellular localization of existing enzyme molecules or slowly changing enzyme content by altering the rate of enzyme synthesis (see Chapter 42).

MANY METABOLIC FUELS ARE INTERCONVERTIBLE

Carbohydrate in excess of requirements for immediate energy-yielding metabolism and formation of glycogen reserves in muscle and liver can readily be used for synthesis of fatty acids, and hence triacylglycerol in both adipose tissue and liver (whence it is exported in very low-density lipoprotein). The rate of lipogenesis in human beings is dependent on the carbohydrate content of the diet and total caloric intake. In Western countries dietary carbohydrates provide ~50% of energy intake. In less-developed countries, carbohydrate may provide 60 to 75% of energy intake. However, the total intake of food is so low that there is little surplus for lipogenesis. A high intake of fat inhibits lipogenesis in adipose tissue and liver. Despite the relatively higher fat intake in Western countries lipogenesis is significant because total caloric intake exceeds energy demand requiring diversion of excess carbohydrate calories to lipogenesis.

Fatty acids (and ketone bodies formed from them) cannot be used for the synthesis of glucose. The reaction of pyruvate dehydrogenase, forming acetyl-CoA, is irreversible, and for every two-carbon unit from acetyl-CoA that enters the citric acid cycle, there is a loss of two carbon atoms as carbon dioxide

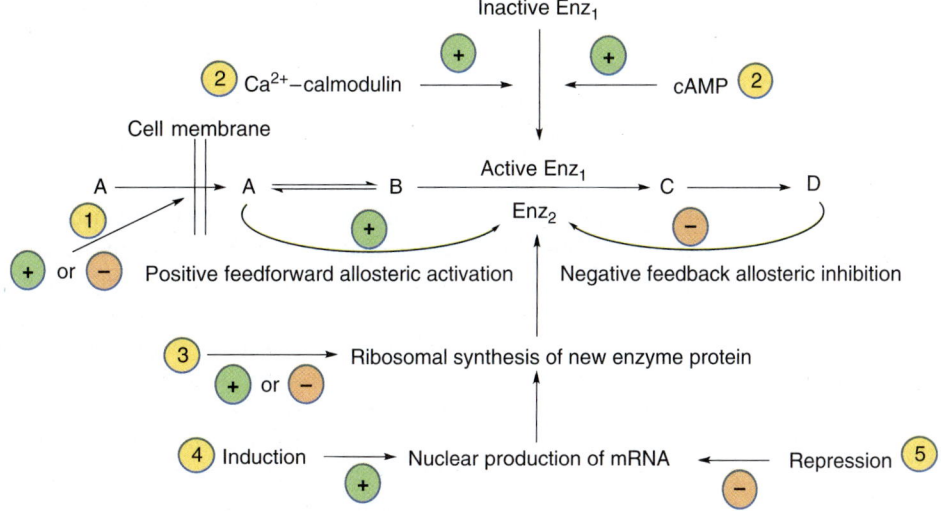

FIGURE 14–9 **Mechanisms of control of an enzyme-catalyzed reaction.** Circled numbers indicate possible sites of action of hormones: ① alteration of membrane permeability; ② conversion of an inactive enzyme to an active enzyme, usually involving phosphorylation/dephosphorylation reactions; ③ alteration of the rate translation of mRNA at the ribosomal level; ④ induction of new mRNA formation; and ⑤ repression of mRNA formation. ① and ② are rapid mechanisms of regulation, whereas ③, ④, and ⑤ are slower.

before oxaloacetate is reformed. This means that acetyl-CoA (and hence any substrates that yield acetyl-CoA) can never be used for gluconeogenesis. The (relatively rare) fatty acids with an odd number of carbon atoms yield propionyl-CoA as the product of the final cycle of β-oxidation. Propionyl-CoA can be a substrate for gluconeogenesis, as can the glycerol released by lipolysis of adipose tissue triacylglycerol reserves.

Most of the amino acids in excess of requirements for protein synthesis (arising from the diet or from tissue protein turnover) yield pyruvate, or four- and five-carbon intermediates of the citric acid cycle (see Chapter 29). Pyruvate can be carboxylated to oxaloacetate, which is the primary substrate for gluconeogenesis, and the other intermediates of the cycle also result in a net increase in the formation of oxaloacetate, which is then available for gluconeogenesis. These amino acids are classified as **glucogenic**. Two amino acids (lysine and leucine) yield only acetyl-CoA on oxidation, and hence cannot be used for gluconeogenesis, and four others (phenylalanine, tyrosine, tryptophan, and isoleucine) give rise to both acetyl-CoA and intermediates that can be used for gluconeogenesis. Those amino acids that give rise to acetyl-CoA are referred to as **ketogenic**. With prolonged fasting and starvation amino acids are mobilized from muscle protein to provide substrates for gluconeogenesis, oxidized by the liver to support liver energy demands, and contribute to synthesis of ketone bodies.

A SUPPLY OF OXIDIZABLE FUEL IS PROVIDED IN BOTH THE FED & FASTING STATES

Glucose Is Always Required by the Central Nervous System & Erythrocytes

Erythrocytes lack mitochondria and hence are wholly reliant on (anaerobic) glycolysis and the pentose phosphate pathway at all times. The brain normally metabolizes glucose but can metabolize ketone bodies. When ketone availability is high such as prolonged fasting, they can meet about 20% of its energy requirements; the remainder must be supplied by glucose. The metabolic changes that occur in the fasting state and starvation serve to preserve plasma glucose for use by the brain and red blood cells, and to provide alternative metabolic (lipids, amino acids) fuels for other tissues (**Figure 14–10**). In pregnancy, the fetus requires a significant amount of glucose, as does the mammary gland for synthesis of lactose during lactation.

In the Fed State, the Exogenous Metabolic Fuels Are Both Oxidized & Stored

In response to a meal typically the caloric intake during the period the food is absorbed exceeds the energy requirements of the organism. The excess calories are stored either as glycogen or lipid. When substrates are oxidized oxygen is consumed

and carbon dioxide is produced. When glucose ($C_6H_{12}O_6$) is oxidized ($C_6H_{12}O_6 + 6O_2 \rightarrow 6CO_2 + 6 H_2O$) for each mole of glucose oxidized a mole of oxygen is consumed and mole of carbon dioxide is released. The molar ratio of CO_2 produced and O_2 consumed is called the respiratory quotient. For carbohydrates this ratio is one. For fatty acid and protein oxidation this ratio is less than one (**Table 14–1**). We can measure this ratio in expired air. This is called the respiratory exchange ratio. This ratio reflects the mixture of substrates being oxidized by all tissues. In a typical person this ratio averages ~0.85 in a 24-hour period for a person on a standard diet. For several hours after a carbohydrate-rich meal, while the products of digestion are being absorbed, there is an abundant supply of carbohydrate. Thus carbohydrate oxidation is the main substrate being oxidized so the respiratory exchange ratio increases toward 1. The process of storing excess calories as glycogen and lipid is an energy requiring process and is called the thermic effect of food, which can account for ~10% of daily energy expenditure. As a person transitions to a fast the rate of glucose oxidation decreases and the rate of fat oxidation increases (this is observed as a decrease in the respiratory exchange ratio toward 0.7 reflecting a shift to fat oxidation; see Table 14–1).

Glucose uptake into muscle and adipose tissue is controlled by **insulin**, which is secreted by the β-islet cells of the pancreas in response to an increased concentration of glucose in the arterial blood. In the fasting state, the glucose transporter of muscle and adipose tissue (GLUT-4) is in intracellular vesicles. An early response to insulin is the migration of these vesicles to the cell surface, where they fuse with the plasma membrane, exposing active glucose transporters. These insulin-sensitive tissues only take up glucose from the bloodstream to any significant extent in the presence of the hormone. As insulin secretion falls in the fasting state, the transporters are internalized, reducing glucose uptake. However, in skeletal muscle, the increase in cytoplasmic calcium ion concentration in response to nerve stimulation and subsequent muscle contraction stimulates the migration of the vesicles to the cell surface and exposure of active glucose transporters whether there is significant insulin stimulation or not. Thus part of the increase in glucose uptake during exercise is independent of an increase in insulin.

The transport capacity for glucose into the liver is high and is independent of insulin, thus transport does not control the rate of glucose uptake in the liver. The liver, however, has an isoenzyme of hexokinase (glucokinase) with a high K_m, so that as the concentration of glucose increases and enters the liver hepatocyte, so does the rate of synthesis of glucose-6-phosphate. Thus, when plasma glucose is elevated in the fed state the liver takes up glucose (see Figure 14–1). If it is in excess of the liver's requirement for energy-yielding metabolism, it is used mainly for synthesis of **glycogen**. In both liver and skeletal muscle, insulin, which increases in response to an increase in glucose acts to amplify glycogen synthesis by stimulating glycogen synthetase and inhibiting glycogen phosphorylase. Some of the additional glucose entering the liver may also be

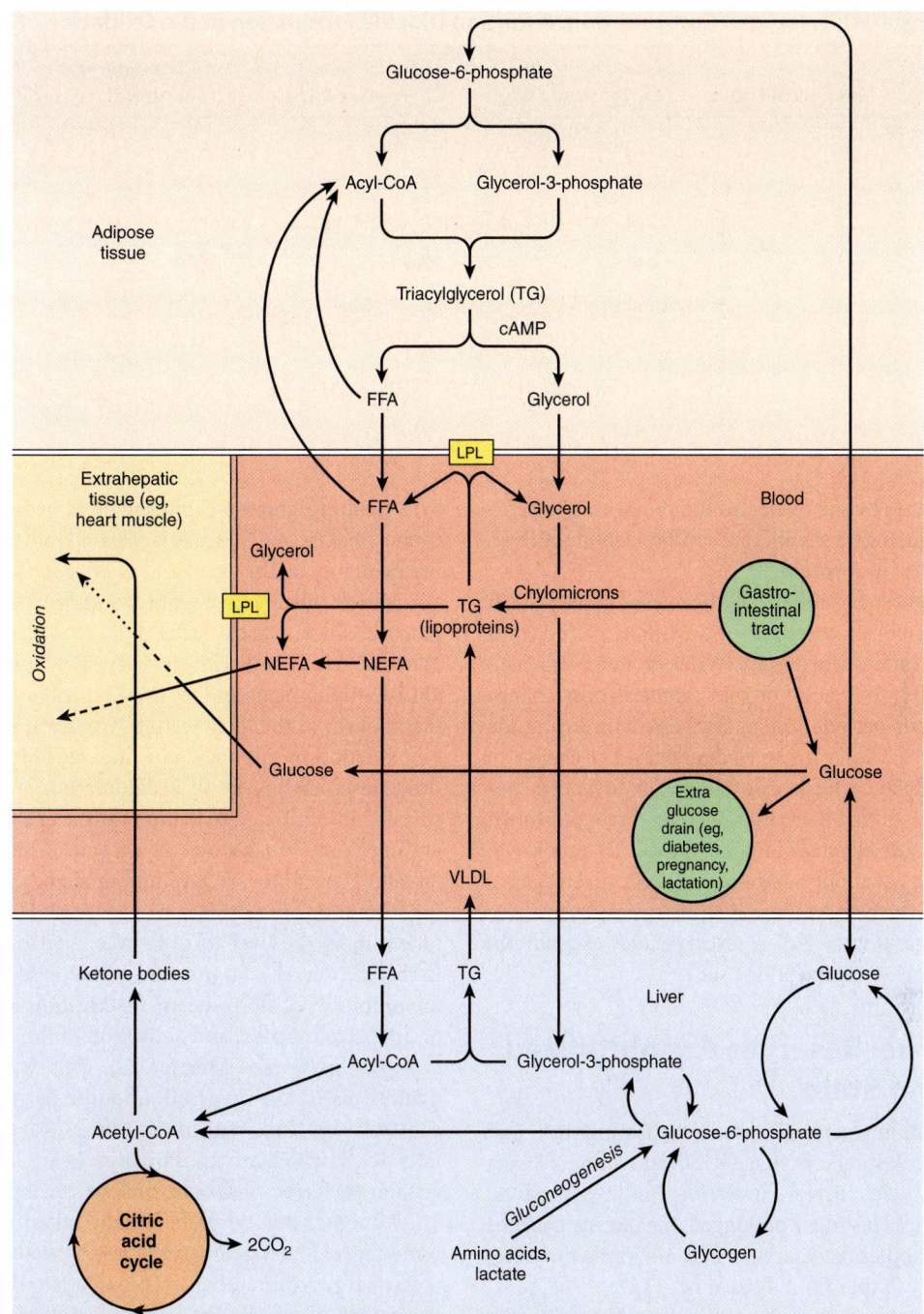

FIGURE 14–10 **Metabolic interrelationships among adipose tissue, the liver, and extrahepatic tissues.** In tissues such as heart, metabolic fuels are oxidized in the following order of preference: fatty acids > ketone bodies > glucose. (LPL, lipoprotein lipase; NEFA, nonesterified fatty acids; VLDL, very low-density lipoproteins.)

used for lipogenesis and hence triacylglycerol synthesis. In adipose tissue, insulin stimulates glucose uptake. Glucose is used to synthesize both the glycerol and fatty acid in triacylglycerol. It inhibits intracellular lipolysis and the release of nonesterified fatty acids by adipose tissue (see Figure 14–1).

The products of lipid digestion enter the circulation as **chylomicrons**, the largest of the plasma lipoproteins, which are especially rich in triacylglycerol (see Chapter 25). In

adipose tissue and skeletal muscle, extracellular lipoprotein lipase is synthesized and activated in response to insulin; the resultant nonesterified fatty acids are largely taken up by the tissue and used for synthesis of triacylglycerol, while the glycerol remains in the bloodstream. It is taken up by the liver and used for gluconeogenesis and glycogen synthesis or lipogenesis. Fatty acids remaining in the bloodstream are taken up by the liver and reesterified. The lipid-depleted chylomicron

TABLE 14–1 Energy Yields, Oxygen Consumption, & Carbon Dioxide Production in the Oxidation of Metabolic Fuels

	Energy Yield (kJ/g)	O_2 Consumed (L/g)	CO_2 Produced (L/g)	RQ (CO_2 Produced/ O_2 Consumed)	Energy (kJ)/L O_2
(glucose) $C_6H_{12}O_6 + 6O_2 \rightarrow 6CO_2 + 6\,H_2O$					
Carbohydrate	16	0.829	0.829	1.00	~20
(albumin) $C_{72}H_{112}N_2O_{22}S + 77O_2 \rightarrow 63CO_2 + 38\,H_2O + SO_3 + 9CO(NH_2)_2$					
Protein	17	0.966	0.782	0.81	~20
(triglyceride) $C_{55}H_{104}O_6 + 78O_2 \rightarrow 55CO_2 + 52\,H_2O$					
Fat	37	2.016	1.427	0.71	~20
(ethanol) $C_2H_5OH + 3O_2 \rightarrow 2CO_2 + 3\,H_2O$					
Alcohol	29	1.429	0.966	0.66	~20

remnants are cleared by the liver, and the remaining triacylglycerol is exported, together with that synthesized in the liver, in **very low-density lipoprotein**.

In healthy weight stable individuals the rates of tissue protein catabolism and anabolism are equal in a 24-hour period, thus whole body protein stores are constant. While protein catabolism is relatively constant, the rate of protein synthesis does change through the 24-hour period. Protein synthesis falls during the fasting period and increases in the feeding period (a change of ~20-25%). It is only in **cachexia** associated with advanced cancer and other diseases that there is an increased rate of protein catabolism. The increased rate of protein synthesis in response to increased availability of amino acids and metabolic fuel is again a response to insulin. Protein synthesis is an energy expensive process; it may account for up to 20% of resting energy expenditure after a meal, but only 9% in the fasting state.

Metabolic Fuel Reserves Are Mobilized in the Fasting State

There is a small fall in plasma glucose in the fasting state, and then little change as fasting is prolonged into starvation. Plasma nonesterified fatty acids increase in fasting, but then rise little more in starvation; as fasting is prolonged, the plasma concentration of ketone bodies (acetoacetate and 3-hydroxybutyrate) increases markedly (**Table 14–2**, **Figure 14–11**).

In the fasting state, as the concentration of glucose in the portal blood coming from the small intestine falls, insulin secretion decreases, and skeletal muscle and adipose tissue take up less glucose. The increase in secretion of **glucagon** by α-cells of the pancreas inhibits glycogen synthetase, and activates glycogen phosphorylase in the liver; mobilizing glycogen stores.

The resulting glucose-6-phosphate is hydrolyzed by glucose-6-phosphatase, and glucose is released into the bloodstream for use primarily by the brain and erythrocytes (see Figure 14–1).

Muscle glycogen cannot contribute directly to plasma glucose, since muscle lacks glucose-6-phosphatase, and the primary use of muscle glycogen is to provide a source of glucose-6-phosphate and pyruvate potentially for energy-yielding metabolism in the muscle itself. However, acetyl-CoA formed by oxidation of fatty acids in muscle inhibits pyruvate dehydrogenase, leading to an accumulation of pyruvate. Most of this is transaminated to alanine, at the expense of amino acids arising from breakdown of muscle protein or released as lactate. The alanine, lactate, and much of the keto acids resulting from this transamination are exported from muscle and are taken up by the liver to support gluconeogenesis. In adipose tissue, the decrease in insulin and increase in glucagon results in inhibition of lipogenesis, inactivation and internalization of lipoprotein lipase, and activation of intracellular hormone-sensitive lipase (see Chapter 25). This leads to release from adipose tissue of increased amounts of glycerol (which is a substrate for gluconeogenesis in the liver) and nonesterified fatty acids, which are used by liver, heart, and skeletal muscle as their preferred metabolic fuel, so sparing glucose.

Although muscle preferentially takes up and metabolizes nonesterified fatty acids in the fasting state, it cannot meet all of its energy requirements by β-oxidation. By contrast, the liver has a greater capacity for β-oxidation than is required to meet its own energy needs, and as fasting becomes more prolonged, it forms more acetyl-CoA than can be oxidized. This acetyl-CoA is used to synthesize the **ketone bodies** (see Chapter 22), which are major metabolic fuels for skeletal and heart muscle and can meet up to 20% of the brain's energy needs in states of

TABLE 14–2 Plasma Concentrations of Metabolic Fuels (mmol/L) in the Fed & Fasting States

	Fed	40-h Fasting	7 Days Starvation
Glucose	5.5	3.6	3.5
Nonesterified fatty acids	0.30	1.15	1.19
Ketone bodies	Negligible	2.9	4.5

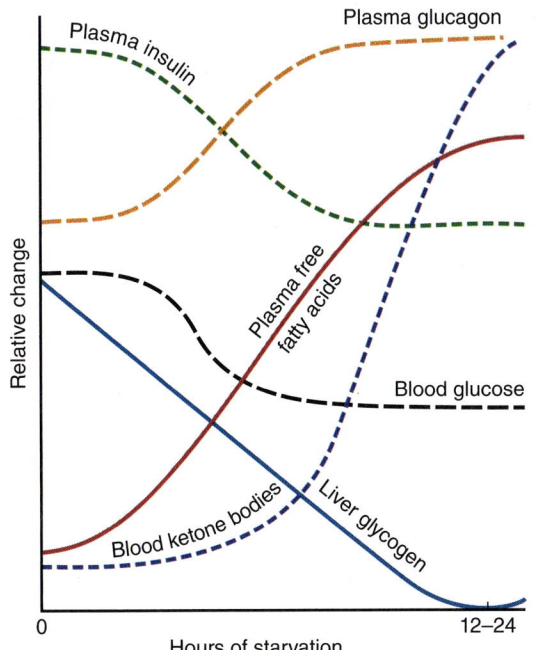

FIGURE 14–11 Relative changes in plasma hormones and metabolic fuels during the onset of starvation.

long-term fasting. In prolonged starvation, glucose may represent less than 10% of whole body energy-yielding metabolism.

Were there no other source of glucose, liver and muscle glycogen would be exhausted after about 18 hours fasting. As fasting becomes more prolonged, an increasing amount of the amino acids released as a result of protein catabolism is utilized in the liver and kidneys for gluconeogenesis (**Table 14–3**).

CLINICAL ASPECTS

In prolonged starvation, as adipose tissue reserves are depleted, there is a considerable increase in the net rate of protein catabolism to provide amino acids, not only as substrates for gluconeogenesis, but also as a metabolic fuel of many tissues. Death results when essential tissue proteins are catabolized and not replaced. In patients with **cachexia** as a result of release of **cytokines** in response to tumors and disease, there is marked increase in the rate of tissue protein catabolism, as well as a considerably increased metabolic rate, so they are in a state of advanced starvation. Again, death results when essential tissue proteins are catabolized and not replaced.

The high demand for glucose by the fetus, and for lactose synthesis in lactation, can lead to ketosis. This may be seen as mild ketosis with hypoglycemia in human beings; in lactating

TABLE 14–3 Summary of the Major Metabolic Features of the Principal Organs

Organ	Major Pathways	Main Substrates	Major Products Exported	Specialist Enzymes
Liver	Glycolysis, gluconeogenesis, lipogenesis, β-oxidation, citric acid cycle, ketogenesis, lipoprotein metabolism, drug metabolism, synthesis of bile salts, urea, uric acid, cholesterol, plasma proteins	Nonesterified fatty acids, glucose (in fed state), lactate, glycerol, fructose, amino acids, alcohol	Glucose, triacylglycerol in VLDL,[a] ketone bodies, urea, uric acid, bile salts, cholesterol, plasma proteins	Glucokinase, glucose-6-phosphatase, glycerol kinase, phosphoenolpyruvate carboxykinase, fructokinase, arginase, HMG-CoA synthase, HMG-CoA lyase, alcohol dehydrogenase
Brain	Glycolysis, citric acid cycle, amino acid metabolism, neurotransmitter synthesis	Glucose, amino acids, ketone bodies in prolonged starvation	Lactate, end products of neurotransmitter metabolism	Those for synthesis and catabolism of neurotransmitters
Heart	β-Oxidation and citric acid cycle	Nonesterified fatty acids, Ketone bodies, lactate, chylomicron and VLDL triacylglycerol, some glucose	—	Lipoprotein lipase, very active electron transport chain
Adipose tissue	Lipogenesis, esterification of fatty acids, lipolysis (in fasting)	Glucose, chylomicron, and VLDL triacylglycerol	Nonesterified fatty acids, glycerol	Lipoprotein lipase, hormone-sensitive lipase, enzymes of the pentose phosphate pathway
Fast twitch muscle	Glycolysis	Glucose, glycogen	Lactate, (alanine and keto acids in fasting)	—
Slow twitch muscle	β-Oxidation and citric acid cycle	Nonesterified fatty acids, ketone bodies, chylomicron, and VLDL triacylglycerol	lactate, alanine	Lipoprotein lipase, very active electron transport chain
Kidney	Gluconeogenesis	Nonesterified fatty acids, lactate, glycerol, glucose	Glucose with long-term fasting	Glycerol kinase, phosphoenolpyruvate carboxykinase
Erythrocytes	Anaerobic glycolysis, pentose phosphate pathway	Glucose	Lactate	Hemoglobin, enzymes of pentose phosphate pathway

[a]VLDL, very low-density lipoprotein.

cattle and in ewes carrying a twin pregnancy, there may be very pronounced ketoacidosis and profound hypoglycemia. Ketogenic diets (low carbohydrate <50 g/day, low protein and high fat) are used to treat intractable (failure to obtain sustained relief with antiepileptic drugs and are not surgical candidates) epilepsy (~30% of all patients with epilepsy). The carbohydrate intake is well below that needed to support the glucose requiring tissues (~6 g/hr × 24 h =144 g/day); thus gluconeogenesis from glucogenic amino acids in the diet partially make up the difference. In addition, because of the very low carbohydrate intake and high rate of fat oxidation they develop a mild ketosis. This provides ketone bodies to the brain to support energy requirements not met by glucose; this decreases overall carbohydrate requirements.

Diabetes (chronic hyperglycemia) affects ~7% of the US adult population. The three common forms of diabetes are Type 1 diabetes mellitus (T1DM), Type II diabetes (T2DM), and gestational diabetes. T1DM is a disorder resulting from a chronic autoimmune destruction of the insulin-producing pancreatic beta cells resulting in a near complete loss of insulin secretion. The predominant form of diabetes T2DM accounts for 90 to 95% of cases, is due to a combination of factors that impair both the sensitivity of tissues to insulin (ie, insulin resistance) and insulin secretion. Gestational diabetes occurs during pregnancy. It affects 7 to 10% of all pregnancies. Inadequate diabetes management can lead to organ dysfunction including blindness, renal and cardiovascular diseases, and in the case of gestational diabetes it can lead to fetal complications. Thus, tissue metabolism can adapt to physiologic events to preserve metabolic health of the organism. This adaptation requires the neuroendocrine systems that controls inter- and intra- organ nutrient homeostasis are intact.

In poorly controlled **diabetes mellitus**, patients may become severely hyperglycemic as a result of lack of or inadequate insulin secretion to restrain hepatic glucose production in the fasted state and impaired stimulation uptake of glucose by hepatic and peripheral tissues in the fed state. The severe hyperglycemia can progress to diabetic ketoacidosis which is a medical emergency. In the absence of insulin to antagonize the actions of glucagon and restrain the delivery of amino acids from muscle to the liver further fuels gluconeogenesis. At the same time, lipolysis in adipose tissue unrestrained by insulin and amplified by the increased sympathetic nervous system activity and glucagon, increases fatty acid delivery to the liver. The relative excess of glucagon, a potent stimulator of ketogenesis, synergizes with the increase in fatty acid supply to markedly increase ketone body formation. Utilization of the ketone bodies in muscle (and other tissues) may be impaired because of the lack of oxaloacetate (all tissues have a requirement for some glucose metabolism to maintain an adequate amount of oxaloacetate for citric acid cycle activity). This will amplify the increase in circulating ketones; ketone bodies (acetoacetate and 3-hydroxybutyrate) are relatively strong acids. Coma results from both the acidosis and the considerably increased osmolality of extracellular fluid and dehydration (mainly as a result of the hyperglycemia, and diuresis resulting from the excretion of glucose and ketone bodies in the urine).

SUMMARY

- The products of digestion provide the tissues with the building blocks for the biosynthesis of complex molecules and also with the fuel for metabolic processes.

- Nearly all products of digestion of carbohydrate, fat, and protein are metabolized to a common metabolite, acetyl-CoA, before oxidation to CO_2 in the citric acid cycle.

- Acetyl-CoA is the precursor for synthesis of long-chain fatty acids, steroids (including cholesterol), and ketone bodies.

- Glucose provides carbon skeletons for the glycerol of triacylglycerols and nonessential amino acids.

- Water-soluble products of digestion are transported directly to the liver via the hepatic portal vein. The liver regulates the concentrations of glucose and amino acids available to other tissues. Lipids and lipid-soluble products of digestion enter the bloodstream from the lymphatic system, and the liver clears the remnants after extrahepatic tissues have taken up fatty acids.

- Pathways are compartmentalized within the cell. Glycolysis, glycogenesis, glycogenolysis, the pentose phosphate pathway, and lipogenesis occur in the cytosol. The mitochondria contain the enzymes of the citric acid cycle and for β-oxidation of fatty acids, as well as the respiratory chain and ATP synthase. The membranes of the endoplasmic reticulum contain the enzymes for a number of other processes, including triacylglycerol synthesis and drug metabolism.

- Metabolic pathways are regulated by rapid mechanisms affecting the activity of existing enzymes, that is, allosteric and covalent modification (often in response to hormone action) and slow mechanisms that affect the synthesis of enzymes.

- Dietary carbohydrate and amino acids in excess of requirements can be used for fatty acid and hence triacylglycerol synthesis.

- In fasting and starvation, glucose must be provided for the brain and red blood cells; in the early fasting state, this is supplied from glycogen reserves. In order to spare glucose, muscle and other tissues do not take up glucose when insulin secretion is low; they utilize fatty acids (and later ketone bodies) as their preferred fuel.

- Adipose tissue releases nonesterified fatty acids in the fasting state. In prolonged fasting and starvation, these are used by the liver for synthesis of ketone bodies, which along with the fatty acids provide the major fuel for muscle.

- Many of the amino acids, arising from the diet or from tissue protein turnover, can be used for gluconeogenesis, as can the glycerol from triacylglycerol.

- Neither fatty acids, arising from the diet or from lipolysis of adipose tissue triacylglycerol, nor ketone bodies, formed from fatty acids in the fasting state, can provide substrates for gluconeogenesis.

REFERENCES

Frayn KN. The glucose-fatty acid cycle: a physiological perspective. Biochem Soc Trans 2003;31(Pt 6):1115-1119.

Hall KD, Heymsfield SB, Kemnitz JW, Klein S, Schoeller DA, Speakman JR. Energy balance and its components: implications for body weight regulation. Am J Clin Nutr 2012;95(4):989-994.

Moore MC, Cherrington AD, Wasserman DH. Regulation of hepatic and peripheral glucose disposal. Best Pract Res Clin Endocrinol Metab 2003;17(3):343-364.

Saccharides (ie, Carbohydrates) of Physiological Significance

Owen P. McGuinness, PhD

<table>
<tr><td>

OBJECTIVES

After studying this chapter, you should be able to:

</td><td>

- Explain what is meant by the glycome, glycobiology, and the science of glycomics.
- Explain what is meant by the terms monosaccharide, disaccharide, oligosaccharide, and polysaccharide.
- Explain the different ways in which the structures of glucose and other monosaccharides can be represented, and describe the various types of isomerism of sugars and the pyranose and furanose ring structures.
- Describe the formation of glycosides and the structures of the important disaccharides and polysaccharides.
- Explain what is meant by the glycemic index of a carbohydrate.
- Describe the roles of saccharides in cell membranes and lipoproteins.

</td></tr>
</table>

BIOMEDICAL IMPORTANCE

Carbohydrates (saccharides is the preferred nomenclature for biochemists) are extremely polar molecules that are widely distributed in plants and animals. They have important structural and metabolic roles. The word carbohydrate comes from the designation of these molecules as carbon hydrate. All saccharides are not simply comprised of C, H, and O or have the proportions $(CH_2O)_n$; thus saccharide is a more inclusive term. Saccharides are also called sugars, but this designation focusses on the sweetness of the molecule and misses the many functions of saccharides beyond sweetness.

In plants, glucose is synthesized from carbon dioxide and water by photosynthesis and stored as starch or used to synthesize the cellulose of the plant cell walls. Animals can synthesize carbohydrates from amino acids, but most are derived ultimately from plants they ingest. **Glucose** is the most important carbohydrate; most dietary carbohydrate is absorbed into the bloodstream as simple sugars (ie, monosaccharide). Glucose is absorbed after the hydrolysis in the intestine of dietary starch and disaccharides. Other simple sugars are absorbed after digestion and then rapidly converted to glucose in the liver.

This was adapted from chapter in 30th edition by David A. Bender, PhD, & Peter A. Mayes, PhD, DSc

Glucose is a major metabolic fuel of mammals (except ruminants) and a universal fuel of the fetus. It is the precursor for synthesis of all the other carbohydrates in the body, including **glycogen** for storage, **ribose** and **deoxyribose** in nucleic acids, **galactose** for synthesis of lactose in milk, in glycolipids, and in combination with protein in glycoproteins (see Chapter 46) and proteoglycans. Sucrose, the most commonly used carbohydrate in cooking is a naturally sweet molecule, but with it comes calories. With the obesity epidemic sweeteners were developed that are 600 to 20,000-fold sweeter than sucrose, so one can use them in substantially less amount. Sweeteners give same sweetness as sucrose, but are considered "calorie free". Diseases associated with carbohydrate metabolism include **diabetes mellitus, galactosemia, glycogen storage diseases, fructose intolerance,** and **lactose intolerance**.

Glycobiology is the study of the roles of saccharides in health and disease. The **glycome** is the entire complement of saccharides of an organism, whether free or in more complex molecules. **Glycomics**, an analogous term to genomics and proteomics, is the comprehensive study of glycomes, including genetic, physiological, pathological, and other aspects.

A very large number of glycoside links can be formed between saccharides. Thus, saccharides are the most abundant biomolecules both due to their reactivity and structural plasticity. For example, three different saccharides (eg, hexoses)

may be linked to each other to form over 1000 different tri-saccharides. The conformations of the sugars in oligosaccharide chains vary depending on their linkages and proximity to other molecules with which the oligosaccharides may interact. Oligosaccharide chains encode **biological information** that depends on their constituent sugars, sequences, and linkages.

SACCHARIDES ARE ALDEHYDE OR KETONE DERIVATIVES OF POLYHYDRIC ALCOHOLS

Carbohydrates (saccharides) are classified as follows:

1. **Monosaccharides** are those sugars that cannot be hydrolyzed into simpler saccharides. They may be classified as **trioses**, **tetroses**, **pentoses**, **hexoses**, or **heptoses**, depending on the number of carbon atoms (3-7), and the location of the carbonyl group C=O. If it is a terminal aldehyde, it is an **aldose**. If it is not terminal, then there is a ketone and is designated as ketose. For trioses there are only two possibilities—glyceraldehyde and dihydroxyacetone. Examples are listed in **Table 15–1**. In addition to aldehydes and ketones, the polyhydric alcohols (sugar alcohols or **polyols**), in which the aldehyde or ketone group has been reduced to an alcohol group, can occur naturally in foods. They are also synthesized by reduction of monosaccharides for use in the manufacture of foods for weight reduction and for diabetics. They are poorly absorbed, and have about half the energy yield of sugars. The biggest side effect is flatulence; bacteria in the intestine ferment the sugar alcohol that was not absorbed.

2. **Disaccharides** are condensation products of two monosaccharide units, for example, lactose, maltose, isomaltose, sucrose, and trehalose.

3. **Oligosaccharides** are condensation products of 3 to 10 monosaccharides. Most are not digested by human enzymes.

4. **Polysaccharides** are condensation products of more than 10 monosaccharide units; examples are the starches and dextrans, which may be linear or branched polymers.

Polysaccharides are sometimes classified as hexosans or pentosans, depending on the constituent monosaccharides (hexoses or pentoses, respectively). In addition to starches and dextrans (which are hexosans), foods contain a wide variety of other polysaccharides that are collectively known as nonstarch polysaccharides; they are not digested by human enzymes, and are the major component of dietary fiber. Examples are cellulose from plant cell walls (a glucose polymer; see Figure 15–13) and inulin, the storage carbohydrate in some plants (a fructose polymer; see Figure 15–13).

BIOMEDICALLY, GLUCOSE IS THE MOST IMPORTANT MONOSACCHARIDE

The Structure of Glucose Can Be Represented in Three Ways

The straight-chain structural formula (aldohexose; **Figure 15–1A**) can account for some of the properties of glucose, but a cyclic structure (a **hemiacetal** formed by reaction between the aldehyde group and a hydroxyl group) is thermodynamically favored and accounts for other properties. The cyclic structure is normally drawn as shown in **Figure 15–1B**, the Haworth projection, in which the molecule is viewed from the side and above the plane of the ring; the bonds nearest to the viewer are bold and thickened, and the hydroxyl groups are above or below the plane of the ring. The hydrogen atoms attached to each carbon are not shown in this figure. The ring is actually in the form of a chair (**Figure 15–1C**).

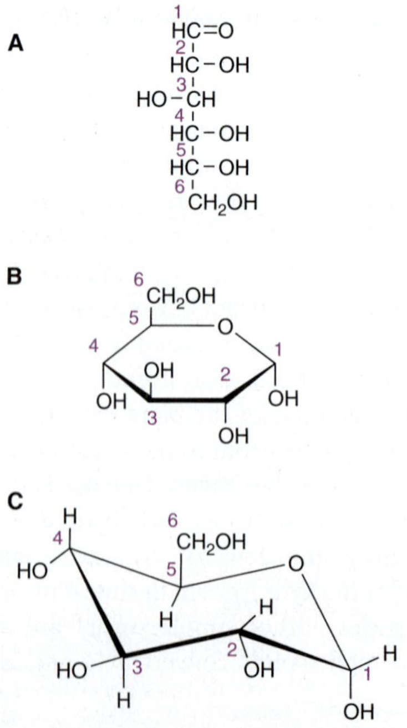

FIGURE 15–1 D-Glucose. (**A**) Straight-chain form. (**B**) α-D-glucose; Haworth projection. (**C**) α-D-glucose; chair form.

TABLE 15–1 Classification of Important Monosaccharides

	Aldoses	Ketoses
Trioses ($C_3H_6O_3$)	Glycerose (glyceraldehyde)	Dihydroxyacetone (dihydroxyacetone)
Tetroses ($C_4H_8O_4$)	Erythrose	Erythrulose
Pentoses ($C_5H_{10}O_5$)	Ribose	Ribulose
Hexoses ($C_6H_{12}O_6$)	Glucose, galactose, mannose	Fructose
Heptoses ($C_7H_{14}O_7$)	—	Sedoheptulose

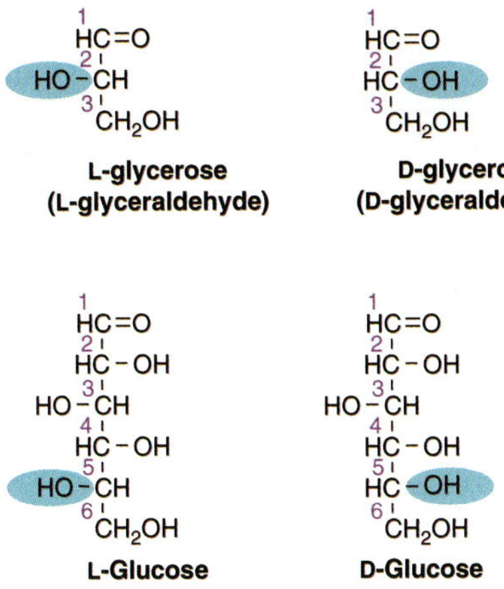

FIGURE 15–2 D- and L-isomerism of glycerose and glucose.

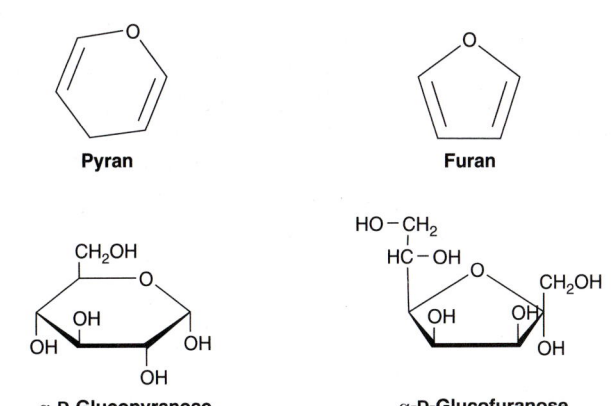

FIGURE 15–3 Pyranose and furanose forms of glucose.

Monosaccharides Exhibit Various Forms of Isomerism

Glucose, with four asymmetric carbon atoms, can form 16 isomers. The more important types of isomerism found with glucose are as follows:

1. **D- and L-isomerism:** The designation of a sugar isomer as the D form or its mirror image as the L form is determined by its spatial relationship to the parent compound of the carbohydrates, the three-carbon sugar glycerose (glyceraldehyde). The L and D forms of this sugar, and of glucose, are shown in **Figure 15–2**. The orientation of the —H and —OH groups around the carbon atom adjacent to the terminal alcohol carbon (carbon 5 in glucose) determines whether the sugar belongs to the D or L series. When the —OH group on this carbon is on the right (as seen in Figure 15–2), the sugar is the D-isomer; when it is on the left, it is the L-isomer. Most of the naturally occurring monosaccharides are D sugars, and the enzymes responsible for their metabolism are specific for this configuration.

2. The presence of asymmetric carbon atoms also confers **optical activity** on the compound. When a beam of plane-polarized light is passed through a solution of an **optical isomer**, it rotates either to the right, dextrorotatory (+), or to the left, levorotatory (−). The direction of rotation of polarized light is independent of the stereochemistry of the sugar, so it may be designated D(−), D(+), L(−), or L(+). For example, the naturally occurring form of fructose is the D(−) isomer. Confusingly, dextrorotatory (+) was at one time called d-, and levorotatory (−) l-. This nomenclature is obsolete, but may sometimes be found; it is unrelated to D- and L-isomerism. In solution, glucose is dextrorotatory, and glucose solutions are sometimes known as **dextrose**.

3. **Pyranose and furanose ring structures:** The ring structures of monosaccharides are similar to the ring structures of either pyran (a six-membered ring) or furan (a five-membered ring) (**Figures 15–3** and **15–4**). For glucose in solution, more than 99% is in the pyranose form.

4. **Alpha- and beta-anomers:** The ring structure of an aldose is a hemiacetal, since it is formed by reaction between an aldehyde and an alcohol group. Similarly, the ring structure of a ketose is a hemiketal. Crystalline glucose is α-D-glucopyranose. The cyclic structure is retained in the solution, but isomerism occurs about position 1, the carbonyl or **anomeric carbon atom**, to give a mixture of α-glucopyranose (38%) and β-glucopyranose (62%). Less than 0.3% is represented by α- and β-anomers of glucofuranose.

5. **Epimers:** Isomers differing as a result of variations in configuration of the —OH and —H on carbon atoms 2, 3, and 4 of glucose are known as epimers. Biologically, the most important epimers of glucose are mannose (epimerized at carbon 2) and galactose (epimerized at carbon 4) (**Figure 15–5**).

6. **Aldose-ketose isomerism:** Fructose has the same molecular formula as glucose but differs in that there is a potential keto group in position 2, the anomeric carbon of fructose, whereas in glucose there is a potential aldehyde group in position 1, the anomeric carbon. Examples of aldose and ketose sugars are shown in **Figures 15–6** and **15–7**, respectively. Chemically, aldoses are reducing compounds, and are sometimes known as reducing sugars. This provides the basis for a simple chemical test for glucose in urine in

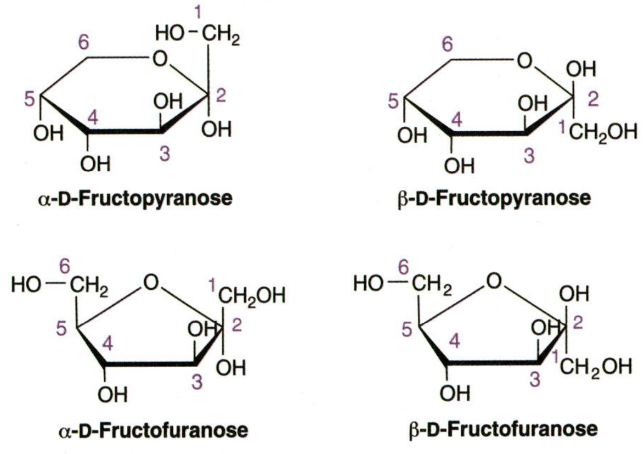

FIGURE 15–4 Pyranose and furanose forms of fructose.

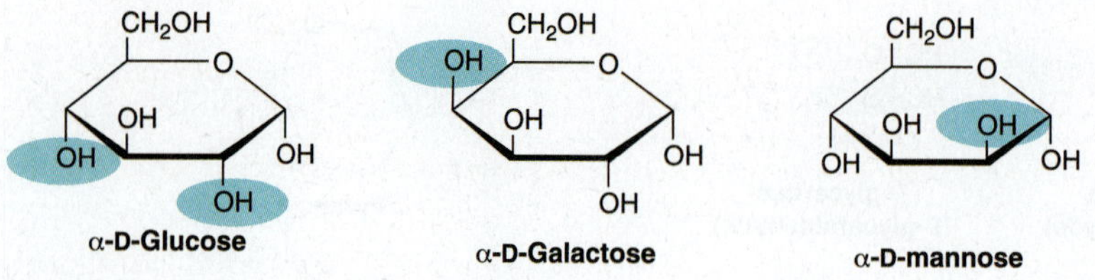

FIGURE 15–5 Epimers of glucose.

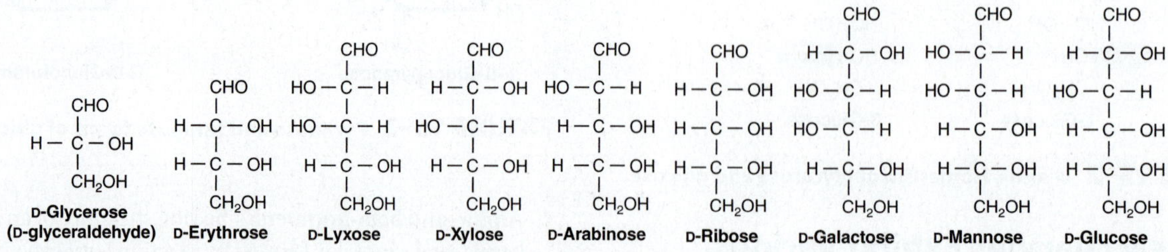

FIGURE 15–6 Examples of aldoses of physiological significance.

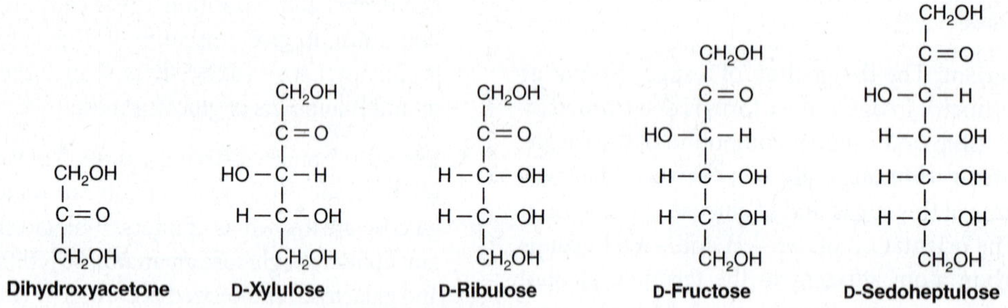

FIGURE 15–7 Examples of ketoses of physiological significance.

poorly controlled diabetes mellitus by reduction of an alkaline copper solution (see Chapter 48).

Many Monosaccharides Are Physiologically Important

Derivatives of trioses, tetroses, and pentoses and of the seven-carbon sugar sedoheptulose are formed as metabolic intermediates in glycolysis (see Chapter 17) and the pentose phosphate pathway (see Chapter 20). Pentoses are important in nucleotides, nucleic acids, and several coenzymes (**Table 15–2**). Glucose, galactose, fructose, and mannose are the physiologically important hexoses (**Table 15–3**). The biochemically important aldoses and ketoses are shown in Figures 15–6 and 15–7, respectively.

In addition, carboxylic acid derivatives of glucose are important, including D-glucuronate (for glucuronide formation and in glycosaminoglycans), its metabolic derivative, L-iduronate (in glycosaminoglycans, **Figure 15–8**) and L-gulonate (an intermediate in the uronic acid pathway; see Figure 20–4).

Saccharides Form Glycosides With Other Compounds & With Each Other

Glycosides are formed by condensation between the hydroxyl group of the anomeric carbon of a monosaccharide, and a second compound that may be another monosaccharide (denoted **glycone**) or, a nonsaccharide group (denoted **aglycone**). If the second group is also a hydroxyl, the *O*-glycosidic bond is an **acetal** link because it results from a reaction between a hemiacetal group (formed from an aldehyde and an —OH group) and another —OH group. If the hemiacetal portion is glucose, the resulting compound is a **glucoside**; if galactose, a **galactoside**; and so on. If the second group is an amine, an *N*-glycosidic bond is formed, for example, between adenine and ribose in nucleotides such as ATP (see Figure 11–4).

Glycosides are widely distributed in nature; the aglycone may be methanol, glycerol, a sterol, a phenol, or a base such as adenine. The glycosides that are important in medicine because of their action on the heart (**cardiac glycosides**), all contain steroids as the aglycone. These include derivatives of digitalis and strophanthus such as **ouabain**, an inhibitor of the

TABLE 15–2 Pentoses of Physiological Importance

Sugar	Source	Biochemical and Clinical Importance
D-Ribose	Nucleic acids and metabolic intermediate	Structural component of nucleic acids and coenzymes, including ATP, NAD(P), and flavin coenzymes
D-Ribulose	Metabolic intermediate	Intermediate in the pentose phosphate pathway
D-Arabinose	Plant gums	Constituent of glycoproteins
D-Xylose	Plant gums, proteoglycans, glycosaminoglycans	Constituent of glycoproteins
L-Xylulose	Metabolic intermediate	Excreted in the urine in essential pentosuria

TABLE 15–3 Hexoses of Physiological Importance

Sugar	Source	Biochemical Importance	Clinical Significance
D-Glucose	Fruit juices, hydrolysis of starch, cane or beet sugar, maltose and lactose	The main metabolic fuel for tissues; "blood sugar"	Excreted in the urine (glucosuria) in poorly controlled diabetes mellitus as a result of hyperglycemia
D-Fructose	Fruit juices, honey, hydrolysis of cane or beet sugar and inulin, enzymic isomerization of glucose syrups for food manufacture	Rapidly metabolized by the liver	Hereditary fructose intolerance leads to accumulation of metabolites of fructose that impair glucose production and induce hypoglycemia
D-Galactose	Hydrolysis of lactose	Readily metabolized to glucose; synthesized in the mammary gland for synthesis of lactose in milk. A constituent of glycolipids and glycoproteins	Hereditary galactosemia as a result of failure to metabolize galactose leads to cataracts
D-Mannose	Hydrolysis of plant mannan gums	Constituent of glycoproteins	

Na^+–K^+-ATPase of cell membranes. Other glycosides include antibiotics such as **streptomycin**.

Deoxy Sugars Lack an Oxygen Atom

Deoxy sugars are those in which one hydroxyl group has been replaced by hydrogen. An example is **deoxyribose** (**Figure 15–9**) in DNA. The deoxy sugar L-fucose (see Figure 15–15) occurs in glycoproteins. In clinical medicine the tissue accumulation of a tracer quantity of 2-deoxyglucose ([18]F 2-fluoro-2-deoxyglucose) is used to detect metabolically active tumors, which have very high rates of glucose uptake. 2-deoxyglucose after being phosphorylated by hexokinase cannot be further metabolized, so it accumulates. This accumulation is detected using positron emission tomography.

Amino Sugars (Hexosamines) Are Components of Glycoproteins, Gangliosides, & Glycosaminoglycans

The amino sugars include D-glucosamine, a constituent of hyaluronic acid (**Figure 15–10**), D-galactosamine (also known as chondrosamine), a constituent of chondroitin, and D-mannosamine. Several **antibiotics** (eg, **erythromycin**) contain amino sugars, which are important for their antibiotic activity.

Maltose, Sucrose, & Lactose Are Important Disaccharides

The disaccharides are sugars composed of two monosaccharide residues linked by a glycoside bond (**Figure 15–11**).

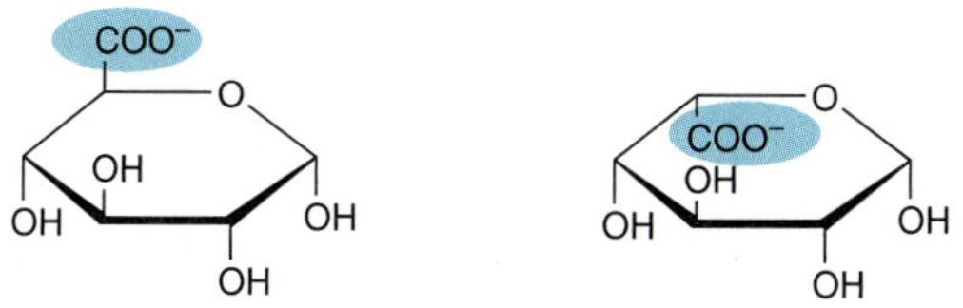

FIGURE 15–8 α-D-**Glucuronate (left) and β-L-iduronate (right).**

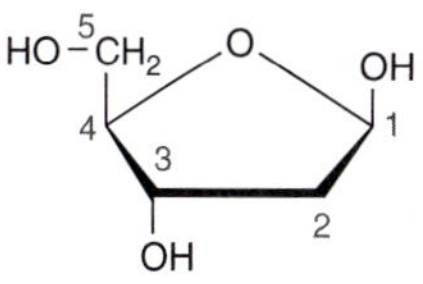

FIGURE 15–9 2-Deoxy-D-ribofuranose (β-form).

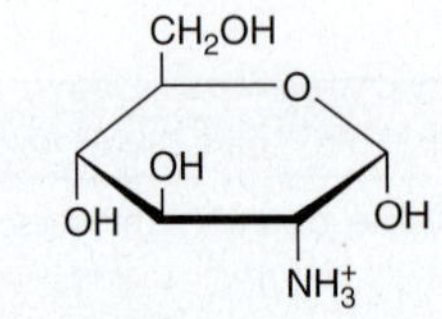

FIGURE 15–10 **Glucosamine (2-amino-D-glucopyranose) (α-form).** Galactosamine is 2-amino-D-galactopyranose. Both glucosamine and galactosamine occur as *N*-acetyl derivatives in complex carbohydrates, for example, glycoproteins.

The physiologically important disaccharides are maltose, sucrose, and lactose (**Table 15–4**). Hydrolysis of sucrose yields a mixture of glucose and fructose called "invert sugar" because fructose is strongly levorotatory and changes (inverts) the weaker dextrorotatory action of sucrose.

POLYSACCHARIDES SERVE STORAGE & STRUCTURAL FUNCTIONS

Polysaccharides include a number of physiologically important carbohydrates. They can provide one of the three functions (1) structural or mechanical protection, (2) energy storage, and (3) bind water to resist dehydration.

Starch is a homopolymer of glucose forming an α-glucosidic chain, called a **glucosan** or **glucan**. It is the most important dietary carbohydrate in cereals, potatoes, legumes, and other vegetables. The two main constituents are **amylose** (13-20%), which has a nonbranching helical structure, and **amylopectin** (80-87%), which consists of branched chains, consisting of 24 to 30 glucose residues with α1 → 4 linkages in the chains and by α1 → 6 linkages at the branch points (**Figure 15–12**).

The extent to which starch in foods is hydrolyzed by amylase is determined by its structure, the degree of crystallization or hydration (the result of cooking), and whether it is enclosed in intact (and indigestible) plant cell walls. Individuals with diabetes or prediabetes have trouble controlling their glucose concentration and are encouraged to eat complex carbohydrates that do not rapidly increase glucose levels on ingestion. The **glycemic index** of a starchy food is based on the extent to which it raises the blood concentration of glucose compared with an equivalent amount of glucose or a reference food such as white bread or boiled rice. An important determinant of the glycemic index is its digestibility. Glycemic index ranges from 1 (or 100%) for starches that are readily hydrolyzed in the small intestine to 0 for those that are not hydrolyzed at all.

Glycogen is the energy storage polysaccharide in animals and is sometimes called animal starch. It is a more highly branched structure than amylopectin, with chains of 12 to 15 α-D-glucopyranose residues (in α1 → 4 glucosidic linkage) with branching by means of α1 → 6 glucosidic bonds. Muscle glycogen granules (β-particles) are spherical and contain up to 60,000 glucose residues; in liver there are similar granules and also rosettes of glycogen granules that appear to be aggregated β-particles. Glycogen can be readily mobilized by the liver to support glucose needs of peripheral tissues.

Inulin is a polysaccharide of fructose (a fructosan) found in tubers and roots of dahlias, artichokes, and dandelions. It is not hydrolyzed by intestinal enzymes, so has no nutritional value.

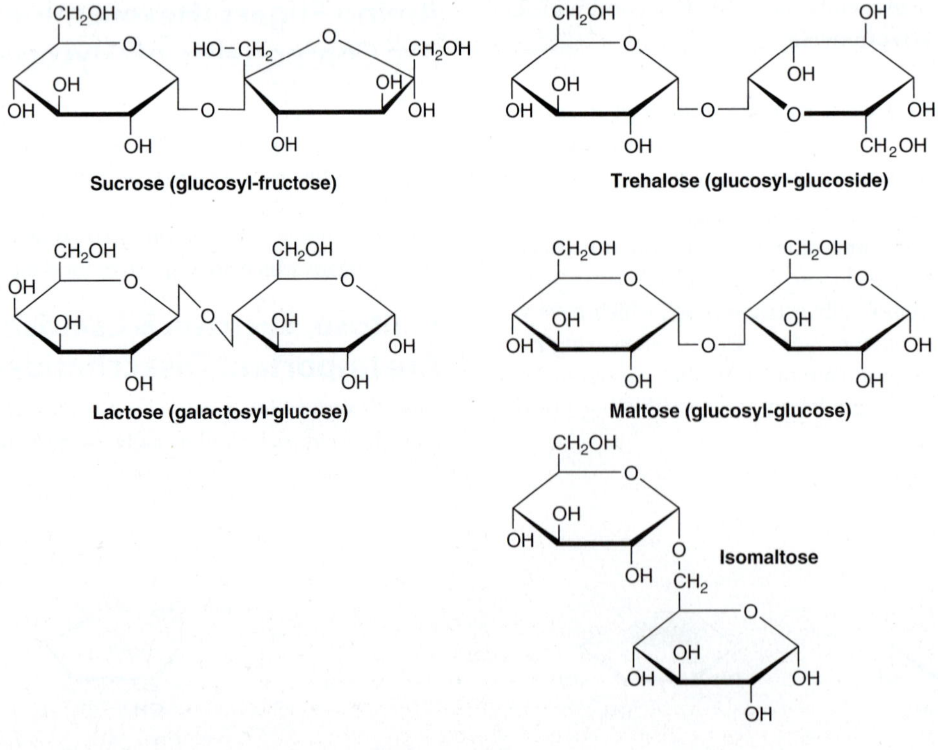

FIGURE 15–11 Structures of nutritionally important disaccharides.

TABLE 15–4 **Disaccharides of Physiological Importance**

Sugar	Composition	Source	Clinical Significance
Sucrose	O-α-D-glucopyranosyl-(1→2)-β-D-fructofuranoside	Cane and beet sugar, sorghum, and some fruits and vegetables	Rare genetic lack of sucrase leads to sucrose intolerance—diarrhea and flatulence
Lactose	O-β-D-galactopyranosyl-(1→4)-β-D-glucopyranose	Milk (and many pharmaceutical preparations as a filler)	Lack of lactase (alactasia) leads to lactose intolerance—diarrhea and flatulence; may be excreted in the urine in pregnancy
Maltose	O-α-D-glucopyranosyl-(1→4)-α-D-glucopyranose	Enzymic hydrolysis of starch (amylase); germinating cereals and malt	
Isomaltose	O-α-D-glucopyranosyl-(1→6)-α-D-glucopyranose	Enzymic hydrolysis of starch (the branch points in amylopectin)	
Lactulose	O-α-D-galactopyranosyl-(1→4)-β-D-fructofuranose	Heated milk (small amounts), mainly synthetic	Not hydrolyzed by intestinal enzymes, but fermented by intestinal bacteria; used as a mild osmotic laxative
Trehalose	O-α-D-glucopyranosyl-(1→1)-α-D-glucopyranoside	Yeasts and fungi; the main sugar of insect hemolymph	

It is used clinically to assess kidney function. When infused it is excreted by the kidney. The ability of the kidney to remove inulin is a reflection of the glomerular filtration rate. **Dextrins** are intermediates in the hydrolysis of starch. **Cellulose** is the chief constituent of plant cell walls. It is insoluble and consists of β-D-glucopyranose units linked by β1 → 4 bonds to form long, straight chains strengthened by cross-linking hydrogen bonds. Mammals lack any enzyme that hydrolyzes the β1 → 4 bonds, and so cannot digest cellulose. It is the major component of dietary fiber. Microorganisms in the gut

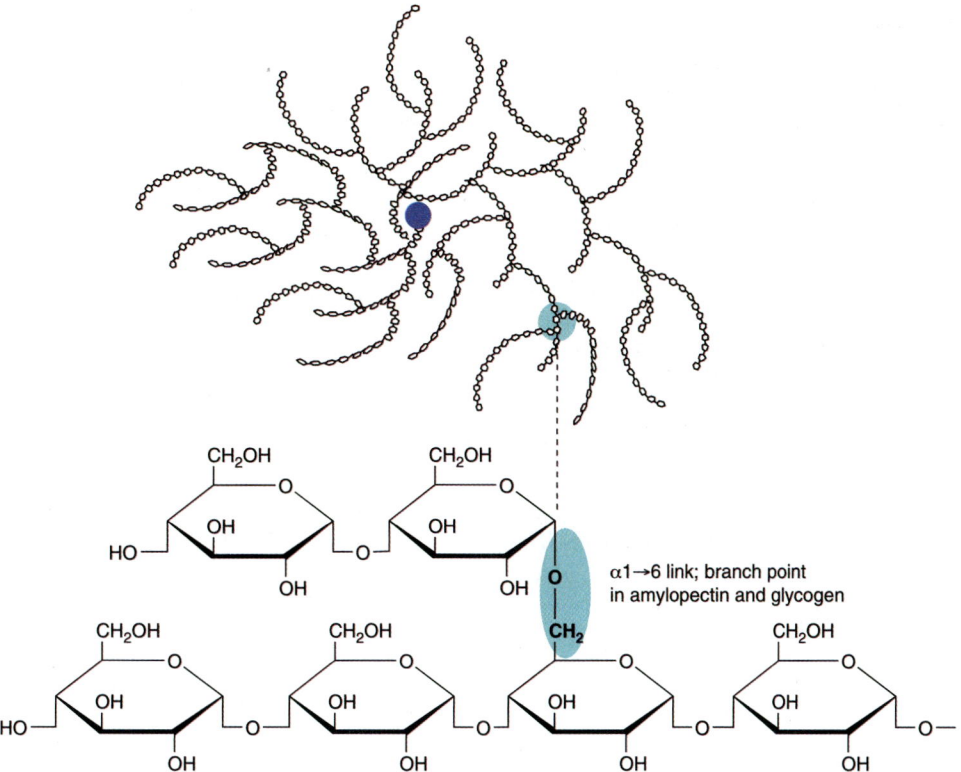

FIGURE 15–12 **The structure of starch and glycogen.** Amylose is a linear polymer of glucose residues linked α1→4, which coils into a helix. Amylopectin and glycogen consist of short chains of glucose residues linked α1→4 with branch points formed by α1→6 glycoside bonds. The glycogen molecule is a sphere ~21 nm in diameter that can be seen in electron micrographs. It has a molecular mass of ~10^7 Da and consists of polysaccharide chains, each containing about 13 glucose residues. The chains are either branched or unbranched and are arranged in 12 concentric layers. The branched chains (each has two branches) are found in the inner layers and the unbranched chains in the outermost layer. The blue dot at the center of the glycogen molecule is glycogenin, the primer molecule for glycogen synthesis.

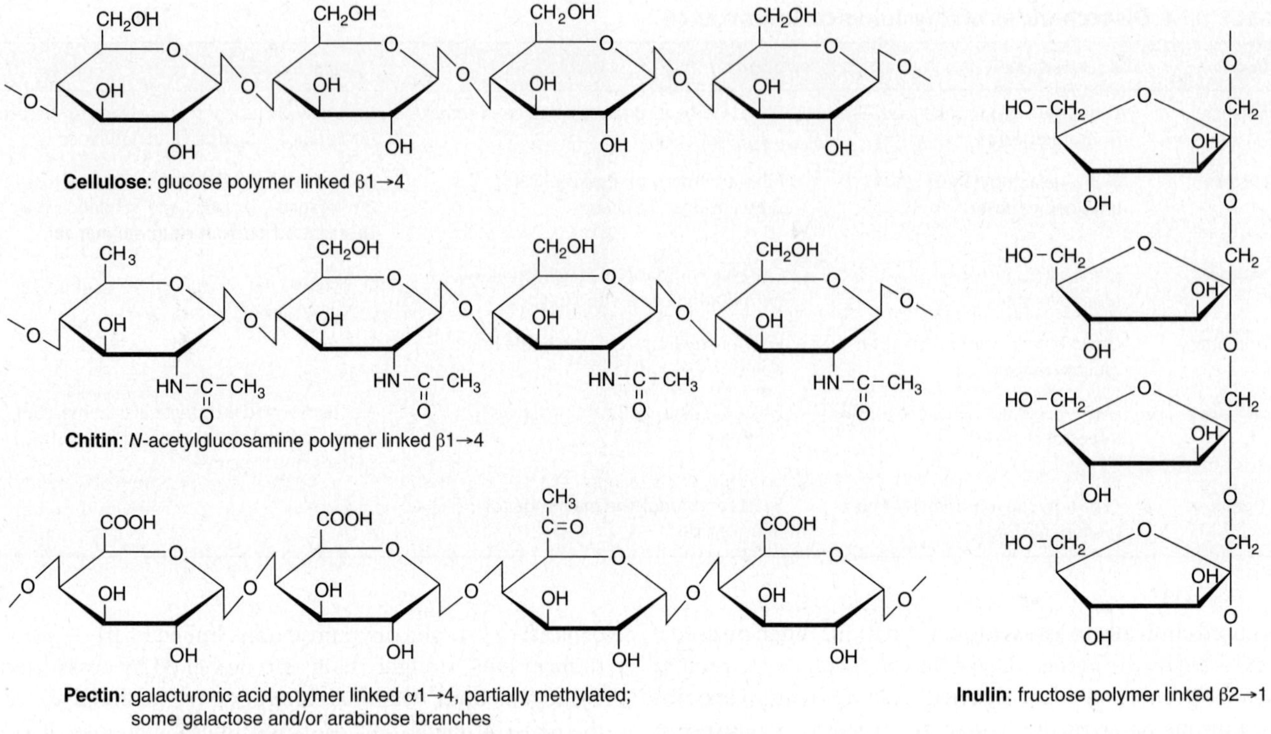

Cellulose: glucose polymer linked β1→4

Chitin: *N*-acetylglucosamine polymer linked β1→4

Pectin: galacturonic acid polymer linked α1→4, partially methylated; some galactose and/or arabinose branches

Inulin: fructose polymer linked β2→1

FIGURE 15–13 **The structures of some important nonstarch polysaccharides.**

of ruminants and other herbivores can hydrolyze the linkage and ferment the products to short-chain fatty acids as a major energy source. Recent data suggest that the availability of these short-chain fatty acids, which is determined by the microbial composition in the intestine, can impact metabolic health in at-risk individuals (diabetes, obesity, irritable bowel disease). By manipulating the microbial composition using prebiotics, one can shift the microbial population and possibly improve metabolic health. There is some bacterial metabolism of cellulose in the human colon. **Chitin** is a structural polysaccharide in the exoskeleton of crustaceans and insects, and also in mushrooms. It consists of *N*-acetyl-D-glucosamine units joined by β1 → 4 glycosidic bonds. **Pectin** occurs in fruits; it is a polymer of galacturonic acid linked α1 → 4, with some galactose an/or arabinose branches, and is partially methylated (**Figure 15–13**).

Glycosaminoglycans (mucopolysaccharides) are complex carbohydrates containing **amino sugars** and **uronic acids**. They may be attached to a protein molecule to form a **proteoglycan**. Proteoglycans provide the ground or packing substance of connective tissue (see Chapter 50). They hold large quantities of water and occupy space because of the large number of —OH groups and negative charges on the molecule, which, by repulsion, keep the carbohydrate chains apart. They serve to cushion or lubricate other structures, such as connective tissue in a joint and cartilage. Examples are **hyaluronic acid** and **chondroitin sulfate. Heparin is an important anticoagulant** (**Figure 15–14**).

Glycoproteins (also known as mucoproteins) are proteins containing branched or unbranched oligosaccharide chains (**Table 15–5**), **including fucose** (**Figure 15–15**). They occur

in cell membranes (see Chapters 40 and 46) and many proteins are glycosylated. The **sialic acids** are *N*- or *O*-acyl derivatives

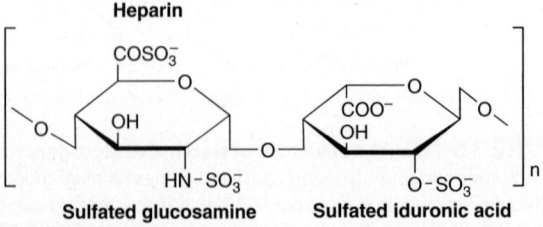

Hyaluronic acid

β-Glucuronic acid   *N*-Acetylglucosamine

Chondroitin 4-sulfate

β-Glucuronic acid   *N*-Acetylgalactosamine sulfate

Heparin

Sulfated glucosamine   Sulfated iduronic acid

FIGURE 15–14 **Structure of some complex polysaccharides and glycosaminoglycans.**

TABLE 15–5 Saccharides Found in Glycoproteins

Hexoses	Mannose (Man), Galactose (Gal)
Acetyl hexosamines	N-Acetylglucosamine (GlcNAc), N-acetylgalactosamine (GalNAc)
Pentoses	Arabinose (Ara), Xylose (Xyl)
Methyl pentose	L-Fucose (Fuc, see Figure 15–15)
Sialic acids	N-Acyl derivatives of neuraminic acid; the predominant sialic acid is N-acetylneuraminic acid (NeuAc, see Figure 15–15)

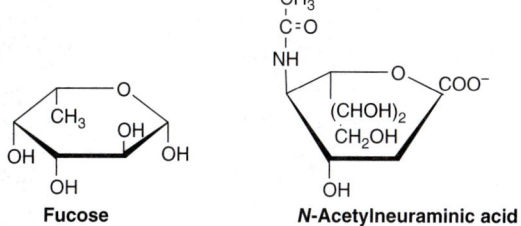

FIGURE 15–15 β-L-Fucose (6-deoxy-β-L-galactose) and *N*-acetylneuraminic acid, a sialic acid.

of neuraminic acid (see Figure 15–15). **Neuraminic acid** is a nine-carbon sugar derived from mannosamine (an epimer of glucosamine) and pyruvate. Sialic acids are constituents of both **glycoproteins** and **gangliosides**.

Proteins can undergo glycation, which is the nonenzymatic addition of a saccharide (eg, glucose) to a protein. Glycation increases as glucose levels increase. A glycated form of hemoglobin (hemoglobin A1c) increases in individuals with diabetes. It is used for the diagnosis of diabetes. It is also monitored during medical treatment of diabetes to determine how well glucose is managed.

CARBOHYDRATES OCCUR IN CELL MEMBRANES & IN LIPOPROTEINS

Approximately 5% of the weight of cell membranes is the carbohydrate part of glycoproteins (see Chapter 46) and glycolipids. Their presence on the outer surface of the plasma membrane (the **glycocalyx**) has been shown with the use of plant **lectins**, proteins that bind specific glycosyl residues. For example, **concanavalin A** binds α-glucosyl and α-mannosyl residues. **Glycophorin** is a major integral membrane glycoprotein of human erythrocytes. It has 130 amino acid residues and spans the lipid membrane, with polypeptide regions outside both the external and internal (cytoplasmic) surfaces. Carbohydrate chains are attached to the amino terminal portion outside the external surface. Carbohydrates are also present in apolipoprotein B of plasma lipoproteins. ABO blood groups are defined by different immunogenic molecules on erythrocytes. ABO blood group antigens are oligosaccharide chains that are attached to proteins. Polymorphisms in the enzyme ABO **glycosyltransferase** determine which single saccharide is placed in the oligosaccharide chain. The blood group A expresses A-transferase that places *N*-acetylgalactosamine. The blood group B expresses the B-transferase which places galactose. Individuals with blood type O have a mutation that produces an inactive ABO glycosyltransferase; no

oligosaccharide is added to the protein. Adding oligosaccharides to cell surface proteins make them an antigen. Individuals with blood type O do not have this modification are thus universal blood donors.

SUMMARY

- The glycome is the entire complement of sugars of an organism, whether free or present in more complex molecules. Glycomics is the study of glycomes, including genetic, physiological, pathological, and other aspects.

- Carbohydrates are major constituents of animal food and animal tissues. They are characterized by the type and number of monosaccharide residues in their molecules.

- Glucose is the most important carbohydrate in mammalian biochemistry because nearly all carbohydrate in food is converted to glucose for metabolism.

- Saccharides have large numbers of stereoisomers because they contain several asymmetric carbon atoms.

- The physiologically important monosaccharides include glucose, the "blood sugar," and ribose, an important constituent of nucleotides and nucleic acids.

- The important disaccharides include maltose (glucosyl–glucose), an intermediate in the digestion of starch; sucrose (glucosyl–fructose), important as a dietary constituent containing fructose; and lactose (galactosyl–glucose), in milk.

- Starch and glycogen are storage polymers of glucose in plants and animals, respectively. Starch is the major metabolic fuel in the diet.

- Complex carbohydrates contain other sugar derivatives such as amino sugars, uronic acids, and sialic acids. They include proteoglycans and glycosaminoglycans, which are associated with structural elements of the tissues, and glycoproteins, which are proteins containing oligosaccharide chains; they are found in many situations including the cell membrane.

- Oligosaccharide chains encode biological information, depending on their constituent sugars and their sequence and linkages.

The Citric Acid Cycle: A Pathway Central to Carbohydrate, Lipid, & Amino Acid Metabolism

Owen P. McGuinness, PhD

OBJECTIVES

After studying this chapter, you should be able to:

- Describe the reactions of the citric acid cycle.
- Identify the reactions that generate reducing equivalents that are oxidized in the mitochondrial electron transport chain to yield ATP.
- Identify the steps that require vitamins in the citric acid cycle.
- Explain how the citric acid cycle provides both a route for catabolism of amino acids and a route for their synthesis.
- Describe the main anaplerotic and cataplerotic pathways that permit replenishment and removal of citric acid cycle intermediates.
- Describe the role of the citric acid cycle in fatty acid synthesis.
- Explain how the activity of the citric acid cycle is controlled by the availability of oxidized cofactors.
- Explain how hyperammonemia can impair citric acid cycle flux.

BIOMEDICAL IMPORTANCE

The citric acid cycle (the Krebs or tricarboxylic acid cycle) is a sequence of reactions in mitochondria that oxidizes the acetyl moiety of acetyl-CoA to CO_2 and couples this to the reduction of coenzymes that are reoxidized in the electron transport chain (see Chapter 13), linked to the formation of ATP.

The citric acid cycle is the final common pathway for the oxidation of carbohydrate, lipid, and protein because glucose, fatty acids, and most amino acids are metabolized to acetyl-CoA or intermediates of the cycle. It also has a central role in gluconeogenesis, lipogenesis, and interconversion of amino acids. Many of these processes occur in most tissues, but liver is the only tissue in which all occur to a significant extent. The repercussions are therefore profound when, for example, large numbers of hepatic cells are damaged as in acute **hepatitis** or replaced by connective tissue (as in **cirrhosis**). The few genetic defects of citric acid cycle enzymes that have been reported are associated with severe neurologic damage as a result of impaired ATP formation in the central nervous system.

Hyperammonemia, as occurs in advanced liver disease, leads to loss of consciousness, coma, and convulsions. The impaired activity of the citric acid cycle leads to reduced formation of ATP. Ammonia both depletes citric acid cycle intermediates (by withdrawing α-ketoglutarate for the formation of glutamate and glutamine) and inhibits the oxidative decarboxylation of α-ketoglutarate.

THE CITRIC ACID CYCLE PROVIDES SUBSTRATES FOR THE RESPIRATORY CHAIN

The cycle starts with reaction between the acetyl moiety of acetyl-CoA and the four-carbon dicarboxylic acid oxaloacetate, forming a six-carbon tricarboxylic acid, citrate. In the subsequent reactions, two molecules of CO_2 are released and oxaloacetate is regenerated (**Figure 16–1**). Only a small quantity of oxaloacetate is needed for the oxidation of a large

This was adapted from chapter in 30th edition by David A. Bender, PhD, & Peter A. Mayes, PhD, DSc

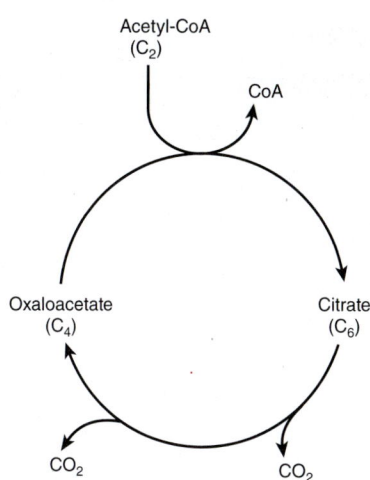

FIGURE 16–1 The citric acid cycle, illustrating the importance of regenerating oxaloacetate to sustain the cycle.

quantity of acetyl-CoA; it can be considered as playing a **catalytic role**, since it is regenerated at the end of the cycle.

The citric acid cycle provides the main pathway for ATP formation linked to the oxidation of metabolic fuels. During the oxidation of acetyl-CoA, coenzymes are reduced, then reoxidized in the respiratory chain, linked to the formation of ATP (oxidative phosphorylation, **Figure 16–2**; see also Chapter 13). This process is **aerobic**, requiring oxygen as the final oxidant of the reduced coenzymes. The enzymes of the citric acid cycle are located in the **mitochondrial matrix**, either free or attached to the inner mitochondrial membrane and the crista membrane. This is where the enzymes and coenzymes of the respiratory chain are also found (see Chapter 13).

To sustain the cycle, the number of carbons entering and leaving must be equal. A two-carbon molecule (Acetyl-CoA) combines with a four-carbon molecule (oxaloacetate) to form a six-carbon molecule (citrate). After one turn of the cycle citrate is converted back to oxaloacetate and two carbons are released as CO_2. If a metabolic pathway adds carbon to the cycle that is not though acetyl-CoA (pyruvate [C3] to oxaloacetate [C4] for gluconeogenesis; see Chapter 19) then there must be an outlet pathway to bring an equal amount of carbon out (oxaloacetate [C4] to phosphoenolpyruvate [C3]) of the cycle. The entry of carbon is called **anaplerosis**. The exiting of carbon is called **cataplerosis**. Thus to sustain the citric acid cycle anaplerosis must equal cataplerosis.

REACTIONS OF THE CITRIC ACID CYCLE GENERATE REDUCING EQUIVALENTS & CO_2

The initial reaction between acetyl-CoA and oxaloacetate (C4) to form citrate (C6) is catalyzed by **citrate synthase**, which forms a carbon–carbon bond between the methyl carbon of acetyl-CoA and the carbonyl carbon of oxaloacetate

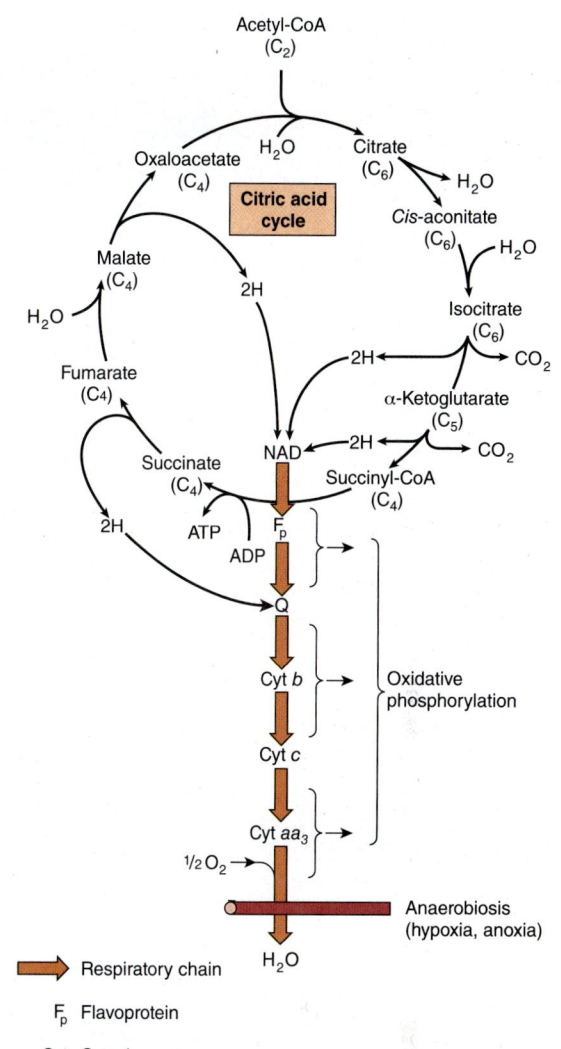

FIGURE 16–2 The citric acid cycle: the major catabolic pathway for acetyl-CoA. Acetyl-CoA, the product of carbohydrate, protein, and lipid catabolism, enters the cycle by forming citrate, and is oxidized to CO_2 with the reduction of coenzymes. Reoxidation of the coenzymes in the respiratory chain leads to phosphorylation of ADP to ATP. For one turn of the cycle, nine ATP are generated via oxidative phosphorylation and one ATP (or GTP) arises at substrate level from the conversion of succinyl-CoA to succinate.

(**Figure 16–3**). The thioester bond of the resultant citryl-CoA is hydrolyzed, releasing citrate and CoASH—an exothermic reaction. The coenzyme A released can be recycled in the conversion of pyruvate to acetyl-CoA by the pyruvate dehydrogenase complex.

Citrate is isomerized to isocitrate by the enzyme **aconitase** (aconitate hydratase); the reaction occurs in two steps: dehydration to *cis*-aconitate and rehydration to isocitrate. Although citrate is a symmetrical molecule, aconitase reacts with citrate asymmetrically, so that the two carbon atoms that are lost in subsequent reactions of the cycle are not those that were added from acetyl-CoA. This asymmetric behavior is the result of **channeling**—transfer of the product of citrate synthase directly onto the active site of aconitase, without

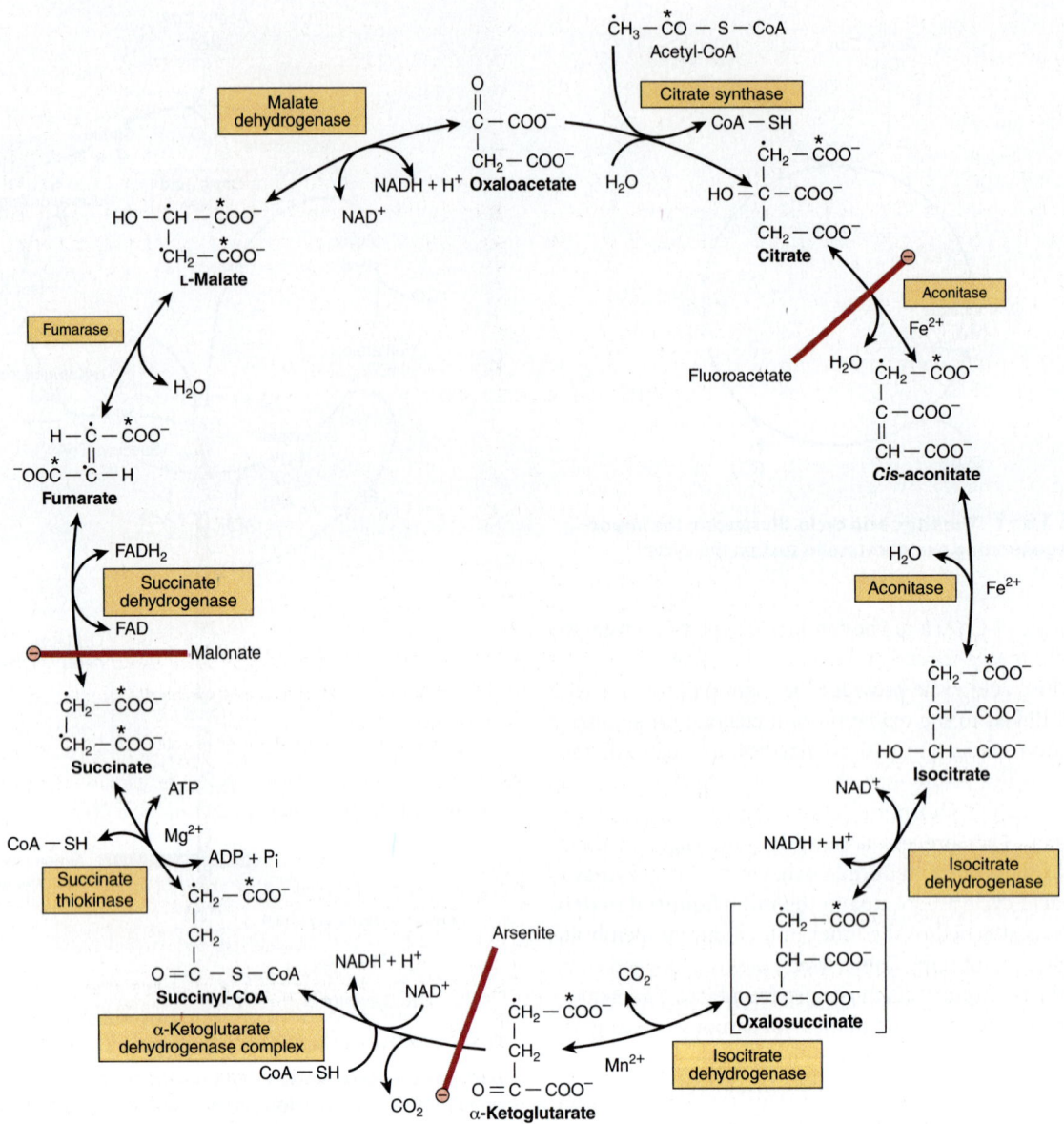

FIGURE 16–3 **The citric acid (Krebs) cycle.** Oxidation of NADH and $FADH_2$ in the respiratory chain leads to the formation of ATP via oxidative phosphorylation. In order to follow the passage of acetyl-CoA through the cycle, the two carbon atoms of the acetyl moiety are shown labeled on the carboxyl carbon (*) and on the methyl carbon (·). Although two carbon atoms are lost as CO_2 in one turn of the cycle, these atoms are not derived from the acetyl-CoA that has immediately entered the cycle, but from that portion of the citrate molecule that was derived from oxaloacetate. However, on completion of a single turn of the cycle, the oxaloacetate that is regenerated is now labeled, which leads to labeled CO_2 being evolved during the second turn of the cycle. Because succinate is a symmetrical compound, "randomization" of label occurs at this step so that all four carbon atoms of oxaloacetate appear to be labeled after one turn of the cycle. During gluconeogenesis, some of the label in oxaloacetate is incorporated into glucose and glycogen (see Figure 20–1). The sites of inhibition (⊖) by fluoroacetate, malonate, and arsenite are indicated.

entering free solution. As fatty acid synthesis uses an anaplerotic pathway (pyruvate carboxylase) to add oxaloacetate to the cycle, this in turn will make extra citrate and thus isocitrate. The excess isocitrate will act as a break on aconitase. Because of channeling, citrate is only available in free solution to be transported from the mitochondria to the cytosol for fatty acid synthesis when aconitase is inhibited by accumulation of its product, isocitrate. The "free" citrate can then be exported out

of the cycle (cataplerosis) to be a source of acetyl-CoA for fatty acid synthesis. This allows citric acid cycle activity to be maintained to generate ATP and reducing equivalents. This energy will be needed to support the energy demanding process of fatty acid synthesis while providing citrate in the cytosol as a source of acetyl-CoA for fatty acid synthesis.

The poison **fluoroacetate** is found in some of the plants, and their consumption can be fatal to grazing animals. Some fluorinated compounds used as anticancer agents and industrial chemicals (including pesticides) are metabolized to fluoroacetate. It is toxic because fluoroacetyl-CoA condenses with oxaloacetate to form fluorocitrate, which inhibits aconitase, causing citrate to accumulate.

Isocitrate undergoes dehydrogenation catalyzed by **isocitrate dehydrogenase** to form, initially, oxalosuccinate, which remains enzyme bound and undergoes decarboxylation to α-ketoglutarate. The decarboxylation requires Mg^{2+} or Mn^{2+} ions. There are three isoenzymes of isocitrate dehydrogenase. One, which uses nicotinamide adenine dinucleotide (NAD^+), is found only in mitochondria. The other two use $NADP^+$ and are found in mitochondria and the cytosol. Respiratory chain–linked oxidation of isocitrate occurs through the NAD^+-dependent enzyme.

α-Ketoglutarate undergoes **oxidative decarboxylation** in a reaction catalyzed by a multienzyme complex similar to that involved in the oxidative decarboxylation of pyruvate (see Figure 17–5). The α-**ketoglutarate dehydrogenase complex** requires the same cofactors as the pyruvate dehydrogenase complex—thiamin diphosphate, lipoate, NAD^+, flavin adenine dinucleotide (FAD), and CoA—and results in the formation of succinyl-CoA. The equilibrium of this reaction is so much in favor of succinyl-CoA formation that it must be considered to be physiologically unidirectional. As in the case of pyruvate oxidation (see Chapter 17), arsenite inhibits the reaction, causing the substrate, α-**ketoglutarate**, to accumulate. High concentrations of ammonia seen in liver disease inhibits α-ketoglutarate dehydrogenase.

Succinyl-CoA is converted to succinate by the enzyme **succinate thiokinase (succinyl-CoA synthetase)**. This is the only example of substrate-level phosphorylation (transfer of a phosphate group bound to the enzyme to GDP or ADP with generation of ATP or GTP) in the citric acid cycle. Tissues in which gluconeogenesis occurs (the liver and kidney) contain two isoenzymes of succinate thiokinase, one specific for GDP and the other for ADP. The GTP formed is used for the decarboxylation of oxaloacetate to phosphoenolpyruvate in gluconeogenesis, and provides a regulatory link between citric acid cycle activity and the withdrawal (cataplerosis) of oxaloacetate for gluconeogenesis. Nongluconeogenic tissues have only the isoenzyme that phosphorylates ADP.

When ketone bodies are being metabolized in extrahepatic tissues, there is an alternative reaction catalyzed by **succinyl-CoA–acetoacetate-CoA transferase (thiophorase)**, involving transfer of CoA from succinyl-CoA to acetoacetate, forming acetoacetyl-CoA and succinate (see Chapter 22).

The onward metabolism of succinate, leading to the regeneration of oxaloacetate, is the same sequence of chemical reactions as occurs in the β-oxidation of fatty acids: dehydrogenation to form a carbon–carbon double bond, addition of water to form a hydroxyl group, and a further dehydrogenation to yield the oxo-group of oxaloacetate.

The first dehydrogenation reaction, forming fumarate, is catalyzed by **succinate dehydrogenase**, which is bound to the inner surface of the inner mitochondrial membrane. The enzyme contains FAD and iron-sulfur (Fe-S) protein, and directly reduces ubiquinone in the electron transport chain. **Fumarase (fumarate hydratase)** catalyzes the addition of water across the double bond of fumarate, yielding malate. Malate is oxidized to oxaloacetate by **malate dehydrogenase**, linked to the reduction of NAD^+ to form NADH. Although the equilibrium of this reaction strongly favors malate, the net flux is to oxaloacetate because oxaloacetate is rapidly being used. Oxaloacetate is used for multiple reactions (form citrate, leave the mitochondria to be a substrate for gluconeogenesis, or to undergo transamination to form aspartate). NADH is reoxidized to NAD by the respiratory chain.

TEN ATP ARE FORMED PER TURN OF THE CITRIC ACID CYCLE

As a result of oxidations catalyzed by the dehydrogenases of the citric acid cycle, three molecules of NADH and one of $FADH_2$ are produced for each molecule of acetyl-CoA catabolized in one turn of the cycle. These reducing equivalents are transferred to the respiratory chain (see Figure 13–3), where reoxidation of each NADH results in formation of ~2.5 ATP, and of $FADH_2$, ~1.5 ATP. In addition, 1 ATP (or GTP) is formed by substrate-level phosphorylation catalyzed by succinate thiokinase.

VITAMINS PLAY KEY ROLES IN THE CITRIC ACID CYCLE

Four of the B vitamins (see Chapter 44) are essential in the citric acid cycle and hence energy-yielding metabolism: **riboflavin**, as FAD, is the cofactor for succinate dehydrogenase; **niacin**, as NAD^+, is the electron acceptor for isocitrate dehydrogenase, α-ketoglutarate dehydrogenase, and malate dehydrogenase; thiamin (**vitamin B_1**), as thiamin diphosphate, is the coenzyme for decarboxylation in the α-ketoglutarate dehydrogenase reaction; and **pantothenic acid**, as part of coenzyme A, is esterified to carboxylic acids to form acetyl-CoA and succinyl-CoA.

THE CITRIC ACID CYCLE PLAYS A PIVOTAL ROLE IN METABOLISM

The citric acid cycle is not only a pathway for oxidation of two carbon units, but is also a major pathway for interconversion of metabolites arising from **transamination** and **deamination** of amino acids (see Chapters 28 and 29), and providing the substrates for **amino acid synthesis** by transamination (see Chapter 27), as well as for **gluconeogenesis** (see Chapter 19) and **fatty acid synthesis** (see Chapter 23). Because it functions in both oxidative and synthetic processes, it is **amphibolic** (**Figure 16–4**).

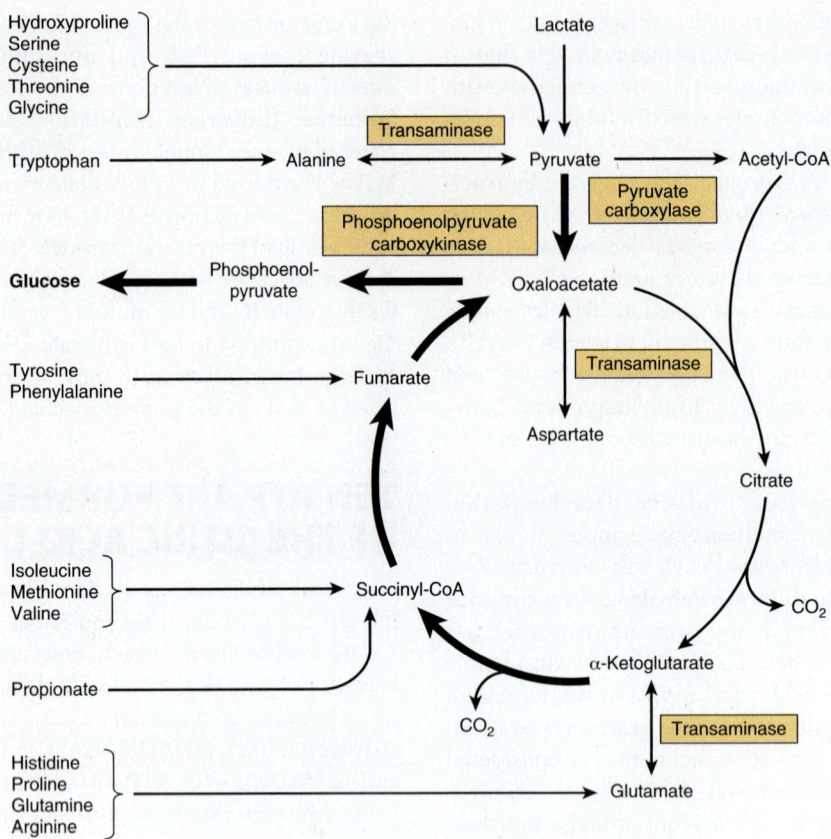

FIGURE 16–4 Involvement of the citric acid cycle in transamination and gluconeogenesis. The bold arrows indicate the main pathway of gluconeogenesis.

The Citric Acid Cycle Takes Part in Gluconeogenesis, Transamination, & Deamination

All the intermediates of the cycle are potentially **glucogenic**, since they can give rise to oxaloacetate, and hence production of glucose (in the liver and kidney, which carry out gluconeogenesis; see Chapter 19). The key enzyme that catalyzes transfer out of the cycle into gluconeogenesis is **phosphoenolpyruvate carboxykinase**, which catalyzes the decarboxylation of oxaloacetate to phosphoenolpyruvate, with GTP acting as the phosphate donor (see Figure 19–1). The GTP required for this reaction is provided by the GDP-dependent isoenzyme of succinate thiokinase. This ensures that oxaloacetate will not be withdrawn from the cycle for gluconeogenesis unless GTP was supplied. Otherwise this would lead to depletion of citric acid cycle intermediates, and hence reduced generation of ATP.

Net transfer into the cycle (ie, anaplerosis) occurs as a result of several reactions. Among the most important of such **anaplerotic** reactions is the formation of oxaloacetate by the carboxylation of pyruvate, catalyzed by **pyruvate carboxylase** (see Figure 16–4). This reaction is important in maintaining an adequate concentration of oxaloacetate for the condensation reaction with acetyl-CoA. If acetyl-CoA accumulates, it acts as both an allosteric activator of pyruvate carboxylase

and an inhibitor of pyruvate dehydrogenase, thereby ensuring a supply of oxaloacetate. Lactate, an important substrate for gluconeogenesis, enters the cycle via oxidation to pyruvate and then carboxylation to oxaloacetate. **Glutamate** and **glutamine** are important anaplerotic substrates. They yield α-ketoglutarate as a result of the reactions catalyzed by glutaminase and glutamate dehydrogenase. Transamination of **aspartate** leads directly to the formation of oxaloacetate, and a variety of compounds that are metabolized can yield **propionyl CoA**, which can be carboxylated and isomerized to succinyl CoA, are also important anaplerotic substrates.

Aminotransferase (transaminase) reactions form pyruvate from alanine, oxaloacetate from aspartate, and α-ketoglutarate from glutamate. Because these reactions are reversible, the cycle also serves as a source of carbon skeletons (ie, cataplerosis) for the synthesis of these amino acids. Other amino acids contribute to gluconeogenesis because their carbon skeletons give rise to citric acid cycle intermediates. Alanine, cysteine, glycine, hydroxyproline, serine, threonine, and tryptophan yield pyruvate; arginine, histidine, glutamine, and proline yield α-ketoglutarate; isoleucine, methionine, and valine yield succinyl-CoA; tyrosine and phenylalanine yield fumarate (see Figure 16–4).

The citric acid cycle itself does not provide a pathway for the complete oxidation of the carbon skeletons of amino acids that give rise to intermediates such as α-ketoglutarate, succinyl CoA, fumarate, and oxaloacetate, because this results in an increase in the amount of oxaloacetate. For complete oxidation to occur a cataplerotic route has to be used. Oxaloacetate must leave the cycle to undergo phosphorylation and carboxylation to phosphoenolpyruvate (at the expense of GTP), then be dephosphorylated to form pyruvate (catalyzed by pyruvate kinase). Pyruvate can then undergo oxidative decarboxylation to acetyl-CoA (catalyzed by pyruvate dehydrogenase).

In ruminants, main metabolic fuel is short-chain fatty acids formed by bacterial fermentation and the formation of propionate. Propionate is the major glucogenic product of rumen fermentation. It enters the cycle at succinyl-CoA via the methylmalonyl-CoA pathway (see Figure 19–2) to form glucose via gluconeogenesis.

The Citric Acid Cycle Takes Part in Fatty Acid Synthesis

Acetyl-CoA, formed from pyruvate by the action of pyruvate dehydrogenase, is the major substrate for long-chain fatty acid synthesis in nonruminants (**Figure 16–5**). (In ruminants, acetyl-CoA is derived directly from acetate.) Pyruvate dehydrogenase is a mitochondrial enzyme, and fatty acid synthesis is in the cytosol; the mitochondrial membrane is impermeable to acetyl-CoA. For acetyl-CoA to be available in the cytosol, citrate is transported from the mitochondrion to the cytosol, then cleaved in a reaction catalyzed by **citrate lyase** (see Figure 16–5). Citrate is only available for transport out of the mitochondrion when aconitase is inhibited by its product and therefore saturated with its substrate, so that citrate cannot be

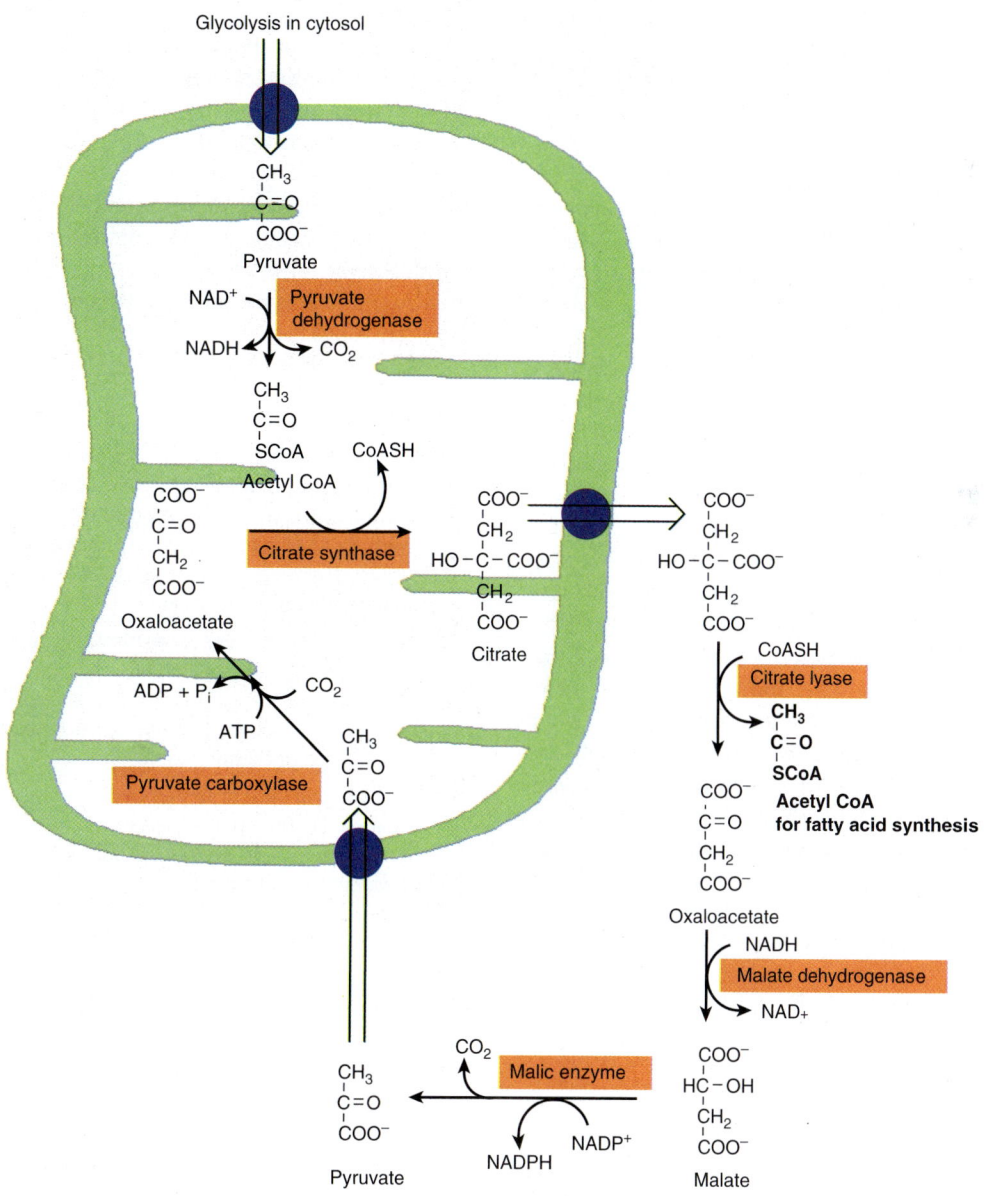

FIGURE 16–5 Participation of the citric acid cycle in provision of cytosolic acetyl-CoA for fatty acid synthesis from glucose. See also Figure 23–5.

channeled directly from citrate synthase onto aconitase. This ensures that citrate is used for fatty acid synthesis only when there is an adequate amount to ensure continued activity of the cycle. The identity and regulation of the many mitochondrial transport systems (eg, malate, citrate, pyruvate) involved in cataplerosis and anaplerosis pathways are poorly understood, but they likely play a critical regulatory role in these pathways.

The oxaloacetate released by citrate lyase cannot reenter the mitochondrion, but is reduced to malate, at the expense of NADH, and the malate undergoes oxidative decarboxylation to pyruvate, reducing $NADP^+$ to NADPH. This reaction, catalyzed by the malic enzyme, is the source of half the NADPH required for fatty acid synthesis (the remainder is provided by the pentose phosphate pathway; see Chapter 20). Pyruvate enters the mitochondrion and is carboxylated to oxaloacetate by pyruvate carboxylase, an ATP-dependent reaction in which the coenzyme is the vitamin biotin.

Regulation of the Citric Acid Cycle Depends Primarily on a Supply of Oxidized Cofactors

In most tissues, where the primary role of the citric acid cycle is in energy-yielding metabolism, **respiratory control** via the respiratory chain and oxidative phosphorylation regulates citric acid cycle activity (see Chapter 13). Thus, cycle activity is immediately dependent on the supply of NAD^+, which in turn, because of the tight coupling between oxidation and phosphorylation, is dependent on the availability of ADP and hence, ultimately on the utilization of ATP in chemical and physical work. Thus as ATP usage increases (eg, muscle contraction) this will generate ADP to help drive the cycle. If ATP demand is low the cycle will slow down to match the demand. In addition, individual enzymes of the cycle are regulated. In general these regulatory events help to couple rapid changes in energy demand with citric acid cycle flux. The main sites for regulation are the nonequilibrium reactions catalyzed by pyruvate dehydrogenase, citrate synthase, isocitrate dehydrogenase, and α-ketoglutarate dehydrogenase. The dehydrogenases are activated by Ca^{2+}, which increases in concentration during contraction of muscle and during secretion by other tissues, when there is increased energy demand. In a tissue such as brain, which is largely dependent on glucose to supply acetyl-CoA, control of the citric acid cycle may occur at pyruvate dehydrogenase.

Several enzymes are also responsive to the energy status as reflected by the [ATP]/[ADP] and [NADH]/[NAD$^+$] ratios.

Thus, there is allosteric inhibition of citrate synthase by ATP and long-chain fatty acyl-CoA. Allosteric activation of mitochondrial NAD-dependent isocitrate dehydrogenase by ADP is counteracted by ATP and NADH. The α-ketoglutarate dehydrogenase complex is regulated in the same way as is pyruvate dehydrogenase (see Figure 17–6). Succinate dehydrogenase is inhibited by oxaloacetate, and the availability of oxaloacetate is controlled by malate dehydrogenase and depends on the [NADH]/[NAD$^+$] ratio. Since the K_m of citrate synthase for oxaloacetate is of the same order of magnitude as the intramitochondrial concentration, it is likely that the concentration of oxaloacetate controls the rate of citrate formation.

Hyperammonemia, as occurs in advanced liver disease and a number of (rare) genetic diseases of amino acid metabolism, leads to loss of consciousness, coma and convulsions, and may be fatal. This is largely because of the withdrawal (cataplerosis) of α-ketoglutarate to form glutamate (catalyzed by glutamate dehydrogenase) and then glutamine (catalyzed by glutamine synthetase) from the citric acid cycle, which is not matched by anaplerosis. It lowers concentrations of all citric acid cycle intermediates, citric acid cycle flux, and the generation of ATP. The equilibrium of glutamate dehydrogenase is finely poised, and the direction of reaction depends on the ratio of NAD$^+$:NADH and the concentration of ammonium ions, which is elevated in liver disease. In addition, ammonia inhibits α-ketoglutarate dehydrogenase, and possibly also pyruvate dehydrogenase further decreasing citric acid cycle flux.

SUMMARY

- The citric acid cycle is the final pathway for the oxidation of carbohydrate, lipid, and protein. Their common end-metabolite, acetyl-CoA, reacts with oxaloacetate to form citrate. By a series of dehydrogenations and decarboxylations, citrate is degraded, reducing coenzymes, releasing two CO_2, and regenerating oxaloacetate.

- The reduced coenzymes are oxidized by the respiratory chain linked to formation of ATP. Thus, the cycle is the major pathway for the formation of ATP. It is located in the matrix of mitochondria adjacent to the enzymes of the respiratory chain and oxidative phosphorylation.

- The citric acid cycle is amphibolic, since in addition to oxidation it is important in the provision of carbon skeletons for gluconeogenesis, acetyl-CoA for fatty acid synthesis, and interconversion of amino acids. For these processes to be sustained they rely on a balance of anaplerosis and cataplerosis in the citric acid cycle.

Glycolysis & the Oxidation of Pyruvate

Owen P. McGuinness, PhD

OBJECTIVES

After studying this chapter, you should be able to:

- Describe the pathway of glycolysis and its control, and explain how glycolysis can operate under anaerobic conditions.
- Describe the reaction of pyruvate dehydrogenase and its regulation.
- Explain how inhibition of pyruvate oxidation leads to lactic acidosis.

BIOMEDICAL IMPORTANCE

Most tissues have at least some requirement for glucose. The brain is predominantly reliant on glucose for its energy needs, except in prolonged fasting where ketone bodies can meet about 20% of its energy needs. Glycolysis is the main pathway through which cells metabolize glucose and other carbohydrates. It occurs in the cytosol of cells and can function either aerobically or anaerobically depending on the availability of oxygen and the activity of the electron transport chain (and hence of the presence of mitochondria). Erythrocytes, which lack mitochondria, are completely reliant on glucose as their metabolic fuel, and metabolize it by anaerobic glycolysis.

The ability of glycolysis to provide ATP in the absence of oxygen allows skeletal muscle to perform at very high levels of work output when oxygen supply is insufficient, and it allows tissues to survive anoxic episodes. However, heart muscle, which is adapted for aerobic performance, has relatively low glycolytic activity and poor survival under conditions of **ischemia**. Diseases in which enzymes of glycolysis (eg, pyruvate kinase) are deficient are mainly seen as **hemolytic anemias** or, if the defect affects skeletal muscle (eg, phosphofructokinase), as **fatigue**. In fast-growing cancer cells, glycolysis proceeds at a high rate, forming large amounts of pyruvate, which is reduced to lactate and released. The lactate is used for gluconeogenesis in the liver (see Chapter 19) and combined with the marked increase in hepatic protein synthesis secondary to the increased delivery of amino acids from the catabolic muscle contributes to the **hypermetabolism** seen in **cancer cachexia**. **Lactic acidosis** can be of two types. Type A is the most common and is due to impaired tissue perfusion or hypoxia (eg, sepsis and hypovolemia). Type B is due to the impaired ability to metabolize lactate (eg, liver disease, thiamine deficiency). Thiamin (vitamin B_1) deficiency impairs the activity of pyruvate dehydrogenase.

GLYCOLYSIS CAN FUNCTION UNDER ANAEROBIC CONDITIONS

Early in the investigations of glycolysis, it was realized that fermentation in yeast was similar to the breakdown of glycogen in muscle. When a muscle contracts under anaerobic conditions, **glycogen disappears** and **lactate appears**. When oxygen is made available, aerobic recovery takes place and lactate is no longer produced. If muscle contraction occurs under aerobic conditions, lactate does not accumulate as pyruvate and NADH, the products of glycolysis. Pyruvate and NADH can enter the mitochondria where they are oxidized further to CO_2, H_2O, and NAD (**Figure 17–1**). When oxygen is in short supply, mitochondrial reoxidation of NADH formed during glycolysis is impaired. As there is a limited supply of NAD to sustain glycolysis the NADH has to be reoxidized. The NADH is reoxidized by reducing pyruvate to lactate. This allows NAD to be available for the early steps in the pathway, so permitting glycolysis to continue. While glycolysis can occur under anaerobic conditions, this has a price. It limits the amount of ATP formed per mole of glucose oxidized, so that much more glucose must be metabolized under anaerobic than aerobic conditions to supply the same quantity of ATP to supply cellular work (**Table 17–1**). In yeast and some other microorganisms, pyruvate formed in anaerobic glycolysis is not reduced to lactate, but is decarboxylated and reduced to form ethanol.

This was adapted from chapter in 30th edition by David A. Bender, PhD, & Peter A. Mayes, PhD, DSc

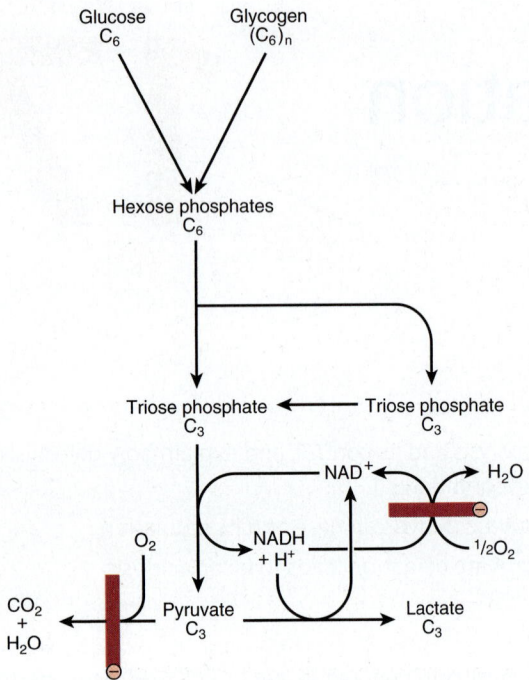

FIGURE 17–1 **Summary of glycolysis.** ⊖, blocked under anaerobic conditions or by absence of mitochondria containing key respiratory enzymes, as in erythrocytes.

GLYCOLYSIS CONSTITUTES THE MAIN PATHWAY OF GLUCOSE UTILIZATION

The overall equation for glycolysis from glucose to lactate is as follows:

$$\text{Glucose} + 2\text{ ADP} + 2\text{ P}_i \rightarrow 2\text{ Lactate} + 2\text{ ATP} + 2\text{ H}_2\text{O}$$

All of the enzymes of glycolysis (**Figure 17–2**) are cytosolic. The general mechanism for the generation of ATP during glycolysis rearranges a phosphorylated monosaccharide glucose-6-phosphate to phosphorylated compounds with a high potential to transfer their phosphate groups. These triosephosphates transfer their phosphate to ADP to form ATP. This process is called substrate-level phosphorylation, as the phosphate is directly donated to ATP from an intermediate in the pathway. This pathway uses ADP and produces ATP. As ADP is limiting, it is required that the generated ATP be used to perform some metabolic work to regenerate the ADP to sustain glycolysis.

After being transported across the plasma membrane by facilitated glucose transporters glucose enters glycolysis by phosphorylation to glucose-6-phosphate, catalyzed by **hexokinase**, using ATP as the phosphate donor. Under physiologic conditions, the phosphorylation of glucose to glucose-6-phosphate can be regarded as irreversible. Hexokinase is inhibited allosterically by its product, glucose-6-phosphate.

TABLE 17–1 **ATP Formation in the Catabolism of Glucose**

Pathway	Reaction Catalyzed by	Method of ATP Formation	ATP per mol of Glucose
Glycolysis	Glyceraldehyde-3-phosphate dehydrogenase	Respiratory chain oxidation of 2 NADH	5[a]
	Phosphoglycerate kinase	Substrate-level phosphorylation	2
	Pyruvate kinase	Substrate-level phosphorylation	2
			9
	Consumption of ATP for reactions of hexokinase and phosphofructokinase		−2
			Net 7
Citric acid cycle	Pyruvate dehydrogenase	Respiratory chain oxidation of 2 NADH	5
	Isocitrate dehydrogenase	Respiratory chain oxidation of 2 NADH	5
	α-Ketoglutarate dehydrogenase	Respiratory chain oxidation of 2 NADH	5
	Succinate thiokinase	Substrate-level phosphorylation	2
	Succinate dehydrogenase	Respiratory chain oxidation of 2 FADH$_2$	3
	Malate dehydrogenase	Respiratory chain oxidation of 2 NADH	5
			Net 25
	Total per mol of glucose under aerobic conditions		32
	Total per mol of glucose under anaerobic conditions		2

[a]This assumes that NADH formed in glycolysis is transported into mitochondria by the malate shuttle (see Figure 13–13). If the glycerophosphate shuttle is used, then only 1.5 ATP will be formed per mol of NADH. There is a considerable advantage in using glycogen rather than glucose for anaerobic glycolysis in muscle, since the product of glycogen phosphorylase is glucose-1-phosphate (see Figure 18–1), which is interconvertible with glucose-6-phosphate. This saves the ATP that would otherwise be used by hexokinase, increasing the net yield of ATP from 2 to 3 per glucose. (Note: The initial formation of glycogen requires ATP so in a net sense glucose first going to glycogen and then coming back out for oxidation via pyruvate is the same. It's just that you have more ATP at a time where ATP is in short supply.

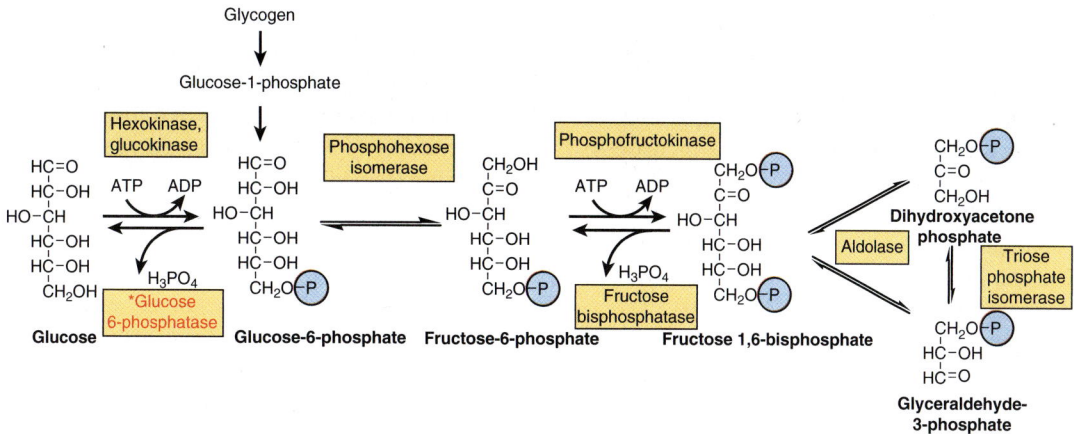

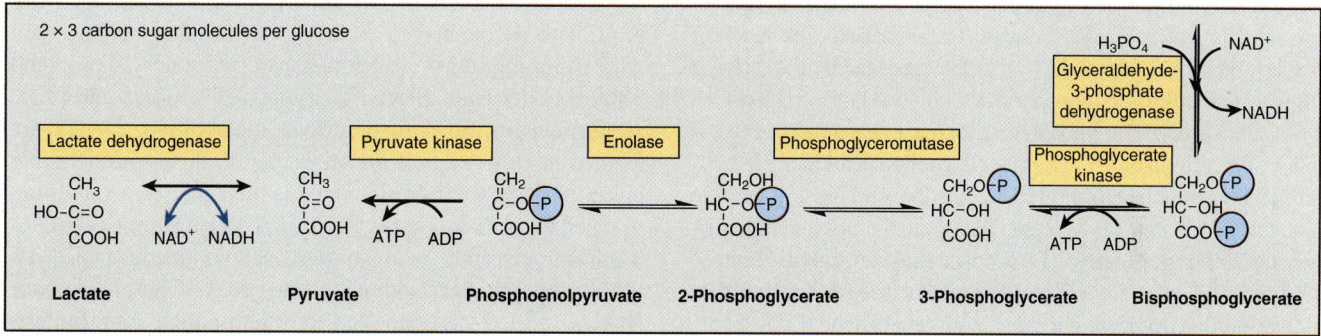

FIGURE 17–2 **The pathway of glycolysis.** (Ⓟ, —PO$_3^{2-}$; P$_i$, HOPO$_3^{2-}$; ⊖, inhibition.) Carbons 1–3 of fructose bisphosphate form dihydroxy-acetone phosphate, and carbons 4–6 form glyceraldehyde-3-phosphate. Glucose-6-phosphatase is expressed *only* in the liver, kidney, and pancreatic islet; it is not expressed in other tissues.

In muscle and adipose tissue, the transport of glucose is stimulated by insulin. In muscle glucose is used for glycolysis (or glycogen synthesis; see Chapter 18). In adipose tissue, it is used for lipogenesis (see Chapter 23). The hexokinase expressed in most tissues has a high affinity (low K_m) for glucose and is allosterically inhibited by its product glucose-6-phosphate. When transport activity is low, transport acts as a barrier to glucose uptake. Thus transport activity is an important determinant of the overall rate of glycolysis. The combined effect of a low transport activity and the high affinity of hexokinase for glucose keeps the intracellular glucose very low. Thus, the concentration of plasma glucose and the cellular ATP demand are the primary determinants of glycolysis in an aerobic environment. In the presence of insulin transport activity increases, thus lowering the barrier to glucose entry. If, for example, transport activity increased 10-fold glycolysis will not increase 10-fold because the consequent increase in glucose-6-phosphate would serve as a brake on hexokinase to limit the overall rate of glycolysis. This effectively shifts the control of glycolysis to pathways (eg, pyruvate oxidation) downstream of hexokinase. In muscle, hexokinase is found bound to the outer mitochondrial membrane. As hexokinase requires ATP, it creates a coupling between hexokinase activity and mitochondrial ATP generation.

In the liver transport activity is not regulated and a different isozyme of hexokinase is expressed (**glucokinase**). **Glucokinase** has a K_m higher than the normal plasma concentration of glucose and it is not inhibited by its product glucose-6-phosphate. Because of the relatively high constitutive glucose transport activity and the low affinity of glucokinase for glucose, the intracellular glucose in the liver is very similar to plasma glucose. The liver can be both a consumer and producer of glucose (feasting vs fasting; see Chapter 14). In contrast to most tissues the liver also expresses glucose-6-phosphatase, which allows the liver in the fasting state to dephosphorylate glucose-6-phosphate generated by the liver and release glucose. In the fed state the function of glucokinase in the liver is to remove glucose from the hepatic portal blood following a meal; this limits the quantity of glucose available to peripheral tissues. Glucokinase in the fasted setting is found inactive in the nucleus bound to a glucokinase regulatory protein. In response to signals during a meal it leaves the nucleus and resides in the cytosol where it is active. Because of the high capacity of glucokinase to phosphorylate glucose and the increase in glucose in the portal vein during a meal, it provides more glucose-6-phosphate than is required for liver glucose oxidation, thus a large fraction is used for glycogen synthesis and a smaller amount is used for lipogenesis.

Glucokinase is also found in pancreatic islet β cells, where it functions to detect changes in concentrations of glucose in the systemic circulation. When glucose is increased more glucose is phosphorylated by glucokinase, increasing glycolysis, and leading to increased formation of ATP. The increase in ATP leads to closure of an ATP sensitive potassium channel, causing membrane depolarization and opening of a voltage-gated calcium channel. The resultant influx of calcium ions leads to fusion of the insulin secretory granules with the cell membrane and the release of insulin.

Glucose-6-phosphate is an important compound at the junction of several metabolic pathways: glycolysis, gluconeogenesis (see Chapter 19), the pentose phosphate pathway (see Chapter 20), glycogenesis, and glycogenolysis (see Chapter 18). In glycolysis, it is converted to fructose-6-phosphate by **phosphohexose isomerase**, which involves an aldose–ketose isomerization. This reaction is followed by another phosphorylation catalyzed by the enzyme **phosphofructokinase** (phosphofructokinase-1) forming fructose 1,6-bisphosphate. The phosphofructokinase reaction is irreversible under physiologic conditions. Phosphofructokinase is both inducible and subject to allosteric regulation, and has a major role in regulating the rate of glycolysis. Note up to this point in the pathway no ATP is generated; ATP is only being consumed. Fructose 1,6-bisphosphate is cleaved by **aldolase** (fructose 1,6-bisphosphate aldolase) into two triose phosphates, glyceraldehyde-3-phosphate and dihydroxyacetone phosphate, which are interconverted by the enzyme **phosphotriose isomerase**.

Glycolysis continues with the oxidation of glyceraldehyde-3-phosphate to 1,3-bisphosphoglycerate and the formation of NADH. The enzyme catalyzing this oxidation, **glyceraldehyde-3-phosphate dehydrogenase**, is NAD dependent. Structurally, it consists of four identical polypeptides (monomers) forming a tetramer. Four —SH groups are present on each polypeptide, derived from cysteine residues within the polypeptide chain. One of the —SH groups is found at the active site of the enzyme (**Figure 17–3**). The substrate initially combines with this —SH group, forming a thiohemiacetal that is oxidized to a thiol ester; the hydrogens removed in this oxidation are transferred to NAD⁺. The thiol ester then undergoes phosphorolysis; inorganic phosphate (P_i) is added, forming 1,3-bisphosphoglycerate and the free —SH group.

In the next reaction, catalyzed by **phosphoglycerate kinase**, phosphate is transferred from 1,3-bisphosphoglycerate onto ADP, forming ATP (substrate-level phosphorylation) and 3-phosphoglycerate. Since two molecules of triose phosphate are formed per molecule of glucose metabolized, 2× ATP are formed in this reaction per molecule of glucose undergoing glycolysis. The toxicity of arsenic is the result of competition of arsenate with inorganic phosphate (P_i) forming 1-arseno-3-phosphoglycerate, which undergoes spontaneous hydrolysis to 3-phosphoglycerate without forming ATP. 3-Phosphoglycerate is isomerized to 2-phosphoglycerate by **phosphoglycerate mutase**. It is likely that 2,3-bisphosphoglycerate (diphosphoglycerate, DPG) is an intermediate in this reaction.

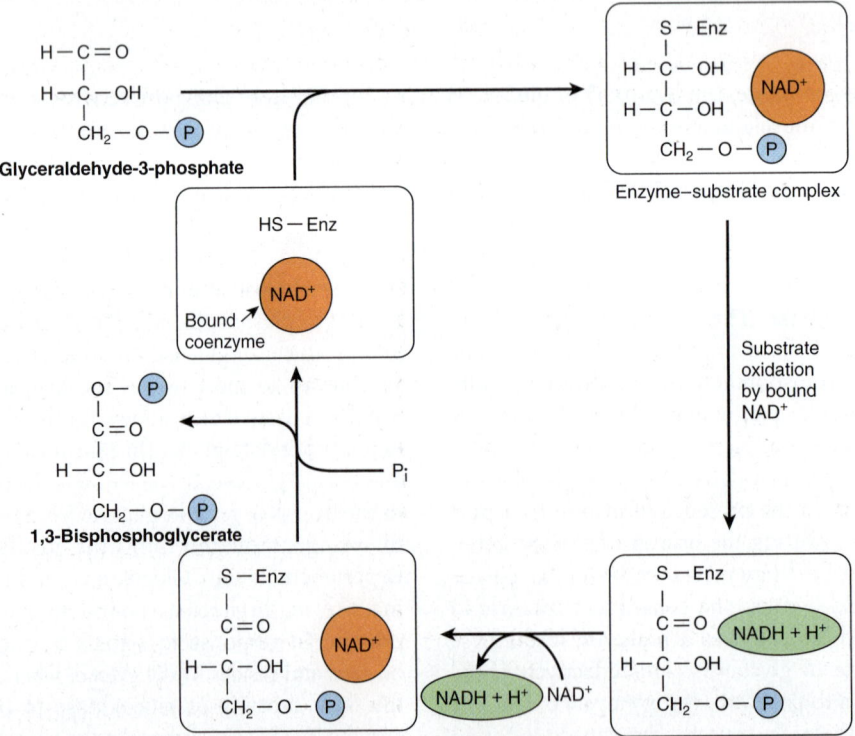

FIGURE 17–3 **Mechanism of oxidation of glyceraldehyde-3-phosphate.** (Enz, glyceraldehyde-3-phosphate dehydrogenase.) The enzyme is inhibited by the —SH poison iodoacetate, which is thus able to inhibit glycolysis. The NADH produced on the enzyme is not so firmly bound to the enzyme as is NAD⁺. Consequently, NADH is easily displaced by another molecule of NAD⁺.

Enolase catalyzes the next step. It involves dehydration, forming phosphoenolpyruvate. Enolase is inhibited by **fluoride**. When blood samples are taken for measurement of glucose, the sample is placed into tubes containing fluoride to inhibit glycolysis and prevent the breakdown of the glucose until the sample is analyzed. Enolase is also dependent on the presence of either Mg^{2+} or Mn^{2+} ions.

The phosphate of phosphoenolpyruvate is transferred to ADP in another substrate-level phosphorylation catalyzed by **pyruvate kinase** to form 2× ATP per molecule of glucose oxidized. The reaction of pyruvate kinase is essentially irreversible under physiologic conditions, partly because of the large free-energy change involved and partly because the immediate product of the enzyme-catalyzed reaction is enolpyruvate, which undergoes spontaneous isomerization to pyruvate, so that the product of the reaction is not available to undergo the reverse reaction.

The availability of oxygen determines which of the two pathways pyruvate follows. Under **anaerobic conditions**, the NADH cannot be reoxidized through the respiratory chain, and pyruvate is reduced to lactate catalyzed by **lactate dehydrogenase**. This permits the oxidization of NADH to form NAD permitting another molecule of glucose to undergo glycolysis. Under **aerobic conditions**, pyruvate is transported into the mitochondria and undergoes oxidative decarboxylation to acetyl-CoA then oxidation to CO_2 in the citric acid cycle (see Chapter 16). The reducing equivalents from the NADH formed in glycolysis are taken up into mitochondria for oxidation via either the malate-aspartate shuttle or the glycerophosphate shuttle (see Chapter 13).

TISSUES THAT FUNCTION UNDER HYPOXIC CONDITIONS OR HAVE INTRISICALLY HIGH RATES OF GLUCOSE OXIDATION PRODUCE LACTATE

In organs where glucose is the major substrate for oxidation, a variable fraction of the glucose uptake can be diverted to lactate depending on the metabolic state of the tissue. In tissues or metabolic states where pyruvate oxidation is absent or impaired, pyruvate is diverted to lactate and released so as to recycle the NAD to support glycolysis and the generation of ATP to support cellular metabolic work. In erythrocytes pyruvate always terminates in lactate. They lack mitochondria and thus cannot oxidize pyruvate. This is true of some skeletal muscle, particularly the white fibers. During states of high-work output they mobilize muscle glycogen stores. Because they lack glucose-6-phosphatase the glucose-6-phosphate mobilized from glycogen traverses down the glycolytic pathway and combined with the relatively low rates of pyruvate oxidation due to lower mitochondrial oxidative capacity, lactate is released. The need for ATP formation may exceed the rate at which oxygen is delivered. In those tissues they rely on anaerobic glycolysis to support metabolism (eg, tumors, retina and renal medulla, skin). Lactate production is also increased in septic shock secondary to both impaired oxygen delivery and shifts on metabolic capacity due to cellular injury. Other tissues like the brain that normally derive much of their energy from glucose oxidation produce some lactate (3-5% of total glycolytic flux) because of the high glycolytic rates. The liver, kidneys, oxidative skeletal muscle, and heart have very high oxidative capacity to oxidize multiple fuels. They normally take up lactate and oxidize it, but can produce it under hypoxic conditions. When lactate production is high, as in vigorous exercise, septic shock, and cancer cachexia, lactate is used in the liver for gluconeogenesis (see Chapter 19), leading to an increase in metabolic rate to provide the ATP and GTP needed. This increase may contribute to the increase in metabolic rate after cessation of vigorous exercise.

Under some states lactate can be directly transported into the mitochondria to generate pyruvate and NADH. This allows for the transfer of reducing equivalents (eg, NADH) from the cytosol into the mitochondrion for the electron transport chain in addition to the glycerophosphate (see Figure 13–12) and malate-aspartate (see Figure 13–13) shuttles.

GLYCOLYSIS IS REGULATED AT THREE NONEQUILIBRIUM REACTIONS

Although most of the reactions of glycolysis are freely reversible, three are markedly exergonic and are considered to be physiologically irreversible. These reactions, catalyzed by **hexokinase** (and glucokinase), **phosphofructokinase**, and **pyruvate kinase**, are the major sites of regulation of glycolysis. Phosphofructokinase is significantly inhibited at normal intracellular concentrations of ATP; as discussed in Chapter 19, this inhibition can be rapidly relieved by 5′ AMP that is formed as ADP begins to accumulate. It is due to a failure to rapidly convert ADP to ATP, signaling the need for an increased rate of glycolysis. Cells that are capable of **gluconeogenesis** (reversing the glycolytic pathway; see Chapter 19) have different enzymes that catalyze reactions to reverse these irreversible steps: glucose-6-phosphatase, fructose 1,6-bisphosphatase and, to reverse the reaction of pyruvate kinase, pyruvate carboxylase, and phosphoenolpyruvate carboxykinase. The reciprocal regulation of phosphofructokinase in glycolysis and fructose 1,6-bisphosphatase in gluconeogenesis is discussed in Chapter 19.

Fructose enters glycolysis by phosphorylation to fructose-1-phosphate, and bypasses the main regulatory steps, resulting in formation of more pyruvate and acetyl-CoA than is required for ATP formation. In addition, increases in fructose-1-P in the liver activate glucokinase (compete for binding of glucokinase with the glucokinase regulatory protein), amplifying hepatic glucose uptake and increasing hepatic lipogenesis and predisposing to hepatic steatosis.

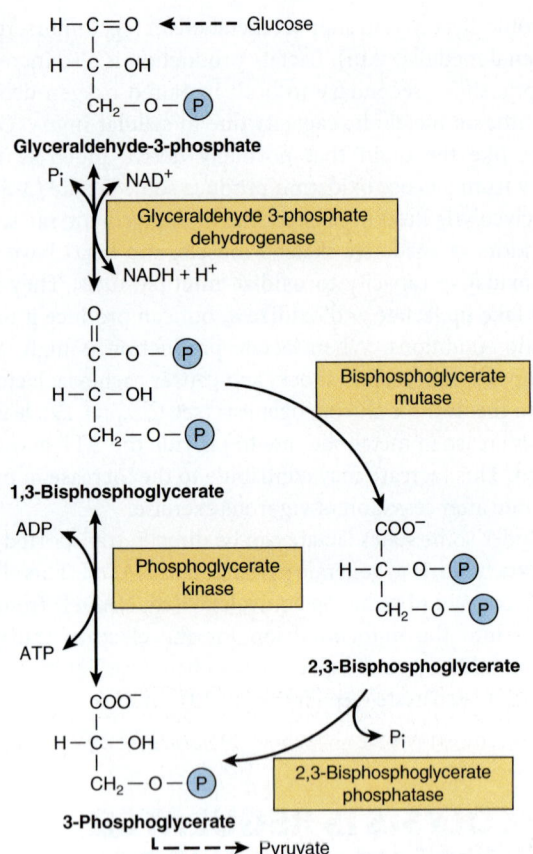

FIGURE 17–4 The 2,3-bisphosphoglycerate pathway in erythrocytes.

In Erythrocytes, the First Site of ATP Formation in Glycolysis May Be Bypassed

In erythrocytes, the reaction catalyzed by **phosphoglycerate kinase** may be bypassed to some extent by the reaction of **bisphosphoglycerate mutase**, which catalyzes the conversion of 1,3-bisphosphoglycerate to 2,3-bisphosphoglycerate, followed by hydrolysis to 3-phosphoglycerate and P_i, catalyzed by **2,3-bisphosphoglycerate phosphatase (Figure 17–4)**. This pathway involves no net yield of ATP from glycolysis, but provides 2,3-bisphosphoglycerate, which binds to hemoglobin, decreasing its affinity for oxygen. This increases the efficiency with which hemoglobin delivers oxygen to tissues (see Chapter 6).

THE OXIDATION OF PYRUVATE TO ACETYL-COA IS THE IRREVERSIBLE ROUTE FROM GLYCOLYSIS TO THE CITRIC ACID CYCLE

Pyruvate is transported into the mitochondrion by a proton symporter. It then undergoes oxidative decarboxylation to acetyl-CoA, catalyzed by a multienzyme complex that is associated

with the inner mitochondrial membrane. This **pyruvate dehydrogenase complex** is analogous to the α-ketoglutarate dehydrogenase complex of the citric acid cycle (see Chapter 16). Pyruvate is decarboxylated by the **pyruvate dehydrogenase** component of the enzyme complex to a hydroxyethyl derivative of the thiazole ring of enzyme-bound **thiamin diphosphate**, which in turn reacts with oxidized lipoamide, the prosthetic group of **dihydrolipoyl transacetylase**, to form acetyl lipoamide (**Figure 17–5**). In thiamin (vitamin B_1; see Chapter 44) deficiency, pyruvate oxidation is impaired, and there is significant (and potentially life-threatening) lactic and pyruvic acidosis. Acetyl lipoamide reacts with coenzyme A to form acetyl-CoA and reduced lipoamide. The reaction is completed when the reduced lipoamide is reoxidized by a flavoprotein, **dihydrolipoyl dehydrogenase**, containing flavin adenine dinucleotide (FAD). Finally, the reduced flavoprotein is oxidized by NAD^+, which in turn transfers reducing equivalents to the respiratory chain. The overall reaction is:

$$\text{Pyruvate} + NAD^+ + CoA \rightarrow \text{Acetyl-CoA} + NADH + H^+ + CO_2$$

The pyruvate dehydrogenase complex consists of a number of polypeptide chains of each of the three component enzymes and the intermediates do not dissociate, but are channeled from one enzyme site to the next. This increases the rate of reaction and prevents side reactions.

Pyruvate Dehydrogenase Is Regulated by End-Product Inhibition & Covalent Modification

Pyruvate dehydrogenase is inhibited by its products, acetyl-CoA and NADH (**Figure 17–6**). It is also regulated by phosphorylation (catalyzed by a kinase) of three serine residues on the pyruvate dehydrogenase component of the multienzyme complex, resulting in decreased activity and by dephosphorylation (catalyzed by a phosphatase) that causes an increase in activity. The kinase is activated by increases in the [ATP]/[ADP], [acetyl-CoA]/[CoA], and [NADH]/[NAD⁺] ratios. Thus, pyruvate dehydrogenase, and therefore glycolysis, is inhibited both when there is adequate ATP (and reduced coenzymes for ATP formation) available, and also when fatty acids are being oxidized. In fasting, when nonesterified fatty acid concentrations increase, there is a decrease in the proportion of the enzyme in the active form, decreasing pyruvate oxidation. In adipose tissue, where glucose provides acetyl-CoA for lipogenesis, the enzyme is activated in response to insulin.

CLINICAL ASPECTS

Inhibition of Pyruvate Metabolism Leads to Lactic Acidosis

Arsenite and mercuric ions react with the —SH groups of lipoic acid and inhibit pyruvate dehydrogenase, as does a **dietary deficiency of thiamin** (see Chapter 44), allowing pyruvate to accumulate. Many alcoholics are thiamin deficient

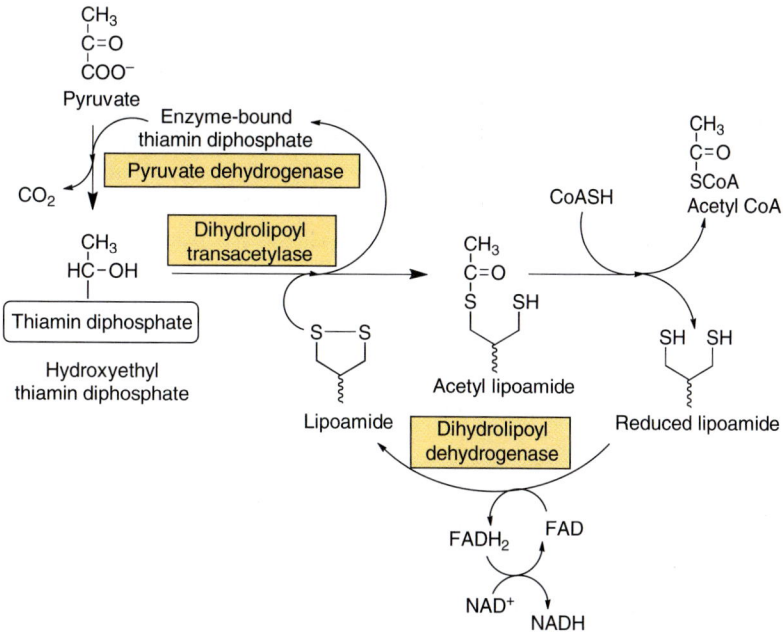

FIGURE 17–5 **Oxidative decarboxylation of pyruvate by the pyruvate dehydrogenase complex.** Lipoic acid is joined by an amide link to a lysine residue of the transacetylase component of the enzyme complex. It forms a long flexible arm, allowing the lipoic acid prosthetic group to rotate sequentially between the active sites of each of the enzymes of the complex. (FAD, flavin adenine dinucleotide; NAD$^+$, nicotinamide adenine dinucleotide.)

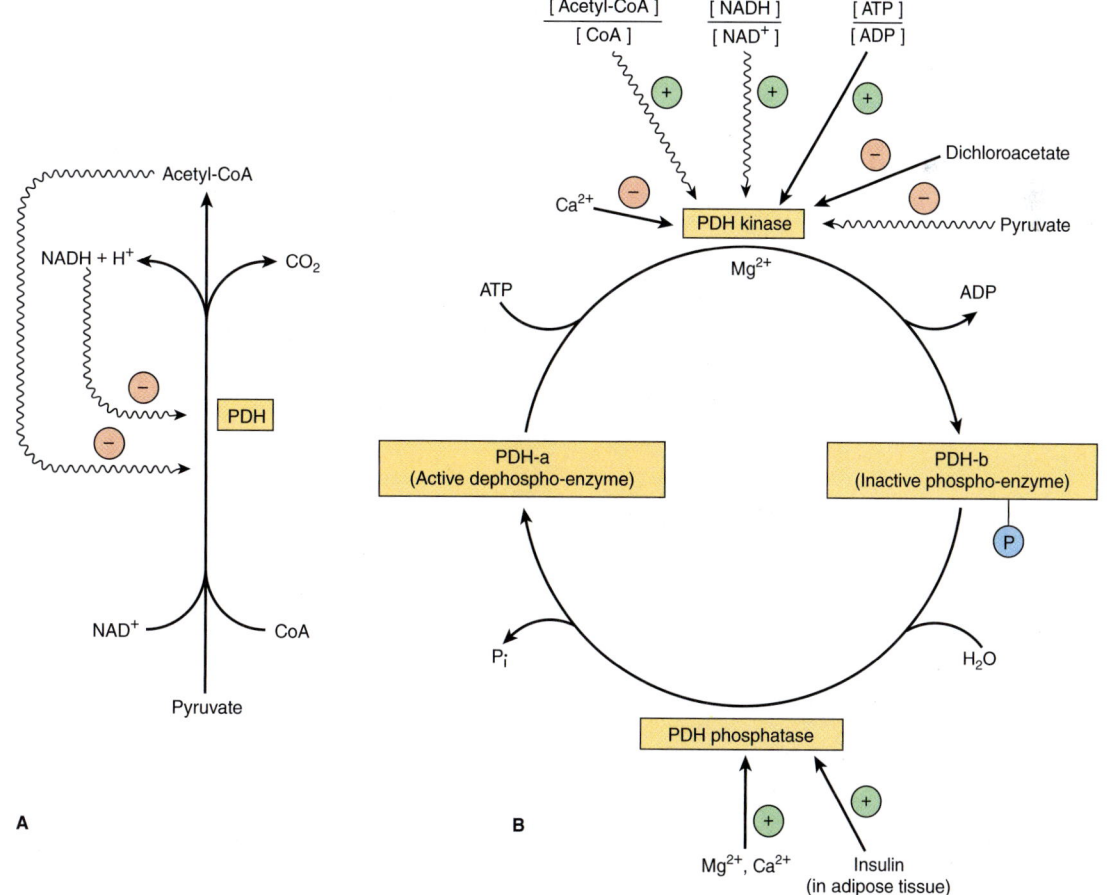

FIGURE 17–6 **Regulation of pyruvate dehydrogenase (PDH).** Arrows with wavy shafts indicate allosteric effects. **(A)** Regulation by end-product inhibition. **(B)** Regulation by interconversion of active and inactive forms.

(both because of a poor diet and also because alcohol inhibits thiamin absorption) and may develop potentially fatal pyruvic and lactic acidosis. Patients with **inherited pyruvate dehydrogenase deficiency**, which can be the result of defects in one or more of the components of the enzyme complex, also present with lactic acidosis, particularly after a glucose load. Because of the dependence of the brain on glucose as a fuel, these metabolic defects commonly cause neurologic disturbances.

Inherited aldolase A deficiency and pyruvate kinase deficiency in erythrocytes cause **hemolytic anemia**. The exercise capacity of patients with **muscle phosphofructokinase deficiency** is low, particularly if they are on high-carbohydrate diets, which requires robust carbohydrate (pyruvate) oxidation.

SUMMARY

- Glycolysis is in the cytosolic pathway in all mammalian cells for the metabolism of glucose (or glycogen) to pyruvate or lactate.

- It can function anaerobically by regenerating oxidized NAD^+ (required in the glyceraldehyde-3-phosphate dehydrogenase reaction), by reducing pyruvate to lactate.

- Lactate is the end product of glycolysis under anaerobic conditions (eg, in exercising muscle) and in erythrocytes, where there are no mitochondria to permit further oxidation of pyruvate.

- Glycolysis is regulated by glucose transport and by three enzymes catalyzing nonequilibrium reactions: hexokinase, phosphofructokinase, and pyruvate kinase.

- In erythrocytes, the first site in glycolysis for generation of ATP may be bypassed, leading to the formation of 2,3-bisphosphoglycerate, which is important in decreasing the affinity of hemoglobin for O_2.

- Pyruvate is oxidized to acetyl-CoA by a multienzyme complex, pyruvate dehydrogenase, which is dependent on the vitamin-derived cofactor thiamin diphosphate.

- Conditions that impair pyruvate metabolism frequently lead to lactic acidosis.

18

Metabolism of Glycogen

Owen P. McGuinness, PhD

OBJECTIVES

After studying this chapter, you should be able to:

- Describe the structure of glycogen and its importance as a carbohydrate reserve.
- Describe the synthesis and breakdown of glycogen and how the processes are regulated in response to hormone action.
- Describe the various types of glycogen storage diseases.

BIOMEDICAL IMPORTANCE

Glycogen is the major storage carbohydrate in animals, corresponding to starch in plants; it is a branched polymer of α-D-glucose (see Figure 15–12). It occurs mainly in liver and muscle, with modest amounts in the brain. The concentration (mg glycogen/g liver) of glycogen in the liver is greater than that of muscle. The total muscle mass of the body is considerably greater than that of the liver. Thus, about three-quarters of total body glycogen is in muscle (**Table 18–1**).

Glycogen in liver and muscle has differing roles in glucose homeostasis. Liver glycogen functions as a reserve to maintain the **blood glucose** concentration in the fasting state. As a reminder (see Chapter 14) the liver makes about 9 grams of glucose per hour after an overnight fast of which 6 g/hr is used by glucose requiring tissues (brain and red blood cells). The total glycogen mass in the liver is ~90 g (**Table 18–1**). Thus, if liver glycogen was the only source of glucose we would have only about 10 hours supply. Of course we have gluconeogenesis (see Chapter 19) as another glucose source in the liver, so glycogen stores do not deplete as rapidly. When we wake up in the morning (assuming we did not have a late night snack!) we will have fasted for ~6 to 8 hours. Our liver glycogen stores will be about 20 g. This amount will depend on when we ate our last meal, the rate of intestinal absorption of the carbohydrate, and composition content of our last meal. If we skip breakfast and lunch, by dinner time (>18 h of fasting) liver glycogen is almost totally depleted. Muscle glycogen provides a readily available source of glucose-1-phosphate for glycolysis within the muscle itself. Muscle glycogen breakdown does not directly yield free glucose because muscle lacks glucose-6-phosphatase. During muscle glycogenolysis, pyruvate formed by glycolysis in muscle can undergo transamination to alanine or reduced to lactate. They can be exported from muscle and used for gluconeogenesis in the liver (see Figure 19–4). After a meal, the replenishment of liver glycogen stores is not solely due to the liver directly taking up glucose and converting it to glycogen. About 50% of the glycogen accretion is from diversion of gluconeogenic carbon (lactate, glucogenic amino acids) into glycogen (called indirect glycogen synthesis; see Chapter 19).

Glycogen storage diseases are a rare (<1:20,000) group of inherited disorders characterized by deficient mobilization of glycogen or deposition of abnormal forms of glycogen, leading to liver damage and muscle weakness; some result in early death. This list has expanded as the specific genes have been identified (not included in the list are Danon and Lafora disease). Glycogen storage diseases are inherited in an autosomal recessive fashion, with the exception of type IX and Danon disease (X-linked recessive).

The highly branched structure of glycogen (see Figure 15–12) provides a large number of sites for glycogenolysis and synthesis. This increases the surface to rapidly release glucose-1-phosphate for muscle glycolytic activity or rapid mobilization of glucose from the liver when glucose demand is high such as during exercise. Endurance athletes adapt muscle metabolism to maintain a slower, more sustained release of glucose-1-phosphate in muscle, to preserve muscle glycogen stores and exercise endurance. Consuming a diet that supplies ample carbohydrates and energy (calories) to meet or exceed daily expenditure (Note: you will not store if you don't eat additional calories to match any change in the training regimen) gradually augments muscle glycogen stores over days and weeks.

This was adapted from the 30th edition by David A. Bender, PhD, & Peter A. Mayes, PhD, DSc

TABLE 18–1 Storage of Carbohydrate in a 70-kg Person

	Percentage of Tissue Weight	Tissue Weight	Body Content (g)
Liver glycogen	5.0	1.8 kg	90
Muscle glycogen	0.7	35 kg	245
Extracellular glucose	0.1	10 L	10

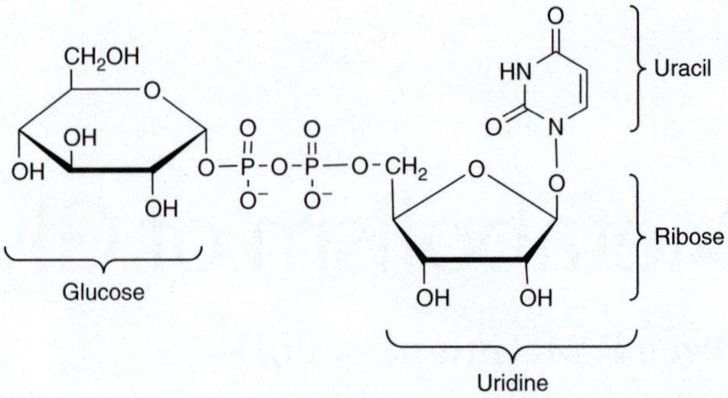

Some endurance athletes practice **carbohydrate loading**—exercise to exhaustion (when muscle glycogen is very low or largely depleted) followed by a high-carbohydrate meal, which results in rapid glycogen synthesis, with fewer branch points than normal and an increase in total stores. Carbohydrate ingestion during exercise prolongs endurance. This is because it helps maintain liver glycogen stores and may spare muscle glycogen especially in fast twitch muscle cells. For each gram of glycogen there is 3 g of water. Thus the early rapid weight loss we see when dieting

FIGURE 18–2 Uridine diphosphate glucose (UDPGlc).

or switching to a low carbohydrate diet (decreases total glycogen stores) in part is due to water and unfortunately not as much fat loss.

We do have small glycogen stores in brain. This is primarily in astrocytes, non-neuronal cells, that support neighboring neurons. This may help explain some of the neurologic symptoms seen in some of the glycogen storage diseases.

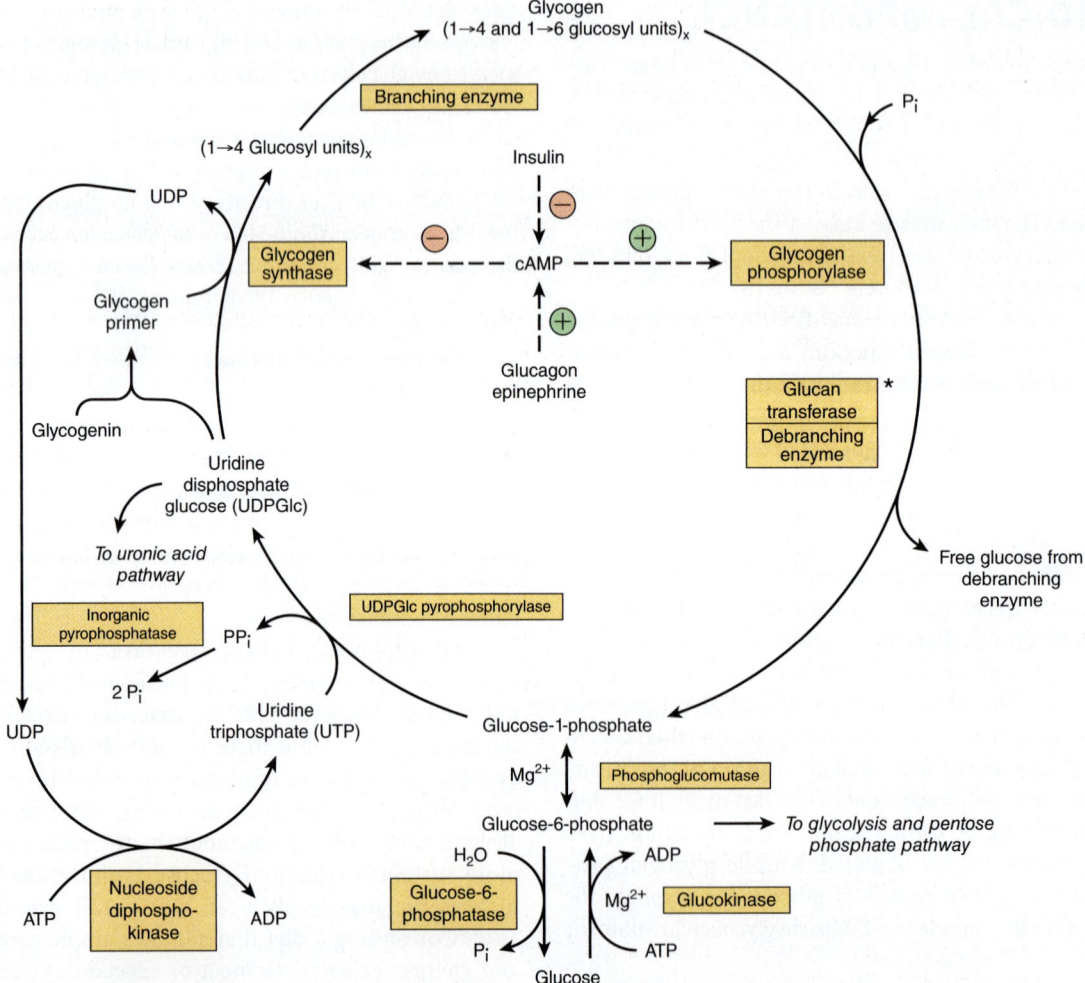

FIGURE 18–1 Pathways of glycogenesis and glycogenolysis in the liver. (⊕, stimulation; ⊖, inhibition.) Insulin decreases the level of cAMP only after it has been raised by glucagon or epinephrine; that is, it antagonizes their action. Glucagon acts on heart muscle but not on skeletal muscle. *Glucan transferase and debranching enzyme appear to be two separate activities of the same enzyme.

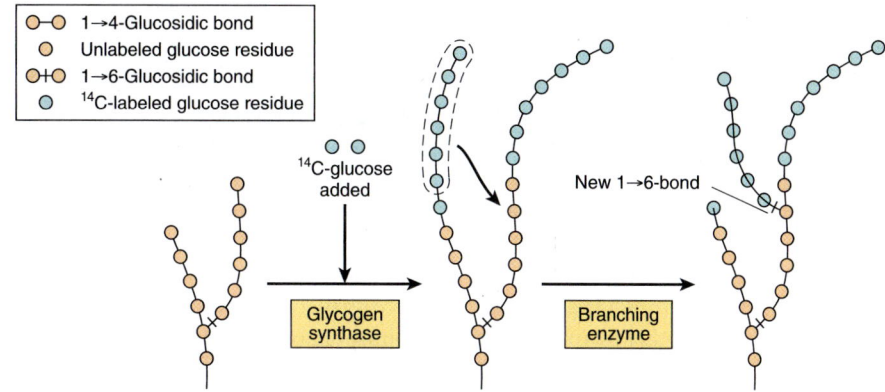

FIGURE 18–3 **The biosynthesis of glycogen.** The mechanism of branching as revealed by feeding ^{14}C-labeled glucose and examining liver glycogen at intervals.

GLYCOGENESIS OCCURS MAINLY IN MUSCLE & LIVER

Glycogen Biosynthesis Involves UDP-Glucose

As in glycolysis, glucose is phosphorylated to glucose-6-phosphate, catalyzed by **hexokinase** in muscle and **glucokinase** in liver (**Figure 18–1**). Glucose-6-phosphate is isomerized to glucose-1-phosphate by **phosphoglucomutase**. The enzyme itself is phosphorylated, and the phosphate group takes part in a reversible reaction in which glucose 1,6-bisphosphate is an intermediate. Next, glucose-1-phosphate reacts with uridine triphosphate (UTP) to form the active nucleotide **uridine diphosphate glucose (UDPGlc)** and pyrophosphate (**Figure 18–2**), catalyzed by **UDPGlc pyrophosphorylase**. The reaction proceeds in the direction of UDPGlc formation because **pyrophosphatase** catalyzes hydrolysis of pyrophosphate to 2× phosphate, so removing one of the reaction products. UDPGlc pyrophosphorylase has a low K_m for glucose-1-phosphate and is present in relatively large amounts, so that it is not a regulatory step in glycogen synthesis.

The initial steps in glycogen synthesis involve the protein **glycogenin**, a 37-kDa protein that is glucosylated on a specific tyrosine residue by UDPGlc. Glycogenin catalyzes the transfer of a further seven glucose residues from UDPGlc, in 1 → 4 linkage, to form a **glycogen primer** that is the substrate for glycogen synthase. The glycogenin remains at the core of the glycogen granule (see Figure 15–12). It is believed that the self-glycosylation of glycogen and the oligosaccharide primer that results is required for glycogen synthesis to occur. One genetic report of a defect in this gene (GSD-IV) primarily impairs heart function. **Glycogen synthase** catalyzes the formation of a glycoside bond between C-1 of the glucose of UDPGlc and C-4 of a terminal glucose residue of glycogen, liberating UDP. The addition of a glucose residue to a preexisting glycogen chain, or "primer," occurs at the nonreducing, outer end of the molecule, so that the branches of the glycogen molecule become elongated as successive 1 → 4 linkages are formed (**Figure 18–3**).

Branching Involves Detachment of Existing Glycogen Chains

When a growing chain is at least 11 glucose residues long, **branching enzyme** transfers a part of the 1 → 4 chain (at least six glucose residues) to a neighboring chain to form a 1 → 6 linkage, establishing a **branch point**. The branches grow by further additions of 1 → 4-glucosyl units and further branching.

GLYCOGENOLYSIS IS NOT THE REVERSE OF GLYCOGENESIS, IT IS A SEPARATE PATHWAY

Glycogen phosphorylase catalyzes the rate-limiting step in glycogenolysis—the phosphorolytic cleavage of the 1 → 4 linkages of glycogen to yield glucose-1-phosphate (**Figure 18–4**). There are different isoenzymes of glycogen phosphorylase in

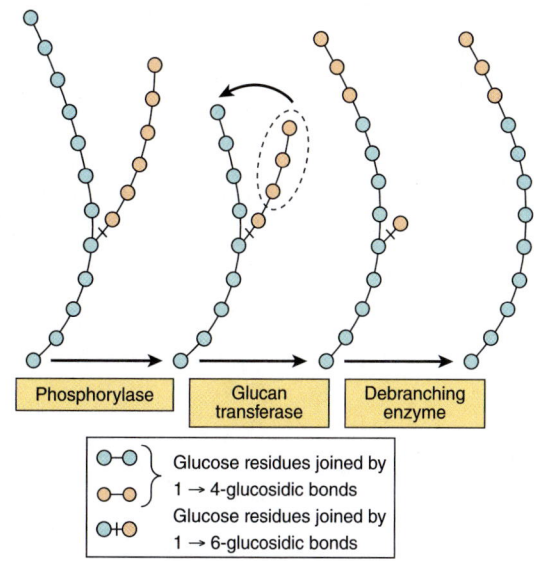

FIGURE 18–4 **Steps in glycogenolysis.**

liver, muscle, and brain, encoded by different genes. Glycogen phosphorylase requires pyridoxal phosphate (see Chapter 44) as its coenzyme. Unlike the reactions of amino acid metabolism (see Chapter 28), in which the aldehyde group of the coenzyme is the reactive group, in phosphorylase the phosphate group is catalytically active.

The terminal glucosyl residues from the outermost chains of the glycogen molecule are removed sequentially until approximately four glucose residues remain on either side of a $1 \rightarrow 6$ branch (see Figure 18–4). The **debranching enzyme** has two catalytic sites in a single polypeptide chain. One is a glucan transferase that transfers a trisaccharide unit from one branch to the other, exposing the $1 \rightarrow 6$ branch point. The other is a 1,6-glycosidase that catalyzes hydrolysis of the $1 \rightarrow 6$-glycoside bond to liberate free glucose. Further phosphorylase action can then proceed. The combined action of phosphorylase and these other enzymes leads to the complete breakdown of glycogen.

The reaction catalyzed by phosphoglucomutase is reversible, so that glucose-1-phosphate derived from hydrolysis of glycogen can form glucose-6-phosphate. In **liver**, but not muscle, **glucose-6-phosphatase** catalyzes hydrolysis of glucose-6-phosphate, yielding glucose that is exported, leading to an increase in the blood glucose concentration. Glucose-6-phosphatase is in the lumen of the smooth endoplasmic reticulum, and genetic defects of the glucose-6-phosphate transporter can cause a variant of type I glycogen storage disease (**Table 18–2**).

TABLE 18–2 Glycogen Storage Diseases

Type	Name	Enzyme Deficiency	Affected Gene	Clinical Features
0	0a	Glycogen synthase	GYS2	Hypoglycemia; hyperketonemia; early death
	0b		GYS1	
I	Ia von Gierke disease	Glucose-6-phosphatase	G6PC	Glycogen accumulation in liver and renal tubule cells; hypoglycemia; lactic acidemia; ketosis; hyperlipemia
	Ib von Gierke Disease	Endoplasmic reticulum glucose-6-phosphate transporter	SLC37A4	
II	Pompe disease	Lysosomal α1 → 4 and α1 → 6 glucosidase (acid maltase)	GAA	Accumulation of glycogen in lysosomes: juvenile-onset variant, muscle hypotonia, death from heart failure by age 2; adult-onset variant, muscle dystrophy
III	Cori/Forbes disease (Type IIa-b)	Liver and muscle debranching enzyme	AGL	Fasting hypoglycemia; hepatomegaly in infancy; accumulation of characteristic branched polysaccharide (limit dextrin); variable muscle weakness
IV	Amylopectinosis, Andersen's disease	Glycogen Branching enzyme	GBE1	Hepatosplenomegaly; accumulation of polysaccharide with few branch points; death from heart or liver failure before age 5
V	McArdle disease	Muscle phosphorylase	PYGM	Poor exercise tolerance; muscle glycogen abnormally high (2.5-4%); blood lactate very low after exercise
VI	Hers disease	Liver phosphorylase	PYGL	Hepatomegaly; accumulation of glycogen in liver; mild hypoglycemia; generally good prognosis
VII	Tarui disease	Muscle and erythrocyte phosphofructokinase 1	PFKM	Poor exercise tolerance; muscle glycogen abnormally high (2.5-4%); blood lactate very low after exercise; also hemolytic anemia
IX	IXa	Liver and muscle phosphorylase kinase	PHKA2	Hepatomegaly; accumulation of glycogen in liver and muscle; mild hypoglycemia; generally good prognosis
	IXb		PHKB	
	IXc		PHKG2	
	IXd		PHKA1	
X		Muscle phosphoglycerate mutase	PGAM2	Hepatomegaly; accumulation of glycogen in liver
XI	Fanconi-Bickel	Glucose transporter 2	SLC2A2	Liver and kidney are affected, weak bones, small for age, renal dysfunction, generally good
XII		Aldolase A	ALDOA	Myopathy, exercise intolerance, hemolytic anemia
XIII		Beta-enolase	ENO3	Exercise intolerance and myalgia
XV		Glycogenin-1	GYG1	Cardiomegaly

Glycogen granules can also be engulfed by **lysosomes**, where acid maltase catalyzes the hydrolysis of glycogen to glucose. This may be especially important in glucose homeostasis in neonates. Genetic lack of lysosomal acid maltase causes type II glycogen storage disease (Pompe disease; see Table 18–2). The lysosomal catabolism of glycogen is under hormonal control.

CYCLIC AMP INTEGRATES THE REGULATION OF GLYCOGENOLYSIS & GLYCOGENESIS

The principal enzymes controlling glycogen metabolism—glycogen phosphorylase and glycogen synthase—are regulated in opposite directions by allosteric mechanisms, subcellular localization, and covalent modification by reversible phosphorylation and dephosphorylation of enzyme protein in response to hormone action (see Chapter 9). Phosphorylation of glycogen phosphorylase increases its activity; phosphorylation of glycogen synthase reduces its activity.

Phosphorylation is increased in response to cyclic AMP (cAMP) (**Figure 18–5**) formed from ATP by **adenylyl cyclase** at the inner surface of cell membranes in response to hormones such as **epinephrine**, **norepinephrine**, and **glucagon**. cAMP is hydrolyzed by **phosphodiesterase**, so terminating hormone action; in liver insulin increases the activity of phosphodiesterase.

Glycogen Phosphorylase Regulation Is Different in Liver & Muscle

In the liver, the role of glycogen is to provide free glucose for export to maintain the blood concentration of glucose; in muscle, the role of glycogen is to provide a source of glucose-6-phosphate for glycolysis in response to the need for ATP for muscle contraction. In both tissues, the enzyme is activated by phosphorylation catalyzed by phosphorylase kinase (to yield phosphorylase a) and inactivated by dephosphorylation catalyzed by phosphoprotein phosphatase (to yield phosphorylase b), in response to hormonal and other signals.

There is instantaneous overriding of this hormonal control. Active phosphorylase a in both tissues is allosterically inhibited by ATP and glucose-6-phosphate; in liver, but not muscle, free glucose is also an inhibitor. Muscle phosphorylase differs from the liver isoenzyme in having a binding site for 5′ AMP (see Figure 18–5), which acts as an allosteric activator of the (inactive) dephosphorylated b-form of the enzyme. 5′ AMP acts as a potent signal of the energy state of the muscle cell; it is formed as the concentration of ADP begins to increase (indicating the need for increased substrate metabolism to permit ATP formation), as a result of the reaction of adenylate kinase: $2 \times ADP \leftrightarrow ATP + 5′ AMP$. The differing regulation of muscle and liver glycogen phosphorylase likely reflects the differing roles of glycogen in liver and muscle.

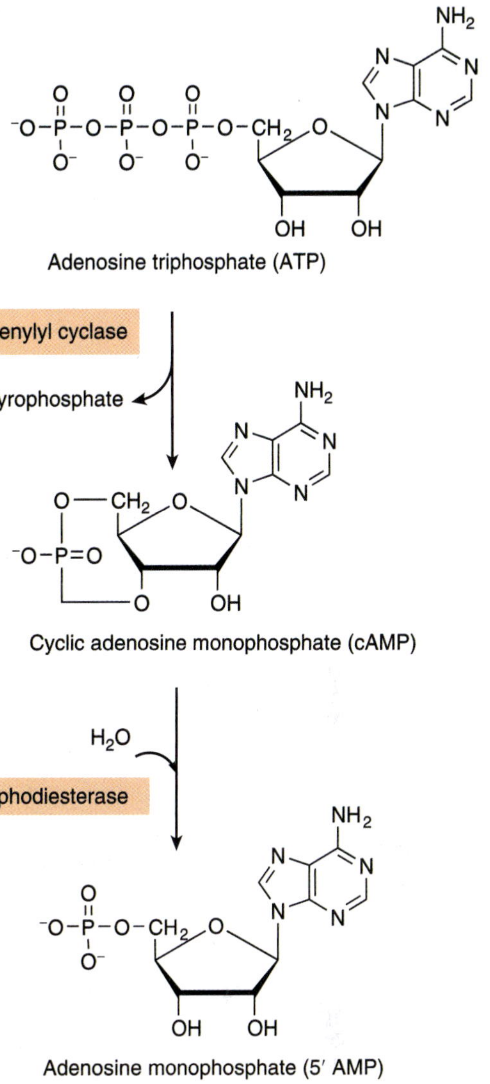

FIGURE 18–5 The formation and hydrolysis of cyclic AMP (3′,5′-adenylic acid, cAMP).

Liver glycogen is released to control plasma glucose. It has to be fast as glucose demands can change rapidly. For example in exercise liver glucose production increases twofold within a few minutes of exercise. Glucose levels would fall if there was a delay. If exogenous glucose is given the liver must rapidly stop making glucose and become a glucose consumer to prevent hyperglycemia and maximize glycogen repletion. In muscle, glycogen is used by the contracting muscle and thus it would be very sensitive to the energy state of the muscle (5′ AMP as well as calcium).

cAMP ACTIVATES GLYCOGEN PHOSPHORYLASE

Phosphorylase kinase is activated in response to cAMP (**Figure 18–6**). Increasing the concentration of cAMP activates **cAMP-dependent protein kinase**, which catalyzes

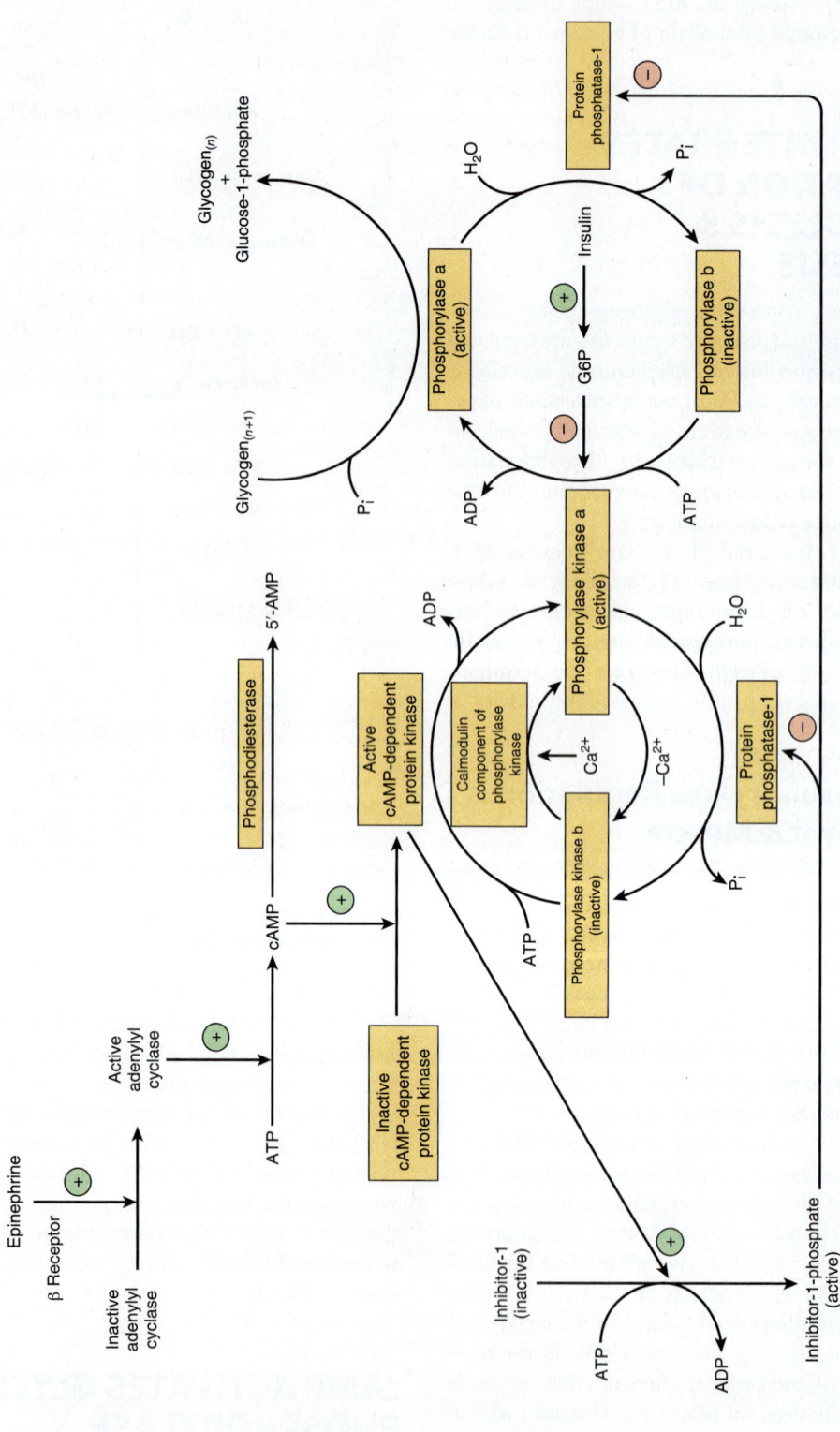

FIGURE 18–6 Control of glycogen phosphorylase in muscle. The sequence of reactions arranged as a cascade allows amplification of the hormonal signal at each step. (G6P, glucose-6-phosphate; *n*, number of glucose residues.)

the phosphorylation by ATP of inactive **phosphorylase kinase b** to active **phosphorylase kinase a**, which in turn, phosphorylates phosphorylase b to phosphorylase a. In the liver, cAMP is formed in response to glucagon, which is secreted in response to falling blood glucose (or exercise). Muscle is insensitive to glucagon; in muscle, the signal for increased cAMP formation is the action of epinephrine, which is secreted in response to fear or fright, when there is a need for increased glycogenolysis to permit rapid muscle activity.

Ca²⁺ Synchronizes the Activation of Glycogen Phosphorylase With Muscle Contraction

Glycogenolysis in muscle increases several 100-fold at the onset of contraction; the same signal (increased cytosolic Ca^{2+} ion concentration) is responsible for initiation of both contraction and glycogenolysis. Muscle phosphorylase kinase, which activates glycogen phosphorylase, is a tetramer of four different subunits, α, β, γ, and δ. The α and β subunits contain serine residues that are phosphorylated by cAMP-dependent protein kinase. The δ subunit is identical to the Ca^{2+}-binding protein **calmodulin** (see Chapter 42) and binds four Ca^{2+}. The binding of Ca^{2+} activates the catalytic site of the γ subunit even while the enzyme is in the dephosphorylated b state; the phosphorylated a form is only fully activated in the presence of high concentrations of Ca^{2+}.

Glycogenolysis in Liver Can Be cAMP-Independent

In the liver, there is cAMP-independent activation of glycogenolysis in response to stimulation of $α_1$ **adrenergic** receptors by norepinephrine (in humans the number of $α_1$ receptors is low). This involves mobilization of Ca^{2+} into the cytosol, followed by the stimulation of a **Ca^{2+}/calmodulin-sensitive phosphorylase kinase**. cAMP-independent glycogenolysis is also activated by vasopressin, oxytocin, and angiotensin II acting either through calcium or the phosphatidylinositol bisphosphate pathway (see Figure 42–10).

Protein Phosphatase-1 Inactivates Glycogen Phosphorylase

Both phosphorylase a and phosphorylase kinase a are dephosphorylated and inactivated by **protein phosphatase-1**. Protein phosphatase-1 is inhibited by a protein, **inhibitor-1**, which is active only after it has been phosphorylated by cAMP-dependent protein kinase. Thus, cAMP controls both the activation and inactivation of phosphorylase (see Figure 18–6). **Insulin** reinforces this effect by inhibiting the activation of phosphorylase b. It does this indirectly by increasing uptake of glucose, leading to increased formation of glucose-6-phosphate, which is an inhibitor of phosphorylase kinase.

The Activities of Glycogen Synthase & Phosphorylase Are Reciprocally Regulated

There are different isoenzymes of glycogen synthase in liver, muscle, and brain. Like phosphorylase, glycogen synthase exists in both phosphorylated and nonphosphorylated states, and the effect of phosphorylation is the reverse of that seen in phosphorylase (**Figure 18–7**). Active **glycogen synthase a** is dephosphorylated and inactive **glycogen synthase b** is phosphorylated.

Six different protein kinases act on glycogen synthase, and there are at least nine different serine residues in the enzyme that can be phosphorylated. Two of the protein kinases are Ca^{2+}/calmodulin dependent (one of these is phosphorylase kinase). Another kinase is cAMP-dependent protein kinase, which allows cAMP-mediated hormone action to inhibit glycogen synthesis synchronously with the activation of glycogenolysis. Insulin also promotes glycogenesis in muscle at the same time as inhibiting glycogenolysis by raising glucose-6-phosphate concentrations, which stimulates the dephosphorylation and activation of glycogen synthase. Dephosphorylation of glycogen synthase b is carried out by protein phosphatase-1, which is under the control of cAMP-dependent protein kinase.

GLYCOGEN METABOLISM IS REGULATED BY A BALANCE IN ACTIVITIES BETWEEN GLYCOGEN SYNTHASE & PHOSPHORYLASE

At the same time, as phosphorylase is activated by a rise in concentration of cAMP (via phosphorylase kinase), glycogen synthase is converted to the inactive form; both effects are mediated via **cAMP-dependent protein kinase** (**Figure 18–8**). Thus, inhibition of glycogenolysis enhances net glycogenesis, and inhibition of glycogenesis enhances net glycogenolysis. Also, the dephosphorylation of phosphorylase a, phosphorylase kinase, and glycogen synthase b is catalyzed by a single enzyme with broad specificity—**protein phosphatase-1**. In turn, protein phosphatase-1 is inhibited by cAMP-dependent protein kinase via inhibitor-1. Insulin ability to inhibit glycogenolysis is via its ability to increase **protein phosphatase-1** and lower cAMP. Thus, glycogenolysis can be terminated and glycogenesis can be stimulated, or vice versa, synchronously, because both processes are dependent on the activity of cAMP-dependent protein kinase and the availability of cAMP. Both phosphorylase kinase and glycogen synthase may be reversibly phosphorylated at more than one site by separate kinases and phosphatases. These secondary phosphorylations modify the sensitivity of the primary sites to phosphorylation and dephosphorylation (**multisite phosphorylation**). They allow increases in glucose and insulin by way of increased glucose-6-phosphate (secondarily to an increase in plasma glucose that activates glucokinase by translocating it from the nucleus to the glycogen granule in the liver), to have effects that act to amplify the effects of insulin to attenuate glycogen mobilization or augment glycogen synthesis (see Figures 18–6 and 18–7).

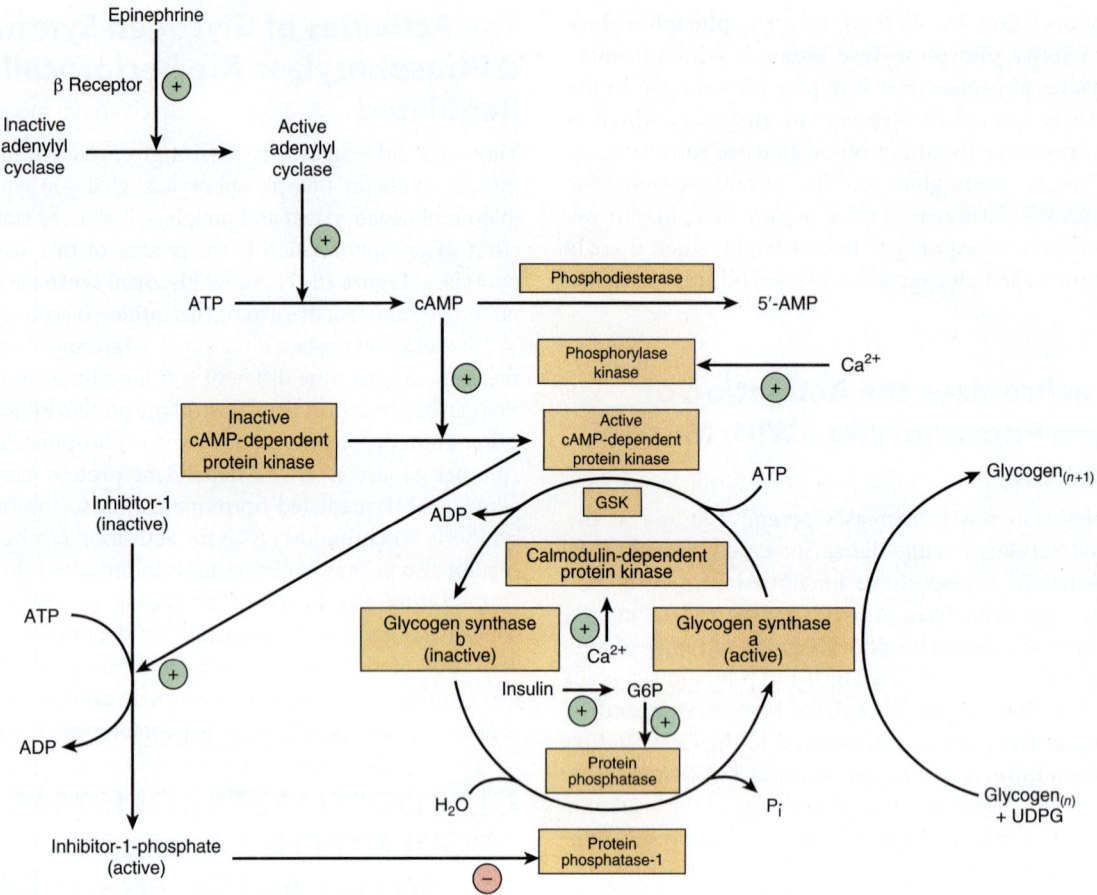

FIGURE 18–7 **Control of glycogen synthase in muscle.** (G6P, glucose-6-phosphate; GSK, glycogen synthase kinase; *n*, number of glucose residues.)

The ability to rapidly fine tune hepatic glycogen metabolism is critical as a failure to rapidly increase hepatic glucose production in response to an increase in glucose demand (ie, exercise) will cause hypoglycemia. It can cause death if severe enough, and even moderate hypoglycemia will prematurely terminate exercise. Gluconeogenesis is a slower process and requires a rise in gluconeogenic substrate supply that helps to sustain glucose production during sustained exercise. In fact, in moderate intensity exercise glucose barely changes because of the tight coupling between liver glycogenolysis, gluconeogenesis, and muscle glucose demand. This occurs because of a rise in glucagon and fall in insulin that drives liver glucose production (remember during exercise glucose uptake in muscle goes up independent of insulin, so even though insulin is falling glucose uptake in exercising muscle increases).

CLINICAL ASPECTS

Hypoglycemia & Diabetes

In individuals' with diabetes the treatments are designed to bring fasting glucose and meal glucose concentration into the normal range to minimize the long-term complications of diabetes. However, despite the best intentions hypoglycemia can

occur. Many individuals with long-standing diabetes develop an unawareness of hypoglycemia. Nonsevere hypoglycemia events usually generate autonomic and/or neuroglycopenic symptoms, which enable the individual to identify the onset, and to treat the falling blood glucose without requiring assistance as it is occurring. If they are unaware the hypoglycemia will get severe enough it will impair their ability to function and will need external assistance. One of the rescue medicines, if the person is unconscious or unable to swallow and ingest a fast-acting (high glycemic index; see Chapter 15) carbohydrate, is glucagon. Glucagon is rapid in onset and is a potent stimulator of glycogenolysis.

Glycogen Storage Diseases Are Inherited

Glycogen storage disease is a generic term to describe a group of inherited disorders characterized by deposition of an abnormal type or quantity of glycogen in tissues, or failure to mobilize glycogen. The principal diseases are summarized in Table 18–2. Glycogen storage diseases are heterogenous and rare. They predominantly affect liver, muscles, heart, and in rare instance brain from infancy to adulthood. Clinical suspicions can lead to diagnosis in primary care and in-patient setting.

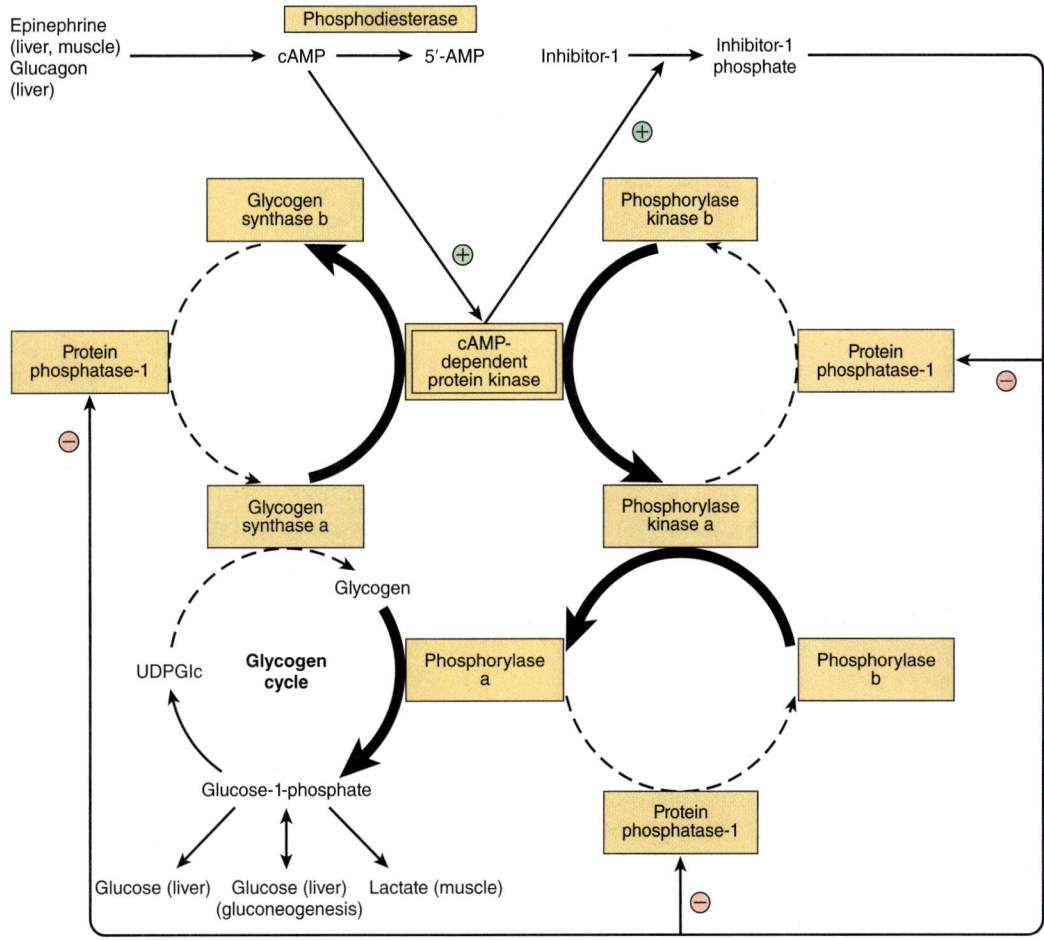

FIGURE 18–8 **Coordinated control of glycogenolysis and glycogenesis by cAMP-dependent protein kinase.** The reactions that lead to glycogenolysis as a result of an increase in cAMP concentrations are shown with bold arrows, and those that are inhibited by activation of protein phosphatase-1 are shown with dashed arrows. The reverse occurs when cAMP concentrations decrease as a result of phosphodiesterase activity, leading to glycogenesis.

Newborn genetic screening is key to detecting glycogen storage disease. Many of the medical crises are preventable with simple nutritional measures. They can limit growth retardation and intellectual disability.

SUMMARY

- Glycogen represents the principal storage carbohydrate in the body, mainly in the liver and muscle.

- In the liver, its major function is to provide glucose for extrahepatic tissues. In muscle, it serves mainly as a source of metabolic fuel for use in muscle. Muscle lacks glucose-6-phosphatase and cannot release free glucose from glycogen.

- Glycogen is synthesized from glucose by the pathway of glycogenesis. It is broken down by a separate pathway, glycogenolysis.

- Cyclic AMP integrates the regulation of glycogenolysis and glycogenesis by promoting the simultaneous activation of phosphorylase and inhibition of glycogen synthase. Insulin acts reciprocally by inhibiting glycogenolysis and stimulating glycogenesis.

- Inherited deficiencies of enzymes of glycogen metabolism in both liver, muscle, and brain cause glycogen storage diseases.

REFERENCES

Ellingwood SS, Cheng A. Biochemical and clinical aspects of glycogen storage diseases. J Endocrinol 2018;238(3):R131-R141. Doi:10.1530/JOE-18-0120.

Murray B, Rosenbloom C. Fundamentals of glycogen metabolism for coaches and athletes. Nutr Rev 2018;76(4):243-259. https://www.ncbi.nlm.nih.gov/pmc/articles/PMC6019055/

Gluconeogenesis & the Control of Blood Glucose

Owen P. McGuinness, PhD

<table>
<tr><td>

O B J E C T I V E S

After studying this chapter, you should be able to:

</td><td>

- Explain the importance of gluconeogenesis in glucose homeostasis.
- Describe the pathway of gluconeogenesis, how irreversible enzymes of glycolysis are bypassed, and how glycolysis and gluconeogenesis are regulated reciprocally.
- Explain how plasma glucose concentration is maintained within narrow limits in the fed and fasting states.

</td></tr>
</table>

BIOMEDICAL IMPORTANCE

Gluconeogenesis is the process of synthesizing glucose from noncarbohydrate precursors. The major substrates are the glucogenic amino acids (see Chapter 29), lactate, glycerol, and propionate. Liver and kidney are the major gluconeogenic tissues. The liver is the primary gluconeogenic organ. While the renal cortex of the kidney may contribute about 10% of whole body gluconeogenesis after a short-term fast (18-24 h), the kidney is not a net source of glucose. This is because the renal medulla is a consumer of glucose. It is only with long-term fasting (~7 days) that the kidney can supply net glucose carbon to contribute to glucose homeostasis. The key gluconeogenic enzymes are expressed in the small intestine. Propionate arising from intestinal bacterial fermentation of carbohydrates is a substrate for gluconeogenesis in enterocytes. The intestine is not a net consumer of lactate and alanine or glycerol, the major substrates for gluconeogenesis. It is a consumer of glucose in the fasting state. Thus, any glucose synthesis that occurs locally is likely metabolized locally.

The rate of hepatic gluconeogenesis is determined by four factors: (1) the availability of gluconeogenic substrates, (2) the capacity of the liver to take up gluconeogenic substrates, (3) the quantity and activity of the gluconeogenic enzymes, and (4) the oxidative capacity of the liver to not only supply the energy to support the energy requiring process of gluconeogenesis but metabolize the nitrogen from the glucogenic amino acids (Ureagenesis; see Chapter 28).

Throughout the 24-hour feeding fasting cycle gluconeogenic precursors are available. In the fasting state adipose

tissue lipolysis releases glycerol. Skeletal muscle releases lactate and gluconeogenic amino acids. As the fast is extended glycerol and amino acids provide an increasing role in supplying carbon for gluconeogenesis. One might think that as a fast progresses gluconeogenesis increases even more. In fact, it does not. This is because the glucose demands of peripheral tissues decrease including the brain. This preserves the vital protein stores. In the fed state gluconeogenic supply does not decrease, in fact it is elevated. The mix of carbon sources is different. Glycerol supply decreases because of a decrease in lipolysis. Lactate supply does not decrease primarily because of the high rates of glycolysis in skeletal muscle. Amino acid supply increases as amino acids derived from dietary protein are directly delivered into the portal vein. During exercise lactate supply from working muscle helps support the increased gluconeogenic demand of exercise.

The transport and uptake of amino acids, glycerol, and lactate are regulated differently. The liver is extremely efficient at taking up glycerol. Greater than 60% of glycerol delivered to the liver is removed on first pass. This fraction remains constant in response to increase or decrease in insulin, glucagon, and epinephrine which are important regulators of gluconeogenesis. In contrast amino acid removal is very sensitive to glucagon and to a lesser extent insulin. Glucagon is a potent stimulator of the transport of glucogenic amino acids by the liver. About 20% of these amino acids are removed on first pass; first pass extraction can increase to more than 60% in the presence of an increase in glucagon. Insulin can stimulate amino acid transport but the response is much smaller than that of glucagon. Lactate uptake by the liver is complex because the liver can produce lactate during high rates of glycogen breakdown. However, in the fasting state the primary driver is the availability of

---

This was adapted from the 30th edition by David A. Bender, PhD & Peter A. Mayes, PhD, DSc

lactate. When lactate is increased along with increases in glucagon the liver can become an efficient consumer of lactate, supporting the gluconeogenic response, for example, as is seen in exercise.

In this chapter we will talk about the gluconeogenic pathways and the sites of regulation. These sites are clearly important in fine-tuning how substrates enter and flow through the gluconeogenic pathway. However, substrate supply and substrate transport can override this regulation by mass action. For example, one might expect that gluconeogenesis decreases after a meal as insulin goes up and glucagon level falls, which should inhibit the gluconeogenic enzymes. However, what is found is gluconeogenesis persists. Delivery and transport of substrates are sustained offsetting the downregulation of the enzymes. However, the liver does not release the gluconeogenic-derived carbon; it diverts it into glycogen to augment meal-derived glycogen synthesis (indirect glycogen synthesis; see Chapter 18).

In disease states associated with hyperglycemia (eg, infection and diabetes) gluconeogenesis is inappropriately increased. In **critically ill patients** in response to injury and infection gluconeogenic supply is very high (elevated lactate, increased lipolysis, increased protein catabolism). Combined with the underlying insulin resistance and high glucagon levels it drives gluconeogenesis and induces **hyperglycemia**, which is associated with poor outcomes. In insulin deficiency (diabetic ketoacidosis) as is seen in Type 1 diabetes (see Chapter 14), in the absence of insulin and very high glucagon the unopposed lipolysis and protein catabolism amplifies the hyperglycemia. Hyperglycemia leads to changes in osmolality of body fluids, impaired blood flow, intracellular acidosis, and increased superoxide radical production (see Chapter 45), resulting in deranged endothelial and immune system function and impaired blood coagulation.

With liver failure gluconeogenesis is impaired and hypoglycemia develops despite the fact that substrate supply is high and there is severe insulin resistance and very high glucagon levels. In this case, the inability to support energy production in the liver (impaired citric acid cycle flux and ureagenesis; see Chapter 16) starves the gluconeogenic pathway of the energy required to support the synthesis of glucose.

GLUCONEOGENESIS INVOLVES GLYCOLYSIS, THE CITRIC ACID CYCLE, PLUS SOME SPECIAL REACTIONS

Thermodynamic Barriers Prevent a Simple Reversal of Glycolysis

Three nonequilibrium reactions in glycolysis (see Chapter 17), catalyzed by hexokinase, phosphofructokinase, and pyruvate kinase, prevent simple reversal of glycolysis for glucose synthesis (**Figure 19–1**). They are circumvented as follows.

Pyruvate & Phosphoenolpyruvate

Reversal of the reaction catalyzed by pyruvate kinase in glycolysis involves two endothermic reactions. Mitochondrial **pyruvate carboxylase** catalyzes the carboxylation of pyruvate to oxaloacetate, an ATP-requiring reaction in which the vitamin biotin is the coenzyme. Biotin binds CO_2 from bicarbonate as carboxybiotin prior to the addition of the CO_2 to pyruvate (see Figure 44–14). The resultant oxaloacetate is reduced to malate, exported from the mitochondrion into the cytosol and then oxidized back to oxaloacetate. A second enzyme, **phosphoenolpyruvate carboxykinase**, catalyzes the decarboxylation and phosphorylation of oxaloacetate to phosphoenolpyruvate using GTP as the phosphate donor. In liver and kidney, the reaction of succinate thiokinase in the citric acid cycle (see Chapter 16) produces GTP (rather than ATP as in other tissues), and this GTP is used for the reaction of phosphoenolpyruvate carboxykinase. This provides a link between citric acid cycle activity and gluconeogenesis, to prevent excessive removal (ie, anaplerosis and cataplerosis have to be equal) of oxaloacetate for gluconeogenesis, which would impair citric acid cycle activity.

Fructose 1,6-Bisphosphate & Fructose-6-Phosphate

The conversion of fructose 1,6-bisphosphate to fructose-6-phosphate, for the reversal of glycolysis, is catalyzed by **fructose 1,6-bisphosphatase**. Its presence determines whether a tissue is capable of synthesizing glucose (or glycogen) not only from pyruvate but also from triose phosphates (eg, glycerol). It is present in liver, kidney, and skeletal muscle, but is probably absent from heart and smooth muscle.

Glucose-6-Phosphate & Glucose

The conversion of glucose-6-phosphate to glucose is catalyzed by **glucose-6-phosphatase**. It is present in liver and kidney (renal cortex), but absent from muscle, which, therefore, cannot export glucose derived from glycogen into the bloodstream.

Glucose-1-Phosphate & Glycogen

The breakdown of glycogen to glucose-1-phosphate is catalyzed by phosphorylase. Glycogen synthesis involves a different pathway via uridine diphosphate glucose and **glycogen synthase** (see Figure 18–1).

The relationships between gluconeogenesis and the glycolytic pathway are shown in Figure 19–1. After transamination or deamination, glucogenic amino acids yield either pyruvate or intermediates of the citric acid cycle. Therefore, the reactions described earlier can account for the conversion of both lactate and glucogenic amino acids to glucose or glycogen.

Propionate is a major precursor of glucose in ruminants; it enters gluconeogenesis via the citric acid cycle. After esterification with CoA, propionyl-CoA is carboxylated to D-methylmalonyl-CoA, catalyzed by **propionyl-CoA carboxylase**, a biotin-dependent enzyme (**Figure 19–2**). **Methylmalonyl-CoA**

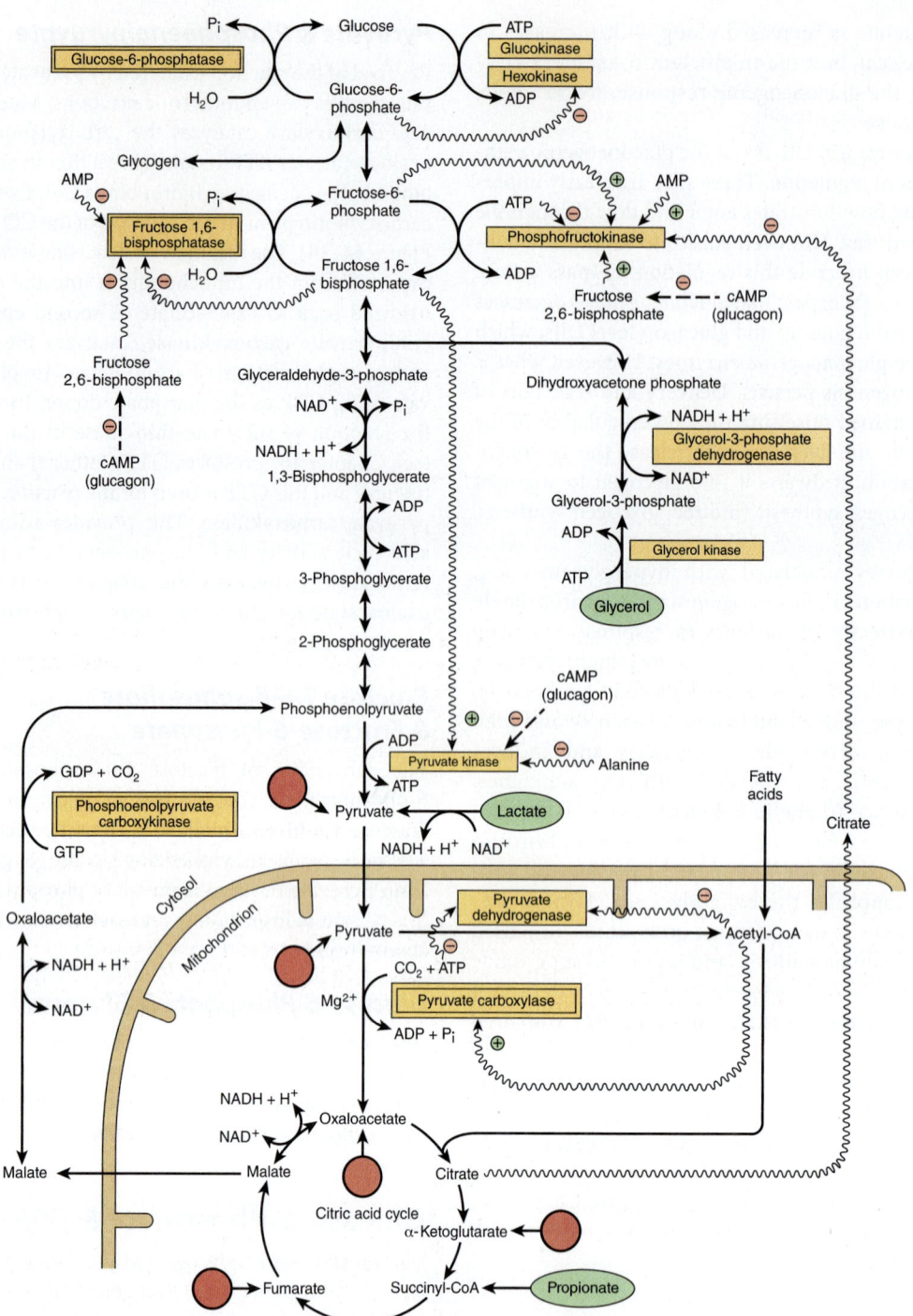

FIGURE 19–1 **Major pathways and regulation of gluconeogenesis and glycolysis in the liver.** Entry points of glucogenic amino acids after transamination are indicated by arrows extended from circles (see also Figure 16–4). The key gluconeogenic enzymes are shown in double-bordered boxes. The ATP required for gluconeogenesis is supplied by the oxidation of fatty acids. Propionate is important only in ruminants. Arrows with wavy shafts signify allosteric effects; dash-shafted arrows, covalent modification by reversible phosphorylation. High concentrations of alanine act as a "gluconeogenic signal" by inhibiting glycolysis at the pyruvate kinase step.

racemase catalyzes the conversion of D-methylmalonyl-CoA to L-methylmalonyl-CoA, which then undergoes isomerization to succinyl-CoA catalyzed by **methylmalonyl-CoA mutase**. In nonruminants, including human beings, propionate arises from the β-oxidation of odd-chain fatty acids that occur in ruminant lipids (see Chapter 22), as well as the oxidation of isoleucine and

the side chain of cholesterol, and is a (relatively minor) substrate for gluconeogenesis. Methylmalonyl-CoA mutase is a vitamin B_{12}-dependent enzyme, and in B_{12} deficiency, methylmalonic acid is excreted in the urine (**methylmalonic aciduria**).

Glycerol is released from adipose tissue as a result of lipolysis of lipoprotein triacylglycerol in the fed state; it may

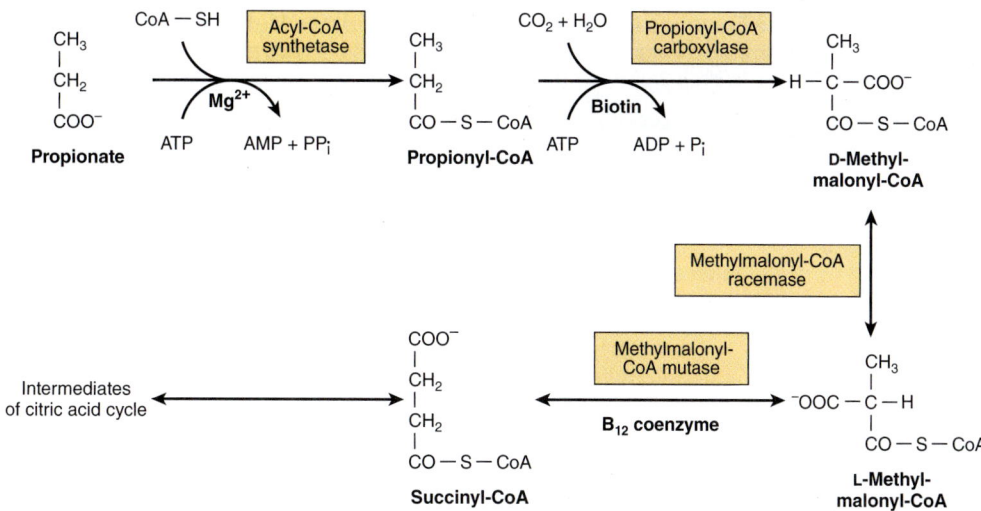

FIGURE 19–2 Metabolism of propionate.

be used for reesterification of free fatty acids to triacylglycerol, or may be a substrate for gluconeogenesis in the liver. In the fasting state, glycerol released from lipolysis of adipose tissue triacylglycerol is used as a substrate for gluconeogenesis in the liver and kidneys.

GLYCOLYSIS & GLUCONEOGENESIS SHARE THE SAME PATHWAY BUT IN OPPOSITE DIRECTIONS, & ARE RECIPROCALLY REGULATED

Changes in the availability of substrates are responsible for most changes in metabolism either directly or indirectly acting via changes in hormone secretion. Three mechanisms are responsible for regulating the activity of enzymes concerned in carbohydrate metabolism: (1) changes in the rate of enzyme synthesis, (2) covalent modification by reversible phosphorylation, and (3) allosteric effects.

Induction & Repression of Key Enzymes Require Several Hours

The changes in enzyme activity in the liver that occur under various metabolic conditions are listed in **Table 19–1**. The enzymes involved catalyze physiologically irreversible nonequilibrium reactions. The effects are generally reinforced because the activity of the enzymes catalyzing the reactions in the opposite direction varies reciprocally (see Figure 19–1). The enzymes involved in the utilization of glucose (ie, those of glycolysis and lipogenesis) become more active when glucose availability is high such as after a meal, and under these conditions the enzymes of gluconeogenesis have relatively low activity. Insulin, secreted in response to increased blood glucose, enhances the synthesis of the key enzymes in glycolysis. It also antagonizes the effect of the glucocorticoids and glucagon-stimulated cAMP, which induce synthesis of the key enzymes of gluconeogenesis.

Covalent Modification by Reversible Phosphorylation Is Rapid

Glucagon and **epinephrine**, hormones that are responsive to a decrease in blood glucose, inhibit glycolysis and stimulate gluconeogenesis in the liver by increasing the concentration of cAMP. This in turn activates cAMP-dependent protein kinase, leading to the phosphorylation and inactivation of **pyruvate kinase**. They also affect the concentration of fructose 2,6-bisphosphate and therefore glycolysis and gluconeogenesis, as described later. In addition, as mentioned glucagon is a potent stimulator of amino acid transport.

Allosteric Modification Is Instantaneous

In gluconeogenesis, pyruvate carboxylase, which catalyzes the synthesis of oxaloacetate from pyruvate, requires acetyl-CoA as an **allosteric activator**. The addition of acetyl-CoA results in a change in the tertiary structure of the protein, lowering the K_m for bicarbonate. This means that as acetyl-CoA is formed from pyruvate, it automatically ensures the provision of oxaloacetate by activating pyruvate carboxylase. The activation of pyruvate carboxylase and the reciprocal inhibition of pyruvate dehydrogenase by acetyl-CoA derived from the oxidation of fatty acids explain the action of fatty acid oxidation in sparing the oxidation of pyruvate (and hence glucose) and stimulating gluconeogenesis. The reciprocal relationship between these two enzymes alters the metabolic fate of pyruvate as the tissue changes from carbohydrate oxidation (glycolysis) to gluconeogenesis during the transition from the fed to fasting state (see Figure 19–1). A major role of fatty acid oxidation in promoting gluconeogenesis is to supply the ATP that is required for glucose synthesis.

TABLE 19–1 Regulatory & Adaptive Enzymes Associated With Carbohydrate Metabolism

	Activity in		Inducer	Repressor	Activator	Inhibitor
	Carbohydrate Feeding	Fasting and Diabetes				
Glycogenolysis, glycolysis, and pyruvate oxidation						
Glycogen synthase	↑	↓			Insulin, glucose-6-phosphate	Glucagon
Hexokinase						Glucose-6-phosphate
Glucokinase	↑	↓	Insulin	Glucagon		
Phosphofructokinase-1	↑	↓	Insulin	Glucagon	5′ AMP, fructose-6-phosphate, fructose 2,6-bisphosphate, P$_i$	Citrate, ATP, glucagon
Pyruvate kinase	↑	↓	Insulin, fructose	Glucagon	Fructose 1,6-bisphosphate, insulin	ATP, alanine, glucagon, norepinephrine
Pyruvate dehydrogenase	↑	↓			CoA, NAD$^+$, insulin, ADP, pyruvate	Acetyl-CoA, NADH, ATP (fatty acids, ketone bodies)
Gluconeogenesis						
Pyruvate carboxylase	↓	↑	Glucocorticoids, glucagon, epinephrine	Insulin	Acetyl-CoA	ADP
Phosphoenolpyruvate carboxykinase	↓	↑	Glucocorticoids, glucagon, epinephrine	Insulin		
Glucose-6-phosphatase	↓	↑	Glucocorticoids, glucagon, epinephrine	Insulin		

Phosphofructokinase (phosphofructokinase-1) occupies a key position in regulating glycolysis and is also subject to feedback control. It is inhibited by citrate and by normal intracellular concentrations of ATP and is activated by 5′ AMP. At the normal intracellular [ATP] the enzyme is about 90% inhibited; this inhibition is reversed by 5′AMP (**Figure 19–3**).

5′ AMP acts as an indicator of the energy status of the cell. The presence of **adenylyl kinase** in liver and many other tissues allows rapid equilibration of the reaction

$$2ADP \leftrightarrow ATP + 5′ AMP$$

Thus, when ATP is used in energy-requiring processes, resulting in the formation of ADP, [AMP] increases. A relatively small decrease in [ATP] causes a several fold increase in [AMP], so that [AMP] acts as a metabolic amplifier of a small change in [ATP], and hence a sensitive signal of the energy state of the cell. The activity of phosphofructokinase-1 is thus regulated in response to the energy status of the cell to control the quantity of carbohydrate undergoing glycolysis prior to its entry into the citric acid cycle. At the same time, AMP

activates glycogen phosphorylase, so increasing glycogenolysis. A consequence of the inhibition of phosphofructokinase-1 by ATP is an accumulation of glucose-6-phosphate, which in turn inhibits further uptake of glucose in extrahepatic tissues by inhibition of hexokinase. Remember glucokinase, which is present in the liver, is not inhibited by glucose-6-phosphate thus allowing for high rates of glucose entry and diversion to glycogen deposition at the same time allow for gluconeogenesis-derived carbon to be diverted to glycogen as well.

Fructose 2,6-Bisphosphate Plays a Unique Role in the Regulation of Glycolysis & Gluconeogenesis in Liver

The most potent positive allosteric activator of phosphofructokinase-1 and inhibitor of fructose 1,6-bisphosphatase in liver is **fructose 2,6-bisphosphate**. It relieves inhibition of phosphofructokinase-1 by ATP and increases the affinity for fructose-6-phosphate. It inhibits fructose 1,6-bisphosphatase by increasing the K_m for fructose 1,6-bisphosphate. Its concentration is under

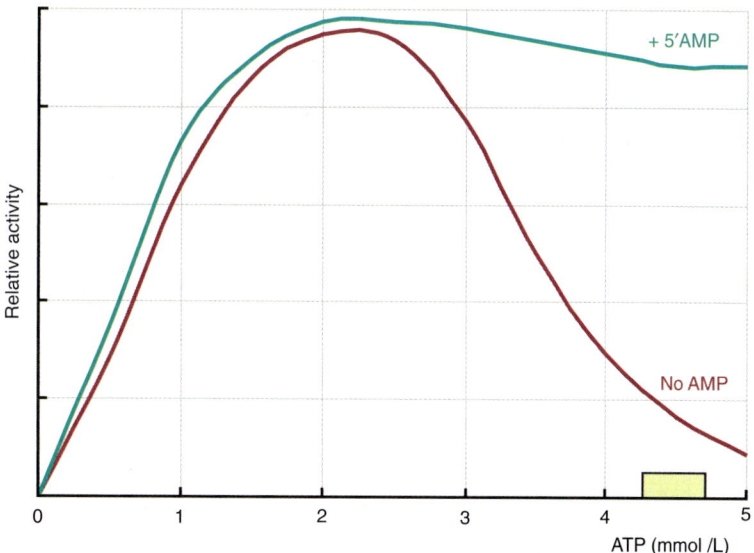

FIGURE 19–3 **The inhibition of phosphofructokinase-1 by ATP and relief of inhibition by AMP.** The yellow bar shows the normal range of the intracellular concentration of ATP.

both substrate (allosteric) and hormonal control (covalent modification) (**Figure 19–4**).

Fructose 2,6-bisphosphate is formed by phosphorylation of fructose-6-phosphate by **phosphofructokinase-2**. The same enzyme protein is also responsible for its breakdown, since it has **fructose 2,6-bisphosphatase** activity. This **bifunctional enzyme** is under the allosteric control of fructose-6-phosphate, which stimulates the kinase and inhibits the phosphatase. Hence, when there is an abundant supply of glucose, the concentration of fructose 2,6-bisphosphate increases, stimulating glycolysis by activating phosphofructokinase-1 and inhibiting fructose 1,6-bisphosphatase. In the fasting state, glucagon stimulates the production of cAMP, activating cAMP-dependent protein kinase, which in turn inactivates phosphofructokinase-2 and activates fructose 2,6-bisphosphatase by phosphorylation. Hence, gluconeogenesis is stimulated by a decrease in the concentration of fructose 2,6-bisphosphate, which inactivates phosphofructokinase-1 and relieves the inhibition of fructose 1,6-bisphosphatase. Xylulose 5-phosphate, an intermediate of the pentose phosphate pathway (see Chapter 20) activates the protein phosphatase that dephosphorylates the bifunctional enzyme, so increasing the formation of fructose 2,6-bisphosphate and increasing the rate of glycolysis. This leads to increased flux through glycolysis and the pentose phosphate pathway and increased fatty acid synthesis (see Chapter 23).

Substrate (Futile) Cycles Allow Fine Tuning & Rapid Response

The control points in glycolysis and glycogen metabolism involve a cycle of phosphorylation and dephosphorylation catalyzed by glucokinase and glucose-6-phosphatase; phosphofructokinase-1 and fructose 1,6-bisphosphatase; pyruvate kinase, pyruvate carboxylase, and phosphoenolpyruvate carboxykinase; and glycogen synthase and phosphorylase.

It would seem obvious that these opposing enzymes are regulated in such a way that when those involved in glycolysis are active, those involved in gluconeogenesis are relatively inactive, since otherwise there would be cycling between phosphorylated and nonphosphorylated intermediates, with net hydrolysis of ATP. In fact in the liver futile cycling of carbon (1-2%) is present at a low rate. The advantage is by having both pathways the liver is allowed to rapidly transition from the fed, fasted, or exercising state. In muscle both phosphofructokinase and fructose 1,6-bisphosphatase have some activity at all times, so that there is indeed even more wasteful substrate cycling. This permits the very rapid increase in the rate of glycolysis necessary for muscle contraction. At rest the rate of phosphofructokinase activity is some 10-fold higher than that of fructose 1,6-bisphosphatase; in anticipation of muscle contraction, the activity of both enzymes increases, fructose 1,6-bisphosphatase 10 times more than phosphofructokinase, maintaining the same net rate of glycolysis. At the start of muscle contraction, the activity of phosphofructokinase increases further, and that of fructose 1,6-bisphosphatase falls, so increasing the net rate of glycolysis (and hence ATP formation) as much as a 1000-fold.

THE BLOOD CONCENTRATION OF GLUCOSE IS REGULATED WITHIN NARROW LIMITS

In the postabsorptive state, the concentration of blood glucose is maintained between 4.5 and 5.5 mmol/L. After the ingestion of a carbohydrate meal, it may rise to 6.5 to 7.2 mmol/L, and in starvation, it may fall to 3.3 to 3.9 mmol/L. A sudden decrease in blood glucose (eg, in response to insulin overdose) causes convulsions, because of the dependence of the brain on a supply of glucose. However, much lower concentrations

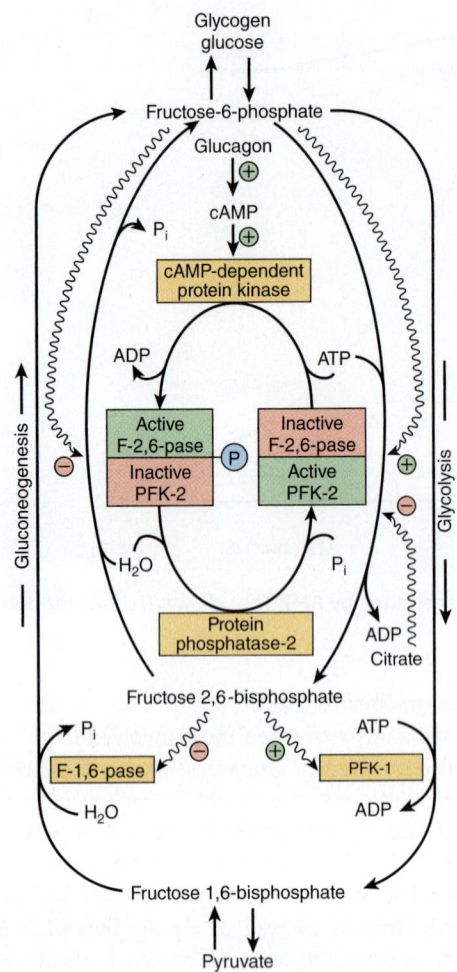

FIGURE 19–4 Control of glycolysis and gluconeogenesis in the liver by fructose 2,6-bisphosphate and the bifunctional enzyme PFK-2/F-2,6-Pase (6-phosphofructo-2-kinase/fructose 2,6-bisphosphatase). (F-1,6-Pase, fructose 1,6-bisphosphatase; PFK-1, phosphofructokinase-1 [6-phosphofructo-1-kinase].) Arrows with wavy shafts indicate allosteric effects.

can be tolerated if hypoglycemia develops slowly enough for adaptation to occur. The blood glucose level in birds is considerably higher (14 mmol/L) and in ruminants considerably lower (~2.2 mmol/L in sheep and 3.3 mmol/L in cattle). These lower normal levels appear to be associated with the fact that ruminants ferment virtually dietary carbohydrate to short-chain fatty acids, and these largely replace glucose as the main metabolic fuel of the tissues in the fed state.

BLOOD GLUCOSE IS DERIVED FROM THE DIET, GLUCONEOGENESIS, & GLYCOGENOLYSIS

The digestible dietary carbohydrates yield glucose, galactose, and fructose that are transported to the liver via the **hepatic portal vein**. Galactose and fructose are readily converted to glucose in the liver (see Chapter 20).

Glucose is formed from two groups of compounds that undergo gluconeogenesis (see Figures 16–4 and 19–1): (1) those that involve a direct net conversion to glucose, including most **amino acids** and **propionate** and (2) those that are the products of the metabolism of glucose in tissues. Thus, **lactate**, formed by glycolysis in skeletal muscle and erythrocytes, is transported to the liver and kidney where it reforms glucose, which again becomes available via the circulation for oxidation in the tissues. This process is known as the **Cori cycle**, or the **lactic acid cycle** (**Figure 19–5**).

In the fasting state, there is a considerable output of alanine from skeletal muscle, far in excess of the amount in the muscle proteins that are being catabolized. It is formed by transamination of pyruvate produced by glycolysis of muscle glycogen, and is exported to the liver, where, after transamination back to pyruvate, it is a substrate for gluconeogenesis. This **glucose-alanine cycle** (see Figure 19–5) provides an indirect way of utilizing muscle glycogen to maintain blood glucose in the fasting state. The glycerol released by adipose tissue is another source of gluconeogenic carbon along with the lactate released by muscle. The ATP required for the hepatic synthesis of glucose from pyruvate (or glycerol) is formed by the oxidation of fatty acids derived from adipose tissue lipolysis. Glucose is also formed from liver glycogen by glycogenolysis (see Chapter 18).

Metabolic & Hormonal Mechanisms Regulate the Concentration of Blood Glucose

The maintenance of a stable blood glucose concentration is one of the most finely regulated of all homeostatic mechanisms, involving the liver, extrahepatic tissues, and several hormones. Liver cells are freely permeable to glucose in either direction (via the GLUT 2 transporter), whereas cells of extrahepatic tissues (apart from pancreatic β-islets) are relatively impermeable, and their unidirectional glucose transporters are regulated by insulin. As a result, uptake from the bloodstream is an important, but not rate limiting under all settings, (see Chapter 14) determinant of the utilization of glucose in extrahepatic tissues. The role of various glucose transporter proteins found in cell membranes is shown in **Table 19–2**.

Glucokinase Is Important in Regulating Blood Glucose After a Meal

Hexokinase has a low K_m for glucose, and in the liver it is saturated with low capacity and acting at a constant rate under all normal conditions. It thus acts to ensure an adequate rate of glycolysis to meet the liver's needs. Glucokinase is an allosteric enzyme with a considerably higher apparent K_m (lower affinity) for glucose, so that its activity increases with increase in the concentration of glucose in the hepatic portal vein (**Figure 19–6**). In the fasting state, glucokinase is located in the nucleus. In response to an increased intracellular concentration of glucose it migrates into the cytosol, mediated by the carbohydrate response element-binding protein (CREBP). It permits hepatic

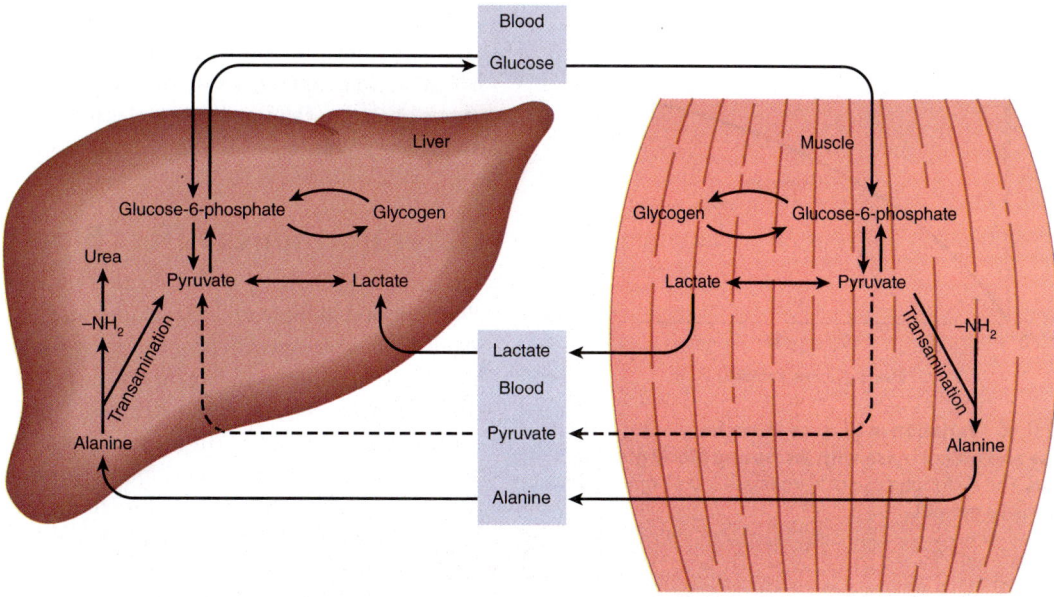

FIGURE 19–5 The lactic acid (Cori cycle) and glucose-alanine cycles.

uptake of large amounts of glucose after a carbohydrate meal, for glycogen and fatty acid synthesis. While the concentration of glucose in the hepatic portal vein may reach 20 mmol/L after a meal, that leaving the liver into the peripheral circulation does not normally exceed 8 to 9 mmol/L. Glucokinase is absent from the liver of ruminants, which have little glucose entering the portal circulation from the intestines.

At normal peripheral blood glucose concentrations (4.5-5.5 mmol/L), the liver is a net producer of glucose. However, as the glucose level rises, the output of glucose ceases, and there is a net uptake (see Figure 14–1).

Insulin & Glucagon Play a Central Role in Regulating Blood Glucose

In addition to the direct effects of hyperglycemia in enhancing the uptake of glucose into the liver, the hormone **insulin** plays a central role in regulating blood glucose. It is produced by the β cells of the islets of Langerhans in the pancreas in response to hyperglycemia. The β-islet cells are freely permeable to glucose via the GLUT 2 transporter, and the glucose is phosphorylated by glucokinase. Therefore, increasing blood glucose increases metabolic flux through glycolysis, the citric acid cycle, and the generation of ATP. The increase in [ATP] inhibits ATP-sensitive K^+ channels, causing depolarization of the cell membrane, which increases Ca^{2+} influx via voltage-sensitive Ca^{2+} channels, stimulating exocytosis of insulin. Thus, the concentration of insulin in the blood parallels that of the blood glucose. Other substances causing release of insulin from the pancreas include amino acids, nonesterified fatty acids, ketone bodies, glucagon, secretin, and the sulfonylurea drugs tolbutamide and glyburide. These drugs are used to stimulate insulin secretion in Type 2 diabetes mellitus via the ATP-sensitive K^+ channels. Drugs that augment glucagon-like-peptide signals increase cyclic-AMP, which

TABLE 19–2 Major Glucose Transporters

	Tissue Location	Functions
Facilitative bidirectional transporters		
GLUT 1	Brain, kidney, colon, placenta, erythrocytes	Glucose uptake
GLUT 2	Liver, pancreatic β cell, small intestine, kidney	Rapid uptake or release of glucose
GLUT 3	Brain, kidney, placenta	Glucose uptake
GLUT 4	Heart and skeletal muscle, adipose tissue	Insulin-stimulated glucose uptake
GLUT 5	Small intestine	Absorption of fructose
Sodium-dependent unidirectional transporter		
SGLT 1	Small intestine and kidney	Active uptake of glucose against a concentration gradient

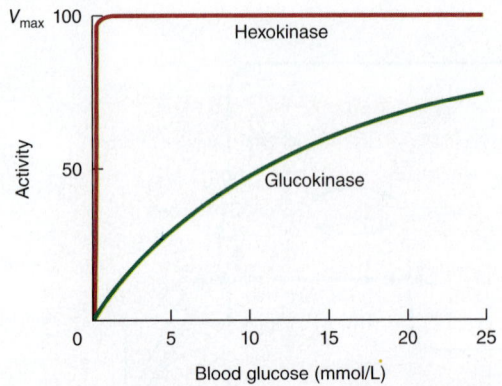

FIGURE 19–6 **Variation in glucose phosphorylating activity of hexokinase and glucokinase with increasing blood glucose concentration.** The K_m for glucose of hexokinase is 0.05 mmol/L and of glucokinase is 10 mmol/L.

potentiate glucose-stimulated insulin secretion. Epinephrine and norepinephrine block the release of insulin. Insulin acts to lower blood glucose immediately by enhancing glucose transport into adipose tissue and muscle by recruitment of glucose transporters (GLUT 4) from the interior of the cell to the plasma membrane. Although it does not affect glucose transport activity in the liver, it directly augments liver glucose uptake and glycogen deposition likely through effects on glucokinase and glycogen synthase and phosphorylase activity. Insulin and other hormones modulate long-term uptake as a result of their actions on transcriptional signals to change an entire enzyme portfolio controlling glycolysis, glycogenesis, and gluconeogenesis (see Chapter 18 and Table 19–1).

Glucagon is the hormone produced by the α cells of the pancreatic islets in response to hypoglycemia. In the liver, it stimulates glycogenolysis by activating glycogen phosphorylase. Unlike epinephrine, glucagon does not have an effect on muscle phosphorylase. Glucagon also enhances gluconeogenesis from amino acids and lactate. In all these actions, glucagon acts via generation of cAMP (see

Table 19–1). Both hepatic glycogenolysis and gluconeogenesis contribute to the **hyperglycemic effect** of glucagon, whose actions oppose those of insulin. Most of the endogenous glucagon (and insulin) is cleared from the circulation by the liver (**Table 19–3**).

Other Hormones Affect Blood Glucose

The **anterior pituitary gland** secretes hormones that tend to elevate blood glucose and therefore antagonize the action of insulin. These are growth hormone, adrenocorticotropic hormone (ACTH), and possibly other "diabetogenic" hormones. Growth hormone secretion is stimulated by hypoglycemia; it decreases glucose uptake in muscle. Some of this effect may be indirect, since it stimulates mobilization of nonesterified fatty acids from adipose tissue, which themselves inhibit glucose utilization. The **glucocorticoids** (11-oxysteroids) are secreted by the adrenal cortex, and are also synthesized in an unregulated manner in adipose tissue. They act to increase gluconeogenesis as a result of enhanced hepatic catabolism of amino acids, due to induction of aminotransferases (and other enzymes such as tryptophan dioxygenase) and key enzymes of gluconeogenesis. In addition, glucocorticoids inhibit the utilization of glucose in extrahepatic tissues. In all these actions, glucocorticoids act in a manner antagonistic to insulin. A number of **cytokines** secreted by macrophages infiltrating adipose tissue also have insulin antagonistic actions; together with glucocorticoids secreted by adipose tissue, this explains the insulin resistance that commonly occurs in obese people.

Epinephrine is secreted by the adrenal medulla because of stressful stimuli (fear, excitement, hemorrhage, hypoxia, hypoglycemia, etc.) and leads to glycogenolysis in liver and muscle owing to stimulation of phosphorylase via generation of cAMP. In muscle, glycogenolysis results in increased glycolysis and lactate release, whereas in liver, it results in the release of glucose into the bloodstream. Epinephrine is a potent stimulator of gluconeogenesis because of the robust increase in substrate supply.

TABLE 19–3 **Tissue Responses to Insulin & Glucagon**

	Liver	Adipose Tissue	Muscle
Increased by insulin	Fatty acid synthesis	Glucose uptake	Glucose uptake
	Glycogen synthesis	Fatty acid synthesis	Glycogen synthesis
	Protein synthesis		Protein synthesis
	Fatty acid synthesis		
Decreased by insulin	Ketogenesis	Lipolysis	
	Gluconeogenesis		
Increased by glucagon	Glycogenolysis	Lipolysis	
	Gluconeogenesis		
	Ketogenesis		

FURTHER CLINICAL ASPECTS

Glucosuria Occurs When the Renal Threshold for Glucose Is Exceeded

When the blood glucose concentration rises above about 10 mmol/L, the kidney also exerts a (passive) regulatory effect. Glucose is continuously filtered by the glomeruli, but is normally completely reabsorbed in the renal tubules by active transport. The capacity of the tubular system to reabsorb glucose is limited to a rate of about 2 mmol/min, and in hyperglycemia (as occurs in poorly controlled diabetes mellitus), the glomerular filtrate may contain more glucose than can be reabsorbed, resulting in **glucosuria** when the **renal threshold** for glucose is exceeded. Thus a common clinical presentation of diabetes is unexplained weight loss and frequent urination. The weight loss is due to the volume loss and subsequent dehydration due to osmotic diuresis in the kidney combined with the loss of glucose calories in the urine.

Hypoglycemia May Occur During Pregnancy & in the Neonate

During pregnancy, fetal glucose consumption increases and there is a risk of maternal, and possibly fetal, hypoglycemia, particularly if there are long intervals between meals or at night. Furthermore, premature and low-birth-weight babies are more susceptible to hypoglycemia, since they have little adipose tissue to provide nonesterified fatty acids. The enzymes of gluconeogenesis may not be fully developed at this time, and gluconeogenesis is anyway dependent on a supply of nonesterified fatty acids for ATP formation. Little glycerol, which would normally be released from adipose tissue, is available for gluconeogenesis.

The Ability to Utilize Glucose May Be Ascertained by Measuring Glucose Tolerance

Glucose tolerance is the ability to regulate the blood glucose concentration after the administration of a test dose of glucose (normally 1 g/kg body weight) (**Figure 19–7**). The normal glucose tolerance is determined by the timing of and quantity of insulin secreted and the ability of tissues to respond to insulin. If a person gains weight, they become insulin resistant. Most people who gain weight do not become glucose intolerant because their beta cells compensate and make additional insulin to maintain normal glucose tolerance. However, their risk of developing diabetes increases. Clinically, the glucose tolerance test is primarily used to detect gestational diabetes. To determine average blood glucose in a person clinicians measure glycosylation status (see Chapter 15) of hemoglobin (HbA1c), which correlates with the person's average glucose over a 2 to 3 month period. If it is (normal <6.0%; prediabetes 6-6.4%; diabetes >6.5%) more than 6.5%, it is diagnostic of diabetes.

Diabetes mellitus (type 1, or insulin-dependent diabetes mellitus [IDDM]) is characterized by impaired glucose tolerance as a result of decreased secretion of insulin because of progressive destruction of pancreatic β-islet cells. Glucose tolerance is also impaired in Type 2 diabetes mellitus (noninsulin-dependent diabetes [NIDDM]) as a result of reduced sensitivity of tissues to insulin action combined with

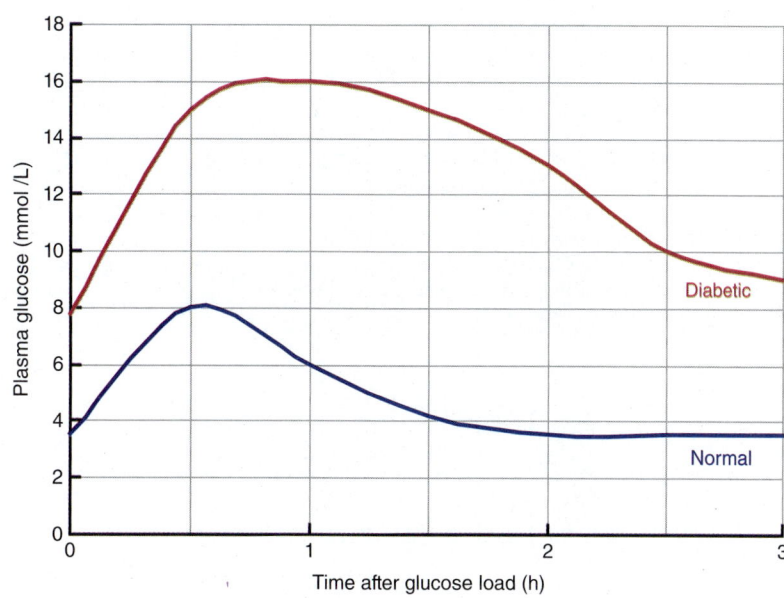

FIGURE 19–7 Glucose tolerance test. Blood glucose curves of a normal and a diabetic person after oral administration of 1 g of glucose/kg body weight. Note the initial raised concentration in the fasting diabetic (>7 mM); so by definition they are diabetic without measuring glucose tolerance. If baseline glucose is in the normal range, a criterion of normal tolerance is the return to the baseline value within 2 hours.

an impaired secretion of insulin. Insulin resistance associated with obesity (and especially abdominal obesity) leading to the development of hyperlipidemia, then atherosclerosis and coronary heart disease, as well as overt diabetes, is known as the **metabolic syndrome**. Impaired glucose tolerance also occurs in conditions where the liver is damaged, in some infections, and in response to some drugs, as well as in conditions that lead to hyperactivity of the pituitary gland or adrenal cortex because hormones secreted by these glands antagonize the action of insulin.

Administration of insulin (as in the treatment of diabetes mellitus) lowers the blood glucose concentration and increases its utilization and storage in the liver and muscle as glycogen. An excess of insulin may cause **hypoglycemia**, resulting in convulsions and even death unless glucose is administered promptly. Hypoglycemia upon fasting can be observed in pituitary or adrenocortical insufficiency. It is due to a decrease in the antagonism to insulin and the lower gluconeogenic capacity of the liver.

Think Otherwise: Very Low Carbohydrate Diets Promote Weight Loss

Very low carbohydrate diets, providing only 20 g per day of carbohydrate or less (compared with a desirable intake of 100-120 g/day), but permitting unlimited consumption of fat and protein, have been promoted as an effective regime for weight loss. Such diets are counter to all advice on a prudent diet composition for health. Since there is a continual demand for glucose, there will be a considerable amount of gluconeogenesis from amino acids; the associated high ATP cost must then be met by oxidation of fatty acids. While this is logical, controlled energy balance studies demonstrate that energy expenditure is in fact not increased. In the end a calorie is a calorie. Weight loss occurs because they eat less calories. The rapid weight loss commonly seem with these diets is most likely due to the fact that low carbohydrate diets decrease glycogen stores, which have 3 g of water per gram of glycogen. The biggest challenge is that staying on these or other extreme diets is nearly impossible.

SUMMARY

- Gluconeogenesis is the process of synthesizing glucose or glycogen from noncarbohydrate precursors. It is of particular importance when carbohydrate is not available from the diet. The main substrates are amino acids, lactate, glycerol, and propionate.

- The pathway of gluconeogenesis in the liver and kidney utilizes those reactions in glycolysis that are reversible plus four additional reactions that circumvent the irreversible nonequilibrium reactions.

- Since glycolysis and gluconeogenesis share the same pathway but operate in opposite directions, their activities are regulated reciprocally.

- The liver regulates the blood glucose concentration after a meal because it contains the high K_m glucokinase that promotes increased hepatic utilization of glucose and is responsive to insulin.

- Insulin is secreted as a direct response to hyperglycemia; it stimulates the liver to store glucose as glycogen and increases uptake of glucose into extrahepatic tissues.

- Glucagon is secreted as a response to hypoglycemia and activates both glycogenolysis and gluconeogenesis in the liver, causing release of glucose into the blood.

The Pentose Phosphate Pathway & Other Pathways of Hexose Metabolism

Owen P. McGuinness, PhD

O B J E C T I V E S

After studying this chapter, you should be able to:

- Describe the pentose phosphate pathway and its roles as a source of NADPH and of ribose for nucleotide synthesis.
- Describe the uronic acid pathway and its importance for synthesis of glucuronic acid for conjugation reactions and (in animals for which it is not a vitamin) vitamin C.
- Describe the metabolism of fructose and the impact of high sugar intake on metabolic disease risk.
- Describe the synthesis and physiologic importance of galactose.
- Explain the consequences of genetic defects of glucose-6-phosphate dehydrogenase (favism), the uronic acid pathway (essential pentosuria), and fructose and galactose metabolism.

BIOMEDICAL IMPORTANCE

The pentose phosphate pathway is an alternative route for the metabolism of glucose. It does not lead to formation of ATP but has two major functions: (1) the formation of **NADPH** for synthesis of fatty acids (see Chapter 23) and steroids (see Chapter 26), and maintaining reduced glutathione for antioxidant activity, and (2) the synthesis of **ribose** for nucleotide and nucleic acid formation (see Chapter 32). Glucose, fructose, and galactose are the main hexoses absorbed from the gastrointestinal tract, derived from dietary starch, sucrose, and lactose, respectively. Fructose and galactose can be converted to glucose, mainly in the liver.

Genetic deficiency of **glucose-6-phosphate dehydrogenase**, the first enzyme of the pentose phosphate pathway, causes acute hemolysis of red blood cells, resulting in **hemolytic anemia**. Glucuronic acid is synthesized from glucose via the **uronic acid pathway**, of minor quantitative importance, but of major significance for the conjugation and excretion of metabolites and foreign chemicals (xenobiotics; see Chapter 47) as **glucuronides**. A deficiency in the pathway leads to the condition of **essential pentosuria**. The lack of one

enzyme of the pathway (gulonolactone oxidase) in primates and some other animals explains why **ascorbic acid** (vitamin C; see Chapter 44) is a dietary requirement for human beings but not most other mammals. Deficiencies in the enzymes of fructose and galactose metabolism lead to metabolic diseases such as **essential fructosuria**, **hereditary fructose intolerance**, **and galactosemia**.

THE PENTOSE PHOSPHATE PATHWAY FORMS NADPH & RIBOSE PHOSPHATE

The pentose phosphate pathway (hexose monophosphate shunt, **Figure 20–1**) is a more complex pathway than glycolysis (see Chapter 17). Three molecules of glucose-6-phosphate give rise to three molecules of CO_2 and three five-carbon sugars. These are rearranged to regenerate two molecules of glucose-6-phosphate and one molecule of the glycolytic intermediate, glyceraldehyde-3-phosphate. Since two molecules of glyceraldehyde-3-phosphate can regenerate glucose-6-phosphate, the pathway can account for the complete oxidation of glucose. If this cycle is repeated glucose will eventually be converted to carbon dioxide and water and NADPH will be generated ($C_6H_{12}O_6 + 12 \text{ NADP}^+ + 6 H_2O \rightarrow 6 CO_2 + 12 H + 12 \text{ NADPH}$).

This was adapted from the 30th edition by David A. Bender, PhD & Peter A. Mayes, PhD, DSc

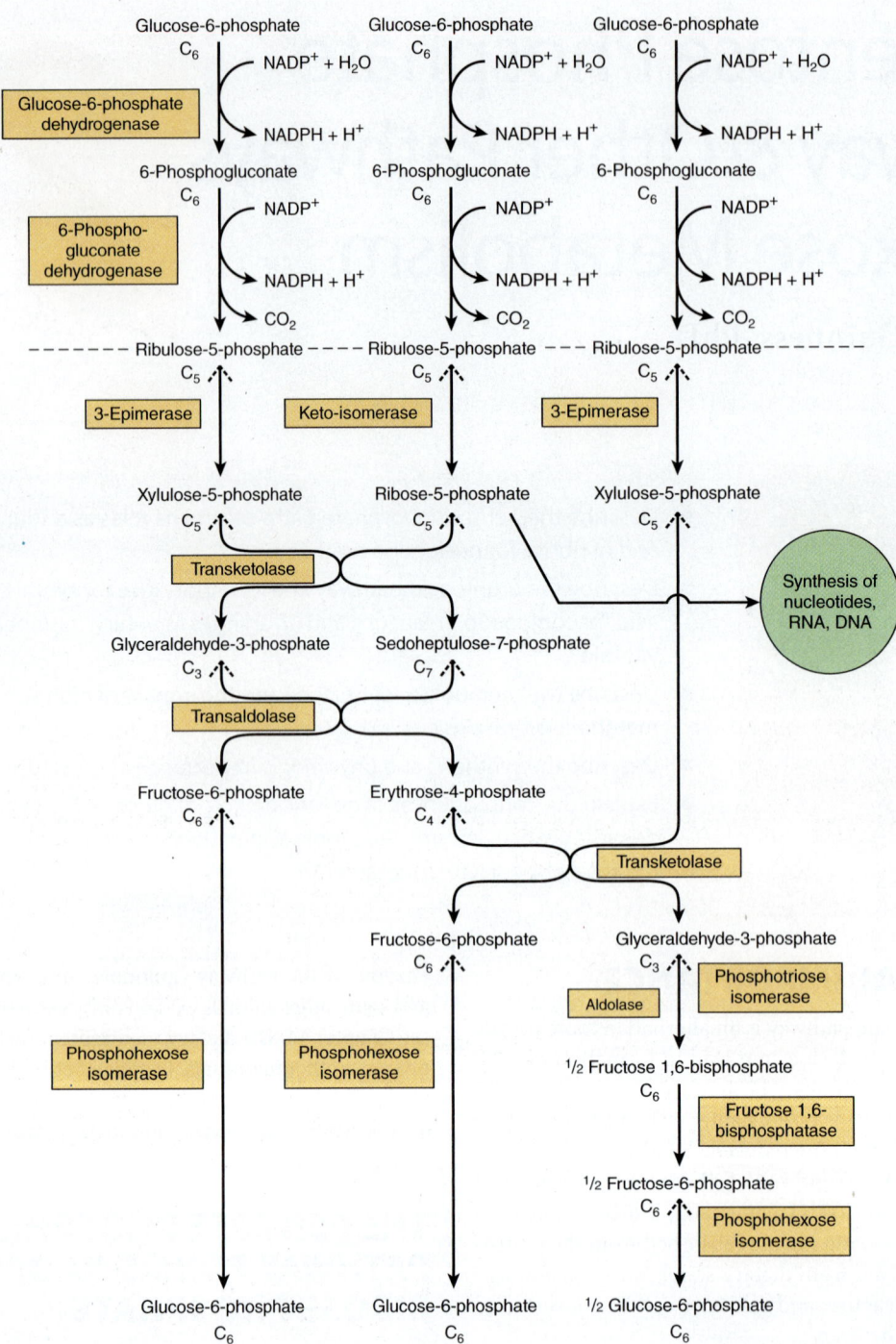

FIGURE 20–1 **Flowchart of pentose phosphate pathway and its connections with the pathway of glycolysis.** The full pathway, as indicated, consists of three interconnected cycles in which glucose-6-phosphate is both substrate and end product. The reactions above the broken line are nonreversible, whereas all reactions under that line are freely reversible apart from that catalyzed by fructose 1,6-bisphosphatase.

REACTIONS OF THE PENTOSE PHOSPHATE PATHWAY OCCUR IN THE CYTOSOL

Like glycolysis, the enzymes of the pentose phosphate pathway are cytosolic. Unlike glycolysis, oxidation is achieved by dehydrogenation using **NADP⁺**, not **NAD⁺**, as the hydrogen acceptor. The sequence of reactions of the pathway may

be divided into two phases: an **irreversible oxidative phase** and a **reversible nonoxidative phase**. In the first phase, glucose-6-phosphate undergoes dehydrogenation and decarboxylation to yield a pentose, ribulose-5-phosphate. In the second phase, ribulose-5-phosphate is converted back to glucose-6-phosphate by a series of reactions involving mainly two enzymes: **transketolase** and **transaldolase** (see Figure 20–1).

The Oxidative Phase Generates NADPH

Dehydrogenation of glucose-6-phosphate to 6-phosphogluconate occurs via the formation of 6-phosphogluconolactone, catalyzed by **glucose-6-phosphate dehydrogenase**, an NADP-dependent enzyme (Figures 20–1 and **20–2**). The hydrolysis of 6-phosphogluconolactone is accomplished by the enzyme **gluconolactone hydrolase**. A second oxidative step is catalyzed by **6-phosphogluconate dehydrogenase**, which also requires NADP$^+$ as hydrogen acceptor. Decarboxylation forms the ketopentose ribulose-5-phosphate.

In the endoplasmic reticulum, an isoenzyme of glucose-6-phosphate dehydrogenase, hexose-6-phosphate dehydrogenase,

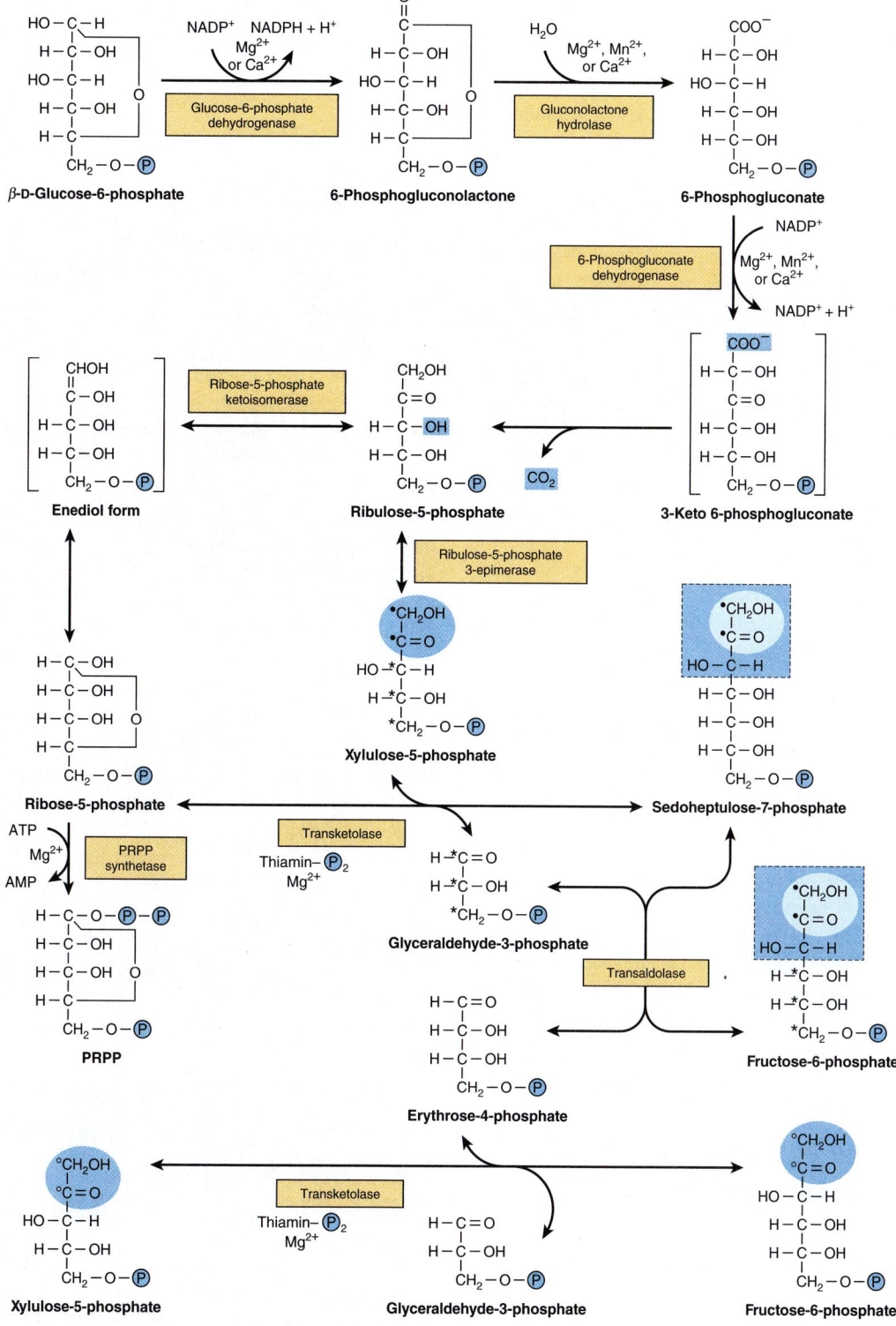

FIGURE 20–2 The pentose phosphate pathway. (P, —PO$_3$$^{2-}$; PRPP, 5-phosphoribosyl 1-pyrophosphate.)

provides NADPH for hydroxylation (mixed function oxidase) reactions, and also for 11-β-hydroxysteroid dehydrogenase-1. This enzyme catalyzes the reduction of (inactive) cortisone to (active) cortisol in liver, the nervous system, and adipose tissue. It is the major source of intracellular cortisol in these tissues and may be important in obesity and the metabolic syndrome.

The Nonoxidative Phase Generates Ribose Precursors

Ribulose-5-phosphate is the substrate for two enzymes. **Ribulose-5-phosphate 3-epimerase** alters the configuration about carbon 3, forming the epimer xylulose 5-phosphate, also a ketopentose. **Ribose-5-phosphate ketoisomerase** converts ribulose-5-phosphate to the corresponding aldopentose, ribose-5-phosphate, which is used for nucleotide and nucleic acid synthesis. **Transketolase** transfers the two-carbon unit comprising carbons 1 and 2 of a ketose onto the aldehyde carbon of an aldose sugar. It therefore affects the conversion of a ketose sugar into an aldose with two carbons less and an aldose sugar into a ketose with two carbons more. The reaction requires Mg^{2+} and **thiamin diphosphate** (vitamin B_1) as coenzyme. Measurement of erythrocyte transketolase and its activation by thiamin diphosphate provides an index of vitamin B_1 nutritional status (see Chapter 44). The two-carbon moiety is transferred as glycolaldehyde bound to thiamin diphosphate. Thus, transketolase catalyzes the transfer of the two-carbon unit from xylulose 5-phosphate to ribose-5-phosphate, producing the seven-carbon ketose sedoheptulose-7-phosphate and the aldose glyceraldehyde-3-phosphate. These two products then undergo transaldolation. **Transaldolase** catalyzes the transfer of a three-carbon dihydroxyacetone moiety (carbons 1–3) from the ketose sedoheptulose-7-phosphate onto the aldose glyceraldehyde-3-phosphate to form the ketose fructose-6-phosphate and the four-carbon aldose erythrose-4-phosphate. Transaldolase has no cofactor, and the reaction proceeds via the intermediate formation of a Schiff base of dihydroxyacetone to the ε-amino group of a lysine residue in the enzyme. In a further reaction catalyzed by **transketolase**, xylulose 5-phosphate serves as a donor of glycolaldehyde. In this case, erythrose-4-phosphate is the acceptor, and the products of the reaction are fructose-6-phosphate and glyceraldehyde-3-phosphate.

In order to oxidize glucose completely to CO_2 via the pentose phosphate pathway, there must be enzymes present in the tissue to convert glyceraldehyde-3-phosphate to glucose-6-phosphate. This involves reversal of glycolysis and the gluconeogenic enzyme **fructose 1,6-bisphosphatase**. In tissues that lack this enzyme, glyceraldehyde-3-phosphate follows the normal pathway of glycolysis to pyruvate.

The Two Major Pathways for the Catabolism of Glucose Have Little in Common

Although glucose-6-phosphate is common to both pathways, the pentose phosphate pathway is markedly different from glycolysis. Oxidation utilizes $NADP^+$ rather than NAD^+, and CO_2, which is not produced in glycolysis, is produced. No ATP is generated in the pentose phosphate pathway, whereas it is a product of glycolysis.

The two pathways are, however, connected. Xylulose 5-phosphate activates the protein phosphatase that dephosphorylates the 6-phosphofructo-2-kinase/fructose 2,6-bisphophatase bifunctional enzyme (see Chapter 17). This activates the kinase and inactivates the phosphatase, leading to increased formation of fructose 2,6-bisphosphate, increased activity of phosphofructokinase-1, and hence increased glycolytic flux. Xylulose 5-phosphate also activates the protein phosphatase that initiates the nuclear translocation and DNA binding of the carbohydrate response element-binding protein, leading to increased synthesis of fatty acids (see Chapter 23) in response to a high-carbohydrate diet. This couples the demand for NADPH for lipogenesis and activation of the lipogenic enzymatic machinery.

Reducing Equivalents Are Generated in Those Tissues Specializing in Reductive Syntheses

The pentose phosphate pathway is active in liver, adipose tissue, adrenal cortex, thyroid, erythrocytes, testis, and lactating mammary gland. Its activity is low in nonlactating mammary gland and skeletal muscle. Those tissues in which the pathway is active use NADPH in reductive syntheses, for example, of fatty acids, steroids, amino acids via glutamate dehydrogenase, and reduced glutathione. The synthesis of glucose-6-phosphate dehydrogenase and 6-phosphogluconate dehydrogenase may also be induced by insulin in the fed state, when lipogenesis increases. NADPH is also used by NADPH oxidase in phagocytes and neutrophils to destroy "respiratory burst" engulfed cells and bacteria using superoxide (see Chapter 54).

Ribose Can Be Synthesized in Virtually All Tissues

Little or no ribose circulates in the bloodstream, so tissues have to synthesize the ribose they require for nucleotide and nucleic acid synthesis using the pentose phosphate pathway (see Figure 20–2). It is not necessary to have a completely functioning pentose phosphate pathway for a tissue to synthesize ribose-5-phosphate. Muscle has only low activity of glucose-6-phosphate dehydrogenase and 6-phosphogluconate dehydrogenase, but, like most other tissues, it is capable of synthesizing ribose-5-phosphate by reversal of the nonoxidative phase of the pentose phosphate pathway utilizing fructose-6-phosphate.

THE PENTOSE PHOSPHATE PATHWAY & GLUTATHIONE PEROXIDASE PROTECT ERYTHROCYTES AGAINST HEMOLYSIS

In red blood cells, the pentose phosphate pathway is the sole source of NADPH for the reduction of oxidized glutathione catalyzed by **glutathione reductase**, a flavoprotein containing

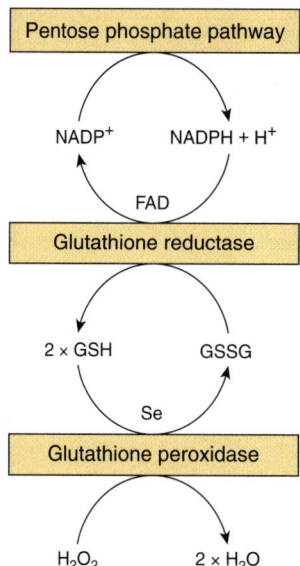

FIGURE 20–3 Role of the pentose phosphate pathway in the glutathione peroxidase reaction of erythrocytes. (GSH, reduced glutathione; GSSG, oxidized glutathione; Se, selenium-containing enzyme.)

flavin adenine dinucleotide (FAD). Reduced glutathione removes H_2O_2 in a reaction catalyzed by **glutathione peroxidase**, an enzyme that contains the **selenium** analog of cysteine (selenocysteine) at the active site (**Figure 20–3**). The reaction is important since accumulation of H_2O_2 may decrease the life span of the erythrocyte by causing oxidative damage to the cell membrane, leading to hemolysis. In other tissues, NADPH can also be generated by the reaction catalyzed by the malic enzyme.

GLUCURONATE, A PRECURSOR OF PROTEOGLYCANS & CONJUGATED GLUCURONIDES, IS A PRODUCT OF THE URONIC ACID PATHWAY

In liver, the **uronic acid pathway** catalyzes the conversion of glucose to glucuronic acid, ascorbic acid (except in human beings and other species for which ascorbate is a vitamin, vitamin C), and pentoses (**Figure 20–4**). It is also an alternative oxidative pathway for glucose that, like the pentose phosphate pathway, does not lead to the formation of ATP. Glucose-6-phosphate is isomerized to glucose-1-phosphate, which then reacts with uridine triphosphate (UTP) to form uridine diphosphate glucose (UDPGlc) in a reaction catalyzed by **UDPGlc pyrophosphorylase**, as occurs in glycogen synthesis (see Chapter 18). UDPGlc is oxidized at carbon 6 by NAD-dependent **UDPGlc dehydrogenase** in a two-step reaction to yield UDP-glucuronate.

UDP-glucuronate is the source of glucuronate for reactions involving its incorporation into proteoglycans (see Chapter 46) or for reaction with substrates such as steroid hormones, bilirubin, and a number of drugs that are excreted in urine or bile as glucuronide conjugates (see Figure 31–13 and Chapter 47).

Glucuronate is reduced to L-gulonate, the direct precursor of **ascorbate** in those animals capable of synthesizing this vitamin, in an NADPH-dependent reaction. In human beings and other primates, as well as guinea pigs, bats, and some birds and fishes, ascorbic acid cannot be synthesized because of the absence of L-**gulonolactone oxidase**. L-Gulonate is oxidized to 3-keto-L-gulonate, which is then decarboxylated to L-xylulose. L-Xylulose is converted to the D-isomer by an NADPH-dependent reduction to xylitol, followed by oxidation in an NAD-dependent reaction to D-xylulose. After conversion to D-xylulose 5-phosphate, it is metabolized via the pentose phosphate pathway.

INGESTION OF LARGE QUANTITIES OF FRUCTOSE HAS PROFOUND METABOLIC CONSEQUENCES

Diets high in sucrose or in high-fructose syrups (HFCS 42 and HFCS55) used in manufactured foods and beverages lead to large amounts of fructose (and glucose) entering the hepatic portal vein. Note that high fructose corn syrup despite the name does not have much more fructose than sucrose (50% fructose). In fact, other dietary sources of sugar have more fructose (eg, apples 73%). The important issue is the total quantity of simple sugars ingested is too high. In 1900 Americans ingested about 15 g/day of fructose (50 kcal/day) primarily from fruits and vegetables, in 2020 it is 77 g/day and in children it is about 81 g/day (~300 kcals/day), a sixfold increase. The American Heart Association recommended keeping the intake below 25 g/day (100 kcal/day) for women and children and 37 g/day (150 kcal/day) for men, as the risk of obesity, hyperuriacidemia, high blood pressure, and diabetes are increased when simple sugar intake is high.

Nearly 90% of the dietary fructose is metabolized by the liver. Fructose undergoes more rapid glycolysis in the liver than does glucose because it bypasses the regulatory step catalyzed by phosphofructokinase (**Figure 20–5**). This allows fructose to flood the pathways in the liver, leading to increased fatty acid synthesis, esterification of fatty acids, and secretion of very-low-density lipoprotein (VLDL), which may raise serum triacylglycerols and ultimately raise LDL cholesterol concentrations. **Fructokinase** in liver, kidney, and intestine catalyzes the phosphorylation of fructose to fructose-1-phosphate. This enzyme does not act on glucose, and, unlike glucokinase, its activity is not affected by fasting or by insulin, which may explain why fructose is cleared from the blood of diabetic patients at a normal rate. Fructose-1-phosphate is cleaved to D-glyceraldehyde and dihydroxyacetone phosphate by **aldolase B**, an enzyme found in the liver, which also functions in glycolysis in the liver by cleaving fructose 1,6-bisphosphate. D-Glyceraldehyde enters glycolysis via phosphorylation to glyceraldehyde-3-phosphate catalyzed by **triokinase**. The two triose phosphates, dihydroxyacetone phosphate and glyceraldehyde-3-phosphate, may either be degraded by glycolysis or may be substrates for

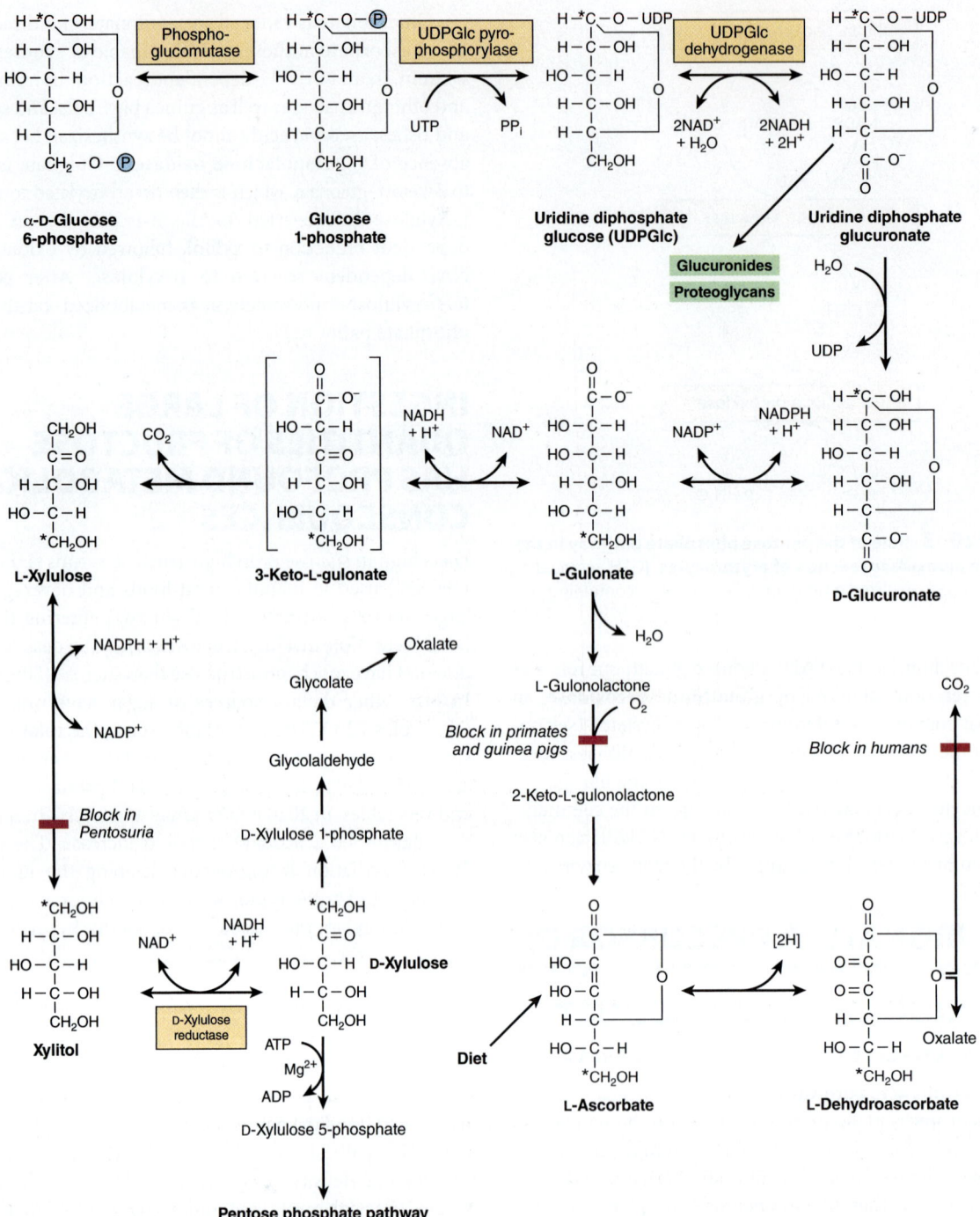

FIGURE 20–4 **Uronic acid pathway.** (*Indicates the fate of carbon 1 of glucose.)

aldolase and hence gluconeogenesis, which is the fate of much of the fructose metabolized in the liver. To amplify the carbohydrate loading effect of fructose, fructose-1-phosphate activates glucokinase and thus amplifies dietary glucose entry into liver. In addition, because of the rapid entry and phosphorylation of fructose the consumption of ATP is very fast causing a rise in ADP and AMP. AMP can be converted to hypoxanthine in the liver and to uric acid (xanthine oxidase) that can cause gout (see Chapter 33).

Extrahepatic tissues generally do not see much fructose. However in those tissues hexokinase catalyzes the phosphorylation of most hexose sugars, including fructose, but glucose inhibits the phosphorylation of fructose since it is a better substrate for hexokinase. Nevertheless, some fructose can be metabolized in adipose tissue and muscle. Fructose is found in seminal plasma and in the fetal circulation of ungulates and whales. Aldose reductase is found in the placenta of the ewe and is responsible for the secretion of sorbitol into the fetal blood. The presence of sorbitol dehydrogenase in the liver, including the fetal liver, is responsible for the conversion of sorbitol into fructose. This pathway is also responsible for the occurrence of fructose in seminal fluid.

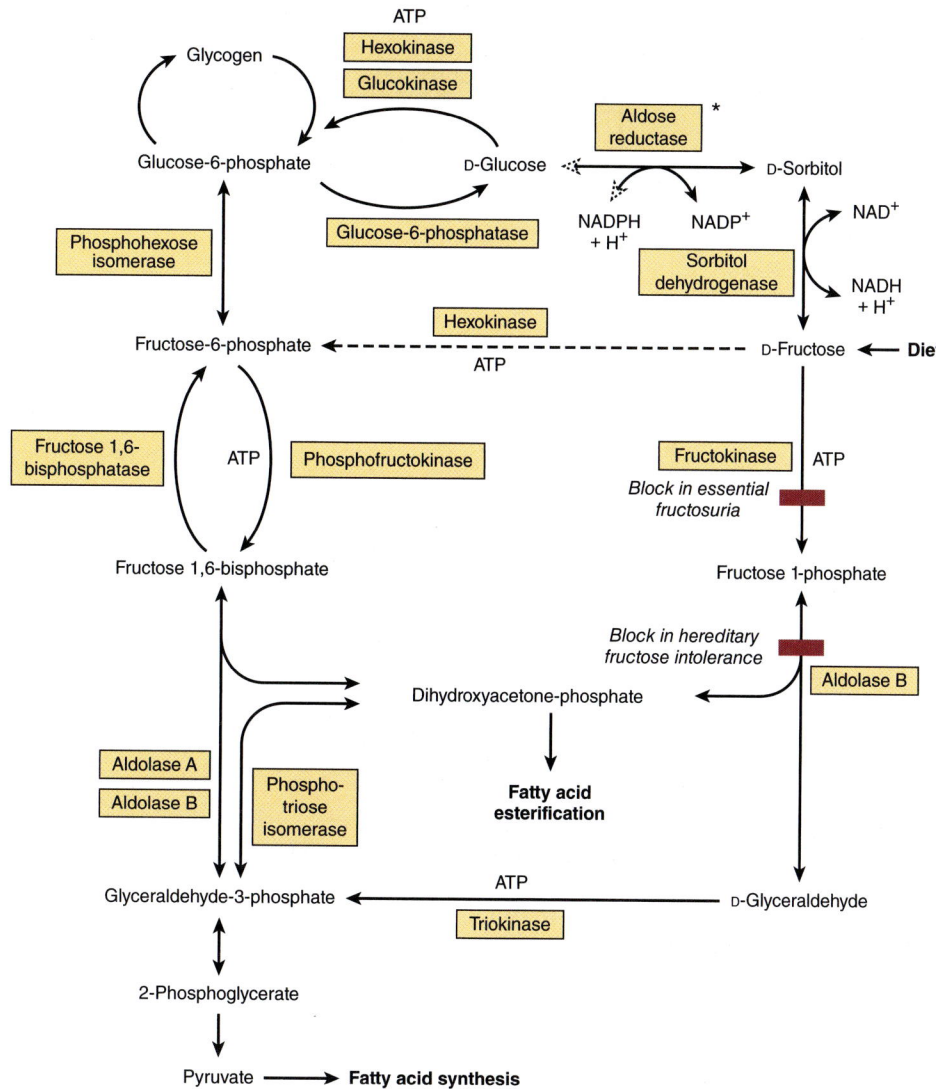

FIGURE 20–5 **Metabolism of fructose.** Aldolase A is found in all tissues, whereas aldolase B is the predominant form in liver. (*Not found in liver.)

GALACTOSE IS NEEDED FOR THE SYNTHESIS OF LACTOSE, GLYCOLIPIDS, PROTEOGLYCANS, & GLYCOPROTEINS

Galactose is derived from intestinal hydrolysis of the disaccharide **lactose**, the sugar found in milk. It is readily converted in the liver to glucose after being transported by GLUT5. So like fructose the majority of dietary galactose is metabolized by the liver. **Galactokinase** catalyzes the phosphorylation of galactose, using ATP as phosphate donor (**Figure 20–6**). Galactose-1-phosphate reacts with UDPGlc to form uridine diphosphate galactose (UDPGal) and glucose-1-phosphate, in a reaction catalyzed by **galactose-1-phosphate uridyl transferase**. The conversion of UDPGal to UDPGlc is catalyzed by **UDPGal 4-epimerase**. The reaction involves oxidation, and then reduction, at carbon 4, with NAD⁺ as a coenzyme. The UDPGlc is then incorporated into glycogen (see Chapter 18).

The epimerase reaction is freely reversible, so glucose can be converted to galactose, and galactose is not a dietary essential. Galactose is required in the body not only for the formation of lactose in lactation but also as a constituent of glycolipids (cerebrosides), proteoglycans, and glycoproteins. In the synthesis of lactose in the mammary gland, UDPGal condenses with glucose to yield lactose, catalyzed by **lactose synthase** (see Figure 20–6).

Glucose Is the Precursor of Amino Sugars (Hexosamines)

Amino sugars are important components of **glycoproteins** (see Chapter 46), of certain **glycosphingolipids** (eg, gangliosides; see Chapter 21), and of glycosaminoglycans (see Chapter 50). The major amino sugars are the hexosamines **glucosamine**, **galactosamine**, and **mannosamine**, and the nine-carbon compound **sialic acid**. The principal sialic acid found in human tissues is *N*-acetylneuraminic acid (NeuAc).

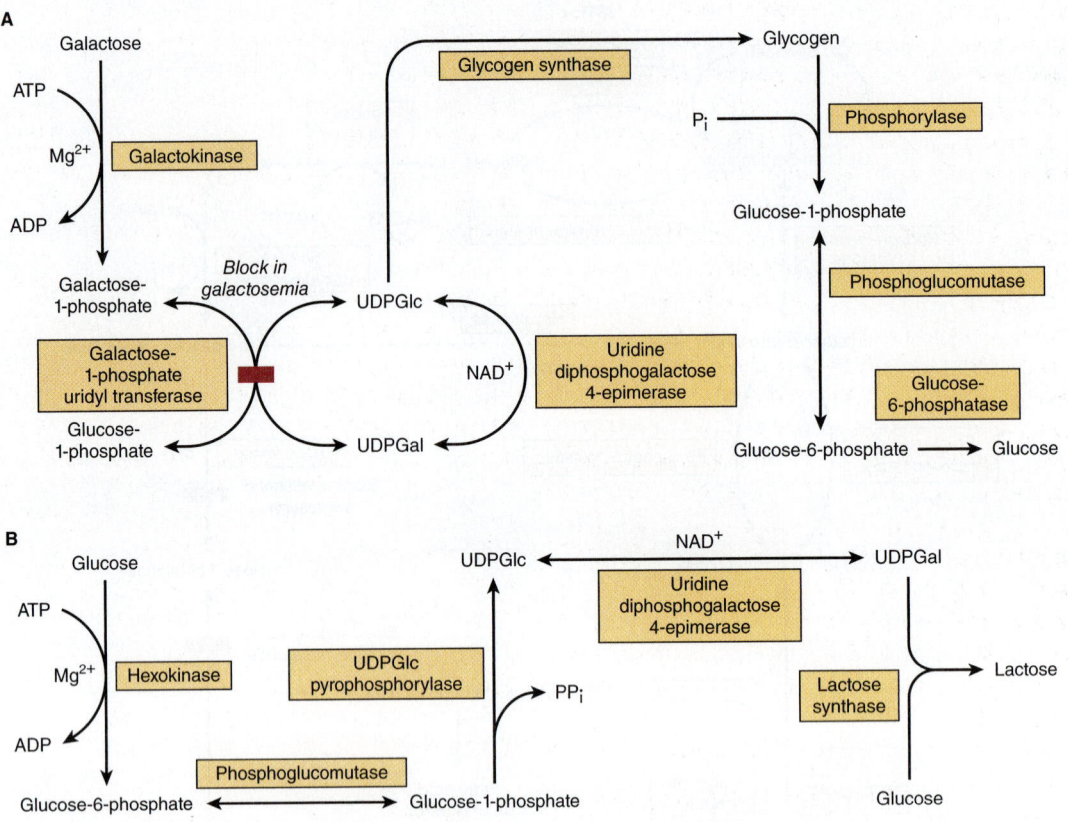

FIGURE 20–6 Pathway of conversion of (A) galactose to glucose in the liver and (B) glucose to lactose in the lactating mammary gland.

A summary of the metabolic interrelationships among the amino sugars is shown in **Figure 20–7**.

CLINICAL ASPECTS

Impairment of the Pentose Phosphate Pathway Leads to Erythrocyte Hemolysis

Genetic defects of glucose-6-phosphate dehydrogenase, with consequent impairment of the generation of NADPH, are common in populations of Mediterranean and Afro-Caribbean origin. The gene is on the X chromosome, so it is mainly males who are affected. Some 400 million people carry a mutated gene for glucose-6-phosphate dehydrogenase, making it the most common genetic defect, but most are asymptomatic. In some populations, glucose-6-phosphatase deficiency is common enough for it to be regarded as a genetic polymorphism. The distribution of mutant genes parallels that of malaria, suggesting that being heterozygous confers resistance against malaria. The defect is manifested as red cell hemolysis (**hemolytic anemia**) when susceptible individuals are subjected to oxidative stress (see Chapter 45) from infection, drugs such as the antimalarial primaquine, and sulfonamides, or when they have eaten fava beans (*Vicia faba*—hence the name of the disease, **favism**).

Many different mutations are known in the gene for glucose-6-phosphate dehydrogenase, leading to two main variants of favism. In the Afro-Caribbean variant, the enzyme is unstable, so that while average red-cell activities are low, it is only the older erythrocytes that are affected by oxidative stress, and the hemolytic crises tend to be self-limiting. By contrast, in the Mediterranean variant the enzyme is stable, but has low activity in all erythrocytes. Hemolytic crises in these people are more severe and can be fatal. Glutathione peroxidase is dependent on a supply of NADPH, which in erythrocytes can only be formed via the pentose phosphate pathway. It reduces organic peroxides and H_2O_2 as part of the body's defense against lipid peroxidation. Measurement of erythrocyte **glutathione reductase**, and its activation by FAD is used to assess **vitamin B$_2$** nutritional status (see Chapter 44).

Disruption of the Uronic Acid Pathway Is Caused by Enzyme Defects & Some Drugs

In the rare benign hereditary condition **essential pentosuria**, considerable quantities of **xylulose** appear in the urine because of a lack of xylulose reductase, the enzyme necessary to reduce xylulose to xylitol. Although pentosuria is benign, with no clinical consequences, xylulose is a reducing sugar and can give false-positive results when urinary glucose is measured using alkaline copper reagents (see Chapter 48). Various drugs increase the rate at which glucose enters the uronic acid pathway. For example, administration of barbital or chlorobutanol

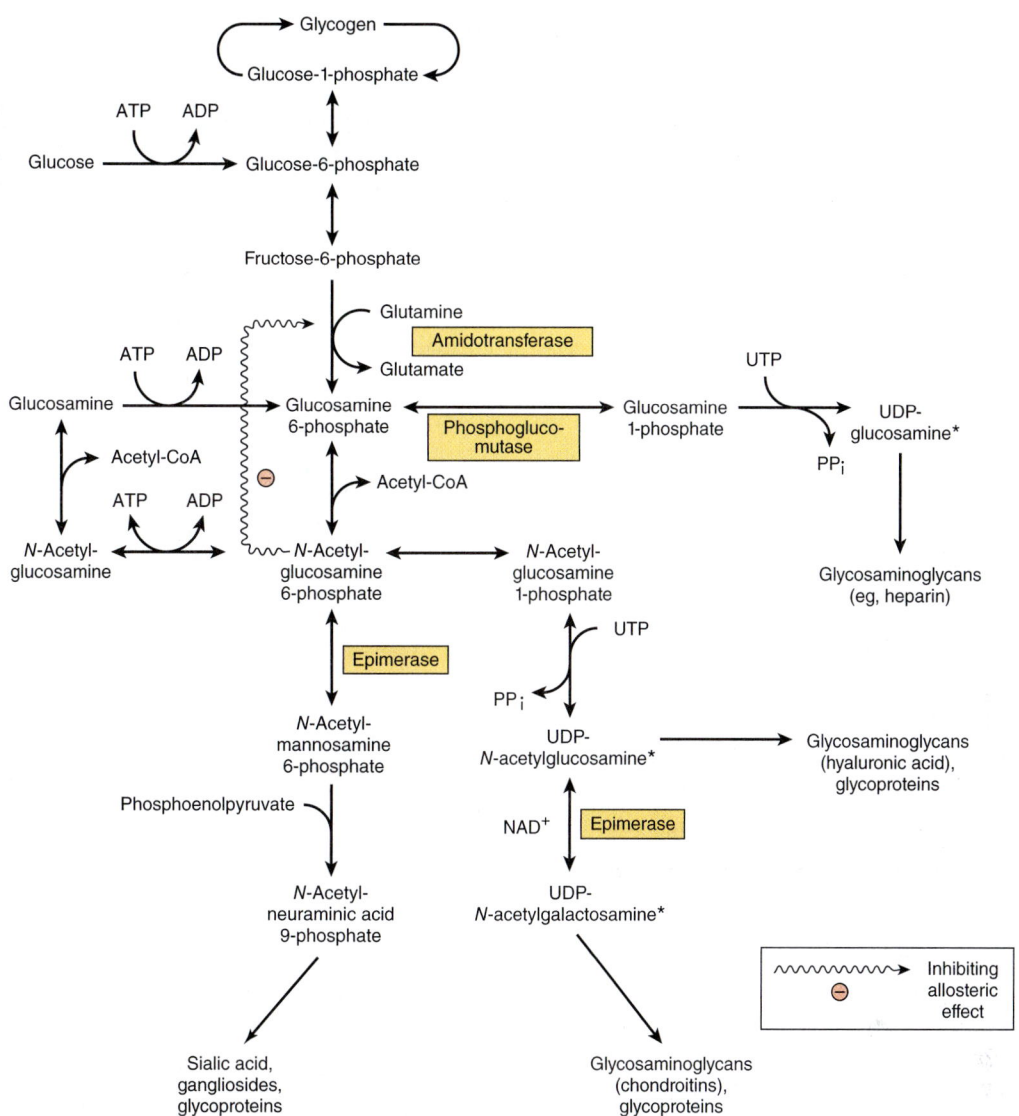

FIGURE 20–7 **Summary of the interrelationships in metabolism of amino sugars.** (*Analogous to UDPGlc.) Other purine or pyrimidine nucleotides may be similarly linked to sugars or amino sugars. Examples are thymidine diphosphate (TDP)-glucosamine and TDP-*N*-acetylglucosamine.

to rats results in a significant increase in the conversion of glucose to glucuronate, L-gulonate, and ascorbate. Aminopyrine and antipyrine increase the excretion of xylulose in pentosuric subjects. Pentosuria also occurs after consumption of relatively large amounts of fruits such as pears that are rich sources of pentoses (**alimentary pentosuria**).

Loading of the Liver With Fructose May Potentiate Hypertriacylglycerolemia, Hypercholesterolemia, & Hyperuricemia

In the liver, fructose increases fatty acid and triacylglycerol synthesis and VLDL secretion, leading to hypertriacylglycerolemia—and increased LDL cholesterol—which can be regarded as potentially atherogenic (see Chapter 26). This is because fructose enters glycolysis via fructokinase, and the resulting fructose-1-phosphate bypasses the regulatory step catalyzed by phosphofructokinase

(see Chapter 17). In addition, acute loading of the liver with fructose, as can occur with intravenous infusion or following very high fructose intakes, causes sequestration of inorganic phosphate in fructose-1-phosphate and diminished ATP synthesis. As a result, there is less inhibition of de novo purine synthesis by ATP, and uric acid formation is increased, causing hyperuricemia, which is the cause of **gout** (see Chapter 33). Since fructose is absorbed from the small intestine by (passive) carrier-mediated diffusion, high oral doses may lead to osmotic diarrhea.

Defects in Fructose Metabolism Cause Disease

A lack of hepatic fructokinase causes **essential fructosuria**, which is a benign and asymptomatic condition. The absence of aldolase B, which cleaves fructose-1-phosphate, leads to

hereditary fructose intolerance, which is characterized by profound hypoglycemia and vomiting after consumption of fructose (or sucrose, which yields fructose on digestion). Diets low in fructose, sorbitol, and sucrose are beneficial for both conditions. One consequence of hereditary fructose intolerance and of a related condition as a result of **fructose 1,6-bisphosphatase deficiency** is fructose-induced **hypoglycemia** despite the presence of high glycogen reserves, because fructose-1-phosphate and 1,6-bisphosphate allosterically inhibit liver glycogen phosphorylase. The sequestration of inorganic phosphate also leads to depletion of ATP and hyperuricemia.

Fructose & Sorbitol in the Lens Are Associated With Diabetic Cataract

Both fructose and sorbitol are found in the lens of the eye in increased concentrations in diabetes mellitus and may be involved in the pathogenesis of **diabetic cataract**. The **sorbitol (polyol) pathway** (not found in liver) is responsible for fructose formation from glucose (see Figure 20–5) and increases in activity as the glucose concentration rises in those tissues that are not insulin sensitive—the lens, peripheral nerves, and renal glomeruli. Glucose is reduced to sorbitol by **aldose reductase**, followed by oxidation of sorbitol to fructose in the presence of NAD+ and sorbitol dehydrogenase (polyol dehydrogenase). Sorbitol does not diffuse through cell membranes, but accumulates, causing osmotic damage. Simultaneously, myoinositol levels fall. In experimental animals, sorbitol accumulation and myoinositol depletion, as well as diabetic cataract, can be prevented by aldose reductase inhibitors. A number of inhibitors are undergoing clinical trials for prevention of adverse effects of diabetes.

Enzyme Deficiencies in the Galactose Pathway Cause Galactosemia

Inability to metabolize galactose occurs in the **galactosemias**, which may be caused by inherited defects of galactokinase, uridyl transferase, or 4-epimerase (see Figure 20–6A),

though deficiency of **uridyl transferase** is best known. Galactose is a substrate for aldose reductase, forming galactitol, which accumulates in the lens of the eye, causing cataract. The condition is more severe if it is the result of a defect in the uridyl transferase since galactose-1-phosphate accumulates and depletes the liver of inorganic phosphate. Ultimately, liver failure and mental deterioration result. In uridyl transferase deficiency, the epimerase is present in adequate amounts, so that the galactosemic individual can still form UDPGal from glucose. This explains how it is possible for normal growth and development of affected children to occur despite the galactose-free diets used to control the symptoms of the disease.

SUMMARY

- The pentose phosphate pathway, present in the cytosol, can account for the complete oxidation of glucose, producing NADPH and CO_2 but no ATP.

- The pathway has an oxidative phase, which is irreversible and generates NADPH, and a nonoxidative phase, which is reversible and provides ribose precursors for nucleotide synthesis. The complete pathway is present mainly in those tissues having a requirement for NADPH for reductive syntheses, for example, lipogenesis or steroidogenesis, whereas the nonoxidative phase is present in all cells requiring ribose.

- In erythrocytes, the pathway has a major function in preventing hemolysis by providing NADPH to maintain glutathione in the reduced state as the substrate for glutathione peroxidase.

- The uronic acid pathway is the source of glucuronic acid for conjugation of many endogenous and exogenous substances before excretion as glucuronides in urine and bile.

- Fructose bypasses the main regulatory step in glycolysis, catalyzed by phosphofructokinase, and stimulates liver glucose uptake, fatty acid synthesis, and hepatic triacylglycerol secretion.

- Galactose is synthesized from glucose in the lactating mammary gland and in other tissues where it is required for the synthesis of glycolipids, proteoglycans, and glycoproteins.

Exam Questions

Section IV – Metabolism of Carbohydrates

1. Which of the following is not an aldose?
   A. Erythrose
   B. Fructose
   C. Galactose
   D. Glucose
   E. Ribose

2. Which of the following is the composition of sucrose?
   A. O-α-D-galactopyranosyl-(1→4)-β-D-glucopyranose
   B. O-α-D-glucopyranosyl-(1→2)-β-D-fructofuranoside
   C. O-α-D-glucopyranosyl-(1→4)-α-D-glucopyranose
   D. O-α-D-glucopyranosyl-(1→1)-α-D-glucopyranoside
   E. O-α-D-glucopyranosyl-(1→6)-α-D-glucopyranose

3. Which of the following is not a pentose?
   A. Fructose
   B. Ribose
   C. Ribulose
   D. Xylose
   E. Xylulose

4. A blood sample is taken from a 50-year-old woman after an overnight fast. Which one of the following will be at a higher concentration than after she had eaten a meal?
   A. Glucose
   B. Insulin
   C. liver glycogen
   D. Nonesterified fatty acids
   E. Triacylglycerol

5. A blood sample is taken from a 25-year-old man after he has eaten three slices of toast and a boiled egg. Which one of the following will be at a higher concentration than if the blood sample had been taken after an overnight fast?
   A. Alanine
   B. Glucagon
   C. Glucose
   D. Ketone bodies
   E. Nonesterified fatty acids

6. A blood sample is taken from a 40-year-old man who has been fasting completely for a week, drinking only water. Which of the following will be at a higher concentration than after a normal overnight fast?
   A. Glucose
   B. Insulin
   C. Ketone bodies
   D. Nonesterified fatty acids
   E. Triacylglycerol

7. Which one of following statements about the fed and fasting metabolic states is correct?
   A. In the fasting state glucagon acts to increase the activity of lipoprotein lipase in adipose tissue.
   B. In the fasting state, glucagon acts to increase the synthesis of glycogen from glucose.
   C. In the fed state insulin acts to increase the breakdown of glycogen to maintain blood glucose.
   D. In the fed state there is decreased secretion of insulin in response to increased glucose in the portal blood.
   E. Ketone bodies are synthesized in liver in the fasting state, and the amount synthesized increases as fasting extends into starvation.

8. Which one of following statements about the fed and fasting metabolic states is correct?
   A. In the fed state muscle can take up glucose for use as a metabolic fuel because glucose transport in muscle is stimulated in response to glucagon.
   B. In the fed state there is increased secretion of glucagon in response to increased glucose in the portal blood.
   C. In the fed state, insulin acts to increase the synthesis of glycogen from glucose.
   D. Plasma glucose is maintained in starvation and prolonged fasting by gluconeogenesis from ketone bodies.
   E. There is an increase in metabolic rate in the fasting state.

9. Which one of following statements about the fed and fasting metabolic states is correct?
   A. In the fasting state liver synthesizes glucose from amino acids.
   B. In the fed state adipose tissue can take up glucose for synthesis of triacylglycerol because glucose transport in adipose tissue is stimulated in response to glucagon.
   C. Ketone bodies are synthesized in muscle in the fasting state, and the amount synthesized increases as fasting extends into starvation.
   D. Ketone bodies provide an alternative fuel for red blood cells in the fasting state.
   E. Plasma glucose is maintained in starvation and prolonged fasting by gluconeogenesis from fatty acids.

10. Which one of following statements about the fed and fasting metabolic states is correct?
    A. In the fasting state adipose tissue synthesizes glucose from the glycerol released by the breakdown of triacylglycerol.
    B. In the fasting state adipose tissue synthesizes ketone bodies.
    C. In the fasting state the main fuel for red blood cells is fatty acids released from adipose tissue.
    D. Ketone bodies provide the main fuel for the central nervous system in the fasting state.
    E. Plasma glucose is maintained in starvation and prolonged fasting by gluconeogenesis in the liver from the amino acids released by the breakdown of muscle protein.

11. Which one of following statements about the fed and fasting metabolic states is correct?
    A. Fatty acids and triacylglycerol are synthesized in the liver in the fasting state.
    B. In the fasting state the main fuel for the central nervous system is fatty acids released from adipose tissue.
    C. In the fasting state the main metabolic fuel for most tissues comes from fatty acids released from adipose tissue.
    D. In the fed state muscle cannot take up glucose for use as a metabolic fuel because glucose transport in muscle is stimulated in response to glucagon.
    E. Plasma glucose is maintained in starvation and prolonged fasting by gluconeogenesis in adipose tissue from the glycerol released from triacylglycerol.

12. A 25-year-old man visits his GP complaining of abdominal cramps and diarrhea after drinking milk. What is the most likely cause of his problem?

    A. Bacterial and yeast overgrowth in the large intestine
    B. Infection with the intestinal parasite Giardia lamblia
    C. Lack of pancreatic amylase
    D. Lack of small intestinal lactase
    E. Lack of small intestinal sucrase-isomaltase

13. Which one of following statements about glycolysis and gluconeogenesis is correct?

    A. All the reactions of glycolysis are freely reversible for gluconeogenesis.
    B. Fructose cannot be used for gluconeogenesis in the liver because it cannot be phosphorylated to fructose-6-phosphate.
    C. Glycolysis can proceed in the absence of oxygen only if pyruvate is formed from lactate in muscle.
    D. Red blood cells only metabolize glucose by anaerobic glycolysis (and the pentose phosphate pathway).
    E. The reverse of glycolysis is the pathway for gluconeogenesis in skeletal muscle.

14. Which one of following statements about the step in glycolysis catalyzed by hexokinase and in gluconeogenesis by glucose-6-phosphatase is correct?

    A. Because hexokinase has a low K_m, its activity in liver increases as the concentration of glucose in the portal blood increases.
    B. Glucose-6-phosphatase is mainly active in muscle in the fasting state.
    C. If hexokinase and glucose-6-phosphatase are both equally active at the same time, there is net formation of ATP from ADP and phosphate.
    D. Liver contains an isoenzyme of hexokinase, glucokinase, which is especially important in the fed state.
    E. Muscle can release glucose into the circulation from its glycogen reserves in the fasting state.

15. Which one of following statements about this step in glycolysis catalyzed by phosphofructokinase and in gluconeogenesis by fructose 1,6-bisphosphatase is correct?

    A. Fructose 1,6-bisphosphatase is mainly active in the liver in the fed state.
    B. Phosphofructokinase is mainly active in the liver in the fasted state.
    C. If phosphofructokinase and fructose 1,6-bisphosphatase are both equally active at the same time, there is a net formation of ATP from ADP and phosphate.
    D. Phosphofructokinase is inhibited by ATP, this is reversed by increases in AMP.
    E. Phosphofructokinase is mainly active in the liver in the fasting state.

16. Which one of the following statements about glucose metabolism in maximum exertion is correct?

    A. Gluconeogenesis from lactate requires less ATP than is formed during anaerobic glycolysis.
    B. In maximum exertion, pyruvate is oxidized to lactate in muscle.
    C. Oxygen debt is caused by the need to exhale carbon dioxide produced in response to acidosis.

D. Oxygen debt reflects the need to replace oxygen that has been used in muscle during vigorous exercise.
E. There is metabolic acidosis as a result of vigorous exercise.

17. Which one of following statements is correct?

    A. Glucose-1-phosphate may be hydrolyzed to yield free glucose in liver.
    B. Glucose-6-phosphate can be formed from glucose, but not from glycogen.
    C. Glucose-6-phosphate cannot be converted to glucose-1-phosphate in liver.
    D. Glucose-6-phosphate is formed from glycogen by the action of the enzyme glycogen phosphorylase.
    E. In liver and red blood cells, glucose-6-phosphate may enter into either glycolysis or the pentose phosphate pathway.

18. Which one of following statements about the pyruvate dehydrogenase multienzyme complex is correct?

    A. In thiamin (vitamin B_1) deficiency, pyruvate formed in muscle cannot be transaminated to alanine.
    B. In thiamin (vitamin B_1) deficiency, pyruvate formed in muscle cannot be carboxylated to oxaloacetate.
    C. The reaction of pyruvate dehydrogenase involves decarboxylation and oxidation of pyruvate, then formation of acetyl-CoA.
    D. The reaction of pyruvate dehydrogenase is readily reversible, so that acetyl-CoA can be used for the synthesis of pyruvate, and hence glucose.
    E. The reaction of pyruvate dehydrogenase leads to the oxidation of NADH to NAD^+, and hence the formation of $\sim 2.5 \times$ ATP per mol of pyruvate oxidized.

19. Which one of following statements about the pentose phosphate pathway is correct?

    A. In favism red blood cells are more susceptible to oxidative stress because of a lack of NADPH for fatty acid synthesis.
    B. People who lack glucose-6-phosphate dehydrogenase cannot synthesize fatty acids because of a lack of NADPH in liver and adipose tissue.
    C. The pentose phosphate pathway is especially important in tissues that are synthesizing fatty acids.
    D. The pentose phosphate pathway is the only source of NADPH for fatty acid synthesis.
    E. The pentose phosphate pathway provides an alternative to glycolysis only in the fasting state.

20. Which one of following statements about glycogen metabolism is correct?

    A. Glycogen is synthesized in the liver in the fed state, then exported to other tissues in low-density lipoproteins.
    B. Glycogen reserves in liver and muscle will meet energy requirements for several days in prolonged fasting.
    C. Liver synthesizes more glycogen when the hepatic portal blood concentration of glucose is high because of the activity of glucokinase in the liver.
    D. Muscle synthesizes glycogen in the fed state because glycogen phosphorylase is activated in response to insulin.
    E. The plasma concentration of glycogen increases in the fed state.

21. Which one of following statements about gluconeogenesis is correct?

    A. Because they form acetyl-CoA, fatty acids can be a substrate for gluconeogenesis.
    B. If oxaloacetate is withdrawn from the citric acid cycle for gluconeogenesis, then it can be replaced by the action of pyruvate dehydrogenase.
    C. The reaction of phosphoenolpyruvate carboxykinase is important to replenish the pool of citric acid cycle intermediates.
    D. The use of GTP as the phosphate donor in the phosphoenolpyruvate carboxykinase reaction provides a link between citric acid cycle activity and gluconeogenesis.
    E. There is a greater yield of ATP in anaerobic glycolysis than the cost for synthesis of glucose from lactate.

22. Which one of following statements about carbohydrate metabolism is correct?

    A. A key step in the biosynthesis of glycogen is the formation of UDP-glucose.
    B. Glycogen can be broken down to glucose-6-phosphate in muscle, which then releases free glucose by the action of the enzyme glucose-6-phosphatase.
    C. Glycogen is stored mainly in the liver and brain.
    D. Insulin inhibits the biosynthesis of glycogen.
    E. Phosphorylase kinase is an enzyme that phosphorylates the enzyme glycogen phosphorylase and thereby decreases glycogen breakdown.

23. Which one of following statements about glycogen metabolism is correct?

    A. Glycogen synthase activity is increased by glucagon.
    B. Glycogen phosphorylase is an enzyme that can be activated by phosphorylation of serine residues.
    C. Glycogen phosphorylase cannot be activated by calcium ions.
    D. cAMP activates glycogen synthesis.
    E. Glycogen phosphorylase breaks the α1-4 glycosidic bonds by hydrolysis.

24. Which one of following statements about glucose metabolism is correct?

    A. Glucagon increases the rate of glycolysis.
    B. Glycolysis requires $NADP^+$.
    C. In glycolysis, glucose is cleaved into two three-carbon compounds.
    D. Substrate-level phosphorylation takes place in the electron transport system.
    E. The main product of glycolysis in red blood cells is pyruvate.

25. Which one of following statements about metabolism of sugars is correct?

    A. Fructokinase phosphorylates fructose to fructose-6-phosphate.
    B. Fructose is an aldose sugar-like glucose.
    C. Fructose transport into cells is insulin dependent.
    D. Galactose is phosphorylated to galactose-1-phosphate by galactokinase.
    E. Sucrose can be biosynthesized from glucose and fructose in the liver.

26. In glycolysis, the conversion of 1 mol of fructose 1,6-bisphosphate to 2 mol of pyruvate results in the formation of:

    A. 1 mol NAD^+ and 2 mol of ATP
    B. 1 mol NADH and 1 mol of ATP
    C. 2 mol NAD^+ and 4 mol of ATP
    D. 2 mol NADH and 2 mol of ATP
    E. 2 mol NADH and 4 mol of ATP

27. Which of the following will provide the main fuel for muscle contraction during short-term maximum exertion?

    A. Muscle glycogen
    B. Muscle reserves of triacylglycerol
    C. Plasma glucose
    D. Plasma nonesterified fatty acids
    E. Triacylglycerol in plasma very-low-density lipoprotein

28. The disaccharide lactulose is not digested, but is fermented by intestinal bacteria, to yield 4 mol of lactate plus four protons. Ammonium (NH_4^+) is in equilibrium with ammonia (NH_3) in the bloodstream. Which of the following best explains how lactulose acts to treat hyperammonemia (elevated blood ammonium concentration)?

    A. Fermentation of lactulose increases the acidy of the bloodstream so that there is more ammonium and less ammonia is available to cross the gut wall.
    B. Fermentation of lactulose results in acidification of the gut contents so that ammonia diffuses from the bloodstream into the gut and is trapped as ammonium that cannot cross back.
    C. Fermentation of lactulose results in acidification of the gut contents so that ammonia produced by intestinal bacteria is trapped as ammonium that cannot diffuse into the bloodstream.
    D. Fermentation of lactulose results in an eightfold increase in the osmolality of the gut contents, so that there is more water for ammonia and ammonium to dissolve in, so that less is absorbed into the bloodstream.
    E. Fermentation of lactulose results in an eightfold increase in the osmolality of the gut contents, so that there is more water for ammonia and ammonium to dissolve in, so that more will diffuse for the bloodstream into the gut.

C H A P T E R

Lipids of Physiologic Significance

21

Kathleen M. Botham, PhD, DSc, & Peter A. Mayes, PhD, DSc

O B J E C T I V E S

After studying this chapter, you should be able to:

■ Define simple and complex lipids and identify the lipid classes in each group.

■ Indicate the structure of saturated and unsaturated fatty acids, explain how the chain length and degree of unsaturation influence their melting point, give examples, and explain the nomenclature.

■ Explain the difference between *cis* and *trans* carbon–carbon double bonds.

■ Describe how eicosanoids are formed by modification of the structure of unsaturated fatty acids; identify the various eicosanoid classes and indicate their functions.

■ Outline the general structure of triacylglycerols and indicate their function.

■ Outline the general structure of phospholipids and glycosphingolipids and indicate the functions of the different classes.

■ Appreciate the importance of cholesterol as the precursor of many biologically important steroids, including steroid hormones, bile acids, and vitamin D.

■ Recognize the cyclic nucleus common to all steroids.

■ Explain why free radicals are damaging to tissues and identify the three stages in the chain reaction of lipid peroxidation that produces them continuously.

■ Describe how antioxidants protect lipids from peroxidation by either inhibiting chain initiation or breaking the chain.

■ Recognize that many lipid molecules are amphipathic, having both hydrophobic and hydrophilic groups in their structure, and explain how this influences their behavior in an aqueous environment and enables certain classes, including phospholipids, sphingolipids, and cholesterol, to form the basic structure of biologic membranes.

BIOMEDICAL IMPORTANCE

The lipids are a heterogeneous group of compounds, including fats, oils, steroids, waxes, and related compounds. They have the common property of being (1) relatively **insoluble in water** and (2) **soluble in nonpolar solvents** such as ether and chloroform. They are important dietary constituents, not only because of the high energy value of fats (see Chapter 22), but also because **essential fatty acids, fat-soluble vitamins,** and other lipophilic **micronutrients** are contained in the fat of natural foods. Dietary supplementation with **long-chain ω3 fatty acids** is believed to have beneficial effects in a number of chronic diseases, including cardiovascular disease, rheumatoid arthritis, and dementia. Fat is stored in **adipose tissue**, where it also serves as a thermal insulator in the subcutaneous tissues and around certain organs. Nonpolar lipids act as **electrical insulators**, allowing rapid propagation of depolarization waves along **myelinated nerves**. Lipids are transported in the blood combined with proteins in **lipoprotein** particles (see Chapters 25 and 26). Lipids have essential roles in nutrition and health, and knowledge of lipid biochemistry is necessary for the understanding of many important biomedical conditions, including **obesity, diabetes mellitus, and atherosclerosis.**

LIPIDS MAY BE SIMPLE, COMPLEX, OR DERIVED

1. **Simple lipids** include fats, oils, and waxes which are esters of fatty acids with various alcohols:
   a. **Fats:** Esters of fatty acids with glycerol
   b. **Oils:** Fats in the liquid state at room temperature
   c. **Waxes:** Esters of fatty acids with higher molecular weight monohydric alcohols

2. **Complex lipids** are esters of fatty acids, which always contain an alcohol and one or more fatty acids, but which also have other groups. They can be divided into three types:
   a. **Phospholipids:** Contain a phosphoric acid residue. They frequently have nitrogen-containing bases (eg, choline) and other substituents. In many phospholipids the alcohol is glycerol (**glycerophospholipids**), but in **sphingophospholipids** it is sphingosine, which contains an amino group.
   b. **Glycolipids (glycosphingolipids):** Contain a fatty acid, sphingosine, and carbohydrate.
   c. **Other complex lipids:** These include lipids such as sulfolipids and amino lipids. Lipoproteins may also be placed in this category.

3. **Derived lipids** are formed from the hydrolysis of both simple and complex lipids. They include **fatty acids**, glycerol, steroids/sterols (including cholesterol), other alcohols, fatty aldehydes, ketone bodies (see Chapter 22), hydrocarbons, lipid-soluble vitamins and micronutrients, and hormones. Some of these (eg, free fatty acids, glycerol) also act as **precursor lipids** in the formation of simple and complex lipids.

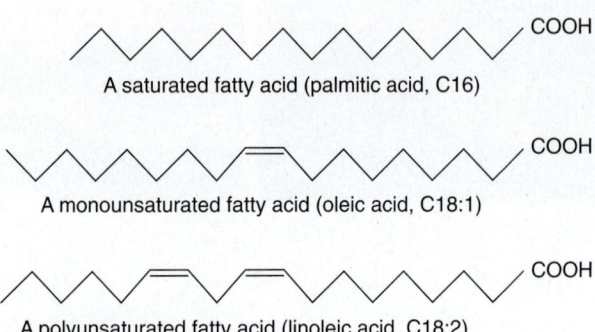

FIGURE 21–1 **Fatty acids.** Examples of a saturated (palmitic acid), monounsaturated (oleic acid), and a polyunsaturated (linoleic acid) fatty acid are shown.

Because they are uncharged, acylglycerols (glycerides), cholesterol, and cholesteryl esters are termed **neutral lipids.**

FATTY ACIDS ARE ALIPHATIC CARBOXYLIC ACIDS

Fatty acids occur in the body mainly as esters in natural fats and oils, but are found in the unesterified form as **free fatty acids**, a transport form in the plasma. Fatty acids that occur in natural fats usually contain an even number of carbon atoms. The chain may be **saturated** (containing no double bonds) or **unsaturated** (containing one or more double bonds) (**Figure 21–1**).

Fatty Acids Are Named After Corresponding Hydrocarbons

The most frequently used systematic nomenclature names the fatty acid after the hydrocarbon with the same number and arrangement of carbon atoms, with **-oic** being substituted for the final **-e** (Genevan system). Saturated acids end in **-anoic**, for example, octanoic acid (C8) (from the hydrocarbon octane), and unsaturated acids with double bonds end in **-enoic**, for example, octadecenoic acid (oleic acid, C18) (from the hydrocarbon octadecane).

Carbon atoms are numbered from the carboxyl carbon (carbon no. 1). The carbon atoms adjacent to the carboxyl carbon (nos. 2, 3, and 4) are also known as the α, β, and γ carbons, respectively, and the terminal methyl carbon is known as the ω- or n-carbon.

Various conventions are used for indicating the number and position of the double bonds (**Figure 21–2**); for example, Δ^9 indicates a double bond on the ninth carbon counting

$$\begin{array}{cccccccccccc} 18 & 17 & 16 & 15 & 14 & 13 & 12 & 11 & 10 & & 9 & & 1 \\ CH_3 & CH_2 & CH_2 & CH_2 & CH_2 & CH_2 & CH_2 & CH_2 & CH & = & CH(CH_2)_7 & COOH \end{array}$$
$$\begin{array}{ccccccccccc} \omega \text{ or } n\text{-}1 & 2 & 3 & 4 & 5 & 6 & 7 & 8 & 9 & 10 & 18 \end{array}$$

FIGURE 21–2 **Nomenclature for number and position of double bonds in unsaturated fatty acids.** Illustrated using oleic acid as an example. In this example, C1 is the Δ^1 carbon and C18 is the ω or *n*-1 carbon. Thus oleic acid may be termed 18:1 ω9 (or *n*-9) or 18:1 Δ^9.

from carbon 1 (the α-carbon); ω9 or n-9 indicates a double bond on the ninth carbon counting from the ω- or n-carbon. In animals, additional double bonds can be introduced only between an existing double bond at the ω9, ω6, or ω3 position and the carboxyl carbon (carbon 1), leading to three series of fatty acids known as the **ω9, ω6, and ω3 families**, respectively.

Saturated Fatty Acids Contain No Double Bonds

Saturated fatty acids may be envisaged as based on acetic acid (CH_3—COOH) as the first member of the series in which CH_2— is progressively added between the terminal CH_3— and —COOH groups. Examples are shown in **Table 21–1**. Other higher members of the series are known to occur, particularly in waxes. A few branched-chain fatty acids have also been isolated from both plant and animal sources.

Unsaturated Fatty Acids Contain One or More Double Bonds

Unsaturated fatty acids (see Figure 21–1, **Table 21–2**, for examples) may be further subdivided as follows:

1. **Monounsaturated** (monoethenoid, monoenoic) acids, containing one double bond.

TABLE 21–1 Saturated Fatty Acids

Common Name	Number of C Atoms	Occurrence
Acetic	2	Major end product of carbohydrate fermentation by rumen organisms
Butyric	4	In certain fats in small amounts (especially butter). An end product of carbohydrate fermentation by rumen organisms[a]
Valeric	5	
Caproic	6	
Lauric	12	Spermaceti, cinnamon, palm kernel, coconut oils, laurels, butter
Myristic	14	Nutmeg, palm kernel, coconut oils, myrtles, butter
Palmitic	16	Common in all animal and plant fats
Stearic	18	Second most common saturated fatty acid in nature after palmitic acid. More abundant in animal than in plant fats.

[a]Also formed in the cecum of herbivores and to a lesser extent in the colon of humans.

TABLE 21–2 Unsaturated Fatty Acids of Physiologic & Nutritional Significance

Number of C Atoms and Number and Position of Common Double Bonds	Family	Common Name	Systematic Name	Occurrence
Monoenoic acids (one double bond)				
16:1;9	ω7	Palmitoleic	cis-9-Hexadecenoic	In nearly all fats
18:1;9	ω9	Oleic	cis-9-Octadecenoic	Possibly the most common fatty acid in natural fats; particularly high in olive oil
18:1;9	ω9	Elaidic	trans-9-Octadecenoic	Hydrogenated and ruminant fats
Dienoic acids (two double bonds)				
18:2;9,12	ω6	Linoleic	all-cis-9,12-Octadecadienoic	Corn, peanut, cottonseed, soy bean, and many plant oils
Trienoic acids (three double bonds)				
18:3;6,9,12	ω6	γ-Linolenic	all-cis-6,9,12-Octadecatrienoic	Some plants, eg, oil of evening primrose, borage oil; minor fatty acid in animals
18:3;9,12,15	ω3	α-Linolenic	all-cis-9,12,15-Octadecatrienoic	Frequently found with linoleic acid but particularly in linseed oil
Tetraenoic acids (four double bonds)				
20:4;5,8,11,14	ω6	Arachidonic	all-cis-5,8,11,14-Eicosatetraenoic	Found in animal fats; important component of phospholipids in animals
Pentaenoic acids (five double bonds)				
20:5;5,8,11,14,17	ω3	Timnodonic	all-cis-5,8,11,14,17-Eicosapentaenoic	Important component of fish oils, eg, cod liver, mackerel, menhaden, salmon oils
Hexaenoic acids (six double bonds)				
22:6;4,7,10,13,16,19	ω3	Cervonic	all-cis-4,7,10,13,16,19-Docosahexaenoic	Fish oils, algal oils, phospholipids in brain

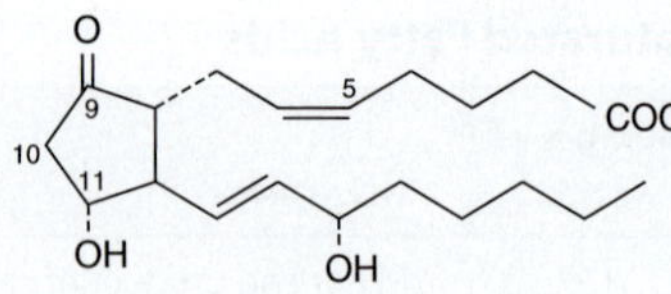

FIGURE 21–3 Prostaglandin E$_2$ (PGE$_2$).

2. **Polyunsaturated** (polyethenoid, polyenoic) acids, containing two or more double bonds.

3. **Eicosanoids:** These compounds, derived from eicosa (20-carbon) polyenoic fatty acids (see Chapter 23), comprise the **prostanoids, leukotrienes (LTs),** and **lipoxins (LXs).** Prostanoids include **prostaglandins (PGs), prostacyclins (PGIs),** and **thromboxanes (TXs).**

Prostaglandins exist in virtually every mammalian tissue, acting as local hormones; they have important physiologic and pharmacologic activities. They are synthesized in vivo by cyclization of the center of the carbon chain of 20-carbon (eicosanoic) polyunsaturated fatty acids (eg, arachidonic acid) to form a cyclopentane ring (**Figure 21–3**). A related series of compounds, the **thromboxanes,** have the cyclopentane ring interrupted with an oxygen atom (oxane ring) (**Figure 21–4**). Three different eicosanoic fatty acids give rise to three groups of eicosanoids characterized by the number of double bonds in the side chains (see Figure 23–12), for example, PG$_1$, PG$_2$, and PG$_3$ (PG, prostaglandin). Different substituent groups attached to the rings give rise to series of prostaglandins and thromboxanes (TX) labeled A, B, etc (see Figure 23–12)—for example, the "E" type of prostaglandin (as in PGE$_2$) has a keto group in position 9, whereas the "F" type has a hydroxyl group in this position. The **leukotrienes** and **lipoxins** (**Figure 21–5**) are a third group of eicosanoid derivatives formed via the **lipoxygenase pathway** (see Figure 23–13). They are characterized by the presence of three or four conjugated double bonds, respectively. Leukotrienes cause bronchoconstriction as well as being potent proinflammatory agents, and play a part in **asthma.**

Most Naturally Occurring Unsaturated Fatty Acids Have *cis* Double Bonds

The carbon chains of saturated fatty acids form a zigzag pattern when extended at low temperatures (see Figure 21–1). At higher temperatures, some bonds rotate, causing chain shortening, which explains why biomembranes become thinner with increases in temperature. Since carbon–carbon double bonds do not allow rotation, a type of **geometric isomerism** occurs in unsaturated fatty acids, termed ***cis–trans* isomerism,** which depends on the orientation of atoms or groups around the axes of their double bonds. If the acyl chains are on the

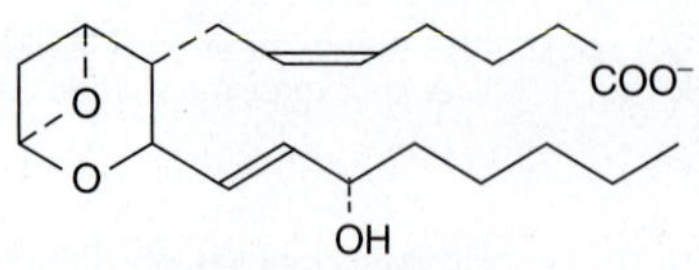

FIGURE 21–4 Thromboxane A$_2$ (TXA$_2$).

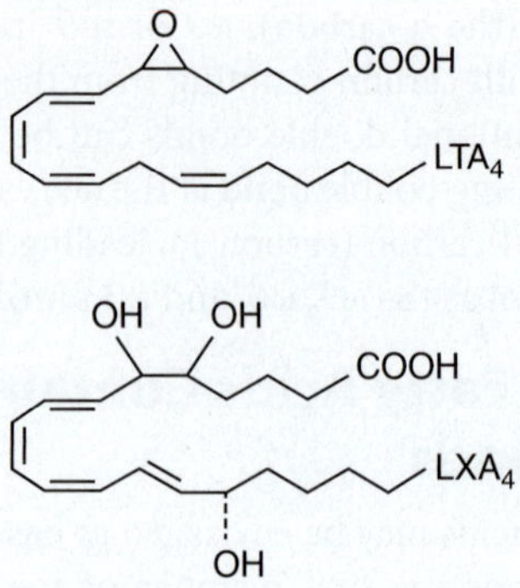

FIGURE 21–5 **Leukotriene and lipoxin structure.** Examples shown are leukotriene A$_4$ (LTA$_4$) and lipoxin A$_4$ (LXA$_4$).

same side of the bond, it is *cis-*, as in oleic acid; if on opposite sides, it is *trans-*, as in elaidic acid, the *trans* isomer of oleic acid (**Figure 21–6**). Double bonds in naturally occurring unsaturated long-chain fatty acids are nearly all in the *cis* configuration, the molecules being "bent" 120° at the double bond. Thus, oleic acid has a V shape, whereas elaidic acid remains "straight." Increase in the number of *cis* double bonds in a fatty acid leads to a variety of possible spatial configurations of the molecule—for example, arachidonic acid, with four *cis* double bonds, is bent into a U shape (**Figure 21–7**). This has profound significance for molecular packing in cell membranes (see Chapter 40) and on the positions occupied by fatty acids in complex lipids such as phospholipids. *Trans* double bonds alter these spatial relationships. Although double bonds in naturally occurring unsaturated fatty acids are almost always in the *cis* configuration, ***trans*-fatty acids** are present in foods such as those derived from ruminant fat (caused by the action of microorganisms in the rumen) and those containing partially hydrogenated vegetable oils (industrial *trans* fats), which are a by-product of the saturation of fatty

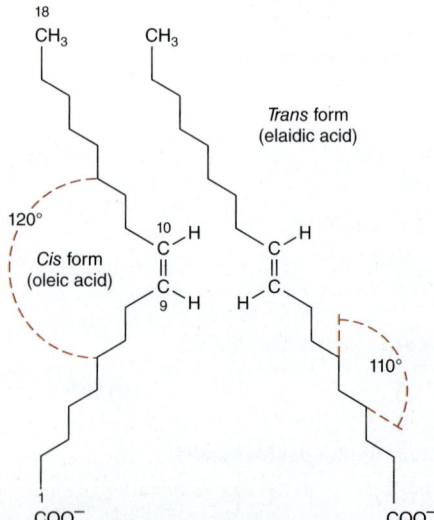

FIGURE 21–6 **Geometric isomerism of Δ⁹, 18:1 fatty acids (oleic and elaidic acids).** There is no rotation around carbon–carbon double bonds. In the *cis* configuration, the acyl chains are on the same side of the bond, while in *trans* form they are on opposite sides.

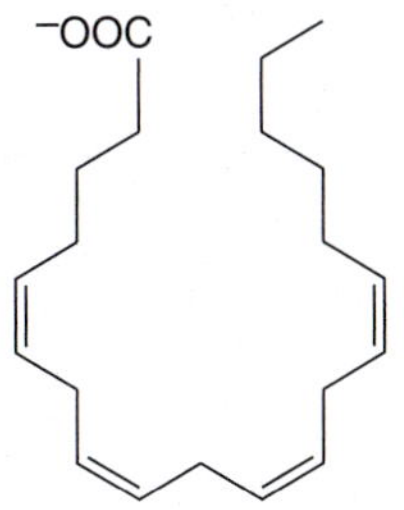

FIGURE 21–7 **Arachidonic acid.** Four double bonds in the *cis* configuration bend the molecule into a U shape.

acids during hydrogenation, or "hardening," of natural oils in the manufacture of margarine. Dietary *trans*-fatty acids, however, are now known to be detrimental to health, being associated with increased risk of diseases including cardiovascular disease, diabetes mellitus, and cancer. For this reason the content of industrial *trans* fats in foods is currently limited to a low level by law in many countries. A plan to eliminate these industrially generated fats completely from the world's food supply was announced by the World Health Organization in 2018.

Physical and Physiologic Properties of Fatty Acids Reflect Chain Length & Degree of Unsaturation

The melting points of even-numbered carbon fatty acids increase with chain length and decrease according to unsaturation. Thus, a triacylglycerol containing three saturated fatty acids of 12 or more carbons is solid at body temperature, whereas if the fatty acid residues are polyunsaturated, it is liquid to below 0°C. In practice, natural acylglycerols contain a mixture of fatty acids tailored to suit their functional roles. For example, membrane lipids, which must be fluid at all environmental temperatures, are more unsaturated than storage lipids. Lipids in tissues that are subject to cooling, for example, during hibernation or in the extremities of animals, are also more unsaturated.

ω3 Fatty Acids Are Anti-Inflammatory & Have Health Benefits

Long-chain ω3 fatty acids such as **α-linolenic (ALA)** (found in plant oils), **eicosapentaenoic (EPA)** (found in fish oil), and **docosahexaenoic (DHA)** (found in fish and algal oils) (see Table 21–2) have anti-inflammatory effects, perhaps due to their promotion of the synthesis of less inflammatory prostaglandins and leukotrienes as compared to ω6 fatty acids (see Figure 23–12). In view of this, their potential use as a therapy in severe chronic disease where inflammation is a contributory cause is under intensive investigation. Current evidence suggests that diets rich in ω3 fatty acids are beneficial, particularly for **cardiovascular disease**, but also for other chronic degenerative diseases such as **cancer**, **rheumatoid arthritis**, **and Alzheimer disease**.

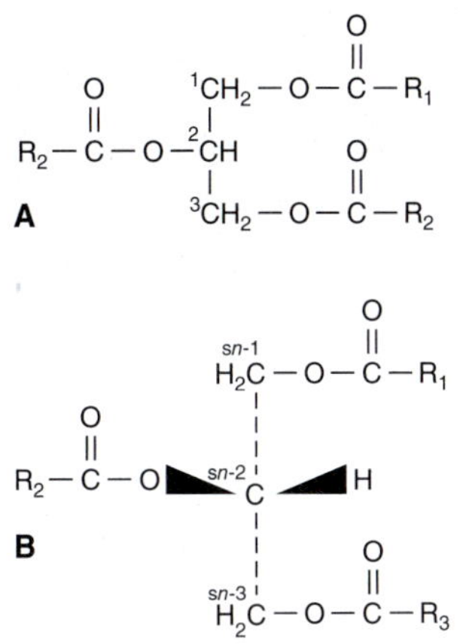

FIGURE 21–8 (A) Triacylglycerol. (B) Projection formula showing triacyl-*sn*-glycerol.

TRIACYLGLYCEROLS (TRIGLYCERIDES)* ARE THE MAIN STORAGE FORMS OF FATTY ACIDS

The triacylglycerols (**Figure 21–8**) are esters of the trihydric alcohol glycerol and fatty acids. Mono- and diacylglycerols, wherein one or two fatty acids are esterified with glycerol, are also found in the tissues. These are of particular significance in the synthesis and hydrolysis of triacylglycerols (see Chapters 24 and 25).

Carbons 1 & 3 of Glycerol Are Not Identical

It is important to realize that carbons 1 and 3 of glycerol are not identical when viewed in three dimensions (shown as a projection formula in Figure 21–8B). To number the carbon atoms of glycerol unambiguously, the -*sn* (stereochemical numbering) system is used. Enzymes readily distinguish between the *sn*-1 and *sn*-3 positions and are nearly always specific for one or the other carbon; for example, glycerol kinase always phosphorylates glycerol on *sn*-3, not *sn*-1 to give glycerol-3-phosphate and not glycerol-1-phosphate (see Figure 24–2).

PHOSPHOLIPIDS ARE THE MAIN LIPID CONSTITUENTS OF MEMBRANES

Many phospholipids are derivatives of **phosphatidic acid** (**Figure 21–9**), in which a glycerol moiety is esterified to a phosphate group and two long-chain fatty acids (glycerophospholipids). Phosphatidic acid is important as an intermediate in the synthesis of triacylglycerols as well as glycerophospholipids

---

*According to the standardized terminology of the International Union of Pure and Applied Chemistry and the International Union of Biochemistry, the monoglycerides, diglycerides, and triglycerides should be designated monoacylglycerols, diacylglycerols, and triacylglycerols, respectively. However, the older terminology is still widely used, particularly in clinical medicine.

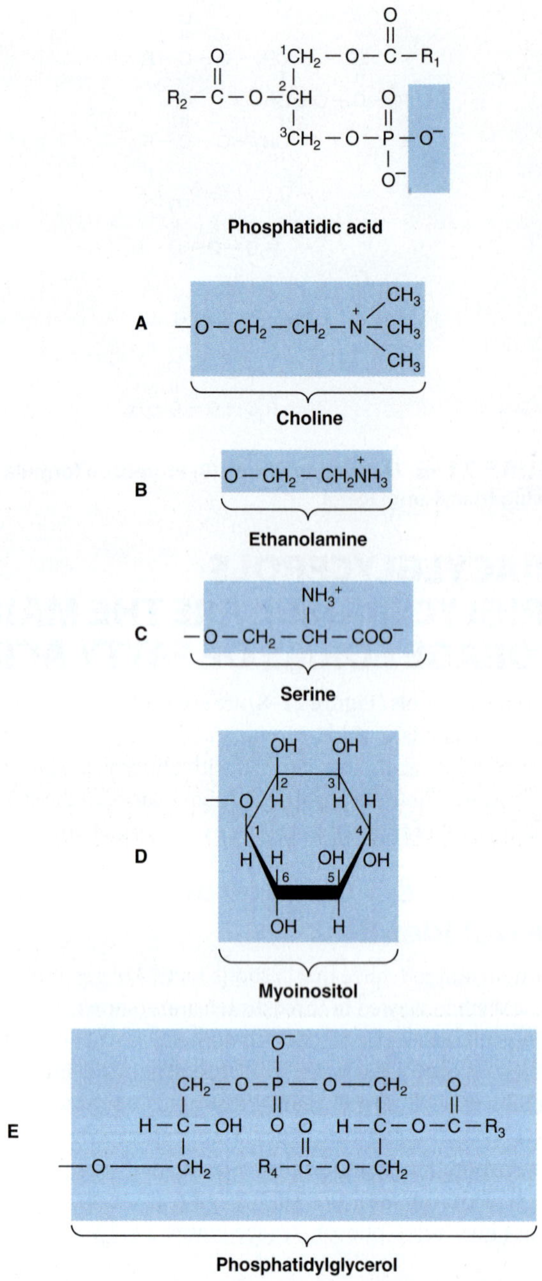

Phosphatidic acid

A   **Choline**

B   **Ethanolamine**

C   **Serine**

D   **Myoinositol**

E   **Phosphatidylglycerol**

FIGURE 21–9 **Phospholipids.** The —O shown shaded in phosphatidic acid is substituted by the substituents A-E to form the phospholipids: **(A)** 3-phosphatidylcholine, **(B)** 3-phosphatidylethanolamine, **(C)** 3-phosphatidylserine, **(D)** 3-phosphatidylinositol, and **(E)** cardiolipin (diphosphatidylglycerol).

(see Figure 24–2) but is not found in any great quantity in tissues. Sphingolipids, such as **sphingomyelin**, in which the phosphate is esterified to **sphingosine**, a complex amino alcohol (**Figure 21–10**), are also important membrane components. Both glycerophospholipids and sphingolipids have two long-chain hydrocarbon tails which are important for their function in forming the lipid bilayer in cell membranes (see Chapter 40), but in the former both are fatty acid chains while in the latter one is a fatty acid and the second is part of the sphingosine molecule (**Figure 21–11**).

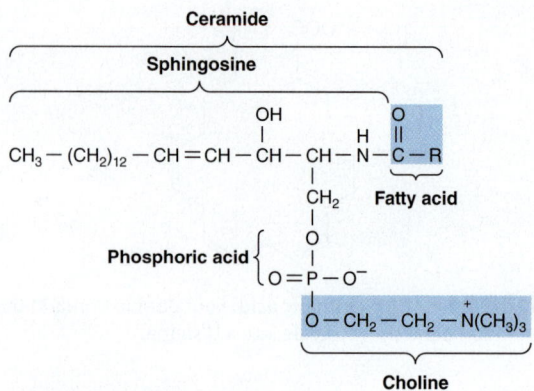

FIGURE 21–10 A sphingomyelin.

Phosphatidylcholines (Lecithins) & Sphingomyelins Are Abundant in Cell Membranes

Glycerophospholipids containing **choline** (see Figure 21–9), (phosphatidylcholines, commonly called **lecithins**) are the most abundant phospholipids of the cell membrane and represent a large proportion of the body's store of choline. Choline is important in nervous transmission, as acetylcholine, and as a store of labile methyl groups. **Dipalmitoyl lecithin** is a very effective surface-active agent and a major constituent of the **surfactant** preventing adherence, due to surface tension, of the inner surfaces of the lungs. Its absence from the lungs of premature infants causes **respiratory distress syndrome**. Most phospholipids have a saturated acyl radical in the *sn*-1 position but an unsaturated radical in the *sn*-2 position of glycerol.

Phosphatidylethanolamine (cephalin) and **phosphatidylserine** (found in most tissues) are also found in cell membranes and differ from phosphatidylcholine only in that ethanolamine or serine, respectively, replaces choline (see Figure 21–9). Phosphatidylserine also plays a role in **apoptosis** (programmed cell death).

Sphingomyelins are found in the outer leaflet of the cell membrane lipid bilayer and are particularly abundant in specialized areas of the plasma membrane known as **lipid rafts** (see Chapter 40). They are also found in large quantities in the **myelin sheath** that surrounds nerve fibers and play a role in **cell signaling** and in **apoptosis**. Sphingomyelins contain no glycerol, and on hydrolysis they yield a fatty acid, phosphoric acid, choline, and sphingosine (see Figure 21–10). The combination of sphingosine plus fatty acid is known as **ceramide** (see Chapter 24), a structure also found in the glycosphingolipids (see next section below).

Phosphatidylinositol Is a Precursor of Second Messengers

The inositol is present in **phosphatidylinositol** as the stereoisomer, myoinositol (see Figure 21–9). Phosphorylated phosphatidylinositols (**phosphoinositides**) are minor components of cell membranes, but play an important part in **cell signaling**

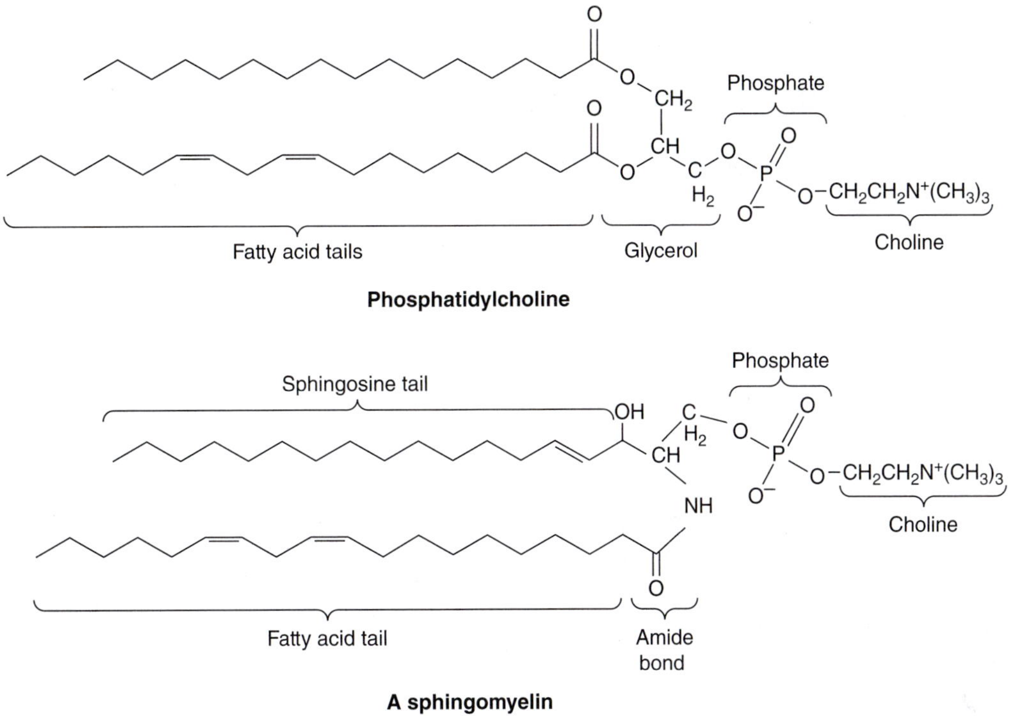

Phosphatidylcholine

A sphingomyelin

FIGURE 21–11 **Comparison of glycerophospholipid and sphingolipid structures.** Both types of phospholipid have two hydrocarbon tails, in glycerophospholipids both are fatty acid chains (a phosphatidylcholine with one saturated and one unsaturated fatty acid is shown) and in sphingolipids one is a fatty acid chain and the other is part of the sphingosine moiety (a sphingomyelin is shown). The two hydrophobic tails and the polar head group are important for the function of these phospholipids in the lipid bilayer in cell membranes (see Chapter 40).

and membrane trafficking. Phosphoinositides may have 1, 2, or 3 phosphate groups attached to the inositol ring. **Phosphatidylinositol 4,5-bisphosphate (PIP$_2$)**, for example, is cleaved into **diacylglycerol** and **inositol trisphosphate** upon stimulation by a suitable hormone agonist, and both of these act as internal signals or second messengers.

Cardiolipin Is a Major Lipid of Mitochondrial Membranes

Phosphatidic acid is a precursor of **phosphatidylglycerol**, which in turn gives rise to **cardiolipin** (see Figure 21–9). This phospholipid is found only in mitochondria and is essential for the mitochondrial function. Decreased cardiolipin levels or alterations in its structure or metabolism cause mitochondrial dysfunction in aging and in pathologic conditions including heart failure, cancer, and Barth syndrome (a rare genetic disease causing cardioskeletal myopathy).

Lysophospholipids Are Intermediates in the Metabolism of Phosphoglycerols

These are glycerophospholipids containing only one acyl radical, for example, **lysophosphatidylcholine (lysolecithin)** (**Figure 21–12**), which is important in the metabolism and interconversion of phospholipids. This compound is also found in oxidized lipoproteins and has been implicated in some of their effects in promoting **atherosclerosis**.

Plasmalogens Occur in Brain & Muscle

These compounds constitute as much as 10 to 30% of the phospholipids of brain and heart. Structurally, the plasmalogens resemble phosphatidylethanolamine but possess an ether link on the *sn*-1 carbon instead of the ester link found in acylglycerols. Typically, the alkyl radical is an unsaturated alcohol (**Figure 21–13**). In some instances, choline, serine, or inositol may be substituted for ethanolamine. Plasmalogens play an important role in the structure of membranes, and are believed to be involved in cell signaling and to act as endogenous antioxidants.

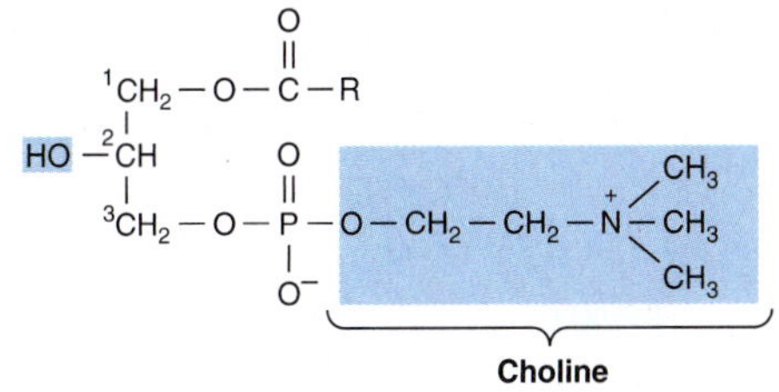

FIGURE 21–12 Lysophosphatidylcholine (lysolecithin).

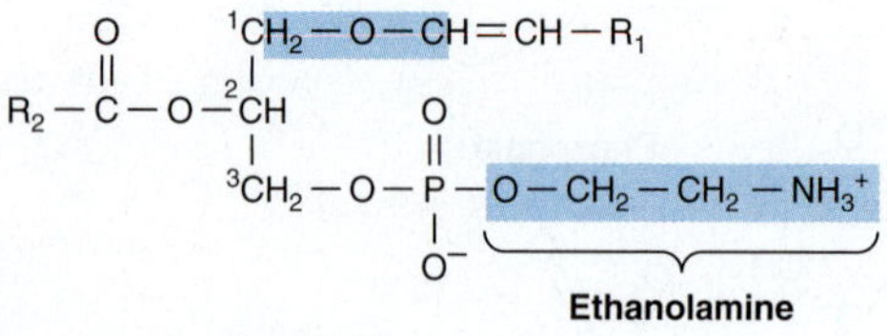

FIGURE 21–13 Plasmalogen.

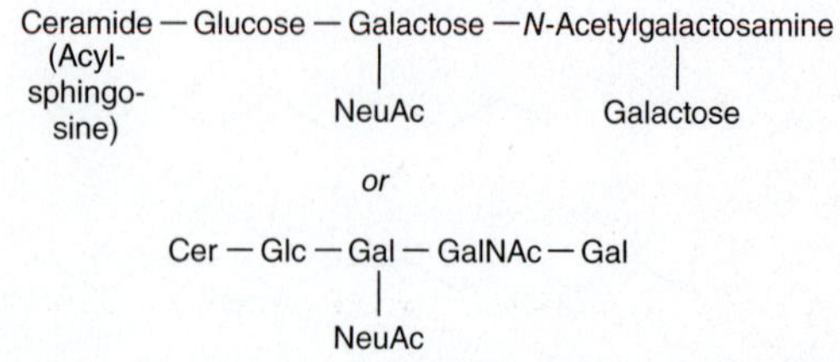

FIGURE 21–15 GM1 ganglioside, a monosialoganglioside, the receptor in human intestine for cholera toxin.

GLYCOLIPIDS (GLYCOSPHINGOLIPIDS) ARE IMPORTANT IN NERVE TISSUES & IN THE CELL MEMBRANE

Glycolipids are lipids with an attached carbohydrate or carbohydrate chain. They are widely distributed in every tissue of the body, particularly in nervous tissue such as brain. They occur particularly in the outer leaflet of the plasma membrane, where they contribute to **cell surface carbohydrates** which form the **glycocalyx** (see Chapter 15).

The major glycolipids found in animal tissues are glycosphingolipids. They contain ceramide and one or more sugars. **Galactosylceramide (Figure 21–14)** is a major glycosphingolipid of brain and other nervous tissue, found in relatively low amounts elsewhere. It contains a number of characteristic C24 fatty acids, for example, cerebronic acid.

Galactosylceramide can be converted to sulfogalactosylceramide (**sulfatide**) which has a sulfo group attached to the O in the three position of galactose and is present in high amounts in **myelin**. **Glucosylceramide** resembles galactosylceramide, but the head group is glucose rather than galactose. It is the predominant simple glycosphingolipid of extraneural tissues, also occurring in the brain in small amounts. **Gangliosides** are complex glycosphingolipids derived from glucosylceramide that contain in addition one or more molecules of **sialic acid**. **Neuraminic acid** (NeuAc; see Chapter 15) is the principal sialic acid found in human tissues. Gangliosides are also present in nervous tissues in high concentration. They function in cell–cell recognition and communication and as receptors for hormones and bacterial toxins such as cholera

toxin. The simplest ganglioside found in tissues is **GM3** (see Figure 24–8), which contains ceramide, one molecule of glucose, one molecule of galactose, and one molecule of NeuAc. In the shorthand nomenclature used, G represents ganglioside; M is a monosialo-containing species; and 3 is a number assigned on the basis of chromatographic migration. **GM1** (**Figure 21–15**), a more complex ganglioside derived from GM3, is of considerable biologic interest, as it is known to be the receptor in human intestine for **cholera toxin**. Other gangliosides can contain anywhere from one to five molecules of sialic acid, giving rise to di-, trisialogangliosides, etc.

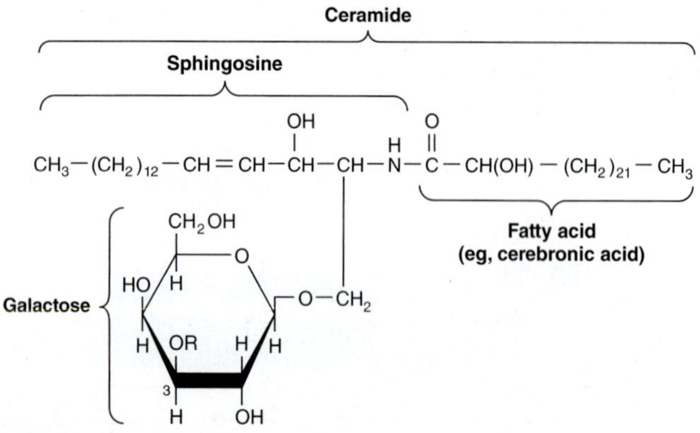

STEROIDS PLAY MANY PHYSIOLOGICALLY IMPORTANT ROLES

Although **cholesterol** is probably best known by most people for its association with **atherosclerosis** and heart disease, it has many essential roles in the body (see Chapter 26). It is the precursor of a large number of equally important **steroids** that include the **bile acids**, **adrenocortical hormones**, **sex hormones**, **vitamin D** (see Chapters 26, 41, 44), and **cardiac glycosides**.

All steroids have a similar cyclic nucleus resembling phenanthrene (rings A, B, and C) to which a cyclopentane ring (D) is attached. The carbon positions on the steroid nucleus are numbered as shown in **Figure 21–16**. It is important to realize that in structural formulas of steroids, a simple hexagonal ring denotes a completely saturated six-carbon ring with all valences satisfied by hydrogen bonds unless shown otherwise; that is, it is not a benzene ring. All double bonds are shown as such. Methyl groups (shown as single bonds unattached at the farther [methyl] end) occur typically at positions 10 and 13 and constitute C atoms 19 and 18. A side chain at position 17 is usual (as in cholesterol). If the compound has one or more hydroxyl groups and no carbonyl or carboxyl groups, it is a **sterol**, and the name terminates in -*ol*.

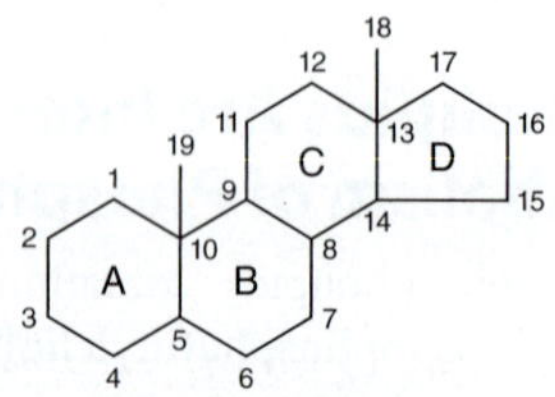

FIGURE 21–14 Structure of galactosylceramide.

FIGURE 21–16 The steroid nucleus.

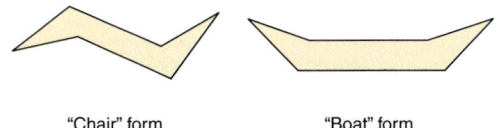

"Chair" form   "Boat" form

FIGURE 21–17 Conformations of stereoisomers of the steroid nucleus.

Because of Asymmetry in the Steroid Molecule, Many Stereoisomers Are Possible

Each of the six-carbon rings of the steroid nucleus is capable of existing in the three-dimensional conformation either of a "chair" or a "boat" (**Figure 21–17**). In naturally occurring steroids, virtually all the rings are in the "chair" form, which is the more stable conformation. With respect to each other, the rings can be either *cis* or *trans* (**Figure 21–18**). The junction between the A and B rings may be *cis* or *trans* in naturally occurring steroids. The junction between B and C is *trans*, as is usually the C/D junction. Bonds attaching substituent groups above the plane of the rings (β bonds) are shown with bold solid lines, whereas those bonds attaching groups below (α bonds) are indicated with broken lines. The A ring of a 5α steroid (ie, the hydrogen at position 5 is in the α configuration) is always *trans* to the B ring, whereas it is *cis* in a 5β steroid (ie, the hydrogen at position 5 is in the β configuration). The methyl groups attached to C10 and C13 are invariably in the β configuration.

Cholesterol Is a Significant Constituent of Many Tissues

Cholesterol (**Figure 21–19**) is widely distributed in all cells of the body but particularly in nervous tissue. It is a major constituent of the plasma membrane (see Chapter 40) and of plasma lipoproteins (see Chapters 25 and 26). It is often found as **cholesteryl ester**, where the hydroxyl group on position 3 is esterified with a long-chain fatty acid. It occurs in animals but not in plants or bacteria.

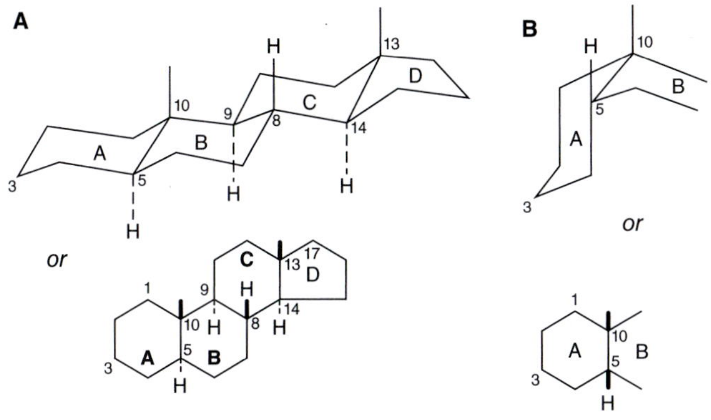

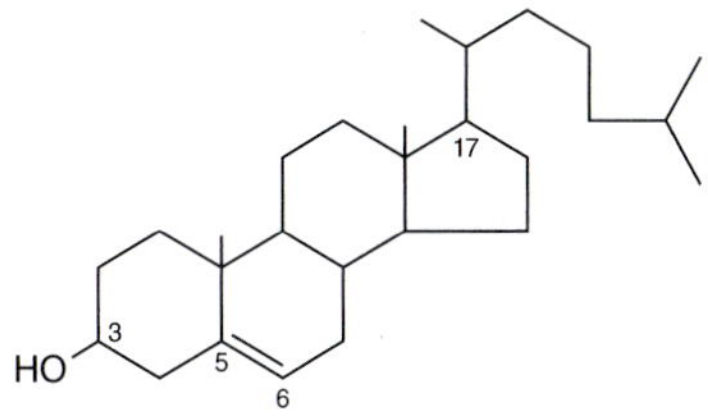

FIGURE 21–19 Cholesterol.

Ergosterol Is a Precursor of Vitamin D

Ergosterol occurs in plants and yeast and is important as a dietary source of vitamin D (see Chapter 44) (**Figure 21–20**). When irradiated with ultraviolet light in the skin, ring B is opened to form vitamin D_2 in a process similar to the one that forms vitamin D_3 from 7-dehydrocholesterol in the skin (see Figure 44–3).

Polyprenoids Share the Same Parent Compound as Cholesterol

Polyprenoids are not steroids but are related to them because they are synthesized, like cholesterol (see Figure 26–2), from five-carbon isoprene units (**Figure 21–21**). They include **ubiquinone** (see Chapter 13), which participates in the respiratory chain in mitochondria, and **dolichols,** long-chain alcohols containing 15 to 23 isoprene units in animal cells (**Figure 21–22**), which take part in glycoprotein synthesis by transferring oligosaccharide residues to asparagine residues of the glycoprotein polypeptide chains as they are formed (see Chapter 46). Plant-derived polyprenoids include rubber, camphor, the fat-soluble vitamins A, D, E, and K, and β-carotene (provitamin A).

LIPID PEROXIDATION IS A SOURCE OF FREE RADICALS

Peroxidation (**auto-oxidation**) of lipids exposed to oxygen-forming peroxides is responsible not only for deterioration of foods (**rancidity**), but also for damage to tissues in vivo, where it may be a cause of cancer, inflammatory diseases, atherosclerosis, and aging. The deleterious effects are considered to be caused by **free radicals**, molecules that have unpaired valence electrons, making them highly reactive. Free radicals containing oxygen (eg, ROO·, RO·, OH·) are termed

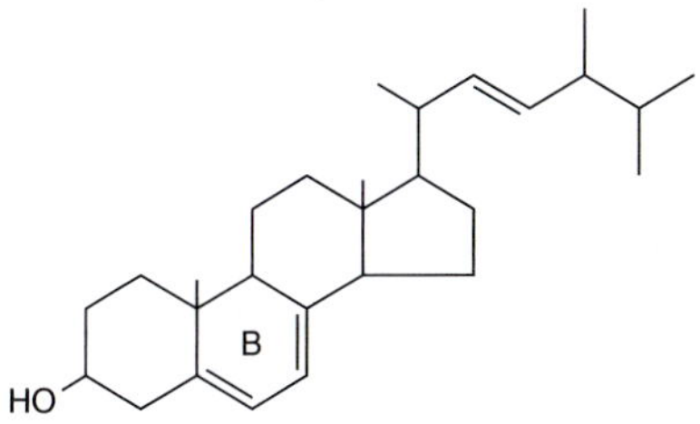

FIGURE 21–18 Generalized steroid nucleus, showing (A) an all-*trans* configuration between adjacent rings and (B) a *cis* configuration between rings A and B.

FIGURE 21–20 Ergosterol.

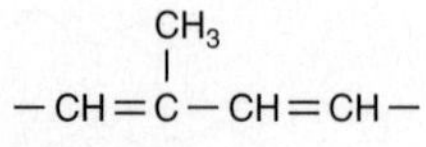

FIGURE 21–21 Isoprene unit.

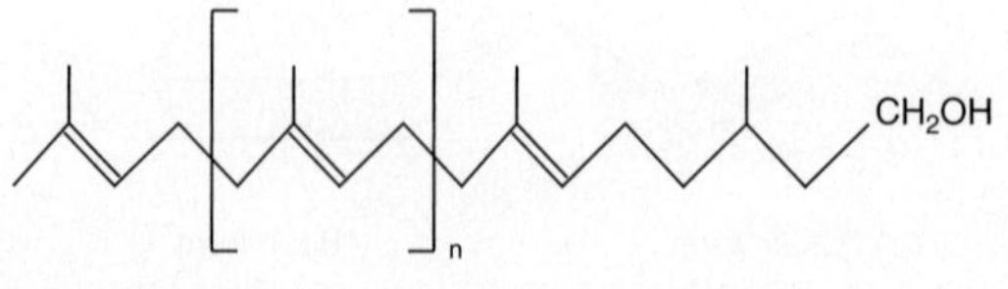

FIGURE 21–22 Dolichol. Major dolichols in humans contain 19 or 20 isoprene units.

reactive oxygen species (ROS). These are produced during peroxide formation from fatty acids containing methylene-interrupted double bonds, that is, those found in the naturally occurring polyunsaturated fatty acids (**Figure 21–23**). **Lipid peroxidation** is a chain reaction in which free radicals formed in the initiation stage in turn generate more (propagation), and thus it has potentially devastating effects. The processes of initiation and propagation can be depicted as follows:

1. **Initiation:** A free radical (X$^{\cdot}$) reacts with a polyunsaturated fatty acid forming the first fatty acid radical (R$^{\cdot 1}$).

   $$X^{\cdot} + R^1H \rightarrow R^{\cdot 1} + XH$$

2. **Propagation:** The unstable fatty acid radical R$^{\cdot 1}$ reacts with oxygen to produce a peroxyl radical (R^{1}OO$^{\cdot}$) which then attacks another fatty acid to form a new fatty acid radical R$^{\cdot 2}$ and a peroxide of R^1. R$^{\cdot 2}$ can then react with fatty acid R^{3}H and so on in a chain reaction. Thus one free radical may be responsible for the peroxidation of a very large number of polyunsaturated fatty acid molecules.

   $$R^{\cdot 1} + O_2 \rightarrow R^1OO^{\cdot}$$
   $$R^1OO^{\cdot} + R^2H \rightarrow R^1OOH + R^{\cdot 2}, \text{ etc.}$$

   The process may be terminated at the third stage.

3. **Termination:** The chain reaction terminates when two radicals react to form a non-radical product (ROOR or RR).

   $$ROO^{\cdot} + ROO^{\cdot} \rightarrow ROOR + O_2$$
   $$ROO^{\cdot} + R^{\cdot} \rightarrow ROOR$$
   $$R^{\cdot} + R^{\cdot} \rightarrow RR$$

Antioxidants are used to control and reduce lipid peroxidation, both by humans in their activities and in nature. Propyl gallate, butylated hydroxyanisole (BHA), and butylated hydroxytoluene (BHT) are antioxidants used as food additives. Naturally occurring antioxidants include vitamin E (tocopherol) (see Chapter 44), which is lipid soluble, and urate and vitamin C, which are water soluble. β-Carotene is an antioxidant at low PO$_2$. Antioxidants fall into two classes: (1) **preventive antioxidants**, which reduce the rate of chain initiation (stage 1 above) and (2) **chain-breaking antioxidants**, which interfere with chain propagation (stage 2 above). Preventive antioxidants include catalase and other peroxidases such as glutathione peroxidase (see Figure 20–3) that react with ROOH; selenium, which is an essential component of glutathione peroxidase and regulates its activity, and chelators of metal ions such as ethylenediaminetetraacetate (EDTA) and diethylenetriaminepentaacetate (DTPA). In vivo, the principal chain-breaking antioxidants are superoxide dismutase, which acts in the aqueous phase to trap superoxide free radicals (O$_2^-$), urate, and vitamin E, which acts in the lipid phase to trap ROO$^{\cdot}$ radicals.

Peroxidation is also catalyzed in vivo by heme compounds and by **lipoxygenases** (see Figure 23–13) found in platelets and leukocytes. Other products of auto-oxidation or enzymic oxidation of physiologic significance include **oxysterols** (formed from cholesterol) and the prostaglandin-like **isoprostanes** (formed from the peroxidation of polyunsaturated fatty acids such as arachidonic acid) which are used as reliable markers of oxidative stress in humans.

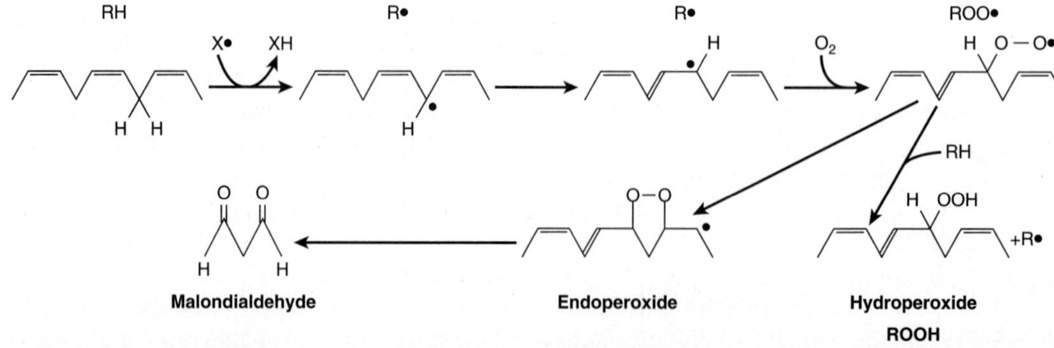

FIGURE 21–23 Lipid peroxidation. The reaction may be initiated by an existing free radical (X$^{\cdot}$), by light, or by metal ions. Malondialdehyde is only formed by fatty acids with three or more double bonds and is used as a measure of lipid peroxidation together with ethane from the terminal two carbons of ω3 fatty acids and pentane from the terminal five carbons of ω6 fatty acids.

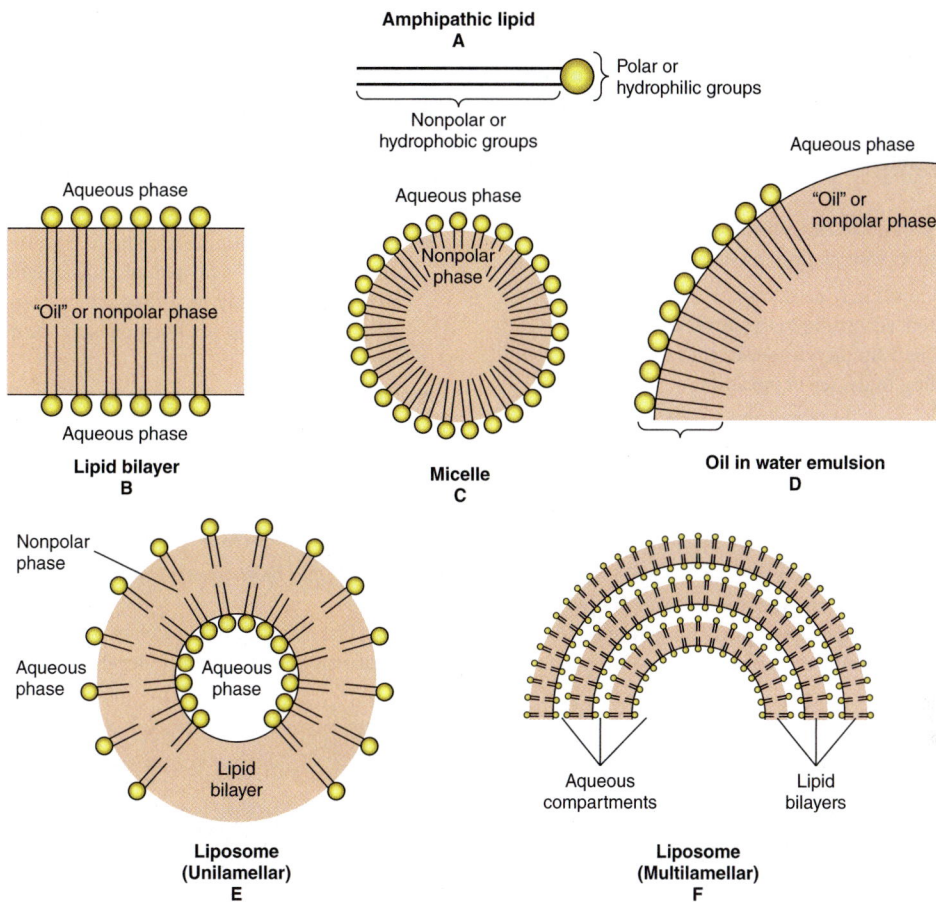

FIGURE 21–24 Formation of lipid membranes, micelles, emulsions, and liposomes from amphipathic lipids, for example, phospholipids.

AMPHIPATHIC LIPIDS SELF-ORIENT AT OIL: WATER INTERFACES

They Form Membranes, Micelles, Liposomes, & Emulsions

In general, lipids are insoluble in water since they contain a predominance of nonpolar (hydrocarbon) groups. However, fatty acids, phospholipids, sphingolipids, bile salts, and, to a lesser extent, cholesterol, contain polar groups. Therefore, a part of the molecule is **hydrophobic**, or water insoluble, and a part is **hydrophilic**, or water soluble. Such molecules are described as **amphipathic** (**Figure 21–24**). They become oriented at oil–water interfaces with the polar group in the water phase and the nonpolar group in the oil phase. A bilayer of such amphipathic lipids is the basic structure in biologic **membranes** (see Chapter 40). When a critical concentration of these lipids is present in an aqueous medium, they form **micelles**, spherical particles in which the hydrophobic tails of the molecules are inside while the hydrophilic heads face the water. **Liposomes** consist of spheres of lipid bilayers that enclose part of the aqueous medium. They may be formed by sonicating an amphipathic lipid in an aqueous medium. Aggregation of bile salts into micelles and the formation of **mixed micelles** with the products of fat digestion are important in facilitating absorption of lipids from the intestine. Liposomes are of potential clinical use—particularly when combined with tissue-specific antibodies—as carriers of drugs in the circulation, targeted to specific organs, for example, in cancer therapy. In addition, they are used for gene transfer into vascular cells and as carriers for topical and transdermal delivery of drugs and cosmetics. **Emulsions** are much larger particles, formed usually by nonpolar lipids in an aqueous medium. These are stabilized by emulsifying agents such as amphipathic lipids (eg, phosphatidylcholine), which form a surface layer separating the main bulk of the nonpolar material from the aqueous phase (see Figure 21–24).

SUMMARY

- Lipids have the common property of being relatively insoluble in water (hydrophobic) but soluble in nonpolar solvents. Amphipathic lipids also contain one or more polar groups, making them suitable as constituents of membranes at lipid-water interfaces.

- Lipids of major physiologic significance include fatty acids and their esters, together with cholesterol and other steroids.

- Long-chain fatty acids may be saturated, monounsaturated, or polyunsaturated, according to the number of double bonds present. Their fluidity decreases with chain length and increases according to degree of unsaturation.

- Eicosanoids are formed from 20-carbon polyunsaturated fatty acids and make up an important group of physiologically and pharmacologically active compounds known as prostaglandins, thromboxanes, leukotrienes, and lipoxins.

- The esters of glycerol are quantitatively the most significant lipids, represented by triacylglycerol ("fat"), a major constituent of some lipoprotein classes and the storage form of lipid in adipose tissue. Glycerophospholipids and sphingolipids are amphipathic lipids and have important roles—as major constituents of membranes and the outer layer of lipoproteins, as surfactant in the lung, as precursors of second messengers, and as constituents of nervous tissue.

- Glycolipids are also important constituents of nervous tissue such as brain and the outer leaflet of the cell membrane, where they contribute to the carbohydrates on the cell surface.

- Cholesterol, an amphipathic lipid, is an important component of membranes. It is the parent molecule from which all other steroids in the body, including major hormones such as the adrenocortical and sex hormones, D vitamins, and bile acids, are synthesized.

- Peroxidation of lipids containing polyunsaturated fatty acids leads to generation of free radicals that damage tissues and cause disease.

REFERENCES

Eljamil AS: *Lipid Biochemistry: For Medical Sciences*. iUniverse, 2015.

Gurr MI, Harwood JL, Frayn KN, et al: *Lipids, Biochemistry, Biotechnology and Health*. Wiley-Blackwell, 2016.

Oxidation of Fatty Acids: Ketogenesis

Kathleen M. Botham, PhD, DSc, & Peter A. Mayes, PhD, DSc

O B J E C T I V E S

After studying this chapter, you should be able to:

- Describe the processes by which fatty acids are transported in the blood, activated and transported into the matrix of the mitochondria for breakdown to obtain energy.
- Outline the β-oxidation pathway by which fatty acids are metabolized to acetyl-CoA and explain how this leads to the production of large quantities of ATP.
- Identify the three compounds termed "ketone bodies" and describe the reactions by which they are formed in liver mitochondria.
- Recognize that ketone bodies are important fuels for extrahepatic tissues and indicate the conditions in which their synthesis and use are favored.
- Indicate the three stages in the metabolism of fatty acids where ketogenesis is regulated.
- Indicate that overproduction of ketone bodies leads to ketosis and, if prolonged, ketoacidosis, and identify pathologic conditions when this occurs.
- Give examples of diseases associated with impaired fatty acid oxidation.

BIOMEDICAL IMPORTANCE

Fatty acids are broken down in mitochondria by oxidation to acetyl-CoA in a process that generates large amounts of energy. When this pathway is proceeding at a high rate, three compounds, **acetoacetate, D-3-hydroxybutyrate**, and **acetone**, known collectively as the **ketone bodies**, are produced by the liver. Acetoacetate and D-3-hydroxybutyrate are used as fuels by extrahepatic tissues in normal metabolism, but overproduction of ketone bodies causes **ketosis**. Increased fatty acid oxidation and consequently ketosis is a characteristic of starvation and of diabetes mellitus. Since ketone bodies are acidic, when they are produced in excess over long periods, as in diabetes, they cause **ketoacidosis**, which is potentially life-threatening. Because gluconeogenesis is dependent on fatty acid oxidation, any impairment in fatty acid oxidation leads to **hypoglycemia**. This occurs in various states of **carnitine deficiency** or deficiency of essential enzymes in fatty acid oxidation, for example, **carnitine palmitoyltransferase**, or inhibition of fatty acid oxidation by poisons, for example, **hypoglycin**.

OXIDATION OF FATTY ACIDS OCCURS IN MITOCHONDRIA

Although acetyl-CoA is both an end point of fatty acid catabolism and the starting substrate for fatty acid synthesis, breakdown is not simply the reverse of the biosynthetic pathway, but an entirely separate process taking place in a different compartment of the cell. The separation of fatty acid oxidation in mitochondria from biosynthesis in the cytosol allows each process to be individually controlled and integrated with tissue requirements. Each step in fatty acid oxidation involves acyl-CoA derivatives, is catalyzed by separate enzymes, utilizes NAD^+ and FAD as coenzymes, and generates ATP. It is an aerobic process, requiring the presence of oxygen.

Fatty Acids Are Transported in the Blood as Free Fatty Acids

Free fatty acids (FFAs)—also called unesterified (UFA) or nonesterified (NEFA) fatty acids (see Chapter 21)—are fatty acids that are in the **unesterified state**. In plasma, longer-chain FFA are combined with **albumin**, and in the cell they are attached to a **fatty acid–binding protein**, so that in fact they are never really "free." Shorter-chain fatty acids are more water soluble and exist as the unionized acid or as a fatty acid anion.

Fatty Acids Are Activated Before Being Catabolized

Fatty acids must first be converted to an active intermediate before they can be catabolized. This is the only step in the complete degradation of a fatty acid that requires energy from ATP. In the presence of ATP and coenzyme A, the enzyme **acyl-CoA synthetase (thiokinase)** catalyzes the conversion of a fatty acid (or FFA) to an "active fatty acid" or **acyl-CoA**, using one high-energy phosphate and forming AMP and PP_i (**Figure 22–1**). The PP_i is hydrolyzed by **inorganic pyrophosphatase** with the loss of a further high-energy phosphate, ensuring that the overall reaction goes to completion. Acyl-CoA synthetases are found on the outer membrane of mitochondria and also in the endoplasmic reticulum and peroxisomes.

Long-Chain Fatty Acids Cross the Inner Mitochondrial Membrane as Carnitine Derivatives

Acyl-CoAs formed as described earlier enter the intermembrane space (see Figure 22–1), but are unable to cross the inner mitochondrial membrane into the matrix where fatty acid breakdown takes place. In the presence of **carnitine** (β-hydroxy-γ-trimethylammonium butyrate), a compound widely distributed in the body and particularly abundant in muscle; however, **carnitine palmitoyltransferase-I**, an enzyme located in the outer mitochondrial membrane, transfers the long-chain acyl group from CoA to carnitine, forming **acylcarnitine** and releasing CoA. Acylcarnitine is able to penetrate the inner membrane and gain access to the β-oxidation system of enzymes via the inner membrane exchange transporter **carnitine-acylcarnitine translocase**. The transporter binds acylcarnitine and transports it across the membrane in exchange for carnitine. The acyl group is then transferred to CoA so that acyl-CoA is reformed and carnitine is liberated. This reaction is catalyzed by **carnitine palmitoyltransferase-II**, which is located on the inside of the inner membrane (see Figure 22–1).

β-OXIDATION OF FATTY ACIDS INVOLVES SUCCESSIVE CLEAVAGE WITH RELEASE OF ACETYL-COA

In the pathway for the oxidation of fatty acids (**Figure 22–2**), two carbons at a time are cleaved from acyl-CoA molecules, starting at the carboxyl end. The chain is broken between the α(2)- and β(3)-carbon atoms—hence the process is termed **β-oxidation**. The two-carbon units formed are acetyl-CoA; thus, palmitoyl(C16)-CoA forms eight acetyl-CoA molecules.

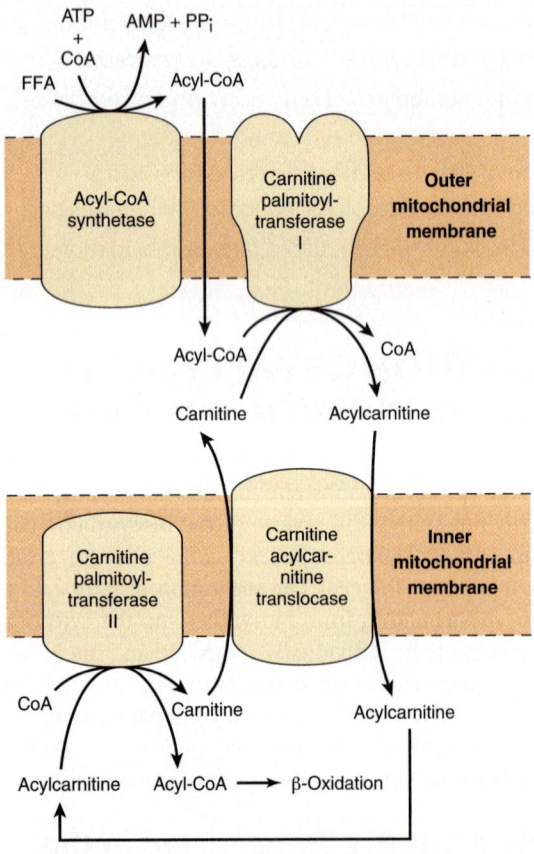

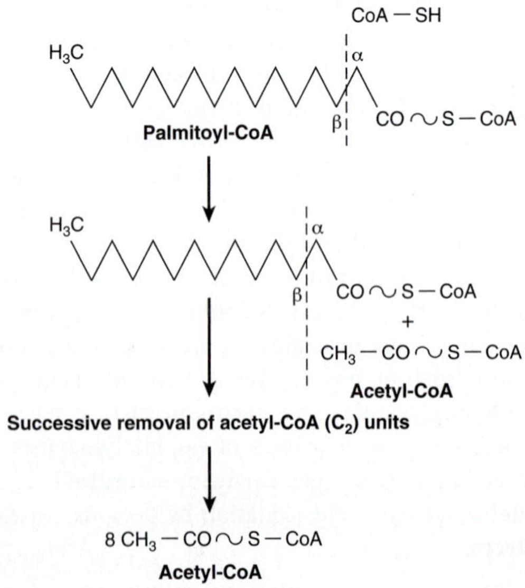

FIGURE 22–1 **Role of carnitine in the transport of long-chain fatty acids through the inner mitochondrial membrane.** Long-chain acyl-CoA formed by acyl-CoA synthetase enters the intermembrane space. For transport across the inner membrane, acyl groups must be transferred from CoA to carnitine by carnitine palmitoyltransferase-I. The acylcarnitine formed is then carried into the matrix by a translocase enzyme in exchange for a free carnitine and acyl-CoA is reformed by carnitine palmitoyltransferase-II.

FIGURE 22–2 Overview of β-oxidation of fatty acids.

The β-Oxidation Cycle Generates FADH₂ & NADH

Several enzymes found in the mitochondrial matrix or inner membrane adjacent to the respiratory chain catalyze the oxidation of acyl-CoA to acetyl-CoA via the β-oxidation pathway. The system proceeds in cyclic fashion which results in the degradation of long fatty acids to acetyl-CoA. In the process, large quantities of the reducing equivalents FADH₂ and NADH are generated and are used to form ATP by oxidative phosphorylation (see Chapter 13) (**Figure 22–3**).

The first step is the removal of two hydrogen atoms from the 2(α)- and 3(β)-carbon atoms, catalyzed by **acyl-CoA dehydrogenase** and requiring flavin adenine dinucleotide (FAD). This results in the formation of Δ²-*trans*-enoyl-CoA and FADH₂. Next, water is added to saturate the double bond and form 3-hydroxyacyl-CoA, catalyzed by Δ²-**enoyl-CoA hydratase**. The 3-hydroxy derivative undergoes further dehydrogenation on the 3-carbon catalyzed by L-**3-hydroxyacyl-CoA dehydrogenase** to form the corresponding 3-ketoacyl-CoA compound. In this case, NAD⁺ is the coenzyme involved. Finally, 3-ketoacyl-CoA is split at the 2,3-position by **thiolase** (3-ketoacyl-CoA-thiolase), forming acetyl-CoA and a new acyl-CoA two carbons shorter than the original acyl-CoA molecule. The shorter acyl-CoA formed in the cleavage reaction reenters the oxidative pathway at reaction 2 (see Figure 22–3). In this way, a long-chain fatty acid with an even number of carbons may be degraded completely to acetyl-CoA (C₂ units). For example, after seven cycles, the C16 fatty acid, palmitate, would be converted to eight acetyl-CoA molecules. Since acetyl-CoA can be oxidized to CO₂ and water via the citric acid cycle (which is also found within the mitochondria), the complete oxidation of fatty acids is achieved.

Fatty acids with an odd number of carbon atoms are oxidized by the pathway of β-oxidation described earlier, producing acetyl-CoA until a three-carbon (propionyl-CoA) residue remains. This compound is converted to succinyl-CoA, a constituent of the citric acid cycle (see Chapter 16). Hence, **the propionyl residue from an odd-chain fatty acid is the only part of a fatty acid that is glucogenic.**

Oxidation of Fatty Acids Produces a Large Quantity of ATP

Each cycle of β-oxidation generates one molecule of FADH₂ and one of NADH. The breakdown of 1 mol of the C16 fatty acid, palmitate, requires seven cycles and produces 8 mol of acetyl-CoA. Oxidation of the reducing equivalents via the respiratory chain leads to the synthesis of 28 mol of ATP (**Table 22–1** and see Chapter 13) and oxidation of acetyl-CoA via the citric acid cycle produces 80 mol of ATP (see Table 22–1 and Chapter 16). The breakdown of 1 mol of palmitate, therefore, yields a gross total of 108 mol of ATP. However, two high-energy phosphates are used in the initial activation step (see Figure 22–3), thus there is a net gain of 106 mol of ATP per mole of palmitate used

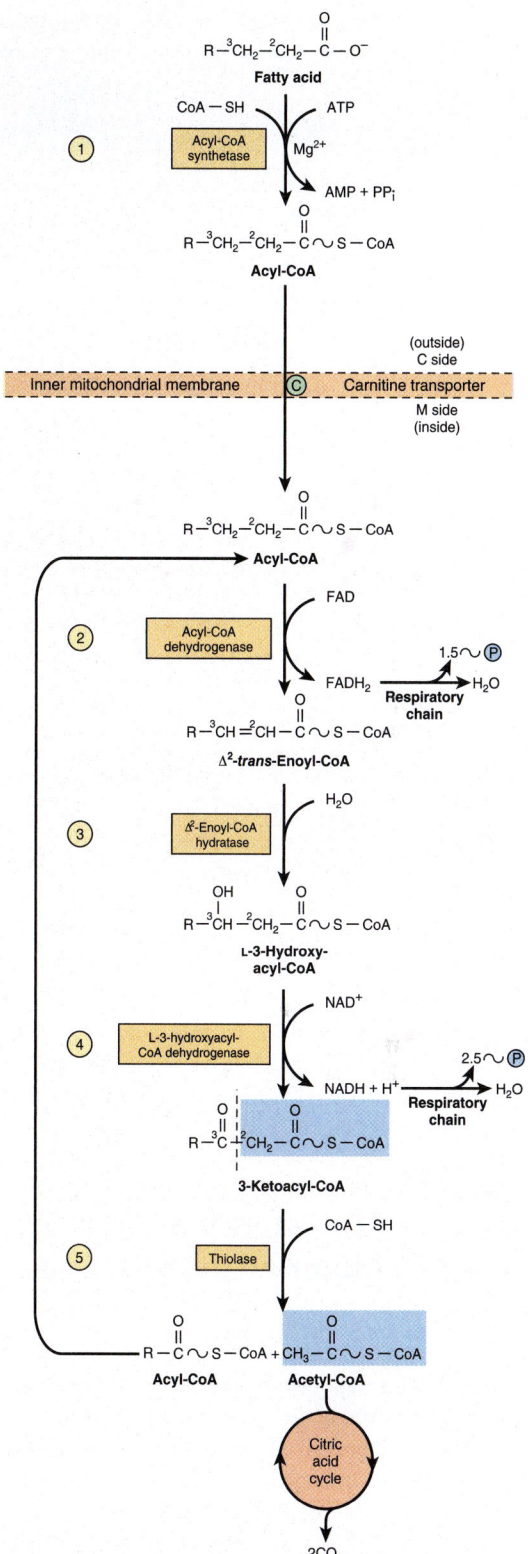

FIGURE 22–3 β-Oxidation of fatty acids. Long-chain acyl-CoA is cycled through reactions ② to ⑤, acetyl-CoA being split off, each cycle, by thiolase (reaction ⑤). When the acyl radical is only four carbon atoms in length, two acetyl-CoA molecules are formed in reaction ⑤.

TABLE 22–1 Generation of ATP From the Complete Oxidation of a C16 Fatty Acid

Step	Product	Amount Product Formed (mol)/mol Palmitate	ATP Formed (mol)/mol Product	Total ATP Formed (mol)/mol Palmitate	ATP Used (mol)/mol Palmitate
Activation		–			2
β-Oxidation	FADH$_2$	7	1.5	10.5	–
β-Oxidation	NADH	7	2.5	17.5	–
Citric acid cycle	Acetyl-CoA	8	10	80	–
	Total ATP formed (mol)/mol palmitate			108	
	Total ATP used (mol)/mol palmitate				2

The table shows how the oxidation of 1 mol of the C16 fatty acid, palmitate, generates 106 mol of ATP (108 formed in total—2 used in the activation step).

(see Table 22–1), or 106 × 30.5* = 3233 kJ. This represents 33% of the free energy of combustion of palmitic acid.

Peroxisomes Oxidize Very-Long-Chain Fatty Acids

A modified form of β-oxidation is found in **peroxisomes** and leads to the breakdown of very-long-chain fatty acids (eg, C20, C22) with the formation of acetyl-CoA and H$_2$O$_2$, which is broken down by catalase (see Chapter 12). This system, however, is not linked directly to phosphorylation and the generation of ATP. The peroxisomal enzymes are induced by high-fat diets and in some species by hypolipidemic drugs such as clofibrate.

Another role of peroxisomal β-oxidation is to shorten the side chain of cholesterol in bile acid formation (see Chapter 26). Peroxisomes also take part in the synthesis of ether glycerolipids (see Chapter 24), cholesterol, and dolichol (see Figure 26–2).

Oxidation of Unsaturated Fatty Acids Occurs by a Modified β-Oxidation Pathway

The CoA esters of unsaturated fatty acids are degraded by the enzymes normally responsible for β-oxidation until there is a *cis* double bond in the Δ^3 or Δ^4 position (**Figure 22–4**). A Δ^3-*cis* compound is isomerized (Δ^3**cis** → Δ^2-*trans*-**enoyl-CoA isomerase**) to the corresponding Δ^2-*trans*-CoA stage of β-oxidation for subsequent hydration and oxidation. Any Δ^4-*cis*-acyl-CoA either remaining, as in the case of linoleic acid (shown in Figure 22–4), or entering the pathway at this point after conversion by acyl-CoA dehydrogenase to Δ^2-*trans*-Δ^4-*cis*-dienoyl-CoA, is then metabolized as indicated in Figure 22–4.

KETOGENESIS OCCURS WHEN THERE IS A HIGH RATE OF FATTY ACID OXIDATION IN THE LIVER

Under metabolic conditions associated with a high rate of fatty acid oxidation, the liver produces considerable quantities of **acetoacetate** and D-**3-hydroxybutyrate** (3-hydroxybutyrate or β-hydroxybutyrate). Acetoacetate continually undergoes spontaneous decarboxylation to yield **acetone**. These three substances are collectively known as the **ketone bodies** (also called acetone bodies or [incorrectly*] "ketones") (**Figure 22–5**). Acetoacetate and 3-hydroxybutyrate are interconverted by the mitochondrial enzyme D-**3-hydroxybutyrate dehydrogenase**; the equilibrium is controlled by the mitochondrial [NAD$^+$]/[NADH] ratio, that is, the **redox state**. The concentration of total ketone bodies in the blood of well-fed mammals does not normally exceed 0.2 mmol/L. However, in ruminants, 3-hydroxybutyrate is formed continuously from butyric acid (a product of ruminal fermentation) in the rumen wall. In nonruminants, the liver appears to be the only organ that adds significant quantities of ketone bodies to the blood. Extrahepatic tissues utilize acetoacetate and 3-hydroxybutyrate as respiratory substrates. Acetone is a waste product which, as it is volatile, can be excreted via the lungs. Because there is active synthesis but little utilization of ketone bodies in the liver, while they are used but not produced in extrahepatic tissues, there is a net flow of the compounds to the extrahepatic tissues (**Figure 22–6**).

Acetoacetyl-CoA Is the Substrate for Ketogenesis

The enzymes responsible for ketone body formation (ketogenesis) are associated mainly with the mitochondria. Acetoacetyl-CoA is formed when two acetyl-CoA molecules produced via fatty acid breakdown condense to form acetoacetyl-CoA

---

*ΔG for the ATP reaction, as explained in Chapter 11.

*The term ketones should not be used as there are ketones in blood that are not ketone bodies, for example, pyruvate and fructose.

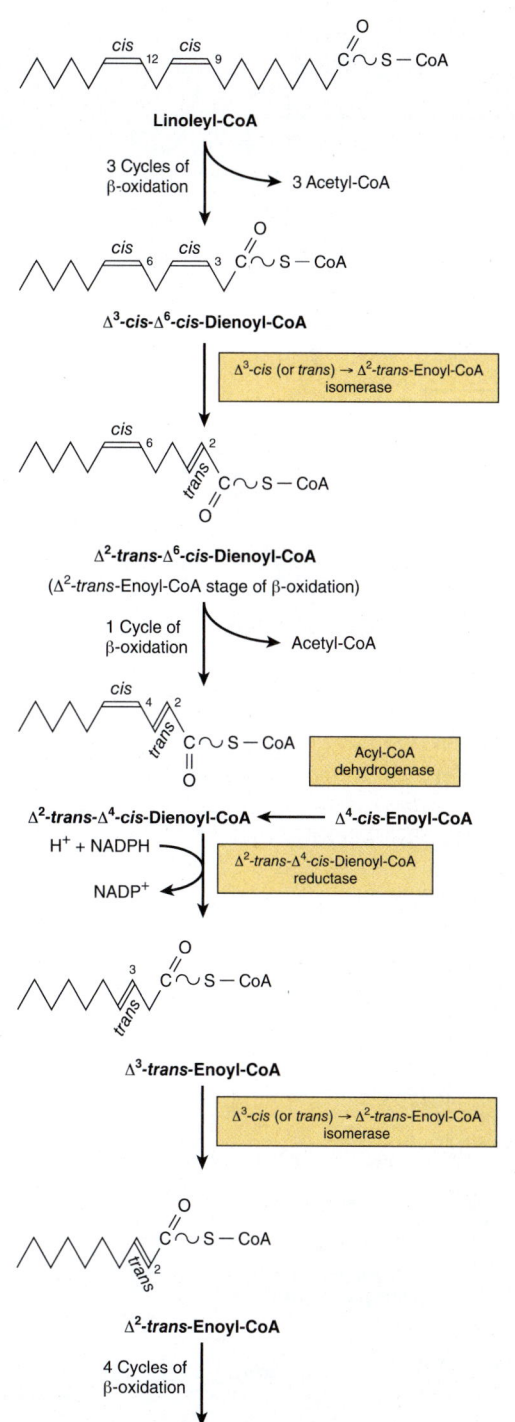

FIGURE 22–4 **Sequence of reactions in the oxidation of unsaturated fatty acids, for example, linoleic acid.** β-Oxidation proceeds as for saturated fatty acids until there is a *cis* double bond in the Δ³ position. This is then isomerized to the corresponding Δ²-*trans* compound allowing one cycle of β-oxidation to proceed, producing the Δ²- *trans*- Δ⁴-*cis* derivative. Δ⁴-*cis*-fatty acids or fatty acids forming Δ⁴-*cis*-enoyl-CoA enter the pathway here. A reduction step forming Δ³-*trans*-enoyl-CoA followed by an isomerization to the Δ²-*trans* form is required to enable β-oxidation to then go to completion. NADPH for the dienoyl-CoA reductase step is supplied by intramitochondrial sources such as glutamate dehydrogenase, isocitrate dehydrogenase, and NAD(P)H transhydrogenase.

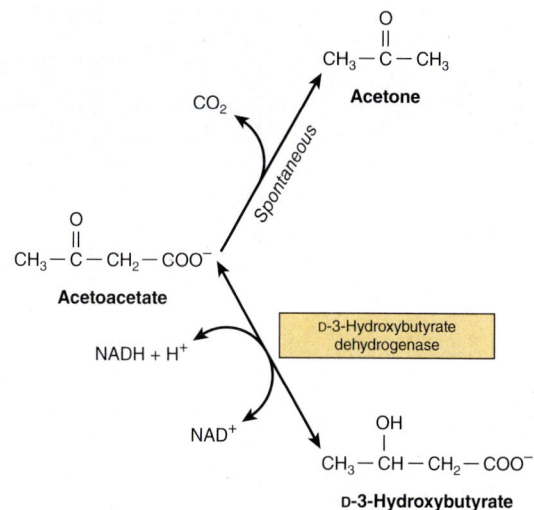

FIGURE 22–5 **Interrelationships of the ketone bodies.** D-3-Hydroxybutyrate dehydrogenase is a mitochondrial enzyme.

by a reversal of the **thiolase** reaction (see Figure 22–3), and may also arise directly from the terminal four carbons of a fatty acid during β-oxidation (**Figure 22–7**). Condensation of acetoacetyl-CoA with another molecule of acetyl-CoA by **3-hydroxy-3-methylglutaryl-CoA (HMG-CoA) synthase** forms **HMG-CoA. HMG-CoA lyase** then causes acetyl-CoA to split off from the HMG-CoA, leaving free acetoacetate. **Both enzymes must be present in mitochondria for ketogenesis to take place.** In mammals, ketone bodies are formed solely in the liver and in the rumen epithelium. 3-Hydroxybutyrate is formed from acetoacetate (see Figure 22–7) and is quantitatively the predominant ketone body present in the blood and urine in ketosis.

Ketone Bodies Serve as a Fuel for Extrahepatic Tissues

While an active enzymatic mechanism produces acetoacetate from acetoacetyl-CoA in the liver, acetoacetate once formed can only be reactivated by linkage to CoA directly in the cytosol, where it is used in a different, much less active pathway as a precursor in cholesterol synthesis (see Chapter 26). This accounts for the net production of ketone bodies by the liver.

In extrahepatic tissues, acetoacetate is activated to acetoacetyl-CoA by **succinyl-CoA-acetoacetate-CoA transferase**. CoA is transferred from succinyl-CoA to form acetoacetyl-CoA (**Figure 22–8**). In a reaction requiring the addition of a CoA, two acetyl-CoA molecules are formed by the splitting of acetoacetyl-CoA by thiolase and these are oxidized in the citric acid cycle. 3-Hydroxybutyrate is utilized by conversion to acetoacetate by the reversal of the reaction by which it is formed in the liver, generating an NADH in the process (see Figure 22–8). Thus, 1 mol of acetoacetate or 3-hydroxbutyrate yields 19 or 21.5 mol of ATP, respectively, by these pathways. If the blood level of ketone bodies rises to a concentration of ~12 mmol/L, the oxidative machinery becomes saturated and

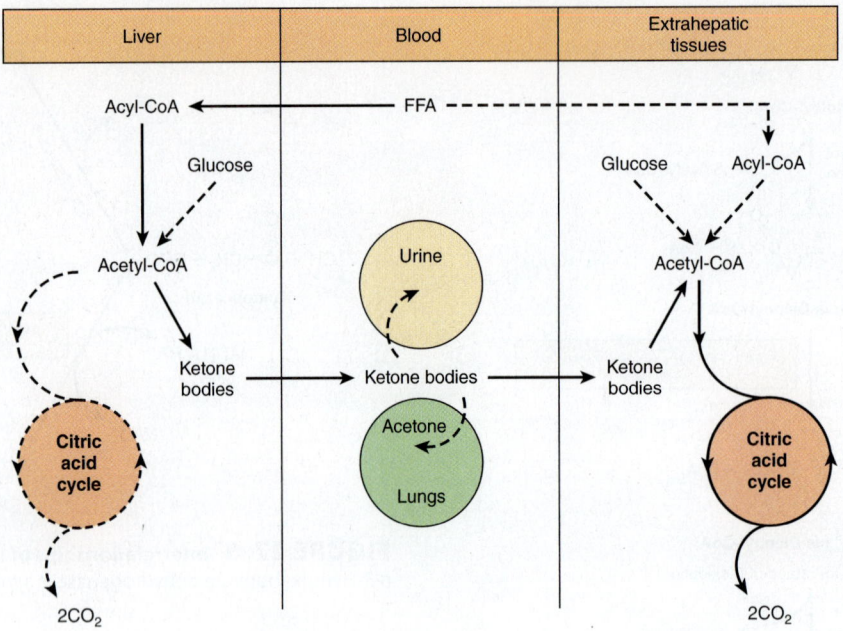

FIGURE 22–6 **Formation, utilization, and excretion of ketone bodies.** (The main pathway is indicated by the solid arrows.)

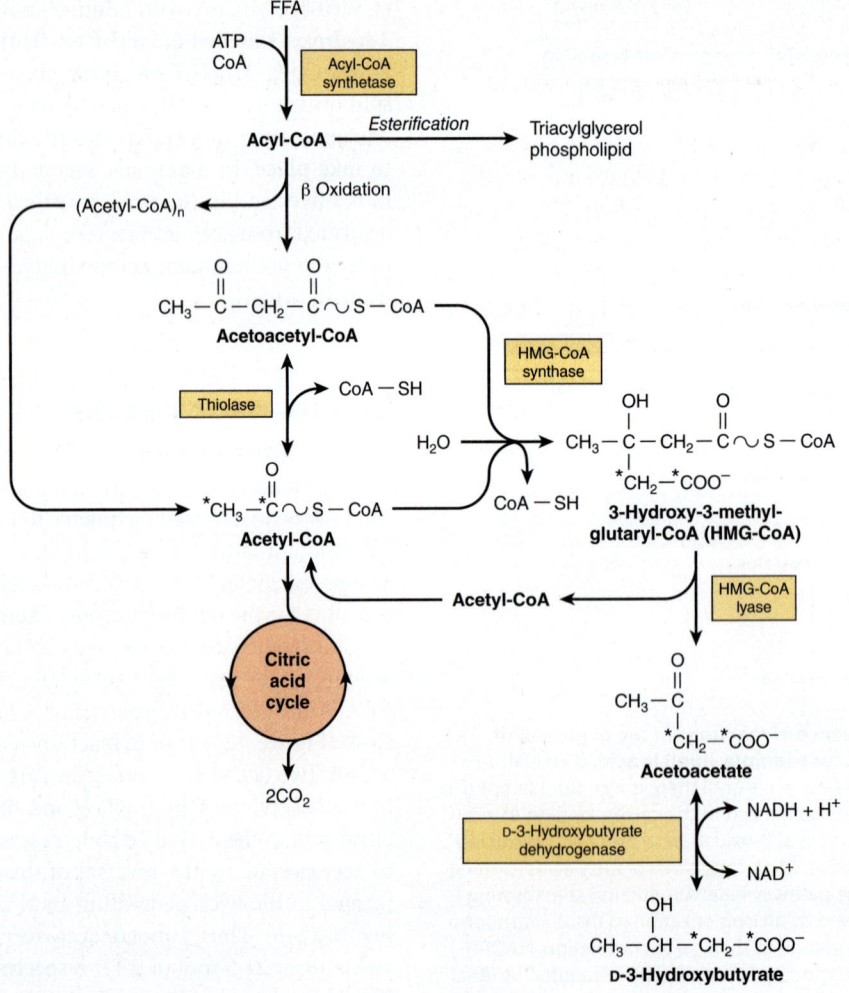

FIGURE 22–7 **Pathways of ketogenesis in the liver.** (FFA, free fatty acids.)

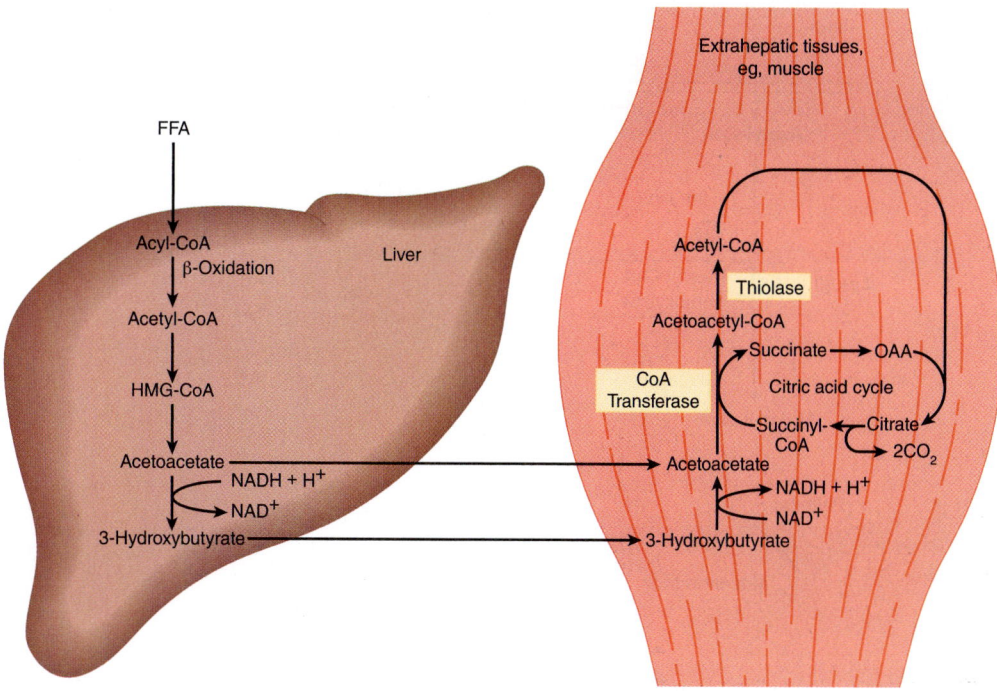

FIGURE 22–8 **Transport of ketone bodies from the liver and pathways of utilization and oxidation in extrahepatic tissues.** CoA transferase, succinyl-CoA-acetoacetate-CoA transferase. The breakdown of acetoacetyl-CoA by thiolase produces two acetyl-CoA molecules and requires the addition of one CoA (not shown).

at this stage, a large proportion of oxygen consumption may be accounted for by their oxidation.

In moderate ketonemia, the loss of ketone bodies via the urine is only a few percent of the total ketone body production and utilization. Since there are renal threshold-like effects (there is not a true threshold) that vary between species and individuals, measurement of the ketonemia, not the ketonuria, is the preferred method of assessing the severity of ketosis.

KETOGENESIS IS REGULATED AT THREE CRUCIAL STEPS

1. Ketosis does not occur in vivo unless there is an increase in the level of circulating FFAs arising from lipolysis of triacylglycerol in adipose tissue. **FFAs are the precursors of ketone bodies in the liver.** Both in fed and in fasting conditions, the liver extracts ~30% of the FFAs passing through it, so that at high concentrations the flux passing into the organ is substantial. **Thus, the factors regulating mobilization of FFA from adipose tissue are important in controlling ketogenesis (Figures 22–9 and 25–8).**

2. After uptake by the liver, FFAs are either **oxidized** to CO_2 or ketone bodies or **esterified** to triacylglycerol and phospholipid (acylglycerols). There is regulation of entry of fatty acids into the oxidative pathway by **carnitine palmitoyltransferase-I** (CPT-I) (see Figure 22–1), and the remainder of the fatty acid taken up is esterified. CPT-I activity is low in the fed state, leading to depression of fatty acid oxidation, and high in starvation, allowing fatty acid oxidation

to increase. **Malonyl-CoA**, the initial intermediate in fatty acid biosynthesis (see Figure 23–1) is a potent inhibitor of CPT-I (**Figure 22–10**). In the fed state, therefore, FFAs

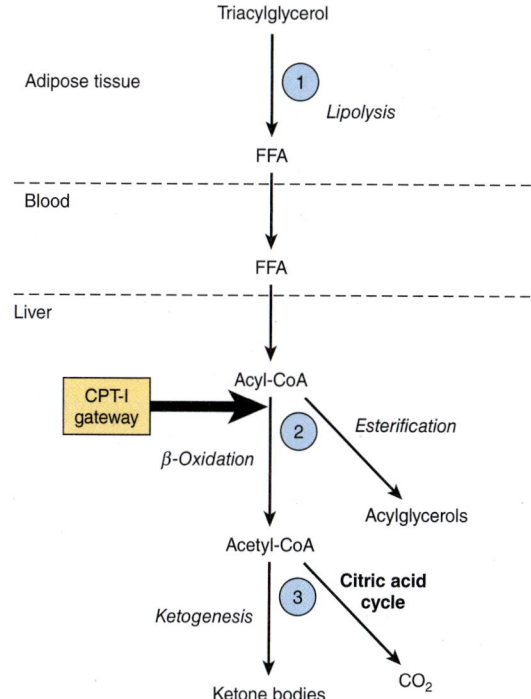

FIGURE 22–9 **Regulation of ketogenesis.** ① to ③ show three crucial steps in the pathway of metabolism of free fatty acids (FFA) that determine the extent of ketogenesis. (CPT-I, carnitine palmitoyltransferase-I.)

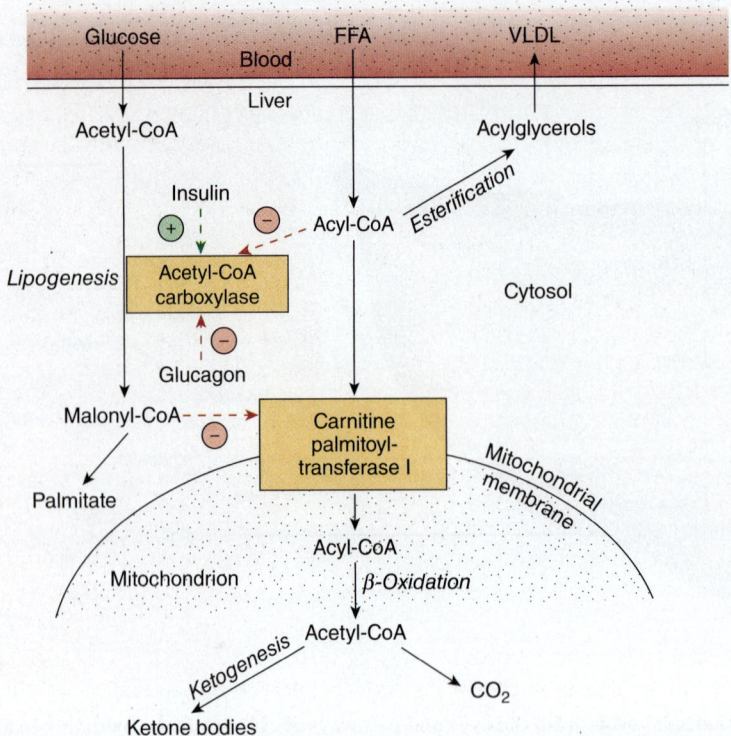

FIGURE 22–10 **Regulation of long-chain fatty acid oxidation in the liver.** (FFA, free fatty acids; VLDL, very-low-density lipoprotein.) Positive (⊕) and negative (⊖) regulatory effects are represented by broken arrows and substrate flow by solid arrows.

enter the liver cell in low concentrations and are nearly all esterified to acylglycerols and transported out of the liver in **very-low-density lipoprotein** (VLDL). However, as the concentration of FFA increases with the onset of starvation, acetyl-CoA carboxylase is inhibited directly by acyl-CoA, and (malonyl-CoA) decreases, releasing the inhibition of CPT-I and allowing more acyl-CoA to be β-oxidized. These events are reinforced in starvation by a decrease in the **(insulin)/(glucagon) ratio**. Thus, β-oxidation from FFA is controlled by the CPT-I gateway into the mitochondria, and the balance of the FFA uptake not oxidized is esterified.

3. In turn, the acetyl-CoA formed in β-oxidation is oxidized in the citric acid cycle, or it enters the pathway of ketogenesis via acetoacetyl-CoA to form ketone bodies. As the level of serum FFA is raised, proportionately more of the acetyl-CoA produced from their breakdown is converted to ketone bodies and less is oxidized via the citric acid cycle to CO_2. The partition of acetyl-CoA between the ketogenic pathway and the pathway of oxidation to CO_2 is regulated so that the total free energy captured in ATP which results from the oxidation of FFA remains constant as their concentration in the serum changes. This may be appreciated when it is realized that complete oxidation of 1 mol of palmitate involves a net production of 106 mol of ATP via β-oxidation and the citric acid cycle (see earlier), whereas only 26 mol of ATP are produced when acetoacetate is the end product and only 16 mol when 3-hydroxybutyrate is the end product. Thus, ketogenesis may be regarded as a mechanism

that allows the liver to oxidize increasing quantities of fatty acids within the constraints of a tightly coupled system of oxidative phosphorylation.

A fall in the concentration of oxaloacetate, particularly within the mitochondria, can impair the ability of the citric acid cycle to metabolize acetyl-CoA and divert fatty acid oxidation toward ketogenesis. Such a fall may occur because of an increase in the (NADH)/(NAD⁺) ratio caused when increased β-oxidation alters the equilibrium between oxaloacetate and malate so that the concentration of oxaloacetate is decreased, and also when gluconeogenesis is elevated due to low blood glucose levels. The activation by acetyl-CoA of pyruvate carboxylase, which catalyzes the conversion of pyruvate to oxaloacetate, partially alleviates this problem, but in conditions such as starvation and untreated diabetes mellitus, ketone bodies are overproduced and cause ketosis.

CLINICAL ASPECTS

Impaired Oxidation of Fatty Acids Gives Rise to Diseases Often Associated With Hypoglycemia

Carnitine deficiency can occur particularly in the newborn—and especially in preterm infants—owing to inadequate biosynthesis or renal leakage. Losses can also occur in hemodialysis. This suggests there may be a vitamin-like dietary requirement for carnitine in some individuals. Symptoms of deficiency include hypoglycemia, which is a consequence of impaired fatty acid oxidation, and lipid accumulation with muscular weakness. Treatment is by oral supplementation with carnitine.

Inherited **CPT-I deficiency** affects only the liver, resulting in reduced fatty acid oxidation and ketogenesis, with hypoglycemia. **CPT-II deficiency** affects primarily skeletal muscle and, when severe, the liver. The sulfonylurea drugs (**glyburide [glibenclamide]** and **tolbutamide**), used in the treatment of Type 2 diabetes mellitus, reduce fatty acid oxidation and, therefore, hyperglycemia by inhibiting CPT-I.

Inherited defects in the enzymes of β-oxidation and ketogenesis also lead to nonketotic hypoglycemia, coma, and fatty liver. Defects have been identified in long- and short-chain 3-hydroxyacyl-CoA dehydrogenase (deficiency of the long-chain enzyme may be a cause of **acute fatty liver of pregnancy**). **3-Ketoacyl-CoA thiolase** and **HMG-CoA lyase deficiency** also affect the degradation of leucine, a ketogenic amino acid (see Chapter 29).

Jamaican vomiting sickness is caused by eating the unripe fruit of the akee tree, which contains the toxin **hypoglycin**. This inactivates medium- and short-chain acyl-CoA dehydrogenase, inhibiting β-oxidation and causing hypoglycemia. **Dicarboxylic aciduria** is characterized by the excretion of C_6—C_{10} ω-dicarboxylic acids and by nonketotic hypoglycemia, and is caused by a lack of mitochondrial **medium-chain acyl-CoA dehydrogenase**. **Refsum disease** is a rare neurologic disorder caused by a metabolic defect that results in the accumulation of phytanic acid, which is found in dairy products and ruminant fat and meat. Phytanic acid is thought to have pathologic effects on membrane function, protein prenylation, and gene expression. **Zellweger (cerebrohepatorenal) syndrome** occurs in individuals with a rare inherited absence of peroxisomes in all tissues. They accumulate C_{26}—C_{38} polyenoic acids in brain tissue and also exhibit a generalized loss of peroxisomal functions. The disease causes severe neurologic symptoms, and most patients die in the first year of life.

Ketoacidosis Results From Prolonged Ketosis

Higher than normal quantities of ketone bodies present in the blood or urine constitute **ketonemia** (hyperketonemia) or **ketonuria**, respectively. The overall condition is called **ketosis**. The basic form of ketosis occurs in **starvation** and involves depletion of available carbohydrate coupled with mobilization of FFA. An exaggeration of this general pattern of metabolism produces the pathologic states found in **diabetes mellitus, the type 2 form of which is increasingly common in Western countries; twin lamb disease**; and **ketosis in lactating cattle**. Nonpathologic forms of ketosis are found under conditions of high-fat feeding and after severe exercise in the postabsorptive state.

Acetoacetic and 3-hydroxybutyric acids are both moderately strong acids and are buffered when present in blood or other tissues. However, their continual excretion in quantity progressively depletes the alkali reserve, causing **ketoacidosis**. This may be fatal in uncontrolled **diabetes mellitus**.

SUMMARY

- Fatty acid oxidation in mitochondria leads to the generation of large quantities of ATP by a process called β-oxidation that cleaves acetyl-CoA units sequentially from fatty acyl chains. The acetyl-CoA is oxidized in the citric acid cycle, generating further ATP.

- The ketone bodies (acetoacetate, 3-hydroxybutyrate, and acetone) are formed in hepatic mitochondria when there is a high rate of fatty acid oxidation. The pathway of ketogenesis involves synthesis and breakdown of HMG-CoA by two key enzymes: HMG-CoA synthase and HMG-CoA lyase.

- Ketone bodies are important fuels in extrahepatic tissues.

- Ketogenesis is regulated at three crucial steps: (1) control of FFA mobilization from adipose tissue; (2) the activity of carnitine palmitoyltransferase-I in liver, which determines the proportion of the fatty acid flux that is oxidized rather than esterified; and (3) partition of acetyl-CoA between the pathway of ketogenesis and the citric acid cycle.

- Diseases associated with impairment of fatty acid oxidation lead to hypoglycemia, fatty infiltration of organs, and hypoketonemia.

- Ketosis is mild in starvation but severe in diabetes mellitus and ruminant ketosis.

REFERENCES

Eljamil AS: *Lipid Biochemistry: For Medical Sciences.* iUniverse, 2015.

Gurr MI, Harwood JL, Frayn KN, et al: *Lipids, Biochemistry, Biotechnology and Health.* Wiley-Blackwell 2016.

Biosynthesis of Fatty Acids & Eicosanoids

Kathleen M. Botham, PhD, DSc, & Peter A. Mayes, PhD, DSc

O B J E C T I V E S

After studying this chapter, you should be able to:

- Describe the reaction catalyzed by acetyl-CoA carboxylase and understand the mechanisms by which its activity is regulated to control the rate of fatty acid synthesis.
- Outline the structure of the fatty acid synthase multienzyme complex, indicating the sequence of enzymes in the two peptide chains of the homodimer.
- Explain how long-chain fatty acids are synthesized by the repeated condensation of two carbon units, with formation of the 16-carbon palmitate being favored in most tissues, and identify the cofactors required.
- Indicate the sources of reducing equivalents (NADPH) for fatty acid synthesis.
- Explain how fatty acid synthesis is regulated by nutritional status and identify other control mechanisms that operate in addition to modulation of the activity of acetyl-CoA carboxylase.
- Identify the nutritionally essential fatty acids and explain why they cannot be formed in the body.
- Explain how polyunsaturated fatty acids are synthesized by desaturase and elongation enzymes.
- Outline the cyclooxygenase and lipoxygenase pathways responsible for the formation of the various classes of eicosanoids.

BIOMEDICAL IMPORTANCE

Fatty acids are synthesized by an **extramitochondrial system**, which is responsible for the complete synthesis of palmitate from acetyl-CoA in the **cytosol**. In most mammals, glucose is the primary substrate for lipogenesis, but in ruminants it is acetate, the main fuel molecule they obtain from the diet. Critical diseases of the pathway have not been reported in humans. However, inhibition of lipogenesis occurs in type 1 (insulin-dependent) **diabetes mellitus**, and variations in the activity of the process affect the nature and extent of **obesity**.

Unsaturated fatty acids in phospholipids of the cell membrane are important in maintaining membrane fluidity (see Chapter 40). A high ratio of polyunsaturated fatty acids to saturated fatty acids (P:S ratio) in the diet is considered to be beneficial in preventing coronary heart disease. Animal tissues have limited capacity for desaturating fatty acids, and require certain dietary polyunsaturated fatty acids derived from plants. These **essential fatty acids** are used to form eicosanoic (C_{20}) fatty acids, which give rise to the **eicosanoids** prostaglandins, thromboxanes, leukotrienes, and lipoxins. Prostaglandins mediate **inflammation**, **pain**, induce **sleep**, and also regulate **blood coagulation** and **reproduction**. **Nonsteroidal anti-inflammatory drugs (NSAIDs)** such as **aspirin and ibuprofen** act by inhibiting prostaglandin synthesis. Leukotrienes have muscle contractant and chemotactic properties and are important in allergic reactions and inflammation.

THE MAIN PATHWAY FOR DE NOVO SYNTHESIS OF FATTY ACIDS (LIPOGENESIS) OCCURS IN THE CYTOSOL

This system is present in many tissues, including liver, kidney, brain, lung, mammary gland, and adipose tissue. Its cofactor requirements include NADPH, ATP, Mn^{2+}, biotin, and HCO_3^- (as a source of CO_2). **Acetyl-CoA** is the immediate substrate, and **free palmitate** is the end product.

Production of Malonyl-CoA Is the Initial & Controlling Step in Fatty Acid Synthesis

The initial step in fatty acid synthesis is the carboxylation of acetyl-CoA to form **malonyl-CoA** by **acetyl-CoA carboxylase**. The reaction requires ATP and the B vitamin **biotin**. Acetyl-CoA carboxylase is a **multienzyme protein** containing biotin carboxylase, and a carboxyl transferase as well as biotin carrier protein (which binds biotin), and a regulatory allosteric site. The reaction takes place in two steps: Step 1 catalyzed by biotin carboxylase results in the carboxylation of biotin and uses ATP and Step 2 catalyzed by carboxyl transferase results in the transfer of the carboxyl group to acetyl-CoA forming the product, malonyl-CoA (**Figure 23–1**). Acetyl-CoA carboxylase has a major role in the regulation of fatty acid synthesis (see following discussion).

The Fatty Acid Synthase Complex Is a Homodimer of Two Polypeptide Chains Containing Six Enzyme Activities & the Acyl Carrier Protein

After the formation of malonyl-CoA, fatty acids are formed by the **fatty acid synthase enzyme complex**. The individual enzymes required for fatty acid synthesis are linked in this multienzyme polypeptide complex that incorporates the **acyl carrier protein (ACP)**, which has a similar function to that of CoA in the β-oxidation pathway (see Chapter 22). It contains the vitamin **pantothenic acid** in the form of 4'-phosphopantetheine (see Figure 44–15). In the primary structure of the protein,

the enzyme domains are believed to be linked in the sequence as shown in **Figure 23–2**. X-ray crystallography of the three-dimensional structure, however, has shown that the complex is a homodimer, with two identical subunits, each containing six enzymes and an ACP, arranged in an X shape (see Figure 23–2). The use of one multienzyme functional unit has the advantages of achieving compartmentalization of the process within the cell without the necessity for permeability barriers, and synthesis of all enzymes in the complex is coordinated since it is encoded by a single gene.

Initially, a priming molecule of acetyl-CoA combines with a cysteine —SH group on the ACP of one monomer of the fatty acid synthase complex (**Figure 23–3**, reaction 1a), while malonyl-CoA combines with the adjacent —SH on the 4'-phosphopantetheine of ACP of the other monomer (reaction 1b). These reactions are catalyzed by **malonyl acetyl transacylase**, to form **acetyl (acyl)-malonyl enzyme**. The acetyl group attacks the methylene group of the malonyl residue, catalyzed by **3-ketoacyl synthase**, and liberates CO_2, forming **3-ketoacyl enzyme** (acetoacetyl enzyme) (reaction 2), freeing the cysteine —SH group. Decarboxylation allows the reaction to go to completion, pulling the whole sequence of reactions in the forward direction. The 3-ketoacyl group is reduced (3-ketoacylreductase), dehydrated (dehydratase), and reduced again (enoyl reductase) (reactions 3-5) to form the corresponding **saturated acyl enzyme** (product of reaction 5). The saturated acyl residue now transfers from the —SH of the 4' phosphopantetheine to the free cysteine —SH group as it is displaced by a new malonyl-CoA. The sequence of reactions is repeated six more times until a saturated 16-carbon acyl radical (palmitoyl) has been assembled. It is liberated from the enzyme complex by the activity of the sixth enzyme in the complex, **thioesterase** (deacylase). The free

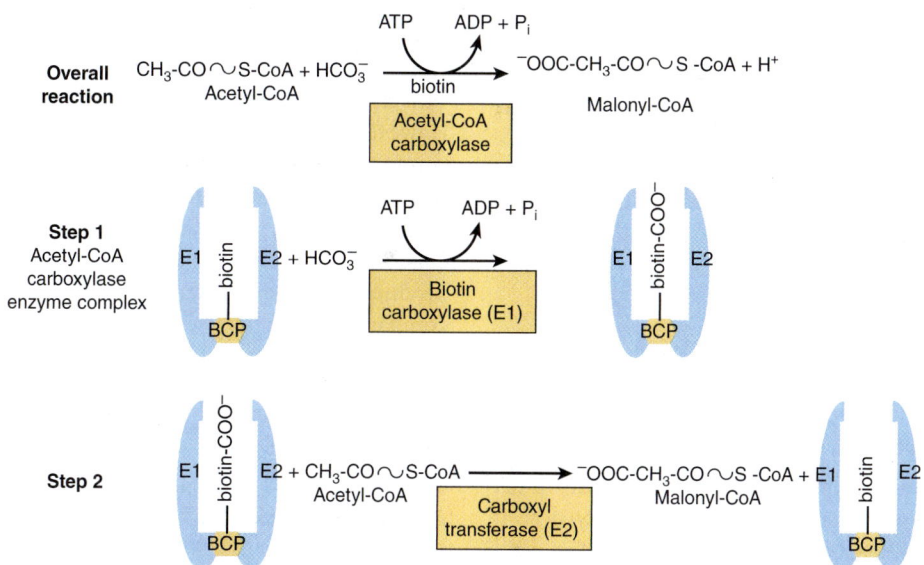

FIGURE 23–1 **Biosynthesis of malonyl-CoA by acetyl-CoA carboxylase.** Acetyl carboxylase is a multienzyme complex containing two enzymes, biotin carboxylase (E1) and a carboxyltransferase (E2), and the biotin carrier protein (BCP). Biotin is covalently linked to the BCP. The reaction proceeds in two steps. In step 1, catalyzed by E1, biotin is carboxylated as it accepts a COO^- group from HCO_3^- and ATP is used. In step 2, catalyzed by E2, the COO^- is transferred to acetyl-CoA forming malonyl-CoA.

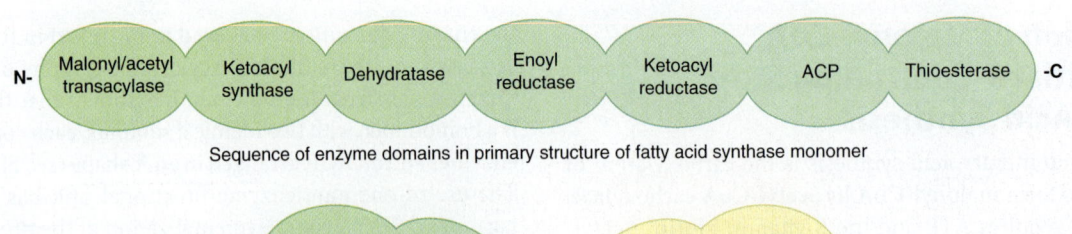

Sequence of enzyme domains in primary structure of fatty acid synthase monomer

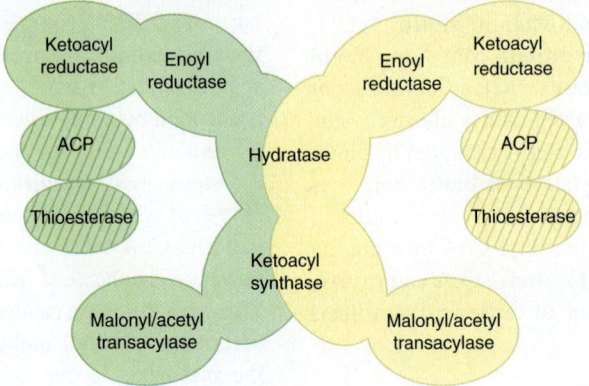

Fatty acid synthase homodimer

FIGURE 23–2 **Fatty acid synthase multienzyme complex.** The complex is a dimer of two identical polypeptide monomers in which six enzymes and the acyl carrier protein (ACP) are linked in the primary structure in the sequence shown. X-ray crystallography of the three-dimensional structure has demonstrated that the two monomers in the complex are arranged in an X-shape.

palmitate must be activated to acyl-CoA before it can proceed via any other metabolic pathway. Its possible fates are esterification into acylglycerols, chain elongation, desaturation, or esterification into cholesteryl ester. In mammary gland, there is a separate thioesterase specific for acyl residues of C_8, C_{10}, or C_{12}, which are subsequently found in milk lipids.

The equation for the overall synthesis of palmitate from acetyl-CoA and malonyl-CoA is

$$CH_3 CO—S—CoA + 7HOOCCHCO—S—CoA + 14NADPH + 14H^+$$
$$\rightarrow CH_3(CH_2)_{14} COOH + 7CO_2 + 6H_2O + 8CoA—SH + 14NADP^+$$

The acetyl-CoA used as a primer forms carbon atoms 15 and 16 of palmitate. The addition of all the subsequent C_2 units is via malonyl-CoA. Propionyl-CoA instead of acetyl-CoA is used as the primer for the synthesis of long-chain fatty acids with an odd number of carbon atoms, which are found particularly in ruminant fat and milk.

The Main Source of NADPH for Lipogenesis Is the Pentose Phosphate Pathway

NADPH is involved as a donor of reducing equivalents in the reduction of the 3-ketoacyl and the 2,3-unsaturated acyl derivatives (see Figure 23–3, reactions 3 and 5). The oxidative reactions of the pentose phosphate pathway (see Chapter 20) are the chief source of the hydrogen required for the synthesis of fatty acids. Significantly, tissues specializing in active lipogenesis—that is, liver, adipose tissue, and the lactating mammary gland—also possess an active pentose phosphate pathway. Moreover,

both metabolic pathways are found in the cytosol of the cell, so there are no membranes or permeability barriers against the transfer of NADPH. Other sources of NADPH include the reaction that converts malate to pyruvate catalyzed by the **NADP malate dehydrogenase (malic enzyme)** (**Figure 23–4**) and the extramitochondrial **isocitrate dehydrogenase** reaction (a substantial source in ruminants).

Acetyl-CoA Is the Principal Building Block of Fatty Acids

Acetyl-CoA is formed from glucose via the oxidation of pyruvate in the matrix of the mitochondria (see Chapter 17). However, as it does not diffuse readily across the mitochondrial membranes, its transport into the cytosol, the principal site of fatty acid synthesis, requires a special mechanism involving **citrate**. After condensation of acetyl-CoA with oxaloacetate in the citric acid cycle within mitochondria, the citrate produced can be translocated into the extramitochondrial compartment via the tricarboxylate transporter, where in the presence of CoA and ATP, it undergoes cleavage to acetyl-CoA and oxaloacetate by **ATP-citrate lyase**, which increases in activity in the well-fed state. The acetyl-CoA is then available for malonyl-CoA formation and synthesis of fatty acids (see Figures 23–1 and 23–3), and the oxaloacetate can form malate via NADH-linked malate dehydrogenase, followed by the generation of NADPH and pyruvate via the malic enzyme. The NADPH becomes available in the cytosol for lipogenesis, and the pyruvate can be used to regenerate acetyl-CoA after transport into the mitochondrion (see Figure 23–4). This pathway is a means of transferring reducing equivalents from extramitochondrial NADH to NADP to form NADPH.

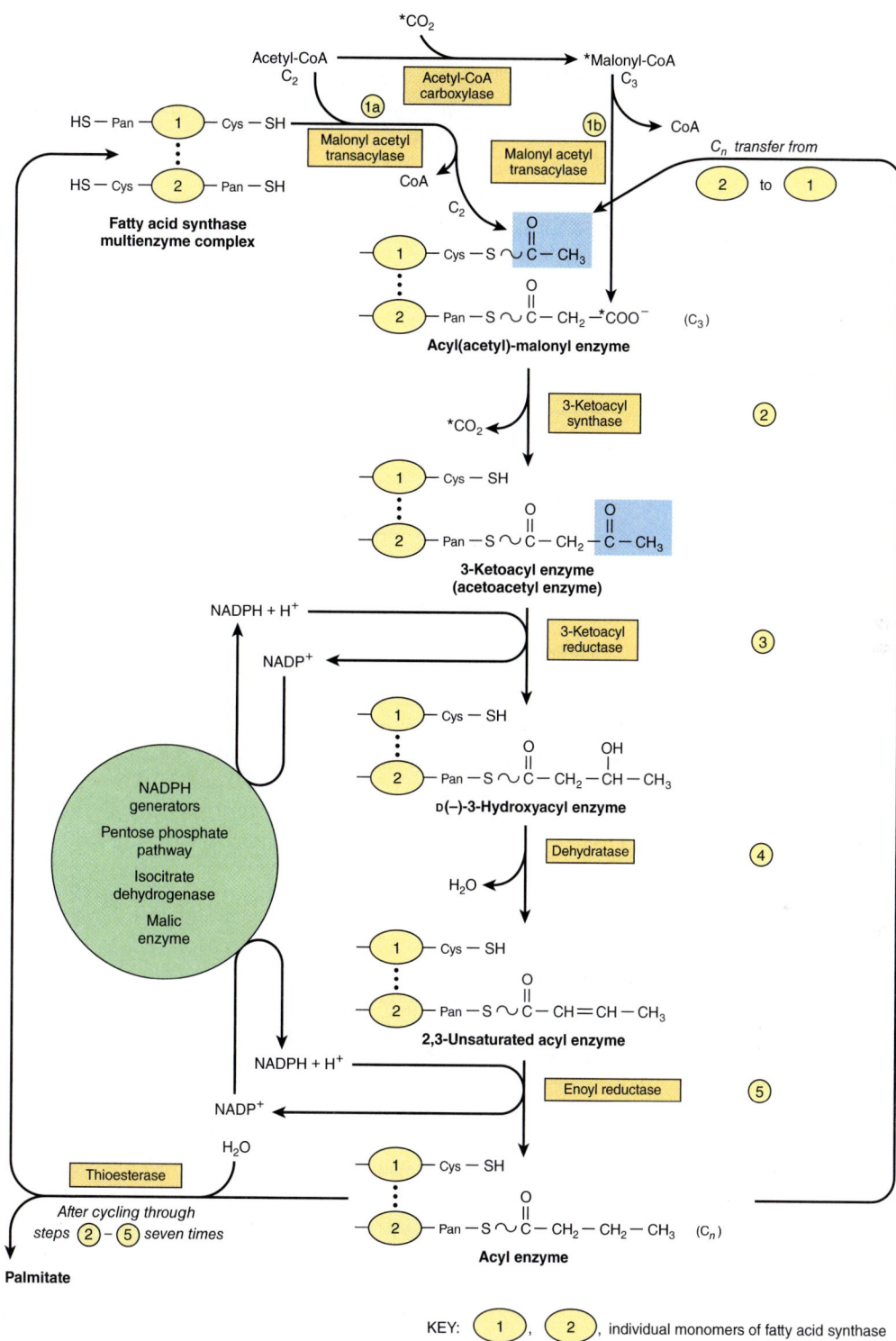

FIGURE 23–3 **Biosynthesis of long-chain fatty acids.** After the initial priming step in which acetyl-CoA is bound to a cysteine-SH group on the fatty acid synthase enzyme (reaction 1a), the addition of a malonyl residue causes the acyl chain to grow by two carbon atoms in each cycle. (Cys, cysteine residue; Pan, 4′-phosphopantetheine.) The blocks highlighted in blue contain the C_2 unit derived from acetyl-CoA initially (as illustrated) and subsequently the C_n unit formed in reaction 5.
*Shows that the carbon in the CO_2 initially incorporated into malonyl-CoA is then released as CO_2 in reaction 2.

As the citrate (tricarboxylate) transporter in the mitochondrial membrane requires malate to exchange with citrate (see Figure 13–10), malate itself can be transported into the mitochondrion, where it is able to reform oxaloacetate. There is little ATP-citrate lyase or malic enzyme in ruminants, probably because in these species acetate (derived from carbohydrate digestion in the rumen and activated to acetyl-CoA extramitochondrially) is the main source of acetyl-CoA.

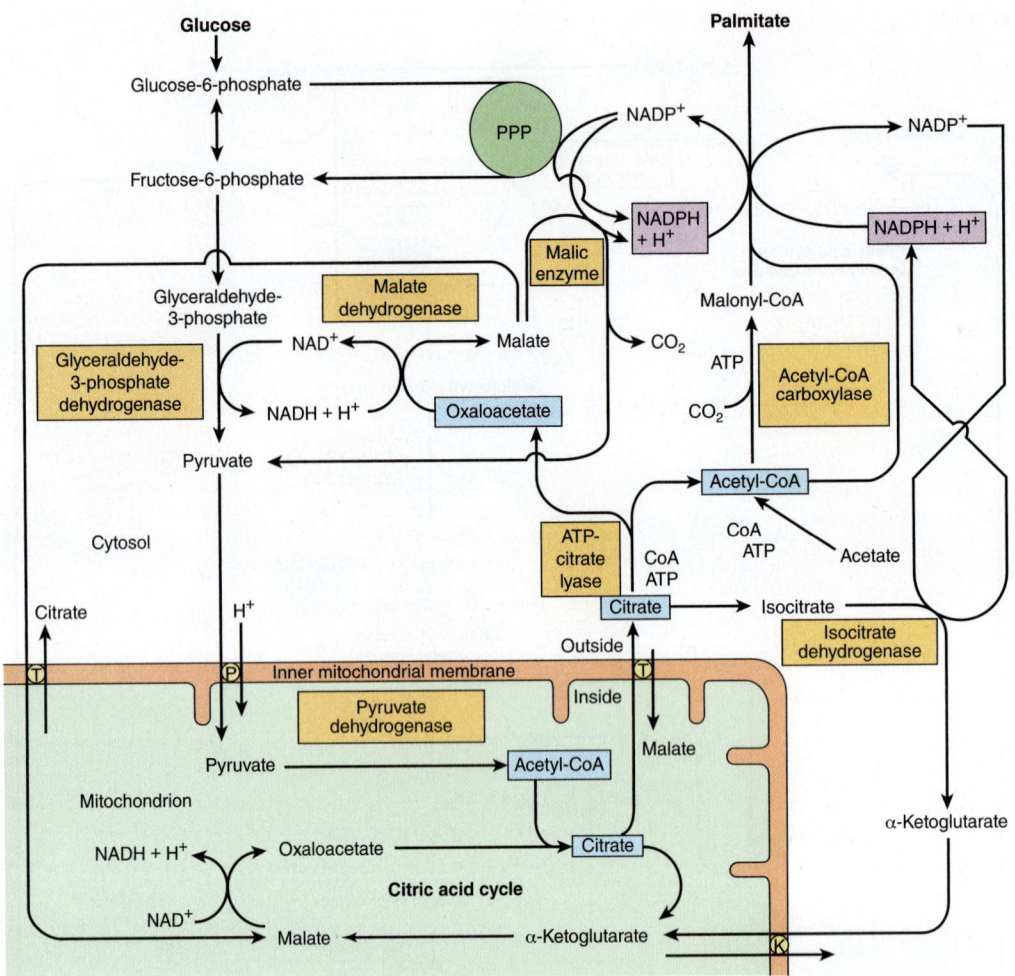

FIGURE 23–4 **The provision of acetyl-CoA and NADPH for lipogenesis.** (K, α-ketoglutarate transporter; P, pyruvate transporter; PPP, pentose phosphate pathway; T, tricarboxylate transporter.)

Elongation of Fatty Acid Chains Occurs in the Endoplasmic Reticulum

This pathway (the "**microsomal system**") elongates saturated and unsaturated fatty acyl-CoAs (from C_{10} upward) by two carbons, using malonyl-CoA as the acetyl donor and NADPH as the reductant, and is catalyzed by the microsomal **fatty acid elongase** system of enzymes (**Figure 23–5**). Elongation of stearyl-CoA in brain increases rapidly during myelination in order to provide C_{22} and C_{24} fatty acids for sphingolipids (see Figures 21–10 and 21–11).

THE NUTRITIONAL STATE REGULATES LIPOGENESIS

Excess carbohydrate is stored as fat in many animals in anticipation of periods of caloric deficiency such as starvation, hibernation, etc., and to provide energy for use between meals in animals, including humans, that take their food at spaced intervals. Lipogenesis converts surplus glucose and intermediates such as pyruvate, lactate, and acetyl-CoA to fat, assisting the anabolic phase of this feeding cycle. The nutritional state of the organism is the main factor regulating the rate of lipogenesis. Thus, the rate is high in the well-fed animal whose diet contains a high proportion of carbohydrate. It is depressed by restricted caloric intake, high-fat diet, or a deficiency of insulin, as in diabetes mellitus. These latter conditions are associated with increased concentrations of plasma-free fatty acids, and an inverse relationship has been demonstrated between hepatic lipogenesis and the concentration of serum-free fatty acids. Lipogenesis is increased when sucrose (a disaccharide consisting of glucose and fructose) is fed instead of glucose because fructose bypasses the phosphofructokinase control point in glycolysis and floods the lipogenic pathway (see Figure 20–5).

SHORT- & LONG-TERM MECHANISMS REGULATE LIPOGENESIS

Long-chain fatty acid synthesis is controlled in the short term by allosteric and covalent modification of enzymes and in the long term by changes in gene expression governing rates of synthesis of enzymes.

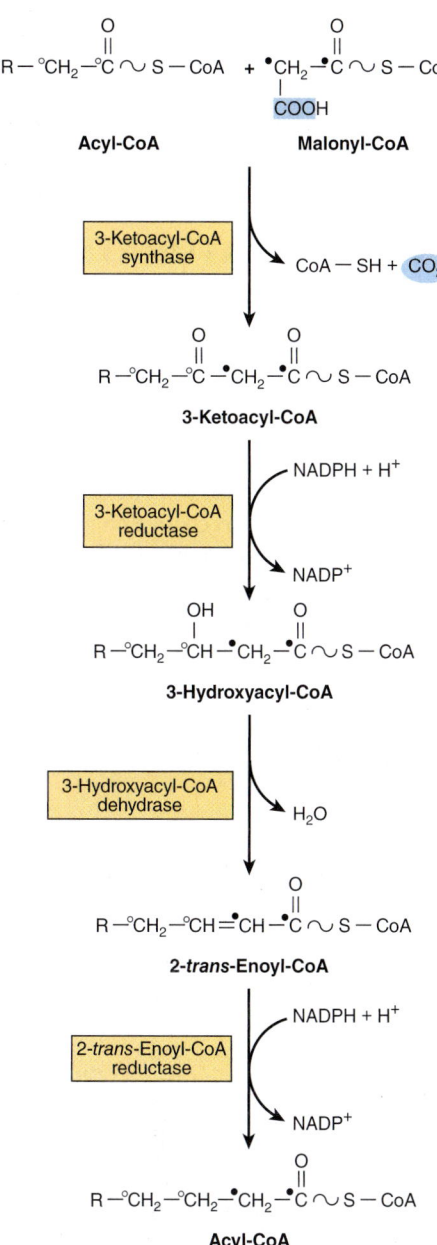

FIGURE 23–5 **Microsomal elongase system for fatty acid chain elongation.** NADH may also be used by the reductases, but NADPH is preferred.

Acetyl-CoA Carboxylase Is the Most Important Enzyme in the Regulation of Lipogenesis

Acetyl-CoA carboxylase is an allosteric enzyme and is activated by **citrate**, which increases in concentration in the well-fed state and is an indicator of a plentiful supply of acetyl-CoA. Citrate promotes the conversion of the enzyme from an inactive dimer (two subunits of the enzyme complex) to an active polymeric form, with a molecular mass of several million. Inactivation is promoted by phosphorylation of the enzyme and by long-chain acyl-CoA molecules, an example of negative feedback inhibition by a product of a reaction

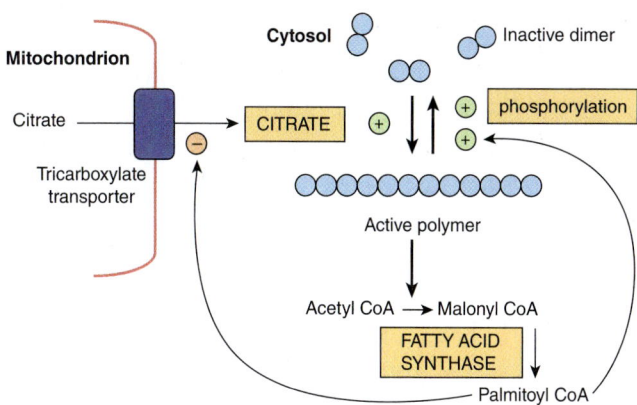

FIGURE 23–6 **Regulation of acetyl-CoA carboxylase.** Acetyl-CoA carboxylase is activated by citrate, which promotes the conversion of the enzyme from an inactive dimer to an active polymeric form. Inactivation is promoted by phosphorylation of the enzyme and by long-chain acyl-CoA molecules such as palmitoyl-CoA. In addition, acyl-CoA inhibits the tricarboxylate transporter, which transports citrate out of mitochondria into the cytosol, thus decreasing the citrate concentration in the cytosol and favoring inactivation of the enzyme.

(**Figure 23–6**). Thus, if acyl-CoA accumulates because it is not esterified quickly enough or because of increased lipolysis or an influx of free fatty acids into the tissue, it will automatically reduce the rate of synthesis of new fatty acid. Acyl-CoA also inhibits the mitochondrial **tricarboxylate transporter**, thus preventing activation of the enzyme by egress of citrate from the mitochondria into the cytosol (see Figure 23–6).

Acetyl-CoA carboxylase is also regulated by hormones such as **glucagon**, **epinephrine**, and **insulin** via changes in its phosphorylation state (details in **Figure 23–7**).

Pyruvate Dehydrogenase Is Also Regulated by Acyl-CoA

Acyl-CoA causes an inhibition of pyruvate dehydrogenase, the enzyme which catalyzes the formation of acetyl-CoA from pyruvate to link glycolysis with the citric acid cycle (see Chapters 16 and 17), by inhibiting the ATP-ADP exchange transporter of the inner mitochondrial membrane. This leads to increased intramitochondrial (ATP)/(ADP) ratios and therefore to conversion of active to inactive pyruvate dehydrogenase (see Figure 17–6), thus regulating the availability of acetyl-CoA for lipogenesis. Furthermore, oxidation of acyl-CoA due to increased levels of free fatty acids may increase the ratios of (acetyl-CoA)/(CoA) and (NADH)/(NAD⁺) in mitochondria, which also inhibits pyruvate dehydrogenase.

Insulin Also Regulates Lipogenesis by Other Mechanisms

Insulin stimulates lipogenesis by several other mechanisms as well as by increasing acetyl-CoA carboxylase activity. It increases the transport of glucose into the cell (eg, in adipose tissue), increasing the availability of both pyruvate (and thus

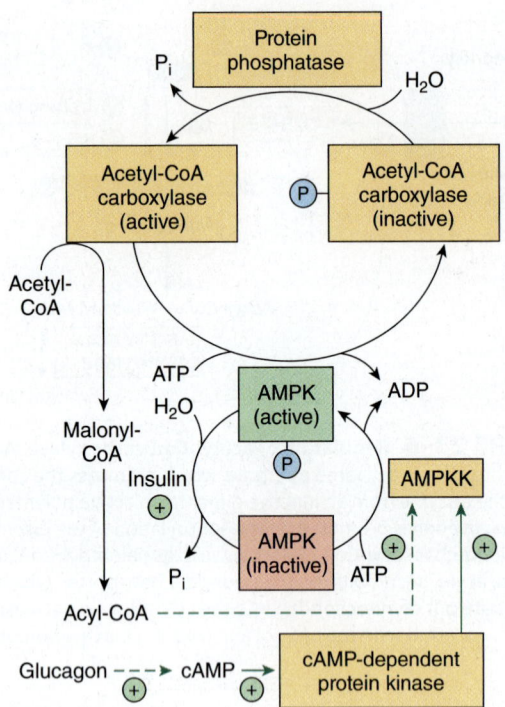

FIGURE 23–7 **Regulation of acetyl-CoA carboxylase by phosphorylation/dephosphorylation.** The enzyme is inactivated by phosphorylation by AMP-activated protein kinase (AMPK), which in turn is phosphorylated and activated by AMP-activated protein kinase kinase (AMPKK). Glucagon and epinephrine increase cAMP, and thus activate this latter enzyme via cAMP-dependent protein kinase. The kinase kinase enzyme is also believed to be activated by acyl-CoA. Insulin activates acetyl-CoA carboxylase via dephosphorylation of AMPK.

acetyl-CoA) for fatty acid synthesis and glycerol-3-phosphate for triacylglycerol synthesis via esterification of the newly formed fatty acids (see Figure 24–2), and also converts the inactive form of pyruvate dehydrogenase to the active form in adipose tissue, although not in liver. Insulin also—by its ability to depress the level of intracellular cAMP—**inhibits lipolysis** in adipose tissue, reducing the concentration of plasma-free fatty acids and, therefore, long-chain acyl-CoA, which are inhibitors of lipogenesis.

The Fatty Acid Synthase Complex & Acetyl-CoA Carboxylase Are Adaptive Enzymes

These enzymes adapt to the body's physiologic needs via changes in gene expression which lead to increases in the total amount of enzyme protein present in the fed state and decreases during intake of a high-fat diet and in conditions such as starvation, and diabetes mellitus. **Insulin** plays an important role, promoting gene expression and induction of enzyme biosynthesis, and **glucagon** (via cAMP) antagonizes this effect. Feeding fats containing polyunsaturated fatty acids coordinately regulates the inhibition of expression of key

enzymes of glycolysis and lipogenesis. These mechanisms for longer-term regulation of lipogenesis take several days to become fully manifested and augment the direct and immediate effect of free fatty acids and hormones such as insulin and glucagon.

SOME POLYUNSATURATED FATTY ACIDS CANNOT BE SYNTHESIZED BY MAMMALS & ARE NUTRITIONALLY ESSENTIAL

Certain long-chain unsaturated fatty acids of metabolic significance in mammals are shown in **Figure 23–8**. Other C_{20}, C_{22}, and C_{24} polyenoic fatty acids may be derived from oleic, linoleic, and α-linolenic acids by chain elongation. Palmitoleic and oleic acids are not essential in the diet because the tissues can introduce a double bond at the Δ^9 position of a saturated fatty acid. **Linoleic and α-linolenic acids** are the only fatty acids known to be essential for the complete nutrition of many species of animals, including humans, and are termed the **nutritionally essential fatty acids**. In humans and most other mammals, **arachidonic acid** can be formed from linoleic acid.

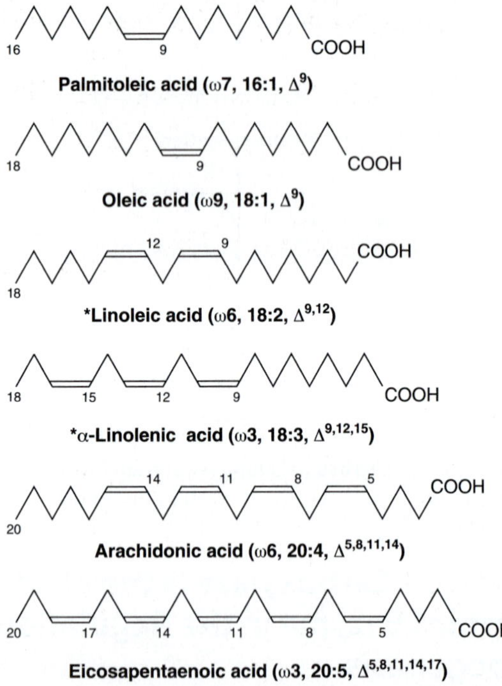

Palmitoleic acid ($\omega7$, $16{:}1$, Δ^9)

Oleic acid ($\omega9$, $18{:}1$, Δ^9)

***Linoleic acid** ($\omega6$, $18{:}2$, $\Delta^{9,12}$)

***α-Linolenic acid** ($\omega3$, $18{:}3$, $\Delta^{9,12,15}$)

Arachidonic acid ($\omega6$, $20{:}4$, $\Delta^{5,8,11,14}$)

Eicosapentaenoic acid ($\omega3$, $20{:}5$, $\Delta^{5,8,11,14,17}$)

FIGURE 23–8 **Structure of some unsaturated fatty acids.** Although the carbon atoms in the molecules are conventionally numbered—that is, numbered from the carboxyl terminal—the ω numbers (eg, ω7 in palmitoleic acid) are calculated from the reverse end (the methyl terminal) of the molecules. The information in parentheses shows, for instance, that α-linolenic acid contains double bonds starting at the third carbon from the methyl terminal, has 18 carbons and 3 double bonds, and has these double bonds at the 9th, 12th, and 15th carbons from the carboxyl terminal. *Nutritionally essential fatty acids in humans.

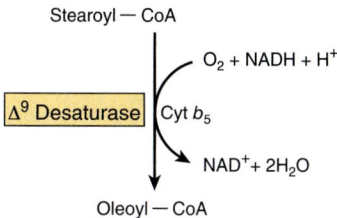

FIGURE 23–9 Microsomal Δ^9 desaturase.

Double bonds can be introduced at the Δ^4, Δ^5, Δ^6, and Δ^9 positions (see Chapter 21) in most animals, but never beyond the Δ^9 position. In contrast, plants are able to synthesize the nutritionally essential fatty acids by introducing double bonds at the Δ^{12} and Δ^{15} positions.

MONOUNSATURATED FATTY ACIDS ARE SYNTHESIZED BY A Δ^9 DESATURASE SYSTEM

Several tissues including the liver are considered to be responsible for the formation of nonessential monounsaturated fatty acids from saturated fatty acids. The first double bond introduced into a saturated fatty acid is nearly always in the Δ^9 position. An enzyme system—Δ^9 **desaturase** (**Figure 23–9**)—in the endoplasmic reticulum catalyzes the conversion of palmitoyl-CoA or stearoyl-CoA to palmitoleoyl-CoA or oleoyl-CoA, respectively.

Oxygen and either NADH or NADPH are necessary for the reaction. The enzymes appear to be similar to a mono-oxygenase system involving cytochrome b_5 (see Chapter 12).

SYNTHESIS OF POLYUNSATURATED FATTY ACIDS INVOLVES DESATURASE & ELONGASE ENZYME SYSTEMS

Additional double bonds introduced into existing monounsaturated fatty acids are always separated from each other by a methylene group (methylene interrupted) except in bacteria. Since animals have a Δ^9 desaturase, they are able to synthesize the $\omega9$ (oleic acid) family of unsaturated fatty acids completely by a combination of chain elongation and desaturation (see Figures 23–9 and **23–10**) after the formation of saturated fatty acids by the pathways described in this chapter. However, as indicated earlier, linoleic ($\omega6$) or α-linolenic ($\omega3$) acids are nutritionally essential, as they are required for the synthesis of the other members of the $\omega6$ or $\omega3$ families (pathways shown in Figure 23–10) and must be supplied in the diet. Linoleic acid is converted to arachidonic acid (20:4 $\omega6$) via **γ-linolenic acid (18:3 $\omega6$)**. The nutritional requirement for arachidonate may thus be dispensed with if there is adequate linoleate in the diet. Cats, however, cannot carry out this conversion owing to the absence of Δ^6 desaturase and must obtain arachidonate in their diet. The desaturation and chain elongation system are greatly diminished in the starving state, in response to

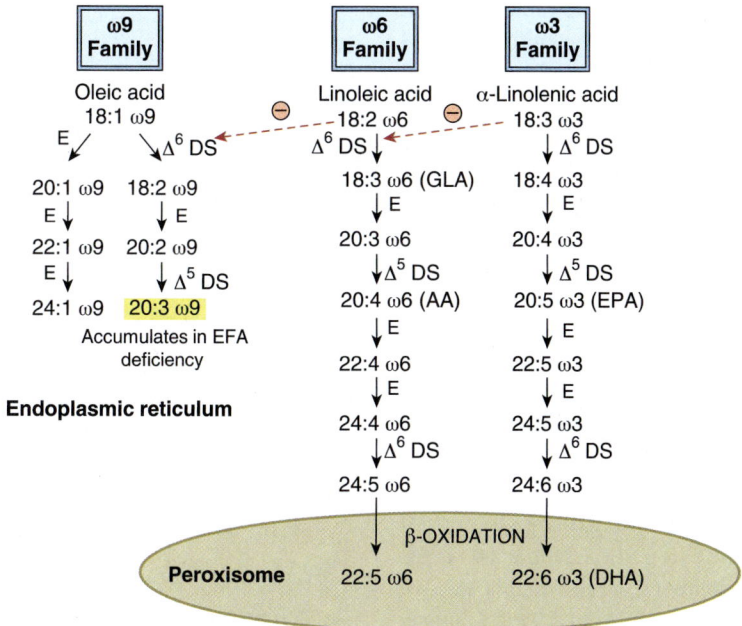

FIGURE 23–10 **Biosynthesis of the $\omega9$, $\omega6$, and $\omega3$ families of polyunsaturated fatty acids.** In animals, the $\omega9$, $\omega6$, and $\omega3$ families of polyunsaturated fatty acids are synthesized in the endoplasmic reticulum from oleic, linoleic, and β-linolenic acids, respectively, by a series of elongation and desaturation reactions. The production of 22:5 $\omega6$ (osbond acid) or 22:6 $\omega3$ (docosahexanoic acid [DHA]), however, requires one cycle of β-oxidation, which takes place inside peroxisomes after the formation of 24:5 $\omega6$ or 24:6 $\omega3$. (AA, arachidonic acid; E, elongase; DS, desaturase; EFA, essential fatty acids; EPA, eicosapentaenoic acid; GLA, γ-linolenic acid; $\ominus$, inhibition.)

glucagon and epinephrine administration, and in the absence of insulin as in type 1 diabetes mellitus.

THE ESSENTIAL FATTY ACIDS (EFA) HAVE IMPORTANT FUNCTIONS IN THE BODY

Rats fed a purified nonlipid diet containing vitamins A and D exhibit a reduced growth rate and reproductive deficiency which may be cured by the addition of **linoleic, α-linolenic,** and **arachidonic acids** to the diet. These fatty acids are found in high concentrations in vegetable oils (see Table 21–2) and in small amounts in animal carcasses. Essential fatty acids are required for prostaglandin, thromboxane, leukotriene, and lipoxin formation (see following discussion). They are found in the structural lipids of the cell, often in the position 2 of phospholipids, and are concerned with the structural integrity of the mitochondrial membrane.

Arachidonic acid is present in membranes and accounts for 5 to 15% of the fatty acids in phospholipids. Docosahexaenoic acid (DHA; ω3, 22:6), which is synthesized to a limited extent from α-linolenic acid or obtained directly from fish oils, is present in high concentrations in retina, cerebral cortex, testis, and sperm. DHA is particularly needed for development of the brain and retina and is supplied via the placenta and milk. Patients with **retinitis pigmentosa** are reported to have low blood levels of DHA. In **essential fatty acid deficiency,** nonessential polyenoic acids of the ω9 family, particularly $\Delta^{5,8,11}$-eicosatrienoic acid (ω9 20:3) (see Figure 23–10), replace the essential fatty acids in phospholipids, other complex lipids, and membranes. The triene:tetraene ratio in plasma lipids can be used to diagnose the extent of essential fatty acid deficiency.

EICOSANOIDS ARE FORMED FROM C$_{20}$ POLYUNSATURATED FATTY ACIDS

Arachidonate and some other C$_{20}$ polyunsaturated fatty acids give rise to **eicosanoids**, physiologically and pharmacologically active compounds known as **prostaglandins (PG), thromboxanes (TX), leukotrienes (LT),** and **lipoxins (LX)** (see Chapter 21). Physiologically, they are considered to act as local hormones functioning through G-protein–linked receptors to elicit their biochemical effects.

There are three groups of eicosanoids that are synthesized from C$_{20}$ eicosanoic acids derived from the essential fatty acids **linoleate** and **α-linolenate,** or directly from dietary arachidonate and eicosapentaenoate (**Figure 23–11**). Arachidonate, which may be obtained from the diet, but is usually derived from the position 2 of phospholipids in the plasma membrane by the action of phospholipase A$_2$ (see Figure 24–5), is the substrate for the synthesis of the PG$_2$, TX$_2$ series (**prostanoids**) by

the **cyclooxygenase pathway**, or the LT$_4$ and LX$_4$ series by the **lipoxygenase pathway,** with the two pathways competing for the arachidonate substrate (see Figure 23–11).

THE CYCLOOXYGENASE PATHWAY IS RESPONSIBLE FOR PROSTANOID SYNTHESIS

Prostanoids (see Chapter 21) are synthesized by the pathway summarized in **Figure 23–12**. In the first reaction, catalyzed by **cyclooxygenase (COX)** (also called **prostaglandin H synthase**), an enzyme that has two activities, a **cyclooxygenase** and **peroxidase,** two molecules of O$_2$ are consumed. COX is present as two isoenzymes, **COX-1** and **COX-2**. The product, an endoperoxide (PGH), is converted to prostaglandins D and E as well as to a thromboxane (TXA$_2$) and prostacyclin (PGI$_2$). Prostanoids are synthesized by many different cell types, but each one produces only one type of prostanoid.

Prostanoids Are Potent, Biologically Active Substances

Thromboxanes are synthesized in platelets and on release cause vasoconstriction and platelet aggregation. Their synthesis is specifically inhibited by low-dose aspirin. **Prostacyclins (PGI$_2$)** are produced by blood vessel walls and are potent inhibitors of platelet aggregation.

Thus, thromboxanes and prostacyclins are antagonistic. PG$_3$ and TX$_3$, formed from eicosapentaenoic acid (EPA), inhibit the release of arachidonate from phospholipids and, therefore, the formation of PG$_2$ and TX$_2$. PGI$_3$ is as potent an antiaggregator of platelets as PGI$_2$, but TXA$_3$ is a weaker aggregator than TXA$_2$, changing the balance of activity and favoring longer clotting times. As little as 1 ng/mL of plasma prostaglandins causes contraction of smooth muscle in animals.

Essential Fatty Acids Do Not Exert All Their Physiologic Effects via Prostaglandin Synthesis

The role of essential fatty acids in membrane formation is unrelated to prostaglandin formation. Prostaglandins do not relieve symptoms of essential fatty acid deficiency, and an essential fatty acid deficiency is not caused by inhibition of prostaglandin synthesis.

Cyclooxygenase Is a "Suicide Enzyme"

"Switching off" of prostaglandin activity is partly achieved by a remarkable property of cyclooxygenase—that of self-catalyzed destruction; that is, it is a "**suicide enzyme**." Furthermore, the inactivation of prostaglandins by **15-hydroxyprostaglandin dehydrogenase** is rapid. Blocking the action of this enzyme

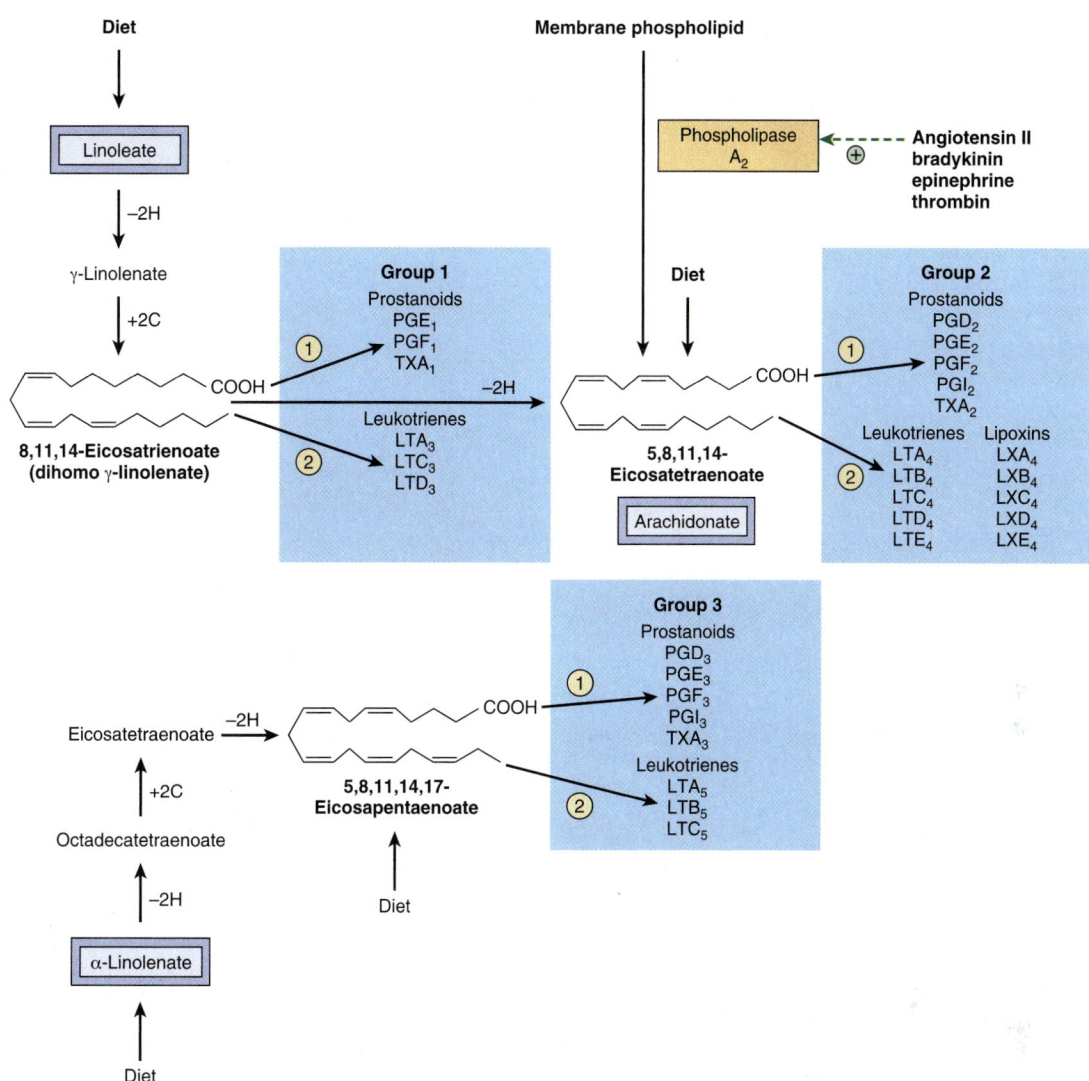

FIGURE 23–11 **The three groups of eicosanoids and their biosynthetic origins.** ①, cyclooxygenase pathway; ②, lipoxygenase pathway; LT, leukotriene; LX, lipoxin; PG, prostaglandin; PGI, prostacyclin; TX, thromboxane. The subscript denotes the total number of double bonds in the molecule and the series to which the compound belongs.

with sulfasalazine or indomethacin can prolong the half-life of prostaglandins in the body.

LEUKOTRIENES & LIPOXINS ARE FORMED BY THE LIPOXYGENASE PATHWAY

The **leukotrienes** are a family of conjugated trienes formed from eicosanoic acids in leukocytes, mastocytoma cells, platelets, and macrophages by the **lipoxygenase pathway** in response to both immunologic and nonimmunologic stimuli. Three different lipoxygenases (dioxygenases) insert oxygen into the 5, 12, and 15 positions of arachidonic acid, giving rise to hydroperoxides (HPETE). Only **5-lipoxygenase** forms leukotrienes (details in **Figure 23–13**). **Lipoxins** are a family

of conjugated tetraenes also arising in leukocytes. They are formed by the combined action of more than one lipoxygenase (see Figure 23–13).

CLINICAL ASPECTS

Essential Fatty Acids Play a Role in the Prevention of Diseases in Humans

Studies have shown that dietary intake of essential fatty acids is associated with a reduction in the development of a number of human diseases including cardiovascular disease, cancer, arthritis, diabetes mellitus, and various neurologic conditions. Various mechanisms are believed to be involved, including changes in gene expression, prostanoid production, and membrane composition.

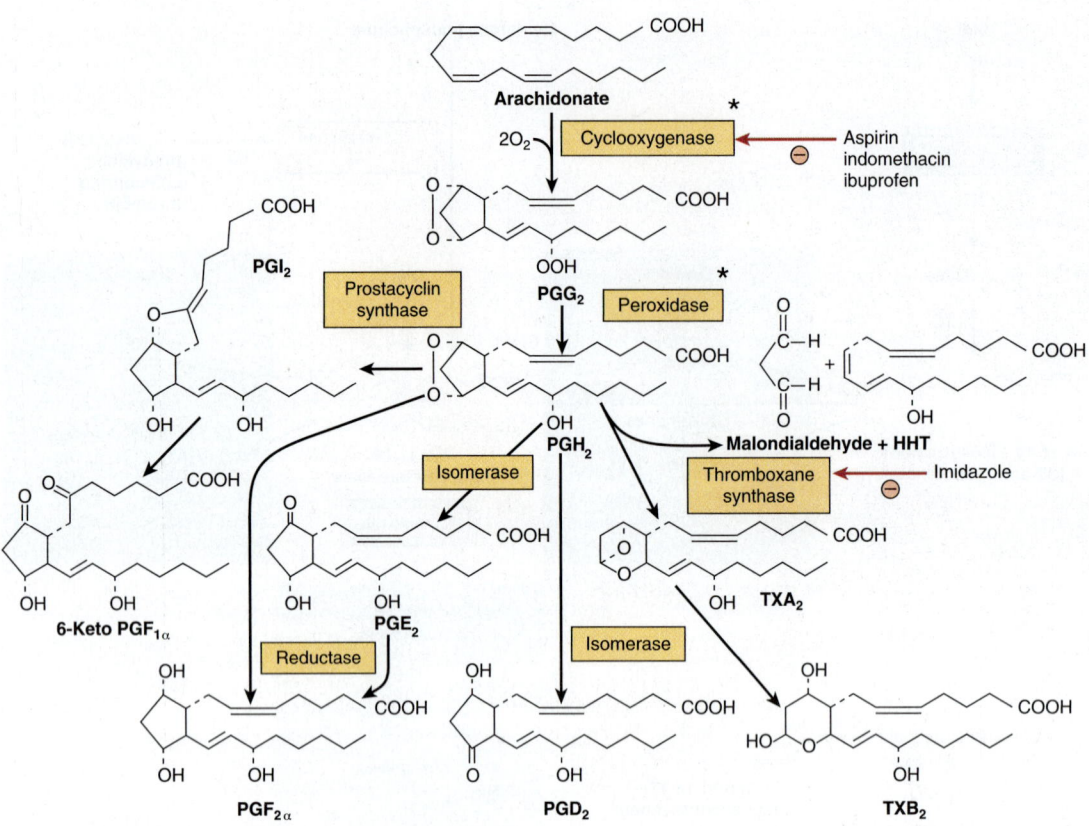

FIGURE 23–12 **Conversion of arachidonic acid to prostaglandins and thromboxanes of series 2.** HHT, hydroxyheptadecatrienoate; PG, prostaglandin; PGI, prostacyclin; TX, thromboxane. *Both of these starred activities are attributed to the cyclooxygenase enzyme (prostaglandin H synthase). Similar conversions occur in prostaglandins and thromboxanes of series 1 and 3.

Symptoms of Essential Fatty Acid Deficiency in Humans Include Skin Lesions & Impairment of Lipid Transport

In adults subsisting on ordinary diets, no signs of essential fatty acid deficiencies have been reported. However, infants receiving formula diets low in fat and patients maintained for long periods exclusively by intravenous nutrition low in essential fatty acids show deficiency symptoms that can be prevented by an essential fatty acid intake of 1 to 2% of the total caloric requirement.

Abnormal Metabolism of Essential Fatty Acids Occurs in Several Diseases

Abnormal metabolism of essential fatty acids, which may be connected with dietary insufficiency, has been noted in cystic fibrosis, acrodermatitis enteropathica, hepatorenal syndrome, Sjögren-Larsson syndrome, multisystem neuronal degeneration, Crohn disease, cirrhosis and alcoholism, and Reye syndrome. Elevated levels of very-long-chain polyenoic acids have been found in the brains of patients with Zellweger syndrome (see Chapter 22). Diets with a high P:S (polyunsaturated:saturated fatty acid) ratio reduce serum cholesterol levels and are considered

to be beneficial in terms of the risk of development of coronary heart disease.

Trans Fatty Acids Are Implicated in Various Disorders

Small amounts of trans-unsaturated fatty acids (see Chapter 21) are found in ruminant fat (eg, butter fat has 2-7%), where they arise from the action of microorganisms in the rumen, but the main source in the human diet is from partially hydrogenated vegetable oils (eg, margarine). Trans fatty acids compete with essential fatty acids and may exacerbate essential fatty acid deficiency. Moreover, they are structurally similar to saturated fatty acids (see Chapter 21) and have comparable effects in the promotion of hypercholesterolemia and atherosclerosis (see Chapter 26).

Nonsteroidal Anti-Inflammatory Drugs Inhibit COX

Aspirin is a nonsteroidal anti-inflammatory drug (NSAID) that inhibits COX-1 and COX-2. Other NSAIDs include **indomethacin** and **ibuprofen**, and these usually inhibit cyclooxygenases by competing with arachidonate. Since inhibition of COX-1 causes the stomach irritation often associated with taking NSAIDs, attempts have been made to develop drugs

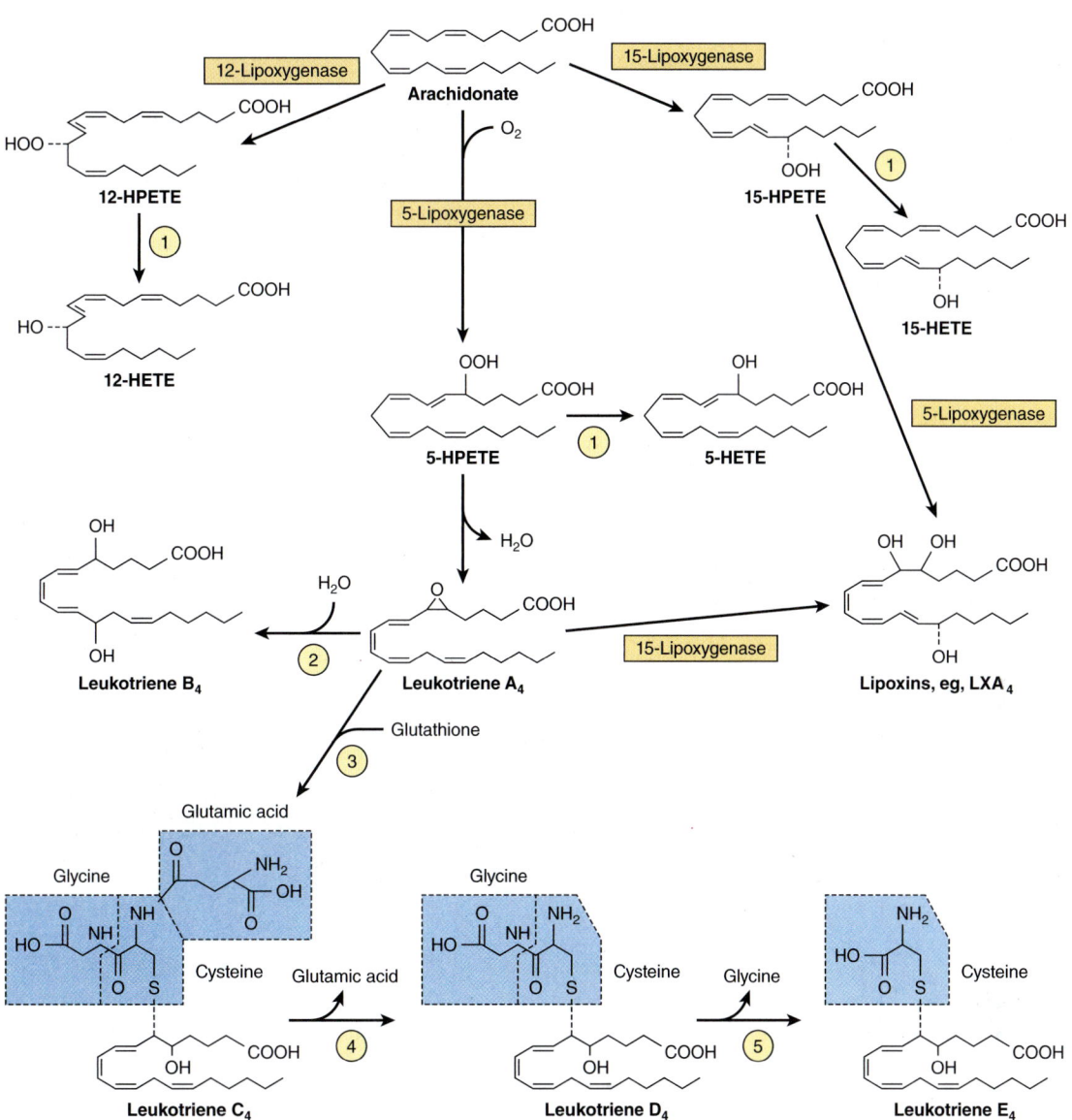

FIGURE 23–13 **Conversion of arachidonic acid to leukotrienes and lipoxins of series 4 via the lipoxygenase pathway.** Some similar conversions occur in series 3 and 5 leukotrienes. ①, peroxidase; ②, leukotriene A₄ epoxide hydrolase; ③, glutathione S-transferase; ④, γ-glutamyltranspeptidase; ⑤, cysteinyl-glycine dipeptidase; HETE, hydroxyeicosatetraenoate; HPETE, hydroperoxyeicosatetraenoate.

that selectively inhibit COX-2 (**coxibs**). Unfortunately, however, the success of this approach has been limited and some coxibs have been withdrawn or suspended from the market due to undesirable side effects and safety issues. Transcription of COX-2—but not of COX-1—is completely inhibited by **anti-inflammatory corticosteroids**.

Prostanoids May Be Used Therapeutically

Potential therapeutic uses of prostanoids include prevention of conception, induction of labor at term, termination of pregnancy, prevention or alleviation of gastric ulcers, control of inflammation and of blood pressure, and relief of asthma and nasal congestion. In addition, PGD_2 is a potent sleep-promoting substance. Prostaglandins increase cAMP in platelets,

thyroid, corpus luteum, fetal bone, adenohypophysis, and lung but reduce cAMP in renal tubule cells and adipose tissue (see Chapter 25).

Leukotrienes & Lipoxins Are Potent Regulators of Many Disease Processes

Slow-reacting substance of anaphylaxis (**SRS-A**) is a mixture of leukotrienes C_4, D_4, and E_4. This mixture of leukotrienes is a potent constrictor of the bronchial airway musculature. These leukotrienes together with **leukotriene B₄** also cause vascular permeability and attraction and activation of leukocytes and are important regulators in many diseases involving inflammatory or immediate hypersensitivity reactions, such as asthma. Leukotrienes are vasoactive, and 5-lipoxygenase has been found in arterial walls. Evidence supports an anti-inflammatory role for lipoxins in vasoactive and immunoregulatory function, for example, as counterregulatory compounds (**chalones**) of the immune response.

SUMMARY

- The synthesis of long-chain fatty acids (lipogenesis) is carried out by two enzyme systems: acetyl-CoA carboxylase and fatty acid synthase.

- The pathway converts acetyl-CoA to palmitate and requires NADPH, ATP, Mn^{2+}, biotin, and pantothenic acid as cofactors.

- Acetyl-CoA carboxylase converts acetyl-CoA to malonyl-CoA, and then fatty acid synthase, a multienzyme complex consisting of two identical polypeptide chains, each containing six separate enzymatic activities and ACP, catalyzes the formation of palmitate from one acetyl-CoA and seven malonyl-CoA molecules.

- Lipogenesis is regulated at the acetyl-CoA carboxylase step by allosteric modifiers, phosphorylation/dephosphorylation, and induction and repression of enzyme synthesis. The enzyme is allosterically activated by citrate and deactivated by long-chain acyl-CoA. Dephosphorylation (eg, by insulin) promotes its activity, while phosphorylation (eg, by glucagon or epinephrine) is inhibitory.

- Biosynthesis of unsaturated long-chain fatty acids is achieved by desaturase and elongase enzymes, which introduce double bonds and lengthen existing acyl chains, respectively.

- Higher animals have Δ^4, Δ^5, Δ^6, and Δ^9 desaturases but cannot insert new double bonds beyond the position 9 of fatty acids. Thus, the essential fatty acids linoleic (ω6) and α-linolenic (ω3) must be obtained from the diet.

- Eicosanoids are derived from C_{20} (eicosanoic) fatty acids synthesized from the essential fatty acids and make up important groups of physiologically and pharmacologically active compounds, including the prostaglandins, thromboxanes, leukotrienes, and lipoxins.

REFERENCES

Eljamil AS: *Lipid Biochemistry: For Medical Sciences*. iUniverse, 2015.

Smith WL, Murphy RC: The eicosanoids: cyclooxygenase, lipoxygenase, and epoxygenase pathways. In *Biochemistry of Lipids, Lipoproteins and Membranes*, 6th ed. Ridgway N, McLeod R (editors). Academic Press, 2015:260-296.

Metabolism of Acylglycerols & Sphingolipids

24

Kathleen M. Botham, PhD, DSc, & Peter A. Mayes, PhD, DSc

O B J E C T I V E S

After studying this chapter, you should be able to:

- Explain that the catabolism of triacylglycerols involves hydrolysis to free fatty acids and glycerol and indicate the fate of these metabolites.
- Indicate that glycerol-3-phosphate is the substrate for the formation of both triacylglycerols and phosphoglycerols and that a branch point at phosphatidate leads to the synthesis of inositol phospholipids and cardiolipin or/and triacylglycerols and other phospholipids.
- Explain that plasmalogens and platelet-activating factor (PAF) are formed by a complex pathway starting from dihydroxyacetone phosphate.
- Illustrate the role of various phospholipases in the degradation and remodeling of phospholipids.
- Explain that ceramide is the precursor from which all sphingolipids are formed.
- Indicate how sphingomyelin and glycosphingolipids are produced by the reaction of ceramide with phosphatidylcholine or sugar residue(s), respectively.
- Identify examples of disease processes caused by defects in phospholipid or sphingolipid synthesis or breakdown.

BIOMEDICAL IMPORTANCE

Acylglycerols constitute the majority of lipids in the body. Triacylglycerols are the major lipids in fat deposits and in food, and their roles in lipid transport and storage and in various diseases such as obesity, diabetes, and hyperlipoproteinemia will be described in subsequent chapters. The amphipathic nature of phospholipids and sphingolipids makes them ideally suitable as the main lipid component of cell membranes.

Phospholipids also take part in the metabolism of many other lipids. Some phospholipids have specialized functions; for example, dipalmitoyl lecithin is a major component of **lung surfactant**, which is lacking in **respiratory distress syndrome** of the newborn. Inositol phospholipids in the cell membrane act as precursors of **hormone second messengers**, and **platelet-activating factor** (PAF), an alkylphospholipid, functions in the mediation of inflammation. Glycosphingolipids, which contain sphingosine and sugar residues as well as a fatty acid, are found in the outer leaflet of the plasma membrane with their oligosaccharide chains facing outward. They form part of the **glycocalyx** of the cell surface and are important (1) in cell adhesion and cell recognition, (2) as receptors for bacterial toxins (eg, the toxin that causes cholera), and (3) as ABO blood group substances. A dozen or so **glycolipid storage diseases** have been described (eg, Gaucher disease and Tay-Sachs disease), each due to a genetic defect in the pathway for glycolipid degradation in the lysosomes.

HYDROLYSIS INITIATES CATABOLISM OF TRIACYLGLYCEROLS

Triacylglycerols must be hydrolyzed by a **lipase** to their constituent fatty acids and glycerol before further catabolism can proceed. Much of this hydrolysis (lipolysis) occurs in adipose tissue with release of free fatty acids into the plasma, where they are found combined with serum albumin (see Figure 25–7). This is followed by free fatty acid uptake into tissues (including liver, heart, kidney, muscle, lung, testis, and adipose tissue, but not readily by brain), where they are oxidized to obtain energy or reesterified. The utilization of glycerol depends on whether such tissues have the enzyme **glycerol kinase**, which is found in significant amounts in liver, kidney, intestine, brown adipose tissue, and the lactating mammary gland.

TRIACYLGLYCEROLS & PHOSPHOGLYCEROLS ARE FORMED BY ACYLATION OF TRIOSE PHOSPHATES

The major pathways of triacylglycerol and phosphoglycerol biosynthesis are outlined in **Figure 24–1**. Important substances such as triacylglycerols, phosphatidylcholine, phosphatidylethanolamine, phosphatidylinositol, and cardiolipin, a constituent of mitochondrial membranes, are formed from **glycerol-3-phosphate**. Significant branch points in the pathway occur at the **phosphatidate** and **diacylglycerol** steps. **Dihydroxyacetone phosphate** is the precursor for phosphoglycerols containing an ether link (—C—O—C—), the best known of which are plasmalogens and PAF. Both glycerol-3-phosphate and dihydroxyacetone phosphate are intermediates in glycolysis, making a very important connection between carbohydrate and lipid metabolism (see Chapter 14).

Both the glycerol and fatty acid components must be activated by ATP before they can be incorporated into acylglycerols. **Glycerol kinase** catalyzes the activation of glycerol to *sn*-glycerol-3-phosphate. If the activity of this enzyme is absent or low, as in muscle or adipose tissue, most of the glycerol-3-phosphate is formed from dihydroxyacetone phosphate by **glycerol-3-phosphate dehydrogenase** (**Figure 24–2**).

Phosphatidate Is the Common Precursor in the Biosynthesis of Triacylglycerols, Many Phosphoglycerols, & Cardiolipin

Biosynthesis of Triacylglycerols

Acyl-CoA, formed by the activation of fatty acids by **acyl-CoA synthetase** (see Chapter 22), combines with glycerol-3-phosphate in two successive reactions catalyzed by **glycerol-3-phosphate acyltransferase** and **1-acylglycerol-3-phosphate acyltransferase**, resulting in the formation of 1,2-diacylglycerol phosphate or

phosphatidate (see Figure 24–2). In the triacylglycerol branch of the pathway, phosphatidate is converted by **phosphatidate phosphohydrolase** (also called **phosphatidate phosphatase** [PAP]) to 1,2-diacylglycerol which is then converted to triacylglycerol by **diacylglycerol acyltransferase (DGAT)**. **Lipins**, a family of three proteins, have phosphatidate phosphohydrolase activity and they also act as transcription factors which regulate the expression of genes involved in lipid metabolism. DGAT catalyzes the only step specific for triacylglycerol synthesis and is thought to be rate limiting in most circumstances. In intestinal mucosa, **monoacylglycerol acyltransferase** converts **monoacylglycerol** to 1,2-diacylglycerol in the **monoacylglycerol pathway**. Most of the activity of these enzymes resides in the endoplasmic reticulum, but some is found in mitochondria. Although phosphatidate phosphohydrolase protein is found mainly in the cytosol, the active form of the enzyme is membrane bound.

Biosynthesis of Phospholipids

The biosynthesis of **phosphatidylcholine** and **phosphatidylethanolamine** also involves the conversion of phosphatidate to 1,2-diacylglycerol (see Figure 24–2). Before the next step, however, choline or ethanolamine must first be activated by phosphorylation by ATP and then linked to CDP. The resulting CDP-choline or CDP-ethanolamine reacts with 1,2-diacylglycerol to form either phosphatidylcholine or phosphatidylethanolamine, respectively (see Figure 24–2, lower left). In the case of **phosphatidylinositol** formation, it is phosphatidate that is linked to CDP, forming CDP-diacylglycerol, which is then linked to inositol by **phosphatidylinositol synthase**. The second messenger, **phosphatidylinositol 4,5 bisphosphate (PIP2)** (see Chapter 21), which regulates essential cell functions including signal transduction and vesicle trafficking, is synthesized by two further kinase catalyzed steps (see Figure 24–2, lower right). **Phosphatidylserine** is formed from phosphatidylethanolamine directly by reaction with serine. Phosphatidylserine may reform phosphatidylethanolamine by decarboxylation (see Figure 24–2, bottom left). An alternative pathway in liver enables phosphatidylethanolamine to give rise directly to phosphatidylcholine by progressive methylation of the ethanolamine residue. Despite these endogenous sources of choline, it is considered to be an essential nutrient for humans and many other mammalian species.

The regulation of triacylglycerol, phosphatidylcholine, and phosphatidylethanolamine biosynthesis is driven by the availability of free fatty acids. Those not required for oxidation are preferentially converted to phospholipids, and when this requirement is satisfied, they are used for triacylglycerol synthesis.

Cardiolipin (diphosphatidylglycerol; see Figure 21–9) is a phospholipid present in mitochondria. It is formed from phosphatidylglycerol, which in turn is synthesized from CDP-diacylglycerol (see Figure 24–2) and glycerol-3-phosphate. Cardiolipin, found in the inner membrane of mitochondria, has a key role in mitochondrial structure and function, and is also thought to be involved in programmed cell death (**apoptosis**).

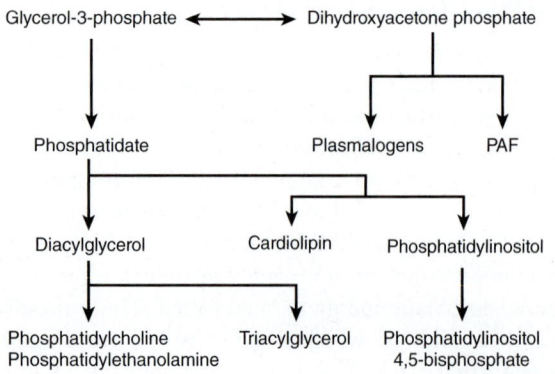

FIGURE 24–1 **Overview of acylglycerol biosynthesis.** (PAF, platelet-activating factor.)

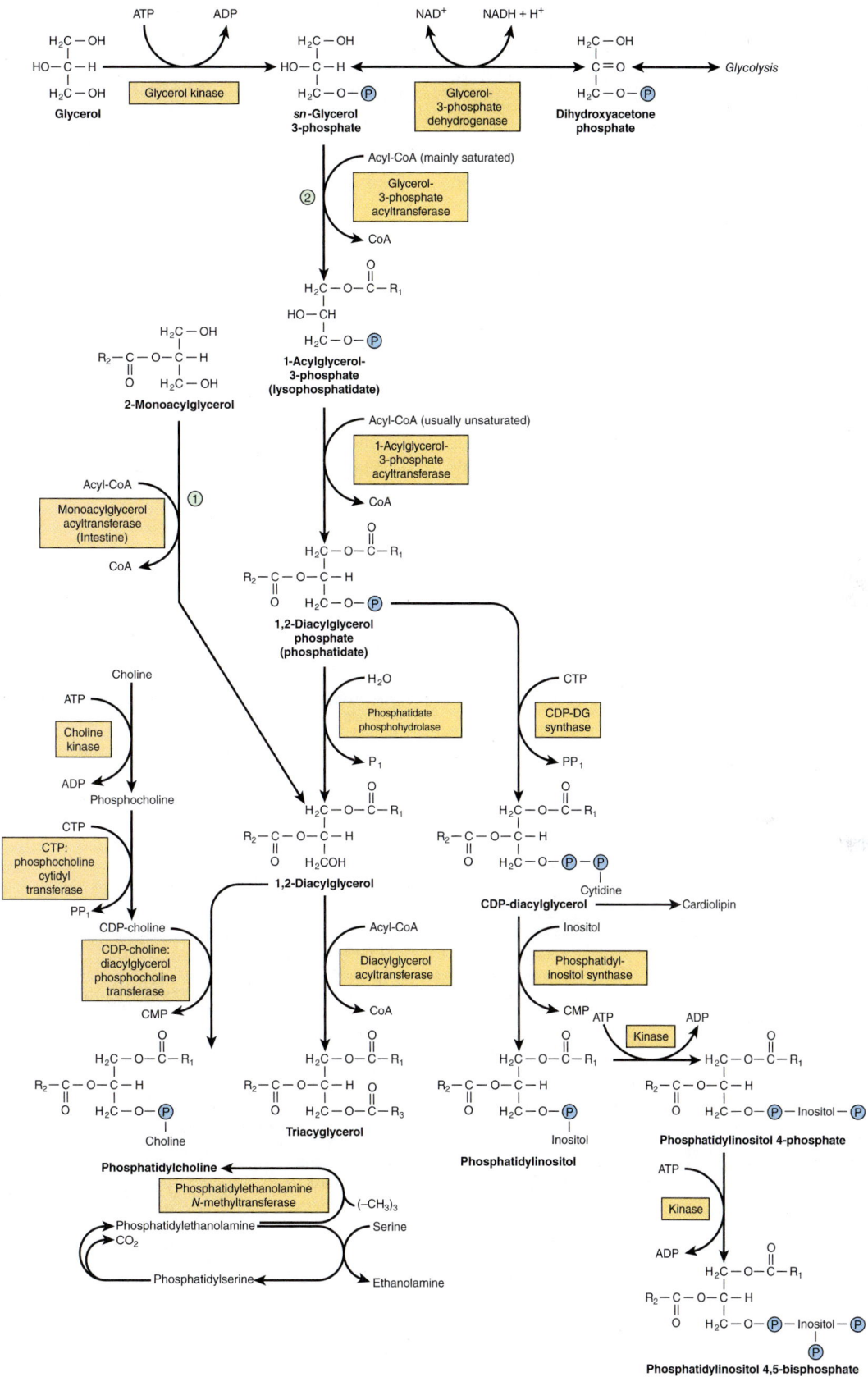

FIGURE 24–2 **Biosynthesis of triacylglycerol and phospholipids.** ①, monoacylglycerol pathway; ②, glycerol phosphate pathway. Phosphatidylethanolamine may be formed from ethanolamine by a pathway similar to that shown for the formation of phosphatidylcholine from choline.

Biosynthesis of Glycerol Ether Phospholipids

In **glycerol ether phospholipids**, one or more of the glycerol carbons is attached to a hydrocarbon chain by an ether linkage rather than an ester bond. **Plasmalogens and PAF** are important examples of this type of lipid. The biosynthetic pathway is located in peroxisomes. The precursor dihydroxyacetone phosphate combines with acyl-CoA to give 1-acyldihydroxyacetone phosphate, and the ether link is formed in the next reaction, producing 1-alkyldihydroxyacetone phosphate, which is then converted to 1-alkylglycerol 3-phosphate (**Figure 24–3**). After further acylation in the 2 position, the resulting 1-alkyl-2-acylglycerol 3-phosphate (analogous to phosphatidate in Figure 24–2) is hydrolyzed to give 1-alkyl-2-acylglycerol. As in the pathway for phospholipid biosynthesis (see Figure 24–2), the next steps in the formation of plasmalogens and PAF require CDP-ethanolamine and CDP-choline, respectively. Plasmalogens, which comprise much of the phospholipid in mitochondria, are produced by desaturation of the resulting 3-phosphoethanolamine derivative (see Figure 24–3), while PAF (1-alkyl-2-acetyl-*sn*-glycerol-3-phosphocholine) is synthesized by acetylation of the corresponding 3-phosphocholine derivative. Production of PAF occurs in many cell types, but particularly leukocytes and endothelial cells. It was recognized initially for its platelet aggregating properties, but is now known to mediate inflammation, and can contribute to allergic reactions and shock.

Phospholipases Allow Degradation & Remodeling of Phosphoglycerols

Although phospholipids are actively degraded, each portion of the molecule turns over at a different rate—for example, the turnover time of the phosphate group is different from that of the 1-acyl group. This is due to the presence of enzymes that allow partial degradation followed by resynthesis (**Figure 24–4**). **Phospholipase A$_2$** catalyzes the hydrolysis of

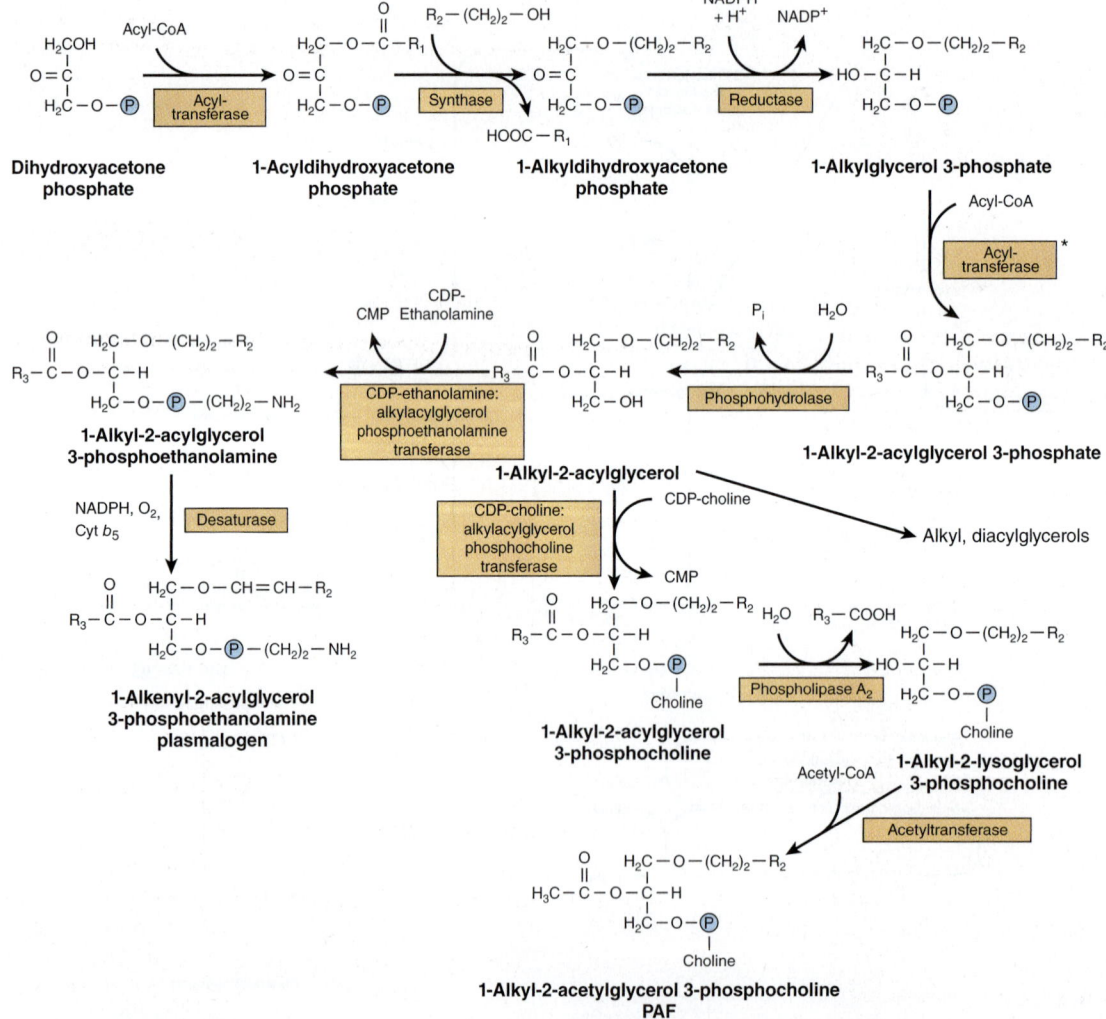

FIGURE 24–3 **Biosynthesis of ether lipids, including plasmalogens, and platelet-activating factor (PAF).** In the de novo pathway for PAF synthesis, acetyl-CoA is incorporated at stage*, avoiding the last two steps in the pathway shown here.

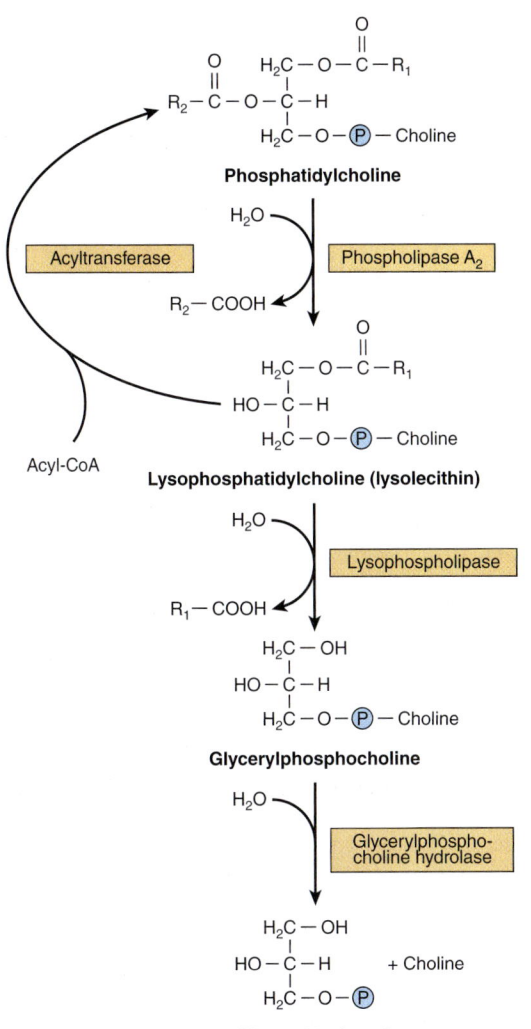

FIGURE 24–4 Metabolism of phosphatidylcholine (lecithin).

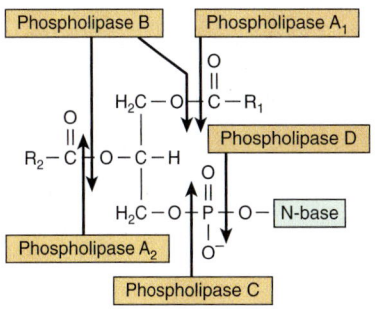

FIGURE 24–5 Sites of the hydrolytic activity of phospholipases on a phospholipid substrate.

glycerophospholipids to form a free fatty acid and lysophospholipid, which in turn may be reacylated by acyl-CoA in the presence of an acyltransferase. Alternatively, lysophospholipid (eg, lysolecithin) is attacked by **lysophospholipase**, forming the corresponding glyceryl phosphoryl base (choline is shown as an example base in Figure 24–4), which may then be split by a hydrolase liberating glycerol-3-phosphate plus base. **Phospholipases A$_1$, A$_2$, B, C,** and **D** attack the bonds indicated in **Figure 24–5**. **Phospholipase A$_2$** is found in pancreatic fluid and snake venom as well as in many types of cells; **phospholipase C** is one of the major toxins secreted by bacteria; and **phospholipase D** is known to be involved in mammalian signal transduction.

Lysophosphatidylcholine, also called lysolecithin, may be formed by an alternative route that involves **lecithin (phosphatidylcholine): cholesterol acyltransferase (LCAT).** This enzyme, found in plasma, catalyzes the transfer of a fatty acid residue from the 2 position of lecithin to cholesterol to form cholesteryl ester and lysolecithin, and is considered to be responsible for much of the cholesteryl ester in plasma lipoproteins (see Chapter 25).

Long-chain saturated fatty acids are found predominantly in the 1 position of phospholipids, whereas the polyunsaturated fatty acids (eg, the precursors of prostaglandins) are incorporated more frequently into the 2 position. The incorporation of fatty acids into phosphatidylcholine occurs in three ways; by complete synthesis of the phospholipid (see Figure 24–4); by transacylation between cholesteryl ester and lysophosphatidylcholine; and by direct acylation of lysophosphatidylcholine by acyl-CoA. Thus, a continuous exchange of the fatty acids is possible, particularly with regard to introducing essential fatty acids (see Chapter 21) into phospholipid molecules.

ALL SPHINGOLIPIDS ARE FORMED FROM CERAMIDE

Ceramide (see Chapter 21) is synthesized in the endoplasmic reticulum from the amino acid serine as shown in **Figure 24–6**. Ceramide is an important signaling molecule (second messenger) regulating pathways including programmed cell death (**apoptosis**), the **cell cycle**, and **cell differentiation** and **senescence**.

Sphingomyelins (see Figure 21–10) are phospholipids and are formed when ceramide reacts with phosphatidylcholine to form sphingomyelin plus diacylglycerol (**Figure 24–7A**). This occurs mainly in the Golgi apparatus and to a lesser extent in the plasma membrane.

Glycosphingolipids Are a Combination of Ceramide With One or More Sugar Residues

The simplest glycosphingolipids (**cerebrosides**) are **galactosylceramide (GalCer)** (see Figure 21–14) and **glucosylceramide (GlcCer)**. GalCer is a major lipid of **myelin**, whereas GlcCer is the major glycosphingolipid of **extraneural tissues** and a precursor of most of the more complex glycosphingolipids. GalCer (**Figure 24–7B**) is formed in a reaction between ceramide and uridine diphosphate galactose (UDPGal) (formed by epimerization from UDPGlc; see Figure 20–6).

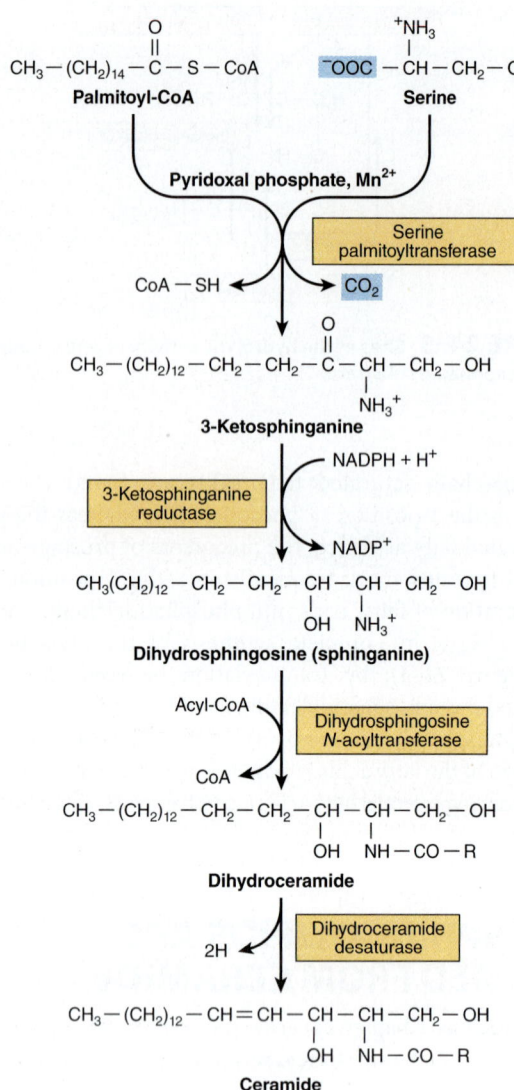

FIGURE 24–6 Biosynthesis of ceramide.

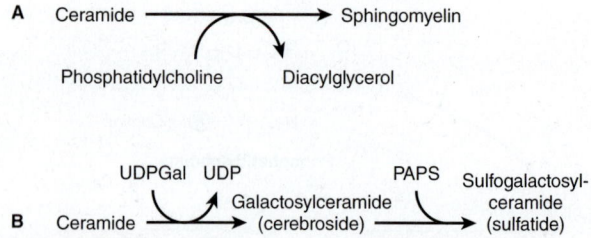

FIGURE 24–7 Biosynthesis of **(A) sphingomyelin, (B) galacto-sylceramide and its sulfo derivative.** (PAPS, "active sulfate," adenosine 3'-phosphate-5'-phosphosulfate.)

and are synthesized from ceramide by the stepwise addition of activated sugars (eg, UDPGlc and UDPGal) and a **sialic acid,** usually *N*-acetylneuraminic acid (**Figure 24–8**). A large number of gangliosides of increasing molecular weight may be formed. Most of the enzymes transferring sugars from nucleotide sugars (glycosyl transferases) are found in the Golgi apparatus. Certain gangliosides function as receptors for bacterial toxins (eg, for **cholera toxin,** which subsequently activates adenylyl cyclase).

Glycosphingolipids are constituents of the outer leaflet of plasma membranes and are important in **cell adhesion, cell recognition,** and signal transduction. Some are antigens, for example, ABO blood group substances.

CLINICAL ASPECTS

Deficiency of Lung Surfactant Causes Respiratory Distress Syndrome

Lung surfactant is composed mainly of lipid with some proteins and carbohydrate and prevents the alveoli from collapsing. The phospholipid **dipalmitoyl-phosphatidylcholine** decreases surface tension at the air-liquid interface and thus greatly reduces the work of breathing, but other surfactant lipid and protein components are also important in surfactant function. Deficiency of lung surfactant in the lungs of many preterm newborns gives

Sulfogalactosylceramide (sulfatide), a component of the myelin sheath, is formed by a further reaction involving 3'-phosphoadenosine-5'-phosphosulfate (PAPS; "active sulfate"). **Gangliosides** are found in cell membranes (see Chapter 40),

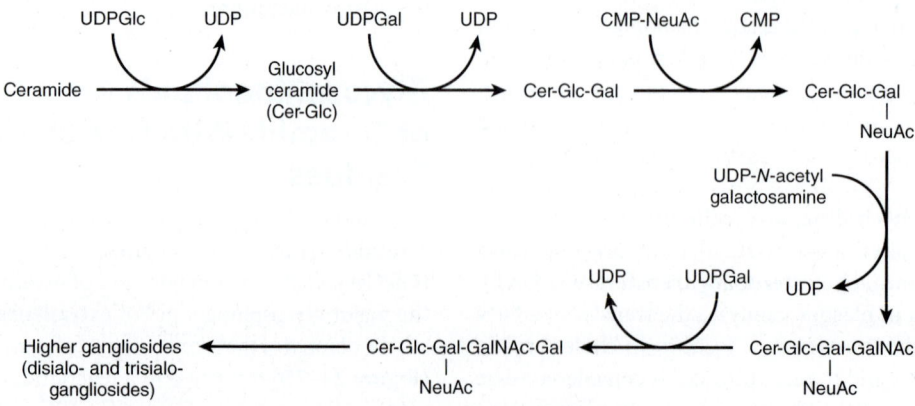

FIGURE 24–8 Biosynthesis of gangliosides. (NeuAc, *N*-acetylneuraminic acid.)

TABLE 24–1 Examples of Sphingolipidoses

Disease	Enzyme Deficiency	Lipid Accumulating	Clinical Symptoms
Tay-Sachs disease	Hexosaminidase A, S	Cer—Glc—Gal(NeuAc)┼GalNAc G_{M2} Ganglioside	Mental retardation, blindness, muscular weakness
Fabry disease	α-Galactosidase	Cer—Glc—Gal—┼Gal Globotriaosylceramide	Skin rash, kidney failure (full symptoms only in males; X-linked recessive)
Metachromatic leukodystrophy	Arylsulfatase A	Cer—Gal—┼OSO$_3$ 3-Sulfogalactosylceramide	Mental retardation and psychologic disturbances in adults; demyelination
Krabbe disease	β-Galactosidase	Cer—┼Gal Galactosylceramide	Mental retardation; myelin almost absent
Gaucher disease	β-Glucosidase	Cer—┼Glc Glucosylceramide	Enlarged liver and spleen, erosion of long bones, mental retardation in infants
Niemann-Pick disease types A, B	Sphingomyelinase	Cer—┼P—choline Sphingomyelin	Enlarged liver and spleen, mental retardation; fatal in early life
Farber disease	Ceramidase	Acyl—┼Sphingosine Ceramide	Hoarseness, dermatitis, skeletal deformation, mental retardation; fatal in early life

Abbreviations: Cer, ceramide; Gal, galactose; Glc, glucose; NeuAc, *N*-acetylneuraminic acid; ┼, site of deficient enzyme reaction.

rise to **infant respiratory distress syndrome (IRDS)**. Administration of either natural or artificial surfactant is of therapeutic benefit.

Phospholipids & Sphingolipids Are Involved in Multiple Sclerosis & Lipidoses

Certain diseases are characterized by abnormal quantities of these lipids in the tissues, often in the nervous system. They may be classified into two groups: (1) true demyelinating diseases and (2) sphingolipidoses.

In **multiple sclerosis**, which is a demyelinating disease, there is loss of both phospholipids (particularly ethanolamine plasmalogen) and of sphingolipids from white matter. Thus, the lipid composition of white matter resembles that of gray matter. The cerebrospinal fluid shows raised phospholipid levels.

The **sphingolipidoses (lipid storage diseases)** are a group of inherited diseases that are caused by a genetic defect in the catabolism of lipids containing sphingosine. They are part of a larger group of lysosomal disorders and exhibit several constant features: (1) complex lipids containing ceramide accumulate in cells, particularly neurons, causing neurodegeneration and shortening the life span. (2) The rate of **synthesis** of the stored lipid is normal. (3) The enzymatic defect is in the **degradation pathway** of sphingolipids in lysosomes. (4) The extent to which the activity of the affected enzyme is decreased is similar in all tissues. There is no effective treatment for many of the diseases, although some success has been achieved with **enzyme replacement therapy** and **bone marrow transplantation** in the treatment of Gaucher and Fabry diseases. Other promising approaches are **substrate deprivation therapy** to inhibit the synthesis of sphingolipids and **chemical chaperone therapy**.

Gene therapy for lysosomal disorders is also currently under investigation. Some examples of the more important lipid storage diseases are shown in **Table 24–1**.

Multiple sulfatase deficiency results in accumulation of sulfogalactosylceramide, steroid sulfates, and proteoglycans owing to a combined deficiency of arylsulfatases A, B, and C and steroid sulfatase. Symptoms include abnormalities in neurologic and metabolic functions, as well as in hearing, sight, and bone. **Metachromatic leukodystrophy** is characterized by a build-up of sulfatides in tissues caused by a defect in arylsulfatase A, and leads to irreversible damage to the myelin sheath.

SUMMARY

- Triacylglycerols and some phosphoglycerols are synthesized by progressive acylation of glycerol-3-phosphate. The pathway bifurcates at phosphatidate, forming inositol phospholipids and cardiolipin on the one hand and triacylglycerol and choline and ethanolamine phospholipids on the other.

- Plasmalogens and PAF are ether phospholipids formed from dihydroxyacetone phosphate.

- Sphingolipids are formed from ceramide (*N*-acylsphingosine). Sphingomyelin is present in membranes of organelles involved in secretory processes (eg, Golgi apparatus). The simplest glycosphingolipids are a combination of ceramide plus a sugar residue (eg, GalCer in myelin). Gangliosides are more complex glycosphingolipids containing more sugar residues plus sialic acid. They are present in the outer layer of the plasma membrane, where they contribute to the glycocalyx and are important as antigens and cell receptors.

- Phospholipids and sphingolipids are involved in several disease processes, including infant respiratory distress syndrome (lack of lung surfactant), multiple sclerosis (demyelination), and sphingolipidoses (inability to break down sphingolipids in lysosomes due to inherited defects in hydrolase enzymes).

REFERENCES

Eljamil AS: *Lipid Biochemistry: For Medical Sciences*. iUniverse, 2015.

Futerman AH: Sphingolipids. In *Biochemistry of Lipids, Lipoproteins and Membranes*, 6th ed. Ridgway N, McLeod R (editors). Academic Press, 2015:297-327.

Ridgway ND: Phospholipid synthesis in mammalian cells. In *Biochemistry of Lipids, Lipoproteins and Membranes*, 6th ed. Academic Press, 2015:210-236.

Lipid Transport & Storage

Kathleen M. Botham, PhD, DSc, & Peter A. Mayes, PhD, DSc

OBJECTIVES

After studying this chapter, you should be able to:

- Identify the four major groups of plasma lipoproteins and the four major lipid classes they carry.
- Illustrate the structure of a lipoprotein particle.
- Indicate the major types of apolipoprotein found in the different lipoprotein classes.
- Explain that triacylglycerol from the diet is carried to the liver in chylomicrons and from the liver to extrahepatic tissues in very-low-density lipoprotein (VLDL), and that these particles are synthesized in intestinal and liver cells, respectively, by similar processes.
- Illustrate the processes by which chylomicrons are metabolized by lipases to form chylomicron remnants, which are then removed from the circulation by the liver.
- Explain how VLDL is metabolized by lipases to intermediate-density lipoprotein (IDL) which may be cleared by the liver or converted to low-density lipoprotein (LDL), which functions to deliver cholesterol from the liver to extrahepatic tissues via the LDL (apoB100, E) receptor.
- Explain how high-density lipoprotein (HDL) is synthesized, indicate the mechanisms by which it accepts cholesterol from extrahepatic tissues and returns it to the liver in reverse cholesterol transport.
- Describe how the liver plays a central role in lipid transport and metabolism and how hepatic VLDL secretion is regulated by the diet and hormones.
- Indicate the roles of LDL and HDL in promoting and retarding, respectively, the development of atherosclerosis.
- Indicate the causes of alcoholic and nonalcoholic fatty liver disease (NAFLD).
- Explain the processes by which fatty acids are released from triacylglycerol stored in adipose tissue.
- Understand the role of brown adipose tissue in the generation of body heat.

BIOMEDICAL IMPORTANCE

Fat absorbed from the diet and lipids synthesized by the liver and adipose tissue must be transported between the various tissues and organs for utilization and storage. Since lipids are insoluble in water, the problem of how to transport them in the aqueous blood plasma is solved by associating nonpolar lipids (triacylglycerol and cholesteryl esters) with amphipathic lipids (phospholipids and cholesterol) and proteins to make **water-miscible lipoproteins**.

In a meal-eating omnivore such as the human, excess calories are ingested in the anabolic phase of the feeding cycle, followed by a period of negative caloric balance when the organism draws on its carbohydrate and fat stores. Lipoproteins mediate this cycle by transporting lipids from the intestines as **chylomicrons**—and from the liver as **very-low-density lipoproteins (VLDL)**—to most tissues for oxidation and to adipose tissue for storage. Lipid is mobilized from adipose tissue as free fatty acids (FFAs) bound to serum albumin. Abnormalities of lipoprotein metabolism cause various **hypo-** or **hyperlipoproteinemias**. The most common of these is in **diabetes mellitus**, where insulin deficiency causes excessive mobilization of FFA and underutilization of chylomicrons and VLDL, leading to **hypertriacylglycerolemia**. Most other pathologic conditions affecting lipid transport are due primarily to inherited defects, some of which cause **hypercholesterolemia** and premature **atherosclerosis** (see Table 26–1). **Obesity**—particularly abdominal obesity—is a risk factor for increased mortality, hypertension, Type 2 diabetes mellitus, hyperlipidemia, hyperglycemia, and various endocrine dysfunctions.

LIPIDS ARE TRANSPORTED IN THE PLASMA AS LIPOPROTEINS

Four Major Lipid Classes Are Present in Lipoproteins

Plasma lipids consist of **triacylglycerols** (16%), **phospholipids** (30%), **cholesterol** (14%), and **cholesteryl esters** (36%) and a much smaller fraction of unesterified long-chain fatty acids (or FFAs) (4%). This latter fraction, the **FFA**, is metabolically the most active of the plasma lipids.

Four Major Groups of Plasma Lipoproteins Have Been Identified

Since fat is less dense than water, the density of a lipoprotein decreases as the proportion of lipid to protein increases (**Table 25–1**). Four major groups of lipoproteins have been identified that are important physiologically and in clinical diagnosis.

These are (1) **chylomicrons**, derived from intestinal absorption of triacylglycerol and other lipids; (2) **VLDL**, derived from the liver for the export of triacylglycerol; (3) **low-density lipoproteins** (LDL), responsible for the transport of cholesterol in humans and representing a final stage in the catabolism of VLDL; and (4) **high-density lipoproteins**, (HDL), involved in reverse cholesterol transport (see later and Chapter 26) and also in VLDL and chylomicron metabolism. Triacylglycerol is the predominant lipid in chylomicrons and VLDL, whereas cholesterol and phospholipid are the predominant lipids in LDL and HDL, respectively (see Table 25–1). Lipoproteins may also be classified according to their electrophoretic properties into α- (HDL), β- (LDL), and **pre-β** (VLDL)-**lipoproteins**.

Lipoproteins Consist of a Nonpolar Core & a Single Surface Layer of Amphipathic Lipids

The **nonpolar lipid core** consists of mainly **triacylglycerol** and **cholesteryl ester** and is surrounded by a **single surface layer** of amphipathic phospholipid and **cholesterol** molecules (**Figure 25–1**). These are oriented so that their polar groups face outward to the aqueous medium, as in the cell membrane (see Chapters 21 and 40). The protein moiety of a lipoprotein is known as an **apolipoprotein** or **apoprotein**, constituting nearly 70% of some HDL and as little as 1% of chylomicrons.

TABLE 25–1 Composition of the Lipoproteins in Plasma of Humans

| Lipoprotein | Source | Diameter (nm) | Density (g/Ml) | Composition | | Main Lipid Components | Apolipoproteins |
				Protein (%)	Lipid (%)		
Chylomicrons	Intestine	90-1000	<0.95	1-2	98-99	Triacylglycerol	A-I, A-II, A-IV[a], B-48, C-I, C-II, C-III, E
Chylomicron remnants	Chylomicrons	45-150	<1.006	6-8	92-94	Triacylglycerol, phospholipids, cholesterol	B-48, E
VLDL	Liver (intestine)	30-90	0.95-1.006	7-10	90-93	Triacylglycerol	B-100, C-I, C-II, C-III
IDL	VLDL	25-35	1.006-1.019	11	89	Triacylglycerol, cholesterol	B-100, E
LDL	VLDL	20-25	1.019-1.063	21	79	Cholesterol	B-100
HDL	Liver, intestine, VLDL, chylomicrons					Phospholipids, cholesterol	A-I, A-II, A-IV, C-I, C-II, C-III, D[b], E
HDL$_1$		20-25	1.019-1.063	32	68		
HDL$_2$		10-20	1.063-1.125	33	67		
HDL$_3$		5-10	1.125-1.210	57	43		
Preβ-HDL[c]		<5	>1.210				A-I
Albumin/free fatty acids	Adipose tissue		>1.281	99	1	Free fatty acids	

[a]Secreted with chylomicrons but transfers to HDL.
[b]Associated with HDL$_2$ and HDL$_3$ subfractions.
[c]Part of a minor fraction known as very-high-density lipoproteins (VHDL).
Abbreviations: HDL, high-density lipoproteins; IDL, intermediate-density lipoproteins; LDL, low-density lipoproteins; VLDL, very-low-density lipoproteins.

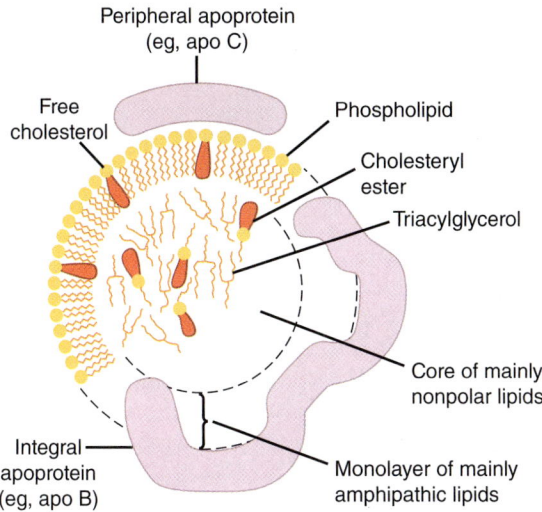

FIGURE 25–1 **Generalized structure of a plasma lipoprotein.** Small amounts of cholesteryl ester and triacylglycerol are found in the surface layer and a little free cholesterol in the core.

The Distribution of Apolipoproteins Characterizes the Lipoprotein

One or more apolipoproteins are present in each lipoprotein. They are usually abbreviated as apo followed by the letters A, B, C, etc. (see Table 25–1). Some apolipoproteins are integral and cannot be removed (eg, apo B), whereas others are bound to the surface and are free to transfer to other lipoproteins (eg, apos C and E). The major apolipoproteins of HDL (α-lipoprotein) are apo As (see Table 25–1). The main apolipoprotein of LDL (β-lipoprotein) is apo B (B-100), which is found also in VLDL. Chylomicrons contain a truncated form of apo B-100, termed apoB-48 as it constitutes 48% of the full-sized molecule. Apo B-48 is synthesized in the intestine, while B-100 is synthesized in the liver. Apo B-100 is one of the longest single polypeptide chains known, having more than 4500 amino acids and a molecular mass of 540,000 Da. To produce Apo B-48, a stop signal is introduced into the mRNA transcript for apo B-100 by an RNA editing enzyme. Apos C-I, C-II, and C-III are smaller polypeptides (molecular mass 7000-9000 Da) freely transferable between several different lipoproteins. Apo E, found in VLDL, HDL, chylomicrons, and chylomicron remnants, is also freely transferable; it accounts for 5 to 10% of total VLDL apolipoproteins in normal subjects.

Apolipoproteins carry out several roles: (1) they can form part of the structure of the lipoprotein, for example, apo B; (2) they are enzyme cofactors, for example, C-II for lipoprotein lipase, A-I for lecithin:cholesterol acyltransferase (LCAT), or enzyme inhibitors, for example, apo A-II and apo C-III for lipoprotein lipase, apo C-I for cholesteryl ester transfer protein (see Chapter 26); and (3) they act as ligands for interaction with lipoprotein receptors in tissues, for example, apo B-100 and apo E for the LDL receptor, apo E for the LDL-receptor–related protein-1 (LRP-1), which recognizes remnant lipoproteins (see later), and apo A-I for the HDL receptor. Apo A-IV

is thought to have a role in chylomicron metabolism and may also act as a regulator of satiety and glucose homeostasis, making it a potential therapeutic target for the treatment of diabetes and obesity, while apo D is believed to be an important factor in human neurodegenerative disorders.

FREE FATTY ACIDS ARE RAPIDLY METABOLIZED

The FFAs (also termed nonesterified fatty acids [NEFAs] or unesterified fatty acids) arise in the plasma from the breakdown of triacylglycerol in adipose tissue or as a result of the action of lipoprotein lipase on the plasma triacylglycerols. They are found **in combination with albumin**, a very effective solubilizer. Levels are low in the fully fed condition and rise to 0.7 to 0.8 mEq/mL in the starved state. In uncontrolled **diabetes mellitus**, the level may rise to as much as 2 mEq/mL.

FFAs are removed from the blood extremely rapidly by the tissues and oxidized (fulfilling 25-50% of energy requirements in starvation) or esterified to form triacylglycerol. In starvation, esterified lipids from the circulation or in the tissues are also oxidized, particularly in heart and skeletal muscle cells, where considerable stores of lipid are found.

The FFA uptake by tissues is related directly to the plasma-FFA concentration, which in turn is determined by the rate of lipolysis in adipose tissue. After dissociation of the fatty acid–albumin complex at the plasma membrane, fatty acids bind to a **membrane fatty acid transport protein** that acts as a transmembrane cotransporter with Na^+. On entering the cytosol, FFAs are bound by intracellular **fatty acid–binding proteins**. The role of these proteins in intracellular transport is thought to be similar to that of serum albumin in extracellular transport of long-chain fatty acids.

TRIACYLGLYCEROL IS TRANSPORTED FROM THE INTESTINES IN CHYLOMICRONS & FROM THE LIVER IN VERY-LOW-DENSITY LIPOPROTEINS

By definition, **chylomicrons** are found in **chyle** formed only by the lymphatic system **draining the intestine**. They are responsible for the transport of all dietary lipids into the circulation. Small quantities of VLDL are also to be found in chyle; however, most **VLDL in the plasma** are of hepatic origin. **They are the vehicles of transport of triacylglycerol from the liver to the extrahepatic tissues.**

There are striking similarities in the mechanisms of formation of chylomicrons by intestinal cells and of VLDL by hepatic parenchymal cells (**Figure 25–2**), perhaps because—apart from the mammary gland—the intestine and liver are the only tissues from which particulate lipid is secreted. Newly secreted or "nascent" chylomicrons and VLDL contain only a

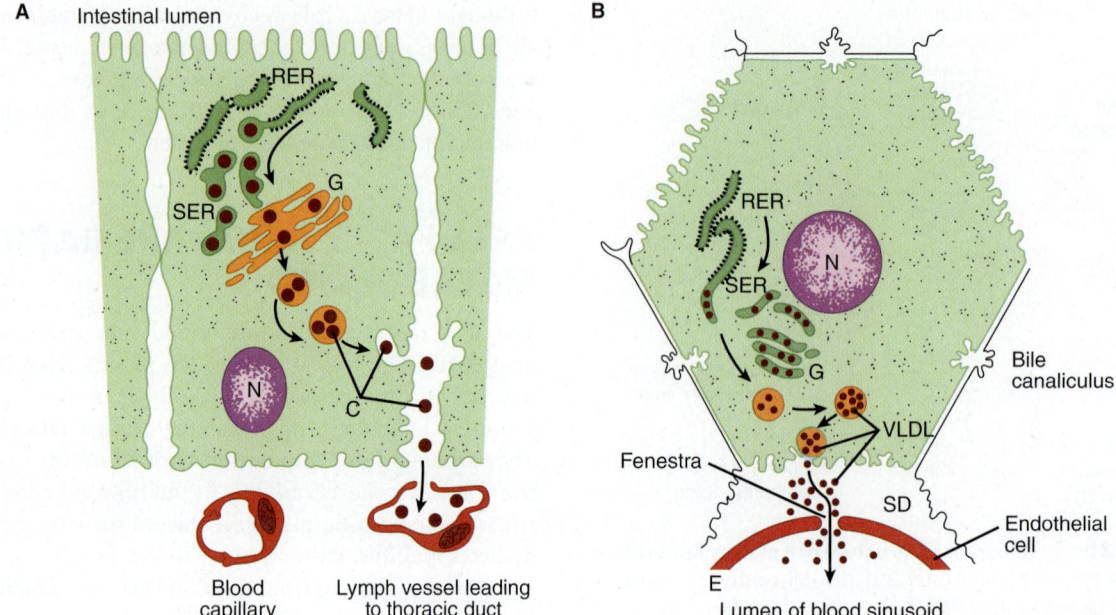

FIGURE 25–2 **The formation and secretion of (A) chylomicrons by an intestinal cell and (B) very-low-density lipoproteins by a hepatic cell.** Apolipoprotein B, synthesized in the rough endoplasmic reticulum, is incorporated into particles with triacylglycerol, cholesterol, and phospholipids in the smooth endoplasmic reticulum. After the addition of carbohydrate residues in the Golgi apparatus, the particles are released from the cell by reverse pinocytosis. Chylomicrons pass into the lymphatic system. VLDL are secreted into the space of Disse and then into the hepatic sinusoids through fenestrae in the endothelial lining. (C, chylomicrons; E, endothelium; G, Golgi apparatus; N, nucleus; RER, rough endoplasmic reticulum; SD, space of Disse, containing blood plasma; SER, smooth endoplasmic reticulum; VLDL, very-low-density lipoprotein.)

small amount of apolipoproteins C and E, and the full complement is acquired from HDL in the circulation (**Figures 25–3** and **25–4**). Apo B, however, is an integral part of the lipoprotein particles. It is incorporated into the particles during their

assembly inside the cells and is essential for chylomicron and VLDL formation. In **abetalipoproteinemia** (a rare disease), lipoproteins containing apo B are not formed and lipid droplets accumulate in the intestine and liver.

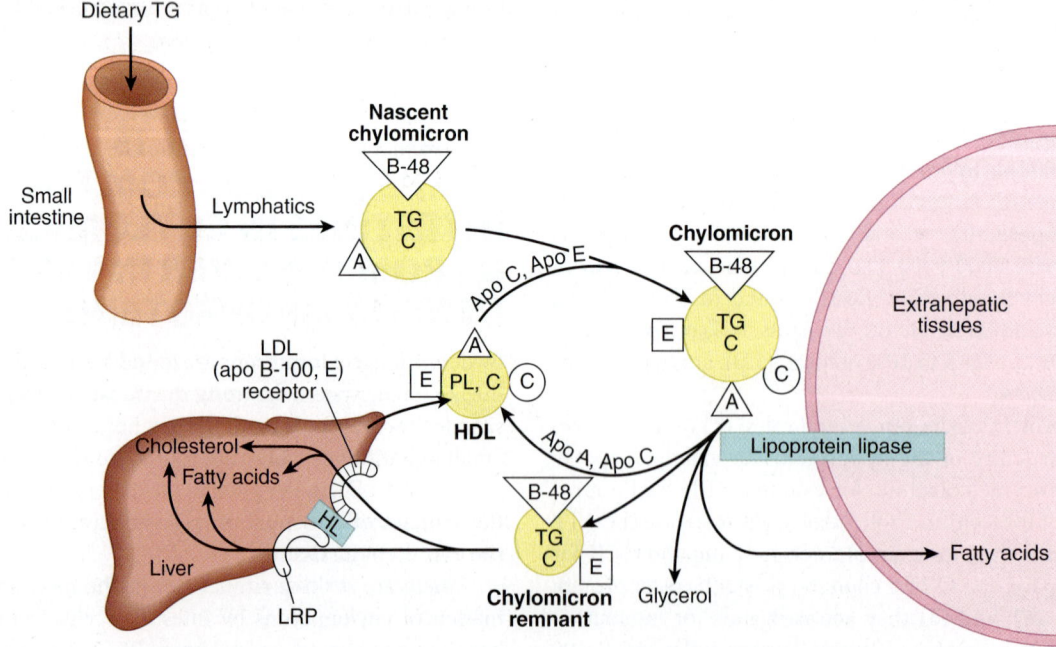

FIGURE 25–3 **Metabolic fate of chylomicrons.** A, apolipoprotein A; B-48, apolipoprotein B-48; C, apolipoprotein C; C, cholesterol and cholesteryl ester; E, apolipoprotein E; HDL, high-density lipoprotein; HL, hepatic lipase; LRP, LDL-receptor–related protein; PL, phospholipid; TG, triacylglycerol. Only the predominant lipids are shown.

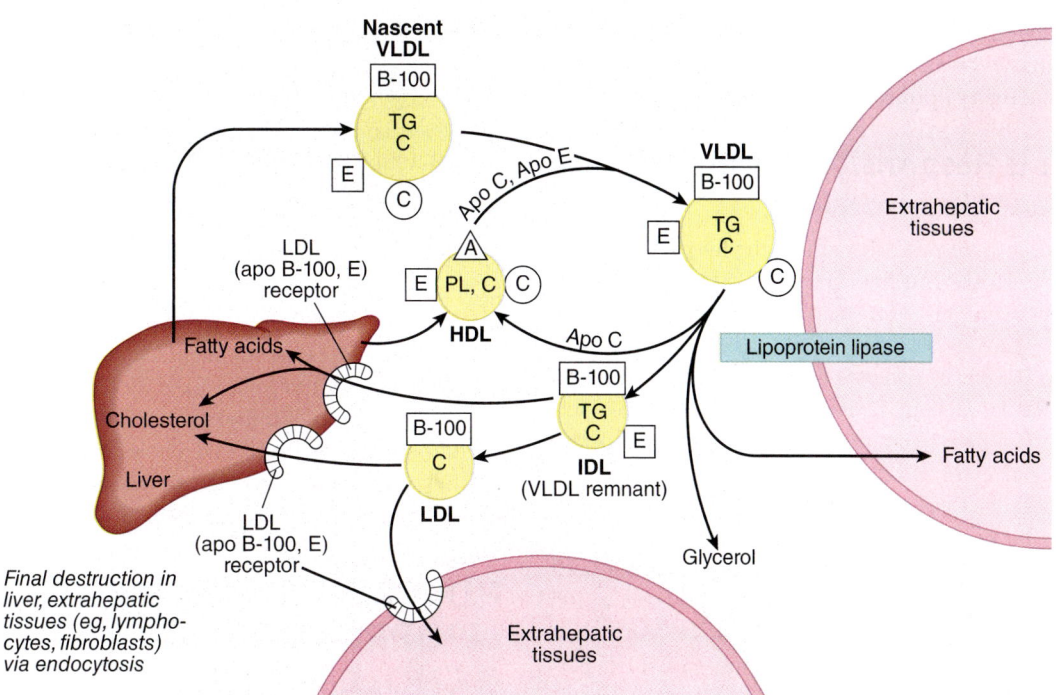

FIGURE 25–4 **Metabolic fate of very-low-density lipoproteins (VLDL) and production of low-density lipoproteins (LDL).** A, apolipoprotein A; B-100, apolipoprotein B-100; C, apolipoprotein C; C, cholesterol and cholesteryl ester; E, apolipoprotein E; HDL, high-density lipoprotein; IDL, intermediate-density lipoprotein; PL, phospholipid; TG, triacylglycerol. Only the predominant lipids are shown. It is possible that some IDL is also metabolized via the low-density lipoprotein receptor–related protein-1 (LRP-1.)

CHYLOMICRONS & VERY-LOW-DENSITY LIPOPROTEINS ARE RAPIDLY CATABOLIZED

The clearance of chylomicrons from the blood is rapid, the half-time of disappearance being under 1 hour in humans. Larger particles are catabolized more quickly than smaller ones. Fatty acids originating from chylomicron triacylglycerol are delivered mainly to adipose tissue, heart, and muscle (80%), while ~20% goes to the liver. However, **the liver does not metabolize native chylomicrons or VLDL significantly**; thus, the fatty acids in the liver must be secondary to their metabolism in extrahepatic tissues.

Triacylglycerols of Chylomicrons & VLDL Are Hydrolyzed by Lipoprotein Lipase to Form Remnant Lipoproteins

Lipoprotein lipase is an enzyme located on the walls of blood capillaries, anchored to the endothelium by negatively charged proteoglycan chains of heparan sulfate. It has been found in heart, adipose tissue, spleen, lung, renal medulla, aorta, diaphragm, and lactating mammary gland, although it is not active in adult liver. It is not normally found in blood; however, following injection of **heparin**, lipoprotein lipase is released from its heparan sulfate–binding sites into the circulation. **Hepatic lipase** is bound to the sinusoidal surface of liver cells and is also released by heparin. This enzyme, however, does not react

readily with chylomicrons or VLDL but is involved in chylomicron remnant and HDL metabolism (see later).

Both **phospholipids** and **apo C-II** are required as cofactors for lipoprotein lipase activity, while **apo A-II** and **apo C-III** act as inhibitors. Hydrolysis takes place while the lipoproteins are attached to the enzyme on the endothelium. Triacylglycerol is hydrolyzed progressively through a diacylglycerol to a monoacylglycerol and finally to FFA plus glycerol. Some of the released FFA return to the circulation, attached to albumin, but the bulk is transported into the tissue (see Figures 25–3 and 25–4). Heart lipoprotein lipase has a low K_m for triacylglycerol, about one-tenth of that for the enzyme in adipose tissue. This enables the delivery of fatty acids from triacylglycerol to be **redirected from adipose tissue to the heart in the starved state** when the plasma triacylglycerol decreases. A similar redirection to the mammary gland occurs during lactation, allowing uptake of lipoprotein triacylglycerol fatty acid for **milk fat** synthesis. The **VLDL receptor** plays an important part in the delivery of fatty acids from VLDL triacylglycerol to adipocytes by binding VLDL and bringing it into close contact with lipoprotein lipase. In adipose tissue, **insulin** enhances lipoprotein lipase synthesis in adipocytes and its translocation to the luminal surface of the capillary endothelium.

Reaction with lipoprotein lipase results in the loss of 70 to 90% of the triacylglycerol of chylomicrons and in the loss of apo C (which returns to HDL) but not apo E, which is retained. The resulting **chylomicron remnant** is about half the diameter of the parent chylomicron and is relatively enriched in cholesterol and cholesteryl esters because of the loss of triacylglycerol (see Figure 25–3). Similar changes occur to VLDL, with the formation of **VLDL remnants** (also called **intermediate-density lipoprotein (IDL)** (see Figure 25–4).

The Liver Is Responsible for the Uptake of Remnant Lipoproteins

Chylomicron remnants are taken up by the liver by receptor-mediated endocytosis, and the cholesteryl esters and triacylglycerols are hydrolyzed and metabolized. Uptake is mediated by **apo E** (see Figure 25–3), via two apo E-dependent receptors, the **LDL (apo B-100, E) receptor** and **LDL receptor–related protein-1 (LRP-1)**. Hepatic lipase has a dual role: (1) it acts as a ligand to facilitate remnant uptake and (2) it hydrolyzes remnant triacylglycerol and phospholipid.

After VLDL has been converted to IDL, the remnant particles may be taken up by the liver directly via the LDL (apo B-100, E) receptor, or they may be further metabolized to LDL in the circulation. Only one molecule of apo B-100 is present in each of these lipoprotein particles, and this is conserved during the transformations. Thus, each LDL particle is derived from a single precursor VLDL particle (see Figure 25–4). In humans, a relatively large proportion of IDL forms LDL, accounting for the increased concentrations of LDL in humans compared with many other mammals.

LDL IS METABOLIZED VIA THE LDL RECEPTOR

The liver and many extrahepatic tissues express the **LDL (apo B-100, E) receptor** (LDL receptor). It is so designated because it is specific for apo B-100 but not B-48, which lacks the carboxyl terminal domain of B-100 containing the LDL receptor ligand, and it also takes up lipoproteins rich in apo E. The LDL receptor is found in many tissues in the body, including the liver, arterial wall, ovary, testes, and adrenocortex. A positive correlation exists between the incidence of **atherosclerosis** and the plasma concentration of LDL cholesterol. The LDL receptor is defective in **familial hypercholesterolemia**, a genetic condition in which blood LDL cholesterol levels are increased, causing premature atherosclerosis (see Table 26–1). For further discussion of the regulation of the LDL receptor, see Chapter 26.

HDL TAKES PART IN BOTH LIPOPROTEIN TRIACYLGLYCEROL & CHOLESTEROL METABOLISM

HDL is synthesized and secreted from both liver and intestine (**Figure 25–5**). However, apo C and apo E are synthesized in the liver and transferred from liver HDL to intestinal HDL

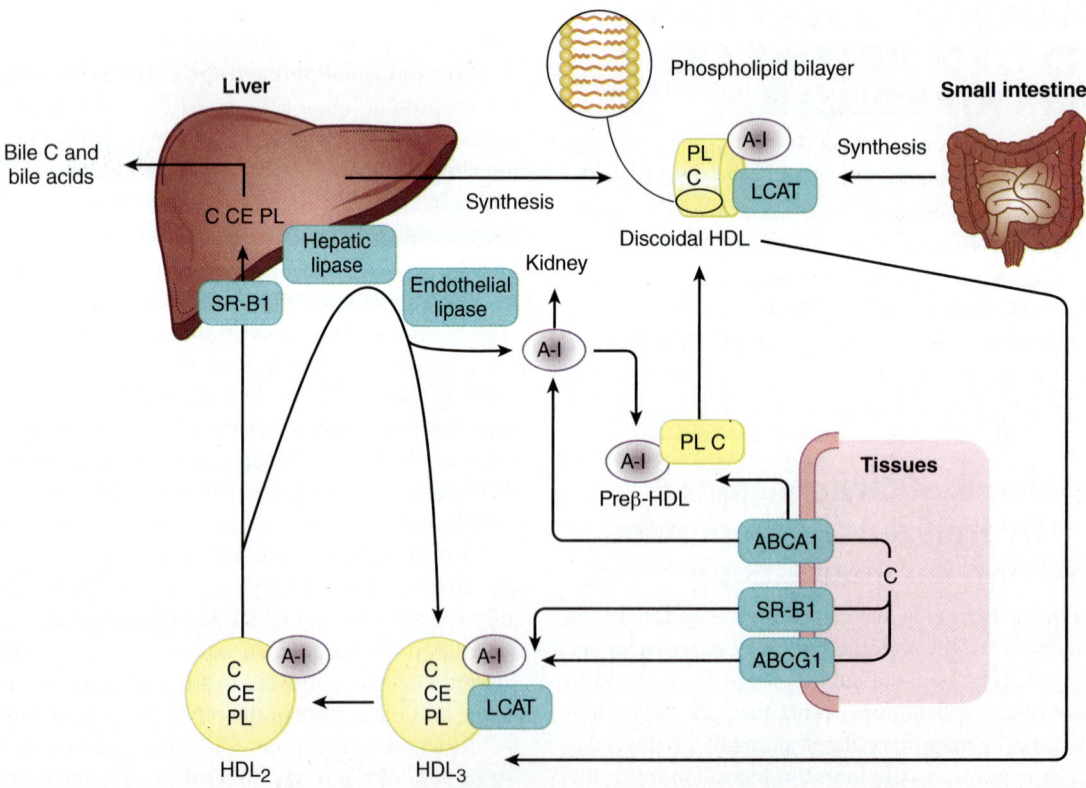

FIGURE 25–5   Metabolism of high-density lipoprotein (HDL) in reverse cholesterol transport. A-I, apolipoprotein A-I; ABCA1, ATP-binding cassette transporter A1; ABCG1, ATP-binding cassette transporter G1; C, cholesterol; CE, cholesteryl ester; LCAT, lecithin:cholesterol acyltransferase; PL, phospholipid; SR-B1, scavenger receptor B1. Preβ-HDL, HDL₂, HDL₃—see Table 25–1. Surplus surface constituents from the action of lipoprotein lipase on chylomicrons and VLDL are another source of preβ-HDL.

when the latter enters the plasma. A major function of HDL is to act as a repository for the apo C and apo E required in the metabolism of chylomicrons and VLDL. Nascent HDL consists of discoid phospholipid bilayers containing apo A and free cholesterol. These lipoproteins are similar to the particles found in the plasma of patients with a deficiency of the plasma enzyme **LCAT** and in the plasma of patients with **obstructive jaundice**. LCAT—and the LCAT activator apo A-I—bind to the discoidal particles, and the surface phospholipid and free cholesterol are converted into cholesteryl esters and lysolecithin (see Chapter 24). The nonpolar cholesteryl esters move into the hydrophobic interior of the bilayer, whereas lysolecithin is transferred to plasma albumin. Thus, a nonpolar core is generated, forming a spherical, pseudomicellar HDL covered by a surface film of polar lipids and apolipoproteins. This aids the removal of excess unesterified cholesterol from lipoproteins and tissues as described in the following discussion.

Reverse Cholesterol Transport Is a Major Function of HDL

The removal of cholesterol from the tissues and its return to the liver for excretion is termed **reverse cholesterol transport**. The **class B scavenger receptor B1 (SR-B1)** is an HDL receptor with a dual role in this process. In the liver, it binds HDL via apo A-I, and cholesteryl ester is selectively delivered to the cells, although the particle itself, including apo A-I, is not taken up. In the extrahepatic tissues, on the other hand, SR-B1 mediates the acceptance of cholesterol effluxed from the cells by HDL, which then transports it to the liver where it is excreted in the bile either as cholesterol or after conversion to bile acids (see Figure 25–5). HDL$_3$, generated from discoidal HDL by the action of LCAT, accepts cholesterol from the tissues via the **SR-B1** and the cholesterol is then esterified by LCAT, increasing the size of the particles to form the less dense HDL$_2$. HDL$_3$ is then reformed, either after selective delivery of cholesteryl ester to the liver via the SR-B1 or by hydrolysis of HDL$_2$ phospholipid and triacylglycerol by hepatic lipase and endothelial lipase. This interchange of HDL$_2$ and HDL$_3$ is called the **HDL cycle** (see Figure 25–5). Free apo A-I is released by these processes and forms **preβ-HDL** after associating with a minimum amount of phospholipid and cholesterol. Surplus apo A-I is destroyed in the kidney. A second important mechanism for reverse cholesterol transport involves the **ATP-binding cassette transporters A1 (ABCA1)** (also called cholesterol efflux regulatory protein) and **G1 (ABCG1)**. These transporters are members of a family of transporter proteins that couple the hydrolysis of ATP to the binding of a substrate, enabling it to be transported across the membrane. ABCG1 mediates the transport of cholesterol from cells to HDL, while ABCA1 preferentially promotes efflux to poorly lipidated particles such as preβ-HDL or apo A-1, which are then converted to HDL$_3$ via discoidal HDL (see Figure 25–5). Preβ-HDL is the most potent form of HDL inducing cholesterol efflux from the tissues.

HDL concentrations vary reciprocally with plasma triacylglycerol concentrations and directly with the activity of lipoprotein lipase. This may be due to surplus surface constituents, for example, phospholipid and apo A-I, being released during hydrolysis of chylomicrons and VLDL and contributing toward the formation of preβ-HDL and discoidal HDL. HDL$_2$ concentrations are **inversely related to the incidence of atherosclerosis**, possibly because they reflect the efficiency of reverse cholesterol transport. HDL$_c$ (HDL$_1$) is found in the blood of diet-induced hypercholesterolemic animals. It is rich in cholesterol, and its sole apolipoprotein is apo E. It appears that all plasma lipoproteins are interrelated components of one or more metabolic cycles that together are responsible for the complex process of plasma lipid transport.

THE LIVER PLAYS A CENTRAL ROLE IN LIPID TRANSPORT & METABOLISM

The liver is responsible for a number of major functions in lipid metabolism including:

1. Facilitation of the digestion and absorption of lipids by the production of **bile** (see Chapter 26).

2. Active **synthesis and oxidation of fatty acids** (see Chapters 22 and 23) and also synthesis of triacylglycerols and phospholipids (see Chapter 24).

3. **Conversion of fatty acids to ketone bodies (ketogenesis)** (see Chapter 22).

4. **Synthesis and metabolism of plasma lipoproteins**.

Hepatic VLDL Secretion Is Related to Dietary & Hormonal Status

The cellular events involved in hepatic VLDL formation and secretion are shown in **Figure 25–6**. VLDL assembly requires the synthesis of apo B-100 and a source of triacylglycerol. Apo B-100 is synthesized on polyribosomes and translocated to the lumen of the endoplasmic reticulum (ER) as it is formed. As the protein enters the lumen, it is lipidated with phospholipid with the aid of the **microsomal triacylglycerol transfer protein (MTP)**, which also facilitates the transfer of triacylglycerol across the ER membrane, and apo B–containing **VLDL2** (or precursor VLDL) particles are formed. The triacylglycerol is derived from lipolysis of cytosolic triacylglycerol lipid droplets (this pool of triacylglycerol is formed using FFA either synthesized *de novo* or taken up from the plasma, see later) followed by reesterification in a pathway requiring phospholipid derivatives and diacylglycerol acyl transferases. Triacylglycerol not used for VLDL1 formation is recycled to the cytosolic droplets. After assembly in the ER, VLDL2 are carried in coat protein II (COPII) vesicles (see Chapter 49) to the golgi, where they fuse with triacylglycerol-rich lipid droplets to produce **VLDL1**. Phosphatidic acid produced by the action of phospholipase D after activation by a small GTP-binding protein called **ADP-ribosylation factor-1 (ARF-1)** is needed for the

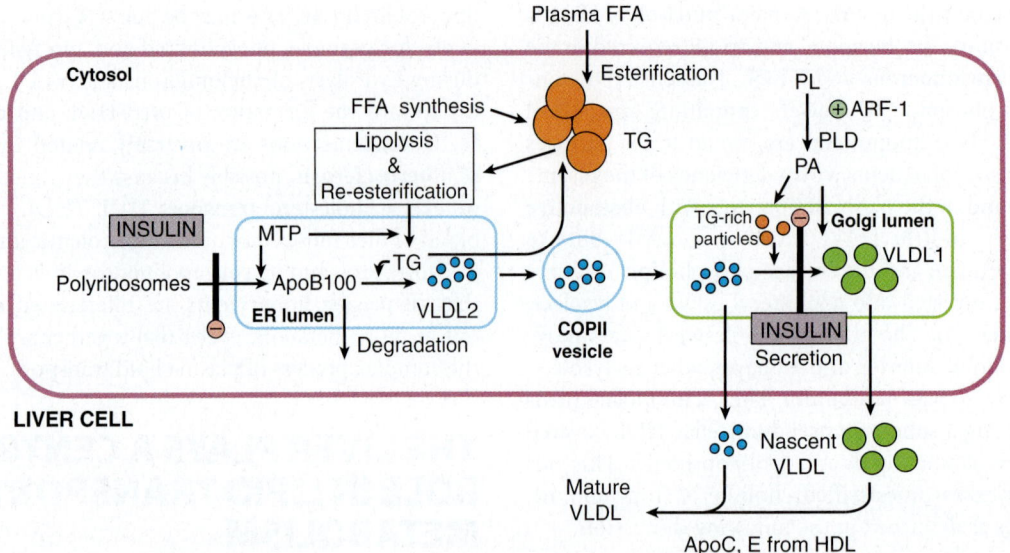

FIGURE 25–6 **The assembly of very-low-density lipoprotein (VLDL) in the liver.** The pathways indicated underlie the events depicted in Figure 25–2. Apo B-100 is synthesized on polyribosomes and is lipidated with PL by MTP as it enters the ER lumen. Any excess is degraded in proteasomes. TG derived from lipolysis of cytosolic lipid droplets followed by resynthesis is transferred into the ER lumen with the aid of MTP and interacts with apo B-100, forming VLDL2. Excess TG is recycled to the cytosolic lipid droplets. VLDL2 are translocated to the golgi in COPII vesicles where they fuse with TG-rich particles to form VLDL1. PA is produced by activation of PLD by ARF-1 and is incorporated into the TG-rich VLDL1 and/or VLDL2. Both VLDL1 and VLDL2 may be secreted into the blood. Insulin inhibits VLDL secretion by inhibiting apo B-100 synthesis and the formation of VLDL1 from VLDL2. (Apo, apolipoprotein; ARF-1, ADP-ribosylation factor-1; FFA, free fatty acids; HDL, high-density lipoproteins; MTP, microsomal triacylglycerol transfer protein; PA, phosphatidic acid; PL, phospholipid; PLD, phospholipase D; TG, triacylglycerol.)

formation of the triacylglycerol—rich particles and/or VLDL2. Although some VLDL2 particles may be secreted without fusion, most particles which leave the cell are in the form of VLDL1. These nascent VLDL then acquire apolipoproteins C and E from HDL in the circulation to become mature VLDL.

Triacylglycerol for VLDL formation is synthesized from FFA. The fatty acids used are derived from two possible sources: (1) de novo synthesis within the liver from **acetyl-CoA** derived mainly from carbohydrate (perhaps not so important in humans) and (2) uptake of **FFA** from the circulation. The first source is predominant in the well-fed condition, when fatty acid synthesis is high and the level of circulating FFAs is low. As triacylglycerol does not normally accumulate in the liver in these conditions, it must be inferred that it is transported from the liver in VLDL as rapidly as it is synthesized. FFAs from the circulation are the main source during starvation, the feeding of high-fat diets, or in diabetes mellitus, when hepatic lipogenesis is inhibited. Factors that enhance both the synthesis of triacylglycerol and the secretion of VLDL by the liver include (1) the fed state rather than the starved state; (2) the feeding of diets high in carbohydrate (particularly if they contain sucrose or fructose), leading to high rates of lipogenesis and esterification of fatty acids; (3) high levels of circulating FFA; (4) ingestion of ethanol; and (5) the presence of high concentrations of insulin and low concentrations of glucagon, which enhance fatty acid synthesis and esterification and inhibit their oxidation.

Insulin suppresses hepatic VLDL secretion by inhibiting both apo B-100 synthesis and the conversion of the smaller VLDL2 into VLDL1 by fusion with bulk triacylglycerol. Some

other factors which are known to inhibit or prevent VLDL assembly in the liver include the antibiotic brefeldin A, which inhibits the action of ARF-1; the sulfonylurea hypoglycemic drug, tolbutamide, dietary ω3 fatty acids (see Chapter 21), and orotic acid, an intermediate in the synthesis of pyrimidines (see Chapter 33), all of which decrease the rate of triacylglycerol lipolysis; and a defect in the *MTP* gene, which lowers MTP levels. The regulation of VLDL formation in the liver is complex and involves interactions between hormonal and dietary factors that are not yet fully understood.

CLINICAL ASPECTS

Imbalance in the Rate of Triacylglycerol Formation & Export Causes Fatty Liver

For a variety of reasons, lipid—mainly as triacylglycerol—can accumulate in the liver. Extensive accumulation causes **fatty liver** and is regarded as a pathologic condition. **Nonalcoholic fatty liver disease (NAFLD)** is the most common liver disorder worldwide. When accumulation of lipid in the liver becomes chronic, inflammatory and fibrotic changes may develop leading to **nonalcoholic steatohepatitis (NASH),** which can progress to liver diseases including **cirrhosis, hepatocarcinoma,** and **liver failure.**

Fatty livers fall into two main categories. The first type is associated with **raised levels of plasma FFA** resulting from mobilization of fat from adipose tissue or from the hydrolysis of lipoprotein triacylglycerol by lipoprotein lipase in extrahepatic tissues. The production of VLDL does not keep pace with the increasing influx and esterification of free fatty acids, allowing triacylglycerol to accumulate, which in turn causes a fatty liver. This occurs during **starvation** and the feeding of **high-fat diets**. The ability to secrete VLDL may also be impaired (eg, in starvation). In uncontrolled **diabetes mellitus, twin lamb disease,** and **ketosis in cattle** fatty infiltration is sufficiently severe to cause visible pallor (fatty appearance) and enlargement of the liver with possible liver dysfunction.

The second type of fatty liver is usually due to a **metabolic block in the production of plasma lipoproteins,** thus allowing triacylglycerol to accumulate. Theoretically, the lesion may be due to (1) a block in apolipoprotein synthesis (or an increase in its degradation before it can be incorporated into VLDL), (2) a block in the synthesis of the lipoprotein from lipid and apolipoprotein, (3) a failure in provision of phospholipids that are found in lipoproteins, or (4) a failure in the secretory mechanism itself.

One type of fatty liver that has been studied extensively in rats is caused by a deficiency of **choline**, which has therefore been called a **lipotropic factor**. **Orotic acid** also causes fatty liver; it is believed to interfere with glycosylation of VLDL, thus inhibiting release, and may also impair the recruitment of triacylglycerol to the particles. A deficiency of vitamin E enhances the hepatic necrosis of the choline deficiency type of fatty liver. Added vitamin E or a source of **selenium** has a protective effect by combating lipid peroxidation. In addition to protein deficiency, essential fatty acid and vitamin deficiencies (eg, linoleic acid, pyridoxine, and pantothenic acid) can cause fatty infiltration of the liver.

Ethanol Also Causes Fatty Liver

Alcoholic fatty liver is the first stage in **alcoholic liver disease (ALD)** which is caused by **alcoholism** and ultimately leads to **cirrhosis**. The fat accumulation in the liver is caused by a combination of impaired fatty acid oxidation and increased lipogenesis, which is thought to be due to changes in the [NADH]/[NAD$^+$] redox potential in the liver, and also to interference with the action of transcription factors regulating the expression of the enzymes involved in the pathways. Oxidation of ethanol by **alcohol dehydrogenase** leads to excess production of NADH, which competes with reducing equivalents from other substrates, including fatty acids, for the respiratory chain. This inhibits their oxidation and causes increased esterification of fatty acids to form triacylglycerol, resulting in the fatty liver. Oxidation of ethanol leads to the formation of acetaldehyde, which is oxidized by **aldehyde dehydrogenase**, producing acetate. The increased (NADH)/(NAD$^+$) ratio also causes increased (lactate)/(pyruvate), resulting in **hyperlacticacidemia**, which decreases excretion of uric acid, aggravating **gout**.

Some metabolism of ethanol takes place via a cytochrome P450–dependent microsomal ethanol oxidizing system (MEOS) involving NADPH and O_2. This system increases in activity in **chronic alcoholism** and may account for the increased metabolic clearance of ethanol in this condition, but may also promote the development of ALD. Ethanol also inhibits the metabolism of some drugs, for example, barbiturates, by competing for cytochrome P450–dependent enzymes.

In some Asian populations and Native Americans, alcohol consumption results in increased adverse reactions to acetaldehyde owing to a genetic defect of mitochondrial aldehyde dehydrogenase.

ADIPOSE TISSUE IS THE MAIN STORE OF TRIACYLGLYCEROL IN THE BODY

Triacylglycerols are stored in adipose tissue in large lipid droplets and are continually undergoing lipolysis (hydrolysis) and reesterification. These two processes are entirely different pathways involving different reactants and enzymes. This allows the processes of esterification or lipolysis to be regulated separately by many nutritional, metabolic, and hormonal factors. The balance between these two processes determines the magnitude of the FFA pool in adipose tissue, which in turn determines the level of FFA circulating in the plasma. Since the latter has most profound effects on the metabolism of other tissues, particularly liver and muscle, the factors operating in adipose tissue that regulate the outflow of FFA exert an influence far beyond the tissue itself. Moreover, since the discovery in the last 20 years that adipose tissue secretes hormones such as **leptin** and **adiponectin**, known as **adipokines**, its role as an endocrine organ has been recognized. Leptin regulates energy homeostasis by stimulating energy use and limiting food intake. If it is lacking, food intake may be uncontrolled, causing obesity. Adiponectin modulates glucose and lipid metabolism in muscle and liver, and enhances the sensitivity of tissues to insulin.

The Provision of Glycerol-3-Phosphate Regulates Esterification: Lipolysis Is Controlled by Hormone-Sensitive Lipase

Triacylglycerol is synthesized from acyl-CoA and glycerol-3-phosphate (see Figure 24–2). Since the enzyme **glycerol kinase** is not expressed in adipose tissue, glycerol cannot be utilized for the provision of glycerol-3-phosphate, which must be supplied from glucose via glycolysis (**Figure 25–7**).

Triacylglycerol undergoes hydrolysis by a **hormone-sensitive lipase** to form FFA and glycerol. This lipase is distinct from lipoprotein lipase, which catalyzes lipoprotein triacylglycerol hydrolysis before its uptake into extrahepatic tissues (see earlier). Since the glycerol cannot be utilized, it enters the blood and is taken up and transported to tissues such as the liver and kidney, which possess an active glycerol kinase. The FFA formed by lipolysis can be reconverted in adipose tissue to acyl-CoA by **acyl-CoA synthetase** and reesterified with glycerol-3-phosphate to form triacylglycerol. Thus, **there is a continuous cycle of lipolysis and reesterification within the**

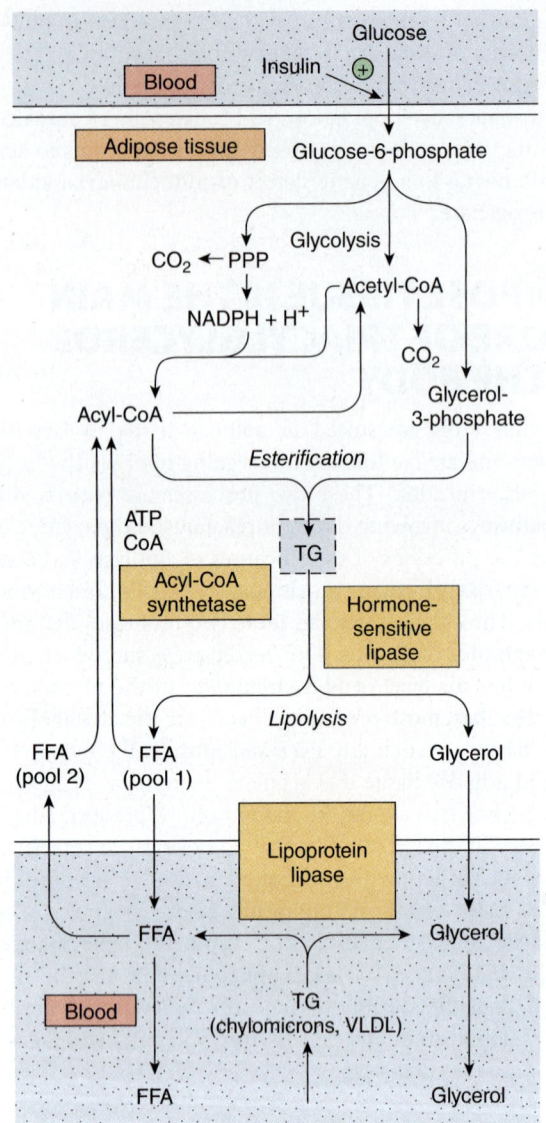

FIGURE 25–7 **Triacylglycerol metabolism in adipose tissue.** Hormone-sensitive lipase is activated by ACTH, TSH, glucagon, epinephrine and norepinephrine, and inhibited by insulin, prostaglandin E_1, and nicotinic acid. Details of the formation of glycerol-3-phosphate from intermediates of glycolysis are shown in Figure 24–2. (FFA, free fatty acids; PPP, pentose phosphate pathway; TG, triacylglycerol; VLDL, very-low-density lipoprotein.)

tissue (see Figure 25–7). However, when the rate of reesterification is not sufficient to match the rate of lipolysis, FFAs accumulate and diffuse into the plasma, where they bind to albumin and raise the concentration of plasma-free fatty acids.

Increased Glucose Metabolism Reduces the Output of FFA

When the utilization of glucose by adipose tissue is increased, the FFA outflow decreases. However, the release of glycerol continues, demonstrating that the effect of glucose is not mediated by reducing the rate of lipolysis. The effect is due to the increased provision of glycerol-3-phosphate, which enhances

esterification of FFA. Glucose can take several pathways in adipose tissue, including oxidation to CO_2 via acetyl-CoA and the citric acid cycle, oxidation in the pentose phosphate pathway, conversion to long-chain fatty acids (via acyl-CoA), and formation of acylglycerol via glycerol-3-phosphate (see Figure 25–7). When glucose utilization is high, a larger proportion of the uptake is oxidized to CO_2 and converted to fatty acids. However, as total glucose utilization decreases, the greater proportion of the glucose is directed to the production of glycerol-3-phosphate for the esterification of acyl-CoA to form triacylglycerol, which helps to minimize the efflux of FFA.

HORMONES REGULATE FAT MOBILIZATION

Adipose Tissue Lipolysis Is Inhibited by Insulin

The rate of release of FFA from adipose tissue is affected by many hormones that influence either the rate of esterification or the rate of lipolysis. **Insulin** inhibits the release of FFA from adipose tissue, which results in a fall in circulating plasma FFA. Insulin also enhances lipogenesis and the synthesis of acylglycerol and increases the oxidation of glucose to CO_2 via the pentose phosphate pathway. All of these effects are dependent on the presence of glucose and can be explained, to a large extent, on the basis of the ability of insulin to enhance the uptake of glucose into adipose cells via the **GLUT 4 transporter**. In addition, insulin increases the activity of the enzymes pyruvate dehydrogenase, acetyl-CoA carboxylase, and glycerol phosphate acyltransferase, reinforcing the effects of increased glucose uptake on the enhancement of fatty acid and acylglycerol synthesis. These three enzymes are regulated in a coordinate manner by phosphorylation–dephosphorylation mechanisms (see Chapters 17, 23, and 24).

Another principal action of insulin in adipose tissue is to inhibit the activity of **hormone-sensitive lipase**, reducing the release not only of FFA but also of glycerol. Adipose tissue is much more sensitive to insulin than many other tissues, and is thus a major site of insulin action in vivo.

Several Hormones Promote Lipolysis

Other hormones accelerate the release of FFA from adipose tissue and raise the plasma-free fatty acid concentration by increasing the rate of lipolysis of the triacylglycerol stores (**Figure 25–8**). These include **epinephrine, norepinephrine, glucagon, adrenocorticotropic hormone (ACTH), thyroid-stimulating hormone (TSH), growth hormone (GH)**. Many of these activate hormone-sensitive lipase. For an optimal effect, most of these lipolytic processes require the presence of **glucocorticoids** and **thyroid hormones**. These hormones act in a **facilitatory** or **permissive** capacity with respect to other lipolytic endocrine factors.

The hormones that act rapidly in promoting lipolysis, that is, catecholamines (epinephrine and norepinephrine), do so by

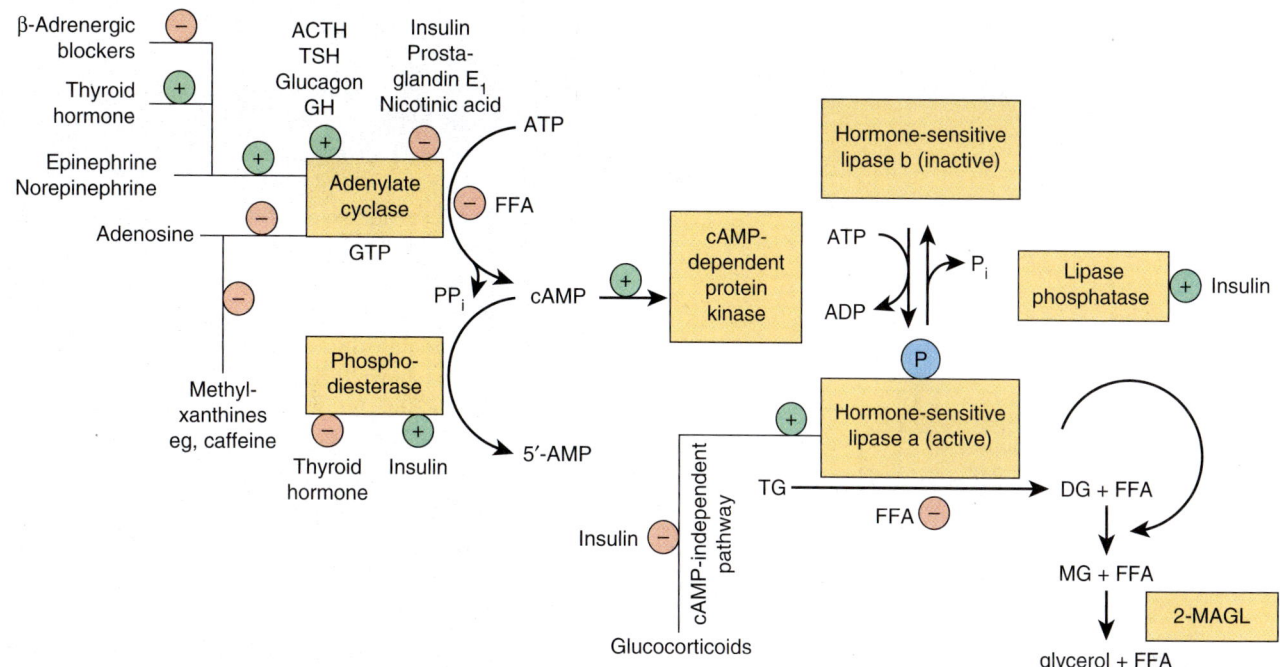

FIGURE 25–8 **Control of adipose tissue lipolysis.** Hormone-sensitive lipase is activated when phosphorylated by cAMP-dependent protein kinase. cAMP is generated by adenylate cyclase and degraded by phosphodiesterase. The lipolytic stimulus is "switched off" by removal of the stimulating hormone; the action of lipase phosphatase; the inhibition of the lipase and adenylyl cyclase by high concentrations of FFA; the inhibition of adenylyl cyclase by adenosine; and the removal of cAMP by the action of phosphodiesterase. The stimulatory effects of epinephrine and norepinephrine on adenylate cyclase are promoted by β-adrenergic blockers and inhibited by thyroid hormone. Similarly, methylxanthines block the inhibition of the enzyme by adenosine. ⊕, positive and ⊖, negative regulatory effects. (2-MAGL, 2-monoacylglycerol lipase; ACTH, adrenocorticotropin hormone; FFA, free fatty acids; GH, growth hormone; TSH, thyroid-stimulating hormone.)

stimulating the activity of **adenylyl cyclase**, the enzyme that converts ATP to cAMP. The mechanism is analogous to that responsible for hormonal stimulation of glycogenolysis (see Chapter 18). cAMP, by stimulating **cAMP-dependent protein kinase**, activates hormone-sensitive lipase. Thus, processes which destroy or preserve cAMP influence lipolysis. cAMP is degraded to 5′-AMP by the enzyme **cyclic 3′,5′-nucleotide phosphodiesterase**. This enzyme is inhibited by methylxanthines such as **caffeine** and **theophylline**. **Insulin** antagonizes the effect of the lipolytic hormones. Lipolysis appears to be more sensitive to changes in concentration of insulin than are glucose utilization and esterification. The antilipolytic effects of insulin, as well as those of nicotinic acid and prostaglandin E₁ which also act in this way, are accounted for by inhibition of the synthesis of cAMP at the adenylyl cyclase site, acting through a G_i protein. Insulin also stimulates the phosphodiesterase and the lipase phosphatase that inactivates hormone-sensitive lipase. The effect of growth hormone in promoting lipolysis is dependent on synthesis of proteins involved in the formation of cAMP. Glucocorticoids promote lipolysis via synthesis of new lipase protein by a cAMP-independent pathway, which may be inhibited by insulin, and also by promoting transcription of genes involved in the cAMP signal cascade. These findings help to explain the role of the pituitary gland and the adrenal cortex in enhancing fat mobilization. The sympathetic nervous system, through liberation of norepinephrine in adipose tissue, plays a central role in the mobilization of FFA.

Thus, the increased lipolysis caused by many of the factors described earlier can be reduced or abolished by denervation of adipose tissue or by ganglionic blockade.

Perilipin Regulates the Balance Between Triacylglycerol Storage & Lipolysis in Adipocytes

Perilipin, a protein involved in the formation of lipid droplets in adipocytes, inhibits lipolysis in basal conditions by preventing access of the lipase enzymes to the stored triacylglycerols. On stimulation with hormones which promote triacylglycerol degradation, however, the protein becomes phosphorylated and changes its conformation, exposing the lipid droplet surface to hormone-sensitive lipase and thus promoting lipolysis. Perilipin, therefore, enables the storage and breakdown of triacylglycerol to be coordinated according to the metabolic needs of the body.

BROWN ADIPOSE TISSUE PROMOTES THERMOGENESIS

Brown adipose tissue (BAT) is a specialized form of adipose tissue involved in metabolism, and in **thermogenesis** (heat generation). Thus, it is extremely active in some species, for example, during arousal from hibernation, in animals exposed

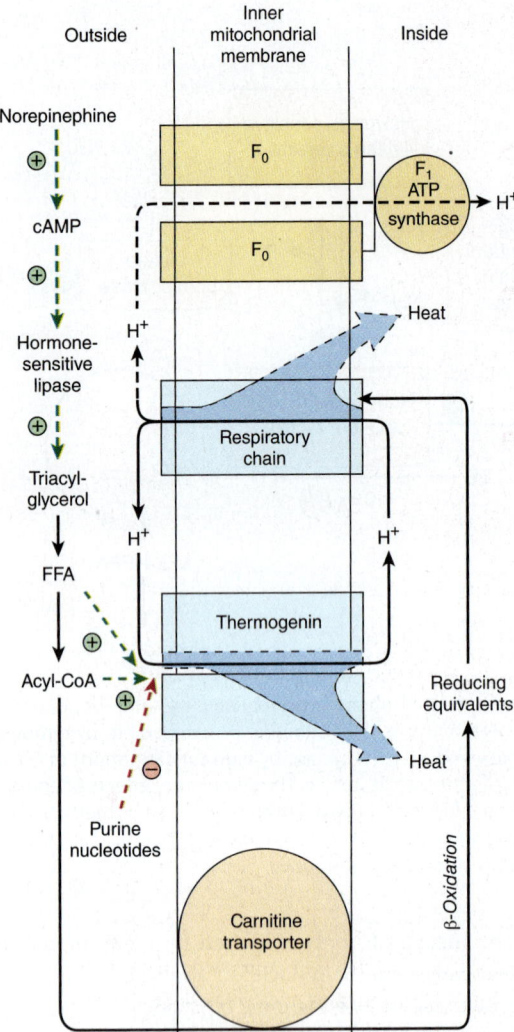

FIGURE 25–9 **Thermogenesis in brown adipose tissue.** Activity of the respiratory chain results in the translocation of protons from the mitochondrial matrix into the intermembrane space (see Figure 13–7). The presence of thermogenin (UCP1) in brown adipose tissue enables the protons to flow back into the matrix without passing through the F₁ ATP synthase and the energy generated is dissipated as heat instead of being captured as ATP. The passage of H⁺ via thermogenin is inhibited by purine nucleotides when brown adipose tissue is unstimulated. Under the influence of norepinephrine, the inhibition is removed by the production of free fatty acids (FFA) and acyl-CoA. Note the dual role of acyl-CoA in both facilitating the action of thermogenin and supplying reducing equivalents for the respiratory chain. ⊕ and ⊖ signify positive or negative regulatory effects.

to cold (nonshivering thermogenesis), and in heat production in the newborn. Though not a prominent tissue in humans, it is present in normal individuals. BAT is characterized by a well-developed blood supply and a high content of mitochondria and cytochromes, but low activity of ATP synthase. Metabolic emphasis is placed on oxidation of both glucose and fatty acids. **Norepinephrine** liberated from sympathetic nerve endings is important in increasing lipolysis in the tissue and

increasing synthesis of lipoprotein lipase to enhance utilization of triacylglycerol-rich lipoproteins from the circulation. Oxidation and phosphorylation are not coupled in mitochondria of this tissue because of the presence of a thermogenic uncoupling protein, thermogenin (also called uncoupling protein 1 [UCP1]) and the phosphorylation that does occur is at the substrate level, for example, at the succinate thiokinase step and in glycolysis. Thus, **oxidation produces much heat, and little free energy is trapped in ATP. Thermogenin** acts as a proton conductance pathway dissipating the electrochemical potential across the mitochondrial membrane (**Figure 25–9**). Recent research has shown that BAT activity is inversely related to body fat content, and thus is a potential target for the treatment of obesity and related metabolic disorders.

SUMMARY

- Since nonpolar lipids are insoluble in water, for transport between the tissues in the aqueous blood plasma they are combined with amphipathic lipids and proteins to make water-miscible lipoproteins.

- Four major groups of lipoproteins are recognized. Chylomicrons transport lipids resulting from digestion and absorption. VLDLs transport triacylglycerol from the liver. LDLs deliver cholesterol to the tissues, and HDLs remove cholesterol from the tissues and return it to the liver for excretion in the process known as reverse cholesterol transport.

- Chylomicrons and VLDL are metabolized by hydrolysis of their triacylglycerol, leaving lipoprotein remnants in the circulation. These are taken up by liver, but some of the remnants (IDL) resulting from VLDL form LDL are taken up by extrahepatic tissues as well as the liver via the LDL receptor.

- Apolipoproteins constitute the protein moiety of lipoproteins. They act as enzyme activators (eg, apo C-II and apo A-I) or as ligands for cell receptors (eg, apo A-I, apo E, and apo B-100).

- Triacylglycerol is the main storage lipid in adipose tissue. Upon mobilization, FFA and glycerol are released. FFAs are an important fuel source.

- Brown adipose tissue is the site of "nonshivering thermogenesis." It is found in hibernating and newborn animals and is present in adult humans. Thermogenesis results from the presence of UCP1 (thermogenin) in the inner mitochondrial membrane.

REFERENCES

Eljamil AS: *Lipid Biochemistry: For Medical Sciences.* iUniverse, 2015.

Francis G: High density lipoproteins: metabolism and protective roles against atherosclerosis. In *Biochemistry of Lipids, Lipoproteins and Membranes*, 6th ed. Ridgway N, McLeod R (editors). Academic Press, 2015:437-459.

McLeod RS, Yao Z: Assembly and secretion of triacylglycerol-rich lipoproteins. In *Biochemistry of Lipids, Lipoproteins and Membranes*, 6th ed. Ridgway N, McLeod R (editors). Academic Press, 2015:460-488.

Cholesterol Synthesis, Transport, & Excretion

Kathleen M. Botham, PhD, DSc, & Peter A. Mayes, PhD, DSc

OBJECTIVES

After studying this chapter, you should be able to:

- Explain the importance of cholesterol as an essential structural component of cell membranes and as a precursor of all other steroids in the body, and indicate its pathologic role in cholesterol gallstone disease and atherosclerosis development.

- Identify the five stages in the biosynthesis of cholesterol from acetyl-CoA.

- Indicate the role of 3-hydroxy-3-methylglutaryl-CoA reductase (HMG-CoA reductase) in controlling the rate of cholesterol synthesis and explain the mechanisms by which its activity is regulated.

- Explain that cholesterol balance in cells is tightly regulated and indicate the factors involved in maintaining the correct balance.

- Explain the role of plasma lipoproteins, including chylomicrons, very-low-density lipoprotein (VLDL), low-density lipoprotein (LDL), and high-density lipoprotein (HDL), in the transport of cholesterol between tissues in the plasma.

- Name the two main primary bile acids found in mammals and outline the pathways by which they are synthesized from cholesterol in the liver.

- Explain the importance of bile acid synthesis not only in the digestion and absorption of fats but also as a major excretory route for cholesterol.

- Indicate how secondary bile acids are produced from primary bile acids by intestinal bacteria.

- Explain what is meant by the "enterohepatic circulation" and why it is important.

- Identify various factors related to plasma cholesterol concentrations that affect the risk of coronary heart disease, including diet and lifestyle and the class of lipoprotein in which it is carried.

- Give examples of inherited and noninherited conditions affecting lipoprotein metabolism that cause hypo- or hyperlipoproteinemia.

BIOMEDICAL IMPORTANCE

Cholesterol is present in tissues and in plasma either as free cholesterol or combined with a long-chain fatty acid as cholesteryl ester, the storage form. In plasma, both forms are transported in lipoproteins (see Chapter 25). Cholesterol is an amphipathic lipid and as such is an essential structural component of membranes, where it is important for the maintenance of the correct permeability and fluidity (see Chapter 40), and of the outer layer of plasma lipoproteins. It is synthesized in many tissues from acetyl-CoA and is the precursor of all other steroids in the body, including **corticosteroids, sex hormones, bile acids,** and **vitamin D**. As a typical product of animal metabolism, cholesterol occurs in foods of animal origin such as egg yolk, meat, liver, and brain. Plasma **low-density lipoprotein (LDL)** is the vehicle that supplies cholesterol and cholesteryl ester to many tissues. Free cholesterol is removed from tissues by plasma **high-density lipoprotein (HDL)** and transported to the liver, where it is eliminated from the body either unchanged or after conversion to bile acids in the process known as **reverse cholesterol transport** (see Chapter 25).

Cholesterol is a major constituent of **gallstones**. However, its chief role in pathologic processes is as a factor in the development of **atherosclerosis** of vital arteries, causing cerebrovascular, coronary, and peripheral vascular disease.

CHOLESTEROL IS BIOSYNTHESIZED FROM ACETYL-CoA

A little more than half the cholesterol of the body arises by synthesis (about 700 mg/d), and the remainder is provided by the average diet. The liver and intestine account for approximately 10% each of total synthesis in humans. Virtually all tissues containing nucleated cells are capable of cholesterol synthesis, which occurs in the endoplasmic reticulum and the cytosolic compartments.

Acetyl-CoA Is the Source of All Carbon Atoms in Cholesterol

Cholesterol is a 27-carbon compound consisting of four rings and a side chain (see Figure 21–19). It is synthesized from acetyl-CoA by a lengthy pathway that may be divided into five steps: (1) synthesis of **mevalonate** from acetyl-CoA (**Figure 26–1**); (2) formation of **isoprenoid units** from mevalonate by loss of CO_2 (**Figure 26–2**); (3) condensation of six isoprenoid units form **squalene** (see Figure 26–2); (4) cyclization of squalene gives rise to the parent steroid, **lanosterol** (**Figure 26–3**); (5) formation of cholesterol from lanosterol (see Figure 26–3).

Step 1—Biosynthesis of Mevalonate: HMG-CoA (3-hydroxy-3-methylglutaryl-CoA) is formed by the reactions used in mitochondria to synthesize ketone bodies (see Figure 22–7). However, since cholesterol synthesis is extramitochondrial, the two pathways are distinct. Initially, two molecules of acetyl-CoA condense to form acetoacetyl-CoA catalyzed by cytosolic **thiolase**. Acetoacetyl-CoA condenses with a further molecule of acetyl-CoA catalyzed by **HMG-CoA synthase** to form HMG-CoA, which is reduced to **mevalonate** by NADPH in a reaction catalyzed by **HMG-CoA reductase**. This last step is the principal regulatory step in the pathway of cholesterol synthesis and is the site of action of the most effective class of cholesterol-lowering drugs, the statins, which are HMG-CoA reductase inhibitors (see Figure 26–1).

Step 2—Formation of Isoprenoid Units: Mevalonate is phosphorylated sequentially using ATP by three kinases, and after decarboxylation (see Figure 26–2) of the active isoprenoid unit, **isopentenyl diphosphate**, is formed.

Step 3—Six Isoprenoid Units Form Squalene: Isopentenyl diphosphate is isomerized by a shift of the double bond to form **dimethylallyl diphosphate**, and then condensed with another molecule of isopentenyl diphosphate to form the 10-carbon intermediate **geranyl diphosphate** (see Figure 26–2). A further condensation with isopentenyl diphosphate forms **farnesyl diphosphate**. Two molecules of farnesyl diphosphate condense

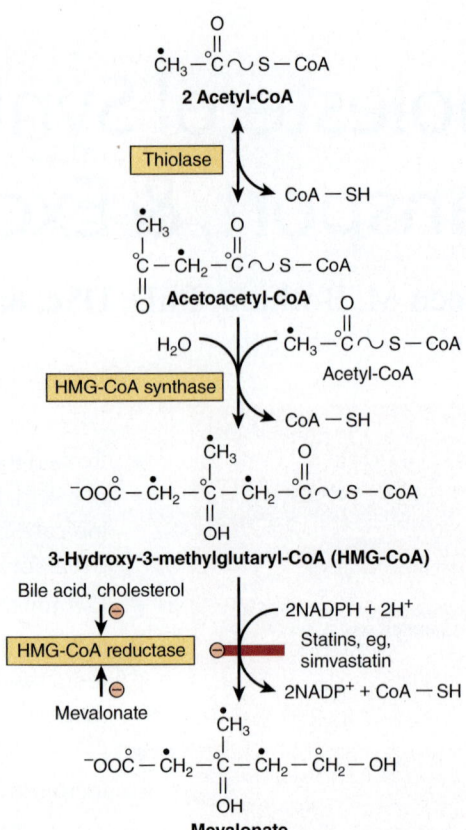

FIGURE 26–1 **Biosynthesis of mevalonate.** HMG-CoA reductase is inhibited by statins. The open and solid circles indicate the fate of each of the carbons in the acetyl moiety of acetyl-CoA.

at the diphosphate end to form **squalene**. Initially, inorganic pyrophosphate is eliminated, forming presqualene diphosphate, which is then reduced by NADPH with elimination of a further inorganic pyrophosphate molecule. Farnesyl diphosphate is also an intermediate in the formation of other important compounds which contain isoprenoid units such as dolichol and ubiquinone (see following discussion and Figure 26–2).

Step 4—Formation of Lanosterol: Squalene can fold into a structure that closely resembles the steroid nucleus (see Figure 26–3). Before ring closure occurs, squalene is converted to squalene 2,3-epoxide by **squalene epoxidase**, a mixed-function oxidase in the endoplasmic reticulum. The methyl group on C_{14} is transferred to C_{13} and that on C_8 to C_{14} as cyclization occurs, catalyzed by **oxidosqualene-lanosterol cyclase**.

Step 5—Formation of Cholesterol: The formation of cholesterol from **lanosterol** takes place in the membranes of the endoplasmic reticulum and involves changes in the steroid nucleus and the side chain (see Figure 26–3). The methyl groups on C_{14} and C_4 are removed to form 14-desmethyl lanosterol and then zymosterol. The double bond at C_8—C_9 is subsequently moved to C_5—C_6 in two steps, forming **desmosterol (24-dehydrocholesterol)**. Finally, the double bond of the side chain is reduced, producing cholesterol.

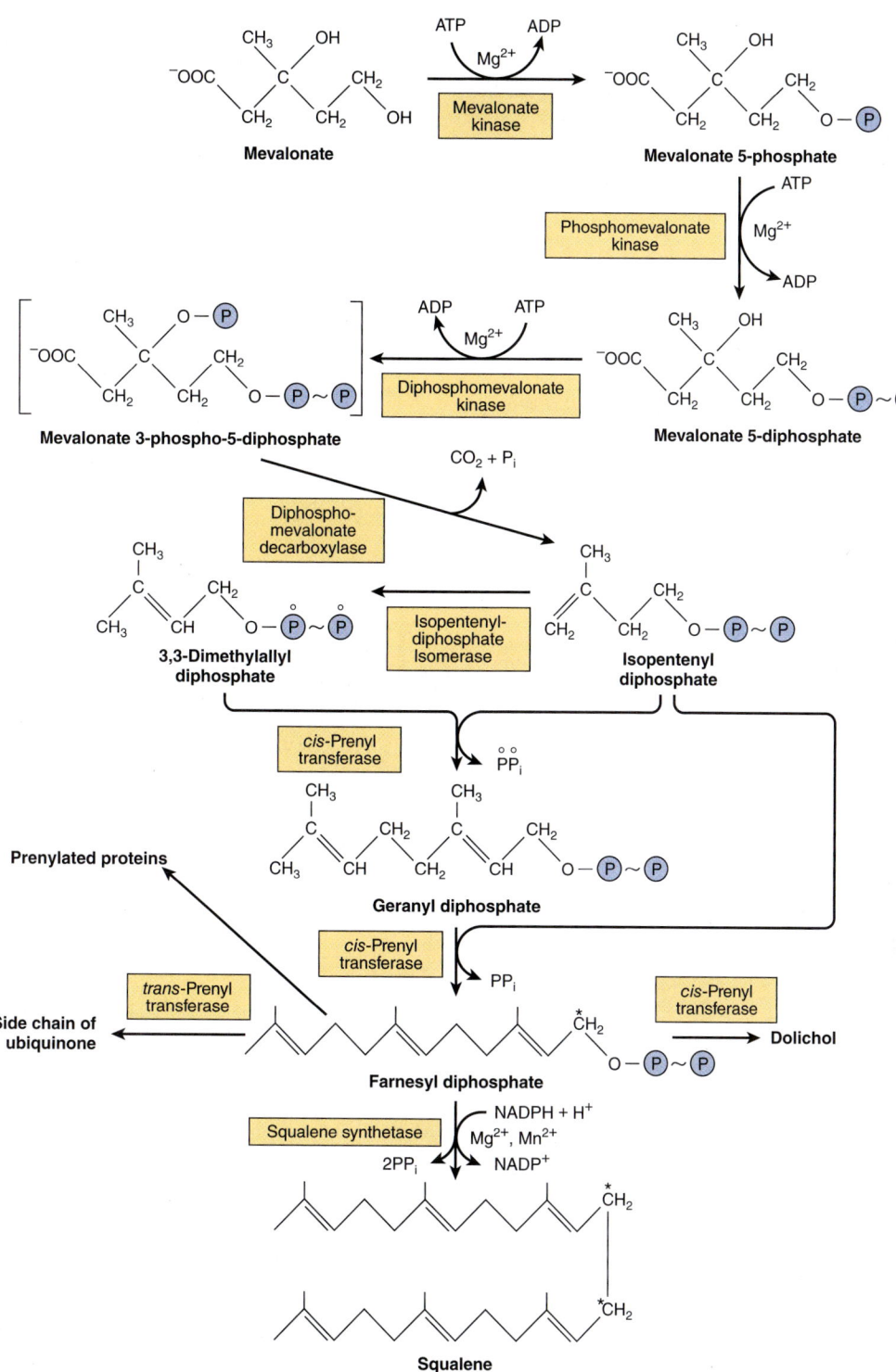

FIGURE 26–2 **Biosynthesis of squalene, ubiquinone, dolichol, and other polyisoprene derivatives.** (HMG, 3-hydroxy-3-methylglutaryl.) A farnesyl residue is present in heme of a cytochrome oxidase. The carbon marked with an asterisk becomes C_{11} or C_{12} in squalene. The open circles indicate the origin of the PP_i eliminated in the formation of geranyl diphosphate. Squalene synthetase is a microsomal enzyme; all other enzymes indicated are soluble cytosolic proteins, and some are found in peroxisomes.

Farnesyl Diphosphate Gives Rise to Dolichol & Ubiquinone

The polyisoprenoids, **dolichol** (see Figure 21–22 and Chapter 46) and **ubiquinone** (see Figure 13–6) are formed from farnesyl diphosphate by the further addition of up to 16 (dolichol) or 3 to 7 (ubiquinone) isopentenyl diphosphate residues (see Figure 26–2). Some **GTP-binding proteins** in the cell membrane are prenylated with farnesyl or geranyl-geranyl (20 carbon) residues. **Protein prenylation** is believed

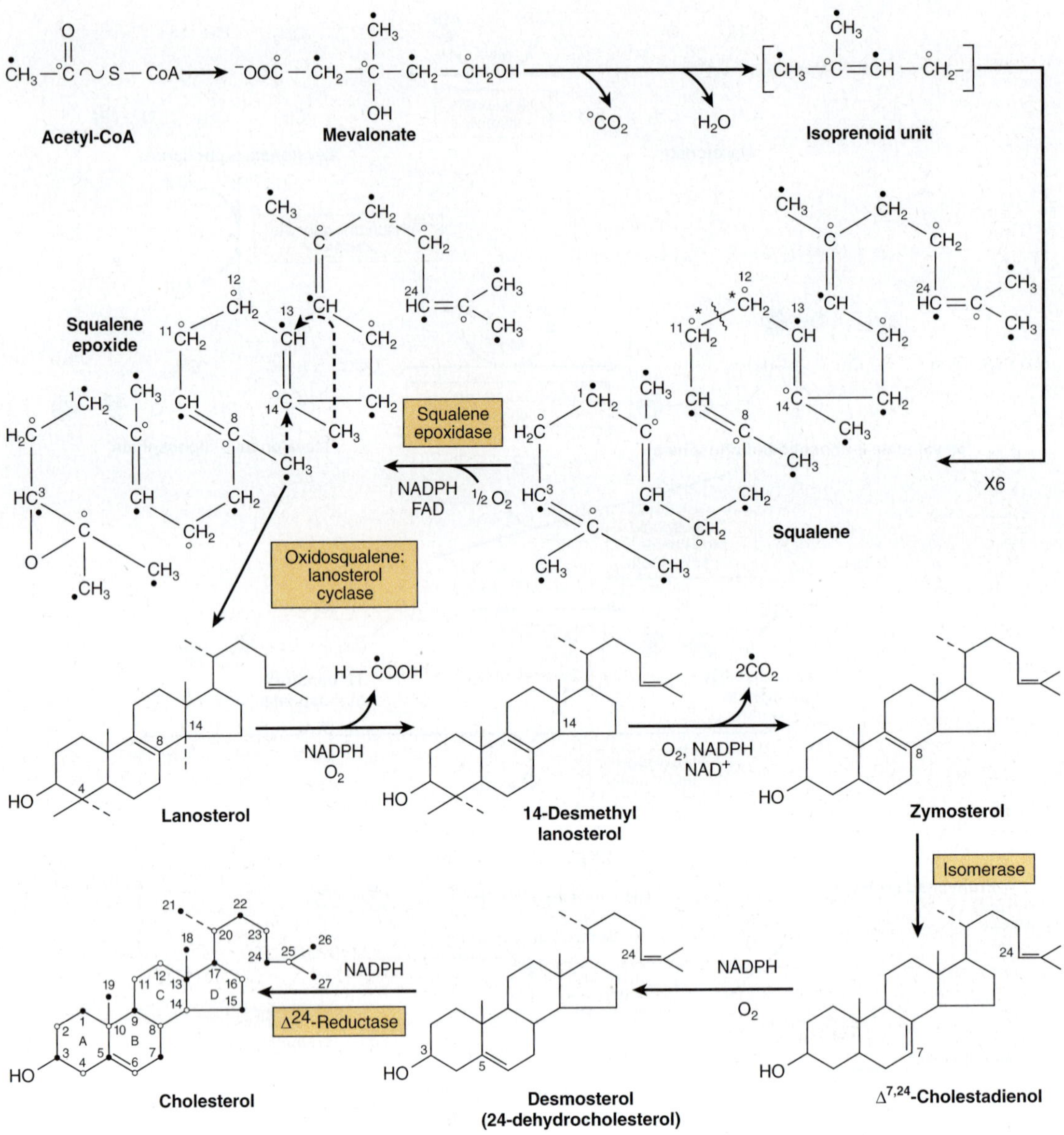

FIGURE 26–3 **Biosynthesis of cholesterol.** The numbered positions are those of the steroid nucleus and the open and solid circles indicate the fate of each of the carbons in the acetyl moiety of acetyl-CoA. (*Refer to labeling of squalene in Figure 26–2.)

to facilitate the anchoring of proteins into lipoid membranes and may also be involved in protein–protein interactions and membrane-associated protein trafficking.

CHOLESTEROL SYNTHESIS IS CONTROLLED BY REGULATION OF HMG-CoA REDUCTASE

Cholesterol synthesis is tightly controlled by regulation at the HMG-CoA reductase step. The activity of the enzyme is inhibited by mevalonate, the immediate product of the reaction, and by cholesterol, the main product of the pathway. Thus, increased intake of cholesterol from the diet leads to a decrease in *de novo* synthesis, especially in the liver.

Regulatory mechanisms include both modulation of the synthesis of enzyme protein and posttranslational modification. Cholesterol and metabolites repress transcription of HMG-CoA reductase mRNA via inhibition of a **sterol regulatory element-binding protein (SREBP)** transcription factor. SREBPs are a family of proteins that regulate the transcription of a range of genes involved in the cellular uptake and metabolism of cholesterol and other lipids. SREBP activation is inhibited by insulin-induced gene **(Insig)**, a protein whose expression, as its name indicates, is induced by insulin and is present in the endoplasmic reticulum. Insig also promotes degradation of HMG-CoA reductase. A **diurnal variation** occurs both in cholesterol synthesis and reductase activity. Short-term changes in enzyme activity, however, are brought about by posttranslational modification (**Figure 26–4**). **Insulin** or

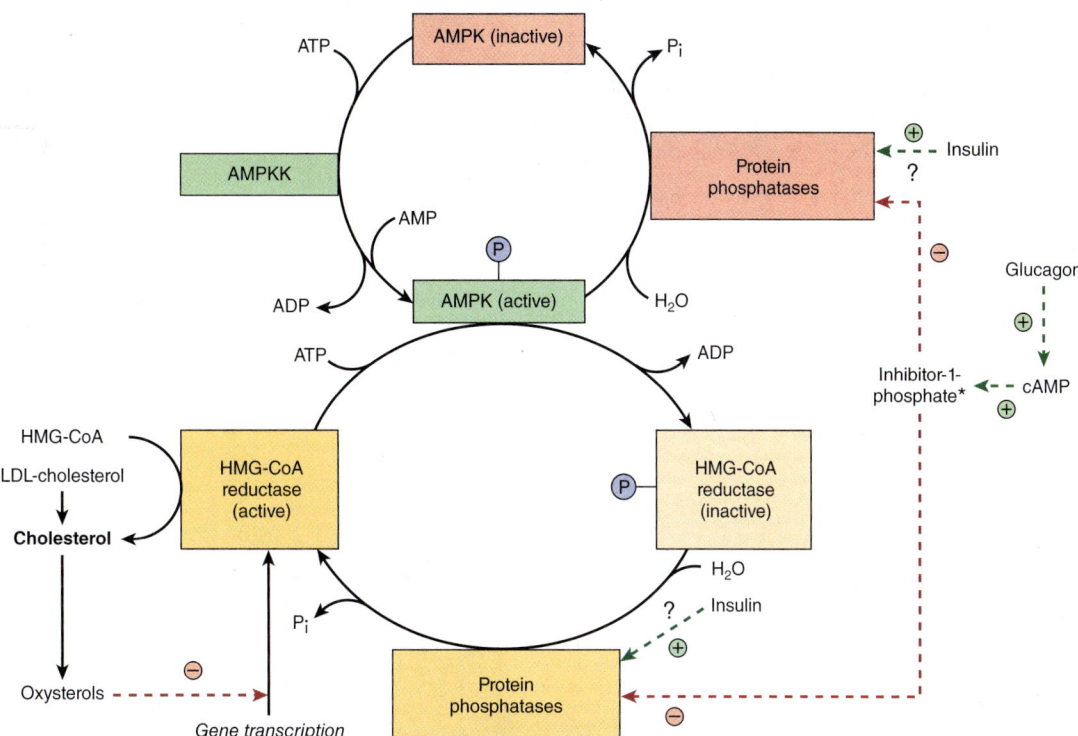

FIGURE 26–4 **Possible posttranslational mechanisms in the regulation of cholesterol synthesis by HMG-CoA reductase.** Insulin has a dominant role compared with glucagon. Stimulatory (⊕) or inhibitory (⊖) effects are shown as dotted green or red arrows, respectively. (AMPK, AMP-activated protein kinase; AMPKK, AMP-activated protein kinase kinase.) *See Figure 18–6.

thyroid hormone increases HMG-CoA reductase activity, whereas **glucagon** or **glucocorticoids** decrease it. Activity is reversibly modified by phosphorylation–dephosphorylation mechanisms, some of which may be cAMP-dependent and therefore immediately responsive to glucagon. **AMP-activated protein kinase (AMPK)** (formerly called HM-CoA reductase kinase) phosphorylates and inactivates HMG-CoA reductase. AMPK is activated via phosphorylation by **AMPK kinase (AMPKK)** and by allosteric modification by AMP.

THE CHOLESTEROL BALANCE IN TISSUES IS TIGHTLY REGULATED

In tissues, a tight balance must be maintained between those factors which increase cholesterol levels and those which decrease them, thus keeping intracellular concentrations constant (**Figure 26–5**). An increase in cell cholesterol may be caused by: (1) Uptake of cholesterol-containing lipoproteins by receptors, for example, the LDL receptor or scavenger receptors such as CD36. (2) Uptake of free cholesterol from cholesterol-rich lipoproteins to the cell membrane. (3) Cholesterol synthesis. (4) Hydrolysis of cholesteryl esters by the enzyme **cholesteryl ester hydrolase**. A decrease may be due to: (1) Efflux of cholesterol from the membrane to HDL via the ABCA1, ABCG1, or SR-B1 (see Figure 25–5). (2) Esterification of cholesterol by **ACAT** (acyl-CoA:cholesterol acyltransferase). (3) Utilization of cholesterol for synthesis of other steroids, such as hormones, in steroidogenic tissues, or bile acids in the liver.

The LDL Receptor Plays an Important Role in the Maintenance of Intracellular Cholesterol Balance

LDL (apo B-100, E) receptors occur on the cell surface in pits that are coated on the cytosolic side of the cell membrane with a protein called **clathrin**. The glycoprotein receptor spans the membrane, the B-100 binding region being at the exposed amino terminal end. After binding, LDL is taken up intact by **endocytosis**. The apoprotein and cholesteryl ester are then hydrolyzed in the lysosomes, and cholesterol is translocated into the cell. The receptors are recycled to the cell surface. This influx of cholesterol inhibits the transcription of the genes encoding HMG-CoA synthase, HMG-CoA reductase, and other enzymes involved in cholesterol synthesis, as well as the LDL receptor itself, via the SREBP pathway, and thus coordinately suppresses cholesterol synthesis and uptake. ACAT activity is also stimulated, promoting cholesterol esterification. In addition, recent research has shown that the protein **proprotein convertase subtilisin/kexin type 9 (PCSK9)** regulates the recycling of the receptor to the cell surface by targeting it for degradation. By these mechanisms, LDL receptor activity on the cell surface is regulated by the cholesterol requirement for membranes, steroid hormones, or bile acid

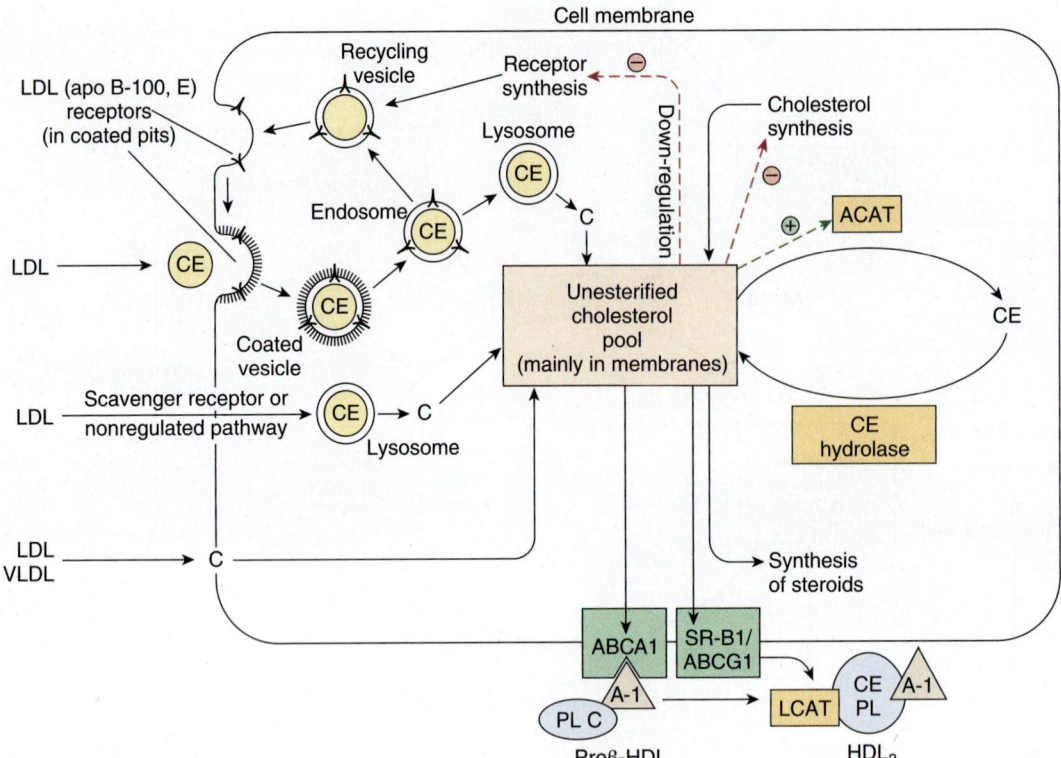

FIGURE 26–5 **Factors affecting cholesterol balance at the cellular level.** Reverse cholesterol transport may be mediated via the ABCA1 transporter protein (with preβ-HDL as the exogenous acceptor) or the SR-B1 or ABCG1 (with HDL$_3$ as the exogenous acceptor). Stimulatory (⊕) or inhibitory (⊖) effects are shown as dotted green or red arrows, respectively. (ACAT, acyl-CoA:cholesterol acyltransferase; A-I, apolipoprotein A-I; C, cholesterol; CE, cholesteryl ester; LCAT, lecithin:cholesterol acyltransferase; LDL, low-density lipoprotein; PL, phospholipid; VLDL, very-low-density lipoprotein.) LDL and HDL are not shown to scale.

synthesis, and the free cholesterol content of the cell is kept within relatively narrow limits (see Figure 26–5).

CHOLESTEROL IS TRANSPORTED BETWEEN TISSUES IN PLASMA LIPOPROTEINS

Cholesterol is transported in plasma in lipoproteins (see Table 25–1), with the greater part in the form of cholesteryl ester (**Figure 26–6**), and in humans the highest proportion is found in LDL. Dietary cholesterol equilibrates with plasma cholesterol in days and with tissue cholesterol in weeks. Cholesteryl ester in the diet is hydrolyzed to cholesterol, which is then absorbed by the intestine together with dietary unesterified cholesterol and other lipids. With cholesterol synthesized in the intestines, it is then incorporated into chylomicrons (see Chapter 25). Of the cholesterol absorbed, 80 to 90% is esterified with long-chain fatty acids in the intestinal mucosa. Ninety-five percent of the chylomicron cholesterol is delivered to the liver in chylomicron remnants, and most of the cholesterol secreted by the liver in very-low-density lipoprotein (VLDL) is retained during the formation of intermediate-density lipoprotein (IDL) and ultimately LDL, which is taken up by the LDL receptor in liver and extrahepatic tissues (see Chapter 25).

Plasma LCAT Is Responsible for Virtually All Plasma Cholesteryl Ester in Humans

Lecithin: cholesterol acyltransferase (LCAT) activity is associated with HDL containing apo A-I. As cholesterol in HDL becomes esterified, it creates a concentration gradient and draws in cholesterol from tissues and from other lipoproteins (see Figures 26–5 and 26–6), thus enabling HDL to function in **reverse cholesterol transport** (see Figure 25–5).

Cholesteryl Ester Transfer Protein Facilitates Transfer of Cholesteryl Ester From HDL to Other Lipoproteins

Cholesteryl ester transfer protein, associated with HDL, is found in plasma of humans and many other species. It facilitates transfer of cholesteryl ester from HDL to VLDL, IDL, and LDL in exchange for triacylglycerol, relieving product inhibition of the LCAT activity in HDL. Thus, in humans, much of the cholesteryl ester formed by LCAT finds its way to the liver via VLDL remnants (IDL) or LDL (see Figure 26–6). The triacylglycerol-enriched HDL$_2$ delivers its cholesterol to the liver in the HDL cycle (see Figure 25–5).

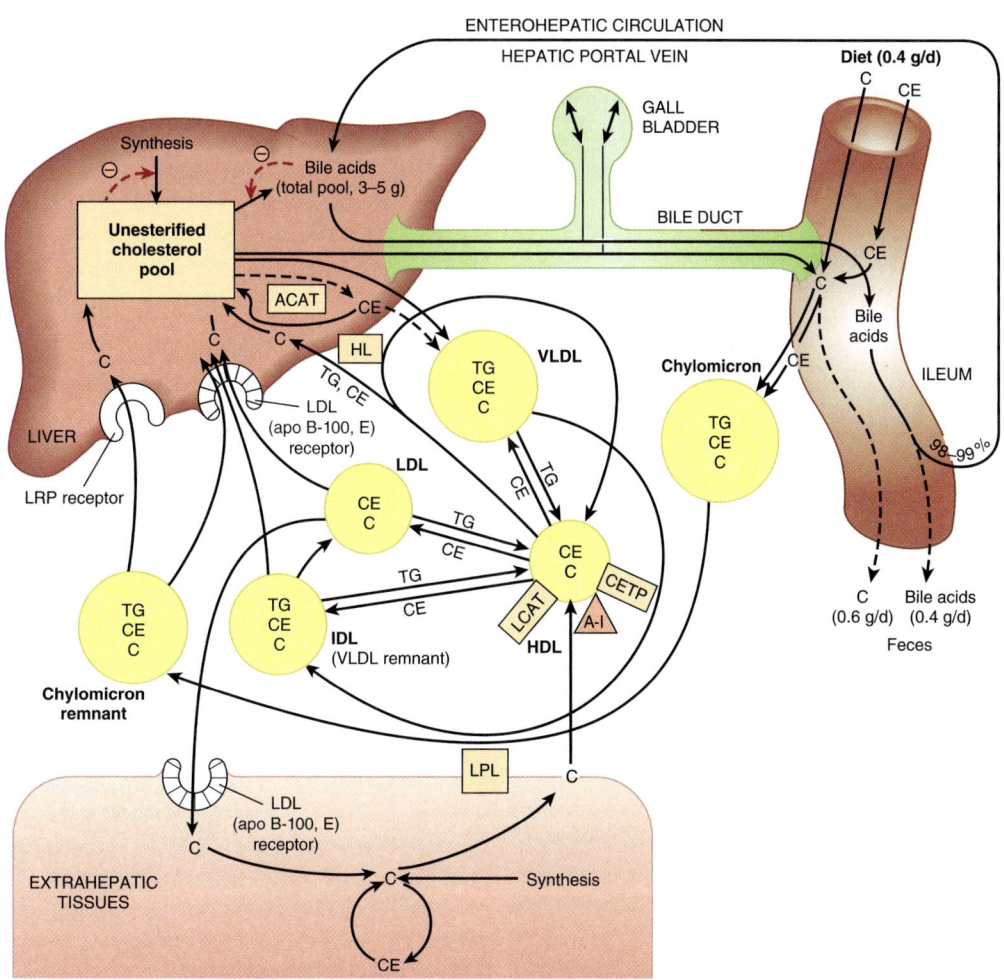

FIGURE 26–6 **Transport of cholesterol between the tissues in humans.** ACAT, acyl-CoA:cholesterol acyltransferase; A-I, apolipoprotein A-I; C, unesterified cholesterol; CE, cholesteryl ester; CETP, cholesteryl ester transfer protein; HDL, high-density lipoprotein; HL, hepatic lipase; IDL, intermediate-density lipoprotein; LCAT, lecithin:cholesterol acyltransferase; LDL, low-density lipoprotein; LPL, lipoprotein lipase; LRP, LDL receptor–related protein-1; TG, triacylglycerol; VLDL, very-low-density lipoprotein.

CHOLESTEROL IS EXCRETED FROM THE BODY IN THE BILE AS CHOLESTEROL OR AS BILE ACIDS

Cholesterol is excreted from the body via the bile either in the unesterified form or after conversion into bile acids in the liver. **Coprostanol** is the principal sterol in the feces; it is formed from cholesterol by the bacteria in the lower intestine.

Bile Acids Are Formed from Cholesterol

The primary bile acids are synthesized in the liver from cholesterol. These are **cholic acid** (found in the largest amount in most mammals) and **chenodeoxycholic acid** (**Figure 26–7**). The 7α-hydroxylation of cholesterol is the first and principal regulatory step in the biosynthesis of bile acids and is catalyzed by **cholesterol 7α-hydroxylase**, a microsomal cytochrome P450 enzyme–designated **CYP7A1** (see Chapter 12). A typical monooxygenase, it requires oxygen, NADPH, and cytochrome

P450. Subsequent hydroxylation steps are also catalyzed by monooxygenases. The pathway of bile acid biosynthesis divides early into one subpathway leading to **cholyl-CoA**, characterized by an extra α-OH group on position 12, and another pathway leading to **chenodeoxycholyl-CoA** (see Figure 26–7). A second pathway in mitochondria involving the 27-hydroxylation of cholesterol by the cytochrome P450 **sterol 27-hydroxylase** (**CYP27A1**) as the first step is responsible for a significant proportion of the primary bile acids synthesized. The primary bile acids (see Figure 26–7) enter the bile as glycine or taurine conjugates. Conjugation takes place in liver peroxisomes. In humans, the ratio of the glycine to the taurine conjugates is normally 3:1. The anions of the bile acids and their glyco- or tauro- conjugates are termed bile salts, and in the alkaline bile (pH 7.6-8.4), they are assumed to be in this form.

Primary bile acids are further metabolized in the intestine by the activity of the intestinal bacteria. Thus, deconjugation and 7α-dehydroxylation occur, producing the **secondary bile acids, deoxycholic acid,** and **lithocholic acid** (see Figure 26–7).

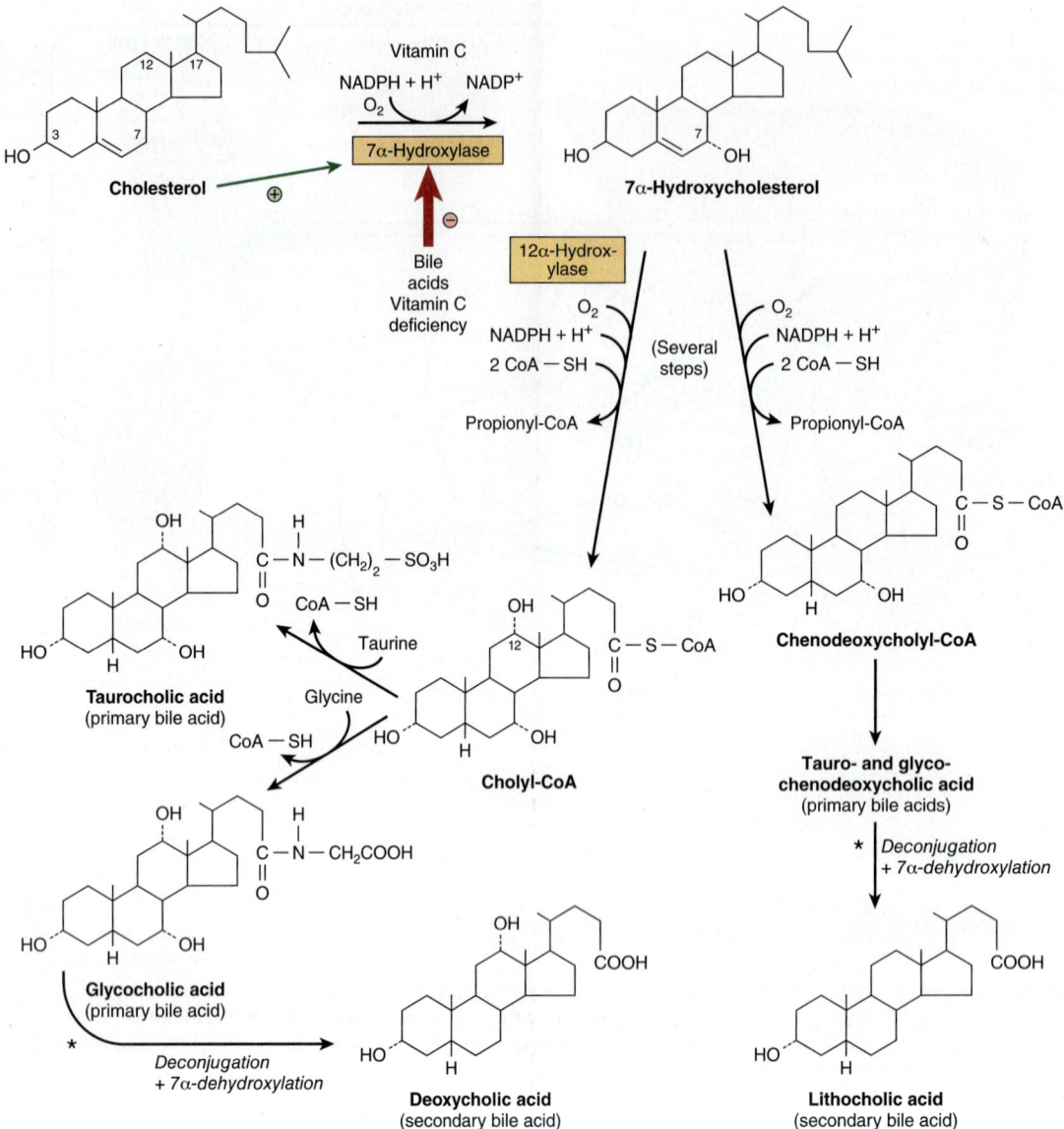

FIGURE 26–7 **Biosynthesis and degradation of bile acids.** A second pathway in mitochondria involves hydroxylation of cholesterol by sterol 27-hydroxylase. *Catalyzed by microbial enzymes.

Most Bile Acids Return to the Liver in the Enterohepatic Circulation

Although products of fat digestion, including cholesterol, are absorbed in the first 100 cm of small intestine, the primary and secondary bile acids are absorbed almost exclusively in the ileum, and 98 to 99% is returned to the liver via the portal circulation. This is known as the **enterohepatic circulation** (see Figure 26–6). However, lithocholic acid, because of its insolubility, is not reabsorbed to any significant extent. Only a small fraction of the bile acids escapes absorption and is therefore eliminated in the feces. Nonetheless, this represents a major pathway for the elimination of cholesterol. Each day the pool of bile acids (about 3-5 g) is cycled through the intestine 6 to 10 times and an amount of bile acid equivalent to that lost in the feces is synthesized from cholesterol, so that a pool of bile

acids of constant size is maintained. This is accomplished by a system of feedback controls.

Bile Acid Synthesis Is Regulated at the CYP7A1 Step

The principal rate-limiting step in the biosynthesis of bile acids is at the **CYP7A1 reaction** (see Figure 26–7). The activity of the enzyme is feedback regulated via the nuclear bile acid–binding receptor, **farnesoid X receptor (FXR)**. When the size of the bile acid pool in the enterohepatic circulation increases, FXR is activated, and transcription of the *CYP7A1* gene is suppressed. Chenodeoxycholic acid is particularly important in activating FXR. CYP7A1 activity is also enhanced by cholesterol of endogenous and dietary origin and regulated by insulin, glucagon, glucocorticoids, and thyroid hormone.

CLINICAL ASPECTS

Serum Cholesterol Is Correlated With the Incidence of Atherosclerosis & Coronary Heart Disease

Atherosclerosis is an inflammatory disease characterized by the deposition of cholesterol and cholesteryl ester from the plasma lipoproteins into the artery wall and is a major cause of heart disease. Elevated plasma cholesterol levels (>5.2 mmol/L) are one of the most important factors in promoting atherosclerosis, but it is now recognized that elevated blood triacylglycerol is also an independent risk factor. Diseases in which there is a prolonged elevation of levels of VLDL, IDL, chylomicron remnants, or LDL in the blood (eg, **diabetes mellitus, lipid nephrosis, hypothyroidism,** and **other conditions of hyperlipidemia**) are often accompanied by premature or more severe atherosclerosis. There is also an inverse relationship between HDL (HDL$_2$) concentrations and coronary heart disease, making the **LDL:HDL cholesterol ratio a good predictive parameter**. This is consistent with the function of HDL in reverse cholesterol transport. Susceptibility to atherosclerosis varies widely among species, and humans are one of the few in which the disease can be induced by diets high in cholesterol.

Diet Can Play an Important Role in Reducing Serum Cholesterol

Hereditary factors play the most important role in determining the serum cholesterol concentrations of individuals; however, dietary and environmental factors also play a part, and the most beneficial of these is the substitution in the diet of **polyunsaturated** and **monounsaturated fatty acids** for saturated fatty acids. Plant oils such as corn oil and sunflower seed oil contain a high proportion of ω6 polyunsaturated fatty acids, while olive oil contains a high concentration of monounsaturated fatty acids. ω3 fatty acids found in fish oils are also beneficial (see Chapter 21). On the other hand, butter fat, beef fat, and palm oil contain a high proportion of saturated fatty acids. Sucrose and fructose have a greater effect in raising blood lipids, particularly triacylglycerols, than do other carbohydrates.

One of the mechanisms by which unsaturated fatty acids lower blood cholesterol levels is by the upregulation of LDL receptors on the cell surface, causing an increase in the catabolic rate of LDL, the main atherogenic lipoprotein. In addition, ω3 fatty acids have anti-inflammatory and triacylglycerol-lowering effects. Saturated fatty acids also cause the formation of smaller VLDL particles that contain relatively more cholesterol, and they are utilized by extrahepatic tissues at a slower rate than are larger particles—tendencies that may be regarded as atherogenic.

Lifestyle Affects the Serum Cholesterol Level

Additional factors considered to play a part in coronary heart disease include **high blood pressure, smoking, male gender,** obesity (particularly abdominal obesity), lack of exercise, and drinking soft as opposed to hard water. Factors associated with elevation of plasma-free fatty acids (FFAs) followed by increased output of triacylglycerol and cholesterol into the circulation in VLDL include **emotional stress** and **coffee drinking**. Premenopausal women appear to be protected against many of these deleterious factors, and this is thought to be related to the beneficial effects of **estrogen**. There is an association between **moderate alcohol consumption** and a lower incidence of coronary heart disease. This may be due to elevation of HDL concentrations resulting from increased synthesis of apo A-I and changes in activity of cholesteryl ester transfer protein. It has been claimed that red wine is particularly beneficial, perhaps because of its content of antioxidants. Regular exercise lowers plasma LDL but raises HDL. Triacylglycerol concentrations are also reduced, due most likely to increased insulin sensitivity, which enhances the expression of lipoprotein lipase.

When Diet Changes Fail, Hypolipidemic Drugs Can Reduce Serum Cholesterol & Triacylglycerol

A family of drugs known as **statins** have proved highly efficacious in lowering plasma cholesterol and preventing heart disease. Statins act by inhibiting HMG-CoA reductase and upregulating LDL receptor activity. Examples currently in use include **atorvastatin, simvastatin, fluvastatin,** and **pravastatin. Ezetimibe** reduces blood cholesterol levels by inhibiting the absorption of cholesterol by the intestine via blockage of uptake by the **Niemann-Pick C-like 1 protein**. Other drugs used include fibrates such as **clofibrate, gemfibrozil,** and **nicotinic acid**, which act mainly to lower plasma triacylglycerols by decreasing the secretion of triacylglycerol and cholesterol-containing VLDL by the liver. Since PCSK9 reduces the number of LDL receptors exposed on the cell membrane, it has the effect of raising blood cholesterol levels, thus drugs that inhibit its activity are potentially antiatherogenic and two such compounds, **alirocumab** and **evolocumab**, have recently been approved for use and others are currently in development.

Primary Disorders of the Plasma Lipoproteins (Dyslipoproteinemias) Are Inherited

Inherited defects in lipoprotein metabolism lead to the primary condition of either **hypo-** or **hyperlipoproteinemia** (**Table 26–1**). For example, **familial hypercholesterolemia (FH)**, causes severe hypercholesterolemia and is associated with premature atherosclerosis. The defect is most often in the gene for the LDL receptor, so that LDL is not cleared from the blood. In addition, diseases such as diabetes mellitus, hypothyroidism, kidney disease (nephrotic syndrome), and atherosclerosis are associated with secondary abnormal lipoprotein patterns that are similar to one or another of the primary inherited conditions. Virtually all of the primary conditions

TABLE 26-1 Primary Disorders of Plasma Lipoproteins (Dyslipoproteinemias)

Name	Defect	Remarks
Hypolipoproteinemias Abetalipoproteinermia	No chylomicrons, VLDL, or LDL are formed because of defect in the loading of apo B with lipid.	Rare; blood acylglycerols low; intestine and liver accumulate acylglycerols. Intestinal malabsorption. Early death avoidable by administration of large doses of fat-soluble vitamins, particularly vitamin E.
Familial α-lipoprotein deficiency Tangier disease Fish-eye disease Apo A-I deficiencies	All have low or near absence of HDL.	Tendency toward hypertriacylglycerolemia as a result of absence of apo C-II, causing inactive LPL. Low LDL levels. Atherosclerosis in the elderly.
Hyperlipoproteinemias Familial lipoprotein lipase deficiency (type I)	Hypertriacylglycerolemia due to deficiency of LPL, abnormal LPL, or apo C-II deficiency causing inactive LPL.	Slow clearance of chylomicrons and VLDL. Low levels of LDL and HDL. No increased risk of coronary disease.
Familial hypercholesterolemia (type IIa)	Defective LDL receptors or mutation in ligand region of apo B-100.	Elevated LDL levels and hypercholesterolemia, resulting in atherosclerosis and coronary disease.
Familial type III hyperlipoproteinemia (broad β-disease, remnant removal disease, familial dysbetalipoproteinemia)	Deficiency in remnant clearance by the liver is due to abnormality in apo E. Patients lack isoforms E3 and E4 and have only E2, which does not react with the E receptor.[a]	Increase in chylomicron and VLDL remnants of density <1.019 (β-VLDL). Causes hypercholesterolemia, xanthomas, and atherosclerosis.
Familial hypertriacylglycerolemia (type IV)	Overproduction of VLDL often associated with glucose intolerance and hyperinsulinemia.	Cholesterol levels rise with the VLDL concentration. LDL and HDL tend to be subnormal. This type of pattern is commonly associated with coronary heart disease, type 2 diabetes mellitus, obesity, alcoholism, and administration of progestational hormones.
Familial hyperalphalipoproteinemia	Increased concentrations of HDL.	A rare condition apparently beneficial to health and longevity.
Hepatic lipase deficiency	Deficiency of the enzyme leads to accumulation of large triacylglycerol-rich HDL and VLDL remnants.	Patients have xanthomas and coronary heart disease.
Familial lecithin:cholesterol acyltransferase (LCAT) deficiency	Absence of LCAT leads to block in reverse cholesterol transport. HDL remains as nascent disks incapable of taking up and esterifying cholesterol.	Plasma concentrations of cholesteryl esters and lysolecithin are low. Present is an abnormal LDL fraction, lipoprotein X, found also in patients with cholestasis. VLDL is abnormal (β-VLDL).
Familial lipoprotein(a) excess	Lp(a) consists of 1 mol of LDL attached to 1 mol of apo(a). Apo(a) shows structural homologies to plasminogen.	Premature coronary heart disease due to atherosclerosis, plus thrombosis due to inhibition of fibrinolysis.

[a]There is an association between patients possessing the apo E4 allele and the incidence of Alzheimer disease. Apparently, apo E4 binds more avidly to β-amyloid found in neuritic plaques.
Abbreviations: LDL, low-density lipoprotein; LPL, lipoprotein lipase; HDL, high-density lipoprotein; VLDL, very-low-density lipoprotein.

are due to a defect at a stage in lipoprotein formation, transport, or degradation (see Figures 25–4, 26–5, and 26–6). Not all of the abnormalities are harmful.

SUMMARY

- Cholesterol is the precursor of all other steroids in the body, for example, corticosteroids, sex hormones, bile acids, and vitamin D. It also plays an important structural role in membranes and in the outer layer of lipoproteins.

- Cholesterol is synthesized in the body entirely from acetyl-CoA. Three molecules of acetyl-CoA form mevalonate via the important regulatory reaction for the pathway, catalyzed by HMG-CoA reductase. Next, a five-carbon isoprenoid unit is formed, and six of these condense to form squalene. Squalene

undergoes cyclization to form the parent steroid lanosterol, which, after the loss of three methyl groups and other changes, forms cholesterol.

- Cholesterol synthesis in the liver is regulated partly by cholesterol in the diet. In tissues, cholesterol balance is maintained between the factors causing gain of cholesterol (eg, synthesis, uptake via the LDL or scavenger receptors) and the factors causing loss of cholesterol (eg, steroid synthesis, cholesteryl ester formation, excretion). The activity of the LDL receptor is modulated by cellular cholesterol levels to achieve this balance. In reverse cholesterol transport, HDL takes up cholesterol from the tissues and LCAT then esterifies it and deposits it in the core of the particles. The cholesteryl ester in HDL is taken up by the liver, either directly or after transfer to VLDL, IDL, or LDL by the action of the cholesteryl ester transfer protein.

■ Excess cholesterol is excreted from the liver in the bile as cholesterol or bile acids. Most (98-99%) of the bile acids are reabsorbed in the ileum and returned to the liver in the enterohepatic circulation.

■ Elevated levels of cholesterol present in VLDL, IDL, or LDL are associated with atherosclerosis, whereas high levels of HDL have a protective effect.

■ Inherited defects in lipoprotein metabolism lead to a primary condition of hypo- or hyperlipoproteinemia. Conditions such as diabetes mellitus, hypothyroidism, kidney disease, and atherosclerosis exhibit secondary abnormal lipoprotein patterns that resemble certain primary conditions.

REFERENCES

Brown AJ, Sharpe LJ: Cholesterol synthesis. In *Biochemistry of Lipids, Lipoproteins and Membranes*, 6th ed. Ridgway N, McLeod R (editors). Academic Press, 2015:328-358.

Dawson PA: Bile acid metabolism. In *Biochemistry of Lipids, Lipoproteins and Membranes*, 6th ed. Ridgway N, McLeod R (editors). Academic Press, 2015:359-390.

Francis G: High density lipoproteins: metabolism and protective roles against atherosclerosis. In *Biochemistry of Lipids, Lipoproteins and Membranes*, 6th ed. Ridgway N, McLeod R (editors). Academic Press, 2015;437-459.

Exam Questions

Section V – Metabolism of Lipids

1. Which one of the following statements concerning fatty acid molecules is CORRECT?
   A. They consist of a carboxylic acid head group attached to a carbohydrate chain.
   B. They are called polyunsaturated when they contain one or more carbon–carbon double bonds.
   C. Their melting points increase with increasing unsaturation.
   D. They almost always have their double bonds in the *cis* configuration when they occur naturally.
   E. They occur in the body mainly in the form of free (nonesterified) fatty acids.

2. Which one of the following is NOT a phospholipid?
   A. Sphingomyelin
   B. Plasmalogen
   C. Cardiolipin
   D. Galactosylceramide
   E. Lysolecithin

3. Which one of the following statements about gangliosides is INCORRECT?
   A. They are derived from galactosylceramide.
   B. They contain one or more molecules of sialic acid.
   C. They are present in nervous tissue in high concentrations.
   D. The ganglioside GM1 is the receptor for cholera toxin in the human intestine.
   E. They function in cell–cell recognition.

4. Which one of the following is a chain-breaking antioxidant?
   A. Glutathione peroxidase
   B. Selenium
   C. Superoxide dismutase
   D. EDTA
   E. Catalase

5. After they are produced from acetyl-CoA in the liver, ketone bodies are mainly used for which one of the following processes?
   A. Excretion as waste products
   B. Energy generation in the liver
   C. Conversion to fatty acids for storage of energy
   D. Generation of energy in the tissues
   E. Generation of energy in red blood cells

6. The subcellular site of the breakdown of long-chain fatty acids to acetyl-CoA via β-oxidation is:
   A. The cytosol
   B. The matrix of the mitochondria
   C. The endoplasmic reticulum
   D. The mitochondrial intermembrane space
   E. The Golgi apparatus

7. Carnitine is needed for fatty acid oxidation BECAUSE:
   A. It is a cofactor for acyl-CoA synthetase, which activates fatty acids for breakdown.
   B. Long-chain acyl-CoA ("activated fatty acids") need to enter the mitochondrial intermembrane space to be oxidized, but cannot cross the outer mitochondrial membrane. Transfer of the acyl group from CoA to carnitine enables translocation to occur.
   C. Acylcarnitine, formed when long-chain acyl groups are transferred from CoA to carnitine is the substrate for the first step in the β-oxidation pathway.
   D. Long-chain acyl-CoA ("activated fatty acids") need to enter the mitochondrial matrix to be oxidized, but cannot cross the inner mitochondrial membrane. Transfer of the acyl group from CoA to carnitine enables translocation to occur.
   E. It prevents the breakdown of long-chain fatty acyl-CoA in the mitochondrial intermembrane space.

8. The breakdown of one molecule of a C16 fully saturated fatty acid (palmitic acid) by β-oxidation lead to the formation of:
   A. 8 $FADH_2$, 8 NADH, and 8 acetyl-CoA molecules
   B. 7 $FADH_2$, 7 NADH, and 7 acetyl-CoA molecules
   C. 8 $FADH_2$, 8 NADH, and 7 acetyl-CoA molecules
   D. 7 $FADH_2$, 8 NADH, and 8 acetyl-CoA molecules
   E. 7 $FADH_2$, 7 NADH, and 8 acetyl-CoA molecules

9. Malonyl-CoA, the first intermediate in fatty acid synthesis, is an important regulator of fatty acid metabolism BECAUSE:
   A. Its formation from acetyl-CoA and bicarbonate by the enzyme acetyl-CoA carboxylase is the main rate-limiting step in fatty acid synthesis.
   B. It prevents entry of fatty acyl groups into the matrix of the mitochondria because it is a potent inhibitor of carnitine palmitoyl transferase-I.
   C. It prevents entry of fatty acyl groups into the matrix of the mitochondria because it is a potent inhibitor of carnitine palmitoyl transferase-II.
   D. It prevents entry of fatty acyl groups into the matrix of the mitochondria because it is a potent inhibitor of carnitine–acylcarnitine translocase.
   E. It inhibits the synthesis of fatty acyl-CoA.

10. α-Linolenic acid is considered to be nutritionally essential in humans BECAUSE:
    A. It is an ω3 fatty acid.
    B. It contains three double bonds.
    C. In humans, double bonds cannot be introduced into fatty acids beyond the Δ9 position.
    D. In humans, double bonds cannot be introduced into fatty acids beyond the Δ12 position.
    E. Human tissues are unable to introduce a double bond in the Δ9 position of fatty acids.

11. Inactivation of acetyl-CoA carboxylase is favored WHEN:
    A. Cytosolic citrate levels are high.
    B. It is in a polymeric form.
    C. Palmitoyl-CoA levels are low.
    D. The tricarboxylate transporter is inhibited.
    E. It is dephosphorylated.

12. Which one of the following eicosanoids is synthesized from linoleic acid via the cyclooxygenase pathway?
    A. Prostaglandin E_1 (PGE_1)
    B. Leukotriene A_3 (LTA_3)
    C. Prostaglandin E_3 (PGE_3)
    D. Lipoxin A_4 (LXA_4)
    E. Thromboxane A_3 (TXA_3)

13. Which one of the following enzymes is inhibited by the nonsteroidal anti-inflammatory drug (NSAID) aspirin?
    A. Lipoxygenase
    B. Prostacyclin synthase
    C. Cyclooxygenase
    D. Thromboxane synthase
    E. Δ6 desaturase

14. Which one of the following is the major product of fatty acid synthase?
    A. Acetyl-CoA
    B. Oleate
    C. Palmitoyl-CoA
    D. Acetoacetate
    E. Palmitate

15. Fatty acids are broken down by repeated removal of two carbon fragments as acetyl-CoA in the β-oxidation cycle, and synthesized by repeated condensation of acetyl-CoAs until a long-chain saturated fatty acid with an even number of carbons is formed. Since fatty acids need to be broken down when energy is in short supply and synthesized when it is plentiful, there are important differences between the two processes which help cells to regulate them efficiently. Which one of the following statements concerning these differences is INCORRECT?
    A. Fatty acid breakdown takes place inside mitochondria, while synthesis occurs in the cytosol.
    B. Fatty acid breakdown uses NAD^+ and produces NADH, while synthesis uses NADPH and produces NADP.
    C. Fatty acyl groups are activated for breakdown using CoA and for synthesis using acyl carrier protein.
    D. Transport across the mitochondrial membrane of fatty acyl groups is required for fatty acid breakdown, but not for synthesis.
    E. Glucagon promotes fatty acid synthesis and inhibits fatty acid breakdown.

16. Hormone-sensitive lipase, the enzyme which mobilizes fatty acids from triacylglycerol stores in adipose tissue is inhibited by:
    A. Glucagon
    B. ACTH
    C. Epinephrine
    D. Vasopressin
    E. Prostaglandin E

17. Which one of the following best describes the action of phospholipase C?
    A. It releases the fatty acyl chain from the *sn*-2 position of a phospholipid.
    B. It cleaves a phospholipid into its phosphate-containing head group and a diacylglycerol.
    C. It releases the head group of a phospholipid, leaving phosphatidic acid.
    D. It releases the fatty acyl chain from the *sn*-1 position of a phospholipid.
    E. It releases the fatty acyl chains from the *sn*-1 and *sn*-2 positions of a phospholipid.

18. Tay-Sachs disease is a lipid storage disease caused by a genetic defect in which one of the following enzymes:
    A. β-Galactosidase
    B. Sphingomyelinase
    C. Ceramidase
    D. Hexosaminidase A
    E. β-Glucosidase

19. Which of the plasma lipoproteins is best described as follows: synthesized in the intestinal mucosa, contains a high concentration of triacylglycerol, and is responsible for the transport of dietary lipids in the circulation?
    A. Chylomicrons
    B. High-density lipoprotein
    C. Intermediate-density lipoprotein
    D. Low-density lipoprotein
    E. Very-low-density lipoprotein

20. Which of the plasma lipoproteins is best described as follows: synthesized in the liver, contains a high concentration of triacylglycerol, and is mainly cleared from the circulation by adipose tissue and muscle?
    A. Chylomicrons
    B. High-density lipoprotein
    C. Intermediate-density lipoprotein
    D. Low-density lipoprotein
    E. Very-low-density lipoprotein

21. Which of the plasma lipoproteins is best described as follows: formed in the circulation by removal of triacylglycerol from very-low-density lipoprotein, contains apo B-100, delivers cholesterol to extrahepatic tissues?
    A. Chylomicrons
    B. High-density lipoprotein
    C. Intermediate-density lipoprotein
    D. Low-density lipoprotein
    E. Very-low-density lipoprotein

22. Which of the following will be elevated in the bloodstream about 2 hours after eating a high-fat meal?
    A. Chylomicrons
    B. High-density lipoprotein
    C. Ketone bodies
    D. Nonesterified fatty acids
    E. Very-low-density lipoprotein

23. Which of the following will be elevated in the bloodstream about 4 hours after eating a high-fat meal?

    A. Low-density lipoprotein
    B. High-density lipoprotein
    C. Ketone bodies
    D. Nonesterified fatty acids
    E. Very-low-density lipoprotein

24. Which one of the following processes is NOT involved in the transfer of cholesterol from extrahepatic tissues and its delivery to the liver for excretion by HDL?

    A. Efflux of cholesterol from tissues to preβ-HDL via ABCA1.
    B. Esterification of cholesterol to cholesteryl ester by LCAT to form HDL_3.
    C. Transfer of cholesteryl ester from HDL to VLDL, IDL, and LDL by the action of cholesteryl ester transfer protein (CETP).
    D. Efflux of cholesterol from tissues to HDL_3 via SR-B1 and ABCG1.
    E. Selective uptake of cholesteryl ester from HDL_2 by the liver via SR-B1.

25. Which one of the following statements concerning chylomicrons is CORRECT?

    A. Chylomicrons are made inside intestinal cells and secreted into lymph, where they acquire apolipoproteins B and C.
    B. The core of chylomicrons contains triacylglycerol and phospholipids.
    C. The enzyme hormone-sensitive lipase acts on chylomicrons to release fatty acids from triacylglycerol when they are bound to the surface of endothelial cells in blood capillaries.
    D. Chylomicron remnants differ from chylomicrons in that they are smaller and contain a lower proportion of triacylglycerol and a higher proportion of cholesterol.
    E. Chylomicrons are taken up by the liver.

26. Which one of the following statements concerning the biosynthesis of cholesterol is CORRECT?

    A. The rate-limiting step is the formation of 3-hydroxy-3-methylglutaryl-CoA (HMG-CoA) by the enzyme HMG-CoA synthase.
    B. Synthesis occurs in the cytosol of the cell.
    C. All the carbon atoms in the cholesterol synthesized originate from acetyl-CoA.
    D. Squalene is the first cyclic intermediate in the pathway.
    E. The initial substrate is mevalonate.

27. The class of drugs called statins have proved very effective against hypercholesterolemia, a major cause of atherosclerosis and associated cardiovascular disease. These drugs reduce plasma cholesterol levels by:

    A. Preventing absorption of cholesterol from the intestine.
    B. Increasing the excretion of cholesterol from the body via conversion to bile acids.

    C. Inhibiting the conversion of 3-hydroxy-3-methylglutaryl-CoA to mevalonate in the pathway for cholesterol biosynthesis.
    D. Increasing the rate of degradation of 3-hydroxy-3-methylglutaryl-CoA reductase.
    E. Stimulating the activity of the LDL receptor in the liver.

28. Which of the following statements about bile acids (or bile salts) is INCORRECT?

    A. Primary bile acids are synthesized in the liver from cholesterol.
    B. Bile acids are needed for the breakdown of fats by pancreatic lipase.
    C. Secondary bile acids are produced by modification of primary bile acids in the liver.
    D. Bile acids facilitate the absorption of the products of lipid digestion in the jejunum.
    E. Bile acids are recirculated between the liver and the small intestine in the enterohepatic circulation.

29. A 35-year-old man with severe hypercholesterolemia has a family history of deaths at a young age from heart disease and stroke. Which of the following genes is likely to be defective?

    A. Apolipoprotein E
    B. The LDL receptor
    C. Lipoprotein lipase
    D. *PCSK9*
    E. LCAT

30. Compounds which inhibit the action of the recently discovered protein, proprotein convertase subtilisin/kexin type 9 (PCSK9), have been identified as a potential antiatherogenic drugs BECAUSE PCSK9:

    A. Decreases the number of LDL receptors exposed at the cell surface, thus LDL uptake is lowered and blood cholesterol levels rise.
    B. Inhibits the binding of apo B to the LDL receptor, thus blocking uptake of the lipoprotein and raising blood cholesterol levels.
    C. Increases the absorption of cholesterol from the intestine.
    D. Prevents the breakdown of cholesterol to bile acids in the liver.
    E. Increases the synthesis and secretion of VLDL in the liver, leading to increased LDL formation in the blood.

Biosynthesis of the Nutritionally Nonessential Amino Acids

Victor W. Rodwell, PhD

OBJECTIVES

After studying this chapter, you should be able to:

- Explain why the absence of certain amino acids that are present in most proteins from the human diet typically is not deleterious to human health.

- Appreciate the distinction between the terms "essential" and "nutritionally essential" amino acids, and identify the amino acids that are nutritionally nonessential.

- Name the intermediates of the citric acid cycle and of glycolysis that are precursors of aspartate, asparagine, glutamate, glutamine, glycine, and serine.

- Illustrate the key role of transaminases in amino acid metabolism.

- Explain the process by which the 4-hydroxyproline and 5-hydroxylysine of proteins such as collagen are formed.

- Describe the clinical presentation of scurvy, and provide a biochemical explanation for why a severe deprivation of vitamin C (ascorbic acid) results in this nutritional disorder.

- Appreciate that, despite the toxicity of selenium, selenocysteine is an essential component of several mammalian proteins.

- Define and outline the reaction catalyzed by a mixed-function oxidase.

- Identify the role of tetrahydrobiopterin in tyrosine biosynthesis.

- Indicate the role of a modified transfer RNA (tRNA) in the cotranslational insertion of selenocysteine into proteins.

BIOMEDICAL IMPORTANCE

Amino acid deficiency states can result if nutritionally essential amino acids are absent from the diet, or are present in inadequate amounts. In certain regions of West Africa, for example, **kwashiorkor** sometimes results when a child is weaned onto a starchy diet poor in protein, while **marasmus** may occur if both caloric intake and specific amino acids are deficient. Patients with short bowel syndrome, if unable to absorb sufficient quantities of calories and nutrients, may suffer from significant nutritional and metabolic abnormalities. Both the nutritional disorder **scurvy**, a dietary deficiency of vitamin C, and specific genetic disorders are associated with an impaired ability of connective tissue to form peptidyl 4-hydroxyproline and peptidyl 5-hydroxylysine. The resulting conformational instability of collagen is accompanied by bleeding gums, swelling joints, poor wound healing, and ultimately in death. **Menkes syndrome**, characterized by kinky hair and growth retardation, results from a dietary deficiency of copper, an essential cofactor for the enzyme lysyl oxidase that functions in formation of the covalent cross-links that strengthen collagen fibers. Genetic disorders of collagen biosynthesis include several forms of **osteogenesis imperfecta**, characterized by fragile bones, and **Ehlers-Danlos syndrome**, a group of connective tissue disorders that result in mobile joints and skin abnormalities due to defects in the genes that encode enzymes, including procollagen-lysine 5-hydroxylase.

NUTRITIONALLY ESSENTIAL & NUTRITIONALLY NONESSENTIAL AMINO ACIDS

While often employed with reference to amino acids, the terms "essential" and "nonessential" are misleading, since for human subjects all 20 common amino acids are essential to ensure health. Of these 20 amino acids, 8 *must* be present in the human diet, and thus are best termed "*nutritionally essential.*" The other 12 "*nutritionally nonessential*" amino acids, while *metabolically* essential, need not be present in the diet (**Table 27–1**). The distinction between these two classes of amino acids was established in the 1930s by feeding human subjects a diet in which purified amino acids replaced protein. Subsequent biochemical investigations ultimately revealed the reactions and intermediates involved in the biosynthesis of all 20 amino acids. Amino acid deficiency disorders are endemic in certain regions of West Africa where diets rely heavily on grains that are poor sources of tryptophan and lysine. These nutritional disorders include **kwashiorkor**, which results when a child is weaned onto a starchy diet poor in protein, and **marasmus**, in which both caloric intake and specific amino acids are deficient.

TABLE 27–1 Amino Acid Requirements of Humans

Nutritionally Essential	Nutritionally Nonessential
Arginine[a]	Alanine
Histidine	Asparagine
Isoleucine	Aspartate
Leucine	Cysteine
Lysine	Glutamate
Methionine	Glutamine
Phenylalanine	Glycine
Threonine	Hydroxyproline[b]
Tryptophan	Hydroxylysine[b]
Valine	Proline
	Serine
	Tyrosine

[a]Nutritionally "semiessential." Synthesized at rates inadequate to support growth of children.
[b]Not necessary for protein synthesis, but is formed during posttranslational processing of collagen.

Biosynthesis of the Nutritionally Essential Amino Acids Involves Lengthy Metabolic Pathways

The existence of nutritional requirements suggests that dependence on an *external* source of a specific nutrient can be of greater survival value than the ability to biosynthesize it. Why? Because if the diet contains ample quantities of a nutrient, retention of the ability to biosynthesize it represents information of *negative* survival value, because ATP and nutrients are not required to synthesize "unnecessary" DNA—even if specific encoded genes are no longer expressed. The number of enzymes required by prokaryotic cells to biosynthesize the nutritionally *essential* amino acids is large relative to the number of enzymes required for the formation of the nutritionally *nonessential* amino acids (**Table 27–2**). This suggests a survival advantage for humans to retain the ability to biosynthesize "easy" amino acids while losing the ability to make others more difficult to biosynthesize. The reactions by which organisms such as plants and bacteria, but not human subjects, form certain amino acids are not discussed. This chapter addresses the reactions and intermediates involved in the biosynthesis of the 12 nutritionally *nonessential* amino acids in human tissues and emphasizes selected medically significant disorders associated with their metabolism.

TABLE 27–2 Enzymes Required for the Synthesis of Amino Acids From Amphibolic Intermediates

Number of Enzymes Required to Synthesize			
Nutritionally Essential		Nutritionally Nonessential	
Arg[a]	7	Ala	1
His	6	Asp	1
Thr	6	Asn[b]	1
Met	5 (4 shared)	Glu	1
Lys	8	Gln[a]	1
Ile	8 (6 shared)	Hyl[c]	1
Val	6 (all shared)	Hyp[d]	1
Leu	7 (5 shared)	Pro[a]	3
Phe	10	Ser	3
Trp	5 (8 shared)	Gly[e]	1
	59 (total)	Cys[f]	2
		Tyr[g]	1
			17 (total)

[a]From Glu.
[b]From Asp.
[c]From Lys.
[d]From Pro.
[e]From Ser.
[f]From Ser plus sulfate.
[g]From Phe.

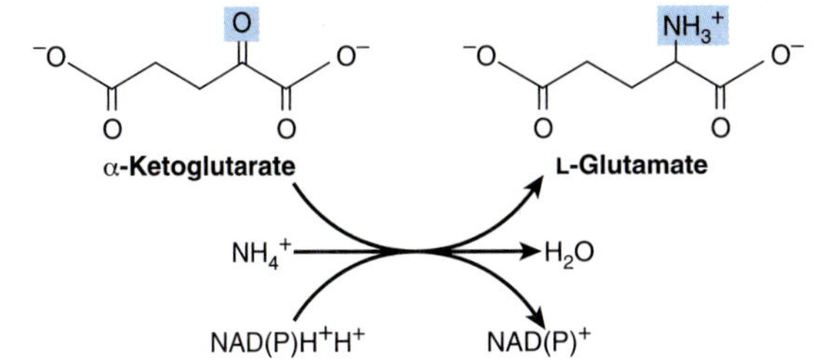

FIGURE 27–1 The reaction catalyzed by glutamate dehydrogenase (EC 1.4.1.3).

BIOSYNTHESIS OF THE NUTRITIONALLY NONESSENTIAL AMINO ACIDS

Glutamate

Glutamate, the precursor of the so-called "glutamate family" of amino acids, is formed by the reductive amidation of the citric acid cycle intermediate α-ketoglutarate, a reaction catalyzed by mitochondrial glutamate dehydrogenase (**Figure 27–1**). The reaction both strongly favors glutamate synthesis and lowers the concentration of cytotoxic ammonium ion.

Glutamine

The amidation of glutamate to glutamine catalyzed by glutamine synthetase (**Figure 27–2**) involves the intermediate formation of γ-glutamyl phosphate (**Figure 27–3**). Following the ordered binding of glutamate and ATP, glutamate attacks the γ-phosphorus of ATP, forming γ-glutamyl phosphate and ADP. NH_4^+ then binds, and uncharged NH_3 attacks γ-glutamyl phosphate. Release of P_i and of a proton from the γ-amino group of the tetrahedral intermediate then allows release of the product, glutamine.

Alanine & Aspartate

Transamination of pyruvate forms alanine (**Figure 27–4**). Similarly, transamination of oxaloacetate forms aspartate.

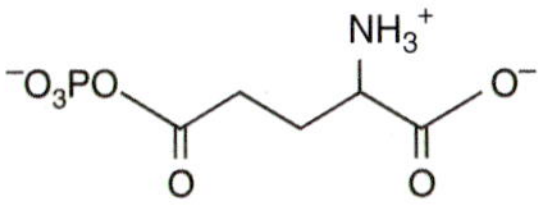

FIGURE 27–3 γ-Glutamyl phosphate.

Glutamate Dehydrogenase, Glutamine Synthetase, & Aminotransferases Play Central Roles in Amino Acid Biosynthesis

The combined action of the enzymes glutamate dehydrogenase, glutamine synthetase, and the aminotransferases (see Figures 27–1, 27–2, and 27–4) results in the incorporation of potentially cytotoxic ammonium ion into nontoxic amino acids.

Asparagine

The conversion of aspartate to asparagine, catalyzed by asparagine synthetase (**Figure 27–5**), resembles the glutamine synthetase reaction (see Figure 27–2), but glutamine, rather than ammonium ion, provides the nitrogen. Bacterial asparagine synthetases can, however, also use ammonium ion. The reaction involves the intermediate formation of aspartyl phosphate (**Figure 27–6**). The coupled hydrolysis of PP_i to P_i by pyrophosphatase, EC 3.6.1.1, ensures that the reaction is strongly favored.

Serine

Oxidation of the α-hydroxyl group of the glycolytic intermediate 3-phosphoglycerate, catalyzed by 3-phosphoglycerate dehydrogenase, converts it to 3-phosphohydroxypyruvate. Transamination and subsequent dephosphorylation then form serine (**Figure 27–7**).

Glycine

Glycine aminotransferases can catalyze the synthesis of glycine from glyoxylate and glutamate or alanine. Unlike most aminotransferase reactions, is strongly favored glycine synthesis. Additional important mammalian routes for glycine formation are from choline (**Figure 27–8**) and from serine (**Figure 27–9**).

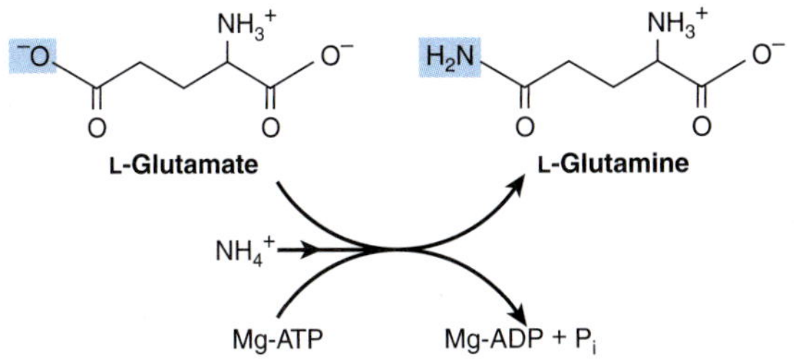

FIGURE 27–2 The reaction catalyzed by glutamine synthetase (EC 6.3.1.2).

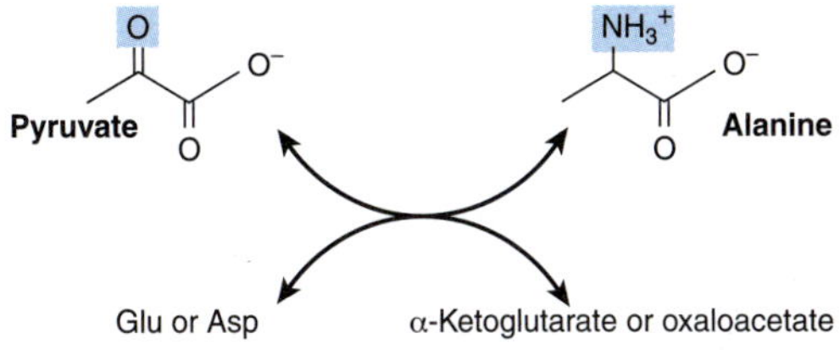

FIGURE 27–4 Formation of alanine by transamination of pyruvate. The amino donor may be glutamate or aspartate. The other product thus is α-ketoglutarate or oxaloacetate.

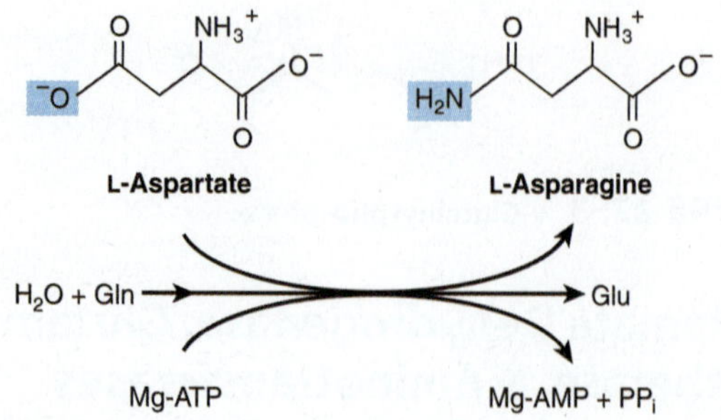

FIGURE 27–5 The reaction catalyzed by asparagine synthetase (EC 6.3.5.4). Note similarities to and differences from the glutamine synthetase reaction (Figure 27–2).

Proline

The initial reaction of proline biosynthesis converts the γ-carboxyl group of glutamate to the mixed acid anhydride of glutamate γ-phosphate (see Figure 27–3). Subsequent reduction forms glutamate γ-semialdehyde, which following spontaneous cyclization is reduced to L-proline (**Figure 27–10**).

Cysteine

While not itself nutritionally essential, cysteine is formed from methionine, which is nutritionally essential. Following conversion of methionine to homocysteine (see Figure 29–18), homocysteine and serine form cystathionine, whose hydrolysis forms cysteine and homoserine (**Figure 27–11**).

Tyrosine

Phenylalanine hydroxylase converts phenylalanine to tyrosine (**Figure 27–12**). If the diet contains adequate quantities of the nutritionally essential amino acid phenylalanine, tyrosine is nutritionally nonessential. However, since the phenylalanine hydroxylase reaction is irreversible, dietary tyrosine cannot replace phenylalanine. Catalysis by this mixed-function oxidase incorporates one atom of O_2 into the *para* position of phenylalanine and reduces the other atom to water. Reducing power, provided as tetrahydrobiopterin, derives ultimately from NADPH.

Hydroxyproline & Hydroxylysine

Hydroxyproline and hydroxylysine occur principally in collagen. Since there is no tRNA for either hydroxylated amino acid, neither dietary hydroxyproline nor dietary hydroxylysine is incorporated during protein synthesis. Peptidyl hydroxyproline and hydroxylysine arise from proline and lysine, but only after these amino acids have been incorporated into peptides. Hydroxylation of peptidyl prolyl and peptidyl lysyl residues, catalyzed by **prolyl hydroxylase** and **lysyl hydroxylase** of skin, skeletal muscle, and

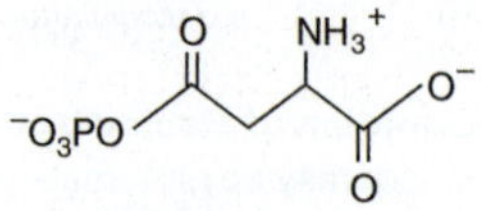

FIGURE 27–6 Aspartyl phosphate.

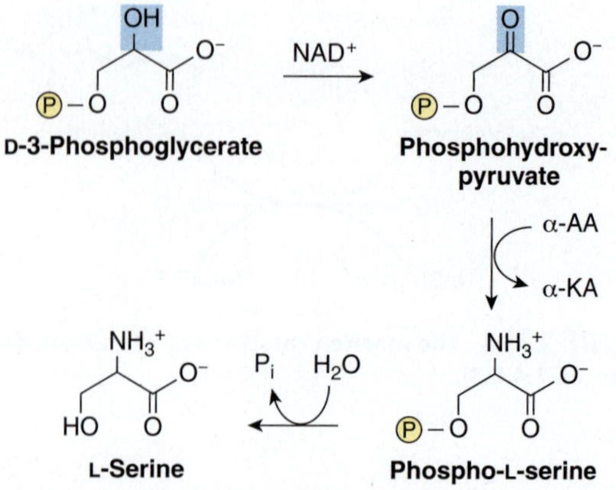

FIGURE 27–7 Serine biosynthesis. Oxidation of 3-phosphoglycerate is catalyzed by 3-phosphoglycerate dehydrogenase (EC 1.1.1.95). Transamination converts phosphohydroxypyruvate to phosphoserine. Hydrolytic removal of the phosphoryl group catalyzed by phosphoserine hydrolase (EC 3.1.3.3) then forms L-serine.

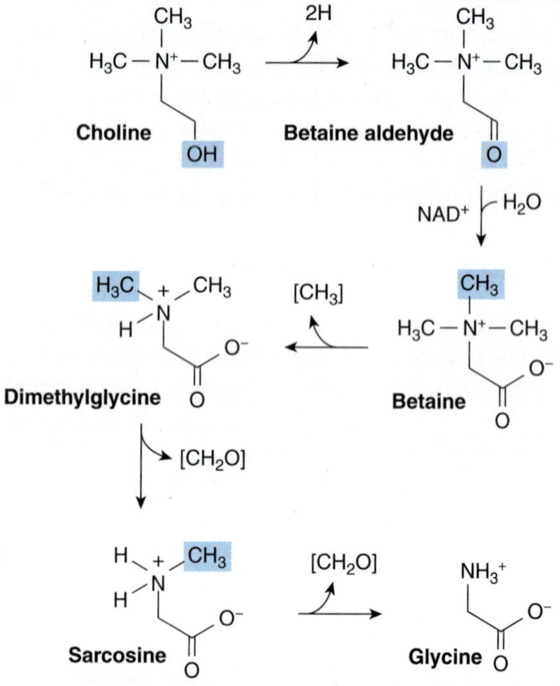

FIGURE 27–8 Formation of glycine from choline. Catalysts include choline dehydrogenase (EC 1.1.3.17), betaine aldehyde dehydrogenase (EC 1.2.1.8), betaine-homocysteine N-methyltransferase (EC 2.1.1.157), sarcosine dehydrogenase (EC 1.5.8.3), and dimethylglycine dehydrogenase (EC 1.5.8.4).

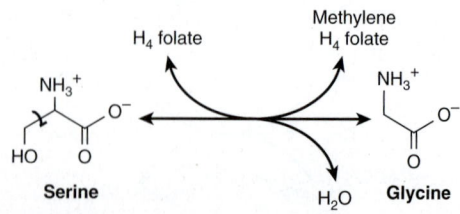

FIGURE 27–9 Interconversion of serine and glycine, catalyzed by serine hydroxymethyltransferase (EC 2.1.2.1). The reaction is freely reversible. (H_4 folate, tetrahydrofolate.)

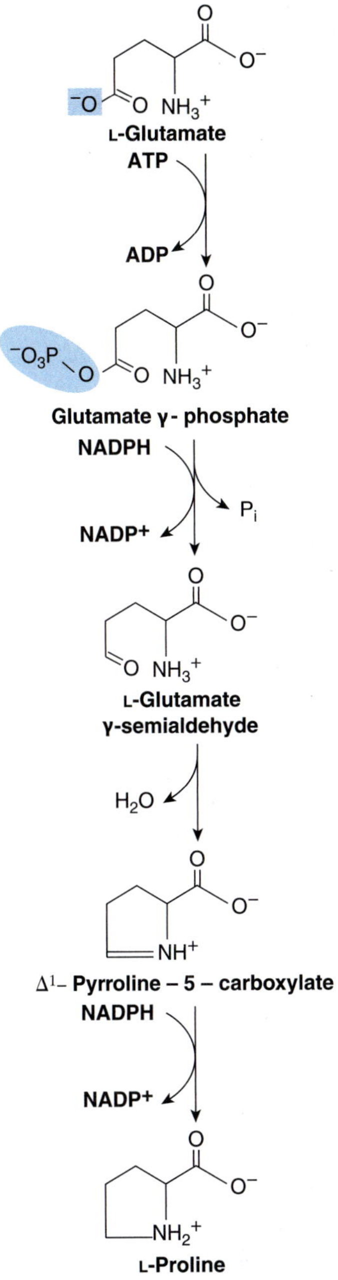

FIGURE 27–10 Biosynthesis of proline from glutamate.
Catalysts for these reactions are glutamate-5-kinase (EC 2.7.2.11),
glutamate-5-semialdehyde dehydrogenase (EC 1.2.1.41), and pyrroline-
5-carboxylate reductase (EC 1.5.1.2). Ring closure of glutamate semialdehyde is spontaneous.

granulating wounds requires, in addition to the substrate, molecular O_2, ascorbate, Fe^{2+}, and α-ketoglutarate (**Figure 27–13**).
For every mole of proline or lysine hydroxylated, one mole of
α-ketoglutarate is decarboxylated to succinate. The hydroxylases
are mixed-function oxidases. One atom of O_2 is incorporated into
proline or lysine, the other into succinate (see Figure 27–13). A
deficiency of the vitamin C required for these two hydroxylases
results in **scurvy**, in which bleeding gums, swelling joints, and
impaired wound healing result from the impaired stability of collagen (see Chapters 5 and 50).

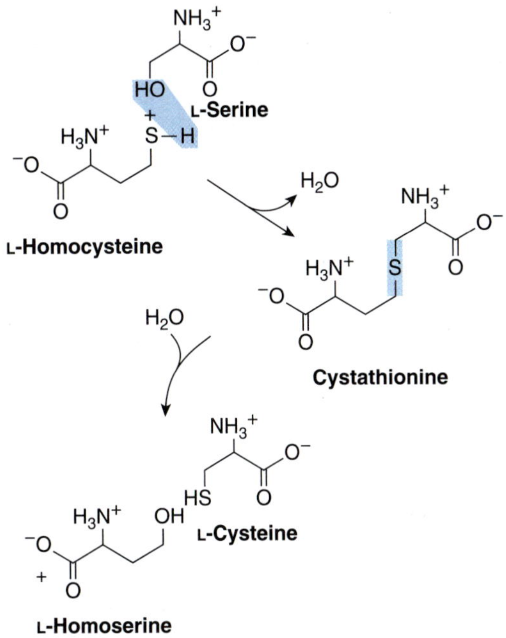

FIGURE 27–11 Conversion of homocysteine and serine to
homoserine and cysteine. The sulfur of cysteine derives from methionine and the carbon skeleton from serine. The catalysts are cystathionine
β-synthase (EC 4.2.1.22) and cystathionine γ-lyase (EC 4.4.1.1).

Valine, Leucine, & Isoleucine

While leucine, valine, and isoleucine are all nutritionally essential
amino acids, tissue aminotransferases reversibly interconvert
all three amino acids and their corresponding α-keto acids.

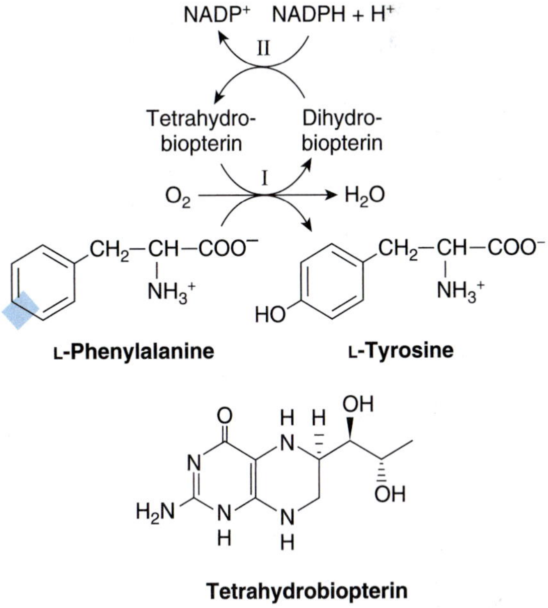

FIGURE 27–12 Conversion of phenylalanine to tyrosine by
phenylalanine hydroxylase (EC 1.14.16.1). Two distinct enzymatic
activities are involved. Activity II catalyzes reduction of dihydrobiopterin by NADPH, and activity I the reduction of O_2 to H_2O and of phenylalanine to tyrosine. This reaction is associated with several defects
of phenylalanine metabolism discussed in Chapter 29.

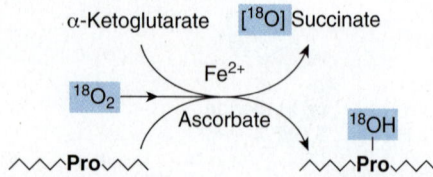

FIGURE 27–13 **Hydroxylation of a proline-rich peptide.** Molecular oxygen is incorporated into both succinate and proline. Procollagen-proline 4-hydroxylase (EC 1.14.11.2) thus is a mixed-function oxidase. Procollagen-lysine 5-hydroxylase (EC 1.14.11.4) catalyzes an analogous reaction.

These α-keto acids thus can replace their corresponding amino acids in the diet.

Selenocysteine, the 21st Amino Acid

While the occurrence of selenocysteine (**Figure 27–14**) in proteins is relatively uncommon, at least 25 human selenoproteins are known. Selenocysteine is present at the active site of several human enzymes that catalyze redox reactions. Examples include thioredoxin reductase, glutathione peroxidase, and the deiodinase that converts thyroxine to triiodothyronine. Where present, selenocysteine participates in the catalytic mechanism of these enzymes. Significantly, the replacement of selenocysteine by cysteine can actually *reduce* catalytic activity. Impairments in human selenoproteins have been implicated in tumorgenesis and atherosclerosis, and are associated with selenium deficiency cardiomyopathy (Keshan disease).

Biosynthesis of selenocysteine requires serine, selenate (SeO_4^{2-}), ATP, a specific tRNA, and several enzymes. Serine provides the carbon skeleton of selenocysteine. Selenophosphate, formed from ATP and selenate (see Figure 27–14), serves as the selenium donor. Unlike 4-hydroxyproline or 5-hydroxylysine, selenocysteine arises *cotranslationally* during its incorporation into peptides. The UGA anticodon of the unusual tRNA called tRNA^{Sec} normally signals STOP. The ability of the protein synthetic apparatus to identify a selenocysteine-specific UGA codon involves the selenocysteine insertion element, a stem-loop structure in the untranslated region of the mRNA. tRNA^{Sec} is first charged with serine by the ligase that charges tRNA^{Ser}.

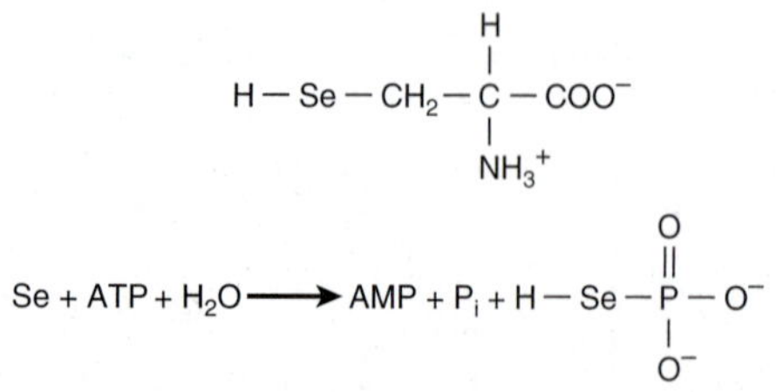

$$H - Se - CH_2 - \underset{\underset{NH_3^+}{|}}{\overset{\overset{H}{|}}{C}} - COO^-$$

$$Se + ATP + H_2O \longrightarrow AMP + P_i + H - Se - \underset{\underset{O^-}{||}}{\overset{\overset{O}{||}}{P}} - O^-$$

FIGURE 27–14 **Selenocysteine (top) and the reaction catalyzed by selenophosphate synthetase (EC 2.7.9.3) (bottom).**

Subsequent replacement of the serine oxygen by selenium involves selenophosphate formed by selenophosphate synthetase (see Figure 27–14). Successive enzyme-catalyzed reactions convert cysteyl-tRNA^{Sec} to aminoacrylyl-tRNA^{Sec} and then to selenocysteyl-tRNA^{Sec}. In the presence of a specific elongation factor that recognizes selenocysteyl-tRNA^{Sec}, selenocysteine can then be incorporated into proteins.

SUMMARY

- All vertebrates can form certain amino acids from amphibolic intermediates or from other dietary amino acids. The intermediates and the amino acids to which they give rise are α-ketoglutarate (Glu, Gln, Pro, Hyp), oxaloacetate (Asp, Asn), and 3-phosphoglycerate (Ser, Gly).

- Cysteine, tyrosine, and hydroxylysine are formed from nutritionally essential amino acids. Serine provides the carbon skeleton and homocysteine the sulfur for cysteine biosynthesis.

- In scurvy, a nutritional disease that results from a deficiency of vitamin C, impaired hydroxylation of peptidyl proline and peptidyl lysine results in a failure to provide the substrates for cross-linking of maturing collagens.

- Phenylalanine hydroxylase converts phenylalanine to tyrosine. Since the reaction catalyzed by this mixed function oxidase is irreversible, tyrosine cannot give rise to phenylalanine.

- Neither dietary hydroxyproline nor hydroxylysine is incorporated into proteins because no codon or tRNA dictates their insertion into peptides.

- Peptidyl hydroxyproline and hydroxylysine are formed by hydroxylation of peptidyl proline or lysine in reactions catalyzed by mixed-function oxidases that require vitamin C as cofactor.

- Selenocysteine, an essential active site residue in several mammalian enzymes, arises by cotranslational insertion from a previously modified tRNA.

REFERENCES

Gladyshev VN, Arner ES, Brigelius FR, et al: Selenoprotein gene nomenclature. J Biol Chem 2016;291:20436.

Kilberg MS: Asparagine synthetase chemotherapy. Annu Rev Biochem 2006;75:629.

Rayman RP: Selenium and human health. Lancet 2012;379;9822;1256.

Ruzzo EK, Capo-Chichi JM, Ben-Zeev B, et al: Deficiency of asparagine synthetase causes congenital microcephaly and a progressive form of encephalopathy. Neuron 2013;80:429.

Stickel F, Inderbitzin D, Candinas D: Role of nutrition in liver transplantation for end-stage chronic liver disease. Nutr Rev 2008;66:47.

Turanov AA, Shchedrina VA, Everley RA, et al: Selenoprotein S is involved in maintenance and transport of multiprotein complexes. Biochem J 2014;462:555.

Catabolism of Proteins & of Amino Acid Nitrogen

Victor W. Rodwell, PhD

O B J E C T I V E S

After studying this chapter, you should be able to:

- Describe protein turnover, indicate the mean rate of protein turnover in healthy individuals, and provide examples of human proteins that are degraded at rates greater than the mean rate.

- Outline the events in protein turnover by both ATP-dependent and ATP-independent pathways, and indicate the roles in protein degradation played by the proteasome, ubiquitin, cell surface receptors, circulating asialoglycoproteins, and lysosomes.

- Indicate how the ultimate end products of nitrogen catabolism in mammals differ from those in birds and fish.

- Illustrate the central roles of transaminases (aminotransferases), of glutamate dehydrogenase, and of glutaminase in human nitrogen metabolism.

- Use structural formulas to represent the reactions that convert NH_3, CO_2, and the amide nitrogen of aspartate into urea, and identify the subcellular locations of the enzymes that catalyze urea biosynthesis.

- Indicate the roles of allosteric regulation and of acetylglutamate in the regulation of the earliest steps in urea biosynthesis.

- Explain why metabolic defects in different enzymes of urea biosynthesis, although distinct at the molecular level, present similar clinical signs and symptoms.

- Describe both the classical approaches and the role of tandem mass spectrometry in screening neonates for inherited metabolic diseases.

BIOMEDICAL IMPORTANCE

In normal adults, nitrogen intake matches nitrogen excreted. Positive nitrogen balance, an excess of ingested over excreted nitrogen, accompanies growth and pregnancy. Negative nitrogen balance, where output exceeds intake, may follow surgery, advanced cancer, and the nutritional disorders kwashiorkor and marasmus. Genetic disorders that result from defects in the genes that encode ubiquitin, ubiquitin ligases, or deubiquitinating enzymes that participate in the degradation of certain proteins include Angelman syndrome, juvenile Parkinson disease, von Hippel-Lindau syndrome, and congenital polycythemia. This chapter describes how the nitrogen of amino acids is converted to urea, and the metabolic disorders that accompany defects in this process. Ammonia, which is highly toxic, arises in humans primarily from the α-amino nitrogen of amino acids. Tissues therefore convert ammonia to the amide nitrogen of the nontoxic amino acid glutamine. Subsequent deamination of glutamine in the liver releases ammonia, which is efficiently converted to urea, which is not toxic. However, if liver function is compromised, as in cirrhosis or hepatitis, elevated blood ammonia levels generate clinical signs and symptoms. Each enzyme of the urea cycle provides examples of metabolic defects and their physiologic consequences. In addition, the urea cycle provides a useful molecular model for the study of other human metabolic defects.

PROTEIN TURNOVER

The continuous degradation and synthesis (turnover) of cellular proteins occur in all forms of life. Each day, humans turn over 1 to 2% of their total body protein, principally muscle protein. High rates of protein degradation occur in tissues that are undergoing structural rearrangement, for example, uterine tissue during pregnancy, skeletal muscle in starvation, and tadpole tail tissue during metamorphosis. While approximately 75% of the amino acids liberated by protein degradation are reutilized, the remaining excess free amino acids are not stored for future use. Amino acids not immediately incorporated into new protein are rapidly degraded. The major portion of the carbon skeletons of the amino acids is converted to amphibolic intermediates, while in humans the amino nitrogen is converted to urea and excreted in the urine.

PROTEASES & PEPTIDASES DEGRADE PROTEINS TO AMINO ACIDS

The relative susceptibility of a protein to degradation is expressed as its **half-life ($t_{1/2}$)**, the time required to lower its concentration to half of its initial value. Half-lives of liver proteins range from under 30 minutes to over 150 hours. Typical "housekeeping" enzymes such as those of glycolysis, have $t_{1/2}$ values of over 100 hours. By contrast, key regulatory enzymes may have $t_{1/2}$ values as low as 0.5 to 2 hours. PEST sequences, regions rich in proline (P), glutamate (E), serine (S), and threonine (T), target some proteins for rapid degradation. Intracellular proteases hydrolyze internal peptide bonds. The resulting peptides are then degraded to amino acids by endopeptidases that hydrolyze internal peptide bonds, and by aminopeptidases and carboxypeptidases that remove amino acids sequentially from the amino- and carboxyl-termini, respectively.

ATP-Independent Degradation

Degradation of blood glycoproteins (see Chapter 46) follows loss of a sialic acid moiety from the nonreducing ends of their oligosaccharide chains. Asialoglycoproteins are then internalized by liver-cell asialoglycoprotein receptors and degraded by lysosomal proteases. Extracellular, membrane-associated, and long-lived intracellular proteins are also degraded in lysosomes by ATP-independent processes.

ATP & Ubiquitin-Dependent Degradation

Degradation of regulatory proteins with short half-lives and of abnormal or misfolded proteins occurs in the cytosol, and requires ATP and **ubiquitin**. Named based on its presence in all eukaryotic cells, ubiquitin is a small (8.5 kDa, 76 residue) polypeptide that targets many intracellular proteins for degradation. The primary structure of ubiquitin is highly conserved. Only 3 of 76 residues differ between yeast and human ubiquitin. **Figure 28–1** illustrates the three-dimensional structure of ubiquitin. Ubiquitin molecules are attached by **non–α-peptide bonds** formed between the carboxyl terminal of ubiquitin and the ε-amino groups of lysyl residues in the target protein (**Figure 28–2**). The residue present at its amino terminus affects whether a protein is ubiquitinated. Amino terminal Met or Ser residues retard, whereas Asp or Arg

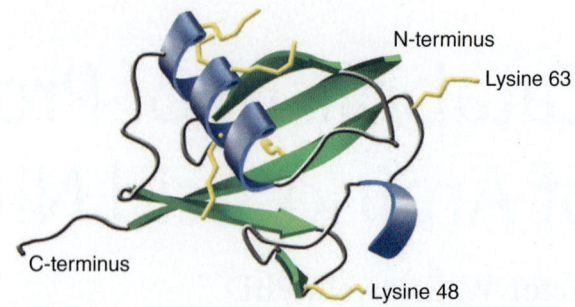

FIGURE 28–1 **Three-dimensional structure of ubiquitin.** Shown are α-helices (blue), β-strands (green), and the R-groups of lysyl residues (orange). Lys48 & Lys63 are sites for attachment of additional ubiquitin molecules during polyubiquitination. (Rogerdodd/Wikipedia)

favor ubiquitination. Attachment of a single ubiquitin molecule to transmembrane proteins alters their subcellular localization and targets them for degradation. Soluble proteins undergo **polyubiquitination**, the ligase-catalyzed attachment of four or more additional ubiquitin molecules (Figure 28–1). Subsequent degradation of ubiquitin-tagged proteins takes place in the **proteasome**, a macromolecule that also is ubiquitous in

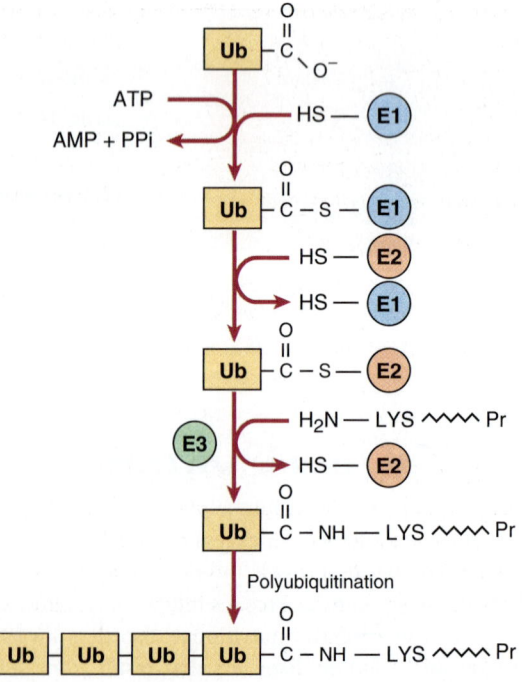

FIGURE 28–2 **Reactions involved in the attachment of ubiquitin (Ub) to proteins.** Three enzymes are involved. E1 is an activating enzyme, E2 a transferase, and E3 a ligase. While depicted as single entities, there are several types of E1, and over 500 types of E2. The terminal COOH of ubiquitin first forms a thioester. The coupled hydrolysis of PP$_i$ by pyrophosphatase ensures that the reaction will proceed readily. A thioester exchange reaction now transfers activated ubiquitin to E2. E3 then catalyzes the transfer of ubiquitin to the ε-amino group of a lysyl residue of the target protein. Additional rounds of ubiquitination result in subsequent polyubiquitination.

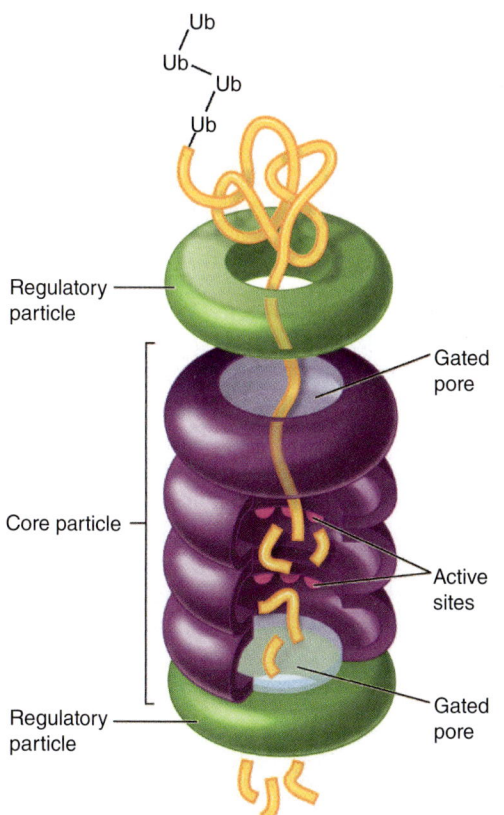

FIGURE 28–3 **Representation of the structure of a proteasome.** The upper ring is gated to permit only polyubiquitinated proteins to enter the proteosome, where immobilized internal proteases degrade them to peptides.

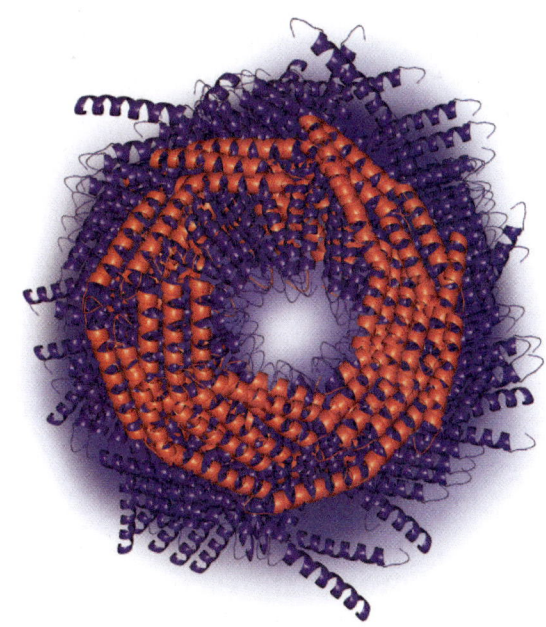

FIGURE 28–4 **An end-on view of a proteasome.** (Thomas Splettstoesser/Wikipedia)

eukaryotic cells. The proteasome consists of a macromolecular, cylindrical complex of proteins, whose stacked rings form a central pore that harbors the active sites of proteolytic enzymes. For degradation, a protein thus must first enter the central pore. Entry into the core is regulated by the two outer rings that recognize polyubiquitinated proteins (**Figures 28–3** and **28–4**).

For the discovery of ubiquitin-mediated protein degradation, Aaron Ciechanover and Avram Hershko of Israel and Irwin Rose of the United States were awarded the 2004 Nobel Prize in Chemistry. Genetic disorders that result from defects in the genes that encode ubiquitin, ubiquitin ligases, or deubiquitinating enzymes include Angelman syndrome, autosomal recessive juvenile Parkinson disease, von Hippel-Lindau syndrome, and congenital polycythemia. For additional aspects of protein degradation and of ubiquitination, including its role in the cell cycle, see Chapters 4 and 35.

INTERORGAN EXCHANGE MAINTAINS CIRCULATING LEVELS OF AMINO ACIDS

The maintenance of steady-state concentrations of circulating plasma amino acids between meals depends on the net balance between release from endogenous protein stores and

utilization by various tissues. Muscle generates over half of the total body pool of free amino acids, and liver is the site of the urea cycle enzymes necessary for disposal of excess nitrogen. Muscle and liver thus play major roles in maintaining circulating amino acid levels.

Figure 28–5 summarizes the postabsorptive state. Free amino acids, particularly alanine and glutamine, are released from muscle into the circulation. Alanine is extracted primarily by the liver, and glutamine is extracted by the gut and the kidney, both of which convert a significant portion to alanine. Glutamine also serves as a source of ammonia for excretion by the kidney. The kidney provides a major source of serine for uptake by peripheral tissues, including liver and muscle.

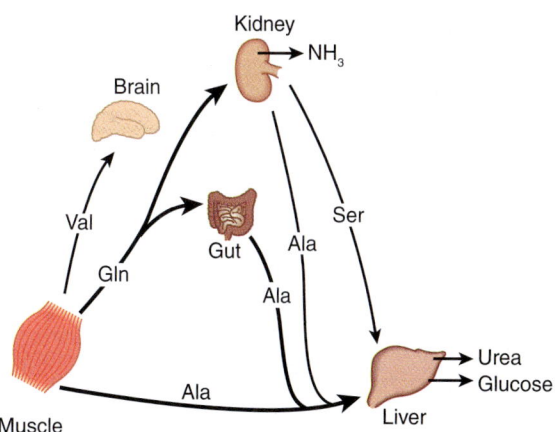

FIGURE 28–5 **Interorgan amino acid exchange in normal postabsorptive humans.** The key role of alanine in amino acid output from muscle and gut and uptake by the liver is shown.

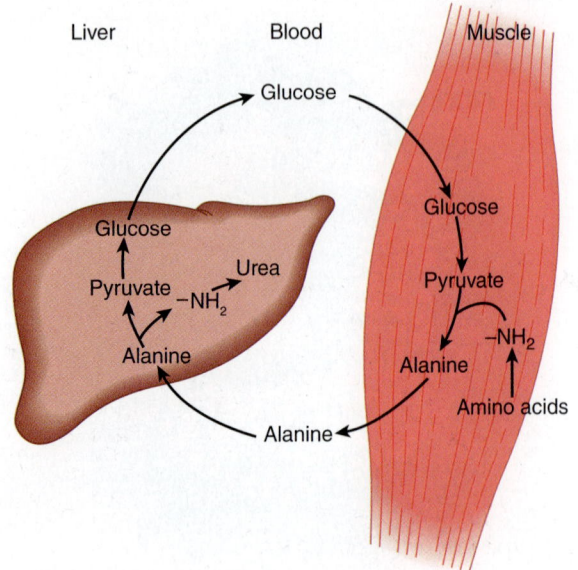

FIGURE 28–6 **The glucose-alanine cycle.** Alanine is synthesized in muscle by transamination of glucose-derived pyruvate, released into the bloodstream, and taken up by the liver. In the liver, the carbon skeleton of alanine is reconverted to glucose and released into the bloodstream, where it is available for uptake by muscle and resynthesis of alanine.

Branched-chain amino acids, particularly valine, are released by muscle and taken up predominantly by the brain.

Alanine is a key **gluconeogenic amino acid** (**Figure 28–6**). The rate of hepatic gluconeogenesis from alanine is far higher than from all other amino acids. The capacity of the liver for gluconeogenesis from alanine does not reach saturation until the alanine concentration reaches 20 to 30 times its normal physiologic level. Following a protein-rich meal, the splanchnic tissues release amino acids (**Figure 28–7**) while the peripheral muscles extract amino acids, in both instances predominantly branched-chain amino acids. Branched-chain amino acids thus serve a special role in nitrogen metabolism. In the fasting

state, they provide the brain with an energy source, and post-prandially they are extracted predominantly by muscle, having been spared by the liver.

ANIMALS CONVERT α-AMINO NITROGEN TO VARIED END PRODUCTS

Depending on their ecological niche and physiology, different animals excrete excess nitrogen as ammonia, uric acid, or urea. The aqueous environment of teleostean fish, which are **ammonotelic** (excrete ammonia), permits them to excrete water continuously to facilitate excretion of ammonia, which is highly toxic. While this approach is appropriate for an aquatic animal, birds must both conserve water and maintain low weight. Birds, which are **uricotelic**, address both problems by excreting nitrogen-rich uric acid (see Figure 33–11) as semisolid guano. Many land animals, including humans, are **ureotelic** and excrete nontoxic, highly water-soluble urea. Since urea is nontoxic to humans, high blood levels in renal disease are a consequence, not a cause, of impaired renal function.

BIOSYNTHESIS OF UREA

Urea biosynthesis occurs in four stages: (1) transamination, (2) oxidative deamination of glutamate, (3) ammonia transport, and (4) reactions of the urea cycle (**Figure 28–8**). The expression in liver of the RNAs for all the enzymes of the urea cycle increases severalfold in starvation, probably secondary to enhanced protein degradation to provide energy.

Transamination Transfers α-Amino Nitrogen to α-Ketoglutarate, Forming Glutamate

Transamination reactions interconvert pairs of α-amino acids and α-keto acids (**Figure 28–9**). Transamination reactions, which are freely reversible, also function in amino acid

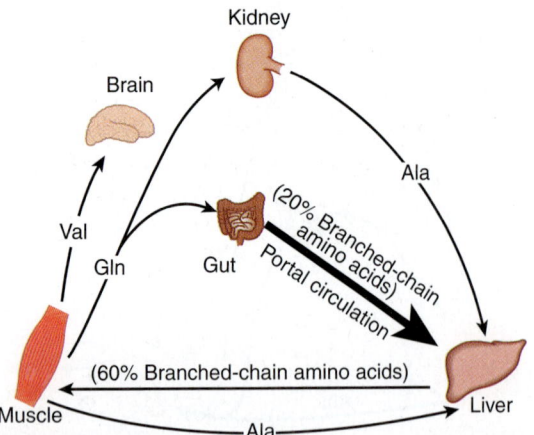

FIGURE 28–7 Summary of amino acid exchange between organs immediately after feeding.

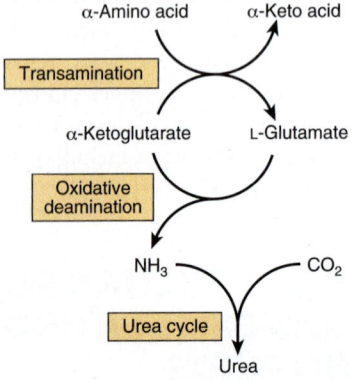

FIGURE 28–8 Overall flow of nitrogen in amino acid catabolism.

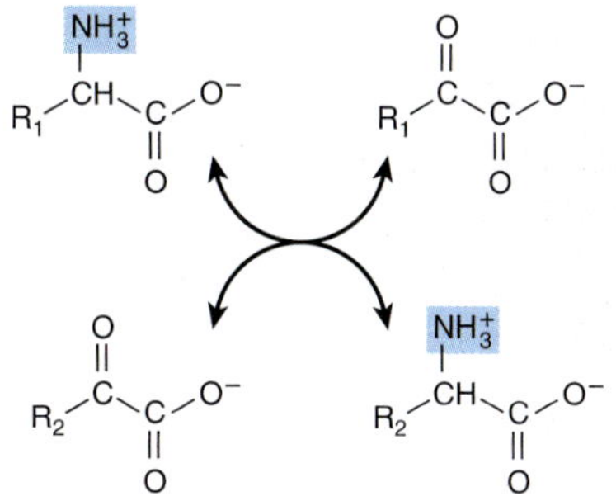

FIGURE 28–9 **Transamination.** The reaction is freely reversible with an equilibrium constant close to unity.

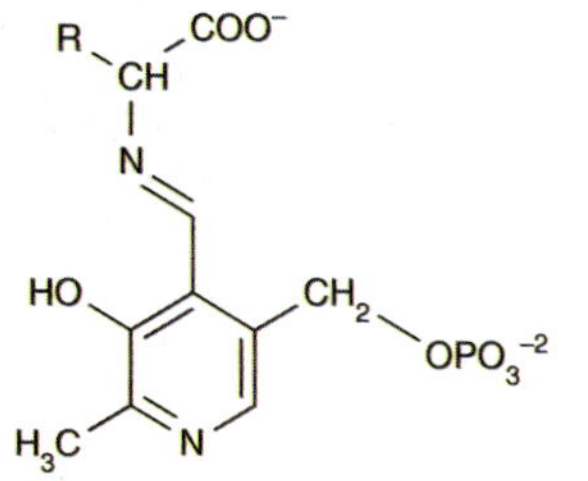

FIGURE 28–11 **Structure of a Schiff base formed between pyridoxal phosphate and an amino acid.**

biosynthesis (see Figure 27–4). All of the common amino acids except lysine, threonine, proline, and hydroxyproline participate in transamination. Transamination is not restricted to α-amino groups. The δ-amino group of ornithine (but not the ε-amino group of lysine) readily undergoes transamination.

Alanine-pyruvate aminotransferase (alanine aminotransferase, EC 2.6.1.2) and glutamate-α-ketoglutarate aminotransferase (glutamate aminotransferase, EC 2.6.1.1) catalyze the transfer of amino groups to pyruvate (forming alanine) or to α-ketoglutarate (forming glutamate).

Each aminotransferase is specific for one pair of substrates, but nonspecific for the other pair. Since alanine is also a substrate for glutamate aminotransferase, the α-amino nitrogen from all amino acids that undergo transamination can be concentrated in glutamate. This is important because L-glutamate is the only amino acid that undergoes oxidative deamination at an appreciable rate in mammalian tissues. The formation of ammonia from α-amino groups thus occurs mainly via the α-amino nitrogen of L-glutamate.

Transamination occurs via a "ping-pong" mechanism characterized by the alternate addition of a substrate and release of a product (**Figure 28–10**). Following removal of its α-amino nitrogen by transamination, the remaining carbon "skeleton" of an amino acid is degraded by pathways discussed in Chapter 29.

Pyridoxal phosphate (PLP), a derivative of vitamin B_6, is present at the catalytic site of all aminotransferases, and plays a key role in catalysis. During transamination, PLP serves as a "carrier" of amino groups. An enzyme-bound Schiff base (**Figure 28–11**) is formed between the oxo group of enzyme-bound PLP and the α-amino group of an α-amino acid. The Schiff base can rearrange in various ways. In transamination, rearrangement forms an α-keto acid and enzyme-bound pyridoxamine phosphate. As noted earlier, certain diseases are

associated with elevated serum levels of aminotransferases (see Table 7–1).

L-GLUTAMATE DEHYDROGENASE OCCUPIES A CENTRAL POSITION IN NITROGEN METABOLISM

Transfer of amino nitrogen to α-ketoglutarate forms L-glutamate. Hepatic **L-glutamate dehydrogenase** (GDH), which can use either NAD+ or NADP+, releases this nitrogen as ammonia (**Figure 28–12**). Conversion of α-amino nitrogen to ammonia by the concerted action of glutamate aminotransferase and GDH is often termed "transdeamination." Liver GDH activity is allosterically inhibited by ATP, GTP, and NADH, and is activated by ADP. The GDH reaction is freely reversible, and also functions in amino acid biosynthesis (see Figure 27–1).

AMINO ACID OXIDASES REMOVE NITROGEN AS AMMONIA

L-Amino acid oxidase of liver and kidney convert an amino acid to an α-imino acid that decomposes to an α-keto acid with release of ammonium ion (**Figure 28–13**). The reduced flavin is reoxidized by molecular oxygen, forming hydrogen peroxide (H_2O_2), which then is split to O_2 and H_2O by **catalase**, EC 1.11.1.6.

Ammonia Intoxication Is Life-Threatening

The ammonia produced by enteric bacteria and absorbed into portal venous blood and the ammonia produced by tissues are rapidly removed from circulation by the liver and converted to urea. Thus, normally, only traces (10-20 µg/dL) are present in peripheral blood. This is essential, since ammonia is toxic

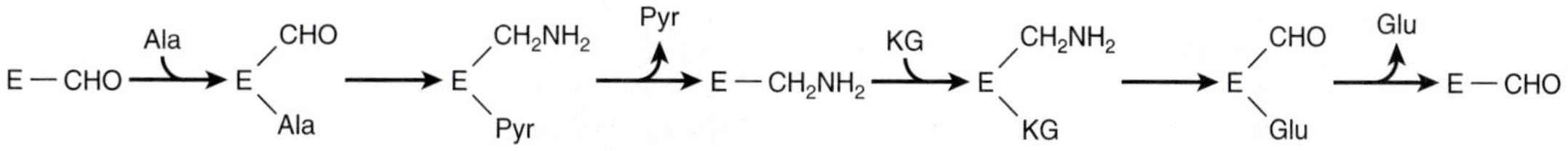

FIGURE 28–10 **"Ping-pong" mechanism for transamination.** E—CHO and E—CH$_2$NH$_2$ represent enzyme-bound pyridoxal phosphate and pyridoxamine phosphate, respectively. (Ala, alanine; Glu, glutamate; KG, α-ketoglutarate; Pyr, pyruvate.)

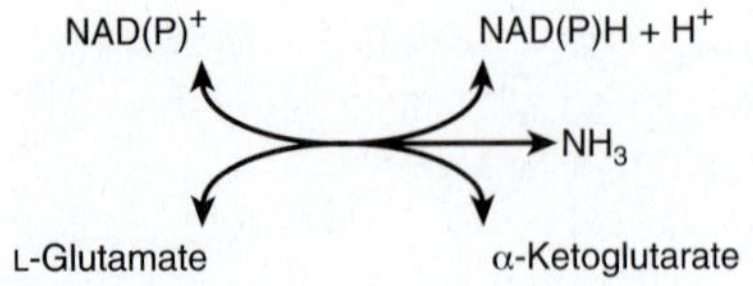

NAD(P)$^+$ NAD(P)H + H$^+$

$\longrightarrow$ NH$_3$

L-Glutamate   α-Ketoglutarate

FIGURE 28–12   **The reaction catalyzed by glutamate dehydrogenase, EC 1.4.1.2.** NAD(P)$^+$ means that either NAD$^+$ or NADP$^+$ can serve as the oxidoreductant. The reaction is reversible, but strongly favors glutamate formation.

to the central nervous system. Should portal blood bypass the liver, systemic blood ammonia may reach toxic levels. This occurs in severely impaired hepatic function or the development of collateral links between the portal and systemic veins in cirrhosis. Symptoms of **ammonia intoxication** include tremor, slurred speech, blurred vision, coma, and ultimately death. Ammonia may be toxic to the brain in part because it reacts with α-ketoglutarate to form glutamate. The resulting depletion of α-ketoglutarate then impairs function of the tricarboxylic acid (TCA) cycle in neurons.

Glutamine Synthetase Fixes Ammonia as Glutamine

Formation of glutamine is catalyzed by mitochondrial **glutamine synthetase** (**Figure 28–14**). Since amide bond synthesis is coupled to the hydrolysis of ATP to ADP and P$_i$, the reaction strongly favors glutamine synthesis. During catalysis, glutamate attacks the γ-phosphoryl group of ATP, forming γ-glutamyl phosphate and ADP. Following deprotonation of NH$_4^+$, NH$_3$ attacks γ-glutamyl phosphate, and glutamine and P$_i$ are released. In addition to providing glutamine to serve as a carrier of nitrogen, carbon and energy between organs (Figure 28–5), glutamine synthetase plays a major role both in ammonia detoxification and in acid–base homeostasis. A rare

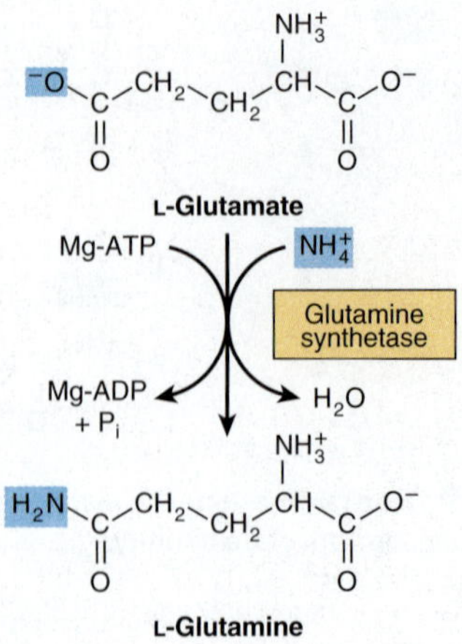

FIGURE 28–14   **Formation of glutamine, catalyzed by glutamine synthetase, EC 6.3.1.2.**

deficiency in neonate glutamine synthetase results in severe brain damage, multiorgan failure, and death.

Glutaminase & Asparaginase Deamidate Glutamine & Asparagine

There are two human isoforms of mitochondrial **glutaminase**, termed liver-type and renal-type glutaminase. Products of different genes, the glutaminases differ with respect to their structure, kinetics, and regulation. Hepatic glutaminase levels rise in response to high protein intake while renal kidney-type glutaminase increases in metabolic acidosis. Hydrolytic release of the amide nitrogen of glutamine as ammonia, catalyzed by glutaminase (**Figure 28–15**), strongly favors glutamate formation. An analogous reaction is catalyzed by L-asparaginase (EC 3.5.1.1). The concerted action of glutamine synthetase and glutaminase thus catalyzes the interconversion of free ammonium ion and glutamine.

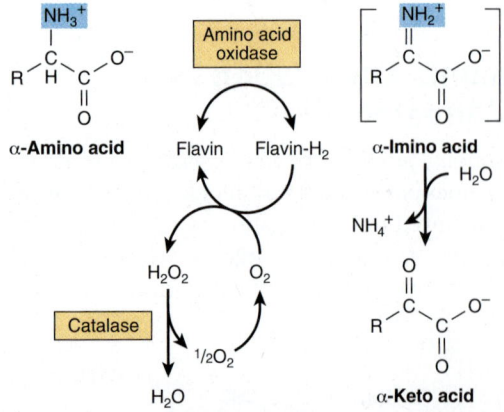

FIGURE 28–13   **Oxidative deamination catalyzed by L-amino acid oxidase (L-α-amino acid:O$_2$ oxidoreductase, EC 1.4.3.2).** The α-imino acid, shown in brackets, is not a stable intermediate.

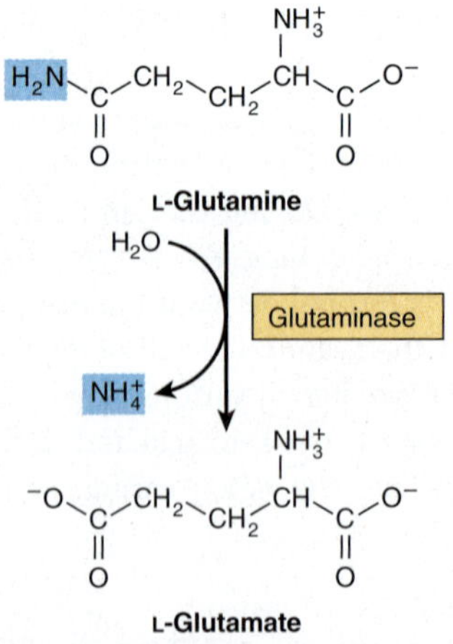

FIGURE 28–15   **The reaction catalyzed by glutaminase, EC 3.5.1.2.** The reaction proceeds essentially irreversibly in the direction of glutamate and NH$_4^+$ formation. Note that the *amide* nitrogen, not the α-amino nitrogen, is removed.

Formation & Secretion of Ammonia Maintains Acid–Base Balance

Excretion into urine of ammonia produced by renal tubular cells facilitates cation conservation and regulation of acid–base balance. Ammonia production from intracellular renal amino acids, especially glutamine, increases in **metabolic acidosis** and decreases in **metabolic alkalosis**.

Urea Is the Major End Product of Nitrogen Catabolism in Humans

Synthesis of 1 mol of urea requires 3 mol of ATP, 1 mol each of ammonium ion and of aspartate, and employs five enzymes (**Figure 28–16**). Of the six participating amino acids, *N*-acetylglutamate functions solely as an enzyme activator. The

others serve as carriers of the atoms that ultimately become urea. The major metabolic role of **ornithine**, **citrulline**, and **argininosuccinate** in mammals is urea synthesis. Urea synthesis is a cyclic process. While ammonium ion, CO_2, ATP, and aspartate are consumed, the ornithine consumed in reaction 2 is regenerated in reaction 5. Thus, there is no net loss or gain of ornithine, citrulline, argininosuccinate, or arginine. As indicated in Figure 28–16, some reactions of urea synthesis occur in the matrix of the mitochondrion, and other reactions in the cytosol.

Carbamoyl Phosphate Synthetase I Initiates Urea Biosynthesis

Condensation of CO_2, ammonia, and ATP to form **carbamoyl phosphate** is catalyzed by mitochondrial **carbamoyl phosphate synthetase I** (EC 6.3.4.16). A cytosolic form of this

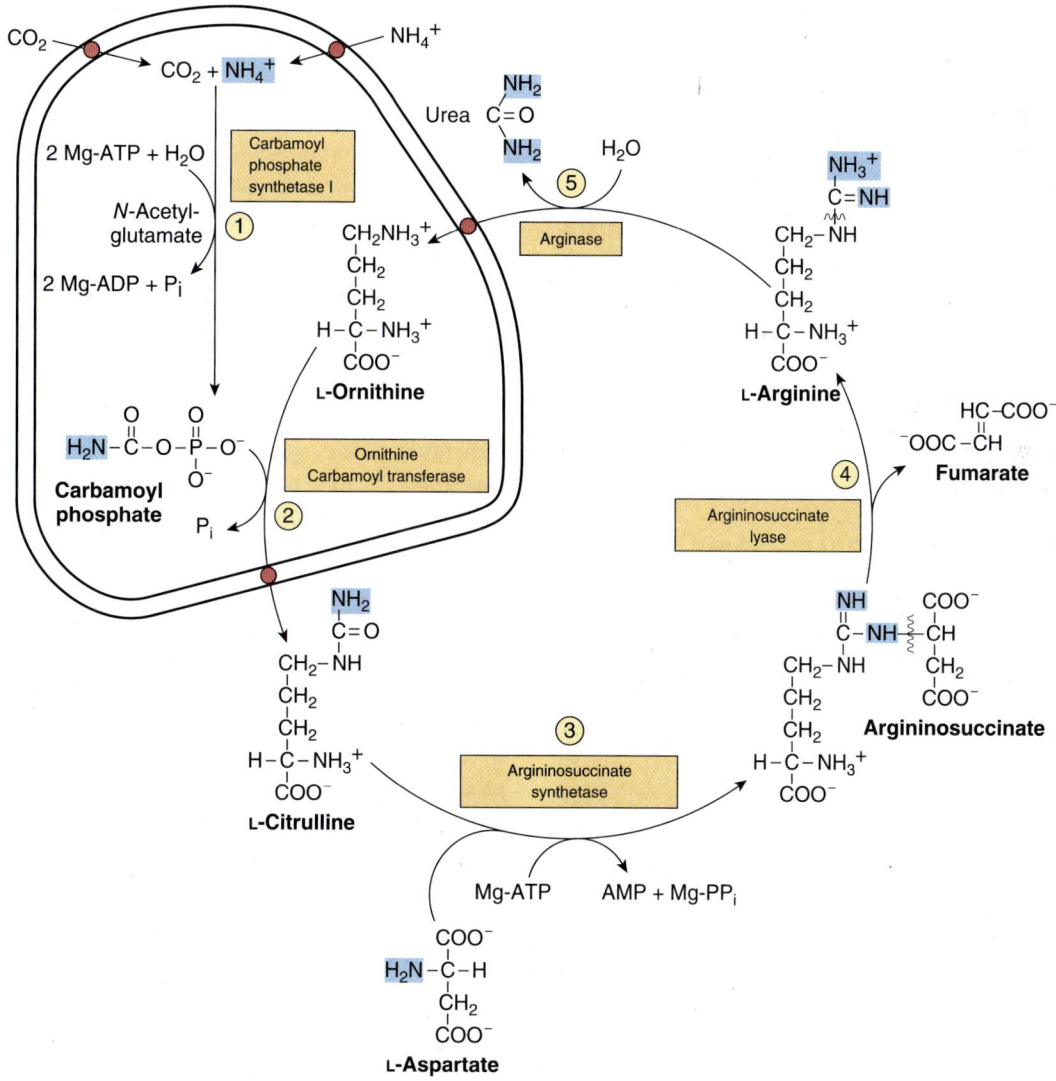

FIGURE 28–16 **Reactions and intermediates of urea biosynthesis.** The nitrogen-containing groups that contribute to the formation of urea are shaded. Reactions ① and ② occur in the matrix of liver mitochondria and reactions ③, ④, and ⑤ in liver cytosol. CO_2 (as bicarbonate), ammonium ion, ornithine, and citrulline enter the mitochondrial matrix via specific carriers (see red dots) present in the inner membrane of liver mitochondria.

enzyme, carbamoyl phosphate synthetase II, uses glutamine rather than ammonia as the nitrogen donor and functions in pyrimidine biosynthesis (see Figure 33–9). The concerted action of glutamate dehydrogenase and carbamoyl phosphate synthetase I thus shuttles amino nitrogen into carbamoyl phosphate, a compound with high group transfer potential.

Carbamoyl phosphate synthetase I, the rate-limiting enzyme of the urea cycle, is active only in the presence of **N-acetylglutamate**, an allosteric activator that enhances the affinity of the synthetase for ATP. Synthesis of 1 mol of carbamoyl phosphate requires 2 mol of ATP. One ATP serves as the phosphoryl donor for formation of the mixed acid anhydride bond of carbamoyl phosphate. The second ATP provides the driving force for synthesis of the amide bond of carbamoyl phosphate. The other products are 2 mol of ADP and 1 mol of P_i (reaction 1, Figure 28–16). The reaction proceeds stepwise. Reaction of bicarbonate with ATP forms carbonyl phosphate and ADP. Ammonia then displaces ADP, forming carbamate and orthophosphate. Phosphorylation of carbamate by the second ATP then forms carbamoyl phosphate.

Carbamoyl Phosphate Plus Ornithine Forms Citrulline

L-Ornithine transcarbamoylase (EC 2.1.3.3) catalyzes transfer of the carbamoyl group of carbamoyl phosphate to ornithine, forming citrulline and orthophosphate (reaction 2, Figure 28–16). While the reaction occurs in the mitochondrial matrix, both the formation of ornithine and the subsequent metabolism of citrulline take place in the cytosol. Entry of ornithine into mitochondria and exodus of citrulline from mitochondria involves the mitochondrial inner membrane carriers ORC1, ORC2, and SLCA25A29 (Figure 28–16).

Citrulline Plus Aspartate Forms Argininosuccinate

Argininosuccinate synthetase (EC 6.3.4.5) links aspartate and citrulline via the amino group of aspartate (reaction 3, Figure 28–16), which provides the second nitrogen of urea. The reaction requires ATP and involves intermediate formation of citrullyl-AMP. Subsequent displacement of AMP by aspartate then forms argininosuccinate.

Cleavage of Argininosuccinate Forms Arginine & Fumarate

Cleavage of argininosuccinate is catalyzed by **argininosuccinate lyase** (EC 4.3.2.1). The reaction proceeds with retention of all three nitrogens in arginine and release of the aspartate skeleton as fumarate (reaction 4, Figure 28–16). Subsequent addition of water to fumarate forms L-malate, whose subsequent NAD⁺-dependent oxidation forms oxaloacetate. These two reactions are analogous to reactions of the citric acid cycle, but are catalyzed by *cytosolic* **fumarase** and **malate dehydrogenase**. Transamination of oxaloacetate by glutamate aminotransferase then reforms aspartate. The carbon skeleton of aspartate-fumarate thus acts as a carrier of the nitrogen of glutamate into a precursor of urea.

Cleavage of Arginine Releases Urea & Reforms Ornithine

Hydrolytic cleavage of the guanidino group of arginine, catalyzed by liver **arginase** (EC 3.5.3.1), releases urea (reaction 5, Figure 28–16). The other product, ornithine, reenters liver mitochondria and participates in additional rounds of urea synthesis. Ornithine and lysine are potent inhibitors of arginase, and compete with arginine. Arginine also serves as the precursor of the potent muscle relaxant nitric oxide (NO) in a Ca^{2+}-dependent reaction catalyzed by NO synthetase.

Carbamoyl Phosphate Synthetase I Is the Pacemaker Enzyme of the Urea Cycle

The activity of carbamoyl phosphate synthetase I is determined by **N-acetylglutamate**, whose steady-state level is dictated by the balance between its rate of synthesis from acetyl-CoA and glutamate and its rate of hydrolysis to acetate and glutamate, reactions catalyzed by N-acetylglutamate synthetase (NAGS) and N-acetylglutamate deacylase (hydrolase), respectively.

Acetyl-CoA + L-glutamate → N-acetyl-L-glutamate + CoASH

N-acetyl-L-glutamate + H_2O → L-glutamate + acetate

Major changes in diet can increase the concentrations of individual urea cycle enzymes 10- to 20-fold. For example, starvation elevates enzyme levels, presumably to cope with the increased production of ammonia that accompanies enhanced starvation-induced degradation of protein.

GENERAL FEATURES OF METABOLIC DISORDERS

The comparatively rare, but well-characterized and medically devastating metabolic disorders associated with the enzymes of urea biosynthesis illustrate the following general principles of inherited metabolic diseases:

1. Similar or identical clinical signs and symptoms can accompany various genetic mutations in a gene that encodes a given enzyme or in enzymes that catalyze successive reactions in a metabolic pathway.

2. Rational therapy is based on an understanding of the relevant biochemical enzyme-catalyzed reactions in both normal and impaired individuals.

3. The identification of intermediates and of ancillary products that accumulate prior to a metabolic block provides the basis for metabolic screening tests that can implicate the reaction that is impaired.

4. Definitive diagnosis involves quantitative assay of the activity of the enzyme suspected to be defective.

5. The DNA sequence of the gene that encodes a given mutant enzyme is compared to that of the wild-type gene to identify the specific mutation(s) that cause the disease.

6. The exponential increase in DNA sequencing of human genes has identified dozens of mutations of an affected gene that are benign or are associated with symptoms of varying severity of a given metabolic disorder.

METABOLIC DISORDERS ARE ASSOCIATED WITH EACH REACTION OF THE UREA CYCLE

Five well-documented diseases represent defects in the biosynthesis of enzymes of the urea cycle. Molecular genetic analysis has pinpointed the loci of mutations associated with each deficiency, each of which exhibits considerable genetic and phenotypic variability (**Table 28–1**).

Urea cycle disorders are characterized by hyperammonemia, encephalopathy, and respiratory alkalosis. Four of the five metabolic diseases, deficiencies of carbamoyl phosphate synthetase I, ornithine carbamoyl transferase, argininosuccinate synthetase, and argininosuccinate lyase, result in the accumulation of precursors of urea, principally ammonia and glutamine. Ammonia intoxication is most severe when the metabolic block occurs at reactions 1 or 2 (Figure 28–16), for if citrulline can be synthesized, some ammonia has already been removed by being covalently linked to an organic metabolite.

Clinical symptoms common to all urea cycle disorders include vomiting, avoidance of high-protein foods, intermittent ataxia, irritability, lethargy, and severe mental retardation. The most dramatic clinical presentation occurs in full-term infants who initially appear normal, then exhibit progressive lethargy, hypothermia, and apnea due to high plasma ammonia levels. The clinical features and treatment of all five disorders are similar. Significant improvement and minimization of brain damage can accompany a low-protein diet ingested as frequent small meals to avoid sudden increases in blood ammonia levels. The goal of dietary therapy is to provide sufficient protein, arginine, and energy to promote growth and development while simultaneously minimizing the metabolic perturbations.

Carbamoyl Phosphate Synthetase I

N-Acetylglutamate is essential for the activity of carbamoyl phosphate synthetase I, EC 6.3.4.16 (reaction 1, Figure 28–16). Defects in carbamoyl phosphate synthetase I are responsible for the relatively rare (estimated frequency 1:62,000) metabolic disease termed "hyperammonemia type 1."

N-Acetylglutamate Synthetase

N-Acetylglutamate synthetase, EC 2.3.1.1 (NAGS), catalyzes the formation from acetyl-CoA and glutamate of the N-acetylglutamate essential for carbamoyl phosphate synthetase I activity.

L-Glutamate + acetyl-CoA → N-acetyl-L-glutamate + CoASH

While the clinical and biochemical features of NAGS deficiency are indistinguishable from those arising from a defect in carbamoyl phosphate synthetase I, a deficiency in NAGS may respond to administered N-acetylglutamate.

Ornithine Permease

The hyperornithinemia, hyperammonemia, and homocitrullinuria (**HHH**) syndrome results from mutation of the *ORC1* gene that encodes the mitochondrial membrane ornithine carrier. The inability to import cytosolic ornithine into the mitochondrial matrix renders the urea cycle inoperable, with consequent hyperammonemia, and hyperornithinemia due to the accompanying accumulation of cytosolic ornithine. In the absence of its normal acceptor (ornithine), mitochondrial carbamoyl phosphate carbamoylates lysine to homocitrulline, resulting in homocitrullinuria.

Ornithine Transcarbamoylase

The X-chromosome–linked deficiency termed "hyperammonemia type 2" reflects a defect in ornithine transcarbamoylase (reaction 2, Figure 28–16). The mothers also exhibit hyperammonemia and an aversion to high-protein foods. Levels

TABLE 28–1 Enzymes of Inherited Metabolic Disorders of the Urea Cycle

Enzyme	Enzyme Catalog Number	OMIM[a] Reference	Figure and Reaction
Carbamoyl-phosphate synthetase 1	6.3.4.16	237300	28-13①
Ornithine carbamoyl transferase	2.1.3.3	311250	28-13②
Argininosuccinate synthetase	6.3.4.5	215700	28-13③
Argininosuccinate lyase	4.3.2.1	608310	28-13④
Arginase	3.5.3.1	608313	28-13⑤

[a]Online Mendelian inheritance in man database: ncbi.nlm.nih.gov/omim/

of glutamine are elevated in blood, cerebrospinal fluid, and urine, probably as a result of enhanced glutamine synthesis in response to elevated levels of tissue ammonia.

Argininosuccinate Synthetase

In addition to patients who lack detectable argininosuccinate synthetase activity (reaction 3, Figure 28–16), 25-fold elevations in K_m for citrulline have been reported. In the resulting citrullinemia, plasma and cerebrospinal fluid citrulline levels are elevated, and 1 to 2 g of citrulline are excreted daily.

Argininosuccinate Lyase

Argininosuccinic aciduria, accompanied by elevated levels of argininosuccinate in blood, cerebrospinal fluid, and urine, is associated with friable, tufted hair (trichorrhexis nodosa). Both early- and late-onset types are known. The metabolic defect is in argininosuccinate lyase (reaction 4, Figure 28–16). Diagnosis by the measurement of erythrocyte argininosuccinate lyase activity can be performed on umbilical cord blood or amniotic fluid cells.

Arginase

Hyperargininemia is an autosomal recessive defect in the gene for arginase (reaction 5, Figure 28–16). Unlike other urea cycle disorders, the first symptoms of hyperargininemia typically do not appear until age 2 to 4 years. Blood and cerebrospinal fluid levels of arginine are elevated. The urinary amino acid pattern, which resembles that of lysine-cystinuria (see Chapter 29), may reflect competition by arginine with lysine and cysteine for reabsorption in the renal tubule.

Analysis of Neonate Blood by Tandem Mass Spectrometry Can Detect Metabolic Diseases

Metabolic diseases caused by the absence or functional impairment of metabolic enzymes can be devastating. Early dietary intervention, however, can in many instances ameliorate the otherwise inevitable dire effects. The early detection of such metabolic diseases is thus is of primary importance. Since the initiation in the United States of newborn screening programs in the 1960s, all states now conduct metabolic screening of newborn infants. The powerful and sensitive technique of **tandem mass spectrometry** (MS) (see Chapter 4) can in a few minutes detect over 40 analytes of significance in the detection of metabolic disorders. Most states employ tandem MS to screen newborns to detect metabolic disorders such as organic acidemias, aminoacidemias, disorders of fatty acid oxidation, and defects in the enzymes of the urea cycle. An article in *Clinical Chemistry* 2006 39:315 reviews the theory of tandem MS, its application to the detection of metabolic disorders, and situations that can yield false positives, and includes a lengthy table of detectable analytes and the relevant metabolic diseases.

Can Metabolic Disorders Be Rectified by Gene or Protein Modification

Despite results in animal models using an adenoviral vector to treat citrullinemia, at present gene therapy provides no effective solution for human subjects. However, direct CRISPR/Cas9-based modification of a defective enzyme can restore functional enzyme activity of cultured human pluripotent stem cells.

SUMMARY

- Human subjects degrade 1 to 2% of their body protein daily at rates that vary widely between proteins and with physiologic state. Key regulatory enzymes often have short half-lives.
- Proteins are degraded by both ATP-dependent and ATP-independent pathways. Ubiquitin targets many intracellular proteins for degradation. Liver cell surface receptors bind and internalize circulating asialoglycoproteins destined for lysosomal degradation.
- Polyubiquitinated proteins are degraded by proteases on the inner surface of a cylindrical macromolecule, the proteasome. Entry into the proteasome is gated by a donut-shaped protein pore that rejects entry to all but polyubiquitinated proteins.
- Fishes excrete highly toxic NH_3 directly. Birds convert NH_3 to uric acid. Higher vertebrates convert NH_3 to urea.
- Transamination channels amino acid nitrogen into glutamate. GDH occupies a central position in nitrogen metabolism.
- Glutamine synthetase converts NH_3 to nontoxic glutamine. Glutaminase releases NH_3 for use in urea synthesis.
- NH_3, CO_2, and the amide nitrogen of aspartate provide the atoms of urea.
- Hepatic urea synthesis takes place in part in the mitochondrial matrix and in part in the cytosol.
- Changes in enzyme levels and allosteric regulation of carbamoyl phosphate synthetase I by N-acetylglutamate regulate urea biosynthesis.
- Metabolic diseases are associated with defects in each enzyme of the urea cycle, of the ORC1 ornithine carrier, and of NAGS.
- The metabolic disorders of urea biosynthesis illustrate six general principles of all metabolic disorders.
- Tandem mass spectrometry is the technique of choice for screening neonates for inherited metabolic diseases.

REFERENCES

Adam S, Almeida MF, Assoun M, et al: Dietary management of urea cycle disorders: European practice. Mol Genet Metab 2013;110:439.

Burgard P, Kölker S, Haege G, et al. Neonatal mortality and outcome at the end of the first year of life in early onset urea cycle disorders. J Inherit Metab Dis. 2016;39:219.

Dwane L, Gallagher WM, Ni Chonghaile T, et al: The emerging role of non-traditional ubiquitination in oncogenic pathways. J Biol Chem 2017;292:3543.

Häberle J, Pauli S, Schmidt E, et al: Mild citrullinemia in caucasians is an allelic variant of argininosuccinate synthetase deficiency (citrullinemia type 1). Mol Genet Metab 2003;80:302.

Jiang YH, Beaudet AL: Human disorders of ubiquitination and proteasomal degradation. Curr Opin Pediatr 2004;16:419.

Monné M, Miniero DV, Dabbabbo L, et al: Mitochondrial transporters for ornithine and related amino acids: a review. Amino Acids 2015;9:1963.

Pal A, Young MA, Donato NJ: Emerging potential of therapeutic targeting of ubiquitin-specific proteases in the treatment of cancer. Cancer Res 2014;14:721.

Pickart CM: Mechanisms underlying ubiquitination. Annu Rev Biochem 2001;70:503.

Sylvestersen KB, Young C, Nielsen ML: Advances in characterizing ubiquitylation sites by mass spectrometry. Curr Opin Chem Biol 2013;17:49.

Waisbren SE, Gropman AL: Improving long term outcomes in urea cycle disorders. J Inherit Metab Dis 2016;39:573.

Catabolism of the Carbon Skeletons of Amino Acids

Victor W. Rodwell, PhD

- Name the principal catabolites of the carbon skeletons of the protein amino acids and the major metabolic fates of these catabolites.

- Write an equation for an aminotransferase (transaminase) reaction and illustrate the role played by the coenzyme.

- Outline the metabolic pathways for each of the protein amino acids, and identify reactions associated with clinically significant metabolic disorders.

- Provide examples of aminoacidurias that arise from defects in glomerular tubular reabsorption, and the consequences of impaired intestinal absorption of tryptophan.

- Explain why metabolic defects in different enzymes of the catabolism of a specific amino acid can be associated with similar clinical signs and symptoms.

- Describe the implications of a metabolic defect in Δ^1-pyrroline-5-carboxylate dehydrogenase for the catabolism of proline and of 4-hydroxyproline.

- Explain how the α-amino nitrogen of proline and of lysine is removed by processes other than transamination.

- Draw analogies between the reactions that participate in the catabolism of fatty acids and of the branched-chain amino acids.

- Identify the specific metabolic defects in hypervalinemia, maple syrup urine disease, intermittent branched-chain ketonuria, isovaleric acidemia, and methylmalonic aciduria.

BIOMEDICAL IMPORTANCE

Chapter 28 described the removal and the metabolic fate of the nitrogen atoms of most of the protein L-α-amino acids. This chapter addresses the metabolic fates of the resulting hydrocarbon skeletons of each of the protein amino acids, the enzymes and intermediates involved, and several associated metabolic diseases or "inborn errors of metabolism." Most disorders of amino acid catabolism are comparatively rare, but if left untreated, they can result in irreversible brain damage and early mortality. Prenatal or early postnatal detection of metabolic disorders and timely initiation of treatment thus are essential. The ability to detect the activities of enzymes in cultured amniotic fluid cells facilitates prenatal diagnosis by amniocentesis. In the United States, all states conduct screening tests of newborns for up to 40 metabolic diseases, which include disorders associated with defects in the catabolism of amino acids. The most reliable screening tests use tandem mass spectrometry to detect, in a few drops of neonate blood, catabolites suggestive of a given metabolic defect, and thereby implicate the absence or lowered activity of one or more specific enzymes.

Mutations either of a gene or of associated regulatory regions of DNA can result either in the failure to synthesize the encoded enzyme, or in synthesis of a partially or completely nonfunctional enzyme. Mutations that affect enzyme activity, those that compromise its three-dimensional structure, or that disrupt its catalytic or regulatory sites, can have severe metabolic consequences. Low catalytic efficiency of a mutant enzyme can result from impaired positioning of residues involved in catalysis, or in binding a substrate, coenzyme, or metal ion. Mutations may also impair the ability of certain enzymes to respond appropriately to the signals that modulate their activity by altering an enzyme's affinity for an allosteric regulator of activity. Since different mutations can have similar effects on any of the above factors, at a molecular level these represent distinct molecular diseases, although various mutations may give rise to the same clinical signs and symptoms. Remediation of metabolic disorders of amino acid metabolism consists primarily of feeding diets low in the amino acid whose catabolism is impaired. Ultimately, however, genetic engineering may be able to permanently correct a given metabolic defect.

AMINO ACIDS ARE CATABOLIZED TO INTERMEDIATES FOR CARBOHYDRATE & LIPID BIOSYNTHESIS

Nutritional studies in the period 1920 to 1940, reinforced and confirmed by studies using isotopically labeled amino acids conducted from 1940 to 1950, established the interconvertibility of the carbon atoms of fat, carbohydrate, and protein. These studies also revealed that all or a portion of the carbon skeleton of every amino acid is convertible either to carbohydrate, fat, or both fat and carbohydrate (**Table 29–1**). **Figure 29–1** outlines overall aspects of these interconversions.

TRANSAMINATION TYPICALLY INITIATES AMINO ACID CATABOLISM

Removal of α-amino nitrogen by transamination, catalyzed by a transaminase (see Figure 28–9), is the first catabolic reaction of most of the protein amino acids. The exceptions are proline, hydroxyproline, threonine, and lysine, whose α-amino groups do not participate in transamination. The hydrocarbon skeletons that remain are then degraded to amphibolic intermediates as outlined in Figure 29–1.

TABLE 29–1 Fate of the Carbon Skeletons of the Protein L-α-Amino Acids

Converted to Amphibolic Intermediates That Form		
Carbohydrate (Glycogenic)	Fat (Ketogenic)	Glycogen and Fat (Glycogenic and Ketogenic)
Ala	Hyp	Leu	Ile
Arg	Met	Lys	Phe
Asp	Pro		Trp
Cys	Ser		Tyr
Glu	Thr		
Gly	Val		
His			

Asparagine & Aspartate Form Oxaloacetate

All four carbons of asparagine and of aspartate form **oxaloacetate** via subsequent reactions catalyzed by **asparaginase** (EC 3.5.1.1) and a **transaminase**.

$$\text{Asparagine} + H_2O \rightarrow \text{Aspartate} + NH_4^+$$
$$\text{Aspartate} + \text{Pyruvate} \rightarrow \text{Alanine} + \text{Oxaloacetate}$$

Glutamine & Glutamate Form α-Ketoglutarate

Successive reactions catalyzed by **glutaminase** (EC 3.5.1.2) and a **transaminase** form **α-ketoglutarate**.

$$\text{Glutamine} + H_2O \rightarrow \text{Glutamate} + NH_4^+$$
$$\text{Glutamate} + \text{Pyruvate} \rightarrow \text{Alanine} + \text{α-Ketoglutarate}$$

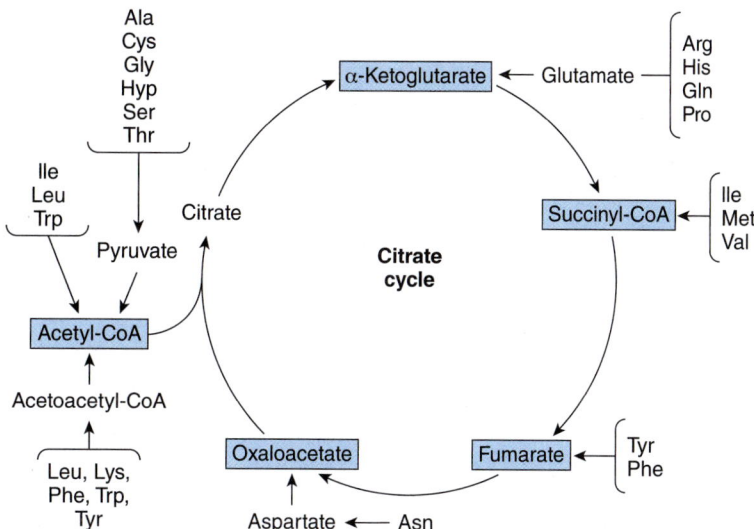

FIGURE 29–1 Overview of the amphibolic intermediates that result from catabolism of the protein amino acids.

While both glutamate and aspartate are substrates for the same transaminase, metabolic defects in transaminases, which fulfill central amphibolic functions, may be incompatible with life. Consequently, no known metabolic defect is associated with these two short catabolic pathways that convert asparagine and glutamine to amphibolic intermediates.

Proline

The catabolism of proline takes place in mitochondria. Since proline does not participate in transamination, its α-amino nitrogen is retained throughout a two-stage oxidation to glutamate. Oxidation to Δ^1-pyrroline-5-carboxylate is catalyzed by proline dehydrogenase, EC 1.5.5.2. Subsequent oxidation to glutamate is catalyzed by Δ^1-pyrroline-5-carboxylate dehydrogenase (also called glutamate-γ-semialdehyde dehydrogenase, EC 1.2.1.88; **Figure 29–2**). There are two metabolic

disorders of proline catabolism. Inherited as autosomal recessive traits, both are consistent with a normal adult life. The metabolic block in **type I hyperprolinemia** is at **proline dehydrogenase**. There is no associated impairment of hydroxyproline catabolism. The metabolic block in **type II hyperprolinemia** is at Δ^1-pyrroline-5-carboxylate dehydrogenase, which also participates in the catabolism of arginine, ornithine, and hydroxyproline (see later). Since proline and hydroxyproline catabolism are affected, both Δ^1-pyrroline-5-carboxylate and Δ^1-pyrroline-3-hydroxy-5-carboxylate (see Figure 29–11) are excreted.

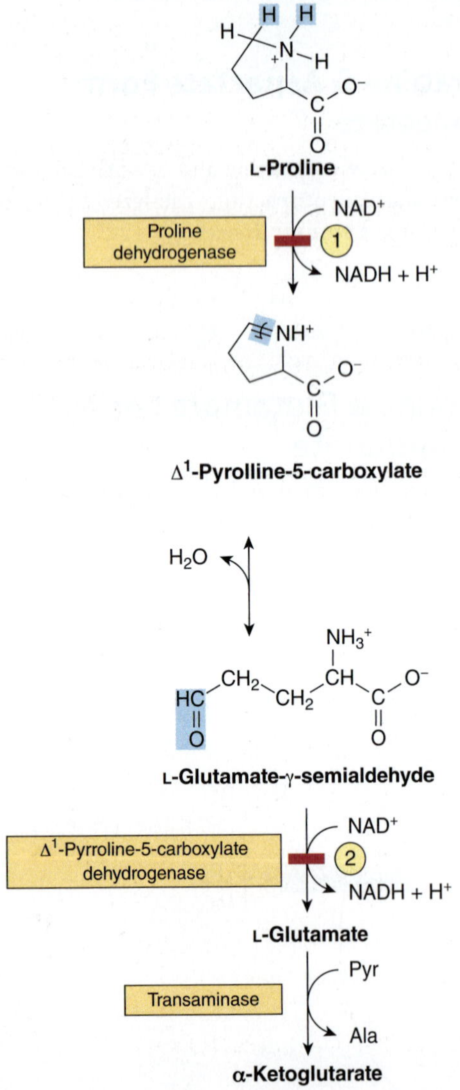

Arginine & Ornithine

The initial reactions in arginine catabolism are conversion to ornithine followed by transamination of ornithine to glutamate-γ-semialdehyde (**Figure 29–3**). Subsequent catabolism of glutamate-γ-semialdehyde to α-ketoglutarate occurs as described for proline (see Figure 29–2). Mutations in **ornithine δ-aminotransferase** (ornithine transaminase, EC 2.6.1.13) elevate plasma and urinary ornithine, and are associated with **gyrate atrophy of the choroid and retina**. Treatment involves restricting dietary arginine. In the **hyperornithinemia–hyperammonemia syndrome**, a defective ORC1 mitochondrial **ornithine-citrulline antiporter** (see Figure 28–16) impairs transport of ornithine into mitochondria, where it participates in urea synthesis.

Histidine

Catabolism of histidine proceeds via urocanate, 4-imidazolone-5-propionate, and N-formiminoglutamate (Figlu). Formimino

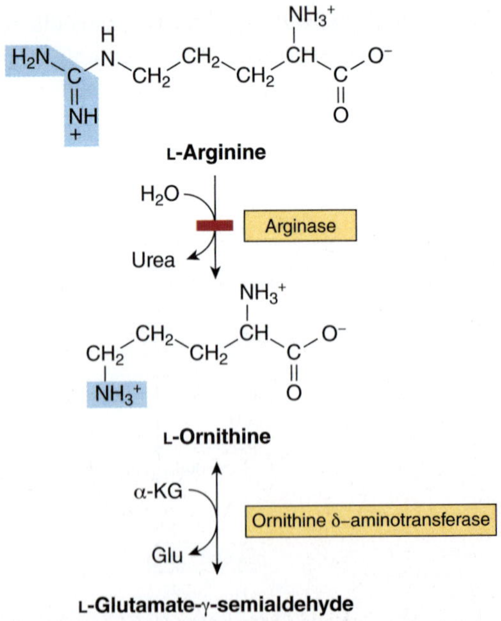

FIGURE 29–2 **Catabolism of proline.** Red bars and circled numerals indicate the locus of the inherited metabolic defects in ① type-I hyperprolinemia and ② type-II hyperprolinemia. In this and subsequent figures, blue highlights emphasize the portions of the molecules that are undergoing chemical change.

FIGURE 29–3 **Catabolism of arginine.** Arginase-catalyzed cleavage of L-arginine forms urea and L-ornithine. This reaction (red bar) represents the site of the inherited metabolic defect in hyperargininemia. Subsequent transamination of ornithine to glutamate-γ-semialdehyde is followed by its oxidation to α-ketoglutarate.

group transfer to tetrahydrofolate forms glutamate, then α-ketoglutarate (**Figure 29–4**). In **folic acid deficiency**, transfer of the formimino group is impaired, and Figlu is excreted. Excretion of Figlu following a dose of histidine thus can be used to detect folic acid deficiency. Benign disorders of histidine catabolism include **histidinemia** and **urocanic aciduria** associated with impaired **histidase** and **urocanase**, respectively.

CATABOLISM OF GLYCINE, SERINE, ALANINE, CYSTEINE, THREONINE, & 4-HYDROXYPROLINE

Glycine

The **glycine cleavage system** of liver mitochondria splits glycine to CO_2 and NH_4^+ and forms N^5,N^{10}-methylene tetrahydrofolate.

$$Glycine + H_4folate + NAD^+ \rightarrow CO_2 + NH_3 + 5,10\text{-}CH_2\text{-}H_4folate + NADH + H^+$$

The glycine cleavage complex (**Figure 29–5**) consists of three enzymes and an "H-protein" that has a covalently attached dihydrolipoyl moiety. Figure 29–5 also illustrates the individual reactions and intermediates in glycine cleavage. In **nonketotic hyperglycinemia**, a rare inborn error of glycine degradation, glycine accumulates in all body tissues including the central nervous system. The defect in **primary hyperoxaluria** is the failure to catabolize glyoxylate formed by the deamination of glycine. Subsequent oxidation of glyoxylate to oxalate results in urolithiasis, nephrocalcinosis, and early mortality from renal failure or hypertension. **Glycinuria** results from a defect in renal tubular reabsorption.

Serine

Following conversion to glycine, catalyzed by glycine hydroxy-methyltransferase (EC 2.1.2.1), serine catabolism merges with that of glycine (**Figure 29–6**).

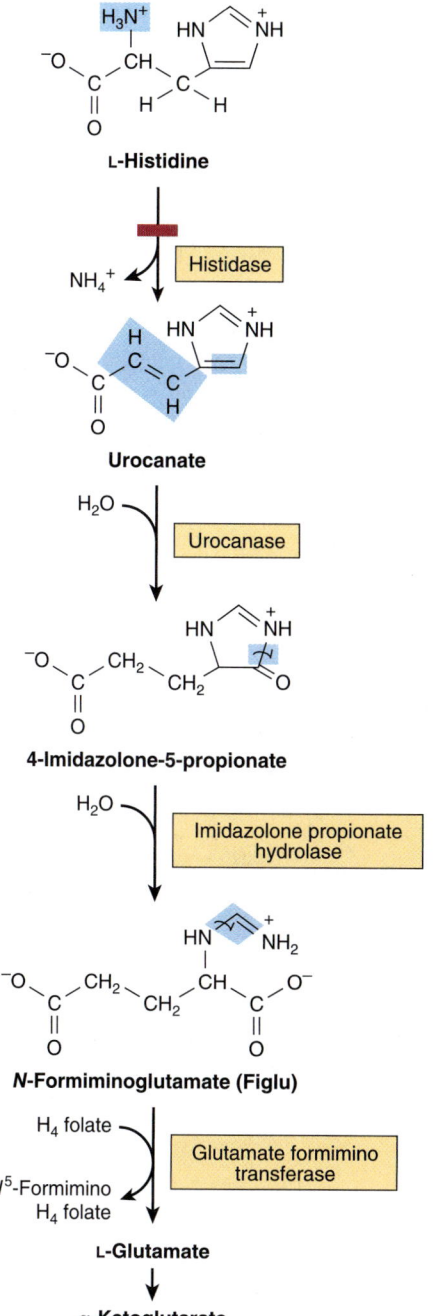

FIGURE 29–4 Catabolism of L-histidine to α-ketoglutarate. (H_4 folate, tetrahydrofolate.) The red bar indicates the site of an inherited metabolic defect.

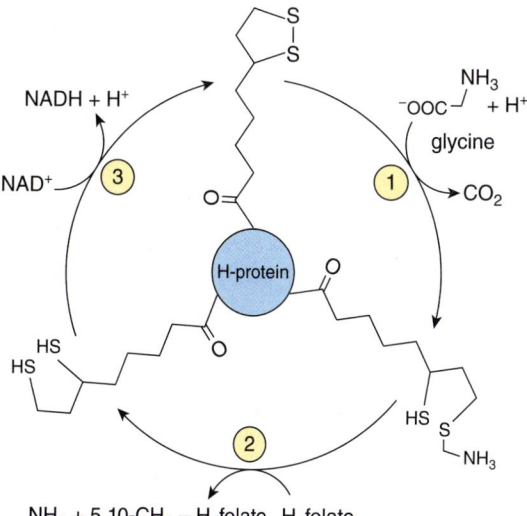

FIGURE 29–5 The glycine cleavage system of liver mitochondria. The glycine cleavage complex consists of three enzymes and an "H-protein" that has covalently attached dihyrolipoate. Catalysts for the numbered reactions are ① glycine dehydrogenase (decarboxylating), ② an ammonia-forming aminomethyltransferase, and ③ dihydrolipoamide dehydrogenase. (H_4 folate, tetrahydrofolate).

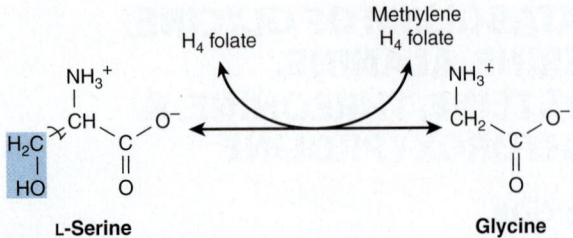

FIGURE 29–6 Interconversion of serine and glycine by glycine hydroxymethyltransferase. (H_4 folate, tetrahydrofolate.)

Alanine

Transamination of α-alanine forms pyruvate. Probably on account of its central role in metabolism, there is no known viable metabolic defect of α-alanine catabolism.

Cystine & Cysteine

Cystine is first reduced to cysteine by **cystine reductase**, EC 1.8.1.6 (**Figure 29–7**). Two different pathways then convert cysteine to pyruvate (**Figure 29–8**). There are numerous abnormalities of cysteine metabolism. Cystine, lysine, arginine, and ornithine are excreted in **cystine-lysinuria (cystinuria)**, a defect in renal reabsorption of these amino acids. Apart from cystine calculi, cystinuria is benign. The mixed disulfide of L-cysteine and L-homocysteine (**Figure 29–9**) excreted by cystinuric patients is more soluble than cystine and reduces formation of cystine calculi.

Several metabolic defects result in vitamin B_6-responsive or vitamin B_6-unresponsive **homocystinurias**. These include a deficiency in the reaction catalyzed by cystathionine β-synthase, EC 4.2.1.22:

$$\text{Serine} + \text{homocysteine} \rightarrow \text{cystathionine} + H_2O$$

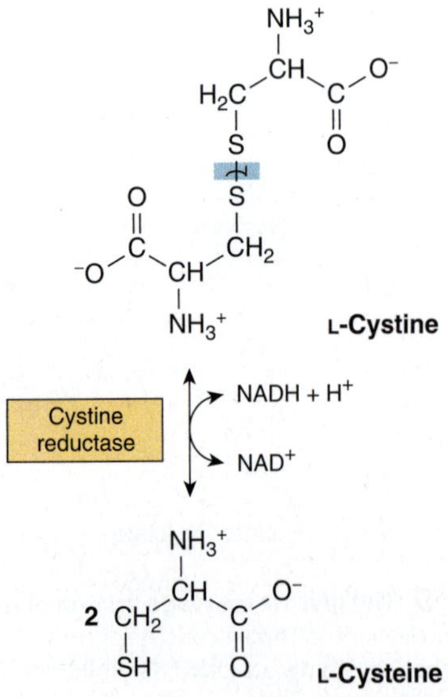

FIGURE 29–7 Reduction of cystine to cysteine by cystine reductase.

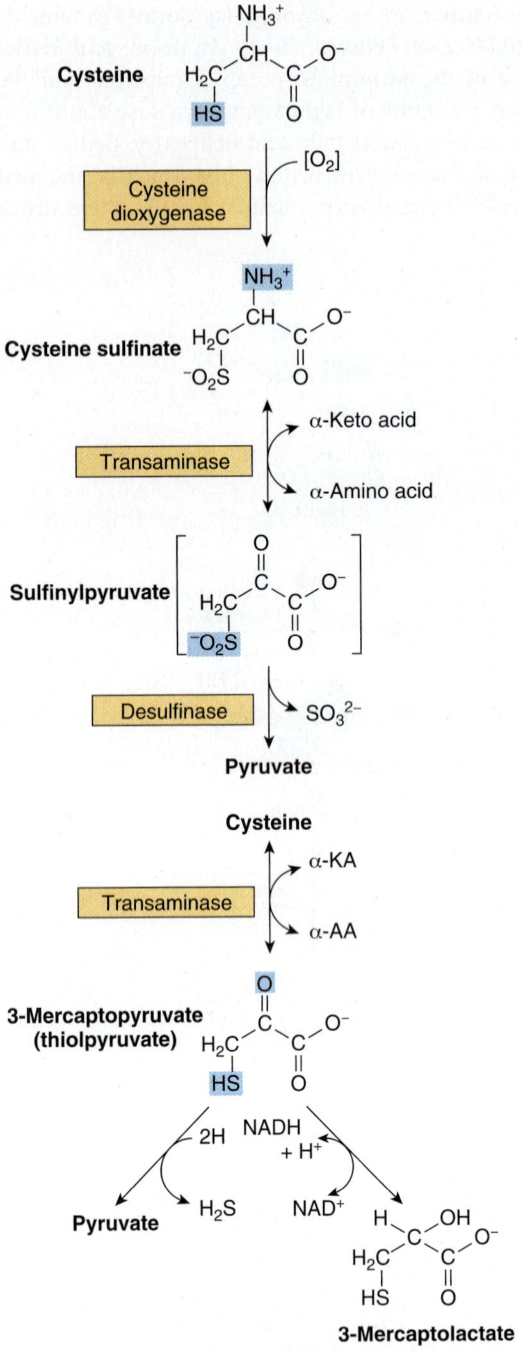

FIGURE 29–8 Two pathways catabolize cysteine: the cysteine sulfinate pathway (*top*) and the 3-mercaptopyruvate pathway (*bottom*).

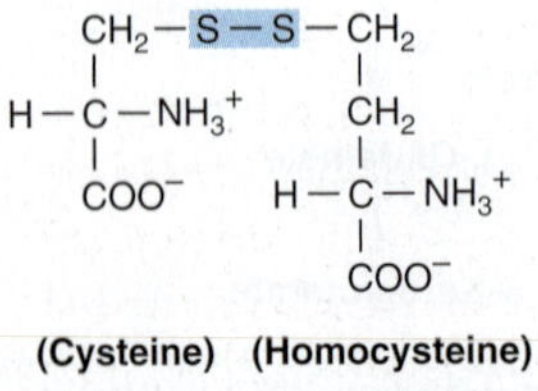

FIGURE 29–9 Structure of the mixed disulfide of cysteine and homocysteine.

Consequences include osteoporosis and mental retardation. Defective carrier-mediated transport of cystine results in **cystinosis (cystine storage disease)** with deposition of cystine crystals in tissues and early mortality from acute renal failure. Epidemiologic and other data link plasma homocysteine levels to cardiovascular risk, but the role of homocysteine as a causal cardiovascular risk factor remains controversial.

Threonine

Threonine aldolase (EC 4.1.2.5) cleaves threonine to glycine and acetaldehyde. Catabolism of glycine is discussed earlier. Oxidation of acetaldehyde to acetate is followed by formation of acetyl-CoA (**Figure 29–10**).

4-Hydroxyproline

Catabolism of 4-hydroxy-L-proline forms, successively, L-Δ^1-pyrroline-3-hydroxy-5-carboxylate, γ-hydroxy-L-glutamate-γ-semialdehyde, erythro-γ-hydroxy-L-glutamate, and α-keto-γ-hydroxyglutarate. An aldol-type cleavage then forms glyoxylate plus pyruvate (**Figure 29–11**). A defect in **4-hydroxyproline dehydrogenase** results in **hyperhydroxyprolinemia**, which is benign. There is no associated impairment of proline catabolism. A defect

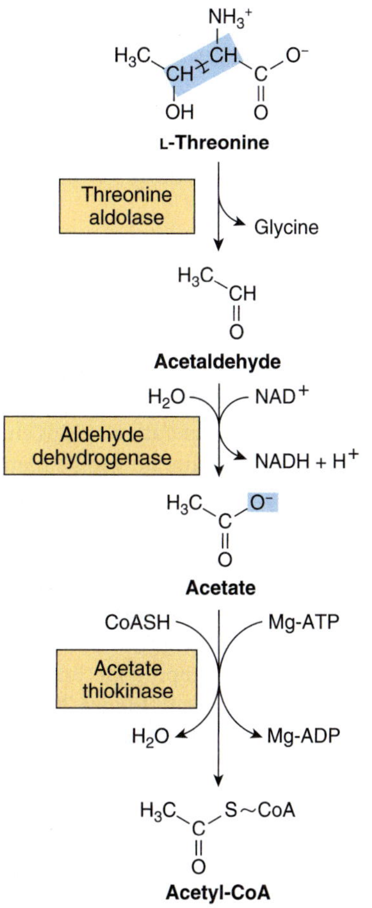

FIGURE 29–10 Intermediates in the conversion of threonine to glycine and acetyl-CoA.

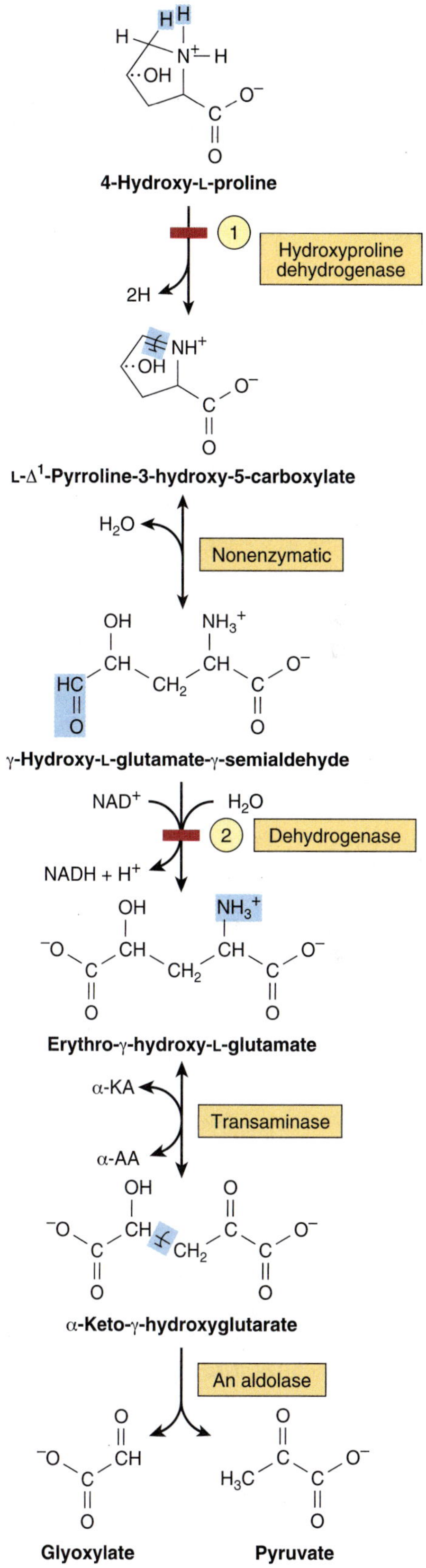

FIGURE 29–11 Intermediates in hydroxyproline catabolism. (α-AA, α-amino acid; α-KA, α-keto acid.) Red bars indicate the sites of the inherited metabolic defects in ① hyperhydroxyprolinemia and ② type II hyperprolinemia.

in **glutamate-γ-semialdehyde dehydrogenase** is accompanied by excretion of Δ^1-pyrroline-3-hydroxy-5-carboxylate.

ADDITIONAL AMINO ACIDS THAT FORM ACETYL-CoA

Tyrosine

Figure 29–12 illustrates the intermediates and enzymes that participate in the catabolism of tyrosine to amphibolic intermediates. Following transamination of tyrosine to *p*-hydroxyphenylpyruvate, successive reactions form homogentisate, maleylacetoacetate, fumarylacetoacetate, fumarate, acetoacetate, and ultimately acetyl-CoA and acetate.

Several metabolic disorders are associated with the tyrosine catabolic pathway. The probable metabolic defect in **type I tyrosinemia (tyrosinosis)** is at **fumarylacetoacetate hydrolase**, EC 3.7.1.12 (reaction 4, see Figure 29–12). Therapy employs a diet low in tyrosine and phenylalanine. Untreated acute and chronic tyrosinosis leads to death from liver failure. Alternate metabolites of tyrosine are also excreted in **type II tyrosinemia (Richner-Hanhart syndrome)**, a defect in **tyrosine aminotransferase** (reaction 1, see Figure 29–12), and in **neonatal tyrosinemia**, due to lowered activity of *p*-hydroxyphenylpyruvate hydroxylase, EC 1.13.11.27 (reaction 2, see Figure 29–12). Therapy employs a diet low in protein.

The metabolic defect in **alkaptonuria** is a defective **homogentisate oxidase** (EC 1.13.11.5), which catalyzes reaction 3 of Figure 29–12. The urine darkens on exposure to air due to oxidation of excreted homogentisate. Late in the disease, there is arthritis and connective tissue pigmentation (ochronosis) due to oxidation of homogentisate to benzoquinone acetate, which polymerizes and binds to connective tissue. First described in the 16th century based on the observation that the urine darkened on exposure to air, alkaptonuria provided the basis for Sir Archibald Garrod's early 20th century classic ideas concerning heritable metabolic disorders. Based on the presence of ochronosis and on chemical evidence, the earliest known case of alkaptonuria is, however, its detection in 1977 in an Egyptian mummy dating from 1500 B.C.!

Phenylalanine

Phenylalanine is first converted to tyrosine (see Figure 27–12). Subsequent reactions are those of tyrosine (see Figure 29–12). **Hyperphenylalaninemias** arise from defects in phenylalanine hydroxylase, EC 1.14.16.1 (**type I, classic phenylketonuria [PKU]**, frequency 1 in 10,000 births), in dihydrobiopterin reductase (**types II and III**), or in dihydrobiopterin biosynthesis (**types IV and V**) (see Figure 27–12). Alternative catabolites are excreted (**Figure 29–13**). A diet low in phenylalanine can prevent the mental retardation of PKU.

DNA probes facilitate prenatal diagnosis of defects in phenylalanine hydroxylase or dihydrobiopterin reductase. Elevated blood phenylalanine may not be detectable until 3 to 4 days postpartum. False positives in premature infants may reflect delayed maturation of enzymes of phenylalanine catabolism. An older and less reliable screening test employs $FeCl_3$ to detect urinary phenylpyruvate. $FeCl_3$ screening for PKU of the urine of newborn infants is compulsory in many countries, but in the United States has been largely supplanted by tandem mass spectrometry.

Lysine

Removal of the **ε-nitrogen** of lysine proceeds via initial formation of **saccharopine** and subsequent reactions that also liberate the **α-nitrogen**. The ultimate product of the carbon skeleton is crotonyl-CoA. Circled numerals refer to the corresponding numbered reactions of **Figure 29–14**. Reactions 1 and 2 convert the Schiff base formed between α-ketoglutarate and the ε-amino group of lysine to L-α-aminoadipate-δ-semialdehyde. Reactions 1 and 2 both are catalyzed by a single bifunctional enzyme, aminoadipate-δ-semialdehyde synthase (EC 1.5.1.8) whose *N*-terminal and *C*-terminal domains contain lysine-α-ketoglutarate reductase and saccharopine dehydrogenase activity, respectively. Reduction of L-α-aminoadipate-δ-semialdehyde to L-α-aminoadipate (reaction 3) is followed by transamination to α-ketoadipate (reaction 4). Conversion to the thioester glutaryl-CoA (reaction 5) is followed by the decarboxylation of glutaryl-CoA to crotonyl-CoA (reaction 6). Reduction of crotonyl-CoA by crotanoyl-CoA reductase, EC 1.3.1.86, forms butanoyl-CoA:

$$\text{Crotonyl-CoA} + \text{NADPH} + \text{H}^+ \rightarrow \text{butanoyl-CoA} + \text{NADP}^+$$

Subsequent reactions are those of fatty acid catabolism (see Chapter 22).

Hyperlysinemia can result from a metabolic defect in either the first or second activity of the bifunctional enzyme aminoadipate-δ-semialdehyde synthase, but only if the defect involves the second activity that is accompanied by elevated levels of blood saccharopine. A metabolic defect at reaction 6 results in an inherited metabolic disease that is associated with striatal and cortical degeneration, and is characterized by elevated concentrations of glutarate and its metabolites glutaconate and 3-hydroxyglutarate. The challenge in clinical management of these metabolic defects is to restrict dietary intake of L-lysine without producing malnutrition.

Tryptophan

Tryptophan is degraded to amphibolic intermediates via the kynurenine-anthranilate pathway (**Figure 29–15**). **Tryptophan 2,3-dioxygenase**, EC 1.13.11.11 (**tryptophan pyrrolase**) opens the indole ring, incorporates molecular oxygen, and forms *N*-formylkynurenine. Tryptophan oxygenase, an iron porphyrin metalloprotein that is inducible in liver by adrenal corticosteroids and by tryptophan, is feedback inhibited by nicotinic acid derivatives, including NADPH. Hydrolytic removal of the formyl group of *N*-formylkynurenine, catalyzed

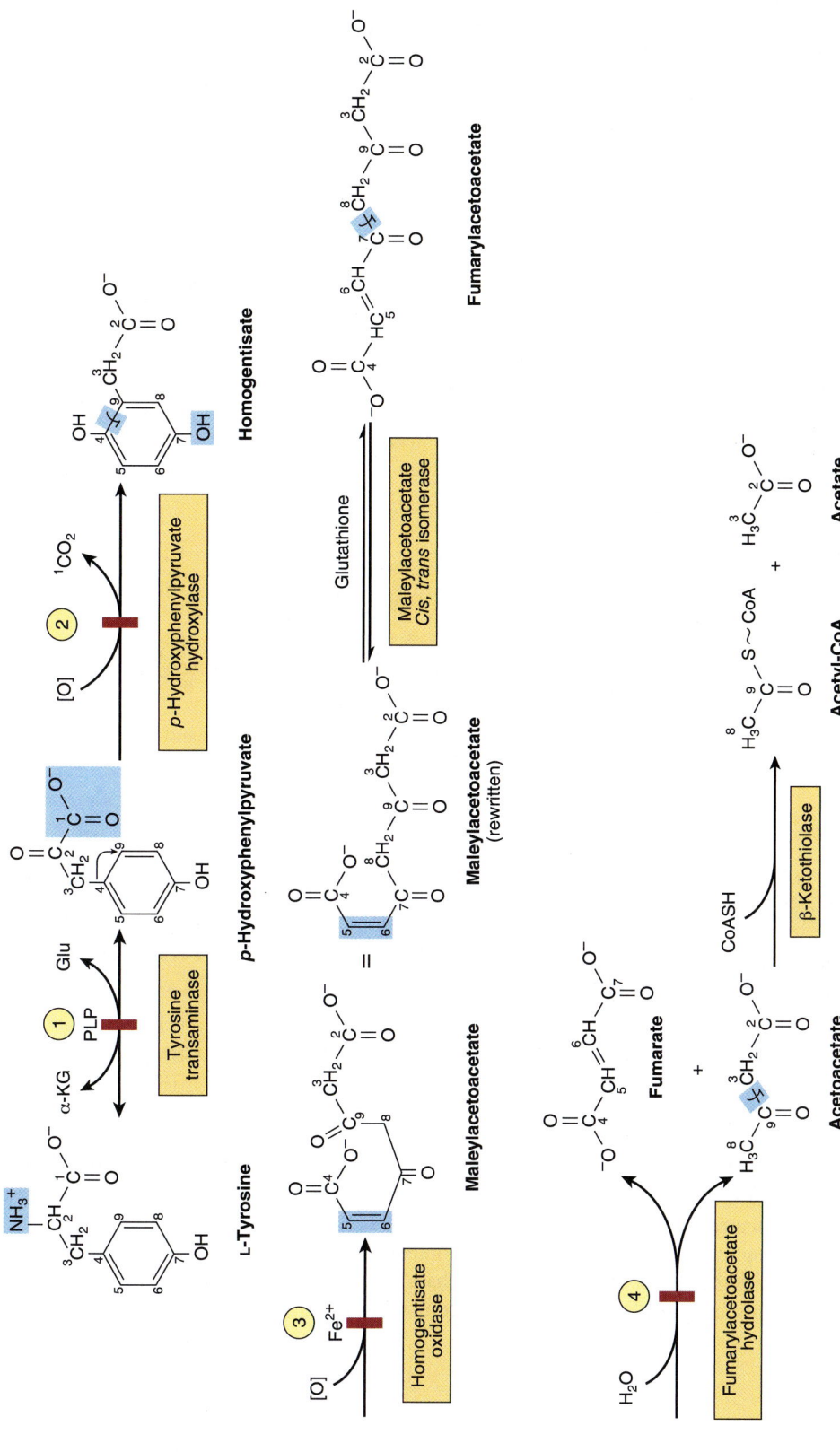

FIGURE 29–12 Intermediates in tyrosine catabolism. Carbons are numbered to emphasize their ultimate fate. (α-KG, α-ketoglutarate; Glu, glutamate; PLP, pyridoxal phosphate.) Red bars indicate the probable sites of the inherited metabolic defects in type II tyrosinemia; neonatal tyrosinemia; ① alkaptonuria; and ② type I tyrosinemia, or tyrosinosis. ③ alkaptonuria; and ④ type I tyrosinemia, or tyrosinosis.

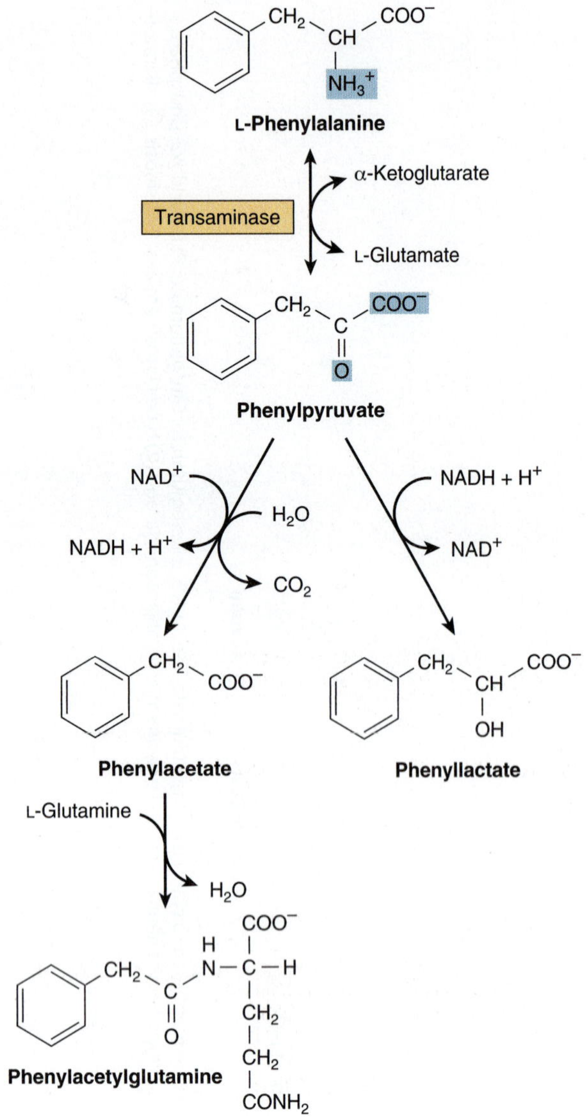

FIGURE 29–13 **Alternative pathways of phenylalanine catabolism in phenylketonuria.** The reactions also occur in normal liver tissue but are of minor significance.

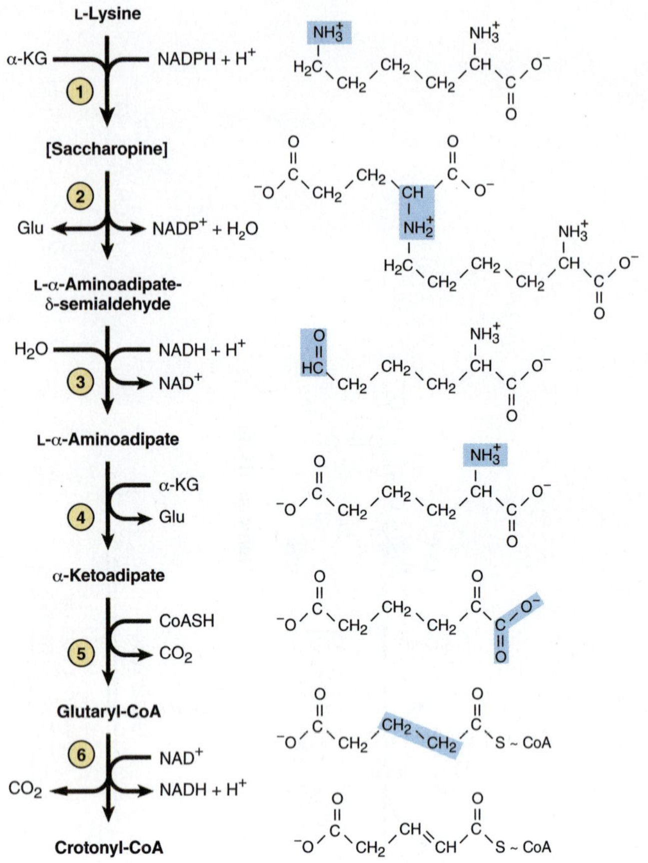

FIGURE 29–14 **Reactions and intermediates in the catabolism of lysine.**

by **kynurenine formylase** (EC 3.5.1.9), produces kynurenine. Since **kynureninase** (EC 3.7.1.3) requires pyridoxal phosphate, excretion of xanthurenate (**Figure 29–16**) in response to a tryptophan load is diagnostic of vitamin B_6 deficiency. **Hartnup disease** reflects impaired intestinal and renal transport of tryptophan and other neutral amino acids. Indole derivatives of unabsorbed tryptophan formed by intestinal bacteria are excreted. The defect limits tryptophan availability for niacin biosynthesis and accounts for the pellagra-like signs and symptoms.

Methionine

Methionine reacts with ATP forming *S*-adenosylmethionine, "active methionine" (**Figure 29–17**). Subsequent reactions form propionyl-CoA (**Figure 29–18**), whose conversion to succinyl-CoA occurs via reactions 2, 3, and 4 of Figure 19–2.

THE INITIAL REACTIONS ARE COMMON TO ALL THREE BRANCHED-CHAIN AMINO ACIDS

Several of the initial reactions of the catabolism of isoleucine, leucine, and valine (**Figure 29–19**) are analogous to reactions of fatty acid catabolism (see Figure 22–3). Following transamination (see Figure 29–19, reaction 1), the carbon skeletons of the resulting α-keto acids undergo oxidative decarboxylation and conversion to coenzyme A thioesters. This multistep process is catalyzed by the **mitochondrial branched-chain α-ketoacid dehydrogenase complex**, whose components are functionally identical to those of the pyruvate dehydrogenase complex (PDH) (see Figure 18–5). Like PDH, the branched-chain α-ketoacid dehydrogenase complex consists of five components.

E1: thiamin pyrophosphate (TPP)-dependent branched-chain α-ketoacid decarboxylase

E2: dihydrolipoyl transacylase (contains lipoamide)

E3: dihydrolipoamide dehydrogenase (contains FAD)

PDH complex kinase (PDK)

PDH complex phosphatase (PDP)

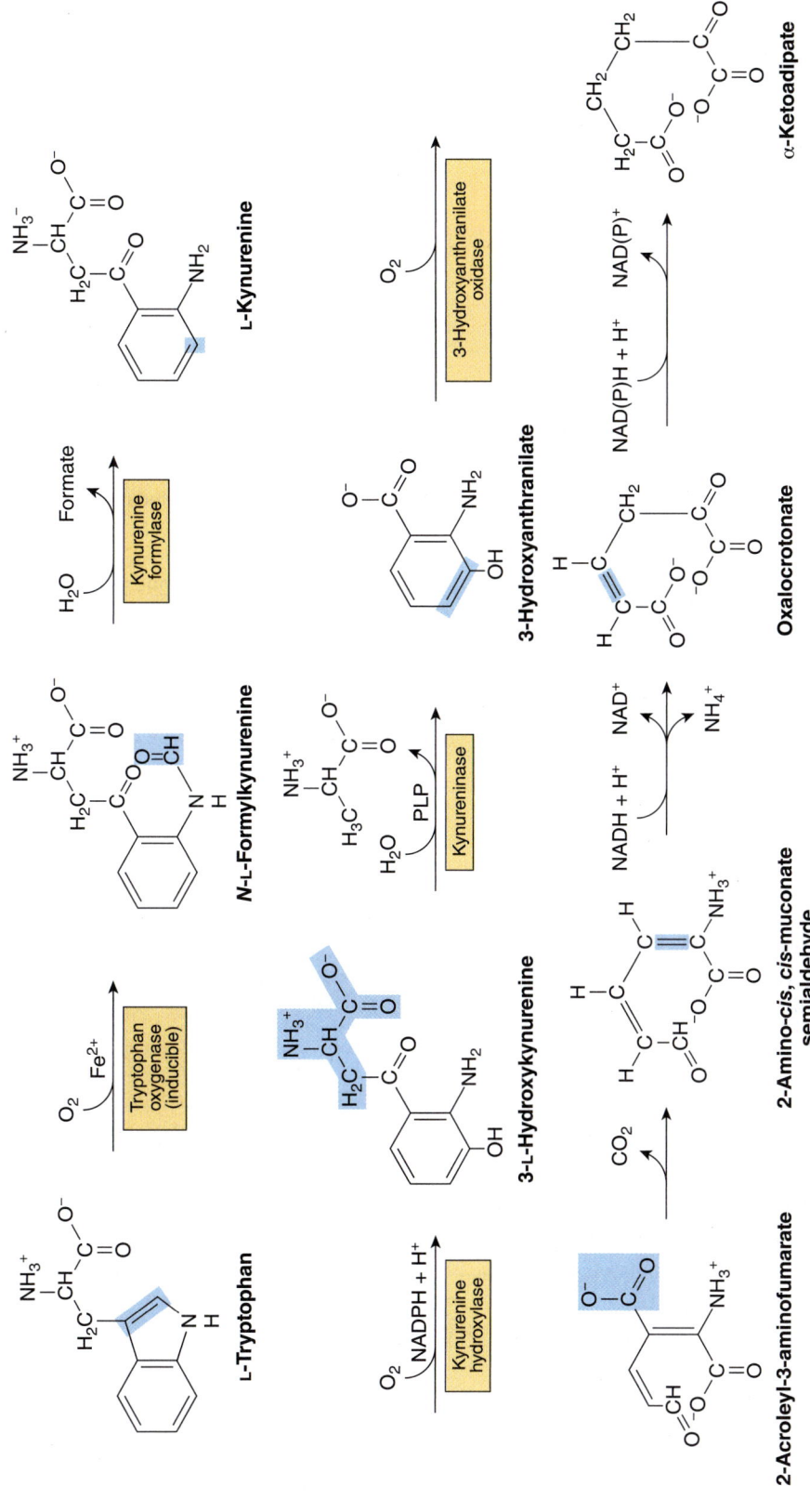

FIGURE 29–15 Reactions and intermediates in the catabolism of tryptophan. (PLP, pyridoxal phosphate.)

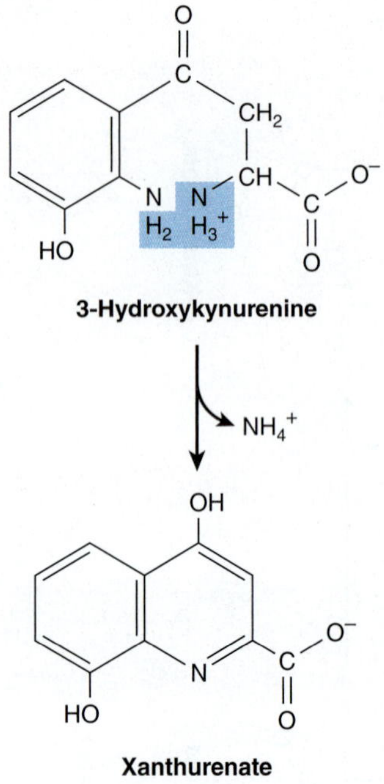

FIGURE 29–16 **Formation of xanthurenate in vitamin B$_6$ deficiency.** Conversion of the tryptophan metabolite 3-hydroxykynurenine to 3-hydroxyanthranilate is impaired (see Figure 29–15). A large portion is therefore converted to xanthurenate.

As for pyruvate dehydrogenase (see Figure 17–6), the PDH complex kinase and PDH complex phosphatase regulate activity of the branched-chain α-ketoacid dehydrogenase complex via phosphorylation (inactivation) and dephosphorylation (activation).

Dehydrogenation of the resulting coenzyme A thioesters (reaction 3, Figure 29–19) proceeds like the dehydrogenation of lipid-derived fatty acyl-CoA thioesters (see Chapter 22). **Figures 29–20, 29–21,** and **29–22** illustrate the subsequent reactions unique for each amino acid skeleton.

METABOLIC DISORDERS OF BRANCHED-CHAIN AMINO ACID CATABOLISM

As the name implies, the odor of urine in **maple syrup urine disease (branched-chain ketonuria,** or **MSUD)** suggests maple syrup, or burnt sugar. The biochemical defect in MSUD involves the **α-ketoacid decarboxylase complex** (reaction 2, Figure 29–19). Plasma and urinary levels of leucine, isoleucine, valine, and their cognate α-keto acids and α-hydroxy acids (reduced α-keto acids) are elevated, but the urinary keto acids derive principally from leucine. Signs and symptoms of MSUD often include ketoacidosis, neurologic derangements, mental retardation, and a maple syrup odor of urine. The mechanism of toxicity is unknown. Early diagnosis by enzymatic analysis is essential to avoid brain damage and early mortality by replacing dietary protein by an amino acid mixture that lacks leucine, isoleucine, and valine.

The molecular genetics of MSUD are heterogeneous. MSUD can result from mutations in the genes that encode E1α, E1β, E2, and E3. Based on the locus affected, genetic subtypes of MSUD are recognized. Type IA MSUD arises from mutations in the *E1α* gene, type IB in the *E1β* gene, type II in the *E2* gene, and type III in the *E3* gene (**Table 29–2**). In **intermittent branched-chain ketonuria,** the α-ketoacid decarboxylase retains some activity, and symptoms occur later in life. In **isovaleric acidemia,** ingestion of protein-rich foods elevates isovalerate, the deacylation product of isovaleryl-CoA. The impaired enzyme in **isovaleric acidemia** is **isovaleryl-CoA dehydrogenase**, EC 1.3.8.4 (reaction 3, Figure 29–19). Vomiting, acidosis, and coma follow ingestion of excess protein. Accumulated isovaleryl-CoA is hydrolyzed to isovalerate and excreted.

Table 29–3 summarizes the metabolic disorders associated with the catabolism of amino acids, and lists the impaired enzyme, its IUB enzyme catalog (EC) number, a cross-reference to a specific figure, and numbered reaction in this text, and a numerical link to the Online Mendelian Inheritance in Man (OMIM) database.

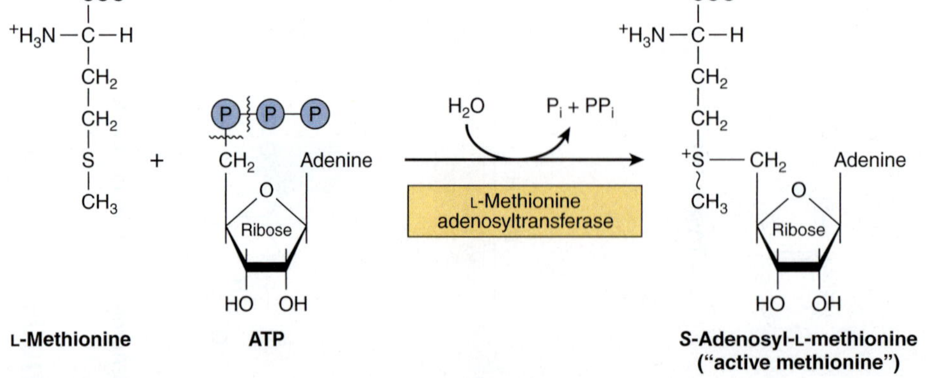

FIGURE 29–17 **Formation of *S*-adenosylmethionine.** ~ CH$_3$ represents the high group transfer potential of "active methionine."

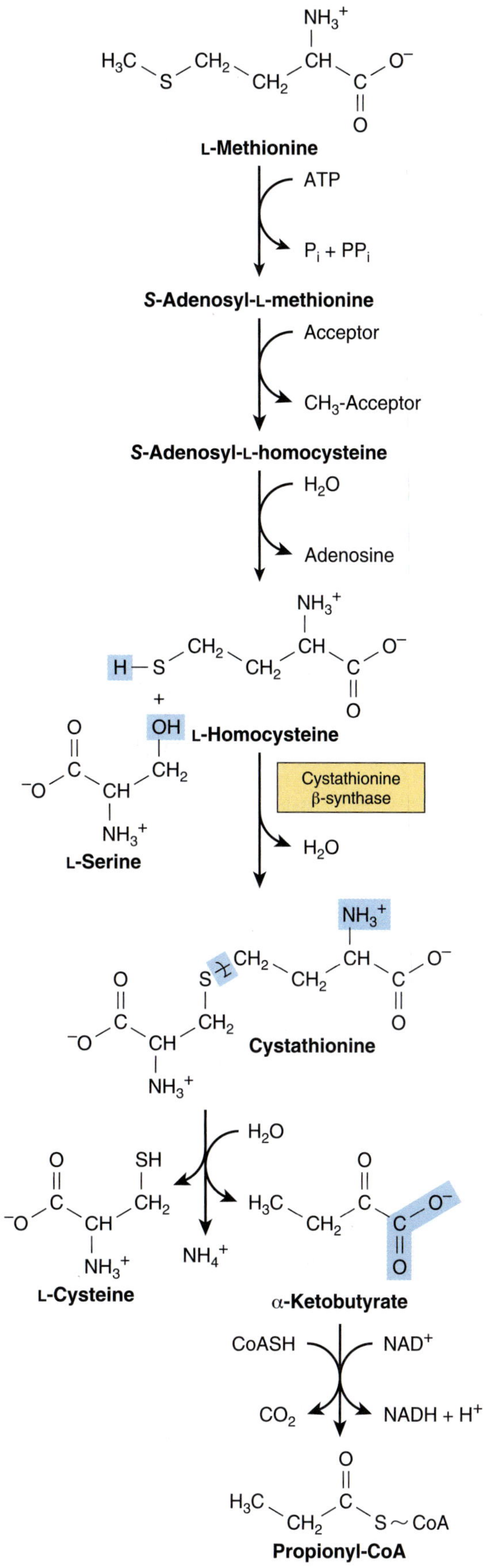

FIGURE 29–18 Conversion of methionine to propionyl-CoA.

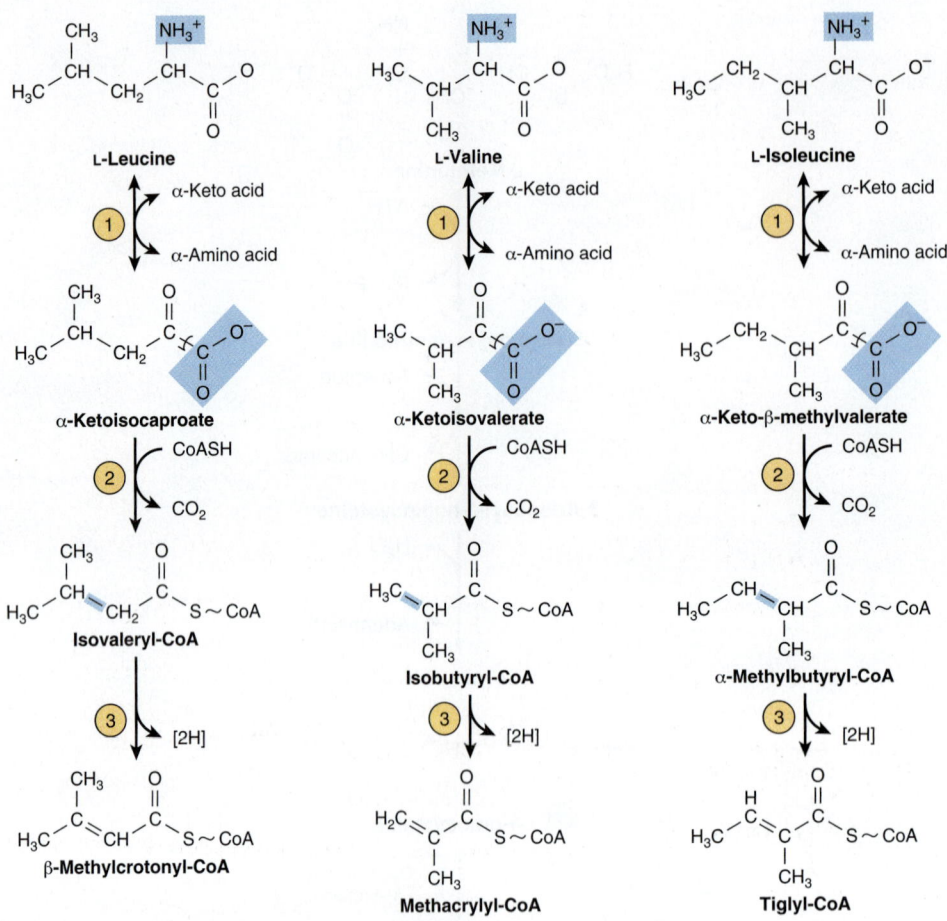

FIGURE 29–19 **The first three reactions in the catabolism of leucine, valine, and isoleucine.** Note the analogy of reactions 2 and 3 to reactions of the catabolism of fatty acids (see Figure 22–3). The analogy to fatty acid catabolism continues, as shown in subsequent figures.

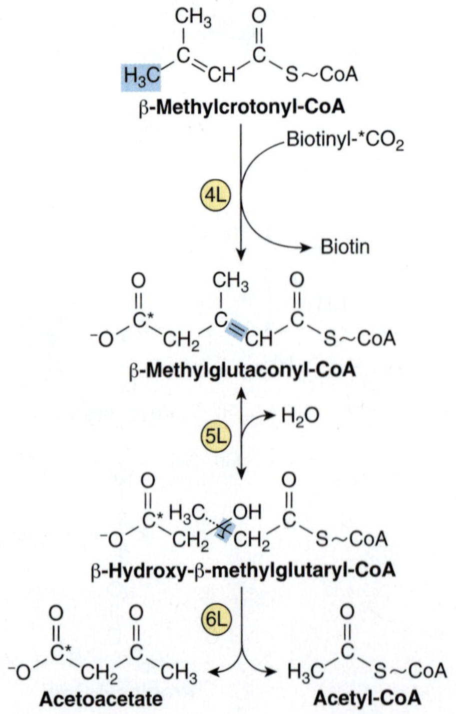

FIGURE 29–20 **Catabolism of the β-methylcrotonyl-CoA formed from L-leucine.** Asterisks indicate carbon atoms derived from CO_2.

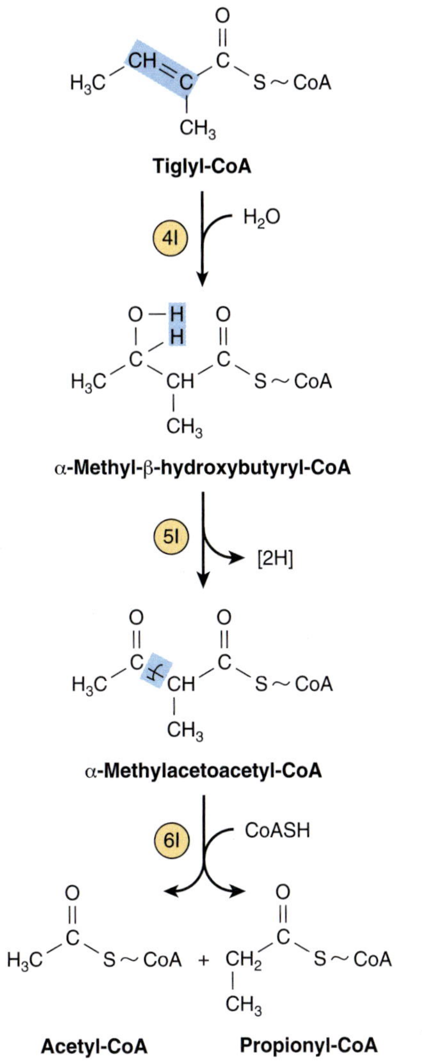

FIGURE 29–21 Subsequent catabolism of the tiglyl-CoA formed from ʟ-isoleucine.

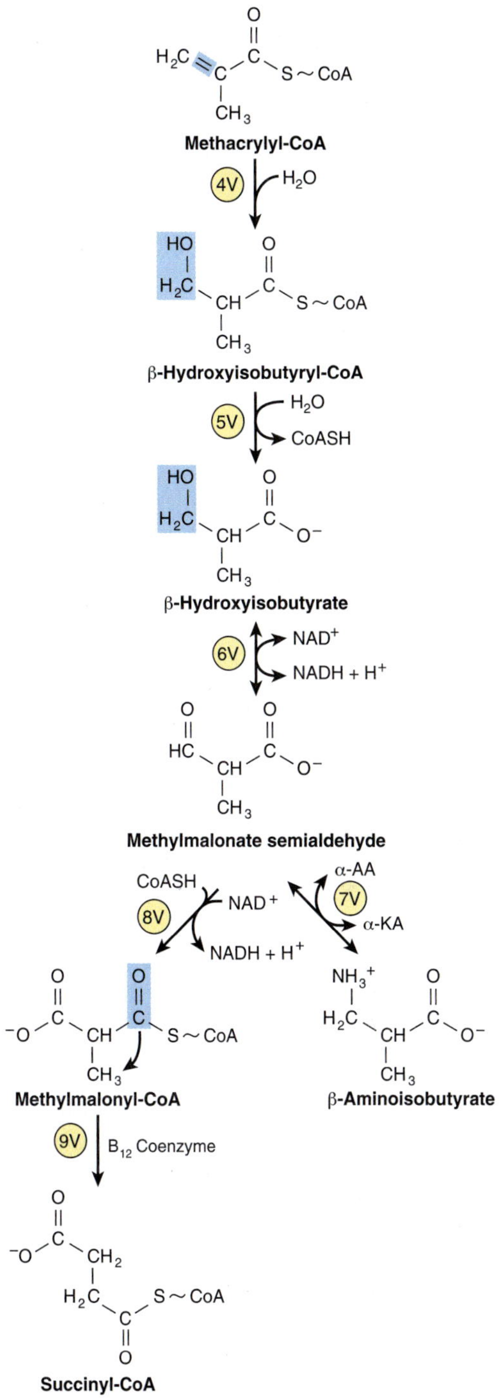

FIGURE 29–22 Subsequent catabolism of the methacrylyl-CoA formed from ʟ-valine (see Figure 29–19). (α-AA, α-amino acid; α-KA, α-keto acid.)

TABLE 29–2 Maple Syrup Urine Disease Can Reflect Impaired Function of Various Components of the α-Ketoacid Decarboxylase Complex

Branched-Chain α-Ketoacid Decarboxylase Component		OMIM[a] Reference	Maple Syrup Urine Disease
E1α	α-Ketoacid decarboxylase	608348	Type 1A
E1β	α-Ketoacid decarboxylase	248611	Type 1B
E2	Dihydrolipoyl transacylase	608770	Type II
E3	Dihydrolipoamide dehydrogenase	238331	Type III

[a]Online Mendelian Inheritance in Man database: ncbi.nlm.nih.gov/omim/.

TABLE 29–3 Metabolic Diseases of Amino Acid Metabolism

Defective Enzyme	Enzyme Catalog Number	OMIM[a] Reference	Major Signs and Symptoms	Figure and Reaction
S-Adenosylhomocysteine hydrolase	3.3.1.1	180960	Hypermethioninemia	29–18③
Arginase	3.5.3.1	207800	Argininemia	29–3①
Cystathionine-β-synthase	4.2.1.22	236200	Homocystinuria	29–18④
Fumarylacetoacetate hydrolase	3.7.1.12	276700	Type I tyrosinemia (tyrosinosis)	29–12④
Histidine ammonia lyase (histidase)	4.3.1.3	609457	Histidinemia & urocanic acidemia	29–4①
Homogentisate oxidase	1.13.11.5	607474	Alkaptonuria. Homogentisate excreted	29–12③
p-Hydroxyphenylpyruvate hydroxylase	1.13.11.27	276710	Neonatal tyrosinemia	29–12③
Isovaleryl-CoA dehydrogenase	1.3.8.4	607036	Isovaleric acidemia	29–19③
Branched chain α-ketoacid decarboxylase complex		248600	Branched-chain ketonuria (MSUD)	29–19①
Methionine adenosyltransferase	2.5.1.6	250850	Hypermethioninemia	29–17①
Ornithine-δ-aminotransferase	2.6.1.13	258870	Ornithemia, gyrate atrophy	29–3②
Phenylalanine hydroxylase	1.14.16.1	261600	Type I (classic) phenylketonuria	27–9①
Proline dehydrogenase	1.5.5.2	606810	Type I hyperprolinemia	29–2①
Δ'-Pyrroline-5-carboxylate dehydrogenase	1.2.1.88	606811	Type II hyperprolinemia & hyper 4-hydroxyprolinemia	29–2②
Saccharopine dehydrogenase	1.5.1.7	268700	Saccharopinuria	29–14②
Tyrosine aminotransferase	2.6.1.5	613018	Type II tyrosinemia	29–12①

[a]Online Mendelian Inheritance in Man database: ncbi.nlm.nih.gov/omim/.

SUMMARY

- Excess amino acids are catabolized to amphibolic intermediates that serve as sources of energy or for the biosynthesis of carbohydrates and lipids.

- Transamination is the most common initial reaction of amino acid catabolism. Subsequent reactions remove any additional nitrogen and restructure hydrocarbon skeletons for conversion to oxaloacetate, α-ketoglutarate, pyruvate, and acetyl-CoA.

- Metabolic diseases associated with glycine catabolism include glycinuria and primary hyperoxaluria.

- Two distinct pathways convert cysteine to pyruvate. Metabolic disorders of cysteine catabolism include cystine-lysinuria, cystine storage disease, and the homocystinurias.

- Threonine catabolism merges with that of glycine after threonine aldolase cleaves threonine to glycine and acetaldehyde.

- Following transamination, the carbon skeleton of tyrosine is degraded to fumarate and acetoacetate. Metabolic diseases of tyrosine catabolism include tyrosinosis, Richner-Hanhart syndrome, neonatal tyrosinemia, and alkaptonuria.

- Metabolic disorders of phenylalanine catabolism include PKU and several hyperphenylalaninemias.

- Neither nitrogen of lysine participates in transamination. The same net effect is, however, achieved by the intermediate formation of saccharopine. Metabolic diseases of lysine catabolism include periodic and persistent forms of hyperlysinemia-ammonemia.

- The catabolism of leucine, valine, and isoleucine presents many analogies to fatty acid catabolism. Metabolic disorders of branched-chain amino acid catabolism include hypervalinemia, maple syrup urine disease, intermittent branched-chain ketonuria, isovaleric acidemia, and methylmalonic aciduria.

REFERENCES

Bliksrud YT, Brodtkorb E, Andresen PA, et al: Tyrosinemia type I, de novo mutation in liver tissue suppressing an inborn splicing defect. J Mol Med 2005;83:406.

Dobrowolski SF, Pey AL, Koch R, et al: Biochemical characterization of mutant phenylalanine hydroxylase enzymes and correlation with clinical presentation in hyperphenylalaninaemic patients. J Inherit Metab Dis 2009;32:10.

Garg U, Dasouki M: Expanded newborn screening of inherited metabolic disorders by tandem mass spectrometry. Clinical and laboratory aspects. Clin Biochem 2006;39:315.

Geng J, Liu A: Heme-dependent dioxygenases in tryptophan oxidation. Arch Biochem Biophys 2014;44:18.

Heldt K, Schwahn B, Marquardt I, et al: Diagnosis of maple syrup urine disease by newborn screening allows early intervention without extraneous detoxification. Mol Genet Metab 2005;84:313.

Houten SM, Te Brinke H, Denis S, et al: Genetic basis of hyperlysinemia. Orphanet J Rare Dis 2013;8:57.

Lamp J, Keyser B, Koeller DM, et al: Glutaric aciduria type 1 metabolites impair the succinate transport from astrocytic to neuronal cells. J Biol Chem 2011;286:17-777.

Mayr JA, Feichtinger RG, Tort F, et al: Lipoic acid biosynthesis defects. J Inherit Metab Dis 2014;37:553.

Nagao M, Tanaka T, Furujo M: Spectrum of mutations associated with methionine adenosyltransferase I/III deficiency among individuals identified during newborn screening in Japan. Mol Genet Metab 2013;110:460.

Stenn FF, Milgram JW, Lee SL, et al: Biochemical identification of homogentisic acid pigment in an ochronotic Egyptian mummy. Science 1977;197:566.

Tondo M, Calpena E, Arriola G, et al: Clinical, biochemical, molecular and therapeutic aspects of 2 new cases of 2-aminoadipic semialdehyde synthase deficiency. Mol Genet Metab 2013;110:231.

Conversion of Amino Acids to Specialized Products

Victor W. Rodwell, PhD

- Cite examples of how amino acids participate in a variety of biosynthetic processes other than protein synthesis.

- Outline how arginine participates in the biosynthesis of creatine, nitric oxide (NO), putrescine, spermine, and spermidine.

- Indicate the contribution of cysteine and of β-alanine to the structure of coenzyme A.

- Discuss the role played by glycine in drug catabolism and excretion.

- Document the role of glycine in the biosynthesis of heme, purines, creatine, and sarcosine.

- Identify the reaction that converts an amino acid to the neurotransmitter histamine.

- Document the role of *S*-adenosylmethionine in metabolism.

- Recognize the structures of tryptophan metabolites serotonin, melatonin, tryptamine, and indole 3-acetate.

- Describe how tyrosine gives rise to norepinephrine and epinephrine.

- Illustrate the key roles of peptidyl serine, threonine, and tyrosine in metabolic regulation and signal transduction pathways.

- Diagram the roles of glycine, arginine, and *S*-adenosylmethionine in the biosynthesis of creatine.

- Explain the role of creatine phosphate in energy homeostasis.

- Illustrate the formation of γ-aminobutyrate (GABA) and the rare metabolic disorders associated with defects in GABA catabolism.

BIOMEDICAL IMPORTANCE

Certain proteins contain amino acids that have been post-translationally modified to permit them to perform specific functions. Examples include the carboxylation of glutamate to form γ-carboxyglutamate, which functions in Ca^{2+} binding, the hydroxylation of proline to form 3- and 4-hydroxyproline in collagen, and the hydroxylation of lysine to 5-hydroxylysine, whose subsequent modification and cross-linking stabilize maturing collagen fibers. In addition to serving as the building blocks for protein synthesis, amino acids serve as precursors of biologic materials as diverse and important as heme, purines, pyrimidines, hormones, neurotransmitters, and biologically active peptides. Histamine plays a central role in many allergic reactions. Neurotransmitters derived from amino acids include γ-aminobutyrate (GABA), 5-hydroxytryptamine (serotonin), dopamine, norepinephrine, and epinephrine. Many drugs used to treat neurologic and psychiatric conditions act by altering the metabolism of these neurotransmitters. Discussed below are the metabolism and metabolic roles of selected α- and non–α-amino acids.

L-α-AMINO ACIDS

Alanine

Alanine serves as a carrier of ammonia and of the carbons of pyruvate from skeletal muscle to liver via the Cori cycle

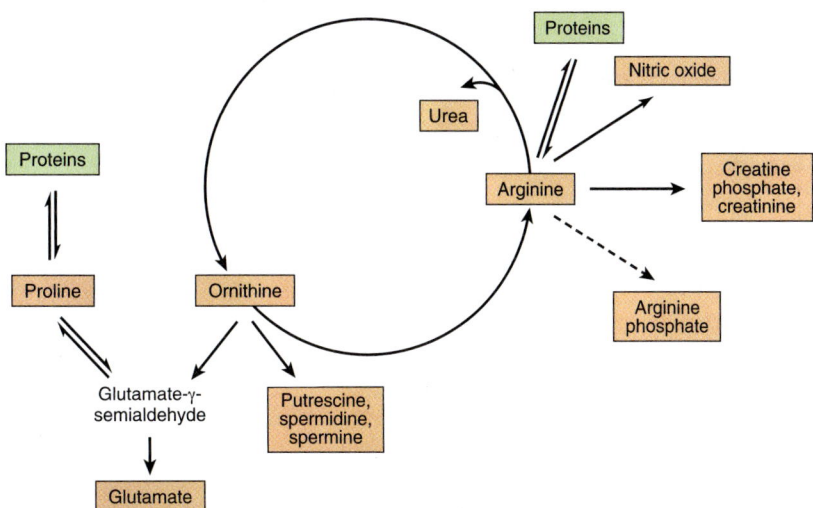

FIGURE 30–1 **Arginine, ornithine, and proline metabolism.** Reactions with solid arrows all occur in mammalian tissues. Putrescine and spermine synthesis occurs in both mammals and bacteria. Arginine phosphate of invertebrate muscle functions as a phosphagen analogous to creatine phosphate of mammalian muscle.

(see Chapters 19 & 28), and together with glycine constitutes a major fraction of the free amino acids in plasma.

Arginine

Figure 30–1 summarizes the metabolic fates of arginine. In addition to serving as a carrier of nitrogen atoms in urea biosynthesis (see Figure 28–16), the guanidino group of arginine is incorporated into creatine, and following conversion to ornithine, its carbon skeleton serves as a precursor of the polyamines putrescine and spermine (see below).

The reaction catalyzed by nitric oxide synthase, EC 1.14.13.39 (**Figure 30–2**), a five-electron oxidoreductase with multiple cofactors, converts one nitrogen of the guanidine group of arginine to nitric oxide, an intercellular signaling molecule that serves as a neurotransmitter, smooth muscle relaxant, and vasodilator (see Chapter 51).

Cysteine

Cysteine participates in the biosynthesis of coenzyme A (see Chapter 44) by reacting with pantothenate to form 4-phosphopantothenoylcysteine. In addition, taurine, formed from cysteine, can displace the coenzyme A moiety of cholyl-CoA to form the bile acid taurocholic acid (see Chapter 26). The conversion of cysteine to taurine involves catalysis by the nonheme Fe^{2+} enzyme cysteine dioxygenase (EC 1.13.11.20), sulfinoalanine decarboxylase (EC 4.1.1.29), and hypotaurine dehydrogenase (EC 1.8.1.3) (**Figure 30–3**).

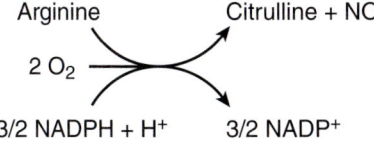

FIGURE 30–2 **The reaction catalyzed by nitric oxide synthase.**

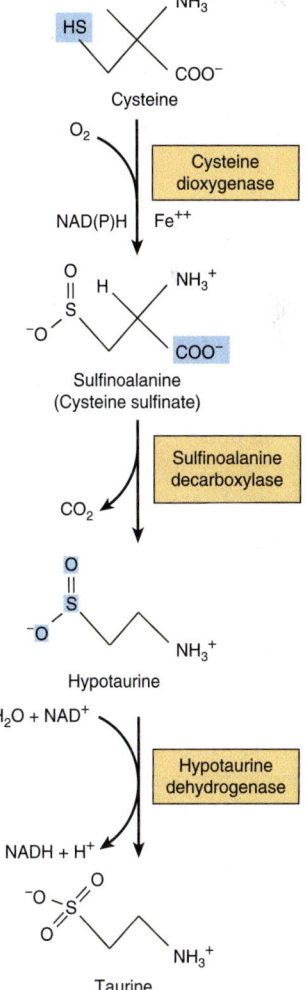

FIGURE 30–3 **Conversion of cysteine to taurine.** The reactions are catalyzed by cysteine dioxygenase, cysteine sulfinate decarboxylase, and hypotaurine decarboxylase, respectively.

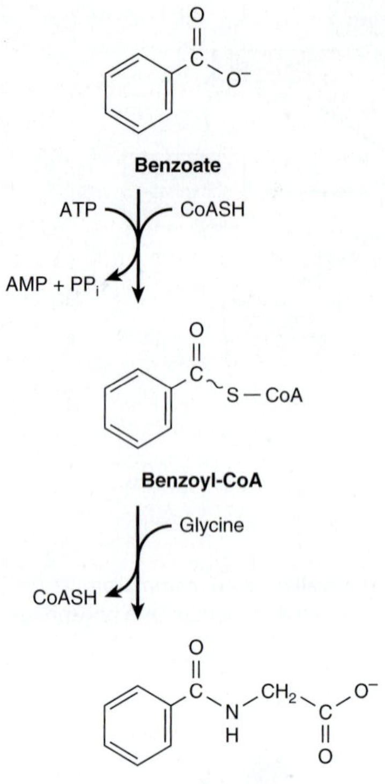

FIGURE 30–4 **Biosynthesis of hippurate.** Analogous reactions occur with many acidic drugs and catabolites.

Glycine

Many relatively apolar metabolites are converted to water-soluble glycine conjugates. An example is the hippuric acid formed from the food additive benzoate (**Figure 30–4**). Many drugs, drug metabolites, and other compounds with carboxyl groups also are conjugated with glycine. This makes them more water soluble and thereby facilitates their excretion in the urine.

Glycine is a component of creatine, and its nitrogen and α-carbon are incorporated into the pyrrole rings and the methylene bridge carbons of heme (see Chapter 31). The entire glycine molecule supplies atoms 4, 5, and 7 of the purine bases (see Figure 33–1).

Histidine

Decarboxylation of histidine to histamine is catalyzed by the pyridoxal 5′-phosphate-dependent enzyme histidine decarboxylase, EC 4.1.1.22 (**Figure 30–5**). A biogenic amine that functions in allergic reactions and gastric secretion, histamine is present in all tissues. Its concentration in the brain hypothalamus varies in accordance with a circadian rhythm.

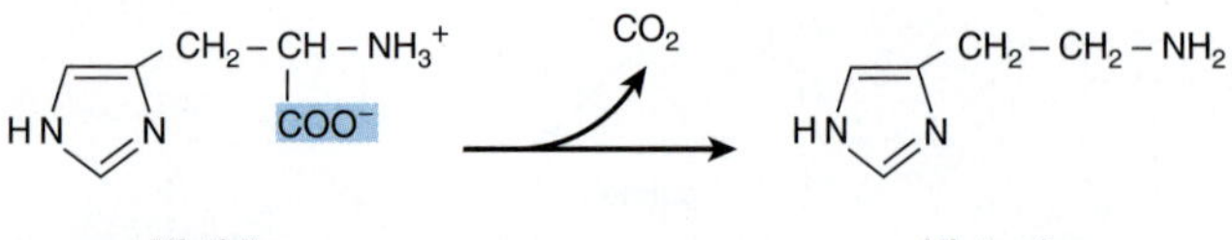

FIGURE 30–5 **The reaction catalyzed by histidine decarboxylase.**

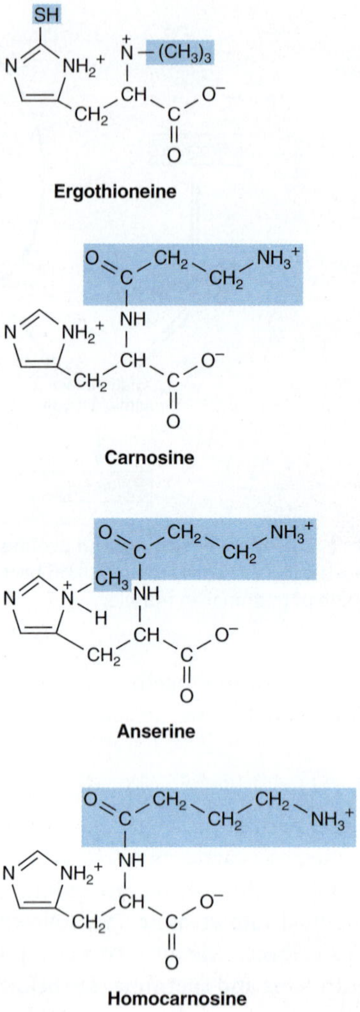

FIGURE 30–6 **Derivatives of histidine.** Colored boxes surround the components not derived from histidine. The SH group of ergothioneine derives from cysteine.

Histidine-containing compounds present in the human body include carnosine, and dietarily derived ergothioneine and anserine (**Figure 30–6**). Carnosine (β-alanyl-histidine) and homocarnosine (γ-aminobutyryl-histidine) are major constituents of excitable tissues, brain, and skeletal muscle. Urinary levels of 3-methylhistidine are unusually low in patients with **Wilson disease**.

Methionine

The major nonprotein fate of methionine is conversion to S-adenosylmethionine, the principal source of methyl groups in the body. Biosynthesis of S-adenosylmethionine from methionine and ATP is catalyzed by methionine adenosyltransferase (MAT), EC 2.5.1.6 (**Figure 30–7**). Human tissues contain three MAT isozymes: MAT-1 and MAT-3 of liver and MAT-2 of nonhepatic tissues. Although **hypermethioninemia** can result from severely decreased hepatic MAT-1 and MAT-3 activity, if there is residual MAT-1 or MAT-3 activity and MAT-2 activity is normal, a high tissue

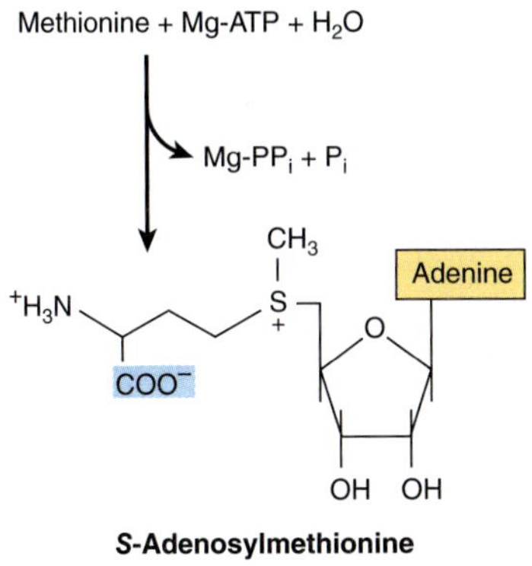

FIGURE 30–7 Biosynthesis of *S*-adenosylmethionine, catalyzed by methionine adenosyltransferase.

concentration of methionine will ensure synthesis of adequate amounts of *S*-adenosylmethionine.

Following decarboxylation of *S*-adenosylmethionine by methionine decarboxylase (EC 4.1.1.57), three carbons and the α-amino group of methionine can be utilized for the biosynthesis of the polyamines **spermine** and **spermidine**. These polyamines function in cell proliferation and growth, are growth factors for cultured mammalian cells, and stabilize intact cells, subcellular organelles, and membranes. Pharmacologic doses of polyamines are hypothermic and hypotensive. Since they bear multiple positive charges, polyamines readily associate with DNA and RNA. **Figure 30–8** summarizes the biosynthesis of polyamines from methionine and ornithine, and **Figure 30–9** the catabolism of polyamines.

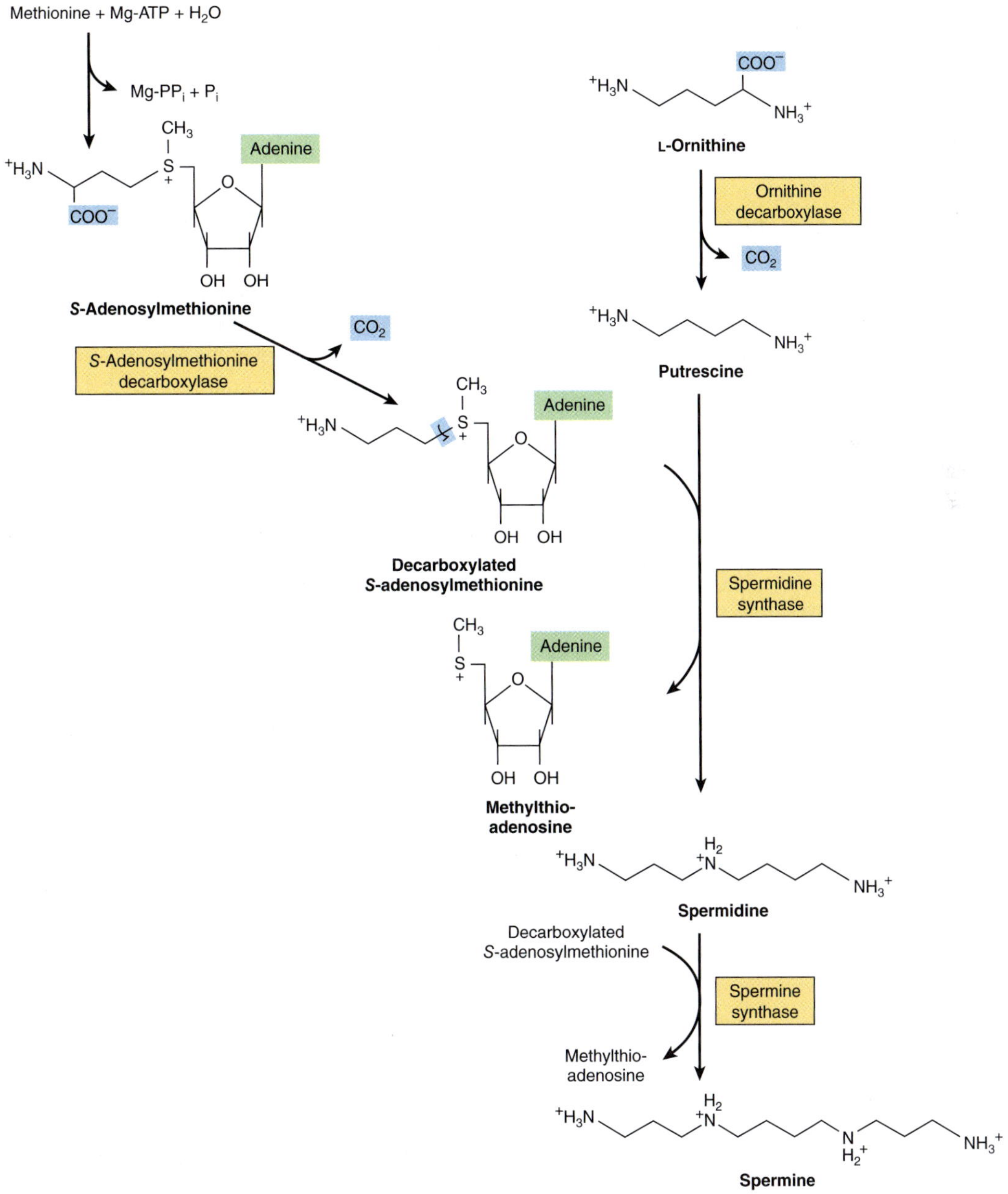

FIGURE 30–8 Intermediates and enzymes that participate in the biosynthesis of spermidine and spermine.

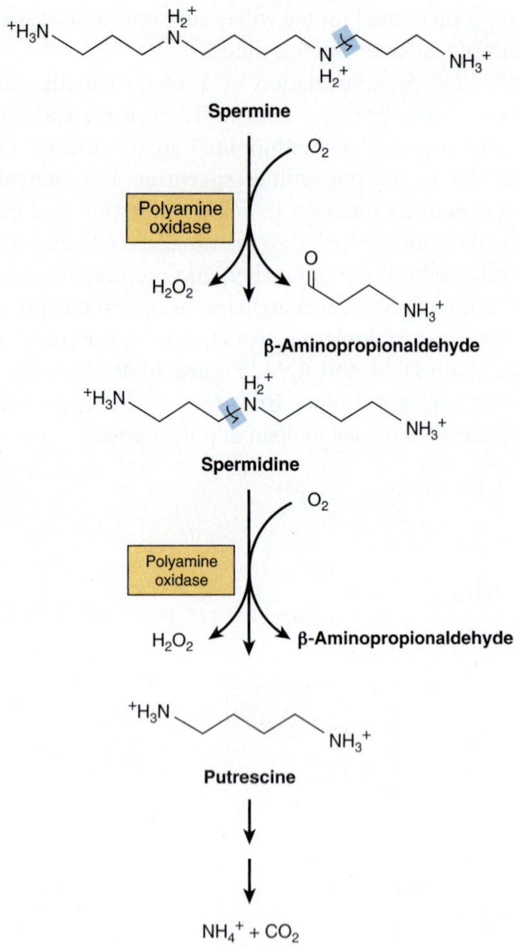

FIGURE 30–9   Catabolism of polyamines.

Serine

Serine participates in the biosynthesis of sphingosine (see Chapter 24), and of purines and pyrimidines, where it provides carbons 2 and 8 of purines and the methyl group of thymine (see Chapter 33). Genetic defects in cystathionine β-synthase (EC 4.2.1.22)

$$\text{Serine} + \text{Homocysteine} \rightarrow \text{Cystathionine} + H_2O$$

a heme protein that catalyzes the pyridoxal 5′-phosphate–dependent condensation of serine with homocysteine to form cystathionine, result in **homocystinuria**. Finally, serine (not cysteine) serves as the precursor of peptidyl selenocysteine (see Chapter 27).

Tryptophan

Following hydroxylation of tryptophan to 5-hydroxytryptophan by liver tryptophan hydroxylase (EC 1.14.16.4), subsequent decarboxylation forms serotonin (5-hydroxytryptamine), a potent vasoconstrictor and stimulator of smooth muscle contraction. Catabolism of serotonin is initiated by deamination to 5-hydroxyindole-3-acetate, a reaction catalyzed by monoamine

oxidase, EC 1.4.3.4 (**Figure 30–10**). The psychic stimulation that follows administration of iproniazid results from its ability to prolong the action of serotonin by inhibiting monoamine oxidase. In carcinoid (argentaffinoma), tumor cells overproduce serotonin. Urinary metabolites of serotonin in patients with carcinoid include N-acetylserotonin glucuronide and the glycine conjugate of 5-hydroxyindoleacetate. Serotonin and 5-methoxytryptamine are metabolized to the corresponding acids by monoamine oxidase. N-Acetylation of serotonin, followed by its O-methylation in the pineal body, forms melatonin. Circulating melatonin is taken up by all tissues, including brain, but is rapidly metabolized by hydroxylation followed by conjugation with sulfate or with glucuronic acid. Kidney tissue, liver tissue, and fecal bacteria all convert tryptophan to tryptamine, then to indole 3-acetate. The principal normal urinary catabolites of tryptophan are 5-hydroxyindoleacetate and indole 3-acetate (Figure 30–10).

Tyrosine

Neural cells convert tyrosine to epinephrine and norepinephrine (**Figure 30–11**). While dopa is also an intermediate in the formation of melanin, different enzymes hydroxylate tyrosine in melanocytes. DOPA decarboxylase (EC 4.1.1.28), a pyridoxal phosphate-dependent enzyme, forms dopamine. Subsequent hydroxylation, catalyzed by dopamine β-oxidase (EC 1.14.17.1), then forms norepinephrine. In the adrenal medulla, phenylethanolamine N-methyltransferase (EC 2.1.1.28) utilizes S-adenosylmethionine to methylate the primary amine of norepinephrine, forming epinephrine (Figure 30–11). Tyrosine is also a precursor of triiodothyronine and thyroxine (see Chapter 41).

Phosphoserine, Phosphothreonine, & Phosphotyrosine

The phosphorylation and dephosphorylation of specific seryl, threonyl, or tyrosyl residues of proteins regulate the activity of certain enzymes of lipid and carbohydrate metabolism and of proteins that participate in signal transduction cascades (see Chapter 42).

Sarcosine (*N*-Methylglycine)

The biosynthesis and catabolism of sarcosine (N-methylglycine) occur in mitochondria. Formation of sarcosine from dimethyl glycine is catalyzed by the flavoprotein dimethyl glycine dehydrogenase EC 1.5.8.4, which requires reduced pteroylpentaglutamate (TPG).

$$\text{Dimethylglycine} + FADH_2 + H_4TPG + H_2O \rightarrow \text{Sarcosine} + N\text{-formyl-TPG}$$

Traces of sarcosine can also arise by methylation of glycine, a reaction catalyzed by glycine N-methyltransferase, EC 2.1.1.20.

$$\text{Glycine} + S\text{-Adenosylmethionine} \rightarrow \text{Sarcosine} + S\text{-Adenosylhomocysteine}$$

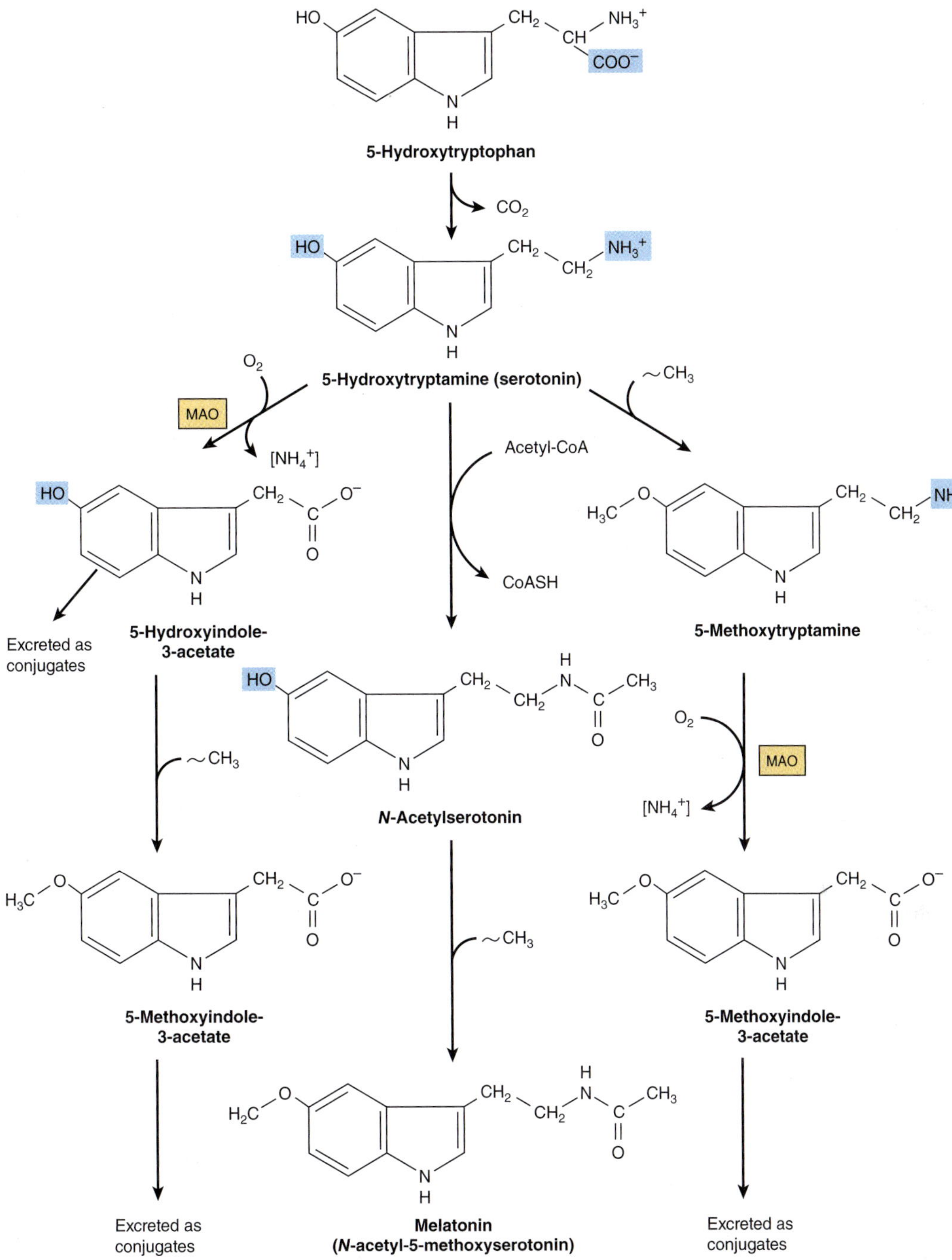

FIGURE 30–10 **Biosynthesis and metabolism of serotonin and melatonin.** ([NH$_4^+$], by transamination; MAO, monoamine oxidase; ~CH$_3$, from S-adenosylmethionine.)

Catabolism of sarcosine to glycine, catalyzed by the flavoprotein sarcosine dehydrogenase EC 1.5.8.3, also requires reduced TPG.

$$\text{Sarcosine} + \text{FAD} + \text{H}_4\text{TPG} + \text{H}_2\text{O} \rightarrow \text{Glycine} + \text{FADH}_2 \\ + \text{N-formyl-TPG}$$

The demethylation reactions that form and degrade sarcosine represent important sources of one-carbon units. FADH$_2$ is reoxidized via the electron transport chain (see Chapter 13).

Creatine & Creatinine

Creatinine is formed in muscle from creatine phosphate by irreversible, nonenzymatic dehydration, and loss of phosphate (**Figure 30–12**). Since the 24-hour urinary excretion of creatinine is proportionate to muscle mass, it provides a measure of whether a complete 24-hour urine specimen has been collected. Glycine, arginine, and methionine all participate in creatine biosynthesis. Synthesis of creatine is completed by methylation of guanidoacetate by S-adenosylmethionine (Figure 30–12).

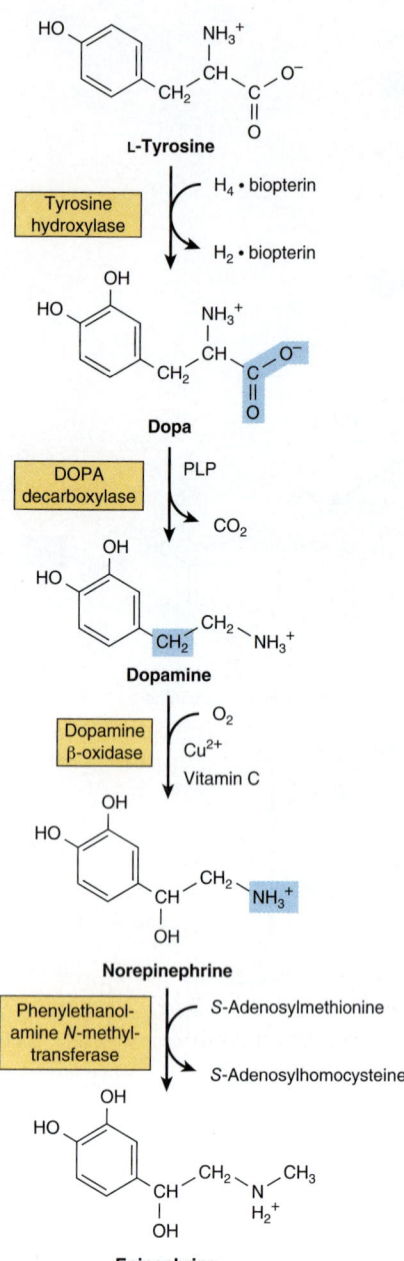

FIGURE 30–11 **Conversion of tyrosine to epinephrine and norepinephrine in neuronal and adrenal cells.** (PLP, pyridoxal phosphate.)

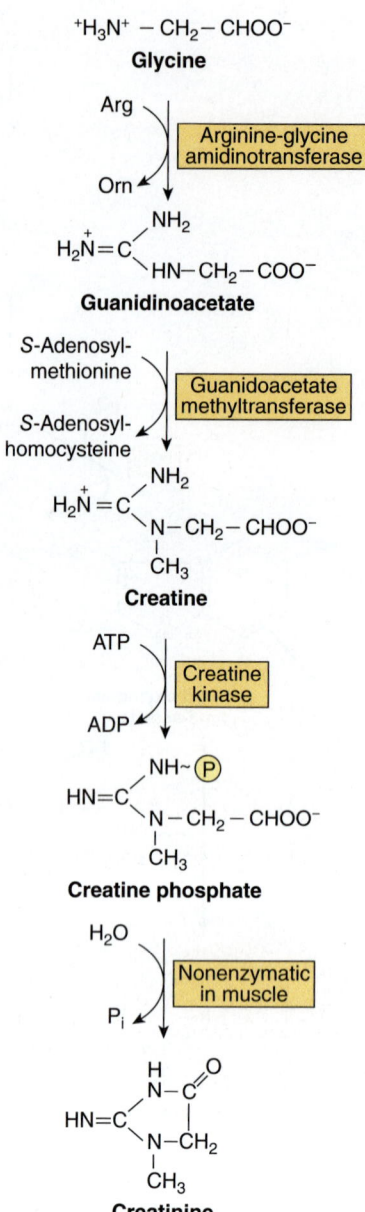

FIGURE 30–12 **Biosynthesis of creatine and creatinine.** Conversion of glycine and the guanidine group of arginine to creatine and creatine phosphate. Also shown is the nonenzymic hydrolysis of creatine phosphate to creatinine.

NON–α-AMINO ACIDS

Non–α-amino acids present in tissues in a free form include β-alanine, β-aminoisobutyrate, and GABA. β-Alanine is also present in combined form in coenzyme A, and in the β-alanyl dipeptides carnosine, anserine, and homocarnosine (see below).

β-Alanine & β-Aminoisobutyrate

β-Alanine and β-aminoisobutyrate are formed during catabolism of the pyrimidines uracil and thymine, respectively (see Figure 33–9). Traces of β-alanine also result from the

hydrolysis of β-alanyl dipeptides by the enzyme carnosinase, EC 3.4.13.20. β-Aminoisobutyrate also arises by transamination of methylmalonate semialdehyde, a catabolite of L-valine (see Figure 29–22).

The initial reaction of β-alanine catabolism is transamination to malonate semialdehyde. Subsequent transfer of coenzyme A from succinyl-CoA forms malonyl-CoA semialdehyde, which is then oxidized to malonyl-CoA and decarboxylated to the amphibolic intermediate acetyl-CoA. Analogous reactions characterize the catabolism of β-aminoisobutyrate. Transamination forms methylmalonate semialdehyde, which is converted to the amphibolic intermediate succinyl-CoA by

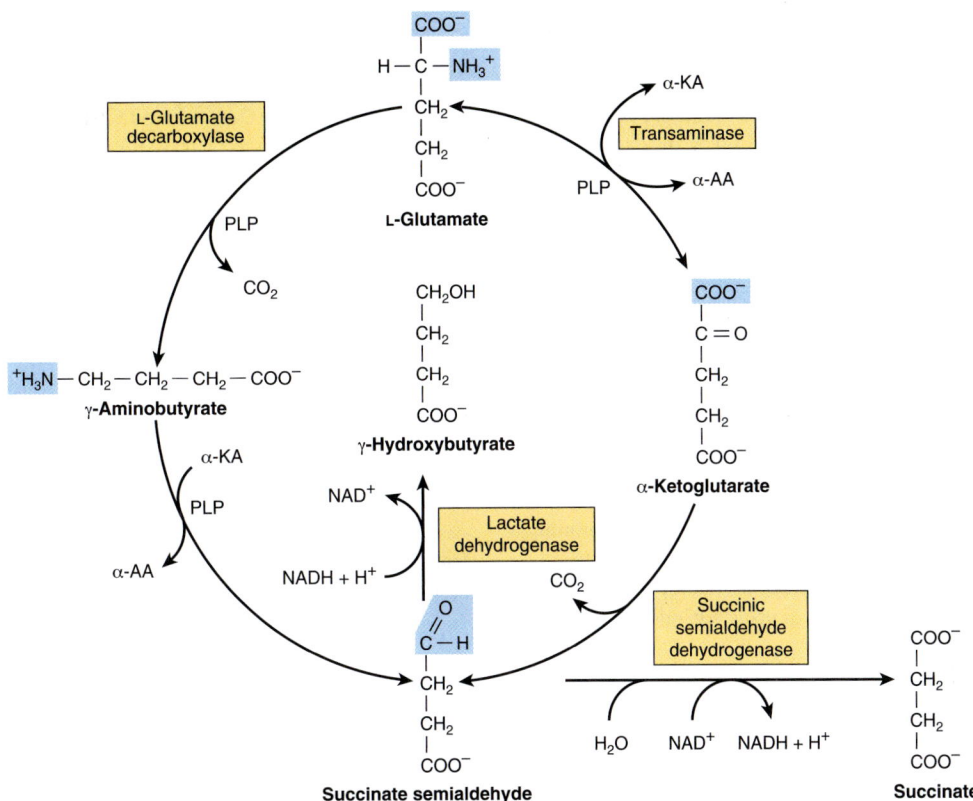

FIGURE 30–13 **Metabolism of γ-aminobutyrate.** (α-AA, α-amino acids; α-KA, α-keto acids; PLP, pyridoxal phosphate.)

reactions 8V and 9V of Figure 29–22. Disorders of β-alanine and β-aminoisobutyrate metabolism arise from defects in enzymes of the pyrimidine catabolic pathway. Principal among these are disorders that result from a total or partial deficiency of dihydropyrimidine dehydrogenase (see Chapter 33).

β-Alanyl Dipeptides

The β-alanyl dipeptides carnosine and anserine (N-methylcarnosine) (Figure 30–6) activate myosin ATPase (EC 3.6.4.1), chelate copper, and enhance copper uptake. β-Alanyl-imidazole buffers the pH of anaerobically contracting skeletal muscle. Biosynthesis of carnosine is catalyzed by carnosine synthetase (EC 6.3.2.11) in a two-stage reaction that involves initial formation of an enzyme-bound acyl-adenylate of β-alanine and subsequent transfer of the β-alanyl moiety to L-histidine.

$$ATP + \beta\text{-Alanine} \rightarrow \beta\text{-Alanyl-AMP} + PP_i$$
$$\beta\text{-Alanyl-AMP} + L\text{-Histidine} \rightarrow Carnosine + AMP$$

Hydrolysis of carnosine to β-alanine and L-histidine is catalyzed by carnosinase. The heritable disorder carnosinase deficiency is characterized by **carnosinuria**.

Homocarnosine (Figure 30–6, present in human brain at higher levels than carnosine, is synthesized in brain tissue by carnosine synthetase. Serum carnosinase does not hydrolyze homocarnosine. **Homocarnosinosis**, a rare genetic disorder, is associated with progressive spastic paraplegia and mental retardation.

γ-Aminobutyrate

GABA functions in brain tissue as an inhibitory neurotransmitter by altering transmembrane potential differences. GABA is formed by decarboxylation of glutamate by L-glutamate decarboxylase, EC 4.1.1.15 (**Figure 30–13**). Transamination of GABA forms succinate semialdehyde, which can be reduced to γ-hydroxybutyrate by L-lactate dehydrogenase, or be oxidized to succinate and thence via the citric acid cycle to CO_2 and H_2O (Figure 30–13). A rare genetic disorder of GABA metabolism involves a defective GABA aminotransferase EC 2.6.1.19, an enzyme that participates in the catabolism of GABA subsequent to its postsynaptic release in brain tissue. Defects in succinic semialdehyde dehydrogenase, EC 1.2.1.24 (Figure 30–13) are responsible for **4-hydroxybutyric aciduria**, a rare metabolic disorder of GABA catabolism characterized by the presence of 4-hydroxybutyrate in urine, plasma, and cerebrospinal fluid (CSF). No present treatment is available for the accompanying mild-to-severe neurologic symptoms.

SUMMARY

■ In addition to serving structural and functional roles in proteins, α-amino acids participate in a wide variety of other biosynthetic processes.

■ Arginine provides the formamidine group of creatine and the nitrogen of NO. Via ornithine, arginine provides the skeleton of the polyamines putrescine, spermine, and spermidine.

- Cysteine provides the thioethanolamine portion of coenzyme A, and following its conversion to taurine, is part of the bile acid taurocholic acid.

- Glycine participates in the biosynthesis of heme, purines, creatine, and N-methylglycine (sarcosine). Many drugs and drug metabolites are excreted as glycine conjugates. This enhances their water solubility for urinary excretion.

- Decarboxylation of histidine forms the neurotransmitter histamine. Histidine compounds present in the human body include ergothioneine, carnosine, and anserine.

- S-Adenosylmethionine, the principal source of methyl groups in metabolism, contributes its carbon skeleton to the biosynthesis of the polyamines spermine and spermidine.

- In addition to its roles in phospholipid and sphingosine biosynthesis, serine provides carbons 2 and 8 of purines and the methyl group of thymine.

- Key tryptophan metabolites include serotonin and melatonin. Kidney and liver tissue, and also fecal bacteria, convert tryptophan to tryptamine and thence to indole 3-acetate. The principal tryptophan catabolites in urine are indole 3-acetate and 5-hydroxyindoleacetate.

- Tyrosine forms norepinephrine and epinephrine, and following iodination the thyroid hormones triiodothyronine and thyroxine.

- The enzyme-catalyzed interconversion of the phospho- and dephospho- forms of peptide-bound serine, threonine, and tyrosine plays key roles in metabolic regulation, including signal transduction.

- Glycine, arginine, and S-adenosylmethionine all participate in the biosynthesis of creatine, which as creatine phosphate serves as a major energy reserve in muscle and brain tissue. Excretion in the urine of its catabolite creatinine is proportionate to muscle mass.

- β-Alanine and β-aminoisobutyrate both are present in tissues as free amino acids. β-Alanine also occurs in bound form in coenzyme A. Catabolism of β-alanine involves stepwise conversion to acetyl-CoA. Analogous reactions catabolize β-aminoisobutyrate to succinyl-CoA. Disorders of β-alanine and β-aminoisobutyrate metabolism arise from defects in enzymes of pyrimidine catabolism.

- Decarboxylation of glutamate forms the inhibitory neurotransmitter GABA. Two rare metabolic disorders are associated with defects in GABA catabolism.

REFERENCES

Allen GF, Land JM, Heales SJ: A new perspective on the treatment of aromatic L-amino acid decarboxylase deficiency. Mol Genet Metab 2009;97:6.

Caine C, Shohat M, Kim JK, et al: A pathogenic S250F missense mutation results in a mouse model of mild aromatic L-amino acid decarboxylase (AADC) deficiency. Hum Mol Genet 2017;26:4406.

Cravedi E, Deniau E, Giannitelli M, et al: Tourette syndrome and other neurodevelopmental disorders: a comprehensive review. Child Adolesc Psychiatry Ment Health 2017;11:59.

Jansen EE, Vogel KR, Salomons GS, et al: Correlation of blood biomarkers with age informs pathomechanisms in succinic semialdehyde dehydrogenase deficiency (SSADHD), a disorder of GABA metabolism. J Inherit Metab Dis 2016;39:795.

Manegold C, Hoffmann GF, Degen I, et al: Aromatic L-amino acid decarboxylase deficiency: clinical features, drug therapy and followup. J Inherit Metab Dis 2009;32:371.

Moinard C, Cynober L, de Bandt JP: Polyamines: metabolism and implications in human diseases. Clin Nutr 2005;24:184.

Montioli R, Dindo M, Giorgetti A, et al: A comprehensive picture of the mutations associated with aromatic amino acid decarboxylase deficiency: from molecular mechanisms to therapy implications. Hum Mol Genet 2014;23:5429.

Pearl PL, Gibson KM, Cortez MA, et al: Succinic semialdehyde dehydrogenase deficiency: lessons from mice and men. J Inherit Metab Dis 2009;32:343.

Schippers KJ, Nichols SA: Deep, dark secrets of melatonin in animal evolution. Cell 2014;159:9.

Wernli C, Finochiaro S, Volken C, et al: Targeted screening of succinic semialdehyde dehydrogenase deficiency (SSADHD) employing an enzymatic assay for γ-hydroxybutyric acid (GHB) in biofluids. Mol Genet Metab Rep. 2016;17:81.

Porphyrins & Bile Pigments

Victor W. Rodwell, PhD

- Write the structural formulas of the two amphibolic intermediates whose condensation initiates heme biosynthesis.
- Identify the enzyme that catalyzes the key regulated step of hepatic heme biosynthesis.
- Explain why, although porphyrinogens and porphyrins both are tetrapyrroles, porphyrins are colored whereas porphyrinogens are colorless.
- Specify the intracellular locations of the enzymes and metabolites of heme biosynthesis.
- Outline the causes and clinical presentations of various porphyrias.
- Identify the roles of heme oxygenase and of UDP-glucosyl transferase in heme catabolism.
- Define jaundice, name some of its causes, and suggest how to determine its biochemical basis.
- Specify the biochemical basis of the clinical laboratory terms "direct bilirubin" and "indirect bilirubin."

BIOMEDICAL IMPORTANCE

The biochemistry of the porphyrins and of the bile pigments are closely related topics. Heme is synthesized from porphyrins and iron, and the products of degradation of heme include the bile pigments and iron. The biochemistry of the porphyrins and of heme is basic to understanding the varied functions of **hemoproteins**, and the **porphyrias**, a group of diseases caused by abnormalities in the pathway of porphyrin biosynthesis. A much more common clinical condition is **jaundice**, a consequence of an elevated level of plasma bilirubin, due either to overproduction of bilirubin or to failure of its excretion. Jaundice occurs in numerous diseases ranging from hemolytic anemias to viral hepatitis and to cancer of the pancreas.

PORPHYRINS

Porphyrins are cyclic compounds formed by the linkage of four **pyrrole rings** through methyne ($=HC-$) bridges (**Figure 31-1**). Various **side chains** can replace the eight numbered hydrogen atoms of the pyrrole rings.

Porphyrins can form complexes with metal ions that form coordinate bonds to the nitrogen atom of each of the four pyrrole rings. Examples include **iron porphyrins** such as the **heme** of hemoglobin and the **magnesium-containing** porphyrin **chlorophyll**, the photosynthetic pigment of plants. Heme proteins are ubiquitous in biology and serve diverse functions including, but not limited to, oxygen transport and storage (eg, hemoglobin and myoglobin) and electron transport (eg, cytochrome c and cytochrome P450). Hemes are **tetrapyrroles**, of which two types, heme b and heme c, predominate (**Figure 31-2**). In heme c the vinyl groups of heme b are replaced by covalent thioether links to an apoprotein, typically via cysteinyl residues. Unlike heme b, heme c thus does not readily dissociate from its apoprotein.

Proteins that contain heme are widely distributed in nature (**Table 31-1**). Vertebrate heme proteins generally bind one mole of heme c per mole, although those of nonvertebrates may bind significantly more heme.

HEME IS SYNTHESIZED FROM SUCCINYL-CoA & GLYCINE

The biosynthesis of heme involves both cytosolic and mitochondrial reactions and intermediates. Heme biosynthesis occurs in most mammalian cells except mature erythrocytes,

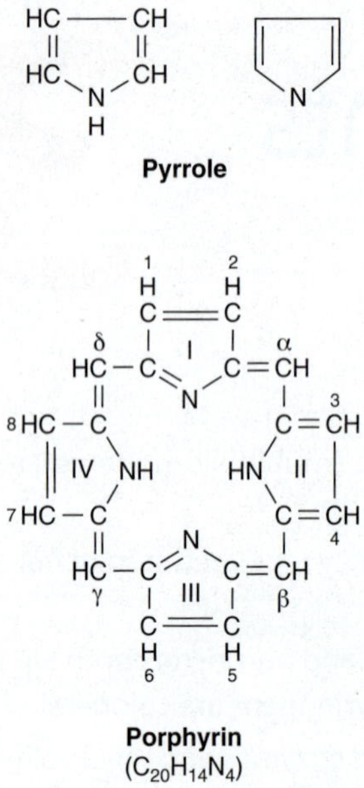

FIGURE 31–1 **The porphyrin molecule.** Rings are labeled I, II, III, and IV. Substituent positions are labeled 1 through 8. The four methyne bridges (=HC—) are labeled α, β, γ, and δ.

which lack mitochondria. Approximately 85% of heme synthesis occurs in erythroid precursor cells in the **bone marrow**, and the majority of the remainder in **hepatocytes**. Heme biosynthesis is initiated by the condensation of succinyl-CoA and glycine in a pyridoxal phosphate-dependent reaction catalyzed by mitochondrial **δ-aminolevulinate synthase** (**ALA synthase**, EC 2.3.1.37).

$$(1) \text{ Succinyl-CoA + glycine} \rightarrow \text{δ-aminolevulinate} + \text{CoA-SH} + CO_2$$

Humans express two isozymes of ALA synthase. ALAS1 is ubiquitously expressed throughout the body, whereas ALAS2 is expressed in erythrocyte precursor cells. Formation of δ-aminolevulinate is rate-limiting for porphyrin biosynthesis in mammalian liver (**Figure 31–3**).

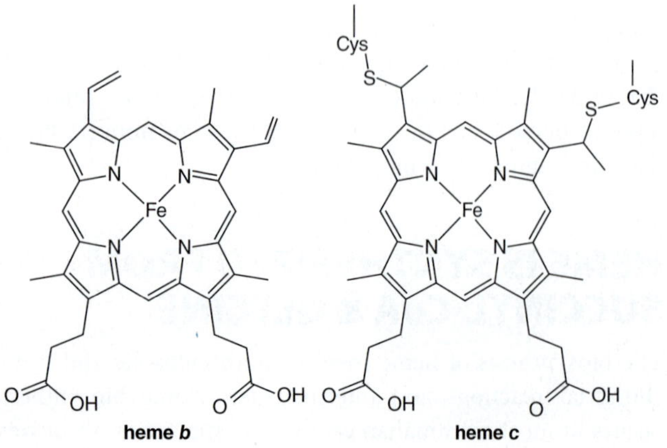

TABLE 31–1 **Examples of Important Heme Proteins[a]**

Protein	Function
Hemoglobin	Transport of oxygen in blood
Myoglobin	Storage of oxygen in muscle
Cytochrome c	Involvement in the electron transport chain
Cytochrome P450	Hydroxylation of xenobiotics
Catalase	Degradation of hydrogen peroxide
Tryptophan pyrrolase	Oxidation of tryptophan

[a]The functions of the above proteins are described in various chapters of this text.

Following the exit of δ-aminolevulinate into the cytosol, the reaction catalyzed by cytosolic **ALA dehydratase** (EC 4.2.1.24; porphobilinogen synthase) condenses two molecules of ALA, forming **porphobilinogen**:

$$(2) \text{ 2 δ-Aminolevulinate} \rightarrow \text{porphobilinogen} + 2 H_2O$$

(**Figure 31–4**). A zinc metalloprotein, ALA dehydratase is sensitive to inhibition by **lead**, as can occur in lead poisoning.

The third reaction, catalyzed by *cytosolic* **hydroxymethylbilane synthase** (uroporphyrinogen I synthase, EC 2.5.1.61) involves head-to-tail condensation of four molecules of porphobilinogen to form the *linear* tetrapyrrole **hydroxymethylbilane** (**Figure 31–5,** *top*):

$$(3) \text{ 4 Porphobilinogen} + H_2O \rightarrow \text{hydroxymethylbilane} + 4 NH_3$$

Subsequent cyclization of hydroxymethylbilane, catalyzed by cytosolic **uroporphyrinogen III synthase**, EC 4.2.1.75:

$$(4) \text{ Hydroxymethylbilane} \rightarrow \text{uroporphyrinogen III} + H_2O$$

forms **uroporphyrinogen III** (Figure 31–5, ***bottom right***). Hydroxymethylbilane can undergo spontaneous cyclization forming **uroporphyrinogen I** (Figure 31–5, ***bottom left***), but under normal conditions, the uroporphyrinogen formed is almost exclusively the type III isomer. The type I isomers of porphyrinogens are, however, formed in excess in certain porphyrias. Since the pyrrole rings of these uroporphyrinogens are connected by **methylene** (—CH₂—) rather than by

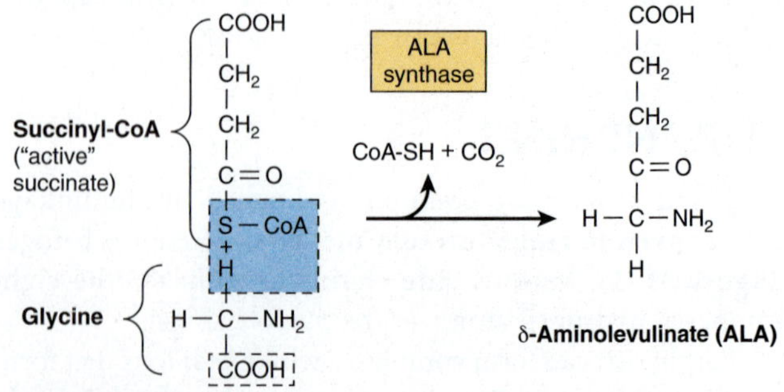

FIGURE 31–2 **Structures of heme *b* and heme *c*.**

FIGURE 31–3 **Synthesis of δ-aminolevulinate (ALA).** This *mitochondrial* reaction is catalyzed by ALA synthase.

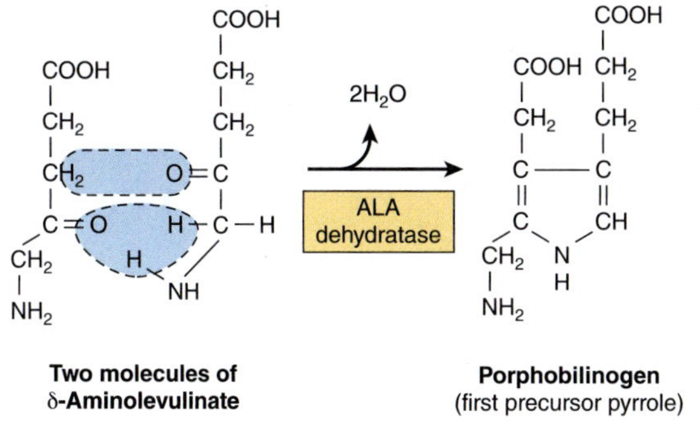

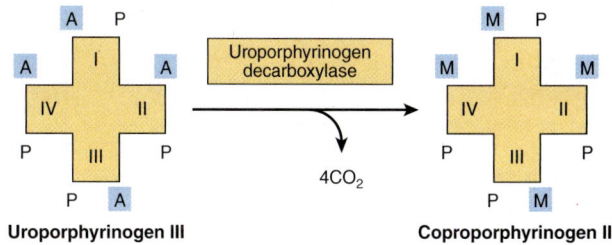

FIGURE 31–6 **Decarboxylation of uroporphyrinogen III to coproporphyrinogen III.** Shown is a representation of the tetrapyrrole to emphasize the conversion of four attached acetyl groups to methyl groups. (A, acetyl; M, methyl; P, propionyl.)

All four acetate moieties of uroporphyrinogen III next undergo decarboxylation to methyl (M) substituents, forming **coproporphyrinogen III** in a cytosolic reaction catalyzed by **uroporphyrinogen decarboxylase**, EC 4.1.1.37 (**Figure 31–6**):

(5) Uroporphyrinogen III → coproporphyrinogen III + 4 CO_2

This decarboxylase can also convert uroporphyrinogen I, if present, to coproporphyrinogen I.

The final three reactions of heme biosynthesis all occur in mitochondria. Coproporphyrinogen III enters the mitochondria and is converted, successively, to **protoporphyrinogen III**, and then to **protoporphyrin III**. These reactions are catalyzed by **coproporphyrinogen oxidase** (EC 1.3.3.3), which decarboxylates and oxidizes the two propionic acid side chains to form **protoporphyrinogen III**:

(6) Coproporphyrinogen III + O_2 + 2 H^+ →
protoporphyrinogen III + 2 CO_2 + 2 H_2O

This oxidase is specific for type III coproporphyrinogen, so type I protoporphyrins generally do not occur in humans. Protoporphyrinogen III is next oxidized to **protoporphyrin III** in a reaction catalyzed by **protoporphyrinogen oxidase**, EC 1.3.3.4:

(7) Protoporphyrinogen III + 3 O_2 → protoporphyrin III
+ 3 H_2O_2

The eighth and final step in heme synthesis involves the incorporation of ferrous iron into protoporphyrin III in a reaction catalyzed by **ferrochelatase** (heme synthase, EC 4.99.1.1), (**Figure 31–7**):

(8) Protoporphyrin III + Fe^{2+} → heme + 2 H^+

Figure 31–8 summarizes the stages of the biosynthesis of the porphyrin derivatives from porphobilinogen. For the above reactions, numbers correspond to those in Figure 31–8 and in **Table 31–2**.

ALA Synthase Is the Key Regulatory Enzyme in Hepatic Biosynthesis of Heme

Unlike ALAS2, which is expressed exclusively in erythrocyte precursor cells, ALAS1 is expressed throughout body tissues.

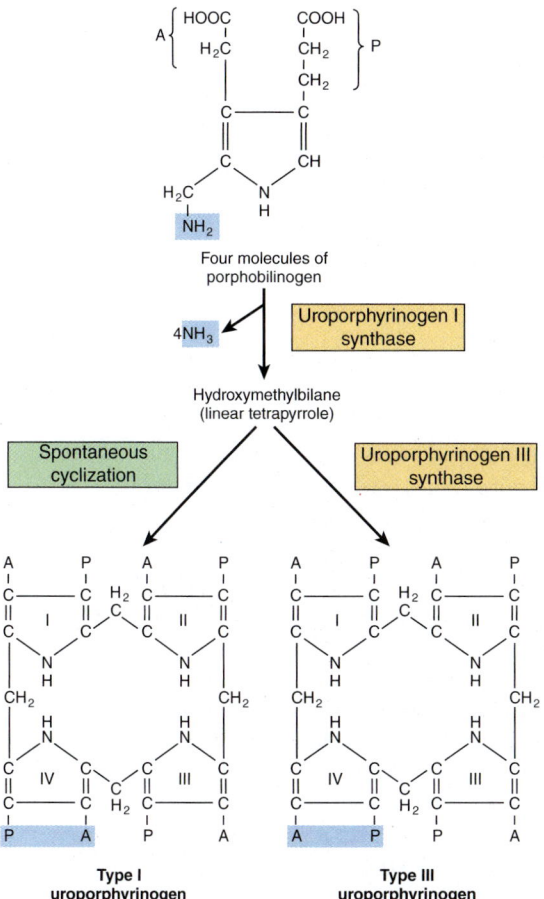

FIGURE 31–4 **Formation of porphobilinogen.** *Cytosolic* porphobilinogen synthase converts two molecules of δ-aminolevulinate to porphybilinogen.

methyne bridges (=HC—), the double bonds do not form a conjugated system. **Porphyrinogens** thus are **colorless**. They are, however, readily auto-oxidized to **colored porphyrins**.

FIGURE 31–5 **Synthesis of hydroxymethylbilane and its subsequent cyclization to porphobilinogen III.** *Cytosolic* hydroxymethylbilane synthase (ALA dehydratase) forms a linear tetrapyrrole, which *cytosolic* uroporphyrinogen synthase cyclizes to form **uroporphyrinogen III**. Notice the asymmetry of the substituents on ring 4, so that the highlighted acetate and propionate substituents are reversed in uroporphyrinogens I and III. (A, acetate [—CH_2COO^-]; P, propionate [—$CH_2CH_2COO^-$].)

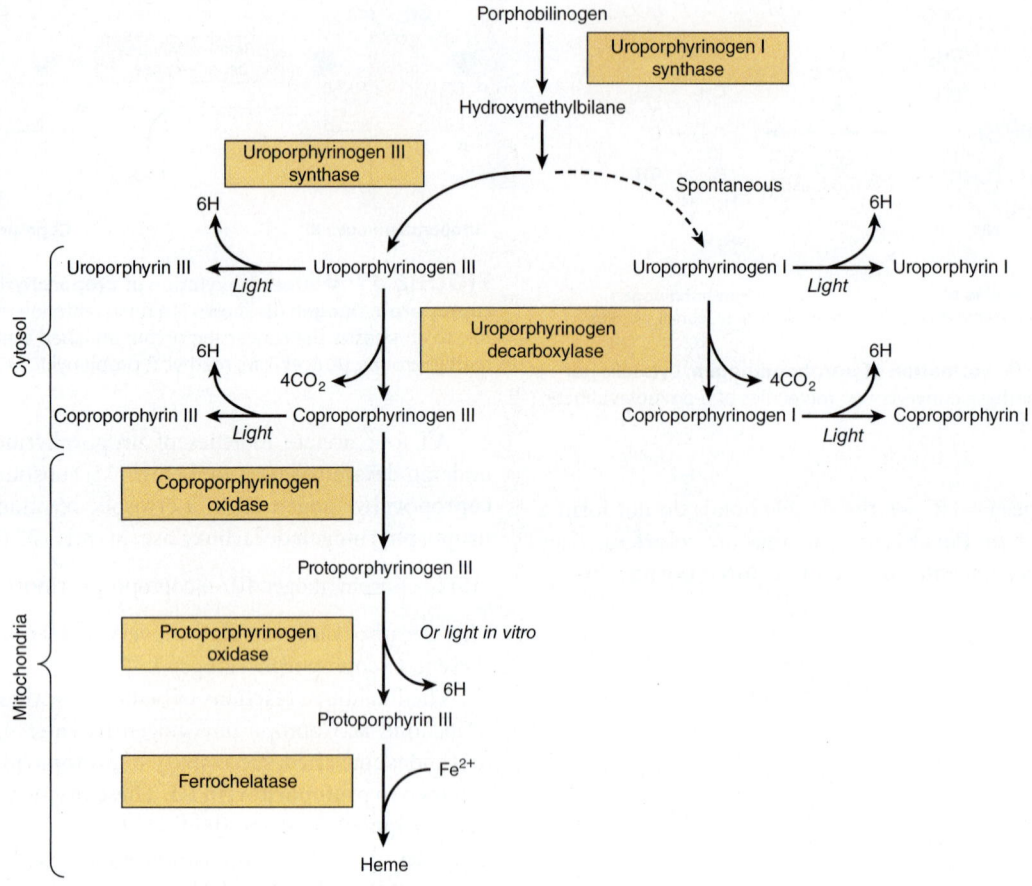

FIGURE 31–7 Biosynthesis from porphobilinogen of the indicated porphyrin derivatives.

The reaction catalyzed by ALA synthase 1 (Figure 31–3) is rate-limiting for biosynthesis of heme in liver. Typically for an enzyme that catalyzes a rate-limiting reaction, ALAS1 has a short half-life. **Heme**, acting through the Erg-1 aporepressor and one of its NAB corepressors, acts as a **negative regulator** of the synthesis of ALAS1 (Figure 31–8). Synthesis of ALAS1 thus increases greatly in the *absence* of heme, but diminishes in its *presence*. Heme also affects translation of ALAS1 and its translocation from its cytosolic site of synthesis into the mitochondrion. Many drugs whose metabolism requires the hemoprotein cytochrome P450 increase cytochrome P450 biosynthesis. The resulting depletion of the intracellular heme pool induces synthesis of ALAS1, and the rate of heme synthesis rises to meet metabolic demand. By contrast, since ALAS2 is not feedback regulated by heme, its biosynthesis is not induced by these drugs.

PORPHYRINS ARE COLORED & FLUORESCE

While **porphyrinogens are colorless**, the various **porphyrins are colored**. The **conjugated double bonds** in the pyrrole rings and linking methylene groups of porphyrins (absent in the porphyrinogens) are responsible for their characteristic absorption and fluorescence spectra. The visible and the ultraviolet spectra of porphyrins and porphyrin derivatives are useful for their identification (**Figure 31–9**). The sharp absorption band **near 400 nm**, a distinguishing feature shared by all porphyrins, is termed the **Soret band** after its discoverer, the French physicist Charles Soret.

Porphyrins dissolved in strong mineral acids or in organic solvents and illuminated by ultraviolet light emit a strong red **fluorescence**, a property often used to detect small amounts of free porphyrins. The photodynamic properties of porphyrins have suggested their possible use in the treatment of certain types of cancer, a procedure called **cancer phototherapy**. Since tumors often take up more porphyrins than do normal tissues, **hematoporphyrin** or related compounds are administered to a patient with an appropriate tumor. The tumor is then exposed to an **argon laser** to excite the porphyrins, producing cytotoxic effects.

Spectrophotometry Is Used to Detect Porphyrins & Their Precursors

Coproporphyrins and uroporphyrins are excreted in increased amounts in the **porphyrias**. When present in urine or feces, they can be separated by extraction with appropriate solvents, then identified and quantified using spectrophotometric methods.

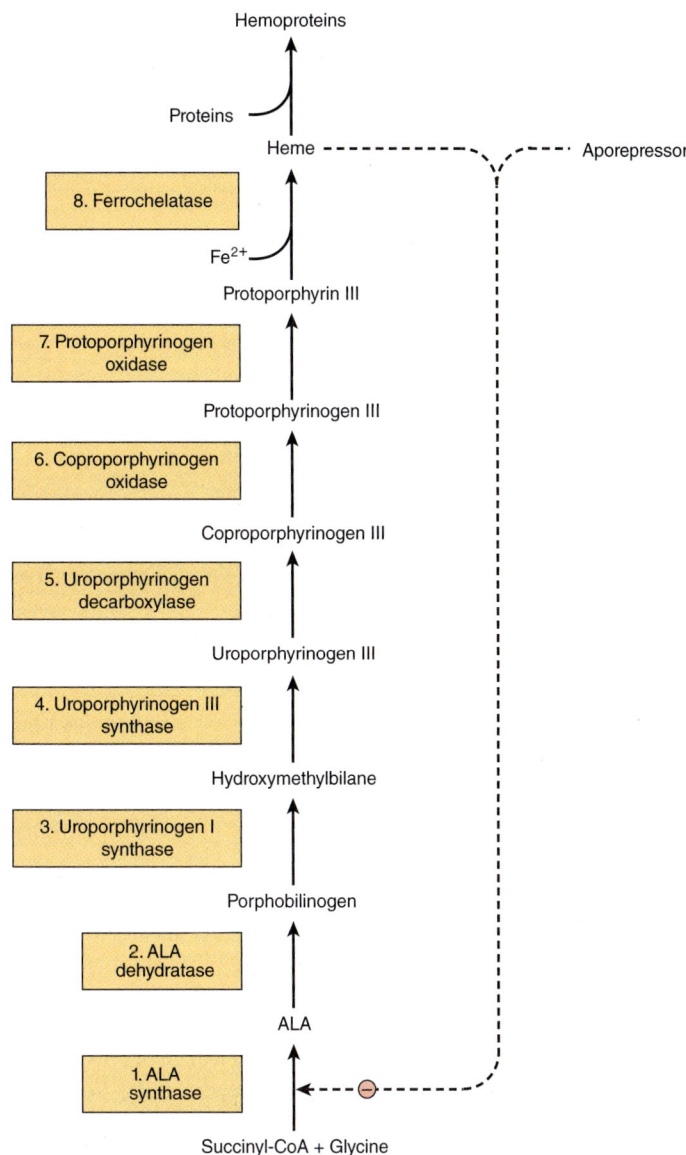

FIGURE 31–8 **Intermediates, enzymes, and regulation of heme synthesis.** The numbers of the enzymes that catalyze the indicated reactions are those used in the accompanying text and in column 1 of **Table 31–2**. Enzymes 1, 6, 7, and 8 are *mitochondrial*, but enzymes 2 to 5 are *cytosolic*. Regulation of hepatic heme synthesis occurs at ALA synthase (ALAS1) by a repression–derepression mechanism mediated by heme and a hypothetical aporepressor (not shown). Mutations in the gene encoding enzyme 1 cause X-linked sideroblastic anemia. Mutations in the genes encoding enzymes 2 to 8 give rise to the porphyrias.

DISORDERS OF HEME BIOSYNTHESIS

Disorders of heme biosynthesis may be genetic or acquired. An example of an acquired defect is lead poisoning. Lead can inactivate ferrochelatase and ALA dehydratase by complexing with essential thiol groups. Signs include elevated levels of protoporphyrin in erythrocytes and elevated urinary levels of ALA and coproporphyrin.

Genetic disorders of heme metabolism and of bilirubin metabolism (see below) share the following features with metabolic disorders of urea biosynthesis (see Chapter 28):

1. Similar or identical clinical signs and symptoms can arise from different mutations in genes that encode either a given enzyme or an enzyme that catalyzes a successive reaction.

2. Rational therapy requires an understanding of the biochemistry of the enzyme-catalyzed reactions in both normal and impaired individuals.

3. Identification of the intermediates and side products that accumulate prior to a metabolic block can provide the basis for metabolic screening tests that can implicate the impaired reaction.

4. Definitive diagnosis involves quantitative assay of the activity of the enzyme(s) suspected to be defective. To this might

TABLE 31–2 Summary of Major Findings in the Porphyrias[a]

Enzyme Involved[b]	Type, Class, and OMIM Number	Major Signs and Symptoms	Results of Laboratory Tests
1. ALA synthase 2 (ALAS2), EC 2.3.1.37	X-linked sideroblasticanemia[c] (erythropoietic) (OMIM 301300)	Anemia	Red cell counts and hemoglobin decreased
2. ALA dehydratase EC 4.2.1.24	ALA dehydratase deficiency (hepatic) (OMIM 125270)	Abdominal pain, neuropsychiatric symptoms	Urinary ALA and coproporphyrin III increased
3. Uroporphyrinogen I synthase[d] EC 2.5.1.61	Acute intermittent porphyria (hepatic) (OMIM 176000)	Abdominal pain, neuropsychiatric symptoms	Urinary ALA and PBG[e] increased
4. Uroporphyrinogen III synthase EC 4.2.1.75	Congenital erythropoietic (erythropoietic) (OMIM 263700)	Photosensitivity	Urinary, fecal, and red cell uroporphyrin I increased
5. Uroporphyrinogen decarboxylase EC 4.1.1.37	Porphyria cutaneatarda (hepatic) (OMIM 176100)	Photosensitivity	Urinary uroporphyrin I increased
6. Coproporphyrinogen oxidase EC 1.3.3.3	Hereditary coproporphyria (hepatic) (OMIM 121300)	Photosensitivity, abdominal pain, neuropsychiatric symptoms	Urinary ALA, PBG, and coproporphyrin III and fecal coproporphyrin III increased
7. Protoporphyrinogen oxidase EC 1.3.3.4	Variegate porphyria (hepatic) (OMIM 176200)	Photosensitivity, abdominal pain, neuropsychiatric symptoms	Urinary ALA, PBG, and coproporphyrin III and fecal protoporphyrin IX increased
8. Ferrochelatase EC 4.99.1.1	Protoporphyria (erythropoietic) (OMIM 177000)	Photosensitivity	Fecal and red cell protoporphyrin IX increased

[a]Only the biochemical findings in the active stages of these diseases are listed. Certain biochemical abnormalities are detectable in the latent stages of some of the above conditions. Conditions 3, 5, and 8 are generally the most prevalent porphyrias. Condition 2 is rare.
[b]The numbering of the enzymes in this Table corresponds to that used in Figure 31–8.
[c]X-linkedsideroblastic anemia is not a porphyria but is included here because ALA synthase is involved.
[d]This enzyme is also called PBG deaminase or hydroxymethylbilane synthase.
[e]PBG = porphyrobilinogen III.
Abbreviations: ALA, δ-aminolevulinic acid; PBG, porphobilinogen.

be added consideration of the as yet incompletely identified factors that facilitate translocation of enzymes and intermediates between cellular compartments.

5. Comparison of the DNA sequence of the gene that encodes a given mutant enzyme to that of the wild-type gene can identify the specific mutation(s) that cause the disease.

The Porphyrias

The signs and symptoms of porphyria result either from a **deficiency** of intermediates beyond the enzymatic block, or

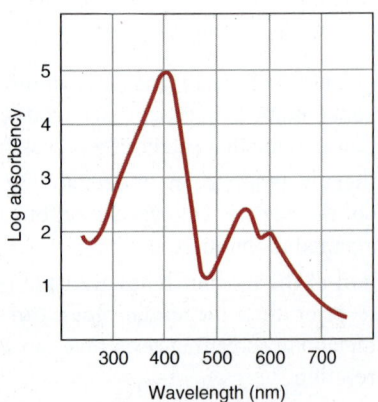

FIGURE 31–9 Absorption spectrum of hematoporphyrin. The spectrum is of a dilute (0.01%) solution of hematoporphyrin in 5% HCl.

from the **accumulation** of metabolites prior to the block. Table 31–2 lists six major types of **porphyria** that reflect low or absent activity of enzymes that catalyze reactions 2 through 8 of Figure 31–8. Possibly due to potential lethality, there is **no** known defect of ALAS1. Individuals with low ALAS2 activity develop anemia, not porphyria (Table 31–2). Porphyria consequent to low activity of ALA dehydratase, termed ALA dehydratase-deficient porphyria, is extremely rare.

Congenital Erythropoietic Porphyria

While most porphyrias are inherited in an **autosomal dominant manner**, congenital erythropoietic porphyria is inherited in a **recessive mode**. The defective enzyme in congenital erythropoietic porphyria is **uroporphyrinogen III synthase** (Figure 31–5, bottom). The photosensitivity and severe disfigurement exhibited by some victims of congenital erythropoietic porphyria has suggested them as prototypes of so-called werewolves.

Acute Intermittent Porphyria

The defective enzyme in acute intermittent porphyria is hydroxymethylbilane synthase (Figure 31–5, bottom). ALA and porphobilinogen accumulate in body tissues and fluids (**Figure 31–10**).

Subsequent Metabolic Blocks

Blocks later in the pathway result in the **accumulation of the porphyrinogens** indicated in Figures 31–8 and 31–10.

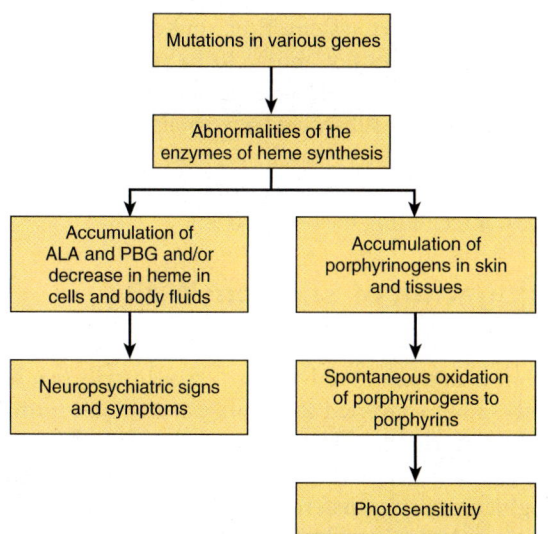

FIGURE 31–10 **Biochemical basis of the major signs and symptoms of the porphyrias.**

Their oxidation to the corresponding porphyrin derivatives cause **photosensitivity** to visible light of about 400-nm wavelength. Possibly as a result of their excitation and reaction with molecular oxygen, the resulting oxygen radicals injure lysosomes and other subcellular organelles, releasing proteolytic enzymes that cause variable degrees of skin damage, including scarring.

CLASSIFICATION OF THE PORPHYRIAS

Porphyrias may be termed **erythropoietic** or **hepatic** based on the organs most affected, typically bone marrow and the liver (Table 31–2). Different and variable levels of heme, toxic precursors, or metabolites probably account for why specific porphyrias differentially affect some cell types and organs. Alternatively, porphyrias may be classified as **acute** or **cutaneous** based on their clinical features. The diagnosis of a specific type of porphyria involves consideration of the clinical and family history, physical examination, and appropriate laboratory tests. Table 31–2 lists the major signs, symptoms, and relevant laboratory findings in the six principal types of porphyria.

Drug-Induced Porphyria

Certain drugs (eg, barbiturates, griseofulvin) induce the production of cytochrome P450. In patients with porphyria, this can precipitate an attack of porphyria by depleting heme levels. The compensating derepression of synthesis of ALAS1 then results in increased levels of potentially harmful heme precursors.

Possible Treatments for Porphyrias

Present treatment of porphyrias is essentially symptomatic: avoiding drugs that induce production of cytochrome P450, ingestion of large amounts of carbohydrate, and administration of hematin to repress ALAS1 synthesis to diminish production of harmful heme precursors. Patients exhibiting photosensitivity benefit from sunscreens and possibly from administered β-carotene, which appears to lessen production of free radicals, decreasing photosensitivity.

CATABOLISM OF HEME PRODUCES BILIRUBIN

Human adults normally destroy about 200 billion erythrocytes per day. A 70-kg human therefore turns over approximately **6 g of hemoglobin** daily. All products are reused. The **globin** is degraded to its constituent amino acids, and the released **iron** enters the iron pool. The iron-free **porphyrin** portion of heme is also degraded, mainly in the reticuloendothelial cells of the liver, spleen, and bone marrow.

The catabolism of heme from all heme proteins takes place in the **microsomal fraction** of cells by **heme oxygenase**, EC 1.14.18.18. Heme oxygenase synthesis is substrate-inducible, and heme also serves both as a substrate and as a cofactor for the reaction. The iron of the heme that reaches heme oxygenase has usually been oxidized to its **ferric form** (**hemin**). Conversion of one mole of heme-Fe^{3+} to biliverdin, carbon monoxide, and Fe^{3+} consumes three moles of O_2, plus seven electrons provided by NADH and NADPH–cytochrome P450 reductase:

$$Fe^{3+}\text{-Heme} + 3\ O_2 + 7\ e^- \rightarrow \text{biliverdin} + CO + Fe^{3+}$$

Despite its high affinity for heme-Fe^{2+} (see Chapter 6), the carbon monoxide produced does not severely inhibit heme oxygenase. Birds and amphibians excrete the green-colored biliverdin directly. In humans, **biliverdin reductase** (EC 1.3.1.24) reduces the central methylene bridge of biliverdin to a methyl group, producing the yellow-pigment **bilirubin** (**Figure 31–11**):

$$\text{Biliverdin} + NADPH + H^+ \rightarrow \text{bilirubin} + NADP^+$$

Since 1 g of hemoglobin yields about 35 mg of bilirubin, **human adults form 250 to 350 mg of bilirubin per day**. This is derived principally from hemoglobin, and also from ineffective erythropoiesis and from catabolism of other heme proteins.

Conversion of heme to bilirubin by reticuloendothelial cells can be observed visually as the purple color of the heme in a **hematoma** slowly converts to the yellow pigment of bilirubin.

Bilirubin Is Transported to the Liver Bound to Serum Albumin

Bilirubin is only sparingly soluble in water. Consequently, it must be bound to serum albumin for transport to the liver. Albumin appears to have both high-affinity and low-affinity sites for bilirubin. The high-affinity site can bind approximately 25 mg of bilirubin/100 mL of plasma. More loosely bound bilirubin can readily be detached and diffused into tissues, and antibiotics and certain other drugs can compete with and displace bilirubin from albumin's high-affinity site.

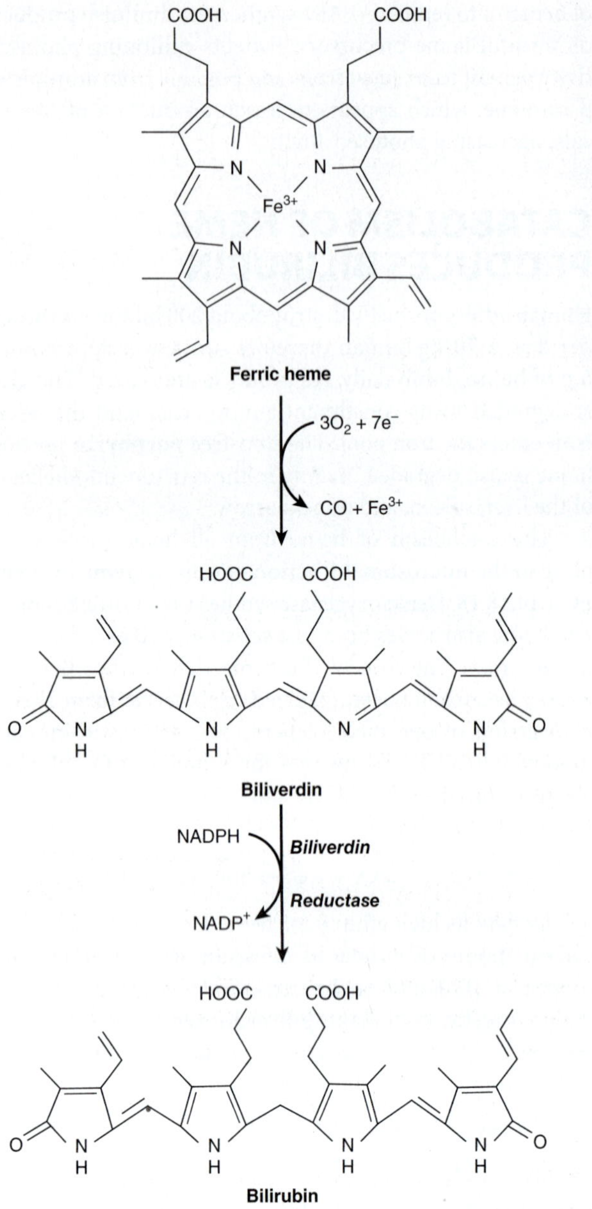

Ferric heme

$$3O_2 + 7e^-$$

$$CO + Fe^{3+}$$

Biliverdin

NADPH

Biliverdin

Reductase

$NADP^+$

Bilirubin

FIGURE 31–11 **Conversion of ferric heme to biliverdin, and then to bilirubin.** (1) Conversion of ferric heme to biliverdin is catalyzed by the heme oxygenase system. (2) Subsequently, biliverdin reductase reduces bilirubin to bilirubin.

Further Metabolism of Bilirubin Occurs Primarily in the Liver

Hepatic catabolism of bilirubin takes place in three stages: uptake by the liver, conjugation with glucuronic acid, and secretion in the bile.

Uptake of Bilirubin by Liver Parenchymal Cells

Bilirubin is removed from albumin and taken up at the sinusoidal surface of hepatocytes by a large capacity, saturable facilitated transport system. Even under pathologic conditions, transport does not appear to be rate-limiting for the metabolism of bilirubin. The net uptake of bilirubin depends on its **removal** by subsequent metabolism. Once internalized, bilirubin binds to cytosolic proteins such as glutathione S-transferase, previously known as a **ligandin**, to prevent bilirubin from reentering the bloodstream.

Conjugation of Bilirubin With Glucuronate

Bilirubin is **nonpolar**, and would persist in cells (eg, bound to lipids) if not converted to a more water-soluble form. Bilirubin is converted to a more **polar** molecule by conjugation with glucuronic acid (**Figure 31–12**). A bilirubin-specific **UDP-glucuronosyltransferase** (EC 2.4.1.17) of the endoplasmic reticulum catalyzes stepwise transfer to bilirubin of two glucosyl moieties from UDP-glucuronate:

Bilirubin + UDP-glucuronate → bilirubin monoglucuronide + UDP

Bilirubin monoglucuronide + UDP-glucuronate → bilirubin diglucuronide + UDP

Secretion of Bilirubin Into the Bile

Secretion of conjugated bilirubin into the bile occurs by an **active transport** mechanism, which probably is rate-limiting for the entire process of hepatic bilirubin metabolism. The protein involved is **a multispecific organic anion transporter (MOAT)** located in the **plasma membrane** of the bile canaliculi. A member of the family of ATP-binding cassette transporters, MOAT transports a number of organic anions. The hepatic transport of conjugated bilirubin into the bile is **inducible** by the same drugs that can induce the conjugation of bilirubin. Conjugation and excretion of bilirubin thus constitute a coordinated functional unit.

Most of the bilirubin excreted in the bile of mammals is bilirubin diglucuronide. Bilirubin UDP-glucuronosyltransferase activity can be **induced** by several drugs, including phenobarbital. However, when bilirubin conjugates exist abnormally in human plasma (eg, in obstructive jaundice), they are predominantly **monoglucuronides**. **Figure 31–13** summarizes

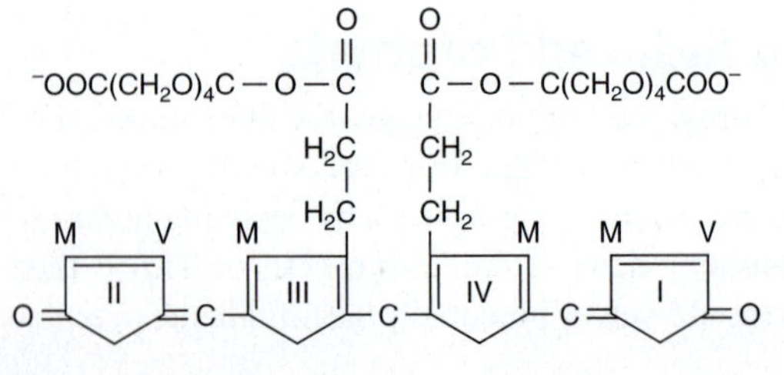

FIGURE 31–12 **Bilirubin diglucuronide.** Glucuronate moieties are attached via ester bonds to the two propionate groups of bilirubin. Clinically, the diglucuronide is also termed "direct reacting" bilirubin.

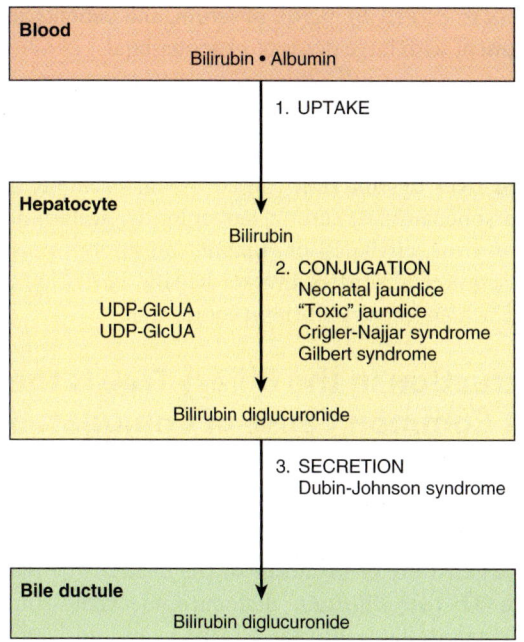

FIGURE 31–13 Diagrammatic representation of the three major processes (uptake, conjugation, and secretion) involved in the transfer of bilirubin from blood to bile. Certain proteins of hepatocytes bind intracellular bilirubin and may prevent its efflux into the bloodstream. The processes affected in certain conditions that cause jaundice are also shown.

the three major processes involved in the transfer of bilirubin from blood to bile. Sites that are affected in a number of conditions causing jaundice are also indicated.

Intestinal Bacteria Reduce Conjugated Bilirubin to Urobilinogen

When conjugated bilirubin reaches the terminal ileum and the large intestine, the glucuronosyl moieties are removed by specific bacterial **β-glucuronidases** (EC 3.2.1.31). Subsequent reduction by the fecal flora forms a group of colorless tetrapyrroles called **urobilinogens**. Small portions of urobilinogens are reabsorbed in the terminal ileum and large intestine and subsequently are reexcreted via the **enterohepatic urobilinogen cycle**. Under abnormal conditions, particularly when excessive bile pigment is formed or when liver disease disrupts this intrahepatic cycle, urobilinogen may also be excreted in the urine. Most of the colorless urobilinogens formed in the colon are **oxidized** there to colored **urobilins** and excreted in the feces. Fecal darkening upon standing in air results from the oxidation of residual urobilinogens to urobilins.

Measurement of Bilirubin in Serum

Quantitation of bilirubin employs a colorimetric method based on the reddish-purple color formed when bilirubin reacts with diazotized sulfanilic acid. An assay conducted in the *absence* of added methanol measures "**direct bilirubin**," which is **bilirubin glucuronide**. An assay conducted in the

presence of added methanol measures **total bilirubin**. The *difference* between total bilirubin and direct bilirubin is known as "**indirect bilirubin**," and is **unconjugated bilirubin**.

HYPERBILIRUBINEMIA CAUSES JAUNDICE

Hyperbilirubinemia, a blood level that exceeds 1 mg of bilirubin per dL (17 μmol/L), may result from **production** of more bilirubin than the normal liver can excrete, or from the failure of a damaged liver to **excrete** normal amounts of bilirubin. In the absence of hepatic damage, **obstruction** of the excretory ducts of the liver prevents the excretion of bilirubin, and will also cause hyperbilirubinemia. In all these situations, when the blood concentration of bilirubin reaches 2 to 2.5 mg/dL, it diffuses into the tissues, which turn yellow, a condition termed **jaundice** or **icterus**.

Occurrence of Unconjugated Bilirubin in Blood

Forms of hyperbilirubinemia include **retention hyperbilirubinemia** due to overproduction of bilirubin, and **regurgitation hyperbilirubinemia** due to reflux into the bloodstream because of biliary obstruction.

Because of its **hydrophobicity**, only *unconjugated* bilirubin can cross the blood–brain barrier into the central nervous system. Encephalopathy due to hyperbilirubinemia (**kernicterus**) thus occurs only with unconjugated bilirubin, as in retention hyperbilirubinemia. Alternatively, because of its water solubility, only *conjugated* bilirubin can appear in urine. Accordingly, **choluric jaundice** (choluria is the presence of bile pigments in the urine) occurs only in regurgitation hyperbilirubinemia, and **acholuric jaundice** occurs only in the presence of an excess of unconjugated bilirubin. **Table 31–3** lists some causes of unconjugated and conjugated hyperbilirubinemia. A moderate hyperbilirubinemia accompanies **hemolytic anemias**.

TABLE 31–3 Some Causes of Unconjugated and Conjugated Hyperbilirubinemia

Unconjugated	Conjugated
Hemolytic anemias	Obstruction of the biliary tree
Neonatal "physiological jaundice"	Dubin–Johnson syndrome
Crigler-Najjar syndromes types I and II	Rotor syndrome
Gilbert syndrome	Liver diseases such as the various types of hepatitis
Toxic hyperbilirubinemia	

These causes are discussed briefly in the text. Common causes of obstruction of the biliary tree are a stone in the common bile duct and cancer of the head of the pancreas. Various liver diseases (eg, the various types of hepatitis) are frequent causes of predominantly conjugated hyperbilirubinemia.

Hyperbilirubinemia is usually modest (< 4 mg bilirubin per dL; < 68 µmol/L) despite extensive hemolysis, due to the high capacity of a healthy liver to metabolize bilirubin.

DISORDERS OF BILIRUBIN METABOLISM

Neonatal "Physiologic Jaundice"

The unconjugated hyperbilirubinemia of neonatal "physiologic jaundice" results from accelerated hemolysis and an immature hepatic system for the uptake, conjugation, and secretion of bilirubin. In this transient condition, bilirubin-glucuronosyltransferase activity, and probably also synthesis of UDP-glucuronate, are reduced. When the plasma concentration of unconjugated bilirubin exceeds that which can be tightly bound by albumin (20–25 mg/dL), bilirubin can penetrate the blood–brain barrier. If left untreated, the resulting **hyperbilirubinemic toxic encephalopathy**, or **kernicterus**, can result in mental retardation. Exposure of jaundiced neonates to blue light (phototherapy) promotes hepatic excretion of unconjugated bilirubin by converting some to derivatives that are excreted in the bile, and phenobarbital, a promoter of bilirubin metabolism, may be administered.

Defects of Bilirubin UDP-Glucuronosyltransferase

Glucuronosyltransferases (EC 2.4.1.17), a family of enzymes with differing substrate specificities, increase the polarity of various drugs and drug metabolites, thereby facilitating their excretion. Mutations in the gene that encodes **bilirubin UDP-glucuronosyltransferase** can result in the encoded enzyme having reduced or absent activity. Syndromes whose clinical presentation reflects the severity of the impairment include Gilbert syndrome and two types of Crigler-Najjar syndrome.

Gilbert Syndrome

Providing that about 30% of the bilirubin UDP-glucuronosyltransferase activity is retained in Gilbert syndrome, the condition is harmless.

Type I Crigler-Najjar Syndrome

The severe congenital jaundice (over 20 mg bilirubin per dL serum) and accompanying brain damage of type I Crigler-Najjar syndrome reflect the complete absence of hepatic UDP-glucuronosyltransferase activity. Phototherapy reduces plasma bilirubin levels somewhat, but phenobarbital has no beneficial effect. The disease is often fatal within the first 15 months of life.

Type II Crigler-Najjar Syndrome

In type II Crigler-Najjar syndrome, some bilirubin UDP-glucuronosyltransferase activity is retained. This condition thus is more benign than the type I syndrome. Serum bilirubin tends not to exceed 20 mg/dL of serum, and patients respond to treatment with large doses of phenobarbital.

Toxic Hyperbilirubinemia

Unconjugated hyperbilirubinemia can result from **toxin-induced liver dysfunction** caused by, for example, chloroform, arsphenamines, carbon tetrachloride, acetaminophen, hepatitis virus, cirrhosis, or *Amanita* mushroom poisoning. These acquired disorders involve hepatic parenchymal cell damage, which impairs bilirubin conjugation.

Obstruction in the Biliary Tree Is the Most Common Cause of Conjugated Hyperbilirubinemia

Conjugated hyperbilirubinemia commonly results from blockage of the hepatic or common bile ducts, most often due to a **gallstone** or to **cancer of the head of the pancreas** (**Figure 31–14**). Bilirubin diglucuronide that cannot be excreted regurgitates into the hepatic veins and lymphatics, conjugated bilirubin appears in the blood and urine (**choluric jaundice**), and the stools typically are a pale color.

The term **cholestatic jaundice** includes both all cases of extrahepatic obstructive jaundice and also conjugated hyperbilirubinemia due to micro-obstruction of intrahepatic biliary ductules by damaged hepatocytes, such as may occur in infectious hepatitis.

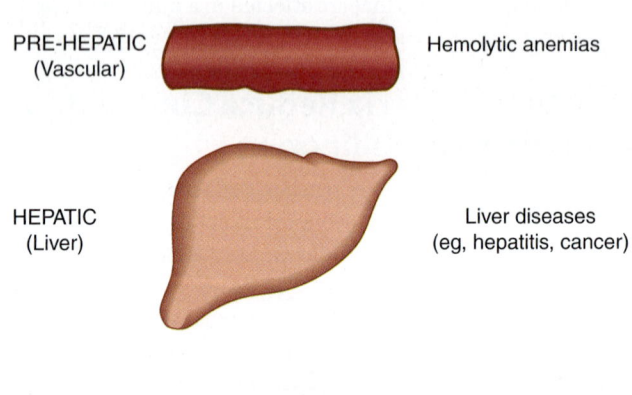

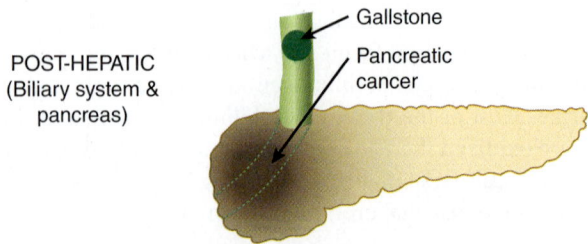

PRE-HEPATIC (Vascular) — Hemolytic anemias

HEPATIC (Liver) — Liver diseases (eg, hepatitis, cancer)

POST-HEPATIC (Biliary system & pancreas) — Gallstone, Pancreatic cancer

FIGURE 31–14 **Major causes of jaundice. Prehepatic jaundice** indicates events in the bloodstream, major causes being various hemolytic anemias. **Hepatic jaundice** arises from hepatitis or other liver diseases (eg, cancer). **Posthepatic jaundice** refers to events in the biliary tree, for which the major causes are obstruction of the common bile duct by a gallstone (biliary calculus) or by cancer of the head of the pancreas.

TABLE 31–4 Laboratory Results in Normal Patients and Patients With Three Different Causes of Jaundice

Condition	Serum Bilirubin	Urine Urobilinogen	Urine Bilirubin	Fecal Urobilinogen
Normal	Direct: 0.1-0.4 mg/dL	0-4 mg/24 h	Absent	40-280 mg/24 h
	Indirect: 0.2-0.7 mg/dL			
Hemolytic anemia	↑Indirect	Increased	Absent	Increased
Hepatitis	↑Direct and indirect	Decreased if micro-obstruction is present	Present if micro-obstruction occurs	Decreased
Obstructive jaundice[a]	↑Direct	Absent	Present	Trace to absent

[a]The most common causes of obstructive (posthepatic) jaundice are cancer of the head of the pancreas and a gallstone lodged in the common bile duct. The presence of bilirubin in the urine is sometimes referred to as choluria—therefore, hepatitis and obstruction of the common bile duct cause choluric jaundice, whereas the jaundice of hemolytic anemia is referred to as acholuric. The laboratory results in patients with hepatitis are variable, depending on the extent of damage to parenchymal cells and the extent of micro-obstruction to bile ductules. Serum levels of **alanine aminotransferase** and **aspartate aminotransferase** are usually markedly elevated in hepatitis, whereas serum levels of **alkaline phosphatase** are elevated in obstructive liver disease.

Dubin-Johnson Syndrome

This benign autosomal recessive disorder consists of **conjugated hyperbilirubinemia** in childhood or during adult life. The hyperbilirubinemia is caused by mutations in the gene encoding the protein involved in the **secretion** of conjugated bilirubin into bile.

Some Conjugated Bilirubin Can Bind Covalently to Albumin

When levels of conjugated bilirubin remain high in plasma, a fraction can bind covalently to albumin. This fraction, termed **δ-bilirubin**, has a **longer half-life** in plasma than does conventional conjugated bilirubin, and remains elevated during recovery from obstructive jaundice. Some patients therefore continue to appear jaundiced even after the circulating conjugated bilirubin level has returned to normal.

Urinary Urobilinogen & Bilirubin Are Clinical Indicators

In **complete obstruction of the bile duct**, bilirubin has no access to the intestine for conversion to urobilinogen, so no urobilinogen is present in the urine. The presence of conjugated bilirubin in the urine without urobilinogen suggests intrahepatic or posthepatic obstructive jaundice.

In **jaundice secondary to hemolysis**, the increased production of bilirubin leads to increased production of **urobilinogen**, which appears in the urine in large amounts. Bilirubin is not usually found in the urine in hemolytic jaundice, so the combination of increased urobilinogen and absence of bilirubin is suggestive of hemolytic jaundice. Increased blood destruction from any cause brings about an increase in urine urobilinogen.

Table 31–4 summarizes laboratory results obtained in patients with jaundice due to prehepatic, hepatic, or posthepatic causes: **hemolytic anemia** (prehepatic), **hepatitis** (hepatic), and **obstruction of the common bile duct** (posthepatic); see Figure 31–14. Laboratory tests on **blood** (evaluation of the possibility of a hemolytic anemia and measurement of prothrombin time) and on **serum** (eg, electrophoresis of proteins; alkaline phosphatase and alanine aminotransferase and aspartate aminotransferase activities) also help to distinguish between prehepatic, hepatic, and posthepatic causes of jaundice.

SUMMARY

- The heme of hemoproteins such as hemoglobin and the cytochromes is an iron-porphyrin complex. Porphyrin consists of four pyrrole rings joined by methyne bridges.

- The eight methyl, vinyl, and propionyl substituents on the four pyrrole rings of heme are arranged in a specific sequence. The metal ion (Fe^{2+} in hemoglobin; Mg^{2+} in chlorophyll) is linked to the four nitrogen atoms of the pyrrole rings.

- Biosynthesis of the heme ring involves eight enzyme-catalyzed reactions, some of which occur in mitochondria, others in cytosol.

- Synthesis of heme commences with the condensation of succinyl-CoA and glycine to form ALA. This reaction is catalyzed by ALAS1, the regulatory enzyme of heme biosynthesis.

- Synthesis of ALAS1 increases in response to a low level of available heme. For example, certain drugs (eg, phenobarbital) indirectly trigger enhanced synthesis of ALAS1 by promoting synthesis of the heme protein cytochrome P450, which thereby depletes the heme pool. By contrast, ALAS2 is not regulated by heme levels, and consequently not by drugs that promote synthesis of cytochrome P450.

- Genetic abnormalities of seven of the eight enzymes of heme biosynthesis result in inherited porphyrias. Erythrocytes and liver are the major sites of expression of the porphyrias. Photosensitivity and neurologic problems are common complaints. Intake of certain toxins (eg, lead) can cause acquired porphyrias. Increased amounts of porphyrins or their precursors can be detected in blood and urine, facilitating diagnosis.

- Catabolism of the heme ring, initiated by the mitochondrial enzyme heme oxygenase, produces the linear tetrapyrrole, biliverdin. Subsequent reduction of biliverdin in the cytosol forms bilirubin.

- Bilirubin binds to albumin for transport from peripheral tissues to the liver, where it is taken up by hepatocytes. The iron of heme is released and reutilized.

- The water solubility of bilirubin is increased by the addition of two moles of the highly polar glucuronosyl moiety, derived from UDP-glucuronate, per mole of bilirubin. Attachment of the glucuronosyl moieties is catalyzed by bilirubin UDP-glucuronosyltransferase, one of a large family of enzymes of differing substrate specificities that increase the polarity of various drugs and drug metabolites, thereby facilitating their excretion.

- Mutations in the encoding gene may result in reduced or absent bilirubin UDP-glucuronosyltransferase activity. Clinical presentations that reflect the severity of the mutation(s) include Gilbert syndrome and two types of Crigler-Najjar syndrome, conditions whose severity depend on the extent of remaining glucuronosyltransferase activity.

- Following secretion of bilirubin from the bile into the gut, bacterial enzymes convert bilirubin to urobilinogen and urobilin, which are excreted in the feces and urine.

- Colorimetric measurement of bilirubin employs the color formed when bilirubin reacts with diazotized sulfanilic acid. Assays conducted in the *absence* of added methanol measure "direct bilirubin" (ie, bilirubin glucuronide). Assays conducted in the *presence* of added methanol measure total bilirubin. The difference between total bilirubin and direct bilirubin, termed "indirect bilirubin," is unconjugated bilirubin.

- Jaundice results from an elevated level of plasma bilirubin. The causes of jaundice can be distinguished as prehepatic (eg, hemolytic anemias), hepatic (eg, hepatitis), or posthepatic (eg, obstruction of the common bile duct). Measurements of plasma total and nonconjugated bilirubin, of urinary urobilinogen and bilirubin, of the activity of certain serum enzymes, and the analysis of stool samples help distinguish between the causes of jaundice.

REFERENCES

Ajioka RS, Phillips JD, Kushner JP: Biosynthesis of heme in mammals. Biochim Biophys Acta 2006;1763:723.

Desnick RJ, Astrin KH: The porphyrias. In *Harrison's Principles of Internal Medicine,* 17th ed. Fauci AS (editor). McGraw-Hill, 2008.

Dufour DR: Liver disease. In *Tietz Textbook of Clinical Chemistry and Molecular Diagnostics,* 4th ed. Burtis CA, Ashwood ER, Bruns DE (editors). Elsevier Saunders, 2006.

Higgins T, Beutler E, Doumas BT: Hemoglobin, iron and bilirubin. In *Tietz Textbook of Clinical Chemistry and Molecular Diagnostics,* 4th ed. Burtis CA, Ashwood ER, Bruns DE (editors). Elsevier Saunders, 2006.

Pratt DS, Kaplan MM: Evaluation of liver function. In *Harrison's Principles of Internal Medicine,* 17th ed. Fauci AS (editor). McGraw-Hill, 2008.

Wolkoff AW: The hyperbilirubinemias. In *Harrison's Principles of Internal Medicine,* 17th ed. Fauci AS (editor). McGraw-Hill, 2008.

Exam Questions

Section VI - Metabolism of Proteins & Amino Acids

1. Select the one of the following statements that is NOT CORRECT:
   A. Δ^1-Pyrroline-5-carboxylate is an intermediate both in the biosynthesis and in the catabolism of L-proline.
   B. Human tissues can form dietarily nonessential amino acids from amphibolic intermediates or from dietarily essential amino acids.
   C. Human liver tissue can form serine from the glycolytic intermediate 3-phosphoglycerate.
   D. The reaction catalyzed by phenylalanine hydroxylase interconverts phenylalanine and tyrosine.
   E. The reducing power of tetrahydrobiopterin derives ultimately from NADPH.

2. Identify the metabolite that does NOT serve as a precursor of a dietarily essential amino acid:
   A. α-Ketoglutarate
   B. 3-Phosphoglycerate
   C. Glutamate
   D. Aspartate
   E. Histamine

3. Select the one of the following statements that is NOT CORRECT:
   A. Selenocysteine is present at the active sites of certain human enzymes.
   B. Selenocysteine is inserted into proteins by a posttranslational process.
   C. Transamination of dietary α-keto acids can replace the dietary essential amino acids leucine, isoleucine, and valine.
   D. Conversion of peptidyl proline to peptidyl-4-hydroxyproline is accompanied by the incorporation of oxygen into succinate.
   E. Serine and glycine are interconverted in a single reaction in which tetrahydrofolate derivatives participate.

4. Select the CORRECT answer:
   The first reaction in the degradation of most of the protein amino acids involves the participation of:
   A. NAD$^+$
   B. Thiamine pyrophosphate (TPP)
   C. Pyridoxal phosphate
   D. FAD
   E. NAD$^+$ and TPP

5. Identify the amino acid that is the major contributor to the transport of nitrogen destined for excretion as urea:
   A. Alanine
   B. Glutamine
   C. Glycine
   D. Lysine
   E. Ornithine

6. Select the one of the following statements that is NOT CORRECT:
   A. Angelman syndrome is associated with a defective ubiquitin E3 ligase.
   B. Following a protein-rich meal, the splanchnic tissues release predominantly branched-chain amino acids. which are taken up by peripheral muscle tissue.
   C. The rate of hepatic gluconeogenesis from glutamine exceeds that of any other amino acid.
   D. The L-α-amino oxidase-catalyzed conversion of an α-amino acid to its corresponding α-keto acid is accompanied by the release of NH_4^+.
   E. Similar or even identical signs and symptoms can be associated with different mutations of the gene that encodes a given enzyme.

7. Select the one of the following statements that is NOT CORRECT:
   A. PEST sequences target some proteins for rapid degradation.
   B. ATP and ubiquitin typically participate in the degradation of membrane-associated proteins and other proteins with long half-lives.
   C. Ubiquitin molecules are attached to target proteins via non-α peptide bonds.
   D. The discoverers of ubiquitin-mediated protein degradation received a Nobel Prize.
   E. Degradation of ubiquitin-tagged proteins takes place in the proteasome, a multi-subunit macromolecule present in all eukaryotes.

8. For metabolic disorders of the urea cycle, which statement is NOT CORRECT:
   A. Ammonia intoxication is most severe when the metabolic block in the urea cycle occurs prior to the reaction catalyzed by argininosuccinate synthase.
   B. Clinical symptoms include mental retardation and the avoidance of protein-rich foods.
   C. Clinical signs can include acidosis.
   D. Aspartate provides the second nitrogen of argininosuccinate.
   E. Dietary management focuses on a low-protein diet ingested as frequent small meals.

9. Select the one of the following statements that is NOT CORRECT:
   A. One metabolic function of glutamine is to sequester nitrogen in a nontoxic form.
   B. Liver glutamate dehydrogenase is allosterically inhibited by ATP and activated by ADP.
   C. Urea is formed both from absorbed ammonia produced by enteric bacteria and from ammonia generated by tissue metabolic activity.
   D. The concerted action of glutamate dehydrogenase and glutamate aminotransferase may be termed transdeamination.
   E. Fumarate generated during biosynthesis of argininosuccinate ultimately forms oxaloacetate in reactions in mitochondria catalyzed successively by fumarase and malate dehydrogenase.

10. Select the one of the following statements that is NOT CORRECT:

    A. Threonine provides the thioethanol moiety for biosynthesis of coenzyme A.
    B. Histamine arises by decarboxylation of histidine.
    C. Ornithine serves as a precursor of both spermine and spermidine.
    D. Serotonin and melatonin are metabolites of tryptophan.
    E. Glycine, arginine, and methionine each contribute atoms for biosynthesis of creatine.

11. Select the one of the following statements that is NOT CORRECT:

    A. Excreted creatinine is a function of muscle mass, and can be used to determine whether a patient has provided a complete 24-hour urine specimen.
    B. Many drugs and drug catabolites are excreted in urine as glycine conjugates.
    C. The major nonprotein metabolic fate of methionine is conversion to S-adenosylmethionine.
    D. The concentration of histamine in brain hypothalamus exhibits a circadian rhythm.
    E. Decarboxylation of glutamine forms the inhibitory neurotransmitter GABA (γ-aminobutyrate).

12. What distinguishes the routes by which each of the following amino acids appears in human proteins?

    5-Hydroxylysine
    γ-Carboxyglutamate
    Selenocysteine

13. What evolutionary advantage might be gained by the fact that certain amino acids are *dietarily* essential for human subjects?

14. What explanation can you offer to explain that metabolic defects that result in the complete absence of the activity of glutamate dehydrogenase have not been detected?

15. Which of the following is NOT a hemoprotein?

    A. Myoglobin
    B. Cytochrome *c*
    C. Catalase
    D. Cytochrome P450
    E. Albumin

16. A 30-year-old man presented at clinic with a history of intermittent abdominal pain and episodes of confusion and psychiatric problems. Laboratory tests revealed increases of urinary δ-aminolevulinate and porphobilinogen. Mutational analysis revealed a mutation in the gene for uroporphyrinogen I synthase (porphobilinogen deaminase). The probable diagnosis was:

    A. Acute intermittent porphyria.
    B. X-linked sideroblastic anemia.
    C. Congenital erythropoietic porphyria.
    D. Porphyria cutanea tarda.
    E. Variegate porphyria.

17. Select the one of the following statements that is NOT CORRECT:

    A. Bilirubin is a cyclic tetrapyrrole.
    B. Albumin-bound bilirubin is transported to the liver.
    C. High levels of bilirubin can cause damage to the brains of newborn infants.
    D. Bilirubin contains methyl and vinyl groups.
    B. Bilirubin does not contain iron.

18. A 62-year-old female presented at clinic with intense jaundice, steadily increasing over the preceding 3 months. She gave a history of severe upper abdominal pain. radiating to the back. and had lost considerable weight. She had noted that her stools had been very pale for some time. Lab tests revealed a very high level of direct bilirubin. and also elevated urinary bilirubin. The plasma level of alanine aminotransferase (ALT) was only slightly elevated, whereas the level of alkaline phosphatase was markedly elevated. Abdominal ultrasonography revealed no evidence of gallstones. Of the following, which is the most likely diagnosis?

    A. Gilbert syndrome
    B. Hemolytic anemia
    C. Type 1 Crigler-Najjar syndrome
    D. Carcinoma of the pancreas
    E. Infectious hepatitis

19. Clinical laboratories typically use diazotized sulfanilic acid to measure serum bilirubin and its derivatives. What is the physical basis that permits the laboratory to report results to the physician in terms of these two forms of bilirubin?

20. What signals the synthesis of heme to take place?

SECTION VII

Structure, Function, & Replication of Informational Macromolecules

CHAPTER 32

Nucleotides

Victor W. Rodwell, PhD

OBJECTIVES

After studying this chapter, you should be able to:

- Write structural formulas to represent the amino- and oxo-tautomers of a purine and of a pyrimidine and state which tautomer predominates under physiologic conditions.
- Reproduce the structural formulas for the principal nucleotides present in DNA and in RNA and the less common nucleotides 5-methylcytosine, 5-hydroxymethylcytosine, and pseudouridine (ψ).
- Represent D-ribose or 2-deoxy-D-ribose linked as either a *syn* or an *anti* conformer to a purine, name the bond between the sugar and the base, and indicate which conformer predominates under most physiologic conditions.
- Number the C and N atoms of a pyrimidine ribonucleoside and of a purine deoxyribonucleoside, including using a primed numeral for C atoms of the sugars.
- Compare the phosphoryl group transfer potential of each phosphoryl group of a nucleoside triphosphate.
- Outline the physiologic roles of the cyclic phosphodiesters cAMP and cGMP.
- Appreciate that polynucleotides are directional macromolecules composed of mononucleotides linked by 3′ → 5′-phosphodiester bonds.
- Be familiar with the abbreviated representations of polynucleotide structures such as pTpGpT or TGCATCA, for which the 5′-end is always shown at the left and all phosphodiester bonds are 3′ → 5′.
- For specific synthetic analogs of purine and pyrimidine bases and their derivatives that have served as anticancer drugs, indicate in what ways these compounds inhibit metabolism.

BIOMEDICAL IMPORTANCE

In addition to serving as precursors of nucleic acids, purine and pyrimidine nucleotides participate in metabolic functions as diverse as energy metabolism, protein synthesis, regulation of enzyme activity, and signal transduction. When linked to vitamins or vitamin derivatives, nucleotides form a portion of many coenzymes. As the principal donors and acceptors of phosphoryl groups in metabolism, nucleoside tri- and diphosphates such as ATP and ADP are the principal players in the energy transductions that accompany metabolic interconversions and oxidative phosphorylation. Linked to sugars or lipids, nucleosides constitute key biosynthetic intermediates. The sugar derivatives UDP-glucose and UDP-galactose participate in sugar interconversions and in the biosynthesis of starch and glycogen. Similarly, nucleoside-lipid derivatives such as CDP-acylglycerol are intermediates in lipid biosynthesis. Roles that nucleotides perform in metabolic regulation include ATP-dependent phosphorylation of key metabolic enzymes, allosteric regulation of enzymes by ATP, ADP, AMP, and CTP, and control by ADP of the rate of oxidative phosphorylation. The cyclic nucleotides cAMP and cGMP serve as the second messengers in hormonally regulated events, and GTP and GDP play key roles in the cascade of events that characterize signal transduction pathways. In addition to the central roles that nucleotides play in metabolism, their medical applications include the use of synthetic purine and pyrimidine analogs that contain halogens, thiols, or additional nitrogen atoms in the chemotherapy of cancer and AIDS, and as suppressors of the immune response during organ transplantation.

CHEMISTRY OF PURINES, PYRIMIDINES, NUCLEOSIDES, & NUCLEOTIDES

Purines & Pyrimidines Are Heterocyclic Compounds

Purines and pyrimidines are aromatic **heterocycles**, cyclic structures that contain, in addition to carbon, other (hetero) atoms such as nitrogen. Note that the smaller pyrimidine molecule has the *longer* name and the larger purine molecule the *shorter* name, and that their six-atom rings are numbered in opposite directions (**Figure 32–1**). Purines or

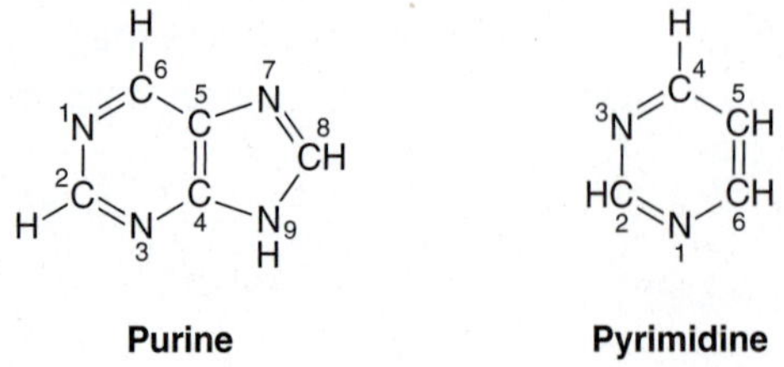

FIGURE 32–2   Tautomerism of the oxo and amino functional groups of purines and pyrimidines.

pyrimidines with an NH_2 group are weak bases (pK_a values 3-4), although the proton present at low pH is associated, not as one might expect with the exocyclic amino group, but with a ring nitrogen, typically N1 of adenine, N7 of guanine, and N3 of cytosine. The planar character of purines and pyrimidines facilitates their close association, or "stacking," that stabilizes double-stranded DNA (see Chapter 34). The oxo and amino groups of purines and pyrimidines exhibit keto–enol and amine–imine **tautomerism** (**Figure 32–2**), although physiologic conditions strongly favor the amino and oxo forms.

Nucleosides Are *N*-Glycosides

Nucleo**sides** are derivatives of purines and pyrimidines that have a sugar linked to a ring nitrogen of a purine or pyrimidine. Numerals with a prime (eg, 2′ or 3′) distinguish atoms of the sugar from those of the heterocycle. The sugar in **ribonucleosides** is D-ribose, and in **deoxyribonucleosides** is 2-deoxy-D-ribose. Both sugars are linked to the heterocycle by an **a-*N*-glycosidic bond**, almost always to the *N*-1 of a pyrimidine or to *N*-9 of a purine (**Figure 32–3**).

Nucleotides Are Phosphorylated Nucleosides

Mononucleo**tides** are nucleo**sides** with a phosphoryl group esterified to a hydroxyl group of the sugar. The 3′- and 5′-nucleotides are nucleosides with a phosphoryl group on the 3′- or 5′-hydroxyl group of the sugar, respectively. Since most nucleotides are 5′-, the prefix "5′-" usually is omitted when naming them. UMP and dAMP thus represent nucleotides with a phosphoryl group on C-5 of the pentose. Additional phosphoryl groups, ligated by **acid anhydride bonds** to the phosphoryl group of a mononucleotide, form **nucleoside diphosphates** and **triphosphates** (**Figure 32–4**).

Heterocylic *N*-Glycosides Exist as *Syn* and *Anti* Conformers

Steric hindrance by the heterocyclic ring blocks free rotation about the β-*N*-glycosidic bond of nucleosides or nucleotides. Both therefore exist as noninterconvertible **syn or anti conformers** (**Figure 32–5**). While both *syn* and *anti* conformers occur in nature, the *anti* conformers predominate.

Table 32–1 lists the major purines and pyrimidines and their nucleoside and nucleotide derivatives. Single-letter abbreviations are used to identify adenine (A), guanine (G), cytosine (C), thymine (T), and uracil (U), whether free or

FIGURE 32–1   **Purine and pyrimidine.** The atoms are numbered according to the international system.

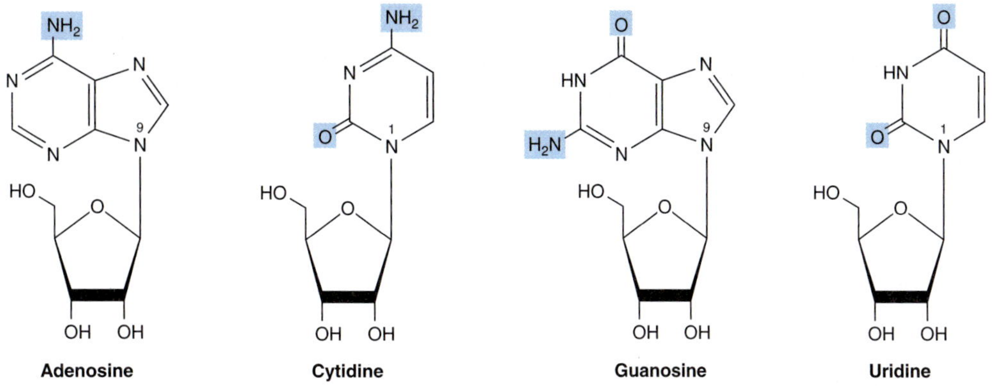

FIGURE 32–3 Ribonucleosides, drawn as the *syn* conformers.

present in nucleosides or nucleotides. The prefix "d" (deoxy) indicates that the sugar is 2′-deoxy-D-ribose (eg, in dATP) (**Figure 32–6**).

Modification of Polynucleotides Can Generate Additional Structures

Small quantities of additional purines and pyrimidines occur in DNA and RNAs. Examples include 5-methylcytosine of bacterial and human DNA, 5-hydroxymethylcytosine of bacterial and viral nucleic acids, and mono- and the di-*N*-methylated adenine and guanine of mammalian messenger RNAs (**Figure 32–7**) that function in oligonucleotide recognition and in regulating the half-lives of RNAs. Free heterocyclic bases include hypoxanthine, xanthine, and uric acid (**Figure 32–8**), intermediates in the catabolism of adenine and guanine (see Chapter 33). Methylated heterocycles of plants include the xanthine derivatives caffeine of coffee, theophylline of tea, and theobromine of cocoa (**Figure 32–9**).

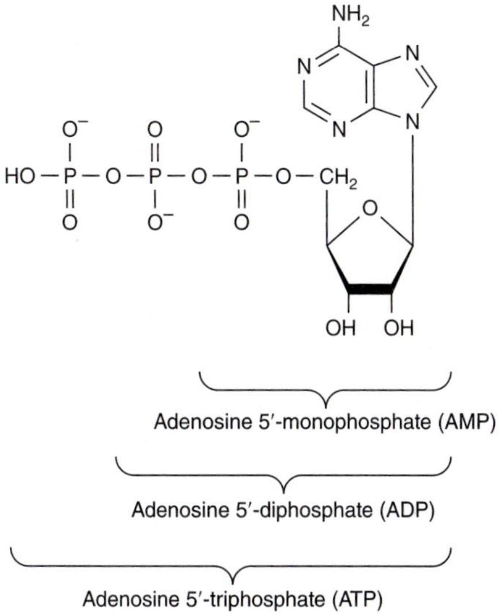

Nucleotides Are Polyfunctional Acids

The primary and secondary phosphoryl groups of nucleosides have pK_a values of about 1.0 and 6.2, respectively. Since purines and pyrimidines are neutral at physiologic pH, nucleotides bear a net negative charge. The pK_a values of the secondary phosphoryl groups are such that they can serve either as proton donors or as proton acceptors at pH values approximately two or more units above or below neutrality.

Nucleotides Absorb Ultraviolet Light

The conjugated double bonds of purine and pyrimidine derivatives absorb ultraviolet light. While their spectra are pH-dependent, at pH 7.0 all the common nucleotides absorb light at wavelengths around 260 nm. The concentration of nucleotides and nucleic acids thus often is expressed in terms of "absorbance at 260 nm." The mutagenic effect of ultraviolet light is due to its absorption by nucleotides in DNA that results in chemical modifications (see Chapter 35).

Nucleotides Serve Diverse Physiologic Functions

In addition to their roles as precursors of nucleic acids, ATP, GTP, UTP, CTP, and their derivatives each serve unique

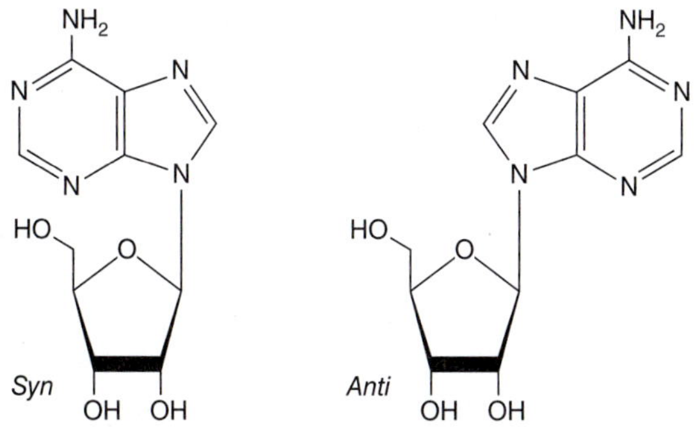

Adenosine 5′-monophosphate (AMP)

Adenosine 5′-diphosphate (ADP)

Adenosine 5′-triphosphate (ATP)

FIGURE 32–4 ATP, its diphosphate, and its monophosphate.

FIGURE 32–5 The *syn* and *anti* conformers of adenosine differ with respect to orientation about the *N*-glycosidic bond.

TABLE 32–1 Purine Bases, Ribonucleosides, and Ribonucleotides

Purine or Pyrimidine	X = H	X = Ribose	X = Ribose Phosphate
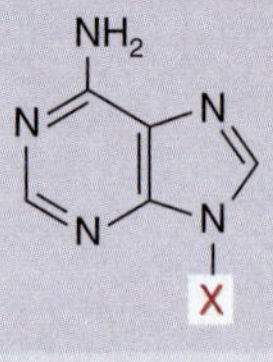	Adenine	Adenosine	Adenosine monophosphate (AMP)
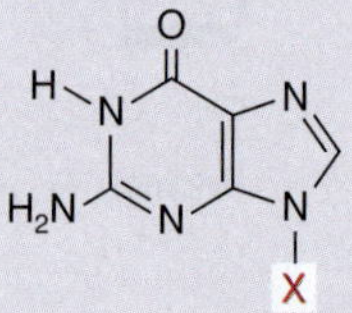	Guanine	Guanosine	Guanosine monophosphate (GMP)
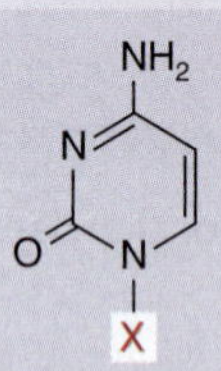	Cytosine	Cytidine	Cytidine monophosphate (CMP)
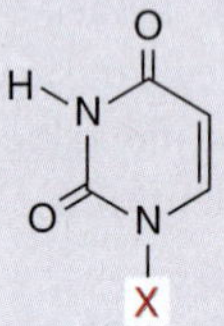	Uracil	Uridine	Uridine monophosphate (UMP)
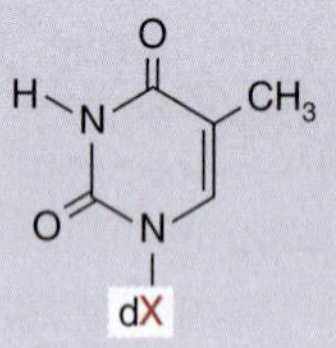	Thymine	Thymidine	Thymidine monophosphate (TMP)

physiologic functions discussed in other chapters. Selected examples include the role of ATP as the principal biologic transducer of free energy, and the second messenger cAMP (**Figure 32–10**). The mean intracellular concentration of ATP, the most abundant free nucleotide in mammalian cells, is about 1 mmol/L. The intracellular concentration of the information carrier cAMP (about 1 nmol/L) is six orders of magnitude below that of ATP. Other examples include adenosine 3′-phosphate-5′-phosphosulfate (**Figure 32–11**), the sulfate donor for sulfated proteoglycans (see Chapter 50) and

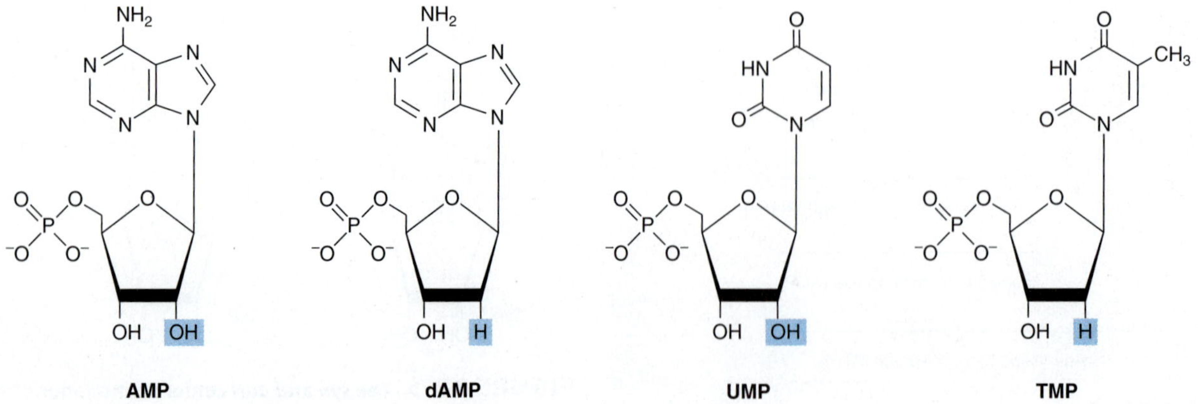

FIGURE 32–6 Structures of AMP, dAMP, UMP, and TMP.

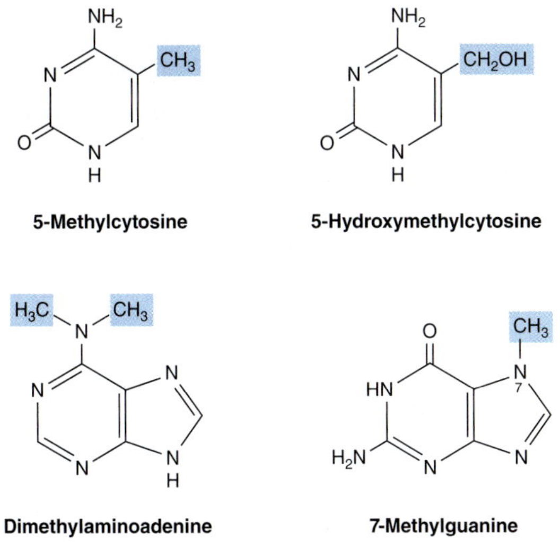

FIGURE 32–7 Four uncommon naturally occurring pyrimidines and purines.

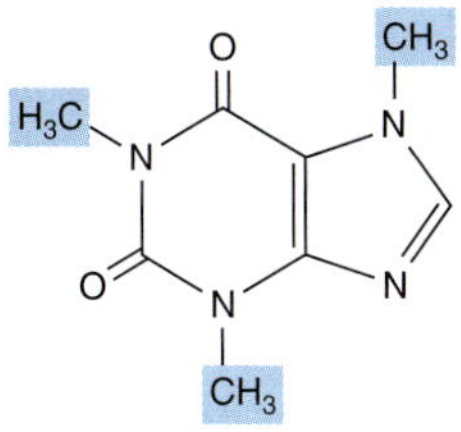

FIGURE 32–9 **Caffeine, a trimethylxanthine.** The dimethylxanthines theobromine and theophylline are similar but lack the methyl group at *N*-1 and at *N*-7, respectively.

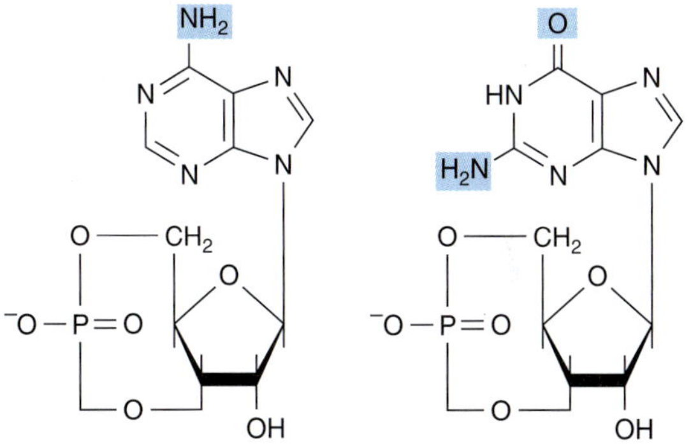

for sulfate conjugates of drugs; and the methyl group donor *S*-adenosylmethionine (**Figure 32–12**). GTP serves as an allosteric regulator and as an energy source for protein synthesis, and cGMP (Figure 32–10) serves as a second messenger in response to nitric oxide (NO) during relaxation of smooth muscle (see Chapter 51).

UDP-sugar derivatives participate in sugar epimerizations and in biosynthesis of glycogen (see Chapter 18), glucosyl disaccharides, and the oligosaccharides of glycoproteins and proteoglycans (see Chapters 46 & 50). UDP-glucuronic acid forms the urinary glucuronide conjugates of bilirubin (see Chapter 31) and of many drugs, including aspirin. CTP participates in biosynthesis of phosphoglycerides, sphingomyelin, and other substituted sphingosines (see Chapter 24). Finally, many coenzymes incorporate nucleotides as well as structures similar to purine and pyrimidine nucleotides (**Table 32–2**).

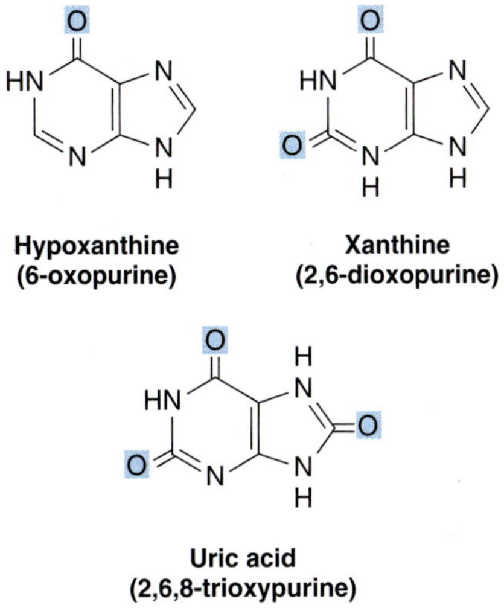

FIGURE 32–10 **cAMP, 3′,5′-cyclic AMP, and cGMP, 3′, 5′-cyclic GMP.**

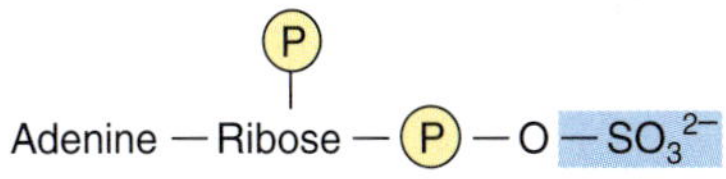

FIGURE 32–11 **Adenosine 3′-phosphate-5′-phosphosulfate.**

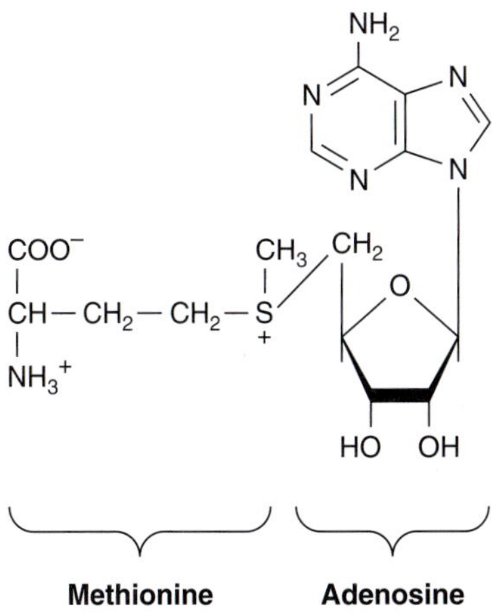

FIGURE 32–8 Structures of hypoxanthine, xanthine, and uric acid, drawn as the oxo tautomers.

FIGURE 32–12 *S*-Adenosylmethionine.

TABLE 32–2 Many Coenzymes and Related Compounds Are Derivatives of Adenosine Monophosphate

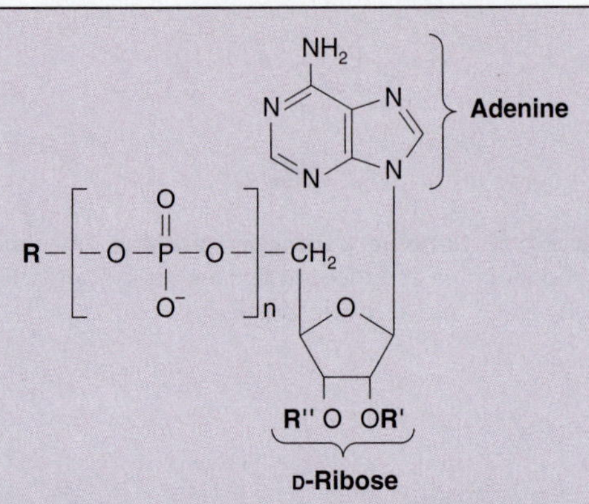

Coenzyme	R	R'	R''	n
Active methionine	Methionine[a]	H	H	0
Amino acid adenylates	Amino acid	H	H	1
Active sulfate	SO_3^{2-}	H	PO_3^{2-}	1
3',5'-Cyclic AMP		H	PO_3^{2-}	1
NAD[b]	Nicotinamide	H	H	2
NADP[b]	Nicotinamide	PO_3^{2-}	H	2
FAD	Riboflavin	H	H	2
Coenzyme A	Pantothenate	H	PO_3^{2-}	2

[a]Replaces phosphoryl group.
[b]R is a vitamin B derivative.

Nucleoside Triphosphates Have High-Group Transfer Potential

Nucleotide triphosphates have two acid anhydride bonds and one ester bond. Relative to esters, acid anhydrides have a high-group transfer potential. $\Delta G^{0\prime}$ for the hydrolysis of each of the two terminal (β and γ) phosphoryl groups of a nucleoside triphosphate is about -7 kcal/mol (-30 kJ/mol). This high-group transfer potential not only permits purine and pyrimidine nucleoside triphosphates to function as group transfer reagents, most commonly of the γ-phosphoryl group, but also on occasion transfer of a nucleotide monophosphate moiety with an accompanying release of PP_i. Cleavage of an acid anhydride bond typically is coupled with a highly endergonic process such as covalent bond synthesis, for example, the polymerization of nucleoside triphosphates to form a nucleic acid (see Chapter 34).

SYNTHETIC NUCLEOTIDE ANALOGS ARE USED IN CHEMOTHERAPY

Synthetic analogs of purines, pyrimidines, nucleosides, and nucleotides modified in the heterocyclic ring or in the sugar moiety have numerous applications in clinical medicine. Their toxic effects reflect either inhibition of enzymes essential for nucleic acid synthesis or their incorporation into nucleic acids with resulting disruption of base pairing. Oncologists employ 5-fluoro- or 5-iodouracil, 3-deoxyuridine, 6-thioguanine and 6-mercaptopurine, 5- or 6-azauridine, 5- or 6-azacytidine, and 8-azaguanine (**Figure 32–13**), which are incorporated into DNA prior to cell division. The purine analog allopurinol, used in treatment of hyperuricemia and gout, inhibits purine

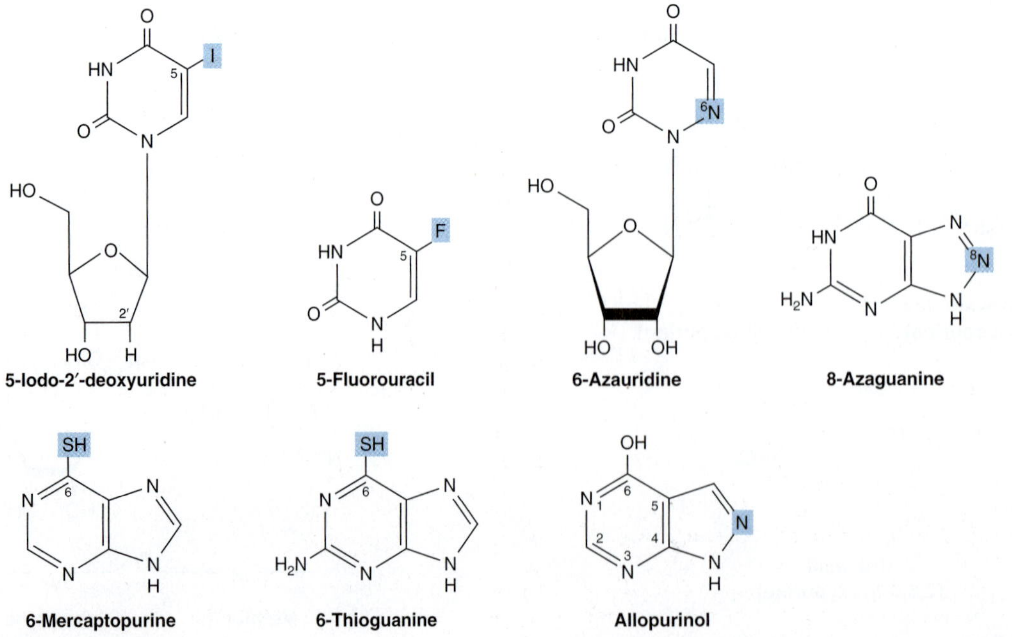

FIGURE 32–13 Selected synthetic pyrimidine and purine analogs.

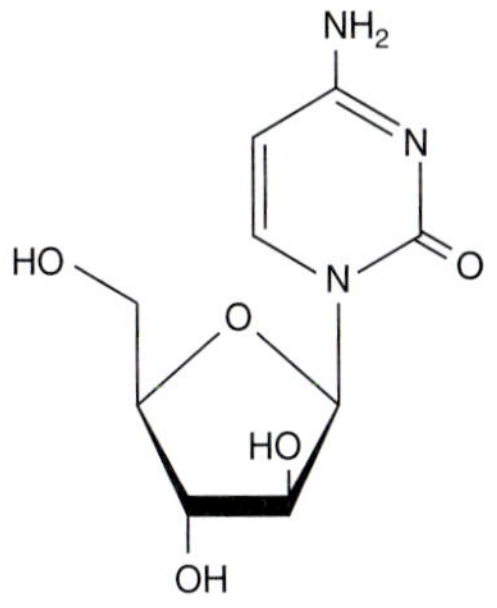

FIGURE 32–14 Cytarabine and azathioprine.

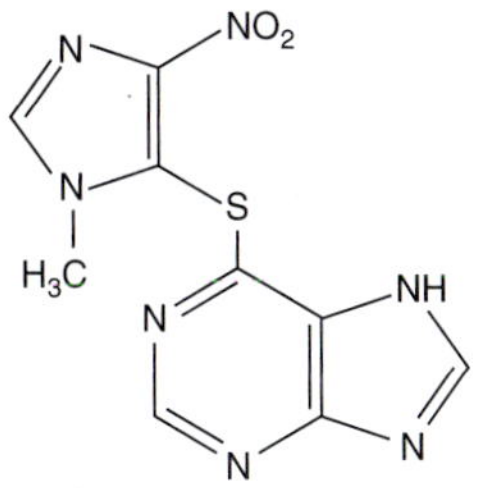

biosynthesis and xanthine oxidase activity. Cytarabine is used in chemotherapy of cancer, and azathioprine, which is catabolized to 6-mercaptopurine, is employed during organ transplantation to suppress immunologic rejection (**Figure 32–14**).

Non-Hydrolyzable Nucleoside Triphosphate Analogs Serve as Research Tools

Synthetic, non-hydrolyzable analogs of nucleoside triphosphates (**Figure 32–15**) allow investigators to distinguish the effects of nucleotides due to phosphoryl transfer from effects mediated by occupancy of allosteric nucleotide-binding sites on regulated enzymes (see Chapter 9).

DNA & RNA ARE POLYNUCLEOTIDES

The 5′-phosphoryl group of a mononucleotide can esterify a second hydroxyl group, forming a **phosphodiester**. Most commonly, this second hydroxyl group is the 3′-OH of the pentose

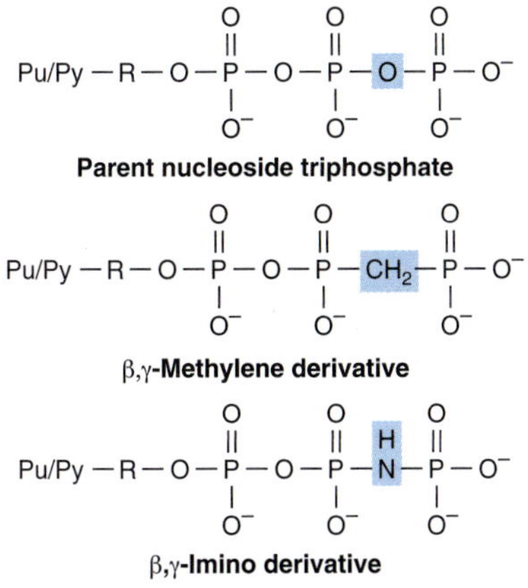

FIGURE 32–15 Synthetic derivatives of nucleoside triphosphates incapable of undergoing hydrolytic release of the terminal phosphoryl group. (Pu/Py, a purine or pyrimidine base; R, ribose or deoxyribose.) Shown are the parent (hydrolyzable) nucleoside triphosphate (**top**) and the unhydrolyzable β-methylene (**center**) and γ-imino derivatives (**bottom**).

of a second nucleotide. This forms a **dinucleotide** in which the pentose moieties are linked by a 3′,5′-phosphodiester bond to form the "backbone" of RNA and DNA. The formation of a dinucleotide may be represented as the elimination of water between two mononucleotides. Biologic formation of dinucleotides does not, however, occur in this way because the reverse reaction, hydrolysis of the phosphodiester bond, is strongly favored on thermodynamic grounds. However, despite an extremely favorable ΔG, in the absence of catalysis by **phosphodiesterases** hydrolysis of the phosphodiester bonds of DNA occurs only over long periods of time. DNA therefore persists for considerable periods, and has been detected even in fossils. RNAs are far less stable than DNA since their 2′-hydroxyl groups (absent from DNA) can function as nucleophiles for the chemical hydrolysis of 3′,5′-phosphodiester bonds.

Posttranslational modification of preformed **polynucleotides** can generate additional structures such as **pseudouridine**, a nucleoside in which D-ribose is linked to C-5 of uracil by a **carbon-to-carbon bond** rather than by the usual β-N-glycosidic bond. The nucleotide pseudouridylic acid (ψ) arises by rearrangement of a UMP of a preformed tRNA. Similarly, methylation by S-adenosylmethionine of a UMP of preformed tRNA forms TMP (thymidine monophosphate), which contains ribose rather than deoxyribose.

Polynucleotides Are Directional Macromolecules

Directional 3′ → 5′ phosphodiester bonds link the monomers of polynucleotides. Since each end of a polynucleotide thus is distinct, we refer to the "5′-end" or the "3′-end" of a polynucleotide. Since the phosphodiester bonds all are 3′ → 5′, the representation pGpGpApTpCpA indicates that the terminal 5′-hydroxyl is phosphorylated. More concisely, the representation GGATC, which shows only the base sequence, is by convention written with the 5′-base (G) at the *left* and the 3′-base (C) at the *right*.

SUMMARY

- Under physiologic conditions, the amino and oxo tautomers of purines, pyrimidines, and their derivatives predominate.

- Nucleic acids contain, in addition to A, G, C, T, and U, traces of 5-methylcytosine, 5-hydroxymethylcytosine, pseudouridine (ψ), and N-methylated heterocycles.

- Most nucleosides contain D-ribose or 2-deoxy-D-ribose linked to N-1 of a pyrimidine or to N-9 of a purine by a β-glycosidic bond whose *syn* conformers predominate.

- A primed numeral indicates the hydroxyl to which the phosphoryl group of the sugars of mononucleotides (eg, 3′-GMP, 5′-dCMP) is attached. Additional phosphoryl groups linked to the first by acid anhydride bonds form nucleoside diphosphates and triphosphates.

- Nucleoside triphosphates have high group transfer potential and participate in covalent bond syntheses. The cyclic phosphodiesters cAMP and cGMP function as intracellular second messengers.

- Mononucleotides linked by $3' \rightarrow 5'$-phosphodiester bonds form polynucleotides, directional macromolecules with distinct 3'- and 5'-ends. When represented as pTpGpT or TGCATCA, the 5'-end is at the left, and all phosphodiester bonds are $3' \rightarrow 5'$.

- Synthetic analogs of purine and pyrimidine bases and their derivatives serve as anticancer drugs either by inhibiting an enzyme of nucleotide biosynthesis or by being incorporated into DNA or RNA.

REFERENCES

Adams RLP, Knowler JT, Leader DP: *The Biochemistry of the Nucleic Acids*, 11th ed. Chapman & Hall, 1992.

Blackburn GM, Gait MJ, Loaks D, et al: *Nucleic Acids in Chemistry and Biology*, 3rd ed., RSC Publishing, 2006.

Pacher P, Nivorozhkin A, Szabo C: Therapeutic effects of xanthine oxidase inhibitors: renaissance half a century after the discovery of allopurinol. Pharmacol Rev 2006;58:87.

Metabolism of Purine & Pyrimidine Nucleotides

Victor W. Rodwell, PhD

- Compare and contrast the roles of dietary nucleic acids and of de novo biosynthesis in the production of purines and pyrimidines destined for polynucleotide biosynthesis.
- Explain why antifolate drugs and analogs of the amino acid glutamine inhibit purine biosynthesis.
- Outline the sequence of reactions that convert inosine monophosphate (IMP), first to AMP and GMP, and subsequently to their corresponding nucleoside triphosphates.
- Describe the formation from ribonucleotides of deoxyribonucleotides (dNTPs).
- Indicate the regulatory role of phosphoribosyl pyrophosphate (PRPP) in hepatic purine biosynthesis and the specific reaction of hepatic purine biosynthesis that is feedback inhibited by AMP and GMP.
- State the relevance of coordinated control of purine and pyrimidine nucleotide biosynthesis.
- Identify the reactions discussed that are inhibited by anticancer drugs.
- Write the structure of the end product of purine catabolism. Comment on its solubility and indicate its role in gout, Lesch-Nyhan syndrome, and von Gierke disease.
- Identify reactions whose impairment leads to modified pathologic signs and symptoms.
- Indicate why there are few clinically significant disorders of pyrimidine catabolism.

BIOMEDICAL IMPORTANCE

Despite a diet that may be rich in nucleoproteins, dietary purines and pyrimidines are not incorporated directly into tissue nucleic acids. Humans synthesize the nucleic acids and their derivatives ATP, NAD$^+$, coenzyme A, etc, from amphibolic intermediates. However, *injected* purine or pyrimidine analogs, including potential anticancer drugs, may nevertheless be incorporated into DNA. The biosyntheses of purine and pyrimidine ribonucleotide triphosphates (NTPs) and dNTPs are precisely regulated events. Coordinated feedback mechanisms ensure their production in appropriate quantities and at times that match varying physiologic demand (eg, cell division). Human diseases that involve abnormalities in purine metabolism include gout, Lesch-Nyhan syndrome, adenosine deaminase deficiency, and purine nucleoside phosphorylase deficiency. Diseases of pyrimidine biosynthesis are rarer, but include orotic acidurias. Unlike the low solubility of uric acid formed by catabolism of purines, the end products of pyrimidine catabolism (carbon dioxide, ammonia, β-alanine, and γ-aminoisobutyrate) are highly water soluble. One genetic disorder of pyrimidine catabolism, β-hydroxybutyric aciduria, is due to total or partial deficiency of the enzyme dihydropyrimidine dehydrogenase. This disorder of pyrimidine catabolism, also known as combined uraciluria-thyminuria, is also a disorder of β-amino acid biosynthesis, since the formation of β-alanine and of β-aminoisobutyrate is impaired. A nongenetic form can be triggered by administration of 5-fluorouracil to patients with low levels of dihydropyrimidine dehydrogenase.

PURINES & PYRIMIDINES ARE DIETARILY NONESSENTIAL

Normal human tissues can synthesize purines and pyrimidines from amphibolic intermediates in quantities and at times appropriate to meet variable physiologic demand. Ingested nucleic acids and nucleotides therefore are dietarily nonessential. Following their degradation in the intestinal tract, the resulting mononucleotides may be absorbed or converted to purine and pyrimidine bases. The purine bases are then oxidized to uric acid, which may be absorbed and excreted in the urine. While little or no dietary purine or pyrimidine is incorporated into tissue nucleic acids, *injected* compounds are incorporated. The incorporation of injected [^{3}H] thymidine into newly synthesized DNA thus can be used to measure the rate of DNA synthesis.

BIOSYNTHESIS OF PURINE NUCLEOTIDES

With the exception of parasitic protozoa, all forms of life synthesize purine and pyrimidine nucleotides. Synthesis from amphibolic intermediates proceeds at controlled rates appropriate for all cellular functions. To achieve homeostasis, intracellular mechanisms sense and regulate the pool sizes of NTPs, which rise during growth or tissue regeneration when cells are rapidly dividing.

Purine and pyrimidine nucleotides are synthesized in vivo at rates consistent with physiologic need. Early investigations of nucleotide biosynthesis first employed birds, and later *Escherichia coli*. Isotopic precursors of uric acid fed to pigeons established the source of each atom of a purine (**Figure 33–1**) and initiated study of the intermediates of purine biosynthesis.

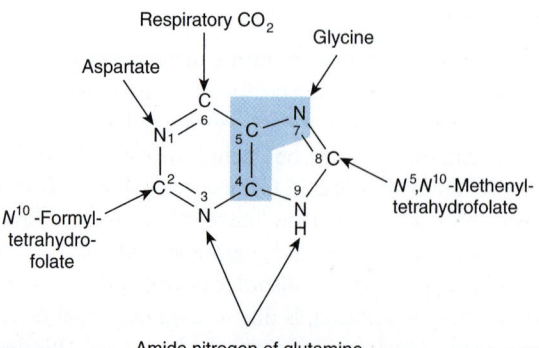

FIGURE 33–1   **Sources of the nitrogen and carbon atoms of the purine ring.** Atoms 4, 5, and 7 (**blue highlight**) derive from glycine.

Avian tissues also served as a source of cloned genes that encode enzymes of purine biosynthesis and the regulatory proteins that control the rate of purine biosynthesis.

The three processes that contribute to purine nucleotide biosynthesis are, in order of decreasing importance:

1. Synthesis from amphibolic intermediates (synthesis de novo)
2. Phosphoribosylation of purines
3. Phosphorylation of purine nucleosides

INOSINE MONOPHOSPHATE (IMP) IS SYNTHESIZED FROM AMPHIBOLIC INTERMEDIATES

The initial reaction of purine biosynthesis, transfer of two phosphoryl groups from ATP to carbon 1 of ribose 5-phosphate forming phosphoribosyl pyrophosphate (PRPP), is catalyzed by PRPP synthetase, EC 2.7.6.1. The end product of the ten subsequent enzyme-catalyzed reactions is IMP (**Figure 33–2**).

Following synthesis of IMP, separate branches lead to AMP and GMP (**Figure 33–3**). Subsequent phosphoryl transfer from ATP converts AMP and GMP to ADP and GDP, respectively. Conversion of GDP to GTP involves a second phosphoryl transfer from ATP, whereas conversion of ADP to ATP is achieved primarily by oxidative phosphorylation (see Chapter 13).

Multifunctional Catalysts Participate in Purine Nucleotide Biosynthesis

In prokaryotes, each reaction of Figure 33–2 is catalyzed by a different polypeptide. By contrast, the enzymes of eukaryotes are polypeptides that possess multiple catalytic activities whose adjacent catalytic sites facilitate channeling of intermediates between sites. Three distinct multifunctional enzymes catalyze reactions ③, ④, and ⑥; reactions ⑦ and ⑧; and reactions ⑩ and ⑪ of Figure 33–2.

Antifolate Drugs & Glutamine Analogs Block Purine Nucleotide Biosynthesis

The carbons added in reactions ④ and ⑩ of Figure 33–2 are contributed by derivatives of tetrahydrofolate. Purine deficiency states, while rare in humans, generally reflect a deficiency of folic acid. Compounds that inhibit formation of tetrahydrofolates and therefore block purine synthesis have been used in cancer chemotherapy. Inhibitory compounds and the reactions they inhibit include **azaserine** (reaction ⑤, Figure 33–2), **diazanorleucine** (reaction ②, Figure 33–2),

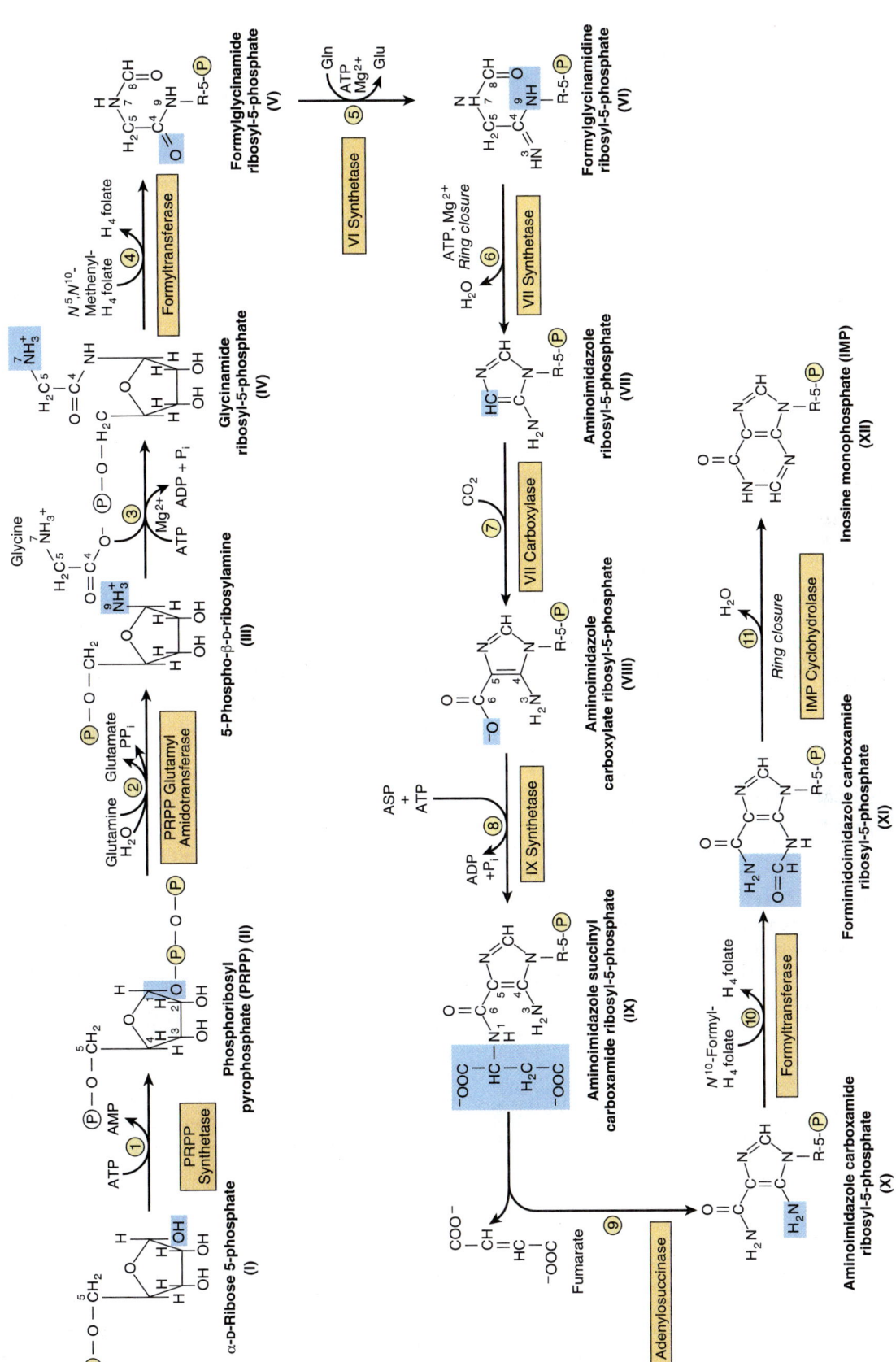

FIGURE 33-2 **Purine biosynthesis from ribose 5-phosphate and ATP.** See the text for explanations. (Ⓟ, PO_3^{2-} or PO_2^-.)

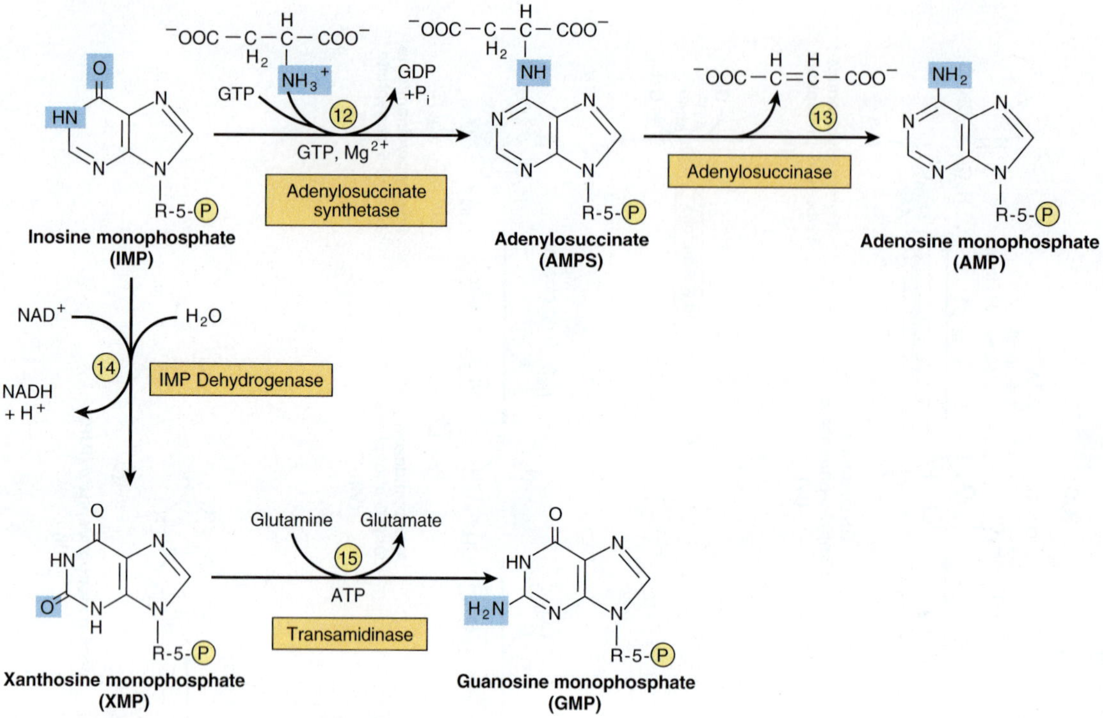

FIGURE 33–3 Conversion of IMP to AMP and GMP.

6-mercaptopurine (reactions ⑬ and ⑭, Figure 33–3), and **mycophenolic acid** (reaction ⑭, Figure 33–3).

"SALVAGE REACTIONS" CONVERT PURINES & THEIR NUCLEOSIDES TO MONONUCLEOTIDES

Conversion of purines, their ribonucleo**sides**, and their deoxyribonucleo**sides** to mononucleo**tides** involves "salvage reactions" that require far less energy than de novo synthesis. The more important mechanism involves phosphoribosylation by PRPP (structure II, Figure 33–2) of a free purine (Pu) to form a purine 5′-mononucleotide (Pu-RP):

$$Pu + PR\text{–}PP \rightarrow Pu\text{–}RP + PP_i$$

Phosphoryl transfer from PRPP catalyzed by adenosine- and hypoxanthine-phosphoribosyl transferases (EC 2.4.2.7 & EC 2.4.2.8, respectively), converts adenine, hypoxanthine, and guanine to their mononucleotides (**Figure 33–4**).

A second salvage mechanism involves phosphoryl transfer from ATP to a purine ribonucleo**side** (Pu-R):

$$Pu\text{-}R + ATP \rightarrow PuR\text{-}P + ADP$$

Phosphorylation of the purine nucleotides, catalyzed by adenosine kinase (EC 2.7.1.20), converts adenosine and deoxyadenosine to AMP and dAMP. Similarly, deoxycytidine kinase (EC 2.7.1.24) phosphorylates deoxycytidine and 2′-deoxyguanosine, forming dCMP and dGMP, respectively.

Liver, the major site of purine nucleotide biosynthesis, provides purines and purine nucleosides for salvage and for utilization by tissues incapable of their biosynthesis. Human brain tissue has a low level of PRPP glutamyl amidotransferase, EC 2.4.2.14 (reaction ②, Figure 33–2) and hence depends in part on exogenous purines. Erythrocytes and polymorphonuclear leukocytes cannot synthesize 5-phosphoribosylamine (structure III, Figure 33–2), and therefore also utilize exogenous purines to form nucleotides.

HEPATIC PURINE BIOSYNTHESIS IS STRINGENTLY REGULATED

AMP & GMP Feedback Regulate PRPP Glutamyl Amidotransferase

Biosynthesis of IMP is energetically expensive. In addition to ATP, glycine, glutamine, aspartate, and reduced tetrahydrofolate derivatives all are consumed. Thus, it is of survival advantage to closely regulate purine biosynthesis in response to varying physiologic need. The overall determinant of the rate of de novo purine nucleotide biosynthesis is the concentration of PRPP. This, in turn, depends on the rate of PRPP synthesis, utilization, degradation, and regulation. The rate of PRPP synthesis depends on the availability of ribose 5-phosphate and on the activity of PRPP synthetase, EC 2.7.6.1 (reaction ② **Figure 33–5**), an enzyme whose activity is feedback inhibited by AMP, ADP, GMP, and GDP. Elevated levels of these nucleoside phosphates thus signal a physiologically appropriate overall decrease in their biosynthesis.

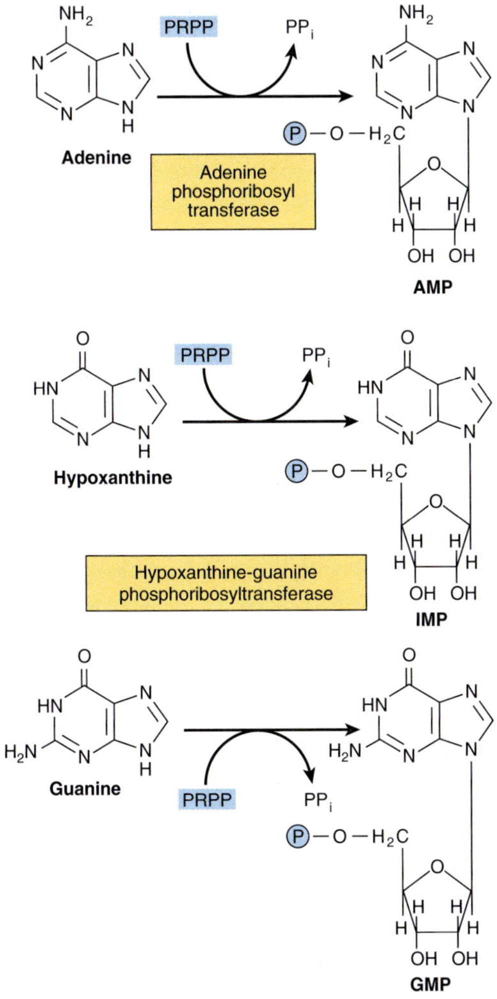

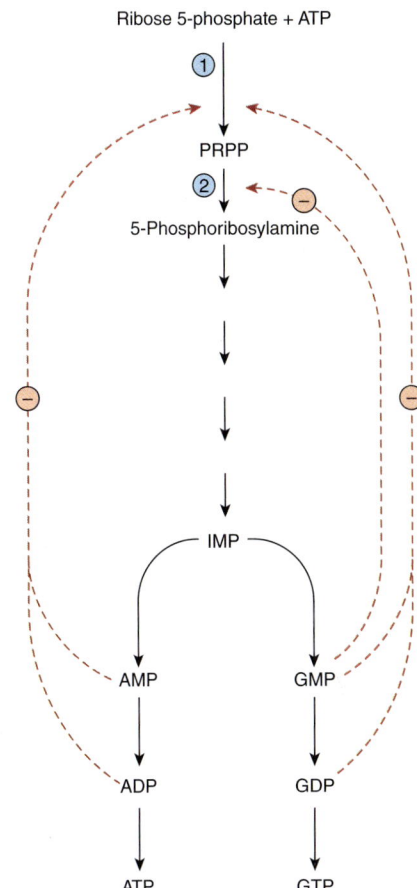

FIGURE 33–5 **Control of the rate of de novo purine nucleotide biosynthesis.** Reactions ① and ② are catalyzed by PRPP synthetase and by PRPP glutamyl amidotransferase, respectively. Solid lines represent chemical flow. Broken red lines represent feedback inhibition by intermediates of the pathway.

FIGURE 33–4 **Phosphoribosylation of adenine, hypoxanthine, and guanine to form AMP, IMP, and GMP, respectively.**

AMP & GMP Feedback Regulate Their Formation From IMP

In addition to regulation at the level of PRPP biosynthesis, additional mechanisms that regulate conversion of IMP to ATP and GTP are summarized in **Figure 33–6**. AMP feedback inhibits adenylosuccinate synthetase, EC 6.3.4.4 (reaction ⑫, Figure 33–3), and GMP inhibits IMP dehydrogenase, EC 1.1.1.205 (reaction ⑭, Figure 33–3). Furthermore, conversion of IMP to adenylosuccinate en route to AMP (reaction ⑫, Figure 33–3) requires GTP, and conversion of xanthinylate (XMP) to GMP requires ATP. This cross-regulation between the pathways of IMP metabolism thus serves to balance the biosynthesis of purine nucleoside triphosphates by decreasing the synthesis of one purine nucleotide when there is a deficiency of the other nucleotide. AMP and GMP also inhibit hypoxanthine-guanine phosphoribosyltransferase, which converts hypoxanthine and guanine to IMP and GMP (Figure 33–4), while GMP also feedback inhibits PRPP glutamyl amidotransferase (reaction ②, Figure 33–2).

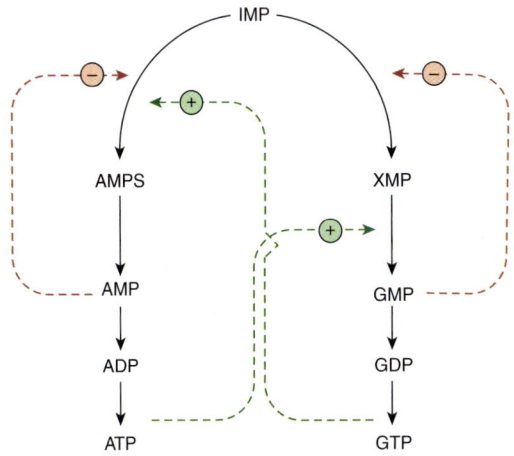

FIGURE 33–6 **Regulation of the conversion of IMP to adenosine nucleotides and guanosine nucleotides.** Solid lines represent chemical flow. Broken green lines represent positive feedback loops ⊕, and broken red lines represent negative feedback loops ⊖. (AMPS, adenylosuccinate; XMP, xanthosine monophosphate; their structures are given in Figure 33–3.)

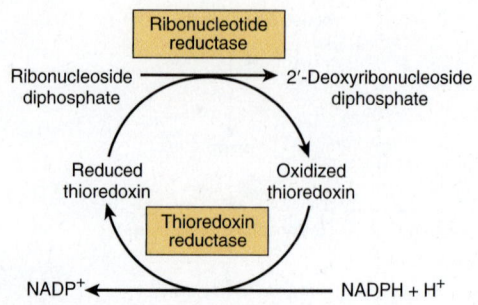

FIGURE 33–7 Reduction of ribonucleoside diphosphates to 2′-deoxyribonucleoside diphosphates.

REDUCTION OF RIBONUCLEOSIDE DIPHOSPHATES FORMS DEOXYRIBONUCLEOSIDE DIPHOSPHATES

Reduction of the 2′-hydroxyl of purine and pyrimidine ribonucleotides, catalyzed by the complex that includes **ribonucleotide reductase**, EC 1.17.4.1 (**Figure 33–7**), provides the deoxyribonucleoside diphosphates (dNDPs) needed for both the synthesis and repair of DNA (see Chapter 35). The enzyme complex is functional only when cells are actively synthesizing DNA. Reduction requires reduced thioredoxin, thioredoxin reductase (EC 1.8.1.9), and NADPH. The immediate reductant, reduced thioredoxin, is produced by NADPH-dependent reduction of oxidized thioredoxin (Figure 33–7). The reduction of ribonucleoside diphosphates (NDPs) to dNDPs is subject to complex regulatory controls that achieve balanced production of dNTPs for synthesis of DNA (**Figure 33–8**).

BIOSYNTHESIS OF PYRIMIDINE NUCLEOTIDES

Figure 33–9 illustrates the intermediates and enzymes of pyrimidine nucleotide biosynthesis. The catalyst for the initial reaction is *cytosolic* carbamoyl phosphate synthetase II (EC 6.3.5.5), a different enzyme from the *mitochondrial* carbamoyl phosphate synthetase I of urea synthesis (see Figure 28–16). Compartmentation thus provides an independent pool of carbamoyl phosphate for each process. Unlike in purine biosynthesis where PRPP serves as a scaffold for assembly of the purine ring (Figure 33–2), PRPP participates in pyrimidine biosynthesis only subsequent to assembly of the pyrimidine ring. As for the biosynthesis of pyrimidines, purine nucleoside biosynthesis is energetically costly.

Multifunctional Proteins Catalyze the Early Reactions of Pyrimidine Biosynthesis

Five of the first six enzyme activities of pyrimidine biosynthesis reside on **multifunctional polypeptides**. CAD, a single

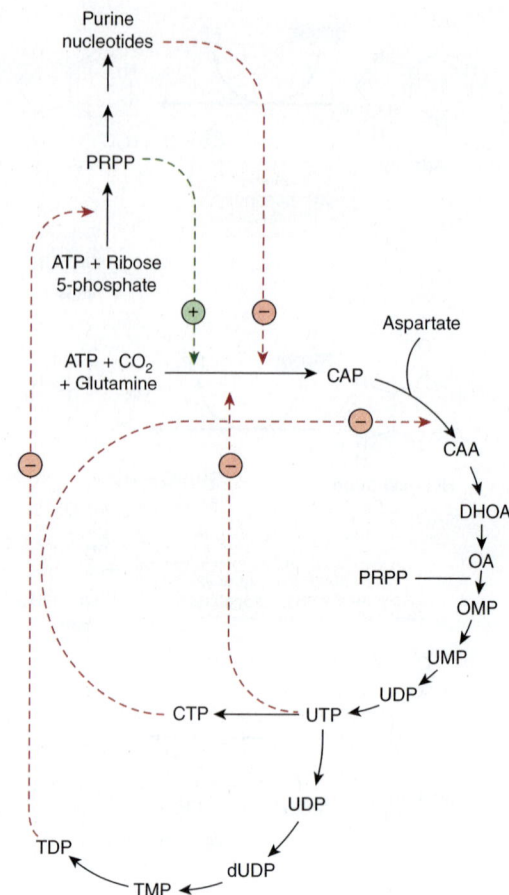

FIGURE 33–8 Regulatory aspects of the biosynthesis of purine and pyrimidine ribonucleotides and reduction to their respective 2′-deoxyribonucleotides. The broken green line represents a positive feedback loop. Broken red lines represent negative feedback loops. Abbreviations are provided for the intermediates in the biosynthesis of pyrimidine nucleotides whose structures are given in Figure 33–9. (CAA, carbamoyl aspartate; DHOA, dihydroorotate; OA, orotic acid; OMP, orotidine monophosphate; PRPP phosphoribosyl pyrophosphate.)

polypeptide named for the the first letters of its enzyme activities, catalyzes the first three reactions of Figure 33–9. A second bifunctional enzyme catalyzes reactions ⑤ and ⑥ of Figure 33–9. The close proximity of multiple active sites on a multifunctional polypeptide facilitates efficient channeling of the intermediates of pyrimidine biosynthesis.

THE DEOXYRIBONUCLEOSIDES OF URACIL & CYTOSINE ARE SALVAGED

Adenine, guanine, and hypoxanthine released during the turnover of nucleic acids, notably messenger RNAs, are reconverted to nucleoside triphosphates via so-called **salvage pathways**. While mammalian cells reutilize few *free* pyrimidines, "salvage reactions" convert the pyrimidine ribonucleo*sides*

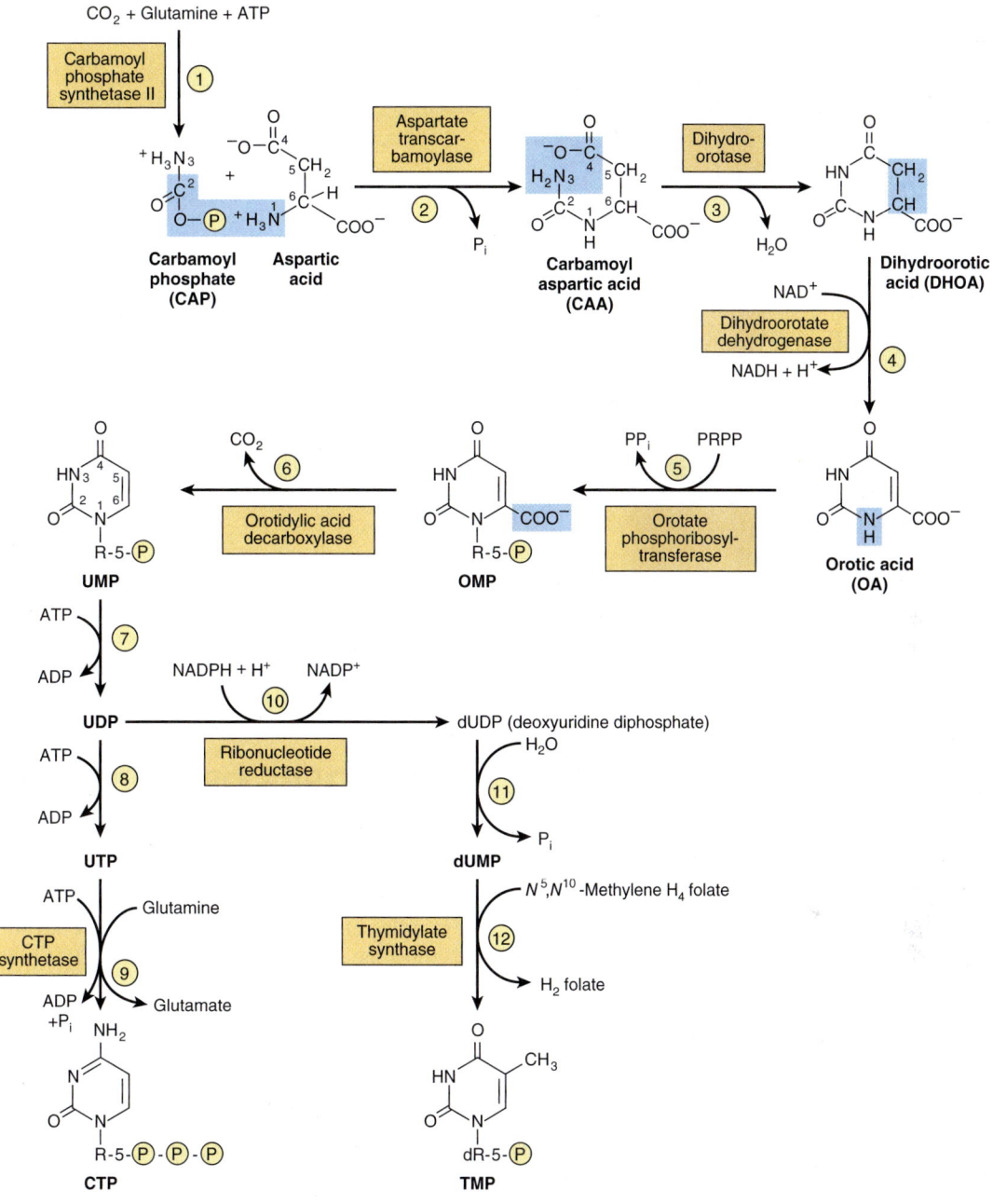

FIGURE 33–9 The biosynthetic pathway for pyrimidine nucleotides.

uridine and cytidine and the pyrimidine deoxyribonucleo*sides* thymidine and deoxycytidine to their respective nucleo*tides*.

$$\text{Guanine} + \text{PRPP} \rightarrow \text{GMP} + \text{PP}_i$$
$$\text{Hypoxanthine} + \text{PRPP} \rightarrow \text{IMP} + \text{PP}_i$$

Phosphoryltransferases (kinases) catalyze transfer of the γ-phosphoryl group of ATP to the diphosphates of the dNDPs 2′-deoxycytidine, 2′-deoxyguanosine, and 2′-deoxyadenosine, converting them to the corresponding nucleoside triphosphates.

$$\text{NDP} + \text{ATP} \rightarrow \text{NTP} + \text{ADP}$$
$$\text{dNDP} + \text{ATP} \rightarrow \text{dNTP} + \text{ADP}$$

Methotrexate Blocks Reduction of Dihydrofolate

The reaction catalyzed by thymidylate synthase, EC 2.1.1.45 (reaction ⑫, Figure 33–9) is the only reaction of pyrimidine nucleotide biosynthesis that requires a tetrahydrofolate derivative. During this reaction, the methylene group of N^5,N^{10}-methylene-tetrahydrofolate is reduced to the methyl group that is transferred to the 5-position of the pyrimidine ring, as well as dihydrofolate, the oxidized form of tetrahydrofolate. For further pyrimidine synthesis to occur, dihydrofolate must be reduced back to tetrahydrofolate. This reduction, catalyzed by dihydrofolate reductase (EC 1.5.1.3), is inhibited by **methotrexate**. Dividing cells, which must generate TMP and dihydrofolate, are especially sensitive to inhibitors of dihydrofolate reductase such as the anticancer drug methotrexate.

Certain Pyrimidine Analogs Are Substrates for Enzymes of Pyrimidine Nucleotide Biosynthesis

Allopurinol and the anticancer drug **5-fluorouracil** (see Figure 32–13) are alternate substrates for orotate phosphoribosyltransferase, EC 2.4.2.10 (reaction ⑤, Figure 33–9). Both drugs are phosphoribosylated, and allopurinol is converted to a nucleotide in which the ribosyl phosphate is attached to N^1 of the pyrimidine ring.

REGULATION OF PYRIMIDINE NUCLEOTIDE BIOSYNTHESIS

Gene Expression & Enzyme Activity Both Are Regulated

CAD represents the primary focus for regulation of pyrimidine biosynthesis. Expression of the CAD gene is regulated at the level both of transcription and of translation. At the level of enzyme activity, the carbamoyl phosphate synthetase II (CPS) activity of CAD is activated by PRPP and is feedback inhibited by UTP. The effect of UTP is, however, abolished by phosphorylation of serine 1406 of CAD.

Purine & Pyrimidine Nucleotide Biosynthesis Are Coordinately Regulated

Purine and pyrimidine biosynthesis parallel one another quantitatively, that is, mole for mole, suggesting coordinated control of their biosynthesis. Several sites of *cross-regulation* characterize the pathways that lead to the biosynthesis of purine and pyrimidine nucleotides. PRPP synthetase (reaction ①, Figure 33–2), which forms a precursor essential for both processes, is feedback inhibited by both purine and pyrimidine nucleotides, as is the conversion of both pyrimidine and purine nucleotides NDPs to NTPs (**Figure 33–10**).

HUMANS CATABOLIZE PURINES TO URIC ACID

Humans convert adenosine and guanosine to uric acid (**Figure 33–11**). Adenosine is first converted to inosine by adenosine deaminase, EC 3.5.4.4. In mammals other than

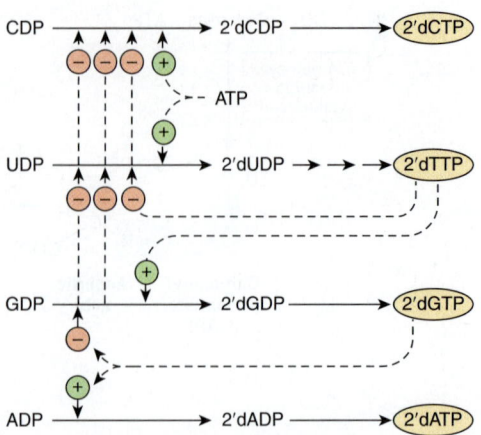

FIGURE 33–10 **Regulation of the conversion of purine and pyrimidine NDPs to NTPs.** Solid lines represent chemical flow. Broken lines indicate targets of positive ⊕ or negative ⊖ feedback inhibition.

higher primates, uricase (EC 1.7.3.3) converts uric acid to the water-soluble product allantoin. However, since humans lack uricase, the end product of purine catabolism in humans is uric acid.

DISORDERS OF PURINE METABOLISM

Various genetic defects in PRPP synthetase (reaction ①, Figure 33–2) present clinically as gout. Each defect—for example, an elevated V_{max}, increased affinity for ribose 5-phosphate, or resistance to feedback inhibition—results in overproduction and overexcretion of purine catabolites. When serum urate levels exceed the solubility limit, sodium urate crystalizes in soft tissues and joints where it causes an inflammatory reaction, **gouty arthritis**. However, most cases of gout reflect abnormalities in renal handling of uric acid.

While purine deficiency states are rare in human subjects, there are numerous genetic disorders of purine catabolism. **Hyperuricemias** may be differentiated based on whether patients excrete normal or excessive quantities of total urates. Some hyperuricemias reflect specific enzyme defects. Others are secondary to diseases such as cancer or psoriasis that enhance tissue turnover.

Lesch-Nyhan Syndrome

The Lesch-Nyhan syndrome, an overproduction hyperuricemia characterized by frequent episodes of uric acid lithiasis and a bizarre syndrome of self-mutilation, reflects a defect in **hypoxanthine-guanine phosphoribosyl transferase**, an enzyme of purine salvage (Figure 33–4). The accompanying rise in intracellular PRPP results in purine overproduction. Mutations that decrease or abolish hypoxanthine-guanine phosphoribosyltransferase activity include deletions, frameshift mutations, base substitutions, and aberrant mRNA splicing.

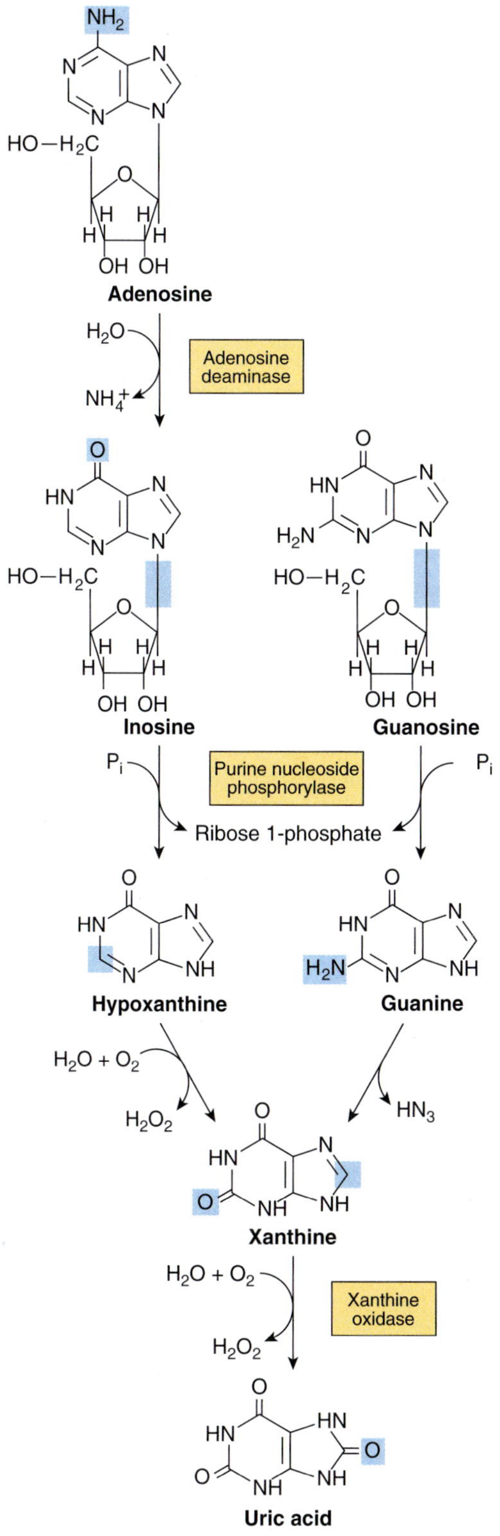

von Gierke Disease

Purine overproduction and hyperuricemia in von Gierke disease (**glucose-6-phosphatase deficiency**) occurs secondary to enhanced generation of the PRPP precursor ribose 5-phosphate. An associated lactic acidosis elevates the renal threshold for urate, elevating total body urates.

Hypouricemia

Hypouricemia and increased excretion of hypoxanthine and xanthine are associated with a deficiency in **xanthine oxidase**, EC 1.17.3.2 (Figure 33–11) due to a genetic defect or to severe liver damage. Patients with a severe enzyme deficiency may exhibit xanthinuria and xanthine lithiasis.

Adenosine Deaminase & Purine Nucleoside Phosphorylase Deficiency

Adenosine deaminase deficiency (Figure 33–11) is associated with an immunodeficiency disease in which both thymus-derived lymphocytes (T cells) and bone marrow–derived lymphocytes (B cells) are sparse and dysfunctional. Patients suffer from severe immunodeficiency. In the absence of enzyme replacement or bone marrow transplantation, infants often succumb to fatal infections. Defective activity of **purine nucleoside phosphorylase** (EC 2.4.2.1) is associated with a severe deficiency of T cells, but apparently normal B-cell function. Immune dysfunctions appear to result from accumulation of dGTP and dATP, which inhibit ribonucleotide reductase and thereby deplete cells of DNA precursors. **Table 33–1** summarizes known disorders of purine metabolism.

PYRIMIDINE CATABOLITES ARE WATER SOLUBLE

Unlike the low solubility products of purine catabolism, catabolism of the pyrimidines forms highly water-soluble products—CO_2, NH_3, β-alanine, and β-aminoisobutyrate (**Figure 33–12**). Excretion of β-aminoisobutyrate increases in leukemia and severe x-ray radiation exposure due to increased destruction of DNA. However, many persons of Chinese or Japanese ancestry routinely excrete β-aminoisobutyrate.

Disorders of β-alanine and β-aminoisobutyrate metabolism arise from defects in enzymes of pyrimidine catabolism. These include **β-hydroxybutyric aciduria**, a disorder due to total or partial deficiency of the enzyme **dihydropyrimidine dehydrogenase**, EC 1.3.1.2 (Figure 33–12). The genetic disease reflects an absence of the enzyme. A disorder of

FIGURE 33–11 Formation of uric acid from purine nucleosides by way of the purine bases hypoxanthine, xanthine, and guanine. Purine deoxyribonucleosides are degraded by the same catabolic pathway and enzymes, all of which exist in the mucosa of the mammalian gastrointestinal tract.

TABLE 33–1 Metabolic Disorders of Purine & Pyrimidine Matabolism

Defective Enzyme	Enzyme Catalog Number	OMIM Reference	Major Signs and Symptoms	Figure and Reaction
Purine Metabolism				
Hypoxanthine-guanine phosphoribosyl transferase	2.4.2.8	308000	Lesch-Nyhan syndrome. Uricemia, self-mutilation	33-4 ②
PRPP synthase	2.7.6.1	311860	Gout; gouty arthritis	33-2 ①
Adenosine deaminase	3.5.4.6	102700	Severely compromised immune system	33-1 ①
Purine nucleoside phosphorylase	2.4.2.1	164050	Autoimmune disorders; benign and opportunistic infections	33-11 ②
Pyrimidine Metabolism				
Dihydropyrimidine dehydrogenase	1.3.1.2	274270	Can develop toxicity to 5-fluorouracil, also a substrate for this dehydrogenase	33-12 ②
Orotate phosphoribosyl transferase and orotidylic acid decarboxylase	2.4.2.10 and 4.1.1.23	258900	Orotic acid aciduria type 1; megaloblastic anemia	33-9 ⑤ and ⑥
Orotidylic acid decarboxylase	4.1.1.23	258920	Orotic acid aciduria type 2	33-9 ⑥

pyrimidine catabolism, known also as combined uraciluria-thyminuria, is also a disorder of β-amino acid metabolism, since the *formation* of β-alanine and of β-aminoisobutyrate is impaired. When caused by a genetic error, there are serious neurologic complications. A nongenetic form is triggered by the administration of the anticancer drug 5-fluorouracil (see Figure 32–13) to patients with low levels of dihydropyrimidine dehydrogenase.

Pseudouridine Is Excreted Unchanged

No human enzyme catalyzes hydrolysis or phosphorolysis of the pseudouridine (ψ) derived from the degradation of RNA molecules. This unusual nucleotide therefore is excreted unchanged in the urine of normal subjects. Pseudouridine was indeed first isolated from human urine (**Figure 33–13**).

OVERPRODUCTION OF PYRIMIDINE CATABOLITES

Since the end products of pyrimidine catabolism are highly water soluble, pyrimidine overproduction results in few clinical signs or symptoms. Table 33–1 lists exceptions. In hyperuricemia associated with severe overproduction of PRPP, there is overproduction of pyrimidine nucleotides and increased excretion of β-alanine. Since N^5,N^{10}-methylene-tetrahydrofolate is required for thymidylate synthesis, disorders of folate and vitamin B_{12} metabolism result in deficiencies of TMP.

Orotic Aciduria

The orotic aciduria that accompanies the **Reye syndrome** probably is a consequence of the inability of severely damaged mitochondria to utilize carbamoyl phosphate, which then becomes available for cytosolic overproduction of orotic acid. **Type I orotic aciduria** reflects a deficiency of both orotate phosphoribosyltransferase (EC 2.1.3.3) and orotidylate decarboxylase, EC 4.1.1.23 (reactions ⑤ and ⑥, Figure 33–9). The rarer **Type II orotic aciduria** is due to a deficiency only of orotidylate decarboxylase (reaction ⑥, Figure 33–9).

Deficiency of a Urea Cycle Enzyme Results in Excretion of Pyrimidine Precursors

Increased excretion of orotic acid, uracil, and uridine accompanies a deficiency in liver mitochondrial ornithine transcarbamoylase (see reaction ②, Figure 28–16). Excess carbamoyl phosphate exits to the cytosol, where it stimulates pyrimidine nucleotide biosynthesis. The resulting mild **orotic aciduria** is increased by high-nitrogen foods.

Drugs May Precipitate Orotic Aciduria

Allopurinol (see Figure 32–13), an alternative substrate for orotate phosphoribosyltransferase (reaction ⑤, Figure 33–9), competes with orotic acid. The resulting nucleotide product also inhibits orotidylate decarboxylase (reaction ⑥, Figure 33–9), resulting in **orotic aciduria** and **orotidinuria**. 6-Azauridine, following conversion to 6-azauridylate, also competitively inhibits orotidylate decarboxylase (reaction ⑥, Figure 33–9), enhancing excretion of orotic acid and orotidine. Four genes that encode urate transporters have been identified. Two of the encoded proteins are localized to the apical membrane of proximal tubular cells.

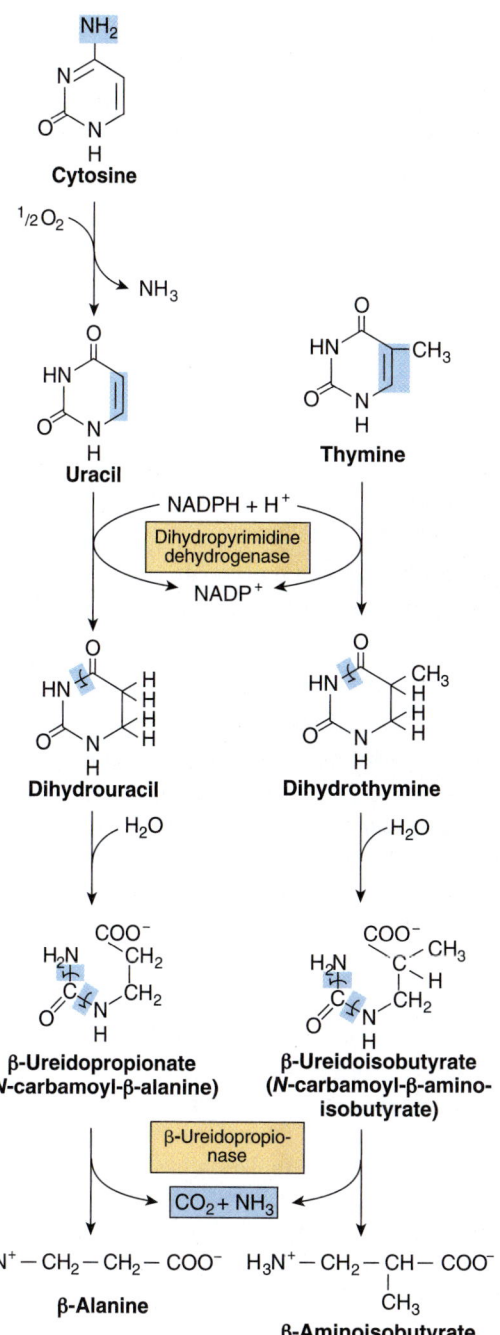

FIGURE 33–12 **Catabolism of pyrimidines.** Hepatic β-ureidopropionase catalyzes the formation of both β-alanine and β-aminoisobutyrate from their pyrimidine precursors.

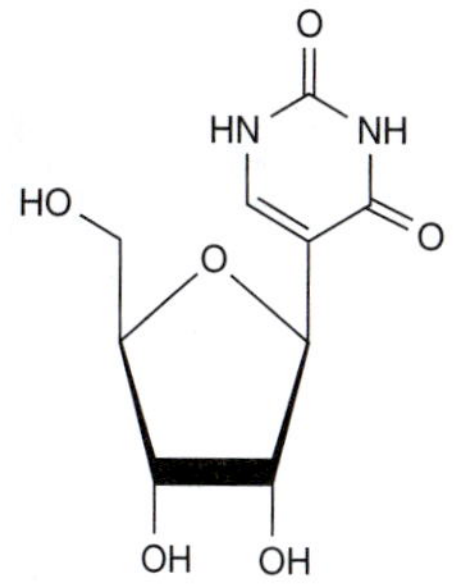

FIGURE 33–13 **Pseudouridine, in which ribose is linked to C5 of uridine.**

- IMP is a precursor both of AMP and GMP. Glutamine provides the 2-amino group of GMP, and aspartate the 6-amino group of AMP.

- Phosphoryl transfer from ATP converts AMP and GMP to ADP and GDP. A second phosphoryl transfer from ATP forms GTP, but ADP is converted to ATP primarily by oxidative phosphorylation.

- Hepatic purine nucleotide biosynthesis is stringently regulated by the pool size of PRPP and by feedback inhibition of PRPP glutamyl amidotransferase by AMP and GMP.

- Coordinated regulation of purine and pyrimidine nucleotide biosynthesis ensures their presence in proportions appropriate for nucleic acid biosynthesis and other metabolic needs.

- Humans catabolize purines to uric acid (pK_a 5.8), present as the relatively insoluble acid at acidic pH or as its more soluble sodium urate salt at a pH near neutrality. Urate crystals are diagnostic of gout. Other disorders of purine catabolism include Lesch-Nyhan syndrome, von Gierke disease, and hypouricemias.

- Since pyrimidine catabolites are water soluble, their overproduction does not result in clinical abnormalities. Excretion of pyrimidine precursors can, however, result from a deficiency of ornithine transcarbamoylase because excess carbamoyl phosphate is available for pyrimidine biosynthesis.

REFERENCES

Brassier A, Ottolenghi C, Boutron A, et al: Dihydrolipoamide dehydrogenase deficiency: a still overlooked cause of recurrent acute liver failure and Reye-like syndrome. Mol Genet Metab 2013;109:28.

Fu R, Jinnah HA: Genotype-phenotype correlations in Lesch-Nyhan disease: moving beyond the gene. J Biol Chem 2012;287:2997.

Fu W, Li Q, Yao J, et al: Protein expression of urate transporters in renal tissue of patients with uric acid nephrolithiasis. Cell Biochem Biophys 2014;70:449.

Moyer RA, John DS: Acute gout precipitated by total parenteral nutrition. J Rheumatol 2003;30:849.

Uehara I, Kimura T, Tanigaki S, et al: Paracellular route is the major urate transport pathway across the blood-placental barrier. Physiol Rep 2014;20:2.

Wu VC, Huang JW, Hsueh PR, et al: Renal hypouricemia is an ominous sign in patients with severe acute respiratory syndrome. Am J Kidney Dis 2005;45:88.

SUMMARY

- Degradation of ingested nucleic acids yields free purines and pyrimidines. Purines and pyrimidines are formed from amphibolic intermediates and thus are dietarily nonessential.

- Several reactions of IMP biosynthesis require folate derivatives and glutamine. Consequently, antifolate drugs and glutamine analogs inhibit purine biosynthesis.

Nucleic Acid Structure & Function

P. Anthony Weil, PhD

O B J E C T I V E S

After studying this chapter, you should be able to:

- Understand the chemical monomeric and polymeric structure of the genetic material, deoxyribonucleic acid, or DNA, which is found primarily within the nucleus of eukaryotic cells, but also in organelles.
- Explain why genomic DNA is double stranded and highly negatively charged.
- Understand the outline of how the genetic information of DNA can be faithfully duplicated via the process of DNA replication.
- Describe how the genetic information of DNA is transcribed, or copied, into myriad, distinct forms of ribonucleic acid (RNA).
- Appreciate that one form of information-rich RNA, the so-called messenger RNA (mRNA), is processed posttranscriptionally, shuttled to the cytoplasm, and then translated into proteins, the molecules that form the structures, shapes, and ultimately functions of individual cells, tissues, and organs.

BIOMEDICAL IMPORTANCE

The discovery that genetic information is coded along the length of a polymeric molecule composed of only four types of monomeric units was one of the major scientific achievements of the 20th century. This polymeric molecule, **deoxyribonucleic acid (DNA)**, is the chemical basis of heredity and is organized into genes, the fundamental units of genetic information. The basic information pathway—that is, DNA, which directs the synthesis of RNA, which in turn both directs and regulates protein synthesis—has been elucidated. Genes do not function autonomously; rather their replication and function are controlled by various gene products, primarily proteins often in collaboration with components of various signal transduction pathways. Knowledge of the structure and function of nucleic acids is essential in understanding genetics and many aspects of pathophysiology as well as the genetic basis of disease.

DNA CONTAINS THE GENETIC INFORMATION

The demonstration that DNA contained the genetic information was first made in 1944 in a series of classic experiments by Avery, MacLeod, and McCarty. They showed that the genetic determination of the character (type) of the capsule of a specific pneumococcus bacterium could be transmitted to another of a different capsular type by introducing purified DNA from the former pneumococcus into the latter. These authors referred to the agent (later shown to be DNA) accomplishing the change as "transforming factor." Subsequently, this type of genetic manipulation has become commonplace. Conceptually similar experiments are now regularly performed utilizing a variety of cells, including human cells and mammalian embryos as recipients and molecularly cloned DNA as the donor of genetic information.

DNA Contains Four Distinct Deoxynucleotides

The chemical nature of the monomeric deoxynucleotide units of DNA—**deoxyadenylate**, **deoxyguanylate**, **deoxycytidylate**, and **thymidylate**—is described in Chapter 32. These monomeric units of DNA are held in polymeric form by 3′,5′-phosphodiester bonds constituting a single strand, as depicted in **Figure 34–1**. The informational content of DNA (the genetic code) resides in the sequence in which these monomers—purine and pyrimidine deoxyribonucleotides—are ordered. The polymer as depicted possesses a polarity; one end has a 5′-hydroxyl or phosphate terminus while the other has a 3′-phosphate or hydroxyl terminus. The importance of this

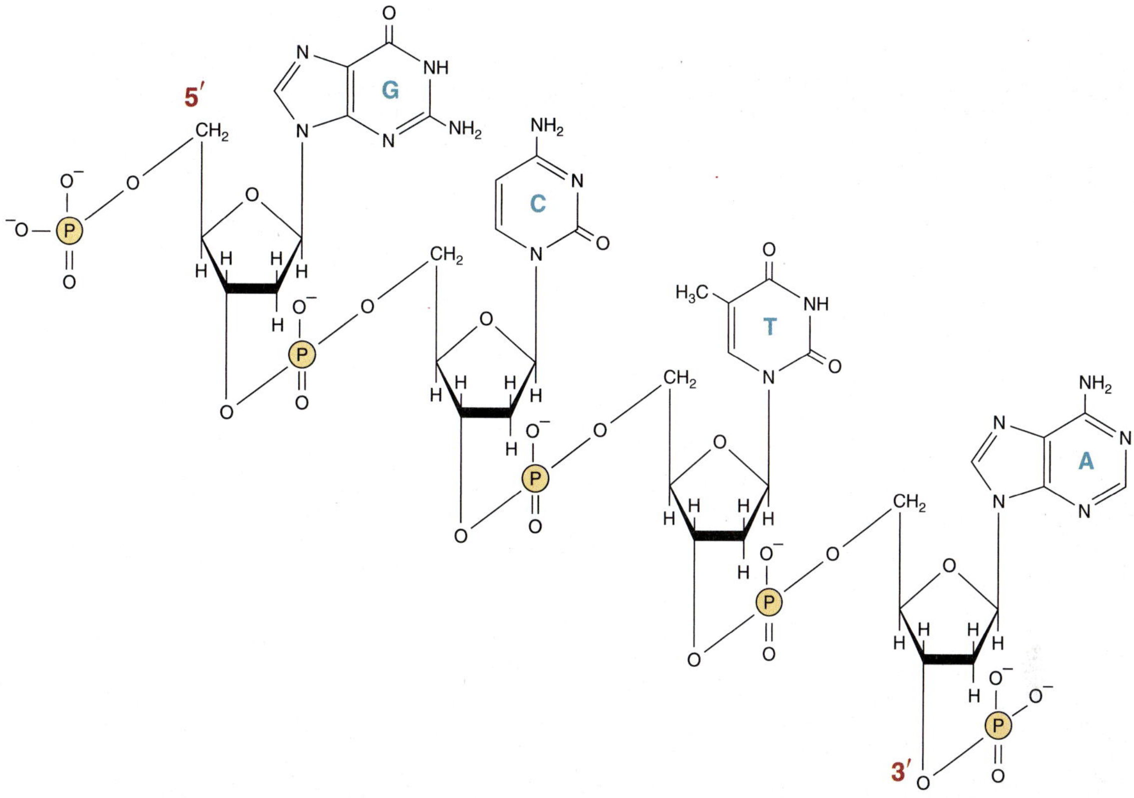

FIGURE 34–1 **A segment of one strand of a DNA molecule in which the purine and pyrimidine bases guanine (G), cytosine (C), thymine (T), and adenine (A) are held together by a phosphodiester backbone between 2′-deoxyribosyl moieties attached to the nucleobases by an *N*-glycosidic bond.** Note that the phosphodiester backbone is negatively charged and has a polarity (ie, a direction). Convention dictates that a single-stranded DNA sequence is written in the 5′ to 3′ direction (ie, pGpCpTpAp, where G, C, T, and A represent the four bases and P represents the interconnecting phosphates).

polarity will become evident. Since the genetic information resides in the order of the monomeric units within the polymers, there must exist a mechanism of reproducing or replicating this specific information with a high degree of fidelity. That requirement, together with x-ray diffraction data from the DNA molecule generated by Franklin, and the observation of Chargaff that in DNA molecules the concentration of deoxyadenosine (A) nucleotides equals that of thymidine (T) nucleotides (A = T), while the concentration of deoxyguanosine (G) nucleotides equals that of deoxycytidine (C) nucleotides (G = C), led Watson, Crick, and Wilkins to propose in the early 1950s a model of a **double-stranded (ds) DNA** molecule. The model they proposed is depicted in **Figure 34–2**. The two strands of this double-stranded helix are held in register by both **hydrogen bonds** (**H-bonds**; see Figure 2–2) between the purine and pyrimidine bases of the respective linear molecules and by **van der Waals** and **hydrophobic interactions** (see Chapter 2 and Figure 2–4) between the stacked adjacent base pairs. The pairings between the purine and pyrimidine nucleotides on the opposite strands are very specific, and are dependent on hydrogen bonding of **A with T** and **G with C** (see Figure 34–2). **A–T and G–C base pairs** are often referred to as **Watson-Crick base pairs**.

This common form of DNA is said to be right handed because as one looks down the double helix, the base residues form a spiral in a clockwise direction. In the double-stranded molecule, restrictions imposed by the rotation about the phosphodiester bond, the favored *anti*-configuration of the glycosidic bond (see Figure 32–5), and the predominant tautomers (see Figure 32–2) of the four bases (A, G, T, and C) allow A to pair only with T, and G only with C, as depicted in **Figure 34–3**. These base-pairing restrictions explain the earlier observation that in a double-stranded DNA molecule the content of A equals that of T and the content of G equals that of C. The two strands of the double-helical molecule, each of which possesses a **polarity**, are **antiparallel**; that is, one strand runs in the 5′ to 3′ direction and the other in the 3′ to 5′ direction. Within a particular gene in the double-stranded DNA molecules, the genetic information resides in the sequence of nucleotides on one strand, the **template strand**. This is the strand of DNA that is copied, or transcribed, during **ribonucleic acid (RNA)** synthesis. It is sometimes referred to as the **noncoding strand**. The opposite strand is considered the **coding strand** because it matches the sequence of the RNA transcript (but containing uracil in place of thymine; see Figure 34–8) that encodes the protein.

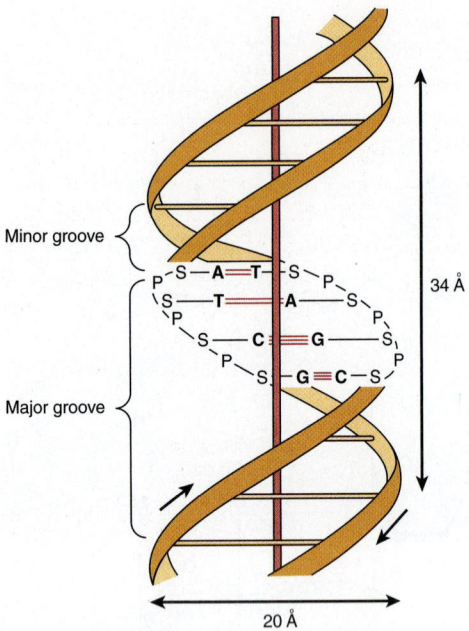

FIGURE 34–2 **A diagrammatic representation of the Watson and Crick model of the double-helical structure of the B form of DNA.** The horizontal arrow indicates the width of the double helix (20 Å), and the vertical arrow indicates the distance spanned by one complete turn of the double helix (34 Å). One turn of B-DNA includes 10 base pairs (bp), so the rise is 3.4 Å per bp. The central axis of the double helix is indicated by the vertical rod. The short arrows designate the polarity of the antiparallel strands. The major and minor grooves are depicted. (A, adenine; C, cytosine; G, guanine; P, phosphate; S, sugar [deoxyribose]; T, thymine.) Hydrogen bonds between A/T and G/C bases indicated by short, red, horizontal lines.

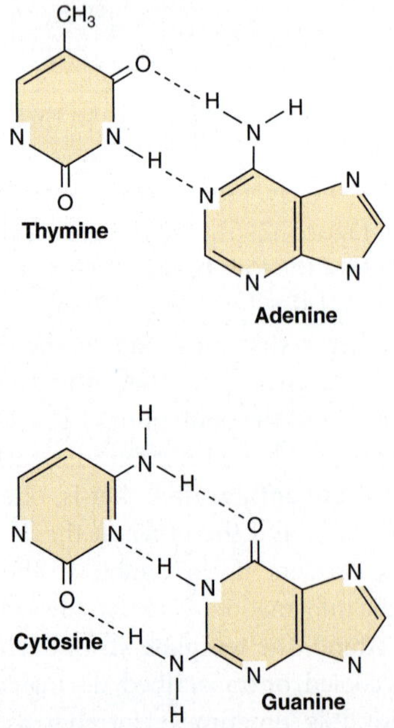

FIGURE 34–3 Classic Watson-Crick DNA base pairing between complementary deoxynucleotides involves the formation of hydrogen bonds. Two such H-bonds form between adenine and thymine, and three H-bonds form between cytidine and guanine. The broken lines represent H-bonds.

The two strands, in which opposing bases are held together by interstrand hydrogen bonds, wind around a central axis in the form of a **double helix**. In the test tube, double-stranded DNA can exist in at least six forms (A-DNA, through E-DNA and Z-DNA). These different forms of DNA differ with regard to intra- and interstrand interactions and involve structural rearrangements within the monomeric units of DNA. The **B form is usually found under physiologic conditions**. A single turn of B form DNA about the long axis of the molecule contains 10 bp. The distance spanned by one turn of B-DNA is 3.4 nm (34 Å). The width (helical diameter) of the double helix in B-DNA is 2 nm (20 Å).

There Are Grooves in the DNA Molecule

Examination of the model depicted in Figure 34–2 reveals a **major groove** and a **minor groove** winding along the molecule parallel to the phosphodiester backbones. In these grooves, proteins often interact specifically with exposed atoms of the nucleotides (via specific hydrophobic and ionic interactions), thereby recognizing and binding to specific nucleotide sequences as well as the unique shapes formed therefrom. Binding usually occurs without disrupting the base pairing of the double-helical DNA molecule. As discussed in Chapters 35, 36, and 38, regulatory proteins that control DNA replication, repair, and recombination as well as the transcription of specific genes occur through such protein-DNA interactions, and thus contribute critically to cellular function.

The Denaturation of DNA Is Used to Analyze Its Structure

As depicted in Figure 34–3, three H-bonds, formed by hydrogen atoms bonded to electronegative N or O atoms, hold the deoxyguanosine nucleotide to the deoxycytidine nucleotide. By contrast, the other canonical base pair, the A–T pair, is held together by two H-bonds. Note that the four DNA nucleotide bases ([dG, dA] purines and [dT, dC] pyrimidines) are flat, planar molecules (see Figure 32–1 and Table 32–1). These two key properties of the nucleotide bases allow them to closely stack within duplex DNA (see Figure 34–2). Moreover, the atoms within the aromatic, heterocyclic bases are highly polarizable and, coupled with the fact that many of the atoms within the bases contain partial charges, allows for the stacked bases to form van der Waals and electrostatic interactions. Collectively these forces are referred to as base-stacking forces or base-stacking interactions. The base-stacking interactions between adjacent G–C (or C–G) base pairs are stronger than A–T (or T–A) base pairs. Thus, overall, G–C-rich DNA sequences are more resistant to denaturation, or strand separation, termed "melting," than are A–T-rich regions of DNA.

DNA Can be Reversibly Denatured & Specifically Renatured, Both in the Test Tube & in Living Cells

In the laboratory **the double-stranded structure of DNA can be separated**, or **denatured** into its two component strands in solution by: increasing temperature, decreasing solution salt concentrations, adding **chaotropic agents, which can form competing H-bonds** with the individual deoxynucleotide bases, or often experimentally, by a combination of all three treatments. Under such conditions not only do the two stacks of bases pull apart, but the bases themselves unstack while remaining connected within the now, two single-stranded polymers, that are connected by their phosphodiester backbones. Concomitant with the denaturation of the DNA molecule into two single strands is an increase in the optical absorbance in the ultraviolet light spectrum (260 nm; see Chapter 32) of the purine and pyrimidine bases of each strand; this phenomenon is referred to as **hyperchromicity** of denaturation. Because of the combined strength of base stacking *and* the H-bonding between the complementary bases in each strand, the double-stranded DNA molecule exhibits properties of a rigid rod. Thus, native double-stranded DNA in solution is an extremely viscous material. However, on denaturation, DNA solutions lose their viscosity.

The strands of a given molecule of double-stranded DNA separate over a temperature range. The midpoint of the measured **DNA denaturation** is called the **melting temperature**, or T_m. The T_m is influenced by the base composition of the DNA and by the salt concentration or other components of the solution (see following discussion). DNA rich in G–C pairs melts at a higher temperature than DNA rich in A–T pairs, due to differences in hydrogen bond content and base stacking, as discussed earlier. A 10-fold increase of monovalent cation concentration significantly increases the T_m by neutralizing the intrinsic interchain repulsion between the highly negatively charged phosphates of the phosphodiester backbone of each DNA strand. For example, an increase of NaCl concentration from 0.01 to 0.1 M increases T_m by 16.6°C. By contrast, chaotropes such as **urea** (NH_2CONH_2; see Figure 28–16) and **formamide** (CH_3NO) can efficiently form H-bonds with the nucleotide bases, which destabilizes H-bonding between bases. Such solution conditions will lower the T_m. Chaotrope addition allows the strands of DNA or complementary DNA–RNA, and intramolecular RNA-RNA hybrids (see following discussion) to be separated at much lower temperatures. Lower temperatures minimize phosphodiester bond breakage and chemical damage to nucleotides that can occur on extended incubation in solution. In living cells, both DNA denaturation and **renaturation** (see following discussion) occurs naturally during the processes of DNA replication, DNA recombination, DNA repair (see Chapter 35), and DNA gene transcription (see Chapter 36). In all of these instances DNA strand separation and renaturation is mediated through the action of specific nucleic acid binding proteins and various enzymes, in combination with thermal and/or chemical energy supplied via ATP hydrolysis.

Renaturation of DNA Requires Precise Base Pair Matching

Importantly, separated strands of DNA will **renature, or reassociate** when appropriate temperature and salt conditions are achieved. Reannealing is often referred to as **renaturation** or

hybridization. The rate of strand reassociation depends on the concentration of the complementary strands. At a given temperature and salt concentration, a particular nucleic acid strand will associate tightly only with its complementary strand. Thus, **renaturation is highly specific**. Indeed, researchers have shown that renatured DNA hybrid molecules with but one base pair mismatch can readily be detected and quantified. Importantly, DNA-DNA, DNA-RNA, and RNA-RNA hybrid molecules will also form under appropriate conditions. For example, DNA will form a perfect double-stranded hybrid molecule with a **complementary DNA (cDNA)** sequence, or with a cognate, **complementary RNA**. When hybridization is combined with various sophisticated analytical techniques scientists can specifically detect, identify, and even determine the sequence of vanishingly small amounts of nucleic acids (both DNA and RNA). An overview of much of the enabling technology of such nucleic acid analyses is described in Chapter 39.

DNA Exists in Relaxed & Supercoiled Forms

In some organisms such as bacteria, bacteriophages, many DNA-containing animal viruses, as well as organelles such as mitochondria (see Figure 35–8), the ends of the DNA molecules are joined to create a closed circle with no covalently free ends. This of course does not destroy the polarity of the molecules, but it eliminates all free 3′ and 5′ hydroxyl and phosphoryl groups. Closed circles exist in relaxed or supercoiled forms. Supercoils are introduced when a closed circle is twisted around its own axis or when a linear piece of duplex DNA, whose ends are fixed, is twisted. This energy-requiring process puts the molecule under torsional stress, and the greater the number of supercoils, the greater the stress or torsion (test this by twisting a rubber band). **Negative supercoils** are formed when the molecule is twisted in the direction opposite from the clockwise turns of the right-handed double helix found in B-DNA. Such DNA is said to be underwound. The energy required to achieve this state is, in a sense, stored in the supercoils. The transition to another form that requires energy is thereby facilitated by the underwinding (see Figure 35–19). One such transition is strand separation, which is a prerequisite for DNA replication and transcription. Supercoiled DNA is therefore a preferred form in biologic systems. Enzymes that catalyze topologic changes of DNA are called **topoisomerases**. Topoisomerases can relax or insert supercoils, using ATP as an energy source. Homologs of this enzyme exist in all organisms and are important targets for cancer chemotherapy. Supercoils can also form within linear DNAs if particular segments of DNA are constrained by interacting tightly with nuclear proteins that establish two boundary sites defining a topologic domain.

DNA PROVIDES A TEMPLATE FOR REPLICATION & TRANSCRIPTION

The genetic information stored in the nucleotide sequence of DNA serves two purposes. It is the source of information for the synthesis of all protein molecules of the cell and organism, and it provides the information inherited by daughter cells or offspring. Both of these functions require that the DNA molecule serve as a template—in the first case for the transcription of the information into RNA and in the second case for the replication of the information into daughter DNA molecules.

When each strand of the double-stranded parental DNA molecule separates from its complement during replication, each independently serves as a template on which a new complementary strand is synthesized (**Figure 34–4**). The two newly formed double-stranded daughter DNA molecules, each containing one strand (but complementary rather than identical) from the parent double-stranded DNA molecule, are then sorted between the two daughter cells during mitosis (**Figure 34–5**). Each daughter cell contains DNA molecules with information identical to that which the parent possessed; yet, in each daughter cell, the DNA molecule of the parent cell has been only semiconserved.

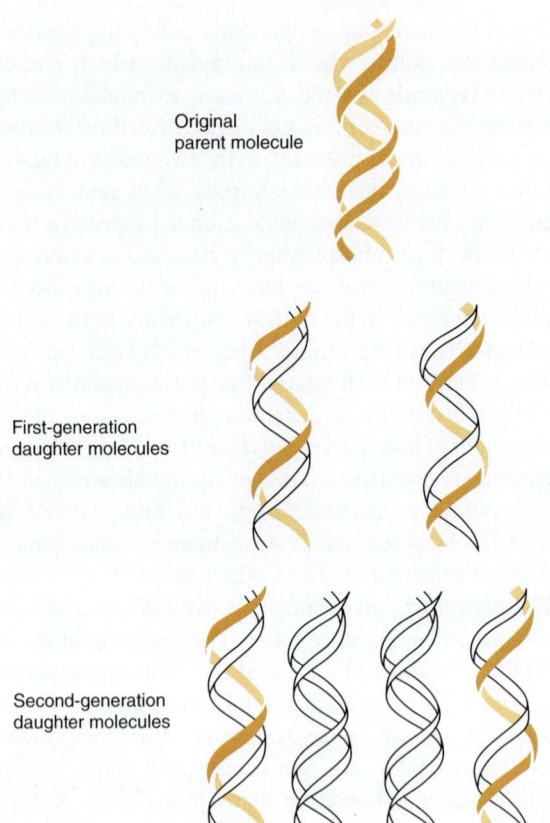

FIGURE 34–5 **DNA replication is semiconservative.** During a round of replication, each of the two strands of DNA is used as a template for synthesis of a new, complementary strand. The semiconservative nature of DNA replication has implications for the biochemical (see Figure 35–16), cytogenetic (see Figure 35–12), and epigenetic control of gene expression (see Figures 38–8 and 38–9).

THE CHEMICAL NATURE OF RNA DIFFERS FROM THAT OF DNA

RNA is a polymer of purine and pyrimidine ribonucleotides linked together by 3',5'-phosphodiester bonds analogous to those in DNA (**Figure 34–6**). Although sharing many features with DNA, RNA possesses several specific differences:

1. In RNA, the sugar moiety to which the phosphates and purine and pyrimidine bases are attached is ribose rather than the 2'-deoxyribose of DNA (see Figures 19–2 and 32–3).

2. The pyrimidine components of RNA can differ from those of DNA. Although RNA contains the ribonucleotides of adenine, guanine, and cytosine, it does not possess thymine except in the rare case mentioned in the following discussion. Instead of thymine, RNA contains the ribonucleotide of uracil.

3. RNA can exist as a single strand, whereas DNA exists as a double-stranded helical molecule. However, given the proper complementary base sequence with opposite polarity, the single strand of RNA—as demonstrated in **Figures 34–7** and 34–11—is capable of folding back on itself like a hairpin

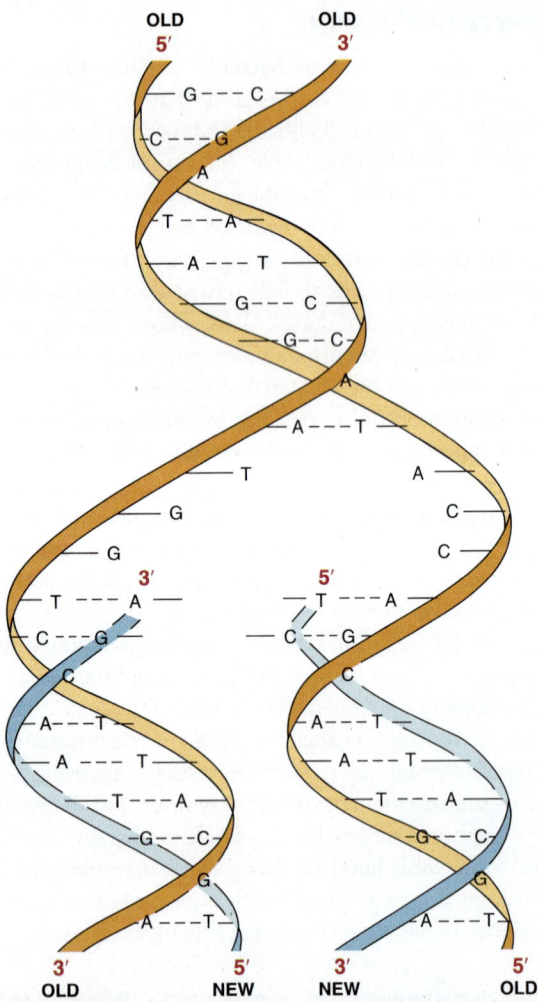

FIGURE 34–4 **DNA synthesis maintains the sequence and structure of the original template DNA.** The double-stranded structure of DNA and the template function of each old parental strand (orange) on which a new complementary daughter strand (blue) is synthesized.

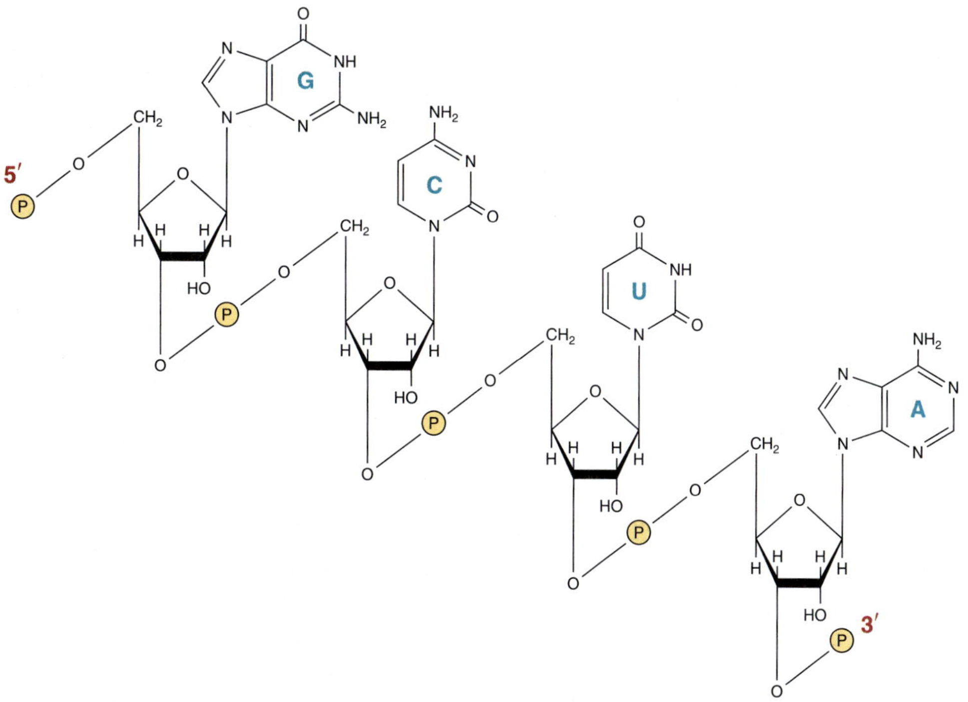

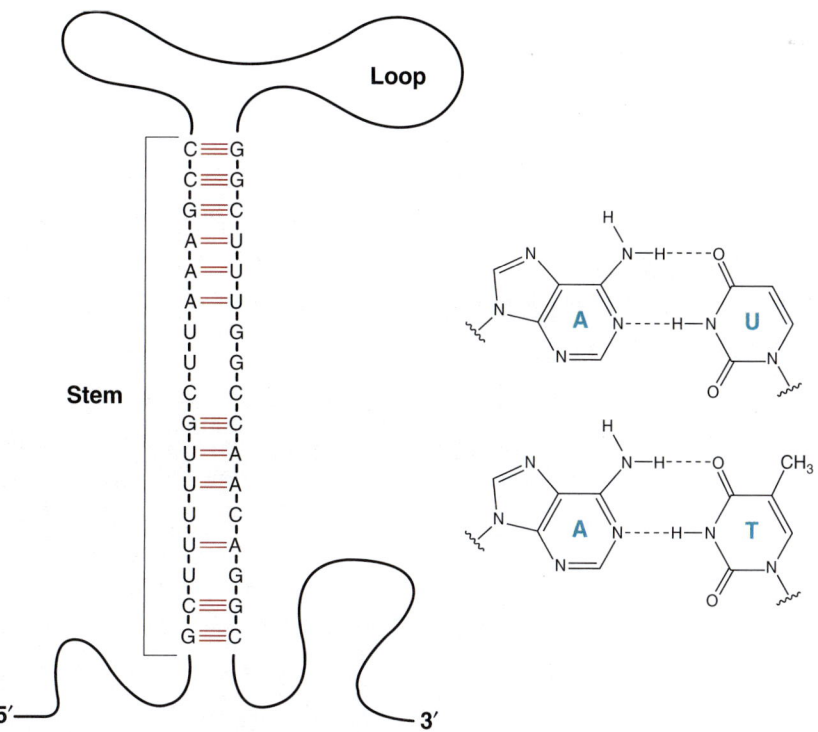

FIGURE 34–6 **A segment of a ribonucleic acid (RNA) molecule in which the purine and pyrimidine bases—guanine (G), cytosine (C), uracil (U), and adenine (A)—are held together by phosphodiester bonds between ribosyl moieties attached to the nucleobases by** ***N*-glycosidic bonds.** Note that negative charge(s) on the phosphodiester backbone are not illustrated (ie, Figure 34–1) and that the polymer has a polarity as indicated by the labeled 3′- and 5′-attached phosphates.

FIGURE 34–7 **RNA Secondary Structure. (Left) Diagrammatic representation of the secondary structure of a hypothetical single-stranded RNA molecule in which a stem loop, or "hairpin", has been formed.** Formation of this structure is dependent on the indicated intramolecular base pairing (colored horizontal lines between complementary bases). Note that in RNA G pairs with C as in DNA, but that A pairs with U. (Right) Schematic of A-U (top) compared to A-T base pairs (bottom).

DNA strands:

Coding → 5′—T G G A A T T G T G A G C G G A T A A C A A T T T C A C A C A G G A A A C A G C T A T G A C C A T G—3′
Template → 3′—A C C T T A A C A C T C G C C T A T T G T T A A A G T G T G T C C T T T G T C G A T A C T G G T A C—5′

RNA transcript: 5′——ppp A U U G U G A G C G G A U A A C A A U U U C A C A C A G G A A A C A G C U A U G A C C A U G 3′

FIGURE 34–8 **The relationship between the sequences of an RNA transcript and its gene, in which the coding and template strands are shown with their polarities.** The RNA transcript with a 5′ to 3′ polarity is complementary to the template strand with its 3′ to 5′ polarity. Note that the sequence in the RNA transcript and its polarity is the same as that in the coding strand, except that the U of the transcript replaces the T of the gene; the initiating nucleotide of RNAs contain a terminal 5-triphosphate (ie, pppA-above).

and thus acquiring double-stranded characteristics: G pairing with C, and A pairing with U. G-C base pairs form three H-bonds and A-U base form only two.

4. Since the RNA molecule is a single strand complementary to only one of the two strands of a gene, its guanine content does not necessarily equal its cytosine content, nor does its adenine content necessarily equal its uracil content.

5. RNA can be hydrolyzed by alkali to 2′, 3′ cyclic diesters of the mononucleotides, compounds that cannot be formed from alkali-treated DNA because of the absence of a 2′-hydroxyl group. The alkali lability of RNA is useful both diagnostically and analytically.

Information within the single strand of RNA is contained in its sequence ("primary structure") of purine and pyrimidine nucleotides within the polymer. The sequence is complementary to the template strand of the gene from which it was transcribed. Because of this complementarity, an RNA molecule can bind specifically via the base-pairing rules to its template DNA strand (**A**–T, **G**–C, **C**–G, **U**–A; RNA base bolded); it will not bind ("hybridize") with the other (coding) strand of its gene. The sequence of the RNA molecule (except for U replacing T) is the same as that of the coding strand of the gene (**Figure 34–8**).

NEARLY ALL THE SEVERAL SPECIES OF STABLE, ABUNDANT RNAs ARE INVOLVED IN SOME ASPECT OF PROTEIN SYNTHESIS

Those cytoplasmic RNA molecules that serve as templates for protein synthesis (ie, that transfer genetic information from DNA to the protein-synthesizing machinery) are designated **messenger RNAs (mRNAs)**. Many other very abundant cytoplasmic RNA molecules (**ribosomal RNAs [rRNAs]**) have structural roles wherein they contribute to the formation and function of ribosomes (the organellar machinery for protein synthesis) or serve as adapter molecules (**transfer RNAs [tRNAs]**) for the translation of RNA information into specific sequences of polymerized amino acids.

Interestingly, some RNA molecules have intrinsic catalytic activity. The activity of these RNA enzymes, or **ribozymes**, often involves the cleavage of a nucleic acid. Two ribozymes are the ribozymes involved in the RNA splicing, and

peptidyl transferase that catalyzes peptide bond formation on the ribosome.

In all eukaryotic cells, there are **small nuclear RNA (snRNA)** species that are not directly involved in protein synthesis but play pivotal roles in RNA processing, particularly mRNA processing. These relatively small molecules vary in size from 90 to about 300 nucleotides (**Table 34–1**). The properties of the several classes of cellular RNAs are detailed in the following discussion.

The genetic material for some animal and plant viruses is RNA rather than DNA. Although some RNA viruses never have their information transcribed into a DNA molecule (eg, Influenza and Coronaviruses like COVID-19), certain animal RNA viruses—specifically, the retroviruses (eg, the human immunodeficiency, or HIV virus)—are transcribed by **viral RNA–dependent DNA polymerase, the so-called reverse transcriptase**, to produce a double-stranded DNA copy of their RNA genome. In many cases, the resulting double-stranded DNA transcript is integrated into the host genome and subsequently serves as a template for gene expression and from which new viral RNA genomes and viral mRNAs can be transcribed and subsequently translated by the host cell machinery into viral proteins. Genomic insertion of such integrating "proviral" DNA molecules can, depending on the site involved, be mutagenic, inactivating a gene or disregulating its expression (see Figure 35–11).

TABLE 34–1 Some of the Species of Small-Stable RNAs Found in Mammalian Cells

Name	Length (Nucleotides)	Molecules per Cell	Localization
U1	165	1×10^6	Nucleoplasm
U2	188	5×10^5	Nucleoplasm
U3	216	3×10^5	Nucleolus
U4	139	1×10^5	Nucleoplasm
U5	118	2×10^5	Nucleoplasm
U6	106	3×10^5	Perichromatin granules
4.5S	95	3×10^5	Nucleus and cytoplasm
7SK	280	5×10^5	Nucleus and cytoplasm

THERE EXIST SEVERAL DISTINCT CLASSES OF RNA

As noted earlier, in all prokaryotic and eukaryotic organisms, four main classes of RNA molecules exist: mRNA, tRNA, rRNA, and small RNAs. Each differs from the others by abundance, size, function, and general stability.

Messenger RNA

This class is the most heterogeneous in abundance, size, and stability; for example, in brewer's yeast, specific mRNAs are present in 100s/cell to, on average, ≤0.1/mRNA/cell in a genetically homogeneous population. As detailed in Chapters 36 and 38, both specific transcriptional and posttranscriptional mechanisms contribute to this large dynamic range in mRNA content. In mammalian cells, specific mRNA abundance likely varies over a 10^4-fold range. All members of this RNA class function as messengers conveying the information in a gene to the protein-synthesizing machinery, where each mRNA serves as a template on which a specific sequence of amino acids is polymerized to form a specific protein molecule, in this case the ultimate gene product (**Figure 34–9**).

Eukaryotic mRNAs have unique chemical characteristics. The 5′ terminus of mRNA is "capped" by a 7-methylguanosine triphosphate that is linked to an adjacent 2′-O-methyl ribonucleoside at its 5′-hydroxyl through the three phosphates (**Figure 34–10**). mRNA molecules frequently contain internal N^6-methyladenine and other 2′-O-ribose-methylated nucleotides. The cap is involved in the recognition of mRNA by the translation machinery, and also helps stabilize the mRNA by preventing the nucleolytic attack by 5′-exoribonucleases. The protein-synthesizing machinery begins translating the mRNA into proteins beginning downstream of the 5′ or capped terminus. At the other end of almost all eukaryotic mRNA molecules, the 3′-hydroxyl terminus has an attached, nongenetically encoded polymer of adenylate residues 20 to 250 nucleotides

in length. The poly(A) "tail" at the 3′-end of mRNAs maintains the intracellular stability of the specific mRNA by preventing the attack of 3′-exoribonucleases and also facilitates translation (see Figure 37–7). Both the mRNA "cap" and "poly(A) tail" are added posttranscriptionally by nontemplate-directed enzymes to mRNA precursor molecules (pre-mRNA). mRNA represents 2 to 5% of total eukaryotic cellular RNA.

In mammalian cells, including cells of humans, the mRNA molecules present in the cytoplasm are not the RNA products immediately synthesized from the DNA template but must be formed by processing from the precursor, or pre-mRNA before entering the cytoplasm. Thus, in mammalian cell nuclei, the immediate products of gene transcription (primary transcripts) are very heterogeneous and can be 10- to 50-fold longer than mature mRNA molecules. As discussed in Chapter 36, pre-mRNA molecules are processed to generate mRNA molecules, which then enter the cytoplasm to serve as templates for protein synthesis.

Transfer RNA

tRNA molecules vary in length from 74 to 95 nucleotides and, like many other RNAs, are also generated by nuclear processing of a precursor molecule (see Chapter 36). The tRNA molecules serve as adapters for the translation of the information in the sequence of nucleotides of the mRNA into specific amino acids. There are at least 20 species of tRNA molecules in every cell, at least one (and often several) corresponding to each of the 20 amino acids required for protein synthesis. Although each specific tRNA differs from the others in its sequence of nucleotides, the tRNA molecules as a class have many features in common. The primary structure—that is, the nucleotide sequence—of all tRNA molecules allows extensive folding and intrastrand complementarity to generate a secondary structure that appears in two dimensions like a cloverleaf (**Figure 34–11**).

All tRNA molecules contain four main double-stranded arms or stems, connected by single-stranded loops named for their respective nucleotide composition or function. The amino acid **acceptor arm** terminates in the nucleotides $CpCpA_{OH}$. As with the mRNA 5′-Cap and 3′ Poly A tail, these three nucleotides (CCA) are added posttranscriptionally, in this case by a specific nucleotidyl transferase enzyme. The tRNA-appropriate amino acid is attached, or "charged," onto the posttranscriptionally added 3′-OH group of the A moiety of the acceptor arm through the action of specific aminoacyl tRNA synthetases (see Figure 37–1). The **D**, **TψC**, and **extra arms** help define a specific tRNA. tRNAs compose roughly 20% of total cellular RNA. Recent work has shown that many tRNAs are specifically cleaved by certain ribonucleases to generate unique subfragments termed **tRNA-derived small RNA (tsRNAs)**. These relatively stable tsRNAs are thought to regulate both transcription and translation (see Chapters 36–38).

Ribosomal RNA

A ribosome is a cytoplasmic nucleoprotein structure that acts as the machinery for the synthesis of proteins from the mRNA

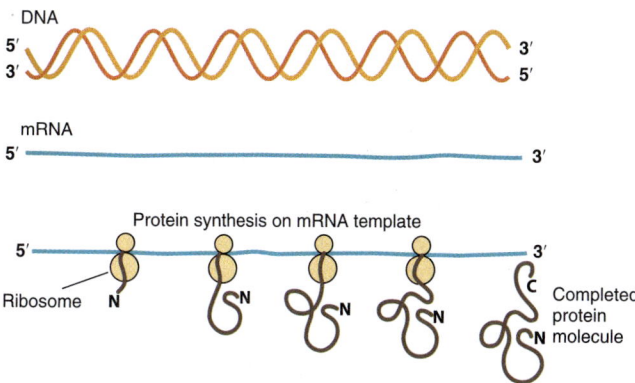

FIGURE 34–9 **The expression of genetic information within DNA into the form of an mRNA transcript with 5′ to 3′ polarity and then into protein with N- to C-polarity is shown.** DNA is transcribed into mRNA that is subsequently translated by ribosomes into a specific protein molecule that exhibits polarity, N-terminus (**N**) to C-terminus (**C**).

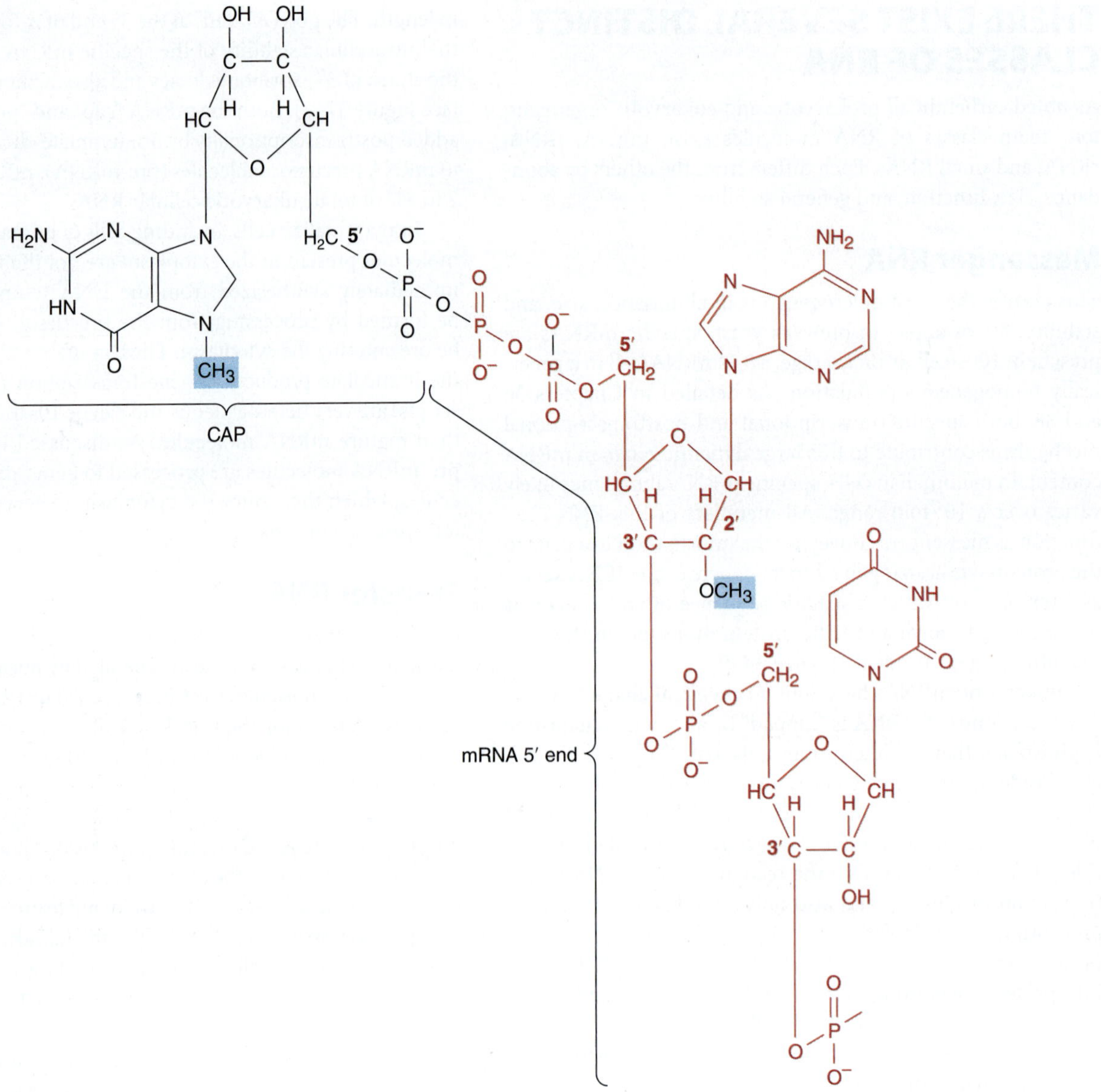

FIGURE 34–10 **The cap structure attached to the 5′ terminal of most eukaryotic messenger RNA molecules.** A 7-methylguanosine triphosphate (black) is attached at the 5′ end of the mRNA (red), which usually also contains a 2′-*O*-methylpurine nucleotide. These modifications (the cap and methyl group) are added after the mRNA is transcribed from DNA. Note that the γ- and β-phosphates of the GTP added to form the cap (black in figure) are lost on cap addition while the γ-phosphate of the initiating nucleotide (here an A-residue; red in figure) is lost during cap addition.

templates. On the ribosomes, the mRNA and tRNA molecules interact to translate the information transcribed from the gene during mRNA synthesis into a specific protein. During periods of active protein synthesis, many ribosomes can be associated with any single mRNA molecule to form an assembly called the **polysome** (see Figure 37–7).

The components of the mammalian ribosome, which has a molecular weight of about 4.2×10^6 and a sedimentation velocity coefficient of 80S (**S** = **Svedberg units**, a parameter sensitive to molecular size and shape) are shown in **Table 34–2**. The mammalian ribosome contains two major nucleoprotein subunits—a larger one with a molecular weight of 2.8×10^6 (60S) and a smaller subunit with a molecular weight of 1.4×10^6 (40S).

The 60S subunit contains a 5S rRNA, a 5.8S rRNA, and a 28S rRNA; there are also more than 50 specific polypeptides. The 40S subunit is smaller and contains a single 18S rRNA and 30 distinct polypeptide chains. All of the rRNA molecules except the 5S rRNA, which is independently transcribed, are processed from a single 45S precursor RNA molecule in the nucleolus (see Chapter 36). rRNAs are highly methylated post-transcriptionally, and are packaged in the nucleolus with the specific ribosomal proteins. In the cytoplasm, the ribosomes remain quite stable and capable of many cycles of translation. The exact functions of the rRNA molecules in the ribosome, over and above their scaffold functions are not yet fully understood. It is clear though that rRNAs are necessary for

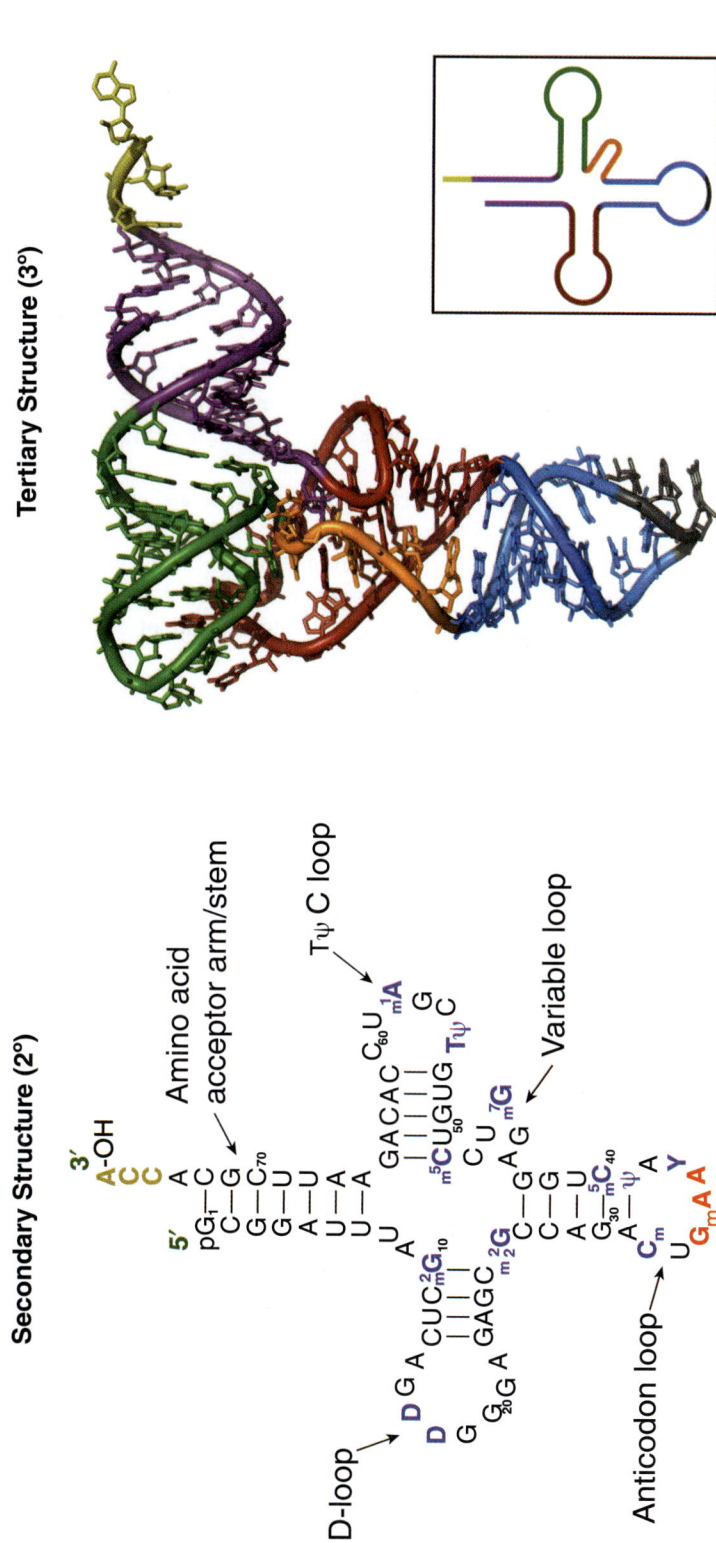

FIGURE 34–11 Structure of a mature, functional tRNA, yeast phenylalanyl-tRNA (tRNA^Phe). Shown are primary (1°), secondary (2°), and tertiary (3°) structures (top, lower left, and lower right, respectively) of tRNA^Phe. Numerals below the 76 nucleotide-long tRNA^Phe primary structure indicate nucleotide numbering from the 5′ (+1) to the 3′ end (+76) of the molecule. Note that the +1 nucleotide contains a 5′ phosphate moiety (P), while the 3′ nucleotide has a free 3′ hydroxyl group (OH). Bases underlined in bold type within the sequence of tRNA^Phe are heavily modified to the nucleotides shown in the 2° structural representation of tRNA^Phe. This structure is often referred to as a "cloverleaf." Some of these nucleotides have noncanonical ribonucleotide names, as represented in the 2° structural model. Within tRNA^Phe nucleotides U_16 and U_17 are modified to D_16; D_17; G_37 to Y_37; U_39 and U_55 to Ψ, and U_54 to T_54 (see following discussion for details). Straight lines between bases within the tRNA secondary structure represent hydrogen bonds formed between bases (A–U; G–C). Note that these regions of secondary structure form with the same strand polarity (ie, 5′ to 3′ and 3′ to 5′) as base-paired regions of DNA. The three bases of the anticodon loop are shown in red. In the case of amino acid–charged tRNAs, an aminoacyl moiety is esterified to the 3′-CCA_OH terminus (brown; in this case the amino acid would be phenylalanine; not shown). Blue type highlights nontraditional nucleotides introduced by posttranslational modification, abbreviated as follows: m^2G = 2-methylguanosine; D = 5,6-dihydrouridine; m^2_2 G = N2-dimethylguanosine; C_m = O2′-methylcytidine; G_m = O2′-methylguanosine; T = 5-methyluridine; Y = wybutosine; Ψ = pseudouridine; m^5C = 5-methylcytidine; m^7G = 7-methylguanosine; m^1A = 1-methyladenosine. Essentially all tRNAs fold into similar, characteristic, tertiary structures (3°) as shown, lower right. The distinct portions of the molecule in 2° (insert) and 3° configurations are color-coded in this image for clarity. tRNA^Phe was the first nucleic acid whose structure was determined by x-ray crystallography. Such distinct three-dimensional tRNA structures bind specifically to important functional sites on both aminoacyl tRNA synthetases and the ribosomes during protein synthesis (see Chapter 37). (Reproduced with permission from Transfer RNA/Wikipedia Commons https://en.wikipedia.org/wiki/Transfer_RNA.)

TABLE 34–2 Components of Mammalian Ribosomes

Component	Mass (MW)	Protein		RNA		
		Number	Mass	Size	Mass	Bases
40S subunit	1.4×10^6	33	7×10^5	18S	7×10^5	1900
60S subunit	2.8×10^6	50	1×10^6	5S	3.5×10^4	120
				5.8S	4.5×10^4	160
				28S	1.6×10^6	4700

Note: The ribosomal subunits are defined according to their sedimentation velocity in Svedberg (**S**) units (40S or 60S). The number of unique proteins and their total mass (MW) and the RNA components of each subunit in size (Svedberg units), mass, and number of bases are listed.

ribosome assembly, and also play key roles in the binding of mRNA to ribosomes and mRNA translation. Recent studies indicate that the large rRNA component performs the peptidyl transferase activity and thus is a ribozyme. The rRNAs (28S + 18S) represent roughly 70% of total cellular RNA.

Small RNA

A large number of discrete, highly conserved small RNA species are found in eukaryotic cells; some are quite stable. Most of these molecules are complexed with proteins to form ribonucleoproteins and are distributed in the nucleus, the cytoplasm, or both. They range in size from 20 to 1000 nucleotides and are present in 100,000 to 1,000,000 copies per cell, collectively representing less than or equal to 5% of cellular RNA.

Small Nuclear RNAs

snRNAs, a subset of the small nuclear RNAs (see Table 34–1), are significantly involved in rRNA and mRNA processing and gene regulation. Of the several snRNAs, U1, U2, U4, U5, and U6 are involved in mRNA splicing, a nuclear process whereby introns are removed from mRNA precursor molecules to generate functional, translatable cytoplasmic mRNAs (see Chapter 36). The U7 snRNA is involved in production of the correct 3′ ends of histone mRNA—which lacks a poly(A) tail. 7SK RNA associates with several proteins to form a ribonucleoprotein complex, termed P-TEFb, that modulates mRNA gene transcription elongation by RNA polymerase II (see Chapter 36).

Large & Small Noncoding Regulatory RNAs: Micro-RNAs (miRNAs), Silencing RNAs (siRNAs), Long Noncoding RNAs (lncRNAs), and Circular RNAs (circRNAs)

One of the most exciting and unanticipated discoveries in the last decade of eukaryotic regulatory biology has been the identification and characterization of regulatory nonprotein coding RNAs (ncRNAs). NcRNAs exist in two general size classes, large (50–1000nt) and small (20–22nt). Regulatory ncRNAs have been described in most eukaryotes (see Chapter 38).

The **small ncRNAs termed miRNAs and siRNAs typically inhibit gene expression** at the level of specific protein production by targeting mRNAs through one of several distinct mechanisms. miRNAs are generated by specific nucleolytic processing of the products of distinct genes/transcription units (see Figure 36–17). miRNA precursors, which are 5′-capped and 3′-polyadenylated, usually range in size from about 500 to 1000 nucleotides.

By contrast, siRNAs are generated by the specific nucleolytic processing of large dsRNAs that are either produced from other endogenous RNAs, or dsRNAs introduced into the cell by, for example, RNA viruses. Both **siRNAs and miRNAs hybridize via the formation of RNA–RNA hybridization to their targeted mRNAs** (see Figure 38–19). To date, hundreds of distinct miRNAs and siRNAs have been described in humans; estimates suggest that there are likely 1000s of human miRNA-encoding genes. Given their exquisite genetic specificity, both miRNAs and siRNAs represent exciting new **potential agents for therapeutic drug development**. In the laboratory siRNAs are frequently used to decrease or "knockdown" specific protein levels (via siRNA homology–directed mRNA degradation), and thus serve as an extremely useful and powerful alternative to gene-knockout technology (see Chapter 39). Indeed, several siRNA-based therapeutic clinical trials are in progress to test the efficacy of these novel molecules as drugs for treating human disease.

Other exciting recent observations in the RNA realm are the identification and characterization of two classes of larger noncoding RNAs, the **circular RNAs (circRNAs)** and the **long noncoding RNAs**, or **lncRNAs**. Many **circRNAs** have recently been discovered and characterized. circRNAs appear to be produced by RNA splicing-type reactions from a wide range of precursor RNAs, both mRNA precursors and nonprotein coding lncRNA precursors (see following discussion for more information on lncRNAs). Though not an abundant class of RNA molecules in most cells, circRNAs have been detected in all eukaryotes tested, and seem particularly abundant in metazoans. While the functions of circRNAs are still being elucidated they seem to be particularly abundant in cells of the nervous system. Similar to lncRNAs, these molecules likely play important roles in cellular biology by regulating gene expression at multiple levels. LncRNAs, which as their name implies, do not code for protein, and range in size from ~300 to 1000s of nucleotides in length. These RNAs are typically transcribed from the large regions of eukaryotic genomes that do not encode for protein (ie, the mRNA encoding genes). In fact, transcriptome analyses indicate that **more than 90% of all eukaryotic genomic DNA may be transcribed** at some level. ncRNAs make up a significant portion of this transcription. ncRNAs play many roles ranging from contributing to structural aspects of chromatin to regulation of mRNA gene transcription by RNA polymerase II. Future work will further characterize this important, newly discovered class of RNA molecules.

Interestingly, bacteria also contain small, heterogeneous regulatory RNAs termed sRNAs. Bacterial sRNAs range in size from 50 to 500 nucleotides, and like eukaryotic mi/si/lncRNAs, control the expression/activity of a large array of distinct genes. sRNAs often repress, but sometimes activate protein synthesis by binding to specific mRNA.

SPECIFIC NUCLEASES DIGEST NUCLEIC ACIDS

Enzymes capable of degrading nucleic acids have been recognized for many years. These nucleases can be classified in several ways. Those that exhibit specificity for DNA are referred to as **deoxyribonucleases**. Those nucleases that specifically hydrolyze RNA are **ribonucleases**. Some nucleases degrade DNA and RNA. Within both of these classes are enzymes capable of cleaving internal phosphodiester bonds to produce either 3′-hydroxyl and 5′-phosphoryl terminals or, 5′-hydroxyl and 3′-phosphoryl terminals. These are referred to as **endonucleases**. Some are capable of hydrolyzing both strands of a **double-stranded** molecule, whereas others can only cleave **single strands** of nucleic acids. Some nucleases can hydrolyze only unpaired single strands, while others are capable of hydrolyzing single strands participating in the formation of a double-stranded molecule. There exist classes of endonucleases that recognize specific sequences in DNA. One class of these DNA cleaving enzymes, the **restriction endonucleases**, also termed restriction enzymes, do so directly by binding specific (usually) contiguous DNA base pairs (typically 4, 5, 6, or 8 bp), and cleaving both strands of DNA, usually DNA within the binding/recognition sequence element. The second class of enzymes, which are ribonucleoprotein complexes, utilizes a "guide RNA" of specific nucleotide sequence that targets a nuclease to cleave distinct DNA or RNA sequences. These are the **CRISPR-Cas** family of enzymes. Both classes of DNA-cleaving enzyme are described in greater detail in Chapter 39.

Some nucleases are capable of hydrolyzing a nucleotide only when it is present at a terminal of a molecule; these are referred to as **exonucleases**. Exonucleases act in one direction ($3′ \rightarrow 5′$ or $5′ \rightarrow 3′$) only. In bacteria, a $3′ \rightarrow 5′$ exonuclease is an integral part of the DNA replication machinery and there serves to edit—or proofread—the most recently added deoxynucleotide for base-pairing errors.

In addition to their roles in nucleic acid metabolism in living cells, the nucleases described here, in concert with a panoply of other nucleic acid synthesizing and modifying enzymes, coupled with nucleic acid cloning and sequencing techniques represent the essential tools of modern molecular genetics and molecular medicine (see Chapter 39).

SUMMARY

- DNA consists of four bases—A, G, C, and T—that are held in linear array by phosphodiester bonds through the 3′ and 5′ positions of adjacent deoxyribose moieties.

- DNA is organized into two strands by the pairing of bases A to T and G to C on complementary strands. These strands form a double helix around a central axis.

- The ~3×10^9 bp of DNA in humans are organized into the haploid complement of 23 chromosomes. The exact sequence of these 3 billion nucleotides defines the uniqueness of each individual.

- DNA provides a template, both for its own replication and thus maintenance of the genotype, and for the transcription of the roughly 25,000 protein coding human genes as well as a large array of nonprotein coding regulatory ncRNAs.

- RNA exists in several different cellular nucleoprotein structures, most of which are directly or indirectly involved in protein synthesis or its regulation. The linear array of nucleotides in RNA consists of A, G, C, and U, and the sugar moiety is ribose.

- The major forms of RNA include mRNA, rRNA, tRNA, snRNAs and regulatory ncRNAs. Certain RNA molecules act as catalysts (ribozymes).

REFERENCES

Ali T, Grote P: Beyond the RNA-dependent function of LncRNA genes. eLife 2020; 9:e60583. doi.org/10.7554/eLife.60583.

Berget SM, Moore C, Sharp PA: Spliced segments at the 5′ terminus of adenovirus 2 late mRNA. Proc Natl Acad Sci. USA 1977;74:3171.

Doudna JA: The promise and challenge of therapeutic genome editing. Nature 2020;578:229.

Goodall GJ, VO Wickramasinghe: RNA in Cancer. Nat Rev Cancer 2020; doi: 10.1038/s41568-020-00306-0.

Herbert A: A Genetic Instruction Code Based on DNA Conformation. Trends Genet 2019; 35:887.

Noller HF: The parable of the caveman and the Ferrari: protein synthesis and the RNA world. Philos Trans R Soc Lond B Biol Sci 2017;372(1716):20160187.

Rich A, Zhang S: Timeline: Z-DNA: the long road to biological function. Nat Rev Genet 2003;4:566.

Watson JD, Crick FH: Molecular structure of nucleic acids: a structure for deoxyribose nucleic acid. Nature 1953;171:737.

DNA Organization, Replication, & Repair

P. Anthony Weil, PhD

O B J E C T I V E S

After studying this chapter, you should be able to:

- Appreciate that the roughly 3×10^9 base pairs of DNA that compose the haploid genome of humans are divided uniquely between 23 linear DNA units, the chromosomes. Humans, being diploid, have 23 pairs of these linear chromosomes: 22 autosomes and two sex chromosomes.

- Understand that human genomic DNA, if extended end-to-end, would be meters in length, yet still fits within the nucleus of the cell, an organelle that is only microns (μ; 10^{-6} meters) in diameter. Such condensation in DNA length, in part, is induced following its association with the highly positively charged histone proteins resulting in the formation of a unique DNA-histone complex termed the nucleosome. Nucleosomes have DNA wrapped around the surface of an octamer of histones.

- Explain that strings of nucleosomes form along the linear sequence of genomic DNA to form chromatin, which itself can be more tightly packaged and condensed, this ultimately leads to the formation of the chromosomes.

- Appreciate that while the chromosomes are the macroscopic functional units for DNA transcription, replication, recombination, gene assortment, and cellular division, it is DNA function at the level of the individual nucleotides that composes regulatory sequences linked to specific genes that are essential for life.

- Describe the steps, phase of the cell cycle, and the molecules responsible for the replication, repair, and recombination of DNA, and understand the negative effects that errors in any of these processes can have on cellular, and thus, organismal integrity and health.

BIOMEDICAL IMPORTANCE[*]

The genetic information in the DNA of a chromosome can be transmitted by exact replication or it can be exchanged by a number of processes, including recombination, transposition, and gene conversion. These processes provide a means of ensuring adaptability and diversity for the organism, but when they go awry, can also result in disease. A number of enzyme systems are involved in DNA replication, alteration, and repair. Mutations are due to a change in the base sequence of DNA and may result from the faulty replication, transposition, or repair of DNA and occur with a frequency of about one in every 10^6 cell divisions. Abnormalities in gene products (either in RNA, protein function, or amount) can be the result of mutations that occur in transcribed protein coding, and nonprotein coding DNA, or nontranscribed regulatory-region DNA. A mutation in a germ cell is transmitted to offspring (so-called vertical transmission of hereditary disease). A number of factors, including viruses, chemicals, ultraviolet light, and ionizing radiation can all increase the rate of mutation.

---

[*]So far as possible, the discussion in this chapter and in Chapters 36, 37, and 38 will focus on mammalian organisms, which are, of course, among the higher eukaryotes. At times, it will be necessary to refer to observations in prokaryotic organisms such as the bacterium *Escherichia coli* and its viruses, or less complex eukaryotic model systems such as the fruit fly *Drosophila*, the nematode roundworm *Caenorhabditis elegans*, or the Brewer's yeast *Saccharomyces cerevisae*. However, in such cases the information will be of a kind that can be readily extrapolated to mammalian organisms.

Mutations often affect somatic cells and so are passed on to successive generations of cells, but only within an organism (ie, horizontally). It is becoming apparent that a number of diseases—and perhaps most cancers—are due to the combined effects of vertical transmission of mutations as well as horizontal transmission of induced mutations and the impact thereof on cellular function.

CHROMATIN IS THE CHROMOSOMAL MATERIAL IN THE NUCLEI OF CELLS OF EUKARYOTIC ORGANISMS

Chromatin consists of very long **double-stranded DNA (dsDNA) molecules** and a nearly equal mass of small basic proteins termed **histones** as well as a smaller amount of **nonhistone proteins** (most of which are acidic and larger than histones) and a small quantity of **RNA**. The nonhistone proteins include enzymes involved in DNA replication and repair, and the proteins involved in RNA synthesis, processing, and transport to the cytoplasm as well as an array of proteins that regulate these various processes. The dsDNA helix in each chromosome has a length that is thousands of times the diameter of the cell nucleus. One purpose of the molecules that comprise chromatin, particularly the histones, is to condense the DNA; however, it is important to note that the histones also integrally participate in gene regulation (see Chapters 36, 38, and 42); indeed, histones contribute importantly to all DNA-directed molecular transactions. Electron microscopic studies of chromatin have demonstrated dense spherical particles called **nucleosomes**, which are approximately 10 nm in diameter and connected by DNA filaments (**Figure 35–1**). Nucleosomes are composed of DNA wound around an octameric complex of histone molecules. Biochemical, biophysical, and X-ray crystallography data all corroborate this model of nucleosome structure.

Histones Are the Most Abundant Chromatin Proteins

Histones are a small family of closely related basic proteins. **H1 histones** are the ones least tightly bound to chromatin (**Figures 35–1**, **35–2**, and **35–3**) and are, therefore, easily removed with a salt solution, after which chromatin becomes more soluble. The organizational unit of this soluble chromatin is the nucleosome. **Nucleosomes contain four major types of histones: H2A, H2B, H3, and H4**. The sequence and structures of all four histones, H2A, H2B, H3, and H4, the so-called core histones that form the nucleosome, have been highly conserved between species, although variants of these four histones exist in various organisms, and are used for specialized purposes. The extreme conservation of core histone sequences implies that the overall functions of histones are identical in all eukaryotes, and that the entire molecule

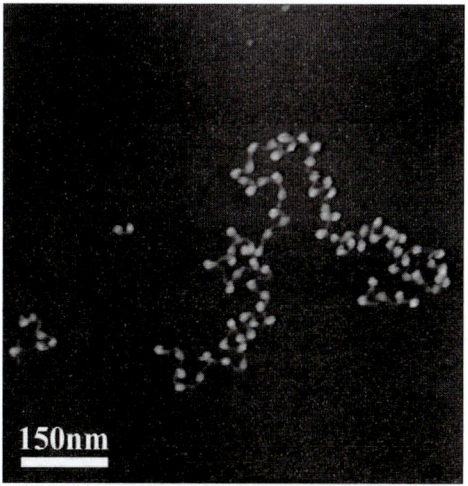

FIGURE 35–1 Electron micrograph of chromatin showing individual nucleosomes (white, ball-shaped) attached to strands of DNA (thin, gray line); see also Figure 35–2. (Reproduced with permission from Shao Z. Probing Nanometer Structures with Atomic Force Microscopy. News Physiol Sci. 1999;14:142-149.)

is involved quite specifically in carrying out these functions. The carboxyl terminal two-thirds of the histone molecules are hydrophobic, while their amino terminal thirds are particularly rich in basic amino acids. **These four core histones are subject to at least six major, or frequent, types of covalent modification or posttranslational modifications (PTMs):** acetylation, methylation, phosphorylation, ADP-ribosylation, monoubiquitylation, and sumoylation. These histone modifications play important roles in chromatin structure and function, as illustrated in **Table 35–1**. Chapters 38 and 42 provide a more detailed discussion of the role of histone PTMs in cellular biology.

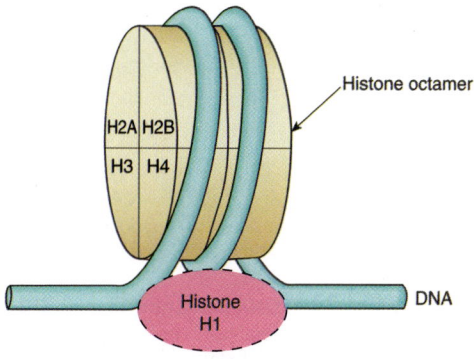

FIGURE 35–2 **Model for the structure of the nucleosome.** DNA is wrapped around the surface of a protein cylinder consisting of two each of histones H2A, H2B, H3, and H4 that form the histone octamer. The ~145 bp of DNA, consisting of 1.75 superhelical turns, are in contact with the histone octamer. The position of histone H1, when it is present, is indicated by the dashed outline at the bottom of the figure. Note that histone H1 interacts with DNA as it enters and exits the nucleosome.

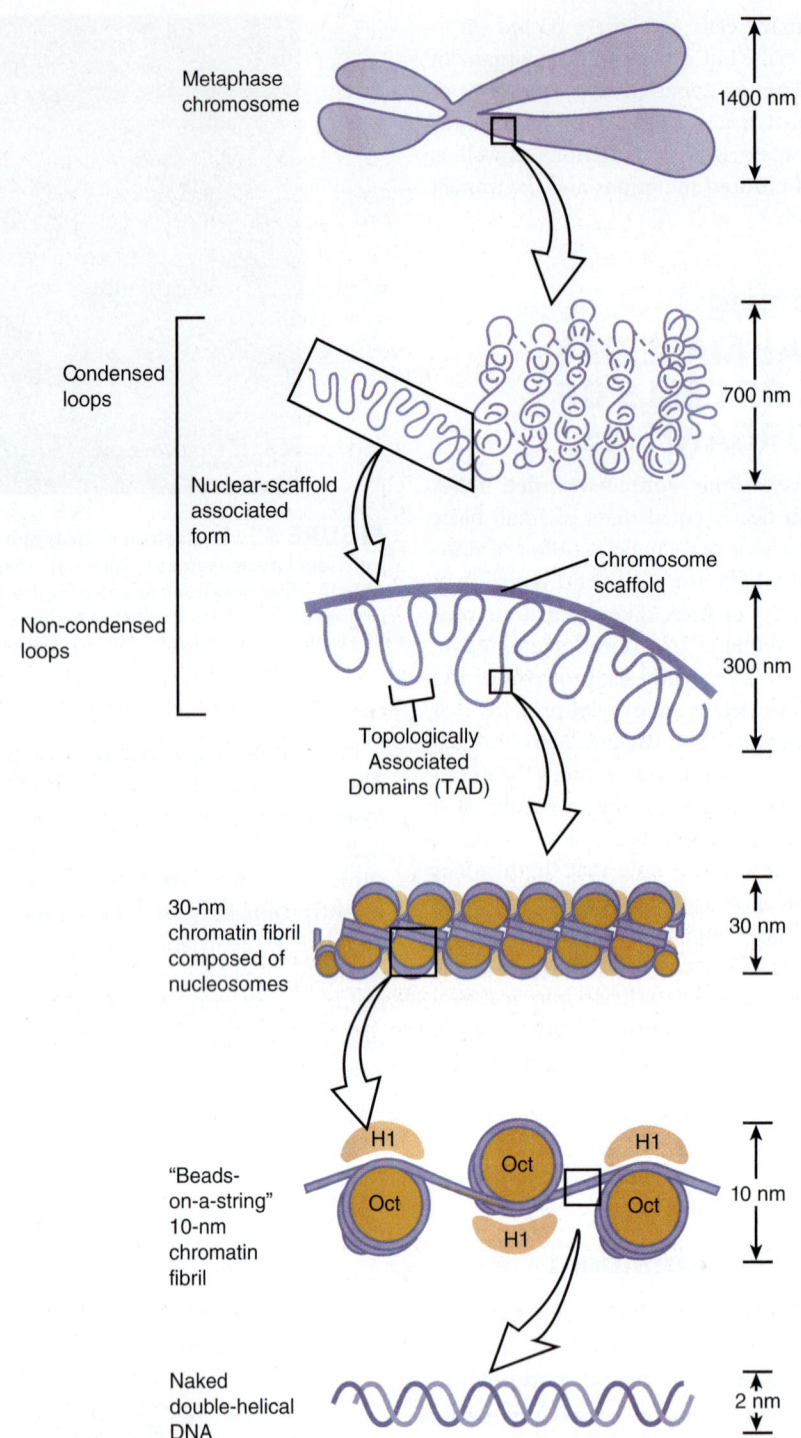

FIGURE 35–3   **Extent of DNA packaging in metaphase chromosomes (top) to noted duplex DNA (bottom).** Chromosomal DNA is packaged and organized at several levels as shown (see Table 35–2). Each phase of condensation or compaction and organization (bottom to top) decreases overall DNA accessibility to an extent that the DNA sequences in metaphase chromosomes are likely almost totally transcriptionally inert. In toto, these five levels of DNA compaction result in nearly a 10^4-fold linear decrease in end-to-end DNA length. Complete condensation and decondensation of the linear DNA in chromosomes occur in the space of just a few hours during the normal replicative cell cycle (see Figure 35–20).

TABLE 35–1 Possible Roles of Posttranslationally Modified Histones

1. Acetylation of histones H3 and H4 is associated with the activation or inactivation of gene transcription.

2. Acetylation of core histones is associated with chromosomal assembly during DNA replication.

3. Phosphorylation of histone H1 is associated with the condensation of chromosomes during the replication cycle.

4. ADP-ribosylation of histones is associated with DNA repair.

5. Methylation of histones is correlated with activation and repression of gene transcription.

6. Monoubiquitylation is associated with gene activation, repression, and heterochromatic gene silencing.

7. Sumoylation of histones (SUMO; small ubiquitin-related modifier) is associated with transcription repression.

8. Replacement of H2A with alternative H2AZ within nucleosomes is associated with transcriptional activation.

9. Alternative acylations of histones (propionylation, butyrylation, crotonylation, succinylation, malonylation and 2-hydroxyisobutyrylation) that likely link histone modifications with intracellular metabolism. These newly discovered modifications of histones correlate with gene activity.

The histones interact with each other in very specific ways. **H3 and H4 form a tetramer** containing two molecules of each $(H3–H4)_2$, while **H2A and H2B form dimers** (H2A–H2B). Under physiologic conditions, these histone oligomers associate to form the **histone octamer** of the composition $(H3–H4)_2$–$(H2A–H2B)_2$.

The Nucleosome Contains Histone & DNA

When the histone octamer is mixed with purified dsDNA under appropriate ionic conditions, the same x-ray diffraction pattern is formed as that observed in freshly isolated chromatin. Biochemical and electron microscopic studies confirm the existence of reconstituted nucleosomes in such in vitro-generated preparations. Furthermore, the reconstitution of nucleosomes from DNA and histones H2A, H2B, H3, and H4 is independent of the organismal or cellular origin of the various components. A result argues that nucleosome formation is an ancient and evolutionarily conserved, fundamental cellular process. Moreover, neither the histone H1, nor the nonhistone proteins are necessary for the reconstitution of the nucleosome core.

In the nucleosome, the DNA is supercoiled in a left-handed helix over the surface of the disk-shaped histone octamer (see Figure 35–2). The majority of core histone proteins interact with the DNA on the inside of the supercoil without protruding, although the amino terminal tails of all the histones are thought to extend outside of this structure and are available for regulatory PTMs (see Table 35–1).

The $(H3–H4)_2$ tetramer itself can confer nucleosome-like properties on DNA and thus has a central role in the formation of the nucleosome. The addition of two H2A–H2B dimers

stabilizes the primary particle and firmly binds two additional half-turns of DNA previously bound only loosely to the $(H3–H4)_2$. Thus, 1.75 superhelical turns of DNA are wrapped around the surface of the histone octamer, protecting 145 to 150 bp of DNA and forming the nucleosome core particle (see Figure 35–2). In chromatin, **core particles are separated by a roughly 30-bp region of DNA termed "linker."** Most of the DNA is in a repeating series of these structures, giving chromatin a repeating "beads-on-a-string" appearance when examined by electron microscopy (see Figure 35–1).

In vivo the assembly of nucleosomes is mediated by one of several nuclear chromatin assembly factors whose functions are facilitated by histone chaperones, a group of proteins that exhibit high affinity for binding histones. As the nucleosome is assembled, histones are released from the histone chaperones. Nucleosomes appear to exhibit preference for certain regions on specific DNA molecules, but the basis for this nonrandom distribution, termed **phasing**, is not yet completely understood. Phasing is likely related both to the relative physical flexibility of particular nucleotide sequences to accommodate the regions of kinking within the nucleosomal supercoil, as well as the presence of other DNA-bound factors that limit the sites of nucleosome deposition.

HIGHER-ORDER STRUCTURES PROVIDE FOR THE COMPACTION OF CHROMATIN

Electron microscopy of chromatin reveals two higher orders of structure—the 10-nm fibril and the 30-nm chromatin fiber—beyond that of the nucleosome itself. The disk-like nucleosome structure has a 10-nm diameter and a height of 5 nm. The **10-nm fibril** consists of nucleosomes arranged with their edges separated by a small distance (30 bp of linker DNA) with their flat faces parallel to the fibril axis (see Figure 35–3). The 10-nm fibril is probably further supercoiled with six or seven nucleosomes per turn to form the **30-nm chromatin fiber** (see Figure 35–3). Each turn of the supercoil is relatively flat, and the faces of the nucleosomes of successive turns would be nearly parallel to each other. H1 histones appear to stabilize the 30-nm fiber, but their position and that of the variable length linker DNA are not clear. It is probable that nucleosomes can form a variety of packed structures. In order to form a mitotic chromosome, the 30-nm fiber must be compacted in length another 100-fold (see following discussion).

In **interphase chromosomes**, chromatin fibers appear to be organized into 30,000 to 100,000 bp **loops or domains** anchored in a **scaffolding**, or supporting matrix structure within the nucleus, the so-called **nuclear matrix**. Within these domains, referred to as **Topologically Associated Domains, or TADs**; see Figure 35–3, some DNA sequences are located nonrandomly. It has been suggested that at least some of the looped domains of chromatin correspond to one or more separate genetic functions, containing both coding and noncoding regions of the cognate gene or genes. This nuclear architecture is dynamic, having important regulatory effects on gene regulation. Indeed, recent data suggest that certain genes or gene regions are mobile within the nucleus, moving obligatorily to discrete loci within the nucleus on activation. Future work in the area of the 3-D organization of nuclear chromatin, and further characterization of regulatory TADs, will determine the exact biological functions and molecular mechanisms responsible (see Chapter 38).

SOME REGIONS OF CHROMATIN ARE "ACTIVE" & OTHERS ARE "INACTIVE"

Generally, every cell of an individual metazoan organism contains the same genetic information. Thus, the differences between cell types within an organism must be explained by differential expression of the common genetic information. Chromatin containing active genes (ie, transcriptionally or potentially transcriptionally active chromatin) has been shown to differ in several important ways from that of inactive regions. The nucleosome structure of active chromatin appears to be altered, sometimes quite extensively, in highly active regions. DNA in active chromatin contains large regions (about 100,000 bases long) that are relatively more **sensitive to digestion by a nuclease** such as DNase I. DNase I makes single-strand cuts in nearly any segment of DNA due to its low-sequence specificity. However, DNase I will only avidly digest DNA that is not protected, or bound by protein. The sensitivity to DNase I of active chromatin regions reflects only a potential for transcription rather than transcription itself, and in several different cellular systems can be correlated with a relative lack of 5-methyldeoxycytidine (meC; see Figure 32–7) in the DNA, and particular histone variants and/or histone PTMs (phosphorylation, acetylation, etc.; see Table 35–1).

Within the large regions of active chromatin there exist shorter stretches of 100 to 300 nucleotides that exhibit an even greater (another 10-fold) sensitivity to DNase I. These **hypersensitive sites** probably result from a structural conformation that favors access of the nuclease to the DNA. These regions are often located immediately upstream from the active gene and are the location of interrupted nucleosomal structure caused by the binding of nonhistone regulatory transcription factor proteins (enhancer-binding transcriptional activator proteins; see Chapters 36 and 38). In many cases, it seems that if a gene is capable of being transcribed, it very often has a DNase-hypersensitive site(s) in the chromatin immediately upstream. As noted earlier, nonhistone regulatory proteins involved in transcription control and those involved in maintaining access to the template strand lead to the formation of hypersensitive sites. Such sites often provide the first clue about the presence and location of a transcription control element.

By contrast, transcriptionally inactive chromatin is densely packed during interphase as observed by electron microscopic studies and is referred to as **heterochromatin**; transcriptionally active chromatin stains less densely and is referred to as **euchromatin**. Generally, euchromatin is replicated earlier than heterochromatin in the mammalian cell cycle (see following discussion). The chromatin in these regions of inactivity is often high in meC content, and histones therein contain relatively lower levels of certain "activating" covalent modifications and higher levels of "repressing" histone PTMs (see Table 35–1).

There are two types of heterochromatin: constitutive and facultative. **Constitutive heterochromatin** is always relatively highly condensed (ie, heterochromatic), and thus essentially always inactive. Such constitutive heterochromatin is found in the regions near the chromosomal centromere and at chromosomal ends (telomeres). **Facultative heterochromatin** is at times condensed, but at other times it is actively transcribed and, thus, uncondensed and appears as euchromatin. Of the two members of the X-chromosome pair in mammalian females, one X chromosome is almost completely inactive transcriptionally and is heterochromatic. However, the heterochromatic X chromosome decondenses during gametogenesis and becomes transcriptionally active during early embryogenesis—thus, it is facultative heterochromatin.

Certain cells of insects, for example, *Chironomus* and *Drosophila*, contain giant chromosomes that have been replicated for multiple cycles without separation of daughter chromatids. These copies of DNA line up side by side in precise register and produce a banded chromosome containing regions of condensed chromatin and lighter bands of more extended chromatin. Transcriptionally active regions of these **polytene chromosomes** are especially decondensed into "puffs" that can be shown to contain the enzymes responsible for transcription and to be the sites of RNA synthesis (**Figure 35–4**). Using highly sensitive fluorescently labeled hybridization probes, specific gene sequences can be mapped, or "painted," within the nuclei of human cells, even without polytene chromosome formation, using **fluorescence in situ hybridization, or FISH** techniques (see Chapter 39).

DNA IS ORGANIZED INTO CHROMOSOMES

In preparation for cell duplication via the cyclical process termed mitosis, cellular DNA content is doubled (see Figure 35–20). During one phase of the mitotic cycle termed metaphase, the duplicated chromosomes condense and can readily be visualized. Condensed **chromosomes** possess a twofold symmetry, with the identical duplicated **sister chromatids** connected at a chromosomal structure termed the **centromere**, the relative position of which is characteristic for a given chromosome (**Figure 35–5**). The centromere is an adenine–thymine (A–T)-rich region containing repeated DNA sequences that range in size from 10^2 (brewers' yeast) to 10^6 (mammals) **base pairs** (**bp**). Metazoan centromeres are bound by nucleosomes containing the special histone H3 variant protein CENP-A and other specific centromere-binding proteins. This complex, called the **kinetochore**, provides the anchor for the mitotic spindle, on which chromosomal segregation occurs during mitosis.

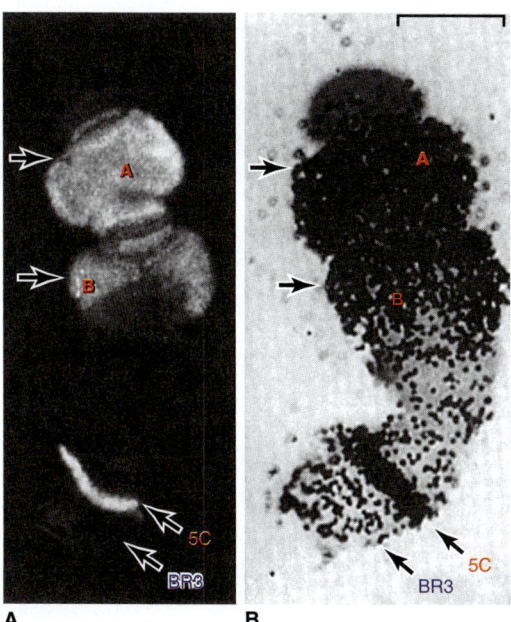

A   **B**

FIGURE 35–4 **Illustration of the tight correlation between the presence of RNA polymerase II (Table 36–2) and messenger RNA synthesis.** A number of genes, labeled A, B (top), and 5C, but not genes at locus (band) BR3 (5C, BR3, bottom) are activated when midge fly *Chironomus tentans* larvae are subjected to heat shock (39°C for 30 minutes). (**A**) Distribution of RNA polymerase II in isolated chromosome IV from the salivary gland (groups of light spots **at arrows**). The enzyme was detected by immunofluorescence using a fluorescently labeled antibody directed against the polymerase. The 5C and BR3 are specific bands of chromosome IV, and the arrows indicate puffs (ie, A, B, 5C). (**B**) Autoradiogram of a chromosome IV that was incubated in ³H-uridine to label the RNA. Note the correspondence of the immunofluorescence and presence of the radioactive RNA (black dots) (ie, A, B, 5C). Bar = 7 μm. (Reproduced with permission from Sass H. RNA polymerase B in polytene chromosomes: immunofluorescent and autoradiographic analysis during stimulated and repressed RNA synthesis. Cell. 1982;28(2):269-278.)

The ends of each chromosome contain structures called **telomeres**. **Telomeres consist of short TG-rich repeats.** Human telomeres have a variable number of repeats of the sequence 5′-TTAGGG-3′, which can extend for several kilobases. **Telomerase**, a multisubunit RNA template-containing complex related to viral RNA-dependent DNA polymerases (reverse transcriptases), is the enzyme responsible for telomere synthesis, and thus for maintaining the length of the telomere. Since telomere shortening has been associated with both malignant transformation (see Chapter 56) and aging (see Chapter 58), this enzyme has become an attractive target for cancer chemotherapy and drug development (see Chapter 56). Each sister chromatid contains one dsDNA molecule. As schematized in Figure 35–3, during interphase, the packing of the DNA molecule is less dense than it is in the condensed chromosome during metaphase. Metaphase chromosomes are nearly completely transcriptionally inactive.

The human haploid genome consists of about 3×10^9 bp and about 1.7×10^7 nucleosomes. Thus, each of the 23 chromatids in the human haploid genome would contain on the average

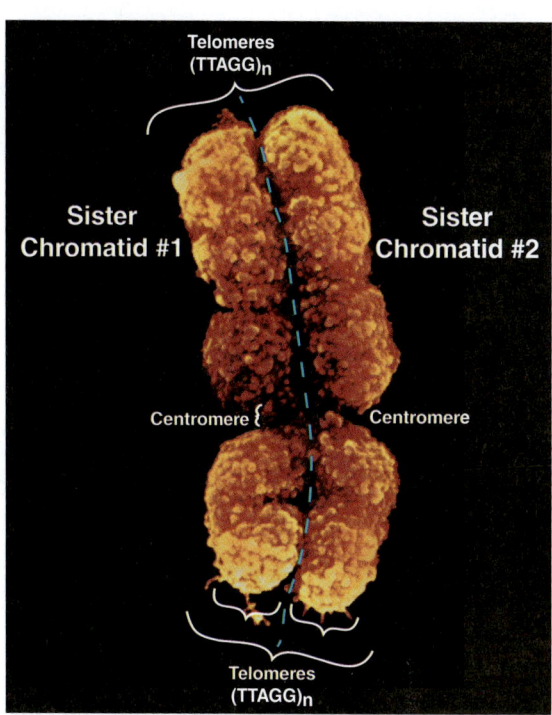

FIGURE 35–5 **The two sister chromatids of mitotic human chromosome 12.** The dashed line demarcates the sister chromatids. The location of the A+T-rich centromeric region connecting sister chromatids is indicated, as are two of the four telomeres residing at the very ends of the chromatids that are attached one to the other at the centromere. (Reproduced with permission from Biophoto Associates/Photo Researchers, Inc.)

1.3×10^8 nucleotides in one dsDNA molecule. Consequently, the length of each DNA molecule must be compressed about 8000-fold to generate the structure of a condensed metaphase chromosome. In metaphase chromosomes, the 30-nm chromatin fibers are also folded into a series of **looped domains**, the proximal portions of which are anchored to the nuclear matrix, likely through interactions with proteins termed **lamins** that constitute integral components of the inner nuclear membrane within the nucleus (see Figures 35–3 and 49–4). The packing ratios of each of the orders of DNA structure are summarized in **Table 35–2**. Though chromosomes are highly compacted, certain transcription proteins have been shown to still be able to access their target DNA sequences. The packaging of nucleoproteins within chromatids is not random, as

TABLE 35–2 **The Packing or Compaction Ratios of Each of the Orders of DNA Structure**

Chromatin Form	Packing Ratio
Naked double-helical DNA	~1.0
10-nm fibril of nucleosomes	7-10
30-nm chromatin fiber of superhelical nucleosomes	40-60
Condensed metaphase chromosome loops	8000

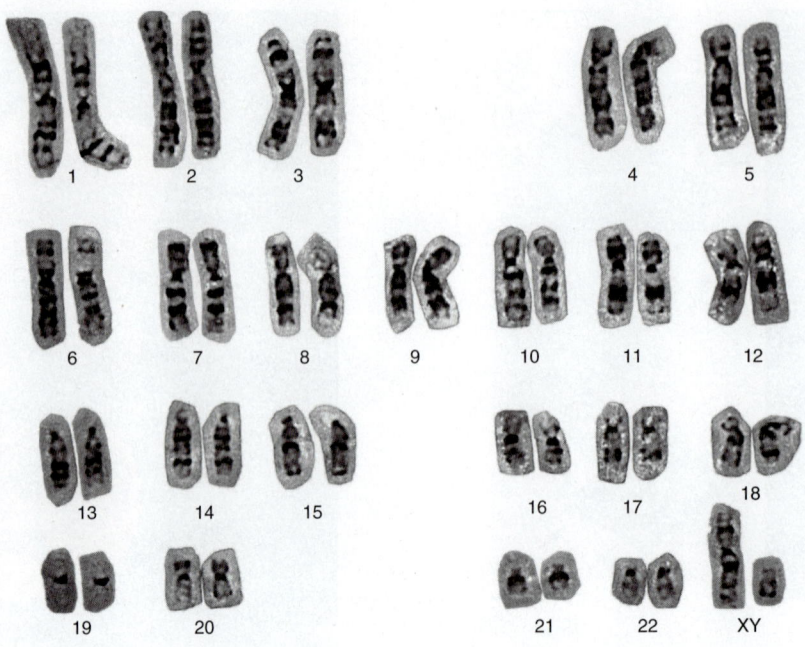

FIGURE 35–6 A human karyotype (of a man with a normal 46,XY constitution), in which the metaphase chromosomes have been stained by the Giemsa method and aligned according to the Paris Convention. (Reproduced with permission from H Lawce and F Conte.)

evidenced by the characteristic patterns observed when chromosomes are stained with specific dyes such as quinacrine or Giemsa stain (**Figure 35–6**).

From individual to individual within a single species, the pattern of staining (banding) of the entire chromosome complement is highly reproducible; nonetheless, it differs significantly between species, even those closely related. Thus, the packaging of the nucleoproteins in chromosomes of higher eukaryotes must in some way be dependent on species-specific characteristics of the DNA molecules.

A combination of specialized staining techniques and high-resolution microscopy has allowed cytogeneticists to quite precisely map many genes to specific regions of mouse and human chromosomes. With the recent elucidation of the human and mouse genome sequences (among others), it has become clear that many of these visual mapping methods were remarkably accurate.

Coding Regions Are Often Interrupted by Intervening Sequences

The protein coding regions of DNA, the transcripts of which ultimately appear in the cytoplasm as single mRNA molecules, are usually interrupted in the eukaryotic genome by large intervening sequences of nonprotein-coding DNA. Accordingly, the **primary transcripts** or **mRNA precursors** (originally termed **hnRNA** because this species of RNA was quite heterogeneous in size [length] and mostly restricted to the nucleus), contain nonprotein coding intervening sequences of RNA that must be removed in a process that also joins together the appropriate protein-coding segments to form the mature mRNA. Most coding sequences for a single mRNA are interrupted in the genome

(and thus in the primary transcript) by at least one—and in some cases as many as 50—noncoding intervening sequences termed **introns**. In most cases, the introns are much longer than the coding regions termed **exons**. The processing of the primary transcript, which involves precise removal of introns and splicing of adjacent exons, is described in Chapter 36.

The function of the intervening sequences, or introns, is not totally clear. However, mRNA precursor molecules can be differentially spliced thereby increasing the number of distinct (yet related) proteins produced by a single gene and its corresponding primary mRNA gene transcript. Introns may also serve to separate functional domains (exons) of coding information in a form that permits genetic rearrangement by recombination to occur more rapidly than if all coding regions for a given genetic function were contiguous. Such an enhanced rate of genetic rearrangement of functional domains might allow more rapid evolution of biologic function. In some instances, other protein-coding or noncoding RNAs are localized within the intronic DNA of certain genes (see Chapter 34). The relationships among chromosomal DNA, gene clusters on the chromosome, the exon–intron structure of genes, and the final mRNA product are illustrated in **Figure 35–7**.

THE EXACT FUNCTION OF MUCH OF THE MAMMALIAN GENOME IS NOT WELL UNDERSTOOD

The haploid genome of each human cell consists of 3.3×10^9 bp of DNA subdivided into 23 chromosomes. The entire haploid genome contains sufficient DNA to code for nearly 1.5 million average-sized protein coding genes (ie, ~2200 bp

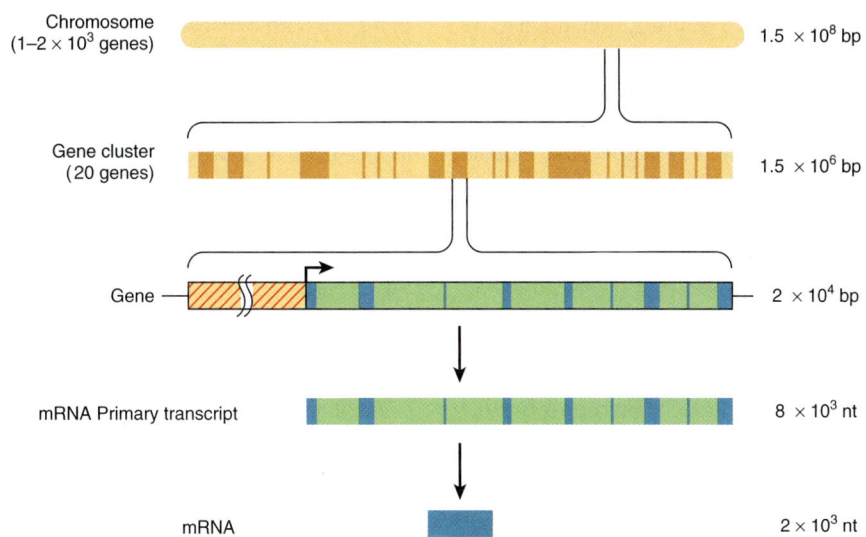

FIGURE 35–7 **The relationship between chromosomal DNA and mRNA.** The human haploid DNA complement of 3×10^9 bp is unequally distributed between 23 chromosomes (see Figure 35–6). Genes are often clustered on these chromosomes. An average gene is 2×10^4 bp in length, including the regulatory region (red-hatched area), which is often located at the 5′ end of the gene. The regulatory region is shown here as being adjacent to the transcription initiation site (bent arrow). Most eukaryotic genes have alternating exons and introns. In this example, there are nine exons (blue-colored areas) and eight introns (green-colored areas). The introns are removed from the primary transcript by processing reactions, and the exons are ligated together in sequence to form the mature mRNA through a process termed RNA splicing. (nt, nucleotides.)

of protein-coding DNA). However, early studies of mutation rates and of the complexities of the genomes of higher organisms suggested that humans have significantly fewer than 100,000 proteins encoded by the ~1% of the human genome that is composed of exonic DNA. Indeed, current estimates based on sequencing of the human genome and the collection of mRNA species produced therefrom suggest there are about 25,000 protein-coding genes in humans. This implies that most genomic DNA is nonprotein coding—that is, its information is never translated into an amino acid sequence of a protein molecule. Certainly, some of the excess DNA sequences serve to regulate the expression of genes during development, differentiation, and adaptation to the environment, either by serving as binding sites for regulatory proteins or by encoding regulatory ncRNAs. Some excess clearly makes up the intervening sequences or introns that split the coding regions of genes, and another portion of the excess appears to be composed of many families of repeated sequences for which clear functions have yet to be defined, though some small RNAs transcribed from these repeats can modulate transcription, either directly by interacting with the transcription machinery or indirectly by affecting the activity of the chromatin template. Interestingly, the ENCODE Project Consortium (see Chapters 10 and 39) has shown that most of the genomic sequence is indeed transcribed in at least some human cell types, albeit at a low level. A large fraction of such transcription appears to generate the lncRNAs (see Chapter 34). Further research will elucidate the role(s) played by such transcripts.

The DNA in an eukaryotic genome can be divided into two broad "sequence classes." These are unique-sequence DNA, or nonrepetitive DNA and repetitive-sequence DNA. In the haploid genome, unique-sequence DNA generally includes the

single copy genes that code for proteins. The repetitive DNA in the haploid genome includes sequences that vary in copy number from 2 to as many as 10^7 copies per cell.

More Than Half the DNA in Eukaryotic Organisms Is in Unique or Nonrepetitive Sequences

This estimation and genome-wide organization of repetitive sequence DNA was based on a variety of techniques, and most recently on direct genomic DNA sequencing. Similar techniques were used to determine the number of protein-encoding genes. In brewers' yeast (*Saccharomyces cerevisiae*, a lower eukaryote), about two-thirds of its 6200 genes are expressed, but only approximately one-fifth are required for viability under laboratory growth conditions. In typical tissues in a higher eukaryote (eg, mammalian liver and kidney), between 10,000 and 15,000 genes are actively expressed. Different combinations of genes are expressed in each tissue of course, and how this is accomplished is one of the major unanswered questions in biology.

In Human DNA, at Least 30% of the Genome Consists of Repetitive Sequences

Repetitive-sequence DNA can be broadly classified as moderately repetitive or as highly repetitive. The highly repetitive sequences consist of 5 to 500 base pair lengths repeated many times in tandem. These sequences are often clustered in centromeres and telomeres of the chromosome and some are present in about 1 to 10 million copies per haploid genome.

The majority of these sequences are transcriptionally inactive and some of these sequences play a structural role in the chromosome (see Figure 35–5 and Chapter 39).

The moderately repetitive sequences, which are defined as being present in numbers of less than 10^6 copies per haploid genome, are not clustered but are interspersed with unique sequences. In many cases, these long interspersed repeats are transcribed by RNA polymerase II and the produced RNAs contain 5′-Cap structures (see Figure 34–10) indistinguishable from those on mRNA. Depending on their length, moderately repetitive sequences are classified as **long interspersed nuclear elements (LINEs)** or **short interspersed nuclear elements (SINEs)**. Both types appear to be **retroposons**; that is, they arose from movement from one location to another **(transposition)** through an RNA intermediate by the action of reverse transcriptase that transcribes an RNA template into DNA. Mammalian genomes contain 20,000 to 50,000 copies of the 6 to 7 kbp LINEs. These represent species-specific families of repeat elements. SINEs are shorter (70-300 bp), and there may be more than 100,000 copies per genome. Of the SINEs in the human genome, one family, the **Alu family**, is present in about 500,000 copies per haploid genome and accounts for ~10% of the human genome. Members of the human Alu family and their closely related analogs in other animals can be transcribed as integral components of mRNA precursors or as discrete RNA molecules, including the well-studied 4.5S RNA and 7S RNA. These particular family members are highly conserved within a species as well as between mammalian species. Components of the short-interspersed repeats, including the members of the Alu family, may be mobile elements, capable of jumping into and out of various sites within the genome (see following discussion). These transposition events can have disastrous results, as exemplified by the insertion of Alu sequences into a gene, which, when so mutated, causes neurofibromatosis. Additionally, Alu B1 and B2 SINE RNAs have been shown to regulate mRNA production at the levels of transcription and mRNA splicing.

Microsatellite Repeat Sequences

One category of repeat sequences exists as both dispersed and grouped tandem arrays. The sequences consist of 2 to 6 bp repeated up to 50 times. These **microsatellite sequences** most commonly are found as dinucleotide repeats of AC on one strand and TG on the opposite strand, but several other forms occur, including CG, AT, and CA. The AC repeat sequences occur at 50,000 to 100,000 locations in the genome. At any locus, the number of these repeats may vary on the two chromosomes, thus providing heterozygosity of the number of copies of a particular microsatellite number in an individual. This is a heritable trait, and because of their number and the ease of detecting them using the **polymerase chain reaction (PCR)** (see Chapter 39), such repeats are useful in constructing genetic linkage maps. Most genes are associated with one or more microsatellite markers, so the relative position of genes on chromosomes can be assessed, as can the association

of a gene with a disease. Using PCR, a large number of family members can be rapidly screened for a certain **microsatellite polymorphism**. The association of a specific polymorphism with a gene in affected family members—and the lack of this association in unaffected members—may be the first clue about the genetic basis of a disease.

Trinucleotide sequences that increase in number (microsatellite instability) can cause disease. The unstable $(CGG)_n$ repeat sequence (n = the number of repeats; in this case CGG) is associated with the fragile X syndrome. Other trinucleotide repeats that undergo dynamic mutation (usually an increase in repeat numbers) are associated with Huntington chorea (CAG), myotonic dystrophy (CTG), spinobulbar muscular atrophy (CAG), and Kennedy disease (CAG). The advent of next-generation, high-throughput DNA sequencing technologies (see Chapter 39) has dramatically impacted both the speed, accuracy, and precision with which scientists and clinicians can analyze human genome structure. Some newly instituted clinical tests involve targeted genomic DNA sequencing prepared either from tissues or serum samples.

ONE PERCENT OF CELLULAR DNA IS IN MITOCHONDRIA

The majority of the polypeptides in mitochondria (about 54 out of 67) are encoded by nuclear genes, while the rest are coded by genes found in mitochondrial (mt) DNA. Human mitochondria contain 2 to 10 copies of a small circular ~16 kbp dsDNA molecule that makes up approximately 1% of total cellular DNA. This mtDNA codes for mt-specific ribosomal and transfer RNAs and for 13 proteins that play key roles in the respiratory chain (see Chapter 13). The linearized structural map of the human mitochondrial genes is shown in **Figure 35–8**. Some of the features of mtDNA are shown in **Table 35–3**.

An important feature of human mitochondrial mtDNA is that—because all mitochondria are contributed by the ovum during zygote formation—it is transmitted by maternal nonmendelian inheritance. Thus, in diseases resulting from mutations of mtDNA, an affected mother would in theory pass the disease to all of her children but only her daughters would transmit the trait. However, in some cases, deletions in mtDNA occur during oogenesis and thus are not inherited from the mother. A number of diseases have now been shown to be due to mutations of mtDNA. These include a variety of myopathies, neurologic disorders, and some forms of diabetes mellitus.

GENETIC MATERIAL CAN BE ALTERED & REARRANGED

An alteration in the sequence of purine and pyrimidine bases in a gene due to a change—a removal or an insertion—of one or more bases may result in an altered gene product or alteration of gene expression if nonprotein coding DNA is involved.

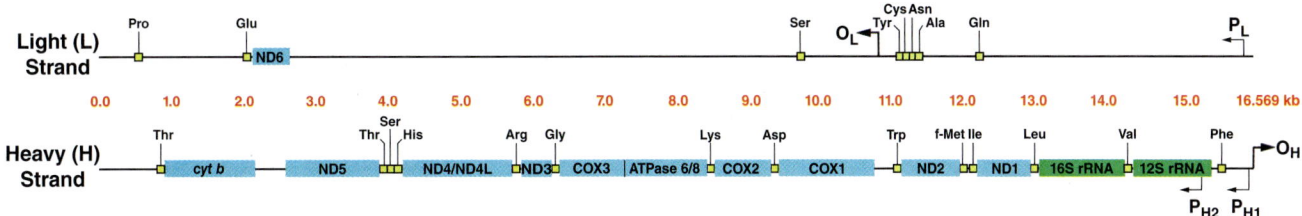

FIGURE 35–8 **Map of human mitochondrial genes.** The maps represent the so-called light (L; upper) and heavy (H; lower) strands of the 16,569 base pair linearized mitochondrial (mt) DNA The maps show the mt genes encoding subunits of NADH-coenzyme Q oxidoreductase (ND1 through ND6), cytochrome c oxidase (COX1 through COX3), cytochrome b (cyt b), ATP synthase (ATPase 6 and 8) and the 12S and 16S mt ribosomal rRNAs. Mt transfer RNA (tRNA) encoding genes are denoted by small yellow boxes and the three-letter code indicating the cognate amino acids which they specify during mt translation. The origin of heavy-strand (O_H), and light-strand (O_L) DNA replication, as well as the promoters for the initiation of heavy-strand (P_{H1} and P_{H2}), and light-strand (P_L) gene transcription are indicated by arrows and letters (see also Table 57–3). Figure generated using *Homo sapiens* mitochondrion, complete genome; Sequence: NCBI Reference NC_012920.1 and annotations therein.

Such **in**sertions or **del**etions are termed **indels**. Indels often result in a **mutation** whose consequences are discussed in detail in Chapter 37.

Chromosomal Recombination Is One Way of Rearranging Genetic Material

Genetic information can be exchanged between similar or homologous chromosomes. The exchange, or **recombination** event, occurs primarily during meiosis in mammalian cells and requires alignment of homologous metaphase chromosomes, an alignment that almost always occurs with great exactness. A process of chromosome (chromatid) crossing over occurs as shown in **Figure 35–9**. This usually results in an equal and reciprocal exchange of genetic information between homologous chromosomes. If the homologous chromosomes possess different alleles (ie, gene/DNA sequence variants) of the same genes, the crossover may produce noticeable and heritable genetic linkage differences. In the rare case where the alignment of homologous chromosomes is not exact, the crossing over, or recombination event, may result in an unequal exchange of information. One chromosome may receive less genetic material and thus a deletion, while the other partner

TABLE 35–3 **Major Features of Human Mitochondrial DNA**

• Is circular, double-stranded, and composed of heavy (H) and light (L) chains or strands
• Contains 16,569 bp

• Encodes 13 protein subunits of the respiratory chain (of a total of about 67)
    Seven subunits of NADH dehydrogenase (complex I)
    Cytochrome b of complex III
    Three subunits of cytochrome oxidase (complex IV)
    Two subunits of ATP synthase

• Encodes large (16S) and small (12S) mt ribosomal RNAs
• Encodes 22 mt tRNA molecules

• Genetic code differs slightly from the standard code
    UGA (standard stop codon) is read as Trp
    AGA and AGG (standard codons for Arg) are read as stop codons

• Contains very few untranslated sequences
• High mutation rate (5-10 times that of nuclear DNA)
• Comparisons of mtDNA sequences provide evidence about evolutionary origins of primates and other species

Data from Harding AE. Neurological disease and mitochondrial genes. Trends Neurosci. 1991;14(4):132-138.

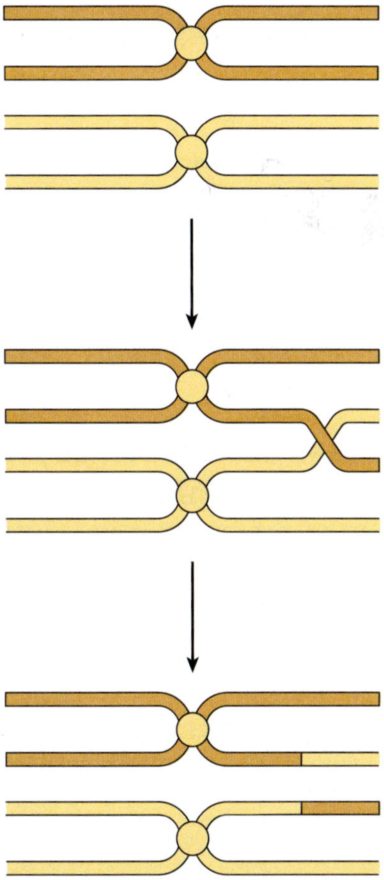

FIGURE 35–9 **The process of crossing over between homologous metaphase chromosomes to generate recombinant chromosomes.** See also Figure 35–12.

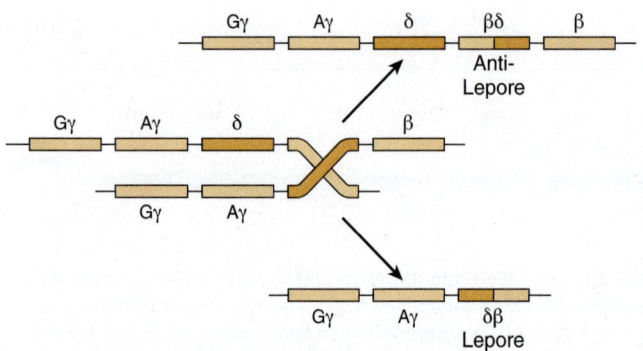

FIGURE 35–10 The process of unequal crossover in the region of the mammalian genome that harbors the structural genes encoding hemoglobins and the generation of the unequal recombinant products hemoglobin delta-beta Lepore and beta-delta anti-Lepore. The examples given show the locations of the crossover regions within amino acid coding regions of the indicated genes (ie, β and δ globin genes). (Modified with permission from Clegg JB, Weatherall DJ: β⁰ Thalassemia: time for a reappraisal? Lancet 1974;304(7873):133-135.)

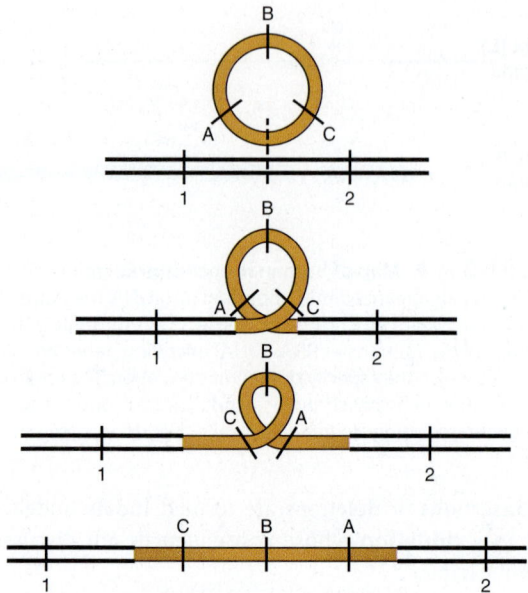

FIGURE 35–11 The integration of a circular genome from a virus (with genes A, B, and C) into the DNA molecule of a host (with genes 1 and 2) and the consequent ordering of the genes.

of the chromosome pair receives more genetic material and thus an insertion or duplication. One well-studied example of unequal crossing that occurs in humans involves the genes encoding hemoglobins. Unequal crossing over results in a human hemoglobinopathy designated Lepore and anti-Lepore (**Figure 35–10**).

The farther apart any two genes are on an individual chromosome, the greater the likelihood of a crossover recombination event. This is the basis for genetic mapping methods. **Unequal crossover** affects tandem arrays of repeated DNAs whether they are related globin genes, as in Figure 35–10, or more abundant repetitive DNA. Unequal crossover through slippage in the base pairing can result in expansion or contraction in the copy number of the repeat family and may contribute to the expansion and fixation of variant members throughout the repeat array.

Some Viruses Chromosomally Integrate Their Genomes in Infected Cells

Some bacterial viruses (bacteriophages) are capable of recombining with the DNA of a bacterial host in such a way that the genetic information of the bacteriophage is incorporated in a linear fashion into the genetic information of the host. This integration, which is a form of recombination, occurs by the mechanism illustrated in **Figure 35–11**. The backbone of the circularized bacteriophage genome is broken, as is that of the DNA molecule of the host; the appropriate ends are resealed with the proper polarity. The bacteriophage DNA is figuratively straightened out ("linearized") as it is integrated into the bacterial DNA molecule—frequently a closed circle as well. The site at which the bacteriophage genome integrates or recombines with the bacterial genome is chosen by one of two mechanisms. If the bacteriophage contains a DNA sequence **homologous** to a sequence in the host DNA molecule, then

a recombination event analogous to that occurring between homologous chromosomes can occur. However, some bacteriophages synthesize proteins that bind specific sites on bacterial chromosomes to a **nonhomologous** site characteristic of the bacteriophage DNA molecule. Integration occurs at the site and is said to be **"site specific."**

Many animal viruses, particularly the oncogenic viruses—either directly or, in the case of RNA viruses such as HIV that causes AIDS, double-stranded DNA copies generated by the action of the viral **RNA-dependent DNA polymerase**, or **reverse transcriptase**—can be integrated into chromosomes of the mammalian cell. Integration of the viral DNA into the genome of the infected cells generally is not "site specific" but does display site preferences. Not surprisingly a subset of such integration events is mutagenic.

Transposition Can Produce Processed Genes

In eukaryotic cells, small DNA elements that clearly are not viruses are capable of transposing themselves in and out of the host genome in ways that affect the function of neighboring DNA sequences. These mobile elements, sometimes called **"jumping DNA,"** or jumping genes, can carry flanking regions of DNA and, therefore, profoundly affect evolution. As mentioned earlier, the Alu family of moderately repeated DNA sequences has structural characteristics similar to the termini of retroviruses, which would account for the ability of the latter to move into and out of the mammalian genome.

Direct evidence for the transposition of other small DNA elements into the human genome has been provided by the discovery of **"processed genes"** for immunoglobulin molecules, α-globin molecules, and many others. These processed genes consist of DNA sequences identical or nearly identical to those of the messenger RNA for the appropriate gene product. That is, the 5′-nontranslated region, the coding region without intron representation, and the 3′ poly(A) tail are all present contiguously. This particular DNA sequence arrangement must have resulted from the reverse transcription of an appropriately processed messenger RNA molecule from which the intron regions had been removed and the poly(A) tail added. The only recognized mechanism that this reverse transcript could have used to integrate into the genome would have been a transposition event. In fact, these "processed genes" have short terminal repeats at each end, as do known transposed sequences in other organisms. In the absence of their transcription and thus genetic selection for function, many of the processed genes have been randomly altered through evolution so that they now contain nonsense codons that preclude their ability to encode a functional, intact protein even if they could be transcribed (see Chapter 37). Thus, such transposed sequences are referred to as "**pseudogenes**."

Gene Conversion Produces Rearrangements

Besides unequal crossover and transposition, a third mechanism can effect rapid changes in the genetic material. Similar sequences on homologous or nonhomologous chromosomes may occasionally pair up and eliminate any mismatched sequences between them. This may lead to the accidental fixation of one variant or another throughout a family of repeated sequences and thereby homogenize the sequences of the members of repetitive DNA families. This process is referred to as **gene conversion**.

Sister Chromatids Exchange

In diploid eukaryotic organisms such as humans, after cells progress through the DNA synthetic, or S phase of the mitotic cell cycle (see Figure 35–20), they contain a tetraploid content of DNA. This is in the form of sister chromatids of chromosome pairs (see Figure 35–6). Each of these sister chromatids contains identical genetic information since each is a product of the semiconservative replication of the original parent DNA molecule of that chromosome. Crossing over can occur between these genetically identical sister chromatids. Of course, these **sister chromatid exchanges** (**Figure 35–12**) have no genetic consequence as long as the exchange is the result of an equal crossover.

Immunoglobulin Genes Rearrange

In mammalian cells, some interesting gene rearrangements occur normally during development and differentiation. For example, the V_L and C_L genes, which encode for the immunoglobulin G (IgG) light-chain variable (V_L) and constant (C_L) portions of the IgG light chain in a single IgG molecule (see Chapters 38, 52), are widely separated in the germ line DNA. In the DNA of a differentiated IgG-producing (plasma) cell,

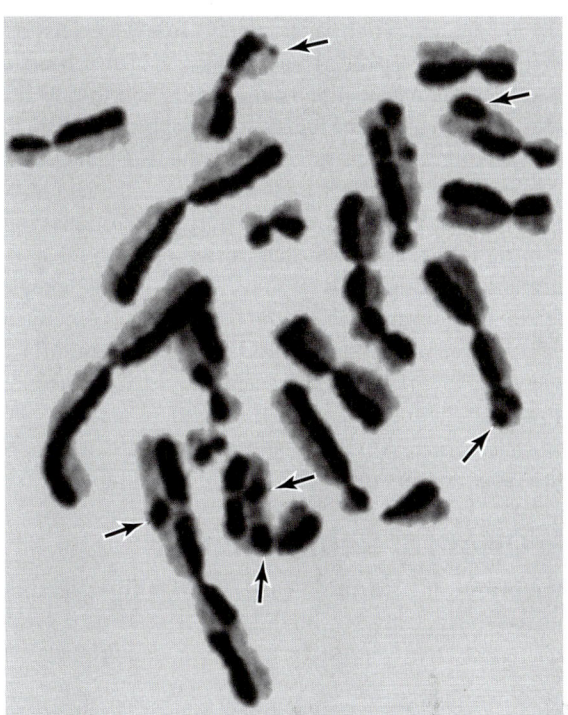

FIGURE 35–12 **Sister chromatid exchanges between human chromosomes.** The exchanges are detectable by Giemsa staining of the chromosomes of cells replicated for two cycles in the presence of **bromodeoxyuridine** (**BrdU**; 5-Bromo-2′-deoxyuridine; see Figure 32–13 with a depiction of 5-Iodo-2′deoxyuridine). The arrows indicate some regions of exchange. DNA synthesized with the thymine analog BrdU appears black in this image. (Reproduced with permission from S Wolff and J Bodycote.)

the same V_L and C_L genes have been moved physically closer, and linked together in the genome within a single transcription unit. However, even then, this rearrangement of DNA during differentiation does not bring the V_L and C_L genes into contiguity in the DNA. Instead, the DNA contains an intron of about 1200 bp at or near the junction of the V and C regions. This intron sequence is transcribed into RNA along with the V_L and C_L exons, and the interspersed, intronic non-IgG sequence information is removed from the RNA during its nuclear processing via mRNA splicing (see Chapters 36 and 38).

DNA SYNTHESIS & REPLICATION ARE RIGIDLY CONTROLLED

The primary function of DNA replication is the provision of progeny with the genetic information possessed by the parent. Thus, the replication of DNA must be complete and carried out in such a way as to maintain genetic stability within the organism and the species. The process of DNA replication is complex and involves many cellular functions and several verification procedures to ensure fidelity in replication. About 30 proteins are involved in the replication of the *Escherichia coli* chromosome; this process is significantly more complex in eukaryotic organisms.

In all cells, replication can occur only from a single-stranded DNA (ssDNA) template. Therefore, mechanisms must exist to target the site of initiation of replication and to unwind the dsDNA in that region. The replication complex must then form. After replication is complete in an area, the parent and daughter strands must reform dsDNA. In eukaryotic cells, an additional step must occur. The dsDNA must reform the chromatin structure, including nucleosomes that existed prior to the onset of replication. Although this entire process is not completely understood in eukaryotic cells, replication has been quite precisely described in prokaryotic cells, and the general principles are the same in both. The major steps are listed in **Table 35–4**, illustrated in **Figure 35–13**, and discussed, in sequence, in following discussion. A number of proteins, most with specific enzymatic action, are involved in this process (**Table 35–5**).

The Origin of Replication

At the **origin of replication (ori)**, there is an association of sequence-specific dsDNA-binding proteins with a series of direct repeat DNA sequences. In *E. coli*, the origin of DNA

TABLE 35–4 Steps Involved in DNA Replication in Eukaryotes

1. Identification of the origins of replication

2. ATP hydrolysis-driven removal of nucleosomes and unwinding of dsDNA to provide an ssDNA template

3. Formation of the replication fork; synthesis of RNA primer

4. Initiation of DNA synthesis and elongation

5. Formation of replication bubbles with ligation of the newly synthesized DNA segments

6. Reconstitution of chromatin structure

synthesis termed oriC is bound by the protein dnaA, which forms a complex consisting of 150 to 250 bp of DNA and multimers of this single-strand DNA-binding protein. This binding event leads to the local denaturation and unwinding of an adjacent A+T-rich region of DNA. Functionally similar **autonomously replicating sequences (ARS) or replicators** have been identified in yeast cells. The ARS contains a somewhat

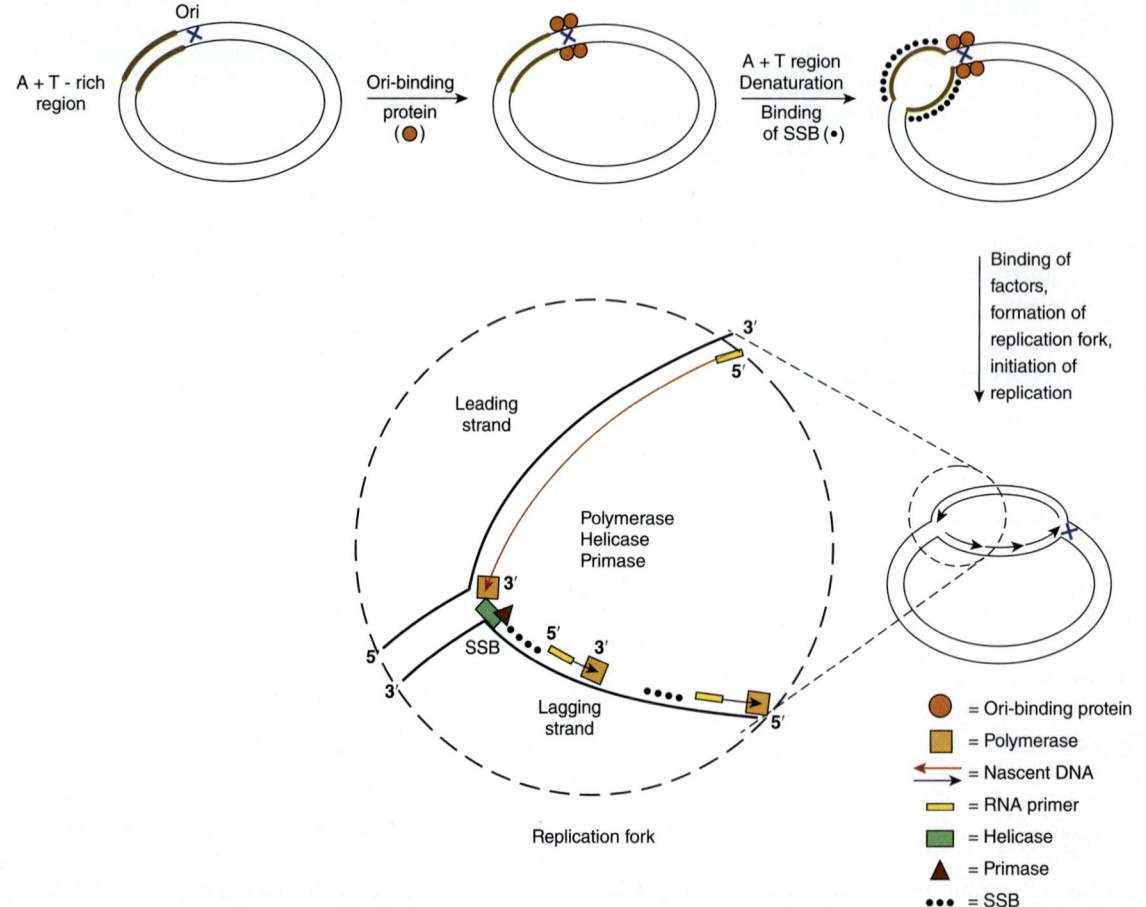

FIGURE 35–13 **Steps involved in DNA replication.** This figure describes DNA replication in an *E. coli* cell, but the general steps are similar in eukaryotes. A specific interaction of a protein (the dnaA protein) to the origin of replication (oriC) results in local unwinding of DNA at an adjacent A+T-rich region. The DNA in this area is maintained in the single-strand conformation (ssDNA) by single-strand-binding proteins (SSBs). This allows a variety of proteins, including helicase, primase, and DNA polymerase, to bind and to initiate DNA synthesis. The replication fork proceeds as DNA synthesis occurs continuously (long red arrow) on the leading strand and discontinuously (short black arrows) on the lagging strand. The nascent DNA is always synthesized in the 5′ to 3′ direction, as DNA polymerases can add a nucleotide only to the 3′ end of a DNA strand.

TABLE 35-5 Classes of Proteins Involved in Replication

Protein	Function
DNA polymerases	Deoxynucleotide polymerization
Helicases	ATP-driven processive unwinding of DNA
Topoisomerases	Relieve torsional strain that results from helicase-induced unwinding
DNA primase	Initiates synthesis of RNA primers
Single-strand binding proteins (SSBs)	Prevent premature reannealing ssDNA strands to form dsDNA
DNA ligase	Seals the single-strand nick between the nascent chain and Okazaki fragments on lagging strand

degenerate 11-bp sequence called the **origin replication element (ORE)**. The ORE binds a set of proteins, analogous to the dnaA protein of *E. coli*, the group of proteins is collectively called the **origin recognition complex (ORC)**. ORC homologs have been found in all eukaryotes examined. The ORE is located adjacent to an approximately 80-bp A+T-rich sequence that is easy to unwind. This is called the **DNA unwinding element (DUE)**. The DUE is the origin of replication in yeast and is bound by the MCM protein complex.

Consensus sequences similar in sequence to oriC or ARS in structure have not been precisely defined in mammalian cells. However, several of the proteins that participate in ori recognition and function have been identified in human cells and appear quite similar to their yeast counterparts in both amino acid sequence and function. This fact argues that at the functional level ORE-like elements exist in humans.

Unwinding of DNA

The interaction of proteins with ori defines the start site of replication and provides a short region of ssDNA essential for initiation of synthesis of the nascent DNA strand. This process requires the formation of a number of protein–protein and protein-DNA interactions. A critical step is provided by a DNA helicase that allows for processive unwinding of DNA. This function is provided by a complex of dnaB helicase and the dnaC protein. Single-stranded DNA-binding proteins (SSBs) stabilize this complex once formed.

Formation of the Replication Fork

A replication fork consists of four components that form in the following order: (1) the DNA helicase unwinds a short segment of the parental duplex DNA; (2) SSBs bind to ssDNA and prevent premature reannealing of ssDNA back into dsDNA; (3) a primase initiates synthesis of an RNA molecule that is essential for priming DNA synthesis; and (4) the DNA polymerase initiates nascent, daughter-strand synthesis on the free 3′-OH of the primer.

The DNA polymerase III enzyme (the *dnaE* gene product in *E. coli*) binds to template DNA as part of a multiprotein complex that consists of several polymerase accessory factors (β', γ, δ, δ', and τ). DNA polymerases only synthesize DNA in the 5′ to 3′ direction, and only one of the several different types of polymerases is involved at the replication fork. Because the DNA strands are antiparallel (see Chapter 34), the polymerase functions asymmetrically. On the **leading (forward) strand**, the DNA is synthesized continuously. On the **lagging (retrograde) strand**, the DNA is synthesized in short (1-5 kb; see Figure 35–16) fragments, the so-called **Okazaki fragments**, so named after the scientist who discovered them. Several Okazaki fragments (up to 1000) must be sequentially synthesized for each replication fork. To ensure that this happens, the helicase acts on the lagging strand to unwind dsDNA in a 5′ to 3′ direction. The helicase associates with the primase to afford the latter proper access to the template. This allows the RNA primer to be made and, in turn, the polymerase to begin replicating the DNA. This is an important reaction sequence since DNA polymerases cannot initiate DNA synthesis de novo. The mobile complex between helicase and primase has been called a **primosome**. As the synthesis of an Okazaki fragment is completed and the polymerase is released, a new primer has been synthesized. The same polymerase molecule remains associated with the replication fork and proceeds to synthesize the next Okazaki fragment.

The DNA Polymerase Complex

A number of different DNA polymerase molecules engage in DNA replication. These share three important properties: (1) **chain elongation**, (2) **processivity**, and (3) **proofreading**. Chain elongation accounts for the rate (in **nucleotides per second**; **nt/s**) at which polymerization (ie, phosphodiester bond formation) occurs. Processivity is an expression of the number of nucleotides added to the nascent chain before the polymerase disengages from the template. The proofreading function identifies copying errors and corrects them. In *E. coli*, DNA polymerase III (pol III) functions at the replication fork. Of all polymerases, it catalyzes the highest rate of chain elongation and is the most processive. It is capable of polymerizing 0.5 Mb of DNA during one cycle on the leading strand. Pol III is a large (>1 MDa), multisubunit protein complex in *E. coli*. DNA pol III associates with the two identical β subunits of the DNA sliding "clamp"; this association dramatically increases pol III-DNA complex stability, processivity (100 to >50,000 nucleotides) and rate of chain elongation (20-50 nt/s) generating the high degree of processivity the enzyme exhibits.

Polymerase I (pol I) and II (pol II) are mostly involved in proofreading and DNA repair. Eukaryotic cells have counterparts for each of these enzymes plus a large number of additional DNA polymerases primarily involved in DNA repair. A comparison is shown in **Table 35–6**.

In mammalian cells, the polymerase is capable of polymerizing at a rate that is somewhat slower than the rate of polymerization of deoxynucleotides by the bacterial DNA polymerase complex. This reduced rate may result from interference by nucleosomes.

TABLE 35–6   **A Comparison of Prokaryotic & Eukaryotic DNA Polymerases**

E. coli	Eukaryotic	Function
I		Gap filling following DNA replication, repair, and recombination
II		DNA proofreading and repair
	β	DNA repair
	γ	Mitochondrial DNA synthesis
III	ε	Processive, leading strand synthesis
DnaG	α	Primase
	δ	Processive, lagging strand synthesis

Initiation & Elongation of DNA Synthesis

The initiation of DNA synthesis (**Figure 35–14**) requires **priming** by a short length of RNA, about 10 to 200 nucleotides long. In *E. coli* primer formation is catalyzed by dnaG (primase), in eukaryotes DNA Pol α synthesizes these RNA primers. Once complete DNA synthesis commences on this short RNA primer. The priming process involves nucleophilic attack by the 3′-hydroxyl group of the RNA primer on the phosphate of the first entering deoxynucleoside triphosphate (N in Figure 35–14) with the splitting off of pyrophosphate; this transition to DNA synthesis is catalyzed by the appropriate DNA polymerases (DNA pol III in *E. coli*; DNA pol δ and ε in eukaryotes). The 3′-hydroxyl group of the recently attached deoxyribonucleoside monophosphate is then free to carry out a **nucleophilic attack** on the next entering deoxyribonucleoside triphosphate (N + 1 in Figure 35–14), again at its α phosphate moiety, with the splitting off of pyrophosphate. Of course, selection of the proper deoxyribonucleotide whose terminal 3′-hydroxyl group is to be attacked is dependent on proper base pairing with the other strand of the DNA molecule according to Watson and Crick base pairing rules (**Figure 35–15**). When an adenine deoxyribonucleoside monophosphoryl moiety is in the template position, a thymidine triphosphate will interact with the dXTP binding site of the DNA polymerase and its α phosphate will be attacked by the 3′-hydroxyl group of the deoxyribonucleoside monophosphoryl most recently added to the polymer. By this stepwise process, the template dictates which deoxyribonucleoside triphosphate is complementary, and by hydrogen bonding, holds it in place while the 3′-hydroxyl group of the growing strand attacks and incorporates the new nucleotide into the polymer. These segments of DNA attached to an RNA primer component are the Okazaki fragments (**Figure 35–16**). In both bacteria and mammals, after many Okazaki fragments are generated, the replication complex begins to remove the RNA primers, to fill in the gaps left by their removal with the proper base-paired deoxynucleotide, and then to seal the fragments of newly synthesized DNA by enzymes referred to as **DNA ligases**.

Replication Exhibits Polarity

As has already been noted, DNA molecules are double stranded and the two strands are antiparallel. The replication of DNA in prokaryotes and eukaryotes occurs on both strands simultaneously. However, an enzyme capable of polymerizing DNA in the 3′ to 5′ direction does not exist in any organism, so that both of the newly replicated DNA strands cannot grow in the same direction simultaneously. Nevertheless, in bacteria the same enzyme does replicate both strands at the same time (in eukaryotes pol ε and pol δ catalyze leading and lagging strand synthesis; see Table 35–6). The single enzyme replicates one strand ("leading strand") in a continuous manner in the 5′ to 3′ direction, with the same overall forward direction. It replicates the other strand ("lagging strand") discontinuously while polymerizing the nucleotides in short spurts of 150 to 250 nucleotides, again in the 5′ to 3′ direction, but at the same time it faces toward the back end of the preceding RNA primer rather than toward the unreplicated portion. This process of **semidiscontinuous DNA synthesis** is shown diagrammatically in Figures 35–13 and 35–16.

Formation of Replication Bubbles

Replication of the circular bacterial chromosome, composed of roughly 5×10^6 bp of DNA proceeds from a single ori. This process is completed in about 30 minutes, a replication rate of 3×10^5 bp/min. The entire mammalian genome replicates in approximately 9 hours, the average period required for formation of a tetraploid genome from a diploid genome in a replicating cell. If a mammalian genome (3×10^9 bp) replicated at the same rate as bacteria (ie, 3×10^5 bp/min) from but a single ori, replication would take over 150 hours! Metazoan organisms get around this problem using two strategies. First, replication is bidirectional. Second, replication proceeds from multiple origins in each chromosome (a total of as many as 100 in humans). Thus, replication occurs in both directions along all of the chromosomes, and both strands are replicated simultaneously. This replication process generates "**replication bubbles**" (**Figure 35–17**).

The multiple ori sites that serve as origins for DNA replication in eukaryotes are poorly defined except in a few animal viruses and in yeast. However, it is clear that initiation is regulated both spatially and temporally, since clusters of adjacent sites initiate replication synchronously. Replication firing, or DNA replication initiation at a replicator/ori, is influenced by a number of distinct properties of chromatin structure that are just beginning to be understood. It is clear, however, that there are more replicators and excess ORC than needed to replicate the mammalian genome within the time of a typical S phase. Therefore, mechanisms for controlling the excess ORC-bound replicators must exist. Understanding the control of the formation and firing of replication complexes is one of the major challenges in this field.

During the replication of DNA, there must be a separation of the two strands to allow each to serve as a template by hydrogen bonding its nucleotide bases to the incoming

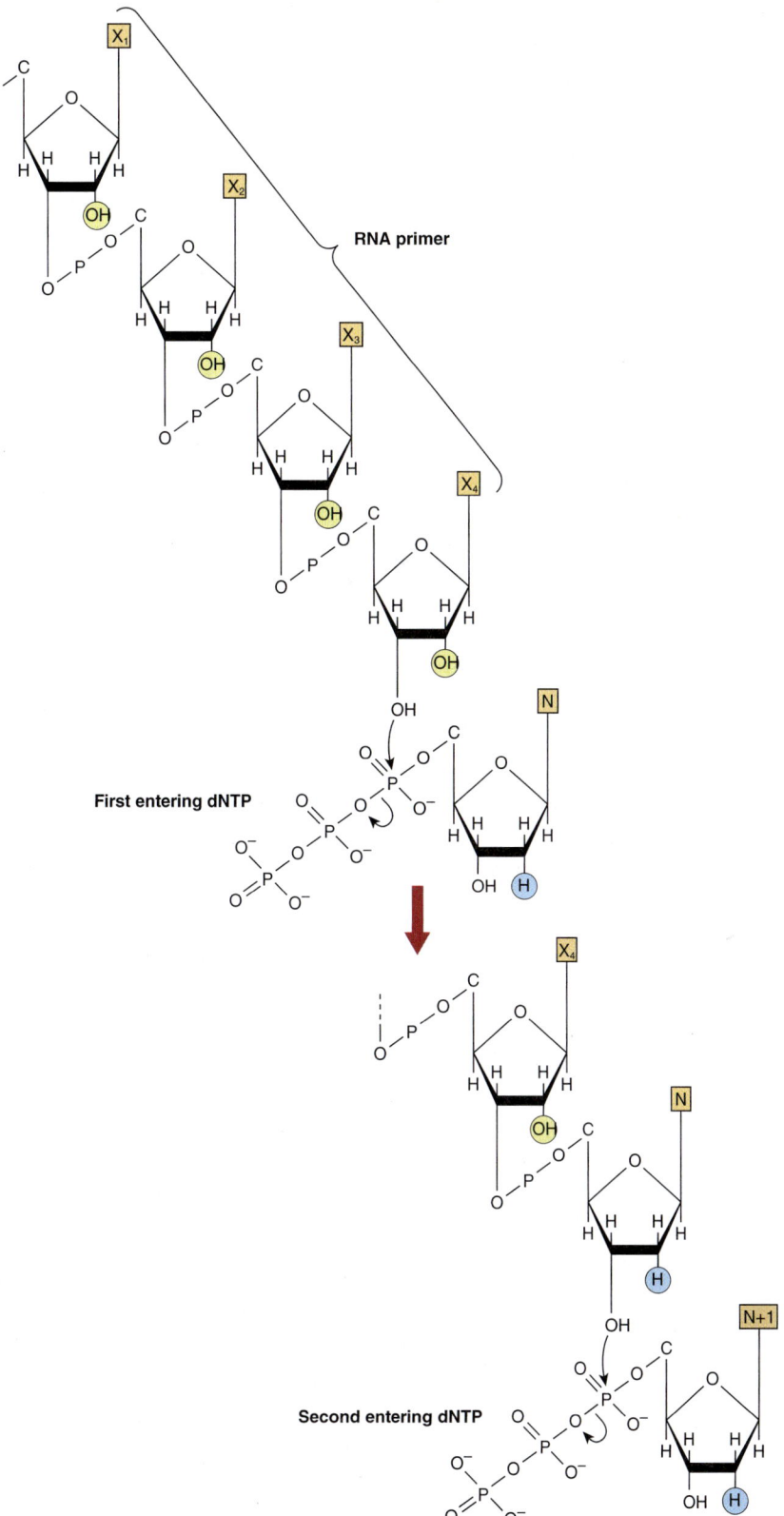

FIGURE 35–14 **The initiation of DNA synthesis upon a primer of RNA and the subsequent attachment of the second deoxyribo-nucleoside triphosphate.** Note the blue highlighting of the 2′-H moiety of deoxyribose and the yellow highlighting of the 2′-OH moiety within the RNA primer.

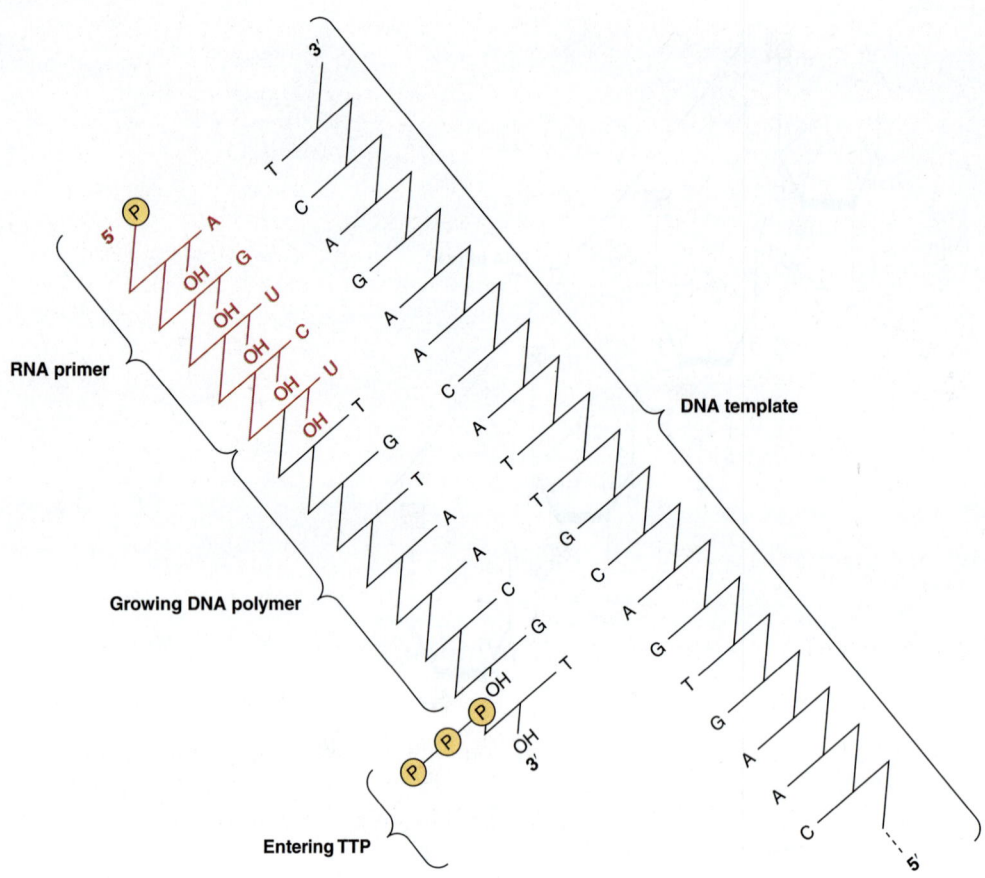

FIGURE 35–15   The RNA-primed synthesis of DNA demonstrating the template function of the complementary strand of parental DNA.

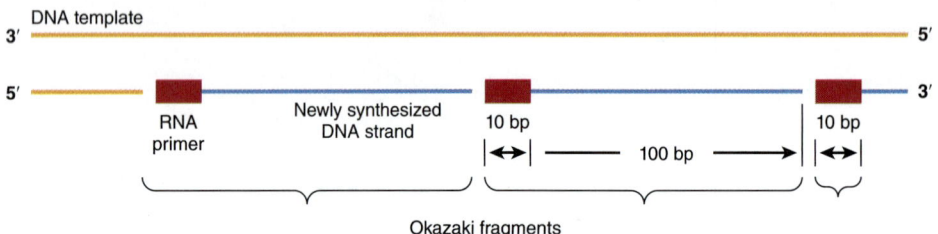

FIGURE 35–16   **The discontinuous polymerization of deoxyribonucleotides on the lagging strand; formation of Okazaki fragments during lagging strand DNA synthesis is illustrated.** Okazaki fragments are 100 to 250 nucleotides long in eukaryotes, 1000 to 2000 nucleotides in prokaryotes.

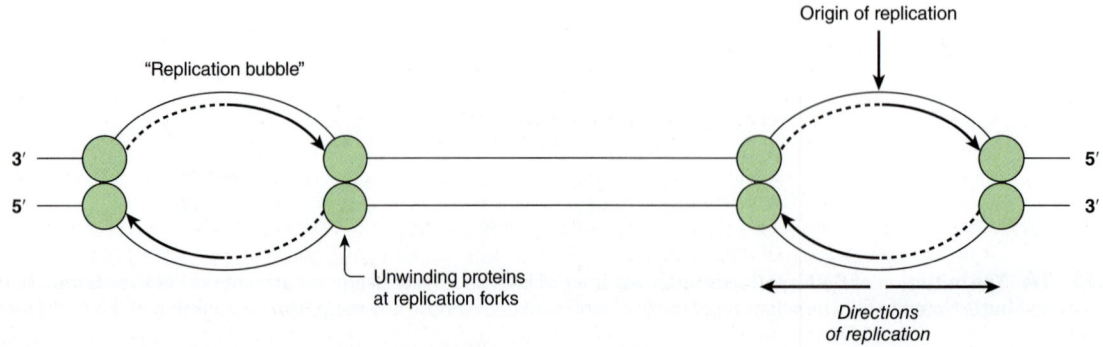

FIGURE 35–17   **The generation of "replication bubbles" during the process of DNA synthesis.** The bidirectional replication and the proposed positions of unwinding proteins at the replication forks are depicted.

deoxynucleoside triphosphate. The separation of the DNA strands is promoted by **SSBs** in *E. coli*, and a protein termed **replication protein A (RPA) in eukaryotes**. These molecules stabilize the single-stranded structure as the replication fork progresses. The stabilizing proteins bind cooperatively and stoichiometrically to the single strands without interfering with the abilities of the nucleotides to serve as templates (see Figure 35–13). In addition to separating the two strands of the double helix, there must be an unwinding of the molecule (once every 10 nucleotide pairs) to allow strand separation. The hexameric dnaB protein complex unwinds DNA in *E. coli*, whereas the hexameric MCM complex unwinds eukaryotic DNA. This unwinding happens in segments adjacent to the replication bubble. To counteract this unwinding, there are multiple "swivels"

interspersed in the DNA molecules of all organisms. The swivel function is provided by specific enzymes that introduce **"nicks" in one strand of the unwinding double helix**, thereby allowing the unwinding process to proceed. The nicks are quickly resealed without requiring energy input, because of the formation of a high-energy covalent bond between the nicked phosphodiester backbone and the nicking-sealing enzyme. The nicking-resealing enzymes are called **DNA topoisomerases**. This process is depicted diagrammatically in **Figure 35–18** and there compared with the ATP-dependent resealing carried out by the DNA ligases (see Table 39–2). Topoisomerases are also capable of unwinding supercoiled DNA. Supercoiled DNA is a higher-ordered structure occurring in circular DNA molecules wrapped around a core, as depicted in Figures 35–2 and **35–19**.

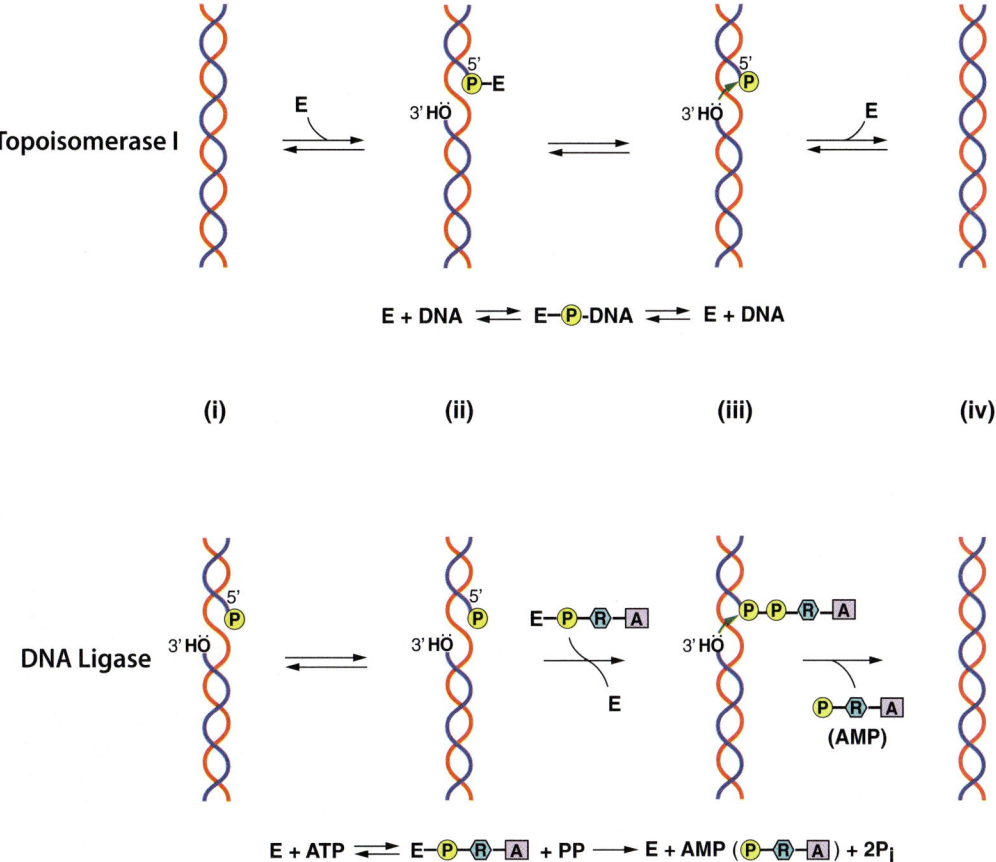

FIGURE 35–18 **Two types of DNA nick-sealing reactions.** Two forms of nick sealing are represented: ATP-independent (top) and ATP-dependent (bottom). Nick sealing processes proceeds in multiple steps: (i)-substrate to → (iv)-product. The enzymes involved are signified by E (top, bottom), while small molecule reactants and products are indicated as Phosphate (P); Pyrophosphate (PP), inorganic Phosphate (Pi) generated from PP by the action of ubiquitous pyrophosphatases, Ribose (R), and Adenine (A). The nick-sealing reaction at the top is catalyzed by DNA topoisomerase I and is ATP-energy independent because the energy for reformation of DNA phosphodiester bonds is stored within the covalent attachment of topoisomerase I to DNA (P-E; top; step ii). Bond reformation is accomplished by the nucleophilic attack of the 3' OH group (green arrow, step iii) to the phosphate of the P-E complex. This reaction releases free topoisomerase I (E) and intact double stranded DNA (step iv). The overall enzyme reaction is schematized at the bottom of the figure (steps i → iv). The nick-sealing reaction catalyzed by DNA ligase (bottom) repairs single strand DNA breaks in the phosphodiester backbone that are a result of DNA replication and/or DNA repair (step i; bottom). The complete DNA ligase reaction requires hydrolysis of two of the high-energy phosphodiester bonds of ATP. The overall reaction scheme of DNA ligase nick-sealing from nick, to enzyme-DNA binding, to enzyme activation that releases Pyrophosphate (PP) to release of free enzyme, AMP and intact DNA is depicted (bottom; as noted in the text, PP is rapidly converted to 2 moles of Pi by the action of ubiquitous pyrophosphatases). The activated ligase (E-P-R-A) reacts with the 5' P at the nick site to form a transient DNA-P-P-R-A complex (note: P-R-A = AMP) that liberates free DNA Ligase Enzyme (E). Nucleophilic attack of the free 3' OH group with the 5'P of the DNA-5'P-AMP complex (green arrow, step iii) reseals the nick and liberates AMP. The overall enzyme reaction converting nicked DNA to intact DNA (E + ATP → E + AMP + 2Pi) is schematized at the bottom of the figure (steps i → iv).

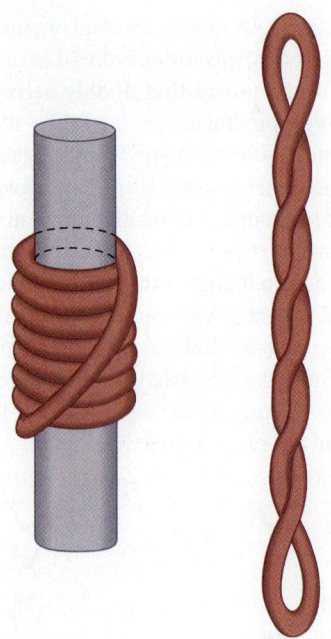

FIGURE 35–19 **Supercoiling of DNA.** A left-handed toroidal (solenoidal) supercoil, at left, will convert to a right-handed inter-wound supercoil, at right, when the cylindric core is removed. Such a transition is analogous to that which occurs when nucleosomes are disrupted by the high salt extraction of histones from chromatin.

There exists in one species of animal viruses (retroviruses) a class of enzymes capable of synthesizing a single-stranded and then a dsDNA molecule from a single-stranded RNA template. This polymerase, termed RNA-dependent DNA polymerase, or "**reverse transcriptase**," first synthesizes a DNA–RNA hybrid molecule utilizing the RNA genome as a template. A specific virus-encoded nuclease, **RNase H**, degrades the hybridized template RNA strand. Subsequently, the remaining DNA strand in turn serves as a template for the viral reverse transcriptase to form a dsDNA molecule containing the genetic information originally present in the RNA genome of the animal virus. The resulting dsDNA can then integrate into the host genome, and from this integrated proviral DNA, viral genes can be expressed via transcription using host cell machinery.

Reconstitution of Chromatin Structure

There is evidence that nuclear organization and chromatin structure are involved in determining the regulation and initiation of DNA synthesis. As noted earlier, the rate of polymerization in eukaryotic cells, which have chromatin and nucleosomes, is slower than that in prokaryotic cells, which lack canonical nucleosomes. It is also clear that chromatin structure must be reformed after replication. Newly replicated DNA is rapidly assembled into nucleosomes, and the preexisting and newly assembled histone octamers are randomly distributed to each arm of the replication fork. These reactions are facilitated through the actions of histone chaperone proteins working in concert with chromatin assembly and remodeling complexes.

DNA Synthesis Occurs During the S Phase of the Cell Cycle

In eukaryotic cells, including human cells, the replication of the DNA genome occurs only at a specified time during the mitotic life cycle of the cell. This period is referred to as the **Synthetic** or **S phase**. S phase is temporally separated from the **Mitotic, or M phase** (where duplicated chromosomes distribute equally to both daughter cells), and by two non-DNA synthetic periods referred to as **gap 1 (G_1)** and **gap 2 (G_2) phases**, which occur before and after the S phase, respectively (**Figure 35–20**). Among other things, the cell prepares for DNA synthesis in G_1, and for mitosis in G_2. The cell regulates DNA synthesis by allowing it to occur only once per cell cycle, and only during S phase, in cells preparing to divide by a mitotic process.

All eukaryotic cells have gene products that govern the transition from one phase of the cell cycle to another. The **cyclins** are a family of proteins whose intracellular concentrations cyclically increase and decrease at specific times during the cell cycle—thus their name. The cyclins activate, at the appropriate time, different **cyclin-dependent protein kinases (CDKs)** that phosphorylate substrates essential for progression through the cell cycle (**Figure 35–21**). For example, cyclin D levels rise in late G_1 phase and allow progression beyond the **start (yeast)** or **restriction point (mammals)**, the point beyond which cells irrevocably proceed into the S or DNA synthesis phase.

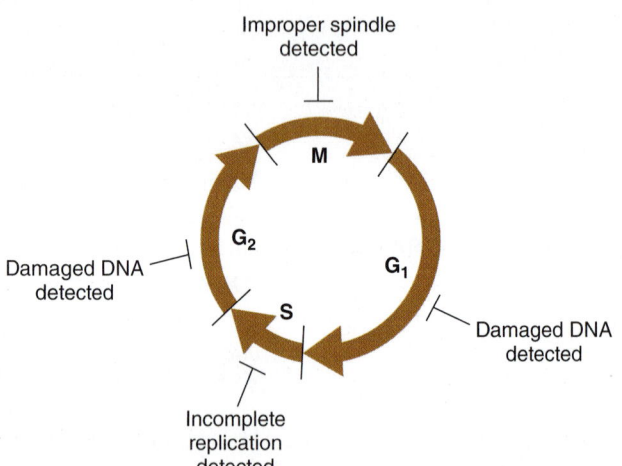

FIGURE 35–20 **Progress through the mammalian cell cycle is continuously monitored via multiple cell-cycle checkpoints.** DNA, chromosome, and chromosome segregation integrity are continuously monitored throughout the cell cycle. If DNA damage is detected in either the G_1 or the G_2 phase of the cell cycle, if the genome is incompletely replicated, or if normal chromosome segregation machinery is incomplete (ie, a defective spindle), cells will not progress through the phase of the cycle in which defects are detected. In some cases, if the damage cannot be repaired, such cells undergo programmed cell death (apoptosis). Note that cells can reversibly leave the cell cycle during G_1 entering a nonreplicative state termed G_0 (not shown, but see Figure 9–8). When appropriate signals/conditions occur, cells reenter G_1 and progress normally through the cell cycle as depicted.

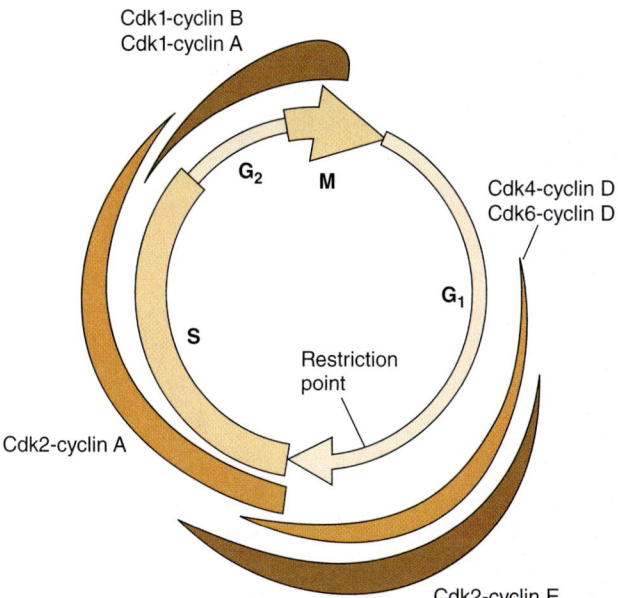

FIGURE 35–21 Schematic illustration of the points during the mammalian cell cycle during which the indicated cyclins and cyclin-dependent kinases are activated. The thickness of the various colored lines is indicative of the extent of activity.

The D cyclins activate CDK4 and CDK6. These two kinases are also synthesized during G_1 in cells undergoing active division. The D cyclins and CDK4 and CDK6 are nuclear proteins that assemble as a complex in late G_1 phase. The cyclin–CDK complex is now an active serine–threonine protein kinase. One substrate for this kinase is the retinoblastoma (Rb) protein. Rb is a cell-cycle regulator because it binds to and inactivates a transcription factor (E2F) necessary for the transcription of certain genes (histone genes, DNA replication proteins, etc.) needed for progression from G_1 to S phase. The phosphorylation of Rb by CDK4 or CDK6 results in the release of E2F from Rb-mediated transcription repression—thus, gene transcription activation ensues and cell-cycle progression takes place.

Other cyclins and CDKs are involved in different aspects of cell-cycle progression (**Table 35–7**). Cyclin E and CDK2 form a complex in late G_1. Cyclin E is rapidly degraded, and the released CDK2 then forms a complex with cyclin A. This sequence is necessary for the initiation of DNA synthesis in S phase. A complex between cyclin B and CDK1 is rate-limiting for the G_2/M transition in eukaryotic cells.

TABLE 35–7 Cyclins and Cyclin-Dependent Kinases Involved in Cell-Cycle Progression

Cyclin	Kinase	Function
D	CDK4, CDK6	Progression past restriction point at G_1/S boundary
E, A	CDK2	Initiation of DNA synthesis in early S phase
B	CDK1	Transition from G_2 to M

Many of the cancer-causing viruses (oncoviruses) and cancer-inducing genes (oncogenes) are capable of alleviating or disrupting the restriction that normally controls the entry of mammalian cells from G_1 into the S phase. From the foregoing, one might have surmised that excessive production of a cyclin, loss of a specific CDK inhibitor (see following discussion), or production or activation of a cyclin/CDK at an inappropriate time might result in abnormal or unrestrained cell division. Similarly, the oncoproteins (or transforming proteins) produced by several DNA viruses target the Rb transcription repressor for inactivation, inducing cell division inappropriately, while inactivation of Rb, itself a tumor suppressor gene, leads to uncontrolled cell growth and tumor formation.

During the S phase, mammalian cells contain greater quantities of DNA polymerase than during the nonsynthetic phases of the cell cycle. Furthermore, those enzymes responsible for formation of the substrates for DNA synthesis—that is, deoxyribonucleoside triphosphates—are also increased in activity, and their expression drops following the synthetic phase until the reappearance of the signal for renewed DNA synthesis. During the S phase, the **nuclear DNA is completely replicated once and only once**. After chromatin has been replicated, it is marked so as to prevent its further replication until it again passes through mitosis. This process is termed replication licensing. The molecular mechanisms for this phenomenon in human cells involve dissociation and/or cyclin–CDK phosphorylation and subsequent degradation of several origin binding proteins that play critical roles in replication complex formation. Consequently, origins fire only once per cell cycle.

In general, a given pair of chromosomes will replicate simultaneously and within a fixed portion of the S phase on every replication. On a chromosome, clusters of replication units replicate coordinately. The nature of the signals that regulate DNA synthesis at these levels is unknown, but the regulation does appear to be an intrinsic property of each individual chromosome that is mediated by the several replication origins contained therein.

All Organisms Contain Elaborate Evolutionarily Conserved Mechanisms to Repair Damaged DNA

Repair of damaged DNA is critical for maintaining genomic integrity and thereby preventing the propagation of mutations, either horizontally (somatic cells), or vertically (germ cells). DNA is subjected to a huge array of chemical, physical, and biologic assaults on a daily basis (**Table 35–8**), hence efficient recognition and repair of DNA lesions is essential. Consequently, eukaryotic cells contain five major DNA repair pathways, each of which contain multiple, sometimes shared proteins; these DNA repair proteins typically have orthologues in prokaryotes. The mechanisms of DNA repair include **nucleotide excision repair (NER); mismatch repair (MMR); base excision repair (BER); homologous recombination (HR);** and **nonhomologous end-joining (NHEJ)** repair pathways (**Figure 35–22**). The experiment of testing the importance of

TABLE 35–8 **Types of Damage to DNA**

I. Single-base alteration
   A. Depurination
   B. Deamination of cytosine to uracil
   C. Deamination of adenine to hypoxanthine
   D. Alkylation of base
   E. Base-analog incorporation

II. Two-base alteration
   A. UV light-induced thymine–thymine (pyrimidine) dimer
   B. Bifunctional alkylating agent cross-linkage

III. Chain breaks
   A. Ionizing radiation
   B. Radioactive disintegration of backbone element
   C. Oxidative free-radical formation

IV. Cross-linkage
   A. Between bases in same or opposite strands
   B. Between DNA and protein molecules (eg, histones)

TABLE 35–9 **Human Diseases of DNA Damage Repair**

Defective Nonhomologous End-Joining Repair (NHEJ)
Severe combined immunodeficiency disease (SCID)
Radiation-sensitive severe combined immunodeficiency disease (RS-SCID)

Defective Homologous Repair (HR)
AT-like disorder (ATLD)
Nijmegen breakage syndrome (NBS)
Bloom syndrome (BS)
Werner syndrome (WS)
Rothmund-Thomson syndrome (RTS)
Breast cancer susceptibility 1 and 2 (BRCA1, BRCA2)

Defective DNA Nucleotide Excision Repair (NER)
Xeroderma pigmentosum (XP)
Cockayne syndrome (CS)
Trichothiodystrophy (TTD)

Defective DNA Base Excision Repair (BER)
MUTYH-associated polyposis (MAP)

Defective DNA Mismatch Repair (MMR)
Hereditary nonpolyposis colorectal cancer (HNPCC)

many of these DNA repair proteins to human biology has been performed by nature—mutations in a large number of these genes lead to human disease (**Table 35–9**). Moreover, systematic gene-directed "knockout" experiments (see Chapter 39) with laboratory mice and cells in culture have clearly ascribed critical genome integrity maintenance functions to these genes as well. In these genetic studies, it was observed that indeed targeted mutations within these genes induce defects in DNA repair while often also dramatically increasing susceptibility to cancer.

One of the most intensively studied mechanisms of DNA repair is the mechanism used to repair DNA **double-strand breaks** (**DSBs**). DSBs are DNA damage events that are highly mutagenic, and will be discussed in some detail here. There are two pathways, HR and NHEJ, that eukaryotic cells utilize to remove DSBs. The choice between the two depends on the phase of the cell cycle (see Figures 35–20 and 35–21) and the exact type of DSB breaks to be repaired (**Table 35–8**). During the G_0/G_1 phases of the cell cycle, DSBs are corrected by the NHEJ pathway, whereas during S,

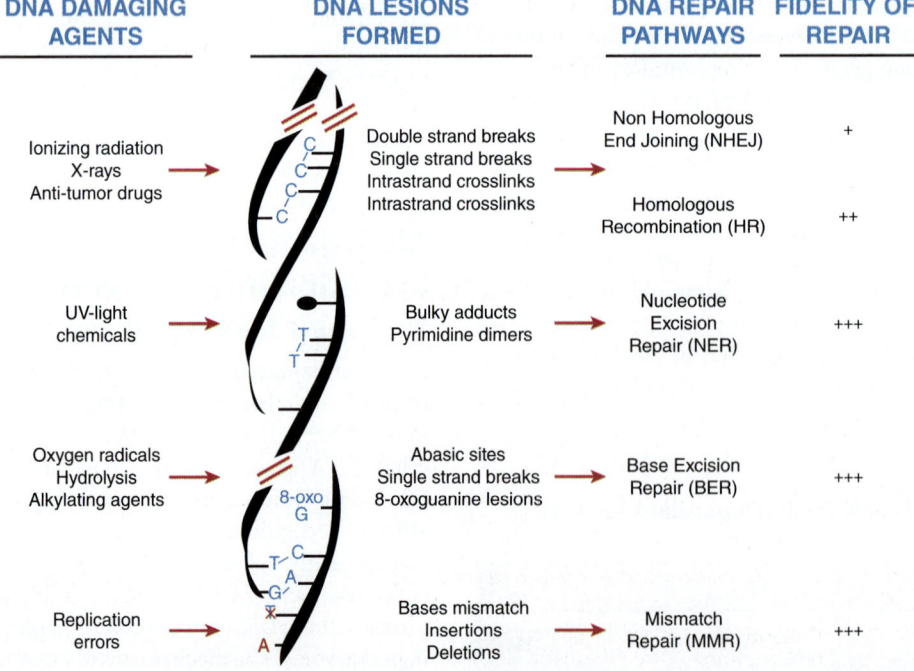

DNA DAMAGING AGENTS	DNA LESIONS FORMED	DNA REPAIR PATHWAYS	FIDELITY OF REPAIR
Ionizing radiation X-rays Anti-tumor drugs	Double strand breaks Single strand breaks Intrastrand crosslinks Intrastrand crosslinks	Non Homologous End Joining (NHEJ)	+
		Homologous Recombination (HR)	++
UV-light chemicals	Bulky adducts Pyrimidine dimers	Nucleotide Excision Repair (NER)	+++
Oxygen radicals Hydrolysis Alkylating agents	Abasic sites Single strand breaks 8-oxoguanine lesions	Base Excision Repair (BER)	+++
Replication errors	Bases mismatch Insertions Deletions	Mismatch Repair (MMR)	+++

FIGURE 35–22 **Mammals use multiple DNA repair pathways of variable accuracy to repair the myriad forms of DNA damage that genomic DNA is subjected to.** Listed are the major types of DNA damaging agents, the DNA lesions so formed (schematized and listed), the DNA repair pathway responsible for repairing the different lesions, and the relative fidelity of these pathways. (Reproduced with permission from Blanpain C, Mohrin M, Sotiropoulou PA, et al: DNA-damage response in tissue-specific and cancer stem cells. Cell Stem Cell. 2011;8(1):16-29.)

G_2, and M phases of the cell cycle HR is utilized. All steps of DNA damage repair are catalyzed by evolutionarily conserved molecules, which include **DNA damage sensors**, **transducers**, and **damage repair mediators**. Collectively, these cascades of proteins participate in the cellular response to DNA damage. Importantly, the ultimate cellular outcomes of DNA damage, and cellular attempts to repair DNA damage, range from **cell-cycle delay** to allow for DNA repair, to **cell-cycle arrest**, to **apoptosis** or **senescence** (**Figure 35–23**; and further detail in

following discussion). The molecules involved in these complex and highly integrated processes range from recognition of damage-specific histone modifications (ie, dimethylated lysine 20 histone H4; H4K20me2) and incorporation of histone isotype variants such as histone **H2AX** into nucleosomes at the site of DNA damage (see Table 35–1), poly ADP ribose polymerase (**PARP**), the MRN protein complex (Mre11-Rad50-NBS1 subunits); to DNA damage-activated kinase recognition/signaling proteins (**ATM** [ataxia telangiectasia, mutated] and

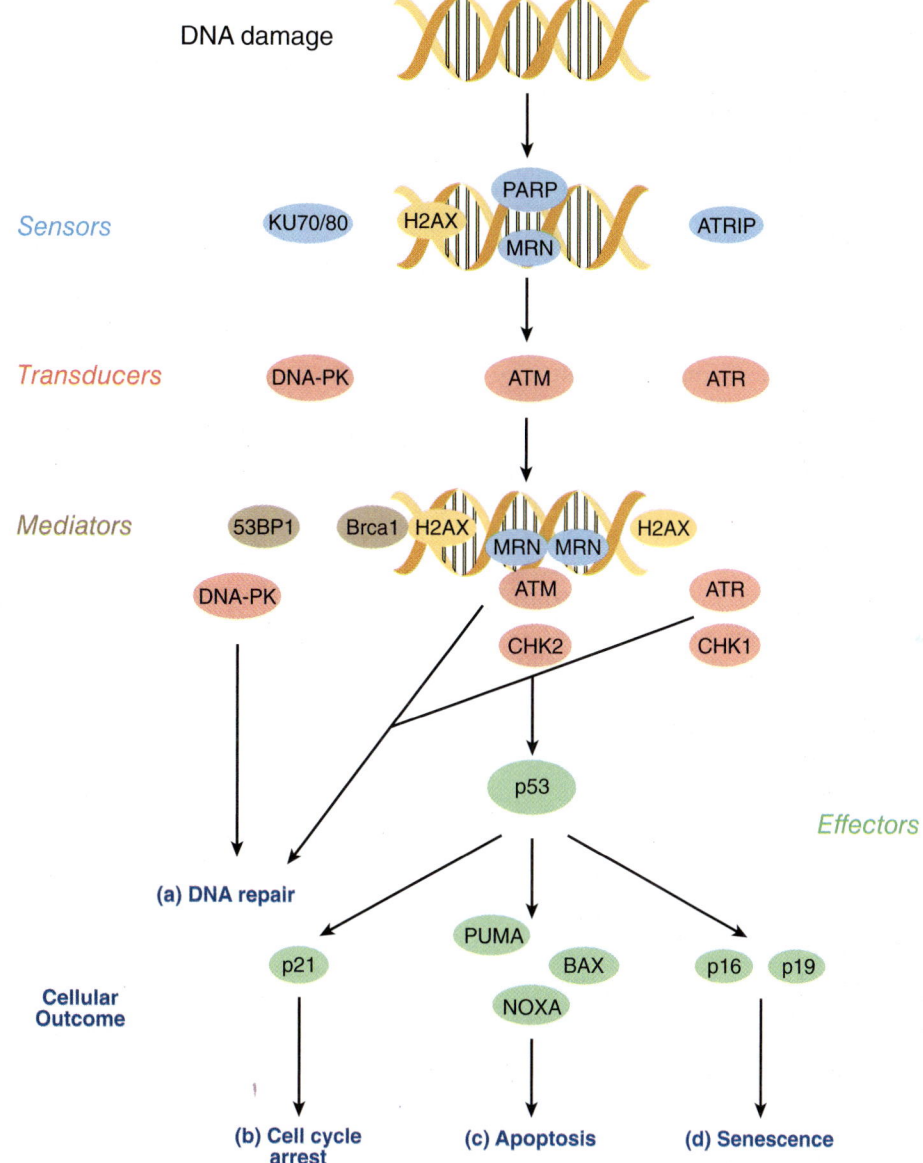

FIGURE 35–23 **The multistep mechanism of DNA double-strand break repair.** Shown top to bottom are the proteins (protein complexes) that identify DSBs in genomic DNA (sensors), transduce, and amplify the recognized DNA damage (transducers and mediators), as well as the molecules that dictate the ultimate outcomes of the DNA damage response (effectors). Damaged DNA can be (a) repaired directly (DNA repair), or, via p53-mediated pathways and depending on the severity of DNA damage and p53-activated genes induced, (b), cells can be arrested in the cell cycle by p21/WAF1 the potent CDK–cyclin complex inhibitor to allow time for extensively damaged DNA to be repaired, or (c), and (d) if the extent of DNA damage is too great to repair, cells can either undergo apoptosis (programmed cell death) or senescence (gradual loss of viability due to cessation of cell division). Both of these processes prevent the cell containing such damaged DNA from ever dividing and hence inducing cancer or other deleterious biologic outcomes. (Reproduced with permission from Blanpain C, Mohrin M, Sotiropoulou PA, et al: DNA-damage response in tissue-specific and cancer stem cells. Cell Stem Cell. 2011;8(1):16-29.)

ATM-related kinase (**ATR**), the multisubunit DNA-dependent protein kinase [**DNA-PK and Ku70/80**], and checkpoint kinases 1 and 2 [**CHK1, CHK2**]). These multiple kinases phosphorylate and consequently modulate the activities of dozens of proteins, such as numerous DNA repair, checkpoint control, and cell-cycle control proteins like CDC25A, B, C, Wee1, p21, p16, and p19 (all cyclin–CDK regulators [see Figure 9–8; and following discussion]; various exo- and endonucleases; DNA single-strand-specific DNA-binding proteins [RPA]; PCNA and specific DNA polymerases [DNA pol δ and η]). Several of these (types of) proteins/enzymes have been discussed earlier in the context of DNA replication. DNA repair and its relationship to cell-cycle control are very active areas of research given their central roles in cell biology and potential for generating and preventing cancer.

DNA & Chromosome Integrity Is Monitored Throughout the Cell Cycle

Given the importance of normal DNA and chromosome function to survival and propagation, it is not surprising that eukaryotic cells have developed elaborate mechanisms to monitor the integrity of the genetic material. As detailed earlier, a number of complex multisubunit enzyme systems have evolved to repair damaged DNA at the nucleotide sequence level. Similarly, DNA mishaps at the chromosome level are also monitored and repaired. As shown in Figure 35–20, both DNA and chromosomal integrity are continuously monitored throughout the cell cycle. The four specific steps at which this monitoring occurs have been termed **checkpoint controls**. If problems are detected at any of these checkpoints, progression through the cycle is interrupted and transit through the cell cycle is halted until the damage is repaired. The molecular mechanisms underlying detection of DNA damage during the G_1 and G_2 phases of the cycle are understood better than those operative during S and M phases.

The **tumor suppressor p53**, a protein of apparent MW 53 kDa on SDS-PAGE, plays a key role in both G_1 and G_2 checkpoint control. Normally a very unstable protein, p53 is a DNA-binding transcription factor, **one of a family of related proteins** (ie, **p53**, **p63**, and **p73**) that is stabilized in response to DNA damage, perhaps by direct p53-DNA interactions. Like the histones discussed earlier, p53 is subject to a panoply of regulatory PTMs, all of which likely modify its multiple biologic activities. Increased levels of p53 activate transcription of an ensemble of genes that collectively serve to delay transit through the cycle. One of these induced proteins, **p21**, **is a potent CDK–cyclin inhibitor (CKI)** that is capable of efficiently inhibiting the action of all CDKs. Clearly, inhibition of CDKs will halt progression through the cell cycle (see Figures 35–20 and 35–21). If DNA damage is too extensive to repair, the affected cells undergo **apoptosis (programmed cell death)** in a p53-dependent fashion. In this case, p53 induces the activation of a collection of genes that induce apoptosis (see Figure 56–9). Cells lacking functional p53 fail to undergo apoptosis in response to high levels of radiation or DNA-active

chemotherapeutic agents. It may come as no surprise, then, that *p53* is one of the most frequently mutated genes in human cancers (see Chapter 56). Indeed, recent genomic sequencing studies of a multitude of tumor DNA samples suggest that over 80% of human cancers carry p53 loss-of-function mutations. Additional research into the mechanisms of checkpoint control will prove invaluable for the development of effective anticancer therapeutic options.

SUMMARY

- DNA in eukaryotic cells is associated with a variety of proteins, resulting in a structure called chromatin.

- Much of the DNA is associated with histone proteins to form a structure called the nucleosome. Nucleosomes are composed of an octamer of histones around which about 150 bp of DNA is wrapped.

- Histones are subject to an extensive array of dynamic covalent modifications that have important regulatory consequences.

- Nucleosomes and higher-order structures formed from them serve to compact the DNA.

- DNA in transcriptionally active regions is relatively more sensitive to nuclease attack in vitro. This property is used as a proxy for a relative increase in DNA accessibility; some regions, so-called hypersensitive sites are exceptionally sensitive and such DNA regions are often found to contain transcription control sites.

- Highly transcriptionally active DNA (genes) is often clustered in regions of each chromosome. Within these regions, genes may be separated by inactive DNA in nucleosomal structures. In many eukaryotic transcription units (ie, the portion of a gene that is copied by RNA polymerase) often consists of protein-coding regions of DNA (exons) interrupted by intervening sequences of nonprotein-coding DNA (introns). This is particularly true for mRNA-encoding genes.

- After transcription, during RNA processing, introns are removed and the exons are ligated together to form the mature RNAs that appear in the cytoplasm; this process is termed RNA splicing.

- DNA in each chromosome is exactly replicated according to the rules of base pairing during the S phase of the cell cycle.

- Each strand of the double helix is replicated simultaneously but by somewhat different mechanisms. A complex of proteins, including DNA polymerase, replicates the leading strand continuously in the 5′ to 3′ direction. The lagging strand is replicated discontinuously, in short pieces of 100 to 250 nucleotides by DNA polymerase synthesizing in the 5′ → 3′ direction.

- DNA replication is initiated at special sites termed origins to generate replication bubbles. Each eukaryotic chromosome contains multiple origins. The entire process takes about 9 hours in a typical human cell and only occurs during the S phase of the cell cycle.

- A variety of mechanisms that employ different enzyme systems repair damaged cellular DNA after exposure of cells to chemical and physical mutagens.

- Normal cells containing DNA that cannot be repaired either undergo senescence or programmed cell death via apoptosis.

REFERENCES

Almannai M, El-Hattab AW, Ali M, Soler-Alfonso C: Clinical trials in mitochondrial disorder, an update. Mol Gen Metab 2020;131:1.

Barnes CE, English DM, Cowley SM: Acetylation & Co: an expanding repertoire of histone acylations regulations chromatin and transcription. Essays Biochem 2019;63:97.

Braunschweig U, Gueroussov S, Plocik AM, Graveley BR, Blencowe BJ: Dynamic integration of splicing within gene regulatory pathways. Cell 2013;152:1252.

Burgers PMJ, Kunkel TA: Eukaryotic DNA replication fork. Annu Rev Biochem 2017;86:417.

Chabot B, Shkreta L: Defective control of pre-messenger RNA splicing in human disease. J Cell Biol 2016;212:13.

Dance A: Researchers peek into chromosomes' 3D structure in unprecedented detail. Proc Nat Acad Sci 2020;117:25186.

Dominguez-Brauer C, Thu KL, Mason, JL, Blaser H, Bray MR, Mak TM: Targeting mitosis in cancer: emerging strategies. Mol Cell 2015;60:524.

Hills SA, Diffley JFX: DNA replication and oncogene-induced replicative stress. Curr Biol 2014;24:R435.

Kunkel TA: Celebrating DNA's repair crew. Cell 2015;163:1301.

Naftelberg S, Schor IE, Ast G, Kornblihtt AR: Regulation of alternative splicing through coupling with transcription and chromatin structure. Ann Rev Biochem 2015;84:165.

Neupert W: Mitochondrial gene expression: a playground of evolutionary tinkering. Ann Rev Biochem 2016;85:65.

Paloozola, KC, Lerner J, Zaret KS: A changing paradigm of transcriptional memory propagation through mitosis Nat Rev Mol Cell Biol 2019;20:55.

Pozo K, Bibb JA: The Emerging role of Cdk5 in cancer. Trends Cancer 2016;2:606.

Sabari BR, Zhang D, Allis CD, Zhao Y: Metabolic regulation of gene expression through histone acylations. Nat Rev Mol Cell Biol 2017;18:90.

Salazar-Roa M, Malumbres M: Fueling the cell division cycle. Trends Cell Biol 2017;27:69.

Smith OK, Aladjem MI: Chromatin structure and replication origins: determinants of chromosome replication and nuclear organization. J Mol Biol 2014;426:3330.

Tang YC, Amon A: Gene copy-number alterations: a cost-benefit analysis. Cell 2013;152:394.

Yao NY, O'Donnell ME: The DNA replication machine: structure and dynamic function. Subcell Biochem 2021;96:233.

Yuan Li H: Molecular mechanisms of eukaryotic origin initiation, replication fork progression, and chromatin maintenance. Biochem J 2020;477:3499.

RNA Synthesis, Processing, & Modification

P. Anthony Weil, PhD

- Describe both the molecules involved and the mechanism of RNA synthesis.
- Describe the major differences between the prokaryotic and eukaryotic transcription machineries.
- Explain how eukaryotic DNA-dependent RNA polymerases, in collaboration with an array of specific accessory factors, can differentially transcribe genomic DNA to produce specific messenger RNA (mRNA) precursor molecules.
- Diagram the critical functional elements of eukaryotic mRNA encoding genes and detail important similarities and differences with their prokaryotic counterparts.
- Describe the key structural elements of eukaryotic mRNA precursors as well as their fully processed forms.
- Appreciate the fact that the majority of mammalian mRNA-encoding genes are interrupted by multiple nonprotein coding sequences termed introns, which are interspersed between protein coding regions termed exons.
- Explain that since intron RNA does not encode protein, the intronic RNA must be specifically and accurately removed in order to generate functional mRNAs from the mRNA precursor molecules in a series of precise molecular events termed RNA splicing.
- Explain the steps and molecules that catalyze mRNA splicing, a process that converts the end-modified precursor molecules into mRNAs that are functional for translation.

BIOMEDICAL IMPORTANCE

The synthesis of an RNA molecule from eukaryotic DNA is a complex process involving one of the group of DNA-dependent RNA polymerase enzymes and a number of additional proteins. The general steps required to synthesize the primary transcript are initiation, elongation, and termination. Most is known about initiation. A number of DNA regions (generally located upstream from the initiation site) and protein factors that bind to these sequences to regulate the initiation of transcription have been identified. Certain RNAs—mRNAs in particular—have very different life spans in a cell. The RNA molecules synthesized in mammalian cells are made as precursor molecules that have to be processed into mature, active RNA. It is important to understand the basic principles of messenger RNA (mRNA) synthesis and metabolism, for modulation of

these processes results in altered rates of protein synthesis, and thus, a variety of cellular phenotypic changes. Such alterations in gene expression allow organisms to adapt to changes in their environment. It is also how differentiated cell structures and functions are established and maintained. Errors or changes in synthesis, processing, splicing, stability, or function of mRNA transcripts can be a cause of disease.

RNA EXISTS IN TWO MAJOR CLASSES

All eukaryotic cells have two major classes of RNA (**Table 36–1**), the **protein coding RNAs**, or **mRNAs**, and various forms of abundant **nonprotein coding RNAs** delineated on the basis of size: the large ribosomal RNAs (**rRNAs**) and long noncoding RNAs

TABLE 36–1 Classes of Eukaryotic RNA

RNA	Distinct Forms	Abundance	Stability
Protein Coding RNAs			
Messenger (mRNA)	$\geq 10^5$ Different species	2-5% of total	Unstable to very stable
Nonprotein Coding RNAs (ncRNAs)			
Large ncRNAs			
Ribosomal (rRNA)	28S, 18S	80% of total	Very stable
lncRNAs	~1000s	~1-2%	Unstable to very stable
Small ncRNAs	5.8S, 5S	~2%	Very stable
Small ribosomal RNAs			
Transfer RNAs (tRNAs)	~60 Different species	~15% of total	Very stable
Small nuclear (snRNA)	~30 Different species	≤1% of total	Very stable
Micro/Silencing (mi/SiRNAs)	100s-1000	<1% of total	Stable
tRNA-derived small fragments (tsRNAs)	100s of different species	<1% of total tRNAs	Stable to unstable

(**lncRNAs**) and small noncoding transfer RNAs (**tRNAs**), the small nuclear RNAs (**snRNAs**) and the micro and silencing RNAs (**miRNAs** and **siRNAs**). mRNAs, rRNAs, and tRNAs are directly involved in protein synthesis while the other RNAs participate in either mRNA splicing (snRNAs) or modulation of gene expression by altering mRNA function (mi/siRNAs) and/or expression (lncRNAs). These RNAs differ in their diversity, stability, and abundance in cells.

RNA IS SYNTHESIZED FROM A DNA TEMPLATE BY RNA POLYMERASES

The processes of DNA and RNA synthesis are similar in that they involve (1) the general steps of initiation, elongation, and termination with 5′–3′ polarity; (2) large, multicomponent initiation and polymerization complexes; and (3) adherence to Watson-Crick base-pairing rules. However, DNA and RNA synthesis do differ in several important ways, including the following: (1) ribonucleotides are used in RNA synthesis rather than deoxyribonucleotides; (2) in RNA U replaces T as the complementary base for A; (3) a primer is not involved in RNA synthesis because RNA polymerases have the intrinsic ability to initiate synthesis de novo; (4) in a given cell only portions of the genome are vigorously transcribed or copied into RNA, at any given time/condition, whereas the entire genome must be copied, once and only once during DNA replication; and (5) there is no highly active, efficient proofreading function during transcription.

The process of synthesizing RNA from a DNA template has been characterized best in prokaryotes. Although in mammalian cells, aspects of the regulation of RNA synthesis and the processing of the RNA transcripts are different from those in prokaryotes, the process of RNA synthesis per se is quite

similar in these two classes of organisms. Indeed, there is significant amino acid sequence conservation between prokaryotic and eukaryotic DNA-dependent RNA polymerases. Consequently, the description of RNA synthesis in prokaryotes, where it is best understood, is applicable to eukaryotes even though the enzymes involved and the regulatory signals, though related, are different.

The Template Strand of DNA Is Transcribed, Or Copied Into RNA

The sequence of ribonucleotides in an RNA molecule is complementary to the sequence of deoxyribonucleotides in one strand of the double-stranded DNA molecule (see Figure 34–8). The strand that is transcribed or copied into an RNA molecule is referred to as the **template strand** of the DNA. The other DNA strand, the **non-template strand**, is frequently referred to as the **coding strand** of that gene. It is called this because, with the exception of T for U changes, it corresponds exactly to the sequence of the mRNA primary transcript, which encodes the (protein) product of the gene. In the case of a double-stranded DNA molecule containing many genes, the template strand for each gene will not necessarily be the same strand of the DNA double helix (**Figure 36–1**). Thus, a given strand of a double-stranded DNA molecule will serve as the template

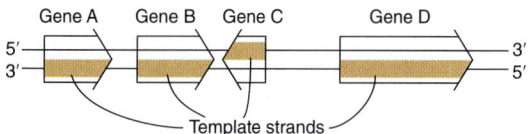

FIGURE 36–1 Genes can be transcribed from both strands of DNA. The arrowheads indicate the direction of transcription (polarity). Note that the template strand is always read in the 3′–5′ direction. The opposite strand is called the coding strand because it is identical (except for T for U changes) to the mRNA transcript (the primary transcript in eukaryotic cells) that encodes the protein product of the gene.

strand for some genes and the coding strand of other genes. Note that the nucleotide sequence of an RNA transcript will be the same (except for U replacing T) as that of the coding strand. The information in the template strand is read out in the 3′–5′ direction. Though not shown in Figure 36–1, there are instances of genes embedded within other genes.

DNA-Dependent RNA Polymerase Binds to a Distinct Site, the Promoter, & Initiates Transcription

DNA-dependent RNA polymerase (RNAP) is the enzyme responsible for the polymerization of ribonucleotides into a sequence complementary to the template strand of the gene (**Figures 36–2** and **36–3**). A key aspect of the process of RNA synthesis is how RNAP identifies and binds template DNA at the correct genomic location. The enzyme attaches at a specific site—the promoter—on the DNA template. This is followed by unwinding of the DNA duplex at, and immediately downstream of the promoter. Once RNAP is specifically bound and has unwound the DNA template at the promoter, initiation of RNA synthesis commences at the **transcription**

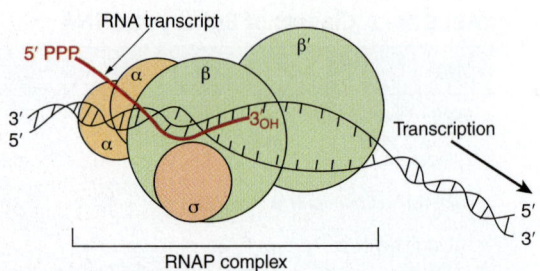

FIGURE 36–2 **RNA polymerase catalyzes the polymerization of ribonucleotides into an RNA sequence that is complementary to the template strand of the gene.** The RNA transcript has the same polarity (5′–3′) as the coding strand but contains U rather than T. Bacterial RNAP consists of a core complex of two β subunits (β and β′) and two α subunits. The holoenzyme form of RNA polymerase contains the σ subunit bound to the α₂ββ′ core assembly. The ω subunit is not shown. The transcription "bubble" is an approximately 20-bp area of melted DNA, and the entire complex covers 30 to 75 bp of DNA depending on the conformation of RNAP.

start site (TSS) with the formation of an initial dinucleotide, the sequence of which is dictated by the DNA sequence of the template strand. The process of phosphodiester bond formation continues until a termination sequence is reached

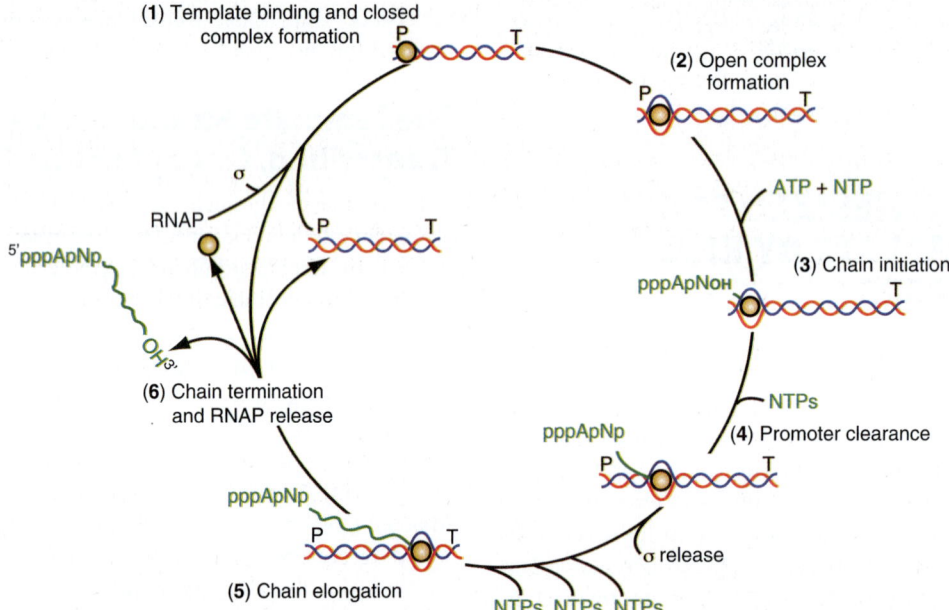

FIGURE 36–3 **The transcription cycle.** The transcription cycle can be described in six steps: **(1) Template binding and closed RNA polymerase-promoter complex formation:** RNAP binds to DNA and then locates a promoter (**P**) DNA sequence element. **(2) Open promoter complex formation:** Once bound to the promoter, RNAP melts the two DNA strands to form an open promoter complex; this complex is also referred to as the preinitiation complex or PIC. Strand separation allows the polymerase to access the coding information in the template strand of DNA. **(3) Chain initiation:** Using the coding information of the template, RNAP catalyzes the coupling of the first base (often a purine) to the second, template-directed ribonucleoside triphosphate to form a dinucleotide (in this example forming the dinucleotide 5′ pppApN_OH 3′). **(4) Promoter clearance:** After RNA chain length reaches ~10 to 20 nt, the polymerase undergoes a conformational change and then is able to move away from the promoter, transcribing down the transcription unit. On many genes σ-factor is released from RNAP at this phase of the transcription cycle. **(5) Chain elongation:** Successive residues are added to the 3′-OH terminus of the nascent RNA molecule until a transcription termination DNA sequence element (**T**) is encountered by RNAP. **(6) Chain termination and RNAP release:** Upon encountering the transcription termination site, RNAP undergoes an additional conformational change that leads to release of the completed RNA chain, the DNA template, and RNAP. Following reformation of holoenzyme (Eσ), RNAP can rebind to DNA beginning the promoter search process and the cycle is repeated. Note that all of the steps in the transcription cycle are facilitated by additional proteins, and indeed are often subjected to regulation by positive- and/or negative-acting factors.

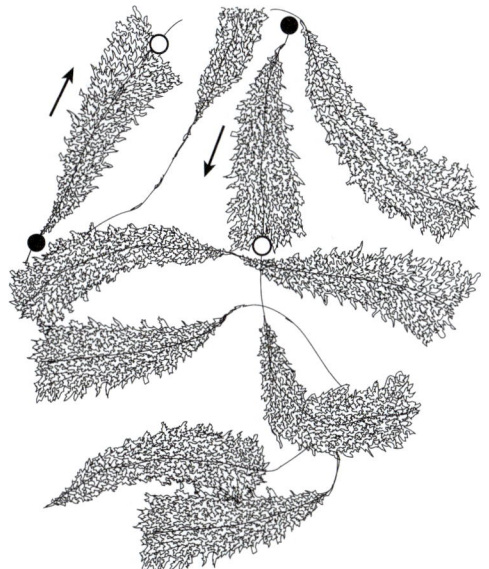

FIGURE 36-4 **Schematic representation of an electron photomicrograph of tandem arrays of amphibian rRNA-encoding genes in the process of being transcribed.** The magnification is about 6000×. Note that the length of the transcripts increases as the RNA polymerase molecules progress along the individual rRNA genes from transcription start sites (filled circles) to transcription termination sites (open circles). RNA polymerase I (not visualized here) is at the base of the nascent rRNA transcripts. Thus, the proximal end of the transcribed gene has short transcripts attached to it, while much longer transcripts are attached to the distal end of the gene. The arrows indicate the direction (5'→3') of transcription.

(see Figure 36–3). A **transcription unit** is defined as that region of DNA that includes the signals for transcription initiation, elongation, and termination. The RNA product, which is synthesized in the 5'–3' direction, is the **primary transcript**. Transcription frequency varies from gene to gene but can be quite high. An electron micrograph of transcription in action is presented in **Figure 36–4**. In prokaryotes, a transcription

unit can represent the product of several contiguous genes, while mammalian cell transcription units usually contain but a single gene. The 5' termini of the primary RNA transcript and the mature cytoplasmic RNA are identical in bacteria. Thus, the bacterial TSS corresponds to the 5' nucleotide of the mRNA. This is designated position +1, as is the corresponding nucleotide in the DNA. The numbering system of the transcription unit increases positively as the sequence proceeds *downstream* from the start site. This convention makes it easy to locate particular regions, or functional elements of transcripts (and their encoded proteins if protein coding), such as intron and exon boundaries; translation start/stop signals etc. The nucleotide in the promoter adjacent to the transcription initiation site in the upstream direction is designated –1, and these negative numbers increase as the sequence proceeds *upstream*, away from the TSS. This +/– numbering system provides a conventional way of defining the location of regulatory elements in a gene (**Figure 36–5**).

The primary transcripts generated by RNA polymerase II— one of the three distinct nuclear DNA-dependent RNA polymerases in eukaryotes—are promptly modified by the addition of 7-methylguanosine triphosphate caps (see Figure 34–10), which persist and eventually appear on the 5' end of mature cytoplasmic mRNA. These caps are necessary for the subsequent processing of the primary transcript to mRNA, for the translation of the mRNA, and for protection of the mRNA against nucleolytic attack by 5'-exonucleases.

Bacterial DNA-Dependent RNA Polymerase Is a Multisubunit Enzyme

The basic DNA-dependent RNA polymerase of the bacterium *Escherichia coli* exists as an approximately 400-kDa core complex consisting of two identical α subunits, two large β and β' subunits, and an ω subunit. The β subunit binds Mg^{2+} ions and composes the catalytic subunit (see Figure 36–2).

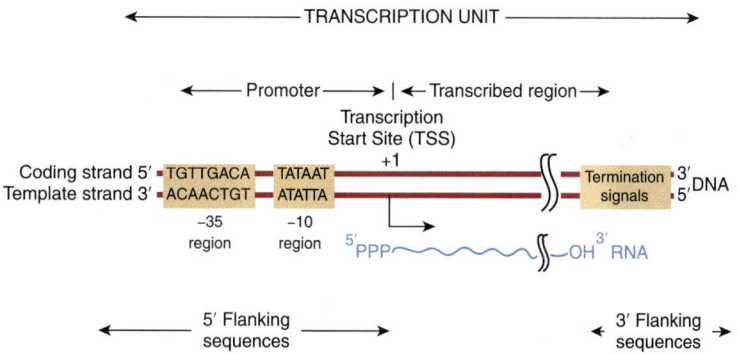

FIGURE 36-5 **Prokaryotic promoters share two regions of highly conserved nucleotide sequence.** These regions are located 35- and 10-bp upstream of the TSS, which is indicated as +1. By convention, all nucleotides upstream of the transcription initiation site (at +1) are numbered in a negative sense and are referred to as 5'-flanking sequences, while sequences downstream of the +1 TSS are numbered in a positive sense. Also, by convention, the promoter DNA regulatory sequence elements such as the –35 and the –10 TATAAT elements are described in the 5'→3' direction and as being on the coding strand. These elements function only in double-stranded DNA. Other transcriptional regulatory elements, however, can often act in a direction independent fashion, and such *cis*-elements are drawn accordingly in any schematic (see Figure 36–8). Note that the transcript produced from this transcription unit has the same polarity or "sense" (ie, 5'→3' orientation) as the coding strand. Termination-determining *cis*-elements reside at the end of the transcription unit (see Figure 36–6 for more detail). By convention, the sequences downstream of the site at which transcription termination occurs are termed 3'-flanking sequences.

The **core RNA polymerase, $\beta\beta'\alpha_2\omega$**, often termed **E**, associates with a specific protein factor (the **sigma [σ] factor**) to form **holoenzyme, $\beta\beta'\alpha_2\sigma\omega$**, or **E$\sigma$**. The genes encoding all these proteins are essential for viability with an exception of ω-encoding gene. The σ subunit enables the core enzyme to recognize and bind the promoter region (see Figure 36–5) to form the **preinitiation complex (PIC)**. There are multiple, distinct σ-factor encoding genes in all bacterial species. Sigma factors have a dual role in the process of promoter recognition: σ association with core RNA polymerase decreases its affinity for nonpromoter DNA, while simultaneously increasing holoenzyme affinity for promoter DNA. Within the bacterial cell these multiple σ-factors compete for interaction with limiting core RNA polymerase (ie, **E**). Each of these unique σ-factors act as a regulatory protein that modifies the promoter recognition specificity of the resulting unique RNA polymerase holoenzyme (ie, $E\sigma_1$, $E\sigma_2$,...). The appearance of different σ-factors and their association with core RNA polymerase to form novel holoenzyme forms $E\sigma_1$, $E\sigma_2$,..., can be correlated temporally with various programs of gene expression in prokaryotic systems such as sporulation, growth in various poor nutrient sources, and the response to heat shock.

Mammalian Cells Possess Three Distinct Nuclear DNA-Dependent RNA Polymerases & a Single DNA-Dependent Mitochondrial RNA Polymerase

Some of the distinguishing properties of mammalian nuclear RNA polymerases are described in **Table 36–2**. Each of these DNA-dependent RNA polymerases is responsible for transcription of different sets of genes. The sizes of the RNA polymerases range from MW 500 kDa to 600 kDa, and the enzymes exhibit more complex subunit profiles than prokaryotic RNA polymerases. They all have two large subunits, which remarkably bear strong sequence and structural similarities to prokaryotic β and β' subunits, and a number of smaller subunits—as many as 14 in the case of RNA pol III. The functions of each of the subunits are not yet fully understood. A peptide toxin from the mushroom *Amanita phalloides*, α-amanitin, is a specific differential inhibitor of the eukaryotic nuclear DNA-dependent RNA polymerases and as such has proved to

be a powerful research tool (**Table 36–2**). α-Amanitin blocks the translocation of RNA polymerase during phosphodiester bond formation. Mitochondria contain a dedicated DNA-dependent RNA polymerase (**mtRNAP**) that is encoded by a nuclear gene. This mtRNAP, in concert with two accessory initiation and one termination factor, is responsible for all mtDNA gene transcription (see Figure 35–8).

RNA SYNTHESIS IS A CYCLICAL PROCESS THAT INVOLVES RNA CHAIN INITIATION, ELONGATION, & TERMINATION

The process of RNA synthesis in bacteria—depicted in Figure 36–3—is cyclical and involves multiple steps. First RNA polymerase holoenzyme (Eσ) must locate and then specifically bind a promoter (**P**; see Figure 36–3). Once the promoter is located, the Eσ–promoter DNA complex undergoes a temperature-dependent conformational change and unwinds, or melts the DNA in and around the TSS (at +1). This complex is termed the **preinitiation complex (PIC)**. Unwinding allows the active site of the Eσ to access the template strand, which of course dictates the sequence of ribonucleotides to be polymerized into RNA. The first DNA template-directed nucleotide (typically, though not always a purine) then associates with the nucleotide-binding site of the enzyme, and in the presence of the next appropriate nucleotide bound to the polymerase, RNAP catalyzes the formation of the first phosphodiester bond, and the nascent chain is now attached to the polymerization site on the β subunit of RNAP. This reaction is termed **initiation**. The nascent dinucleotide retains the 5′-triphosphate of the initiating nucleotide (in Figure 36–3 this happens to be ATP).

RNA polymerase continues to incorporate nucleotides +3 to ~+10, at which point the polymerase undergoes another conformational change and moves away from the promoter; this reaction is termed **promoter clearance**. On many genes the σ–factor dissociates from the $\beta\beta'\alpha_2$ assembly at this point. The **elongation phase** then commences, and here the nascent RNA molecule grows 5′–>3′ as consecutive NTP incorporation steps continue cyclically, antiparallel to the template. The enzyme polymerizes the ribonucleotides in the specific sequence dictated by the template strand and interpreted by Watson-Crick base-pairing rules. **Pyrophosphate (PP_i)** is released following each cycle of polymerization. As for DNA synthesis, the liberated PP_i is rapidly degraded to two molecules of **inorganic phosphate (P_i)** by ubiquitous **pyrophosphatases**, thereby providing irreversibility on the overall synthetic reaction. The decision to stay at the promoter in a poised or stalled state, or transition to elongation can be an important regulatory step in both prokaryotic and eukaryotic mRNA gene transcription.

As the **elongation** complex containing RNA polymerase progresses along the DNA molecule, **DNA unwinding** must occur in order to provide access for the appropriate base pairing to the nucleotides of the coding strand. The extent

TABLE 36–2 **Nomenclature & Properties of Mammalian Nuclear DNA-Dependent RNA Polymerases**

Form of RNA Polymerase	Sensitivity to α-Amanitin	Major Transcription Products
I	Insensitive	rRNA
II	High sensitivity	mRNA, lncRNA, miRNA, snRNA
III	Intermediate sensitivity	tRNA, 5s rRNA, some snRNAs

of this transcription bubble (ie, DNA unwinding) is constant throughout transcription and has been estimated to be about 20 bp per polymerase molecule (see Figures 36–2; 36–3). Thus, the size of the unwound DNA region is dictated by the polymerase and is independent of the DNA sequence in the complex. RNA polymerase has an intrinsic "unwindase" activity that opens the DNA helix (see PIC formation earlier). The fact that the DNA double helix must unwind, and the strands separate at least transiently for transcription implies some temporary disruption of the nucleosome structure of eukaryotic cells. Topoisomerase both precedes and follows the progressing RNA polymerase to prevent the formation of superhelical tension that would serve to increase the energy required to unwind the template DNA ahead of RNAP.

Termination of the synthesis of RNA in bacteria is signaled by sequences in the template DNA (see Figure 36–3; **T**) and sequences within the transcript. On many genes RNAP alone efficiently terminates transcription. However, on a subset of genes a **termination protein** termed **rho (ρ) factor** is required to mediate transcription termination. After termination both free core RNAP (E) and product RNA dissociate from the DNA template. The resulting free core RNAP (E) is able to associate with σ-factor to reform Eσ, and re-enter the transcription cycle. In eukaryotic cells, termination is less well understood; however, the proteins catalyzing RNA processing, termination, and polyadenylation all appear to load onto RNA polymerase II soon after initiation. In both prokaryotes and eukaryotes, multiple RNA polymerase molecules typically transcribe the same template strand of a gene simultaneously, but the process is phased and spaced in such a way that at any one moment each is transcribing a different portion of the DNA sequence (see Figures 36–1 and 36–4).

THE FIDELITY & FREQUENCY OF TRANSCRIPTION IS CONTROLLED BY PROTEINS BOUND TO CERTAIN DNA SEQUENCES

Biochemical and genetic analyses of the DNA sequence of specific genes has allowed the recognition of a number of sequences important in gene transcription. From the large number of bacterial genes studied, it is possible to construct consensus models of transcription initiation and termination signals.

The question, "How does RNAP find the correct site to initiate transcription?" is not trivial when the complexity of the genome is considered. *Escherichia coli* has about 4×10^3 transcription initiation sites (ie, gene promoters) within the 4.2×10^6 bp genome. The situation is even more complex in humans, where as many as 150,000 distinct transcription initiation sites are distributed throughout 3×10^9 bp of DNA. Bacterial RNAP can bind, with low affinity, to many regions of DNA, but it scans the DNA sequence—at a rate of more than or equal to 10^3 bp/s—until it recognizes certain specific regions of DNA to which it binds with higher affinity. These

regions are termed promoters, and it is the base-specific association of RNAP with promoters that ensures accurate initiation of transcription. Perhaps not surprisingly then, the promoter recognition-utilization process is a critical target for regulation of transcription in both bacteria and humans.

Bacterial Promoters Are Relatively Simple

Bacterial promoters are approximately 40 nucleotides (40 bp or four turns of the DNA double helix) in length, a region small enough to be covered by an *E. coli* RNA holopolymerase molecule (Eσ). In a consensus promoter, there are two short conserved sequence elements. Approximately 35-bp upstream of the TSS there is a consensus sequence of eight nucleotide pairs (consensus: 5′-TGTTGACA-3′) to which the RNAP binds to form the so **closed promoter complex**. More proximal to the TSS—about 10 nucleotides upstream—is a six-nucleotide-pair A+T-rich sequence (consensus: 5′-TATAAT-3′). These conserved sequence elements together comprise the promoter, and are shown schematically in Figure 36–5. The latter sequence has a lower melting temperature because of its lack of GC nucleotide pairs. Thus, this so-called "**TATA box**" is thought to ease the dissociation of the two DNA strands so that RNA polymerase bound to the promoter region can have access to the nucleotide sequence of its immediately downstream template strand. Once the process of strand separation occurs, the combination of RNA polymerase plus promoter is called the **open complex**. Other bacteria have slightly different consensus sequences in their promoters, but all generally have two components to the promoter; these tend to be in the same position relative to the TSS, and in all cases the sequences between the two promoter elements have no similarity but still provide critical spacing functions that facilitate recognition of −35 and −10 sequences by the cognate RNA polymerase holoenzyme. Within a bacterial cell, different sets of genes are often coordinately regulated. One important way that this is accomplished is through the fact that these coregulated genes share particular −35 and −10 promoter sequences. These unique sets of promoters are recognized by different σ-factors bound to core RNA polymerase as intimated earlier (ie, $E\sigma_1$, $E\sigma_2$,...).

Bacterial RNAP alone has the intrinsic ability to specifically terminate transcription on about 50% of cellular genes. On the remaining bacterial genes the accessory **ρ (rho) termination factor** is required. Proposed mechanisms for ρ-independent and ρ-dependent transcription termination events are presented in **Figure 36–6**. The majority of eukaryotic mRNA gene transcription events require multiple accessory transcription factors.

As discussed in detail in Chapter 38, bacterial gene transcription is controlled through the action of repressor and activator proteins. These proteins typically bind to unique and specific DNA sequences that lie adjacent to promoters. These repressors and activators affect the ability of the RNA polymerase to bind promoter DNA and/or form open complexes.

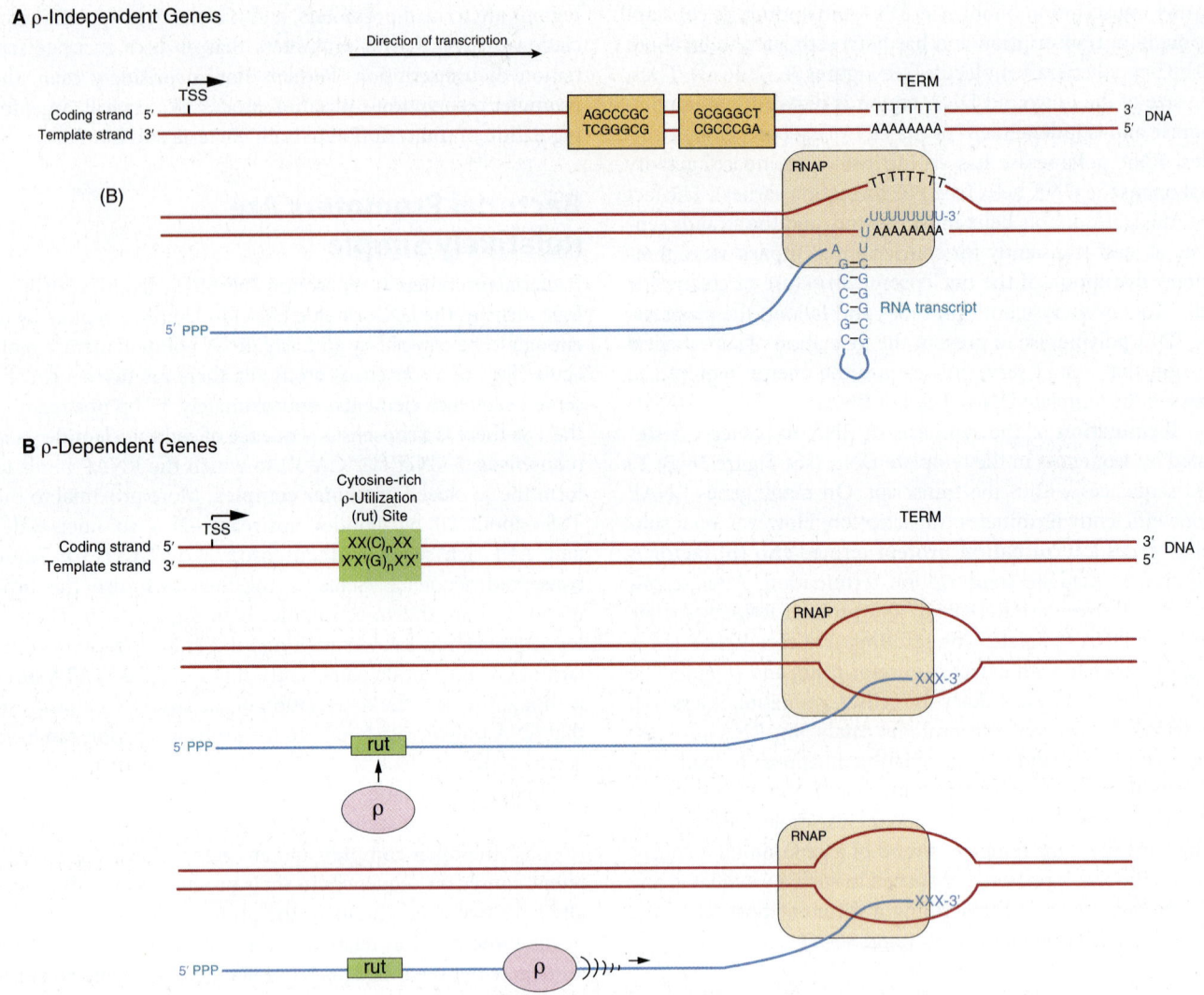

FIGURE 36–6 **Two major mechanisms of transcription termination in bacteria.** **(A)** Bacterial RNAP can directly terminate transcription following the recognition of both specific RNA and DNA signals within transcripts/transcription units. In such situations, the transcription termination signal contains an inverted, hyphenated repeat (the two boxed areas) followed by a stretch of A nucleotides in the template strand (here, bottom strand). The inverted repeat-containing sequences, when transcribed into RNA, can generate a secondary structure like that present in the RNA transcript shown. Formation of this RNA hairpin causes RNA polymerase to pause and on recognition of the polyA sequence in the template strand induces transcription chain termination. **(B)** In cases where genes do not contain the two cis-elements noted earlier, a novel guanine-rich element in the DNA, and an accessory transcription factor, the ρ-protein, together serve to facilitate transcription termination. The transcript contains a run of C-residues that serve as a binding site for the ρ-factor, itself a hexameric ATPase. When present in the transcript, this element, termed the ρ-factor utilization site, or rut, is directly recognized by ρ-factor. On rut element binding, its intrinsic ATPase activity is activated and ρ-factor translocates 5′ to 3′ on the transcript until ρ encounters the transcribing RNA polymerase. ρ-factor-RNAP interaction induces transcription termination and DNA, RNA, and protein dissociation.

The net effect is to stimulate or inhibit PIC formation and transcription initiation—consequently blocking or enhancing specific RNA synthesis.

Eukaryotic Promoters Are More Complex

There are two types of TSS proximal signals in DNA that control transcription in eukaryotic cells. One of these, the **promoter**, defines where transcription initiation will occur on the DNA template; the other set of DNA signals serve to either stimulate and repress transcription, and thus contribute to the control of how frequently transcription occurs. For example, in the thymidine kinase (*tk*) gene of the herpes simplex virus (HSV), which utilizes transcription factors of its mammalian host for its early gene expression program, there is a single unique TSS, and accurate transcription initiation from this site depends on a nucleotide sequence located about 25 nucleotides upstream from the start site (ie, at −25) (**Figure 36–7**). This region has the sequence of **TATAAAAG** and bears remarkable similarity to the functionally related **TATA box** that is located about 10-bp upstream from the prokaryotic

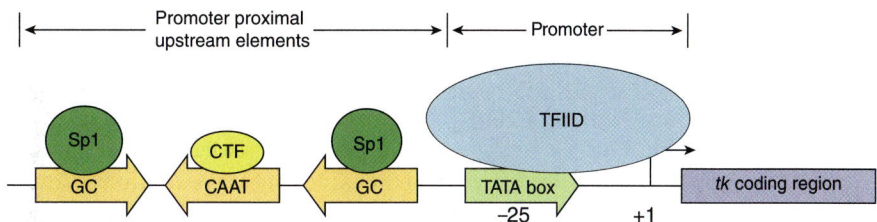

FIGURE 36–7 **Transcription elements and binding factors in the herpes simplex virus thymidine kinase *(tk)* gene.** DNA-dependent RNA polymerase II (not shown) binds to the region encompassing the TATA box (which is shown here bound by transcription factor TFIID) and TSS at +1 (see also Figure 36–9) to form a multicomponent PIC capable of initiating transcription at a single nucleotide (+1 TSS). The frequency of this event is increased by the presence of upstream *cis*-acting elements (the GC and CAAT boxes) located either near to the promoter (promoter proximal) or distant from the promoter (distal elements; see Figure 36–8). Proximal and distal DNA *cis*-elements are bound by *trans*-acting transcriptional activating factors, in this example Sp1 and CTF (also called C/EBP, NF1, NFY) bound to proximal elements. Most proximal and distal *cis*-elements can function independently of orientation (arrows; see also Figure 36-8).

RNA TSS (see Figure 36–5). Mutation of the TATA box markedly reduces transcription of the HSV *tk* gene. Other cellular genes contain this consensus *cis*-active element (**Figures 36–7** and **36–8**). The TATA box is located 25 bp upstream from the TSS in mammalian genes that contain it. The consensus sequence for a TATA box is TATAAA, though numerous variations have been characterized. The human TATA box is bound by the 34-kDa **TATA-binding protein (TBP)**, a subunit in an essential multisubunit complex termed TFIID. The non-TBP subunits of TFIID are proteins called **TBP-associated factors (TAFs)**. Binding of the TBP-TAF TFIID complex to the TATA box sequence is thought to represent a first step in the formation of the transcription complex on the promoter.

A large number of eukaryotic mRNA-encoding genes lack a consensus TATA box. In such instances, additional DNA *cis*-elements, an **initiator (Inr) sequence** and/or the **downstream promoter element (DPE)** among others, direct the RNA polymerase II transcription machinery to the promoter and serve to direct RNA pol II to start transcription from the correct site. The hexameric Inr element spans the start site (from −3 to +5) and consists of the general consensus sequence TCA^{+1} G/T T T/C (A^{+1} indicates the first nucleotide transcribed, ie, TSS). The proteins that bind to Inr in order to direct pol II binding include TFIID. Promoters that have both a TATA box and an Inr may be "stronger," or more frequently transcribed, than those that have just one of these elements. The DPE has the consensus sequence A/GGA/TCGTG and is localized about 25-bp downstream of the +1 TSS. Like the Inr, DPE sequences are also bound by the TAF subunits of TFIID. In a survey of hundreds of thousands of eukaryotic protein

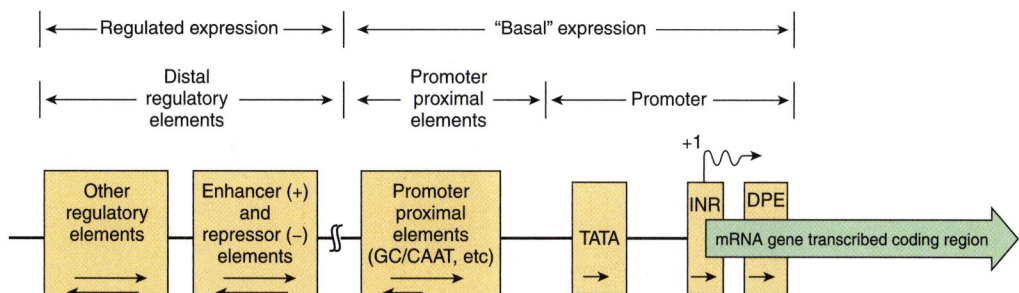

FIGURE 36–8 **Schematic showing the transcription control regions in a hypothetical mRNA-producing eukaryotic gene transcribed by RNA polymerase II.** Such a gene can be divided into its coding and regulatory regions, as defined by the transcription start site (arrow; +1). The coding region contains the DNA sequence that is transcribed into mRNA, which is ultimately translated into protein, typically after extensive mRNA processing via splicing (see Figures 36–12 to 36–16). The regulatory region consists of two classes of elements. One is responsible for ensuring basal expression. The "promoter," often composed of the TATA box and/or Inr and/or DPE elements (see Table 36–3), directs the RNA polymerase II transcription machinery to the correct site (fidelity). However, in certain genes that lack a consensus TATA, the so-called TATA-less promoters, an initiator (Inr) and/or DPE element(s) may direct the polymerase to this site. Another component, the upstream elements, specifies the frequency of initiation; such elements can either be proximal (50–200 bp) or distal (1000–10^5 bp) to the promoter as shown. Among the best studied of the proximal elements is the CAAT box, but several other elements (bound by the transactivator proteins Sp1, NF1, AP1, etc.; Table 36–3) may be used in various genes. The distal elements enhance or repress expression, several of which mediate the response to various signals, including hormones, heat shock, heavy metals, and chemicals. Tissue-specific expression also involves specific sequences of this sort. The orientation dependence of all the elements is indicated by the arrows within the boxes. For example, the proximal promoter elements (TATA box, INR, DPE) must be in the 5′→3′ orientation, while the proximal upstream elements often work best in the 5′→3′ orientation but, most can be reversed. The locations of some elements are not fixed with respect to the transcription start site. Indeed, some elements responsible for regulated expression can be located interspersed with the upstream elements or can be located downstream from the start site, within, or even downstream of the regulated gene itself.

coding genes, roughly 30% contained a TATA box and Inr, 25% contained Inr and DPE, 15% contained all three elements, whereas ~30% contained just the Inr.

Sequences generally, though not always, just upstream from the start site contribute importantly to how frequently transcription occurs. Not surprisingly, mutations in these regions reduce the frequency of transcription initiation 10-fold to 20-fold. Typical of these DNA elements are the GC and CAAT boxes, so named because of the DNA sequences involved. As illustrated in Figure 36–7, each of these DNA elements are bound by a specific protein, Sp1 in the case of the GC box and CTF by the CAAT box; both bind through their distinct **DNA-binding domains (DBD;** see Chapter 38). The frequency of transcription initiation is a consequence of these protein-DNA interactions and complex interactions between particular domains of the transcription factors (distinct from the DBD domains—so-called **activation domains; AD;** see Chapter 38) and the rest of the transcription machinery (RNA polymerase II, the **basal, or general factors, GTFs, TFIIA, B, D, E, F, H,** and other coregulatory factors such as mediator, chromatin remodelers, and chromatin modifying factors). (**Figures 36–9** and **36–10;** Chapter 38.) The protein-DNA interactions at the TATA box involving RNA polymerase II and other components of the basal transcription machinery ensures the fidelity of initiation.

Together, the promoter plus promoter-proximal *cis*-active upstream elements confer fidelity and modulate the frequency of initiation on a gene respectively. The TATA box has a particularly rigid requirement for both position and orientation. As with bacterial promoters, single-base changes in any of these *cis*-elements can have dramatic effects on function by reducing the binding affinity of the cognate *trans*-factors (either TFIID/TBP or Sp1, CTF, and similar factors). The spacing of the TATA box, Inr, and DPE is also critical.

A third class of sequence elements also increase or decrease the rate of transcription of eukaryotic genes. These elements are called either **enhancers** or **repressors/silencers,** depending on how they affect transcription. They have been found in a variety of locations, both upstream and downstream of the TSS, and even within the transcribed protein coding portions of some genes. Enhancers and silencers can exert their effects when located thousands or even many tens of thousands of bases away from transcription units located on the same chromosome. Surprisingly, enhancers and silencers can function in an orientation-independent fashion. Literally, hundreds of these elements have been described. In some cases, the sequence requirements for binding are rigidly constrained; in others, considerable sequence variation is allowed. Some sequences bind only a single protein; however, the majority of these regulatory sequences are bound by several different proteins. As suggested by the discussion of chromatin folding in the previous chapter (see Figure 35–3), communication between drastically separated regions of genomic chromatin regularly occur. Indeed, recent studies indicate that specific patterns of long-range chromatin–chromatin interactions play key roles in the regulation of gene transcription in eukaryotes. Such interactions are likely mediated through the actions of specific nucleosomal, specific chromosomal structural/regulatory proteins, as well as transfactor-transfactor interactions. Thus, interactions between the many transfactors binding to promoter distal and proximal *cis*-elements, regulate transcription in response to a vast array of biologic signals.

Specific Signals Regulate Transcription Termination

The signals for the termination of transcription by eukaryotic RNA polymerase II are only now being fully understood. Termination signals exist relatively far downstream of the protein coding sequences of eukaryotic genes. For example, the transcription termination signal for mouse β-globin gene transcription (the first eukaryotic gene so dissected) occurs at several positions 1000 to 2000 bases beyond the site at which the mRNA poly(A) tail (see later) will eventually be added.

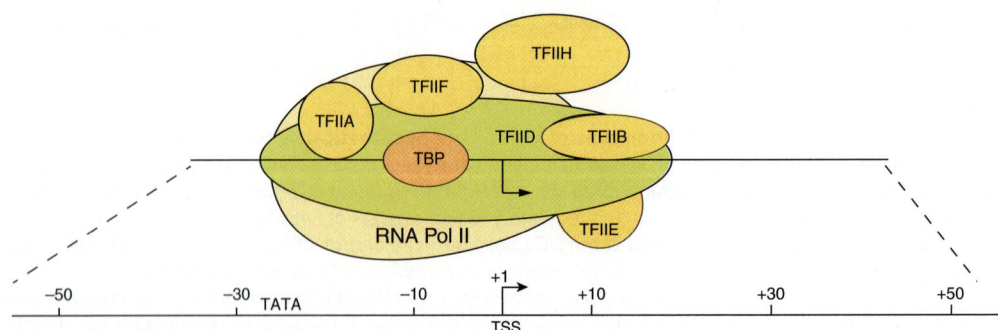

FIGURE 36–9 **The eukaryotic basal transcription complex.** Formation of the basal transcription complex begins when TFIID binds, via its TATA binding protein (TBP) subunit and several of its 14 TBP-associated factor (TAF) subunits, to the TATA box. TFIID then directs the assembly of several other components by protein-DNA and protein–protein interactions: TFIIA, B, E, F, H, and polymerase II (pol II). The entire complex spans DNA from about positions −30 to +30 relative to the TSS at +1 (marked by bent arrow). The atomic level, X-ray–derived structures of RNA polymerase II alone and of the TBP subunit of TFIID bound to TATA promoter DNA in the presence of either TFIIB or TFIIA have all been solved at 3-Å resolution. The structures of mammalian and yeast basal transcription complexes (aka preinitiation complexes, or PICs) have also recently been determined at 10-Å resolution by electron microscopy. Thus, the molecular structures of the transcription machinery in action are beginning to be elucidated. Much of this structural information is consistent with the models presented here.

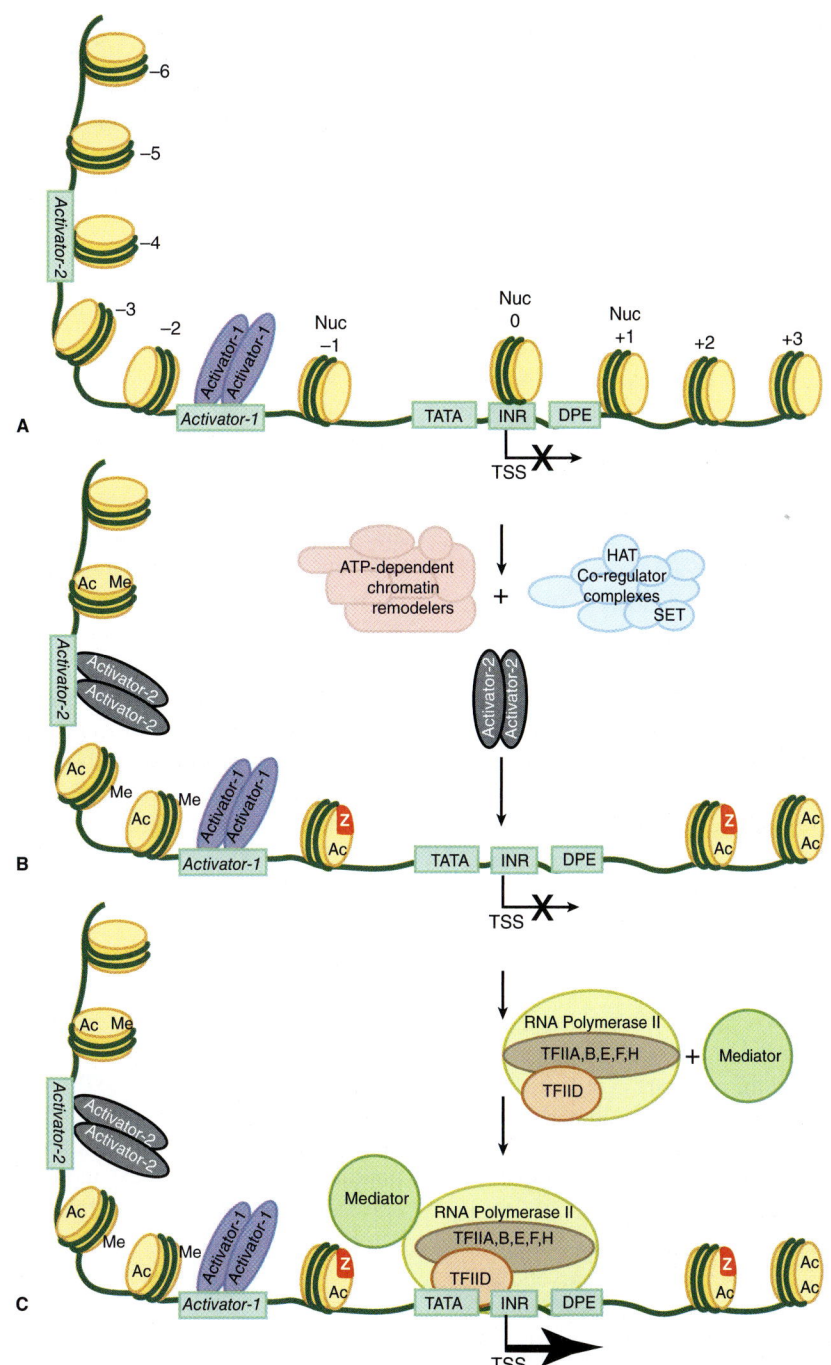

FIGURE 36–10 **Nucleosome covalent modifications, remodeling, and eviction by chromatin-active coregulators modulate PIC formation and transcription.** Shown in (**A**), is an inactive mRNA encoding gene (see **X** over the +1 transcription start site/TSS) with a single dimeric transcription factor (Activator-1; violet ovals) bound to its cognate enhancer binding site (*Activator-1*). This particular enhancer element was nucleosome-free and hence available for interaction with its cognate activator binding protein. However, this gene is still inactive (X over TSS) due to the fact that a portion of its enhancer (in this illustration the enhancer is bipartite and composed of *Activator-1* and *Activator-2*, DNA-binding sites) *and* the promoter are covered by nucleosomes. Recall that nucleosomes have ~150 bp of DNA wound around the histone octamer. Hence, the single nucleosome over the promoter will occlude access of the transcription machinery (pol II + GTFs) to the TATA, Inr and/or DPE promoter elements. (**B**) Enhancer DNA-bound Activator-1 interacts with any of a number of distinct ATP-dependent chromatin remodelers and chromatin-modifying coregulator complexes. These coregulators together have the ability to both move, or remodel (ie, change the octameric histone content, and/or remove nucleosomes) through the action of various ATP-dependent remodelers as well as to covalently modify nucleosomal histones using intrinsic acetylases (HAT; resulting in acetylation [Ac]) and methylases (SET; resulting in methylation [Me], among other PTMs; Table 35–1) carried by subunits of these complexes. (**C**) The resulting changes in nucleosome position and nucleosome occupancy (ie, nucleosomes −4, 0 and +1), composition (nucleosome −1 and nucleosome +2; replacement of nucleosomal H2A with histone H2AX[Z]) thus allows for the binding of the second Activator-2 dimer to *Activator-2* DNA sequences, which ultimately allows for the binding of the transcription machinery (TFIIA, B, D, E, F, H; polymerase II and mediator) to the promoter (TATA-INR-DPE) and the formation of an active PIC, which leads to activated transcription (large arrow at TSS).

Multiple accessory factors participate in this process, hence eukaryotic mRNA gene transcript formation is much more involved than in bacteria. However, it is known that formation of the mRNA 3′ terminus, which is generated posttranscriptionally, is coupled to events or structures formed at the time and site of initiation. Moreover, mRNA formation, and in this case mRNA 3′-end formation, depends on a special structure present on the C-terminus of the largest subunit of **RNA polymerase II, the carboxy-terminal domain (CTD)**, and this process appears to involve at least two steps as follows. After RNA polymerase II has traversed the region of the transcription unit encoding the 3′ end of the transcript, RNA endonucleases cleave the primary transcript at a position about 15 bases 3′ of the consensus sequence **AAUAAA that serves in eukaryotic transcripts as a cleavage and polyadenylation signal**. Finally, this newly formed 3′ terminal is polyadenylated in the nucleoplasm, as described later.

THE EUKARYOTIC TRANSCRIPTION MACHINERY IS COMPLEX

A collection of 42 proteins comprise the three nuclear DNA-dependent RNA polymerase enzymes that transcribe the eukaryotic nuclear genome. These enzymes, termed RNA polymerase I, II, and III transcribe information contained in the template strand of DNA into RNA. Eukaryotic RNAP's I, II, and III contain 12 to 16 subunits. Each major class of eukaryotic RNA polymerase transcribes a unique general class of genes: RNA Pol I transcribes the genes encoding rRNAs, Pol II transcribes mRNA-encoding genes, and Pol III transcribes small RNA-encoding genes. As in all other life forms, these polymerases must recognize a specific site in the promoter in order to initiate transcription at the proper nucleotide. However, in stark contrast to the situation described earlier for prokaryotes where purified holoenzyme (ie, $\beta\beta'\alpha_2\rho$) can specifically bind, transcribe, and terminate transcription on purified DNA, purified eukaryotic RNAs are unable to discriminate between promoter sequences and nonpromoter DNA in the test tube. All eukaryotic RNA polymerases require other proteins known as **general transcription factors (GTFs)** in order to catalyze specific transcription. The 12-subunit RNA polymerase II requires six GTFs: **TFIIA, TFIIB, TFIID (or TBP), TFIIE, TFIIF, and TFIIH** (a total of 33 additional polypeptides) to both facilitate promoter-specific binding of the enzyme, and formation of a functional PIC capable of specific mRNA gene transcription in vitro. RNA polymerases I and III require their own unique set of polymerase-specific GTFs. Complicating an already complex situation, within the cell the transcription machinery (RNA polymerase II, the six GTFs) and activator proteins interact with another set of proteins—termed **coactivators** (also known as **coregulators**). Coregulators bridge between enhancer DNA-bound activator proteins and the transcription machinery, and in so doing modulate the rate of transcription through various molecular mechanisms (described in detail in Chapter 38).

Formation of the Pol II Transcription Complex

In bacteria, a σ-factor–polymerase holoenzyme complex, Eσ, selectively and directly binds to promoter DNA to form the PIC (see Figure 36–3). The situation is much more complex in the case of eukaryotic gene transcription. mRNA-encoding genes, which are transcribed by pol II, are described as an example. In the case of pol II–transcribed genes, the function of σ-factors is assumed by a number of proteins. Thus, accurate, specific PIC formation requires both RNA Pol II *and* the six GTFs (TFIIA, TFIIB, TFIID, TFIIE, TFIIF, TFIIH). These GTFs serve to promote RNA polymerase II transcription on essentially all genes. As noted earlier, a number of the GTFs are composed of multiple subunits. The TFIID complex (**TFIID consists of 15 subunits, the TATA-box binding protein-TBP**, and 14 **TBP associated factors; TAFs**) binds to the TATA box promoter element through its TBP and TAF subunits. TFIID is the only GTF that is independently capable of specific, high-affinity binding to promoter DNA.

TBP binds to the TATA box in the minor groove of DNA (most transcription factors bind in the major groove; see Figure 38–15) and causes an approximately 100-degree bend of the DNA helix. This bending is thought to facilitate the interaction of TFIID TAF subunits with other components of the transcription initiation complex, the multicomponent eukaryotic promoter, and possibly with factors bound to upstream elements. Although initially defined as a component solely required for transcription of pol II gene promoters, TBP, by virtue of its association with distinct, polymerase-specific sets of TAFs, is also an important component of pol I and pol III transcription initiation complexes even if they do not contain TATA boxes.

The binding of TFIID marks a specific promoter for transcription. Of several subsequent in vitro steps, the first is the binding of TFIIA, then TFIIB to the TFIID-promoter complex. Binding of TFIID, TFIIA, and TFIIB results in a stable multiprotein-DNA complex, one that is more precisely located and more tightly bound at the transcription initiation site. This complex attracts and tethers the pol II and TFIIF complex to the promoter. Addition of TFIIE and TFIIH are the final steps in the assembly of the PIC. Each of these binding events extends the size of the complex so that finally about 60 bp (from −30 to +30 relative to the +1 TSS) are covered (see Figure 36–9). The PIC is now complete and capable of basal transcription initiated from the correct +1 template-directed nucleotide. In genes that lack a TATA box, the same factors are required. In such cases, the Inr and/or DPE serve to position the complex for accurate initiation of transcription (see Figure 36–8).

Promoter Accessibility & Hence PIC Formation Is Often Modulated by Nucleosomes

On certain eukaryotic genes, the transcription machinery (pol II, etc.) cannot access promoter sequences (ie, TATA–INR–DPE) because these essential promoter elements are wrapped up in nucleosomes (see Figures 35–2, 35–3, and **36–10**). Only after transcription factors bind to enhancer DNA upstream of the promoter and recruit chromatin remodeling and modifying coregulatory factors, such as the Swi/Snf, SRC-1, p300/CBP (see Chapter 42), P/CAF or other factors, are the repressing nucleosomes removed (see Figure 36–10). Once the remodeled promoter is "open" following nucleosome eviction, GTFs and RNA polymerase II can bind and initiate mRNA gene transcription. Note that the binding of transactivator and coregulator protein complexes can be sensitive to, and/or directly control the composition and/or covalent modification status of the DNA *and* the histones within the nucleosomes in and around the promoter and enhancer, and thereby increase or decrease the ability of all the other components required for PIC formation to interact with a particular gene. This so-called **epigenetic code of DNA, histone and protein modifications** contributes importantly to gene transcription control. Indeed, mutations in proteins that catalyze (code writers), remove (code erasers), or differentially bind (code readers) modified DNA and/or histones can lead to human disease.

Phosphorylation Activates Pol II

Eukaryotic pol II is more complex than bacterial RNAP (Eσ) as it consists of 12 subunits. Surprisingly though, the two largest subunits of pol II (MW 220 and 150 kDa) are partially homologous in sequence (and structure) to the bacterial β′ and β subunits. The most striking difference between eukaryotic pol II and its prokaryotic counterpart in that pol II has a series of **heptad repeats with consensus sequence Tyr-Ser-Pro-Thr-Ser-Pro-Ser (YSPTSPS)** at the carboxyl terminus of the largest pol II subunit, the so-called **C-terminal domain (CTD)**. The length of this CTD can vary from organism to organism; for example, there are 26 repeated units in yeast and 52 units in mammals. The CTD is a substrate for several enzymes (kinases, phosphatases, prolyl isomerases, glycosylases); phosphorylation of the CTD was the first CTD PTM discovered. Among other proteins the kinase subunit of TFIIH can modify the CTD. Covalently modified CTD is the binding site, or platform, for a wide array of proteins, and it has been shown to interact with many transcription regulators, mRNA modifying and processing enzymes, and nuclear transport proteins. The association of these factors with the CTD of RNA polymerase II (and other components of the basal machinery) thus serves to couple transcription initiation with mRNA capping, splicing, 3′–end formation, mRNA quality control, and transport to the cytoplasm. Pol II polymerization is activated when phosphorylated on the Ser and Thr residues and displays reduced activity when the CTD is dephosphorylated. CTD phosphorylation/dephosphorylation is critical for promoter clearance, elongation, termination, and even appropriate mRNA processing. Genetically modified Pol II lacking the CTD tail is incapable of activating transcription in vitro, while cells expressing pol II lacking the CTD rapidly lose viability. These results underscore the importance of this domain to mRNA biogenesis and metabolism.

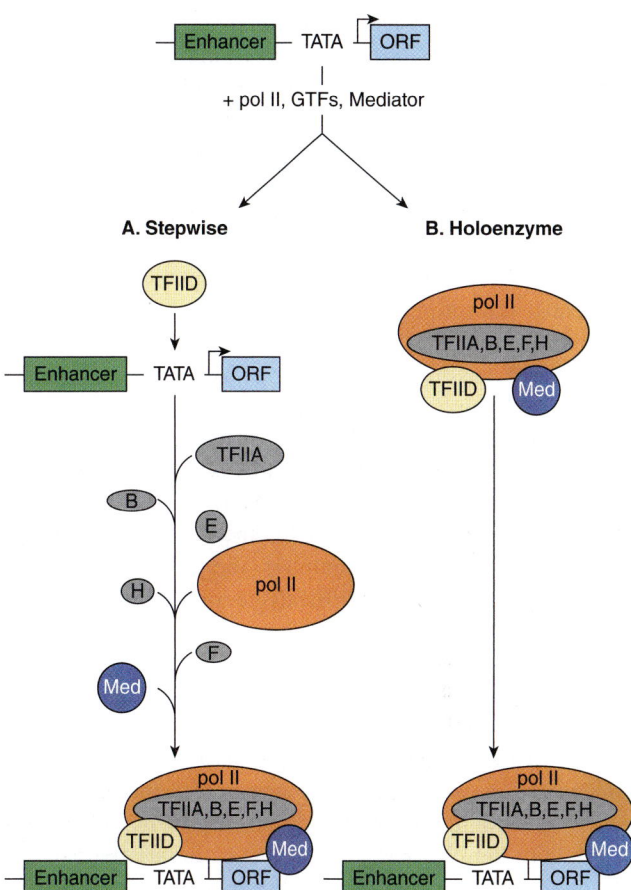

FIGURE 36–11 **Models for the formation of an RNA polymerase II preinitiation complex.** Shown at top is a typical mRNA gene transcription unit: enhancer-promoter (TATA)-TSS (bent arrow) and ORF (open reading frame; sequences encoding the protein) within the transcribed region. PICs have been shown to form by at least two distinct mechanisms in vitro: **(A)** the stepwise binding of GTFs, pol II, and mediator (Med), or **(B)** by the binding of a single multiprotein complex composed of pol II, Med, and the six GTFs. DNA-binding transactivator proteins specifically bind enhancers and in part facilitate PIC formation (or PIC function) by binding directly to the TFIID-TAF subunits, or Med subunits of mediator (not shown, see Figure 36–10), or other components of the transcription machinery. The molecular mechanism(s) by which such protein–protein interactions stimulate transcription remain a subject of intense investigation.

Pol II can associate with other proteins termed **mediator** or **Med** proteins to form a complex sometimes referred to as the pol II holoenzyme; this complex can form on the promoter or in solution prior to PIC formation. The Med proteins (more than 30 proteins; Med1-Med31) are essential for appropriate regulation of pol II transcription by serving myriad roles, both activating and repressing transcription. Thus, mediator, like TFIID is a transcriptional coregulator (**Figure 36–11**).

The Role of Transcription Activators & Coregulators

TFIID was originally considered to be a single protein, TBP. However, several pieces of evidence led to the discovery that TFIID is actually a complex consisting of TBP and the 14 TAFs.

The first evidence that TFIID was more complex than just the TBP molecules came from the observation that TBP binds to a 10-bp segment of DNA, immediately over the TATA box of the gene, whereas native holo-TFIID covers a 35-bp or larger region (see Figure 36–9). Second, purified recombinant *E. coli*–expressed TBP has a molecular mass of 20 to 40 kDa (depending on the species), whereas the native TFIID complex has a mass of about 1000 kDa. Finally, and perhaps most importantly, TBP supports basal transcription but not the augmented transcription provided by certain activators, for example, Sp1 bound to the GC box. TFIID, on the other hand, supports both basal and enhanced transcription by Sp1, Oct1, AP1, CTF, ATF, etc. (**Table 36–3**). The TAFs are essential for this activator-enhanced transcription. There are likely several forms of TFIID in metazoans that differ slightly in their complement of TAFs. Thus, different combinations of TAFs with TBP—or one of several recently discovered TBP-like factors (TLFs)—bind to different promoters, and recent reports suggest that this may account for the tissue- or cell-selective gene activation noted in various promoters and for the different strengths of certain promoters. TAFs, since they are required for the action of activators, are often called coactivators or coregulators. There are thus three classes of transcription factors involved in the regulation of pol II genes: pol II and GTFs, coregulators, and DNA-binding activator or repressor proteins (**Table 36–4**). How these classes of proteins interact to govern both the site and frequency of transcription is a question of central importance and active investigation. It is currently thought that coregulators both act as a bridge between

TABLE 36–4 **Three Classes of Transcription Factors Involved in mRNA Gene Transcription**

General Mechanisms	Specific Components
Basal components	RNA polymerase II, TBP, TFIIA, B, D, E, F, and H
Coregulators	TAFs (TBP + TAFs) = TFIID; certain genes
	Mediator, Meds
	Chromatin modifiers (pCAF, p300/CBP)
	Chromatin remodelers (Swi/Snf)
Activators	SP1, ATF, CTF, AP1, etc.

the DNA-binding transactivators and pol II/GTFs, and act to modify chromatin.

Two Models Can Explain the Assembly of the Pol II Preinitiation Complex

The formation of the PIC described earlier is based on the sequential addition of purified components as observed through in vitro experiments. An essential feature of this model is that PIC assembly takes place on a DNA template where the transcription proteins all have ready access to DNA. Accordingly, transcription activators, which have autonomous DNA binding and activation domains (DBDs and ADs; see Chapter 38), are thought to function by stimulating PIC formation. Here the TAF or mediator complexes are viewed as bridging factors that communicate between the upstream-bound activators, and the GTFs and pol II. This view assumes that there is **stepwise assembly** of the PIC—promoted by various interactions between activators, coactivators, and PIC components, and is illustrated in panel A of Figure 36–11. This model is supported by observations that many of these proteins can indeed bind to one another in vitro.

Recent evidence suggests that there is another possible mechanism of PIC formation and thus transcription regulation. First, large preassembled complexes of GTFs and pol II are found in cell extracts, and these complexes can associate with the promoter in a single step. Second, the rate of transcription achieved when activators are added to limiting concentrations of pol II holoenzyme can be matched by artificially increasing the concentration of pol II and GTFs in the absence of activators. Thus, at least in vitro, one can establish conditions where activators are not in themselves absolutely essential for PIC formation. These observations led to the "**recruitment hypothesis,**" which has now been tested experimentally. Simply stated, the role of activators and some coactivators may be solely to recruit a preformed holoenzyme–GTF complex to the promoter. The requirement for an activation domain is circumvented when either a component of TFIID or the pol II holoenzyme is artificially tethered, using recombinant DNA techniques, to the DBD of an activator. This anchoring, through the DBD component of the activator molecule, leads to a transcriptionally competent structure, and there is

TABLE 36–3 **Some of the Mammalian RNA Polymerase II Transcription Control Elements, Their Consensus Sequences, & the Factors That Bind to Them**

Element	Consensus Sequence	Factor
TATA box	TATAAA	TBP/TFIID
Inr	T/cT/cANT/AT/cT/c	TFIID
DPE	A/gGA/tCGTG	TFIID
CAAT box	CCAATC	C/EBP*, NF-Y*
GC box	GGGCGG	Sp1*
E-box	CAACTGAC	Myo D
κB motif	GGGACTTTCC	NF-κB
Ig octamer	ATGCAAAT	Oct1, 2, 4, 6*
AP1	TGAG/cTC/AA	Jun, Fos, ATF*
Serum response	GATGCCCATA	SRF
Heat shock	(NGAAN)₃	HSF

Note: All elements listed are written 5′ to 3′ and only the top strand of the duplex element is shown. A complete list would include hundreds of examples. The asterisks signify that there are several members of this family. Nucleotides separated by a / indicate that either of two nucleotides can be at that position (ie, T/C in the Inr first position implies that either T or C can occupy that position). N implies any of the four DNA bases A, G, C, or T can occupy that particular position in the indicated *cis*-element.

no further requirement for the activation domain of the activator. In this view, the role of activation domains is to direct preformed holoenzyme–GTF complexes to the promoter; they do not assist in PIC assembly (see panel B, Figure 36–11). In this model, the efficiency of the recruitment process directly determines the rate of transcription at a given promoter. Future studies will continue to test these and other models of the molecular mechanisms of eukaryotic gene transcription regulation (see also Chapter 38).

RNA MOLECULES ARE EXTENSIVELY PROCESSED BEFORE THEY BECOME FUNCTIONAL

In prokaryotic organisms, the primary transcripts of mRNA-encoding genes begin to serve as translation templates even before their transcription has been completed. This can occur because the site of transcription is not compartmentalized into the nucleus as it is in eukaryotic organisms. Thus, transcription and translation are coupled in prokaryotic cells. Consequently, prokaryotic mRNAs are subjected to little processing prior to carrying out their intended function in protein synthesis. Indeed, appropriate regulation of some genes (eg, the *Trp* operon) relies on this coupling of transcription and translation. Prokaryotic rRNA and tRNA molecules are transcribed in units considerably longer than the ultimate molecule. In fact, many of the tRNA transcription units encode more than one tRNA molecule. Thus, in prokaryotes, the processing of these rRNA and tRNA precursor molecules is required for the generation of the mature functional molecules.

Nearly all eukaryotic RNA primary transcripts undergo extensive processing between the time they are synthesized and the time at which they serve their ultimate function, whether it be as mRNA, miRNAs, or as a component of the translation machinery such as rRNA or tRNA. Processing occurs primarily within the nucleus. The processes of transcription, RNA processing, and even RNA transport from the nucleus are highly coordinated. Indeed, a transcriptional coactivator termed SAGA in yeasts and P/CAF in human cells (see Figure 36–10) is thought to link transcription activation to RNA processing by recruiting a second complex termed **transcription export** (TREX) to transcription elongation, splicing, and nuclear export. TREX represents a likely molecular link between transcription elongation complexes, the RNA splicing machinery, and nuclear export (**Figure 36–12**). This coupling is thought to dramatically increase both the fidelity and rate of processing and movement of mRNA to the cytoplasm for translation.

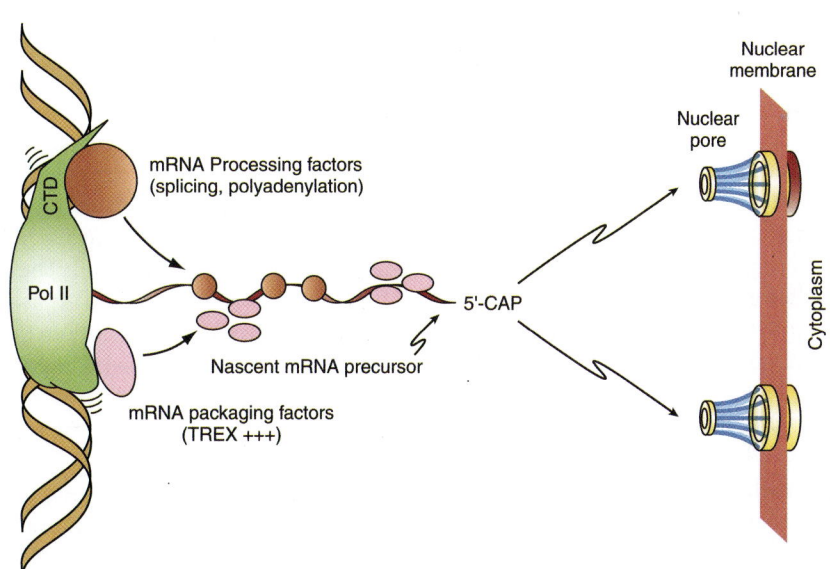

FIGURE 36–12 **RNA polymerase II–mediated mRNA gene transcription is cotranscriptionally coupled to RNA processing and transport.** Shown (left) is RNA pol II actively transcribing an mRNA encoding gene (elongation top to bottom of figure). RNA processing factors (ie, SR/RRM-motif-containing splicing factors as well as polyadenylation and termination factors) interact with the C-terminal domain (CTD, composed of multiple copies of a heptapeptide with consensus sequence –YSPTSPS-) of pol II, while mRNA packaging and transport factors such as TREX complex (pink ovals) are recruited to the nascent mRNA primary transcript, either through direct pol II interactions as shown or, through interactions with SR/splicing factors (brown circles) resident on the nascent mRNA precursor molecules. Note that the CTD is not drawn to scale. The evolutionarily conserved CTD of the Rpb1 subunit of pol II is in reality 5 to 10 times the length of the polymerase due to its many prolines and consequent unstructured nature, and thus a significant docking site for RNA processing and transport proteins among many other critical mRNA metabolizing activities (ie, mRNA polyadenylation and transcription termination factors). In both cases, nascent mRNA chains are thought to be more rapidly and accurately processed due to the immediate recruitment of these many factors to the growing mRNA (precursor) chain. Following appropriate mRNA processing, the mature mRNA is delivered to the nuclear pores (Figures 36–17, 49–4) dotting the nuclear membrane, where, upon transport through the pores, the mRNAs can be engaged by ribosomes and translated into protein.

The Coding Portions (Exons) of Most Eukaryotic mRNA Encoding Genes Are Interrupted by Introns

The RNA sequences that appear in mature RNAs are termed **exons**. In mRNA encoding genes these exons are often interrupted by long sequences of DNA that neither appear in mature mRNA, nor contribute to the genetic information ultimately translated into the amino acid sequence of a protein molecule (see Chapter 35). In fact, these sequences often interrupt the coding region of protein-encoding genes. These **intervening sequences**, or **introns**, exist within most but not all mRNA encoding genes of higher eukaryotes. Human mRNA-encoding gene exons average ~150 nt, while introns are much more heterogeneous, ranging from 10 to 30,000 nucleotides in length. The largest characterized intron in humans is 1,100,000 nt and resides within the 1,586,329 bp *KCNIP4* gene (Genebank Record: NC000004.12), which encodes the potassium voltage-gated channel interacting protein 4. Intron RNA sequences are cleaved out of pre-mRNA transcripts, and the exons of the transcript are appropriately spliced together in the nucleus before the resulting mRNA molecule is transported to the cytoplasm for translation (**Figures 36–13** and **36–14**).

Introns Are Removed & Exons Are Spliced Together

Several different splicing reaction mechanisms for intron removal have been described. The one most frequently used in eukaryotic cells is described later. Although the sequences of nucleotides in the introns of the various eukaryotic transcripts—and even those within a single transcript—are quite heterogeneous, there are reasonably conserved sequences at each of the two exons–intron (splice) junctions and at the branch site, which is located 20 to 40 nucleotides upstream from the 3′–splice site (see consensus sequences in Figure 36–14). A special multi-component complex, the **spliceosome**, is involved in converting the primary transcript into mRNA. Spliceosomes consist of the primary transcript, five snRNAs (U1, U2, U4, U5, and U6), and more than 60 proteins, many of which contain conserved **RRM** (**RNA recognition**) and **SR** (**serine–arginine**) protein **motifs**. Collectively, the five snRNAs and RRM/SR-containing proteins form a small nuclear ribonucleoprotein termed an **snRNP complex**. It is likely that this penta-snRNP spliceosome forms prior to interaction with mRNA precursors. snRNPs are thought to position the exon and intron RNA segments for the necessary

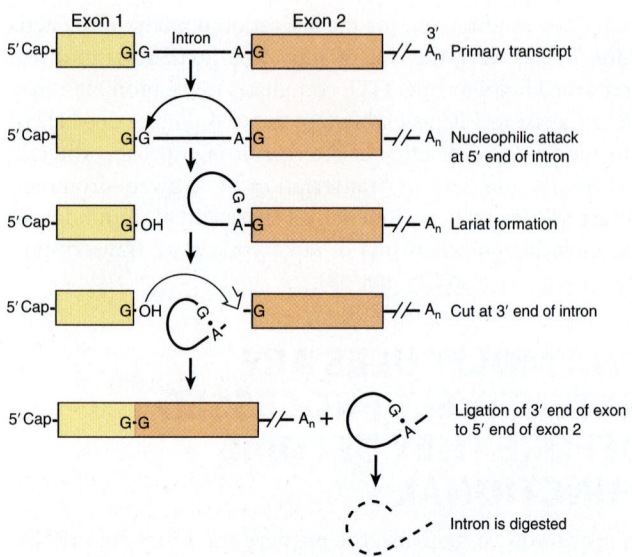

FIGURE 36–13 **The processing of the primary transcript to mRNA.** In this hypothetical transcript, the 5′ (**left**) end of the intron is cut (→) and a structure resembling a lariat forms between the **G** at the 5′ end of the intron and an **A** near the 3′ end, in the consensus sequence UACUAAC. This sequence is called the branch site, and it is the 3′ most **A** that forms the 5′–2′ bond with the G. The 3′ (**right**) end of the intron is then cut (⇓). This releases the lariat, which is digested, and exon 1 is joined to exon 2 at G residues.

splicing reactions. The splicing reaction starts with a cut at the junction of the 5′ exon (donor on left) and intron (see Figure 36–13). This is accomplished by a nucleophilic attack by an adenylyl residue in the branch point sequence located just upstream from the 3′ end of this intron. The free 5′ terminus then forms a loop or lariat structure that is linked by an unusual 5′–2′ phosphodiester bond to the reactive A in the PyNPyPyPuAPy branch site sequence (see Figure 36–14). This adenylyl residue is typically located 20 to 30 nucleotides upstream from the 3′ end of the intron being removed. The branch site identifies the 3′–splice site. A second cut is made at the junction of the intron with the 3′ exon (donor on right). In this second transesterification reaction, the 3′ hydroxyl of the upstream exon attacks the 5′ phosphate at the downstream exon–intron boundary and the lariat structure containing the intron is released and hydrolyzed. The 5′ and 3′ exons are ligated to form a continuous sequence.

The snRNAs and associated proteins are required for formation of the various structures and intermediates. U1 within the snRNP complex binds first by base pairing to the 5′ exon–intron boundary. U2 within the snRNP complex then

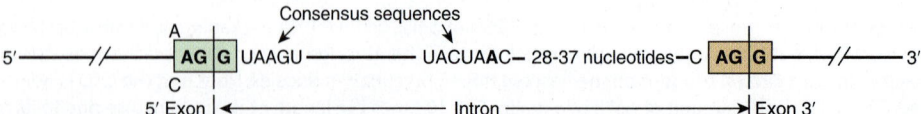

FIGURE 36–14 **Consensus sequences at splice junctions.** The 5′ (donor; **left**) and 3′ (acceptor; **right**) consensus sequences are shown. Also shown is the yeast consensus sequence (UACU**AA**C) for the branch site. In mammalian cells, this consensus sequence is PyNPyPyPuAPy, where Py is a pyrimidine, Pu is a purine, and N is any nucleotide. The site of branch formation (ie, Figure 36–13) is located 20 to 40 nucleotides upstream from the 3′–splice site.

binds by base pairing to the branch site, and this exposes the nucleophilic A residue. U4/U5/U6 within the snRNP complex mediates an ATP-dependent protein-mediated unwinding that results in disruption of the base-paired U4–U6 complex with the release of U4. U6 is then able to interact first with U2, then with U1. These interactions serve to approximate the 5′–splice site, the branch point with its reactive A, and the 3′–splice site. Alignment is enhanced by U5. This process also results in the formation of the loop or lariat structure. The two ends are cleaved by the U2–U6 within the snRNP complex. It is important to note that RNA (in concert with two specifically bound Mg^{2+} ions) serves as the catalytic agent. This sequence of events is then repeated in genes containing multiple introns. In such cases, a definite pattern is followed for each gene, though the introns are not necessarily removed in sequence—1, then 2, then 3, etc. Recent high-resolution **cryoelectron microscopy** (**cryo-EM**) studies have elucidated the structures of various states of the spliceosome in action.

Alternative Splicing Provides for Production of Different mRNAs From a Single mRNA Primary Transcript, Thereby Increasing the Genetic Potential of an Organism

The processing of mRNA molecules is a site for regulation of gene expression. Alternative patterns of mRNA splicing result from tissue-specific adaptive and developmental control mechanisms. Interestingly, recent studies suggest that alternative splicing is controlled, at least in part, through chromatin epigenetic marks (ie, Table 35–1). This form of coupling of transcription and mRNA processing may either be kinetic and/or mediated through interactions between specific histone PTMs and alternative splicing factors that can load onto nascent mRNA gene transcripts during the process of transcription (see Figures 36–11 and 36–12).

As mentioned earlier, the sequence of exon–intron splicing events generally follows a hierarchical order for a given gene. The fact that very complex RNA structures are formed during splicing—and that a number of snRNAs and proteins are involved—affords numerous possibilities for a change of this order and for the generation of different mRNAs. Similarly, the use of alternative termination-cleavage polyadenylation sites also results in mRNA variability. Some schematic examples of these processes, all of which have been well documented in a large array of organisms, including humans, are shown in **Figure 36–15**.

Not surprisingly, defects in mRNA splicing can cause disease. One of the first examples of the critical importance of accurate splicing was the discovery that one form of β-thalassemia, a disease in which the β-globin gene of hemoglobin is severely under-expressed, results from a nucleotide change at an exon–intron junction. This mutation precludes removal of the intron, altering the translational reading frame of β-globin mRNA, thereby blocking β-chain protein production,

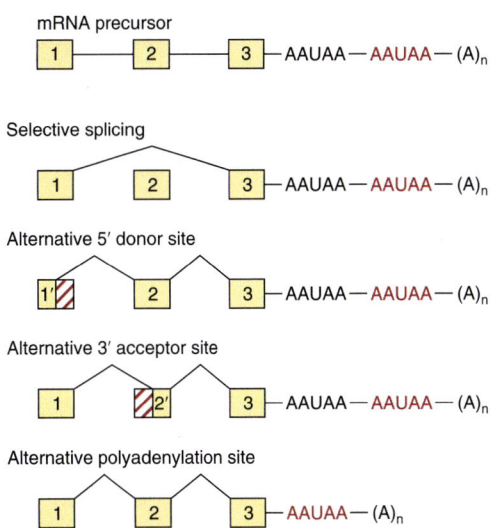

FIGURE 36–15 **Mechanisms of alternative processing of mRNA precursors.** This form of mRNA processing involves the selective inclusion or exclusion of exons, the use of alternative 5′–donor or 3′–acceptor sites, and the use of different polyadenylation sites, and dramatically increases the differential protein coding potential of the genome.

and hence hemoglobin. Note that all of these mechanisms represent control of gene expression post- or after transcription. Such mutations function in the **cis-configuration—that is only the mutated allele**, carrying the mutated β-globin gene will be defective in β-globin protein production. Not surprisingly, mutations in a/the gene that encodes the proteins that catalyze splicing and/or polyadenylation can also cause defects in splicing and RNA metabolism, and hence disease. **Because these altered proteins will affect the formation/function of many, or all mRNAs, such variants are said to act in trans**.

Alternative Promoter Utilization Also Provides a Form of Regulation

Tissue-specific regulation of gene expression can be provided by alternative splicing, as noted earlier, by regulatory DNA control elements in the promoter, or by the use of alternative promoters. The glucokinase (*GK*) gene consists of 10 exons interrupted by 9 introns. The sequence of exons 2 to 10 is identical in liver and pancreatic β cells, the primary tissues in which GK protein is expressed. Expression of the *GK* gene is regulated very differently—by two different promoters—in these two tissues. The liver promoter and exon 1L are located near exons 2 to 10; exon 1L is ligated directly to exon 2. By contrast, the pancreatic β-cell promoter is located about 30-kbp upstream. In this case, the 3′ boundary of exon 1β is ligated to the 5′ boundary of exon 2. The liver promoter and exon 1L are excluded and removed during the splicing reaction (**Figure 36–16**). The existence of multiple distinct promoters allows for cell- and tissue-specific expression patterns of a particular gene (mRNA). In the case of *GK*, insulin and cAMP (see Chapter 42) control *GK* transcription in liver, while glucose controls *GK* expression

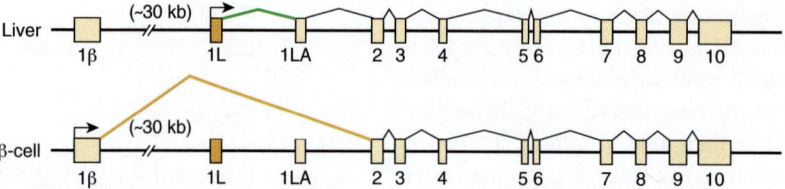

FIGURE 36–16 **Alternative promoter use in the liver and pancreatic β-cell glucokinase (*GK*) genes.** Differential regulation of the glucokinase gene is accomplished by the use of tissue-specific promoters. The β-cell *GK* gene promoter and exon 1β are located about 30-kbp upstream from the liver promoter and exon 1L. Each promoter has a unique structure and is regulated differently. Exons 2 to 10 are identical in the two genes, and the GK proteins encoded by the liver and β-cell mRNAs have identical kinetic properties.

in β cells. Moreover, as noted earlier, such variation in spliced mRNAs can also change the protein coding protein potential of these mRNAs.

Both Ribosomal RNAs & Most Transfer RNAs Are Processed From Larger Precursors

In mammalian cells three rRNA molecules (28S, 18S, 5.8S) are transcribed as part of a single large 45S precursor molecule. The precursor is subsequently processed in the nucleolus to provide these three RNA components for the ribosome subunits found in the cytoplasm. The rRNA genes are located in the nucleoli of mammalian cells. Hundreds of copies of these genes are present in every cell. This large number of genes is needed to synthesize sufficient copies of each type of rRNA to form the 10^7 ribosomes required for every round of cell duplication. Whereas a single mRNA molecule may be copied into 10^5 protein molecules, providing a large amplification, the rRNAs are end products. This lack of amplification requires both a large number of genes and a high transcription rate, typically synchronized with cell growth rate. Similarly, tRNAs are often synthesized as precursors, with extra sequences both 5′ and 3′ of the sequences comprising the mature tRNA. A small fraction of tRNAs contain introns. As with rRNA-encoding genes, tRNA-encoding genes are also vigorously transcribed.

RNAs CAN BE EXTENSIVELY MODIFIED

As introduced in the description of tRNAs (see Figure 34–11), essentially all are covalently modified after transcription. It is clear that at least some of these modifications are regulatory.

Messenger RNA Is Extensively Modified Both Internally & at 5′ & 3′ Termini

Eukaryotic mRNAs contain a **7-methylguanosine cap structure** at their 5′ terminus (see Figure 34–10), and most have a **poly(A) tail** at the 3′ terminus. The cap structure is added to the 5′ end of the newly transcribed mRNA precursor in the nucleus very soon after synthesis and prior to transport of the mRNA molecule to the cytoplasm. The 5′ cap of the RNA transcript is required both for efficient translation initiation

(see Figure 37–7) and protection of the 5′ end of mRNA from attack by 5′ → 3′ exonucleases. The secondary methylations of mRNA molecules, those on the 2′-hydroxy and the N^7 of adenylyl residues, occur after the mRNA molecule has appeared in the cytoplasm.

Poly(A) tails are added to the 3′ end of mRNA molecules in a posttranscriptional processing step. The mRNA is first cleaved about 20 nucleotides downstream from an AAUAA recognition sequence. Another enzyme, **poly(A) polymerase**, adds a poly(A) tail which is subsequently extended to as many as 200 A residues. The poly(A) tail both protects the 3′ end of mRNA from 3′ → 5′ exonuclease attack and facilitates translation (see Figure 37–7). The presence or absence of the poly(A) tail does not determine whether a precursor molecule in the nucleus appears in the cytoplasm, because all poly(A)-tailed nuclear mRNA molecules do not contribute to cytoplasmic mRNA, nor do all cytoplasmic mRNA molecules contain poly(A) tails (histone mRNAs are most notable in this regard). Following nuclear transport, cytoplasmic enzymes in mammalian cells can both add and remove adenylyl residues from the poly(A) tails; this process has been associated with an alteration of mRNA stability and translatability.

The size of some cytoplasmic mRNA molecules, even after the poly(A) tail is removed, is still considerably greater than the size required to code for the specific protein for which it is a template, often by a factor of 2 or 3. The extra nucleotides occur in **untranslated (nonprotein coding) exonic regions** both 5′ and 3′ of the coding region; the longest untranslated sequences are usually at the 3′ end. The **5′ UTR and 3′ UTR** sequences have been implicated in RNA processing, transport, storage, degradation, and translation; each of these reactions potentially contributes additional levels of control of gene expression. Some of these posttranscriptional events involving mRNAs occur in cytoplasmic organelles termed P bodies (see Chapter 37).

Recent research has shown that eukaryotic mRNAs are subject to extensive and dynamic posttranscriptional modifications. The best characterized of these base modifications is N^6-methyladenosine, in which the amino group attached to the C_6 position of adenine is methylated (see Figures 32–1 and 32–3). As described for histone-specific posttranslational modifications and functions (see Table 35–1; Chapter 38) and regulatory phosphorylation of protein/function-activity (see Chapter 42), distinct families of enzymes/proteins that attach ("writers"), and bind ("readers"), and remove ("erasers") the methyl group of N^6-methyl adenosine have been described and characterized. Genetic deletion experiments wherein the genes encoding these writers/readers/erasers are mutated all show that the N^6-methyladenosine mRNA "mark" contributes importantly to cell viability and function. Future work will uncover the extent to which mRNA posttranscriptional modifications support human cell function and impact disease.

Micro-RNAs Are Derived From Large Primary Transcripts Through Specific Nucleolytic Processing

The majority of miRNAs are transcribed by RNA pol II into **primary transcripts** termed **pri-miRNAs**. pri-miRNAs are 5′-capped and 3′-polyadenylated (**Figure 36–17**). pri-miRNAs are synthesized from transcription units encoding one or several

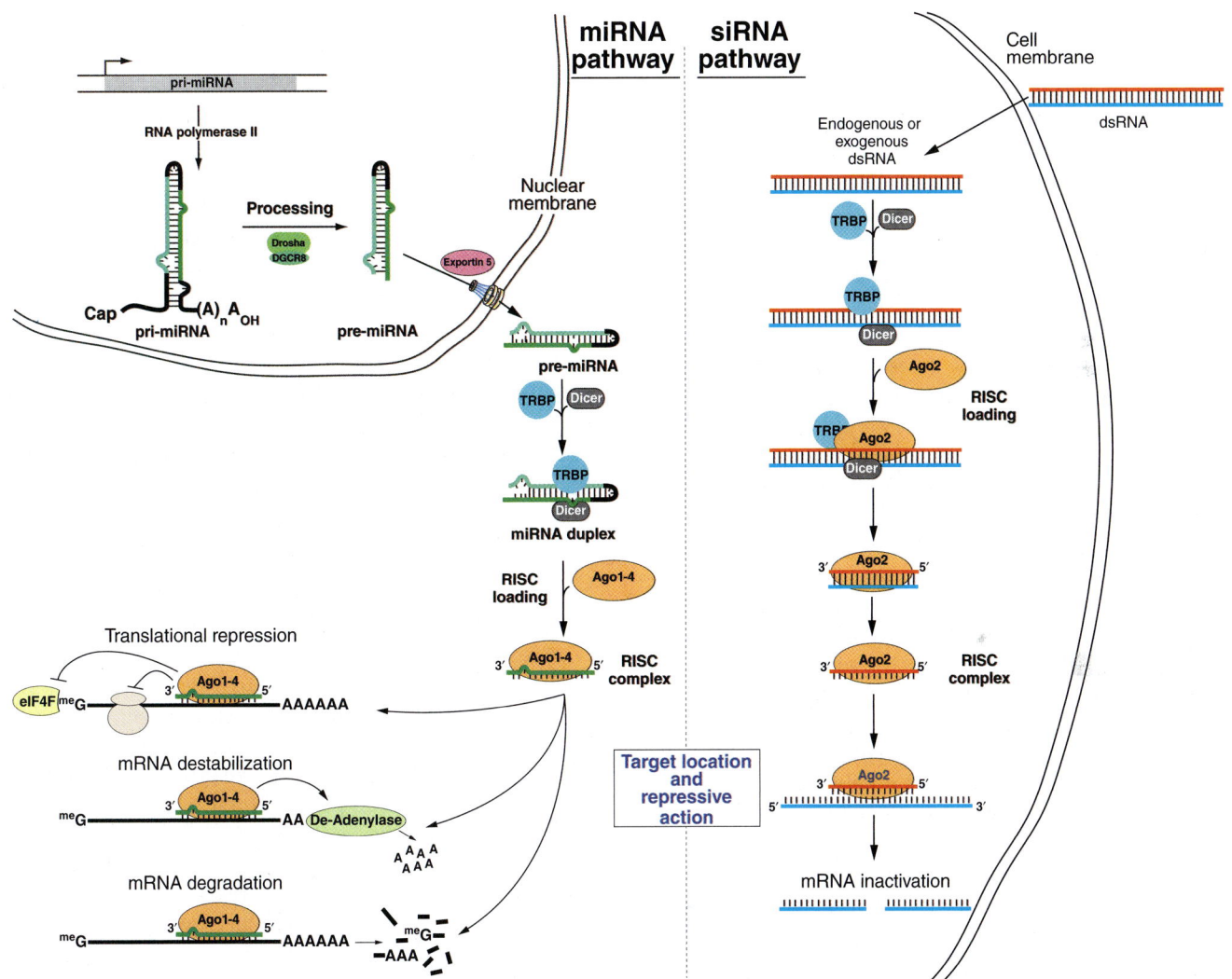

FIGURE 36–17 Biogenesis of micro (mi) and silencing (si)RNAs. (Left) miRNA encoding genes are transcribed by RNA pol II into a primary miRNA (pri-miRNA), which is 5′-capped and polyadenylated as is typical of mRNA coding primary transcripts. This pri-miRNA is subjected to processing within the nucleus by the action of the Drosha-DGCR8 nuclease, which trims sequences from both 5′ and 3′ ends to generate the pre-miRNA. This partially processed double-stranded RNA is transported through the nuclear pore by exportin-5. The cytoplasmic pre-miRNA is then trimmed further by the action of the heterodimeric nuclease termed Dicer (TRBP-Dicer), to form the 21–22 nt miRNA duplex. One of the two resulting 21–22 nucleotide-long RNA strands is selected, the duplex unwound, and the selected strand loaded into the RNA-induced silencing complex, or RISC complex containing one of several Ago proteins (Ago1, 2, 3, or 4) and other important accessory proteins, thereby generating the mature, functional miRNA. Following target mRNA location and sequence-specific miRNA–mRNA annealing, the functional miRNA can modulate mRNA function by one of three mechanisms: translational repression, mRNA destabilization by mRNA deadenylation, or mRNA degradation. **(Right)** The siRNA pathway generates functional siRNAs from large double-stranded RNAs that are formed either intracellularly by RNA–RNA hybridization (inter- or intramolecular) or from extracellular sources such as RNA viruses. These viral dsRNAs are again processed to ~22 nt siRNA dsRNA segments via the heterodimeric Dicer nuclease, loaded into the Ago2-containing RISC complex, one strand is then selected to generate the siRNA that locates target RNA sequences via sequence-specific siRNA–RNA annealing. This ternary target RNA–siRNA–Ago2 complex induces RNA cleavage, which inactivates the target RNA.

distinct miRNAs; these transcription units are either located independently in the genome, or within the intronic DNA of other genes. Given this organization miRNA-encoding genes must therefore minimally possess a distinct promoter, coding region, and polyadenylation/termination signals. pri-miRNAs have extensive 2° structure, and this intramolecular structure is maintained following processing by the **Drosha-DGCR8 nuclease**; the portion containing the RNA hairpin is preserved, transported through the nuclear pore via the action of exportin 5, and once in the cytoplasm, further processed by the heterodimeric **dicer nuclease-TRBP complex** to a **21 or 22-mer**. Ultimately, one of the two strands is selected for loading into the **RNA-induced silencing complex (RISC)**, which is composed of one of four **Argonaute proteins (Ago 1→4)**, to form a mature, functional 21–22 nt single-stranded miRNA. siRNAs are produced similarly. Once in the RISC complex, miRNAs can modulate mRNA function by one of three mechanisms: (a) promoting mRNA degradation directly; (b) stimulating CCR4/NOT complex-mediated poly(A) tail degradation; or (c) inhibition of translation by targeting the 5′-methyl cap binding translation factor eIF4 (see Figures 37–7 and 37–8) or the ribosome directly. Recent data suggest that at least some regulatory miRNA-encoding genes may be linked, and hence coevolve with their target genes.

RNA Editing Alters mRNA Sequence After Transcription

The central dogma states that for a given gene and gene product there is a linear relationship between the coding sequence in DNA, the mRNA sequence, and the protein sequence (see Figure 35–7). Changes in the DNA sequence should be reflected in a change in the mRNA sequence and, depending on codon usage, in protein sequence. However, exceptions to this dogma have been documented. Coding information can be changed at the mRNA level by **RNA editing**. In such cases, the coding sequence of the mRNA differs from that in the cognate DNA. An example is the apolipoprotein B (*apoB*) gene and mRNA. In liver, the single *apoB* gene is transcribed into an mRNA that directs the synthesis of a 100-kDa protein, apoB100. In the intestine, the same gene directs the synthesis of the identical mRNA primary transcript; however, a cytidine deaminase converts a CAA codon in the mRNA to UAA at a single specific site. Rather than encoding glutamine, this codon becomes a termination signal (see Table 37–1), and hence production of a truncated 48-kDa protein (apoB48) results. ApoB100 and apoB48 have different functions in the two organs. A growing number of other examples include a glutamine to arginine change in the glutamate receptor and several changes in trypanosome mitochondrial mRNAs, generally involving the addition or deletion of uridine. Current estimates suggest that perhaps 0.01% of mRNAs are edited in this fashion. Recently, editing of miRNAs has been described suggesting that these two forms of posttranscriptional control mechanisms could cooperatively contribute to gene regulation. Enzymes that catalyze RNA editing have been identified and

characterized extensively. Scientists have been able to harness these enzymes to "correct" mutated mRNAs derived from disease-causing gene variants using cell culture models. These exciting new results offer yet another molecular genetic technology to potentially mitigate, or cure human disease.

Transfer RNA Is Extensively Processed & Modified

As described in Chapter 34 and detailed in Chapter 37, tRNA molecules serve as adapter molecules for the translation of mRNA into protein sequences. tRNAs contain many modifications of the standard bases A, U, G, and C, including methylation, reduction, deamination, and rearranged glycosidic bonds. Further posttranscriptional modification of the tRNA molecules includes nucleotide alkylations and the attachment of the characteristic $CpCpA_{OH}$ terminal at the 3′ end of the molecule by a nucleotidyl transferase. The 3′ OH of the A ribose is the point of attachment for the specific amino acid that is to enter the polymerization reaction of protein synthesis. The methylation of mammalian tRNA precursors probably occurs in the nucleus, whereas the cleavage and attachment of $CpCpA_{OH}$ are cytoplasmic functions; the terminal nucleotides turn over more rapidly than do the tRNA molecules themselves. Specific aminoacyl tRNA synthetases within the cytoplasm of mammalian cells are required for the attachment of the different amino acids to the $CpCpA_{OH}$ residues (see Chapter 37).

RNA CAN ACT AS A CATALYST

In addition to the catalytic action served by the snRNAs in the formation of mRNA, several other enzymatic functions have been attributed to RNA. **Ribozymes** are RNA molecules with catalytic activity. These ribozyme activities generally involve transesterification reactions, and most are concerned with RNA metabolism (splicing and endoribonuclease). Recently, a rRNA component has been implicated in hydrolyzing an aminoacyl ester and thus to play a central role in peptide bond function (peptidyl transferases; see Chapter 37). Collectively, these observations, made using RNA molecules derived from the organelles from plants, yeast, viruses, and higher eukaryotic cells, show that RNA can act as an enzyme, and have revolutionized thinking about enzyme action and the origin of life itself.

SUMMARY

- RNA is synthesized from a DNA template by the enzyme DNA-dependent RNA polymerase.

- While bacteria contain but a single RNA polymerase ($\beta\beta\alpha_2\sigma\omega$), there are three distinct nuclear DNA-dependent RNA polymerases in mammals: RNA polymerases I, II, and III. These enzymes catalyze the transcription of rRNA (pol I), mRNA/mi/siRNAs/lncRNAs (pol II), and tRNA and 5S rRNA (pol III) encoding genes.

- RNA polymerases interact with unique *cis*-active regions of genes, termed promoters, in order to form PICs capable of initiation. In eukaryotes, the process of pol II PIC formation requires, in addition to polymerase, multiple general transcription factors (GTFs), TFIIA, B, D, E, F, and H. Distinct sets of GTFs are utilized by RNA Polymerases I and III.

■ Eukaryotic PIC formation can occur on accessible promoters either stepwise—by the sequential, ordered interactions of GTFs and RNA polymerase with DNA promoters—or in one step by the recognition of the promoter by a preformed GTF-RNA polymerase holoenzyme complex.

■ Transcription exhibits three phases: initiation, elongation, and termination. All are dependent on distinct DNA *cis*-elements and can be modulated by distinct *trans*-acting protein factors.

■ The presence of nucleosomes can either enhance or occlude the binding of both transfactors and the transcription machinery to their cognate DNA *cis*-elements, thereby modulating transcription.

■ Most eukaryotic RNAs are synthesized as precursors that contain excess sequences that are removed prior to the generation of mature, functional RNA. These processing reactions provide additional potential steps for regulation of gene expression.

■ Eukaryotic mRNA synthesis results in a pre-mRNA precursor that contains extensive amounts of excess RNA (introns) that must be precisely removed by RNA splicing to generate functional, translatable mRNA composed of exonic coding and 5′ and 3′ noncoding sequences.

■ All steps—from changes in DNA template, sequence, and accessibility in chromatin, to RNA stability and translatability—are subject to modulation and hence are potential control sites for eukaryotic gene regulation.

REFERENCES

Chen H, Pugh BF: What do transcription factors interact with? J Mol Biol 2021;Feb 20:166883.

Davis MC, Kesthely CA, Franklin EA, MacLellan SR: The essential activities of the bacterial sigma factor. Can J Microbiol 2017;63:89-99.

Decker KB, Hinton DM: Transcription regulation at the core: similarities among bacterial, archaeal, and eukaryotic RNA polymerases. Ann Rev Microbiol 2013;67:113-139.

Eaton JD, West S: Termination of transcription by RNA polymerase II: BOOM! Trends Genet 2020;36:664-675.

Elkon R, Ugalde AP, Agami R: Alternative cleavage and polyadenylation: extent, regulation and function. Nat Rev Gen 2013;14:496-506.

Hillen HS, Teminkov D, Cramer P: Structural basis of mitochondrial transcription. Nat Struc Mol Biol 2018; 25:754-765.

Kugel JF, Goodrich JA: Finding the start site: redefining the human initiator element. Genes Dev 2017;31:1-2.

Lee Y, Rio DC: Mechanisms and regulation of alternative pre-mRNA splicing. Ann Rev Biochem 2015;84:291-323.

Luse DS, Parida M, Spector BM, Nilson KA, Price DH: A unified view of the sequence and functional organization of the human RNA polymerase II promoter. Nuc Acids Res 2020;48:7767-7785.

Miguel-Escalada I, Pasquali L, Ferrer J: Transcriptional enhancers: functional insights and role in human disease. Curr Opin Genet Dev 2015;33:71-76.

Niederriter AR, Varshney A, Parker SC, Martin DM: Super enhancers in cancers, complex disease, and developmental disorders. Genes 2015;6:1183-1200.

Nogales E, Louder RK, He Y: Cryo-EM in the study of challenging systems: the human transcription pre-initiation complex. Curr Opin Struc Biol 2016;40:120-127.

Osman S, Cramer P: Structural biology of RNA polymerase II transcription: 20 years on. Ann Rev Cell Dev Biol 2020;3611-3634.

Rambout X, Maquat LE: The nuclear cap-binding complex as choreographer of gene transcription and pre-mRNA processing. Genes Dev 2020;34:1113-1127.

Roeder RG: 50⁺ years of eukaryotic transcription: an expanding universe of factors and mechanisms. Nat Struc Mol Biol 2019:26:783-791.

Takizawa Y, Binshtein E, Erwin AL, Pyburn TM, Mittendorf KF, Ohi MD: While the revolution will not be crystallized, biochemistry reigns supreme. Prot Sci 2017;26:69-81.

Venkatesh S, Workman JL: Histone exchange, chromatin structure and the regulation of transcription. Nat Rev Mol Cell Biol 2015;16:178-189.

Wan R, Bai R, Zan X, Shi Y: How is precursor messenger RNA spliced by the spliceosome? Ann Rev Biochem 2020;89:333-358.

Zhang Q, Lenardo MJ, Baltimore D: 30 years of NF-KB: a blossoming of relevance to human pathobiology. Cell 2017;168:37-57.

37

Protein Synthesis & the Genetic Code

P. Anthony Weil, PhD

O B J E C T I V E S

After studying this chapter, you should be able to:

- Understand that the genetic code is a three-letter nucleotide code, which is contained within the linear array of the exon DNA (composed of triplets of A, G, C, and T) of protein coding genes, and that this three-letter code is translated into mRNA (composed of triplets of A, G, C, and U) to specify the linear order of amino acid addition during protein synthesis via the process of translation.

- Appreciate that the universal genetic code is degenerate, unambiguous, nonoverlapping, and punctuation free.

- Explain that the genetic code is composed of 64 codons, 61 of which encode amino acids while 3 induce the termination of protein synthesis.

- Describe how the transfer RNAs (tRNAs) serve as the ultimate informational agents that decode the genetic code of messenger RNAs (mRNAs).

- Understand the mechanism of the energy-intensive process of protein synthesis that involves the steps of initiation, elongation, and termination, occurs on RNA-protein complexes termed ribosomes.

- Appreciate that protein synthesis, like DNA replication and transcription, is precisely controlled through the action of multiple accessory factors that are responsive to multiple extra- and intracellular regulatory signaling inputs.

BIOMEDICAL IMPORTANCE

The letters A, G, T, and C correspond to the nucleotides found in DNA. Within the protein-coding genes, these nucleotides are organized into three-letter code words called **codons**, and the collection of these codons, once transcribed into mRNA, make up the **genetic code**. It was impossible to understand protein synthesis—or to explain the molecular effects of gene mutations—before the genetic code was deciphered. The code provides a foundation for explaining the way in which protein defects may cause genetic disease and for the diagnosis and possible treatment of these disorders. In addition, the pathophysiology of many viral infections is related to the ability of these infectious agents to disrupt host cell protein synthesis. Many antibacterial drugs are effective because they selectively disrupt protein synthesis in the invading bacterial cell but do not affect protein synthesis in eukaryotic cells.

GENETIC INFORMATION FLOWS FROM DNA TO RNA TO PROTEIN

The genetic information within the nucleotide sequence of DNA is transcribed in the nucleus into the specific nucleotide sequence of an mRNA molecule. The sequence of nucleotides in the RNA transcript is complementary to the nucleotide sequence of the template strand of its gene in accordance with Watson-Crick base-pairing rules. Several different classes of RNA combine to direct the synthesis of proteins.

In prokaryotes there is a linear correspondence between the gene, the messenger RNA transcribed from the gene, and the polypeptide product. The situation is more complicated in higher eukaryotic cells where the primary transcript is much larger than the mature mRNA. The large mRNA precursors contain coding regions (exons) that will form the mature mRNA and long intervening sequences (introns) that separate the exons. The mRNA is processed within the nucleus, and the introns, which typically make up much more of this RNA than the exons, are removed. Exons are spliced together to form mature mRNAs, which are transported to the cytoplasm, where they are translated into protein (see Chapter 36).

The cell must possess the machinery necessary to translate information accurately and efficiently from the nucleotide sequence of an mRNA into the sequence of amino acids of the corresponding specific protein. Clarification of our understanding of this process, which is termed **translation**, awaited deciphering of the genetic code. It was realized early that mRNA molecules themselves have no affinity for amino acids and, therefore, that the translation of the information in the mRNA nucleotide sequence into the amino acid sequence of a protein requires an intermediate adapter molecule. This adapter molecule must recognize a specific nucleotide sequence on the one hand as well as a specific amino acid on the other. With such an adapter molecule, the cell can direct a specific amino acid into the proper sequential position of a protein during its synthesis as dictated by the nucleotide sequence of the specific mRNA. The functional groups of the amino acids do not themselves actually come into contact with the mRNA template.

THE NUCLEOTIDE SEQUENCE OF AN mRNA MOLECULE CONTAINS A SERIES OF CODONS THAT SPECIFY THE AMINO ACID SEQUENCE OF THE ENCODED PROTEIN

Twenty different amino acids are required for the synthesis of the cellular complement of proteins; thus, there must be at least 20 distinct codons that make up the genetic code. Since there are only four different nucleotides in mRNA, each codon must consist of more than a single purine or pyrimidine nucleotide. Codons consisting of two nucleotides each could provide for only 16 (ie, 4^2) distinct codons, whereas codons of three nucleotides could provide 64 (4^3) specific codons.

It is now known that each codon consists of a sequence of three nucleotides; that is, **it is a triplet code (Table 37–1)**. The initial deciphering of the **genetic code** depended heavily on in vitro synthesis of nucleotide polymers, particularly triplets in repeated sequence. These synthetic triplet ribonucleotides were used as mRNAs to program protein synthesis in the test tube, which allowed investigators to deduce the genetic code.

THE GENETIC CODE IS DEGENERATE, UNAMBIGUOUS, NONOVERLAPPING, WITHOUT PUNCTUATION, & UNIVERSAL

Three of the 64 possible codons do not code for specific amino acids; these have been termed **nonsense codons**. Nonsense codons are utilized in the cell as **translation termination**

TABLE 37–1 The Genetic Codea (Codon Assignments in Mammalian Messenger RNAs)

First Nucleotide	Second Nucleotide				Third Nucleotide
	U	C	A	G	
U	Phe	Ser	Tyr	Cys	U
	Phe	Ser	Tyr	Cys	C
	Leu	Ser	Term	Termb	A
	Leu	Ser	Term	Trp	G
C	Leu	Pro	His	Arg	U
	Leu	Pro	His	Arg	C
	Leu	Pro	Gln	Arg	A
	Leu	Pro	Gln	Arg	G
A	Ile	Thr	Asn	Ser	U
	Ile	Thr	Asn	Ser	C
	Ilea	Thr	Lys	Argb	A
	Met	Thr	Lys	Argb	G
G	Val	Ala	Asp	Gly	U
	Val	Ala	Asp	Gly	C
	Val	Ala	Glu	Gly	A
	Val	Ala	Glu	Gly	G

aThe terms first, second, and third nucleotide refer to the individual nucleotides of a triplet codon read 5′-3′, left to right. A, adenine nucleotide; C, cytosine nucleotide; G, guanine nucleotide; Term, chain terminator codon; U, uridine nucleotide. AUG, which codes for Met, serves as the initiator codon in mammalian cells and also encodes for internal methionines in a protein. (Abbreviations of amino acids are explained in Chapter 3.)
bIn mammalian mitochondria, AUA codes for Met and UGA for Trp, and AGA and AGG serve as chain terminators.

signals by specifying where the polymerization of amino acids into a protein molecule is to stop. The remaining 61 codons code for the 20 naturally occurring amino acids (see Table 37–1). Thus, there is "**degeneracy**" in the genetic code—that is, multiple codons decode the same amino acid. Some amino acids are encoded by several codons; for example, six different codons, UCU, UCC, UCA, UCG, AGU, and AGC all specify serine. Other amino acids, such as methionine and tryptophan, have a single codon. In general, the third nucleotide in a codon is less important than the first two in determining the specific amino acid to be incorporated, and this accounts for most of the degeneracy of the code. However, for any specific codon, only a single amino acid is specified; with rare exceptions, the genetic code is **unambiguous**—that is, given a specific codon, only a single amino acid is indicated. The distinction between **ambiguity** and **degeneracy** is an important concept.

The unambiguous but degenerate code can be explained in molecular terms. The recognition of specific codons in the mRNA by the tRNA adapter molecules is dependent on the tRNA **anticodon region** and specific base-pairing rules that dictate tRNA–mRNA codon binding. Each tRNA molecule contains a specific sequence, complementary to a codon, which is termed its anticodon. For a given codon in the mRNA, only a single species of tRNA molecule possesses the proper anticodon. Since each tRNA molecule can be charged with only one specific amino acid, each codon therefore specifies only one amino acid. However, some tRNA molecules can utilize the anticodon to recognize more than one codon. With few exceptions, given a specific codon, only a specific amino acid will be incorporated—although, given a specific amino acid, more than one codon may be used.

As discussed in the following section, the reading of the genetic code during the process of protein synthesis does not involve any overlap of codons. Thus, the genetic code is **nonoverlapping**. Furthermore, once the reading is commenced at a specific start codon, there is **no punctuation** between codons, and the message is read in a continuing sequence of nucleotide triplets until a translation stop codon is reached.

Until recently, the genetic code was thought to be universal. It has now been shown that the set of tRNA molecules in mitochondria (which contain their own separate and distinct translation machinery) from lower and higher eukaryotes, including humans, reads four codons differently from the tRNA molecules in the cytoplasm of even the same cells. As noted in a footnote to Table 37–1, in mammalian mitochondria the codon AUA is read as Met, and UGA codes for Trp. In addition, in mitochondria, the codons AGA and AGG are read as stop or chain terminator codons rather than as Arg. As a result of these organelle-specific changes in genetic code, mitochondria require only 22 tRNA molecules (see Figure 35–8 for the location of these genes in mtDNA) to read their genetic code, whereas the cytoplasmic translation system possesses a full complement of 31 tRNA species. These exceptions noted, the genetic code is **universal**. The frequency of use of each amino acid codon varies considerably between species and among different tissues within a species. The specific tRNA levels generally mirror these codon usage biases. Thus, a particular abundantly used codon is decoded by a similarly abundant-specific tRNA which recognizes that particular codon. Tables of **codon usage** are quite accurate now that many genomes have been sequenced and such information is vital for large-scale production of proteins for therapeutic purposes (ie, insulin, erythropoietin). Such proteins are often produced in nonhuman cells using recombinant DNA technology (see Chapter 39).

TABLE 37–2 Features of the Genetic Code

- Degenerate
- Unambiguous
- Nonoverlapping
- Not punctuated
- Universal

A summary of the main features of the genetic code are listed in **Table 37–2**.

AT LEAST ONE SPECIES OF tRNA EXISTS FOR EACH OF THE 20 AMINO ACIDS

tRNA molecules have extraordinarily similar functions and three-dimensional structures. The adapter function of the tRNA molecules requires the charging of each specific tRNA with its specific amino acid. Since there is no affinity of nucleic acids for specific functional groups of amino acids, this recognition must be carried out by a protein molecule capable of recognizing both a specific tRNA molecule and a specific amino acid. At least 20 specific enzymes are required for these specific recognition functions and for the proper attachment of the 20 amino acids to specific tRNA molecules. This energy requiring process of recognition and attachment, **tRNA amino acid charging**, proceeds in two steps and is catalyzed by one enzyme for each of the 20 amino acids. These enzymes are termed **aminoacyl-tRNA synthetases**. They form an activated intermediate of aminoacyl-AMP–enzyme complex (**Figure 37–1**). The specific aminoacyl-AMP–enzyme complex then recognizes a specific tRNA to which it attaches the aminoacyl moiety at the 3′-hydroxyl adenosyl terminal. The charging reactions have an error rate of less than 10^{-4} and so are quite accurate. The amino acid remains attached to its specific tRNA in an ester linkage until it is incorporated at a specific position during the synthesis of a polypeptide on the ribosome.

The regions of the tRNA molecule referred to in Chapter 34 (and illustrated in Figure 34–11) now become important. The **ribothymidine pseudouridine cytidine (TψC) arm** is involved

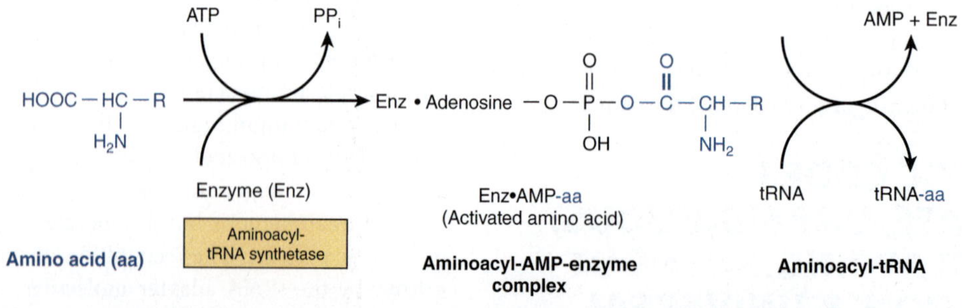

FIGURE 37–1 Formation of aminoacyl-tRNA. A two-step reaction, involving the enzyme aminoacyl-tRNA synthetase, results in the formation of aminoacyl-tRNA. The first reaction involves the formation of an AMP-amino acid–enzyme complex. This activated amino acid is next transferred to the corresponding tRNA molecule. The AMP and enzyme are released, and the latter can be reutilized. The charging reactions have an error rate (ie, esterifying the incorrect amino acid on tRNA^{XXX}) of less than one mischarging event out of 10^4 amino acid charging events.

in binding of the aminoacyl-tRNA to the ribosomal surface at the site of protein synthesis. The **tRNA dihydrouridine (D) arm** is one of the sites important for the proper recognition of a given tRNA species by its proper aminoacyl-tRNA synthetase. The **tRNA acceptor arm**, located at the 3′-hydroxyl adenosyl terminal, is the site of attachment of the specific amino acid.

The anticodon region (arm) consists of seven nucleotides, and it recognizes the three-letter codon in mRNA (**Figure 37–2**). The sequence read from the 3′ to 5′ direction in that anticodon loop consists of a variable base (N)–modified purine (Pu˙)–XYZ (the anticodon)–pyrimidine (Py)–pyrimidine (Py)-5′. Note that this direction of reading the anticodon is 3′–5′, whereas the genetic code in Table 37–1 is read 5′–3′, since the codon and the anticodon loop of the mRNA and tRNA molecules, respectively, are antiparallel in their complementarity just like all other intermolecular interactions between nucleic acid strands.

The degeneracy of the genetic code resides mostly in the last nucleotide of the codon triplet, suggesting that the base pairing between this last nucleotide and the corresponding nucleotide of the anticodon is not strictly by the Watson-Crick rule. This is called **wobble**; the pairing of the codon and anticodon can "wobble" at this specific nucleotide-to-nucleotide pairing site. For example, the two codons for arginine, AGA and AGG, can bind to the same anticodon having an uracil at its 5′ end (UCU). Similarly, three codons for glycine—GGU, GGC, and GGA—can form a base pair from one anticodon, 3′ CCI 5′, because **inosine (I)** can base pair with U, C, and A. Inosine is generated by deamination of adenine (see Figure 33–2 for structure).

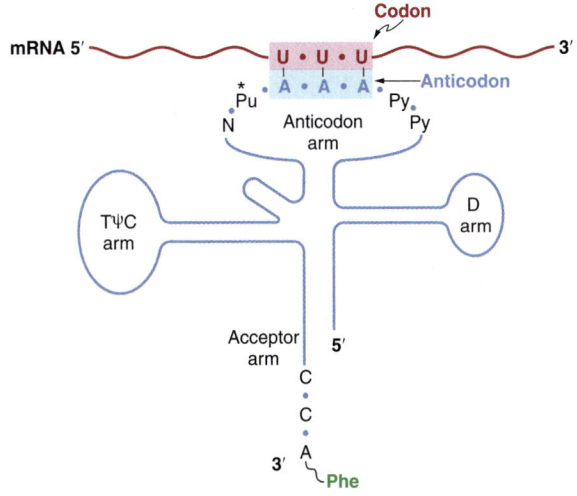

FIGURE 37–2 Recognition of the codon by the anticodon. One of the codons for phenylalanine is UUU. tRNA charged with phenylalanine (Phe) has the complementary sequence AAA; hence, it forms a base-pair complex with the codon. The anticodon region (arm) typically consists of a sequence of seven nucleotides: variable (N), modified purine (Pu*), X, Y, Z (here, A A A), and two pyrimidines (Py) in the 3′ to 5′ direction. Note the antiparallel mode of interaction between mRNA and tRNA.

MUTATIONS RESULT WHEN CHANGES OCCUR IN THE NUCLEOTIDE SEQUENCE

Although the initial change may not occur in the template strand of the double-stranded DNA molecule for that gene, after replication, daughter DNA molecules with mutations in the template strand will segregate and appear in the population of organisms.

Some Mutations Occur by Base Substitution

Single-base changes (**point mutations**) may be **transitions** or **transversions**. In the former, a given pyrimidine is changed to the other pyrimidine or a given purine is changed to the other purine. Transversions are changes from a purine to either of the two pyrimidines or the change of a pyrimidine into either of the two purines, as shown in **Figure 37–3**.

When the DNA nucleotide sequence of a protein-coding gene containing the mutation is transcribed into an mRNA molecule, then the RNA molecule will of course possess the base change at the corresponding location.

Single-base changes in the mRNA may have one of several effects when translated into protein:

1. There may be no detectable effect because of the degeneracy of the code; such mutations are often referred to as **silent mutations**. This would be most likely if the changed base in the mRNA molecule were to be at the third nucleotide of a codon. Because of wobble, the translation of a codon is least sensitive to a change at the third position.

2. A **missense effect** will occur when a different amino acid is incorporated at the corresponding site in the protein molecule. This mistaken amino acid—or missense, depending on its location in the specific protein—might be acceptable, partially acceptable, or unacceptable to the function of that protein molecule. From a careful examination of the genetic code, one can conclude that most single-base changes would result in the replacement of one amino acid by another with rather similar functional groups. This is an effective genetic "buffering" mechanism to avoid drastic change in the physical properties of a protein molecule. If an acceptable missense effect occurs, the resulting protein molecule may not be distinguishable from the normal one.

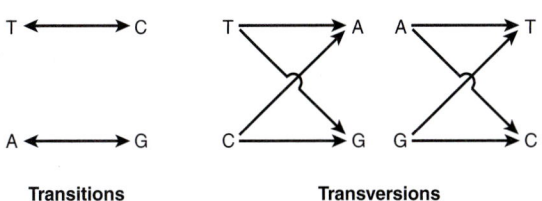

FIGURE 37–3 Diagrammatic representation of transition and transversion mutations.

A partially acceptable missense will result in a protein molecule with partial but abnormal function. If an unacceptable missense effect occurs, then the protein molecule will not be capable of functioning normally.

3. A **nonsense** codon may appear that would then result in the **premature termination** of translation and the production of only a fragment of the intended protein molecule. The probability is high that a prematurely terminated protein molecule or peptide fragment will not function in its normal role. Examples of the different types of mutations, and their effects on the coding potential of mRNA are presented in **Figures 37–4** and **37–5**.

Frameshift Mutations Result From Deletion or Insertion of Nucleotides in DNA That Generates Altered mRNAs

The deletion of a single nucleotide from the coding strand of a gene results in an altered reading frame in the mRNA. The machinery translating the mRNA does not recognize that a base was missing, since there is no punctuation in the reading of codons. Thus, a major alteration in the sequence of polymerized amino acids, as depicted in Example 1, Figure 37–5, results. Altering the reading frame results in a garbled translation of the mRNA distal to the single nucleotide deletion. Not only is the sequence of amino acids distal to this deletion garbled, but reading of the message can also result in the appearance of a nonsense codon and thus the production of a

polypeptide both garbled and prematurely terminated (Example 3, Figure 37–5).

If three nucleotides or a multiple of three nucleotides are deleted from a coding region, translation of the corresponding mRNA will generate a protein that is missing the corresponding number of amino acids (Example 2, Figure 37–5). Because the reading frame is a triplet, the reading phase will not be disturbed for those codons distal to the deletion. If, however, deletion of one or two nucleotides occurs just prior to or within the normal termination codon (nonsense codon), the reading of the normal termination signal is disturbed. Such a deletion might result in reading through the now "mutated" termination signal until another nonsense codon is encountered (not shown here).

Insertions of one or two or nonmultiples of three nucleotides into a gene result in an mRNA in which the reading frame is distorted on translation, and the same effects that occur with deletions are reflected in the mRNA translation. This may result in garbled amino acid sequences distal to the insertion and the generation of a **nonsense codon** at, or distal to the insertion, or perhaps reading through the normal termination codon. Following a deletion in a gene, an insertion (or vice versa) can reestablish the proper reading frame (Example 4, Figure 37–5). The corresponding mRNA, when translated, would contain a garbled amino acid sequence between the insertion and deletion. Beyond the reestablishment of the reading frame, the amino acid sequence would be correct. One can imagine that different combinations of **insertions or deletions** (ie, **indels**), or of both, would result in

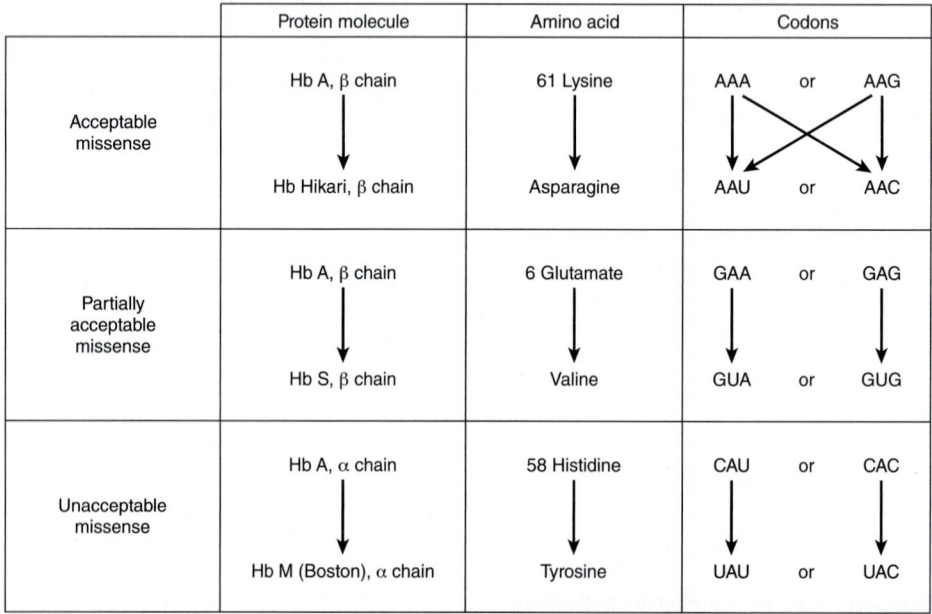

FIGURE 37–4 **Examples of three types of missense mutations resulting in abnormal hemoglobin chains.** The amino acid alterations and possible alterations in the respective codons are indicated. The hemoglobin Hikari β-chain mutation has apparently normal physiologic properties but is electrophoretically altered. Hemoglobin S has a β-chain mutation and partial function; hemoglobin S binds oxygen but precipitates when deoxygenated; this causes red blood cells to sickle, and represents the cellular and molecular basis of sickle cell disease (see Figure 6–13). Hemoglobin M Boston, an α-chain mutation, permits the oxidation of the heme ferrous iron to the ferric state and so will not bind oxygen at all.

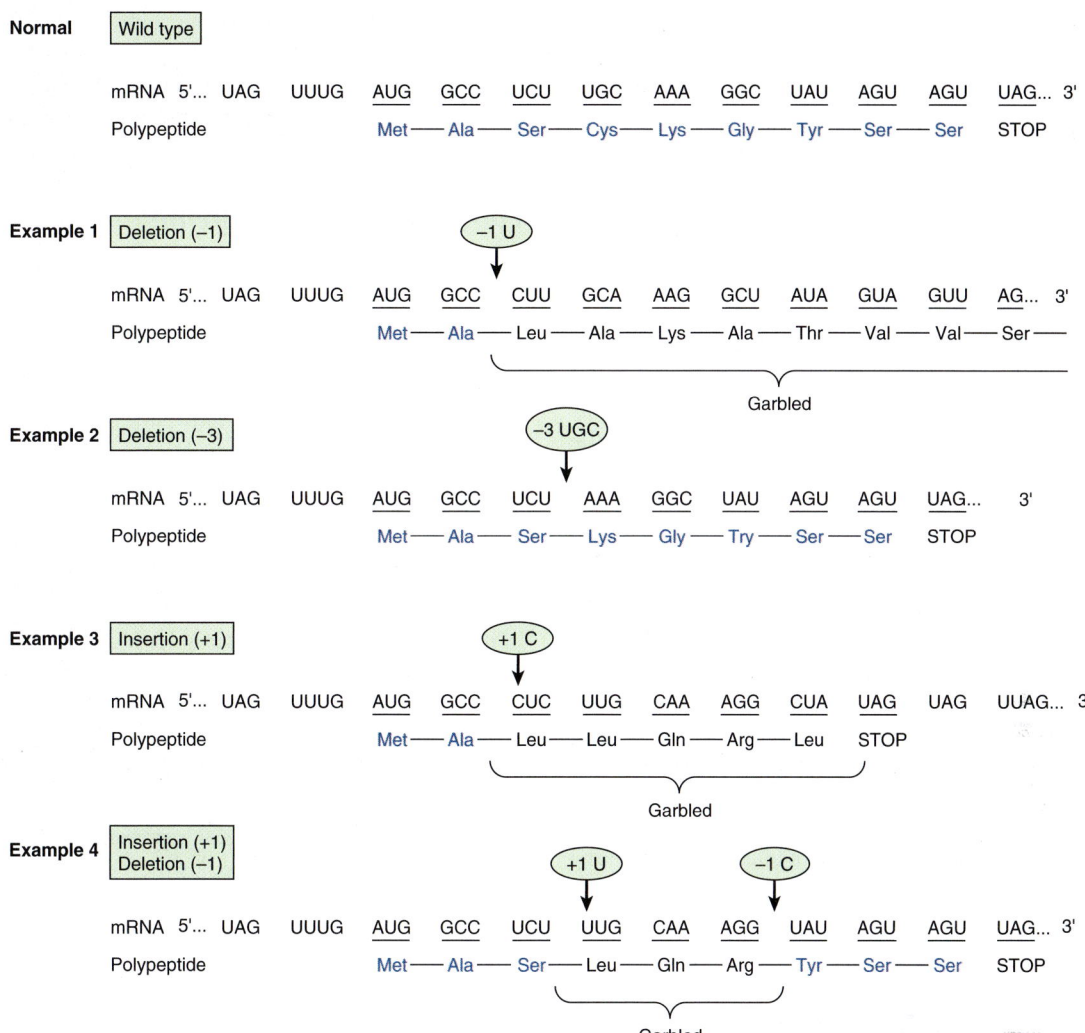

FIGURE 37–5 Examples of the effects of deletions and insertions in a gene on the sequence of the mRNA transcript and of the polypeptide chain translated therefrom. The arrows indicate the sites of deletions or insertions, and the numbers in the ovals indicate the number of nucleotide residues deleted or inserted. Colored type indicates the correct amino acids in the correct order.

formation of a protein wherein a portion is abnormal, but this portion is surrounded by the normal amino acid sequences. Such phenomena have been demonstrated convincingly in a number of human diseases.

Suppressor Mutations Can Counteract Some of the Effects of Missense, Nonsense, & Frameshift Mutations

The discussion of the altered protein products of gene mutations is based on the presence of normally functioning tRNA molecules. However, in prokaryotic and lower eukaryotic organisms, abnormally functioning tRNA molecules have been discovered that are themselves the results of mutations. Some of these abnormal tRNA molecules are capable of binding to and decoding altered codons, thereby suppressing the effects of mutations in distinct mutated mRNA-encoding structural genes. These **suppressor tRNA molecules**, usually

formed as a result of alterations in their anticodon regions, are capable of suppressing certain missense mutations, nonsense mutations, and frameshift mutations. However, since the suppressor tRNA molecules are not capable of distinguishing between a normal codon and one resulting from a gene mutation, their presence in the cell usually results in decreased viability. For instance, the nonsense suppressor tRNA molecules can suppress the normal termination signals to allow a read-through when it is not desirable. Frameshift suppressor tRNA molecules may read a normal codon plus a component of a juxtaposed codon to provide a frameshift, also when it is not desirable. Suppressor tRNA molecules may exist in mammalian cells, since read-through of translation has on occasion been observed. In the laboratory context, such suppressor tRNAs, coupled with mutated variants of aminoacyl-tRNA synthetases, can be utilized to incorporate unnatural amino acids into defined locations within altered genes that carry engineered nonsense mutations. The resulting labeled proteins can be used for in vivo and in vitro cross-linking and biophysical studies. This new tool adds significantly to biologists interested in studying the mechanisms of a wide range of biologic processes.

LIKE TRANSCRIPTION, PROTEIN SYNTHESIS CAN BE DESCRIBED IN THREE PHASES: INITIATION, ELONGATION, & TERMINATION

The general structural characteristics of ribosomes are discussed in Chapter 34. These particulate entities serve as the machinery on which the mRNA nucleotide sequence is translated into the sequence of amino acids of the specified protein. The translation of the mRNA commences near its 5′ end with the formation of the corresponding amino terminus of the protein molecule. The message is decoded from 5′ to 3′, concluding with the formation of the carboxyl terminus of the protein. Again, the concept of **polarity** is apparent. As described in Chapter 36, the transcription of a gene into the corresponding mRNA or its precursor first forms the 5′ end of the RNA molecule. In prokaryotes, this allows for the beginning of mRNA translation before the transcription of the gene is completed. In eukaryotic organisms, the process of transcription is a nuclear one, while mRNA translation occurs in the cytoplasm, precluding simultaneous transcription and translation in eukaryotic organisms and enabling the processing necessary to generate mature mRNA from the primary transcript.

Initiation Involves Several Protein-RNA Complexes

Initiation of eukaryotic protein synthesis requires that an mRNA molecule be selected for translation by a ribosome (**Figure 37–6**). Once the mRNA binds to the ribosome, the ribosome must locate the initiation codon thereby setting the correct reading frame on the mRNA, and translation begins. This process involves tRNA, rRNA, mRNA, and at least 10 **eukaryotic initiation factors (eIFs)**, some of which have multiple (three to eight) subunits. Also involved are GTP, ATP, and amino acids. Initiation can be divided into three steps, all of which are obligatorily preceded by dissociation of the 80S ribosome into its constituent 40S and 80S subunits: (1) binding of a **ternary complex consisting of the initiator methionyl-tRNA (met-tRNA^i), GTP, and eIF-2** to the 40S ribosome to form the **43S preinitiation complex**; (2) binding of mRNA to the 40S preinitiation complex to form the **48S initiation complex**; and (3) combination of the 48S initiation complex with the 60S ribosomal subunit to form the **80S initiation complex**.

Ribosomal Dissociation

Prior to initiation, 80S ribosomes dissociate into component 40S and 60S subunits during translation termination (see following discussion). Dissociation allows these components to participate in subsequent rounds of translation. Two initiation factors, **eIF-3, eIF-1,** and **eIF-1A**, bind to the newly dissociated 40S ribosomal subunit. Binding of these three eIFs delay reassociation of the 40S subunit with the 60S subunit, allowing other translation initiation factors to associate with the 40S subunit.

Formation of the 43S Preinitiation Complex

The first step of translation initiation involves the binding of GTP by eIF-2. This binary complex then binds to **methionyl tRNA^i**, a tRNA specifically involved in binding to the initiation codon AUG. It is important to note that there are two tRNAs for methionine. One specifies methionine for the initiator codon, the other for internal methionines. Each has a unique nucleotide sequence; both are aminoacylated by the same methionyl-tRNA synthetase. The GTP-eIF-2-tRNA^i ternary complex binds to the 40S ribosomal subunit to form the 43S preinitiation complex. The ternary complex–40S subunit complex is stabilized by eIF-3 and eIF-1A and the subsequent binding of **eIF5**.

eIF-2 is one of two control points for protein synthesis initiation in eukaryotic cells. eIF-2 consists of α, β, and γ subunits. **eIF-2α is phosphorylated** (on serine 51) by at least **four different protein kinases (HCR, PKR, PERK, and GCN2)** that are activated when a cell is under stress and when the energy expenditure required for protein synthesis would be deleterious. Such conditions include amino acid or glucose starvation, virus infection, intracellular presence of large quantities of misfolded proteins (endoplasmic reticulum [ER] stress), serum deprivation (for cells in culture), hyperosmolality, and heat shock. PKR is particularly interesting in this regard. This kinase is activated by viruses and provides a host defense mechanism that decreases protein synthesis, including viral protein synthesis, thereby inhibiting viral replication. Phosphorylated eIF-2α binds tightly to and inactivates the GTP–GDP recycling protein eIF-2B, thus preventing formation of the 43S preinitiation complex and blocking protein synthesis.

Formation of the 48S Initiation Complex

As described in Chapter 36, the 5′ termini of mRNA molecules in eukaryotic cells are "capped." The ^{7me}G-cap facilitates the binding of mRNA to the 43S preinitiation complex. A **cap-binding protein complex, eIF-4F (4F)**, which consists of **eIF-4E (4E)** and the **eIF-4G (4G)-eIF-4A (4A) complex**, binds to the cap through the 4E protein. Subsequently **eIF-4B (4B)** binds and reduces the complex secondary structure of the 5′ end of the mRNA through its ATP-dependent helicase activity. The association of mRNA with the 43S preinitiation complex to form the 48S initiation complex requires ATP hydrolysis. eIF-3 is a key protein because it binds with high affinity to the 4G component of 4F, and links this complex to the 40S ribosomal subunit. Following association of the 43S preinitiation complex with the mRNA cap, and reduction ("melting") of the secondary structure near the 5′ end of the mRNA through the action of the 4B helicase and ATP, the complex translocates 5′ → 3′ and scans the mRNA for a suitable initiation codon. Generally, this is the 5′-most AUG, but the precise initiation

STEPS IN THE REFORMATION OF THE 80S INITIATION COMPLEX

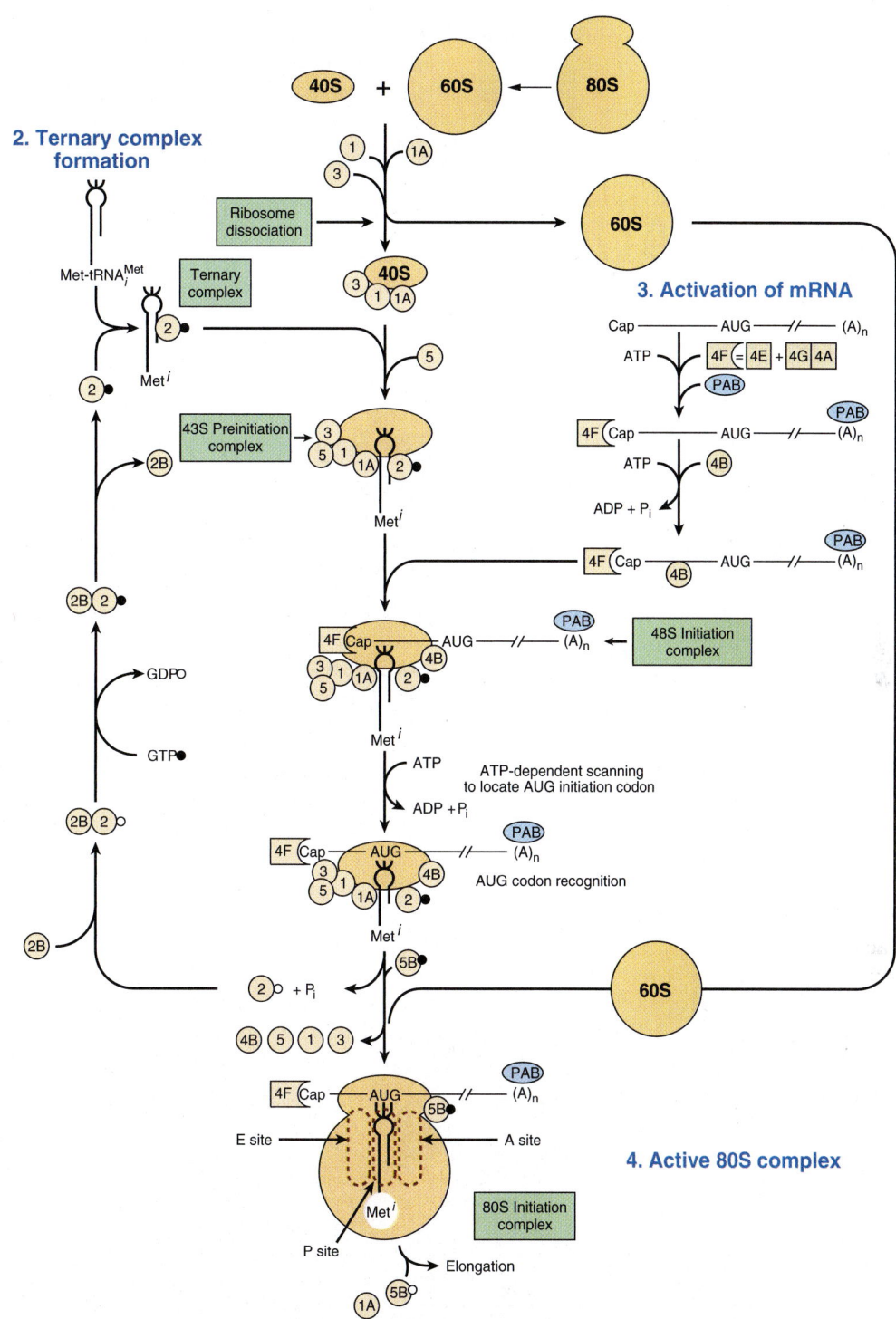

FIGURE 37–6 **Diagrammatic representation of the initiation phase of protein synthesis on a eukaryotic mRNA.** Eukaryotic mRNAs contain a 5′ ^{7me}G-cap (Cap) and 3′ poly(A) terminal [$(A)_n$] as shown. Translation preinitiation complex formation proceeds in several steps: (1) Dissociation of the 80S complex to component 40S and 60S subunits, a process facilitated by binding of factors eIF1, eIF1A, and eIF3 to the ribosomal 40S subunit (top). (2) Formation of the 43S preinitiation complex, a ternary complex consisting of met-tRNA$_i$ and GTP-bound to the initiation factor eIF-2 (eIF-2-GTP; left). This complex is then bound by the eIF5 initiation factor forming the complete 43S preinitiation complex. (3) Activation of 5′-capped mRNA and formation of the 48S initiation complex. mRNA is bound via its 5′-cap by eIF4F (composed of eIF4E, eIF4G, and eIF4A factors) and 3′ Poly(A) tail by Poly A binding (PAB) protein forming the 48S initiation complex. ATP hydrolysis-dependent 5′ to 3′ mRNA scanning enables location of the initiation codon AUG, which is then bound by met-tRNA$_i$. (4) Following addition of GTP-bound eI5B and dissociation of eIF1, eIF2-GDP, eIF3, and eIF5, formation of the 80S initiation complex occurs when a recycled 60S ribosomal subunit joins the 48S complex. This reaction positions the initiator met-tRNA$_i$ within the P-Site of the active 80S initiation complex formation induces dissociation of eIF1A and GDP-bound eIF5B (see text for details). This complex is now competent for translation initiation. (GTP, •; GDP, ○); the various initiation factors appear in abbreviated form as circles or squares, for example, eIF-3, (③), eIF-4F, (4F), ([4F]). 4•F is a complex consisting of 4E and 4A bound to 4G (see Figure 37–7). Note that the "circular" structure of mRNA illustrated in Figure 37–7 is thought to be the actual form of mRNA on which steps 1 to 4 actually occur.

codon is determined by so-called **Kozak consensus sequences** that surround the **AUG** initiation codon:

$$-3 \quad\quad -2 \quad\quad +1 \quad\quad +4$$

$$\text{GCCPu } ^A\!/_C \text{ C AUG G}$$

Role of the Poly(A) Tail in Initiation

Biochemical and genetic experiments have revealed that the **3′ poly(A) tail** and the **poly(A) binding protein**, **PAB**, are both required for efficient initiation of protein synthesis. Further studies showed that the poly(A) tail stimulates recruitment of the 40S ribosomal subunit to the mRNA through a complex set of interactions. PAB (**Figure 37–7**) bound to the poly(A) tail, interacts with eIF-4G, and 4E subunits of cap-bound eIF-4F to form a circular structure that helps direct the 40S ribosomal subunit to the 5′ end of the mRNA and also likely stabilizes mRNAs from exonucleolytic degradation. This helps explain how the cap and poly(A) tail structures have a synergistic effect on protein synthesis. Indeed, differential protein–protein interactions between general and specific mRNA translational repressors and eIF-4E result in m^{7}G cap-dependent translation control (**Figure 37–8**).

Formation of the 80S Initiation Complex

The binding of the 60S ribosomal subunit to the 48S initiation complex involves hydrolysis of the GTP bound to eIF-2 by **eIF-5**. This reaction results in release of the initiation factors bound to the 48S initiation complex (these factors then are recycled) and the rapid association of the 40S and 60S subunits to form the 80S ribosome. At this point, the met-tRNA^i is on the P site of the ribosome, ready for the elongation cycle to commence.

The Regulation of eIF-4E Controls the Rate of Initiation

The 4F complex is particularly important in controlling the rate of protein translation. As described earlier, 4F is a complex consisting of 4E, which binds to the m^{7}G cap structure at the 5′ end of the mRNA, and 4G, which serves as a scaffolding protein. In addition to binding 4E, 4G binds to eIF-3, which links the complex to the 40S ribosomal subunit. It also binds 4A and 4B, the ATPase-helicase complex that helps unwind the RNA (see Figure 37–8).

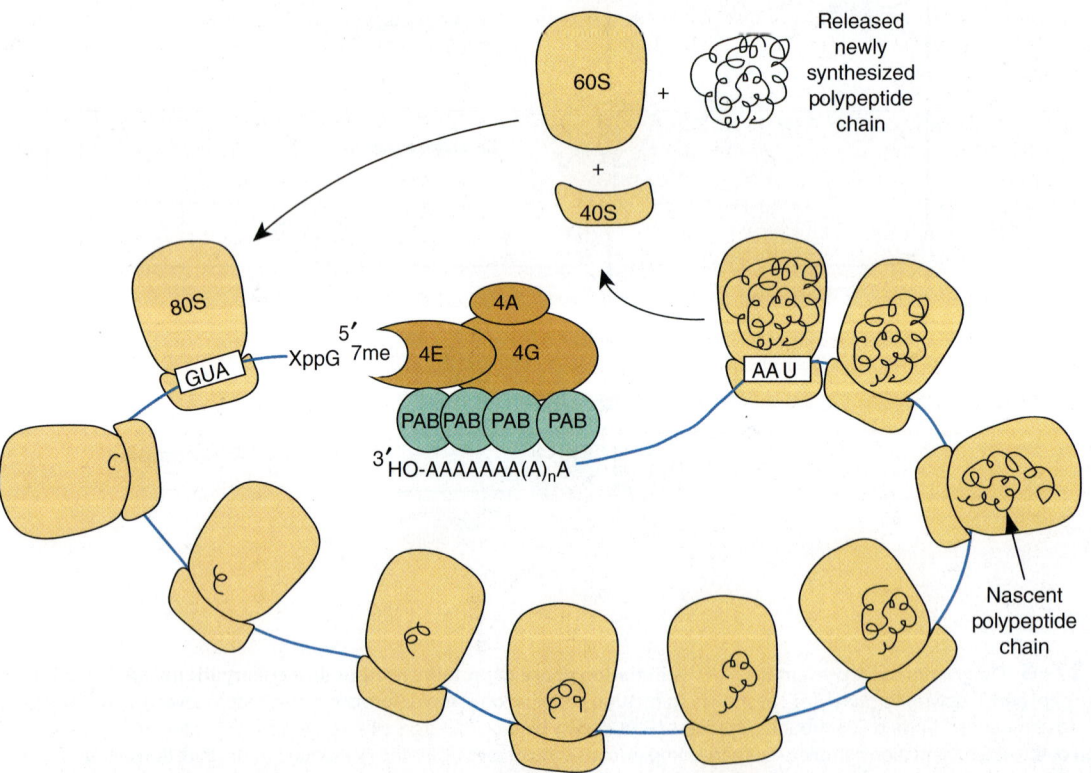

FIGURE 37–7 **Schematic illustrating the circularization of mRNA through protein–protein interactions between ^{7me}G cap-bound eIF4F and poly(A) tail-bound poly(A) binding protein.** eIF4F, composed of eIF4A, 4E, and 4G subunits binds the mRNA 5′-^{7me}G "Cap" (^{7me}GpppX-) upstream of the translation initiation codon (AUG) with high affinity. The eIF4G subunit of the complex also binds poly(A) binding protein (PAB) with high affinity. Since PAB is bound tightly to the mRNA 3′-poly(A) tail (5′-(X)$_n$A(A)$_n$ AAAAAAA$_{OH}$ 3′), circularization results. Shown are multiple 80S ribosomes that are in the process of translating the circularized mRNA into protein (black curlicues), forming a polysome. On encountering a termination codon (here UAA), translation termination occurs leading to release of the newly translated protein and dissociation of the 80S ribosome into 60S, 40S subunits. Dissociated ribosomal subunits can recycle through another round of translation (see Figures 37–6 and 37–10).

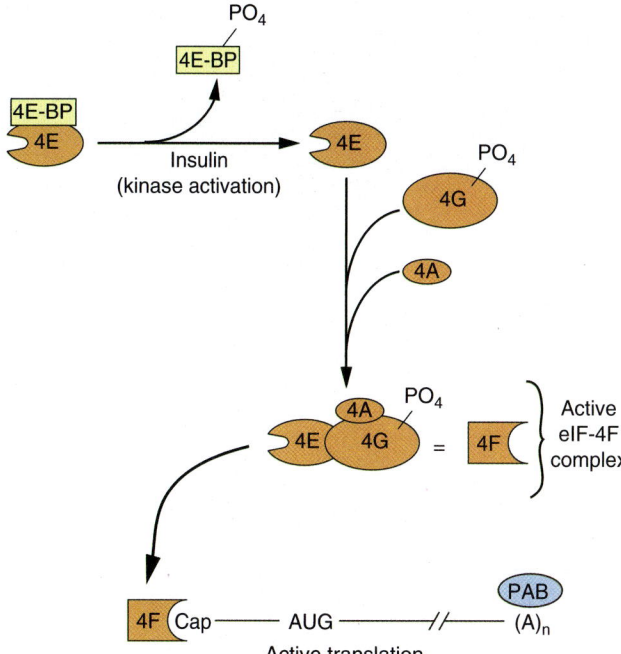

FIGURE 37–8 **Activation of eIF-4E by insulin and formation of the cap-binding eIF-4F complex.** The 4F-cap mRNA complex is depicted as in Figures 37–6 and 37–7. The 4F complex consists of eIF-4E (4E), eIF-4A (4A), and eIF-4G (4G). 4E is inactive when bound by one of a family of binding proteins (4E-BPs). Insulin and mitogenic growth polypeptides, or growth factors (eg, IGF-1, PDGF, interleukin-2, and angiotensin II) activate the PI3 kinase/AKT kinase signaling pathways, which activate the mTOR kinase; this results in the phosphorylation of 4E-BP (see Figure 42–8). Phosphorylated 4E-BP dissociates from 4E, and the latter is then able to form the 4F complex and bind to the mRNA cap. These growth polypeptides also induce phosphorylation of 4G itself by the mTOR and MAP kinase pathways (see Chapter 42). Phosphorylated 4F binds much more avidly to the cap than does nonphosphorylated 4F, which stimulates 48S initiation complex formation and hence translation.

4E is responsible for recognition of the mRNA cap structure, a rate-limiting step in translation. This process is further regulated by phosphorylation (see Figure 37–8). Insulin and mitogenic growth factors result in the phosphorylation of 4E on Ser209 (or Thr210). Phosphorylated 4E binds to the cap much more avidly than does the nonphosphorylated form, thus enhancing the rate of initiation. Components of the MAP kinase, PI3K, mTOR, RAS, and S6 kinase signaling pathways (see Figure 42–8) can all, under appropriate conditions, be involved in these regulatory phosphorylation reactions.

The activity of 4E is modulated in a second way, and this also involves phosphorylation; a set of proteins bind to and inactivate 4E. These proteins include **4E-BP1** (**BP1**, also known as **PHAS-1**) and the closely related proteins **4E-BP2** and **4E-BP3**. BP1 binds with high affinity to 4E. The 4E-BP1 association prevents 4E from binding to 4G (to form 4F). Since this interaction is essential for the binding of 4F to the ribosomal 40S subunit and for correctly positioning it on the capped mRNA, BP-1 effectively inhibits translation initiation.

Insulin and other growth factors result in the phosphorylation of BP-1 at seven unique sites. Phosphorylation of BP-1

results in its dissociation from 4E, and it cannot rebind until critical sites are dephosphorylated. These effects on the activation of 4E explain in part how insulin causes a marked posttranscriptional increase of protein synthesis in liver, adipose, and muscle tissue.

Elongation Is Also a Multistep, Accessory Factor-Facilitated Process

Elongation is a cyclic process on the ribosome in which one amino acid at a time is added to the nascent peptide chain (**Figure 37–9**). The peptide sequence is determined by the order of the codons in the mRNA. Elongation involves several steps catalyzed by proteins called **elongation factors** (**EFs**). These steps are (1) binding of aminoacyl-tRNA to the A site, (2) peptide bond formation, (3) translocation of the ribosome on the mRNA, and (4) expulsion of the deacylated tRNA from the P- and E-sites.

Binding of Aminoacyl-tRNA to the A Site

In the complete 80S ribosome formed during the process of initiation, both the **A site** (**aminoacyl or acceptor site**) and **E site** (**deacylated tRNA exit site**) are free (see Figure 37–6). The binding of the appropriate aminoacyl-tRNA in the A site requires proper codon recognition. **Elongation factor 1A (EF1A)** forms a ternary complex with GTP and the entering aminoacyl-tRNA (see Figure 37–9). This complex then allows the correct aminoacyl-tRNA to enter the A site with the release of EF1A-GDP and phosphate. GTP hydrolysis is catalyzed by an active site on the ribosome; hydrolysis induces a conformational change in the ribosome concomitantly increasing affinity for the tRNA. As shown in Figure 37–9, EF1A-GDP then recycles to EF1A-GTP with the aid of other soluble protein factors and GTP.

Peptide Bond Formation

The α-amino group of the new aminoacyl-tRNA in the A site carries out a nucleophilic attack on the esterified carboxyl group of the **peptidyl-tRNA** occupying the **P site** (**peptidyl or polypeptide site**). At initiation, this site is occupied by the initiator met-tRNA[i]. This reaction is catalyzed by a **peptidyl transferase**, a component of the 28S RNA of the 60S ribosomal subunit. This is another example of ribozyme activity and indicates an important—and previously unsuspected—direct role for RNA in protein synthesis (**Table 37–3**). Because the amino acid on the aminoacyl-tRNA is already "activated," no further energy source is required for this reaction. The reaction results in attachment of the growing peptide chain to the tRNA in the A site.

Translocation

The now deacylated tRNA is attached by its anticodon to the P site at one end and by its open 3′ CCA tail to the E site on the large ribosomal subunit (middle portion of Figure 37–9). At this point, **elongation factor 2 (EF2)** binds to and displaces the peptidyl tRNA from the A site to the P site. In turn, the

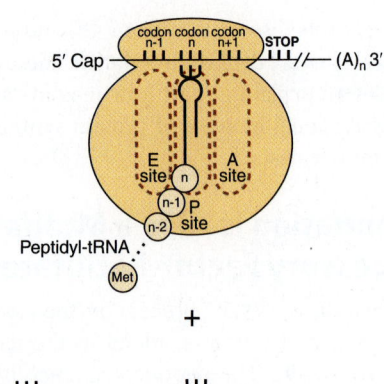

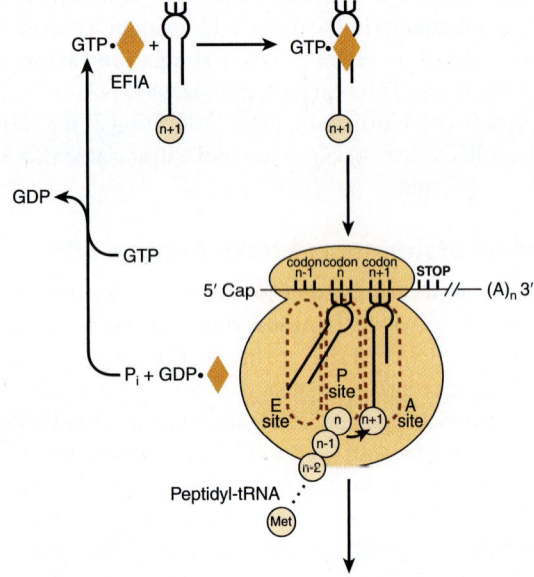

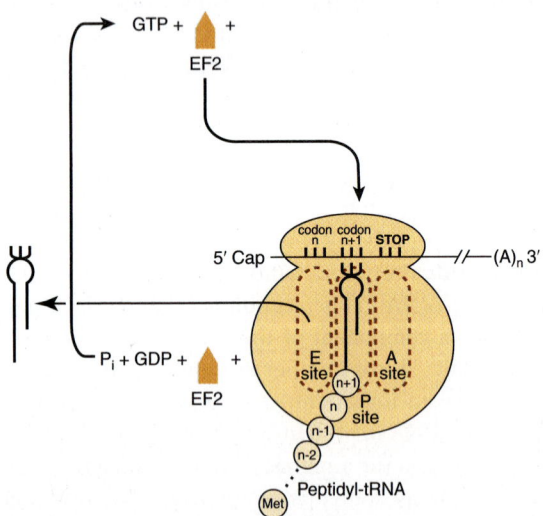

FIGURE 37–9 **Diagrammatic representation of the peptide elongation process of protein synthesis.** The small circles labeled n − 1, n, n + 1, etc., represent the amino acid residues of the newly formed protein molecule (in N-terminal to C-terminal orientation) and the corresponding codons in the mRNA. EFIA and EF2 represent elongation factors 1 and 2, respectively. The peptidyl-tRNA, aminoacyl-tRNA, and exit sites on the ribosome are represented by P site, A site, and E site, respectively.

TABLE 37–3 **Evidence That rRNA Is a Peptidyl Transferase**

- Ribosomes can make peptide bonds (albeit inefficiently) even when proteins are removed or inactivated.
- Certain parts of the rRNA sequence are highly conserved in all species.
- These conserved regions are on the surface of the RNA molecule.
- RNA can be catalytic in many other chemical reactions.
- Mutations that result in antibiotic resistance at the level of protein synthesis are more often found in rRNA than in the protein components of the ribosome.
- X-ray crystal structure of large subunit bound to tRNAs suggests detailed mechanism.

deacylated tRNA is on the E site, from which it leaves the ribosome. The EF2-GTP complex is hydrolyzed to EF2-GDP, effectively moving the mRNA forward by one codon and leaving the A site open for occupancy by another ternary complex of amino acid tRNA–EF1A-GTP and another cycle of elongation.

The charging of the tRNA molecule with the aminoacyl moiety requires the hydrolysis of an ATP to an AMP, equivalent to the hydrolysis of two ATPs to two ADPs and phosphates. The entry of the aminoacyl-tRNA into the A site results in the hydrolysis of one GTP to GDP. Translocation of the newly formed peptidyl-tRNA in the A site into the P site by EF2 similarly results in hydrolysis of GTP to GDP and phosphate. Thus, the energy requirements for the formation of one peptide bond include the equivalent of the hydrolysis of two ATP molecules to ADP and of two GTP molecules to GDP, or the hydrolysis of four high-energy phosphate bonds. A eukaryotic ribosome can incorporate as many as six amino acids per second; prokaryotic ribosomes incorporate as many as 18 per second. Thus, the energy requiring process of peptide synthesis occurs with great speed and accuracy until a termination codon is reached.

Termination Occurs When a Stop Codon Is Recognized

In comparison to initiation and elongation, termination is a relatively simple process (**Figure 37–10**). After multiple cycles of elongation culminating in polymerization of the specific amino acids into a protein molecule, the stop or terminating codon of mRNA (UAA, UAG, UGA) appears in the A site. Normally, there is no tRNA with an anticodon capable of recognizing such a termination signal. **Releasing factor 1 (RF1)** recognizes that a stop codon resides in the A site (see Figure 37–10). RF1 is bound by a complex consisting of **releasing factor 3 (RF3)** with bound GTP. This complex, with the peptidyl transferase, promotes hydrolysis of the bond between the peptide and the tRNA occupying the P site. Thus, a water molecule rather than an amino acid is added. This hydrolysis releases the protein and the tRNA from the P site. On hydrolysis and release, the 80S ribosome dissociates into its 40S and 60S subunits, which are then recycled (see Figure 37–7). Therefore, the releasing factors are proteins that hydrolyze the peptidyl-tRNA bond when a stop codon occupies the A site. The mRNA is then released from the ribosome, which dissociates into its component 40S and 60S subunits, and another cycle can be repeated (see Figure 37–6).

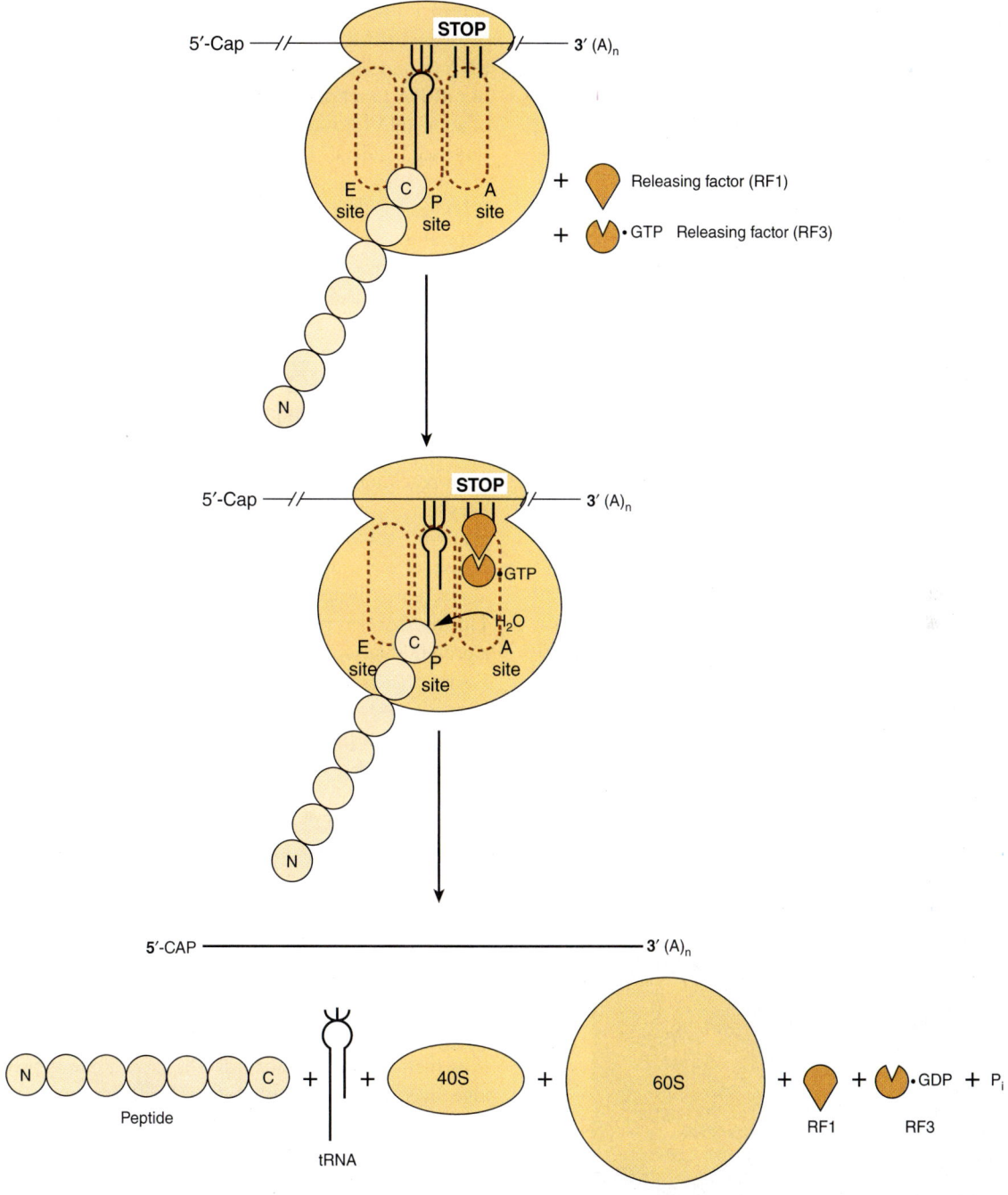

FIGURE 37–10 Diagrammatic representation of the termination process of protein synthesis. The 60S ribosomal peptidyl-tRNA, aminoacyl-tRNA, and exit sites are indicated as P site, A site, and E site, respectively. The termination (stop) codon is indicated by the three vertical bars and STOP. Releasing factor RF1 binds to the stop codon in the A site. Releasing factor RF3, with bound GTP, binds to RF1. Hydrolysis of the peptidyl-tRNA complex is shown by the entry of water (H_2O); arrow. N and C indicate the amino- and carboxy-terminal amino acids of the nascent polypeptide chain, respectively, and illustrate the polarity of protein synthesis. Termination results in release of the mRNA, the newly synthesized protein (N- and C-termini; N, C), free tRNA, 40S and 60S subunits, as well as RF1, GDP-bound RF3, and inorganic P_i, as shown at bottom.

Polysomes Are Assemblies of Ribosomes

Many ribosomes can translate the same mRNA molecule simultaneously. Because of their relatively large size, the ribosome particles cannot attach to an mRNA any closer than 35 nucleotides apart. **Multiple ribosomes on the same mRNA molecule** form a polyribosome, or "**polysome**" (see Figure 37–7). In an unrestricted system, the number of ribosomes attached to an mRNA (and thus the size of polyribosomes) correlates positively with the length of the mRNA molecule.

Polyribosomes actively synthesizing proteins can exist as free particles in the cellular cytoplasm or may be attached to sheets of membranous cytoplasmic structures referred to as the **endoplasmic reticulum (ER)**. Attachment of the particulate polyribosomes to the ER is responsible for its "rough" appearance as seen by electron microscopy. The proteins synthesized by the attached polyribosomes are extruded into the cisternal space between the sheets of rough ER and are exported from there. Some of the protein products of the rough ER are packaged by the Golgi apparatus for eventual export (see Figures 49–2, 49–6). The polyribosomal particles free in the cytosol are responsible for the synthesis of proteins required for intracellular functions.

Nontranslating mRNAs Can Form Ribonucleoprotein Particles That Accumulate in Cytoplasmic Organelles Termed P Bodies or Stress Granules

mRNAs, bound by specific packaging proteins and exported from the nucleus as **ribonucleoproteins particles (mRNPs)** sometimes do not immediately associate with ribosomes to be translated. Alternatively, under certain conditions where translation is slowed or stopped (cellular stress or developmental cues among other signals/conditions) select, untranslated mRNAs can associate with a range of specific RNAs and proteins to form **P bodies** and the related **stress granules (SG)**. These structures are biomolecular condensates composed of interacting RNAs and proteins. P bodies can readily be visualized via immunohistochemistry using appropriate antibodies (**Figure 37–11**). These cytoplasmic structures are related to similar small mRNA-containing granules found in neurons and certain maternal cells. Overall P bodies (and SGs) are thought to contribute importantly to mRNA metabolism. Over 35 distinct proteins have been suggested to reside exclusively or extensively

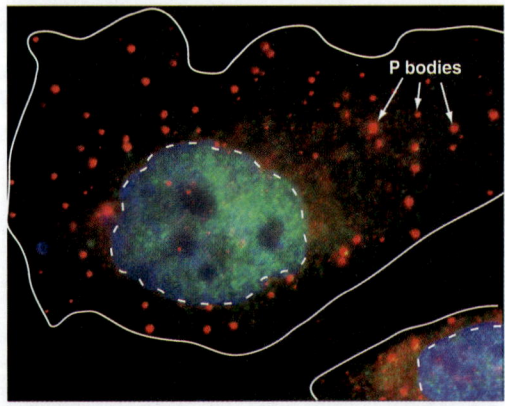

FIGURE 37–11 **The P body is a cytoplasmic structure involved in mRNA metabolism.** Shown is a photomicrograph of two mammalian cells in which a single distinct protein constituent of the P body has been visualized using the cognate-specific fluorescently labeled antibody. P bodies appear as small red circles of varying size throughout the cytoplasm. The cell plasma membranes are indicated by a solid white line, nuclei by a dashed line. Nuclei were counterstained using a fluorescent dye with different fluorescence excitation/emission spectra from the labeled antibody used to identify P bodies; the nuclear stain intercalates between the DNA base pairs and appears as blue/green. Modified from http://www.mcb.arizona.edu/parker/WHAT/what.htm. (Reproduced with permission from Dr. Roy Parker.)

within P bodies. These proteins range from mRNA binding proteins to mRNA decapping enzymes, RNA helicases, and RNA exonucleases (5′-3′ and 3′-5′), to components involved in miRNA function and mRNA quality control. Incorporation of an mRNA into such complexes is not an unequivocal "death sentence." Indeed, though the mechanisms are not yet fully understood, certain mRNAs appear to be temporarily stored in P bodies (or SGs) and then retrieved and utilized for protein translation. This molecular behavior suggests that the cytoplasmic functions of mRNA (translation and degradation) are controlled, at least in part, by the dynamic interaction of mRNA with polysomes and P body/SG protein/enzyme constituents.

The Machinery of Protein Synthesis Can Respond to Environmental Threats

Ferritin, an iron-binding protein, prevents ionized iron (Fe^{2+}) from reaching toxic levels within cells. Elemental iron stimulates ferritin synthesis by causing the release of a cytoplasmic protein that binds to a specific region in the 5′ nontranslated region of ferritin mRNA. Disruption of this protein-mRNA interaction activates ferritin mRNA and results in its translation. This mechanism provides for rapid control of the synthesis of a protein that sequesters Fe^{2+}, a potentially toxic molecule (see Figures 52–7 and 52–8). Similarly, environmental stress and starvation inhibit the positive roles of mTOR (see Figures 37–8 and 42–8) on promoting activation of eIF-4F and 48S complex formation.

Many Viruses Co-opt the Host Cell Protein Synthesis Machinery

The protein synthesis machinery can also be modified in deleterious ways. Viruses replicate by using host cell processes, including those involved in protein synthesis. Some viral mRNAs are translated much more efficiently than those of the host cell (eg, encephalomyocarditis virus). Others, such as reovirus and vesicular stomatitis virus, replicate efficiently, and thus their very abundant mRNAs have a competitive advantage over host cell mRNAs for limited translation factors. Other viruses inhibit host cell protein synthesis by preventing the association of mRNA with the 40S ribosome.

Poliovirus and other picornaviruses gain a selective advantage by disrupting the function of the 4F complex. The mRNAs of these viruses do not have a cap structure to direct the binding of the 40S ribosomal subunit (see earlier). Instead, the 40S ribosomal subunit contacts an **internal ribosomal entry site (IRES)** in a reaction that requires 4G but not 4E. The virus gains a selective advantage by having a protease that attacks 4G and removes the amino terminal 4E binding site. Now the 4E-4G complex (4F) cannot form, so the 40S ribosomal subunit cannot be directed to host capped mRNAs, abolishing host cell protein synthesis. The 4G fragment can direct binding of the 40S ribosomal subunit to IRES-containing mRNAs, so viral mRNA translation is very efficient (**Figure 37–12**). These viruses also promote the dephosphorylation of BP1 (PHAS-1), thereby decreasing cap (4E)-dependent translation (see Figure 37–8).

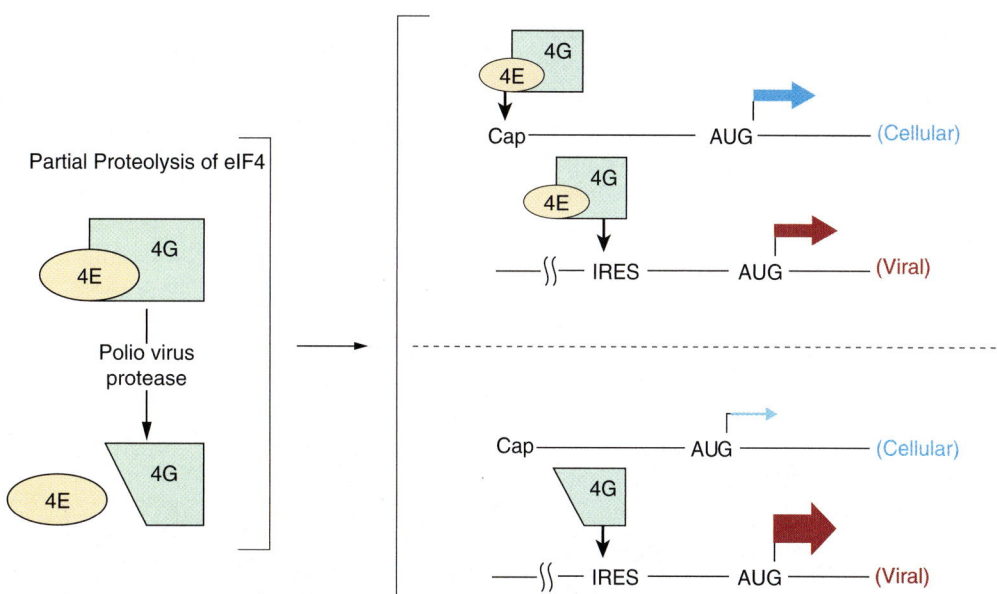

FIGURE 37–12 **Picornaviruses disrupt the 4F complex.** The 4E-4G complex (4F) directs the 40S ribosomal subunit to the typical capped mRNA (see text). However, 4G alone is sufficient for targeting the 40S subunit to the internal ribosomal entry site (IRES) of certain viral mRNAs. To gain selective advantage, some viruses (eg, poliovirus) express a protease that cleaves the 4E binding site from the amino terminal end of 4G. This truncated 4G can direct the 40S ribosomal subunit to mRNAs that have an IRES but not to those that have a cap (ie, host cell mRNAs). The widths of the arrows indicate the rate of translation initiation from the AUG codon in each example. Other viruses utilize distinct processes to effect selective initiation of translation on their cognate viral mRNAs via IRES elements.

POSTTRANSLATIONAL PROCESSING AFFECTS THE ACTIVITY OF MANY PROTEINS

Some animal viruses, notably HIV, poliovirus, and hepatitis virus, synthesize long polycistronic proteins from one long mRNA molecule. The viral protein molecules translated from these long mRNAs are subsequently cleaved at defined sites to provide the several specific viral proteins required for viral function. In animal cells, many cellular proteins are synthesized from the mRNA template as a precursor molecule, which then must be modified to achieve the active protein. The prototype is insulin, a small protein having two polypeptide chains with interchain and intrachain disulfide bridges. The molecule is synthesized as a single chain precursor, or **prohormone**, which folds to allow the formation of specific, intra- or intermolecular disulfide bridges (see intramolecular disulfide bonds in insulin; Figure 41–12). A specific protease then clips out the segment that connects the two chains which form the functional insulin molecule (see Figure 41–12).

Many other peptides are synthesized as precursor proproteins that require modifications before attaining biologic activity. Many of the posttranslational modifications involve the removal of amino terminal amino acid residues by specific aminopeptidases (see Figure 41–14). By contrast, collagen, an abundant protein in the extracellular spaces of higher eukaryotes, is synthesized as procollagen. Three procollagen polypeptide molecules, frequently not identical in sequence, align themselves in a particular way that is dependent on the existence of specific amino terminal peptides (see Figure 5–11).

Specific enzymes then carry out hydroxylations and oxidations of specific amino acid residues within the procollagen molecules to provide cross-links for greater stability. Amino terminal peptides are cleaved off the molecule to form the final product—a strong, insoluble collagen molecule. Many other posttranslational modifications of proteins occur. Covalent modification by acetylation, phosphorylation, methylation, ubiquitylation, and glycosylation is common (see Chapter 5; Table 35–1).

MANY ANTIBIOTICS WORK BY SELECTIVELY INHIBITING PROTEIN SYNTHESIS IN BACTERIA

Ribosomes in bacteria and in the mitochondria of higher eukaryotic cells differ from the mammalian ribosome described in Chapter 34. The bacterial ribosome is smaller (70S vs 80S) and has a different, somewhat simpler complement of RNA and protein molecules. This difference can be exploited for clinical purposes because many effective antibiotics interact specifically with the proteins and RNAs of prokaryotic ribosomes and thus only inhibit bacterial protein synthesis. This results in growth arrest (ie, bacteriostatic action) or death (ie, bactericidal action) of the bacterium. The most useful members of this class of **antibiotics** (eg, **tetracyclines, lincomycin, erythromycin**, and **chloramphenicol**) do not interact with components of eukaryotic ribosomes and thus are not toxic to eukaryotes. Tetracycline prevents the binding of aminoacyl-tRNAs to the bacterial ribosome A site. Chloramphenicol works by binding to 23S rRNA, which is interesting in view of the newly appreciated role of rRNA in peptide bond formation through its peptidyl transferase activity. It should be mentioned that the close similarity between prokaryotic and mitochondrial ribosomes can lead to complications in the use of some antibiotics.

Other antibiotics inhibit protein synthesis on all ribosomes (**puromycin**) or only on those of eukaryotic cells (**cycloheximide**). Puromycin (**Figure 37–13**) is a structural analog of tyrosinyl-tRNA. Puromycin is incorporated via the A site on the ribosome into the carboxyl terminal position of a peptide but causes the premature release of the polypeptide. Puromycin, as a tyrosinyl-tRNA analog, effectively inhibits protein synthesis in both prokaryotes and eukaryotes. Cycloheximide inhibits peptidyl transferase in the 60S ribosomal subunit in eukaryotes, presumably by binding to an rRNA component.

Diphtheria toxin, an exotoxin of *Corynebacterium diphtheriae* infected with a specific lysogenic phage, catalyzes the ADP-ribosylation of EF-2 on the unique amino acid diphthamide (a posttranslationally modified version of histidine) in mammalian cells. This modification inactivates EF-2 and thereby specifically inhibits mammalian protein synthesis. Many animals (eg, mice) are resistant to diphtheria toxin. This resistance is due to inability of diphtheria toxin to cross the cell membrane rather than to insensitivity of mouse EF-2 to diphtheria toxin-catalyzed ADP-ribosylation by NAD.

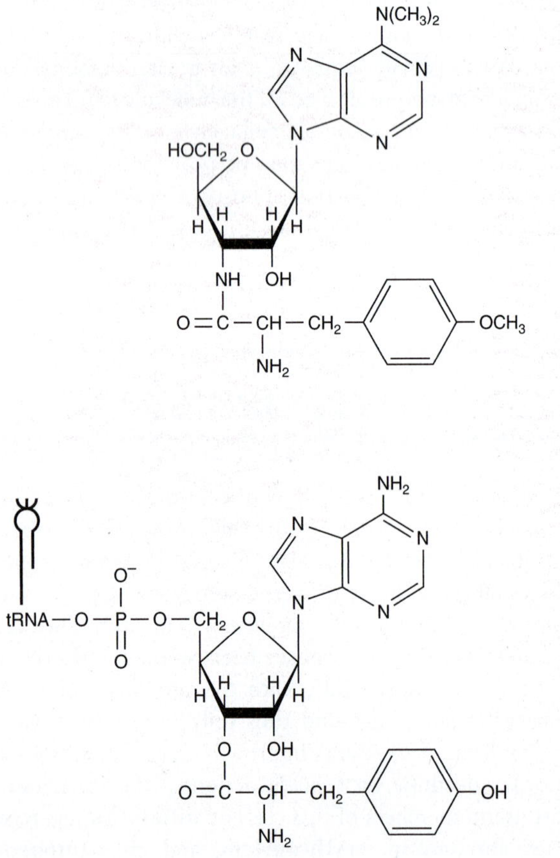

Ricin, an extremely toxic molecule isolated from the castor bean, inactivates eukaryotic 28S ribosomal RNA by catalyzing the N-glycolytic cleavage or removal of a single adenine.

Many of these compounds—puromycin and cycloheximide in particular—are not clinically useful but have been important in the laboratory as a tool to elucidate the role of protein synthesis in the regulation of metabolic processes, particularly enzyme induction by hormones.

SUMMARY

- The flow of genetic information generally follows the sequence DNA → RNA → protein.

- Ribosomal RNA (rRNA), transfer RNA (tRNA), and messenger RNA (mRNA) are directly involved in protein synthesis.

- The information in mRNA is a continuous linear array of codons, each of which is three nucleotides long.

- The mRNA is read continuously from a start (AUG) to termination (UAA, UAG, UGA) codon.

- The open reading frame, or ORF, of the mRNA is the series of contiguous codons (AUG to STOP), each codon specifying a certain amino acid, which determines the precise amino acid sequence of the protein.

- Protein synthesis, like DNA and RNA synthesis, follows the 5′ to 3′ polarity of mRNA and can be divided into three processes: initiation, elongation, and termination.

- Mutant proteins arise when base substitutions result in codons that specify a different amino acid at a given position, when a stop codon results in a truncated protein, or when base additions or deletions alter the reading frame, so different codons are read.

- A variety of compounds, including several antibiotics, inhibit protein synthesis by affecting one or more of the steps involved in protein synthesis.

FIGURE 37–13 **The comparative structures of the antibiotic puromycin (top) and the 3′ terminal portion of tyrosinyl-tRNA (bottom).**

REFERENCES

Aitken CE, Lorch R: A mechanistic overview of translation in eukaryotes. Nat Struc Mol Biol 2012;19:568-576.

Crick FH, Barnett L, Brenner S, et al: General nature of the genetic code for proteins. Nature 1961;192:1227-1232.

Frank J: Whither ribosome structure and dynamics? (A perspective). J Mol Biology 2016;428:3565-3569.

Hernandez G, Osnaya VG, Perez-Martinez X: Conservation and Variability of the AUG initiation codon context in eukaryotes. Trends Biochem Sci 2019;44:1009-1021.

Hinnebusch AG: The scanning mechanism of eukaryotic translation initiation. Ann Rev Biochem 2014;83:779-812.

Hinnebusch AG, Ivanov IP, Sonenberg N: Translational control by 5′-untranslated regions of eukaryotic mRNAs. Science 2016;352:1413-1416.

Jackson R, Hellen CUT, Pestova TV: The mechanism of eukaryotic translation initiation and principles of its regulation. Nat Rev Mol Cell Biol 2010;10:113-127.

Jain S, Parker R: The discovery and analysis of P bodies. Adv Exp Med Biol 2013;768:23-43.

Kozak M: Structural features in eukaryotic mRNAs that modulate the initiation of translation. J Biol Chem 1991;266:1986-1970.

Liu CC, Schultz PG: Adding new chemistries to the genetic code. Annu Rev Biochem 2010;79:413-444.

Moore PB, Steitz TA: The roles of RNA in the synthesis of protein. Cold Spring Harb Perspect Biol 2011;3:a003780.

Sonenberg N, Hinnebusch AG: Regulation of translation initiation in eukaryotes: mechanisms and biological targets. Cell 2009;136:731-745.

Tauber D, Tauber G, Parker R: Mechanisms and Regulation of RNA Condensation in RNP Granule Formation. Trends Biochem Sci 2020;45:764-778.

Thompson SR: Tricks an IRES uses to enslave ribosomes. Trends Microbiol 2012;20:558-566.

Torre D de la, Chin JW: Reprogramming the genetic code. Nat Rev Genetics 2021;22:169-184.

Wang Q, Parrish AR, Wang L: Expanding the genetic code for biological studies. Chem Biol 2009;16:323-336.

Weatherall DJ: Thalassemia: the long road from bedside to genome. Nat Rev Genet 2004;5:625-631.

Wilson DN: Ribosome-targeting antibiotics and mechanisms of bacterial resistance. Nat Rev Microbiol 2013;12:35-48.

Regulation of Gene Expression

P. Anthony Weil, PhD

O B J E C T I V E S

*After studying this chapter,
you should be able to:*

- Explain that the many steps involved in the vectorial processes of gene expression: targeted modulation of gene copy number, gene rearrangement, transcription, mRNA processing and transport from the nucleus, translation, protein subcellular compartmentalization, posttranslational modification, and degradation are all subject to regulatory control, both positive and negative. Changes in any, or multiple of these processes, can increase or decrease the amount and/or activity of the cognate gene product.

- Appreciate that DNA-binding transcription factors, proteins that bind to specific DNA sequences that are physically linked to their target transcriptional promoter elements, can either activate or repress gene transcription.

- Recognize that DNA-binding transcription factors are often modular proteins composed of structurally and functionally distinct domains. These transcription factors can directly or indirectly control messenger RNA (mRNA) gene transcription, either through contacts with RNA polymerase and its cofactors, or through interactions with coregulators that modulate nucleosome occupancy, position, structure, composition, and histone covalent modifications.

- Understand that nucleosome-directed regulatory events typically increase or decrease the accessibility of the underlying DNA such as enhancer or promoter sequences, although some nucleosome modifications can also create new binding sites for other coregulators.

- Describe how the processes of gene transcription, RNA processing, and nuclear export of RNA are all coupled.

- Describe the phenomenon of epigenetic gene regulation and how such processes occur at the molecular level.

BIOMEDICAL IMPORTANCE

Organisms alter expression of genes in response to genetic developmental cues or programs, environmental challenges, or disease, by modulating the amount, the spatial, and/or the temporal patterns of gene expression. The mechanisms controlling gene expression have been studied in detail and often involve changes in gene transcription. Control of transcription ultimately results from changes in the mode of interaction of specific regulatory molecules, usually proteins, with various regions of DNA in the regulated gene. Such interactions can either have a positive or negative effect on transcription. Transcription control can result in tissue-specific gene expression,

and gene regulation can be influenced by a range of physiologic, biologic, environmental, and pharmacologic agents.

In addition to transcription level controls, gene expression can also be modulated by gene amplification, gene rearrangement, posttranscriptional modifications, RNA stabilization, translational control, protein modification, protein compartmentalization, and protein stabilization or degradation. Many of the mechanisms that control gene expression are utilized to respond to developmental cues, growth factors, hormones, environmental agents, and therapeutic drugs. Dysregulation of gene expression can lead to human disease. Thus, a molecular understanding of these processes will lead to development of therapeutics that can alter pathophysiologic mechanisms or inhibit the function or arrest the growth of pathogenic organisms.

REGULATED EXPRESSION OF GENES IS REQUIRED FOR DEVELOPMENT, DIFFERENTIATION, & ADAPTATION

The genetic information present in each normal somatic cell of a metazoan organism is practically identical. The genetically reproducible, hardwired exceptions are found in those few cells that have amplified or rearranged genes in order to perform specialized cellular functions. Of course, in various disease states chromosome integrity is altered (ie, cancer; see Figure 56–11) sometimes even at the whole chromosome level (eg, trisomy 21, that causes Down syndrome). Expression of the genetic information must be regulated during ontogeny and differentiation of the organism and its cellular components. Furthermore, in order for the organism to adapt to its environment and to conserve energy and nutrients, the expression of genetic information must be cued to extrinsic signals and respond only when necessary. As organisms have evolved, more sophisticated regulatory mechanisms have appeared which provide the organism and its cells with the responsiveness necessary for survival in a complex environment. Mammalian cells possess about 1000 times more genetic information than does the bacterium *Escherichia coli*. Much of this additional genetic information is likely involved in regulation of gene expression during the differentiation of tissues and biologic processes in the multicellular organism and in ensuring that the organism can respond to complex environmental challenges.

In simple terms, there are only two types of gene regulation: **positive regulation** and **negative regulation** (Table 38–1). When the expression of genetic information is quantitatively increased by the presence of a specific regulatory element, regulation is said to be positive; when the expression of genetic information is diminished by the presence of a specific regulatory element, regulation is said to be negative. The element or molecule mediating negative regulation is said to be a **negative regulator**, a **silencer** or **repressor**; while the element mediating positive regulation is a **positive regulator**, an **enhancer** or **activator**. However, a **double negative** has the effect of acting as a positive. Thus, an effector that inhibits the function of a negative regulator will appear to bring about a positive regulation. Many regulated systems that appear to be induced are in fact **derepressed** at the molecular level. (See Chapter 9 for additional discussion of these terms.)

BIOLOGIC SYSTEMS EXHIBIT THREE TYPES OF TEMPORAL RESPONSES TO A REGULATORY SIGNAL

Figure 38–1 depicts the extent or amount of gene expression in three types of temporal responses to an inducing signal. A **type A response** is characterized by an increased extent of gene expression that is dependent on the continued presence of the inducing signal. When the inducing signal is removed, the amount of gene expression diminishes to its basal level, but the amount repeatedly increases in response to

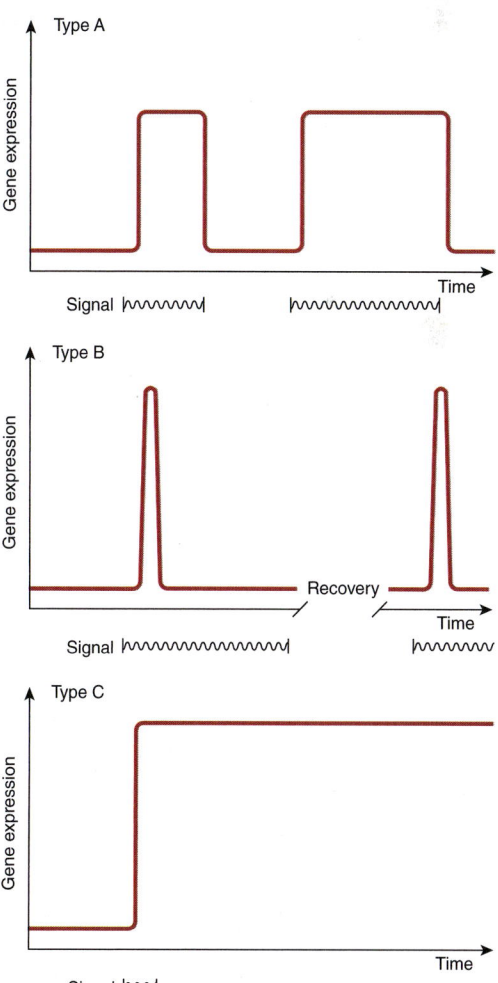

FIGURE 38–1 Diagrammatic representations of the responses of the extent of expression of a gene to specific regulatory signals as a function of time.

TABLE 38–1 **Effects of Positive & Negative Regulation on Gene Expression**

	Rate of Gene Expression	
	Negative Regulation	Positive Regulation
Regulator present	Decreased	Increased
Regulator absent	Increased	Decreased

the reappearance of the specific signal. This type of response is commonly observed in prokaryotes in response to sudden changes of the intracellular concentration of a nutrient. It is also observed in many higher organisms after exposure to inducers such as hormones, nutrients, or growth factors (see Chapter 42).

A **type B response** exhibits an increased amount of gene expression that is transient even in the continued presence of the regulatory signal. After the regulatory signal has terminated and the cell has been allowed to recover, a second transient response to a subsequent regulatory signal may be observed. This phenomenon of response-desensitization recovery characterizes the action of many pharmacologic agents, but it is also a feature of many naturally occurring processes. This type of response commonly occurs during development of an organism, when only the transient appearance of a specific gene product is required although the signal persists.

The **type C response** pattern exhibits, in response to the regulatory signal, an increased extent of gene expression that persists indefinitely even after termination of the signal. The signal acts as a trigger in this pattern. Once expression of the gene is initiated in the cell, it cannot be terminated even in the daughter cells; it is therefore an irreversible and inherited alteration. This type of response typically occurs during the development of differentiated function in a tissue or organ.

Simple Unicellular & Multicellular Organisms Serve as Valuable Models for the Study of Gene Expression in Human Cells

Analysis of the regulation of gene expression in prokaryotic cells helped establish the principle that information flows from the gene to a messenger RNA to a specific protein molecule. These studies were aided by the advanced genetic analyses that could be performed in prokaryotic and lower eukaryotic organisms such as the baker's yeast, *Saccharomyces cerevisiae*, and the fruit fly, *Drosophila melanogaster*, among others. In recent years, the principles established in these studies, coupled with a variety of physical, optical, biochemical, informatic, and molecular biological techniques, have led to remarkable progress in the analysis of gene regulation in higher eukaryotic organisms, including humans. In this chapter, the initial discussion will center on prokaryotic systems. The impressive genetic studies will not be described, but the physiology of gene expression will be discussed. However, nearly all of the conclusions about this physiology have been derived from genetic studies and confirmed by molecular biological and biochemical experiments.

Some Features of Prokaryotic Gene Expression Are Unique

Before the physiology of gene expression can be explained, a few specialized genetic and regulatory terms must be defined for prokaryotic systems. In prokaryotes, the genes involved in a metabolic pathway are often present in a linear array called an **operon**, for example, the *lac* operon. An operon can be regulated by a single promoter or regulatory region. The **cistron** is the smallest unit of genetic expression. A single mRNA that encodes more than one separately translated protein is referred to as a **polycistronic mRNA**. For example, the polycistronic *lac* operon mRNA is translated into three separate proteins (see following discussion). Operons and polycistronic mRNAs are common in bacteria but not in eukaryotes.

An **inducible gene** is one whose expression increases in response to an **inducer** or **activator**, a specific positive regulatory signal. In general, inducible genes have relatively low basal rates of transcription. By contrast, genes with high basal rates of transcription are often subject to downregulation by repressors.

The expression of some genes is **constitutive**, meaning that they are expressed at a reasonably constant rate and not known to be subject to extensive regulation. Such genes are often referred to as **housekeeping genes**. As a result of mutation, some inducible gene products become constitutively expressed. A mutation resulting in constitutive expression of what was formerly a regulated gene is called a **constitutive mutation**.

Analysis of Lactose Metabolism in *E. coli* Led to the Discovery of the Basic Principles of Gene Transcription Activation & Repression

Jacob and Monod in 1961 described their **operon model** in a classic paper. Their hypothesis was to a large extent based on observations on the regulation of lactose metabolism by the intestinal bacterium *E. coli*. The molecular mechanisms responsible for the regulation of the genes involved in the metabolism of lactose are now among the best-understood in any organism. β-Galactosidase hydrolyzes the β-galactoside lactose to galactose and glucose. The gene encoding β-galactosidase (*lacZ*) is clustered with the genes encoding lactose permease (*lacY*) and thiogalactoside transacetylase (*lacA*). The genes encoding these three enzymes, along with the *lac* promoter and *lac* operator (a regulatory region), and the *lacI* gene encoding the LacI repressor are physically linked and constitute the *lac* **operon** as depicted in **Figure 38–2**.

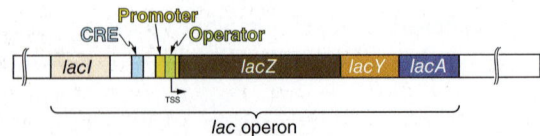

FIGURE 38–2 The positional relationships of the protein coding and regulatory elements of the ~6kbp *lac* operon. *lacZ* encodes β-galactosidase, *lacY* encodes a permease, and *lacA* encodes a transacetylase. *lacI* encodes the *lac* operon repressor protein. Also shown is the transcription start site for *lac* operon transcription (TSS). Note that the binding site for the LacI protein (ie, lac repressor)—the *lac* operator (Operator)—overlaps the *lac* promoter. Immediately upstream of the *lac* operon promoter is the binding site (CRE) for the cAMP-binding protein, CAP, the positive regulator of *lac* operon transcription. See Figure 38–3 for more detail.

This genetic arrangement of the *lac* operon allows for **coordinate expression** of the three enzymes concerned with lactose metabolism. Each of the linked operon genes is transcribed into one large polycistronic mRNA molecule that contains multiple independent translation start (AUG) and stop (UAA) codons for each of the three cistrons. Thus, each protein is translated separately, and they are not processed from a single large precursor protein.

It is now conventional to consider that a gene includes regulatory sequences as well as the region that encodes the primary transcript. Although there are many historical exceptions, a gene is generally italicized in lower case and the encoded protein, when abbreviated, is expressed in roman type with the first letter capitalized. For example, the gene *lacI* encodes the repressor protein LacI. When *E. coli* is presented with lactose or some specific lactose analogs under appropriate nonrepressing conditions (eg, high concentrations of lactose, no or very low glucose in media; see following discussion), the expression of the activities of β-galactosidase, galactoside permease, and thiogalactoside transacetylase is increased 100-fold to 1000-fold. This is a type A response, as depicted in Figure 38–1. The kinetics of induction can be quite rapid; *lac*-specific mRNAs are fully induced within ~5 minutes after addition of lactose to a culture; β-galactosidase protein is maximal within 10 minutes. Under fully induced conditions, there can be up to 5000 β-galactosidase molecules per cell, an amount about 1000 times greater than the basal, uninduced level. Upon removal of the signal, that is, the inducer, the synthesis of these three enzymes declines.

When *E. coli* is exposed to both lactose and glucose as sources of carbon, the cells first metabolize the glucose and then temporarily stop growing until the genes of the *lac* operon become induced to provide the ability to metabolize lactose as a usable energy source. Although lactose is present from the beginning of the bacterial growth phase, the cell does not induce those enzymes necessary for catabolism of lactose until glucose has been exhausted. This phenomenon was first thought to be attributable to repression of the *lac* operon by some catabolite of glucose; hence, it was termed catabolite repression. It is now known that catabolite repression is in fact mediated by a **catabolite activator protein (CAP)** in conjunction **3′,5′ cyclic Adenosine monophosphate (cAMP;** see Figure 18–5). This protein is also referred to as the cAMP regulatory protein (CRP). The expression of many inducible enzyme systems or operons in *E. coli* and other prokaryotes is sensitive to catabolite repression, as discussed in following discussion.

The physiology of induction of the *lac* operon is well understood at the molecular level (**Figure 38–3**). Expression of the normal *lacI* gene of the *lac* operon is constitutive; it is expressed at a constant rate, resulting in formation of the subunits of the **lac repressor**. Four identical subunits with molecular weights of 38,000 assemble into a tetrameric Lac repressor molecule. The LacI repressor protein molecule, the product of *lacI*, has a very high affinity (dissociation constant, K_d about 10^{-13} mol/L) for the operator locus. The **operator locus** is a region of double-stranded DNA that exhibits a twofold rotational symmetry and an inverted palindrome (indicated by arrows about the dotted axis) in a region that is 21-bp long, shown as follows:

$$5' - \text{\textbf{AATTGT} GAG C GATAACAATT}$$
$$3' - \text{TTAACACTCG C CTAT \textbf{TGTTAA}}$$

At any one time, only two of the four subunits of the repressor appear to bind to the operator; within the 21-base-pair operator region nearly every base of each base pair is involved in LacI recognition and binding. Binding occurs mostly in the **major groove** without interrupting the base-paired, double-helical nature of the operator DNA. The **operator locus** (ie, LacI binding site) is between the **promoter**, the site where the DNA-dependent RNA polymerase attaches to commence transcription, and the transcription initiation site of the *lacZ* **gene**, the structural gene for β-galactosidase (see Figures 38–2 and 36–3). When bound to the operator locus, the LacI repressor molecule prevents transcription of the distal structural genes, *lacZ*, *lacY*, and *lacA* by interfering with the binding of RNA polymerase to the promoter; RNA polymerase and LacI repressor cannot be effectively bound to the *lac* operon at the same time. Thus, the LacI repressor molecule is a **negative regulator**, and in its presence (and in the absence of inducer; see following discussion), expression from the *lacZ*, *lacY*, and *lacA* genes is very, very low. There are normally about 30 repressor tetramer molecules in the cell, a concentration (3×10^{-8} mol/L) of tetramer sufficient to effect, at any given time, more than 95% occupancy of the one *lac* operator element in a bacterium, thus ensuring low (but not zero) basal *lac* operon gene transcription in the absence of inducing signals.

A lactose analog that is capable of inducing the *lac* operon while not itself serving as a substrate for β-galactosidase is an example of a **gratuitous inducer**. An example is **isopropyl-thiogalactoside (IPTG)**. The addition of lactose or of a gratuitous inducer such as IPTG to bacteria growing on a poorly utilized carbon source (such as succinate) results in prompt induction of the *lac* operon enzymes. Small amounts of the gratuitous inducer or of lactose are able to enter the cell even in the absence of permease. The LacI repressor molecules—both those attached to the operator loci and those free in the cytosol—have a high affinity for the inducer. Binding of the inducer to repressor molecule induces a conformational change in the structure of the repressor that causes a decrease in operator DNA occupancy because its affinity for the operator is now 10^4 times lower (K_d about 10^{-9} mol/L) than that of LacI in the absence of IPTG. DNA-dependent RNA polymerase can now more efficiently compete with LacI and bind to the promoter (ie, Figures 36–3 and 36–8), and transcription will begin, although this process is relatively inefficient (see following discussion). In such a manner, **an inducer derepresses the *lac* operon** and allows transcription of the genes encoding β-galactosidase, galactoside permease, and thiogalactoside transacetylase. Translation of the

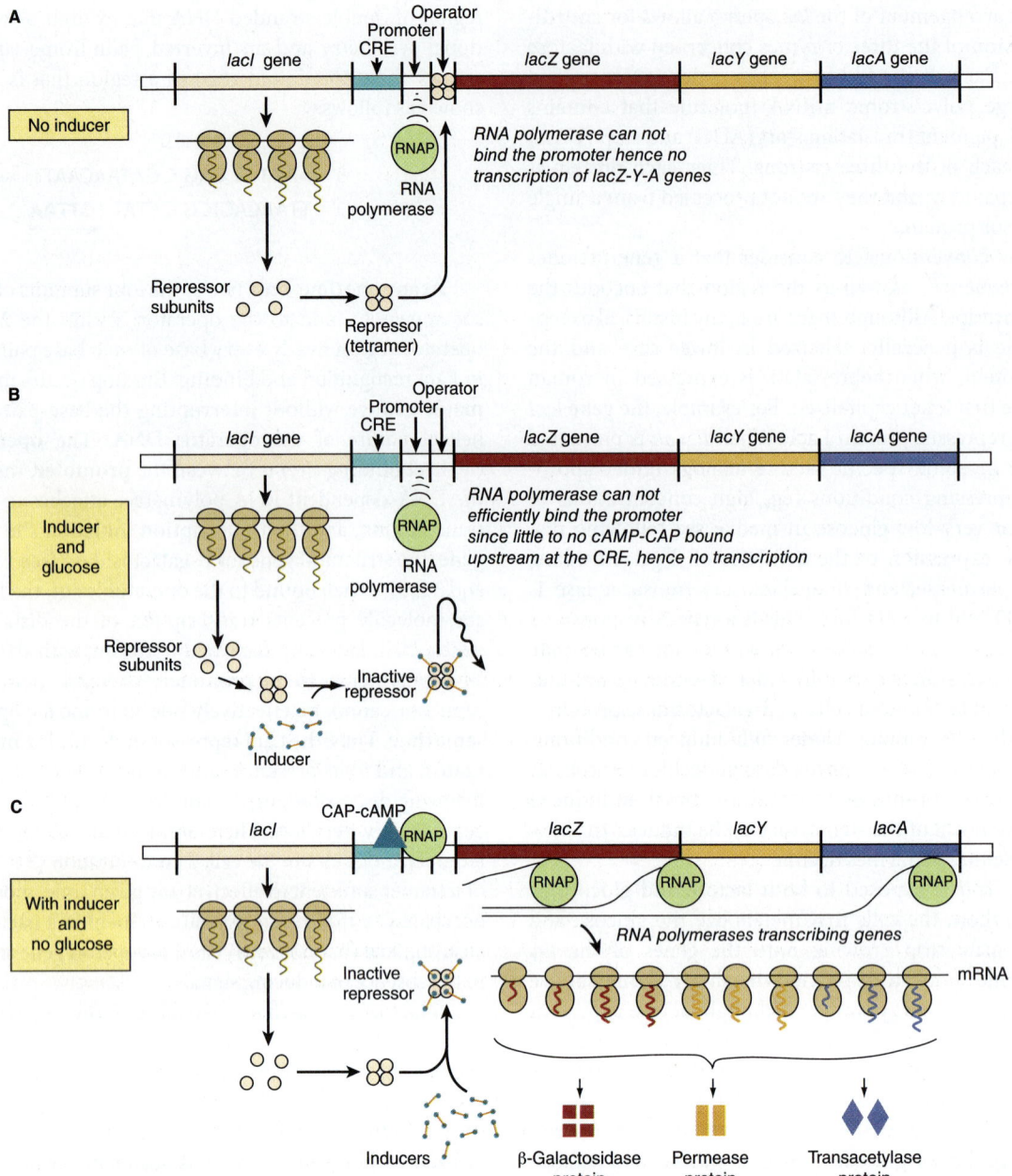

FIGURE 38–3   **The mechanism of repression, derepression, and activation of the *lac* operon.** When no inducer is present (**A**), the constitutively synthesized *lacI* gene products form a repressor tetramer that binds to the operator. Repressor-operator binding prevents the binding of RNA polymerase and consequently prevents transcription of the *lacZ, lacY,* and *lacA* genes into a polycistronic mRNA. When inducer is present, but glucose is also present in the culture medium (**B**), the tetrameric repressor molecules are conformationally altered by inducer, and cannot efficiently bind to the operator locus (affinity of binding reduced >1000-fold). However, RNA polymerase will not efficiently bind the promoter and initiate transcription because positive protein–protein interactions between CRE-bound CAP protein and RNA polymerase fail to occur; thus, the *lac* operon is not efficiently transcribed. However, when inducer is present, *and* glucose is depleted from the medium (**C**), adenylyl cyclase is activated and cAMP is produced. This cAMP binds with high affinity to its binding protein the cyclic AMP activator protein, or CAP. The CAP-cAMP complex binds to its recognition sequence (CRE, the cAMP response element) at *lac* operon nucleotide coordinate −50. Direct protein–protein contacts between the CRE-bound CAP and the RNA polymerase increases promoter binding more than 20-fold; hence RNAP will efficiently transcribe the *lac* operon and the polycistronic *lacZ-lacY-lacA* mRNA molecule formed can be translated into the corresponding protein molecules β-galactosidase, permease, and transacetylase as shown. This protein production enables cellular catabolism of lactose as the sole carbon source for growth.

polycistronic mRNA can occur even before transcription is completed. Derepression of the *lac* operon allows the cell to synthesize the enzymes necessary to catabolize lactose as an energy source. Based on the physiology just described, IPTG-induced expression of transfected plasmids bearing the *lac* operator–promoter ligated to appropriate bioengineered constructs is commonly used to express mammalian recombinant proteins in *E. coli*.

In order for the RNA polymerase to form a PIC at the promoter site most efficiently, the cAMP-CAP complex must also be present in the cell. By an independent mechanism, the bacterium accumulates cAMP only when it is starved for a source of carbon. In the presence of glucose—or of glycerol in concentrations sufficient for growth—the bacteria will lack sufficient cAMP to bind to CAP because glucose inhibits adenylyl cyclase, the enzyme that converts ATP to cAMP (see Chapter 42). Thus, in the presence of glucose or glycerol, cAMP-saturated CAP is lacking, so that the DNA-dependent RNA polymerase cannot initiate transcription of the *lac* operon at the maximal rate. However, in the presence of the CAP-cAMP complex, which binds to **CAP Response Element (CRE)** DNA just upstream of the promoter site, transcription occurs at maximal levels (see Figure 38–3). Studies indicate that a region of CAP directly contacts the RNA polymerase (RNAP) α subunit, and these protein–protein interactions facilitate the binding of RNAP to the promoter. Thus, the CAP-cAMP regulator is acting as a **positive regulator** because its presence is required for optimal gene expression. The *lac* operon is therefore controlled by two distinct, ligand-modulated DNA-binding *trans*-factors; one that acts positively (cAMP-CRP complex) to facilitate productive binding of RNA polymerase to the promoter and one that acts negatively (LacI repressor) that antagonizes RNA polymerase promoter binding. Maximal activity of the *lac* operon occurs when glucose levels are low (high cAMP with CAP activation) and lactose is present (LacI is prevented from binding to the operator) as shown in Figure 38–3, panel C.

With the above information in hand, it becomes relatively to predict the effects of mutations in various components of the *lac*-system upon *lac* operon expression. When the *lacI* gene has been mutated so that its product, LacI, is not capable of binding to operator DNA, the organism will exhibit **constitutive expression** of the *lac* operon. In a contrary manner, an organism with a *lacI* gene mutation that produces a LacI protein which prevents the binding of lactose or other small molecule inducer to the repressor will remain repressed even in the presence of the inducer molecule, because such ligands cannot bind to the repressor on the operator locus in order to derepress the operon. Similarly, bacteria harboring mutations in their *lac* operator locus such that the operator sequence will not bind a normal repressor molecule will constitutively express the *lac* operon genes. Mechanisms of positive and negative regulation comparable to those described here for the *lac* system have been observed in eukaryotic cells (see following discussion).

The Genetic Switch of Bacteriophage Lambda (λ) Provides Another Paradigm for Understanding the Role of Conditional Regulatory Protein-DNA Interactions in Transcriptional Control in Eukaryotic Cells

Like some eukaryotic viruses (eg, herpes simplex virus and HIV), certain bacterial viruses can either reside in a dormant state within the host chromosomes or can replicate within the bacterium and eventually lead to lysis and killing of the bacterial host. Some *E. coli* harbor such a "temperate" virus, bacteriophage lambda (λ). When lambda infects an organism of that species, it injects its 48,490-bp, double-stranded, linear DNA genome into the cell (**Figure 38–4**). Depending on the nutritional state of the cell, the lambda DNA will either **integrate** into the host genome (**lysogenic pathway**) and remain dormant until activated (see following discussion), or it will commence **replicating** until it has made about 100 copies of complete, protein-packaged virus, at which point it causes lysis of its host (**lytic pathway**). The newly generated virus particles can then infect other susceptible host cells. Poor growth conditions favor lysogeny while good growth conditions promote the lytic pathway of lambda growth.

When integrated into the host genome in its dormant state, lambda will remain in that state until activated by exposure of its bacterial host to DNA-damaging agents. In response to such a noxious stimulus, the dormant bacteriophage becomes "induced" and begins to transcribe and subsequently translate those genes of its own genome that are necessary for its excision from the host chromosome, its DNA replication, and the synthesis of its protein coat and lysis enzymes. This event acts like a trigger or type C (see Figure 38–1) response; that is, once dormant lambda has committed itself to induction, there is no turning back until the cell is lysed and the replicated bacteriophage released. This switch from a dormant or **prophage state** to a **lytic infection** is well understood at the genetic and molecular levels and will be described in detail here; though less well understood at the molecular level, HIV and herpes viruses can behave similarly, transitioning from dormant to active states in infected humans.

The lytic/lysogenic genetic switching event in lambda is centered around an 80-bp region in its double-stranded DNA genome referred to as the "right operator" (O_R) (**Figure 38–5A**). The **right operator** is flanked on its left side by the gene for the lambda repressor protein, *cI*, and on its right side by the gene encoding another regulatory protein, the *cro* gene. When lambda is in its prophage state—that is, integrated into the host genome—the *cI* repressor gene is the *only* lambda gene that is expressed. When the bacteriophage is undergoing lytic growth, the *cI* repressor gene is not expressed, but the *cro* gene—as well as many other lambda genes—is expressed. Thus, when the *cI* repressor gene is on, the *cro* gene is off, and when the *cro* gene is on, the *cI* repressor gene is off. As we shall

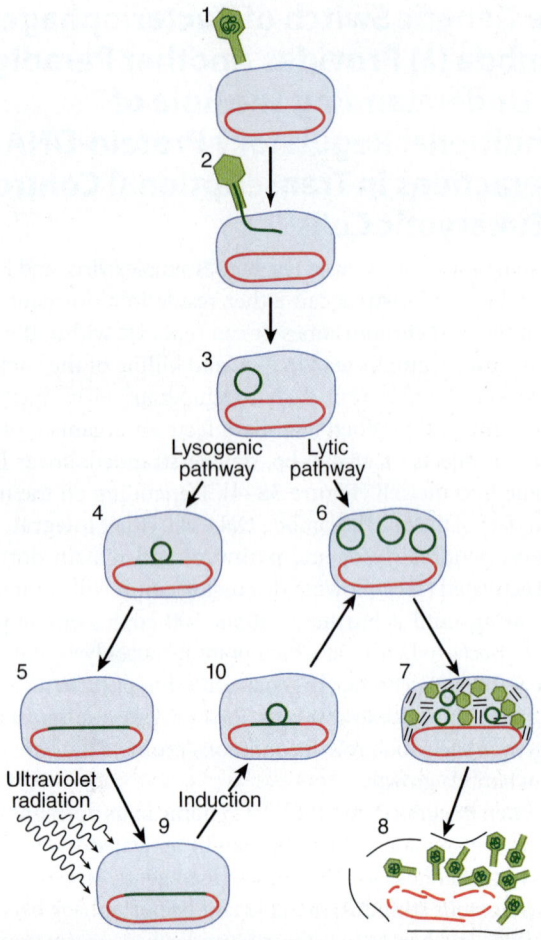

FIGURE 38–4   **Alternate lytic and lysogenic lifestyles of bacteriophage lambda.** Infection of the bacterium *E. coli* by phage lambda begins when a virus particle attaches itself to specific receptors on the bacterial cell surface (**1**) and injects its DNA (dark green line) into the cell (**2**), where the phage genome then circularizes (**3**). Infection can take either of two courses depending on which two sets of viral genes is turned on. In the lysogenic pathway, the viral DNA becomes integrated into the bacterial chromosome (**red**) (**4, 5**), where it is replicated passively as part of the bacterial DNA during *E. coli* cell division. This dormant, bacterial genome-integrated virus is called a prophage, and the cell that harbors is called a lysogen. In the alternative, lytic mode of infection, the viral DNA excises from the *E. coli* chromosome and replicates itself (**6**) in order to direct the synthesis of viral proteins; black lines (**7**). About 100 new virus particles (green hexagons) are formed. The proliferating viruses induce lysis of the cell (**8**). A prophage can be "induced" by a DNA damaging agent such as ultraviolet radiation (**9**). The inducing agent throws a switch (see text and Figure 38–5; the λ "molecular switch."), so that a different set of viral genes is turned on. Viral DNA loops out and is excised from the *E. coli* chromosome (**10**) and replicates; the virus then proceeds along the lytic pathway.

see, these two genes regulate each other's expression and thus, ultimately, the decision between lytic and lysogenic growth of lambda. This decision between repressor gene transcription and *cro* gene transcription is a paradigmatic example of a molecular transcriptional switch.

The 80-bp lambda right operator, O_R, can be subdivided into three discrete, evenly spaced, 17-bp *cis*-active DNA elements that represent the binding sites for either of two bacteriophage lambda regulatory proteins. Importantly, the nucleotide sequences of these three tandemly arranged sites are similar but not identical (**Figure 38–5B**). The three related *cis*-elements, termed operators O_R1, O_R2, and O_R3, can be bound by either cI or cro proteins. However, the relative affinities of cI and cro for each of the sites vary, and this differential binding affinity is central to the appropriate operation of the lambda phage lytic or lysogenic "molecular switch." The DNA region between the *cro* and repressor genes also contains two promoter sequences that direct the binding of RNA polymerase in a specified orientation, where it commences transcribing adjacent genes. One promoter directs RNA polymerase to transcribe in the rightward direction and, thus, to transcribe *cro* and other distal genes, while the other promoter directs the transcription of the *cI* repressor gene in the leftward direction (see Figure 38–5B).

The product of the *cI* repressor gene, the 236-amino-acid **λ cI repressor protein** is a **two-domain** molecule with **amino terminal DNA-binding domain (DBD)** and **carboxyl-terminal dimerization domain**. Association of one repressor protein with another forms a dimer. cI repressor **dimers** bind to operator DNA much more tightly than do monomers (**Figure 38–6A** to **38–6C**).

The product of the *cro* gene, the 66-amino-acid, 9-kDa **cro protein**, has a single domain but also binds the operator DNA more tightly as a dimer (**Figure 38–6D**). The cro protein's single domain mediates both operator binding and dimerization.

In a lysogenic bacterium—that is, a bacterium containing an integrated, dormant lambda prophage—the lambda repressor dimer binds preferentially to O_R1 but in so doing, by a cooperative interaction, enhances the binding (by a factor of 10) of another repressor dimer to O_R2 (**Figure 38–7**). The affinity of repressor for O_R3 is the least of the three operator subregions. The binding of repressor to O_R1 has two major effects. First, occupancy of O_R1 by repressor blocks the binding of RNA polymerase to the rightward promoter and in that way prevents expression of *cro*. Second, as mentioned earlier, repressor dimer bound to O_R1 enhances the binding of repressor dimer to O_R2. The binding of repressor to O_R2 has the important added effect of enhancing the binding of RNA polymerase to the leftward promoter that overlaps O_R3 and thereby enhances transcription and subsequent expression of the repressor gene. This enhancement of transcription is mediated through direct protein–protein interactions between promoter-bound RNA polymerase and O_R2-bound repressor, much as described earlier for CAP protein and RNA polymerase on the *lac* operon. Thus, the λ cI protein is both a negative regulator, by preventing transcription of *cro*, and a positive regulator, by enhancing transcription of its own gene, *cI*. This dual effect of repressor is responsible for the stable state of the dormant lambda bacteriophage; not only does the repressor prevent expression of the genes necessary for lysis, but it also promotes expression of itself to stabilize this state

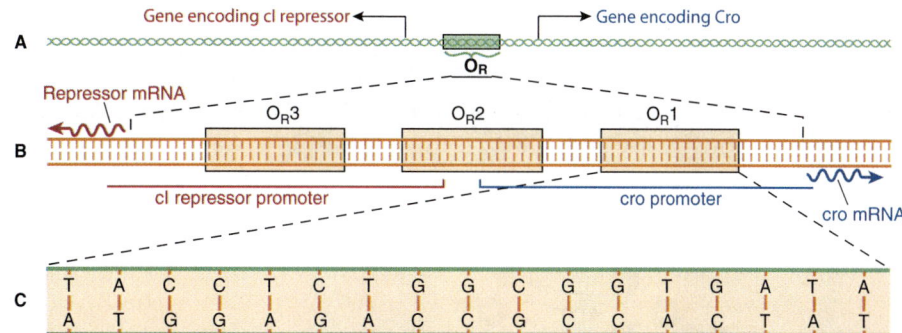

FIGURE 38–5 **Genetic organization of the lambda lifestyle "molecular switch."** Right operator (O_R) is shown in increasing detail in this series of drawings. The operator is a region of the viral DNA some 80-bp long (**A**). To its left lies the gene encoding lambda repressor (*cl*), to its right the gene *(cro)* encoding the regulator protein Cro. When the operator region is enlarged (**B**), it is seen to include three subregions termed operators: O_R1, O_R2, and O_R3, each 17-bp long. These three DNA elements are recognition sites to which both λ cl repressor and Cro proteins can bind. The recognition sites overlap two divergent promoters—sequences of bases to which RNA polymerase binds in order to transcribe these genes into mRNA (wavy lines) that are translated into protein. Site O_R1 is enlarged (**C**) to show its base sequence. (Reproduced with permission from Alan D. Iselin, artist.)

of differentiation. In the event that intracellular repressor protein concentration becomes very high, the excess repressor will bind to O_R3 and by so doing diminish transcription of the repressor gene from the leftward promoter, by blocking RNAP binding to the cl promoter, until the repressor concentration drops and repressor dissociates from O_R3. Similar examples of repressor proteins also having the ability to activate transcription have been observed in eukaryotes.

With such a stable, repressive, cl-mediated, lysogenic state, one might wonder how the lytic cycle could ever be entered. However, this process does occur quite efficiently. When a DNA-damaging signal, such as ultraviolet light, strikes the lysogenic host bacterium, fragments of single-stranded DNA are generated that activate a specific co-protease coded by a bacterial gene and referred to as *recA* (see Figure 38–7). The activated

recA protease hydrolyzes the portion of the repressor protein that connects the amino-terminal and carboxyl-terminal domains of that molecule (see Figure 38–6A). Such cleavage of the repressor domains causes the repressor dimers to dissociate, which in turn causes dissociation of the repressor molecules from O_R2 and eventually from O_R1. The effects of removal of repressor from O_R1 and O_R2 are predictable. RNA polymerase immediately has access to the rightward promoter and commences transcribing the *cro* gene, while simultaneously the enhancing effect of the repressor at O_R2 on leftward transcription is lost as well (see Figure 38–7).

The resulting newly synthesized cro protein also binds to the operator region as a dimer, but as noted earlier, its order of preference is opposite to that of repressor (see Figure 38–7). That is, cro binds most tightly to O_R3, but there is no cooperative

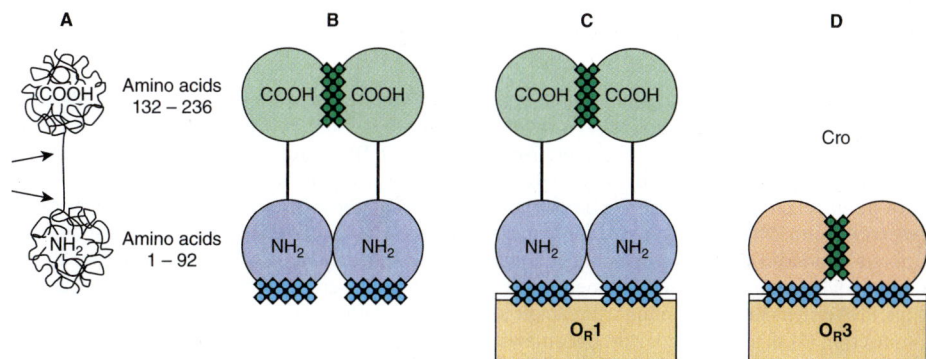

FIGURE 38–6 **Schematic molecular structures of lambda regulatory proteins cl and Cro.** (**A**) The lambda repressor protein is a 236-amino-acid polypeptide. The chain folds itself into a dumbbell shape with two substructures: an amino terminal (NH_2) domain and a carboxyl-terminal (COOH) domain. The two domains are linked by a region of the chain that is less structured and susceptible to cleavage by proteases (indicated by the two arrows). (**B**) Single repressor molecules (monomers) tend to reversibly associate to form dimers. A dimer is held together mainly by contact between the carboxyl-terminal domains (green hatching). (**C**) cl repressor dimers bind to (and can dissociate from) the recognition sites in the operator region; they display differential affinities for the three operator sites, $O_R1 > O_R2 > O_R3$. The DNA-binding domains (DBD) of the repressor molecule that makes contact with DNA (blue hatching). (**D**) Cro is a single globular protein that contains both a DNA binding domain (blue hatching) and a cro-cro dimerization domain, which promotes binding of cro-cro dimers to target operator DNA. It is important that cro exhibits the highest affinity for O_R3, opposite the sequence binding preference of the cl protein. (Reproduced with permission from Alan D. Iselin, artist.)

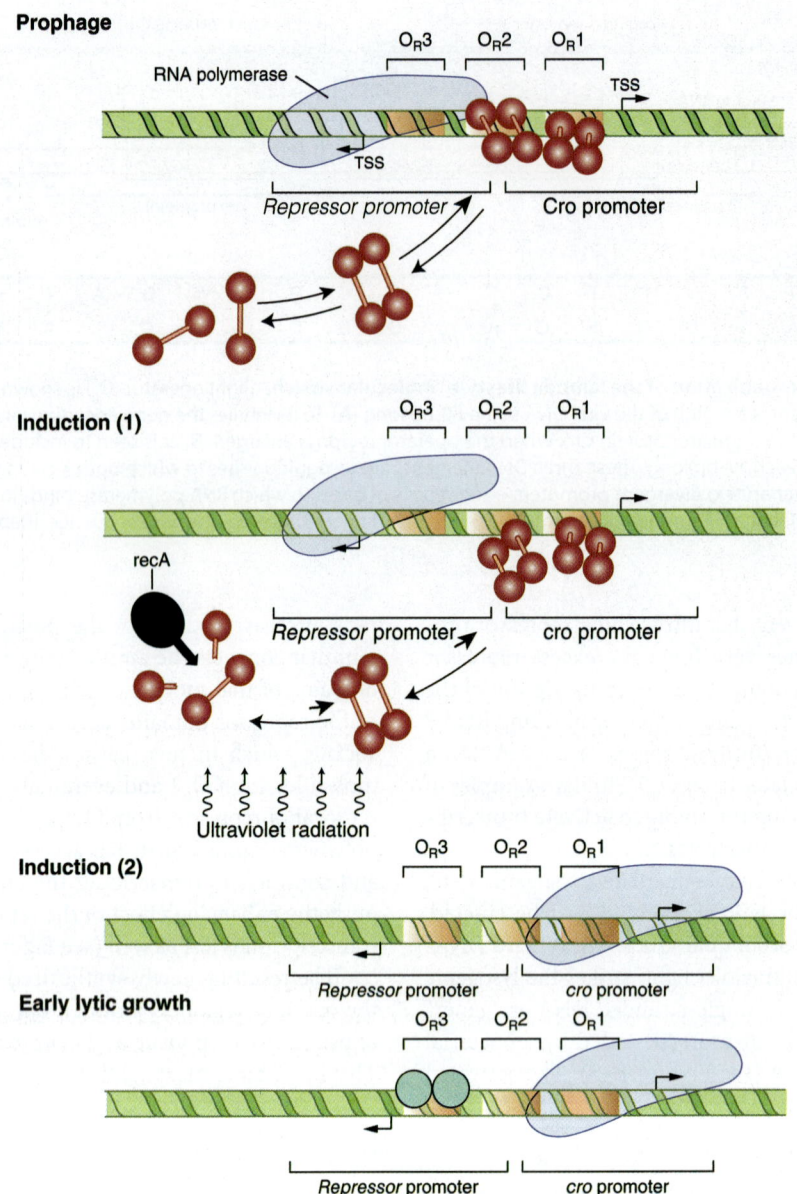

FIGURE 38–7 **Configuration of the lytic/lysogenic switch is shown at four stages of the lambda phage "life" cycle.** The lysogenic pathway (in which the virus remains dormant as a prophage) is selected when a repressor dimer binds to O_R1, thereby making it likely that O_R2 will be bound immediately by another dimer due to the cooperative nature of cI-O_R DNA binding. In the prophage (**top**), the repressor dimers bound at O_R1 and O_R2 prevent RNA polymerase from binding to the rightward *cro* promoter and so block the synthesis of cro (negative control). Simultaneously these DNA-bound cI proteins enhance the binding of polymerase to the leftward promoter (positive control), with the result that the repressor gene is transcribed into RNA (initiation at *cI* gene transcription start site; TSS) and more repressor is synthesized, maintaining the lysogenic state. The prophage is induced (**middle**) when ultraviolet radiation activates the protease recA, which cleaves cI repressor monomers. **Induction** (**1**) The equilibrium of free monomers, free cI dimers, and bound dimers is thereby shifted by mass action, and cI dimers thus dissociate from the operator sites. RNA polymerase is no longer stimulated to bind to the leftward promoter, so that repressor is no longer synthesized. As induction proceeds, **Induction** (**2**) all the operator sites become vacant, thus polymerase can bind to the rightward promoter and cro is synthesized (*cro* TSS shown). During early lytic growth, a single cro dimer binds to O_R3 (light blue shaded circles), the site for which it has the highest affinity thereby occluding the *cI* promoter. Consequently, RNA polymerase cannot bind to the leftward promoter, but the rightward promoter remains accessible. Polymerase continues to bind there, transcribing *cro* and other early lytic genes. Lytic growth ensues (**bottom**). (Reproduced with permission from Alan D. Iselin, artist.)

effect of cro at O_R3 on the binding of cro to O_R2. At increasingly higher concentrations of cro, the protein will bind to O_R2 and eventually to O_R1.

Importantly, occupancy of O_R3 by cro immediately turns off transcription from the leftward *cI* promoter and in that way prevents any further expression of the cI repressor gene. The molecular switch is thus completely "thrown" in the lytic direction. The *cro* gene is now expressed, and the repressor gene is fully turned off. This event is irreversible, and the expression of other lambda genes begins as part of the lytic cycle. When cro repressor concentration becomes quite high, it will eventually occupy O_R1 and in so doing reduce the expression of its own gene, a process that is necessary in order to drive transcription of the genes needed for the final stages of the lytic cycle.

The three-dimensional structures of cro and of the λ cI repressor protein have been determined by x-ray crystallography, and models for their binding and driving the above-described molecular and genetic events have been formulated and tested. Both bind DNA using helix-turn-helix DBD motifs (see following discussion). Along with regulation of the expression of the *lac* operon, the λ molecular switch described here provides arguably the best understanding of the molecular events involved in gene transcription activation and repression.

Detailed analysis of the λ repressor led to the important concept that transcription regulatory proteins have several functional domains. For example, lambda repressor binds to DNA with high affinity. Repressor monomers form dimers that cooperatively interact with each other, these proteins can interact with RNA polymerase, to enhance or block promoter binding or RNAP open complex formation (see Figure 36–3). The protein-DNA interface and the three protein–protein interfaces all involve separate and distinct domains of the two molecules. As will be noted later (see Figure 38–19), this is a characteristic that is typical of most molecules that regulate transcription.

SPECIAL FEATURES ARE INVOLVED IN REGULATION OF EUKARYOTIC GENE TRANSCRIPTION

Most of the DNA in prokaryotic cells is organized into genes, and since the DNA is not compacted with nucleosomal histones bacterial genomes have the potential to be transcribed if appropriate positive and negative *trans*-factors are present in a given cell in an active form. A very different situation exists in eukaryotic cells for two major reasons: first, in human cells relatively little of the total DNA is organized into mRNA-encoding genes and their associated regulatory regions. The function of the extra DNA is being actively investigated (ie, Chapter 39; the ENCODE Projects). Secondly, as described in Chapter 35, the DNA in eukaryotic cells is extensively folded and packed into the protein-DNA complex called chromatin. Histones are an important part of this complex since they both form the structures known as nucleosomes (see Chapter 35) and also factor significantly into gene regulatory mechanisms as outlined in following discussion.

The Chromatin Template Contributes Importantly to Eukaryotic Gene Transcription Control

Chromatin structure provides an additional level of control of gene transcription. As discussed in Chapter 35, large regions of chromatin are transcriptionally inactive while others are either active or potentially active. With few exceptions, each cell contains the same complement of genes; hence, the development of specialized organs, tissues, and cells, and their function in the intact organism depend on the differential expression of genes.

Some of this differential expression is achieved by having different regions of chromatin available for transcription in cells from various tissues. For example, the DNA containing the β-globin gene cluster is in **"active" chromatin** in the reticulocyte but in **"inactive" chromatin** in muscle cells. All the factors involved in the determination of active chromatin have not been elucidated. The presence of nucleosomes and of complexes of histones and DNA (see Chapter 35) certainly provides a barrier against the ready association of most transcription factors with specific DNA regions. The dynamics of the formation and disruption of nucleosome structure are therefore an important part of eukaryotic gene regulation.

Histone covalent modification, also dubbed **the histone code**, is an important determinant of gene activity. Histones are subjected to a wide range of specific posttranslational modifications (see Table 35–1). These modifications are dynamic and reversible. Histone acetylation and deacetylation are best understood. The surprising discovery that histone acetylase and other enzymatic activities are associated with the coregulators involved in regulation of gene transcription (see Chapter 42 for specific examples) has provided a new concept of gene regulation. Acetylation is known to occur on lysine residues in the amino terminal tails of histone molecules, and has been consistently correlated with either active transcription, or alternatively, transcriptional potential. Histone acetylation reduces the positive charge of these tails and likely contributes to a decrease in the binding affinity of histones for the negatively charged DNA. Moreover, such covalent modification of the histones creates new binding, or docking sites for additional proteins such as ATP-dependent chromatin remodeling complexes that contain subunits that carry structural domains that specifically bind to histones that have been subjected to coregulator-deposited PTMs. These complexes can increase accessibility of adjacent DNA sequences by removing or otherwise altering nucleosomal histones. Together then coregulators (chromatin modifiers and chromatin remodelers), working in conjunction, can "open up" gene promoters and regulatory regions, facilitating binding of other *trans*-factors such as transcriptional activator proteins, RNA polymerase II and the GTFs (see Figures 36–10 and 36–11). Histone deacetylation catalyzed by transcriptional corepressors would have the opposite effect. Different proteins with specific acetylase and deacetylase activities are associated with various components of the transcription apparatus. The proteins that catalyze histone PTMs are sometimes referred to as **"code writers"** while the proteins that recognize, bind, and thus interpret these histone PTMs are termed **"code readers"** while the enzymes that remove histone PTMs are called **"code erasers."** (The analogy to signal transduction, with its kinases, phosphatases, and phospho-amino acid binding proteins should be apparent—see Chapter 42.) Collectively then, histone PTMs represent a very dynamic, potentially information-rich source of regulatory information. The exact rules and mechanisms defining the specificity of these various processes are under investigation. Some specific examples are illustrated in Chapter 42. A variety of commercial enterprises are working to develop drugs that specifically alter the activity of the proteins that orchestrate the presence and composition of the histone code, whose relevant PTMs continue to grow at a rapid pace (compare **Table 38–2** with Table 35–1).

In addition to the histone code and its effects on all DNA-mediated reactions, the **methylation of deoxycytidine residues, 5^{me}C,** (in the sequence 5′-^{me}CpG-3′) in DNA has important effects on chromatin, some of which lead to a decrease in gene transcription. For example, in mouse liver, only the unmethylated ribosomal RNA encoding genes can be expressed, and there is evidence that many animal viruses are not transcribed when their DNA is methylated. Acute demethylation of 5^{me}C residues in specific regions of steroid hormone inducible genes has been associated with an increased rate of transcription of the gene. However, as with many histone PTMs, it is not yet possible to generalize that methylated DNA

is transcriptionally inactive, that all inactive chromatin is methylated, or that active DNA is not methylated.

Finally, the binding of specific transcription factors to cognate DNA elements may result in disruption of nucleosomal structure. Most eukaryotic genes have multiple protein-binding DNA elements. The serial binding of transcription factors to these elements—in a combinatorial fashion—may either directly disrupt the structure of the nucleosome, prevent its reformation, or recruit, via protein–protein interactions, multiprotein coregulator complexes that have the ability to covalently modify and/or remodel nucleosomes. These reactions result in chromatin-level structural changes that in the end increase or decrease DNA accessibility to other factors and the transcription machinery.

Eukaryotic DNA that is in an "active" region of chromatin can be transcribed. As in prokaryotic cells, a **promoter** dictates where the RNA polymerase will initiate transcription, but the promoter in mammalian cells (see Chapter 36) is more complex. Additional complexity is added by elements or factors that enhance or repress transcription, define tissue-specific expression, and modulate the actions of many effector molecules. Finally, recent results suggest that gene activation

TABLE 38–2 Summary of Novel Histone PTMs (2011 to 2020)

Histone PTM	Reaction	Donor Precursor	Writer	Eraser	Function	Physiological Relevance
Glutarylation	Acylation	Glutarate	Kat2a, intramolecular catalysis	Sirt7	Nucleosome destabilization, permissive transcription	Glutaric acidemia
Lactylation	Acylation	Lactate	p300		Permissive transcription	Macrophage response, hypoxia
Benzoylation	Acylation	Benzoate		Sirt2	Permissive transcription	Sodium benzoate treatment
S-palmitoylation	S-acylation	Palmitic acid				Cell signaling
O-palmitoylation	O-acylation	Palmitic acid	Lpcat1		Reduced transcription	Cell signaling
Serotonylation	Transamidation	Serotonin	Tgm2		Permissive transcription	Neuronal differentiation
Dopaminylation	Transmidation	Dopamine	Tgm2		Altered transcription	Drug-seeking behaviors
5-Hydroxylysine	Hydroxylation	2-Oxoglutarate	Jmjd6			Testes, development
Glycation	Maillard	Methylglyoxal, monosaccharides	Nonenzymatic	DJ-1	Altered nucleosome stability	Breast cancer, hyperglycemia
4-Oxononanoylation	Ketoamide adduction	4-Oxo-2-nonenal	Nonenzymatic	Sirt2	Nucleosome destabilization	Lipid peroxidation
Acrolein adduct	Michael addition	Acrolein	Nonenzymatic		Nucleosome destabilization	Cigarette smoke, lipid peroxidation
S-glutathionylation	Disulfide formation	Glutathione	Nonenzymatic		Nucleosome destabilization	Aging
Homocysteinylation	Thiolation	Homocysteine thyolactone	Nonenzymatic		Reduced transcription	Hyperhomocysteinemia

Modified with permission from Chan JC, Maze I. Nothing Is Yet Set in (Hi)stone: Novel Post-Translational Modifications Regulating Chromatin Function, Trends Biochem Sci. 2020;45(10):829-844.

and repression might occur when particular genes move into or out of different subnuclear compartments or locations wherein variable amounts of transcription proteins and RNA either promote or disrupt biomolecular condensate formation that stimulate or inhibit transcription.

Epigenetic Mechanisms Contribute Importantly to the Control of Gene Transcription

The molecules and regulatory biology described earlier contributes importantly to transcriptional regulation. Indeed, in recent years the role of covalent modification of DNA and histone (and nonhistone) proteins and the newly discovered ncRNAs has received tremendous attention in the field of gene regulation research, particularly through investigation into how such chemical modifications and/or molecules stably alter gene expression patterns without altering the underlying DNA

gene sequence. This field of study has been termed **epigenetics**. As mentioned in Chapter 35, one aspect of these mechanisms, PTMs of histones has been dubbed the **histone code** or histone epigenetic code. The term "epigenetics" means "above genetics" and refers to the fact that these regulatory mechanisms do not change the underlying regulated DNA sequence, but rather simply the expression patterns, or function, of this DNA. Epigenetic mechanisms play key roles in the establishment, maintenance, and reversibility of transcriptional states. A key feature of epigenetic mechanisms is that the controlled transcriptional on/off states can be maintained through multiple rounds of cell division. This observation indicates that there must be robust, biochemically based mechanisms to maintain and stably propagate these epigenetic states.

Two forms of epigenetic signals, *cis*- and *trans*-epigenetic signals, can be described; these are schematically illustrated in **Figure 38–8**. A simple *trans*-signaling event composed of positive transcriptional feedback mediated by an abundant,

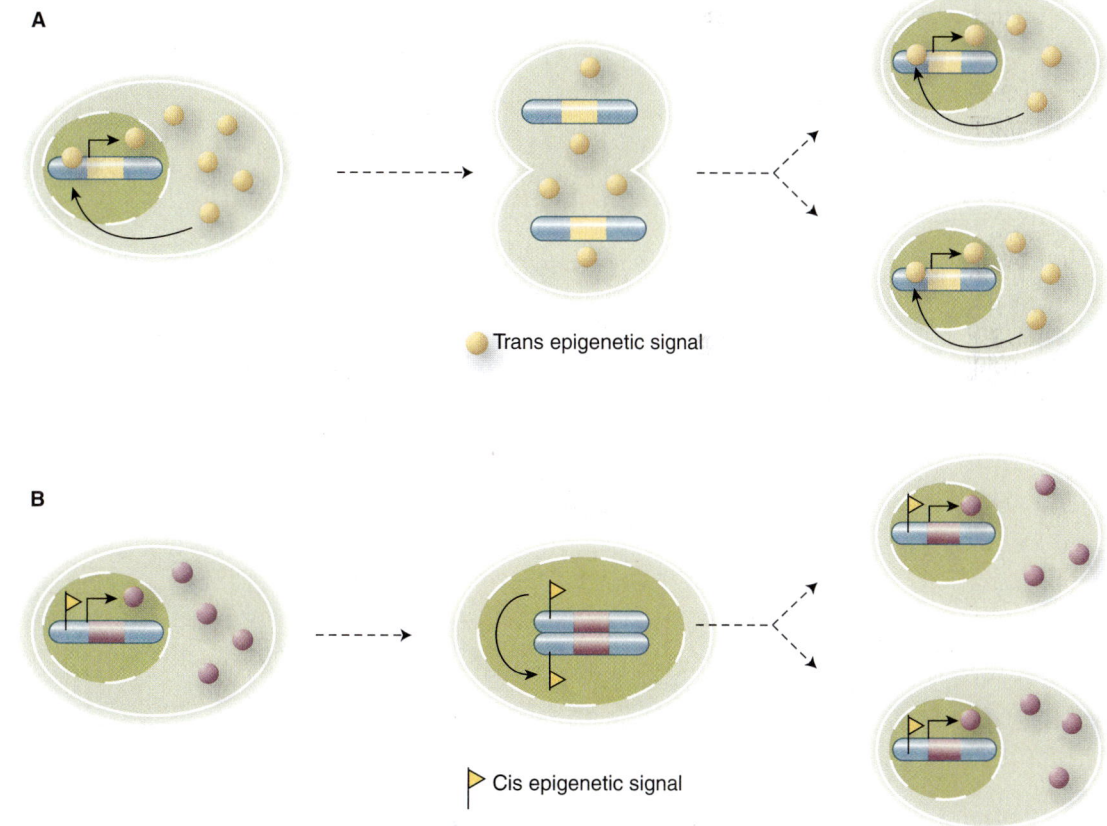

FIGURE 38–8 *cis*- and *trans*-epigenetic signals. (**A**) An example of an epigenetic signal that acts in trans. A DNA-binding transactivator protein (yellow circle) is transcribed from its cognate gene (yellow bar) located on a particular chromosome (blue). The expressed protein is freely diffusible between nuclear and cytoplasmic compartments. Note that excess transactivator reenters the nucleus following cell division, binds to its own gene, and activates transcription in both daughter cells. This cycle re-establishes the positive feedback loop that was in effect prior to cell division, and thereby enforces stable expression of this transcriptional activator protein in both cells. (**B**) A *cis*-epigenetic signal; a gene (pink) located on a particular chromosome (blue) carries a *cis*-epigenetic signal (small yellow flag) within the regulatory region upstream of the pink gene transcription unit. In this case, the epigenetic signal is associated with active gene transcription and subsequent gene product production (pink circles). During DNA replication, the newly replicated chromatid serves as a template that both elicits, and templates, the introduction of the same epigenetic signal, or mark, on the newly synthesized, unmarked chromatid. Consequently, both daughter cells contain the pink gene in a similarly *cis*-epigenetically marked state, which ensures expression in an identical fashion in both cells. See text for more detail.

diffusible transactivator that efficiently partitions roughly equally between mother and daughter cell at each division is depicted in Figure 38–8A. So long as the indicated, transcription factor is expressed at a sufficient level to allow all subsequent daughter cells to inherit the *trans*-epigenetic signal (transcription factor), such cells will have the cellular or molecular phenotype dictated by the other target genes of this transcriptional activator. Shown in Figure 38–8 panel B is an example of how a *cis*-epigenetic signal (here as a specific ᵐᵉCpG methylation mark) can be stably propagated to the two daughter cells following cell division. The hemi-methylated (ie, only one of the two DNA strands is 5ᵐᵉC-modified) DNA mark generated during DNA replication directs the methylation of the newly replicated strand through the action of ubiquitous maintenance

DNA methylases. Thus, the original 5ᵐᵉC methylation mark ultimately results in both DNA daughter strands having the complete *cis*-epigenetic mark.

Both *cis*- and *trans*-epigenetic signals can result in stable and hereditable expression states, and therefore generally represent type C gene expression responses (ie, Figure 38–1). However, it is important to note that both states can be reversed if either the *trans*- or *cis*-epigenetic signals are removed by, for example, extinguishing the expression of the enforcing transcription factor (*trans*-signal) or by completely removing a DNA *cis*-epigenetic signal (via DNA demethylation). Enzymes have been described that can remove both protein PTMs and 5ᵐᵉC modifications.

Stable transmission of epigenetic on/off states can be affected by multiple molecular mechanisms. Shown in **Figure 38–9** are

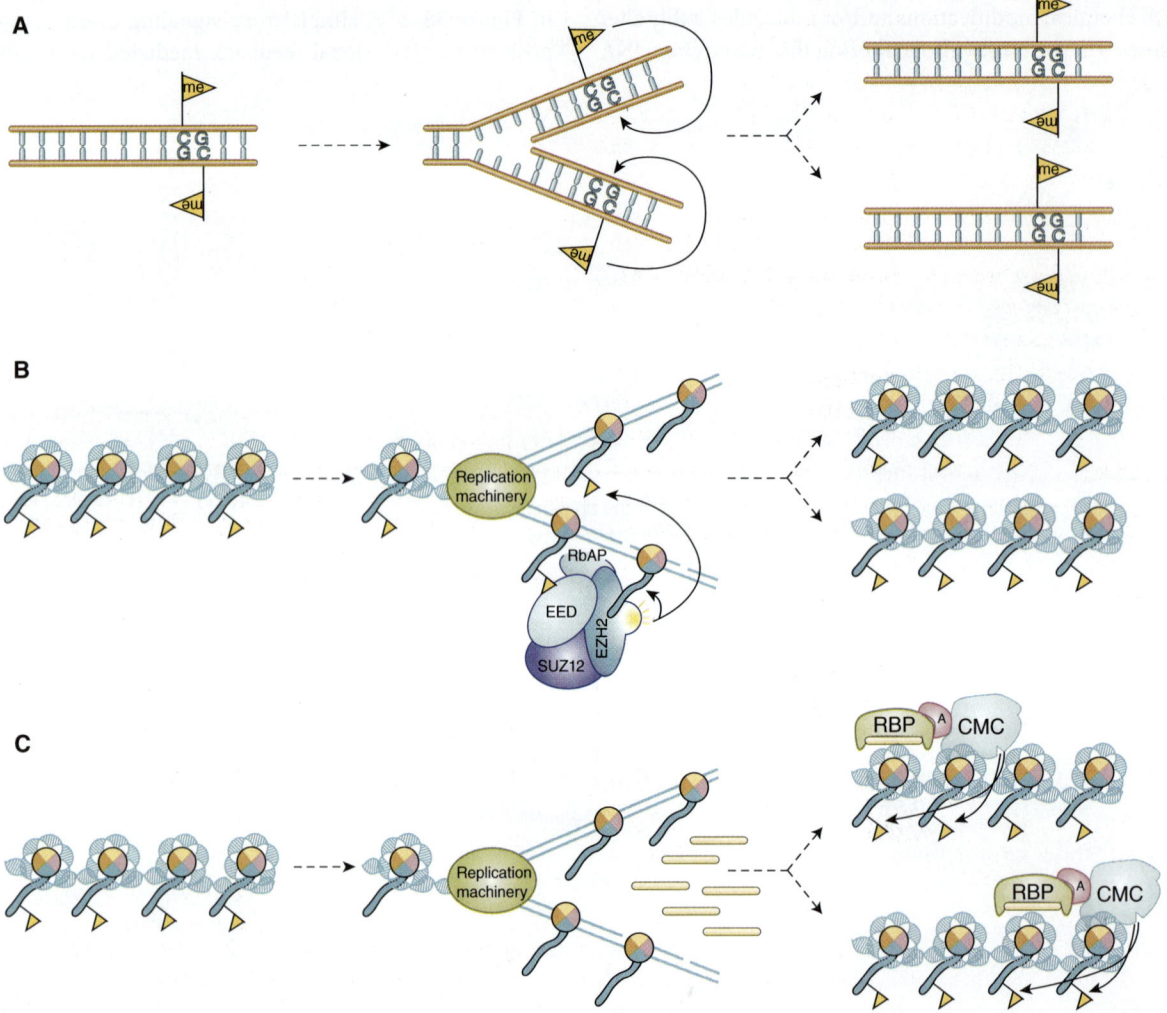

FIGURE 38–9 **Mechanisms for the transmission and propagation of epigenetic signals following a round of DNA replication.** (**A**) Propagation of a 5ᵐᵉC signal (yellow flag; see Figure 38–8B). (**B**) Propagation of a histone PTM epigenetic signal (H3K27me) that is mediated through the action of the PRC2, a chromatin modifying complex, or CMC. PRC2 is composed of EED, EZH2 histone methylase, RbAP, and SUZ12 subunits. Note that in this context, PRC2 is both a histone code reader (via the methylated histone–binding domain in EED) and histone code writer (via the SET domain histone methylase within EZH2). Location-specific deposition of the histone PTM *cis*-epigenetic signal is targeted by the recognition of the H3K27me marks in preexisting nucleosomal histones (yellow flag). (**C**) Another example of the transmission of a histone epigenetic signal (yellow flag) except here signal-targeting is mediated through the action of small ncRNAs that work in concert with an RNA-binding protein (RBP), an adaptor (A) protein, and a CMC. See text for more detail. (Reproduced with permission from Bonasio R, Tu S, Reinberg D. Molecular signals of epigenetic states. Science. 2010;330(6004):612-616.)

three ways by which *cis*-epigenetic marks can be propagated through a round of DNA replication. The first example of epigenetic mark transmission involves the propagation of DNA 5^meC marks, and occurs as described in Figure 38–8. The second example of epigenetic state transmission illustrates how a nucleosomal histone PTM (in this example, Lysine K-27 trimethylated histone H3; H3K27me3) can be propagated. In this example immediately following DNA replication, both H3K27me3-marked and H3-unmarked nucleosomes randomly reform on both daughter DNA strands. The **polycomb repressive complex 2 (PRC2)**, composed of EED-SUZ12-EZH2 and RbAP subunits, binds to the nucleosome containing the preexisting H3K27me3 mark via the EED subunit. Binding of PRC2 to this histone mark stimulates the methylase activity of the EZH2 subunit of PRC2, which results in the local methylation of nucleosomal H3. Histone H3 methylation thus causes the full, stable transmission of the H3K27me3 epigenetic mark to both chromatids. Finally, locus/sequence-specific targeting of nucleosomal histone epigenetic *cis*-signals can be attained through the action of lncRNAs as depicted in Figure 38–9, panel C. Here a specific ncRNA interacts with target DNA sequences and the resulting RNA–DNA complex is recognized by RBP, an RNA-binding protein. Then, likely through a specific adaptor protein (A), the RNA-DNA-RBP complex recruits a **chromatin modifying complex (CMC)** that locally modifies nucleosomal histones. Again, this mechanism leads to the transmission of a stable epigenetic mark.

Additional work will be required to establish the complete molecular details of epigenetic processes, determine how ubiquitously these mechanisms operate, identify the full complement of molecules involved, and genes controlled. Epigenetic signals are critically important to gene regulation as evidenced by the fact that mutations and/or overexpression of many of the molecules that contribute to epigenetic control lead to human disease.

Certain DNA Elements Enhance or Repress Transcription of Eukaryotic Genes

In addition to gross changes in chromatin affecting transcriptional activity, certain DNA elements facilitate or enhance initiation at the promoter and hence are termed **enhancers**. Enhancer elements, which typically contain multiple binding sites for transactivator proteins, differ from the promoter in notable ways. Enhancers can exert their positive influence on transcription even when separated by tens of thousands of base pairs from a promoter; enhancers work when oriented in either direction; and enhancers can work upstream (5') or downstream (3') from the promoter, or even when embedded within the transcription unit of a gene. Experimentally, enhancers can be shown to be promiscuous, in that they can stimulate transcription of any promoter in their vicinity, and may act on more than one promoter. The viral SV40 enhancer can exert an influence on, for example, the transcription of β-globin by increasing its transcription 200-fold in cells containing both

the SV40 enhancer and the β-globin gene on the same plasmid (see following discussion and **Figure 38–10**); in this case the SV40 enhancer-β-globin reporter gene was constructed using recombinant DNA technology—see Chapter 39. The enhancer element does not produce a product that in turn acts on the promoter, since it is active only when it exists within the same DNA molecule as the promoter (ie, in *cis*, or physically linked to). Enhancer-binding proteins are responsible for this effect. The exact mechanism(s) by which these transcription activators work is subject to intensive investigation. Enhancer-binding *trans*-factors, some of which are cell-type specific, while others are ubiquitously expressed, have been shown to interact

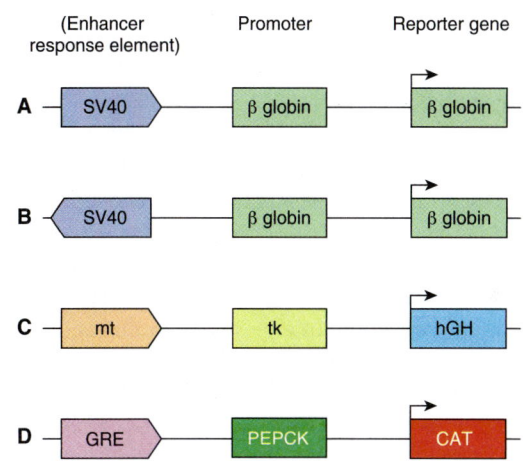

FIGURE 38–10 **A schematic illustrating the methods used to study the organization and action of enhancers and other *cis-acting* regulatory elements.** These model chimeric genes, all constructed by recombinant DNA techniques in vitro (see Chapter 39), consist of a reporter gene that encodes a protein that can be readily assayed, and that is not normally produced in the cells to be studied, a promoter that ensures accurate initiation of transcription, and the indicated enhancer (regulatory response) elements. In all cases, high-level transcription from the indicated chimeras depends on the presence of enhancers, which stimulate transcription ≥100-fold over basal transcriptional levels (ie, transcription of the same chimeric genes containing just promoters fused to the indicated reporter genes). Examples (**A**) and (**B**) illustrate the fact that enhancers (eg, here SV40) work in either orientation and upon a heterologous promoter. Example (**C**) illustrates that the metallothionein (mt) regulatory element (which under the influence of cadmium or zinc induces transcription of the endogenous *mt* gene and hence the metal-binding mt protein) will work through the *herpes simplex* virus (HSV) thymidine kinase (*tk*) gene promoter to enhance transcription of the human growth hormone (*hGH*) reporter gene. In a separate experiment, this engineered genetic construct was introduced into the male pronuclei of single-cell mouse embryos and the embryos placed into the uterus of a surrogate mother to develop as transgenic animals. Offspring have been generated under these conditions, and in some the addition of zinc ions to their drinking water effects an increase in growth hormone expression in liver. In this case, these transgenic animals have responded to the high levels of growth hormone by becoming twice as large as their normal litter mates. Example (**D**) illustrates that a glucocorticoid response element (GRE) enhancer will work through homologous (*PEPCK* gene) or heterologous gene promoters (not shown; ie, HSV *tk* promoter, SV40 promoter, β-globin promoter, etc) to drive expression of the chloramphenicol acetyltransferase (*CAT*) reporter gene.

with a plethora of other transcription proteins. These interactions include chromatin-modifying coactivators, mediator, as well as the individual components of the basal RNA polymerase II transcription machinery. Ultimately, transfactor-enhancer DNA-binding events result in an increase in the binding and/or activity of the basal transcription machinery on the linked promoter. Enhancer elements and associated binding proteins often convey nuclease hypersensitivity to those regions where they reside (see Chapter 35). Recently, while analyzing regulatory sequences that control cellular identity (and other genes essential for cell function) in mammalian genomes investigators have identified large tandem clusters of various enhancer elements in tandem arrays. These sequence elements have been termed super-enhancers. Not surprisingly, the cis-linked genes modulated by super-enhancers are highly expressed. It is highly likely that such super-enhancers contribute importantly to the formation of the biomolecular condensates described earlier. A summary of the properties of enhancers is presented in **Table 38–2**.

One of the best-understood mammalian enhancer systems is that of the β-interferon gene. This gene is induced upon viral infection of mammalian cells. One goal of the cell, once virally infected, is to attempt to mount an antiviral response—if not to save the infected cell, then to help to save the entire organism from viral infection. Interferon production is one mechanism by which this is accomplished. This family of proteins is secreted by virally infected cells. Secreted interferon interacts with neighboring cells to cause an inhibition of viral replication by a variety of mechanisms, thereby limiting the extent of viral infection. The enhancer element controlling induction of the β-interferon gene, which is located between nucleotides −110 and −45 relative to the transcription start site (+1), is well characterized. This enhancer consists of four distinct clustered cis-elements, each of which is bound by unique trans-factors. One cis-element is bound by the transacting factor NF-κB (see Figures 42–10 and 42–13), one by a member of the interferon regulatory factor (IRF) family of transactivator factors, and a third by the heterodimeric leucine zipper factor ATF-2/c-Jun (see following discussion). The fourth factor is the ubiquitous, abundant architectural transcription factor known as HMG I(Y). Upon binding to its A + T-rich binding sites, HMG I(Y) induces a significant bend in the DNA. There are four such HMG I(Y) binding sites interspersed throughout the enhancer. It is believed that these sites play a key role in facilitating the

formation of a unique 3D structure in concert with the aforementioned three trans-factors, by inducing a series of critically spaced DNA bends. Consequently, HMG I(Y) likely induces the cooperative formation of a unique, stereospecific structure within which all four factors are active when viral infection signals are sensed by the cell. The putative structure formed by the cooperative assembly of these four factors has been termed the β-interferon enhanceosome (**Figure 38–11**),

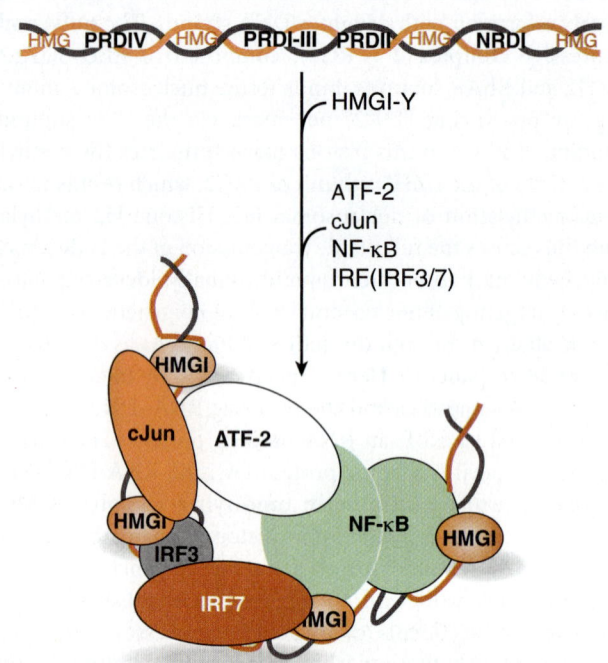

FIGURE 38–11 **Formation and putative structure of the enhanceosome formed on the human β-interferon gene enhancer.** Diagrammatically represented at the top is the distribution of the multiple cis-elements (HMG, PRDIV, PRDI-III, PRDII, NRDI) composing the β-interferon gene enhancer. The intact enhancer mediates transcriptional induction of the β-interferon gene (*IFNB1*) over 100-fold upon virus infection of human cells. The cis-elements of this modular enhancer represent the binding sites for the trans-factors HMG I(Y), cJun-ATF-2, IRF3-IRF7, and NF-κB, respectively. The factors interact with these DNA elements in an obligatory, ordered, and highly cooperative fashion as indicated by the arrow. Initial binding of four HMG I(Y) proteins induces sharp DNA bends in the enhancer, causing the entire 70- to 80-bp region to assume a high level of curvature. This curvature is integral to the subsequent highly cooperative binding of the other trans-factors since bending enables the DNA-bound factors to make critical direct protein–protein interactions that both contribute to the formation and stability of the enhanceosome and generate a unique 3D surface that serves to recruit chromatin-modifying coregulators that carry enzymatic activities (eg, Swi/Snf: ATPase, chromatin remodeler and P/CAF: histone acetyltransferase) as well as the general transcription machinery (RNA polymerase II and GTFs). Although four of the five cis-elements (PRDIV, PRDI-III, PRDII, NRDI) independently can modestly stimulate (~10-fold) transcription of a reporter gene in transfected cells (see Figures 38–10 and 38–12), all five cis-elements, in appropriate order, are required to form an enhancer that can appropriately stimulate transcription of *IFNB1* (ie, ≥100-fold) in response to viral infection of a human cell. This distinction indicates a strict requirement for appropriate enhanceosome architecture for efficient trans-activation. Similar enhanceosomes, involving distinct cis- and trans-factors and coregulators, are proposed to form on many other mammalian genes.

TABLE 38–3 **Summary of the Properties of Enhancers**

- Work when located both short and long distances from target promoter
- Work when upstream or downstream from the promoter
- Work when oriented in either direction
- Work when embedded within target promoter
- Can work with homologous or heterologous promoters
- Work by binding one or more proteins
- Can be composed of one to a few binding elements or many multiples of activation elements (super enhancers)
- Work by recruiting chromatin-modifying coregulatory complexes
- Work by facilitating binding and/or function of the basal transcription complex at the cis-linked promoter

so named because of its proposed structural similarity to the nucleosome, which is also a unique three-dimensional protein-DNA structure that wraps DNA about a core assembly of proteins (see Figures 35–1 and 35–2). The enhanceosome, once formed, induces a large increase in β-interferon gene transcription upon virus infection. Thus, it is thought that it is not simply the protein occupancy of the linearly apposed *cis*-element sites that induces β-interferon gene transcription—rather, it is the formation of the enhanceosome proper that provides appropriate surfaces and 3-dimensional organization for the efficient recruitment of coactivators that results in the enhanced formation of the PIC on the *cis*-linked promoter and thus transcription activation.

cis-Acting DNA elements that decrease the expression of specific genes are termed **silencers**. Silencers have also been identified in a number of eukaryotic genes. However, because fewer of these elements have been intensively studied, it is not possible to formulate accurate generalizations about their mechanism of action. That said, it is clear that as for gene activation, chromatin level covalent modifications of histones, and other proteins, by silencer-recruited repressors and co-recruited multisubunit corepressors likely play central roles in these regulatory events.

Tissue-Specific Expression May Result From Either the Action of Enhancers or Repressors, or a Combination of Both *cis*-Acting Regulatory Elements

Most genes are now recognized to harbor enhancer elements in various locations relative to their coding regions. In addition to being able to enhance gene transcription, some of these enhancer elements clearly possess the ability to do so in a tissue-specific manner. By fusing known or suspected tissue-specific enhancers or silencers to reporter genes (see following discussion) and introducing these chimeric enhancer-reporter constructs via microinjection into single-cell embryos, one can create a transgenic animal (see Chapter 39), and rigorously test whether a given test enhancer or silencer truly modulates expression in a cell- or tissue-specific fashion. This **transgenic animal** approach has proved useful in studying tissue-specific gene expression.

Reporter Genes Are Used to Define Enhancers & Other Regulatory Elements That Modulate Gene Expression

By ligating regions of DNA suspected of harboring regulatory sequences to various reporter genes (the **reporter** or **chimeric gene approach**) (**Figures 38–10, 38–12,** and **38–13**), one can determine which regions in the vicinity of structural genes have an influence on their expression. Pieces of DNA thought to harbor regulatory elements, often identified by bioinformatic sequence alignments, are ligated to a suitable reporter

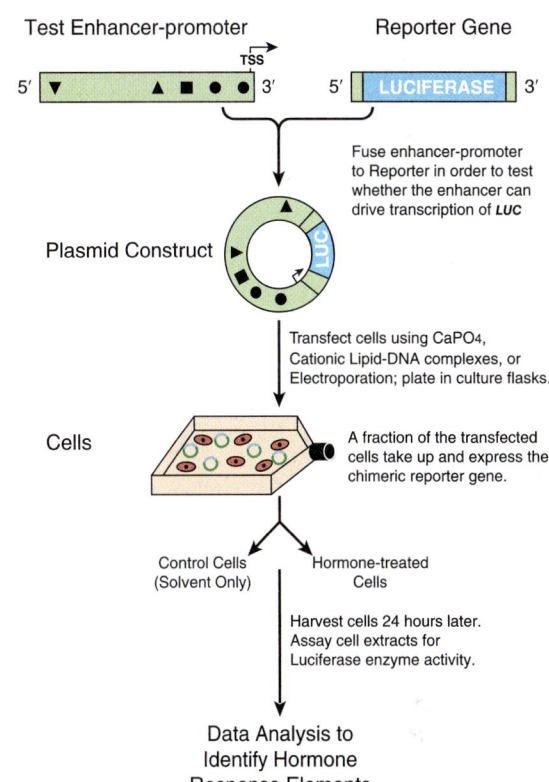

FIGURE 38–12 **The use of reporter genes to define DNA regulatory elements.** A DNA fragment bearing regulatory *cis*-elements (triangles, square, circles in diagram) from the gene in question—in this example, approximately 2 kb of 5′-flanking DNA and cognate promoter—is ligated into a plasmid vector that contains a suitable reporter gene—in this case, the enzyme firefly luciferase, abbreviated LUC. As noted in Figure 38–10 in such experiments, the reporter cannot be present endogenously in the cells transfected. Consequently, any detection of these activities in a cell extract means that the cell was successfully transfected by the plasmid. Not shown here, but typically one cotransfects an additional reporter such as *Renilla* luciferase to serve as a transfection efficiency control. Assay conditions for the firefly and *Renilla* luciferases are different, hence the two activities can be independently sequentially assayed using the same cell extract. An increase of firefly luciferase activity over the basal level, for example, after addition of one or more hormones, means that the region of DNA inserted into the reporter gene plasmid contains functional hormone response elements (HRE). Progressively shorter pieces of DNA, regions with internal deletions, or regions with point mutations can be constructed and inserted upstream of the reporter gene to pinpoint the response element (Figure 38–13). One caveat of this approach is that the transfected plasmid DNAs likely do not form "classical" chromatin structures.

gene and introduced into a host cell (see Figure 38–12). Expression of the reporter gene will be increased if the DNA contains a particular enhancer. For example, addition of different hormones to separate cultures will increase expression of the reporter gene if the DNA contains a particular **hormone response DNA element** (**HRE**) (see Figure 38–13; see also Chapter 42). The location of the element can be pinpointed by using progressively shorter pieces of DNA, deletions, or and/or point mutations (see Figure 38–13).

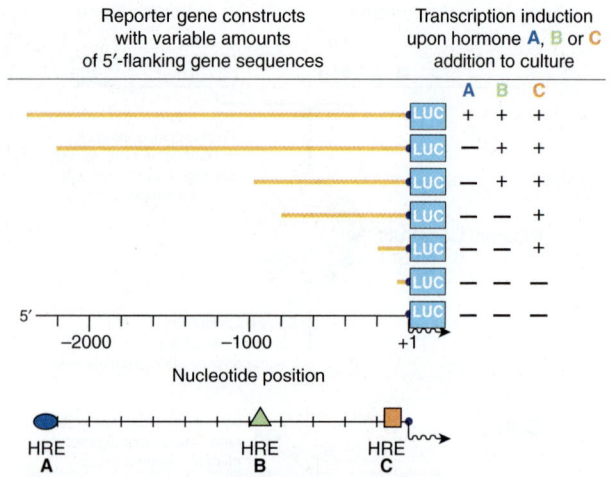

FIGURE 38–13 **Mapping distinct hormone response elements (HREs) (A), (B), and (C) using the reporter gene–transfection approach.** A family of reporter genes, constructed as described in Figures 38–10 and 38–12, can be transfected individually into recipient cells. By analyzing when certain hormone responses are lost in comparison to the 5′ deletion end point, specific hormone-response enhancer elements can be located and defined, ultimately with nucleotide-level precision (see summary, bottom).

This strategy, typically performed using transfected cells in culture (ie, cells induced to take up exogenous DNAs), has led to the identification of hundreds of enhancers, silencers/repressors such as tissue-specific elements, and hormone, heavy metal, and drug-response elements. The activity of a gene at any moment reflects the interaction of these numerous *cis*-acting DNA elements with their respective *trans*-acting factors. Overall, transcriptional output is determined by the balance of positive and negative signaling to the transcription machinery. The challenge now is to figure out exactly how this regulation occurs at the molecular level so that we might ultimately have the ability to modulate gene transcription therapeutically.

Combinations of DNA Elements & Associated Proteins Provide Diversity in Responses

Prokaryotic genes are often regulated in an on-off manner in response to simple environmental cues. Some eukaryotic genes are regulated in the simple on-off manner, but the process in most genes, especially in mammals, is much more complicated. Signals representing a number of complex environmental stimuli may converge on a single gene. The response of the gene to these signals can have several physiologic characteristics. First, the response may extend over a considerable range. This is accomplished by having additive and synergistic positive responses counterbalanced by negative or repressing effects. In some cases, either the positive or the negative response can be dominant. Also required is a mechanism whereby an effector, such as a hormone, can activate some genes in a cell while repressing others and leaving

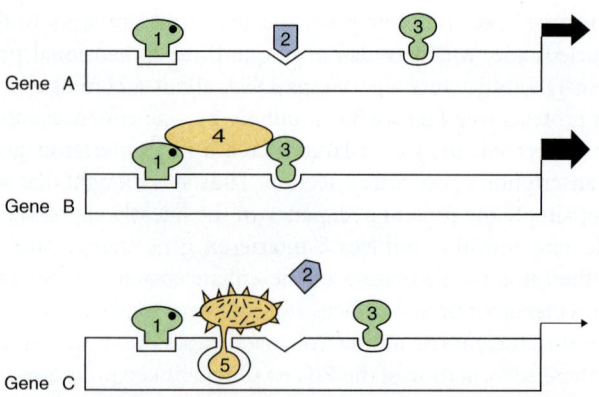

FIGURE 38–14 **Combinations of DNA elements and proteins provide diversity in the response of a gene.** Gene A is activated (the width of the arrow, right, indicates the extent) by the combination of transcriptional activator proteins 1, 2, and 3 (with coactivators, as shown in Figures 36–10 and 38–11). Gene B is activated, in this case more effectively, by the combination of factors 1, 3, and 4; note that transcription factor 4 does not contact DNA directly in this example. The activators could form a linear bridge that links the basal machinery to the promoter, or alternatively, this could be accomplished by DNA looping, or 3D structure formation (ie, Figure 38–11). Regardless, the purpose is to direct the basal transcription machinery to the promoter. Gene C is inactivated by the combination of transcription factors 1, 5, and 3; in this case, factor 5 is shown to preclude the essential binding of factor 2 to DNA, as occurs in example A. If activator 1 promotes cooperative binding of repressor protein 5, and if activator 1 binding requires a ligand (solid dot), it can be seen how the ligand could activate one gene in a cell (gene A) and repress another (gene C) in the same cell.

still others unaffected. When all of these processes are coupled with tissue-specific element factors, considerable flexibility is afforded. These physiologic variables obviously require an arrangement much more complicated than an on-off switch. The collection and organization of DNA elements in a promoter specifies—via associated factors—how a given gene will respond, and how long a particular response is maintained. Some simple examples are illustrated in **Figure 38–14**.

Transcription Domains Can Be Defined by Variable 3-D Localization Within the Cell

The large number of genes in eukaryotic cells and the complex arrays of transcription regulatory factors present an organizational problem. Why are some genes available for transcription in a given cell whereas others are not? If enhancers can regulate several genes from tens of kilobase distances and are not obligatorily position- and orientation-dependent, how are they prevented from triggering transcription of all *cis*-linked genes in the vicinity? Part of the solution to these problems is arrived at by having the chromatin arranged in functional units that restrict patterns of gene expression. This may be achieved by having the chromatin form a structure with the nuclear matrix or other physical entity or compartment within the nucleus. On a macro-scale, the 4D Nucleome Project, an international consortium of scientists studying nuclear function, aims to analyze nuclear genome structure and dynamics to gain insights into how these features map onto transcriptional activity. On a gene-by-gene scale, a molecular mechanism to focus enhancer-driven transcription is provided by **insulators**. These DNA elements, also in association with one or more specific proteins, prevent an enhancer from acting on a promoter on the other side of an insulator in another transcription domain. Insulators thus serve as transcriptional **boundary elements**. In the globin gene cluster, and many other genes, enhancer and promoter sequences are brought into physical contact via specific DNA looping events, often involving boundary elements. The rules controlling such chromosome looping are currently under intense study.

SEVERAL STRUCTURAL MOTIFS COMPOSE THE DNA-BINDING DOMAINS OF REGULATORY TRANSCRIPTION FACTOR PROTEINS

The specificity involved in the control of transcription requires that regulatory proteins bind with high affinity and specificity to the correct region of DNA. Three unique motifs—the **helix-turn-helix (HTH)**, the **zinc finger (ZF)**, and the **leucine zipper (LZ)**—account for many of these specific protein-DNA interactions. Examples of proteins containing these motifs are given in **Table 38–4**.

Comparison of the binding activities of the proteins that contain these motifs leads to several important generalizations.

1. Binding must be of high affinity to the specific site, and of low affinity to all other DNA.

2. Small regions of the protein make direct contact with DNA; the rest of the protein, in addition to providing the

TABLE 38–4 Examples of Transcription Factors That Contain Various DNA-Binding Motifs

Binding Motif	Organism	Regulatory Protein
Helix-turn-helix	*E. coli*	lac repressor, CAP
	Phage	λcI, cro, and 434 repressors
	Mammals	Homeobox proteins Pit-1, Oct1, Oct2
Zinc finger	*E. coli*	Gene 32 protein
	Yeast	Gal4
	Drosophila	Serendipity, hunchback
	Xenopus	TFIIIA
	Mammals	Steroid receptor family, Sp1
Leucine zipper	Yeast	GCN4
	Mammals	C/EBP, fos, Jun, Fra-1, CRE binding protein (CREB), *c-myc*, *n-myc*, *l-myc*

trans-activation domains, may be involved in the dimerization of monomers of the binding protein, may provide a contact surface for the formation of heterodimers, may provide one or more ligand-binding sites, or may provide surfaces for interaction with coactivators, corepressors, or the transcription machinery.

3. The protein-DNA interactions made by these proteins are maintained by hydrogen bonds, ionic interactions, and van der Waals forces.

4. The motifs found in these proteins are class-specific; their presence in a protein of unknown function suggests that the protein may bind to DNA.

5. Proteins with the helix-turn-helix or leucine zipper motifs form dimers, and their respective DNA-binding sites are symmetric palindromes. In proteins with the zinc finger motif, the binding site is repeated two to nine times. These features allow for cooperative interactions between binding sites and enhance the degree and affinity of binding.

The Helix-Turn-Helix Motif

The first motif described was the **helix-turn-helix**. Analysis of the 3D structure of the lambda cro transcription regulator revealed that each monomer consists of three antiparallel β sheets and three α helices (**Figure 38–15**). The dimer forms by association of the antiparallel β_3 sheets. The α_3 helices form the DNA recognition surface, and the rest of the molecule appears to be involved in stabilizing these structures. The average diameter of an α helix is 1.2 nm, which is the approximate width of the major groove in the B form of DNA.

The DNA recognition domain of each cro monomer interacts with 5 bp and the dimer-binding sites span 3.4 nm, allowing it to fit into successive half turns of the major groove on the same surface of DNA (see Figure 38–15). X-ray analyses of the λ cI repressor, CAP (the cAMP receptor protein of *E. coli*), tryptophan repressor, and phage 434 repressor, all also display this dimeric helix-turn-helix structure, which is also present in many eukaryotic DNA-binding proteins (see Table 38–4).

The Zinc Finger Motif

The **zinc finger** was the second DNA-binding motif whose atomic structure was elucidated. It was known that the eukaryotic protein involved, a positive regulator of 5S RNA gene transcription termed TFIIIA, required zinc for activity. Structural and biophysical analyses revealed that each TFIIIA molecule contains nine zinc ions in a repeating coordination complex formed by closely spaced cysteine–cysteine residues followed 12 to 13 amino acids later by a histidine–histidine pair (**Figure 38–16**). In some instances—notably the steroid–thyroid nuclear hormone receptor family—the His–His doublet is replaced by a second Cys–Cys pair. The zinc finger motifs of the protein lie on one face of the DNA helix, with successive fingers alternatively positioned in one turn in the major groove. As is the case with the recognition domain in

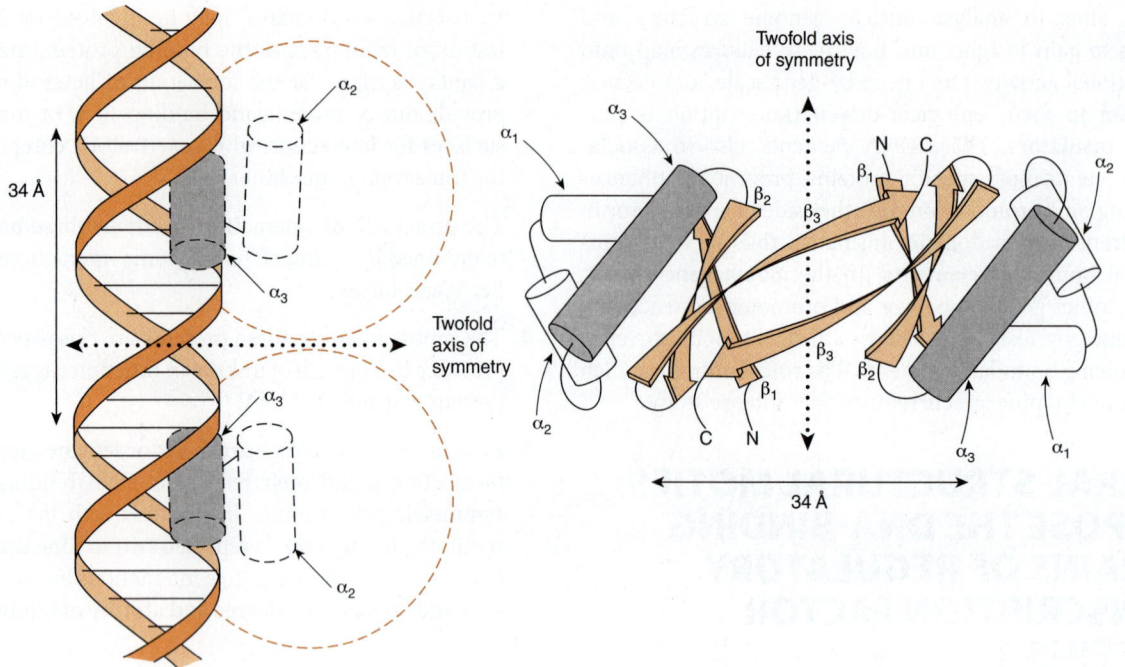

FIGURE 38–15   **A schematic representation of the 3D structure of Cro protein and its binding to DNA by its helix-turn-helix motif (left).** The cro monomer consists of three antiparallel β sheets (β$_1$-β$_3$) and three α helices (α$_1$-α$_3$). The helix-turn-helix (HTH) motif is formed because the α$_3$ and α$_2$ helices are held at about 90° to each other by a turn of four amino acids. The α$_3$ helix of cro is the DNA recognition surface (**shaded**). Two monomers associate through interactions between the two antiparallel β$_3$ sheets to form a dimer that has a twofold axis of symmetry (**right**). A cro dimer binds to DNA through its α$_3$ helices, each of which contacts about 5 bp on the same face of the major groove (see Figures 34–2 and 38–6). The distance between comparable points on the two DNA α helices is 34 Å, the distance required for one complete turn of the double helix. (Reproduced with permission from B Mathews.)

the helix-turn-helix protein, each TFIIIA zinc finger contacts about 5 bp of DNA. The importance of this motif in the action of steroid hormones is underscored by an "experiment of nature." A single amino acid mutation in either of the two zinc fingers of the 1,25(OH)$_2$-D$_3$ receptor protein results in resistance to the action of this hormone and the clinical syndrome of rickets.

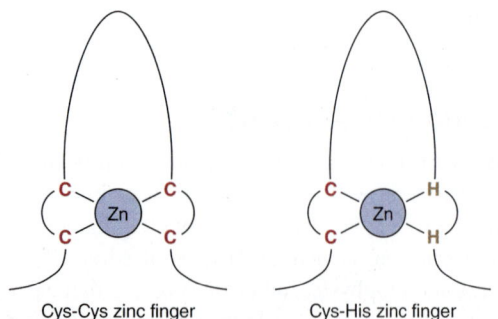

Cys-Cys zinc finger   Cys-His zinc finger

FIGURE 38–16   **Zinc fingers are a series of repeated domains (two to nine) in which each is centered on a tetrahedral coordination with zinc.** In the case of the DNA-binding transcription factor TFIIIA, the coordination is provided by a pair of cysteine residues (C) separated by 12 to 13 amino acids from a pair of histidine (H) residues. In other zinc finger proteins, the second pair also consists of C residues. Zinc fingers bind in the major groove, where adjacent Zn-fingers make contact with 5 bp of DNA along the same face of the helix.

The Leucine Zipper Motif

Analysis of a 30-amino-acid sequence in the carboxyl-terminal region of the mammalian enhancer-binding protein C/EBP revealed a novel structure, the **leucine zipper motif**. As illustrated in **Figure 38–17**, this region of the protein forms an α helix in which there is a periodic repeat of leucine residues at every seventh position. This occurs for eight helical turns and four leucine repeats. Similar structures have been found in a number of other proteins associated with the regulation of transcription in all eukaryotes tested. This structure allows two identical or nonidentical monomers (eg, Jun–Jun or Fos–Jun) to "zip together" in a coiled coil and form a tight dimeric complex (see Figure 38–17). This protein–protein interaction serves to enhance the association of the separate DBDs with their target DNA sites (see Figure 38–17).

THE DNA BINDING & TRANSACTIVATION DOMAINS OF MOST REGULATORY PROTEINS ARE SEPARATE

DNA binding could result in a general conformational change that allows the bound protein to activate transcription, alternatively these two functions could be served by separate and independent domains. Domain swap experiments suggest

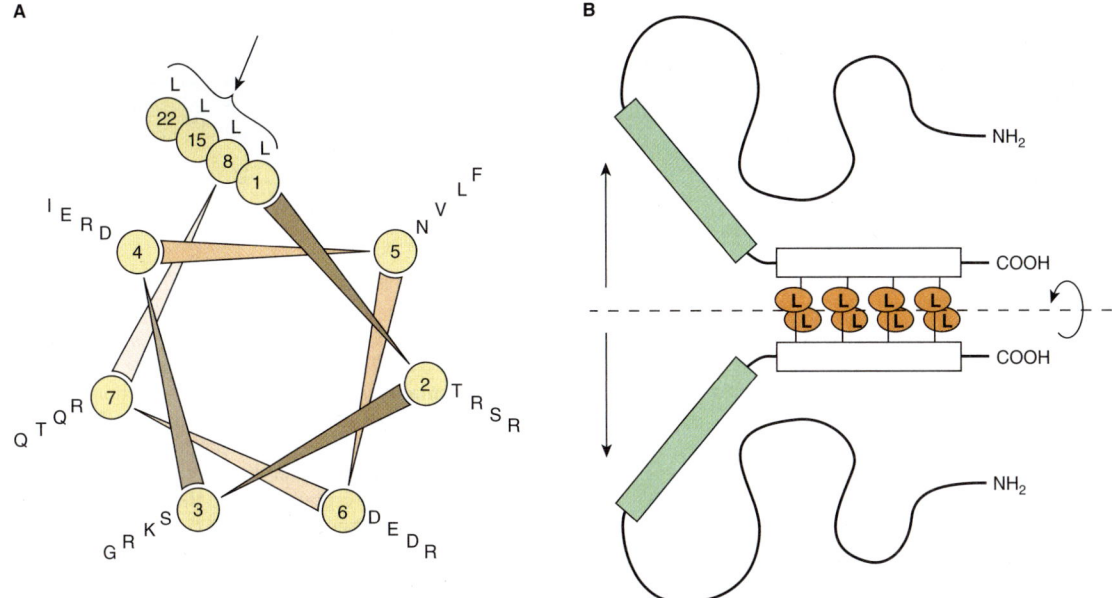

FIGURE 38–17 **The leucine zipper motif.** (**A**) Shown is a helical wheel analysis of a carboxyl-terminal portion of the DNA-binding protein C/EBP (see Table 36–3). The amino acid sequence is displayed end-to-end down the axis of a schematic α helix (see Figures 5–2 to 5–4). The helical wheel consists of seven spokes that correspond to the seven amino acids that comprise every two turns of the α helix. Note that leucine residues (L) occur at every seventh position (in this schematic C/EBP amino acid residues 1, 8, 15, 22; see arrow). Other proteins with "leucine zippers" have a similar helical wheel pattern. (**B**) A schematic model of the DNA-binding domain of C/EBP. Two identical C/EBP polypeptide chains are held in dimer formation by the leucine zipper domain of each polypeptide (denoted by the white rectangles and attached orange-shaded ovals). This association is required to hold the DNA-binding domains of each polypeptide (the green-shaded rectangles) in the proper conformation and register for DNA binding. (Reproduced with permission from S McKnight.)

that the latter is typically the case. The critical tests to address this question was first performed in the yeast *Saccharomyces cerevisiae.*

The yeast *GAL1* gene is a member of a group of genes involved in galactose metabolism. Transcription of *GAL1* is controlled by the DNA-binding transactivator protein Gal4. Gal4 binds to a 17bp enhancer element (termed upstream activator sequence, or UAS in yeast) located upstream of the *Gal1* promoter via its N-terminal DNA binding domain (DBD; Gal4 amino acids 1-73). Gal4 activates *GAL1* transcription through a C-terminal 34 aa activation domain (AD). To systematically test the contributions of the Gal4 AD and DBD sequences to *GAL1* gene transcription activation, and to ask whether the Gal4 DBD uniquely contributed to such transcription activation, a series of domain swap experiments were performed (**Figure 38–18**). The amino terminal 73-amino-acid DBD of Gal4 was removed and replaced with the DBD of LexA, an *E. coli* DNA-binding protein. This domain swap resulted in a chimeric molecule (lexA DBD-Gal4 AD) that did not bind to the *GAL1* UAS, and did not activate transcription of the *GAL1* gene as might be expected (see Figure 38–18). If, however, the *lexA* operator—the DNA sequence normally bound by the LexA DBD—was inserted upstream of the promoter of the *GAL* gene to replace the normal *GAL1* enhancer, the hybrid LexA-Gal4 AD fusion protein bound to this chimeric gene (at the substituted *lexA* operator) and activated transcription of *GAL1*. This general experiment has been repeated many times with a

large array of different DBDs and ADs. Collectively, such data demonstrates that the DBDs and ADs of many transcription factors can function independently.

The hierarchy involved in assembling gene transcription-activating complexes includes proteins that bind DNA and transactivate; others that form protein–protein complexes which bridge DNA-binding proteins to transactivating proteins; and others that form protein–protein complexes with components of coregulators or the basal transcription apparatus. A given protein may thus have several modular surfaces or domains that serve different functions (**Figure 38–19**). As described in Chapter 36, the primary purpose of these molecules is to facilitate the assembly and/or activity of the basal transcription apparatus on the *cis*-linked promoter. Not shown here, but DNA-binding repressor proteins are organized similarly with separable DBDs and **silencing domains, SDs**.

GENE REGULATION IN PROKARYOTES & EUKARYOTES DIFFERS IN OTHER IMPORTANT RESPECTS

In addition to transcription, eukaryotic cells employ a variety of other mechanisms to regulate gene expression (**Table 38–5**). Many more steps, especially in RNA processing, are involved

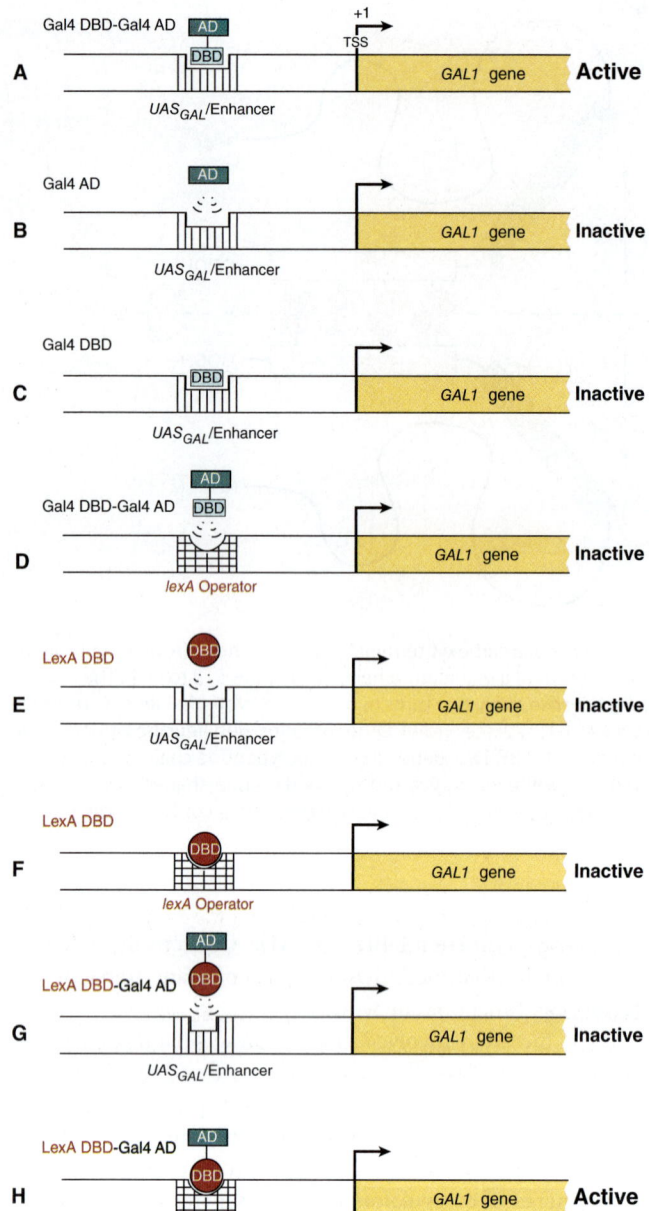

FIGURE 38–18 **Domain-swap experiments demonstrate the independent nature of DNA-binding and transcription activation domains.** The yeast *GAL1* gene contains an upstream activating sequence/enhancer (*UAS*$_{GAL}$/Enhancer) that is bound by the multi-domain DNA binding regulatory transcriptional activator protein Gal4. Gal4, like the lambda cI protein is modular, and contains an N-terminal DNA binding domain (DBD) and a C-terminal activation domain (AD). When the intact Gal4 transcription factor binds the *GAL1 UAS*$_{GAL}$ enhancer, activation of *GAL1* gene transcription ensues [(**A**); **Active**]. Control experiments demonstrate that all three *GAL1*-gene specific components [ie. cis- and trans-active components: *UAS*$_{GAL}$ DNA enhancer, Gal4 DBD and Gal4 AD) are required for active transcription of the natural *GAL1* gene, as expected [(**B**), (**C**), (**D**), (**E**), (**F**)-all **Inactive**]. A chimeric protein, in which the DBD of Gal4 is replaced with the DBD of the *E coli*-specific operator DNA binding protein LexA fails to stimulate *GAL1* transcription because the LexA DBD cannot bind to the *UAS*$_{GAL}$/Enhancer [(**G**); **Inactive**]. By contrast, the LexA DBD-Gal4 AD fusion protein does activate *GAL1* transcription when the *LexA* operator (the natural target for the LexA DBD) is inserted upstream of the *GAL1* promoter region, replacing the normal *UAS*$_{GAL}$/Enhancer [(**H**); **Active**].

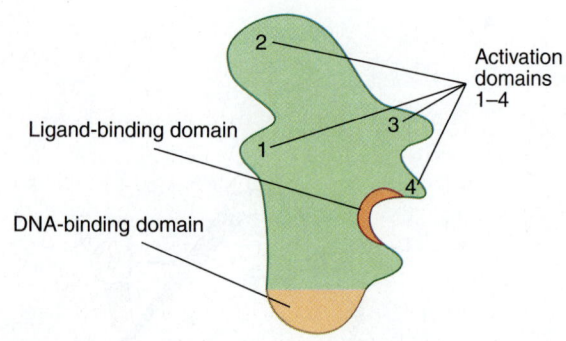

FIGURE 38–19 **Proteins that regulate transcription have several domains.** This hypothetical transcription factor has a DBD that is distinct from a ligand-binding domain (LBD) and several activation domains (ADs) (1-4). Other proteins may lack the DBD or LBD and all may have variable numbers of domains that contact other proteins, including coregulators and those of the basal transcription complex (see also Chapters 41 and 42).

in the expression of eukaryotic genes than of prokaryotic genes, and these steps provide additional sites for regulatory influences that cannot exist in prokaryotes. These RNA processing steps in eukaryotes, described in detail in Chapter 36, include capping of the 5′ ends of primary transcripts, addition of a polyadenylate tail to the 3′ ends of transcripts, mRNA internal base modifications and excision of intron regions to generate spliced exons in the mature mRNA molecule. To date, analyses of eukaryotic gene expression provide evidence that regulation occurs at the level of transcription, nuclear RNA processing, nuclear transport, mRNA stability, and translation. In addition, gene amplification and rearrangement influence gene expression.

Owing to the advent of recombinant DNA technology and high throughput DNA, RNA and protein sequencing methods and other new genetic tools (see Chapter 39), much progress has been made in recent years in our understanding of eukaryotic gene expression. However, because most eukaryotic organisms contain so much more genetic information than do prokaryotes, and because manipulation of their genes is more difficult, molecular aspects of eukaryotic gene regulation are less well understood than the examples discussed earlier in this chapter. This section briefly describes a few different types of eukaryotic gene regulation.

TABLE 38–5 Gene Expression Is Regulated by Transcription & in Numerous Other Ways in Eukaryotic Cells

- Gene amplification
- Gene rearrangement
- RNA processing
- Alternate mRNA splicing
- Transport of mRNA from nucleus to cytoplasm
- Regulation of mRNA stability
- mRNA compartmentalization
- Translational control
- ncRNA silencing and activation

ncRNAs Modulate Gene Expression by Altering mRNA Function

As noted in Chapter 35, the recently discovered class of ubiquitous large and small eukaryotic (non–protein-coding) ncRNAs contribute importantly to the control of gene expression. The mechanism of action of the small miRNA and siRNAs are best understood. These ~22 nucleotide RNAs regulate the function/expression of specific mRNAs by either inhibiting translation or inducing mRNA degradation via different mechanisms; in a very few cases miRNAs have been shown to stimulate mRNA function. miRNA action can result in dramatic changes in protein production and hence gene expression. These small ncRNAs have been implicated in numerous human diseases such as heart disease, cancer, muscle wasting, viral infection, and diabetes.

miRNAs and siRNAs, like the DNA-binding transcription factors described in detail earlier, are transactive, and once synthesized and appropriately processed, interact with specific proteins and bind target mRNAs (see Figure 36–17). Binding of miRNAs to mRNA targets is directed by normal base-pairing rules. In general, if miRNA–mRNA base pairing has one or more mismatches, translation of the cognate "target" mRNA is inhibited, whereas if miRNA–mRNA base pairing is perfect over all 22 nucleotides, the corresponding mRNA is degraded.

Given the tremendous and ever-growing importance of miRNAs, many scientists and biotechnology companies are actively studying miRNA biogenesis, transport, and function in hopes of curing human disease. Time will tell the magnitude and universality of ncRNA-mediated gene regulation.

Eukaryotic Genes Can Be Amplified or Rearranged During Development or in Response to Drugs

During early development of metazoans, there is an abrupt increase in the need for specific molecules such as ribosomal RNA and messenger RNA molecules for proteins that make up specific cell or tissue types. One way to increase the rate at which such molecules can be formed is to increase the number of genes available for transcription of these specific molecules. Among the repetitive DNA sequences within the genome are hundreds of copies of ribosomal RNA-encoding genes. These genes preexist repetitively in the DNA of the gametes and thus are transmitted in high copy numbers from generation to generation. In some specific organisms such as the fruit fly (*Drosophila*), there occurs during oogenesis an amplification of a few preexisting genes such as those for the chorion (eggshell) proteins. Subsequently, these amplified genes, presumably generated by a process of repeated initiations during DNA synthesis, provide multiple sites for gene transcription (see Figures 36–4 and **38–20**). The dark side of specific gene amplification is the fact that in human cancer cells drug resistance can develop upon extended therapeutic treatment due to the amplification and increased expression of genes that encode proteins that either degrade, or pump drugs from target cells.

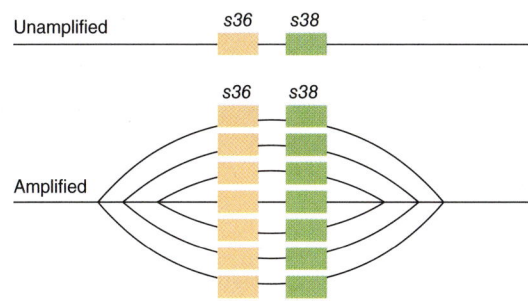

FIGURE 38–20 Schematic representation of the amplification of chorion protein-encoding genes *s36* and *s38*. (Reproduced with permission from Chisholm R: Gene amplification during development, Trends Biochem Sci 1982;7(5):161-162.)

As noted in Chapter 36, the coding sequences responsible for the generation of specific protein molecules are frequently not contiguous in the mammalian genome. In the case of antibody encoding genes, this is particularly true. As described in detail in Chapter 52, immunoglobulins are composed of two polypeptides, the so-called heavy (about 50 kDa) and light (about 25 kDa) chains. The mRNAs encoding these two protein subunits are encoded by gene sequences that are subjected to extensive DNA sequence–coding changes. These DNA coding changes are integral to generating the requisite recognition diversity central to appropriate immune function.

IgG heavy- and light-chain mRNAs are encoded by several different segments that are tandemly repeated in the germline. Thus, for example, the IgG light chain consists of variable (V_L), joining (J_L), and constant (C_L) domains or segments. For particular subsets of IgG light chains, there are roughly 300 tandemly repeated V_L gene coding segments, 5 tandemly arranged J_L coding sequences, and roughly 10 C_L gene coding segments. All of these multiple, distinct coding sequences are located in the same region of the same chromosome, and each type of coding segment (V_L, J_L, and C_L) is tandemly repeated in head-to-tail fashion within the segment repeat region. By having multiple V_L, J_L, and C_L segments to choose from, an immune cell has a greater repertoire of sequences to work with to develop both immunologic flexibility and specificity. However, a given functional IgG light-chain transcription unit—like all other "normal" mammalian transcription units—contains only the coding sequences for a single protein. Thus, before a particular IgG light chain can be expressed, *single* V_L, J_L, and C_L coding sequences must be recombined to generate a *single*, contiguous transcription unit excluding the multiple nonutilized segments (ie, the other approximately 300 unused V_L segments, the other 4 unused J_L segments, and the other 9 unused C_L segments). This deletion of unused genetic information is accomplished by selective DNA recombination that removes the unwanted coding DNA while retaining the required coding sequences: one V_L, one J_L, and one C_L sequence. (V_L sequences are subjected to additional point mutagenesis to generate even more variability—hence the name.) The newly recombined sequences thus form a single

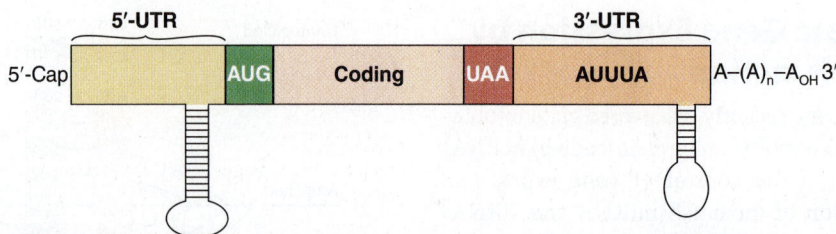

FIGURE 38–21 **Structure of a typical eukaryotic mRNA showing elements that are involved in regulating mRNA stability.** The typical eukaryotic mRNA has a 5′–noncoding sequence (NCS), or untranslated exonic region (5′ UTR), a coding region, and a 3′–exonic untranslated NCS region (3′ UTR). Essentially all mRNAs are capped at the 5′ end, and most have a 100 to 200 nt polyadenylate sequence at their 3′ end. The 5′ cap and 3′ poly(A) tail protect the mRNA against exonuclease attack and are bound by specific proteins that interact to facilitate translation (see Figure 37–7). Stem-loop structures in the 5′ and 3′ NCS, and the AU-rich region in the 3′ NCS represent the binding sites for specific proteins that modulate mRNA stability.

transcription unit that is competent for RNA polymerase II–mediated transcription into a single monocistronic IgG light chain-encoding mRNA. These multiple IgG-gene recombination and mutation events are generated in a single clonal population of B-cells. Although the IgG genes represent one of the best-studied instances of directed DNA rearrangement modulating gene expression, other cases of gene regulatory DNA rearrangement have been described.

Alternative RNA Processing Is Another Control Mechanism

In addition to affecting the efficiency of promoter utilization, eukaryotic cells employ alternative RNA processing to control gene expression. This can result when alternative promoters, intron–exon splice sites, or polyadenylation sites are used. Occasionally, heterogeneity within a cell results, but more commonly the same primary transcript is processed differently in different tissues. A few examples of each of these types of regulation are presented later.

The use of **alternative transcription start sites** results in a different 5′ exon on mRNAs encoding mouse amylase and myosin light chain, rat glucokinase, and *Drosophila* alcohol dehydrogenase and actin. **Alternative polyadenylation sites** in the μ immunoglobulin heavy-chain primary transcript result in mRNAs that are either 2700 bases long (μ_m) or 2400 bases long (μ_s). This results in a different carboxyl-terminal region of the encoded proteins such that the μ_m protein remains attached to the membrane of the B lymphocyte and the μ_s immunoglobulin is secreted. **Alternative splicing and processing** results in the formation of seven unique α-tropomyosin mRNAs in seven different tissues. It is not yet fully understood how these processing-splicing decisions are made or exactly how these steps can be regulated.

Regulation of Messenger RNA Stability Provides Another Control Mechanism

Although most mRNAs in mammalian cells are very stable (half-lives measured in hours), some turn over very rapidly (half-lives of 10-30 minutes). In certain instances, mRNA stability is subject to regulation. This has important implications

since there is usually a direct relationship between mRNA amount and the translation of that mRNA into its cognate protein. Changes in the stability of a specific mRNA can therefore have major effects on biologic processes.

Messenger RNAs exist in the cytoplasm as **ribonucleoprotein particles (RNPs)**. Some of these proteins protect the mRNA from digestion by nucleases, while others may under certain conditions promote nuclease attack. It is thought that mRNAs are stabilized or destabilized by the interaction of proteins with these various structures or sequences. Certain effectors, such as hormones, may regulate mRNA stability by increasing or decreasing the amount of these mRNA-binding proteins.

It is known that the ends of mRNA molecules are involved in mRNA stability (**Figure 38–21**). The 5′–cap structure in eukaryotic mRNA prevents attack by 5′ exonucleases, and the poly(A) tail minimizes the action of 3′ exonucleases. In mRNA molecules with those structures, it is presumed that a single endonucleolytic cut allows exonucleases to attack and digest the entire molecule. Other structures (sequences) in the 5′–untranslated region (5′ UTR), the coding region, and the 3′ UTR are thought to promote or prevent this initial endonucleolytic action (see Figure 38–21).

Thus, it is clear that a number of mechanisms are used to regulate mRNA stability and hence function—just as several mechanisms are used to regulate the synthesis of mRNA, and as detailed in Chapter 37, mRNA translation. Coordinate regulation of these processes confers on the cell remarkable adaptability.

SUMMARY

- The genetic constitutions of metazoan somatic cells are nearly all identical.
- Phenotype (tissue or cell specificity) is dictated by differences in gene expression of the cellular complement of genes.
- Alterations in gene expression allow a cell to adapt to environmental changes, developmental cues, and physiologic signals.
- Gene expression can be controlled at multiple levels by changes in transcription, mRNA processing, localization, and stability or translation. Gene amplification and rearrangements also influence gene expression.

- Transcription controls operate at the level of protein-DNA and protein–protein interactions. These interactions display protein domain modularity and high specificity.

- Many different structural classes of DBDs have been identified in transcription factors.

- Chromatin and DNA modifications contribute importantly in eukaryotic transcription control by modulating DNA accessibility and specifying recruitment of specific coactivators and corepressors to target genes.

- Several epigenetic mechanisms for gene control have been described and the molecular mechanisms through which these processes operate are being elucidated at the molecular level.

- ncRNAs modulate gene expression. The short miRNAs and siRNAs modulate mRNA translation and stability.

REFERENCES

Ambrosi C, Manzo M, Baubec T: Dynamics and context-dependent roles of DNA methylation. J Mol Biol 2017;429(10):1459-1475.

Brandao HB, Gabriele M, Hansen AS: Tracking and interpreting long-range chromatin interactions with super-resolution live-cell imaging. Curr Opin Cell Biol 2021;70:18-26.

Browning DF, Busby SJ: Local and global regulation of transcription initiation in bacteria. Nat Rev Microbiol 2016;14:638-650.

Chan JC, Maze I: Nothing is yet set in (Hi)stone: novel post-transcription modifications regulating chromatin function. Trends Biochem Sci 2020;45:829-844.

Chen H, Pugh BF: What do transcription factors interact with? J Mol Biol 2021; doi.org/10.1016/.mb.2021.166883.

Dekker J, Belmont AS, Guttman M, et al: The 4D nucleome project. Nature 2017;549:219-278.

Henniger E, Oksuz O, Shrinivas K, et al: RNA-mediated feedback control of transcriptional condensates. Cell 2021;184:207-225.

Hnisz D, Abraham BJ, Lee TI, et al: Super-enhancers in the control of cell identity and disease. Cell 2013;155:934-947.

Jacob F, Monod J: Genetic regulatory mechanisms in protein synthesis. J Mol Biol 1961;3:318-356.

Jaeger MG, Winter GE: Fast-acting chemical tools to delineate causality in transcriptional control. Mol Cell 2021;81:1617-1630.

Klug A: The discovery of zinc fingers and their applications in gene regulation and genome manipulation. Annu Rev Biochem 2010;79:213-231.

Manning KS, Cooper TA: The roles of RNA processing in translating genotype to phenotype. Nat Rev Mol Cell Biol 2017;18:102-114.

Narita T, Ito S, Higashiima Y, et al: Enhancers are activated by p300/CPB activity-dependent PIC assembly, RNAPII recruitment, and pause release. Mol Cell 2021;81:1-17.

Ptashne M: *A Genetic Switch*, 2nd ed. Cell Press and Blackwell Scientific Publications, 1992.

Pugh BF: A preoccupied position on nucleosomes. Nat Struct Mol Biol 2010;17:923.

Roeder RG: 50+ years of eukaryotic transcription: an expanding universe of factors and mechanisms. Nat Struc Mol Biol 2019;26:783-791.

Rossi M, Kuntala PK, Lai WKM, et al: A high-resolution protein architecture of the budding yeast genome. Nature 2021;592:309-315.

Schmitt AM, Chang HY: Long noncoding RNAs in cancer pathways. Cancer Cell 2016;29:452-463.

Schwartzman O, Tanay A: Single-cell epigenomics: techniques and emerging applications. Nat Rev Genet 2015;16:716-726.

Scotti MM, Swanson MS: RNA mis-splicing in disease. Nat Rev Genet 2016;17:19-32.

Shao Q, Trinh JT, Zeng L: High-resolution studies of lysis-lysogeny decision-making in bacteriophage lambda. J Biol Chem 2019;294:3343-3349.

Tee WW, Reinberg D: Chromatin features and the epigenetic regulation of pluripotency states in ESCs. Development 2014;141:2376-2390.

Tian B, Manley JL: Alternative polyadenylation of mRNA precursors. Nat Rev Mol Cell Biol 2017;18:18-30.

Wang Z, Cairns MJ, Yan J: Super-enhancers in transcriptional regulation and genome organization. Nuc Acids Res 2019:11481-11496.

Wang Z, Cui M, Shah AM, et al: Cell-type-specific gene regulatory networks underlying murine neonatal heart regeneration at single-cell resolution. Cell Reports 2020;33:108472.

Zaborowska J, Egloff S, Murphy S: The pol II CTD: new twists in the tail. Nat Struct Mol Biol 2016;23:771-777.

Molecular Genetics, Recombinant DNA, & Genomic Technology

P. Anthony Weil, PhD

O B J E C T I V E S

After studying this chapter, you should be able to:

- Understand the basic procedures and methods involved in recombinant DNA technology and genetic engineering.
- Appreciate the rationale behind the methods used to synthesize, analyze, and sequence DNA and RNA.
- Be able to describe how to identify and quantify individual proteins, as well as proteins bound to specific sequences of genomic DNA or RNA.

BIOMEDICAL IMPORTANCE*

The development of recombinant DNA techniques, high-density DNA microarrays, high-throughput screening, low-cost genome-scale DNA and RNA sequencing, high sensitivity mass spectrometry-based protein identification and sequencing, and other molecular genetic methodologies has revolutionized biology. Collectively these powerful new technologies are having an increasing impact on clinical medicine. Although much has been learned about human genetic disease from pedigree analysis and study of the affected, mutated proteins, in many cases where the specific genetic defect was unknown, these approaches could not be used. Fortunately, these new technologies circumvent these limitations by going directly to cellular DNA, RNA, and protein molecules for such information. Manipulation of a DNA sequence and the construction of chimeric molecules—so-called genetic engineering—provide a means of studying how a specific segment of DNA controls cellular function. New biochemical and molecular genetic tools allow investigators to query and manipulate genomic sequences as well as to examine the entire complement of cellular RNA, protein, and protein PTM status at the molecular level in small tissue samples, and even single cells.

Understanding molecular genetics technology is important for several reasons: (1) It offers a rational approach to understanding the molecular basis of disease. For example, familial hypercholesterolemia, sickle cell disease, the thalassemias, cystic fibrosis, muscular dystrophy as well as more complex multifactorial diseases like vascular and heart disease, Alzheimer disease,

cancer, obesity, and diabetes. (2) Human proteins can be produced in abundance for therapy (eg, insulin, growth hormone, tissue plasminogen activator). (3) Proteins or nucleic acids for preparation of vaccines (eg, hepatitis B, COVID-19), and for diagnostic testing (eg, Ebola and AIDS tests) can be readily obtained. (4) This technology is used both to diagnose existing diseases as well as to predict the risk of developing a given disease and individual responses to pharmacologic therapeutics—so called **personalized medicine**. (5) Special techniques have led to remarkable advances in forensic medicine, which have allowed for the molecular diagnostic analysis of DNA from single cells. (6) Finally, in extremely well understood diseases, potentially curative gene therapy for diseases caused by a single-gene deficiency such as sickle cell disease, the thalassemias, adenosine deaminase deficiency, and others are being devised.

RECOMBINANT DNA TECHNOLOGY INVOLVES ISOLATION & MANIPULATION OF DNA TO MAKE CHIMERIC MOLECULES

Isolation and manipulation of DNA, including end-to-end joining of sequences from very different sources to make chimeric molecules (eg, molecules containing both human and bacterial DNA sequences in a sequence-independent fashion; gene fragments containing discrete functional elements, or complete genes), is the essence of recombinant DNA research. This involves several unique techniques and reagents.

---

*See glossary of terms at the end of this chapter.

Restriction Enzymes Cleave DNA Chains at Specific Locations

Certain endonucleases—enzymes that cut DNA at specific DNA sequences within the molecule—(as opposed to exonucleases, which processively digest from the ends of DNA molecules in a primarily sequence-independent fashion)—are a key tool in recombinant DNA research. These enzymes were termed **restriction enzymes**, or **REs**, because their presence in a given bacterium restricted, or prevented the growth of certain bacterial viruses (bacteriophages). Restriction enzymes cut DNA of any source into unique pieces in a sequence-specific manner—in contrast to most other enzymatic, chemical, or physical methods, which break DNA randomly. These defensive enzymes (hundreds have been discovered) protect the host bacterial DNA from the DNA genome of foreign organisms (primarily infective phages) by specifically inactivating the invading phage DNA by digestion. The viral RNA-inducible interferon system (see Chapter 38; Figure 38–11) provides the same sort of molecular defense against RNA viruses in mammalian cells. However, restriction endonucleases are present only in cells that also have a companion enzyme that site-specifically methylates the DNA of the bacterial host, thereby rendering it noncleavable by that particular restriction enzyme. Thus, sequence-specific DNA methylases and sequence-specific restriction endonucleases that target the exact same sites always exist in pairs in a given bacterium.

Restriction enzymes are named after the bacterium from which they are isolated. For example, EcoRI is from *Escherichia coli*, and BamHI is from *Bacillus amyloliquefaciens* (**Table 39–1**). The first three letters in the restriction enzyme name consist of the first letter of the genus (*E*) and the first two letters of the species (*co*) in the case of the restriction enzyme EcoRI derived from *E. coli* strain R. These designations may be followed by a strain designation (*R*) and a roman numeral (*I*) to indicate the order of discovery (eg, EcoRI and EcoRII). Each enzyme recognizes and cleaves a specific double-stranded DNA sequence that is typically 4- to 8-bp long. These DNA cuts result in **blunt ends** (eg, HpaI) or overlapping (**sticky or cohesive**) **ends** (eg, BamHI) (**Figure 39–1**), depending on the mechanism used by the enzyme. Sticky ends are particularly useful in constructing hybrid or chimeric DNA molecules (see later). If the four nucleotides are distributed randomly in a given DNA molecule, one can calculate how frequently a given enzyme will cut a length of DNA. For each position in the DNA molecule, there are four possibilities (A, C, G, and T); therefore, a restriction enzyme that recognizes a 4-bp sequence cuts DNA, on average, once every 256 bp (4^4), whereas another enzyme that recognizes a 6-bp sequence cuts once every 4096 bp (4^6). A given piece of DNA has a characteristic linear array of sites for the various enzymes dictated by the linear sequence of its bases; hence, a **restriction map** can be constructed. When DNA is digested with a particular enzyme, the cut ends of all the fragments have the same

TABLE 39–1 Selected Restriction Endonucleases & Their Sequence Specificities

Endonuclease	Sequence Recognized Cleavage Sites Shown	Bacterial Source
BamHI	↓ G GATCC CCTAC C ↑	*Bacillus amyloliquefaciens H*
BglII	↓ A GATCT TCTAG A ↑	*Bacillus globigii*
EcoRI	↓ G AATTC CTTAA C ↑	*Escherichia coli RY13*
EcoRII	↓ CCTGG GGACC ↑	*Escherichia coli R245*
HindIII	↓ A AGCTT TTCGA A ↑	*Haemophilus influenzae R_d*
HhaI	↓ GCG C C GCG ↑	*Haemophilus haemolyticus*
HpaI	↓ GTT AAC CAA TTC ↑	*Haemophilus Parainfluenza*
MstII	↓ CC TnAGG GGAnT CC ↑	*Microcoleus strain*
PstI	↓ CTGCA G G ACGTC ↑	*Providencia stuartii 164*
TaqI	↓ T CGA AGC T ↑	*Thermus aquaticus YTI*

Abbreviations: A, adenine; C, cytosine; G, guanine; T, thymine. Arrows show the site of cleavage; depending on the site, the ends of the resulting cleaved double-stranded DNA are termed sticky ends (*BamHI*) or blunt ends (*HpaI*). The length of the recognition sequence can be 4 bp (*TaqI*), 5 bp (*EcoRII*), 6 bp (*EcoRI*), or 7 bp (*MstII*) or longer. By convention, these are written in the 5′ to 3′ direction for the upper strand of each recognition sequence, and the lower strand is shown with the opposite (ie, 3′-5′) polarity. Note that most recognition sequences are palindromes (ie, the sequence reads the same in opposite directions on the two strands). A residue designated n means that any nucleotide is permitted.

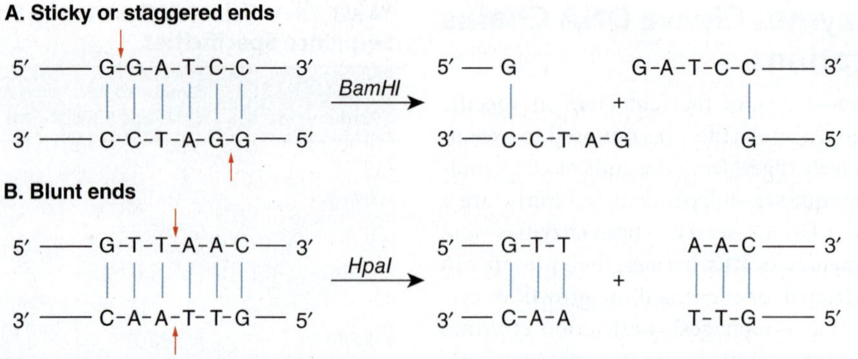

A. Sticky or staggered ends

```
5' —— G-G-A-T-C-C —— 3'              5' — G          G-A-T-C-C —— 3'
      | | | | | |          BamHI         |              |
3' —— C-C-T-A-G-G —— 5'     ──────►   3' — C-C-T-A-G  +     G —— 5'
```

B. Blunt ends

```
5' —— G-T-T-A-A-C —— 3'              5' — G-T-T        A-A-C —— 3'
      | | | | | |          HpaI         | | |          | | |
3' —— C-A-A-T-T-G —— 5'     ──────►   3' — C-A-A    +   T-T-G —— 5'
```

FIGURE 39–1   **Results of restriction endonuclease digestion.** Digestion with a restriction endonuclease (arrows) can result in the formation of DNA fragments with sticky, or cohesive, ends (**A**), or blunt ends (**B**); phosphodiester backbone, black lines; interstrand hydrogen bonds between purine and pyrimidine bases, blue. Generating fragments whose ends have particular structures (ie, blunt, cohesive) is an important consideration in devising cloning strategies.

DNA sequence. The fragments produced can be isolated by electrophoresis on agarose or polyacrylamide gels (see the discussion of blot transfer, later); this is an essential step in DNA cloning as well as various DNA analyses, and a major use of these enzymes.

A number of other enzymes that act on DNA and RNA are an important part of recombinant DNA technology. Many of these are referred to in this and other chapters (**Table 39–2**).

Restriction Enzymes, Endonucleases, Recombinases, Thermostable DNA Polymerases, DNA Synthesizers & DNA Ligase Are Used to Engineer & Prepare Chimeric DNA Molecules

Sticky, or complementary cohesive-end ligation of DNA fragments is technically easy, but some special techniques are often

TABLE 39–2   Some of the Enzymes Used in Recombinant DNA Research

Enzyme	Reaction	Primary Use
Phosphatases	Dephosphorylates 5′ ends of RNA and DNA	Removal of 5′-PO$_4$ groups prior to kinase labeling; also used to prevent self-ligation
DNA ligase	Catalyzes bonds between DNA molecules	Joining of DNA molecules
DNA polymerase I	Synthesizes double-stranded DNA from single-stranded DNA	Synthesis of double-stranded cDNA; DNA labeling and nick translation; generation of blunt ends from sticky ends
Thermostable DNA polymerases	Synthesize DNA at elevated temperatures (60°-80°C)	Polymerase chain reaction (DNA synthesis), mutagenesis
DNase I	Under appropriate conditions, produces single-stranded nicks in DNA	Nick translation; mapping of hypersensitive sites; mapping protein-DNA interactions
Exonuclease III	Removes nucleotides from 3′ ends of DNA	DNA sequencing; ChIP-Exo, mapping of DNA-protein interactions
λ Exonuclease	Removes nucleotides from 5′ ends of DNA	DNA sequencing, mapping of DNA-protein interactions
Polynucleotide kinase	Transfers terminal phosphate (γ position) from ATP to 5′-OH groups of DNA or RNA	^{32}P end-labeling of DNA or RNA
Reverse transcriptase	Synthesizes DNA from RNA template	Synthesis of cDNA from mRNA; RNA (5′ end) mapping studies
RNAse H	Degrades the RNA portion of a DNA–RNA hybrid	Synthesis of cDNA from mRNA
S1 nuclease	Degrades single-stranded DNA	Removal of "hairpin" in synthesis of cDNA; RNA mapping studies (both 5′ and 3′ ends)
Terminal transferase	Adds nucleotides to the 3′ ends of DNA	Homopolymer tailing
Recombinases (CRE, INT, FLP)	Catalyze site-specific recombination between DNA containing homologous target sequences	Generation of specific chimeric DNA molecules, work both in vitro and in vivo
CRISPER-Cas9/C2c2	RNA-targeted DNA-, or RNA-directed nuclease	Genome editing, and with variations, modulation of gene expression at DNA and RNA levels

required to overcome problems inherent in this approach. Sticky ends of a vector may reconnect with themselves, with no net gain of DNA. Sticky ends of fragments also anneal so that heterogeneous tandem inserts can form. Also, sticky-end sites may not be available or in a convenient position. To alleviate these problems, an enzyme that generates blunt ends can be used. Blunt ends can be ligated directly; however, ligation is not directional. To circumvent this problem new DNA ends of specific sequence can be added by direct blunt-end ligation using the bacteriophage T4 enzyme DNA ligase. Alternatively, convenient RE recognition sites can be added to a DNA fragment through the use of polymerase chain reaction (PCR) amplification via thermostable DNA polymerases or, alternatively through direct DNA synthesis on a DNA synthesizer (see Figure 39–7).

As an adjunct to the use of restriction endonucleases to combine and engineer DNA fragments, scientists have begun utilizing recombinases such as bacterial lox P sites, which are recognized by the CRE recombinase, bacteriophage λ *att* sites recognized by the λ phage encoded INT protein or yeast *FRT* sites recognized by the yeast Flp recombinase. These recombinase systems all catalyze specific incorporation of two DNA fragments that carry the appropriate recognition sequences and carry out homologous recombination (see Figure 35–9) between the relevant recognition sites. A novel DNA editing/ gene regulatory system termed **CRISPR-Cas9 (clustered regularly interspersed short palindromic repeats–associated gene 9)** first discovered in 2012, has revolutionized genomic DNA studies. The CRISPR system, found in many bacteria, represents a form of acquired, or adaptive immunity (see Chapters 52 and 54) to prevent reinfection of a bacterium by

specific bacteriophages. CRISPR complements the system of restriction endonucleases and methylases described earlier. CRISPR uses RNA-based targeting to bring the Cas9 nuclease to foreign (or any complementary) DNA. Within bacteria this CRISPR-RNA-Cas9 complex then degrades and inactivates the targeted DNA. The CRISPR system has been adapted for use in eukaryotic cells, including human cells, where it has been shown to be an RNA-directed site-specific nuclease just as it is in bacteria. Variations on the use of CRISPR allow for gene deletion, gene editing, gene visualization, and even modulation of gene transcription. Thus, CRISPR has added an exciting new, highly efficient, and very specific technology to the toolbox of methods for the manipulation of DNA and genetic analysis of mammalian cells. The basic aspects of CRISPR-Cas9 function are outlined in **Figure 39–2**.

The similarities of the CRISPR-Cas RNA-directed targeting and gene inactivation method and mi/siRNA-mediated repression of expression in higher eukaryotes are notable. Both methodologies are being actively pursued for experimental and therapeutic purposes. Interestingly, a variant of the CRISPR-Cas system, C2c2, has been shown to site-specifically cleave RNA. This discovery paves the way for potential specific alteration of mRNA/ncRNA levels in cells absent the ethical and technical challenges inherent in genome editing with the CRISPR-Cas9 system.

Cloning Amplifies DNA

A **clone** is a large population of identical molecules, cells, or organisms that arise from a common ancestor. Molecular cloning allows for the production of a large number of

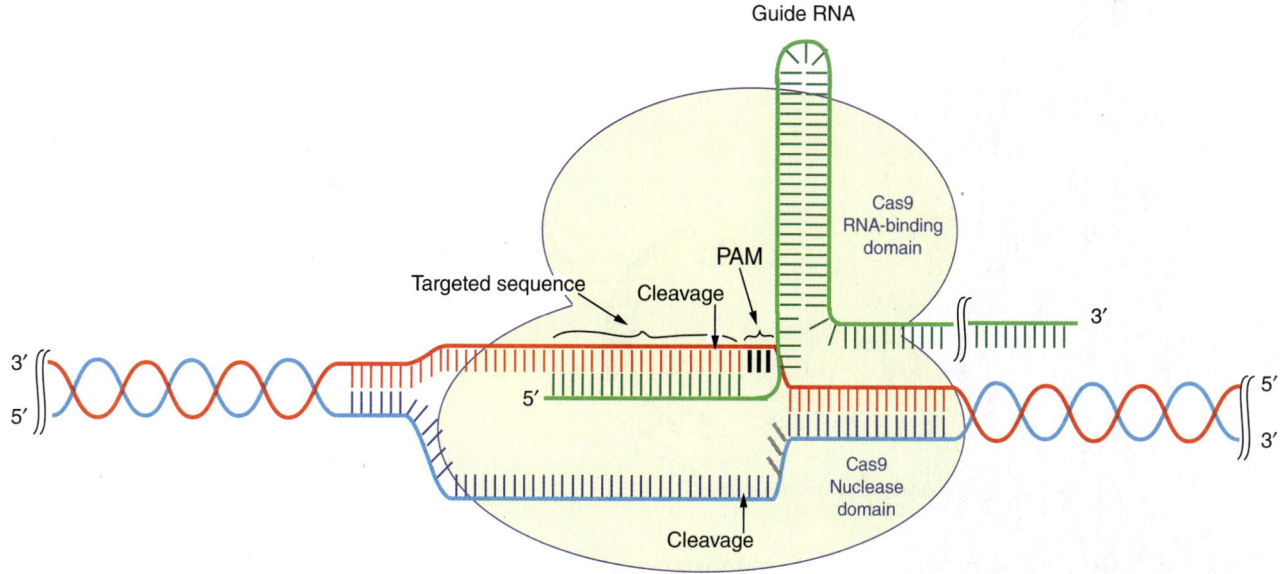

FIGURE 39–2 Overview of CRISPR-Cas9. Shown is the two-domain CRISPR-Cas9 nuclease protein bound to target genomic DNA (red, blue) and specific guide RNA (green), which through base complementarity (20 nts) locates its genomic target, which is adjacent to a short protospacer adjacent motif, or PAM. The guide RNA binding, and nuclease domains are labeled. Once specifically localized, the two distinct Cas9 nuclease active centers cleave both strands of the targeted genomic DNA (cleavage; arrows) immediately downstream of the PAM, which results in DNA double-strand break. Subsequent DNA repair by cellular activities (see Chapter 35) can introduce mutations thereby inactivating the targeted gene. Variations on the use of CRISPR-Cas9 are numerous and allow for the sculpting of the structure and expression of genomic DNA.

identical DNA molecules, which can then be characterized or used for other purposes. This technique is based on the fact that chimeric or hybrid DNA molecules can be constructed in **cloning vectors**—typically bacterial plasmids, phages, or **cosmids (hybrid plasmids that also contain specific phage sequences)**—which then continue to replicate clonally in a single host cell under their own control systems. In this way, the chimeric DNA is amplified. The general procedure is illustrated in **Figure 39–3**.

Bacterial **plasmids** are small, circular, duplex DNA molecules whose natural function is to confer antibiotic resistance to the host cell. Plasmids have several properties that make them extremely useful as cloning vectors. They exist as single or multiple copies within the bacterium and replicate independently from the bacterial DNA as **episomes** (ie, **a genome above or outside the bacterial genome**) while using primarily the host replication machinery. The complete DNA sequence of thousands of plasmids is known; hence, the precise location of restriction enzyme cleavage sites for inserting foreign DNA is available. Plasmids are smaller than the host chromosome and are therefore easily biochemically separated from the latter, and the desired plasmid-inserted DNA can be readily

removed by cutting the plasmid with the enzyme specific for the restriction site into which the original piece of DNA was inserted.

Phages (bacterial viruses) often have linear DNA genomes into which foreign DNA can be inserted at unique restriction enzyme sites. The resulting chimeric DNA is collected after the phage proceeds through its lytic cycle and produces mature, infective phage particles. A major advantage of phage vectors is that while plasmids accept DNA pieces up to about 10-kb long, phages can readily accept DNA fragments up to ~20-kb long. The ultimate insert size is imposed by the amount of DNA that can be packed into the phage head during virus propagation.

Larger fragments of DNA can be cloned in cosmids, DNA cloning vectors that combine the best features of plasmids and phages. Cosmids are plasmids that contain the DNA sequences, so-called **cohesive end sites (cos sites)**, required for packaging lambda DNA into the phage particle. These vectors grow in the plasmid form in bacteria, but since much of the unnecessary lambda DNA has been removed, more chimeric DNA can be packaged into the particle head. Cosmids can carry inserts of chimeric DNA that are 35- to 50-kb long. Even larger pieces of DNA can be incorporated into **bacterial**

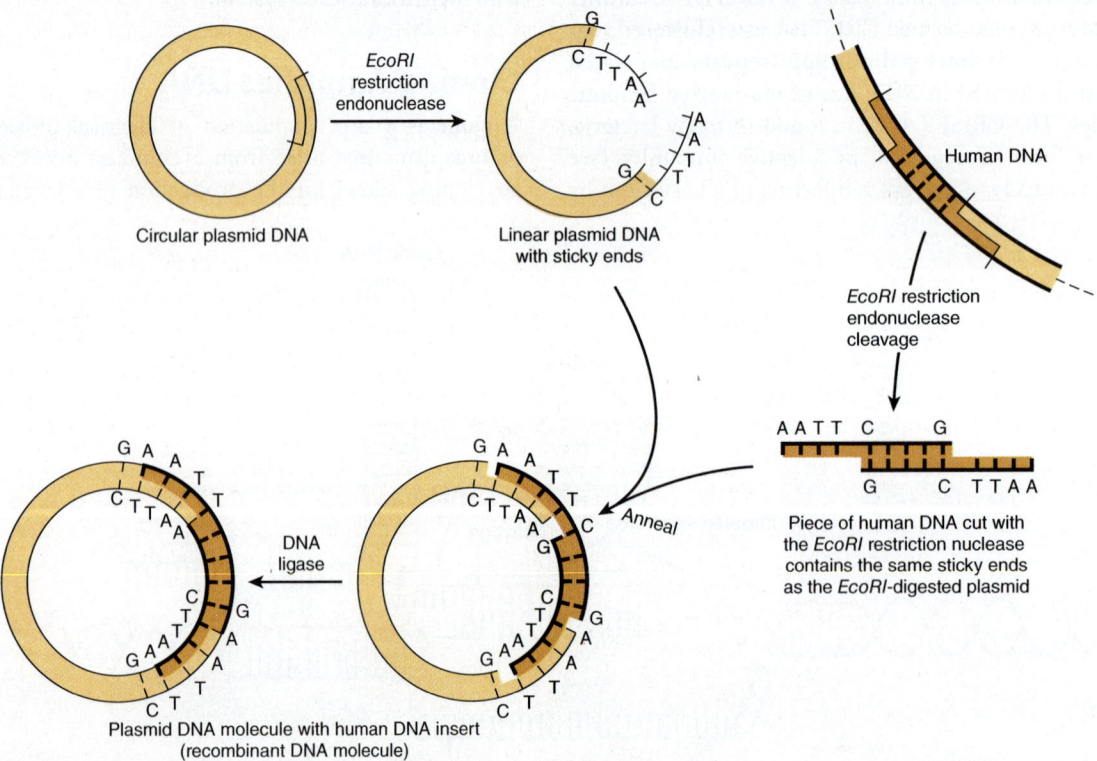

FIGURE 39–3 **Use of restriction endonucleases to make new recombinant or chimeric DNA molecules.** When inserted back into a bacterial cell (by the process called DNA-mediated transformation), typically only a single plasmid is taken up by a single cell, and the plasmid DNA replicates clonally, not only itself, but also the physically linked new DNA insert. Since recombining the sticky ends, as indicated, typically regenerates the same DNA sequence recognized by the original restriction enzyme, the cloned DNA insert can be cleanly cut back out of the recombinant plasmid circle with this endonuclease. Alternatively, the insert sequences can be specifically amplified from the purified chimeric plasmid DNA by PCR (see Figure 39–7). If a mixture of all of the DNA pieces created by treatment of total human DNA with a single restriction nuclease is used as the source of human DNA, a million or so different types of recombinant DNA molecules can be obtained, each pure in its own bacterial clone.

TABLE 39–3 **Cloning Capacities of Common Cloning Vectors**

Vector	DNA Insert Size (kb)
Plasmid pUC19	0.01-10
Lambda charon 4A	10-20
Cosmids	35-50
BAC, P1	50-250
YAC	500-3000

artificial chromosome (BAC), **yeast artificial chromosome (YAC)**, or *E. coli* bacteriophage **P1-derived artificial chromosome (PAC)** vectors. These vectors will accept and propagate DNA inserts of several hundred kilobases or more, and have largely replaced the plasmid, phage, and cosmid vectors for some cloning and eukaryotic gene mapping/expression applications. A comparison of these vectors is shown in **Table 39–3**.

Because insertion of DNA into a functional region of the vector will interfere with the action of this region, care must be taken not to interrupt an essential function of the vector. This concept can be exploited, however, to provide a powerful double positive/negative selection technique. For example, a common early plasmid vector **pBR322** has genes conferring resistance to both **tetracycline (Tet)** and **ampicillin (Amp)**, that is, **Tet**^r- and **Amp**^r-resistant growth, respectively. A single *PstI* restriction enzyme site within the Amp resistance gene

is commonly used as the insertion site for a piece of foreign DNA. In addition to having sticky ends (see Table 39–1 and Figure 39–1), the DNA inserted at this site disrupts the ORF of the β-lactamse-encoding *bla* gene. β-Lactamase, a secreted enzyme degrades and inactivates ampicillin. A bacterium carrying a plasmid with an inserted DNA fragment in the *bla* gene will be Amp-sensitive (Amp^s). Thus, cells carrying the parental plasmid, which provides resistance to both antibiotics, can be readily distinguished via replica plating, and separated from cells carrying the chimeric plasmid, which is resistant only to tetracycline (**Figure 39–4**). YACs contain selection, replication, and segregation functions that work in both bacteria and yeast cells and therefore can be propagated in either organism.

In addition to the vectors described in Table 39–3 that are designed primarily for propagation in bacterial cells, vectors for mammalian cell propagation and insert gene-(cDNA)/protein expression have also been developed. These vectors are all based on various eukaryotic viruses that are composed of RNA or DNA genomes. Notable examples of such **viral vectors** are those utilizing **adenoviral (Ad)**, or **adenovirus-associated viral (AAV)** (DNA-based) and **retroviral** (RNA based) genomes. Though somewhat limited in the size of DNA sequences that can be inserted, such **mammalian viral cloning vectors** make up for this shortcoming because they will efficiently infect a wide range of different cell types. For this reason, various mammalian viral vectors, some with both positive and negative selection genes (*aka* selectable "markers" as noted earlier for pBR322) are being used for both laboratory experiments and human **gene therapy**.

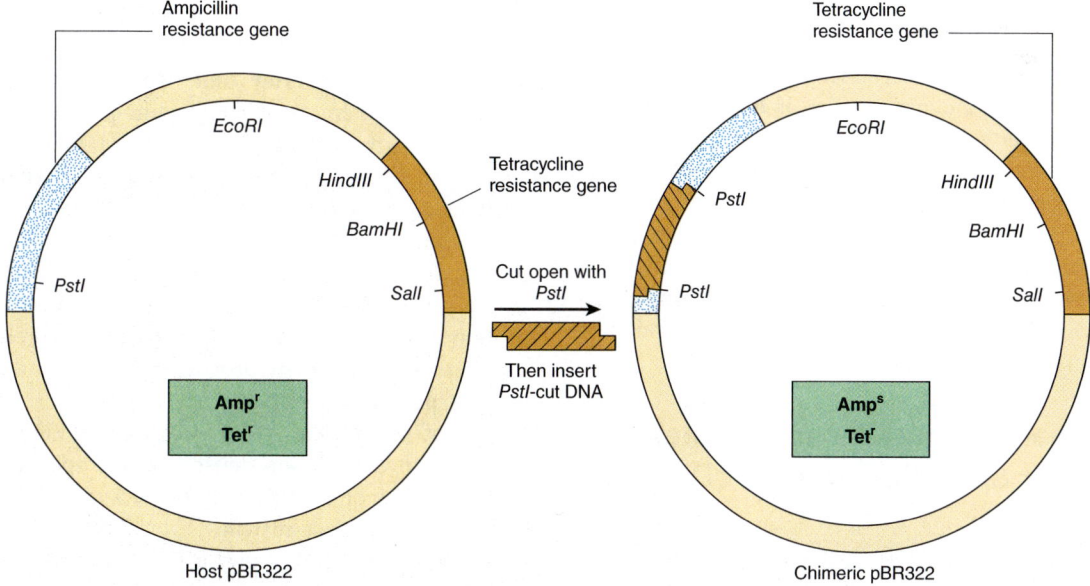

FIGURE 39–4 **A method of screening recombinants for inserted DNA fragments.** Using the plasmid pBR322, a piece of DNA is inserted into the unique *PstI* site. This insertion disrupts the codon reading frame for the gene coding for a protein that provides ampicillin resistance to the host bacterium. Hence, cells carrying the chimeric plasmid will no longer grow/survive when cultured in liquid, or plated on a substrate medium that contains this antibiotic. The differential sensitivity to tetracycline and ampicillin can therefore be used to distinguish clones of plasmid that contain an insert. A similar scheme relying on production of an in-frame fusion of a newly inserted DNA producing a peptide fragment capable of complementing an inactive, N-terminally truncated form of the enzyme β-galactosidase, a component of the *lac* operon (see Figure 38–2) allows for blue–white colony formation on agar plates containing a dye hydrolyzable by β-galactosidase. β-galactosidase–positive colonies are blue; such colonies contain plasmids in which a DNA was successfully inserted.

A Library Is a Collection of Recombinant Clones

The combination of restriction enzymes and various cloning vectors allows the entire genome of an organism to be individually packed, in small segments, into a vector. A collection of these different recombinant clones is called a library. A **genomic library** is prepared from the total DNA of a cell line or tissue that has been fragmented using either restriction endonucleases, or shearing and adaptor ligation to the resulting fragments. A **cDNA library** comprises complementary DNA copies of the population of mRNAs in a tissue. Genomic DNA libraries are often prepared by performing **partial digestion of total DNA** with a restriction enzyme that cuts DNA frequently (eg, a four-base cutter such as *Taq*I). The idea is to generate rather large fragments so that most genes will be left intact. The BAC, YAC, and P1 vectors are preferred since they can accept very large fragments of DNA and thus offer a better chance of isolating an intact eukaryotic mRNA-encoding gene on a single DNA fragment.

A vector in which the protein coded by the gene introduced by recombinant DNA technology is actually synthesized is known as an **expression vector**. Such vectors are now commonly used to detect specific cDNA molecules in libraries and to produce proteins by genetic engineering techniques. These vectors are specially constructed to contain very active inducible promoters, proper in-phase translation initiation codons, transcription and translation termination signals, and appropriate protein processing signals, if needed. Some expression vectors even contain genes that code for protease inhibitors, so that the final yield of product is enhanced. Interestingly, as the cost of synthetic DNA synthesis has dropped, many investigators now often simply synthesize an entire cDNA (gene) of interest (in 100–150 nt segments) incorporating the codon preferences of the host used for expression in order to maximize protein production. New efficiencies in synthetic DNA synthesis now allow for the de novo synthesis of complete genes and even (small) genomes, obviating, in part, the need to generate and carefully probe complex genomic libraries to identify/obtain specific gene sequences. These advances usher in new and exciting possibilities in synthetic biology while concomitantly introducing potential ethical conundrums.

Probes Search Libraries or Complex Samples for Specific Genes or cDNA Molecules

A variety of molecules can be used to "probe" libraries in search of a specific gene or cDNA molecule or to define and quantitate DNA or RNA separated by electrophoresis through various gels. Probes are generally molecules of DNA or RNA labeled with a ^{32}P-containing nucleotide—or fluorescently labeled nucleotide(s) (more commonly now). Importantly, neither modification (^{32}P- or fluorescent-label) affects the hybridization properties of the resulting labeled nucleic acid

probes. The probe must recognize a complementary sequence to be effective. A cDNA synthesized from a specific mRNA (or a synthetic oligonucleotide) can be used to screen either a cDNA library for a longer cDNA or a genomic library for a complementary sequence in the coding region of a gene. cDNA/oligonucleotide/cRNA probes are used to detect DNA fragments on Southern blot transfers and to detect and quantitate RNA on Northern blot transfers (see later). PCR methods can be used for both tasks (see later).

Blotting & Probing Techniques Allow Visualization of Specific Target Molecules

Visualization of a specific DNA or RNA fragment (or protein, see later) among the many thousands of nontarget molecules in a complex sample requires the convergence of a number of techniques, collectively termed **blot transfer**. **Figure 39–5** illustrates the **Southern** (DNA), **Northern** (RNA), and **Western** (protein) blot transfer procedures. The first technique is named for the person who devised the technique, Edward Southern; the other names began as laboratory jargon, but are now accepted terms. These procedures are useful in determining how many copies of a gene are in a given tissue or whether there are any alterations in a gene (deletions, insertions, or rearrangements) because the requisite, initial gel electrophoresis step separates the molecules on the basis of size. Occasionally, if a specific base is changed and a restriction site is altered, these procedures can detect a point mutation. The Northern and Western blot transfer techniques are used to size and quantify specific RNA and protein molecules, respectively. A fourth technique, the **Southwestern** or **overlay blot**, which examines protein–nucleic acid interactions or protein–protein interactions, respectively are variants of the Southern/Northern/Western blotting methods (not shown). In these last two techniques, proteins are separated by electrophoresis, blotted to a membrane, renatured, and analyzed for an interaction with a particular DNA, or RNA sequence, or protein by incubation with a specific labeled nucleic acid probe (Southwestern) or protein probe (overlay assay) using either a labeled protein, or alternatively protein–protein interactions are detected using a specific antibody.

All of the nucleic acid–based hybridization procedures discussed in this section depend on the specific base-pairing properties of complementary nucleic acid strands (see Chapter 34). Perfect matches hybridize readily and withstand high temperatures and/or low ionic strength buffer in the hybridization and washing reactions. Less than perfect matches do not tolerate such **stringent conditions** (ie, elevated temperatures and low-salt concentrations); thus, hybridization either never occurs or is disrupted during the washing step. Hybridization conditions capable of detecting just a single base-pair (bp) mismatch between probe and target have been devised.

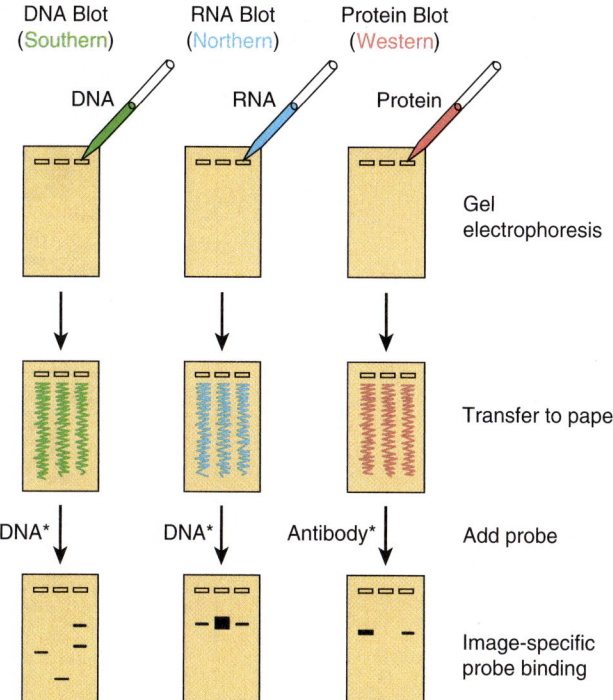

DNA Blot (Southern)   RNA Blot (Northern)   Protein Blot (Western)

DNA   RNA   Protein

Gel electrophoresis

Transfer to paper

DNA*   DNA*   Antibody*   Add probe

Image-specific probe binding

FIGURE 39–5 The blot transfer procedure. In a Southern, or DNA blot transfer, DNA isolated from a cell line or tissue is digested with one or more restriction enzymes. This mixture is pipetted into a well in an agarose or polyacrylamide gel and exposed to a direct electrical current. DNA, being negatively charged, migrates toward the anode; the smaller fragments move the most rapidly. After a suitable time, the DNA within the gel is denatured by exposure to mild alkali and transferred, via capillary action (or electrotransfer—not shown), to nitrocellulose or nylon paper, resulting in an exact replica of the pattern on the gel, using the blotting technique devised by Southern blot. The DNA is bound to the paper by exposure to heat or UV, and the paper is then exposed to the labeled DNA probe, which hybridizes to complementary strands on the filter. After thorough washing, the paper is exposed to x-ray film or an imaging screen, which is developed to reveal several specific bands corresponding to the DNA fragment(s) that were recognized (hybridized to) by the sequences in the DNA probe. The RNA, or Northern, blot is conceptually similar. RNA is subjected to electrophoresis before blot transfer. This requires some different steps from those of DNA transfer, primarily to ensure that the RNA remains intact, and is generally somewhat more difficult. In the protein, or Western, blot, proteins are electrophoresed and transferred to special paper that avidly binds proteins and then probed with a specific antibody or other probe molecule. (Asterisks signify labeled probes, either radioactive or fluorescent.) In the case of Southwestern blotting (see the text; not shown), a protein blot similar to that shown above under "Western" is exposed to labeled nucleic acid, and protein–nucleic acid complexes formed are detected by autoradiography or imaging.

Manual & Automated Techniques Are Available to Determine the Sequence of DNA

The segments of specific DNA molecules obtained by recombinant DNA technology can be analyzed to determine their nucleotide sequence. DNA sequencing depends on having a population of identical DNA molecules. This requirement can be satisfied by cloning the fragment of interest, either using the techniques described earlier, or by using PCR methods (see later). The **manual enzymatic Sanger method** employs specific dideoxynucleotides that terminate DNA strand synthesis catalyzed by DNA polymerase, at specific nucleotides as the strand is synthesized on purified single-stranded template DNA. The reactions are adjusted so that a population of DNA fragments representing termination at every nucleotide is obtained. By having a radioactive label incorporated at the termination site, one can separate the fragments according to size using high-resolution polyacrylamide gel electrophoresis. The resulting gel is vacuum dried onto filter paper and used to expose an x-ray film or more commonly a special imaging plate, which allows the visualization of the labeled DNA fragments generated during sequencing. These images are read in order to give the DNA sequence as shown in **Figure 39–6**, which is an example of first-generation, Sanger DNA sequencing. So-called next-generation sequencing (NGS), or second-generation techniques are automated, but still utilized DNA polymerase and utilize four different fluorescent labels, one representing each nucleotide. Each incorporated labeled deoxynucleotide emits a specific signal on excitation by a laser beam of a particular wavelength that is measured by sensitive detectors, and these signals can be recorded by a computer. The newest DNA sequencing machines use fluorescently labeled nucleotides but detect incorporation using microscopic optics. Such sequencing machines have reduced the cost of DNA sequencing by orders of magnitude, and the ensuing cost reductions, and massive parallelization of second-generation sequencing have ushered in the era of personalized genome sequencing. Third-generation DNA sequencing, though still in full development, enables real-time single-molecule sequencing of very long DNA and RNA molecules. There are currently two distinct third-generation technologies. One approach still uses DNA polymerase (PacBio) while the other utilizes changes in ion flow that occur when single-stranded nucleic acid molecules pass through nanometer scale nanopores (Oxford Nanopore Technologies [ONT]). These two new technologies can read the sequence of nucleic acids that range in sizes up to 100 to 900 kilobases. Such third-generation sequencers promise to yet again revolutionize the field of nucleic acid sequencing.

Oligonucleotide Synthesis Is Now Routine

Automated chemical synthesis of moderately long oligonucleotides (100+ nucleotides) of exact sequence is now a routine laboratory procedure. Each synthetic cycle takes but minutes such that very large dsDNA molecules can be made by synthesizing relatively short segments that are annealed and then ligated to one another. As mentioned earlier for DNA sequencing, the process has been miniaturized and can be significantly parallelized to allow the synthesis of hundreds to thousands of defined sequence oligonucleotides simultaneously.

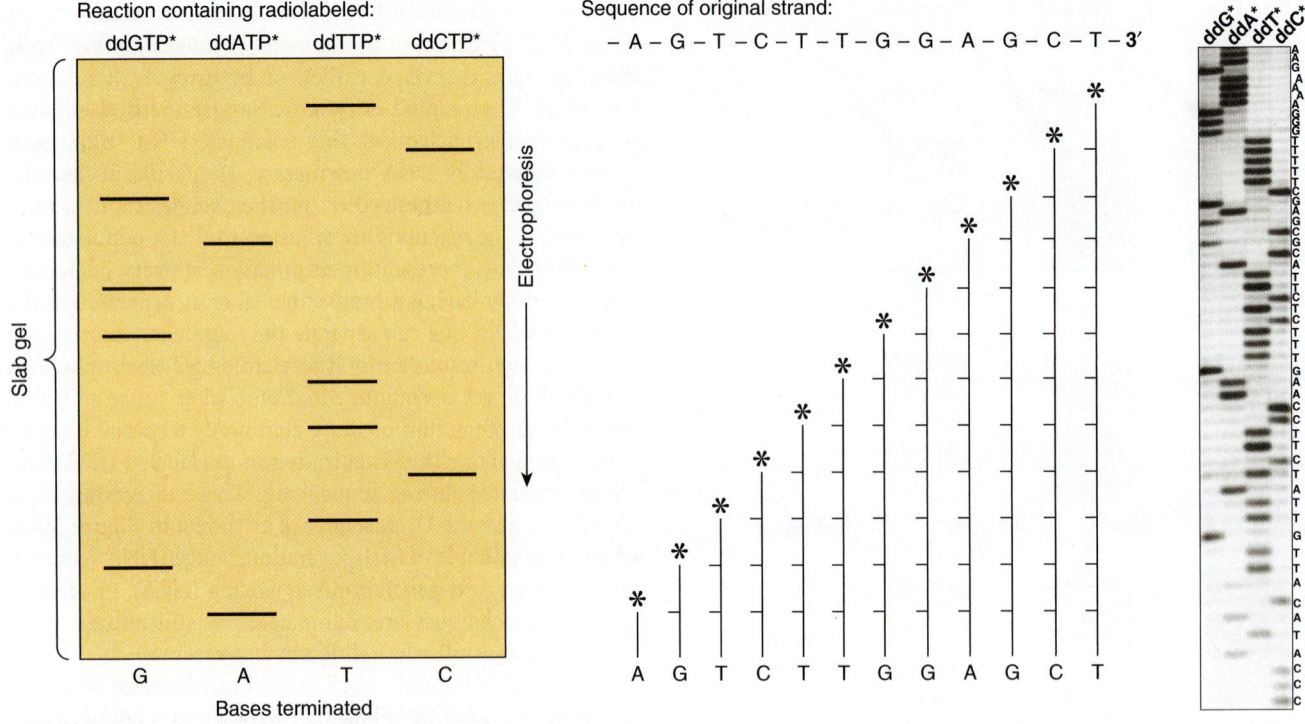

FIGURE 39–6 **Sequencing of DNA by the chain termination method devised by Sanger.** The ladder-like arrays represent, from bottom to top, all of the successively longer fragments of the original DNA strand. Knowing which specific dideoxynucleotide reaction was conducted to produce each mixture of fragments, one can determine the sequence of nucleotides from the unlabeled end toward the labeled end (*) by reading up the gel. The base-pairing rules of Watson and Crick (A–T, G–C) dictate the sequence of the other (complementary) strand. (Asterisks signify site of radiolabeling.) Schematically shown **(left, middle)** are the terminated synthesis products of a hypothetical fragment of DNA, sequence listed **(middle, top)**. An autoradiogram **(right)** of an actual set of DNA sequencing reactions that utilized the four ^{32}P-labeled dideoxynucleotides indicated at the top of the scanned autoradiogram (dideoxy(dd)G, ddA, ddT, ddC). Electrophoresis was from top to bottom. The deduced DNA sequence is listed on the right side of the gel. Note the log-linear relationship between distance of migration (ie, top to bottom of gel) and DNA fragment length. Current state-of-the-art DNA sequencers no longer utilize gel electrophoresis for fractionation of labeled synthesis products. Moreover, in the majority of NGS sequencing platforms, synthesis is followed by monitoring incorporation of the four fluorescently labeled dXTPs.

Oligonucleotides are now indispensable for DNA sequencing, library screening, protein-DNA binding assays, the PCR (see later), site-directed mutagenesis, complete synthetic gene synthesis, as well as complete (bacterial) genome synthesis, and numerous other applications.

The Polymerase Chain Reaction (PCR) Method Amplifies DNA Sequences

PCR is a method of amplifying a target sequence of DNA. The development of PCR has revolutionized the ways in which both DNA and RNA can be studied. PCR provides a sensitive, selective, and extremely rapid means of amplifying any desired sequence of DNA. Specificity is based on the use of two oligonucleotide primers that hybridize to complementary sequences on opposite strands of DNA and flank the target sequence (**Figure 39–7**). The DNA sample is first heat denatured (>90°C) to separate the two strands of the template DNA containing the target sequence; the primers, added in excess, are allowed to anneal to the DNA (typically at 50-75°C) in order to generate the required template-primer complex. Subsequently, each strand is copied by a special DNA polymerase, starting at the primer sites in the presence of all four dXTPs. The two DNA strands each serve as a template for the synthesis of new DNA from the two primers. Repeated cycles of heat denaturation, annealing of the primers to their complementary sequences, and extension of the annealed primers with DNA polymerase result in the exponential amplification of DNA segments of defined length. Product DNA doubles with each PCR cycle. DNA synthesis is catalyzed by a heat-stable DNA polymerase purified from one of a number of different thermophilic bacteria, organisms that grow at 70° to 80°C. Thermostable DNA polymerases can withstand multiple, short incubations at more than 90°, temperatures required to completely denature DNA. These thermostable DNA polymerases have made automation of PCR possible.

DNA sequences as short as 50 to 100 bp and as long as 10 kb can readily be amplified by PCR. Twenty cycles provide an amplification of 10^6 (ie, 2^{20}) and 30 cycles, 10^9 (2^{30}). Each cycle takes less than or equal to 5 to 10 minutes so that even large DNA molecules can be amplified fairly rapidly. Because of subtle differences in DNA sequence with each new PCR target, the exact conditions for amplification must be empirically optimized. The workhorse PCR technique is central to

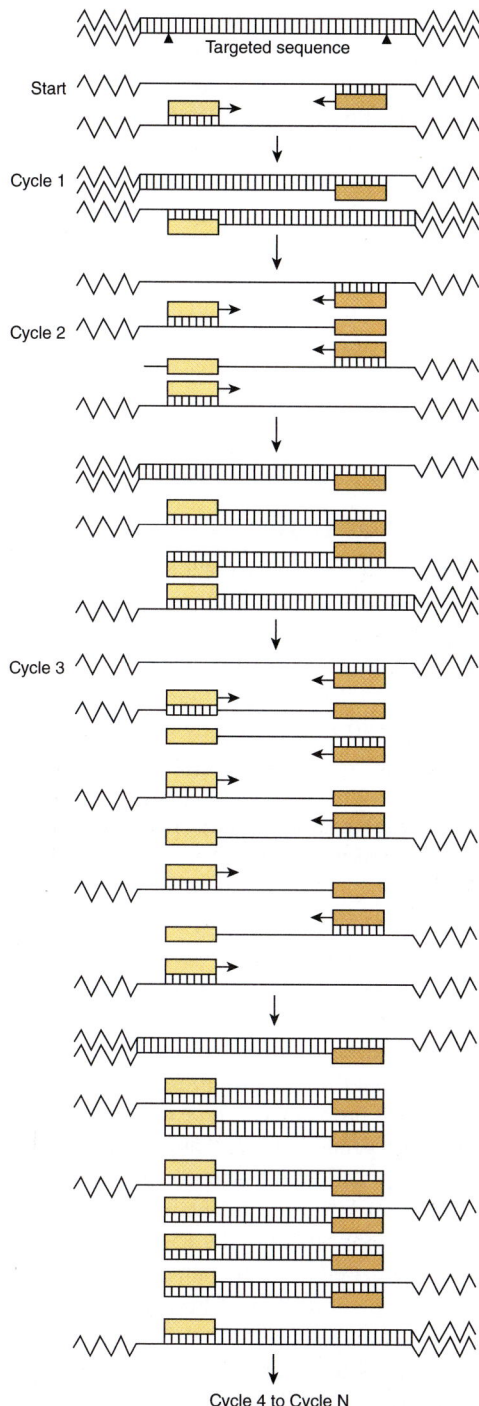

FIGURE 39–7 **The polymerase chain reaction technique is used to amplify specific gene sequences.** Double-stranded DNA is heated to separate it into individual strands. These bind two distinct primers that are directed at specific sequences on opposite strands and that define the segment to be amplified. DNA polymerase extends the primers in each direction and synthesizes two strands complementary to the original two. This cycle is repeated 30 or more times, giving an amplified product of defined length and sequence. Note that the four dXTPs and the two primers are present in excess to minimize the possibility that these components are limiting for polymerization/amplification. It is important to note though that as cycle number increases incorporation rates can drop, and mutation/error rates can increase.

many DNA/RNA sequencing technologies. The PCR method allows the DNA in a single cell, hair follicle, or spermatozoon to be amplified and analyzed. Thus, the applications of PCR to forensic medicine are obvious. PCR is also used to: (1) detect and quantify infectious agents, especially latent viruses; (2) make accurate genetic diagnoses by producing DNA fragments for subsequent DNA sequence analyses to identify mutations; (3) detect allelic polymorphisms ranging from single base pair changes to large and small indels and gene amplification; (4) establish precise tissue types for transplants; and (5) study evolution, using DNA from archeological samples (6) for quantitative RNA analyses after RNA copying and mRNA quantitation by the so-called RT-PCR method (cDNA copies of mRNA generated by a retroviral reverse transcriptase) or (7) to score in vivo protein-DNA occupancy using chromatin immunoprecipitation assays (see later). New uses of PCR are developed every year.

PRACTICAL APPLICATIONS OF RECOMBINANT DNA TECHNOLOGY ARE NUMEROUS

The isolation of a specific (ca 1000 bp) mRNA-encoding gene from an entire human genome requires a technique that will discriminate one part in 1,000,000. The identification of a regulatory region that may be only 10 bp in length requires a sensitivity of one part in 300,000,000; a disease such as sickle cell anemia is caused by a single base change, or one part in 3,000,000,000. DNA technology is powerful enough to accomplish all these things.

Proteins Can Be Produced for Research, Diagnosis, & Commerce

A practical goal of recombinant DNA research is the production of materials for biomedical applications. This technology has two distinct merits: (1) it can supply large amounts of material that could not be obtained by conventional purification methods (eg, interferon, tissue plasminogen activating factor, etc.); and (2) it can provide human proteins (eg, insulin and growth hormone). The advantages in both cases are obvious. Although the primary aim is to supply products—generally proteins—for treatment (insulin) and diagnosis (AIDS testing) of human and other animal diseases and for disease prevention (hepatitis B; COVID-19 vaccines), there are other potential commercial applications, especially in agriculture. An example of the latter is the attempt to engineer plants that are more resistant to drought or temperature extremes, more efficient at fixing nitrogen, or that produce seeds containing the complete complement of essential amino acids (rice, wheat, corn, etc.).

Direct Sequencing of Genomic DNA & Exomes

As noted earlier, recent advances in DNA-sequencing technology, the so-called NGS, or high-throughput sequencing (HTS) platforms, have dramatically reduced the per base cost of DNA sequencing. The initial sequence of the human genome costs roughly $350,000,000 (US). The cost of sequencing of the same 3×10^9 bp diploid human genome using the new NGS platforms is estimated to be less than 0.03% of the original. Thus, human genome sequencing for <u>less than or equal to</u> $1000 (US), and exomes for <u>less than or equal to</u> $500 (US) is readily available. This dramatic reduction in cost has stimulated various international initiatives to sequence the entire genomes/exomes of hundreds of thousands of individuals of various racial and ethnic backgrounds in order to determine the true extent of DNA/genome polymorphisms present within the human population. The resulting cornucopia of genetic information, and the ever-decreasing cost of genomic DNA sequencing is dramatically increasing our ability to diagnose and, ultimately treat human disease. Obviously, when personal genome sequencing does become commonplace, dramatic changes in the practice of medicine will result because therapies will ultimately be custom tailored to the exact genetic makeup of each individual.

Gene Therapy & Stem Cell Biology

Diseases caused by deficiency of a single-gene product are all theoretically amenable to replacement therapy. The strategy is to clone a normal copy of the relevant gene (eg, the gene that codes for adenosine deaminase) into a vector that will readily be taken up and incorporated into the genome of a host cell. Bone marrow precursor cells are being investigated for this purpose because they presumably will resettle in the marrow and replicate there. The introduced gene would begin to direct the expression of its protein product, and this would correct the deficiency in the host cell.

As an alternative to "replacing" defective genes to cure human disease, many scientists are investigating the feasibility of identifying and characterizing pluripotent stem cells that have the ability to differentiate into any cell type in the body. Recent results in this field have shown that adult human somatic cells can readily be converted into apparent **induced pluripotent stem cells (iPSCs)** by transfection with cDNAs encoding a handful of DNA-binding transcription factors. These and other new developments in the fields of gene therapy and stem cell biology promise exciting new potential therapies for curing human disease. Finally, generating iPSCs from diseased patient cells also offer the opportunity to create authentic human cell–based models for laboratory studies of the molecular basis of human disease.

Transgenic Animals

The somatic cell gene replacement therapy described earlier would obviously not be passed on to offspring. Other strategies to alter germ cell lines have been devised but have been tested only in experimental animals. A certain percentage of genes injected into a fertilized mouse ovum will be incorporated into the genome and found in both somatic and germ cells. Hundreds of transgenic animals have been established, and these are useful for analysis of tissue-specific effects on gene expression and effects of overproduction of gene products (eg, those from the growth hormone gene or oncogenes)

and in discovering genes involved in development, a process that heretofore has been difficult to study in mammals. The transgenic approach has been used to correct a genetic deficiency in mice. Fertilized ova obtained from mice with genetic hypogonadism were injected with DNA containing the coding sequence for the gonadotropin-releasing hormone (GnRH) precursor protein. This gene was expressed and regulated normally in the hypothalamus of a certain number of the resultant mice, and these animals were in all respects normal. Their offspring also showed no evidence of GnRH deficiency. This is, therefore, evidence of somatic cell expression of the transgene and of its maintenance in germ cells.

Targeted Gene Regulation by Disruption or "Knockout", "Knockin", Editing, & Controlled Expression

Various technical advances have allowed for the precise, targeted modification of mammalian genes. The exact methods used for genetically engineering mammalian genomes have evolved from tedious, low-efficiency methods based on positive and negative drug selections and homologous recombination (Knockout/Knockin) to the recently described CRISPR-Cas9 system described earlier. The goal of all of these methods is ultimately to generate a family of genetic variants of a gene of interest that are (a) a null, or complete loss-of-function allele; (b) recessive, loss-of-function alleles; and (c) ideally, dominant gain-of-function alleles. These genetic alterations are generated in pluripotent stem cells, which ultimately allow for introduction and propagation in whole model organisms (fruit fly, fish, worms, rodents, etc.). Having all three of these genetic variants allows one to precisely dissect the mechanism of action of any gene. However, complicating genetic analyses of many genes is the fact that their function(s) are essential for viability. To get around this problem, cell-type or tissue-specific genetic variants must be generated. This hurdle has been overcome through the use of cell- and tissue-specific enhancers that can drive the conditional (ie, experimentally controlled) expression of the targeting recombinases (ie, CRE-lox) and/or nuclease (CRISPR-Cas9) that generate the altered genes, either null or loss- or gain-of-function alleles. Alternatively, selected loss-of-function alleles can be generated through equivalent siRNA expression to knock down production of a specific gene product. Collectively, these methods allow for sophisticated genetic and biochemical tests of gene function and allow scientists to probe the structure and function of mammalian genes in physiologic contexts. Tremendous molecular mechanistic insights have been, and will continue to be, obtained into the molecular etiology of human disease through these and other biochemical approaches. Exciting times lie ahead!

RNA & Protein Profiling, & Protein-DNA Interaction Mapping

The "-omic" revolution of the last decade has culminated in the determination of the complete nucleotide sequence of many thousands of genomes, including those of budding and fission yeasts, numerous bacteria, the fruit fly, the worm *Caenorhabditis elegans*, plants, the mouse, rat, chicken, monkey, and, most notably, humans. Additional genomes are being sequenced at an accelerating pace. The availability of all of this DNA sequence information, coupled with engineering advances, has led to the development of several revolutionary methodologies, most of which are based on second- and third-generation sequencing platforms. In the case of automated DNA sequencing, mRNAs are converted to cDNAs using retroviral RNA-dependent DNA polymerases to reverse transcribe mRNAs to DNAs. The resulting cDNAs are amplified and directly sequenced; this method is termed **RNA-Seq**. These methods allow for the quantitative description of the entire transcriptome. Recent reports in the literature have used RNA-Seq to describe the transcriptome of single cells, and when coupled with high-sensitivity ribosome profiling (see later) and mass spectrometry–based proteomics (see later), confidently define gene expression profiles at the mRNA and protein levels.

Recent methodologic advances (**PRO-Seq, Precision Run-On sequencing**, and **NET-Seq, native elongating transcript sequencing**) allow for sequencing of RNA within elongating RNA polymerase-DNA-RNA ternary complexes, thereby allowing nucleotide-level descriptions, genome-wide, of active transcription in living cells. A parallel method termed **ribosome profiling** allows investigators to use high-throughput DNA sequencing to determine both the identity and number of cellular mRNAs in the process of being actively translated—thereby defining the cellular proteome. Transcriptome information allows one to quantitatively *predict* the collection of proteins that might be expressed in a particular cell, tissue, or organ in normal and disease states based on the mRNAs present in those cells, while ribosome profiling allows for the quantitative measurement of the *actual* cellular proteome, particularly when coupled with newer high sensitivity mass spectrometry to analyze protein content and PTM status of cellular proteomes (see later).

Complementing the very high-throughput, genome-wide expression profiling methods described earlier is the development of methods to map the location, or occupancy of specific proteins bound to discrete DNA sequences within living cells. This method, illustrated in **Figure 39–8**, is termed **chromatin immunoprecipitation** (**ChIP**). Proteins are cross-linked in situ in cells or tissues, the cellular chromatin is isolated, sheared, and specific cross-linked protein-DNA complexes purified using antibodies that recognize a particular protein, or protein isoform. DNA bound to this protein is recovered and amplified using PCR and analyzed using gel electrophoresis imaging, microarray hybridization (**ChIP-chip**), or direct sequencing. There are two versions of the DNA sequencing assay readout. In the first, the immunopurified DNA is directly subjected to NGS/high-throughput DNA sequencing (**ChIP-Seq**); in the second version, the immunopurified, cross-linked protein-DNA complex is treated with exonucleases to remove cross-linked DNA sequences that are not in intimate contact with the protein of interest; this is termed **ChIP-Exo**. Collectively ChIP-chip and ChIP-Seq methods allow investigators to identify the locations of a single protein genome-wide throughout all the chromosomes. ChIP-Exo has the added advantage of allowing investigators to map in vivo protein occupancy at single nucleotide-level resolution. Finally, methods for high-sensitivity, high-throughput mass spectrometry of metabolites (**metabolomics**), various small molecules (lipids, **lipidomics**; carbohydrates, **glycomics, etc.**), and complex protein samples (**proteomics**) have been developed. Newer mass spectrometry methods allow scientists to identify thousands of proteins in complex samples extracted from very small numbers of cells (<1 g). Such analyses can now be used to quantify the relative amounts of proteins in two samples, as well as the level of certain PTMs, such as phosphorylation, acetylation etc.; and with the use of specific antibodies, define specific protein–protein interactions. This critical information tells investigators which of the many mRNAs detected in transcriptome mapping studies are being translated into protein, generally the ultimate dictator of cellular/tissue/organismal phenotypes.

New genetic means for identifying protein–protein interactions and protein function have also been devised. Systematic genome-wide gene expression knockdown using siRNAs, synthetic lethal genetic interaction screens, or most recently CRISPR-Cas9 knockdown have been used to assess the contribution of individual genes to a variety of processes in model systems (yeast, worms, and flies) and mammalian cells (human and mouse). Specific network mappings of protein–protein interactions, on a genome-wide basis, have been identified using high-throughput variants of the **two-hybrid interaction** test (**Figure 39–9**). This simple yet powerful method can be performed in bacteria, yeast, or metazoan cells, and allows for the detection of specific protein–protein interactions in living cells. Reconstruction experiments indicate that protein–protein interactions with binding affinities of K_d ~10^{-6} mol/L or tighter can readily be detected with this method. Together, these technologies provide powerful tools with which to dissect the intricacies of human biology.

SYSTEMS BIOLOGY AIMS TO INTEGRATE THE FLOOD OF –OMIC DATA IN ORDER TO DECIPHER FUNDAMENTAL BIOLOGIC REGULATORY PRINCIPLES

Microarray techniques, high-throughput genomic DNA sequencing, RNA-Seq, ribosome profiling, mass spectrometry, ChIP-Seq/ChIP-Exo, genome-wide two-hybrid screens, genetic knockdown, and synthetic lethal screens coupled with mass spectrometric protein and metabolite identification experiments have led to the generation of enormous amounts of data. Appropriate data management, analysis, and interpretation of the deluge of information forthcoming from such studies have relied on the application of statistical methods

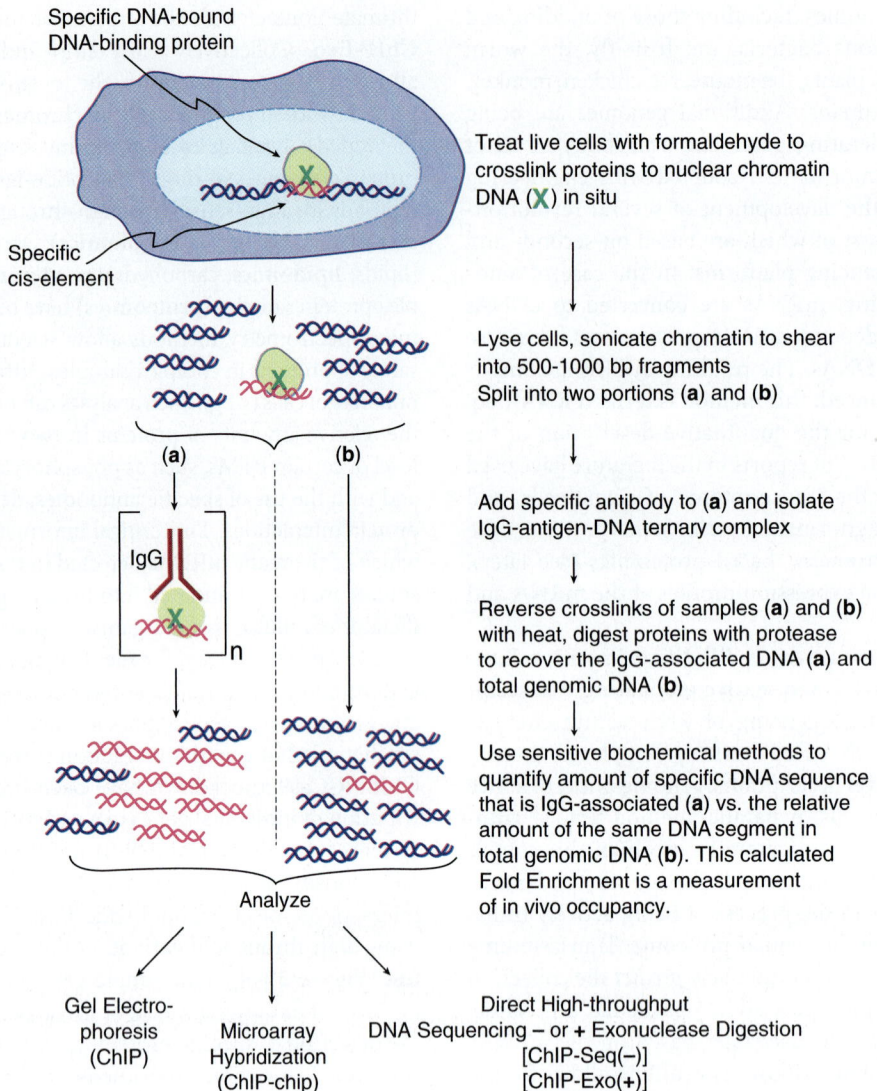

FIGURE 39-8 **Outline of the chromatin immunoprecipitation (ChIP) technique.** This method allows for the precise localization of a particular protein (or modified protein if an appropriate antibody is available; for example, **n** = anti-phosphorylated or acetylated histones, transcription factors, etc.) on a particular sequence element in living cells. Depending on the method used to analyze the immunopurified DNA, quantitative or semiquantitative information, at near nucleotide level resolution, can be obtained. Protein-DNA occupancy can be scored genome-wide in two ways. First, by ChIP-chip, a method that uses a hybridization readout. In ChIP-chip total genomic DNA is labeled with one particular fluorophore and the immunopurified DNA is labeled with a spectrally distinct fluorophore. These differentially labeled DNAs are mixed and hybridized to microarray "chips" (microscope slides) that contain specific DNA fragments, or more commonly now, synthetic oligonucleotide 50 to 70 nucleotides long. These gene-specific oligonucleotides are deposited and covalently attached at predetermined, known X,Y position/ coordinates on the slide. The labeled DNAs are hybridized, the slides washed and hybridization to each gene-specific oligonucleotide probe is scored using differential laser scanning and sensitive photodetection at micron resolution. The hybridization signal intensities are quantified, and the ratio of IP DNA/genomic DNA signals is used to score occupancy levels. The second method, termed ChIP-Seq, directly sequences immuno-purified DNAs using NGS sequencing methods. Two variants of ChIP-Seq are shown: "standard" ChIP-Seq and ChIP-Exo. These two approaches differ in their ability to resolve and map the locations of the bound protein on genomic DNA. Standard ChIP-Seq resolution is 50 to 100 nt reso-lution, while ChIP-Exo has near single nt level resolution. Both approaches rely on efficient bioinformatic algorithms to deal with the very large datasets that are generated. ChIP-chip and ChIP-Seq techniques provide a (semi-) quantitative measure of in vivo protein occupancy. Though not schematized here, similar methods termed RIP (RNA immunoprecipitation) or CLIP (cross-linking protein-RNA and immunoprecipitation), which differ primarily in the method of protein-RNA cross-linking, can score the in vivo binding of specific proteins to specific RNA species (typically mRNAs, though any RNA species can be analyzed by these techniques).

and novel algorithms for analyzing or "mining" and visual-izing such huge datasets. This has led to the development of the field of **bioinformatics** (see also Chapter 11). These new technologies, coupled with the flood of experimental data, have further led to the development of the field of **systems**

biology, a discipline whose goal is to quantitatively analyze, and integrate this flood of biologically important information in order to gain novel insights into biology and its pathologic manifestations. Future work at the intersection of bioinfor-matics, engineering, biophysics, genetics, transcript, protein,

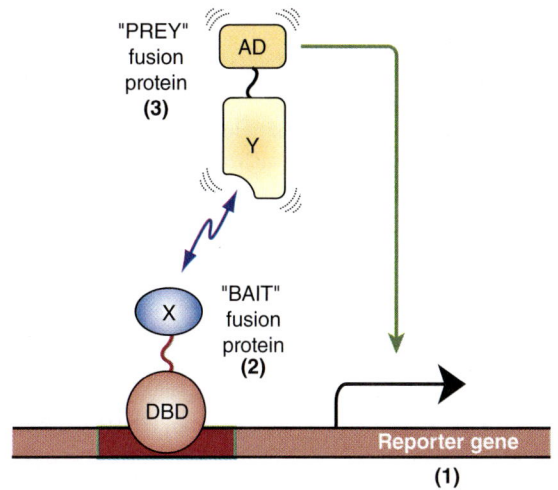

FIGURE 39–9 **Overview of two hybrid system for identifying and characterizing protein–protein interactions.** Shown are the basic components and operation of the two hybrid systems, originally devised by Fields and Song (Nature 340:245-246 [1989]) to function in the baker's yeast system. **(1)** A reporter gene, either a selectable marker (ie, a gene conferring prototrophic growth on selective media, or producing an enzyme for which a colony colorimetric assay exists, such as β-galactosidase) that is expressed only when a transcription factor binds upstream to a *cis*-linked enhancer (dark red bar). **(2)** A "bait" fusion protein **(DBD-X)** produced from a chimeric gene expressing a modular DNA-binding domain **(DBD**; often derived from the yeast Gal4 protein or the bacterial Lex A protein, both high-affinity, high-specificity DNA-binding proteins) fused in-frame to a protein of interest, here X. In two hybrid experiments, one is testing whether any protein can interact with protein X. Prey protein X may be fused in its entirety or often alternatively just a portion of protein X is expressed in-frame with the DBD. **(3)** A "prey" protein **(Y-AD)**, which represents a fusion of a specific protein fused in-frame to a transcriptional activation domain **(AD**; often derived from either the *herpes simplex* virus VP16 activator protein or the yeast GAL4 protein). This system serves as a useful test of protein–protein interactions between proteins X and Y because in the absence of a functional transactivator binding to the indicated enhancer, no transcription of the reporter gene occurs (ie, see Figure 38–16). Thus, one observes transcription only if protein X–protein Y interaction occurs, thereby bringing a functional AD to the *cis*-linked transcription unit, in this case activating transcription of the reporter gene. In this scenario, protein DBD-X alone fails to activate reporter transcription because the X-domain fused to the DBD does not contain an AD. Similarly, protein Y-AD alone fails to activate reporter gene transcription because it lacks a DBD to target the Y-AD protein to the enhancer-promoter-reporter gene. Only when both proteins are expressed in a single cell and bind the enhancer and, via DBD-X–Y-AD protein–protein interactions, regenerate a functional transactivator binary "protein," does reporter gene transcription result in activation and mRNA synthesis (green line from AD to reporter gene).

protein PTM profiling, and systems biology will revolutionize our understanding of physiology and medicine and ultimately human health.

SUMMARY

- In DNA cloning, a particular segment of DNA is either directly synthesized, or is removed from its normal environment using PCR or one of many DNA-directed endonucleases. Such DNA is then ligated into a vector in which the DNA segment can be individually amplified and produced in abundance.

- Manipulation of the DNA to change its sequence, so-called genetic engineering, is a key element in cloning (eg, the construction of chimeric molecules) and can also be used to study the function of a certain fragment of DNA and to analyze how genes are regulated.

- A variety of very sensitive techniques can now be applied to the isolation and characterization of genes and to the quantitation of gene products in both static (ie, equilibrium) and dynamic (kinetic) modes. These methods allow for the identification of the genes responsible for diseases, and for the study of how faulty genes/gene regulation causes disease.

- Mammalian genomes can now be precisely engineered to knockin (add/replace a gene), knockout (delete or inactivate), and/or actively and conditionally manipulate specific genes using miRNAs and SiRNAs, and novel genome editing enzymes (recombinases) and enzyme-RNA systems (CRISPR-Cas).

REFERENCES

Baltimore D, Berg P, Botchan M, et al.: Biotechnology. A prudent path forward for genomic engineering and germline gene modification. Science 2015;348:36-38.

Boyd JM: De novo assembly of plasmids using yeast recombinational cloning. Methods Mol Biol 2016;1373:33-41.

Brosh R, Laurent JM, Ordoñez R, et al.: A versatile platform for locus-scale genome rewriting and verification. Proc Natl Acad Sci USA 2021;118:e2023952118. doi: 10.1073/pnas.2023952118.

Churchman LS, Weissman JS: Nascent transcript sequencing visualizes transcription at nucleotide resolution. Nature 2011;469:368-373.

Datlinger P, Rendeiro AF, Boenke T, et al.: Ultra-high-throughput single-cell RNA sequencing and perturbation screening with combinatorial fluidic indexing. Nat Methods 2021;18:635–642.

Di Blasi R, Zouein A, Ellis T, Ceroni F: Genetic toolkits to design and build mammalian synthetic systems. Trends Biotechnol 2021;S0167-7799(20)30332-2. doi: 10.1016/j.tibtech.2020.12.007.

Doudna JA: The promise and challenge of therapeutic genome editing. Nature 2020;578:229-236.

Gandhi TK, Zhong J, Mathivanan S, et al.: Analysis of the human protein interactome and comparison with yeast, worm and fly interaction datasets. Nat Genet 2006;38:285-293.

Gibson DG, Glass JI, Lartigue C, et al.: Creation of a bacterial cell controlled by a chemically synthesized genome. Science 2010;329:52-56.

Green MR, Sambrook J: *Molecular Cloning: A Laboratory Manual,* 4th ed. Cold Spring Harbor Laboratory Press, 2012.

Ingolia TN: Ribosome footprint profiling of translation throughout the genome. Cell 2016;165:22-33.

Jain M, Koren S, Miga KH, et al.: Nanopore sequencing and assembly of a human genome with ultra-long reads. Nat Biotechnol 2018;36:338-345.

Kohman RE, Kunjapur AM, Hysolli E, Wang Y, Church GM: From designing the molecules of life to designing life: future applications derived from advances in DNA technologies. Angew Chem Int Ed Engl 2018;57:4313-4328.

Kwak H, Fuda NJ, Core LJ, Lis JT: Precise maps of RNA polymerase reveal how promoters direct initiation and pausing. Science 2013;339:950-953.

Liu SJ, Horlbeck MA, Cho SW, et al.: CRISPRi-based genome-scale identification of functional long noncoding RNA loci in human cells. Science 2017;355:355(6320).

Midha MK, Wu M, Chiu K-P: Long-read sequencing in deciphering human genetics to a greater depth. Hum Genet 2019;138:1201-1215.

Mitchell LA, McCulloch LH, Pinglay S, et al.: De novo assembly and delivery to mouse cells of a 101 kb functional human gene. Genetics 2021;218:iyab038. doi: 10.1093/genetics/iyab038.

Moufarrej MN, Wong RJ, Shaw GM, Stevenson DK, Quake SR: Investigating pregnancy and its complications using circulating cell-free RNA in women's blood during gestation. Front Pediatr 2020;8:605219. doi: 10.3389/fped.2020.605219.

Myers RM, Stamatoyannopoulos J, Snyder M, et al.: A user's guide to the encyclopedia of DNA elements (ENCODE). PLoS Biol 2011;9:e1001046.

Nurk S, Koren S, Rhie A, Rautiainen M, Bzikadze AV, et al.: The complete sequence of a human genome. Science 2022; 376:44–53.

Ohno M, Ando T, Priest DG, Taniguchi Y: Hi-CO: 3D genome structure analysis with nucleosome resolution. Nat Protoc 2021;May 28. doi: 10.1038/s41596-021-00543-z. Epub ahead of print. PMID: 34050337.

Rossi MJ, Kuntala PK, Lai WKM, et al.: A high-resolution protein architecture of the budding yeast genome. Nature 2021 Apr;592(7853):309-314. doi: 10.1038/s41586-021-03314-8. Epub 2021 Mar 10. PMID: 33692541; PMCID: PMC8035251.

Shendure J, Balasubramanian S, Church GM, et al.: DNA sequencing at 40: past, present and future. Nature 2017;550: 345-353.

Shivram H, Cress BF, Knott GJ, Doudna JA: Controlling and enhancing CRISPR systems. Nat Chem Biol 2021; 17: 10-19.

Suwinski P, Ong C, Ling MHT, Poh YM, Khan AM, Ong HS: Advancing personalized medicine through the application of whole exome sequencing and big data analytics. Front Genet 2019;10:49. doi: 10.3389/fgene.2019.00049.

Takahashi K, Tanabe K, Ohnuki M, et al.: Induction of pluripotent stem cells from adult human fibroblasts by defined factors. Cell 2007;131:861-872.

Turnbull C, Scott RH, Thomas E, et al.: 100 000 Genomes Project. The 100 000 Genomes Project: bringing whole genome sequencing to the NHS. Brit Med J 2018 Apr 24;361:k1687. doi: 10.1136/bmj.k1687.

Van Dijk EL, Jaszczyszyn Y, Naquin D, Thermes C: The third revolution in sequencing technology. Trends Genet 2018;34: 666-681.

Wang L, Wheeler DA: Genomic sequencing for cancer diagnosis and therapy. Annu Rev Med 2014;65:33-48.

Watson JF, García-Nafría J: *In vivo* DNA assembly using common laboratory bacteria: A re-emerging tool to simplify molecular cloning. J Biol Chem 2019;294:15271-15281.

Weatherall DJ: A journey in science: early lessons from the hemoglobin field. Mol Med 2014;20:478-485.

Wheeler DA, Srinivasan M, Egholm M, et al.: The complete genome of an individual by massively parallel DNA sequencing. Nature 2008;452:872-876.

Workman RE, Tang AD, Tang PS, et al.: Nanopore native RNA sequencing of a human poly(A) transcriptome. Nat Methods 2019;16:1297-1305.

GLOSSARY

ARS: Autonomously replicating sequence; the origin of replication in yeast.

Autoradiography: The detection of radioactive molecules (DNA, RNA, and protein) by visualization of their effects on photographic or x-ray film.

Bacteriophage: A virus that infects a bacterium.

Blunt-ended DNA: Two strands of a DNA duplex having ends that are flush with each other.

CAGE: Cap analysis of gene expression. A method that allows the selective capture, amplification, cloning, and sequencing of mRNAs via the 5′-cap structure.

cDNA: A single-stranded DNA molecule that is complementary to an mRNA molecule and is synthesized from it by the action of reverse transcriptase.

Chimeric molecule: A molecule (DNA, RNA, protein) containing sequences derived from two different species.

ChIP, chromatin immunoprecipitation: A technique that allows for the determination of the exact localization of a particular protein, or protein isoform, on any particular genomic location in a living cell. The method is based on cross-linking of living cells, cell disruption, DNA fragmentation, and immunoprecipitation with specific antibodies that purify the cognate protein cross-linked to DNA. Cross-links are reversed, associated DNA purified, and specific sequences that are purified are measured using any of several different methods (see following next three glossary entries).

ChIP-chip, chromatin immunoprecipitation assayed via a microarray chip hybridization readout: A hybridization-based method that uses chromatin immunoprecipitation (ChIP) techniques to map, genome-wide, the in vivo sites of binding of specific proteins within chromatin in living cells. Sequence binding is determined by annealing fluorescently labeled ChIP DNA samples to microarrays (array).

ChIP-Exo, chromatin immunoprecipitation assayed via a NGS/deep sequencing readout after treatment of immunoprecipitated protein-DNA complexes with exonucleases: A variation on ChIP-Seq (see next entry) that allows nucleotide-level precision in the mapping and description of DNA *cis*-elements bound by a particular protein.

ChIP-Seq, chromatin immunoprecipitation assayed via a NGS sequencing readout: Genomic DNA-binding location in a ChIP determined by high-throughput sequencing, rather than hybridization to microarrays.

CLIP: a method that uses UV cross-linking to induce covalent attachment of distinct proteins to specific RNAs in vivo: RNAs that are protein bound can subsequently be purified from cell lysates by immunoprecipitation and subsequent sequencing. A recently developed variant of CLIP, termed PAR-CLIP uses photoactivable nucleotides to enhance cross-linking efficiency.

Clone: A large number of organisms, cells, or molecules that are identical with a single parental organism, cell, or molecule.

Copy number variation (CNV): Change in the copy number of specific genomic regions of DNA between two or more individuals. CNVs can be as large as 10^6 bp of DNA and as small as a few bp. CNVs also include insertions or deletions (indels).

Cosmid: A plasmid into which the DNA sequences from bacteriophage lambda that are necessary for the packaging of DNA (λ *cos* sites) have been inserted; this permits the plasmid DNA to be packaged in vitro.

CRISPER/Cas9: A prokaryotic "immune system" conferring resistance to external genes from bacteriophage. This system provides a bacterial version of acquired immunity. CRISPR (clustered regularly interspaced short palindromic repeats) spacer-derived RNA combines with the Cas9 nuclease to target and specifically cleave the DNA of invading phage, thereby inactivating these invading genomes. This effectively protects the bacterium from productive phage infection and lysis. A variant of CRISPR-Cas9 that utilizes related proteins and guide RNAs to target specific RNAs for degradation is catalyzed by the C2c2 bacterial system.

ENCODE project: Encyclopedia of DNA elements project; an effort of multiple laboratories throughout the world to provide a detailed, biochemically informative representation of the human genome using high-throughput sequencing methods to identify and catalog the functional elements within the human genome. modENCODE is a parallel initiative to perform similar analyses on model organisms (yeasts, worms, flies, etc.).

Endonuclease: An enzyme that cleaves internal bonds in DNA or RNA.

Epigenetic code: The patterns of modification of chromosomal DNA (ie, cytosine methylation) and nucleosomal histone posttranslational modifications. These changes in modification status can lead to dramatic alterations in gene expression. Notably though, the actual underlying DNA sequence involved does not change.

Excinuclease: The excision nuclease involved in nucleotide exchange repair of DNA.

Exome: The nucleotide sequence of the entire complement of mRNA exons expressed in a particular cell, tissue, organ, or organism. The exome differs from the transcriptome that represents the entire collection of genome transcripts.

Exon: The sequence of a gene that is represented (expressed) as mRNA (or other mature, fully processed RNA).

Exonuclease: An enzyme that cleaves nucleotides from either the 3′ or 5′ ends of DNA or RNA.

Fingerprinting: The use of RFLPs or repeat sequence DNA to establish a unique pattern of DNA fragments for an individual.

FISH: Fluorescence in situ hybridization, a method used to map the location of specific DNA sequences within fixed nuclei.

Footprinting: DNA (or RNA; see Ribosome profiling later) with protein bound is resistant to digestion by DNase (or RNase) enzymes. When a sequencing reaction is performed using such DNA (or RNA), a protected area, representing the "footprint" of the bound protein, will be detected because nucleases are unable to cleave the DNA (or RNA) directly bound by the protein.

Hairpin: A double-helical stretch formed by base pairing between neighboring complementary sequences of a single strand of DNA or RNA.

Hybridization: The specific reassociation of complementary strands of nucleic acids (DNA with DNA, DNA with RNA, or RNA with RNA).

Insert: An additional length of base pairs in DNA, generally introduced by the techniques of recombinant DNA technology.

Intron: The sequence of an mRNA-encoding gene that is transcribed but excised before translation. tRNA genes can also contain introns.

Library: A collection of cloned fragments that represents, in aggregate, the entire genome. Libraries may be either genomic DNA (in which both introns and exons are represented) or cDNA (in which only exons are represented).

Ligation: The enzyme-catalyzed joining in phosphodiester linkage of two stretches of DNA or RNA into one; the respective enzymes are DNA and RNA ligases.

LINES: Long interspersed repeat sequences.

Microsatellite polymorphism: Heterozygosity of a certain microsatellite repeat in an individual.

Microsatellite repeat sequences: Dispersed or group repeat sequences of 2 to 5 bp repeated up to 50 times. May occur at 50 to 100 thousand locations in the genome.

miRNAs: MicroRNAs, 21 to 22 nucleotide long RNA species derived primarily from RNA polymerase II transcription units, via RNA processing. These RNAs play crucial roles in gene regulation by altering mRNA function.

NET-seq, native elongating sequencing: Genome-wide analysis of eukaryotic mRNA nascent chain 3′ ends mapped at nucleotide-level resolution. RNA polymerase II elongation complexes are captured by immunopurification with anti-Pol II antibodies and nascent RNAs containing a free 3′ OH group are tagged via ligation with an RNA linker and subsequently amplified by PCR and subjected to NGS sequencing.

Nick translation: A technique for labeling DNA based on the ability of the DNA polymerase I from *E. coli* to degrade a strand of DNA that has been nicked and then to resynthesize the strand; if a radioactive nucleoside triphosphate is employed, the resynthesized strand becomes labeled and can be used as a radioactive probe.

Northern blot: A method for transferring RNA from an agarose or polyacrylamide gel to a nitrocellulose or nylon filter, on which the RNA can be detected by a suitable probe by base-specific hybridization.

Oligonucleotide: A short, defined sequence of nucleotides joined together in the typical phosphodiester linkage.

Ori: The origin of DNA replication.

PAC: A high-capacity (70-95 kb) cloning vector based on the lytic *E. coli* bacteriophage P1 that replicates in bacteria as an extrachromosomal element.

Palindrome: A sequence of duplex DNA that is the same when the two strands are read in opposite directions.

Plasmid: A small, extrachromosomal, circular molecule of DNA, or episome, that replicates independently of the host DNA.

Polymerase chain reaction (PCR): An enzymatic method for the repeated copying (and thus amplification) of the two strands of DNA that make up a particular gene sequence.

Primosome: The mobile complex of helicase and primase that is involved in DNA replication.

Probe: A molecule used to detect the presence of a specific fragment of DNA or RNA in, for instance, a bacterial colony that is formed from a genetic library or during analysis by blot transfer techniques; common probes are cDNA molecules, synthetic oligodeoxynucleotides of defined sequence, or antibodies to specific proteins.

PRO-Seq, precision run on sequencing: A method where nascent transcripts are specifically captured and sequenced using NGS sequencing techniques. This method allows for the mapping of the location of active transcription complexes genome-wide.

Proteome: The entire collection of expressed proteins in an organism.

Pseudogene: An inactive segment of DNA arising by mutation of a parental active gene; typically generated by transposition of a cDNA copy of an mRNA.

Recombinant DNA: The altered DNA that results from the insertion of a sequence of deoxynucleotides not previously present into an existing molecule of DNA by enzymatic or chemical means.

Replica plating: A standard bacteriological method wherein 1 to 10 (or more) agar-containing petri dishes that contain various selective growth compounds are inoculated with the bacterial colonies from a primary, or master plate, thereby reproducing all of the spatial relationships of the original colonies. Transfer from original to replica is usually accomplished using a sterile velvet cloth.

Restriction enzyme: A DNA endonuclease that causes cleavage of both strands of DNA at highly specific sites dictated by the base sequence.

Reverse transcription: RNA-directed synthesis of DNA catalyzed by reverse transcriptase.

Ribosome profiling: A variant of footprinting that allows isolation of small mRNA fragments protected from nuclease digestion by actively translating ribosomes. Allows for the quantitative estimation of the amount and rates of synthesis of the proteome of cells.

RNA-Seq: A method where cellular RNA populations are converted, via linker ligation and PCR into cDNAs that are then subjected to deep sequencing to determine the complete sequence of essentially all RNAs in the preparation.

RIP: An RNA immunoprecipitation method, performed like ChIP, which is used to score specific binding of a protein to a specific RNA in vivo. RIP uses formaldehyde cross-linking to induce covalent attachment of proteins to RNA (see also CLIP/PAR-CLIP).

RT-PCR: A method used to quantitate mRNA levels that relies on a first step of cDNA copying of mRNAs catalyzed by reverse transcriptase prior to PCR amplification and quantitation.

Signal: The end product observed when a specific sequence of DNA or RNA is detected by autoradiography or some other method. Hybridization with a complementary radioactive polynucleotide (eg, by Southern or Northern blotting) is commonly used to generate the signal.

SINES: Short interspersed repeat sequences.

siRNAs: Silencing RNAs, 21 to 25 nt in length generated by selective nucleolytic degradation of double-stranded RNAs of cellular or viral origin. siRNAs anneal to various specific sites within target in RNAs leading to mRNA degradation, hence gene "knockdown."

SNP: Single-nucleotide polymorphism. Refers to the fact that single-nucleotide genetic variation in genome sequence exists at discrete loci throughout the chromosomes. Measurement of allelic SNP differences is useful for gene mapping studies.

snRNA: Small nuclear RNA. This family of RNAs is best known for its role in mRNA processing.

Southern blot: A method for transferring DNA from an agarose gel to nitrocellulose filter, on which the DNA can be detected by a suitable probe (eg, complementary DNA or RNA).

Southwestern blot: A method for detecting protein-DNA interactions by applying a labeled DNA probe to a transfer membrane that contains a renatured protein.

Spliceosome: The macromolecular complex responsible for precursor mRNA splicing. The spliceosome consists of at least five small nuclear RNAs (snRNA; U1, U2, U4, U5, and U6) and many proteins.

Splicing: The removal of introns from RNA accompanied by the joining of its exons.

Sticky-ended DNA: Complementary single strands of DNA that protrude from opposite ends of a DNA duplex or from the ends of different duplex molecules (see also Blunt-ended DNA, earlier).

Tandem: Used to describe multiple copies of the same sequence (eg, DNA) that lie adjacent to one another.

Terminal transferase: An enzyme that adds nucleotides of one type (eg, deoxyadenonucleotidyl residues) to the 3′ end of DNA strands.

Transcription: Template DNA-directed synthesis of nucleic acids, typically DNA-directed synthesis of RNA.

Transcriptome: The entire collection of expressed RNAs in a cell, tissue, organ, or organism; includes both mRNAs and ncRNAs.

Transgenic: Describing the introduction of new DNA into germ cells by its injection into the nucleus of the ovum.

Translation: Synthesis of protein using mRNA as template.

Vector: A plasmid or bacteriophage into which foreign DNA can be introduced for the purposes of cloning.

Western blot: A method for transferring protein to a nitrocellulose filter, on which the protein can be detected by a suitable probe (eg, an antibody).

Exam Questions

Section VII – Structure, Function, & Replication of Informational Macromolecules

1. Which of the following statements about β,γ-methylene and β,γ-imino derivatives of purine and pyrimidine triphosphates is CORRECT?

   A. They are potential anticancer drugs.
   B. They are precursors of B vitamins.
   C. They readily undergo hydrolytic removal of the terminal phosphate.
   D. They can be used to implicate involvement of nucleotide triphosphates by effects other than phosphoryl transfer.
   E. They serve as polynucleotide precursors.

2. Which of the following statements about nucleotide structures is NOT CORRECT?

   A. Nucleotides are polyfunctional acids.
   B. Caffeine and theobromine differ structurally solely with respect to the number of methyl groups attached to their ring nitrogens.
   C. A purine is a heterocyclic aromatic molecule composed of a pyrimidine ring fused to an imidazole ring.
   D. NAD^+, FMN, S-adenosylmethionine, and coenzyme A all are derivatives of ribonucleotides.
   E. 3′,5′-Cyclic AMP and 3′,5′-cyclic GMP (cAMP and cGMP) serve as second messengers in human physiology.

3. Which of the following statements about purine nucleotide metabolism is NOT CORRECT?

   A. An early step in purine biosynthesis is the formation of PRPP (phosphoribosyl-1-pyrophosphate).
   B. Inosine monophosphate (IMP) is a precursor of both AMP and GMP.
   C. Orotic acid is an intermediate in pyrimidine nucleotide biosynthesis.
   D. Humans catabolize uridine and pseudouridine by analogous reactions.
   E. Ribonucleotide reductase converts nucleoside diphosphates to the corresponding deoxyribonucleoside diphosphates.

4. Which of the following statements is NOT CORRECT?

   A. Metabolic disorders are only infrequently associated with defects in the catabolism of purines.
   B. Immune dysfunctions are associated both with a defective adenosine deaminase and with a defective purine nucleoside phosphorylase.
   C. The Lesch-Nyhan syndrome reflects a defect in hypoxanthine-guanine phosphoribosyl transferase.
   D. Xanthine lithiasis can be due to a severe defect in xanthine oxidase.
   E. Hyperuricemia can result from conditions such as cancer characterized by enhanced tissue turnover.

5. Which of the following components are found in DNA? **Choose the most complete answer**.

   A. A phosphate group, adenine, and ribose
   B. A phosphate group, guanine, and deoxyribose
   C. Cytosine and ribose
   D. Thymine and deoxyribose
   E. A phosphate group and adenine

6. The backbone of a DNA molecule consists of which of the following?

   A. Alternating sugars and nitrogenous bases
   B. Nitrogenous bases alone
   C. Phosphate groups alone
   D. Alternating phosphate and sugar groups
   E. Five carbon sugars alone

7. Which of the following is the interconnecting bond that connects the nucleotides of RNA and DNA ?

   A. *N*-glycosidic bonds
   B. 3′-5′ phosphodiester linkages
   C. Phosphomonoesters
   D. -2′ phosphodiester linkages
   E. Peptide nucleic acid bonds

8. Which component of the DNA duplex causes the molecule to have a net negative charge at physiologic pH?

   A. Deoxyribose
   B. Ribose
   C. Phosphate groups
   D. Chlorine ion
   E. Adenine

9. Which molecular feature listed causes duplex DNA to exhibit a near-constant width along its long axis?

   A. A purine nitrogenous base always pairs with another purine nitrogenous base.
   B. A pyrimidine nitrogenous base always pairs with another pyrimidine nitrogenous base.
   C. A pyrimidine nitrogenous base always pairs with a purine nitrogenous base.
   D. Repulsion between phosphate groups keeps the strands a uniform distance apart.
   E. Attraction between phosphate groups keeps the strands a uniform distance apart.

10. The model for DNA replication first proposed by Watson and Crick posited that every newly replicated double-stranded daughter duplex DNA molecule.

    A. Was composed of the two strands from the parent DNA molecule
    B. Contained solely the two newly synthesized strands of DNA
    C. Contained two strands that are random mixtures of new and old DNA within each strand
    D. Was composed of one strand derived from the original parental DNA duplex and one strand that was newly synthesized
    E. Was composed of nucleotide sequences completely distinct from either parental DNA strand

11. Name the mechanism through which RNAs are synthesized from DNA.
    A. Replicational duplication
    B. Translation
    C. Translesion repair
    D. Transesterification
    E. Transcription

12. Which of the forces or interactions listed below play the predominant role in driving RNA secondary and tertiary structure formation?
    A. Hydrophilic repulsion
    B. Formation of complementary base pair regions
    C. Hydrophobic interaction
    D. van der Waals interactions
    E. Salt bridge formation

13. Name the enzyme that synthesizes RNA from a double-stranded DNA template.
    A. RNA-dependent RNA polymerase
    B. DNA-dependent RNA convertase
    C. RNA-dependent replicase
    D. DNA-dependent RNA polymerase
    E. Reverse transcriptase

14. Define the most notable characteristic difference with regard to gene expression between eukaryotes and prokaryotes.
    A. Ribosomal RNA nucleotide lengths
    B. Mitochondria
    C. Lysosomes and peroxisomes
    D. Sequestration of the genomic material in the nucleus
    E. Chlorophyll

15. Which entry below correctly describes the approximate number of bp of DNA_____, which is separated into _____ chromosomes in a typical diploid human cell in a nonreplicating state?
    A. 64 billion, 23
    B. 6.4 trillion, 46
    C. 23 billion, 64
    D. 64 billion, 46
    E. 6.4 billion, 46

16. What is the approximate number of base pairs associated with a single nucleosome?
    A. 146
    B. 292
    C. 73
    D. 1460
    E. 900

17. All but one of the following histones are found located within the superhelix formed between DNA and the histone octamer; which of the following is this histone?
    A. Histone H2B
    B. Histone H3
    C. Histone H1
    D. Histone H3
    E. Histone H4

18. Chromatin can be broadly defined as active and repressed. Which of the following is termed as a subclass of chromatin that is specifically inactivated at certain times within an organism's life and/or in particular sets of differentiated cells?
    A. Constitutive euchromatin
    B. Facultative heterochromatin
    C. Euchromatin
    D. Constitutive heterochromatin

19. Which of the following hypothesizes that the physical and functional status of a certain region of genomic chromatin is dependent on the patterns of specific histone posttranslational modifications (PTMs), and/or DNA methylation status?
    A. Morse code
    B. PTM hypothesis
    C. Nuclear body hypothesis
    D. Epigenetic code
    E. Genetic code

20. What is the name of the unusual repeated stretch of DNA localized at the tips of all eukaryotic chromosomes?
    A. Kinetochore
    B. Telomere
    C. Centriole
    D. Chromomere
    E. Micromere

21. Given that DNA polymerases are unable to synthesize DNA without a primer, what molecule serves as the primer for these enzymes during DNA replication?
    A. Five carbon sugars
    B. Deoxyribose alone
    C. A short RNA molecule
    D. Proteins with free hydroxyl groups
    E. Phosphomonoester

22. Which of the following terms is used for the discontinuous DNA replication that occurs during replication is catalyzed via the production of small DNA segments?
    A. Okazaki fragments
    B. Toshihiro pieces
    C. Onishi oligonucleotides
    D. Crick strands
    E. Watson fragments

23. What molecule or force supplies the energy that drives the relief of mechanical strain by DNA gyrase?
    A. Pyrimidine to purine conversion
    B. Hydrolysis of GTP
    C. Hydrolysis of ATP
    D. Glycolysis
    E. A proton gradient molecule or force

24. What is the name of the phase of the cell cycle between the conclusion of cell division and the beginning of DNA synthesis?
    A. G_1
    B. S
    C. G_2
    D. M
    E. G_0

25. At what stage of the cell cycle are key protein kinases, like cyclin-dependent kinase, activated?
    A. Right before mitosis
    B. At the beginning of S phase
    C. Near the end of G_1 phase
    D. At the end of the G_2 phase
    E. All of the above

26. What disease is often associated with a breakdown of a cell's ability to regulate/control its own division?
    A. Kidney disease
    B. Cancer
    C. Emphysema
    D. Diabetes
    E. Heart disease

27. What is the molecular mechanism that is responsible for the quick decrease in the Cdk activity that leads to exit from the M phase and the entry into G_1?
    A. Drop in mitotic cyclin concentration
    B. Decreased G_1 cyclin concentration
    C. Rise in G_2 cyclin concentration
    D. Rise in mitotic cyclin concentration
    E. Rise in G_1 cyclin concentration

28. Which of the following is the site to which RNA polymerase binds on the DNA template prior to the initiation of transcription?
    A. Intron/exon junction
    B. Open reading frame DNA
    C. Terminator
    D. Initiator methionine codon
    E. Promoter

29. The large eukaryotic rRNA genes, such as 18S and 28S RNA-encoding genes, are transcribed by which of the following RNA polymerases?
    A. RNA polymerase III
    B. RNA-dependent RNA polymerase δ
    C. RNA polymerase I
    D. RNA polymerase II
    E. Mitochondrial RNA polymerase

30. Eukaryotic RNA polymerases all have a requirement for a large variety of accessory proteins to enable them to bind promoters and form physiologically relevant transcription complexes. What are these proteins termed as?
    A. Basal or general transcription factors
    B. Activators
    C. Accessory factors
    D. Elongation factors
    E. Facilitator polypeptides

31. The DNA segment from which the primary transcript is copied or transcribed is called which of the following?
    A. Coding strand
    B. Initiator methionine domain
    C. Translation unit
    D. Transcriptome
    E. Initial codon

32. What class of DNA sequences are the eukaryotic genes that encode rRNAs?
    A. Single-copy DNA
    B. Highly repetitive DNA
    C. Moderately repetitive DNA
    D. Mixed sequence DNA

33. How the modifications to the nucleotides of the pre-tRNAs, pre-rRNAs, and pre-mRNAs occur?
    A. Postprandially
    B. Postmitotically
    C. Pretranscriptionally
    D. Posttranscriptionally
    E. Prematurely

34. RNA polymerase II promoters are located on which side of the transcription unit?
    A. Internal
    B. 3′ downstream
    C. Nearest to the C-terminus
    D. Nearest to the N-terminus
    E. 5′ upstream

35. With regard to eukaryotic mRNAs, which one of the following is not a normal property of mRNAs?
    A. Eukaryotic mRNAs have special modifications at their 5′ (cap) and 3′ (poly(A) tail) termini.
    B. Are attached to ribosomes when they are translated.
    C. They are found in the cytoplasm within peroxisomes.
    D. Most have a significant noncoding segment that does not direct assembly of amino acids.
    E. Contain continuous nucleotide sequences that encode a particular polypeptide.

36. Which of the following is the bond connecting the initiation nucleotide of the mRNA with the 5me-G Cap structure?
    A. 3′-5′ phosphodiester bridge
    B. 5′-5′ triphosphate bridge
    C. 3′-3′ triphosphate bridge
    D. 3′-5′ triphosphate bridge
    E. 5′-3′ triphosphate bridge

37. What sequence feature of mature mRNAs listed below is thought to protect mRNAs from degradation?
    A. Special posttranslational modifications
    B. 3′ Poly(C)$_n$ tail
    C. 5me-G Cap
    D. Introns
    E. Lariat structures

38. What could the consequences of inaccurate mRNA splicing be for the RNA?
    A. A single base error at a splice junction will cause a large deletion.
    B. A single base error at a splice junction will cause a large insertion.
    C. A single base error at a splice junction will cause a large inversion.
    D. C and E.
    E. A single base error at a splice junction will change the reading frame and result in mRNA mistranslation.

39. What is the macromolecular complex that associates with introns during mRNA splicing?

    A. Splicer
    B. Dicer
    C. Nuclear body
    D. Spliceosome
    E. Slicer

40. What reaction does reverse transcriptase catalyze?

    A. Translation of RNA to DNA
    B. Transcription of DNA to RNA
    C. Conversion of ribonucleotides into deoxyribonucleotides
    D. Transcription of RNA to DNA
    E. Conversion of a ribonucleotide to deoxynucleotides in the DNA double helix

41. RNAi or dsRNA-mediated RNA interference mediates which of the following?

    A. RNA ligation
    B. RNA silencing
    C. RNA inversion
    D. RNA restoration
    E. RNA quelling

42. While the genetic code has 64 codons, there are only 20 naturally occurring amino acids. Consequently, some amino acids are encoded by more than one codon. This feature of the genetic code is an illustration of the genetic code being which of the following?

    A. Degenerative
    B. Duplicative
    C. Nonoverlapping
    D. Overlapping
    E. Redundant

43. The genetic code contains how many termination codons?

    A. 3
    B. 21
    C. 61
    D. 64
    E. 20

44. If a tRNA has the sequence 5′-CAU-3′, what codon would it recognize (ignore wobble base pairing).

    A. 3′-UAC-5′
    B. 3′-AUG-5′
    C. 5′-ATG-3′
    D. 5′-AUC-3′
    E. 5′-AUG-3′

45. What is on the 3′ end of all functional, mature tRNAs?

    A. The cloverleaf loop
    B. The anticodon
    C. The sequence CCA
    D. The codon

46. Most aminoacyl-tRNA synthetases possess an activity that is shared with DNA polymerases. This activity is a _____ function.

    A. Proofreading
    B. Hydrogenase
    C. Proteolytic
    D. Helicase
    E. Endonucleolytic

47. Which of the following is the CORRECT order for the three distinct phases of protein synthesis?

    A. Initiation, termination, elongation
    B. Termination, initiation, elongation
    C. Initiation, elongation, termination
    D. Elongation, initiation, termination
    E. Elongation, termination, initiation

48. Which amino acid is the initiating amino acid for essentially all proteins?

    A. Cysteine
    B. Threonine
    C. Tryptophan
    D. Methionine
    E. Glutamic acid

49. The initiator tRNA is placed within the active 80S complex at which of the three-canonical ribosomal "sites" during protein synthesis?

    A. E site
    B. I site
    C. P site
    D. A site
    E. Releasing factor binding site

50. Name the enzyme that forms the peptide bond during protein synthesis and define its chemical composition.

    A. Pepsynthase, protein
    B. Peptidyl transferase, RNA
    C. Peptidase, glycolipid
    D. Peptidyl transferase, protein
    E. GTPase, glycopeptide

51. What is the term used for mutations in the middle of an open reading frame that create a stop codon?

    A. Frameshift mutation
    B. Missense mutation
    C. No-nonsense mutation
    D. Point mutation
    E. Nonsense mutation

52. What is the directionality of polypeptide synthesis?

    A. C-terminal to N-terminal direction
    B. N-terminal to 3′ direction
    C. N-terminal to C-terminal direction
    D. 3′ to 5′ direction
    E. 5′ to 3′ direction

53. Which of the following *cis*-acting elements typically resides adjacent to or overlaps with many prokaryotic promoters?

    A. Regulatory gene
    B. Structural gene(s)
    C. Repressor
    D. Operator
    E. Terminator

54. What is the term applied to a segment of a bacterial chromosome where genes for the enzymes of a particular metabolic pathway are clustered and subject to coordinate control?

    A. Operon
    B. Operator
    C. Promoter
    D. Terminal controller
    E. Origin

55. What is the term applied to the complete collection of proteins present in a particular cell type?

    A. Genome
    B. Peptide collection
    C. Transcriptome
    D. Translatome
    E. Proteome

56. How does nucleosome formation on genomic DNA affect the initiation and/or elongation phases of transcription?

    A. Nucleosomes inhibit access of enzymes involved in all phases of transcription.
    B. Nucleosomes recruit histone and DNA-modifying enzymes, and the actions of these recruited enzymes affect the access of transcription proteins to DNA.
    C. Nucleosomes induce DNA degradation where the DNA contacts the histones.
    D. Nucleosomes have no significant effect on transcription.

57. Which types of molecules interact with eukaryotic mRNA gene core promoter sites to facilitate the association of RNA polymerase II?

    A. Termination factors
    B. Sequence-specific transcription factors (transactivators)
    C. Elongation factors
    D. GTPases
    E. General, or basal transcription factors (ie, the GTFs)

58. Most eukaryotic transcription factors contain at least two domains, each of which mediate different aspects of transcription factor function. Which of the following are these domains?

    A. RNA-binding domain and repression domain
    B. Activation domain and repression domain
    C. DNA-binding domain and activation domain
    D. DNA-binding domain and ligand-binding domain
    E. RNA-binding domain and the activation domain

59. Transcription factors bound at enhancers stimulate the initiation of transcription at the *cis*-linked core promoter through the action of intermediaries are termed as which of the following?

    A. Coactivators
    B. Cotranscription proteins
    C. Corepressors
    D. Receptors
    E. Coordinators

60. What reactions among transcription proteins greatly expand the diversity of regulatory factors that can be generated from a small number of polypeptides?

    A. Recombination
    B. Homodimerization

    C. Heterozygosity
    D. Heterodimerization
    E. Trimerization

61. The gene region containing the TATA box and extending to the transcription start site (TSS) is often termed which of the following?

    A. Polymerase home
    B. Initiator
    C. Start selector
    D. Core promoter
    E. Operator

62. Which of the following possible mechanisms for how enhancers can stimulate transcription from great distances are currently thought to be CORRECT?

    A. Enhancers can reversibly excise the intervening DNA between enhancers and promoters.
    B. RNA polymerase II binds avidly to enhancer sequences.
    C. Enhancers unwind DNA.
    D. Enhancers can search through DNA and bind directly to the associated core promoter.
    E. Enhancers and core promoters are brought into close proximity through DNA loop formation mediated by DNA-binding proteins.

63. Which of the following histone amino acids are typically acetylated?

    A. Lysine
    B. Arginine
    C. Asparagine
    D. Histidine
    E. Leucine

64. Place the following steps in order; what are the steps that occur sequentially during a transcription activation event following the binding of a transcriptional activator to its cognate activator–binding site on genomic DNA.

    1. The chromatin remodeling complex binds to the core histones at the target region.
    2. The combined actions of the various molecular complexes increase promoter accessibility to the transcriptional machinery.
    3. The activator recruits a coactivator to a region of chromatin targeted for transcription.
    4. Transcriptional machinery assembles at the site where transcription will be initiated.
    5. The coactivator acetylates the core histones of nearby nucleosomes.

    A. 1 – 2 – 3 – 4 – 5
    B. 3 – 1 – 5 – 2 – 4
    C. 3 – 5 – 1 – 2 – 4
    D. 5 – 3 – 1 – 2 – 4
    E. 3 – 5 – 1 – 4 – 2

65. What strategy in transcription factor research allows for the simultaneous identification of all of the genomic sites bound by a given transcription factor under a given set of physiologic conditions?

    A. Systematic deletion mapping
    B. DNAse I sensitivity
    C. Chromatin immunoprecipitation-sequencing (ChIP-seq)
    D. FISH
    E. Fluorescence lifetime imaging microscopy

66. Which sequences extend between the 5′–methylguanosine cap present on eukaryotic mRNAs to the AUG initiation codon?

    A. Stop codon
    B. Last exon
    C. Last intron
    D. 3′ UTR
    E. 5′ UTR

67. Which of the following features of eukaryotic mRNA contribute importantly to message half-life?

    A. 5′–UTR sequences
    B. The promoter
    C. The operator
    D. 3′ UTR and poly(A) tail
    E. The first intron

CHAPTER

40

Membranes: Structure & Function

P. Anthony Weil, PhD

BIOMEDICAL IMPORTANCE

Membranes are dynamic, highly fluid structures consisting of a lipid bilayer and associated proteins. **Plasma membranes** form closed compartments around the cytoplasm to define cell boundaries. The plasma membrane has **selective permeabilities** and acts as a barrier, thereby maintaining differences in composition between the inside and outside of the cell. Selective membrane molecular permeability is generated through the action of specific **transporters** and **ion channels**. The plasma membrane also exchanges material with the extracellular environment by **exocytosis** and **endocytosis**, and there are special areas of membrane structure—**gap junctions**—through which adjacent cells may communicate by exchanging material. In addition, the plasma membrane plays key roles in **cell–cell interactions** and in **transmembrane signaling**.

Membranes also form **specialized compartments** within the cell. Such intracellular membranes help **shape** many of the morphologically distinguishable cellular structures (organelles), for example, mitochondria, endoplasmic reticulum (ER), Golgi, secretory granules, lysosomes, and the nucleus. Membranes localize **enzymes**, function as integral elements in **excitation-response coupling**, and provide sites of **energy transduction**, such as in photosynthesis in plants (chloroplasts) and oxidative phosphorylation (mitochondria).

Changes in membrane components can affect water balance and ion flux, and therefore many processes within the cell. Specific deficiencies or alterations of certain membrane components (eg, caused by mutations in genes encoding membrane proteins) lead to a variety of **diseases** (see Table 40–7). In short, normal cellular function critically depends on normal membranes.

467

MAINTENANCE OF A NORMAL INTRA- & EXTRACELLULAR ENVIRONMENT IS FUNDAMENTAL TO LIFE

Life originated in an aqueous environment; enzyme reactions, cellular, and subcellular processes have therefore evolved to work in this milieu, encapsulated within a cell.

The Body's Internal Water Is Compartmentalized

Water makes up about **60%** of the lean body mass of the human body and is distributed in two large compartments.

Intracellular Fluid (ICF)

This compartment constitutes **two-thirds** of total body water and provides a specialized environment for the cell to (1) make, store, and utilize energy; (2) to repair itself; (3) to replicate; and (4) to perform cell-specific functions.

Extracellular Fluid (ECF)

This compartment contains about **one-third** of total body water and is distributed between the plasma and interstitial compartments. The extracellular fluid is a bidirectional **delivery system**. The ECF brings cells nutrients (eg, glucose, fatty acids, and amino acids), oxygen, various ions and trace minerals, and a variety of regulatory molecules (hormones) that coordinate the functions of widely separated cells. Similarly, extracellular fluid **removes** CO_2, waste products, and toxic or detoxified materials from the immediate cellular environment.

The Ionic Compositions of Intracellular & Extracellular Fluids Differ Greatly

As illustrated in **Table 40–1**, the **internal environment** is rich in K^+ and Mg^{2+}, and phosphate is its major inorganic anion. The cytosol of cells contains a high concentration of protein that acts as a major intracellular buffer. **Extracellular fluid** is characterized by high Na^+ and Ca^{2+} content, and Cl^- is the major anion. These ionic differences are maintained due to various membranes found in cells. These membranes have unique lipid and protein compositions. A fraction of the protein constituents of membrane proteins are specialized to generate and maintain the differential ionic compositions of the extra- and intracellular compartments.

MEMBRANES ARE COMPLEX STRUCTURES COMPOSED OF LIPIDS, PROTEINS, & CARBOHYDRATE-CONTAINING MOLECULES

We shall mainly discuss the membranes present in eukaryotic cells, although many of the principles described also apply to the membranes of prokaryotes. The various cellular membranes have different lipid and protein compositions. The ratio of protein to lipid in different membranes is presented in **Figure 40–1**, and is responsible for the many divergent functions of cellular organelles. Membranes are sheet-like enclosed structures consisting of an asymmetric lipid bilayer with distinct inner and outer surfaces or leaflets. These structures and surfaces are protein-studded, sheet-like, noncovalent

TABLE 40–1 **Comparison of the Mean Concentrations of Various Molecules Outside & Inside a Mammalian Cell**

Substance	Extracellular Fluid	Intracellular Fluid
Na^+	140 mmol/L	10 mmol/L
K^+	4 mmol/L	140 mmol/L
Ca^{2+} (free)	2.5 mmol/L	0.1 μmol/L
Mg^{2+}	1.5 mmol/L	30 mmol/L
Cl^-	100 mmol/L	4 mmol/L
HCO_3^-	27 mmol/L	10 mmol/L
PO_4^{3-}	2 mmol/L	60 mmol/L
Glucose	5.5 mmol/L	0-1 mmol/L
Protein	2 g/dL	16 g/dL

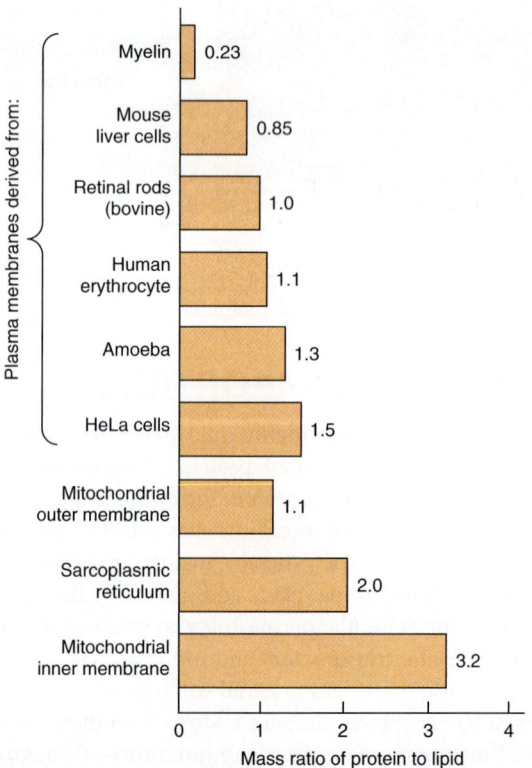

FIGURE 40–1 **Membrane protein content is highly variable.** The amount of proteins equals or exceeds the quantity of lipid in nearly all membranes. The outstanding exception is myelin, an electrical insulator found on many nerve fibers.

assemblies that form spontaneously in aqueous environments due to the amphipathic nature of lipids and the proteins contained within the membrane.

The Major Lipids in Mammalian Membranes Are Phospholipids, Glycosphingolipids & Cholesterol

Phospholipids

Of the two major phospholipid classes present in membranes, **phosphoglycerides** are the more common and consist of a glycerol-phosphate backbone to which are attached two fatty acids in ester linkages and an alcohol (**Figure 40–2**). The **fatty acid** constituents are usually even-numbered carbon molecules, most commonly containing 16 or 18 carbons. They are unbranched and can be saturated or unsaturated with one or more double bonds. The simplest phosphoglyceride is **phosphatidic acid**, a 1,2-diacylglycerol 3-phosphate, a key intermediate in the formation of other phosphoglycerides (see Chapter 24). In most phosphoglycerides present in membranes, the 3-phosphate is esterified to an **alcohol** such as choline, ethanolamine, glycerol, inositol, or serine (see Chapter 21). Phosphatidylcholine is generally the major phosphoglyceride by mass in the membranes of human cells.

The second major class of phospholipids comprises **sphingomyelin** (see Figure 21–11), a phospholipid that contains a sphingosine rather than a glycerol backbone. A fatty acid is attached by an amide linkage to the amino group of sphingosine, forming **ceramide**. When the primary hydroxyl group of sphingosine is esterified to phosphorylcholine, sphingomyelin is formed. As the name suggests, sphingomyelin is prominent in myelin sheaths.

Glycosphingolipids

The **glycosphingolipids (GSLs)** are sugar-containing lipids built on a backbone of **ceramide**. GSLs include **galactosyl-** and **glucosyl-ceramides** (cerebrosides) and the **gangliosides** (see structures in Chapter 21), and are mainly located in the plasma membranes of cells, displaying their sugar components to the exterior of the cell.

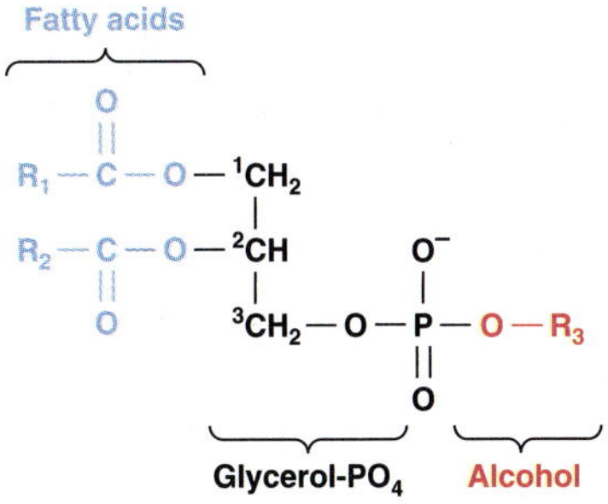

Sterols

The most common sterol in the membranes of animal cells is **cholesterol** (see Chapter 21). The majority of cholesterol resides within **plasma membranes**, but smaller amounts are found within mitochondrial, Golgi complex, and nuclear membranes. Cholesterol intercalates among the phospholipids of the membrane, with its hydrophilic hydroxyl group at the aqueous interface and the remainder of the molecule buried within the lipid bilayer leaflet. From a nutritional viewpoint, it is important to know that cholesterol is not present in plants.

Lipids can be separated from one another and quantified by techniques such as column, thin-layer, and gas-liquid chromatography and their structures established by mass spectrometry and other techniques (see Chapter 4).

Membrane Lipids Are Amphipathic

All major lipids in membranes contain both hydrophobic and hydrophilic regions and are therefore termed **amphipathic**. If the hydrophobic region were separated from the rest of the molecule, it would be insoluble in water but soluble in organic solvents. Conversely, if the hydrophilic region were separated from the rest of the molecule, it would be insoluble in organic solvents but soluble in water. The amphipathic nature of a phospholipid is represented in **Figure 40–3** and also Figure 21–24. Thus, the **polar head groups** of the phospholipids and the hydroxyl group of cholesterol interface with the

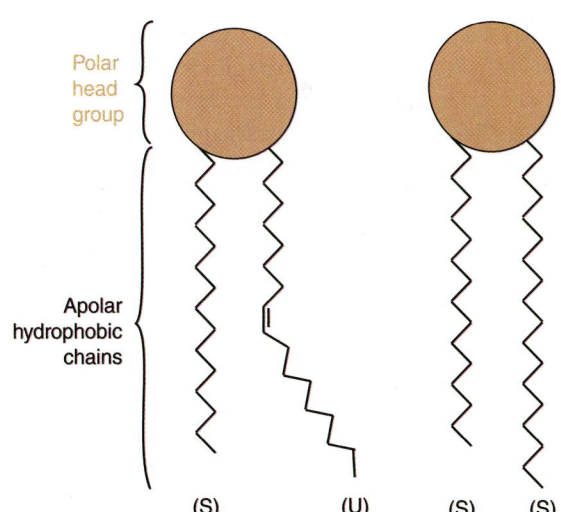

FIGURE 40–3 **Diagrammatic representation of a phospholipid or other membrane lipid.** The polar head group is hydrophilic, and the hydrocarbon tails are hydrophobic or lipophilic. The fatty acids in the tails are saturated (**S**) or unsaturated (**U**); the former is usually attached to carbon 1 of glycerol and the latter to carbon 2 (see Figure 40–2). Note the kink in the tail of the unsaturated fatty acid (U), which is important in conferring increased membrane fluidity.

The **S-U** phospholipid on the left contains the C16 saturated lipid palmitic acid, and the monounsaturated C18 lipid cis-oleic acid; both are esterified to glycerol (see Figure 40–2). The **S-S** phospholipid schematized on the right contains the C16 saturated lipid palmitic acid and the saturated C18 lipid, stearic acid.

FIGURE 40–2 **A phosphoglyceride showing the fatty acids (R₁ and R₂), glycerol, and a phosphorylated alcohol component.** Saturated fatty acids are usually attached to carbon 1 of glycerol, and unsaturated fatty acids to carbon 2. In phosphatidic acid, R_3 is hydrogen.

aqueous environment; a similar situation applies to the **sugar moieties** of the GSLs (see later).

Saturated fatty acids form relatively straight tails, whereas unsaturated fatty acids, which generally exist in the cis form in membranes, form "kinked" tails (see Figure 40–3; see also Figures 21–1, 21–6). As the number of double bonds within the lipid side chains increase, the number of kinks in the tails increases. As a consequence, the membrane lipids become less tightly packed and the membrane more fluid. The problem caused by the presence of **trans fatty acids** in membrane lipids is described in Chapter 21.

Detergents are amphipathic molecules that are important in biochemistry as well as in the household. The molecular structure of a detergent is not unlike that of a phospholipid. Certain detergents are widely used to **solubilize** and purify membrane proteins. The hydrophobic end of the detergent binds to hydrophobic regions of the proteins, displacing most of their bound lipids. The polar end of the detergent is free, bringing the proteins into solution as detergent-protein complexes, usually also containing some residual lipids.

Membrane Lipids Form Bilayers

The amphipathic character of phospholipids suggests that the two regions of the molecule have incompatible solubilities. However, in a solvent such as water, phospholipids spontaneously organize themselves into **micelles** (**Figure 40–4** and Figure 21–24), an assembly that thermodynamically satisfies the solubility requirements of the two chemically distinct regions of these molecules. Within the micelle the hydrophobic regions of the amphipathic phospholipids are shielded from water, while the hydrophilic polar groups are immersed in the aqueous environment. Micelles are usually relatively small in size (eg, ~200 nm) and consequently are limited in their potential to form membranes. Detergents commonly form micelles.

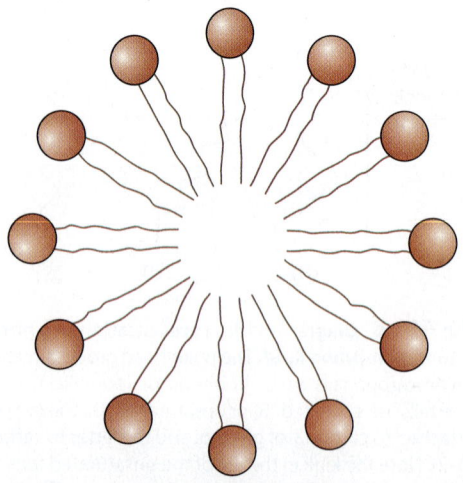

FIGURE 40–4 **Diagrammatic cross-section of a micelle.** The polar head groups are bathed in water, whereas the hydrophobic hydrocarbon tails are surrounded by other hydrocarbons and thereby protected from water. Micelles are relatively small (compared with lipid bilayers) spherical structures.

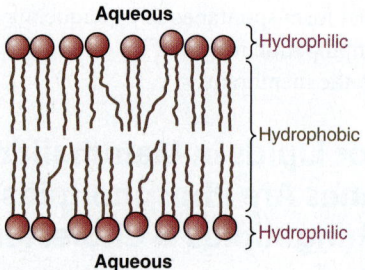

FIGURE 40–5 **Diagram of a section of a bilayer membrane formed from phospholipids.** The unsaturated fatty acid tails are kinked and lead to more spacing between the polar head groups, and hence to more room for movement. This in turn results in increased membrane fluidity.

Phospholipids and similar amphipathic molecules can form another structure, the **bimolecular lipid bilayer**, which also satisfies the thermodynamic requirements of amphipathic molecules in an aqueous environment. Bilayers are the key structures in biologic membranes. Bilayers exist as sheets wherein the hydrophobic regions of the phospholipids are sequestered from the aqueous environment, while the hydrophilic, charged portions are exposed to water (**Figure 40–5** and Figure 21–24). The ends or edges of the bilayer sheet can be eliminated by folding the sheet back on itself to form an enclosed vesicle with no edges. The closed bilayer provides one of the most essential properties of membranes. The lipid bilayer is **impermeable to most water-soluble molecules** since such charged molecules would be insoluble in the hydrophobic core of the bilayer. The **self-assembly of lipid bilayers** is driven by the **hydrophobic effect**, which describes the tendency of nonpolar molecules to self-associate in an aqueous environment, while in the process excluding H_2O. When lipid molecules come together in a bilayer, the entropy of the surrounding solvent molecules increases due to the release of immobilized water.

Two questions arise from consideration of the information described earlier. First, how many biologically important molecules are **lipid-soluble** and can therefore readily enter the cell? Gases such as oxygen, CO_2, and nitrogen—small molecules with little interaction with solvents—readily diffuse through the hydrophobic regions of the membrane. The **permeability coefficients** of several ions and a number of other molecules in a lipid bilayer are shown in **Figure 40–6**. The electrolytes Na^+, K^+, and Cl^- cross the bilayer much more slowly than water. In general, the permeability coefficients of small molecules in a lipid bilayer **correlate with their solubilities in nonpolar solvents**. For instance, **steroids** more readily traverse the lipid bilayer compared with electrolytes. The high permeability coefficient of **water** itself is surprising, but is partly explained by its small size and relative lack of charge. Many **drugs** are hydrophobic and can readily cross membranes and enter cells.

The second question concerns **non–lipid-soluble molecules**. How are the transmembrane concentration gradients for these molecules maintained? The answer is that **membranes**

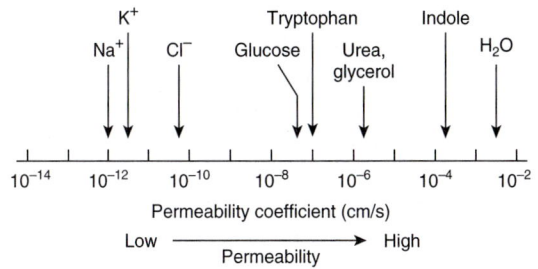

FIGURE 40–6 **Permeability coefficients of water, some ions, and other small molecules in lipid bilayer membranes.** The permeability coefficient is a measure of the ability of a molecule to diffuse across a permeability barrier. Molecules that move rapidly through a given membrane are said to have a high permeability coefficient.

contain proteins, many of which span the lipid bilayer. These proteins either form **channels** for the movement of ions and small molecules or serve as **transporters** for molecules that otherwise could not readily traverse the lipid bilayer (membrane). The nature, properties, and structures of membrane channels and transporters are described later.

Membrane Proteins Are Associated With the Lipid Bilayer

Membrane **phospholipids** act as a solvent for membrane proteins, creating an environment in which the latter can function. As described in Chapter 5, the α-**helical structure of proteins** minimizes the hydrophilic character of the peptide bonds themselves. Thus, proteins can be amphipathic and form an integral part of the membrane by having hydrophilic regions protruding at the inside and outside faces of the membrane but connected by a hydrophobic region traversing the hydrophobic core of the bilayer. In fact, those portions of membrane proteins that traverse membranes do contain substantial numbers of hydrophobic amino acids and almost invariably have a high α-helical content. For most membranes, a stretch of ~20 amino acids in an α-helical configuration will span the lipid bilayer (see Figure 5–2).

It is possible to calculate whether a particular sequence of amino acids present in a protein is consistent with a **transmembrane location**. This can be done by consulting a table that lists the hydrophobicities of each of the 20 common amino acids and the free energy values for their transfer from the interior of a membrane to water. Hydrophobic amino acids have positive values; polar amino acids have negative values. The total free energy values for transferring successive sequences of 20 amino acids in the protein are plotted, yielding a so-called **hydropathy plot**. Values of over 20 kcal mol⁻¹ are consistent with—but do not prove—the interpretation that the hydrophobic sequence is a transmembrane segment.

Another aspect of the interaction of lipids and proteins is that some proteins are anchored to one leaflet of the bilayer by covalent linkages to certain lipids; this process is termed **protein lipidation**. Lipidation can occur at protein termini (N- or C-) or internally. Common protein lipidation events are C-terminal protein **isoprenylation, cholesterylation,** and **glycophosphatidylinositol** (**GPI**; see Figure 46–1); N-terminal protein **myristoylation** and internal cysteine **S-prenylation** and **S-acylation**. Such lipidation only occurs on a specific subset of proteins and typically plays key roles in their biology.

Different Membranes Have Different Protein Compositions

The **number of different proteins** in a membrane varies from less than a dozen very abundant proteins in the sarcoplasmic reticulum of muscle cells to hundreds in plasma membranes. Proteins are the **major functional molecules** of membranes and consist of **enzymes, pumps** and **transporters, channels, structural components, antigens** (eg, for histocompatibility), and **receptors** for various molecules. Because every type of membrane possesses a different complement of proteins, there is no such thing as a typical membrane structure. The enzymes associated with several different membranes are shown in **Table 40–2**.

Membranes Are Dynamic Structures

Membranes and their components are dynamic structures. Membrane lipids and proteins undergo turnover, just as they do in other compartments of the cell. Different lipids have different turnover rates, and the turnover rates of individual species of membrane proteins may vary widely. In some instances, the membrane itself can turn over even more rapidly than any of its constituents. This is discussed in more detail in the section on endocytosis.

Another indicator of the dynamic nature of membranes is that a variety of studies have shown that lipids and certain proteins exhibit **lateral diffusion** in the plane of their membranes. Many nonmobile proteins do not exhibit lateral diffusion because they are anchored to the underlying actin cytoskeleton. By contrast, the **transverse** movement of lipids

TABLE 40–2 **Enzymatic Markers of Different Membranes**[a]

Membrane	Enzyme
Plasma	5′-Nucleotidase
	Adenylyl cyclase
	Na⁺-K⁺-ATPase
Endoplasmic reticulum	Glucose-6-phosphatase
Golgi apparatus	
Cis	GlcNAc transferase I
Medial	Golgi mannosidase II
Trans	Galactosyl transferase
Trans Golgi network	Sialyltransferase
Inner mitochondrial membrane	ATP synthase

[a]Membranes contain many proteins, some of which have enzymatic activity. Some of these enzymes are located only in certain membranes and can therefore be used as markers to follow the purification of these membranes.

across the membrane (**flip-flop**) is extremely slow (see later), and does not appear to occur at an appreciable rate in the case of membrane proteins.

Membranes Are Asymmetric Structures

Proteins have unique orientations in membranes, making the **outside surfaces different from the inside surfaces**. An **inside-outside asymmetry** is also provided by the external location of the carbohydrates attached to membrane proteins. In addition, specific proteins are located exclusively on the outsides or insides of membranes.

There are also **regional heterogeneities** in membranes. Some, such as occur at the villous borders of mucosal cells, are almost macroscopically visible. Others, such as those at gap junctions, tight junctions, and synapses, occupy much smaller regions of the membrane and generate correspondingly smaller local asymmetries.

There is also inside-outside **asymmetry of the phospholipids**. The **choline-containing phospholipids** (phosphatidylcholine and sphingomyelin) are located mainly in the **outer leaflet**; the **aminophospholipids** (phosphatidylserine and phosphatidylethanolamine) are preferentially located in the **inner leaflet**. Obviously, if this **lipid asymmetry** is to exist at all, there must be **limited transverse mobility**, or "flip-flop" the membrane phospholipids. In fact, phospholipids in synthetic bilayers exhibit an extraordinarily slow rate of flip-flop; the half-life of the asymmetry in these synthetic bilayers is on the order of several weeks.

The mechanisms involved in the lipid asymmetry are not well understood. The enzymes involved in the synthesis of phospholipids are located on the cytoplasmic side of microsomal membrane vesicles. Translocases (**flippases**) exist that transfer certain phospholipids (eg, phosphatidylcholine) from the inner to the outer leaflet. Specific proteins that preferentially bind individual phospholipids also appear to be present in the two leaflets; thus, lipid binding also contributes to the asymmetric distribution of specific lipid molecules. In addition, **phospholipid exchange proteins** recognize certain phospholipids and transfer them from one membrane (eg, the **ER**) to others (eg, mitochondrial and peroxisomal). A related issue is how lipids enter membranes. This has not been studied as intensively as the topic of how proteins enter membranes (see Chapter 49) and knowledge is still relatively meager. Many membrane lipids are synthesized in the ER. At least three pathways have been recognized: (1) transport from the ER in vesicles, which then transfer the contained lipids to the recipient membrane; (2) entry via direct contact of one membrane (eg, the ER) with another, facilitated by specific proteins; and (3) transport via the phospholipid exchange proteins (also known as lipid transfer proteins) mentioned earlier, which only exchanges lipids, but does not cause net transfer.

There is further asymmetry with regard to GSLs and **glycoproteins**; the sugar moieties of these molecules all protrude outward from the plasma membrane and are absent from its inner face.

Membranes Contain Integral & Peripheral Proteins

It is useful to classify membrane proteins into two types: **integral** and **peripheral** (**Figure 40–7**). Most membrane proteins fall into the **integral class**, meaning that they interact extensively with the phospholipids and **require the use of**

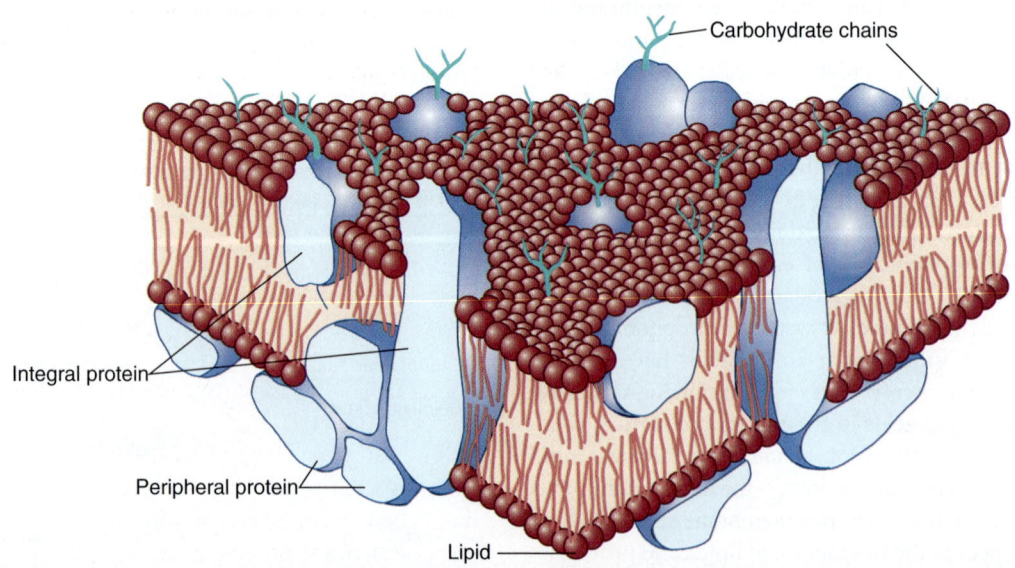

FIGURE 40–7 **The fluid mosaic model of membrane structure.** The membrane consists of a bimolecular lipid layer with proteins inserted in it or bound to either surface. Integral membrane proteins are firmly embedded in the lipid layers. Some of these proteins completely span the bilayer and are called transmembrane proteins, while others are embedded in either the outer or inner leaflet of the lipid bilayer. Loosely bound to the outer or inner surface of the membrane are the peripheral proteins. Many of the proteins and all the glycolipids have externally exposed oligosaccharide carbohydrate chains. (Reproduced with permission from Mescher AL: *Junqueira's Basic Histology Text and Atlas*, 16th ed. New York, NY: McGraw Hil; 2021.)

detergents for their solubilization. Also, they generally span the bilayer as a bundle of α-helical transmembrane segments. Integral proteins are usually globular and are themselves amphipathic. They consist of two hydrophilic ends separated by an intervening hydrophobic region that traverses the hydrophobic core of the bilayer. As the structures of integral membrane proteins were being elucidated, it became apparent that certain ones (eg, transporter molecules, ion channels, various receptors, and G proteins) span the bilayer many times, whereas other simple membrane proteins (eg, glycophorin A) span the membrane only once (see Figures 42–4 and 52–5). Integral proteins are asymmetrically distributed across the membrane bilayer. This asymmetric orientation is conferred at the time of their insertion in the lipid bilayer during biosynthesis in the ER. The molecular mechanisms involved in insertion of proteins into membranes and the topic of membrane assembly are discussed in Chapter 49.

Peripheral proteins do not interact directly with the hydrophobic cores of the phospholipids in the bilayer and thus **do not require use of detergents** for their release. They are bound to the hydrophilic regions of specific integral proteins and head groups of phospholipids and can be released from them by treatment with salt solutions of high ionic strength. For example, ankyrin, a peripheral protein, is bound to the inner aspect of the integral protein "band 3" of the erythrocyte membrane. Spectrin, a cytoskeletal structure within the erythrocyte, is in turn bound to ankyrin and thereby plays an important role in maintenance of the biconcave shape of the erythrocyte (see Figure 53–6).

ARTIFICIAL MEMBRANES MODEL MEMBRANE FUNCTION

Artificial membrane systems can be prepared by appropriate techniques. These systems generally consist of mixtures of one or more phospholipids of natural or synthetic origin that have been treated by using **mild sonication** to induce the formation of spherical vesicles in which the lipids form a bilayer. Such vesicles, surrounded by a lipid bilayer with an aqueous interior, are termed **liposomes** (see Figure 21–24).

The advantages and uses of artificial membrane systems for the biochemical study of membrane function are as follows:

1. The **lipid content** of the membranes can be varied, allowing systematic examination of the effects of varying lipid composition on certain functions.

2. Purified **membrane proteins** or enzymes can be incorporated into these vesicles in order to assess what factors (eg, specific lipids or ancillary proteins) the proteins require to reconstitute their function.

3. The **environment** of these systems can be rigidly controlled and systematically varied (eg, ion concentrations and ligands).

4. When liposomes are formed, they can be made to **entrap** certain compounds within the vesicle such as drugs and isolated genes. There is interest in using liposomes to distribute drugs to certain tissues, and if components (eg, antibodies to certain cell surface molecules) could be incorporated into liposomes so that they would be targeted to specific tissues or tumors, the therapeutic impact would be considerable. DNA entrapped inside liposomes appears to be less sensitive to attack by nucleases; this approach may prove useful in attempts at **gene therapy**.

THE FLUID MOSAIC MODEL OF MEMBRANE STRUCTURE IS WIDELY ACCEPTED

The **fluid mosaic model** of membrane structure proposed in 1972 by Singer and Nicolson (see Figure 40–7) is now widely accepted. The model is often likened to integral membrane protein "icebergs" floating in a sea of (predominantly) fluid phospholipid molecules. Early evidence for the model was the finding that well characterized, fluorescently labeled integral membrane proteins could be seen microscopically to rapidly and randomly redistribute within the plasma membrane of a hybrid cell formed by the artificial fusion of two different (mouse and human) parent cells (one labeled the other not). It has subsequently been demonstrated that **phospholipids** undergo even more rapid lateral diffusion with subsequent redistribution within the plane of the membrane. Measurements indicate that within the plane of the membrane, one molecule of phospholipid can move several micrometers per second.

The **phase changes**—and thus the **fluidity** of membranes—are largely dependent on the lipid composition of the membrane. In a lipid bilayer, the hydrophobic chains of the fatty acids can be highly aligned or ordered to provide a rather stiff structure. As the temperature increases, the hydrophobic side chains undergo a **transition** from the **ordered state** (more gel-like or crystalline phase) to a **disordered** one, taking on a more liquid-like or fluid arrangement. The temperature at which membrane structure undergoes the transition from ordered to disordered (ie, melts) is called the "**transition temperature**" (T_m). Longer and more saturated fatty acid chains interact more strongly with each other via their extended hydrocarbon chains and thus cause higher values of T_m—that is, higher temperatures are required to increase the fluidity of the bilayer. On the other hand, **unsaturated bonds** that exist **in the cis configuration** tend to increase the fluidity of a bilayer by decreasing compactness of the side chain packing without diminishing hydrophobicity (see Figures 40–3 and 40–5). The phospholipids of cellular membranes generally contain at least one unsaturated fatty acid with at least one cis double bond.

Cholesterol acts as a buffer to modify the fluidity of membranes. At temperatures below the T_m, it interferes with the interaction of the hydrocarbon tails of fatty acids and thus increases fluidity. At temperatures above the T_m, it limits disorder because it is more rigid than the hydrocarbon tails of the fatty acids and cannot move in the membrane to the same extent, thus limiting, or "buffering" membrane fluidity.

The fluidity of a membrane significantly affects its functions. As membrane fluidity increases, so does its permeability to water and other small hydrophilic molecules. The lateral mobility of integral proteins increases as the fluidity of the membrane increases. If the active site of an integral protein involved in a given function is exclusively in its hydrophilic regions, changing lipid fluidity will probably have little effect on the activity of the protein; however, if the protein is involved in a transport function in which transport components span the membrane, lipid-phase effects may significantly alter transport rate. The insulin receptor (see Figure 42–8) is an excellent example of altered function with changes in fluidity. As the concentration of unsaturated fatty acids in the membrane is increased (by growing cultured cells in a medium rich in such molecules), fluidity increases. Increased fluidity alters the receptor such that it binds insulin more effectively. At normal body temperature (37°C), the lipid bilayer is in a fluid state. Underscoring the importance of membrane fluidity, it has been shown that bacteria can modify the composition of their membrane lipids to adapt to changes in temperature.

Lipid Rafts, Caveolae, & Tight Junctions Are Specialized Features of Plasma Membranes

Plasma membranes contain **certain specialized structures** whose biochemical natures have been investigated in some detail.

Lipid rafts are specialized areas of the **exoplasmic (outer) leaflet** of the lipid bilayer enriched in cholesterol, sphingolipids, and certain proteins (**Figure 40–8**). They have been hypothesized to be involved in signal transduction and other processes. It is thought that clustering certain components of signaling systems closely together may increase the efficiency of their function.

Caveolae may derive from lipid rafts. Many, if not all, contain the protein **caveolin-1**, which may be involved in

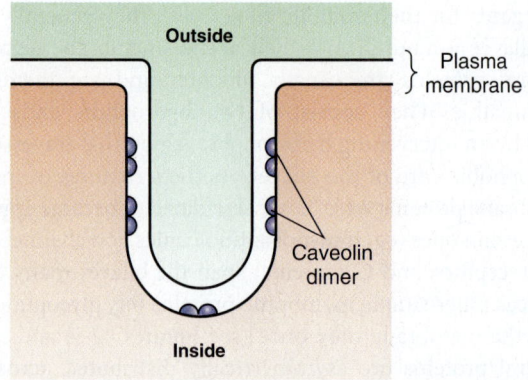

FIGURE 40–9 **Schematic diagram of a caveola.** A caveola is an invagination in the plasma membrane. The protein caveolin appears to play an important role in the formation of caveolae and occurs as a dimer. Each caveolin monomer is anchored to the inner leaflet of the plasma membrane by three palmitoyl molecules (not shown).

their formation from rafts. Caveolae are observable by electron microscopy as flask, or tube-shaped indentations of the cell membrane into the cytosol (**Figure 40–9**), and likely play a role in endocytosis (cellular uptake of various components). Proteins detected in caveolae include various components of the signal transduction system (eg, the insulin receptor and some G proteins; see Chapter 42), the folate receptor, and endothelial nitric oxide synthase (eNOS). Caveolae and lipid rafts are active areas of research, and ideas concerning them and their roles in various biologic processes are rapidly evolving.

Tight junctions are other structures found in surface membranes. They are often located below the apical surfaces of epithelial cells and prevent the diffusion of macromolecules between cells. They are composed of various proteins, including occludin, various claudins, and junctional adhesion molecules.

Yet other specialized structures found in surface membranes include **desmosomes, adherens junctions**, and **microvilli**; their chemical natures and functions are not discussed here. The nature of **gap junctions** is described later.

MEMBRANE SELECTIVITY ALLOWS ADJUSTMENTS OF CELL COMPOSITION & FUNCTION

If the plasma membrane is relatively impermeable, how do most molecules enter a cell? How is selectivity of this movement established? Answers to such questions are important in understanding how cells adjust to a constantly changing extracellular environment. Metazoan organisms also must have means of communicating between adjacent and distant cells, so that complex biologic processes can be coordinated. These signals must arrive at and be transmitted by the membrane, or they must be generated as a consequence of some interaction with the membrane. Some of the major mechanisms used to accomplish these different objectives are listed in **Table 40–3.**

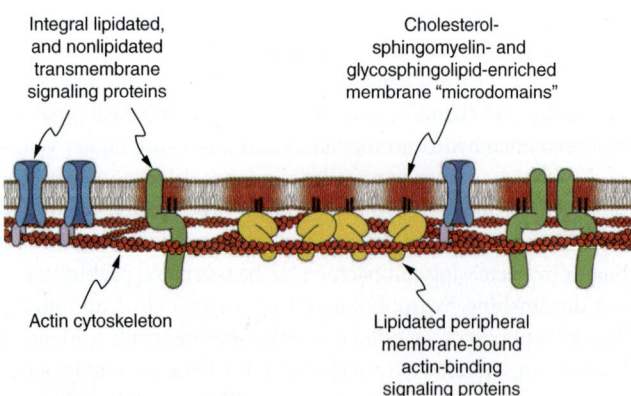

FIGURE 40–8 **Schematic diagram of a lipid raft.** Shown in schematic form are multiple lipid rafts (red membrane shading) that represent localized microdomains rich in the indicated lipids and signaling proteins (blue, green, yellow). Lipid rafts are stabilized through interactions (direct and indirect) with the actin cytoskeleton (red bihelical chains; see Figure 51–3). (Reproduced with permission from Owen DM, Magenau A, Williamson D, et al: The lipid raft hypothesis revisited—new insights on raft composition and function from super-resolution fluorescence microscopy, Bioessays. 2012;34(9):739-747.)

TABLE 40–3 Transfer of Material & Information Across Membranes

Cross-membrane movement of small molecules
Diffusion (passive and facilitated)
Active transport
Cross-membrane movement of large molecules
Endocytosis
Exocytosis
Signal transmission across membranes
Cell surface receptors
1. Signal transduction (eg, glucagon → cAMP)
2. Signal internalization (coupled with endocytosis, eg, the LDL receptor)
Movement to intracellular receptors (steroid hormones; a form of diffusion)
Intercellular contact and communication
Passive (simple) diffusion is the flow of solute from a higher to a lower concentration due to random thermal movement
Facilitated diffusion is passive transport of a solute from a higher concentration to a lower concentration, mediated by a specific protein transporter
Active transport is transport of a solute across a membrane in the direction of increasing concentration, and thus requires energy (frequently derived from the hydrolysis of ATP); a specific transporter (pump) is involved
Extracellular microvesicle and exosome secretion and uptake

The other terms used in this table are explained later in this chapter or elsewhere in this text.

Passive Diffusion Involving Transporters & Ion Channels Moves Many Small Molecules Across Membranes

Molecules can **passively** traverse the bilayer down electrochemical gradients by **simple diffusion** or by **facilitated diffusion**. This spontaneous movement toward equilibrium contrasts with **active transport**, which **requires energy** because it constitutes movement against an electrochemical gradient. **Figure 40–10** provides a schematic representation of these mechanisms.

Simple diffusion is the passive flow of a solute from a higher to a lower concentration due to random thermal movement. By contrast, **facilitated diffusion** is passive transport of a solute from a higher to a lower concentration mediated by a specific protein transporter. **Active transport** is vectorial movement of a solute across a membrane against a concentration gradient, and thus requires energy (frequently derived from the hydrolysis of ATP); a specific transporter (**pump**) is involved.

As mentioned earlier in this chapter, some solutes such as gases can enter the cell by diffusing down an electrochemical gradient across the membrane and do not require metabolic energy. **Simple diffusion** of a solute across the membrane is limited by three factors: (1) the thermal agitation of that specific molecule; (2) the concentration gradient across the membrane; and (3) the solubility of that solute (the permeability coefficient, Figure 40–6) in the hydrophobic core of the

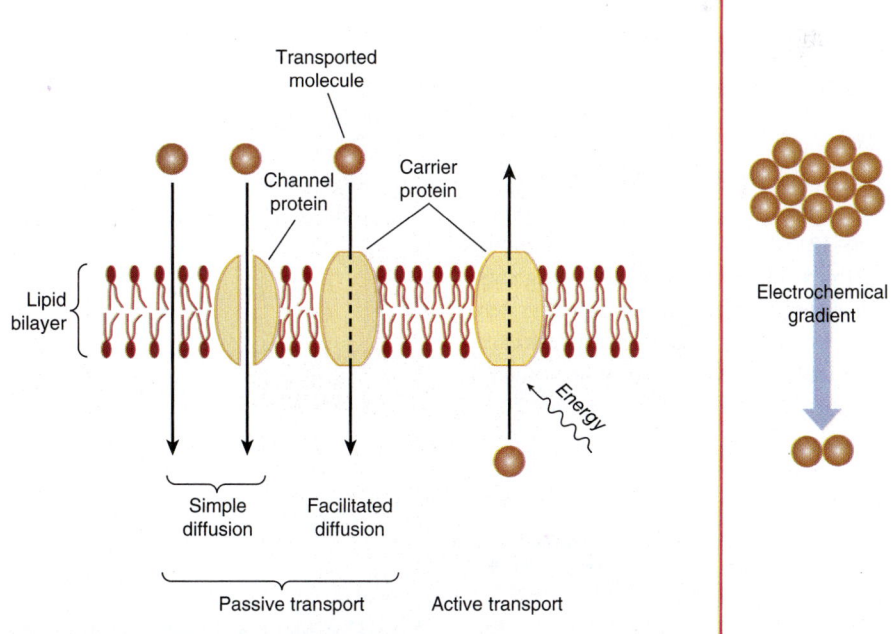

FIGURE 40–10 **Many small, uncharged molecules pass freely through the lipid bilayer by simple diffusion.** Larger uncharged molecules, and some small uncharged molecules, are transferred by specific carrier proteins (transporters) or through channels or pores. Passive transport is always down an electrochemical gradient (shown schematically, right), toward equilibrium. Active transport is against an electrochemical gradient and requires an input of energy, whereas passive transport does not. (Reproduced with permission from Alberts B, Bray D, Lewis J, et al: *Molecular Biology of the Cell*. New York, NY: Garland; 1983.)

membrane bilayer. Solubility is inversely proportional to the number of hydrogen bonds that must be broken in order for a solute in the external aqueous phase to become incorporated in the hydrophobic bilayer. Electrolytes, poorly soluble in lipid, do not form hydrogen bonds with water, but they do acquire a shell of water from hydration by electrostatic interaction. The size of the shell is directly proportional to the charge density of the electrolyte. Electrolytes with a large charge density have a larger shell of hydration and thus a slower diffusion rate. Na^+, for example, has a higher charge density than K^+. Hydrated Na^+ is therefore larger than hydrated K^+; hence, the latter tends to move more easily through the membrane.

The following affect **net diffusion** of a substance: (1) concentration gradient across the membrane—solutes move from high to low concentration; (2) electrical potential across the membrane: solutes move toward the solution that has the opposite charge. The inside of the cell usually has a net negative charge; (3) permeability coefficient of the substance for the membrane; (4) hydrostatic pressure gradient across the membrane: increased pressure will increase the rate and force of the collision between the molecules and the membrane; and (5) temperature, since increased temperature will increase particle motion and thus increase the frequency of collisions between external particles and the membrane.

Facilitated diffusion involves either certain transporters or ion channels (**Figure 40–11**). Active transport is mediated by other transporters most of which are ATP-driven. A multitude of transporters and channels exist in biologic membranes that route the entry of ions into and out of cells. **Table 40–4** summarizes some important differences between transporters and ion channels.

Transporters Are Specific Proteins Involved in Facilitated Diffusion & Also Active Transport

Transport systems can be described in a functional sense according to the number of molecules moved and the direction of movement (**Figure 40–12**) or according to whether movement is toward or away from equilibrium. The following **classification** depends primarily on the former. A **uniport** system moves one type of molecule bidirectionally. In **cotransport** systems, the transfer of one solute depends on the stoichiometric

TABLE 40–4 Comparison of Transporters & Ion Channels

Transporters	Ion Channels
Bind solute and undergo conformational changes, transferring the solute across the membrane	Form pores in membranes
Involved in passive (facilitated diffusion) and active transport	Involved only in passive transport
Transport is significantly slower than via ion channels	Transport is significantly faster than via transporters

Note: Transporters are also known as carriers or permeases. Active transporters are often called pumps.

simultaneous or sequential transfer of another solute. A **symport** moves two solutes in the same direction. Examples are the proton-sugar transporter in bacteria and the Na^+-sugar transporters (for glucose and certain other sugars) and Na^+-amino acid transporters in mammalian cells. **Antiport** systems move two molecules in opposite directions (eg, Na^+ in and Ca^{2+} out).

Hydrophilic molecules that cannot pass freely through the lipid bilayer membrane do so passively by **facilitated diffusion** or by **active transport**. Passive transport is driven by the transmembrane gradient of substrate. Active transport always occurs against an electrical or chemical gradient, and so it requires energy, usually in the form of ATP. Both types of transport involve **specific carrier proteins** (transporters) and both show **specificity** for ions, sugars, and amino acids. Passive and active transports resemble a substrate-enzyme interaction. Points of resemblance of both to enzyme action are as follows: (1) There is a specific binding site for the solute. (2) The carrier is saturable, so it has a maximum rate of transport (V_{max}; **Figure 40–13**). (3) There is a binding constant (K_m) for the solute, and so the whole system has a K_m (Figure 40–13). (4) Structurally similar competitive inhibitors block transport. Transporters are thus like enzymes, but generally do not modify their substrates.

Cotransporters use the gradient of one substrate created by active transport to drive the movement of the other substrate.

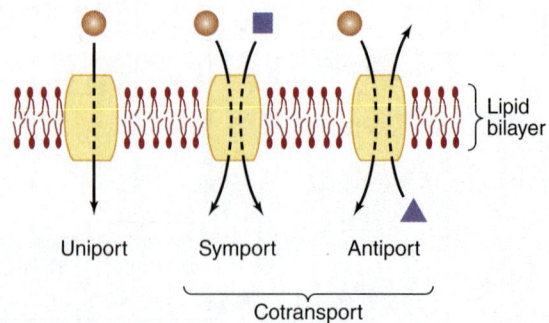

FIGURE 40–12 Schematic representation of types of transport systems. Transporters can be classified with regard to the direction of movement and whether one or more unique molecules are moved. A uniport can also allow movement in the opposite direction, depending on the concentrations inside and outside a cell of the molecule transported. (Reproduced with permission from Alberts B, Bray D, Lewis J, et al: *Molecular Biology of the Cell*. New York, NY: Garland; 1983.)

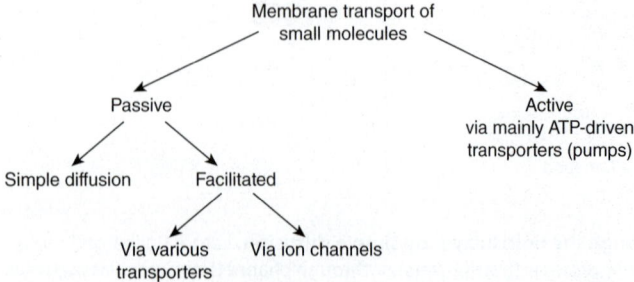

FIGURE 40–11 A schematic diagram of the two types of membrane transport of small molecules.

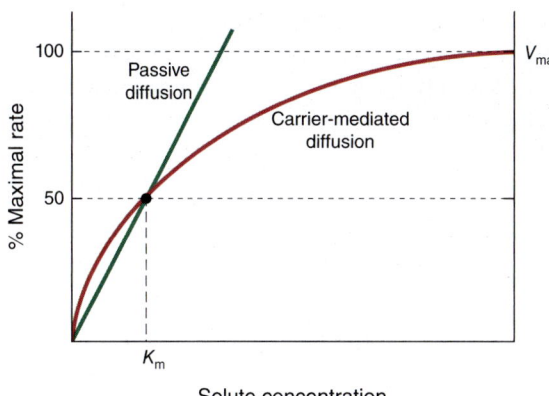

FIGURE 40–13 **A comparison of the kinetics of carrier-mediated (facilitated) diffusion with passive diffusion.** The rate of movement in the latter is directly proportionate to solute concentration, whereas the process is saturable when carriers are involved. The concentration at half-maximal velocity is equal to the binding constant (K_m) of the carrier for the solute. (V_{max}, maximal rate.)

The Na^+ gradient produced by the Na^+-K^+-ATPase is used to drive the transport of a number of important metabolites. The ATPase is a very important example of **primary transport**, while the Na^+-dependent systems are examples of **secondary transport** that rely on the gradient produced by another system. Thus, inhibition of the Na^+-K^+-ATPase in cells also blocks the Na^+-dependent uptake of substances like glucose.

Facilitated Diffusion Is Mediated by a Variety of Specific Transporters

Some specific solutes diffuse down electrochemical gradients across membranes more rapidly than might be expected from their size, charge, or partition coefficient. This is because specific transporters are involved. This **facilitated diffusion** exhibits properties distinct from those of simple diffusion. The rate of facilitated diffusion, a uniport system, can be saturated; that is, the number of sites involved in diffusion of the specific solutes appears finite. Many facilitated diffusion systems are stereospecific but, like simple diffusion, are driven by the transmembrane electrochemical gradient.

A **"ping-pong" mechanism** (**Figure 40–14**) helps to explain facilitated diffusion. In this model, the carrier protein exists in two principal conformations. In the **"ping"** state, it is exposed to high concentrations of solute, and molecules of the solute bind to specific sites on the carrier protein. Binding induces a conformational change that exposes the carrier to a lower concentration of solute (**"pong"** state). This process is completely reversible, and net flux across the membrane depends on the concentration gradient. The rate at which solutes enter a cell by facilitated diffusion is determined by (1) the concentration gradient across the membrane; (2) the amount of carrier available (this is a key control step); (3) the affinity of the solute-carrier interaction; (4) the rapidity of the conformational change for both the loaded and the unloaded carrier.

Hormones can regulate facilitated diffusion by changing the number of transporters available. Insulin via a complex signaling pathway increases glucose transport in fat and muscle by recruiting **glucose transporters (GLUT)** from an intracellular reservoir. Insulin also enhances amino acid transport in liver and other tissues. One of the coordinated actions of glucocorticoid hormones is to enhance transport of amino acids into liver, where the amino acids then serve as a substrate for gluconeogenesis. Growth hormone increases amino acid transport in all cells, and estrogens do this in the uterus. There are at least five different carrier systems for amino acids in animal cells. Each is specific for a group of closely related amino acids, and most operate as Na^+-symport systems (see Figure 40–12).

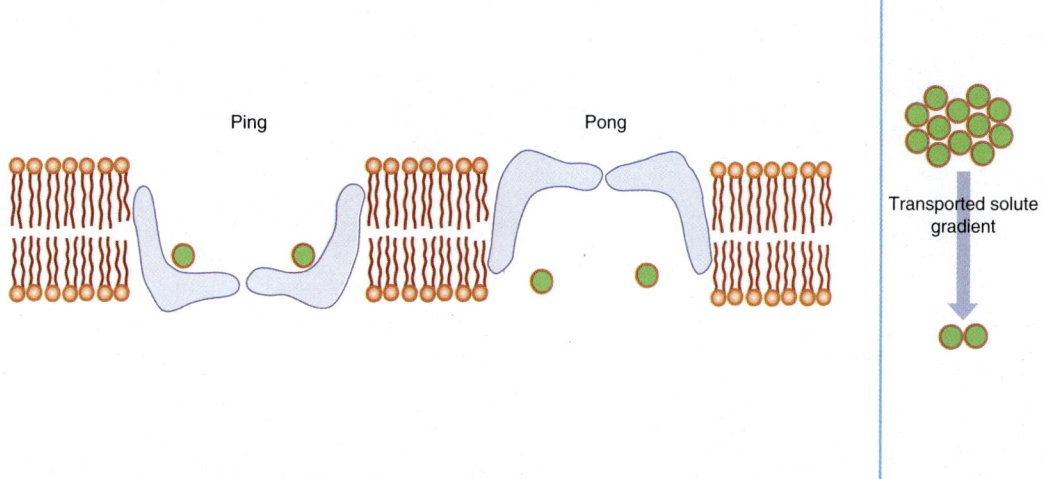

FIGURE 40–14 **The "ping-pong" model of facilitated diffusion.** A protein carrier (blue structure) in the lipid bilayer associates with a solute in high concentration on one side of the membrane. A conformational change ensues ("ping" to "pong"), and the solute is discharged on the side favoring the new equilibrium (solute concentration gradient shown schematically, right). The empty carrier then reverts to the original conformation ("pong" to "ping") to complete the cycle.

Ion Channels Are Transmembrane Proteins That Allow the Selective Entry of Various Ions

Natural membranes contain transmembrane channels, pore-like structures composed of proteins that constitute selective **ion channels**. Cation-conductive channels have an average diameter of about 5 to 8 nm. The **permeability** of a channel depends on the size, the extent of hydration, and the extent of charge density on the ion. Specific channels for Na^+, K^+, Ca^{2+}, and Cl^- have been identified. The functional α subunit of a Na^+ channel is schematically illustrated in **Figure 40–15**. The α subunit is composed of four domains (I-IV) each of which is formed by six contiguous transmembrane α-helices; each of these domains is connected by variable-length intra- and extracellular loops. The amino- and carboxy termini of the α subunit are located in the cytoplasm. The actual pore in the channel through which Na^+ ions pass is formed by interactions between the four domains, generating a tertiary structure by interactions between the four sets of 5,6 α-helices of domains I to IV. Na^+ channels are often **voltage sensitive** or **gated**; the voltage sensor of the channel is formed through the interaction domain I to IV, the four α-helices-4 formed when domains I to IV interact. This ~5 to 8 nm pore constitutes the center of the tertiary channel structure.

Ion channels are very **selective**, in most cases permitting the passage of only one type of ion (Na^+, Ca^{2+}, etc.). The **selectivity filter** of K^+ channels is made up of a ring of carbonyl groups donated by the subunits. The carbonyls displace bound water from the ion, and thus restrict its size to appropriate precise dimensions for passage through the channel.

Many variations on the structural theme described earlier for the Na^+ channel have been described. However, all ion channels are basically made up of transmembrane subunits that come together to form a central pore through which ions pass selectively.

The membranes of nerve cells contain well-studied ion channels that are responsible for the generation of action potentials. The activity of some of these channels is controlled by neurotransmitters; hence, channel activity can be regulated.

Ion channels are open transiently and thus are "**gated**." Gates can be controlled by opening or closing. In **ligand-gated channels**, a specific molecule binds to a receptor and opens the channel. **Voltage-gated channels** open (or close) in response to changes in membrane potential. **Mechanically gated** channels respond to mechanical stimuli (pressure and touch). Some properties of ion channels are listed in **Tables** 40–4 and **40–5**.

Detailed Studies of a K⁺ Channel & of a Voltage-Gated Channel Have Yielded Major Insights Into Their Actions

There are at least four features of ion channels that must be elucidated: (1) their overall structures; (2) how they conduct ions so rapidly; (3) their selectivity; and (4) their gating properties. As described later, considerable progress in tackling these difficult problems has been made.

The **K⁺ channel** (KvAP) is an integral membrane protein composed of four identical subunits, each with two transmembrane segments, creating an inverted "**V**"-**like** structure (**Figure 40–16**). The part of the channels that confers ion

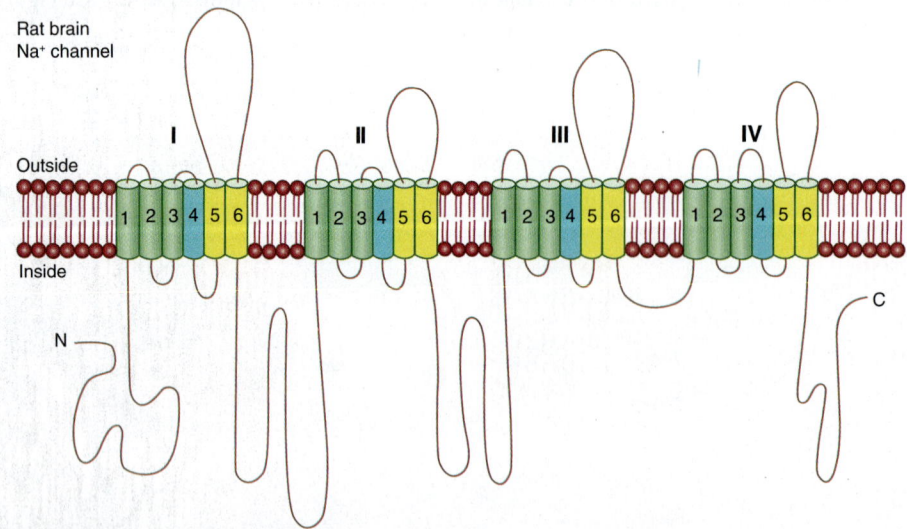

FIGURE 40–15 **Diagrammatic representation of the structures of an ion channel (a Na⁺ channel of rat brain).** The Roman numerals indicate the four domains (I-IV) of the Na⁺ channel α subunit. The α-helical transmembrane domains of each domain are numbered 1 to 6. The four blue-shaded subunits in the different domains represent the voltage-sensing portion of the α subunit. The actual pore through which the ions (Na⁺) pass is not shown, but is formed by apposition of the 5 and 6 transmembrane α-helices of domains I to IV (*colored yellow*). The specific areas of the subunits involved in the opening and closing of the channel are also not indicated. (Reproduced with permission from Catterall WA. Structure and function of voltage-sensitive ion channels. *Science.* 1988;242(4875):50-61.)

TABLE 40–5 Some Properties of Ion Channels

- They are composed of transmembrane protein subunits.
- Most are highly selective for one ion; a few are nonselective.
- They allow impermeable ions to cross membranes at rates approaching diffusion limits.
- They can permit ion fluxes of 10^6-10^7/s.
- Their activities are regulated.
- The main types are voltage-gated, ligand-gated, and mechanically gated.
- They are usually highly conserved across species.
- Most cells have a variety of Na^+, K^+, Ca^{2+}, and Cl^- channels.
- Mutations in genes encoding them can cause specific diseases.
- Their activities are affected by certain drugs.

selectivity (the **selectivity filter**) measures 12-Å long (a relatively short length of the membrane, so K^+ does not have far to travel in the membrane) and is situated at the wide end of the inverted "**V**." The large, water-filled cavity and helical dipoles shown in Figure 40–16 help overcome the relatively large electrostatic energy barrier for a cation to cross the membrane. The selectivity filter is lined with carbonyl oxygen atoms (contributed by a TVGYG sequence), providing a number of sites with which K^+ can interact. K^+ ions, which dehydrate as they enter the narrow selectivity filter, fit with proper coordination into the filter, but Na^+ is too small to interact with the carbonyl oxygen atoms in correct alignment and is rejected. Two K^+ ions, when close to each other in the filter, repel one another. This repulsion overcomes interactions between K^+ and the surrounding protein molecule and allows very rapid conduction of K^+ with high selectivity.

Other studies on a voltage-gated ion channel (HvAP) in *Aeropyrum pernix* have revealed many features of its voltage-sensing and voltage-gating mechanisms. This channel is made up of four subunits, each with six transmembrane segments. One of the six segments (S4 and part of S3) is the voltage sensor. It behaves like a **charged paddle** (**Figure 40–17**), in that it can move through the interior of the membrane transferring four positive charges (due to four Arg residues in each subunit) from one membrane surface to the other in response

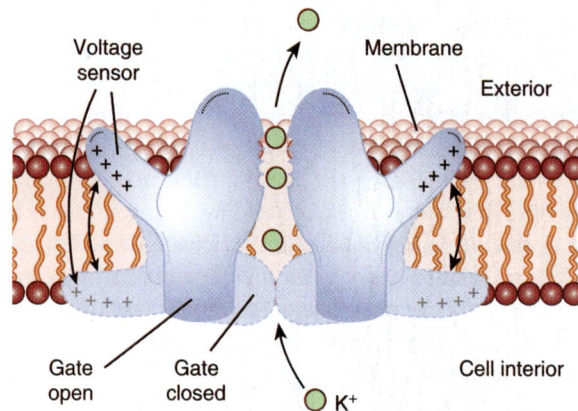

FIGURE 40–17 Schematic diagram of the voltage-gated K^+ channel of *Aeropyrum pernix*. The voltage sensors behave like charged paddles that move through the interior of the membrane. Four voltage sensors (only two are shown here) are linked mechanically to the gate of the channel. Each sensor has four positive charges contributed by arginine residues.

to changes in voltage. There are four voltage sensors in each channel, linked to the gate. The gate part of the channel is constructed from S6 helices (one from each of the subunits). Movements of this part of the channel in response to changing voltage effectively close the channel or reopen it, in the latter case allowing a current of ions to cross the membrane.

Ionophores Are Molecules That Act as Membrane Shuttles for Various Ions

Certain microbes synthesize small cyclic organic molecules, **ionophores, such as valinomycin** that function as shuttles for the movement of ions (K^+ in the case of valinomycin) across membranes. Ionophores contain hydrophilic centers that are surrounded by peripheral hydrophobic regions. Specific ions bind within the hydrophilic center of the molecule, which then diffuses through the membrane efficiently delivering the ion in question to the cytosol. Other ionophores (the polypeptide antibiotic **gramicidin**) fold up to form hollow channels through which ions can traverse the membrane.

Microbial toxins such as **diphtheria toxin** and activated **serum complement components** can produce large pores in cellular membranes and thereby provide macromolecules with direct access to the internal milieu. The toxin α-**hemolysin** (produced by certain species of *Streptococcus*) consists of seven subunits that come together to form a β-barrel that allows metabolites like ATP to leak out of cells, resulting in cell lysis.

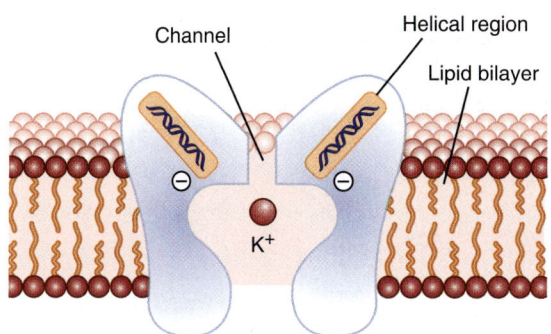

FIGURE 40–16 Schematic diagram of the structure of a K^+ channel (KvAP) from *Streptomyces lividans*. A single K^+ is shown in a large aqueous cavity inside the membrane interior. Two helical regions of the channel protein are oriented with their carboxylate ends pointing to where the K^+ is located. The channel is lined by carboxyl oxygen.

Aquaporins Are Proteins That Form Water Channels in Certain Membranes

In certain cells (eg, red cells and cells of the collecting ductules of the kidney), the movement of water by simple diffusion is augmented by movement through **water channels**. These channels are composed of tetrameric transmembrane proteins named **aquaporins**. At least 10 distinct aquaporins (AP-1 to AP-10) have been identified. Crystallographic and other studies have revealed how these channels permit passage of water but exclude passage of ions and protons. In essence, the pores are too narrow to permit passage of ions. Protons are excluded by the fact that the oxygen atom of water binds to two asparagine residues lining the channel, making the water unavailable to participate in an H^+ relay, and thus preventing entry of protons. Mutations in the gene encoding AP-2 have been shown to be the cause of one type of **nephrogenic diabetes insipidus**, a condition in which there is an inability to concentrate urine.

ACTIVE TRANSPORT SYSTEMS REQUIRE A SOURCE OF ENERGY

The process of active transport differs from diffusion in that molecules are transported against concentration gradients; hence, energy is required. This energy can come from the hydrolysis of ATP, from electron movement, or from light. The **maintenance of electrochemical** gradients **in biologic systems is so important that it consumes approximately 30% of the total energy expenditure in the cell**.

As shown in **Table 40–6, four major classes of ATP-driven active transporters** (**P, F, V,** and **ABC** transporters) have been recognized. The nomenclature is explained in the legend to the table. The first example of the P class, the Na⁺-K⁺-ATPase, is discussed later. The Ca²⁺ ATPase of muscle is discussed in Chapter 51. The second class is referred to as F-type. The most important example of this class is the mt ATP synthase, described in Chapter 13. V-type active transporters pump protons into lysosomes and other structures. ABC transporters include the **CFTR** protein, a chloride channel involved in the causation of cystic fibrosis (described later in this chapter and in Chapter 58). Another important member of this class is the multidrug-resistance-1 protein (**MDR-1 protein**). This transporter

will pump a variety of drugs, including many anticancer agents, out of cells. It is a very important cause of cancer cells exhibiting resistance to chemotherapy, although many other mechanisms are also implicated (see Chapter 56).

The Na⁺-K⁺-ATPase of the Plasma Membrane Is a Key Enzyme in Regulating Intracellular Concentrations of Na⁺ & K⁺

As shown in Table 40–1, cells maintain a low intracellular Na⁺ concentration and a high intracellular K⁺ concentration, along with a net negative electrical potential inside. The pump that maintains these ionic gradients is an ATPase that is activated by Na⁺ and K⁺ (**Na⁺-K⁺-ATPase**). The Na⁺-K⁺-ATPase pumps three Na⁺ out and two K⁺ into cells (**Figure 40–18**). This pump is an integral membrane protein that contains a transmembrane domain allowing the passage of ions, and cytosolic domains that couple ATP hydrolysis to transport. There are catalytic centers for both ATP and Na⁺ on the cytoplasmic (inner) side of the plasma membrane (PM), while there are K⁺–binding sites located on the extracellular side of the membrane. Phosphorylation by ATP induces a conformational change in the protein leading to the transfer of three Na⁺ ions from the inner to the outer side of the plasma membrane. Two molecules of K⁺ bind to sites on the protein on the external surface of the cell membrane, resulting in dephosphorylation of the protein and transfer of the K⁺ ions across the membrane to the interior. Thus, three Na⁺ ions are transported out for every two K⁺ ions entering. This differential ion transport creates a charge imbalance between the inside and the outside of the cell, making the cell interior more negative (an **electrogenic** effect). Two clinically important cardiac drugs **ouabain** and **digitalis**, inhibit the Na⁺-K⁺-ATPase by binding to the extracellular domain. This enzyme can consume significant amounts of cellular ATP energy. The Na⁺-K⁺-ATPase can be coupled to various other transporters, such as those involved in transport of glucose (see later).

TABLE 40–6   Major Types of ATP-Driven Active Transporters

Type	Example With Subcellular Location
P-type	Ca²⁺ ATPase (SR); Na⁺-K⁺-ATPase (PM)
F-type	mt ATP synthase of oxidative phosphorylation
V-type	The ATPase that pumps protons into lysosomes and synaptic vesicles
ABC transporter	CFTR protein (PM); MDR-1 protein (PM)

Abbreviations: CFTR, cystic fibrosis transmembrane regulator protein, a Cl⁻ transporter, and the protein implicated in the causation of cystic fibrosis (see later in this chapter); MDR-1 protein (multidrug-resistance-1 protein), a protein that pumps many chemotherapeutic agents out of cancer cells and is thus an important contributor to the resistance of certain cancer cells to treatment; mt, mitochondrial; PM, plasma membrane; SR, sarcoplasmic reticulum of muscle.
P (in P-type) signifies phosphorylation (these proteins autophosphorylate).
F (in F-type) signifies energy coupling factors.
V (in V-type) signifies vacuolar.
ABC signifies ATP-binding cassette transporter (all have two nucleotide-binding domains and two transmembrane segments).

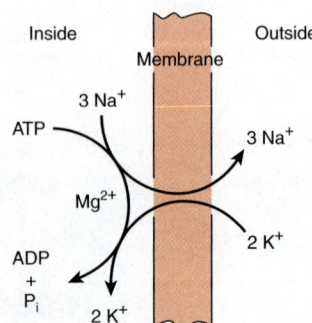

FIGURE 40–18   Stoichiometry of the Na⁺-K⁺-ATPase pump. This pump moves three Na⁺ ions from inside the cell to the outside and brings two K⁺ ions from the outside to the inside for every molecule of ATP hydrolyzed to ADP by the membrane-associated ATPase. Ouabain and other cardiac glycosides inhibit this pump by acting on the extracellular surface of the membrane. (Reproduced with permission from R Post.)

TRANSMISSION OF NERVE IMPULSES INVOLVES ION CHANNELS & PUMPS

The membrane enclosing **neuronal cells** maintains an asymmetry of inside-outside voltage (electrical potential) and is also **electrically excitable** due to the presence of voltage-gated channels. When appropriately stimulated by a chemical signal mediated by a specific synaptic membrane receptor (see discussion of the transmission of biochemical signals, later), channels in the membrane are opened to allow the rapid influx of Na^+ or Ca^{2+} (with or without the efflux of K^+), so that the voltage difference rapidly collapses, and that segment of the membrane is **depolarized**. However, as a result of the action of the ion pumps in the membrane, the gradient is quickly restored.

When large areas of the membrane are **depolarized** in this manner, the electrochemical disturbance propagates in wave-like form down the membrane, generating a **nerve impulse**. **Myelin sheets**, formed by Schwann cells, wrap around nerve fibers and provide an **electrical insulator** that surrounds most of the nerve and greatly speeds up the propagation of the wave (signal) by allowing ions to flow in and out of the membrane only where the membrane is free of the insulation (at the **nodes of Ranvier**). The myelin membrane has a very high lipid content that accounts for its excellent insulating property. Relatively few proteins are found in the myelin membrane; those present appear to hold together multiple membrane bilayers to form the hydrophobic insulating structure that is impermeable to ions and water. Certain diseases, for example, **multiple sclerosis** and the **Guillain-Barré syndrome**, are characterized by demyelination and impaired nerve conduction.

TRANSPORT OF GLUCOSE INVOLVES SEVERAL MECHANISMS

A discussion of the transport of glucose summarizes many of the points discussed earlier. Glucose must enter cells as the first step in energy utilization. A number of different glucose transporters (GLUTs) are involved, varying in different tissues (see Table 19–2). In adipocytes and skeletal muscle, glucose enters by a specific transport system (GLUT4) that is enhanced by insulin. Changes in transport are primarily due to alterations of V_{max} (presumably from more or fewer transporters), but changes in K_m may also be involved.

Glucose transport in the small intestine involves some different aspects of the principles of transport discussed earlier. Glucose and Na^+ bind to different sites on a **Na^+-glucose symporter** located at the **apical surface**. Na^+ moves into the cell down its electrochemical gradient and "drags" glucose with it (**Figure 40–19**). Therefore, the greater the Na^+ gradient, the more glucose enters; and if Na^+ in extracellular fluid is low, glucose transport stops. To maintain a steep Na^+ gradient, this Na^+-glucose symporter is dependent on gradients generated by the Na^+-K^+-ATPase, which maintains a low intracellular

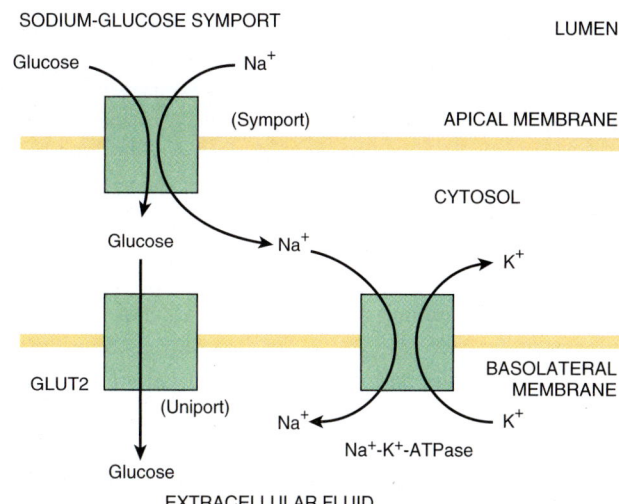

FIGURE 40–19 **The transcellular movement of glucose in an intestinal cell.** Glucose follows Na^+ across the luminal epithelial membrane. The Na^+ gradient that drives this symport is established by Na^+-K^+ exchange, which occurs at the basolateral membrane facing the extracellular fluid compartment via the action of the Na^+-K^+-ATPase. Glucose at high concentration within the cell moves "downhill" into the extracellular fluid by facilitated diffusion (a uniport mechanism), via GLUT2 (a glucose transporter, see Table 19–2). The sodium-glucose symport actually carries 2 Na^+ for each glucose.

Na^+ concentration. Similar mechanisms are used to transport other sugars as well as amino acids across the apical lumen in polarized cells such as are found in the intestine and kidney. The transcellular movement of glucose in this case involves one additional component, a uniport (see Figure 40–19) that allows the glucose accumulated within the cell to move across the **basolateral membrane** and involves **a glucose uniporter** (GLUT2).

The treatment of severe cases of **diarrhea** (such as is found in cholera) makes use of the above information. In **cholera** (see Chapter 57), massive amounts of fluid can be passed as watery stools in a very short time, resulting in severe dehydration and possibly death. **Oral rehydration therapy**, consisting primarily of **NaCl and glucose**, has been developed by the World Health Organization (WHO). The transport of glucose and Na^+ across the intestinal epithelium forces (via osmosis) movement of water from the lumen of the gut into intestinal cells, resulting in rehydration. Glucose alone or NaCl alone would not be effective.

CELLS TRANSPORT CERTAIN MACROMOLECULES ACROSS THE PLASMA MEMBRANE BY ENDOCYTOSIS & EXOCYTOSIS

The process by which cells take up large molecules is called **endocytosis**. Some of these molecules, when hydrolyzed inside the cell, **yield nutrients** (eg, polysaccharides, proteins, and polynucleotides). Endocytosis also provides a mechanism for **regulating** the content of certain membrane components, hormone receptors being a case in point. Endocytosis can be used to learn more about how cells function. DNA from one cell type can be used to transfect a different cell and alter the latter's function or phenotype. A specific gene is often employed in these experiments, and this provides a unique way to study and analyze the regulation of that gene. **DNA transfection** depends on endocytosis, which is responsible for the entry of DNA into the cell. Such experiments commonly use calcium phosphate since Ca^{2+} stimulates endocytosis and precipitates DNA, which makes the DNA a better object for endocytosis (see Chapter 39). Cells also **release macromolecules** by **exocytosis**. Endocytosis and exocytosis both involve vesicle formation with or from the plasma membrane.

Endocytosis Involves Ingestion of Parts of the Plasma Membrane

Almost all eukaryotic cells are continuously recycling parts of their plasma membranes. Endocytotic vesicles are generated when segments of the plasma membrane invaginate, enclosing a small volume of extracellular fluid and its contents. The vesicle then pinches off as the fusion of plasma membranes seals the neck of the vesicle at the original site of invagination (**Figure 40–20**). The bilayer lipid membrane, or **vesicle** so generated, then fuses with other membrane structures and

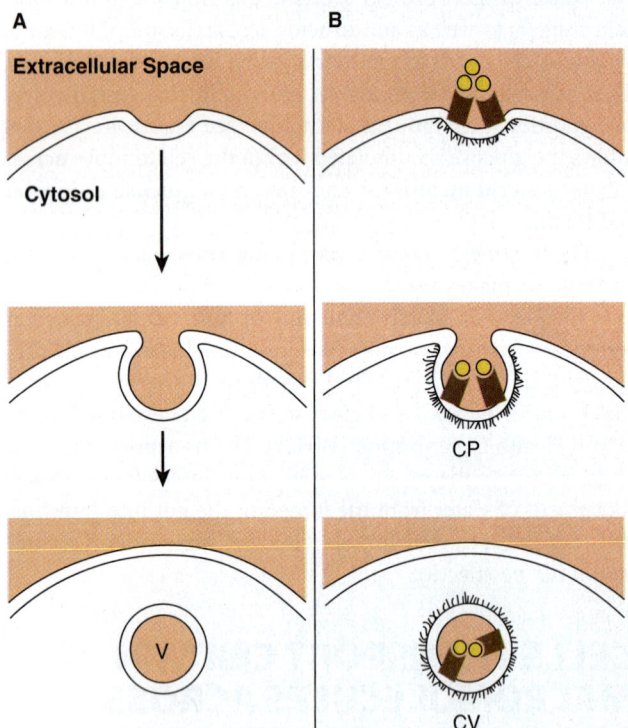

FIGURE 40–20 **Two types of pinocytosis.** An endocytotic vesicle (V) forms as a result of invagination of a portion of the plasma membrane. Fluid-phase pinocytosis (**A**) is random and nondirected. Absorptive (receptor-mediated endocytosis) (**B**) is selective and occurs in coated pits (CP) lined with the protein clathrin (the fuzzy material). Targeting is provided by receptors (*brown symbols*) specific for a variety of molecules. This results in the formation of an internalized clathrin-coated vesicle (CV).

thus achieves the transport of its contents to other cellular compartments or even back to the cell exterior. Most endocytotic vesicles fuse with **primary lysosomes** to form **secondary lysosomes**, which contain hydrolytic enzymes and are therefore specialized organelles for intracellular disposal. The macromolecular contents are digested to yield amino acids, simple sugars, or nucleotides, which are transported out of the vesicles to be (re)used by the cell. Endocytosis requires (1) energy, usually from the hydrolysis of ATP; (2) Ca^{2+}; and (3) cytoskeletal elements in the cell (see Chapter 51).

There are two **general types of endocytosis**. **Phagocytosis** occurs only in specialized cells such as macrophages and granulocytes. Phagocytosis involves the ingestion of large particles such as viruses, bacteria, cells, or cellular debris. Macrophages are extremely active in this regard and may ingest 25% of their volume per hour. In so doing, a macrophage may internalize 3% of its plasma membrane each minute or the entire membrane every 30 minutes.

Pinocytosis ("cell drinking") is a property of all cells and leads to the cellular uptake of fluid and fluid contents. There are two types of pinocytosis. **Fluid-phase pinocytosis** is a nonselective process in which the uptake of a solute by formation of small vesicles is simply proportionate to its concentration in the surrounding extracellular fluid. The formation of these vesicles is an extremely active process. Fibroblasts, for example, internalize their plasma membrane at about one-third the rate of macrophages. This process occurs more rapidly than membranes are made. The surface area and volume of a cell do not change much, so membranes must be replaced by exocytosis or by being recycled as fast as they are removed by endocytosis.

The other type of pinocytosis, **absorptive pinocytosis** or **receptor-mediated endocytosis**, is primarily responsible for the uptake of specific macromolecules for which there are binding sites on the plasma membrane. These high-affinity receptors permit the selective concentration of ligands from the medium, minimize the uptake of fluid or soluble unbound macromolecules, and markedly increase the rate at which specific molecules enter the cell. The vesicles formed during absorptive pinocytosis are derived from invaginations (pits) that are coated on the cytoplasmic side with a filamentous material and are appropriately named **coated pits**. In many systems, the protein **clathrin** is the filamentous material. It has a three-limbed structure (called a **triskelion**), with each limb being made up of one light and one heavy chain of clathrin. The polymerization of clathrin into a vesicle is directed by **assembly particles**, composed of four **adapter proteins**. These interact with certain amino acid sequences in the receptors that become cargo, ensuring selectivity of uptake. The lipid **phosphatidylinositol 4,5-bisphosphate** (PIP$_2$) (see Chapter 21) also plays an important role in vesicle assembly. In addition, the protein **dynamin**, which both binds and hydrolyzes GTP, is necessary for the pinching off of clathrin-coated vesicles from the cell surface. Coated pits may constitute as much as 2% of the surface of some cells. Other aspects of vesicles are discussed in Chapter 49.

As an example, **the low-density lipoprotein (LDL)** molecule and its receptor (see Chapter 25) are internalized by means of coated pits containing the LDL receptor. Endocytotic vesicles containing the LDL-bound LDL receptor complex fuse to lysosomes in the cell. The receptor is released and recycled back to the cell surface membrane, but the apoprotein of LDL is degraded and the cholesteryl esters metabolized. Synthesis of the LDL receptor is regulated by secondary or tertiary consequences of pinocytosis, for example, by metabolic products—such as cholesterol—released during the degradation of LDL. Disorders of the LDL receptor and its internalization are medically important and are discussed in Chapters 25 and 26.

Absorptive pinocytosis of **extracellular glycoproteins** requires that the glycoproteins carry specific carbohydrate recognition signals. These recognition signals are bound by membrane receptor molecules that play a role analogous to that of the LDL receptor. A **galactosyl receptor** on the surface of hepatocytes is instrumental in the absorptive pinocytosis of **asialoglycoproteins** from the circulation (see Chapter 46). **Acid hydrolases** taken up by absorptive pinocytosis in fibroblasts are recognized by their **mannose-6-phosphate** moieties. Interestingly, the mannose-6-phosphate moiety also seems to play an important role in the intracellular targeting of the acid hydrolases to the lysosomes of the cells in which they are synthesized.

There is a problematic side to receptor-mediated endocytosis in that **viruses** which cause such diseases as hepatitis (affecting liver cells), poliomyelitis (affecting motor neurons), AIDS (affecting T cells), and COVID-19 (affecting lung and other cell types), initiate their infectious cycles by entering cells via this mechanism. **Iron toxicity** also begins with excessive uptake due to endocytosis.

Exocytosis Releases Certain Macromolecules From Cells

Most cells release macromolecules to the exterior by **exocytosis**. This process is also involved in membrane remodeling, when the components synthesized in the ER and Golgi are carried in vesicles that fuse with the plasma membrane. The signal for this "classical exocytosis" (see later) is often a hormone which, when it binds to a cell surface receptor, induces a local and transient change in Ca^{2+} concentration. Ca^{2+} triggers exocytosis. **Figure 40–21** provides a comparison of the mechanisms of exocytosis and endocytosis.

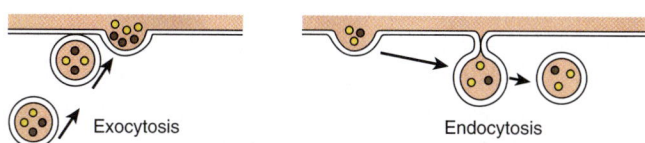

FIGURE 40–21 **A comparison of the mechanisms of endocytosis and exocytosis.** Exocytosis involves the contact of two inside-surface (cytoplasmic side) monolayers, whereas endocytosis results from the contact of two outer-surface monolayers.

Molecules released by this mode of exocytosis have at least three fates: (1) they are membrane proteins and remain associated with the cell surface; (2) they can become part of the extracellular matrix, for example, collagen and glycosaminoglycans; (3) they can enter extracellular fluid and signal other cells. Insulin, parathyroid hormone, and the catecholamines are all packaged in granules and processed within cells, to be released on appropriate stimulation.

VARIOUS SIGNALS CAN BE TRANSMITTED ACROSS MEMBRANES

Specific biochemical signals such as neurotransmitters, hormones, and immunoglobulins bind to integral transmembrane receptor proteins via their exposed extracellular domains, thereby transmitting information through the membranes to the cytoplasm. This process, called **transmembrane signaling** or **signal transduction**, involves the generation of a number of second messenger signaling molecules, including cyclic nucleotides, calcium, phosphoinositides, and diacylglycerol (see Chapter 42). Many of the steps involve phosphorylation of receptors and downstream proteins.

GAP JUNCTIONS ALLOW DIRECT FLOW OF MOLECULES FROM ONE CELL TO ANOTHER

Gap junctions are structures that permit direct transfer of small molecules (up to ~1200 Da) from one cell to its neighbor. Gap junctions are composed of a family of proteins called **connexins** that form a bihexagonal structure consisting of 12 such proteins. Six connexins form a connexin hemichannel and join to a similar structure in a neighboring cell to make a complete, membrane-spanning **connexon channel** (**Figure 40–22**). One gap junction contains several connexons. Different connexins are found in different tissues. Mutations in genes encoding connexins have been found to be associated with a number of conditions, including cardiovascular abnormalities, one type of deafness, and the X-linked form of Charcot-Marie-Tooth disease (a demyelinating neurologic disorder).

EXTRACELLULAR VESICLES (EXOSOMES) REPRESENT A NOVEL, & PREVIOUSLY UNDERAPPRECIATED MECHANISM OF CELL–CELL COMMUNICATION

In the last decade, a class of small, heterogeneous, secreted vesicles, broadly termed **extracellular vesicles**, have been identified and characterized. These extracellular vesicles have been implicated as a new and important mediator of

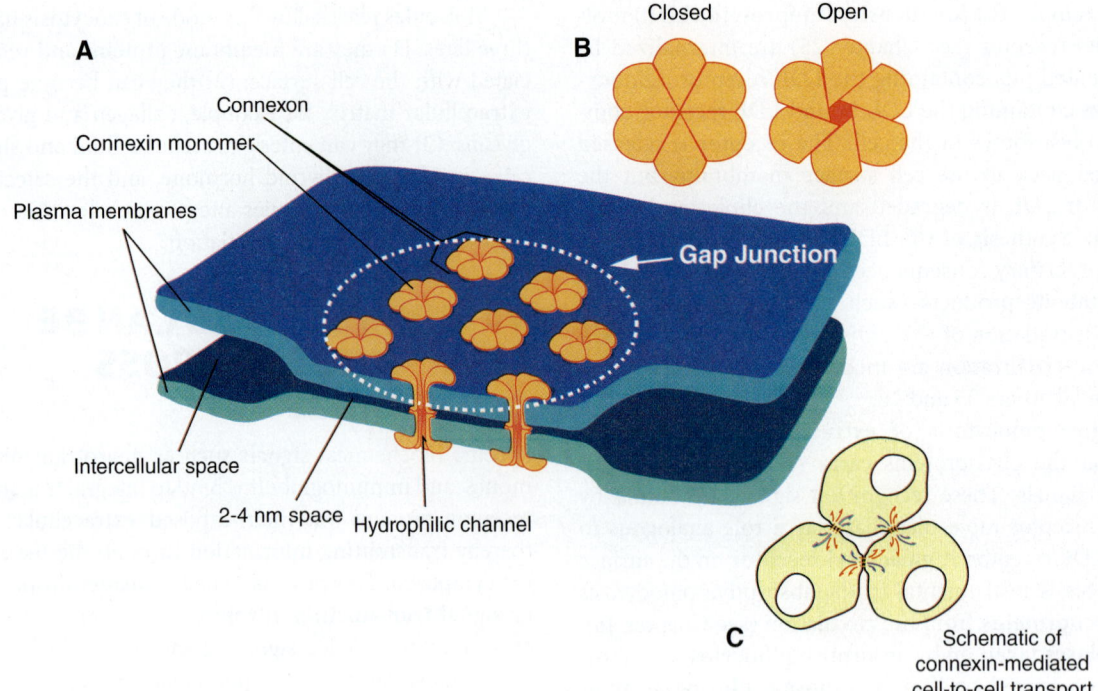

FIGURE 40–22 **Schematic diagram of a gap junction.** Shown schematically are (**A**) the relationships between cells containing connexin; (**B**) open and closed complete connexin channels; and (**C**) the flow of molecules (*blue, red arrows*) between a group of three cells. One connexon is made from two hemiconnexons. Each hemiconnexon is made from six connexin molecules. Small solutes are able to diffuse through the central channel when open, thereby providing a direct mechanism of cell–cell communication. Note that connexins connect cells that are within 2 to 4 nm of each other. (Image source: http://upload.wikimedia.org/wikipedia/commons/b/b7/Gap_cell_junction-en.svg).

cell–cell communication that likely contribute importantly to both normal and pathologic physiology. These vesicles, enclosed by a lipid bilayer, are somewhat heterogeneous in size (30-2000 nm diameter), and are generated by at least two distinct mechanisms (**Figure 40–23**): **microvesicles** are generated by budding from the plasma membrane of a **source cell**, while **exosomes** are generated from the **multivesicular body** (**MVB**), a component of the endocytic membrane trafficking system described earlier (see Figures 40–20 and 40–21). Exosomes are secreted from the source cell on fusion of the MVB with the plasma membrane. In both cases, the released extracellular vesicles (exosomes and microvesicles) ultimately fuse to their **target cell** to deliver a distinct "payload." Unfortunately, given the recent discovery of extracellular vesicles, the exact names and terms used to describe these vesicles, their cargos, and relevant source and target cells vary in the biomedical literature. Moreover, the terms "microvesicle" and "exosome" are often lumped together as simply "exosomes."

Vesicle content varies from source cell to source cell and has even been reported to be different from the same source cell grown under different conditions. Vesicle payloads can include a variety of cytoplasmic and nuclear proteins, membrane-bound proteins ranging from channels to receptors, major histocompatibility complex (MHC) molecules, lipid raft-interacting proteins, DNA, mRNA, large and small ncRNAs, as well as small protein and bioactive small molecules (see Figure 40–23). Given the rich and broad diversity of

vesicle/exosome contents, it is not surprising that these structures have been implicated in a very broad range of biology and diseases. Moreover, given their membrane protein content and the fact that extracellular vesicles appear to target specific recipient cells, the potential value of exosomes as therapeutic delivery systems is receiving significant interest and attention in the pharmaceutical and biotechnology industries. Future work will determine whether this new and exciting area of biomedical research on extracellular vesicles generates efficacious therapeutics.

MUTATIONS AFFECTING MEMBRANE PROTEINS CAUSE DISEASES

In view of the fact that membranes are located in so many organelles and are involved in so many processes, it is not surprising that mutations affecting their protein constituents should result in many diseases or disorders. While some mutations directly affect the function of membrane proteins, the majority cause protein misfolding that impair membrane trafficking at any of a number of steps from their site of synthesis in the ER to the plasma membrane or other intracellular sites/organelles (see Chapter 49). Examples of diseases or disorders due to abnormalities in membrane proteins are listed in **Table 40–7**. These mainly reflect mutations in proteins of the plasma membrane, with one affecting lysosomal function (I-cell disease).

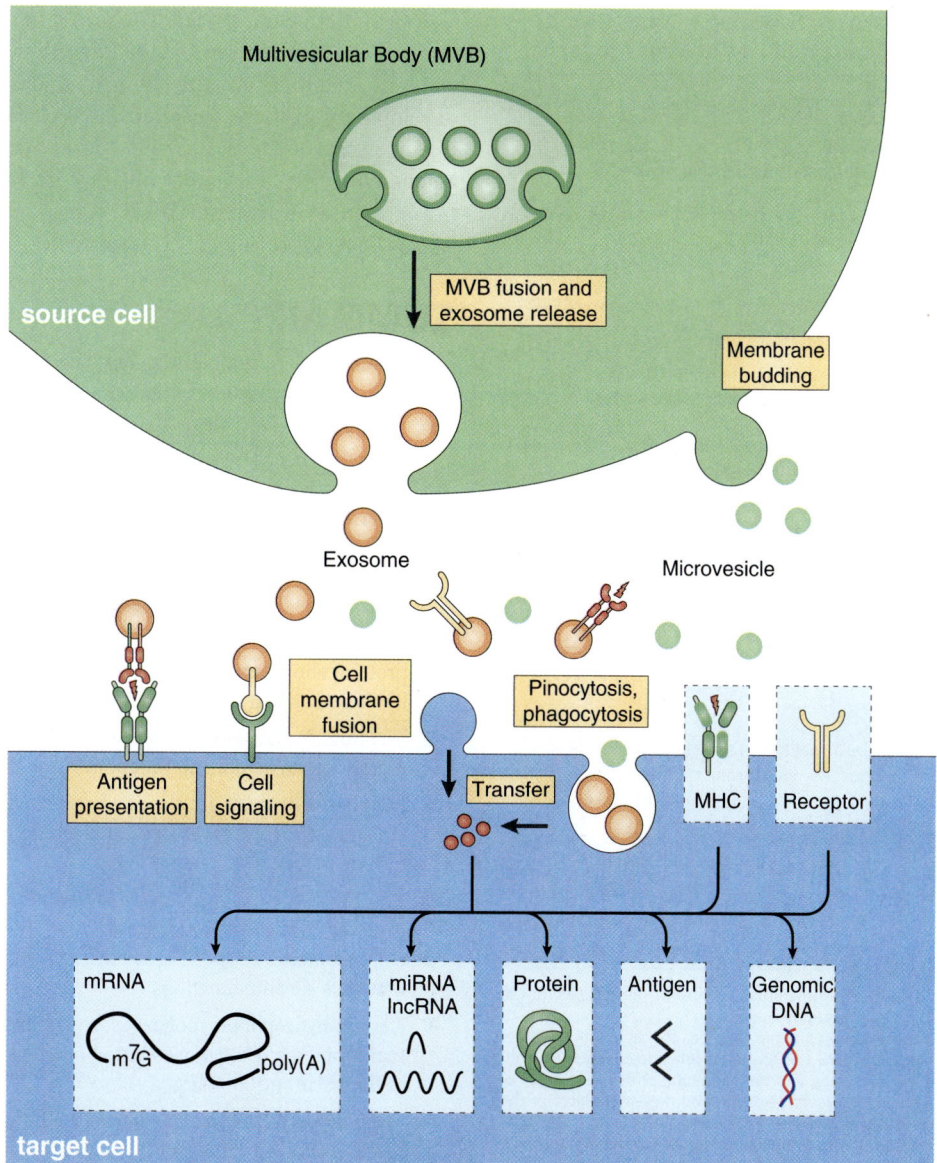

FIGURE 40–23 **Cell–cell communication via extracellular vesicles.** Shown are the proposed mechanisms for the formation and production of exosomes and microvesicles via exocytosis (exosome) and membrane budding (microvesicle) from a **source** cell. Vesicles produced in the multivesicular body (MVB) can be exocytosed following fusion with the plasma membrane as shown, or budded into the extracellular space. All of these processes involve the collection of proteins, lipids, and signaling molecules previously implicated in exocytosis and budding (not shown). Once released from the source cell, the resulting exosomes and/or microvesicles locate their **target** cell, and following the types of vesicle-target cell interactions shown, release their contents (see black arrows within the target cell). Different vesicles have been shown to contain RNA (mRNA, miRNA, lncRNA; see Chapter 36) and DNA, specific bioactive proteins and lipids; antigens; and biologically active small molecules. Importantly, extracellular vesicles have been shown to have both positive and negative biologic effects on target cells in both normal and pathologic states.

Proteins in plasma membranes can be classified as receptors, transporters, ion channels, enzymes, and structural components. Members of all of these classes are often glycosylated, so that mutations affecting this process may alter their function (see Chapter 46). Mutations in receptors can cause defects in transmembrane signaling, a common occurrence in cancer (see Chapter 56). Many genetic diseases or disorders have been ascribed to mutations affecting various proteins involved in the transport of amino acids, sugars, lipids, urate, anions, cations, water, and vitamins across the plasma membrane.

Mutations in genes encoding proteins in other membrane-bound compartments can also have harmful consequences. For example, mutations in genes encoding mitochondrial membrane proteins involved in oxidative phosphorylation can cause neurologic and other problems (eg, **Leber hereditary optic neuropathy [LHON]**, a condition in which some success with gene therapy has been reported).

Membrane proteins can also be affected by conditions other than mutations. Formation of autoantibodies to the acetylcholine receptor in skeletal muscle causes myasthenia gravis.

TABLE 40–7 Some Diseases or Pathologic States Resulting From or Attributed to Abnormalities of Membranes[a]

Disease	Abnormality
Achondroplasia (OMIM 100800)	Mutations in the gene encoding the fibroblast growth factor receptor 3
Familial hypercholesterolemia (OMIM 143890)	Mutations in the gene encoding the LDL receptor
Cystic fibrosis (OMIM 219700)	Mutations in the gene encoding the CFTR protein, a Cl⁻ transporter
Congenital long QT syndrome (OMIM 192500)	Mutations in genes encoding ion channels in the heart
Wilson disease (OMIM 277900)	Mutations in the gene encoding a copper-dependent ATPase
I-cell disease (OMIM 252500)	Mutations in the gene encoding GlcNAc phosphotransferase, resulting in absence of the Man 6-P signal for lysosomal localization of certain hydrolases
Hereditary spherocytosis (OMIM 182900)	Mutations in the genes encoding spectrin or other structural proteins in the red cell membrane
Metastasis of cancer cells	Abnormalities in the oligosaccharide chains of membrane glycoproteins and glycolipids are thought to be of importance
Paroxysmal nocturnal hemoglobinuria (OMIM 311770)	Mutation resulting in deficient attachment of the GPI anchor (see Chapter 46) to certain proteins of the red cell membrane

Abbreviations: CFTR, cystic fibrosis transmembrane regulator protein; GPI, glycosylphosphatidylinositol; LDL, low-density lipoprotein.
[a]The disorders listed are discussed further in other chapters. The table lists examples of mutations affecting two receptors, one transporter, several ion channels (ie, congenital long QT syndrome), two enzymes, and one structural protein. Examples of altered or defective glycosylation of glycoproteins are also presented. Most of the conditions listed involve the plasma membrane.
See also: Ammendolia DA, Bement WM, Brumell JH: Plasma membrane integrity: implications for health and disease. BMC Biology 2021:19:71.

Ischemia can quickly affect the integrity of various ion channels in membranes. Overexpression of P-glycoprotein (MDR-1), a plasma membrane–localized drug pump, results in multidrug resistance (MDR) in cancer cells. Abnormalities of membrane constituents other than proteins can also be harmful. With regard to lipids, excess of cholesterol (eg, in familial hypercholesterolemia), of lysophospholipid (eg, after bites by certain snakes, whose venom contains phospholipases), or of GSLs (eg, in a sphingolipidosis), can all affect membrane structure and hence function.

Cystic Fibrosis Is Due to Mutations in the Gene Encoding CFTR, a Chloride Transporter

Cystic fibrosis (CF) is a recessive genetic disorder prevalent among whites in North America and certain parts of northern Europe. CF is characterized by chronic bacterial infections of the airways and sinuses, fat maldigestion due to pancreatic exocrine insufficiency, infertility in males due to abnormal development of the vas deferens, and elevated levels of chloride in sweat (>60 mmol/L). It is now known that mutations in a gene encoding a protein named **cystic fibrosis transmembrane regulator protein (CFTR)** is responsible for CF. CFTR is a cyclic AMP-regulated Cl⁻ transporter.

SUMMARY

■ Membranes are complex dynamic structures composed of lipids, proteins, and carbohydrate-containing molecules.

■ The basic structure of all membranes is the lipid bilayer. This bilayer is formed by two sheets of phospholipids in which the hydrophilic polar head groups are directed away from each other and are exposed to the aqueous environment on the outer and inner surfaces of the membrane. The hydrophobic nonpolar tails of these molecules are oriented toward each other, in the direction of the center of the membrane.

■ Membranes are very dynamic structures. Lipids and certain proteins show rapid lateral diffusion. Flip-flop is very slow for lipids and almost nonexistent for proteins.

■ The fluid mosaic model forms a useful basis for thinking about membrane structure and function.

■ Membrane proteins are classified as integral if they are firmly embedded in the bilayer and as peripheral if they are attached to the outer or inner membrane surface.

■ The 20 or so membranes in a mammalian cell have different compositions and functions and they define essential compartments, or specialized environments, within the cell that have specific functions.

■ Certain hydrophobic molecules freely diffuse across membranes, but the movement of others is restricted because of their size and/or charge.

■ Various passive and active (usually ATP-dependent) mechanisms are employed to maintain gradients of many different molecules across different membranes.

■ Certain solutes, for example, glucose, enter cells by facilitated diffusion along a downhill gradient from high to low concentration using specific carrier proteins (transporters).

■ The major ATP-driven pumps are classified as P (phosphorylated), F (energy factors), V (vacuolar), and ABC transporters.

■ Ligand- or voltage-gated ion channels are often employed to move charged molecules (Na^+, K^+, Ca^{2+}, etc.) across membranes down their electrochemical gradients.

■ Large molecules can enter or leave cells through mechanisms such as endocytosis or exocytosis. These processes often require binding of the molecule to a receptor, which affords specificity to the process.

■ Extracellular vesicles, termed exosomes, also allow direct movement of macromolecules from cell to cell via small vesicles. Exosome payloads can include specific lipids, proteins (receptors, channels, signaling proteins), DNA, RNAs, and small bioactive molecules.

■ Mutations that affect the structure of membrane proteins may cause diseases.

REFERENCES

Ammendolia DA, Bement WM, Brumell JH: Plasma membrane integrity: implications for health and disease. BMC Biol 2021;19:7. doi.org/10.1186/s12915-021-00972-y.

Boulanger CM, Loyer X, Rautou PE, Amabile N: Extracellular vesicles in coronary artery disease. Nat Rev Cardiol 2017;14(5):259-272.

Busija AR, Patel HH, Insel PA: Caveolins and cavins in the trafficking, maturation, and degradation of caveolae: implications for cell physiology. Am Physiol Cell Physiol 2017;312:C459-C477.

Caprioli RM: Imaging Mass Spectrometry: A Perspective. J Biomol Tech 2019;30:7-11.

Dingian T, Futerman AH: The fine-tuning of cell membrane lipid bilayers accentuates their compositional complexity. Bioessays 2021;43:e2100021.

Doherty GJ, McMahon HT: Mechanisms of endocytosis. Annu Rev Biochem 2009;78:857-902.

Fujimoto T, Parmryd I: Interleaflet coupling, pinning, and leaflet asymmetry-major players in plasma membrane nanodomain formation. Front Cell Dev Biol 2017;4:155.

Harkewicz R, Dennis EA: Applications of mass spectrometry to lipids and membranes. Ann Rev Biochem 2011;80:301-325.

Jeppesen DK, Fenix AM, Franklin JL, et al.: Reassessment of exosome composition. Cell 2019;177:428-445.

Kefauver JM, Ward AB, Patapoutian A: Discoveries in structure and physiology of mechanically activated ion channels. Nature 2020;587:567-576.

Longo N: Inherited defects of membrane transport. In *Harrison's Principles of Internal Medicine*, 17th ed. Fauci AS, et al (editors). McGraw-Hill, 2008.

Mittelbrunn M, Sánchez-Madrid F: Intercellular communication: diverse structures for exchange of genetic information. Nat Rev Mol Cell Biol 2012;13:328-335.

Nakagawa T: Structures of the AMPA receptor in complex with its auxiliary subunit cornichon. Science 2019;366:1259-1263.

Nicolson GL: The Fluid-Mosaic Model of Membrane Structure: still relevant to understanding the structure, function and dynamics of biological membranes after more than 40 years. Biochim Biophys Acta 2014;1838:1451-1466.

Raposo G, Stoorvogel W: Extracellular vesicles: exosomes, microvesicles, and friends. J Cell Biol 2013;200:373-383.

Regen SL: The origin of lipid rafts. Biochemistry 2020;59:4617-4621.

Singer SJ: Some early history of membrane molecular biology. Annu Rev Physiol 2004;66:1-27.

Spielberg DR, Clancy JP: Cystic fibrosis and its management through established and emerging therapies. Annu Rev Genomics Hum Genet 2016;17:155-175.

Stone MB, Shelby SA, Veatch SL: Super-resolution microscopy: shedding light on the cellular plasma membrane. Chem Rev 2017;17(11):7457-7477.

Vance DE, Vance J (editors): *Biochemistry of Lipids, Lipoproteins and Membranes*, 5th ed. Elsevier, 2008.

Voelker DR: Genetic and biochemical analysis of non-vesicular lipid traffic. Annu Rev Biochem 2009;78:827-856.

Züllig T, Köfeler HC: High resolution mass spectrometry in lipidomics. Mass Spectrom Rev 2021;40:162-176.

The Diversity of the Endocrine System

P. Anthony Weil, PhD

ACTH	Adrenocorticotropic hormone	**IGF-I**	Insulin-like growth factor I
ANF	Atrial natriuretic factor	**LH**	Luteotropic hormone
cAMP	Cyclic adenosine monophosphate	**LPH**	Lipotropin
CBG	Corticosteroid-binding globulin	**MIT**	Monoiodotyrosine
CG	Chorionic gonadotropin	**MSH**	Melanocyte-stimulating hormone
cGMP	Cyclic guanosine monophosphate	**OHSD**	Hydroxysteroid dehydrogenase
CLIP	Corticotropin-like intermediate lobe peptide	**PNMT**	Phenylethanolamine-*N*-methyltransferase
DBH	Dopamine β-hydroxylase	**POMC**	Pro-opiomelanocortin
DHEA	Dehydroepiandrosterone	**PRL**	Prolactin
DHT	Dihydrotestosterone	**SHBG**	Sex hormone–binding globulin
DIT	Diiodotyrosine	**StAR**	Steroidogenic acute regulatory (protein)
DOC	Deoxycorticosterone	**TBG**	Thyroxine-binding globulin
EGF	Epidermal growth factor	**TEBG**	Testosterone-estrogen–binding globulin
FSH	Follicle-stimulating hormone	**TRH**	Thyrotropin-releasing hormone
GH	Growth hormone	**TSH**	Thyrotropin-stimulating hormone

BIOMEDICAL IMPORTANCE

The survival of multicellular organisms depends on their ability to adapt to a constantly changing environment. Intercellular communication mechanisms are necessary requirements for this adaptation. The nervous system and the endocrine system provide this intercellular, organism-wide communication. The nervous system was originally viewed as providing a fixed communication system, whereas the endocrine system supplied hormones, which are mobile messages. In fact, there is a remarkable convergence of these regulatory systems. For example, neural regulation of the endocrine system is important in the production and secretion of some hormones; many neurotransmitters resemble hormones in their synthesis, transport, and mechanism of action; and many hormones are synthesized in the nervous system. The word "hormone" is derived from a Greek term that means to arouse to activity. As classically defined, a hormone is a substance that is synthesized in one organ and transported by the circulatory system to act on another tissue. However, this original description is too restrictive because hormones can act on adjacent cells (paracrine action) and on the cell in which they were synthesized (autocrine action) without entering the systemic circulation. A diverse array of hormones—each with distinctive mechanisms of action and properties of biosynthesis, storage, secretion, transport, and metabolism—has evolved to provide homeostatic responses. This biochemical diversity is the topic of this chapter.

THE TARGET CELL CONCEPT

There are over 200 types of differentiated cells in humans. Only a few produce hormones, but virtually all of the 75 trillion cells in a human are targets of one or more of the 50+ known hormones. The concept of the target cell is a useful way of looking at hormone action. It was thought that hormones affected a single cell type—or only a few kinds of cells—and that a hormone elicited a unique biochemical or physiologic action. We now know that a given hormone can affect several different cell types; that more than one hormone can affect a given cell type; and that hormones can exert many different effects in one cell or in different cells. With the discovery of specific cell surface and intracellular hormone receptors, the definition of a target has been expanded to include any cell in which the hormone (ligand) binds to its receptor, whether or not a biochemical or physiologic response has yet been determined.

Several factors determine the response of a target cell to a hormone. These can be thought of in two general ways: (1) as factors that affect the concentration of the hormone at the target cell (**Table 41–1**) and (2) as factors that affect the actual response of the target cell to the hormone (**Table 41–2**).

HORMONE RECEPTORS ARE OF CENTRAL IMPORTANCE

Receptors Discriminate Precisely

One of the major challenges faced in making the hormone-based communication system work is illustrated in **Figure 41–1**. Hormones are present at very low concentrations in the extracellular fluid, generally in the femto- to nanomolar range (10^{-15}-10^{-9} mol/L). This concentration is much lower than that of the many structurally similar molecules (sterols, amino acids, peptides, and proteins) and other molecules that circulate at concentrations in the micro- to millimolar (10^{-6}-10^{-3} mol/L) range. Target cells, therefore, must

TABLE 41–1 Determinants of the Concentration of a Hormone at the Target Cell

The rate of synthesis and secretion of the hormones.
The proximity of the target cell to the hormone source (dilution effect).
The affinity (dissociation constant; K_d) of the hormone with specific plasma transport proteins (if any).
The conversion of inactive or suboptimally active forms of the hormone into the fully active form.
The rate of clearance of hormone from plasma, by other tissues, or by degradation, metabolism, or excretion.

TABLE 41–2 Determinants of the Target Cell Response

The number, relative activity, and state of occupancy of the specific receptors on the plasma membrane or in the cytoplasm or nucleus.
The metabolism (activation or inactivation) of the hormone in the target cell.
The presence of other factors within the cell that are necessary for the hormone response.
Up- or downregulation of the receptor consequent to the interaction with its ligand.
Postreceptor desensitization of the cell, including downregulation of the receptor.

distinguish not only between different hormones present in small amounts but also between a given hormone and the 10^6- to 10^9-fold excess of other similar molecules. This high degree of discrimination is provided by cell-associated recognition molecules called receptors. Hormones initiate their biologic effects by binding to hormone-specific receptors, and since any effective control system also must provide a means of stopping a response, hormone-induced actions generally, but not always, terminate when the effector dissociates from the receptor (see Figure 38–1; Type A response).

A target cell is defined by its ability to selectively bind a given hormone to its cognate receptor. Several biochemical features of this interaction are important in order for hormone-receptor interactions to be physiologically relevant: (1) binding should be specific, that is, displaceable by agonist or antagonist; (2) binding should be saturable; and (3) binding should occur within the concentration range of the expected biologic response.

Both Recognition & Coupling Domains Occur on Receptors

All receptors have at least two functional domains. A recognition domain binds the hormone ligand and a second region

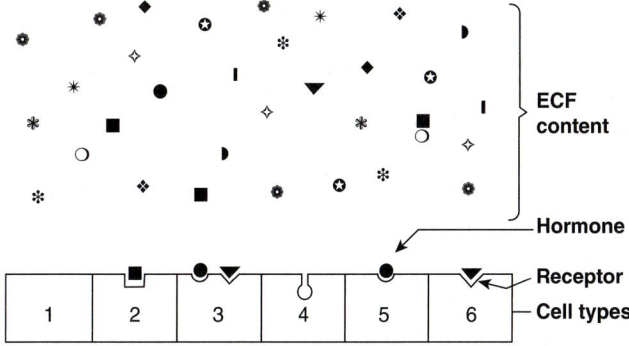

FIGURE 41–1 Specificity and selectivity of hormone receptors. Many different molecules circulate in the extracellular fluid (ECF), but only a few are recognized by hormone receptors. Receptors must select these molecules from among high concentrations of the other molecules. This simplified drawing shows that a cell may have no hormone receptors (Cell type 1), have one receptor (Cell types 2+5+6), have receptors for several hormones (Cell type 3), or have a receptor but no hormone in the vicinity (Cell type 4).

generates a signal that couples hormone recognition to some intracellular function. This coupling, or signal transduction, occurs in two general ways. Polypeptide and protein hormones and the catecholamines bind to receptors located in the plasma membrane and thereby generate a signal that regulates various intracellular functions, often by changing the activity of an enzyme. By contrast, the lipophilic steroid, retinoid, and thyroid hormones interact with intracellular receptors, and it is this ligand-receptor complex within the nucleus that directly provides the signal, generally to specific genes whose rate of transcription is thereby affected.

The domains responsible for hormone recognition and signal generation have been identified in the protein polypeptide and catecholamine hormone receptors. Like many other DNA-binding transcription factors, the steroid, thyroid, and retinoid hormone receptors have several functional domains: one site binds the hormone; another binds to specific DNA regions; a third is involved in the interaction with various coregulator proteins that result in the activation (or repression) of gene transcription; and a fourth region may specify binding to one or more other proteins that influence the intracellular trafficking of the receptor (see Figure 38–19).

The dual functions of binding and coupling ultimately define a receptor, and it is the coupling of hormone binding to signal transduction, the so-called **receptor-effector coupling**—that provides the first step in amplification of the hormonal response. This dual purpose also distinguishes the target cell receptor from the plasma carrier proteins that bind hormone but do not generate a signal (see Table 41–6).

Receptors Are Proteins

Several classes of peptide hormone receptors have been defined. For example, the insulin receptor is a heterotetramer composed of two copies of two different protein subunits ($\alpha_2\beta_2$) linked by multiple disulfide bonds in which the extracellular α subunit binds insulin and the membrane-spanning β subunit transduces the signal through the tyrosine protein kinase domain located in the cytoplasmic portion of this polypeptide. The receptors for **insulin-like growth factor I (IGF-I)** and **epidermal growth factor (EGF)** are generally similar in structure to the insulin receptor. The **growth hormone (GH)** and **prolactin (PRL)** receptors also span the plasma membrane of target cells but do not contain intrinsic protein kinase activity. Ligand binding to these receptors, however, results in the association and activation of a completely different protein kinase signaling pathway, the Jak-Stat pathway. Polypeptide hormone and catecholamine receptors, which transduce signals by altering the rate of production of **cAMP** through **G-proteins**, which are guanosine nucleotide-binding proteins that are characterized by the presence of seven membrane-spanning domains. Protein kinase activation and the generation of cyclic AMP (cAMP, 3′5′-adenylic acid; see Figure 18–5) is a downstream action of this class of receptor (see Chapter 42 for further details).

A comparison of several different steroid receptors with thyroid hormone receptors revealed a remarkable conservation of the amino acid sequence in certain regions, particularly in the DNA-binding domains. This observation led to the realization that receptors of the steroid and thyroid type are members of a large superfamily of nuclear receptors. Many related members of this family currently have no known ligand and thus are called orphan receptors. The nuclear receptor superfamily plays a critical role in the regulation of gene transcription by hormones, as described in Chapter 42.

HORMONES CAN BE CLASSIFIED IN SEVERAL WAYS

Hormones can be classified according to chemical composition, solubility properties, location of receptors, and the nature of the signal used to mediate hormonal action within the cell. A classification based on the last two properties is illustrated in **Table 41–3**, and general features of each group are illustrated in **Table 41–4**.

The hormones in group I are lipophilic. After secretion, these hormones associate with plasma transport or carrier proteins, a process that circumvents the problem of solubility while prolonging the plasma half-life of the hormone. The relative percentages of bound and free hormone are determined by the amount, binding affinity, and binding capacity of the transport protein. The free hormone, which is the biologically active form, readily traverses the lipophilic plasma membrane of all cells and encounters receptors in either the cytosol or nucleus of target cells. The ligand-receptor complex is the intracellular messenger in this group.

The second major group consists of water-soluble hormones that bind to specific receptors spanning the plasma membrane of the target cell. Hormones that bind to these surface receptors of cells communicate with intracellular metabolic processes through intermediary molecules called **second messengers** (the hormone itself is the first messenger), which are generated as a consequence of the ligand-receptor interaction. The second messenger concept arose from an observation that epinephrine binds to the plasma membrane of certain cells and increases intracellular cAMP. This was followed by a series of experiments in which cAMP was found to mediate the effects of many hormones. Hormones that employ this mechanism are shown in group II.A of Table 41–3. Atrial natriuretic factor (ANF) uses cGMP as its second messenger (group II.B). Several hormones, many of which were previously thought to affect cAMP, appear to use ionic calcium (Ca^{2+}) or metabolites of complex phosphoinositides (or both) as the intracellular second messenger signal, and are shown in group II.C of the table. The intracellular messenger for group II.D is a protein kinase–phosphatase cascade; several have been identified, and a given hormone may use more than one kinase cascade. A few hormones fit into more than one category, and assignments change as new information is discovered.

TABLE 41–3 Classification of Hormones by Mechanism of Action

I. *Hormones that bind to intracellular receptors*
　Androgens
　Calcitriol (1,25[OH]$_2$-D$_3$)
　Estrogens
　Glucocorticoids
　Mineralocorticoids
　Progestins
　Retinoic acid
　Thyroid hormones (T$_3$ and T$_4$)
II. *Hormones that bind to cell surface receptors*
　A. The second messenger is cAMP
　　α$_2$-Adrenergic catecholamines
　　β-Adrenergic catecholamines
　　Adrenocorticotropic hormone (ACTH)
　　Antidiuretic hormone (vasopressin)
　　Calcitonin
　　Chorionic gonadotropin (CG)
　　Corticotropin-releasing hormone
　　Follicle-stimulating hormone (FSH)
　　Glucagon
　　Lipotropin (LPH)
　　Luteinizing hormone (LH)
　　Melanocyte-stimulating hormone (MSH)
　　Parathyroid hormone (PTH)
　　Somatostatin
　　Thyroid-stimulating hormone (TSH)
　B. The second messenger is cGMP
　　Atrial natriuretic factor
　　Nitric oxide
　C. The second messenger is calcium or phosphatidylinositols (or both)
　　Acetylcholine (muscarinic)
　　α$_1$-Adrenergic catecholamines
　　Angiotensin II
　　Antidiuretic hormone (vasopressin)
　　Cholecystokinin
　　Gastrin
　　Gonadotropin-releasing hormone
　　Oxytocin
　　Platelet-derived growth factor (PDGF)
　　Substance P
　　Thyrotropin-releasing hormone (TRH)
　D. The second messenger is a kinase or phosphatase cascade
　　Adiponectin
　　Chorionic somatomammotropin
　　Epidermal growth factor (EGF)
　　Erythropoietin (EPO)
　　Fibroblast growth factor (FGF)
　　Growth hormone (GH)
　　Insulin
　　Insulin-like growth factors I and II
　　Leptin
　　Nerve growth factor (NGF)
　　Platelet-derived growth factor
　　Prolactin

TABLE 41–4 General Features of Hormone Classes

	Group I	Group II
Types	Steroids, iodothyronines, calcitriol, retinoids	Polypeptides, proteins, glycoproteins, catecholamines
Solubility	Lipophilic	Hydrophilic
Transport proteins	Yes	No
Plasma half-life	Long (hours to days)	Short (minutes)
Receptor	Intracellular	Plasma membrane
Mediator	Receptor-hormone complex	cAMP, cGMP, Ca^{2+}, metabolites of complex phosphinositols, kinase cascades

adrenal (glucocorticoids and mineralocorticoids), and the pituitary (TSH, FSH, LH, GH, PRL, ACTH). Some organs are designed to perform two distinct but closely related functions. For example, the ovaries produce mature oocytes and the reproductive hormones estradiol and progesterone. The testes produce mature spermatozoa and testosterone. Hormones are also produced in specialized cells within other organs such as the small intestine (glucagon-like peptide), thyroid (calcitonin), and kidney (angiotensin II). Finally, the synthesis of some hormones requires the parenchymal cells of more than one organ—for example, the skin, liver, and kidney are required for the production of 1,25(OH)$_2$-D$_3$ (calcitriol). Examples of this diversity in the approach to hormone synthesis, each of which has evolved to fulfill a specific purpose, are discussed later.

Hormones Are Chemically Diverse

Hormones are synthesized from a wide variety of chemical building blocks. A large series is derived from cholesterol. These include the glucocorticoids, mineralocorticoids, androgens, estrogens, progestins, and 1,25(OH)$_2$-D$_3$ (**Figure 41–2**). In some cases, a steroid hormone is the precursor molecule for another hormone. For example, progesterone is a hormone in its own right but is also a precursor in the formation of glucocorticoids, mineralocorticoids, testosterone, and estrogens. Testosterone is an obligatory intermediate in the biosynthesis of estradiol and in the formation of dihydrotestosterone (DHT). In these examples, described in detail in the following discussion, the final product is determined by the cell type and the associated set of enzymes in which the precursor exists.

The amino acid tyrosine is the starting point in the synthesis of both the catecholamines and thyroid hormones tetraiodothyronine (thyroxine; T$_4$) and triiodothyronine (T$_3$) (see Figure 41–2). T$_3$ and T$_4$ are unique in that they require the addition of iodine (as I$^-$) for bioactivity. Since dietary iodine is very scarce in many parts of the world, an intricate mechanism for accumulating and retaining I$^-$ has evolved (described later, Figure 41–11).

DIVERSITY OF THE ENDOCRINE SYSTEM

Hormones Are Synthesized in a Variety of Cellular Arrangements

Hormones are synthesized in discrete organs designed solely for this specific purpose, such as the thyroid (triiodothyronine),

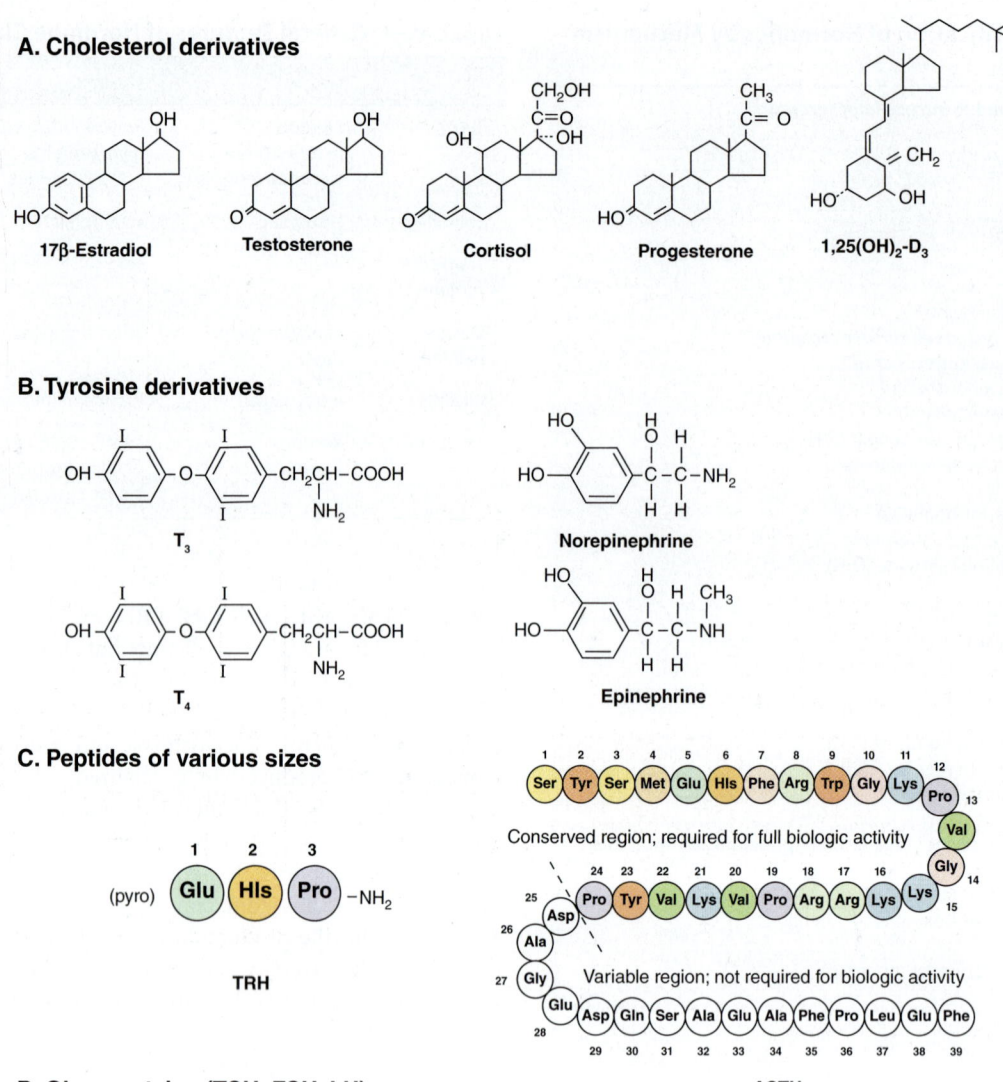

A. Cholesterol derivatives

17β-Estradiol   Testosterone   Cortisol   Progesterone   1,25(OH)$_2$-D$_3$

B. Tyrosine derivatives

T$_3$ Norepinephrine

T$_4$ Epinephrine

C. Peptides of various sizes

TRH

Conserved region; required for full biologic activity

Variable region; not required for biologic activity

ACTH

D. Glycoproteins (TSH, FSH, LH)

common α subunits

unique β subunits

FIGURE 41–2 **Chemical diversity of hormones. (A)** cholesterol derivatives; **(B)** tyrosine derivatives; **(C)** peptides of various sizes; note: pyroglutamic acid (pyro) is a cyclized variant of glutamic acid in which side chain carboxyl and free amino groups cyclize to form a lactam. **(D)** glycoproteins (TSH, FSH, and LH) with common α subunits and unique β subunits.

Many hormones are polypeptides or glycoproteins. These range in size from the small thyrotropin-releasing hormone (TRH), a tripeptide, to single-chain polypeptides like adreno-corticotropic hormone (ACTH; 39 amino acids), parathyroid hormone (PTH; 84 amino acids), and growth hormone (GH; 191 amino acids) (see Figure 41–2). Insulin is an A-B chain heterodimer of 21 and 30 amino acids, respectively. Follicle-stimulating hormone (FSH), luteinizing hormone (LH), thyroid-stimulating hormone (TSH), and chorionic gonadotropin (CG) are glycoprotein hormones of αβ heterodimeric structure. The α chain is identical in all of these hormones, and distinct β chains impart hormone uniqueness. These hormones have a molecular mass in the range of 25 to 30 kDa depending on the degree of glycosylation and the length of the β chain.

Hormones Are Synthesized & Modified for Full Activity in a Variety of Ways

Some hormones are synthesized in final form and secreted immediately. Included in this class are hormones derived from cholesterol. Some, such as the catecholamines, are synthesized in final form and stored in the producing cells, while others, like insulin, are synthesized as precursor molecules in the producing cell, and then are processed and secreted upon a physiologic cue (plasma glucose concentrations). Finally, still others are converted to active forms from precursor molecules in the peripheral tissues (T$_3$ and DHT). All of these examples are discussed in more detail in the following discussion.

MANY HORMONES ARE MADE FROM CHOLESTEROL

Adrenal Steroidogenesis

The adrenal steroid hormones are synthesized from cholesterol, which is mostly derived from the plasma, but a small portion is synthesized in situ from acetyl-CoA via mevalonate and squalene. Much of the cholesterol in the adrenal is esterified and stored in cytoplasmic lipid droplets. Upon stimulation of the adrenal by ACTH, an esterase is activated, and the free cholesterol formed is transported into the mitochondrion, where a **cytochrome P450 side chain cleavage enzyme (P450scc)** converts cholesterol to pregnenolone. Cleavage of the side chain involves sequential hydroxylations, first at C_{22} and then at C_{20}, followed by side chain cleavage (removal of the six-carbon fragment isocaproaldehyde) to give the 21-carbon steroid (**Figure 41–3**, top). An ACTH-dependent **steroidogenic acute regulatory (StAR)** protein is essential for the transport of cholesterol to P450scc in the inner mitochondrial membrane.

All mammalian steroid hormones are formed from cholesterol via pregnenolone through a series of reactions that occur in either the mitochondria or endoplasmic reticulum of the producing cell. Hydroxylases that require molecular oxygen and NADPH are essential, and dehydrogenases, an isomerase, and a lyase reaction are also necessary for certain steps. There is cellular specificity in adrenal steroidogenesis. For instance, 18-hydroxylase and 19-hydroxysteroid dehydrogenases, which are required for aldosterone synthesis, are found only in the zona glomerulosa cells (the outer region of the adrenal cortex), so that the biosynthesis of this mineralocorticoid is confined to this region. A schematic representation of the pathways involved in the synthesis of the three major classes of adrenal steroids is presented in **Figure 41–4**. The enzymes are shown in the rectangular boxes, and the modifications at each step are shaded.

Mineralocorticoid Synthesis

Synthesis of aldosterone follows the mineralocorticoid pathway and occurs in the zona glomerulosa. Pregnenolone is converted to progesterone by the action of two smooth endoplasmic reticulum enzymes, **3β-hydroxysteroid dehydrogenase (3β-OHSD)** and **$\Delta^{5,4}$-isomerase.** Progesterone is hydroxylated at the C_{21} position to form 11-deoxycorticosterone (DOC), which is an active (Na$^+$-retaining) mineralocorticoid. The next hydroxylation, at C_{11}, produces corticosterone, which has glucocorticoid activity and is a weak mineralocorticoid (it has <5% of the potency of aldosterone). In some species (eg, rodents), it is the most potent glucocorticoid. C_{21} hydroxylation is necessary for both mineralocorticoid and glucocorticoid activity, but most steroids with a C_{17} hydroxyl group have more glucocorticoid and less mineralocorticoid action. In the zona glomerulosa, which does not have the smooth endoplasmic reticulum enzyme 17α-hydroxylase, a mitochondrial 18-hydroxylase is present. The **18-hydroxylase (aldosterone**

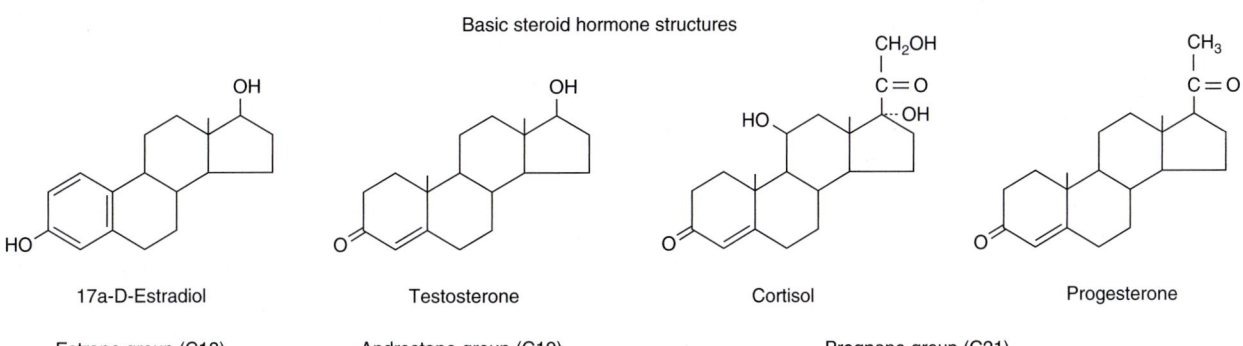

FIGURE 41–3 **Cholesterol side chain cleavage and basic steroid hormone structures.** The basic sterol rings are identified by the letters A to D. The carbon atoms are numbered 1 to 21, starting with the A ring (see Figure 26–3).

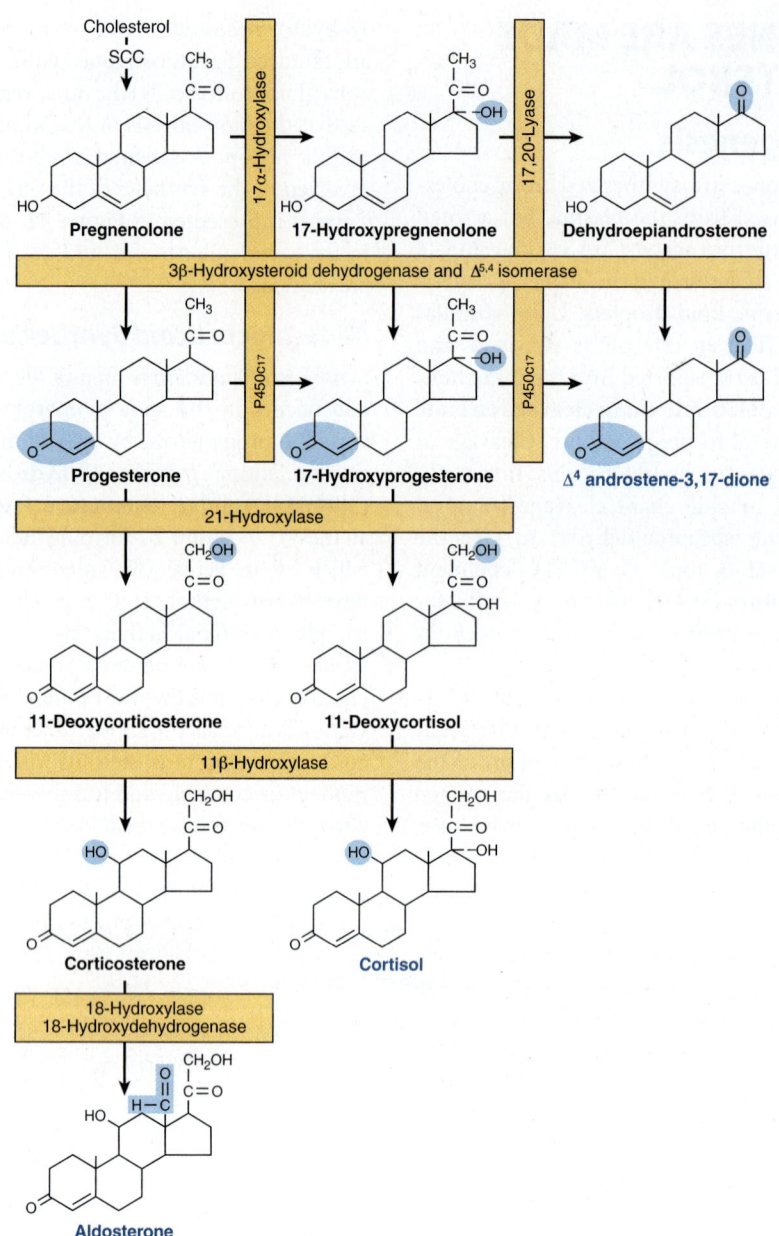

FIGURE 41–4 Pathways involved in the synthesis of the three major classes of adrenal steroids (mineralocorticoids, glucocorticoids, and androgens). Enzymes are shown in the rectangular boxes, and the modifications at each step are shaded. Note that the 17α-hydroxylase and 17,20-lyase activities are both part of one enzyme, designated P450c17. (Reproduced with permission from DeGroot LJ: Endocrinology, vol 2. Philadelphia, PA: Grune & Stratton; 1979.)

synthase) acts on corticosterone to form 18-hydroxycorticosterone, which is changed to aldosterone by conversion of the 18-alcohol to an aldehyde. This unique distribution of enzymes and the special regulation of the zona glomerulosa by K^+ and angiotensin II have led some investigators to suggest that, in addition to the adrenal being two glands, the adrenal cortex is actually two separate organs.

Glucocorticoid Synthesis

Cortisol synthesis requires three hydroxylases located in the fasciculata and reticularis zones of the adrenal cortex that act sequentially on the C_{17}, C_{21}, and C_{11} positions. The first two reactions are rapid, while C_{11} hydroxylation is relatively slow. If the C_{11} position is hydroxylated first, the action of **17α-hydroxylase** is impeded, and the mineralocorticoid pathway is followed (forming corticosterone or aldosterone, depending on the cell type). 17α-Hydroxylase is a smooth endoplasmic reticulum enzyme that acts on either progesterone or, more commonly, pregnenolone. 17α-Hydroxyprogesterone is hydroxylated at C_{21} to form 11-deoxycortisol, which is then hydroxylated at C_{11} to form cortisol, the most potent natural glucocorticoid hormone in humans. 21-Hydroxylase is a smooth endoplasmic reticulum enzyme, whereas 11β-hydroxylase is a mitochondrial enzyme. Steroidogenesis thus involves the repeated shuttling of substrates into and out of the mitochondria.

Androgen Synthesis

The major androgen or androgen precursor produced by the adrenal cortex is dehydroepiandrosterone (DHEA). Most 17-hydroxypregnenolone follows the glucocorticoid pathway, but a small fraction is subjected to oxidative fission and removal of the two-carbon side chain through the action of 17,20-lyase. The lyase activity is actually part of the same enzyme (P450c17) that catalyzes 17α-hydroxylation. This is therefore a **dual-function protein**. The lyase activity is important in both the adrenals and the gonads and acts exclusively on 17α-hydroxy-containing molecules. Adrenal androgen production increases markedly if glucocorticoid biosynthesis is impeded by the lack of one of the hydroxylases (**adrenogenital syndrome**). DHEA is really a prohormone since the actions of 3β-OHSD and $\Delta^{5,4}$-isomerase convert the weak androgen DHEA into the more potent **androstenedione**. Small amounts of androstenedione are also formed in the adrenal by the action of the lyase on 17α-hydroxyprogesterone. Reduction of androstenedione at the C_{17} position results in the formation of **testosterone**, the most potent adrenal androgen. Small amounts of testosterone are produced in the adrenal by this mechanism, but most of this conversion occurs in the testes.

Testicular Steroidogenesis

Testicular androgens are synthesized in the interstitial tissue by the Leydig cells. The immediate precursor of the gonadal steroids, as for the adrenal steroids, is cholesterol. The rate-limiting step, as in the adrenal, is delivery of cholesterol to the inner membrane of the mitochondria by the transport protein StAR. Once in the proper location, cholesterol is acted upon by the side chain cleavage enzyme P450scc. The conversion of cholesterol to pregnenolone is identical in adrenal, ovary, and testis. In the latter two tissues, however, the reaction is promoted by LH rather than ACTH.

The conversion of pregnenolone to testosterone requires the action of five enzyme activities contained in three proteins: (1) 3β-hydroxysteroid dehydrogenase (3β-OHSD) and $\Delta^{5,4}$-isomerase; (2) 17α-hydroxylase and 17,20-lyase; and (3) 17β-hydroxysteroid dehydrogenase (17β-OHSD). This sequence, referred to as the **progesterone (or Δ^4) pathway**, is shown on the right side of **Figure 41–5**. Pregnenolone can also be converted to testosterone by the **dehydroepiandrosterone (or Δ^5) pathway**, which is illustrated on the left side of Figure 41–5. The **$\Delta 5$** route appears to be most used in human testes.

The five enzyme activities are localized in the microsomal fraction in rat testes, and there is a close functional association between the activities of 3β-OHSD and $\Delta^{5,4}$-isomerase and between those of a 17α-hydroxylase and 17,20-lyase. These enzyme pairs, both contained in a single protein, are shown in the general reaction sequence in Figure 41–5.

DHT Is Formed From Testosterone in Peripheral Tissues

Testosterone is metabolized by two pathways. One involves oxidation at the 17-position, and the other involves reduction of the A ring double bond and the 3-ketone. Metabolism by the first pathway occurs in many tissues, including liver, and produces 17-ketosteroids that are generally inactive or less active than the parent compound. Metabolism by the second pathway, which is less efficient, occurs primarily in target tissues and produces the potent metabolite DHT.

The most significant metabolic product of testosterone is DHT, since in many tissues, including prostate, external genitalia, and some areas of the skin, this is the active form of the hormone. The plasma content of DHT in the adult male is about one-tenth that of testosterone, and ~400 μg of DHT is produced daily as compared with about 5 mg of testosterone. About 50 to 100 μg of DHT are secreted by the testes. The rest is produced peripherally from testosterone in a reaction catalyzed by the NADPH-dependent **5α-reductase** (**Figure 41–6**). Testosterone can thus be considered a prohormone since it is converted into a much more potent compound (DHT) and since most of this conversion occurs outside the testes. Some estradiol is formed from the peripheral aromatization of testosterone, particularly in males.

Ovarian Steroidogenesis

The estrogens are a family of hormones synthesized in a variety of tissues. 17β-Estradiol is the primary estrogen of ovarian origin. In some species, estrone, synthesized in numerous tissues, is more abundant. In pregnancy, relatively more estriol is produced, and this comes from the placenta. The general pathway and the subcellular localization of the enzymes involved in the early steps of estradiol synthesis are the same as those involved in androgen biosynthesis. Features unique to the ovary are illustrated in **Figure 41–7**.

Estrogens are formed by the aromatization of androgens in a complex process that involves three hydroxylation steps, each of which requires O_2 and NADPH. The **aromatase enzyme complex** is thought to include a P450 monooxygenase. Estradiol is formed if the substrate of this enzyme complex is testosterone, whereas estrone results from the aromatization of androstenedione.

The cellular source of the various ovarian steroids has been difficult to unravel, but a transfer of substrates between two cell types is involved. Theca cells are the source of androstenedione and testosterone. These are converted by the aromatase enzyme in granulosa cells to estrone and estradiol, respectively. Progesterone, a precursor for all steroid hormones, is produced and secreted by the corpus luteum as an end-product hormone because these cells do not contain the enzymes necessary to convert progesterone to other steroid hormones (**Figure 41–8**).

Significant amounts of estrogens are produced by the peripheral aromatization of androgens. In human males, the peripheral

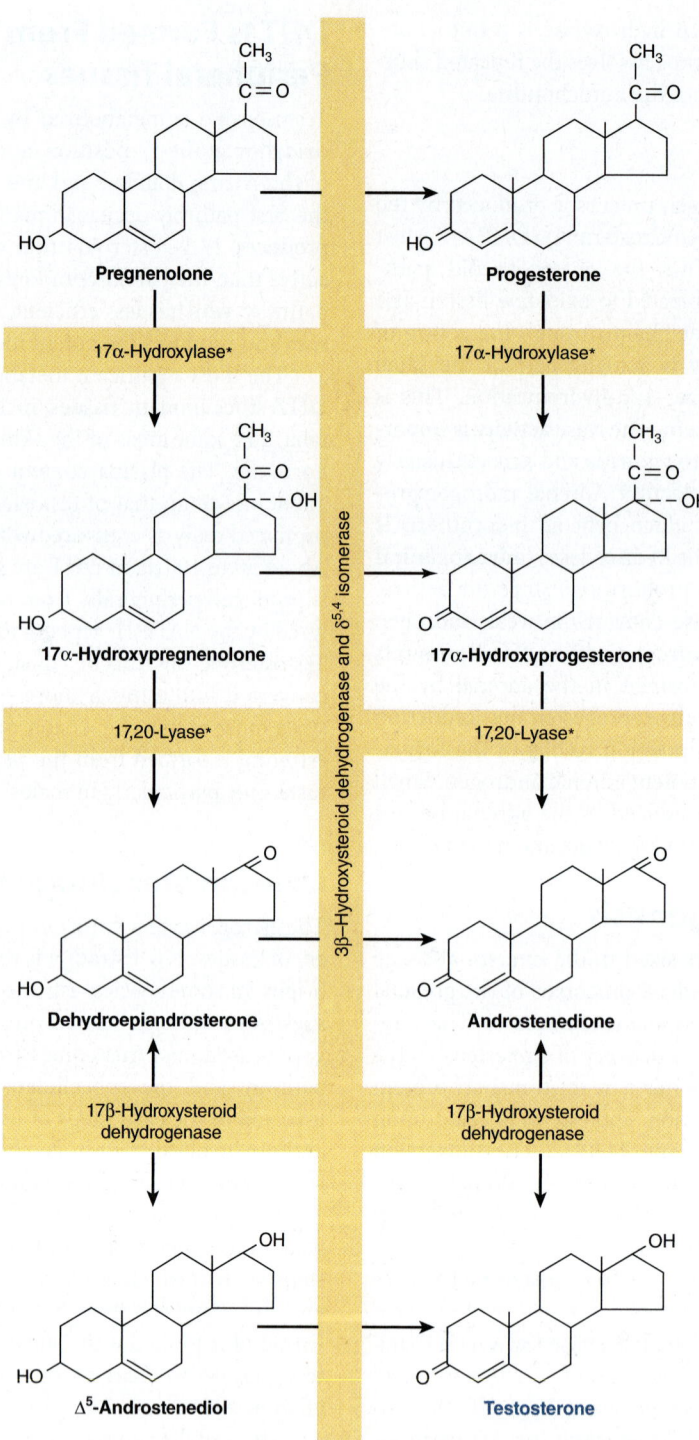

FIGURE 41–5 **Pathways of testosterone biosynthesis.** The pathway on the left side of the figure is called the Δ^5 or dehydroepiandrosterone pathway; the pathway on the right side is called the Δ^4 or progesterone pathway. The asterisk indicates that the 17α-hydroxylase and 17,20-lyase activities reside in a single protein, P450c17.

aromatization of testosterone to estradiol (E_2) accounts for 80% of the production of the latter. In females, adrenal androgens are important substrates since as much as 50% of the E_2 produced during pregnancy comes from the aromatization of androgens. Finally, conversion of androstenedione to estrone is the major source of estrogens in postmenopausal women. Aromatase activity is present in adipose cells and also in liver, skin, and other tissues. Increased activity of this enzyme may contribute to the "estrogenization" that characterizes such diseases as cirrhosis of the liver, hyperthyroidism, aging, and obesity. Aromatase inhibitors show promise as therapeutic agents in breast cancer and possibly in other female reproductive tract malignancies.

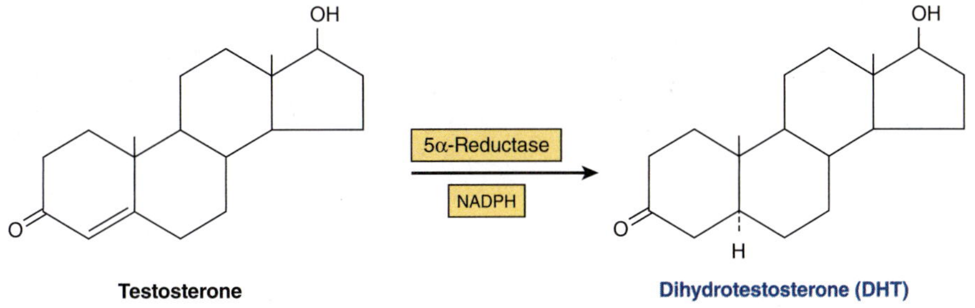

FIGURE 41–6 Dihydrotestosterone is formed from testosterone through action of the enzyme 5α-reductase.

1,25(OH)$_2$-D$_3$ (Calcitriol) Is Synthesized From a Cholesterol Derivative

1,25(OH)$_2$-D$_3$ is produced by a complex series of enzymatic reactions that involve the plasma transport of precursor molecules to a number of different tissues (**Figure 41–9**). One of these precursors is vitamin D—really not a vitamin, but this common name persists. The active molecule, 1,25(OH)$_2$-D$_3$, is transported to other organs where it activates biologic processes in a manner similar to that employed by the steroid hormones.

Skin

Small amounts of the precursor for 1,25(OH)$_2$-D$_3$ synthesis are present in food (fish liver oil and egg yolk), but most of the precursor for 1,25(OH)$_2$-D$_3$ synthesis is produced in the malpighian layer of the epidermis from 7-dehydrocholesterol in an ultraviolet light-mediated, nonenzymatic **photolysis** reaction. The extent of this conversion is related directly to the intensity of the exposure and inversely to the extent of pigmentation in the skin. There is an age-related loss of 7-dehydrocholesterol

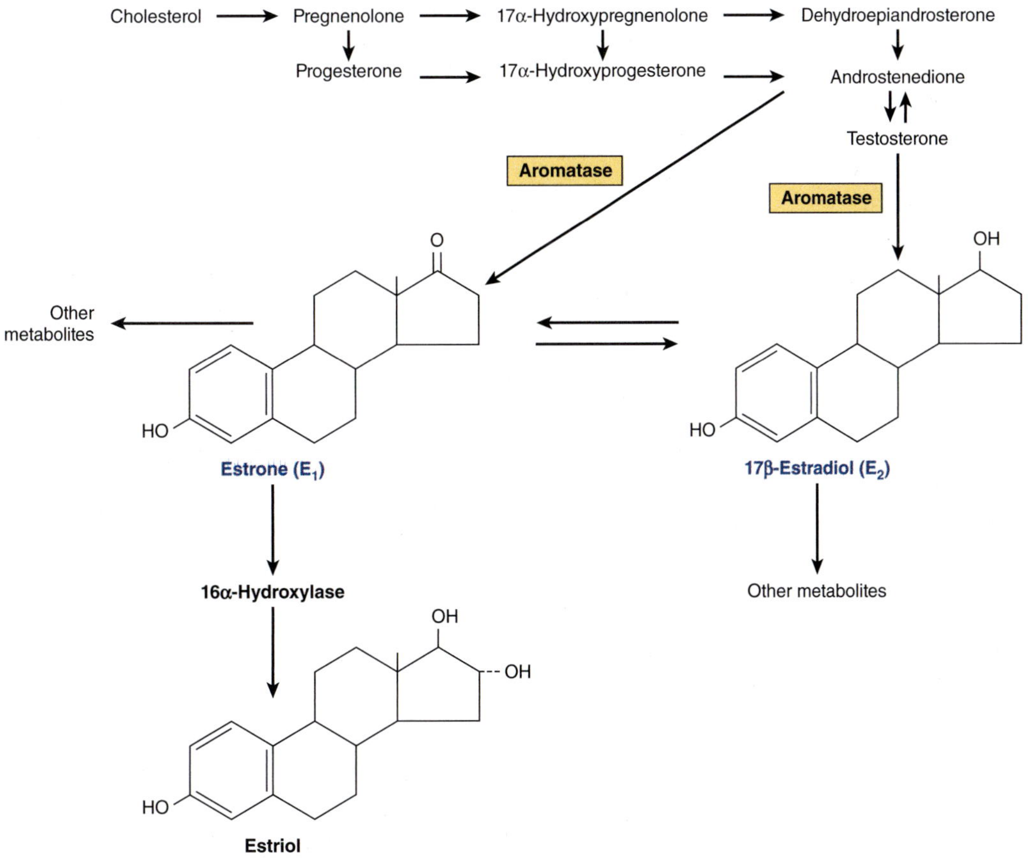

FIGURE 41–7 **Biosynthesis of estrogens.** (Reproduced with permission from Barrett KE, Barman SM, Brooks HL, et al: Ganong's *Review of Medical Physiology*, 26th ed. New York, NY: McGraw Hill; 2019.)

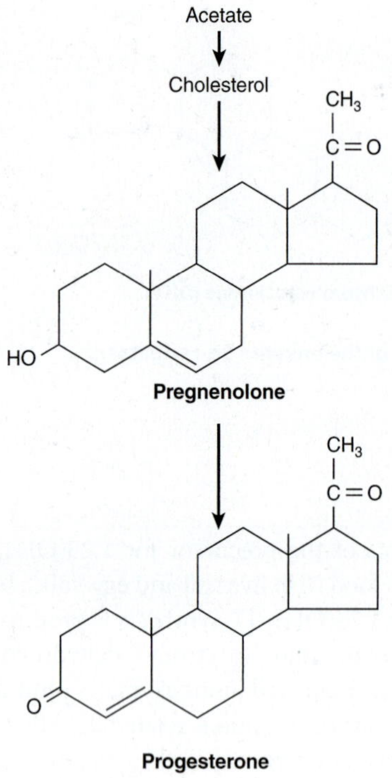

FIGURE 41–8 **Biosynthesis of progesterone in the corpus luteum.**

in the epidermis that may be related to the negative calcium balance associated with old age.

Liver

A specific transport protein called the **vitamin D–binding protein** binds vitamin D_3 and its metabolites and moves vitamin D_3 from the skin or intestine to the liver, where it undergoes 25-hydroxylation, the first obligatory reaction in the production of $1,25(OH)_2$-D_3. 25-Hydroxylation occurs in the endoplasmic reticulum in a reaction that requires magnesium, NADPH, molecular oxygen, and an uncharacterized cytoplasmic factor. Two enzymes are involved: an NADPH-dependent cytochrome P450 reductase and a cytochrome P450. This reaction is not regulated, and it also occurs with low efficiency in kidney and intestine. The $25(OH)_2$-D_3 enters the circulation, where it is the major form of vitamin D found in plasma, and is transported to the kidney by the vitamin D–binding protein.

Kidney

$25(OH)_2$-D_3 is a weak agonist and must be modified by hydroxylation at position C_1 for full biologic activity. This is accomplished in mitochondria of the renal proximal convoluted tubule by a three-component monooxygenase reaction that requires NADPH, Mg^{2+}, molecular oxygen, and at least

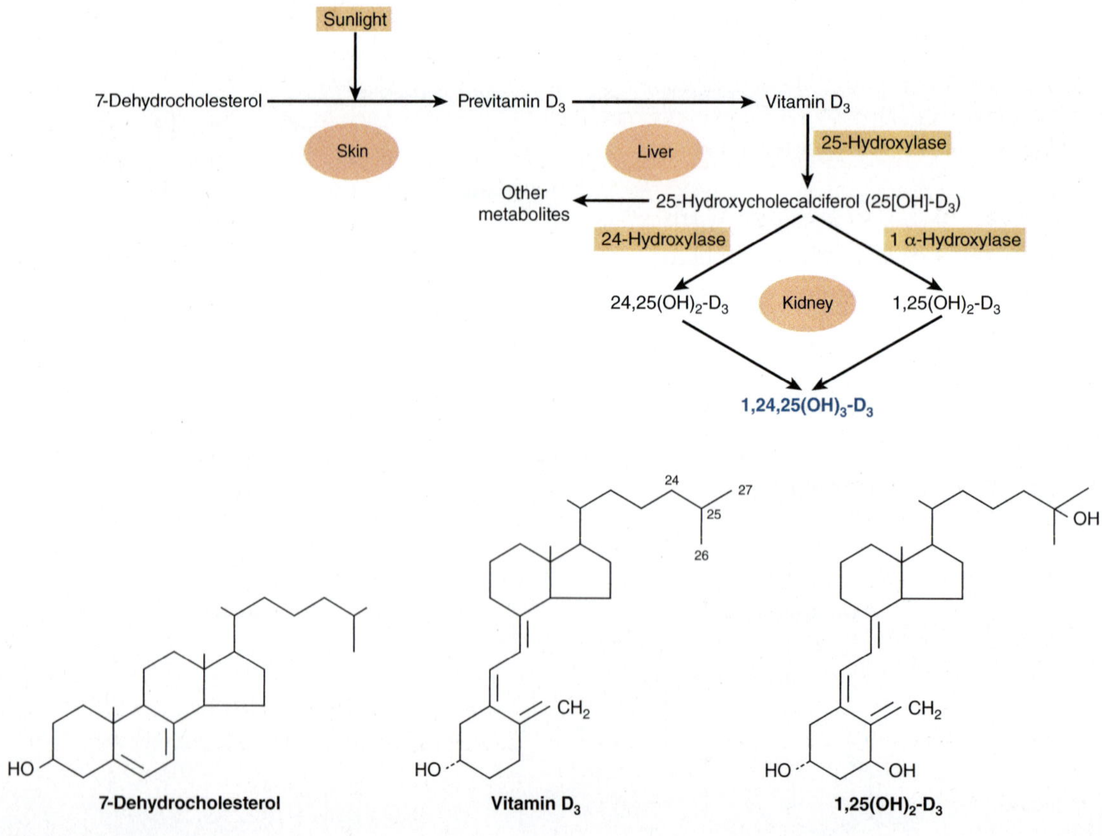

FIGURE 41–9 **Formation and hydroxylation of vitamin D_3.** 25-Hydroxylation takes place in the liver, and the other hydroxylations occur in the kidneys. $25,26(OH)_2$-D_3 and $1,25,26(OH)_3$-D_3 are probably formed as well. The structures of 7-dehydrocholesterol, vitamin D_3, and $1,25(OH)_2$-D_3 are also shown. (Modified with permission from Barrett KE, Barman SM, Brooks HL, et al: Ganong's *Review of Medical Physiology*, 26th ed. New York, NY: McGraw Hill; 2019.)

three enzymes: (1) a flavoprotein, renal ferredoxin reductase; (2) an iron-sulfur protein, renal ferredoxin; and (3) cytochrome P450. This system produces $1,25(OH)_2$-D_3, which is the most potent naturally occurring metabolite of vitamin D.

CATECHOLAMINES & THYROID HORMONES ARE MADE FROM TYROSINE

Catecholamines Are Synthesized in Final Form & Stored in Secretion Granules

Three amines—dopamine, norepinephrine, and epinephrine—are synthesized from tyrosine in the chromaffin cells of the adrenal medulla. The major product of the adrenal medulla is epinephrine. This compound constitutes about 80% of the catecholamines in the medulla, and it is not made in extramedullary tissue. In contrast, most of the norepinephrine present in organs innervated by sympathetic nerves is made in situ (about 80% of the total), and most of the rest is made in other nerve endings and reaches the target sites via the circulation. Epinephrine and norepinephrine may be produced and stored in different cells in the adrenal medulla and other chromaffin tissues.

The conversion of tyrosine to epinephrine requires four sequential steps: (1) ring hydroxylation; (2) decarboxylation; (3) side chain hydroxylation to form norepinephrine; and (4) *N*-methylation to form epinephrine. The biosynthetic pathway and the enzymes involved are illustrated in **Figure 41–10**.

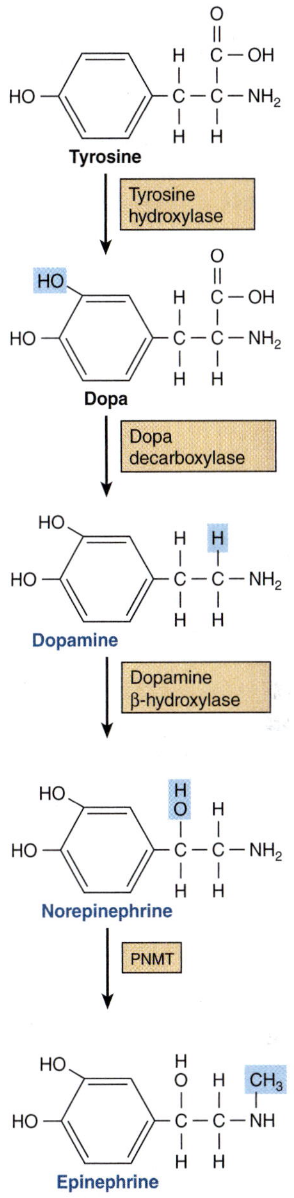

Tyrosine Hydroxylase Is Rate-Limiting for Catecholamine Biosynthesis

Tyrosine is the immediate precursor of catecholamines, and **tyrosine hydroxylase** is the rate-limiting enzyme in catecholamine biosynthesis. Tyrosine hydroxylase is found in both soluble and particle-bound forms only in tissues that synthesize catecholamines; it functions as an oxidoreductase, with tetrahydropteridine as a cofactor, to convert L-tyrosine to L-dihydroxyphenylalanine (L-**dopa**). As the rate-limiting enzyme, tyrosine hydroxylase is regulated in a variety of ways. The most important mechanism involves feedback inhibition by the catecholamines, which compete with the enzyme for the pteridine cofactor. Catecholamines cannot cross the blood–brain barrier; hence, in the brain they must be synthesized locally. In certain central nervous system diseases (eg, Parkinson disease), there is a local deficiency of dopamine synthesis. L-Dopa, the precursor of dopamine, readily crosses the blood–brain barrier and so is an important agent in the treatment of Parkinson disease.

Dopa Decarboxylase Is Present in All Tissues

This soluble enzyme requires pyridoxal phosphate for the conversion of L-dopa to 3,4-dihydroxyphenylethylamine

FIGURE 41–10 **Biosynthesis of catecholamines.** (PNMT, phenylethanolamine-*N*-methyltransferase.)

(**dopamine**). Compounds that resemble L-dopa, such as α-methyldopa, are competitive inhibitors of this reaction. α-Methyldopa is effective in treating some kinds of hypertension.

Dopamine β-Hydroxylase (DBH) Catalyzes the Conversion of Dopamine to Norepinephrine

DBH is a monooxygenase and uses ascorbate as an electron donor, copper at the active site, and fumarate as modulator. DBH is in the particulate fraction of the medullary cells, probably in the secretion granule; thus, the conversion of dopamine to **norepinephrine** occurs in this organelle.

Phenylethanolamine-N-Methyltransferase (PNMT) Catalyzes the Production of Epinephrine

PNMT catalyzes the *N*-methylation of norepinephrine to form **epinephrine** in the epinephrine-forming cells of the adrenal medulla. Since PNMT is soluble, it is assumed that norepinephrine-to-epinephrine conversion occurs in the cytoplasm. The synthesis of PNMT is induced by glucocorticoid hormones that reach the medulla via the intra-adrenal portal system. This special system provides for a 100-fold steroid concentration gradient over systemic arterial blood, and this high intra-adrenal concentration appears to be necessary for the induction of PNMT.

T_3 & T_4 Illustrate the Diversity in Hormone Synthesis

The formation of **triiodothyronine (T_3)** and **tetraiodothyronine (thyroxine; T_4)** (see Figure 41–2) illustrates many of the principles of diversity discussed in this chapter. These hormones require a rare element (iodine) for bioactivity; they are synthesized as part of a very large precursor molecule (thyroglobulin); they are stored in an intracellular reservoir (colloid); and there is peripheral conversion of T_4 to T_3, which is a much more active hormone.

The thyroid hormones T_3 and T_4 are unique in that iodine (as iodide) is an essential component of both. In most parts of the world, iodine is a scarce component of soil, and for that reason there is little in food. A complex mechanism has evolved to acquire and retain this crucial element and to convert it into a form suitable for incorporation into organic compounds. At the same time, the thyroid must synthesize thyronine from tyrosine, and this synthesis takes place in thyroglobulin (**Figure 41–11**).

Thyroglobulin is the precursor of T_4 and T_3. It is a large iodinated, glycosylated protein with a molecular mass of 660 kDa. Carbohydrate accounts for 8 to 10% of the weight of thyroglobulin and iodide for about 0.2 to 1%, depending on the iodine content in the diet. Thyroglobulin is composed of two large subunits. It contains 115 tyrosine residues, each of which is a potential site of iodination. About 70% of the iodide in thyroglobulin exists in the inactive precursors, **monoiodotyrosine (MIT)** and **diiodotyrosine (DIT)**, while 30% is in the **iodothyronyl residues**, T_4 and T_3. When iodine supplies are sufficient, the T_4:T_3 ratio is about 7:1. In **iodine deficiency**, this ratio decreases, as does the DIT:MIT ratio. Thyroglobulin, a large molecule of about 5000 amino acids, provides the conformation required for tyrosyl coupling and iodide organification necessary in the formation of the diaminoacid thyroid hormones. It is synthesized in the basal portion of the cell and moves to the lumen, where it is a storage form of T_3 and T_4 in the colloid; several weeks' supply of these hormones exist in the normal thyroid. Within minutes after stimulation of the thyroid by TSH, colloid reenters the cell and there is a marked increase of phagolysosome activity. Various acid proteases and peptidases hydrolyze the thyroglobulin into its constituent amino acids, including T_4 and T_3, which are discharged into the extracellular space (see Figure 41–11). Thyroglobulin is thus a very large prohormone.

Iodide Metabolism Involves Several Discrete Steps

The thyroid is able to concentrate I^- against a strong electrochemical gradient. This is an energy-dependent process and is linked to the Na^+-K^+-ATPase–dependent thyroidal I^- transporter. The ratio of iodide in thyroid to iodide in serum (T:S ratio) is a reflection of the activity of this transporter. This activity is primarily controlled by TSH and ranges from 500:1 in animals chronically stimulated with TSH to 5:1 or less in hypophysectomized animals (no TSH). The T:S ratio in humans on a normal iodine diet is about 25:1.

The thyroid is the only tissue that can oxidize I^- to a higher valence state, an obligatory step in I^- organification and thyroid hormone biosynthesis. This step involves a heme-containing peroxidase and occurs at the luminal surface of the follicular cell. Thyroperoxidase, a tetrameric protein with a molecular mass of 60 kDa, requires hydrogen peroxide (H_2O_2) as an oxidizing agent. The H_2O_2 is produced by an NADPH-dependent enzyme resembling cytochrome *c* reductase. A number of compounds inhibit I^- oxidation and therefore its subsequent incorporation into MIT and DIT. The most important of these are the thiourea drugs. They are used as antithyroid drugs because of their ability to inhibit thyroid hormone biosynthesis at this step. Once iodination occurs, the iodine does not readily leave the thyroid. Free tyrosine can be iodinated, but it is not incorporated into proteins since no tRNA recognizes iodinated tyrosine.

The coupling of two DIT molecules to form T_4—or of an MIT and DIT to form T_3—occurs within the thyroglobulin molecule. A separate coupling enzyme has not been found, and since this is an oxidative process, it is assumed that the same thyroperoxidase catalyzes this reaction by stimulating free radical formation of iodotyrosine. This hypothesis is supported by the observation that the same drugs which inhibit I^- oxidation also inhibit coupling. The formed thyroid hormones remain as integral parts of thyroglobulin until the latter is degraded, as described earlier.

A deiodinase removes I^- from the inactive mono and diiodothyronine molecules in the thyroid. This mechanism provides a substantial amount of the I^- used in T_3 and T_4 biosynthesis. A peripheral deiodinase in target tissues such as pituitary, kidney, and liver selectively removes I^- from the 5' position of T_4 to make T_3 (see Figure 41–2), which is a much more active molecule. In this sense, T_4 can be thought of as a prohormone, though it does have some intrinsic activity.

Several Hormones Are Made From Larger Peptide Precursors

Formation of the critical disulfide bridges in insulin requires that this hormone be first synthesized as part of a larger

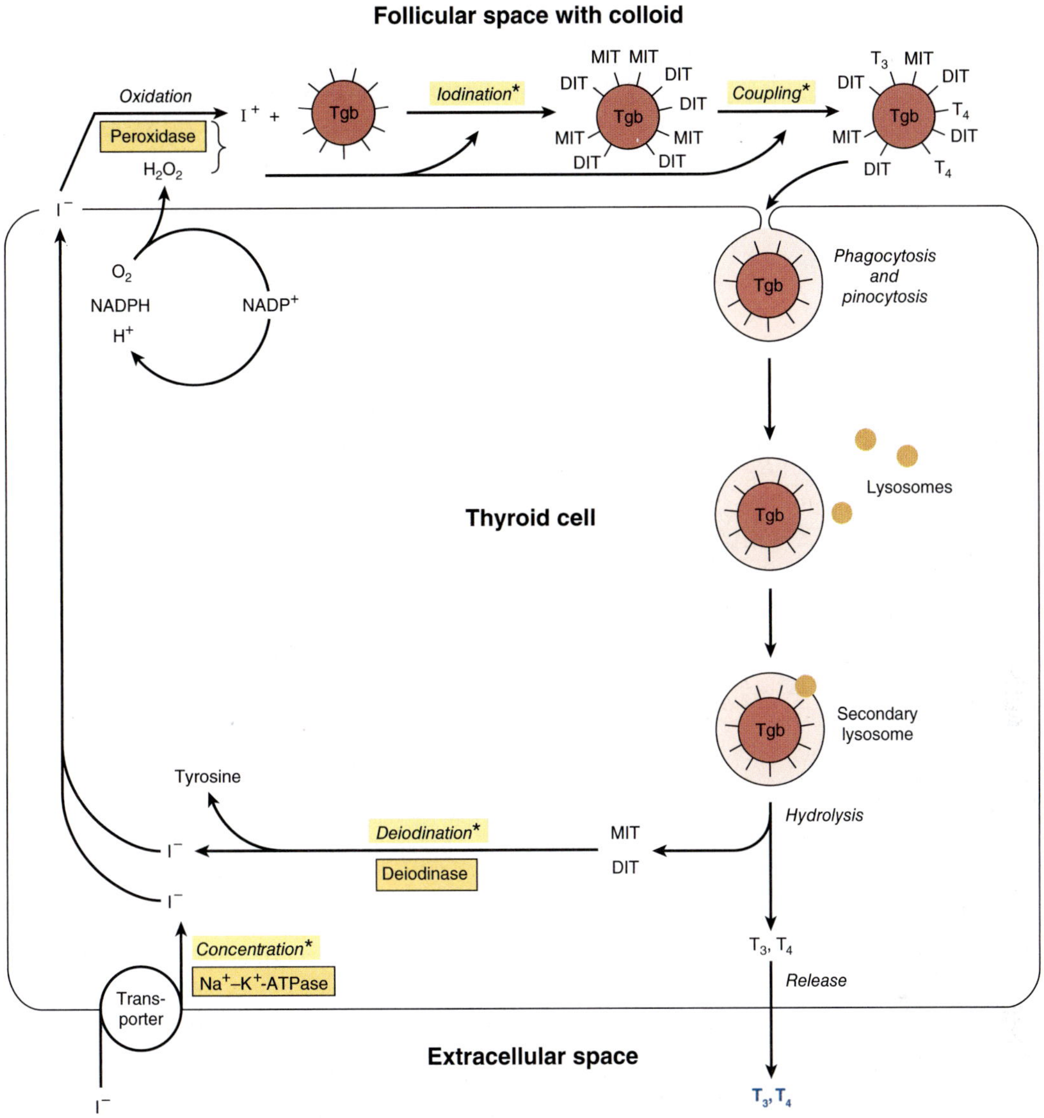

Follicular space with colloid

FIGURE 41–11 **Model of iodide metabolism in the thyroid follicle.** A follicular cell is shown facing the follicular lumen (**top**) and the extracellular space (**bottom**). Iodide enters the thyroid primarily through a transporter (bottom left). Thyroid hormone synthesis occurs in the follicular space through a series of reactions, many of which are peroxidase-mediated. Thyroid hormones, stored in the colloid in the follicular space, are released from thyroglobulin by hydrolysis inside the thyroid cell. (DIT, diiodotyrosine; MIT, monoiodotyrosine; Tgb, thyroglobulin; T_3, triiodothyronine; T_4, tetraiodothyronine; T_3 and T_4 structures are shown in Figure 41–2B.) Asterisks indicate steps or processes where inherited enzyme deficiencies cause congenital goiter and often result in hypothyroidism.

precursor molecule, proinsulin. This is conceptually similar to the example of the thyroid hormones, which can only be formed in the context of a much larger molecule. Several other hormones are synthesized as parts of large precursor molecules, not because of some special structural requirement but rather as a mechanism for controlling the available amount of the active hormone. PTH and angiotensin II are examples of this type of regulation. Another interesting example is the pro-opiomelanocortin (POMC) protein, which can be processed into many different hormones in a tissue-specific manner. These examples are discussed in detail in the following discussion.

Insulin Is Synthesized as a Preprohormone & Modified Within the β Cell

Insulin has an AB heterodimeric structure with one intrachain (A6-A11) and two interchain disulfide bridges (A7-B7 and A20-B19) (**Figure 41–12**). The A and B chains could be synthesized in the laboratory, but attempts at a biochemical synthesis of the mature insulin molecule yielded very poor results. The reason for this became apparent when it was discovered that insulin is synthesized as a **preprohormone** (molecular weight ~11,500), which is the prototype for peptides that are

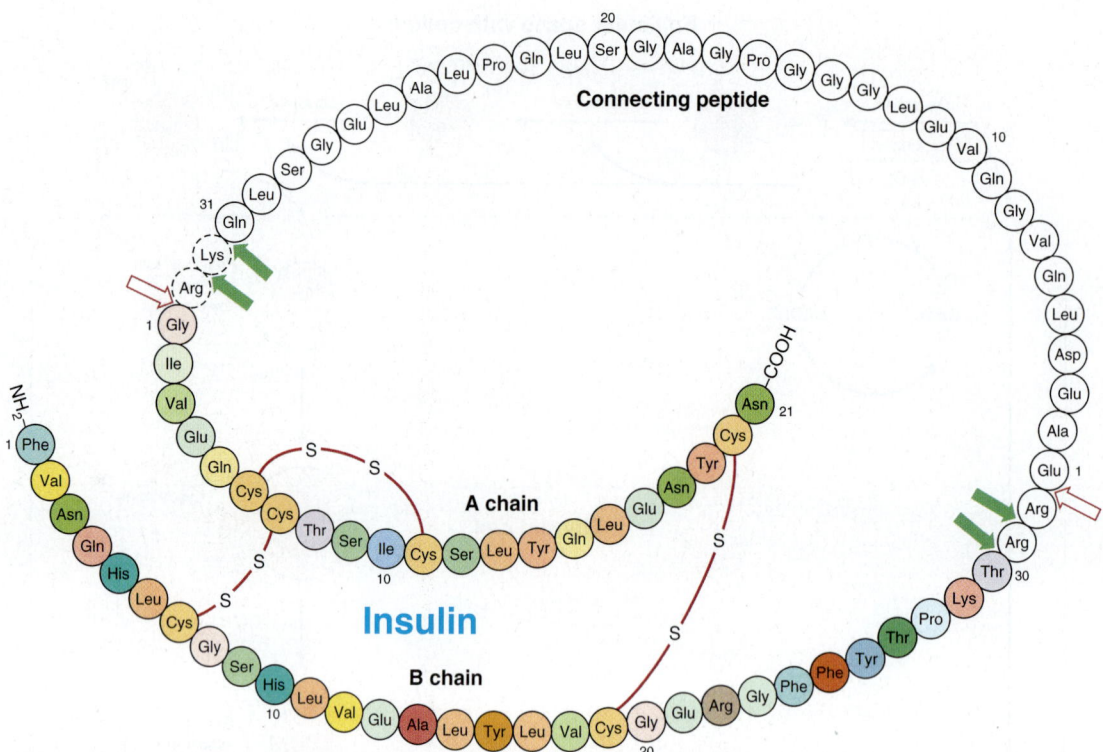

FIGURE 41–12 **Structure of human proinsulin.** Insulin and C-peptide molecules are connected at two sites by peptide bonds. An initial cleavage by a trypsin-like enzyme (*red open arrows*) followed by several cleavages by a carboxypeptidase-like enzyme (*green solid arrows*) results in the production of the heterodimeric (AB) insulin molecule (*colored*), which is held together by two intrapeptide cysteine disulfide bonds; insulin C-peptide (*white*).

processed from larger precursor molecules. The hydrophobic 23-amino-acid pre-, or leader, sequence directs the molecule into the cisternae of the endoplasmic reticulum and then is removed. This results in the 9000-MW proinsulin molecule, which provides the conformation necessary for the proper and efficient formation of the disulfide bridges. As shown in Figure 41–12, the sequence of proinsulin, starting from the amino terminus, is B chain—connecting (C) peptide—A chain. The proinsulin molecule undergoes a series of site-specific peptide cleavages that result in the formation of equimolar amounts of mature insulin and C-peptide. These enzymatic cleavages are summarized in Figure 41–12.

PTH Is Secreted as an 84-Amino-Acid Peptide

The immediate precursor of **parathyroid (PTH)** is **proPTH**, which differs from the native 84-amino-acid hormone by having a highly basic hexapeptide amino terminal extension. The primary gene product and the immediate precursor for proPTH is the 115-amino-acid **preproPTH**. This differs from proPTH by having an additional 25-amino-acid NH$_2$-terminal extension that is common with the other leader or signal sequences characteristic of secreted proteins; it is predominantly hydrophobic in nature. The complete structure of pre-proPTH and the sequences of proPTH and PTH are illustrated in **Figure 41–13**. PTH$_{1–34}$ has full biologic activity, and the region 25 to 34 is primarily responsible for receptor binding.

The biosynthesis of PTH and its subsequent secretion are regulated by the plasma ionized calcium (Ca^{2+}) concentration through a complex process. An acute decrease of Ca^{2+} results in a marked increase of PTH mRNA, and this is followed by an increased rate of PTH synthesis and secretion. However, about 80 to 90% of the proPTH synthesized cannot be accounted for as intact PTH in cells or in the incubation medium of experimental systems. This finding led to the conclusion that most of the proPTH synthesized is quickly degraded. It was later discovered that this rate of degradation decreases when Ca^{2+} concentrations are low, and it increases when Ca^{2+} concentrations are high. A Ca^{2+} receptor on the surface of the parathyroid cell mediates these effects. Very specific fragments of PTH are generated during its proteolytic digestion (see Figure 41–13). A number of proteolytic enzymes, including cathepsins B and D, have been identified in parathyroid tissue. Cathepsin B cleaves PTH into two fragments: PTH$_{1–36}$ and PTH$_{37–84}$. PTH$_{37–84}$ is not further degraded; however, PTH$_{1–36}$ is rapidly and progressively cleaved into di- and tripeptides. Most of the proteolysis of PTH occurs within the gland, but a number of studies confirm that PTH, once secreted, is proteolytically degraded in other tissues, especially the liver, by similar mechanisms.

Angiotensin II Is Also Synthesized From a Large Precursor

The renin-angiotensin system is involved in the regulation of blood pressure and electrolyte metabolism (through production

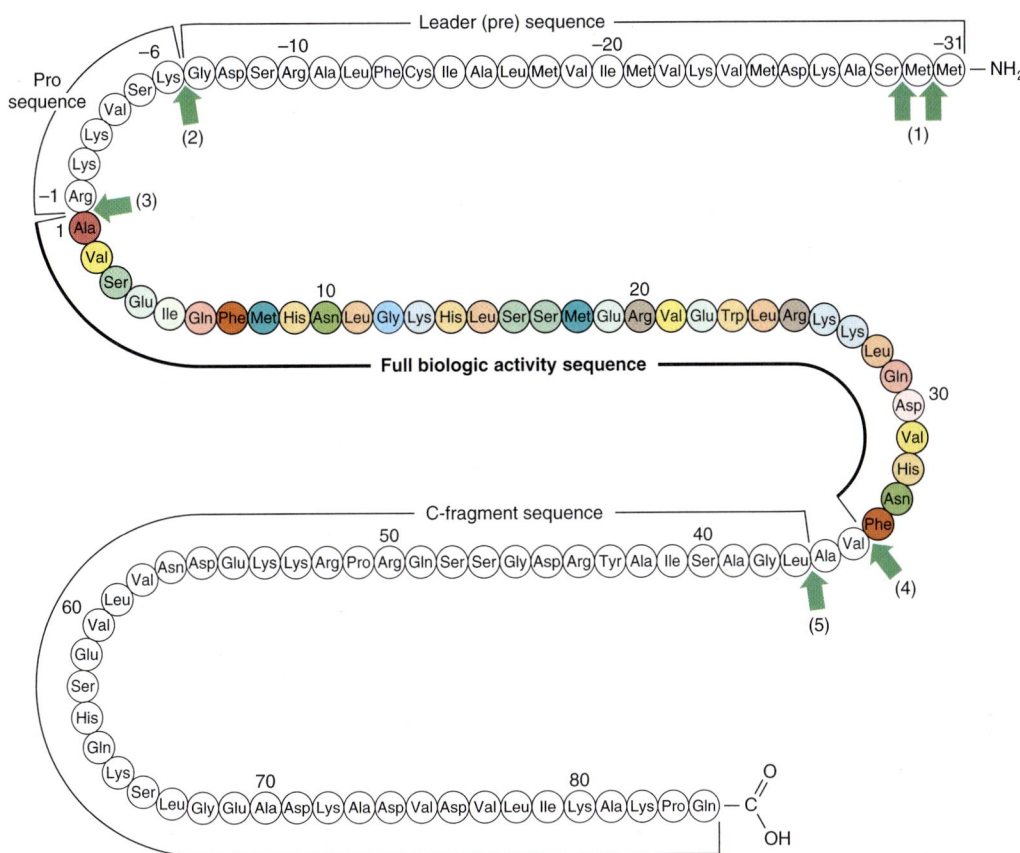

FIGURE 41–13 **Structure of bovine preproparathyroid hormone.** The green arrows indicate sites of cleavage by processing enzymes in the parathyroid gland and in the liver after secretion of the hormone (numbered 1 to 5). The biologically active region of the molecule (colored) is flanked by sequence not required for activity on target receptors. (Reproduced with permission from Habener JF. Recent advances in parathyroid hormone research. Clin Biochem. 1981;14(5):223-229.)

of aldosterone). The primary hormone involved in these processes is angiotensin II, an octapeptide made from angiotensinogen (**Figure 41–14**). Angiotensinogen, a large α_2-globulin made in liver, is the substrate for renin, an enzyme produced in the juxtaglomerular cells of the renal afferent arteriole. The position of these cells makes them particularly sensitive to blood pressure changes, and many of the physiologic regulators of renin release act through renal baroreceptors. The juxtaglomerular cells are also sensitive to changes of Na^+ and Cl^- concentration in the renal tubular fluid; therefore, any combination of factors that decreases fluid volume (dehydration, decreased blood pressure, fluid, or blood loss) or decreases NaCl concentration stimulates renin release. Renal sympathetic nerves that terminate in the juxtaglomerular cells mediate the central nervous system and postural effects on renin release independently of the baroreceptor and salt effects, a mechanism that involves the β-adrenergic receptor. Renin acts on the substrate angiotensinogen to produce the decapeptide angiotensin I.

Angiotensin-converting enzyme (ACE), a glycoprotein found in lung, endothelial cells, and plasma, removes two carboxyl terminal amino acids from the decapeptide angiotensin I to form angiotensin II in a step that is not thought to be rate-limiting. Various nonapeptide analogs of angiotensin I

and other compounds act as competitive inhibitors of converting enzyme and are used to treat renin-dependent hypertension. These are referred to as ACE inhibitors. Angiotensin II increases blood pressure by causing vasoconstriction of the arteriole and is a very potent vasoactive substance. It inhibits renin release from the juxtaglomerular cells and is a potent stimulator of aldosterone production. This results in Na^+ retention, volume expansion, and increased blood pressure.

In some species, angiotensin II is converted to the heptapeptide angiotensin III (see Figure 41–14), an equally potent stimulator of aldosterone production. In humans, the plasma level of angiotensin II is four times greater than that of angiotensin III, so most effects are exerted by the octapeptide. Angiotensins II and III are rapidly inactivated by angiotensinases.

Angiotensin II binds to specific adrenal cortex glomerulosa cell receptors. The hormone-receptor interaction does not activate adenylyl cyclase, and cAMP does not appear to mediate the action of this hormone. The actions of angiotensin II, which are to stimulate the conversion of cholesterol to pregnenolone and of corticosterone to 18-hydroxycorticosterone and aldosterone, may involve changes in the concentration of intracellular calcium and of phospholipid metabolites by mechanisms similar to those described in Chapter 42.

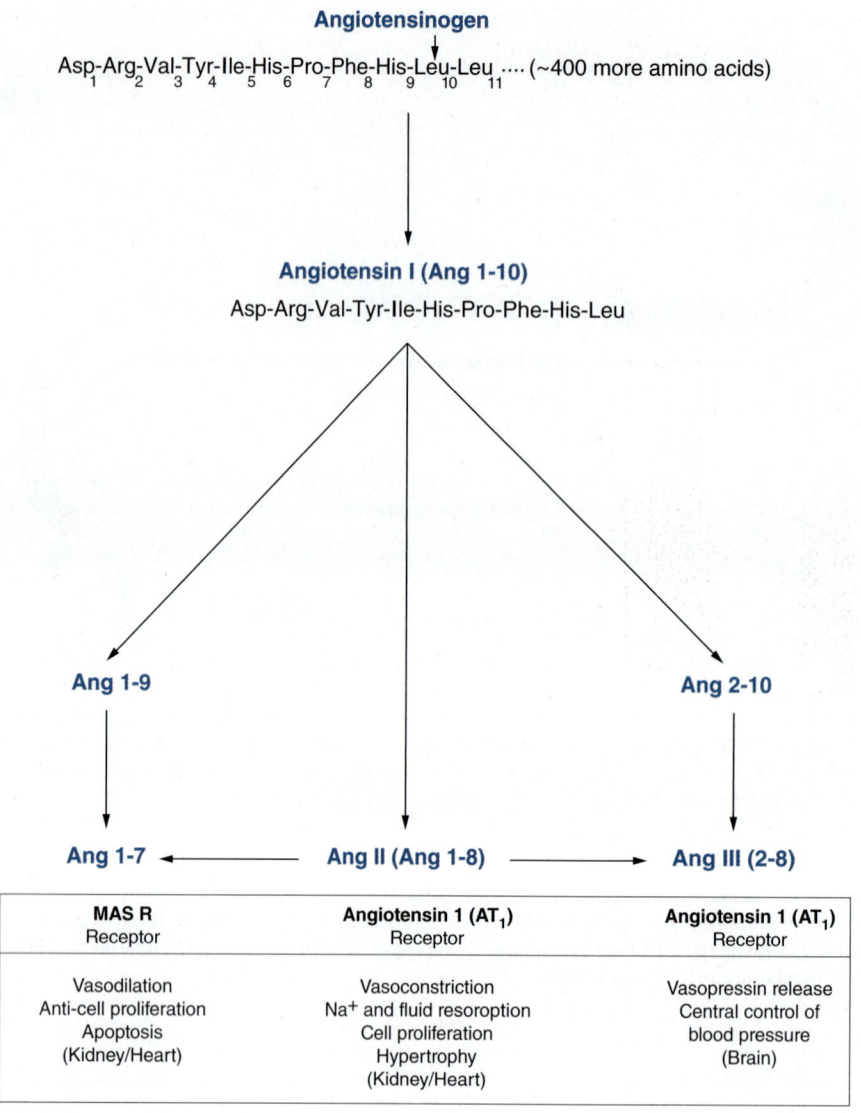

Angiotensinogen

Asp-Arg-Val-Tyr-Ile-His-Pro-Phe-His-Leu-Leu .... (~400 more amino acids)

Angiotensin I (Ang 1-10)

Asp-Arg-Val-Tyr-Ile-His-Pro-Phe-His-Leu

Ang 1-9

Ang 2-10

Ang 1-7 ← **Ang II (Ang 1-8)** → **Ang III (2-8)**

MAS R Receptor	**Angiotensin 1 (AT₁)** Receptor	**Angiotensin 1 (AT₁)** Receptor
Vasodilation Anti-cell proliferation Apoptosis (Kidney/Heart)	Vasoconstriction Na⁺ and fluid resorption Cell proliferation Hypertrophy (Kidney/Heart)	Vasopressin release Central control of blood pressure (Brain)

FIGURE 41–14 **Formation, metabolism, and selected physiologic activities of angiotensins.** The three most biologically active forms of angiotensin (Ang), Ang 1-7, Ang 1-8 (Ang II), and Ang 2-8 (Ang III) are shown. The numbers represent the amino acids present in each Ang are numbered relative to the sequence of Ang 1-10 (Ang I). All Ang forms are derived by proteolysis catalyzed by a number of different proteases. Initial processing of the 400+ amino acid long angiotensinogen precursor is catalyzed by renin, while several of the other proteolytic events are catalyzed by angiotensin-converting enzyme 1 (ACE1), or ACE2. The receptors bound by the different Ang forms, as well as physiologic outcomes of receptor binding are shown (**bottom**).

Complex Processing Generates the Pro-Opiomelanocortin (POMC) Peptide Family

The POMC family consists of peptides that act as hormones (ACTH, LPH, MSH) and others that may serve as neurotransmitters or neuromodulators (endorphins) (**Figure 41–15**). POMC is synthesized as a pre-proprecursor molecule of 267 amino acids and is differentially processed in various regions of the pituitary and other tissues (see later).

The POMC gene is expressed in the anterior and intermediate lobes of the pituitary. The most conserved sequences between species are within the amino terminal fragment, the ACTH region, and the β-endorphin region. POMC or related

products are found in several other vertebrate tissues, including the brain, placenta, gastrointestinal tract, reproductive tract, lung, and lymphocytes.

The POMC protein is processed differently in the anterior lobe than in the intermediate lobe. The intermediate lobe of the pituitary is rudimentary in adult humans, but it is active in human fetuses and in pregnant women during late gestation and is also active in many animal species. Processing of the human POMC protein in the peripheral tissues (gut, placenta, and male reproductive tract) begins with the removal of a 26 amino acid signal peptide (see Chapter 49). Removal of the POMC signal peptide generates the 241 amino acid POMC protein. There are three polypeptide groups generated from the POMC precursor: (1) ACTH, which can give

Pre-pro-opiomelanocortin:

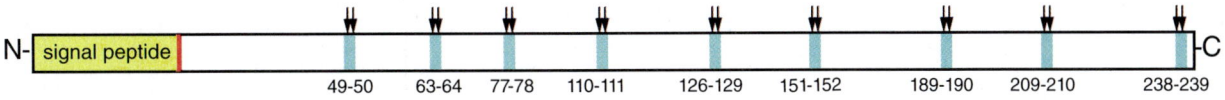

pro-opiomelanocortin:

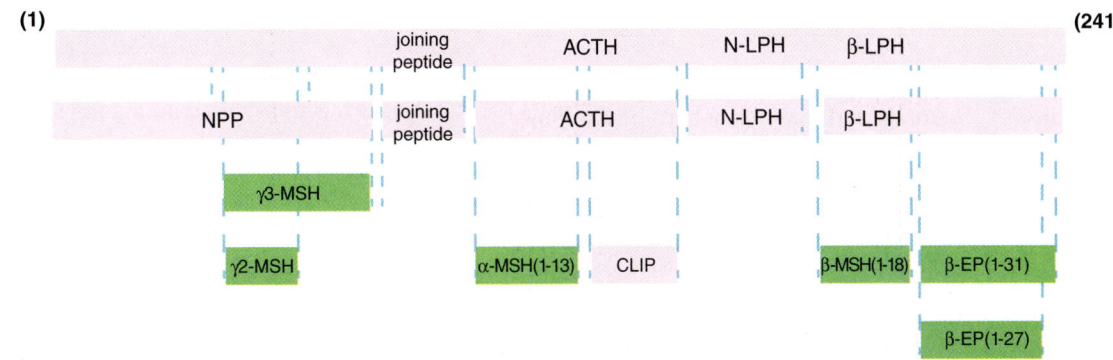

FIGURE 41-15 **Products of pre-pro-opiomelanocortin (POMC) cleavage.** Schematic showing the structure of human pre-pro-opiomelanocortin (*top*) with N-terminal signal sequence (yellow) and eight di-basic and one tetra-basic amino acid clusters (ie, R, K residues; amino acid coordinates within the 241 aa pro-opiomelanocortin polypeptide, *blue*), which represent the nine potential sites of trypsin-like enzyme cleavage (*vertical arrows*) of pro-opiomelanocortin that generate the biologically active polypeptide cleavage products in the hypothalamus shown below (green/pink, labeled). Abbreviations: ACTH, adrenocorticotropic hormone; CLIP, corticotropin-like intermediate lobe peptide; EP, Endorphin; LPH, lipotropin; MSH, melanocyte-stimulating hormone. Numbers in parentheses indicate the amino acid length of the indicated processed polypeptides.

rise to α-MSH(1-13) and corticotropin-like intermediate lobe peptide (CLIP); (2) β-lipotropin (β-LPH), which can yield β-MSH(1-18), and two β-endorphins (β-EPs): β-EP (1-31) and β-EP(1-27); and (3) a large N-terminal polypeptide (NPP), which generates several forms of γ-MSH (γ2-MSH and γ2-MSH) that differ slightly in length (see Figure 41-15). The diversity of these products is generated via specific proteolysis at the locations of multiple dibasic amino acid clusters that reside along the length of the POMC molecule (see Figure 41-15). Each of the peptides mentioned is preceded by either Lys-Arg, Arg-Lys, Arg-Arg, or Lys-Lys residues. In addition, many of these small polypeptide hormones are subject to extensive, additional, tissue-specific modifications that significantly affect their activities (phosphorylation, acetylation, glycosylation, and amidation).

Mutations of the α-MSH receptor are linked to a common, early-onset form of obesity. This observation has directed additional attention to the POMC peptide hormones in recent years.

THERE IS SIGNIFICANT VARIATION IN THE STORAGE & SECRETION OF HORMONES

As mentioned earlier, the steroid hormones and $1,25(OH)_2$-D_3 are synthesized in their final active form. They are also secreted as they are made, and thus there is no intracellular reservoir

of these hormones. The catecholamines, also synthesized in active form, are stored in granules in the chromaffin cells in the adrenal medulla. In response to appropriate neural stimulation, these granules are released from the cell through exocytosis, and the catecholamines are released into the circulation. A several-hour reserve supply of catecholamines exists in the chromaffin cells.

PTH also exists in storage vesicles. As much as 80 to 90% of the proPTH synthesized is degraded before it enters this final storage compartment, especially when Ca^{2+} levels are high in the parathyroid cell (see earlier). PTH is secreted when Ca^{2+} is low in the parathyroid cells, which contain a several-hour supply of the hormone.

The human pancreas secretes about 40 to 50 units of insulin daily; this represents about 15 to 20% of the hormone stored in the β cells. Insulin and the C-peptide (see Figure 41-12) are normally secreted in equimolar amounts. Stimuli such as glucose, which provokes insulin secretion, therefore trigger the processing of proinsulin to insulin as an essential part of the secretory response.

A several-week supply of T_3 and T_4 exists in the thyroglobulin that is stored in colloid in the lumen of the thyroid follicles. These hormones can be released upon stimulation by TSH. This is the most exaggerated example of a prohormone, as a molecule containing ~5000 amino acids must be first synthesized, then degraded, to supply a few molecules of the active hormones T_4 and T_3.

TABLE 41–5 Diversity in the Storage of Hormones

Hormone	Supply Stored in Cell
Steroids and 1,25(OH)$_2$-D$_3$	None
Catecholamines and PTH	Hours
Insulin	Days
T$_3$ and T$_4$	Weeks

TABLE 41–7 Comparison of T$_4$ and T$_3$ in Plasma

Total Hormone (µg/dL)	Free Hormone			$t_{1/2}$ in Blood (days)
	Percentage of Total	ng/dL	Molarity	
T$_4$ 8	0.03	~2.24	3.0×10^{-11}	6.5
T$_3$ 0.15	0.3	~0.4	0.6×10^{-11}	1.5

The diversity in storage and secretion of hormones is illustrated in **Table 41–5**.

SOME HORMONES HAVE PLASMA TRANSPORT PROTEINS

The class I hormones are hydrophobic in chemical nature and thus are not very soluble in plasma. These hormones, principally the steroids and thyroid hormones, have specialized plasma transport proteins that serve several purposes. First, these proteins circumvent the solubility problem and thereby deliver the hormone to the target cell. They also provide a circulating reservoir of the hormone that can be substantial, as in the case of the thyroid hormones. Hormones, when bound to the transport proteins, cannot be metabolized, thereby prolonging their plasma half-life ($t_{1/2}$). The binding affinity of a given hormone to its transporter determines the bound versus free ratio of the hormone. This is important because only the free form of a hormone is biologically active. In general, the concentration of free hormone in plasma is very low, in the range of 10^{-15} to 10^{-9} mol/L. It is important to distinguish between plasma transport proteins and hormone receptors. Both bind hormones but with very different characteristics (**Table 41–6**).

The hydrophilic hormones—generally class II and of peptide structure—are freely soluble in plasma and do not require transport proteins. Hormones such as insulin, growth hormone, ACTH, and TSH circulate in the free, active form and have very short plasma half-lives. A notable exception is

TABLE 41–6 Comparison of Receptors With Transport Proteins

Feature	Receptors	Transport Proteins
Concentration	Very low (thousands/cell)	High (billions/µL)
Binding affinity (K_d)	High (pmol/L to nmol/L range)	Low (µmol/L range)
Binding specificity	Very high	Low
Saturability	Yes	No
Reversibility	Yes	Yes
Signal transduction	Yes	No

IGF-I, which is transported bound to members of a family of binding proteins.

Thyroid Hormones Are Transported by Thyroid-Binding Globulin

Many of the principles discussed above are illustrated in a discussion of thyroid-binding proteins. One-half to two-thirds of T$_4$ and T$_3$ in the body is in an extrathyroidal reservoir. Most of this circulates in bound form, that is, bound to a specific binding protein, **thyroxine-binding globulin (TBG)**. TBG, a glycoprotein with a molecular mass of 50 kDa, binds T$_4$ and T$_3$ and has the capacity to bind 20 µg/dL of plasma. Under normal circumstances, TBG binds—noncovalently—nearly all of the T$_4$ and T$_3$ in plasma, and it binds T$_4$ with greater affinity than T$_3$ (**Table 41–7**). The plasma half-life of T$_4$ is correspondingly four to five times that of T$_3$. The small, unbound (free) fraction is responsible for the biologic activity. Thus, in spite of the great difference in total amount, the free fraction of T$_4$ approximates that of T$_3$, and given that T$_3$ is intrinsically more active than T$_4$, most biologic activity is attributed to T$_3$. TBG does not bind any other hormones.

Glucocorticoids Are Transported by Corticosteroid-Binding Globulin

Hydrocortisone (cortisol) also circulates in plasma in protein-bound and free forms. The main plasma binding protein is an α-globulin called **transcortin**, or **corticosteroid-binding globulin (CBG)**. CBG is produced in the liver, and its synthesis, like that of TBG, is increased by estrogens. CBG binds most of the hormone when plasma cortisol levels are within the normal range; much smaller amounts of cortisol are bound to albumin. The affinity of binding helps determine the biologic half-lives of various glucocorticoids. Cortisol binds tightly to CBG and has a $t_{1/2}$ of 1.5 to 2 hours, while corticosterone, which binds less tightly, has a $t_{1/2}$ of less than 1 hour (**Table 41–8**). The unbound (free) cortisol constitutes ~8% of the total and represents the biologically active fraction. Binding to CBG is not restricted to glucocorticoids. Deoxycorticosterone and progesterone interact with CBG with sufficient affinity to compete for cortisol binding. Aldosterone, the most potent natural mineralocorticoid, does not have a specific plasma transport protein. Gonadal steroids bind very weakly to CBG (see Table 41–8).

TABLE 41–8 Approximate Affinities of Steroids for Serum-Binding Proteins

	SHBG[a]	CBG[a]
Dihydrotestosterone	1	>100
Testosterone	2	>100
Estradiol	5	>10
Estrone	>10	>100
Progesterone	>100	~2
Cortisol	>100	~3
Corticosterone	>100	~5

[a]Affinity expressed as K_d (nmol/L).
Abbreviations: CBG, corticosteroid-binding globulin; SHBG, sex hormone–binding globulin.

Gonadal Steroids Are Transported by Sex Hormone–Binding Globulin

Most mammals, humans included, have a plasma β-globulin that binds testosterone with specificity, relatively high affinity, and limited capacity (see Table 41–8). This protein, usually called **sex hormone–binding globulin (SHBG)** or testosterone-estrogen–binding globulin (TEBG), is produced in the liver. Its production is increased by estrogens (women have twice the serum concentration of SHBG as men), certain types of liver disease, and hyperthyroidism; it is decreased by androgens, advancing age, and hypothyroidism. Many of these conditions also affect the production of CBG and TBG. Since SHBG and albumin bind 97 to 99% of circulating testosterone, only a small fraction of the hormone in circulation is in the free (biologically active) form. The primary function of SHBG may be to restrict the free concentration of testosterone in the serum. Testosterone binds to SHBG with higher affinity than does estradiol (see Table 41–8). Therefore, a change in the level of SHBG causes a greater change in the free testosterone level than in the free estradiol level.

Estrogens are bound to SHBG and progestins to CBG. SHBG binds estradiol about five times less avidly than it binds testosterone or DHT, while progesterone and cortisol have little affinity for this protein (see Table 41–8). In contrast, progesterone and cortisol bind with nearly equal affinity to CBG, which in turn has little affinity for estradiol and even less for testosterone, DHT, or estrone.

These binding proteins also provide a circulating reservoir of hormone, and because of the relatively large binding capacity, they probably buffer against sudden changes in the plasma level. Because the metabolic clearance rates of these steroids are inversely related to the affinity of their binding to SHBG, estrone is cleared more rapidly than estradiol, which in turn is cleared more rapidly than testosterone or DHT.

SUMMARY

- The presence of a specific receptor defines the target cells for a given hormone.

- Receptors are proteins that bind specific hormones and generate an intracellular signal (receptor-effector coupling).

- Some hormones have intracellular receptors; others bind to receptors on the plasma membrane.

- Hormones are synthesized from a number of precursor molecules, including cholesterol, tyrosine per se, and all the constituent amino acids of peptides and proteins.

- A number of modification processes alter the activity of hormones. For example, many hormones are synthesized from larger precursor molecules.

- The complement of enzymes in a particular cell type allows for the production of a specific class of steroid hormone.

- Most of the lipid-soluble hormones are bound to rather specific plasma transport proteins.

REFERENCES

Bain DL, Heneghan AF, Connaghan-Jones KD, Miura MT: Nuclear receptor structure: implications for function. Annu Rev Physiol 2007;69:201-220.

Bartalina L: Thyroid hormone-binding proteins: update 1994. Endocr Rev 1994;13:140-142.

Cristina Casals-Casas C, Desvergne B: Endocrine disruptors: from endocrine to metabolic disruption. Annu Rev Physiol 2011;73:135-162.

DeLuca HR: The vitamin D story: a collaborative effort of basic science and clinical medicine. FASEB J 1988;2:224-236.

Douglass J, Civelli O, Herbert E: Polyprotein gene expression: generation of diversity of neuroendocrine peptides. Annu Rev Biochem 1984;53:665-715.

Fan W, Atkins AR, Yu RT, et al: Road to exercise mimetics: targeting nuclear receptor in skeletal muscle. J Mol Endocrinol 2013;51:T87-T100.

Farooqi IS, O'Rahilly S: Monogenic obesity in humans. Annu Rev Med 2005;56:443-458.

Hah N, Kraus WL: Hormone-regulated transcriptomes: lessons learned from estrogen signaling pathways in breast cancer cells. Mol Cell Endocrinol 2014;382:652-664.

Kirwin P, Kay RG, Brouwers B, et al: Quantitative mass spectrometry for human melancortin peptides in vitro and in vivo suggests prominent roles for β-MSH and desacetyl α-MSH in energy homeostasis. 2018;17:82-97. doi: 10.1016/j.molmet.2018.08.006.

Miller WL: Molecular biology of steroid hormone biosynthesis. Endocr Rev 1988;9:295-318.

Russell DW, Wilson JD: Steroid 5 alpha-reductase: two genes/two enzymes. Annu Rev Biochem 1994;63:25-61.

Steiner DF, Smeekens SP, Ohagi S, et al: The new enzymology of precursor processing endoproteases. J Biol Chem 1992;267:23435-23438.

Taguchi A, White M: Insulin-like signaling, nutrient homeostasis, and life span. Annu Rev Physiol 2008;70:191-212.

Weikum ER, Knuesel MT, Ortlund EA, Yamamoto KR: Glucocorticoid receptor control of transcription: precision and plasticity via allostery. Nat Rev Mol Cell Biol 2017;18:159-174.

Xu Y, O'Malley BW, Elmquist JK: Brain nuclear receptors and body weight regulation. J Clin Invest 2017;127:1172-1180.

Hormone Action & Signal Transduction

42

P. Anthony Weil, PhD

O B J E C T I V E S

After studying this chapter, you should be able to:

- Explain the roles of stimulus, hormone release, signal generation, and effector response in hormone-regulated physiologic processes.
- Describe the role of receptors and guanosine nucleotide-binding G-proteins in hormone signal transduction, particularly with regard to the generation of second messengers.
- Appreciate the complex patterns of signal transduction pathway cross-talk in mediating complicated physiologic outputs.
- Understand the key roles that protein-ligand, protein–protein, protein posttranslational modification, and protein-DNA interactions play in mediating hormone-directed physiologic processes.
- Detail why hormone-modulated receptors, second messengers, and associated signaling molecules represent a rich source of potential drug target development given their key roles in the regulation of physiology.

BIOMEDICAL IMPORTANCE

The homeostatic adaptations an organism makes to a constantly changing environment are in large part accomplished through alterations of the activity and amount of proteins. Hormones provide a major means of facilitating these changes. A hormone-receptor interaction results in generation of an amplified intracellular signal that can either regulate the activity of a select set of genes that alters the amounts of certain proteins in the target cell or affect the activity of specific proteins, including enzymes, transporters, or channel proteins. Signals can influence the location of proteins in the cell, and often affects general processes such as protein synthesis, cell growth, and replication, via their effects on gene expression. Other signaling molecules—including cytokines, interleukins, growth factors, and metabolites—use some of the same general mechanisms and signal transduction pathways. Excessive, deficient, or inappropriate production and release of hormones and the other regulatory signaling molecules are major causes of disease. Many pharmacotherapeutic agents are aimed at correcting or otherwise influencing the pathways discussed in this chapter.

HORMONES TRANSDUCE SIGNALS TO AFFECT HOMEOSTATIC MECHANISMS

The general steps involved in producing a coordinated response to a particular stimulus are illustrated in **Figure 42–1**. The stimulus can be a challenge or a threat to the organism, to an organ, or to the integrity of a single cell within that organism. Recognition of the stimulus is the first step in the adaptive response. At the organismic level, this generally involves the nervous system and the special senses (sight, hearing, pain, smell, and touch). At the organ, tissue, or cellular level, recognition involves physicochemical factors such as pH, O_2 tension, temperature, nutrient supply, noxious metabolites, and osmolarity. Appropriate recognition of such stimuli results in the release of one or more hormones that will govern generation of the necessary adaptive response. For purposes of this discussion, the hormones are categorized as described in Table 41–4, that is, based on the location of their specific cellular receptors and the type of signals generated. Group I hormones interact with an intracellular receptor and group II hormones with

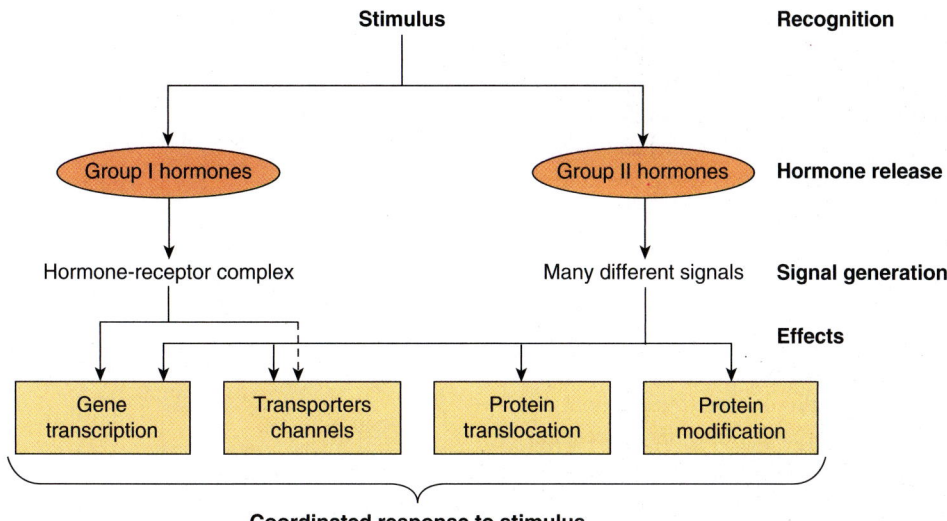

FIGURE 42–1 **Hormonal involvement in responses to a stimulus.** Physiologic needs, or challenges to the integrity of the organism, elicit a response that includes the release of one or more hormones. These hormones generate signals at or within target cells and these signals regulate a variety of biologic processes that provide for a coordinated response to the stimulus or challenge. See Figure 42–8 for a specific example.

receptor recognition sites located on the extracellular surface of the plasma membrane of target cells. The cytokines, interleukins, and growth factors should also be considered in this latter category. These molecules, of critical importance in homeostatic adaptation, are hormones in the sense that they are produced in specific cells, have the equivalent of autocrine, paracrine, and endocrine actions, bind to cell surface receptors, and activate many of the same signal transduction pathways employed by the more traditional group II hormones.

SIGNAL GENERATION

The Ligand-Receptor Complex Is the Signal for Group I Hormones

The lipophilic group I hormones diffuse through the plasma membrane of all cells but only encounter their specific, high-affinity intracellular receptors in target cells. These receptors can be located in the cytoplasm or in the nucleus of target cells. The hormone-receptor complex first undergoes an **activation reaction**. As shown in **Figure 42–2**, receptor activation occurs by at least two mechanisms. For example, glucocorticoids diffuse across the plasma membrane and encounter their cognate receptor in the cytoplasm of target cells. Ligand-receptor binding results in a conformational change in the receptor leading to the dissociation of the heat shock protein 90 (hsp90 chaperone; Chapters 5, 49). This step is necessary for subsequent nuclear translocation of the glucocorticoid receptor (GR). The GR receptor also contains a nuclear localization sequence (NLS) that is now free to facilitate translocation from cytoplasm to nucleus. The activated receptor moves into the nucleus (Figure 42–2) and binds with high affinity to a specific DNA sequence called the **hormone response element (HRE)**. In the case of the GR,

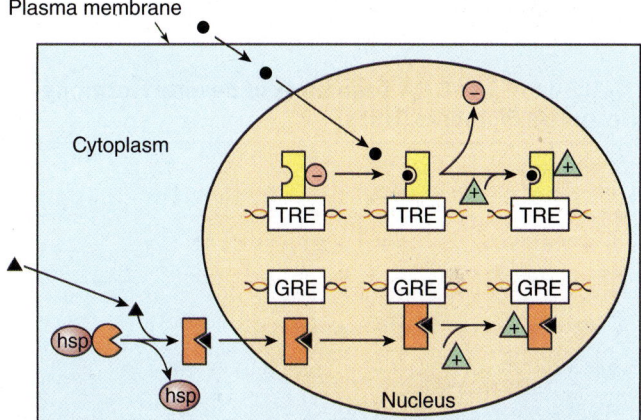

FIGURE 42–2 **Regulation of gene expression by two different group I hormones, thyroid hormone and glucocorticoids.** The hydrophobic steroid hormones readily gain access to the cytoplasmic compartment of target cells by diffusion through the plasma membrane. Glucocorticoid hormones (solid triangles) encounter their cognate receptor, the glucocorticoid receptor (GR) in the cytoplasm where GR exists in a complex with a chaperone protein, heat shock protein 90 (hsp). Ligand binding causes dissociation of hsp90 and a conformational change of the receptor. The receptor-ligand complex then traverses the nuclear membrane and binds to DNA with specificity and high affinity at a glucocorticoid response element (GRE). This event affects the architecture of a number of transcription coregulators (green triangles), and enhanced transcription ensues. By contrast, thyroid hormones and retinoic acid (*black circle*) directly enter the nucleus, where their cognate heterodimeric (TR-RXR; see Figure 42–12) receptors are already bound to the appropriate response elements with an associated transcription corepressor complex (*red circles*). Binding of hormones occurs, which again induces conformational changes in the receptor leading to dissociation of the corepressor complex from the receptor, thereby allowing an activator complex, consisting of the TR-TRE and coactivator, to assemble. The gene is then actively transcribed.

this is a **glucocorticoid response element,** or **GRE**. Consensus sequences for various HREs are shown in **Table 42–1**. The liganded DNA-bound receptor serves as a high-affinity binding target for one or more transcriptional coactivator proteins, and accelerated gene transcription typically ensues when this occurs. By contrast, certain hormones such as the thyroid hormones and retinoids diffuse from the extracellular fluid across the plasma membrane and go directly into the nucleus. In this case, the cognate receptor is already bound to the HRE (the thyroid hormone response element [TRE], in this example). However, this DNA-bound receptor fails to activate transcription because it exists in complex with a corepressor. Indeed, this receptor-corepressor complex serves as a tonic repressor of gene transcription. Association of ligand with these receptors results in dissociation of the corepressor(s). The liganded receptor is now capable of binding one or more coactivators with high affinity, resulting in the recruitment of RNA polymerase II and the GTFs and activation of gene transcription as noted earlier for the GR-GRE complex. The relationship of hormone receptors to other nuclear receptors and to coregulators is discussed in more detail later.

By selectively affecting gene transcription and the consequent production of appropriate target mRNAs, the amounts of specific proteins are changed and metabolic processes are

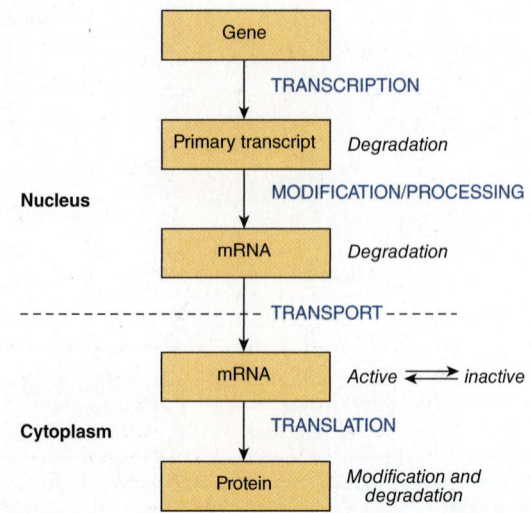

FIGURE 42–3 The "information pathway." Information flows from the gene to the primary transcript to mRNA to protein. Hormones can affect any of the steps involved and can affect the rates of processing, degradation, or modification of the various products.

influenced. The influence of each of these hormones is quite specific; generally, a given hormone directly affects less than 1% of the genes, mRNA, or proteins in a target cell; sometimes only a few genes are affected. The nuclear actions of steroid, thyroid, and retinoid hormones are quite well defined. Most evidence suggests that these hormones exert their dominant effect on modulating gene transcription, but they—and many of the hormones in the other classes discussed later—can act at any step of the "information pathway," as illustrated in **Figure 42–3**, to control specific gene expression and, ultimately, a biologic response. Direct actions of steroids in the cytoplasm and on various organelles and membranes have also been described. Recently, microRNAs and lncRNAs have been implicated in mediating some of the diverse actions of hormones.

GROUP II (PEPTIDE & CATECHOLAMINE) HORMONES HAVE MEMBRANE RECEPTORS & USE INTRACELLULAR MESSENGERS

Many hormones are water soluble, have no transport proteins (and therefore have a short plasma half-life), and initiate a response by binding to a receptor located in the plasma membrane (see Tables 41–3 and 41–4). The mechanism of action of this group of hormones can best be discussed in terms of the **intracellular signals** they generate. These signals include **cAMP** (cyclic AMP; 3′,5′-adenylic acid; see Figure 18–5), a nucleotide derived from ATP through the action of adenylyl cyclase; **cGMP**, a nucleotide formed from GTP by guanylyl cyclase; **Ca²⁺**; and **phosphatidylinositides**; such small molecules are **termed second messengers** as their synthesis is triggered by the presence of the primary hormone (molecule)

TABLE 42–1 The DNA Sequences of Several Hormone Response Elements (HREs)[a]

Hormone or Effector	HRE	DNA Sequence
Glucocorticoids	GRE	GGTACA NNN TGTTCT ◄——— ———►
Progestins	PRE	
Mineralocorticoids	MRE	
Androgens	ARE	
Estrogens	ERE	AGGTCA — TGACCT ◄——— ———►
Thyroid hormone	TRE	AGGTCA N_{(1-5)} AGGTCA ———► ———►
Retinoic acid	RARE	
Vitamin D	VDRE	
Camp	CRE	TGACGTCA

[a]Letters indicate nucleotide; N means any one of the four can be used in that position. The arrows pointing in opposite directions illustrate the slightly imperfect inverted palindromes present in many HREs; in some cases, these are called "half binding sites," or half-sites, because each binds one monomer of the receptor. The GRE, PRE, MRE, and ARE consist of the same DNA sequence. Specificity is likely conferred by the intracellular concentration of the ligand or hormone receptor, by flanking DNA sequences not included in the consensus, or by other accessory elements. A second group of HREs includes those for thyroid hormones, estrogens, retinoic acid, and vitamin D. These HREs are similar except for the orientation and spacing between the half palindromes. Spacing determines the hormone specificity. VDRE (N = 3), TRE (N = 4), and RARE (N = 5) bind to direct repeats rather than to inverted repeats. Another member of the steroid receptor superfamily, the retinoid X receptor (RXR), forms heterodimers with VDR, TR, and RARE, and these constitute the functional forms of these DNA-binding transacting factors. cAMP affects gene transcription through the CRE.

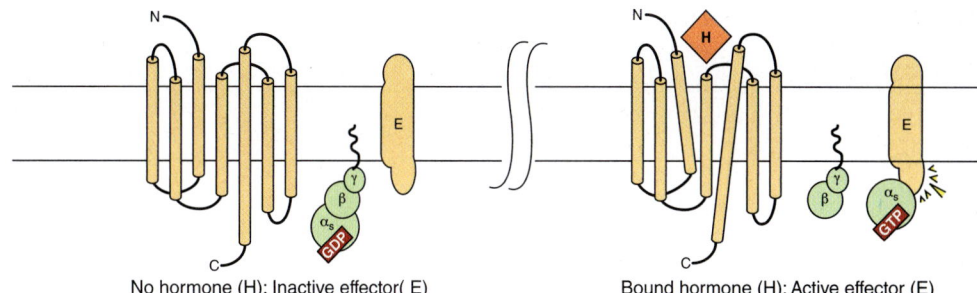

No hormone (H): Inactive effector(E)   Bound hormone (H): Active effector (E)

FIGURE 42–4 **Components of the hormone receptor–G-protein effector system.** Receptors that couple to effectors through G-proteins, the G-protein–coupled receptors (GPCRs), typically have seven α-helical membrane-spanning domains (here shown as long cylinders). In the absence of hormone (**left**), the heterotrimeric G-protein complex (α,β,γ) is in an inactive guanosine diphosphate (GDP)-bound form and is probably not associated with the receptor. This complex is anchored to the plasma membrane through prenylated groups on the βγ subunits (**wavy lines**) and perhaps by myristoylated groups on α subunits (not shown). On binding of hormone (H) to the receptor, there are conformational changes within the receptor (as indicated by the tilted membrane-spanning domains) and association of the G-protein complex with the rearranged receptor—this then activates the G-protein complex. This activation results from the exchange of GDP with guanosine triphosphate (GTP) on the α subunit, after which α and βγ dissociate. The α subunit binds to and activates the effector (E). E can be adenylyl cyclase, Ca^{2+}, Na^+, or Cl^- channels ($α_s$), or it could be a K^+ channel ($α_i$), phospholipase Cβ ($α_q$), or cGMP phosphodiesterase ($α_t$); see Table 42–3. The βγ subunit can also have direct actions on E. (Reproduced with permission from Becker KL: *Principles and Practice of Endocrinology and Metabolism*, 3rd ed. Philadelphia, PA: Lippincott; 2001.)

binding its receptor. Many of these second messengers affect gene transcription, as described in the previous paragraph; but they also influence a variety of other biologic processes, as shown in Figure 42–3, but see also Figures 42–6 and 42–8.

G-Protein–Coupled Receptors

Many of the group II hormones bind to receptors that couple to effectors through a **guanine-nucleotide binding protein (G-protein)** intermediary. These receptors typically have seven α-helical hydrophobic plasma membrane-spanning domains, here illustrated by the seven interconnected cylinders extending through the lipid bilayer in **Figure 42–4**. Receptors of this class, which signal through G-proteins, are known as **G-protein–coupled receptors (GPCRs)**. To date, hundreds of *GPCR* genes have been identified, and represent the largest family of cell surface receptors in humans. Not surprisingly, a wide variety of responses are mediated by the GPCRs.

cAMP Is the Intracellular Signal for Many Responses

Cyclic AMP was the first intracellular second messenger signal identified in mammalian cells. Several components comprise a system for the generation, degradation, and action of cAMP (**Table 42–2**).

Adenylyl Cyclase

Different peptide hormones can either stimulate (s) or inhibit (i) the production of cAMP from adenylyl cyclase through the action of the G-proteins. G-proteins are encoded by at least 10 different genes (**Table 42–3**). Two parallel systems, a stimulatory (s) one and an inhibitory (i) one, converge on a catalytic molecule (C). Each consists of a receptor, R_s or R_i, and a regulatory G-protein complex termed G_s and G_i. Both G_s and G_i are

heterotrimeric G-proteins composed of α, β, and γ subunits. Since the α subunit in G_s differs from that in G_i, the proteins, which are distinct gene products, are designated $α_s$ and $α_i$. The α subunits bind guanine nucleotides. The β and γ subunits are likely always associated (βγ) and appear to function predominantly as a heterodimer. The binding of a hormone to R_s or R_i results in a receptor-mediated activation of G-protein, which entails the exchange of GDP by GTP on α and the concomitant dissociation of βγ from α.

TABLE 42–2 Subclassification of Group II.A Hormones

Hormones That Stimulate Adenylyl Cyclase (H_s)	Hormones That Inhibit Adenylyl Cyclase (H_i)
ACTH	Acetylcholine
ADH	$α_2$-Adrenergics
β-Adrenergics	Angiotensin II
Calcitonin	Somatostatin
CRH	
FSH	
Glucagon	
hCG	
LH	
LPH	
MSH	
PTH	
TSH	

Abbreviations: ACTH, adrenocorticotropic hormone; ADH, antidiuretic hormone; CRH, corticotropin-releasing hormone; FSH, follicle-stimulating hormone; hCG, human chorionic gonadotropin; LH, luteinizing hormone; LPH, lipotropin; MSH, melanocyte-stimulating hormone; PTH, parathyroid hormone; TSH, thyroid-stimulating hormone.

TABLE 42–3 Classes & Functions of Selected G-Proteins[a]

Class or Type	Stimulus	Effector	Effect
Gs			
α_s	Glucagon, β-adrenergics	↑Adenylyl cyclase	Glyconeogenesis, lipolysis, glycogenolysis
		↑Cardiac Ca^{2+}, Cl^-, and Na^+ channels	Olfaction
α_{olf}	Odorant	↑Adenylyl cyclase	
G_i			
$\alpha_{i-1,2,3}$	Acetylcholine, α_2-adrenergics	↓Adenylyl cyclase	Slowed heart rate
		↑Potassium channels	
	M_2 cholinergics	↓Calcium channels	
α_o	Opioids, endorphins	↑Potassium channels	Neuronal electrical activity
α_t	Light	↑Cgmp phosphodiesterase	Vision
G_q			
α_q	M_1 cholinergics		
	α_1-Adrenergics	↑Phospholipase C-β1	↓Muscle contraction
α_{11}	α_1-Adrenergics	↑Phospholipase C-β2	↓Blood pressure
G_{12}			
α_{12}	Thrombin	Rho	Cell shape changes

[a]The four major classes or families of mammalian G-proteins (G_s, G_i, G_q, and G_{12}) are based on protein sequence conservation. Representative members of each are shown, along with known stimuli, effectors, and well-defined biologic effects. Nine isoforms of adenylyl cyclase have been identified (isoforms I-IX). All isoforms are stimulated by α_s; α_i isoforms inhibit types V and VI, and α_o inhibits types I and V. At least 16 different α subunits have been identified.
Reproduced with permission from Becker KL: *Principles and Practice of Endocrinology and Metabolism*, 3rd ed. Philadelphia, PA: Lippincott; 2001.

The α_s protein has intrinsic GTPase activity. The active form, α_s-GTP, is inactivated on hydrolysis of the GTP to GDP; the **trimeric G_s complex (αβγ)** is then reformed and is ready for another cycle of activation. **Cholera** and **pertussis toxins** catalyze the **ADP ribosylation** of α_s and α_{i-2} (see Table 42–3), respectively. In the case of α_s, this modification disrupts the intrinsic GTPase activity; thus, α_s cannot reassociate with βγ and is therefore irreversibly activated. ADP ribosylation of α_{i-2} prevents the dissociation of α_{i-2} from βγ, and free α_{i-2} thus cannot be formed. α_s activity in such cells is therefore unopposed.

There is a large family of G-proteins, and these are part of the superfamily of GTPases. The G-protein family is classified according to sequence homology into four subfamilies, as illustrated in Table 42–3. There are 21 α, 5 β, and 8 γ subunit genes. Various combinations of these subunits provide a large number of possible αβγ complexes.

The α subunits and the βγ complex have actions independent of those on adenylyl cyclase (see Figure 42–4 and Table 42–3). Some forms of α_i stimulate K^+ channels and inhibit Ca^{2+} channels, and some α_s molecules have the opposite effects. Members of the G_q family activate the phospholipase C group of enzymes. The βγ complexes have been associated with K^+ channel stimulation and phospholipase C activation. G proteins are involved in many important biologic processes in addition to hormone action. Notable examples include olfaction (α_{OLF}) and vision (α_t). Some examples are listed in

Table 42–3. GPCRs are implicated in a number of diseases and are major targets for pharmaceutical agents.

Protein Kinase

As discussed in Chapter 38, in prokaryotic cells, cAMP binds to a specific protein called cAMP activator protein (CAP) that binds directly to DNA and influences gene expression. By contrast, in eukaryotic cells, cAMP binds to a protein kinase called **protein kinase A (PKA)**, a heterotetrameric molecule consisting of two regulatory subunits (R) that inhibit the activity of the two catalytic subunits (C) when bound as a tetrameric complex. cAMP binding to the R_2C_2 tetramer results in the following reaction:

$$4cAMP + R_2C_2 \rightleftharpoons R_2\text{-}4cAMP + 2C$$

The R_2C_2 complex has no enzymatic activity, but the binding of cAMP to the R subunit induces dissociation of the R-C complex, thereby activating the latter (**Figure 42–5**). The active C subunit catalyzes the transfer of the γ phosphate of ATP to a serine or threonine residue in a variety of proteins. The consensus PKA phosphorylation sites are -ArgArg/Lys-X-Ser/Thr- and -Arg-Lys-X-X-Ser-, where X can be any amino acid.

Historically protein kinase activities were described as being "cAMP-dependent" or "cAMP-independent." This classification has changed, as protein phosphorylation is now recognized as being a major and ubiquitous regulatory mechanism.

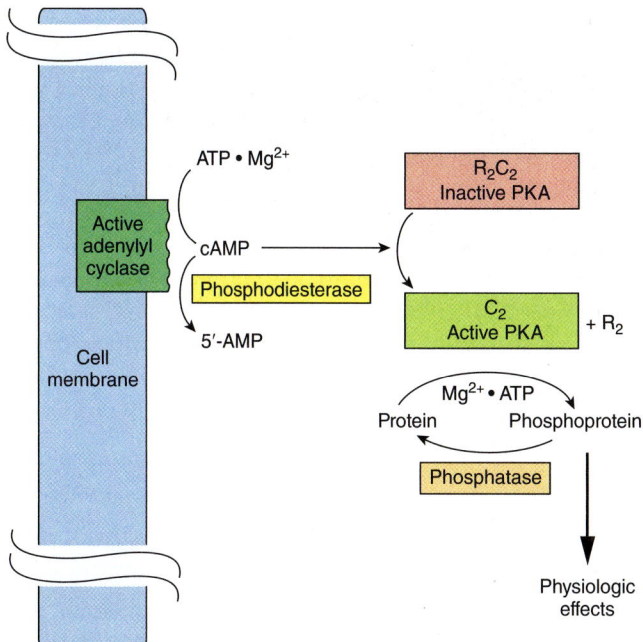

FIGURE 42–5 **Hormonal regulation of cellular processes through cAMP-dependent protein kinase (PKA).** PKA exists in an inactive form as an R_2C_2 heterotetramer consisting of two regulatory (R) and two catalytic (C) subunits. The cAMP generated by the action of adenylyl cyclase (activated as shown in Figure 42–4) binds to the regulatory subunit of PKA. This results in dissociation of the regulatory and catalytic subunits and activation of the latter. The active catalytic subunits phosphorylate a number of target proteins on serine and threonine residues. Phosphatases remove phosphate from these residues and thus terminate the physiologic response. A phosphodiesterase can also terminate the response by converting cAMP to 5′-AMP.

Several hundred protein kinases have now been described. These kinases are related in sequence and structure within the catalytic domain, but each is a unique molecule with considerable variability with respect to subunit composition, molecular weight, autophosphorylation, K_m for ATP, and substrate specificity. Both kinase and protein phosphatase activities can be targeted by interaction with specific kinase-binding proteins. In the case of PKA, such targeting proteins are termed **A kinase anchoring proteins**, or **AKAPs**. AKAPs serve as scaffolds, which localize PKA near to substrates thereby focusing PKA activity toward physiologic substrates and facilitating spatiotemporal biologic regulation while also allowing for common, shared proteins to elicit specific physiologic responses. Multiple AKAPs have been described and importantly they can bind PKA and other kinases as well as phosphatases, phosphodiesterases (which hydrolyze cAMP), and protein kinase substrates. The multifunctionality of AKAPs facilitates signaling localization, rate (production and destruction of signals), specificity, and dynamics.

Phosphoproteins

The effects of cAMP in eukaryotic cells are all thought to be mediated by protein phosphorylation and dephosphorylation, principally on serine and threonine residues. The control of any of the effects of cAMP, including such diverse processes

as steroidogenesis, secretion, ion transport, carbohydrate and fat metabolism, enzyme induction, gene regulation, synaptic transmission, and cell growth and replication, could be conferred by a specific protein kinase, by a specific phosphatase, or by specific substrates for phosphorylation. The array of specific substrates contributes critically to defining a target tissue, and are involved in defining the extent of a particular response within a given cell. For example, the effects of cAMP on gene transcription are mediated by the **cyclic AMP response element binding protein (CREB)**. When CREB binds to a **cAMP responsive DNA enhancer element (CRE)** (see Table 42–1) in its nonphosphorylated state, it is a weak activator of transcription. However, when phosphorylated by PKA at key amino acids, CREB binds the coactivator **CREB-binding protein CBP/p300** (see later) and as a result is a much more potent transcriptional activator. CBP and the related p300 contain **histone acetyltransferase activities (HATs)**, and hence serve as chromatin-active transcriptional coregulators (see Chapters 36, 38). Interestingly, CBP/p300 can also acetylate certain transcription factors thereby stimulating their ability to bind DNA and modulate transcription.

Phosphodiesterases

Actions caused by hormones that increase cAMP concentration can be terminated in a number of ways, including the hydrolysis of cAMP to 5′-AMP by phosphodiesterases (see Figure 42–5). The presence of these hydrolytic enzymes ensures a rapid turnover of the signal (cAMP) and hence a rapid termination of the biologic process once the hormonal stimulus is removed. There are at least 11 known members of the phosphodiesterase family of enzymes. These are subject to regulation by their substrates, cAMP and cGMP; by hormones; and by intracellular messengers such as calcium, probably acting through calmodulin. Inhibitors of phosphodiesterase, most notably methylated xanthine derivatives such as caffeine, increase intracellular cAMP and mimic or prolong the actions of hormones through this signal.

Phosphoprotein Phosphatases

Given the importance of protein phosphorylation, it is not surprising that regulation of the protein dephosphorylation reaction is another important control mechanism (see Figure 42–5). The phosphoprotein phosphatases are themselves subject to regulation by phosphorylation-dephosphorylation reactions and by a variety of other mechanisms, such as protein–protein interactions. In fact, the substrate specificity of the phosphoserine-phosphothreonine phosphatases may be dictated by distinct regulatory subunits whose binding is regulated hormonally. One of the best-studied roles of regulation by the dephosphorylation of proteins is that of glycogen metabolism in muscle (see Figures 18–6 to 18–8). Two major types of phosphoserine-phosphothreonine phosphatases have been described. Type I preferentially dephosphorylates the β subunit of phosphorylase kinase, whereas type II dephosphorylates the α subunit. Type I phosphatase is implicated in the regulation of glycogen synthase, phosphorylase, and phosphorylase kinase. This phosphatase is itself regulated by phosphorylation of certain of its subunits, and these reactions are reversed by the action of one of the type II phosphatases. In addition, two heat-stable protein inhibitors regulate type I phosphatase activity. Inhibitor-1 is phosphorylated and activated by cAMP-dependent protein kinases, and inhibitor-2, which may be a subunit of the inactive phosphatase, is also phosphorylated, possibly by glycogen synthase kinase-3. Phosphatases that target phosphotyrosine are also important in signal transduction (see Figure 42–8).

cGMP Is Also an Intracellular Signal

Cyclic GMP is made from GTP by the enzyme guanylyl cyclase, which exists in soluble and membrane-bound forms. Each of these enzyme forms has unique physiologic properties. The atriopeptins, a family of peptides produced in cardiac atrial tissues, cause natriuresis, diuresis, vasodilation, and inhibition of aldosterone secretion. These peptides (eg, atrial natriuretic factor) bind to and activate the membrane-bound form of guanylyl cyclase. Activation of guanylyl cyclase results in an increase of cGMP by as much as 50-fold in some cases, and this is thought to mediate the effects mentioned earlier. Other evidence links cGMP to vasodilation. A series of compounds, including nitroprusside, nitroglycerin, nitric oxide, sodium nitrite, and sodium azide, all cause smooth muscle relaxation and are potent vasodilators. These agents increase cGMP by activating the soluble form of guanylyl cyclase, and inhibitors of cGMP phosphodiesterase (eg, the drug sildenafil [Viagra]) enhance and prolong these responses. The increased cGMP activates cGMP-dependent protein kinase (PKG), which in turn phosphorylates a number of smooth muscle proteins. Presumably, this is involved in relaxation of smooth muscle and vasodilation.

Several Hormones Act Through Calcium or Phosphatidylinositols

Ionized calcium, Ca^{2+}, is an important regulator of a variety of cellular processes, including muscle contraction, stimulus-secretion coupling, the blood clotting cascade, enzyme activity, and membrane excitability. Ca^{2+} is also an intracellular messenger of hormone action.

Calcium Metabolism

The extracellular Ca^{2+} concentration is ~5 mmol/L and is very rigidly controlled. Although substantial amounts of calcium are associated with intracellular organelles such as mitochondria and the endoplasmic reticulum, the intracellular concentration of free or ionized calcium (Ca^{2+}) is very low: 0.05 to 10 μmol/L. In spite of this large concentration gradient and a favorable transmembrane electrical gradient, Ca^{2+} is restrained from entering the cell. A considerable amount of energy is expended to ensure that the intracellular Ca^{2+} is controlled, as a prolonged elevation of Ca^{2+} in the cell is very toxic. A Na^+/Ca^{2+} exchange mechanism that has a high-capacity but low-affinity pumps Ca^{2+} out of cells. There also is a Ca^{2+}/proton

ATPase-dependent pump that extrudes Ca^{2+} in exchange for H^+. This pump has a high affinity for Ca^{2+} but a low capacity and is probably responsible for fine-tuning cytosolic Ca^{2+}. Furthermore, Ca^{2+}-ATPases pump Ca^{2+} from the cytosol to the lumen of the endoplasmic reticulum. There are three ways of changing cytosolic Ca^{2+} levels: (1) Certain hormones (class II.C, Table 41–3) by binding to receptors that are themselves Ca^{2+} channels, enhance membrane permeability to Ca^{2+}, and thereby increase Ca^{2+} influx. (2) Hormones also indirectly promote Ca^{2+} influx by modulating the membrane potential at the plasma membrane. Membrane depolarization opens voltage-gated Ca^{2+} channels and allows for Ca^{2+} influx. (3) Ca^{2+} can be mobilized from the endoplasmic reticulum, and possibly from mitochondrial pools.

Calmodulin

An important observation linking Ca^{2+} to hormone action involved the definition of the intracellular targets of Ca^{2+} action. The discovery of a Ca^{2+}-dependent regulator of phosphodiesterase activity provided the basis for a broad understanding of how Ca^{2+} and cAMP interact within cells. The calcium-dependent regulatory protein is calmodulin, a 17-kDa protein that is homologous to the muscle protein troponin C in structure and function. Calmodulin has four Ca^{2+}-binding sites, and full occupancy of these sites leads to a marked conformational change, which allows calmodulin to activate enzymes and ion channels. The interaction of Ca^{2+} with calmodulin (with the resultant change of activity of the latter) is conceptually similar to the binding of cAMP to PKA and the subsequent activation of this molecule. Calmodulin can be one of numerous subunits of complex proteins and is particularly involved in regulating various kinases and enzymes of cyclic nucleotide generation and degradation. A partial list of the enzymes regulated directly or indirectly by Ca^{2+}, probably through calmodulin, is presented in **Table 42–4**.

In addition to its effects on enzymes and ion transport, Ca^{2+}/calmodulin regulates the activity of many structural elements in cells. These include the actin-myosin complex of smooth muscle, which is under β-adrenergic control, and various microfilament-mediated processes in noncontractile cells, including cell motility, cell conformation changes, mitosis, granule release, and endocytosis.

TABLE 42–4 Some Enzymes & Proteins Regulated by Calcium or Calmodulin

- Adenylyl cyclase
- Ca^{2+}-dependent protein kinases
- Ca^{2+}-Mg^{2+}-ATPase
- Ca^{2+}-phospholipid–dependent protein kinase
- Cyclic nucleotide phosphodiesterase
- Some cytoskeletal proteins
- Some ion channels (eg, I-type calcium channels)
- Nitric oxide synthase
- Phosphorylase kinase
- Phosphoprotein phosphatase 2B
- Some receptors (eg, NMDA-type glutamate receptor)

Abbreviation: NDMA, *N*-methyl-ᴅ-aspartate receptor.

Calcium Is a Mediator of Hormone Action

A role for Ca^{2+} in hormone action is suggested by the observations that the effect of many hormones is (1) blunted by Ca^{2+}-free media or when intracellular calcium is depleted; (2) can be mimicked by agents that increase cytosolic Ca^{2+}, such as the Ca^{2+} ionophore A23187; and (3) influences cellular calcium flux. Again, the regulation of glycogen metabolism in liver (by vasopressin and β-adrenergic catecholamines; see Figures 18–6 and 18–7) Is instructive. A number of critical metabolic enzymes are regulated by Ca^{2+}, phosphorylation, or both. These include glycogen synthase, pyruvate kinase, pyruvate carboxylase, glycerol-3-phosphate dehydrogenase, and pyruvate dehydrogenase among others (see Figure 19–1).

Phosphatidylinositide Metabolism Affects Ca^{2+}-Dependent Hormone Action

Some signal must provide communication between the hormone receptor on the plasma membrane and the intracellular Ca^{2+} reservoirs. This is accomplished by products of phosphatidylinositol metabolism. Cell surface receptors such as those for acetylcholine, antidiuretic hormone, and $α_1$-type catecholamines are, when occupied by their respective ligands, potent activators of phospholipase C. Receptor binding and activation of phospholipase C are coupled by the G_q isoforms (see Table 42–3 and **Figure 42–6**). Phospholipase C catalyzes the hydrolysis of phosphatidylinositol 4,5-bisphosphate to **inositol trisphosphate (IP$_3$)** and **1,2-diacylglycerol (diacylglcerol, DAG)** (**Figure 42–7**). Diacylglycerol is itself capable of activating **protein kinase C (PKC)**, the activity of which also depends on Ca^{2+} (see Chapter 21 and Figures, 24–1, 24–2, and 55–1). IP$_3$, by interacting with a specific intracellular receptor, is an effective releaser of Ca^{2+} from intracellular storage sites in the endoplasmic reticulum. Thus, the hydrolysis of phosphatidylinositol 4,5-bisphosphate leads to activation of PKC and promotes an increase of cytoplasmic Ca^{2+}. As shown in Figure 42–4, the activation of G-proteins can also have a direct action on Ca^{2+} channels. The resulting elevations of cytosolic Ca^{2+} activate Ca^{2+}-calmodulin–dependent kinases and many other Ca^{2+}-calmodulin–dependent enzymes.

Steroidogenic agents—including ACTH and cAMP in the adrenal cortex; angiotensin II, K$^+$, serotonin, ACTH, and

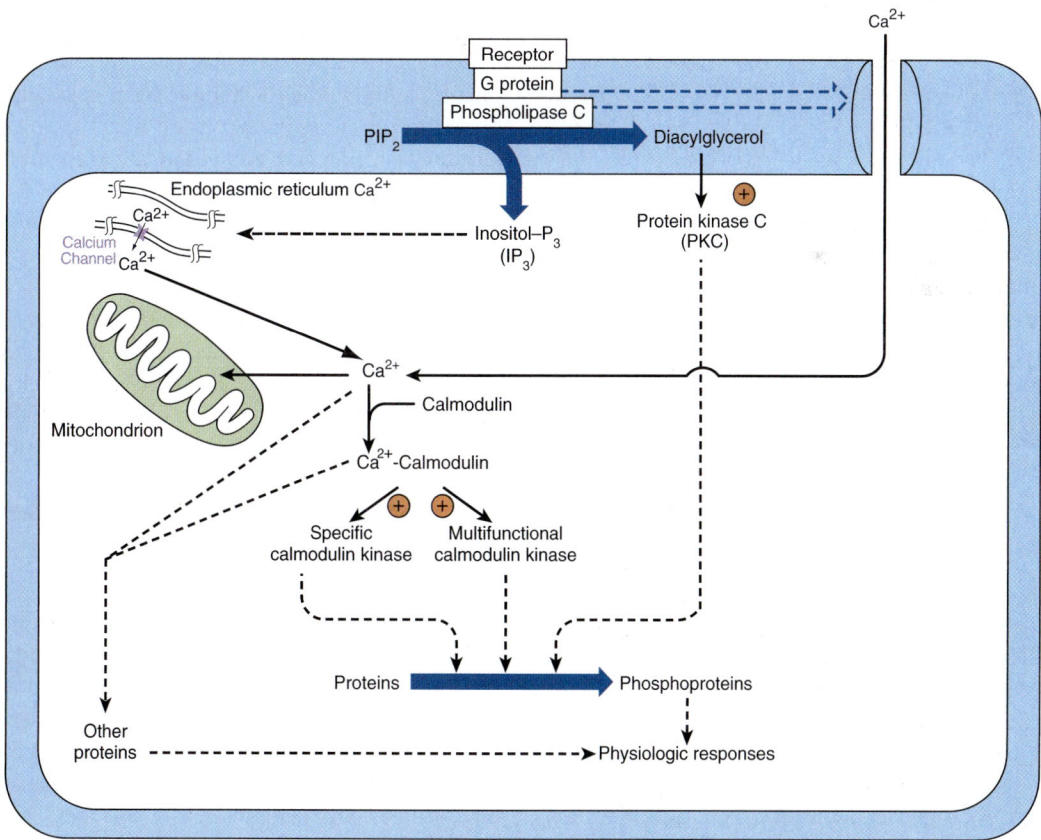

FIGURE 42–6 **Certain hormone-receptor interactions result in the activation of phospholipase C (PLC).** PLC activation appears to involve a specific G-protein, which also may activate a calcium channel. Phospholipase C generates inositol trisphosphate (IP$_3$) from PIP$_2$ (phosphoinositol 4,5-bisphosphate; see Figure 42–7), which liberates stored intracellular Ca^{2+}, and diacylglycerol (DAG), a potent activator of protein kinase C (PKC). In this scheme, the activated PKC phosphorylates specific substrates, which then alter physiologic processes. Likewise, the Ca^{2+}-calmodulin complex can activate specific kinases, two of which are shown here. These actions result in phosphorylation of substrates, and this leads to altered physiologic responses. This figure also shows that Ca^{2+} can enter cells through voltage- or ligand-gated Ca^{2+} channels. The intracellular [Ca^{2+}] is also regulated through storage and release by the mitochondria and endoplasmic reticulum. (Reproduced with permission from JH Exton.)

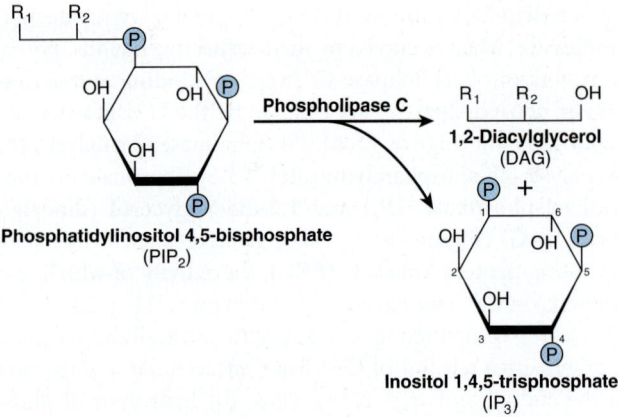

FIGURE 42–7   **Phospholipase C cleaves PIP$_2$ into diacylglycerol and inositol trisphosphate.** R$_1$ generally is stearate, and R$_2$ is usually arachidonate. IP$_3$ can be dephosphorylated (to the inactive I-1,4-P$_2$) or phosphorylated (to the potentially active I-1,3,4,5-P$_4$).

cAMP in the zona glomerulosa of the adrenal; LH in the ovary; and LH and cAMP in the Leydig cells of the testes—have been associated with increased amounts of phosphatidic acid, phosphatidylinositol, and polyphosphoinositides (see Chapter 21) in the respective target tissues. Several other examples could be cited.

The roles that Ca^{2+} and polyphosphoinositide breakdown products might play in hormone action are presented in Figure 42–6. In this scheme, the activated protein kinase C can phosphorylate specific substrates, which then alter physiologic processes. Likewise, the Ca^{2+}-calmodulin complex can activate specific kinases. These then modify substrates and thereby alter physiologic responses.

Some Hormones Act Through a Protein Kinase Cascade

Single protein kinases such as PKA, PKC, and Ca^{2+}-**calmodulin-dependent kinases (CaMKs)**, which result in the phosphorylation of serine and threonine residues in target proteins, play a very important role in hormone action. The discovery that the epidermal growth factor (EGF) receptor contains an intrinsic tyrosine kinase activity that is activated by the binding of the ligand EGF was an important breakthrough. The insulin and insulin-like growth factor 1 (IGF-1) receptors also contain intrinsic ligand-activated tyrosine kinase activity. Several receptors—generally those involved in binding ligands that control cellular growth, differentiation, and the inflammatory response—either have intrinsic tyrosine kinase activity or are tightly associated with proteins that are tyrosine kinases. Another distinguishing feature of this class of hormone action is that these kinases preferentially phosphorylate tyrosine residues, and tyrosine phosphorylation is infrequent (<0.03% of total amino acid phosphorylation) in mammalian cells. A third distinguishing feature is that the ligand-receptor interaction that results in a tyrosine phosphorylation event initiates a cascade that may involve several protein kinases, phosphatases, and other regulatory proteins.

Insulin Transmits Signals by Several Kinase Cascades

The **insulin, EGF,** and **IGF-1 receptors** have intrinsic protein tyrosine kinase activities located in their cytoplasmic domains. These activities are stimulated when their ligands bind to the cognate receptor. The receptors are then autophosphorylated on tyrosine residues, and this phosphorylation initiates a complex series of events (summarized in simplified fashion in **Figure 42–8**). The phosphorylated insulin receptor next phosphorylates **insulin receptor substrates** (there are at least four of these molecules, called **IRS 1-4**) on tyrosine residues. Phosphorylated IRS binds to the **Src homology 2 (SH2)** domains of a variety of proteins that are directly involved in mediating different effects of insulin. One of these proteins, PI-3 kinase, links insulin receptor activation to insulin action through activation of a number of molecules, including the kinase **phosphoinositide-dependent kinase 1 (PDK1)**. This enzyme propagates the signal through several other kinases, including **PKB** (also known as **AKT**), **SKG,** and **aPKC** (see legend to Figure 42–8 for definitions and expanded abbreviations). An alternative pathway downstream from PDK1 involves **p70S6K** and perhaps other as yet unidentified kinases. A second major pathway involves **mTOR**. This enzyme is directly regulated by amino acid levels and insulin and is essential for p70S6K activity. The mTOR-signaling system provides a distinction between the PKB and p70S6K branches downstream from PKD1. These pathways are involved in protein translocation, enzyme activation, and the regulation, by insulin, of genes involved in metabolism (see Figure 42–8). Another SH2 domain–containing protein is **GRB2**, which binds to IRS-1 and links tyrosine phosphorylation to several proteins, the result of which is activation of a cascade of threonine and serine kinases. A pathway showing how this insulin-receptor interaction activates the mitogen-activated protein kinase (**MAPK**) pathway and the anabolic effects of insulin is illustrated in Figure 42–8. The exact roles of many of these docking proteins, kinases, and phosphatases are actively being studied.

The JAK/STAT Pathway Is Used by Hormones and Cytokines

Tyrosine kinase activation can also initiate a phosphorylation and dephosphorylation cascade that involves the action of several other protein kinases and the counterbalancing actions of phosphatases. Two mechanisms are employed to initiate this cascade. Some hormones, such as growth hormone, prolactin, erythropoietin, and the cytokines, initiate their action by activating a tyrosine kinase, but this activity is not an integral part of the hormone receptor. The hormone-receptor interaction promotes binding and activation of **cytoplasmic protein tyrosine kinases,** such as the **Janus kinases 1 and 2, JAK1,** or **JAK2,** or **tyrosine kinase 2, TYK2.**

These kinases phosphorylate one or more cytoplasmic proteins, which then associate with other docking proteins through binding to SH2 domains. One such interaction results in the activation of a family of cytosolic proteins called **STATs,** or

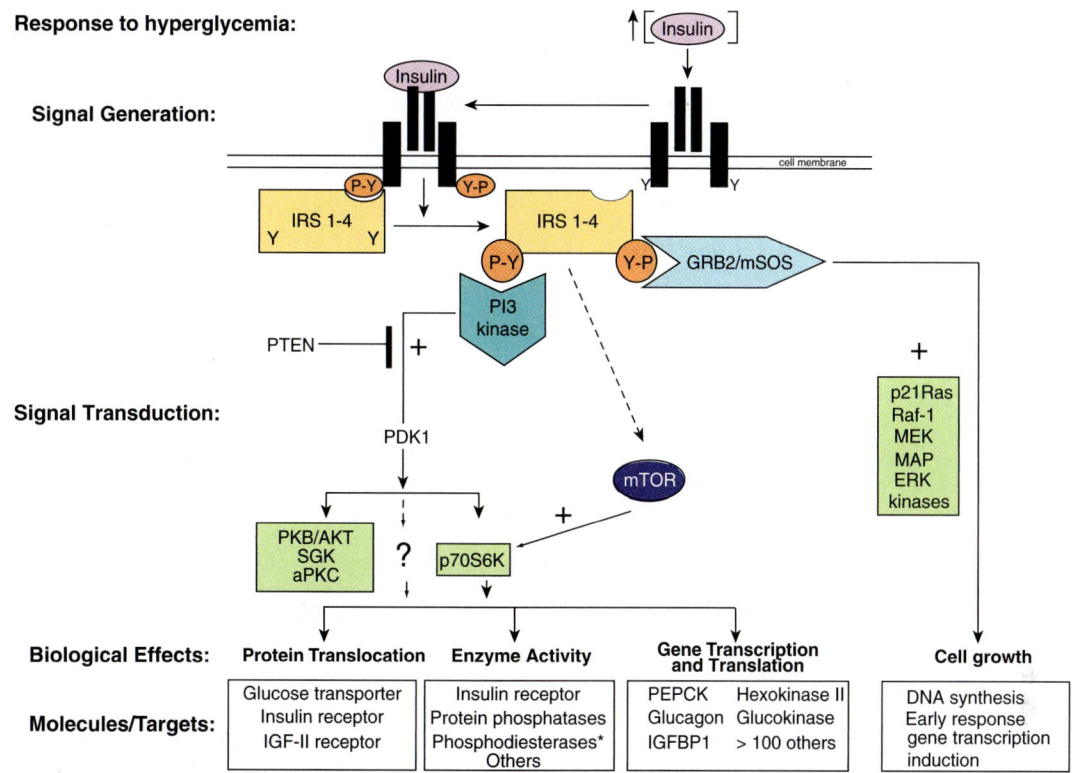

FIGURE 42–8 Insulin signaling pathways. The insulin signaling pathways provide an excellent example of the "recognition -> hormone release -> signal generation -> effects" paradigm outlined in Figure 42–1. Insulin is released into the bloodstream from pancreatic β cells in response to hyperglycemia. Binding of insulin to a target cell-specific plasma membrane heterotetrameric insulin receptor (IR) results in a cascade of intracellular events. First, the intrinsic tyrosine kinase activity of the insulin receptor is activated, and marks the initial event. Receptor activation results in increased tyrosine phosphorylation (conversion of specific Y residues to Y-P) within the receptor. One or more of the insulin receptor substrate (IRS) molecules (IRS 1-4) then bind to the tyrosine-phosphorylated receptor and themselves are specifically tyrosine phosphorylated. IRS proteins interact with the activated IR via N-terminal PH (pleckstrin homology) and phosphotyrosine binding (PTB) domains. IR-docked IRS proteins are tyrosine phosphorylated and the resulting P-Y residues form the docking sites for several additional signaling proteins (ie, PI-3 kinase, GRB2, and mTOR). GRB2 and PI-3K bind to IRS P-Y residues via their SH (*Src* homology) domains; binding to IRS-Y-P residues leads to activation of the activity of many intracellular signaling molecules such as GTPases, protein kinases, and lipid kinases, all of which play key roles in certain metabolic actions of insulin. The two best-described pathways are shown. In detail, phosphorylation of an IRS molecule results in docking and activation of the lipid kinase, PI-3 kinase; PI-3K generates novel inositol lipids that act as "second messenger" molecules. These, in turn, activate PDK1 and then a variety of downstream signaling molecules, including protein kinase B (PKB/AKT), SGK, and aPKC. An alternative pathway involves the activation of p70S6K and perhaps other as yet unidentified kinases. Next, additional phosphorylation of IRS results in docking of GRB2/mSOS and activation of the small GTPase, p21Ras, which initiates a protein kinase cascade that activates Raf-1, MEK, and the p42/p44 MAP kinase isoforms. These protein kinases are important in the regulation of proliferation and differentiation of many cell types. The mTOR pathway provides an alternative way of activating p70S6K and is involved in nutrient signaling as well as insulin action. Each of the indicated signaling cascades may influence different biologic processes, as shown (Protein Translocation, Protein/Enzyme Activity, Gene Transcription/Translation, Cell Growth). All of the phosphorylation events are reversible through the action of specific phosphatases. As an example, the lipid phosphatase PTEN dephosphorylates the product of the PI-3 kinase reaction, thereby antagonizing the pathway and terminating the signal. Representative effects of major actions of insulin are shown in each of the boxes (bottom). The asterisk after phosphodiesterase indicates that insulin indirectly affects the activity of many enzymes by activating phosphodiesterases and reducing intracellular cAMP levels. (aPKC, atypical protein kinase C; GRB2, growth factor receptor binding protein 2; IGFBP, insulin-like growth factor binding protein; IRS 1-4, insulin receptor substrate isoforms 1-4; MAP kinase, mitogen-activated protein kinase; MEK, MAP kinase and ERK kinase; mSOS, mammalian son of sevenless; mTOR, mammalian target of rapamycin; p70S6K, p70 ribosomal protein S6 kinase; PDK1, phosphoinositide-dependent kinase; PI-3 kinase, phosphatidylinositol 3-kinase; PKB, protein kinase B; PTEN, phosphatase and tensin homolog deleted on chromosome 10; SGK, serum and glucocorticoid-regulated kinase.)

signal transducers and activators of transcription. The phosphorylated STAT protein dimerizes and translocates into the nucleus, binds to a specific DNA element such as the interferon response element (IRE), and activates transcription. This is illustrated in **Figure 42–9**. Other SH2 docking events may result in the activation of PI-3 kinase, the MAP kinase pathway (through SHC or GRB2), or G-protein–mediated activation of phospholipase C (PLCγ) with the attendant production of diacylglycerol

and activation of protein kinase C. It is apparent that there is a potential for "cross-talk" when different hormones activate these various signal transduction pathways.

The NF-κB Pathway Is Regulated by Glucocorticoids

The DNA-binding transcription factor **NF-κB** is a heterodimeric complex typically composed of two subunits termed

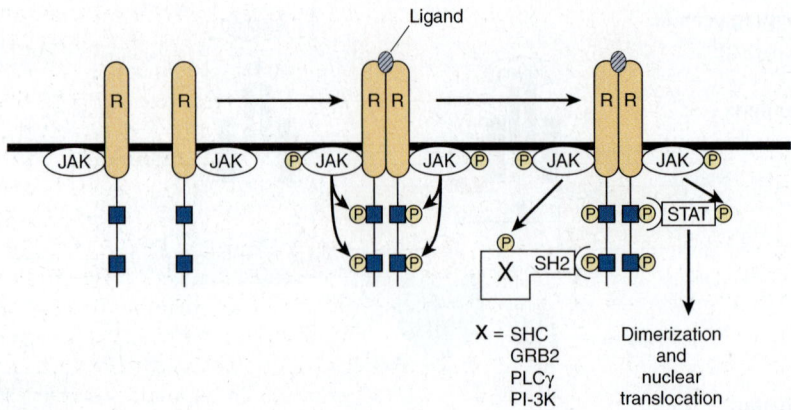

FIGURE 42–9 Initiation of signal transduction by receptors linked to Jak kinases. The receptors (R) that bind prolactin, growth hormone, interferons, and cytokines lack endogenous tyrosine kinase. On ligand binding, these receptors dimerize and an associated, though inactive protein kinase (JAK1, JAK2, or TYK) is phosphorylated. Phospho-JAK is now activated, and proceeds to phosphorylate the receptor on tyrosine residues. The STAT proteins associate with the phosphorylated receptor and then are themselves phosphorylated by JAK-P. The phosphorylated STAT protein, STAT-P dimerizes, translocates to the nucleus, binds to specific DNA elements, and regulates transcription. The phosphotyrosine residues of the receptor also bind to several SH2 domain-containing proteins (X-SH2), which result in activation of the MAP kinase pathway (through SHC or GRB2), PLCγ, or PI-3 kinase.

p50 and **p65** (**Figure 42–10**). Normally NF-κβ is sequestered in the cytoplasm in a transcriptionally inactive form by interactions with members of the **IκB** (inhibitor of NF-κβ) family of proteins. Extracellular stimuli such as proinflammatory cytokines, reactive oxygen species, and mitogens lead to activation of the **IKK** (IκB kinase) **complex**, which is a heterohexameric structure consisting of α, β, and γ subunits. Activated IKK phosphorylates IκB on two serine residues.

This phosphorylation targets IκB for polyubiquitylation and subsequent degradation by the proteasome. Following IκB degradation, free NF-κB translocates to the nucleus, where it binds to a number of gene enhancers and activates transcription, particularly of genes involved in the **inflammatory response**. Transcriptional regulation by NF-κB is mediated by a variety of coactivators such as CREB-binding protein (CBP), as described later (see Figure 42–13).

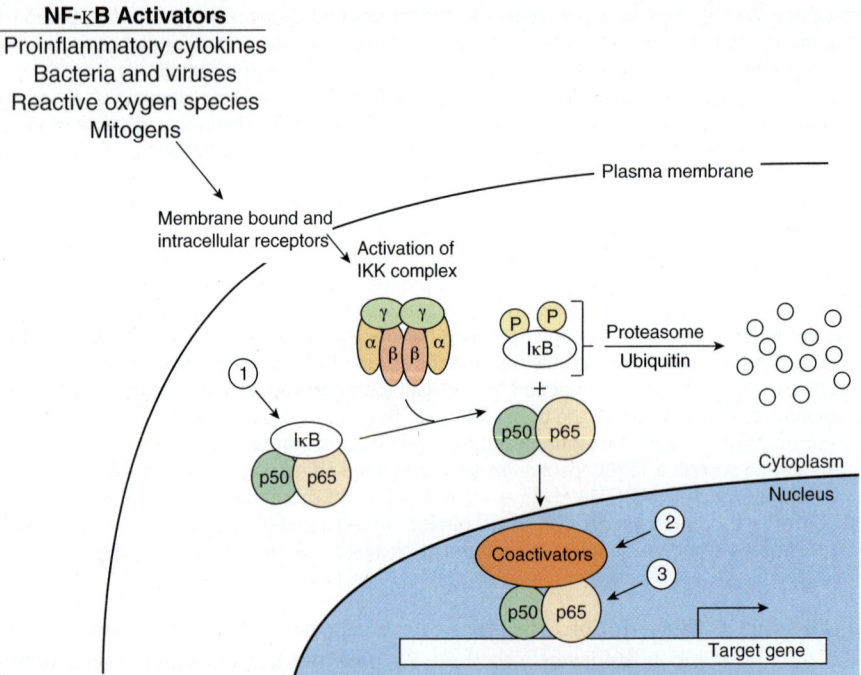

FIGURE 42–10 Regulation of the NF-κB pathway. NF-κB consists of two subunits, p50 and p65, which when present in the nucleus regulates transcription of the multitude of genes important for the inflammatory response. NF-κB is restricted from entering the nucleus by IκB, an inhibitor of NF-κB. IκB binds to—and masks—the nuclear localization signal of NF-κB. This cytoplasmic protein is phosphorylated by an IKK complex which is activated by cytokines, reactive oxygen species, and mitogens. Phosphorylated IκB can be ubiquitylated and degraded, thus releasing its hold on NF-κB, and allowing for nuclear translocation. Glucocorticoids, potent anti-inflammatory agents, are thought to affect at least three steps in this process (1, 2, 3), as described in the text.

Glucocorticoid hormones are therapeutically useful agents for the treatment of a variety of inflammatory and immune diseases. Their anti-inflammatory and immunomodulatory actions are explained in part by the inhibition of NF-κB and its subsequent actions. Evidence for three mechanisms for the inhibition of NF-κB by glucocorticoids has been described: (1) glucocorticoids increase IκB mRNA, which leads to an increase of IκB protein and more efficient sequestration of NF-κB in the cytoplasm. (2) The GR competes with NF-κB for binding to coactivators. (3) The GR directly binds to the p65 subunit of NF-κB and inhibits its activation (see Figure 42–10).

HORMONES CAN INFLUENCE SPECIFIC BIOLOGIC EFFECTS BY MODULATING TRANSCRIPTION

The signals generated as described earlier have to be translated into an action that allows the cell to effectively adapt to a physiologic challenge (see Figure 42–1). Much of this adaptation is accomplished through alterations in the rates of transcription of specific genes. Many different observations have led to the current view of how hormones affect transcription. Some of these are as follows: (1) actively transcribed genes are in regions of "open" chromatin (experimentally defined as relatively high susceptibility to DNase I digestion, and containing certain histone PTMs or "histone marks"), which allows for the access of transcription factors to DNA. (2) Genes have regulatory regions, and transcription factors bind to these to modulate the frequency of transcription (initiation, promoter clearance, and elongation; see Chapters 36, 38). (3) The hormone-receptor complex can be one of these transcription factors. The DNA sequence to which receptors bind is called a **HRE** (see Table 42–1 for examples). (4) Alternatively, other hormone-generated signals can modify the location, amount, or activity of transcription factors and thereby influence binding to the regulatory or response element. (5) Members of a large superfamily of nuclear receptors act with—or in a manner analogous to—the hormone receptors described earlier. (6) These nuclear receptors interact with another large group of coregulatory molecules to effect changes in the transcription of specific genes.

Several HREs Have Been Defined

HREs resemble enhancer elements in that they are not strictly dependent on position and location or orientation. They generally are found within a few hundred nucleotides upstream (5′) of the transcription initiation site, but they may be located within the coding region of the gene, in introns. HREs were defined by the strategy illustrated in Figure 38–11. The consensus sequences illustrated in Table 42–1 were arrived at through analysis of many genes regulated by a given hormone using simple, heterologous reporter systems (see Figure 38–10) coupled with genome-wide chromatin immunoprecipitation and bioinformatic analyses (see Chapter 39). Although these simple HREs bind the hormone-receptor complex more avidly than surrounding DNA—or DNA from an unrelated

source—and confer hormone responsiveness to a reporter gene, it soon became apparent that the regulatory circuitry of natural genes must be much more complicated. Glucocorticoids, progestins, mineralocorticoids, and androgens have vastly different physiologic actions. How could the specificity required for these effects be achieved through regulation of gene expression by the same HRE (see Table 42–1)? Questions like this have led to experiments which have allowed for elaboration of a more complex model of transcription regulation by the steroid hormone receptor family of proteins. For example, in the vast majority of cellular genes, the HRE is found associated with other specific regulatory DNA elements (and associated binding proteins); these associations are mandatory for optimal function. The extensive sequence similarity noted between steroid hormone receptors, particularly in their DNA-binding domains (DBD), led to discovery of the **nuclear receptor superfamily** of proteins. These—and a large number of **coregulator proteins**—allow for a wide variety of DNA–protein and protein–protein interactions and the specificity necessary for highly regulated physiologic control. A schematic of such an assembly is illustrated in **Figure 42–11**.

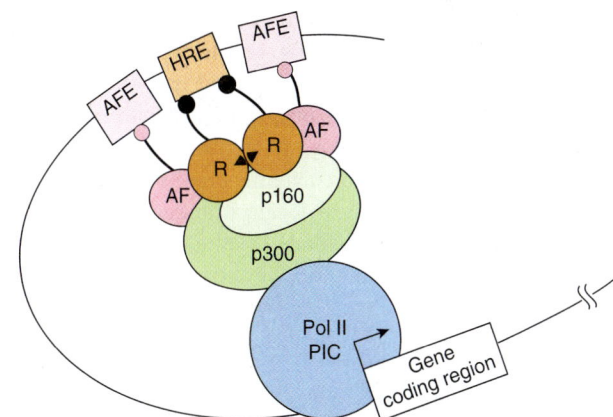

FIGURE 42–11   The hormone response transcriptional activation unit. The hormone response transcription unit is an assembly of DNA elements and complementary, cognate DNA-bound proteins that interact, through protein–protein interactions, with a number of coactivator or corepressor molecules. An essential component is the hormone response element that is bound by the ligand (▲)-bound receptor (R). Also important are the accessory factor elements (AFEs) with bound transcription factors. More than two dozen of these accessory factors (AFs), which are often members of the nuclear receptor superfamily, have been linked to hormone effects on transcription. The AFs can interact with each other, with the liganded nuclear receptors, or with coregulators. These components communicate with the basal transcription machinery, forming the polymerase II PIC (ie, RNAP II and GTFs; see Figure 36–10) through a coregulator complex that can consist of one or more members of the p160, corepressor, mediator-related, or CBP/p300 families (see Table 42–6). Recall (see Chapters 36, 38) that many of the transcription coregulators carry intrinsic enzymatic activities that covalently modify the DNA, transcription proteins, and the histones present in the nucleosomes (not shown here) in and around the enhancer (HRE, AFE) and promoter. Collectively the hormone, hormone receptor, chromatin, DNA and transcription machinery integrate and process hormone signals to regulate transcription in a physiologic fashion.

There Is a Large Family of Nuclear Receptor Proteins

The nuclear receptor superfamily consists of a diverse set of transcription factors that were discovered because of a sequence similarity in their DBDs. This family, now with more than 50 members, includes the nuclear hormone receptors discussed earlier, a number of other receptors whose ligands were discovered after the receptors were identified, and many putative, or orphan receptors, for which the regulatory physiologic ligand has yet to be discovered.

These nuclear receptors have several common structural features (**Figure 42–12**). All have a centrally located DBD that allows the receptor to bind with high affinity to its cognate HRE. The DBD contains two zinc finger binding motifs (see Figure 38–14) that direct binding either as homodimers, as heterodimers (usually with a retinoid X receptor [RXR] partner), or as monomers. The target response element consists of one or two DNA half-site consensus sequences arranged as an inverted or direct repeat. The spacing between the latter helps determine binding specificity. Thus, in general, a direct repeat with three, four, or five nucleotide spacer regions specifies the binding of the vitamin D, thyroid, and retinoic acid receptors, respectively, to the same consensus response element (Table 42–1). A multifunctional **ligand-binding domain (LBD)** is located in the carboxyl terminal half of the receptor. The LBD binds hormones or metabolites with selectivity and thus specifies a particular biologic response. The LBD also contains domains that mediate the binding of heat shock proteins, dimerization, nuclear localization, and transactivation. The latter function is facilitated by the carboxyl-terminal transcription activation function, or **activation domain/AD**

(**AF-2 domain**), which forms a surface required for the interaction with coactivators. A highly variable **hinge region** separates the DBD from the LBD. This region provides flexibility to the receptor, so it can assume different DNA-binding conformations. Finally, there is a highly variable amino-terminal region that contains another AD referred to as **AF-1**. The AF-1 AD likely provides for distinct physiologic functions through the binding of different coregulator proteins. This region of the receptor, through the use of different promoters, alternative splice sites, and multiple translation initiation sites, provides for receptor isoforms that share DBD and LBD identity but exert different physiologic responses because of the association of various coregulators with this variable amino terminal AF-1 AD.

It is possible to sort this large number of receptors into groups in a variety of ways. Here, they are discussed according to the way they bind to their respective DNA elements (see Figure 42–12). Classic hormone receptors for glucocorticoids (GR), mineralocorticoids (MR), estrogens (ER), androgens (AR), and progestins (PR) bind as homodimers to inverted repeat sequences. Other hormone receptors such as thyroid (TR), retinoic acid (RAR), and vitamin D (VDR) and receptors that bind various metabolite ligands such as PPAR α, β, and γ, FXR, LXR, PXR, and CAR bind as heterodimers, with RXR as a partner, to direct repeat sequences (see Figure 42–12 and **Table 42–5**). Another group of orphan receptors that as yet have no known ligand bind as homodimers or monomers to direct repeat sequences.

As illustrated in Table 42–5, the discovery of the nuclear receptor superfamily has led to an important understanding of how a variety of metabolites and xenobiotics regulate gene expression and thus the metabolism, detoxification, and

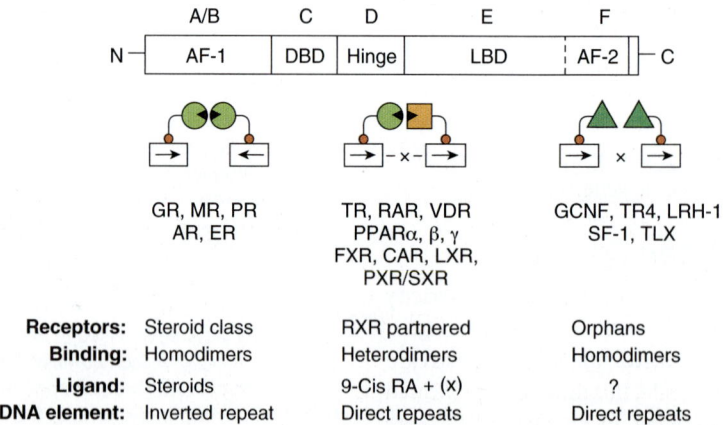

Receptors:	Steroid class	RXR partnered	Orphans
Binding:	Homodimers	Heterodimers	Homodimers
Ligand:	Steroids	9-Cis RA + (x)	?
DNA element:	Inverted repeat	Direct repeats	Direct repeats

FIGURE 42–12   The nuclear receptor superfamily. Members of this family are divided into six structural domains (A-F). Domain A/B is also called AF-1, or the modulator region, because it contains an AD and is involved in activating transcription. The C domain consists of the DNA-binding domain (DBD). The D region contains the hinge, which provides flexibility between the DBD and the ligand-binding domain (LBD, region E). The C-terminal part of region E contains AF-2, another AD that also contributes importantly to activation of transcription. The F region is less well defined. The functions of these domains are discussed in more detail in the text. Receptors with known ligands, such as the steroid hormones, bind as homodimers on inverted repeat half-sites. Other receptors form heterodimers with the partner RXR on direct repeat elements. There can be nucleotide spacers of one to five bases between these direct repeats (DR1-5; see Table 42–1 for more details). Another class of receptors for which ligands have not been definitively determined (orphan receptors) bind as homodimers to direct repeats and occasionally as monomers to a single half-site.

TABLE 42–5 Nuclear Receptors With Special Ligands[a]

Receptor		Partner	Ligand	Process Affected
Peroxisome	PPAR$_\alpha$	RXR (DR1)	Fatty acids	Peroxisome proliferation
Proliferator-activated	PPAR$_\beta$		Fatty acids	
	PPAR$_\gamma$		Fatty acids	Lipid and carbohydrate metabolism
			Eicosanoids, thiazolidinediones	
Farnesoid X	FXR	RXR (DR4)	Farnesol, bile acids	Bile acid metabolism
Liver X	LXR	RXR (DR4)	Oxysterols	Cholesterol metabolism
Xenobiotic X	CAR	RXR (DR5)	Androstanes, phenobarbital, xenobiotics	Protection against certain drugs, toxic metabolites, and xenobiotics
	PXR	RXR (DR3)	Pregnanes	
			Xenobiotics	

[a]Many members of the nuclear receptor superfamily were discovered by "homology" cloning, and the corresponding ligands were subsequently identified. These ligands are not hormones in the classic sense, but they do have a similar function in that they activate specific members of the nuclear receptor superfamily. The receptors described here form heterodimers with RXR and have variable nucleotide sequences separating the direct repeat binding elements (DR1-5). These receptors regulate a variety of genes encoding cytochrome p450s (CYP), cytosolic binding proteins, and ATP-binding cassette (ABC) transporters to influence metabolism and protect cells against drugs and noxious agents.

elimination of normal body products and exogenous agents such as pharmaceuticals. Not surprisingly, this area is a fertile field for investigation of new therapeutic interventions.

A Large Number of Nuclear Receptor Coregulators Also Participate in Regulating Transcription

Chromatin remodeling (histone modifications, DNA methylation, nucleosome repositioning/remodeling/displacement) transcription factor modification by various enzyme activities, and the communication between the nuclear receptors and the basal transcription apparatus are accomplished by protein–protein interactions with one or more of a class of coregulator molecules. The number of these coregulator molecules now exceeds 100, not counting species variations and splice variants. The first of these to be described was the **CREB-binding protein, CBP**. CBP, through an amino-terminal domain, binds to phosphorylated serine 137 of CREB and mediates transactivation in response to cAMP. It thus is described as a coactivator. CBP and its close relative, p300, interact directly or indirectly with a number of DNA-binding transcription factors, including **activator protein-1 (AP-1)**, **STATs**, **nuclear receptors**, and **CREB** (**Figure 42–13**). CBP/p300 also binds

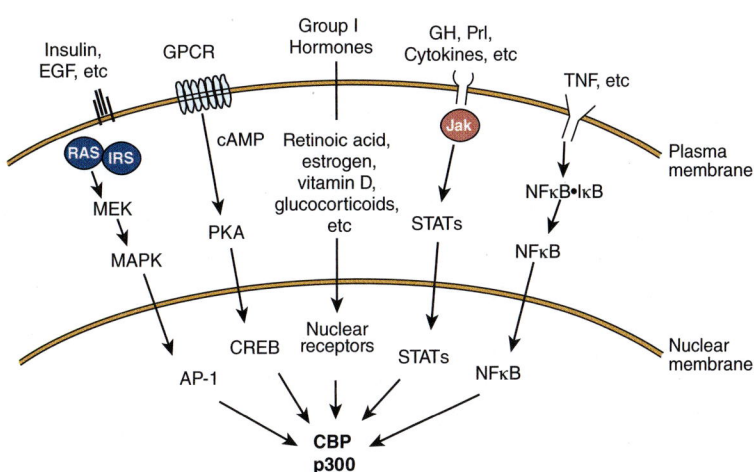

FIGURE 42–13 Several signal transduction pathways converge on CBP/p300. Many ligands that associate with membrane or nuclear receptors eventually converge on CBP/p300. Several different signal transduction pathways are illustrated. (EGF, epidermal growth factor; GH, growth hormone; Prl, prolactin; TNF, tumor necrosis factor; other abbreviations are expanded in the text.)

to the p160 family of coactivators described later and to a number of other proteins, including the p90^{rsk} protein kinase and RNA helicase A. It is important to note, as mentioned earlier, that **CBP/p300** also has **intrinsic histone acetyltransferase (HAT) activity**. Some of the many actions of CBP/p300, which appear to depend on intrinsic enzyme activities and its ability to serve as a scaffold for the binding of other proteins, are illustrated in Figure 42–11. Other coregulators serve similar functions.

Several other families of coactivator molecules have been described. Members of the **p160 family of coactivators**, all of about 160 kDa, include (1) SRC-1 and NCoA-1; (2) GRIP 1, TIF2, and NCoA-2; and (3) p/CIP, ACTR, AIB1, RAC3, and TRAM-1 (**Table 42–6**). The different names for members within a subfamily often represent species variations or minor splice variants. There is about 35% amino acid identity between members of the different subfamilies. The p160 coactivators

TABLE 42–6 Some Mammalian Coregulator Proteins

I. 300-kDa family of coactivators	
A. CBP	CREB-binding protein
B. p300	Protein of 300 kDa
II. 160-kDa family of coactivators	
A. SRC-1,2,3	Steroid receptor coactivator 1, 2, and 3
NCoA-1	Nuclear receptor coactivator 1
B. TIF2	Transcriptional intermediary factor 2
GRIP1	Glucocorticoid receptor–interacting protein
NCoA-2	Nuclear receptor coactivator 2
C. p/CIP	p300/CBP cointegrator-associated protein 1
ACTR	Activator of the thyroid and retinoic acid receptors
AIB	Amplified in breast cancer
RAC3	Receptor-associated coactivator 3
TRAM-1	TR activator molecule 1
III. Corepressors	
A. NCoR	Nuclear receptor corepressor
B. SMRT	Silencing mediator for RXR and TR
IV. Mediator subunits	
A. TRAPs	Thyroid hormone receptor–associated proteins
B. DRIPs	Vitamin D receptor–interacting proteins
C. ARC	Activator-recruited cofactor

share several properties. They (1) bind nuclear receptors in an agonist- and AF-2 AD-dependent manner, (2) have a conserved amino terminal basic helix-loop-helix (bHLH) motif (see Chapter 38), (3) have a weak carboxyl-terminal transactivation domain and a stronger amino-terminal transactivation domain in a region that is required for CBP-p160 interaction, (4) contain at least three of the **LXXLL motifs** required for protein–protein interaction with other coactivators, and (5) often have HAT activity. The role of HAT is particularly interesting, as mutations of the HAT domain disable many of these transcription factors. Current thinking holds that these HAT activities acetylate histones, which facilitate the remodeling of chromatin into a transcription-efficient environment. Histone acetylation/deacetylation thus plays a critical role in gene expression. Finally, it is important to note that other protein substrates for HAT-mediated acetylation, such as DNA-binding transcription activators and other coregulators have been reported. Such nonhistone PTM events likely also factor importantly into the overall regulatory response.

A small number of proteins, including **NCoR** and **SMRT**, comprise the **corepressor family**. They function, at least in part, as described in Figure 42–2. Another family includes the TRAPs, DRIPs, and ARC (see Table 42–6). These proteins represent subunits of the mediator (see Chapter 36) and range in size from 80 to 240 kDa and are thought to link the nuclear receptor-coactivator complex to RNA polymerase II and the other components of the basal transcription apparatus.

The exact role of these coactivators is presently under intensive investigation. Many of these proteins have intrinsic enzymatic activities. This is particularly interesting in view of the fact that acetylation, phosphorylation, methylation, sumoylation, and ubiquitination—as well as proteolysis and cellular translocation—have been proposed to alter the activity of some of these coregulators and their targets.

It appears that certain combinations of coregulators—and thus different combinations of activators and inhibitors—are responsible for specific ligand-induced actions through various receptors. Furthermore, these interactions on a given promoter are dynamic. In some cases, complexes consisting of over 45 transcription factors have been observed on a single gene.

SUMMARY

- Hormones, cytokines, interleukins, and growth factors use a variety of signaling mechanisms to facilitate cellular adaptive responses.

- The ligand-receptor complex serves as the initial signal for members of the nuclear receptor family.

- Class II peptide/protein and catecholamine hormones, which bind to cell surface receptors, generate a variety of intracellular signals. These include cAMP, cGMP, Ca^{2+}, phosphatidylinositides, and protein kinase cascades.

- Many hormone responses are accomplished through alterations in the rate of transcription of specific genes.

- The nuclear receptor superfamily of proteins plays a central role in the regulation of gene transcription.

- DNA-binding nuclear receptors, which may have hormones, metabolites, or drugs as ligands, bind to specific HREs as homodimers or as heterodimers with RXR.

- Another large family of coregulator proteins remodel chromatin, modify other transcription factors, and bridge the nuclear receptors to the basal transcription apparatus.

REFERENCES

Arvanitakis L, Geras-Raaka E, Gershengorn MC: Constitutively signaling G-protein-coupled receptors and human disease. Trends Endocrinol Metab 1998;9:27.

Auphan N, Didonato JA, Rosette C, Helmberg A, Karin M: Immunosuppression by glucocorticoids: inhibition of NF-KB activity through induction of IKB synthesis. Science 1995;270:286-290.

Beene DL, Scott JD: A-kinase anchoring proteins take shape. Curr Opin in Cell Biol 2007;19:192.

Bucko PJ, Scott JD: Drugs that regulate local cell signaling: AKAP targeting as a therapeutic option. Ann Rev Pharmacol Toxical 2021;61:361-379.

Cheung E, Kraus WL: Genomic analyses of hormone signaling and gene regulation. Annu Rev Physiol 2010;72:191-218.

Cohen P, Spiegelman BM: Cell biology of fat storage. Mol Biol Cell 2016;27:2523-2527.

Dasgupta S, Lonard DM, O'Malley BW: Nuclear receptor coactivators: master regulators of human health and disease. Annu Rev Med 2014;65:279-292.

Evans RM, Mangelsdorf DJ: Nuclear receptors. RXR, and the big bang: Cell 2014;157:255-266.

Fan W, Evans RM: Exercise mimetics: impact on health and performance. Cell Metab 2017;25:242-247.

Fasouli ES, Katsaatoni E: JAK-STAT in early hematopoiesis and leukemia. Front Cell Dev Biol 2021; 9:669363.doi:10.3389/fcell2021.669363.

Lazar MA: Maturing the nuclear receptor family. J Clin Invest 2017;127:1123-1125.

Métivier R, Reid G, Gannon F: Transcription in four dimensions: nuclear receptor-directed initiation of gene expression. EMBO Rep 2006;7:161.

O'Malley B: Coregulators: from whence came these "master genes." Mol Endocrinol 2007;21:1009.

Ravnskjaer K, Madiraju A, Montminy M: Role of the cAMP pathway in glucose and lipid metabolism. Handb Exp Pharmacol 2016;233:29-49.

Szwarc MM, Lydon JP, O'Malley BW: Steroid receptor coactivators as therapeutic targets in the female reproductive system. J Steroid Biochem Mol Biol 2015;154:32-38.

Wang Z, Schaffer NE, Kliewer SA, Mangelsdorf DJ: Nuclear receptors: emerging drug targets for parasitic diseases. J Clin Invest 2017;127:1165-1171.

Weikum ER, Knuesel MT, Ortlund EA, Yamamoto KR: Glucocorticoid receptor control of transcription: precision and plasticity via allostery. Nat Rev Mol Cell Biol 2017;18:159-174.

Weis WI, Kobilka BK: The molecular basis of G protein-coupled receptor activation. Ann Rev Biochem 2018;87:897-919.

Welboren W-J, van Driel MA, Janssen-Megens EM, et al: ChIP-Seq of ERα and RNA polymerase II defines genes differentially responding to ligands. EMBO J 2009;28:1418-1428.

Zhang H, Cao X, Tang M, et al: A subcellular map of the human kinome. eLife 2021;10:e64943.

Zhang Q, Lenardo MJ, Baltimore D: 30 years of NF-κB: a blossoming of relevance to human pathobiology. Cell 2017;168:37-57.

Exam Questions

Section VIII – Biochemistry of Extracellular & Intracellular Communication

1. Regarding membrane lipids, select the one FALSE answer.
   A. The major phospholipid by mass in human membranes is generally phosphatidylcholine.
   B. Glycolipids are located on the inner and outer leaflets of the plasma membrane.
   C. Phosphatidic acid is a precursor of phosphatidylserine, but not of sphingomyelin.
   D. Phosphatidylcholine and phosphatidylethanolamine are located primarily on the outer leaflet of the plasma membrane.
   E. The flip-flop of phospholipids in membranes is very slow.

2. Regarding membrane proteins, select the one FALSE answer.
   A. Because of steric considerations, α-helices cannot exist in membranes.
   B. A hydropathy plot helps one to estimate whether a segment of a protein is predominantly hydrophobic or hydrophilic.
   C. Certain proteins are anchored to the outer leaflet of plasma membranes via glycophosphatidylinositol (GPI) structures.
   D. Adenylyl cyclase is a marker enzyme for the plasma membrane.
   E. Myelin has a very high content of lipid compared with protein.

3. Regarding membrane transport, select the one FALSE statement.
   A. Potassium has a lower charge density than sodium and tends to move more quickly through membranes than does sodium.
   B. The flow of ions through ion channels is an example of passive transport.
   C. Facilitated diffusion requires a protein transporter.
   D. Inhibition of the Na^+-K^+-ATPase will inhibit sodium-dependent uptake of glucose in intestinal cells.
   E. Insulin, by recruiting glucose transporters to the plasma membrane, increases uptake of glucose in fat cells but not in muscle.

4. Regarding the Na^+-K^+-ATPase, select the one FALSE statement.
   A. Its action maintains the high intracellular concentration of sodium compared with potassium.
   B. It can use as much as 30% of the total ATP expenditure of a cell.
   C. It is inhibited by digitalis, a drug that is useful in certain cardiac conditions.
   D. It is located in the plasma membrane of cells.
   E. Phosphorylation is involved in its mechanism of action, leading to its classification as a P-type ATP-driven active transporter.

5. What molecules enable cells to respond to a specific extracellular signaling molecule?
   A. Specific receptor carbohydrates localized to the inner plasma membrane surface
   B. Plasma lipid bilayer
   C. Ion channels
   D. Receptors that specifically recognize and bind that particular messenger molecule
   E. Intact nuclear membranes

6. Indicate the term generally applied to the extracellular messenger molecules that bind to transmembrane receptor proteins.
   A. Competitive inhibitor
   B. Ligand
   C. Scatchard curve
   D. Substrate
   E. Key

7. In autocrine signaling:
   A. Messenger molecules reach their target cells via passage through bloodstream.
   B. Messenger molecules travel only short distances through the extracellular space to cells that are in close proximity to the cell that is generating the message.
   C. The cell producing the messenger expresses receptors on its surface that can respond to that messenger.
   D. The messenger molecules are usually rapidly degraded and hence can only work over short distances.

8. Regardless of how a signal is initiated, the ligand-binding event is propagated via second messengers or protein recruitment. What is the ultimate, or final biochemical outcome of these binding events?
   A. A protein in the middle of an intracellular signaling pathway is activated.
   B. A protein at the bottom of an intracellular signaling pathway is activated.
   C. A protein at the top of an extracellular signaling pathway is activated.
   D. A protein at the top of an intracellular signaling pathway is deactivated.
   E. A protein at the top of an intracellular signaling pathway is activated.

9. What features of the nuclear receptor superfamily suggest that these proteins have evolved from a common ancestor?
   A. They all bind the same ligand with high affinity.
   B. They all function within the nucleus.
   C. They are all subject to regulatory phosphorylation.
   D. They all contain regions of high amino acid sequence similarity/identity.
   E. They all bind DNA.

10. What effect does degradation of receptor-ligand complexes after internalization have on the ability of a cell to respond if immediately re-exposed to the same hormone?
    A. The cellular response is attenuated due to a decrease in cellular receptor number.
    B. Cellular response is enhanced due to reduced receptor-ligand competition.
    C. The cellular response is unchanged to subsequent stimuli.

D. Cell hormone response is now bimodal; enhanced for a short time and thereafter inactivated.

11. Typically, what is the first reaction after most receptor protein-tyrosine kinases (RTKs) bind their ligand?

    A. Receptor trimerization
    B. Receptor degradation
    C. Receptor denaturation
    D. Receptor dissociation
    E. Receptor dimerization

12. Where is the kinase catalytic domain of the receptor protein-tyrosine kinases found?

    A. On the extracellular surface of the receptor, immediately adjacent to the ligand-binding domain.
    B. On the cytoplasmic domain of the receptor.
    C. On an independent protein that rapidly binds the receptor on ligand binding.
    D. Within the transmembrane spanning portion of the receptor.

13. The subunits of the heterotrimeric G-proteins are called the __, __, and__ subunits.

    A. α, β, and χ
    B. α, β, and δ
    C. α, γ, and δ
    D. α, β, and γ
    E. γ, δ, and η

14. Of the receptors listed below, which can directly conduct a flow of ions across the plasma membrane when bound to their cognate ligand?

    A. Receptor tyrosine kinases (RTKs)
    B. G-protein–coupled receptors (GPCRs)
    C. G-protein gamma α subunit.
    D. Steroid hormone receptors
    E. Ligand-gated channels

15. Which of the following is NOT a natural ligand that binds to G-protein–coupled receptors?

    A. Hormones
    B. Steroid hormones
    C. Chemoattractants
    D. Opium derivatives
    E. Neurotransmitters

16. Place the events of signaling listed below in the CORRECT order.

    1. G-protein binds to activated receptor forming a receptor–G-protein complex
    2. Release of GDP by the G-protein
    3. Change in conformation of the cytoplasmic loops of the receptor
    4. Binding of GTP by the G-protein
    5. Increase in the affinity of the receptor for a G-protein on the cytoplasmic surface of the membrane
    6. Binding of a hormone or neurotransmitter to a G-protein–coupled receptor
    7. Conformational shift in the α subunit of the G-protein
    A. 6 – 3 – 5 – 1 – 2 – 4 – 7
    B. 6 – 5 – 4 – 1 – 7 – 2 – 3

C. 6 – 3 – 5 – 1 – 7 – 2 – 4
D. 6 – 7 – 3 – 5 – 1 – 2 – 4
E. 6 – 3 – 5 – 4 – 7 – 2 – 1

17. Which heterotrimeric G-proteins couple receptors to adenylyl cyclase via the activation of GTP-bound G_α subunits?

    A. G_s family
    B. G_q family
    C. G_i family
    D. $G_{12/13}$ family
    E. G_x family

18. What must happen in order to prevent overstimulation by a hormone?

    A. Hormones must be degraded.
    B. G-proteins must be recycled and then degraded.
    C. Receptors must be blocked from continuing to activate G-proteins.
    D. Receptors must dimerize.

19. Which of the following hormones termed the "flight-or-fight" hormone is secreted by the adrenal medulla?

    A. Epinephrine
    B. Oxytocin
    C. Insulin
    D. Glucagon
    E. Somatostatin

20. Which hormone is secreted by α cells in the pancreas in response to low blood glucose levels?

    A. Insulin
    B. Glucagon
    C. Estradiol
    D. Epinephrine
    E. Somatostatin

21. In liver cells, the expression of genes encoding gluconeogenic enzymes such as phosphoenolpyruvate carboxykinase (PEPCK) is induced in response to which of the following molecules?

    A. cGMP
    B. Insulin
    C. ATP
    D. cAMP
    E. Cholesterol

22. What happens to protein kinase A (PKA) following the binding of cAMP?

    A. The regulatory subunits of PKA dissociate, thereby activating the catalytic subunits.
    B. PKA catalytic subunits then bind to two regulatory subunits, thereby activating the catalytic subunits.
    C. The inhibitory regulatory subunits dissociate from the catalytic subunits, completely inactivating the enzyme.
    D. The stimulatory regulatory subunits dissociate from the catalytic subunits, inhibiting the enzyme.
    E. Phosphodiesterase binds to the catalytic subunits, which results in enzyme inactivation.

C H A P T E R

43

Nutrition, Digestion, & Absorption

David A. Bender, PhD

OBJECTIVES

After studying this chapter, you should be able to:

- Describe the digestion and absorption of carbohydrates, lipids, proteins, vitamins, and minerals.
- Explain how energy requirements can be measured and estimated, and how measuring the respiratory quotient permits estimation of the mix of metabolic fuels being oxidized.
- Describe the consequences of undernutrition: marasmus, cachexia, and kwashiorkor.
- Explain how protein requirements are determined and why more of some proteins than others are required to maintain nitrogen balance.

BIOMEDICAL IMPORTANCE

In addition to water, the diet must provide metabolic fuels (mainly carbohydrates and lipids), protein (for growth and turnover of tissue proteins, as well as a source of metabolic fuel), fiber (for bulk in the intestinal lumen), minerals (containing elements with specific metabolic functions), and vitamins and essential fatty acids (organic compounds needed in smaller amounts for other metabolic and physiological functions). The polysaccharides, triacylglycerols, and proteins that make up the bulk of the diet must be hydrolyzed to their constituent monosaccharides, fatty acids, and amino acids, respectively, before absorption and utilization. Minerals and vitamins must be released from the complex matrix of food before they can be absorbed and utilized.

Globally, **undernutrition** is widespread, leading to impaired growth, defective immune system, and reduced work capacity. By contrast, in developed countries, and increasingly in developing countries, there is excessive food consumption, leading to obesity, and the development of diabetes, cardiovascular disease, and some cancers. Worldwide,

there are more overweight and obese people than undernourished people. Deficiencies of vitamin A, iron, and iodine pose major health concerns in many countries, and deficiencies of other vitamins and minerals are a major cause of ill health. In developed countries nutrient deficiency is rare, although there are vulnerable sections of the population at risk. Intakes of minerals and vitamins that are adequate to prevent deficiency may be inadequate to promote optimum health and longevity.

Excessive secretion of gastric acid, associated with *Helicobacter pylori* infection, can result in the development of gastric and duodenal **ulcers**; small changes in the composition of bile can result in crystallization of cholesterol as **gallstones**; failure of exocrine pancreatic secretion (as in **cystic fibrosis**) leads to undernutrition and steatorrhea. **Lactose intolerance** is the result of lactase deficiency, leading to diarrhea and intestinal discomfort when lactose is consumed. Absorption of intact peptides that stimulate antibody responses causes **allergic reactions**; **celiac disease** is an allergic reaction to wheat gluten.

DIGESTION & ABSORPTION OF CARBOHYDRATES

The digestion of carbohydrates liberates oligosaccharides, which are further hydrolyzed to free mono- and disaccharides. The increase in blood glucose after a test dose of a carbohydrate compared with that after an equivalent amount of glucose (as glucose or from a reference starchy food) is known as the **glycemic index**. Glucose and galactose have an index of 1 (or 100%), as do lactose, maltose, isomaltose, and trehalose, which give rise to these monosaccharides on hydrolysis. Fructose and the sugar alcohols are absorbed less rapidly and have a lower glycemic index, as does sucrose. The glycemic index of starch varies between near 1 (or 100%) and near 0 as a result of variable rates of hydrolysis, and that of nonstarch polysaccharides (see Figure 15–13) is 0. Foods that have a low glycemic index are considered to be more beneficial since they cause less fluctuation in insulin secretion. Resistant starch and nonstarch polysaccharides provide substrates for bacterial fermentation in the large intestine, and the resultant butyrate and other short-chain fatty acids provide a significant source of fuel for intestinal enterocytes. There is evidence that butyrate also has antiproliferative activity, and so provides protection against colorectal cancer.

Amylases Catalyze the Hydrolysis of Starch

The hydrolysis of starch is catalyzed by salivary and pancreatic amylases, which catalyze random hydrolysis of $\alpha(1 \rightarrow 4)$ glycoside bonds, yielding dextrins, then a mixture of glucose, maltose, and maltotriose and small branched dextrins (from the branchpoints in amylopectin, Figure 15–12).

Disaccharidases Are Brush Border Enzymes

The disaccharidases, maltase, sucrase-isomaltase (a bifunctional enzyme catalyzing hydrolysis of sucrose and isomaltose), lactase, and trehalase are located on the brush border of the intestinal mucosal cells, where the resultant monosaccharides and those directly ingested are absorbed. Congenital deficiency of lactase occurs rarely in infants, leading to lactose intolerance and failure to thrive when fed on breast milk or normal infant formula. Congenital deficiency of sucrase-isomaltase occurs among the Inuit, leading to sucrose intolerance, with persistent diarrhea and failure to thrive when the diet contains sucrose.

In most mammals, and most human beings, lactase activity begins to fall after weaning and is almost completely lost by late adolescence, leading to **lactose intolerance**. Lactose remains in the intestinal lumen, where it is a substrate for bacterial fermentation to lactate, resulting in abdominal discomfort and diarrhea after consumption of relatively large amounts. In two population groups, people of north European origin and nomadic tribes of sub-Saharan Africa and Arabia, lactase persists after weaning and into adult life. Marine mammals secrete a high-fat milk that contains no carbohydrate, and their pups lack lactase.

There Are Two Separate Mechanisms for the Absorption of Monosaccharides in the Small Intestine

Glucose and galactose are absorbed by a sodium-dependent process. They are carried by the same transport protein (SGLT 1) and compete with each other for intestinal absorption (**Figure 43–1**). Other monosaccharides are absorbed by carrier-mediated diffusion. Because they are not actively transported, fructose and sugar alcohols are only absorbed down their concentration gradient, and after a moderately high intake, some may remain in the intestinal lumen, acting as a substrate for bacterial fermentation. Large intakes of fructose and sugar alcohols can lead to osmotic diarrhea.

DIGESTION & ABSORPTION OF LIPIDS

The major lipids in the diet are triacylglycerols and, to a lesser extent, phospholipids. These are hydrophobic molecules and have to be hydrolyzed and emulsified to very small droplets

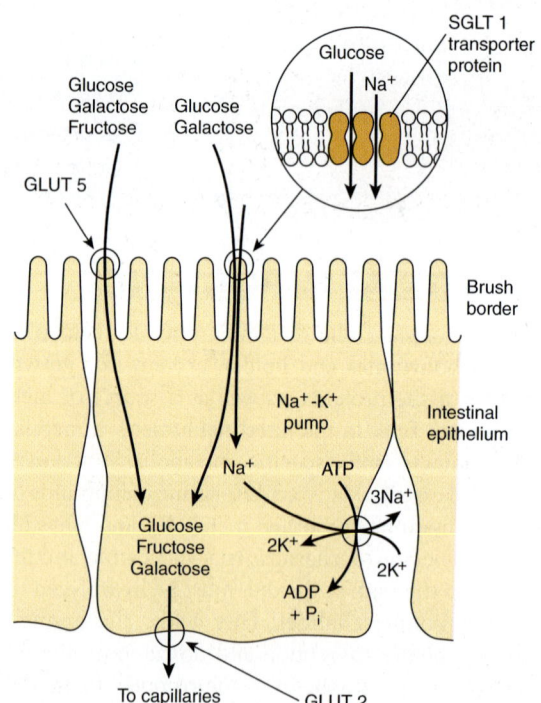

FIGURE 43–1 Transport of glucose, fructose, and galactose across the intestinal epithelium. The SGLT 1 transporter is coupled to the Na⁺-K⁺ pump, allowing glucose and galactose to be transported against their concentration gradients. The GLUT 5 Na⁺-independent facilitative transporter allows fructose, as well as glucose and galactose, to be transported down their concentration gradients. Exit from the cell for all sugars is via the GLUT 2 facilitative transporter.

(micelles, 4-6 nm in diameter) before they can be absorbed. The fat-soluble vitamins, A, D, E, and K, and a variety of other lipids (including cholesterol and carotenes) are absorbed dissolved in the lipid micelles. Absorption of carotenes and fat-soluble vitamins is impaired on a very low-fat diet.

Hydrolysis of triacylglycerols is initiated by lingual and gastric lipases, which attack the *sn*-3 ester bond forming 1,2-diacylglycerols and free fatty acids, which act as emulsifying agents. Pancreatic lipase is secreted into the small intestine and requires a further pancreatic protein, colipase, for activity. It is specific for the primary ester links—that is, positions 1 and 3 in triacylglycerols—resulting in 2-monoacylglycerols and free fatty acids as the major end products of luminal triacylglycerol digestion. Inhibitors of pancreatic lipase are used to inhibit triacylglycerol hydrolysis in the treatment of severe obesity. Pancreatic esterase in the intestinal lumen hydrolyzes monoacylglycerols, but they are poor substrates, and only ~25% of ingested triacylglycerol is completely hydrolyzed to glycerol and fatty acids before absorption (**Figure 43–2**). Bile salts, formed in the liver and secreted in the bile, permit emulsification of the products of lipid digestion into micelles together with dietary phospholipids and cholesterol secreted in the bile (about 2 g/d) as well as dietary cholesterol (about 0.5 g/d). Micelles are less than 1 μm in diameter, and soluble, so they allow the products of digestion, including the fat-soluble vitamins, being transported through the aqueous environment of the intestinal lumen to come into close contact with the brush border of the mucosal cells, allowing uptake into the epithelium. The bile salts remain in the intestinal lumen, where most are absorbed from the ileum into the **enterohepatic circulation** (see Chapter 26).

Within the intestinal epithelium, 1-monoacylglycerols are hydrolyzed to fatty acids and glycerol and 2-monoacylglycerols are reacylated to triacylglycerols via the **monoacylglycerol pathway**. Glycerol released in the intestinal lumen is absorbed into the hepatic portal vein; glycerol released within the epithelium is reutilized for triacylglycerol synthesis via the normal phosphatidic acid pathway (see Chapter 24). Long-chain fatty acids are esterified to triacylglycerol in the mucosal cells and together with the other products of lipid digestion, secreted as chylomicrons into the lymphatics, entering the bloodstream via the thoracic duct (see Chapter 25). Short- and medium-chain fatty acids are mainly absorbed into the hepatic portal vein as free fatty acids.

Cholesterol is absorbed dissolved in lipid micelles and is mainly esterified in the intestinal mucosa before being incorporated into chylomicrons. Plant sterols and stanols (in which the B ring is saturated) compete with cholesterol for esterification, but are poor substrates, so that there is an increased amount of unesterified cholesterol in the mucosal cells. Unesterified cholesterol and other sterols are actively transported out of the mucosal cells into the intestinal lumen. This means that plant sterols and stanols effectively inhibit the absorption of not only dietary cholesterol, but also the larger amount that is secreted in the bile, so lowering the whole body cholesterol content, and hence the plasma cholesterol concentration.

DIGESTION & ABSORPTION OF PROTEINS

Native proteins are resistant to digestion because few peptide bonds are accessible to the proteolytic enzymes without prior denaturation by heat in cooking and by the action of gastric acid.

Several Groups of Enzymes Catalyze the Digestion of Proteins

There are two main classes of proteolytic digestive enzymes (**proteases**), with different specificities for the amino acids forming the peptide bond to be hydrolyzed. **Endopeptidases** hydrolyze peptide bonds between specific amino acids throughout the molecule. They are the first enzymes to act, yielding a larger number of smaller fragments. Pepsin in the gastric juice catalyzes hydrolysis of peptide bonds adjacent to amino acids with bulky side-chains (aromatic and branched-chain amino acids and methionine). Trypsin, chymotrypsin, and elastase are secreted into the small intestine by the pancreas. Trypsin catalyzes hydrolysis of lysine and arginine amides, chymotrypsin amides of aromatic amino acids, and elastase amides of small neutral aliphatic amino acids. **Exopeptidases** catalyze the hydrolysis of peptide bonds, one at a time, from the ends of peptides. **Carboxypeptidases**, secreted in the pancreatic juice, release amino acids from the free carboxyl terminal; **aminopeptidases**, secreted by the intestinal mucosal cells, release amino acids from the amino terminal. **Dipeptidases** and **tripeptidases** in the brush border of intestinal mucosal cells catalyze the hydrolysis of di- and tripeptides, which are not substrates for amino- and carboxypeptidases.

The proteases are secreted as inactive **zymogens**; the active site of the enzyme is masked by a small region of the peptide chain that is removed by hydrolysis of a specific peptide bond. Pepsinogen is activated to pepsin by gastric acid and by activated pepsin. In the small intestine, trypsinogen, the precursor of trypsin, is activated by enteropeptidase, which is secreted by the duodenal epithelial cells; trypsin can then activate chymotrypsinogen to chymotrypsin, proelastase to elastase, procarboxypeptidase to carboxypeptidase, and proaminopeptidase to aminopeptidase.

Free Amino Acids & Small Peptides Are Absorbed by Different Mechanisms

The end product of the action of endopeptidases and exopeptidases is a mixture of free amino acids, di- and tripeptides, and oligopeptides, all of which are absorbed. Free amino acids are absorbed across the intestinal mucosa by sodium-dependent active transport. There are several different amino acid transporters, with specificity for the nature of the amino acid side chain (large or small, neutral, acidic, or basic). The amino acids carried by any one transporter compete with each other for absorption and tissue uptake. Dipeptides and tripeptides enter the brush border of the intestinal mucosal cells, where they are hydrolyzed

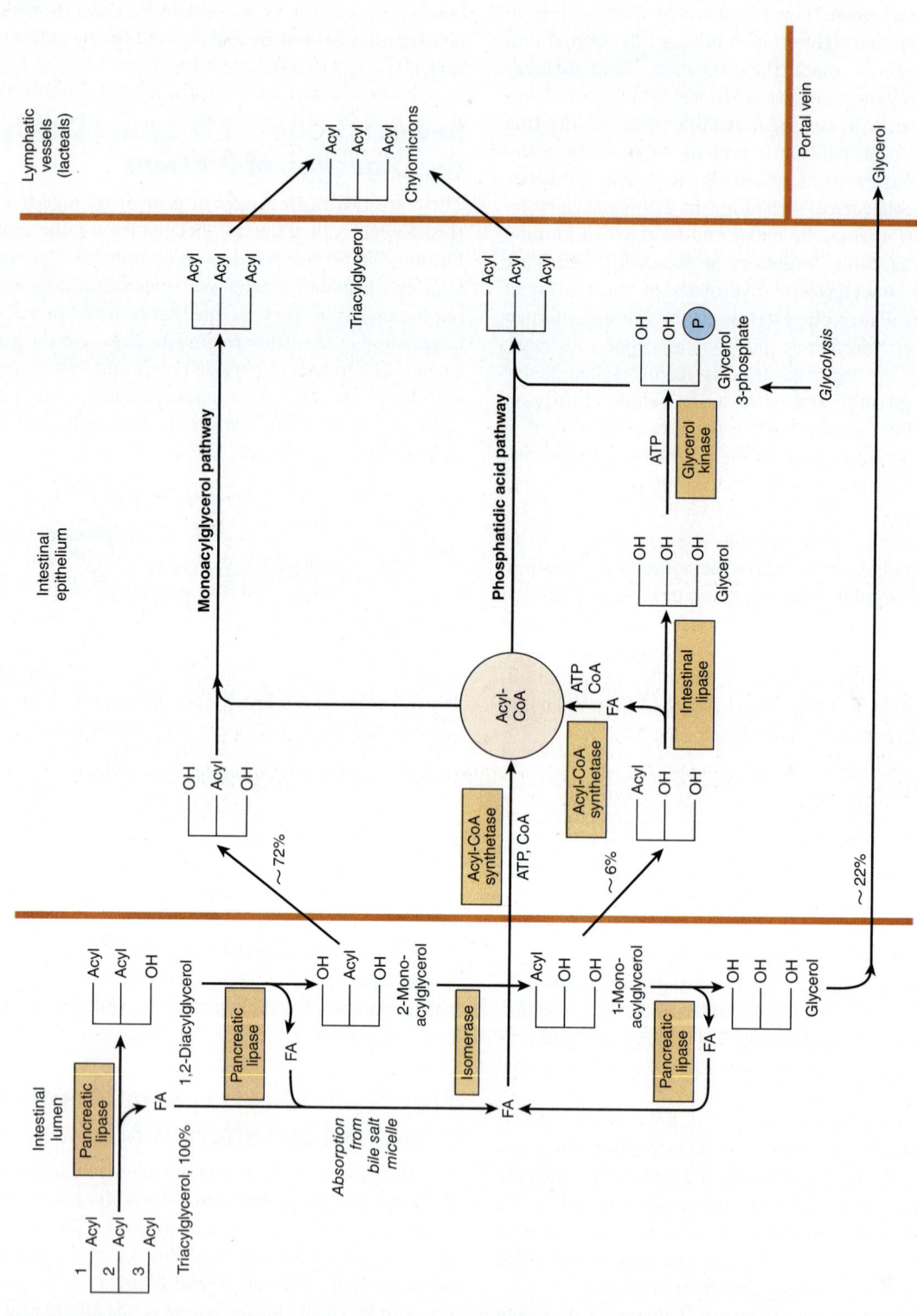

FIGURE 43–2 Digestion and absorption of triacylglycerols. The values given for percentage uptake may vary widely but indicate the relative importance of the three routes shown.

to free amino acids, which are then transported into the hepatic portal vein. Relatively large peptides may be absorbed intact, either by uptake into mucosal epithelial cells (transcellular) or by passing between epithelial cells (paracellular). Many such peptides are large enough to stimulate antibody formation—this is the basis of **allergic reactions** to foods.

DIGESTION & ABSORPTION OF VITAMINS & MINERALS

Vitamins and minerals are released from food during digestion, although this is not complete, and the availability of vitamins and minerals depends on the type of food and, especially for minerals, the presence of chelating compounds. The fat-soluble vitamins are absorbed in the lipid micelles that are the result of fat digestion; water-soluble vitamins and most mineral salts are absorbed from the small intestine either by active transport or by carrier-mediated diffusion followed by binding to intracellular proteins to achieve concentrative uptake. Vitamin B_{12} absorption requires a specific transport protein, intrinsic factor (see Chapter 44); calcium absorption is dependent on vitamin D; zinc absorption requires a zinc-binding ligand secreted by the exocrine pancreas, and the absorption of iron is limited (see following text).

Calcium Absorption Is Dependent on Vitamin D

In addition to its role in regulating calcium homeostasis, vitamin D is required for the intestinal absorption of calcium. Synthesis of the intracellular calcium-binding protein, **calbindin**, required for calcium absorption, is induced by vitamin D. Vitamin D also acts to recruit calcium transporters to the cell surface, so increasing calcium absorption rapidly—a process that is independent of new protein synthesis.

Phytic acid (inositol hexaphosphate) in cereals binds calcium in the intestinal lumen, preventing its absorption. Other minerals, including zinc, are also chelated by phytate. This is mainly a problem among people who consume large amounts of unleavened whole-wheat products; yeast contains an enzyme, **phytase**, that dephosphorylates phytate, so rendering it inactive. High concentrations of fatty acids in the intestinal lumen, as a result of impaired fat absorption, can also reduce calcium absorption by forming insoluble calcium salts; a high intake of oxalate can sometimes cause deficiency since calcium oxalate is insoluble.

Iron Absorption Is Limited and Strictly Controlled, but Enhanced by Vitamin C and Alcohol

Although iron deficiency is a common problem in both developed and developing countries, about 10% of the population are genetically at risk of iron overload (**hemochromatosis**), and in order to reduce the risk of adverse effects of nonenzymic generation of free radicals by iron salts, absorption is strictly regulated. Inorganic iron is transported into the mucosal cell by a proton-linked divalent metal ion transporter, and accumulated intracellularly by binding to **ferritin**. Iron leaves the mucosal cell via a transport protein ferroportin, but only if there is free **transferrin** in plasma to bind to. Once transferrin is saturated with iron, any that has accumulated in the mucosal cells is lost when the cells are shed. Expression of the ferroportin gene (and possibly also that for the divalent metal ion transporter) is downregulated by hepcidin, a peptide secreted by the liver when body iron reserves are adequate. In response to hypoxia, anemia, or hemorrhage, the synthesis of hepcidin is reduced, leading to increased synthesis of ferroportin and increased iron absorption (**Figure 43–3**). As a result of this

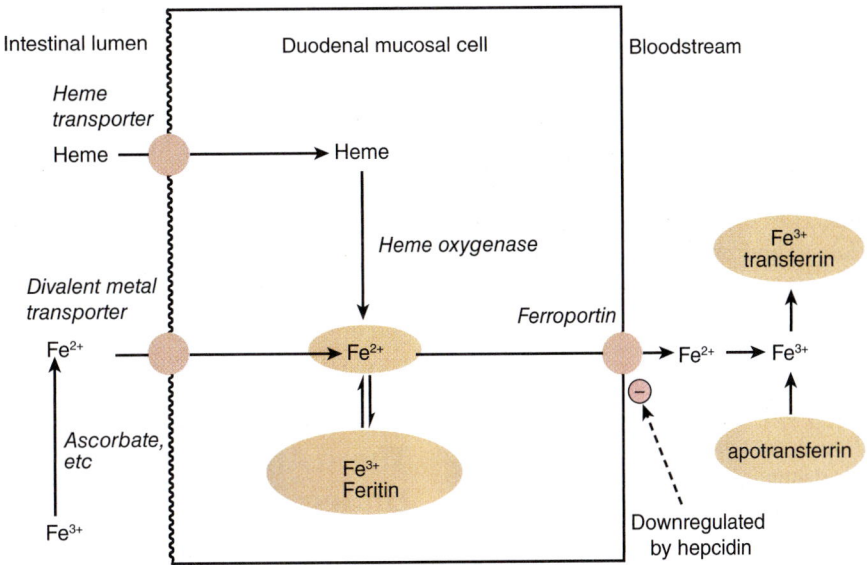

FIGURE 43–3 **Absorption of iron.** Hepcidin secreted by the liver downregulates synthesis of ferroportin and limits iron absorption.

mucosal barrier, only ~10% of dietary iron is absorbed, and only 1 to 5% from many plant foods.

Inorganic iron is absorbed in the Fe^{2+} (reduced) state, and hence, the presence of reducing agents enhances absorption. The most effective compound is **vitamin C**, and while intakes of 40 to 80 mg of vitamin C per day are more than adequate to meet requirements, an intake of 25 to 50 mg per meal enhances iron absorption, especially when iron salts are used to treat iron deficiency anemia. Alcohol and fructose also enhance iron absorption. Heme iron from meat is absorbed separately and is considerably more available than inorganic iron. However, the absorption of both inorganic and heme iron is impaired by calcium—a glass of milk with a meal significantly reduces iron availability.

ENERGY BALANCE: OVER- & UNDERNUTRITION

After the provision of water, the body's first requirement is for metabolic fuels—fats, carbohydrates, and amino acids from proteins. Food intake in excess of energy expenditure leads to **obesity**, while intake less than expenditure leads to emaciation and wasting, **marasmus**, and **kwashiorkor**. Both obesity and severe undernutrition are associated with increased mortality. The body mass index = weight (in kg)/height² (in m) is commonly used as a way of expressing relative obesity; a desirable range is between 20 and 25.

Energy Requirements Are Estimated by Measurement of Energy Expenditure

Energy expenditure can be determined directly by measuring heat output from the body, but is normally estimated indirectly from the consumption of oxygen. There is an energy expenditure of ~20 kJ/L of oxygen consumed, regardless of whether the fuel being metabolized is carbohydrate, fat, or protein (see Table 14–1).

Measurement of the ratio of the volume of carbon dioxide produced: volume of oxygen consumed (**respiratory quotient [RQ]**) is an indication of the mixture of metabolic fuels being oxidized (see Table 14–1).

A more recent technique permits estimation of total energy expenditure over a period of 1 to 3 weeks, using dual isotopically labeled water, $^2H_2^{18}O$. ^2H is lost from the body only in water, while ^{18}O is lost in both water and carbon dioxide; the difference in the rate of loss of the two labels permits estimation of total carbon dioxide production, and hence oxygen consumption and energy expenditure (**Figure 43–4**).

Basal metabolic rate (BMR) is the energy expenditure by the body when at rest, but not asleep, under controlled conditions of thermal neutrality, measured about 12 hours after the last meal. BMR depends on weight, age, and gender. **Total energy expenditure** depends on the BMR, the energy required for physical activity, and the energy cost of synthesizing reserves in the fed state. It is therefore possible to estimate an individual's energy requirement from body weight,

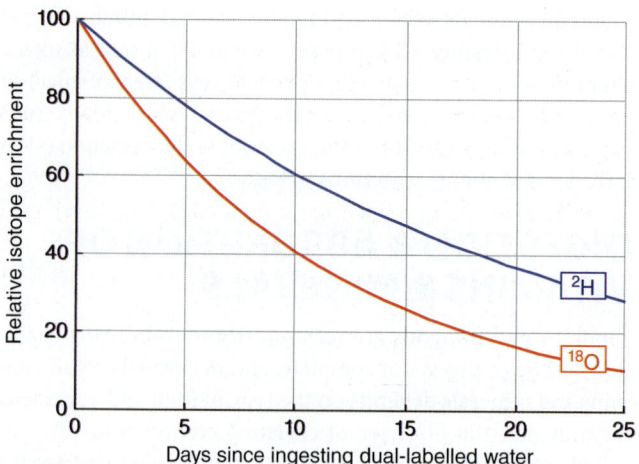

FIGURE 43–4   Dual isotopically labeled water for estimation of energy expenditure.

age, gender, and level of physical activity. Body weight affects BMR because there is a greater amount of active tissue in a larger body. The decrease in BMR with increasing age, even when body weight remains constant, is the result of muscle tissue replacement by adipose tissue, which is metabolically less active. Similarly, women have a significantly lower BMR than do men of the same body weight and age because women's bodies contain proportionally more adipose tissue.

Energy Requirements Increase With Activity

The most useful way of expressing the energy cost of physical activities is as a multiple of BMR. This is known as the **physical activity ratio (PAR)** or **metabolic equivalent of the task (MET)**. Sedentary activities use only about 1.1 to 1.2 × BMR. By contrast, vigorous exertion, such as climbing stairs, cross-country walking uphill, etc, may use 6 to 8 × BMR. The overall **physical activity level (PAL)** is the sum of the PAR of different activities, multiplied by the time taken for that activity, divided by 24 hours.

Ten Percent of the Energy Yield of a Meal May Be Expended in Forming Reserves

There is a considerable increase in metabolic rate after a meal (**diet-induced thermogenesis**). A small part of this is the energy cost of secreting digestive enzymes and of active transport of the products of digestion; the major part is the energy cost of synthesizing reserves of glycogen, triacylglycerol, and protein.

There Are Two Extreme Forms of Undernutrition

Marasmus can occur in both adults and children and occurs in vulnerable groups of all populations. **Kwashiorkor** affects only children and has been reported only in developing countries.

The distinguishing feature of kwashiorkor is that there is fluid retention, leading to edema, and fatty infiltration of the liver. Marasmus is a state of extreme emaciation; it is the outcome of prolonged negative energy balance. Not only have the body's fat reserves been exhausted, but there is wastage of muscle as well, and as the condition progresses, there is loss of protein from the heart, liver, and kidneys. The amino acids released by the catabolism of tissue proteins are used as a source of metabolic fuel and as substrates for gluconeogenesis to maintain a supply of glucose for the brain and red blood cells (see Chapter 20). As a result of the reduced synthesis of proteins, there is impaired immune response and increased risk of infection. Impairment of cell proliferation in the intestinal mucosa occurs, resulting in reduction in the surface area of the intestinal mucosa, and reduction in the absorption of such nutrients as are available.

Patients With Advanced Cancer and AIDS Are Often Malnourished

Patients with advanced cancer, HIV infection and AIDS, and a number of other chronic diseases are frequently undernourished, a condition called **cachexia**. Physically, they show all the signs of marasmus, but there is considerably more loss of body protein than occurs in starvation. Unlike marasmus, where protein synthesis is reduced, but catabolism is unaffected, in cachexia the secretion of cytokines in response to infection and cancer increases the catabolism of tissue protein by the ATP-dependent ubiquitin-proteasome pathway, so increasing energy expenditure. Patients are **hypermetabolic**, that is, they have a considerably increased BMR. In addition to activation of the ubiquitin-proteasome pathway of protein catabolism, three other factors are involved. Many tumors metabolize glucose anaerobically to release lactate. This is then used for gluconeogenesis in the liver, which is energy consuming with a net cost of six ATP for each mol of glucose cycled (see Figure 19–4). There is increased stimulation of mitochondrial **uncoupling proteins** by **cytokines** leading to thermogenesis and increased oxidation of metabolic fuels. **Futile cycling of lipids** occurs because hormone-sensitive lipase is activated by a proteoglycan secreted by tumors, resulting in liberation of fatty acids from adipose tissue and ATP-expensive reesterification to triacylglycerols in the liver, which are exported in very-low-density lipoprotein (VLDL).

Kwashiorkor Affects Undernourished Children

In addition to the wasting of muscle tissue, loss of intestinal mucosa and impaired immune responses seen in marasmus, children with **kwashiorkor** show a number of characteristic features. The defining feature is **edema**, associated with a decreased concentration of plasma proteins. In addition, there is enlargement of the liver as a result of accumulation of fat. It was formerly believed that the cause of kwashiorkor was a lack of protein, with a more or less adequate energy intake;

however, analysis of the diets of affected children shows that this is not so. Protein deficiency leads to stunting of growth, and children with kwashiorkor are less stunted than those with marasmus. Furthermore, the edema begins to improve early in treatment, when the child is still receiving a low-protein diet.

Very commonly, an infection precipitates kwashiorkor. Superimposed on general food deficiency, there is probably a deficiency of antioxidant nutrients such as zinc, copper, carotene, and vitamins C and E. The **respiratory burst** in response to infection leads to the production of oxygen and halogen **free radicals** as part of the cytotoxic action of stimulated macrophages. This added oxidant stress triggers the development of kwashiorkor.

PROTEIN & AMINO ACID REQUIREMENTS

Protein Requirements Can Be Determined by Measuring Nitrogen Balance

The state of protein nutrition can be determined by measuring the dietary intake and output of nitrogenous compounds from the body. Although nucleic acids also contain nitrogen, protein is the major dietary source of nitrogen and measurement of total nitrogen intake gives a good estimate of protein intake (mg N × 6.25 = mg protein, as N is 16% of most proteins). The output of N from the body is mainly in urea and smaller quantities of other compounds in urine, undigested protein (including digestive enzymes and shed intestinal mucosal cells) in feces; significant amounts may also be lost in sweat and shed skin. The difference between intake and output of nitrogenous compounds is known as **nitrogen balance**. Three states can be defined. In a healthy adult, there is nitrogen **equilibrium**; intake equals output, and there is no change in the total body content of protein. In a growing child, a pregnant woman, or a person in recovery from protein loss, the excretion of nitrogenous compounds is less than the dietary intake and there is net retention of nitrogen in the body as protein—**positive nitrogen balance**. In response to trauma or infection, or if the intake of protein is inadequate to meet requirements, there is net loss of protein nitrogen from the body—**negative nitrogen balance**. Except when replacing protein losses, nitrogen equilibrium can be maintained at any level of protein intake above requirements. A high intake of protein does not lead to positive nitrogen balance; although it increases the rate of protein synthesis, it also increases the rate of protein catabolism, so that nitrogen equilibrium is maintained, albeit with a higher rate of protein turnover. Both protein synthesis and catabolism are ATP expensive, and this increased rate of protein turnover explains the increased diet-induced thermogenesis seen in people consuming a high-protein diet.

The continual catabolism of tissue proteins creates the requirement for dietary protein, even in an adult who is not growing; although some of the amino acids released can be reutilized, much is used for gluconeogenesis in the fasting state. Nitrogen balance studies show that the average daily requirement is 0.66 g of protein per kg body weight (giving a reference intake of 0.825 g of protein/kg body weight, allowing for individual variation); ~55 g/d, or 8 to 9% of energy intake. Average intakes of protein in developed countries are of the order of 80 to 100 g/d, that is, 14 to 15% of energy intake. Because body protein increases in growing children, they have a proportionally greater requirement than adults and should be in positive nitrogen balance. Even so, the need is relatively small compared with the requirement for protein turnover. In some countries, protein intake is inadequate to meet these requirements, resulting in stunting of growth. There is little or no evidence that athletes and body builders require large amounts of protein; simply consuming more of a normal diet providing about 14% of energy from protein will provide more than enough protein for increased muscle protein synthesis—the main requirement is for an increased energy intake to permit increased protein synthesis.

There Is a Loss of Body Protein in Response to Trauma & Infection

One of the metabolic reactions to a major trauma, such as a burn, a broken limb, or surgery, is an increase in the net catabolism of tissue proteins, both in response to cytokines and glucocorticoid hormones, and as a result of excessive utilization of threonine and cysteine in the synthesis of **acute-phase proteins**. As much as 6 to 7% of the total body protein may be lost over 10 days. Prolonged bed rest results in considerable loss of protein because of atrophy of muscles. Protein catabolism may be increased in response to cytokines, and without the stimulus of exercise it is not completely replaced. Lost protein is replaced during **convalescence**, when there is positive nitrogen balance. Again, as in the case of athletes, a normal diet is adequate to permit this replacement protein synthesis.

The Requirement Is Not Just for Protein, but for Specific Amino Acids

Not all proteins are nutritionally equivalent. More of some is needed to maintain nitrogen balance than others because different proteins contain different amounts of the various amino acids. The body's requirement is for amino acids in the correct proportions to replace tissue proteins. The amino acids can be divided into two groups: **essential** and **nonessential**. There are nine essential or indispensable amino acids, which cannot be synthesized in the body: histidine, isoleucine, leucine, lysine, methionine, phenylalanine, threonine, tryptophan, and valine. If one of these is lacking or inadequate, then regardless of the total intake of protein, it will not be possible to maintain

nitrogen balance since there will not be enough of that amino acid for protein synthesis.

Two amino acids, cysteine and tyrosine, can be synthesized in the body, but only from essential amino acid precursors—cysteine from methionine and tyrosine from phenylalanine. The dietary intakes of cysteine and tyrosine thus affect the requirements for methionine and phenylalanine. The remaining 11 amino acids in proteins are considered to be nonessential or dispensable since they can be synthesized as long as there is enough total protein in the diet. If one of these amino acids is omitted from the diet, nitrogen balance can still be maintained. However, only three amino acids, alanine, aspartate, and glutamate, can be considered to be truly dispensable; they are synthesized by transamination of common metabolic intermediates (pyruvate, oxaloacetate, and ketoglutarate, respectively). The remaining amino acids are considered as nonessential, but under some circumstances the requirement may outstrip the capacity for their synthesis.

SUMMARY

- Digestion involves hydrolyzing food molecules into smaller molecules for absorption through the gastrointestinal epithelium. Polysaccharides are absorbed as monosaccharides, triacylglycerols as 2-monoacylglycerols, fatty acids and glycerol, and proteins as amino acids and small peptides.

- Digestive disorders arise as a result of (1) enzyme deficiency, for example, lactase and sucrase; (2) malabsorption, for example, of glucose and galactose as a result of defects in the Na^+-glucose cotransporter (SGLT 1); (3) absorption of unhydrolyzed polypeptides leading to immune responses, for example, as in celiac disease; and (4) precipitation of cholesterol from bile as gallstones.

- In addition to water, the diet must provide metabolic fuels (carbohydrate and fat) for body growth and activity, protein for synthesis of tissue proteins, fiber for bulk in the intestinal contents, minerals for specific metabolic functions (see Chapter 44), polyunsaturated fatty acids of the $n-3$ and $n-6$ families, and vitamins—organic compounds needed in small amounts for other essential functions (see Chapter 44).

- Undernutrition occurs in two extreme forms: marasmus, in adults and children, and kwashiorkor in children. Chronic illness can also lead to undernutrition (cachexia) as a result of hypermetabolism.

- Overnutrition leads to excess energy intake and is associated with chronic noncommunicable diseases such as obesity, type 2 diabetes, atherosclerosis, cancer, and hypertension.

- Twenty different amino acids are required for protein synthesis, of which nine are essential in the human diet. The quantity of protein required can be determined by studies of nitrogen balance and is affected by protein quality—the amounts of essential amino acids present in dietary proteins compared with the amounts required for tissue protein synthesis.

Micronutrients: Vitamins & Minerals

David A. Bender, PhD

BIOMEDICAL IMPORTANCE

Vitamins are a group of organic nutrients, required in small quantities for a variety of biochemical functions that, generally, cannot be synthesized by the body and must therefore be supplied in the diet.

The lipid-soluble vitamins are hydrophobic compounds that can be absorbed efficiently only when there is normal fat absorption. Like other lipids, they are transported in the blood in lipoproteins or attached to specific binding proteins. They have diverse functions—for example, vitamin A, vision and cell differentiation; vitamin D, calcium and phosphate metabolism, and cell differentiation; vitamin E, antioxidant; and vitamin K, blood clotting. As well as dietary inadequacy, conditions affecting the digestion and absorption of the lipid-soluble vitamins, such as a very-low-fat diet, steatorrhea, and disorders of the biliary system, can all lead to deficiency syndromes, including night blindness and xerophthalmia (vitamin A); rickets in young children and osteomalacia in adults (vitamin D); neurologic disorders and hemolytic anemia of the newborn (vitamin E); and hemorrhagic disease of the newborn (vitamin K). Toxicity can result from excessive intake of vitamins A and D. Vitamin A and the carotenes (many of which are precursors of vitamin A), and vitamin E are antioxidants (see Chapter 45) and have possible roles in prevention of atherosclerosis and cancer, although in excess they may also act as damaging pro-oxidants.

The water-soluble vitamins are vitamins B and C, folic acid, biotin, and pantothenic acid; they function mainly as enzyme cofactors. Folic acid acts as a carrier of one-carbon units. Deficiency of a single vitamin of the B complex is rare since poor diets are most often associated with **multiple deficiency states**. Nevertheless, specific syndromes are characteristic of deficiencies of individual vitamins, for example, beriberi (thiamin); cheilosis, glossitis, seborrhea (riboflavin); pellagra (niacin); megaloblastic anemia, methylmalonic aciduria, and pernicious anemia (vitamin B_{12}); megaloblastic anemia (folic acid); and scurvy (vitamin C).

Inorganic mineral elements that have a function in the body must be provided in the diet. When the intake is insufficient, deficiency signs may arise, for example, anemia (iron), and cretinism and goiter (iodine). Excessive intakes may be toxic.

The Determination of Micronutrient Requirements Depends on the Criteria of Adequacy Chosen

For any nutrient, there is a range of intakes between that which is clearly inadequate, leading to **clinical deficiency disease**, and that which is so much in excess of the body's metabolic capacity that there may be signs of **toxicity**. Between these two extremes is a level of intake that is adequate for normal health and the maintenance of metabolic integrity. Requirements are determined in depletion/repletion studies, in which people are deprived of the nutrient until there is a metabolic change, then repleted with the nutrient until the abnormality is normalized. Individuals do not all have the same requirement for nutrients, even when calculated on the basis of body size or energy expenditure. There is a range of individual requirements of up to 25% around the mean. Therefore, in order to assess the adequacy of diets, it is necessary to set a reference level of intake high enough to ensure that no one either suffers from deficiency or is at risk of toxicity. If it is assumed that individual requirements are distributed in a statistically normal fashion around the observed mean requirement, then a range of ±2 × the standard deviation (SD) around the mean includes the requirements of 95% of the population. Reference or recommended intakes are therefore set at the average requirement plus 2 × SD, and so meet or exceed the requirements of 97.5% of the population.

Reference and recommended intakes of vitamins and minerals published by different national and international authorities differ because of different interpretations of the available data, and the availability of new experimental data in more recent publications.

THE VITAMINS ARE A DISPARATE GROUP OF COMPOUNDS WITH A VARIETY OF METABOLIC FUNCTIONS

A vitamin is defined as an organic compound that is required in the diet in small amounts for the maintenance of normal metabolic integrity. Deficiency causes a specific disease, which is cured or prevented only by restoring the vitamin to the diet (**Table 44–1**). However, **vitamin D**, which is formed in the skin from 7-dehydrocholesterol on exposure to sunlight, and **niacin**, which can be formed from the essential amino acid tryptophan, do not strictly comply with this definition.

TABLE 44–1 **The Vitamins**

Vitamin		Functions	Deficiency Disease
Lipid-soluble			
A	Retinol, β-carotene	Visual pigments in the retina; regulation of gene expression and cell differentiation (β-carotene is an antioxidant)	Night blindness, xerophthalmia; keratinization of skin
D	Calciferol	Maintenance of calcium balance; enhances intestinal absorption of Ca^{2+} and mobilizes bone mineral; regulation of gene expression and cell differentiation	Rickets = poor mineralization of bone in children; osteomalacia = bone demineralization in adults
E	Tocopherols, tocotrienols	Antioxidant, especially in cell membranes; roles in cell signaling	Extremely rare—serious neurologic dysfunction
K	Phylloquinone; menaquinones	Coenzyme in formation of γ-carboxyglutamate in enzymes of blood clotting and bone matrix	Impaired blood clotting, hemorrhagic disease
Water-soluble			
B_1	Thiamin	Coenzyme in pyruvate and α-ketoglutarate dehydrogenases, and transketolase; regulates Cl^- channel in nerve conduction	Peripheral nerve damage (beriberi) or central nervous system lesions (Wernicke-Korsakoff syndrome)
B_2	Riboflavin	Coenzyme in oxidation and reduction reactions (FAD and FMN); prosthetic group of flavoproteins	Lesions of corner of mouth, lips, and tongue, seborrheic dermatitis
Niacin	Nicotinic acid, nicotinamide	Coenzyme in oxidation and reduction reactions, functional part of NAD and NADP; role in intracellular calcium regulation and cell signaling	Pellagra—photosensitive dermatitis, depressive psychosis
B_6	Pyridoxine, pyridoxal, pyridoxamine	Coenzyme in transamination and decarboxylation of amino acids and glycogen phosphorylase; modulation of steroid hormone action	Disorders of amino acid metabolism, convulsions
	Folic acid	Coenzyme in transfer of one-carbon fragments	Megaloblastic anemia
B_{12}	Cobalamin	Coenzyme in transfer of one-carbon fragments and metabolism of folic acid	Pernicious anemia = megaloblastic anemia with degeneration of the spinal cord
	Pantothenic acid	Functional part of CoA and acyl carrier protein: fatty acid synthesis and metabolism	Peripheral nerve damage (nutritional melalgia or "burning foot syndrome")
H	Biotin	Coenzyme in carboxylation reactions in gluconeogenesis and fatty acid synthesis; role in regulation of cell cycle	Impaired fat and carbohydrate metabolism, dermatitis
C	Ascorbic acid	Coenzyme in hydroxylation of proline and lysine in collagen synthesis; antioxidant; enhances absorption of iron	Scurvy—impaired wound healing, loss of dental cement, subcutaneous hemorrhage

LIPID-SOLUBLE VITAMINS

TWO GROUPS OF COMPOUNDS HAVE VITAMIN A ACTIVITY

Retinoids comprise **retinol, retinaldehyde**, and **retinoic acid** (preformed vitamin A, found only in foods of animal origin); carotenoids, found in plants, are a variety of carotenes and related compounds; many are precursors of vitamin A, as they can be cleaved to yield retinaldehyde, then retinol and retinoic acid (**Figure 44–1**). The α-, β-, and γ-carotenes and crypto-xanthin are quantitatively the most important provitamin A carotenoids. β-Carotene and other provitamin A carotenoids are cleaved in the intestinal mucosa by carotene dioxygenase, yielding retinaldehyde, which is reduced to retinol, esterified and secreted in chylomicrons together with esters formed from dietary retinol. The intestinal activity of carotene dioxygenase is low, so that a relatively large proportion of ingested β-carotene may appear in the circulation unchanged. There are two isoenzymes of carotene dioxygenase. One catalyzes cleavage of the central bond of β-carotene; the other catalyzes asymmetric cleavage leading to the formation of 8′-, 10′-, and 12′-apo-carotenals, which are oxidized to retinoic acid, but cannot be used as sources of retinol or retinaldehyde.

Although it would appear that one molecule of β-carotene should yield two of retinol, this is not so in practice; 6 μg of β-carotene is equivalent to 1 μg of preformed retinol. The total amount of vitamin A in foods is therefore expressed as micrograms of retinol equivalents = μg preformed vitamin A + 1/6 × μg β-carotene + 1/12 × μg other provitamin A carotenoids. Before pure vitamin A was available for chemical analysis, the vitamin A content of foods was determined by biological assay and the results expressed as international units (IU). 1 IU = 0.3 μg retinol; 1 μg retinol = 3.33 IU. Although obsolete, IU is sometimes still used in food labeling. The term **retinol**

activity equivalent takes account of the incomplete absorption and metabolism of carotenoids; 1 RAE = 1 μg all-*trans*-retinol, 12 μg β-carotene, 24 μg α-carotene or β-cryptoxanthin. On this basis, 1 IU of vitamin A activity is equal to 3.6 μg β-carotene or 7.2 μg of other provitamin A carotenoids.

Vitamin A Has a Function in Vision

In the retina, retinaldehyde functions as the prosthetic group of the light-sensitive opsin proteins, forming **rhodopsin** (in rods) and **iodopsin** (in cones). Any one cone cell contains only one type of opsin and is sensitive to only one color. In the pigment epithelium of the retina, all-*trans*-retinol is isomerized to 11-*cis*-retinol and oxidized to 11-*cis*-retinaldehyde. This reacts with a lysine residue in opsin, forming the holoprotein rhodopsin. As shown in **Figure 44–2**, the absorption of light by rhodopsin causes isomerization of the retinaldehyde from 11-*cis* to all-*trans*, and a conformational change in opsin. This results in the release of retinaldehyde from the protein, and the initiation of a nerve impulse. The formation of the initial excited form of rhodopsin, bathorhodopsin, occurs within picoseconds of illumination. There is then a series of conformational changes leading to the formation of metarhodopsin II, which initiates a guanine nucleotide amplification cascade and then a nerve impulse. The final step is hydrolysis to release all-*trans*-retinaldehyde and opsin. The key to initiation of the visual cycle is the availability of 11-*cis*-retinaldehyde, and hence vitamin A. In deficiency, both the time taken to adapt to darkness and the ability to see in poor light are impaired.

Retinoic Acid Has a Role in the Regulation of Gene Expression and Tissue Differentiation

A major role of vitamin A is the control of cell differentiation and turnover. All-*trans*-retinoic acid and 9-*cis*-retinoic acid (Figure 44–1) regulate growth, development, and tissue differentiation; they have different actions in different tissues. Like the thyroid and steroid hormones and vitamin D, retinoic acid binds to nuclear receptors that bind to response elements of DNA and regulate the transcription of specific genes. There are two families of nuclear retinoid receptors: the retinoic acid receptors (RAR) bind all-*trans*-retinoic acid or 9-*cis*-retinoic acid, and the retinoid X receptors (RXR) bind 9-*cis*-retinoic acid. RXRs also form hetero dimers with vitamin D, thyroid, and other nuclear acting hormone receptors. Deficiency of vitamin A impairs vitamin D and thyroid hormone function because of lack of 9-*cis*-retinoic acid to form active receptor dimers. Unoccupied RXRs form dimers with occupied vitamin D and thyroid hormone receptors, but not only are these unable to activate gene expression, they may repress it. Consequently vitamin A deficiency has a more severe effect on vitamin D and thyroid hormone function than simply interfering with gene expression. Excessive vitamin A also impairs vitamin D and thyroid hormone function, because of formation of RXR homodimers, meaning that there are not enough

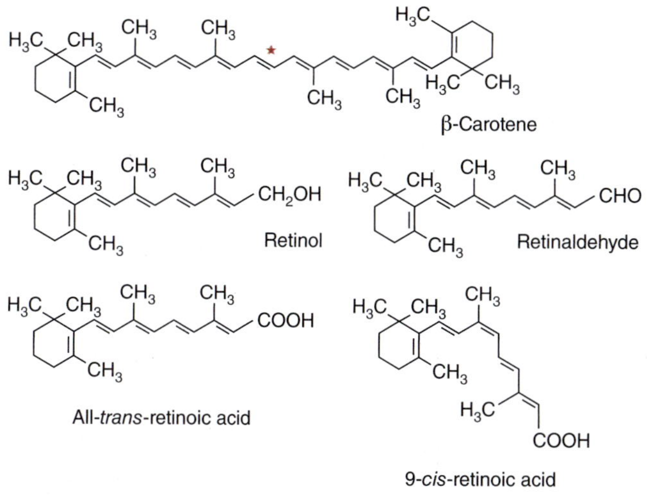

FIGURE 44–1 **β-Carotene and the major vitamin A vitamers.** Asterisk shows the site of symmetrical cleavage of β-carotene by carotene dioxygenase, to yield retinaldehyde.

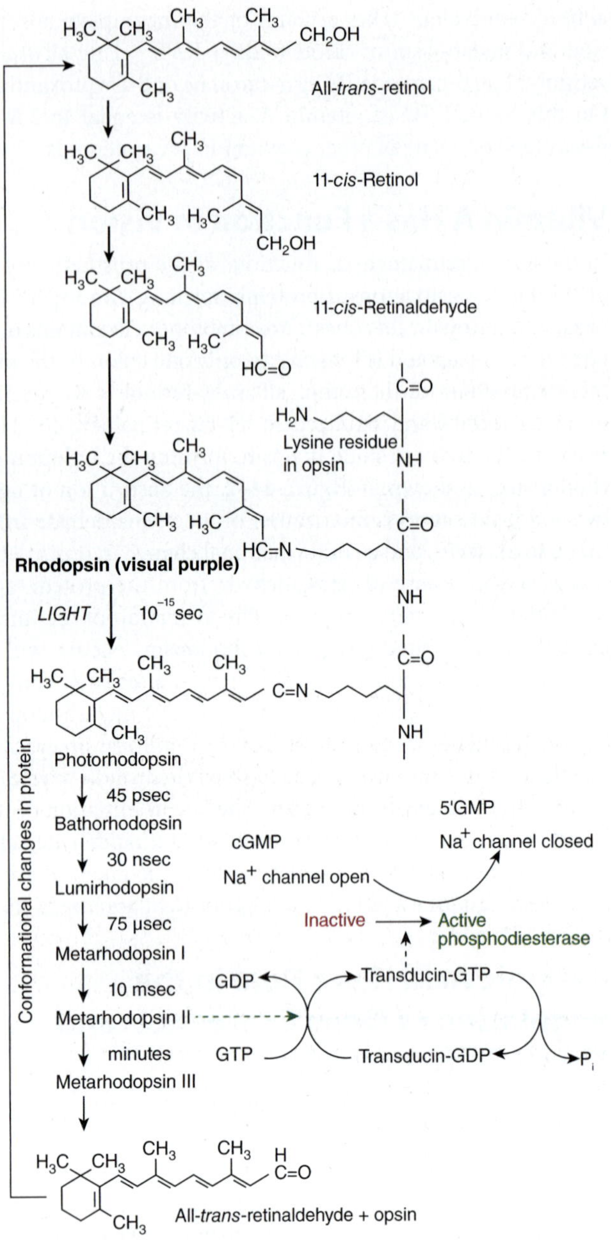

FIGURE 44–2 The role of retinaldehyde in the visual cycle.

RXR available to form heterodimers with the vitamin D and thyroid hormone receptors.

Vitamin A Deficiency Is a Major Public Health Problem Worldwide

Vitamin A deficiency is the most important preventable cause of blindness. The earliest sign of deficiency is a loss of sensitivity to green light, followed by impairment to adapt to dim light, then night blindness, an inability to see in the dark. More prolonged deficiency leads to **xerophthalmia**: keratinization of the cornea, and blindness. Vitamin A also has an important role in differentiation of immune system cells, and even mild deficiency leads to increased susceptibility to infectious diseases. The synthesis of retinol-binding protein, which is required to transport the vitamin in the bloodstream, is reduced in response to infection (it is a negative **acute phase protein**), decreasing the circulating concentration of the vitamin, and further impairing immune responses.

Vitamin A Is Toxic in Excess

Humans possess only a limited capacity to metabolize vitamin A, and excessive intakes lead to accumulation beyond the capacity of intracellular binding proteins; unbound vitamin A causes membrane lysis and tissue damage. Symptoms of toxicity affect the central nervous system (headache, nausea, ataxia, and anorexia, all associated with increased cerebrospinal fluid pressure); the liver (hepatomegaly with histological changes and hyperlipidemia); calcium homeostasis (thickening of the long bones, hypercalcemia, and calcification of soft tissues); and the skin (excessive dryness, desquamation, and alopecia).

VITAMIN D IS REALLY A HORMONE

Vitamin D is not strictly a vitamin since it can be synthesized in the skin, and under most conditions this is the major source of the vitamin. Only when sunlight exposure is inadequate is a dietary source required. Vitamin D's main function is to regulate calcium absorption and homeostasis; most of its actions are mediated by way of nuclear receptors that regulate gene expression. It also has a role in regulating cell proliferation and differentiation. There is evidence that intakes considerably higher than are required to maintain calcium homeostasis reduce the risk of insulin resistance, obesity, and metabolic syndrome, as well as various cancers. Deficiency, leading to rickets in children and osteomalacia in adults, continues to be a problem in northern latitudes, where sunlight exposure is inadequate.

Vitamin D Is Synthesized in the Skin

7-Dehydrocholesterol (an intermediate in the synthesis of cholesterol that accumulates in the skin) undergoes a nonenzymic reaction on exposure to ultraviolet light, yielding previtamin D (**Figure 44–3**). This undergoes a further reaction over a period of hours to form cholecalciferol, which is absorbed into the bloodstream. In temperate climates, the plasma concentration of vitamin D is highest at the end of summer and lowest at the end of winter. Beyond latitudes about 40° north or south, there is very little ultraviolet radiation of the appropriate wavelength in winter.

Vitamin D Is Metabolized to the Active Metabolite, Calcitriol, in Liver & Kidney

Cholecalciferol, either synthesized in the skin or from food, undergoes two hydroxylations to yield the active metabolite, 1,25-dihydroxyvitamin D or calcitriol (**Figure 44–4**). Ergocalciferol from fortified foods undergoes similar hydroxylation to

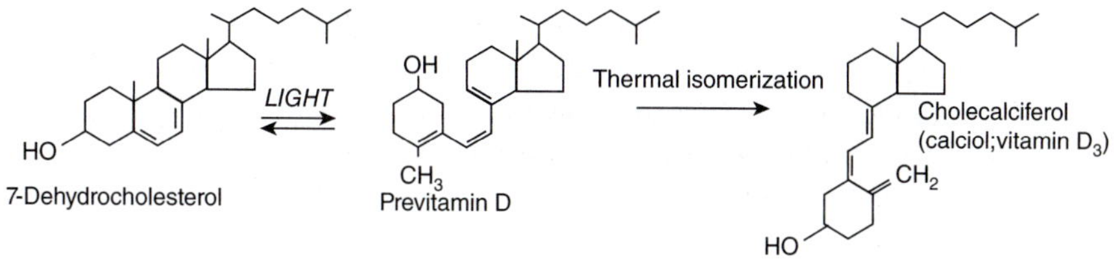

FIGURE 44–3 The synthesis of vitamin D in the skin.

yield ercalcitriol. In the liver, cholecalciferol is hydroxylated to form the 25-hydroxy derivative, calcidiol. This is released into the circulation bound to a vitamin D–binding globulin, which is the main storage form of the vitamin. In the kidney, calcidiol undergoes either 1-hydroxylation to yield the active metabolite 1,25-dihydroxyvitamin D (calcitriol), or 24-hydroxylation to yield a probably inactive metabolite, 24,25-dihydroxyvitamin D (24-hydroxycalcidiol). Some tissues, other than those that are involved in calcium homeostasis, take up calcidiol from the circulation and synthesize calcitriol that acts within that cell.

Vitamin D Metabolism Is Both Regulated by and Regulates Calcium Homeostasis

The main function of vitamin D is in the control of calcium homeostasis, and in turn, vitamin D metabolism is regulated by factors that respond to plasma concentrations of calcium and phosphate. Calcitriol acts to reduce its own synthesis by inducing the 24-hydroxylase and repressing the 1-hydroxylase in the kidney. The principal function of vitamin D is to maintain the plasma calcium concentration. Calcitriol achieves this in three ways: it increases intestinal absorption of calcium; it reduces excretion of calcium (by stimulating reabsorption in the distal renal tubules); and it mobilizes bone mineral. In addition, calcitriol is involved

in insulin secretion, synthesis and secretion of parathyroid and thyroid hormones, inhibition of production of interleukin by activated T-lymphocytes and of immunoglobulin by activated B-lymphocytes, differentiation of monocyte precursor cells, and modulation of cell proliferation. In most of these actions, it acts like a steroid hormone, binding to nuclear receptors and enhancing gene expression, although it also has a rapid effect to mobilize calcium transporters in the intestinal mucosa.

Higher Intakes of Vitamin D May Be Beneficial

There is growing evidence that higher vitamin D status is protective against various cancers, including prostate and colorectal cancer, and also against prediabetes and metabolic syndrome. Desirable levels of intake may be considerably higher than current reference intakes, and certainly could not be met from unfortified foods. While increased sunlight exposure would meet the need, it carries the risk of developing skin cancer.

Vitamin D Deficiency Affects Children & Adults

In the vitamin D deficiency disease **rickets**, the bones of children are undermineralized as a result of poor absorption of

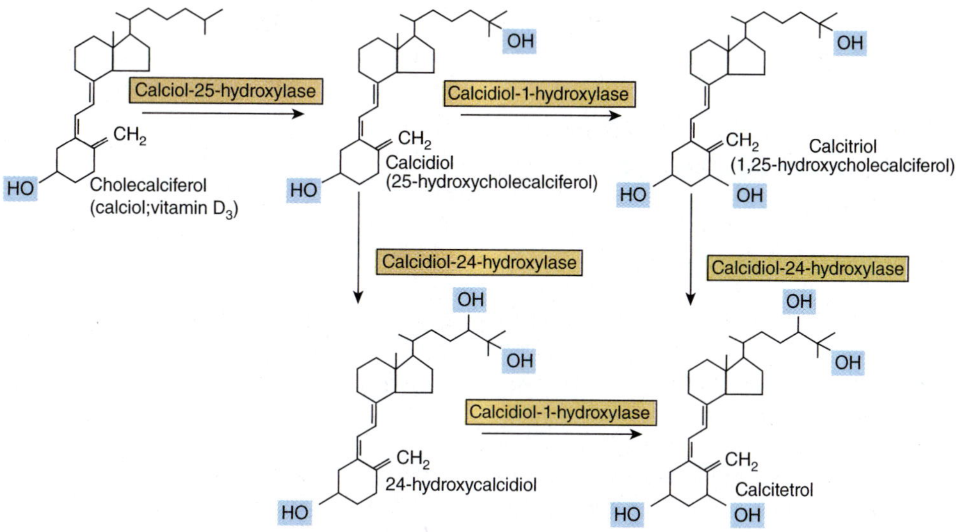

FIGURE 44–4 Metabolism of vitamin D.

calcium. Similar problems occur as a result of deficiency during the adolescent growth spurt. **Osteomalacia** in adults results from the demineralization of bone, especially in women who have little exposure to sunlight, especially after several pregnancies. Although vitamin D is essential for prevention and treatment of osteomalacia in the elderly, there is less evidence that it is beneficial in treating **osteoporosis**.

Vitamin D Is Toxic in Excess

Some infants are sensitive to intakes of vitamin D as low as 50 µg/d, resulting in an elevated plasma concentration of calcium. This can lead to contraction of blood vessels, high blood pressure, and **calcinosis**—the calcification of soft tissues. In at least a few cases, hypercalcemia in response to low intakes of vitamin D is due to genetic defects of calcidiol 24-hydroxylase, the enzyme that leads to inactivation of the vitamin. Although excess dietary vitamin D is toxic, excessive exposure to sunlight does not lead to vitamin D poisoning, because there is a limited capacity to form the precursor, 7-dehydrocholesterol, and prolonged exposure of previtamin D to sunlight leads to formation of inactive compounds.

VITAMIN E DOES NOT HAVE A PRECISELY DEFINED METABOLIC FUNCTION

No unequivocal unique function for vitamin E has been defined. It acts as a lipid-soluble **antioxidant** in cell membranes, where many of its functions can be provided by synthetic antioxidants, and is important in maintaining the fluidity of cell membranes. It also has a (relatively poorly defined) role in cell signaling. Vitamin E is the generic descriptor for two families of compounds, the **tocopherols** and the **tocotrienols** (**Figure 44–5**). The different vitamers have different biologic potency; the most active is D-α-tocopherol, and it is usual to express vitamin E intake in terms of milligrams D-α-tocopherol equivalents. Synthetic DL-α-tocopherol does not have the same biologic potency as the naturally occurring compound.

Vitamin E Is the Major Lipid-Soluble Antioxidant in Cell Membranes & Plasma Lipoproteins

The main function of vitamin E is as a chain-breaking, free radical–trapping antioxidant in cell membranes and plasma lipoproteins by reacting with the lipid peroxide radicals formed by peroxidation of polyunsaturated fatty acids (see Chapter 45). The tocopheroxyl radical is relatively unreactive, and ultimately forms nonradical compounds. Commonly, the tocopheroxyl radical is reduced back to tocopherol by reaction with vitamin C from plasma. The resultant, stable, monodehydroascorbate radical then undergoes enzymic or nonenzymic reaction to yield ascorbate and dehydroascorbate, neither of which is a radical.

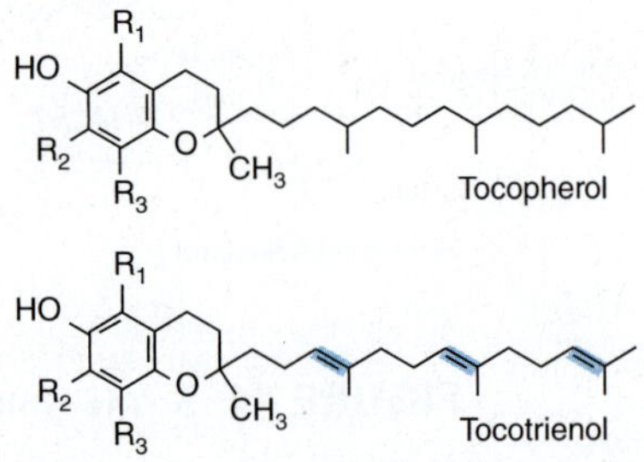

FIGURE 44–5 **Vitamin E vitamers.** In α-tocopherol and tocotrienol R_1, R_2, and R_3 are all –CH_3 groups. In the β-vitamers R_2 is H, in the γ-vitamers R_1 is H, and in the δ-vitamers R_1 and R_2 are both H.

Vitamin E Deficiency

In experimental animals, vitamin E deficiency results in resorption of fetuses and testicular atrophy. Dietary deficiency of vitamin E in human beings is unknown, although patients with severe fat malabsorption, cystic fibrosis, and some forms of chronic liver disease suffer deficiency because they are unable to absorb or transport the vitamin, leading to nerve and muscle membrane damage. Premature infants are born with inadequate reserves of the vitamin. The erythrocyte membranes are abnormally fragile as a result of lipid peroxidation, leading to hemolytic anemia.

VITAMIN K IS REQUIRED FOR SYNTHESIS OF BLOOD-CLOTTING PROTEINS

Vitamin K was discovered as a result of investigations into the cause of a bleeding disorder, hemorrhagic (sweet clover) disease of cattle and of chickens fed on a fat-free diet. The missing factor in the diet of the chickens was vitamin K, while the cattle feed contained **dicumarol**, an antagonist of the vitamin. Antagonists of vitamin K are used to reduce blood coagulation in patients at risk of thrombosis; the most widely used is **warfarin**.

Three compounds have the biological activity of vitamin K (**Figure 44–6**): **phylloquinone**, the normal dietary source, found in green vegetables; **menaquinones**, synthesized by intestinal bacteria, with differing lengths of side chain; and **menadione** and menadiol diacetate, synthetic compounds that can be metabolized to phylloquinone. Menaquinones are absorbed to some extent, but it is not clear to what extent they are biologically active as it is possible to induce signs of vitamin K deficiency simply by feeding a phylloquinone-deficient diet, without inhibiting intestinal bacterial action.

Vitamin K Is the Coenzyme for Carboxylation of Glutamate in Postsynthetic Modification of Calcium-Binding Proteins

Vitamin K is the cofactor for the carboxylation of glutamate residues in the postsynthetic modification of proteins to form the unusual amino acid γ-carboxyglutamate (GLA) (**Figure 44–7**).

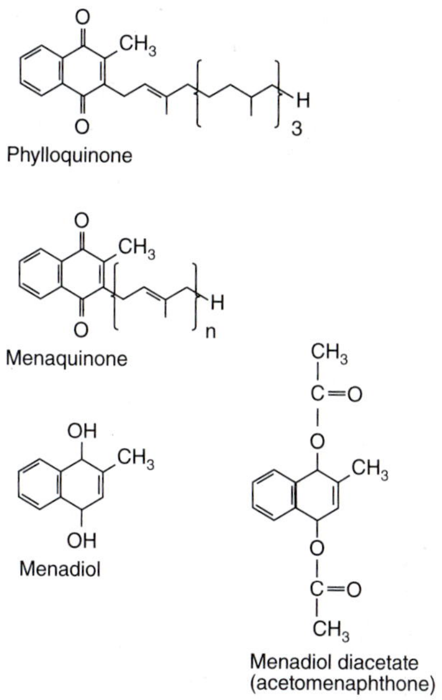

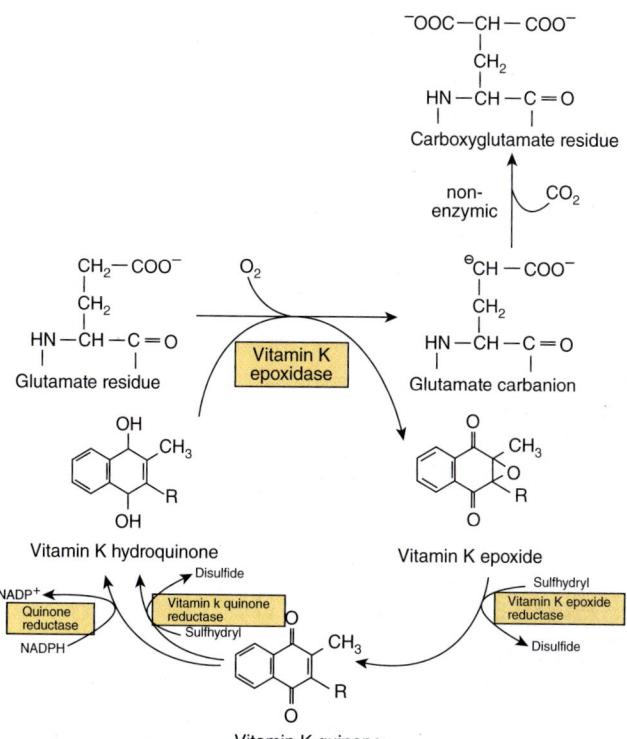

FIGURE 44–6 **The vitamin K vitamers.** Menadiol (or menadione) and menadiol diacetate are synthetic compounds that are converted to menaquinone in the liver.

Initially, vitamin K hydroquinone is oxidized to the epoxide, which activates a glutamate residue in the protein substrate to a carbanion, which reacts nonenzymically with carbon dioxide to form γ-carboxyglutamate. Vitamin K epoxide is reduced to the quinone by a warfarin-sensitive reductase, and the quinone

is reduced to the active hydroquinone by either the same warfarin-sensitive reductase or a warfarin-insensitive quinone reductase. In the presence of warfarin, vitamin K epoxide cannot be reduced, but accumulates and is excreted. If enough vitamin K (as the quinone) is provided in the diet, it can be reduced to the active hydroquinone by the warfarin-insensitive enzyme, and carboxylation can continue, with stoichiometric utilization of vitamin K and excretion of the epoxide. A high dose of vitamin K is the antidote to an overdose of warfarin.

Prothrombin and several other proteins of the blood-clotting system (factors VII, IX, and X, and proteins C and S, Chapter 52) each contain 4 to 6 γ-carboxyglutamate residues. γ-Carboxyglutamate chelates calcium ions, and so permits the binding of the blood-clotting proteins to membranes. In vitamin K deficiency, or in the presence of warfarin, an abnormal precursor of prothrombin (preprothrombin) containing little or no γ-carboxyglutamate, and incapable of chelating calcium, is released into the circulation.

Vitamin K Is Also Important in Synthesis of Bone & Other Calcium-Binding Proteins

A number of other proteins undergo the same vitamin K–dependent carboxylation of glutamate to γ-carboxyglutamate, including osteocalcin and the matrix Gla protein in bone, nephrocalcin in kidney and the product of the growth arrest specific gene *Gas6*, which is involved in both the regulation of differentiation and development in the nervous system, and control of apoptosis in other tissues. All of these γ-carboxyglutamate–containing proteins bind calcium, which causes a conformational change so that they interact with membrane phospholipids. The release into the circulation of osteocalcin provides an index of vitamin D status.

WATER-SOLUBLE VITAMINS

VITAMIN B₁ (THIAMIN) HAS A KEY ROLE IN CARBOHYDRATE METABOLISM

Thiamin has a central role in energy-yielding metabolism, and especially the metabolism of carbohydrates (**Figure 44–8**). **Thiamin diphosphate** is the coenzyme for three multienzyme complexes that catalyze oxidative decarboxylation reactions: pyruvate dehydrogenase in carbohydrate metabolism (see Chapter 17); α-ketoglutarate dehydrogenase in the citric acid cycle (see Chapter 16); and the branched-chain keto acid

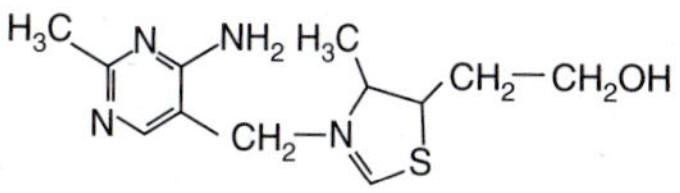

FIGURE 44–7 The role of vitamin K in the synthesis of γ-carboxyglutamate.

FIGURE 44–8 Thiamin.

dehydrogenase involved in the metabolism of leucine, isoleucine, and valine (see Chapter 29). In each case, the thiamin diphosphate provides a reactive carbon on the thiazole moiety that forms a carbanion, which then adds to the carbonyl group, for example, pyruvate. The addition compound is then decarboxylated, eliminating CO_2. Thiamin diphosphate is also the coenzyme for transketolase, in the pentose phosphate pathway (see Chapter 20).

Thiamin triphosphate has a role in nerve conduction; it phosphorylates, and so activates, a chloride channel in the nerve membrane.

Thiamin Deficiency Affects the Nervous System & the Heart

Thiamin deficiency can result in three distinct syndromes: a chronic peripheral neuritis, **beriberi**, which may or may not be associated with **heart failure** and **edema**; acute pernicious (fulminating) beriberi (Shoshin beriberi), in which heart failure and metabolic abnormalities predominate, without peripheral neuritis; and **Wernicke encephalopathy** with **Korsakoff psychosis**, which is associated especially with alcohol and narcotic abuse. The role of thiamin diphosphate in pyruvate dehydrogenase means that in deficiency there is impaired conversion of pyruvate to acetyl-CoA. In subjects on a relatively high carbohydrate diet, this results in increased plasma concentrations of lactate and pyruvate, which may cause life-threatening **lactic acidosis**.

Thiamin Nutritional Status Can Be Assessed by Erythrocyte Transketolase Activation

The activation of apotransketolase (the enzyme protein) in erythrocyte lysate by thiamin diphosphate added in vitro has become the accepted index of thiamin nutritional status.

VITAMIN B₂ (RIBOFLAVIN) HAS A CENTRAL ROLE IN ENERGY-YIELDING METABOLISM

Riboflavin provides the reactive moieties of the coenzymes **flavin mononucleotide (FMN)** and **flavin adenine dinucleotide (FAD)** (see Figure 12–2). FMN is formed by ATP-dependent phosphorylation of riboflavin; FAD is synthesized by further reaction with ATP in which the AMP moiety is transferred onto FMN. The main dietary sources of riboflavin are milk and dairy products. In addition, because of its intense yellow color, riboflavin is widely used as a food additive.

Flavin Coenzymes Are Electron Carriers in Oxidoreduction Reactions

These include the mitochondrial respiratory chain, key enzymes in fatty acid and amino acid oxidation, and the citric acid cycle. Reoxidation of the reduced flavin in oxygenases

and mixed-function oxidases proceeds by way of formation of the flavin radical and flavin hydroperoxide, with the intermediate generation of superoxide and perhydroxyl radicals and hydrogen peroxide. Because of this, flavin oxidases make a significant contribution to the total oxidant stress in the body (see Chapter 45).

Riboflavin Deficiency Is Widespread but Not Fatal

Although riboflavin is centrally involved in lipid and carbohydrate metabolism, and deficiency occurs in many countries, it is not fatal, because there is very efficient conservation of tissue riboflavin. Riboflavin released by the catabolism of enzymes is rapidly incorporated into newly synthesized enzymes. Deficiency is characterized by cheilosis, desquamation and inflammation of the tongue, and seborrheic dermatitis. Riboflavin nutritional status is assessed by measurement of the activation of erythrocyte glutathione reductase by FAD added in vitro.

NIACIN IS NOT STRICTLY A VITAMIN

Niacin was discovered as a nutrient during studies of **pellagra**. It is not strictly a vitamin since it can be synthesized in the body from the essential amino acid tryptophan. Two compounds, **nicotinic acid** and **nicotinamide**, have the biological activity of niacin; its metabolic function is as the nicotinamide ring of the coenzymes **NAD** and **NADP** in oxidation/reduction reactions (see Figures 7–2 and 12–4). Some 60 mg of tryptophan is equivalent to 1 mg of dietary niacin. The niacin content of foods is expressed as:

$$\text{mg niacin equivalents} = \text{mg preformed niacin} + 1/60 \times \text{mg tryptophan}$$

Since most of the niacin in cereals is biologically unavailable, this is discounted.

NAD Is the Source of ADP-Ribose

In addition to its coenzyme role, NAD is the source of ADP-ribose for the **ADP-ribosylation** of proteins and polyADP-ribosylation of nucleoproteins involved in the **DNA repair mechanism**. Cyclic ADP-ribose and nicotinic acid adenine dinucleotide, formed from NAD, act to increase intracellular calcium in response to neurotransmitters and hormones.

Pellagra Is Caused by Deficiency of Tryptophan & Niacin

Pellagra is characterized by a photosensitive dermatitis. As the condition progresses, there is depressive psychosis and possibly diarrhea. Untreated pellagra is fatal. Although the nutritional etiology of pellagra is well established, and either tryptophan or niacin prevents or cures the disease, additional factors, including deficiency of riboflavin or vitamin B$_6$, both of which are required for synthesis of nicotinamide from tryptophan, may be important. In most outbreaks of pellagra, twice as many women as men are affected, probably the result of inhibition of tryptophan metabolism by estrogen metabolites.

Pellagra Can Occur as a Result of Disease Despite an Adequate Intake of Tryptophan & Niacin

A number of genetic diseases that result in defects of tryptophan metabolism are associated with the development of pellagra, despite an apparently adequate intake of both tryptophan and niacin. **Hartnup disease** is a rare genetic condition in which there is a defect of the membrane transport mechanism for tryptophan, resulting in large losses as a result of intestinal malabsorption and failure of renal reabsorption. In **carcinoid syndrome**, there is metastasis of a primary liver tumor of enterochromaffin cells, which synthesize 5-hydroxytryptamine. Overproduction of 5-hydroxytryptamine may account for as much as 60% of the body's tryptophan metabolism, causing pellagra because of the diversion away from NAD synthesis.

Niacin Is Toxic in Excess

Nicotinic acid has been used to treat hyperlipidemia when of the order of 1 to 6 g/d are required, causing dilation of blood vessels and flushing, along with skin irritation. Intakes of both nicotinic acid and nicotinamide in excess of 500 mg/d cause liver damage.

VITAMIN B$_6$ IS IMPORTANT IN AMINO ACID & GLYCOGEN METABOLISM & IN STEROID HORMONE ACTION

Six compounds have vitamin B$_6$ activity (**Figure 44–9**): **pyridoxine, pyridoxal, pyridoxamine**, and their 5′-phosphates. The active coenzyme is pyridoxal 5′-phosphate. Some 80% of the body's total vitamin B$_6$ is pyridoxal phosphate in muscle, mostly associated with glycogen phosphorylase. This is not available in deficiency, but is released in starvation, when glycogen reserves become depleted, and is then available, especially to liver and kidney, to meet increased requirement for gluconeogenesis from amino acids.

Vitamin B$_6$ Has Several Roles in Metabolism

Pyridoxal phosphate is a coenzyme for many enzymes involved in amino acid metabolism, especially transamination and decarboxylation. It is also the cofactor of glycogen phosphorylase, where the phosphate group is catalytically important. In addition, it is important in steroid hormone

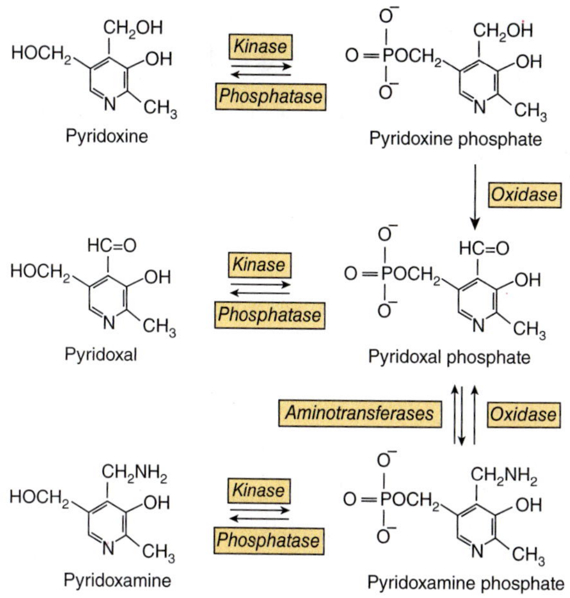

FIGURE 44–9 Interconversion of the vitamin B$_6$ vitamers.

action. Pyridoxal phosphate removes the hormone-receptor complex from DNA binding, terminating the action of the hormones. In vitamin B$_6$ deficiency, there is increased sensitivity to the actions of low concentrations of estrogens, androgens, cortisol, and vitamin D.

Vitamin B$_6$ Deficiency Is Rare

Although clinical deficiency disease is rare, there is evidence that a significant proportion of the population has marginal vitamin B$_6$ status. Moderate deficiency results in abnormalities of tryptophan and methionine metabolism. Increased sensitivity to steroid hormone action may be important in the development of **hormone-dependent cancer** of the breast, uterus, and prostate, and vitamin B$_6$ status may affect the prognosis.

Vitamin B$_6$ Status Is Assessed by Assaying Erythrocyte Transaminases

The most widely used method of assessing vitamin B$_6$ status is by the activation of erythrocyte transaminases by pyridoxal phosphate added in vitro, expressed as the activation coefficient. Measurement of plasma concentrations of the vitamin is also used.

In Excess, Vitamin B$_6$ Causes Sensory Neuropathy

The development of sensory neuropathy has been reported in patients taking 2 to 7 g of pyridoxine per day for a variety of reasons. There was some residual damage after withdrawal of these high doses; other reports suggest that intakes in excess of 100 to 200 mg/d are associated with neurological damage.

VITAMIN B₁₂ IS FOUND ONLY IN FOODS OF ANIMAL ORIGIN

The term "vitamin B_{12}" is used as a generic descriptor for the **cobalamins**—those **corrinoids** (cobalt-containing compounds possessing the corrin ring) having the biological activity of the vitamin (**Figure 44–10**). Some corrinoids that are growth factors for microorganisms not only have no vitamin B_{12} activity, but may also be antimetabolites of the vitamin. Although it is synthesized exclusively by microorganisms, for practical purposes vitamin B_{12} is found only in foods of animal origin, there being no plant sources of this vitamin. This means that strict vegetarians (vegans) are at risk of developing B_{12} deficiency. The small amounts of the vitamin formed by bacteria on the surface of fruits may be adequate to meet requirements, but preparations of vitamin B_{12} made by bacterial fermentation are available.

Vitamin B₁₂ Absorption Requires Two Binding Proteins

Vitamin B_{12} is absorbed bound to **intrinsic factor**, a small glycoprotein secreted by the parietal cells of the gastric mucosa. Gastric acid and pepsin release the vitamin from protein binding in food and make it available to bind to **cobalophilin**, a binding protein secreted in the saliva. In the duodenum, cobalophilin is hydrolyzed, releasing the vitamin for binding to intrinsic factor. **Pancreatic insufficiency** can therefore be a factor in the development of vitamin B_{12} deficiency, resulting in the excretion of cobalophilin-bound vitamin B_{12}. Intrinsic factor binds only the active vitamin B_{12} vitamers and not other corrinoids. Vitamin B_{12} is absorbed from the distal third of the ileum via receptors that bind the intrinsic factor–vitamin B_{12} complex, but not free intrinsic factor or free vitamin. There is considerable enterohepatic circulation of vitamin B_{12}, with excretion in the bile, then reabsorption after binding to intrinsic factor in the ileum.

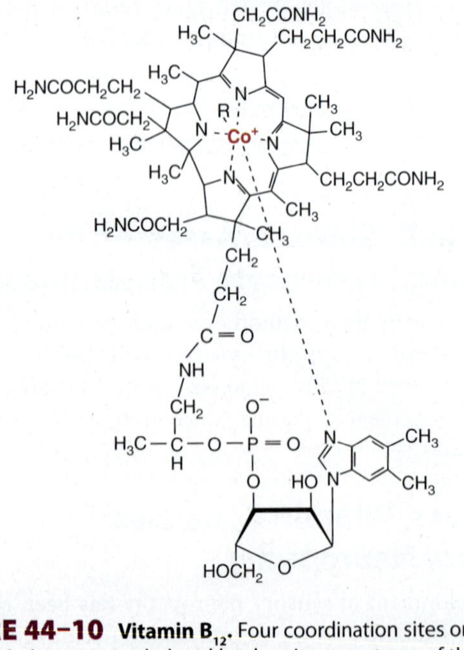

There Are Two Vitamin B₁₂–Dependent Enzymes

Methylmalonyl-CoA mutase, and **methionine synthase** (**Figure 44–11**) are vitamin B_{12}–dependent enzymes. Methylmalonyl-CoA is formed as an intermediate in the catabolism of valine as well as by the carboxylation of propionyl-CoA arising from the catabolism of isoleucine, cholesterol, and fatty acids with an odd number of carbon atoms, or propionate, a major product of microbial fermentation ruminants. Methylmalonyl-CoA undergoes a vitamin B_{12}–dependent rearrangement to succinyl-CoA, catalyzed by methylmalonyl-CoA mutase (see Figure 19–2). The activity of this enzyme is greatly reduced in vitamin B_{12} deficiency, leading to an accumulation of methylmalonyl-CoA and urinary excretion of methylmalonic acid, which provides a means of assessing vitamin B_{12} nutritional status.

Vitamin B₁₂ Deficiency Causes Pernicious Anemia

Pernicious anemia arises when vitamin B_{12} deficiency impairs the metabolism of folic acid, leading to functional folate deficiency that disturbs erythropoiesis, causing immature precursors of erythrocytes to be released into the circulation (megaloblastic anemia). The most common cause of pernicious anemia is failure of the absorption of vitamin B_{12} rather than dietary deficiency. This can be the result of failure of intrinsic factor secretion caused by autoimmune disease affecting parietal cells or from production of anti-intrinsic factor antibodies. There is irreversible degeneration of the spinal cord in pernicious anemia, as a result of failure of methylation of one arginine residue in myelin basic protein. This is the

FIGURE 44–10 **Vitamin B₁₂.** Four coordination sites on the central cobalt atom are chelated by the nitrogen atoms of the corrin ring, and one by the nitrogen of the dimethylbenzimidazole nucleotide. The sixth coordination site may be occupied by CN⁻ (cyanocobalamin), OH⁻ (hydroxocobalamin), H₂O (aquocobalamin, —CH₃ (methyl cobalamin), or 5′-deoxyadenosine (adenosylcobalamin).

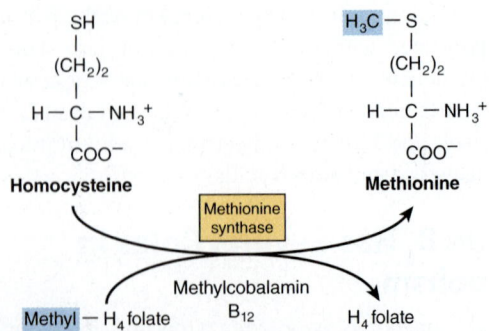

FIGURE 44–11 **Homocysteine and the folate trap.** Vitamin B₁₂ deficiency leads to impairment of methionine synthase, resulting in accumulation of homocysteine and trapping folate as methyltetrahydrofolate.

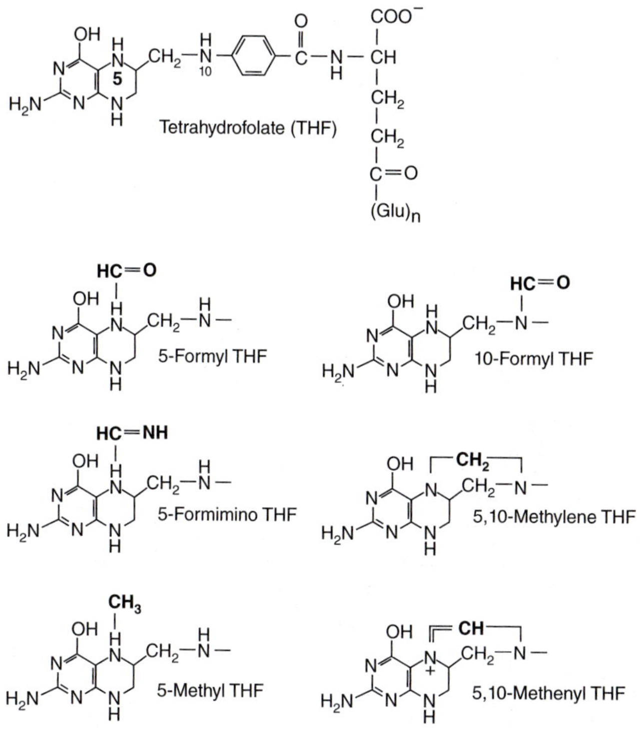

FIGURE 44–12 Tetrahydrofolic acid and the one-carbon substituted folates.

result of methionine deficiency in the central nervous system, rather than secondary folate deficiency.

THERE ARE MULTIPLE FORMS OF FOLATE IN THE DIET

The active form of folic acid (pteroyl glutamate) is tetrahydrofolate (**Figure 44–12**). The folates in foods may have up to seven additional glutamate residues linked by γ-peptide bonds. In addition, all of the one-carbon substituted folates in Figure 44–12 may also be present in foods. The extent to which the different forms of folate can be absorbed varies, and folate intakes are calculated as dietary folate equivalents—the sum of μg food folates + 1.7 × μg of folic acid (used in food enrichment).

Tetrahydrofolate Is a Carrier of One-Carbon Units

Tetrahydrofolate can carry one-carbon fragments attached to *N*-5 (formyl, formimino, or methyl groups), *N*-10 (formyl) or bridging *N*-5–*N*-10 (methylene or methenyl groups). 5-Formyl-tetrahydrofolate (known as **folinic acid**) is more stable than folate and is therefore used pharmaceutically. The synthetic (racemic) form is called **leucovorin**. The major point of entry for one-carbon fragments into substituted folates is methylenetetrahydrofolate (**Figure 44–13**), which is formed by the reaction of glycine, serine, and choline with tetrahydrofolate. Serine is the most important source of substituted folates for biosynthetic reactions, and the activity of serine hydroxymethyltransferase is regulated by the state of folate substitution and the availability of folate. The reaction is reversible, and in liver it can form serine from glycine as a substrate for gluconeogenesis. Methylene-, methenyl-, and 10-formyltetrahydrofolates are interconvertible. When one-carbon folates are not required, the oxidation of formyltetrahydrofolate to yield carbon dioxide provides a means of maintaining a pool of free folate.

Inhibitors of Folate Metabolism Provide Cancer Chemotherapy, Antibacterial, & Antimalarial Drugs

The methylation of deoxyuridine monophosphate (dUMP) to thymidine monophosphate (TMP), catalyzed by thymidylate synthase, is essential for the synthesis of DNA. The one-carbon fragment of methylene-tetrahydrofolate is reduced to a methyl group with release of dihydrofolate, which is then reduced back to tetrahydrofolate by **dihydrofolate reductase**. Thymidylate synthase and dihydrofolate reductase are especially active in tissues with a high rate of cell division. **Methotrexate**, an analog of 10-methyltetrahydrofolate, which inhibits dihydrofolate reductase, is used in cancer chemotherapy. The dihydrofolate reductases of some bacteria and parasites differ from the human enzyme; enabling inhibitors of these enzymes can be used as antibacterial (eg, **trimethoprim**) and antimalarial drugs (eg, **pyrimethamine**).

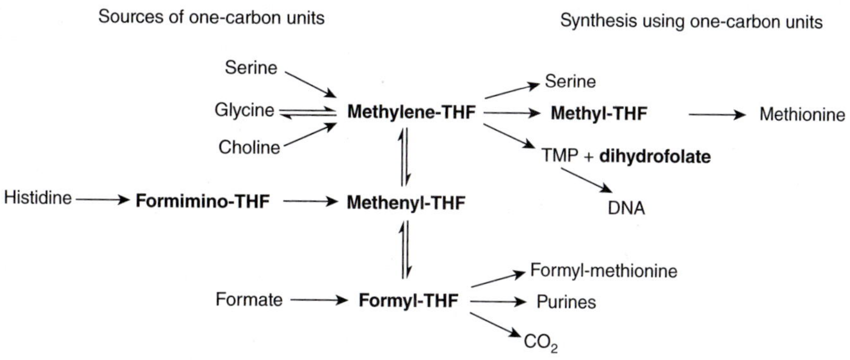

FIGURE 44–13 Sources and utilization of one-carbon substituted folates.

Vitamin B$_{12}$ Deficiency Causes Functional Folate Deficiency—the "Folate Trap"

When acting as a methyl donor, S-adenosyl methionine forms homocysteine, which may be remethylated using methyltetrahydrofolate by methionine synthase, a vitamin B$_{12}$–dependent enzyme (Figure 44–11). As the reduction of methylenetetrahydrofolate to methyltetrahydrofolate is irreversible. Since the major source of tetrahydrofolate for tissues is methyltetrahydrofolate, the role of methionine synthase is vital, and provides a link between the functions of folate and vitamin B$_{12}$. Impairment of methionine synthase in vitamin B$_{12}$ deficiency results in the accumulation of methyltetrahydrofolate that cannot be used—the "folate trap." There is therefore functional deficiency of folate, secondary to the deficiency of vitamin B$_{12}$.

Folate Deficiency Causes Megaloblastic Anemia

Deficiency of folic acid itself or deficiency of vitamin B$_{12}$, which leads to functional folic acid deficiency, affects cells that are dividing rapidly because they have a large requirement for thymidine for DNA synthesis. Clinically, this affects the bone marrow, leading to megaloblastic anemia.

Folic Acid Supplements Reduce the Risk of Neural Tube Defects & Hyperhomocysteinemia, & May Reduce the Incidence of Cardiovascular Disease & Some Cancers

Supplements of 400 µg/d of folic acid begun before conception result in a significant reduction in the incidence of **spina bifida** and other **neural tube defects**. Because of this, there is mandatory enrichment of flour with folic acid in many countries. Elevated blood homocysteine is a significant risk factor for **atherosclerosis, thrombosis**, and **hypertension**. The condition is the result of an impaired ability to form methyltetrahydrofolate by methylenetetrahydrofolate reductase, causing functional folate deficiency, resulting in failure to remethylate homocysteine to methionine. People with an abnormal variant of methylenetetrahydrofolate reductase that occurs in 5 to 10% of the population do not develop hyperhomocysteinemia if they have a relatively high intake of folate. A number of placebo-controlled trials of supplements of folate (commonly together with vitamins B$_6$ and B$_{12}$) have shown the expected lowering of plasma homocysteine, but apart from reduced incidence of stroke there has been no effect on death from cardiovascular disease.

There is also evidence that low folate status results in impaired methylation of CpG islands in DNA, which is a factor in the development of colorectal and other cancers. A number of studies suggest that folic acid supplementation or food enrichment may reduce the risk of developing some cancers. However, there is also some evidence that folate supplements increase the rate of transformation of preneoplastic colorectal polyps into cancers, so that people with such polyps may be at increased risk of developing colorectal cancer if they have a high folate intake.

Folic Acid Enrichment of Foods May Put Some People at Risk

Folic acid supplements will rectify the megaloblastic anemia of vitamin B$_{12}$ deficiency but not the irreversible nerve damage. A high intake of folic acid can thus mask vitamin B$_{12}$ deficiency. This is especially a problem for elderly people, since atrophic gastritis that develops with increasing age leads to failure of gastric acid secretion, and hence failure to release vitamin B$_{12}$ from dietary proteins. Because of this, although many countries have adopted mandatory enrichment of flour with folic acid to prevent neural tube defects, others have not. There is also antagonism between folic acid and some anticonvulsants used in the treatment of epilepsy, and, as noted above, there is some evidence that folate supplements may increase the risk of developing colorectal cancer among people with preneoplastic colorectal polyps.

DIETARY BIOTIN DEFICIENCY IS UNKNOWN

The structures of biotin, biocytin, and carboxybiotin (the active metabolic intermediate) are shown in **Figure 44–14**. Biotin is widely distributed in many foods as biocytin (ε-amino-biotinyl lysine), which is released on proteolysis. It is synthesized by intestinal flora in excess of requirements. Deficiency is unknown, except among people maintained for many months on total parenteral nutrition, and a very small number who eat abnormally large amounts of uncooked egg white, which contains avidin, a protein that binds biotin and renders it unavailable for absorption.

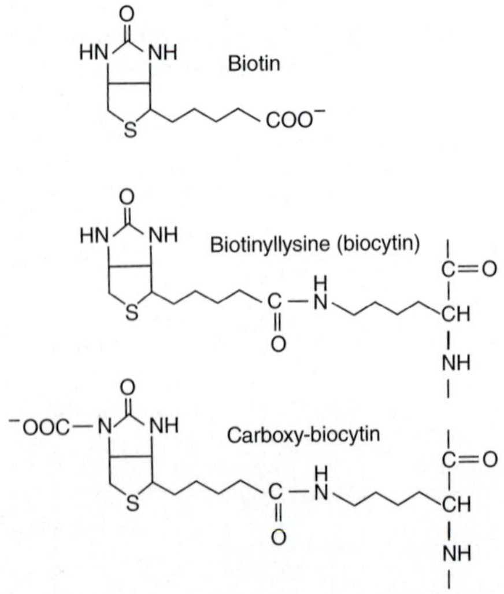

FIGURE 44–14 Biotin, biocytin, and carboxybiocytin.

Biotin Is a Coenzyme of Carboxylase Enzymes

Biotin functions to transfer carbon dioxide in a small number of enzymes: acetyl-CoA carboxylase (see Figure 23–1), pyruvate carboxylase (see Figure 19–1), propionyl-CoA carboxylase (see Figure 19–2), and methylcrotonyl-CoA carboxylase. A holocarboxylase synthetase catalyzes the transfer of biotin onto a lysine residue of the apoenzyme to form the biocytin residue of the holoenzyme. The reactive intermediate is 1-N-carboxybiocytin, formed from bicarbonate in an ATP-dependent reaction. The carboxyl group is then transferred to the substrate for carboxylation.

Biotin also has a role in regulation of the cell cycle, acting via the biotinylation of key nuclear proteins.

AS PART OF COENZYME A & ACP, PANTOTHENIC ACID ACTS AS A CARRIER OF ACYL GROUPS

Pantothenic acid has a central role in acyl group metabolism when acting as the pantetheine functional moiety of coenzyme A (CoA) or acyl carrier protein (ACP) (**Figure 44–15**). The pantetheine moiety is formed after combination of pantothenate with cysteine, which provides the—SH prosthetic group of CoA and ACP. CoA takes part in reactions of the citric acid cycle (see Chapter 16), fatty acid oxidation (see Chapter 22), acetylations, and cholesterol synthesis (see Chapter 26). ACP participates in fatty acid synthesis (see Chapter 23). The vitamin is widely distributed in all foodstuffs, and deficiency has not been unequivocally reported in humans except in specific depletion studies.

ASCORBIC ACID IS A VITAMIN FOR ONLY SOME SPECIES

Vitamin C (**Figure 44–16**) is a vitamin for human beings and other primates, the guinea pig, bats, passeriform birds, and

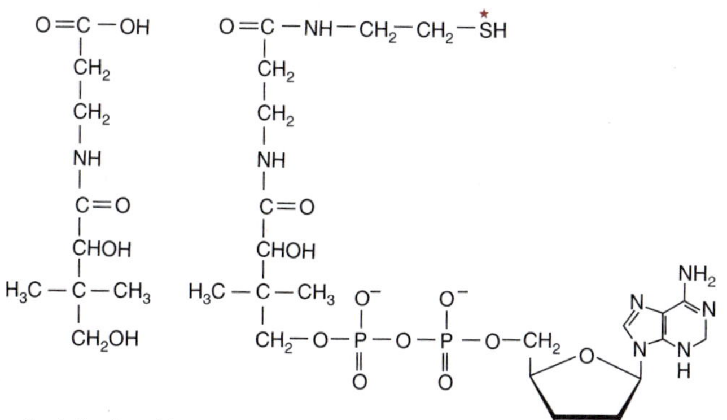

FIGURE 44–15 Pantothenic acid and coenzyme A. Asterisk shows site of acylation by fatty acids.

FIGURE 44–16 Vitamin C.

most fishes and invertebrates; other animals synthesize it as an intermediate in the uronic acid pathway of glucose metabolism (see Figure 20–4). In those species for which it is a vitamin, gulonolactone oxidase is absent. Both ascorbic acid and dehydroascorbic acid have vitamin activity.

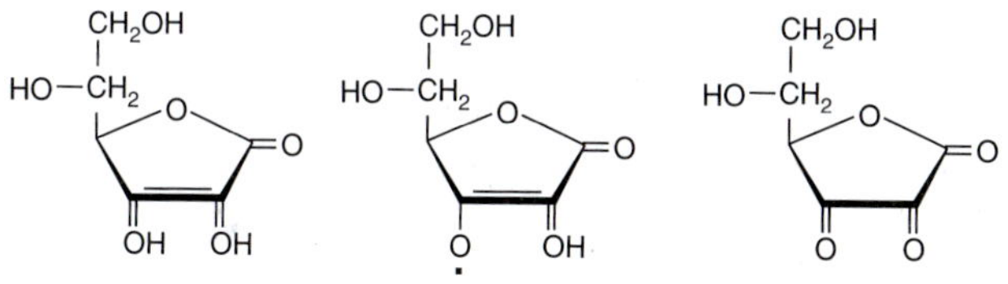

Vitamin C Is the Coenzyme for Two Groups of Hydroxylases

Ascorbic acid has specific roles in the copper-containing hydroxylases and the α-ketoglutarate–linked iron-containing hydroxylases. It also increases the activity of a number of other enzymes in vitro, although this is a nonspecific reducing action. In addition, it has a number of nonenzymic effects as a result of its action as a reducing agent and oxygen radical quencher (see Chapter 45).

Dopamine β-hydroxylase is a copper-containing enzyme involved in the synthesis of the catecholamines (norepinephrine and epinephrine), from tyrosine in the adrenal medulla and central nervous system. During hydroxylation the Cu^+ is oxidized to Cu^{2+}; reduction back to Cu^+ specifically requires ascorbate, which is oxidized to monodehydroascorbate.

A number of peptide hormones have a carboxy-terminal amide that is derived from a terminal glycine residue. This glycine is hydroxylated on the α-carbon by a copper-containing enzyme, **peptidylglycine hydroxylase**, which, again, requires ascorbate for reduction of Cu^{2+}.

A number of iron-containing, ascorbate-requiring hydroxylases share a common reaction mechanism, in which hydroxylation of the substrate is linked to oxidative decarboxylation of α-ketoglutarate. Many of these enzymes are involved in the modification of precursor proteins. **Proline** and **lysine hydroxylases** are required for the postsynthetic modification of **procollagen** to **collagen**, and proline hydroxylase is also required in formation of **osteocalcin** and the C1q component of **complement**. Aspartate β-hydroxylase is required for the postsynthetic modification of the precursor of protein C, the vitamin K–dependent protease that hydrolyzes activated factor V in the blood-clotting cascade (see Chapter 52). Trimethyllysine and γ-butyrobetaine hydroxylases are required for the synthesis of carnitine. In these enzymes ascorbate is required to reduce the iron prosthetic group after accidental oxidation during reaction; it is neither consumed stoichiometrically with the substrates nor does it have a simple catalytic role.

Vitamin C Deficiency Causes Scurvy

Signs of vitamin C deficiency include skin changes, fragility of blood capillaries, gum decay, tooth loss, and bone fracture, which can be attributed to impaired collagen synthesis, and psychological changes that can be attributed to impaired synthesis of catecholamines.

There May Be Benefits From Higher Intakes of Vitamin C

At intakes above about 100 mg/d, the body's capacity to metabolize vitamin C is saturated, and any further intake is excreted in the urine. However, in addition to its other roles, vitamin C enhances the absorption of inorganic iron, and this depends on the presence of the vitamin in the gut. Therefore, increased intake may be beneficial, and it is frequently prescribed together with iron supplements to treat iron deficiency anemia. There is very little good evidence that high doses of vitamin C prevent the common cold, although they may reduce the duration and severity of symptoms.

MINERALS ARE REQUIRED FOR BOTH PHYSIOLOGIC & BIOCHEMICAL FUNCTIONS

Many of the essential minerals (**Table 44–2**) are widely distributed in foods, and most people eating a mixed diet are likely to receive adequate intakes. The amounts required vary from grams per day for sodium and calcium, through milligrams per day (eg, iron and zinc), to micrograms per day for the trace elements. In general, mineral deficiencies occur when foods come from one region where the soil may be deficient in some minerals (eg, iodine and selenium, deficiencies of both of which occur in many areas of the world). When

TABLE 44–2 Classification of Minerals According to Their Function

Function	Mineral
Structural function	Calcium, magnesium, phosphate
Involved in membrane function	Sodium, potassium
Function as prosthetic groups in enzymes	Cobalt, copper, iron, molybdenum, selenium, zinc
Regulatory role or role in hormone action	Calcium, chromium, iodine, magnesium, manganese, sodium, potassium
Known to be essential, but function unknown	Silicon, vanadium, nickel, tin
Have effects in the body, but essentiality is not established	Fluoride, lithium
May occur in foods and known to be toxic in excess	Aluminum, arsenic, antimony, boron, bromine, cadmium, cesium, germanium, lead, mercury, silver, strontium

foods come from a variety of regions, mineral deficiency is less likely to occur. Iron deficiency is an important problem worldwide, because if iron losses from the body are relatively high (eg, from heavy menstrual blood loss or intestinal parasites), it is difficult to achieve an adequate intake to replace losses. However, 10% of the population (and more in some areas) are genetically at risk of iron overload, leading to formation of free radicals as a result of nonenzymic reactions of iron ions in free solution when the capacity of iron-binding proteins has been exceeded. Foods grown on soil containing high levels of selenium cause toxicity, and excessive intakes of sodium cause hypertension in susceptible people.

SUMMARY

- Vitamins are organic nutrients with essential metabolic functions that are required in small amounts in the diet because they cannot be synthesized by the body. The lipid-soluble vitamins (A, D, E, and K) are hydrophobic molecules requiring normal fat absorption for their absorption.

- Vitamin A (retinol), present in meat, and the provitamin (β-carotene), found in plants, form retinaldehyde, utilized in vision, and retinoic acid, which acts in the control of gene expression.

- Vitamin D is a steroid prohormone yielding the active hormone calcitriol, which regulates calcium and phosphate metabolism; deficiency leads to rickets and osteomalacia. It has a role in controlling cell differentiation and insulin secretion.

- Vitamin E (tocopherol) is the most important lipid-soluble antioxidant in the body, acting in the lipid phase of membranes protecting against the effects of free radicals.

- Vitamin K functions as the cofactor of a carboxylase that acts on glutamate residues in the protein precursors of clotting factors and as well as osteocalcin and matrix 1a protein in bone, enabling them to chelate calcium.

- The water-soluble vitamins act as enzyme cofactors. Thiamin is a cofactor in oxidative decarboxylation of α-keto acids and of transketolase in the pentose phosphate pathway. Riboflavin and niacin are important cofactors in oxidoreduction reactions, present in flavoprotein enzymes and in NAD and NADP, respectively.

- Pantothenic acid is present in coenzyme A and acyl carrier protein, which act as carriers of acyl groups in metabolic reactions.

- Vitamin B_6 as pyridoxal phosphate is the coenzyme for several enzymes of amino acid metabolism, including the transaminases, and of glycogen phosphorylase. Biotin is the coenzyme for several carboxylase enzymes.

- Vitamin B_{12} and folate provide one-carbon residues for DNA synthesis and other reactions; deficiency results in megaloblastic anemia.

- Vitamin C is a water-soluble antioxidant that maintains vitamin E and many metal cofactors in the reduced state.

- Inorganic mineral elements that have a function in the body must be provided in the diet. When intake is insufficient, deficiency may develop, and excessive intakes may be toxic.

Free Radicals & Antioxidant Nutrients

David A. Bender, PhD

OBJECTIVES

After studying this chapter, you should be able to:

- Describe the damage caused to DNA, lipids, and proteins by free radicals, and the diseases associated with radical damage.
- Describe the main sources of oxygen radicals in the body.
- Describe the mechanisms and dietary factors that protect against radical damage.
- Explain how antioxidants can also act as pro-oxidants, and why intervention trials of antioxidant nutrients have generally yielded disappointing results.

BIOMEDICAL IMPORTANCE

Free radicals are formed in the body under normal conditions. They cause damage to nucleic acids, proteins, and lipids in cell membranes and plasma lipoproteins. This can cause cancer, atherosclerosis and coronary artery disease, and autoimmune diseases. Epidemiological and laboratory studies have identified a number of protective antioxidant nutrients: selenium, vitamins C and E, β-carotene, and other carotenoids, and a variety of polyphenolic compounds derived from plant foods. Many people take supplements of one or more antioxidant nutrients. However, intervention trials show little benefit of antioxidant supplements except among people who were initially deficient, and many trials of β-carotene and vitamin E have shown increased mortality among those taking the supplements.

Free Radical Reactions Are Self-Perpetuating Chain Reactions

Free radicals are highly reactive molecular species with an unpaired electron; they persist for only a very short time (of the order of 10^{-9} to 10^{-12} seconds) before they collide with another molecule and either abstract or donate an electron in order to achieve stability. In so doing, they generate a new radical from the molecule with which they collided. The main way in which a free radical can be quenched, so terminating this chain reaction, is if two radicals react together, when the unpaired electrons can become paired in one or other of the parent molecules. This is a rare occurrence, because of the very short half-life of an individual radical and the very low concentrations of radicals in tissues.

The most damaging radicals in biological systems are oxygen radicals (sometimes called reactive oxygen species)—especially superoxide, $^{\bullet}O_2^-$, hydroxyl, $^{\bullet}OH$, and perhydroxyl, $^{\bullet}O_2H$. Tissue damage caused by oxygen radicals is often called oxidative damage, and factors that protect against oxygen radical damage are known as antioxidants.

Radicals Can Damage DNA, Lipids, & Proteins

Interaction of radicals with bases in DNA can lead to chemical changes that, if not repaired (see Chapter 35), may be inherited in daughter cells. Radical damage to unsaturated fatty acids in cell membranes and plasma lipoproteins leads to the formation of lipid peroxides, precursors of highly reactive dialdehydes that can chemically modify proteins and nucleic acid bases. Proteins are also subject to direct chemical modification by interaction with radicals. Oxidative damage to tyrosine residues in proteins can lead to the formation of dihydroxyphenylalanine that can undergo nonenzymic reactions leading to further formation of oxygen radicals (**Figure 45–1**).

The total body radical burden can be estimated by measuring the products of lipid peroxidation. Lipid peroxides can be measured by the ferrous oxidation in xylenol orange (FOX) assay. Under acidic conditions, they oxidize Fe^{2+} to Fe^{3+}, which forms a chromophore with xylenol orange. The dialdehydes formed from lipid peroxides can be measured by reaction with thiobarbituric acid, with which they form a red fluorescent adduct—the results of this assay are generally reported as total thiobarbituric acid reactive substances, TBARS. Peroxidation of n-6 polyunsaturated fatty acids leads to the formation of

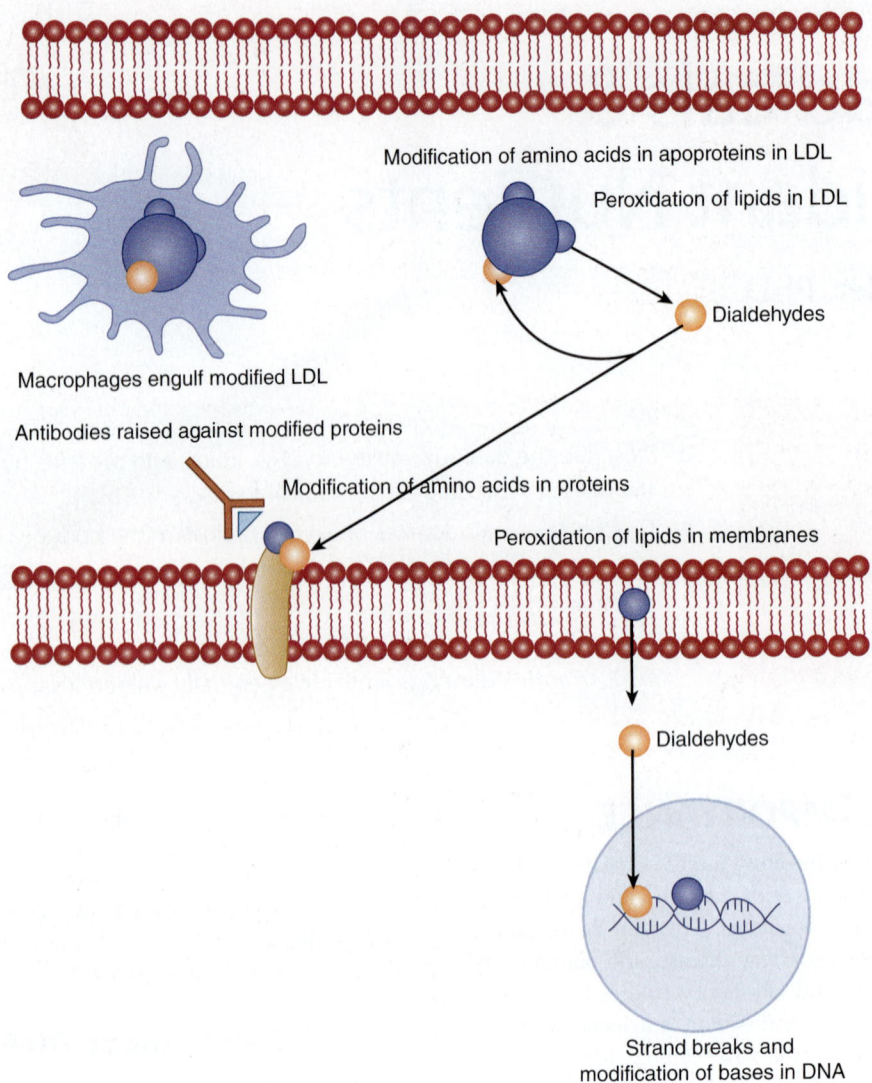

FIGURE 45–1 **Tissue damage by radicals.**

pentane, and of n-3 polyunsaturated fatty acids to ethane, both of which can be measured in exhaled air.

Radical Damage May Cause Mutations, Cancer, Autoimmune Disease, & Atherosclerosis

Radical damage to the DNA of germline cells in ovaries and testes can lead to heritable mutations; in somatic cells, the result may be initiation of cancer. The dialdehydes formed as a result of radical-induced lipid peroxidation in cell membranes can also modify bases in DNA.

Chemical modification of amino acids in proteins, either by direct radical action or as a result of reaction with the products of radical-induced lipid peroxidation, leads to proteins that are recognized as nonself by the immune system. The resultant antibodies may also cross-react with normal tissue proteins, so initiating autoimmune disease.

Chemical modification of the proteins or lipids in plasma low-density lipoprotein (LDL) leads to abnormal LDL that is not recognized by the liver LDL receptors, and so is not cleared by the liver. Instead, the modified LDL is taken up by macrophage scavenger receptors. Lipid-engorged macrophages infiltrate under blood vessel endothelium (especially when there is already some damage to the endothelium), where they die as a consequence of the toxic levels of unesterified cholesterol they have accumulated. This leads to the development of atherosclerotic plaques, which, in extreme cases, can more or less completely occlude a blood vessel.

There Are Multiple Sources of Oxygen Radicals in the Body

Ionizing radiation (x-rays and UV) can lyse water, leading to the formation of hydroxyl radicals. Transition metal ions, including Cu^+, Co^{2+}, Ni^{2+}, and Fe^{2+} can react nonenzymically with oxygen or hydrogen peroxide, again leading to the

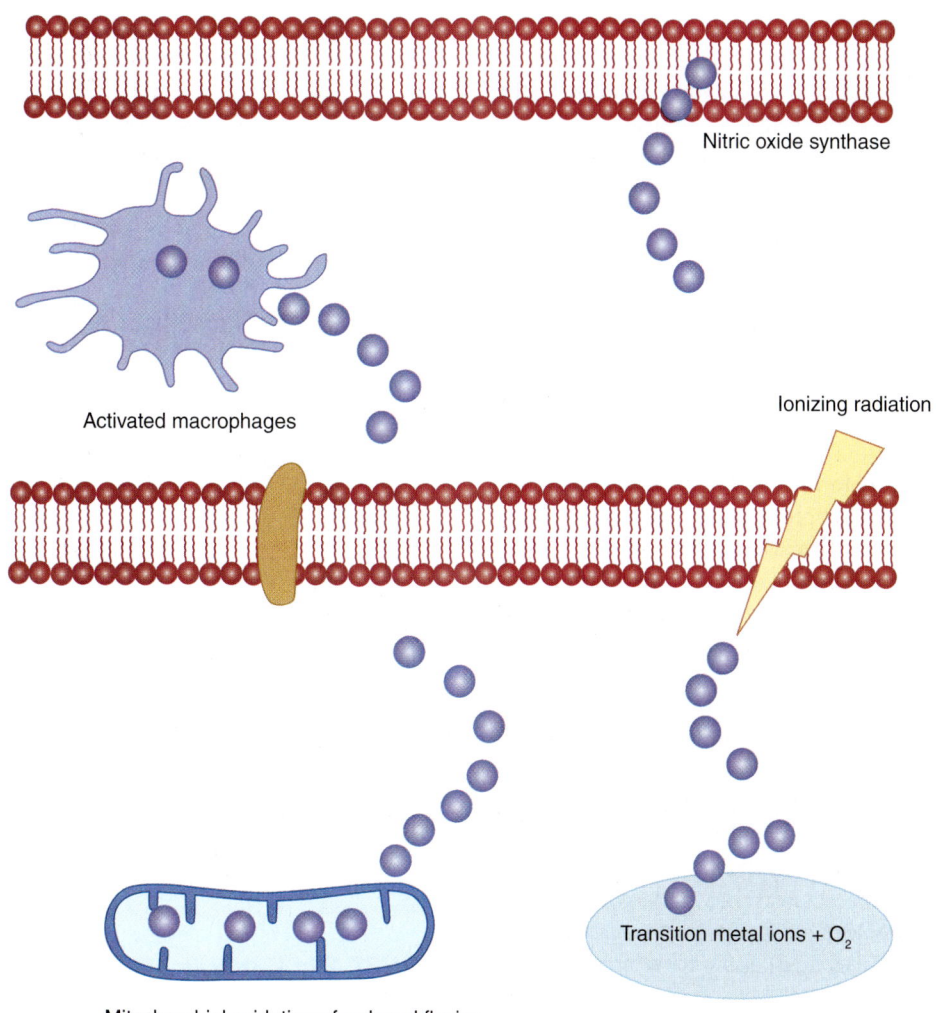

FIGURE 45–2 Sources of radicals.

formation of hydroxyl radicals. Nitric oxide (an important compound in cell signaling, originally described as the endothelium-derived relaxation factor) is itself a radical, and, more importantly, can react with superoxide to yield peroxynitrite, which decays to form hydroxyl radicals (**Figure 45–2**).

The respiratory burst of activated macrophages (see Chapter 53) is characterized by the increased utilization of glucose via the pentose phosphate pathway (see Chapter 20) to reduce $NADP^+$ to NADPH, and increased utilization of oxygen to oxidize NADPH to produce cytotoxic oxygen (and halogen) radicals that kill phagocytosed microorganisms. The respiratory burst oxidase (NADPH oxidase) is a flavoprotein that reduces oxygen to superoxide:

$$NADPH + 2O_2 \rightarrow NADP^+ + 2{}^{\bullet}O_2^- + 2H^+$$

Plasma markers of radical damage to lipids increase considerably in response to even a mild infection.

The oxidation of reduced flavin coenzymes in the mitochondrial (see Chapter 13) and microsomal electron transport chains proceeds through a series of steps in which the flavin semiquinone radical is stabilized by the protein to which it is bound, and forms oxygen radicals as transient intermediates. Because of the unpredictable nature of these radical intermediates, there is considerable "leakage" of radicals, and some 3 to 5% of the daily consumption of 30 mol of oxygen by an adult human being is converted to singlet oxygen, hydrogen peroxide, and superoxide, perhydroxyl, and hydroxyl radicals, rather than undergoing complete reduction to water. This results in daily production of about 1.5 mol of reactive oxygen species.

There Are Various Mechanisms of Protection Against Radical Damage

The metal ions that undergo nonenzymic reactions to form oxygen radicals are not normally free in solution, but are bound to either the proteins for which they provide a prosthetic group, or to specific transport and storage proteins, so that they are unreactive. Iron is bound to transferrin, ferritin, and hemosiderin, copper to ceruloplasmin, and other metal ions are bound to metallothionein. This binding to transport proteins that are too large to be filtered in the kidneys also prevents loss of metal ions in the urine.

Superoxide is produced both accidentally and also as the reactive oxygen species required for a number of enzyme-catalyzed reactions. A family of superoxide dismutases catalyze the reaction between superoxide and protons to yield oxygen and hydrogen peroxide:

$$\cdot O_2^- + 2H^+ \rightarrow H_2O_2$$

The hydrogen peroxide is then removed by catalase and various peroxidases: $2H_2O_2 \rightarrow 2H_2O + O_2$. Most enzymes that produce and require superoxide are contained in the peroxisomes, together with superoxide dismutase, catalase, and peroxidases.

The peroxides that are formed by radical damage to lipids in membranes and plasma lipoproteins are reduced to hydroxy fatty acids by glutathione peroxidase, a selenium-dependent enzyme (hence the importance of adequate selenium intake to maximize antioxidant activity), and oxidized glutathione, which is reduced by NADPH-dependent glutathione reductase (see Figure 20–3). Lipid peroxides are also reduced to fatty acids by reaction with vitamin E, forming the tocopheroxyl radical, which is relatively stable, since the unpaired electron can be located in any one of three positions in the molecule (**Figure 45–3**). The tocopheroxyl radical persists long enough to undergo reduction back to tocopherol by reaction with vitamin C at the surface of the cell or lipoprotein. The resultant monodehydroascorbate radical then undergoes enzymic reduction back to ascorbate or a nonenzymic reaction of 2 mol of monodehydroascorbate to yield 1 mol each of ascorbate and dehydroascorbate.

Ascorbate, uric acid, and a variety of polyphenols derived from plant foods act as water-soluble radical-trapping antioxidants, forming relatively stable radicals that persist long enough to undergo reaction to nonradical products. Ubiquinone and carotenes similarly act as lipid-soluble radical-trapping antioxidants in membranes and plasma lipoproteins.

The Antioxidant Paradox— Antioxidants Can Also Be Pro-Oxidants

Although ascorbate is an antioxidant, reacting with superoxide and hydroxyl to yield monodehydroascorbate and hydrogen peroxide or water, it can also be a source of superoxide radicals by reaction with oxygen, and hydroxyl radicals by reaction with Cu^{2+} ions (**Table 45–1**). However, these pro-oxidant actions require relatively high concentrations of ascorbate, which are unlikely to be reached in tissues, since once the plasma concentration of ascorbate reaches about 30 mmol/L, the renal threshold is reached, and at intakes above about 100 to 120 mg/d the vitamin is excreted in the urine quantitatively with intake.

A considerable body of epidemiological evidence suggests that carotene is protective against lung and other cancers. However, two major intervention trials in the 1990s showed an

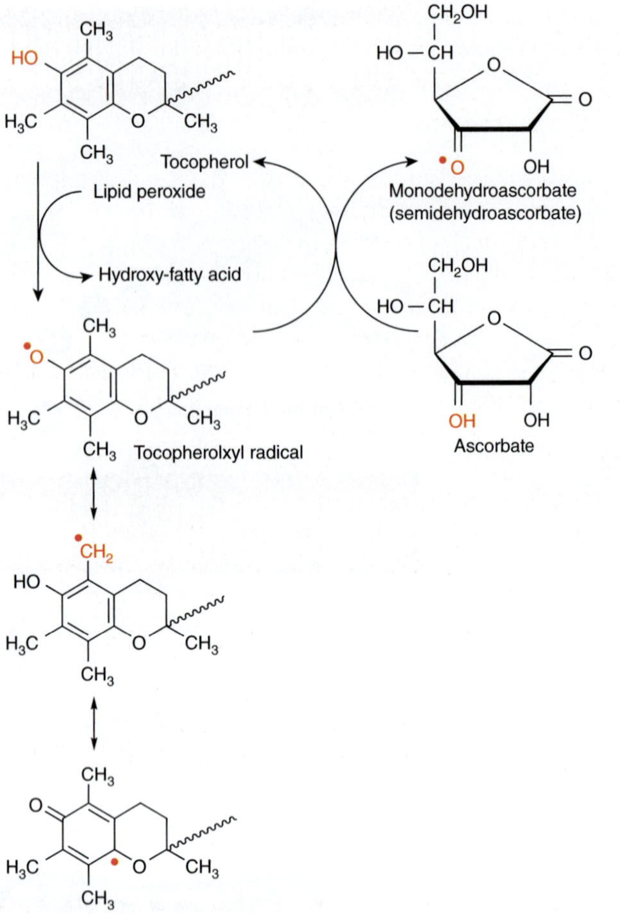

FIGURE 45–3 The roles of vitamins E and C in reducing lipid peroxides, and stabilization of the tocopheroxyl radical by delocalization of the unpaired electron.

increase in death from lung (and other) cancer among people who were given supplements of β-carotene. The problem is that although β-carotene is indeed a radical-trapping antioxidant under conditions of low partial pressure of oxygen, as in most tissues, at high partial pressures of oxygen (as in the lungs) and especially in high concentrations, β-carotene is an autocatalytic pro-oxidant, and hence can initiate radical damage to lipids and proteins.

TABLE 45–1 Antioxidant and Pro-Oxidant Roles of Vitamin C

Antioxidant roles:
Ascorbate + $\cdot O_2^-$ → H_2O_2 + monodehydroascorbate; catalase and peroxidases catalyze the reaction: $2H_2O_2 \rightarrow 2H_2O + O_2$
Ascorbate + $\cdot OH$ → H_2O + monodehydroascorbate
Pro-oxidant roles:
Ascorbate + O_2 → $\cdot O_2^-$ + monodehydroascorbate
Ascorbate + Cu^{2+} → Cu^+ + monodehydroascorbate
$Cu^+ + H_2O_2 \rightarrow Cu^{2+} + OH^- + \cdot OH$

Epidemiological evidence also suggests that vitamin E is protective against atherosclerosis and cardiovascular disease. However, meta-analysis of intervention trials with vitamin E shows increased mortality among those taking (high dose) supplements. These trials have all used α-tocopherol, and it is possible that the other vitamers of vitamin E that are present in foods, but not the supplements, may be important. In vitro, plasma lipoproteins form lower levels of cholesterol ester hydroperoxide when incubated with sources of low concentrations of perhydroxyl radicals when vitamin E has been removed than when it is present. The problem seems to be that vitamin E acts as an antioxidant by forming a stable radical that persists long enough to undergo metabolism to nonradical products. This means that the radical also persists long enough to penetrate deeper in to the lipoprotein, causing further radical damage, rather than interacting with a water-soluble antioxidant at the surface of the lipoprotein.

Nitric oxide and other radicals are important in cell signaling, and especially in signaling for programmed cell death (apoptosis) of cells that have suffered DNA and other damage. It is likely that high concentrations of antioxidants, rather than protecting against tissue damage, may quench the signaling radicals, and so permit the continued survival of damaged cells, so increasing, rather than decreasing, the risk of cancer development.

SUMMARY

- Free radicals are highly reactive molecular species that contain an unpaired electron. They can react with, and modify, proteins, nucleic acids, and fatty acids in cell membranes and plasma lipoproteins.

- Radical damage to lipids and proteins in plasma lipoproteins is a factor in the development of atherosclerosis and coronary artery disease; radical damage to nucleic acids may induce heritable mutations and cancer; radical damage to proteins may lead to the development of autoimmune diseases.

- Oxygen radicals arise as a result of exposure to ionizing radiation, nonenzymic reactions of transition metal ions, the respiratory burst of activated macrophages, and the normal oxidation of reduced flavin coenzymes.

- Protection against radical damage is afforded by enzymes that remove superoxide ions and hydrogen peroxide, enzymic reduction of lipid peroxides linked to oxidation of glutathione, nonenzymic reaction of lipid peroxides with vitamin E, and reaction of radicals with compounds such as vitamins C and E, carotene, ubiquinone, uric acid, and dietary polyphenols that form relatively stable radicals that persist long enough to undergo reaction to nonradical products.

- Except in people who were initially deficient, intervention trials of vitamin E and β-carotene have generally shown increased mortality among those taking the supplements. β-Carotene is only an antioxidant at low concentrations of oxygen; at higher concentrations of oxygen it is an autocatalytic pro-oxidant. Vitamin E forms a stable radical that is capable of either undergoing reaction with water-soluble antioxidants or penetrating further into lipoproteins and tissues, so increasing radical damage.

- Radicals are important in cell signaling. However, it is likely that high concentrations of antioxidants, so far from protecting against tissue damage, may quench the signaling radicals, and so permit the continued survival of damaged cells, so increasing, rather than decreasing, the risk of cancer development.

Glycoproteins

David A. Bender, PhD

OBJECTIVES	
After studying this chapter, you should be able to:	▪ Explain the importance of glycoproteins in health and disease.
	▪ Describe the principal sugars found in glycoproteins.
	▪ Describe the major classes of glycoproteins (*N*-linked, *O*-linked, and GPI-linked).
	▪ Describe the major features of the pathways of biosynthesis and degradation of glycoproteins.
	▪ Explain how many microorganisms, such as influenza virus, attach to cell surfaces via sugar chains.

BIOMEDICAL IMPORTANCE

Glycoproteins are proteins that contain oligosaccharide chains (glycans) covalently bound to amino acids; **glycosylation** (the enzymic attachment of sugars) is the most frequent posttranslational modification of proteins. Many proteins also undergo reversible glycosylation with a single sugar (*N*-acetylglucosamine) bound to a serine or threonine residue that is also a site for reversible phosphorylation. This is an important mechanism of metabolic regulation. Nonenzymic attachment of sugars to proteins can also occur, and is referred to as **glycation**. This process can have serious pathologic consequences (eg, in poorly controlled diabetes mellitus).

Glycoproteins are one class of **glycoconjugate** or **complex carbohydrate**—molecules containing one or more carbohydrate chains covalently linked to protein (to form **glycoproteins** or **proteoglycans**; see Chapter 50) or lipid (to form **glycolipids**; see Chapter 21). Almost all **plasma proteins**, and many **peptide hormones**, are glycoproteins, as are a number of **blood group substances** (others are glycosphingolipids). Many **cell membrane proteins** (see Chapter 40) contain substantial amounts of carbohydrate, and many are anchored to the lipid bilayer by a glycan chain. Evidence is accumulating that alterations in the structures of glycoproteins and other glycoconjugates on the surface of cancer cells are important in metastasis.

GLYCOPROTEINS OCCUR WIDELY & PERFORM NUMEROUS FUNCTIONS

Glycoproteins occur in most organisms, from bacteria to human beings. Many viruses also contain glycoproteins, some of which play key roles in viral attachment to host cells. Glycoproteins have a wide range of functions (**Table 46–1**); their carbohydrate content ranges from 1 to over 85% by weight. The glycan structures of glycoproteins change in response to signals involved in cell differentiation, normal physiology, and neoplastic transformation. This is the result of different expression patterns of glycosyltransferases. **Table 46–2** lists some of the major functions of the glycan chains of glycoproteins.

OLIGOSACCHARIDE CHAINS ENCODE BIOLOGICAL INFORMATION

The biological information contained in the sequence and linkages of sugars in glycans differs from that in DNA, RNA, and proteins in one important respect; it is secondary rather than primary information. The pattern of glycosylation of a given protein depends on the pattern of expression of the various **glycosyltransferases** in the cells that are involved in glycoprotein synthesis, the affinity of the different glycosyltransferases for their carbohydrate substrates, and the relative availability of the different carbohydrate substrates. Because of this most glycoproteins display **microheterogeneity** something that complicates their analysis. Not all of the glycan chains of a given glycoprotein are complete; some are truncated.

The information from the sugars is expressed via interactions between the glycans and proteins such as **lectins** (see following text) or other molecules. These interactions lead to changes in cellular activity.

TABLE 46–1 Some Functions Served by Glycoproteins

Function	Glycoproteins
Structural molecule	Collagens
Lubricant and protective agent	Mucins
Transport molecule	Transferrin, ceruloplasmin
Immunological molecule	Immunoglobulins, histocompatibility antigens
Hormone	Chorionic gonadotropin, thyroid-stimulating hormone (TSH)
Enzyme	Various, eg, alkaline phosphatase
Cell attachment–recognition site	Various proteins involved in cell–cell (eg, sperm-oocyte), virus-cell, bacterium-cell, and hormone-cell interactions
Antifreeze	Plasma proteins of cold-water fish
Interact with specific carbohydrates	Lectins, selectins (cell adhesion lectins), antibodies
Receptor	Cell surface proteins involved in hormone and drug action
Regulate folding of proteins that are exported from the cell	Calnexin, calreticulin
Regulation of differentiation and development	Notch and its analogs, key proteins in development
Hemostasis (and thrombosis)	Specific glycoproteins on the surface membranes of platelets

EIGHT SUGARS PREDOMINATE IN HUMAN GLYCOPROTEINS

Only eight monosaccharides are commonly found in glycoproteins (**Table 46–3** and Chapter 15). *N*-acetylneuraminic acid (NeuAc) is usually found at the termini of oligosaccharide chains, attached to subterminal galactose (Gal) or

TABLE 46–2 Some Functions of the Oligosaccharide Chains of Glycoproteins

- Change physicochemical properties of the protein such as solubility, viscosity, charge, conformation, denaturation.
- Provide binding sites for various molecules, as well as bacteria, viruses, and some parasites.
- Provide cell surface recognition signals.
- Protect against proteolysis.
- Ensure correct folding of proteins that are exported from the cell and target incorrectly folded proteins for transport from the endoplasmic reticulum back to the cytosol for catabolism.
- Protect peptide hormones and other plasma proteins against clearance by the liver.
- Permit anchoring of extracellular proteins in the cell membrane, and of intracellular proteins inside subcellular organelles such as endoplasmic reticulum and Golgi.
- Direct intracellular migration, sorting and secretion of proteins.
- Affect embryonic development and cell and tissue differentiation.
- May affect sites of metastases selected by cancer cells.

N-acetylgalactosamine (GalNAc) residues. The other sugars are usually found in more internal positions. **Sulfate** is often found in glycoproteins, usually attached to galactose, *N*-acetylglucosamine, or *N*-acetylgalactosamine.

As in most biosynthetic reactions, it is not the free or phosphorylated sugar that is the substrate for glycoprotein synthesis, but the corresponding **sugar nucleotide** (see Figure 18–2); some contain UDP and others guanosine diphosphate (GDP) or cytidine monophosphate (CMP).

LECTINS CAN BE USED TO PURIFY GLYCOPROTEINS & TO INVESTIGATE THEIR FUNCTIONS

Lectins are **carbohydrate-binding proteins** that agglutinate cells or precipitate glycoconjugates; a number of lectins are themselves glycoproteins. Immunoglobulins that react with sugars are not considered to be lectins. Lectins contain at least two sugar-binding sites; proteins with only a single sugar-binding site will not agglutinate cells or precipitate glycoconjugates.

Lectins were first discovered in plants and microorganisms, but many lectins of animal origin are now known, including the mammalian **asialoglycoprotein receptor**. Many peptide hormones, and most plasma proteins are glycoproteins. Treatment of the protein with neuraminidase removes the terminal *N*-acetylneuraminic acid moiety, exposing the subterminal galactose residue. This asialoglycoprotein is cleared from the circulation very much faster than the intact glycoprotein. Liver cells contain an asialoglycoprotein receptor that recognizes the galactose moiety of many desialylated plasma proteins, leading to their endocytosis and catabolism.

Plant lectins were formerly called **phytohemagglutinins**, because of their ability to agglutinate red blood cells by reacting with the cell surface glycoproteins. Undenatured lectins in undercooked legumes can lead to stripping of the intestinal mucosa by agglutinating mucosal cells.

Lectins are used to purify glycoproteins, as tools for probing the glycoprotein profiles of cell surfaces, and as reagents for generating mutant cells deficient in certain enzymes involved in the biosynthesis of oligosaccharide chains.

THERE ARE THREE MAJOR CLASSES OF GLYCOPROTEINS

Glycoproteins can be divided into three main groups, based on the nature of the linkage between the polypeptide and oligosaccharide chains (**Figure 46–1**); there are other minor classes of glycoprotein:

1. Those containing an **O-glycosidic linkage** (*O*-linked), involving the hydroxyl side chain of serine or threonine (and sometimes also tyrosine) and a sugar such as *N*-acetylgalactosamine (GalNAc-Ser[Thr]).

2. Those containing an **N-glycosidic linkage** (*N*-linked), involving the amide nitrogen of asparagine and *N*-acetylglucosamine (GlcNAc-Asn).

TABLE 46–3 **The Principal Sugars Found in Human Glycoproteins[a]**

Sugar	Type	Abbreviation	Sugar Nucleotide	Comments
Galactose	Hexose	Gal	UDP-Gal	Often found subterminal to NeuAc in N-linked glycoproteins. Also, found in the core trisaccharide of proteoglycans.
Glucose	Hexose	Glc	UDP-Glc	Present during the biosynthesis of N-linked glycoproteins but not usually present in mature glycoproteins. Present in some clotting factors.
Mannose	Hexose	Man	GDP-Man	Common sugar in N-linked glycoproteins.
N-Acetylneuraminic acid	Sialic acid (nine C atoms)	NeuAc	CMP-NeuAc	Often the terminal sugar in both N- and O-linked glycoproteins. Other types of sialic acid are also found, but NeuAc is the major species found in humans. Acetyl groups may also occur as O-acetyl species as well as N-acetyl.
Fucose	Deoxyhexose	Fuc	GDP-Fuc	May be external in both N- and O-linked glycoproteins or internal, linked to the GlcNAc residue attached to Asn in N-linked species. Can also occur internally attached to the OH of Ser (eg, in t-PA and certain clotting factors).
N-Acetylgalactosamine	Aminohexose	GalNAc	UDP-GalNAc	Present in both N- and O-linked glycoproteins.
N-Acetylglucosamine	Aminohexose	GlcNAc	UDP-GlcNAc	The sugar attached to the polypeptide chain via Asn in N-linked glycoproteins; also found at other sites in the oligosaccharides of these proteins. Many nuclear proteins have GlcNAc attached to the OH of Ser or Thr as a single sugar.
Xylose	Pentose	Xyl	UDP-Xyl	Xyl is attached to the OH of Ser in many proteoglycans. Xyl in turn is attached to two Gal residues, forming a link trisaccharide. Xyl is also found in t-PA and certain clotting factors.

[a]The structures of these sugars are illustrated in Chapter 15.

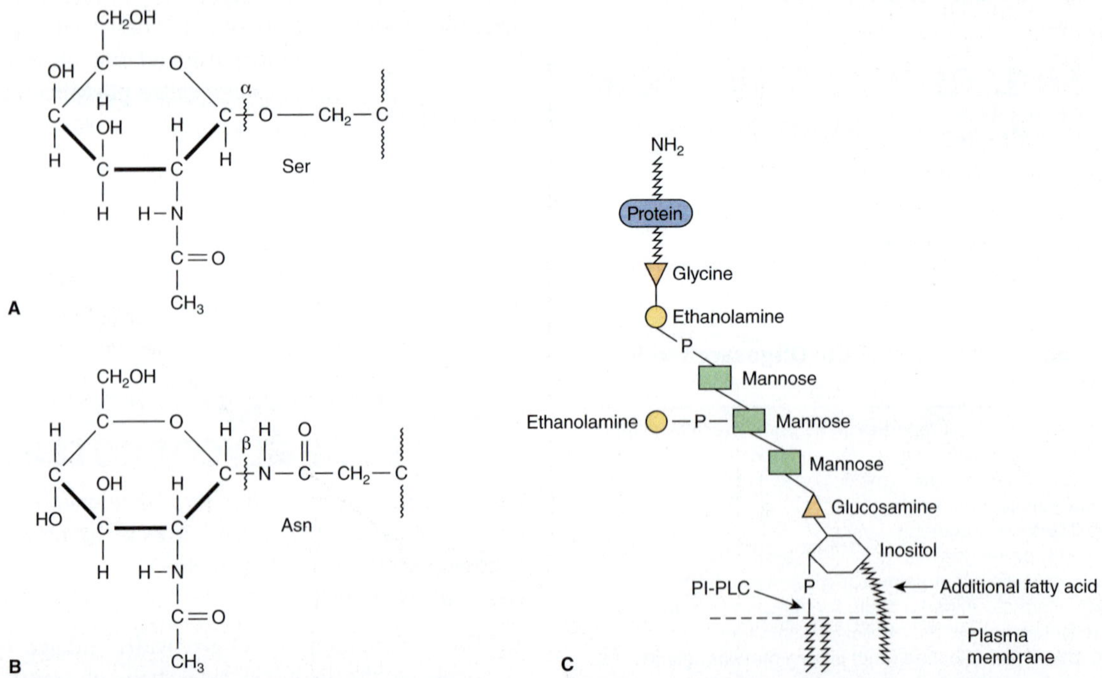

FIGURE 46–1 **The three main types of glycoprotein: (A) an O-linkage (N-acetylgalactosamine to serine), (B) an N-linkage (N-acetylglucosamine to asparagine), and (C) a glycosylphosphatidylinositol (GPI) linkage.** The GPI structure shown is that linking acetylcholinesterase to the plasma membrane of the human red blood cell. The site of action of PI-phospholipase C (PI-PLC), which releases the enzyme from membrane binding is indicated. This particular GPI contains an extra fatty acid attached to inositol and also an extra phosphoryl-ethanolamine moiety attached to the central of the mannose residue. Variations found among different GPI structures include the identity of the carboxyl-terminal amino acid, the molecules attached to the mannose residues, and the precise nature of the lipid moiety.

3. Those linked to the carboxyl-terminal amino acid of a protein via a phosphorylethanolamine moiety joined to an oligosaccharide (glycan), which in turn is linked via glucosamine to phosphatidylinositol (PI). These are **glycosylphosphatidylinositol-anchored (GPI-anchored)** glycoproteins. Among other functions, they are involved in directing glycoproteins to the apical or basolateral areas of the plasma membrane of polarized epithelial cells (see Chapter 40 and below).

The number of oligosaccharide chains attached to one protein can vary from 1 to 30 or more, with the sugar chains ranging from one or two residues in length to much larger structures. The glycan chain may be linear or branched. Many proteins contain more than one type of sugar chain; for instance, **glycophorin**, an important red cell membrane glycoprotein (see Chapter 53), contains both O- and N-linked oligosaccharides.

GLYCOPROTEINS CONTAIN SEVERAL TYPES OF O-GLYCOSIDIC LINKAGES

At least four subclasses of O-glycosidic linkages are found in human glycoproteins:

1. The **N-acetylgalactosamine-Ser(Thr)** linkage shown in Figure 46–1 is the predominant linkage. Usually a galactose or an N-acetylneuraminic acid residue is attached to the N-acetylgalactosamine, but many variations in the sugar compositions and lengths of such oligosaccharide chains are found. This type of linkage is found in **mucins** (see below).

2. **Proteoglycans** contain a galactose-galactose-xylose trisaccharide (the so-called link trisaccharide) attached to serine.

3. **Collagens** (see Chapter 50) contain a **galactose-hydroxylysine** linkage.

4. Many **nuclear and cytosolic proteins** contain side chains consisting of a single N-acetylglucosamine attached to a serine or threonine residue.

Mucins Have a High Content of O-Linked Oligosaccharides & Exhibit Repeating Amino Acid Sequences

Mucins are glycoproteins that are highly resistant to proteolysis because the density of oligosaccharide chains makes it difficult for **proteases** to access the polypeptide chain. They help to **lubricate** and form a **protective physical barrier** on epithelial surfaces.

Mucins have two distinctive characteristics: a high content of **O-linked oligosaccharides** (the carbohydrate content is generally more than 50%); and the presence of **variable numbers of tandem repeats (VNTRs)** of peptide sequence in the center of the polypeptide chain, to which the O-glycan chains are attached in clusters. These tandem repeat sequences are rich in serine, threonine, and proline; indeed, up to 60% of the dietary requirement for threonine can be accounted for by the synthesis of mucins. Although O-glycans predominate, mucins often also contain a number of N-glycan chains.

Both **secretory** and **membrane-bound** mucins occur. **Mucus** secreted by the gastrointestinal, respiratory, and reproductive tracts is a solution containing about 5% mucins. Secretory mucins generally have an oligomeric structure, with monomers linked by disulfide bonds, and hence a very high molecular mass. Mucus has a high viscosity and often forms a gel because of its content of mucins. The high content of O-glycans confers an extended structure. This is partly explained by steric interactions between the N-acetylgalactosamine moieties and adjacent amino acids, resulting in a chain-stiffening effect, so that the conformation of mucins often become rigid rods. Intermolecular noncovalent interactions between sugars on neighboring glycan chains contribute to gel formation. The high content of **N-acetylneuraminic acid** and **sulfate** residues found in many mucins gives them a negative charge.

Membrane-bound mucins participate in **cell–cell interactions**. They may also mask cell surface antigens. Many cancer cells form large amounts of mucins that mask surface antigens and protect the cancer cells from immune surveillance. Mucins also carry cancer-specific peptide and carbohydrate epitopes. Some of these have been used to stimulate an immune response against cancer cells.

O-Linked Glycoproteins Are Synthesized by Sequential Addition of Sugars from Sugar Nucleotides

Because most glycoproteins are membrane-bound or secreted, their mRNA is usually translated on membrane-bound polyribosomes (see Chapter 37). The glycan chains are built up by the sequential donation of sugars from sugar nucleotides, catalyzed by **glycoprotein glycosyltransferases**. Because many glycosylation reactions occur within the lumen of the Golgi apparatus, there are **carrier systems** (permeases and transporters) to transport sugar nucleotides (UDP-galactose, GDP-mannose, and CMP-N-acetylneuraminic acid) across the Golgi membrane. These are **antiporter** systems in which the influx of one molecule of sugar nucleotide is balanced by the efflux of one molecule of the corresponding nucleotide (UMP, GMP, or CMP).

Humans possess 41 different types of glycoprotein glycosyltransferases. Families of glycosyltransferases are named for the sugar nucleotide donor, and subfamilies on the basis of the linkage formed between the sugar and the acceptor substrate; transfer may occur with retention or inversion of the conformation at C-1 of the sugar. Binding of the sugar nucleotide to the enzyme causes a conformational change in the enzyme that permits binding of the acceptor substrate. Glycosyltransferases show a high degree of specificity for the acceptor substrate, typically acting only on the product of the preceding reaction. The different stages in glycan formation, and hence the different glycosyltransferases, are located in different regions of the Golgi, which provides for spatial separation of the steps in the process. Not all of the glycan chains of a given glycoprotein are complete; some are truncated, leading to microheterogeneity. No consensus sequence is known to determine which serine and threonine residues are glycosylated, but the first sugar moiety incorporated is usually *N*-acetylgalactosamine.

N-LINKED GLYCOPROTEINS CONTAIN AN ASPARAGINE-*N*-ACETYLGLUCOSAMINE LINKAGE

N-Linked glycoproteins are the predominant class of glycoproteins, including both **membrane-bound** and **circulating** glycoproteins. They are distinguished by the presence of the asparagine—*N*-acetylglucosamine linkage (**Figure 46–1**). There are three major classes of *N*-linked oligosaccharides: **complex, high-mannose**, and **hybrid**. All three have the same core pentasaccharide, $Man_3GlcNAc_2$, bound to asparagine, but differ in their outer branches (**Figure 46–2**).

Complex oligosaccharides contain two, three, four, or five outer branches. The oligosaccharide branches are often referred to as **antennae**, so that bi-, tri-, tetra-, and penta-antennary structures may all be found. They generally contain terminal *N*-acetylneuraminic acid residues and underlying galactose and *N*-acetylglucosamine residues, the latter often

constituting the disaccharide *N*-acetyllactosamine. Repeating **N*-acetyllactosamine units—[Galβ1–3/4GlcNAcβ1–3]*n* (poly-*N*-acetyllactosaminoglycans)—are often found on *N*-linked glycan chains. I/i blood group substances belong to this class. A bewildering number of chains of the complex type exist, and that indicated in Figure 46–2 is only one of many. Other complex chains may terminate in galactose or fucose.

High-mannose oligosaccharides typically have two to six additional mannose residues linked to the pentasaccharide core. Hybrid molecules contain features of both of the other classes.

The Biosynthesis of *N*-Linked Glycoproteins Involves Dolichol-P-P-Oligosaccharide

The synthesis of all *N*-linked glycoproteins begins with the synthesis of a branched oligosaccharide attached to **dolichol pyrophosphate** (**Figure 46–3**) on the cytosolic side of the endoplasmic reticulum membrane, which is then translocated to the lumen of the endoplasmic reticulum, where it undergoes further glycosylation, before the oligosaccharide chain is transferred by an oligosaccharyltransferase onto an asparagine residue of the acceptor apoglycoprotein as it enters the endoplasmic reticulum during synthesis on membrane-bound polyribosomes. This is thus a cotranslational modification. In many of the *N*-linked glycoproteins there is a consensus

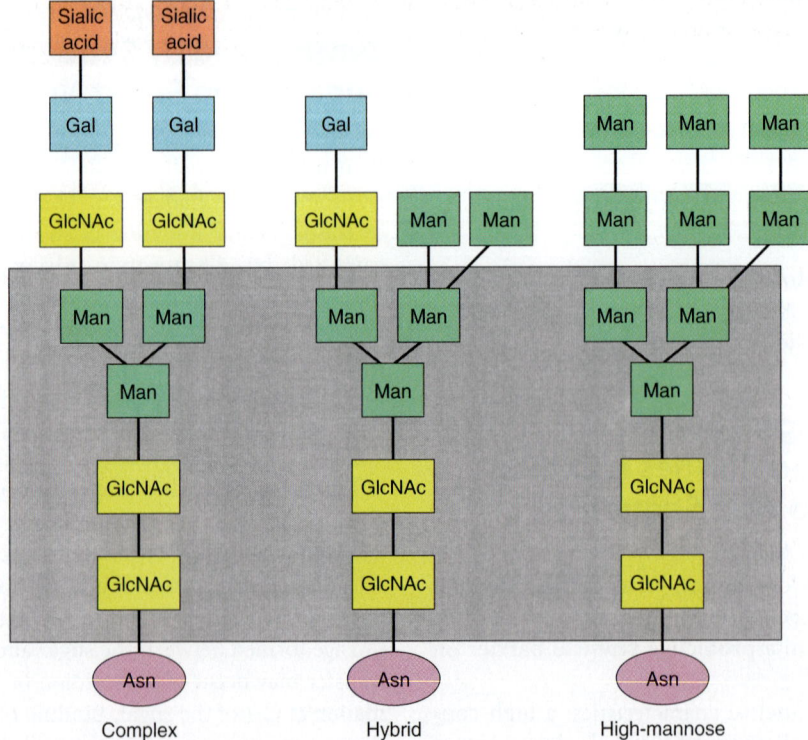

FIGURE 46–2 **Structures of the major types of asparagine-linked oligosaccharides.** The boxed area encloses the pentasaccharide core common to all *N*-linked glycoproteins.

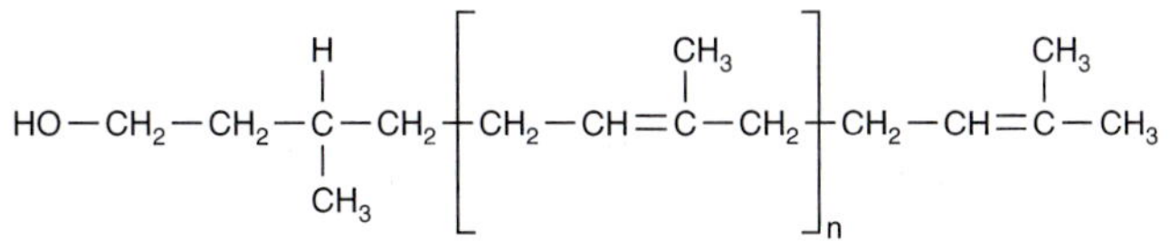

$$HO-CH_2-CH_2-\underset{\underset{CH_3}{|}}{\overset{\overset{H}{|}}{C}}-CH_2\left[CH_2-CH=\underset{CH_3}{\overset{CH_3}{C}}-CH_2\right]_n CH_2-CH=\underset{CH_3}{\overset{CH_3}{C}}-CH_3$$

FIGURE 46–3 **The structure of dolichol.** The group within the brackets is an isoprene unit (*n* = 17-20 isoprenoid units).

sequence of Asn-X-Ser/Thr (where X = any amino acid other than proline) that determines the site of glycosylation; in others there is no clear consensus sequence for glycosylation.

As shown in **Figure 46–4**, the first step is a reaction between UDP-*N*-acetylglucosamine and dolichol phosphate, forming *N*-acetylglucosamine dolichol pyrophosphate. A second *N*-acetylglucosamine is added from UDP-*N*-acetylglucosamine, followed by the addition of five molecules of mannose from GDP-mannose. The dolichol pyrophosphate oligosaccharide is then translocated into the lumen of the endoplasmic reticulum, and further mannose and glucose molecules are added, to form the final dolichol pyrophosphate oligosaccharide, using dolichol phosphate mannose and dolichol phosphate glucose as the donors. The dolichol pyrophosphate oligosaccharide is then transferred onto the acceptor asparagine residue of the nascent protein chain.

To form **high-mannose** chains, the glucose and some of the peripheral mannose residues are removed by glycosidases. To form an oligosaccharide chain of the **complex type**, the glucose residues and four of the mannose residues are removed by glycosidases in the endoplasmic reticulum and Golgi, then *N*-acetylglucosamine, galactose, and *N*-acetylneuraminic acid are added in reactions catalyzed by glycosyltransferases in the Golgi apparatus. **Hybrid chains** are formed by partial

processing, forming complex chains on one arm and mannose units on the other arm.

Glycoproteins & Calnexin Ensure Correct Folding of Proteins in the Endoplasmic Reticulum

Calnexin is a chaperone protein in the endoplasmic reticulum membrane; binding to calnexin prevents a glycoprotein from aggregating. It is a lectin, recognizing specific carbohydrate sequences in the glycan chain of the glycoprotein. Incorrectly folded glycoproteins undergo partial deglycosylation, and are targeted to undergo transport from the endoplasmic reticulum back to the cytosol for catabolism.

Calnexin binds to glycoproteins that possess a monoglycosylated core from which the terminal glucose residue has been removed, leaving only the innermost glucose attached. Calnexin and the bound glycoprotein form a complex with **ERp57**, a homolog of protein disulfide isomerase, which catalyzes disulfide bond interchange, facilitating correct folding. The bound glycoprotein is released from its complex with calnexin-ERp57 when the sole remaining glucose is hydrolyzed by a glucosidase and is then available for secretion if it is correctly folded. If not, a **glucosyltransferase**

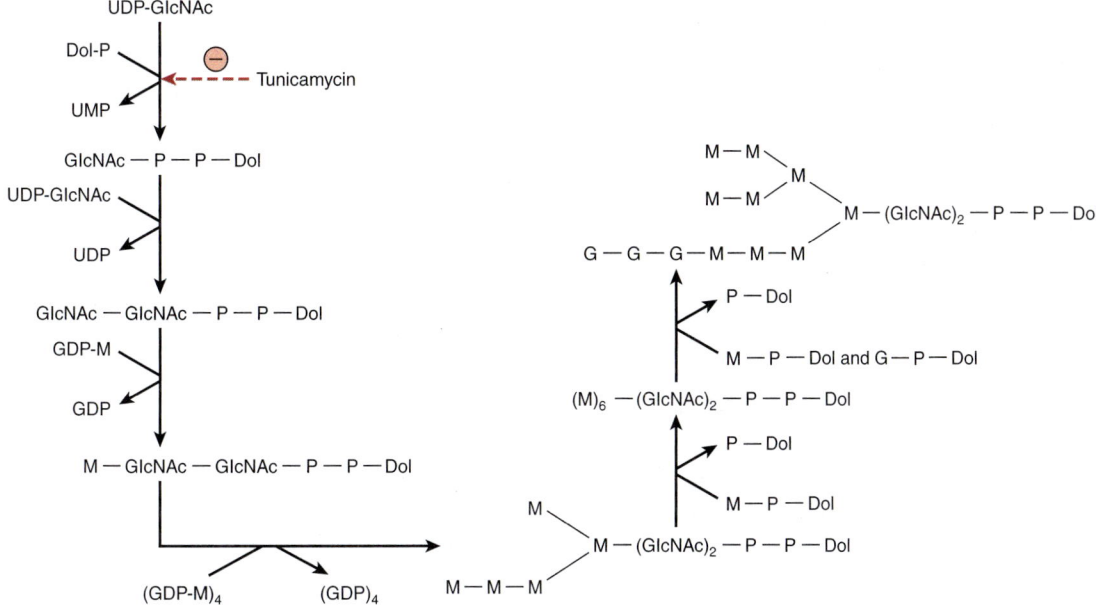

FIGURE 46–4 **Pathway of biosynthesis of dolichol pyrophosphate oligosaccharide.** Note that the first five internal mannose residues are donated by GDP-mannose, whereas the more external mannose residues and the glucose residues are donated by dolichol-P-mannose and dolichol-P-glucose. (Dol, dolichol; GDP, guanosine diphosphate; P, phosphate; UDP, uridine diphosphate; UMP, uridine monophosphate.)

recognizes and reglucosylates the misfolded glycoprotein, which rebinds to the calnexin-Erp57 complex. If it is now correctly folded, the glycoprotein is again deglucosylated and secreted. If it is not capable of correct folding, it is translocated out of the endoplasmic reticulum into the cytosol for catabolism. The glucosyltransferase senses the folding of the glycoprotein and only reglucosylates misfolded proteins. The soluble endoplasmic reticulum protein **calreticulin** performs a similar function to that of calnexin.

Several Factors Regulate the Glycosylation of Glycoproteins

The glycosylation of glycoproteins involves a large number of enzymes; about 1% of the human genome codes for genes that are involved with protein glycosylation. These include at least 10 distinct *N*-acetylglucosamine transferases, and multiple isoenzymes of the other glycosyltransferases. Controlling factors in the first stage of *N*-linked glycoprotein biosynthesis (assembly and transfer of the dolichol pyrophosphate oligosaccharide) include not only the availability of the sugar nucleotides, but also the presence of suitable acceptor sites in proteins, the tissue concentration of dolichol phosphate, and the activity of the relevant oligosaccharide: protein transferase.

Various **cancer cells** are found to synthesize different oligosaccharide chains from those made in normal cells (eg, greater branching). This is due to cancer cells expressing different patterns of glycosyltransferases from those in normal cells, as a result of specific gene activation or repression. The differences in oligosaccharide chains could affect adhesive interactions between cancer cells and the normal parent tissue cells, contributing to metastasis.

SOME PROTEINS ARE ANCHORED TO THE PLASMA MEMBRANE BY GLYCOPHOSPHATIDYL-INOSITOL STRUCTURES

Membrane-bound proteins that are anchored to the lipid bilayer by a glycophosphatidylinositol (GPI) tail (Figure 46–1) constitute the third major class of glycoproteins. The GPI linkage is the most common way by which various proteins are anchored to cell membranes.

The proteins are anchored to the outer leaflet of the plasma membrane or the inner (luminal) leaflet of the membrane in secretory vesicles by the fatty acids of phosphatidylinositol. The phosphatidylinositol is linked via *N*-acetylglucosamine to a glycan chain containing a variety of sugars, including mannose and glucosamine. In turn, the oligosaccharide chain is linked via phosphorylethanolamine in an amide linkage to the carboxyl-terminal amino acid of the protein. Additional constituents are found in many GPI structures; for example, that shown in Figure 46–1 contains an extra phosphorylethanolamine attached to the middle of the three mannose moieties of the glycan and an extra fatty acid attached to glucosamine.

There are three functions of this GPI linkage:

1. The GPI anchor allows greatly enhanced **mobility** of a protein in the plasma membrane compared with that of a protein that employ transmembrane domains. The GPI anchor is attached only to the outer leaflet of the lipid bilayer, so that it is freer to diffuse than a protein anchored through both layers of the membrane. Enhanced mobility may be important in facilitating rapid responses to stimuli.

2. Some GPI anchors may connect with **signal transduction** pathways, enabling proteins that lack a transmembrane domain to nevertheless serve as receptors for hormones and other cell surface signals.

3. GPI structures can **target** proteins to apical or basolateral domains of the plasma membrane of polarized epithelial cells.

The GPI anchor is preformed in the endoplasmic reticulum, and is then attached to the protein after ribosomal synthesis is complete. The primary translation products of GPI-anchored proteins have not only an amino-terminal signal sequence that directs them into the endoplasmic reticulum during synthesis, but also a carboxy-terminal hydrophobic domain that acts as the signal for attachment of the GPI anchor. The first stage in synthesis of the GPI anchor is insertion of the fatty acids of phosphatidylinositol into the luminal face of the endoplasmic reticulum membrane, followed by glycosylation, starting with esterification of *N*-acetyl-glucosamine to the phosphate group of phosphatidylinositol. A terminal phosphoethanolamine moiety is added to the completed glycan chain. The hydrophobic carboxy-terminal domain of the protein is displaced by the amino group of ethanolamine in the transamidation reaction that forms the amide linkage between the GPI anchor and an aspartate residue in the protein.

SOME PROTEINS UNDERGO RAPIDLY REVERSIBLE GLYCOSYLATION

Many proteins, including nuclear pore proteins, proteins of the cytoskeleton, transcription factors, and proteins associated with chromatin, as well as nuclear oncogene proteins and tumor suppressor proteins, undergo rapidly reversible *O*-glycosylation with a single sugar moiety, *N*-acetylglucosamine. The serine and threonine sites of glycosylation are the same as those of phosphorylation of these proteins, and glycosylation and phosphorylation occur reciprocally in response to cellular signaling.

The *O*-linked *N*-acetylglucosamine transferase that catalyzes this glycosylation uses UDP-*N*-acetylglucosamine as the sugar donor, and has phosphatase activity, so can directly replace a serine or threonine phosphate with *N*-acetylglucosamine. There is no absolute consensus sequence for the enzyme, but about half the sites that are subject to reciprocal glycosylation and phosphorylation are Pro-Val-Ser. This *O*-linked *N*-acetylglucosamine transferase is activated by phosphorylation in response to insulin action, and the *N*-acetylglucosamine is removed by *N*-acetylglucosaminidase, leaving the site available for phosphorylation.

Both the activity and peptide specificity of *O*-linked *N*-acetylglucosamine transferase depend on the concentration of UDP-*N*-acetylglucosamine. Depending on cell type, up to 2 to 5% of glucose metabolism flows by way of the hexosamine pathway leading to *N*-acetylglucosamine formation, giving the *O*-linked *N*-acetylglucosamine transferase a role in nutrient sensing in the cell. Excessive *O*-glycosylation with *N*-acetylglucosamine (and hence reduced phosphorylation) of target proteins is implicated in **insulin resistance** and glucose toxicity in **diabetes mellitus**, as well as neurodegenerative diseases.

ADVANCED GLYCATION END-PRODUCTS (AGEs) ARE IMPORTANT IN CAUSING TISSUE DAMAGE IN DIABETES MELLITUS

Glycation is the nonenzymic attachment of sugars (mainly glucose) to amino groups of proteins (and also to other molecules including DNA and lipids), unlike **glycosylation**, which is enzyme-catalyzed. Initially, glucose forms a **Schiff base** to the amino terminal of the protein, which then undergoes the **Amadori rearrangement** to yield **ketoamines (Figure 46–5)**, and further reactions to yield **advanced glycation end-products (AGEs)**. The overall series of reactions is known as the **Maillard reaction**, which is involved in the **browning** of certain foodstuffs during storage or heating, and provides much of the flavor of some foods.

Advanced glycation end-products induce **tissue damage** in poorly controlled **diabetes mellitus**. When the blood glucose concentration is consistently elevated, there is increased glycation of proteins. Glycation of collagen and other proteins in the extracellular matrix alters their properties (eg, increasing the **cross-linking of collagen**). Cross-linking can lead to

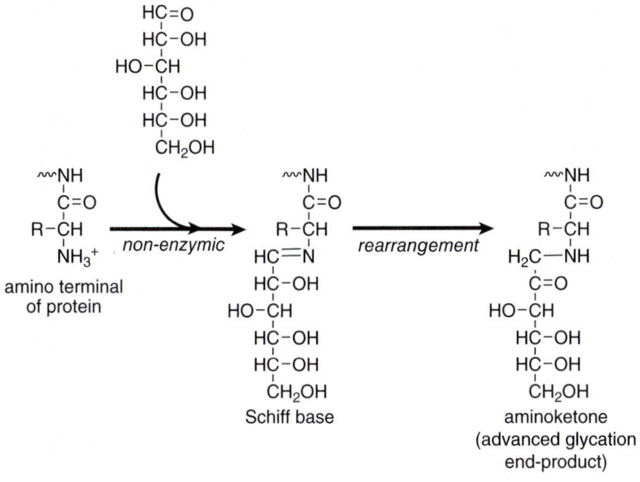

accumulation of various plasma proteins in the walls of blood vessels; in particular, accumulation of low-density lipoprotein (**LDL**) can contribute to **atherogenesis**. AGEs appear to be involved in both **microvascular** and **macrovascular** damage in diabetes mellitus. Endothelial cells and macrophages have AGE receptors on their surfaces; uptake of glycated proteins by these receptors can activate the transcription factor **NF-kB** (see Chapter 52), generating a variety of **cytokines** and **proinflammatory molecules**.

Nonenzymic glycation of **hemoglobin A** present in red blood cells leads to the formation of HbA_{1c}. It occurs normally to a small extent, but is increased in diabetic patients with poor glycemic control, whose blood glucose concentration is consistently elevated. As discussed in Chapter 6, measurement of HbA_{1c} has become a very important part of the **management of patients with diabetes mellitus**.

GLYCOPROTEINS ARE INVOLVED IN MANY BIOLOGICAL PROCESSES & IN MANY DISEASES

As listed in Table 46–1, glycoproteins have many different functions, including transport molecules, immunological molecules, and hormones. They are also important in fertilization and inflammation; a number of diseases are due to defects in the synthesis and catabolism of glycoproteins.

Glycoproteins Are Important in Fertilization

To reach the plasma membrane of an oocyte, a sperm has to traverse the **zona pellucida (ZP)**, a thick, transparent, envelope that surrounds the oocyte. The glycoprotein ZP3 is an *O*-linked glycoprotein that functions as a sperm receptor. A protein on the sperm surface interacts with the oligosaccharide chains of ZP3. By transmembrane signaling, this interaction induces the **acrosomal reaction**, in which enzymes such as proteases and hyaluronidase along with other contents of the acrosome of the sperm are released. Liberation of these enzymes permits the sperm to pass through the zona pellucida and reach the plasma membrane of the oocyte. Another glycoprotein, PH-30, is important in binding of the sperm plasma membrane to that of the oocyte, and the subsequent fusion of the two membranes, enabling the sperm to enter and fertilize the oocyte.

Selectins Play Key Roles in Inflammation & in Lymphocyte Homing

Leukocytes play important roles in many inflammatory and immunological processes; the first steps are interactions between circulating leukocytes and **endothelial cells** prior to passage of the leukocytes out of the circulation. Leukocytes and endothelial cells contain cell surface lectins, called **selectins**, which participate in intercellular adhesion. Selectins are single-chain Ca^{2+}-binding transmembrane proteins; the

FIGURE 46–5 Formation of advanced glycation end-products from glucose.

amino terminals contain the lectin domain, which is involved in binding to specific carbohydrate ligands.

Interactions between selectins on the neutrophil cell surface and glycoproteins on the endothelial cell trap the neutrophils temporarily, so that they roll over the endothelial surface. During this the neutrophils are activated, undergo a change in shape, and now adhere firmly to the endothelium. This adhesion is the result of interactions between **integrins** (see Chapter 53) on the neutrophils and immunoglobulin-related proteins on the endothelial cells. After adhesion, the neutrophils insert pseudopodia into the junctions between endothelial cells, squeeze through these junctions, cross the basement membrane, and are then free to migrate in the extravascular space.

Selectins bind **sialylated and fucosylated oligosaccharides**. Sulfated lipids (see Chapter 21) may also be ligands. Administration of compounds or monoclonal antibodies that block selectin-ligand interactions may be therapeutically useful to inhibit inflammatory responses. **Cancer cells** often have selectin ligands on their surfaces, which may have a role in the invasion and metastasis of cancer cells.

Abnormalities in the Synthesis of Glycoproteins Underlie Certain Diseases

Leukocyte adhesion deficiency II is a rare condition caused by mutations affecting the activity of a Golgi-located GDP-fucose transporter. The absence of fucosylated ligands for selectins leads to a marked decrease in neutrophil rolling. Patients suffer life-threatening recurrent bacterial infections, and also psychomotor and mental retardation. The condition may respond to oral fucose.

Paroxysmal nocturnal hemoglobinuria is an acquired mild anemia characterized by the presence of hemoglobin in urine due to hemolysis of red cells, particularly during sleep, reflecting a slight drop in plasma pH during sleep, which increases susceptibility to lysis by the complement system (see Chapter 52). The condition is due to the acquisition by hematopoietic cells of somatic mutations in the gene coding for the enzyme that links glucosamine to phosphatidylinositol in the GPI structure. This leads to a deficiency of proteins that are anchored to the red cell membrane by GPI linkages. Two proteins, **decay accelerating factor** and **CD59**, normally interact with components of the complement system to prevent the hemolysis. When they are deficient, the complement system acts on the red cell membrane to cause hemolysis.

Some of the **congenital muscular dystrophies** are the result of defects in the synthesis of glycans in the protein α-dystroglycan. This protein protrudes from the surface membrane of muscle cells and interacts with laminin-2 (merosin) in the basal lamina. If the glycans of α-dystroglycan are not correctly formed (as a result of mutations in genes encoding some glycosyltransferases), this results in defective interaction of α-DG with laminin.

Rheumatoid arthritis is associated with an alteration in the glycosylation of circulating immunoglobulin G (IgG) molecules (see Chapter 52), such that they lack galactose in their Fc regions and terminate in N-acetylglucosamine. **Mannose-binding protein**, a lectin synthesized by liver cells and secreted into the circulation, binds mannose, N-acetylglucosamine, and some other sugars. It can thus bind agalactosyl IgG molecules, which subsequently activate the complement system, contributing to chronic inflammation in the synovial membranes of joints.

Mannose-binding protein can also bind sugars when they are present on the surfaces of bacteria, fungi, and viruses, preparing these pathogens for opsonization or for destruction by the complement system. This is an example of **innate immunity**, which does not involve immunoglobulins or T lymphocytes. Deficiency of this protein in young infants as a result of mutation renders them susceptible to recurrent infections.

Inclusion Cell (I-Cell) Disease Results From Faulty Targeting of Lysosomal Enzymes

Mannose 6-phosphate serves to target enzymes into the lysosome. I-cell disease is a rare condition characterized by severe progressive psychomotor retardation and a variety of physical signs, with death often occurring in the first decade of life. Cells from patients with I-cell disease lack almost all of the normal lysosomal enzymes; the lysosomes thus accumulate many different types of undegraded molecules, forming inclusion bodies. The patients' plasma contains very high activities of lysosomal enzymes, suggesting that the enzymes are synthesized but fail to reach their proper intracellular destination and are instead secreted. Lysosomal enzymes from normal individuals carry the mannose 6-phosphate recognition marker, cells from patients with I-cell disease lack the Golgi-located N-acetylglucosamine phosphotransferase. Two lectins act as **mannose 6-phosphate receptor proteins**; both function in the intracellular sorting of lysosomal enzymes into clathrin-coated vesicles in the Golgi. These vesicles then leave the Golgi and fuse with a prelysosomal compartment.

Genetic Deficiencies of Glycoprotein Lysosomal Hydrolases Cause Diseases Such as α-Mannosidosis

Turnover of glycoproteins involves the catabolism of the oligosaccharide chains by a number of lysosomal hydrolases, including α-neuraminidase, β-galactosidase, β-hexosaminidase, α- and β-mannosidases, α-N-acetylgalactosaminidase, α-fucosidase, endo-β-N-acetylglucosaminidase, and aspartylglucosaminidase. Genetic defects of these enzymes result in abnormal degradation of glycoproteins. The accumulation in tissues of partially degraded glycoproteins leads to various diseases. Among the best recognized of these are mannosidosis, fucosidosis, sialidosis, aspartylglucosaminuria, and Schindler disease, due respectively to deficiencies of α-mannosidase, α-fucosidase, α-neuraminidase, aspartylglucosaminidase, and α-N-acetylgalactosaminidase.

GLYCANS ARE INVOLVED IN THE BINDING OF VIRUSES, BACTERIA, & SOME PARASITES TO HUMAN CELLS

A feature of glycans that explains many of their biological actions is that they bind specifically to proteins and other glycans. One reflection of this is their ability to bind some viruses, bacteria, and parasites.

Influenza virus A binds to cell surface glycoprotein receptor molecules containing N-acetylneuraminic acid via a **hemagglutinin** protein. The virus also has a **neuraminidase** that plays a key role in allowing elution of newly synthesized progeny from infected cells. If this process is inhibited, spread of these viruses is markedly diminished. Inhibitors of this enzyme (eg, zanamivir, oseltamivir) are now available for use in treating patients with influenza. Influenza viruses are classified according to the type of hemagglutinin (H) and neuraminidase (N) that they possess. There are at least 16 types of hemagglutinin and 9 types of neuraminidase. Thus, **avian influenza virus** is classified as **H5N1**.

Human immunodeficiency virus type 1 (**HIV-1**), the cause of AIDS, attaches to cells via one of the surface glycoproteins gp 120 and uses another surface glycoprotein (gp 41) to fuse with the host cell membrane. **Antibodies** to gp 120 develop during infection by HIV-1, and there has been interest in using the isolated protein as a vaccine. One major problem with this approach is that the structure of gp 120 can change relatively rapidly due to mutations, allowing the virus to escape from the neutralizing activity of antibodies directed against it.

Helicobacter pylori is the major cause of **peptic ulcers**. It binds to at least two different glycans present on the surfaces of epithelial cells in the stomach allowing it to establish a stable attachment site to the stomach lining. Similarly, many bacteria that cause **diarrhea** attach to surface cells of the intestinal mucosa via glycans present in glycoproteins or glycolipids. The attachment of the malarial parasite *Plasmodium falciparum* to human cells is mediated by a GPI present on the surface of the parasite.

SUMMARY

- Glycoproteins are widely distributed proteins with diverse functions that contain one or more covalently linked carbohydrate chains.

- The carbohydrate content of glycoproteins ranges from 1 to more than 85% by weight. They may be simple or very complex in structure. Eight sugars are mainly found in the sugar chains of human glycoproteins: xylose, fucose, galactose, glucose, mannose, N-acetylgalactosamine, N-acetylglucosamine, and N-acetylneuraminic acid.

- Some of the oligosaccharide chains of glycoproteins encode biological information; they are also important in modulating the solubility and viscosity of glycoproteins, in protecting them against proteolysis, and in their biologic actions.

- Glycosidases hydrolyze specific linkages in oligosaccharides.

- Lectins are carbohydrate-binding proteins involved in cell adhesion and many other processes.

- The major classes of glycoproteins are O-linked (involving serine or threonine), N-linked (involving the amide group of asparagine), and GPI-linked.

- Mucins are a class of O-linked glycoproteins that are distributed on the surfaces of epithelial cells of the respiratory, gastrointestinal, and reproductive tracts.

- The endoplasmic reticulum and Golgi apparatus play a major role in glycosylation reactions involved in the biosynthesis of glycoproteins.

- The oligosaccharide chains of O-linked glycoproteins are synthesized by the stepwise addition of sugars donated by sugar nucleotides in reactions catalyzed by glycoprotein glycosyltransferases.

- The synthesis of N-linked glycoproteins involves a dolichol-P-P-oligosaccharide and various glycotransferases and glycosidases. Depending on the enzymes and precursor proteins in a tissue, the resulting N-linked oligosaccharides may be simple, complex, or of the high-mannose type.

- Glycoproteins are implicated in many biological processes, including fertilization and inflammation.

- There are a number of diseases involving abnormalities in the synthesis and degradation of glycoproteins. Glycoproteins are also involved in many other diseases, including influenza, AIDS, rheumatoid arthritis, cystic fibrosis, and peptic ulcer.

47

Metabolism of Xenobiotics

David A. Bender, PhD & Robert K. Murray, MD, PhD

O B J E C T I V E S

After studying this chapter, you should be able to:

- Describe the two phases of xenobiotic metabolism, the first involving mainly hydroxylation reactions catalyzed by cytochromes P450 and the second conjugation reactions.
- Describe the metabolic importance of glutathione.
- Describe how xenobiotics can have toxic, immunological, and carcinogenic effects.

BIOMEDICAL IMPORTANCE

We are exposed to a wide variety of compunds that are foreign to the body (**xenobiotics**, from the Greek *xenos* = foreign); naturally occurring compounds in plant foods, and synthetic compounds in medicines, food additives, and environmental pollutants. Knowledge of the metabolism of xenobiotics is essential for an understanding of pharmacology, therapeutics, and toxicology. Many of the xenobiotics in plant foods have potentially beneficial effects (eg, acting as antioxidants, Chapter 45).

Understanding the mechanisms involved in xenobiotic metabolism will permit the development of transgenic microorganisms and plants containing genes that encode enzymes that can be used to render potentially hazardous pollutants harmless. Similarly, transgenic organisms may be used for biosynthesis of drugs and other chemicals.

WE ENCOUNTER MANY XENOBIOTICS THAT MUST BE METABOLIZED BEFORE BEING EXCRETED

The main xenobiotics of medical relevance are **drugs, chemical carcinogens**, naturally occurring compounds in plant foods, and a wide variety of compounds that have found their way into our environment, such as polychlorinated biphenyls (PCBs), and insecticides and other pesticides. Most of these compounds are metabolized, mainly in the liver. While the metabolism of xenobiotics is generally considered to be

a process of detoxification, sometimes the metabolism of compounds that are themselves inert or harmless renders them biologically active. This may be desirable, as in the activation of a prodrug to the active compound, or it may be undesirable, as in the formation of a carcinogen or mutagen from an inert precursor.

The metabolism of xenobiotics occurs in two phases. In **phase 1**, the major reaction involved is **hydroxylation**, catalyzed by the **monooxygenases** of the **cytochrome P450** family. In addition to hydroxylation, these enzymes catalyze a wide range of other reactions, including deamination, dehalogenation, desulfuration, epoxidation, peroxygenation, and reduction. Reactions involving hydrolysis (eg, catalyzed by esterases) and other, non-P450–catalyzed, reactions also occur in phase 1.

Phase 1 metabolism renders compounds more reactive by introducing groups that can be conjugated with glucuronic acid, sulfate, acetate, glutathione, or amino acids in phase 2 metabolism. This produces **polar compounds** that are water-soluble and can therefore be excreted in urine or bile.

In some cases, phase 1 metabolic reactions convert xenobiotics from **inactive** to **biologically active** compounds. In these instances, the original xenobiotics are referred to as **prodrugs** or **procarcinogens**. Sometimes, additional phase 1 reactions (eg, further hydroxylation reactions) convert these active compounds into less active or inactive forms prior to conjugation. In other cases, it is the conjugation reactions that convert the active products of phase 1 reactions to inactive compounds, which are excreted.

ISOFORMS OF CYTOCHROME P450 HYDROXYLATE A WIDE VARIETY OF XENOBIOTICS IN PHASE 1 METABOLISM

The main reaction involved in phase 1 metabolism is **hydroxylation**, catalyzed by a family of monoxygenase enzymes known as **cytochromes P450**. There are at least 57 genes encoding cytochrome P450 in the human genome. Cytochrome P450 is a heme enzyme. It is so named because it was originally discovered when it was noted that preparations of microsomes (fragments of the endoplasmic reticulum) that had been chemically reduced and then exposed to carbon monoxide had an absorption peak at 450 nm.

At least half of the common drugs that we ingest are metabolized by isoforms of cytochrome P450. They also act on steroid hormones, carcinogens, and pollutants. In addition, cytochromes P450 are important in the metabolism of a number of physiological compounds—for example, the synthesis of steroid hormones (see Chapter 26) and the conversion of vitamin D to its active metabolite, calcitriol (see Chapter 44).

The overall reaction catalyzed by cytochrome P450 is:

$$RH + O_2 + NADPH + H^+ \rightarrow R\text{-}OH + H_2O + NADP$$

The reaction mechanism is shown in Figure 12–6. **NADPH-cytochrome P450 reductase** catalyzes the transfer of electrons from NADPH to cytochrome P450. Reduced cytochrome P450 catalyzes the **reductive activation of molecular oxygen**, one atom of which becomes the hydroxyl group in the substrate and the other is reduced to water. **Cytochrome b_5**, another hemoprotein found in the membranes of the smooth endoplasmic reticulum (see Chapter 12), may be involved as an electron donor in some cases.

Isoforms of Cytochrome P450 Form a Superfamily of Heme-Containing Enzymes

There is a **systematic nomenclature** for the cytochromes P450 and their genes, based on their amino acid sequence homology. The root CYP denotes a cytochrome P450. This is followed by number designating the **family**; cytochromes P450 are included in the same family if they exhibit 40% or more amino acid sequence identity. This is followed by a capital letter indicating the **subfamily**; P450s are in the same subfamily if they exhibit more than 55% sequence identity. The **individual** P450s are then assigned numbers in their subfamily. Thus, CYP1A1 denotes a cytochrome P450 that is a member of family 1 and subfamily A and is the first individual member of that subfamily. The nomenclature for the **genes** encoding cytochromes P450 is the same, except that italics are used; thus, the gene encoding CYP1A1 is *CYP1A1*.

The major families of cytochromes P450 involved in drug metabolism are CYP1 (with 3 subfamilies), CYP2

(13 subfamilies), and CYP3 (1 subfamily). The various cytochromes P450 have overlapping substrate specificities, so that a very broad range of xenobiotics can be metabolized by one or other of the enzymes.

Cytochromes P450 are present in greatest amount in **liver cells** and enterocytes. In liver and most other tissues, they are present mainly in the **membranes of the smooth endoplasmic reticulum**, which constitute part of the **microsomal fraction** when tissue is subjected to subcellular fractionation. In hepatic microsomes, cytochromes P450 can comprise as much as 20% of the total protein. In the **adrenal gland**, where they are involved in cholesterol and steroid hormone biosynthesis, they are found in **mitochondria** as well as in the endoplasmic reticulum (see Chapters 26 and 41).

Overlapping Specificity of Cytochromes P450 Explains Interactions Between Drugs and Between Drugs and Nutrients

Most isoforms of cytochrome P450 are **inducible**. For instance, the administration of phenobarbital or other drugs causes hypertrophy of the smooth endoplasmic reticulum and a three- to fourfold increase in the amount of cytochrome P450 within a few days. In most cases this involves increased transcription to mRNA. However, in some cases, induction involves stabilization of mRNA or the enzyme protein itself, or increased translation of existing mRNA.

Induction of cytochrome P450 underlies **drug interactions**, when the effects of one drug are altered by administration of another. For example, the anticoagulant **warfarin** is metabolized by **CYP2C9**, which is induced by phenobarbital. Induction of CYP2C9 will increase the metabolism of warfarin, so reducing its efficacy, so that the dose must be increased. CYP2E1 catabolizes the metabolism of some widely used solvents and compounds found in tobacco smoke, many of which are **procarcinogens**; it is induced by **ethanol**, so increasing the risk of carcinogenicity.

Naturally occurring compounds in foods may also affect cytochrome P450. Grapefruit contains a variety of furanocoumarins, which inhibit cytochrome P450 and so affect the metabolism of many drugs. Some drugs are activated by cytochrome P450, so that grapefruit will reduce their activity; others are inactivated by cytochrome P450, so that grapefruit increases their activity. Drugs that are affected include statins, omeprazole, antihistamines, and benzodiazepine antidepressants.

Polymorphism of cytochromes P450 may explain much of the variation in drug responses by different patients; variants with low catalytic activity will lead to slower metabolism of the substrate, and hence prolonged drug action. **CYP2A6** is involved in the metabolism of **nicotine** to conitine. Three *CYP2A6* alleles have been identified: a wild type and two inactive alleles. Individuals with the null alleles, who have impaired metabolism of nicotine, are apparently protected against becoming tobacco-dependent smokers. These individuals smoke less, presumably because their blood and brain concentrations of nicotine remain elevated longer than those of individuals with the wild-type allele. It has been speculated that inhibiting CYP2A6 may provide a novel way to help smoking cessation.

CONJUGATION REACTIONS IN PHASE 2 METABOLISM PREPARE XENOBIOTICS FOR EXCRETION

In phase 1 reactions, xenobiotics are converted to more polar, hydroxylated derivatives. In phase 2 reactions, these derivatives are conjugated with molecules such as glucuronic acid, sulfate, or glutathione. This renders them even more water-soluble, facilitating their eventual excretion in the urine or bile.

Glucuronidation Is the Most Frequent Conjugation Reaction

The **glucuronidation** of bilirubin is discussed in Chapter 31; xenobiotics are glucuronidated in the same way, using UDP-glucuronic acid (see Figure 20–4) along with a variety of glucuronosyltransferases, present in both the endoplasmic reticulum and cytosol. Molecules such as 2-acetylaminofluorene (a carcinogen), aniline, benzoic acid, meprobamate (a tranquilizer), phenol, and many steroid hormones are excreted as glucuronides. The glucuronide may be attached to oxygen, nitrogen, or sulfur groups of the substrates.

Some Alcohols, Arylamines, and Phenols Are Sulfated

The **sulfate donor** in these and other biologic and sulfation reactions (eg, sulfation of steroids, glycosaminoglycans, glycolipids, and glycoproteins) is "active sulfate"—**adenosine 3′-phosphate-5′-phosphosulfate** (**PAPS**; see Chapter 24).

Glutathione Is Required for Conjugation of Electrophilic Compounds

The tripeptide glutathione (γ-glutamylcysteinylglycine) is important in the phase II metabolism of electrophilic compounds, forming glutathione S-conjugates that are excreted in urine and bile. The reaction catalyzed by glutathione S-transferases is:

$$R + GSH \rightarrow R\text{--}S\text{--}G$$

where R is an electrophilic compound.

There are four classes of cytosolic glutathione S-transferase and two classes of membrane-bound microsomal enzymes, as well as a structurally distinct kappa class that is found in mitochondria and peroxisomes. Glutathione S-transferases exist as homo- or heterodimers constructed from at least seven different types of subunit, and different subunits are induced by different xenobiotics.

Because glutathione S-transferases also bind a number of ligands that are not substrates, including bilirubin, steroid hormones, and some carcinogens and their metabolites, they are sometimes known as **ligandin**. Glutathione S-transferase binds bilirubin at a site distinct from the catalytic site. The complex transits from the bloodstream to the liver, then to the endoplasmic reticulum for conjugation of bilirubin with glucuronic acid, and excretion in the bile (see Chapter 31). Binding of carcinogens sequesters them, so preventing their actions on DNA.

The liver exhibits a very high level of glutathione S-transferase activity; in vitro the entire pool of glutathione can be depleted within minutes on exposure to a xenobiotic substrate. The activity of glutathione S-transferase is upregulated in many tumors, leading to resistance to chemotherapy.

Glutathione conjugates may be transported out of the liver, where they are substrates for extracellular γ-glutamyltranspeptidase and some dipeptidases. The resultant cysteine S-conjugates are taken up by other tissues (especially the kidney) and N-acetylated to yield mercapturic acids (N-acetylcysteine S-conjugates), which are excreted in the urine. Some hepatic glutathione S-conjugates enter the bile canniculi, where they are broken down to cysteine S-conjugates that are then taken up into the liver for N-acetylation, and re-excreted in the bile.

In addition to its role in phase 2 metabolism, glutathione has a number of other roles in metabolism:

1. It provides the reductant for the reduction of **hydrogen peroxide** to water in the reaction catalyzed by glutathione peroxidase (see Figure 20–3).

2. It is an important **intracellular reductant and antioxidant**, helping to maintain essential SH groups of enzymes in their reduced state.

3. A metabolic cycle involving GSH as a carrier has been implicated in the **transport of some amino acids** across membranes in the kidney. The first reaction of the cycle is catalyzed by γ-**glutamyltransferase** (**GGT**):

$$\text{Amino acid} + \text{GSH} \rightarrow \gamma\text{-Glutamyl amino acid} + \text{Cysteinylglycine}$$

This reaction transfers amino acids across the plasma membrane as dipeptides, which are subsequently hydrolyzed, releasing the amino acid and glutamate and the GSH being resynthesized from cysteinylglycine. GGT is present in the plasma membrane of renal tubular cells and bile ductule cells, and in the endoplasmic reticulum of hepatocytes. It is released into the blood from hepatic cells in various hepatobiliary diseases, providing an early indication of liver damage.

OTHER REACTIONS ARE ALSO INVOLVED IN PHASE 2 METABOLISM

The two most important reactions other than conjugation are acetylation and methylation.

The reaction of **acetylation** is:

$$X + Acetyl\text{-}CoA \rightarrow Acetyl\text{-}X + CoA$$

where X represents a xenobiotic or its metabolite. As for other acetylation reactions, **acetyl-CoA** is the acetyl donor. These reactions are catalyzed by **acetyltransferases** present in the cytosol of various tissues, particularly liver. The drug **isoniazid**, used in the treatment of tuberculosis, is subject to acetylation. There is polymorphism of acetyltransferases, resulting in individuals who are classified as **slow or fast acetylators**. Slow acetylators are more subject to the toxic effects of isoniazid because the drug persists for longer.

Some xenobiotics are subject to **methylation** by methyltransferases, employing S-adenosylmethionine (see Figure 29–17) as the methyl donor.

RESPONSES TO XENOBIOTICS INCLUDE TOXIC, IMMUNOLOGICAL, & CARCINOGENIC EFFECTS

There are very few xenobiotics, including drugs, that do not have some toxic effects if the dose is large enough. The **toxic effects of xenobiotics** cover a wide spectrum, but can be considered under three general headings:

1. Covalent binding of xenobiotic metabolites to macromolecules including **DNA**, **RNA**, and **protein** can lead to cell injury (**cytotoxicity**), which can be severe enough to result in cell death. In response to damage to DNA, the **DNA repair mechanism** of the cell is activated. Part of this involves the transfer of multiple ADP-ribose units onto DNA-binding proteins, catalyzed by poly(ADP-ribose polymerase). The source of ADP-ribose is NAD, and in response to severe DNA damage there is considerable depletion of NAD. In turn, this leads to severely impaired ATP formation, and cell death.

2. The reactive metabolite of a xenobiotic may bind to a protein, acting as a hapten, and altering its **antigenicity**. On its own it will not stimulate antibody production, but does so when bound to a protein. The resultant antibodies react not only with the modified protein but also with the unmodified protein, so potentially initiating **autoimmune disease**.

3. Reactions of some activated xenobiotics with **DNA** are important in **chemical carcinogenesis**. Some chemicals (eg, benzo[α]pyrene) require activation by cytochrome P450 in the endoplasmic reticulum to become carcinogenic (they are therefore called **indirect carcinogens**). The activities of the xenobiotic-metabolizing enzymes present in the endoplasmic reticulum thus help to determine whether such compounds become carcinogenic or are "detoxified."

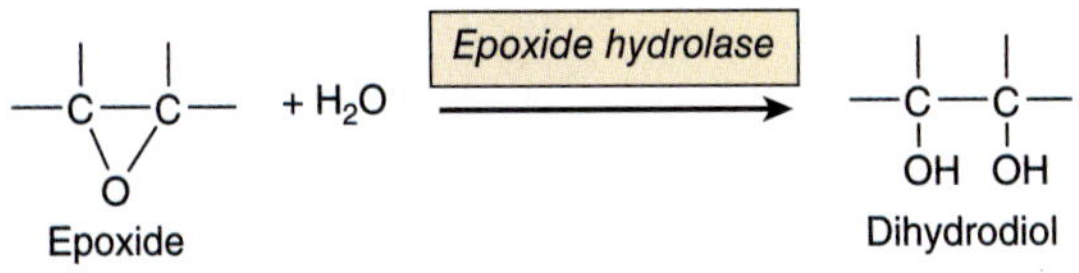

FIGURE 47–1 The reaction of epoxide hydrolase.

The enzyme **epoxide hydrolase**, located in the membranes of the endoplasmic reticulum, can protect against some carcinogens. The products of the action of cytochrome P450 on some procarcinogen substrates are **epoxides**. Epoxides are highly reactive and, hence, potentially mutagenic or carcinogenic. As shown in **Figure 47–1**, the hydrolase catalyzes hydrolysis of epoxides to much less reactive dihydrodiols.

SUMMARY

- Xenobiotics are chemical compounds foreign to the body, including drugs, food additives, and environmental pollutants, as well as naturally occurring compounds in plant foods.

- Xenobiotics are metabolized in two phases. The major reaction of phase 1 is hydroxylation catalyzed by a variety of monooxygenases, known as the cytochromes P450. In phase 2, the hydroxylated species are conjugated with a variety of hydrophilic compounds such as glucuronic acid, sulfate, or glutathione. The combined operation of these two phases converts lipophilic compounds into water-soluble compounds that can be excreted in urine or bile.

- Cytochromes P450 catalyze reactions that introduce one atom of oxygen derived from molecular oxygen into the substrate, yielding a hydroxylated product, and the other into water. NADPH and NADPH cytochrome P450 reductase are involved in the reaction mechanism.

- Cytochromes P450 are hemoproteins and generally have a wide substrate specificity, acting on many exogenous and endogenous substrates. At least 57 cytochrome P450 genes are found in the human genome.

- Cytochromes P450 are generally located in the endoplasmic reticulum of cells, especially in the liver.

- Many cytochromes P450 are inducible. This has important implications for interactions between drugs.

- Phase 2 conjugation reactions are catalyzed by enzymes such as glucuronyltransferases, sulfotransferases, and glutathione S-transferases, using UDP-glucuronic acid, PAPS (active sulfate), and glutathione, respectively, as donors.

- Glutathione not only plays an important role in phase 2 reactions, but also is an intracellular reducing agent.

- Xenobiotics can produce a variety of biological effects, including toxicity, immunological reactions, and cancer.

Clinical Biochemistry

David A. Bender, PhD

OBJECTIVES

After studying this chapter, you should be able to:

- Explain the importance of laboratory tests in clinical and veterinary medicine.
- Explain what is meant by the reference range for the results of a test.
- Explain the difference between the precision and accuracy of an assay method, and explain the sensitivity and specificity of an assay method.
- Explain what is meant by the sensitivity, specificity, and predictive value of a laboratory test.
- List techniques that are commonly used in a diagnostic lab carrying out biochemical tests and explain the principle of each method.
- Explain why high plasma concentrations of certain enzymes are considered to be indicators of tissue damage.
- Describe in outline the different requirements for measuring an enzyme in a plasma sample and using an enzyme to measure an analyte.

THE IMPORTANCE OF LABORATORY TESTS IN MEDICINE

Various laboratory tests are an essential part of medicine and veterinary practice. Biochemical tests can be used for screening for disease, for confirmation (or otherwise) of a diagnosis made on clinical examination, for monitoring progression of a disease and the outcome of treatment. Blood and urine samples are most commonly used; occasionally feces, saliva, or cerebrospinal fluid (CSF) may be analyzed, and on rare occasions, tissue biopsy samples. Most of our knowledge and understanding of the underlying causes of metabolic diseases and of the effects of disease on metabolism has come from analysis of metabolites in blood and urine, and from measurement of enzymes in blood. In turn, that knowledge has permitted advances in the treatment of disease and the development of more effective drugs.

Advances in technology mean that many tests that were formerly carried out only in specialist laboratories can now be performed at the bedside, in the doctor's office, or veterinary practice, sometimes even at home by patients themselves, with automated machines or "dipsticks" that are simple to use. Other tests are still conducted in hospital laboratories or by private clinical chemistry laboratories, with samples sent in by the referring physician. Some tests that are less commonly requested and may be technically more demanding are performed only in specialist centers. These often involve specialist techniques to study rare (and sometimes newly discovered) metabolic diseases. In addition, testing of samples from athletes and race horses for performance-enhancing drugs and other banned substances is normally carried out in only a limited number of specially licensed laboratories.

CAUSES OF ABNORMALITIES IN LEVELS OF ANALYTES MEASURED IN THE LABORATORY

A great many different conditions can lead to abnormalities of the results of laboratory tests. Tissue injury that results in damage to cell membranes and an increase in the permeability of the plasma membrane leads to leakage of intracellular material into the bloodstream (eg, leakage of creatine kinase MB into the bloodstream following a myocardial infarction). In other cases, the synthesis of proteins and hormones is increased or decreased (eg, C-reactive protein in inflammatory states, or hormones in endocrine disorders). Kidney and liver failure lead to the accumulation of a number of compounds (eg, creatinine and bilirubin, respectively) in the blood, due to an inability of the organ concerned to excrete or metabolize the compound concerned.

THE REFERENCE RANGE

For any compound that is measured (an **analyte**), there is a range of values around the average or mean that can be considered to be normal. This is the result of biological variations between individuals. In addition, day-to-day or week-to-week variations can occur in the results for the same individual. Therefore, the first step in establishing any new laboratory test for screening for, or diagnosis of, disease, or monitoring treatment, is to determine the range of results in a population of healthy people. For some tests, this will also mean determining the normal ranges of analytes in people of different ages. The normal range of some analytes will differ between men and women, and there may be differences between different ethnic groups to be considered as well.

If the results obtained for a target healthy population group (depending on age, gender, and perhaps ethnicity) are statistically normally distributed (ie, the results show a symmetrical gaussian distribution around the mean), then the acceptable or normal range is taken to be ± 2x standard deviation around the mean. This range includes 95% of the target population, and is known as the reference range. Values outside the reference range are considered to be abnormal, meriting further investigation. If the results from the healthy population are not statistically normally distributed, but skewed, then further statistical analysis is required before the 95% reference range can be established.

For some tests, the results from different laboratories will differ, usually because they use different methods of measurement. Each laboratory establishes its own set of reference ranges for the analyses it performs. Some laboratories report the results as the value; others report results as the number of standard deviations away from the mean—the so-called Z-score. This allows the physician to see how far from the mean the result is—in other words, how abnormal it is. Sometimes the results will be reported as 5 or 10 (or more) times above the upper limit of normal.

The use of the 95% range as the reference range has an unfortunate consequence. By chance, 5% of the "normal" results will be outside the reference range. This first became apparent in the 1970s, when multichannel analysers were developed that were capable of determining 20 or more analytes in each sample. Almost every sample gave one result that was outside the reference range, but if the same person gave a sample a few days later, that apparently abnormal result was now within the reference range, although by chance the result for another analyte might now be outside the reference range.

VALIDITY OF LABORATORY RESULTS

Diagnostic laboratories are subject to inspection and regulatory procedures to assess the validity of their results and ensure **quality control** of their reports. Such measures will ensure that the value of the concentration, activity, or amount of a substance in a specimen reported represents the best value obtainable with the method, reagents, and instruments used.

In establishing a new test, or a new method, four questions have to be answered:

1. **How precise is the method?** This is a measure of the reproducibility of the method. If the same sample is analyzed many times over, how much variation will be seen in the results obtained? **Figure 48–1** illustrates this. In this example, one set of results is much more precise than the other (there is a difference between the two in the spread of results around the mean), even though they yield the same mean result. Precision is not absolute, but subject to variations inherent in the complexity of the method used, the stability of reagents, the sophistication of the equipment used for the assay, and the skill of the technicians involved.

2. **How accurate is the result?** This is a measure of how close the result is to the true value. **Figure 48–2** shows the results of assays by two different methods or by the same method but in two different laboratories. Both have similar precision, but their mean values are very different. It is not possible to say from this information which laboratory is correct (and this is part of the reason why laboratories establish their own reference ranges). There are a number of national or regional quality control schemes in which all participating laboratories are sent the same (pooled) blood or urine sample. Each laboratory measures the various analytes in the pooled sample. The results obtained by all laboratories are plotted as a distribution curve. The mean of these values is calculated and considered to be the "true value." Such a quality control scheme allows each participating laboratory to determine how close its results are to the "true value."

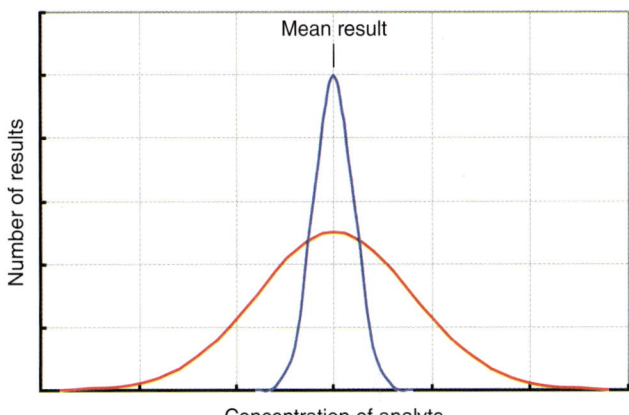

FIGURE 48–1 Precision of an analytical method. The graph shows the results of an analyte measured multiple times in the same sample, either by two different analytical methods or by the same method in two different laboratories. In both cases, the mean result is the same. However, one method or laboratory, shown in blue, has a low scatter of results, and hence a low standard deviation, and high precision, while the other, shown in red, has a high scatter of results, a high standard deviation, and low precision.

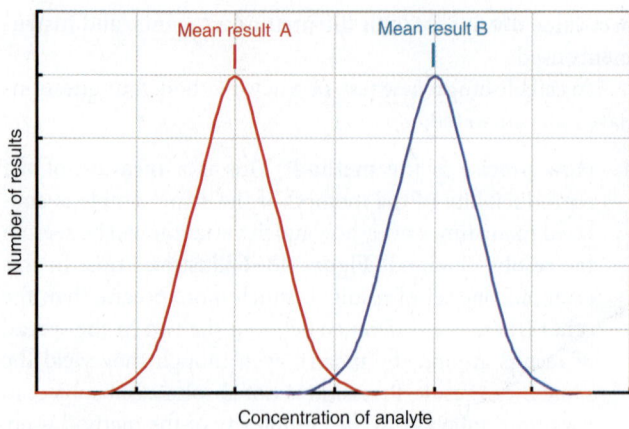

FIGURE 48–2 **Accuracy of an analytical method.** Two different analytical methods, performed on multiple samples, or the same method performed in two different laboratories, with the same scatter of results, and hence the same standard deviation and the same precision. However, the mean values of analytes obtained for the two methods or laboratories are very different; it is not possible to tell which result is closer to the true value.

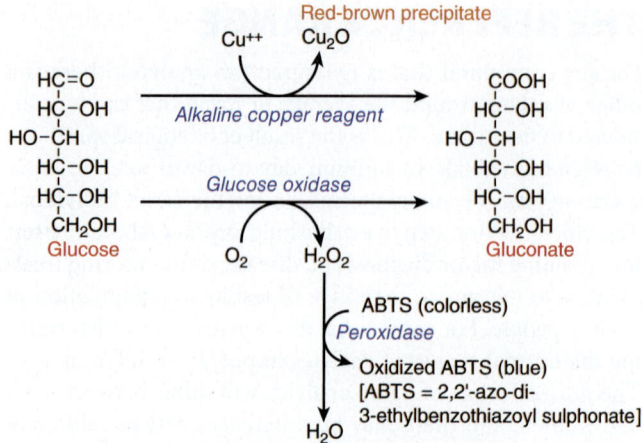

FIGURE 48–3 **Specificity of an analytical method.** Measurement of blood glucose by two methods. Chemical reduction of Cu^{2+} in alkaline solution will detect not only glucose, but any other reducing sugar and other substances such as vitamin C. Enzymic oxidation of glucose using glucose oxidase is a specific reaction; no other compound will be oxidized and contribute to the value obtained.

3. **How sensitive is the method?** In other words, how little of the analyte can be determined reliably? What is the lower limit of reliable detection? This is obviously important when results below the reference range are clinically significant, or when samples are being analyzed for narcotics or performance-enhancing substances that are banned in competitive sport.

4. **How specific is the method?** This question deals with the issue of confidence that the assay is actually measuring the analyte of interest. For example, the now obsolete method of measuring glucose in blood or urine used an alkaline copper (Cu^{2+}) solution, which was reduced to Cu^{+} by glucose. However, the presence of other reducing compounds in urine or blood, such as xylose or vitamin C, will result in a falsely high value. Modern methods of measuring glucose depend on the enzyme glucose oxidase, which only reacts with glucose, and so is highly specific. One of the products of the action of glucose oxidase on glucose is hydrogen peroxide; the second step in the assay is to reduce this hydrogen peroxide to water and oxygen, using a peroxidase along with a colorless compound that turns blue when it is oxidized by the oxygen produced. However, high concentrations of vitamin C, as would be seen with patients taking vitamin supplements, reduce the dye back to its colorless form, giving a low or false-negative result (**Figure 48–3**).

ASSESSMENT OF CLINICAL VALIDITY OF A LABORATORY TEST

The above four criteria must be established for each analytical method. In addition, the **clinical value** of the test has to be established by taking into consideration its sensitivity, specificity, and positive and negative predictive values (**Table 48–1**). Here, unfortunately, the same two terms, sensitivity and specificity, are used, but with very different meanings from those used in establishing the analytical method.

The **sensitivity** of a test is the **percentage of positive test results in patients with the disease ("true positive")**. For example, the test for phenylketonuria is highly sensitive; a positive result is obtained in all who have the disease (100% sensitivity). By contrast, the carcinoembryonic antigen (CEA) test for carcinoma of the colon has lower sensitivity; only 72% of those with carcinoma of the colon test positive when the disease is extensive, and only 20% with early disease.

The **specificity** of a test is the **percentage of negative test results among people who do not have the disease**. The test for phenylketonuria is highly specific; 99.9% of normal individuals give a negative result; only 0.1% gives a false-positive result. By contrast, the CEA test has a variable specificity; about 3% of nonsmoking individuals give a false-positive result (97% specificity), whereas 20% of smokers give a false-positive result (80% specificity).

The sensitivity and specificity of a test are inversely related to each other. If the cutoff point is set too high, then very few healthy people will give a false-positive result, but many people with the disease may give a false-negative result. The sensitivity will thus be low, but the specificity will be high. Conversely, if the cutoff point is too low, then almost all people with the disease will be detected (the test will have a high sensitivity), but more disease-free people may give a false-positive result (the test will have a low specificity).

The **predictive value of a positive test** (positive predictive value) is the percentage of positive results that are true positives. Similarly, the **predictive value of a negative test** (negative predictive value) is the percentage of negative results that are true

TABLE 48–1 Sensitivity, Specificity, and Positive and Negative Predictive Values of a Laboratory Test

What is the result of the test?		Does the Patient Have the Disease?	
		Yes	**No**
	Positive	True positive (a)	False positive (b)
	Negative	False negative (c)	True negative (d)
Sensitivity	=	$\dfrac{\text{True positive (a)} \times 100}{\text{Number of patients who have the disease (a + c)}}$	
Specificity	=	$\dfrac{\text{True negative (d)} \times 100}{\text{Number of patients who do not have the disease (b + d)}}$	
Positive predictive value	=	$\dfrac{\text{True positive (a)} \times 100}{\text{Number of patients who have a positive test (a + b)}}$	
Negative predictive value	=	$\dfrac{\text{True negative (d)} \times 100}{\text{Number of patients who have a negative test (c + d)}}$	

negatives. This is related to the prevalence of the disease. For example, in a group of patients in a urology ward, the prevalence of renal disease is higher than in the general population. In this group, the serum concentration of creatinine will have a higher predictive value than in the general population. Formulae for calculating sensitivity, specificity, and predictive values of a diagnostic test are shown in Table 48–1.

SAMPLES FOR ANALYSIS

The usual samples for analysis are blood and urine. Blood is collected into tubes with or without an anticoagulant, depending on whether plasma or serum is required. Less commonly, samples of saliva, CSF, or feces may be used.

There is a difference between measurement of an analyte in a blood sample and in urine. The concentration of an analyte in blood reflects levels at the time the sample was taken, whereas a urine sample represents the cumulative excretion of the analyte over a period of time. A further difference is that it is usual to report results of blood tests as amount of analyte (or enzyme activity) per milliliter or liter of blood (or plasma or serum). Reporting the concentration of the analyte in urine in the same way is not useful, since urine volume depends very largely on fluid intake. In some cases, the patient is asked to provide a complete 24-hour urine sample; this is a tedious procedure, and it is difficult to know whether there really has been a complete 24-hour collection. Alternatively, the concentration of the analyte is reported per mol of creatinine. Creatinine excretion is reasonably constant from day to day for any one individual, but varies between individuals because it depends mainly on muscle mass; creatinine is formed nonenzymically from creatine and creatine phosphate, most of which is in skeletal muscle.

Apart from measurement of blood gases, for which arterial samples are required, blood samples are usually of venous blood. Blood glucose is often measured in capillary blood from a finger prick. Some analyses use whole blood; others require either serum or plasma. For a serum sample, the blood is allowed to clot, then the red cells and fibrin clot are removed by centrifugation. For a plasma sample, the blood is collected into a tube containing an anticoagulant, and the red cells are removed by centrifugation. The difference between serum and plasma is that plasma contains prothrombin and the other clotting factors, including fibrinogen, while serum does not. Different anticoagulants (citrate, EDTA or oxalate, all of which chelate calcium and so inhibit coagulation) are used for collection of plasma samples, depending on the assay to be performed. Heparin, which acts by activating antithrombin III, is also used. For measurement of blood glucose, potassium fluoride is added, as an inhibitor of glycolysis by red blood cells.

TECHNIQUES USED IN CLINICAL CHEMISTRY

Most routine clinical chemistry reactions involve linking a chemical or enzymic reaction to the development of a colored product, or chromophore, that is measured by **absorption spectrophotometry**. Different chromophores absorb light at different wavelengths; the energy of the absorbed light excites electrons to an unstable orbital. The absorbance of light at a specific wavelength in the visible or ultraviolet range is directly proportional to the concentration of the colored end-product, and hence to the concentration of the analyte in the sample. Although at one time such analyses were performed manually, nowadays most assays are automated, and a single instrument can carry out multiple assays on a single sample.

In absorption spectrophotometry, the excited electrons return to their basal state in a series of small quantum jumps, emitting the energy absorbed as heat. For some compounds the electrons return to a lower energy state in a single quantum jump, emitting light of a higher wavelength (lower energy) than the exciting light. This is fluorescence, and the technique is known as **fluorescence spectrophotometry** or

spectrophotofluorimetry. The sample is illuminated with light of a specific wavelength, and the light emitted is measured, at right angles to the direction of the illuminating wavelength. Again, the intensity of the fluorescence is proportional to the concentration of the fluorophore, and hence the concentration of the analyte. Fluorimetry permits both greater specificity and sensitivity of the assay. The specificity is greater than for absorption spectrophotometry because both the exciting wavelength and the emitted wavelength are specific for the fluorophore, while for absorption spectrophotometry there is only one wavelength to be set, that of the light that is absorbed. Fluorimetry is more sensitive because it is easier to detect the emission than the absorption of small amounts of light than.

Increasingly, especially in research and specialist centers, multiple analytes are measured in the same sample using **high-pressure liquid chromatography** to separate analytes, followed by colorimetric, fluorimetric, or electrochemical detection, or linked to mass spectrometry to identify compounds. Such methods form the basis of **metabolomics**, the study of a whole array of metabolites in a single sample, and **metabonomics**, the study of changes in analytes in response to a drug or experimental treatment of some kind.

Historically, **electrolytes** such as sodium and potassium were measured by **flame photometry**, measuring the light emitted when the ion was introduced into a clear flame. Sodium gives a yellow flame and potassium a purple one. Nowadays these and other ions are measured using **ion-specific electrodes**. In some cases, metal ions are measured by **atomic absorption spectrometry**. Here the sample is introduced into a flame, and illuminated at a specific wavelength. The light energy absorbed excites electrons to an unstable orbital, and the absorption of light is directly proportional to the concentration of the element in the sample, as is the case for absorption spectrophotometry.

Enzymes in Clinical Chemistry

Enzymes are important in clinical chemistry in three different ways: to measure analytes in a sample; to measure the activity of enzymes themselves in a sample; and as a test of vitamin nutritional status.

Using an enzyme to measure the concentration of an analyte confers a high degree of specificity on the assay, since in general an enzyme will act on only a single substrate, or a small range of closely related substrates, while a simple chemical reaction may well respond to a variety of (possibly unrelated) analytes. For example, as shown in Figure 48–3, a variety of reducing compounds will react with an alkaline copper reagent to give a false-positive result for glucose, whereas the enzymic assay using glucose oxidase will only give a positive result for glucose, and not other reducing compounds.

When an enzyme is used to detect an analyte, the limiting factor in the assay must be the analyte itself; the other enzyme substrates must be present in excess. More importantly, the concentration of the analyte in the sample must be adjusted to be below the K_m of the enzyme, so that there is a large change

in the rate of reaction when the concentration of the analyte changes (region A in **Figure 48–4**).

When cells are damaged or die, their contents leak out into the bloodstream. Measurement of enzymes in plasma can therefore be used to detect tissue damage; information is obtained from the pattern of enzymes (and tissue-specific isoenzymes) released. The extent of the increase in enzyme activity in plasma above the normal range often indicates the severity of tissue damage. When an assay is to determine the activity of an enzyme in plasma, the limiting factor must be the enzyme itself. The concentration of substrate added must be considerably in excess of the K_m of the enzyme, so that the enzyme is acting at or near V_{max}, and even a relatively large change in the concentration of substrate does not have a significant effect on the rate of reaction (region B in Figure 48–4). In practice, this means that the concentration of substrate added is about 20-fold higher than the K_m of the enzyme.

If an enzyme has a vitamin-derived coenzyme that is essential for activity, then measurement of the activity of the enzyme in red blood cells with and without added coenzyme can be used as an index of vitamin nutritional status. This provides an indication of functional nutritional status, while measurement of the vitamin and its metabolites may reflect recent intake rather than physiological adequacy. The underlying assumption is that red blood cells have to compete with other tissues in the body for what may be a limited supply of the coenzyme. Therefore, the extent to which the red cell enzyme

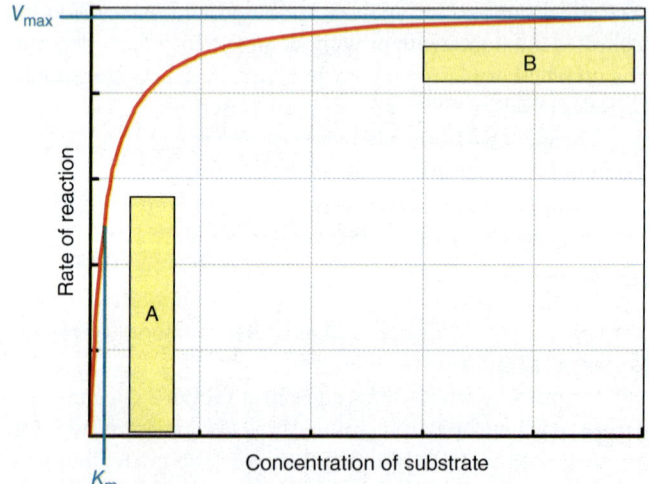

FIGURE 48–4 Use of enzymes to measure analytes and measurement of enzyme activity in biologic samples. At concentrations of substrate (analyte) at or below the K_m of the enzyme (region A in the graph), there is a very sharp increase in the rate of reaction with a small change in the concentration of analyte, so the enzyme-linked assay has greatest sensitivity over this range of concentration. At concentrations of substrate considerably above the K_m of the enzyme, as the enzyme is approaching V_{max}, (region B in the graph), it is the amount of enzyme in the sample that is limiting for the rate of formation of product, so that this is the appropriate range of substrate concentration to use for measurement of enzyme activity in a biological sample.

is saturated with its coenzyme will reflect the availability of the coenzyme over a period of time corresponding to the half-life of red cells. The assay consists of incubating two samples of the red cell lysate: one that has been preincubated with, and one without, addition of the coenzyme. Then substrate is added to both, and the activity of the enzyme is measured. In the sample preincubated without addition of the coenzyme, only that enzyme that had coenzyme bound (the holoenzyme) will be active. In the sample that was preincubated with the coenzyme, any apoenzyme (inactive enzyme protein without bound coenzyme) will have been activated to the holoenzyme. There is, therefore, always either no change in enzyme activity on addition of coenzyme, indicating complete saturation of the enzyme with coenzyme, or an increase in activity, reflecting the activation of the apoenzyme by added coenzyme. The results are reported as an enzyme **activation coefficient** (the ratio of activity in the sample preincubated with coenzyme: that without). Reference ranges for the activation coefficient are established in the same way as for any other test. Such enzyme activation assays are available for thiamin (vitamin B_1, using red cell transketolase), riboflavin (vitamin B_2, using red cell glutathione reductase), and vitamin B_6 (using one or the other of the red cell transaminases).

Competitive Ligand-Binding Assays and Immunoassays

If there is a protein that will bind the analyte, and the bound and free analyte (ligand) can be separated and measured, then it is possible to devise an assay for the analyte. Perhaps the simplest such ligand-binding assay is that for the hormone cortisol, which is transported in the bloodstream bound to a specific cortisol-binding globulin. It is easy to prepare a plasma sample containing the binding globulin that has been stripped of its ligand (cortisol) by incubation with aluminum oxide or charcoal. This is done using a relatively large pooled plasma sample, and provides binding globulin for a large number of assays. The hormone is extracted from each sample to be analyzed, using an organic solvent, evaporated to dryness, then dissolved in ethanol and a suitable buffer, with the addition of a trace amount of highly radioactive hormone. Each sample is then incubated with the binding globulin at 37°C, then cooled to 4°C. Charcoal is added to adsorb the unbound ligand, and rapidly removed by centrifugation. The radioactivity in the supernatant is measured. This gives a measure of bound ligand, and is expressed as a percentage of the total radioactivity added to each sample. A standard curve is constructed using known amounts of hormone, so that the concentration of hormone in the samples can be determined.

A wide variety of additional hormones and other analytes can be measured in the same way, by raising either monoclonal antibodies or polyclonal antisera against the analyte, for example, by injecting the analyte covalently bound to a protein into an animal. The antiserum against a hormone raised in a single rabbit can be used for many thousands of assays. Each batch of antiserum must, of course, be tested for its specificity

for the hormone (ensuring that it does not also bind related hormones, especially a problem with steroid hormones), and for its sensitivity. When the binding protein is an antibody or antiserum, the assay is usually called a **radioimmunoassay**.

In a variant of the competitive binding assay, the antibody is bound covalently to the surface of beads. It is then easy to separate the bound and free ligand simply by washing the beads with ice-cold buffer, leaving the bound ligand attached to the beads for measurement of bound radioactivity. Alternatively, the antibody may be bound covalently to the surface of the test tube, or to each well in a multiwell plate. After incubation, a sample of the incubation medium is taken for measurement of the radioactivity that is not bound.

Increasingly, in order to minimize exposure to radioactive materials, fluorescently labeled ligand or antibody is used. A further development is the **sandwich assay**, in which two different antibodies against the ligand are used, each of which binds to a different region (epitope) of the analyte. The first antibody is covalently bound to the surface of each well of a multiple well plate, and the sample is added and incubated. After removal of the incubation medium and washing each well, the second antibody is added, sandwiching the analyte between the two antibodies. The second antibody can be labeled with a radioactive isotope or a fluorophore, thus permitting measurement of the bound second antibody, and hence bound ligand. In most cases, the second antibody is labeled with an enzyme. The amount of bound second antibody, and hence bound ligand, is determined by measuring the activity of the enzyme bound to the walls of each well of the plate, after washing to remove unbound second antibody and adding the enzyme substrate. This is the **enzyme-linked immunosorbent assay (ELISA)**.

Dry Chemistry Dipsticks

For a number of assays, the enzymes or antibodies and reagents can be combined on a plastic strip. For measurement of blood glucose, a finger-prick blood sample is placed on the test strip that contains glucose and the reagents shown in Figure 48–3. The intensity of the blue color formed, and hence the concentration of glucose, is measured using a handheld device called the glucometer. This provides a simple and reliable method to estimate glucose at the bedside in a hospital ward, a doctor's clinic or even at home. For urine testing, several different assays can be included as separate pellets on a plastic stick called a dipstick—for example, to detect or semiquantitatively estimate levels of glucose, ketone bodies, protein, and several other analytes at the same time. Similar dipsticks are available to detect human chorionic gonadotropin (hCG) in urine, as a home pregnancy test.

Screening Neonates for Inborn Errors of Metabolism

Many of the inborn errors of metabolism can lead to very severe mental retardation if treatment is not initiated early enough. For conditions such as phenylketonuria and maple syrup urine disease, dietary restriction of the amino acids that are not metabolized normally (phenylalanine in phenylketonuria; the branched-chain amino acids leucine, isoleucine, and valine in maple syrup urine disease) is essential for management of the condition. Therefore, it is usual in most developed countries to screen neonates for such conditions. The concentration of the offending amino acid(s) is measured in a blood sample that is normally taken a week after birth, when the enzymes that are affected in the disease should have reached full expression. Most commonly, a capillary blood sample is taken by heel prick, and is blotted onto absorbent paper to be sent to the laboratory for analysis.

The first such screening test for an inborn error of metabolism was the Guthrie bacterial inhibition test. A disk from the paper containing the blood sample is laid onto an agar plate that has been seeded with a phenylalanine-requiring strain of *Bacillus subtilis*, together with a competitive inhibitor of phenylalanine uptake into the bacteria (β-thienylalanine) at such a concentration that it will compete with phenylalanine at levels normally found in blood, so that the bacteria will not grow. If the concentration of phenylalanine is more than that usually found in blood, it will be taken up by the bacteria despite the inhibitor, and the bacteria will form visible colonies on the agar.

In most centers, the bacterial inhibition test has been superseded by chromatographic techniques that permit the detection of a variety of abnormal metabolites, and hence the detection of a variety of different inborn errors of metabolism.

ORGAN FUNCTION TESTS

Tests that provide information on the functioning of particular organs are often grouped together as organ function tests. Such grouped tests include tests of kidney, liver, and thyroid function.

Tests of Kidney Function

A complete **urinalysis** includes assessment of the physical and chemical characteristics of urine. Physical characteristics to be assessed include urine volume (this requires a timed urine sample, usually 24 hours), odor, color, appearance (clear or turbid), specific gravity, and pH. Protein, glucose, blood, ketone bodies, bile salts, and bile pigments are abnormal constituents of urine that appear in different disease conditions.

Urea and **creatinine** are excreted in urine; their serum concentrations can be used as markers of renal function because the serum concentration increases as renal function deteriorates. Creatinine is a better marker of renal function than urea because its blood concentration is not significantly affected by nonrenal factors, thus making it a specific indicator of renal function; a number of factors affect blood urea concentration.

Normally, less than 150 mg of protein, and less than 30 mg of albumin, is excreted in urine per 24 hours. This is below the limit of detection by routine tests. Presence of protein in excess of this is referred to as **proteinuria** and is a sign of renal disease. The most common cause of proteinuria is loss of

integrity of the glomerular basement membrane (glomerular proteinuria), as seen in nephrotic syndrome and diabetic nephropathy. The major protein found in glomerular proteinuria is albumin. **Microalbuminuria** is defined as the presence of 30 to 300 mg of albumin in a 24-hour urine sample. It is an early marker of renal damage in diabetes mellitus.

Even though serum creatinine is a marker of renal function, a significant increase in its blood concentration is seen only when there has been about a 50% decline in the glomerular filtration rate (GFR). Measurement of serum creatinine is therefore a test with poor sensitivity. Measurement of **creatinine clearance** gives an estimate of the GFR, and so can be used to detect the early stages of renal failure. **Clearance** is the volume of plasma from which a compound is completely cleared by the kidney in unit time. It is calculated by the formula:

$$\text{Clearance (mL/min)} = (U \times V)/P$$

where U is concentration of the measured analyte in a timed sample of urine (usually 24 hours); P is plasma concentration of the analyte; and V is volume of urine produced per minute (calculated by dividing the volume of urine collected over 24 hours by $[24 \times 60]$).

A compound that is useful for measurement of renal clearance has a fairly constant blood concentration, is excreted only in urine, is freely filtered at the glomerulus, and is neither reabsorbed nor secreted by the renal tubules. Although creatinine clearance is commonly measured, it overestimates GFR because it is secreted by the renal tubules to a small extent. Inulin clearance satisfies all the criteria essential for a compound to be used in clearance tests. However, unlike creatinine, inulin is an exogenous compound that has to be infused intravenously at a constant rate.

Liver Function Tests

Liver function tests (LFTs) are a group of tests that help in diagnosis, assessing prognosis and monitoring therapy of liver disease. Each test assesses a specific aspect of liver function. An increase in **serum bilirubin** occurs due to many causes, and results in **jaundice**. In obstruction of the bile duct (obstructive jaundice), it is mainly conjugated bilirubin that increases. In hepatocellular disease, both conjugated and unconjugated bilirubin are elevated, reflecting the inability of the liver to take up, conjugate, and excrete bilirubin into the bile (see Chapter 31). Total serum protein and albumin levels are low in chronic liver diseases, such as cirrhosis. Prothrombin time (see Chapter 55) may be prolonged in acute disorders of the liver because of impaired synthesis of coagulation factors.

The activities of serum alanine (ALT) and aspartate (AST) aminotransferases (see Chapter 28) are significantly elevated several days before onset of jaundice in acute viral hepatitis. ALT is considered to be more specific for liver disease than AST, because AST is elevated in cases of cardiac or skeletal muscle injury while ALT is not. Serum alkaline phosphatase activity is elevated in obstructive jaundice. A high activity of serum alkaline phosphatase is also seen in bone disease.

Thyroid Function Tests

The thyroid gland secretes the thyroid hormones—thyroxine or tetraiodothyronine (T_4) and triiodothyronine (T_3). Diseases associated with increased or decreased synthesis of thyroid hormones (hyperthyroidism and hypothyroidism, respectively) occur commonly. A clinical diagnosis of a thyroid disorder is confirmed by measurement of serum thyroid-stimulating hormone (thyrotropin, TSH) and free thyroxine and triiodothyronine. The concentration of total serum thyroxine can be affected by changes in the concentration of thyroid-binding globulin in the absence of thyroid disease. Total thyroxine is seldom measured nowadays, because assays to measure free thyroxine are available.

Adrenal Function Tests

A clinical diagnosis of adrenal hyperfunction (Cushing syndrome) or hypofunction (Addison disease) is confirmed by adrenal function tests. Secretion of cortisol from the adrenal gland shows diurnal variation; serum cortisol is highest during the early morning hours and lowest around midnight. Loss of this diurnal variation is one of the earliest signs of adrenal hyperfunction. Measurement of serum cortisol in blood samples drawn at midnight and 8 AM is, therefore, useful as a test. A diagnosis of adrenal hyperfunction is confirmed by demonstration of failure of suppression of the early morning concentration of cortisol following the administration of 1 mg dexamethasone (a potent synthetic glucocorticoid) at midnight; this is the **dexamethasone suppression test**.

Markers of Cardiovascular Risk and Myocardial Infarction

As discussed in Chapter 25, plasma total cholesterol, and especially the ratio of LDL:HDL cholesterol provides an index of the risk of developing atherosclerosis. The plasma lipoproteins were originally separated by centrifugation, hence their classification by density. Later methods involved separation by electrophoresis. Nowadays, total plasma cholesterol is measured, then lipoproteins containing apoprotein B (see Table 25–1) are precipitated using a divalent cation, allowing measurement of the cholesterol associated with high-density lipoprotein (HDL).

An electrocardiogram may not always show typical changes following a myocardial infarction. In such a situation, elevation in serum levels of cardiac troponin or creatine kinase MB isoenzyme provides confirmation of the occurrence of a myocardial infarction, as both of these proteins are specific to cardiac muscle.

SUMMARY

- Laboratory tests can provide useful information for diagnosis and treatment of disease as well as providing information about normal metabolism and the pathology of disease.

- The reference range of an analyte is the range ± 2 × standard deviations around the mean value for the population group under consideration. Values outside this reference range are suggestive of an abnormality that merits further investigation.

- The precision of an analytical method is a measure of its reproducibility; the accuracy of a method is a measure of how close the result is to the true value.

- The sensitivity of an analytical method is a measure of how little of the analyte can be detected. The specificity is the extent to which other compounds present in the sample may give a false-positive result.

- The sensitivity of a test refers to the percentage of patients with the disease who will give a positive result. The specificity of a test is the percentage of patients without the disease who will give a negative result.

- Samples for analysis are usually blood and urine, although saliva, feces, and CSF may also be used. Blood samples may be collected in tubes containing an anticoagulant (for plasma samples) or without (for serum samples).

- Many laboratory tests rely on production of a colored or fluorescent product that can be measured by absorption spectrophotometry or fluorimetry.

- Many compounds can be measured by high-pressure liquid chromatography, sometimes in conjunction with mass spectrometry. The measurement of a large number of analytes in a sample is the basis of metabolomics, and of metabonomics, which is the effect of a disease, drug, or other treatment on metabolism.

- Enzymes may be used to provide sensitive and specific assay methods for analytes. In this case, there must be an excess of the other substrate(s) in the sample, so that the limiting factor is the concentration of the analyte in the sample.

- Many enzymes are released into the bloodstream from dying or damaged cells in disease, and their measurement can give useful diagnostic and prognostic information. In order to determine the activity of an enzyme in a sample, there must be an excess of substrate, so that the limiting factor is the amount of enzyme present.

- Many analytes (and especially hormones) are measured by competitive binding assays, using either a naturally occurring binding protein or an antiserum or monoclonal antibody to bind the ligand. Trace amounts of high specific activity radioactive ligand, or fluorescently labelled ligand or binding protein are used.

REFERENCES

Lab Tests Online: www.labtestsonline.org (A comprehensive web site provided by the American Association of Clinical Chemists that provides accurate information on many laboratory tests).

MedlinePlus: http://www.nlm.nih.gov/medlineplus/encyclopedia.html (The A.D.A.M. Medical Encyclopedia includes over 4000 articles about diseases, lab tests and other matters).

Exam Questions

Section IX – Special Topics (A)

1. Which of the following will be elevated in the bloodstream about 1 to 2 hours after eating a high-fat meal?

   A. Chylomicrons
   B. High density lipoprotein
   C. Ketone bodies
   D. Nonesterified fatty acids
   E. Very low-density lipoprotein

2. Which of the following will be elevated in the bloodstream about 4 to 5 hours after eating a high-fat meal?

   A. Chylomicrons
   B. High-density lipoprotein
   C. Ketone bodies
   D. Nonesterified fatty acids
   E. Very low-density lipoprotein

3. Which of the following is the best definition of glycemic index?

   A. The increase in the blood concentration of glucagon after consuming the food compared with that after an equivalent amount of white bread.
   B. The increase in the blood concentration of glucose after consuming the food.
   C. The increase in the blood concentration of glucose after consuming the food compared with that after an equivalent amount of white bread.
   D. The increase in the blood concentration of insulin after consuming the food.
   E. The increase in the blood concentration of insulin after consuming the food compared with that after an equivalent amount of white bread.

4. Which of the following will have the lowest glycemic index?

   A. A baked apple
   B. A baked potato
   C. An uncooked apple
   D. An uncooked potato
   E. Apple juice

5. Which of the following will have the highest glycemic index?

   A. A baked apple
   B. A baked potato
   C. An uncooked apple
   D. An uncooked potato
   E. Apple juice

6. Which one of the following statements concerning chylomicrons is CORRECT?

   A. Chylomicrons are made inside intestinal cells and secreted into lymph, where they acquire apolipoproteins B and C.
   B. The core of chylomicrons contains triacylglycerol and phospholipids.
   C. The enzyme hormone-sensitive lipase acts on chylomicrons to release fatty acids from triacylglycerol when they are bound to the surface of endothelial cells in blood capillaries.

   D. Chylomicron remnants differ from chylomicrons in that they are smaller and contain a lower proportion of triacylglycerol.
   E. Chylomicrons are taken up by the liver.

7. Plant sterols and stanols inhibit the absorption of cholesterol from the gastrointestinal tract. Which of the following best describes how they act?

   A. They are incorporated into chylomicrons in place of cholesterol.
   B. They compete with cholesterol for esterification in the intestinal lumen, so that less cholesterol is esterified.
   C. They compete with cholesterol for esterification in the mucosal cell, and unesterified cholesterol is actively transported out of the cell into the intestinal lumen.
   D. They compete with cholesterol for esterification in the mucosal cell, and unesterified cholesterol is not incorporated into chylomicrons.
   E. They displace cholesterol from lipid micelles, so that it is not available for absorption.

8. Which one of following statements about energy metabolism is CORRECT?

   A. Adipose tissue does not contribute to basal metabolic rate (BMR).
   B. Physical activity level (PAL) is the sum of physical activity ratios for different activities throughout the day, multiplied by the time spent in each activity, expressed as a multiple of BMR.
   C. Physical activity ratio (PAR) is the energy cost of physical activity throughout the day.
   D. Resting metabolic rate (RMR) is the energy expenditure of the body when asleep.
   E. The energy cost of physical activity can be determined by measuring respiratory quotient (RQ) production during the activity.

9. A patient with metastatic colorectal cancer has lost 6 kg of body weight over the last month. Which of the following is the best explanation for her weight loss?

   A. Because of the tumour she is oedematous.
   B. Chemotherapy has caused nausea and loss of appetite.
   C. Her basal metabolic rate has fallen as a result of protein catabolism caused by tumour necrosis factor and other cytokines.
   D. Her basal metabolic rate (BMR) has increased as a result of anaerobic glycolysis in the tumour and the energy cost of gluconeogenesis from the resultant lactate in her liver.
   E. The tumour has a very high energy requirement for cell proliferation.

10. A 5-year-old child arriving at a refugee center in East Africa is stunted in growth (only 89% of expected height for age) but not oedematous. Would you consider him to be:

    A. Suffering from kwashiorkor
    B. Suffering from marasmic kwashiorkor
    C. Suffering from marasmus
    D. Suffering from undernutrition
    E. Underfed but not considered to be clinically malnourished

11. A 5-year-old child arriving at a refugee center in East Africa is stunted in growth (only 55% of expected height for age) but not oedematous. Would you consider him to be:

    A. Suffering from kwashiorkor
    B. Suffering from marasmic kwashiorkor
    C. Suffering from marasmus
    D. Suffering from undernutrition
    E. Underfed but not considered to be clinically malnourished

12. Which of the following is the definition of nitrogen balance?

    A. Protein intake as a percentage of total energy intake
    B. The difference between protein intake and excretion of nitrogenous compounds
    C. The ratio of excretion of nitrogenous compounds/protein intake
    D. The ratio of protein intake/excretion of nitrogenous compounds
    E. The sum of protein intake and excretion of nitrogenous compounds

13. Which one of following statements about nitrogen balance is CORRECT?

    A. If the intake of protein is greater than requirements, there will always be positive nitrogen balance.
    B. In nitrogen equilibrium, the excretion of nitrogenous metabolites is greater than the dietary intake of nitrogenous compounds.
    C. In positive nitrogen balance, the excretion of nitrogenous metabolites is less than the dietary intake of nitrogenous compounds.
    D. Nitrogen balance is the ratio of intake of nitrogenous compounds/output of nitrogenous metabolites from the body.
    E. Positive nitrogen balance means that there is a net loss of protein from the body.

14. In a series of experiments to determine amino acid requirements, healthy young adult volunteers were fed mixtures of amino acids as their sole protein source. Which of the following mixtures would lead to negative nitrogen balance (assuming that all other amino acids are provided in adequate amounts)?

    A. One lacking alanine, glycine, and tyrosine
    B. One lacking arginine, glycine, and cysteine
    C. One lacking asparagine, glutamine, and cysteine
    D. One lacking lysine, glycine, and tyrosine
    E. One lacking proline, alanine, and glutamate

15. Which of the following vitamins provides the cofactor for reduction reactions in fatty acid synthesis?

    A. Folate
    B. Niacin
    C. Riboflavin
    D. Thiamin
    E. Vitamin B_6

16. Deficiency of which one of these vitamins is a major cause of blindness worldwide?

    A. Vitamin A
    B. Vitamin B_{12}
    C. Vitamin B_6
    D. Vitamin D
    E. Vitamin K

17. Deficiency of which one of these vitamins may lead to megaloblastic anaemia?

    A. Vitamin B_6
    B. Vitamin B_{12}
    C. Vitamin D
    D. Vitamin E
    E. Vitamin K

18. Which one of the following criteria of vitamin adequacy can be defined as "There are no signs of deficiency under normal conditions, but any trauma or stress reveals the precarious state of the body reserves and may precipitate clinical signs"?

    A. Abnormal response to a metabolic load
    B. Clinical deficiency disease
    C. Covert deficiency
    D. Incomplete saturation of body reserves
    E. Subclinical deficiency

19. Which one of the following criteria of vitamin adequacy can be defined as metabolic abnormalities under normal conditions?

    A. Abnormal response to a metabolic load
    B. Clinical deficiency disease
    C. Covert deficiency
    D. Incomplete saturation of body reserves
    E. Subclinical deficiency

20. Which of the following is the best definition of the reference nutrient intake (RNI) or recommended daily amount (RDA), of a vitamin or mineral?

    A. One standard deviation above the average requirement of the population group under consideration
    B. One standard deviation below the average requirement of the population group under consideration
    C. The average requirement of the population group under consideration
    D. Two standard deviations above the average requirement of the population group under consideration
    E. Two standard deviations below the average requirement of the population group under consideration

21. What percentage of the population will have met their requirement for a vitamin or mineral if their intake is equal to the RNI or RDA?

    A. 2.5%
    B. 5%
    C. 50%
    D. 95%
    E. 97.5%

22. What percentage of the population will have met their requirement for a vitamin or mineral if their intake is equal to the lower reference nutrient intake (LRNI)?
    A. 2.5%
    B. 5%
    C. 50%
    D. 95%
    E. 97.5%

23. What percentage of the population will have met their requirement for a vitamin or mineral if their intake is equal to the average requirement?
    A. 2.5%
    B. 5%
    C. 50%
    D. 95%
    E. 97.5%

24. For a person whose intake of a vitamin or mineral is equal to the average requirement, what is the probability that this level of intake is adequate to meet his/her individual requirement?
    A. 2.5%
    B. 5%
    C. 50%
    D. 95%
    E. 97.5%

25. For a person whose intake of a vitamin or mineral is equal to the LRNI, what is the probability that this level of intake is adequate to meet his/her individual requirement?
    A. 2.5%
    B. 5%
    C. 50%
    D. 95%
    E. 97.5%

26. For a person whose intake of a vitamin or mineral is equal to the RNI, what is the probability that this level of intake is adequate to meet his/her individual requirement?
    A. 2.5%
    B. 5%
    C. 50%
    D. 95%
    E. 97.5%

27. Which one of the following is NOT a source of oxygen radicals?
    A. Action of superoxide dismutase
    B. Activation of macrophages
    C. Nonenzymic reactions of transition metal ions
    D. Reaction of β-carotene with oxygen
    E. Ultraviolet radiation

28. Which one of the following provides protection against oxygen radical damage to tissues?
    A. Action of superoxide dismutase
    B. Activation of macrophages
    C. Nonenzymic reactions of transition metal ions
    D. Reaction of β-carotene with oxygen
    E. Ultraviolet radiation

29. Which one of the following is NOT the result of oxygen radical action?
    A. Activation of macrophages
    B. Modification of bases in DNA
    C. Oxidation of amino acids in apoproteins of LDL
    D. Peroxidation of unsaturated fatty acids in membranes
    E. Strand breaks in DNA

30. Which of the following types of oxygen radical damage may lead to the development of autoimmune thyroid disease?
    A. Chemical modification of DNA bases in somatic cells
    B. Chemical modification of DNA in germ-line cells
    C. Oxidation of amino acids in cell membrane proteins
    D. Oxidation of amino acids in mitochondrial proteins
    E. Oxidation of unsaturated fatty acids in plasma lipoproteins

31. Which of the following types of oxygen radical damage may lead to the development of atherosclerosis and coronary heart disease?
    A. Chemical modification of DNA bases in somatic cells
    B. Chemical modification of DNA in germ-line cells
    C. Oxidation of amino acids in cell membrane proteins
    D. Oxidation of amino acids in mitochondrial proteins
    E. Oxidation of unsaturated fatty acids in plasma lipoproteins

32. Which of the following types of oxygen radical damage may lead to the development of cancer?
    A. Chemical modification of DNA bases in somatic cells
    B. Chemical modification of DNA in germ-line cells
    C. Oxidation of amino acids in cell membrane proteins
    D. Oxidation of amino acids in mitochondrial proteins
    E. Oxidation of unsaturated fatty acids in plasma lipoproteins

33. Which of the following types of oxygen radical damage may lead to the development of hereditary mutations?
    A. Chemical modification of DNA bases in somatic cells
    B. Chemical modification of DNA in germ-line cells
    C. Oxidation of amino acids in cell membrane proteins
    D. Oxidation of amino acids in mitochondrial proteins
    E. Oxidation of unsaturated fatty acids in plasma lipoproteins

34. Which one of the following best explains the antioxidant action of vitamin E?
    A. It forms a stable radical that can be reduced back to active vitamin E by reaction with vitamin C.
    B. It is a radical, so that when it reacts with another radical, a nonradical product is formed.
    C. It is converted to a stable radical by reaction with vitamin C.
    D. It is lipid soluble and can react with free radicals in the blood plasma resulting from nitric oxide (NO) formation by vascular endothelium.
    E. Oxidized vitamin E can be reduced back to active vitamin E by reaction with glutathione and glutathione peroxidase.

35. Which of the following best describes the glycome?
    A. The DNA coding for glycosyltransferases
    B. The full complement of all carbohydrates in the body
    C. The full complement of free sugars in cells and tissues
    D. The full complement of glycoproteins and glycolipids in the body
    E. The full complement of glycosyltransferases in the body

36. Which of the following methods CANNOT be used to determine the structures of glycoproteins?
    A. Carbohydrate microarrays
    B. Degradation using endo- and exoglycosidases
    C. Genome analysis
    D. Mass spectrometry
    E. Sepharose-lectin chromatography

37. Which of the following is NOT a function of glycoproteins?
    A. Anchoring proteins at the cell surface
    B. Protecting plasma proteins against clearance by the liver
    C. Providing a transport system for folate into cells
    D. Providing a transport system for uptake of low-density lipoprotein into the liver
    E. Providing cell surface recognition signals

38. Which of the following is NOT a constituent of glycoproteins?
    A. Fucose
    B. Galactose
    C. Glucose
    D. Mannose
    E. Sucrose

39. Which of the following is used as a sugar donor in the synthesis of the common pentasaccharide of N-linked glycoproteins?
    A. CMP-N-acetylneuraminic acid
    B. Dolichol pyrophosphate N-acetylglucosamine
    C. Dolichol pyrophosphate-mannose
    D. GDP-fucose
    E. UDP-N-acetylglucosamine

40. Which of the following is NOT used as a sugar donor in the synthesis of N-linked glycoproteins in the endoplasmic reticulum?
    A. Dolichol pyrophosphate fructose
    B. Dolichol pyrophosphate galactose
    C. Dolichol pyrophosphate mannose
    D. Dolichol pyrophosphate N-acetylglucosamine
    E. Dolichol pyrophosphate N-acetylneuraminic acid

41. Which of the following best describes the attachment of the common pentapeptide to the apoprotein in synthesis of an N-linked glycoprotein?
    A. Direct glycation of the amino terminal amino acid of the peptide
    B. Displacement of the amino terminal region of the peptide in a transamidation reaction
    C. Displacement of the amino terminal region of the peptide in a transamination reaction
    D. Displacement of the carboxy terminal region of the peptide in a transamidation reaction
    E. Displacement of the carboxy terminal region of the peptide in a transamination reaction

42. Which of the following is NOT a glycoprotein?
    A. Collagen
    B. Immunoglobulin G
    C. Serum albumin
    D. Thyroid-stimulating hormone
    E. Transferrin

43. Which one of the following statements is INCORRECT?
    A. Calnexin ensures the correct folding of glycoproteins in the endoplasmic reticulum.
    B. Dolichol-pyrophosphate oligosaccharide donates all of the sugars found in N-linked glycoproteins.
    C. Mucins contain predominantly O-linked glycans.
    D. N-Acetylneuraminic acid is commonly found at the termini of N-linked sugar chains of glycoproteins.
    E. O-linked sugar chains of glycoproteins are built up by the stepwise addition of sugars from sugar nucleotides.

44. Which of the following is NOT an activity of cytochrome P450?
    A. Activation of vitamin D
    B. Hydroxylation of steroid hormone precursors
    C. Hydroxylation of xenobiotics
    D. Hydroxylation of retinoic acid
    E. Methylation of xenobiotics

45. Which of the following best describes the reaction of a cytochrome P450?
    A. $RH + O_2 + NADP^+ \rightarrow R\text{-}OH + H_2O + NADPH$
    B. $RH + O_2 + NAD^+ \rightarrow R\text{-}OH + H_2O + NADH$
    C. $RH + O_2 + NADPH \rightarrow R\text{-}OH + H_2O + NADP^+$
    D. $RH + O_2 + NADPH \rightarrow R\text{-}OH + H_2O_2 + NADP^+$
    E. $RH + O_2 + NADH \rightarrow R\text{-}OH + H_2O + NAD^+$

46. Which of the following is the preferred lipid component of the cytochrome P450 system?
    A. Dolichol phosphate
    B. Phosphatidylcholine
    C. Phosphatidylethanolamine
    D. Phosphatidylinositol
    E. Phosphatidylserine

47. Which of the following best describes the drug interactions between phenobarbital and warfarin?
    A. Phenobarbital induces CYP2C9, and this results in decreased catabolism of warfarin.
    B. Phenobarbital induces CYP2C9, and this results in increased catabolism of warfarin.
    C. Phenobarbital represses CYP2C9, and this results in increased catabolism of warfarin.
    D. Warfarin induces CYP2C9, and this results in decreased catabolism of phenobarbital.
    E. Warfarin induces CYP2C9, and this results in increased catabolism of phenobarbital.

48. Which of the following best describes the effects of polymorphisms of CYP2A6?

    A. People with the active allele are less likely to become tobacco-dependent smokers because this cytochrome inactivates nicotine to cotinine.
    B. People with the inactive (null) allele are less likely to become tobacco-dependent smokers because this cytochrome inactivates nicotine to cotinine.
    C. People with the inactive (null) allele are less likely to become tobacco-dependent smokers because this cytochrome activates nicotine to cotinine.
    D. People with the inactive (null) allele are more likely to become tobacco-dependent smokers because this cytochrome inactivates nicotine to cotinine.
    E. People with the inactive (null) allele are more likely to become tobacco-dependent smokers because this cytochrome activates nicotine to cotinine.

49. Which of the following is NOT a function of glutathione?

    A. Coenzyme for the reduction of hydrogen peroxide
    B. Conjugation of bilirubin
    C. Conjugation of some products of phase I metabolism of xenobiotics
    D. Transport of amino acids across cell membranes
    E. Transport of bilirubin in the bloodstream

50. Which of the following best describes the reference range for a laboratory test?

    A. A range $\pm 1 \times$ standard deviation around the mean value
    B. A range $\pm 1.5 \times$ standard deviation around the mean value
    C. A range $\pm 2 \times$ standard deviation around the mean value
    D. A range $\pm 2.5 \times$ standard deviation around the mean value
    E. A range $\pm 3 \times$ standard deviation around the mean value

51. Which of the following statements about laboratory tests is INCORRECT?

    A. The predictive value of a test is the extent to which it will correctly predict whether or not a person has the disease.
    B. The sensitivity and specificity of a test are inversely related.
    C. The sensitivity of a test is a measure of how many people with the disease will give a positive result.
    D. The specificity of a test is a measure of how many people with the disease will give a positive result.
    E. The specificity of a test is a measure of how many people without the disease will give a negative result.

52. Which of the following is CORRECT when an enzyme is used to measure an analyte in a blood sample?

    A. The concentration of substrate must be about 20-times the K_m of the enzyme.
    B. The concentration of substrate must be equal to the K_m of the enzyme.
    C. The concentration of substrate must be equal to or lower than the K_m of the enzyme.
    D. The concentration of the substrate in the assay is not important.
    E. The concentration of substrate must be about 1/20th of the K_m of the enzyme.

53. Which of the following is CORRECT when an enzyme is being measured in a blood sample?

    A. The concentration of substrate must be about 20 times the K_m of the enzyme.
    B. The concentration of substrate must be equal to the K_m of the enzyme.
    C. The concentration of substrate must be equal to or lower than the K_m of the enzyme.
    D. The concentration of the substrate in the assay is not important.
    E. The concentration of substrate must be about 1/20th of the K_m of the enzyme.

54. Which of the following best explains the use of enzyme activation assays to assess vitamin nutritional status?

    A. Adding the vitamin-derived cofactor to the incubation converts previously inactive apoenzyme into active holoenzyme.
    B. Adding the vitamin-derived cofactor to the incubation converts previously inactive holoenzyme into active apoenzyme.
    C. Adding the vitamin-derived cofactor to the incubation converts previously active holoenzyme into inactive apoenzyme.
    D. Adding the vitamin-derived cofactor to the incubation converts previously active apoenzyme into inactive holoenzyme.
    E. Adding the vitamin-derived cofactor to the incubation leads to a reduction in enzyme activity.

55. Which of the following would be used to prepare serum from a blood sample?

    A. A plain tube
    B. A tube containing citrate
    C. A tube containing EDTA
    D. A tube containing oxalate
    E. An evacuated tube to exclude oxygen

56. Which of the following would be used to take a blood sample for blood gas analysis?

    A. A plain tube
    B. A tube containing citrate
    C. A tube containing EDTA
    D. A tube containing oxalate
    E. An evacuated tube to exclude oxygen

57. Which of the following best explains the difference between creatinine clearance and inulin clearance as tests of renal function?

    A. Creatinine clearance is higher than inulin clearance because creatinine is actively secreted in the distal renal tubules.
    B. Creatinine clearance is higher than inulin clearance because inulin is actively secreted in the proximal renal tubules.
    C. Creatinine clearance is higher than inulin clearance because inulin is actively secreted in the distal renal tubules.
    D. Creatinine clearance is lower than inulin clearance because creatinine is actively secreted in the distal renal tubules.
    E. Creatinine clearance is lower than inulin clearance because inulin is not completely filtered at the glomerulus.

C H A P T E R

Intracellular Traffic & Sorting of Proteins

49

Kathleen M. Botham, PhD, DSc, &
Robert K. Murray, MD, PhD

O B J E C T I V E S

After studying this chapter,
you should be able to:

- Indicate that many proteins are targeted by signal sequences to their correct destinations and that the Golgi apparatus plays an important role in sorting proteins.

- Recognize that specialized signals are involved in sorting proteins to mitochondria, the nucleus, and to peroxisomes.

- Explain that N-terminal signal peptides play a key role in directing newly synthesized proteins into the lumen of the endoplasmic reticulum (ER).

- Explain how chaperones prevent faulty folding of other proteins, how misfolded proteins are disposed of, and how the ER acts as a quality control compartment.

- Explain the role of ubiquitin as a key molecule in protein degradation.

- Recognize the important role of transport vesicles in intracellular transport.

- Indicate that many diseases result from mutations in genes encoding proteins involved in intracellular transport.

BIOMEDICAL IMPORTANCE

Proteins are synthesized on polyribosomes, but perform varied functions at different subcellular locations, including the cytosol, specific organelles, or membranes. Yet others are destined for export. Thus, there is considerable **intracellular traffic of proteins**. As recognized by Blobel in 1970, to enable proteins to attain their proper locations, their structure contains a signal or coding sequence that **targets** them appropriately. This led to the identification of numerous specific signals (**Table 49–1**), and also the recognition that **certain diseases** result from mutations that adversely affect these signals. This chapter will discuss the sorting and intracellular traffic of proteins and briefly consider some of the disorders that result when abnormalities occur.

MANY PROTEINS ARE TARGETED BY SIGNAL SEQUENCES TO THEIR CORRECT DESTINATIONS

The protein biosynthetic pathways in cells can be considered to be **one large sorting system**. Many proteins carry **signals** (usually but not always specific sequences of amino acids known as targeting sequences) that direct them to their specific subcellular destinations (see Table 49–1); these signals are a fundamental component of the sorting system. Usually, the signal sequences are recognized by and interact with complementary areas of other proteins that serve as receptors.

TABLE 49–1   Sequences or Molecules That Direct Proteins to Specific Organelles

Targeting Sequence or Compound	Organelle Targeted
N-terminal signal peptide	ER
Carboxyl-terminal KDEL sequence (Lys-Asp-Glu-Leu) in ER-resident proteins in COPI vesicles	Lumen of ER
Diacidic sequences (eg, Asp-X-Glu) in membrane proteins in COPII vesicles	Golgi membranes
Amino terminal sequence (20-50 residues)	Mitochondrial matrix
NLS (eg, Pro$_2$-Lys$_3$-Arg-Lys-Val)	Nucleus
PTS (eg, Ser-Lys-Leu [SKL])	Peroxisome
Mannose 6-phosphate	Lysosome

Abbreviations: ER, endoplasmic reticulum; NLS, nuclear localization signal; PTS, peroxisomal-matrix targeting sequence.

A major sorting decision is made early in protein biosynthesis, when specific proteins are synthesized either on **cytosolic** (free) or **membrane-bound polyribosomes** (see Chapter 37). The **signal hypothesis** was proposed by Blobel and Sabatini in 1971 partly to explain the distinction between free and membrane-bound polyribosomes. They proposed that proteins synthesized on membrane-bound polyribosomes contain an **N-terminal signal peptide**, which causes them to become attached to the membranes of the endoplasmic reticulum (ER), and facilitates protein transfer into the ER lumen. On the other hand, proteins synthesized on free polyribosomes lack the signal peptide and retain free movement in the cytosol. An important aspect of the signal hypothesis is that **all ribosomes have the same structure**, and that the distinction between membrane-bound and free ribosomes depends solely on the former carrying proteins that have signal peptides. Because many membrane proteins are synthesized on membrane-bound polyribosomes, the signal hypothesis plays an important role in **concepts of membrane assembly**. ER regions containing attached polyribosomes are called the **rough ER (RER)**, and the distinction between the two types of ribosomes results in two branches of the protein-sorting pathway, called the **cytosolic branch** and the **RER branch** (**Figure 49–1**).

Proteins synthesized by cytosolic polyribosomes are directed to mitochondria, nuclei, and peroxisomes by specific signals, or remain in the cytosol if they lack a signal. Any protein that contains a targeting sequence that is subsequently removed is designated as a **preprotein**. In some cases, a second peptide is also removed, and in that event the original, newly synthesized protein is known as a **preproprotein** (eg, preproalbumin; see Chapter 52).

Proteins synthesized and sorted in the **RER branch** (see Figure 49–1) include many destined for various membranes (eg, of the ER, Golgi apparatus [GA], plasma membrane [PM]), and also lysosomal enzymes. In addition, proteins for **export from the cell via exocytosis** (secretion) are synthesized via this route. These various proteins may thus reside in the membranes or lumen of the ER, or follow the major

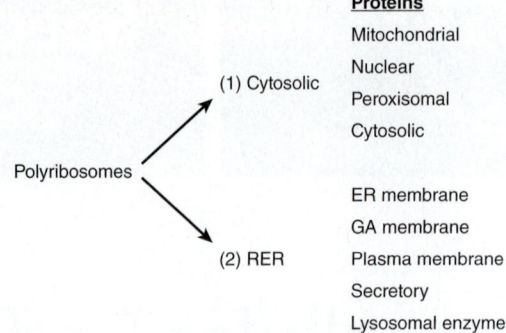

FIGURE 49–1   The two branches of protein sorting. Proteins are synthesized on cytosolic (free) polyribosomes or membrane-bound polyribosomes in the rough endoplasmic reticulum. Mitochondrial proteins encoded by nuclear genes are derived from the cytosolic pathway. (ER; endoplasmic reticulum; GA, Golgi apparatus; RER, rough endoplasmic reticulum.)

transport route of intracellular proteins to the GA. In the **secretory** or **exocytotic pathway**, proteins are transported from the ER → GA → PM and then released into the external environment. Secretion may be **constitutive**, meaning that transport occurs continuously, or **regulated**, where transport is switched on and off as required. Proteins destined for the GA, the PM, certain other sites, or for constitutive secretion are carried in **transport vesicles** (**Figure 49–2**) (see also following discussion). Other proteins which are subject to **regulated secretion** are carried in **secretory vesicles** (see Figure 49–2). These are particularly prominent in the pancreas and certain other glands. Passage of enzymes to the lysosomes using the mannose 6-phosphate signal is described in Chapter 46.

The Golgi Apparatus Is Involved in Glycosylation & Sorting of Proteins

The **GA** plays two major roles in protein synthesis. First, it is involved in the **processing of the oligosaccharide chains** of membrane and other N-linked glycoproteins and also contains enzymes involved in O-glycosylation (see Chapter 46). Second, it is involved in the **sorting** of various proteins prior to their delivery to their appropriate intracellular destinations. **The GA consists of *cis*- (facing the ER), medial and *trans*-cisternae (membrane stacks), and the *trans*-Golgi network (TGN)** (see Figure 49–2). All parts of the GA participate in oligosaccharide chain processing, whereas the TGN is particularly involved in protein sorting, and is very rich in vesicles.

Chaperones Are Proteins That Stabilize Unfolded or Partially Folded Proteins

Molecular chaperones are proteins which **stabilize unfolded or partially folded intermediates**, allowing them time to fold covalently and preventing inappropriate interactions, thus combating the formation of nonfunctional structures. Three families of these proteins are known, Hsp70, Hsp 90, and Hsp33. Most chaperones exhibit **ATPase activity** and bind ADP and ATP. This activity is important for their effect on protein folding.

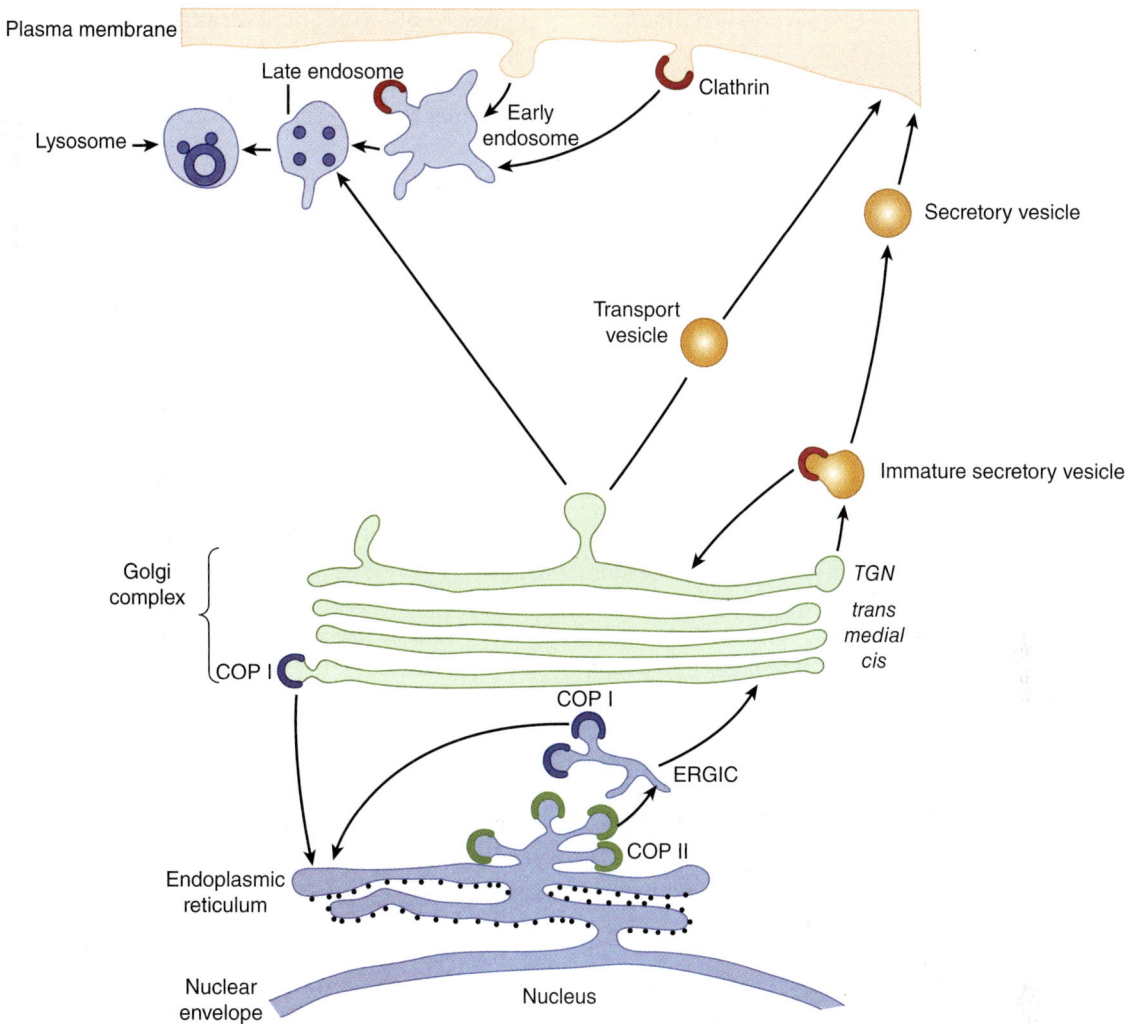

FIGURE 49–2 **The rough ER branch of protein sorting.** Newly synthesized proteins are inserted into the ER membrane or lumen from membrane-bound polyribosomes (small black circles studding the cytosolic face of the ER). Proteins that are transported out of the ER are carried in COPII vesicles to the *cis*-Golgi (anterograde transport). Proteins move through the Golgi as the cisternae (membrane sac-like structures) mature. In the *trans*-Golgi network (TGN), the exit side of the Golgi, proteins are segregated and sorted. For regulated secretion, proteins accumulate in secretory vesicles, while proteins destined for insertion in the plasma membrane for constitutive secretion are carried to the cell surface in transport vesicles. Clathrin-coated vesicles are involved in endocytosis, carrying cargo to late endosomes and to lysosomes. Mannose 6-phosphate (not shown; see Chapter 46) acts as a signal for transporting enzymes to lysosomes. COPI vesicles transport protein from GA to the ER (retrograde transport) and may be involved in some intra-Golgi transport. Cargo normally passes through the ER-Golgi intermediate complex (ERGIC) compartment to the GA. (Reproduced with permission from E Degen.)

The ADP-chaperone complex often has a high affinity for the unfolded protein, which, when bound, stimulates the replacement of ADP with ATP. The ATP-chaperone complex, in turn, releases segments of the protein that have folded properly, and the **cycle** involving ADP and ATP binding is repeated until the protein is released. Chaperones are required for the correct targeting of proteins to their subcellular locations. A number of important properties of these proteins are listed in **Table 49–2**.

Chaperonins are a second type of chaperone which enable the correct folding of denatured proteins. They also differ from chaperones in that they are much larger, being oligomers with a molecular weight of 800 kDa as compared to monomers of 70 to 100 kDa. The structure of the bacterial chaperonin **GroEL/GroES** has been studied in detail. It has two ring-like structures, each composed of seven identical subunits, which

form barrel-like shape, and again ATP is involved in its action. Its structure allows it to sequester an unfolded protein away from other proteins, giving it time and suitable conditions to fold properly. The heat shock protein **Hsp60** is the equivalent of GroEL in eukaryotes.

THE CYTOSOLIC PROTEIN SORTING BRANCH DIRECTS PROTEINS TO SUBCELLULAR ORGANELLES

Proteins synthesized via the cytosolic sorting branch either have no targeting signal, if they are destined for the cytosol, or they contain a signal which enables them to be taken up

TABLE 49–2 Some Properties of Chaperone Proteins

- Present in a wide range of species from bacteria to humans
- Many are so-called heat shock proteins (Hsp), which are induced by stressors including heat, oxidants, toxins, free radicals, and viruses
- Some are inducible by conditions that cause unfolding of newly synthesized proteins (eg, elevated temperature and various chemicals)
- They bind to predominantly hydrophobic regions of unfolded proteins and prevent their aggregation
- They act in part as a quality control or editing mechanism for detecting misfolded or otherwise defective proteins
- Most chaperones show associated ATPase activity, with ATP or ADP being involved in the protein-chaperone interaction
- Found in various cellular compartments such as cytosol, mitochondria, and the lumen of the endoplasmic reticulum

into the correct subcellular organelle. Specific uptake signals direct proteins to the mitochondria, nucleus, and peroxisomes (see Table 49–1). Since protein synthesis is complete before transport occurs, these processes are termed posttranslational translocation.

Specific mechanisms for the import of proteins into mitochondria, the nucleus, peroxisomes, and the ER are discussed in the following discussion, but in general, there are usually three stages, recognition, translocation, and maturation. Energy is required for translocation and chaperone proteins play a part in keeping the protein unfolded and in pulling of the protein through the membrane. Folding and processing of the newly translocated protein is carried out by other proteins within the organelle.

Most Mitochondrial Proteins Are Imported

Mitochondria contain many proteins. Thirteen polypeptides (mostly membrane components of the electron transport chain) are encoded by the **mitochondrial (mt) genome** and synthesized in that organelle using its own protein-synthesizing system. However, the vast majority (at least several hundreds) are encoded by **nuclear genes**, and synthesized outside the mitochondria on **cytosolic polyribosomes**, and must be imported. Most progress has been made in the study of proteins present in **the mitochondrial matrix**, such as the ATP synthase subunits (see Chapter 13). Only the pathway of import of matrix proteins will be discussed in any detail here.

Matrix proteins must pass from cytosolic polyribosomes through the **outer** and **inner mitochondrial membranes** to reach their destination. Passage through the two membranes is called **translocation**. They have an amino terminal leader sequence (**presequence**), about 20 to 50 amino acids in length (see Table 49–1), which is amphipathic and contains many hydrophobic and positively charged amino acids (eg, Lys or Arg). The presequence is equivalent to a signal peptide mediating attachment of polyribosomes to membranes of the ER (see later), but in this instance it **targets proteins to the matrix**. Some general features of the passage of a protein from the cytosol to the mitochondrial matrix are shown in **Figure 49–3**.

Translocation occurs **posttranslationally**, after the matrix preproteins are released from the cytosolic polyribosomes. Interactions with a number of cytosolic proteins that act as **chaperones** (see later) and as **targeting factors** occur prior to translocation.

Two distinct **translocation complexes** are situated in the outer and inner mitochondrial membranes, referred to (respectively) as **translocase of the outer membrane (TOM)** and translocase of the inner membrane (**TIM**). Each complex has been analyzed and found to be composed of a number of proteins, some of which act as **receptors** (eg, **Tom20/22**) for the incoming proteins and others as **components** (eg, **Tom40**) **of the transmembrane pores** through which these proteins must pass. Proteins must be in the **unfolded state** to pass through the complexes, and this is made possible by **ATP-dependent binding to several chaperone proteins** including **Hsp70** (see Figure 49–3). In mitochondria, chaperones are involved in translocation, sorting, folding, assembly, and degradation of imported proteins. A **proton-motive force** across the inner membrane is required for import; it is made up of the **electric potential** across the membrane (inside negative) and the **pH gradient** (see Chapter 13). The positively charged leader sequence may be helped through the membrane by the negative charge in the matrix. In addition, close apposition at **contact sites** between the outer and inner membranes is necessary for translocation to occur.

The presequence is split off in the matrix by a **matrix protease**. Contact with **other chaperones** present in the matrix is essential to complete the overall process of import. Interaction with mt-Hsp70 (mt = mitochondrial; Hsp = heat shock protein; 70 = ~70 kDa) ensures proper import into the matrix and prevents misfolding or aggregation, while interaction with the mt-Hsp60–Hsp10 system ensures proper folding. The interactions of imported proteins with the above chaperones require **hydrolysis of ATP** to drive them.

The above describes the major pathway of proteins destined for the mitochondrial matrix. However, certain proteins insert into the **outer mitochondrial membrane** facilitated by the TOM complex. Others remain in the **intermembrane space**, and some insert into the **inner membrane**. Yet others proceed into the matrix and then return to the inner membrane or intermembrane space. A number of proteins contain two signaling sequences—one to enter the mitochondrial matrix and the other to mediate subsequent relocation (eg, into the inner membrane). Certain mitochondrial proteins do not contain presequences (eg, cytochrome *c*, which locates in the intermembrane space), and others contain **internal presequences**. Overall, proteins employ a variety of mechanisms and routes to attain their final destinations in mitochondria.

Transport of Macromolecules in & out of the Nucleus Involves Localization Signals

It has been estimated that more than a million macromolecules per minute are transported between the nucleus and

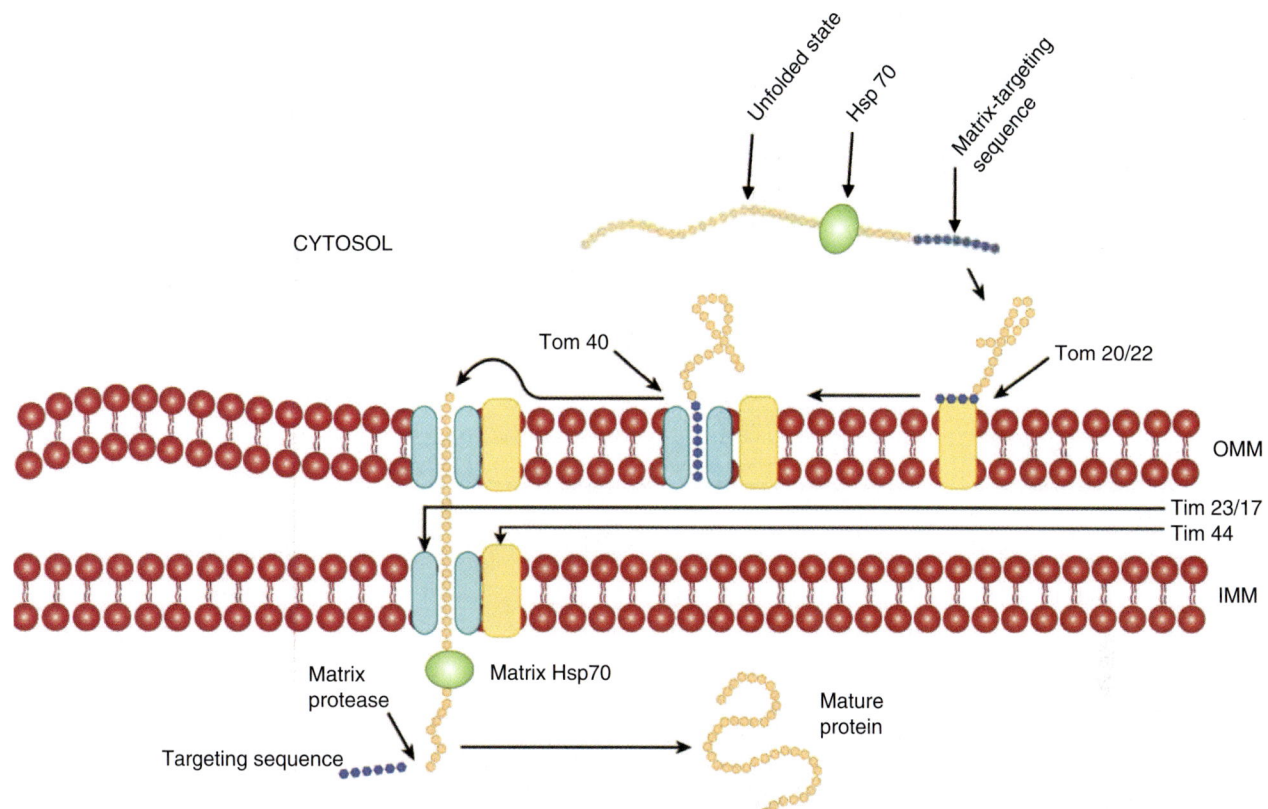

FIGURE 49–3 **Entry of a protein into the mitochondrial matrix.** After synthesis on cytosolic polyribosomes, an unfolded protein containing a matrix-targeting sequence interacts with the cytosolic chaperone Hsp70 and then with mitochondria (mt) via the receptor translocase of the outer membrane (Tom) 20/22. Transfer to the import channel Tom 40 followed by translocation across the outer membrane is the next step. Transport across the inner mt membrane occurs via a complex comprising Tim (translocase of the inner membrane) 23 and Tim 17 proteins. On the inside of the inner mt membrane, the protein interacts with the matrix chaperone Hsp 70, which in turn interacts with membrane protein Tim 44. The hydrolysis of ATP by mt Hsp70 probably helps drive the translocation, as does the electronegative interior of the matrix. The targeting sequence is subsequently cleaved by the matrix protease, and the imported protein assumes its final shape, sometimes with the prior aid of interaction with an mt chaperonin. At the site of translocation, the outer and inner mt membranes are in close contact. OMM, outer mitochondrial membrane; IMM, inner mitochondrial membrane.

the cytoplasm in an active eukaryotic cell. These macromolecules include histones, ribosomal proteins and ribosomal subunits, transcription factors, and mRNA molecules. The transport is bidirectional and occurs through the **nuclear pore complexes** (NPCs). These are complex structures with a mass approximately 15 times that of a ribosome and are composed of aggregates of about 30 different proteins. The minimal diameter of an NPC is approximately 9 nm. Molecules smaller than about 40 kDa can pass through the channel of the NPC by **diffusion**, but **special translocation mechanisms** exist for larger molecules.

Here we shall mainly describe current knowledge about the **nuclear import** of certain macromolecules. The general picture that has emerged is that proteins to be imported (cargo molecules) carry a **nuclear localization signal (NLS)**. One example of an NLS is the amino acid sequence $(Pro)_2$-$(Lys)_3$-Arg-Lys-Val (see Table 49–1), which is markedly rich in basic residues. A cargo molecule containing an NLS interacts with **importin**, a soluble protein which has two subunits, termed α and β. For the transport of proteins into the nucleus, importin

usually functions as a heterodimer, importin α/β. The α subunit binds to the NLS and the complex **docks** transiently at the NPC via interaction with the β subunit. Another family of proteins called **Ran** plays a critical regulatory role in the interaction of the complex with the NPC and in its translocation through the NPC. Ran proteins are small monomeric nuclear **GTPases** and, like other GTPases, exist in either GTP-bound or GDP-bound states. They are themselves regulated by **guanine nucleotide exchange factors (GEFs)**, which are located in the nucleus, and Ran **GTPase-accelerating proteins (GAPs)**, which are predominantly cytoplasmic. The GTP-bound state of Ran is favored in the nucleus and the GDP-bound state in the cytoplasm. The conformations and activities of Ran molecules vary depending on whether GTP or GDP is bound to them (the GTP-bound state is active; see discussion of G-proteins in Chapter 42). The **asymmetry** between nucleus and cytoplasm—with respect to which of these two nucleotides is bound to Ran molecules—is thought to be crucial in understanding the roles of Ran in transferring complexes unidirectionally across the NPC. When **cargo molecules** are

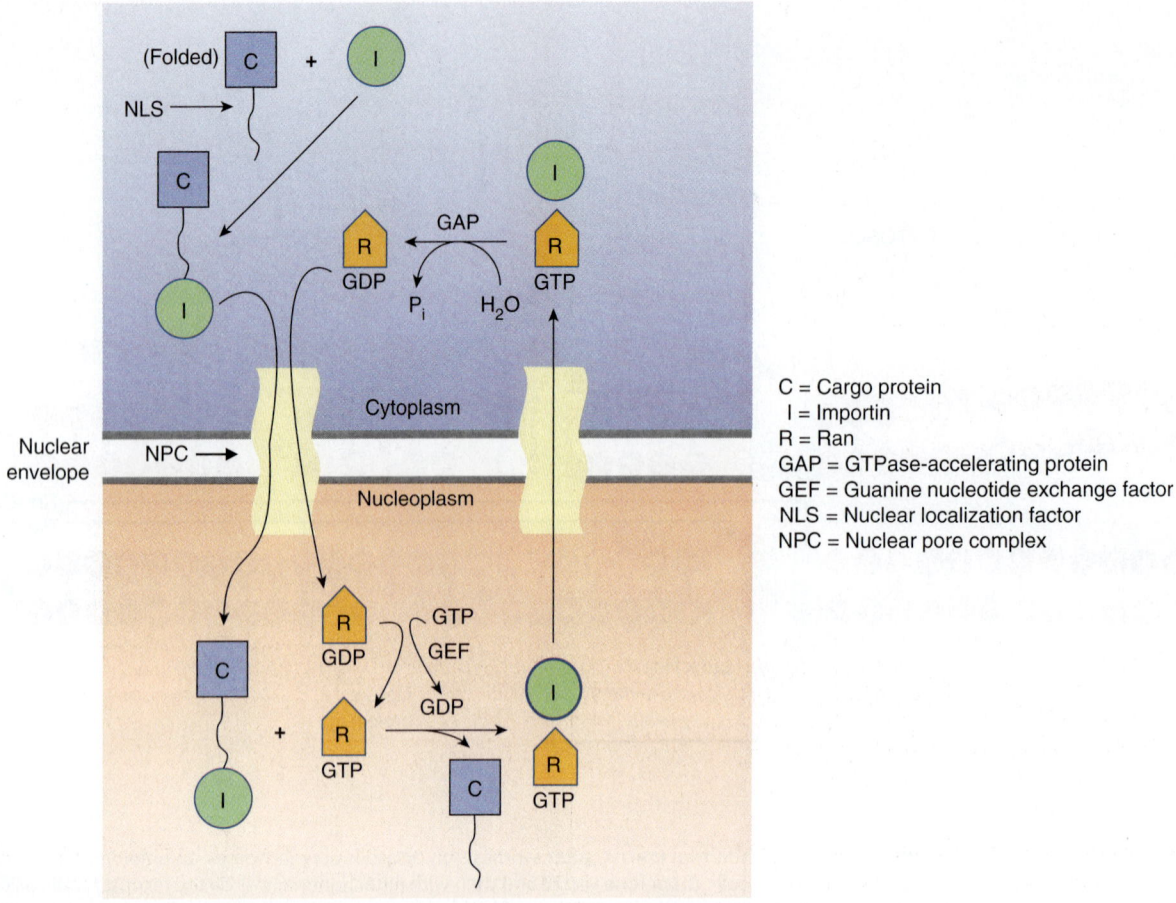

FIGURE 49–4 **The entry of a protein into the nucleoplasm.** A cargo molecule (C) in the cytoplasm interacts via its nuclear localization signal (NLS) to form a complex with importin (I). This complex binds to Ran (R)·GDP and traverses the nuclear pore complex (NPC) into the nucleoplasm. In the nucleoplasm, Ran·GDP is converted to Ran·GTP by guanine nuclear exchange factor (GEF), causing a conformational change in Ran which releases the cargo molecule. The I-Ran·GTP complex then leaves the nucleoplasm via the NPC to return to the cytoplasm. Here I is released by the action of GTPase-accelerating protein (GAP), which converts GTP to GDP, enabling it to bind to another C. The Ran·GTP is the active form of the complex, with the Ran·GDP form is inactive. Directionality is believed to be conferred on the overall process by the dissociation of Ran·GTP in the cytoplasm.

released inside the nucleus, the **importins recirculate to the cytoplasm** to be used again. **Figure 49–4** summarizes some of the principal features in the above process.

Proteins similar to importins, referred to as **exportins**, are involved in the export of many macromolecules (various proteins, tRNA molecules, ribosomal subunits, and certain mRNA molecules) from the nucleus. Cargo molecules for export carry **nuclear export signals (NESs)**. Ran proteins are involved in this process also, and it is now established that the processes of import and export share a number of common features. The family of importins and exportins are referred to as **karyopherins**.

Another system is involved in the translocation of the majority of **mRNA molecules**. These are exported from the nucleus to the cytoplasm as messenger ribonucleoprotein (mRNP) complexes attached to a protein termed **mRNP exporter** that carries mRNP molecules through the NPC. Ran is not involved. This system appears to use the hydrolysis of **ATP** by an RNA helicase (Dbp5) to drive translocation.

Other **small monomeric GTPases** (eg, ARF, Rab, Ras, and Rho) are important in various cellular processes such as vesicle formation and transport (ARF and Rab; see later), certain growth and differentiation processes (Ras), and formation of the actin cytoskeleton (Rho). A process involving GTP and GDP is also crucial in the transport of proteins across the membrane of the ER (see later).

Proteins Imported Into Peroxisomes Carry Unique Targeting Sequences

The **peroxisome** is an important organelle involved in aspects of the metabolism of many molecules, including fatty acids and other lipids (eg, plasmalogens, cholesterol, and bile acids), purines, amino acids, and hydrogen peroxide. The peroxisome is bounded by a single membrane and contains more than 50 enzymes; catalase and urate oxidase are marker enzymes for this organelle. Its proteins are **synthesized on cytosolic polyribosomes** and fold prior to import. The pathways of import of a number of its proteins and enzymes have been studied, some being **matrix components (Figure 49–5)** and others **membrane components**. At least two **peroxisomal–matrix**

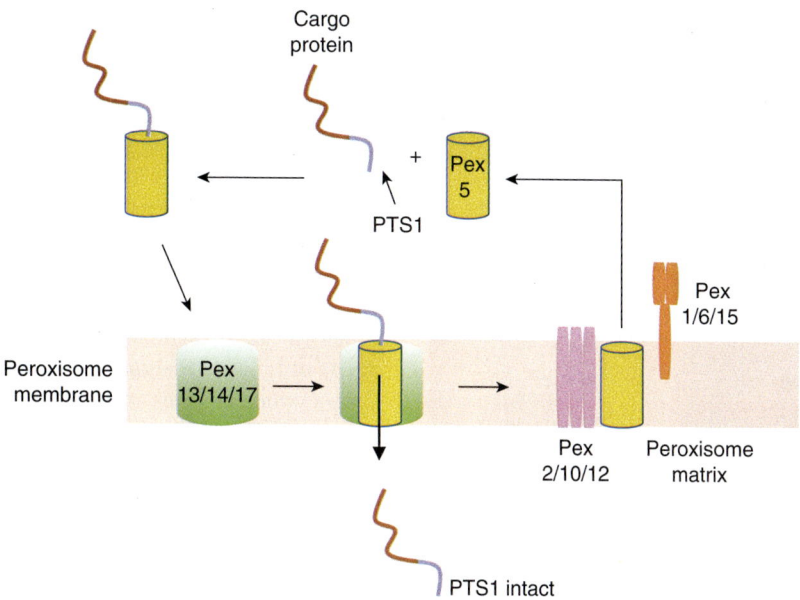

FIGURE 49–5 **Entry of a protein into the peroxisomal matrix.** The protein for import into the matrix (eg, catalase) is synthesized on cytosolic polyribosomes, assumes its folded shape prior to import, and contains a C-terminal peroxisomal-targeting sequence (PTS1) which docks with the cytosolic receptor protein Pex5 and the complex interacts with a receptor on the peroxisomal membrane consisting of Pex13, 14, and 17 (Pex13/14/17). A transient pore is then formed in the membrane and the protein is translocated into the peroxisome matrix. Pex 5 remains in the membrane and is recycled to the cytosol with the aid of complexes of Pex 2, 10, and 12 (Pex2/10/12) and 1, 6, and 15 (Pex 1/6/15). The protein retains PTS1 in the matrix.

targeting sequences (PTSs) have been discovered. One, **PTS1**, is a tripeptide (ie, Ser-Lys-Leu [SKL], but variations of this sequence have been detected) located at the carboxyl terminal of a number of matrix proteins, including catalase. Proteins containing PTS1 sequences **form complexes** with a cytosolic receptor protein (**Pex5**) which then dock with a peroxisomal membrane receptor complex containing **Pex13, 14, and 17**, and the cargo protein is translocated into the lumen. The Pex5 receptor remains in the membrane and is recycled to the cytosol by a process involving other Pexs, including complexes of **Pex2, 10, and 12** and **Pex1, 6, and 15**. A second peroxisome matrix targeting sequence, **PTS2**, is a nine amino acid sequence at the N-terminus and has been found in at least four matrix proteins (eg, thiolase). Proteins containing PTS2 sequences are known to complex with the receptor protein **Pex7** rather than Pex5, but the exact translocation mechanism is less well understood than for PTS1. The PTS1 and PTS2 signals are not cleaved after entry into the matrix.

Most **peroxisomal membrane proteins** have been found to contain neither of the two targeting sequences, but seem to contain others. The import system can handle **intact oligomers** (eg, tetrameric catalase). Import of **matrix proteins** requires **ATP**, whereas import of **membrane proteins** does **not**.

Most Cases of Zellweger Syndrome Are due to Mutations in Genes Involved in the Biogenesis of Peroxisomes

Interest in import of proteins into peroxisomes has been stimulated by studies on **Zellweger syndrome**. This condition is apparent at birth and is characterized by **profound neurologic impairment**, with victims often dying within a year. The number of peroxisomes can vary from being almost normal to being virtually absent in some patients. Biochemical findings include an accumulation of very-long-chain fatty acids, abnormalities of the synthesis of bile acids, and a marked reduction of plasmalogens. The condition is usually caused by **mutations** in genes encoding certain proteins, such as the PEX family of genes (also called **peroxins**), involved in various steps of **peroxisome biogenesis** (eg, the import of proteins described earlier and in Figure 49–5), or in genes encoding certain peroxisomal enzymes themselves. Two closely related conditions are **neonatal adrenoleukodystrophy** and **infantile Refsum disease**. Zellweger syndrome and these two conditions represent a **spectrum** of overlapping features, with Zellweger syndrome being the **most severe** (many proteins affected) and infantile Refsum disease the least severe (only one or a few proteins affected). **Table 49–3** lists these and related conditions.

PROTEINS SORTED VIA THE ROUGH ER BRANCH HAVE N-TERMINAL SIGNAL PEPTIDES

As indicated earlier, the **RER branch** is the second of the two branches involved in the synthesis and sorting of proteins. In this branch, proteins have **N-terminal signal peptides** and are synthesized on **membrane-bound polyribosomes**. They are usually **translocated into the lumen** of the RER prior to further sorting (see Figure 49–2). Certain membrane proteins,

TABLE 49–3 Disorders due to Peroxisomal Abnormalities

	OMIM Number[a]
Zellweger syndrome	214100
Neonatal adrenoleukodystrophy	202370
Infantile Refsum disease	266510
Pipecolic acidemia	239400
Rhizomelic chondrodysplasia punctata (RCDP) type 1	215100
X-linked Adrenoleukodystrophy	300100
Acyl-CoA oxidase deficiency	264470
D-bifunctional protein deficiency	261515
Hyperoxaluria type 1	259900
Acatalasemia	115500
Glutaryl-CoA oxidase deficiency	231690
Heimler syndrome 1	234580

[a]OMIM, *Online Mendelian Inheritance in Man.* Each number specifies a reference in which information regarding each of the above conditions can be found.
Modified with permission from Seashore MR, Wappner RS: *Genetics in Primary Care & Clinical Medicine.* Stamford, CT: Appleton & Lange; 1996.

however, are transferred directly into the membrane of the ER without reaching its lumen.

Some characteristics of N-terminal signal peptides are summarized in **Table 49–4**.

There is much **evidence to support** the signal hypothesis, confirming that N-terminal signal peptides are involved in the process of protein translocation across the ER membrane. For example, mutant proteins containing altered signal peptides in which hydrophobic amino acids are replaced by hydrophilic ones are not inserted into the lumen of the ER. On the other hand, nonmembrane proteins (eg, α-globin) to which signal peptides have been attached by genetic engineering can be inserted into the lumen of the ER, or even secreted.

Translocation of Proteins to the Endoplasmic Reticulum May Be Cotranslational or Posttranslational

Most nascent proteins are transferred across the ER membrane into the lumen by the **cotranslational pathway**, so called because the process occurs during ongoing protein synthesis.

TABLE 49–4 Some Properties of Signal Peptides Directing Proteins to the ER

- Usually, but not always, located at the amino terminal
- Contain approximately 12-35 amino acids
- Methionine is usually the amino-terminal amino acid
- Contain a central cluster (~6-12) of hydrophobic amino acids
- The region near the N-terminus usually carries a net positive charge
- The amino acid residue at the cleavage site is variable, but residues −1 and −3 relative to the cleavage site must be small and neutral

The process of elongation of the remaining portion of the protein being synthesized probably facilitates passage of the nascent protein across the lipid bilayer. It is important that proteins be kept in an **unfolded state** prior to entering the conducting channel—otherwise, they may not be able to gain access to the channel. The pathway involves a number of specialized proteins, including the **signal recognition particle (SRP)**, the **SRP receptor (SRP-R)**, and the **translocon**. The translocon consists of three membrane proteins (the Sec61 complex) that form a **protein-conducting channel** in the ER membrane through which the newly synthesized protein may pass. The channel **opens only when a signal peptide is present**. Closure of the channel when proteins are not being translocated prevents ions such as calcium and other molecules leaking through it, causing cell dysfunction. The process proceeds in five steps summarized as follows and in **Figure 49–6**.

Step 1: The signal sequence emerges from the ribosome and binds to the **SRP**. The SRP contains **six proteins** associated with an RNA molecule, and each of these plays a role (eg, binding of another molecule) in its function. This step temporarily stops further elongation of the polypeptide chain (elongation arrest) after some 70 amino acids have been polymerized.

Step 2: The SRP-ribosome-nascent protein complex travels to the ER membrane, where it binds to the **SRP-R**, an ER membrane protein composed of two subunits. The α subunit (SRP-Rα) binds the SRP complex and the membrane-spanning β subunit (SRP-Rβ) anchors SRP-Rα in the ER membrane. The SRP guides the complex to the SRP-R, preventing premature expulsion of the growing polypeptide into the cytosol.

Step 3: The SRP is released, translation resumes, the ribosome binds to the **translocon (Sec 61 complex)**, and the signal peptide inserts into the channel in the translocon. SRP and both subunits of the SRP-R bind **GTP**, this enables their interaction, resulting in the hydrolysis of GTP.

Step 4: The signal peptide induces opening of the channel in the translocon, by causing the plug (shown at the bottom on the translocon in Figure 49–6) to move. The growing polypeptide is then fully translocated across the membrane, driven by its ongoing synthesis.

Step 5: Cleavage of the signal peptide by **signal peptidase** occurs, and the fully translocated polypeptide/protein is released into the lumen of the ER. The signal peptide is degraded by proteases. Ribosomes are released from the ER membrane and dissociate into their two types of subunits.

Secretory proteins and **soluble proteins destined for organelles distal to the ER** completely traverse the membrane bilayer and are discharged into the lumen. Many secretory proteins are N-glycosylated. *N*-**Glycan chains**, if present, are added by the enzyme **oligosaccharide:protein transferase** (see Chapter 46) as these proteins traverse the inner part of the ER membrane—a process called **cotranslational**

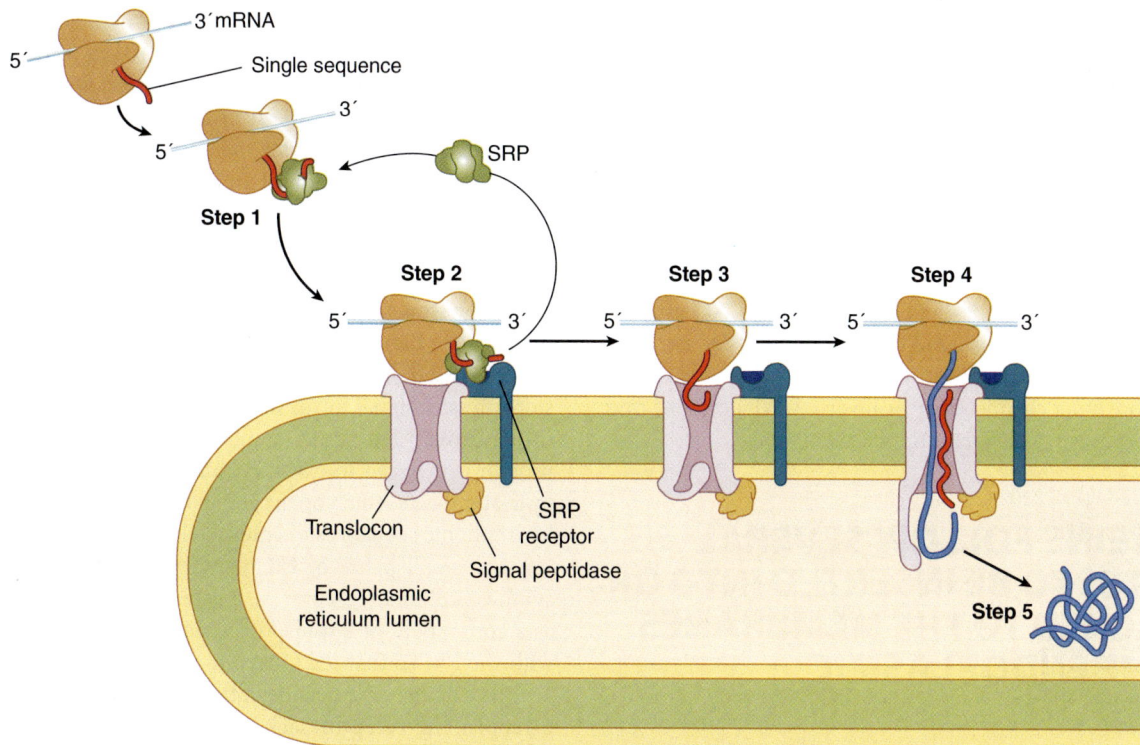

FIGURE 49–6 **Cotranslational targeting of secretory proteins to the ER. Step 1:** As the signal sequence emerges from the ribosome, it is recognized and bound by the signal recognition particle (SRP) and translation is arrested. **Step 2:** The SRP escorts the complex to the ER membrane where it binds to the SRP receptor. **Step 3:** The SRP is released, the ribosome binds to the translocon, translation resumes, and the signal sequence is inserted into the membrane channel. **Step 4:** The signal sequence opens the translocon and the growing polypeptide chain is translocated across the membrane. **Step 5:** Cleavage of the signal sequence by signal peptidase releases the polypeptide into the lumen of the ER. (Reproduced with permission from Cooper GM, Hausman RE: *The Cell: A Molecular Approach*, 6th ed. Sunderland, MA: Sinauer Associates, Inc, 2013.)

glycosylation. Subsequently, these glycoproteins are found in the **lumen of the Golgi apparatus**, where further changes in glycan chains occur (see Chapter 46) prior to intracellular distribution or secretion.

In contrast, proteins destined to be embedded in **membranes of the ER** or in **other membranes** along the secretory pathway only **partially translocate** across the ER membrane (steps 1–4). They are able to insert into the ER membrane by lateral transfer through the wall of the translocon (see later).

Posttranslational translocation of proteins to the ER does occur in eukaryotes, although it is less common than the cotranslational route. The process (**Figure 49–7**) involves the Sec61 translocon complex, the **Sec62/Sec63 complex** which is also membrane bound, and chaperone proteins of the Hsp70 family. Some of these prevent the protein folding in the cytosol, but one of them, **binding immunoglobulin protein (BiP)** (also known as GRP78 or Hsp70), is inside the ER lumen. The protein to be translocated initially binds to the translocon, and cytosolic chaperones are released. The leading end of the peptide then binds to BiP in the lumen. ATP bound to BiP interacts with Sec62/63, ATP is hydrolyzed to ADP providing energy to move the protein forwards, while the bound BiP-ADP prevents its moving backward into the cytosol. It can then be pulled through by sequential binding of BiP molecules and ATP hydrolysis. When the entire protein has entered the

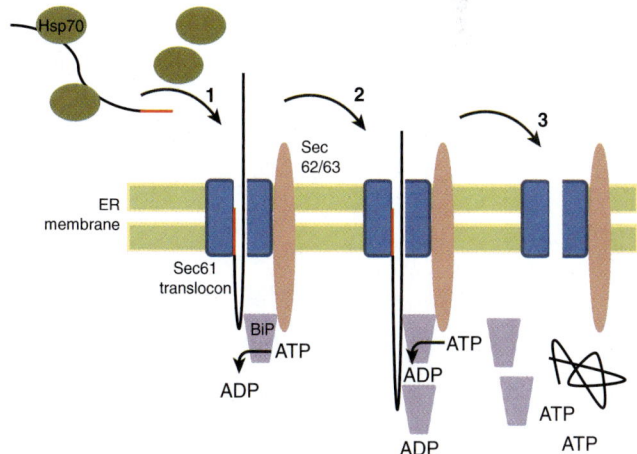

FIGURE 49–7 **Posttranslational translocation of proteins into the ER.** 1. Proteins synthesized in the cytosol are prevented from folding by chaperone proteins such as members of the Hsp70 family. The N-terminal signal sequence inserts into the Sec61 translocon complex and the cytosolic chaperones are released. Binding immunoglobulin protein (BiP) interacts with the protein and the Sec62/63 complex and the bound ATP is hydrolyzed to ADP. 2. The protein is prevented from moving back into the cytosol by the bound BiP, and successive binding of BiP and ATP hydrolysis pulls the protein into the lumen. 3. When the whole protein is inside, ADP is exchanged for ATP and BiP is released.

lumen, ADP is exchanged for ATP, allowing BiP to be released. In addition to its function in protein sorting to the ER lumen, BiP **promotes proper folding by preventing aggregation** and will temporarily bind abnormally folded immunoglobulin heavy chains and many other proteins, preventing them from leaving the ER.

There is evidence that the ER membrane is involved in **retrograde transport** of various molecules from the ER lumen **to the cytosol**. These molecules include unfolded or misfolded glycoproteins, glycopeptides, and oligosaccharides. At least some of these molecules are **degraded in proteasomes** (see later). The involvement of the translocon in retrotranslocation is not clear; one or more other channels may be involved. Whatever the case, there is **two-way traffic** across the ER membrane.

PROTEINS FOLLOW SEVERAL ROUTES TO BE INSERTED INTO OR ATTACHED TO THE MEMBRANES OF THE ENDOPLASMIC RETICULUM

The routes that proteins follow to be inserted into the membranes of the ER include cotranslational insertion; posttranslational insertion; retention in the GA followed by retrieval to the ER; and retrograde transport from the GA.

Cotranslational Insertion Requires Stop Transfer Sequences or Internal Insertion Sequences

Figure 49–8 shows a variety of ways in which proteins are distributed in membranes. In particular, the **amino termini** of certain proteins (eg, the low-density lipoprotein [LDL] receptor)

can be seen to be on the extracytoplasmic face, whereas for other proteins (eg, the asialoglycoprotein receptor) the **carboxyl termini** are on this face. These dispositions are explained by the initial biosynthetic events at the ER membrane. Proteins like the **LDL receptor** enter the ER membrane in a manner analogous to a secretory protein (see Figure 49–6); they partly traverse the ER membrane, the signal peptide is cleaved, and their amino terminal protrudes into the lumen. However, this type of protein contains a highly hydrophobic segment which acts as a **halt- or stop-transfer signal** and causes its retention in the membrane (**Figure 49–9**). This sequence has its N-terminal end in the ER lumen and the C-terminal in the cytosol; the stop-transfer signal forms the single transmembrane segment of the protein and is its membrane-anchoring domain. The protein is believed to exit the translocon into the membrane by a lateral gate which opens and closes continuously allowing hydrophobic sequences to enter the lipid bilayer.

The small patch of ER membrane in which the newly synthesized LDL receptor is located subsequently buds off as a component of a transport vesicle which eventually fuses with the PM so that the C-terminal faces the cytosol and the N-terminal faces the outside of the cell. In contrast, the **asialoglycoprotein receptor** lacks a cleavable N-terminal signal peptide, but possesses an **internal insertion sequence**, which inserts into the membrane but is not cleaved. This acts as an anchor, and its C-terminus is extruded through the membrane into the ER lumen. Cytochrome P450 is anchored in a similar way, but its N-terminal, rather than C-terminal, is extruded into the lumen. The more complex disposition of a **transmembrane transporter** (eg, for glucose) which may cross the membrane up to 12 times, can be explained by the fact that alternating transmembrane α-helices act as uncleaved insertion sequences and as halt-transfer signals, respectively. Each pair of helical segments is inserted as a hairpin. Sequences that determine the structure of a protein in a membrane are called

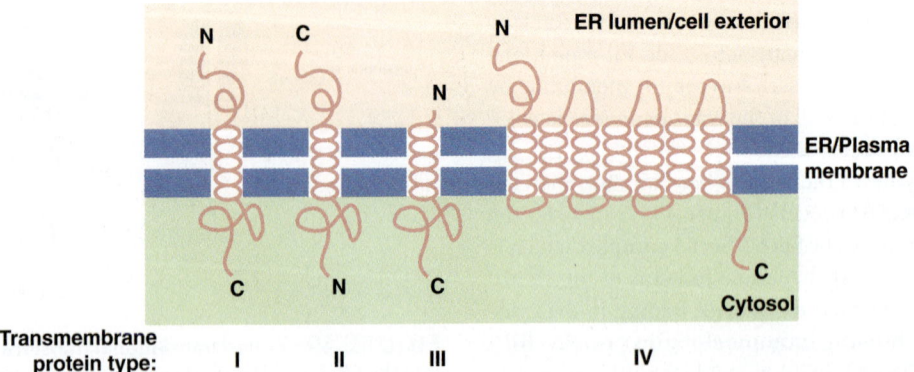

FIGURE 49–8 Variations in the way in which proteins are inserted into membranes. This schematic representation illustrates a number of possible orientations. The orientations form initially in the ER membrane, but are retained when vesicles bud off and fuse with the plasma membrane, so that the terminal initially facing the ER lumen always faces the outside of the cell. Type I transmembrane proteins (eg, the LDL receptor and influenza hemagglutinin) cross the membrane once and have their amino termini in the ER lumen/cell exterior. Type II transmembrane proteins (eg, the asialoglycoprotein and transferrin receptors) also cross the membrane once, but have their C-termini in the ER lumen/cell exterior. Type III transmembrane proteins (eg, cytochrome P450, an ER membrane protein) have a disposition similar to type I proteins, but do not contain a cleavable signal peptide. Type IV transmembrane proteins (eg, G-protein–coupled receptors and glucose transporters) cross the membrane a number of times (7 times for the former and 12 times for the latter); they are also called polytopic membrane proteins. (C, carboxyl terminal; N, amino terminal.)

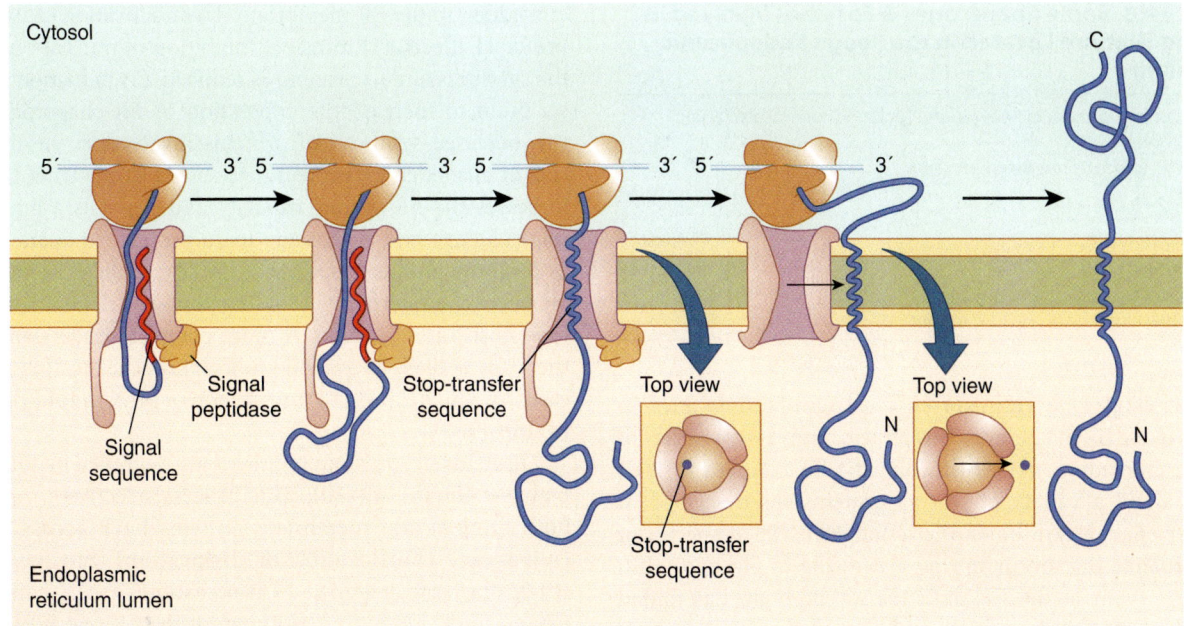

FIGURE 49–9 **Insertion of a membrane protein with a stop-transfer signal into the ER membrane.** The protein enters the membrane in a similar way to a secretory protein (Figure 49–6), the signal peptide is cleaved as the polypeptide chain crosses the membrane, so the amino terminus of the polypeptide chain is exposed in the ER lumen. However, translocation of the polypeptide chain across the membrane is halted when the translocon recognizes a transmembrane stop-transfer sequence which is highly hydrophobic. The protein then exits the translocon channel via a lateral gate and become anchored in the ER membrane. Continued translation results in a membrane-spanning protein with its carboxy terminus on the cytosolic side. (Reproduced with permission from Cooper GM, Hausman RE: *The Cell: A Molecular Approach*, 6th ed. Sunderland, MA: Sinauer Associates, Inc, 2013.)

topogenic sequences. The LDL receptor, asialoglycoprotein receptor, and glucose transporter are examples of types I, II, and IV transmembrane proteins and are found in the PM, while cytochrome P450 is an example of a type III protein which remains in the ER membrane (see Figure 49–8).

Some Proteins Are Synthesized on Free Polyribosomes & Attach to the Endoplasmic Reticulum Membrane Posttranslationally

Proteins may enter the ER membrane posttranslationally through the lateral gate in the translocon in a similar way to cotranslationally sorted molecules. An example is **cytochrome b_5**, which appears to enter the ER membrane subsequent to translation, assisted by several chaperones.

Other Routes Include Retention in the GA With Retrieval to the ER & Also Retrograde Transport From the GA

A number of proteins possess the amino acid sequence **KDEL** (Lys-Asp-Glu-Leu) at their carboxyl terminal (see Table 49–1). KDEL-containing proteins first travel to the **GA** in **vesicles coated with coat protein II (COPII)** (see later). This process is known as **anterograde vesicular transport**. In the GA, they interact with a specific KDEL receptor protein, which retains them transiently. They then **return to the ER in vesicles coated**

with coat protein I (**COPI, retrograde vesicular transport**), where they dissociate from the receptor, and are thus retrieved. HDEL sequences (H = histidine) serve a similar purpose. The above processes result in net localization of certain soluble proteins to the ER lumen.

Certain other **non-KDEL–containing proteins** also pass to the Golgi and then return, by retrograde vesicular transport, to the ER to be inserted therein. These include vesicle components that must be recycled, as well as certain ER membrane proteins. These proteins often possess a C-terminal signal located in the cytosol rich in basic residues.

Thus, proteins reach the ER membrane by a **variety of routes**, and similar pathways are likely to be used for other membranes (eg, the mitochondrial membranes and the PM). Precise targeting sequences have been identified in some instances (eg, KDEL sequences).

The topic of membrane biogenesis is discussed further later in this chapter.

THE ER FUNCTIONS AS THE QUALITY CONTROL COMPARTMENT OF THE CELL

After entering the ER, newly synthesized proteins fold with the assistance of chaperones and folding enzymes, and their folding status is monitored by chaperones and also enzymes (**Table 49–5**).

TABLE 49–5 Some Chaperones & Enzymes Involved in Folding That Are Located in the Rough Endoplasmic Reticulum

- BiP (immunoglobulin heavy-chain binding protein) (also called GRP78 or Hsp70)
- GRP94 (glucose-regulated protein)
- GRP170
- Hsp47
- Calnexin
- Calreticulin
- PDI (protein disulfide isomerase)
- PPI (peptidyl prolyl *cis-trans* isomerase)

The chaperone **calnexin** is a calcium-binding protein located in the ER membrane. This protein binds a wide variety of proteins, including major histocompatibility complex (MHC) antigens and a variety of plasma proteins. As described in Chapter 46, calnexin binds the monoglucosylated species of glycoproteins that occur during processing of glycoproteins, retaining them in the ER until the glycoprotein has folded properly. **Calreticulin**, which is also a calcium-binding protein, has properties similar to those of calnexin, but it is not membrane-bound. In addition to chaperones, two enzymes in the ER lumen are concerned with proper folding of proteins. **Protein disulfide isomerase (PDI)** promotes **rapid formation** and reshuffling of disulfide bonds until the correct set is achieved. **Peptidyl prolyl isomerase (PPI)** accelerates folding of proline-containing proteins by catalyzing the cis–trans isomerization of X-Pro bonds, where X is any amino acid residue.

Misfolded or incompletely folded proteins interact with chaperones, which retain them in the ER and prevent them from being exported to their final destinations. If such interactions continue for a prolonged period of time, harmful buildup of misfolded proteins is avoided by **ER-associated degradation (ERAD)**. In a number of genetic diseases, such as cystic fibrosis, misfolded proteins are retained in the ER, and in some cases, the retained proteins still exhibit some functional activity. As discussed later in this chapter, drugs that will interact with such proteins and promote their correct folding and export out of the ER are currently under investigation.

MISFOLDED PROTEINS UNDERGO ENDOPLASMIC RETICULUM–ASSOCIATED DEGRADATION

Maintenance of **homeostasis in the ER** is important for normal cell function. Perturbation of the unique environment within the lumen of the ER (eg, by changes in ER Ca^{2+}, alterations of redox status, exposure to various toxins or some viruses), can lead to reduced protein folding capacity and the accumulation of misfolded proteins. The accumulation of misfolded proteins in the ER is referred to as **ER stress**. The **unfolded protein response (UPR)** is a mechanism within cells which senses the levels of misfolded proteins and activates intracellular signaling mechanisms to restore ER homeostasis. The UPR is initiated by **ER stress sensors**, which are transmembrane proteins embedded in the ER membrane. Their activation causes three principal effects: (1) transient inhibition of translation so that the synthesis of new proteins is reduced, (2) induction of transcription to increase the expression of ER chaperones, and (3) increased synthesis of proteins involved in the degradation of misfolded ER proteins (discussed later). Thus, the UPR increases the ER folding capacity and prevents a buildup of unproductive and potentially toxic protein products, as well as promoting other responses to restore cellular homeostasis. However, if impairment of folding persists, cell death pathways (apoptosis) are activated. A more complete understanding of the UPR is likely to provide new approaches to treating diseases in which ER stress and defective protein folding occur (**Table 49–6**).

Proteins that misfold in the ER are degraded by the ERAD pathway (**Figure 49–10**). This process selectively transports both luminal and membrane proteins **back across the ER (retrograde translocation or dislocation)** into the cytosol where they are degraded in **proteasomes** (see Chapter 28). **Chaperones** present in the lumen of the ER (eg, BiP) help to target misfolded proteins to proteasomes.

The assembly of the retrotranslocation translocon comprising a number of proteins is initiated by recognition of

TABLE 49–6 Some Conformational Diseases That Are Caused by Abnormalities in Intracellular Transport of Specific Proteins & Enzymes due to Mutations

Disease	Affected Protein
α_1-Antitrypsin deficiency with liver disease	α_1-Antitrypsin
Chediak-Higashi syndrome	Lysosomal trafficking regulator
Combined deficiency of factors V and VIII	ERGIC53, a mannose-binding lectin
Cystic fibrosis	CFTR
Diabetes mellitus (some cases)	Insulin receptor (α-subunit)
Familial hypercholesterolemia, autosomal dominant	LDL receptor
Gaucher disease	β-Glucosidase
Hemophilia A and B	Factors VIII and IX
Hereditary hemochromatosis	HFE
Hermansky-Pudlak syndrome	AP-3 adaptor complex β3A subunit
I-cell disease	*N*-acetylglucosamine 1-phosphotransferase
Lowe oculocerebrorenal syndrome	PIP_2 5-phosphatase
Tay-Sachs disease	β-Hexosaminidase
von Willebrand disease	von Willebrand factor

Abbreviations: LDL, low-density lipoprotein; PIP_2, phosphatidylinositol 4,5-bisphosphate. Data from Schröder M, Kaufman RJ. The mammalian unfolded protein response. Annu Rev Biochem. 2005;74:739-789 and Olkkonen VM, Ikonen E. Genetic defects of intracellular-membrane transport. N Engl J Med. 2000;343(15):1095–1104.

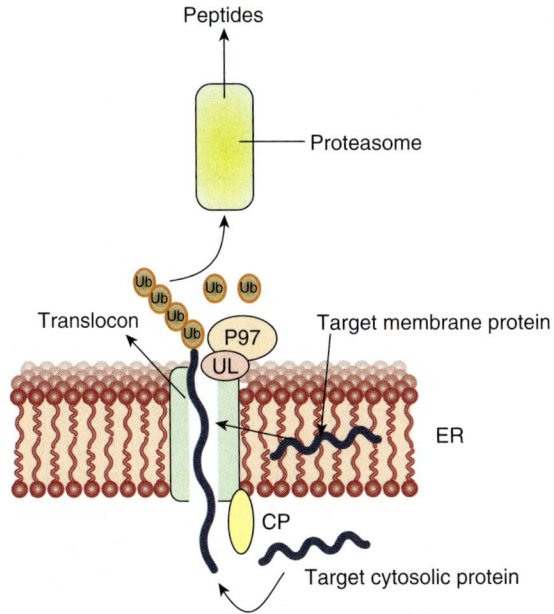

FIGURE 49–10 **Simplified scheme of the events in ERAD.** An ER target protein which is misfolded either in the lumen or in the membrane undergoes retrograde transport into the cytosol via a translocon. Chaperone proteins (CP) target misfolded proteins for retrotranslocation. As the protein enters the cytosol, it is ubiquitinated by ubiquitin ligases (UL), pulled out of the membrane by P97, an ATPase, and delivered to the proteasome. Inside the proteasome it is degraded to small peptides that may have several fates after exit.

misfolded proteins by chaperones and adaptor proteins. As retrotranslocation occurs, misfolded proteins are polyubiquinated (see Chaper 28) by ubiquitin ligases on the cytosolic side, and are then pulled from the membrane by **P97** (also called Vcp), an **ATPase** and delivered to the proteasome for degradation (see Figure 49–10).

Ubiquitin Is a Key Molecule in Protein Degradation

There are two major pathways of protein degradation in eukaryotes. One involves **lysosomal proteases** and does not require ATP, but the major pathway involves **ubiquitin** and is ATP-dependent. The ubiquitin pathway is particularly associated with **disposal of misfolded proteins and regulatory enzymes that have short half-lives**. Ubiquitin is known to be involved in diverse important physiologic processes including **cell cycle regulation** (degradation of cyclins), **DNA repair, inflammation** and **the immune response** (see Chapter 52), **muscle wasting, viral infections**, and **many others**. Ubiquitin is a small (76 amino acids), highly conserved protein that tags various proteins for degradation in proteasomes. The mechanism of attachment of ubiquitin to a target protein (eg, a misfolded form of cystic fibrosis transmembrane conductance regulator [CFTR], the protein involved in the causation of cystic fibrosis; see Chapter 40) is shown in Figure 28–2 and the process is described in detail in Chapter 28.

Ubiquitinated Proteins Are Degraded in Proteasomes

Polyubiquitinated target proteins enter **proteasomes** located in the cytosol. Proteasomes are protein complexes with a relatively **large cylindrical structure** and are composed of four rings with a hollow **core** containing the protease active sites, and one or two **caps** or **regulatory particles** that recognize the polyubiquinated substrates and initiate degradation (see Figures 28–3 and 28–4). Target proteins are unfolded by ATPases present in the proteasome caps and pass into the core to be degraded to small peptides, which then exit the proteasome to be further degraded by cytosolic peptidases. Both normally and abnormally folded proteins are substrates for the proteasome, which can hydrolyze a wide variety of peptide bonds. Liberated ubiquitin molecules are recycled. The proteasome plays an important role in **presenting small peptides** produced by **degradation of various viruses** and other molecules to **MHC class I molecules**, a key step in antigen presentation to T lymphocytes.

TRANSPORT VESICLES ARE KEY PLAYERS IN INTRACELLULAR PROTEIN TRAFFIC

Proteins that are synthesized on membrane-bound polyribosomes and are destined for the Golgi or PM reach these sites inside **transport vesicles**. As indicated in **Table 49–7**, there are a number of different types of vesicles. There may be other types of vesicles still to be discovered.

Each vesicle has its own set of coat proteins. **Clathrin** is used in vesicles destined for exocytosis (see discussions of the LDL receptor in Chapters 25 and 26) and in some of those carrying cargo to lysosomes. This protein consists of three interlocking spirals, which interact to form a lattice around the vesicle. Coat proteins I and II, rather than clathrin, are

TABLE 49–7 **Some Types of Vesicles & Their Functions**

Vesicle	Function
COPI	Involved in intra-GA transport and retrograde transport from the GA to the ER
COPII	Involved in export from the ER to either ERGIC or the GA
Clathrin	Involved in transport in post-GA locations including the PM, TGN, and endosomes
Secretory vesicles	Involved in regulated secretion from organs such as the pancreas (eg, secretion of insulin)
Vesicles from the TGN to the PM	They carry proteins to the PM and are also involved in constitutive secretion

Abbreviations: ER, endoplasmic reticulum; ERGIC, ER-GA intermediate compartment; GA, Golgi apparatus; PM, plasma membrane; TGN, *trans*-Golgi network (the network of membranes located on the side of the Golgi apparatus distal to the ER).
Note: Each vesicle has its own set of coat proteins. Clathrin is associated with various adaptor proteins forming different types of clathrin vesicles which have different intracellular targets.

associated with COPI and COPII vesicles, which are involved in **retrograde transport** (from the Golgi to the ER) and **anterograde transport** (from the ER to the Golgi), respectively. Transport and secretory vesicles carrying cargo from the Golgi to the plasma membrane are also clathrin-free. Here we focus mainly on COPII, COPI, and clathrin-coated vesicles. Each type has a different complement of proteins in its coat. For the sake of clarity, the non–clathrin-coated vesicles are referred to in this text as **transport vesicles**. The principles concerning assembly of these different types are generally similar, although some details of assembly for COPI and clathrin-coated vesicles differ from those for COPII (see following discussion).

Transport Vesicle Formation and Function Involves SNAREs & Other Factors

Vesicles lie at the heart of intracellular transport of many proteins. **Genetic approaches** and **cell-free systems** have been used to elucidate the mechanisms of their formation and transport. The overall process is complex, and involves a variety of cytosolic and membrane proteins, GTP, ATP, and accessory factors. **Budding, tethering, docking,** and **membrane fusion** are key steps in the life cycles of vesicles, with the GTP-binding proteins, **Sar1, ARF,** and **Rab** acting as **molecular switches.** Sar1 is the protein involved in step 1 of formation of COPII vesicles, whereas ARF is involved in the formation of COPI and clathrin-coated vesicles. The functions of the various proteins involved in vesicle processing and the abbreviations used are shown in **Table 49–8.**

There are common general steps in transport vesicle formation, vesicle targeting, and fusion with a target membrane, irrespective of the membrane the vesicle forms from or its intracellular destination. The nature of the coat proteins, GTPases and targeting factors differ depending on where the vesicle forms from and its eventual destination. Anterograde transport from the ER to the Golgi involving COPII vesicles is the best studied example. The process can be considered to occur in eight steps (**Figure 49–11**). The basic concept is that each transport vesicle is loaded with specific cargo and also one or more **v-SNARE** proteins that direct targeting. Each target membrane bears one or more **complementary t-SNARE proteins** with which v-SNAREs interact, mediating SNARE protein-dependent vesicle–membrane fusion. In addition, **Rab proteins** also help direct the vesicles to specific membranes and their tethering at a target membrane.

Step 1: Budding is initiated when the GTPase **Sar1** is activated by binding of GTP in exchange for GDP via the action of the GEF **Sec12p** (see Table 49–9), switching it from a soluble to a membrane bound form via conformational change which exposes a hydrophobic tail. This enables it to become embedded in the ER membrane to form a focal point for vesicle assembly.

Step 2: For **bud formation,** various **coat proteins** bind to **Sar1·GTP.** In turn, membrane cargo proteins bind to the coat proteins either **directly** or via **intermediary proteins** that attach to coat proteins, and they then become enclosed in their appropriate vesicles. Soluble cargo proteins bind to receptor regions inside the vesicles. A number of **signal sequences** on cargo molecules have been identified (see Table 49–1). For example, KDEL sequences direct certain proteins in retrograde flow from the Golgi to the ER in COPI vesicles. Diacidic sequences (eg, Asp-X-Glu, X = any amino acid) and short hydrophobic sequences on membrane proteins destined for the Golgi membrane are involved in interactions with coat proteins of COPII vesicles. However, not all cargo molecules have a sorting signal. Some highly abundant secretory proteins travel to various cellular destinations in transport vesicles by **bulk flow;** that is, they enter into transport vesicles at the same concentration that they occur in the organelle. However, it appears that most proteins are actively sorted (concentrated) into transport vesicles and bulk flow is used by only a select group of cargo proteins. Additional coat proteins are assembled to **complete bud formation.** Coat proteins promote budding, contribute to the curvature of buds, and also help sort proteins.

Step 3: The **bud pinches off,** completing formation of the coated vesicle. The curvature of the ER membrane and protein–protein and protein-lipid interactions in the bud facilitate pinching off from ER exit sites. Vesicles move through cells along **microtubules** or along **actin filaments.**

Step 4: Coat disassembly, or **uncoating** (involving **dissociation** of **Sar1** and the **shell** of coat proteins) follows **hydrolysis of bound GTP to GDP** by Sar1, promoted by a specific coat protein. Sar1 thus plays key roles in both assembly and dissociation of the coat proteins. **Uncoating** is necessary for fusion to occur.

TABLE 49–8 Some Factors Involved in the Formation of Non–Clathrin-Coated Vesicles & Their Transport

- ARF: ADP-ribosylation factor, a GTPase involved in formation of COPI and also clathrin-coated vesicles.
- Coat proteins: A family of proteins found in coated vesicles. Different transport vesicles have different complements of coat proteins.
- NSF: *N*-ethylmaleimide-sensitive factor, an ATPase.
- Sar1: A GTPase that plays a key role in assembly of COPII vesicles.
- Sec12p: A guanine nucleotide exchange factor (GEF) that interconverts Sar1·GDP and Sar1·GTP.
- α-SNAP: Soluble NSF attachment protein. Along with NSF, this protein is involved in dissociation of SNARE complexes.
- SNARE: SNAP receptor. SNAREs are key molecules in the fusion of vesicles with acceptor membranes.
- t-SNARE: Target SNARE.
- v-SNARE: Vesicle SNARE.
- Rab proteins: A family of Ras-related proteins (monomeric GTPases). They are active when GTP is bound. Different Rab molecules dock different vesicles to acceptor membranes.
- Rab effector proteins: A family of proteins that interact with Rab molecules; some act to tether vesicles to acceptor membranes.

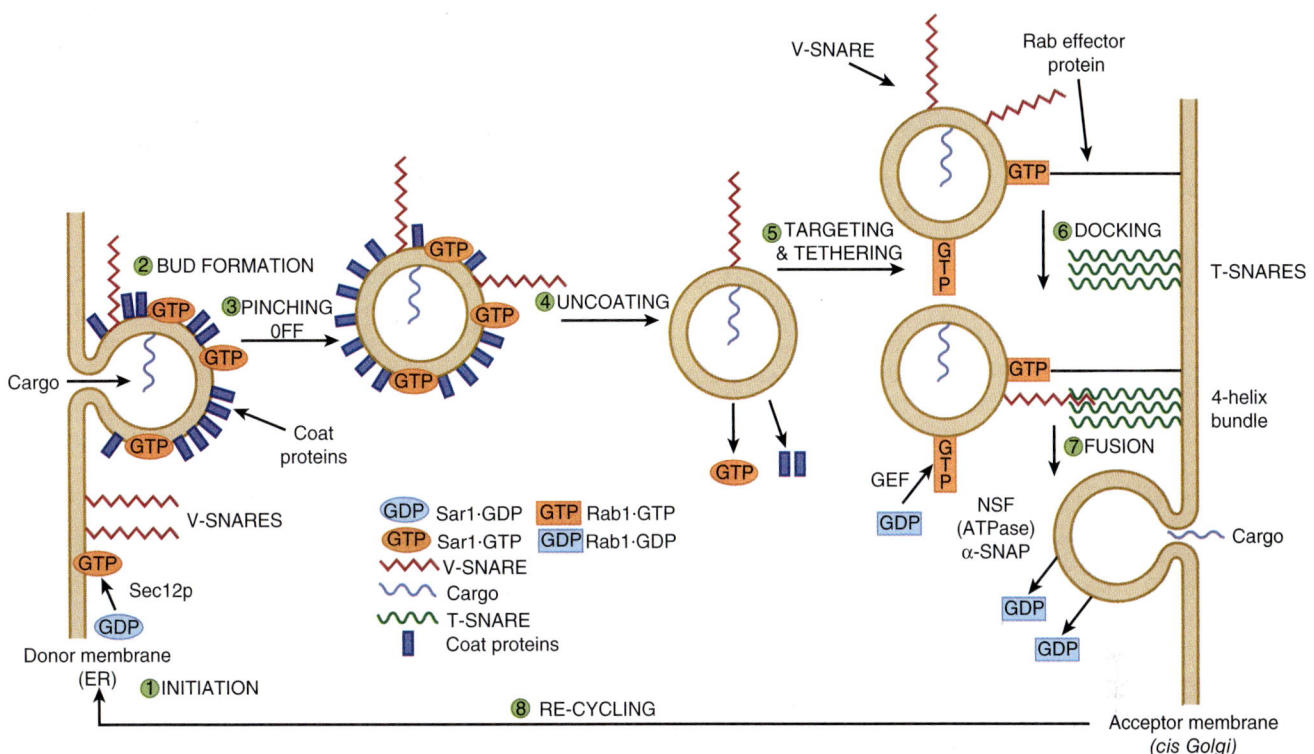

FIGURE 49–11 **Anterograde transport involving COPII vesicles. Step 1:** Sar1 is activated when GDP exchanged for GTP and it becomes embedded in the ER membrane to form a focal point for bud formation. **Step 2:** Coat proteins bind to Sar1·GTP and cargo proteins become enclosed inside the vesicles. **Step 3:** The bud pinches off, formatting a complete coated vesicle. Vesicles move through cells along microtubules or actin filaments. **Step 4:** The vesicle is uncoated when bound GTP is hydrolyzed to GDP by Sar1. **Step 5:** Rab molecules are attached to vesicles after switching of Rab·GDP to Rab·GTP, a specific GEF (see Table 49–9). Rab effector proteins on target membranes bind to Rab·GTP, tethering the vesicles to the target membrane. **Step 6:** v-SNAREs pair with cognate t-SNAREs in the target membrane to form a four-helix bundle which docks the vesicles and initiates fusion. **Step 7:** When the v- and t-SNARES are closely aligned, the vesicle fuses with the membrane and the contents are released. GTP is then hydrolyzed to GDP, and the Rab·GDP molecules are released into the cytosol. An ATPase (NSF) and α-SNAP (see **Table 49–9**) dissociate the four-helix bundle between the v- and t-SNARES so that they can be reused. **Step 8:** Rab and SNARE proteins are recycled for further rounds of vesicle fusion.

Step 5: **Vesicle targeting** is achieved by attachment of **Rab** molecules to vesicles. Rabs are a family of Ras-like proteins required in several steps of intracellular protein transport and also in regulated secretion and endocytosis. They are **small monomeric GTPases** that attach to the cytosolic faces of budding vesicles in the **GTP-bound state** and are also present on acceptor membranes. Rab·GDP molecules in the cytosol are switched to Rab·GTP molecules by a specific GEF (see Table 49–9). **Rab effector proteins** on acceptor membranes bind to Rab·GTP, but not Rab GDP molecules, thus **tethering** the vesicles to the membranes.

Step 6: **v-SNAREs pair with cognate t-SNAREs** in the target membrane to **dock** the vesicles and initiate fusion. Generally, one v-SNARE in the vesicle pairs with three t-SNAREs on the acceptor membrane to form a tight **four-helix bundle**. In **synaptic vesicles** one v-SNARE is called **synaptobrevin**. **Botulinum B toxin**, one of the most lethal toxins known and the most serious cause of food poisoning, contains a **protease** that binds **synaptobrevin**, thus **inhibiting release of acetylcholine** at the neuromuscular junction and often proving fatal.

Step 7: **Fusion** of the vesicle with the acceptor membrane occurs once the v- and t-SNARES are closely aligned. After vesicle fusion and release of contents, GTP is hydrolyzed to GDP, and the Rab·GDP molecules are released into the cytosol. When a SNARE on one membrane interacts with a SNARE on another membrane, linking the two, this is referred to as a **trans-SNARE complex** or a **SNARE pin**. Interactions of SNARES on the same membrane form a

TABLE 49–9 **Some Major Features of Membrane Assembly**

- Lipids and proteins are inserted independently into membranes.
- Individual membrane lipids and proteins turn over independently and at different rates.
- Topogenic sequences (eg, signal [amino terminal or internal] and stop-transfer) are important in determining the insertion and disposition of proteins in membranes.
- Membrane proteins inside transport vesicles bud off the endoplasmic reticulum on their way to the Golgi; final sorting of many membrane proteins occurs in the *trans*-Golgi network.
- Specific sorting sequences guide proteins to particular organelles such as lysosomes, peroxisomes, and mitochondria.

cis-SNARE complex. In order to dissociate the four-helix bundle between the v- and t-SNARES so that they can be reused, two additional proteins are required. These are an **ATPase** (NSF) and **a-SNAP** (see Table 49–9). NSF hydrolyzes ATP and the energy released dissociates the four-helix bundle making the SNARE proteins available for another round of membrane fusion.

Step 8: Certain components, such as the Rab and SNARE proteins, are **recycled** for subsequent rounds of vesicle fusion.

During the above cycle, SNARES, tethering proteins, Rab, and other proteins all collaborate to deliver a vesicle and its contents to the appropriate site.

Some Transport Vesicles Travel via the *Trans*-Golgi Network

Proteins found in the **apical** or **basolateral** areas of the PMs of polarized epithelial cells may be transported to these sites in various ways; in **transport vesicles** budding from the *trans*-Golgi network, with different Rab proteins directing some vesicles to apical regions and others to basolateral regions; via initial direction to the basolateral membrane, followed by endocytosis and transport across the cell by **transcytosis** to the apical region; or via a process involving the **glycosylphosphatidylinositol (GPI) anchor** described in Chapter 46. This structure is also often present in **lipid rafts** (see Chapter 40).

Once proteins in the secretory pathway reach the *cis*-Golgi from the ER, they may be retained in vesicles for travel through the Golgi to the *trans*-Golgi, or they may cross by a process called **cisternal maturation,** in which the **cisternae (the flattened membrane disks of the Golgi that bud off from the ER)** move and transform into one another, or perhaps in some cases **diffusion** via intracisternal connections. In this model, vesicular elements from the ER fuse with one another to help form the *cis*-Golgi, which in turn can move forward toward the *trans*-Golgi. COPI vesicles return Golgi enzymes (eg, glycosyltransferases) back from distal (*trans*-) cisternae of the Golgi to more proximal (*cis*) cisternae.

The fungal metabolite **brefeldin A inhibits the formation of COPI vesicles by preventing GTP from binding to ARF.** In its presence, the GA appears to **collapse into the ER.** Brefeldin A has proven to be a useful tool for examining some aspects of Golgi structure and function.

Some Proteins Undergo Further Processing While Inside Vesicles

Some proteins are subjected to further processing **by proteolysis** while inside either transport or secretory vesicles. For example, **albumin** is synthesized by hepatocytes as **preproalbumin** (see Chapter 52). Its signal peptide is removed, converting it to **proalbumin**. In turn, proalbumin, while inside secretory vesicles, is converted to **albumin** by action of **furin** (**Figure 49–12**). This enzyme cleaves a hexapeptide from proalbumin immediately C-terminal to a dibasic amino acid site (ArgArg). The resulting mature albumin is secreted into the plasma. Hormones such as **insulin** (see Chapter 41) are subjected to similar proteolytic cleavages while inside secretory vesicles.

THE ASSEMBLY OF MEMBRANES IS COMPLEX

There are a number of different types of cell membranes, ranging from the PM which separates the cell contents from the external environment to the internal membranes of subcellular organelles, such a mitochondria and the ER. Although the general lipid bilayer structure is similar in all membranes, they differ in their specific protein and lipid content and each type has its own specific features (see Chapter 40). No satisfactory scheme describing the assembly of any one of these membranes is currently available. Vesicular transport and the way in which various proteins are initially inserted into the membrane of the ER have been discussed earlier. Some general points about membrane assembly are addressed in the following discussion.

Asymmetry of Both Proteins & Lipids Is Maintained During Membrane Assembly

Vesicles formed from membranes of the ER and GA, either naturally or pinched off by homogenization, exhibit **transverse asymmetries** of both lipid and protein. These **asymmetries are maintained** during fusion of transport vesicles with the PM. The **inside** of the vesicles after fusion becomes the **outside of the plasma membrane**, and the cytoplasmic side of the vesicles remains the cytoplasmic side of the membrane (**Figure 49–13**). The enzymes responsible for the synthesis

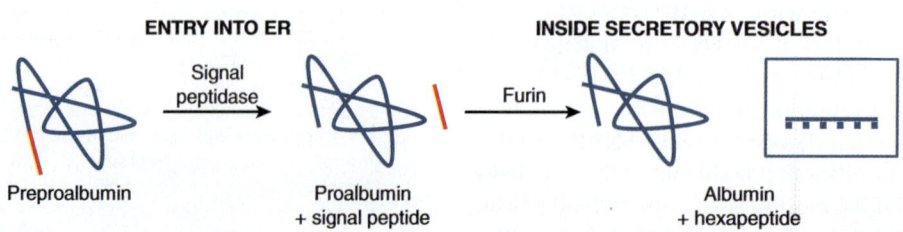

ENTRY INTO ER   INSIDE SECRETORY VESICLES

Preproalbumin   →(Signal peptidase)→   Proalbumin + signal peptide   →(Furin)→   Albumin + hexapeptide

FIGURE 49–12   Processing of preproalbumin albumin. The signal peptide is removed from preproalbumin as it moves into the ER. Furin cleaves proalbumin at the C-terminal end of a basic dipeptide (ArgArg) while the protein is inside the secretory vesicle. The mature albumin is secreted into the plasma.

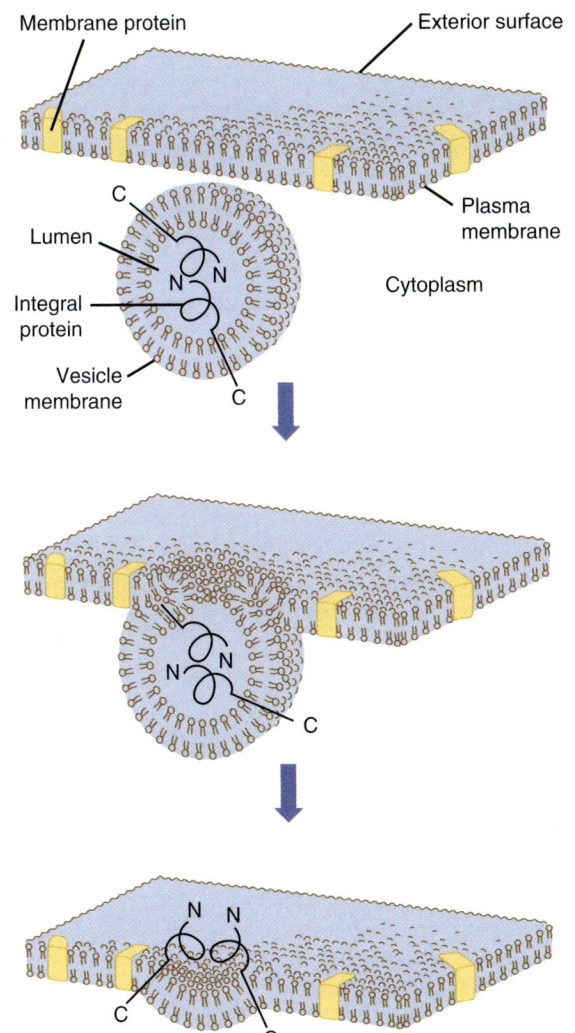

Membrane protein

Exterior surface

Plasma membrane

Lumen

Integral protein

Vesicle membrane

Cytoplasm

FIGURE 49–13 **Fusion of a vesicle with the plasma membrane preserves the orientation of any integral proteins embedded in the vesicle bilayer.** Initially, the amino terminal of the protein faces the lumen, or inner cavity, of such a vesicle. After fusion, the amino terminal is on the exterior surface of the plasma membrane. The lumen of a vesicle and the outside of the cell are topologically equivalent.

of phospholipids, the major class of lipids in membranes (see Chapter 40) reside in the cytoplasmic surface of the cisternae (the sac-like structures) of the ER. Phospholipids are synthesized at that site, and it is thought that they self-assemble into thermodynamically stable bimolecular layers, thereby expanding the membrane and perhaps promoting the detachment of so-called **lipid vesicles** from it. It has been proposed that these vesicles travel to other sites, donating their lipids to other membranes. **Phospholipid exchange proteins** are cytosolic proteins that take up phospholipids from one membrane and release them to another are believed to play a role in regulating the specific lipid composition of various membranes.

It should be noted that the **lipid compositions** of the ER, Golgi, and PM differ, the latter two membranes containing

higher amounts of **cholesterol, sphingomyelin, and glycosphingolipids**, and **less phosphoglycerides** than does the ER. Sphingolipids pack more densely in membranes than do phosphoglycerides. These differences affect the structures and functions of membranes. For example, the **thickness of the bilayer** of the Golgi and plasma membrane is greater than that of the ER, which affects which particular transmembrane proteins are found in these organelles. Also, **lipid rafts** (see Chapter 40) are believed to be formed in the Golgi.

Lipids & Proteins Undergo Turnover at Different Rates in Different Membranes

It has been shown that the half-lives of the lipids of the ER membranes are generally shorter than those of its proteins, so that the **turnover rates of lipids and proteins are independent**. Indeed, different lipids have been found to have different half-lives. Furthermore, the half-lives of the proteins of these membranes vary widely, some exhibiting short (hours) and others long (days) half-lives. Thus, individual lipids and proteins of the ER membranes appear to be inserted into it relatively independently and this is believed to be the case for many other membranes.

The biogenesis of membranes is thus a complex process about which much remains to be learned. One indication of the complexity involved is to consider the number of **posttranslational modifications** that membrane proteins may be subjected prior to attaining their mature state. These may include disulfide formation, proteolysis, assembly into multimers, glycosylation, addition of a glycophosphatidylinositol (GPI) anchor, sulfation on tyrosine or carbohydrate moieties, phosphorylation, acylation, and prenylation. Nevertheless, significant progress has been made; **Table 49–9** summarizes some of the major features of membrane assembly that have emerged to date.

Various Disorders Result From Mutations in Genes Encoding Proteins Involved in Intracellular Transport

Some disorders reflecting abnormal **peroxisomal** function and abnormalities of protein synthesis in the **ER** and of the synthesis of **lysosomal proteins** have been listed earlier in this chapter (see Tables 49–3 and 49–6, respectively). Many other mutations affecting folding of proteins and their intracellular transport to various organelles have been reported, including neurodegenerative disorders such as Alzheimer's disease, Huntington's disease, and Parkinson's disease. The elucidation of the causes of these various **conformational disorders** has contributed significantly to our understanding of **molecular pathology**. The term "**diseases of proteostasis deficiency**" has also been applied to diseases due to misfolding of proteins. Proteostasis is a composite word derived from protein homeostasis. Normal proteostasis is due to a balance of many factors, such as synthesis, folding, trafficking, aggregation, and normal degradation. If any one of these is disturbed (eg, by mutation, aging, cell stress, or injury), a variety of disorders can occur, depending on the particular proteins involved.

Hsp90 is a chaperone protein which plays a part in the processing of 300 to 400 proteins, including some that are involved in cancer development, and it was identified as a target for antitumor therapy in the 1990s. A number of Hsp90 inhibitors have undergone clinical trials since then, but none so far have been approved for use by the FDA, mainly because of toxicity problems. The search for an Hsp90 inhibitor suitable for clinical use, however, is continuing. The chaperone Hsp70 is another potential anticancer target currently under investigation, since it is expressed much more strongly in tumor cells than in normal cells. Small drug molecules that act as chemical chaperones have also been shown to prevent misfolding and restore protein function.

SUMMARY

- Many proteins are targeted to their destinations by signal sequences. A major sorting decision is made when proteins are partitioned between cytosolic (or free) and membrane-bound polyribosomes by virtue of the absence or presence of an N-terminal signal peptide.

- Proteins synthesized on cytosolic polyribosomes are targeted by specific signal sequences to mitochondria, nuclei, peroxisomes, and the ER. Proteins which lack a signal remain in the cytosol.

- Proteins synthesized on membrane-bound polyribosomes initially enter the ER membrane or lumen, and many are ultimately destined for other membranes including the plasma membrane and that of the Golgi, for lysosomes and for secretion via exocytosis.

- Many glycosylation reactions occur in compartments of the Golgi, and proteins are further sorted in the TGN.

- Molecular chaperones stabilize unfolded or partially folded proteins. Chaperones are required for the correct targeting of proteins to their subcellular locations.

- In posttranslational translocation, proteins are transported to their target organelles after their synthesis is complete. Proteins destined for mitochondria, the nucleus, and peroxisomes follow this route, as well as a minority of proteins targeted to the ER.

- Most proteins enter the ER lumen by the cotranslational pathway, where translocation occurs during ongoing protein synthesis.

- Proteins embedded in the ER membrane may inserted cotranslationally, posttranslationally, or after transport to the Golgi (anterograde transport), transient retention, and return to the ER (retrograde transport).

- Harmful buildup of misfolded proteins triggers the UPR and they are degraded via the ERAD pathway. Proteins are tagged for degradation by the addition of a number of ubiquitin molecules and then enter the cytosol where they are broken down in proteasomes.

- Different types of transport vesicles are coated with different proteins. Clathrin-coated vesicles are destined for exocytosis and lysosomes, while coat proteins I and II are associated with COPI and COPII vesicles, which are responsible for retrograde and anterograde transport, respectively.

- Transport vesicle processing is complex and requires many protein factors. Budding from the donor membrane is followed by movement through the cytosol, tethering, docking, and fusion with the target membrane.

- Certain proteins (eg, precursors of albumin and insulin) are subjected to proteolysis while inside transport vesicles, producing the mature proteins.

- Small GTPases (eg, Ran, Rab) and GEFs play key roles in many aspects of intracellular trafficking.

- Vesicles formed from membranes of the ER and Golgi are asymmetrical in both lipid and protein content. The asymmetry is maintained during assembly and during fusion of transport vesicles with the PM, so that the inside of the vesicles after fusion becomes the outside of the PM, and the cytoplasmic side of the vesicles remains facing the cytosol.

- Lipids and proteins are inserted into membranes independently and turn over at different rates. Exact details of the assembly process remain to be established.

- The molecular chaperones Hsp90 and Hsp70 are under intensive investigation as possible targets for anticancer therapy.

REFERENCES

Alberts B, Johnson A, Lewis J, et al: *Molecular Biology of the Cell*, 6th ed. Garland Science, 2014. (An excellent textbook of cell biology, with comprehensive coverage of trafficking and sorting.)

Cooper GM, Hausman RE: *The Cell: A Molecular Approach*, 8th ed. Palgrave, 2019. (An excellent textbook of cell biology, with comprehensive coverage of trafficking and sorting.)

The Extracellular Matrix

Kathleen M. Botham, PhD, DSc, &
Robert K. Murray, MD, PhD

OBJECTIVES

After studying this chapter, you should be able to:

- Indicate the importance of the extracellular matrix (ECM) and its components in health and disease.

- Describe the structural and functional properties of collagen and elastin, the major proteins of the ECM.

- Indicate the major features of fibrillin, fibronectin, and laminin, other important proteins of the ECM.

- Describe the properties and general features of the synthesis and degradation of glycosaminoglycans and proteoglycans, and their contributions to the ECM.

- Give a brief account of the major biochemical features of bone and cartilage.

BIOMEDICAL IMPORTANCE

Most mammalian cells are located in tissues where they are surrounded by a complex **extracellular matrix** (**ECM**) often referred to as "**connective tissue**," which protects the organs and also provides elasticity where required (eg, in blood vessels, lungs, and skin). The ECM contains three major classes of biomolecules: **structural proteins**, for example, **collagen, elastin,** and **fibrillin**; certain **specialized proteins** such as **fibronectin** and **laminin**, which form a mesh of fibers that are embedded in the third class, **proteoglycans**. The ECM is involved in many processes, both normal and pathologic, for example, it plays important roles in development, in inflammatory states, and in the spread of cancer cells. Certain components of the ECM play a part in both **rheumatoid arthritis** and **osteoarthritis**. Several diseases (eg, osteogenesis imperfecta and a number of types of the Ehlers-Danlos syndrome) are due to genetic disturbances of the synthesis of collagen, a major ECM component. Specific components of proteoglycans (the glycosaminoglycans; GAGs) are affected in the group of genetic disorders known as the **mucopolysaccharidoses**. Changes occur in the ECM during the **aging process**. This chapter describes the basic biochemistry of the three major classes of biomolecules found in the ECM and illustrates their biomedical significance. Major biochemical features of two specialized forms of ECM—bone and cartilage—and of a number of diseases involving them are also briefly considered.

COLLAGEN IS THE MOST ABUNDANT PROTEIN IN THE ANIMAL WORLD

Collagen, the major component of most connective tissues, constitutes approximately 25% of the protein of mammals. It provides an extracellular framework for all metazoan animals and exists in virtually every animal tissue. Twenty-nine types of collagens made up of over 30 distinct polypeptide chains (each encoded by a separate gene) have been identified in human tissues (**Table 50–1**). Although several of these are present only in small proportions, they may play important roles in determining the physical properties of specific tissues. In addition, a number of proteins (eg, the C1q component of the complement system, pulmonary surfactant proteins SPA and SPD) that are not classified as collagens have collagen-like domains in their structures; these proteins are sometimes referred to as "noncollagen collagens."

COLLAGENS HAVE A TRIPLE HELIX STRUCTURE

All collagen types have a **triple helical structure** made up of three polypeptide chain subunits (**α chains**). In some collagens, the entire molecule is a triple helix, whereas in others only a fraction of the structure may be in this form.

TABLE 50–1 Types of Collagen & Their Tissue Distribution

Type	Distribution
I	Noncartilaginous connective tissues, including bone, tendon, skin
II	Cartilage, vitreous humor
III	Extensible connective tissues, including skin, lung, vascular system
IV	Basement membranes
V	Minor component in tissues containing collagen I
VI	Muscle and most connective tissues
VII	Dermal-epidermal junction
VIII	Endothelium and other tissues
IX	Tissues containing collagen II
X	Hypertrophic cartilage
XI	Tissues containing collagen II
XII	Tissues containing collagen I
XIII	Many tissues, including neuromuscular junctions and skin
XIV	Tissues containing collagen I
XV	Associated with collagens close to basement membranes in many tissues including in eye, muscle, microvessels
XVI	Many tissues
XVII	Epithelia, skin hemidesmosomes
XVIII	Associated with collagens close to basement membranes, close structural homologue of XV
XIX	Rare, basement membranes, rhabdomyosarcoma cells
XX	Many tissues, particularly corneal epithelium
XXI	Many tissues
XXII	Tissue junctions, including cartilage-synovial fluid, hair follicle–dermis
XXIII	Limited in tissues, mainly transmembrane and shed forms
XXIV	Developing cornea and bone
XXV	Brain
XXVI	Testis, ovary
XXVII	Embryonic cartilage and other developing tissues, cartilage in adults
XXVIII	Basement membrane around Schwann cells
XXIX	Epidermis

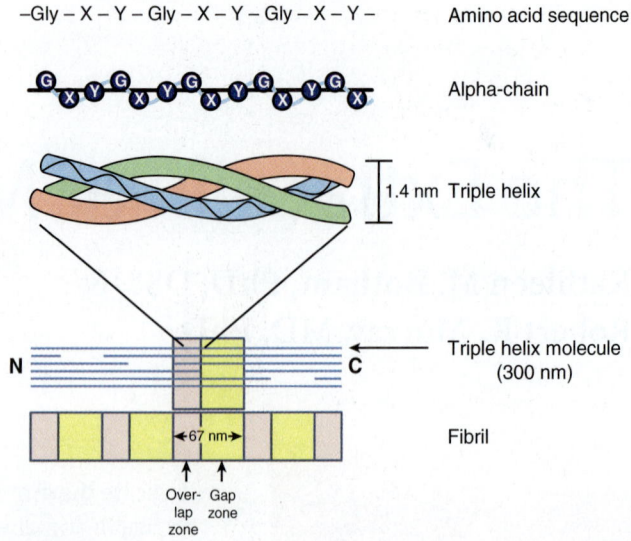

FIGURE 50–1 **Molecular features of collagen structure from the primary sequence to the fibril.** Each individual polypeptide chain is twisted into a left-handed helix of three residues (Gly-X-Y) per turn, and three of these chains are then wound into a right-handed superhelix. The triple helices are then assembled into a quarter-staggered alignment to form fibrils. This arrangement leads to areas where there is complete overlap of the molecules alternating with areas where there is a gap, giving the fibrils a regular banded appearance.

Mature **collagen type I** belongs to the former type; each polypeptide subunit is twisted into a left-handed helix of three residues per turn forming an α chain. Three of these are then wound into a **right-handed triple-** or **superhelix**, forming a rod-like molecule 1.4 nm in diameter and about 300-nm long (**Figure 50–1**). **Glycine** residues occur at every third position

of the triple helical portion of the α chain. This is necessary because glycine is the only amino acid small enough to be accommodated in the limited space available in the central core of the triple helix. This **repeating structure**, represented as (Gly-X-Y)n, is an absolute requirement for the formation of the triple helix. While X and Y can be any other amino acids, the X positions are often proline and the Y positions are often hydroxyproline. Proline and hydroxyproline confer **rigidity** on the collagen molecule. **Hydroxyproline** is formed by the post-translational hydroxylation of peptide-bound proline residues catalyzed by the enzyme **prolyl hydroxylase**, whose cofactors are **ascorbic acid** (vitamin C) and α-ketoglutarate. Lysines in the Y position may also be posttranslationally modified to hydroxylysine through the action of **lysyl hydroxylase**, an enzyme with similar cofactors to prolyl hydroxylase. Some of these hydroxylysines may be further modified by the addition of galactose or galactosyl-glucose through an **O-glycosidic linkage** (see Chapter 46), a glycosylation site that is unique to collagen.

Some collagen types form long rod-like fibers in tissues. These are assembled by lateral association of these triple helical units into **fibrils** (10-300 nm in diameter) in a **"quarter-staggered" alignment** such that each is displaced longitudinally from its neighbor by slightly less than one-quarter of its length (see Figure 50–1). Fibrils, in turn, associate into thicker **fibers** (1-20 μm in diameter). Because the quarter staggered alignment results in regularly spaced gaps between the triple helical molecules in the array, fibers have a banded appearance in connective tissues. In some tissues, for example tendons, fibers associate into even larger bundles, which may have a diameter of up to 500 μm. Collagen fibers are further stabilized by the formation of **covalent cross-links**, both within and between the triple helical units. These cross-links form through the action of **lysyl oxidase**, a copper-dependent enzyme that oxidatively deaminates the ε-amino groups of certain lysine and hydroxylysine residues, yielding reactive aldehydes. Such aldehydes can form aldol condensation products with other lysine- or hydroxylysine-derived aldehydes or form Schiff bases with the ε-amino groups of unoxidized lysines or hydroxylysines. These reactions, after further chemical rearrangements, result in the stable covalent cross-links that are important for the tensile strength of the fibers. Histidine may also be involved in certain cross-links.

The main fibril-forming collagens in skin and bone and in cartilage, respectively, are types **I** and **II**, although other collagens also adopt this structure. In addition, however, there are many nonfibril-forming collagens and their structures and functions are described briefly in the following section.

Some Collagen Types Do Not Form Fibrils

Several collagen types do not form fibrils in tissues (**Figure 50–2**). They are characterized by interruptions of the triple helix with stretches of protein lacking Gly-X-Y repeat sequences. Thus, areas of globular structure are interspersed in the triple helical structure. **Network-like collagens** such as type IV form networks in basement membranes; **fibril-associated collagens with interrupted triple helices (FACITs)**, as their name indicates, have interruptions in the triple helical domains; **beaded filaments** consist of long chains of collagen molecules which have a regular beaded appearance; collagen VII forms the main part of **anchoring fibrils** in epithelial tissues; **transmembrane collagens** have short intracellular N-terminal domains and extracellular domains with long interrupted triple helices; **multiplexins** are collagens with multiple triple helix domains and interruptions.

Collagen Undergoes Extensive Posttranslational Modifications

Newly synthesized collagen undergoes extensive **posttranslational modification** before becoming part of a mature extracellular collagen fiber (**Table 50–2**). Like most secreted proteins, collagen is synthesized on ribosomes in a precursor form, **preprocollagen**, which contains a leader or signal sequence that directs the polypeptide chain into the lumen of the endoplasmic reticulum (ER) (see Chapter 49). As it enters the ER, this leader sequence is removed enzymatically. **Hydroxylation** of proline and lysine residues and **glycosylation** of hydroxylysines in the **procollagen** molecule also take place at this site. The procollagen molecule contains polypeptide extensions (**extension peptides**) of 20 to 35 kDa at both its amino- and carboxyl-terminal ends, which are not present in mature collagen. Both extension peptides contain cysteine residues. While the amino terminal propeptide forms only intrachain disulfide bonds, the carboxyl-terminal propeptides form both intrachain and interchain disulfide bonds. Formation of these disulfide bonds assists in the **registration** of the three collagen molecules to form the triple helix, winding from the carboxyl-terminal end. After formation of the triple helix, no further hydroxylation of proline or lysine or glycosylation of hydroxylysines can take place. **Self-assembly** is a cardinal principle in the biosynthesis of collagen.

Following **secretion** from the cell by way of the Golgi apparatus, extracellular enzymes called **procollagen aminoproteinase** and **procollagen carboxyproteinase** remove the extension peptides at the amino- and carboxyl-terminal ends, respectively, forming the monomeric units of collagen, termed **tropocollagen**. Cleavage of the propeptides may occur within crypts or folds in the cell membrane. Once the propeptides are removed, the tropocollagen molecules, containing approximately 1000 amino acids per α chain, **spontaneously assemble** into collagen fibers. These are further stabilized by the formation of **inter- and intrachain cross-links** through the action of lysyl oxidase, as described previously.

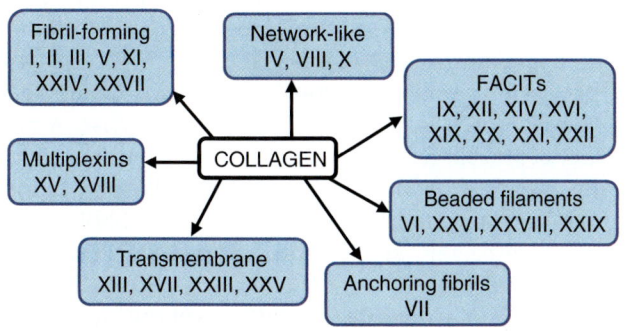

FIGURE 50–2 **Classification of collagens according to the structures they form.** FACIT, fibril-associated collagen with interrupted triple helices; multiplexin, multiple triple helix domains and interruptions.

TABLE 50–2 **Order & Location of Processing of the Fibrillar Collagen Precursor**

Intracellular
1. Cleavage of signal peptide
2. Hydroxylation of prolyl residues and some lysyl residues; glycosylation of some hydroxylysyl residues
3. Formation of intrachain and interchain S–S bonds in extension peptides
4. Formation of triple helix

Extracellular
1. Cleavage of amino- and carboxyl-terminal propeptides
2. Assembly of collagen fibers in quarter-staggered alignment
3. Oxidative deamination of η-amino groups of lysyl and hydroxylysyl residues to aldehydes
4. Formation of intra- and interchain cross-links via Schiff bases and aldol condensation products

The same cells that secrete collagen also secrete **fibronectin**, a large glycoprotein present on cell surfaces, in the ECM, and in blood (see later). Fibronectin binds collagen fibers during aggregation and alters the kinetics of fiber formation in the pericellular matrix. Associated with fibronectin and procollagen in this matrix are the **proteoglycans** heparan sulfate and chondroitin sulfate (see later). In fact, **type IX collagen**, a minor collagen type from cartilage, contains an attached glycosaminoglycan chain. Such interactions may serve to regulate the formation of collagen fibers and to determine their orientation in tissues.

Once formed, collagen is relatively **metabolically stable**. However, its breakdown is increased during starvation and various inflammatory states. Excessive production of collagen occurs in a number of conditions, for example, hepatic cirrhosis.

A Number of Genetic & Deficiency Diseases Result From Abnormalities in the Synthesis of Collagen

More than 30 genes encode the collagens, and they are designated according to the procollagen type and their constituent α chains, called proα chains. Collagens may be homotrimeric, containing three identical proα chains, or heterotrimeric, where the proα chains are different. For example, type I collagen is heterotrimeric, containing two proα1(I) and one proα2(I) chains (the arabic number refers to the proα chain, and the roman numeral in parentheses indicates the collagen type), while type II collagen is homotrimeric, having three proα1(II) chains. Collagen genes have the prefix *COL* followed by the type in arabic numerals, then an A and the number of the proα chain they encode. Thus, *COL1A1* and *COL1A2* are the genes for the proα1 and 2 chains of type I collagen, *COL2A1* is the gene for the proα1 chain of type II collagen, and so on.

The pathway of collagen biosynthesis is complex, involving at least eight enzyme-catalyzed posttranslational steps. Thus, it is not surprising that a number of diseases (**Table 50–3**) are due to **mutations in collagen genes** or in **genes encoding some of the enzymes** involved in these posttranslational modifications. Diseases affecting bone (eg, osteogenesis imperfecta) and cartilage (eg, the chondrodysplasias) will be discussed later in this chapter.

Ehlers-Danlos syndrome (formerly called Cutis hyperelastica), comprises a group of inherited disorders whose principal clinical features are hyperextensibility of the skin, abnormal tissue fragility, and increased joint mobility. The clinical picture is variable, reflecting underlying extensive genetic heterogeneity. A number of forms of the disease caused by genetic defects in proteins involved in the synthesis and assembly of various collagens types are known. In 2017 the Villefranche classification of six subtypes, first published in 1997, was superceded by the 2017 International Classification of Ehlers Danlos Syndromes which identifies 13 subtypes, distinguished by their phenotype and the underlying

TABLE 50–3 Diseases Caused by Mutations in Collagen Genes or by Deficiencies in the Activities of Enzymes Involved in the Posttranslational Biosynthesis of Collagen[a]

Gene or Enzyme Affected	Disease[a]
COL1A1, COL1A2	Osteogenesis imperfecta type 1[b]
	Osteoporosis
	Arthrochalasia Ehlers-Danlos syndrome (aEDS)
COL2A1	Severe chondrodysplasia
	Osteoarthritis
COL3A1	Vascular Ehlers-Danlos syndrome (vEDS)
COL4A3, COL4A4, COL4A5	Alport syndrome (autosomal and X-linked)
COL7A1	Epidermolysis bullosa, dystrophic
COL10A1	Schmid metaphyseal chondrodysplasia
COL5A1, COL5A2, COL1A1	Classical Ehlers-Danlos syndrome (cEDS)
Tenascin XB (*TNXB*)	Classical-like Ehlers-Danlos syndrome (clEDS)
Lysyl hydroxylase	Kyphoscoliotic Ehlers-Danlos syndrome (kEDS)
ADAM metallopeptidase with thrombospondin type 1 motif (*ADAMTS2*) (also called procollagen *N*-proteinase)	Dermatosparaxis Ehlers-Danlos syndrome (dEDS)
ATP7A (a copper-transporting ATPase)	Menkes disease[c]

[a]Genetic linkage to collagen genes has been shown to a few other conditions not listed here.
[b]Eight different types of osteogenesis imperfecta are recognized, but most cases are caused by mutations in the *COL1A1* and *COL1A2* genes.
[c]Secondary to a deficiency of copper (see Chapter 52).

molecular defect (**Table 50–4**). The **hypermobility, vascular and classical** subtypes are more common, while the other ten are rare. The vascular subtype is the most serious because of its tendency for spontaneous rupture of arteries or the bowel, reflecting abnormalities in type III collagen. Patients with kyphoscoliosis exhibit progressive curvature of the spine (scoliosis) due to a deficiency of lysyl hydroxylase. A deficiency of procollagen *N*-proteinase (ADAM metallopeptidase with thrombospondin type 1 motif [*ADAMTS2*]), causing formation of abnormal thin, irregular collagen fibrils, results in dermatosparaxis, manifested by marked fragile and sagging skin.

The **Alport syndrome** (hereditary nephritis) is the name given to a number of genetic disorders (both X-linked and autosomal) affecting **type IV** collagen, a network-like collagen which forms part of the structure of the basement membranes of the renal glomeruli, inner ear, and eye (see discussion of

TABLE 50–4 The 2017 Ehlers-Danlos Syndrome International Classification of Subtypes[a]

Subtype Name	Defect in	Incidence	Clinical Signs
Hypermobile EDS (hEDS)	Unknown	1:5000-20,000	Joint hypermobility, musculoskeletal abnormalities
Classical EDS (cEDS)	Type V collagen Type I collagen (rare)	1:20,000-40,000	Skin hyperextensibility, generalized joint hypermobility
Vascular EDS (vEDS)	Type III collagen Type I collagen (rare)	1:100,000	Fragile blood vessels and organs, thin and translucent skin, easy bruising
Kyphoscoliotic EDS (kEDS)	Lysyl hydroxylase	<1:100,000	Curvature of the spine (scoliosis), severe muscle weakness, generalized joint hypermobility
Arthrochalasia EDS (aEDS)	Type I collagen	<40 cases	Severe joint hypermobility and dislocation of both hips, skin hyperextensibility
Dermatosparaxis EDS (dEDS)	ADAM metallopeptidase with thrombospondin type 1 motif (ADAMTS2)[b]	<1:1000,000	Very fragile and sagging skin
Classical-like EDS (clEDS)	Tenascin XB[c]	<1:1000,000	Skin hyperextensibility, generalized joint hypermobility, easily bruised skin
Cardiac-valvular EDS (cvEDS)	Type I collagen	<1:1000,000	Severe progressive cardiac valvular disorders, skin hyperextensibility, joint hypermobility
Brittle cornea syndrome (BCS)	Zinc finger protein	<1:1000,000	Thin or ruptured cornea
Spondylodysplastic EDS (spEDS)	β-1,4 galactosyltranferase Zinc transporter	<1:1000,000	Short stature, muscle weakness, bowing of limbs
Musculo-contractural EDS (mcEDS)	Carbohydrate sulfotransferase Dermatan sulfate epimerase	<1:1000,000	Multiple muscle and joint contractures (permanent tightening of muscles and nearby tissues, hypermobility of distal joints)
Myopathic EDS (mEDS)	Type XII collagen	<1:1000,000	Muscle weakness or atrophy that improves with age, joint contractures, hypermobility of distal joints
Periodontal EDS (pEDS)	Complement component	Unknown, but very rare	Early onset of severe and intractable gum disease

[a]Data from Malfait F, Francomano C, Byers P, et al: The 2017 international classification of the Ehlers-Danlos syndromes, Am J Med Genet C Semin Med Genet. 2017;175(1):8-26.
[b]Also called procollagen *N*-proteinase
[c]A glycoprotein expressed in connective tissues such as skin, joints, and muscles.

laminin, later). Mutations in several genes encoding type IV collagen fibers have been demonstrated. The main presenting sign is hematuria, accompanied by ocular lesions and hearing loss, and patients may eventually develop end-stage renal disease. Electron microscopy reveals characteristic abnormalities of the structure of the basement membrane and lamina densa.

In **epidermolysis bullosa**, the skin breaks and blisters as a result of minor trauma. The dystrophic form of the disease is due to mutations in *COL7A1*, which affect the structure of **type VII** collagen. This collagen forms delicate fibrils that anchor the basal lamina to collagen fibrils in the dermis. These anchoring fibrils have been shown to be markedly reduced in dystrophic epidermolysis bullosa, probably resulting in the blistering. Epidermolysis bullosa simplex, another variant, is due to mutations in keratins 5 and 14 (see Chapter 51).

Although **Scurvy** affects the structure of collagen, it is caused by **a dietary deficiency of ascorbic acid** (vitamin C) (see Chapter 44), not a genetic abnormality. Its major signs are bleeding gums, subcutaneous hemorrhages, and poor wound healing. These signs reflect defective synthesis of collagen due to reduced activity of the enzymes **prolyl and lysyl hydroxylases**, both of which require ascorbic acid as a cofactor and are involved in posttranslational modifications which give collagen molecules rigidity.

In **Menkes disease**, deficiency of copper resulting from mutations in *ATP7A*, the gene for a copper-transporting ATPase (also called Menkes protein) causes defective cross-linking of collagen and elastin by the copper-dependent enzyme lysyl oxidase. (Menkes disease is discussed in Chapter 52.)

ELASTIN CONFERS EXTENSIBILITY & RECOIL ON LUNG, BLOOD VESSELS, & LIGAMENTS

Elastin is a connective tissue protein that is responsible for properties of extensibility and elastic recoil in tissues. Although not as widespread as collagen, elastin is present in large amounts in tissues that require these physical properties, for example, lung, large arterial blood vessels, and some elastic ligaments. Smaller quantities of elastin are also found in skin, ear cartilage, and several other tissues. In contrast to collagen, only one genetic type of elastin is known, although variants arise by alternative splicing (see Chapter 36) of the hnRNA for elastin. Elastin is synthesized as a soluble monomer of ~70 kDa called **tropoelastin**. Some of the prolines of tropoelastin are hydroxylated to **hydroxyproline** by prolyl hydroxylase, though hydroxylysine and glycosylated hydroxylysine are not present. Unlike collagen, tropoelastin is not synthesized in a proform with extension peptides. Furthermore, elastin does not contain repeat Gly-X-Y sequences, triple helical structure, or carbohydrate moieties.

After secretion from the cell, certain lysyl residues of tropoelastin are oxidatively deaminated to aldehydes by **lysyl oxidase**, the same enzyme involved in this process in collagen. However, the major cross-links formed in elastin are the **desmosines**, which result from the condensation of three of these lysine-derived aldehydes with an unmodified lysine to form a tetrafunctional cross-link unique to elastin. Once cross-linked in its mature, extracellular form, elastin is highly insoluble, **extremely stable**, and has a very low-turnover rate. The various random coil conformations present in its structure permit the protein to stretch and subsequently recoil during the performance of its physiologic functions.

Table 50–5 summarizes the main differences between collagen and elastin.

Deletions in the elastin gene (located at chromosome 7q11.23) have been found in approximately 90% of subjects with the **Williams-Beuren syndrome**, a developmental disorder affecting connective tissue and the central nervous system. The mutations affect the synthesis of elastin, and probably play a causative role in the **supravalvular aortic stenosis** often found in this condition. Fragmentation or, alternatively, a decrease of elastin is found in conditions such as **pulmonary emphysema, cutis laxa, and aging of the skin**.

TABLE 50–5 Major Differences Between Collagen & Elastin

Collagen	Elastin
1. Many different genetic types	One genetic type
2. Triple helix	No triple helix; random coil conformations permitting stretching
3. (Gly-X-Y)$_n$ repeating structure	No (Gly-X-Y)$_n$ repeating structure
4. Presence of hydroxylysine	No hydroxylysine
5. Carbohydrate-containing	No carbohydrate
6. Intramolecular aldol cross-links	Intramolecular desmosine cross-links
7. Presence of extension peptides during biosynthesis	No extension peptides present during biosynthesis

FIBRILLINS ARE STRUCTURAL COMPONENTS OF MICROFIBRILS

Microfibrils are fine fiber-like strands 10 to 12 nm in diameter which provide a **scaffold** for the deposition of elastin in the ECM. **Fibrillins** are large glycoproteins (about 350 kDa) that are major structural component of these fibers. They are secreted (subsequent to a proteolytic cleavage) into the ECM by fibroblasts and become incorporated into the insoluble microfibrils. **Fibrillin-1** is the main fibrillin present, but fibrillins-2 and 3 have also been identified in humans, and fibrillin-2 is thought to be important in deposition of microfibrils early in development. Other proteins including **microfibril-associated glycoproteins (MAGPs)**, **fibulins**, and **members of the ADAMTS family** are also associated with microfibrils. Fibrillin microfibrils are found in elastic fibers and also in elastin-free bundles in the eye, kidney, and tendons.

Marfan Syndrome Is Caused by Mutations in the Gene for Fibrillin-1

Marfan syndrome is a relatively prevalent inherited disease affecting connective tissue; it is inherited as an autosomal dominant trait. It affects the **eyes** (eg, causing dislocation of the lens, known as ectopia lentis), the **skeletal system** (most patients are tall and exhibit long digits [arachnodactyly] and hyperextensibility of the joints), and the **cardiovascular system** (eg, causing weakness of the aortic media, leading to dilation of the ascending aorta). Abraham Lincoln may have had this condition. Most cases are caused by mutations in the gene (on chromosome 15) for fibrillin-1. This results in abnormal fibrillin and/or lower amounts being deposited in the ECM. Since the mutation interferes with the normal binding of fibrillin-1 to the cytokine transforming growth factor β (TGF-β), Marfan syndrome causes disturbances in TGF-β signaling. This has led to the development of therapies for the condition using drugs that antagonize TGF-β (eg, the angiotensin II receptor antagonist, Losartan).

Mutations in the fibrillin-1 gene have also been identified as the cause of **acromicric dysplasia**, which is characterized by short stature, skin thickening, and stiff joints. **Congenital contractural arachnodactyly** is associated with a mutation in the gene for fibrillin-2. The probable sequence of events leading to Marfan syndrome is summarized in **Figure 50–3**.

FIBRONECTIN IS INVOLVED IN CELL ADHESION & MIGRATION

Fibronectin is a major glycoprotein of the ECM, also found in a soluble form in plasma. It consists of two identical subunits, each of about 230 kDa, joined by two disulfide bridges near their carboxyl terminals. The gene encoding fibronectin is very large, containing some 50 exons; the RNA produced by its transcription is subject to considerable alternative splicing, and as many as 20 different mRNAs have been detected

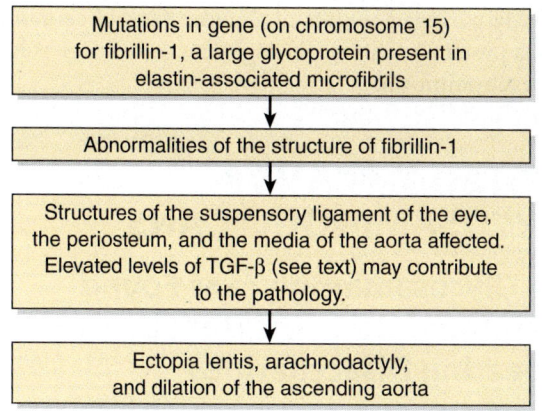

FIGURE 50–3 Probable sequence of events in the causation of the major signs exhibited by patients with Marfan syndrome.

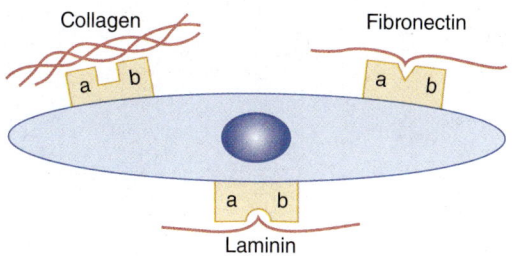

FIGURE 50–5 Schematic representation of a cell interactions with major proteins of the ECM. a and b indicate α- and β-polypeptide chains of integrins.

in various tissues. Fibronectin contains three types of repeating motifs (I, II, and III), which are organized into functional **domains** (at least seven); functions of these domains include binding fibronectin (thus enabling molecules of the protein to interact) **heparin** (see later), fibrin, collagen, and cell surfaces (**Figure 50–4**). Fibronectin contains an **Arg-Gly-Asp (RGD) sequence** that binds to cells via a transmembrane receptor protein which belongs to the **integrin** class of proteins (see Chapter 55). This sequence is shared by a number of other proteins present in the ECM that bind to integrins present in cell plasma membranes, and its presence in synthetic peptides enables them to inhibit the binding of fibronectin to cells. **Figure 50–5** illustrates the interaction of collagen, fibronectin, and laminin, all major proteins of the ECM, with a typical cell (eg, fibroblast) present in the matrix.

The fibronectin integrin receptor interacts indirectly with **actin** microfilaments (see Chapter 51) present in the cytosol via a number of proteins, collectively known as **attachment proteins**; these include **talin, vinculin, α-actinin,** and **paxillin** (**Figure 50–6**). Such large protein complexes form **focal adhesions** which not only anchor cells in the ECM, but also relay signals from the exterior which influence cell behavior. Thus, the interaction of fibronectin with its receptor provides one route whereby the **outside of the cell can communicate with the inside**. Fibronectin is also involved in **cell migration**, as it provides a binding site for cells and thus helps them to steer their way through the ECM. The amount of fibronectin around

many **transformed cells** is sharply reduced, partly explaining their faulty interaction with the ECM.

LAMININ IS A MAJOR PROTEIN COMPONENT OF BASAL LAMINAS

Basal laminas are specialized areas of the ECM that surround epithelial and some other cells (eg, muscle cells). **Laminin** (a glycoprotein of about 850 kDa and 70-nm length) consists of three distinct elongated polypeptide chains (α, β, and γ chains, each of which have genetic variants) linked together to form a complex, elongated shape (see Figure 51–11, **which shows laminin-2, also called merosin**). In basal laminas, laminin forms networks which are attached to type IV collagen by **nidogen** (previously called entactin), a glycoprotein containing an RGD sequence, and the heparan sulfate proteoglycan, **perlecan**. The collagen interacts with laminin (rather than directly with the cell surface), which in turn interacts with integrins or other proteins, such as **dystroglycans** (see Chapter 51) thus anchoring the lamina to the cells (**Figure 50–7**).

In the **renal glomerulus**, the basal lamina consists of two separate sheets of cells (one endothelial and one epithelial), each disposed on opposite sides of the lamina; these three layers make up the **glomerular membrane**. This relatively thick basal lamina has an important role in **glomerular filtration**. The glomerular membrane allows small molecules, such as **inulin** (5.2 kDa), to pass through as easily as water. On the other hand, large molecules, including most plasma proteins, are not able to pass through. The reason for this is two-fold: (1) The **pores** in the glomerular membrane are about 8-nm across, thus molecules larger than this are unable to pass through.

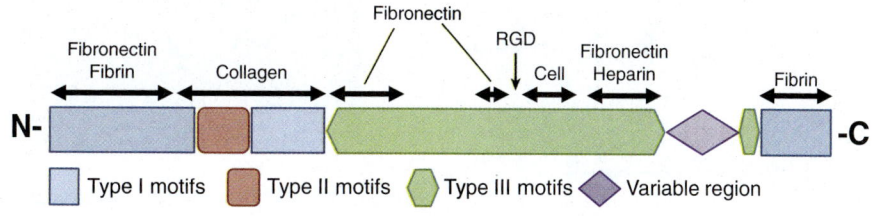

FIGURE 50–4 Structure of the fibronectin monomer. Fibronectin is a dimer joined by disulfide bridges (not shown) near the carboxyl terminals of the monomers. Each monomer consists mainly of repeating motifs of type I, II, or III and has a number of protein-binding domains. Four bind fibronectin and there are also domains for collagen, heparin, fibrin, and cell binding. The approximate location of the Arg-Gly-Asp (RGD) sequence of fibronectin, which interacts with a variety of fibronectin integrin receptors on cell surfaces, is indicated by the arrow.

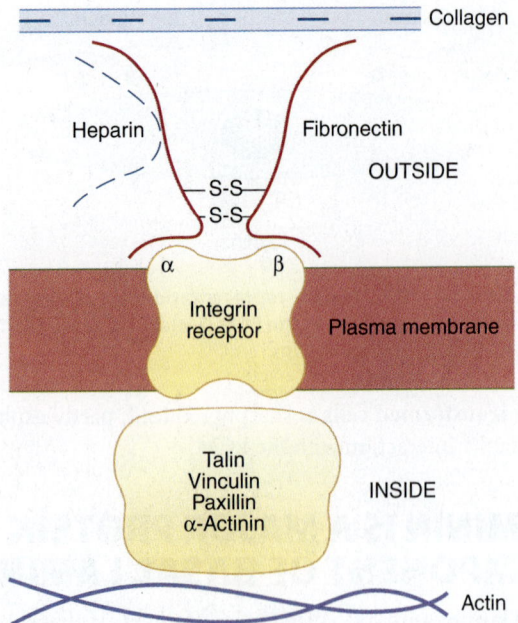

FIGURE 50–6 **Schematic representation of fibronectin interacting with actin in the cytosol via an integrin fibronectin receptor.** The fibronectin dimer on the outside of the plasma membrane binds to the membrane-spanning integrin receptor via its Arg-Gly-Asp (RGD) sequences. On the cytosolic side, integrin interacts with attachment proteins including talin, vinculin, α-actinin, and paxillin (shown as a complex) which interact with actin microfilaments, thus indirectly linking fibronectin in the ECM with actin in the cell cytosol.

(2) Although some plasma proteins, such as albumin, are smaller than this pore size, they are prevented from passing through easily by the **negative charges** of heparan sulfate and of certain sialic acid–containing glycoproteins present in the lamina which repel most plasma proteins, which are negatively charged at the pH of blood. The normal structure of the glomerulus may be severely damaged in certain types of **glomerulonephritis** (eg, caused by antibodies directed against various components of the glomerular membrane). This affects the pores and the amounts and dispositions of the negatively charged macromolecules referred to earlier, and

relatively massive amounts of albumin (and of certain other plasma proteins) can pass through into the urine, resulting in severe **albuminuria**.

PROTEOGLYCANS & GLYCOSAMINOGLYCANS

The Glycosaminoglycans Found in Proteoglycans Are Built Up of Repeating Disaccharides

Proteoglycans, proteins that contain covalently linked **glycosaminoglycans (GAGs)** (see also Chapters 15 and 46), are a major component of the ECM. At least 30 have been characterized, for example, syndecan, betaglycan, serglycin, perlecan, aggrecan, versican, decorin, biglycan, and fibromodulin. A proteoglycan consists of a **core protein** bound covalently to GAGs, and these units form large complexes with other components of the ECM, such as hyaluronic acid or collagen. **Figures 50–8** and **50–9** show the general structure of such complexes. They are very large with an overall structure resembling that of a bottlebrush. The example shown in

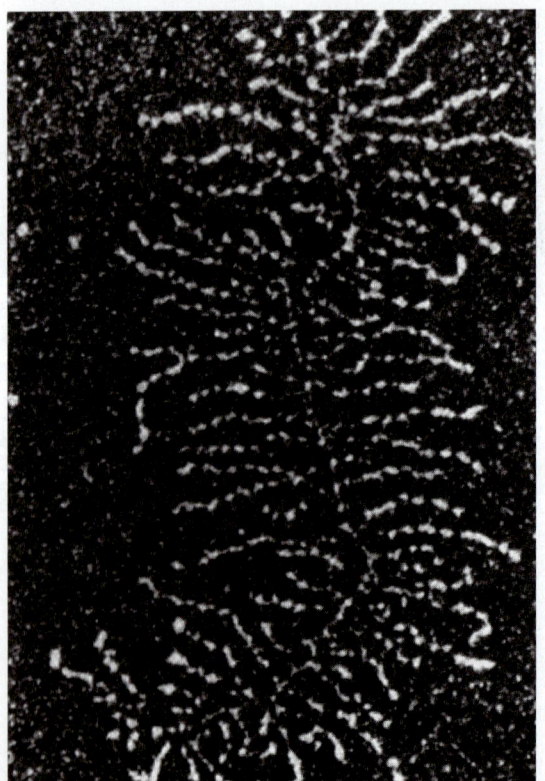

FIGURE 50–8 **Darkfield electron micrograph of a proteoglycan aggregate.** The proteoglycan subunits and filamentous backbone are particularly well extended in this image. (Reproduced with permission from Rosenberg L, Hellmann W, Kleinschmidt AK. Electron microscopic studies of proteoglycan aggregates from bovine articular cartilage, J Biol Chem. 1975;250(5):1877–1883.)

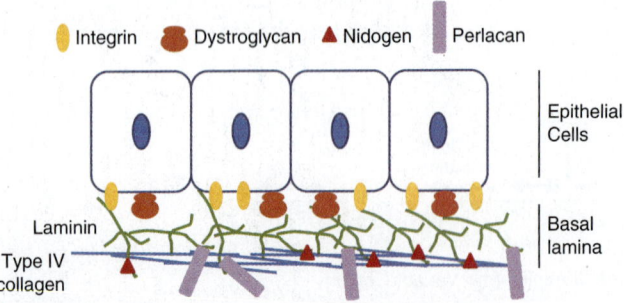

FIGURE 50–7 **Structure of the basal lamina.** Laminin is attached to type IV collagen via nidogen and perlecan (forming the basal lamina) and to the epithelial cell layer via integrins and dystroglycans.

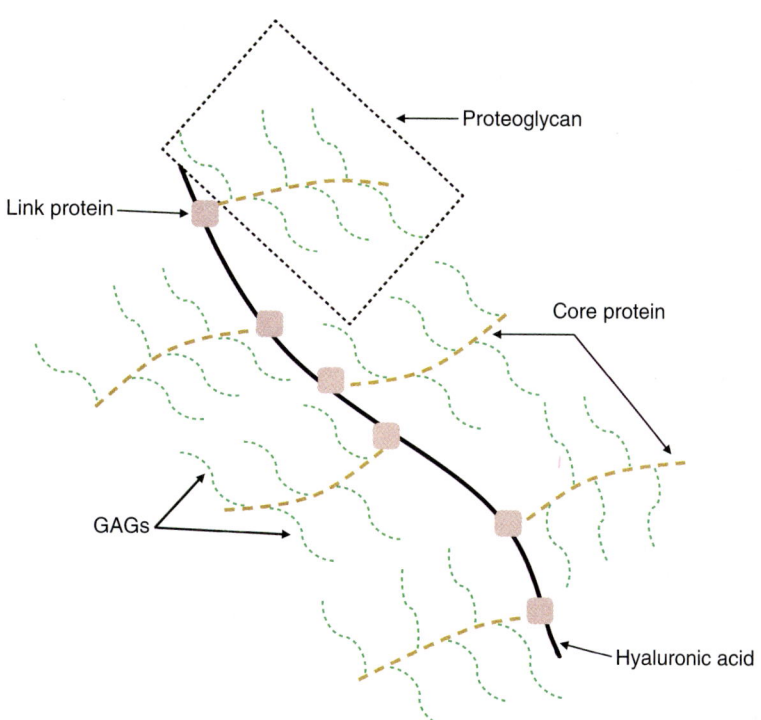

FIGURE 50–9 **Schematic representation of a proteoglycan complex.** In this example, proteoglycans are attached via noncovalent bonds to link proteins which, in turn, bond noncovalently to a long strand of the glycosaminoglycan (GAG), hyaluronic acid.

Figure 50–9 contains a long strand of hyaluronic acid (one type of GAG) (see Chapter 15) to which link proteins are attached **noncovalently**. In turn, the link proteins interact noncovalently with core protein molecules from which chains of other GAGs (eg, keratan sulfate and chondroitin sulfate) project. Proteoglycans vary in tissue distribution, nature of the core protein, attached GAGs, and their function. The amount of **carbohydrate** in a proteoglycan is usually much greater than that found in a glycoprotein, and may comprise up to 95% of its weight.

GAGs are unbranched polysaccharides made up of repeating disaccharides, one component of which is always an **amino sugar** (hence, the name GAG), either D-glucosamine or D-galactosamine. The other component of the repeating disaccharide (except in the case of keratan sulfate) is a **uronic acid**, either D-glucuronic acid (GlcUA) or its 5′-epimer, L-iduronic acid (IdUA). There are at least seven GAGs: **hyaluronic acid (hyaluronan), chondroitin sulfate, keratan sulfates I and II, heparin, heparan sulfate,** and **dermatan sulfate**. With the exception of hyaluronic acid, all the GAGs contain **sulfate groups**, either as *O*-esters or as *N*-sulfate (in heparin and heparan sulfate). Hyaluronic acid is also exceptional because it can exist as a polysaccharide in the ECM, with no covalent attachment to protein, as the definition of a proteoglycan given above specifies. Both GAGs and proteoglycans have proved difficult to work with, partly because of their complexity. However, since they are major components of the ECM and have a number of important biologic roles as well as being involved in a number of disease processes, interest in them has increased greatly in recent years.

Biosynthesis of Glycosaminoglycans Involves Attachment to Core Proteins, Chain Elongation, & Chain Termination

Attachment to Core Proteins

The linkage between GAGs and their core proteins is generally one of the following three types:

1. An *O*-**glycosidic bond** between **xylose** (Xyl) and **a serine residue** on the protein, a bond that is unique to proteoglycans. This linkage is formed by transfer of a Xyl residue to serine from UDP-xylose. Two residues of galactose (Gal) are then added to the Xyl residue, forming a **link trisaccharide**, Gal-Gal-Xyl. Further chain growth of the GAG occurs on the terminal Gal.

2. An *O*-**glycosidic bond** between *N*-**acetylgalactosamine (GalNAc)** and **serine (Ser) or threonine (Thr)** (see Figure 46–1A), present in keratan sulfate. This bond is formed by donation to Ser or Thr of a GalNAc residue, employing UDP-GalNAc as its donor.

3. An *N*-**glycosylamine bond** between *N*-acetylglucosamine (**GlcNAc**) and the amide nitrogen of **asparagine (Asn)**, as found in *N*-linked glycoproteins (see Figure 46–1B). Its synthesis is believed to involve dolichol-PP oligosaccharide.

The synthesis of the core proteins occurs in the **endoplasmic reticulum**, and formation of at least some of the above linkages also occurs there. Most of the later steps in the biosynthesis of GAG chains and their subsequent modifications occur in the **Golgi apparatus**.

Chain Elongation

Appropriate **nucleotide sugars** and highly specific Golgi-located **glycosyltransferases** are employed to synthesize the oligosaccharide chains of GAGs. The "**one enzyme, one linkage**" relationship appears to hold here, as in the case of certain types of linkages found in glycoproteins. The enzyme systems involved in chain elongation are capable of high-fidelity reproduction of complex GAGs (see also Chapter 46).

Chain Termination

This appears to result from (1) **sulfation**, particularly at certain positions of the sugars, and (2) the **progression** of the growing GAG chain away from the membrane site where catalysis occurs.

Further Modifications

After formation of the GAG chain, **numerous chemical modifications** occur, such as the introduction of sulfate groups onto GalNAc and other moieties and the epimerization of GlcUA to IdUA residues. The enzymes catalyzing sulfation are designated **sulfotransferases** and use **3′-phosphoadenosine-5′-phosphosulfate** (PAPS, also called active sulfate) (see Chapter 32) as the sulfate donor. These Golgi-located enzymes are highly specific, and distinct enzymes catalyze sulfation at different positions (eg, carbons 2, 3, 4, and 6) on the acceptor sugars. An **epimerase** catalyzes conversions of glucuronyl to iduronyl residues.

Proteoglycans Are Important in the Structural Organization of the Extracellular Matrix

Proteoglycans are found in **every tissue** of the body, mainly in the ECM or ground substance, transparent, colorless material of the ECM which fills spaces between cells and largely consists of proteoglycans. Here they are associated with each other and also with the other major structural components of the matrix, collagen, and elastin, in specific ways. Some proteoglycans bind to collagen and others to elastin. These interactions are important in determining the structural organization of the matrix. Some proteoglycans (eg, decorin) can also **bind growth factors** such as TGF-β, modulating their effects on cells. In addition, some of them interact with certain **adhesive proteins** such as fibronectin and laminin (see earlier), also found in the matrix. The GAGs present in the proteoglycans are **polyanions** and hence bind polycations and cations such as Na⁺ and K⁺. This latter ability attracts water by osmotic pressure into the ECM and contributes to its turgor. GAGs also **gel** at relatively low concentrations. Because of the long, extended nature of the polysaccharide chains of GAGs and their ability to gel, the proteoglycans can act as **sieves**, restricting the passage of large macromolecules into the ECM but allowing relatively free diffusion of small molecules. Again, because of their extended structures and the huge macromolecular aggregates they often form, they occupy a **large volume** of the matrix relative to proteins.

Various Glycosaminoglycans Exhibit Differences in Structure & Have Characteristic Distributions and Diverse Functions

The seven GAGs named above differ from each other in a number of the following properties: amino sugar composition, uronic acid composition, linkages between these components, chain length of the disaccharides, the presence or absence of sulfate groups and their positions of attachment to the constituent sugars, the nature of the core proteins to which they are attached, the nature of the linkage to core protein, their tissue and subcellular distribution, and their biologic functions.

The structure (**Figure 50–10**), distribution, and functions of each of the GAGs will now be briefly discussed. The major features of the seven GAGs are summarized in **Table 50–6**.

Hyaluronic Acid

Hyaluronic acid consists of an unbranched chain of repeating disaccharide units containing glucuronic acid (GlcUA) and GlcNAc. It is present in bacteria and is found in the ECM of nearly all animal tissues, but is especially high in concentration in highly hydrated types such as skin and umbilical cord, and in bone, cartilage, joints (synovial fluid) and in vitreous humor in the eye, as well as in embryonic tissues. It is thought to play an important role in permitting **cell migration** during morphogenesis and wound repair. Its ability to attract water into the ECM triggers loosening of the matrix, aiding this process. The high concentrations of hyaluronic acid together with chondroitin sulfates present in **cartilage** contribute to its compressibility (see later).

Chondroitin Sulfates (Chondroitin 4-Sulfate & Chondroitin 6-Sulfate)

Proteoglycans linked to **chondroitin sulfate** by the Xyl-Ser O-glycosidic bond are prominent components of **cartilage** (see later). The repeating disaccharide is similar to that found in hyaluronic acid, containing GlcUA but with **GalNAc** replacing GlcNAc. The GalNAc is substituted with **sulfate** at either its 4′ or its 6′ position, with approximately one sulfate being present per disaccharide unit. Chondroitin sulfates have an important role in maintaining the structure of the ECM. They are located at sites of calcification in endochondral **bone** as well as in cartilage. They are also found in high amounts in the ECM of the central nervous system and, in addition to their structural function, are thought to act as signaling molecules in the prevention of the repair of nerve endings after injury.

Keratan Sulfates I & II

As shown in Figure 50–10, the keratan sulfates consist of repeating **Gal-GlcNAc** disaccharide units containing **sulfate** attached to the 6′ position of GlcNAc or occasionally of Gal. **Keratan sulfate I** was originally isolated from the **cornea**,

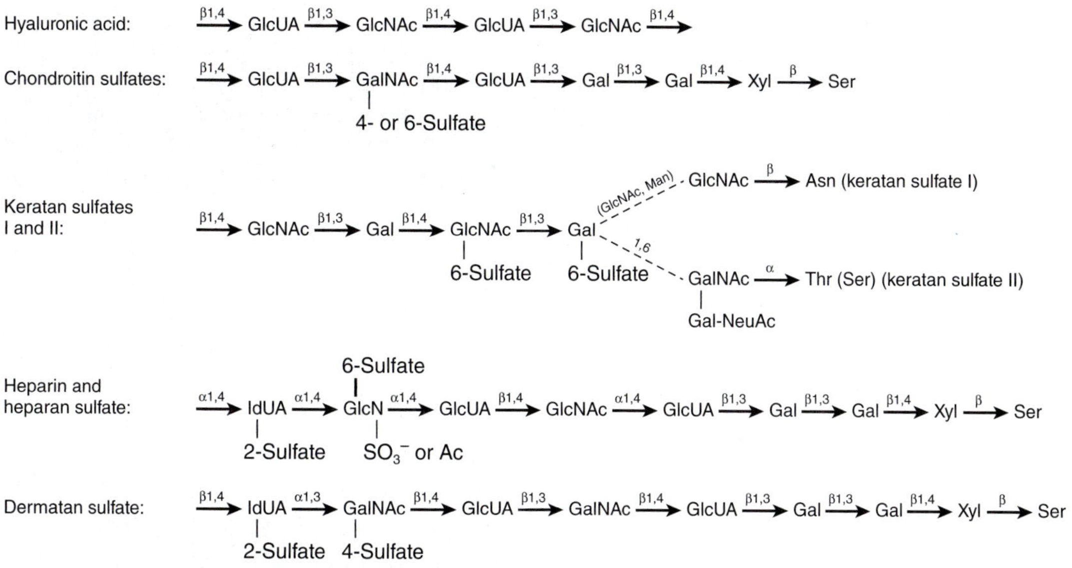

FIGURE 50–10 **Structures of glycosaminoglycans and their attachments to core proteins.** (Ac, acetyl; Asn, L-asparagine; Gal, D-galactose; GalN, D-galactosamine; GalNAc, N-acetyl D-galactosamine; GlcN, D-glucosamine; GlcNAc, N-acetyl D-glucosamine; GlcUA, D-glucuronic acid; IdUA, L-iduronic acid; Man, D-mannose; NeuAc, N-acetylneuraminic acid; Ser, L-serine; Thr, L-threonine; Xyl, L-xylose.) The summary structures are qualitative representations only and do not reflect, for example, the uronic acid composition of hybrid glycosaminoglycans such as heparin and dermatan sulfate, which contain both L-iduronic and D-glucuronic acid. Hyaluronic acid has no covalent attachment to protein. Chondroitin sulfates, heparin, heparan sulfate, and dermatan sulfate attach to a Ser on the core protein via the Gal-Gal-Xyl link trisaccharide. Keratan sulfate I links to a core protein Asn via GlcNAc and Keratan sulfate II to a Ser (or Thr) via GalNAc.

while **keratan sulfate II** came from cartilage. The two GAGs I or II, are classified based on their different linkages to the core protein (see Figure 50–10). In the eye, they lie between collagen fibrils and play a critical role in corneal transparency. Changes in proteoglycan composition found in corneal scars disappear when the cornea heals.

Heparin

The repeating disaccharide **heparin** contains **glucosamine** (GlcN) and either of the two uronic acids (GlcUA or iduronic acid [IdUA]) (**Figure 50–11**). Most of the amino groups of the GlcN residues are **N-sulfated**, but a few are acetylated (GlcNAc). The GlcN also carries a sulfate attached to carbon 6.

TABLE 50–6 Properties of Glycosaminoglycans

GAG	Sugars	Sulfate[a]	Protein Linkage	Location
Hyaluronic acid	GlcNAc, GLcUA	–	None	Skin, synovial fluid, bone, cartilage, vitreous humor, embryonic tissues
Chondroitin sulfate	GalNAc, GlcUA	GalNAc	Xyl-Ser; associated with HA via link proteins	Cartilage, bone, CNS
Keratan sulfate I and II	GlcNAc, Gal	GlcNAc	GlcNAc-Asn (KS I) GalNAc-Thr (KS II)	Cornea, cartilage, loose connective tissue
Heparin	Gln, IdUA	GlcN GlcN IdUA	Ser	Mast cells, liver, lung, skin
Heparan sulfate	GlcN, GlcUA	GlcN	Xyl-Ser	Skin, kidney basement membrane
Dermatan sulfate	GalNAc, IdUA, (GlcUA)	GalNAc IdUA	Xyl-Ser	Skin, wide distribution

[a]The sulfate is attached to various positions of the sugars indicated (see Figure 50–10). Note that all GAGs except the keratan sulfates contain a uronic acid which may be glucuronic or iduronic acid.

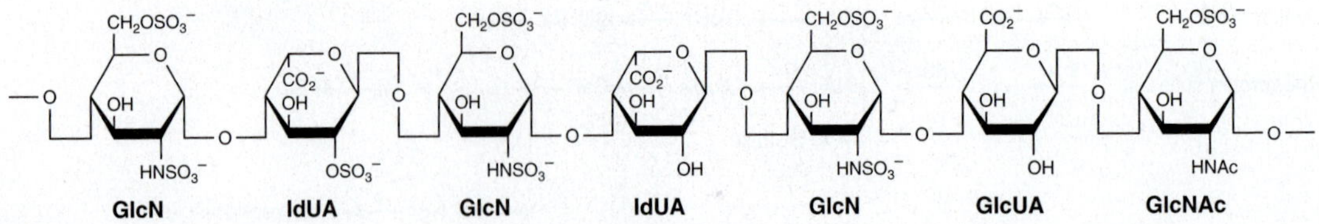

FIGURE 50–11 Structure of heparin. Structural features typical of heparin are shown. Each repeating disaccharide contains glucosamine (GlcN) and either D-glucuronic (GlcUA) or L-iduronic acid (IdUA). A few GlcN residues are acetylated (GlcNAc). The sequence of variously substituted repeating disaccharide units has been arbitrarily selected. Non-O-sulfated or 3-O-sulfated glucosamine residues may also occur.

The vast majority of the uronic acid residues are **IdUA**. Initially, all of the uronic acids are GlcUA, but a 5′-epimerase converts approximately 90% of the GlcUA residues to IdUA after the polysaccharide chain is formed. The protein molecule of the heparin proteoglycan is unique, consisting exclusively of serine and glycine residues. Approximately two-thirds of the serine residues contain GAG chains, usually of 5 to 15 kDa but occasionally much larger. Heparin is found in the granules of **mast cells** and also in liver, lung, and skin. It is an important **anticoagulant**. It is released into the blood from capillary walls by the action of lipoprotein lipase and it binds with factors IX and XI, but its most important interaction is with **plasma antithrombin** (discussed in Chapter 55).

Heparan Sulfate

This molecule is present in a proteoglycan found on many extracellular **cell surfaces**. It contains **GlcN** with fewer N-sulfates than heparin, and, unlike heparin, its predominant uronic acid is **GlcUA**. **Heparan sulfates** are associated with the plasma membrane of cells, with their core proteins actually spanning that membrane. In this, they may act as **receptors** and may also participate in the mediation of the **cell growth** and **cell–cell communication**. The attachment of cells to their substratum in culture is mediated at least in part by heparan sulfate. This proteoglycan is also found in the **basement membrane of the kidney** along with type IV collagen and laminin (see earlier), where it plays a major role in determining the charge selectiveness of glomerular filtration.

Dermatan Sulfate

This substance is widely distributed in animal tissues. Its structure is similar to that of chondroitin sulfate, except that in place of a GlcUA in β-1,3 linkage to GalNAc it contains an **IdUA** in an α-1,3 linkage to **GalNAc**. Formation of the IdUA occurs, as in heparin and heparan sulfate, by 5′-epimerization of GlcUA. Because this is regulated by the degree of sulfation and because sulfation is incomplete, dermatan sulfate contains **both** IdUA-GalNAc and GlcUA-GalNAc disaccharides. **Dermatan sulfate** has a widespread distribution in tissues, and is the main GAG in skin. Evidence suggests it may play a part in blood coagulation, wound repair, and resistance to infection.

Proteoglycans are also found in **intracellular locations** such as the nucleus where they are thought to have a regulatory role in functions such as cell proliferation and transport

of molecules between the nucleus and the cytosol. The various functions of GAGs are summarized in **Table 50–7**.

Deficiencies of Enzymes That Degrade Glycosaminoglycans Result in Mucopolysaccharidoses

Both **exo-** and **endoglycosidases** degrade GAGs. Like most other biomolecules, GAGs are subject to **turnover**, being both synthesized and degraded. In adult tissues, GAGs generally exhibit relatively **slow** turnover, their half-lives being days to weeks.

Understanding of the degradative pathways for GAGs, as in the case of glycoproteins (see Chapter 46) and glycosphingolipids (see Chapter 24), has been greatly aided by elucidation of the specific enzyme deficiencies that occur in certain **inborn errors of metabolism**. When GAGs are involved, these inborn errors are called **mucopolysaccharidoses (MPSs)** (Table 50–8).

Degradation of GAGs is carried out by a battery of **lysosomal hydrolases**. These include **endoglycosidases, exoglycosidases,** and **sulfatases**, generally acting in sequence. The **MPSs** (see Table 50–8) share a common mechanism of causation involving a mutation in a gene encoding a lysosomal

TABLE 50–7 Some Functions of Glycosaminoglycans & Proteoglycans

- Act as structural components of the ECM
- Have specific interactions with collagen, elastin, fibronectin, laminin, and other proteins such as growth factors
- As polyanions, bind polycations and cations
- Contribute to the characteristic turgor of various tissues
- Act as sieves in the ECM
- Facilitate cell migration (HA)
- Have role in compressibility of cartilage in weight bearing (HA, CS)
- Play role in corneal transparency (KS I and DS)
- Have structural role in sclera (DS)
- Act as anticoagulant (heparin)
- Are components of plasma membranes, where they may act as receptors and participate in cell adhesion and cell–cell interactions (eg, HS)
- Determine charge selectivity of renal glomerulus (HS)
- Are components of synaptic and other vesicles (eg, HS)
- Have a role in nuclear functions such as cell proliferation and transport of molecules between the nucleus and the cytosol

Abbreviations: CS, chondroitin sulfate; DS, dermatan sulfate; ECM, extracellular matrix; HA, hyaluronic acid; HS, heparan sulfate; KS I, keratan sulfate I.

TABLE 50–8 **The Mucopolysaccharidoses**

Disease Name	Abbreviation[a]	Enzyme Defective	GAG(s) Affected	Symptoms
Hurler-, Scheie- Hurler-Scheie syndrome	MPS I	α-L-Iduronidase	Dermatan sulfate, heparan sulfate	Mental retardation, coarse facial features, hepatosplenomegaly, cloudy cornea
Hunter syndrome	MPS II	Iduronate sulfatase	Dermatan sulfate, heparan sulfate	Mental retardation
Sanfilippo syndrome A	MPS IIIA	Heparan sulfate-*N*-sulfatase[b]	Heparan sulfate	Delay in development, motor dysfunction
Sanfilippo syndrome B	MPS IIIB	α-*N*-Acetylglucosaminidase	Heparan sulfate	As MPS IIIA
Sanfilippo syndrome C	MPS IIIC	α-Glucosaminide *N*-acetyltransferase	Heparan sulfate	As MPS IIIA
Sanfilippo syndrome D	MPS IIID	*N*-Acetylglucosamine 6-sulfatase	Heparan sulfate	As MPS IIIA
Morquio syndrome A	MPS IVA	Galactosamine 6-sulfatase	Keratan sulfate, chondroitin 6-sulfate	Skeletal dysplasia, short stature
Morquio syndrome B	MPS IVB	β-Galactosidase	Keratan sulfate	As MPS IVA
Maroteaux-Lamy syndrome	MPS VI	*N*-Acetylgalactosamine 4-sulfatase[c]	Dermatan sulfate	Curvature of the spine, short stature, skeletal dysplasia, cardiac defects
Sly syndrome	MPS VII	β-Glucuronidase	Dermatan sulfate, heparan sulfate, chondroitin 4-sulfate, chondroitin 6-sulfate	Skeletal dysplasia, short stature, hepatomegaly, cloudy cornea
Natowicz syndrome	MPS IX	Hyaluronidase	Hyaluronic acid	Joint pain, short stature

[a]The terms MPS V and MPS VIII are no longer used.
[b]Also called sulfaminidase.
[c]Also called arylsulfatase B.

hydroxylase responsible for the degradation of one or more GAGs. This leads to a defect in the enzyme and the accumulation of the substrate GAGs in various tissues, including the liver, spleen, bone, skin, and the central nervous system. The diseases are usually inherited in an **autosomal recessive** manner, with **Hurler** and **Hunter syndromes** being perhaps the most widely studied. None is common. In general, these conditions are chronic and progressive and affect multiple organs. Many patients exhibit organomegaly (eg, hepato- and splenomegaly); severe abnormalities in the development of cartilage and bone; abnormal facial appearance; and mental retardation. In addition, defects in hearing, vision, and the cardiovascular system may be present. Diagnostic tests include analysis of GAGs in urine or tissue biopsy samples; assay of suspected defective enzymes in white blood cells, fibroblasts or serum; and test for specific genes. Prenatal diagnosis is now sometimes possible using amniotic fluid cells or chorionic villus biopsy samples. In some cases, **a family history** of a mucopolysaccharidosis is obtained.

The term "**mucolipidosis**" was introduced to denote diseases that combined features common to both mucopolysaccharidoses and sphingolipidoses (see Chapter 24). In **sialidosis** (mucolipidosis I, ML-I), various oligosaccharides derived from glycoproteins and certain gangliosides accumulate in tissues. **I-cell disease** (ML-II) and **pseudo-Hurler polydystrophy** (MLIII) are described in Chapter 46. The term "mucolipidosis"

is retained because it is still in relatively widespread clinical usage, but it is not appropriate for these two latter diseases since the mechanism of their causation involves **mislocation** of certain lysosomal enzymes. Genetic defects of the catabolism of the oligosaccharide chains of glycoproteins (eg, mannosidosis, fucosidosis) are also described in Chapter 46. Most of these defects are characterized by increased excretion of various fragments of glycoproteins in the urine, which accumulate because of the metabolic block, as in the case of the mucolipidoses.

Hyaluronidase is one important enzyme involved in the catabolism of both hyaluronic acid and chondroitin sulfate. It is a widely distributed endoglycosidase that cleaves hexosaminidic linkages. From hyaluronic acid, the enzyme generates a tetrasaccharide with the structure (GlcUAβ-1, 3-GlcNAc-β-1,4)$_2$, which can be degraded further by a β-glucuronidase and β-*N*-acetylhexosaminidase. A genetic defect in hyaluronidase causes MPS IX (Natowicz syndrome), a lysosomal storage disorder in which hyaluronic acid accumulates in the joints.

Proteoglycans Are Associated With Major Diseases & With Aging

Hyaluronic acid may be important in permitting **tumor cells to migrate** through the ECM. Tumor cells can induce fibroblasts to synthesize greatly increased amounts of this GAG, thereby facilitating their own spread. Some tumor cells have less heparan sulfate at their surfaces, and this may play a role in the **lack of adhesiveness** that these cells display.

The intima of the **arterial wall** contains hyaluronic acid and chondroitin sulfate, dermatan sulfate, and heparan sulfate proteoglycans. Of these proteoglycans, dermatan sulfate binds plasma low-density lipoproteins. In addition, dermatan sulfate appears to be the major GAG synthesized by arterial smooth muscle cells. Because these cells proliferate in **atherosclerotic lesions** in arteries, dermatan sulfate may play an important role in development of the atherosclerotic plaque.

In various types of **arthritis**, proteoglycans may act as **autoantigens**, thus contributing to the pathologic features of these conditions. The amount of chondroitin sulfate in cartilage diminishes with age, whereas the amounts of keratan sulfate and hyaluronic acid increase. These changes may contribute to the development of **osteoarthritis**, as may increased activity of the enzyme aggrecanase, which acts to degrade aggrecan. Changes in the amounts of certain GAGs in the skin help to account for its characteristic alterations with **aging**.

In the past few years, it has become clear that in addition to their structural role in the ECM, proteoglycans function as signaling molecules which influence cell behavior, and they are now believed to play a part in diverse diseases such as fibrosis, cardiovascular disease, and cancer.

BONE IS A MINERALIZED CONNECTIVE TISSUE

Bone contains both **organic** and **inorganic** material. The **organic** matter is mainly **protein**. The principal proteins of bone are listed in **Table 50–9**; **type I collagen** is the major protein, comprising 90 to 95% of the organic material. Type V

collagen is also present in small amounts, as are a number of noncollagen proteins, some of which are relatively specific to bone. These are now believed to play an active part of the mineralization process. The **inorganic** or mineral component is mainly crystalline **hydroxyapatite**—$Ca_{10}(PO_4)_6(OH)_2$—along with sodium, magnesium, carbonate, and fluoride; approximately 99% of the body's calcium is contained in bone (see Chapter 44). Hydroxyapatite confers on bone the strength and resilience required for its physiologic roles.

Bone is a **dynamic structure** that undergoes continuing cycles of remodeling, consisting of resorption (demineralization) followed by deposition of new bone tissue (mineralization). This remodeling permits bone to adapt to both physical (eg, increases in weight bearing) and hormonal signals.

The major cell types involved in bone resorption and deposition are **osteoclasts** and **osteoblasts**, respectively (**Figure 50–12**). **Osteocytes** are found in mature bone and are also involved in the maintenance of the bone matrix. They are descended from osteoblasts and are very long-lived, with an average half-life of 25 years.

Osteoclasts are multinucleated cells derived from pluripotent hematopoietic stem cells. Osteoclasts possess an apical membrane domain, exhibiting a ruffled border that plays a key role in bone resorption (**Figure 50–13**). A special proton-translocating **ATPase** expels protons across the ruffled border into the resorption area, which is the microenvironment of low pH shown in the figure. This lowers the local pH to 4.0 or less, thus increasing the solubility of hydroxyapatite and helping its breakdown into Ca^{2+}, H_3PO_4 and H_2CO_3, and water, thus allowing demineralization to occur. Lysosomal acid proteases such as cathepsins are also released to digest the now accessible matrix proteins. **Osteoblasts**—mononuclear cells derived from pluripotent mesenchymal precursors—synthesize most of the proteins found in bone (see Table 50–9) as well as various growth factors and cytokines. They are responsible for the

TABLE 50–9 **The Principal Proteins Found in Bone**[a]

Proteins	Comments
Collagens	
Collagen type I	Approximately 90% of total bone protein. Composed of two α1(I) and one α2(I) chains.
Collagen type V	Minor component.
Noncollagen proteins	
Plasma proteins	Mixture of various plasma proteins.
Proteoglycans[b] CS-PG I (biglycan)	Contains two GAG chains; found in other tissues.
CS-PG II (decorin)	Contains one GAG chain; found in other tissues.
CS-PG III	Bone-specific.
Bone SPARC[c] protein (osteonectin)	Not bone-specific.
Osteocalcin (bone Gla protein)	Contains γ-carboxyglutamate (Gla) residues that bind to hydroxyapatite. Bone-specific.
Osteopontin	Not bone-specific. Glycosylated and phosphorylated.
Bone sialoprotein	Bone-specific. Heavily glycosylated, and sulfated on tyrosine.
Bone morphogenetic proteins (BMPs)	A family (at least 20) of secreted proteins with a variety of actions on bone; many induce ectopic bone growth.
Osteoprotegerin	Inhibits osteoclastogenesis

[a]Noncollagen proteins are involved in the regulation of the mineralization process. A number of other proteins are also present in bone, including a tyrosine-rich acidic matrix protein (TRAMP), some growth factors (eg, TGF-β), and enzymes involved in collagen synthesis (eg, lysyl oxidase).
[b]CS-PG, chondroitin sulfate–proteoglycan; these are similar to the dermatan sulfate PGs (DS-PGs) of cartilage.
[c]SPARC, secreted protein acidic and rich in cysteine.

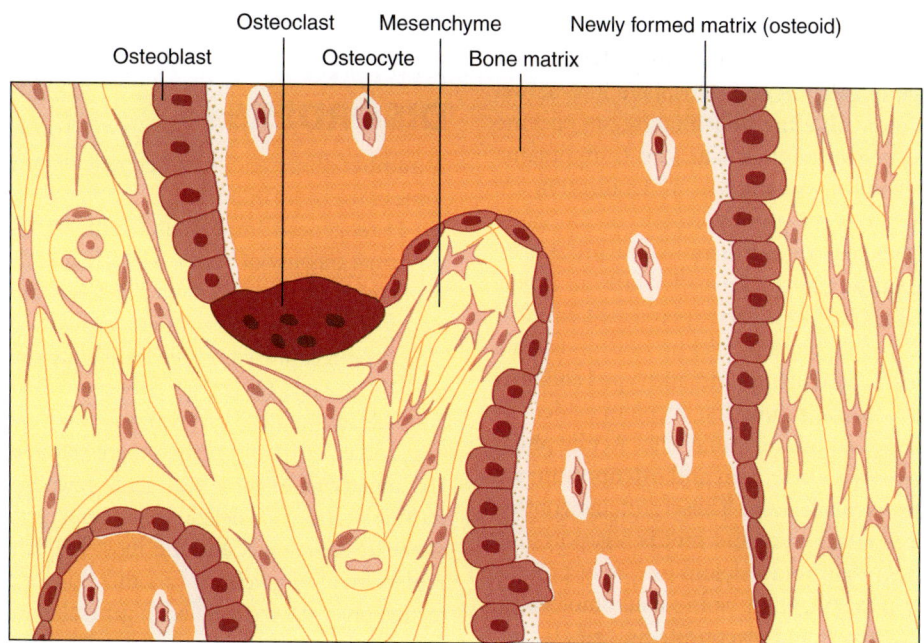

FIGURE 50–12 **Schematic illustration of the major cells present in the membranous bone.** Osteoblasts (lighter color) synthesize type I collagen, which forms a matrix that traps cells. As this occurs, osteoblasts gradually differentiate to become osteocytes. (Reproduced with permission from Mescher AL: Junqueira's *Basic Histology Text and Atlas*, 16th ed. New York, NY: McGraw Hill; 2021.)

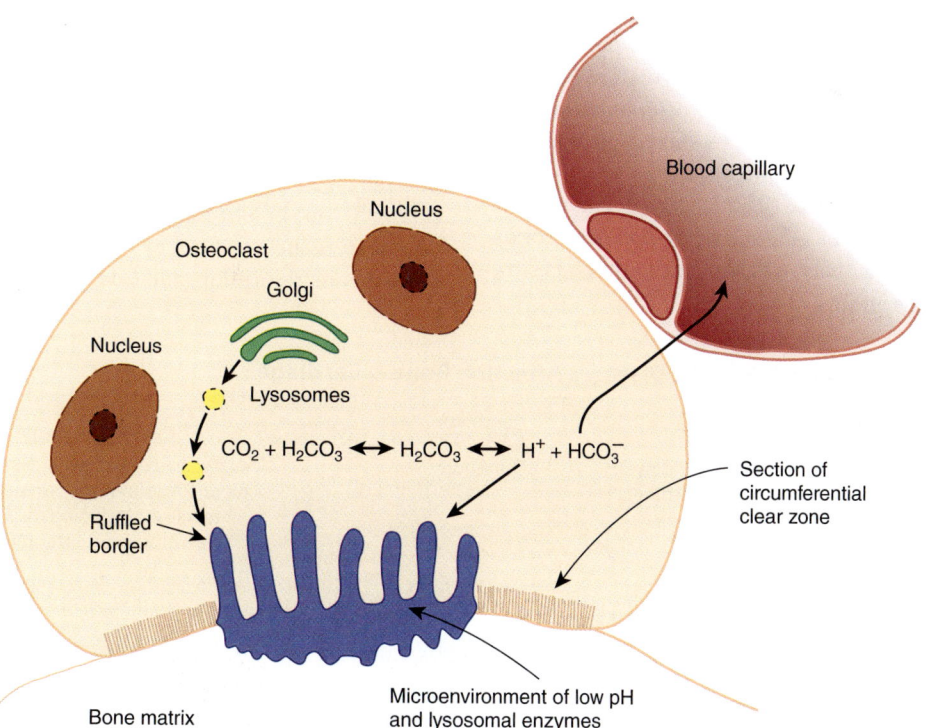

FIGURE 50–13 **Schematic illustration of the role of the osteoclast in bone resorption.** Lysosomal enzymes and hydrogen ions are released into the confined microenvironment created by the attachment between the bone matrix and the peripheral clear zone of the osteoclast. The acidification of this confined space facilitates the dissolution of calcium phosphate from bone and is the optimal pH for the activity of lysosomal hydrolases. The bone matrix is thus removed, and the products of bone resorption are taken up into the cytoplasm of the osteoclast, probably digested further, and transferred into capillaries. The production of H^+ from CO_2 and H_2CO_3 is facilitated by carbonic anhydrase II. (Reproduced with permission from Mescher AL: Junqueira's *Basic Histology Text and Atlas*, 16th ed. New York, NY: McGraw Hill; 2021.)

deposition of the new bone matrix (osteoid) and its subsequent mineralization. Osteoblasts **control mineralization** by regulating the passage of calcium and phosphate ions across their surface membranes. **Alkaline phosphatase**, an enzyme in the cell membrane, generates phosphate ions from organic phosphates. The mechanisms involved in mineralization are not fully understood, however, the hydrolysis of inorganic pyrophosphate (PP_i), which is known to inhibit the process, by **tissue nonspecific alkaline phosphatase (TNAP)** (an isoenzyme of alkaline phosphatase) is known to play a part. In addition, matrix vesicles, which contain calcium and phosphate and bud off from the osteoblast membrane, and **type I collagen** have been implicated. Mineralization first becomes evident in the gaps between successive collagen molecules. **Acidic phosphoproteins**, such as **bone sialoprotein and osteopontin**, are thought to act as sites of nucleation. These proteins contain RGD sequences for cell attachment and motifs (eg, poly-Asp and poly-Glu stretches) that bind calcium and may provide an initial scaffold for mineralization. Some macromolecules, such as certain proteoglycans and glycoproteins, can also act as **inhibitors** of nucleation.

Bone consists of two types of tissue, **trabecular (also called cancellous or spongy) bone** is found at the end of long bones close to the joints and is less dense than **compact (or cortical) bone**, which, as well as being denser, is harder and stronger. It forms the cortex (or outer shell) of most bones and accounts for about 80% of the weight of the human skeleton. It is estimated that approximately 4% of compact bone and 20% of trabecular bone is **renewed annually** in the typical healthy adult. Many factors are involved in the **regulation of bone metabolism**. Some of which **stimulate osteoblast activity** (eg, parathyroid hormone and 1,25-dihydroxycholecalciferol; see Chapter 44) to promote mineralization, while others **inhibit** it (eg, corticosteroids). Parathyroid hormone and 1,25-dihydroxycholecalciferol also stimulate bone resorbtion by increasing osteoclast activity, whereas calcitonin and estrogens have the opposite effect.

BONE IS AFFECTED BY MANY METABOLIC & GENETIC DISORDERS

A number of metabolic and genetic disorders affect bone, and some of the more important examples are listed in **Table 50–10**.

Osteogenesis imperfecta (brittle bones) is characterized by abnormal fragility of bones. The sclera of the eye is often abnormally thin and translucent and may appear blue owing to a deficiency of connective tissue. **Eight types** (I-VIII) of this condition have been recognized. Types I to IV are caused by mutations in the *COL1A1* or *COL1A2* genes or both. Type I is mild, but type II is severe and infants born with the condition usually do not survive, and types III and IV are progressive and/or deforming. Over 100 mutations in these two genes have been documented and include partial gene deletions and duplications. Other mutations affect RNA splicing, and the most frequent type results in the **replacement of glycine** by another bulkier amino acid, affecting formation of the triple helix. In general, these mutations result in decreased expression of collagen or in structurally abnormal pro chains that assemble into **abnormal fibrils**, weakening the overall structure of bone. When one abnormal chain is present, it may interact with two normal chains, but folding may be prevented, resulting in enzymatic degradation of all of the chains. This is called "**procollagen suicide**" and is an example of a dominant negative mutation, a result often seen when a protein consists of multiple different subunits. Types V to VIII are less common and are caused by mutations in the genes for proteins involved in bone mineralization other than collagen.

Osteopetrosis (marble bone disease), characterized by **increased bone density**, is a rare condition characterized by inability to resorb bone. It is due to mutations in the gene (located on chromosome 8q22) encoding **carbonic anhydrase II (CA II)**, one of four isozymes of carbonic anhydrase present in human tissues. Deficiency of CA II in osteoclasts prevents normal bone resorption, and osteopetrosis results.

TABLE 50–10 **Some Metabolic & Genetic Diseases Affecting Bone & Cartilage**

Condition	Causes	Condition	Causes
Dwarfism	Often deficiency of growth hormone, but many other causes	Osteoporosis	Age-related, estrogen deficiency following menopause, mutations in genes affecting bone metabolism,[a] including the vitamin D receptor (*VDR*), estrogen receptor-α (*ER-α*), and *COL1A1*
Rickets	Deficiency of vitamin D in childhood	Osteoarthritis	Age-related cartilage degeneration, mutations in various genes[a] including *VDR*, *ER-α*, and *COL2A1*
Osteomalacia	Deficiency of vitamin D in adults	Chondrodysplasias	Mutations in *COL2A1*
Hyperparathyroidism	Excess parathyroid hormone causing bone resorption	Pfeiffer, Jackson-Weiss, and Crouzon syndromes[b]	Mutations in the gene for fibroblast growth factor receptor (FGFR) 1 and/or 2
Osteogenesis imperfecta	Mutations in *COL1A1* and *COL1A2* affecting the synthesis and structure of collagen	Achondroplasia and thanatophoric dysplasia[c]	Mutation in the gene for FGFR3

[a]Only a small number of cases.
[b]In the Pfeiffer, Jackson-Weiss, and Crouzon syndromes, there is premature fusion of some bones in the skull (craniosynostosis).
[c]Thanatophoric dysplasia in the most common neonatal lethal skeletal dysplasia.

Osteoporosis is a generalized progressive reduction in bone tissue mass per unit volume, caused by an imbalance between bone resorption and deposition, and leads to skeletal weakness. The primary type 1 condition commonly occurs in women after the menopause. This is thought to be mainly due to lack of estrogen, which leads to increased bone resorption accompanied by decreased bone mineralization. Primary type 2 or senile osteoporosis occurs in both sexes post 75 years, although is more prevalent in women (ratio 2:1 female:male). The ratio of **mineral** to **organic elements** is unchanged in the remaining normal bone. Fractures of various bones, such as the head of the femur, occur very easily and represent a huge burden to both the affected patients and to society in general.

THE MAJOR COMPONENTS OF CARTILAGE ARE TYPE II COLLAGEN & CERTAIN PROTEOGLYCANS

There are three types of cartilage. The major type is **hyaline (articular) cartilage** and its principal proteins are listed in **Table 50–11**. **Type II collagen** is the major protein component (**Figure 50–14**), and a number of other minor types of collagen are also present. In addition to these components, the second type, **elastic cartilage**, contains elastin, and the third, **fibroelastic cartilage**, contains type I collagen. Cartilage contains a number of **proteoglycans**, which play an important role in its compressibility. **Aggrecan** (about 2×10^3 kDa) is the major proteoglycan. As shown in **Figure 50–15**, it has a very complex structure, containing several GAGs (hyaluronic acid, chondroitin sulfate, and keratan sulfate) and both link and core proteins. The core protein contains three domains: A,

TABLE 50–11 The Principal Proteins Found in Cartilage

Proteins	Comments
Collagen proteins	
Collagen type II	90-98% of total hyaline cartilage collagen. Composed of three α1(II) chains.
Collagens V, VI, IX, X, XI	Type IX cross-links to type II collagen. Type XI may help control diameter of type II fibrils.
Noncollagen proteins	
Cartilage oligomeric matrix protein (COMP)	An important structural component of cartilage. Regulates cell movement and attachment.
Aggrecan	The major proteoglycan of cartilage.
DS-PG I (biglycan)ᵃ	Similar to CS-PG I of bone.
DS-PG II (decorin)	Similar to CS-PG II of bone.
Chondronectin	Promotes chondrocyte attachment to type II collagen

ᵃThe core proteins of the proteoglycans DS-PG I and DS-PG II are homologous to those of CS-PG I and CS-PG II found in bone, except that instead the GlcUA in a β-1,3 linkage to GalNAc found in CS-PG, DS-PG contains an IdUA in an α-1,3 linkage to GalNAc.A possible explanation is that osteoblasts lack the epimerase required to convert glucuronic acid to iduronic acid.
Abbreviations: CS-PG, chondroitin sulfate proteoglycans; DS-PG, dermatan sulfate proteoglycans.

B, and C. Hyaluronic acid binds noncovalently to domain A of the core protein as well as to the link protein, which stabilizes the hyaluronate–core protein interactions. The keratan sulfate chains are located in domain B, whereas the chondroitin sulfate chains are located in domain C; both of these types of GAGs are bound covalently to the core protein. The core protein also contains both O- and N-linked oligosaccharide chains.

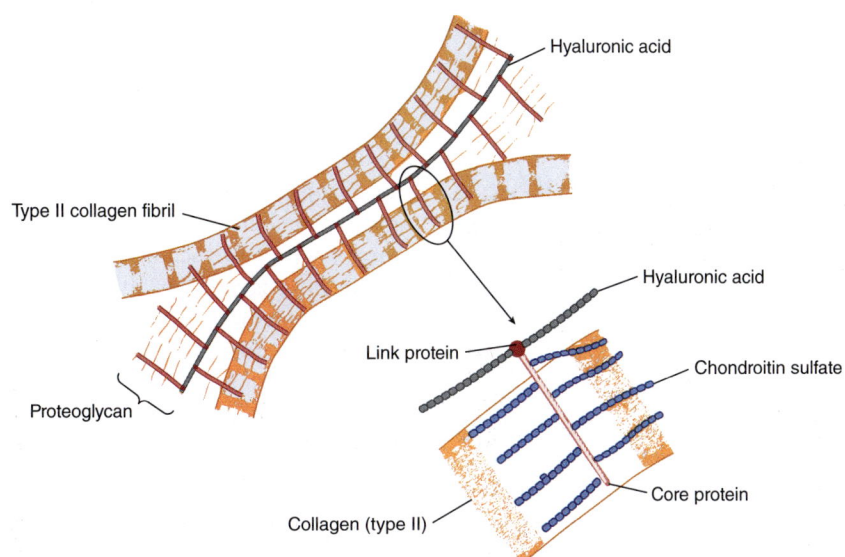

FIGURE 50–14 **Schematic representation of the molecular organization in the cartilage matrix.** Link proteins noncovalently bind the core protein (red) of proteoglycans to the linear hyaluronic acid molecules (gray). The chondroitin sulfate side chains of the proteoglycan electrostatically bind to the collagen fibrils, forming a cross-linked matrix. The oval outlines the area enlarged in the lower part of the figure. (Reproduced with permission from Mescher AL: Junqueira's *Basic Histology Text and Atlas*, 16th ed. New York, NY: McGraw Hill; 2021.)

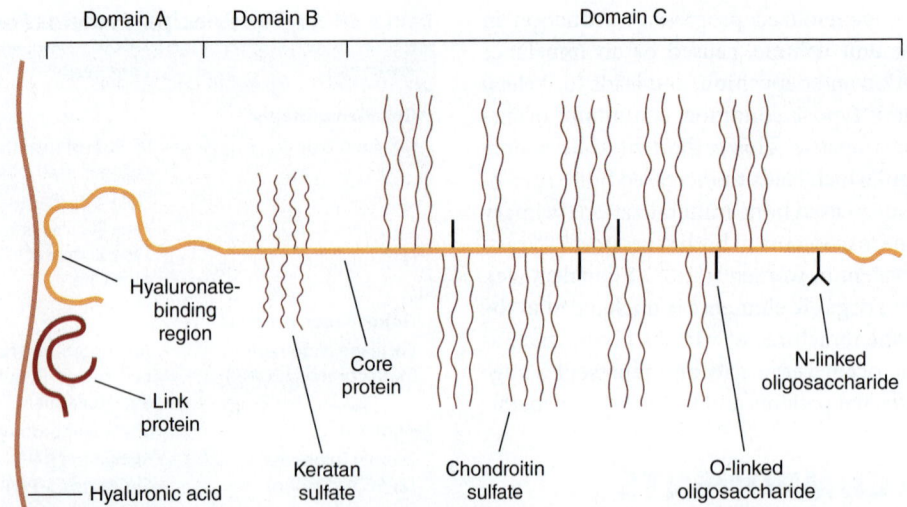

FIGURE 50–15 **Schematic diagram of aggrecan.** A strand of hyaluronic acid is shown on the left. The core protein (about 210 kDa) has three major domains. Domain A, at its amino-terminal end, interacts with approximately five repeating disaccharides in hyaluronate. The link protein interacts with both hyaluronate and domain A, stabilizing their interactions. Approximately 30 keratan sulfate chains are attached, via GalNAc-Ser linkages, to domain B. Domain C contains about 100 chondroitin sulfate chains attached via Gal-Gal-Xyl-Ser linkages and about 40 O-linked oligosaccharide chains. One or more N-linked glycan chains are also found near the carboxyl terminal of the core protein. (Reproduced with permission from Moran LA, Scrimgeour KG, Horton HR, et al: *Biochemistry*, 2nd ed. Upper Saddle River, NJ: Pearson Hall; 1994.)

The other proteoglycans found in cartilage have simpler structures than aggrecan. **Chondronectin** is involved in the attachment of type II collagen to chondrocytes (the cells in cartilage).

Cartilage is an avascular tissue and obtains most of its nutrients from synovial fluid. It exhibits slow but continuous **turnover**. Various **proteases** (eg, collagenases and stromelysin) synthesized by chondrocytes can **degrade collagen** and the other proteins found in cartilage. Interleukin-1 (IL-1) and tumor necrosis factor α (TNF-α) appear to stimulate the production of such proteases, whereas TGF-β and insulin-like growth factor 1 (IGF-I) generally exert an anabolic influence on the tissue.

CHONDRODYSPLASIAS ARE CAUSED BY MUTATIONS IN GENES ENCODING TYPE II COLLAGEN & FIBROBLAST GROWTH FACTOR RECEPTORS

Chondrodysplasias are a mixed group of hereditary disorders affecting cartilage. They are manifested by short-limbed dwarfism and numerous skeletal deformities. A few are due to a variety of mutations in the *COL2A1* gene, leading to abnormal forms of type II collagen. One example is the **Stickler syndrome**, manifested by degeneration of the joint cartilage and of the vitreous body of the eye.

The best known of the chondrodysplasias is **achondroplasia**, the most common cause of **short-limbed dwarfism**. Affected individuals have short limbs, normal trunk size, macrocephaly, and a variety of other skeletal abnormalities.

The condition is often inherited as an autosomal dominant trait, but many cases are due to new mutations. Achondroplasia is not a collagen disorder but is due to mutations in the gene encoding **fibroblast growth factor receptor 3 (FGFR3)**. **Fibroblast growth factors** are a family of more than 20 proteins that affect the growth and differentiation of cells of mesenchymal and neuroectodermal origin. Their **receptors** are transmembrane proteins and form a subgroup of four in the family of receptor tyrosine kinases. FGFR3 is one member of this subgroup and mediates the actions of FGF3 on cartilage. In almost all cases of achondroplasia that have been investigated, the mutations were found to involve nucleotide 1138 and resulted in substitution of arginine for glycine (residue number 380) in the transmembrane domain of the protein, rendering it inactive. No such mutation was found in unaffected individuals.

Other mutations in the same gene can result in **hypochondroplasia**, **thanatophoric dysplasia** (types I and II) (other forms of short-limbed dwarfism), and the **SADDAN phenotype** (severe achondroplasia with developmental delay and acanthosis nigricans [the latter is a brown to black hyperpigmentation of the skin]).

As indicated in Table 50–10, **other skeletal dysplasias** (including certain craniosynostosis syndromes) are also due to mutations in genes encoding FGF receptors. Another type of skeletal dysplasia, **diastrophic dysplasia** has been found to be due to mutation in a sulfate transporter.

SUMMARY

- The major components of the ECM are the structural proteins collagen, elastin, and fibrillin-1, a number of specialized proteins (eg, fibronectin and laminin), and various proteoglycans.

- Collagen is the most abundant protein in the animal kingdom; approximately 29 types have been identified. All collagens contain greater or lesser stretches of triple helix and the repeating structure (Gly-X-Y)n.

- The biosynthesis of collagen is complex, featuring many posttranslational events, including hydroxylation of proline and lysine.

- Diseases associated with impaired synthesis of collagen include scurvy, osteogenesis imperfecta, Ehlers-Danlos syndrome (thirteen subtypes), and Menkes disease.

- Elastin confers extensibility and elastic recoil on tissues. Elastin lacks hydroxylysine, Gly-X-Y sequences, triple helical structure, and sugars, but contains desmosine and isodesmosine cross-links not found in collagen.

- Fibrillin-1 is located in microfibrils. Mutations in the gene encoding fibrillin-1 cause Marfan syndrome. The cytokine TGF-β appears to contribute to the cardiovascular pathology.

- The GAGs are made up of repeating disaccharides containing a uronic acid (glucuronic or iduronic) or hexose (galactose) and a hexosamine (galactosamine or glucosamine). Sulfate is also frequently present.

- The major GAGs are hyaluronic acid, chondroitin sulfate, keratan sulfates I and II, heparin, heparan sulfate, and dermatan sulfate.

- The GAGs are synthesized by the sequential actions of specific enzymes (glycosyltransferases, epimerases, sulfotransferases, etc) and are degraded by the sequential action of lysosomal hydrolases. Genetic deficiencies of the latter result in mucopolysaccharidoses (eg, the Hurler syndrome).

- GAGs occur in tissues bound to various proteins (link proteins and core proteins), constituting proteoglycans. These structures are often of very high molecular weight and serve many functions in tissues.

- Many components of the ECM bind to integrins, proteins found on the cell surface, and this constitutes one pathway by which the cell exterior can communicate with the interior.

- Bone and cartilage are specialized forms of the ECM. Collagen I and hydroxyapatite are the major constituents of bone. Collagen II and certain proteoglycans are major constituents of cartilage.

- A number of heritable diseases of bone (eg, osteogenesis imperfecta) and of cartilage (eg, the chondrodystrophies) are caused by mutations in the genes for collagen and proteins involved in bone mineralization and cartilage formation.

REFERENCES

Kasper DL, Fauci AS, Hauser SL et al: *Harrison's Principles of Internal Medicine*, 19th ed. McGraw Hill Education, 2017.

Theocharis AD, Manou D, Karamanos NK. The extracellular matrix as a multitasking player in disease. The FEBS J 2019;286:2830.

Muscle & the Cytoskeleton

January D. Haile, PhD, & Peter J. Kennelly, PhD

O B J E C T I V E S

After studying this chapter, you should be able to:

- Describe the differences between actin-based and myosin-based regulation of muscle contraction.

- Outline the composition and organization of the thick and thin filaments in striated muscle tissues.

- Describe the pivotal role of Ca^{2+} in the initiation of both contraction and relaxation in muscle.

- List the various channels, pumps, and exchangers involved in regulating intracellular Ca^{2+} levels in various types of muscle.

- List the major energy sources for regenerating ATP in muscle tissue.

- Identify the preferred energy sources for fast and slow twitch fibers.

- Understand the molecular bases of malignant hyperthermia, Duchenne and Becker muscular dystrophies, and inherited cardiomyopathies.

- Explain how nitric oxide (NO) induces relaxation of vascular smooth muscle.

- Know the general structures and functions of the major components of the cytoskeleton, namely microfilaments, microtubules, and intermediate filaments.

- Explain the role of mutations in the gene encoding lamin A and lamin C in Hutchinson-Gilford progeria syndrome (**progeria**).

BIOMEDICAL IMPORTANCE

Highly specialized muscle cells (myocytes) utilize a contractile apparatus consisting of physically juxtaposed **actin** and **myosin** polymers, or fibrils. Muscle contraction is controlled by signal transduction pathways in which the second messenger **Ca^{2+}** plays a pivotal role. Muscle tissue is subject to a variety of pathologic conditions, including hereditary diseases, malignant hyperthermia, and heart failure. Hereditary diseases include Duchenne-type MS, which is often treated with glucocorticoids. **Heart failure** is a very common medical condition, with a variety of causes. Diagnosis and therapeutic intervention require an understanding of the biochemistry of heart muscle. **Vasodilators**—such as nitroglycerin, used in the treatment of angina pectoris—act by increasing the formation of nitrous oxide (NO).

The **cytoskeleton**, a network of polymeric protein fibers and molecular motors, maintains the shape, integrity, and internal organization of all mammalian cells. It is a structural and mechanical network that mediates motional processes such as cytokinesis, endocytosis, exocytosis, secretion, phagocytosis, and diapedesis. Several pathogenic microorganisms, including *Yersinia*, *Salmonella enterica*, *Listeria monocytogenes*, and *Shigella*, attack or co-opt the cytoskeleton of the infected host as an integral feature of their mechanisms of virulence.

MUSCLE IS A STRUCTURALLY & FUNCTIONALLY SPECIALIZED TISSUE

Three Types of Muscle Exist: Skeletal, Cardiac, & Smooth

Muscle is a highly specialized tissue configured to convert the chemical energy of ATP potential into mechanical energy on a mass, macroscopic scale. Three types of muscle are found in vertebrates. **Skeletal** and **cardiac** muscle display a striated appearance, a consequence of the parallel alignment of the fibrils of their contractile apparatus. **Smooth** muscle is devoid of striations, a consequence of the (more) random orientation of its contractile fibrils. Mechanically, these differences in orientation mean that cardiac and skeletal muscle contract and exert force in only one dimension, like a coil spring, while contracting smooth muscle contracts exerts force and shortens in all directions, like the polymer skin of an inflated balloon. Skeletal muscle also is under conscious or **voluntary** nervous control, while both cardiac and smooth muscle work in an unconscious, **involuntary** manner.

The Sarcomere Is the Functional Unit of Muscle

Striated muscle is composed of multinucleated muscle fiber cells, which may extend the entire length of the muscle, surrounded by an electrically excitable plasma membrane, the **sarcolemma**. Within each individual muscle fiber cell oriented longitudinally along its length are bundles of parallel **myofibrils**, consisting of interdigitated, overlapping thick and thin filaments, embedded in intracellular fluid termed **sarcoplasm**. When a **myofibril** is examined by electron microscopy, alternating dark and light bands can be observed (**Figure 51–1**).

These bands are referred to as **A** and **I bands**, respectively. In the A band, the thin filaments (dark bands) are arranged around the thick filament (myosin) as a secondary hexagonal array. Each thin filament lies symmetrically between three thick filaments (**Figure 51–2**, center, mid cross section), and each thick filament is surrounded symmetrically by six thin filaments. The central region of the A band, a seemingly less dense region known as the **H band**, consists entirely of thick filaments. The I band (light bands) corresponds to a zone containing only thin filaments, and is bisected by the dense and narrow **Z line**, which consists of a complex network of polypeptides anchoring the thin filaments together (see Figure 51–2).

The **sarcomere** is defined as the region between two Z lines (see Figures 51–1 and 51–2) and is repeated along the axis of a fibril at distances of 1500 to 2300 nm depending on the state of contraction. Most muscle fiber cells are aligned so that their sarcomeres are in parallel register (see Figure 51–1).

Thin & Thick Filaments Slide Past Each Other During Contraction

The sliding filament model was largely based on careful morphologic observations on resting, extended, and contracting muscle. When muscle contracts, the H zones and the I bands shorten (see legend to Figure 51–2). Given that the thin and thick filaments remain intact, it was concluded that the interdigitated filaments slide past one another during contraction. As the tension developed during muscle contraction is proportionate to the degree to which the filaments overlapped,

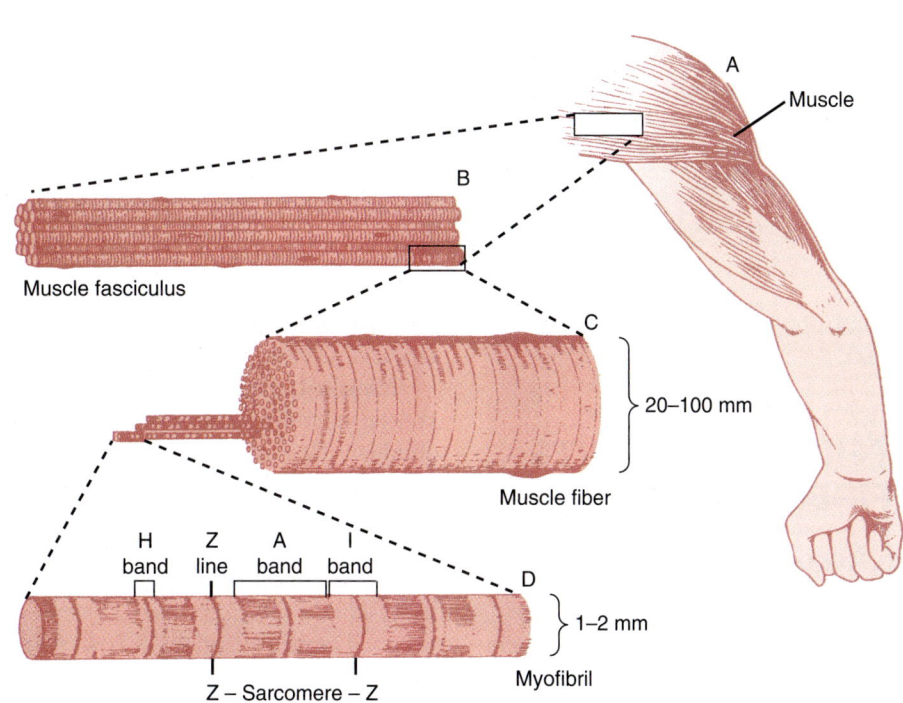

FIGURE 51–1 **The structure of voluntary muscle.** The sarcomere is the region between the Z lines. (Reproduced with permission from Bloom W, Fawcett DW: A Textbook of Histology, 11th ed. Philadelphia, PA: Saunders; 1986. Drawing by Sylvia Colard Keene.)

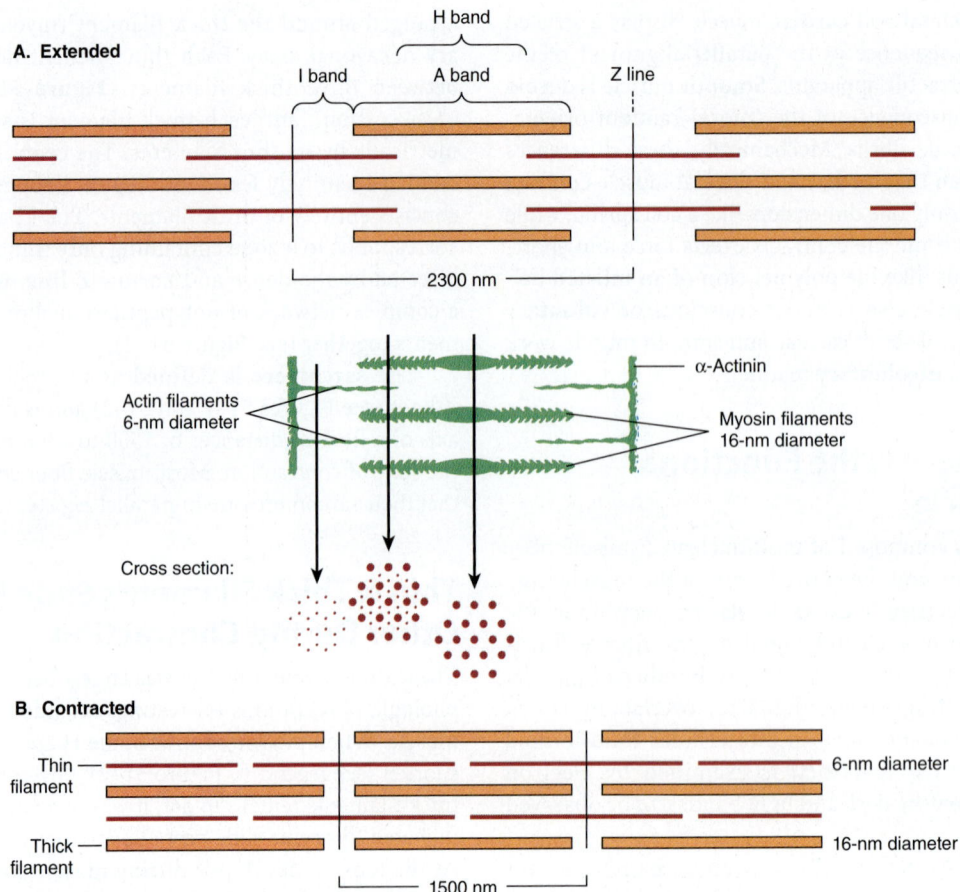

A. Extended

H band

I band   A band   Z line

2300 nm

α-Actinin

Actin filaments
6-nm diameter

Myosin filaments
16-nm diameter

Cross section:

B. Contracted

Thin
filament   6-nm diameter

Thick
filament   16-nm diameter

1500 nm

FIGURE 51–2 **Arrangement of filaments in striated muscle. (A)** Extended. The positions of the I, A, and H bands in the extended state are shown. The thin filaments partly overlap the ends of the thick filaments, and the thin filaments are shown anchored in the Z lines (often called Z disks). In the lower part of Figure 51–2A, "arrowheads," pointing in opposite directions, are shown emanating from the myosin (thick) filaments. Four actin (thin) filaments are shown attached to two Z lines via α-actinin. The central region of the three myosin filaments, free of arrowheads, is called the M band (not labeled). Cross sections through the M bands, through an area where myosin and actin filaments overlap and through an area in which solely actin filaments are present, are shown. **(B)** Contracted. The actin filaments are seen to have slipped along the sides of the myosin fibers toward each other. The lengths of the thick filaments (indicated by the A bands) and the thin filaments (distance between Z lines and the adjacent edges of the H bands) have not changed. However, the lengths of the sarcomeres have been reduced (from 2300 to 1500 nm), and the lengths of the H and I bands are also reduced because of the overlap between the thick and thin filaments. These morphologic observations provided part of the basis for the sliding filament model of muscle contraction.

it was apparent that contraction involved a dynamic interaction between the filaments.

MAJOR PROTEIN COMPONENTS OF MUSCLE FIBERS

Actin & Myosin Account for 75% of the Protein Mass in Muscle

Actin and myosin contribute 20% and 55% of the protein mass in muscle, respectively. The monomeric form of actin, **G-actin** (43 kDa; G, globular) polymerizes in the presence of Mg^{2+} to form an insoluble double helical filament called F-actin (**Figure 51–3**). The **F-actin** fiber, also known as the

thin filament, is 6- to 7-nm thick and has a structure that repeats every 35.5 nm.

Myosin-I, which exists in cells as a monomer, links cytoskeletal microfilaments to the plasma membrane. The predominant form of myosin in contractile tissues is **myosin II** (≈ 460 kDa), hereafter referred to simply as myosin, an asymmetric hexamer. The hexamer consists of one pair of **heavy (H) chains** (≈ 200 kDa), and two pairs of **light (L) chains** (≈ 20 kDa), referred to as essential and regulatory. The two heavy chains are intertwined, forming an extended helical tail, each capped by a globular head domain to which the light chains associate (**Figure 51–4**). Myosin possesses low, but detectable levels of **ATPase** activity in vitro: the catalysis of the hydrolysis of ATP by water to form ADP and P_i. Low levels of ATPase activity are a common feature of enzymes that

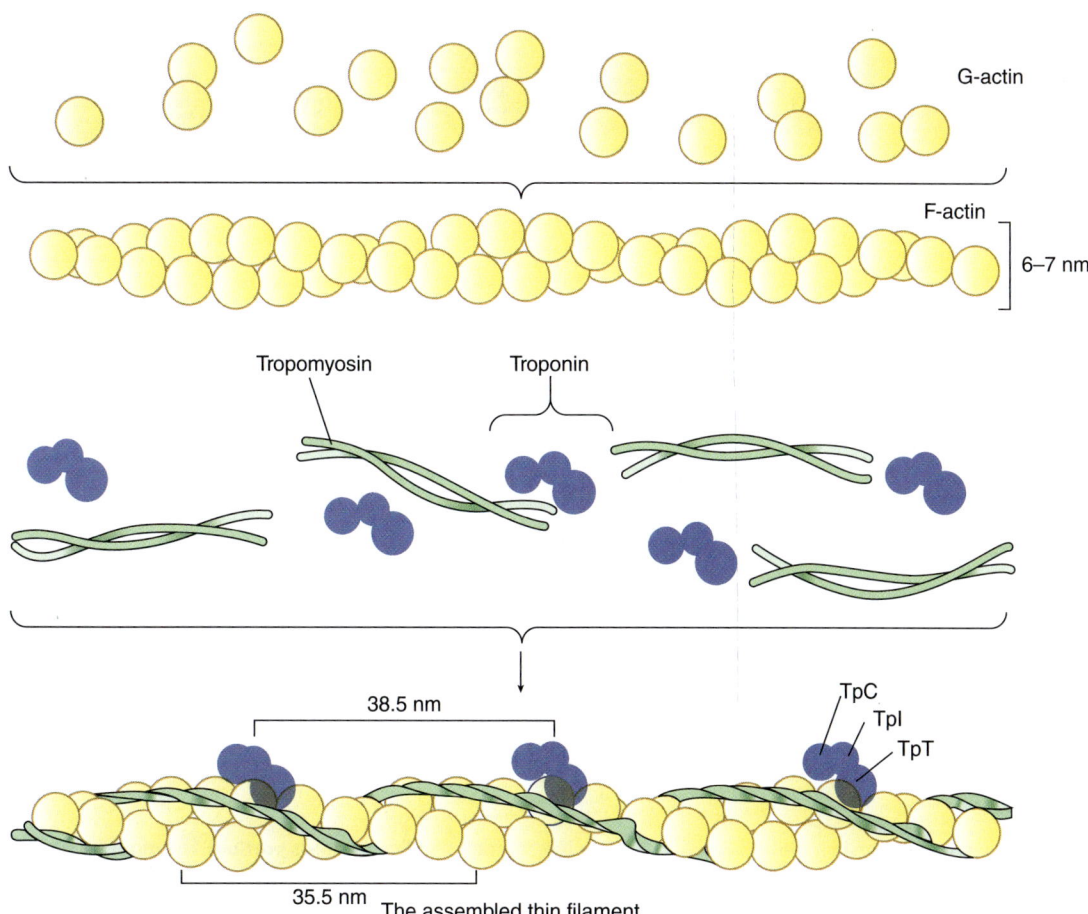

FIGURE 51-3 **Schematic representation of the thin filament, showing the spatial configuration of its three major protein components: actin, troponin, and tropomyosin.** The upper panel shows individual molecules of G-actin. The middle panel shows actin monomers assembled into F-actin. Individual molecules of tropomyosin (two strands wound around one another) and of troponin (made up of its three subunits) are also shown. The lower panel shows the assembled thin filament, consisting of F-actin, tropomyosin, and the three subunits of troponin (TpC, TpI, and TpT).

employ ATP as a substrate. When skeletal muscle myosin is complexed to actin to form **actomyosin** complex, its ATPase activity greatly increases.

The Structural & Functional Organization of Myosin Was Mapped by Limited Proteolysis

Limited proteolysis with **trypsin** yielded two myosin fragments called light and heavy **meromyosin** (HMM, ≈ 340 kDa). Light meromyosin (LMM) consists of the aggregated α-helical fibers from the tail of myosin (see Figure 51–4). It does not hydrolyze ATP or bind to F-actin. By contrast, HMM is a soluble protein that possesses both fibrous and globular regions (see Figure 51–4). HMM exhibits ATPase activity and binds to F-actin. Limited proteolysis of HMM with the protease papain cleaves it into two subfragments: S-1 globular region (≈115 kDa) and S-2 fibrous region. Only the **S-1** fragment, exhibits ATPase activity and binds both actin and myosin light chains (see Figure 51–4).

Tropomyosin & the Troponins Are Key Components of the Thin Filaments in Striated Muscle

In striated muscle, there are two other proteins that are minor in terms of their mass but important in terms of their function. **Tropomyosin**, present in all muscle and muscle-like structures, is a fibrous molecule that consists of two chains, α and β. The chains attach to F-actin in the groove between its filaments (see Figure 51–3). The **troponin complex** is unique to striated muscle and consists of three polypeptides. **Troponin T** (TpT) binds to tropomyosin as well as to the other two troponin components. **Troponin I** (TpI) inhibits the F-actin–myosin interaction and also binds to the other components of troponin. **Troponin C** (TpC) is a calcium-binding polypeptide that is structurally and functionally analogous to **calmodulin**, an important calcium-binding protein widely distributed in nature. Up to four calcium ions can bind per molecule of troponin C or calmodulin.

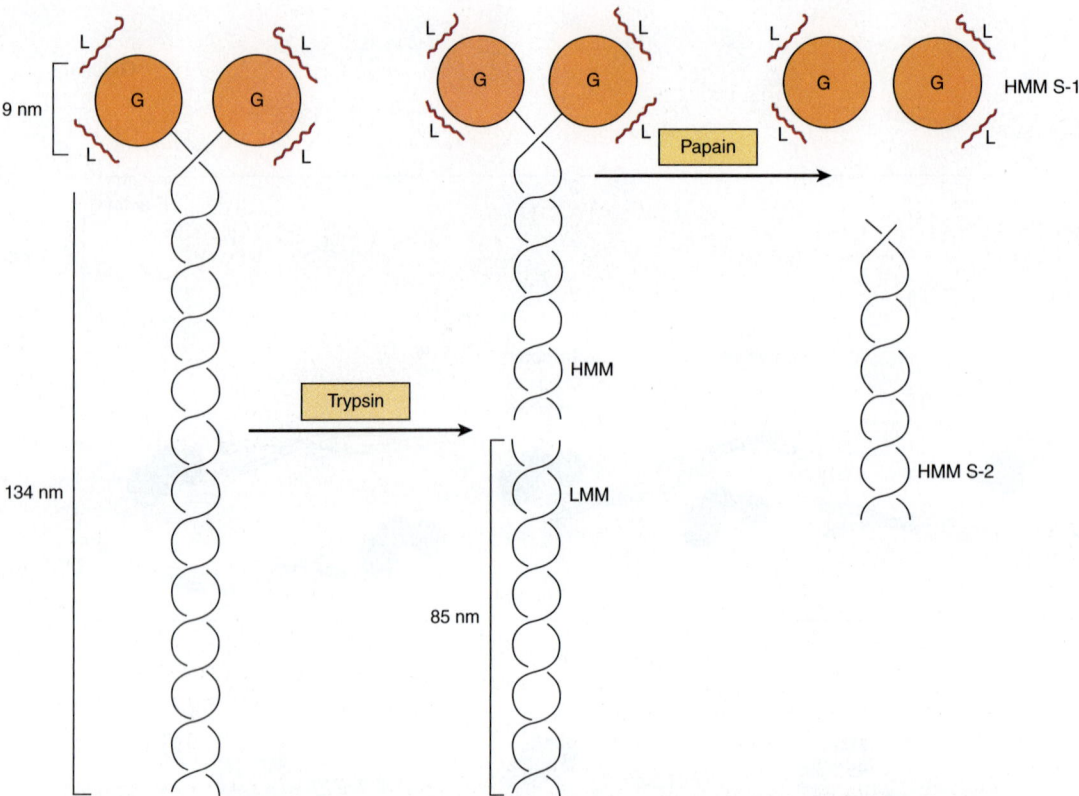

FIGURE 51–4 Diagram of a myosin molecule showing the two intertwined α-helices (fibrous portion), the globular region or head (G), the light chains (L), and the effects of proteolytic cleavage by trypsin and papain. The globular head region, which contains an actin-binding site and an L chain-binding site, attaches to the remainder of the myosin molecule.

MUSCLE CONVERTS CHEMICAL ENERGY TO MECHANICAL ENERGY

Muscle contraction occurs when thick filaments translocate along adjacent thin filaments by a process analogous to climbing a rope hand over hand. The hands in this instance are the S-1 domains of the myosin head, which ascend via a repeated cycle of attachment, an ATP-powered conformational change or **power stroke**, and detachment. Although individually minute in terms of both distance covered and the energy unleashed, when multiplied by the $1–2 \times 10^{18}$ myosin molecules in a human bicep, both great force and rapid motion can be generated.

A schematic diagram of the cycle of events involved in each power stroke is presented in **Figure 51–5**.

At rest, the S-1 head of myosin contains bound ADP and P_i, remnants of the last prior power stroke. On stimulation by the second messenger Ca^{2+} (see later), actin becomes accessible to the S-1 head of myosin which finds it and binds it, thereby forming a **cross-bridge** linking the thick and thin filaments. As depicted in Figure 51–2, the clusters of cross-bridges (abbreviated as arrowheads with the remainder of the thin filaments not shown) at ends of a thick filament have opposite polarities. The two polar regions of the thick filament are separated by a 150-nm segment (the M band, not labeled in the figure) which does not participate in cross-bridge formation.

Once the actin:myosin:ADP:P_i cross-bridge complex is formed, the bound P_i is released. This is followed by release of the bound ADP, triggering a large conformational change in the head of myosin in relation to its tail (**Figure 51–6**), the power stroke, that pulls the actin about 10 nm toward the center of the sarcomere. This nucleotide-free actin-myosin complex represents its low-energy state.

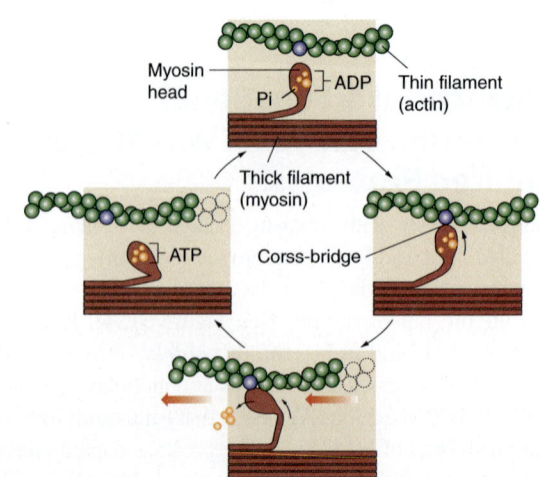

FIGURE 51–5 The hydrolysis of ATP drives the cyclic association and dissociation of actin and myosin in five reactions described in the text.

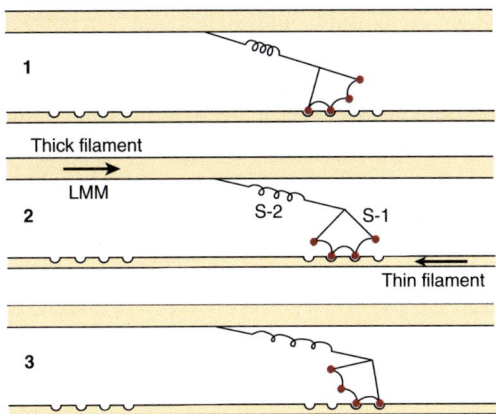

FIGURE 51–6 **Representation of the active cross-bridges between thick and thin filaments.** HE Huxley proposed that the force involved in muscular contraction originates in a tendency for the myosin head (S-1) to rotate relative to the thin filament and is transmitted to the thick filament by the S-2 portion of the myosin molecule acting as an inextensible link. Flexible points at each end of S-2 permit S-1 to rotate and allow for variations in the separation between filaments. This figure is based on HE Huxley's proposal, and also incorporates elastic (the coils in the S-2 portion) and stepwise-shortening elements (depicted here as four sites of interaction between the S-1 portion and the thin filament). The strengths of binding of the attached sites are higher in position 2 than in position 1 and higher in position 3 than position 2. The myosin head can be detached from position 3 with the utilization of a molecule of ATP; this is the predominant process during shortening. The myosin head is seen to vary in its position from about 90° to about 45°, as indicated in the text. (S-1, myosin head; S-2, portion of the myosin molecule; LMM, light meromyosin) (see legend to Figure 49–4).

Next, a molecule of ATP binds to the S-1 myosin head, which drastically lowers the affinity of the myosin head for actin. Actin is thus released, or detached. Finally, in the **relaxation phase** of muscle contraction, the S-1 head of myosin hydrolyzes ATP to ADP and P_i, but these products remain bound. The hydrolysis of ATP to ADP-P_i-myosin raises myosin into a so-called high-energy conformation. The myosin:ADP:P_i complex now stands ready to engage in another cycle and move another 10 nm along the thin filaments, provided Ca^{2+} levels remain elevated.

Calculations have indicated that the thermodynamic efficiency of muscle contraction is about 50%, which compares quite favorably to the 20% or less of the internal combustion engine. It should also be noted that, while hydrolysis of ATP ultimately powers the cycle, the release of ADP provides the immediate driver for the conformational change in S-1 that translates chemical energy into the mechanical power stroke. **Rigor mortis**, the stiffening of the body that occurs postmortem, occurs because the resulting fall in intracellular ATP levels prevents the dissociation of the myosin S-1 head from actin.

CONTRACTION IS ORCHESTRATED BY THE SECOND MESSENGER Ca²⁺

The contraction of muscles from all sources occurs by the general mechanism described earlier. However, the manner in which contraction is regulated differs among the various types of muscle tissue. Striated muscle relies on **actin-based** regulation, while regulation in smooth muscle is **myosin-based**. However, regardless of whether regulation is actin- or myosin-based, the second messenger Ca^{2+} plays a central role in the initiation and control of contraction.

Actin-Based Regulation Occurs in Skeletal Muscle

The contractile apparatus in skeletal muscle is regulated by the **troponin system**, which is bound to tropomyosin and F-actin in the thin filament (see Figure 51–3). In resting muscle, TpI prevents binding of the myosin head to its F-actin attachment site either by altering the conformation of F-actin via the tropomyosin molecules or by simply rolling tropomyosin into a position that directly blocks the sites on F-actin to which the myosin heads attach. Binding of Ca^{2+} to TpC relieves inhibition by TpI, permitting the power stroke cycle to proceed. When Ca^{2+} levels fall, the TpC:Ca^{2+} complex dissociates, allowing TpI to reestablish its inhibitory block on the binding of the myosin S-1 head to F-actin. The thick and thin filaments detach from one another, allowing the muscle to relax.

Myosin-Based Regulation Occurs in Smooth Muscle

Unlike skeletal muscle, **smooth muscles** contract along multiple axes. Thus, although they have molecular structures similar to those in striated muscle, they lack striations because their sarcomeres are not aligned. Smooth muscles contain α-actinin and tropomyosin molecules, as do skeletal muscles, but they lack the troponin system. In smooth muscle, it is the light chains that prevent the binding of the myosin head to F-actin in the resting state. This inhibitory block is released by the phosphorylation of the regulatory, or type 2, light chains in the myosin S-1 head by myosin light-chain kinase. The activity of myosin light-chain kinase is controlled by the binding of calmodulin (**Figure 51–7**).

Calmodulin is a small, (≈ 17 kDa), ubiquitous protein that contains four EF-hand motifs, each of which can bind one molecule of calcium. Binding of calcium to all four of these sites triggers a conformational change that enables the $(Ca^{2+})_4$:calmodulin complex to bind to and activate various intracellular target enzymes, including myosin light-chain kinase and Ca^{2+}-ATPase (see later). Ca^{2+} binding to calmodulin exhibits strong positive cooperativity, enabling calmodulin to act as a Ca^{2+}-sensitive molecular trigger. When Ca^{2+} levels fall, the balance of activity between myosin light-chain kinases and phosphatases shifts in favor of the latter, leading to the dephosphorylation of regulatory light chains, the detachment of myosin S-1 domains from actin, and muscle relaxation.

Rho kinase provides a Ca^{2+}-independent pathway for initiating contraction. Rho kinase phosphorylates myosin's regulatory light chains, as well as phosphorylating the phosphatase that dephosphorylates the light chains. Since phosphorylation of the phosphatase inhibits its activity, this further shifts the

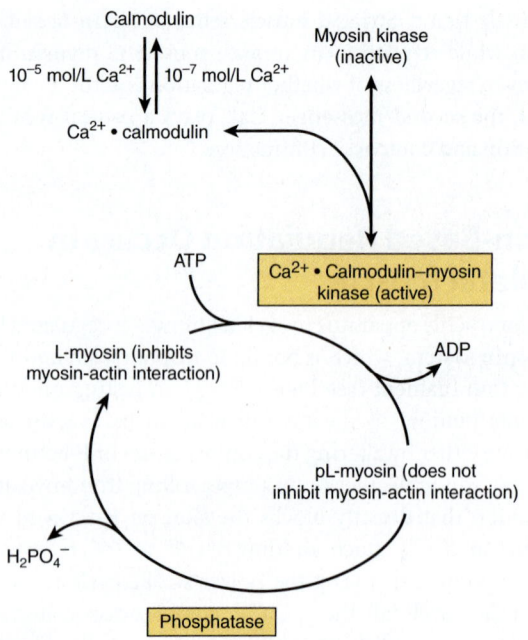

FIGURE 51–7 **Regulation of smooth muscle contraction by Ca²⁺.** The pL-myosin is the phosphorylated light chain of myosin and L-myosin is the dephosphorylated light chain.

balance between the kinases and phosphatases that act on myosin in favor of the former. **Table 51–1** summarizes and compares the regulation of actin-myosin interactions (activation of myosin ATPase) in striated and smooth muscles. Phosphorylation of Rho kinase by the cAMP-dependent protein kinase may play a role in the dampening effect of this second messenger on smooth muscle contraction.

Caldesmon (87 kDa), another regulatory protein, is ubiquitous in smooth muscle and also present in nonmuscle tissue.

At low [Ca²⁺], caldesmon binds to tropomyosin and actin, preventing the interaction of actin with myosin, thereby keeping muscle in a relaxed state. At higher [Ca²⁺], (Ca²⁺)₄:calmodulin binds caldesmon, releasing it from actin. Now actin is free to bind myosin and contraction can occur. Caldesmon is also subject to covalent modification by phosphorylation-dephosphorylation (Chapter 9); when phosphorylated, it cannot bind actin, again freeing the latter to interact with myosin.

The Sarcoplasmic Reticulum Regulates Intracellular Levels of Ca²⁺ in Skeletal Muscle

When striated muscle is in the relaxed state, the Ca²⁺ needed to initiate contraction is kept stored, ready for release into the sarcoplasm, in the **sarcoplasmic reticulum (SR)**, a network of fibrous sacks. The SR is linked to the sarcomeres by the transverse channels of the T-tubule system. In resting muscle, the [Ca²⁺] in the sarcoplasm typically ranges between 10^{-8} and 10^{-7} mol/L. This low resting concentration is maintained by the basal activity of the Ca²⁺ ATPase (**Figure 51–8**), a Ca²⁺-activated calcium pump that uses the energy of hydrolysis of ATP to move calcium ions from regions of low to high concentration. Once inside the SR, Ca²⁺ is bound by **calsequestrin**, a specific Ca²⁺-binding protein.

On release of Ca²⁺ into the sarcoplasm, (Ca²⁺)₄:calmodulin binds to and activates the Ca²⁺-dependent ATPase. This immediately sets into motion the export of Ca²⁺ from the sarcoplasm back into the SR, enabling the muscle to rapidly relax and ready for future contraction. If the concentration of ATP in the sarcoplasm falls dramatically (eg, by excessive usage during the cycle of contraction-relaxation or by diminished formation, such as might occur in ischemia), the Ca²⁺-ATPase will cease pumping and calcium ion levels will remain high.

TABLE 51–1 **Actin-Myosin Interactions in Striated & Smooth Muscle**

	Striated Muscle	Smooth Muscle (and Nonmuscle Cells)
Proteins of muscle filaments	Actin Myosin Tropomyosin Troponin (Tpl, TpT, TpC)	Actin Myosin[a] Tropomyosin
Spontaneous interaction of F-actin and myosin alone (spontaneous activation of myosin ATPase by F-actin)	Yes	No
Inhibitor of F-actin–myosin interaction (inhibitor of F-actin-dependent activation of ATPase)	Troponin system (Tpl)	Unphosphorylated myosin light chain
Contraction activated by	Ca²⁺	Ca²⁺
Direct effect of Ca²⁺	4Ca²⁺ bind to TpC	4Ca²⁺ bind to calmodulin
Effect of protein-bound Ca²⁺	TpC · 4Ca²⁺ antagonizes Tpl inhibition of F-actin–myosin interaction (allows F-actin activation of ATPase)	Calmodulin · 4Ca²⁺ activates myosin light-chain kinase that phosphorylates myosin p-light chain. The phosphorylated p-light chain no longer inhibits F-actin–myosin interaction (allows F-actin activation of ATPase)

[a]Light chains of myosin are different in striated and smooth muscles.

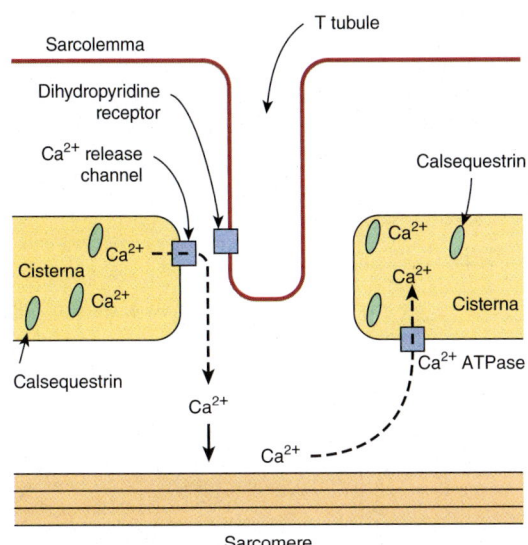

FIGURE 51–8 **Diagram of the relationships among the sarcolemma (plasma membrane), a T tubule, and two cisternae of the SR of skeletal muscle (not to scale).** The T tubule extends inward from the sarcolemma. A wave of depolarization, initiated by a nerve impulse, is transmitted from the sarcolemma down the T tubule. It is then conveyed to the Ca^{2+} release channel (RYR), perhaps by interaction between it and the dihydropyridine receptor (slow Ca^{2+} voltage channel), which lie in proximity to one other. Release of Ca^{2+} from the Ca^{2+} release channel into the cytosol initiates contraction. Subsequently, Ca^{2+} is pumped back into the cisternae of the SR by the Ca^{2+} ATPase (Ca^{2+} pump) and stored there, in part bound to calsequestrin.

When the sarcolemma is excited by a **nerve impulse**, the excitable membranes of the T-tubule system become depolarized, opening the **voltage-gated** Ca^{2+} release channels (homotetramer, ≈565 kDa per subunit) in the nearby SR. Ca^{2+} rapidly floods into the sarcoplasm, raising the concentration nearly 100-fold, to 10^{-5} mol/L, where it binds to troponin C and calmodulin to initiate contraction. **Ryanodine**, a plant alkaloid, binds to and modulates the activity a voltage-gated Ca^{2+} channel also known as the ryanodine receptor (RYR). The two isoforms, RYR1 and RYR2, present in skeletal muscle and heart muscle, respectively. RYR2 is also present in brain tissue. RYR is in the proximity of the **dihydropyridine receptor**, a voltage-gated Ca^{2+} channel of the transverse tubule system, (see Figure 51–8).

Mutations in the Gene Encoding the Ca^{2+} Release Channel Are Cause of Human Malignant Hyperthermia

Exposure to certain anesthetics (eg, halothane) and depolarizing skeletal muscle relaxants (eg, succinylcholine) can cause **malignant hyperthermia (MH)** in some genetically predisposed patients. Symptoms of MH include rigid skeletal muscles, hypermetabolism, and high fever. To prevent other complications or death, anesthetic is stopped immediately and

the drug **dantrolene** is administered. Dantrolene is a skeletal muscle relaxant that acts to inhibit release of Ca^{2+} from the SR into the cytosol.

The causes of MH in humans may involve mutations in the genes encoding the Ca^{2+} release channel, calsequestrin-1, the dihydropyridine receptor, or in the RYR receptor that foster high cytosolic Ca^{2+} levels. Mutations in the *RYR1* gene are also associated with **central core disease**. This is a rare myopathy presenting in infancy with hypotonia and proximal muscle weakness. Electron microscopy reveals an absence of mitochondria in the center of many type I muscle fibers. Damage to mitochondria induced by high intracellular levels of Ca^{2+} secondary to abnormal functioning of *RYR1* appears to be responsible for the morphologic findings.

CARDIAC MUSCLE RESEMBLES SKELETAL MUSCLE IN MANY RESPECTS

Regulation of Contraction in Cardiac Muscle Is Also Actin-Based

Cardiac muscle, like skeletal muscle, is **striated**. It also is regulated by a similar actin-myosin-tropomyosin-troponin system. Unlike skeletal muscle, cardiac muscle exhibits **intrinsic rhythmicity**, and individual myocytes communicate with each other because of its syncytial nature. The **T-tubular system** (see later) is more developed in cardiac muscle, whereas the SR is less extensive and consequently the intracellular supply of Ca^{2+} for contraction is lower. Cardiac muscle thus relies on **extracellular Ca^{2+}** for contraction; if isolated cardiac muscle is deprived of Ca^{2+}, it ceases to beat within approximately 1 minute, whereas skeletal muscle can continue to contract without an extracellular source of Ca^{2+} for a longer period. **Cyclic AMP (cAMP)** plays a more prominent role in cardiac muscle than it does in skeletal muscle. cAMP modulates intracellular levels of Ca^{2+} through the activation of the protein kinases that phosphorylate various transport proteins in the sarcolemma and SR. The protein kinases also target the troponin-tropomyosin regulatory complex, affecting its responsiveness to intracellular Ca^{2+}. There is a rough correlation between the phosphorylation of TpI and the increased contraction of cardiac muscle induced by catecholamines. This may account for the **inotropic effects** (increased contractility) of β-adrenergic compounds on the heart. Some differences among skeletal, cardiac, and smooth muscle are summarized in **Table 51–2**.

Ca^{2+} Enters Cardiomyocytes via Ca^{2+} Channels & Leaves via the Na^{+}-Ca^{2+} Exchanger & the Ca^{2+} ATPase

Extracellular Ca^{2+} enters cardiomyocytes via highly selective channels. The major portal of entry is the L-type

TABLE 51–2 Some Differences Among Skeletal, Cardiac, & Smooth Muscle

Skeletal Muscle	Cardiac Muscle	Smooth Muscle
1. Striated	1. Striated	1. Nonstriated
2. No syncytium	2. Syncytial	2. Syncytial
3. Small T tubules	3. Large T tubules	3. Generally rudimentary T tubules
4. Sarcoplasmic reticulum well developed and Ca^{2+} pump acts rapidly.	4. Sarcoplasmic reticulum present and Ca^{2+} pump acts relatively rapidly.	4. Sarcoplasmic reticulum often rudimentary and Ca^{2+} pump acts slowly.
5. Plasmalemma contains few hormone receptors.	5. Plasmalemma contains a variety of receptors (eg, α- and β-adrenergic).	5. Plasmalemma contains a variety of receptors (eg, α- and β-adrenergic).
6. Nerve impulse initiates contraction.	6. Has intrinsic rhythmicity.	6. Contraction initiated by nerve impulses, hormones, etc.
7. Extracellular fluid Ca^{2+} not important for contraction.	7. Extracellular fluid Ca^{2+} important for contraction.	7. Extracellular fluid Ca^{2+} important for contraction.
8. Troponin system present.	8. Troponin system present.	8. Lacks troponin system; uses regulatory head of myosin.
9. Caldesmon not involved.	9. Caldesmon not involved.	9. Caldesmon is important regulatory protein.
10. Very rapid cycling of the cross-bridges.	10. Relatively rapid cycling of the cross-bridges.	10. Slow cycling of the cross-bridges permits slow, prolonged contraction and less utilization of ATP.

(long-duration current, large conductance) or **slow Ca^{2+}** channel, which is voltage-gated, opening during depolarization and closing when the action potential declines. These channels are equivalent to the dihydropyridine receptors of skeletal muscle (see Figure 51–8). Slow Ca^{2+} channels are **regulated** by cAMP-dependent protein kinases (stimulatory) and cGMP-dependent protein kinases (inhibitory), and can be inhibited by so-called calcium channel blockers (eg, verapamil). **Fast** (or T, transient) Ca^{2+} channels are also present in the plasmalemma, though in much lower numbers; they probably contribute to the early phase of increase of myoplasmic $[Ca^{2+}]$.

When the $[Ca^{2+}]$ in the myoplasm increases, it triggers the opening of the Ca^{2+} release channel of the SR. This Ca^{2+}-induced Ca^{2+} release from the SR stores accounts for approximately 90% of the Ca^{2+} that enters a stimulated cardiomyocyte. However, the 10% that enters from the cytoplasm is vitally important, as it serves as a trigger for mobilization of Ca^{2+} from the SR.

The Na^+-Ca^{2+} exchanger serves as the principal route of egress of Ca^{2+} from cardiomyocytes. Na^+ and Ca^{2+} are exchanged at a ratio of 3:1, with movement of Na^+ into the cell from the plasma providing the energy needed to move Ca^{2+} into the plasma against a concentration gradient. The Na^+-Ca^{2+} exchange contributes to relaxation, but may run in the reverse direction during excitation. Consequently, anything that causes intracellular $[Na^+]$ to rise will secondarily cause intracellular $[Ca^{2+}]$ to rise as well, causing more forceful contraction (**positive inotropic effect**). **Digitalis** promotes the inflow of Ca^{2+} via the Ca^{2+}-Na^+ exchanger by inhibiting the sarcolemmal Na^+-K^+-ATPase, reducing the rate of Na^+ exit by this route, thereby increasing intracellular $[Na^+]$. The resulting increase in intracellular $[Ca^{2+}]$ enhances the force of cardiac

contraction to the benefit of a patient experiencing heart failure (**Figure 51–9**).

In contrast to skeletal muscle, the sarcolemmal **Ca^{2+} ATPase** is a to be a minor contributor to Ca^{2+} egress as compared with the Ca^{2+}-Na^+ exchanger. Ion channels are important in skeletal muscle, as well as in cardiac muscle (see Chapter 40). Mutations in genes encoding ion channels are responsible for a number of relatively rare conditions affecting muscle. These and other diseases due to mutations of ion channels have been termed channelopathies; some are listed in **Table 51–3**.

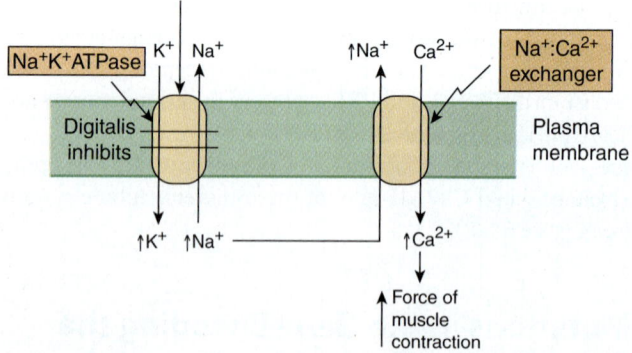

FIGURE 51–9 **Scheme of how the drug digitalis (used in the treatment of certain cases of heart failure) increases cardiac contraction.** Digitalis inhibits the Na^+-K^+ ATPase (see Chapter 40). This results in less Na^+ being pumped out of the cardiac myocyte and leads to an increase of the intracellular $[Na^+]$. In turn, this stimulates the Na^+-Ca^{2+} exchanger so that more Na^+ is exchanged outward, and more Ca^{2+} enters the myocyte. The resulting increased intracellular $[Ca^{2+}]$ increases the force of muscular contraction.

TABLE 51–3 Some Disorders (Channelopathies) due to Mutations in Genes Encoding Polypeptide Constituents of Ion Channels

Disorder[a]	Ion Channel and Major Organs Involved
Central core disease (OMIM 117000)	Ca^{2+} release channel (RYR1), skeletal muscle
Hyperkalemic periodic paralysis (OMIM 170500)	Sodium channel, skeletal muscle
Hypokalemic periodic paralysis (OMIM 170400)	Slow Ca^{2+} voltage channel (DHPR), skeletal muscle
Malignant hyperthermia (OMIM 145600)	Ca^{2+} release channel (RYR1), skeletal muscle
Myotonia congenita (OMIM 160800)	Chloride channel, skeletal muscle

[a]Other channelopathies include the long QT syndrome (OMIM 192500); pseudoaldosteronism (Liddle syndrome, OMIM 177200); persistent hyperinsulinemic hypoglycemia of infancy (OMIM 601820); hereditary X-linked recessive type II nephrolithiasis of infancy (Dent syndrome, OMIM 300009); and generalized myotonia, recessive (Becker disease, OMIM 255700). The term "myotonia" signifies any condition in which muscles do not relax after contraction.
Data from Ackerman MJ, Clapham DE. Ion channels—basic science and clinical disease, N Engl J Med. 1997;336(22):1575–1586.

MUSCLE CONTRACTION REQUIRES LARGE QUANTITIES OF ATP

The amount of ATP in skeletal muscle is only sufficient to provide energy for contraction for a few seconds. Consequently, muscle cells have developed multiple mechanisms to regenerate the ATP needed to sustain the contraction-relaxation cycle: (1) by glycolysis, using blood glucose or muscle glycogen, (2) by oxidative phosphorylation, (3) from creatine phosphate, and (4) from two molecules of ADP in a reaction catalyzed by adenylyl kinase (**Figure 51–10**).

Skeletal Muscle Contains Large Quantities of Glycogen

The sarcoplasm of skeletal muscle contains large stores of **glycogen**, located in granules close to the I bands. The release of glucose from glycogen is dependent on a specific muscle **glycogen phosphorylase** (see Chapter 18), which can be activated by Ca^{2+}, epinephrine, and AMP. Ca^{2+} also activates phosphorylase b kinase, allowing glucose mobilization to begin with the initiation of muscle contraction.

Under Aerobic Conditions, Muscle Generates ATP Mainly by Oxidative Phosphorylation

Glucose, derived from the blood glucose or from endogenous glycogen, along with fatty acids, derived from the triacylglycerols of adipose tissue, are the principal substrates used for aerobic metabolism in muscle. Oxidative phosphorylation is the primary mechanism of ATP production in aerobic respiration. Therefore, contracting muscles have a high demand for oxygen. Some muscles, which can be distinguished by their red color, contain the oxygen storage protein myoglobin (see Chapter 6) to protect against oxygen shortfalls.

Creatine Phosphate Constitutes a Major Energy Reserve in Muscle

Creatine kinase, a muscle-specific enzyme with clinical utility in the detection of acute or chronic diseases of muscle,

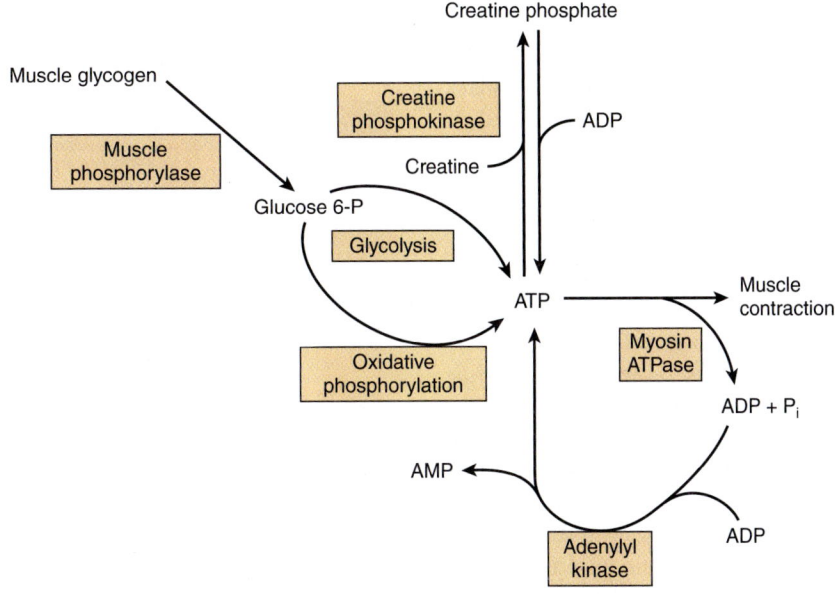

FIGURE 51–10 The multiple sources of ATP in muscle.

catalyzes the synthesis of creatine phosphate from creatine and ATP (see Figure 51–10). The equilibrium constant for this reaction is near 1. Hence, when ATP levels are high, formation of creatine phosphate is favored. However, when ATP levels drop, the equilibrium shifts in favor of ATP synthesis at the expense of stored creatine phosphate. Creatine phosphate thus provides a readily available source for a high-energy phosphate group that can be used to regenerate ATP from ADP.

Adenylyl Kinase Serves as a Reserve of Last Resort

ATP synthase (see Figure 13–8) is the primary vehicle for regenerating ATP in living cells. However, it can only synthesize ATP from ADP. In order to regenerate ATP from AMP, the latter must first be phosphorylated to form ADP, a process catalyzed by the enzyme adenylyl kinase using ATP as phosphodonor. This enzyme thus can generate two molecules of ADP from one molecule each of AMP and ATP. When other means for regenerating ATP become exhausted and nucleotide triphosphate level plummet, the equilibrium shifts in favor of synthesizing ATP at the expense of ADP. It is important to note that the by-product of this reaction is AMP. Hence, this is only a temporary expedient as the cell can only cannibalize their total adenine nucleotide pool for so long.

SKELETAL MUSCLE CONTAINS SLOW (RED) & FAST (WHITE) TWITCH FIBERS

Different types of fibers have been detected in skeletal muscle. One classification subdivides them into type I (slow twitch), type IIA (fast twitch-oxidative), and type IIB (fast twitch-glycolytic). For the sake of simplicity, we shall consider only two types: type I (slow twitch, oxidative) and type II (fast twitch, glycolytic). The **type I** fibers contain myoglobin and mitochondria; their metabolism is aerobic, and they maintain relatively sustained contractions. The white **type II** fibers lack myoglobin and contain few mitochondria: they derive their energy from anaerobic glycolysis and perform contractions of relatively short duration.

The proportion of these two types of fibers varies among the muscles of the body, depending on function and training. For example, the number of type I fibers in certain leg muscles increases in athletes training for marathons, whereas the number of type II fibers increases in sprinters. Type I fibers rely heavily on aerobic metabolism to regenerate ATP, while type II fibers rely on oxygen-independent sources such as creatine phosphate. For this reason, many endurance athletes will engage in carbohydrate loading, the ingestion of meals that predominantly consist of foods with a high starch content, in an effort to increase the size of their glycogen stores, while some sprinters will attempt to increase the pool of creatine phosphate in their muscles by ingesting creatine as a dietary supplement.

MUSCLE TISSUES ARE THE TARGET OF SEVERAL GENETIC DISORDERS

Inherited Cardiomyopathies Can Arise From Disorders of Cardiac Energy Metabolism or Abnormal Myocardial Proteins

A **cardiomyopathy** is any structural or functional abnormality of the ventricular myocardium. These abnormalities can arise from a number of causes, many of them hereditary. As shown in **Table 51–4**, the causes of inherited cardiomyopathies fall into two broad classes: (1) disorders of cardiac energy metabolism, mainly reflecting mutations in genes encoding enzymes or proteins involved in fatty acid oxidation (a major source of energy for the myocardium) and oxidative phosphorylation; (2) mutations in genes encoding proteins involved in or affecting myocardial contraction, such as myosin, tropomyosin, the troponins, and cardiac myosin-binding protein C.

Mutations in the Cardiac β-Myosin Heavy-Chain Gene Are One Cause of Familial Hypertrophic Cardiomyopathy

Familial hypertrophic cardiomyopathy is one of the most commonly encountered hereditary cardiac diseases. Patients exhibit hypertrophy—often massive—of one or both ventricles, starting early in life. Most cases are transmitted in an autosomal dominant manner; the rest are sporadic. The root cause of this condition is any one of several **missense mutations** in the gene encoding the **β-myosin heavy chain** that leads to the replacement of a highly conserved amino acid with some other residue. The substitutions cluster in the head

TABLE 51–4 Biochemical Causes of Inherited Cardiomyopathies[a]

Cause	Proteins or Process Affected
Inborn errors of fatty acid oxidation	Carnitine entry into cells and mitochondria
	Certain enzymes of fatty acid oxidation
Disorders of mitochondrial oxidative phosphorylation	Proteins encoded by mitochondrial genes
	Proteins encoded by nuclear genes
Abnormalities of myocardial contractile and structural proteins	β-Myosin heavy chains, troponin, tropomyosin, dystrophin

[a]Mutations (eg, point mutations, or in some cases deletions) in the genes (nuclear or mitochondrial) encoding various proteins, enzymes, or tRNA molecules are the fundamental causes of the inherited cardiomyopathies. Some conditions are mild, whereas others are severe and may be part of a syndrome affecting other tissues.
Data from Kelly DP, Strauss AW. Inherited cardiomyopathies, N Engl J Med. 1994; 330(13):913–919.

and head-rod regions of the myosin heavy chain. One hypothesis is that these heavy chain variants ("poison polypeptides") cause formation of abnormal myofibrils, eventually resulting in compensatory hypertrophy.

Patients with familial hypertrophic cardiomyopathy can show great variation in clinical picture. This in part reflects genetic heterogeneity as it appears that mutations in a number of other genes (eg, those encoding cardiac actin, tropomyosin, cardiac troponins I and T, essential and regulatory myosin light chains, cardiac myosin-binding protein C, titin, and mitochondrial tRNA-glycine and tRNA-isoleucine) may also cause familial hypertrophic cardiomyopathy. Patients harboring mutations that are predicted to alter the charge character of the affected amino acid side chain exhibit a significantly shorter life expectancy than patients in whom the mutation produced no alteration in charge.

Mutations in the genes encoding dystrophin, muscle LIM protein (so called because it was found to contain a cysteine-rich domain originally detected in three proteins: Lin-II, Isl-1, and Mec-3), the cyclic AMP response-element binding protein (CREB), desmin, and lamin have each been implicated in the causation of **dilated cardiomyopathy**. The first two proteins help organize the contractile apparatus of cardiac muscle cells, while CREB regulates the expression of several genes within these cells.

Mutations in the Gene Encoding Dystrophin Cause Duchenne Muscular Dystrophy

Other protein components of the contractile apparatus include titin, the world's largest known protein whose role is to anchor the ends of myofibrils, nebulin, α-actinin, desmin, and dystrophin. Among these proteins, **dystrophin** is of particular biomedical interest. Mutations in the gene that encodes it play a causative role in **Duchenne muscular dystrophy** and **Becker muscular dystrophy** and have been implicated in **dilated cardiomyopathy** (see earlier). Dystrophin bridges the actin cytoskeleton to the extracellular matrix at the interior face of the plasma membrane. Formation of this link is necessary for assembly of the synaptic junction. Duchenne muscular dystrophy appears to result from the inability of mutationally altered forms of dystrophin to support formation of functionally competent synaptic junctions. Similarly, mutations in genes encoding the glycosyltransferases that modify **α-dystroglycan** or those encoding polypeptide components of the **sarcoglycan** complex (**Figure 51–11**) are responsible for certain other congenital forms of muscular dystrophy such as **limb-girdle**.

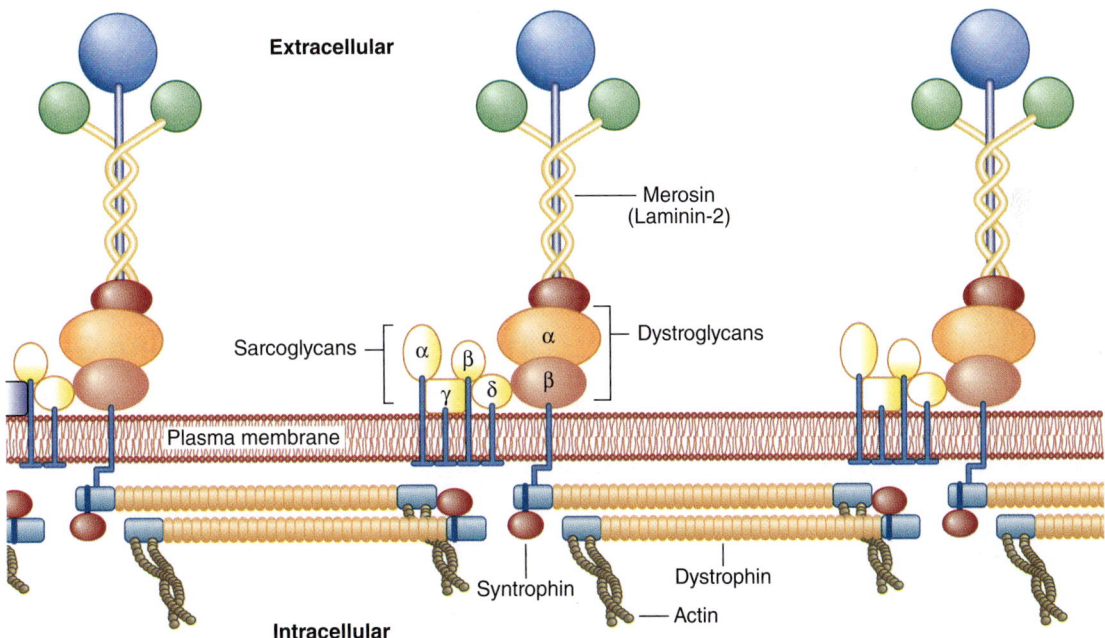

FIGURE 51–11 **Organization of dystrophin and other proteins in relation to the plasma membrane of muscle cells.** Dystrophin is part of a large oligomeric complex associated with several other protein complexes. The dystroglycan complex consists of α-dystroglycan, which associates with the basal lamina protein merosin (also named laminin-2, see Chapter 50), and α-dystroglycan, which binds α-dystroglycan and dystrophin. Syntrophin binds to the carboxyl terminal of dystrophin. The sarcoglycan complex consists of four transmembrane proteins: α-, β-, γ-, and δ-sarcoglycan. The function of the sarcoglycan complex and the nature of the interactions within the complex and between it and the other complexes are not clear. The sarcoglycan complex is formed only in striated muscle, and its subunits preferentially associate with each other, suggesting that the complex may function as a single unit. Mutations in the gene encoding dystrophin cause Duchenne and Becker muscular dystrophies. Mutations in the genes encoding the various sarcoglycans have been shown to be responsible for limb-girdle dystrophies (eg, OMIM 604286) and mutations in genes encoding other muscle proteins cause other types of muscular dystrophy. Mutations in genes encoding certain glycosyltransferases involved in the synthesis of the glycan chains of α-dystroglycan are responsible for certain congenital muscular dystrophies. (Reproduced with permission from Duggan DJ, Gorospe JR, Fanin M, et al: Mutations in the sarcoglycan genes in patients with myopathy, N Engl J Med. 1997;336(9):618–624.)

Nitric Oxide Relaxes the Smooth Muscle of Blood Vessels

Acetylcholine triggers the relaxation of the smooth muscle of blood vessels. On binding to its cell surface receptors on the endothelial cells surrounding vascular smooth muscle cells, acetylcholine triggers the activation of associated phospholipases on the interior surface of the plasma membrane. Phospholipases hydrolyze and release the polyphosphorylated head groups, particularly 3,4,5-triphosphoinositol, from phosphatidylinositol, a quantitatively minor but functionally important phospholipid component of the plasma membrane. These polyphosphoinositol second messengers initiate the release of Ca^{2+} into the cytoplasm of these vascular epithelial cells, which in turn triggers the release of **endothelium-derived relaxing factor (EDRF)**, which diffuses into the adjacent smooth muscle where it causes subsequent relaxation. EDRF was identified as **NO, nitrous oxide**, which has a half-life of only ~3-4 s in tissues.

NO synthase, a Ca^{2+}-activated enzyme found in the cytosol, catalyzes the five-electron oxidation of a guanidino nitrogen in the side chain of arginine, yielding citrulline and NO (**Figure 51–12**). This complex reaction utilizes NADPH and four redox-active prosthetic groups: FAD, FMN, heme, and tetrahydrobiopterin. On diffusing into the surrounding vascular smooth muscle cells, NO binds to the heme moiety of a soluble guanylyl cyclase, activating the enzyme and elevating the intracellular levels of the second messenger 3′,5′-cyclic GMP (cGMP). This in turn stimulates the activities of certain cGMP-dependent protein kinases, which phosphorylate specific muscle proteins, causing relaxation.

NO can also be formed from **nitrite**, derived from the metabolism of vasodilators such as glyceryl trinitrate, also known as nitroglycerin, which is commonly administered to treat angina. Another important cardiovascular effect of NO is the inhibition of platelet aggregation, a consequence of the increased synthesis of cGMP.

SKELETAL MUSCLE CONSTITUTES THE MAJOR RESERVE OF PROTEIN IN THE BODY

In humans, **skeletal muscle proteins** are the major nonfat source of stored energy. This explains the large losses of muscle mass, particularly in adults, that accompany prolonged caloric undernutrition (see Chapter 28).

THE CYTOSKELETON PERFORMS MULTIPLE CELLULAR FUNCTIONS

All cells perform mechanical work, albeit of lower physical magnitude than muscle cells. These include self-propulsion, cytokinesis, endocytosis, exocytosis, and phagocytosis. Like muscle cells, the core of this machinery is comprised of filamentous protein polymers called the **cytoskeleton**. The cytoskeleton's more quantitatively prominent filamentous structures include **actin filaments** (also known as microfilaments), **microtubules**, and **intermediate filaments**.

Nonmuscle Cells Contain Microfilaments Containing G-Actin

G-actin is present in most if not all cells of the body. Under physiologic conditions, G-actin monomers spontaneously polymerize to form double helical **F-actin** filaments (7-9.5 nm in diameter), similar to those found in muscle. The actin microfilaments of the cytoskeleton generally exist as bundles within a tangled-appearing meshwork. These prominent bundles, which just underlie the plasma membranes of many cells, are referred to as **stress fibers**. Stress fibers disappear as cell motility increases or on malignant transformation of cells.

Cytoskeletal microfilaments are composed of β-actin and γ-actin. Although not organized as in muscle, actin filaments in nonmuscle cells can interact with **myosin** to cause cellular movements.

Microtubules Contain α- & β-Tubulins

Microtubules, an integral component of the cellular cytoskeleton, consist of cytoplasmic tubes, 25 nm in diameter that are often of extreme length (**Figure 51–13**). Each cylindrical tube is comprised of 13 longitudinally arranged protofilaments

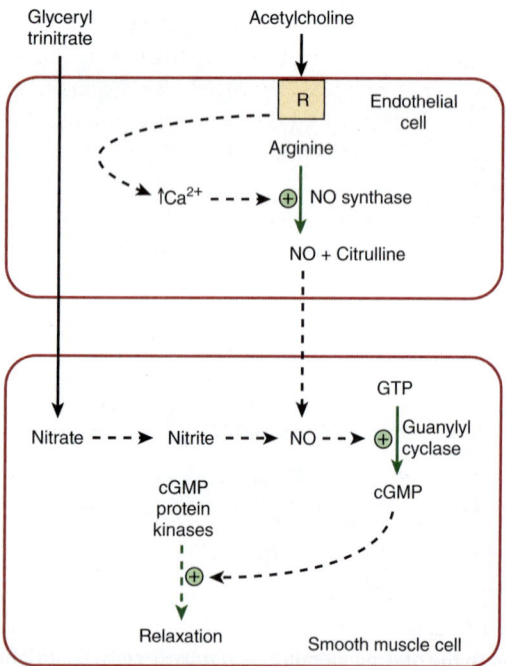

FIGURE 51–12 **Diagram showing formation in an endothelial cell of nitric oxide (NO) from arginine in a reaction catalyzed by NO synthase.** Interaction of an agonist (eg, acetylcholine) with a receptor (R) leads to intracellular release of Ca^{2+} induced by inositol trisphosphate generated by the phosphoinositide pathway, resulting in activation of NO synthase. The NO subsequently diffuses into adjacent smooth muscle, where it leads to activation of guanylyl cyclase, formation of cGMP, stimulation of cGMP protein kinases, and subsequent relaxation. The vasodilator nitroglycerin is shown entering the smooth muscle cell, where its metabolism also leads to formation of NO.

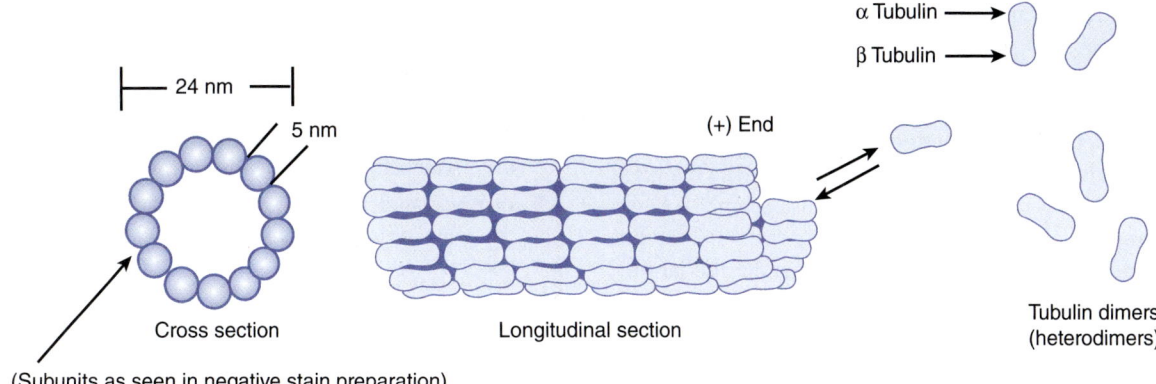

α Tubulin

β Tubulin

(+) End

Cross section

Longitudinal section

(Subunits as seen in negative stain preparation)

Tubulin dimers (heterodimers)

FIGURE 51–13 **Schematic representation of microtubules.** At left and center are shown drawings of microtubules as seen in the electron microscope following fixation with tannic acid in glutaraldehyde. The unstained tubulin subunits are delineated by the dense tannic acid. Cross sections of tubules reveal a ring of 13 subunits of dimers arranged in a spiral. Changes in microtubule length are due to the addition or loss of individual tubulin subunits. Characteristic arrangements of microtubules (not shown here) are found in centrioles, basal bodies, cilia, and flagellae. (Reproduced with permission from Junqueira LC, Carneiro J, Kelley RO: *Basic Histology*, 7th ed. New York, NY: Appleton & Lange; 1992.)

comprised of **α-tubulin** and **β-tubulin**, closely related proteins of ≈50 kDa molecular mass. Assembly starts with the formation of tubulin dimers that assemble into protofilaments that associate with one another in parallel to form sheets and eventually cylinders. **GTP** is required for assembly. A microtubule-organizing center, located around a pair of centrioles, nucleates the growth of new microtubules. A third species of tubulin, **γ-tubulin**, appears to play an important role in this process.

Microtubules are necessary for **mitotic spindle** formation and function. The latter include chromosome segregation during cell division, intracellular movement of endocytic and exocytic vesicles, and structural components of **cilia** and **flagella**. Microtubules also maintain the structure of **axons** and **dendrites** as well as participate in the axoplasmic flow of material along neuronal processes. A variety of **microtubule-associated proteins [MAPs]**, one of which is **tau,** play important roles in microtubule assembly and stabilization.

Microtubules exist in a state of dynamic instability, constantly assembling and disassembling. They exhibit **polarity** (plus and minus ends); a feature important for growth from centrioles and directing intracellular movement. For instance, in axonal transport, the protein **kinesin**, with a myosin-like ATPase activity, uses hydrolysis of ATP to move vesicles down the axon toward the positive end of the microtubular formation. Flow of materials in the opposite direction, toward the negative end, is powered by **cytosolic dynein**, another protein with ATPase activity. Similarly, **axonemal dyneins** power ciliary and flagellar movement while **dynamin**, which uses GTP instead of ATP, is involved in endocytosis. Kinesins, dyneins, dynamin, and myosins are referred to as **molecular motors**.

An absence of dynein in cilia and flagella results in immotile cilia and flagella, leads to **Kartagener syndrome** (OMIM 244400), which can cause male sterility, situs inversus, and chronic respiratory infections. Mutations in genes affecting the synthesis of dynein have been detected in individuals with this syndrome. Pharmaceuticals such as **colchicine** (used for treatment of acute gouty arthritis), **vinblastine** (a vinca alkaloid used

for treating certain types of cancer), **paclitaxel** (Taxol) (effective against ovarian cancer), and **griseofulvin** (an antifungal agent) also disrupt microtubule assembly or disassembly.

Intermediate Filaments Differ From Microfilaments & Microtubules

An intracellular fibrous system exists comprised of filaments with an axial periodicity of 21 nm and a diameter of 8 to 10 nm that is intermediate between that of microfilaments (6 nm) and microtubules (23 nm). At least four classes of **intermediate filaments** are found, as indicated in **Table 51–5**. Each is made up

TABLE 51–5 Classes of Intermediate Filaments of Eukaryotic Cells & Their Distributions

Proteins	Molecular Mass (kDa)	Distributions
Lamins		
A, B, and C	65-75	Nuclear lamina
Keratins		
Type I (acidic)	40-60	Epithelial cells, hair, nails
Type II (basic)	50-70	As for type I (acidic)
Vimentin-like		
Vimentin	54	Various mesenchymal cells
Desmin	53	Muscle
Glial fibrillary acid protein	50	Glial cells
Peripherin	66	Neurons
Neurofilaments		
Low (L), medium (M), and high (H)[a]	60-130	Neurons

[a]Refers to their molecular masses.
Note: Intermediate filaments have an approximate diameter of 10 nm and have various functions. For example, keratins are distributed widely in epithelial cells and adhere via adapter proteins to desmosomes and hemidesmosomes. Lamins provide support for the nuclear membrane.

of elongated, fibrous molecules with a central rod domain, an amino-terminal head, and a carboxyl-terminal tail. These subunits assemble in a helical a manner to form repeating tetrameric units to form rope-like fibrils. Intermediate filaments are important structural components of cells that serve as **relatively stable** components of the cytoskeleton. Intermediate filaments are not undergoing rapid assembly and disassembly and not disappearing during mitosis, as do actin and many microtubular filaments. An important exception to this are the **lamins**, which, subsequent to phosphorylation, disassemble at mitosis and reappear when it terminates. **Lamins** form a meshwork positioned in apposition to the inner nuclear membrane. In addition, lamin A is an important component of the structural scaffolding that maintains the integrity of the cell nucleus.

Hutchinson-Gilford progeria syndrome (**progeria**, OMIM 176670) is caused by mutations in the *LMNA* gene that encodes **lamin A** and, via alternative splicing (see Chapter 36), **lamin C**. In progeria, a farnesylated form (see Figure 26–2 for the structure of farnesyl) of prelamin A accumulates in this condition, because the site at which the farnesylated portion of lamin A is normally cleaved by proteases has been altered by mutation. It appears that the accumulation of farnesylated prelamin A destabilizes nuclei, altering their shape, somehow predisposing victims to **manifest signs of premature aging**. Experiments in mice have indicated that administration of a farnesyltransferase inhibitor may ameliorate the development of misshapen nuclei. Children affected by this condition often die in their teens of atherosclerosis.

Keratins form a large family consisting of about 30 members. As indicated in Table 51–5, two major types of keratins are found that assemble into **heterodimers** made up of one member of each class. **Vimentins** are widely distributed in mesodermal cells. Desmin, glial fibrillary acidic protein, and peripherin are related to them. All members of the vimentin-like family can copolymerize with each other.

Intermediate filaments in nerve cells, called neurofilaments, are classified as low, medium, and high on the basis of their molecular masses. The **distribution of intermediate filaments** in normal and abnormal (eg, cancer) cells can be studied by the use of immunofluorescent techniques. Pathologists utilize antibodies against specific intermediate filaments to determine the origin of certain dedifferentiated malignant tumors.

A number of **skin diseases**, mainly characterized by blistering, have been found to be due to mutations in genes encoding **various keratins**. Two of these disorders are epidermolysis bullosa simplex (OMIM 131800) and epidermolytic palmoplantar keratoderma (OMIM 144200). The **blistering** found in these disorders probably reflects a diminished capacity of various layers of the skin to resist mechanical stresses due to abnormalities in the keratin structure.

SUMMARY

- The myofibrils of skeletal muscle consist of thick, myosin-based and thin, actin-based filaments.
- During contraction, these interdigitating filaments slide past one another with cross-bridges between myosin and actin generating and sustaining tension.

- Muscle contraction is driven by the cyclic attachment, conformational shift, and detachment of massive numbers of myosin head domains to adjacent actin fibrils.
- The hydrolysis of ATP is used to drive movement of the filaments. ATP binds to myosin heads and is hydrolyzed to ADP and P_i by the ATPase activity of the actomyosin complex.
- In striated muscle, the contractile apparatus is held in check by the troponin complex (troponins T, I, and C) until inhibition is relieved by the binding of Ca^{2+} to troponin C.
- In smooth muscle, the contractile apparatus is held in check by the regulatory light chains in myosin. This block is relieved when the regulatory light chains are phosphorylated by a $(Ca^{2+})_4$:calmodulin-activated protein kinase, myosin light-chain kinase.
- In skeletal muscle, the SR regulates distribution of Ca^{2+} to the sarcomeres, whereas in cardiac and smooth muscle the inflow of Ca^{2+} occurs via Ca^{2+} channels in the sarcolemma.
- Ca^{2+} not only initiates contraction, it also activates the calcium efflux systems that will bring contraction to a close.
- Many cases of malignant hyperthermia in humans are due to mutations in the gene encoding the Ca^{2+} release channel (RYR1).
- Some cases of familial hypertrophic cardiomyopathy are due to missense mutations in the gene coding for the β-myosin heavy chain. Mutations in genes encoding a number of other proteins have also been detected.
- NO synthase catalyzes the formation of NO, a regulator of vascular smooth muscle.
- Duchenne-type muscular dystrophy is a hereditary disease with mutations in the gene encoding the protein dystrophin.
- Two major types of muscle fibers are found in humans: white (anaerobic) and red (aerobic).
- Nonmuscle cells contain an internal network of fibers called the cytoskeleton that provides the mechanical apparatus needed to maintain and change cell shape, provide cell motility, phagocytosis, etc.
- The cytoskeleton is comprised of a variety of filaments that include actin-based microfilaments, α-tubulin and β-tubulin containing microtubules, and lamin-, keratin-, and vimentin-containing intermediate filaments.
- Mutations in the gene encoding lamin A cause progeria, a condition characterized by symptoms resembling premature aging.
- Mutations in genes for certain keratins cause a number of skin diseases.
- The cytoskeleton provides a scaffold for a variety of molecular motors, such as kinesin and dynein, that participate in vesicle transport, axonal flow, flagellar motion, and morphologic changes in cell shape.

REFERENCES

Blanchoin L, Bouiemaa-Paterski R, Sykes C, Plastino J: Actin dynamics, architecture, and mechanics in cell motility. Physiol Rev 2014;94:235.

Goldman RD, Pollard TD (editors): *The Cytoskeleton*. Cold Spring Harbor Lab Press, 2017.

Khalil RA (editor): *Vascular Pharmacology: Cytoskeleton and Extracellular Matrix.* Academic Press, 2018.

Kull FJ, Endow SA: Force generation by kinesin and myosin cytoskeleton proteins. J Cell Sci 2013;126:9.

Rall JA: *Mechanism of Muscle Contraction.* Springer, 2014.

Sequeira V, Nijenkamp LL, Regan JA, van der Velden J: The physiological role of cardiac cytoskeleton and its alterations in heart failure. Biochim Biophys Acta 2014;1838:700.

Smith DA: *The Sliding Filament Theory of Muscle Contraction.* Springer, 2018.

Szent-Gyorgyi AG: The early history of the biochemistry of muscle contraction. J Gen Physiol 2004;123:631.

Plasma Proteins & Immunoglobulins

Peter J. Kennelly, PhD

- List the major functions of blood.
- Describe the principal functions of serum albumin.
- Explain how haptoglobin protects the kidney against formation of damaging iron precipitates.
- Describe the roles of ferritin, transferrin, and ceruloplasmin in iron homeostasis.
- Describe the mechanism by which transferrin, transferrin receptors, and HFE protein regulate the synthesis of hepcidin.
- Explain how iron homeostasis can be perturbed by dietary deficiencies or certain disorders.
- Describe the general structures and functions of the five classes of immunoglobulins.
- Explain how up to a million different immunoglobulins can be generated utilizing fewer than 150 human genes.
- Describe the activation and mode of action of the complement system.
- Compare and contrast the adaptive and innate immune systems.
- Define the term lectin.
- Outline the key differences between polyclonal and monoclonal antibodies.
- Describe the salient features of autoimmune and immunodeficiency disorders.

BIOMEDICAL IMPORTANCE

The proteins that circulate in blood plasma play important roles in human physiology. **Albumins** facilitate the transit of fatty acids, steroid hormones, and other ligands between tissues, while **transferrin** aids the uptake and distribution of iron. Circulating **fibrinogen** serves as a readily mobilized building block of the fibrin mesh that provides the foundation of the clots used to seal injured vessels. Formation of these clots is triggered by a cascade of latent proteases, or **zymogens**, called blood coagulation factors. Plasma also contains several proteins that function as inhibitors of proteolytic enzymes. **Antithrombin** helps confine the formation of clots to the vicinity of a wound, while α_1-antiproteinase and α_2-macroglobulin shield healthy tissues from the proteases that destroy invading pathogens and remove dead or defective cells. Circulating immunoglobulins called **antibodies** form the front line of the body's immune system.

Perturbances in the production of plasma proteins can have serious health consequences. Deficiencies in key components of the blood clotting cascade can result in excessive bruising and bleeding (**hemophilia**). Persons lacking plasma ceruloplasmin, the body's primary carrier of copper, are subject to hepatolenticular degeneration (Wilson disease), while emphysema is associated with a genetic deficiency in the production of circulating α_1-antiproteinase. More than one in every 30 residents of North America suffer from an **autoimmune disorder**, such as type 1 diabetes, asthma, and rheumatoid arthritis, resulting from the aberrant production immunoglobulins (**Table 52–1**). **Immunocompromised** individuals are rendered vulnerable to infection by viral or microbial invaders by insufficiencies in the production of protective antibodies. Such insufficiencies may result from infection by pathogens such as the **human immunodeficiency virus** (HIV) or from the administration of immunosuppressant drugs. While the root causes of plasma protein–related diseases such as hemophilia

TABLE 52–1 Prevalence of Selected Autoimmune Diseases Among U.S. Population

Autoimmune Disease	Mean Prevalence Rate (per 100,000)	Percentage Female
Graves disease/hyperthyroidism	1152	88
Rheumatoid arthritis	860	75
Thyroiditis/hypothyroidism	792	95
Vitiligo	400	52
Type 1 diabetes	192	48
Pernicious anemia	151	67
Multiple sclerosis	58	64
Primary glomerulonephritis	40	32
Systemic lupus erythematosus	24	88
IgA glomerulonephritis	23	67
Sjogren syndrome	14	94
Myasthenia gravis	5	73
Addison's disease	5	93
Scleroderma	4	92

Data from Jacobson DL, Gange SJ, Rose NR, et al: Epidemiology and estimated population burden of selected autoimmune diseases in the United States, Clin Immunol Immunopathol. 1997;84(3):223–243.

TABLE 52–2 Major Functions of Blood

1. **Respiration**—transport of oxygen from the lungs to the tissues and of CO_2 from the tissues to the lungs
2. **Nutrition**—transport of absorbed food materials
3. **Excretion**—transport of metabolic waste to the kidneys, lungs, skin, and intestines for removal
4. Maintenance of the normal **acid–base balance** in the body
5. Regulation of **water balance** through the effects of blood on the exchange of water between the circulating fluid and the tissue fluid
6. Regulation of **body temperature** by the distribution of body heat
7. **Defense** against infection by the white blood cells and circulating antibodies
8. Transport of **hormones** and regulation of metabolism
9. Transport of **metabolites**
10. **Coagulation**

are relatively straightforward, others—in particular many autoimmune disorders—arise due to the cryptic interplay of complex genetic, dietary, nutritional, environmental, and medical factors.

THE BLOOD HAS MANY FUNCTIONS

As the primary avenue by which tissues are connected to each other and the surrounding environment, the blood that circulates throughout our body performs a variety of functions. These include delivering nutrients and oxygen, removing waste products, conveying hormones, and defending against infectious microorganisms (**Table 52–2**). These myriad functions are carried out by a diverse set of components that include cellular entities such as red blood cells, platelets, and leukocytes (see Chapters 53 and 54), and the water, electrolytes, metabolites, nutrients, proteins, and hormones that comprise the **plasma**.

PLASMA CONTAINS A COMPLEX MIXTURE OF PROTEINS

Early scientists classified the proteins found in plasma into three groups, **fibrinogen**, **albumin**, and **globulins**, on the basis of their relative solubility in solutions to which water-miscible organic solvents such as ethanol or salting out agents such as

ammonium sulfate had been added. Using electrophoresis in a **cellulose acetate** matrix, clinical scientists subsequently were able to separate the salt-soluble serum protein fraction into five major components designated albumin and the α_1-, α_2-, β-, and γ-**globulins** (**Figure 52–1**). Plasma proteins tend to be rich in disulfide bonds and frequently contain bound carbohydrate (**glycoproteins**) or lipid (**lipoproteins**). The relative dimensions and molecular masses of several plasma proteins are shown in **Figure 52–2**.

Plasma Proteins Help Determine the Distribution of Fluid Between Blood & Tissues

The aggregate concentration of the proteins present in human plasma typically falls in the range of 7 to 7.5 g/dL. The resulting **osmotic pressure** (oncotic pressure) is approximately 25 mm Hg. Since the **hydrostatic pressure** in the arterioles is approximately 37 mm Hg, with an interstitial (tissue) pressure of 1 mm Hg opposing it, a net outward force of about 11 mm Hg drives fluid from the plasma into the interstitial spaces. By contrast, the hydrostatic pressure in venules is about 17 mm Hg; thus, a net force of about 9 mm Hg drives water from tissues back into the circulation. The earlier pressures are often referred to as **Starling forces**. If the concentration of plasma proteins is markedly diminished (eg, due to severe protein malnutrition), fluid will cease flowing back into the intravascular compartment and begin to accumulate in the extravascular tissue spaces, a condition known as **edema**.

Most Plasma Proteins Are Synthesized in the Liver

Roughly 70 to 80% of all plasma proteins are synthesized in the liver. These include albumin, fibrinogen, transferrin, and most components of the complement and blood

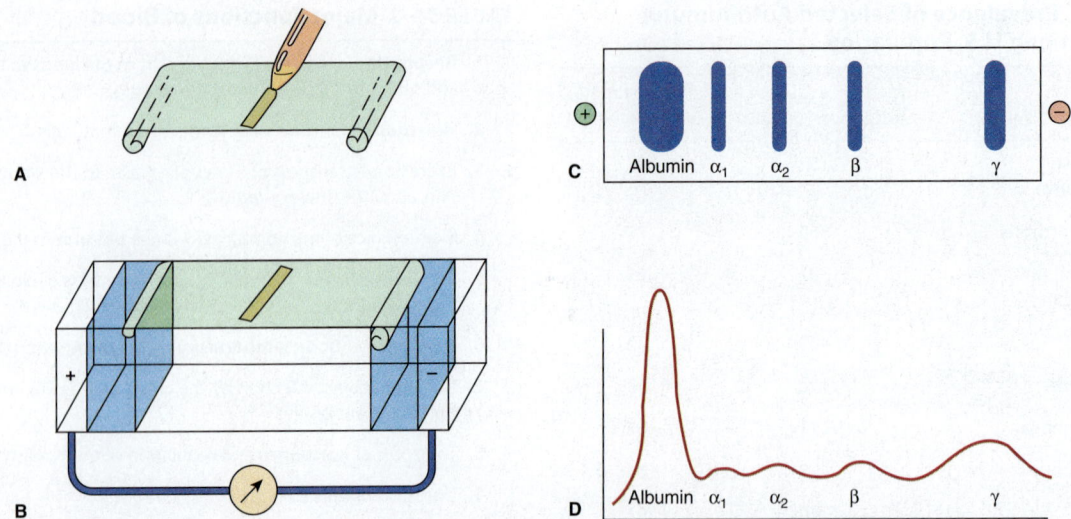

FIGURE 52–1 **Technique of cellulose acetate zone electrophoresis. (A)** A small amount of serum or other fluid is applied to a cellulose acetate strip. **(B)** Electrophoresis in electrolyte buffer is performed. **(C)** Staining enables separated bands of protein to be visualized. **(D)** Densitometer scanning reveals the relative mobilities of albumin, α_1-globulin, β_2-globulin, β-globulin, and γ-globulin. (Reproduced with permission from Parslow TG, Stites DP, Terr AI, et al: *Medical Immunology*, 10th ed. New York, NY: McGraw Hill; 2001.)

coagulation cascades. Two prominent exceptions are von Willebrand factor, which is synthesized in the vascular endothelium, and the γ-globulins, which are synthesized in the lymphocytes. Most plasma proteins are covalently modified by the addition of either N- or O-linked oligosaccharide chains, or both (see Chapter 46). Albumin is the major exception. These oligosaccharide chains fulfill a variety of functions (see Table 46–2). Loss of terminal sialic acid residues accelerates clearance of plasma glycoproteins from the circulation.

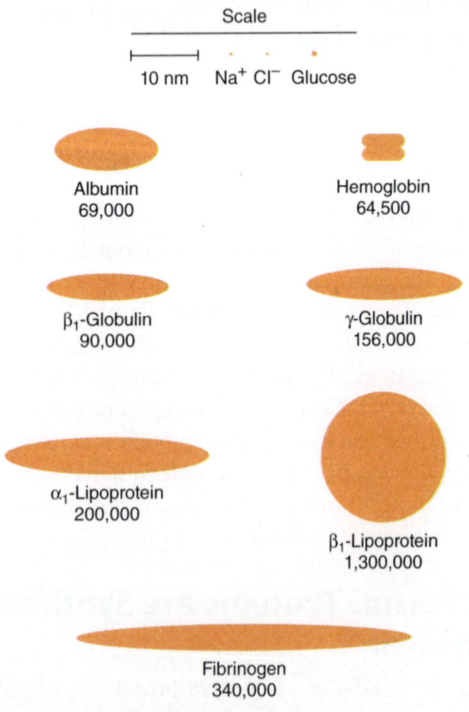

As is the case for other proteins destined for secretion from a cell, the genes for plasma proteins code for an amino-terminal **signal sequence** that targets them to the endoplasmic reticulum. As this leader sequence emerges from the ribosome, it binds to a transmembrane protein complex in the endoplasmic reticulum called the **signal recognition particle**. The emerging polypeptide chain is pulled through the signal recognition particle into the lumen of the endoplasmic reticulum and its leader sequence cleaved off by an associated **signal peptidase** (see Chapter 49). The newly synthesized proteins then move from the rough endoplasmic membrane via the smooth endoplasmic membrane and Golgi apparatus intro secretory vesicles, during which time they are subject to various posttranslational modifications (proteolysis, glycosylation, phosphorylation, etc). Ultimately, the mature proteins are released via the secretory vesicles into the plasma.

Many Plasma Proteins Exhibit Polymorphism

A **polymorphism** is a mendelian or monogenic trait that exists in the population in two or more phenotypes, neither of which is rare (ie, it occurs with frequency of at least 1-2%). Most polymorphisms are innocuous. While the ABO blood group substances (see Chapter 53) are perhaps the best-known example of a human polymorphism, other human plasma proteins exhibit polymorphism including α_1-antitrypsin, haptoglobin, transferrin, ceruloplasmin, and immunoglobulins.

Scale

|———| 10 nm Na^+ Cl^- Glucose

Albumin
69,000

Hemoglobin
64,500

β_1-Globulin
90,000

γ-Globulin
156,000

α_1-Lipoprotein
200,000

β_1-Lipoprotein
1,300,000

Fibrinogen
340,000

FIGURE 52–2 **Relative dimensions and approximate molecular masses of protein molecules in the blood.**

Each Plasma Protein Has a Characteristic Half-Life in the Circulation

The **half-life** of a plasma protein is the time required for 50% of the molecules present at any given moment to be degraded or otherwise cleared from the blood. Clearance permits the continuous replacement of older, possibly damaged, protein molecules by newly-synthesized ones, a process called **turnover.** During normal turnover, the total concentration of these proteins will remain constant as the countervailing processes of synthesis and clearance reach a **steady state.**

In certain diseases, the half-life of a protein may be markedly altered. For instance, in some gastrointestinal diseases such as regional ileitis (Crohn disease), considerable amounts of plasma proteins, including albumin, may be lost into the bowel through the inflamed intestinal mucosa. The half-life of albumin in these subjects may be reduced from its normal 20 days to as little as 1 day, a condition referred to as a **protein-losing gastroenteropathy.**

ALBUMIN IS THE MOST ABUNDANT PROTEIN IN HUMAN PLASMA

The liver synthesizes approximately 12 g of albumin per day, representing about 25% of total hepatic protein synthesis and half its secreted protein. About 40% of the body's albumin circulates in the plasma, where it accounts for roughly three-fifths of total plasma protein by weight (3.4-4.7 g/dL). The remainder resides in the extracellular space. Because of its relatively high concentration, albumin is thought to contribute 75 to 80% of the **osmotic pressure** of human plasma. Like most other secreted proteins, albumin is initially synthesized as a **preproprotein.** Its **signal peptide** is removed as it passes into the cisternae of the rough endoplasmic reticulum. Subsequently, a second **hexapeptide** is cleaved from the new N-terminus as it moves along the secretory pathway.

Mature human albumin consists of a single polypeptide chain, 585 amino acids in length, that is organized into three functional domains. Its ellipsoidal conformation is stabilized by a total of 17 intrachain disulfide bonds. A major role of albumin is to bind to and transport numerous **ligands.** These include free fatty acids (FFA), calcium, certain steroid hormones, bilirubin, copper, and tryptophan. A variety of drugs, including sulfonamides, penicillin G, dicumarol, and aspirin, also bind to albumin; a finding with important pharmacologic implications. Preparations of human albumin have been widely used in the treatment of burns and of hemorrhagic shock.

Some humans suffer from genetic mutations that impair their ability to synthesize albumin. Individuals whose plasma is completely devoid of albumin are said to exhibit **analbuminemia.** Surprisingly, persons suffering from analbuminemia display only moderate edema. Depressed synthesis of albumin also occurs in a variety of diseases, particularly those of the liver, which manifests as a decrease in the ratio of albumin to globulins (decreased albumin-globulin ratio). The synthesis of albumin decreases relatively early in conditions of protein malnutrition, such as **kwashiorkor.**

THE LEVELS OF CERTAIN PLASMA PROTEINS INCREASE DURING INFLAMMATION OR FOLLOWING TISSUE DAMAGE

Table 52–3 summarizes the functions of many of the plasma proteins. **Acute-phase proteins** such as **C-reactive protein** (CRP, so named because it reacts with the C polysaccharide of pneumococci), α_1-antiproteinase, haptoglobin, α_1-acid

TABLE 52–3 Some Functions of Plasma Proteins

Function	Plasma Proteins
Antiproteases	Antichymotrypsin
	α_1-Antitrypsin (α_1-antiproteinase)
	α_2-Macroglobulin
	Antithrombin
Blood clotting	Various coagulation factors, fibrinogen
Enzymes	Function in blood, for example, coagulation factors, cholinesterase
	Leakage from cells or tissues, eg, aminotransferases
Hormones	Erythropoietin[a]
Immune defense	Immunoglobulins, complement proteins, and β_2-macroglobulin
Involvement in inflammatory responses	Acute phase response proteins (eg, C-reactive protein, α_1-acid glycoprotein [orosomucoid])
Oncofetal	α_1-Fetoprotein (AFP)
Transport or binding proteins	Albumin (various ligands, including bilirubin, free fatty acids, ions [Ca²⁺], metals [eg, Cu²⁺, Zn²⁺], metheme, steroids, other hormones, and a variety of drugs)
	Corticosteroid-binding globulin (transcortin) (binds cortisol)
	Haptoglobin (binds extracorpuscular hemoglobin)
	Lipoproteins (chylomicrons, VLDL, LDL, HDL)
	Hemopexin (binds heme)
	Retinol-binding protein (binds retinol)
	Sex-hormone-binding globulin (binds testosterone, estradiol)
	Thyroid-binding globulin (binds T_4, T_3)
	Transferrin (transport iron)
	Transthyretin (formerly prealbumin; binds T_4 and forms a complex, with retinol-binding protein)

[a]Various other protein hormones circulate in the blood but are not usually designated as plasma proteins. Similarly, ferritin is also found in plasma in small amounts, but it too is not usually characterized as a plasma protein.

glycoprotein, and fibrinogen are believed to play a role in the body's response to inflammation. For example, C-reactive protein stimulates the complement pathway (see later), while α_1-antitrypsin neutralizes certain of the proteases released during acute inflammation.

During chronic inflammatory states and in patients with cancer the levels of acute-phase proteins may increase from 1.5 to as much as 1000-fold (in the case of CRP). Hence, CRP is used as a biomarker of tissue injury, infection, and inflammation. **Interleukin 1 (IL-1)**, a polypeptide released from mononuclear phagocytic cells, is the principal stimulator of acute-phase protein synthesis by hepatocytes. However, additional molecules such as IL-6 also participate.

The small proteins that facilitate cell–cell communication between the components of the immune system, such as interferons, ILs, and tumor necrosis factors, are called **cytokines**. Cytokines can be both autocrine and paracrine in nature. One of the primary targets of IL-1 and IL-6 is **nuclear factor kappa-B (NFκB)**, a transcription factor that regulates the expression of the genes encoding many cytokines, chemokines, growth factors, and cell adhesion molecules. NFκB, a heterodimer comprised of a 50- and a 65-kDa polypeptide, normally resides in the cytosol as an inactive complex with a second protein, NFκB inhibitor-α, also known as **IκBα**. Inflammation, injury, or radiation stimulates the phosphorylation of IκBα, which in turn targets it for ubiquitination and degradation. Once freed from its inhibitory partner, active NFκB translocates to the nucleus where it stimulates transcription of its target genes.

HAPTOGLOBIN PROTECTS THE KIDNEYS

Iron in Senescent Erythrocytes Is Recycled by Macrophages

Erythrocytes normally have a half-life of approximately 28 days, a turnover rate that requires the catabolism of ≈200 billion erythrocytes per day. Senescent or damaged erythrocytes are phagocytosed in the spleen and liver by macrophages of the reticuloendothelial system (RES). Within these macrophages, hemoglobin-derived heme groups are oxidized to biliverdin by the enzyme **heme oxygenase** (see Figure 31–13), releasing carbon monoxide and iron. The liberated iron is exported from phagocytic vesicles in the macrophage by **NRAMP 1** (natural resistance–associated macrophage protein 1), a transporter homologous to DMT1. Iron is subsequently secreted into the circulation by the transmembrane protein ferroportin (**Figure 52–3**). Thus, ferroportin plays a central role in both iron absorption by the intestine and iron secretion from macrophages.

In the blood, Fe^{2+} is oxidized to Fe^{3+} in a reaction catalyzed by the ferrioxidase **ceruloplasmin** (see later), a copper-containing plasma enzyme synthesized by liver. Once oxidized, Fe^{3+} is then bound by transferrin in blood. The iron released from macrophages in this way (about 25 mg/d) is recycled, thereby reducing the need for intestinal iron absorption, which averages only 1 to 2 mg/d.

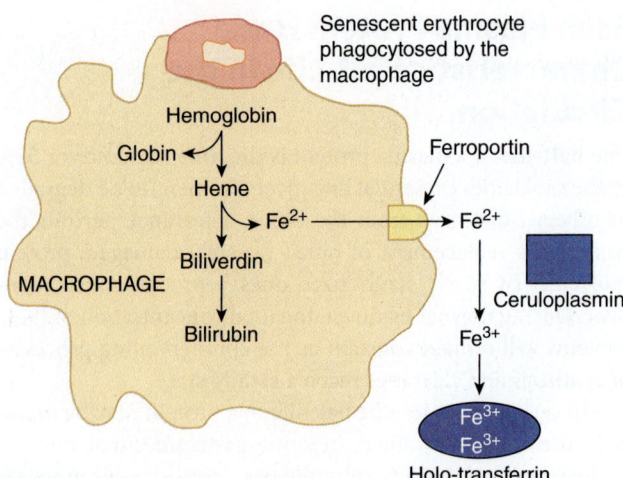

FIGURE 52–3   Recycling of iron in macrophages. Senescent erythrocytes are phagocytosed by macrophages. Hemoglobin is degraded and iron is released from heme by the action of the enzyme heme oxygenase. Ferrous iron is then transported out of the macrophage via ferroportin (Fp). In the plasma, it is oxidized to the ferric form by ceruloplasmin before binding to transferrin (Tf). Iron circulates in blood tightly bound to Tf.

Haptoglobin Scavenges Hemoglobin That Has Escaped Recycling

During the course of red blood cell turnover, approximately 10% of an erythrocyte's hemoglobin escapes into the circulation. This free, **extracorpuscular** hemoglobin is sufficiently small at ≈65 kDa to pass through the glomerulus of the kidney into the tubules, where it tends to form damaging precipitates. **Haptoglobin** (Hp) is a plasma glycoprotein that binds extracorpuscular hemoglobin (Hb), forming a tight noncovalent complex (Hb-Hp). The large size of the resulting Hb-Hp complex (≥ 155 kDa) prevents its passage through the glomerulus, thereby protecting the kidney from the formation of harmful precipitates and reducing the loss of the iron associated with extracorpuscular hemoglobin. Normally, a deciliter of human plasma contains sufficient haptoglobin to bind 40 to 180 mg of hemoglobin.

Human haptoglobin exists in **three polymorphic forms**, known as Hp 1-1, Hp 2-1, and Hp 2-2, that reflect the patterns of inheritance of two genes, designated Hp^1 and Hp^2. Homozygotes synthesize Hp 1-1 or Hp 2-2, respectively, while Hp 2-1 is synthesized by heterozygotes. Other plasma proteins bind free **heme** rather than hemoglobin. They include hemopexin and **albumin**, the latter of which binds metheme (ferric heme) to form methemalbumin. Methemalbumin subsequently transfers this metheme to hemopexin.

Haptoglobin Can Serve as a Diagnostic Indicator

In situations where hemoglobin is constantly being released from red blood cells, such as occurs in hemolytic anemias, the level of haptoglobin can fall dramatically. This decrease reflects the marked difference in the half-lives of free haptoglobin, approximately 5 days, and the Hb-Hp complex, approximately 90 minutes. In some cancer patients the level of **haptoglobin-related protein**, a homologue of haptoglobin also present in plasma, becomes elevated. However, its significance is not yet understood.

IRON IS STRICTLY CONSERVED

Iron is a key constituent of many human proteins, including hemoglobin, myoglobin, the cytochrome P450 group of enzymes, numerous components of the electron transport chain, and ribonucleotide reductase, which catalyzes the conversion of ribonucleotides into deoxyribonucleotides. Body iron, which is distributed as shown in **Table 52–4**, is highly conserved. A healthy adult loses only about 1 to 1.5 mg (< 0.05%) of their 3 to 4 g of body iron each day. However, an adult premenopausal female can experience iron deficiency due to blood loss during menstruation.

Enterocytes can absorb dietary iron in its free, ferrous Fe^{2+}, form, or as heme. Absorption of non-heme iron by enterocytes of the proximal duodenum is a highly regulated process (**Figure 52–4**). The transfer of iron across the apical membrane of the enterocytes is mediated via the **divalent metal transporter 1 (DMT1** or **SLC11A2)**, a transporter that also conveys Mn^{2+}, Co^{2+}, Zn^{2+}, Cu^{2+}, and Pb^{2+}. Since DMT1

TABLE 52–4 Distribution of Iron in a 70-kg Adult Male[a]

Transferrin	3-4 mg
Hemoglobin in red blood cells	2500 mg
In myoglobin and various enzymes	300 mg
In stores (ferritin)	1000 mg
Absorption	1 mg/d
Losses	1 mg/d

[a]In an adult female of similar weight, the amount in stores would generally be less (100-400 mg) and the losses would be greater (1.5-2 mg/d).

is specific for divalent metal ions, free ferric iron (Fe^{3+}) must be converted to its ferrous form (Fe^{2+}) by ingested reducing agents such as vitamin C, or enzymatically by a brush border membrane-bound ferrireductase, **duodenal cytochrome b (Dcytb)**. Once absorbed, heme-bound iron is released by the enzymatic action of heme oxygenase (see Chapter 31).

On entering the enterocytes, iron can either be stored bound to the iron-storage protein **ferritin** or transferred across the basolateral membrane by the iron exporter protein **ferroportin**, also known as **iron-regulated protein 1 (IREG1** or **SLC40A1)**. In the plasma, iron is transported in the Fe^{3+} form bound to the transport protein, **transferrin**. **Hephaestin**, a

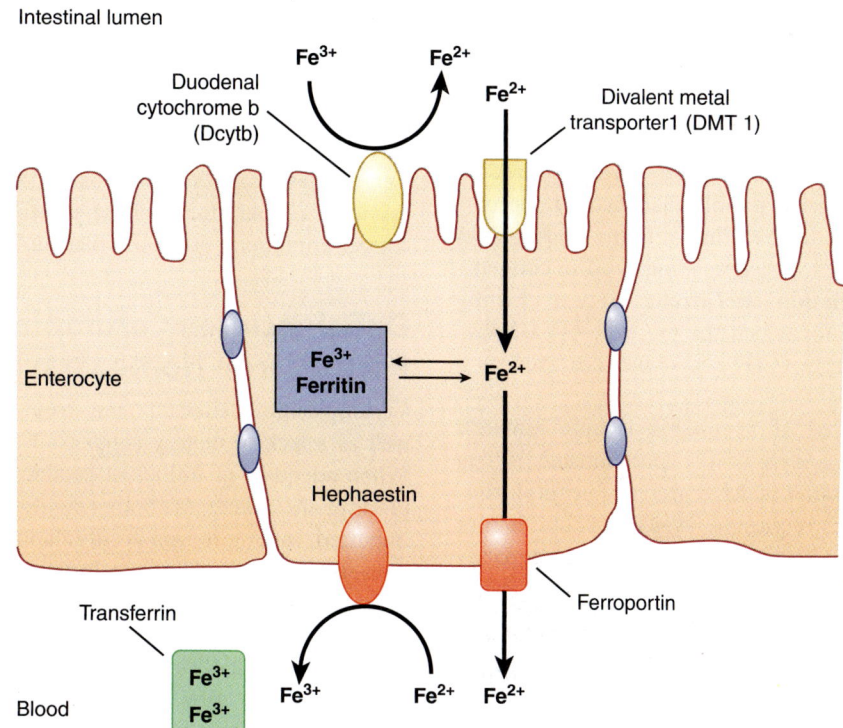

FIGURE 52–4 Non-heme iron transport in enterocytes. Ferric iron is reduced to the ferrous form by a luminal ferrireductase, duodenal cytochrome b (Dcytb). Ferrous iron is transported into the enterocyte via divalent metal transporter1 (DMT1). Within the enterocyte, iron is either stored as ferritin, or transported out of the cell by ferroportin (Fp). Ferrous iron is oxidized to its ferric form by hephaestin. The ferric iron is then bound by transferrin for transport by the blood to various sites in the body.

copper-containing ferroxidase homologous to ceruloplasmin, oxidizes Fe^{2+} to Fe^{3+} prior to export. Any excess ferritin-bound iron remaining in the enterocytes is disposed of when they are sloughed off into the gut lumen.

Ferritin Can Bind Thousands of Fe^{3+} Atoms

The human body can typically store up to 1 g of iron, the vast majority of which is bound to **ferritin**. Ferritin (MW 440 kDa) forms a hollow ball composed of two-dozen ≈19 to 21 kDa polypeptide subunits that can encapsulate as many as 3000 to 4500 ferric atoms. Ferritin subunits may be of the H (heavy) or the L (light) type. The H-subunit possesses ferroxidase activity, which is required for iron-loading of ferritin. The L-subunit is proposed to play a role in ferritin nucleation and stabilization. Normally, a small amount of ferritin is present in human plasma (50-200 µg/dL) proportionate to the total stores of iron in the body. While it is not known whether the ferritin in plasma is derived from damaged cells or secretion from healthy ones, plasma ferritin levels provide a useful indicator of body iron stores. **Hemosiderin**, a partly degraded form of ferritin, also may appear in tissues under conditions of iron overload (**hemosiderosis**).

Transferrin Shuttles Iron to Where It Is Needed

The toxicity of free iron is largely a consequence of its ability to induce the formation of damaging reactive oxygen species (**Figure 52–5**). Living organisms protect themselves against iron's potential toxicity by employing specialized storage and transport proteins, and transporting iron in its less reactive, Fe^{3+}, state. In humans, Fe^{3+} is transported through the circulation bound to **transferrin (Tf)**, a glycoprotein synthesized by the liver. This β_1-globulin has a molecular mass of approximately 76 kDa and contains two high-affinity binding sites for Fe^{3+}. Glycosylation of transferrin is impaired in **congenital disorders of glycosylation** (see Chapter 46) or in **chronic alcoholism**, a condition for which the presence of **carbohydrate-deficient transferrin (CDT)** is sometimes used as a biomarker.

The concentration of Tf in plasma is approximately 300 mg/dL, sufficient to carry a total of approximately 300 µg of iron per deciliter of plasma, which represents the **total iron-binding capacity (TIBC)** of plasma. Typically, about 30% of the iron-binding sites in transferrin are occupied. Occupancy

$$Fe^{2+} + H_2O_2 \longrightarrow Fe^{3+} + OH^{\cdot} + OH^-$$

FIGURE 52–5   The Fenton reaction. Free iron is extremely toxic as it can catalyze the formation of hydroxyl radical ($OH^{\cdot}$) from hydrogen peroxide (see also Chapter 53). The hydroxyl radical is a transient but highly reactive species and that oxidize cellular macromolecules resulting in tissue damage.

can decrease to less than 16% during severe iron deficiency and may increase to more than 45% in iron overload conditions.

The Transferrin Cycle Facilitates Cellular Uptake of Iron

For the delivery of transported iron, the recipient cell must bind circulating transferrin via a cell surface receptor, the **transferrin receptor 1 (TfR1)**. The receptor-transferrin complex is then internalized by **receptor-mediated endocytosis** (see Chapter 25) into late endosomes. The bound iron is then released as late endosomes become acidified and is transported to the cytoplasm by DMT1. Still bound to the transferrin receptor, the iron-free or **apo**transferrin receptor (apoTf) is returned by the endosome to the cell surface, where it dissociates from the receptor and reenters the plasma ready to bind more iron. This recycling process is called the **transferrin cycle (Figure 52–6)**.

While TfR1 is found on the surface of most cells, the homologous **transferrin receptor 2 (TfR2)** is encountered primarily on the surface of hepatocytes and the crypt cells of the small intestine. The affinity of TfR2 for transferrin is much lower than that of TfR1, optimizing the former as a sensor, rather than an importer, for iron.

Oxidation by Ceruloplasmin Is a Key Feature of the Iron Cycle

Macrophages play a key role in the turnover of red blood cells. Following phagocytosis and digestion via lysosomal hydrolases, the free iron is expelled largely in the ferrous, Fe^{2+}, state. In order to be recovered via the transferrin cycle, this iron must be oxidized to the ferric, Fe^{3+}, state by the ferroxidase **ceruloplasmin**, a 160-kDa α_2-globulin synthesized in the liver. Ceruloplasm carries out this oxidation using six, catalytically essential, copper atoms whose presence renders it the major copper-containing protein in plasma.

Deficiencies in Ceruloplasmin Perturb Iron Homeostasis

Ceruloplasmin deficiency can arise from genetic causes as well as a lack of dietary copper, an essential micronutrient. When adequate quantities of catalytically functional ceruloplasmin are lacking, the body's ability to recycle Fe^{2+} becomes impaired, leading to iron accumulation in tissues. While persons suffering from **hypoceruloplasmenia**, a genetically heritable condition in which ceruloplasmin levels are roughly 50% of normal, generally display no clinical abnormalities, genetic mutations that abolish the ferroxidase activity of ceruloplasmin, **aceruloplasminemia**, can have severe physiologic consequences. If left untreated, the progressive accumulation of iron in pancreatic islet cells and basal ganglia eventually leads to the development of insulin-dependent diabetes and neurologic degeneration that may manifest as dementia, dysarthria, and dystonia.

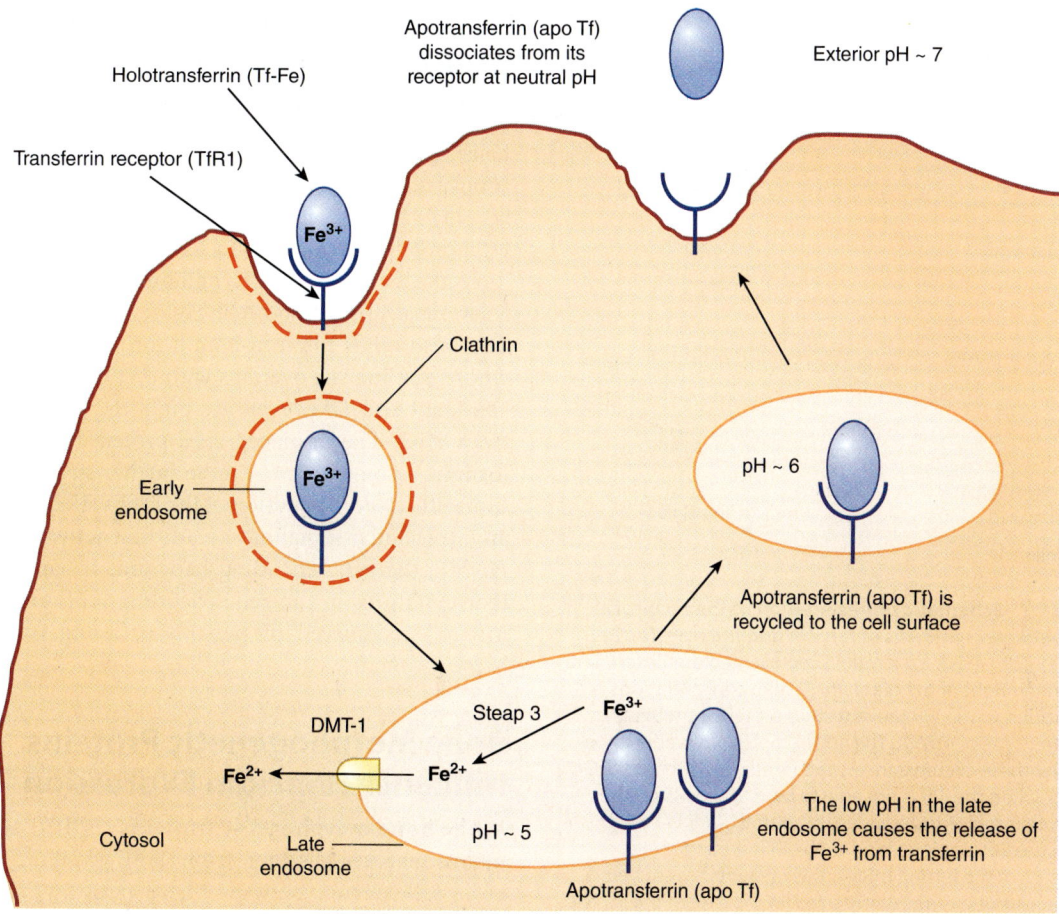

FIGURE 52–6 **The transferrin cycle.** Holotransferrin (Tf-Fe) binds to transferrin receptor 1 (TfR1) present in clathrin-coated pits on the cell surface. The TfR1–Tf-Fe complex is endocytosed and the endocytic vesicles fuse to form early endosomes. The early endosomes mature in to late endosomes, which have a low internal pH. These acidic conditions cause the release of iron from transferrin. The resulting apotransferrin (apoTf) remains bound to TfR1. Ferric iron is converted to its ferrous form by the ferrireductase, Steap 3, and is then transported to the cytosol via DMT1. The TfR1-apoTf complex then is recycled back to the cell surface. At the cell surface, apoTf is released from TFR1. TfR1 then binds to new Tf-Fe. This completes the transferrin cycle.

Ceruloplasmin Levels Decrease in Wilson Disease

In **Wilson disease**, a mutation in the gene for a **copper-binding P-type ATPase** (ATP7B protein) blocks the excretion of excess copper in the bile. As a consequence, copper accumulates in the liver, brain, kidney, and red blood cells. Paradoxically, rising levels of copper within the liver apparently interferes with the incorporation of this metal into newly synthesized ceruloplasmin polypeptides (apoceruloplasm) leading to a fall in plasma ceruloplasmin levels. If left untreated, patients suffering from this form of **copper toxicosis** may develop hemolytic anemia or chronic liver disease (cirrhosis and hepatitis), while the accumulation of copper in the basal ganglia can lead to neurologic symptoms. Wilson disease can be treated by limiting the dietary intake of copper while concurrently depleting any excess copper already present by the regular administration of **penicillamine**, a copper chelating agent that is excreted in the urine.

INTRACELLULAR IRON HOMEOSTASIS IS TIGHTLY REGULATED

Synthesis of TfR1 & Ferritin Are Reciprocally Regulated

Changes in intracellular iron levels influence the synthesis of both TfR1 and ferritin. When iron is low, the rate of TfR1 synthesis increases while that of ferritin declines. The opposite occurs when iron is abundant and tissue needs have been sated. Control is exerted through the binding of Fe^{2+} to **iron regulatory proteins (IRPs)** **1** and **2**, cytoplasmic isoforms of the tricarboxylic TCA enzyme aconitase, by hairpin loops structures called **iron response elements (IREs).** The IREs are located in the 5′ and 3′ untranslated regions (UTRs) of the mRNAs coding for ferritin and TfR1, respectively (**Figure 52–7**). Binding of iron-free IRP at the 3′ UTR of the mRNA for TfR1 stabilizes it,

(A) Ferritin mRNA

(i) ↑Fe : IRP does not bind to IRE

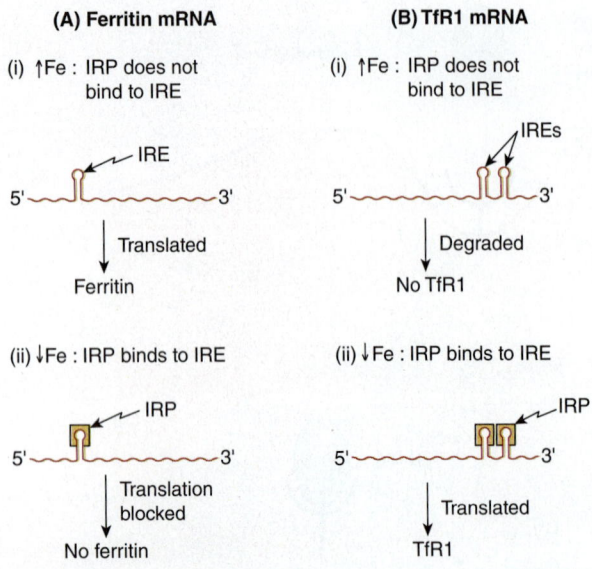

(B) TfR1 mRNA

(i) ↑Fe : IRP does not bind to IRE

FIGURE 52–7 **Schematic representation of the reciprocal relationship between synthesis of ferritin and the transferrin receptor (TfR1).** The mRNA for ferritin is represented on the left, and that for TfR1 on the right of the diagram. At high concentrations of iron, the iron bound to the IRP prevents that protein from binding the IREs on either type of mRNA. The mRNA for ferritin is able to be translated under these circumstances, and ferritin is synthesized. On the other hand, when the IRP is not able to bind to the IRE on the mRNA for TfR1, that mRNA is degraded. In contrast, at low concentrations of iron the IRP is able to bind to the IREs on both types of mRNA. In the case of the ferritin mRNA, this prevents it from being translated. Hence ferritin is not synthesized. In the case of the mRNA for TfR1, binding of the IRP prevents that mRNA from being degraded, enabling it to be translated and TfR1 to be synthesized. IRE, iron response element; IRP, iron regulatory protein.

thereby increasing TfR1 synthesis, while binding of an IRP to the IRE located at the 5′ UTR of ferritin mRNA blocks translation. When iron levels rise, Fe^{2+} binds to IRP, completing the assembly of a 4Fe-4S cluster that triggers dissociation of the protein from these DNA hairpins. Once free of IRP, the mRNA encoding ferritin is available to undergo translation. Concurrently, in its IRP-free state the mRNA encoding TfR1 is subject rapid degradation, thereby slowing TfR1 synthesis.

Hepcidin Is the Chief Regulator of Systemic Iron Homeostasis

The 25-amino acid peptide **hepcidin** plays a central role in iron homeostasis. Synthesized in the liver as an 84-amino acid precursor (prohepcidin), hepcidin binds to the cellular iron exporter, **ferroportin**, triggering the latter's internalization and degradation. The consequent decrease in ferroportin produces a "mucosal block" that lowers iron absorption in the intestine and depresses recycling of the iron liberated as red blood cells turnover (**Figure 52–8**). Together, these result in a reduction in circulating iron levels (hypoferremia) as well as reduced placental iron transfer during pregnancy. When plasma iron levels are high, synthesis of hepcidin by the liver increases, reducing both iron absorption and recycling.

Hepcidin Expression Is Influenced by Iron, Erythropoiesis, Inflammation, & Hypoxia

Liver cells monitor iron levels using one of two "iron-sensing complexes" consisting of a homodimer of either **TfR1** and **TfR2** bound to a third transmembrane protein, **HFE** (**Figure 52–9**). HFE protein is a major histocompatibility (MHC) class 1–like molecule that binds β_2-**microglobulin** (a component of class I MHC molecules, not shown in Figure 52–9) and, normally, TfR1. TfR1 also binds the iron-bound form of transferrin (Tf-Fe), whose binding site overlaps with that for HFE. When iron is abundant and TfFe levels are high, the latter displaces HFE from TfR1. The displaced HFE protein then binds to TfR2, forming a complex that can be further stabilized by association with Tf-Fe. Formation of the HFE-TfR2 complex triggers an intracellular signaling cascade that activates expression of *HAMP*, the gene encoding hepcidin. It has been observed that mutations in the gene encoding HFE protein are commonly encountered in persons suffering from hereditary hemochromatosis.

Bone Morphogenetic Proteins Influence Hepcidin Expression

While **bone morphogenic proteins** (BMPs) act by mechanisms that are distinct from HFE protein, considerable cross-talk occurs between these pathways. For example, the binding affinity of the cell surface receptor for BMP (BMPR) is augmented when BMPR is associated with the coreceptor, **hemojuvelin** (HJV). The activation of the BMPR-HJV complex triggers the phosphorylation of intracellular signaling proteins called **SMADs**, which stimulates the transcription of the gene that codes for hepcidin (see Figure 52–9).

Inflammatory & Erythropoietic Signals Regulate Hepcidin Levels

During an inflammatory response, synthesis of hepcidin is induced by small secreted proteins called cytokines. Binding of cytokines such as **interleukin 6 (IL-6)** to their cell surface receptor activates expression of the gene encoding hepcidin via the JAK-STAT (Janus kinase—signal transducer and activator of transcription) pathway (see Figure 52–9). Inflammation-associated cytokines also are thought to trigger the increase in hepcidin levels that accompanies **anemia of inflammation (AI)**. AI manifests as a microcytic, hypochromic anemia that is refractory to iron supplementation.

Hepcidin expression decreases during hypoxia or β-thalassemia. The former is mediated by erythropoietin, whose synthesis is controlled by hypoxia-inducible transcription factors 1 and 2 (HIF-1 and HIF-2). In β-thalassemia, the expression of hepcidin is inhibited by **growth differentiation factor 15 (GDF15)** and **twisted gastrulation 1 (TWSG1)**, which are secreted by erythroblasts.

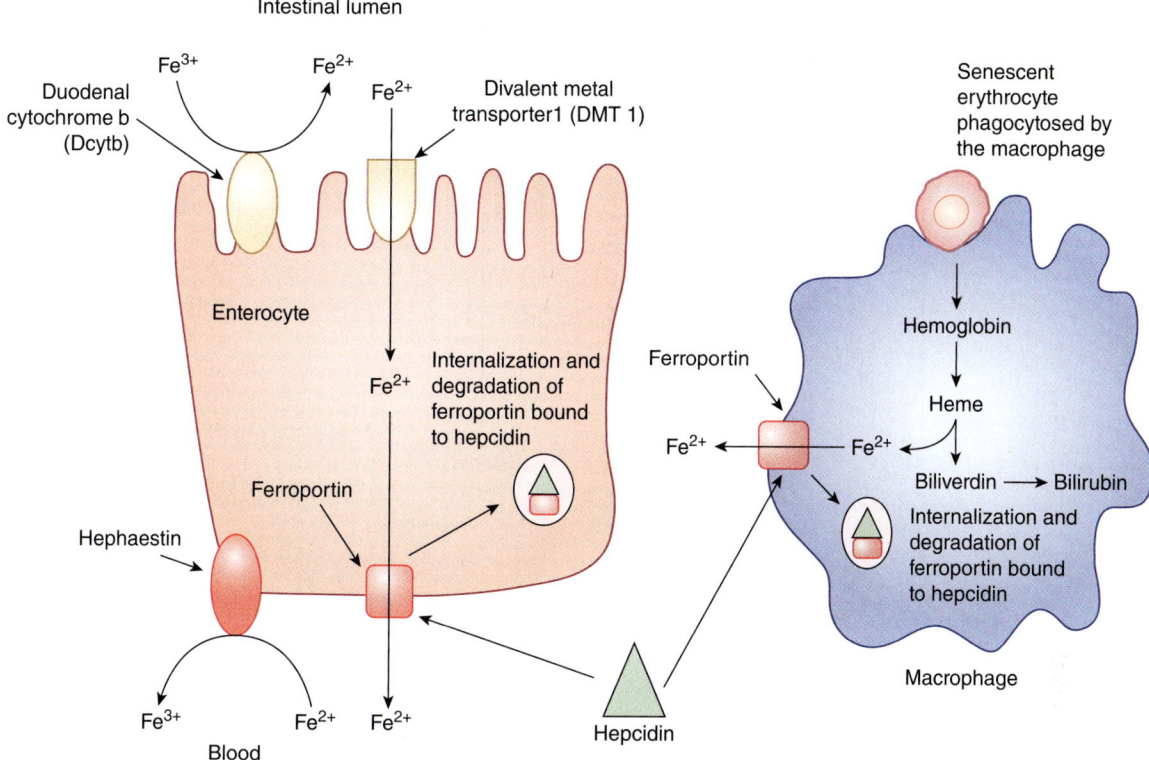

FIGURE 52–8 **Role of hepcidin in systemic iron regulation.** Hepcidin binds to and triggers the internalization and degradation of ferroportin expressed on the surface of enterocytes and macrophages. This decreases iron absorption from the intestine and inhibits iron release from macrophages, leading to hypoferremia.

IRON DEFICIENCY & ANEMIA ARE COMMON WORLDWIDE

Iron deficiency is extremely common in many parts of the world, especially in developing countries. Major causes of iron insufficiency include dietary deficiency, malabsorption, gastrointestinal bleeding, and episodic blood loss such as from menstruation. Persistent iron deficiency can lead to the progressive depletion of bodily iron stores. If the level of transferrin saturation falls to 20% or below, hemoglobin synthesis will be impaired, resulting in **iron-deficient erythropoiesis**. Over time hemoglobin levels in blood will gradually fall, resulting in **iron-deficiency anemia**, which manifests as a **hypochromic, microcytic blood picture** that is accompanied by fatigue, pallor, and reduced exercise capacity.

During iron deficient anemia, erythrocytes exhibit increased levels of surface transferrin receptor-1 as well as deficits in the ferrochelatase-catalyzed incorporation of iron into protoporphyrin IX. Partial proteolysis of cell surface transferrin receptors releases increased quantities of **soluble transferrin receptor (sTfR) protein** into the plasma. This increase, along with the accumulation of **red-cell protoporphyrin,** serve as diagnostic biomarkers for iron deficiency anemia since chronic inflammation does not affect the level of erythrocyte transferrin receptors, and hence sTfR. **Table 52–5**

lists several clinical biomarkers utilized to detect and monitor the progression of iron deficiency anemia.

Hereditary Hemochromatosis Is Characterized by Iron Overload

Hemochromatosis or iron overload, is characterized by **hemisoderosis,** the accumulation of stainable concentrations of iron in tissues. Hereditary hyperabsorption of iron by the intestines can be caused by mutations in the gene encoding homeostatic iron regulator protein (*HFE*) or, less frequently, hepcidin, TfR2, HJV, or ferroportin (**Table 52–6**). Secondary iron overload is usually associated with ineffective erythropoiesis, as seen in the thalassemia syndromes.

SERUM INHIBITORS PREVENT INDISCRIMINATE PROTEOLYSIS

Proteases are essential participants in tissue remodeling, blood clotting, elimination of old or diseased cells, destruction of invading pathogens, and other physiologic functions. Left unchecked, however, proteolytic enzymes released into the blood by secretion or tissue damage can harm healthy tissue. Protection from indiscriminate proteolysis involves a battery of serum proteins that inhibit, and thereby limit the scope of, protease action.

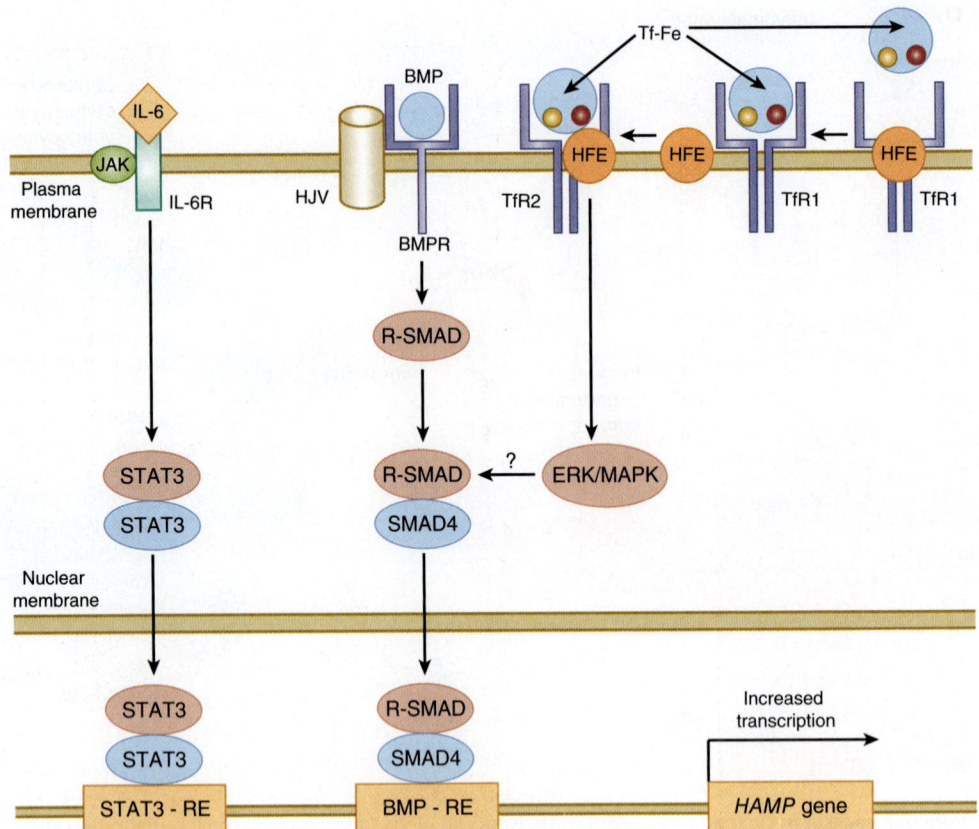

FIGURE 52–9 **Regulation of hepcidin gene expression.** Tf-Fe (holotransferrin) competes with HFE for binding to TFR1. High levels of Tf-Fe displace HFE from its binding site on TfR1. Displaced HFE binds to TfR2 along with Tf-Fe to signal via the ERK/MAPK pathway to induce hepcidin expression. BMP binds to its receptor BMPR and HJV (coreceptor) to activate R-SMAD. R-SMAD dimerizes with SMAD4, then translocates to the nucleus where it binds to the BMP-RE, resulting in transcriptional activation of hepcidin as shown. IL-6, which is a biomarker of inflammation, binds to its cell surface receptor and activates the JAK-STAT pathway. STAT3 translocates to the nucleus where it binds to its response element (STAT-RE) on the hepcidin gene to induce it. BMP, bone morphogenetic protein; BMPR, bone morphogenetic protein receptor; BMP-RE, BMP response element; ERK-MAPK, extracellular signal-regulated kinase/mitogen-activated protein kinase; *HAMP*, gene encoding hepcidin antimicrobial peptide (hepcidin); HJV, hemojuvelin; IL-6, interleukin 6; IL-6R, interleukin 6 receptor; JAK, Janus-associated kinase; SMAD, Sma and MAD (mothers against decapentaplegic)-related protein; STAT, signal transduction and activator of transcription; STAT3-RE, STAT 3 response element; TfR1, transferrin receptor 1; TfR2, transferrin receptor 2.

TABLE 52–5 **Changes in Various Laboratory Tests Used to Assess Iron-Deficiency Anemia**

Parameter	Normal	Negative Iron Balance	Iron-Deficient Erythropoiesis	Iron-Deficiency Anemia
Serum ferritin (μg/dL)	50-200	Decreased <20	Decreased <15	Decreased <15
Total iron binding capacity (TIBC) (μg/dL)	300-360	Slightly increased >360	Increased >380	Increased >400
Serum iron (μg/dL)	50-150	Normal	Decreased <50	Decreased <30
Transferrin saturation (%)	30-50	Normal	Decreased <20	Decreased <10
RBC protoporphyrin (μg/L)	30-50	Normal	Increased	Increased
Soluble transferrin receptor (μg/L)	4-9	Increased	Increased	Increased
RBC morphology	Normal	Normal	Normal	Microcytic Hypochromic

Data from Hillman RS, Finch CA: The Red Cell Manual, 7th ed. Philadelphia, PA: FA Davis and Co; 1996.

TABLE 52–6 Iron Overload Conditions

Hereditary Hemochromatosis
• HFE-related hemochromatosis (type 1)
• Non-HFE-related hemochromatosis
• H_o Juvenile hemochromatosis (type 2)
• Hepcidin mutation (type 2A)
• Hemojuvelin mutation (type 2B)
• H_o Transferrin receptor 2 mutation (type 3)
• H_o Ferroportin mutation (type 4)
Secondary Hemochromatosis
• Anemia Characterized by ineffective erythropoiesis (eg, thalassemia major)
• Repeated blood transfusions
• Parenteral iron therapy
• Dietary iron overload (Bantu siderosis)
Miscellaneous Conditions Associated with Iron Overload
• Alcoholic liver disease
• Nonalcoholic steatohepatitis
• Hepatitis C infection

Deficiency of α₁-Antiproteinase Is Associated With Emphysema & Liver Disease

α1-Antiproteinase, a 394-residue glycoprotein is the principal **serine protease inhibitor (serpin)** in human plasma. Formerly called α₁-antitrypsin, α₁-antiproteinase inactivates trypsin, elastase, and other serine proteases by forming a covalent complex with them. α₁-Antiproteinase, which constitutes >90% of the α₁-albumin fraction in plasma, is synthesized by hepatocytes and macrophages. At least 75 polymorphic forms of this serpin, or Pi, exist. The major genotype is MM, whose phenotypic product is PiMM. A deficiency in α₁-antiproteinase plays a role in some cases (~ 5%) of emphysema, particularly in subjects with the ZZ genotype (who synthesize PiZ) and in PiSZ heterozygotes, both of whom secrete lower levels of serpins than normal individuals.

Oxidation of Met₃₅₈ Inactivates α₁-Antiproteinase

In the lungs, oxidation of a key methionine residue, Met₃₅₈, located in the protease-binding domain renders α₁-antiproteinase unable to covalently bind and neutralize serine proteases. Individuals who use tobacco products or are regularly exposed to the smoke generated from burning fossil fuels are especially prone to oxidation of Met₃₅₈. Unchecked by this key inhibitor, proteolytic activity in the lungs can contribute to the development of emphysema, particularly for patients possessing low levels of α₁-antiproteinase (eg, PiZZ phenotype). Intravenous administration of serpins (augmentation therapy) has been used as an adjunct in the treatment of emphysema patients that exhibit α₁-antiproteinase deficiency.

Individuals deficient in α₁-antiproteinase are also at greater risk of lung damage from the accumulation of polymorphonuclear white blood cells in the lung that occurs during pneumonia and other respiratory infections. Deficiency of α₁-antiproteinase is also implicated in **α₁-antitrypsin deficiency liver disease**, a form of cirrhosis that afflicts persons possessing the ZZ phenotype. In these individuals, a mutation leading to the substitution of Glu₃₄₂ by lysine produces a form of α₁-antiproteinase that is prone to aggregation in the cisternae of the endoplasmic reticulum in hepatic cells.

α₂-Macroglobulin Neutralizes Proteases & Targets Cytokines to Tissues

α₂-Macroglobulin, a member of the thioester plasma protein family, comprises 8 to 10% of the total plasma protein in humans. This homotetrameric glycoprotein, which is synthesized by monocytes, hepatocytes, and astrocytes, is the most abundant member of a group of homologous plasma proteins that include complement proteins C3 and C4. α₂-Macroglobulin mediates the inhibition and clearance of a broad spectrum of truant proteases by a "Venus flytrap" mechanism that utilizes a 35-residue "bait domain" and an internal cyclic thioester linking a cysteine and a glutamine residue (**Figure 52–10**). Cleavage of the bait domain triggers a massive conformational change that results in the envelopment of the attacking protease. The reactive thioester then reacts with the protease to covalently link the two proteins. In addition, this conformational change exposes a sequence in α₂-macroglobulin that is recognized by the cell surface receptors responsible for binding and clearing the α₂-macroglobulin-protease complex from the plasma.

In addition to serving as the plasma's predominant broad-spectrum, or **panproteinase**, inhibitor, α₂-macroglobulin also binds to and transports approximately 10% of the **zinc** carried by plasma (the remainder being transported by albumin) as well as cytokines such as platelet-derived growth factor and transforming growth factor β. α₂-Macroglobulin routes these bound effectors toward particular tissues or cells. Once taken up, the complex dissociates, thereby freeing the cytokines to exert their modulatory effects on cell growth and function.

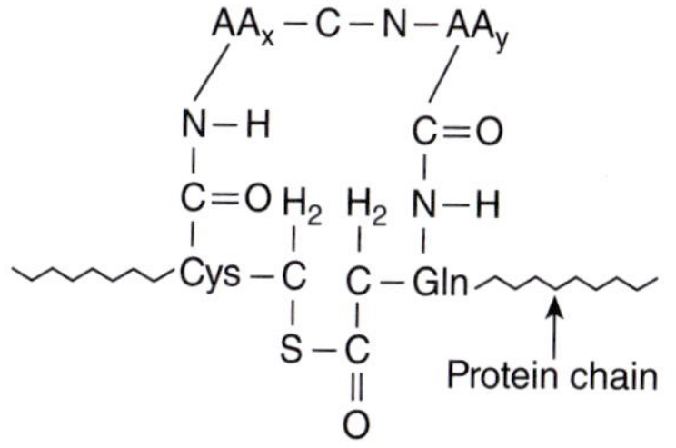

FIGURE 52–10 An internal cyclic thiol ester bond, as present in α₂-macroglobulin. AAₓ and AAᵧ are neighboring amino acids to cysteine and glutamine.

DEPOSITION OF PLASMA PROTEINS IN TISSUES LEADS TO AMYLOIDOSIS

Amyloidosis refers to an impairment of tissue function that results from the accumulation of insoluble protein aggregates in the interstitial spaces between cells. The term is a misnomer, as it was originally thought that the fibrils were starch-like in nature. Actually, the fibrils are made up primarily of proteolytic fragments from plasma proteins. The conformation of these fragment is extremely rich in β-pleated sheet. They generally also contain a **P component** derived from a C-reactive protein-like protein called serum amyloid P component.

Structural abnormalities or overproduction of more than 20 different proteins have been implicated in various types of amyloidosis. Primary amyloidosis (**Table 52–7**) typically is caused by a monoclonal plasma cell disorder that leads to the accumulation of protein fragments derived from immunoglobulin **light chains** (see later). **Secondary** amyloidosis results from an accumulation of fragments of serum amyloid A (SAA) associated with chronic infections or cancer. In these instances, elevated levels of inflammatory cytokines stimulate the liver to synthesize SAA, which leads to a concomitant rise in its proteolytic degradation products. **Familial amyloidosis** results from accumulation of mutated forms of certain plasma proteins such as **transthyretin** (see Table 52–3). Over 80 mutationally altered forms of this protein have been identified. In addition, because patients undergoing long-term hemodialysis are unable to clear β_2-**microglobulin**, which is retained by dialysis membranes, they are at greater risk of amyloidosis as the levels of this plasma protein increase in the blood.

PLASMA IMMUNOGLOBULINS DEFEND AGAINST INVADERS

The major humoral components of the body's immune system include **B lymphocytes (B cells)**, **T lymphocytes (T cells)**, and **the innate immune system**. B lymphocytes are mainly derived from bone marrow cells, while T lymphocytes originate from the thymus. B cells are responsible for the synthesis of circulating antibodies, also known as **immunoglobulins**, while **T cells** are involved in a variety of important cell-mediated immunologic processes. The latter include defense against malignant cells and many viruses as well as hypersensitivity reactions and graft rejection. B and T cells respond in an **adaptive** manner, developing a targeted response for each invader encountered. The **innate immune system** defends against infection in a nonspecific manner. It contains a variety of cells such as phagocytes, neutrophils, natural killer cells, and others that will be discussed in Chapter 54.

Immunoglobulins Are Comprised of Multiple Polypeptide Chains

Immunoglobulins are oligomeric glycoproteins composed of heavy (H) or light (L) chains, designations based on their rate of migration during SDS-polyacrylamide gel electrophoresis. Human immunoglobulins can be grouped into five classes abbreviated as IgA, IgD, IgE, IgG, and IgM (**Table 52–8**). The most abundant of the five, IgG, consists of two identical light chains (23 kDa) linked together by a network of disulfide bonds to each other as well as two identical heavy chains (53-75 kDa). The respective biologic functions of each class are summarized in **Table 52–9**.

Each class of immunoglobulin contains a different H chain isoform: α (IgA), δ (IgD), ε (IgE), γ (IgG), and μ (IgM). The γ-type H chains of IgG are organized into an amino-terminal variable region (V_H) and three **constant regions (C_H1, C_H2, C_H3)**. The μ and ε chains each have four C_H domains rather than the usual three. Some immunoglobulins such as IgG exist only as the basic tetramer, with the various chains arranged in the Y-shaped configuration shown in **Figure 52-11**. Others such as IgA and IgM can form higher oligomers comprised of two, three (IgA), or five (IgM) copies of the core tetrameric unit (**Figure 52-12**).

The IgG light chain can be subdivided into a C-terminal **constant region (C_L)** and amino-terminal **variable region** (V_L). There are two general types of light chains, **kappa (κ)** and **lambda (λ)**, which differ in the structure of their C_L regions. A given immunoglobulin molecule always contains either two κ or two λ light chains—never a mixture of a κ and a λ, with the former predominating in humans.

Each IgG molecule binds to its target molecules, or **antigens**, at a specific site, or **epitope** (sometimes called an antigenic determinant). Typical epitopes may consist of a short polysaccharide or amino acid sequence, or a specific three-dimensional arrangement of amino acids or sugars drawn from disparate parts of the antigen's primary sequence. Each IgG molecule contains two identical antigen binding sites, which are located near the tips of the Y, making this immunoglobulin bivalent—able to bind two antigen molecules simultaneously. These antigen binding sites are comprised of V_H and V_L domains arranged together as two antiparallel sheets.

Because the region between the C_H1 and C_H2 domains can be readily cleaved by pepsin or papain (see Figure 52–11), it is

TABLE 52–7 A Classification of Amyloidosis

Type	Protein Implicated
Primary	Principally light chains of immunoglobulins
Secondary	Serum amyloid A (SAA)
Familial	Transthyretin; also rarely apolipoprotein A-1, cystain C, fibrinogen, gelsolin, lysozyme
Alzheimer's disease	Amyloid β peptide (see Chapter 57, case no. 2)
Dialysis-related	β_2-microglobulin

Note: Proteins other than these listed have also been implicated in amyloidosis.

TABLE 52–8 Properties of Human Immunoglobulins

Property	IgG	IgA	IgM	IgD	IgE
Percentage of total immunoglobulin in serum (approximate)	75	15	9	0.2	0.004
Serum concentration (mg/dL) (approximate)	1000	200	120	3	0.05
Sedimentation coefficient	7S	7S or 11S[a]	19S	7S	8S
Molecular weight (x1000)	150	170 or 400[a]	900	180	190
Structure	Monomer	Monomer or dimer	Monomer or pentamer	Monomer	Monomer
H-chain symbol	γ	α	μ	δ	ε
Complement fixation	+	—	+	—	—
Transplacental passage	+	—	—	?	—
Mediation of allergic responses	—	—	—	—	+
Found in secretions	—	+	—	—	—
Opsonization	+	—	—[b]	—	—
Antigen receptor on B cell	—	—	+	?	—
Polymeric form contains J chain	—	+	+	—	—

[a]The 11S form is found in secretions (eg, saliva, milk, and tears) and fluids of the respiratory, intestinal, and genital tracts.
[b]IgM opsonizes indirectly by activating complement. This produces C3b, which is an opsonin.
Reproduced with permission from Levinson W, Jawetz E: *Medical Microbiology and Immunology*, 7th ed. New York, NY: McGraw Hill; 2002.

referred to as the hinge region. The hinge region confers **flexibility** to the Fab arms, which facilitates binding to separate antigen molecules. If IgGs recognizing more than one epitope are present, large antibody-antigen clusters can form which are readily recognized and disposed by phagocytic leukocytes. Cluster formation is often demonstrated in the laboratory by the formation of erythrocyte **rosettes**.

The Constant Regions Determine Class-Specific Effector Functions

The **constant regions** of the immunoglobulin molecules, particularly C_H2 and C_H3 (and C_H4 of IgM and IgE) on the Fc fragment, are responsible for the **class-specific effector functions** of the different immunoglobulin molecules (see

Table 52–9, bottom part), such as complement fixation or transplacental passage.

Hypervariable Regions Confer Binding Specificity

Within the variable regions of the L and H chains are a handful of **hypervariable regions**, which align together on the surface of the immunoglobulin as a projecting loop that serves as the antigen binding site. Each hypervariable region is made up of short (5-10 residue) islands interspersed within the relatively invariable **framework** or **complementarity-determining regions (CDRs)** (**Figure 52–13**).

The essence of antigen-antibody interactions is **mutual complementarity** between the surfaces of CDRs and epitopes

TABLE 52–9 Major Functions of Immunoglobulins

Immunoglobulin	Major Functions
IgG	Main antibody in the secondary response. Opsonizes bacteria, making them easier to phagocytose. Fixes complement, which enhances bacterial killing. Neutralizes bacterial toxins and viruses. Crosses the placenta.
IgA	Secretory IgA prevents attachment of bacteria and viruses to mucous membranes. Does not fix complement.
IgM	Produced in the primary response to an antigen. Fixes complement. Does not cross the placenta. Antigen receptor on the surface of B cells.
IgD	Found on the surfaces of B cells where it acts as a receptor for antigen.
IgE	Mediates immediate hypersensitivity by causing release of mediators from mast cells and basophils upon exposure to antigen (allergen). Defends against worm infections by causing release of enzymes from eosinophils. Does not fix complement. Main host defense against helminthic infections.

Reproduced with permission from Levinson W, Jawetz E: *Medical Microbiology and Immunology*, 7th ed. New York, NY: McGraw Hill; 2002.

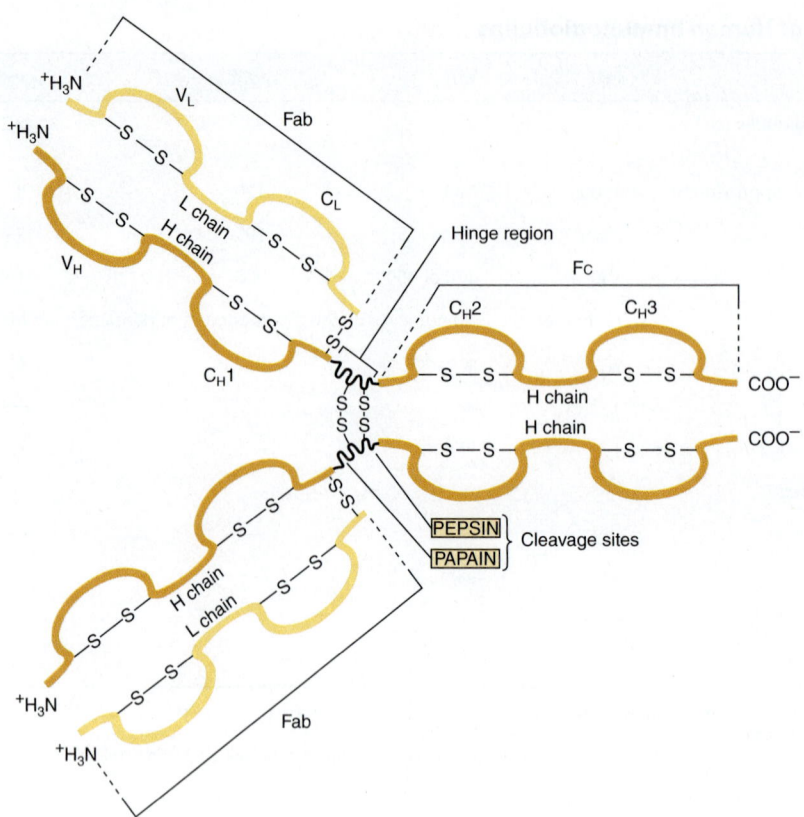

FIGURE 52–11   Structure of IgG. The molecule consists of two light (L) chains and two heavy (H) chains. Each light chain consists of a variable (V_L) and a constant (C_L) region. Each heavy chain consists of a variable region (V_H) and a constant region that is divided into three domains (C_H1, C_H2, and C_H3). The C_H2 domain contains the complement-binding site and the C_H3 domain contains a site that attaches to receptors on neutrophils and macrophages. The antigen-binding site is formed by the hypervariable regions of both the light and heavy chains, which are located in the variable regions of these chains (see Figure 52–13). The light and heavy chains are linked by disulfide bonds, and the heavy chains are also linked to each other by disulfide bonds. (Reproduced with permission from Parslow TG, Stites DP, Terr AI, et al: *Medical Immunology*, 10th ed. New York, NY: McGraw Hill; 2001.)

that involve multiple **noncovalent** interactions such as hydrogen bonding, salt bridges, hydrophobic interactions, and van der Waals forces (see Chapter 2). The exceptional binding affinity and specificity of immunoglobulins for their target antigens derives from the ability of activated B cells to uniquely configure the **hypervariable regions** of the immunoglobulin light and heavy chains to complement a specific epitope.

Antibody Diversity Depends on Gene Rearrangements

The human genome contains fewer than 150 immunoglobulin genes. Nevertheless, each person is capable of synthesizing perhaps 1 million different antibodies, each specific for a unique antigen. Clearly, immunoglobulin expression does not follow the "one gene, one protein" paradigm. Instead, immunoglobulin diversity is generated by **combinatorial mechanisms** based on mixing and rearranging a finite pool of genetic information in multiple ways (see Chapters 35 and 38).

Antibody diversity arises, in part, from the distribution of the coding sequence for each immunoglobulin chain among multiple genes. Each light chain is the product of three or more separate structural genes that code for the **variable region** (V_L), **joining region** (**J**) (bearing no relationship to the J chain of IgA or IgM), and **constant region** (C_L). Similarly, each heavy chain is the product of at least **four** different genes that code for a **variable region** (V_H), a **diversity region** (**D**), a **joining region** (**J**), and a **constant region** (C_H). Each gene is present in the human genome in several versions that can be assembled in a multiplicity of combinations.

Diversity is further augmented through the action of the **activation-induced cytidine deaminase** (AID). By catalyzing the conversion of cytidine to uracil, AID massively increases the frequency of mutation of immunoglobulin *V* genes. AID-generated mutations are **somatic** in nature, unique to the differentiated cell where they occurred rather than to a germline cell. Consequently, activation of AID can generate unique subpopulations of B cells that harbor distinct mutations of their *V* genes, causing each to synthesize immunoglobulins of differing antigen specificity. In some pathologic states, the mutagenic action of AID can lead to the generation of **autoantibodies** that target the body's endogenous components, a phenomenon termed **autoimmunity**.

A third mechanism for synthesizing antibodies that target novel antigens is **junctional diversity**. This refers to the addition or deletion of random numbers of nucleotides that takes

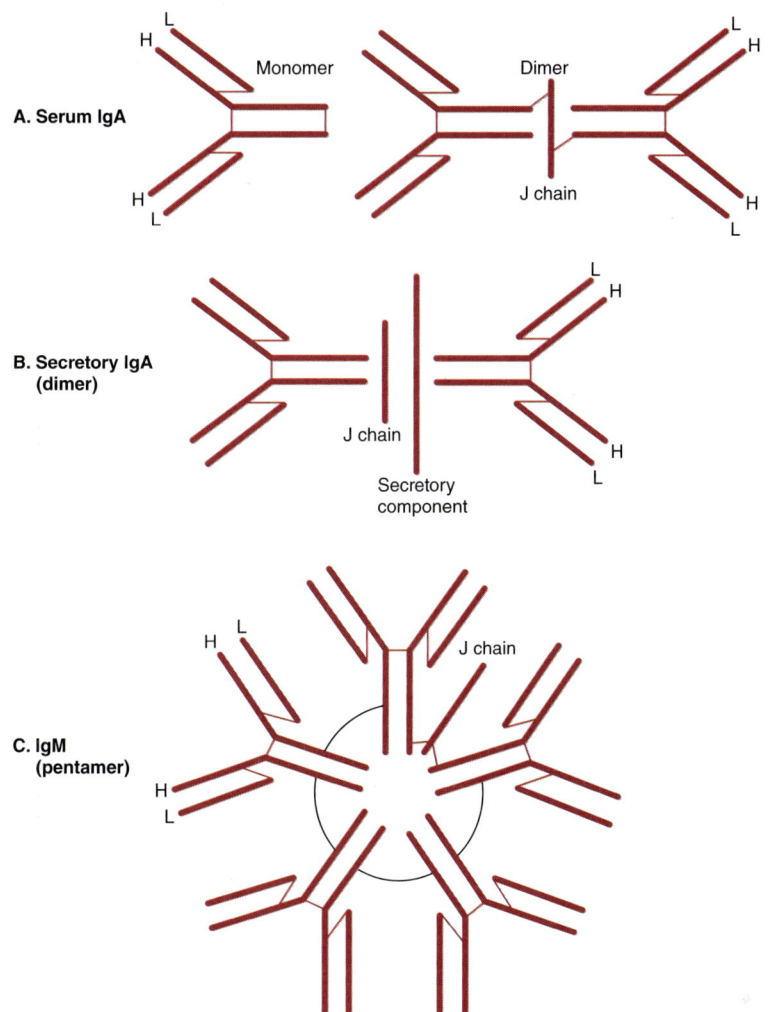

FIGURE 52–12 **Schematic representation of serum IgA, secretory IgA, and IgM.** Both IgA and IgM have a J chain, but only secretory IgA has a secretory component. Polypeptide chains are represented by thick lines; disulfide bonds linking different polypeptide chains are represented by thin lines. (Reproduced with permission from Parslow TG, Stites DP, Terr AI, et al: *Medical Immunology*, 10th ed. New York, NY: McGraw Hill; 2001.)

place when certain gene segments are joined together. As is the case with AID, the mutations generated by junctional diversity are somatic in nature.

Class (Isotype) Switching Occurs During Immune Responses

In most humoral immune responses, antibodies of different classes are generated that target the same epitope. Each class appears in a specific chronologic order following exposure to an immunogen (immunizing antigen). For instance, the appearance of antibodies of the IgM class normally precedes that of the IgG class. The transition from the synthesis of one class to another is called **class** or **isotype switching**. Switching involves combining a given immunoglobulin light chain with different heavy chains. Whereas a newly synthesized light chain will initially be mated with a μ chain to generate a specific IgM molecule, over time the same antigen-specific light chain will be mated with a γ chain to generate an IgG whose V_H region

and, consequently, antigen specificity will be identical to that of the μ chain of the preceding IgM molecule. Combining this light chain with an α heavy chain, in turn, forms an IgA molecule with identical antigen specificity. Immunoglobulin molecules that possess identical hypervariable and variable domains (and epitope specificity) are said to share a common **idiotype**.

Monoclonal Antibodies Are an Important Research Tool

Antibodies have emerged as a major tool in biomedical research, diagnosis, and treatment. Originally, the production of antibodies against a selected antigen required that the antigen be injected into a host animal, such as a rabbit or goat, and serum containing plasma immunoglobulins that included (hopefully) antibodies against the antigen of interest obtained. When an antigen is injected into an animal, a mixture of B cells is induced to synthesize antibodies directed against epitopes on the antigen. The population of antibodies

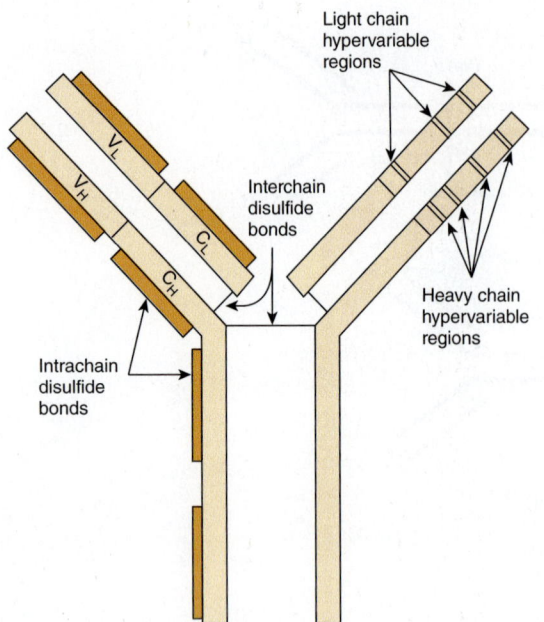

FIGURE 52–13 **Schematic model of an IgG molecule showing approximate positions of the hypervariable regions in heavy and light chains.** The antigen-binding site is formed by these hypervariable regions. The hypervariable regions are also called complementarity-determining regions (CDRs). (Reproduced with permission from Parslow TG, Stites DP, Terr AI, et al: *Medical Immunology*, 10th ed. New York, NY: McGraw Hill; 2001.)

produced is heterogeneous or **polyclonal** in nature. Moreover, unless subjected to costly affinity purification, the serum will contain all the antibodies produced by the host animal, not thus those against the injected laboratory antigen.

Homogenous **monoclonal** antibodies that target, not just a single antigen, but a single epitope on its surface can be generated by isolating B cells from the spleen of a mouse (or another suitable animal) that has been previously injected with antigen. The cultured B cells are fused with mouse **myeloma cells** to generate an immortalized **hybridoma** cell line that secretes a single, monoclonal antibody. These antibodies are then screened to identify hybridoma lines that secrete a monoclonal antibody specific for the antigen or even the epitope of choice.

For therapeutic applications, such as fighting infections by the coronavirus, the monoclonal antibodies produced by murine cell lines are **humanized.** Using genetic engineering, the variable regions of the murine antibodies are inserted into the appropriate sites in a human immunoglobulin molecule. Humanizing antibodies in this manner markedly reduces their **immunogenicity**, diminishing the chances of triggering an anaphylactic reaction.

THE COMPLEMENT SYSTEM ALSO PROTECTS AGAINST INFECTION

Immunoglobulins form the core of the body's **adaptive immune system**, a name that reflects its ability to generate antibodies against novel infectious agents. The **innate** immune system derives its name from the fact that the number, function, and specificity of its components are fixed and remain constant throughout life. The **complement system** forms the humoral arm of the innate immune system. Since it can be activated by exposure to antibody-antigen complexes, it can be considered to function as a support for or "complement" of the adaptive immune system.

The complement system displays features reminiscent of the clot-inducing coagulation cascade. Both consist of sets of circulating zymogens (proproteins) that remain catalytically dormant until activated by proteolytic cleavage. These proproteins, called **complement factors**, are synthesized by a variety of cell types that include hepatocytes, macrophages, monocytes, and intestinal endothelial cells. As is the case for clotting factors, most complement factors are proproteases (see Chapter 9) that, on activation, target other components of the system, thereby generating a series or **cascade** of proteolytic activation events that produce one or more protective end products (eg, fibrin).

The **classical pathway** for activating the complex system is triggered when an antibody-antigen complex binds to and stimulates the protease activity of factor **C1**. Activated C1 then cleaves complement factor C2 to form two smaller proteins, C2a and C2b, as well as cleaving complement factor C4 to form C4a and C4b (**Figure 52–14**). Two of the proteolytic fragments, C2a and C4b, then associate together to form a new protease, the C3 convertase, which cleaves complement factor C3 into C3a and C3b. C3a now binds with the C2a:C4b heterodimer to form a heterotrimeric complex, the C5 convertase, that cleaves complement factor C5 into C5a and C5b. The C5b protein then combines with complement factors C6, C7, C8, and C9 to form the **membrane attack complex** (MAC). MAC kills bacterial invaders by binding to and opening a pore in their plasma membrane. Following lysis, the bacterial remains are destroyed by phagocytic macrophages. Meanwhile, the C3a and C5a proteins serve as chemoattractants that recruit leukocytes to the site of infection and stimulate an inflammatory response.

Targeting of the MAC to invading bacteria is facilitated by the presence of thioester bonds in C3 and C4. Like the thioester bond in the plasma protease inhibitor α_2-macroglobulin, these highly reactive bonds becomes exposed as a result of the conformational change that accompanies proteolytic activation. In the case of C3 and C4, the thioester reacts with the hydroxyl groups of the bacteria's surface polysaccharides, covalently anchoring the C5 convertase complex of which they are a part to the targeted pathogen. This enables the remaining components of the MAC to be formed in close proximity to the bacterial membrane, facilitating assembly.

Activation can also be triggered via the **lectin pathway**. Bacterial polysaccharides bind to a complement factor called **mannose-binding lectin** (MBL) or **mannan-binding protein** (MBP). The lectin-polysaccharide complex then recruits and activates C4 (see Figure 52–14). The term **lectin** refers to any protein that binds polysaccharides. Most lectins are highly selective. MBL is specific for the mannose-containing carbohydrate moieties (**mannans**) of glycoproteins and **lipopolysaccharides** present on the surface of gram-positive bacteria, some viruses,

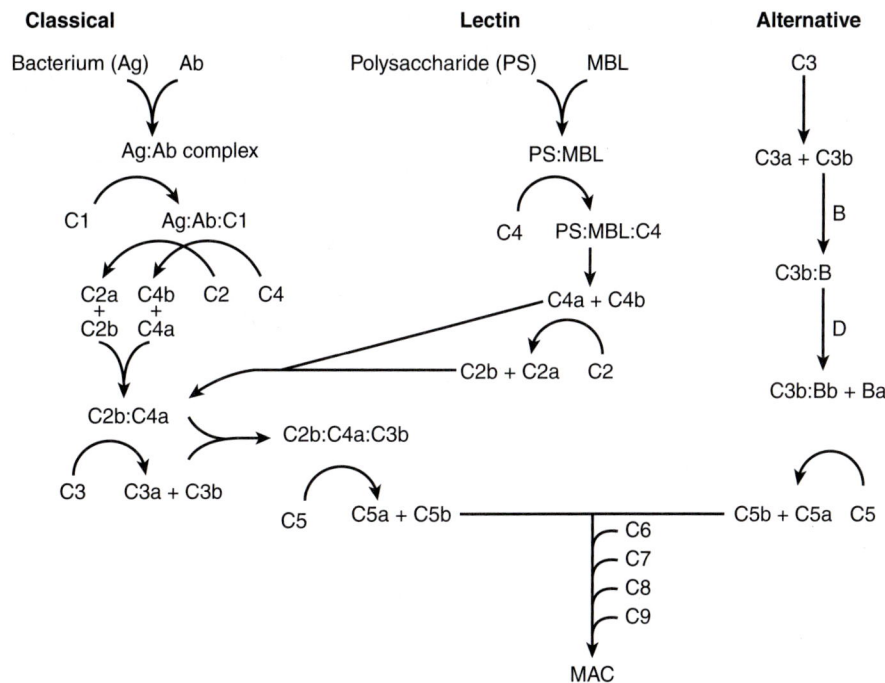

FIGURE 52–14 **The complement cascade.** Activation of the complement system can occur via three different mechanisms, referred to as the classical, lectin, and alternative pathways. Shown are the major components involved in each pathway, the products formed by proteolytic cleavage of the inactive proproteins, and the major complexes formed. Colons are used to indicate association in a complex.

and several fungi. On binding to the polysaccharide-MBL complex, C4 undergoes autoproteolysis, releasing C4a and C4b. In addition, C4 cleaves C2 into C2a and C2b. The remainder of cascade mirrors that for the classical pathway.

MBL circulates as large, ≈400 to 700 kDa, multivalent complexes consisting of four or more copies of a homotrimeric core unit, each of which contains three identical ≈30-kDa subunits. The polypeptides in this core are linked by the intertwining of their collagen-like domains. The carbohydrate binding-domain reside in their globular heads. To form MBL, four or more homotrimers become covalently linked by disulfide bonds between their amino-terminal tail regions. These form a "stalk" from which the C-terminal carbohydrate-binding heads project in a branched arrangement resembling that of an immunoglobulin (**Figure 52–15**).

The complement system also can be activated by the **alternative pathway**, in which C3 is activated by direct chemical hydrolysis, a process sometimes referred to as "ticking over." In the alternative pathway, C3b binds complement factor B, forming a C3b:B complex that is then cleaved by complex factor D. The resulting C3b:Bb complex possesses C5 convertase activity.

DYSFUNCTIONS OF THE IMMUNE SYSTEM CONTRIBUTE TO MANY PATHOLOGIC CONDITIONS

Dysfunctions of the innate and adaptive immune systems can have serious physiologic consequences. Deficits in the production of immunoglobulins or complement factors can leave the

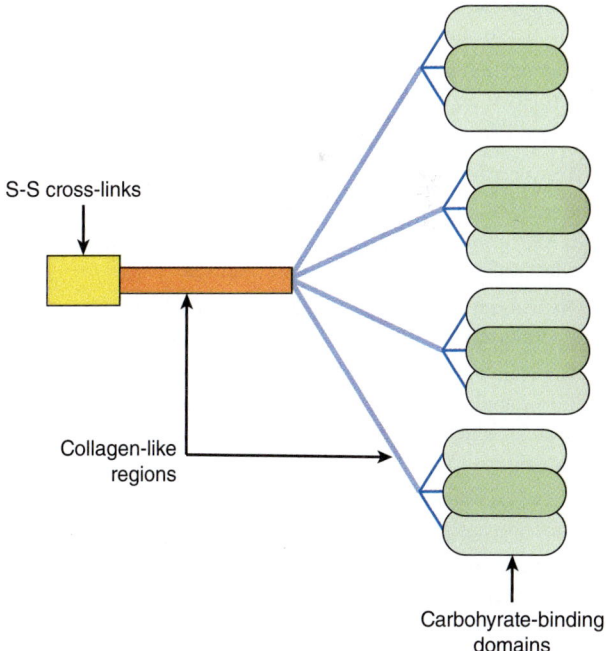

FIGURE 52–15 **Schematic representation of mannose-binding lectin (MBL).** Shown is a schematic diagram of a MBL comprised of four sets of MBL homotrimers. The carbohydrate-binding domains are colored. The intertwined collagen-like binding domains for each trimer are shown in blue. The stalk region, where the amino-terminal portions of the homotrimers of collagen-like domains associate together, is colored orange and yellow, with yellow marking the region where the S—S cross-links that stabilize the tetramer of homotrimers are located.

affected, **immunocompromised** individual susceptible to the occurrence and spread of bacterial, fungal, or viral infections. Many factors can contribute to a depression in the effectiveness of the immune system. These include genetic abnormalities (eg, **agammaglobulinemia**, in which production of IgG is markedly affected), toxins, viral infections, malnutrition, neoplastic transformation, or treatment with immunosuppressant drugs.

Overproduction and precocious activation of the immune and complement systems can also be deleterious. Failure to differentiate host cells from a foreign invader can trigger an **autoimmune response** in which the body's immune system attacks its own tissues and organs. The resulting damage may be cumulative, such as occurs in rheumatoid arthritis and multiple sclerosis, or acute, such as the complete destruction of pancreatic islet cells that occurs in type 1 diabetes. In North America, autoimmune disorders affect three in every hundred persons.

Table 52–1 lists several of the more commonly encountered autoimmune disorders.

SUMMARY

- Most plasma proteins are synthesized in the liver. The majority are glycosylated.
- Albumin accounts for roughly 60%, by mass, of the protein content of plasma. As such, it is the principal determinant of intravascular osmotic pressure.
- Albumin also binds to and transports fatty acids, bilirubin, metal ions, and certain drugs.
- Haptoglobin binds extracorpuscular hemoglobin to prevent the formation of damaging precipitates in the tubules.
- Ferritin binds to and stores ferric iron inside cells.
- Transferrin transports iron to the sites where it is required.
- Ceruloplasmin, the major copper-containing protein in plasma, is a ferroxidase that plays a key role in recycling the iron released when senescent red blood cells are destroyed.
- Hepcidin regulates iron homeostasis by blocking internalization of the cellular iron export protein, ferrocidin.
- Hepcidin expression is stimulated when binding of transferrin-iron complexes to type 1 transferrin receptors displaces HFE protein, which then binds to and activates type 2 transferrin receptors.
- Hereditary hemochromatosis is a genetic disease involving excessive absorption of iron.
- α_1-Antitrypsin is the major serine protease inhibitor of plasma. Genetically produced deficiencies of this protein can lead to emphysema and liver disease.

- α_2-Macroglobulin is a major plasma protein that neutralizes many proteases and targets select cytokines to specific organs.
- Using their adaptive immune system, humans can synthesize immunoglobulins that specifically target a million or more different antigens.
- The core structure of the immunoglobulins is a tetramer consisting of two light and two heavy chains arranged in a "Y" configuration.
- Synthesis of diverse antibodies from a limited set of genes is made possible by the combination, rearrangement, and somatic mutation of immunoglobulin genes.
- Hybridoma cells can provide monoclonal antibodies for laboratory and clinical use.
- The complement system is generally activated by complexes formed between infecting microbes and protective antibodies or between mannose-rich polysaccharides on the pathogen's surface and mannose-binding protein.
- In the complement system, the components from which the membrane attack complex is assembled are produced by a series of proteolytic cleavage events that transform dormant zymogens into active proteases.
- Autoimmune disorders result when the immune system attacks our body's own tissues.

REFERENCES

Craig WY, Ledue TB, Ritchie RF: *Plasma Proteins: Clinical Utility and Interpretation.* Foundation for Blood Research, 2008.

Crichton R (editor): *Iron Metabolism, from Molecular Mechanics to Clinical Consequences,* 4th ed. Wiley, 2016.

Garred P, Tenner AJ, Molines TE: Therapeutic targeting of the complement system: From rare diseases to pandemics. Pharmacol Rev 2021;73:792.

Hentz MW, Muckenthaler MU, Gali B, et al: Two to tango: regulation of mammalian iron metabolism. Cell 2010;142:24.

Nimmerjahn F, Ravetch JV: *Fc Mediated Activity of Antibodies.* Springer, 2020.

Reese AR: *The Antibody Molecule: From Antitoxins to Therapeutic Antibodies.* Oxford University Press, 2015.

Roumenina LT (editor): *The Complement System. Innovative Diagnostic and Research Protocols,* Springer, 2021.

Schaller H, Gerber S, Kaempfer U, et al: *Human Blood Plasma Proteins: Structure and Function.* Wiley, 2008.

Smith SA, Travers RJ, Morrissey JH: How it all starts: Initiation of the clotting cascade. Critic Rev Biochem Mol Biol 2015;50:326.

Williams NA, Kivimaki M, Langenberg C, et al: Plasma protein patterns as comprehensive indicators of health. Nature Med 2019;25:1851.

Red Blood Cells

Peter J. Kennelly, PhD

O B J E C T I V E S

After studying this chapter, you should be able to:

- Understand the concept of stem cells and their importance.
- Explain why red blood cells are reliant on glucose for energy.
- Describe the roles of erythropoietin, thrombopoietin, and other cytokines in the production of red blood cells and platelets.
- Describe the enzyme systems that protect heme iron from oxidation and reduce methemoglobin.
- Identify the major components of the erythrocyte cytoskeleton.
- Summarize the causes of the major disorders affecting red blood cells.
- Describe the major function of erythrocyte band 3 protein.
- Explain the biochemical bases of the ABO blood group substances.
- List the major components contained in the dense granules and α-granules in platelets.
- Describe the molecular bases of immune thrombocytopenic purpura and von Willebrand disease.

BIOMEDICAL IMPORTANCE

The evolution of a diverse array of freely circulating blood cells was critical to the development of animal life. The packaging of hemoglobin and carbonic anhydrase inside specialized cells called **erythrocytes** greatly amplified the capacity of circulating blood to carry oxygen to and carbon dioxide away from peripheral tissues. **Anemia**, a deficiency in the level of circulating hemoglobin (<120-130 g/L), compromises health by reducing the ability of the blood to supply tissues with adequate levels of oxygen. Anemia can arise from a variety of causes that include genetic abnormalities (eg, sickle cell trait, pernicious anemia), excessive bleeding, insufficiencies of dietary iron or vitamin B_{12}, or the lysis of red blood cells by invading pathogens (eg, malaria). **Platelets** help staunch the outflow of blood from damaged tissues. Deficits in platelet number or function increase a patient's vulnerability to hemorrhage by reducing the speed of formation and structural integrity of protective clots. As is the case for anemia, a low platelet count, known as **thrombocytopenia**, can be triggered by a range of factors that include bacterial infection, some medications including sulfa-containing antibiotics, or autoimmune reactions such as

idiopathic thrombocytopenic purpura. Other pathophysiologic syndromes, such as **von Willebrand disease** and **Glanzmann thrombasthenia**, are caused by genetic mutations that impair platelet adherence or aggregation rather than their abundance.

RED BLOOD CELLS DERIVE FROM HEMATOPOIETIC STEM CELLS

Both red blood cells and platelets turn over at a relatively high rate. Hence, replacements are constantly being produced from precursor **stem cells**. Stem cells are considered to exist in an undifferentiated state. They possess a unique capacity both to produce unaltered daughter cells (**self-renewal**) and to generate a diverse range of specialized cell types (**potency**). **Totipotent** stem cells are capable of producing all the cells in an organism, while **unipotent** stem cells can produce only one type of differentiated cell of a closely related family. **Pluripotent** stem cells can differentiate into cells of any of the three germ layers. **Multipotent** stem cells are restricted to forming cells that belong to a single family or class. In addition, stem cells can also be classified as **embryonic** or **adult**, the latter of which are more limited in their capacity to differentiate.

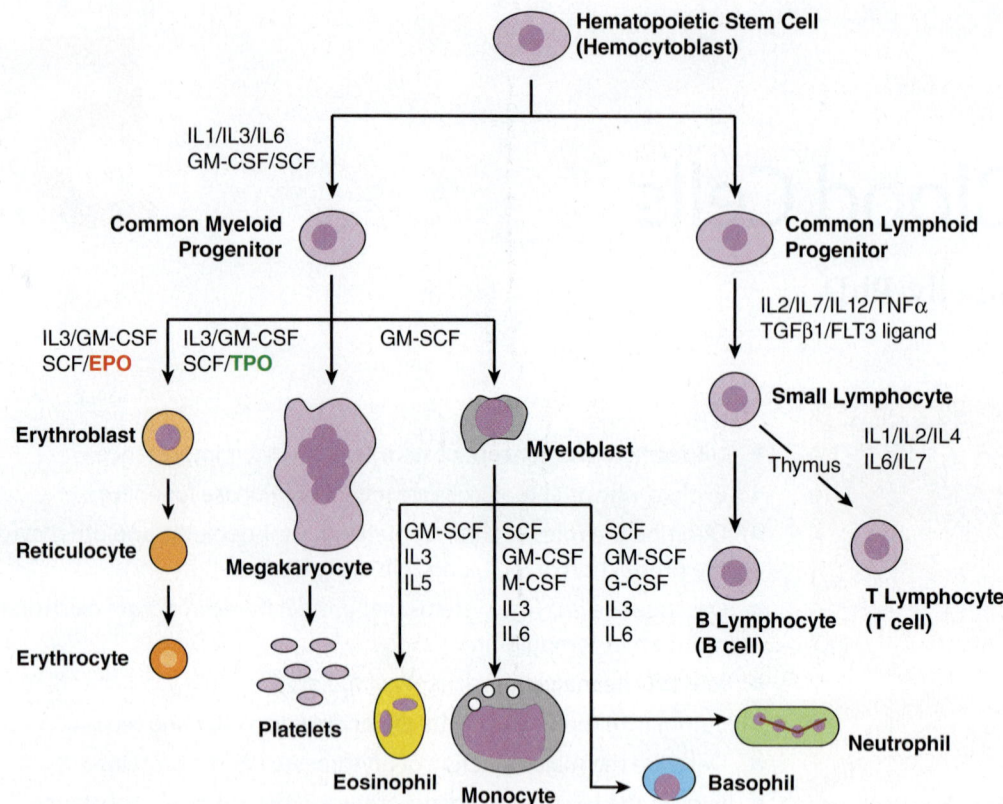

FIGURE 53–1 **Hematopoiesis.** Shown is a simplified and heavily abbreviated scheme indicating the paths by which hematopoietic stem cells differentiate to produce many of the more quantitatively prominent red and white blood cells. Only selected developmental intermediates are shown. The names for each cell type are indicated in **bold type.** Cell nuclei are shown in **purple.** Each arrow summarizes a multistage transition. The hormones and cytokines that stimulate each transition are listed next to the arrows. **EPO,** erythropoietin; FLT3 ligand, FMS-like tyrosine kinase 3 ligand; G-CSF, granulocyte colony-stimulating factor; GM-CSF, granulocyte-macrophage colony-stimulating factor; IL, interleukin; M-CSF, macrophage colony-stimulating factor; SCF, stem cell factor; TGF-β_1, transforming growth factor β_1; TNF-α, tumor necrosis factor α; **TPO,** thrombopoietin.

Differentiation of hematopoietic stem cells is regulated by a set of secreted glycoproteins called **cytokines.** **Stem cell factor** and several **colony-stimulating factors** collaborate with interleukins 1, 3, and 6 to stimulate the proliferation of hematopoietic stem cells in the bone marrow and induce their differentiation into one of several myeloid cell types (**Figure 53–1**). **Erythropoietin** or **thrombopoietin** then directs the further differentiation of these myeloid progenitor cells into erythrocytes or platelets, respectively.

RED BLOOD CELLS ARE HIGHLY SPECIALIZED

Mature Erythrocytes Are Devoid of Internal Organelles

The structure and composition of red blood cells reflect their highly specialized function: to deliver the maximum quantity of oxygen to tissues and aid in the removal of carbon dioxide, a waste product of cellular respiration, and urea. The interior of a red blood cell contains a massive amount of hemoglobin, roughly one-third by weight (30-34 g/dL for an adult). This extraordinary hemoglobin concentration is achieved, in part,

by dispensing with the intracellular organelles normally found in eukaryotic cells (eg, nucleus, lysosome, Golgi apparatus, mitochondria). As a consequence, mature **enucleated** red blood cells are unable to reproduce.

Red blood cells possess an extensive cytoskeletal network responsible for maintaining their biconcave configuration (**Figure 53–2**). Their unusual shape enhances the exchange of

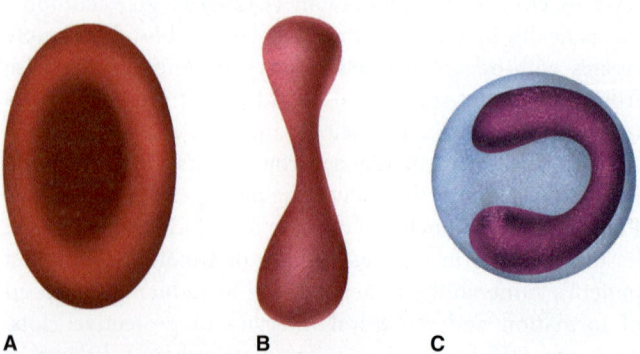

FIGURE 53–2 **Red blood cells are shaped like biconcave disks.** Shown are drawings of **(A)** a red blood cell, **(B)** a section through a red blood cell illustrating its biconcave shape, and **(C)** a red blood cell folded for passage through a narrow capillary.

oxygen and carbon dioxide between erythrocytes and tissues in two ways. First, their disc-like configuration possesses a much higher ratio of surface area to volume than more spherical forms. Second, it enables red blood cells to fold over and squeeze through small capillaries, whose diameter is less than that of the erythrocyte itself. Both factors reduce the distance gas molecules must diffuse to and from the rapidly moving (up to 2 mm/s) erythrocytes.

Erythrocytes Generate ATP Exclusively via Glycolysis

Mature red blood cells lack mitochondria and their contents, including ATP synthase and the enzymes of the tricarboxylic acid cycle (TCA), electron transport chain, and β-oxidation pathway. They are therefore incapable of utilizing fatty acids or ketone bodies as metabolic fuel and therefore are completely reliant on glycolysis to generate ATP. Glucose enters red blood cells by **facilitated diffusion** (see Chapter 40), a process mediated by **glucose transporter 1 (GLUT1)**, also known as glucose permease (**Table 53–1**).

The glycolytic pathway in red blood cells also possesses a unique branch, or shunt, whose purpose is to isomerize 1,3-bisphosphoglycerate (1,3-BPG) to **2,3-bisphosphoglycerate** (2,3-BPG). 2,3-BPG binds to and stabilizes hemoglobin in the T-state (see Chapter 6). Conversion of 1,3-BPG to 2,3-BPG is catalyzed by 2,3-BPG mutase, a bifunctional enzyme that also catalyzes the hydrolysis of 2,3-BPG to the glycolytic intermediate 3-phosphoglycerate. A second enzyme, multiple inositol polyphosphate phosphatase, also catalyzes the hydrolysis of 2,3-BPG, in this case to the glycolytic intermediate 2-phosphoglycerate. The activities of these enzymes are sensitive to pH, which ensures that 2,3-BPG levels rise and fall at the appropriate times as the red blood cells travel the circulatory system.

Various aspects of the metabolism of the red cell, several of which are discussed in other chapters, are summarized in **Table 53–2**.

TABLE 53–1 Some Properties of the Glucose Transporter of the Membrane of the Red Blood Cell (GLUT1)

- It accounts for ~2% of the protein of the membrane of the RBC.
- It exhibits specificity for glucose and related d-hexoses (l-hexoses are not transported).
- The transporter functions at ~75% of its V_{max} at the physiologic concentration of blood glucose, is saturable, and can be inhibited by certain analogs of glucose.
- It is a member of a family of homologous glucose transporters found in mammalian tissues.
- It is not dependent upon insulin, unlike the corresponding carrier in muscle and adipose tissue.
- Its 492 amino acid sequence has been determined.
- It transports glucose when inserted into artificial liposomes.
- It is estimated to contain 12 transmembrane helical segments.
- It functions by generating a gated pore in the membrane to permit passage of glucose; the pore is conformationally dependent on the presence of glucose and can oscillate rapidly (~900 times/s).

TABLE 53–2 Important Aspects of the Metabolism of the Red Blood Cell

- The RBC is highly dependent upon glucose as its energy source, for which its membrane contains high-affinity glucose transporters.
- Glycolysis, producing lactate, is the mode of production of ATP.
- Because RBCs lack mitochondria there is no production of ATP by oxidative phosphorylation.
- The RBC has a variety of transporters that maintain ionic and water balance.
- Production of 2,3-bisphosphoglycerate by reactions closely associated with glycolysis is important in regulating the ability of Hb to transport oxygen.
- The pentose phosphate pathway of the RBC metabolizes about 5%-10% of the total flux of glucose) and produces NADPH. Hemolytic anemia due to a deficiency of the activity of glucose- 6-phosphate dehydrogenase is common.
- Reduced glutathione (GSH) is important in the metabolism of the RBC, in part to counteract the action of potentially toxic peroxides. The RBC can synthesize GSH and the NADPH required to return oxidized glutathione (G-S-S-G) to the reduced state GSH.
- The iron of Hb must be maintained in the ferrous state. Ferric iron is reduced to the ferrous state by the action of an NADH-dependent methemoglobin reductase system involving cytochrome b_5 reductase and cytochrome b_5.
- While biosynthesis of glycogen, fatty acids, protein, and nucleic acids does not occur in the RBC, some lipids (eg, cholesterol) in the red cell membrane can exchange with corresponding plasma lipids.
- The RBC contains certain enzymes of nucleotide metabolism (eg, adenosine deaminase, pyrimidine nucleotidase, and adenylyl kinase). Deficiencies of these enzymes are involved in some cases of hemolytic anemia.
- When RBCs reach the end of their lifespan, the globin is degraded to amino acids (which are reutilized in the body), the iron is released from heme and reutilized, and the tetrapyrrole component of heme is converted to bilirubin, which is mainly excreted into the bowel via the bile.

Carbonic Anhydrase Facilitates CO$_2$ Transport

Every molecule of O_2 consumed results in the formation of corresponding number of CO_2 molecules, which must be disposed. Like oxygen, the solubility of carbon dioxide in aqueous solution is much too low to accommodate more than a few percent of the CO_2 produced by metabolically active tissues. However, the solubility of the hydrated form of CO_2, carbonic acid (H_2CO_3), and its protonic dissociation product, bicarbonate (HCO_3^-), is relatively high. The presence of high levels of the enzyme **carbonic anhydrase** (see Figure 6–11) enables erythrocytes both to absorb waste CO_2 by catalyzing its rapid conversion to carbonic acid, and to reverse this process in order to facilitate its expulsion from the lungs. While red blood cells carry some CO_2 in the form of hemoglobin-bound carbamates (see Chapter 6), the majority, ≈80%, is carried as dissolved carbonic acid and bicarbonate.

RED BLOOD CELLS MUST BE CONTINUALLY REPLACED

About Two Million New Red Blood Cells Enter the Circulation Each Second

The **120-day lifespan** of a normal red blood cell requires that nearly 1% of the 20 to 30 trillion erythrocytes in a typical individual must be replaced daily. This equates to a rate of production of ~ 2 million new red blood cells per second. When initially formed, differentiated red blood cells retain a portion of the ribosomes, endoplasmic reticulum, mitochondria, etc. that were present in their nucleated precursors. Consequently, during the ≈ 24 hours required to complete maturation to erythrocytes, these nascent red blood cells, called **reticulocytes**, retain the capacity to synthesize polypeptides (eg, globin) under the direction of vestigial mRNA molecules.

In rare cases, genetic mutations that lead to an impairment of ribosome function, called **ribomyopathies**, can result in red blood cell hypoplasia. **Diamond-Blackfan anemia** is caused by mutations in the gene encoding the ribosomal processing protein RPS19. **5q-syndrome**, which presents a similar clinical picture, is caused by mutations that lead to an insufficiency of ribosomal protein RPS14.

Erythropoietin Regulates Production of Red Blood Cells

The initial stages of **erythropoiesis**, the production of red blood cells, are modulated by stem cell factor, colony-stimulating factors, and interleukins 1, 3, and 6. Commitment of myeloid progenitor cells to differentiate into erythrocytes is largely dependent on **erythropoietin (EPO)**, a glycoprotein of 166 amino acids (molecular mass ≈ 34 kDa). Glycosylation occurs at four sites. Glycosylation at two of these sites is essential for physiologic function. EPO is synthesized mainly by the kidney, which releases it into the bloodstream in response to hypoxia. On reaching the bone marrow, EPO stimulates red blood cell progenitors via a transmembrane receptor. Binding of EPO to its receptor causes the latter to dimerize. The transition to a dimeric state then activates associated molecules of the Jak2 protein-tyrosine kinase.

Erythropoietin is administered therapeutically to treat anemias arising from chronic kidney failure, disorders of hematopoietic stem cells (**myelodysplasia**), or from the collateral effects of chemical and radiologic treatments for cancer. Currently, pharmaceutical EPO is obtained via recombinant expression in Chinese Hamster Ovary cell lines, which are commonly used to produce therapeutic proteins because of their capacity to closely mimic the glycosylation patterns found in their naturally produced counterparts. Unfortunately, some recipients of recurring doses of recombinant EPO develop neutralizing antibodies that in turn causes the development of pure red blood cell aplasia. Scientists are working to develop less immunogenic variants of or biomimetic substitutes for current versions of recombinant EPO.

OXIDATION OF HEME IRON COMPROMISES OXYGEN TRANSPORT

Cytochrome b_5 Reductase Reduces Methemoglobin

The ferrous, Fe^{2+}, iron atoms in hemoglobin are susceptible to oxidation by **reactive oxygen species (ROS)**. Hemoglobin in which one or more heme irons has been oxidized to the ferric (Fe^{3+}) state is called **methemoglobin**. Ferric hemes do not bind oxygen, which not only reduces the number of O_2-binding sites, but can interfere with the cooperative interactions between the subunits of the hemoglobin tetramer (see Chapter 6). The ability to rescue methemoglobin by reducing ferric iron is thus of great physiologic importance. In red blood cells, methemoglobin reduction is catalyzed by the NADH–cytochrome b_5 methemoglobin reductase system. The first component of the system, a flavoprotein named **cytochrome b_5 reductase** (also known as methemoglobin reductase), transfers electrons from **NADH** to the second component, **cytochrome b_5**:

$$\text{Cyt } b_{5ox} + \text{NADH} \rightarrow \text{Cyt } b_{5red} + \text{NAD}^+$$

Reduced cytochrome b_5 then transfers the electrons to methemoglobin, reducing Fe^{3+} back to the Fe^{2+} state, which restores hemoglobin to its fully functional state:

$$\text{Hb} - Fe^{3+} + \text{Cyt } b_{5red} \rightarrow \text{Hb} - Fe^{2+} + \text{Cyt } b_{5ox}$$

The ultimate source of the electrons used to reduce methemoglobin is glycolysis, where NAD^+ is reduced to NADH by the action of glyceraldehyde-3-phosphate dehydrogenase. The efficiency of this system is such that only trace quantities of methemoglobin are normally present in erythrocytes.

Methemoglobinemia Is Inherited or Acquired

Methemoglobinemia, the abnormal accumulation of methemoglobin, can arise from genetic abnormalities (inherited methemoglobinemia) or from the ingestion of certain substances (acquired methemoglobinemia) such as nitrates, aniline, or sulfonamide antibiotics (**Table 53–3**). Affected patients often exhibit bluish discoloration of the skin and mucous membranes (cyanosis). The inherited form most commonly arises from a deficiency in the quantity or activity of **cytochrome b_5 reductase**, although mutations that affect the properties of cytochrome b_5 itself have also been encountered. In rare instances, methemoglobinemia can result from mutations in hemoglobin itself, such as those affecting the proximal and distal histidine residues (see Figure 6–3), that render it more susceptible to oxidation. Collectively referred to as hemoglobin M (HbM), these include HbM_{Iwate}, in which His87 in the α subunit is replaced by Tyr; $HbM_{Hyde Park}$, in which His92 of the β subunit is replaced by Tyr; HbM_{Boston}, in which His58 in the α subunits of hemoglobin is replaced by Tyr; and $HbM_{Saskatoon}$, in which His63 in the β subunit is replaced by Tyr. One exception to this pattern is $HbM_{Milwaukee-1}$, in which Val67 of the β subunit is replaced by Glu. All known carriers of HbM are heterozygotes.

Superoxide Dismutase, Catalase, & Glutathione Protect Blood Cells From Oxidative Damage

The radical anion **superoxide**, $O_2^{\overline{\cdot}}$, is generated in red blood cells by the autoxidation of hemoglobin to methemoglobin.

TABLE 53–3 Summary of the Causes of Some Important Disorders Affecting Red Blood Cells

Disorder	Sole or Major Cause
Iron deficiency anemia	Inadequate intake or exerssive loss of iron
Methemoglobinemia	Intake of excess oxidants (various chemicals and drugs)
	Genetic deficiency in the NADH-dependent methemoglobin reductase system (OMIM 250800)
	Inheritance of HbM (OMIM 141900)
Sickle cell anemia (OMIM 603903)	Sequence of condon 6 of the β chain changed from GAG in the normal gene to GTC in the sickle cell gene, resulting in substitution of valine for glutamic acid
α-Thalassemias (OMIM 141800)	Mutations in the α-globin genes, mainly unequal crossing-over and large deletions and less commonly nonsense and frameshift mutations
β-Thalassemias (OMIM 141900)	A very wide variety of mutations in the β-globin gene, including deletions, nonsense and frameshift mutations, and others affecting every aspect of its structure (eg, splice sites, promoter mutants)
Megaloblastic anemias	Deficiency of vitamin B12. Decreased absorpotion of B_{12}, often due to a deficiency of intrinsic factor, normally secreted by gastric parietal cells
	Deficiency of folic acid. Decreased intake, defective absorption, or increased demand (eg, in pregnancy) for folate
Hereditary spherocytosis[a] (OMIM 182900)	Deficiencies in the amount or in the structure of α- or β-spectrin, band 3, or band 4.1.
Glucose-6-phosphate dehydrogenase (G6PD) deficiency[a] (OMIM 305900)	A variety of mutations in the gene (X-linked) for G6PD, mostly single-point mutations
Pyruvate kinase (PK) deficiency[a] (OMIM 266200)	A variety of mutations in the gene for the R (red cell) isozyme of PK
Parosysmal nocturnal hemoglobinuria[a] (OMIM 311770)	Mutations in the PIG-A gene, affecting synthesis of GPI-anchored proteins

[a]OMIM members apply only to disorders with a genetic basis.

This potent ROS can react with and damage proteins, lipids, nucleotides, and other biomolecules (see Chapter 58). Approximately 3% of the hemoglobin of human blood undergoes auto-oxidation each day. In addition, oxidation of the iron storage protein ferritin by superoxide can result in the release of free Fe^{2+} and the subsequent iron-catalyzed generation of OH· (see Figure 58–2). Superoxide may thus produce the tissue damage that occurs in persons suffering from abnormally high levels of iron in the body, known as iron overload. Iron overload is characteristic of individuals suffering from **hereditary hemochromatosis**, a genetic condition that causes the body to absorb excessive quantities of dietary iron. Another endogenous source of superoxide is the enzyme **NADPH-hemoprotein reductase** (a cytochrome P450 reductase, see Chapter 12), which also catalyzes the reduction of the Fe^{3+} in methemoglobin to Fe^{2+}.

Deficiency of Glucose-6-Phosphate Dehydrogenase Is an Important Cause of Hemolytic Anemia

The limited suite of metabolic pathways present in red blood cells renders them completely reliant on the **pentose phosphate pathway** (see Chapter 20) or, to be more specific, the X-linked enzyme **glucose-6-phosphate dehydrogenase**, for the reduction of NADP+ to NADPH. Glucose-6-phosphate dehydrogenase deficiency is the most common of all **enzymopathies** (diseases caused by abnormalities of enzymes). More than 400 million people are estimated to carry one of the over 140 genetic variants of glucose-6-phosphate dehydrogenase. This deficiency is most common among natives of tropical Africa (and their African-American descendants), the Mediterranean, and certain parts of Asia.

Individuals harboring this deficiency are vulnerable to hemolytic anemia, a consequence of their inability to generate sufficient NADPH to maintain glutathione, a key intracellular antioxidant, in the reduced state (**Figure 53–3**). A deficiency in glucose-6-phosphate dehydrogenase renders red blood cells hypersensitive to oxidative stress and the formation of **Heinz bodies**, insoluble aggregates consisting of hemoglobin molecules whose —SH groups have become oxidized. Like sickle cell trait, the persistence of these genetic variants has been attributed to their potential to confer enhanced resistance to malaria.

Hemolytic Anemias Can Be Caused by Extrinsic, Intrinsic, or Membrane-Specific Factors

Hemolytic anemia can be triggered by factors other than a deficiency in glucose-6-phosphate dehydrogenase (**Figure 53–4**). **Extrinsic** causes (beyond the erythrocyte membrane) include **hypersplenism**, in which the enlargement of the spleen causes red blood cells to become sequestered within this organ.

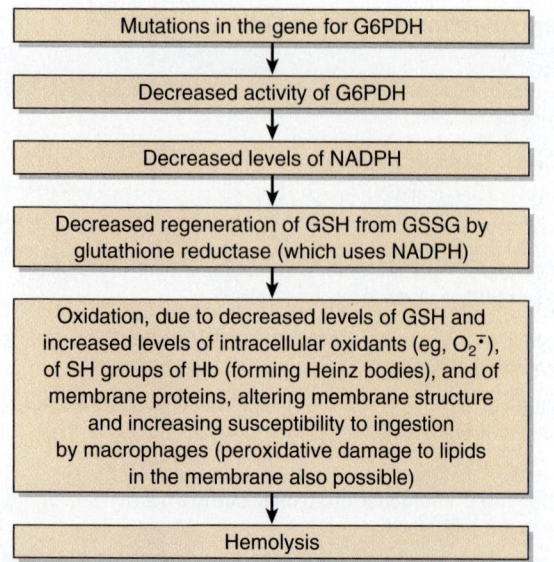

FIGURE 53–3 Summary of probable events causing hemolytic anemia due to deficiency of the activity of glucose-6-phosphate dehydrogenase (OMIM 305900).

Erythrocytes also can lyse if attacked by incompatible antibodies present in intravenously administered plasma or blood (eg, **transfusion reaction**). Immunologic incompatibilities may arise when an Rh⁺ fetus is carried by an Rh⁻ mother (**Rh disease**) or as a consequence of an autoimmune disorder (eg, **warm** or **cold antibody hemolytic anemias**). Some infectious and toxic agents, such as the proteases and phospholipases found in many insect and reptile venoms, act by directly undermining the structural integrity of the erythrocyte membrane. Similarly, some infectious bacteria, including certain strains of *Escherichia coli* and clostridia, secrete factors that attack and lyse red blood cell membranes. These factors, collectively referred to as **hemolysins,** can be composed of proteins, lipids, or a combination thereof. Parasitic infections

(eg, the plasmodia causing malaria) are also a major cause of hemolytic anemias in certain geographic areas.

The root cause of many hemolytic anemias is intracellular, also referred to as **intrinsic**. Gluocse-6-phosphate dehydrogenase deficiency falls into this category. Defects in the composition or structure of hemoglobin, called **hemoglobinopathies**, constitute the second major intrinsic cause of hemolysis. Most hemoglobinopathies, such as sickle cell anemia and the various thalassemias (see Chapter 6), are genetic in nature. In rare cases, hemolytic anemia can arise from an insufficiency in the enzyme **pyruvate kinase**. The resulting impairment of glycolysis reduces the production of the ATP required to power the export of excess water and Na⁺. The resulting osmotic pressure (see later) can compromise and potentially overwhelm the integrity of the erythrocyte membrane.

Mutations that affect the cytoskeletal proteins responsible for maintaining the erythrocyte's biconcave shape and resistance to osmotic pressure are classified as **membrane-specific** causes of hemolytic anemia (see later). The most important of these are **hereditary spherocytosis** and **hereditary elliptocytosis**, which arise from abnormalities in the amount or structure of the cytoskeletal protein spectrin. Defects may also occur in the synthesis of the glycophosphatidylinositol groups that anchor certain proteins, such as acetylcholinesterase and decay-accelerating factor, to the surface of the erythrocyte membrane, as is the case in **paroxysmal nocturnal hemoglobinuria** (see Chapter 46).

THE RED BLOOD CELL MEMBRANE

Early analyses by SDS-PAGE of the polypeptides present in red blood cells revealed 10 major proteins (**Figure 53–5**). These proteins were initially assigned numeric designators based on their migration on SDS-PAGE. Thus, the polypeptide with the

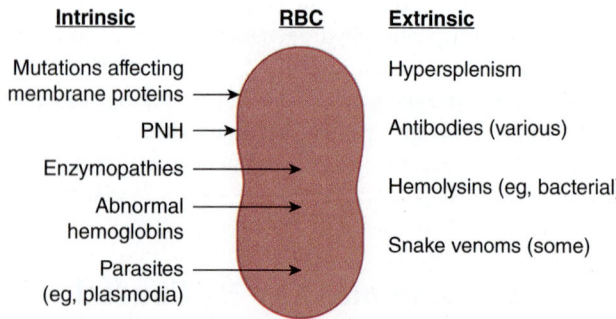

FIGURE 53–4 Schematic diagram of some causes of hemolytic anemias. Extrinsic causes include hypersplenism, various antibodies, certain bacterial hemolysins, and some snake venoms. Causes intrinsic to the red cells include mutations that affect the structures of membrane proteins (eg, in hereditary spherocytosis and hereditary elliptocytosis), paroxysmal nocturnal hemoglobinuria (PNH, see Chapter 47), enzymopathies, abnormal hemoglobins, and certain parasites (eg, plasmodia causing malaria).

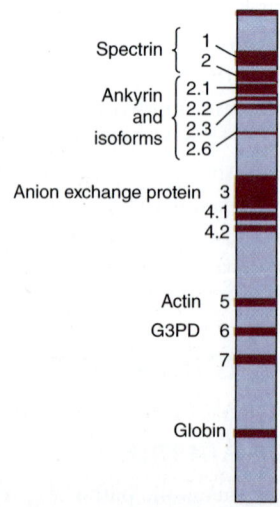

FIGURE 53–5 Major membrane proteins of the human red blood cell. Proteins separated by SDS-PAGE were detected by staining with Coomassie blue dye. (Reproduced with permission from Beck WS, Tepper RI: *Hematology*, 5th ed. Cambridge, MA: The MIT Press; 1991.)

TABLE 53–4 **Principal Proteins of the Red Cell Membrane**

Band Number[a]	Protein	Integral (I) or Peripheral (P)	Approximated Molecular Mass (kDa)
1	Spectrin (α)	P	240
2	Spectrin (β)	P	220
2.1	Ankyrin	P	210
2.2	Ankyrin	P	195
2.3	Ankyrin	P	175
2.6	Ankyrin	P	145
3	Anion exchange protein	I	100
4.1	Unnamed	P	80
5	Actin	P	43
6	Glyceraldehyde-3-phosphate dehydrogenase	P	35
7	Troomyosin	P	29
8	Unnamed	P	23
	Glycophorins A, B, and C	I	31, 23, and 28

[a]The band number refers to the relative rates of migration on SDS-PAGE (see Figure 53–5). A number of other components (eg, 4.2 and 4.9) are not listed.
Data from Scriver CR, Beaudet AL, Valle D, et al *The Metabolic basis of inherited Disease*, 8th ed. New York, NY: McGraw Hill; 2001.

highest molecular mass, which migrates slowest, was designated band 1 protein, also known as **spectrin** (**Table 53–4**). As illustrated by **Figure 53–6**, certain of these proteins are glycosylated. Several span the membrane bilayer (integral membrane proteins), while others associate with its surface, generally via protein–protein interactions (peripheral membrane proteins).

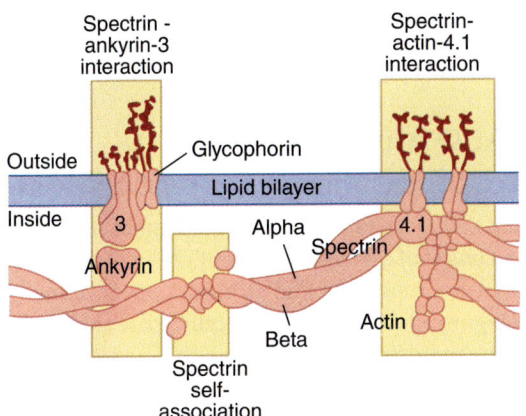

FIGURE 53–6 **Interactions of cytoskeletal proteins with each other and with certain integral proteins of the membrane of the red blood cell.** (Reproduced with permission from Beck WS, Tepper RI: *Hematology*, 5th ed. Cambridge, MA: The MIT Press; 1991.)

The Red Blood Cell Membrane Contains Anion Exchange Protein & the Glycophorins

Band 3 protein is a transmembrane glycoprotein whose polypeptide chain crosses the lipid bilayer 14 times. Band 3 protein is oriented with its carboxyl-terminal end projecting from the external surface of the erythrocyte membrane and its amino-terminal end from the cytosolic face. The principal function of this dimeric **anion exchange protein** is to provide a channel through the membrane via which chloride and bicarbonate anions generated from the hydration of CO_2 can be exchanged. At the tissues, bicarbonate enters erythrocytes in exchange for chloride. At the lungs, where carbon dioxide is exhaled, this process is reversed. Band 3 protein's amino-terminal end also serves as an anchoring point for several other red blood cell proteins, including band 4.1 and 4.2 proteins, ankyrin, hemoglobin, and several glycolytic enzymes.

Glycophorins A, **B**, and **C** are transmembrane proteins. Their membrane spanning domain consists of a single 23-amino acid arranged in the form of an α-helix. The most abundant form, **glycophorin A**, is comprised of a 131-amino acid polypeptide possessing 16 covalently-bound oligosaccharide chains. The majority of these are bound to the side chain hydroxyl groups of serine and threonine residues. These O-linked oligosaccharides account for roughly 60% of glycophorin A's total mass and nearly 90% of the sialic acid residues exposed on the surface of the red cell membrane. The glycoprotein's carboxyl-terminal end extends into the cytosol, where it binds to band 4.1 protein, which in turn is bound to spectrin. **Genetic polymorphisms** affecting the glycosylation of glycophorin A serve as the basis of the MN blood group system (see later). Intriguingly, individuals whose red cells lack glycophorin A exhibit no adverse effects. However, glycophorin A serves as the site of binding for several viral and bacterial pathogens, including influenza virus and the malaria parasite, *Plasmodium falciparum*.

Spectrin, Ankyrin, & Other Peripheral Membrane Proteins Help Determine the Shape & Flexibility of the Red Blood Cell

In order to maximize the efficiency of gas exchange, red blood cells must possess the structural strength to maintain their biconcave shape, yet remain sufficiently flexible to squeeze through peripheral capillaries and the sinusoids of the spleen. The inherently fluid and deformable foundation of the plasma membrane, the lipid bilayer, is molded to produce the erythrocyte's characteristic biconcave shape by a strong but flexible internal network of **cytoskeletal proteins** (see Figure 53–6).

Spectrin is the most abundant protein of the erythrocyte cytoskeleton. It is composed of two polypeptides, each more than 2100 residues in length: spectrin 1 (α chain) and spectrin 2 (β chain). The α and β chains of each spectrin dimer intertwine in an antiparallel orientation to form a highly extended structural unit ≈100 nm in length. Normally, two spectrin dimers self-associate head-to-head to form an approximately 200-nm long heterotetramer that is linked to the inner surface of the plasma membrane (and is bridged to other spectrin tetramers) via ankyrin, actin, and band 4.1 protein. The result is an internal mesh, the cytoskeleton, that is strong enough to maintain cell shape and resist swelling due to osmotic pressure, yet flexible enough to allow the erythrocyte to fold when needed.

Band 2 protein, better known as **ankyrin** is a pyramid-shaped protein that binds tightly to both spectrin and band 3 protein, an integral membrane protein that anchors spectrin to the membrane. Ankyrin is sensitive to proteolysis, accounting for the appearance of bands 2.2, 2.3, and 2.6, all of which are derived from intact ankyrin, which forms band 2.1.

Actin (band 5 protein) is a 42-kDa protein that can exist in two different conformations. The globular form, which is monomeric, is known as G-actin. When G-actin transitions into its filamentous, or F- form, the F-actin monomers rapidly polymerize into an extended double helical microfilament. These F-actin microfilaments bind to spectrin as well as **band 4.1 protein**. Band 4.1 protein is globular protein in shape and binds tightly to a site near the actin-binding domain in the tail of spectrin to form a protein 4.1-spectrin-actin ternary complex. Protein 4.1 also binds to the integral membrane proteins glycophorin A and glycophorin C, as well as certain phospholipids, thereby anchoring the ternary complex to the membrane.

Other, less quantitatively prominent proteins of the erythrocyte cytoskeleton include band 4.9 protein, adducin, and tropomyosin.

Spectrin Abnormalities Can Cause Hereditary Spherocytosis & Elliptocytosis

Hereditary spherocytosis, a genetic disease transmitted as an autosomal dominant, is characterized by hemolytic anemia and splenomegaly resulting from the presence of spherocytes (spherical red blood cells) in the peripheral blood. These spherocytes are less deformable and more prone to destruction in the spleen than normal erythrocytes, greatly shortening their life in the circulation. This condition, which affects about 1 in 5000 persons of Northern European ancestry, is caused by a deficiency in the amount or abnormalities in the structure of spectrin or, less frequently, ankyrin or band 3, 4.1, or 4.2 proteins. Loss of these proteins or impairments of their capacity to associate with other cytoskeletal components weakens the links that anchor the erythrocyte membrane to the cytoskeleton, allowing the erythrocyte to swell into a spherical shape. The anemia associated with hereditary spherocytosis is generally relieved by surgical removal of the patient's spleen (splenectomy).

Hereditary elliptocytosis can be readily distinguished from hereditary spherocytosis by virtue of the fact that the affected red blood cells assume an elliptic shape. This condition, which affects 1 in 2500 persons of Northern European descent and is even more frequent among populations from regions where malaria is common, results from genetic abnormalities that affect spectrin or, less frequently, band 4.1 protein or glycophorin C.

THE BIOCHEMICAL BASIS OF THE ABO SYSTEM

Approximately 30 human blood group systems have been recognized, the best known of which are the **ABO**, **Rh** (Rhesus), and **MN** systems. The term "**blood group**" refers to a defined set of red blood cell antigens, or blood group substances, controlled by a genetic locus having a variable number of alleles (eg, A, B, and O in the ABO system). The term "**blood type**" refers to the antigenic phenotype, usually recognized by the use of appropriate antibodies.

The ABO System Is of Crucial Importance in Blood Transfusion

The ABO system was discovered by Landsteiner in 1900 while investigating the basis of compatible and incompatible transfusions in humans. The membranes of the erythrocytes of most individuals contain one blood group substance, either type A, B, AB, or O. Individuals of type A have anti-B antibodies in their plasma that will agglutinate the erythrocytes in type B or type AB blood. Individuals of type B have anti-A antibodies that will agglutinate type A or type AB erythrocytes. Type AB blood has neither anti-A nor anti-B antibodies, and has been designated the **universal recipient**. Type O blood has neither A nor B antigens, and has been designated the **universal donor**. The above description has been simplified considerably, as further subgroups exist such as A_1 and A_2. The genes responsible for production of the ABO substances, which are located on the long arm of chromosome 9, fall into three genotypes, or **alleles**, two of which are codominant (A and B) and the third (O) recessive; these ultimately determine which of the four phenotypic products is synthesized: the A, B, AB, and O substances.

The ABO Antigens Are Glycosphingolipids & Glycoproteins

The ABO antigens or, as they are sometimes referred to, substances consist of a set of complex oligosaccharides located on the surface of most cells and as a component of many secretions (**Figure 53–7**). In red blood cells, these antigens are generally anchored to the cell surface by covalent attachment to a membrane lipid, forming a type of molecule known as a **glycolipid.** In secretions the same oligosaccharides are anchored to proteins, forming **glycoproteins.** Their presence in secretions is determined by a gene designated *Se* (for secretor), which codes for a specific **fucosyl (Fuc) transferase** in secretory organs, such as the exocrine glands. This gene is normally silent

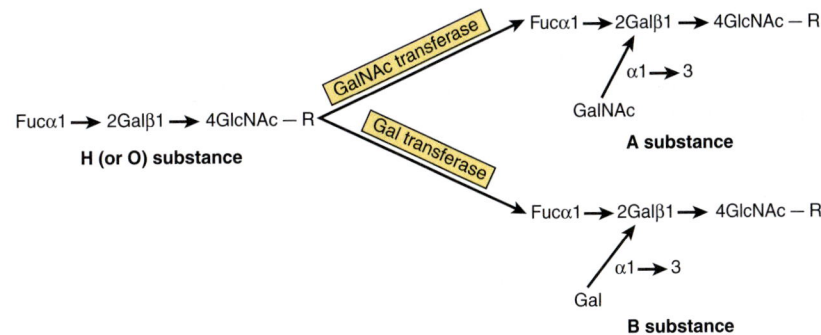

FIGURE 53–7 **Diagrammatic representation of the structures of the H, A, and B blood group substances.** R represents a long complex oligosaccharide chain, joined either to ceramide where the substances are glycosphingolipids, or to the polypeptide backbone of a protein via a serine or threonine residue where the substances are glycoproteins. Note that the blood group substances are biantennary; that is, they have two arms, formed at a branch point (not indicated) between the GlcNAc—R, and only one arm of the branch is shown. Thus, the H, A, and B substances each contains two of their respective short oligosaccharide chains shown above. The AB substance contains one type A chain and one type B chain.

in other cells. Individuals of *SeSe* or *Sese* genotypes secrete either or both A and B antigens whereas individuals of the *sese* genotype do not. However, red blood cells of both genotypes can express the A and B glycolipid antigens.

The *A* Gene Encodes a GalNAc Transferase, the *B* Gene a Gal Transferase, & the *O* Gene an Inactive Product

H substance, the blood group substance found in persons of type O, is the precursor of both the A and B substances (see Figure 53–7). It is formed by the action of a **fucosyltransferase,** coded for by the H locus, that catalyzes the addition of an α1 → 2 linked fucose onto the terminal Gal residue of its precursor:

$$GDP - Fuc + Gal - \beta - R \rightarrow Fuc - \alpha1, 2 - Gal - \beta - R + GDP$$
$$\text{Precursor} \qquad\qquad\qquad \text{H substance}$$

The *A* gene encodes a UDP-GalNAc-specific GalNAc transferase that adds a GalNAc to H substance, forming A substance. The *B* gene encodes a UDP-Gal-specific Gal transferase that adds the Gal residue to H substance to form B substance. Individuals of **type AB** possess both enzymes, and thus synthesize two oligosaccharide chains (see Figure 53–7), one with a terminal GalNAc and the other with a terminal Gal.

Anti-A antibodies recognize the additional GalNAc residue present in the A substance, and anti-B antibodies recognize the additional Gal residue found in the B substance. Type O individuals harbor a frameshift mutation in the gene encoding the terminal glycosyltransferase that results in the production of an inactive protein. Thus, H substance thus constitutes the O antigen.

An *h* allele arises when a mutation in the portion of H locus that codes for the fucosyltransferase yields an inactive enzyme. While individuals of the heterozygous H*h* genotype still can synthesize adequate levels of H substance, *hh* homozygous individuals cannot. Since H substance is the precursor for A and B substances, all individuals carrying

the *hh* genotype will have red blood cells of type O regardless of whether or not they express one of both of the A and B terminal glycosyltransferase(s). This is referred to as the Bombay phenotype (O*h*).

PLATELETS

Platelets Contain Mitochondria, But Lack a Nucleus

When megakaryocytes, the progenitors of red blood cells, are exposed to **thrombopoietin** they may fragment and form platelets (see Figure 53–1). Like red blood cells, platelets lack a nucleus, but unlike erythrocytes they possess mitochondria, lysozymes, and a tubular network that forms an **open canalicular system.** This honeycomb of channels increases the surface area of the platelets, which are spheroidal at rest, thereby facilitating the secretion of various endocrine and coagulation factors (see Chapter 55). These factors are stored inside the platelets within densely packed secretory vesicles, called granules. **Dense granules** contain Ca^{2+}, ADP, and serotonin, while **α-granules** contain fibrinogen, fibronectin, platelet-derived growth factor, von Willebrand factor, or other coagulation factors ready for release on receipt of an appropriate stimulus (see Chapter 55). Under normal circumstances, these small (2-μm diameter), enucleated cells circulate at a density of 2 to 4×10^5 platelets per milliliter of blood. While platelets derive the majority of their energy from metabolizing glucose, their mitochondria enables them to generate ATP via the β-oxidation of fatty acids.

Platelet Disorders Compromise Hemostasis

Abnormalities in platelet number or function can have serious physiologic consequences. For example, in **acute coronary syndrome,** the formation of enlarged, hyperreactive platelets results in an increased risk of **thrombosis,** the unregulated formation of blood clots that obstruct the circulation. The presence of larger than normal platelets also correlates with an increased frequency of myocardial infarction.

Immune thrombocytopenic purpura is caused by the formation of antibodies against ones own platelets. This autoimmune disorder is marked by depressed platelet counts (**thrombocytopenia**). When a platelet's surface becomes decorated with antibodies, it is subject to clearance by splenic macrophages. In some instances, platelet autoantibodies will bind to differentiating megakaryocytes, depressing platelet production. Thrombocytopenia also can occur when persons who are homozygous for a mutant variant of glycoprotein IIb/IIIa in which the leucine 33 is replaced by proline receive blood from a donor that is homo- or heterozygous for the wild-type form of this major platelet antigen. Exposure to the donor's platelets triggers the production of **alloantibodies** that attack not only the donated platelets, but the patient's endogenous platelets as well. In **neonatal alloimmune thrombocytopenia**, which affects roughly 1 in 200 term pregnancies, antibodies from the mother's circulation cross the placental barrier and attack platelets in the fetus' circulatory system. Thrombocytopenia also can be induced by drugs such as tamoxifen, ibuprofen, vancomycin, and many sulfonamides.

The symptoms of **hemolytic-uremic syndrome**, a disease of infants characterized by progressive kidney failure, include both thrombocytopenia and hemolytic anemia. By contrast, the abnormal bleeding associated with **von Willebrand disease** is caused by a genetic defect that compromises the ability of platelets to adhere to the endothelium, rather than a deficit in platelet number. Other bleeding disorders resulting from defects in platelet adherence include **Bernard-Soulier syndrome** (genetically inherited deficiency in glycoprotein 1b), and **Glanzmann thrombasthenia** (genetically inherited deficiency in the glycoprotein IIb/IIIa complex).

RECOMBINANT DNA TECHNOLOGY HAS HAD A PROFOUND IMPACT ON HEMATOLOGY

The bases of the **thalassemias** and of many **disorders of coagulation** (see Chapter 55) have been greatly clarified by investigations utilizing gene cloning and DNA sequencing, while the study of oncogenes and chromosomal translocations has advanced our understanding of the **leukemias**. As discussed earlier, recombinant DNA technology has made available therapeutically useful quantities of **erythropoietin** and **other growth factors**.

The first pathophysiologic condition to be treated by gene therapy was a deficiency of **adenosine deaminase**. Lymphocytes are particularly sensitive to deficits in this enzyme. In 1990, Dr. William French Anderson introduced a new copy of the gene, carried on a retroviral vector, into a 4-year-old girl suffering from severe combined immunodeficiency (bubble boy disease). Although the patient is still required to take medications, the replacement gene has remained stable into adulthood.

SUMMARY

- Major causes of anemia include blood loss, deficiencies of iron, folate, and vitamin B_{12}, and various factors that cause hemolysis.
- The shape of the red blood cell contributes to the efficiency of gas exchange and to its ability to undergo deformation when passing through capillaries.
- The production of red cells and platelets is regulated by erythropoietin, thrombopoietin, and other cytokines.
- Mature red blood cells, which lack internal organelles, are dependent on glycolysis to generate ATP.
- 2,3-Bisphosphoglycerate mutase catalyzes the isomerization of the glycolytic intermediate 1,3-bisphosphoglycerate to form 2,3-bisphosphoglycerate, which stabilizes T-state hemoglobin.
- Methemoglobin is unable to transport oxygen.
- Cytochrome b_5 reductase reduces the Fe^{3+} of methemoglobin to Fe^{2+}, restoring function.
- The red cell contains a battery of cytosolic enzymes—superoxide dismutase, catalase, and glutathione peroxidase—that catalyze the neutralization of reactive oxygen species.
- Deficiencies in the quantity or the activity of glucose-6-phosphate dehydrogenase, which produces NADPH, constitute a major cause of hemolytic anemia.
- Cytoskeletal proteins such as spectrin, ankyrin, and actin interacting with integral membrane proteins underlie the flexible biconcave shape of red blood cells.
- Deficiencies or defects of spectrin can lead to hereditary spherocytosis and hereditary elliptocytosis, both causes of hemolytic anemia.
- Band 4.1 protein facilitates the exchange of bicarbonate and chloride ions across erythrocyte membranes.
- The ABO blood group substances of the red cell membrane are complex glycosphingolipids. The immunodominant sugar of A substance is N-acetylgalactosamine, whereas that of B substance is galactose. O substance contains neither of these sugar residues.
- Platelets are small, enucleated fragments of larger precursor cells called megakaryocytes.
- When activated, platelets release effector molecules and fibrinogen stored in secretory granules.
- von Willebrand disease, a bleeding disorder, is caused by a genetic mutation that impairs the ability of platelets to adhere.

REFERENCES

Alkrimi J, George L: *Medical Diagnosis by Analysis of Blood Cell Images.* Lambert Academic Publishing, 2014.

Bain BJ: *Blood Cells. A Practical Guide,* 5th ed. Wiley Blackwell, 2015.

Bresnick E (ed): *Hematopoiesis.* Curr Topics Develop Biol vol 118, Academic Press, 2017.

Dzierzak E, Philipsen S: Erythropoiesis: development and differentiation. Cold Spring Harb Perspect Med 2013;3:a011601.

Hoffman R, Benz EJ Jr, Silberstein LE, et al (editors): *Hematology: Basic Principles and Practice.* Elsevier, 2017.

Ledford H: CRISPR gene therapy shows promise against blood diseases. Nature 2020;588:383.

Michelson A, Cattaneo M, Frelinger A, Newman P: *Platelets,* 4th ed. Academic Press, 2019.

Randolph TR: Hemoglobinopathies (Structural defects in hemoglobin). *Rodak's Hematology,* 6th ed. Keohane EM, Otto CN, Walenga JM (editors). Elsevier, 2020.

Smyth SS, Whiteheart S, Italiano JE Jr, Coller BS: Platelet morphology, biochemistry, and function. In: *Williams Hematology,* 8th ed. Kaushansky K, Lichtman MA, Beutler E, et al (editors). McGraw-Hill, 2010;1735.

Susanstad T, Fuangthon M, Tharakaraman K, et al: Modified recombinant human erythropoietin with potentially reduced immunogenicity. Sci Rep 2021;11:1491.

White Blood Cells

Peter J. Kennelly, PhD

- Describe how white blood cells work in concert to combat infection and to trigger an inflammatory response.
- List the basic steps in the elimination of infectious microorganisms by phagocytosis.
- Describe the role of chemotaxis in leukocyte function.
- List the key components found within the granules of phagocytes and basophils and describe their primary functions.
- List the reactive oxygen species produced during the respiratory burst.
- Explain the basis for the physiologic effects caused by defects in the NADPH oxidase system.
- Explain the molecular basis of type 1 leukocyte adhesion deficiency.
- Describe how neutrophils and eosinophils entrap parasites using neutrophil extracellular traps (NETs).
- Describe the role of the helper T cells in the production of new antibodies.
- Define the term cytokine and describe the key characteristics of interleukins, interferons, prostaglandins, and leukotrienes.

BIOMEDICAL IMPORTANCE

White blood cells, or **leukocytes**, serve as key sentries and potent defenders against invading pathogens. **Neutrophils**, the most abundant type of white blood cell, ingest and destroy invading bacteria and fungi by a process known as **phagocytosis**, while **eosinophils** phagocytize larger parasites. Circulating **monocytes** migrate from the bloodstream to diseased tissues, where they differentiate into phagocytic **macrophages**. **Granulocytes** such as **basophils** and **mast cells** release stored effectors that attract additional leukocytes to the site of infection and trigger an inflammatory response. **B lymphocytes** generate and release protective antibodies with the assistance of **T lymphocytes**. Other lymphocytes, such as **cytotoxic T cells** and **natural killer cells**, target virally infected and malignantly transformed host cells.

Malignant neoplasms of blood-forming tissues, called **leukemias**, can lead to the uncontrolled production of one or more of the major classes of white blood cells. The hyperactivation of granulocytes during an allergic response can, in extreme cases, lead to **anaphylaxis** and death. **Leukopenia**, a depression in the production of white blood cells, can result from physical injury or infection of the bone marrow, chemotherapy, ionizing radiation, infection by the **Epstein-Barr virus** (mononucleosis), an autoimmune response (**lupus**), or the displacement of bone marrow cells by fibrous tissues (**myelofibrosis**). The resulting deficit in the levels of circulating leukocytes can leave the affected individual **immunocompromised**, that is vulnerable to infection.

DEFENSE AGAINST INFECTION REQUIRES MULTIPLE CELL TYPES

The white blood cells, or **leukocytes**, are key participants in the **acute inflammatory response**, a multicomponent process that defends the body against infectious organisms and ameliorates the impact of tissue infection or morbidity. **Lymphocytes** produce protective antibodies that target foreign invaders and tag them for elimination. **Basophils** secrete hematologic effectors such as histamines (**Figure 54–1**) that facilitate the accumulation of fluid within infected or damaged

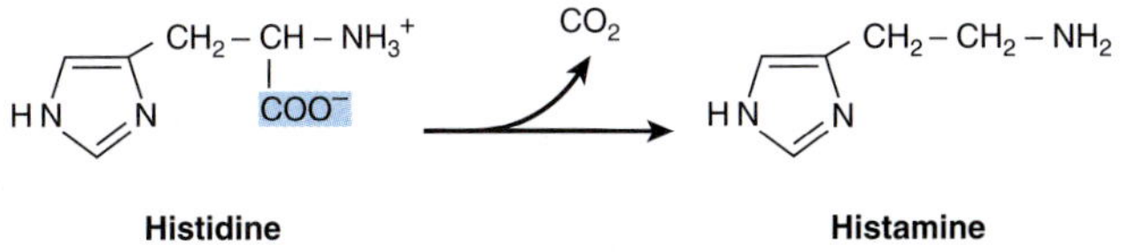

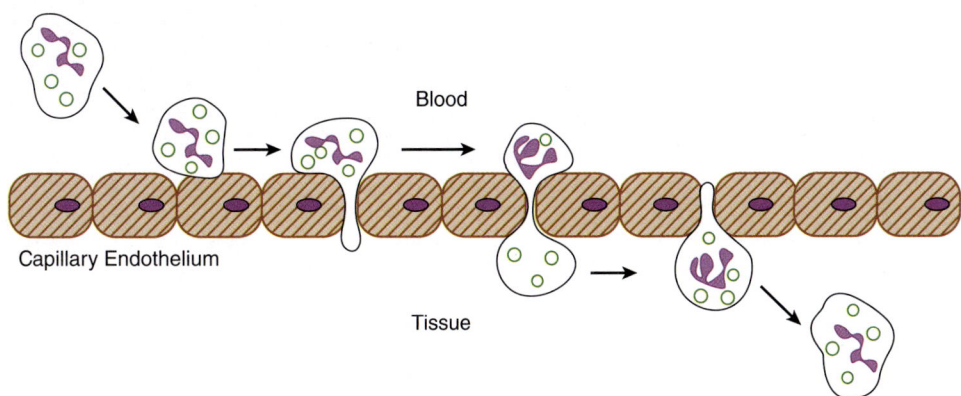

FIGURE 54–1 Structures of histidine and its decarboxylation product, histamine.

tissues as well as chemokines that attract additional **neutrophils**. Once activated, neutrophils encapsulate invading bacteria within membrane vesicles (**phagocytosis**) and destroy them using a combination of hydrolytic enzymes, reactive oxygen species (ROS), and antimicrobial peptides. Circulating **monocytes** migrate into the tissues where, on stimulation, they transform into phagocytic **macrophages**, which ingest and destroy infected and damaged host cells.

Leukocytes, unlike red blood cells and platelets, possess a full complement of internal organelles. However, the nuclei of many leukocytes exhibit marked deviations from the compact, spherical organelle typical of most eukaryotic cells. In monocytes, for example, the nuclei are unusually large and noticeably irregular in shape while in neutrophils, eosinophils, and other **polymorphonuclear leukocytes** they segment into multiple lobes.

MULTIPLE EFFECTORS REGULATE THE PRODUCTION OF WHITE BLOOD CELLS

Most white blood cells turn over rapidly and thus must be continually replaced. The lifetime of a circulating myeloid leukocyte, for example, ranges from a few hours to a few days, while most lymphocytes persist for only a few weeks in the blood. A notable exception to this pattern is **memory lymphocytes**, which may live for several years. The production of monocytes and granulocytes proceeds via the formation of a common myeloid progenitor, while differentiation of hematopoietic stem cells into lymphocytes proceeds via the formation of a

common lymphoid progenitor (see Figure 53–1). The proliferation of hematopoietic stem cells and the determination of their ultimate fate is controlled by the concerted influences of multiple effector molecules. Stem cell growth factor, granulocyte-macrophage colony-stimulating factor, and interleukins 5 and 6, for example, stimulate the production of granulocytes (neutrophils, eosinophils, basophils) and monocytes, a process that proceeds via the formation of myeloid progenitor cells. Tumor necrosis factor α, transforming growth factor β₁, and interleukins 2 and 7 promote the formation of lymphoid progenitor cells and their eventual maturation into B and T lymphocytes.

LEUKOCYTES ARE MOTILE

Leukocytes Migrate in Response to Chemical Signals

Leukocytes can be found throughout the body, migrating from the blood to sites of injury or infection in response to chemical signals, a process referred to as **chemotaxis**. Migration out of the circulation takes place via **diapedesis**, an amoeboid mechanism involving the cytoskeleton-mediated contortion of the cell that begins with the extension of a thin pseudopod between the cells of the capillary epithelium (**Figure 54–2**). Once the pseudopod becomes anchored on the other side, the contents of the cell are squeezed through the projection by cytoskeletal proteins, filling the distal end of the pseudopod to form a new, translocated cell body. Once inside the tissues, locomotion proceeds via a similar, stepwise amoeboid mechanism.

Chemotaxis Is Mediated by G-Protein–Coupled Receptors

Leukocytes are attracted to tissues by **chemotactic factors** such as chemokines, complement fragment C5a, small peptides derived from bacteria (eg, N-formyl-methionyl-leucyl-phenylalanine), and several leukotrienes. These factors bind

FIGURE 54–2 Diapedesis. Shown, from left to right, are the major steps in diapedesis, the process by which neutrophils and other leukocytes traverse the capillary wall, whose cells are shown in red, in response to chemotactic signals. Cell nuclei are shown in purple and granules in green.

to one of several cell surface receptors that share similar transmembrane domains comprised of seven membrane-spanning α-helices. Because these receptors all are closely coupled with one or more heterotrimeric guanosine nucleotide-binding proteins (**G-proteins**), they are often referred to as G-protein–coupled receptors. On ligand binding, a signal transduction cascade is initiated in which G-proteins activate **phospholipase C**, which hydrolyses phosphatidylinositol 4,5-bisphosphate to produce diacylglycerols and the water-soluble second messenger **inositol 1,4,5-triphosphate** (IP_3). The surge in IP_3 triggers the release of Ca^{2+} into the cytoplasm. In neutrophils, the appearance of cytoplasmic Ca^{2+} activates the components of the actin-myosin cytoskeleton responsible for effecting cell migration and granule secretion. Diacylglycerol, together with Ca^{2+}, stimulates protein kinase C and induces its translocation from the cytosol to the plasma membrane, where it catalyzes the **phosphorylation** of various proteins, including some involved in triggering the respiratory burst (see later).

Chemokines Are Stabilized by Disulfide Bonds

Chemokines are small, generally 6 to 10 kDa, proteins secreted by activated white blood cells in order to attract additional neutrophils to a site of infection or injury. Chemokines can be divided into four subclasses based on the number and spacing of the cysteine residues that participate in the disulfide bonds that stabilize the protein's conformation. Type C chemokines are characterized by a single intrachain disulfide bond linking a pair of conserved cysteine residues. In addition to this conserved disulfide bond, the other three recognized chemokine groups possess a second disulfide bond (**Figure 54–3**). In type

CC chemokines, one of the additional cysteine residues lies adjacent to the first of the universally conserved residues. In types CXC and CX_3C, these cysteines are separated by one and three intervening amino acid residues, respectively. CX_3C chemokines, the largest of the four types of cytokines, are covalently modified by glycosylation at several sites in their longer C-terminus.

Integrins Facilitate Diapedesis

The adhesion of leukocytes to vascular endothelial cells is mediated by transmembrane glycoproteins of the **integrin** and **selectin** families (see the discussion of **selectins** in Chapter 46). **Integrins** consist of noncovalently associated α and β subunits, each of which consists of an extracellular, transmembrane, and intracellular segment. The extracellular segments bind to various extracellular matrix proteins that possess Arg-Gly-Asp sequences, while the intracellular domains bind to cytoskeletal components such as actin and vinculin. Their ability to bridge the exterior and interior of a cell enables integrins to link leukocyte responses (eg, movement and phagocytosis) to changes in the surrounding environment. Some integrins of specific interest with regard to neutrophils are listed in **Table 54–1**.

Type 1 leukocyte adhesion deficiency is caused by the lack of the $β_2$ subunit (also designated CD18) of LFA-1 and of two related integrins found in neutrophils and macrophages, Mac-1 (CD11b/CD18) and p150,95 (CD11c/CD18). This impairs the ability of the affected leukocytes to adhere to vascular endothelial cells, impeding diapedesis. Fewer white blood cells enter their infected tissues; thereby lowering the resistance of the affected individuals to bacterial and fungal infections.

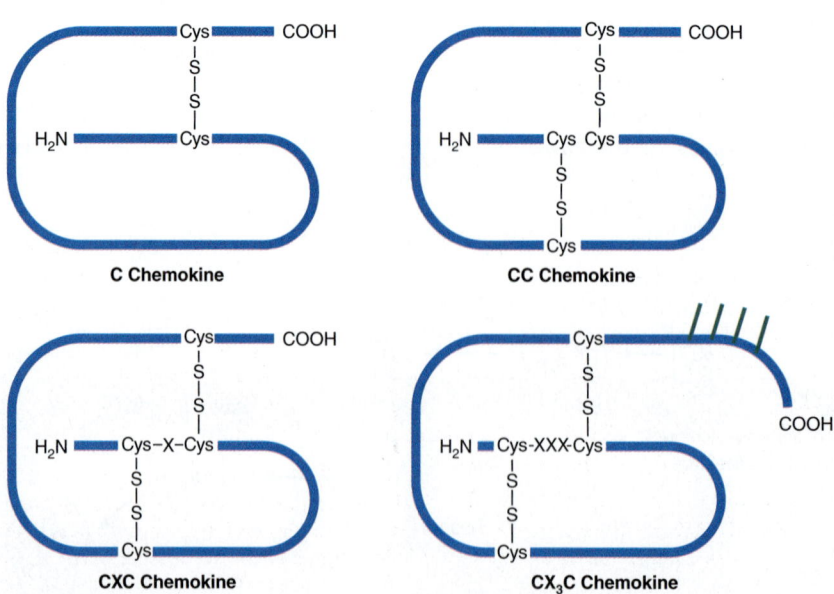

FIGURE 54–3 Chemokines. This figure depicts the key structural features of type C, CC, CXC, and CX_3C chemokines. The polypeptide chains are depicted in blue with their amino and carboxy termini marked by H_2N and COOH, respectively. Key cysteine residues are denoted as Cys, conserved disulfide bonds at S-S, and spacer amino acids for types CXC and CX_3C using X. Bound carbohydrate is depicted in green.

TABLE 54–1 Principal Integrins of White Blood Cells & Platelets[a]

Integrin	Cell	Subunit	Ligand	Function
VLA-1 (CD49a)	WBCs, others	α1β1	Collagen, laminin	Cell-ECM adhension
VLA-5 (CD49e)	WBCs, others	α5β1	Fibronectin	Cell-ECM adhesion
VLA-6 (CD49f)	WBCs, others	α6β1	Laminin	Cell-ECM adhesion
LFA-1 (CD11a)	WBCs	αLβ2	ICAM-1	Adhesion of WBCs
Glycoprotein IIb/IIIa	Platelets	αIIbβ3	ICAM-2	
Fibrinogen, fibronectin, von Willebrand factor	Platelet adhesion and aggregation			

[a]CD, cluster of differentiation; ECM, extracellular matrix; ICAM, intercellular adhesion molecule; LFA-1, lymphocyte function-associated antigen 1; VLA, very late antigen.
Note: A deficiency of LFA-1 and related integrins is found in type I leukocyte adhesion deficiency (OMIM 116920). A deficiency of platelet glycoprotein IIb/IIIb complex is found in Glazmann thrombasthenia (OMIM 273800), a condition characterized by a history of bleeding, a normal platelet count, and abnormal clot retraction. These findings illustrate how fundamental knowledge of cell surface adhesion proteins is shedding light on the causation of a number of diseases.

INVADING MICROBES & INFECTED CELLS ARE DISPOSED BY PHAGOCYTOSIS

Phagocytes Ingest Target Cells

White blood cells typically destroy invading microorganisms via **phagocytosis (Figure 54–4)**. While phagocytic leukocytes possess cell surface receptors against bacterial lipopolysaccharides or peptidoglycans, in most cases they recognize infective pathogens indirectly, by the presence of antibodies or complement factors that have previously adhered to their surface (see Chapter 52). The process of tagging an invader with protective proteins to facilitate recognition by phagocytic leukocytes is called **opsonization**.

Pathogen binding triggers dramatic alterations in the shape of the phagocyte, which proceeds to envelop the target cell until it is encased within an internalized membrane vesicle called a **phagosome** (phagolysosome). The encapsulated invader is then destroyed using a combination of hydrolytic enzymes (eg, lysozyme, proteases), antimicrobial peptides (defensins), and reactive oxygen species. This arsenal of toxins and degradatory enzymes are stored in cytoplasmic vesicles known as **granules**, which fuse with the phagosome (**Table 54–2**). Since these granules can be observed under a microscope, the cells that harbor them are referred to as **granulocytes**. Eventually, after digestion of the microbial invader and absorption of their component sugars, amino acids, etc., the phagosome migrates to and fuses with the plasma membrane of the white blood cell, expelling any remaining debris.

The components of this debris, which include fragments of proteins, oligosaccharides, lipopolysaccharides, peptidoglycans, and polynucleotides, provide an important source of antigens for stimulating the production of new antibodies. **Helper T cells** and other leukocytes absorb these materials via endocytosis (see Figure 40–21), then route them to the cell surface in association with a membrane protein called the **major histocompatibility complex (MHC)**. The MHC serves as a scaffold for presenting potential antigens to surrounding lymphocytes in a form likely to stimulate the production of new antibodies.

The three principal classes of phagocytic leukocytes are **neutrophils**, **eosinophils**, and **macrophages**. Neutrophils, which comprise roughly 60% of the white blood cells present

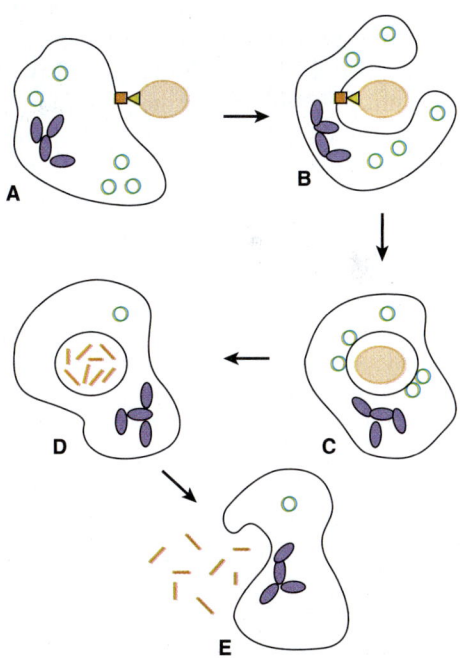

FIGURE 54–4 Phagocytosis. This figure depicts the destruction of an opsonized microorganism, shaded in ORANGE, by a neutrophil via phagocytosis. The multilobed nucleus of the neutrophil is shown in purple, secretory granules in green. The presence of an antibody or complement tag is indicated by a yellow triangle, with the corresponding cell surface receptor as a bright orange square. Cellular debris from the microorganism is represented as orange line segments. **(A)** The neutrophil binds an antigen molecule on the opsonized microbe via a receptor. **(B)** The neutrophil envelops the microbe. **(C)** Secretory granules fuse with the newly internalized phagosome, delivering their contents. **(D)** Granule-derived enzymes and cytotoxins destroy the microorganism. **(E)** The phagosome then fuses with the cell membrane, expelling any remaining debris.

TABLE 54–2 Enzymes & Proteins of the Granules of Phagocytic Leukocytes

Enzyme or Protein	Reaction Catalyzed or Function	Comment
Myeloperoxidase (MPO)	$H_2O_2 + X^-$ (halide) $+ H^+ \rightarrow HOX + H_2O$ where $X^- = Cl^-$, HOX = hypochlorous acid	Responsible for the green color of pus Genetic deficiency can cause recurrent infections
NADPH oxidase	$2O_2 + NADPH \rightarrow 2O_2^{--} + NADP + H^+$	Key component of the respiratory burst Deficient in chronic granulomatous disease
Lysozyme	Hydrolyzes link between N-acetylmuramic acid and N-acetyl-D-glucosamine found in certain bacterial cell walls	Abundant in macrophages. Hydrolyzes bacterial peptidoglycans
Defensins	Basic antibiotic peptides of 20-33 amino acids	Apparently kill bacteria by causing membrane damage
Lactoferrin	Iron-binding protein	May inhibit growth of certain bacteria by binding iron and may be involved in regulation of proliferation of myleoid cells
Elastase Collagenase Gelatinase Cathepsin G	Proteases	Abundant in phagocytes; Breakdown protein components of infectious organisms; Generate fragments for antigen presentation

in the circulation, phagocytize bacteria and small eukaryotic microorganisms such as fungi. The less numerous **eosinophils**, which make up 2 to 3% of the leukocytes in the blood, ingest larger eukaryotic microorganisms such as **paramecia**. Macrophages are derived from monocytes, which comprise about 5% of the leukocytes in the blood. Monocytes migrate from the bloodstream into tissues throughout the body where, on receipt of a stimulus, they differentiate to form **macrophages**. While macrophages can ingest invading microbes, the signature function of these large phagocytes is to destroy human host cells that have been compromised by infection, malignant transformation, or programmed cell death, also known as **apoptosis**. These functionally compromised cells are recognized by the appearance of aberrant proteins and oligosaccharides on their surface. Precocious activation of macrophages is associated with the etiology of many degenerative diseases such as osteoporosis, atherosclerosis, arthritis, and cystic fibrosis. They also can facilitate the metastasis of cancer cells.

Phagocytic Leukocytes Generate Reactive Oxygen Species During the Respiratory Burst

Phagocytes employ **ROS** such as, superoxide, H_2O_2, hydroxyl radical, and HOCl (hypochlorous acid) as major components of the chemical and enzymatic arsenal used to destroy ingested cells. Production of ROS begins shortly (15-60 seconds) after internalization of an encapsulated cell, using O_2 and electrons derived from NADPH. The accompanying surge in oxygen consumption has been termed the **respiratory burst**. Producing the large quantities of NADPH required via the pentose phosphate pathway (see Chapter 20) is facilitated by the phagocyte's heavy reliance on aerobic glycolysis to generate ATP, a consequence of the low number of mitochondria contained within them.

The first step in the formation of microbicidal ROS during the respiratory burst is the synthesis of superoxide, which is catalyzed by the **NADPH oxidase system**. Catalysis proceeds via a two-step mechanism, the reduction of molecular oxygen to form superoxide (see Table 54–2):

$$2O_2 + NADPH + H+ \rightarrow 2O_2^{\bullet} + NADP^+ + H$$

followed by the spontaneous dismutation of **hydrogen peroxide** from two molecules of superoxide:

$$O_2^{\bullet-} + O_2^{\bullet-} + 2H^+ \rightarrow H_2O_2 + O_2$$

The NADPH oxidase system is comprised of **cytochrome b_{558}**, a plasma membrane–associated heterodimer, and two cytoplasmic polypeptides of 47 and 67 kDa. On activation, the cytoplasmic peptides are recruited to the plasma membrane where they associate with cytochrome b_{558} to form the active complex. Flux through the pentose phosphate cycle, the cell's primary source of NADPH, also increases markedly during phagocytosis. The cell is protected from any superoxide that may escape from the phagosomes by **superoxide dismutase**, which catalyzes the disproportionation of two superoxide radical anions into one molecule each of H_2O_2 and O_2. The hydrogen peroxide can be used as a substrate for myeloperoxidase (see later) or disposed of by the action of glutathione peroxidase or catalase.

Myeloperoxidase Catalyzes the Production of Chlorinated Oxidants

The formation of hypohalous acids during the respiratory burst is catalyzed by the enzyme **myeloperoxidase**.

$$H_2O_2 + X^- + H^+ \xrightarrow{\text{MYELOPEROXIDASE}} HOX + H_2O$$

($X^- = Cl^-$, Br^-, I^- or SCN^-; HOX = hypohalous acid)

Present in large amounts in neutrophil granules, this enzyme catalyzes the oxidization of Cl^- and other halides by H_2O_2, which yields **HOCl** and other hypohalous acids. HOCl, the active ingredient of household liquid bleach, is a powerful oxidant that is highly microbicidal. When used to sterilize wounds, it reacts with any primary or secondary amines present to produce various nitrogen-chlorine derivatives. While **chloramines** are less powerful oxidants than HOCl, their microbicidal effects are much less likely to cause collateral damage to surrounding tissue.

Mutations Affecting the NADPH Oxidase System Cause Chronic Granulomatous Disease

Chronic granulomatous disease is caused by functionally deleterious mutations in the genes encoding any of the four polypeptides of the NADPH oxidase system that impair its ability to produce ROS. This impairment, in turn, undermines the ability of phagocytic leukocytes to kill ingested pathogens. Although relatively uncommon, persons suffering from this condition experience recurrent infections. In an attempt to wall off persistent infections, the body develops chronic inflammatory lesions called granulomas in the skin, lungs, and lymph nodes. In some cases, relief can be provided by the administration of gamma interferon in an effort to increase transcription of the 91-kDa component of cytochrome b_{558}.

NEUTROPHILS & EOSINOPHILS EMPLOY NETS TO ENTRAP PARASITES

In addition to ingesting small microorganisms such as bacteria by phagocytosis, neutrophils and eosinophils can assist in the elimination of larger invaders by trapping them within webs called **neutrophil extracellular traps** or **NETs (Figure 54–5)**. The strands for these NETs are composed of polynucleotide strands generated by the dispersal, or **decondensation**, of a neutrophil's chromosomal DNA. This process involves rupture of the nuclear membrane and the disruption of the favorable charge–charge interactions that normally stabilize chromatin. Dissolution of histone-polynucleotide complexes is promoted by the enzyme **peptidylarginine deiminase**, which catalyzes the deimination of the strongly basic side chains of arginine residues to form neutral citrulline residues (**Figure 54–6**). Some chromatin proteins do remain associated with the DNA, forming cross-links between the polynucleotide strands. Granule membranes also rupture at this time, releasing their contents into the cytoplasm where they can bind to the decondensing polynucleotide strands, decorating the DNA with granule-derived proteases, antimicrobial peptides, and other factors. Eventually, the neutrophils lyse, unleashing their NETs on invading parasites in order to immobilize them and hinder their spread.

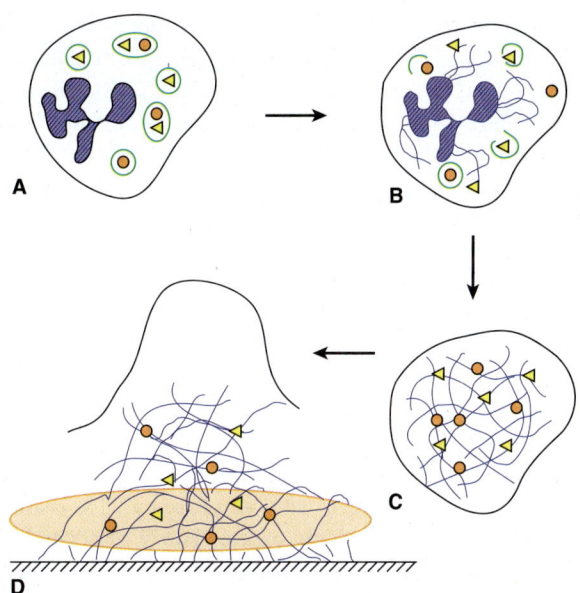

FIGURE 54–5 Trapping parasites using NETs. The figure depicts the basic stages in the formation and deployment of a DNA-based web by a neutrophil or eosinophil to trap a parasitic microorganism. **(A)** Resting neutrophil. The multilobed nucleus is shown in hatched purple, intracellular granules in green, and granule enzymes and cytotoxins as orange circles and yellow triangles. **(B)** On stimulation, the membranes encasing the nucleus and granules rupture, releasing enzymes, cytotoxins, and strands of DNA (purple) from decondensing chromosomes. **(C)** The DNA strands form a mesh that fills the interior of the cell to which some granule-derived proteins adhere. **(D)** The neutrophil lyses, releasing its DNA-protein web, which entraps the parasite (orange) against surface of the epithelium (hatched).

PHAGOCYTE-DERIVED PROTEASES CAN DAMAGE HEALTHY CELLS

Macrophages and other phagocytes produce numerous proteases (see Table 54–2), several of which can hydrolyze elastin, collagen, and other proteins present in the extracellular matrix. Although it is normal for small amounts of elastase

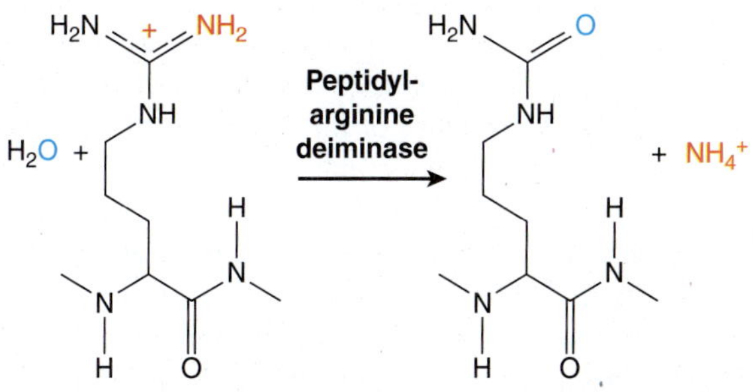

FIGURE 54–6 Citrullination. The enzyme peptidyl arginine deiminase displaces one of the imino groups (red) on the side chain of arginine by an oxygen atom (blue) derived from water. The net result is to replace a positive charge provided by the protonated arginine side chain by an amide, which is neutral.

and other proteinases leak into tissues throughout the body, their activities are normally kept in check by **antiproteinases** present in plasma and the extracellular fluid (see Chapter 52). One of these, α_2-**macroglobulin**, attracts a wide range of proteases with a "bait" region. Cleavage of a peptide bond in the bait region springs a conformational change that traps the protease responsible in a noncovalent complex with the antiproteinase, blocking further proteolytic assaults.

Disruption of the balance between proteases and antiproteinases can have deleterious effects. For example, when inadequate drainage leads to the accumulation of large numbers of neutrophils, considerable **tissue damage** can result. In the lungs, a genetic defect that renders elastase insensitive to α_1-antiproteinase inhibitor (α_1-antitrypsin) and other antiproteinases contributes significantly to the pulmonary tissue damage associated with emphysema. The elevated levels of chlorinated oxidants such as HOCl associated with inflammation can activate several of the proteinases listed in Table 54–2, while simultaneously inactivating several countervailing antiproteinases. In addition, these inhibitory proteins can themselves be degraded by proteases. For example, tissue inhibitor of metalloproteinases and α_1-antichymotrypsin can be hydrolyzed by elastase while α_1-antiproteinase inhibitor can be hydrolyzed by collagenase and gelatinase.

LEUKOCYTES COMMUNICATE USING SECRETED EFFECTORS

The development of the immune and inflammatory responses by injured or infected tissues requires the coordinated action of leukocytes and other cells. Much of this coordination is accomplished by secreting a diverse set of small (<25 kDa) proteins, termed **cytokines**, that includes interleukins, interferons, and chemokines.

The more than three dozen known **interleukins** derive their name from the cells in which they are synthesized and from which they are secreted. They are generally designated by the class abbreviation **IL** followed by an identifying number, for example, IL1, IL3, IL22. The **interferons (IFN)**, on the other hand, derive their name from their ability to inhibit, or interfere, with the replication of infecting viruses. Approximately 10 distinct families of interferons have been identified in animals to date. **Chemokines** attract and activate migrating leukocytes to a site of injury or infection. Most cytokines are glycosylated. In general, they stimulate both the type of leukocytes from which they are secreted (**autocrine signaling**) as well as other types of leukocytes (**paracrine signaling**). Historically, cytokines have been distinguished from hormones by their close association with immunity and inflammation.

Leukocytes also secrete lipid mediators, called **eicosanoids**, produced by the oxidation of arachidonic acid (see Chapter 15). These lipid mediators fall into two broad classes, **leukotrienes** and **prostaglandins**. Leukotrienes are characterized by the presence of a set of three conjugated carbon–carbon double bonds. Several incorporate the amino acid cysteine into their structure. Prostaglandins, which were first isolated from the prostate gland, contain 20 carbon atoms and are distinguished by their common five-membered ring.

Histamine (see Figure 54–1), which is synthesized by decarboxylating the amino acid histidine, is secreted in large amounts by activated **basophils** and **mast cells**. Histamine works with other hematologic factors, such as heparin and eicosanoids, to maintain blood flow to a site of injury or infection and stimulate the accumulation of fluid (edema). The accumulated fluid at the site of inflammation facilitates leukocyte migration, thereby enhancing the immune response.

LYMPHOCYTES PRODUCE PROTECTIVE ANTIBODIES

Lymphocytes make up approximately 30% of the leukocytes present in the blood. Their capacity to produce novel protective antibodies against newly encountered antigens (see Chapter 53) makes them the cornerstone of the body's **adaptive immune system**. The classification of lymphocytes into B and T types originally was based on the identity of the tissues in which each form completed their maturation. In avian species, the B lymphocytes (B cells) are processed in the **bursa of Fabricius**. The B cells in humans, which lack this organ, mature in the bone marrow. Maturation of **T lymphocytes (T cells)** takes place in the **thymus**. B lymphocytes secrete the soluble antibodies present in bodily humors, for example, plasma and interstitial fluids, and are therefore said to confer **humoral immunity**.

Lymphocytes that have not yet been stimulated to produce immunoglobulins are said to be **naïve**. Synthesis of a new antibody by a naive lymphocyte can be triggered in several ways. Glycoproteins, lipopolysaccharides, or peptidoglycans from an infecting microorganism can bind to their corresponding receptor on the lymphocyte's surface. Alternatively, the lymphocyte can be activated by encountering an antigen that has been presented on the surface of another white blood cell in association with the MHC. **Plasma cells** such as macrophages, neutrophils, and phagocytic lymphocytes all possess the capacity to display or present fragments of macromolecules derived from phagocytized microbes. **Helper T cells** also can present antigens on their surface, including material derived from debris ejected by phagocytes that they have ingested by endocytosis. Helper T cells also serve as "cellular switchboards," coordinating the immune response by receiving, processing, and sending signals from and to other components of the immune system.

Cytotoxic T cells recognize proteins that appear on the surface of host cells as a consequence of viral infection or oncogenic transformation. Once bound, they induce the lysis of the target cell using perforins, proteins that form channels in the plasma membrane, and proteases called granzymes, which mimic the action of the cathepsin proteases that normally trigger programmed cell death (**apoptosis**). **Natural killer cells** resemble cytotoxic T cells, but contain granules holding additional toxic chemicals to aid in their attack.

SUMMARY

- The elimination of infectious microorganisms involves the cumulative actions of multiple types of leukocytes, including lymphocytes, phagocytes, and basophils.

- White blood cells communicate using secreted effector molecules such as chemokines, prostaglandins, leukotrienes, interleukins, and interferons.

- Leukocytes migrate from the blood to the tissues in response to specific chemical attractants, a process termed chemotaxis.

- The amoeboid migration of leukocytes from blood to the tissues relies on cytoskeletally mediated cell flexibility and deformation.

- Basophils secrete histamine and heparin, which facilitate the migration of leukocytes by inducing fluid accumulation at a site of infection or injury.

- Integrins mediate the adhesion of white blood cells to the vascular endothelium, the first step in migration toward infected tissues.

- Phagocytes internalize invading microorganisms inside membrane vesicles called phagosomes.

- Once inside the phagosomes, foreign microorganisms are destroyed using a combination of reactive oxygen species (the respiratory burst), hydrolytic enzymes, and cytotoxic peptides.

- Mutations in proteins of the NADPH oxidase system cause chronic granulomatous disease.

- Neutrophils and eosinophils defend against large parasites by immobilizing them within webs comprised of strands of decondensed chromosomal DNA.

- Decondensation of chromosomal DNA is facilitated by citrullination of arginine side chains on histones.

- Lymphocytes produce protective immunoglobulins (antibodies) that confer humoral immunity.

- Phagocytes and helper T cells stimulate the production of new antibodies by presenting fragments of pathogen-derived macromolecules on their surface.

- Cytotoxic T cells and natural killer cells recognize and destroy host cells that display cell surface proteins and glycolipids characteristic of viral infection or malignant transformation.

REFERENCES

Adkis M, Burgler S, Crameri R, et al: Interleukins, from 1 to 37, and interferon-γ: receptors, functions, and roles in diseases. J Allergy Clin Immunol 2011;127:701.

Gulati G, Caro J: *Blood Cells: Morphology and Clinical Relevance*, 2nd ed. American Society of Clinical Oncology, 2014.

Henderson GI: *Leukocytes: Biology, Classification and Role in Disease*. Novoc Sci Publ 2012.

Hillman R, Ault K, Rinder H: *Hematology in Clinical Practice*. 5th ed. McGraw Hill, 2010.

Kloc M (editor): *Macrophages. Origin, Function and Biointervention*. Springer, 2017.

Masopust D, Ahmed R (editors): *T-Cell Memory*. Cold Spring Harbor Press, 2021.

Mayadas TN, Cullere X, Lowell CA: The multifaceted functions of neutrophils. Annu Rev Pathol 2014;9:181.

Melo RCN, Dvorak AM, Weller PF: *Eosinophil Ultrastructure. Atlas of Eosinophil Cell Biology and Pathology*. Elsevier, 2021.

Nordenfelt P, Tapper H: Phagosome dynamics during phagocytosis by neutrophils. J Leukocyte Biol 2011;90:271.

Ribatti D: *The Mast Cell. A Multifunctional Effector Cell*. Springer, 2019.

Wynn TA, Chawla A, Pollard JW: Macrophage development in development, homeostasis and disease. Nature 2013;496:445.

Exam Questions

Section X – Special Topics (B)

1. Briefly describe the mode of action of nitroglycerin, a common agent for treating angina.

2. Patients being treated for heart failure oftentimes exhibit decreased expression and defective regulation of SERCA2a, the principal Ca^{2+}-ATPase of the sarcoplasmic reticulum. Explain how defects in this protein might contribute to deterioration in cardiac function.

3. Select the one of the following that is NOT CORRECT:
   A. The troponin system regulates contraction of smooth muscle.
   B. Muscle contraction takes place via a sliding filament mechanism.
   C. Myosin light-chain kinase phosphorylates the regulatory light chains in the myosin head domain.
   D. F-actin is formed via the polymerization of G-actin.
   E. Ca^{2+} both activates muscle contraction and stimulates its own removal by activating the Ca^{2+} ATPase.

4. A patient anesthetized using a halothane compound exhibits a marked rise in body temperature, a behavior indicative of malignant hyperthermia (HT). Select the one of the following statements that is NOT CORRECT:
   A. MH can arise from mutations that alter the amino acid sequence of the Na^+-K^+-ATPase.
   B. MH can arise from mutations that alter the amino acid sequence of the ryanodine-sensitive Ca^{2+} release channel.
   C. The muscle rigidity that occurs during MH is triggered by the presence of high concentrations of Ca^{2+} in the cytoplasm.
   D. MH can arise from mutations that alter the amino acid sequence of the voltage-gated, slow K-type Ca^{2+} channel.
   E. MH can be treated by intravenous administration of dantrolene, which inhibits release of Ca^{2+} from the sarcoplasmic reticulum into the cytosol.

5. Select the one of the following statements that is NOT CORRECT:
   A. Fast-twitch fibers rely heavily on creatine phosphate to regenerate ATP.
   B. Slow-twitch fibers appear red because they contain hemoglobin.
   C. Fast-twitch fibers contain relatively few mitochondria.
   D. Marathoners try to increase the quantity of glycogen in their muscles by eating carbohydrate-rich meals before an event (carbo loading).
   E. Skeletal muscle serves as the major reserve of protein in the body.

6. Select the one of the following that is NOT a feature of the contractile cycle in striated muscle:
   A. Binding of Ca^{2+} to troponin C uncovers the myosin-binding sites on actin.
   B. The power stroke is initiated by the release of P_i from the actin-myosin-ADP-P_i complex.

   C. Release of ADP from the actin-myosin-ADP complex is accompanied by a large change in the conformation of myosin's head domain (relative to its tail domain).
   D. The binding of ATP by myosin increases its affinity for actin.
   E. Rigor mortis results for the inability of actin to release from the actin-myosin complex when cells are deficient in ATP.

7. Select the one of the following that does NOT serve as a major energy reserve for replenishing ATP in muscle tissue:
   A. Glycogen
   B. Creatine phosphate
   C. ADP (in conjunction with adenylyl kinase)
   D. Fatty acids
   E. Epinephrine

8. Select the one of the following statements that is NOT CORRECT:
   A. The drugs colchicine and vinblastine inhibit microtubule assembly.
   B. Mutations affecting keratin can lead to blistering.
   C. Mutations in the gene encoding lamin A and lamin C cause progeria (accelerated aging).
   D. α- and β-tubulin are the major components of stress fibers.
   E. Molecular motors such as dynein, kinesis, and dynamin power ciliary movement, vesicle transport, and endocytosis.

9. Select the one of the following statements that is NOT CORRECT:
   A. The major function of the Ca^{2+} channels in cardiomyocytes is to admit extracellular calcium ions into the cell in order to trigger Ca^{2+}-induced Ca^{2+} release from the SR.
   B. Digitalis increases the forcefulness of cardiac contractions by raising the level of intracellular Na^+.
   C. Certain types of muscular dystrophies are caused by mutations in enzymes called glycosyltransferases.
   D. Dantrolene relaxes skeletal muscle by inhibiting the release of Ca^{2+} from the SR.
   E. In the SR, Ca^{2+} is bound to a specific Ca^{2+}-binding protein called calmodulin.

10. Describe the role of haptoglobin in the protection of the kidneys from the potentially damaging effects of extracorpuscular hemoglobin.

11. Briefly describe how activation of cytidine deaminase helps generate immunoglobulins with unique antigen-binding sites.

12. Select the one of the following statements that is NOT CORRECT:
    A. Interleukin 1 stimulates the production of acute-phase proteins.
    B. Iron must be reduced to the ferrous (Fe^{2+}) state in order to be recovered via the transferrin cycle.
    C. Many complement proteins are zymogens.
    D. The type 2 transferrin receptor (TfR2) functions primarily as an iron sensor.
    E. Mannose-binding lectin binds carbohydrate groups present on the surface of invading bacteria.

13. Select the one of the following statements that is NOT CORRECT:

    A. Albumin is synthesized as a proprotein.
    B. Albumin is stabilized by multiple intrachain disulfide bonds.
    C. Albumin is a glycoprotein.
    D. Albumin facilitates the movement of fatty acids through the circulation.
    E. Albumin is the major determinant of plasma osmotic pressure.

14. Select the one of the following statements that is NOT CORRECT:

    A. Wilson disease can be treated using copper chelators such as penicillamine.
    B. Wilson disease is characterized by copper toxicosis (abnormally high levels of copper).
    C. Wilson disease is caused by mutations in the gene encoding ceruloplasmin.
    D. Albumin facilitates the movement of sulfonamide drugs through the circulation.
    E. Albumin can be lost from the body if the intestinal mucosa becomes inflamed.

15. You encounter a 50-year-old woman in the clinic who is pale and tired. You suspect that she is suffering from iron-deficiency anemia and prescribe a series of laboratory tests. Select the one of the following potential test outcomes that would NOT be consistent with your provisional diagnosis.

    A. Lower than normal levels of red cell protoporphyrin
    B. Increased saturation of transferrin
    C. Increased expression of TfR
    D. Increased levels of plasma hepcidin
    E. Decreased levels of hemoglobin

16. Select the one of the following that is NOT a potential cause of amyloidosis:

    A. Accumulation of β_2-macroglobulin
    B. Deposition of fragments derived from immunoglobulin light chains
    C. Accumulation of degradation products of serum amyloid A
    D. Presence of mutationally altered forms of transthyretin
    E. Amylase deficiency

17. Select the one of the following statements that is NOT CORRECT:

    A. All immunoglobulins contain at least two heavy-chain polypeptides and two light-chain polypeptides.
    B. Immunoglobulin polypeptide chains are linked together by disulfide bonds.
    C. Immunoglobulins are multivalent.
    D. Immunoglobulins are glycosylated.
    E. Immunoglobulins are primary components of the body's innate immune system.

18. Explain the linkage how a deficiency in glucose-6-phosphate dehydrogenase within erythrocytes can lead to hemolytic anemia.

19. Select the one of the following statements that is NOT CORRECT:

    A. The high surface area of biconcave red blood cells facilitates gas exchange.
    B. Hereditary elliptocytosis can be caused by defects in or a deficiency of spectrin.

    C. The diameter of red blood cells exceeds that of many peripheral capillaries.
    D. Protein 4.1 helps link the erythrocyte cytoskeleton to proteins in the cell's plasma membrane.
    E. In order to pass through narrow capillaries, red blood cells must be squeezed into a compact, spherical shape.

20. Select the one of the following statements that is NOT CORRECT:

    A. Red blood cells contain high levels of superoxide dismutase.
    B. A and B substances are formed by the addition of fucose and N-acetylglucosamine, respectively, to H substance.
    C. Platelets generate ATP exclusively via glycolysis.
    D. Mature red blood cells are devoid of internal organelles.
    E. Erythrocyte membranes contain high levels of the Band 3 anion exchange protein.

21. Select the one of the following statements that is NOT CORRECT:

    A. Erythropoietin stimulates the formation of red blood cells from hematopoietic stem cells.
    B. Multipotent stem cells are able to differentiate into cells of a closely related type.
    C. Carbonic anhydrase increases the capacity of red blood cells to transport CO_2.
    D. GLUT1 mediates the active transport of glucose into erythrocytes.
    E. Hypoxia stimulates the production of erythropoietin by the kidneys.

22. A patient recently exposed to aniline displays bluish discoloration of their skin and mucous membranes. Select a plausible provisional diagnosis from the following list:

    A. Methemoglobinemia
    B. Hereditary hemochromatosis
    C. 5q-syndrome
    D. Immune thrombocytopenic purpura
    E. Glanzmann thrombasthenia

23. Select the one of the following statements that is NOT CORRECT:

    A. The accumulation of fluid at a site of infection (edema) facilitates leukocyte migration.
    B. Type 1 leukocyte adhesion deficiency is caused by a lack of the β_2 subunit of an integrin designated LFA-1.
    C. The components of the complement cascade circulate though the plasma as inactive zymogens.
    D. Leukocytes are recruited to a site of infection by chemotaxis toward the sources of epinephrine.
    E. Neutrophils can trap large pathogens in NETS constructed, in part, from strands of chromosomal DNA.

24. Select the one of the following statements that is NOT CORRECT:

    A. Interleukins are key mediators of leukocyte production.
    B. Lymphocytes produce protective antibodies.
    C. Monocytes can be found in tissues throughout the body.
    D. The hematologic factor histamine is synthesized by the deamination of the amino acid histidine.
    E. The term polymorphonuclear refers to leukocytes possessing a segmented nucleus.

25. Select the one of the following statements that is NOT CORRECT:

    A. Phagocytes destroy ingested bacteria using reactive oxygen species and hydrolytic enzymes.
    B. Chronic granulomatous disease is caused by a deficiency in myeloperoxidase activity.
    C. NADPH serves as the primary source of electrons for generating ROS during the oxidative burst.
    D. Neutrophils aid in the elimination of some parasites by enmeshing them in NETs formed from their chromosomal DNA.
    E. Chemokines are stabilized by the formation of intrachain disulfide bonds.

26. Select the one of the following statements that is NOT CORRECT:

    A. Activated leukocytes secrete lipid mediators called interferons.
    B. Neutrophils facilitate production of protective antibodies by presenting fragments of phagocytized microbes on their surface in association with the major histocompatibility complex (MHC).
    C. Cytotoxic T cells use perforins to lyse infected cells.
    D. Soluble antibodies are released into the plasma primarily by B lymphocytes.
    E. Emphysema can arise from the action of elastase and other granule-derived proteases on pulmonary tissue.

27. Select the one FALSE statement:

    A. The great majority of mitochondrial proteins are encoded by the nuclear genome.
    B. Ran proteins, like ARF and Ras proteins, are monomeric GTPases.
    C. One cause of Refsum disease is mutations in genes encoding peroxisomal proteins.
    D. Peroxisomal proteins are synthesized on cytosolic polyribosomes.
    E. Import of proteins into mitochondria involves proteins known as importins.

28. Select the one FALSE statement:

    A. *N*-terminal signal peptides directing nascent proteins to the ER membrane contain a hydrophobic sequence.
    B. Posttranslational translocation of proteins to the ER does not occur in mammalian species.
    C. The SRP contains one RNA species.
    D. *N*-glycosylation is catalyzed by oligosaccharide: protein transferase.
    E. Type I membrane proteins have their *N*-termini facing the lumen of the ER.

29. Select the one FALSE statement:

    A. Chaperones often exhibit ATPase activity.
    B. Protein disulfide isomerase and peptidyl prolyl isomerase are enzymes involved in helping proteins fold properly.
    C. Ubiquitin is a small protein involved in protein degradation by lysosomes.
    D. Mitochondria contain chaperones.
    E. Retrotranslocation across the ER membrane is involved in helping dispose of misfolded proteins.

30. Select the one FALSE statement:

    A. Rab is a small GTPase involved in vesicle targeting.
    B. COPII vesicles are involved in anterograde transport of cargo from the ER to the ERGIC or Golgi apparatus.
    C. Brefeldin A prevents GTP binding to ARF, and thus inhibits formation of COPI vesicles.
    D. Botulinum toxin B acts by cleaving synaptobrevin, inhibiting release of acetylcholine at the neuromuscular junction.
    E. Furin converts preproalbumin to proalbumin.

31. Which one of the following types of protein does NOT act as a GTPase?

    A. ADP ribosylation factor (ARF)
    B. Rab proteins
    C. *N*-ethylmaleimide-sensitive factor (NSF)
    D. Sar1
    E. Ran proteins

32. Select the one FALSE statement:

    A. Collagen has a triple helical structure, forming a right-hand superhelix.
    B. Proline and hydroxyproline confer rigidity on collagen.
    C. Collagen contains one or more *O*-glycosidic linkages.
    D. Collagen lacks cross-links.
    E. Deficiency of vitamin C impairs the action of prolyl and lysyl hydroxylases.

33. Select the one FALSE statement:

    A. Elastin contains hydroxyproline, but not hydroxylysine.
    B. Elastin contains cross-links formed by desmosines.
    C. No genetic diseases due to abnormalities of elastin have as yet been identified.
    D. Unlike collagen, there is only one gene encoding elastin.
    E. Elastin does not contain any sugar molecules.

34. Select the one FALSE statement:

    A. Marfan syndrome is due to mutations in the gene encoding fibrillin-1, a major constituent of microfibrils.
    B. All subtypes of Ehlers-Danlos syndrome are due to mutations affecting the genes encoding the various types of collagen.
    C. Laminin is found in renal glomeruli along with entactin, type IV collagen, and heparin or heparan sulfate.
    D. Mutations affecting type IV collagen can cause serious renal disease.
    E. Mutations in the collagen *1A1* gene can cause osteogenesis imperfecta.

35. Select the one FALSE statement:

    A. Most but not all GAGs contain an amino sugar and a uronic acid.
    B. All GAGs are sulfated.
    C. GAGs are built up by the actions of glycosyltransferases using sugars donated by nucleotide sugars.
    D. Glucuronic acid can be converted to iduronic acid by an epimerase.
    E. The proteoglycan aggrecan contains hyaluronic acid, keratan sulfate, and chondroitin sulfate.

36. A male infant is failing to thrive and, on examination, is noted to have hepatomegaly and splenomegaly, among other findings. Urinalysis reveals the presence of both dermatan sulfate and heparan sulfate. You suspect the patient has Hurler syndrome. From the following list, select the enzyme that you would wish to have assayed to support your diagnosis.

   A. β-Glucuronidase
   B. β-Galactosidase
   C. α-L-Iduronidase
   D. α-N-Acetylglucosaminidase
   E. Neuraminidase

37. You see a child in clinic who is well below average height. You note that the child has short limbs, normal trunk size, macrocephaly, and a variety of other skeletal abnormalities. You suspect that the child has achondroplasia. Select from the following list the test that would best confirm your diagnosis:

   A. Measurement of growth hormone
   B. Assays for enzymes involved in the metabolism of GAGs
   C. Tests for urinary mucopolysaccharides
   D. Gene tests for abnormalities of the fibroblast growth factor receptor 3 (FGFR3)
   E. Gene tests for abnormalities of growth hormone

C H A P T E R

55

Hemostasis & Thrombosis

Peter L. Gross, MD, MSc, FRCP(C),
P. Anthony Weil, PhD, & Margaret L. Rand, PhD

OBJECTIVES

After studying this chapter, you should be able to:

- Understand the significance of hemostasis and thrombosis in health and disease.
- Outline the steps leading to platelet aggregation.
- Identify antiplatelet drugs and their mode of inhibition.
- Outline the pathways of coagulation that result in the formation of fibrin.
- Identify the vitamin K–dependent coagulation factors.
- Provide examples of genetic disorders that lead to bleeding.
- Describe the process of fibrinolysis.

BIOMEDICAL IMPORTANCE

Fundamental aspects of platelet biology are described in this chapter as well as basic aspects of the proteins of the blood coagulation system and of fibrinolysis. Hemorrhagic and thrombotic states can cause serious medical emergencies, and thromboses in the coronary and cerebral arteries are major causes of death in many parts of the world. Rational management of these conditions requires a clear understanding of the bases of platelet activation, blood coagulation, and fibrinolysis.

HEMOSTASIS & THROMBOSIS HAVE THREE COMMON PHASES

Hemostasis is the cessation of bleeding from a cut or severed vessel, whereas **thrombosis** occurs when the endothelium lining blood vessels is damaged or removed (eg, on rupture of an atherosclerotic plaque). These processes involve blood vessels, platelet aggregation, and plasma proteins that cause formation or dissolution of platelet aggregates and fibrin.

In hemostasis, there is initial vasoconstriction of the injured vessel, causing diminished blood flow distal to the injury. Then, hemostasis and thrombosis share **three phases**:

1. Formation of a loose and temporary **platelet aggregate** at the site of injury. Platelets bind to collagen at the site of vessel wall injury, form thromboxane A_2 (TxA_2), and release ADP, which activate other platelets flowing by the vicinity of the injury. (The mechanism of platelet activation is described later.) Thrombin, formed during coagulation at the same site, causes further platelet activation. On activation, platelets change shape and, in the presence of fibrinogen and/or von Willebrand factor, aggregate to form the hemostatic plug (in hemostasis) or thrombus (in thrombosis).

2. Formation of a **fibrin mesh** that binds to the platelet aggregate forming a more stable hemostatic plug or thrombus.

3. Partial or complete **dissolution** of the hemostatic plug or thrombus by plasmin.

There Are Three Types of Thrombi

Three types of thrombi or clots are distinguished. All three contain **fibrin** in various proportions:

1. The **white thrombus** is composed of platelets and fibrin and is relatively poor in erythrocytes. It forms at the site of an injury or abnormal vessel wall, particularly in areas where blood flow is rapid (arteries).

2. The **red thrombus** consists primarily of red cells and fibrin. It morphologically resembles the clot formed in a test tube and may form in vivo in areas of retarded blood flow or stasis (eg, veins) with or without vascular injury, or it may form at a site of injury or in an abnormal vessel in conjunction with an initiating platelet plug.

3. A third type is **fibrin deposits** in very small blood vessels or capillaries.

We shall first describe some of the aspects of the involvement of platelets and blood vessel walls in the overall process. Then, we shall describe the coagulation pathway leading to the formation of fibrin. This separation of platelets and clotting factors is artificial since both play intimate and often mutually interdependent roles in hemostasis and thrombosis; however, this strategy facilitates description of the overall processes involved.

Platelet Aggregation Requires Outside-In and Inside-Out Transmembrane Signaling

Platelets normally circulate in an unstimulated disk-shaped form. **During hemostasis or thrombosis, platelets become activated** and help **form hemostatic plugs or thrombi (Figure 55–1)**. Three major steps are involved: (1) adhesion to exposed collagen in blood vessels; (2) release (exocytosis) of the contents of their storage granules; and (3) aggregation.

Platelets adhere to collagen via specific receptors on the platelet surface, including the glycoprotein complexes GPIa-IIa ($\alpha 2\beta 1$ integrin; see Chapter 54) and GPIb-IX-V, and GPVI. The binding of GPIb-IX-V to collagen is mediated via von Willebrand factor; this interaction is especially important in platelet adherence to the subendothelium under conditions of high shear stress that occur in small vessels and partially stenosed arteries.

Platelets that are bound to collagen change shape and spread out on the subendothelium. These adherent platelets release the contents of their **storage granules** (the dense granules and the alpha granules); some of the molecules released amplify the responses to vessel wall injury. Granule release is also stimulated by thrombin.

Thrombin, formed from the coagulation cascade (described later), is the most potent activator of platelets and initiates activation by interacting with its receptors **protease-activated receptor-1 (PAR-1)**, **PAR-4**, and **GPIb-IX-V** on the platelet plasma membrane (see Figure 55–1A). The further events leading to platelet activation on binding to PAR-1 and PAR-4 are examples of **outside-in transmembrane signaling**, in which

a chemical messenger outside the cell generates effector molecules inside the cell (see Chapter 42). In this instance, thrombin acts as the external chemical messenger (stimulus or agonist). The interaction of thrombin with its G-protein–coupled receptors PAR-1 and PAR-4 stimulates the activity of an intracellular **phospholipase Cβ (PLCβ)**. This enzyme hydrolyzes the membrane phospholipid **phosphatidylinositol 4,5-bisphosphate (PIP_2, a polyphosphoinositide)** to form the two internal effector molecules (**1,2-diacylglycerol [DAG]** and **1,4,5-inositol trisphosphate [IP_3]**; see Figures 42–6 and 42–7).

Hydrolysis of PIP_2 is also involved in the action of many hormones and drugs. DAG stimulates **protein kinase C (PKC)**, which phosphorylates the protein **pleckstrin** (47 kDa). This results in aggregation and release of the contents of the storage granules. **ADP** released from dense granules can also activate platelets via its specific G-protein–coupled receptors (see Figure 55–1A; $P2Y_1$, $P2Y_{12}$), resulting in aggregation of additional platelets. IP_3 causes release of Ca^{2+} into the cytosol mainly from the dense tubular system (or residual smooth endoplasmic reticulum from the megakaryocyte), which then interacts with calmodulin and myosin light-chain kinase, leading to phosphorylation of the light chains of myosin. These chains then interact with actin, causing changes of platelet shape.

Collagen-induced activation of a platelet **cytosolic phospholipase A$_2$ (cPLA$_2$)** by increased levels of intracellular Ca^{2+} results in liberation of arachidonic acid from platelet membrane phospholipids, leading to the formation of thromboxane A2 (**TxA$_2$**; see Chapters 21 and 23). TxA$_2$, in turn, by binding to its specific G-protein–coupled receptor (**TP**), can further activate PLCβ, promoting platelet aggregation (see Figure 55–1A).

All of the **aggregating agents**, including thrombin, collagen, ADP, TxA$_2$, and others such as platelet-activating factor, via an **inside-out signaling pathway**, modify the platelet surface **glycoprotein complex GPIIb-IIIa** ($\alpha IIb\beta 3$ integrin; see Chapter 54) so that the receptor has a higher affinity for **fibrinogen** or **von Willebrand factor** (see Figure 55–1B). Molecules of divalent fibrinogen or multivalent von Willebrand factor then link adjacent activated platelets to each other, forming a platelet aggregate. von Willebrand factor–mediated platelet aggregation occurs under conditions of high shear stress. Agents such as epinephrine, serotonin, and vasopressin, exert synergistic effects with other aggregating agents.

Activated platelets, besides forming a platelet aggregate, accelerate the activation of the coagulation factor X and prothrombin by exposing the procoagulant anionic phospholipid phosphatidylserine on their membrane surface (see following section for more detail). Polyphosphate, released from the dense granules, accelerates factor V activation and also accelerates factor XI activation by thrombin.

Endothelial Cells Synthesize Prostacyclin & Other Compounds That Affect Clotting & Thrombosis

The **endothelial cells** in the walls of blood vessels make important contributions to the overall regulation of hemostasis and

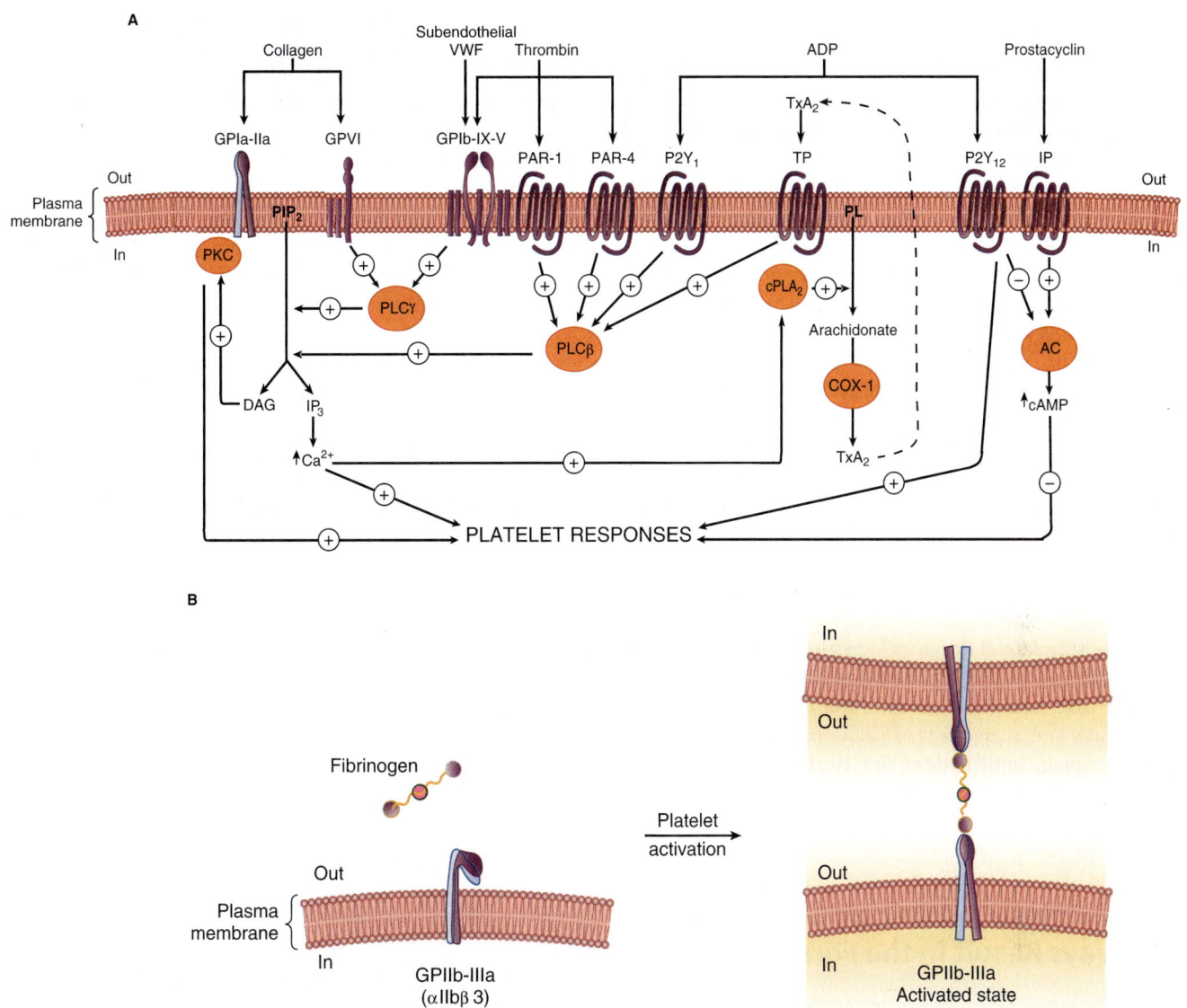

FIGURE 55–1 **(A) Diagrammatic representation of platelet activation by collagen, thrombin, thromboxane A₂ and ADP, and inhibition by prostacyclin.** The external environment (Out), the plasma membrane, and the inside of a platelet (In) are depicted from top to bottom. Platelet responses include, depending on the agonist, change of platelet shape, release of the contents of the storage granules, and aggregation. (AC, adenylyl cyclase; cAMP, cyclic AMP; COX-1, cyclooxgenase-1; cPLA₂, cytosolic phospholipase A₂; DAG, 1,2-diacylglycerol; GP, glycoprotein; IP, prostacyclin receptor; IP₃, inositol 1,4,5-trisphosphate; P2Y₁, P2Y₁₂, purinoceptors; PAR, protease activated receptor; PIP₂, phosphatidylinositol 4,5-bisphosphate; PKC, protein kinase C; PL, phospholipid; PLCβ, phospholipase Cβ; PLCγ, phospholipase Cγ; TP, thromboxane A₂ receptor; TxA₂, thromboxane A₂; VWF, von Willebrand factor.) The G-proteins that are involved are not shown. **(B) Diagrammatic representation of platelet aggregation mediated by fibrinogen binding to activated GPIIb-IIIa molecules on adjacent platelets.** Signaling events initiated by all aggregating agents transform GPIIb-IIIa from its resting state to an activated form that can bind divalent fibrinogen or, at the high shear that occurs in small vessels, multivalent von Willebrand factor.

thrombosis. As described in Chapter 23, these cells synthesize the prostanoid **prostacyclin** (PGI₂), a potent inhibitor of platelet aggregation. Prostacyclin acts by stimulating the activity of adenylyl cyclase in the surface membranes of platelets via its G protein–coupled receptor (see Figure 55–1A). The resulting increase of intraplatelet **cAMP** opposes the increase in the level of intracellular Ca²⁺ produced by IP₃ and thus inhibits platelet activation. This is in contrast with the effect of the prostanoid TxA₂ that is formed by activated platelets, which is that of promoting aggregation. Endothelial cells play other roles in the regulation of thrombosis. For instance, these cells possess a surface-expressed ecto-**ADPase**, which hydrolyzes ADP, and thus opposes its aggregating effect on platelets. In addition, these cells express proteoglycans such as **heparan sulfate**, an anticoagulant, and they also release **plasminogen activators**, which may help dissolve thrombi. **Nitric oxide** (endothelium-derived relaxing factor), another potent platelet inhibitor, is discussed in Chapter 51.

Aspirin Is One of Several Effective Antiplatelet Drugs

Antiplatelet drugs inhibit platelet responses. The most used antiplatelet drug is **aspirin (acetylsalicylic acid)**, which irreversibly acetylates and thus inhibits the platelet **cyclooxygenase (COX-1)** involved in formation of TxA_2 (see Chapter 21); TxA_2 is a potent aggregator of platelets and also a vasoconstrictor. Platelets are very sensitive to aspirin; as little as 30 mg of aspirin/day (one regular aspirin tablet contains 325 mg) effectively eliminates the synthesis of TxA_2 by platelets. Aspirin also inhibits production of prostaglandin I_2 (PGI_2), which opposes platelet aggregation and is a vasodilator, by endothelial cells, but unlike platelets, these cells regenerate cyclooxygenase within few hours. Thus, the overall balance between TxA_2 and PGI_2 can be shifted in favor of the latter, opposing platelet aggregation. Indications for treatment with aspirin include management of acute coronary syndromes (angina, myocardial infarction), acute stroke syndromes (transient ischemic attacks, acute stroke), severe carotid artery stenosis, and primary prevention of these and other atherothrombotic diseases.

Other **antiplatelet drugs** include specific inhibitors of the $P2Y_{12}$ receptor for ADP (eg, **clopidogrel, prasugrel, ticagrelor,** and **cangrelor**) and of the thrombin receptor PAR-1 (vorapaxar), and antagonists of ligand binding to GPIIb-IIIa (eg, **abciximab, eptifibatide,** and **tirofiban**) that interfere with fibrinogen and von Willebrand factor binding and thus platelet aggregation.

Both Extrinsic & Intrinsic Coagulation Pathways Result in the Formation of Fibrin

Two coagulation pathways lead to **fibrin clot** formation: the **extrinsic** and the **intrinsic** pathways. Contrary to previous thought, these pathways are not independent. However, this artificial distinction is retained in the following text to facilitate their description.

Initiation of fibrin clot formation in response to **tissue injury** is carried out by the **extrinsic pathway.** The **intrinsic pathway** can be activated by negatively charged surfaces **in vitro,** for example, glass. Both pathways lead to the proteolytic conversion of **prothrombin** to **thrombin.** Thrombin catalyzes cleavage of **fibrinogen** to initiate **fibrin** clot formation. The extrinsic and intrinsic pathways are complex and involve many different proteins (**Figures 55-2** and **55-3; Table 55-1**). The coagulation factors are another example of multidomain proteins sharing conserved domains (see Figure 5-9). In general, as shown in **Table 55-2**, these proteins can be classified into **five types:** (1) zymogens of serine-dependent proteases that are activated during the process of coagulation; (2) cofactors; (3) fibrinogen; (4) a zymogen of a transglutaminase that covalently cross-links fibrin and stabilizes the fibrin clot; and (5) regulatory and other proteins.

The Extrinsic Pathway Leads to Activation of Factor X

The **extrinsic pathway** involves tissue factor, factors VII and X, and Ca^{2+} and results in the production of factor Xa (by convention, activated clotting factors are referred to by use of the suffix a). The extrinsic pathway is initiated at the site of **tissue injury** with the exposure of **tissue factor** (TF; see Figure 55–2), located in the subendothelium and on activated monocytes. TF interacts with and activates **factor VII** (53 kDa, a zymogen-containing vitamin K–dependent γ-carboxyglutamate [Gla] residues; see Chapter 44), synthesized in the liver. It should be noted that in the Gla-containing zymogens (factors II, VII, IX, and X), the Gla residues in the amino-terminal regions of the molecules serve as high-affinity binding sites for Ca^{2+}. TF acts as a cofactor for **factor VIIa,** enhancing its enzymatic activity to activate **factor X** (56 kDa). The reaction by which **factor X** is activated requires the assembly of the **extrinsic tenase complex (Ca^{2+}-TF-factor VIIa)** formed on

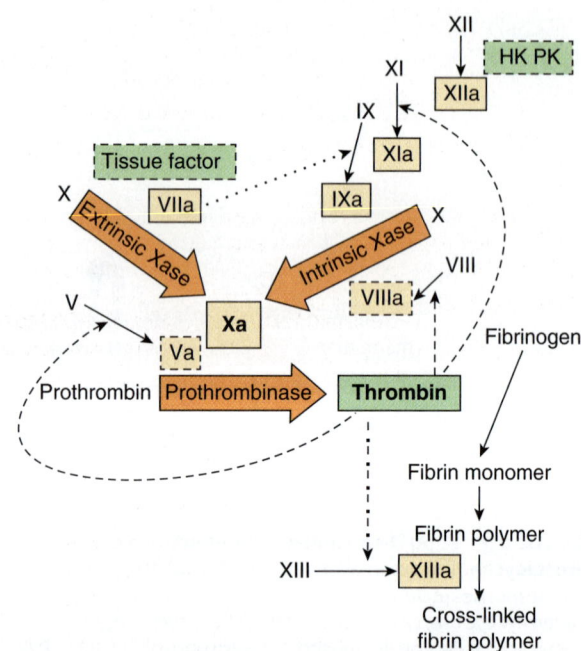

FIGURE 55–2 The pathways of blood coagulation, with the extrinsic pathway indicated at the top left and the intrinsic pathway at the top right. The pathways converge in the formation of active factor X (ie, factor Xa) and culminate in the formation of cross-linked fibrin (lower right). Complexes of tissue factor and factor VIIa activate not only factor X (extrinsic Xase [tenase]) but also factor IX in the intrinsic pathway (dotted arrow). In addition, thrombin feedback activates at the sites indicated (dashed arrows) and also activates factor XIII to factor XIIIa (dashed-dotted arrow) and activates factor VII to factor VIIa (not shown). The three predominant complexes, extrinsic Xase, intrinsic Xase, and prothrombinase, are indicated within the large arrows; these reactions require anionic procoagulant phospholipid membrane and calcium. Activated proteases are in solid-outlined boxes; active cofactors are in dash-outlined boxes; and inactive factors are not in boxes. (HK, high-molecular-weight kininogen; PK, prekallikrein.)

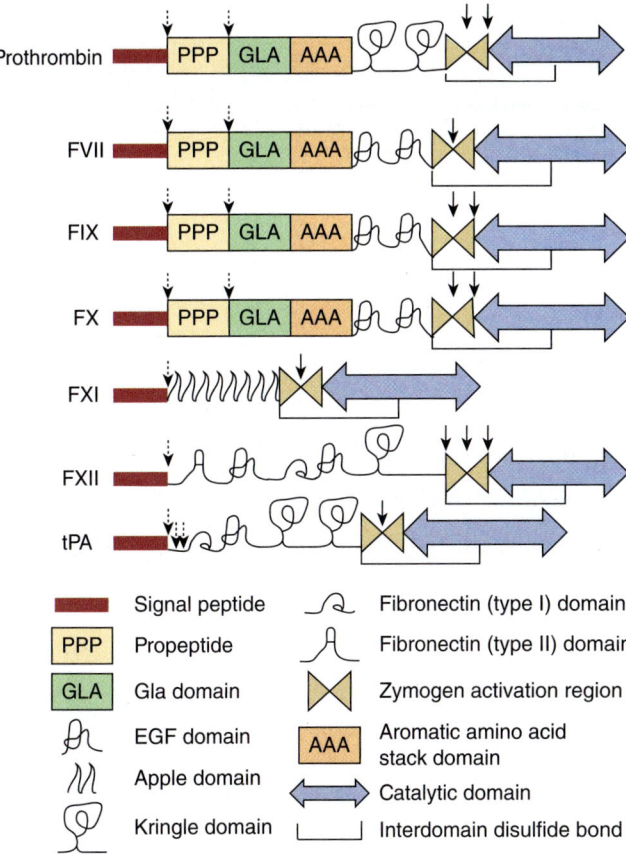

Signal peptide

PPP Propeptide

GLA Gla domain

EGF domain

Apple domain

Kringle domain

Fibronectin (type I) domain

Fibronectin (type II) domain

Zymogen activation region

AAA Aromatic amino acid stack domain

Catalytic domain

Interdomain disulfide bond

FIGURE 55–3 **The structural domains of selected proteins involved in coagulation and fibrinolysis.** Shared domains are a result of gene duplication and exon shuffling that contributed to the molecular evolution of the coagulation system. The domains are as identified at the bottom of the figure and include signal peptide, propeptide, Gla (γ-carboxyglutamate) domain, epidermal growth factor (EGF) domain, apple domain, kringle domain, fibronectin (types I and II) domain, the zymogen activation region, aromatic amino acid stack, and the catalytic domain. Interdomain disulfide bonds are indicated, but numerous intradomain disulfide bonds are not. Sites of proteolytic cleavage in synthesis or activation are indicated by arrows (dashed and solid, respectively). (FVII, factor VII; FIX, factor IX; FX, factor X, FXI; factor XI; FXII, factor XII; t-PA, tissue plasminogen activator.)

Factor	Common Name(s)	
I	Fibrinogen	These factors are usually referred to by their common names
II	Prothrombin	
III	Tissue factor	
IV	Ca²⁺	Ca²⁺ is usually not referred to as a coagulation factor
V	Proaccelerin, labile factor, accelerator (Ac-) globulin	
VII[a]	Proconvertin, serum prothrombin conversion accelerator (SPCA), cothromboplastin	
VIII	Antihemophilic factor A, antihemophilic globulin (AHG)	
IX	Antihemophilic factor B, Christmas factor, plasma thromboplastin component (PTC)	
X	Stuart-Prower factor	
XI	Plasma thromboplastin antecedent (PTA)	
XII	Hageman factor	
XIII	Fibrin stabilizing factor (FSF), fibrinoligase	

[a]There is no factor VI.

Note: The numbers indicate the order in which the factors have been discovered and bear no relationship to the order in which they act.

The Intrinsic Pathway Also Leads to Activation of Factor X

The formation of **factor Xa** is the major site where the intrinsic and extrinsic pathways converge (see Figure 55–2). The **intrinsic pathway** (see Figure 55–2) involves factors XII, XI, IX, VIII, and X, as well as prekallikrein, high-molecular-weight (HMW) kininogen, Ca²⁺, and cell surface exposed phosphatidylserine. This pathway results in the production of **factor Xa** by the intrinsic tenase complex (see later for composition), in which factor IXa serves as the serine protease and factor VIIIa as the cofactor. As noted earlier, the activation of **factor X provides an important link between the intrinsic and extrinsic pathways**.

The **intrinsic pathway** can be initiated by "contact" in which prekallikrein, HMW kininogen, factors XII and XI are exposed to a negatively charged activating surface. In vivo, polymers of phosphates, such as extracellular DNA, RNA, and polyphosphate (macromolecules available only following cell damage) may serve as this negatively charged activating surface. Kaolin, a highly negatively charged hydrated aluminum silicate, can be used for in vitro tests as an initiator of the intrinsic pathway. When the components of the contact phase assemble on the activating surface, factor XII is activated to **factor XIIa** on proteolysis by kallikrein. This factor XIIa, generated by kallikrein, cleaves prekallikrein to generate more kallikrein, setting up a positive feedback activation loop. Factor XIIa, once formed, activates **factor XI to XIa** and also releases **bradykinin** (a peptide with potent vasodilator action) from HMW kininogen.

a cell membrane surface exposing the procoagulant anionic aminophospholipid phosphatidylserine. Factor VIIa cleaves an Arg-Ile bond in factor X to produce the two-chain serine protease, **factor Xa**. TF and factor VIIa also activate factor IX in the intrinsic pathway. Indeed, **the formation of complexes between membrane-bound TF and factor VIIa is now considered to be the key process involved in initiation of blood coagulation in vivo**.

Tissue factor pathway inhibitor (TFPI) is a major physiologic inhibitor of coagulation. TFPI is a protein that circulates in the blood where it directly inhibits factor Xa by binding to the enzyme near its active site. This factor Xa-TFPI complex then inhibits the factor VIIa-TF complex.

TABLE 55–2   The Functions of the Proteins Involved in Blood Coagulation

Zymogens of Serine Proteases	
Factor XII	Binds to negatively charged surface, eg, kaolin, glass; activated by high-molecular-weight kininogen and kallikrein
Factor XI	Activated by factor XIIa
Factor IX	Activated by factors XIa and VIIa
Factor VII	Activated by factor VIIa and thrombin
Factor X	Activated on the surface of activated platelets by tenase complex (Ca^{2+}, factors VIIIa and IXa) and by the extrinsic tenase complex (Ca^{2+}, tissue factor, and factor VIIa)
Prothrombin (factor II)	Activated on the surface of activated platelets by prothrombinase complex (Ca^{2+}, factors Va and Xa) to form thrombin (factors II, VII, IX, and X are Gla-containing zymogens) (Gla = γ-carboxyglutamate)
Cofactors	
Factor VIII	Activated by thrombin; factor VIIIa is a cofactor in the activation of factor X by factor IXa
Factor V	Activated by thrombin; factor Va is a cofactor in the activation of prothrombin by factor Xa
Tissue factor (factor III)	A glycoprotein located in the subendothelium and expressed on activated monocytes to act as a cofactor for factor VIIa
Fibrinogen	
Factor I	Cleaved by thrombin to form fibrin clot
Zymogen of a Thiol-Dependent Transglutaminase	
Factor XIII	Activated by thrombin; stabilizes fibrin clot by covalent cross-linking
Regulatory and Other Proteins	
Protein C	Activated to the serine protease activated protein C (APC) by thrombin bound to thrombomodulin which then degrades factors VIIIa and Va
Protein S	Acts as a cofactor of protein C; both proteins contain Gla (γ-carboxyglutamate) residues
Thrombomodulin	Protein on the surface of endothelial cells; binds thrombin, which then activates protein C

In the presence of Ca^{2+}, factor XIa activates factor IX (55 kDa, a Gla-containing zymogen), to the serine protease, **factor IXa**. This, in turn, also cleaves an Arg-Ile bond in factor X to produce **factor Xa**. This latter reaction requires the assembly of components, called **the intrinsic tenase complex**, **composed of Ca^{2+}-factor VIIIa-factor IXa**, which forms on procoagulant membrane surfaces expressing phosphatidylserine, often activated platelets. (Note that this phospholipid is normally on the inner leaflet side of the plasma membrane bilayer of resting, nonactivated platelets.)

Factor VIII (330 kDa), a circulating glycoprotein, is not a protease precursor but a cofactor precursor that, when activated, serves as a receptor on the platelet surface for factors IXa and X. Factor VIII is activated by minute quantities of thrombin to form **factor VIIIa**, which is in turn inactivated on further cleavage by thrombin-mediated activated protein C (see later).

The role of the **initial steps of the intrinsic pathway** in initiating coagulation has been called into question because patients with a hereditary deficiency of factor XII, prekallikrein or HMW kininogen do not exhibit bleeding problems. Similarly, patients with a deficiency of factor XI may not have bleeding problems. In thrombosis, deficiencies in the intrinsic pathway are protective. The intrinsic pathway largely serves to **amplify factor Xa** and ultimately **thrombin formation**,

through feedback mechanisms (see later). The intrinsic pathway may also be important in **fibrinolysis** (see later) since kallikrein, and factors XIIa and XIa can cleave plasminogen, and kallikrein can activate single-chain urokinase.

Factor Xa Leads to Activation of Prothrombin to Thrombin

Factor Xa, produced by either the extrinsic or the intrinsic pathway, activates **prothrombin** (factor II) to **thrombin** (factor IIa) (see Figure 55–2; Table 55–1).

The activation of prothrombin, like that of factor X, occurs on a procoagulant membrane surface and requires the assembly of a **prothrombinase complex**, consisting of Ca^{2+}, and factors Va and Xa. The assembly of the prothrombinase complex, like that of the tenase complex, takes place on the phosphatidylserine-exposing membrane surface, often activated platelets.

Factor V (330 kDa) is synthesized in the liver, spleen, and kidney and is found in platelets as well as in plasma. Factor Va functions as a cofactor in the prothrombinase complex in a manner similar to that of factor VIIIa in the tenase complex. Factor V is activated to **factor Va** by traces of thrombin, and binds specifically to a procoagulant membrane (often that of platelets) **(Figure 55–4)** and forms a complex with factor Xa and

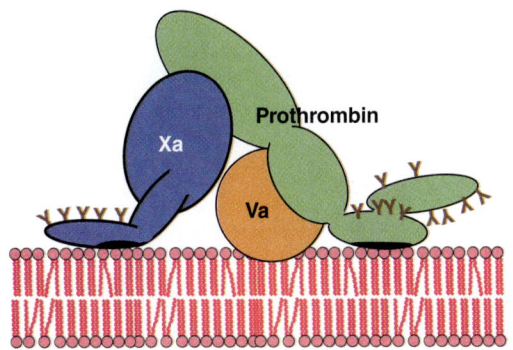

FIGURE 55–4 **Schematic representation of the prothrombinase complex bound to the procoagulant plasma membrane.** The prothrombinase complex contains factors Va, Xa, and prothrombin. A central theme in blood coagulation is the assembly of protein complexes, that is, the tenase complexes and the prothrombinase complex, in a Ca²⁺-dependent fashion, on membrane surfaces on which phosphatidylserine is exposed. The catalytic efficiency of zymogen activation is increased by many orders of magnitude by the membrane-bound complexes. γ-Carboxyglutamate residues (indicated by **Y**) on vitamin K–dependent proteins bind calcium and contribute to the exposure of membrane-binding sites on these proteins (black ovals, Xa, and prothrombinase).

prothrombin. It is subsequently inactivated by activated protein C (see later), thereby providing a means of limiting the activation of prothrombin to thrombin. **Prothrombin** (72 kDa; see Figure 55–4) is a single-chain glycoprotein synthesized in the liver. The amino-terminal region of prothrombin (see Figure 55–3) contains 10 Gla residues, and the serine-dependent active protease site is in the catalytic domain close to the carboxyl-terminal region of the molecule. On binding to the complex of factors Va and Xa on the procoagulant membrane (see Figure 55–4), prothrombin is cleaved by factor Xa at

two sites to generate the active, two-chain thrombin molecule, which is then released from the membrane surface.

Conversion of Fibrinogen to Fibrin Is Catalyzed by Thrombin

Thrombin, produced by the prothrombinase complex, in addition to having a potent stimulatory effect on platelets (see earlier), **converts fibrinogen to fibrin** (see Figure 55–2). **Fibrinogen** (factor I, 340 kDa; see Figure 55–2 and **Figure 55–5**; Tables 55–1 and 55–2) is an abundant (3 mg/mL) soluble plasma glycoprotein that consists of a dimer of three polypeptide chains, $(A\alpha, B\beta, \gamma)_2$, that is covalently linked by 29 disulfide bonds. The $B\beta$ and γ chains contain asparagine-linked complex oligosaccharides (see Chapter 46). All three chains are synthesized in the liver; the three genes encoding these proteins are on the same chromosome where their expression is coordinately regulated in humans. The amino-terminal regions of the six chains are held in close proximity by a number of disulfide bonds (a subset is shown in Figure 55–5), while the carboxyl-terminal regions are spread apart. Thus, the fibrinogen molecule has a trinodular, elongated structure with a central E domain that is linked to lateral D domains via coiled coil regions (Figure 55–5 and **Figure 55–6A**). The **N-terminal A and B** portions of the $A\alpha$ and $B\beta$ chains are termed **fibrinopeptide A (FPA)** and **fibrinopeptide B (FPB)**, respectively; these domains are highly negatively charged as a result of an abundance of aspartate and glutamate residues (see later). The negative charges contribute to the solubility of fibrinogen in plasma and importantly also serve to prevent aggregation by causing electrostatic repulsion between fibrinogen molecules.

Thrombin (34 kDa), the serine protease formed by the prothrombinase complex, hydrolyzes the four Arg-Gly bonds

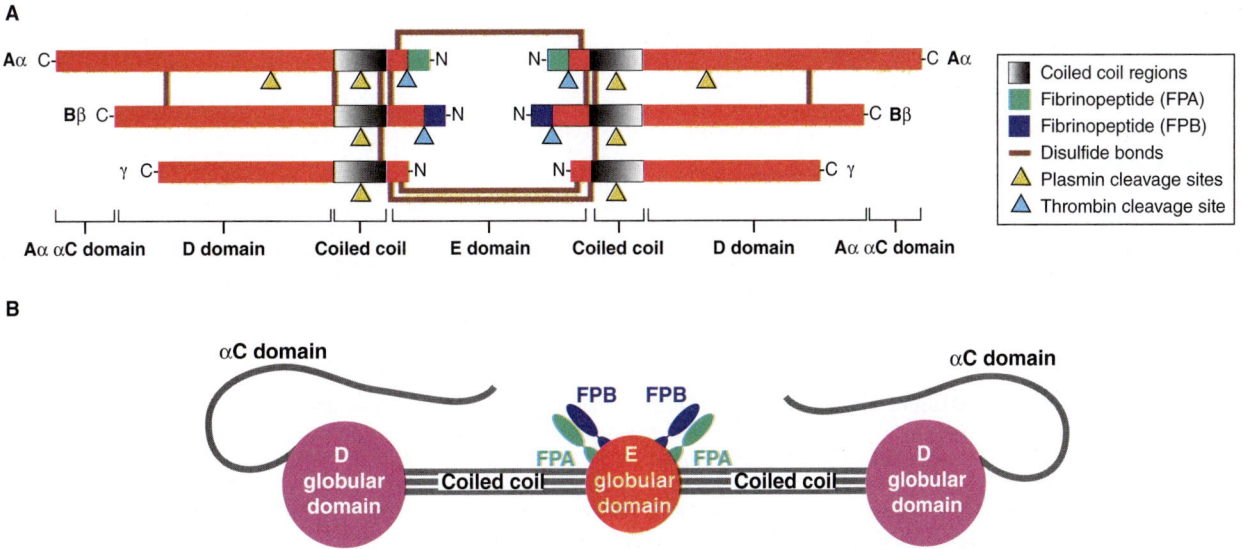

FIGURE 55–5 **Diagrammatic representation of fibrinogen. (A)** Fibrinogen is a dimeric molecule, with each half composed of three polypeptide chains: **A**α, **B**β, and γ. Disulfide bonds join together the chains and the two halves of the molecule. **(B)** Fibrinogen forms a trinodular structure with a central E domain linked via coiled coil regions to two lateral D domains each of which contains a flexible Aα chain αC domain. The thrombin-cleaved regulatory peptides fibrinopeptide A (FPA) and fibrinopeptide B (FPB) reside within the E nodule as shown.

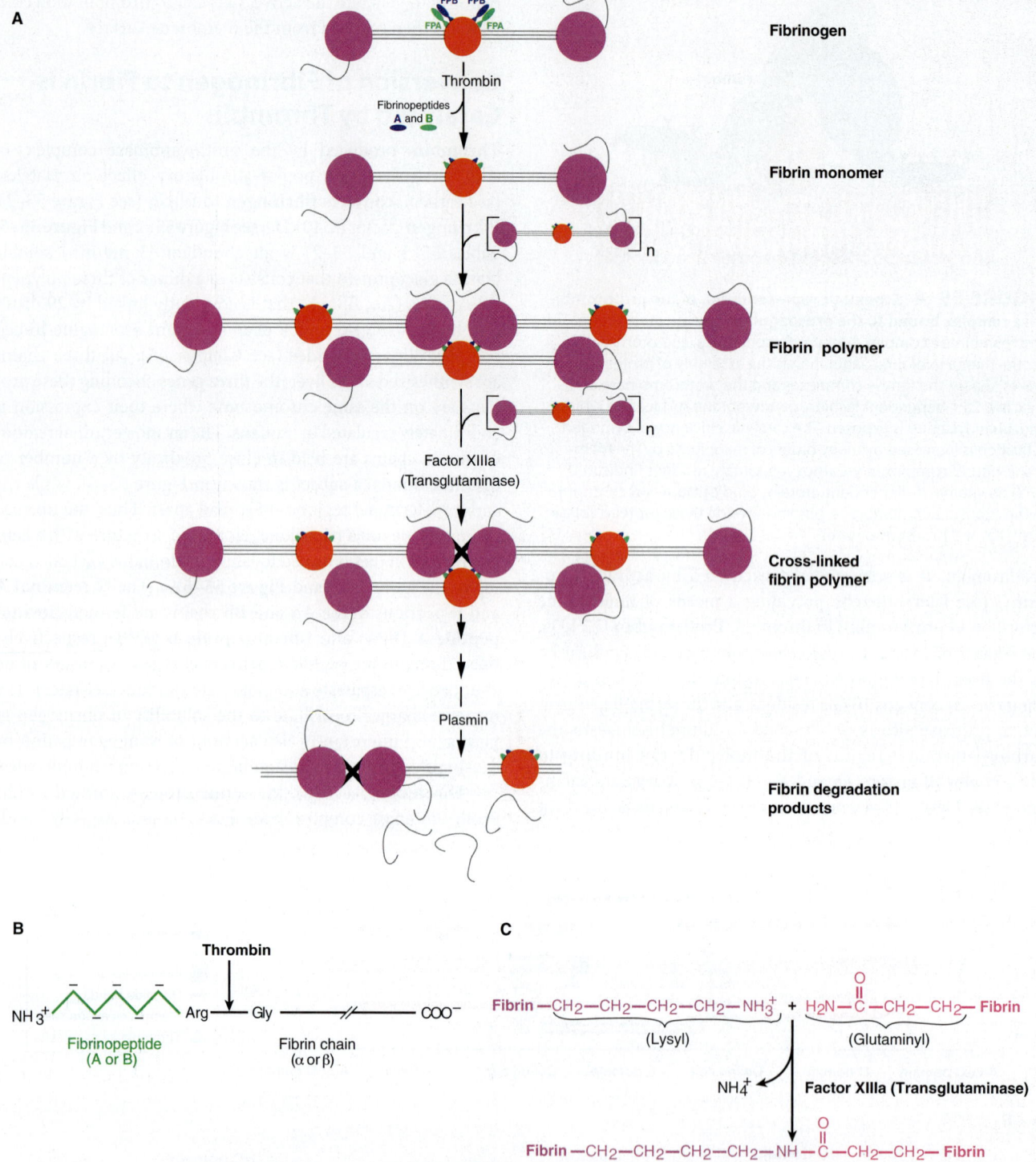

FIGURE 55–6 Fibrin polymerization and degradation. (A) The formation of fibrin monomer via cleavage of fibrinopeptide A (FPA) and fibrinopeptide B (FPB) from fibrinogen by thrombin, the spontaneous polymerization of fibrin monomers to dimers and higher oligomers, followed by the stabilization of fibrin oligomers by factor XIIIa-mediated covalent cross-linking of adjacent fibrin monomers. Finally (bottom) is illustrated the degradation of fibrin polymers into soluble degradation products by plasmin digestion, which leads to clot dissolution. **(B)** Thrombin cleavage site of the **A**α and **B**β chains of fibrinogen to yield **FPA/FPB** (left; green) and the α and β chains of fibrin monomer (right; black). **(C)** Schematic of factor XIIIa (transglutaminase)-mediated cross-linking of fibrin molecules.

between the N-terminal fibrinopeptides and the α and β portions of the **Aα** and **Bβ** chains of fibrinogen (Figure 55–6A, B). The release of FPA and FPB by thrombin generates **fibrin monomer**, which has the subunit structure $(\alpha, \beta, \gamma)_2$. Since FPA and FPB contain only 16 and 14 residues, respectively, the fibrin molecule retains 98% of the residues present in fibrinogen. The removal of the fibrinopeptides exposes binding sites within the E domain of fibrin monomers that specifically interact with complementary domains within the D domains of other fibrin monomers. In this way, fibrin monomers spontaneously polymerize in a half-staggered pattern to form long strands (protofibrils) (see Figure 55–6A). Although insoluble, this initial fibrin clot is unstable, held together only by the noncovalent association of fibrin monomers.

In addition to converting fibrinogen to fibrin, thrombin also activates **factor XIII to factor XIIIa**. Factor XIIIa is a highly specific **transglutaminase** that covalently cross-links γ chains and, more slowly, α chains of fibrin molecules by forming peptide bonds between the amide groups of glutamine and the ε-amino groups of lysine residues (see Figure 55–6C). Such cross-linking yields a more stable fibrin clot with increased resistance to proteolysis. This fibrin mesh serves to stabilize the hemostatic plug or thrombus.

Levels of Circulating Thrombin Are Carefully Controlled

Once active thrombin is formed in the course of hemostasis or thrombosis, its concentration must be carefully controlled to prevent further fibrin formation or platelet activation. This is achieved in **two ways**. Thrombin circulates as its inactive precursor, prothrombin, which is activated as a consequence of a cascade of enzymatic reactions, each converting an inactive zymogen to an active enzyme and leading finally to the conversion of prothrombin to thrombin (see Figure 55–2). At each point in the cascade, **feedback mechanisms** produce a delicate balance of activation and inhibition. The concentration of factor XII in plasma is approximately 30 μg/mL, while that of fibrinogen is 3 mg/mL, with intermediate clotting factors increasing in concentration as one proceeds down the cascade; these facts illustrate that the clotting cascade provides **amplification**. The second means of controlling thrombin activity is **the inactivation of any formed thrombin** by **circulating inhibitors**, the most important of which is antithrombin (see later).

The Activity of Antithrombin, an Inhibitor of Thrombin, Is Increased by Heparin

Four naturally occurring **thrombin inhibitors** exist in normal plasma. The most important is **antithrombin**, which contributes approximately 75% of the antithrombin activity. Antithrombin can also inhibit the activities of factors IXa, Xa, XIa, XIIa, and VIIa complexed with tissue factor. α_2-**macroglobulin** contributes most of the remainder of the antithrombin activity,

with **heparin cofactor II** and α_1-**antitrypsin** acting as minor inhibitors under physiologic conditions.

The endogenous activity of antithrombin is greatly potentiated by the presence of sulfated glycosaminoglycans (heparans) (see Chapter 50). Heparans bind to a specific cationic site of antithrombin, which induces a conformational change that promotes binding of antithrombin to thrombin and factor Xa, as well as to its other substrates. This mechanism is the basis for the use of **heparin**, a derivatized heparan, in clinical medicine to inhibit coagulation. The anticoagulant effects of heparin can be antagonized by strongly cationic polypeptides such as **protamine,** which bind strongly to heparin, thus inhibiting the binding of heparin to antithrombin.

Low-molecular-weight heparins (LMWHs), derived from enzymatic or chemical cleavage of unfractionated heparin, have more clinical use. They can be administered subcutaneously at home, have greater bioavailability than unfractionated heparin, and do not need frequent laboratory monitoring.

Individuals with **inherited deficiencies of antithrombin** are prone to develop venous thrombosis, providing evidence that antithrombin has a physiologic function and that the coagulation system in humans is normally in a dynamic state.

Thrombin is involved in an additional regulatory mechanism that operates in coagulation. It combines with **thrombomodulin**, a glycoprotein present on the surfaces of endothelial cells. The complex activates **protein C** on the **endothelial protein C receptor**. In combination with **protein S**, **activated protein C (APC)** degrades factors Va and VIIIa, thereby limiting their actions in coagulation (see Table 55–2). A genetic deficiency of either protein C or protein S can cause venous thrombosis. Furthermore, patients with **factor V Leiden** (which has a glutamine residue in place of an arginine at position 506) have an increased risk of venous thrombotic disease because activated factor V Leiden is resistant to inactivation by APC; this condition is also termed APC resistance.

Coumarin Anticoagulants Inhibit the Vitamin K–Dependent Carboxylation of Factors II, VII, IX, & X

The **coumarin drugs** (eg, warfarin), which are used as anticoagulants, inhibit the vitamin K–dependent carboxylation of Glu to Gla residues (see Chapter 44) in the amino-terminal regions of factors II, VII, IX, and X and also proteins C and S. These proteins, all of which are synthesized in the liver, are dependent on the Ca^{2+}-binding properties of the Gla residues for their normal function in the coagulation pathways. **Coumarins inhibit the reduction of the quinone derivatives of vitamin K to the active hydroquinone forms** (see Chapter 44). Thus, the administration of vitamin K will bypass the coumarin-induced inhibition and allow the posttranslational modification of carboxylation to occur. Reversal of coumarin inhibition by vitamin K requires 12 to 24 hours, whereas reversal of the anticoagulant effects of heparin by protamine is almost instantaneous. Coumarin reversal is achieved faster by infusing normal coagulation factors.

Heparin and **warfarin** are used in the treatment of thrombotic and thromboembolic conditions, such as deep vein thrombosis and pulmonary embolism, and the prevention of stroke in patients with the heart rhythm abnormality atrial fibrillation. Heparin is administered first, because of its prompt onset of action, whereas warfarin takes several days to reach full effect. Their effect is not well predictable by dosage, and thus because of the risk of producing hemorrhage, these drugs are closely monitored by use of appropriate tests of coagulation (see later).

New oral inhibitors of thrombin (eg, dabigatran) or of factor Xa (eg, rivaroxaban, apixaban, edoxaban, and others) are also used in the prevention and treatment of thrombotic conditions. These drugs are advantageous because their effect is predictable based on the dose, and some do not require routine monitoring by laboratory tests. Specific reversal agents, such as antibodies or decoy molecules, are approved, or in various stages of development. Agents that target the "intrinsic" factors are being evaluated as antithrombotics.

Fibrin Clots Are Dissolved by Plasmin

As stated earlier, the coagulation system is normally in a state of dynamic equilibrium in which fibrin clots are constantly being laid down and dissolved. This latter process is termed **fibrinolysis**. **Plasmin**, the serine protease mainly responsible for degrading fibrin and fibrinogen, circulates in the form of its inactive zymogen, **plasminogen** (90 kDa), and any small amounts of plasmin that are formed in the fluid phase under physiologic conditions are rapidly inactivated by the fast-acting plasmin inhibitor, α_2-antiplasmin. Plasminogen binds to fibrin and thus becomes incorporated in clots as they are produced; since plasmin that is formed when bound to fibrin is protected from α_2-antiplasmin, it remains active. **Activators of plasminogen** of various types are found in most body tissues, and all cleave the same Arg-Val bond in plasminogen to produce the disulfide bridge-linked two-chain serine protease, plasmin (**Figure 55–7**). The **specificity of plasmin for fibrin** is another mechanism to regulate fibrinolysis. Via one of its kringle domains, plasmin(ogen) specifically binds lysine residues on fibrin and so is increasingly incorporated into the fibrin mesh

as it cleaves it. (Kringle domains [see Figure 55–3] are common protein motifs of about 100 amino acid residues in length; they have a characteristic covalent structure defined by a pattern of three disulfide bonds.) Thus, the carboxypeptidase **activated thrombin-activatable fibrinolysis inhibitor (TAFIa)** (see Figure 55–7), which removes terminal lysines from fibrin, also inhibits fibrinolysis. Thrombin activates TAFI to TAFIa, thereby inhibiting fibrinolysis during clot formation.

Tissue plasminogen activator (t-PA) (see Figures 55–3 and 55–7) is a serine protease that is released into the circulation from vascular endothelium under conditions of injury or stress and is catalytically inactive unless bound to fibrin. On binding to fibrin, t-PA cleaves plasminogen within the clot to generate plasmin, which in turn digests the fibrin to form soluble degradation products and thus dissolves the clot. Neither plasmin nor the plasminogen activator can remain bound to these degradation products, and so they are released into the fluid phase where they are inactivated by their natural inhibitors. Prourokinase is the precursor of a second activator of plasminogen, **urokinase**. Originally isolated from urine, it is now known that urokinase is synthesized by various cell types including monocytes and macrophages, fibroblasts, and epithelial cells. The main action of urokinase appears to be the degradation of extracellular matrix. Figure 55–7 indicates the sites of action of five proteins that influence the formation and action of plasmin.

Recombinant t-PA & Streptokinase Are Used as Clot Busters

t-PA, marketed as **alteplase**, is produced by recombinant DNA methods. It is used therapeutically as a fibrinolytic agent, as is **streptokinase**, an enzyme secreted by a number of streptococcal bacterial strains. However, the latter is less selective than t-PA, activating plasminogen in the fluid phase (where it can degrade circulating fibrinogen) as well as plasminogen that is bound to a fibrin clot. The amount of plasmin produced by therapeutic doses of streptokinase may exceed the capacity of the circulating α_2-antiplasmin, causing fibrinogen as well as fibrin to be degraded and resulting in the bleeding often encountered during fibrinolytic therapy. Because of its relative

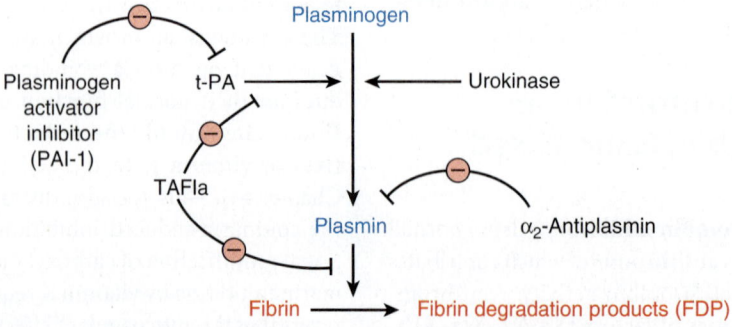

FIGURE 55–7 **Initiation of fibrinolysis by the activation of plasminogen to plasmin.** Scheme of sites and modes of action of tissue plasminogen activator (t-PA), urokinase, plasminogen activator inhibitor, α_2-antiplasmin, and thrombin-activatable fibrinolysis inhibitor (TAFIa).

selectivity for degrading fibrin, recombinant t-PA has been widely used to restore the patency of coronary arteries following thrombosis. If administered early enough, before irreversible damage of heart muscle occurs (about 6 hours after onset of thrombosis), t-PA can significantly reduce the mortality rate from myocardial damage following coronary thrombosis. Streptokinase has also been widely used in the treatment of coronary thrombosis, but has the disadvantage of being antigenic. t-PA has also been used in the treatment of ischemic stroke, peripheral arterial occlusion, deep vein thrombosis, and pulmonary embolism. The fibrinolysis inhibitor tranexamic acid, a lysine analogue, is an inhibitor of plasmin activity used therapeutically to treat bleeding.

There are a number of disorders, including cancer and sepsis, in which the **concentrations of plasminogen activators increase**. In addition, the **antiplasmin activities** contributed by α_1-antitrypsin and α_2-antiplasmin may be impaired in diseases such as cirrhosis. Since certain bacterial proteins, such as streptokinase, are capable of activating plasminogen, they may be responsible for the diffuse hemorrhage sometimes observed in patients with disseminated bacterial infections.

There Are Several Hereditary Bleeding Disorders, Including Hemophilia A

Inherited deficiencies of the clotting system that result in bleeding are found in humans. The most common is deficiency of factor VIII, causing **hemophilia A**, which is an X chromosome-linked disease. **Hemophilia B**, also X chromosome-linked, is due to a deficiency of factor IX and has recently been identified as the form of hemophilia that played a major role in the history of the royal families of Europe. The clinical features of hemophilia A and B are almost identical, but these two diseases can be readily distinguished by specific assays for the two factors.

The **gene for human factor VIII, F8,** one of the largest known genes, is 186 kb in length, containing 26 exons that encode a protein of 2332 amino acids, while the **gene for human factor IX, F9,** is much smaller, 33 kb in length, containing 8 exons that encode a protein of 415 amino acids. A variety of variants in *F8* and *F9* have been detected leading to diminished activities of the factor VIII and IX proteins; these include partial gene deletions and point and missense variant. Prenatal diagnosis by DNA analysis after chorionic villus sampling or amniocentesis is now possible (see Chapter 39).

Treatment for patients with hemophilia A consisted initially, in the 1960s, of intravenous administration of cryoprecipitate (enriched in factor VIII) prepared from individual donors; in the 1970s, lyophilized factor VIII or factor IX concentrates, prepared from very large plasma pools, became available for treatment of hemophilia A and B patients, respectively. In the 1990s, factors VIII and IX prepared by recombinant DNA technology (see Chapter 39) were introduced. Such preparations are free of contaminating viruses (eg, hepatitis A, B, C, or HIV-1) found in human plasma, but are expensive; their use will increase as the cost of production decreases. Longer-acting

recombinant factors with extended half-lives in the circulation are now in use. A novel, nonfactor replacement, therapy for hemophilia A that is administered subcutaneously is a bispecific antibody to factor X and factor IXa that brings these two proteins together as factor VIIIa would. Gene therapy for hemophilia is presently in late-phase clinical investigations.

The most common hereditary bleeding disorder is **von Willebrand disease**, with a prevalence of up to 1% of the population. It results from a deficiency or defect in **von Willebrand factor**, a large multimeric glycoprotein that is secreted by endothelial cells and platelets into the plasma, where it stabilizes factor VIII. von Willebrand factor also promotes platelet adhesion at sites of vessel wall injury and platelet aggregation under conditions of high shear (see earlier).

Laboratory Tests Measure Platelet Aggregation, Coagulation, & Thrombolysis

A number of **laboratory tests** are available to **measure the phases of hemostasis** described earlier. The tests include **platelet count**, **bleeding time/closure time**, **platelet aggregation**, **activated partial thromboplastin time (aPTT or PTT)**, **prothrombin time (PT)**, **thrombin time (TT)**, **concentration of fibrinogen**, **fibrin clot stability**, and **measurement of fibrin degradation products**. The **platelet count** quantitates the number of platelets. The **skin bleeding time** is an overall test of platelet and vessel wall function, while the **closure time** measured using the high shear Platelet Function Analyzer PFA-100/200 is an in vitro test of platelet-related hemostasis. **Platelet aggregation** measures responses to specific aggregating agents. aPTT is a measure of the intrinsic pathway and PT of the extrinsic pathway, with aPTT being used to monitor heparin therapy and PT to measure the effectiveness of warfarin. The reader is referred to a textbook of hematology for a discussion of these tests.

SUMMARY

- Hemostasis and thrombosis are complex processes involving platelets, coagulation factors, and blood vessels.

- Thrombin and other agents cause platelet aggregation, which involves a variety of biochemical and morphologic events. Stimulation of phospholipase C and the polyphosphoinositide pathway is a key event in platelet activation, but other processes are also involved.

- Aspirin is an important antiplatelet drug that acts by inhibiting production of thromboxane A_2.

- Many coagulation factors are zymogens of serine proteases, becoming activated, then inactivated during the overall process.

- Both extrinsic and intrinsic pathways of coagulation exist, the former initiated in vivo by tissue factor. The pathways converge at factor Xa, ultimately resulting in thrombin-catalyzed conversion of fibrinogen to fibrin, which is strengthened by covalent cross-linking, catalyzed by factor XIIIa.

- Genetic disorders that lead to bleeding occur; the principal disorders involve factor VIII (hemophilia A), factor IX (hemophilia B), and von Willebrand factor (von Willebrand disease).

- Antithrombin is an important natural inhibitor of coagulation; genetic deficiency of this protein can result in thrombosis.

- Factors II, VII, IX, and X and proteins C and S require vitamin K–dependent γ-carboxylation of specific glutamate residues to function in coagulation. This carboxylation process can be inhibited by the anticoagulant warfarin.

- Fibrin is dissolved by plasmin. Plasmin exists as an inactive precursor, plasminogen, which can be activated by tissue plasminogen activator (t-PA). t-PA is widely used clinically to treat early thrombosis in the coronary and cerebral arteries.

REFERENCES

Hoffman R, Benz EJ Jr, Silberstein LR, et al (editors): *Hematology: Basic Principles and Practice*, 7th ed. Elsevier Saunders, 2017.

Israels SJ (editor): *Mechanisms in Hematology*, 4th ed. Core Health Sciences Inc, 2011. (This text has many excellent illustrations of basic mechanisms in hematology.)

Jameson JL, Fauci AS, Kasper DL, et al (editors): *Harrison's Principles of Internal Medicine*, 20th ed. McGraw-Hill, 2018.

Marder VJ, Aird WC, Bennett JS, et al (editors): *Hemostasis and Thrombosis: Basic Principles and Clinical Practice*, 6th ed. Lippincott Williams & Wilkins, 2013.

Michelson AD, Cattaneo M, Frelinger A, Newman P (editors): *Platelets*, 4th ed. Elsevier, 2019.

Cancer: An Overview

Molly Jacob, MD, PhD, MNASc, FRCPath,
Joe Varghese, MD, PhD, & P. Anthony Weil, PhD

OBJECTIVES

After studying this chapter, you should be able to:

- Present an overview of carcinogenesis and important biochemical and genetic features of cancer cells.
- Describe the important characteristics of cancer cells that distinguish them from normal cells.
- Describe important properties of oncogenes and tumor suppressor genes and their role in carcinogenesis.
- Briefly describe the concepts of genomic instability, aneuploidy, and angiogenesis in tumor formation and growth.
- Discuss the relevance of tumor markers.
- Outline how understanding of the biology of cancer has led to the development of various new therapies.

BIOMEDICAL IMPORTANCE

Cancers constitute the **second most common cause of death**, after cardiovascular disease, in many countries. Approximately 10 million people around the world die from cancer each year, and this figure is projected to increase. Humans of all ages develop cancer, and a wide variety of organs are affected. Worldwide, the main types of cancer accounting for mortality are those involving the lung, stomach, colon, rectum, liver, and breast. Other types of cancers that lead to death include cervical, esophageal, and prostate cancers. Skin cancers are also very common, but apart from melanomas, they are generally not as aggressive as those mentioned earlier. The **incidence of many cancers increases with age**. Hence, as people live longer, many more will develop the disease. Hereditary factors play a role in some types of tumors. Apart from great individual suffering caused by the disease, the economic burden to society is immense.

SOME GENERAL COMMENTS ON NEOPLASMS

A neoplasm or tumor refers to any abnormal new growth of tissue. It may be benign or malignant in nature. The term "cancer" is usually associated with malignant tumors. Tumors can arise in any organ in the body and result in different clinical features, depending on the location of the growth.

Cancer cells are characterized by certain **key properties**: (1) ability for **rapid proliferation**; (2) **increased genomic mutations** at the level of the nucleotide, small and large insertions and deletions (indels), and gross chromosomal rearrangements, duplications and loss; (3) **loss of contact inhibition** in culture *in vitro*; (4) **invasion of local tissues** and **spread to other parts of the body** (**metastasis**); (5) **self-sufficiency in growth signals** while **developing insensitivity to antigrowth signals**; (6) **ability to stimulate local angiogenesis**; and (7) are often able to **evade apoptosis**. These properties are characteristic of cells of **malignant tumors**. Metastasis of tumor cells is generally responsible for the deaths of patients who have cancer. By contrast, cells of **benign tumors** also show diminished control of growth, but do not invade local tissue or spread to other parts of the body. These points are summarized in **Figure 56–1** and amplified on in **Figure 56–2**.

Central issues in oncology (the study of cancer) include elucidation of biochemical and genetic mechanisms that underlie the uncontrolled growth of cancer cells and their ability to invade and metastasize, and development of successful treatments that will destroy cancer cells, while causing minimal damage to normal cells. Considerable progress has been made in understanding the basic nature of cancer cells, and it is now generally accepted that though mutations in key genes contribute

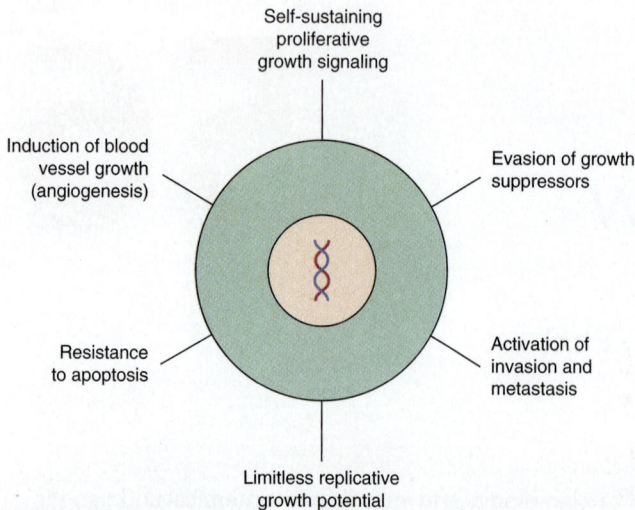

FIGURE 56–1 **Major biological features of cancer cells.** Other key properties of cancer cells are shown in Figure 56–2.

significantly to malignancies, particularly at the initiation phase of oncogenesis, other factors are also implicated in the development of malignant phenotypes. Organismal immunologic status and tissue microenvironment are two such factors. Recent work has shown that the microenvironment of host

and tumor cells, and the interactions between them contribute to the pathogenesis of malignancies. However, many aspects of the behavior of cancer cells, in particular their ability to metastasize, have yet to be fully explained. In addition, despite improvements in treatment of certain types of cancers, therapies are still often unsuccessful. This chapter aims to introduce the reader to key concepts of cancer biology. A **glossary** at the end of this chapter defines many of the terms used herein.

FUNDAMENTAL FEATURES OF CARCINOGENESIS

Nonlethal genetic damage is thought to be the initiating event in carcinogenesis. There are several broad classes of genes, which when mutated cause either gain- or loss-of-function or misregulation of cellular processes, and can thus result in the development of a tumor. Such genes are **proto-oncogenes**, **tumor suppressor genes**, **genes** involved in **DNA synthesis** and **repair**, genes involved in **chromosome segregation**, genes regulating **apoptosis**, or those involved in **evasion of immune surveillance**. Mutations in cancer cells fall into two categories: **driver mutations** and **passenger mutations**. Driver mutations are mutations in genes from one (or more) of the classes of genes listed earlier; **only driver mutations transform normal cells to cancer cells**. By contrast, passenger mutations are

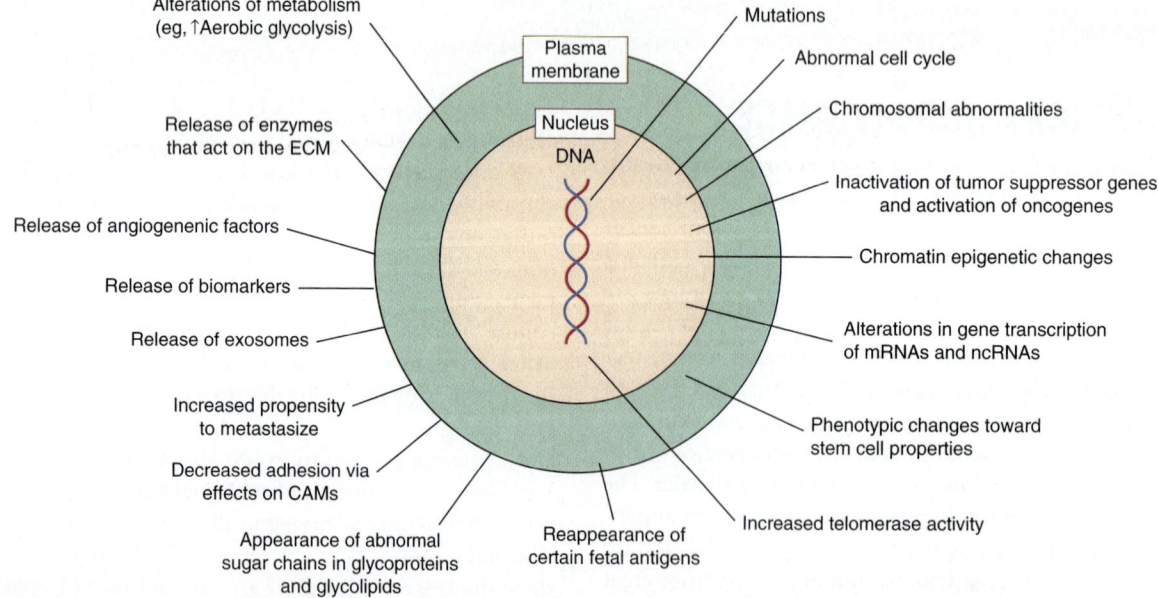

FIGURE 56–2 **Biochemical and genetic changes that occur in human cancer cells.** Many changes, in addition to those indicated in Figure 56–1, can be observed in cancer cells; only some of these are shown here. The roles of mutations in activating oncogenes and inactivating tumor suppressor genes are discussed in the text. Abnormalities in cell cycle progression and of chromosome and chromatin structure, including aneuploidy, are common in cancer cells. Alterations of expression of specific mRNAs and regulatory ncRNAs have been reported, and the relationship of stem cells to cancer cells is a very active area of research. Telomerase activity is often elevated in cancer cells. Tumors sometimes synthesize certain fetal antigens, which may be measurable in the blood. Changes in plasma membrane constituents (eg, alteration of the sugar chains of various glycoproteins—some of which are cell adhesion molecules—and glycolipids) have been detected in many studies, and may be of importance in relation to decreased cell adhesion and metastasis. Various molecules are released from cancer cells, in either soluble or membrane-bound vesicular forms (exosomes), and can be detected in the blood or extracellular fluid; these include metabolites, lipids, carbohydrates, proteins, and nucleic acids. Angiogenic factors and various proteinases are also released by some tumors. Many changes in metabolism have been observed; for example, cancer cells often exhibit a high rate of aerobic glycolysis. (CAM, cell adhesion molecule; ECM, extracellular matrix.)

mutations in genes found in cancer cells **that neither cause cancer, nor affect tumor progression**.

Cancer is of **clonal origin**, with a single abnormal cell (often with many genetic alterations) multiplying to become a mass of cells that forms a tumor. The cellular milieu of a tumor, termed the **tumor microenvironment (TME)**, significantly affects tumor formation. The exact nature of these influences may vary with cell types involved, cell-to-cell interactions, and the presence of factors such as paracrine signals, local hypoxia, and local pro-inflammatory responses. Thus, **carcinogenesis is a multistep process** that ultimately transforms normal cells into malignant cells. Tumor formation can take from a few, to tens of years, to develop to macroscopic levels.

CAUSES OF GENETIC DAMAGE

The nonlethal genetic damage causing cancer can be due to either **acquired** or **inherited, or both types of mutations**. Those that are acquired result either from errors in DNA replication or DNA repair (termed **replication mutations [R]**), or from exposure to environmental carcinogens (termed **environmental mutations [E]**) (radiation, chemicals, and viruses; Figures 35–22, 35–23 and see following discussion). The third major class of oncogenic mutations is termed **hereditary mutations (H)**; inherited mutations are derived from one or both parents. Such hereditary abnormalities result in a number of **familial conditions** that predispose to development of a cancer that is hereditary in nature. These mutations are found in specific genes (eg, tumor suppressor genes; DNA repair genes, cell cycle control genes, etc; see Chapter 35) present in germ cells, and are discussed later in this chapter. Mutations of the R, E, and H variety collectively cause the majority of human cancers, though the exact percentage of a particular type of cancer caused by these three types of mutations varies.

Spontaneous mutations, some of which may predispose to the development of cancer, occur at a frequency of approximately 10^{-7} to 10^{-6} per cell per generation. This rate is greater in tissues with a high rate of proliferation, a dynamic that can increase the generation of cancer cells from affected parent cells. **Oxidative stress** (see Chapter 45), generated as a result of enhanced production of **reactive oxygen species (ROS)**, may also be a factor in increasing mutation rates.

RADIATION, CHEMICALS, & CERTAIN VIRUSES ARE MAJOR KNOWN CAUSES OF CANCER

In general, there are three classes of environmental carcinogens that can induce tumor formation: **radiation, chemicals**, and certain **oncogenic viruses (Figure 56–3)**. The first two carcinogens cause mutations in DNA, while viruses generally act either by introducing novel genes, or modulating the expression of growth-regulatory cellular genes in normal cells.

We shall only briefly describe how radiation, chemicals, and oncogenic viruses cause cancer because many of the specific

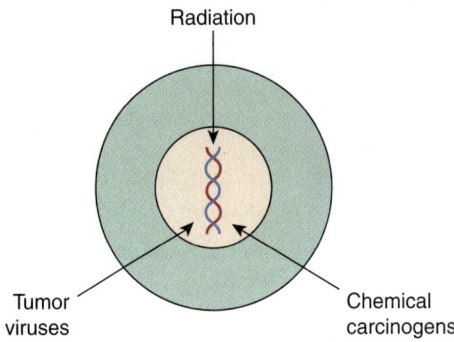

FIGURE 56–3 Radiation, chemical carcinogens and certain viruses can cause cancer by damaging chromosomal DNA.

lesions, and molecules involved in their repair, have been presented in Chapter 35.

Radiation Can Be Carcinogenic

Ultraviolet (UV) rays, x-rays, and **γ-rays** are mutagenic and carcinogenic. Extensive studies have shown that these agents damage DNA in a number of ways (**Table 56–1**; see also Table 35–8, and Figure 35–22). Mutations in DNA, if not corrected, are thought to underlie the carcinogenic effect of **radiation**, although the exact pathways are still under investigation. Additionally, x-rays and γ-rays can induce formation of **reactive oxygen species (ROS)**, which (as mentioned earlier) are also mutagenic and probably contribute to the carcinogenic effects of radiation.

Exposure to UV radiation commonly occurs due to exposure to sunlight, which is its main source. Ample evidence shows that UV radiation is linked to cancers of the skin. The risk of developing a skin cancer due to ultraviolet radiation increases with increased frequency and intensity of exposure, and decreases with increasing melanin content of skin.

As detailed in Chapter 35, DNA damage produced by environmental agents is usually removed by DNA repair mechanisms. Not surprisingly then, given the mutagenic nature of DNA damage, individuals who have an inherited inability to efficiently repair DNA have increased risk of developing a malignancy (see Table 35–9).

Many Chemicals Are Carcinogenic

A wide variety of chemical compounds are carcinogenic (**Table 56–2**).

Most **chemical carcinogens** are thought to **covalently modify DNA,** thereby forming a wide range of **nucleotide adducts** (see Table 35–8). Depending on the extent of damage to

TABLE 56–1 Some Types of DNA Damage Caused by Radiation

- Formation of pyrimidine dimers
- Formation of apurinic or apyrimidinic sites by elimination of corresponding bases
- Formation of single- or double-strand breaks or cross-linking of DNA strands

TABLE 56–2 **Some Chemical Carcinogens**

Class	Compound
Polycyclic aromatic hydrocarbons	Benzo[a]pyrene, dimethylbenzanthracene
Aromatic amines	2-Acetylaminofluorene, *N*-methyl-4-aminoazobenzene (MAB)
Nitrosamines	Dimethylnitrosamine, diethylnitrosamine
Various drugs	Alkylating agents (eg, cyclophosphamide), diethylstilbestrol
Naturally occurring compounds	Aflatoxin B_1

Note: As listed above, some drugs used as chemotherapeutic agents (eg, cyclophosphamide) can be carcinogenic.

DNA and its repair by DNA repair systems (see Figure 35–22), a variety of mutations in DNA can result from exposure of humans to chemical carcinogens, some of which contribute to the development of cancer.

Some chemicals interact directly with DNA (eg, mechlorethamine and β-propiolactone), but others, termed **procarcinogens**, require conversion (by enzyme action) to become **ultimate carcinogens**. Most ultimate carcinogens are **electrophiles** (molecules deficient in electrons; see Chapter 2) and readily attack nucleophilic (electron-rich) groups in DNA. Conversion of chemicals to ultimate carcinogens is principally due to the actions of various species of the enzyme **cytochrome P450**, which is located in the endoplasmic reticulum (ER) (see Chapters 12, 47). This fact is applied in the Ames assay (see later), in which an aliquot of post-mitochondrial supernatant (containing ER) is added to the assay system as a source of cytochrome P450 enzymes.

Chemical carcinogenesis involves two stages—initiation and **promotion**. Initiation is the stage where chemical exposure of DNA results in genomic DNA damage, some of which goes unrepaired. Unrepaired DNA changes are a necessary initial event for a cell to become cancerous. Promotion is the stage at which such an "initiated" cell begins to grow and proliferate abnormally. The cumulative effect of these two events can result in a neoplasm.

Chemical carcinogens can be identified by testing their ability to induce mutations. A simple way to do this is by using the **Ames assay** (**Figure 56–4**). This simple test, which detects mutations in the bacterium *Salmonella typhimurium* caused by chemicals, has proven very valuable for screening potential carcinogens. A modification of the Ames test is to add an aliquot of mammalian ER to the assay, to make it possible to identify procarcinogens. Very few, if any, compounds that have tested negative in the Ames test have been shown to cause tumors in animals. However, animal testing is required to unambiguously show that a chemical is carcinogenic.

It should be noted that compounds that alter epigenetic factors (such as DNA methylation and/or histone modifications [posttranslational changes]; see Chapters 35, 38) that might predispose to cancer, would not be identifiable by the Ames test, as they are not mutagenic.

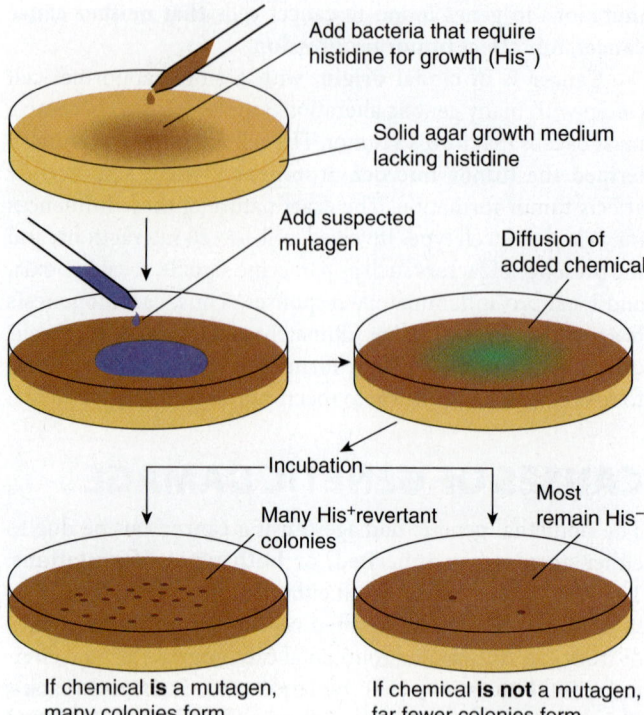

FIGURE 56–4 **The Ames assay to screen for mutagens.** The bacteria that are added (as shown) lack the ability to synthesize histidine (hence designated His⁻). They are therefore unable to grow in a medium that lacks histidine. Hence, they will not form colonies in the medium shown, as it does not contain histidine. If the chemical to be tested is mutagenic, it is likely that it may induce mutations in the bacteria that cause the *HIS* gene to become functional, resulting in synthesis of histidine (which will enable growth of colonies). Such cells/colonies are called revertants, because the non-functional gene in cells added at the start has been mutated (ie, reverted from being non-functional to being functional) (Reproduced with permission from Nester EW, Anderson DG, Roberts CE, et al: *Microbiology: A Human Perspective*, 5th ed. New York, NY: McGraw Hill; 2007.)

Certain Human Cancers Are Caused by Viruses

The study of **tumor viruses** has contributed significantly to the understanding of cancer. For example, discovery of both oncogenes and tumor suppressor genes (see later) emerged from studies of oncogenic viruses. Both DNA and RNA viruses have been identified as being able to cause cancers in humans (**Table 56–3**). The details of how each of these viruses causes cancer will not be described here. In general, the process involves incorporation (ie, integration) of genetic material of viruses into the genome of the host cell. In the case of RNA viruses, this would occur after reverse transcription of the viral RNA to viral DNA. (An exception to this rule is Hepatitis C virus [HCV] which is an RNA virus that is oncogenic but does not contain/utilize a viral reverse transcriptase.) Such integration of viral DNA (called the provirus) into host cell DNA results in various effects such as **deregulation of the cell cycle, inhibition of apoptosis**, and **abnormalities of cell-signaling pathways** (see Chapters 35, 42). All these

TABLE 56–3 Some Viruses That Cause or Are Associated With Human Cancers

Virus	Genome	Cancer
Epstein-Barr virus	DNA	Burkitt lymphoma, nasopharyngeal cancer, B-cell lymphoma
Hepatitis B	DNA	Hepatocellular carcinoma
Hepatitis C	RNA	Hepatocellular carcinoma
Human herpesvirus 8 (HHV-8)	DNA	Kaposi sarcoma
Human papilloma viruses (types 16 and 18)	DNA	Cancer of the cervix
Human T-cell leukemia virus type 1	RNA	Adult T-cell leukemia

events are discussed later in this chapter. **DNA viruses** often act by **downregulation** of the expression, and/or function of tumor suppressor genes *P53* and *RB* (see later) and their protein products. RNA viruses often carry oncogenes in their genomes; how oncogenes act to cause malignancy is discussed later.

ONCOGENES & TUMOR SUPPRESSOR GENES PLAY KEY ROLES IN CAUSING CANCER

Over the past several decades, major advances have been made in understanding how cancer cells develop and grow. Two key findings were the discoveries of **oncogenes** and **tumor suppressor genes**. These discoveries pointed to specific molecular mechanisms through which cell growth and division could be dysregulated, resulting in abnormal growth.

Oncogenes Are Derived From Cellular Genes Termed Proto-Oncogenes, Which Encode a Wide Variety of Proteins That Stimulate Cell Growth

An **oncogene** can be defined as an altered gene, the product of which acts in a **dominant** manner to accelerate cell growth or cell division. Oncogenes are generated by "activation" of normal cellular genes, termed **proto-oncogenes**, which encode growth-stimulating proteins. Such activation can be affected through any of several different mechanisms (**Table 56–4**).

Table 56–4 lists an example of a point mutation occurring in the *RAS* oncogene. *RAS* encodes a small GTPase. Loss of the GTPase activity of this G-protein results in chronic stimulation of the activity of adenylyl cyclase and the MAP kinase pathway, leading to cell proliferation (recall that G-proteins are active when complexed with GTP and inactive when the bound GTP is hydrolyzed to GDP by their intrinsic GTPase activity; see Chapter 42). Another way an oncogene can be activated

TABLE 56–4 Mechanisms by Which Oncogenes Are Activated

Mechanism	Explanation
Mutation	A classic example is a point mutation of the *RAS* oncogene. This results in the gene product, a small G-protein (RAS), in which the intrinsic GTPase activity of the protein is lost. Consequently, cells are subjected to constitutively active signaling via stimulation of adenylyl cyclase and the mitogen-activated protein (MAP) kinase pathway.
Promoter insertion	Insertion of a viral promoter sequence near an oncogene activates it.
Enhancer insertion	Insertion of a viral enhancer sequence near an oncogene activates it.
Chromosomal translocation	A piece of one chromosome is split off and is joined to another, resulting in activation of an oncogene at the site where insertion occurs. Classic examples of such translocations include those seen in Burkitt lymphoma (see Figure 56–5) and in the Philadelphia chromosome in chronic myelocytic leukemia (see the Glossary).
Gene amplification[a]	Abnormal multiplication of a gene occurs, resulting in many copies. This can occur with oncogenes and genes involved in drug resistance.

[a]Regions of gene amplification may be recognized as homogeneously stained regions (HSRs) on chromosomes, or as double minute chromosomes (which are extrachromosomal in location).

is via insertion of an enhancer and/or a strong promoter upstream of a protein-coding gene, resulting in increased transcription (and hence protein expression) of the cognate gene. **Figure 56–5A** illustrates how integration of a retroviral provirus (ie, the reverse transcriptase-generated dsDNA copy of the RNA genome of a tumor virus such as **Rous sarcoma virus, RSV**) can act as an enhancer/promoter to activate *MYC*, a neighboring host gene. Overproduction of the oncogenic transcription factor MYC activates the transcription of many genes, cell cycle–regulatory genes among them. Such Myc-induced overexpression of proteins stimulates unregulated cell proliferation.

Chromosomal translocations are found quite frequently in cancer cells; literally hundreds of different examples have been documented. The translocation found in cases of Burkitt lymphoma is illustrated in **Figure 56–5B**. The overall effect of this translocation is that the *MYC* gene comes under the influence of the powerful IgG heavy chain-encoding gene enhancer, resulting in increased transcription of the *MYC* gene, and subsequent increased cell proliferation.

Yet another mechanism of oncogene activation is via **gene amplification** (see Figure 38–20), a phenomenon that occurs quite commonly in various cancers. What happens is that multiple copies of an oncogene are formed, which results in increased production of the growth-promoting protein encoded by that gene.

Activated oncogenes promote cancer through a variety of mechanisms; summarized in **Figure 56–6**. The protein products

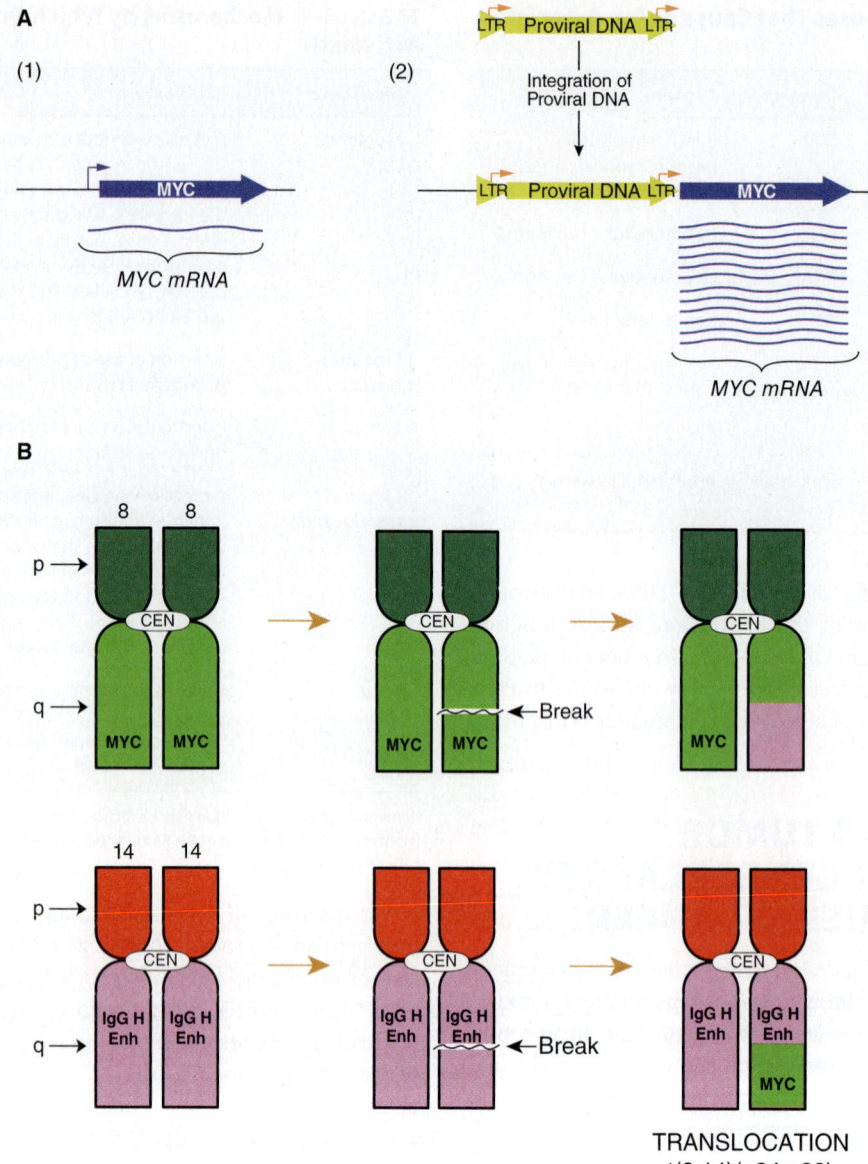

FIGURE 56–5 **(A) Schematic representation of how promoter insertion may activate a proto-oncogene.** (1) Schematic representation of a *MYC* gene on a chromosome. (2) An avian leukemia virus has integrated in the chromosome in its pro-viral form (a DNA copy of its RNA genome), as shown. This has occurred adjacent to the *MYC* gene. The long terminal repeats (LTR) of the proviral DNA serve as strong enhancer and promoter sequences (see Chapter 36). They now reside just upstream of the *MYC* gene and therefore activate *MYC*, resulting in vigorous transcription of the *MYC* mRNA. For simplicity, only one strand of DNA is depicted and other details have been omitted. **(B) Schematic representation of the reciprocal translocation involved in Burkitt lymphoma.** The chromosomes involved are 8 and 14. The letters p and q refer to the short and long arms of the chromosomes respectively, while the white oval, labeled CEN, is the centromere. Only a portion of each chromosome is shown. A segment from the end of the q arm of chromosome 8 breaks off and translocates to chromosome 14. The reverse process moves a small segment from the q arm of chromosome 14 to chromosome 8; these translocations involve regions q24 (chromosome 8) and q32 (chromosome 14). The *MYC* gene is contained in the small piece of chromosome 8 that was transferred to chromosome 14. Thus, *MYC* is placed adjacent to the genes transcribing the heavy chains of immunoglobulin molecules, and *MYC* gene transcription is activated by the potent IgG H gene enhancer. Many other translocations have been identified, with perhaps the best known being that involved in formation of the Philadelphia chromosome seen in chronic myeloid leukemia (see the Glossary).

of activated oncogenes affect cell signaling pathways, where they may act as a growth factor, a receptor for a growth factor, a G-protein, or as a downstream signaling molecule. Other oncoproteins act to alter transcription of genes crucial for oncogenesis or to deregulate the cell cycle. Still other oncoproteins affect

cell–cell interactions, or the process of apoptosis. Collectively, these mechanisms explain many of the major features of cancer cells shown in Figure 56–1, such as their limitless replicative potential, their constitutively activated signaling pathways, their ability to invade and spread, and their evasion of apoptosis.

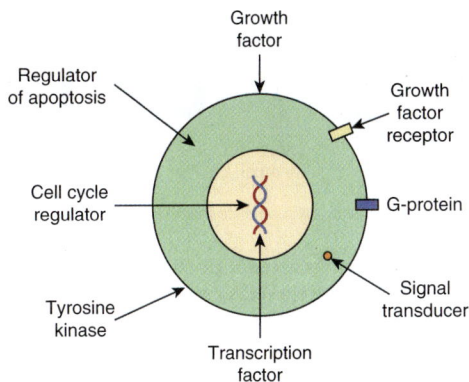

FIGURE 56–6 **Examples of ways by which oncoproteins work.** Shown are examples of various proteins encoded by oncogenes (oncoproteins). These are listed below with the corresponding oncogene given in parentheses along with its OMIM number (see glossary for description of OMIM). A growth factor, fibroblast growth factor 3 (*INT2*, 164950); a growth factor receptor, epidermal growth factor receptor [EGFR] (*HER1*, 131550); a G-protein (*H-RAS-1*, 190020); a signal transducer (*BRAF*, 164757); a transcription factor (*MYC*, 190080); a tyrosine kinase involved in cell–cell adhesion (*SRC*, 190090); a cell cycle regulator (*PRAD*, 168461); a regulator of apoptosis (*BCL2*, 151430).

TABLE 56–5 **Differences Between Oncogenes & Tumor Suppressor Genes**

Oncogenes	Tumor Suppressor Genes
Mutation in one of the two alleles is sufficient to produce oncogenic effects.	Mutations in both alleles are required to produce oncogenic effects.
Oncogenic effect is due to gain-of-function of a protein that stimulates cell growth and proliferation	Oncogenic effect is due to loss-of-function of a protein that inhibits cell growth and proliferation.
Mutations in oncogenes arise in somatic cells, and hence are not inherited.	Mutations in tumor suppressor genes may be present in somatic or germ cells (and hence, may be inherited)
Usually do not show tissue preference with respect to the type of cancers in which they are found	Often demonstrate strong tissue preference with respect to the type of cancers in which they are found (eg, mutations in the *RB* gene result in retinoblastoma)

Data from Levine AJ. The p53 tumor-suppressor gene. N Engl J Med. 1992;326(20): 1350-1352.

Certain **tumor viruses** (eg, retroviruses, papovaviruses) **contain oncogenes**. It was the study of the **Rous Sarcoma virus (RSV)** retrovirus, that first revealed the existence of oncogenes. Further study showed that many retroviral oncogenes were derived from normal cellular genes, the so-called proto-oncogenes, which the tumor viruses had acquired during their passage through host cells.

Tumor Suppressor Genes Act to Inhibit Cell Growth & Division

A **tumor suppressor gene** produces a protein that normally suppresses cell growth or cell division. When such a gene is altered by mutation, the inhibitory effect of its product is lost or diminished. Thus, loss-of-function of a tumor suppressor gene leads to increased cell growth or division. As first suggested by AG Knudson based on his studies of the inheritance of retinoblastomas in 1971, it is now known that both copies of the *RB* tumor suppressor gene must be mutated for the RB protein to lose its inhibitory effect on cell growth regulation. The loss-of-function retinoblastoma gene allele, rb^-, is recessive to a wild-type *RB* copy of the gene.

A useful distinction has been made between **gatekeeper** and **caretaker** functions of tumor suppressor genes. Gatekeeper genes (products) control cell proliferation, and include mainly genes that act to regulate the cell cycle and apoptosis. By contrast, caretaker gene products are concerned with preserving the integrity of the genome, and include proteins that are involved in recognizing and correcting DNA damage, and in maintaining chromosomal integrity during cell division.

Many oncogenes and tumor suppressor genes have now been identified. Only a few are mentioned here. Important differences between oncogenes and tumor suppressor genes are listed in **Table 56–5**. **Table 56–6** lists some properties of two oncogenes (*MYC* and *RAS*), and tumor suppressor genes (*P53* and *RB*; see Chapter 35) that have been studied most intensively.

Studies of the Development of Colorectal Cancers Have Illuminated the Involvement of Specific Oncogenes & Tumor Suppressor Genes

Many types of tumors have been analyzed for genetic changes. One of the earliest studied and most informative areas in this respect has been analysis of the **development of colorectal cancer** by Vogelstein and colleagues. Their work, and that of others, has shown the involvement of various oncogenes and tumor suppressor genes in human cancer. These workers analyzed the sequence and expression of various oncogenes, tumor suppressor genes, and other relevant genes in samples of **normal colonic epithelium**, of **dysplastic epithelium** (a preneoplastic condition characterized by abnormal development of epithelium), various stages of **adenomatous polyps**, and **adenocarcinomas**. Some of their key findings are summarized in **Figure 56–7**. It can be seen that certain genes are mutated at relatively specific stages of the sequence of events shown. The functions of the various genes identified are listed in **Table 56–7**. The overall sequence of changes can vary from that shown, and other genes may also be involved. Similar studies have been performed on a number of other human tumors, revealing different patterns of activation of oncogenes, and mutations of tumor suppressor genes. Further mutations in these and other genes are involved in **tumor progression**,

TABLE 56–6 Properties of Some Important Oncogenes & Tumor Suppressor Genes

Name	Properties
MYC	An oncogene (OMIM 190080) that encodes a transcription factor that can alter transcription of many key cellular regulatory genes. MYC is involved in stimulation of cell growth, cell cycle progression, and DNA replication. It is mutated in a variety of tumors.
P53	A tumor suppressor gene (OMIM 191170) that is activated in response to various stimuli that produce DNA damage. Activation of p53 induces cell cycle arrest, apoptosis, senescence, and DNA repair. It is also involved in some aspects of regulation of cellular metabolism. It has, therefore, been named "the guardian of the genome." p53 is mutated in about 50% of human tumors.
RAS	A family of oncogenes encoding small G-proteins, specifically GTPases that were initially identified as the transforming genes present in certain murine sarcoma viruses. Important members of the family are K-RAS (Kirsten), H-RAS (Harvey) (OMIM 190020), and N-RAS (neuroblastoma). Persistent activation of these genes due to mutations contributes to the development of a variety of cancers.
RB	A tumor suppressor gene (OMIM 180200) encoding the Rb protein. Rb, a transcriptional repressor, regulates the cell cycle by binding to the activating transcription factor E2F. Rb represses the transcription of various genes involved in the transition of the G_1 phase of the cell cycle to the S phase. Mutation of the RB gene results in retinoblastoma, but it is also involved in the genesis of certain other tumors (see Chapter 35).

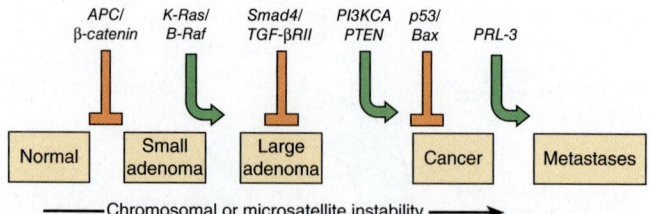

FIGURE 56–7 Multistep genetic changes associated with the development of colorectal cancers. Mutations in the *APC* gene initiate the formation of adenomas. One sequence of mutations in an oncogene and in various tumor suppressor genes that can result in further progression to large adenomas and cancer is indicated. Patients with familial adenomatous polyposis (OMIM 175100) inherit mutations in the *APC* gene and develop numerous dysplastic aberrant crypt foci (ACF), some of which progress as they acquire the other mutations indicated in the figure. The tumors from patients with hereditary nonpolyposis colon cancer (OMIM 120435) go through a similar, though not identical, series of mutations; mutations in the DNA mismatch repair system (see Chapter 35) speed up this process. K-RAS, BRAF, and PI3KCA are oncogenes; the other genes listed are tumor suppressor genes. The sequence of events shown here is not invariable in the development of all colorectal cancers. A variety of other genetic alterations have been described in a small fraction of advanced colorectal cancers. These may be responsible for the heterogeneity of biologic and clinical properties observed among different cases. Instability of chromosomes and microsatellites (see Chapter 35) occurs in many tumors, and likely involves mutations in a considerable number of genes. (Modified with permission from Valle DL, Antonarakis S, Ballabio A, et al: The Online Metabolic & Molecular Bases of Inherited Disease. New York, NY: McGraw Hill; 2019.)

TABLE 56–7 Some Genes Associated With Colorectal Carcinogenesis

Gene[a]	Action of Encoded Protein
APC (OMIM 611731)	Antagonizes WNT[b] signaling; if mutated, WNT signaling is enhanced, stimulating cell growth.
β-CATENIN (OMIM 116806)	Encodes β-catenin, a protein present in adherens junctions, which are important for the integrity of epithelial tissues. It is an integral part of the WNT signaling pathway.
K-RAS (OMIM 601599)	Involved in tyrosine kinase signaling, especially in the mitogen-activated protein (MAP) kinase pathway.
BRAF (OMIM 164757)	A serine/threonine kinase that acts in concert with Ras to activate the MAP kinase pathway.
SMAD4 (OMIM 600993)	Involved in intracellular signaling by transforming growth factor β (TGF-β[c]).
TGF-βRII	Acts as a receptor for TGF-β.
PI3KCA (OMIM 171834)	Acts as a catalytic subunit of phosphatidylinositol 3-kinase (PI3K).
PTEN (OMIM 601728)	A protein tyrosine phosphatase which acts as an important inhibitor of signaling via the PI3K-Akt pathway.
P53 (OMIM 191170)	The product, p53, is induced in response to DNA damage and is also a transcription factor for many genes involved in cell division (see Chapter 35, Table 56–10).
BAX (OMIM 600040)	Acts to induce cell death (activator of apoptosis).
PRL3 (OMIM 606449)	A protein tyrosine phosphatase involved in cell cycle regulation.

Abbreviations: *APC*, adenomatous polyposis coli gene; *BAX*, encodes BCL2-associated X protein (BCL2 is a repressor of apoptosis); *BRAF*, the human homolog of an avian protooncogene; *K-RAS*, Kirsten–Ras-associated gene; *PI3KCA*, encodes the catalytic subunit of phosphatidylinositol 3-kinase; *PRL3*, encodes a protein tyrosine phosphatase with homology to PRL1, another protein tyrosine phosphatase found in regenerating liver; *PTEN*, encodes a protein tyrosine phosphatase and tensin homolog; *P53*, encodes p53, a polypeptide of molecular mass ~53 kDa; *SMAD4*, the homolog of a gene found in *Drosophila*.
[a]*K-RAS*, *BRAF*, and *PI3KCA* are oncogenes; the other genes listed are either tumor suppressor genes or genes the products of which are associated with the actions of the products of tumor suppressor genes.
[b]The WNT family of secreted glycoproteins is involved in a variety of developmental processes.
[c]TGF-β is a polypeptide (a growth factor) that regulates proliferation and differentiation in many cell types.
Note: The various genes listed are either oncogenes, tumor suppressor genes, or genes the products of which are closely associated with the products of these two types of genes. The cumulative effects of mutations in the genes listed are to drive colonic epithelial cells to proliferate and eventually become cancerous. They achieve this mainly via effects on various signaling pathways that affect cellular proliferation. Other genes and proteins not listed here are also involved. This table and Figure 56–7 vividly illustrate the importance of a detailed knowledge of cell signaling for understanding the genesis of cancer.

a phenomenon whereby clones of tumor cells that grow rapidly and are able to metastasize predominate. Thus, tumors may contain a variety of cells with different genotypes, making it difficult to treat successfully.

Several conclusions can be drawn from these results and those from other similar studies. First, **cancer is truly a genetic disease**, but in a somewhat different sense from the normal meaning of the phrase, insofar as many of the gene alterations are due to somatic mutations. Second, **carcinogenesis is a multistep process**. It is estimated that in most cases a minimum of five to six genes must be mutated for cancer to occur. Third, additional subsequent mutations are thought to confer selective advantages on certain clones of cells, some of which acquire the **ability to metastasize** (see later). Finally, many of the genes implicated in colorectal carcinogenesis and other types of cancers are involved in cell signaling events, showing again the **central role that alterations in signaling play in the development of cancer**.

GROWTH FACTORS & ABNORMALITIES OF THEIR RECEPTORS & SIGNALING PATHWAYS PLAY MAJOR ROLES IN CANCER DEVELOPMENT

There Are Many Growth Factors

A large variety of polypeptide growth factors that affect human tissues and cells have been identified. Some are listed in **Table 56–8**. Here, we focus mostly on their relationship with cancer.

Growth factors can act in an **endocrine**, **paracrine**, or **autocrine** manner (see Chapter 41) and affect a wide variety

TABLE 56–8 Some Polypeptide Growth Factors

Growth Factor	Function
Epidermal growth factor (EGF)	Stimulates growth of many epidermal and epithelial cells
Erythropoietin (EPO)	Regulates development of early erythropoietic cells
Fibroblast growth factors (FGFs)	Promote proliferation of many different cells
Interleukins	Interleukins exert a variety of effects on cells of the immune system
Nerve growth factor (NGF)	Trophic effect on certain neurons
Platelet-derived growth factor (PDGF)	Stimulates growth of mesenchymal and glial cells
Transforming growth factor α (TGF-α)	Similar to EGF
Transforming growth factor β (TGF-β)	Exerts both stimulatory and inhibitory effects on certain cells

Note: Many other growth factors have also been identified. Growth factors may be produced by a variety of cells, or may have mainly one source. Many different interleukins have now been isolated; along with the interferons and some other proteins/polypeptides, they are referred to as cytokines.

of cells, to produce a **mitogenic response** (see Chapter 42). Recall (see Chapter 53) that growth factors play important roles in the growth and differentiation of hematopoietic cells.

Growth inhibitory factors also exist. For example, **transforming growth factor β (TGF-β)** exerts inhibitory effects on the growth of certain cells. Thus, chronic exposure to either increased amounts of a growth factor, or to decreased amounts of a growth inhibitory factor, can alter the balance of cellular growth.

Growth Factors Work via Specific Receptors & Transmembrane Signaling to Affect the Activities of Specific Genes

Growth factors produce their effects by interacting with **specific receptors** on cell surfaces, initiating **various signaling events** (see Chapter 42). Genes encoding receptors for growth factors have been identified and characterized. These receptors generally have short membrane-spanning segments and external and cytoplasmic domains (see Chapters 40, 42). A number of these receptors (eg, receptors for **epidermal growth factor [EGF]**, **insulin, insulin-like growth factor [IGF-I]**, and **platelet-derived growth factor [PDGF]**) have **tyrosine kinase** activity. The kinase activity, located within the cytoplasmic domains of these receptors, causes autophosphorylation of the cognate receptor protein and also phosphorylates certain other proteins when the receptors are bound by their target ligand.

Consideration of platelet-derived growth factor, **PDGF**, illustrates how a particular growth factor brings about its effects. Interaction of PDGF with its receptor stimulates the activity of **phospholipase C (PLC)**. PLC splits **phosphatidylinositol bisphosphate (PIP$_2$)** (found in biologic cell membranes) into **inositol trisphosphate (IP$_3$)** and **diacylglycerol (DAG)** (see Figure 42–6). Elevations in IP$_3$ levels result in increased levels of intracellular Ca^{2+}, while DAG activates **protein kinase C (PKC)**. Hydrolysis of DAG may release **arachidonic acid**, which can stimulate production of **prostaglandins** and **leukotrienes**, each of which has various biologic effects. Exposure of target cells to PDGF can result in rapid (minutes to 1–2 hours) activation of expression of genes encoding certain cellular proto-oncogenes (eg, *MYC* and *FOS*) that participate in stimulation of mitosis through their effects on the cell cycle (see later). The bottom line is that growth factors interact with specific receptors to stimulate specific signaling pathways that serve to increase or decrease the amounts and/or activities of various critical proteins that affect cell division.

MICRO-RNAS ARE KEY PLAYERS IN CARCINOGENESIS & TUMOR METASTASIS

Micro-RNAs (miRNA), discovered in 1993, are non-protein-coding RNAs (ncRNA) that are ~22 nucleotides long (see Chapters 34 and 36). miRNAs are expressed in different tissue and cell types and are known to be involved in the regulation of gene expression by decreasing translation of specific mRNA, or decreasing mRNA stability (see Figure 36–17).

The expression of miRNA has been found to be dysregulated in many cancers. Such misregulation is thought to have a causal role in the pathogenesis of such cancers. Some miRNAs are oncogenic (referred to as oncomiRs) and are overexpressed in cancerous tissue. On the other hand, tumor-suppressive miRNAs that counteract oncogenic characteristics are found to be under-expressed in some cancers; the similarity to oncogenes and tumor suppressors (described previously) is apt. Examples of oncogenic microRNAs include the miR-17-92 polycistronic miRNA encoding gene cluster (implicated in cancers of the lung, breast, pancreas, colon, etc.); miR-21 (in cancers of the lung, breast, etc.); and miR-155 (in lung cancer and lymphoma). Examples of tumor suppressing microRNAs include let-7 (dysregulated in ovarian and lung cancer), miR-34 (implicated in various types of cancers), miR-15, and miR-16 (both involved in chronic lymphocytic leukemia).

MicroRNAs have also been shown to influence extrinsic factors that modulate tumors. These include interactions between the immune system and cancer cells, interactions of stromal cells, effects on oncoviruses, etc. It is likely that the overall protein expression resulting from specific types and amounts of miRNAs determines whether the net effect of a specific miRNA is oncogenic or tumor suppressive.

The use of miRNAs in clinical applications is gaining ground. miRNAs are being investigated as biomarkers for cancer diagnosis, prognosis, and classification. These small RNAs also hold promise in therapy, where oligonucleotides are used to enhance expression of tumor suppressor miRNAs in cells, or antisense oligonucleotides are employed to counter the action of oncogenic miRNAs. Several such miRNA-targeted therapeutics have been developed, some of which have reached clinical trials. However, many challenges remain. For example, Phase I clinical trials using miR-34 for hepatocellular carcinoma showed promising therapeutic effects but had to be stopped due to severe adverse effects. Nevertheless, other efforts, such as the use of miR-16 for the treatment of malignant mesotheliomas and non-small cell lung cancers, have met with early success in Phase I trials.

Another important clinical application of miRNAs lies in their ability to potentiate therapeutic responses during anticancer chemotherapy and radiotherapy. Thus, the use of miRNAs as specific targets for treatment and/or for therapeutic supplementation in the management of cancer appears highly promising. However, as with all gene-based therapies, effective delivery of stable oligonucleotides into appropriate target cells is extremely challenging. Nevertheless, tremendous advances in scientific and pharmaceutical developments are anticipated in this area in the years ahead, bringing to clinical practice a group of novel therapeutic agents for the treatment of cancers.

EXTRACELLULAR VESICLES & CANCER

Extracellular vesicles (EVs), also referred to as **exosomes** and **microvesicles**, are a group of small vesicles that are enclosed by a lipid bilayer that are released by most eukaryotic cells,

both normal and diseased (see Figure 40–23). These vesicles vary from one another in their size and the way in which they are formed. Secreted EVs contain a variety of biomolecules such as lipids, proteins, and nucleic acids, and can be delivered to target cells through a range of mechanisms such that they can produce autocrine, paracrine, and even endocrine effects. EVs often contain noncoding RNAs (ncRNA) (comprising miRNAs, long non-coding RNAs [lncRNAs], and circular RNAs [circRNAs]; see Chapters 34, 36). Thus, EVs represent a novel mechanism of cellular communication.

Exosomes have been shown to transfer information to target cells by mechanisms such as receptor-ligand binding, direct fusion of the vesicle with the plasma membrane of recipient cells, or by endocytosis. There is now growing evidence to show that **tumor-derived exosomes** (**TEX**s) play an important role in the pathogenesis of tumors. Transfer of ncRNAs in this way has been implicated in the pathogenesis of cancer (through laboratory studies that have demonstrated that EVs can modulate expression of crucial genes that are involved in tumorigenesis). The effects produced by ncRNAs on their target cells include control of cell proliferation, induction of angiogenesis, alterations in the TME and tumor immunity, and even induction of drug resistance. EVs have also been shown to act on distant organs, thereby affecting metastasis.

Exosomes are found in several body fluids (serum, plasma, urine, etc.) and so have the potential to serve as diagnostic or prognostic markers of cancers. They may also be useful as therapeutics. Since they can readily cross biologic membranes, they may be used as effective delivery systems for anticancer drugs (including miRNA or anti-miRNA). Exosomes thus represent an exciting new area of oncology. However, additional experimentation will be required in order to make the diagnostic and therapeutic potential of exosomes a clinical reality.

EPIGENETIC MECHANISMS ARE INVOLVED IN CANCER

There is strong evidence that epigenetic mechanisms (see Chapter 36) are involved in the causation of cancer. Such mechanisms produce nonmutational changes that affect regulation of gene expression. For example, methylation of specific cytosine bases in particular genes has been implicated in turning off the transcription of certain other genes. Multiple changes in the normal pattern of methylation status of cytosine residues in specific genes have been detected in cancer cells. Histone posttranslational modification status (ie, acetylation, ADP-ribosylation, methylation, phosphorylation, sumoylation, and ubiquitylation; see Chapters 35 and 38) have been found in cancer cells. Similarly, mutations that affect the structure and/or activity of the nucleosome remodeling complexes that are involved in chromatin remodeling can also affect gene transcription. Indeed, several of the components of the SWI/SNF complexes likely act as tumor suppressor genes. Consistent with these observations that show that chromatin and/or DNA modification status is important in oncogenesis, it has recently been shown that a significant fraction of certain human tumor types contain mutations in histones; **such histone variants are termed oncohistones**. Oncohistones contribute importantly to oncogenesis by changing normal epigenetic modification patterns that are essential for specific gene transcription. Importantly, oncohistones can have dramatic effects on the functions of multiple chromatin remodelers and, hence, gene regulatory programs.

A matter of particular interest regarding the role of epigenetic changes in oncogenic gene expression is that many of the protein and DNA modifications responsible are reversible. In this regard, 5-azadeoxycytidine and decitabine are inhibitors of DNA methyltransferases (DNMTs) (see also Figure 32–13), while valproic acid and vorinostat act to inhibit histone deacetylases (HDACs). Both of these agents have been used to treat certain types of leukemias and lymphomas and are thought to work by de-repressing the transcription of certain critical growth-regulatory genes, such as tumor suppressors. Epigenetic mechanisms are also involved in many aspects of resistance to anticancer drugs. For example, resistance to temozolomide (TMZ), the standard chemotherapeutic agent for glioblastoma multiforme, is linked to promotor methylation of the DNA repair enzyme, O^6-methylguanine-DNA methyltransferase (MGMT). Increased activity of MGMT (by decreased promotor methylation) rescues cancer cells from TMZ-induced cell death by accelerating repair of DNA damage induced by the drug. Detection of MGMT promotor methylation patterns is now being used to classify newly diagnosed glioblastomas and to predict response to TMZ. Synthesis and study of additional specific inhibitors that act at DNA and protein/enzyme epigenetic levels are active areas of investigation. Finally, the increasing use of various –omic technologies to identify, quantify, and score epigenetic changes and their effects on gene expression in many types of cancers is adding considerably to knowledge in this important area of research. Two notable examples in this regard are the efforts and results of **TCGA** (**The Cancer Genome Atlas**) and **PCAWG** (**Pan-Cancer Analysis of Whole Genomes**) cancer biology consortia.

A NUMBER OF CANCERS DISPLAY A HEREDITARY PREDISPOSITION

It has been known for many years that certain cancers have a hereditary basis. Depending on the specific cancer type, it has been estimated that 5 to 15% of cancers have a hereditary etiology. The discovery of oncogenes and tumor suppressor genes has allowed investigation of the molecular basis of this phenomenon. Many types of hereditary cancer have now been recognized; a few of these are listed in **Table 56–9**. In a number of cases, where a hereditary syndrome is suspected, appropriate genetic screening of individuals and families has allowed early interventions to be made. Some young women, who have inherited either a mutated *BRCA1* or *BRCA2* gene, have opted for prophylactic mastectomies to prevent cancer of the breast occurring in later life. Even without prior knowledge regarding a familial predisposition to a particular type of cancer, screening for colorectal cancer can be readily performed. Commercial analyses of home, self-sample-collected fecal samples can be analyzed for the presence of intestinal cell DNA carrying mutations that predispose to colorectal cancer (see Figure 56–7).

ABNORMALITIES OF THE CELL CYCLE ARE UBIQUITOUS IN CANCER CELLS

Knowledge of **the cell cycle** is necessary for understanding the molecular mechanisms involved in the development of many types of cancer. It is also of importance because many

TABLE 56–9 Some Hereditary Cancer Conditions

Condition	Gene	Major Function	Major Clinical Feature
Adenomatous polyposis of the colon (OMIM 175100)	APC	See Table 56–7	Development of early-onset adenomatous polyps, which are immediate precursors of colorectal cancers
Breast cancer 1, early onset (OMIM 113705)	BRCA1	DNA repair	About 5% of women with breast cancer in North America carry mutations in this gene or in *BRCA2*. Also substantially increases risk of ovarian cancer
Breast cancer 2, early onset (OMIM 600185)	BRCA2	DNA repair	As stated above for *BRCA1*. Mutations in this gene also increase the risk of ovarian cancer, but to a lesser extent
Hereditary nonpolyposis cancer, type I (OMIM 120435)	MSH2	DNA mismatch repair	Early onset of colorectal cancers
Li-Fraumeni syndrome (OMIM 151623)	P53	See Table 56–6	A rare syndrome involving cancers at different sites, developing at an early age
Neurofibromatosis, type 1 (OMIM 162200)	NF1	Encodes neurofibromin	Presentation varies from presence of a few café au lait spots to development of thousands of neurofibromas
Retinoblastoma (MIM 180200)	RB1	See Table 56–6	Hereditary or sporadic retinoblastoma[a]

[a]In hereditary retinoblastoma, one allele is mutated in the germ line, requiring only one subsequent mutation for a tumor to form. In sporadic retinoblastoma, neither allele is mutated at birth, so that subsequent mutations in both alleles are required for a tumor to develop.
Many other hereditary cancer conditions have also been identified.

anticancer drugs act only against cells that are dividing, or are in a certain phase of the cycle.

Basic aspects of the cell cycle were described in Chapter 35. As shown in Figure 35–20, the cycle has four phases: G_1, S, G_2, and M. If cells are not in one of these phases, they are said to be in the G_0 phase and are termed quiescent. Cells can be recruited back into the cycle from G_0 by various influences (eg, certain growth factors). Cancer cells usually have a shorter generation time than normal cells, and there are fewer of them in the quiescent G_0 phase.

The roles of various **cyclins**, **cyclin-dependent kinases** (**CDKs**), and a number of other important molecules that affect the cell cycle (eg, the protein products of the *RB* and *P53* genes) were described in Chapter 35. The points in the cycle at which some of these molecules act have been indicated in Figures 35–20, 35–21, and Table 35–7.

Because a major property of cancer cells is uncontrolled growth, many aspects of their cell cycle have been studied in considerable depth. A few important results are mentioned here. A variety of mutations that affect cyclins and CDKs have been reported. Many products of proto-oncogenes and tumor suppressor genes play important roles in regulating the normal cell cycle. A wide variety of mutations have been found in these types of genes, including *RAS*, *MYC*, *RB*, and *P53* (which are among the most studied, see later).

For example, as discussed in Chapter 35, the protein product of the *RB* gene is a regulator of the cell cycle. It acts by binding to a transcription factor E2F, thereby blocking transcription of genes required for progression of the cell from G1 to S phase. Mutation-induced loss of RB protein function, thus, removes this element of control of the cell cycle.

When damage to DNA occurs (by radiation or chemicals), the p53 protein increases in amount and activates transcription of genes that delay transit through the cycle. If the damage is too severe to repair, p53 activates genes that cause apoptosis (see later, and Figure 35–23). If p53 is absent or inactive due to mutations, apoptosis does not occur and cells with damaged DNA persist, potentially becoming progenitors of cancer cells.

GENOMIC INSTABILITY & ANEUPLOIDY ARE IMPORTANT CHARACTERISTICS OF CANCER CELLS

As referred earlier and also later in this chapter, cancer cell genomes carry many mutations. One possible explanation for their **genomic instability** is that they have **a mutator phenotype**. The idea of mutator phenotypes was originally postulated by Loeb and colleagues, who argued that such phenotypes were caused by cancer cells having acquired mutations in genes involved in DNA replication and DNA repair, thus allowing mutations to accumulate. The concept was later expanded to include mutations that affect chromosomal segregation, DNA damage surveillance, and processes

such as apoptosis. Increasing evidence suggests that collectively these various mechanisms, termed errors in genome replicative functions, contribute to a very large proportion of cancers.

The term genomic instability is frequently used to refer to two abnormalities seen in many cancer cells: **microsatellite instability** (**MSI**) and **chromosomal instability** (**CIN**). MSI was described briefly in Chapter 35. MSI involves expansion or contraction of microsatellite DNA sequences, usually due to abnormalities of mismatch repair or replication slippage. CIN occurs more often than MSI, and the two are often mutually exclusive. CIN refers to gain or loss of chromosomes caused by abnormalities of chromosomal segregation during mitosis.

Another area of interest with regard to CIN is **copy number variation** (**CNV**) (see Glossary; Chapter 39). Associations of various CNVs with many types of cancers have been identified, and their precise roles in carcinogenesis are under investigation.

An important aspect of CIN is **aneuploidy**, a very common feature of solid tumors. Aneuploidy exists when the chromosomal number of a cell is not a multiple of the haploid number. The degree of aneuploidy often correlates with a poor prognosis suggesting that abnormalities of chromosomal segregation contribute to tumor progression by increasing genetic diversity. Indeed, some scientists believe that aneuploidy is a fundamental aspect of cancer.

Much research is aimed at determining the basis of CIN and aneuploidy. As shown in **Figure 56–8**, a number of different processes are involved in normal chromosomal segregation. Each process is complex, and involves various organelles and many individual proteins. A textbook of cell biology should be consulted for details of the process of chromosomal segregation and cell division. Studies are in progress to compare these processes in normal and tumor cells, and to determine which of the differences detected may contribute to CIN and aneuploidy. One hope of this line of research is that it might be possible to develop drugs that diminish or even prevent CIN and aneuploidy.

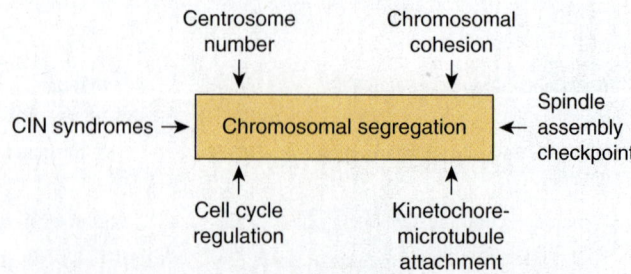

FIGURE 56–8 **Factors involved in chromosomal segregation that are relevant to understanding chromosomal instability (CIN) and aneuploidy.** CIN syndromes include Bloom syndrome (OMIM 210900) and others. (Data from Thompson SL, Bakhoum SF, Compton DA. Mechanisms of chromosomal instability, Curr Biol. 2010;20(6):R285-R295.)

MANY CANCER CELLS DISPLAY ELEVATED LEVELS OF TELOMERASE ACTIVITY

There has been considerable interest in the involvement of telomeres in a number of diseases and also in aging (see Chapters 35 and 57). With respect to cancer, when tumor cells divide rapidly their telomeres often shorten. Such shortened telomeres have been implicated as a risk factor for many, but not all, solid tumors. Indeed, **telomere length** has strong predictive value with regard to gauging the progression of chronic inflammatory diseases, such as **ulcerative colitis** and **Barrett esophagus**, to cancer. Abnormalities of telomere structure and function can contribute to CIN (see earlier). The activity of **telomerase**, the main enzyme involved in synthesizing (and thus maintaining the length of) telomeres, is very low in normal somatic cells, but is frequently elevated in cancer cells, providing a mechanism for overcoming telomere shortening. Hence, inhibiting telomerase activity represents an attractive target of cancer chemotherapy, as most cancer cells (in contrast to normal somatic cells) exhibit elevated activity of this enzyme. However, any such inhibitor could also adversely affect normal stem cells (a ubiquitous class of essential cells found in most tissue types and that require telomerase activity for their regenerative function). This is a major limitation to such an approach. However, Imetelstat (GRN163L) is one such agent that has reached clinical trials and has shown promising results in the treatment of several tumors, including glioblastoma, lung and ovarian cancers. It is likely that additional structural analysis of the telomerase enzyme, coupled with larger-scale studies of small molecule inhibitors, will serve to identify more potent compounds with therapeutic anticancer efficacy.

CANCER CELLS HAVE ABNORMALITIES OF APOPTOSIS THAT PROLONG THEIR PROLIFERATIVE CAPACITY

Apoptosis, also known as **programmed cell death (PCD)**, is a genetically directed program, which when activated, causes cell death. The main proteins involved in apoptosis are proteolytic enzymes termed **caspases** that normally exist in inactive forms, the **procaspases**. The name **caspase (cysteine-dependent aspartate-directed protease)** is derived from the fact that these enzymes utilize cysteine at their active sites to cleave target polypeptide chains at a point immediately after aspartic acid residues. About 15 human caspases are known, although not all participate in apoptosis. When the caspases involved in apoptosis are activated (mainly caspases 2, 3, 6, 7, 8, 9, and 10), they result in a **cascade** of events that ultimately kills cells by digesting various proteins and other biomolecules (compare with the coagulation cascade, Chapter 55). The **upstream caspases** (eg, 2, 8, and 10) at the beginning of the cascade are often called **initiators**, and those downstream at the end

of the pathway (caspases 3, 6, and 7) are called **effectors** or **executioners**. A **caspase-activated DNase (CAD)** fragments/digests nuclear DNA, thereby producing a characteristic DNA laddering pattern that is readily detected by native gel electrophoresis. Histological features of apoptosis include condensation of chromatin, changes of nuclear shape, and membrane blebbing. The cells that die as a result of these effects are rapidly disposed of by phagocytic activity, thereby avoiding development of an inflammatory reaction.

Apoptosis differs from **necrosis**, a pathologic form of cell death that is not genetically programmed. Necrosis occurs on exposure of cells or tissues to external agents, such as certain chemicals and extreme heat (eg, burns). Various hydrolytic enzymes (proteases, phospholipases, nucleases, etc.) are involved in the process of necrosis. Release of cell contents from dying cells can cause local inflammation (unlike apoptosis).

The overall process of apoptosis is complex and tightly regulated. The apoptotic regulatory pathway involves proteins that act as receptors and adapters, procaspases and caspases, and pro- and antiapoptotic factors. There are two major pathways, **extrinsic** (death receptor) and **intrinsic**, with **mitochondria** being an important participant in the latter pathway. **Figure 56–9** shows a simplified diagram of some of the key events in apoptosis.

Major features of the **apoptotic death receptor pathway** are shown on the left-hand side of the figure. **External signals** initiating apoptosis include **tumor necrosis factor α (TNF-α)** and Fas ligand. A number of death receptors have been identified. These receptors are transmembrane proteins, some of which interact with **adapter proteins** (such as **FADD [Fas-associated protein with death domain]**). These complexes, in turn, interact with **procaspase-8**, resulting in its conversion to **caspase-8** (an initiator). **Caspase-3** (an effector) is activated via a series of further reactions. Caspase 3 digests important structural proteins such as lamin (associated with nuclear condensation), various cytoskeletal proteins, and enzymes involved in DNA repair, causing cell death.

Regulation of this pathway occurs at several levels. **FLIP (FLICE [FADD-like IL-1β-converting enzyme]-inhibitory protein)** is an antiapoptotic regulator that inhibits the conversion of procaspase-8 to its active form. **Inhibitors of apoptosis (IAPs)** inhibit the conversion of procaspase-3 (and procaspase-9; see later) to its active form. These effects can be overcome by the protein **SMAC** (second mitochondria-derived activator of caspase), which is released from mitochondria.

The **mitochondrial pathway** can be initiated by exposure to stimuli such as reactive oxygen species (ROS) and DNA damage. Initiation of this pathway results in the formation of pores in the outer mitochondrial membrane, through which **cytochrome C** escapes into the cytoplasm. In the cytoplasm, cytochrome C interacts with **APAF-1** (apoptotic peptidase-activating factor-1), **procaspase-9**, and ATP to form a multiprotein complex known as the **apoptosome**. As a result of this interaction, procaspase-9 is converted to its active form and, in turn, acts on **procaspase-3** to produce **caspase-3**.

Activation of the *p53* gene upregulates transcription of *BAX*. The **BAX** protein is proapoptotic, by causing loss of

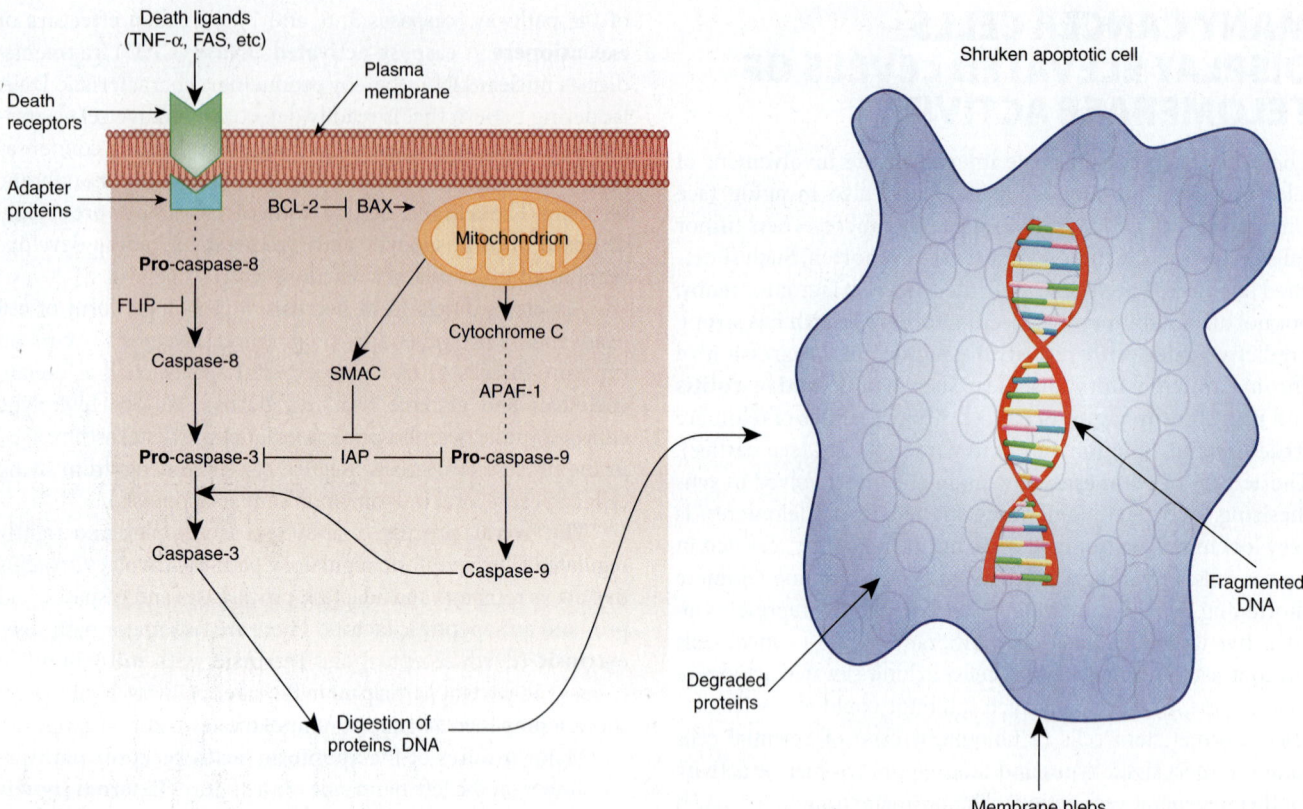

FIGURE 56–9   Simplified scheme of apoptosis. The major molecular events in the extrinsic pathway are shown on the left: Death signals include TNF-α and FAS (present on the surface of lymphocytes and some other cells). The signals (ligands) interact with specific death receptors (there are a number of them). The activated receptor then interacts with an adapter protein (FADD is one of a number of them), and then forms a complex with procaspase 8. (The complex is indicated by the interrupted line between the adapter protein and procaspase-8 in the figure). Through a series of further steps, active caspase-3 is formed, which is a major effector (executioner) of cell damage. Regulation of the extrinsic pathway can occur due to the inhibitory effect of FLIP on the conversion of procaspase-8 to caspase-8, and also the inhibitory effect of IAP on procaspase-3. On the right are the major cellular events in the intrinsic (mitochondrial) pathway. Various cell stresses affect the permeability of the mitochondrial outer membrane, resulting in efflux of cytochrome c into the cytoplasm. This forms a multiprotein complex with APAF-1 and procaspase-9, called an apoptosome. Through these interactions, procaspase-9 is converted to caspase-9. This, in turn, can act on procaspase-3 to convert it to its active form. Regulation of the intrinsic pathway can occur at the level of BAX, which facilitates increasing mitochondrial permeability permitting efflux of cytochrome c, and is thus pro-apoptotic. BCL-2 opposes this effect of BAX and is thus antiapoptotic. IAP also inhibits procaspase-9; this effect of IAP can be overcome by SMAC. (APAF-1, apoptotic protease activating factor-1; BAX, BCL-2–associated X protein; BCL-2, B-cell CLL/lymphoma 2 (CLL represents chronic lymphatic leukemia); FADD, FAS-associated via death domain; FAS, FAS antigen; FLICE, FADD-like ICE; FLIP, FLICE inhibitory protein; IAP, inhibitor of apoptosis proteins; ICE, interleukin-1-β convertase; SMAC, second mitochondria-derived activator of caspase; —| signifies opposes the action of; arrows indicate activation.

mitochondrial membrane potential, thus helping initiate the mitochondrial apoptotic pathway. On the other hand, **BCL-2** inhibits this loss of membrane potential, and is thus antiapoptotic. IAPs inhibit conversion of procaspase-9 to caspase-9; SMAC can overcome this. Note that the death pathway uses caspase-8 as an initiator, whereas the mitochondrial pathway uses caspase-9. These two pathways can interact. In addition, there are also other pathways of apoptosis, which are not discussed here.

Cancer Cells Evade Apoptosis

Cancer cells have developed mechanisms to evade apoptosis, and thus continue to grow and divide. In general, these mechanisms involve mutations that cause loss-of-function of proteins that are proapoptotic, or from overexpression of antiapoptotic genes. One such example concerns loss-of-function

of the *P53* gene, perhaps the most commonly mutated gene in cancers. Resultant loss of upregulation of transcription of proapoptotic *BAX* (see earlier) in p53-deficient cells shifts the balance in favor of antiapoptotic proteins. Overexpression of many antiapoptotic genes is a frequent finding in cancers. The consequent evasion of apoptosis favors the continuing growth of cancer cells. Attempts are being made to develop drugs that will specifically turn on apoptosis in cancer cells. Most of these agents are small molecules that inhibit the BCL-2 family of antiapoptotic proteins. Several chemo- and immunotherapeutic agents in current use induce apoptosis in cancer cells via the extrinsic pathway. The possibility of potentiating the antitumor effects of these drugs by combining them with apoptosis-inducing agents is being explored.

As indicated earlier, apoptosis is a complex, highly regulated pathway with numerous components, many of which are not

TABLE 56–10 Summary of Some Important Features of Apoptosis

- It involves a genetically programmed series of events and differs from necrosis, which is a result of direct cellular damage.
- The series of cellular events involved is a cascade, similar to the process of blood coagulation.
- It is characterized by cell shrinking, membrane blebbing, absence of inflammation, and a distinct electrophoresis pattern (laddering) of degraded DNA.
- Many caspases (proteinases) are involved; some are initiators and others act as effectors (executioners).
- Apoptosis can occur either by extrinsic (death receptor–mediated) or intrinsic (mitochondrial) pathways.
- FAS and other receptors are involved in the extrinsic pathway of apoptosis.
- Cellular stress and other factors activate the intrinsic (mitochondrial) pathway of apoptosis; release of cytochrome c into the cytoplasm is an important event in this pathway.
- Apoptosis is regulated by a balance between inhibitors (antiapoptotic factors) and activators (pro-apoptotic factors) of the process.
- Acquired mutations found in cancer cells enable them to evade apoptosis, thus promoting cell proliferation.

mentioned here in this abbreviated account. Apoptosis is also involved in various developmental and physiologic processes. It may seem paradoxical, but regulated cell death is as important in maintaining health, as in formation of new cells. In addition to cancer, apoptosis is implicated in other diseases, including certain autoimmune and chronic neurologic disorders, such as Alzheimer disease and Parkinson disease, where **excessive cell death** (rather than excessive growth) is a feature. **Table 56–10** summarizes some of the principal features of apoptosis.

Pro-Inflammatory and Tumor-Promoting Effects of Necrosis

Unlike apoptosis, necrosis of tissue results in release of intracellular contents into the surrounding microenvironment. These include molecules that serve as mediators of the pro-inflammatory response, resulting in the infiltration of tissue by immune inflammatory cells. It has been shown that such cells can have active tumor-promoting effects. Immune inflammatory cells have been reported to promote angiogenesis, cell proliferation, and invasiveness. Necrosis, which appears to counter the proliferative tendency of cancerous cells, may paradoxically benefit tumorigenesis. Thus, developing tumors appear to gain by tolerance of some degree of cell necrosis, because this results in recruitment of inflammatory cells that ultimately supply the tumor cells with growth-promoting factors via angiogenesis.

THE TUMOR MICROENVIRONMENT PLAYS A CRITICAL ROLE IN CANCER DEVELOPMENT, METASTASIS, & RESPONSE TO TREATMENT

The tumor microenvironment (TME) has been shown to play a vital role in tumor biology. It is involved in the pathogenesis of a tumor, its progression, and how it responds to treatment.

In addition to cancer cells, the TME consists of a variety of non-cancer cells and components of the extracellular matrix. These include cells of the immune system (T and B lymphocytes, natural killer [NK] cells, and macrophages) and mesenchymal cells (stem cells, myofibroblasts, endothelial cells, and adipocytes). Complex intercellular and stromal interactions occurring in the TME play important roles in cancer cell proliferation, survival, spread (metastasis), and response to treatment. Exosomes produced from cells within the TME signal between cancer cells and surrounding cells, and have been implicated in tumor metastasis and drug resistance.

Although a variety of immune cells infiltrate tumors, their antitumor effector functions are dampened by tumor-derived signals. Furthermore, T lymphocytes and macrophages are reprogrammed (eg, by sustained nuclear factor kappa B [NF-κB] activation; see Figure 42–10) to promote tumor growth and survival. This reprograming can allow cancer cells to escape immune surveillance and also co-opts the immune system to accelerate cancer progression. A greater understanding of the role of immune cells in the TME in driving carcinogenesis has opened up a new and important avenue of cancer treatment called immunotherapy (discussed in more detail later in the chapter).

Mesenchymal cells, such as the myofibroblasts and mesenchymal stem cells, are known to facilitate formation of cancer stem cell niches, thus aiding cancer stem cell survival and proliferation. Other mesenchymal cells, such as endothelial cells, respond to paracrine signals in the TME such as **vascular endothelial growth factor** (**VEGF**), by promoting tumor angiogenesis and metastasis (see also following discussion). Adipocytes in the TME secrete a number of growth factors that support tumor growth. In addition, the extracellular matrix of the TME also contributes to tumor progression by facilitating the formation of cancer stem cell niches, aiding tumor invasion and metastasis. **Figure 56–10** shows the typical constituents of the TME.

Overall, it is important to note that tumors are not simply a collection of cancer cells. They are comprised of a variety of different types of cells, and consist of both transformed and nontransformed cells. Understanding the complex interactions between these cells and their microenvironment is an important aspect of ongoing cancer research.

CANCER CELLS EXHIBIT ALTERED METABOLIC PROGRAMMING

In order to survive, and ultimately grow and proliferate to form a tumor mass, tumor cells must develop the ability to procure all necessary nutrients from typically hypoxic and nutrient-poor environments. Given these immutable facts, it is not surprising that multiple transcriptomic (RNA-seq) studies have shown that genes that code for proteins involved in nutrient capture, uptake, and metabolism are often mutated, and/or over-/underexpressed in different types of tumors. These observations have reinvigorated research into metabolism in general, and in cancer cells in particular.

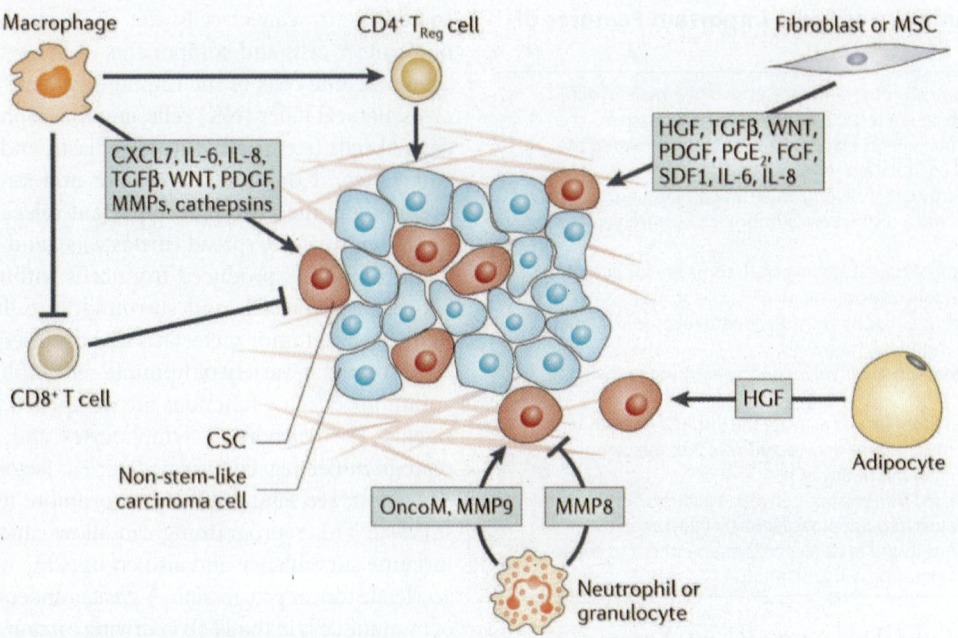

FIGURE 56–10 **The tumor microenvironment contributes critically to tumor cell growth.** Each cell type depicted affects the tumor by secreting factors that stimulate the formation of cancer stem cells (CSCs) and helping maintain the already formed CSCs in the stem cell state. Summarized here are some of the major cell types and their secreted factors, listed by cell type, that are known to have an impact on CSCs and cancer cells. (CXCL7, CXC-chemokine ligand 7; FGF, fibroblast growth factor; HGF, hepatocyte growth factor; IL-6, interleukin 6; MMP, matrix metalloproteinase; MSC, mesenchymal stem cell; OncoM, oncostatin M; PDGF, platelet-derived growth factor; PGE$_2$, prostaglandin E$_2$; SDF1, stromal cell-derived factor 1; TGF-β, transforming growth factor β.) (Reproduced with permission from Pattabiraman DR, Weinberg RA. Tackling the cancer stem cells—what challenges do they pose? *Nat Rev Drug Discov.* 2014;13(7):497-451.)

Glucose and the amino acid glutamine are two of the most abundant metabolites in plasma. Together, they account for much of the carbon and nitrogen metabolism in human cells. In 1924, the biochemist Otto Warburg and his colleagues made the discovery that cancer cells take up large amounts of glucose and metabolize it via glycolysis to lactic acid, even in the presence of oxygen. This observation was termed the **Warburg effect**. Based on these data, Warburg proposed two hypotheses: first, that such an increased ratio of anaerobic glycolysis to aerobic respiration in cancer cells was likely to be due to defects in the mitochondrial respiratory chain; and second, that enhanced rates of glycolysis enabled cancer cells to preferentially proliferate in an environment of reduced oxygen tension often seen in tumors. Furthermore, Warburg argued that the switch from aerobic to anaerobic glucose metabolism was a/the driver of tumorigenesis.

It is now thought that the reprogramming of mitochondrial respiration typically observed in tumor cells is a direct effect of at least two kinds of influences, rather than to overt defects in mitochondria. The first of these is the self-sustaining signaling by growth factors that characterizes cancer cells, which causes increased cellular proliferation (Figures 56–1 and 56–2). The second of these are the genetic changes in specific metabolic enzyme-encoding genes, metabolite transporters, and other related genes. These genetic alterations include preferential expression of certain mRNA splice variants, amplification of particular enzyme-encoding genes, as well as altered catalytic efficiencies and specificities and metabolic

products. Collectively, the resulting changes elicit important metabolic rewiring as well as epigenetic changes in the activity of the transcription machinery (ie, DNA and protein methylation, acetylation, and other posttranslational modifications) that lead to more efficient cellular anabolism and alteration of the TME, which *in toto* facilitates tumor cell proliferation.

An example of tumor cell–specific metabolic rewiring can be illustrated by the example of pyruvate kinase. There are two isozymes, PKM1 and PKM2, which are encoded by the glycolytic pyruvate kinase muscle gene, *PKM*. These are generated by alternative splicing (see Chapter 38). Unlike PKM1, which is expressed constitutively in normal cells, PKM2 is often more highly expressed in cancer cells. More likely to be important though is the fact that PKM2 exists in either a dimeric form that exhibits very low catalytic activity, or a tetrameric form, with high catalytic activity. PKM2 in cancer cells is most often in the low-activity dimeric form, which results in the accumulation of glycolytic intermediates. These intermediates allow cancer cells to synthesize macromolecules that support rapid proliferation (as proposed in the original Warburg hypothesis). Such **metabolic enzyme reprogramming** ultimately leads to less shuttling of glucose-derived chemical energy into the production of ATP (**Figure 56–11**), with a concomitant shunting of such energy into building up the cellular biomass of proteins, lipids, nucleic acids, etc. These macromolecules are essential for cell proliferation (in this case, proliferation of cancer cells). Collectively, these observations help explain why a high rate of glycolysis confers a selective advantage

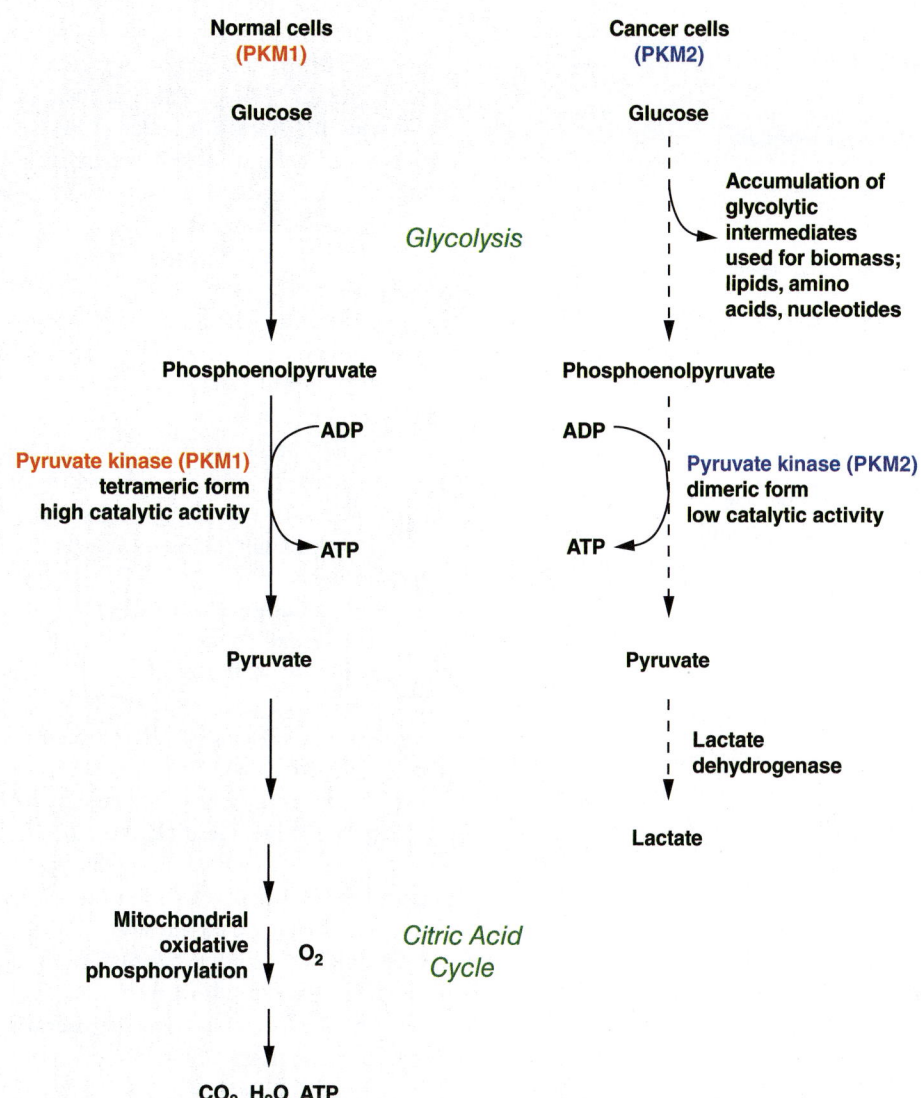

FIGURE 56–11 **Pyruvate kinase isozymes and glycolysis in normal and in cancer cells.** In normal cells, the major source of ATP is oxidative phosphorylation. Some ATP is obtained from glycolysis. The major pyruvate kinase (PK) isozyme in normal cells is PKM1. In cancer cells, anaerobic glycolysis is prominent, lactic acid is produced via the action of lactate dehydrogenase (LDH) and production of ATP from oxidative phosphorylation is diminished (not shown in the figure; see Chapters 14 and 16). In cancer cells, PKM2 is the major PK isozyme. For complex reasons not as yet fully understood, this change of isozyme profile in cancer cells is associated with decreased net production of ATP from glycolysis, but increased use of metabolites to build up biomass.

on tumor cells. Based on such observations, a current promising approach is to analyze blood and urine samples to look for alterations in metabolite profiles that may help detect cancer at an early stage and to classify them. Such metabolic profiling can be achieved by combining mass spectrometry (MS) with nuclear magnetic resonance (NMR) spectroscopy. While MS can analyze thousands of metabolites in a sample with high sensitivity, NMR spectroscopy can accurately quantitate these metabolites in a highly reproducible manner.

Solid tumors typically have localized areas of **poor blood supply** and, as noted earlier, preferentially utilize glycolytic metabolism and thus secrete lactic acid into the TME, which leads to local **acidosis**. It has been postulated that local TME acidosis allows tumor cells to invade tissues easily. **Hypoxia**, or low oxygen tension (partial pressure of O_2; pO_2) in areas of tumors with poor blood supply stimulates the expression of **hypoxia-inducible factor-1** (**HIF-1**). This transcription factor, which is activated by low partial pressure of oxygen, upregulates (among other actions) transcription of at least eight genes that control synthesis of enzymes involved in glycolysis.

The pH and pO_2 in tumors are important factors that affect responses to anticancer drugs and other treatments. For example, the anticancer efficacy of radiation treatment of cancers is significantly lower in hypoxic conditions. Chemicals have been developed to inhibit glycolysis in tumor cells, and perhaps selectively kill them (**Table 56–11**). These include **3-bromopyruvate** (an inhibitor of hexokinase-2) and **2-deoxy-D-glucose** (an inhibitor of hexokinase-1). Another compound,

TABLE 56–11　Some Compounds That Inhibit Glycolysis & Have Been Found to Display Variable Anticancer Activity

Compound	Enzyme Inhibited
3-Bromopyruvate	Hexokinase II
2-Deoxy-D-glucose	Hexokinase I
Dichloroacetate	Pyruvate dehydrogenase kinase (PDH)
Iodoacetate	Glyceraldehyde-phosphate dehydrogenase

Note: The rationale for development of these agents is that glycolysis is usually much more active in tumor cells, so that inhibiting it may damage them more than normal cells. Inhibition of PDH kinase results in stimulation of PDH, diverting pyruvate away from glycolysis.

dichloroacetate (DCA), inhibits the activity of pyruvate dehydrogenase kinase, and thus stimulates the activity of pyruvate dehydrogenase (see Chapter 17), diverting substrate from glycolysis into the citric acid cycle. However, so far, none of these have been found to be clinically useful.

STEM CELLS IN CANCER

Stem cells were discussed briefly in Chapters 39 and 53. Much work is currently ongoing to investigate the role of stem cells in cancer. **Cancer stem cells (CSC)** are believed to harbor mutations that, either by themselves or in combination with further mutations, make these cells potentially cancerous. CSCs can be detected by the use of specific surface markers, or other techniques. It appears that surrounding tissues (eg, components of the extracellular matrix of the TME) significantly influence the behavior of these cells (see Figure 56–10). An important concept driving some of the research in this area is the belief that one of the reasons that cancer chemotherapy is often not successful is that **a pool of localized and/or disseminated cancer stem cells exist** that are not susceptible to conventional chemotherapy. Reasons for this include the facts that many stem cells are relatively dormant, have active DNA repair systems (see Figure 35–23), express drug transporters that can expel anticancer drugs, and are often resistant to apoptosis.

Evidence is accumulating that cancer stem cells do indeed play key roles in many types of neoplasia. If so, development of therapies with high specificity for killing these stem cells will be of considerable value.

TUMORS OFTEN STIMULATE ANGIOGENESIS

As with normal cells, tissues and organs, tumor cells need an adequate blood supply to provide nutrients for their survival. Both tumor cells and cells in tissues surrounding tumors have been found to **secrete angiogenic factors** that stimulate the growth of new blood vessels. There has been much interest in tumor angiogenesis, because inhibiting this process represents a potential means of killing tumor cells.

The growth of blood vessels that supply tumor cells can be stimulated by hypoxia and other factors. As mentioned earlier, hypoxia induces HIF-1, which in turn increases levels of vascular endothelial growth factor (VEGF), a family of proteins that serve as major stimulators of angiogenesis. VEGF proteins interact with specific tyrosine kinase receptors on endothelial and lymphatic cells. These interactions activate intracellular signaling pathways that cause upregulation of the NF-κB pathway (see Chapter 42), resulting in proliferation of endothelial cells and formation of new blood vessels. Blood vessels supplying tumors are not normal; their structure is often disorganized, with lower than normal levels of integrity. Consequently, they are often leakier than normal blood vessels. Additional factors other than VEGFs, such as **angiopoietin**, **β-fibroblast growth factor (β-FGF)**, **TGF-β**, and **placental growth factor (PlGF)**, also stimulate angiogenesis. Not surprisingly there are other factors that inhibit blood vessel growth (eg, **arresten** and **endostatin**).

Monoclonal antibodies (mAbs) to one form of VEGF have been developed (eg, bevacizumab or Avastatin) and have been used in the treatment of certain types of cancer (eg, colon and breast). These mAbs bind to VEGF and block it from acting, presumably by blocking VEGF from interacting with the VEGF receptor. These therapeutic mAbs were found to increase patient survival, but most patients eventually relapsed. As with many antineoplastic therapies, it is now believed that these mAbs are best used in combination with other anticancer therapies. Monoclonal antibodies to other growth factors that stimulate angiogenesis are also being developed and are in clinical trials, as are small molecule inhibitors of angiogenesis. Inhibitors of angiogenesis are also useful in other conditions, such as "wet," or age-related, **macular degeneration** and **diabetic retinopathy**, in which proliferation of blood vessels is a feature.

METASTASIS IS THE MOST SERIOUS ASPECT OF CANCER

It has been estimated that about **85% of the mortality** associated with cancer results from **metastasis**. Spread of cancer is usually via lymphatics or blood vessels. Metastasis is a complex process, the molecular mechanisms of which are only now beginning to be understood.

Figure 56–12 presents a simplified scheme of metastasis. The earliest event is **detachment** of tumor cells from the primary tumor. The cells can then gain access to the circulation (or lymphatics), a process termed **intravasation**. Once in the circulation, they tend to **arrest** in the nearest small capillary bed. In that site, they **extravasate** and **migrate** through the neighboring **extracellular matrix (ECM)**, before finding a site to settle. Thereafter, if they survive host defense mechanisms, extravasated cells proliferate and tumors can grow at variable rates. To ensure growth, metastatic cells need an adequate blood supply, as discussed earlier.

Many studies have shown that cancer cells have an abnormal complement of proteins on their surfaces. These changes

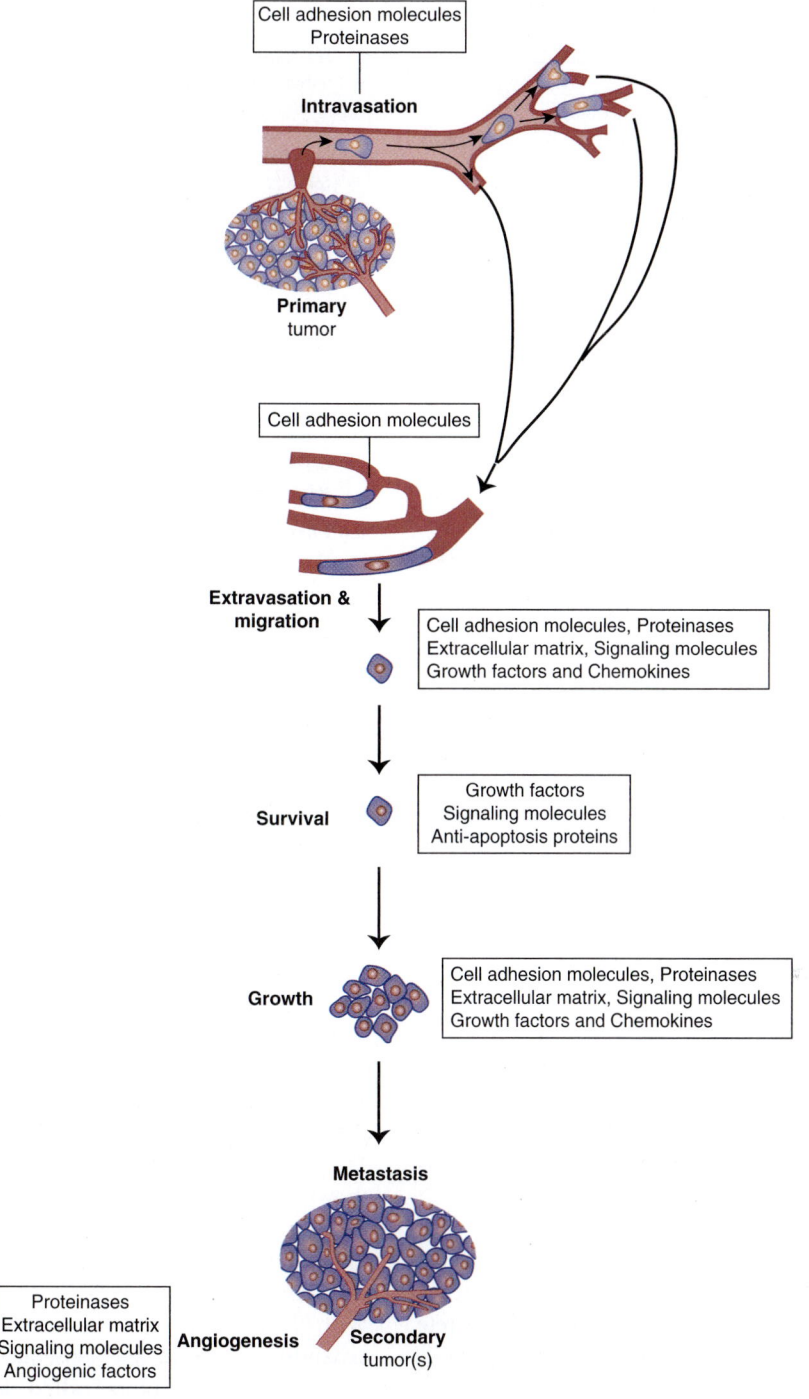

FIGURE 56–12 **Simplified scheme of metastasis.** Schematic representation of the sequence of steps in metastasis, indicating some of the factors believed to be involved.

may permit decreased cell adhesion and allow individual cancer cells to detach more readily from the parent tumor. Molecules on cell surfaces involved in cell adhesion are called **cell adhesion molecules**, or **CAMs** (**Table 56–12**). Decreases in the amounts of the CAM **E-cadherin**, a molecule of major importance in the adhesion of many normal cells, may help to account for the decreased adhesiveness of many cancer cells. Many studies have shown changes in the oligosaccharide chains of cell surface glycoproteins (see Figure 40–7), which

occur due to altered activities of various glycosyltransferases (see Chapter 46). One such important change is an increase in the activity of GlcNAc transferase V. This enzyme catalyzes transfer of GlcNAc to a growing oligosaccharide chain, forming a β1-6 linkage and allowing further growth of the chain. It has been proposed that such elongated chains participate in an altered glycan lattice at the cell surface. This may cause structural reorganization of receptors and other molecules, perhaps predisposing to the spread of cancer cells.

TABLE 56–12 Some Important Cell Adhesion Molecules (CAMs)

- Cadherins
- ICAMS, Intercellular adhesion molecules
- Integrins
- Selectins

Note: CAMs may be homophilic or heterophilic. Homophilic CAMs interact with identical molecules on neighboring cells, whereas heterophilic CAMs interact with different molecules. Cadherins are homophilic, selectins and integrins are heterophilic, and Ig CAMs may be either. Integrins are discussed briefly in Chapter 54, and selectins in Chapter 46.

Another important property of many cancer cells is that they can release various **proteinases** into the ECM. Of the four major classes of proteinases (serine, cysteine, aspartate, and metalloproteinases), in cancer, particular interest has focused on the **matrix metalloproteinases** (**MMPs**). The MMPs constitute a large family of metal-dependent (usually zinc) enzymes. A number of studies have shown increased activity of MMPs, such as MMP-2 and MMP-9 (also known as gelatinases), in tumors. These enzymes are capable of degrading proteins, such as collagen, in the basement membrane and in the ECM, thereby facilitating the spread of tumor cells. Inhibitors of these enzymes have been developed, but so far these have not exhibited much clinical success.

Another factor that allows increased movement of cancer cells is the molecular process of **epithelial-to-mesenchymal transition**, or **EMT**. The EMT involves a change of cell morphology and function from epithelial to mesenchymal type, likely induced by the cellular environment. Mesenchymal cell types have more actin filaments, permitting increased movement, an essential property of cells that metastasize.

The ECM also plays an important role in metastasis. There is evidence of communication by signaling mechanisms between cancer cells and those of the ECM. The types of cells in the ECM can also affect metastasis. As mentioned earlier, proteinases that degrade proteins in the ECM can facilitate spread of cancer cells. In addition, the ECM contains various growth factors that can influence tumor behavior.

On their travels, tumor cells are exposed to various cells of the immune system (such as T cells, NK cells, and macrophages; see Chapter 54), and must be able to survive exposure to them. Some of these immune surveillance cells secrete various **chemokines**, small proteins that can attract various cells such as leukocytes, sometimes causing an inflammatory response to tumor cells.

It has been estimated that significantly fewer than 1 in 10,000 cancer cells may have the genetic capacity to successfully colonize (metastasize). Certain tumor cells show a predilection to metastasize to specific organs (eg, prostate tumor cells to bone; breast tumor cells to bone and brain). It is likely that specific cell surface molecules are involved in this tropism.

Various studies have shown that certain genes enhance metastasis, whereas others act as metastasis suppressor genes. Determining exactly how these genes work is the subject of intense investigation. **Table 56–13** summarizes some important points regarding metastasis.

TABLE 56–13 Important Features of Metastasis

- An epithelial-to-mesenchymal cell transition is often found in cancers, allowing detachment and spread of potentially metastatic cells.
- Metastasis is relatively inefficient (only about 1:10,000 tumor cells may have the genetic potential to colonize).
- Metastatic cells must evade various cells of the immune system to survive.
- Changes in cell surface molecules (eg, CAMs and others) are involved in the process.
- Increased proteinase activity (eg, of MMP-2 and MMP-9) facilitates invasion.
- The existence of metastasis-enhancer and suppressor genes has been shown.
- Some cancer cells metastasize preferentially to specific organs.
- Metastasis-gene signatures may be detected by transcriptome/exome analysis; such transcriptome information can be of prognostic value, potentially allowing for personalized therapeutic treatment.

Abbreviations: CAM, cell adhesion molecule; MMP, matrix metalloproteinase.

THERE ARE MANY IMMUNOLOGIC ASPECTS OF CANCER

Tumor immunology is a broad area of study. Hence, due to the extensive breadth of this field, it can only be dealt with here briefly. It is likely that the normal decline in immune responsiveness that accompanies **aging** plays a role in the increased incidence of cancer in older people. A longstanding hope of molecular oncologists and physicians has been to harness the exquisite specificity of the immune system to selectively kill cancer cells. There are many ongoing clinical trials investigating this possibility. These studies involve the use of antibodies, vaccines, and various types of T cells that have been genetically manipulated to increase their ability to kill neoplastic cells. One of the methods proven to be effective is the use of antibodies against certain T-lymphocyte surface proteins. For example, antibodies developed against cytotoxic T-lymphocyte antigen 4 (anti-CTLA-4) or programmed death 1 (anti-PD1) have been shown to "remove the brakes" on these cells, thus setting them free to attack cancer cells.

Other strategies using modified T cells have proven to be effective as well. This approach is termed **CAR-T cell therapy**. With CAR-T, a patient's own T cells are harvested, grown, and expanded in culture, and subsequently engineered *in vitro* to express specific receptors (**called chimeric antigen receptors or CAR**) that can bind to antigens specifically expressed on tumor cells. Following reinfusion into the same patient, these genetically modified CAR–T cells bind and kill the targeted lymphoblastic leukemia cells. This approach is especially valuable in cases in which conventional chemotherapy had failed. Many other CAR-T cell therapies are under development and is likely that this approach will revolutionize treatment for many cancers for which treatment options are limited. The major advantage of immunotherapy is that it has a broad spectrum of action and therefore can be used against a wide variety of cancers. In addition, resistance is less likely to occur with this form of treatment. It is hoped that immunotherapy will be the fourth major weapon against cancer, after surgery, radiotherapy, and chemotherapy.

Chronic inflammation involves aspects of immune function. There is evidence that it can **predispose to cancer**; for example, the incidence of colorectal cancer is much higher in individuals who have had long-standing ulcerative colitis than in those without this condition. Some inflammatory cells produce relatively large amounts of **reactive oxygen species (ROS)**, which can cause damage to DNA and other macromolecules (see Chapter 57), and perhaps contribute to oncogenesis. It has also been reported that **low doses of aspirin** may lower the risk of development of colorectal cancer, perhaps via its anti-inflammatory action.

Cancer: Its Relationship to Inflammation & Obesity

The association between inflammation and cancer is now well established. Inflammation is known to be a critical component of tumorigenesis. That said, the exact mechanisms linking inflammation and cancer are poorly understood. Examples of possible molecules that are involved in induction of an inflammatory process include nuclear factor kappa B (NF-κB) and signal transducer and activator of transcription 3 (STAT3). NF-κB is a transcription factor that induces expression of proteins that are involved in proinflammatory, proliferative, and reparative processes. Activation of NF-κB has been shown to occur in tumors in response to inflammatory stimuli or oncogenic mutations (see Chapter 42). Signaling via STAT3 is activated by interleukin 6 (IL-6), a pro-inflammatory cytokine that activates Janus kinase (JAK)-STAT signaling and its downstream effects (see Chapter 42). Such events are thought to be responsible for driving many hallmark features of cancer. In addition, the "**inflammasome**," a multi-protein complex that acts as a sensor of cellular damage, is another potential candidate that mediates inflammation. Activation of inflammasomes leads to secretion of **pro-inflammatory cytokines**, such as **IL-1β** and **IL-18**, both of which have been implicated in tumorigenesis. There is a large body of evidence to implicate other inflammatory mediators in the development of tumors.

Obesity is associated with low-grade inflammation. Visceral adipose tissue is considered an important source of proinflammatory cytokines. It is now known that the microenvironment surrounding tumor cells influences tumorigenesis (see earlier). Inflammatory cells in the TME are considered to play a crucial role in the process. Obesity has been found to mediate and exacerbate dysfunctional changes in the microenvironment; this has been shown to occur both in normal tissue and tumors. Such changes include alterations in factors that may be endocrine, metabolic, or inflammatory in nature. By contrast, caloric restriction has been shown to inhibit tumorigenesis (and even longevity; Chapter 57) in experimental models. Many cellular pathways, such as those involving growth factor signaling, inflammation, cellular homeostasis, and the TME are affected by such caloric restriction. These observations suggest that such targets may be considered for prevention of cancer in humans.

TUMOR BIOMARKERS CAN BE MEASURED IN SAMPLES OF BLOOD & OTHER BODY FLUIDS

Biochemical tests are often helpful in the management of patients with cancer (eg, some patients with advanced cancers may have elevated levels of plasma calcium, which can cause serious problems if not attended to). Many cancers are associated with the abnormal production of enzymes, proteins, and hormones that can be measured in plasma or serum. These molecules are known as **tumor biomarkers**. Some of them are listed in **Table 56–14**.

However, significant elevations of some of the biomarkers listed in Table 56–14 also occur in a variety of **non-cancerous conditions**. For example, elevations of the level of **prostate-specific antigen** (PSA), a glycoprotein synthesized by prostate cells, occur not only in patients with cancer of the prostate, but also in patients with prostatitis and **benign prostatic hyperplasia** (BPH). Similarly, elevations of **carcinoembryonic antigen** (CEA) are found not only in patients with various types of cancer, but also in heavy smokers and people with ulcerative colitis and cirrhosis. The fact that elevations of tumor biomarkers are usually not specific for cancer has meant that measurements of most of them are not used primarily for diagnosis of cancer. Their main uses have been in following the effectiveness of treatments and in detecting early recurrences. As with other laboratory tests (see Chapter 48), the clinical picture must be considered when interpreting the results of measurements of tumor biomarkers.

It is hoped that ongoing -**omic** analyses of body fluids and accessible cancer cells (in blood, serum, and biopsy samples) will provide new **prognostic tumor biomarkers** of increased

TABLE 56–14 Some Useful Tumor Biomarkers Measurable in Blood

Tumor Biomarker	Associated Cancer
α-Fetoprotein (AFP)	Hepatocellular carcinoma, germ cell tumor
Calcitonin (CT)	Thyroid (medullary carcinoma)
Carcinoembryonic antigen (CEA)	Colon, lung, breast, pancreas, ovary
Human chorionic gonadotropin (hCG)	Trophoblastic disease, germ cell tumor
Monoclonal immunoglobulin	Myeloma
Prostate-specific antigen (PSA)	Prostate
CA-125	Ovary
CA 19-9	Pancreas

Note: Most of these tumor biomarkers are also elevated in the blood of patients with noncancerous diseases. For example, CEA is elevated in a variety of noncancerous gastrointestinal disorders, and PSA is elevated in prostatitis and benign prostatic hyperplasia. This is why interpretation of elevated results of tumor markers must be made with caution and their main uses are to follow effectiveness of treatments and to detect recurrences. There are a number of other tumor biomarkers that are widely used.

sensitivity and specificity, and those capable of alerting physicians to the presence of cancers at an early stage of their development. A new class of tumor markers that is receiving a lot of attention is called **circulating tumor DNA (ctDNA)**. ctDNA refers to DNA derived from cancers, which can be isolated from blood. Since DNA from different cancers harbor characteristic mutations, identification of these mutations can indicate the presence and type of cancer, even when conventional biopsies are not possible. It is hoped that these tests (called liquid biopsies) can help in early diagnosis of cancers, suggest personalized therapies, and aid in early detection of relapses. Note the similarity in basic approach to analysis of stool samples to detect genetic changes in DNA from sloughed intestinal epithelial cells to diagnose colon cancer as described earlier.

Transcriptome and whole genome sequencing analyses (see Chapter 39 and following discussion) of cancer cells have revealed a plethora of potentially very useful biomarkers of oncogenesis. These methods are also useful in more accurately subclassifying tumors (so-called "personalized medicine"; see Chapter 39) in order to provide more accurate diagnoses and guide more efficacious, targeted modes of therapy. Such molecular diagnostic methods are becoming the standard of care for a select subset of cancers.

DETAILED GENETIC ANALYSES OF TUMOR CELLS IS PROVIDING NEW INSIGHTS INTO CANCER

Since the completion of the Human Genome Project, the technology of large-scale DNA sequencing and bioinformatic analyses and interpretation of sequence data has advanced considerably. Large-scale DNA sequencing has become both faster and cheaper with the widespread availability of various high-throughput DNA sequencing technologies (see Chapter 39). These advances have allowed for analyses of DNA sequences of a large number of different types of tumors. Recently, worldwide teams of scientists affiliated with **The Cancer Genome Atlas (TCGA)**, and the **Pan-Cancer Analysis of Whole Genomes Consortium** (PCAWG) have published and organized the results of integrated DNA- and RNA-sequencing analyses where both DNA genomes and RNA exomes (the sequence and quantification of the complete collection of RNA species expressed in cells) of thousands of tumors and matched noncancerous cells derived from dozens of different types of tissues. These scientists are using this information to uncover important new information regarding the mechanisms of oncogenesis. The resulting **comprehensive catalogs** of data from the PCAWG and TCGA consortia (and other efforts) regarding specific types, numbers, and effects of specific gene mutations in different cancers is revolutionizing **diagnostic testing** and the development of **custom-tailored therapy**.

CRISPR (clustered regularly interspaced short palindromic repeats), a system originally discovered as a form of adaptive immunity in bacteria, has been transformed into a versatile and precise gene-editing tool (see Figure 39–2). CRISPR is a protein-RNA complex composed of a **guide RNA (gRNA)** and an endonuclease called Cas (Cas9 is commonly used). Inside a cell, the gRNA targets Cas to a specific segment of DNA that is complementary to its sequence. On binding, Cas can either degrade the segment of DNA (thus knocking out a gene, for example), or edit it precisely to incorporate desired changes in the DNA sequence. CRISPR technology has been used to develop new cancer treatments. Recently, the CRISPR approach was used to engineer T cells (called NYCE T cells), in which three genes were precisely edited, thus enabling it to better recognize and kill tumor cells that expressed a protein, NY-ESO-1.

As noted earlier, tumors often exhibit extreme heterogeneity, such that they consist of subpopulations of cells that are genetically and phenotypically distinct from one another (see Figure 56–10). It is now possible to isolate and sequence the nucleic acids of single cells obtained from tumors (**single-cell sequencing**), in order to understand the genetic landscape of a particular tumor. Such understanding is important to allow for the development of multimodal treatment strategies that efficiently target all the subpopulations in a given tumor.

Overall, it is expected that information obtained from these new technologies will dramatically impact the next stages of cancer genomics and help in development of methods that allow for earlier diagnosis of cancer, identification of critical genetic changes that drive cancer progression and, ultimately, personalized therapy for cancer in individual patients. Such an individualized approach to cancer diagnosis and treatment is termed **precision oncology**.

KNOWLEDGE OF MECHANISMS INVOLVED IN CARCINOGENESIS HAS LED TO THE DEVELOPMENT OF NEW THERAPIES

One of the great hopes of cancer research is that insights into fundamental biochemical and genetic mechanisms involved in carcinogenesis will lead to new and better therapies. This has already occurred to a certain extent and it is hoped that ongoing developments will accelerate this process.

Classical chemotherapeutic drugs include alkylating agents, platinum complexes, antimetabolites, and mitotic spindle poisons, among other classes of chemical compounds. These agents will not be discussed here. Within the classes of drugs developed more recently are inhibitors of signal transduction (including tyrosine kinase inhibitors), monoclonal antibodies directed to various target molecules, inhibitors of hormone receptors, drugs that affect differentiation, anti-angiogenesis agents, and biologic response modifiers. Examples of each of these are listed in **Table 56–15**.

The finding of widespread defects in signaling mechanisms in cancer cells, and in particular the detection of mutations in **tyrosine kinases**, has led to the development of inhibitors of these enzymes. The most dramatic success has probably been the introduction of imatinib (marketed as Gleevec) for the

TABLE 56–15 Some Anticancer Agents That Are Based on Recent Advances in Knowledge of Cancer Biology

Class	Example	Used to Treat
Inhibitors of signal transduction	Imatinib, an inhibitor of tyrosine kinase	Chronic myelocytic leukemia
Monoclonal antibodies	Trastuzumab, an mAb to the HER2/Neu receptor	Late-stage breast cancer
Antiangiogenesis agents	Bevacizumab, an mAb to VEGF A	Colon and breast cancers
Antihormonal agents	Tamoxifen, antagonist of the estrogen receptor	Breast cancer
Affect differentiation	All-trans retinoic acid (ATRA), targets the retinoic acid receptor on promyelocytic leukemia cells, causing them to differentiate	Promyelocytic leukemia
Affect epigenetic changes	5-Azadeoxycytidine inhibits DNA methyltransferases; SAHA inhibits histone deacetylases	Certain leukemias, cutaneous T-cell lymphoma

Abbreviations: Chronic myelocytic leukemia; mAb, monoclonal antibody; SAHA, suberoylanilide hydroxamic acid (Vorinostat); VEGF A, vascular endothelial growth factor A.
Note: In some cases, the above agents may have been replaced by other more effective agents. Some of the agents listed above are also used to treat conditions other than those listed.

treatment of **chronic myelocytic leukemia (CML)**. Imatinib is an orally administered drug that inhibits the tyrosine kinase formed due to the *ABL-BCR* chromosomal translocation that is involved in the genesis of CML. It is an ATP analog that competitively binds to the ATP-binding pocket of the kinase. Treatment with this drug has produced complete remission of disease in many patients. Other tyrosine kinase inhibitors include erlotinib and gefitinib, which inhibit the epidermal growth factor receptor (EGFR). EGFR is overexpressed in certain lung (eg, non–small cell cancers) and breast cancers, resulting in aberrant (constitutive) signaling. It is important to appreciate that the design of such drugs requires detailed structural knowledge, such as that provided by x-ray crystallography, NMR studies, and model building of the molecules being targeted. Another class of useful drugs is **monoclonal antibodies** to various molecules exposed on the surfaces of neoplastic cells (see discussion earlier regarding anti-VEGF mAb). A few of these mAbs that are clinically useful as therapeutic agents are listed in Table 56–15.

Other approaches to treating cancer that are in use or being developed, but are not listed in Table 56–15, include various types of **gene therapy** (including siRNAs, see Chapter 34), **immunotherapy** (described earlier), oncolytic viruses (viruses that preferentially infect and replicate in tumor cells, ultimately leading to tumor cell death via several mechanisms), **inhibitors of steroid hormone receptors, aromatase inhibitors** (see Chapter 41) (for some breast, ovarian, and prostate cancers), **telomerase inhibitors**, applications of **nanotechnology** (eg, nanoshells and other nanoparticles), **phototherapy** (see Chapter 31), and drugs that **selectively target cancer stem cells**.

It is important to appreciate that anticancer drugs have side effects, like all other therapeutic agents. Sometimes, these are severe. Moreover, resistance to many drugs can develop after variable periods of time, as a result of therapy-driven/selected genetic changes in tumor cells.

The study of mechanisms by which cancer cells develop resistance to drugs is an important area of research. Cancer cells use a number of strategies to develop drug resistance (see summary, **Table 56–16**). The overall thrust of drug development for cancer therapy is to use new information that emerges from studies of basic immunology, biochemistry, and cellular, molecular, and cancer biology to develop safer and

TABLE 56–16 Mechanisms by Which Cancer Cells Can Develop Drug Resistance

Mechanism of Drug Resistance	Example
Increased drug efflux from the cell	Overexpression of transport proteins, such as multidrug resistance proteins (MDRs) (eg, *P*-glycoprotein or MDR1), causes efflux of cancer chemotherapeutic drugs, such as taxanes, topoisomerase inhibitors, and antimetabolites.
Decreased drug activation	Decreased conversion of prodrugs (eg, 5-fluorouracil) to their active forms due to downregulation of enzymes that catalyze their activation.
Drug inactivation	Platinum drugs (cisplatin and carboplatin) are inactivated by conjugation with glutathione.
Increased drug target expression	Increased expression of thymidylate synthase, the target of antimetabolites such as 5-fluorouracil.
Dysfunctional apoptosis	Overexpression of anti-apoptotic proteins such as the BCL2 family of proteins, and decreased expression of pro-apoptotic proteins such as BAX and BAK.
Activation of prosurvival signaling	Activation of epidermal growth factor receptor (EGFR)-mediated signaling in response to various chemotherapeutic drugs.
Modification of tumor microenvironment	Increased expression of integrins (proteins that attach cells to the extracellular matrix), which inhibits apoptosis and alters drug targets.

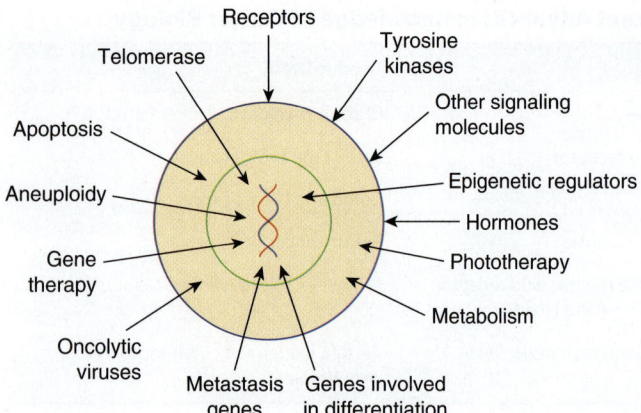

FIGURE 56–13 Examples of targets for anticancer drugs and some emerging therapies, both of which have developed from relatively recent research. Not shown in the figure are anti-angiogenic agents, applications of nanotechnology, therapies directed against cancer stem cells, and immunologic approaches. Most of the targets and therapies indicated are discussed briefly in the text.

more effective agents. Intensive research over the past several decades has resulted in an improved understanding of genetic alterations that underlie the development of specific types of cancer. This knowledge has led to a shift from the use of broad-spectrum cytotoxic drugs to therapies that are specifically designed to target individual tumors. Currently, a major area of research is to identify specific driver mutations, that is, the genetic changes that play critical roles in development of tumors (see discussion of colorectal cancer earlier). **Molecular profiling of cancer** in individual patients allows oncologists to choose the most appropriate drug or treatment modality that targets the molecular abnormality in each tumor, while simultaneously allowing for monitoring the efficacy of such treatment(s) over time. Such **personalized anticancer therapy** has been shown to significantly improve clinical response to drugs and increase survival in various types of cancer. An understanding of individuals' genetic differences in metabolism of anticancer drugs may also help personalize anticancer treatments.

Figure 56–13 summarizes some of the targets for drug therapy and some emerging therapies that have developed from studies of fundamental aspects of cancer.

MANY CANCERS CAN BE PREVENTED

The tremendous individual suffering caused by cancer, and the heavy economic burden it imposes on society, makes it important to adopt measures to **prevent the development of cancer**. **Modifiable risk factors** have been linked to a wide variety of cancers. It is likely that a significant number of all cancers in developed countries may be prevented, if the measures summarized in **Table 56–17** are introduced on a population-wide basis.

The use of tobacco, in its various forms, continues to be a major cause of cancer affecting the lungs, mouth, larynx,

TABLE 56–17 Public Health Measures That Might Prevent a Significant Number of Cancers if Introduced on a Population-Wide Basis

- Reduced tobacco use
- Increased physical activity
- Weight control
- Improved diet
- Limited alcohol intake
- Use of safer sex practices
- Routine cancer screening tests
- Avoidance of excess exposure to the sun

Data from Stein CJ, Colditz GA. Modifiable risk factors for cancer. Br J Cancer. 2004; 90(2):299-303.

esophagus, and stomach. A sustained public education campaign regarding the adverse effects of tobacco use has resulted in significant decreases in the incidence of cancers associated with tobacco use. Vaccines against the human papillomavirus (HPV) (known to be associated with cancer of the cervix) and hepatitis B virus (HBV) (associated with hepatocellular carcinoma) have been found to be effective in decreasing the incidence of cancers caused by these viruses.

Chemoprevention, which refers to the use of drugs to prevent the development of cancer, has been found to be effective in certain types of cancer. For example, the use of estrogen receptor modulators (like tamoxifen) has been shown to decrease the incidence of breast cancer in high-risk women by about 50%. Similarly, the use of finasteride (a drug that inhibits the enzyme, 5α-reductase, that converts testosterone to dihydrotestosterone), has been associated with reduced incidence of prostate cancer. It has also been shown that long-term use of aspirin, which is commonly prescribed as an antiplatelet agent, is associated with decreased incidence of colon cancer.

In some instances, identification of genetic risk factors for cancer is opening up possibilities for newer strategies in cancer prevention. For example, women who have mutations in the breast-cancer associated genes, *BRCA1* and *BRCA2*, can undergo prophylactic mastectomy (surgery for removal of the breast) in order to mitigate any future risk of developing cancer.

Overall, the rapid progression of research in cancer biology is not only opening up newer ways to treat cancer, but also to decrease and/or prevent the occurrence of the disease in the first place.

SUMMARY

- Cancer is caused by mutations in genes that control cell growth and duplication, cell death (apoptosis), and cell-cell interactions (eg, cell adhesion). Other important aspects of cancer are defects in cell signaling pathways, stimulation of angiogenesis, aneuploidy, and changes in cellular metabolism and cell microenvironment.

- A majority of cancers are likely due to replicative errors (as broadly defined) that affect somatic cells. However, a number of cancers have been identified that are caused by hereditary or environmental factors.

- Major classes of genes involved in cancer are oncogenes, tumor suppressor genes, and genes encoding proteins important for DNA synthesis and repair.

- Mutations affecting genes directing the synthesis and expression of microRNAs are implicated in oncogenesis.

- Epigenetic changes that alter gene expression are increasingly being recognized in cancer (and in other diseases); one reason for interest in epigenetics is that epigenetic "marks" are potentially reversible by drugs.

- Mechanisms of metastasis are being explored intensively; the discovery of metastasis enhancer and suppressor genes, among other findings, may lead to new therapies.

- Apoptosis (programmed cell death) plays important roles in oncogenesis. Cancer cells acquire mutations that permit them to evade apoptosis, thus prolonging viability and enabling their continued replication.

- Cancer cells display various alterations of metabolism, nutrient capture, and utilization.

- The development of cancer is a multistep process involving genetic, epigenetic, and microenvironmental changes that confer selective advantages on clones of cells, some of which eventually acquire the ability to metastasize successfully. Because of the diversity of mutations involved, it is likely that no two tumors have identical genomes.

- Metastasis (the spread of cancer to distant locations) is associated with changes in expression of cell adhesion molecules and modification of the extracellular matrix, which allows cancer cells to detach from the primary tumor and migrate to distant locations.

- It is likely that extracellular vesicles (exosomes) released by cancer cells play important roles in cancer progression and metastasis.

- Tumor biomarkers may help in both the early diagnosis of cancer and for monitoring response to treatment and in detecting recurrences.

- Diagnostic whole genome-, exome-, and circulating tumor-derived DNA sequencing is now able to reveal important driver and passenger mutations present in many types of cancers and is becoming a powerful complement to existing treatments.

- Advances in understanding the molecular biology of cancer cells have led to the introduction of a number of new therapies, with others in the pipeline.

- Several strategies have proven useful in preventing cancer. These include modification of risk factors (like decrease in tobacco use in its various forms), vaccination against tumor-causing viruses (such as HPV and HBV), use of drugs (anti-estrogens in breast cancer; anti-androgens in prostate cancer), and risk-modifying surgery (mastectomy in women with mutations in *BRCA1* and *BRCA2*).

REFERENCES

Aravanis AM, Lee M, Klausner RD: Next-generation sequencing of circulating tumor DNA for early cancer detection. Cell 2017;168:571-574.

Bassiouni R, Gibbs, LD, Craig DW, Carpten JD, McEachron TA: Applicability of spatial transcriptional profiling to cancer research. Mol Cell (2021);81:1631-1639.

Borrebaick CAK: Precision diagnostics: moving towards protein biomarker signatures of clinical utility in cancer. Nat Rev Cancer 2017;17:199-204.

Breast Cancer Association Consortium, Dorling L, Carvalho S, Allen J, González-Neira, A, et al: Breast Cancer Risk Genes–Association Analysis in More than 113,000 Women. N Engl J Med 2021;384:428-439.

Brown KA: Metabolic pathways in obesity-related breast cancer. Nat Rev Endocrinol 2021 Apr 29. doi: 10.1038/s41574-021-00487-0.

Cenik BK, Shilatifard A: COMPASS and SWI/SNF complexes in development and disease. Nat Rev Genet 2021;22:38-58.

Chakravarti D, LaBella KA, DePinho RA: Telomeres: history, health, and hallmarks of aging. Cell 2021;184:306-322.

Dai, J, Su, Y, Zhong, S, et al: Exosomes: key players in cancer and potential therapeutic strategy. Sig Transduct Target Ther 2020;5:145.

Dawson MA: The cancer epigenome: concepts, challenges, and therapeutic opportunities. Science 2017;355:1147-1152.

Degasperi A, Zou X, Amarante TD, et al: Substitution mutational signatures in whole-genome–sequenced cancers in the UK population. Science 2022;376. DOI: 10.1126/science.abl9283.

Guterres AN, Villanueva J: Targeting telomerase for cancer therapy. Oncogene 2020;39:5811-5824.

Hahn WC, Bader S, Braun TP, et al: An expanded universe of cancer targets. Cell 2021;184:1142-1155.

Hanahan D, Weinberg RA: Hallmarks of cancer: the next generation. Cell 2011;144:646-674.

Hu D, Shilatifard A: Epigenetics of hematopoiesis and hematological malignancies. Genes Dev 2016;30:2012-2041.

Huang A, Garraway LA, Ashworth A, Weber B: Synthetic lethality as an engine for cancer drug target discovery. Nat Rev Drug Discov 2020;19:23-38.

Ju YS: A large-scale snapshot of intratumor heterogeneity in human cancer. Cancer Cell 2021;39:463-465.

Ling H, Fabbri M, Calin GA: MicroRNAs and other non-coding RNAs as targets for anticancer drug development. Nat Rev Drug Discov 2013;12:847-865.

Ma P, Pan Y, Li W, et al: Extracellular vesicles-mediated noncoding RNAs transfer in cancer. J Hematol Oncol 2017;10:57.

Martinez P, Blasco MA: Telomere-driven diseases and telomere-targeting therapies. J Cell Biol 2017;216:875-887.

Mattox AK, Bettegowda C, Zhou S, Papadopoulos N, Kinzler KW, Vogelstein B: Applications of liquid biopsies for cancer. Sci Transl Med 2019;11:eaay1984.

Morgan MAJ, Shilatifard A: Reevaluating the roles of histone-modifying enzymes and their associated chromatin modifications in transcriptional regulation. Nat Genet 2020;52:1271-1281.

Nacev BA, Feng L, Bagert JD, et al: The expanding landscape of 'oncohistone' mutations in human cancers. Nature 2019;567:473-478.

Otto T, Sicinski P: Cell cycle proteins as promising targets in cancer therapy. Nat Rev Cancer 2017;17:79-92.

Pavalova NN, Thompson B: The emerging hallmarks of cancer metabolism. Cell Metab 2016;23:27-47.

Piunti A, Shilatifard A: The roles of Polycomb repressive complexes in mammalian development and cancer. Nat Rev Mol Cell Biol 2021;22:326-345.

Reiter JG, Baretti M, Gerold JM, et al: An analysis of genetic heterogeneity in untreated cancers. Nat Rev Cancer 2019;19:639-650.

Shi J, Kantoff PW, Wooster R, Farokhzad OC: Cancer nanomedicine: progress, challenges and opportunities. Nat Rev Cancer 2017; 17:20-37.

Suski JM, Braun M, Strmiska V, Sicinski P: Targeting cell-cycle machinery in cancer. Cancer Cell 2021;39:1-15.

Tomasetti C, Li L, Vogelstein B: Stem cell divisions, somatic mutations, cancer etiology, and cancer prevention Science 2017;355:1330-1334.

Weinberg R: *The Biology of Cancer*, 2nd ed. Garland Science, 2013.

Zhang L, Zhang, Y, Hu X: Targeting the transcription cycle and RNA processing in cancer treatment. Curr Opin Pharmacol 2021;58:69-75.

Zhang Z, Lu M, Qin Y, Gao W, Tao L, Su W, Zhong J: Neoantigen: A New Breakthrough in Tumor Immunotherapy. Front Immunol 2021;12:672356.

USEFUL WEB SITES

American Cancer Society. http://www.cancer.org

National Cancer Institute-The Cancer Genome Atlas (TCGA) Program. https://www.cancer.gov/about-nci/organization/ccg/research/structural-genomics/tcga

International Cancer Genome Consortium: Pan-Cancer Analysis of Whole Genomes (PCAWG) study. https://dcc.icgc.org/pcawg

OMIM: Online Mendelian Inheritance in Man—an Online catalog of human genes and genetic disorders https://www.omim.org

GLOSSARY

Adenomatous polyp: A benign tumor of epithelial origin that has the potential to become a carcinoma. Adenomas are often polypoid. A polyp is a growth that protrudes from a mucous membrane; most are benign, but some polyps can become malignant.

Ames assay: An assay system devised by Dr. Bruce Ames that uses a specially designed *Salmonella typhimurium* strain to detect mutagens. Most carcinogens are mutagens, but if mutagenicity of a chemical is detected, ideally such chemical compounds should be tested for carcinogenicity by animal testing.

Aneuploidy: Refers to any condition in which the chromosome number of a cell is not an exact multiple of the basic haploid number. Aneuploidy is found in many tumor cells and may play a fundamental role in the development of cancer.

Angiogenesis: The formation of new blood vessels. Angiogenesis is often active around tumor cells, ensuring that they obtain an adequate blood supply. A number of growth factors are secreted by tumors and surrounding cells (eg, vascular endothelial growth factor, or VEGF) and are involved in this process.

Apoptosis: Cell death due to activation of a genetic program that causes fragmentation of cellular DNA and other changes. Caspases play a central role in the process. Many positive and negative regulators affect it. The protein p53 induces apoptosis as a response to cell DNA damage. Most cancer cells exhibit abnormalities of apoptosis, due to various mutations that help to ensure their prolonged survival.

Benign tumor: A mass of abnormal proliferating cells, which are noninvasive and do not metastasize.

Biologic response modifiers: Molecules produced by the body or in the laboratory, which when administered to patients alter the body's response to infection, inflammation, and other processes. Examples include monoclonal antibodies, cytokines, interleukins, interferons, and growth factors.

Bloom syndrome: One of the chromosomal instability (CIN) syndromes. Because of mutations in a DNA helicase, subjects are sensitive to DNA damage and are at increased risk of developing tumors.

Burkitt lymphoma: This is a B-cell lymphoma, endemic in parts of Africa, where it mainly affects the jaw and facial bones. A reciprocal translocation, involving the *c-MYC* gene on chromosome 8 and the immunoglobulin heavy-chain gene on chromosome 14, is a characteristic driver of this disease.

Cancer: A tumor consisting of malignant cells

Cancer stem cell: A cell within a tumor that has the capacity to self-renew and to give rise to the heterogeneous lineages of cancer cells found in the tumor.

Carcinogen: Any agent (eg, a chemical or a physical agent) capable of causing cells to become cancerous.

Carcinoma: A malignant growth of epithelial origin. A cancer of glandular origin or showing glandular features is usually designated as an adenocarcinoma.

Caspases: Proteolytic enzymes that play a central role in apoptosis, but are also involved in other processes. About 15 types are present in humans. Caspases hydrolyze peptide bonds on the C-terminal side of aspartate residues.

Cell cycle: The various events pertaining to cell division that occur as a cell goes from one mitosis to another.

Centriole: An array of microtubules that is paired and found in the center of a centrosome. (Also see **Centrosome**.)

Centromere: The constricted region of a mitotic chromosome where chromatids are joined together. It is in close proximity to the kinetochore. Abnormalities of centromeres may contribute to CIN. (Also see **Kinetochore**.)

Centrosome: A centrally located organelle that is the primary microtubule-organizing center of a cell. It acts as the spindle pole during cell division. Both centrosome number and intracellular location within the cell during cell division are critical for accurate cell duplication.

Chromatid: A single chromosome.

Chromatin remodeling: This involves conformational, or covalent changes of nucleosomes, brought about by the actions of multiprotein complexes (such as the SW1/SNF complex). These changes can alter gene transcription (turning it on or off, depending on specific conditions). Mutations affecting different proteins of these complexes are often found in cancer cells. (See also **Epigenetics**.)

Chromosomal instability (CIN): Gain or loss of whole chromosomes or segments thereof caused by abnormalities of chromosome segregation during mitosis. (See also **Genome instability** and **Microsatellite instability**.) There are a number of disorders that are named CIN syndromes because they are associated with chromosomal abnormalities. An increased incidence of various cancers is found in these conditions.

Chromosomal passenger complex (CPC): A complex of proteins that plays a key role in regulating mitosis by participating in chromosome alignment and spindle assembly. Mutations affecting CPC proteins may contribute to CIN and aneuploidy.

Chromosomal translocation: When part of one chromosome becomes fused to another, often causing activation of a gene at the site. The Philadelphia chromosome (see later) is one of many examples of a chromosomal translocation involved in the causation of cancer (see also Burkitt lymphoma earlier).

Clone: All the cells of a clone are derived from one parent cell.

Copy number variations (CNVs): Variations (because of duplications or deletions) among individuals as to the number of copies they have of particular genes or noncoding DNA. CNVs been identified for many genes and non-protein-coding sequences; some are causative/associated with various diseases, including certain types of cancer.

Driver/passenger mutation: A mutation in a gene that either helps cause cancer or accelerates it. Mutations found in tumors that do not cause cancer or its progression are called passenger mutations.

Epigenetic: Refers to changes of gene expression without changes in the sequence of bases in DNA. Factors causing epigenetic changes include methylation of bases in DNA, posttranslational modifications of histones, and chromatin remodeling.

FAS receptor: A receptor that initiates apoptosis when it binds its ligand, FAS. FAS is a protein present on the surface of activated natural killer cells, cytotoxic T lymphocytes, and other cells.

Gatekeeper gene: Genes that code for proteins that control or inhibit cell growth. Such genes are generally referred to as tumor suppressor genes (eg, *P53* and *RB*). Mutation in such genes often initiates the cascade of events that cause oncogenesis.

Genome instability: This refers to a number of alterations of the genome, of which the two principal ones are CIN and microsatellite instability. In general, it reflects the fact that the genomes of cancer cells are more susceptible to mutations than those of normal cells, in part due to impairment of DNA repair systems.

Growth factors: A variety of polypeptides secreted by many normal and tumor cells. These molecules act via autocrine (affects the cells that produce the growth factor), paracrine (affects neighboring cells), or endocrine (travels in the blood to affect distant cells) modes. Growth factors stimulate proliferation of target cells via interactions with specific receptors. They also have many other biologic properties.

Hypoxia-inducible factors (HIFs): A family of transcription factors important in directing cellular responses to varying levels of oxygen. HIFs are made up of different oxygen-regulated α subunits and a common constitutive β subunit. At physiologic levels of oxygen, the α subunit undergoes rapid degradation, initiated by prolyl hydroxylases. HIFs have various functions; for example, HIF-1 upregulates various genes encoding enzymes of glycolysis, and also the expression of vascular endothelial growth factor (VEGF).

Kinetochore: A structure that forms on each mitotic chromosome adjacent to the centromere. Mutations affecting the structures of its component proteins could contribute to causing CIN. (See also **Centromere**.)

Leukemias: A variety of malignant diseases in which various white blood cells (eg, myeloblasts, lymphoblasts, etc) proliferate in an unrestrained manner. Leukemias may be acute or chronic.

Loss of heterozygosity (LOH): LOH occurs when there is loss of the normal allele (often encoding a tumor suppressor gene) from a pair of heterozygous chromosomes, allowing the effects of the defective allele to be manifest clinically.

Lymphoma: A group of neoplasms arising in the reticuloendothelial and lymphatic systems. Major members of the group are Hodgkin and non-Hodgkin lymphomas.

Malignant cells: They are cancer cells, which have the ability to grow in an unrestrained manner, to invade, and to spread (metastasize) to other parts of the body.

Metastasis: The ability of cancer cells to spread to distant parts of the body and grow there.

Microsatellite instability (MSI): Expansion or contraction of short tandem repeats (microsatellites) due to replication slippage, abnormalities of mismatch repair or of homologous recombination.

Nanotechnology: The development and application of devices that are only a few nanometers in size (10^{-9} m = 1 nm). Some are being applied to cancer therapy.

Necrosis: Cell death induced by chemicals or tissue injury. During necrosis, a variety of hydrolytic enzymes are released that subsequently digest cellular molecules. Necrosis is not a genetically programmed process, as is apoptosis. Affected cells usually lyse releasing their contents, causing local inflammation.

Neoplasm: Any new growth of tissue, benign or malignant.

Oncogene: A mutated cellular gene (ie, proto-oncogene), the gene product of which is involved in the transformation of a normal cell to a cancer cell.

Oncology: The area of medical science that concerns itself with all aspects of cancer (causes, diagnosis, treatment, etc).

OMIM: Online Mendelian Inheritance in Man: An authoritative and comprehensive compendium of human genes and genetic phenotypes.

Philadelphia chromosome: A chromosome formed by a reciprocal translocation between chromosomes 9 and 22. This chromosomal translocation is the cause of chronic myeloid leukemia (CML). To form the abnormal Philadelphia chromosome, part of the *BCR* (breakpoint cluster region) gene of chromosome 22 fuses with part of the *ABL* gene (encodes a tyrosine kinase) of chromosome 9, directing the synthesis of a chimeric protein that has unregulated tyrosine kinase activity that drives cell proliferation. The activity of this kinase is inhibited by the drug imatinib (Gleevec), which has been successfully used to treat CML. (See also **Chromosomal translocation**.)

Procarcinogen: A chemical that becomes a carcinogen when it undergoes metabolism in the body.

Proto-oncogene: A normal cellular gene, which when mutated can give rise to a product that stimulates the growth of cells, contributing to the development of cancer.

Replication slippage: A process in which, because of misalignment of DNA strands where repeat sequences occur, DNA polymerase pauses and dissociates, resulting in deletions or insertions of repeat sequences.

Retinoblastoma: A rare tumor of the retina. Mutation of the *RB* tumor suppressor gene plays a key role in its development. Patients with hereditary retinoblastomas have inherited one mutated copy of the *RB* gene, and only need a mutation in the other copy of the gene to develop the tumor. Patients with sporadic retinoblastomas are born with two normal copies, and require two mutations to inactivate the gene.

Rous sarcoma virus (RSV): An RNA tumor virus that causes sarcomas in chickens. It was discovered in 1911 by Peyton Rous. RSV is a retrovirus that uses a reverse transcriptase in its replication. The resultant DNA copy of its genome subsequently integrates into the host cell genome. RSV has been widely used in studies of cancer, which has led to many important findings regarding the molecular bases of oncogenic transformation.

Sarcoma: A malignant tumor of mesenchymal origin (eg, from cells of the extracellular matrix or other sources).

Telomeres: Structures at the ends of chromosome that contain multiple repeats of specific hexanucleotide DNA sequences. The telomeres of normal cells shorten on repeated cell division, which may result in cell death. The enzyme telomerase replicates telomeres and is often expressed in cancer cells, helping them to evade cell death. Telomerase levels are extremely low in normal somatic cells.

Transformation: The process by which normal cells in tissue culture become changed to abnormal cells (eg, by oncogenic viruses or chemicals), some of which may be malignant.

Translocation: The displacement of one part of a chromosome to a different chromosome or to a different part of the same chromosome. Classic examples are the translocation found in Burkitt lymphoma (see earlier) and the translocation between chromosomes 9 and 22, which causes the appearance of the Philadelphia chromosome found in chronic myelogenous leukemia. Translocations have been found in many cancer cells.

Tumor: Any new growth of tissue, which may be benign or malignant.

Tumor suppressor gene: A gene the protein product of which normally restrains cell growth, but when its activity is lost or reduced by mutation contributes to the development of a cancer cell.

The Biochemistry of Aging

Peter J. Kennelly, PhD

- Describe the essential features of wear and tear theories of aging.
- List four or more common environmental constituents known to damage biologic macromolecules such as proteins and DNA.
- Explain what makes nucleotide bases especially vulnerable to damage.
- Describe the most physiologically important difference between mitochondrial and nuclear genomes.
- Describe the oxidative theory of aging.
- List the primary sources of reactive oxygen species (ROS) in humans.
- Describe the basic tenets of the mitochondrial theory of aging
- Describe three mechanisms by which cells prevent or repair damage inflicted by ROS.
- Describe the basic tenets of metabolic theories of aging.
- Explain the mechanism of the telomere "countdown clock."
- Outline our current understanding of the genetic contribution to aging.
- Explain the benefits of model organisms to biomedical research.

BIOMEDICAL IMPORTANCE

Consider the various stages in the lifespan of *Homo sapiens*. Infancy and childhood are characterized by continual growth in height and body mass. Basic motor and intellectual skills develop: walking, language, etc. Infancy and childhood also represent a period of vulnerability wherein a youngster is dependent on adults for water, food, shelter, protection, and instruction. Adolescence witnesses a final burst of growth in the body's skeletal framework. More importantly, a series of dramatic developmental changes occur—an accumulation of muscle mass, maturation of the gonads and brain, and the emergence of secondary sex characteristics—that transform a child into an independent and reproductively capable adult. Adulthood, the longest stage, is a period devoid of dramatic physical growth or developmental change. With the notable exception of pregnancy in females, it is not unusual for adults to maintain the same body weight, overall appearance, and general level of activity for two or three decades or more.

Barring fatal illness or injury, the onset of the final stage of life, old age, is signaled by a resurgence of physical and physiologic change. Muscle and bone mass progressively decrease.

Hair begins to thin and lose its pigmentation. Skin loses its suppleness and accumulates blemishes. Attention span and recall decline. Eventually, inevitably, life itself comes to an end as essential bodily functions decline.

Understanding the underlying causes and instigating triggers of aging and the changes that accompany it is of great biomedical importance. Hutchison-Gilford, Werner, and Down syndrome are three human genetic diseases whose pathologies include an acceleration of many of the physiologic events associated with aging. Slowing or preventing some of the degenerative processes that cause or accompany aging such as arthritis, osteoporosis, Alzheimer and Parkinson's diseases, can render the later stages of life more vital, productive, and fulfilling. Coopting the factors responsible for triggering cell death may enable physicians to selectively destroy harmful tumors, polyps, and cysts.

LIFESPAN VERSUS LONGEVITY

From Paleolithic to Medieval times the average life expectancy of a newborn baby oscillated over a range of 25 to 35 years. Beginning with the Renaissance, however, this number gradually

717

TABLE 57–1 Average Life Expectancy by Decade, USA

| Sample Period | Average Life Expectancy (Years) | |
	From Birth	If Survived to Age 5
1900-1902	49.24	59.98
1909-1911	51.49	61.21
1919-1921	56.40	62.99
1929-1931	59.20	64.29
1939-1941	63.62	67.49
1949-1951	68.07	70.54
1959-1961	69.89	72.04
1969-1971	70.75	72.43
1979-1981	73.88	75.00
1989-1991	75.37	76.22
1999-2001	76.83	77.47

Data from Arias E, Curtin LR, Wei R, Anderson RN. U.S. decennial life tables for 1999-2001, United States life tables. *Natl Vital Stat Rep.* 2008;57(1):1-36.

increased such that, by the beginning of the 20th century, the average life expectancy of persons born in developing countries reached the mid-40s. Today, 100 years later, the current world average is 67 years, and that for developed nations is approaching 80. These dramatic increases have led to speculation about how long this trend might be expected to continue. Can future generations expect to routinely live past the century mark? Is it possible that human beings possess the potential, with proper care and maintenance, to live indefinitely?

Unfortunately, such extrapolations are unlikely to be realized because they are based on a misunderstanding of the term **life expectancy**. Life expectancy is calculated by averaging over all births. Hence, it is dramatically influenced by infant mortality rates. While the life expectancy of a Roman child was 25 years, if one calculated the expected lifespan only for those persons who survived infancy, which we will refer to as **longevity**, the average nearly doubled to 48. When one factors out the dramatic decline in infant mortality rates that has taken place over the past century and a half, the predicted longevity of a 5-year-old child in the United States has increased from 70.5 in 1950 to 77.5 years in 2000 (**Table 57–1**). Is there some sort of upper limit to the lifespan of a properly nourished, protected, well-maintained human being?

AGING & MORTALITY: NONSPECIFIC OR PROGRAMMED PROCESSES?

Are aging and death non-determinant or **stochastic** processes wherein living creatures inevitably reach a tipping point after a lifetime's accumulation of damage from disease, injury, and simple wear and tear? Alternatively, are aging and death genetically programmed processes analogous to puberty that have

evolved through a process of natural selection? In all likelihood, aging and death are multifactorial processes to which both stochastic and programmed factors contribute.

WEAR & TEAR THEORIES OF AGING

Some researchers theorize that aging and mortality are the inevitable outcomes of a lifetime's accumulation of damage from injury, disease, and exposure to deleterious environmental constituents such as ultraviolet light. These theories note that while repair and turnover mechanisms exist to restore or replace many types of damaged molecules, they are less than perfect. Hence, some damage inevitably leaks through—damage that will accumulate over time, particularly among those cell populations that undergo little, if any, turnover (**Table 57–2**). Ironically, many prominent sources of molecular and cellular damage also are essential for terrestrial life: water, oxygen, and sunlight.

Hydrolytic Reactions Can Damage Proteins & Nucleotides

Water is a relatively weak nucleophile. However, because of its ubiquity and high concentration (>55 M; see Chapter 2), even this weak nucleophile will occasionally react with susceptible targets inside the cell. These targets include the amide bonds that form polypeptide chains and, especially, the side chain amides of the amino acids asparagine and glutamine. Hydrolysis transforms asparagine and glutamine residues into aspartate and glutamate residues, respectively, thereby introducing an acidic and potentially negatively charged group in place of a pH- and charge-neutral amide (**Figure 57–1A** and **B**).

Both the bonds linking nucleotide bases to the deoxyribose-phosphodiester backbone of DNA and the amino groups projecting from the heterocyclic aromatic rings of the nucleotide bases cytosine, adenine, and guanine are also susceptible to hydrolytic attack. In the latter case, the amino group is replaced by a carbonyl oxygen to form uracil, hypoxanthine, and

TABLE 57–2 Time Required for All of the Average Cells of This Type to Be Replaced

Tissue or Cell Type	Turnover
Intestinal epithelium	34 h[a]
Epidermis	39 d[b]
Leukocyte	<1 y[c]
Adipocytes	9.8 y[c]
Intercostal skeletal muscle	15.2 y[c]
Cardiomyocytes	≥100 y[c]

Data from:
[a]Potten CS, Kellett M, Rew DA, et al: Proliferation in human gastrointestinal epithelium using bromodeoxyuridine in vivo. Gut 1992;33:524.
[b]Weinstein GD, McCullough JL, Ross P: Cell proliferation in normal epidermis. J Invest Dermatol 1984;82:623.
[c]Spalding KL, Arner E, Westermark PO, et al: Dynamics of fat cell turnover in humans. Nature 2008;453:783.

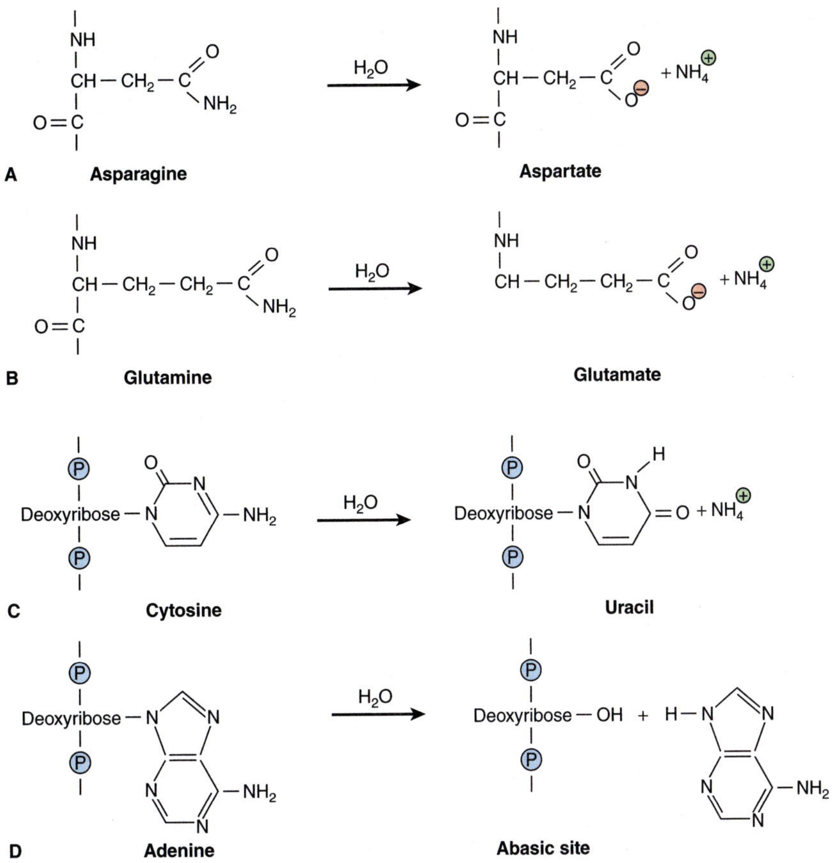

FIGURE 57–1 **Examples of hydrolytic damage to biologic macromolecules.** Shown are a few of the ways in which water can react with and chemically alter proteins and DNA: (**A**) Net substitution of aspartic acid via hydrolytic deamidation of the neutral side chain of asparagine. (**B**) Net substitution of glutamic acid via hydrolytic deamidation of the neutral side chain of glutamine. (**C**) Net mutation of cytosine to uracil by water. (**D**) Formation of an abasic site in DNA via hydrolytic cleavage of a ribose-base bond.

xanthine, respectively (**Figure 57–1C**). In the former, the nucleotide base is completely eliminated, leaving a gap in the sequence (**Figure 57–1D**). The elimination or alteration of nucleotide bases in DNA is of potentially much greater biomedical significance than those affecting proteins. If the affected gene encodes a protein, if left unrepaired (see Chapter 35) every subsequent copy of the protein will be altered, whereas direct damage affects only a single copy of a protein. If the mutant cell divides, the alteration will be passed on to its progeny, amplifying its impact.

Other biologically relevant bonds that are susceptible to hydrolysis include the ester linkages that connect fatty acids to their cognate glycerolipids, the glucosidic bonds that link the monosaccharide units of carbohydrates, and the phosphodiester bonds that hold polynucleotides together and link the head groups of phospholipids to their diacylglycerol partners. In most instances, with the notable exception of polynucleotide chain breaks, the products of these reactions appear to be biologically innocuous.

Respiration Generates Reactive Oxygen Species

Numerous biologic processes require enzyme-catalyzed oxidation of organic molecules by molecular oxygen (O_2). These processes include the hydroxylation of proline and lysine side chains in collagen (see Chapter 5), the detoxification of xenobiotics by cytochrome P450 (see Chapter 47), the degradation of purine nucleotides to uric acid (see Chapter 33), the reoxidation of the prosthetic groups in the flavin-containing enzymes that catalyze oxidative decarboxylation (eg, the pyruvate dehydrogenase complex; see Chapter 17) and other redox reactions (eg, amino acid oxidases; see Chapter 28), and the generation of the chemiosmotic gradient in mitochondria by the electron transport chain (see Chapter 13). Redox enzymes frequently employ prosthetic groups such as flavin nucleotides, iron-sulfur centers, or heme-bound metal ions (see Chapters 12 and 13) to assist in generating and stabilizing the free radical and oxyanion intermediates formed during these processes.

Occasionally, these highly reactive intermediates escape to form of ROS such as superoxide and hydrogen peroxide inside the cell (**Figure 57–2A**). The most common of these sources is the electron transport chain, whose high levels of electron flux and structural complexity render it vulnerable to "leakage" of ROS. In addition, many mammalian cells synthesize and release nitric oxide (NO•), a free radical containing second messenger that promotes vasodilation and muscle relaxation in the cardiovascular system (see Chapter 55).

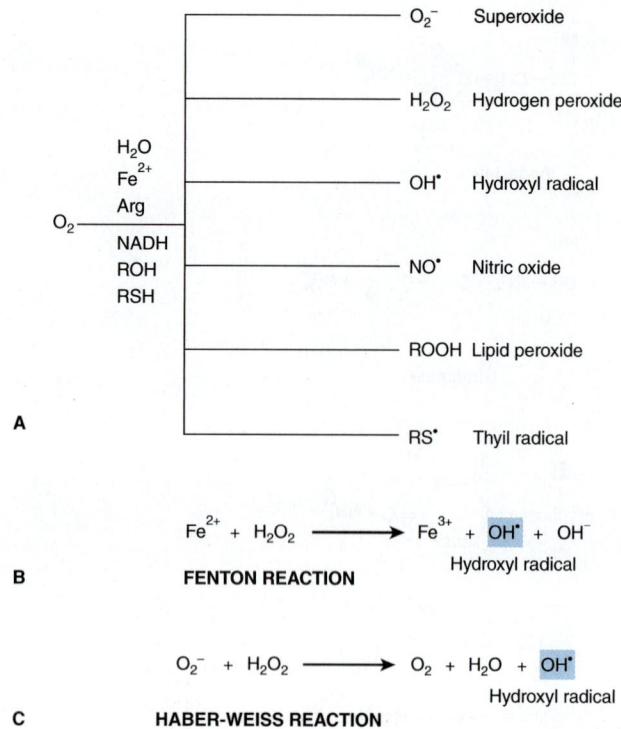

FIGURE 57–2 **Reactive oxygen species (ROS) are toxic by-products of life in an aerobic environment. (A)** Many types of ROS are encountered in living cells. **(B)** Generation of hydroxyl radical via the Fenton reaction. **(C)** Generation of hydroxyl radical by the Haber-Weiss reaction.

Reactive Oxygen Species Are Chemically Prolific

The extremely high reactivity of ROS makes them extremely dangerous. ROS can react with and chemically alter virtually any organic compound, including proteins, nucleic acids, and lipids. ROS display a particularly strong tendency to form **adducts**—products arising from the combination of precursors—with biologic compounds that contain multiple double bonds such as nucleotide bases and polyunsaturated fatty acids (**Figure 57–3**). Adducts formed with nucleotide bases can be especially dangerous because of their potential, if uncorrected, to cause mutation generating errors during DNA replication.

The ease with which exposure to warm air can turn household butter rancid is a testament to the reactivity of unsaturated fats, those containing one or more carbon–carbon double bonds (see Chapter 23), with ROS. The resulting peroxidation of lipids can lead to the formation of cross-linked lipid-lipid and lipid-protein adducts that may compromise membrane fluidity and integrity. In mitochondria, a loss of membrane integrity can potentially allow protons to leak across, thereby reducing the efficiency of ATP generation and increasing the production of deleterious ROS. Accumulated damage to mitochondrial membranes may eventually lead to the efflux of cytochrome c, an inducer of programmed cell death or **apoptosis** (see later).

Chain Reactions Amplify the Destructiveness of ROS

The destructiveness inherent in the high reactivity of many of ROS, particularly those containing an unpaired electron, is exacerbated by their capacity to participate in chain reactions in which the product of the reaction may include, not just a damaged biomolecule, but another free radical species capable of producing additional damage with the accompanying formation of yet another free radical by-product. This chain of events will continue until a free radical by-product is able to acquire an unpaired electron from another ROS molecule or a redox protectant such as reduced glutathione. Alternatively, the ROS may be eliminated by the catalytic action of a member of the cell's arsenal of dedicated antioxidant enzymes (see Chapters 12 and 53).

The reactivity, and hence destructiveness, of individual ROS varies. Hydrogen peroxide, for example, is less reactive than superoxide, which in turn is less reactive than hydroxyl radical (OH·). Two pathways exist by which the highly toxic hydroxyl radical can be generated from less destructive ROS in living organisms. One is through the Fenton reaction, in which H_2O_2 reacts with ferrous (+2) iron to generate a hydroxyl radical along with a hydroxide ion and ferric (+3) iron (see Figure 57–2B). The ferric iron, in turn, can be reduced back to the ferrous (+2) state by other hydrogen peroxide molecules, furthering the production of additional hydroxyl radicals via the Fenton reaction. Hydroxyl radical can also be generated when superoxide and hydrogen peroxide disproportionate via the Haber-Weiss reaction (see Figure 57–2C).

THE MITOCHONDRIAL THEORY OF AGING

Mitochondria Are Major Sources of ROS

In 1956, Denham Harmon proposed the free radical theory of aging. It had been suggested that the similarities between the toxic side effects produced on exposure to hyperbaric oxygen or ionizing radiation could be traced to a common source, their shared propensity for generating ROS. This proposition dovetailed nicely with Harmon's own observation that lifespan was inversely related to metabolic rate and, by extrapolation, respiration. Since all animals respire, free radical theory postulates that aging is a reflection of the cumulative biomolecular damage that results from the continual and inescapable production of ROS.

The dominant natural source of cellular ROS is the occasional escape or leakage of electrons from the electron transport chain. As damage to the components of this pathway would be expected to increase the rate of leakage, it is not difficult to envision the emergence of a vicious cycle in which some initial insult to the mitochondria would lead to an increased rate of ROS production that would, in turn, cause further damage. Accumulating damage to mitochondrial components could, in turn, adversely affect the production of ATP, perhaps to the point where it undermines cell vitality and function.

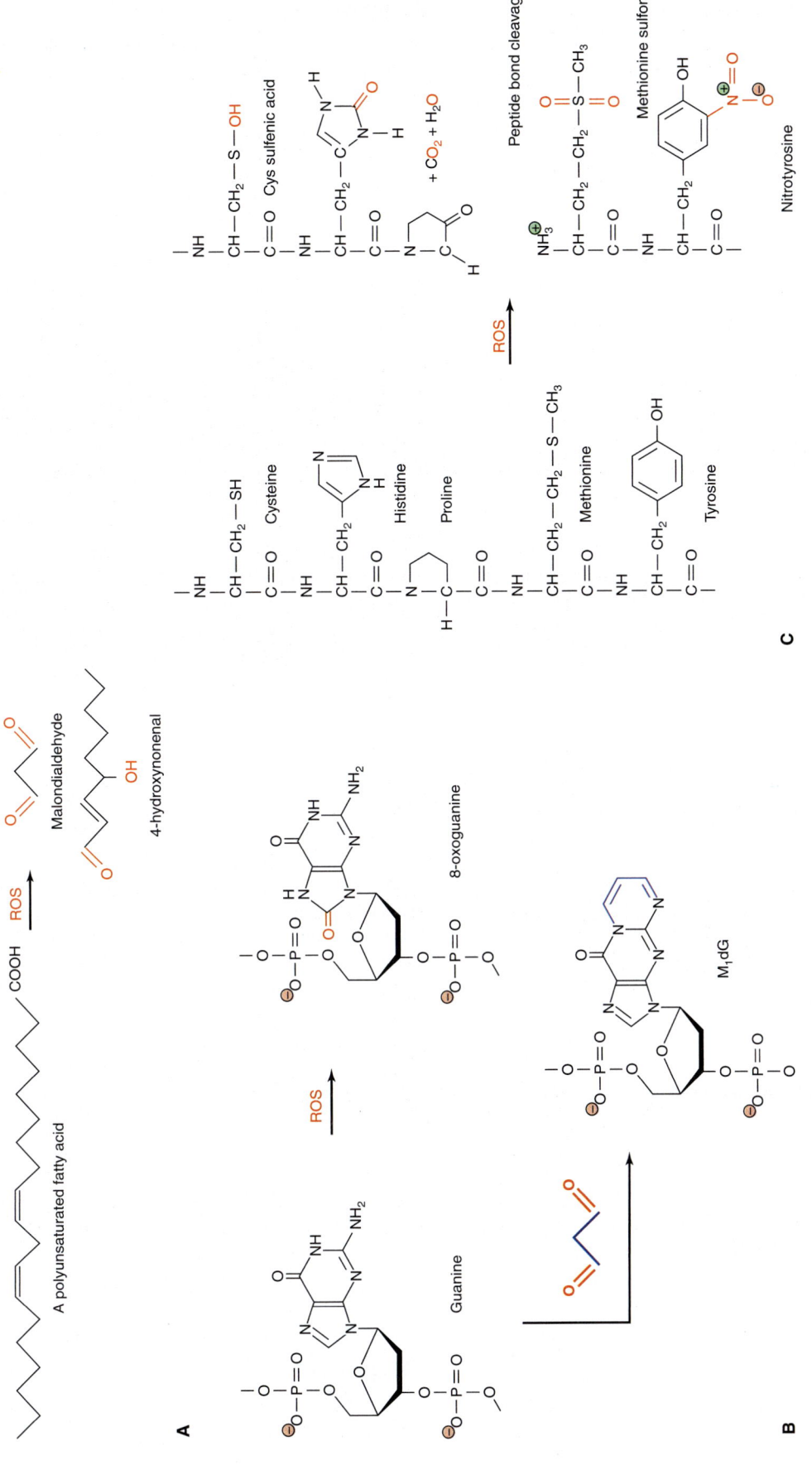

FIGURE 57-3 ROS react directly and indirectly with a wide range of biologic molecules. (A) Peroxidation of unsaturated lipids generates reactive products such as malondialdehyde and 4-hydroxynonenal. **(B)** Guanine can be directly oxidized by ROS to produce 8-oxoguanine or form an adduct, M_1dG, with the ROS product malondialdehyde. **(C)** Common reactions of proteins with ROS, including oxidation of amino acid side chains and cleavage of peptide bonds. Oxygen atoms derived from ROS are marked in red. Carbon atoms derived from malondialdehyde in M_1dG are colored blue. The complete chemical name for M_1dG is 3-(2-Deoxy-D-*erythro*-pentofuranosyl)pyrimido(1,2-α)purin-10(3*H*)-one.

Mitochondria Are Unable to Repair Their Endogenous DNA

The mitochondrion's endogenous genome also plays a role in this proposed redox damage cycle. The mitochondrial genome is a much reduced, vestigial remnant of the genome of the ancient bacterium that was the precursor of the current organelle. It is presumed that the exchange of metabolites between primitive eukaryotes and neighboring bacteria developed into a mutually dependency. The ultimate outcome of this intimate metabolic integration was **endosymbiosis**, the internalization of the smaller bacterium by its eukaryotic partner. Over time most, but not all, of the genes contained in the genome of the internalized bacterium's were either eliminated or were transferred to the nuclear DNA of the eukaryotic host. Today, the genome of the human mitochondrion encodes two ribosomal RNAs (one for each subunit), 22 tRNAs, several of the polypeptide components of complexes I, III, and IV of the electron transport chain, and parts of F_1, F_0 ATPase (**Table 57–3**). The mitochondrial genome lacks genes encoding the types of surveillance and repair enzymes responsible for maintaining the integrity of nuclear DNA. Hence, when mutations occur, including those that adversely affect components of the electron transport chain, they become a permanent feature of a mitochondrion's genome. Not only do these mutations and their impacts persist, additional mutations can also accrue over time that further compromise electron transport chain function. One can envision a vicious cycle in which an ROS-generated mutation to the mitochondrial genome leads to the increased leakage of ROS from the electron chain that in turn cause further DNA damage and still higher levels of ROS leakage.

While the mitochondrial hypothesis is no longer viewed as a comprehensive, unifying explanation for all of the changes that are associated with human aging and its comorbidities, this organelle remains a likely contributor. It has been observed that levels of the key electron carrier, NAD(H), decline with age. In addition, the key part played by mitochondria in programmed cell death (apoptosis) can be viewed as offering circumstantial support for mitochondria as drivers of aging and morbidity.

TABLE 57–3 Genes Encoded by the Genome of Human Mitochondria

rRNA	12S, 16S rRNA
tRNA	22 tRNAs (2 for Leu and Ser)
Subunits of NADH-ubiquinone oxidoreductase (Complex I, >40 total)	ND 1-6, ND 4L
Subunits of ubiquinol-cytochrome c oxidoreductase (Complex III, 11 total)	Cytochrome b
Subunits of cytochrome oxidase (Complex IV, 13 total)	COX I, COX II, COX III
Subunits of the F_1, F_0 ATPase (ATP synthase, 12 total)	ATPase 6, ATPase 8

Mitochondria Are Key Participants in Apoptosis

Apoptosis imbues higher organisms with the ability to selectively eliminate cells that are rendered superfluous by developmental changes, such as those that continually take place during embryogenesis, or which have been damaged beyond repair. During developmental tissue remodeling, the apoptotic cell death program is triggered by receptor-mediated signals. In the case of damaged cells, ROS, viral dsRNA, DNA damage, or heat shock may act as triggers. These signals induce the opening of the permeability transition pore complex embedded in the mitochondrial outer membrane, which allows cytochrome c, a small ($\approx$12.5 kDa), soluble electron carrier protein, to escape into the cytoplasm. Here, cytochrome c provides the core around which the **apoptosome**, a multiprotein complex, coalesces. Assembly of the apoptosome initiates a cascade of proteolytic activation events targeting the proenzyme forms of the caspases, a family of cysteine proteases. The terminal caspases, numbers 3 and 7, break down structural proteins in the cytoplasm and chromatin proteins in the nucleus; events that lead to the death and, eventually, phagocytosis of the morbid cell. Many researchers are working to find ways to exploit the presence of this intrinsic, receptor-mediated cell death pathway as a means for selectively eliminating cancer and other malicious cells.

Ultraviolet Radiation Can Be Extremely Damaging

The term **ultraviolet (UV) radiation** refers to those wavelengths of light that lie immediately below the blue end of the visible spectrum. While the human eye cannot detect them, they are strongly absorbed by organic compounds possessing aromatic rings or multiple, conjugated double bonds. These include the nucleotide bases of DNA and RNA; the side chains of phenylalanine, tyrosine, and tryptophan; polyunsaturated fatty acids; heme groups; and numerous cofactors and coenzymes including flavins, cyanocobalamin, etc. Absorption of this short wavelength, high-energy light can cause the rupture of covalent bonds in proteins, DNA, and RNA; the formation of thymine dimers in DNA (**Figure 57–4**); cross-linking of proteins; and the generation of free radicals. While UV radiation does not penetrate beyond the first few layers of epidermal cells, the high efficiency with which it is absorbed can lead to the rapid accumulation of damage by the skin. Because the nucleotide bases of DNA and RNA are particularly effective at absorbing UV radiation, it is highly mutagenic. Thus, prolonged exposure to intense sunlight can lead to the accumulation of multiple DNA lesions that can overwhelm a cell's intrinsic repair capacity, leading to the development of myelomas—some of which aggressively proliferate if left untreated.

Protein Glycation Often Leads to the Formation of Damaging Cross-Links

When amine groups in proteins and nucleotides are exposed to a reducing sugar such as glucose, they may form an adduct

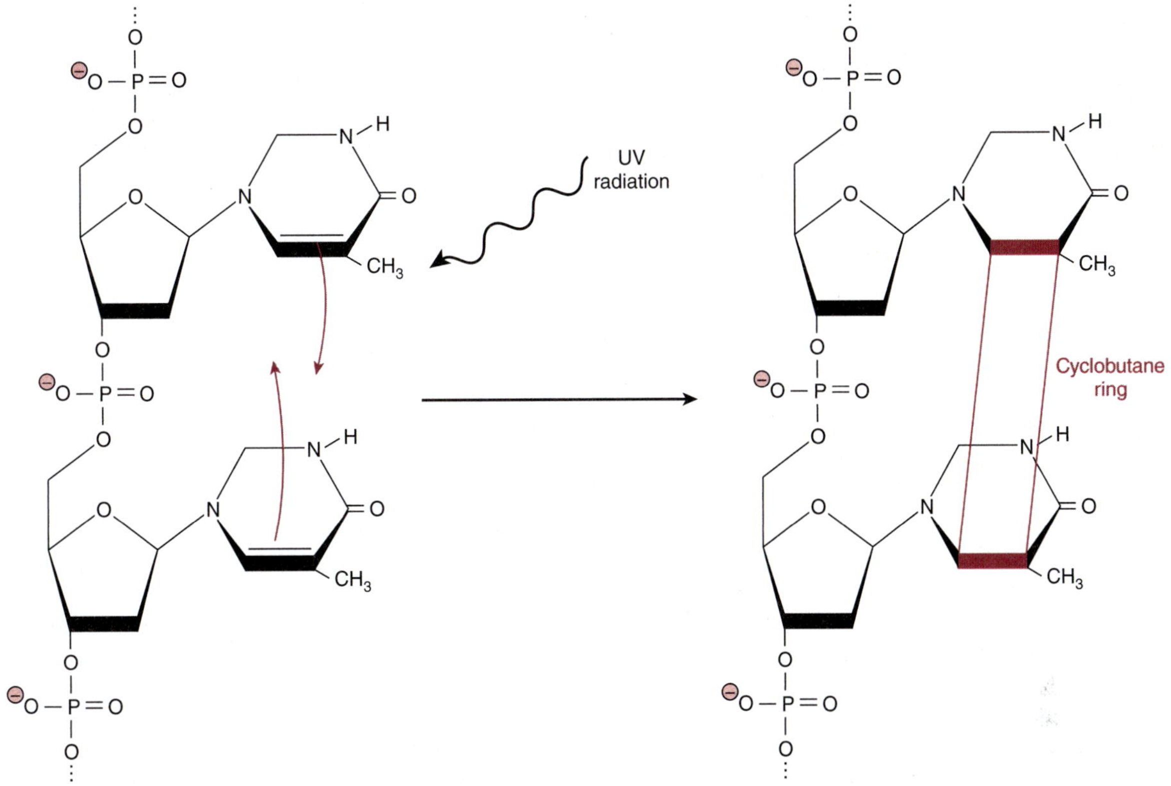

FIGURE 57-4 **Formation of a thymine dimer following excitation by UV light.** When consecutive thymine bases are stacked one above the other in a DNA double helix, absorption of UV light can lead to the formation of a cyclobutane ring (red, not to scale) covalently linking the two bases together to form a thymine dimer.

by a process called **glycation**. The initial step in this process is the formation of a Schiff base between the aldehyde or ketone group of the sugar and amine groups on proteins and other macromolecules. Over time, glycated proteins undergo a series of rearrangements to form **Amadori** products, which contain a conjugated carbon–carbon double bond that can react with the amino group on a neighboring macromolecule (**Figure 57-5**). The net result is the formation of a covalent cross-link between two proteins or other biologic macromolecules. These same macromolecules can, in turn, be glycated further, extending the network of cross-links to include other macromolecules. These cross-linked aggregates are sometimes called **advanced glycation end products** or **AGEs**.

The physiologic impact of protein glycation can be especially pronounced when long-lived proteins such as collagen or β-crystallins are affected. Their persistence enhances the opportunity for multiple glycation and subsequent cross-linking events to occur. In blood vessels, the accumulation of cross-links in the collagen network of vascular endothelial cells can lead to the progressive loss of elasticity and thickening of the basement membrane, both of which potentiate plaque formation. The end result is an increasing workload for heart. In the eye, the accumulation of aggregated proteins compromises the opacity of the lens, eventually manifesting itself as cataracts. Diabetics, whose ability to control blood glucose levels is impaired, are particularly susceptible to the

formation of AGEs. In fact, the glycation of hemoglobin and serum albumin are used as biomarkers for the diagnosis of diabetes.

MOLECULAR REPAIR MECHANISMS COMBAT WEAR & TEAR

Enzymatic & Chemical Mechanisms Intercept Damaging ROS

A corollary to the wear and tear theory of aging is that longevity reflects the robustness and durability of an individual's molecular protection, repair, and replacement mechanisms. For example, fruit flies that have been genetically altered to express elevated levels of superoxide dismutase exhibit significantly extended lifespans relative to wild-type.

In the cytoplasm, the cysteine-containing tripeptide glutathione acts as a chemical redox protectant that reacts directly with ROS to generate less reactive compounds such as water. Oxidized glutathione, which consists of two tripeptides linked by an S-S bond, is then enzymatically reduced to maintain the pool of protectant (see Chapter 53). Glutathione can also reduce cysteine sulfenic acids and aberrant disulfides on proteins and form adducts with toxic xenobiotics

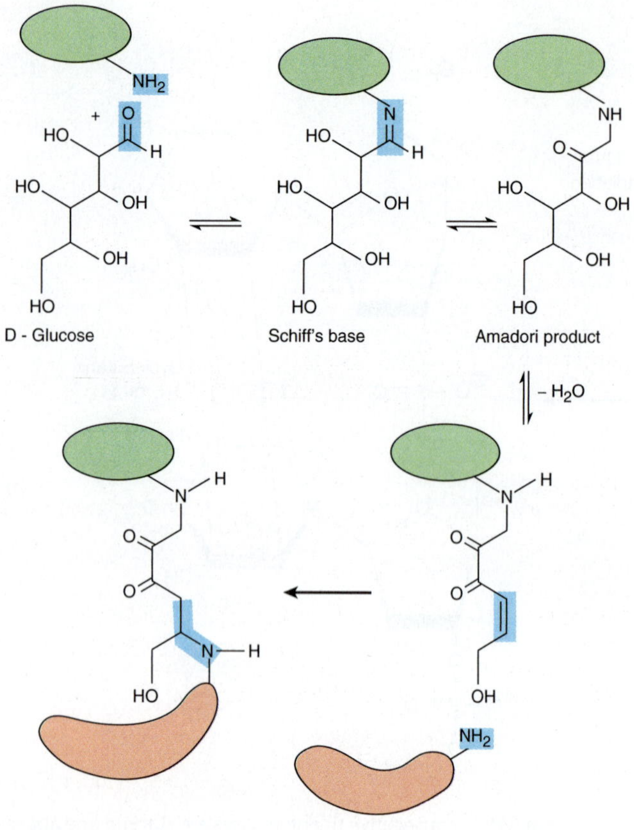

FIGURE 57–5 Protein glycation can lead to the formation of protein–protein cross-links. Shown are the sequence of reactions that generate the Amadori product on the surface of the protein marked in green, and the subsequent formation of a protein–protein cross-link via an amino group on the surface of a second, red, protein.

(see Chapter 47). Ascorbic acid and vitamin E also possess antioxidant properties, which accounts for the popularity of diets that emphasize foods or supplements rich in these compounds to combat ROS and, hopefully, slow aging.

The Integrity of DNA Is Maintained by Proofreading & Repair Mechanisms

In addition to the prophylactic measures mentioned earlier, living organisms possess a limited capacity to replace or repair damaged macromolecules. The largest suite of repair enzymes is the set devoted to maintaining the integrity of the nuclear (but not the mitochondrial) genome. This is to be expected given DNA's pivotal role in inheritance, its vulnerability to chemical assault and UV radiation, and the fact that—by contrast to almost every other macromolecule—each human cell contains only one or two copies of each chromosome.

Maintaining the integrity of the genome begins at replication, where careful proofreading is performed to ensure that the new genome formed during **somatic** cell division faithfully replicates the template that directed its synthesis. The term somatic refers to the differentiated cells that comprise the body of an organism. In addition, most living organisms possess an cadre of enzymes whose role is to inspect and correct

aberrations that either escaped replicative proofreading or were subsequently generated through the action of water (double-strand breaks, depurination, and cytosine deamidation), UV radiation (thymine dimers and strand breaks), or exposure to chemical modifiers (adduct formation). This multilayered system is composed of mismatch repair enzymes, nucleotide excision repair enzymes, and base excision repair enzymes as well as the Ku system for repairing double-strand breaks in the phosphodiester backbone (see Chapter 35). As a last resort, cells harboring damaging mutations are subject to removal by apoptosis.

Despite the many precautions taken to identify and correct errors, some mutations inevitably slip through. Indeed, a low, but finite, frequency of mutation is necessary in order to generate the genetic variability that furthers evolution. The **somatic mutation theory of aging** proposes that these mutations also serve as drivers of the aging process. Simply put, the accumulation of mutated somatic cells must inevitably lead to compromised biologic function that manifests itself, at least in part, as the physical changes we associate with aging.

Some Types of Protein Damage Can Be Repaired

In contrast to DNA, a cell's capacity to repair damage to other biomolecules is relatively limited. For the most part, cells appear to rely on routine turnover, wherein the individual members of the global population of a given biomolecule are degraded and replaced by new synthesis on a continuing, or constitutive, basis (see Chapter 9) to remove damaged lipids, carbohydrates, and proteins. However, some proteins, particularly the fibrous proteins found in tendons, ligaments, bones, matrix, etc, undergo little, if any, turnover. These long-lived proteins tend to accumulate damage over the years, which contributes to the loss of elasticity in vascular tissues and joints, loss of lens opacity, etc. The most prominent mechanisms for repairing damaged proteins target the oxidized side chain sulfur atoms of cysteine and methionine, and the isoaspartyl groups formed when a peptide bond shifts from an α- to a side chain carboxyl group.

The side chain sulfhydryl group of cysteine frequently plays important catalytic, regulatory, redox, and structural roles (eg, cysteine disulfides, Fe-S centers) in proteins. However, both the sulfhydryl group of cysteine and the sulfur ether of methionine are extremely vulnerable to oxidation (see Figure 57–3C). Cysteine disulfides, cysteine sulfenic acids, and methionine sulfoxide can be reduced by disulfide reductases and methionine sulfoxide reductases, all of which use NADPH as electron donor, or via direct reaction with reduced glutathione. Unfortunately, the reduction potentials of glutathione and NADPH are only sufficient to reduce the lowest oxidation states of these sulfur atoms: cysteine disulfides or sulfenic acids and methionine sulfoxide. Hence, cysteine sulfinic acid, cysteine sulfonic acid, and methionine sulfone are refractory to reduction by available biochemical means.

Aspartic acid possesses the precise geometry needed to bring its side chain carboxyl group into close proximity with

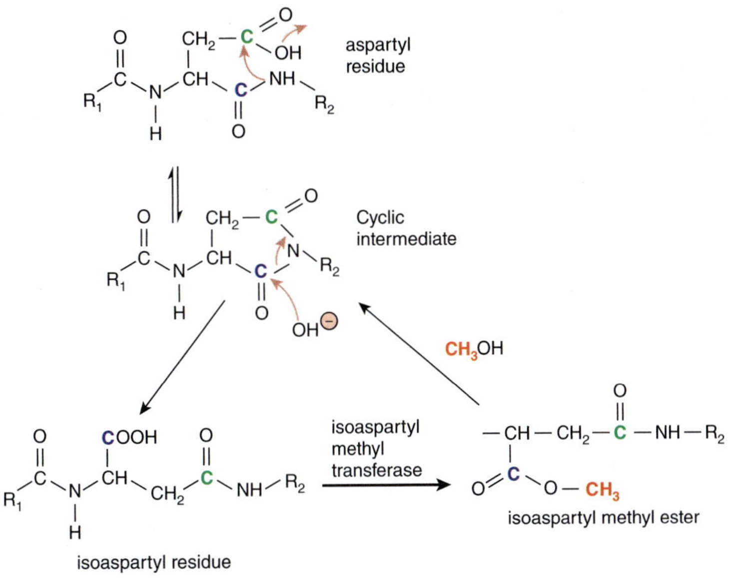

FIGURE 57–6 **Formation of an isoaspartyl linkage in a polypeptide backbone and its repair via the intervention of isoaspartyl methyltransferase.** Shown is the sequence of chemical and enzyme-catalyzed reactions that lead to formation of an isoaspartyl linkage and restoration of a normal peptide linkage. The carbons corresponding to the α and side chain carboxylic acid groups in aspartic acid are colored blue and green, respectively. Red arrows denote the routes of nucleophilic attack during the cyclization and hydrolysis reactions. The methyl group added by isoaspartyl methyltransferase is colored red.

the peptide bond involving its α-carboxyl group, at which point they may react to generate a cyclic diamide. This intermediate can reopen in one of two ways to either reform the original peptide bond or one which the side chain carboxyl now forms part of the protein's peptide backbone, forming an isoaspartyl residue (**Figure 57–6**). Restoration of the normal peptide bond linkage is facilitated by the methylation of the α-carboxyl by isoaspartyl methyltransferase. Formation of the methyl ester introduces a leaving group that potentiates the reformation of the cyclic diamide, which can then reopen to form the normal peptide bond linkage (see Figure 57–6).

Aggregated Proteins Are Highly Refractory to Degradation or Repair

Modifications to a protein's composition or conformation that cause it to adhere to other protein molecules can lead to the formation of toxic aggregates, sometimes called **amyloid**. Such aggregates are a hallmark of several neurodegenerative diseases, including Parkinson, Alzheimer, Huntington disease, spinocerebellar ataxias, and the transmissible spongiform encephalopathies. The toxic effects of these insoluble aggregates are exacerbated by their persistence, as they are generally refractory to degradation by the proteases normally responsible for protein turnover.

AGING AS A PREPROGRAMMED PROCESS

While molecular wear and tear undoubtedly contributes to aging, several observations suggest a prominent role for programmed, deterministic mechanisms. Female menopause, for example, provides an example of an age-associated physiologic change that is genetically programmed and hormonally controlled. The following paragraphs describe several current theories regarding deterministic, programmed mechanisms for controlling aging and death.

Metabolic Theories of Aging: "The Brighter the Candle, the Quicker It Burns"

One of the many variants of this famous quote attributed to the ancient Chinese philosopher Lao Tzu summarizes the salient features of **metabolic theories of aging**. Its origins can be traced to the observation that the larger members of the animal kingdom tend to live longer than the smaller ones (**Table 57–4**). Reasoning that the causal basis for this correlation derived from in something connected with size, rather than size itself, scientists focused their attention on the organ most frequently associated with life and vitality—the heart. In general, the resting heart rate of small animals such as hummingbirds, 250 beats per minute, tends to be higher than those of large animals such as whales, 10 to 30 beats per minute. Estimates of the total number of times each vertebrate animal's heart beat over the course of a lifetime exhibited an amazing convergence on 1×10^9.

Is the vertebrate heart physically or genetically limited to 1 billion beats? A more nuanced variation of this **heartbeat hypothesis** was put forward by Raymond Pearl in the 1920s. Pearl's **metabolic** or **rate of living hypothesis** posited that an individual's lifespan was reciprocally linked to their basal metabolic rate. It was calculated that every vertebrate animal expends a similar amount of total metabolic energy *per unit*

TABLE 57–4 **Lifespan Versus Body Mass for Several Mammals**

Species	Approximate Mass (kg)	Mean Expectation of Life at Maturity (years)
White-footed mouse	0.02	0.28
Deer mouse	0.02	0.43
Bank vole	0.025	0.48
Eastern chipmunk	0.1	1.63
American pika	0.13	2.33
Golden mantled grd. squirrel	0.155	2.12
Red squirrel	0.189	2.45
Belding's ground squirrel	0.25	1.78
Uinta ground squirrel	0.35	1.72
Eastern gray squirrel	0.6	2.17
Arctic ground squirrel	0.7	1.71
Eastern cottontail	1.25	1.48
Striped skunk	2.25	1.90
American badger	7.15	2.33
North Arnerican river otter	7.2	3.79
Bobcat	7.5	2.48
North American beaver	18	1.52
Impala	44	4.80
Bighorn sheep	55	5.48
Wild boar	85	1.91
Warthog	87	2.82
Nilgiri tahr	100	4.71
Blue wildebeest	165	4.79
Red deer stag	175	4.90
Waterbuck	200	5.87
Burchell's zebra	270	7.95
African buffalo	490	4.82
Hippopotamus	2390	16.40
African elephant	4000	19.10

Data from Millar JS, Zammuto: Life histories of mammals: an analysis of life tables, Ecology 1983;64(4):631-635.

body mass, 7×10^5 J/g, over their lifetime. While intuitively appealing, identification of a mechanistic link between lifespan and energy expenditure or metabolic rate has proven elusive. Adherents of the mitochondrial theory of aging suggest that what is being "counted" is not heartbeats or energy, but the ROS that are the by-product of respiration. Over time, the continued generation of energy and related consumption of O_2 leads to the accumulation of ROS-induced damage to DNA, proteins, and lipids until, eventually, a universally conserved tipping point is reached. Cells experiencing caloric deficits adjust (reprogram) their metabolic pathways to utilize available resources in a more efficient manner that concomitantly decreases the yield of collateral ROS.

Telomeres: A Molecular Countdown Clock?

A second school of thought holds that the putative countdown clock that controls aging and lifespan does not monitor heartbeats, energy, or ROS. Rather, it uses **telomeres** to track the number of times each somatic cell divides.

Unlike the closed circular DNA of bacterial genomes, the genomic DNA of eukaryotes is linear. If left unprotected, the exposed ends of these linear polynucleotides would be available to participate in potentially deleterious genetic recombination events. Telomeres cap the ends of eukaryotic chromosomes with a few hundred GT-rich hexanucleotide repeats. These caps also provide a source dispensable DNA to accommodate the wastage that occurs when linear chromosomes are replicated.

This wastage is a consequence of the fact that all DNA polymerases work unidirectionally, 3′ to 5′ (see Chapter 35). When replicating linear double-stranded DNA via discontinuous 3′ to 5′ synthesis, the 5′ end of each new strand will generally commence roughly 100 bp short of the extreme 5′ end of the template strand. Consequently, each time a dividing cell replicates its genome, its chromosomes become shorter (**Figure 57–7**). The telomeres provide an expendable source of DNA whose gradual loss is of little **immediate** consequence to the cell. However, once the supply of telomere DNA is exhausted, roughly 100 cell divisions for humans, mitosis ceases, causing the somatic cell to enter a state of **replicative senescence**. Hence, as our bodies age, they progressively lose the capacity to replace lost or damaged cells.

Organisms are able to generate progeny that contain full-length telomeres through the intervention of the enzyme **telomerase**. Telomerase is a ribonucleoprotein that is expressed in stem cells and most cancer cells, but not in somatic cells. Using an RNA template, telomerase adds GT-rich hexanucleotide repeat sequences to the ends of linear DNA molecules that restore their telomere caps to full length. In the laboratory, the operation of the proposed telomere clock has been demonstrated using genetically engineered somatic cells that express telomerase. As predicted by the telomere clock hypothesis, the continual restoration of their telomere caps enabled the genetically engineered cells continued to divide in culture long after wild-type control cells entered replicative senescence.

Kenyon Used a Model Organism to Discover the First Aging Genes

Many advances in biomedical science are the product of research that uses a variety of so-called model organisms as

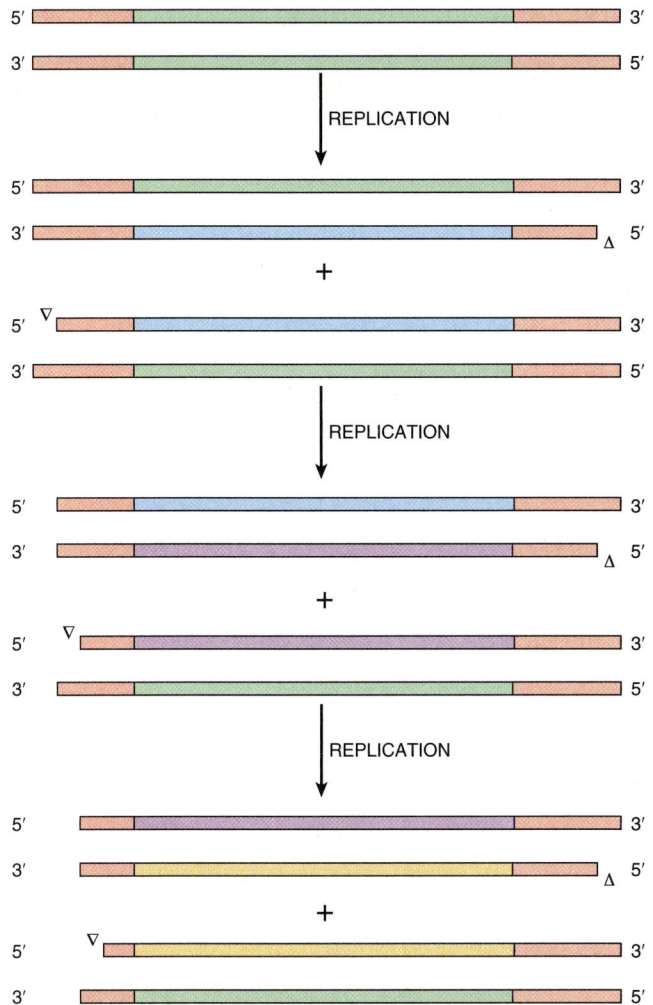

FIGURE 57–7 **The telomeres at the ends of eukaryotic chromosomes progressively shorten with each cycle of replication.** Shown is a schematic diagram of the linear DNA of a eukaryotic chromosome (green) containing telomeres at each end (red). During the first replication, new DNA strands are synthesized (green) using the original chromosome as template. For simplicity, the next two replication cycles (purple, yellow) show the fate of only the lower of the two nucleotide products from the preceding replicative cycle. Open arrowheads denote the site of incomplete strand synthesis. The model assumes that the single strand overhangs at the ends of each chromosome are trimmed at the completion of each cycle of cell division. Note the progressive shortening of the telomere repeats.

their test subjects. The fruit fly, *Drosophila melanogaster*, has yielded a rich harvest of information concerning the genes that guide cellular differentiation and organ development. Baker's yeast and the African clawed frog, *Xenopus laevis*, have served as the workhorses for dissecting the signal transduction circuitry that orchestrates the cell division cycle. A variety of cultured mammalian cell lines serve as surrogates for adipocytes, kidney cells, tumors, dendrites, etc. While at first glance it would appear that many of these model systems share little in common with humans, each possesses unique attributes that render them convenient vehicles for addressing particular problems or investigating specific cellular or molecular processes.

Caenorhabditis elegans is a worm that has served as an important model for the study of developmental biology. *Caenorhabditis elegans* adults contain only 959 cells. Moreover, this worm is transparent and grows rapidly, enabling the developmental program that generates each cell in a mature adult to be traced back to the fertilized egg. In the early 1990s, Cynthia Kenyon and colleagues observed that worms carrying mutations in the gene encoding an insulin receptor-like molecule, *daf-2*, lived 70% longer than their wild-type counterparts. Equally important, the behavior of adult mutant worms resembled that of young wild-type *C. elegans* for much of this period. This illustrates an important characteristic of a *bona fide* "aging gene." It must do more than simply prolong existence. It must also delay some of the physiologic changes associated with aging.

Investigation of further aging genes indicate they code for either one of a small set of transcription factors that include PHA-4 or DAF-16, which presumably control expression of aging critical genes, or signaling proteins such as DAF-2 and mTOR that probably activate PHA-4, DAF-16, etc, in response to specific environmental signals. Much remains to be learned about the extent to which aging is controlled by genetically programmed events, and how these genes and their products interact with nutritional and other factors that influence vitality and longevity.

WHY WOULD EVOLUTION SELECT FOR LIMITED LIFESPANS?

The idea that animals would have evolved mechanisms designed specifically to limit their lifespan would appear, at first glance, to be counterintuitive. If the driving force behind evolution is the selection for traits that enhance fitness and survival, should not this translate into an ever-increasing life expectancy? However, while extending lifespan may represent a desirable trait from the point of view of the individual, it does not necessarily follow that this applies to a population or species as a whole. A genetically programmed limit on lifespan could benefit the group by eliminating the drain on available resources imposed by members no longer actively involved in the production, development, and training of offspring. Indeed, the current three-generation lifespan can be rationalized as providing time: (a) for newborns to develop into reproductively active young adults, (b) for these young adults to produce and nurture their offspring, and (c) for older adults to serve as a source of guidance and assistance to young adults facing the challenges of childbirth and childrearing.

SUMMARY

■ Aging and longevity are controlled via the complex interplay between random and deterministic factors that include genetic programming, environmental stresses, lifestyle, cellular countdown clocks, and molecular repair processes.

■ Wear and tear theories hypothesize that the aging results from the accumulation of biomolecular damage over time.

- Water, oxygen, and light are essential for life, but possess an intrinsic capacity to damage biologic macromolecules.

- ROS are continually generated as a by-product of aerobic metabolism, particularly by the electron transport chain.

- The deleterious effects of ROS are often amplified by free radical chain reactions.

- The reactivity of their aromatic ring systems and ability to absorb UV light render the nucleotide bases of DNA particularly vulnerable to UV or chemical damage.

- Missing or damaged nucleotide bases provide a potential source of deleterious gene mutations.

- Their critical roles in energy production and apoptosis, along with their endogenous genome, render mitochondria a central player in many theories of aging and death.

- Because the telomere caps at the ends of our chromosomes progressively shorten with each division of a somatic cell, they are hypothesized to serve as a molecular countdown clock.

- Model organisms provide useful vehicles for investigating biologic processes.

- Mutation of the *daf-2* gene in *C. elegans* extends the lifespan of worms by 70%.

- Evolutionary selection of a limited lifespan may optimize the vitality of the population rather than that of its individual members.

REFERENCES

Arias E, Curtin LR, Wei R, et al: U.S. decennial life tables for 1999–2001, United States life tables. Natl Vital Stat Rep 2008;57:1.

Haas RH: Mitochondrial dysfunction in aging and diseases of aging. Biology (Basel) 2019;8:48.

Myakotnykh VS, Berezina DA, Borokova TA, Gavrilov IV: Comparative biochemistry of aging process in men and women. Adv Gerontol 2015;5:72.

Niedernhofer LJ, Gurkar AU, Wang Y, et al: Genomic instability and aging. Ann Rev Biochem 2018;87:295.

Rhoads TW, Anderson RM: Taking the long view on metabolism. Science 2021;373:738.

Speakman JR: Body size, energy metabolism and lifespan. J Exp Biol 2005;208:1717.

Whittemore K, Vera E, Martinez-Nevado E, et al: Telomere shortening rate predicts life span. Proc Natl Acad Sci USA 2019;116:15122.

Zhang S, Li F, Zhou T, et al: *Caenorhabditis elegans* as a useful model for studying aging mutations. Front Endocrinol 2020;11:554994.

Biochemical Case Histories

David A. Bender, PhD

O B J E C T I V E S

After studying this book, you should be able to:

■ Use your knowledge to explain the underlying biochemical defects in diseases.

INTRODUCTION

In this final chapter, nine case histories are presented as open-ended problems for you to solve, based on what you have learnt from studying this book. No solutions are provided, and there is no discussion of the cases; all that you need to know in order to explain the problems is available elsewhere in this book.

In many cases, the patient's clinical chemistry results are presented together with reference ranges. These may differ from problem to problem, because, as discussed in Chapter 48, reference ranges from different laboratories may differ.

CASE 1

The patient is a 5-year-old boy, who was born, at term, after an uneventful pregnancy. He was a sickly infant, and did not grow well. On a number of occasions his mother noted that he appeared drowsy, or even comatose, and said that there was a "chemical, alcohol-like" smell on his breath, and in his urine. The GP suspected diabetes mellitus, and sent him to the Middlesex Hospital in London for a glucose tolerance test. The results are shown in **Figure 58–1**.

Blood samples were also taken for measurement of insulin at zero time and 1 hour after the glucose load. At this time, a new method of measuring insulin was being developed, radio-immunoassay (see Chapter 48), and therefore both this and the conventional biological assay were used. The biological method of measuring insulin is by its ability to stimulate the uptake and metabolism of glucose in rat muscle in vitro; this can be performed relatively simply by measuring the radioactivity in $^{14}CO_2$ after incubating duplicate samples of the muscle with [^{14}C]glucose, with and without the sample containing insulin. The results are shown in **Table 58–1**.

As a part of their studies of the new radioimmunoassay for insulin, the team at the Middlesex Hospital performed gel exclusion chromatography of a pooled sample of normal serum,

and determined insulin in the fractions eluted from the columns both by radioimmunoassay (graph A in **Figure 58–2**) and by stimulation of glucose oxidation (graph B). Three molecular mass markers were used; they eluted as follows: M_r 9000 in fraction 10, M_r 6000 in fraction 23, and M_r 4500 in fraction 27.

The investigators also measured insulin in the fractions eluted from the chromatography column after treatment of each fraction with trypsin. The results are shown in graph C.

After seeing the results of these studies, they subjected the same-pooled serum sample to brief treatment with trypsin, and performed gel exclusion chromatography on the product. Again, they measured insulin by radioimmunoassay (graph D) and biologic assay (graph E).

Since these studies in the 1960s, the gene for human insulin has been cloned. Although insulin consists of two peptide chains, 21- and 30-amino acids long, respectively, these are

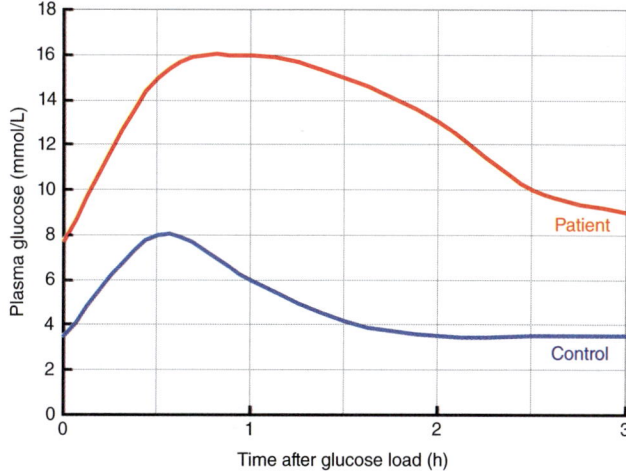

FIGURE 58–1 Plasma glucose in the patient and a control subject after a test dose of glucose.

TABLE 58–1 Serum Insulin (mU/L) Measured by Biological Assay and Radioimmunoassay

	Baseline (Fasting) Blood Sample		1 Hour After Glucose Load	
	Patient (Case 1)	Control Subjects	Patient (Case 1)	Control Subjects
Biological assay	0.8	6 ± 2	5	40 ± 11
Radioimmunoassay	10	6 ± 2	50	40 ± 11

coded for by a single gene, which has a total of 330 base pairs between the initiator and stop codons. As you would expect for a secreted protein, there is a signal sequence coding for 24 amino acids at the 5′ end of the gene.

What does this information suggest about the processes that occur in the synthesis of insulin?

What is likely to be the underlying biochemical basis of the patient's problem?

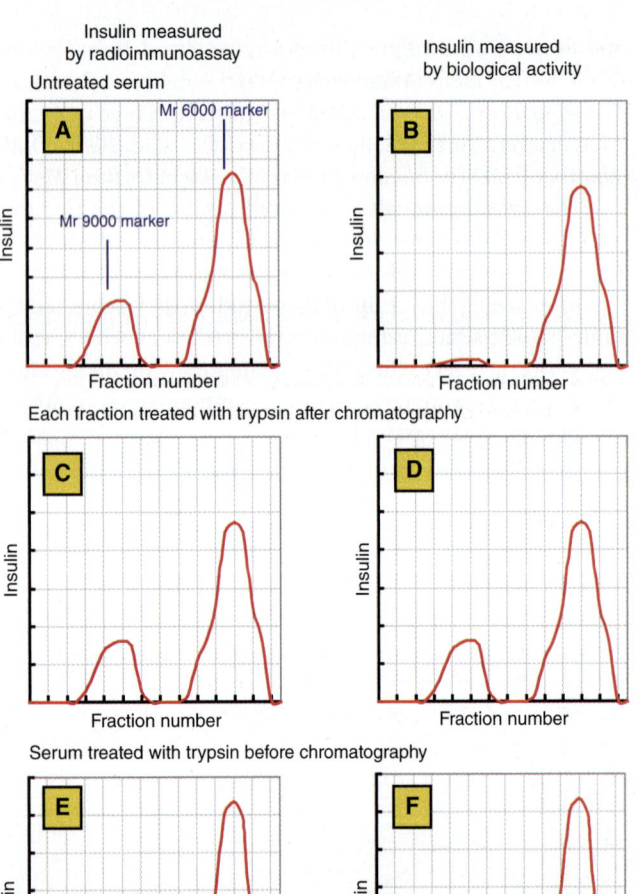

FIGURE 58–2 Insulin measured by radioimmunoassay and biological assay before and after treatment of plasma samples with trypsin.

CASE 2

The patient is a 50-year-old man, 174-cm tall, and weighing 105 kg. He is an engineer, and works on secondment in one of the strict Islamic states in the Gulf, where alcohol is prohibited. At the beginning of August, he returned home for his annual leave. According to his family, he behaved as he usually did when on leave, consuming a great deal of alcohol and refusing meals. He was known to be drinking 2 L of whiskey, two or three bottles of wine, and a dozen or more cans of beer each day; his only solid food consisted of sweets and biscuits.

On September 1st, he was admitted to the emergency department, semiconscious, and with a rapid respiration rate (40/minute). His blood pressure was 90/60 and his pulse rate was 136/minute. His temperature was normal (37.1°C). Emergency blood gas analysis revealed severe acidosis: pH 7.02 and base excess −23; Po_2 91 mm Hg and Pco_2 10 mm Hg. He was transferred to intensive care and given intravenous bicarbonate.

His pulse rate remained high, and his blood pressure low, so emergency cardiac catheterization was performed; this revealed a cardiac output of 23 L/min (normal 4-6). A chest x-ray shows significant cardiac enlargement.

Table 58–2 shows the clinical chemistry results from a plasma sample taken shortly after he was admitted.

What is the likely biochemical basis of the patient's problem, which led to his emergency hospitalization?

What additional test(s) might you request to confirm your assumption?

What emergency treatment would you suggest?

TABLE 58–2 Clinical Chemistry Results for the Patient in Case 2 on Admission. All Values Are mmol/L

	Patient in Case 2	Reference Range
Glucose	10.6	3.5-5
Sodium	142	131-151
Potassium	3.9	3.4-5.2
Chloride	91	100-110
Bicarbonate	5	21-29
Lactate	18.9	0.9-2.7
Pyruvate	2.5	0.1-0.2

TABLE 58–3 The Effect of Incubation With 330 µmol/L Acetylphenylhydrazine on Erythrocyte Glutathione

	Patient in Case 3		Control Subjects	
	GSH (mmol/L)	GSSG (µmol/L)	GSH (mmol/L)	GSSG (µmol/L)
Initial	1.61	400	2.01 ± 0.29	4.2 ± 0.61
+ Acetylphenylhydrazine	0.28	1540	1.82 ± 0.24	190 ± 28

CASE 3

The patient is an African American recruit to the army. He was given the antimalarial drug primaquine, and suffered a delayed reaction with kidney pain, dark urine, and low red blood cell counts that led to anemia and weakness. Centrifugation of a blood sample showed a low hematocrit, and the plasma was red colored.

Similar acute hemolytic attacks have been observed, predominantly in men of Afro-Caribbean origin, in response to primaquine and a variety of other drugs, including dapsone, the antipyretic acetylphenylhydrazine, the antibacterial bactrim/septrin, sulfonamides, and sulfones, whose only common feature is that they all undergo cyclic nonenzymic reactions in the presence of oxygen to produce hydrogen peroxide and a variety of oxygen radicals that can cause oxidative damage to membrane lipids, leading to hemolysis. Moderately severe infection can also precipitate a hemolytic crisis in susceptible people.

One way of screening for sensitivity to primaquine is based on the observation that the glutathione concentration of erythrocytes from sensitive subjects is somewhat lower than that of control subjects, and falls considerably on incubation with acetylphenylhydrazine.

Glutathione (GSH) is a tripeptide, γ-glutamyl-cysteinylglycine, which readily undergoes oxidation to form a disulphide-linked hexapeptide, oxidized glutathione, generally abbreviated to GSSG. **Table 58–3** shows the concentrations of GSH and GSSG in red cells from the patient and 10 control subjects, before and after incubation with acetylphenylhydrazine.

How much GSH is oxidized per mol of acetylphenylhydrazine added?

The reported K_m of glutathione reductase for GSSG is 65 µmol/L and for NADPH is 8.5 µmol/L. Erythrocyte lysates were incubated with a saturating concentration of GSSG (1 mmol/L) and either NADPH or NADH (100 µmol/L). Each incubation contained the hemolysate from 0.5 mL packed cells (**Table 58–4**).

TABLE 58–4 Glutathione Reductase, µmol Product Formed/min

	Patient in Case 3	Control Subjects
NADPH	0.64	0.63 ± 0.06
NADH	0.01	0.01 ± 0.001

Since none of the red cell lysates showed any significant activity with NADH, it is unlikely that there is any transhydrogenase activity in erythrocytes reducing NADP⁺ to NADPH at the expense of NADH. This raises the problem of identifying the source of NADPH in erythrocytes.

The dye methylene blue will oxidize NADPH; the reduced dye then undergoes nonenzymic oxidation in air, so the addition of a relatively small amount of methylene blue will effectively deplete NADPH, and would be expected to stimulate any pathway that reduces NADP⁺ to NADPH.

Erythrocytes from control subjects were incubated with 10 mmol/L [^{14}C]glucose with or without the addition of methylene blue; all six possible positional isomers of [^{14}C]glucose were used, and the radioactivity in (lactate + pyruvate) was determined after thin layer chromatography of the incubation mixture. Each incubation contained 1 mL of erythrocytes in a total incubation volume of 2 mL (**Table 58–5**).

In further studies, only the formation of $^{14}CO_2$ from [^{14}C-1]glucose was measured, with the addition of:

- Sodium ascorbate (which undergoes a nonenzymic reaction in air to produce H_2O_2)
- Acetylphenylhydrazine (which is known to precipitate hemolysis in sensitive subjects, and depletes reduced glutathione)
- Methylene blue (which oxidizes NADPH)

The incubations were repeated with N-ethylmaleimide, which undergoes a nonenzymic reaction with the —SH group of reduced glutathione, and thus depletes the cell of total glutathione. The results are shown in **Table 58–6**.

Further studies showed that the patient's red blood cells contained only about 20% of the normal activity of glucose-6-phosphate dehydrogenase (see Chapter 20). In order to investigate why his enzyme activity was so low, a sample of his red blood cells was incubated at 45°C for 60 minutes, then cooled to 30°C and the activity of glucose-6-phosphate dehydrogenase was determined. After the preincubation at 45°C, his red cells showed only 60% of their initial activity. By contrast, red cells from control subjects retained 90% of their initial activity after preincubation at 45°C for 60 minutes.

What conclusions can you draw from these results?

CASE 4

The patient is a 10-year-old Maltese boy. On his birthday his aunt gave him a pie made from fava beans (a local delicacy), and that evening he suffered kidney pain, and

TABLE 58–5 Production of [^{14}C]lactate, Pyruvate, and CO_2 by 1 mL Erythrocytes From Control Subjects Incubated for 1 Hour With 10 mmol/L [^{14}C]glucose at 10 µCi/mmol

	Control		+ Methylene Blue	
	Lactate + Pyruvate	CO_2	Lactate + Pyruvate	CO_2
[^{14}C-1]glucose	12680 ± 110	1410 ± 15	1830 ± 20	12260 ± 130
[^{14}C-2]glucose	14080 ± 120	ND	14120 ± 120	ND
[^{14}C-3]glucose	14100 ± 120	ND	14090 ± 120	ND
[^{14}C-4]glucose	14060 ± 120	ND	14080 ± 120	ND
[^{14}C-5]glucose	14120 ± 120	ND	14060 ± 120	ND
[^{14}C-6]glucose	14090 ± 110	ND	14100 ± 120	ND

Abbreviation: ND, not detectable, ie, below the limits of detection.
Figures show dpm, mean ± sd for 5 replicate incubations.

passed dark urine. A blood film showed a low red blood cell count and the plasma was red colored. This problem is not uncommon in Malta, and indeed several of his classmates (all boys) have died when an acute crisis has been precipitated by eating fava beans, or after a moderate fever associated with an infection.

Further studies showed that his erythrocyte glucose-6-phosphate dehydrogenase activity was only 10% of normal and had a very high K_m for NADP$^+$. Unlike the patient in Case 3, his red blood cell enzyme was as stable to incubation at 45°C as that from control subjects.

What conclusions can you draw from these observations?

CASE 5

The patient is a 28-week-old baby girl. She was admitted to the emergency department in a coma, having suffered a convulsion after feeding. She had a mild infection and slight fever at the time. Since birth she had been a sickly child, and had frequently vomited and become drowsy after feeding. She was bottle-fed and at one time, cows' milk allergy was suspected, although the problems persisted when she was fed on soya milk.

On admission, she was mildly hypoglycemic, ketotic, and her plasma pH was 7.29. Analysis of a blood sample showed normal levels of insulin, but considerable hyperammonemia (plasma ammonium ion concentration 500 µmol/L; reference range 40-80 µmol/L). She responded well to intravenous glucose infusion and rectal infusion of lactulose, regaining consciousness. She had poor muscle tone.

A liver biopsy sample was taken, and the activities of the enzymes of urea synthesis (see Chapter 28) were determined, and compared with activities in postmortem liver samples from six infants of the same age. The results are shown in **Table 58–7**. She remained well on a high-carbohydrate, low-protein diet for several days, although the poor muscle tone and muscle weakness persisted. A second liver biopsy sample was taken after 4 days and the activity of the enzymes determined again.

TABLE 58–6 Production of $^{14}CO_2$ by 1 mL Erythrocytes From Control Subjects Incubated for 1 Hour With 10 mmol/L [^{14}C-1]glucose at 10 µCi/mmol

Additions	Control	+ N-Ethylmaleimide
None	1410 ± 70	670 ± 30
Ascorbate	8665 ± 300	2133 ± 200
Acetylphenylhydrazine	7740 ± 320	4955 ± 325
Methylene blue	12230 ± 500	11265 ± 450

Figures show dpm, mean ± sd for 5 replicate incubations.

TABLE 58–7 Activity of Enzymes of the Urea Synthesis Cycle in Liver Biopsy Samples From the Patient in Case 5 on Admission and After 4 Days on a High Carbohydrate, Low-Protein Diet, Compared With Activities in Postmortem Samples From 6 Infants of the Same Age

	µmol Product Formed/min/mg Protein		
	Patient		
	On Admission	After 4 Days	Control Subjects
Carbamoyl phosphate synthetase	0.337	1.45	1.30 ± 0.40
Ornithine carbamoyltransferase	29.0	28.6	18.1 ± 4.9
Argininosuccinate synthetase	0.852	0.75	0.49 ± 0.09
Argininosuccinase	1.19	0.95	0.64 ± 0.15
Arginase	183	175	152 ± 56

TABLE 58–8 Metabolism of [^{13}C]propionate

	Patient in Case 5	Mother	Father	Control Subjects
Percent recovered in $^{13}CO_2$ over 3 h	1.01	32.6	33.5	65 ± 5
dpm fixed/ mg fibroblast protein/30 min	5.0	230	265	561 ± 45

Her very low-protein diet was continued, but in order to ensure an adequate supply of essential amino acids for growth, she was fed a mixture of the keto acids of threonine, methionine, leucine, isoleucine, and valine. After each feed she again became abnormally drowsy and markedly ketotic, with significant acidosis. Her plasma ammonium ion concentration was within the normal range, and a glucose tolerance test was normal, with a normal increase in insulin secretion after glucose load.

High-pressure liquid chromatography of her plasma revealed an abnormally high concentration of propionic acid (24 μmol/L; reference range 0.7-3 μmol /L). Urine analysis showed considerable excretion of methylcitrate (1.1 μmol/mg creatinine), which is not normally detectable. She was also excreting a significant amount of short-chain acyl carnitine (mainly propionyl carnitine)—28.6 μmol/ 24 hours, compared with a reference range of 5.7 + 3.5 μmol/ 24 hours.

The metabolism of a test dose of [^{13}C]propionate given by intravenous infusion was determined in the patient, her parents, and a group of control subjects; skin fibroblasts were cultured and the activity of propionyl-CoA carboxylase was determined by incubation with propionate and $NaH^{14}CO_3$, followed by acidification and measurement of the radioactivity in products. The results are shown in **Table 58–8**.

The results of measuring carnitine in the first liver biopsy sample and in a muscle biopsy sample gave the results shown in **Table 58–9**.

What conclusions can you draw from these results?

Can you explain the biochemical basis of the patient's condition?

CASE 6

The patient is a 9-month-old girl, the second child of unrelated parents. She was born at term after an uneventful pregnancy, weighing 3.4 kg and was breast-fed, with gradual introduction of solids from 3 months of age onward. Her mother reported that, while she liked cheese, meat, and fish, she frequently became irritable and grizzly after meals, and became lethargic, drowsy, and "floppy" after eating relatively large amounts of protein-rich foods. Her urine had a curious odor, described by her mother as being "cat-like," on such occasions.

At 9 months of age she was admitted to the emergency department in a coma, and suffering convulsions. She had been unwell for the last 3 days, with a slight fever, and for the last 12 hours had been refusing all food and drink. At this time, she weighed 8.8 kg, and her body length was 70.5 cm.

Emergency blood tests revealed moderate acidosis (pH 7.25) and severe hypoglycemia (glucose < 1 mmol/L); a dipstick test for plasma ketone bodies was negative. A blood sample was taken for full clinical chemistry tests, and she was given intravenous glucose. Within a short time, she recovered consciousness. The results of the blood tests are shown in **Table 58–10**.

She remained in hospital for several weeks, while further tests were performed. She was generally well through this time, but became drowsy and severely hypoglycemic, and hyperventilated, if she was deprived of food for more than about 8 to 9 hours. Her muscle tone was poor, and she was very weak, with considerably less strength (eg, in pushing her arms or legs against the pediatrician's hand) than would be expected for a girl of her age.

On one occasion her blood glucose was monitored at 30-minute intervals over 3 hours from waking, without being fed. It fell from 3.4 mmol/L on waking to 1.3 mmol/L 3 hours later. She was deprived of breakfast again the next day, and again blood glucose was measured at 30-minute intervals for 3 hours during which she received an intravenous infusion of β-hydroxybutyrate (50 μmol/min/kg body weight). During the infusion of β-hydroxybutyrate, her plasma glucose remained between 3.3 and 3.5 mmol/L.

At no time were ketone bodies detected in her urine, and there was no evidence of any abnormal excretion of amino acids. However, a number of abnormal organic acids were detected in her urine, including relatively large amounts of

TABLE 58–9 Liver and Muscle Carnitine

	Liver		Muscle	
μmol/g Weight Tissue	Patient in Case 5	Control Subjects	Patient in Case 5	Control Subjects
Total carnitine	0.23	0.83 ± 0.26	1.56	2.29 ± 0.75
Free carnitine	0.05	0.41 ± 0.17	0.29	1.62 ± 0.67
Short-chain acylcarnitine	0.16	0.37 ± 0.20	1.16	0.58 ± 0.32
Long-chain acylcarnitine	0.01	0.05 ± 0.02	0.11	0.09 ± 0.03

TABLE 58–10 Clinical Chemistry Results for a Plasma Sample From the Patient in Case 6 on Admission & Reference Range for 24-Hour Fasting

	Patient in Case 6	Reference Range
Glucose, mmol/L	0.22	4-5
pH	7-25	7-35-7-45
Bicarbonate, mmol/L	11	21-29
Ammonium, µmol/L	120	<50
Ketone bodies, mmol/L	undetectable	2.5-3.5
Nonesterified fatty acids, mmol/L	2	1.0-1.2
Insulin, mU/L	5	5-35
Glucagon, ng/mL	140	130-160

3-hydroxy-3-methylglutaric and 3-hydroxy-3-methylglutaconic acids. The excretion of these two acids increased considerably under two conditions:

1. When she was fed a relatively high-protein meal (she became grizzly, lethargic, and irritable). A blood sample taken after such a meal showed significant hyperammonemia (130 µmol/L), but normal glucose (5.5 mmol/L).

2. When she was fasted for more than the normal overnight fast, with or without the infusion of β-hydroxybutyrate.

One obvious metabolic precursor of 3-hydroxy-3-methylglutaric acid is 3-hydroxy-3-methylglutaryl–CoA (HMG-CoA), which is normally cleaved to yield acetoacetate and acetyl-CoA by the enzyme hydroxymethyglutaryl-CoA lyase (see Chapter 22). Therefore, the activity of this enzyme was determined in leukocytes from blood samples from the patient and her parents. The results are shown in **Table 58–11**.

Analysis of her urine also revealed considerable excretion of carnitine, as shown in **Table 58–12**.

What is the likely biochemical basis of the patient's problem? To what extent can you account for her various metabolic problems from the information you are given?

What dietary manipulation(s) would be likely to maintain her in good health, and prevent further emergency hospital admissions?

TABLE 58–11 Leukocyte HMG-CoA Lyase Activity, nmol Product Formed/min/g Protein

Patient in Case 6	1.7
Mother	10.2
Father	11.4
Control subjects	19.7 ± 2.0

TABLE 58–12 Urinary Excretion of Carnitine, nmol/mg Creatinine

	Patient in Case 6	Reference Range
Total carnitine	680	125 ± 75
Free carnitine	31	51 ± 40
Acyl carnitine	649	74 ± 40

CASE 7

The patient is a 9-month-old boy, the second child of unrelated parents; his brother is 5 years old, fit, and healthy. He was born at full term after an uneventful pregnancy, weighing 3.4 kg (the 50th centile), and developed normally until he was 6 months old, after when he showed some retardation of development. He also developed a fine scaly skin rash about this time, and his hair, which had been normal, became thin and sparse.

At 9 months of age, he was admitted to the emergency department in a coma; the results of clinical chemistry tests on a plasma sample are shown in **Table 58–13**.

The acidosis was treated by intravenous infusion of bicarbonate, and he recovered consciousness. Over the next few days he continued to show signs of acidosis (rapid respiration), and even after a meal ketone bodies were present in his urine. His plasma lactate, pyruvate, and ketone bodies remained high; plasma glucose was in the low normal range, and his plasma insulin was normal both in the fasting state and after an oral glucose load.

Urine analysis revealed the presence of significant amounts of a number of organic acids that are not normally excreted in the urine, including:

- Lactate, pyruvate, and alanine
- Propionate, hydroxypropionate, and propionyl glycine
- Methylcitrate
- Tiglate and tiglylglycine
- 3-Methyl crotonate, 3-methylcrotonylglycine, and 3-hydroxyisovalerate

TABLE 58–13 Clinical Chemistry Results for a Plasma Sample From the Patient in Case 7 on Admission & Reference Range for 24-Hour Fasting

	Patient in Case 7	Reference Range
Glucose, mmol/L	3.3	3.5-5.5
pH	6.9	7-35-7-45
Bicarbonate, mmol/L	2.0	21-25
Ketone bodies, mmol/L	21	1-2.5
Lactate, mmol/L	7-3	0.5-2.2
Pyruvate, mmol/L	0.31	<0.15

His skin rash and hair loss were reminiscent of the signs of biotin deficiency (see Chapter 44), as caused by excessive consumption of uncooked egg white. However, his mother said that he did not eat raw or undercooked eggs at all, although he was fond of hard-boiled eggs and yeast extract (which are rich sources of biotin). His plasma biotin was 0.2 nmol/L (normal > 0.8 nmol/L), and he excreted a significant amount of biotin in the form of biocytin (see Figure 44–14) and small biocytin-containing peptides, which are not normally detectable in urine.

He was treated with 5 mg of biotin per day. After 3 days the abnormal organic acids were no longer detectable in his urine, and his plasma lactate, pyruvate, and ketone bodies had returned to normal, although his excretion of biocytin and biocytin-containing peptides increased. At this stage he was discharged from hospital, with a supply of biotin tablets. After 3 weeks his skin rash began to clear, and his hair loss ceased.

Three months later, at a regular out-patient visit, it was decided to cease the biotin supplements. Within a week, the abnormal organic acids were again detected in his urine, and he was treated with varying doses of biotin until the organic aciduria ceased. This was achieved at an intake of 150 µg/d (compared with the reference intake of 10-20 µg/d for an infant under 2 years old).

He has continued to take 150 µg of biotin daily, and has remained in good health for the last 4 years.

Can you account for the biochemical basis of the patient's problem?

CASE 8

The patient is a 4-year-old girl, the only child of nonconsanguineous parents, born at term after an uneventful pregnancy. At 14 months of age she was admitted to hospital with a 1-day history of persistent vomiting, rapid shallow respiration, and dehydration. On admission, her respiration rate was 60/minute and her pulse 178/minute. The first column in **Table 58–14** shows the results of clinical chemistry tests at that time. She responded rapidly to intravenous bicarbonate and a single intramuscular injection of insulin.

The results of a glucose tolerance test 3 days after admission were normal, and her plasma insulin response to an oral glucose load was within the normal range. She was discharged from hospital 7 days after admission, apparently fit and well. The second column in Table 58–14 shows the results of clinical chemistry tests taken shortly before her discharge.

She was readmitted to hospital at 16, 25, 31, and 48 months of age, suffering from restlessness, unsteady gait, rapid shallow respiration, persistent vomiting, and dehydration. Each time the crisis was preceded by a common childhood illness and decreased appetite, and she responded well to intravenous fluids and bicarbonate. A number of milder episodes were treated at home by oral fluid and bicarbonate.

During her admission at age 25 months, a skin biopsy was taken, fibroblasts were cultured, and the mitochondrial enzyme activities shown in **Table 58–15** were determined.

TABLE 58–14 Clinical Chemistry Results for Plasma & Urine Samples From the Patient in Case 8 on Admission and Again 1 Week Later

	Acute Admission	1 Week Later	Reference Range
Plasma			
Glucose, mmol/L	14	5.1	3.5-5.5
Sodium, mmol/L	132	137	135-145
Chloride, mmol/L	111	105	100-106
Bicarbonate, mmol/L	1.5	20	21-25
Urea, mmol/L	4.1	4.9	2.9-8.9
Lactate, mmol/L	7-3	5.5	0.5-2.2
Pyruvate, mmol/L	0.31	0.25	<0.15
Alanine, mmol/L	–	852	99-313
Aspartate, mmol/L	–	Undetectable	3-11
pH	6.89	7-36	7-35–7-45
Urine			
Lactate, mg/g creatinine	–	1.48	<0.1
Ketone bodies, using Ketostix	Very high	Negative	Negative

Can you explain the biochemical basis of the patient's condition?

CASE 9

The patient is a 5-year-old boy who is diabetic. There is a family history of diabetes, which strongly suggests a dominant pattern of inheritance. He secretes a significant amount of insulin, although less than normal subjects,

TABLE 58–15 Activities of Mitochondrial Enzymes From Cultured Skin Fibroblasts (nmol product formed/min/mg protein)

	Patient in Case 8	Control Subjects
Citrate synthase	32.8	76.3 ± 15.1
Cytochrome c reduction by NADH	11.6	16.7 ± 4.6
Cytochrome c reduction by succinate	9.43	12.3 ± 3.2
Cytochrome oxidase	37-7	50.3 ± 11.6
NADH dehydrogenase	633	910 ± 169
Pyruvate carboxylase	0.03	1.62 ± 0.39
Pyruvate dehydrogenase	1.86	1.72 ± 0.35
Succinate oxidase	190	210 ± 30

TABLE 58–16 Secretion of Insulin (µg/minute/incubation) by Rabbit Pancreas In Vitro

	Control	+ Mannoheptulose
3.3 mmol/L glucose	3.5	3.5
16.6 mmol/L glucose	12.5	3.5

Data from Coore HG, Randle PJ. Regulation of insulin secretion studied with pieces of rabbit pancreas incubated in vitro. Biochem J. 1964;93(1):66–78.

suggesting that the problem is not type 1 diabetes. Unlike type 2 diabetes, this condition develops in early childhood, and is generally referred to as maturity-onset diabetes of the young (MODY).

The results of studies of the secretion of insulin by rabbit pancreas incubated in vitro with two concentrations of glucose, with and without the addition of the 7-carbon sugar mannoheptulose, which is an inhibitor of the phosphorylation of glucose to glucose-6-phosphate are shown in **Table 58–16**.

Two enzymes catalyze the formation of glucose-6-phosphate from glucose (see Chapter 17):

- Hexokinase is expressed in all tissues; it has a K_m for glucose of ~0.15 mmol/L.
- Glucokinase is expressed only in liver and the β cells of the pancreas; it has a K_m for glucose of ~20 mmol/L.

The normal range of plasma glucose is between 3.5 and 5 mmol/L, rising in peripheral blood to 8 to 10 mmol/L after a moderately high intake of glucose. After a meal, the concentration of glucose in the portal blood, coming from the small intestine to the liver, may be considerably higher than this.

What effect will changes in the plasma concentration of glucose have on the rate of formation of glucose-6-phosphate catalyzed by hexokinase?

What effect will changes in the plasma concentration of glucose have on the rate of formation of glucose-6-phosphate catalyzed by glucokinase?

What is the importance of glucokinase in the liver?

Froguel and coworkers (1993) reported studies of the glucokinase gene in a number of families affected by MODY, and also in unaffected families. They published a list of 16 variants of the glucokinase gene, shown in **Table 58–17**. All their patients with MODY had an abnormality of the gene.

What are the amino acid changes associated with each mutation in the gene?

TABLE 58–17 Mutations in the Glucokinase Gene

Codon	Nucleotide Change	Amino Acid Change	Effect
4	GAC ⇒ AAC	?	None
10	GCC ⇒ GCT	?	None
70	GAA ⇒ AAA	?	MODY
98	CAG ⇒ TAG	?	MODY
116	ACC ⇒ ACT	?	None
175	GGA ⇒ AGA	?	MODY
182	GTG ⇒ ATG	?	MODY
186	CGA ⇒ TGA	?	MODY
203	GTG ⇒ GCG	?	MODY
228	ACG ⇒ ATG	?	MODY
261	GGG ⇒ AGG	?	MODY
279	GAG ⇒ TAG	?	MODY
300	GAG ⇒ AAG	?	MODY
300	GAG ⇒ CAG	?	MODY
309	CTC ⇒ CCC	?	MODY
414	AAG ⇒ GAG	?	MODY

Data from Froguel P, Zouali H, Vionnet N, et al. Familial hyperglycemia due to mutations in glucokinase. Definition of a subtype of diabetes mellitus, N Engl J Med. 1993;328(10):697–702.

Why do the mutations affecting codons 4, 10, and 116 have no effect on the people involved?

What conclusions can you draw from this information?

The same authors also studied the secretion of insulin in response to glucose infusion in patients with MODY and control subjects. They were given an intravenous infusion of glucose; the rate of infusion was varied so as to maintain a constant plasma concentration of glucose of 10 mmol/L. Their plasma concentrations of glucose and insulin were measured before and after 60 minutes of glucose infusion; the results are shown in **Table 58–18**.

What conclusions can you draw from this information about the probable role of glucokinase in the β cells of the pancreas?

Can you deduce the way in which the β cells of the pancreas sense an increase in plasma glucose and signal the secretion of insulin?

TABLE 58–18 Plasma Concentrations of Glucose and Insulin Before and After 60 Minutes of Glucose Infusion

	Plasma Glucose (mmol/L)		Insulin (mU/L)	
	Patients With MODY	Control Subjects	Patients With MODY	Control Subjects
Fasting	7-0 ± 0.4	5.1 ± 0.3	5 ± 2	6 ± 2
60-min glucose infusion	Maintained at 10 mmol/L by varying rate of infusion		12 ± 7	40 ± 11

Data from Froguel P, Zouali H, Vionnet N, et al. Familial hyperglycemia due to mutations in glucokinase. Definition of a subtype of diabetes mellitus, N Engl J Med. 1993; 328(10):697–702.

Exam Questions

Section XI – Special Topics (C)

1. Which one of the following statements regarding the blood coagulation pathways is NOT CORRECT?

   A. The components of the extrinsic Xase (tenase) complex are factor VIIa, tissue factor, Ca^{2+}, and factor X.

   B. The components of the intrinsic Xase (tenase) complex are factors IXa and VIIIa, Ca^{2+}, and factor X.

   C. The components of the prothrombinase complex are factors Xa and Va, Ca^{2+}, and factor II (prothrombin).

   D. The extrinsic and intrinsic Xase complexes and prothrombinase complex require anionic procoagulant phosphatidylserine on low-density lipoprotein (LDL) for their assembly.

   E. Fibrin formed by cleavage of fibrinogen by thrombin is covalently cross-linked by the action of factor XIIIa, which itself is formed by the action of thrombin on factor XIII.

2. On which one of the following coagulation factors does a patient taking warfarin for his thrombotic disorder have decreased Gla (γ-carboxyglutamate) residues?

   A. Tissue factor

   B. Factor XI

   C. Factor V

   D. Factor II (prothrombin)

   E. Fibrinogen

3. A 65-year-old man suffers a myocardial infarction and is given tissue plasminogen activator within 6 hours of onset of the thrombosis to achieve which one of the following?

   A. Prevent activation of the extrinsic pathway of coagulation

   B. Inhibit thrombin

   C. Enhance degradation of factors VIIIa and Va

   D. Enhance fibrinolysis

   E. Inhibit platelet aggregation

4. Which one of the following statements regarding platelet activation in hemostasis and thrombosis is NOT CORRECT?

   A. Platelets adhere directly to subendothelial collagen via GPIa-IIa and GPVI, while binding of GPIb-IX-V is mediated via von Willebrand factor.

   B. The aggregating agent thromboxane A_2 is formed from arachidonic acid liberated from platelet membrane phospholipids by the action of phospholipase A_2.

   C. The aggregating agent ADP is released from the dense granules of activated platelets.

   D. The aggregating agent thrombin activates intracellular phospholipase Cβ, which forms the internal effector molecules 1,2-diacylglycerol and 1,4,5-inositol trisphosphate from the membrane phospholipid phosphatidylinositol 4,5-bisphosphate.

   E. The ADP receptors, the thromboxane A_2 receptor, the thrombin PAR-1 and PAR-4 receptors, and the fibrinogen GPIIb-IIIa receptor are all examples of G-protein–coupled receptors.

5. A 15-year-old adolescent girl presented at clinic with bruises on her lower extremities. Of the following, which is *least likely* to explain the bleeding signs exhibited by this individual?

   A. Hemophilia A

   B. von Willebrand disease

   C. A low platelet count

   D. Aspirin ingestion

   E. A platelet disorder with absence of storage granules

6. Regarding chemical carcinogenesis, select the one FALSE statement:

   A. Approximately 80% of human cancers may be due to environmental factors.

   B. In general, chemical carcinogens interact noncovalently with DNA.

   C. Some chemicals are converted to carcinogens by enzymes, usually cytochrome P450 species.

   D. Most ultimate carcinogens are electrophiles and attack nucleophilic groups in DNA.

   E. The Ames assay is a useful test for screening chemicals for mutagenicity; however, animal testing is required to show that a chemical is carcinogenic.

7. Regarding viral carcinogenesis, select the one FALSE statement:

   A. Approximately 15% of human cancers may be caused by viruses.

   B. Only RNA viruses are known to be carcinogens.

   C. RNA viruses causing or associated with tumors include hepatitis C virus.

   D. Retroviruses possess reverse transcriptase, which copies RNA to DNA.

   E. Tumor viruses act by deregulating the cell cycle, inhibiting apoptosis, and interfering with normal cell signaling processes.

8. Regarding oncogenes and tumor suppressor genes, select the one FALSE statement:

   A. Both copies of a tumor suppressor gene must be mutated for its product to lose its activity.

   B. Mutation of an oncogene occurs in somatic cells and is not inherited.

   C. The product of an oncogene shows a gain of function that signals cell division.

   D. *RB* and *P53* are tumor suppressor genes; *MYC* and *RAS* are oncogenes.

   E. Mutation of one tumor suppressor gene or one oncogene is thought to be sufficient to cause cancer.

9. Regarding growth factors, select the one FALSE statement:
   A. They include a large number of polypeptides, most of which stimulate cell growth.
   B. Growth factors can act in an endocrine, paracrine, or autocrine manner.
   C. Certain growth factors, such as TGF-β, can act in a growth inhibitory manner.
   D. Some receptors for growth factors have tyrosine kinase activity; mutations of these receptors occur in cancer cells.
   E. PDGF stimulates phospholipase A_2, which hydrolyzes PIP_2 to form DAG and IP_3, both of which are second messengers.

10. Regarding the cell cycle, select the one FALSE statement:
    A. Cells transiting the cell cycle can reside within any of the five phases of the cell cycle (ie, G_1, G_0, S, G_2, and M).
    B. Cancer cells usually have a shorter generation time than normal cells and there are fewer of them in G_0 phase.
    C. A variety of mutations in cyclins and CDKs have been reported in cancer cells.
    D. RB is a cell cycle regulator, where it binds to transcription factor E2F, thus allowing progression of the cell from G_1 to S phase.
    E. When damage to DNA occurs, p53 increases in amount and activates transcription of genes that delay transit through the cycle.

11. Regarding chromosomes and genomic instability, select the one FALSE statement:
    A. Cancer cells may have a mutator phenotype, which means that they have mutations in genes that affect DNA replication and repair, chromosomal segregation, DNA damage surveillance, and apoptosis.
    B. Chromosomal instability refers to gain or loss of chromosomes caused by abnormalities of chromosomal segregation during mitosis.
    C. Microsatellite instability involves expansion or contraction of microsatellites due to abnormalities of nucleotide excision repair.
    D. Aneuploidy (when the chromosomal number of a cell is not a multiple of the haploid number) is a common feature of tumor cells.
    E. Abnormalities of chromosome cohesion and of kinetochore-microtubule attachment may contribute to chromosomal instability and aneuploidy.

12. Select the one FALSE statement:
    A. The activity of telomerase is frequently elevated in cancer cells.
    B. A number of cancers have a strong hereditary predisposition or susceptibility; these include Li-Fraumeni syndrome and retinoblastoma.
    C. The products of *BRCA1* and *BRCA2* (responsible for hereditary breast cancer types I and II) appear to be involved in DNA repair.
    D. Tumor cells usually exhibit a high rate of anaerobic glycolysis; this may be at least partly explained by the presence in many tumor cells of the PK-2 isozyme, which is associated with lesser production of ATP and possibly increased use of metabolites to build up biomass.
    E. Dichloroacetate, a compound found to display some anticancer activity, inhibits pyruvate carboxylase, and thus diverts pyruvate away from glycolysis.

13. Select the one FALSE statement:
    A. Whole-genome and exome sequencing is revealing important new information about the numbers and types of mutations in cancer cells.
    B. Abnormalities of epigenetic mechanisms, such as demethylation of cytosine residues, abnormal modification of histones, and aberrant chromatin remodeling are being increasingly detected in cancer cells.
    C. Persistence of cancer stem cells (which are often relatively dormant and have active DNA repair systems) may help to explain some of the shortcomings of chemotherapy.
    D. Angiogenin is an inhibitor of angiogenesis.
    E. Chronic inflammation, possibly via increased production of reactive oxygen species, predisposes to development of certain types of cancer.

14. Regarding apoptosis, select the one FALSE statement:
    A. Apoptosis can be initiated by the interaction of certain ligands with specific receptors on cell surface.
    B. Cell stress and other factors activate the mitochondrial pathway of apoptosis; release of cytochrome P450 into the cytoplasm is an important event in this pathway.
    C. A distinct pattern of fragments of DNA is found in apoptotic cells; it is caused by caspase-activated DNase.
    D. Caspase 3 digests cell proteins such as lamin, certain cytoskeletal proteins, and various enzymes, leading to cell death.
    E. Cancer cells have acquired various mutations that allow them to evade apoptosis, prolonging their existence.

15. Select the one FALSE statement:
    A. Proteins involved in cell adhesion include cadherins, integrins, and selectins.
    B. Decreased amounts of E-cadherin on the surfaces of cancer cells may help account for the decreased adhesiveness shown by tumor cells.
    C. Increased activity of GlcNAc transferase V in cancer cells may lead to an altered glycan lattice at the cell surface, perhaps predisposing to their spread.
    D. Cancer cells secrete metalloproteinases that degrade proteins in the ECM and facilitate their spread.
    E. All tumor cells have the genetic capacity to colonize.

16. The number of enzymes dedicated to repairing hydrolytic, oxidative, and photochemical damage to polynucleotides such as DNA is much greater than the number devoted to repairing damaged proteins. Identify the statement from the list below that is INCONSISTENT with this observation:
    A. Polynucleotides absorb ultraviolet light more efficiently than do proteins.
    B. Proteins contain sulfur, an element that is susceptible to oxidation.
    C. In general, proteins turn over more frequently than does DNA.
    D. Mutations in a structural gene have the potential to alter the proteins they encode as well as the DNA itself.
    E. If left uncorrected, genome mutations will be passed on to succeeding generations.

17. Which of the following is NOT a feature of the mitochondrial hypothesis of aging?

   A. Reactive oxygen species are generated as a by-product by the electron transport chain.
   B. Mitochondria lack the capacity to repair damaged DNA.
   C. Many of the complexes in the electron transport chain are constructed from a mixture nuclearly encoded and mitochondrially encoded subunits.
   D. Damaged mitochondria form protease-resistant aggregates.
   E. Damaged mitochondria can trigger apoptosis—programmed cell death.

18. Which of the following is NOT a component of the cell's suite of damage repair and prevention agents?

   A. Superoxide dismutase
   B. Glutathione
   C. Isoaspartyl methyltransferase
   D. Catalase
   E. Caspase 7

19. Select the one of the following statements that describes an aspect of the metabolic theory of aging:

   A. Elevated levels of plasma glucose promote the formation of cross-linked protein aggregates.
   B. Damage from ROS is multiplied by the tendency of oxygen radicals to multiply via chain reactions.
   C. Calorically restricted diets promote lower and more efficient metabolic activity.
   D. Blood flow to the heart muscle becomes restricted over time due to the cholesterol-induced formation of arterial plaques.
   E. Vigorous physical activity correlates with the loss of STEM cells.

20. Select the one of the following statements that is NOT CORRECT:

   A. Telomeres prevent genetic recombination by capping the ends of linear DNA molecules.
   B. Aging genes can be distinguished by their impact on an organism's lifespan.
   C. The short lifespan of *Caenorhabditis elegans* renders them an attractive model organism for studying aging.
   D. Telomere shortening is a consequence of the discontinuous nature of the process by which the "lagging strand" is synthesized during chromosome replication.
   E. Telomerase activity is high in both STEM cells and in many cancer cells.

The Answer Bank

Section I – Proteins: Structure & Function

1. **B.**

2. **D.**

3. That fermentation required intact cells was disproved by the discovery that a cell-free yeast extract could convert sugar to ethanol and carbon dioxide. This discovery led to the identification of the intermediates, enzymes, and cofactors of fermentation and glycolysis.

4. Fermentation ceased over time, but resumed when inorganic orthophosphate was added. This led to the isolation of phosphorylated intermediates. Other experiments using heated yeast extract led to the discovery of ATP, ADP, and NAD.

5. Preparations used to identify metabolites and enzymes included perfused liver, liver slices, and tissue homogenates fractionated by centrifugation.

6. Radioactive^{14}C, ^{3}H, and ^{32}P facilitated the isolation of intermediates of carbohydrate, lipid, nucleotide, and amino acid metabolism and enabled precursor product relationships between intermediates to be tracked.

7. Garrod proposal that alkaptonuria, albinism, cystinuria, and pentosuria resulted from "inborn errors of metabolism" led to the field of biochemical genetics.

8. Regulation of cholesterol biosynthesis illustrates the link between biochemistry and genetics. Cell surface receptors internalize plasma cholesterol, which then regulates cholesterol biosynthesis. Defective receptors result in extreme hypercholesterolemia.

9. Key model organisms include yeast, slime mold, fruit fly, and a small round worm, each with a short generation time and readily mutated.

10. **D.** Hydrocarbons are water insoluble.

11. **A.** Of the protein amino acids, only phenylalanine, tyrosine, and tryptophan absorb light at 280 nm.

12. **D.** When present in solution at a pH equal to their pK_a only half of the molecules of a monofunctional weak acid (eg, ammonium ion or acetic acid) are in the charged state. Maximal mobility will occur either at a pH 3 or more pH units below the pK_a for ammonium ion, or at a pH 3 or more pH units above the pK_a for acetic acid.

13. **C.** At its pI an amino acid has an equal number of positive and negative charges, but has no *net* overall charge.

14. **C.** The Edman technique involves successive derivatization and removal of N-terminal residues.

15. Self-association in an aqueous environment as a large droplet minimizes the surface area in contact with water, and hence the number of water molecules whose degrees of rotational freedom are restricted.

16. Strong bases and acids dissociate essentially completely in water, NaOH as Na^+ and OH^-. By contrast, a weak acid such as pyruvic acid dissociates only partially in solution.

17. **E.** Tandem mass spectrometry can separate complex mixtures of peptides.

18. **E.** Many proteins undergo posttranslational processing, for example, insulin, which is synthesized as a single polypeptide which subsequent proteolysis converts to two polypeptide chains linked by disulfide bonds.

19. pI is the pH at which a molecule bears no *net* charge. In this example, the pI is a pH midway between the third and fourth pK_a values: pI = (6.3 + 7.7)/2 = 7.0. As pH is adjusted from acidic to basic, net charge will change successively as follows: +3, +2, +1, 0, −1, −2, −3.

20. All of the protein amino acids are *essential* since all are required for protein synthesis, but "nutritionally essential" amino acids (10 for humans) are those which an organism cannot synthesize. Many vitamins are "dietarily essential," although vitamin C is *dietarily essential* only for humans, catfish, and certain other organisms.

21. **D.** Gene arrays, also termed DNA chips or DNA arrays, contain multiple DNA probes with differing sequences bound at known locations on a solid support. Hybridization of complementary DNA or RNA probes at particular locations provides information about their nucleic acid composition.

22. **D.** A hydrogen bond interaction involves the residue in fourth place along the helix.

23. **E.** Prions contain no nucleic acid, just protein. Prion diseases therefore are transmitted by protein without involvement of DNA or RNA.

24. Unlike pK_2 (6.82) of phosphoric acid, the other two dissociating groups of phosphoric acid cannot serve as effective buffers at physiologic pH because they are either completely dissociated or predominantly protonated at pH 7.

25. A: Carboxyl groups (pK_1 through pK_3) and amino groups (pK_a through pK_7)
    B: Minus one
    C: Plus 0.5
    D: Toward the cathode

26. To act as an effective buffer, a compound should have a pK_a no less than 0.5 pH units removed from the desired pH, and be present in sufficient quantity.

27. Carboxylation of a glutamyl residue forms γ-carboxyglutamate, a potent chelator of Ca^{++} required for blood clotting and clot dissolution. 4-Hydroxyproline and 5-hydroxylysine are present in several structural proteins.

28. (a) Copper is an essential prosthetic group for the amine oxidase that converts lysine to the hydroxylysine that participates in formation of covalent crosslinks that strengthen collagen.

29. (b) Ascorbic acid is essential for proline hydroxylase to convert proline to the hydroxyproline, which provides interchain hydrogen bonds that stabilize the collagen triple helix.

30. Signal sequences target proteins to specific subcellular locations in the cell, or for secretion.

Section II – Enzymes: Kinetics, Mechanism, Regulation, & Role of Transition Metals

1. Carbonic anhydrase catalyzes the hydration of carbon dioxide to form carbonic acid. A portion of this weak acid, in turn, dissociates to produce bicarbonate and a proton. As the concentration of carbon dioxide falls, carbonic acid is broken down to form carbon dioxide and water. To compensate for the loss of

carbonic acid, bicarbonate and protons recombine to restore equilibrium, leading to a net drop in [H$^+$] and a rise in pH.

2. **D.**	7. **B.**	12. **B.**	17. **B.**
3. **E.**	8. **C.**	13. **B.**	18. **D.**
4. **B.**	9. **A.**	14. **C.**	19. **A**
5. **A.**	10. **D.**	15. **D.**	
6. **E.**	11. **E.**	16. **A.**	

Section III – Bioenergetics

1. **A.** A reaction with a negative ΔG is exergonic; it proceeds spontaneously and free energy is released.
2. **E.** In an exergonic reaction ΔG is negative and in an endergonic reaction it is positive. When ΔG is zero, the reaction is at equilibrium.
3. **B.** When the reactants are present in concentrations of 1.0 mol/L, ΔG^0 is the standard free-energy change. For biochemical reaction, the pH (7.0) is also defined and this is ΔG$^{0'}$.
4. **D.** ATP contains two high-energy phosphate bonds and is needed to drive endergonic reactions. It is not stored in the body and in the presence of uncouplers its synthesis is blocked.
5. **A.** Reduced cytochrome c is oxidized by cytochrome c oxidase (complex IV of the respiratory chain), with the concomitant reduction of molecular oxygen to two molecules of water.
6. **E.** Cytochrome oxidase is not a dehydrogenase, although all other cytochromes are classed as such.
7. **B.** Although Cytochromes p450 are located mainly in the endoplasmic reticulum, they are found in mitochondria in some tissues.
8. **D.** Oxidation of one molecule of NADH via the respiratory chain generates 2.5 molecules of ATP in total. One is formed via complex I, 1 via complex II and 0.5 via complex IV.
9. **C.** 1.5 molecules of ATP are formed in total as FADH$_2$ is oxidized, 1 via complex II and 0.5 via complex IV.
10. **E.** Oligomycin blocks oxidation and ATP synthesis as it prevents the flow of electrons back into the mitochondrial matrix through ATP synthase.
11. **A.** Uncouplers allow electrons to reenter the mitochondrial matrix without passing through ATP synthase.
12. **E.** In the presence of an uncoupler, the energy released as electrons flow into the mitochondrial matrix is not captured as ATP and is dissipated as heat.
13. **C.** Thermogenin is a physiological uncoupler found in brown adipose tissue. Its function is to generate body heat.
14. **D.** Three ATP molecules are generated for each revolution of the ATP synthase molecule.
15. **B.** The electrochemical potential difference across the inner mitochondrial membrane caused by electron transport must be negative on the matrix side so that protons are forced to reenter via the ATP synthase to discharge the gradient.

Section IV – Metabolism of Carbohydrates

1. **B.**	8. **C.**	15. **D.**	22. **A.**
2. **B.**	9. **A.**	16. **E.**	23. **B.**
3. **A.**	10. **E.**	17. **E.**	24. **C.**
4. **D.**	11. **C.**	18. **C.**	25. **D.**
5. **C.**	12. **D.**	19. **C.**	26. **E.**
6. **C.**	13. **D.**	20. **C.**	27. **A.**
7. **E.**	14. **D.**	21. **D.**	28. **B.**

Section V – Metabolism of Lipids

1. **D.**
2. **D.**
3. **A.** Gangliosides are derived from glucosylceramide.
4. **C.** A, B, D, and E are classed as preventive antioxidants as they act by reducing the rate of chain initiation.
5. **D.**
6. **B.**
7. **D.** Long-chain fatty acids are activated by coupling to CoA, but fatty acyl CoA cannot cross the inner mitochondrial membrane. After transfer of the acyl group from CoA to carnitine by carnitine palmitoyl transferase (CPT)-I, acylcarnitine is carried across by carnitine-acylcarnitine translocase in exchange for a carnitine. Inside the matrix, CPT-II transfers the acyl group back to CoA and carnitine is taken back into the intermembrane space by the translocase enzyme.
8. **E.** The breakdown of palmitic acid (C16) requires seven cycles of β-oxidation each producing 1 FADH$_2$ and 1 NADH molecule and results in the formation of eight 2C acetyl CoA molecules.
9. **B.** When the action of carnitine palmitoyl transferase-I is inhibited by malonyl CoA, fatty acyl groups are unable to enter the matrix of the mitochondria where their breakdown by β-oxidation takes place.
10. **C.** Humans (and most mammals) do not possess enzymes able to introduce a double bond into fatty acids beyond Δ9.
11. **D.** Inhibition of the tricarboxylic acid transporter causes levels of citrate in the cytosol to decrease and favors inactivation of the enzyme.
12. **A.**
13. **C.**
14. **E.**
15. **E.** Glucagon is released when blood glucose levels are low. In this situation, fatty acids are broken down for energy and fatty acid synthesis is inhibited.
16. **E.** Glucagon, ACTH, epinephrine and vasopressin promote activation of the enzyme.
17. **B.**
18. **D.**
19. **A.** Chylomicrons are triacylglycerol-rich lipoproteins synthesized in the intestinal mucosa using fat from the diet and secreted into lymph.
20. **E.** VLDL is synthesized and secreted by the liver, and adipose tissue and muscle take up the fatty acids released by the action of lipoprotein lipase.
21. **D.** Very low-density lipoprotein secreted by the liver is converted to intermediate-density lipoprotein and then to low-density lipoprotein (LDL) by the action of lipases and the transfer of cholesterol and proteins from high-density lipoprotein. LDL delivers cholesterol to extrahepatic tissues and is also cleared by the liver.
22. **A.** Chylomicrons are synthesized in the intestine and secreted into lymph after a fat meal.
23. **E.** Chylomicrons and their remnants are cleared from the circulation rapidly after a meal, and the secretion of very low-density lipoprotein by the liver then increases. Ketone bodies and non-esterified fatty acids are elevated in the fasting state.
24. **C.** When cholesteryl ester is transferred from HDL to other lipoproteins by the action of CETP it is ultimately delivered to the liver in VLDL, IDL, or LDL.

25. **D.** Chylomicrons are metabolized by lipoprotein lipase when bound to the surface of endothelial cells. This process releases fatty acids from triacylglycerol which are then taken up by the tissues. The resulting smaller, cholesterol-enriched chylomicron remnant particles are released into the circulation and cleared by the liver.

26. **C.** Cholesterol is synthesized in the endoplasmic reticulum from acetyl CoA. The rate-limiting step is the formation of mevalonate from 3-hydroxy 3-methylglutaryl-CoA by HMG CoA reductase, and lanosterol is the first cyclic intermediate.

27. **C.**

28. **C.** Secondary bile acids are produced by the modification of primary bile acids in the intestine.

29. **B.** If the LDL receptor is defective, LDL is not cleared from the blood, causing severe hypercholesterolemia.

30. **A.** PCSK9 regulates the recycling of LDL receptors to the cell surface after endocytosis has taken place. Inhibition of PCSK9 activity, therefore, increases the number of LDL receptor molecules on the cell surface, leading to an increased rate of clearance and lower blood cholesterol levels.

Section VI – Metabolism of Proteins & Amino Acids

1. **D.** Phenylalanine hydroxylase catalyzes a functionally irreversible reaction, and thus cannot convert tyrosine to phenylalanine.

2. **E.** Histamine is a catabolite, not a precursor, of histidine.

3. **B.** The insertion of selenocysteine into a peptide occurs during, not subsequent to translation.

4. **C.** Pyridoxal-dependent transamination is the first reaction in degradation of all the common amino acids except threonine, lysine, proline, and hydroxyproline.

5. **B.** Glutamine.

6. **C.** The carbon skeleton of alanine contributes the most to hepatic gluconeogenesis.

7. **B.** ATP and ubiquitin participate in the degradation of nonmembrane-associated proteins and proteins with *short* half-lives.

8. **C.** Due to the failure to incorporate NH_4^+ into urea, clinical signs of metabolic disorders of the urea cycle include *alkalosis*, not acidosis.

9. **E.** *Cytosolic* fumarase and *cytosolic* malate dehydrogenase convert fumarase to oxaloacetate following a *cytosolic* reaction of the urea cycle. The *mitochondrial* fumarase and malate dehydrogenase function in the TCA cycle, not urea biosynthesis.

10. **A.** Serine, not threonine, provides the thioethanol moiety of coenzyme A.

11. **E.** Decarboxylation of *glutamate*, not gluta*mine* forms GABA.

12. 5-Hydroxylysine and γ-carboxyglutamate represent examples of posttranslational modification of peptidyl lysyl and peptidyl glutamyl residues, respectively. By contrast, selenocysteine is incorporated into proteins cotranslationally, in the same way as the other 20 common protein amino acids. The process is complex, and involves the unusual tRNA termed tRNA^sec.

13. Biosynthesis of the amino acids that are dietarily essential for humans requires multiple reactions. Since human diets typically contain adequate amounts of these amino acids, loss of the genes that can encode these "unnecessary" enzymes and the lack of need to expend the energy required to copy them provide an evolutionary advantage.

14. Since glutamate dehydrogenase plays multiple central roles in metabolism, its complete absence would unquestionably be fatal.

15. **E.** Albumin is not a hemoprotein. In cases of hemolytic anemia, albumin can bind some metheme, but unlike the other proteins listed, albumin is not a hemoprotein.

16. **A.** Acute intermittent porphyria is due to mutations in the gene for uroporphyrin I synthase.

17. **A.** Bilirubin is a *linear* tetrapyrrole.

18. **D.** The severe jaundice, upper abdominal pain, and weight loss plus the laboratory results indicating an obstructive type of jaundice are consistent with cancer of the pancreas.

19. The assay takes advantage of the different water solubility of unconjugated and conjugated bilirubin. Two assays are conducted, one in the absence and a second in the presence of an organic solvent, typically methanol. The highly polar glucuronic acid groups of conjugated bilirubin convey water solubility that ensures that it will react with the colorimetric reagent even in the absence of any added organic solvent. Data from an assay conducted in the *absence* of added methanol, termed "direct bilirubin," is bilirubin glucuronide. A second assay conducted in the *presence* of added methanol measures *total* bilirubin, that is, both conjugated and unconjugated bilirubin. The *difference* between total bilirubin and direct bilirubin, reported as "indirect bilirubin," is *unconjugated* bilirubin.

20. The biosynthesis of heme from succinyl-CoA and glycine occurs only when the availability of free iron signals the potential for synthesis of heme. Regulation targets the first enzyme of the pathway, Δ-aminolevulinate synthase (ALA synthase) rather than a subsequent reaction. This conserves energy by avoiding wasting of a coenzyme A thioester.

Section VII – Structure, Function, & Replication of Informational Macromolecules

1. **D.** β,γ-Methylene and β,γ-imino purine pyrimidine triphosphates do not readily release the terminal phosphate by hydrolysis or by phosphoryl group transfer.

2. **D.**

3. **E.** Pseudouridine is excreted unchanged in human urine. Its presence there is not indicative of pathology.

4. **A.** Metabolic disorders are infrequently associated with defects in pyrimidine catabolism, which forms water-soluble products.

5. **B.**	21. **C.**	37. **C.**	53. **D.**
6. **D.**	22. **A.**	38. **E.**	54. **A.**
7. **B.**	23. **C.**	39. **D.**	55. **E.**
8. **C.**	24. **A.**	40. **D.**	56. **A.**
9. **C.**	25. **E.**	41. **B.**	57. **E.**
10. **D.**	26. **B.**	42. **A.**	58. **C.**
11. **E.**	27. **A.**	43. **A.**	59. **A.**
12. **B.**	28. **E.**	44. **E.**	60. **D.**
13. **D.**	29. **C.**	45. **C.**	61. **D.**
14. **D.**	30. **A.**	46. **A.**	62. **E.**
15. **E.**	31. **A.**	47. **C.**	63. **A.**
16. **A.**	32. **C.**	48. **D.**	64. **C.**
17. **C.**	33. **D.**	49. **C.**	65. **C.**
18. **B.**	34. **E.**	50. **B.**	66. **E.**
19. **D.**	35. **C.**	51. **E.**	67. **D.**
20. **B.**	36. **B.**	52. **C.**	

Section VIII – Biochemistry of Extracellular & Intracellular Communication

1. **B.** Glycolipids are located on the outer leaflet.
2. **A.** Alpha helices are major constituents of membrane proteins.
3. **E.** Insulin also increases glucose uptake in muscle.
4. **A.** Its action maintains the high intracellular concentration of potassium compared with sodium.

5. **D.**	10. **A.**	15. **B.**	20. **B.**
6. **B.**	11. **E.**	16. **C.**	21. **D.**
7. **C.**	12. **B.**	17. **A.**	22. **A.**
8. **B.**	13. **D.**	18. **C.**	
9. **D.**	14. **E.**	19. **A.**	

Section IX – Special Topics (A)

1. **A.**	16. **A.**	31. **E.**	46. **B.**
2. **E.**	17. **B.**	32. **A.**	47. **B.**
3. **C.**	18. **C.**	33. **B.**	48. **B.**
4. **D.**	19. **E.**	34. **A.**	49. **B.**
5. **E.**	20. **D.**	35. **B.**	50. **C.**
6. **D.**	21. **E.**	36. **C.**	51. **D.**
7. **C.**	22. **A.**	37. **D.**	52. **C.**
8. **B.**	23. **C.**	38. **E.**	53. **A.**
9. **D.**	24. **C.**	39. **E.**	54. **A.**
10. **E.**	25. **A.**	40. **A.**	55. **A.**
11. **C.**	26. **E.**	41. **D.**	56. **E.**
12. **B.**	27. **A.**	42. **C.**	57. **A.**
13. **C.**	28. **A.**	43. **B.**	
14. **D.**	29. **A.**	44. **E.**	
15. **B.**	30. **C.**	45. **C.**	

Section X – Special Topics (B)

1. Within the body, hydrolysis of nitroglycerin releases nitrate ions that can be reduced by mitochondrial aldehyde dehydrogenase to generate nitric oxide (NO), a potent vasodilator.
2. The contractile cycle of cardiac muscle is controlled by oscillations in the level of cytosolic Ca^{2+}. If the reuptake of Ca^{2+} by the sarcoplasmic reticulum is slowed sufficiently by a deficiency in SERCA2a activity, cardiac myocytes will be unable to clear this second messenger from their cytoplasm prior to the onset of the next cycle of excitation. The persistence of high basal levels of cytosolic Ca^{2+} will lead to both a reduction in the amplitude of the contractile cycle and the progressive uncoupling of the excitation-contraction cycle.

3. **A.**	5. **B.**	7. **E.**	9. **E.**
4. **A.**	6. **B.**	8. **D.**	

10. Haptoglobin binds extracorpuscular hemoglobin, forming a complex that is too large to pass through the glomerulus into kidney tubules.
11. The production of new antibodies with unique antigen-binding properties is reliant on the recombination and mutation of the DNA encoding the hypervariable regions of the immunoglobulin heavy and light chains. Cytidine deaminase introduces genetic mutations by catalyzing the hydrolysis of cytosine bases present in DNA to uracil.

12. **B.**	14. **C.**	16. **E.**
13. **C.**	15. **B.**	17. **E.**

18. Erythrocytes deficient in glucose-6-phosphate are rendered extremely vulnerable to destruction by reactive oxygen species resulting from a lack of reduced glutathione, an important agent for protecting against oxidative stress. This is a consequence of their reliance on this enzyme to generate a plentiful supply of the NADPH used by glutathione reductase.

19. **E.**	21. **D.**	23. **D.**	25. **B.**
20. **C.**	22. **A.**	24. **D.**	26. **A.**

27. **E.** Importins are involved in the import of proteins into the nucleus.
28. **B.** Some mammalian proteins are known to be translocated posttranslationally.
29. **C.** Ubiquitin tags proteins for degradation by proteasomes.
30. **E.** Furin converts proalbumin to albumin.
31. **C.** NSF is an ATPase.
32. **D.** Cross-links are an important feature of collagen structure.
33. **C.** Deletions in the elastin gene have been identified as responsible for many cases of Williams-Beuren syndrome.
34. **B.** Ehlers-Danlos syndrome subtypes kyphoscoliosis and dermatosparaxis are caused by defects in noncollagen genes.
35. **B.** Hyaluronic acid (hyaluronan) is not sulfated.
36. **C.** Hurler syndrome is caused by a deficiency of α-L-iduronidase.
37. **D.** Achondroplasia is caused by mutations in the *FGFR3* gene.

Section XI – Special Topics (C)

1. **D.**
2. **D.** Of the listed proteins, only factor II is a vitamin K-dependent coagulation factor.
3. **D.**
4. **E.** GPIIb-IIIa (integrin αIIbβ3) is not a G protein–coupled receptor.
5. **A.** Hemophilia A, being an X chromosome-linked disease, is very unlikely to occur in a female.
6. **B.** Most chemical carcinogens interact covalently with DNA.
7. **B.** Certain DNA viruses are also known to be carcinogenic.
8. **E.** Mutations in approximately 5 to 6 of these two types of cancer promoting or suppressor genes are thought to be necessary for carcinogenesis.
9. **E.** PDGF stimulates phospholipase C, not phospholipase A.
10. **D.** Binding of RB to E2F blocks progression of the cell from G_1 to S phase.
11. **C.** Microsatellite instability is caused by abnormalities of mismatch repair.
12. **E.** Dichloroacetate inhibits pyruvate dehydrogenase kinase.
13. **D.** Angiogenin is an inhibitor of angiogenesis.
14. **B.** Cytochrome C is released from mitochondria.
15. **E.** Only about 1 in 10,000 cancer may have the capacity to colonize.
16. **B.**
17. **D.**
18. **D.**
19. **C.**
20. **B.**

Index

Note: Page numbers followed by *f* indicate figures; and page numbers followed by *t* indicate tables.

A

A (acceptor) site, 413, 414*f*
A band, 619, 619*f*, 620*f*
A blood group substances, 661, 661*f*
A cyclins, 379, 379*f*, 379*t*
A gene, 661
A kinase anchoring proteins (AKAPs), 513
AAV (adenovirus-associated viral) genomes, 449
ABCA1 (ATP-binding cassette transporter A1), 252*f*, 253
ABCG1 (ATP-binding cassette transporter G1), 252*f*, 253
abciximab, 680
abetalipoproteinemia, 250, 268*t*
ABO system, 660–661, 661*f*
absorption
  biomedical importance of, 527
  of carbohydrates, 528, 528*f*
  of lipids, 528–529, 530*f*
  of proteins, 529, 531
  of vitamin B$_{12}$, 105, 544
  of vitamins and minerals, 531–532, 531*f*
absorption spectrophotometry, 571–572
absorptive pinocytosis, 482, 482*f*
ACAT (acyl-CoA:cholesterol acyltransferase), 263
acceptor (A) site, 413, 414*f*
acceptor arm of tRNA, 355, 357*f*, 407, 407*f*
accuracy, of laboratory test, 569–570, 570*f*
ACE (angiotensin-converting enzyme), 504
aceruloplasminemia, 640
acetal link, 150
acetic acid, 13*t*, 207*t*
acetoacetate, 217, 220, 221*f*
acetoacetyl-CoA
  in ketogenesis, 221–222, 222*f*
  in mevalonate synthesis, 260, 260*f*
acetone, 217, 220, 221*f*
acetyl (acyl)-malonyl enzyme, in lipogenesis, 227, 228*f*, 229*f*
acetylation
  enzyme regulation by, 92–93
  of histones, 361, 363*t*
  mass spectrometry detection of, 29*t*
  of xenobiotics, 566–567
acetylcholine, synaptobrevin and, 595

acetyl-CoA
  cholesterol synthesis from, 260, 260*f*, 262*f*
  in citric acid cycle, 156–157, 157*f*, 158*f*, 161–162, 161*f*
  in glycolysis and gluconeogenesis regulation, 183
  in lipogenesis, 226–229, 229*f*, 230*f*, 231
  in metabolic pathways, 135–136, 135*f*, 136*f*, 141–142
  phenylalanine formation of, 296, 298*f*
  pyruvate oxidation to, 168, 169*f*
  threonine conversion to, 295, 295*f*
  tyrosine conversion to, 296, 297*f*
  in VLDL formation, 254, 254*f*
  in xenobiotic metabolism, 567
acetyl-CoA carboxylase
  in lipogenesis, 227*f*, 229*f*, 231, 231*f*
  regulation of, 93, 93*t*
N-acetylgalactosamine (GalNAc), 555, 556*t*, 557
N-acetylglucosamine
  in glycoproteins, 556*t*
  rapidly reversible glycosylation with, 560–561
O-linked *N*-acetylglucosamine transferase, 560–561
N-acetylglutamate
  in urea cycle disorders, 287
  in urea synthesis, 285*f*, 286
N-acetylglutamate deacylase, in urea cycle, 286
N-acetylglutamate synthetase (NAGS)
  defects in, 287
  in urea cycle, 286
N-acetylneuraminic acid, in gangliosides, 244, 244*f*
N-acetylneuraminic acid (NeuAc), 555, 556*t*
acetylsalicylic acid. *See* aspirin
acetyltransferases, 93, 567
acholuric jaundice, 323
achondroplasia, 486*t*, 616
acid anhydride bonds, 330
acid anhydrides, group transfer potential for, 334
acid hydrolases, 483
acid maltase, glycogen hydrolysis by, 175
acid–base balance, ammonia role in, 285

acid–base catalysis, enzymatic, 62, 63*f*
acidic phosphoproteins, 614
acidity, 6
acidosis
  ammonia role in, 285
  in cancer cells, 705
acids
  conjugate, 11
  as proton donors, 10
  as protonated species, 11
  strength of, 13, 13*t*
  strong, 10–11
  transition metals as, 97
  weak (*See* weak acids)
aciduria
  dicarboxylic, 225
  orotic, 346, 346*t*
aconitase, in citric acid cycle, 157, 158*f*
ACP. *See* acyl carrier protein
acrolein adduct, 430*t*
acromicric dysplasia, 604
acrosomal reaction, 561
actin, 618
  F-actin, 620, 621*f*
  G-actin, 620, 621*f*, 630
  in muscle contraction, 620–621, 621*f*, 622*f*
  in red blood cells, 659*t*, 660
  striated muscle regulation by, 622
actin (thin) filaments, 594, 619–620, 620*f*, 621*f*, 630
activated partial thromboplastin time (aPTT, PTT), 687
activated protein C (APC), 685
activation coefficient, 573
activation domains (AD), 392, 439, 440*f*, 520
activation energy (E_{act}), 73, 73*f*, 75
activation reaction, 509
activation-induced cytidine deaminase, 648
activators
  in gene expression, 421–422
  transcription, 395–396
active chromatin, 364, 365*f*, 429
active site, 61–62, 62*f*
  activation energy lowering by, 75
  allosteric site compared with, 8

active transport, 475, 475*f*, 475*t*
ATP in, 480, 480*t*
of bilirubin, 322
transporters involved in, 476–477, 476*f*,
476*t*, 477*f*
actomyosin, 621
acute coronary syndrome, 661
acute fatty liver of pregnancy, 225
acute inflammatory response, 664
acute intermittent porphyria, 320, 320*t*,
321*f*
acute-phase proteins, 534, 637, 637*t*
acyl carrier protein (ACP), 227, 227*f*, 228*f*
pantothenic acid in, 547, 547*f*
acyl enzyme, 227, 229*f*
S-acylation, 471
acylcarnitine, 218, 218*f*
acyl-CoA
formation of, 218, 218*f*
pyruvate dehydrogenase regulation by,
231
in triacylglycerol synthesis, 240
acyl-CoA dehydrogenase, 118
in fatty acid oxidation, 219, 219*f*
medium-chain, deficiency of, 225
acyl-CoA synthetase (thiokinase)
in fatty acid activation, 218, 218*f*
mitochondrial compartmentalization
of, 122
in triacylglycerol synthesis, 240, 255,
256*f*
acyl-CoA:cholesterol acyltransferase
(ACAT), 263
acyl-enzyme intermediate, 64, 64*f*
acylglycerol metabolism
acylation of triose phosphates, 240, 240*f*
biomedical importance of, 239
ceramide in, 243–244, 244*f*
clinical aspects of, 244–245, 245*t*
hydrolysis, 239
phosphatidate in, 240–242, 241*f*, 242*f*
phospholipases in, 242–243, 243*f*
1-acylglycerol-3-phosphate
acyltransferase, 240, 241*f*
acylglycerols, 206
biomedical importance of, 239
synthesis of, 240–243, 240*f*, 241*f*
AD (activation domains), 392, 439, 440*f*,
520
Ad (adenoviral) genomes, 449
adapter proteins, 482, 701
adaptive immune system, 646, 650, 670
adducts, 720
adenine, 332*t*
base pairing in DNA, 349, 350*f*
base pairing in RNA, 352–354, 353*f*
salvage pathways of, 342–343
adenine nucleotide transporter, of
mitochondria, 128, 128*f*

adenocarcinoma, colorectal, 695–697,
696*f*, 696*t*
adenomatous polyps, colorectal, 695–697,
696*f*, 696*t*
adenosine
derivatives of, 332*t*
in S-adenosylmethionine, 333*f*
syn and *anti* conformers of, 331*f*
in uric acid formation, 344, 345*f*
adenosine deaminase, deficiency of, 337,
345, 345*f*, 346*t*, 444, 662
adenosine diphosphate (ADP)
creatine phosphate shuttle for, 130, 130*f*
energy capture and transfer by,
111–112, 111*t*, 112*f*, 113*f*, 125–126
phosphate cycles of, 113, 113*f*
in platelet aggregation, 678, 679*f*
respiratory control by, 126–127, 127*t*
structure of, 112*f*, 331*f*
adenosine monophosphate (AMP), 332*t*,
333*f*
cyclic (*See* cyclic AMP)
energy capture and transfer by, 111–112,
111*t*, 112*f*, 113*f*
IMP conversion to, 338, 340*f*
feedback regulation of, 341, 341*f*
phosphate cycles of, 113, 113*f*
PRPP glutamyl amidotransferase
regulated by, 340, 341*f*
structure of, 112*f*, 331*f*, 332*f*
adenosine triphosphate (ATP)
in active transport, 480, 480*f*
as allosteric effector, 90
biomedical importance of, 109
as cellular energy currency, 112–114,
112*f*, 113*f*, 125–126
citric acid cycle generation of, 157–158,
157*f*, 159
control of supply of, 126–127, 127*t*
in coupling of endergonic and exergonic
reactions, 113
creatine phosphate shuttle for,
130, 130*f*
energy capture and transfer by, 111–112,
111*t*, 112*f*, 113*f*, 125–126
erythrocyte generation of, 655, 655*t*
fatty acid oxidation production of,
219–220, 219*f*, 220*t*
free energy of hydrolysis of, 111–112,
111*t*, 112*f*
in glycolysis and gluconeogenesis
regulation, 184, 185*f*
glycolysis production of, 163–164, 164*f*,
165*f*, 166, 167, 168*f*
in muscle contraction, 622–623, 622*f*,
627–628, 627*f*
oxidative phosphorylation in synthesis
of, 122, 125–126, 126*f*
phosphate cycles of, 113, 113*f*

protein degradation dependent on,
280–281, 280*f*, 281*f*
in purine synthesis, 338
structure of, 111*f*, 112*f*, 331*f*
as transducer of free energy, 332
S-adenosylmethionine, 333, 333*f*, 334*t*
biosynthesis of, 298, 300*f*, 301*f*,
308–309, 309*f*
adenoviral (Ad) genomes, 449
adenovirus-associated viral (AAV)
genomes, 449
adenylyl cyclase, 175, 175*f*, 176*f*
cAMP intracellular signal and, 511–512,
511*t*
in lipolysis, 257, 257*f*
adenylyl kinase (myokinase), 113, 184
mitochondrial compartmentalization of,
122
in muscle contraction, 628
adherens junctions, 474
adhesion, glycoprotein role in, 561–562
adipocytes, 257
turnover of, 718*t*
adipokines, 255
adiponectin, 255
adipose tissue, 206, 239
brown, 257–258, 258*f*
in fasting state, 134–135, 134*f*, 144–145,
144*t*
in fed state, 134–135, 134*f*, 142–144,
143*f*, 144*t*
glucose uptake by, 142–143, 143*f*
glucose utilization by, 256
hormones and lipolysis in, 257, 257*f*
insulin and, 256
lipase, 255–256, 256*f*
lipogenesis and, 257
metabolic features of, 145*t*
perilipin in, 257
triacylglycerol storage in, 255
ADP. *See* adenosine diphosphate
ADPase, in hemostasis and thrombosis,
679
ADP-chaperone complex, 583
ADP-ribosylation
enzyme regulation by, 90, 92
of histones, 361, 363*t*
NAD⁺ as source for, 542
ADP-ribosylation factor-1 (ARF-1),
253–254, 254*f*
adrenal function tests, 575
adrenal gland, cytochromes P450 in, 565
adrenal steroidogenesis
androgen synthesis, 494*f*, 495
glucocorticoid synthesis, 494–495,
494*f*
mineralocorticoid synthesis, 493–494,
494*f*
overview of, 493, 493*f*

adrenergic receptors, glycogenolysis activation by, 176f, 177

adrenocortical hormones, cholesterol as precursor for, 212

adrenocorticotropic hormone (ACTH), 256, 257f

adrenogenital syndrome, 495

adrenoleukodystrophy, neonatal, 587, 588t

advanced glycation end-products (AGEs), 561, 561f, 723

aerobic conditions, 163, 164t, 167, 627

aerobic processes, 157

affinity chromatography
    protein and peptide purification with, 27
    recombinant fusion protein purification with, 68–69, 69f

agammaglobulinemia, 652

aggrecan, 615, 615t, 616f

aggregated proteins, toxic effects of, 725

aggregates, protein, 42

aggregating agents, in platelet aggregation, 678, 679f

aggregation prevention, 590

aging
    biomedical importance of, 717
    cancer and, 708
    extracellular matrix and, 599
    metabolic theories of, 725–726, 726t
    mortality and, 718
    as preprogrammed process, 725–727
    proteoglycans and, 611–612
    somatic mutation theory of, 724
    telomerase and, 365
    wear and tear theories of (See wear and tear theories of aging)

aging genes
    model organisms of, 726–727
    transcription factors, 727

aglycone, 150

AHSP (α-hemoglobin-stabilizing protein), 43

AIDS
    glycan role in, 563
    undernutrition in, 533

AKAPs (A kinase anchoring proteins), 513

ALA. See δ-aminolevulinate; α-linolenic acid

alanine, 16t
    carbon skeleton catabolism of, 294
    circulating plasma levels of, 281–282, 281f, 282f
    metabolic pathways of, 144
    specialized products of, 306–307
    synthesis of, 275, 275f
    transamination reactions forming, 282–283, 283f

β-alanine, biosynthesis of, 312–313

alanine aminotransferase (ALT)
    diagnostic use of, 67, 67t, 574
    transamination reaction of, 283, 283f

β-alanyl dipeptides, biosynthesis of, 313

albumin, 599, 634, 635, 636f, 637, 638
    bilirubin binding to, 321, 325
    free fatty acids in combination with, 218, 247, 248t, 249
    synthesis of, 596, 596f

albuminuria, 599

alcohol dehydrogenase in fatty liver, 255

alcoholic liver disease (ALD), 255

alcoholism
    case study, 730
    cirrhosis and, 255
    transferrin glycosylation in, 640

alcohols
    carbohydrates as aldehyde or ketone derivatives of, 148, 148t
    energy yields from, 144t
    hydrogen bonding of, 7, 7f
    iron absorption and, 531–532
    sulfation of, 566

ALD (alcoholic liver disease), 255

aldehyde dehydrogenase, 116
    in fatty liver, 255

aldehyde oxidase, molybdenum in, 104

aldolase
    deficiency of, 170
    in glycolysis, 165f, 166

aldolase B, 195, 197f
    deficiency of, 199–200

aldose reductase, 200

aldose-ketose isomerism, 149–150, 150f

aldoses, 148, 148t, 149, 150f

aldosterone synthase, 493–494

alimentary pentosuria, 199

Alisporivir, 43

alkaline phosphatase, 67t, 614

alkalosis, ammonia effects on, 285

alkaptonuria, 296, 297f

alkylphospholipid, 239

allergic reactions, to foods, 527, 531

allopurinol, 334, 334f, 344, 346

allosteric activator, 183

allosteric effectors/modifiers, 88–89, 89f, 141

allosteric properties, of hemoglobin, 52–56, 52f, 53f, 54f, 55f

allosteric regulation
    of enzymes, 88–89, 89f, 141, 141f
    in glycolysis and gluconeogenesis regulation, 183–184
    of hexokinase, 140

alpha-anomers, 149

α helix
    in myoglobin, 50, 50f
    as secondary structure unit, 36–37, 36f, 38f

α-actinin, 605, 606f

Alport syndrome, 602t

ALT. See alanine aminotransferase

alteplase, 686

alternative pathway, 651

alternative polyadenylation, 442

alternative transcription start site, 442

altitude, physiologic adaptations to, 56

Alu family, 368, 370

Alzheimer disease
    ω3 fatty acids and, 209
    pathologic protein conformations in, 43
    respiratory chain disorders in, 130

Amadori products, 723

ambiguity and genetic code, 405

Ames assay, 692, 692f

amino acid oxidases, 116, 283–284, 284f

amino acids
    hydrolysis of peptide bonds, 718–719, 719f
    missense mutations and, 407–408, 408f
    nonsense codons and, 408, 409f
    transporter/carrier systems for, 476
    tRNA association with, 406–407, 406f, 407f

D-amino acids, free, 19

D-α-amino acids, biomedical importance of, 15

L-α-amino acids
    absorption of, 529, 531
    biomedical importance of, 15
    biosynthetic conversions of
        alanine, 306–307
        arginine, 307, 307f, 311, 312f
        biomedical importance of, 306
        cysteine, 307, 307f
        glycine, 308, 308f, 310–311, 312f
        histidine, 308, 308f
        methionine, 308–309, 309f, 310f, 311, 312f
        phosphoserine, phosphothreonine, and phosphotyrosine, 310
        serine, 310
        tryptophan, 310, 311f
        tyrosine, 310, 312f
    catabolism of
        alanine, 294
        amino acid oxidases in, 283–284, 284f
        ammonia from, 279, 282–275, 284f
        arginine and ornithine, 292, 292f
        asparagine and aspartate, 291
        biomedical importance of, 279, 290–291
        to carbohydrate and lipid biosynthesis intermediates, 291, 291f, 291t

L-α-amino acids, catabolism of (*Cont.*):
  carbon skeleton fates in, 291, 291*f*, 291*t*
  cystine and cysteine, 294–295, 294*f*
  disorders of, 286–288, 287*t*
  end products of, 282, 285, 285*f*
  glutamate dehydrogenase in, 283, 284*f*
  glutaminase and asparaginase in, 284, 284*f*
  glutamine and glutamate, 291–292
  glutamine synthetase in, 284, 284*f*
  glycine, 293, 294*f*
  histidine, 292–293, 293*f*
  hydroxyproline, 295–296, 295*f*
  initiation of, 291
  isoleucine, leucine, and valine, 298, 300*f*, 302*f*, 303*f*
  lysine, 296, 298*f*
  metabolic diseases of, 290–291, 300, 304*t*
  methionine, 298, 300*f*, 301*f*
  phenylalanine, 296, 298*f*
  proline, 292, 292*f*
  rate of, 279–280
  serine, 293, 294*f*
  threonine, 295, 295*f*
  transamination reactions in, 282–283, 283*f*, 291–293
  tryptophan, 296–298, 299*f*, 300*f*
  tyrosine, 296, 297*f*
  urea synthesis in, 282–283, 282*f*, 283*f*, 285–286, 285*f*
chemical reactions of, 21–22
circulating plasma levels of, 281–282, 281*f*, 282*f*
conservation of catalytic, 64
deamination of, 137
deficiency of, 273–274
determination of sequences of
  Edman reaction for, 27*f*, 28–29
  genomics and, 29
  mass spectrometry for, 29–31, 29*t*, 30*f*, 31*f*
  molecular biology techniques for, 29
  proteomics and, 31–32
  purification techniques for, 24–27, 26*f*, 27*f*, 28*f*
  Sanger's work in, 27–28
essential, 15, 136, 274, 274*t*, 534
extraterrestrial, 18–19
functional groups of, 19–21, 20*f*, 20*f*, 21*t*
genetic code specification of, 16, 16*t*–17*t*, 18*t*, 405, 405*t*
glucogenic, 142, 282, 282*f*
glucose derived from, 186
for health maintenance, 3
hydrophobic and hydrophilic, 18*t*
ketogenic, 142
as metabolic fuel, 134, 142

metabolic pathways of, 135–136, 135*f*, 136*f*, 137*f*
  in fasting state, 144, 144*t*
  in fed state, 143–144, 143*f*, 144*t*
  at organ and cellular level, 136–139, 138*f*
  at subcellular level, 139
metabolic roles of, 19
net charge of, 19–20, 20*f*
nonessential, 136, 274*t*, 275–278, 534
nutritional requirements for, 533–534
peptide bonds between, 22
pI of, 20
pK_a of, 20, 21*t*
pK_a values of, 16*t*–17*t*
posttranslational modifications of, 16, 18, 18*f*
properties of, 16–19, 16*t*–17*t*, 18*f*, 18*t*, 19*t*
protease and peptidase generation of, 280–281, 280*f*, 281*f*
protein and peptide primary structures and, 22
selenocysteine as 21st, 18, 18*f*
solubility of, 20*f*, 21
stereochemistry of, 18
synthesis of
  alanine and aspartate, 275, 275*f*
  asparagine, 275, 276*f*
  biomedical importance of, 273–274
  citric acid cycle role in, 159–161, 160*f*
  cysteine, 276, 277*f*
  glutamate, 275, 275*f*
  glutamine, 275, 275*f*
  glycine, 275, 276*f*
  hydroxyproline and hydroxylysine, 276, 278*f*
  lengthy pathways of, 274, 274*t*
  proline, 276, 277*f*
  selenocysteine, 278, 278*f*
  serine, 275, 276*f*
  tyrosine, 276, 277*f*
  valine, leucine, and isoleucine, 277–278
toxic plant, 19, 19*t*
transport in liver, 180–181
ultraviolet light absorption by, 21, 22*f*
unusual, 22, 22*f*
β-amino acids, biosynthetic conversions of
alanine and aminoisobutyrate, 312–313
alanyl dipeptides, 313
GABA, 313, 313*f*
amino sugars, 151, 152*f*
  synthesis of, 197, 199*f*
aminoacyl residues, 22
aminoacyl site, aminoacyl-tRNA binding to, 413, 414*f*

aminoacyl-tRNA in protein synthesis, 406, 406*f*, 413, 414*f*
aminoacyl-tRNA synthetases, 406, 406*f*
aminoadipate-δ-semialdehyde synthase, 296, 298*f*
γ-aminobutyrate (GABA), biosynthesis of, 313, 313*f*
β-aminoisobutyrate, biosynthesis of, 312–313
δ-aminolevulinate (ALA)
  heavy metal inactivation of, 98
  synthesis of, 316, 316*f*, 319*f*
δ-aminolevulinate (ALA) dehydratase, 316, 317*f*, 319*f*, 320*t*
δ-aminolevulinate (ALA) synthase, 316*f*, 317–318, 319*f*, 320*t*
aminopeptidases, 529
aminophospholipids, membrane asymmetry and, 472
aminopyrine, uronic acid pathway effects of, 199
aminotransferases, 160, 275, 275*f*
  diagnostic use of, 67, 67*t*, 574
  transamination reactions of, 282–283, 283*f*, 291–293
  vitamin B$_6$ in, 283, 283*f*, 543
ammonia
  from amino acid degradation, 279, 282–275, 284*f*
  citric acid cycle and, 156, 162
  intoxication with, 283–284
  nitrogen excretion as, 282
  in urea cycle disorders, 286–288, 287*t*
ammonium ion, pK_a of, 13*t*
ammonotelic animals, 282
amobarbital, 122, 127
AMP. *See* adenosine monophosphate
AMP-activated protein kinase, 93
AMP-activated protein kinase (AMPK), 263, 263*f*
amphibolic pathways, 133
  citric acid cycle as, 159–161, 160*f*
amphipathic helices, 37
amphipathic lipids
  in lipoproteins, 248, 249*f*
  in membranes, 215, 215*f*, 469–470, 469*f*
amphipathic molecules, 8, 8*f*
ampicillin (Amp) resistance genes, 449
AMPK (AMP-activated protein kinase), 263, 263*f*
amplification
  in cancer, 693
  in fibrin clot formation, 685
amylase
  diagnostic use of, 67*t*
  starch hydrolysis by, 528
amyloid, 725
β-amyloid, in Alzheimer disease, 43
amyloidosis, 646, 646*t*

amylopectin, 152, 153*f*
amylopectinosis (Andersen's disease), 174*t*
amylose, 152, 153*f*
anabolic pathways, 133
anabolism, 110
anaerobic conditions, glycolysis in, 163–164, 164*f*, 164*t*, 167, 704–705, 705*f*
analbuminemia, 637
analytes, 569
anaphylaxis, slow-reacting substance of, 237
anaplerosis, 157
anaplerotic substrates, 160–161
Andersen's disease (amylopectinosis), 174*t*
androgens
    steroidogenesis
        ovarian, 495–496, 497*f*, 498*f*
        testicular, 495, 496*f*
    synthesis of, 494*f*, 495
androstenedione, 495, 496*f*
anemias, 56
    causes of, 653, 657*t*
    definition, 653
    Diamond-Blackfan, 656
    hemolytic (*See* hemolytic anemias)
    of inflammation, 642
    iron deficiency, 643, 644*t*, 657*t*
    megaloblastic, 546, 657*t*
    pernicious, 544–545
aneuploidy, in cancer, 700, 700*f*
ANF (atrial natriuretic factor), 490, 491*t*
angiogenesis, in cancer, 706
angiotensin II, 503–504, 504*f*
angiotensin-converting enzyme (ACE), 503
angiotensin-converting enzyme inhibitors, 503
angiotensinogen, 503, 504*f*
anion exchange protein, 659
ankyrin, 659*t*, 660
anomeric carbon atom, 149
anserine, 308, 308*f*, 313
antennae, of glycoproteins, 558
anterior pituitary gland, hormones of, 188
anterograde transport (COPII), 594, 595*f*
anterograde vesicular transport, 591
anti conformers, 330–331, 331*f*
antibiotics
    amino acids in, 15
    amino sugars in, 151
    bacterial protein synthesis and, 417–418, 418*f*
    folate metabolism inhibitors as, 545
    resistance to, 83, 102, 103*f*
antibodies, 634. *See also* immunoglobulins
    monoclonal, 649–650
anticancer agents. *See* chemotherapy

anticoagulant drugs, mechanism of, 685–686
anticodon region of tRNA, 405–406, 407*f*
antifolate drugs, purine nucleotide synthesis affected by, 338, 340
antigenicity, xenobiotic alteration of, 567
antigens, 646
anti-Lepore, 370, 370*f*
antimalarial drugs, folate metabolism inhibitors as, 545
antimycin A, 127
antioxidants, 120
    biomedical importance of, 549
    classes of, 214
    pro-oxidant paradox of, 552–553, 552*t*
    protective mechanisms of, 551–552, 552*f*
    vitamin E as, 540
antiparallel loops, mRNA and tRNA, 407
antiparallel strands, DNA, 349, 350*f*
antiparallel β sheet, 37, 37*f*
antiplatelet drugs, 680
antiport systems, 476, 476*f*
antiporter systems, for sugar nucleotides in glycoprotein synthesis, 557–558
α₁-antiproteinase, 645
antiproteinase inhibitor, 670
antiproteinases, 670
antipyrine, uronic acid pathway effects of, 199
antithrombin, 634, 637*t*, 685
α₁-antitrypsin, 670, 685
APC (activated protein C), 685
apo A-I, 248*t*, 249
    alcohol consumption and, 267
    deficiencies of, 268*t*
    in HDL metabolism, 252*f*, 253
apo A-II, 248*t*, 249, 251
apo A-IV, 248*t*, 249
apo B-48, 248*t*, 249
apo B-100, 248*t*, 249
    in LDL metabolism, 251*f*, 252
    regulation of, 263
apo B-100 receptor
    in LDL metabolism, 252
    regulation of, 263
apo C-I, 248*t*, 249
apo C-II, 248*t*, 249, 251
apo C-III, 248*t*, 249, 251
apo D, 248*t*, 249
apo E, 248*t*, 249, 252
apo E receptor
    in chylomicron remnant uptake, 251*f*, 252
    in LDL metabolism, 252
    regulation of, 263
apolipoprotein E, in Alzheimer disease, 43
apolipoproteins/apoproteins, distribution of, 248*t*, 249
apomyoglobin, 51, 51*f*

apoptosis, 720
    cancer evasion of, 689–690, 701–702, 702*f*, 703*t*
    cardiolipin in, 240
    ceramide in, 243
    DNA damage and, 381–382, 381*f*
    phosphatidylcholines in, 210
apoptosome, 701, 722
apoptotic cell death program, 722
aprotic acids, 97
aPTT (activated partial thromboplastin time), 687
aquaporins, 479–480
aqueous solutions. *See* water
arabinose, 150*f*, 151*t*
arachidonic acid/arachidonate, 207*t*, 209*f*
    eicosanoid formation and, 234, 235*f*, 236*f*, 237*f*
    for essential fatty acid deficiency, 234
    nutritionally essential, 232, 232*f*
ARF-1 (ADP-ribosylation factor-1), 253–254, 254*f*
Arg-Gly-Asp (RGD) sequence, 605
arginase
    defects in, 288
    in urea synthesis, 285*f*, 286
arginine, 17*t*
    carbon skeleton catabolism of, 292, 292*f*
    specialized products of, 307, 307*f*, 311, 312*f*
    stabilization of, 20, 20*f*
    in urea synthesis, 285*f*, 286
arginine phosphate, in energy conservation and capture, 113
argininosuccinate, in urea synthesis, 285–286, 285*f*
argininosuccinate lyase
    defects in, 287*t*, 288
    in urea synthesis, 285*f*, 286
argininosuccinate synthetase
    defects in, 287*t*, 288
    in urea synthesis, 285*f*, 286
Argonaute proteins, 402
aromatase enzyme complex, 495–496, 497*f*
ARS (autonomously replicating sequences), 372–373
arsenic
    in glycolysis pathway, 166
    toxicity of, 97–98
arsenite, 159
arthritis. *See also* osteoarthritis
    gouty, 344, 346*t*
    proteoglycans in, 612
    rheumatoid (*See* rheumatoid arthritis)
artificial membranes, 473
arylamines, sulfation of, 566
ascorbate, 195
ascorbic acid. *See* vitamin C

asialoglycoprotein receptors, 555, 590–591, 590*f*

asialoglycoproteins, 483

a-SNAP, 594*t*, 598

asparaginase, in amino acid catabolism, 284, 284*f*, 291

asparagine, 17*t*
  carbon skeleton catabolism of, 291
  deamination of, 284, 284*f*
  synthesis of, 275, 276*f*

asparagine synthetase, 275, 276*f*

asparagine-*N*-acetylglucosamine linkage, in glycoproteins, 556*f*, 558, 558*f*

aspartate
  carbon skeleton catabolism of, 291
  in citric acid cycle, 159, 160*f*
  synthesis of, 275, 275*f*
  in urea synthesis, 285*f*, 286

aspartate aminotransferase (AST), diagnostic use of, 67, 574

aspartate transcarbamoylase (ATCase), 90
  in pyrimidine synthesis, 343*f*

aspartic acid, 17*t*
  pH effects on, 20, 20*f*

aspartic proteases, acid-base catalysis of, 63, 63*f*

aspirin (acetylsalicylic acid)
  antiplatelet effects of, 680
  cancer development and, 709, 712
  cyclooxygenase affected by, 236–237
  prostaglandin synthesis affected by, 226

assembly particles, 482

AST (aspartate aminotransferase), diagnostic use of, 67, 574

asthma, leukotrienes in, 208

asymmetry
  inside-outside, 472
  of lipid and proteins in membrane assembly, 596–597, 597*f*
  of membranes, 472
  Ran molecules and, 585

ataxia-telangiectasia, mutated (ATM), 381–382, 381*f*

ATCase (aspartate transcarbamoylase), 90
  in pyrimidine synthesis, 343*f*

atherosclerosis
  advanced glycation end-products in, 561
  cholesterol and, 212, 260, 267
  folic acid supplements for, 546
  free radical damage causing, 550
  HDL and, 253, 267
  LDL plasma concentration and, 252, 267
  lipids and, 206, 211
  lysophosphatidylcholine and, 211
  premature, 248, 252
  trans fatty acids and, 236–237

ATM (ataxia-telangiectasia, mutated), 381–382, 381*f*

ATM-related kinase (ATR), 381–382, 381*f*

atomic absorption spectrometry, 572

atorvastatin, 267

ATP. *See* adenosine triphosphate

ATP synthase
  mitochondrial compartmentalization of, 122
  rotary motor function of, 125, 126*f*

ATP/ADP cycle, 112, 113*f*

ATPase, 480, 480*f*
  chaperones exhibiting activity of, 582–583
  in osteoclasts, 612

ATP-binding cassette transporter A1 (ABCA1), 252*f*, 253

ATP-binding cassette transporter G1 (ABCG1), 252*f*, 253

ATP-chaperone complex, 583

ATP-citrate lyase, in lipogenesis, 228, 229*f*

ATP-driven active transporters, 480, 480*t*

ATR (ATM-related kinase), 381–382, 381*f*

atractyloside, 127, 128*f*

atrial natriuretic factor (ANF), 490, 491*t*

attachment proteins, 605, 606*f*

autoantibodies, 648

autocrine signaling, 670

autoimmune disease
  free radical damage causing, 550
  xenobiotic role in, 567

autoimmune disorder, 634, 635*t*

autoimmune response, 652

autoimmunity, 648

autonomously replicating sequences (ARS), 372–373

autotrophic organisms, 111

avian influenza virus, glycans in binding of, 563

axial ratios, 35

axonemal dyneins, 631

5- or 6-azacytidine, 334

5-azadeoxycytidine, 699

8-azaguanine, 334, 334*f*

azaserine, 338, 339*f*

azathioprine, 335, 335*f*

5- or 6-azauridine, 334, 334*f*, 346

B

B blood group substances, 661, 661*f*

B cyclins, 379, 379*f*, 379*t*

B form of DNA, 350, 350*f*

B gene, 661

B lymphocytes, 646, 670

BAC (bacterial artificial chromosome) vector, 448–449, 449*t*

bacitracin, 15

bacteria
  antibiotic and protein synthesis in, 417–418, 418*f*
  bilirubin reduction by intestinal, 323
  glycans in binding of, 563
  transcription cycle in, 387–388, 387*f*
  transcription in, 389–390, 390*f*

bacterial artificial chromosome (BAC) vector, 448–449, 449*t*

bacterial DNA-dependent RNA polymerase, 387–388, 387*f*

bacterial lox P sites, 447

bacterial plasmids, 448, 448*f*

bacterial promoters, in transcription, 389–390, 390*f*

bacteriophage lambda
  att sites of, 447
  as paradigm for protein-DNA interactions, 425–430, 426*f*, 427*f*, 428*f*

balanced chemical equations, 72

BamHI, 445, 445*t*, 446*f*

BAPN (γ-glutamyl-β-aminopropionitrile), 19, 19*t*

barbital, uronic acid pathway effects of, 198–199

barbiturates, respiratory chain inhibition by, 122, 127

barrel-like structures, 583

Barth syndrome, 211

basal laminas, 605

basal metabolic rate (BMR), 532

base excision repair (BER), 379, 380*f*, 380*t*

base pairing in DNA, 349, 350*f*, 351

base pairs (bp), 364

base substitution, mutations occurring by, 407–408, 407*f*, 408*f*, 409*f*

bases
  conjugate, 11
  as proton acceptors, 10
  strong, 10–11
  as unprotonated species, 11
  weak (*See* weak bases)

base-stacking forces, 350

basophils, 664–665, 670

BAT (brown adipose tissue), 127–128, 257–258, 258*f*

BAX, 701–702, 702*f*

BCL-2, in apoptosis, 702, 702*f*

Becker muscular dystrophy, 629

bends, in proteins, 37, 38*f*

benzoate, metabolism of, 308, 308*f*

benzoylation, 430*t*

BER (base excision repair), 379, 380*f*, 380*t*

beriberi, 542

Bernard-Soulier syndrome, 662

beta-anomers, 149

β barrels, 37, 38*f*

β sheets, as secondary structure unit, 37, 37f, 38f

β turns, 37, 38f

β-oxidation of fatty acids, 135–136, 218–220, 218f, 219f, 220t
    ketogenesis regulation and, 223–224, 223f, 224f
    modified, 220, 221f

BgIII, 445t

BHA (butylated hydroxyanisole), 214

BHT (butylated hydroxytoluene), 214

Bi-Bi reactions, 82, 82f, 83f

bicarbonate, carbon dioxide conversion to, 54, 55f

bicarbonate in extracellular and intracellular fluid, 468t

bifunctional enzymes, 185

bilayers, lipid, 470–471, 470f
    membrane proteins and, 471
    thickness of, 597

bile, bilirubin secretion into, 322–323, 323f

bile acids (salts), 212, 259, 265, 265f

bile duct obstruction, hyperbilirubinemia caused by, 323f, 323t, 324–325, 325t

bile pigments, 315

bile salts, 529

biliary tree obstruction, hyperbilirubinemia caused by, 323f, 323t, 324–325, 325t

bilirubin
    conjugation of, 322, 322f, 323f
    elevated levels of, 315, 323–324, 323t
    excretion of, 322–323, 323f
    from heme catabolism, 321–323, 322f, 323f
    intestinal bacteria reduction of, 323
    metabolic disorders of, 324–325, 324f, 325t
    serum measurement of, 323, 325, 325t, 574
    structure of, 321, 322f
    transport of, 322

δ-bilirubin, 325

bilirubin diglucuronide, 322, 322f

bilirubin UDP-glucuronosyltransferase, 322, 324

biliverdin, 321, 322f

biliverdin reductase, 321, 322f

bimolecular lipid bilayer, 470

binding change mechanism, of ATP synthase, 125, 126f

binding constant, Michaelis constant as approximation of, 78–79

binding domains, 38, 39f

binding immunoglobulin protein (BiP), 589–590, 589f

biochemical case histories
    alcoholism, 730

biotin, 734–735
    carnitine, 732–733
    diabetes mellitus, 729–730, 735–736
    glucose-6-phosphate dehydrogenase, 731–732
    glutathione, 731

biochemistry, 1–4, 2f, 3f
    clinical (See laboratory tests)

bioenergetics
    ATP as cellular energy currency, 112–114, 112f, 113f, 125–126
    biomedical importance of, 109
    coupling of endergonic and exergonic reactions, 110–111, 110f, 111f, 113
    energy capture and transfer by high-energy phosphates, 111–112, 111t, 113f, 125–126
    free energy in biologic systems, 109–110

bioengineering, 3

bioethics, 3

bioinformatics, 3, 32

biomarkers
    of cardiovascular risk, 575
    enzymes used as, 67, 67f
    tumor, 709–710, 709t

biophysics, 3

biopolymers, enzymatic synthesis of, 9

biotechnology, 3

biotin
    deficiency of, 536t, 546, 734–735
    functions of, 536t, 547
    in malonyl-CoA synthesis, 227, 227f
    structure of, 546, 546f

BiP (binding immunoglobulin protein), 589–590, 589f

1,3-bisphosphoglycerate, in glycolysis, 165f, 166, 166f

2,3-bisphosphoglycerate (2,3-BPG), 655
    hemoglobin binding of, 52f, 55–56, 55f

bisphosphoglycerate mutase, 168, 168f

2,3-bisphosphoglycerate phosphatase, 168, 168f

2,3-bisphosphoglycerate synthase/2-phosphatase (BPGM), 55–56

bleeding disorders, 686

bleeding time, 687

bleomycin, 15

blindness, vitamin A deficiency causing, 538

blistering, 632

blood. See also plasma proteins
    bilirubin in, 323–324, 323t, 325t
    functions of, 635, 635t
    metabolism integration by, 137–139, 138f

blood cells
    derivation from hematopoietic stem cells, 653–654, 654f
    red (See red blood cells)
    white (See white blood cells)

blood clots
    formation of, 91 (See also fibrin clot formation)
    vitamin K role in, 540–541, 541f

blood coagulation. See coagulation, blood

blood glucose
    biomedical importance of, 180–181
    clinical aspects of, 189–190, 189f
    in fasting state, 145f
    free fatty acids and, 256, 256f
    glucose oxidase in measurement of, 67
    glucose tolerance measurement of, 189–190, 189f
    glycated hemoglobin in measurement of, 57, 561
    glycogen in maintenance of, 171–172
    red blood cells and, 655
    regulation of
        glucokinase role in, 186–187, 188f
        hormonal mechanisms, 186–188, 187t, 188t
        metabolic mechanisms, 186, 187t
        narrow limits of, 185–186
        renal threshold for, 189
        sources of, 186–187, 187f

blood group, 660–661, 661f

blood samples, 571

blood transfusion, ABO system importance in, 660

blood type, 661

blood vessels, nitric oxide affecting, 630, 630f

blot transfer techniques, 450, 451f

blunt end ligation, 446–447

blunt ends, 445, 446f

BMPs (bone morphogenetic proteins), 642

BMR (basal metabolic rate), 532

Bohr effect, 55, 55f

bond energies, of biologically significant atoms, 7t

bonds, biomolecule stabilization by, 7, 7t

bone
    components of, 612–614, 612t, 620f
    metabolic and genetic disorders of, 614–615, 614t
    vitamin K role in, 541

bone marrow
    heme synthesis in, 316
    transplantation of, 245

bone morphogenetic proteins (BMPs), 642

botulinum B toxin, 595

boundary elements, 437

bovine preproparathyroid hormone, 503f

bp (base pairs), 364

2,3-BPG. See 2,3-bisphosphoglycerate

BPGM (2,3-bisphosphoglycerate synthase/2-phosphatase), 55–56

bradykinin, 681

brain
  in fasting state, 134–135, 134*f*, 144–145, 144*t*
  in fed state, 134–135, 134*f*, 142–144, 143*f*, 144*t*
  glucose requirement of, 142
  metabolic features of, 145*t*
branch point, of glycogen, 173
branched-chain amino acids, carbon skeleton catabolism of, 298, 300, 302*f*, 303*f*
branched-chain ketonuria, 300, 303*f*
branching enzyme, in glycogenesis, 172*f*, 173, 173*f*
brefeldin A, 254, 596
broad β-disease, 268*t*
3-bromopyruvate, 705, 706*t*
brown adipose tissue (BAT), 127–128, 257–258, 258*f*
brush border enzymes, 528
budding of vesicles, 594, 595*f*
buffering, 13, 13*t*
buffers
  Henderson-Hasselbalch equation describing, 12, 13*t*
  weak acids acting as, 13, 13*t*
bulk flow, in transport vesicles, 594
bursa of Fabricius, 670
butylated hydroxyanisole (BHA), 214
butylated hydroxytoluene (BHT), 214
butyric acid, 207*t*, 220

C
C2c2, 446*t*, 447
C_{20} polyunsaturated acids, eicosanoids formed from, 234, 235*f*, 236*f*
CA (carbonic anhydrase), 54, 55*f*, 655
CA II (carbonic anhydrase II), 614
Ca^{2+}. *See* calcium
Ca^{2+}/calmodulin-sensitive phosphorylase kinase, glycogenolysis activation by, 177
cachexia, 144, 145, 163, 533
CAD (caspase-activated DNase), 701
caffeine, 331, 333*f*
  hormonal regulation of lipolysis and, 257
calbindin, 531
calcidiol, 539, 539*f*
calciferol. *See* vitamin D
calcinosis, 540
calcitriol (1,25(OH)$_2$-D$_3$), 538–539, 539*f*
  biosynthesis of, 497–499
  storage of, 505, 506*t*
calcium (Ca^{2+})
  absorption of, 531
  in cardiac muscle contraction, 625–626, 626*f*, 627*t*
  enzyme and protein regulation by, 514, 514*t*

in extracellular fluid, 468, 468*t*
glycogen phosphorylase activation by, 177
as hormone action mediator, 515
human requirement for, 97, 97*t*
in intracellular fluid, 468, 468*t*
in malignant hyperthermia, 625
metabolism of, 514
in muscle contraction, 618, 623–625, 624*f*, 624*t*, 625*f*
phosphatidylinositide metabolism and, 515–516, 515*f*, 516*f*
sarcoplasmic reticulum and, 624–625, 625*f*
as second messenger, 89, 490, 491*t*, 510
vitamin D regulation of, 539
calcium channels, in cardiac muscle, 625–626, 626*f*, 627*t*
calcium-binding proteins, vitamin K role in, 540–541, 541*f*
caldesmon, 624
calmodulin, 177, 514, 514*t*, 621, 623
calnexin, 559–560, 592
calreticulin, 560, 592
calsequestrin, 624, 625*f*
cAMP. *See* cyclic AMP
3′,5′-cAMP, as second messenger, 89
cAMP-dependent protein kinase
  glycogen regulation by, 175, 176*f*, 177, 179*f*
  structure of, 39*f*
CAMs (cell adhesion molecules), 707, 708*t*
cancer. *See also* chemotherapy
  angiogenesis in, 706
  apoptosis in, 689–690, 701–702–694, 702*f*, 703*t*
  biomarkers of, 709–710, 709*t*
  biomedical importance of, 689
  carcinogenesis in, 690–691, 697
  causes of (*See* carcinogens)
  cell cycle abnormalities in, 699–700
  cell features in, 689, 690*f*
  colorectal, 380*t*, 695–697, 696*f*, 696*t*
  copy number variation in, 700*f*, 701
  epigenetic mechanisms involved in, 698–699
  extracellular vesicles and, 698
  folate metabolism inhibition in treatment of, 545
  folic acid supplements for, 546
  free radical damage causing, 550
  genetic analyses of, 710
  genetic damage in, 690*f*, 691
  genomic instability and aneuploidy in, 700, 700*f*
  glycoproteins in, 560
  growth factor role in, 697, 697*t*
  hereditary, 691, 697, 699, 699*t*

hyperbilirubinemia caused by, 323*f*, 323*t*, 324
hypermetabolism in, 163
immunologic aspects of, 708–709
incidence of, 689
metabolism in, 703–706, 705*f*, 706*t*
metastatic process in, 706–708, 707*f*, 708*t*
micro-RNA in, 698–699
molecular profiling of, 712
neoplasms and, 689–690, 690*f*
new therapies for, 710–712, 711*t*, 712*f*
ω3 fatty acids and, 209
oncogene role in, 693–695, 693*t*, 694*f*, 695*f*, 695*t*, 696*f*, 696*t*
prevention of, 712, 712*t*
pro-oxidant paradox and, 552–553
selectins in, 562
signaling pathways in, 697, 697*t*
stem cells in, 706
telomerase activity in, 701
tumor microenvironment in, 691, 703, 704*f*
tumor suppressor gene role in, 687*t*, 693*t*, 694*f*, 695–697, 695*f*, 695*t*, 696*f*, 696*t*
undernutrition in, 533
vitamin B$_6$ and, 543
xenobiotic role in, 564, 567, 567*f*
cancer cells
  biochemical and genetic changes in, 690*f*
  cyclins and, 379
  membrane abnormalities and, 486*t*
cancer phototherapy, 318
cap, of mRNA, 355, 356*f*
CAP (catabolite activator protein), 423
CAP Response Element (CRE), 425
caproic acid, 207*t*
carbamates, hemoglobin transport of, 55, 55*f*
carbamoyl phosphate
  excess, 346
  in urea synthesis, 285–286, 285*f*
carbamoyl phosphate synthetase I
  defects in, 287, 287*t*
  in urea synthesis, 285–286, 285*f*
carbamoyl phosphate synthetase II, in pyrimidine synthesis, 342, 343*f*, 344
carbohydrate loading, 172
carbohydrate-deficient transferrin, 640
carbohydrates
  as aldehyde or ketone derivatives of polyhydric alcohols, 148, 148*t*
  biological information encoded in, 148
  biomedical importance of, 147–148
  blood glucose derived from, 186–187, 187*f*

in cell membranes, 154
cell surface, glycolipids and, 212
classification of, 148, 148*t*
diets low in, 190
digestion and absorption of, 528, 528*f*
disaccharides (*See* disaccharides)
diseases associated with metabolism of, 147
energy yields from, 144*t*
glycoproteins (*See* glycoproteins)
in lipoproteins, 155
metabolic pathways of, 135, 135*f*, 136*f*
  in fasting state, 144–145
  in fed state, 142–144, 143*f*
  at organ and cellular level, 137–139, 138*f*
  vitamin B₁ role in, 541–542
monosaccharides (*See* monosaccharides)
polysaccharides (*See* polysaccharides)
carbon dioxide (CO₂)
  carbonic anhydrase and, 655
  citric acid cycle production of, 157–159, 158*f*
  hemoglobin transport of, 54, 55, 55*f*
carbon monoxide
  heme binding of, 51, 51*f*
  respiratory chain inhibition by, 116, 122, 127
carbonic acid
  carbon dioxide conversion to, 54, 55*f*
  pK_a of, 13*t*
carbonic anhydrase (CA), 54, 55*f*, 655
carbonic anhydrase II (CA II), 614
γ-carboxyglutamate, synthesis of, 540–541, 541*f*
γ-carboxyglutamic acid, 18, 18*f*
carboxyl transferase, 227, 227*f*
carboxylases, biotin coenzyme of, 547
carboxypeptidase A, active site of, 62*f*
carboxypeptidases, 529
carboxy-terminal domain (CTD), 394, 426
carcinoembryonic antigen (CEA), 709
carcinogenesis, 690–691, 697
carcinogens
  chemical, 564, 567, 567*f*, 691–692, 691*f*, 692*f*, 692*t*
  radiation, 691, 691*f*
  viruses, 692–693, 693*t*, 694*f*, 712
carcinoid syndrome, 543
cardiac glycosides, 150, 212
cardiac muscle, 619
  calcium channels in, 625–626, 626*f*, 627*t*
  contraction regulation in, 625, 626*t*
cardiac troponins, 67
cardiolipin, 210*f*, 211

mitochondrial compartmentalization of, 122
synthesis of, 240, 240*f*, 241*f*
cardiomyocytes, turnover rate of, 718*t*
cardiomyopathies, 628–629, 628*t*
cardiovascular disease
  biomarkers of, 575
  folic acid supplements for, 546
  lipids and, 209
  ω3 fatty acids and, 209
caretaker tumor suppressor genes, 695
cargo proteins/molecules, 585–586, 586*f*, 594
carnitine
  deficiency of, 217, 224–225, 733–734
  in fatty acid transport, 218, 218*f*
  liver and muscle, 733*t*
  urinary excretion of, 734*t*
carnitine palmitoyltransferase, 217
carnitine palmitoyltransferase-I (CPT-I)
  deficiency of, 225
  in fatty acid transport, 218, 218*f*
  in ketogenesis regulation, 223, 223*f*
carnitine palmitoyltransferase-II (CPT-II)
  deficiency of, 225
  in fatty acid transport, 218, 218*f*
carnitine-acylcarnitine translocase, 218, 218*f*
carnosinase, deficiency of, 313
carnosine, biosynthesis of, 308, 308*f*, 313
carnosinuria, 313
β-carotene, 537, 537*f*
  as antioxidant, 214
  as pro-oxidant, 552
carrier proteins, 476–477, 477*f*
carrier systems, for sugar nucleotides in glycoprotein synthesis, 557–558
CAR-T cell therapy, 708–709
cartilage, 608, 615–616, 615*f*, 615*t*, 616*f*
caspase-activated DNase (CAD), 701
caspases, 701–702, 702*f*
catabolic pathways, 133
catabolism, 110
  respiratory chain capture of energy from, 125–126
catabolite activator protein (CAP), 423
catalase, 118, 283, 552, 656
  as antioxidant, 214
catalysis (enzymatic)
  acid-base, 62, 63*f*
  active site role in, 61–62, 62*f*
  biomedical importance of, 59
  conformational changes in, 63, 63*f*
  covalent, 62, 62*f*, 64*f*, 65*f*
  enzyme detection by, 65–66, 65*f*, 66*f*
  free-energy changes of, 72–73, 73*f*
  by isozymes, 64–65
  kinetics of (*See* enzyme kinetics)

prosthetic group, cofactor, and coenzyme roles in, 60–61, 61*f*
  by proximity, 62
  residues conserved in, 64
  site-directed mutagenesis in study of, 69
  specificity of, 60, 60*f*
  by strain, 62
catalysts
  enzymes as, 9, 60, 60*f*
  oxaloacetate as, 156–157, 157*f*
  ribozymes as, 69–70
  RNA as, 402
catalytic constant (k_{cat}), 78
catalytic efficiency (k_{cat}/K_m), 78
catalytic site. *See* active site
cataplerosis, 157
cataract, diabetic, 200
catecholamine oxidases, copper in, 103, 104*f*
catecholamines
  storage of, 505, 506*t*
  synthesis of, 499–500, 499*f*
caveolae, 474, 474*f*
caveolin-1, 474, 474*f*
CBG (corticosteroid-binding globulin), 506, 507*t*
CBP (CREB-binding protein), 521–522, 522*t*
CBP/p300 and signal transduction pathways, 521–522, 521*f*, 522*t*
CD59, 562
CDK-cyclin inhibitor (CKI), 382
CDKs. *See* cyclin-dependent kinases
cDNA library, 450
CEA (carcinoembryonic antigen), 709
celiac disease, 527
cell, macromolecule transport in, 481–483, 482*f*, 483*f*, 485*f*
cell adhesion, glycosphingolipids in, 244
cell adhesion molecules (CAMs), 707, 708*t*
cell cycle
  cancer abnormalities in, 699–700
  delay or arrest of, 381, 381*f*
  phases of, 378, 378*f*
  regulation of, 593
  S phase of, DNA synthesis during, 378–379, 378*f*, 379*f*, 379*t*
cell death. *See* apoptosis
cell division, checkpoints in, 94, 94*f*
cell growth, 610
cell membranes
  carbohydrates in, 155
  vitamin E in, 540
cell migration, 605, 608
cell recognition, glycosphingolipids in, 244
cell signaling
  phosphatidylcholines in, 210–211
  phosphatidylinositols in, 210–211

cell-cell communication
with gap junctions, 483, 484f
heparan sulfate and, 610
cell-cell interactions, 467
cell-mediated immunity, 646
cellulose, 153–154, 160f
cellulose acetate zone electrophoresis, 635,
636f
central core disease, 625, 627t
central nervous system (CNS), glucose
requirement of, 142
centromere, 364, 365f
cephalin (phosphatidylethanolamine),
210, 210f
membrane asymmetry and, 472
synthesis of, 240, 240f, 241f
ceramide, 210, 212f
accumulation of, 245
in membranes, 469
in sphingolipid synthesis, 243–244, 244f
synthesis of, 243, 244f
cerebrohepatorenal (Zellweger) syndrome,
225, 587, 588t
cerebrosides, 243
ceruloplasmin, 638, 638f, 640–641
diagnostic use of, 67t
cervonic acid, 207t
CF (cystic fibrosis), 68, 486, 486t, 527, 592t
cGMP. See cyclic GMP
chain elongation, 386f, 388–389. See also
elongation
chain initiation, 386f, 388. See also
initiation
chain termination, 386f, 389. See also
termination
chain termination method, 451, 452f
chain-breaking antioxidants, 214
chalones, 238
channeling, in citric acid cycle, 157–158
channelopathies, 626, 627t
channels, amphipathic helices in, 37
chaotropic agents, 351
chaperones, 42
ATP-dependent protein binding to, 584
histone, 363
misfolded proteins and, 592–593, 592t,
593f
properties of, 584t
in protein folding, 582–583
in protein sorting, 583f, 584
in quality control, 592, 592t
in translocation, 584, 585f
chaperonins, 42, 583
charged groups, noncovalent interactions
of, 8
charged paddle, 479, 479f
charge-relay network, 64, 64f
charging, in protein synthesis, 406, 406f
checkpoint controls, 382

checkpoint kinases 1 and 2 (CHK1 and
CHK2), 381f, 382
checkpoints, in cell division, 94, 94f
Chediak-Higashi syndrome, 592t
chemical carcinogenesis, 567, 567f
chemical carcinogens, 564, 567, 567f,
691–692, 691f, 692f, 692t
chemical chaperone therapy, 245
chemical equations
balancing of, 72
kinetic order of, 74
chemical mechanisms and reactive oxygen
species, 723–724
chemical potential, 109
chemical reactions
catalysis of (See catalysis)
coupling of, 110–111, 110f, 111f, 113
equilibrium of, 72, 74–75
free-energy changes of (See free energy)
mechanisms of, 72
multisubstrate, 82, 82f, 83f
rate-limiting, 87
rates of (See rate of reaction)
reversibility of, 72
transition states of, 72–73, 73f
chemiosmotic theory, 125, 126f, 128
chemokines, 666, 666f, 670
chemotaxis, 665–666
chemotherapy
folate metabolism inhibitors, 545
new therapies, 710–712, 711t, 712f
resistance to, 711–712, 711t
synthetic nucleotide analogs in,
334–335, 334f, 335f
chenodeoxycholic acid, 265, 266f
chenodeoxycholyl-CoA, 265, 266f
children, kwashiorkor in, 533
chimeric antigen receptors, 708
chimeric gene approach, 433f, 435, 435f,
436f
chimeric molecules
purpose of, 444
restriction enzymes for
DNA chain cleavage with, 445–446,
445t, 446f, 446t
preparation with, 446–447
ChIP (chromatin immunoprecipitation),
455, 456f
ChIP-chip, 455
ChIP-Exo, 455
ChIP-Seq, 455
chitin, 154, 154f
CHK1 and CHK2 (checkpoint kinases 1
and 2), 381f, 382
chloramines, 669
chloride (Cl⁻)
in extracellular and intracellular fluid,
468, 468t
permeability coefficient of, 471f

chlorinated oxidants, production of, 670
chlorobutanol, uronic acid pathway effects
of, 198–199
chlorophyll, 315
cholecalciferol. See vitamin D₃
cholera
glucose transport in treatment of, 481
toxin, 212, 212f, 244, 512
cholestatic jaundice, 324
cholesterol
absorption of, 529
in bile acid synthesis, 265–266, 266f
biomedical importance of, 259
clinical aspects of, 267, 268t
dietary, 260, 267
excretion of, 265–266, 266f
fructose effects on, 195, 197f, 199
function of, 212
in high-density lipoproteins in,
252–253, 252f
HMG-CoA reductase regulation and,
87
hormone synthesis from, 212, 259, 491,
492f, 493–499, 493f
in lipoprotein, 247–248, 249f
in membranes, 469, 597
fluid mosaic model and, 473–474
metabolic pathways of, 135–136, 136f
plasma levels of
atherosclerosis and coronary heart
disease and, 212, 260, 267
dietary changes and, 267
drug therapy and, 267
lifestyle changes and, 267
synthesis of
acetyl-CoA pathway for, 251–252,
260f, 261f, 262f
diurnal variation in, 262
HMG-CoA reductase regulation of,
262–263, 263f
in tissues, 213, 213f
factors influencing balance of,
263–264, 264f
transport of
with lipoproteins, 263–264, 265f
reverse, 252f, 253, 260, 264, 264f, 267
cholesterol 7α-hydroxylase, 265, 266f
cholesteryl ester hydrolase, 263
cholesteryl ester transfer protein, 249, 264,
265f, 267
cholesteryl esters, 213, 263–264, 265f
in HDL metabolism, 252f, 253
in lipoprotein core, 247, 248, 249f
cholesterylation, 471
cholic acid, 265, 266f
choline, 210, 210f, 211f
deficiency of, fatty liver and, 255
membrane asymmetry and, 472
choluric jaundice, 323–324

cholyl-CoA, in bile acid synthesis, 265, 266*f*

chondrodysplasias, 616

chondroitin sulfates
in cartilage, 615*f*
in extracellular matrix, 608
in mucopolysaccharidoses, 611, 611*t*
properties of, 609*t*
structure of, 154, 154*f*, 607, 609*f*

chondronectin, 615*t*, 616

chorion protein genes, 441, 441*f*

choroid, gyrate atrophy of, 292

chromatids, 364–366, 365*f*, 365*t*, 366*f*, 371, 371*f*

chromatin
active regions of, 364, 365*f*, 429
components of, 361, 361*f*
covalent modifications of, 92
gene expression and template for, 429–430
higher order structure and compaction of, 362*f*, 363
inactive regions of, 364, 429
physiologic roles of, 105
reconstitution in DNA replication, 378

chromatin fibers, 362*f*, 363

chromatin fibrils, 362*f*, 363

chromatin immunoprecipitation (ChIP), 455, 456*f*

chromatin modifying complex (CMC), 433

chromatography
affinity, 27
recombinant fusion protein purification with, 68–69, 69*f*
column, 25, 26*f*
high-pressure liquid, 25, 26*f*, 572
hydrophobic interaction, 26
ion-exchange, 26
protein and peptide purification with, 24–27, 26*f*
size-exclusion, 26, 26*f*

chromium
human requirement for, 97, 97*t*, 105
multivalent states of, 97, 98*f*, 98*t*
toxicity of, 99*t*

chromosomal instability (CIN), 700, 700*f*

chromosomal translocations, in cancer, 693, 694*f*

chromosomes
coding regions of, 366, 367*f*
gene conversion, 371
gene localization to, 453
integration of, 370, 370*f*
integrity of, monitoring, 382
interphase, 363
metaphase, 362*f*, 364, 365*t*, 366*f*
organization of, 364–366, 365*f*, 365*t*, 366*f*

polytene, 364
recombination of, 369–370, 369*f*, 370*f*
scaffolding in, 362*f*, 363
sister chromatid exchange, 371, 371*f*
transposition, 370–371

chronic granulomatous disease, 669

chronic myelocytic leukemia (CML), 711

chyle, 249

chylomicron remnants, 248*t*, 249
formation of, 251–252
liver uptake of, 250*f*, 252

chylomicrons, 137, 138*f*, 143–144
apolipoproteins of, 249
composition of, 248, 248*t*
in lipid transport, 247
metabolism of, 250*f*, 251–252
in triacylglycerol transport, 249–250, 250*f*, 251*f*

chymotrypsin, 529
activation of, 90, 91*f*
covalent catalysis of, 63–64, 64*f*

cI repressor gene, 425, 427*f*

cI repressor protein, 426, 427*f*

CIN (chromosomal instability), 700, 700*f*

circular RNAs (circRNAs), 358–359

circulating tumor DNA (ctDNA), 710

circulation
enterohepatic, 529
metabolism integration by, 137–139, 138*f*

cirrhosis, 156, 236, 254, 255

cis, trans-isomerization, of X-Pro peptide bonds, 42–43, 43*f*

cis-prenyltransferase, 261*f*

cisternal maturation, 596

cistron, 422

citrate
in citric acid cycle, 156–157, 157*f*, 158*f*
in lipogenesis regulation, 228, 231, 231*f*

citrate lyase, 93*t*, 161–162, 161*f*

citrate synthase
in citric acid cycle, 157, 158*f*
regulation of, 162

citric acid, 13*t*

citric acid cycle, 135, 135*f*, 136*f*
ATP generation by, 156–157, 157*f*, 159
biomedical importance of, 156
carbon dioxide produced by, 157–159, 158*f*
in energy conservation and capture, 113
glycolysis route to, 168, 169*f*
mitochondrial location of, 139, 140*f*, 156
pivotal metabolic role of, 159–162, 160*f*, 161*f*
reducing equivalents produced by, 157–159, 158*f*
regulation of, 162

respiratory chain substrates provided by, 156–157, 157*f*
at subcellular level, 139, 140*f*
vitamins essential to, 159

citrullination, 669, 669*f*

citrulline, in urea synthesis, 285–286, 285*f*

citrullinemia, 288

CJD (Creutzfeldt-Jakob disease), 43

CKI (CDK-cyclin inhibitor), 382

Cl⁻. *See* chloride

class B scavenger receptor B1 (SR-B1), 252*f*, 253

classic pathway of complement activation, 650

class-specific effector functions, 647

clathrin, 263, 482, 482*f*, 593–594, 593*t*

clinical biochemistry. *See* laboratory tests

clinical value, of laboratory tests, 570–571, 571*t*

clofibrate, 267

cloning, 447–449, 448*f*, 449*f*, 449*t*

cloning vectors, 448, 448*f*, 449*t*

clopidogrel, 680

closed complex, 386*f*, 389

closure time, 687

clotting. *See* fibrin clot formation

CMC (chromatin modifying complex), 433

CML (chronic myelocytic leukemia), 711

CMP (cytidine monophosphate), 332*t*

CNS (central nervous system), glucose requirement of, 142

CNV (copy number variation), 700, 700*f*

CO₂. *See* carbon dioxide

coactivators, transcription, 394–396, 396*t*

coagulation, blood
disorders of, 662
prostaglandins in, 226

coagulation factors, 680, 681*f*, 681*t*, 682*t*
deficiencies of, 686
vitamin K-dependent carboxylation of, 685–686

coagulation system. *See also* hemostasis
anticoagulant drug effects on, 685–686
biomedical importance of, 677
endothelial cell actions in, 678, 680
fibrin clot formation in (*See* fibrin clot formation)
inherited deficiencies of, 686
laboratory tests measuring, 687
platelet aggregation in, 678, 679*f*
thrombi types in, 678
thrombin inhibition in, 685–686
thrombin regulation in, 685

coat proteins, 591, 594–595, 594*t*, 595*f*

coated pits, 482

cobalamin. *See* vitamin B₁₂

cobalophilin, 544

cobalt
  human requirement for, 97, 97*t*
  multivalent states of, 97, 98*f*, 98*t*
  physiologic roles of, 102–103
  toxicity of, 99*t*
cobamide, 61
code erasers, 429–430
code readers, 429–430
code writers, 429–430
coding regions, 366, 367*f*
coding strand
  in DNA replication, 349, 354*f*
  in RNA synthesis, 385–386, 390*f*
codons, 16, 404
  amino acid sequence and, 405, 405*t*
  anticodon recognition of, 406–407, 407*f*
  nonsense, 405, 408–409, 409*f*
  stop, 414, 415*f*
  tables of usage of, 406
coenzyme A, 61
  cysteine conversion to, 307
  pantothenic acid in, 547, 547*f*
coenzyme Q (Q, ubiquinone), 213
  in cholesterol synthesis, 251–252, 261*f*
  in respiratory chain, 122, 123*f*, 124, 124*f*, 125*f*
coenzymes, 60–61, 61*f*
  laboratory tests for, 572–573
  nicotinamide oxidation and reduction, 117, 117*f*
  nucleotide derivatives, 333, 334*t*
cofactors, 60–61, 61*f*
cognate receptor, 509–510, 509*f*
cohesive end (cos) sites, 448
colchicine, 631
collagen, 417
  abundance of, 599
  in bone, 612, 612*t*
  in cartilage, 615, 615*f*, 615*t*
  chondrodysplasias, 616
  classification of, 601, 601*f*
  cross-linking of, 561
  defects in, 273–274
  diseases caused by abnormalities in, 602*t*, 608–610
  elastin compared with, 604*t*
  fibril formation by, 601
  fibronectin interaction with, 604, 605*f*, 606*f*
  as fibrous protein, 44
  nutritional and genetic disorders of, 44
  O-glycosidic linkages in, 557
  osteogenesis imperfecta and, 614, 614*t*
  platelet adhesion to, 678, 679*f*
  posttranslational modification of, 601–602, 601*t*
  posttranslational processing of, 43–44, 44*f*
  structure of, 599–602, 600*f*, 601*f*
  synthesis of, 44

  triple helix structure of, 44, 44*f*
  types of, 600*t*
  vitamin C effects in, 547
collision theory, 73
colony-stimulating factor, 654
colorectal cancer
  development of, 695–697, 696*f*, 696*t*
  mismatch repair in, 380*t*
column chromatography, protein and peptide purification with, 25, 26*f*
combinatorial chemistry, 65
common lymphoid progenitor, 665
common myeloid progenitor, 665
compact bone, 614
compartmentation
  metabolic regulation with, 86–87
  of mitochondria, 122, 122*f*
competitive inhibitors, 79–81, 80*f*, 81*f*
competitive ligand-binding assays, laboratory tests using, 573
complement, 547
complement system (cascade), 91, 650–651, 651*f*
complementarity
  of DNA, 350*f*, 351, 450
  of RNA, 352–354, 354*f*, 355
complementarity-determining regions, 647–648
complementary DNA (cDNA) library, 450
complex carbohydrates, glycoproteins as, 554
Complex I. *See* NADH-Q oxidoreductase
Complex II. *See* succinate-Q reductase
Complex III. *See* Q-cytochrome *c* oxidoreductase
Complex IV. *See* cytochrome *c* oxidase
complex lipids, 206
complex oligosaccharides, in glycoproteins, 558–559, 558*f*
complexation, of transition metals, 99–100, 99*f*, 100*f*, 101*t*
compulsory-order reactions, 82, 82*f*
concanavalin A, 155
conformational disorders, 597–598
conformations
  of enzymes, 63, 63*f*
  hemoglobin changes in, 53–54, 53*f*, 54*f*
  of proteins
    biomedical importance of, 33
    configuration *versus*, 33–34
    native, 42
    pathologic diseases of, 43
    peptide bond restrictions on, 35–36, 35*f*
congenital contractural arachnodactyly, 604
congenital disorders of glycosylation, 640
congenital erythropoietic porphyria, 320, 320*t*

congenital forms of muscular dystrophy, 629
congenital long QT syndrome, 486*t*
congenital muscular dystrophies, glycoprotein synthesis defects in, 562
conjugate acid, 11
conjugate base, 11
conjugated bilirubin, 322, 322*f*, 323*f*
conjugated double bonds, in porphyrins, 318, 320*f*
conjugated hyperbilirubinemia, 323*f*, 323*t*, 324–325
conjugation reactions, of xenobiotics, 566
connective tissue, 599. *See also* extracellular matrix
connexin, 483, 484*f*
connexon channel, 483, 484*f*
consensus sequences, 394, 398*f*
  Kozak, 412
constant region, immunoglobulins, 646, 647
constitutive gene expression, 422, 425
constitutive heterochromatin, 364
constitutive mutation, 422
constitutive proteins, 87
constitutive secretion, 582
contact sites, 584
contraction, muscle. *See* muscle contraction
convalescence, protein replacement during, 534
Coomassie Blue dye, 27, 28*f*
cooperative binding
  of hemoglobins, 53
  Hill equation for, 79, 79*f*
coordinate expression, 423
COPI vesicles, 591, 593*t*, 594, 596
COPII vesicles, 591, 593*t*, 594, 595*f*
copper, 641
  absorption of, 105
  in Complex IV, 124, 124*f*
  in cytochrome oxidase, 103
  deficiency of, 44, 264
  human requirement for, 97, 97*t*
  multivalent states of, 97, 98*f*, 98*t*
  physiologic roles of, 103, 104*f*
  toxicity of, 99*t*, 641
coproporphyrinogen III, in heme synthesis, 317, 317*f*, 318*f*, 319*f*
coproporphyrinogen oxidase, 317, 318*f*, 319*f*, 320*t*
coprostanol (coprosterol), 264
copy number variation (CNV), 700, 700*f*
coregulators, transcription, 394–396, 396*t*, 521–522, 521*f*, 522*t*
corepressor family, 522, 522*t*
Cori cycle, glucose derived from, 186, 187*f*

Cori-Forbes disease, 174t
cornea, 608
coronary (ischemic) heart disease, 267.
    See also atherosclerosis
corrinoids, 544
corticosteroid-binding globulin (CBG),
    506, 507t
corticosteroids, 259
cortisol
    cortisone reduction to, 193–194
    laboratory tests for, 573, 575
    plasma transport of, 506, 507t
    synthesis of, 492f, 493f, 494–495
cortisone, cortisol synthesis from, 193–194
cos (cohesive end) sites, 448
cosmids, 448–449, 449t
cotranslational glycosylation, 588–589
cotranslational insertion, 590–591, 590f,
    591f
cotranslational pathway, 588–589, 589f
cotransport systems, 476, 476f
Coulomb's law, 7
coumarin drugs, 685–686
coupled enzyme assays, 66, 66f
coupling
    of endergonic and exergonic reactions,
        110–111, 110f, 111f, 113
    of oxidation and phosphorylation in
        respiratory chain, 126–127, 127t
covalent bonds
    biomolecule stabilization by, 7, 7t
    in collagen, 44
    membrane lipid-protein interaction
        and, 471
covalent catalysis
    activation energy lowering by, 75
    enzymatic, 62, 62f, 64f, 65f
covalent cross-links, 601
covalent modifications. *See also*
        phosphorylation
    carcinogens causing, 692
    enzyme regulation by, 90–93, 91f, 92f,
        93t
    in glycolysis and gluconeogenesis
        regulation, 183
    of histones, 361, 363t, 429–430, 430t
    mass spectrometry detection of, 29, 29t
    in protein posttranslational processing,
        43–44
    of pyruvate dehydrogenase, 168, 169f
    reversibility of, 90
COX-1 (cyclooxygenase), 234, 680
COX-2, 234
coxibs, 236–237
cPLA2 (cytosolic phospholipase A₂), 678,
    679f
CPT-I. *See* carnitine palmitoyltransferase-I
CPT-II. *See* carnitine
        palmitoyltransferase-II

CRE (CAP Response Element), 425
C-reactive protein (CRP), 637, 637t
creatine, biosynthesis of, 311, 312f
creatine kinase, 627–628
    in creatine phosphate shuttle, 130, 130f
    diagnostic use of, 67, 67t
    mitochondrial compartmentalization
        of, 122
creatine phosphate
    in energy conservation and capture,
        113, 113f
    in muscle contraction, 626f, 627–628
creatine phosphate shuttle, 130, 130f
creatinine
    biosynthesis of, 311, 312f
    laboratory tests for, 574
creatinine clearance, 574
CREB (cyclic AMP response element
        binding protein), 513
CREB-binding protein (CBP), 521–522,
    522t
Creutzfeldt-Jakob disease (CJD), 43
Crigler-Najjar syndrome, 324
CRISPR-Cas, 359
CRISPR-Cas9, 446t, 447, 447f, 454, 710
cro gene, 425, 427f
Cro protein, 426, 427f
Crohn disease, 236
cross-bridge complex, 622, 623f
crossing-over, in chromosomal
        recombination, 369–370, 369f, 370f
crotonyl-CoA, lysine formation of, 296,
    298f
CRP (C-reactive protein), 637, 637t
cryo-electron microscopy, protein
        structures solved by, 41
crystallography, protein structures solved
        by, 40–41
CTD (carboxy-terminal domain), 394, 426
ctDNA (circulating tumor DNA), 710
C-terminal binding domain, 38
CTP (cytidine triphosphate), 90, 333
cubilin, 105
cutaneous porphyrias, 320t, 321
cyanide, respiratory chain inhibition by,
        116, 122, 127
cyclic 3′,5′-nucleotide phosphodiesterase,
        in lipolysis, 257
cyclic AMP (cAMP), 332, 333f, 334t
    adenylyl cyclase and, 511–512, 511t
    formation and hydrolysis of, 175, 175f
    glycogen regulation by, 175–177, 175f,
        176f, 178f, 179f
    in hemostasis and thrombosis, 679
    in muscle contraction, 625
    in operon model, 425
    phosphodiesterases and, 513
    phosphoprotein phosphatases and,
        513–514

phosphoproteins and, 513
    protein kinases and, 512–513, 513f
    as second messenger, 490, 491t, 510
cyclic AMP regulatory protein (catabolite
        gene activator protein), 423
cyclic AMP response element binding
        protein activator protein (CREB),
        513
cyclic AMP-dependent protein kinase, in
        lipolysis, 257, 257f
cyclic GMP (cGMP), 333f
    as second messenger, 333, 491, 491t,
        510, 514
    synthesis of, 514
cyclin-dependent kinases (CDKs)
    in cancer, 700
    in cell cycle, 378, 379f, 379t
    inhibition of, 382
cyclins, 378–379, 379f, 379t
    in cancer, 700
cycloheximide, 418
cyclooxygenase (COX-1), 234, 680
cyclooxygenase pathway, 234, 235f, 236f
cyclophilins. *See* proline-*cis,*
        trans-isomerases
cyclosporine, 43
CYP2A6, 565
CYP2C9, 565
CYP27A1 (sterol 27-hydroxylase),
        265, 266f
cystathionine β-synthase, 294, 310
cysteine, 16t
    carbon skeleton catabolism of, 294–295,
        294f
    specialized products of, 307, 307f
    synthesis of, 276, 277f
cysteine sulfinate pathway, 294, 294f
cystic fibrosis (CF), 68, 486, 486t,
        527, 592t
cystic fibrosis transmembrane regulator
        protein, 486
    degradation of, 593
cystine, carbon skeleton catabolism of,
        294–295, 294f
cystine reductase, 294, 294f
cystinosis, 295
cystinuria, 294
cytarabine, 335, 335f
cytidine, 331f, 332t, 343
cytidine monophosphate (CMP), 332t
cytidine triphosphate (CTP), 90, 333
cytochrome aa_3, 116
cytochrome b_5, 119, 565, 591, 656
cytochrome b_5 reductase, 656
cytochrome b_{558}, 668
cytochrome b_H, in respiratory chain, 124,
        125f
cytochrome b_L, in respiratory chain, 124,
        125f

cytochrome *c*
  in apoptosis, 701
  iron in, 100
  in respiratory chain, 122, 123*f*, 124, 124*f*, 125*f*
cytochrome *c* oxidase (Complex IV), 122, 123*f*, 124–125, 124*f*
  deficiency of, 130
cytochrome c_1, in respiratory chain, 124, 125*f*
cytochrome oxidase, 116
  transition metals in, 103
cytochrome P450 side chain cleavage enzyme (P450scc), 493, 493*f*
cytochrome P450 system, 115
  carcinogen activation by, 692
  in drug metabolism, 119–120, 119*f*, 120*f*, 564–566
  inducible nature of, 565–558
  iron in, 102
  membrane insertion, 591
  microsomal ethanol oxidizing system, 255
  as monooxygenases, 119–120, 119*f*, 120*f*
  nomenclature of, 565
  polymorphism of, 565
  sterol 27-hydroxylase, 265
  xenobiotic hydroxylation by, 564–566
cytochromes, as dehydrogenases, 118
cytokines, 638, 645, 654, 670
  advanced glycation end-product activation of, 561
  blood glucose regulation by, 188
  cachexia resulting from, 145, 533
  in cancer development, 709
cytoplasmic protein tyrosine kinases, 516–517, 518*f*
cytosine, 332*t*, 347*f*
  base pairing in DNA, 349, 350*f*
  base pairing in RNA, 352–354, 353*f*
  deoxyribonucleosides of, in pyrimidine synthesis, 342–343
  salvage pathways of, 343
cytoskeleton
  biomedical importance of, 618
  cellular functions of, 630–632, 630*f*, 631*t*
cytosol
  fatty acid synthesis in, 217
  glycolysis in, 139, 140*f*
  lipogenesis in, 226–230, 229*f*, 231, 231*f*
  pentose phosphate pathway in, 191–192
  pyrimidine synthesis in, 342
cytosolic branch, of protein sorting, 582*f*
  in cell nucleus, 583–587, 583*f*, 586*f*
  in mitochondria, 584, 585*f*
  in peroxisomes, 586–587, 587*f*
  signals in, 582–583

cytosolic dynein, 631
cytosolic phospholipase A_2 (cPLA_2), 678, 679*f*
cytotoxic T cells, 664, 670
cytotoxicity, of xenobiotics, 567, 567*f*

D

D arm of tRNA, 355, 357*f*, 407, 407*f*
D cyclins
  cancer and, 379
  in cell cycle, 378–379, 379*f*, 379*t*
DAG. *See* diacylglycerol; 1,2-diacylglycerol
damage repair mediators, 381, 381*f*
dAMP, 332*f*
dantrolene, 625
DBDs. *See* DNA-binding domains
DBH (dopamine β-hydroxylase), 499, 499*f*
DCA (dichloroacetate), 706, 706*t*
Dcytb (duodenal cytochrome b), 105, 639
deacetylases, 93
deacetylation, enzyme regulation by, 92–93
deacylase (thioesterase), 227, 228*f*, 229*f*
deamination, 136
  of amino acids, 137, 137*f*
  citric acid cycle role in, 159–161, 160*f*
  in nitrogen metabolism, 283–284, 284*f*
death receptor pathway, of apoptosis, 701–702, 702*f*
debranching enzyme, in glycogenolysis, 173, 173*f*
decarboxylation, oxidative. *See* oxidative decarboxylation
decay accelerating factor, 562
decitabine, 699
decondensation, 669
defensins, 668*t*
degeneracy, of genetic code, 405, 405*t*
dehydratase, 229*f*
7-dehydrocholesterol, 538, 539*f*
dehydroepiandrosterone pathway, 495, 496*f*
dehydrogenases
  coupled enzyme assays using, 66, 66*f*
  cytochromes as, 118
  detection of NAD(P)$^+$-dependent, 66, 66*f*
  flavin groups in, 118
  nicotinamide coenzymes of, 117, 117*f*
  oxidation-reduction reactions of, 116–118, 117*f*
dehydrogenation, hydrogenation coupling to, 110, 111*f*
denaturation, of DNA
  conditions causing, 350–351
  hyperchromicity of, 351
  protein refolding after, 42
  structure analysis and, 350
  temperature increases causing, 75, 351
dense granules, 661

deoxy sugars, 151, 151*f*
deoxyadenylate, 348–349
deoxycholic acid, synthesis of, 265
deoxycytidine residues, methylation of, 430
deoxycytidylate, 348–349
2-deoxy-D-glucose, 705, 706*t*
2-deoxy-D-ribose, 330
deoxyguanylate, 348–349
deoxynucleotides, 348–350, 349*f*
deoxyribonucleases (DNase), 359
  active chromatin and, 363
deoxyribonucleic acid. *See* DNA
deoxyribonucleoside diphosphates (dNDPs), 342, 342*f*
deoxyribonucleosides, 330, 342–343
deoxyribose, 147, 151, 151*f*
3-deoxyuridine, 334
dephosphorylation, enzyme regulation by, 92–94, 92*f*, 93*t*
depolarization
  in nerve impulse transmission, 481
  nonpolar lipids and, 206
derepressed systems, 421
dermatan sulfate, 599, 609*f*, 609*t*, 610
Δ^9-desaturase, 100, 233, 233*f*
desmosines, 604
desmosomes, 474
desmosterol, in cholesterol synthesis, 260, 262*f*
detergents, 470
detoxification, cytochrome P450 roles in, 119–120, 119*f*, 120*f*, 564–566
developmental biology, 727
dexamethasone suppression test, 575
dextrins, 153
dextrose, 149
DGAT (diacylglycerol acyltransferase), 240, 241*f*
DHA (docosahexaenoic acid), 209, 234
DHT. *See* dihydrotestosterone
diabetes insipidus, nephrogenic, 480
diabetes mellitus, 592*t*
  advanced glycation end-products in, 561, 561*f*
  case study, 729–730, 735–736
  excessive gluconeogenesis in, 181
  fatty liver and, 255
  forms of, 146
  free fatty acid levels in, 249
  glucose tolerance in, 189–190, 189*f*
  glycated hemoglobin in management of, 57, 561
  hypoglycemia and, 178
  insulin impairments in, 135, 146
  ketosis/ketoacidosis in, 225
  lipid transport and storage disorders, 247–248
  lipids and, 206, 209

lipogenesis and, 226, 230
respiratory chain disorders in, 130
diabetic cataract, 200
diacylglycerol acyltransferase (DGAT), 240, 241*f*
diacylglycerol (DAG), 211, 666
  formation of, 240*f*, 244*f*
  protein kinase C and, 515, 515*f*
1,2-diacylglycerol (DAG), in platelet aggregation, 678, 679*f*
diagnostic enzymology, 66–68, 67*f*, 67*t*, 572–573, 572*f*
diagnostic testing. *See* laboratory tests
α,γ-diaminobutyric acid, 19, 19*t*
Diamond-Blackfan anemia, 656
diapedesis, 665, 665*f*
diarrhea, 481, 563
diazanorleucine, 338, 339*f*
dicarboxylic aciduria, 225
dicer nuclease-TRBP complex, 402
dichloroacetate (DCA), 706, 706*t*
dicopper center, 103, 104*f*
dicumarol, 540
dielectric constant, 7
diet
  amino acids obtained only through, 15, 136, 274, 274*t*, 534
  biomedical importance of, 527
  blood glucose derived from, 186, 187*f*
  cholesterol levels affected by, 260, 264, 267
  hepatic VLDL secretion and, 253–254, 254*f*
  high fructose, 195–196, 197*f*, 199
  high-fat, fatty liver and, 255
  low carbohydrate, 190
  multiple deficiency states in, 535
  requirements
    energy, 133–134, 532–533, 532*f*
    protein and amino acid, 533–534
    transition metals, 97, 97*t*
    vitamins and minerals, 535–536
diethylenetriaminepentaacetate (DTPA), 214
diet-induced thermogenesis, 532
diffusion
  facilitated (*See* facilitated diffusion)
  net, 476
  simple, 475*f*, 475*t*, 476*f*, 483
diffusion-limited enzymes, 78
digestion
  biomedical importance of, 527
  of carbohydrates, 528, 528*f*
  of lipids, 528–529, 530*f*
  metabolism of major products of, 135–136, 135*f*, 136*f*, 138*f*
  proproteins in, 90
  of proteins, 529, 531
  of vitamins and minerals, 531–532, 531*f*

digitalis, 480, 626, 626*f*
dihydrobiopterin reductase, 296, 298*f*
dihydrofolate, methotrexate and, 343–344
dihydrofolate reductase, 545
dihydrolipoyl dehydrogenase, 118, 168, 169*f*
dihydrolipoyl transacetylase, 168, 169*f*
dihydroorotic acid, 343*f*
dihydropyridine receptor, 625, 625*f*
dihydropyrimidine dehydrogenase deficiency, 337, 345–346, 346*t*, 347*f*
dihydrotestosterone (DHT), 491
  plasma transport of, 507, 507*t*
  synthesis of, 495, 497*f*
dihydroxyacetone phosphate, in glycolysis, 165*f*, 166, 240, 241*f*, 242*f*
1,25-dihydroxyvitamin D, 538–539, 539*f*
diiodotyrosine (DIT), 500, 501*f*
diiron center, of hemerythrin, 100, 101*f*
dilated cardiomyopathy, 629
dimercaprol, 127
dimeric proteins, 39
dimers
  Cro protein, 426, 427*f*
  histone, 361, 361*f*, 363
  lambda repressor (cI) protein, 426, 427*f*
dimethylallyl diphosphate, in cholesterol synthesis, 260, 261*f*
dimethylaminoadenine, 333*f*
2,4-dinitrophenol, 127
dinucleotide, 335
dioxygenase, 119
dipalmitoyl lecithin, 210, 239
dipeptidases, 529
diphosphomevalonate decarboxylase, 261*f*
diphosphomevalonate kinase, 261*f*
diphtheria toxin, 418, 479
dipoles, 6–7, 7*f*
dipsticks, 573
direct bilirubin, 323
disaccharidases, 528
disaccharides, 148
  digestion of, 528, 528*f*
  of physiological importance, 151–152, 152*f*, 153*t*
disease
  autoimmune, 634, 635*t*
  biochemical processes underlying, 3
  biochemistry interrelationship with, 2, 2*f*
  conformational, 592*t*
  of DNA damage repair, 379, 380*t*
  genetic (*See* genetic diseases)
  membranes and, 467, 486*t*
  protein sorting and, 581
  proteoglycans and, 611–612
  of proteostasis deficiency, 597–598

red blood cells and, 653, 657*t*
  of skin, 3
dislocation, 592
dispensable amino acids, 136
dissociation, 9–10
dissociation constant (*K*)
  definition of, 10
  of water, 10
dissociation constant (*K*$_a$), of weak acids and bases, 11
dissociation constant (*K*$_d$), Michaelis constant as approximation of, 78–79
disulfide bonds
  in chemokines, 666, 666*f*
  formation of, 42
  in protein tertiary and quaternary structures, 40
DIT (diiodotyrosine), 500, 501*f*
diurnal variation, in cholesterol synthesis, 262
divalent metal ion transport protein (DMT-1), 105
divalent metal transporter, 639, 639*f*
Dixon plot, 81, 81*f*
DMT-1 (divalent metal ion transport protein), 105
DNA
  base pairing in, 349, 350*f*
    matching of for renaturation, 351
  biomedical importance of, 348
  carcinogen reactions with, 692
  in chromatin, 362*f*
  chromosomal, 362*f*, 365*f*, 365*t*, 366, 366*f*
  cloning of, 447–449, 448*f*, 449*f*, 449*t*
  coding regions of, 366, 367*f*
  complementarity of, 351, 450
  damage to
    double-strand break, 380, 380*f*, 381*f*
    types of, 379, 380*t*
  double-helix structure of, 8–9, 349–350, 350*f*
  enhancer elements, 433–434, 433*f*, 434*f*, 434*t*
  epigenetic code of, 395
  free radical damage to, 549–550, 550*f*
  genetic information contained in, 348–351
  in genetic information flow, 404–405
  immunoglobulin in rearrangements of, 371
  integrity monitoring of, 382
  mitochondrial, 368, 369*f*, 369*t*
  mutations in, 360–361, 378–382
    (*See also* mutations, genetic)
  noncovalent interactions in, 8–9
  in nucleosomes, 361, 361*f*, 362*f*, 363
  partial digestion of, 450
  as polynucleotides, 335
  protein identification using, 29

DNA (*Cont.*):
    recombinant (*See* recombinant DNA technology)
    relationship to mRNA, 367*f*
    relaxed form of, 351
    renaturation of, 351
    repair of, 593
        study of, 379–382
        types of, 379–382, 380*f*, 380*t*
    repetitive-sequence, 367
    replication and synthesis of, 351–352, 352*f*
        biomedical importance of, 360–361
        control of, 371–372, 372*f*, 372*t*
        DNA polymerase complex in, 372*f*, 373, 373*t*
        epigenetic signals during, 431–433, 431*f*
        initiation of, 374, 375*f*, 376*f*
        origin of, 372–373, 372*f*
        polarity of, 374, 376*f*
        reconstitution of chromatin structure in, 378
        repair during, 379–382, 380*t*
        replication bubble formation and, 374, 376*f*, 377*f*, 378
        replication fork formation and, 372*f*, 373, 376*f*
        ribonucleoside diphosphate reduction and, 342
        RNA primer in, 372*f*, 372*t*, 374, 375*f*, 376*f*
        RNA synthesis compared with, 385–386
        in S phase of cell cycle, 378–379, 378*f*, 379*f*, 379*t*
        semiconservative nature of, 352, 352*f*
        semidiscontinuous, 372*f*, 374, 376*f*
        steps in, 372*f*, 372*t*
        unwinding of, 373
    restriction enzyme cleavage of, 445–446, 445*t*, 446*f*, 446*t*
    in RNA synthesis (*See* RNA, synthesis of)
    sequencing of, 451, 452*f*
        direct, 453–454
    silencer elements, 435
    structure of
        components of, 348–349, 349*f*
        denaturation in analysis of, 350–351
        double-helical, 349–350, 350*f*
        forms of, 351
        grooves in, 350, 350*f*
    supercoiled, 351, 377, 378*f*
    topoisomerases of, 351
    transcription of, 351–352 (*See also* RNA, synthesis of)
    transposition of, 370–371
    unique-sequence (nonrepetitive), 367
    unwinding of, 388–389
    xenobiotic reactions with, 567, 567*f*

DNA blot transfer procedure, 450, 451*f*
DNA chips, 32
DNA damage sensors, 381, 381*f*
DNA elements
    combinations of, 435, 436*f*
    gene expression enhancement and repression by, 433–435, 433*f*, 434*f*, 434*t*
    locus control regions and insulators, 436–437
    reporter genes and, 435–436, 435*f*, 436*f*
    tissue-specific expression of, 435
DNA helicase, 372*f*, 373, 373*t*
DNA lesions, 722
DNA ligase, 373*t*, 374
    in chimeric DNA molecule preparation, 446–447
    in recombinant DNA technology, 446*t*
DNA methylases, sequence-specific, 445
DNA methyltransferases (DNMTs), inhibitors of, 699
DNA polymerases
    prokaryotic and eukaryotic, 374*t*
    in recombinant DNA technology, 446*t*, 452, 453*f*
    in replication, 372*f*, 373, 373*t*
DNA primase, 372*f*, 373*t*
DNA regulatory elements, 434–435, 434*f*
DNA repair mechanism, 567
DNA topoisomerases, 377, 377*f*
DNA transfection, endocytosis in, 482
DNA unwinding element (DUE), 373
DNA-binding domains (DBDs), 392
    of cI repressor protein, 426
    of regulatory transcription factor proteins, 437*t*
        helix-turn-helix motif, 437, 438*f*
        leucine zipper motif, 438, 439*f*
        zinc finger motif, 437–438, 438*f*
    transactivation domain separation from, 438–439, 440*f*
DNA-dependent protein kinase (DNA-PK), 381*f*, 382
DNA-dependent RNA polymerases
    initiation by, 386–387, 386*f*, 387*f*
    in operon model, 423, 424*f*, 425
DNA-protein interactions
    bacteriophage lambda as paradigm for, 425–430, 426*f*, 427*f*, 428*f*
    mapping of, 454–455
DNase. *See* deoxyribonucleases
DNase I
    active chromatin and, 363
    in recombinant DNA technology, 446*t*
dNDPs (deoxyribonucleoside diphosphates), 342, 342*f*
DNMTs (DNA methyltransferases), inhibitors of, 699

docking, in protein sorting, 585, 594–595
docosahexaenoic acid (DHA), 209, 234
dolichol, 213, 214*f*
    in cholesterol synthesis, 251–252, 261*f*
dolichol pyrophosphate, in glycoprotein synthesis, 558–559, 559*f*
domains
    albumin, 637
    chromatin, 362*f*, 363, 365–366
    DNA-binding (*See* DNA-binding domains)
    of hormone receptors, 489–490
    protein, 38, 39*f*, 40*f*
L-dopa, 499, 499*f*
dopa decarboxylase, in catecholamine biosynthesis, 499, 499*f*
dopamine, 499, 499*f. See also* catecholamines
dopamine β-hydroxylase
    copper in, 103
    vitamin C as coenzyme for, 547
dopamine β-hydroxylase (DBH), 499, 499*f*
dopaminylation, 430*t*
double displacement reactions, 82, 82*f*
double helix, of DNA structure, 8–9, 349–350, 350*f*
double membrane, of mitochondria, 122, 122*f*
double negative, 421
double reciprocal plot, 77–78, 78*f*
    for Bi-Bi reactions, 82, 83*f*
    inhibitor evaluation with, 80, 80*f*, 81*f*
double-strand breaks (DSBs), 380–381, 380*f*, 381*f*
double-stranded DNA (dsDNA), 349, 361, 371, 385–386
downstream promoter element (DPE), 391
driver mutations, 690, 712
Drosha-DGCR8 nuclease, 402
drug development
    enzyme kinetics role in, 83
    high-throughput screening for, 65–66
    rate-limiting enzymes as targets of, 87
    RNA targets for, 359
    suicide inhibitors in, 81–82
    telomerase as target for, 365
drug interactions, cytochrome P450 role in, 565
drug metabolism, 83
    acetylation and methylation, 566–567
    conjugation, 566
    cytochrome P450 role in, 119–120, 119*f*, 120*f*, 564–566
    hydroxylation, 564–566
drug resistance, 441
    in cancer, 711–712, 711*t*
drug-induced porphyria, 321
dry chemistry dipsticks, 573
DSBs (double-strand breaks), 380–381, 380*f*, 381*f*

dsDNA (double-stranded DNA), 349, 361, 371, 385–386

DTPA (diethylenetriaminepentaacetate), 214

dual-function proteins, 495

Dubin-Johnson syndrome, 325

Duchenne muscular dystrophy, 629

DUE (DNA unwinding element), 373

duodenal cytochrome b (Dcytb), 105, 639

dynamin, 482, 631

dysbetalipoproteinemia, familial, 268t

dyslipoproteinemias, 267, 268t

dysplasia
   acromicric, 604
   colorectal, 695–697, 696f, 696t
   geleophysic, 604
   thanatophoric, 616

α-dystroglycan, 629, 629f

dystroglycans, 605, 606f

dystrophin, 629, 629f

E

E. coli, lactose metabolism and operon hypothesis in, 422–423, 422f, 424f, 425

E. coli bacteriophage P1-based (PAC) vector, 449, 449t

E cyclins, 379, 379f, 379t

E (exit) site, of 80S ribosome, 413, 414f

E'_0. *See* redox potential

E3 ligases, 88

E_{act} (activation energy), 73, 73f, 75

E-cadherin, in cancer, 707, 708t

ECF (extracellular fluid), 468, 468t

ECM. *See* extracellular matrix

*Eco*RI, 445, 445t, 448f

*Eco*RII, 445t

edema
   kwashiorkor causing, 533
   plasma protein concentration and, 635
   vitamin B_1 deficiency in, 542

Edman reaction, protein and peptide sequencing with, 27f, 28–29

Edman reagent, 27f, 28

EDTA (ethylenediaminetetraacetate), 214

EF1A (elongation factor 1A), 413, 414f

EF2 (elongation factor 2), 413–414, 414f

EFA. *See* essential fatty acids

effectors, 88
   of apoptosis, 701–702, 702f

EFs (elongation factors), 413–414, 414f, 414t

EGF. *See* epidermal growth factor

Ehlers-Danlos syndrome, 44, 274, 599, 602t, 603t

eicosanoids, 208, 234, 235f, 670

eicosapentaenoic acid (EPA), 209, 232f, 234

eIF-1A, in protein synthesis, 410, 411f

eIF-2, in protein synthesis, 410, 411f

eIF-3, in protein synthesis, 410, 411f

eIF-4F complex, in protein synthesis, 410, 411f, 412–413, 412f

eIFs, in protein synthesis, 410, 411f

elaidic acid, 207t, 208, 208f

elastase, 529
   covalent catalysis of, 64

elastic cartilage, 615, 615t

elastin, 603–604, 604t, 615

electric potential, 584

electrical insulators
   myelin sheets as, 481
   nonpolar lipids as, 206

electrogenic effect, 480

electrolytes
   lipid bilayer and, 470
   measurement of, 572

electron microscopy (EM), protein structures solved by, 41

electron movement, in active transport, 480

electron transport, in respiratory chain, 122, 123f, 124–125, 124f, 126f

electronic resonance, in amino acids, 19, 19f

electron-transferring flavoprotein (ETF), 118

electrophiles, 9
   carcinogens as, 692
   in enzymatic catalysis, 61

electrophoresis, 635. *See also* gel electrophoresis

electrospray ionization, 30–31, 31f

electrostatic interactions, 8, 350

ELISAs (enzyme-linked immunosorbent assays), 66, 573

elliptocytosis, hereditary, 658, 660

elongase, 230, 231f
   in polyunsaturated fatty acid synthesis, 233–234, 233f

elongation
   in protein synthesis, 413–414, 414f, 414t
   in proteoglycan synthesis, 608
   in RNA synthesis, 386f, 388–389

elongation arrest, 588, 589f

elongation chain, in fatty acid synthesis, 230, 231f

elongation factor 1A (EF1A), 413, 414f

elongation factor 2 (EF2), 413–414, 414f

elongation factors (EFs), 413–414, 414f, 414t

EM (electron microscopy), protein structures solved by, 41

emaciation, 134

emphysema, 645

EMT (epithelial-to-mesenchymal transition), 707f, 708

emulsions, amphipathic lipids forming, 215, 215f

encephalopathy
   lactic acidosis with, 122
   vitamin B_1 deficiency in, 542

endergonic reactions, exergonic reaction coupling to, 110–111, 110f, 111f, 113

endocrine system, 488. *See also* hormone receptors; hormones

endocytosis, 467, 475t
   LDL in, 263
   of macromolecules, 481–483, 482f, 483f, 485f

endoglycosidases, 610–611

endonucleases, 359, 445. *See also* restriction enzymes
   in chimeric DNA molecule preparation, 446–447

endopeptidases, 529

endoplasmic reticulum (ER), 139
   brefeldin A and, 596
   calnexin in, 559–560
   cytochromes P450 in, 565
   exocytosis in, 483
   fatty acid chain elongation in, 230, 231f
   misfolded proteins degradation in, 592–593, 592t, 593f
   protein attachment to, 591
   protein sorting signals in, 582, 582t
   in protein synthesis, 416
   quality control of, 591–592
   rough (*See* rough endoplasmic reticulum)
   signal hypothesis of polyribosome binding to, 582, 582f, 587–590, 588t, 589f

endoplasmic reticulum-associated degradation (ERAD), 592–593, 592t, 593f

endosymbiosis, 722

endothelial cells
   in hemostasis and thrombosis, 678–679, 680
   selectins of, 561–562

endothelium-derived relaxing factor, 630

energy. *See also* bioenergetics; free energy
   for active transport, 480, 480t
   human requirement for, 133–134, 532–533, 532f
   transduction in membranes, 467

energy balance, 109, 532–533, 532f

energy capture, 112

energy conservation, 112

energy expenditure, 532, 532f

energy of activation. *See* activation energy

energy-linked transhydrogenase, mitochondrial, 129

enhancer response element, 433f
enhancer-binding proteins, 433–434
enhancers, in gene expression, 421
    β-interferon gene, 434–435, 434f
    properties of, 433–435, 434t
    study of, 433–434, 433f
    tissue-specific expression of, 435
enolase, in glycolysis, 165f, 167
enoyl reductase, 228f, 229f
Δ²-enoyl-CoA hydratase, 219, 219f, 220
entactin, 605
enterohepatic circulation, 265f, 266, 529
enterohepatic urobilinogen cycle, 323
enteropeptidase, 90, 91f
enthalpy, 110
entropy, 110
environmental mutations, 691
enzymatic assays
    clinical diagnosis using, 66–68, 67f, 67t,
        572–573, 572f
    coupled, 66, 66f
    in drug development, 83
    high-throughput screening, 65–66
    immunoassays, 66
    initial velocity measurements with, 76,
        76f
    single-molecule, 65, 65f
    spectrophotometric, 66, 66f
enzyme inhibitors, transition state
        analogs, 62
enzyme kinetics
    activation energy, 72–73, 73f, 75
    allosteric effects, 88–89
    balanced chemical equations, 72
    biomedical importance of, 71–72
    catalytic constant and catalytic
        efficiency, 78
    catalytic lowering of activation energy
        barriers, 75
    in drug development, 83
    equilibrium constant, 72, 74–75
    free-energy changes, 72–73, 73f
    Hill equation, 79, 79f
    inhibition, 79–82, 80f, 81f
    initial velocity, 76–77, 76f
    kinetic order, 74
    Lineweaver-Burk plots, 77–78, 78f, 80,
        80f, 81f, 82, 83f
    Michaelis-Menten equation, 76–79, 76f,
        78f, 82, 83f
    multisubstrate reactions, 82, 82f, 83f
    reaction rates (See rate of reaction)
    transition states, 72–73, 73f
enzyme-linked immunosorbent assays
        (ELISAs), 66, 573
enzymes
    active site of, 61–62, 62f
    activity of, 78, 572–573, 572f
    bifunctional, 185

biomedical importance of, 59
as catalysts, 9, 60, 60f
classification of, 60
clinical diagnosis using, 66–68, 67f, 67t,
        572–573, 572f
coenzymes of, 60–61, 61f
cofactors of, 60–61, 61f
conformational changes of, 63, 63f
conservation of catalytic residues in,
        64
degradation of
    rate of, 280
    regulation of, 87–88
detection of, 65–66, 65f, 66f
digestive, 529
in DNA repair, 360
as drug targets, 83
group transfer reactions of, 9
half-lives of, 280
heavy metal inactivation of, 98
homologous, 64
inhibitors of (See inhibitors)
isozymes, 64–65
kinetics of (See enzyme kinetics)
mechanisms of action of, 59
    acid-base catalysis, 62, 63f
    covalent catalysis, 62, 62f, 64f, 65f
    proximity, 62
    site-directed mutagenesis in study
        of, 69
    strain, 62
membranes in localization of, 467
of neutrophils, 667, 668t
nucleases, 359
prosthetic groups of, 60–61, 61f
rate-limiting, 87
reactive oxygen species and,
        723–724
recombinant DNA in studies of, 68–69,
        69f
regulation of
    active or passive, 86, 86f
    allosteric, 88–89, 141, 141f
    biomedical importance of, 85–86
    compartmentation role in, 86–87
    control networks of, 94, 94f
    covalent modifications in, 90–93, 92f,
        93t
    hormones in, 141, 141f
    multiple mechanisms of, 88
    proenzymes in, 90, 92f
    quantities, 87–88
    rate-limiting enzymes as targets of, 87
    second messengers in, 89
regulatory, 139
replacement therapy for, 245
RNA, 354
specificity of, 60, 60f
synthesis of, regulation of, 87

enzyme-substrate (ES) complex, stability of,
        61
enzymopathy, 657
eosinophils, 664, 667, 669, 669f
EPA (eicosapentaenoic acid), 209, 232f,
        234
epidermal growth factor (EGF)
    in insulin signal generation, 516, 517f
    receptor for, 490, 711
epidermis, 718t
epidermolysis bullosa, 602t, 603, 632
epidermolytic palmoplantar keratoderma,
        632
epigenetic code of DNA, 395
epigenetic mechanisms, control of gene
        transcription and, 431–433, 431f,
        432f
epigenetic signals, 431–433, 431f, 432f
epigenetics, 86
    in cancer development, 698–699
    histone code as, 92
epilepsy, 146
epimerases, 608
epimers, 149, 150f
epinephrine. See also catecholamines
    blood glucose regulation by, 188
    glycogen regulation by, 175, 176f
    in glycolysis and gluconeogenesis
        regulation, 183
    in lipogenesis regulation, 231, 232f
    in lipolysis, 256, 257f
    synthesis of, 492f, 499f, 500
    tyrosine conversion to, 310, 312f
episomes, 448
epithelial-to-mesenchymal transition
        (EMT), 707f, 708
epitopes, 37, 646
EPO (erythropoietin), 654, 656
epoxide hydrolase, 567, 567f
epoxides, 567, 567f
eptifibatide, 680
equilibrium
    of chemical reactions, 72
    of nitrogen in healthy humans, 533
equilibrium constant (K_{eq})
    enzymatic effects on, 75
    Gibbs free-energy change and, 72, 110
    as ratio of rate constants, 74–75
ER. See endoplasmic reticulum
ERAD (endoplasmic reticulum-associated
        degradation), 592–593, 592t, 593f
ergosterol, 213, 213f
ergothioneine, 308, 308f
erlotinib, 711
ERp57, 559
erythrocytes. See also red blood cells
    in fasting state, 135, 144–145
    in fed state, 135, 142–144, 143f
    glucose requirement of, 142

glycolysis in, 168, 168*f*, 655, 655*t*
metabolic features of, 145*t*
pentose phosphate pathway and
    glutathione peroxidase in, 195, 195*f*
    impairment in, 198
erythromycin, 151
erythropoiesis
    initial stages of, 656
    iron-deficient, 643
erythropoietic porphyrias, 320*t*, 321
erythropoietin (EPO), 654, 656
erythrose-4-phosphate, in pentose
    phosphate pathway, 192*f*, 193*f*, 194
ES (enzyme-substrate) complex, stability
    of, 61
Escherichia coli, lactose metabolism in,
    operon hypothesis and, 422–423,
    422*f*, 424*f*, 425
Escherichia coli bacteriophage P1-based
    (PAC) vector, 449, 449*t*
essential amino acids, 15, 136, 274, 274*t*,
    534
essential fatty acids (EFA), 206, 226
    abnormal metabolism of, 236
    deficiency of, 234
    physiologic effects of, 234
    polyunsaturated, 232, 232*f*
    prostaglandin production and, 234
essential fructosuria, 199
essential pentosuria, 191, 198
estradiol, 491, 492*f*, 493*f*, 507, 507*t*
estriol, 495, 497*f*
estrogen receptor modulators, 712
estrogens
    amino acid transport and, 478
    cholesterol levels and, 267
    synthesis of, 495–496, 497*f*, 498*f*
estrone, 495–496, 497*f*, 507, 507*t*
ETF (electron-transferring flavoprotein),
    118
ethanol
    cytochrome P450 induction by, 565
    fatty liver and, 255
ethanolamine, 211, 211*f*
ether lipids, biosynthesis of, 242*f*
ethylenediaminetetraacetate (EDTA),
    214
euchromatin, 364
eukaryotic gene expression
    alternative RNA processing and, 442
    amplification, 441–442, 441*f*
    bacteriophage lambda as paradigm for
        protein-DNA interactions,
        425–430, 426*f*, 427*f*, 428*f*
    chromatin template in, 429–430
    DNA enhancer elements and, 433–435,
        433*f*, 434*f*, 434*t*
    epigenetic mechanisms, 431–433, 431*f*,
        432*f*

locus control regions and insulators in,
    436–437
methods of, 439–440, 440*t*
as model for study, 422
mRNA stability and, 442, 442*f*
ncRNA alteration of mRNA function
    in, 441
prokaryotes compared with, 439–440,
    441*f*, 442*f*
response diversity in, 435, 436*f*
special features of, 429
tissue-specific expression of, 435
eukaryotic initiation factors (eIFs), in
    protein synthesis, 410, 411*f*
eukaryotic promoters, in transcription,
    390–392, 390*f*, 391*f*, 392*f*, 393*f*, 394
eukaryotic transcription complex, 391*f*
    components of, 394
    nucleosomes and, 393*f*, 395–396
    phosphorylation for RNA polymerase II
        activation, 393*f*, 395–396
    PIC assembly, 395*f*, 396–397
    RNA polymerase II formation, 394
    transcription activators and
        coregulators in, 395–396, 396*t*
evolution
    lifespan and, 727
    RNA world hypothesis of, 69–70
evolutionary conservation, of catalytic
    residues, 64
EVs (extracellular vesicles), 483–484, 485*f*,
    698
exchange diffusion systems, 128
exchange transporters, of mitochondria,
    128–130, 128*f*, 129*f*, 130*f*
excitation-response coupling, 467
executioners, of apoptosis, 701–702, 702*f*
exergonic reactions, endergonic reaction
    coupling to, 110–111, 110*f*, 111*f*,
    113
exit (E) site, of 80S ribosome, 413, 414*f*
exocytosis, 467, 475*t*, 481–483, 483*f*, 485*f*,
    582
exocytotic (secretory) pathway, 582
exoglycosidases, 610–611
exons, 366, 367*f*
    description of, 398
    processing of, 397*f*, 398*f*, 404–405
    splicing of, 397*f*, 398–399, 398*f*
exonucleases, 359, 445
    in recombinant DNA technology, 446*t*
exopeptidases, 529
exosomes, 483–484, 485*f*
    cancer and, 698–699
exportins, 586
expression vector, 450
extra arm, of tRNA, 355, 357*f*
extracellular environment, membranes in
    maintenance of, 468, 468*t*

extracellular fluid (ECF), 468, 468*t*
extracellular matrix (ECM). *See also* bone
    aging process and, 599
    biomedical importance of, 599
    collagen (*See* collagen)
    elastin, 603–604, 604*t*
    fibrillins, 604, 605*f*
    fibronectin, 604–605, 605*f*, 606*f*
    glycosaminoglycans (*See*
        glycosaminoglycans)
    laminin, 605–606, 606*f*
    proteoglycans (*See* proteoglycans)
extracellular vesicles (EVs), 483–484, 485*f*,
    698
extramitochondrial system, fatty acid
    synthesis in, 226
extravasation, of cancer cells, 706, 707*f*
extrinsic pathway, 680–681, 680*f*, 681*f*,
    681*t*, 682*t*
extrinsic tenase complex, 680–681
ezetimibe, for hypercholesterolemia, 267

F
FAB (fast atom bombardment), 30–31, 31*f*
Fabry disease, 245*t*
facilitated diffusion, 475*f*, 475*t*, 483
    for glucose (*See* glucose transporters)
    hormone regulation of, 477
    ping-pong mechanism of, 477, 477*f*
    transporters involved in, 476–477, 476*f*,
        476*t*, 477*f*
facilitated transport, of bilirubin, 322
FACITs (fibril-associated collagens with
    interrupted triple helices), 601
factor I. *See* fibrinogen
factor II. *See* prothrombin
factor III. *See* tissue factor
factor IX, 681*t*, 682, 682*t*
    coumarin drugs and, 685–686
    deficiency of, 687
factor V, 681*t*, 682–683, 682*t*, 683*f*
factor V Leiden, 685
factor VII, 680, 681*t*, 682*t*, 685–686
factor VIII, 681*t*, 682, 682*t*
    deficiency of, 687
factor X, 681*t*, 682*t*
    coumarin drugs and, 685–686
    extrinsic pathway activation of,
        680–681, 680*f*, 681*f*
    intrinsic pathway activation of,
        680–682, 680*f*, 681*f*
    prothrombin activation by, 681–682,
        683*f*
factor Xa, 682–683
factor XI, 681*t*, 682, 682*t*
factor XII, 681*t*, 682, 682*t*
factor XIIa, 681–682, 683*f*
factor XIII, 681*t*, 682*t*, 685
facultative heterochromatin, 364

FAD. *See* flavin adenine dinucleotide

FADH$_2$, fatty acid oxidation generation of, 219, 219*f*

false-positive result, 570

familial amyloidosis, 646

familial hypercholesterolemia (FH), 252, 267, 268*t*, 486, 486*t*, 592*t*

familial hypertrophic cardiomyopathy, 628–629

Fanconi-Bickel disease, 174*t*

Farber disease, 245*t*

farnesoid X receptor (FXR), 266, 521*t*

farnesyl diphosphate, in cholesterol/polyisoprenoid synthesis, 251–252, 261*f*

fast atom bombardment (FAB), 30–31, 31*f*

fast twitch fibers, 145*t*, 628

fasting state, metabolic fuels in, 134–135, 134*f*, 144–145, 144*t*, 145*f*

fatal infantile mitochondrial myopathy and renal dysfunction, 130

fatigue, glycolysis defects causing, 163

fats, 144*t*, 206. *See also* lipids

fatty acid chains, elongation of, 230, 231*f*

fatty acid elongase system, 230, 231*f*, 233–234, 233*f*

fatty acid oxidase, 219*f*

fatty acid synthase complex, 227–228, 227*f*, 228*f*, 229*f*, 232

fatty acid synthesis
  biomedical importance of, 226
  citric acid cycle role in, 159–161, 161*f*
  in cytosol, 226–230, 227*f*, 228*f*
  extramitochondrial, 226
  fructose effects on, 195–196, 197*f*, 199

fatty acid synthetase, 78

fatty acid transport protein, membrane, 249

fatty acid-binding protein, 218, 249

fatty acids
  absorption of, 529
  activation of, 218, 218*f*
  anti-inflammatory, 209
  eicosanoids formed from, 234, 235*f*
  essential (*See* essential fatty acids)
  free (*See* free fatty acids)
  for health maintenance, 3
  long-chain ω3, 206, 209
  in membranes, 469, 469*f*
  as metabolic fuel, 134–135, 141–142
  metabolic pathways of, 135–136, 135*f*, 136*f*
    in fasting state, 144, 144*t*, 145*f*
    in fed state, 143–144, 143*f*, 144*t*
    at organ and cellular level, 137–139, 138*f*
  monounsaturated (*See* monounsaturated fatty acids)

nomenclature of, 206–207, 206*f*

oxidation of (*See also* ketogenesis)
  acetyl-CoA release and, 218–220, 218*f*, 219*f*, 220*t*
  biomedical importance of, 217
  clinical aspects of, 224–225
  hypoglycemia caused by impairment of, 224–225
  impairment of, 224–225

physical/physiologic properties of, 209

polyunsaturated (*See* polyunsaturated fatty acids)

saturated, 206, 206*f*, 207, 207*t* (*See also* saturated fatty acids)

trans, 208–209, 236

transport of, carnitine in, 218, 218*f*

triacylglycerols as storage form of, 209, 209*f*

unsaturated (*See* unsaturated fatty acids)

fatty liver
  alcoholism and, 255
  nonalcoholic fatty liver disease, 254
  nonalcoholic steatohepatitis, 254
  of pregnancy, 225
  triacylglycerol metabolism imbalance and, 254–255
  types of, 255

favism, 198

fed state, metabolic fuels in, 134–135, 134*f*, 142–144, 143*f*

feedback effectors, 88

feedback inhibition, of enzymes, 88–89, 89*f*, 93

feedback regulation, 141
  of thrombin, 685

Fenton reaction, 640, 640*f*

fermentation, by cell-free extract of yeast, 1–2

ferric iron
  in heme, 56, 99–100
  from heme catabolism, 321

ferritin, 416, 531, 531*f*, 639–640, 639*t*, 641–642, 642*f*

ferrochelatase, 317, 318*f*, 319*f*, 320*t*

ferroportin, 639, 642

ferrous iron, in heme, 50–51, 50*f*, 51*f*, 99–100

fertilization, glycoprotein role in, 561

Fe-S clusters, 100, 102*f*

Fe-S (iron-sulfur) proteins, in respiratory chain complexes, 122–124, 123*f*

fetal hemoglobin (HbF), 52–53, 53*f*, 56

α-fetoprotein, 637*t*

FFAs. *See* free fatty acids

FGFR3 (fibroblast growth factor receptor 3), 616

FH (familial hypercholesterolemia), 252, 267, 268*t*, 486, 486*t*, 592*t*

fibril-associated collagens with interrupted triple helices (FACITs), 601

fibrillins, 604, 605*f*

fibrils, 601, 601*f*, 614

fibrin, 678

fibrin clot formation
  amplification in, 685
  extrinsic pathway, 680–681, 680*f*, 681*f*, 681*t*, 682*t*
  fibrinogen conversion to fibrin, 683–685, 683*f*, 684*f*
  inherited deficiencies of, 686
  intrinsic pathway, 680–682, 680*f*, 681*f*, 681*t*, 682*t*
  plasmin dissolution of, 686, 686*f*
  prothrombin conversion to thrombin, 682–683, 683*f*

fibrin deposits, 678

fibrin monomer, 685

fibrinogen (factor I), 634, 635, 636*f*, 680, 681*t*, 682*t*
  conversion to fibrin, 683–685, 683*f*, 684*f*
  in platelet aggregation, 678, 679*f*

fibrinolysis, 686, 686*f*
  intrinsic pathway in, 681–682

fibrinopeptide A (FPA), 683, 684*f*

fibrinopeptide B (FPB), 683, 684*f*

fibroblast growth factor receptor 3 (FGFR3), 616

fibroblasts, pinocytosis by, 483

fibroelastic cartilage, 615, 615*t*

fibronectin, 604–605, 605*f*, 606*f*

fibrous proteins, 35
  collagen as, 44

finasteride, 712

first law of thermodynamics, 109–110

first-order rate constant, 74

fish-eye disease, 268*t*

flame photometry, laboratory tests using, 572

flavin adenine dinucleotide (FAD), 334*t*
  in citric acid cycle, 158*f*, 159
  in cytochromes P450, 119, 119*f*
  in dehydrogenases, 118
  in oxidases, 116, 117*f*
  in respiratory chain complexes, 122
  vitamin B$_2$ in synthesis of, 542

flavin mononucleotide (FMN)
  in cytochromes P450, 119, 119*f*
  in dehydrogenases, 118
  in oxidases, 116, 117*f*
  in respiratory chain complexes, 122
  vitamin B$_2$ in synthesis of, 542

flavoproteins
  cytochromes P450, 119–120, 119*f*
  dehydrogenases, 118
  oxidases, 116, 117*f*
  in respiratory chain complexes, 122

FLIP, in apoptosis, 701, 702*f*

flip-flop, phospholipid, membrane asymmetry and, 472
flippases, 472
fluid mosaic model, 473–474, 474f
fluidity, membrane, 473–474
fluid-phase pinocytosis, 482, 482f
fluorescence, of porphyrins, 318, 320f
fluorescence spectrophotometry, laboratory tests using, 571–572
fluoride, enolase inhibition by, 165f, 167
1-fluoro-2,4-dinitrobenzene, 28
fluoroacetate, 158–159, 158f
5-fluorouracil, 83, 334, 334f, 344, 346t
fluvastatin, 267
flux, metabolite, 86, 86f, 93, 141, 141f
flux-generating reactions, of metabolic pathways, 139–141
FMN. See flavin mononucleotide
focal adhesions, 605
folate trap, 544f, 546
folding
    protein, 42–43, 43f
    of proteins, calnexin role in, 559–560
folic acid
    coenzymes derived from, 61
    deficiency of, 293, 546
    functions of, 536t, 545, 545f
    inhibition of, 545
    structure of, 545, 545f
    supplements of, 546
    toxicity of, 546
formamide, 351
four-helix bundle, 595
Fourier synthesis, 41
FPA (fibrinopeptide A), 683, 684f
FPB (fibrinopeptide B), 683, 684f
frameshift mutations, 408–409, 409f
free energy
    ATP as cellular currency of, 112–114, 112f, 113f, 125–126
    of ATP hydrolysis, 111–112, 111t, 112f
    chemical reaction changes in, 72–73, 73f, 109–110
    enzymatic effects on, 75
    high-energy phosphate capture and transfer of, 111–112, 111t, 113f, 125–126
    metabolic regulation and, 85–86, 86f
    obtainment of, 111
    redox potential and, 115–116, 116t
    as useful energy in biologic systems, 109–110
free fatty acids (FFAs), 206, 217–218
    adipose tissue and, 255
    in fatty liver, 255
    glucose metabolism and, 256
    insulin and, 247, 256
    ketogenesis regulation and, 223–224, 223f, 224f

lipid mobilization and, 247
    lipogenesis affected by, 230, 231f
    in lipoproteins, 248, 248t
    metabolism of, 249
    in VLDL formation, 254, 254f
free polyribosomes, protein synthesis on, 591
free radical theory of aging, 720
free radicals
    antioxidant paradox and, 552–553, 552t
    biomedical importance of, 549
    damage caused by, 549–550, 550f
    lipid peroxidation producing, 213–214, 214f
    mechanisms protecting against, 551–552, 552f
    self-perpetuating chain reactions of, 549
    sources of, 550–551, 551f
    theory of aging, 720
fructokinase, 195, 197f, 199
    deficiency of, 199
fructose
    absorption of, 528, 528f
    defective metabolism of, 147, 199–200
    in diabetic cataract, 200
    in glycolysis pathway, 167
    ingestion of large quantities of, 195–196, 197f, 199
    physiological importance of, 151t
    structure of, 149, 149f
fructose 1,6-bisphosphatase
    deficiency of, 199–200
    in gluconeogenesis, 181, 182f
    in pentose phosphate pathway, 192f, 193f, 194
fructose 1,6-bisphosphate
    in gluconeogenesis, 181, 182f
    in glycolysis, 165f, 166
fructose 2,6-bisphosphatase, 184–185, 186f
    covalent catalysis of, 64, 65f
fructose 2,6-bisphosphate, in glycolysis and gluconeogenesis regulation, 184–185, 186f
fructose-6-phosphate
    in gluconeogenesis, 181, 182f
    in glycolysis, 165f, 166
    in pentose phosphate pathway, 192f, 193f, 194
fructosuria, 199–200
fucose, 154, 155f, 556t
fucosyl (Fuc) transferase, 660
fumarase (fumarate hydratase), in citric acid cycle, 158f, 159
fumarate
    in citric acid cycle, 158f, 159
    in urea synthesis, 285f, 286
fumarylacetoacetate hydrolase, 296, 297f

functional groups
    of amino acids, 19–21, 20f, 20f, 21t
    weak acids acting as, 11–12
furanose ring structures, 149, 149f
furin, 596, 596f
fusion of vesicles, 596
fusion proteins, 68–69, 69f
futile cycles
    in glycolysis and gluconeogenesis regulation, 185
    of lipids, 533
FXR (farnesoid X receptor), 266, 521t

G
ΔG. See Gibbs free-energy change
ΔG⁰. See standard free-energy change
G1 and G2 phase, 378, 378f, 379f
G6PD. See glucose-6-phosphate dehydrogenase
GA. See Golgi apparatus
GABA (γ-aminobutyrate), biosynthesis of, 313, 313f
GAGs. See glycosaminoglycans
Gal transferase, 661
GAL1 enhancer, 439, 440f
galactokinase, 197, 198f
galactosamine, 151, 152f
    synthesis of, 197, 199f
galactose, 147, 150, 150f
    absorption of, 528, 528f
    in glycoproteins, 555, 556t
    impaired metabolism of, 200
    metabolism of, 197, 198f
    physiological importance of, 151t
galactose-1-phosphate uridyl transferase, 197, 198f
galactosemias, 147, 200
galactoside, 150
galactosyl receptor, 483
galactosylceramide, 212, 212f, 243, 244f, 245t, 469
gallium, toxicity of, 98
gallstones, 260, 527
    hyperbilirubinemia caused by, 323f, 323t, 324
GalNAc (N-acetylgalactosamine), 555, 556t, 557
GalNAc transferase, 661
γ-rays, carcinogenic effect of, 691, 691t
gangliosides, 155, 212
    amino sugars in, 151, 152f
    in membranes, 469
    synthesis of, 244, 244f
gap 1 and 2 phase, 378, 378f, 379f
gap junctions, 467, 483, 484f
GAPs (GTPase-accelerating proteins), 585
gastroenteropathy, protein-losing, 637
gated ion channels, 478–479, 479f

gatekeeper tumor suppressor genes, 695
Gaucher disease, 245*t*, 592*t*
GDF15 (growth differentiation factor 15), 642
GDH. *See* glutamate dehydrogenase
GDP, 594
gefitinib, 711
GEFs (guanine nucleotide exchange factors), 585, 586*f*
gel electrophoresis
　polyacrylamide gel, 27, 28*f*
　protein and peptide purification with, 27, 28*f*
geleophysic dysplasia, 604
gel-filtration chromatography. *See* size-exclusion chromatography
gemfibrozil, 267
gene amplification, in cancer, 693
gene arrays, 32
gene conversion, 371
gene expression
　constitutive, 422, 425
　miRNA and siRNA inhibition of, 358–359
　in pyrimidine nucleotide synthesis regulation, 344
　retinoic acid role in, 537–538
　silencing of, 92
gene expression regulation
　alternative RNA processing and, 442
　amplification, 441–442, 441*f*
　biomedical importance of, 420–421
　DNA-binding domains of regulatory transcription factor proteins, 437*t*
　　helix-turn-helix motif, 437, 438*f*
　　leucine zipper motif, 438, 439*f*
　　zinc finger motif, 437–438, 438*f*
　domain separation in, 438–439, 440*f*
　eukaryotic
　　bacteriophage lambda as paradigm for protein-DNA interactions, 425–430, 426*f*, 427*f*, 428*f*
　　chromatin template in, 429–430
　　DNA enhancer elements and, 433–435, 433*f*, 434*f*, 434*t*
　　epigenetic mechanisms, 431–433, 431*f*, 432*f*
　　locus control regions and insulators in, 436–437
　　prokaryotes compared with, 439–440, 440*t*, 441*f*, 442*f*
　　reporter genes, 435–436, 435*f*, 436*f*
　　response diversity in, 435, 436*f*
　　special features of, 429
　　tissue-specific expression of, 435
　models of, 422
　mRNA stability and, 442, 442*f*
　ncRNA alteration of mRNA function in, 441

operon model of, 422–423, 422*f*, 424*f*, 425
　positive and negative, 421, 421*t*
　prokaryotic
　　eukaryotes compared with, 439–440, 440*t*, 441*f*, 442*f*
　　on-off manner of, 435
　　uniqueness of, 422
　RNA alternative processing for, 442
　targeted, 454
　temporal responses to, 421–422, 421*f*
gene therapy, 4, 454
　artificial membranes and, 473
　biomedical importance of, 444
　for lipid storage disorders, 245
　mammalian viral vectors for, 449
　for metabolic disorders, 288
general acid-base catalysis, 62
general transcription factors (GTFs), 391*f*, 392
　activators and coregulators and, 395–396
　in PIC formation, 396–397
　in RNA polymerase II formation, 394
genes
　alteration of, 369*f*, 370*f*, 378–382
　biomedical importance of, 348
　cancer, 690–691, 693–697, 693*t*, 694*f*, 695*f*, 695*t*, 696*f*, 696*t*
　chromosome localization of, 453
　of cytochromes P450, 565
　housekeeping, 422
　immunoglobulin, DNA rearrangement and, 370–371
　inducible, 422
　knockout/knockin, 454
　mapping of, 366
　processed, 370–371
　targeted disruption of, 454
genetic code, 348. *See also* DNA
　L-α-amino acids specified by, 16, 16*t*–17*t*, 18*t*
　biomedical importance of, 404
　features of, 405–406, 406*t*
　triplet codes in, 405, 405*t*
genetic diseases/disorders
　amino acid metabolism defects, 286–288, 287*t*, 290–291, 300, 304*t*
　of bone, 614–615, 614*t*
　cancer, 691, 697, 699, 699*t*
　clotting system deficiencies, 686
　collagen defects in, 44
　enzymes for diagnosis of, 68
　fructose metabolism deficiencies, 199–200
　galactosemias, 200
　gene therapy for, 454
　glycogen storage diseases, 171, 174, 174*t*, 178–179

glycoprotein lysosomal hydrolase deficiencies, 562–563
　hemoglobin mutations, 56, 57*f*
　hemolytic anemia, 163, 170, 191, 198
　membrane protein mutations, 484–486, 486*t*
　of muscles, 628–629, 628*t*, 629*f*
　porphyrias, 315, 318, 320–321, 320*t*, 321*f*
　pyruvate dehydrogenase deficiency, 170
　respiratory chain, 130
　urea cycle defects, 286–288, 287*t*
genetic engineering, 444. *See also* recombinant DNA technology
genetic information, flow of, 404–405
genetic mutations. *See* mutations, genetic
Genevan system, for fatty acid nomenclature, 206
genome
　function of, 366–367
　microsatellite repeat sequences in, 368
　nonrepetitive sequences in, 367
　repetitive sequences in, 367–368
　targeted gene regulation in, 454
genomic instability, in cancer, 700, 700*f*
genomic library, 450
genomic technology. *See* recombinant DNA technology
genomics, protein identification using, 29
geometric isomerism, of unsaturated fatty acids, 208–209, 208*f*
geranyl diphosphate, in cholesterol synthesis, 260, 261*f*
gestational diabetes, 146
GGT. *See* γ-glutamyltransferase
GH. *See* growth hormone
Gibbs free-energy change (ΔG), 72
　of ATP hydrolysis, 111–112, 111*t*, 112*f*
　in biologic systems, 109–110
　enzyme effects on, 75
　equilibrium constant and, 72, 110
　metabolic regulation and, 85–86, 86*f*
　redox potential and, 115–116, 116*t*
Gilbert syndrome, 324
GK (glucokinase) gene, 399–400, 400*f*
Glanzmann thrombasthenia, 662
GlcN. *See* glucosamine
glibenclamide, 225
β-globin gene, 433, 433*f*
globular proteins, 35
globulins, 635
glomerular filtration, 605–606
glomerular membrane, 605–606
glomerulonephritis, 606
glucagon, 135, 144, 145*f*
　blood glucose regulation by, 188*t*, 197
　in cholesterol synthesis, 263, 263*f*
　glycogen regulation by, 175

in glycolysis and gluconeogenesis regulation, 183
for hypoglycemia, 178
in lipogenesis regulation, 231, 232f
in lipolysis, 256
glucagon/insulin ratio, in ketogenesis regulation, 224
glucan, 152
glucan transferase, in glycogenolysis, 173f, 174
β-glucocerebrosidase, diagnostic use of, 67t
glucocorticoids
blood glucose regulation by, 188
in cholesterol synthesis, 263, 263f
in lipolysis, 256, 257f
NF-κB pathway regulation by, 517–518, 518f
plasma transport of, 506, 507t
regulation of gene expression by, 509, 509f
synthesis of, 494–495, 494f
therapeutic use of, 519
glucogenic amino acids, 142, 146, 282, 282f
glucogenic intermediates, 160
glucokinase, 165–166, 172f, 173, 736
blood glucose regulation by, 186–187, 188f
gene mutation of, 736t
glucokinase (GK) gene, 399–400, 400f
gluconeogenesis, 134, 137, 167
biomedical importance of, 180–181
blood glucose derived from, 186, 187f
citric acid cycle role in, 159–161, 160f
clinical aspects of, 189–190, 189f
during exercise, 178
Gibbs free-energy change of, 87
induction and repression of enzymes catalyzing, 183, 184t
low carbohydrate diets and, 190
propionate metabolism and, 181–182, 183f
rate of, 180
regulation of, 183–185, 184t, 185f, 186f
reversal of glycolysis steps in, 181, 182f
gluconolactone hydrolase, in pentose phosphate pathway, 192f, 193, 193f
glucosamine (GlcN), 151, 152f, 197, 199f, 609–610
glucosan, 152
glucose
absorption of, 528, 528f
amino sugar synthesis from, 197, 199f
biomedical importance of, 147–148, 151t
blood levels of (See blood glucose)
breakdown of (See glycolysis)
in cancer cells, 704, 705f, 706t
CNS and erythrocyte requirement for, 142
in extracellular and intracellular fluid, 468t

fatty acid synthesis from, 161–162, 161f
galactose conversion to, 197, 198f
in glycoproteins, 556t
insulin regulation of, 142–144, 165
isomerism of, 149–150, 149f, 150f
as metabolic fuel, 134–135, 134f, 142
metabolic pathways of, 135, 135f, 136f
in fasting state, 144–145, 144t
in fed state, 142–144, 143f, 144t
at organ and cellular level, 137–139, 138f
as most important monosaccharide, 147, 148
oxidation of, lactate production and, 167
in pentose phosphate pathway (See pentose phosphate pathway)
permeability coefficient of, 471f
structure of, 148, 148f
synthesis of (See gluconeogenesis)
transport of, 481, 481f
glucose oxidase, plasma glucose concentration measurement using, 68
glucose tolerance, 189–190, 189f
glucose transporters (GLUT), 186, 187t
in blood glucose regulation, 256, 256f
in erythrocytes, 655, 655t
insulin and, 478, 481, 481f
glucose uniporter (GLUT2), 481, 481f
glucose-1-phosphate
gluconeogenesis and, 181–183, 182f
glycogen release of, 171, 173–174, 173f
in glycogenesis, 172f, 173
glucose-6-phosphatase
deficiency of, 345
in gluconeogenesis, 181, 182f
in glycogenolysis, 174
glucose-6-phosphate
in gluconeogenesis, 181, 182f
in glycogenesis, 172f, 173
in glycolysis, 164–166, 165f
as inhibitor of hexokinase, 140
in pentose phosphate pathway, 191–194, 192f, 193f
in uronic acid pathway, 195, 196f
glucose-6-phosphate dehydrogenase (G6PD)
deficiency of, 191, 198, 657–658, 657t, 658f, 731–732
in pentose phosphate pathway, 192–193, 192f, 193f
glucose-alanine cycle, 186, 187f
glucoside, 150
glucosuria, 189
glucosylceramide, 212, 243, 244f, 469
glucuronate, 150, 151f
bilirubin conjugation with, 322, 322f
uronic acid pathway production of, 195, 196f

glucuronic acid, uronic acid pathway formation of, 191
β-glucuronidases, 323
glucuronidation, of xenobiotics, 566
glucuronide conjugates, glucuronate production for, 195, 196f
glucuronides, 191
GLUT. See glucose transporters
GLUT2 (glucose uniporter), 481, 481f
GLUT4 transporter, 256
glutamate
carbon skeleton catabolism of, 291–292
carboxylation of, 540–541, 541f
in citric acid cycle, 160, 160f
synthesis of, 275, 275f
transamination reactions forming, 282–283, 283f
glutamate aminotransferase, transamination reaction of, 283, 283f
glutamate dehydrogenase (GDH)
in amino acid synthesis, 275, 275f
in nitrogen metabolism, 283, 284f
glutamate-γ-semialdehyde dehydrogenase, 295f, 296
glutamic acid, 17t
glutaminase, 284, 284f
in amino acid catabolism, 291–292
glutamine, 17t
ammonia fixation as, 284, 284f
carbon skeleton catabolism of, 291–292
circulating plasma levels of, 281–282, 281f, 282f
in citric acid cycle, 160, 160f
deamination of, 284, 284f
synthesis of, 275, 275f
glutamine analogs, purine nucleotide synthesis affected by, 338, 340
glutamine synthetase, 275, 275f
ammonia fixation by, 284, 284f
γ-glutamyltransferase (GGT), 566
diagnostic use of, 67t
γ-glutamyl-β-aminopropionitrile (BAPN), 19, 19t
glutaric acid, pK_a of, 13t
glutarylation, 430t
glutathione, 22, 22f, 566, 723, 731, 731t
glutathione peroxidase, 214, 552
hemolysis protection from, 195, 195f, 198
selenium in, 118, 195, 195f
glutathione reductase, 194–195, 195f, 198, 731, 731t
glutathione S-transferase (GST)
fusion protein purification with, 68–69, 69f
in xenobiotic metabolism, 566
S-glutathionylation, 430t
glyburide, 225
N-glycan chains, 588–589

glycans, 554, 555t, 558–559, 558f
in virus, bacteria, and parasite binding, 563
glycated hemoglobin (HbA$_{1c}$), 57, 561
glycation, 430t, 554
in diabetes mellitus, 57, 561, 561f
glycemic index, 152, 528
glyceraldehyde-3-phosphate
in glycolysis, 165f, 166, 166f
in pentose phosphate pathway, 191–194, 192f, 193f
glyceraldehyde-3-phosphate dehydrogenase, in glycolysis, 165f, 166, 166f
glycerol, 206, 209
absorption of, 529
in gluconeogenesis, 182–183
permeability coefficient of, 471f
glycerol ether phospholipids, synthesis of, 242, 242f
glycerol kinase, 239, 240, 241f, 255
glycerol moiety, 135
glycerol phosphate acyltransferase, mitochondrial compartmentalization of, 122
glycerol phosphate pathway, 241f
glycerol-3-phosphate
acylglycerol biosynthesis and, 240, 241f
triacylglycerol esterification and, 255–256, 256f
glycerol-3-phosphate acyltransferase, 240, 241f
glycerol-3-phosphate dehydrogenase, 240, 241f
glycerophosphate shuttle, 129, 129f
glycerophospholipids, 206, 209–210, 211f
glycerose, 149, 149f
glycine, 16t
carbon skeleton catabolism of, 293, 294f
heme biosynthesis from, 315–317, 316f, 317f, 318f, 319f, 320t
specialized products of, 308, 308f, 310–311, 312f
synthesis of, 275, 276f
threonine conversion to, 295, 295f
glycine cleavage complex, 293, 293f
glycine residues, 600
glycinuria, 293
glycobiology, 147
glycocalyx, 155, 212, 239
glycochenodeoxycholic acid, 266f
glycocholic acid, 266f
glycoconjugate carbohydrates, glycoproteins as, 554
glycogen, 147
biomedical importance of, 171–172
breakdown of, 163, 164f, 164t
clinical aspects of, 178–179
gluconeogenesis and, 181–183, 182f

glycogenesis reaction pathway for, 137, 172f, 173–175, 173f
glycogenolysis reaction pathway for, 172f, 173–175, 173f, 174t
liver synthesis of, 142–143
as metabolic fuel, 134–135, 141
metabolic pathways of, 135, 137–139, 138f
in muscle contraction, 627, 627f
regulation of, 175–177, 175f, 176f, 178f, 179f
storage function of, 152, 153f, 171, 172t
glycogen phosphorylase, 627
in glycogenolysis, 173–175, 173f
regulation of, 93t, 175–177, 176f, 177–178, 179f
glycogen primer, 173
glycogen storage diseases, 147, 171, 174, 174t, 178–179
glycogen synthase
gluconeogenesis and, 181–183, 182f
in glycogenesis, 172f, 173, 173f
regulation of, 93t, 175–178, 178f, 179f
glycogen synthase a, 177
glycogen synthase b, 177
glycogenesis, 137
reaction pathway of, 172f, 173–175, 173f
regulation of, 175–177, 175f, 176f, 178f, 179f
glycogenin, in glycogenesis, 172f, 173
glycogenolysis, 137
blood glucose derived from, 186, 187f
induction and repression of enzymes catalyzing, 183, 184t
reaction pathway of, 172f, 173–175, 173f, 174t
regulation of, 175–177, 175f, 176f, 178f, 179f
glycolipid storage diseases, 239
glycolipids, 206, 212, 212f
galactose production for, 197, 198f
glycolysis, 135, 136f
in anaerobic conditions, 163–164, 164f, 164t
biomedical importance of, 163
in cancer cells, 704, 705f, 706t
clinical aspects of, 168–170
dihydroxyacetone phosphate in, 240, 241f, 242f
in energy conservation and capture, 113
in erythrocytes, 168, 168f, 655, 655t
flux-generating reaction in, 139–141
Gibbs free-energy change of, 87
gluconeogenesis regulation and, 183–185, 184t, 185f, 186f
induction and repression of enzymes catalyzing, 183, 184t
pentose phosphate pathway connections with, 192f, 194

pyruvate oxidation after, 168, 169f
reactions of, 163–168, 165f, 166f
regulation of, 167–168
at subcellular level, 139, 140f
thermodynamic barriers to reversal of, 181, 182f
glycome, 147
glycomics, 3, 147, 455
glycone, 150
glycophorins, 155, 557, 659, 659t
glycophosphatidylinositol, 471
glycoprotein complex GPIIb-IIIa, in platelet aggregation, 678, 679f
glycoprotein glycosyltransferases, 557–558
glycoproteins, 35, 154–155, 155t, 635, 660
amino sugars in, 151, 152f, 197
biologic information encoded in, 554
biomedical importance of, 554
classes of, 555, 556f, 557
degradation of, 280
in diabetes mellitus, 561, 561f
disorders involving, 562–563
extracellular, absorptive pinocytosis of, 483
functions of, 554, 555t, 561–562
galactose production for, 197, 198f
as hormone precursors, 492, 492f
membrane asymmetry and, 472
microfibril-associated, 604
monosaccharides commonly found in, 555, 556t
N-glycosidic linkages in, 555, 556f, 558–559, 558f, 559f
O-glycosidic linkages in, 555, 556f, 557–558
oligosaccharide chains in, 554, 555t, 558–559, 558f, 563
on plasma membrane, 560
protein folding role of, 559–560
purification of, 555
rapidly reversible, 560–561
regulation of, 560
synthesis of, 557–558, 559f
in virus, bacteria, and parasite binding, 563
glycosaminoglycans (GAGs), 154, 154f
amino sugars in, 151, 152f, 197
components of, 606–607, 606f, 607f
functions of, 610t
mucopolysaccharidoses and, 610–611, 611t
properties of, 609t
structures of, 608–610, 609f
synthesis of, 606–607
1,6-glycosidase, in glycogenolysis, 173f, 174
glycosides, 150–151
N-glycosides
nucleosides as, 330, 331f

syn and *anti* conformers of, 330–331, 331*f*, 332*t*

N-glycosidic linkages, in glycoproteins, 555, 556*f*, 558–559, 558*f*, 559*f*

O-glycosidic linkages, 600, 607
in glycoproteins, 555, 556*f*, 557–558

glycosphingolipids (GSLs), 206, 212, 212*f*, 239, 486, 660
amino sugars in, 197
membrane asymmetry and, 472, 597
in membranes, 469
synthesis of, 243–244, 244*f*

N-glycosylamine bond, 607

glycosylation, 554. *See also* glycoproteins
congenital disorders of, 640
cotranslational, 588–589
in diabetes mellitus, 561
of hydroxylysines, 601
mass spectrometry detection of, 29*t*
rapidly reversible, 560–561
regulation of, 560

glycosylphosphatidylinositol (GPI), 596

glycosylphosphatidylinositol-anchored (GPI-anchored) glycoproteins, 556*f*, 557, 560

glycosyltransferases, 554, 608

GM1 ganglioside, 212, 212*f*

GM3 ganglioside, 212

GMP, 332*t*
cyclic (*See* cyclic GMP)
IMP conversion to, 338, 340*f*, 341, 341*f*
PRPP glutamyl amidotransferase regulated by, 340, 341*f*

Golgi apparatus (GA)
brefeldin A and, 596
exocytosis in, 483
in membrane synthesis, 582
in protein sorting, 582, 583*f*, 591
protein sorting signals in, 582, 582*t*
in protein synthesis, 416, 582
proteins destined for membrane of, 582, 588–589
retrograde transport from, 591
in VLDL formation, 250*f*

gonadal steroids, transport of, 507, 507*t*

gout, 199, 255, 344, 346*t*

gouty arthritis, 344, 346*t*

gp 120, 563

GPCRs (G-protein-coupled receptors), 511, 511*f*, 665–666

GPI (glycosylphosphatidylinositol), 596

GPI-anchored (glycosylphosphatidylinositol-anchored) glycoproteins, 556*f*, 557, 560

GPIb-IX-V, in platelet aggregation, 678, 679*f*

G-protein-coupled receptors (GPCRs), 511, 511*f*, 665–666

G-proteins (guanine-binding protein), 511, 666
classes and functions of, 512*t*
family of, 512

gramicidin, 15, 479

granules, 667

α-granules, 661

granulocytes, 664, 667

gratuitous inducers, 423

griseofulvin, 631

gRNA (guide RNA), 710

GroEL, 583

group transfer potential, 111, 114
of nucleoside triphosphates, 334

group transfer reactions, 9, 62, 62*f*

growth, energy requirements for, 133

growth differentiation factor 15 (GDF15), 642

growth factors, in cancer, 697, 697*t*

growth hormone (GH)
amino acid transport and, 478
blood glucose regulation by, 188
in lipolysis, 256, 257*f*
receptors for, 490

growth inhibitory factors, in cancer, 697

GSLs. *See* glycosphingolipids

GST. *See* glutathione S-transferase

GTFs. *See* general transcription factors

GTP, 333, 594, 596

GTPase-accelerating proteins (GAPs), 585

GTPases, 585, 595

GTP-binding proteins, 261

guanine, 332*t*
base pairing in DNA, 349, 350*f*
base pairing in RNA, 352–354, 353*f*
ROS oxidation of, 721*f*
salvage pathways of, 342–343

guanine nucleotide exchange factors (GEFs), 585, 586*f*

guanine-binding protein. *See* G-proteins

guanosine, 331*f*, 332*t*
salvage pathways of, 342–343
in uric acid formation, 344, 345*f*

guanosine diphosphate. *See* GDP

guanosine monophosphate. *See* GMP

guanosine triphosphate. *See* GTP

guide RNA (gRNA), 710

Guillain-Barré syndrome, 481

L-gulonolactone oxidase, 195

Guthrie bacterial inhibition test, 574

H

H bands, 619, 619*f*, 620*f*

H blood group substances, 661, 661*f*

H1 histones, 361, 361*f*, 362*f*, 363, 363*t*

H2A histones, 361–363, 361*f*, 363

H2AX histone, 381

H2B histones, 361–363, 361*f*, 363

H3 histones, 361–363, 361*f*, 363, 363*t*

H4 histones, 361–363, 361*f*, 363, 363*t*

hairpin, 352, 353*f*

half-life ($t_{1/2}$)
of plasma proteins, 637
of proteins, 280

halt-transfer signal, 590, 591*f*

haptocorrin, 105

haptoglobin, 638–639, 638*f*

haptoglobin-related protein, 639

Hartnup disease, 298, 299*f*, 543

HAT (histone acetyltransferase activity), 521–522

Haworth projection, 148, 148*f*

Hb Hikari, 408*f*

HbA_{1c} (glycated hemoglobin), 57, 561

HbF (fetal hemoglobin), 52–53, 53*f*, 56

HbM (hemoglobin M), 56, 408*f*

HbS (hemoglobin S), 56, 57*f*, 408*f*

HBV (hepatitis B virus), 712

HDACs (histone deacetylases), 93, 699

HDL. *See* high-density lipoproteins

HDL cycle, 252*f*, 253

HDL receptor, 249, 252*f*, 253

health, 2–4, 2*f*, 3*f*

heart
in fasting state, 134–135, 144–145
in fed state, 134–135, 142–144, 143*f*
metabolic features of, 145*t*

heart failure, 618
cardiolipin in, 211
vitamin B_1 deficiency in, 542

heartbeat hypothesis, 725–726

heat, respiratory chain production of, 127

heat-shock proteins (hsp), as chaperones, 42, 584

heavy metals, toxicity of, 97–98

Heinz bodies, 657

helicases, 372*f*, 373, 373*t*

Helicobacter pylori, glycans in binding of, 563

helix-loop-helix motifs, 37

helix-turn-helix motif, 437, 438*f*

helper T cells, 667, 670

hemagglutinin, influenza, 563

hematoma, 321

hematopoietic stem cells, blood cells derivation from, 653–654, 654*f*

hematoporphyrin, 318

heme
binding of, 638
biomedical importance of, 315
biosynthesis of, 315–318, 316*f*, 317*f*, 318*f*, 319*f*, 320*t*
disorders of, 319–321, 320*t*, 321*f*
regulation of, 317–318
catabolism of, 321–323, 322*f*, 323*f*
in catalase, 118
in Complex IV, 124, 124*f*
in hemoglobin, 50–51, 50*f*, 51*f*, 56

heme (*Cont.*):
    histidines in, 50, 50*f*, 56
    iron in, 50–51, 50*f*, 51*f*, 56, 100–102
    in oxidases, 116
    oxidation of, 656–658, 657*t*, 658*f*
    structure of, 50, 50*f*, 315, 316*f*
heme oxygenase, 321, 322*f*, 638
hemerythrin, diiron center of, 100, 101*f*
hemiacetals, 148, 148*f*, 150
hemiconnexin, 484*f*
hemin, 321
hemochromatosis, 531
    hereditary, 643, 645*t*, 657
hemocyanin, copper in, 103
hemoglobin
    2,3-bisphosphoglycerate binding of, 52*f*,
        55–56, 55*f*
    allosteric properties of, 52–56, 52*f*, 53*f*,
        54*f*, 55*f*
    biomedical implications of, 56–57
    biomedical importance of, 49–50
    carbon dioxide transport by, 54, 55*f*
    conformational changes upon
        oxygenation of, 53–54, 53*f*, 54*f*
    cooperative binding of, 53
    extracorpuscular, haptoglobin binding
        of, 638, 638*f*
    glycated, 57, 561
    heme in, 50–51, 50*f*, 51*f*, 56
    high altitude levels of, 56
    hindered environment of, 51, 51*f*
    molecular mass of, 636*f*
    mutations in, 56, 57*f*, 408*f*
    oxygen dissociation curve of, 51–52,
        51*f*
    P_{50} values of, 53, 53*f*
    pathologic conformations of, 43
    proton binding of, 54–55, 55*f*
    quaternary structure of, 52–56, 52*f*, 53*f*,
        54*f*, 55*f*
    relaxed and taut states of, 51–52
    salt bridges in, 53–54
    secondary and tertiary structures of,
        52–53
    subunits of, 52–53, 52*f*
    tetrameric structure of, 52–53, 52*f*
    turnover of, 321
hemoglobin (Hb) Hikari, 408*f*
hemoglobin M (HbM), 56, 408*f*
hemoglobin S (HbS), 56, 57*f*, 408*f*
hemoglobinopathies, 56, 57*f*, 658
hemoglobinopathy, 370, 370*f*
α-hemoglobin-stabilizing protein (AHSP),
        43
hemoglobinuria, paroxysmal nocturnal,
        486*t*, 657*t*, 658
hemojuvelin (HJV), 642
α-hemolysin, 479
hemolysins, 658

hemolysis
    pentose phosphate pathway and
        glutathione peroxidase protection
        against, 194–195, 195*f*
    pentose phosphate pathway impairment
        leading to, 198
hemolytic anemias, 163, 170, 191
    causes of, 658, 658*f*
    hyperbilirubinemia with, 323, 325, 325*t*
    pentose phosphate pathway impairment
        in, 198
hemolytic-uremic syndrome, 662
hemopexin, 638, 638*f*
hemophilia, 634
hemophilia A, 687
hemophilia B, 687
hemoproteins
    biomedical importance of, 315
    examples of, 315, 316*t*
hemosiderin, 640
hemosiderosis, 640, 643
hemostasis, 661–662
    biomedical importance of, 677
    coagulation (*See* coagulation system)
    fibrinolysis (*See* fibrinolysis)
    laboratory tests measuring, 687
    phases of, 677
    thrombin regulation in, 685
hemostatic plugs, formation of, 678, 679*f*
Henderson-Hasselbalch equation, 12, 13*f*
heparan sulfate, 607, 609*f*, 609*t*, 610
    in hemostasis and thrombosis, 679
heparin, 154, 154*f*, 604, 607, 609*t*, 685–686
    lipoprotein and hepatic lipases and, 251
    structure of, 609–610, 609*f*, 610*f*
heparin cofactor II, 685
hepatic lipase
    in chylomicron remnant uptake,
        251–252, 252*f*
    deficiency of, 268*t*
    metabolism by, 251
hepatic porphyrias, 320*t*, 321
hepatic portal vein, 186
hepatic purine biosynthesis
    AMP and GMP formation regulation
        in, 341, 341*f*
    PRPP glutamyl amidotransferase
        regulation in, 340
hepatitis, 156
    hyperbilirubinemia caused by, 325, 325*t*
hepatitis B virus (HBV), 712
hepatocarcinoma, 254
hepatocytes
    heme synthesis in, 316
    pinocytosis by, 483
hepatolenticular degeneration (Wilson
        disease), 308, 486*t*, 634, 641
hepcidin, 531–532, 531*f*, 642, 643*f*, 644*f*
hepcidin system, 99

hephaestin, 639–640
heptoses, 148, 148*t*
hereditary bleeding disorders, 687
hereditary cancers, 691, 697, 699, 699*t*
hereditary elliptocytosis, 658, 660
hereditary fructose intolerance, 199–200
hereditary hemochromatosis, 643, 645*t*,
        657
hereditary mutations, 691
hereditary nonpolyposis colon cancer, 380*t*
hereditary spherocytosis, 486*t*, 657*t*, 658,
        660
Hermansky-Pudlak syndrome, 592*t*
herpes simplex virus, 390*f*
Hers disease, 174*t*
heterochromatin, 364
heterodimers, 39
heterogeneous regulatory RNAs (sRNAs),
        359
heterotrophic organisms, 111
hexapeptide, in albumin synthesis, 637
hexokinase, 60, 736
    blood glucose concentration and,
        186–187, 188*f*
    coupled enzyme assay for, 66*f*
    as flux-generating step in glycolysis,
        139–141
    in fructose metabolism, 196, 197*f*
    in glycogenesis, 173
    in glycolysis, 164–165, 165*f*
    regulation of, 167
    reversal of reaction catalyzed by, 181,
        182*f*
hexosamines, 151, 152*f*, 155*t*
    synthesis of, 197, 199*f*
hexose monophosphate shunt. *See* pentose
        phosphate pathway
hexose-6-phosphate dehydrogenase,
        193–194
hexoses, 148, 148*t*
    of physiological importance, 150, 151*t*
HFS (high-fructose syrups), 195, 197*f*, 199
HGP (Human Genome Project), 3–4, 3*f*
*Hha*I, 445*t*
HHH (hyperornithinemia,
        hyperammonemia, and
        homocitrullinuria) syndrome, 287
HIF-1 (hypoxia-inducible factor-1), 705
high altitude, physiologic adaptations to,
        56
high-density lipoproteins (HDL), 248, 248*t*
    apolipoproteins of, 249
    atherosclerosis and, 253, 267
    cholesterol removal by, 259
    familial deficiency of, 268*t*
    LDL ratio to, 267
    metabolism of, 252–253, 252*f*
    receptor for, 249, 252*f*, 253
    synthesis of, 252, 252*f*

high-density microarray technology, 455
high-energy phosphates
    creatine phosphate shuttle for, 130, 130f
    cycles of, 113, 113f
    energy capture and transfer by,
        111–112, 111t, 113f, 125–126
    oxidative phosphorylation in synthesis
        of, 122, 125–126, 126f
high-fructose syrups (HFS), 195, 197f, 199
high-mannose oligosaccharides, in
    glycoproteins, 558–559, 558f
high-pressure liquid chromatography
    (HPLC)
    laboratory tests using, 572
    protein and peptide purification with,
        25, 26f
high-throughput sequencing (HTS),
    65–66, 453–454
Hill coefficient (n), 79, 79f
Hill equation, 79, 79f
HindIII, 445t
hippuric acid, glycine conversion to, 308,
    308f
histamine, 308, 308f, 665f, 670
histidase, 293
histidine, 17t, 665f
    carbon skeleton catabolism of, 292–293,
        293f
    fusion protein purification with, 68–69
    in heme, 50, 50f, 56
    resonance hybrids of, 20
    specialized products of, 308, 308f
    stabilization of, 20f
histidinemia, 293
histone acetyltransferase activity (HAT),
    513, 521–522
histone chaperones, 363
histone code. See histones, modifications
    of
histone deacetylases (HDACs), 93, 699
histone dimer, 361, 361f, 363
histone epigenetic code, 431–433
histone octamer, 361f, 362f, 363
histone tetramer, 361, 361f, 363
histones, 361, 361f
    modifications of, 92, 361, 363, 363t,
        394–395, 429–430, 430t
HIV. See human immunodeficiency virus
HIV protease, acid-base catalysis of, 63,
    63f
HJV (hemojuvelin), 642
HMG-CoA. See
    3-hydroxy-3-methylglutaryl-CoA
homeostasis, 85
    in ER, 592
    hormone signal transduction and,
        508–509, 509f
    intracellular iron, 641–642, 642f, 643f,
        644f

homeostatic adaptations, 508
L-homoarginine, 19, 19t
homocarnosine, biosynthesis of, 308, 308f,
    313
homocarnosinosis, 313
homocysteine, 294, 294f
homocysteinylation, 430t
homocystinuria, 294–295, 310
homodimers, 39
homogentisate dioxygenase, 119
homogentisate oxidase, 296, 297f
homologous recombination (HR), 379,
    380f, 380t
homologous sequence, DNA, 370
homologs, protein, 64
homology, protein classification based on,
    35
homology modeling, protein structures
    solved by, 41–42
homopolymer tailing, 446t
hormone receptor-G-protein effector
    system, 511f
hormone receptors
    classification of, 490, 491t
    as proteins, 490
    recognition and coupling domains of,
        489–490
    specificity and selectivity of, 489, 489f
hormone response elements (HRE)
    DNA sequences of, 510t
    gene expression and, 435–436, 435f,
        436f
    in signal generation, 509, 510f
    types of, 519
hormone response transcription unit,
    519f
hormone synthesis
    active form and, 492
    adrenal steroidogenesis, 493–495, 493f
    androgen synthesis, 494f, 495
    angiotensin II, 503–504, 504f
    calcitriol, 497–499
    catecholamines, 499–500, 499f
    cellular arrangements of, 491
    from cholesterol, 212, 259, 491, 492f,
        493–499, 493f
    dihydrotestosterone, 491, 495, 497f
    glucocorticoids, 494–495, 494f
    insulin, 501–502, 502f
    iodide metabolism and, 500
    mineralocorticoids, 493–494, 494f
    ovarian steroidogenesis, 495–496, 497f,
        498f
    parathyroid, 502, 503f
    peptide precursors for, 500–501
    POMC family, 504–505, 505f
    testicular steroidogenesis, 495, 496f
    tetraiodothyronine, 500, 501f
    triiodothyronine, 500, 501f

hormone-dependent cancer, vitamin B_6
    and, 543
hormones
    active form of, 492
    adenylyl cyclase and, 511t
    of adipose tissue, 255
    amino acid precursors of, 19
    biomedical importance of, 508
    in blood glucose regulation, 186–188,
        187t, 188t
    cell surface receptor binding of, 490,
        491t
    chemical diversity of, 491–492, 492f
    definition, 488
    enzyme regulation by, 141, 141f
    facilitated diffusion regulation by, 477
    features of, 491t
    hepatic VLDL secretion and, 253–254,
        254f
    intracellular receptor binding of, 490,
        491t
    laboratory tests for, 573
    lipogenesis regulation by, 256–257
    lipolysis regulation by, 255–256, 257f
    lipophilic, 490, 491t
    peptide, 554
    plasma transport proteins of, 506, 506t,
        507t
    precursor molecule for, 491–492, 492f
    receptor interaction of, 508–509,
        509f
    as second messengers (See second
        messengers)
    signal generation by (See signal
        generation)
    signal transduction by, 508–509, 509f
    stimulus response of, 508–509, 509f
    storage and secretion of, 505, 506t
    transcription modulation by, 519–522,
        519f, 521f, 521t, 522t
    vitamin D as, 538–539, 539f
    water-soluble, 490, 491t
housekeeping genes, 422
HpaI, 445t, 446f
HPLC. See high-pressure liquid
    chromatography
HPV (human papillomavirus), 712
HR (homologous recombination), 379,
    380f, 380t
HRE. See hormone response elements
hsp (heat-shock proteins), as chaperones,
    42, 584
Hsp60, 583
Hsp70, 584, 590
Hsp90, 598
HTS (high-throughput sequencing),
    65–66, 453–454
human genes, localization of, 453
Human Genome Project (HGP), 3–4, 3f

human immunodeficiency virus (HIV)
  glycans in binding of, 563
  undernutrition caused by, 533
human papillomavirus (HPV), 712
human β-interferon gene enhancer,
    434–435, 434*f*
humoral immunity, 670
Hunter syndrome, 611, 611*t*
Hurler syndrome, 611, 611*t*
Hutchinson-Gilford progeria syndrome,
    632
hyaline cartilage, 615, 615*t*
hyaluronic acid, 151, 154, 154*f*, 607, 608,
    609*f*, 609*t*, 615, 615*f*, 616*f*
hyaluronidase, 611
hybrid oligosaccharides, in glycoproteins,
    558–559, 558*f*
hybridization, 351
hybridomas, 650
hydrogen bonds
  in collagen, 44
  definition of, 7
  in DNA, 349, 350*f*
  hydrophobic interactions and, 8
  in protein tertiary and quaternary
    structures, 40
  water molecule formation of, 7, 7*f*
  in α helices, 36, 36*f*
  in β sheets, 37, 37*f*
hydrogen ions
  reaction rate response to, 75–76, 76*f*
  in water, 10–11, 13*f*, 13*t*
hydrogen peroxide, 720
  amino acid oxidase generation of,
    283
  glutathione peroxidase removal of, 118,
    195, 195*f*
  reduction of, 118
hydrogen sulfide, respiratory chain
    inhibition by, 116, 127
hydrogenation, dehydrogenation coupling
    to, 110, 111*f*
hydrolases, 60
  cholesteryl ester, 263
hydrolysis
  of ATP, 111–112, 111*t*, 112*f*
  of bound GTP to GDP, 594
  as group transfer reaction, 9
  of triacylglycerols, 239
  of water, 9
hydronium ions, 10
hydropathy plot, 471
hydroperoxidases, 116, 118
hydroperoxides, 235, 237*f*
hydrophilic portion of lipid molecule, 215,
    215*f*
hydrophobic domains, 38
hydrophobic effect, in lipid bilayer self-
    assembly, 470

hydrophobic interaction chromatography,
    protein and peptide purification
    with, 26
hydrophobic interactions, 8, 8*f*, 349
  in protein tertiary and quaternary
    structures, 40
hydrophobic portion of lipid molecule,
    215, 215*f*
hydrostatic pressure, 635
hydroxide ions, 10
3-hydroxy-3-methylglutaryl-CoA
    (HMG-CoA)
  in cholesterol synthesis, 260, 261*f*
  in ketogenesis, 221, 222*f*
  in mevalonate synthesis, 260, 261*f*
3-hydroxy-3-methylglutaryl-CoA
    (HMG-CoA) lyase
  deficiency of, 225
  in ketogenesis, 221, 222*f*
3-hydroxy-3-methylglutaryl-CoA
    (HMG-CoA) reductase
  cholesterol synthesis controlled by, 260,
    261*f*, 262–263, 263*f*
  drugs targeting, 87
  in mevalonate synthesis, 260, 260*f*
  regulation of, 80, 93*t*
3-hydroxy-3-methylglutaryl-CoA
    (HMG-CoA) reductase kinase,
    regulation of, 93*t*
3-hydroxy-3-methylglutaryl-CoA
    (HMG-CoA) synthase
  in cholesterol synthesis, 260, 261*f*
  in ketogenesis, 221, 222*f*
  in mevalonate synthesis, 260, 260*f*
L-3-hydroxyacyl-CoA dehydrogenase,
    219, 219*f*
3-hydroxyanthranilate dioxygenase, 119
hydroxyapatite, 612
D-3-hydroxybutyrate, 217, 220, 221*f*
D-3-hydroxybutyrate dehydrogenase, 220,
    221*f*
4-hydroxybutyric aciduria, 313
β-hydroxybutyric aciduria, 337, 345–346,
    347*f*
7α-hydroxylase, sterol, 265, 266*f*
17α-hydroxylase, 494, 496*f*
18-hydroxylase, 493–494
27-hydroxylase, sterol, 265, 266*f*
hydroxylase cycle, 119, 120*f*
hydroxylases, 119
  in cortisol synthesis, 494–495
  vitamin C as coenzyme for, 547
hydroxylation
  mass spectrometry detection of, 29*t*
  of proline, 601
  of xenobiotics, 564–566
hydroxylysine, synthesis of, 276, 278*f*
5-hydroxylysine, 18, 18*f*, 430*t*
hydroxylysines, 601

hydroxymethylbilane, in heme synthesis,
    316, 317*f*, 318*f*, 319*f*
hydroxymethylbilane synthase, 316, 317*f*,
    318*f*, 319*f*
  deficiency in, 320, 320*t*
5-hydroxymethylcytosine, 331, 333*f*
hydroxyproline, 600, 604
  carbon skeleton catabolism of, 295–296,
    295*f*
  synthesis of, 276, 278*f*
4-hydroxyproline, 18, 18*f*
4-hydroxyproline dehydrogenase,
    295–296, 295*f*
15-hydroxyprostaglandin dehydrogenase,
    234–235
19-hydroxysteroid, 493
β-hydroxy-γ-trimethylammonium
    butyrate. *See* carnitine
hyperalphalipoproteinemia, familial,
    268*t*
hyperammonemia, 156, 162
  urea cycle disorders causing, 286–288,
    287*t*
hyperbilirubinemia, 315, 323–324, 323*t*
  metabolic disorders underlying,
    324–325, 324*f*, 325*t*
hyperbilirubinemic toxic encephalopathy,
    323–324
hypercholesterolemia, 236, 248, 252, 267,
    268*t*, 486, 486*t*, 592*t*
  fructose role in, 195, 197*f*, 199
hyperchromicity of denaturation, 351
hyperglycemia, 18, 146
hyperglycinemia, nonketotic, 293
hyperhomocysteinemia, folic acid
    supplements for, 546
hyperhydroxyprolinemia, 295*f*, 296
hyperkalemic periodic paralysis, 627*t*
hyperketonemia, 225
hyperlacticacidemia, 255
hyperlipidemia, 267
hyperlipoproteinemias, 247, 267, 268*t*
hyperlysinemia, 296
hypermetabolism, disease causing, 163,
    533
hypermethioninemia, 308–309
hyperornithinemia, hyperammonemia,
    and homocitrullinuria (HHH)
    syndrome, 287
hyperornithinemia–hyperammonemia
    syndrome, 292
hyperoxaluria, 293
hyperphenylalaninemias, 296, 298*f*
hyperprolinemia, 292, 292*f*, 295*f*, 296
hypersensitive sites, chromatin, 364
hypersplenism, 657
hypertension, folic acid supplements for,
    546
hyperthermia, malignant, 625, 627*t*

hypertriacylglycerolemia, 247, 268t
    fructose role in, 197f, 199
hypertrophic cardiomyopathy, familial,
    628–629
hyperuricemia, 344
    fructose role in, 199
hypervariable regions, immunoglobulins,
    647–648
hypoceruloplasmenia, 640
hypoglycemia, 178, 181, 190
    fatty acid oxidation and, 217, 224–225
    fructose-induced, 200
    during pregnancy and in neonates, 189
hypoglycin, 217, 225
hypokalemic periodic paralysis, 627t
hypolipidemic drugs, 267
hypolipoproteinemia, 247, 267, 268t
hypouricemia, 345
hypoxanthine, 331, 333f
    salvage pathways of, 342–343
hypoxanthine-guanine phosphoribosyl
    transferase, 344–345, 346t
hypoxia, in cancer cells, 705
hypoxia-inducible factor-1 (HIF-1), 705
hypoxic conditions, glycolysis in, 164f,
    164t, 167

I
I bands, 619, 619f, 620f
IAPs (inhibitors of apoptosis), 701,
    702f
ibuprofen, 226, 236
IC$_{50}$, 81
I-cell disease, 484, 486t, 562, 592t
ICF (intracellular fluid), 468, 468t
icterus. See jaundice
idiotypes, 649
IDL (intermediate-density lipoproteins),
    248t, 252, 264, 267
iduronate, 150, 151f
IEF (isoelectric focusing), protein and
    peptide purification with, 27, 27f
IgA, 647t, 649f
IgD, 647t
IgE, 647t
IGF-1. See insulin-like growth factor 1
IgG, 647t, 648f
IgM, 647t, 649f
IL-1 (interleukin-1), 638
IL-6 (interleukin-6), 642
imatinib, 710–711, 711t
immune response
    cancer escape from, 703, 704f
    glycoproteins in, 561–562
immune system
    adaptive, 646, 650, 670
    dysfunctions of, 651–652
    innate, 646, 650
immune thrombocytopenic purpura, 662

immunoassays
    enzyme-linked, 66, 573
    laboratory tests using, 573
immunocompromised state, 652, 664
immunogenicity, 650
immunoglobulin genes
    amplification and, 441–442
    DNA rearrangement and, 370–371
immunoglobulin light chains, 371, 646,
    649
immunoglobulins, 634, 637t
    binding specificity of, 647–648, 650f
    class switching and, 649
    class-specific effector functions of, 647
    composition of, 646–647, 647t, 648f, 649f
    diversity of, 648–649
immunotherapy, for cancer, 708–709
IMP. See inosine monophosphate
importins, 585–586, 586f
inactive chromatin, 364, 429
inborn errors of metabolism
    of amino acid metabolism, 286–288,
        287t, 290–291, 300, 304t
    early studies on, 2
    gene or protein modification for, 288
    general features of, 286–287
    mucopolysaccharidoses, 610–611, 611t
    neonate screening for, 573–574
    tandem mass spectrometry in screening
        for, 288, 290
    urea cycle defects, 286–288, 287t
inclusion cell (I-cell) disease, 484, 486t,
    562, 592t
indels, 369, 408–409
indicator metabolites, 88–89
indirect bilirubin, 323
indirect carcinogens, 567
indispensable amino acids. See essential
    amino acids
indole, permeability coefficient of, 471f
indomethacin, cyclooxygenases affected
    by, 236
induced fit model, 63, 63f
induced pluripotent stem cells (iPSCs),
    454
inducers, 87
    gratuitous, 423
    in regulation of gene expression, 422
inducible gene, 422
induction, of enzyme synthesis, 87
infant respiratory distress syndrome
    (IRDS), 244–245
infantile Refsum disease, 587, 588t
infectious disease
    enzymes for diagnosis of, 68
    glycoprotein role in, 563
    kwashiorkor after, 533
    protein loss in, 534
    undernutrition in, 533

inflammasome, 709
inflammation
    advanced glycation end-products in,
        561
    cancer development and, 709
    glycoprotein role in, 561–562
    plasma proteins in, 637–638, 637t
    prostaglandins in, 226
inflammatory response, 518, 664
influenza virus A, glycans in binding of,
    563
information pathway, 510, 510f
inhibitor-1, 177
inhibitors
    of angiogenesis, 706
    competitive, 79–81, 80f, 81f
    of DNA methyltransferases, 699
    drugs acting as, 83
    feedback, 88–89
    of folic acid metabolism, 545
    of histone deacetylases, 699
    irreversible, 81
    kinetic analysis of, 79–82, 80f, 81f
    mechanism-based, 79, 81–82
    noncompetitive, 80–81, 80f, 81f
    of platelet activation, 679f, 680
    of pyruvate dehydrogenase, 168, 169f
    of respiratory chain, 116, 122, 127–128,
        127f, 128f
    of thrombin, 685–686
    tightly bound, 81
    of tyrosine kinases, 710–711, 711t
inhibitors of apoptosis (IAPs), 701, 702f
initial velocity (v_i), 76–79, 76f, 78f, 79f,
    82, 83f
initiation
    in DNA synthesis, 374, 375f, 376f
    in protein synthesis, 410–413, 411f
    in RNA synthesis, 386–387, 386f, 387f
43S initiation complex, in protein
    synthesis, 410, 411f
48S initiation complex, in protein
    synthesis, 410, 411f, 412
80S initiation complex, in protein
    synthesis, 410, 411f, 412
initiation complexes, in protein synthesis,
    410, 411f
initiator methionyl-tRNA, 410, 411f
initiator sequence (Inr), 391
initiators, of apoptosis, 701–702, 702f
innate immune system, 646, 650
innate immunity, glycoproteins in, 562
inner membrane, of mitochondria, 122,
    122f, 123f
    selective permeability of, 128–130, 128f,
        129f, 130f
inner mitochondrial membrane, 584
inorganic elements, human requirement
    for, 97, 97t

inorganic pyrophosphate (PPi), 113–114, 114*f*
inosine monophosphate (IMP)
    conversion to AMP and GMP, 338, 340*f*, 341, 341*f*
    synthesis of, 338, 339*f*, 340*f*
inositol phospholipids, 239
inositol trisphosphate, 211
1,4,5-inositol trisphosphate (IP3)
    in chemotaxis, 666
    in platelet aggregation, 678, 679*f*
Inr (initiator sequence), 391
insertions of nucleotides, 408–409, 409*f*
inside-out transmembrane signaling, in platelet aggregation, 678, 679*f*
inside-outside asymmetry, membrane, 472
Insig (insulin-induced gene), 262
insulators
    in gene expression, 436–437
    nonpolar lipids as, 206
insulin
    adipose tissue metabolism affected by, 256
    biological assay to measure, 729, 730*t*
    blood glucose regulation by, 187–188, 188*t*
    in cholesterol synthesis, 262–263, 263*f*
    in diabetes mellitus, 135, 146
    free fatty acids affected by, 247, 256
    glucose regulation by, 140, 142–144, 165–166
    in glucose transport, 477
    glycogen regulation by, 177
    in glycolysis and gluconeogenesis regulation, 183
    kinase cascade signal transmission by, 516, 517*f*
    in lipogenesis regulation, 231–232, 232*f*
    in lipolysis regulation, 232, 256, 256*f*, 257*f*
    protein synthesis initiation by, 413, 413*f*
    radioimmunoassay to measure, 729, 730*f*, 730*t*
    receptor for, 490
    secretion by rabbit pancreas, 736*t*
    sequencing of, 27–28
    storage of, 505, 506*t*
    synthesis of, 501–502, 502*f*
    VLDL secretion and, 254
insulin receptor substrates (IRS), 516, 517*f*
insulin resistance, glycoprotein role in, 561
insulin/glucagon ratio, in ketogenesis regulation, 224
insulin-induced gene (Insig), 262
insulin-like growth factor 1 (IGF-1), 490
    in insulin signal generation, 516, 517*f*
integral membrane proteins, 35
integral proteins, 472–473, 472*f*

integration, chromosomal, 370, 370*f*
integrins, 562, 666, 667*t*
intercellular communication, 488.
    See also endocrine system
intercostal skeletal muscle cells, turnover of, 718*t*
β-interferon gene, 434–435, 434*f*
interferons, 670
interleukin-1 (IL-1), 638
interleukin-6 (IL-6), 642
interleukins, 670, 709
intermediate filaments, 631–632, 631*t*
intermediate-density lipoproteins (IDL), 248*t*, 252, 264, 267
intermembrane mitochondrial space, proteins in, 584
intermembrane space, of mitochondria, 122, 122*f*
intermittent branched-chain ketonuria, 300
internal ribosomal entry site (IRES), 416, 417*f*
interphase chromosomes, 363
intervening sequences. *See* introns
intestinal epithelium turnover, 718*t*
intestines. *See also* absorption
    in fasting state, 134*f*
    in fed state, 134*f*
    monosaccharide absorption in, 528, 528*f*
    transition metal absorption in, 105
intracellular environment, membranes in maintenance of, 468, 468*t*
intracellular fluid (ICF), 468, 468*t*
intracellular membranes, 467
intracellular signals, 510. *See also* second messengers
intracellular traffic. *See also* protein sorting
    disorders of, 588*t*, 597–598
    of proteins, 581
    transport vesicles in, 593–596, 594*t*, 595*f*
intravasation, of cancer cells, 706, 707*f*
intrinsic factor, 105, 544
intrinsic pathway, 680–682, 680*f*, 681*f*, 681*t*, 682*t*
intrinsic tenase complex, 382
introns (intervening sequences), 366, 367*f*, 371
    description of, 398
    processing of, 397*f*, 398*f*, 404–405
    removal from primary transcript, 397–399, 397*f*, 398*f*
    removal of, 397–399, 397*f*, 398*f*
inulin, 152–153, 154*f*
invert sugar, 152
iodine (iodide)
    deficiency of, 500

human requirement for, 97, 97*t*
    metabolism of, 500
5-iodo-2′-deoxyuridine, 334*f*
iodoacetate, 166*f*
iodopsin, 537
iodothyronyl residues, 500
5-iodouracil, 334
ion channels, 467, 476*t*
    function of, 478, 478*f*
    K⁺ channel, 478–479, 479*f*
    mutations and, 484–486
    nerve impulses and, 481
    properties of, 479*t*
    selectivity of, 478
ion product, 10
ion transport, in mitochondria, 130
ion trap, 31
ion-exchange chromatography, protein and peptide purification with, 26
ionization
    of amino acids, 19–21, 20*f*, 21*t*
    of water, 9–10
ionizing radiation, DNA damage caused by, 380*f*, 380*t*
ionophores, 479
    mitochondrial, 129
ion-specific electrodes, 572
IP₃. *See* 1,4,5-inositol trisphosphate
iproniazid, 310
iPSCs (induced pluripotent stem cells), 454
IPTG (isopropylthiogalactoside), 423
IRDS (infant respiratory distress syndrome), 244–245
IREG1 (iron-regulated protein 1), 639
IRES (internal ribosomal entry site), 416, 417*f*
iron
    absorption of, 105, 531–532, 531*f*
    conservation of, 639–640, 639*f*, 639*t*
    in cytochrome oxidase, 100
    in cytochromes P450, 119, 119*f*
    deficiency of, 643, 644*t*
    heavy metal displacement of, 98
    in heme, 50–51, 50*f*, 51*f*, 56, 100–102
    from heme catabolism, 321
    human requirement for, 97, 97*t*
    intracellular homeostasis, 641–642, 642*f*, 643*f*, 644*f*
    macrophage recycling of, 638, 638*f*
    multivalent states of, 97, 98*f*, 98*t*
    overload, 640, 643, 645*t*
    oxidation of heme, 656–658, 657*t*, 658*f*
    physiologic roles of, 100, 101*f*, 102, 102*f*
    regulation of, 98
    in sulfite oxidase, 104, 104*f*
    toxicity of, 99*t*
    transferrin cycle, 640, 641*f*
iron deficiency anemia, 643, 644*t*, 657*t*
iron porphyrins, 315. *See also* heme

iron response elements, 641–642
iron toxicity, 483
iron-regulated protein 1 (IREG1), 639
iron-sulfur (Fe-S) proteins, in respiratory chain complexes, 122–124, 123*f*
irreversible inhibitors, 81
IRS (insulin receptor substrates), 516, 517*f*
ischemia, 163, 486
isoaspartyl linkage in polypeptide backbone, 725*f*
isoaspartyl methyltransferase, 725*f*
isocitrate, in citric acid cycle, 158*f*, 159
isocitrate dehydrogenase
    in citric acid cycle, 158*f*, 159
    in NADPH production, 228, 229*f*, 230*f*
    regulation of, 162
isoelectric focusing (IEF), protein and peptide purification with, 27, 27*f*
isoelectric pH (pI), 20
isoleucine, 16*t*
    carbon skeleton catabolism of, 298, 300, 302*f*, 303*f*
    synthesis of, 277–278
isomaltose, 152*f*, 153*t*
isomerases, 60
isomerism
    geometric, of unsaturated fatty acids, 208–209, 208*f*
    of monosaccharides, 149–150, 149*f*, 150*f*
isomerization, of folding proteins, 42–43, 43*f*
isomorphous displacement, 40
isoniazid, 567
isopentenyl diphosphate, in cholesterol synthesis, 260, 261*f*
isopentenyl diphosphate isomerase, 261*f*
isoprene units, polyprenoids synthesized from, 213, 214*f*
isoprenoid units, 260, 261*f*, 262*f*
isopropylthiogalactoside (IPTG), 423
isoprostanes, 214
isothermic systems, 109
isotypes, 649
isovaleric acidemia, 300
isovaleryl-CoA dehydrogenase, deficiency of, 300, 302*f*
isozymes, 64–65
IκBα (nuclear factor kappa-B inhibitor α), 638

J
Jak/STAT pathway, 516–517, 518*f*
Jamaican vomiting sickness, 225
jaundice
    hyperbilirubinemia causing, 315, 323–324, 323*t*
    laboratory tests for, 574

metabolic disorders underlying, 324–325, 324*f*, 325*t*
    obstructive, 253
"jumping DNA," 370
junctional diversity, 648–649

K
k. See rate constant
K. See dissociation constant
K⁺. *See* potassium
K⁺ channel (KvAP), 478–479, 479*f*
Kartagener syndrome, 631
karyopherins, 586
karyotype, 366*f*
*k*cat. *See* catalytic constant
*k*cat/*K*m. *See* catalytic efficiency
KDEL-containing proteins, 582*t*, 591
*K*eq. *See* equilibrium constant
keratan sulfate I, 607, 608, 609*f*, 609*t*
keratan sulfate II, 607, 609, 609*f*, 609*t*
keratins, 631*t*, 632
kernicterus, 323
α-ketoacid decarboxylase complex, deficiency of, 300, 302*f*, 303*f*
ketoacidosis, 145–146, 217, 225
3-ketoacyl enzyme, 227, 229*f*
3-ketoacyl reductase, 229*f*
3-ketoacyl synthase, 227, 229*f*
3-ketoacyl-CoA thiolase deficiency, 225
ketoamines, in diabetes mellitus, 561, 561*f*
ketogenesis, 139
    high rates of fatty acid oxidation and, 220–223, 221*f*, 222*f*, 223*f*
    HMG-CoA in, 221, 222*f*
    regulation of, 223–224, 223*f*, 224*f*
ketogenic amino acids, 142
α-ketoglutarate
    arginine and ornithine formation of, 292, 292*f*
    in citric acid cycle, 158*f*, 159
    glutamine and glutamate formation of, 291–292
    histidine formation of, 292–293, 293*f*
    in transamination reactions, 282–283, 283*f*
α-ketoglutarate dehydrogenase complex
    in citric acid cycle, 158*f*, 159
    regulation of, 162
ketone bodies, 217, 220, 221*f*
    brain use of, 142
    free fatty acids as precursors of, 223, 223*f*
    as fuel for extrahepatic tissues, 221–223, 223*f*
    metabolic pathways of, 136, 136*f*, 139, 142
    in fasting state, 144–145, 144*t*, 145*f*
ketonemia, 223, 225

ketonuria, 225, 300, 303*t*
ketoses, 148, 148*t*, 149, 150*f*
ketosis, 145–146, 217, 221, 223, 225, 255
kidney
    basement membrane of, 610
    blood glucose threshold of, 189
    calcitriol synthesis in, 498–499, 498*f*
    metabolic features of, 145*t*
    vitamin D metabolism in, 538–539, 539*f*
kidney function tests, 574
kinesin, 631
kinetic order, 74
kinetic theory, 73
kinetics, enzyme. *See* enzyme kinetics
kinetochore, 364
*K*m. *See* Michaelis constant
Korsakoff psychosis, 542
Kozak consensus sequences, 412
Krabbe disease, 245*t*
Krebs cycle. *See* citric acid cycle
Ku70/80, 381*f*, 382
KvAP (voltage-gated K⁺ channel), 478–479, 479*f*
kwashiorkor, 273, 532–533
kynureninase, 298, 299*f*
kynurenine formylase, 298, 299*f*
kynurenine-anthranilate pathway, 296–298, 299*f*, 300*f*

L
laboratory tests
    biomedical importance of, 568
    for cancer, 709–710, 709*t*
    causes of abnormal results of, 568
    clinical value of, 570–571, 571*t*
    enzymes used in, 66–68, 67*f*, 67*t*, 572–573, 572*f*
    for hemostasis, 687
    of organ function, 574–575
    reference range of, 569
    samples for, 571
    techniques used in, 571–574, 572*f*
    validity of, 569–570, 569*f*, 570*f*, 571*t*
lac operon, 422–423, 422*f*, 424*f*, 425
lac repressor, 422–423, 424*f*
lacA gene, 422–423, 422*f*, 424*f*
lacI gene, 422–423, 422*f*, 424*f*, 425
β-lactamase, 83, 102, 103*f*
lactase, 59, 528
lactate
    in citric acid cycle, 160, 160*f*
    glucose derived from, 186, 187*f*
    glycolysis production of, 163, 164*f*, 164*t*, 167
    uptake by liver, 180–181
lactate dehydrogenase
    diagnostic use of, 67, 67*f*, 67*t*
    pyruvate reduction by, 167
    structure of, 38, 39*f*

lactation, glucose requirement of, 142
lactic acid, pK_a of, 13t
lactic acid cycle, glucose derived from, 186, 187f
lactic acidosis, 122, 163, 168–170
    vitamin B$_1$ deficiency in, 542
lactoferrin, 668t
lactose
    galactose conversion to, 197, 198f
    metabolism of, operon hypothesis and, 422–423, 422f, 424f, 425
    physiological importance of, 151–152, 152f, 153t
lactose intolerance, 147, 527
lactose synthase, 197, 198f
lactulose, 153f
lactylation, 430t
lacY gene, 422–423, 422f, 424f
lacZ gene, 422–423, 422f, 424f
lagging (retrograde) strand, in DNA replication, 372f, 373, 376f
lambda repressor (cI) gene, 425–426, 427f
lambda repressor (cI) protein, 426, 427f
laminin, 605–606, 606f
lamins, 365, 631t, 632
lanosterol, in cholesterol synthesis, 260, 261f, 262f
lathyrism, 15, 19, 19t
lauric acid, 207t
LBD (ligand-binding domain), 520, 520f
LCAT. See lecithin:cholesterol acyltransferase
LCRs (locus control regions), 436–437
LDL. See low-density lipoproteins
LDL receptor, 249
    in chylomicron remnant uptake, 251f, 252
    in cotranslational insertion, 590–591, 590f
    in maintenance of intracellular cholesterol balance, 263–264
LDL-receptor-related protein-1 (LRP-1), 249, 250f, 252
lead, toxicity of, 97–98, 316, 319
leading (forward) strand, in DNA replication, 372f, 373
Leber hereditary optic neuropathy, 485
lecithin:cholesterol acyltransferase (LCAT), 264
    apolipoprotein role in, 249
    familial deficiency of, 268t
    in HDL metabolism, 252f, 253
    in phosphoglycerol degradation and remodeling, 243
lecithins, 210, 210f
lectin pathway, 650, 651f
lectins, 155
    glycoprotein purification with, 555

mannose 6-phosphate receptor proteins, 562
Lepore, 370, 370f
leptin, 255
Lesch-Nyhan syndrome, 337, 344–345, 346t
leucine, 16t
    carbon skeleton catabolism of, 298, 300, 302f, 303f
    synthesis of, 277–278
leucine zipper motif, 438, 439f
leucovorin, 545
leukemias, 662, 664
leukocyte adhesion deficiency II, 562
leukocytes
    communication through effectors, 670
    in infection defense, 664–665
    motility of, 665–666, 665f, 666f, 667t
    multiple types of, 665f
    selectins of, 561–562
    turnover of, 718t
leukodystrophy, metachromatic, 245t
leukopenia, 664
leukotriene A$_4$, 208f
leukotriene B$_4$, 237
leukotrienes (LTs), 208, 208f, 226, 670
    clinical significance of, 234, 237–238
    lipoxygenase pathway in formation of, 235, 235f, 237f
Lewis acids, transition metals as, 97
library, 450
life expectancy, 718, 718t
lifespan
    body mass versus, 725–726, 726t
    evolution and, 727
    versus longevity, 717–718
lifestyle changes, cholesterol levels affected by, 267
ligand-binding assays, laboratory tests using, 573
ligand-binding domain (LBD), 520, 520f
ligand-gated channels, 478
ligandin, 322, 566
ligand-receptor complex, 509–510, 509f, 510f, 510t
ligases, 60
    DNA, 373t, 374
    ubiquitin, 593, 593f
light, as energy source in active transport, 480
light chains, immunoglobulin, 371, 646, 649
LINEs (long interspersed nuclear elements), 367–368
Lineweaver-Burk plot, 77–78, 78f
    for Bi-Bi reactions, 82, 83f
    inhibitor evaluation with, 80, 80f, 81f
linker DNA, 361f, 363

linoleic acid/linoleate, 206f, 207t
    in essential fatty acid deficiency, 234
    nutritionally essential, 232, 232f
    oxidation of, 221f
    synthesis of, 232, 233f
γ-linolenic acid, 207t, 233
α-linolenic acid (ALA), 207t, 209
    for essential fatty acid deficiency, 234
    nutritionally essential, 232, 232f
    synthesis of, 233–234, 233f
lipases
    diagnostic use of, 67t
    insulin and, 256
    triacylglycerol hydrolysis of, 529
    in triacylglycerol metabolism, 239, 255–256, 256f
lipid bilayer, 470–471, 470f
lipid core, of lipoprotein, 248, 249f
lipid droplets, 255, 257
lipid rafts, 210, 474, 474f, 596, 597
lipidomics, 3, 455
lipidoses (lipid storage disorders), 245, 245t
lipids. See also fatty acids; glycolipids; phospholipids; steroids; triacylglycerols
    amphipathic, 215, 215f
    biomedical importance of, 206
    classification of, 206
    complex, 206
    derived, 206
    digestion and absorption of, 528–529, 530f
    disorders associated with abnormalities of, 486
    free radical damage to, 549–550, 550f
    futile cycling of, 533
    in membranes (See also membrane lipids)
        amphipathic, 469–470, 469f
        artificial, 473
        asymmetry and, 472, 596–597, 597f
        protein association with, 471
        protein ratio to, 468, 468f
    as metabolic fuel, 134, 141
    metabolic pathways of, 135–136, 135f, 136f
        in fasting state, 144
        in fed state, 143–144, 143f
        at organ and cellular level, 137–139, 138f
    metabolism of, in liver, 253–254, 254f
    neutral, 206
    peroxidation of, 213–214, 214f, 720, 721f
    precursor, 206
    simple, 206
    solubility of, 206
    transport and storage of

adipose tissue and, 255–256, 256f
    biomedical importance of, 247–248
    brown adipose tissue and, 257–258, 258f
    clinical aspects of, 254–255
    fatty acid deficiency and, 236
    lipolysis and, 257, 257f
    as lipoproteins, 248, 248t, 249f
    liver in, 253–254, 254f
    perilipin and, 257
lipid-soluble molecules, 470
lipid-soluble vitamins, 535, 536t
lipins, 240, 241f
lipoamide, 168, 169f
lipogenesis, 137
    acetyl-CoA for, 228–229, 229f, 230f
    adipose tissue and, 257–258
    ethanol and, 255
    fatty acid synthase complex in, 227–228, 227f, 228f
    main pathway for, 226–230
    malonyl-CoA production in, 227, 227f
    NADPH for, 228, 230f
    regulation of
        enzymes in, 227–228, 231, 231f, 232f
        nutritional state in, 230
        short- and long-term mechanisms in, 230–232, 231f, 232f
lipolysis, 135, 137
    hormone regulation of, 255–256, 257f
    insulin inhibition of, 232
    lipase in, 255–256, 256f
    of triacylglycerol, 239
lipophilic hormones, 490, 491t
lipoprotein lipase, 137
    apolipoprotein role in, 249
    familial deficiency of, 268t
    hydrolysis by, 251, 251f
lipoprotein(a) excess, familial, 268t
lipoproteins, 35, 137, 138f, 635, 637t
    apolipoproteins in, 248t, 249
    carbohydrates in, 155
    in cholesterol transport, 263–264, 265f
    classification of, 248, 248t
    deficiency of, fatty liver and, 255
    disorders of, 267, 268t
    formation of, 247
    fructose effects on, 195, 197f, 199
    remnant, 248t, 251f, 252
    structure of, 248, 249f
    vitamin E in, 540
α-lipoproteins. See high-density lipoproteins
β-lipoproteins. See low-density lipoproteins
liposomes
    amphipathic lipids forming, 215, 215f
    artificial membranes and, 473
lipotropic factor, 255

lipoxin A₄, 208f
lipoxins (LXs), 208, 208f, 226, 234, 235
    clinical significance of, 237–238
    lipoxygenase pathway in formation of, 235, 235f, 237f
lipoxygenase, 235, 237f
    reactive species produced by, 214
5-lipoxygenase, 235, 237, 237f
lipoxygenase pathway, 208, 234, 235, 235f, 237f
liquid chromatography
    laboratory tests using, 572
    protein and peptide purification with, 25, 26f
lithocholic acid, synthesis of, 265, 266f
liver
    amino acid levels and, 281–282, 281f, 282f
    bilirubin metabolism of, 321–323, 322f, 323f
    blood glucose regulation by, 186, 187t
    calcitriol synthesis in, 498, 498f
    carbohydrate metabolism enzyme induction and repression in, 183, 184t
    cirrhosis of, 254, 255
    citric acid cycle in, 156
    cytochromes P450 in, 565
    failure of, 254
    in fasting state, 134–135, 134f, 144–145, 145f
    fatty (See fatty liver)
    in fed state, 134–135, 134f, 142–144, 143f
    fructose 2,6-bisphosphate in, 184–185, 186f
    fructose effects in, 195–196, 197f, 199
    glucose uptake by, 142–143, 143f, 165
    glycerol uptake by, 180–181
    glycogen in, 171, 172t
    glycogen phosphorylase regulation in, 175
    glycogenesis in, 172f, 173–175, 173f, 180–181
    glycogenolysis in, 173, 177
    heme synthesis in, 316–318
    metabolic features of, 137–139, 138f, 145t
    metabolism in
        chylomicrons and VLDL, 251
        fatty acid oxidation and ketogenesis, 220–223, 221f, 222f, 223f
        of lipids, 253
    plasma protein synthesis in, 635–636
    remnant lipoprotein uptake by, 250f, 252
    vitamin D metabolism in, 538–539, 539f
    VLDL formation and secretion in, 253–254, 254f
liver function tests, 574

LMWHs (low-molecular-weight heparins), 685
lncRNAs (long noncoding RNAs), 358–359, 384–385, 385t
locus control regions (LCRs), 436–437
long interspersed nuclear elements (LINEs), 367–368
long noncoding RNAs (lncRNAs), 358–359, 384–385, 385t
long-chain ω3 fatty acids, 207, 209
longevity versus lifespan, 717–718
looped domains, chromatin, 362f, 363, 365
loops, in proteins, 37, 38f
low-density lipoproteins (LDL), 248, 248t
    advanced glycation end-products and, 561
    apolipoproteins of, 249
    atherosclerosis and, 252, 267
    cholesterol supply with, 259
    endocytosis of, 483
    HDL ratio to, 267
    metabolism of, 251f, 252
    receptors for (See LDL receptor)
    regulation of, 263
Lowe oculocerebrorenal syndrome, 592t
low-energy phosphates, 111
low-molecular-weight heparins (LMWHs), 685
LRP-1 (LDL-receptor-related protein-1), 249, 250f, 252
LTs. See leukotrienes
lung surfactant, 210, 239, 244–245
LXs. See lipoxins
LXXLL motifs, 522
lyases, 60
lymphocyte homing, glycoprotein role in, 561–562
lymphocytes, 665, 670
lymphoid progenitor cells, 665
lysine, 17t, 296, 298f
lysine acetyltransferases, 93
lysine hydroxylase, vitamin C as coenzyme for, 547
lysogenic pathway, 425, 426f
lysolecithin, 211, 211f, 243, 243f, 252f, 253
lysophosphatidylcholine, 211, 211f, 243, 243f
lysophospholipase, 243, 243f
lysophospholipids, 211, 211f, 486
lysosomal degradation pathway, 245
lysosomal enzymes in I-cell disease, 484, 486t
lysosomal hydrolases, glycoprotein, 562–563
lysosomal proteases, in protein degradation, 592–593
lysosomal proteins, 597–598

lysosomes
  in endocytosis, 482
  faulty targeting of enzymes into, 562
  glycogen hydrolysis by, 175
  protein sorting signals in, 582, 582t
lysozyme, 38f, 668t
lysyl hydroxylases, 44, 276, 600, 603, 603t
lysyl oxidase, 44, 601, 604
  copper in, 103
lytic pathway, 425, 426f
lytic/lysogenic genetic switching event,
    425–426, 427f

M
M phase, 378, 378f, 379f
mAbs (monoclonal antibodies), 649–650,
    706, 711, 711t
MAC (membrane attack complex), 650
α₂-macroglobulin, 645, 645f, 670, 685
macromolecules
  cell nucleus transport of, 583–587, 583f,
      586f
  cellular transport of, 481–483, 482f, 483f
  mitochondria import of, 584, 585f
macrophages, 664, 665, 668, 670
  iron recycling by, 638, 638f
magnesium (Mg²⁺)
  in extracellular and intracellular fluid,
      468, 468t
  porphyrins containing, 315
Maillard reaction, 561, 561f
major groove, in DNA, 350, 350f
  helix-turn-helix motif and, 437, 438f
  operon model and, 423
major histocompatibility complex (MHC),
    592, 593, 667
major sorting decision, 582
malaria, glycan role in, 563
malate
  in citric acid cycle, 158f, 159
  in lipogenesis, 228, 230f
malate dehydrogenase, in citric acid cycle,
    158f, 159
malate shuttle, 129–130, 129f
MALDI (matrix-assisted laser desorption
    and ionization), 30–31, 31f
malic enzyme, in NADPH production,
    228, 229f, 230f
malignancy. See cancer
malignant hyperthermia (MH), 625, 627t
malonate, 80, 127
malonyl acetyl transacylase, 227, 227f,
    229f
malonyl-CoA
  CPT-I regulation by, 223–224, 224f
  in fatty acid synthesis, 227, 227f
maltose, physiological importance of,
    151–152, 152f, 153t
mammalian coregulator proteins, 522t

mammalian DNA-dependent RNA
    polymerases, 388, 388t
mammalian viral cloning vectors, 449
manganese
  human requirement for, 97, 97t
  multivalent states of, 97, 98f, 98t
  physiologic roles of, 102
  toxicity of, 99t
mannan-binding protein (MBP), 650–651
mannosamine, 151, 197, 199f
mannose, 150, 150f, 151t, 556t
mannose 6-phosphate, 483, 562, 582t
mannose-binding lectin (MBL), 650–651,
    651f
mannose-binding protein, 562
mannosidosis, 562–563
manual enzymatic Sanger method, 451,
    452f
MAPK (mitogen-activated protein kinase)
    pathway, 516, 517f
maple syrup urine disease (MSUD), 300,
    303t
marasmus, 109, 273, 532–533
Marfan syndrome, 604, 605f
Maroteaux-Lamy syndrome, 611t
mass spectrometry (MS)
  metabolic disease detection with, 288
  protein analysis using, 29–31, 29t, 30f,
      31f
  proteomics with, 455
  spectrometer configurations used for,
      29–31, 30f, 31f
  tandem, 31, 288, 290
  volatilization methods for, 30–31, 31f
mast cells, 610, 664, 670
MAT (methionine adenosyltransferase),
    308, 309f
matrix, mitochondrial, 122, 122f, 157
matrix metalloproteinases (MMPs), 708
matrix protease, 584
matrix proteins, 584, 586–587
matrix-assisted laser desorption and
    ionization (MALDI), 30–31, 31f
maximal velocity (Vmax), 76–79, 76f, 78f,
    79f
  allosteric effects on, 88
  in Bi-Bi reactions, 82, 83f
  inhibitor effects on, 80–81, 80f, 81f
MBL (mannose-binding lectin), 650–651,
    651f
MBP (mannan-binding protein),
    650–651
McCardle disease, 174t
MDR (multidrug resistance), 486
MDR-1 protein (multidrug-resistance-1
    protein), 480
mechanically gated channels, 478
mechanism-based inhibitors, 79, 81–82
mediator (Med) proteins, 395, 395f

medicine
  biochemical research impacts on, 2–3
  biochemistry importance to, 1
  biochemistry interrelationship with, 2, 2f
  Human Genome Project impact on,
      3–4, 3f
  personalized, 444
medium-chain acyl-CoA dehydrogenase,
    deficiency of, 225
megaloblastic anemias, 546, 657t
MELAS (mitochondrial encephalopathy,
    lactic acidosis, and stroke), 130
melatonin, tryptophan conversion to, 310,
    311f
melting (transition) temperature (Tm),
    351, 473
membrane attack complex (MAC), 650
membrane fatty acid transport protein,
    249
membrane fusion, 594
membrane lipids
  bilayers of, 470–471, 470f
  cholesterol in, 469
  glycosphingolipids in, 469
  phospholipids in, 209–211, 210f, 469,
      469f
  sterols in, 469
  turnover of, 597
membrane proteins
  differences in, 471, 471t
  integral, 35, 472–473, 472f
  lipid bilayer and, 471
  mutations of, diseases caused by,
      484–485, 486t
  peripheral, 472–473, 472f
  structure of, dynamic, 471–472
  turnover of, 597
membrane-bound polyribosomes, 587
membranes
  artificial, 473
  assembly of, 582, 595t, 596–598, 597f
  asymmetry of, 468, 472, 596–597, 597f
  biomedical importance of, 467
  depolarization of, in nerve impulse
      transmission, 481
  fluidity affecting, 473–474
  gap junctions, 467, 483, 484f
  Golgi apparatus in synthesis of, 582
  intracellular, 468
  mitochondrial, 122, 122f, 123f, 584, 585f
    selective permeability of, 128–130,
        128f, 129f, 130f
  phase changes of, 473–474
  of red blood cells, 658–660, 658f, 659f,
      659t
  selectivity of, 475t (See also transporters
      and transport systems)
    aquaporins, 479–480
    facilitated diffusion, 477, 477f

ion channels, 478, 478f, 479f, 479t
ionophores, 479
mechanisms of, 483
passive diffusion, 475f, 475t, 476f, 476t, 483
transporters, 476–477, 476f, 477f
signal transmission across, 467, 475t, 483
structure of
asymmetry and, 472
fluid mosaic model of, 473–474, 474f
lipids in, 215, 215f, 469–471, 469f, 470f
protein:lipid ratio in, 468, 468f
menadione, 536t, 540–541, 541f
menaquinones, 536t, 540–541, 541f
Menkes syndrome, 44, 264, 603
MEOS (microsomal ethanol oxidizing system), 255
6-mercaptopurine, 334f, 335, 339f, 340
3-mercaptopyruvate pathway, 294, 294f
mercury, toxicity of, 97–98
meromyosin, 621
merosin, 605, 629f
messenger RNA (mRNA), 354, 384–385, 385t
alternative splicing and, 398f, 399
amplification of, 441–442, 441f
chemical characteristics of, 355, 356f
codon assignments in, 404–406, 405t
editing of, 402
function of, 355, 355f
gene expression modulation and, 439–440
in genetic information flow, 404–405
modification of, 400–401
nontranslating, 416, 416f
nucleotide sequence of, 405, 405t
mutations caused by changes in, 408–409, 409f
polycistronic, 422
precursors, 355, 366
processing of, 397f, 398–399, 398f
in protein synthesis
elongation, 413–414, 413f
initiation, 410, 411f, 412–413, 412f
termination, 414
relationship to chromosomal DNA, 367f
stability of, 442, 442f
transcription start site and, 386–387
translocation of, 586
MET (metabolic equivalent), 532
metabolic acidosis, ammonia role in, 285
metabolic alkalosis, ammonia effects on, 285
metabolic disorders
of amino acid metabolism, 286–288, 287t, 290–291, 300, 304t
of bilirubin metabolism, 324–325, 324f, 325t

gene or protein modification for, 288
general features of, 286–287
neonate screening for, 573–574
tandem mass spectrometry in screening for, 288, 290
urea cycle defects, 286–288, 287t
metabolic equivalent (MET), 532
metabolic fuels
in fasting state, 134–135, 134f, 144–145, 145f
in fed state, 134–135, 134f, 142–144, 143f
human intake of, 133–134
interconvertible nature of, 141–142
processing of, 135–136, 135f, 136f, 138f
metabolic rate, 109, 167, 532
metabolic regulation
active or passive, 86, 86f
allosteric, 88–89, 141, 141f
biomedical importance of, 85–86
compartmentation role in, 85–86
control networks of, 94, 94f
covalent modifications in, 90–93, 92f, 93t
of enzyme quantity, 87–88
flux-generating reactions in, 139–141
hormones in, 141, 141f
multiple mechanisms of, 88
nonequilibrium reactions as control points in, 139
proenzymes in, 90, 92f
rate-limiting enzymes as targets of, 87
second messengers in, 89
metabolic syndrome, 190
metabolic theories of aging, 725–726, 726t
metabolism, 110
amino acid roles in, 19
biomedical importance of, 133–134
in cancer cells, 703–706, 705f, 706t
citric acid cycle as pivotal pathway in, 159–162, 160f, 161f
clinical aspects of, 145–146
CNS and erythrocyte glucose requirements and, 142
drug (See drug metabolism)
energy requirements for, 133–134
in fasting state, 134–135, 134f, 144–145, 145f
in fed state, 134–135, 134f, 142–144, 143f
free energy changes of (See bioenergetics)
fructose effects on, 195–196, 197f, 199
group transfer reactions in, 9
inborn errors of, 2
interconvertible fuels in, 141–142
normal, 133
nucleophiles and electrophiles in, 9
organ and cellular level of, 136–139, 138f

pathways for processing products of digestion, 135–136, 135f, 136f, 138f
regulation of (See metabolic regulation)
regulation of metabolite flux through, 139–141, 141f
subcellular level of, 139, 140f
metabolites, flux of, 86, 86f, 93, 139–141, 141f
metabolomics, 3, 455, 572
metabonomics, 572
metachromatic leukodystrophy, 245t
metal ions. See also transition metals
in enzyme prosthetic groups, 61
heavy metal displacement of, 97–98
metal-activated enzymes, 61
metalloenzymes, 61
metalloproteins, 35
physiologic roles of, 100–105, 101t, 102f, 103f, 104f
transition metals in, 99–100, 99f, 100f, 101t
metaphase chromosomes, 364, 365t, 366f
metastasis
mechanisms of, 706–707, 707f, 708t
membrane abnormalities and, 486t
tumor microenvironment role in, 703, 704f
meteorites, extraterrestrial amino acids in, 18–19
methacrylyl-CoA, catabolism of, 302f, 303f
methane monooxygenase, iron in, 101–102
methemoglobin, 56, 656
methemoglobin reductase, 56
methemoglobinemia, 656, 657t
methionine, 17t
carbon skeleton catabolism of, 298, 300f, 301f
in S-adenosylmethionine, 333f, 334t
specialized products of, 308–309, 309f, 310f, 311, 312f
methionine adenosyltransferase (MAT), 308, 309f
methionine synthase, vitamin B_{12} dependency of, 544, 544f, 546
methionyl-tRNA (met-tRNA), 410, 411f
methotrexate, 343–344, 545
N^6-methyladenosine, 400–401
L-β-methylaminoalanine, 19, 19t
methylation
of deoxycytidine residues, 430–431
enzyme regulation by, 92
of histones, 361, 363t
mass spectrometry detection of, 29t
of xenobiotics, 566–567
β-methylcrotonyl-CoA, catabolism of, 302f
5-methylcytosine, 331, 333f

methylene bridges, in porphyrinogens, 316–317

N-methylglycine, 310–311

7-methylguanine, 333*f*

7-methylguanosine cap structure, mRNA, 400

3-methylhistidine, 308

methylmalonic aciduria, 182, 183*f*

methylmalonyl-CoA mutase, 182, 183*f*, 544

methylmalonyl-CoA racemase, 181–182, 183*f*

methylxanthines, 257

methyne bridges, in heme, 50, 50*f*

met-tRNA (methionyl-tRNA), 410, 411*f*

mevalonate, synthesis of, 260, 260*f*, 261*f*

mevalonate kinase, 261*f*

Mg²⁺. *See* magnesium

MH (malignant hyperthermia), 625, 627*t*

MHC (major histocompatibility complex), 592, 667

MI. *See* myocardial infarction

micelles, 215, 215*f*, 470, 470*f*, 529

Michaelis constant (K_m), 76*f*, 77–79, 78*f*
  allosteric effects on, 88
  in Bi-Bi reactions, 82, 83*f*
  cellular substrate concentrations and, 86
  of glutathione reductase, 731
  inhibitor effects on, 80–81, 80*f*, 81*f*

Michaelis-Menten equation, 76–79, 76*f*, 78*f*
  for Bi-Bi reactions, 82, 83*f*

microalbuminuria, 574

microbial toxins, 479

microcystin, 15

microfibrils, 604

microfilaments, 630

β₂-microglobulin, 642, 644*f*, 646

microheterogeneity, of glycoproteins, 554

micronutrients. *See also* minerals; vitamins
  biomedical importance of, 535
  lipid storage of, 206
  nutritional requirements for, 535–536
  toxic levels of, 535
  transition metals, 97, 97*t*

micro-RNAs (miRNAs), 358–359, 384–385, 385*t*, 401–402
  in cancer, 698–699
  gene expression modulation by, 439–440
  transcription of, 400–401, 401*f*

microsatellite instability, 368
  in cancer, 700, 700*f*

microsatellite polymorphisms, 368

microsatellite repeat sequences, 368

microscopy, protein structures solved by, 41

microsomal elongase system, 230, 231*f*

microsomal ethanol oxidizing system (MEOS), 255

microsomal triacylglycerol transfer protein (MTP), 253

microtubule-associated proteins, 630–631

microtubules, 594, 630–631, 631*f*

microvesicles, 484
  cancer and, 698

microvilli, 474

mild sonication, 473

milk fat synthesis, 251

mineralocorticoids, synthesis of, 493–494, 494*f*

minerals
  biomedical importance of, 535
  digestion and absorption of, 531–532, 531*f*
  functions of, 548, 556*t*
  for health maintenance, 3
  nutritional requirements for, 535–536

minor groove, in DNA, 350, 350*f*

MIPP (multiple inositol polyphosphate phosphatase), 56

miRNAs. *See* micro-RNAs

misfolded proteins, endoplasmic reticulum degradation of, 592–593, 592*t*, 593*f*

mismatch repair (MMR), 379, 380*f*, 380*t*

missense mutations, 407–408, 408*f*, 409*f*, 628

MIT (monoiodotyrosine), 500, 501*f*

mitochondria
  apoptosis in, 722
  β-oxidation in, 218–220, 218*f*, 219*f*, 220*t*
  citric acid cycle in, 139, 140*f*, 157
  cytochromes P450 in, 565
  enzyme activity of, 735*t*
  exchange transporters of, 128–130, 128*f*, 129*f*, 130*f*
  fatty acid oxidation in, 217–218, 218*f*
  genes encoded by genome of, 722*t*
  genetic disorders involving, 130
  membranes and compartments of, 122, 122*f*
  protein sorting signals in, 582, 582*t*
  protein synthesis and import by, 584, 585*f*
  redox damage of, 722
  respiratory chain complexes in, 122, 123*f*

mitochondrial branched-chain α-ketoacid dehydrogenase complex, 298, 300, 302*f*

mitochondrial DNA, 368, 369*f*, 369*t*

mitochondrial encephalopathy, lactic acidosis, and stroke (MELAS), 130

mitochondrial genome, 584

mitochondrial glycerol-3-phosphate dehydrogenase, 118

mitochondrial matrix, 584, 585*f*

mitochondrial pathway, of apoptosis, 701–702, 702*f*

mitogen-activated protein kinase (MAPK) pathway, 516, 517*f*

mitotic phase, 378, 378*f*, 379*f*

mitotic spindle, 631

mixed-function oxidases, 119

MMPs (matrix metalloproteinases), 708

MMR (mismatch repair), 379, 380*f*, 380*t*

MOAT (multispecific organic anion transporter), 322

modeling, of protein structures, 41–42

molecular biology, protein and peptide sequencing in, 29

molecular chaperones. *See* chaperones

molecular diagnostic tests, 4

molecular docking programs, protein structures solved by, 41–42

molecular dynamics, protein structures solved by, 41

molecular genetics. *See* recombinant DNA technology

molecular modeling, of protein structures, 41–42

molecular pathology, 597–598

molecular profiling, of cancer, 712

molecular repair mechanisms, wear and tear theory of aging
  aggregated proteins, 725
  enzymatic and chemical mechanisms, 723–724
  proofreading and repair mechanisms, 724
  protein damage, 724–725, 725*f*

molecular replacement, 41

molecular switches, 594

molybdenum
  absorption of, 105
  human requirement for, 97, 97*t*
  multivalent states of, 97, 98*f*, 98*t*
  physiologic roles of, 103–104, 104*f*
  toxicity of, 99*t*

molybdopterin, 99, 100*f*, 103–104

monoacylglycerol acyltransferase, 240, 241*f*

monoacylglycerol pathway, 240, 241*f*, 529

2-monoacylglycerols, 241*f*

monoamine oxidase, 310, 311*f*

monoclonal antibodies (mAbs), 649–650, 706, 711, 711*t*

monocytes, 664, 665

monoiodotyrosine (MIT), 500, 501*f*

monomeric proteins, 39

mononucleotides, 330
  "salvage" reactions and, 340, 341*f*

monooxygenases
  cytochromes P450 as, 119–120, 119*f*, 120*f*
  xenobiotic hydroxylation by, 564–566

monosaccharides, 148, 148*t*
   absorption of, 528, 528*f*
   amino sugars, 151, 152*f*
   deoxy sugars, 151, 151*f*
   glucose as most important, 148 (*See also* glucose)
   in glycoproteins, 555, 556*t*
   glycosides formed from, 150–151
   isomerism of, 149–150, 149*f*, 150*f*
   of physiological importance, 150, 151*f*, 151*t*, 152*f*
monoubiquitylation, of histones, 361, 363*t*
monounsaturated fatty acids, 207, 207*t*
   dietary, cholesterol levels affected by, 267
   synthesis of, 233, 233*f*
Morquio syndrome, 611*t*
mortality and aging, 718
mRNA. *See* messenger RNA
mRNP exporter, 586
mRNPs (ribonucleoprotein particles), 416, 416*f*, 442
MS. *See* mass spectrometry
MS². *See* tandem MS
MS-MS. *See* tandem MS
MstII, 445*t*
MSUD (maple syrup urine disease), 300, 303*t*
MTP (microsomal triacylglycerol transfer protein), 253
mucins, 557
mucolipidosis, 611
mucopolysaccharidoses, 599, 610–611, 611*t*
mucoproteins. *See* glycoproteins
mucus, 557
multidimensional protein identification technology (MudPIT), 32
multidrug resistance (MDR), 486
multidrug-resistance-1 protein (MDR-1 protein), 480
multiple inositol polyphosphate phosphatase (MIPP), 56
multiple sclerosis, 245, 481
multiple sulfatase deficiency, 245
multiplexins, 601
multipotent stem cells, 653
multisite phosphorylation, 177
multispecific organic anion transporter (MOAT), 322
muscle
   actin and myosin in, 620–621, 621*f*, 622*f*
   amino acid levels and, 281–282, 281*f*, 282*f*
   cardiac, 625–626, 626*f*, 626*t*
   in fasting state, 134–135, 134*f*, 144–145
   in fed state, 134–135, 134*f*, 142–144, 143*f*
   genetic disorders of, 628–629, 628*t*, 629*f*
   glucose uptake by, 142–143, 143*f*

glycogen in, 171, 172*t*
   glycogen phosphorylase regulation in, 175, 176*f*
   glycogen synthase regulation in, 178*f*
   glycogenesis in, 172*f*, 173–175, 173*f*
   lactate production by, 167
   metabolic features of, 145*t*
   ribose synthesis in, 194
   structure of, 618–620, 619*f*, 620*f*
   types of, 618–619
muscle contraction
   ATP in, 622–623, 622*f*, 627–628, 627*f*
   calcium role in, 618, 623–625, 624*f*, 624*t*, 625*f*
   energy conversion in, 622–623, 622*f*, 623*f*
   glycogen phosphorylase activation with, 177
   nitric oxide in, 630, 630*f*
   relaxation phase of, 623
   thin and thick filaments in, 619–620, 620*f*
muscular dystrophy, 629
   glycoprotein synthesis defects in, 562
mutations, genetic, 360–361
   Alu family and, 368, 370
   base substitution, 407–408, 407*f*, 408*f*, 409*f*
   cancer-causing, 691, 710, 712
   constitutive, 422
   frameshift, 409–410, 409*f*
   free radical damage causing, 550
   gene conversion and, 371
   indels and, 379
   integration and, 370, 370*f*
   of membrane proteins, diseases caused by, 485, 486*t*
   missense, 407–408, 408*f*, 409*f*, 628
   nonsense, 408–409, 409*f*
   point, 407–408, 407*f*, 408*f*, 409*f*
   premature termination, 408, 409*f*
   recombination and, 369*f*, 370*f*, 379
   silent, 407
   sister chromatid exchanges and, 371, 371*f*
   suppressor, 409–410
   transition, 407, 407*f*
   transposition and, 370–371
   transversion, 407, 407*f*
   xenobiotic role in, 567, 567*f*
   Zellweger syndrome and, 587
mutator phenotype, 700, 700*f*
MYC gene, in cancer, 693, 694*f*, 696*t*
mycophenolic acid, 339*f*, 340
myelin
   galactosylceramide in, 244
   sulfatide in, 212
myelin sheath, 210
myelin sheets, 481

myelinated nerves, depolarization waves along, 206
myelodysplasia, 656
myeloid progenitor cells, 665
myeloperoxidase, 668–669, 668*t*
   iron in, 102
myocardial infarction (MI)
   biomarkers of, 575
   serum enzyme analysis after, 67*f*, 68
myofibrils, 619, 619*f*
myoglobin
   α helices in, 50, 50*f*
   biomedical implications of, 56
   biomedical importance of, 49–50
   heme in, 50–51, 50*f*, 51*f*
   hindered environment of, 51, 51*f*
   iron in, 99–100
   oxygen dissociation curve of, 51–52, 51*f*
   secondary and tertiary structures of, 52–53
myoglobinuria, 56
myokinase. *See* adenylyl kinase
myopathy, lactic acidosis with, 122
myosin, 618
   in muscle contraction, 620–621, 621*f*, 622*f*
   smooth muscle contraction regulation by, 623–624, 624*f*, 624*t*
   structure and function of, 620–621, 621*f*, 622*f*
myosin (thick) filaments, 619–620, 620*f*
myotonia congenita, 627*t*
myristic acid, 207*t*
myristoylation, 471
   mass spectrometry detection of, 29*t*

N
n (Hill coefficient), 79, 79*f*
Na⁺. *See* sodium
NAD⁺. *See* nicotinamide adenine dinucleotide
NAD-linked dehydrogenases, 117
NADH
   fatty acid oxidation production of, 219, 219*f*
   in fatty liver, 255
NADH dehydrogenase, 118
NADH-Q oxidoreductase (Complex I), 122, 123*f*, 124, 124*f*
   deficiency of, 130
NADP⁺. *See* nicotinamide adenine dinucleotide phosphate
NADP malate dehydrogenase, 228, 229*f*, 230*f*
NADP-linked dehydrogenases, 117
NADPH, for lipogenesis, 228, 229*f*, 230*f*
NADPH oxidase, 668, 668*t*, 669
NADPH-cytochrome P450 reductase, 565

NAFLD (nonalcoholic fatty liver disease), 254
NAGS. *See N*-acetylglutamate synthetase
Na⁺-K⁺-ATPase, 480, 480*f*, 481*f*
nanotechnology, 4, 65, 65*f*
NASH (nonalcoholic steatohepatitis), 254
native conformation, protein, 42
Natowicz syndrome, 611*t*
natural killer cells, 664, 670
ncRNAs. *See* noncoding RNAs
NDP (nucleoside diphosphate) kinases, 114
necrosis, 701
negative nitrogen balance, 533
negative predictive value, 570–571, 571*t*
negative regulators, of gene expression, 421, 421*t*, 423, 425, 426–427
negative supercoils, DNA, 351
neonatal adrenoleukodystrophy, 587, 588*t*
neonatal alloimmune thrombocytopenia, 662
neonates
    hypoglycemia in, 189
    inborn errors of metabolism screening in, 573–574
    jaundice in, 324
    metabolic disease screening in, 288
    tyrosinemia in, 296, 297*f*
neoplasms, 689–690, 690*f*
nephrogenic diabetes insipidus, 480
NER (nucleotide excision-repair), 379, 380*f*, 380*t*
nerve impulses, 481
NESs (nuclear export signals), 586
net diffusion, 476
NET-Seq, 455
NeuAc (*N*-acetylneuraminic acid), 555, 556*t*
neural tube defects, folic acid supplements for, 546
neuraminic acid, 155, 212
neuraminidase, influenza, 563
neurodegenerative diseases, prion, 43
neurofilaments, 631*t*, 632
neurolathyrism, 19, 19*t*
neurologic impairment, profound, 587
neurons, membranes of
    impulses transmitted along, 481
    ion channels in, 478*f*
neuropathy, vitamin B₆ causing, 543
neurotransmitters, amino acids as, 19
neutral lipids, 206
neutrophil extracellular traps, 669, 669*f*
neutrophils, 664, 665
    enzymes and proteins of, 667, 668*t*
    integrins in, 666, 667*t*
    myeloperoxidase in, 669
    in parasite entrapment, 669, 669*f*
newborns, hemoglobins of, 53, 53*f*

next-generation sequencing (NGS), 453–454
NFκB (nuclear factor kappa-B), 561, 638
NFκB (nuclear factor kappa-B) inhibitor α, 638
NF-κB pathway, 517–518, 518*f*
NGS (next-generation sequencing), 453–454
NHEJ (nonhomologous end-joining), 379, 380*f*, 380*t*
niacin, 536
    citric acid cycle need for, 159
    deficiency of, 536*t*, 542–543
    functions of, 536*t*, 542
    nicotinamide coenzymes formed from, 117
    toxicity of, 543
nick translation, 446*t*
nickel
    human requirement for, 97, 97*t*
    multivalent states of, 97, 98*f*, 98*t*
    physiologic roles of, 103
    toxicity of, 99*t*
    in urease, 101*t*, 103
nicks and nick-sealing, in DNA replication, 377, 377*f*
nicotinamide, 61, 61*f*, 542
nicotinamide adenine dinucleotide (NAD⁺)
    in citric acid cycle, 158*f*, 159, 162
    as coenzyme, 334*t*
    cytochrome P450 use of, 119, 119*f*
    dehydrogenase use of, 117, 117*f*
    in fatty liver, 255
    metabolic compartmentation using, 87
    niacin in synthesis of, 542
    spectrophotometric assays using, 66, 66*f*
    substrate shuttles in oxidation of, 129–130, 129*f*
nicotinamide adenine dinucleotide phosphate (NADP⁺)
    as coenzyme, 334*t*
    cytochrome P450 use of, 119, 119*f*
    dehydrogenase use of, 117, 117*f*
    intramitochondrial source of, 129
    metabolic compartmentation using, 87
    niacin in synthesis of, 542
    pentose phosphate pathway formation of, 191–194, 192*f*, 193*f*
    spectrophotometric assays using, 66, 66*f*
nicotine, cytochrome P450 metabolism of, 565–566
nicotinic acid, 542
    as hypolipidemic drug, 267
nidogen, 605
Niemann-Pick C-like 1 protein, 267

Niemann-Pick disease, 245*t*
nitric oxide (NO), 630, 630*f*
    arginine conversion to, 307, 307*f*
    in hemostasis and thrombosis, 679
nitric oxide synthases, 307, 307*f*, 630
nitrite, 630
nitrogen, metabolism of
    amino acid oxidases in, 283–284, 284*f*
    ammonia from, 279, 282–275, 284*f*
    disorders of, 286–288, 287*t*
    excreted forms of, 282, 285, 285*f*
    glutamate dehydrogenase in, 283, 284*f*
    glutaminase and asparaginase in, 284, 284*f*
    glutamine synthetase in, 284, 284*f*
    transamination reactions in, 282–283, 283*f*, 291
    urea biosynthesis in, 282–283, 282*f*, 283*f*, 285–286, 285*f*
nitrogen balance, 279, 533–534
nitrogenase, 100
nitroglycerin, 618
p-nitrophenyl phosphate (*p*NPP), 66
NLS (nuclear localization signal), 582*t*, 585, 586*f*
NMP (nucleoside monophosphate) kinases, 114
NMR (nuclear magnetic resonance) spectroscopy, protein structures solved by, 41
NO. *See* nitric oxide
nodes of Ranvier, 481
nodularin, 15
nonalcoholic fatty liver disease (NAFLD), 254
nonalcoholic steatohepatitis (NASH), 254
non-clathrin-coated vesicles. *See* transport vesicles
noncoding RNAs (ncRNAs), 358–359. *See also* long noncoding RNAs; micro-RNAs; ribosomal RNA; silencing RNAs; small nuclear RNA; transfer RNA
    in cancer, 698–699
    gene expression modulation by, 441
noncoding strand, 349, 354*f*
noncompetitive inhibitors, 80–81, 80*f*, 81*f*
noncovalent assemblies, in membranes, 469
noncovalent forces
    biomolecule stabilization by, 7–8, 8*f*
    peptide conformations and, 23
    in protein tertiary and quaternary structures, 40
nonequilibrium reactions
    in glycolysis regulation, 167–168
    regulation of metabolic pathways with, 139

nonessential amino acids, 136, 274*t*, 275–278, 534

nonesterified fatty acids. *See* free fatty acids

nonheme iron proteins. *See* iron-sulfur (Fe-S) proteins

nonhistone proteins, 361

nonhomologous end-joining (NHEJ), 379, 380*f*, 380*t*

nonhomologous sequence, DNA, 370

nonketotic hyperglycinemia, 293

non-lipid-soluble molecules, 470–471

nonoverlapping of genetic code, 406

nonpolar groups, noncovalent interactions of, 8

nonpolar lipid core, of lipoprotein, 248, 249*f*

nonrepetitive (unique-sequence) DNA, 367

nonsense codons, 405, 408–409, 409*f*

nonsense mutations, 408–409, 409*f*

nonsteroidal anti-inflammatory drugs
cyclooxygenase affected by, 236–237
prostaglandins synthesis and, 226, 236–237

nontemplate strand DNA, 385–386

non-α-peptide bonds, between ubiquitin and proteins, 280, 280*f*

norepinephrine. *See also* catecholamines
glycogen regulation by, 175
in lipolysis, 256, 257*f*
synthesis of, 492*f*, 499, 499*f*
in thermogenesis, 258, 258*f*
tyrosine conversion to, 310, 312*f*

Northern RNA blot procedure, 450, 451*f*

Northern (RNA-DNA) blotting, 341

NPCs (nuclear pore complexes), 585

*p*NPP (*p*-nitrophenyl phosphate), 66

N-terminal binding domain, 38

nuclear export signals (NESs), 586

nuclear factor kappa-B (NFκB), 561, 638

nuclear genes, proteins encoded by, 584

nuclear localization signal (NLS), 582*t*, 585, 586*f*

nuclear magnetic resonance (NMR) spectroscopy, protein structures solved by, 41

nuclear pore complexes (NPCs), 585

nuclear receptor coregulators, 521–522, 521*f*, 522*t*

nuclear receptor superfamily, 520–522, 520*f*, 521*t*

nuclear receptors, 490

nucleases, 9
active chromatin and, 364
nucleic acid digestion by, 359

nucleic acids
biomedical importance of, 348
nuclease digestion of, 359

nucleolytic processing, of RNA, 401–402, 401*f*

nucleophiles
definition of, 9
in enzymatic catalysis, 61
water as, 9–10

nucleophilic attack, 374, 375*f*

nucleoproteins, packing of, 365, 365*t*, 366*f*

nucleoside diphosphate (NDP) kinases, 114

nucleoside diphosphates, chemistry of, 330, 331*f*

nucleoside monophosphate (NMP) kinases, 114

nucleoside triphosphates
chemistry of, 330, 331*f*
group transfer potential of, 334
nonhydrolyzable analogs of, 335, 335*f*

nucleosides, 331–333, 332*t*
biomedical importance of, 330
chemistry of, 330, 331*f*
triphosphates, 334

nucleosomes, 361, 361*f*, 362*f*, 363
transcription machinery and, 393*f*, 394–395

nucleotide adducts, carcinogens forming, 691–692

nucleotide excision-repair (NER), 379, 380*f*, 380*t*

nucleotides, 332*t*
biomedical importance of, 330
biosynthesis of, purines, 338, 338*f*
chemistry of, 330, 331*f*
dietarily nonessential, 338
frameshift mutations of, 408–409, 409*f*
metabolism of, biomedical importance of, 337–338
mutations caused by changes in, 407–409, 407*f*, 408*f*, 409*f*
physiologic functions of, 331–333, 333*f*, 334*t*
as polyfunctional acids, 331
polynucleotides, 335
substitution mutations of, 407–408, 407*f*, 408*f*, 409*f*
synthetic analogs of, in chemotherapy, 334–335, 334*f*, 335*f*
triplet codes of, 405, 405*t*
ultraviolet light absorption by, 331

nucleus of cell
macromolecule transport in, 583–587, 583*f*, 586*f*
protein sorting signals in, 582, 582*t*

nutrigenomics, 3

nutrition
antioxidants and (*See* antioxidants)
biochemical research impacts on, 2–3
biomedical importance of, 527

lipogenesis regulated by, 230
requirements
energy, 133–134, 532–533, 532*f*
protein and amino acid, 533–534
transition metals, 97, 97*t*
vitamins and minerals, 535–536

nutritional disorders
amino acid deficiency, 15, 136, 274, 274*t*, 534
collagen defects in, 44
multiple deficiency states in, 535

nutritionally essential fatty acids, 232, 232*f*
abnormal metabolism of, 236
deficiency of, 234, 235–236

O

O gene, 661

obesity, 109, 134, 248, 527, 532
cancer development and, 709
lipids and, 206
lipogenesis and, 226

obstructive jaundice, 253

octamers, histone, 361, 361*f*, 362*f*, 363

oculocerebrorenal syndrome, 592*t*

β-ODAP (β-*N*-oxalyl-ʟ-α, β-diaminopropionic acid), 19, 19*t*

oils, 206

Okazaki fragments, 373, 373*t*, 376*f*

oleic acid, 206, 206*f*, 207*t*, 208*f*
nutritionally essential, 232, 232*f*
synthesis of, 233–234, 233*f*

oleoyl-CoA, 233, 233*f*

oligomers, import of by peroxisomes, 586–587

oligomycin, 127

oligonucleotide, synthesis of, 451–452

oligosaccharide processing, 582

oligosaccharide:protein transferase, 588–589

oligosaccharides, 148, 554, 555*t*, 558–559, 558*f*, 563

ω3 fatty acids
long-chain, 206, 209
synthesis of, 233–234, 233*f*
VLDL secretion and, 254

ω6 fatty acids, 207, 209
synthesis of, 233–234, 233*f*

ω9 fatty acids, 207
synthesis of, 233–234, 233*f*

OMP (orotidine monophosphate), 343*f*

oncogenes, 693–695, 693*t*, 694*f*, 695*f*, 695*t*, 696*f*, 696*t*
cyclins and, 379
early studies on, 2

oncology, 689, 710

oncoproteins, Rb protein and, 379

oncotic (osmotic) pressure, 635, 637

oncoviruses, cyclins and, 379

open complex, 386*f*, 389

operator locus, 422f, 423
operon, 422–423, 422f, 424f, 425
operon model, 422–423–412, 422f, 424f, 425
opsonization, 667
optical activity, 149
optical isomers, of monosaccharides, 149–150
oral rehydration therapy, 481
ORC (origin replication complex), 373
ORE (origin replication element), 373
organ function tests, 574–575
organometallic complexes, transition metals in, 99–100, 99f, 100f, 101t
organs, metabolic pathways in, 137–139, 138f
origin of replication (ori), 372–373, 372f
origin replication complex (ORC), 373
origin replication element (ORE), 373
ornithine
    arginine conversion to, 307, 307f
    carbon skeleton catabolism of, 292, 292f
    in urea synthesis, 285–286, 285f
ornithine permease, defects in, 287
ornithine transaminase, 292, 292f
ornithine transcarbamoylase
    defects in, 287–288
    deficiency of, 346
    in urea synthesis, 285f, 286
ornithine-citrulline antiporter, defect in, 292
orotate phosphoribosyltransferase, 343f, 344, 346, 346t
orotic acid, 254, 255, 343f
orotic aciduria, 346, 346t
orotidine monophosphate (OMP), 343f
orotidinuria, 346
orotidylate decarboxylase, 346
orotidylic acid decarboxylase, 346t
osmotic (oncotic) pressure, 635, 637
osteoarthritis, 599, 612
osteoblasts, 612, 620f
osteocalcin, 547
osteoclasts, 612, 620f
osteocytes, 612
osteogenesis imperfecta, 44, 274, 614, 614t
osteomalacia, 540
osteopetrosis, 614
osteopontin, 614
osteoporosis, 540, 614t, 615
ouabain, 150, 480
outer membrane, of mitochondria, 122, 122f
outer mitochondrial membrane, 584
outside-in transmembrane signaling, in platelet aggregation, 678, 679f
ovarian steroidogenesis, 495–496, 497f, 498f
overlay blot procedure, 450, 451f

overnutrition, 532–533, 532f. See also obesity
oxaloacetate, 139
    asparagine and aspartate formation of, 291
    in citric acid cycle, 156–159, 157f, 158f, 160f
    in gluconeogenesis, 181, 182f
    in lipogenesis, 228–229, 230f
β-N-oxalyl-L-α, β-diaminopropionic acid (β-ODAP), 19, 19t
oxidases, 116, 116f, 117f
oxidation
    β- (See β-oxidation)
    biomedical importance of, 115
    definition of, 115
    dehydrogenase reactions, 116–118, 117f
    endergonic reaction coupling to, 110–111, 110f, 111f
    of fatty acids (See also ketogenesis)
        acetyl-CoA release and, 218–220, 218f, 219f, 220t
        clinical aspects of, 224–225
        hypoglycemia caused by impairment of, 224–225
        ketogenesis regulation and, 223–224, 223f, 224f
        in mitochondria, 217–218, 218f
        modified, 220
    of heme iron, 656–658, 657t, 658f
    hydroperoxidase reactions, 118
    of iodide, 500
    oxidase reactions, 116, 116f, 117f
    oxygenase reactions, 115–116, 119–120, 119f, 120f
    in pentose phosphate pathway, 192f, 193–194, 193f
    phosphorylation coupling to, 126–127, 127t
    pyruvate (See pyruvate oxidation)
    redox potential and, 115–116, 116t
    of reducing equivalents by respiratory chain, 122–125, 123f, 124f, 125f
    substrate shuttles mediating, 129–130, 129f
    superoxide dismutase reactions, 120
    of transition metals, 97, 98f, 98t
oxidative damage. See reactive oxygen species
oxidative decarboxylation
    of amino acid oxidases, 283–284, 284f
    in citric acid cycle, 158f, 159
    of pyruvate, 168, 169f
oxidative phosphorylation, 627
    biomedical importance of, 122
    clinical aspects of, 130
    control of, 126–127, 127t
    in energy conservation and capture, 112

high-energy phosphate synthesis by, 122, 125–126, 126f
    inhibition of, 122
    in mitochondrial compartments, 122, 122f
    uncoupling of, 127–128
oxidoreductases, 60, 116
    structure of, 38, 39f
oxidosqualene:lanosterol cyclase, 260, 262f
4-oxononanolyation, 430t
oxygen
    Complex IV reduction of, 124–125
    heme binding of, 50–51, 50f, 51f
    hemoglobin affinities for, 53, 53f
    hemoglobin binding of, 53–54, 53f, 54f
    toxicity of, 120 (See also reactive oxygen species)
    transport and storage of, 49–50
        biomedical implications of, 56–57
        heme and ferrous iron roles in, 50–51, 50f, 51f
        hemoglobin and myoglobin suitability for, 51–52, 51f
        hemoglobin conformational changes during, 52–56, 52f, 53f, 54f, 55f
        hemoglobin mutations impacting, 56, 57f
        high altitude changes in, 56
oxygen dissociation curve
    of hemoglobin, 51–52, 51f
    of myoglobin, 51–52, 51f
oxygen radicals. See reactive oxygen species
oxygen tension, in cancer cells, 705
oxygenases, 115–116, 119–120, 119f, 120f
oxysterols, 214

P
P bodies, 416, 416f, 441
P component, in amyloidosis, 646
P (peptidyl) site, of 80S ribosome, 413, 414f
p21 protein, 382
P_{50}, 53, 53f
P53 gene, 696t, 700, 701–702
p53 protein, 382
p70S6K, 516, 517f
p97, 593
p160 family of coactivators, 522, 522t
P450scc (cytochrome P450 side chain cleavage enzyme), 493, 493f
PAB (poly(A) binding protein), 412, 412f, 413f
PAC (E. coli bacteriophage P1-based) vector, 449, 449t
paclitaxel, 631
paddle, charged, 479, 479f
PAF. See platelet-activating factor

PAGE (polyacrylamide gel electrophoresis), 27, 28f
pain, prostaglandins in, 226
PAL (physical activity level), 532
palmitate, 226, 229f, 230f
palmitic acid, 206f, 207t
palmitoleic acid, 207t, 232, 232f
palmitoleoyl-CoA, 233
palmitoylation, mass spectrometry detection of, 29t
O-palmitoylation, 430t
S-palmitoylation, 430t
palmitoyl-CoA, 233
Pan-Cancer Analysis of Whole Genomes (PCAWG), 699, 710
pancreatic cancer, hyperbilirubinemia caused by, 323f, 323t, 324
pancreatic insufficiency, vitamin B₁₂ deficiency in, 544
pancreatic islet β cells, 166
pancreatic lipase, 529
pancreatitis, protease activation in, 90
panproteinase, 645
pantothenic acid, 227
    citric acid cycle need for, 159
    coenzymes derived from, 61, 547
    deficiency of, 536t
    functions of, 536t, 547, 547f
PAP (phosphatidate phosphatase), 240, 241f
PAPS (3′-phosphoadenosine-5′-phosphosulfate), 332–333, 333f, 334t, 566, 608
PAR (physical activity ratio), 532
PAR-1 (protease-activated receptor-1), 678, 679f
PAR-4 (protease-activated receptor-4), 678, 679f
paracrine signaling, 670
parallel β sheet, 37, 37f
paramecia, 668
parasites, glycans in binding of, 563
parathyroid hormone (PTH)
    storage of, 505, 506t
    synthesis of, 502, 503f
paroxysmal nocturnal hemoglobinuria, 486t, 562, 657t, 658
PARP (poly ADP ribose polymerase), 381, 381f
partial proteolysis. See selective proteolysis
passenger mutations, 690–691
passive diffusion, 475f, 475t, 476f, 477f, 483
paxillin, 605, 606f
pBR322, 449, 449f
PCAWG (Pan-Cancer Analysis of Whole Genomes), 699, 710
P-cluster, 100
PCR. See polymerase chain reaction

PCSK9 (proprotein convertase subtilisin/kexin type 9), 263
PDGF (platelet-derived growth factor), in cancer, 697, 697t
PDH. See pyruvate dehydrogenase
PDI (protein disulfide isomerase), 592
PDK1 (phosphoinositide-dependent kinase 1), 516, 517f
pectin, 154, 154f
pellagra, 542–543
penicillamine, for Wilson disease, 641
penicillin, 83
pentose phosphate pathway, 135, 136f
    biomedical importance of, 191
    cellular location of, 192
    glycolysis connections with, 192f, 194
    hemolysis protection from, 194–195, 195f
    impairment of, 198
    irreversible oxidative phase of, 192f, 193–194, 193f
    NADPH produced by, for lipogenesis, 228, 229f, 230f
    reducing equivalents generated by, 194
    reversible nonoxidative phase of, 192, 192f, 193f, 194
pentoses, 148, 148t, 150, 151t, 155t
pentosuria, 191, 198
pepsin, 529
pepsinogen, 90, 529
peptic ulcers, 563
peptidases, proteins degradation by, 280–281, 280f, 281f
peptide bonds
    between amino acids, 22
    formation of, 413
    hydrolysis of, 718–719, 719f
    partial double-bond character of, 22, 23f
    secondary structure restrictions from, 35–36, 35f
    between ubiquitin and proteins, 280, 280f
peptide hormone receptors, 490
peptide hormones, glycoproteins as, 554
peptides
    absorption of, 529, 531
    L-α-amino acids in, 16, 16t–17t
    biomedical importance of, 15
    conformations of, noncovalent forces affecting, 23
    as hormone precursors, 492, 492f
    as polyelectrolytes, 23
    as precursors in hormone synthesis, 500–501
    purification of
        chromatographic techniques, 24–27, 27f, 28f
        gel electrophoresis, 27, 27f, 28f
        isoelectric focusing, 27, 27f

sequencing of (See protein sequencing)
structure drawings for, 22
structure of, primary (See primary structure)
unusual amino acids in, 22, 22f
volatilization of, 30–31, 31f
peptidyl (P) site, of 80S ribosome, 413, 414f
peptidyl prolyl isomerase (PPI), 592
peptidyl transferase, 413, 414t
peptidylarginine deiminase, 669, 669f
peptidylglycine hydroxylase, vitamin C as coenzyme for, 547
peptidyl-tRNA, 413–414, 414f, 414t
perilipin, 257
periodic table, 98f
peripheral proteins, 472–473, 472f
perlecan, 605
permeability coefficients, of substances in lipid bilayer, 470, 471f
pernicious anemia, 544–545
peroxidases, 118, 234, 552
peroxidation, of lipids, 213–214, 214f, 720, 721f
peroxides
    lipid, 549–550, 550f
    mechanisms protecting against, 552, 552f
    reduction of, 118
peroxins, 586–587
peroxisomal-matrix targeting sequences (PTSs), 582t, 586–587, 587f
peroxisomes, 118
    biogenesis of, 586–587
    disorders due to abnormalities of, 588t, 597–598
    in fatty acid oxidation, 220
    in polyunsaturated fatty acid synthesis, 233f
    protein sorting signals in, 582, 582t
    proteins imported into, 586–587, 587f
    in transcription regulation, 521t
    in Zellweger syndrome, 225, 587, 588t
personalized medicine, 444, 712
pertussis toxin, 512
PGE₂ (prostaglandin E₂), 208, 208f
PGI₂ (prostacyclin), 678, 679f, 680
PGI₂ (prostaglandin I₂), 234
PGIs. See prostacyclins
PGs. See prostaglandins
pH
    amino acid properties and, 20, 20f
    biomedical importance of, 6
    buffering and, 13, 13t
    in cancer cells, 705
    definition and calculation of, 10–11
    isoelectric, 20
    pKₐ relationship to, 11–12
    reaction rate response to, 75–76, 76f
    of water, 10
pH gradient, 584

phages, in recombinant DNA technology, 448, 449t

phagocytic cells, 668

phagocytosis, 482, 664, 665, 667–668, 667f, 668t

phagosomes, 667

pharmacogenomics, 3

phasing, of nucleosome, 363

phenobarbital, 565

phenols, sulfation of, 566

phenyl isothiocyanate, 27f, 28

phenylalanine, 17t
    carbon skeleton catabolism of, 296, 298f
    ultraviolet light absorption by, 21, 22f

phenylalanine hydroxylase, 276, 277f, 296, 298f

phenylethanolamine-N-methyltransferase (PNMT), 499f, 500

phenylketonuria (PKU), 296, 298f

phi angle, 35, 35f

phosphagens, in energy conservation and capture, 113

phosphatase cascade, 490, 491t

phosphatases, in recombinant DNA technology, 446t

phosphate transporter, of mitochondria, 128, 128f

phosphates
    creatine phosphate shuttle for, 130, 130f
    energy capture and transfer by, 111–112, 111t, 113f, 125–126
    in extracellular and intracellular fluid, 468t
    oxidative phosphorylation in synthesis of, 122, 125–126, 126f
    in RNA synthesis, 388

phosphatidate, 240–242, 240f, 241f, 242f

phosphatidate phosphatase (PAP), 240, 241f

phosphatidate phosphohydrolase, 240, 241f

phosphatidic acid, 209–210, 210f, 469, 469f

phosphatidylcholines, 210, 210f
    membrane asymmetry and, 472
    metabolism of, 243f
    synthesis of, 240, 240f, 241f

phosphatidylethanolamine (cephalin), 210, 210f
    membrane asymmetry and, 472
    synthesis of, 240, 240f, 241f

phosphatidylglycerol, 210f, 211

phosphatidylinositides
    calcium-dependent hormone action and, 515–516, 515f, 516f
    as second messenger, 210–211, 210f, 490, 491t, 510

phosphatidylinositol 4,5-bisphosphate (PiP$_2$), 211, 240

in absorptive pinocytosis, 482
    phospholipase C cleavage of, 515, 516f
    in platelet aggregation, 678, 679f

phosphatidylinositol synthase, 240

phosphatidylinositols
    in membranes, 210–211, 210f
    synthesis of, 240, 240f, 241f

phosphatidylserine, 210, 210f, 240, 240f, 241f
    membrane asymmetry and, 472

3'-phosphoadenosine-5'-phosphosulfate (PAPS), 332–333, 333f, 334t, 566, 608

phosphodiesterases, 175, 175f, 176f, 335, 513

phosphoenolpyruvate
    in gluconeogenesis, 181, 182f
    in glycolysis, 165f, 167

phosphoenolpyruvate carboxykinase, 160
    in gluconeogenesis, 181, 182f

phosphofructokinase
    deficiency of, 170
    in glycolysis, 165f, 166
    regulation of, 167, 184, 185f
    reversal of reaction catalyzed by, 181, 182f

phosphofructokinase-2, 185, 186f

phosphoglucomutase, in glycogenesis, 172f, 173

6-phosphogluconate dehydrogenase, 192f, 193, 193f

6-phosphogluconolactone, 192f, 193, 193f

2-phosphoglycerate, in glycolysis, 165f, 166

3-phosphoglycerate, in glycolysis, 165f, 166

3-phosphoglycerate dehydrogenase, allosteric regulation of, 88, 89f

phosphoglycerate kinase
    in erythrocytes, 168
    in glycolysis, 165f, 166

phosphoglycerate mutase, in glycolysis, 165f, 166

phosphoglycerides, in membranes, 469, 469f, 597

phosphoglycerols
    degradation and remodeling of, 242–243, 243f
    lysophospholipids in metabolism of, 211, 211f
    synthesis of, 240–242, 240f

phosphohexose isomerase, in glycolysis, 165f, 166

phosphoinositide-dependent kinase 1 (PDK1), 516, 517f

phosphoinositides, 210–211, 490, 491t

phospholipase A$_1$, 243, 243f

phospholipase A$_2$, 242–243, 242f, 243f

phospholipase B, 243, 243f

phospholipase C (PLC), 666
    activation and hormone-receptor interactions of, 515, 515f

in phosphoglycerol degradation and remodeling, 243, 243f
    PiP2 cleavage by, 515, 516f

phospholipase Cβ (PLCβ), in platelet aggregation, 678, 679f

phospholipase D, 243, 243f

phospholipases, in phosphoglycerol degradation and remodeling, 242–243, 243f

phospholipid bilayer, noncovalent interactions in, 8

phospholipid exchange proteins, 472

phospholipids, 206
    clinical aspects of, 245
    digestion and absorption of, 528–529, 530f
    glycerol ether, synthesis of, 242, 242f
    in lipoprotein lipase activity, 251
    in lipoproteins, 247, 248
    in membranes, 209–211, 210f, 469, 469f, 471, 596–597
        asymmetry of, 472, 596–597, 597f
    in multiple sclerosis, 245
    as second messenger precursors, 239
    synthesis of, 240–242, 241f

phosphomevalonate kinase, 261f

phosphoprotein phosphatases, 513–514

phosphoproteins, 513

phosphoribosyl pyrophosphate (PRPP)
    in Lesch-Nyhan syndrome, 344–345
    in purine synthesis, 338, 339f, 340
    in pyrimidine synthesis, 342, 342f, 343f

phosphoric acid, pK$_a$ of, 13t

phosphorolysis, as group transfer reaction, 9

phosphorylase kinase a, 176f, 177, 179f

phosphorylase kinase b, 93t, 176f, 177, 179f

phosphorylation
    in energy conservation and capture, 112
    enzyme regulation by, 92–94, 92f, 93t
    in glycogen regulation, 175–177, 176f, 177–178, 179f
    in glycolysis and gluconeogenesis regulation, 183
    of histones, 361, 363t
    mass spectrometry detection of, 29t
    multisite, 177
    oxidation coupling to, 126–127, 127t
    oxidative (See oxidative phosphorylation)
    of pyruvate dehydrogenase, 168, 169f
    respiratory chain level, 126
    in RNA polymerase II activation, 395, 395f
    substrate level, 126

phosphoserine, 310

phosphothreonine, 310

phosphotriose isomerase, in glycolysis, 165f, 166

phosphotyrosine, 310

photolysis, 497–498

photosensitivity, porphyrias causing, 321

phylloquinone, 536t, 540–541, 541f

physical activity, energy requirements for, 532

physical activity level (PAL), 532

physical activity ratio (PAR), 532

physiologic jaundice, 324

phytanic acid, Refsum disease caused by accumulation of, 225

phytase, 531

phytic acid, 531

phytohemagglutinins, 555

pI (isoelectric pH), 27

PIC. See preinitiation complex

ping-pong mechanism, in facilitated diffusion, 477, 477f

ping-pong reactions, 82, 82f, 83f
    in covalent catalysis, 62, 62f
    transamination, 62, 62f, 283, 283f

pinocytosis, 482–483, 482f

PiP₂. See phosphatidylinositol 4,5-bisphosphate

pituitary gland, hormones of, 188

PK. See pyruvate kinase

pKₐ
    of amino acids, 20, 21t
    of L-α-amino acids, 16t–17t
    environmental influence on, 21, 21t
    medium properties affecting, 13
    molecular structure effects on, 13, 13t
    of weak acids, 11–12, 13f, 13t, 20

PKA (protein kinase A), 512–513, 513f

PKC. See protein kinase C

PKU (phenylketonuria), 296, 298f

plants, toxic amino acids of, 19, 19t

plasma, 635–636
    analysis of enzymes in, 66–67, 67f, 67t

plasma glucose concentration. See blood glucose

plasma membrane, 467. See also membranes
    asymmetry of, 596–597, 597f
    cholesterol in, 469
    glycoproteins anchored to, 560
    glycosphingolipids in, 469
    mutations in, diseases caused by, 485, 486t
    specialized structures of, 474, 474f

plasma proteins, 137
    deposition of, 646–647, 646t
    electrophoresis for analysis of, 635
    fluid distribution and, 635
    functions of, 637t
    glycoproteins as, 554
    half-life of, 637
    haptoglobin, 638–639, 638f
    immunoglobulins, 646–650, 647t, 648f, 649f, 650f

    in inflammation, 637–638, 637t
    polymorphism of, 636
    synthesis of, 137, 635–636

plasma samples, 571

plasma transport proteins, 506, 507t

plasmalogens, 211, 212f
    biosynthesis of, 240f, 242, 242f

plasmids, 448, 448f, 449f

plasmin, 686, 686f

plasminogen, 686, 686f

plasminogen activators, in hemostasis and thrombosis, 679

Plasmodium falciparum, glycans in binding of, 563

platelet aggregation, 678, 679f
    laboratory tests measuring, 687

platelet count, 687

platelet-activating factor (PAF), 239
    synthesis of, 240f, 242, 242f

platelet-derived growth factor (PDGF), in cancer, 697, 697t

platelets, 653, 661–662, 667t
    inhibitors of activation of, 679f, 680

PLC. See phospholipase C

PLCβ (phospholipase Cβ), in platelet aggregation, 678, 679f

pleckstrin, in platelet aggregation, 678, 679f

PLP (pyridoxal phosphate), 283, 283f, 543

pluripotent stem cells, induced, 454, 653

PNMT (phenylethanolamine-N-methyltransferase), 499f, 500

pOH, 11

point mutations, 407–408, 407f, 408f, 409f

poisons, as respiratory chain inhibitors, 116, 122, 127–128, 127f, 128f

polar groups
    hydrogen bonding by, 7, 7f
    noncovalent interactions of, 8

polar head groups, 469–470

polarity
    of DNA, 349, 350f
    of DNA synthesis, 374, 376f
    of microtubules, 631
    of protein synthesis, 410

poly ADP ribose polymerase (PARP), 381, 381f

poly(A) binding protein (PAB), 412, 412f, 413f

poly(A) polymerase, 400

poly(A) tail of mRNA, 355, 400, 412, 412f

polyacrylamide gel electrophoresis (PAGE), 27, 28f

polyadenylation, alternative, 442

polyamines, biosynthesis and catabolism of, 309, 309f, 310f

polycistronic mRNA, 422

polycomb repressive complex 2 (PRC2), 433

polycythemia, hemoglobin mutations leading to, 56

polyelectrolytes, peptides as, 23

polyhydric alcohols, carbohydrates as aldehyde or ketone derivatives of, 148, 148t

polyisoprenoids, in cholesterol synthesis, 251–252, 261f

polymerase chain reaction (PCR), 452–453, 453f
    medical applications of, 68
    in microsatellite repeat sequence detection, 368

polymerases
    DNA (See DNA polymerases)
    RNA (See RNA polymerase II; RNA polymerases)

polymerization, of hemoglobin S, 56, 57f

polymorphisms
    microsatellite, 368
    plasma protein, 636

polymorphonuclear leukocytes, 665

polynucleotide kinase, 446t

polynucleotides
    as directional macromolecules, 335
    as DNA and RNA, 335
    functions of, 331
    modification of, 331, 333f
    posttranslational modification of, 335

polyol pathway, 200

polyols, 148

polyprenoids, 213, 214f

polyps, colorectal, 695–697, 696f, 696t

polyribosomes, 356, 412f, 415–416
    membrane-bound, 587–588
    protein synthesis on, 582, 582f, 583f, 584, 586–587, 591
    signal hypothesis of binding of, 582, 582f

polysaccharides, 148
    storage and structural functions of, 152–155, 153f, 154f, 155t

polysomes. See polyribosomes

polytene chromosomes, 364, 365f

polyubiquitination, 280

polyunsaturated fatty acids, 206f, 207t, 208
    dietary, cholesterol levels affected by, 267
    eicosanoids formed from, 234, 235f
    essential, 232, 232f
    synthesis of, 233–234, 233f

POMC peptide family, 504–505, 505f

Pompe disease, 174t, 175

porphobilinogen, in heme synthesis, 316, 317f, 318f, 319f

porphobilinogen synthase, 316, 317f, 318f, 319f

porphyrias, 315, 320, 321f
    classification of, 320t, 321
    drug-induced, 321
    spectrophotometric assessment of, 318

porphyrinogens
  accumulation of, 320–321
  colorless nature of, 318
  in heme synthesis, 316–317, 317f, 318f, 319f
porphyrins
  biomedical importance of, 315
  biosynthesis of, 315–318, 316f, 317f, 318f, 319f, 320t
    regulation of, 317–318
  catabolism of, 321–323, 322f, 323f
  color and fluorescence of, 318, 320f
  structure of, 315, 316f
positive cooperativity, 79
positive nitrogen balance, 533
positive predictive value, of laboratory tests, 570–571, 571t
positive regulators, of gene expression, 421, 421t, 425, 426–427
posttranslational modifications (PTMs)
  of amino acids, 16, 18, 18f
  biomedical importance of, 33
  of collagen, 601–602, 601t
  of histones, 361, 363t, 429–430, 430t
  mass spectrometry detection of, 29, 29t
  of membrane proteins, 597
  polynucleotides, 335
  in protein maturation, 43–44, 44f
posttranslational processing, 417
posttranslational translocation, 584, 589–590, 589f
potassium (K+)
  in extracellular and intracellular fluid, 468, 468t
  permeability coefficient of, 471f
potassium (K+) ion channels, 166, 478–479, 479f
power stroke, 622, 623f
PPi. See inorganic pyrophosphate; pyrophosphate
PPI (peptidyl prolyl isomerase), 592
prasugrel, 680
pravastatin, 267
PRC2 (polycomb repressive complex 2), 433
precision, of laboratory test, 569, 569f
precision oncology, 710
predictive value, of laboratory tests, 570–571, 571t
pregnancy
  fatty liver of, 225
  glucose requirement of, 142
  hypoglycemia during, 189
pregnenolone, 493f, 495, 496f, 498f
preinitiation complex (PIC), 388, 392f
  assembly of, 395f, 396–397
  nucleosomes and, 393f, 394–395
  in protein synthesis, 410–412, 411f
premature termination, 408, 409f

S-prenylation, 471
preproalbumin, 596, 596f
preprocollagen, 601
preprohormone, 501–502
preproprotein, 637
preproPTH, 502, 503f
preproteins, 582
presequence, 584
preventive antioxidants, 214
preventive medicine, biochemical research impacts on, 2–3
primaquine, 731
primary hyperoxaluria, 293
primary lysosomes, 482
primary structure
  amino acids and, 22
  of collagen, 44, 44f
  determination of
    Edman reaction for, 27f, 28–29
    genomics and, 29
    mass spectrometry for, 29–31, 29t, 30f, 31f
    molecular biology techniques for, 29
    proteomics and, 21–32
    purification techniques for, 24–27, 26f, 27f, 28f
    Sanger's work in, 27–28
  as order of protein structure, 35
primary structure, of RNA, 352–354, 357f
primary transcripts, 366, 386–387, 401–402, 401f
primary transport, 477
primases, DNA, 372f, 373t
pri-miRNAs, 401–402, 401f
primosome, 373
prion diseases, pathologic protein conformations in, 43
PRL (prolactin), receptors for, 490
proalbumin, 596, 596f
probing techniques, 450, 451f
procarcinogens, 564, 567, 567f, 692
procaspases, 701–702, 702f
processed genes, 370–371
prochymotrypsin, activation of, 92f
procollagen, 44, 417, 547, 601
procollagen aminoproteinase, 601
procollagen carboxyproteinase, 601
procollagen suicide, 614
prodrugs, 83
  activation of, 564
product inhibition studies, 82
products
  in balanced chemical equations, 72
  enzyme, 60
proenzymes, enzyme regulation using, 90, 91f
progeria, 632
progesterone, 492f, 493f

plasma transport of, 506, 507t
  synthesis of, 495, 496f, 498f
programmed cell death. See apoptosis
prohormones, 417
proinflammatory cytokines, 709
proinsulin, 501–502, 502f
prokaryotic gene expression
  eukaryotes compared with, 439–440, 440t, 441f, 442f
  as model for study, 422
  on-off manner of, 435
  unique features of, 422
prolactin (PRL), receptors for, 490
proline, 17t
  carbon skeleton catabolism of, 292, 292f
  in collagen, 600, 604
  synthesis of, 276, 277f
proline dehydrogenase, deficiency of, 292, 292f
proline hydroxylase, vitamin C as coenzyme for, 547
proline-cis, trans-isomerases, 42–43, 43f
prolyl hydroxylase, 44, 276, 600, 604
promoter clearance, 388
promoter recognition specificity, 388
promoter site, in operon model, 422f, 423
promoters, in transcription, 386–387
  alternative, 399–400, 399f, 400f
  bacterial, 389–390, 390f
  eukaryotic, 390–392, 390f, 391f, 392f, 393f, 394, 430
pro-opiomelanocortin (POMC) peptide family, 504–505, 505f
pro-oxidants, antioxidants as, 552–553, 552t
prophage state, 425
propionate
  glucose derived from, 186
  metabolism of, 181–183, 182f, 183f, 733t
propionyl-CoA
  in citric acid cycle, 160, 160f
  in fatty acid oxidation, 219
  methionine conversion to, 298, 300f, 301f
propionyl-CoA carboxylase, 181–182, 183f
proprotein convertase subtilisin/kexin type 9 (PCSK9), 263
proproteins, 44, 90, 417
proPTH, 502, 503f
propyl gallate, as antioxidant/food preservative, 214
PRO-Seq, 455
prostacyclin (PGI2), 679, 679f, 680
prostacyclins (PGIs), 208
  clinical significance of, 237
  synthesis of, 234, 236f
prostaglandin E2 (PGE2), 208, 208f
prostaglandin H synthase, 234
prostaglandin I2 (PGI2), 234

prostaglandins (PGs), 208, 208f, 226, 234, 670
    cyclooxygenase pathway in synthesis of, 234–235, 235f, 236f
prostanoids, 208
    clinical significance of, 234, 237
    cyclooxygenase pathway in synthesis of, 234–235, 235f, 236f
prostate-specific antigen (PSA), 709
prostatic acid phosphatase, diagnostic use of, 67
prosthetic groups
    of enzymes, 60–61, 61f
    flavins, 116, 117f, 118, 119f
    heme (See heme)
    protoheme, 118
protamine, 685
protease-activated receptor-1 (PAR-1), 678, 679f
protease-activated receptor-4 (PAR-4), 678, 679f
proteases, 9
    as digestive enzymes, 529
    enzyme regulation using, 90, 92f
    α_2-macroglobulin and, 645, 645f
    matrix, 584
    phagocyte-derived, 669–670
    in protein degradation, 593
    proteins degradation by, 280–281, 280f, 281f
    synaptobrevin and, 595
26S proteasome, 87
proteasomes
    protein degradation by, 87–88, 280–281, 280f, 281f
    protein degradation in, 590, 592–593, 593f
    structure of, 280–281, 281f
protein blot procedure, 450, 451f
protein C, 685
protein coding RNAs. See messenger RNA
protein damage, repairing of, 724–725, 725f
Protein Data Bank, 39
protein degradation, ubiquitin in, 593, 593f
protein disulfide isomerase (PDI), 42, 592
protein folding
    chaperones and, 582–583
    misfolded, 592–593, 592t, 593f
protein genes s36 and s38, chorion, 441f
protein kinase A (PKA), 512–513, 513f
protein kinase C (PKC), 515, 515f
    in platelet aggregation, 678, 679f
protein kinases
    cAMP and, 512–513, 513f
    in hormonal regulation of lipolysis, 257, 257f

hormone signaling and, 516–519, 517f, 518f
    in initiation of protein synthesis, 410
    phosphorylation by, 92–94, 92f, 93t
    as second messengers, 490, 491t
protein lipidation, 471
protein misfolding
    endoplasmic reticulum accumulation of, 592–593, 592t
    ubiquitination of, 593, 593f
protein phosphatase-1, 176f, 177, 179f
protein phosphatases, dephosphorylation by, 92–94, 92f, 93t
protein prenylation, 261
protein S, 685
protein sequencing
    Edman reaction for, 27f, 28–29
    genomics and, 29
    mass spectrometry for, 29–31, 29t, 30f, 31f
    molecular biology techniques for, 29
    proteomics and, 31–32
    purification techniques for, 24–27, 26f, 27f, 28f
    Sanger's work in, 27–28
protein sorting
    branches of, 583, 583f
    cell nucleus and, 583–587, 583f, 586f
    cotranslational insertion and, 590–591, 590f, 591f
    disorders of, 597–598
    Golgi apparatus in, 582, 583f, 592–593
    importins and exportins in, 585–586, 586f
    KDEL amino acid sequence and, 582t, 591
    membrane assembly and, 595t, 596–598, 597f
    misfolded proteins, 592–593, 592t, 593f
    mitochondria and, 584, 585f
    peroxisomes and, 586–587, 587f
    peroxisomes/peroxisome disorders and, 587, 588t
    quality control in, 592–593, 592t
    retrograde transport and, 591
    rough ER branch of, 582, 582f, 583f
    signal hypothesis of polyribosome binding, 582, 582f, 587–590, 588t, 589f
    signal sequences and, 581–583, 582t, 590f, 636
    transport vesicles and, 593–596, 594t, 595f
protein synthesis, 139
    antibiotic inhibition of, 417–418, 418f
    biomedical importance of, 404
    elongation in, 413–414, 414f, 414t
    environmental threats and, 416

genetic information flow of, 404–405
    in Golgi apparatus, 582, 583f
    initiation of, 410–413, 411f, 412f
    by mitochondria, 582t, 584
    modular principles in, 35
    mRNA and, 355, 355f, 405, 405t
    mutations and, 407–409, 407f, 408f, 409f
    polysomes in, 416, 582, 582f
    posttranslational processing and, 417
    rate of, 279–280, 412–413, 413f
    with recombinant DNA technology, 453
    regulation of, 87
    reticulocytes and, 656
    RNA and, 354
    termination of, 414, 415f
    translocation and, 413–414
    tRNA and, 406–407, 406f, 407f
    virus replication and, 416, 417f
protein turnover, 87, 279–280
    in membranes, 597
proteinases, of cancer cells, 708
protein-conducting channel, 588
protein-DNA interactions
    bacteriophage lambda as paradigm for, 425–430, 426f, 427f, 428f
    mapping of, 454–455
protein-losing gastroenteropathy, 637
protein-protein cross-links and protein glycation, 724f
protein-protein interactions, identification of, 455, 457f
protein-RNA complexes, in initiation, 410, 411f, 412
proteins
    acute phase, 534, 637, 637t
    aggregates of, 42
    L-α-amino acids in, 16, 16t–17t, 18
    asymmetry of, in membrane assembly, 596–597, 597f
    biomedical importance of, 24
    classification of, 35
    configuration of, 33–34
    conformations of, 33–34, 35f, 42–43
    constitutive, 87
    degradation of
        amino acid levels and, 281–282, 281f, 282f
        ATP- and ubiquitin-dependent, 280–281, 280f, 281f
        ATP-independent, 280
        biomedical importance of, 279
        disorders of, 286–288, 287t
        end products of, 282, 285, 285f
        protease and peptidase role in, 280–281, 280f, 281f
        rate of, 279–280
        regulation of, 87–88

proteins (*Cont.*):
    denaturation of, 42
    diets deficient in, 273–274
    digestion and absorption of, 529, 531
    domains of, 38, 39*f*, 40*f*
    dual-function, 495
    energy yields from, 144*t*
    in extracellular and intracellular fluid, 468, 468*t*
    folding of, 42–43, 43*f*
        calnexin role in, 559–560
    free radical damage to, 549–550, 550*f*
    function of, bioinformatics and, 32
    fusion, 68–69, 69*f*
    in genetic information flow, 404–405
    glycoproteins (*See* glycoproteins)
    half-lives of, 280
    homologous, 64
    hormone receptors as, 490
    identification of
        genomics role in, 29
        mass spectrometry role in, 29–31, 29*t*, 30*f*, 31*f*
        proteomics and, 31–32
    import of
        by cell nucleus, 583–587, 586*f*
        by mitochondria, 584
        by peroxisomes, 586–587, 587*f*
    intracellular traffic of, 582
    ion channels as transmembrane, 478, 478*f*, 479*f*, 479*t*
    life cycle of, 24, 25*f*
    lysosomal, 597–598
    maturation of
        folding, 42–43, 43*f*
        posttranslational processing, 43–44, 44*f*
    in membranes (*See also* membrane proteins)
        in artificial, 473
        lipid ratio to, 468, 468*f*
    as metabolic fuel, 134–135, 141–142
    metabolic pathways of, 135*f*, 136, 136*f*, 138*f*
        in fasting state, 144
        in fed state, 143*f*, 144
        at organ and cellular level, 136–139, 138*f*
    misfolded, 592–593, 592*t*, 593*f*
    modifications of, 394–395
    of neutrophils, 667, 667*t*
    noncovalent interactions in, 8
    nutritional requirements for, 533–534
    phospholipid exchange, 472
    plasma (*See* plasma proteins)
    plasma transport, 506, 507*t*
    posttranslational processing of, 417
    prenylation of, 261
    profiling of, 454–455

purification of
    chromatographic techniques, 24–27, 27*f*, 28*f*
    gel electrophoresis, 27, 27*f*, 28*f*
    isoelectric focusing, 27, 27*f*
ROS reactions with, 721*f*
sequences or molecules that direct, 581–582, 582*t*
sequencing of (*See* protein sequencing)
structure of
    biomedical importance of, 33
    biophysical techniques for solving of, 40–42
    conformation *versus* configuration, 33–34
    folding into, 42–43, 43*f*
    gross characteristics, 35
    modular nature of, 35
    orders of, 35
    pathologic perturbations of, 43
    posttranslational processing of, 43–44, 44*f*
    primary (*See* primary structure)
    quaternary (*See* quaternary structure)
    secondary (*See* secondary structure)
    tertiary (*See* tertiary structure)
    volatilization of, 30–31, 31*f*
proteinuria, 574
proteoglycans, 154
    accumulation of, 245
    in cartilage, 615*f*, 615*t*, 616
    components of, 606–607, 606*f*, 607*f*
    disease and aging with, 611–612
    functions of, 608, 610*t*
    galactose production for, 197, 198*f*
    glucuronate production for, 195, 196*f*
    O-glycosidic linkages in, 557
proteolysis, 596
    enzyme regulation by, 90–91, 92*f*
proteome, 31–32
proteomics, 3, 455
    bioinformatics and, 32
    challenges in, 32
    goal of, 31–32
proteostasis deficiency, diseases of, 597–598
prothrombin (factor II), 680, 681*f*, 681*t*, 682*t*
    coumarin drugs and, 685–686
    factor Xa activation of, 682–683, 683*f*
prothrombinase complex, 682, 683*f*
protic acids, 97
protoheme, 118
proton acceptors, bases as, 10
proton donors, acids as, 10
proton gradient, of respiratory chain, 125, 126*f*
proton motive force, of respiratory chain, 125, 126*f*

proton pump, respiratory chain action as, 122–125, 123*f*, 124*f*, 125*f*
proton shuttle, 64, 64*f*
proton-motive force, 584
protons
    hemoglobin binding of, 54–55, 55*f*
    in water, 9–10
proto-oncogenes, 690, 693, 694*f*
protoporphyrin III, in heme synthesis, 317, 318*f*, 319*f*
protoporphyrinogen III, in heme synthesis, 317, 318*f*, 319*f*
protoporphyrinogen oxidase, 317, 318*f*, 319*f*, 320*t*
proximity, enzymatic catalysis by, 62
PRPP. *See* phosphoribosyl pyrophosphate
PRPP glutamyl amidotransferase, in purine synthesis, 340, 341*f*
PRPP synthetase
    defect in, gout caused by, 344, 346*t*
    in purine synthesis, 340, 341*f*, 344, 344*f*
    in pyrimidine synthesis, 344, 344*f*
PSA (prostate-specific antigen), 709
pseudo-first-order conditions, 74, 76
pseudogenes, 361
pseudouridine, 346, 347*f*
psi angle, 35, 35*f*
PstI, 445*t*
PstI site, insertion of DNA at, 449, 449*f*
pterins, 99, 100*f*
PTH. *See* parathyroid hormone
PTSs (peroxisomal-matrix targeting sequences), 582*t*, 586–587, 587*f*
PTT (activated partial thromboplastin time), 687
"puffs," polytene chromosome, 364, 365*f*
pumps, 475*f*, 481
punctuation of genetic code, 406
purification
    of glycoproteins, 555
    of proteins and peptides
        chromatographic techniques, 24–27, 27*f*, 28*f*
        gel electrophoresis, 27, 27*f*, 28*f*
        isoelectric focusing, 27, 27*f*
purine nucleoside phosphorylase deficiency, 337, 345, 346*t*
purines
    biomedical importance of, 330
    chemistry of, 330, 330*f*
    derivatives of, 330–331, 332*t*
    dietary nonessential, 338
    in DNA, 348–349, 349*f*
    gout and, 344
    metabolism of
        biomedical importance of, 337–338
        disorders of, 344–345, 346*t*
    in RNA, 352, 353*f*
    serine conversion to, 310

substitution mutation of, 407–408, 407*f*, 408*f*, 409*f*

synthesis of, 338, 338*f*, 339*f*, 340, 340*f*
  pyrimidine synthesis coordinated with, 344, 344*f*
    regulation of, 342, 342*f*, 344, 344*f*
  synthetic analogs of, 334*f*
  ultraviolet light absorption by, 331

puromycin, 418, 418*f*

purple acid phosphatases, iron in, 102

pyranose ring structures, 149, 149*f*

pyridoxal. *See* vitamin B$_6$

pyridoxal phosphate (PLP), 283, 283*f*, 543

pyridoxamine. *See* vitamin B$_6$

pyridoxine. *See* vitamin B$_6$

pyrimethamine, 545

pyrimidines
  biomedical importance of, 330
  chemistry of, 330, 330*f*
  derivatives of, 330–331, 332*t*
  dietary nonessential, 338
  in DNA, 348–349, 349*f*
  metabolism of, 345*f*
    biomedical importance of, 337–338
    diseases caused by catabolite overproduction and, 346, 346*t*
    water-soluble metabolites of, 345–346, 347*f*
  precursors of, deficiency of, 346
  in RNA, 352, 353*f*
  serine conversion to, 310
  substitution mutation of, 407–408, 407*f*, 408*f*, 409*f*
  synthesis of, 338, 342, 343*f*
    analogs in, 344
    catalysts in, 342, 343*f*
    purine synthesis coordinated with, 344, 344*f*
    regulation of, 342, 342*f*, 344, 344*f*
  synthetic analogs of, 334*f*
  ultraviolet light absorption by, 331

pyrophosphatase
  in fatty acid activation, 218
  in glycogenesis, 172*f*, 173
  in RNA synthesis, 388

pyrophosphate (PPi)
  in fatty acid activation, 218, 218*f*
  in glycogenesis, 172*f*, 173
  in RNA synthesis, 388

pyrrole rings, of porphyrins, 315, 316*f*

Δ^1-pyrroline-5-carboxylate dehydrogenase, 292, 292*f*

pyruvate, 142, 144
  in citric acid cycle, 159–161, 160*f*
  in gluconeogenesis, 181, 182*f*
  glycolysis production of, 163, 164*f*, 164*t*, 165*f*, 166–167
  inhibition of metabolism of, 168

in transamination reactions, 282–283, 283*f*

pyruvate carboxylase, 160–161, 160*f*
  in gluconeogenesis, 181, 182*f*
  regulation of, 183

pyruvate dehydrogenase (PDH)
  deficiency of, 170
  heavy metal inactivation of, 98
  in pyruvate oxidation, 168, 169*f*
  regulation of, 93*t*, 162, 168, 169*f*
    acyl-CoA in, 231

pyruvate kinase (PK)
  in cancer cells, 704, 705*f*
  deficiency of, 168–170, 657*t*, 658
  in glycolysis, 165*f*, 166
  regulation of, 167, 183
  reversal of reaction catalyzed by, 181, 182*f*

pyruvate oxidation, 168, 169*f*
  induction and repression of enzymes catalyzing, 183, 184*t*

Q

Q. *See* coenzyme Q

Q cycle, 124, 125*f*

Q$_{10}$ (temperature coefficient), 75

Q-cytochrome *c* oxidoreductase (Complex III), 122, 123*f*, 124, 124*f*, 125*f*

QT interval, congenitally long, 486*t*

quadrupole mass spectrometers, 29, 30*f*

quality control, for laboratory tests, 569

quaternary structure, 37–38, 39*f*, 40*f*
  of hemoglobin, 52–56, 52*f*, 53*f*, 54*f*, 55*f*
  as order of protein structure, 35
  schematic diagrams of, 39
  stabilizing factors in, 40

R

α-R groups, of amino acids, 21–22

R (relaxed) state, of hemoglobin, 52, 53*f*, 54*f*, 55

Rab molecules, 595

Rab proteins, 594–596, 594*t*

radiation
  carcinogenic effect of, 691, 691*f*, 691*t*
  DNA damage caused by, 380*f*, 380*t*

radioimmunoassays, laboratory tests using, 573

Ramachandran plot, 35*f*

Ran proteins, 585

rancidity, peroxidation causing, 213

random-order reactions, 82, 82*f*

RAS oncogene, 693, 693*t*, 696*t*

rate constant (*k*), 74–75

rate of reaction
  activation energy and, 72–73, 73*f*, 75
  enzyme effects on, 75
  Gibbs free-energy change and, 72

hydrogen ion concentration effects on, 75–76, 76*f*

initial velocity, 76–79, 76*f*, 78*f*, 79*f*, 82, 83*f*

maximal velocity (*See* maximal velocity)

Michaelis constant (*See* Michaelis constant)

substrate concentration effects on, 74, 76–77, 76*f*, 77*f*

temperature effects on, 73, 74*f*, 75

rate-limiting reactions, 87

RB gene, 696*t*, 700

Rb (retinoblastoma) protein, 379

reactants. *See* substrates

reactive oxygen species (ROS), 656, 719–720
  antioxidant paradox and, 552–553, 552*t*
  in cancer development, 691, 709
  chain reactions and, 720
  damage caused by, 549–550, 550*f*
  enzymatic and chemical mechanisms intercept damaging, 723–724
  heavy metal formation of, 98
  hydroperoxidase reactions with, 118
  in lipid peroxidation, 214, 214*f*
  mechanisms protecting against, 551–552, 552*f*
  reaction with biological molecules, 721*f*
  respiratory burst generation of, 668
  self-perpetuating chain reactions of, 549
  sources of, 550–551, 551*f*
  superoxide dismutase reactions with, 120
  as toxic byproducts of life, 720*f*

receptor-corepressor complex, 510

receptor-effector coupling, 490

receptor-mediated endocytosis, 482–483, 482*f*

recombinant DNA technology
  applications of
    protein production, 453
    targeted gene regulation, 454
  biomedical importance of, 444
  blotting and probing techniques for, 450, 451*f*
  chimeric molecules in, 444
  cloning in, 447–449, 448*f*, 449*f*, 449*t*
  DNA sequencing, 451, 452*f*
  enzyme studies using, 68–69, 69*f*
  genomic library for, 450
  hematology and, 662
  library probe for, 450
  oligonucleotide synthesis, 451–452
  polymerase chain reaction method, 452–453, 453*f*
  restriction enzymes in
    for chimeric DNA molecule preparation, 446–447
    DNA chain cleavage with, 445–446, 446*f*
    selection of, 445–446, 445*t*, 446*t*

recombinant fusion proteins, 68–69, 69f
recombinant t-PA, 686–687
recombinases, 446t, 447
recombination, chromosomal, 369f, 370f, 379–382
recruitment hypothesis, 396
red blood cells
   ABO system, 660–661, 661f
   biomedical importance of, 653
   CO_2 transport by, 655
   derivation from hematopoietic stem cells, 653–654, 654f
   diseases affecting, 653, 657t
   functions of, 654–655, 654f, 655t
   glycolysis of, 655, 655t
   lifespan of, 655–656
   membrane of, 658–660, 658f, 659f, 659t
   replacement of, 655–656
red thrombus, 678
redox potential (E'_0)
   free energy changes and, 115–116, 116t
   of transition metals in organometallic complexes, 100
redox reactions
   biomedical importance of, 115
   free energy changes of, 115–116, 116t
   iron participation in, 100–102
redox state, 220
reducing equivalents
   citric acid cycle production of, 157–159, 158f
   pentose phosphate pathway generation of, 194
   respiratory chain oxidation of, 122–125, 123f, 124f, 125f
5α-reductase, 495, 497f
reduction
   definition of, 115
   dehydrogenase reactions, 116–118, 117f
   hydroperoxidase reactions, 118
   of oxygen by Complex IV, 124–125
   redox potential and, 115–116, 116t
reference range, of laboratory tests, 569
Refsum disease, 225, 587, 588t
regional heterogeneities, 472
regulated secretion, 582
regulation. See metabolic regulation
regulatory domains, 38
regulatory enzymes, of metabolic pathways, 139
regulatory transcription factor proteins, DNA-binding domains of, 437t
   helix-turn-helix motif, 437, 438f
   leucine zipper motif, 438, 439f
   zinc finger motif, 437–438, 438f
relaxation phase of muscle contraction, 623
relaxed form of DNA, 351

relaxed (R) state, of hemoglobin, 52, 53f, 54f, 55
releasing factor 1 (RF1), 413–414, 415f
releasing factor 3 (RF3), 414, 415f
remnant removal disease, 268t
renal glomerulus, 605
renaturation, of DNA, 351
rennin, 59
repair mechanisms and proofreading for DNA, 724
repetitive-sequence DNA, 367–368
replication. See DNA, replication and synthesis of
replication bubbles, 374, 376f, 377f, 378
replication fork, 372f, 373, 376f
replication mutations, 691
replication protein A (RPA), 377
replicative senescence, 726
reporter gene approach, 433f, 435–436, 435f, 436f
repression, of enzyme synthesis, 87
repressor gene, lambda (cI), 425–426, 427f
repressor protein, lambda (cI), 426, 427f
repressors in gene expression, 421–423
reproduction, prostaglandins in, 226
RER. See rough endoplasmic reticulum
REs. See restriction enzymes
respiration, 115
respiratory burst, 533, 668
   free radicals from, 550
respiratory chain
   biomedical importance of, 122
   catabolic energy captured by, 125–126
   citric acid cycle production of substrates for, 156–157, 157f
   clinical aspects of, 130
   coenzyme Q in, 122, 123f, 124, 124f, 125f
   complexes of, 122–125, 123f, 124f, 125f
   control of, 126–127, 127t
   cytochrome c in, 122, 123f, 124, 124f, 125f
   dehydrogenase role in, 116
   electron transport in, 122, 123f, 124–125, 124f, 126f
   flavoproteins and iron-sulfur proteins in, 122–124, 123f
   inhibition of, 116, 122, 127–128, 127f, 128f
   in mitochondrial compartments, 122, 122f
   mitochondrial exchange transporters and, 128–130, 128f, 129f, 130f
   oxidation of reducing equivalents by, 122–125, 123f, 124f, 125f
   oxygen reduction in, 124–125
   proton gradient of, 125, 126f
   proton pump action of, 122–125, 123f, 124f, 125f, 126f
respiratory chain level phosphorylation, 126
respiratory control, 110, 162

respiratory distress syndrome, surfactant deficiency and, 210, 239, 244–245
respiratory quotient (RQ), 532
restriction endonucleases
   clinical diagnosis using, 68
   direct observation of, 65f
restriction enzymes (REs), 359
   in chimeric DNA molecule preparation, 446–447
   DNA chain cleavage with, 445–446, 446f, 446t
   selection of, 445, 445t
restriction fragment length (RFLPs), 68
restriction map, 445
restriction sites, 68
reticulocytes, 656
retina, gyrate atrophy of, 292
retinaldehyde, 537, 537f
retinitis pigmentosa, essential fatty acid deficiency and, 234
retinoblastoma protein (Rb protein), 379
retinoic acid, 537–538, 537f
retinol. See vitamin A
retinol activity equivalent, 537
retrograde transport (COPI), 594
retrograde vesicular transport, 591
retroposons, 367
retrotranslocation, 592–593, 593f
retroviral genomes, 449
retroviruses, reverse transcriptases in, 354, 378
reverse cholesterol transport, 252f, 253, 260, 264, 264f, 267
reverse transcriptase/reverse transcription, 354, 370, 446t
Reye syndrome, 236
   orotic aciduria in, 346
RF1 (releasing factor 1), 413–414, 415f
RF3 (releasing factor 3), 414, 415f
RFLP (restriction fragment length polymorphisms), 68
RGD (Arg-Gly-Asp) sequence, 605
rheumatoid arthritis, 599
   glycoprotein synthesis defects in, 562
   ω3 fatty acids and, 209
rho (ρ) factor, 389
rho kinase, 623–624
rhodopsin, 537, 538f
ribbon diagrams, 39
riboflavin. See vitamin B₂
riboflavin-linked dehydrogenases, 118
ribomyopathies, 656
ribonucleases, 359
ribonucleic acid. See RNA
ribonucleoprotein particles (mRNPs), 416, 416f, 442
ribonucleoside diphosphates, 342, 342f
ribonucleosides, 330, 331f

ribonucleotide reductase
    gallium in, 98
    iron in, 100
    reduction of, 342, 342f
ribose, 147, 150f, 151t
    metabolic pathways of, 135
    in nucleosides, 330, 331f
    pentose phosphate pathway formation
        of, 191–194, 192f, 193f
ribose 5-phosphate, in purine synthesis,
    338, 339f, 340, 341f
ribose-5-phosphate ketoisomerase, in
    pentose phosphate pathway, 192f,
    193f, 194
ribosomal dissociation, in protein
    synthesis, 410
ribosomal RNA (rRNA), 354, 384–385, 385t
    amplification of, 441–442, 441f
    function of, 356
    as peptidyl transferase, 413, 414t
    precursors for, 400
    in protein synthesis
        elongation, 413–414, 413f
        initiation, 410, 411f, 412
    structure of, 355–356, 358t
ribosomes, 139, 356, 358t
    bacterial, 417–418
    profiling of, 455
    in protein synthesis, 410
    as ribozyme, 69
    structure of, 582
ribothymidine pseudouridine cytidine
    (TψC) arm of tRNA, 347, 355,
    357f, 406–407, 407f
ribozymes, 66–67, 354, 402
ribulose, 150f, 151t
ribulose-5-phosphate, in pentose
    phosphate pathway, 192–194, 192f,
    193, 193f
ribulose-5-phosphate 3-epimerase, in
    pentose phosphate pathway, 192f,
    193f, 194
Richner-Hanhart syndrome, 296, 297f
ricin, 418
rickets, 539–540
Rieske iron centers, 100, 102f, 124, 125f
right operator, 425–426, 427f
right-handed triple- or superhelix, 600, 600f
rigor mortis, 623
RISC (RNA-induced silencing complex),
    402
RNA
    biomedical importance of, 348, 384
    catalysis by, 69–70, 402
    in chromatin, 361
    circular, 358–359
    classes/species of, 355–359, 384–385,
        385t
    complementarity of, 352–354, 354f, 355

in genetic information flow, 404–405
long noncoding, 358–359, 384–385, 385t
messenger (See messenger RNA)
micro (See micro-RNAs)
modification of
    of mRNA, at 5′ and 3′ ends, 401–402
    mRNA sequence alteration, 402
    pri-miRNA, 401–402, 401f
    tRNA processing and, 402
mutations caused by changes in,
    407–409, 407f, 408f, 409f, 422
noncoding (See noncoding RNAs)
as polynucleotides, 335
processing of
    after transcription, 397
    alternative, gene expression and, 442
    alternative promoter utilization in,
        399–400, 399f, 400f
    alternative splicing, 398f, 399
    in gene expression regulation,
        439–440, 440t
    of rRNAs and tRNAs, 400
    sequences to remove, 397, 397f, 398f
profiling of, 454–455
in protein synthesis, 354
ribosomal (See ribosomal RNA)
silencing (See silencing RNAs)
small, 358–359
small, heterogeneous regulatory, 359
small nuclear (See small nuclear RNA)
splicing of, 397f, 398–399, 398f
    alternative, 398f, 399
structure of, 352–354, 353f, 354f, 356f,
    357f
synthesis of, 349 (See also transcription)
    cyclical process of, 388–389
    DNA synthesis compared with, 385
    DNA template strand for, 385–386,
        385f
    RNA polymerase in, 386–387, 386f,
        387f
transfer (See transfer RNA)
in viruses, 354
RNA blot procedure, 450, 451f
RNA editing, 402
RNA polymerase II
    activators and coregulators and,
        395–396, 396t
    formation of, 392–394
    phosphorylation activation of, 395, 395f
    RNA processing and, 397, 397f, 398f
    termination signals for, 392–394
    in transcription, 390f, 391–392, 391f,
        392f, 393f
RNA polymerases (RNAP)
    bacterial, 387–388, 387f
    initiation by, 386–387, 386f, 387f
    mammalian, 388, 388t
    in operon model, 423, 424f, 425

RNA primer, in DNA synthesis, 372f, 372t,
    374, 375f, 376f
RNA recognition motifs (RRM), 398
"RNA World" hypothesis, 69–70
RNA-dependent DNA polymerase, 370
RNA-induced silencing complex (RISC),
    402
RNAP. See RNA polymerases
RNA-RNA hybridization, 358
RNAse H, 446t
RNA-Seq, 455
ROS. See reactive oxygen species
rosettes, erythrocyte, 647
Rossmann fold, 38
rough endoplasmic reticulum (RER)
    binding to, 583f, 587–590, 588t, 589f
    in protein sorting, 582, 582f, 593f
    protein synthesis and, 416
    routes of protein insertion into, 589f,
        591, 591f
rough ER branch, of protein sorting, 582,
    582f, 583f
    cotranslational pathway, 588–589,
        589f
    N-terminal signal peptides in, 587–588,
        588t
    posttranslational translocation,
        589–590, 589f
Rous sarcoma virus (RSV), 693, 695
RPA (replication protein A), 377
RQ (respiratory quotient), 532
RRM (RNA recognition motifs), 398
rRNA. See ribosomal RNA
RSV (Rous sarcoma virus), 693, 695
ryanodine, 625

S
S phase of cell cycle, DNA synthesis
    during, 378–379, 378f, 379f, 379t
S1 nuclease, in recombinant DNA
    technology, 446t
S_{50}, 79, 79f
saccharides. See carbohydrates
SADDAN phenotype, 616
StAR (steroidogenic acute regulatory)
    protein, 493, 493f
Starling forces, 635
Stickler syndrome, 616
Stokes radii, 26
salt bridges, 8
    in hemoglobins, 53–54
    in protein tertiary and quaternary
        structures, 40
"salvage" reactions
    in purine synthesis, 340, 341f
    in pyrimidine synthesis, 342–343
samples, for laboratory analysis, 571
sandwich assay, laboratory tests using, 573
Sanfilippo syndrome, 611t

Sanger, Frederick, sequencing work of, 27–28
Sanger reagent, 28
sarcoglycan, 629, 629f
sarcolemma, 619
sarcomere, 619, 619f
sarcoplasm, 619
sarcoplasmic reticulum, 624–625, 625f
sarcosine, 310–311
saturated acyl enzyme, 227, 229f
saturated fatty acids, 206, 206f, 207, 207t
    in membranes, 469–470, 469f
saturation kinetics, 79
scaffolding, 362f, 363
scavenger receptor B1, 252f, 253
screening, for new drugs, 83
scurvy, 44, 264, 277, 548, 603
SDs (silencing domains), 439, 440f
SDS-PAGE (sodium dodecyl sulfate polyacrylamide gel electrophoresis), protein and peptide purification with, 27, 28f
Sec12p, 594
Sec61 complex, 588
Sec62/Sec63 complex, 589
second law of thermodynamics, 110
second messengers, 490, 491t, 510–511.
        See also signal generation
    cGMP as, 333
    enzyme regulation by, 89
    phosphatidylinositide as, 210–211, 210f
    phospholipids as, 239
secondary lysosomes, 482
secondary structure
    α helices, 36–37, 36f, 38f
    β sheets, 37, 37f, 38f
    of collagen, 44, 44f
    folding into, 42
    loops and bends, 37, 38f
    of myoglobin and hemoglobin, 52–53
    as order of protein structure, 35
    peptide bond restrictions on, 35–36, 35f
    of RNA, 352–354, 353f, 357f
secondary transport, 477
second-order rate constant, 74
secretory (exocytotic) pathway, 582
secretory proteins, 588–589
secretory vesicles, 582, 583f
selectins, 561–562
selective permeability
    of inner mitochondrial membrane, 128–130, 128f, 129f, 130f
    membrane, 467, 478, 478f
selective proteolysis, enzyme regulation by, 90–91, 91f
selectivity filter, 478, 479
selenium, 214, 255
    in glutathione peroxidase, 118, 195, 195f
    human requirement for, 97, 97t

selenocysteine, 18, 18f
    synthesis of, 278, 278f
selenophosphate synthetase, 278, 278f
self-assembly
    of collagen, 601
    of lipid bilayer, 470
self-association, of hydrophobic molecules in aqueous solutions, 8
semidiscontinuous DNA synthesis, 372f, 374, 376f
semiquinone, 124, 125f
senescence, DNA damage and, 381, 381f
sensitivity, of laboratory test, 570, 571t
sensory neuropathy, vitamin B6 causing, 543
sequence classes, of genome, 367
sequence-specific DNA methylases, 445
sequencing
    of DNA, 451
    of proteins and peptides (See protein sequencing)
sequential reactions, 82, 82f
serine, 16t
    carbon skeleton catabolism of, 293, 294f
    as feedback inhibitor, 88, 89f
    specialized products of, 310
    synthesis of, 275, 276f
serine protease inhibitor, 645
serine proteases, covalent catalysis of, 63–64, 64f
serine-arginine motifs (SR), 398
serotonin, tryptophan conversion to, 310, 311f
serotonylation, 430t
serpin, 645
serum amyloid P component, 646
serum bilirubin, 323, 325, 325t, 574
serum complement components, 479
serum enzyme analysis, 66–67, 67f, 67t
serum samples, 571
sex hormones, cholesterol as precursor for, 212, 259
sex-hormone-binding globulin (SHBG), 507, 507t
SGLT1 transporter, 528, 528f
SH2 (Src homology 2), 516, 517f
SHBG (sex-hormone-binding globulin), 507, 507t
short interspersed nuclear elements (SINEs), 367–368
sialic acids, 154f, 155, 155t, 212
    in gangliosides, 244, 244f
    synthesis of, 197, 199f
sialidosis, 611
sickle cell anemia/disease
    hemoglobin mutations in, 56, 57f, 408f, 657t
    technology for, 444

signal generation
    in group I hormones, 509–510, 509f, 510f, 510t
    in group II hormones
        calcium, 515
        cAMP, 511–514, 511t, 512t, 513f
        cGMP, 514
        G-protein-coupled receptors, 511, 511f
        intracellular, 510–511
        phosphatidylinositide, 515–516, 515f, 516f
        protein kinases, 516–519, 517f, 518f
signal hypothesis of polyribosome binding, 582, 582f, 587–590, 588t, 589f
signal peptidase, 588, 589f, 636
signal peptide, 582
    of albumin, 637
    in protein sorting, 582f, 584, 585f, 587–590, 589f, 590f
signal recognition particle (SRP), 588–589, 589f, 636
signal recognition particle (SRP) receptor, 588–589, 589f
signal sequences, 581–583, 582t, 590f, 636
signal transducers and activators of transcription (STATs), 516–517, 518f
signal transduction
    across membranes, 467, 475t, 483
    CBP/p300 and, 521–522, 521f
    of hormones, 508–509, 509f
silencers, of gene expression, 92, 421
    DNA elements, 435
    tissue-specific expression of, 435
silencing domains (SDs), 439, 440f
silencing RNAs (siRNAs), 358, 384–385, 385t
    gene expression modulation by, 439–440
    synthesis of, 401f, 402
silent mutations, 407, 407f
simple competitive inhibition, 80, 80f
simple diffusion, 475f, 475t, 476f, 483
simple lipids, 206
simple noncompetitive inhibition, 80–81, 81f
simvastatin, 267
SINEs (short interspersed nuclear elements), 367–368
single quadrupole mass spectrometers, 29, 30f
single-cell sequencing, in cancer research, 710
single-displacement reactions, 82, 82f
single-molecule enzymology, 65, 65f
single-strand binding proteins (SSBs), 372f, 373, 373t

single-stranded DNA (ssDNA), 371
siRNA-miRNA complexes, 358
siRNAs. *See* silencing RNAs
sirtuins, 93
sister chromatids, 364, 365*f*, 371, 371*f*
site-directed mutagenesis, enzyme studies using, 69
site-specific integration, 370
size-exclusion chromatography, 26, 26*f*
Sjögren-Larsson syndrome, 236
skeletal muscle, 619, 626*t*
    actin-based regulation in, 622
    glucose uptake by, 142–143, 143*f*
    glycogen, supplies of, 627
    lactate production by, 167
    malignant hyperthermia of, 625, 627*t*
    metabolic role of, 139
    as protein reserve, 630
    sarcoplasmic reticulum in, 624–625, 625*f*
    twitch fibers of, 628
skin
    calcitriol synthesis in, 497–498, 498*f*
    diseases of, 632
    essential fatty acid deficiency and, 236
    vitamin D synthesis in, 538, 539*f*
skin bleeding time, 687
sleep, prostaglandins in, 226
slow twitch fibers, 145*t*, 628
slow-reacting substance of anaphylaxis, 237
Sly syndrome, 611*t*
SMAC, in apoptosis, 701, 702*f*
small, heterogeneous regulatory RNAs (sRNAs), 359
small nuclear RNA (snRNA), 354, 354*t*, 358, 384–385, 385*t*
    in alternative splicing, 399
    in RNA processing, 398–399
small RNA, 358–359
small ubiquitin-like modifier (SUMO) protein, 92
smooth muscle, 619, 626*t*
    actin-myosin interactions in, 623–624, 624*t*
    regulation of contraction by myosin, 623–624, 624*f*, 624*t*
SNAP (soluble NSF attachment factor) proteins, 594*t*, 595*f*, 596
SNARE pin, 595
SNARE proteins, 594–596, 594*t*, 595*f*
snRNA. *See* small nuclear RNA
sodium dodecyl sulfate polyacrylamide gel electrophoresis (SDS-PAGE), protein and peptide purification with, 27, 28*f*
sodium (Na⁺)
    in extracellular and intracellular fluid, 468, 468*t*
    permeability coefficient of, 471*f*

sodium-potassium pump (Na⁺-K⁺-ATPase), 480, 480*f*, 481*f*
SODs. *See* superoxide dismutases
solubility
    of amino acids, 20*f*, 21
    of lipids, 206
soluble NSF attachment factor (SNAP) proteins, 594*t*, 595*f*, 596
soluble proteins, 35, 588–589
solvent, water as ideal, 6–7, 7*f*
somatic mutation theory of aging, 724
sonication, mild, 473
sorbitol, in diabetic cataract, 200
sorbitol dehydrogenase, 196, 197*f*
sorbitol pathway, 200
Soret band, 318, 320*f*
Southern DNA blot transfer procedure, 450, 451*f*
Southwestern overlay blot procedure, 450, 451*f*
specialized compartments, 467
specific acid-base catalysis, 62
specific activity, 78
specificity
    of enzymes, 60, 60*f*
    of hormone receptors, 489, 489*f*
    of laboratory test, 570, 570*f*, 571*t*
spectrin, 659–660, 659*f*, 659*t*
spectrophotofluorimetry, 571–572
spectrophotometry
    enzymatic assays using, 66, 66*f*
    laboratory tests using, 571–572
    of porphyrins, 318, 320*f*
spectroscopy, protein structures solved by, 41
spermidine, 309, 309*f*, 310*f*
spermine, 309, 309*f*, 310*f*
spherocytosis, hereditary, 486*t*, 657*t*, 658, 660
sphingolipidoses, 245, 245*t*
sphingolipids, 210, 211*f*, 239
    clinical aspects of, 245, 245*t*
    formation of, 243–244, 244*f*
    in multiple sclerosis, 245
sphingomyelins, 210, 210*f*
    membrane asymmetry and, 472, 596–597
    in membranes, 469
    synthesis of, 243–244, 244*f*
sphingophospholipids, 206
sphingosine, 210, 210*f*
    serine conversion to, 310
spina bifida, folic acid supplements for, 546
spliceosome, 398
splicing of mRNA, 397*f*, 398–399, 398*f*, 442
spontaneous assembly, 601
spontaneous chemical reactions, 72
spontaneous mutations, 691

squalene, in cholesterol synthesis, 260, 261*f*, 262*f*
squalene epoxidase, in cholesterol synthesis, 260, 262*f*
squalene synthetase, 261*f*
SR (serine-arginine motifs), 398
SR-B1 (scavenger receptor B1), 252*f*, 253
Src homology 2 (SH2), 516, 517*f*
SREBP (sterol regulatory element-binding protein), 262
sRNAs (small, heterogeneous regulatory RNAs), 359
SRP receptor (SRP-R), 588–589, 589*f*
SRP (signal recognition particle), 588–589, 589*f*, 636
SSBs (single-strand binding proteins), 372*f*, 373, 373*t*
ssDNA (single-stranded DNA), 371
staggered ends, 445, 446*f*
standard free-energy change (ΔG⁰)
    of ATP hydrolysis, 111–112, 111*t*, 112*f*
    in biologic systems, 110
starch, 152, 153*f*
    hydrolysis of, 528
starvation, 109
    fatty liver and, 255
    ketosis in, 225
    metabolic pathways in, 144–145
    plasma hormones and metabolic fuels during, 145*f*
    triacylglycerol redirection and, 251
statin drugs, 87, 260, 260*f*, 267
STATs (signal transducers and activators of transcription), 516–517, 518*f*
stearic acid, 207*t*
stearoyl-CoA, 233, 233*f*
stem cell biology, 4, 454
stem cell factor, 654
stem cells
    in cancer, 706
    hematopoietic, 653–654, 654*f*
    induced pluripotent, 454
step-wise assembly, 396
stereochemical (-*sn*) numbering system, 209, 209*f*
stereochemistry, of L-α-amino acids, 18
stereospecificity, of enzymes, 60, 60*f*
steroid sulfates, 245
steroidogenesis
    ovarian, 495–496, 497*f*, 498*f*
    testicular, 495, 496*f*
steroidogenic acute regulatory (StAR) protein, 493, 493*f*
steroids
    cytochrome P450 in metabolism of, 119–120, 119*f*, 120*f*
    lipid bilayer and, 470
    metabolic pathways of, 136, 136*f*
    nucleus of, 212, 212*f*

plasma transport of, 507, 507*t*
as precursor molecule, 491, 492*f*
receptors for, 490
roles of, 212
stereoisomers of, 213, 213*f*
storage of, 505, 506*t*
vitamin B$_6$ role in actions of, 543
sterol 27-hydroxylase (CYP27A1), 265, 266*f*
sterol regulatory element-binding protein (SREBP), 262
sterols, 212, 469
sticky end ligation, 446–447
sticky ends, 445, 446*f*
sticky patch, in hemoglobin S, 56, 57*f*
stochastic processes, mortality and aging as, 718
stoichiometry, 72
stop codon, 414, 415*f*
stop-transfer signal, 590, 591*f*
storage granules, of platelets, 678
strain, enzymatic catalysis by, 62
streptokinase, 68, 686–687
streptomycin, 151
striated muscle, 620*f*, 623, 624*t*, 625
strong acids, dissociation of, 10–11
strong bases, dissociation of, 10–11
structural proteins, 599. *See also* collagen; elastin; fibrillins
substrate analogs, 80, 80*f*
substrate cycles, in glycolysis and gluconeogenesis regulation, 185
substrate deprivation therapy, 245
substrate level phosphorylation, 126, 164
substrate shuttles, mitochondrial, 129–130, 129*f*
substrates, 60
    anaplerotic, 160–161
    in balanced chemical equations, 72
    cellular concentrations of, 86
    conformational changes induced by, 63, 63*f*
    multiple, 82, 82*f*, 83*f*
    pairs of, 129, 129*f*
    prosthetic group, cofactor, and coenzyme interactions with, 61, 61*f*
    reaction rate response to concentrations of, 74, 76–77, 76*f*, 77*f*
    regulation of metabolic pathways with, 139
    for respiratory chain substrates, 156–157, 157*f*
succinate dehydrogenase, 118
    in citric acid cycle, 158*f*, 159
    inhibition of, 80, 80*f*
succinate thiokinase (succinyl-CoA synthetase), 158*f*, 159, 258, 258*f*
succinate-Q reductase (Complex II), 122, 123*f*, 124, 124*f*
succinic acid, pK_a of, 13*t*

succinic semialdehyde dehydrogenase, defects in, 313, 313*f*
succinyl-CoA
    in citric acid cycle, 158*f*, 159
    heme biosynthesis from, 315–318, 316*f*, 317*f*, 318*f*, 319*f*, 320*t*
succinyl-CoA synthetase (succinate thiokinase), 158*f*, 159, 258, 258*f*
succinyl-CoA-acetoacetate-CoA transferase (thiophorase), 221, 223*f*
    in citric acid cycle, 158*f*, 159
sucrose, 147, 151–152, 152*f*, 153*t*
sugar nucleotides, in glycoprotein synthesis, 557–558
sugars, in amphipathic lipids, 470
"suicide enzyme," cyclooxygenase as, 234–235
suicide inhibitors, 81–82
sulfate, in glycoproteins, 555
sulfatide, 212
sulfation, of xenobiotics, 566
sulfite oxidase
    deficiency of, 104
    transition metals in, 104, 104*f*
sulfogalactosylceramide, 212
    accumulation of, 245
    synthesis of, 244
sulfonylurea drugs, 225
sulfotransferases, 608
SUMO (small ubiquitin-like modifier) protein, 92
sumoylation, of histones, 361, 363*t*
supercoils, DNA, 351, 377, 378*f*
superoxide, 656–657
superoxide anion-free radical, 120
superoxide dismutases (SODs), 214, 656–657
    copper in, 103
    gallium toxicity to, 98
    oxygen toxicity protection by, 120, 552
    transition metals in, 102–103
supersecondary structures, 37
suppressor mutations, 409–410
suppressor tRNA, 409
supravalvular aortic stenosis, 604
surfactant, lung, 239
    deficiency of, 210, 244–245
SV40 enhancer, 433, 433*f*
Svedberg units, 356
symport systems, 476, 476*f*
syn conformers, 330, 331*f*
synaptic vesicles, 595
synaptobrevin, 595
5q-syndrome, 656
synthetic biology, 4
systems biology, 4, 455–457

T
$t_{1/2}$. *See* half-life
T lymphocytes, 646, 664, 670
T state. *See* taut state
T$_3$. *See* triiodothyronine
T$_4$ (thyroxine). *See* tetraiodothyronine
TAFIa (thrombin-activatable fibrinolysis inhibitor), 686, 686*f*
TAFs (TBP-associated factors), 391, 391*f*, 394
talin, 605, 606*f*
tamoxifen, 712
tandem MS (MS-MS, MS^2)
    metabolic disease detection with, 288, 290
    protein analysis using, 31
Tangier disease, 268*t*
TaqI, 445*t*
target cell, 484, 489, 489*t*
targeted gene regulation, 454
Tarui disease, 174*t*
TATA box, in transcription control
    in bacteria, 389–390, 390*f*
    in eukaryotes, 390–392, 390*f*, 391*f*
    TBP binding with, 394
TATA-binding protein (TBP), 391
    TATA box binding of, 394
taurine, cysteine conversion to, 307, 307*f*
taurochenodeoxycholic acid, 266*f*
taurocholic acid, 266*f*
taut (T) state, of hemoglobin, 52, 53*f*, 54*f*
    2,3-bisphosphoglycerate stabilization of, 55–56, 55*f*
    proton binding to, 54–55
tautomerism, 330, 330*f*
Tay-Sachs disease, 245*t*, 592*t*
TBG (thyroxine-binding globulin), 506
TBP. *See* TATA-binding protein
TBP-associated factors (TAFs), 391, 391*f*, 394
TCGA (The Cancer Genome Atlas), 699, 710
TEBG (testosterone-estrogen-binding globulin), 507
telomerase, 365, 701, 726
telomeres, 365, 365*f*, 726, 727*f*
temozolomide, 699
temperature
    in fluid mosaic model of membrane structure, 473
    melting (transition), 351, 473
    reaction rate response to, 73, 74*f*, 75
temperature coefficient (Q$_{10}$), 75
template binding, in transcription, 386*f*
template strand DNA
    in DNA replication, 349, 354, 354*f*
    in RNA synthesis, 385–386, 385*f*, 390*f*
N-terminal signal peptide, 582, 582*f*, 587–590, 588*t*, 589*f*, 590*f*
terminal transferase, 446*t*

termination
    of protein synthesis, 414, 415f
    in proteoglycan synthesis, 608
    in transcription cycle, 386f, 387, 392–394
termination protein, 389
termination signals, of translation,
    405, 405t
tertiary structure, 37–38, 39f, 40f
    of collagen, 44, 44f
    of myoglobin and hemoglobin, 52–53
    as order of protein structure, 35
    schematic diagrams of, 39
    stabilizing factors in, 40
testicular steroidogenesis, 495, 496f
testosterone
    metabolism of, 495, 497f
    plasma transport of, 507, 507t
    synthesis of, 491, 492f, 493f, 495, 496f
testosterone-estrogen-binding globulin
    (TEBG), 507
tethering, 594–595
tetracycline (Tet) resistance genes, 449
tetrahedral geometry, of water molecules,
    6, 7f
tetrahedral transition state intermediates,
    63, 63f
tetrahydrofolate, 545, 545f
tetraiodothyronine (thyroxine; T₄)
    plasma transport of, 506, 506t
    storage of, 505, 506t
    synthesis of, 491, 492f, 500, 501f
tetramers, histone, 361, 361f, 363
tetrapyrroles, in heme, 50, 50f, 315, 316f
tetroses, 148, 148t, 150
TEXs (tumor-derived exosomes), 698
TF (tissue factor), 680–681, 681t, 682t
TFIIA, 391f, 392, 392f, 393f, 394
TFIIB, 391f, 392, 392f, 393f, 394
TFIID, 391f, 392, 392f, 393f
    components of, 395–396
    in RNA polymerase II formation, 394
TFIIE, 391f, 392, 392f, 393f, 394
TFIIF, 391f, 392, 392f, 393f, 394
TFIIH, 391f, 392, 392f, 393f
    CTD modification by, 395
    in RNA polymerase II formation, 394
TFPI (tissue factor pathway inhibitor), 681
TGN (trans-Golgi network), 582, 583f, 596
thalassemias, 57, 662
α-thalassemias, 657t
β-thalassemias, 657t
    pathologic protein conformations in, 43
thanatophoric dysplasia, 616
The Cancer Genome Atlas (TCGA), 699,
    710
theobromine, 331, 333f
theophylline, 257, 331, 333f
thermodynamics, of biologic systems,
    109–110

thermogenesis, 257–258, 258f
    diet-induced, 532
thermogenin, 127–128, 258, 258f
thiamin. See vitamin B₁
thiamin diphosphate, 168, 169f, 541–542
thick (myosin) filaments, 619–620, 620f
thin (actin) filaments, 594, 619–620, 620f,
    621f, 630
thioesterase (deacylase), 227, 228f, 229f
6-thioguanine, 334, 334f
thiokinase. See acyl-CoA synthetase;
    succinate thiokinase
thiolase, 219, 219f, 221
    in mevalonate synthesis, 260, 260f
thiophorase (succinyl-CoA-acetoacetate-
    CoA transferase), 221, 223f
    in citric acid cycle, 158f, 159
thioredoxin, 342, 342f
thioredoxin reductase, 342, 342f
threonine, 16t
    carbon skeleton catabolism of, 295, 295f
thrombin
    in fibrin clot formation, 680
    fibrinogen conversion to fibrin by,
        683–685, 683f, 684f
    inhibition of, 685
    in platelet aggregation, 678, 679f
    prothrombin conversion to, 682–683,
        683f
    regulation of, 685
thrombin-activatable fibrinolysis inhibitor
    (TAFIa), 686, 686f
thrombocytopenia, 653, 662
thrombomodulin, 685
thrombopoietin, 654
thrombosis, 661–662
    alteplase and streptokinase treatment
        of, 686–687
    anticoagulant drug effects on, 685–686
    biomedical importance of, 677
    endothelial cell actions in, 678, 680
    fibrin clot formation in (See fibrin clot
        formation)
    folic acid supplements for, 546
    laboratory tests for, 687
    phases of, 677
    platelet aggregation in, 678, 679f
    thrombi types in, 678
    thrombin inhibition in, 685–686
    thrombin regulation in, 685
thromboxane A₂ (TxA₂), 208f, 234
    inhibition of, 680
    in platelet aggregation, 678, 679f
thromboxanes (TXs), 208, 208f, 226
    cyclooxygenase pathway in formation
        of, 234, 236f
thrombus
    formation of, 678, 679f
    types of, 678

thymidine, 332t
    salvage pathways of, 343
thymidine monophosphate (TMP), 332f,
    332t
thymidylate, 348–349
thymine, 332t, 347f
    base pairing in DNA, 349, 350f
    salvage pathways of, 343
thymine dimer formation and UV light,
    722, 723f
thyroglobulin, 500
thyroid function tests, 575
thyroid hormones
    in cholesterol synthesis, 262–263, 263f
    in lipolysis, 256, 257f
    plasma transport of, 506, 506t
    receptors for, 490
    regulation of gene expression by, 509,
        509f
    storage of, 505, 506t
    synthesis of, 500, 501f
thyroxine (T₄). See tetraiodothyronine
thyroxine-binding globulin (TBG), 506
ticagrelor, 680
tight junctions, 474
tiglyl-CoA, catabolism of, 302f, 303f
TIM (translocase of the inner membrane),
    584
time-of-flight (TOF) mass spectrometers,
    29–30
timnodonic acid, 207t
tirofiban, 680
tissue factor (TF), 680–681, 681t, 682t
tissue factor pathway inhibitor (TFPI), 681
tissue injury
    fibrin clot formation in, 680–681, 680f,
        681f, 681t, 682t
    myoglobinuria after, 56
    serum enzyme analysis after, 67–68, 67f
tissue nonspecific alkaline phosphatase
    (TNAP), 614
tissue plasminogen activator (tPA), 68,
    686–687, 686f
tissues
    differentiation of, retinoic acid role in,
        537–538
    in fasting state, 134–135, 144–145
    in fed state, 134–135, 142–144, 143f
    glucose uptake by, 142–143, 143f
    hypoxic, 167
titration curve, for weak acids, 12, 13f
T_m. See transition (melting) temperature
TME (tumor microenvironment), 691,
    703, 704f
TMP (thymidine monophosphate), 332f, 332t
TNAP (tissue nonspecific alkaline
    phosphatase), 614
tobacco, 712
tocopherols. See vitamin E

tocotrienols. *See* vitamin E

TOF (time-of-flight) mass spectrometers, 29–30

tolbutamide, 225, 254

TOM (translocase of the outer membrane), 584

topogenic sequences, 591

topoisomerases, DNA, 351, 373t, 377, 377f

total bilirubin, 323

total energy expenditure, 532

total iron-binding capacity, 640

totipotent stem cells, 653

toxic hyperbilirubinemia, 324

toxicity
    of ammonia, 283–284
    bilirubin, 323–324
    of folic acid, 546
    heavy metal, 97–98
    lead, 97–98, 316, 319
    micronutrient, 535
    niacin, 543
    oxygen (*See* reactive oxygen species)
    of plant amino acids, 19, 19t
    of transition metals, 98–99, 99t
    vitamin A, 538
    vitamin B_6, 543
    vitamin D, 540
    of xenobiotics, 567, 567f

toxins
    botulinum B, 595
    cholera, 212, 212f, 244, 512
    diphtheria, 418, 479
    microbial, 479
    pertussis, 512

tPA (tissue plasminogen activator), 68, 686–687, 686f

TψC arm. *See* ribothymidine pseudouridine cytidine arm

trabecular bone, 614

trans fatty acids, 208–209, 236, 470

transactivator proteins, 432f, 433

transaldolase, in pentose phosphate pathway, 192–194, 192f, 193f

transaminase
    in amino acid catabolism, 291–292
    vitamin B_6 assessment with, 543

transamination
    in amino acid metabolism, 136, 137f, 282–283, 283f, 291–293
    citric acid cycle role in, 159–161, 160f
    ping-pong mechanism for, 62, 62f, 283, 283f

transcortin, 506, 507t

transcription, 351–352
    activators and coregulators in control of, 395–396, 396t
    bacterial promoters in, 389–390, 390f

eukaryotic promoters in, 390–392, 390f, 391f, 392–394, 392f, 393f

eukaryotic transcription complex, 394–397

    hormone modulation of, 519f, 520–522, 521f, 521t, 522t

    initiation of, 386–387, 386f, 387f

    promoters in, 386–387
        alternative, 399–400, 399f, 400f

    protein-DNA interactions in regulation of, 425–430, 426f, 427f, 428f

    reverse in retroviruses, 354, 370, 378

    in RNA synthesis, 349

    termination of, 392–394

transcription complex, eukaryotic. *See* eukaryotic transcription complex

transcription control elements, 396t

transcription export (TREX), 397, 397f

transcription factors, 396t, 727
    for enzyme synthesis, 87

transcription start site (TSS), 386–387, 387f
    alternative, 442

transcription unit, 387

transcriptome information, 455

transcriptomics, 3

transcytosis, 596

transdeamination, 283, 284f

transducers, 381, 381f

transfected cells in culture, 435f, 436, 436f

transfer RNA (tRNA), 354, 384–385, 385t
    amino acid association with, 406–407, 406f, 407f
    anticodon region of, 405–406, 407f
    function of, 355
    processing of
        modification and, 402
        precursors for, 400
    in protein synthesis
        elongation, 413–414, 413f
        initiation, 410, 411f, 412–413
        termination, 414
    structure of, 355, 357f
    suppressor, 409

transferases, 60

transferrin, 531, 531f, 634, 639–640, 639t, 641–642

transferrin cycle, 640, 641f

transferrin receptor, 640, 641–642, 642f

transgenic animal approach, 435

transgenic animals, 454

trans-Golgi network (TGN), 582, 583f, 596

transhydrogenase, mitochondrial, 129

transition (melting) temperature (T_m), 351, 473

transition metals
    absorption and transport of, 105
    biomedical importance of, 96

heavy metal toxicity and, 97–98

human requirement for, 97, 97t

Lewis acid properties of, 97

multivalent states of, 97, 98f, 98t, 99–100

    in organometallic complexes, 99–100, 99f, 100f, 101t

    oxidation of, 97, 98f, 98t

    physiologic roles of, 100–105, 101t, 102f, 103f, 104f

    toxicity of, 98–99, 99t

transition mutations, 407, 407f

transition state
    of chemical reactions, 72–73, 73f
    enzymatic stabilization of, 75

transition state analogs, 62, 79

transition state intermediates, 62, 73, 73f
    acyl-enzyme, 64, 64f
    tetrahedral, 63, 63f

transition state stabilization, in enzyme catalysis, 62

transketolase
    in pentose phosphate pathway, 192–194, 192f, 193f
    vitamin B_1 assessment with, 542

translation, 405. *See also* protein synthesis

translation termination signals, 405, 405t

translocase of the inner membrane (TIM), 584

translocase of the outer membrane (TOM), 584

translocation
    in cancer, 693, 694f
    into cell nucleus, 584–586, 586f
    of mRNA, 585–586
    of proteins, 584

translocation, in protein synthesis, 414

translocation complexes, 584

translocon, 588

transmembrane collagens, 601

transmembrane proteins
    amino acid sequence of, 471
    ion channels as, 478–479, 478f, 479f, 479t

transmembrane signaling, 467, 475t, 483
    in cancer, 697, 697t
    in platelet aggregation, 678, 679f

transport vesicles
    coat proteins of, 594
    definition of, 593
    model of, 594–596, 594t, 595f
    proteins in, 582, 583f
    in *trans*-Golgi network, 596

transporters and transport systems
    aquaporins, 479–480
    ATP in, 480, 480t
    ATP-binding cassette, 252f, 253

ATP-driven active, 480, 480*t*
facilitated diffusion, 475*f*, 475*t*, 476*f*, 483
  hormone regulation of, 477
    ping-pong mechanism of, 477, 477*f*
    transporters involved in, 476–477, 476*f*, 476*t*, 477*f*
  glucose (*See* glucose transporters)
  ion channels and, 476*t*
  ionophores, 479
  membrane fatty acid, 249
  of membranes, 467, 471, 477
  of mitochondria, 128–130, 128*f*, 129*f*, 130*f*
  passive diffusion, 475*f*, 475*t*, 476*f*, 483
  plasma transport proteins, 506, 506*t*, 507*t*
  reverse cholesterol, 252*f*, 253, 260, 264, 264*f*, 267
  simple diffusion, 475*f*, 475*t*, 476*f*, 483
  tricarboxylate, 231
transposition, 370–371
  retroposons and, 367
trans-prenyltransferase, 261*f*
transthyretin, 646
transverse asymmetry, 596–597
transverse movement, of lipids across membrane, 471–472
transversion mutations, 407, 407*f*
trauma, protein loss in, 534
trehalose, 152*f*, 153*t*
TREX (transcription export), 397, 397*f*
triacylglycerols (triglycerides), 209, 209*f*
  biomedical importance of, 239
  digestion and absorption of, 528–529, 530*f*
  drugs for reduction of serum levels of, 267
  in lipoprotein core, 248, 249*f*
  as metabolic fuel, 134, 141
  metabolic pathways of, 135–136, 136*f*, 137, 138*f*
    in fed state, 143–144, 143*f*
  metabolism of
    in adipose tissue, 255–256, 256*f*
    fatty liver and, 254–255
    hepatic, 253–254, 254*f*
    high-density lipoproteins in, 252–253, 252*f*
    hydrolysis in, 239
  perilipin and, 257
  synthesis of, 139, 240–242, 240*f*, 241*f*
    fructose effects on, 197*f*, 199
  transport of, 249–250, 250*f*, 251*f*
tricarboxylate transporter, 231, 231*f*
tricarboxylic acid cycle. *See* citric acid cycle
triglycerides. *See* triacylglycerols
triiodothyronine (T₃)
  plasma transport of, 506, 506*t*

storage of, 505, 506*t*
  synthesis of, 491, 492*f*, 500, 501*f*
trimeric G complex, 512
trimethoprim, 545
trimethylxanthine, 331, 333*f*
trinucleotide repeats, 368
triokinase, 195, 197*f*
triose phosphate isomerase, structure of, 38*f*
trioses, 148, 148*t*
  of physiological importance, 150
tripeptidases, 529
triple helix, of collagen structure, 44, 44*f*
triplet codes, 405, 405*t*
triskelion, 482
tRNA. *See* transfer RNA
tRNA amino acid charging, 406, 406*f*
tRNA-derived small RNAs (tsRNAs), 355
tropocollagen, 601
tropoelastin, 604
tropomyosin, 621, 621*f*
troponin complex, 621
troponins, 621, 621*f*
  cardiac, 68
true-positive result, 570
trypsin, 529
  covalent catalysis of, 64
  for cystic fibrosis, 68
  limited proteolysis of, 621
trypsinogen, 529
tryptophan, 17*t*
  carbon skeleton catabolism of, 296–298, 299*f*, 300*f*
  deficiency of, 542–543
  permeability coefficient of, 471*f*
  specialized products of, 310, 311*f*
  ultraviolet light absorption by, 21, 22*f*
tryptophan 2,3-dioxygenase (tryptophan pyrrolase), 296–298, 299*f*
L-tryptophan dioxygenase, 119
t-SNARE proteins, 594–596, 595*f*
TSS. *See* transcription start site
T-tubular system, 625
α-tubulin, 631, 631*f*
β-tubulin, 631, 631*f*
γ-tubulin, 631
tumor biomarkers, 709–710, 709*t*
tumor microenvironment (TME), 691, 703, 704*f*
tumor progression, 695–697
tumor suppressor genes, 693*t*, 694*f*, 695–697, 695*f*, 695*t*, 696*f*, 696*t*
  early studies on, 2
  p53, 382
tumor viruses, 692–693, 693*t*, 694*f*, 712
tumor-derived exosomes (TEXs), 698

tumors
  benign, 689
  malignant (*See* cancer)
turnover, protein, 87, 279–280
turnover number, 78
turns, in proteins, 37, 38*f*
twin lamb disease, 225, 255
twisted gastrulation 1 (TWSG1), 642
two-dimensional electrophoresis, protein purity assessment with, 27, 27*f*
two-hybrid interaction test, 455, 457*f*
TWSG1 (twisted gastrulation 1), 642
TxA₂. *See* thromboxane A₂
TXs. *See* thromboxanes
type 1 leukocyte adhesion deficiency, 66
type A response, in gene expression, 421–422, 421*f*
type B response, in gene expression, 421*f*, 422
type C response, in gene expression, 421*f*, 422
type I collagen, 612, 612*t*
type I hyperprolinemia, 292, 292*f*
type I tyrosinemia, 296, 297*f*
type II collagen, 615, 615*f*
type II hyperprolinemia, 292, 292*f*, 295*f*, 296
type II tyrosinemia, 296, 297*f*
type IV collagen, 610
type V collagen, 612
type VII collagen, 603
tyrosinase, copper in, 103
tyrosine, 16*t*–17*t*
  carbon skeleton catabolism of, 296, 297*f*
  hormone synthesis from, 491, 492*f*, 499–500, 499*f*, 501*f*
  specialized products of, 310, 312*f*
  synthesis of, 276, 277*f*
  ultraviolet light absorption by, 21, 22*f*
tyrosine aminotransferase, 296, 297*f*
tyrosine hydroxylase, 499, 499*f*
tyrosine kinases, inhibitors of, 710–711
tyrosinosis, 296, 297*f*
tyrosinyl-tRNA, 418, 418*f*

U
UAS (upstream activator sequence), 439
ubiquinone. *See* coenzyme Q
ubiquitin
  protein degradation dependent on, 280–281, 280*f*, 281*f*
  structure of, 280, 280*f*
ubiquitin and protein degradation, 593, 593*f*
ubiquitin-proteasome pathway, enzyme degradation by, 87–88, 280–281, 280*f*, 281*f*
UCP1 (uncoupling protein 1), 258, 258*f*
UDPGal 4-epimerase, 197, 198*f*

UDPGal (uridine diphosphate galactose), 197, 198*f*
UDPGlc. *See* uridine diphosphate glucose
UDPGlc dehydrogenase, in uronic acid pathway, 195, 196*f*
UDPGlc pyrophosphorylase
  in glycogenesis, 172*f*, 173
  in uronic acid pathway, 195, 196*f*
ulcers, 527, 563
ultimate carcinogens, 692
ultraviolet light (UV), 722, 723*f*
  amino acid absorption of, 21, 22*f*
  carcinogenic effect of, 691, 691*t*
  DNA damage caused by, 380*f*, 380*t*
  nucleotide absorption of, 331
UMP (uridine monophosphate), 332*f*, 332*t*
unambiguity, of genetic code, 405, 405*t*
unconjugated hyperbilirubinemia, 323*t*, 324
uncouplers, 127–128, 533
uncoupling protein 1 (UCP1), 258, 258*f*
undernutrition, 527, 532–533, 532*f*
unequal crossover, 370, 370*f*
unesterified fatty acids. *See* free fatty acids
unfolded protein response (UPR), 592
Union of Biochemistry (IUB), enzyme nomenclature system of, 60
uniport systems, 476, 476*f*
unipotent stem cells, 653
unique-sequence (nonrepetitive) DNA, 367
universality of genetic code, 405*t*, 406
unsaturated fatty acids, 206, 206*f*, 207–208, 207*t*
  cis double bonds in, 208–209, 208*f*
  dietary, cholesterol levels affected by, 267
  eicosanoids formed from, 226, 234, 235*f*, 236*f*
  essential, 232, 232*f*
    abnormal metabolism of, 236
    deficiency of, 234
    prostaglandin production and, 226
  in membranes, 469–470, 469*f*, 470*f*
  oxidation of, 220, 221*f*
  structures of, 232*f*
  synthesis of, 233–234, 233*f*
UPR (unfolded protein response), 592
upstream activator sequence (UAS), 439
uracil, 332*t*, 347*f*
  base pairing in RNA, 352–354, 353*f*
  deoxyribonucleosides of, in pyrimidine synthesis, 342–343
  in RNA synthesis, 385–386
uraciluria-thyminuria, 337–338, 346
urate, as antioxidant, 214

urea
  in DNA denaturation, 351
  laboratory tests for, 574
  metabolic pathways of, 137, 142–143
  nitrogen excretion as, 282, 285, 285*f*
  permeability coefficient of, 471*f*
  synthesis of, 282–283, 282*f*, 283*f*, 285–286, 285*f*
    active enzymes, 732*t*
    defects of, 286–288, 287*t*
    regulation of, 286
urease, transition metals in, 100, 101*t*
ureotelic animals, 282
uric acid, 331, 333*f*
  fructose effects on, 199
  nitrogen excretion as, 282
  purine catabolism to, 344, 345*f*
uricase, 344
uricemia, 346*t*
uricotelic animals, 282
uridine, 331*f*, 332*t*, 343
uridine diphosphate galactose (UDPGal), 197, 198*f*
uridine diphosphate glucose (UDPGlc)
  in glycogenesis, 172*f*, 173
  in uronic acid pathway, 195, 196*f*
uridine monophosphate (UMP), 332*f*, 332*t*
uridyl transferase, deficiency of, 200
urinalysis, 574
urine
  bilirubin in, 323, 325, 325*t*
  glucose in, 189
  myoglobin in, 56
  urobilinogen in, 325, 325*t*
  xylulose in, 198
urine samples, 571
urobilinogen, 323, 325, 325*t*
urobilins, 323
urocanic aciduria, 293
urokinase, 686, 686*f*
uronic acid, 607
uronic acid pathway
  deficiency of, 191
  disruption of, 198–199
  reactions of, 195, 196*f*
uronic acids, in glycosaminoglycans, 154
uroporphyrinogen decarboxylase, 317, 317*f*, 318*f*, 319*f*, 320*t*
uroporphyrinogen I, in heme synthesis, 316, 317*f*, 318*f*, 319*f*
uroporphyrinogen I synthase, 316, 317*f*, 318*f*, 319*f*
  deficiency in, 320, 320*t*
uroporphyrinogen III, in heme synthesis, 316–317, 317*f*, 318*f*, 319*f*

uroporphyrinogen III synthase, 316, 317*f*, 318*f*, 319*f*
  deficiency in, 320, 320*t*
UV. *See* ultraviolet light

V
vaccines, anticancer, 712
valence states, of transition metals, 97, 98*f*, 98*t*, 100
valeric acid, 207*t*
validity, of laboratory tests, 569–570, 569*f*, 570*f*, 571*t*
valine, 16*t*
  carbon skeleton catabolism of, 298, 300, 302*f*, 303*f*
  synthesis of, 277–278
valinomycin, 129, 479
valproic acid, 699
van der Waals forces, 8, 9*f*, 349
vanadium
  absorption of, 105
  human requirement for, 97, 97*t*
  multivalent states of, 97, 98*f*, 98*t*
  physiologic roles of, 105
variable numbers of tandem repeats (VNTRs), 557
variant form of Creutzfeldt-Jakob disease (vCJD), 43
vascular endothelial growth factor (VEGF), 706
vasodilators, 618, 630, 630*f*
vCJD (variant form of Creutzfeldt-Jakob disease), 43
vector cloning, 448–449, 449*t*
VEGF (vascular endothelial growth factor), 706
velocity, enzyme. *See* initial velocity; maximal velocity
very-low-density lipoproteins (VLDLs), 139, 144, 248*t*
  atherosclerosis and, 267
  fatty liver and, 255
  fructose effects on, 195–196, 197*f*, 199
  hepatic secretion of, 253–254, 254*f*
  in ketogenesis regulation, 224
  in lipid transport, 247–248
  metabolism of, 251–252, 251*f*
  remnants, 252, 264
  in triacylglycerol transport, 249–250, 250*f*, 251*f*
vesicles
  in cell-cell communication, 483–484, 485*f*
  coat proteins and, 591, 594–595, 594*t*, 595*f*
  in endocytosis, 482
  extracellular, 483–484, 485*f*
  processing within, 596

secretory, 582, 583*f*
synaptic, 595
targeting of, 595, 595*f*
transport (*See* transport vesicles)
types and functions, 593*t*
v_i. *See* initial velocity
Villefranche classification, 602
vimentins, 631*t*, 632
vinblastine, 631
vinculin, 605, 606*f*
viral RNA-dependent DNA polymerase, 354
viral SV40 enhancer, 433, 433*f*
viruses
    chromosomal integration with, 370, 370*f*
    cyclophilins in, 43
    glycans in binding of, 563
    host cell protein synthesis by, 416, 417*f*
    receptor-mediated endocytosis and, 483
    RNA in, 354
    tumor, 692–693, 693*t*, 694*f*, 712
vision, vitamin A function in, 537, 538*f*
vitamin A (retinol)
    deficiency of, 536*t*, 538
    functions of, 536*t*, 537–538, 538*f*
    structure of, 537, 537*f*
    toxicity of, 538
vitamin B complex
    citric acid cycle need for, 159
    prosthetic groups, cofactors, and
        coenzymes derived from, 61, 61*f*
vitamin B_1 (thiamin)
    citric acid cycle need for, 159
    coenzymes derived from, 61
    deficiency of, 168–170, 536*t*, 542
    functions of, 536*t*, 541–542
    pentose phosphate pathway need for, 194
    structure of, 541, 541*f*
vitamin B_2 (riboflavin)
    citric acid cycle need for, 159
    coenzymes derived from, 61
    deficiency of, 42, 536*t*, 542
    flavin groups formed from, 116, 118
    functions of, 536*t*, 542
    measurement of, 198
vitamin B_6 (pyridoxine, pyridoxal, pyridoxamine)
    in aminotransferases, 283, 283*f*, 543
    deficiency of, 298, 300*f*, 536*t*, 543
    functions of, 536*t*, 543
    structure of, 543, 543*f*
    toxicity of, 543
vitamin B_{12} (cobalamin)
    absorption of, 105, 544
    cobalt in, 102–103
    deficiency of, 536*t*, 544–545, 544*f*

functions of, 536*t*, 544, 544*f*
    structure of, 544, 544*f*
vitamin C (ascorbic acid)
    as antioxidant, 214, 552, 552*f*
    in bile acid synthesis, 266*f*
    in collagen synthesis, 44
    deficiency of, 44, 264, 277, 536*t*, 548, 603
    functions of, 536*t*, 547
    higher intakes of, 548
    human requirement for, 191, 195
    iron absorption and, 531–532
    as pro-oxidant, 552, 552*t*
    structure of, 547, 547*f*
vitamin D (calciferol), 536
    as calcitriol precursor, 497
    calcium absorption and, 531
    in calcium homeostasis, 539
    cholesterol as precursor for, 259
    deficiency of, 536*t*, 539–540
    ergosterol as precursor for, 213, 213*f*
    functions of, 536*t*
    hormone nature of, 538–539, 539*f*
    synthesis and metabolism of, 538–539, 539*f*
    toxicity of, 540
vitamin D_3 (cholecalciferol), 538, 539*f*
    as antioxidant, 213
    formation and hydroxylation of, 497–499, 498*f*
vitamin D-binding protein, 498
vitamin E (tocopherols, tocotrienols)
    as antioxidant, 214, 540, 552, 552*f*
    deficiency of, 536*t*, 540
    fatty liver and, 255
    functions of, 536*t*, 540
    as pro-oxidant, 553
    structure of, 540, 540*f*
vitamin K
    deficiency of, 536*t*, 541
    functions of, 536*t*, 540–541, 541*f*
    structure of, 540–541, 541*f*
vitamin K-dependent carboxylation, of
        coagulation factors, 685–686
vitamins
    biomedical importance of, 535
    classification of, 535, 536*t*
    digestion and absorption of, 531–532
    for health maintenance, 3
    laboratory tests for, 572–573
    lipid- (fat-) soluble, 206
    nutritional requirements for, 535–536
VLDL receptor, 249, 251
VLDLs. *See* very-low-density lipoproteins
V_{max}. *See* maximal velocity
VNTRs (variable numbers of tandem
        repeats), 557

voltage-gated channels, 478–479, 479*f*, 625
voltage-gated K^+ channel (HvAP), 478–479, 479*f*
von Gierke disease, 174*t*, 345
von Willebrand disease, 592*t*, 662, 687
von Willebrand factor, 687
    in platelet aggregation, 678, 679*f*
vorinostat, 699
v-SNARE proteins, 594–596, 595*f*

W
Warburg effect, 704, 705*f*, 706*t*
warfarin, 540–541
    drug interactions of, 565
    mechanism of, 685–686
water
    as biologic solvent, 6–7, 7*f*
    biomedical importance of, 6
    biomolecule interactions with, 7–8, 7*t*, 8*f*
    body compartmentalization of, 468, 468*t*
    dipole formation by, 6–7, 7*f*
    dissociation of, 9–10
    for health maintenance, 3
    hydrogen bonding of, 7, 7*f*
    hydrogen ions in, 10–11, 13*f*, 13*t*
    hydrolysis reactions of, 9
    as nucleophile, 9–10
    permeability coefficient of, 471*f*
    pH of, 10
    respiratory chain production of, 124–125
    tetrahedral geometry of, 6, 7*f*
water channels, 479–480
water-miscible lipoproteins, 247
water-soluble hormones, 490, 491*t*
water-soluble molecules, 470
water-soluble vitamins, 535, 536*t*
Watson-Crick base pairing, 349, 350*f*
waxes, 206
weak acids
    amino acids as, 19–21, 20*f*, 21*t*
    buffering actions of, 13, 13*f*
    dissociation of, 11–12
    as functional groups, 11–12
    Henderson-Hasselbalch equation
        describing behavior of, 12, 13*f*
    medium properties affecting, 13
    molecular structure effects on, 13, 13*t*
    pK_a of, 11–13, 13*f*, 13*t*, 20
weak bases
    buffering actions of, 13, 13*f*
    dissociation of, 11–12

wear and tear theories of aging
   free radicals, 720
   hydrolytic reactions, 718–719, 719f
   mitochondria and, 720–723
   molecular repair mechanisms and,
      723–725
   protein glycation, 722–723, 724f
   reactive oxygen species, 719–720, 720f,
      721f
   ultraviolet radiation, 722, 723f
weight loss, low carbohydrate diets for, 190
Wernicke encephalopathy, 542
Western protein blot procedure, 450, 451f
white blood cells
   biomedical importance of, 664
   in infection defense, 664–665
   integrins in, 666, 667t
   multiple types of, 665f
   phagocytosis, 482, 664, 665, 667–668,
      667f, 668t
   production regulation of, 665
white thrombus, 678
Williams-Beuren syndrome, 604
Wilson disease, 308, 486t, 634, 641
wobble, 407

X
xanthine, 331, 333f
xanthine oxidase, 116
   deficiency of, hypouricemia and,
      345
   molybdenum in, 103–104
xanthurenate, 298, 300f
X-chromosome pair, 364
xenobiotics
   biomedical importance of, 564
   metabolism of
      acetylation and methylation,
         566–567
      conjugation, 566
      hydroxylation by cytochrome P450,
         564–566
   toxic, immunologic, and carcinogenic
      effects of, 567, 567f
   types of, 564
xerophthalmia, 538
x-ray crystallography, 40–41
x-rays, carcinogenic effect of, 691, 691t
xylose, 150f, 151t, 556t
xylulose, 150f, 151t, 199
xylulose 5-phosphate, 194

Y
yeast, fermentation by cell-free extract of,
   1–2
yeast artificial chromosome (YAC) vector,
   449, 449t
yeast Flp recombinase, 447
yeast FRT sites, 447

Z
Z line, 619, 619f, 620f
Zellweger (cerebrohepatorenal) syndrome,
   225, 587, 588t
zinc
   absorption of, 105
   human requirement for, 97, 97t
   multivalent states of, 97, 98f, 98t
   physiologic roles of, 102, 103f
   toxicity of, 99t
zinc finger, 102
   structure of, 99f
zinc finger motif, 437–438, 438f
zona pellucida, 561
zwitterions, 20
zymogens, 90, 91f, 634
   protease secretion as, 529